SPACE SCIENCE SERIES

Tom Gehrels, General Editor

Planets, Stars and Nebulae, Studied with Photopolarimetry
Tom Gehrels, editor, 1974, 1133 pages

Jupiter
Tom Gehrels, editor, 1976, 1254 pages

Planetary Satellites
Joseph A. Burns, editor, 1977, 598 pages

Protostars and Planets
Tom Gehrels, editor, 1978, 756 pages

Asteroids
Tom Gehrels, editor, 1979, 1181 pages

Comets
Laurel L. Wilkening, editor, 1982, 766 pages

Satellites of Jupiter
David Morrison, editor, 1982, 972 pages

Venus
D. M. Hunten, L. Colin, T. M. Donahue and V. I. Moroz, editors, 1983, 1143 pages

Saturn
Tom Gehrels and Mildred S. Matthews, editors, 1984, 968 pages

Planetary Rings
Richard Greenberg and André Brahic, editors, 1984, 784 pages

Protostars and Planets II
David C. Black and Mildred S. Matthews, editors, 1985, 1293 pages

Satellites
Joseph A. Burns and Mildred S. Matthews, editors, 1986, 1021 pages

The Galaxy and the Solar System
Roman Smoluchowski, John N. Bahcall and Mildred S. Matthews, editors, 1986, 485 pages

Meteorites and the Early Solar System
John F. Kerridge and Mildred S. Matthews, editors, 1988, 1269 pages

Mercury
Faith Vilas, Clark R. Chapman and Mildred S. Matthews, editors, 1988, 794 pages

Origin and Evolution of Planetary and Satellite Atmospheres
S. K. Atreya, J. B. Pollack and M. S. Matthews, editors, 1989, 881 pages

Asteroids II
Richard P. Binzel, Tom Gehrels and Mildred S. Matthews, editors, 1989, 1258 pages

Uranus
Jay T. Bergstralh, Ellis D. Miner and Mildred S. Matthews, editors, 1991, 1076 pages

The Sun in Time
C. P. Sonett, M. S. Giampapa and M. S. Matthews, editors, 1991, 996 pages

Solar Interior and Atmosphere
A. N. Cox, W. C. Livingston and M. S. Matthews, editors, 1991, 1414 pages

Mars
H. H. Kieffer, B. M. Jakosky, C. W. Snyder and M. S. Matthews, editors, 1992, 1536 pages

Protostars and Planets III
E. H. Levy and J. I. Lunine, editors, 1993, 1596 pages

Resources of Near-Earth Space
J. S. Lewis, M. S. Matthews and M. L. Guerrieri, editors, 1993, 977 pages

Hazards Due to Comets and Asteroids
T. Gehrels, editor, 1994, 1300 pages

Neptune and Triton
D. P. Cruikshank, editor, 1995, 1247 pages

Cosmic Winds and the Heliosphere
J. R. Jokipii, C. P. Sonett and M. S. Giampapa, editors, 1997, 1052 pages

Venus II—Geology, Geophysics, Atmosphere, and Solar Wind Environment
S. W. Bougher, D. M. Hunten and R. J. Phillips, editors, 1997, 1376 pages

VENUS II

GEOLOGY, GEOPHYSICS, ATMOSPHERE, AND SOLAR WIND ENVIRONMENT

VENUS II

GEOLOGY, GEOPHYSICS, ATMOSPHERE, AND SOLAR WIND ENVIRONMENT

S.W. Bougher
D.M. Hunten
R.J. Phillips

Editors

With the editorial assistance of
M. S. Matthews, A. S. Ruskin,
and M. L. Guerrieri

With 100 collaborating authors

THE UNIVERSITY OF ARIZONA PRESS
TUCSON

About the front cover:

Surveying Venus. A VEGA balloon drifts at the base of the Venus cloud layer. (The two VEGA balloons radioed data from altitudes of about 54 km [middle cloud layer] to as low as 50 km [near the base of the lower cloud layer].) This imaginary view (perhaps after transmission termination, additional helium loss, and further sinking) shows one of the balloons with a view of the surface below, based on Magellan imagery of tectonic faults and scattered impact craters. (Painting by William K. Hartmann)

About the back cover:

Sappas Mons and Maat Mons (P-40176). Seen in the foreground of this computer-generated, three-dimensional perspective view of the surface of Venus is the volcano Sapas Mons—400 km wide—located at 8°N 188°E, on the western edge of Atlas Regio. Sapas Mons is 1.5 km high, with a peak that sits at an elevation of 4.5 km above the planet's mean elevation. In the background on the horizon is Maat Mons, the largest shield volcano on Venus. Its peak rises to an elevation of 8 km above the planet's mean elevation. Lava flows extend to the base of Sapas for hundreds of kilometers across the plains shown in the foreground. In this image, the topographic relief is exaggerated approximately 10 times. (The image was produced by the Solar System Visualization Project and the Magellan Science Team at the Jet Propulsion Laboratory Multimission Image Processing Laboratory.)

The University of Arizona Press
© 1997 The Arizona Board of Regents
First printing
All rights reserved
∞ This book is printed on acid-free, archival-quality paper.
Manufactured in the United States of America.
02 01 00 99 98 97   6 5 4 3 2 1
Library of Congress Cataloging-in-Publication Data
Venus II—geology, geophysics, atmosphere, and solar wind environment.
 / S.W. Bougher, D.M. Hunten, R.J. Phillips, editors ; with the
editorial assistance of M.S. Matthews, A.S. Ruskin, and M.L.
Guerrieri.
      p. cm. — (Space science series)
   Includes bibliographical references and index.
   ISBN 0-8165-1830-0 (cloth)
   1. Venus (Planet)   2. Venus (Planet)—Databases.   3. Astrophysics.
4. CD-ROMs.   I. Bougher, S. W. (Stephen Wesley, 1955– .
II. Hunten, Donald M.   III. Phillips, R. J. (Roger J.), 1940– .
IV. Series.
QB621.V463   1997
559.9'22—dc21   97-28602
                                                                 CIP
British Library Cataloguing-in-Publication Data
A catalogue record for this book is available from the British Library.

VENUS II—GEOLOGY, GEOPHYSICS, ATMOSPHERE, AND SOLAR WIND ENVIRONMENT

CONTENTS

COLLABORATING AUTHORS

PREFACE—Goals, Objectives, and Scope

PART I—Solar Wind Environment

THE VENUS ATMOSPHERE AND IONOSPHERE AND THEIR
   INTERACTION WITH THE SOLAR WIND: AN OVERVIEW   3
T. M. Donahue and C. T. Russell

GLOBAL MODELS OF THE SOLAR WIND INTERACTION
   WITH VENUS   33
*J. G. Luhmann, S. H. Brecht, J. R. Spreiter, S. S. Stahara,
R. S. Steinolfson and A. F. Nagy*

MAGNETOHYDRODYNAMIC PROCESSES: MAGNETIC FIELDS
   IN THE IONOSPHERE OF VENUS   61
T. E. Cravens, H. Shinagawa and J. G. Luhmann

PLASMA WAVE PHENOMENA AT VENUS   95
J. D. Huba and R. J. Strangeway

EVIDENCE FOR VENUS LIGHTNING   125
J. M. Grebowsky, R. J. Strangeway and D. M. Hunten

PART II—Upper Atmosphere (≥ 100 km)

IONOSPHERE: SOLAR CYCLE VARIATIONS   161
J. L. Fox and A. J. Kliore

IONOSPHERE: ENERGETICS   189
A. F. Nagy and T. E. Cravens

SOLAR ACTIVITY BEHAVIOR OF THE THERMOSPHERE   225
*W. T. Kasprzak, G. M. Keating, N. C. Hsu, A. I. F. Stewart,
W. B. Colwell and S. W. Bougher*

UPPER ATMOSPHERE DYNAMICS: GLOBAL CIRCULATION
    AND GRAVITY WAVES   259
  S. W. Bougher, M. J. Alexander and H. G. Mayr

PART III—Lower Atmosphere (≤100 km)

MONITORING OF MESOSPHERIC STRUCTURE
    AND DYNAMICS   295
  *E. Lellouch, T. Clancy, D. Crisp, A. J. Kliore, D. Titov
  and S. W. Bougher*

NEAR-INFRARED SOUNDING OF THE LOWER ATMOSPHERE
    OF VENUS   325
  F. W. Taylor, D. Crisp and B. Bézard

THE THERMAL BALANCE OF THE VENUS ATMOSPHERE   353
  D. Crisp and D. Titov

ION/NEUTRAL ESCAPE OF HYDROGEN AND DEUTERIUM:
    EVOLUTION OF WATER   385
  *T. M. Donahue, D. H. Grinspoon, R. E. Hartle
  and R. R. Hodges, Jr.*

CHEMISTRY OF LOWER ATMOSPHERE AND CLOUDS   415
  *L. W. Esposito, J.-L. Bertaux, V. Krasnopolsky, V. I. Moroz
  and L. V. Zasova*

THE GENERAL CIRCULATION OF THE VENUS ATMOSPHERE:
    AN ASSESSMENT   459
  *P. J. Gierasch, R. M. Goody, R. E. Young, D. Crisp, C. Edwards,
  R. Kahn, D. Rider, A. Del Genio, R. Greeley, A. Hou, C. B. Leovy,
  D. McCleese and M. Newman*

PART IV—Surface Processes

REMOTE SENSING OF SURFACE PROCESSES   503
  B. A. Campbell, R. E. Arvidson, M. K. Shepard and R. A. Brackett

SURFACE SCATTERING AND DIELECTRICAL PROPERTIES   527
  *G. H. Pettengill, B. A. Campbell, D. B. Campbell
  and R. A. Simpson*

AEOLIAN PROCESSES AND FEATURES ON VENUS   547
  *R. Greeley, K. C. Bender, R. S. Saunders, G. Schubert
  and C. M. Weitz*

CONTENTS

GEOCHEMISTRY OF SURFACE–ATMOSPHERE INTERACTIONS
    ON VENUS   591
    B. Fegley, Jr., G. Klingelhöfer, K. Lodders and T. Widemann

ROCK WEATHERING ON THE SURFACE OF VENUS   637
    J. A. Wood

PART V—Geologic Structure

PHYSIOGRAPHY, GEOMORPHIC/GEOLOGIC MAPPING AND
    STRATIGRAPHY OF VENUS   667
    K. L. Tanaka, D. A. Senske, M. Price and R. L. Kirk

PART VI—Volcanism

VOLCANOES AND CENTERS OF VOLCANISM ON VENUS   697
    L. S. Crumpler, J. C. Aubele, D. A. Senske, S. T. Keddie,
    K. P. Magee and J. W. Head

CHANNELS AND VALLEYS   757
    V. R. Baker, G. Komatsu, V. C. Gulick and T. J. Parker

PART VII—Tectonism

TECTONIC OVERVIEW AND SYNTHESIS   797
    V. L. Hansen, J. J. Willis and W. B. Banerdt

LARGE VOLCANIC RISES ON VENUS   845
    S. E. Smrekar, W. S. Kiefer and E. R. Stofan

ISHTAR TERRA   879
    W. M. Kaula, A. Lenardic, D. L. Bindschadler
    and J. Arkani-Hamed

PLAINS TECTONICS ON VENUS   901
    W. B. Banerdt, G. E. McGill and M. T. Zuber

CORONAE ON VENUS: MORPHOLOGY AND ORIGIN   931
    E. R. Stofan, V. E. Hamilton, D. M. Janes and S. E. Smrekar

PART VIII—Impact Cratering

CRATERING ON VENUS: MODELS AND OBSERVATIONS   969
    W. B. McKinnon, K. J. Zahnle, B. A. Ivanov and H. J. Melosh

MORPHOLOGY AND MORPHOMETRY OF IMPACT CRATERS 1015
R. R. Herrick, V. L. Sharpton, M. C. Malin, S. N. Lyons and K. Feely

THE RESURFACING HISTORY OF VENUS 1047
A. T. Basilevsky, J. W. Head, G. G. Schaber and R. G. Strom

PART IX—Geodynamics

VENUSIAN SPIN DYNAMICS 1087
C. F. Yoder

THE VENUS GRAVITY FIELD AND OTHER GEODETIC PARAMETERS 1125
W. L. Sjogren, W. B. Banerdt, P. W. Chodas, A. S. Konopliv, G. Balmino, J. P. Barriot, J. Arkani-Hamed, T. R. Colvin, and M. E. Davies

LITHOSPHERIC MECHANICS AND DYNAMICS OF VENUS 1163
R. J. Phillips, C. L. Johnson, S. L. Mackwell, P. Morgan, D. T. Sandwell and M. T. Zuber

THE CRUST OF VENUS 1205
R. E. Grimm and P. C. Hess

MANTLE CONVECTION AND THE THERMAL EVOLUTION OF VENUS 1245
G. Schubert, V. S. Solomatov, P. J. Tackley and D. L. Turcotte

APPENDIX 1289

GLOSSARY 1295

ACKNOWLEDGMENTS 1329

INDEX 1331

COLLABORATING AUTHORS

Alexander, M. J., 259
Arkani-Hamed, J., 879, 1125
Arvidson, R. E., 503
Aubele, J. C., 697
Baker, V. R., 757
Balmino, G., 1125
Banerdt, W. B., 797, 901, 1125
Barriot, J. P., 1125
Basilevsky, A. T., 1047
Bender, K. C., 547
Bertaux, J.-L., 415
Bézard, B., 325
Bindschadler, D. L., 879
Bougher, S. W., xi, 225, 259, 295
Brackett, R. A., 503
Brecht, S. H., 33
Campbell, B. A., 503, 527
Campbell, D. B., 527
Chodas, P. W., 1125
Clancy, T., 295
Colvin, T. R., 1125
Colwell, W. B., 225
Cravens, T. E., 61, 189
Crisp, D., 295, 325, 353, 459
Crumpler, L. S., 697
Davies, M. E., 1125
Del Genio, A., 459
Donahue, T. M., 3, 385
Esposito, E. W., 415

Edwards, C., 459
Feely, K., 1015
Fegley, Jr., B., 591
Fox, J. L., 161
Gierasch, P. J., 459
Goody, R. M., 459
Grebowsky, J. M., 125
Greeley, R., 459, 547
Grimm, R. E., 1205
Grinspoon, D. H., 385
Gulick, V. C., 757
Hamilton, V. E., 931
Hansen, V. L., 797
Hartle, R. E., 385
Head, J. W., 697, 1047
Herrick, R. R., 1015
Hess, P. C., 1205
Hodges, Jr., R. R., 385
Hou, A., 459
Hsu, N. C., 225
Huba, J. D., 95
Hunten, D. M., xi, 125
Ivanov, B. A., 969
Janes, D. M., 931
Johnson, C. L., 1163
Kahn, R., 459
Kasprzak, W. T., 225
Kaula, W. M., 879
Keating, G. M., 225

Keddie, S. T., 697
Kiefer, W. S., 845
Kirk, R. L., 667
Klingelhöfer, G., 591
Kliore, A. J., 161, 295
Komatsu, G., 757
Konopliv, A. S., 1125
Krasnopolsky, V., 415
Lellouch, E., 295
Lenardic, A., 879
Leovy, C. B., 459
Lodders, K., 591
Luhmann, J. G., 33, 61
Lyon, S. N., 1015
Mackwell, S. L., 1163
Magee, K. P., 697
Malin, M. C., 1015
Mayr, H. G., 259
McCleese, D., 459
McGill, G. E., 901
McKinnon, W. B., 969
Melosh, H. J., 969
Morgan, P., 1163
Moroz, V. I., 415
Nagy, A. F., 33, 189
Newman, M., 459
Parker, T. J., 757
Pettengill, G. H., 527
Phillips, R. J., xi, 1163
Price, M., 667
Rider, D., 459
Russell, C. T., 3
Sandwell, D. T., 1163
Saunders, R. S., 547
Schaber, G. G., 1047
Schubert, G., 547, 1245
Senske, D. A., 667, 697
Sharpton, V. L., 1015

Shepard, M. K., 503
Shinagawa, H., 61
Simpson, R. A., 527
Sjogren, W. L., 1125
Smrekar, S. E., 845, 931
Solomatov, V. S., 1245
Spreiter, J. R., 33
Stahara, S. S., 33
Steinolfson, R. S., 33
Stewart, A. I. F., 225
Stofan, E. R., 845, 931
Strangeway, R. J., 95, 125
Strom, R. G., 1047
Tackley, P. J., 1245
Tanaka, K. L., 667
Taylor, F. W., 325
Titov, D., 295, 353
Turcotte, D. L., 1245
Weitz, C. M., 547
Widemann, T. 591
Willis, J. J., 797
Wood, J. A., 637
Yoder, C. F., 1087
Young, R. E., 459
Zahnle, K. J., 969
Zasova, L. V., 415
Zuber, M. T., 901, 1163

PREFACE

With the demise of the Pioneer Venus Orbiter in October 1992, the Galileo Venus flyby in February 1990, and the final orbit of Magellan in October 1994, we find ourselves at the culmination of a period of exciting Venus reconnaisance and exploration. The resulting scientific studies are unprecedented in their detail for any planet except the Earth. Consequently, this *Venus II* book is motivated by our need to re-evaluate our initial assessment of Venus in light of these and other spacecraft missions and groundbased campaigns conducted over the past 30 years.

A special Venus II scientific meeting was convened in Tucson, Arizona, January 4–7, 1995, in order to review what is known about the geology, interior structure, atmosphere, and solar wind environment of Venus and to study the many intriguing questions and competing theories that remain. The content of this book is an outgrowth of that meeting. The resulting themes of the *Venus II* book are threefold: (1) to summarize those aspects of Venus that are known with reasonable confidence; (2) to elucidate key points where interpretations diverge and the resulting implications; and (3) to identify future measurements of high priority. The diversity of the 36 chapters contained herein clearly reflects the interdisciplinary nature of Venus science, and also attests to the breadth of backgrounds and skills that have contributed to Venus science over many years. Several spacecraft missions are represented, including the Veneras, Vega 1 and 2, Pioneer Venus, the Galileo flyby, and Magellan. In addition, various groundbased observational campaigns are described and their results presented.

The attached CD-ROM is a supplement to the *Venus II* book. It is a collaboration of the Data Distribution Lab of the Jet Propulsion Laboratory and the University of Arizona Press. This CD-ROM incorporates text, graphics, software, and various digital data products from selected *Venus II* book authors. A multimedia interface is provided in order to navigate the text and the extensive data products incorporated. A citation convention for these data products has been implemented in order to connect the book text and the digital products on the CD-ROM (CDPxCyFz, -Az, -Mz, -Tz refer to Part "x", Chapter "y", and Figure, Appendix, Movie and Table "z", respectively). We believe that this CD-ROM is an excellent digital resource that suitably complements the *Venus II* book chapters while also providing non-discipline users a method to access Venus data and general Venus information.

The production of this book was partially supported by the National

Aeronautics and Space Administration through grants from the Magellan Project Office and the Pioneer Project Office. We wish to thank Pete Kahn of the Jet Propulsion Laboratory for coordinating the compilation of the *Venus II* CD-ROM. We are grateful to the University of Arizona Press, publishers of this book in their Space Science Series, and to Tom Gehrels, Series Consultant. Above all, we express our deepest thanks to Amy Schumann Ruskin, Mildred Matthews and Mary Guerrieri, whose experience and hard work made the Venus II project possible.

Finally, as editors, it is our hope that future Venus exploration and science will continue to build upon the data and ideas presented in this *Venus II* book and CD-ROM. Furthermore, it is our sincere belief that ongoing Mars exploration and science will benefit greatly from this new "Venusian perspective."

Stephen W. Bougher
Donald M. Hunten
Roger J. Phillips

December 1996

PART I
Solar Wind Environment

THE VENUS ATMOSPHERE AND IONOSPHERE AND THEIR INTERACTION WITH THE SOLAR WIND: AN OVERVIEW

T. M. DONAHUE
University of Michigan

and

C. T. RUSSELL
University of California at Los Angeles

This chapter summarizes the state of our knowledge of the atmosphere and ionosphere of Venus and their interaction with the solar wind. Its purpose is to set the stage for the chapters that follow. It also briefly identifies the principal gaps in our present knowledge and the measures that might be taken to fill them in.

I. INTRODUCTION: PROPERTIES OF THE ATMOSPHERE

Venus is a very warm, very dry planet with a dense carbon dioxide atmosphere. The average temperature at the mean planetary radius of 6051.5 km is 737 K. The atmosphere there exerts a pressure of 95.0 bar and has a lapse rate of 8.06 K km^{-1}. Its density is 66.47 kg m^{-3}. It consists mainly of CO_2, with a mixing ratio of 0.965±0.008 and of N_2 at 0.035±0.008. The mean molecular weight of this atmosphere is thus 43.33±0.15 kg per mole. The carbon content of the CO_2 per g of planet is $(2.67\pm0.30)\times10^{-5}$ g. This is to be compared with a terrestrial inventory of $(1.5$ to $4.5)\times10^{-5}$ g g^{-1} of carbon, where the lower figure is crustal carbon, mainly carbonate, and the higher figure includes mantle carbon. A similar exercise for nitrogen would compare (2.49 ± 0.30) μg g^{-1} in the atmosphere of Venus with 0.666 μg g^{-1}, 0.78 μg g^{-1}, and 1 μg g^{-1} in Earth's atmosphere, crust and mantle, respectively. CO_2, which is in the atmosphere of hot, desiccated Venus as a gas, has been converted on Earth, in the presence of abundant water, to carbonate in reaction with rocks such as wollastonite.

II. NONRADIOGENIC NOBLE GASES

This rough parity in carbon and nitrogen, two important constituents of volatile compounds on the two planets, does not extend to other volatiles, such as the nonradiogenic noble gases. The elemental abundances of these gases, the isotopic ratios of ^{20}Ne to ^{22}Ne, ^{36}Ar to ^{38}Ar, and the ratio of abundances on Venus and Earth, normalized to the planetary mass, are shown in Table I.

TABLE I

Elemental Abundances, Isotopic Ratios and Ratios of Abundances on Venus and Earth

Element	Venusian Abundance (ppm)	Venus/Earth Ratio	Isotopes	Sun	Isotopic Ratio Venus	Earth
^{20}Ne	7	21	^{20}Ne/^{22}Ne	13	11.8±0.7	10
^{36}Ar	30	70	^{36}Ar/^{38}Ar	5.35	5.55±0.6	5.21
Kr	0.0047	3				
Xe	<0.0040	<35				

The abundances of these elements on Venus, Earth and Mars and typical C3V and CI meteorites are shown in Figs. 1 and 2. The great excess in elemental neon and argon on Venus and the very different relative abundance patterns on the three planets have challenged explanation since they were revealed by probe missions in the 1970s. But, recently, remarkable progress has been made in the construction of scenarios for planetary genesis that provide a plausible explanation for these noble gas abundances. These contemplate that, as the planets grew from planetesimals, their interiors received similar endowments of volatiles from their accreting cores. In the late stage of accretion they are supposed to have received an extra surficial contribution of volatiles from icy planetesimals arriving from the outer solar system. Earth and Venus also received a veneer of material rich in carbon and nitrogen during the last stages of accretion. Initially, Earth and Venus are supposed to have had almost identical supplies of the noble gases with relative abundances like solar gases. Neon, an exception, is supposed to have been depleted before the planetary material was formed. The hydrogen-dominated early upper atmospheres of these water-rich planets were driven away in a supersonic blowoff powered by 450-fold enhanced EUV solar radiation. The outflowing hydrogen carried away the noble gases with it. As the strength of the solar EUV radiation decreased, the outflow rate eventually declined to a level at which it could not carry xenon away. From that time on, the Xe abundances remained fixed in the two atmospheres at today's levels. Earth, but not Venus, suffered a singular event: a collision with a giant planetesimal, probably the one that resulted in the formation of the Moon. This impact dramatically eroded the atmosphere, noble gases included. This is supposed to be the reason that terrestrial abundances of noble gases are so low compared to those of Venus. A subsequent period of degassing of Ne, Ar and Kr, but not Xe, from the interior is supposed to have occurred during the final stages of fractionation on Earth, but not on Venus. The entire process took about 300 Myr to complete on Venus. The final abundances of the noble gases are those shown in Figs. 1 and 2. Blowoff fractionates isotopes as well as elements. The fractionation achieved by the model reproduces the terrestrial and Martian abundances very well (Pepin 1991,1995; Jakosky et al. 1994).

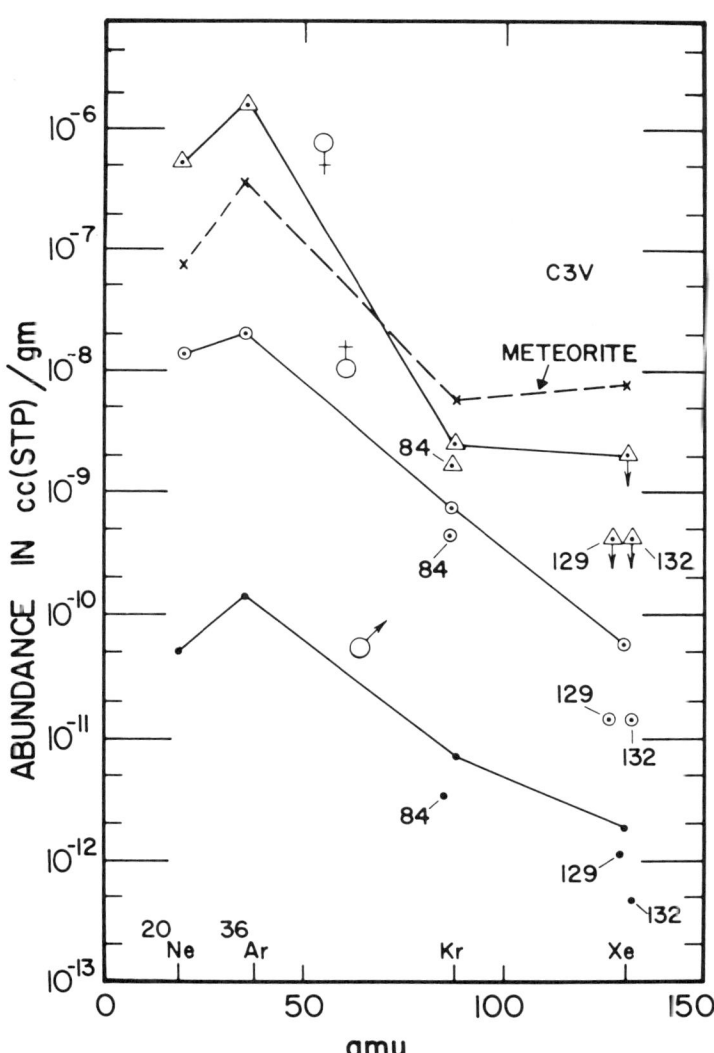

Figure 1. Abundances of noble gases on Venus, Earth, Mars and in a 3CV carbonaceous meteorite (figure from Donahue and Pollack 1983).

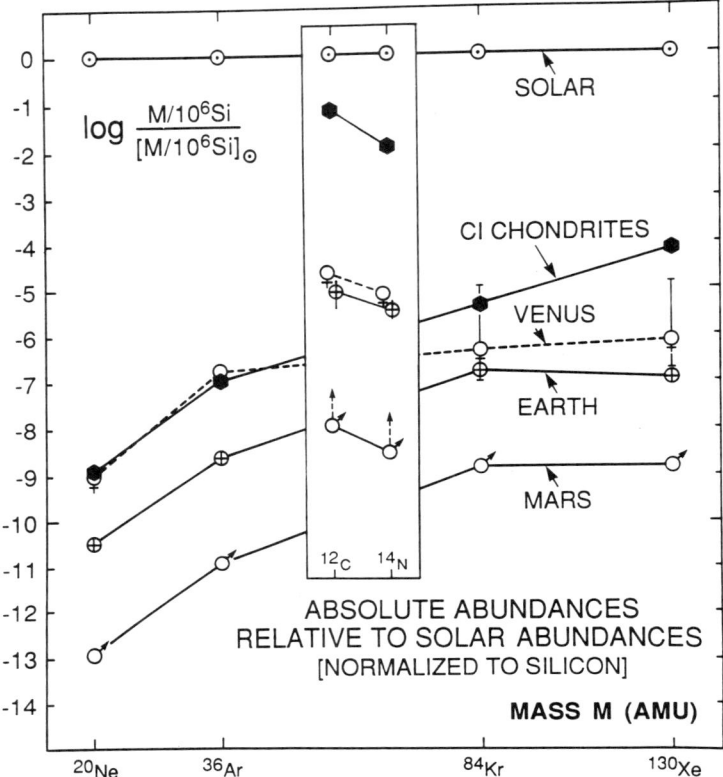

Figure 2. Absolute noble gas, carbon and nitrogen abundances in planetary atmospheres and in CI meteorites relative to the corresponding solar ratios (figure from Pepin 1991).

Crucial for testing this model is a determination of the Xe elemental abundance and a measurement of the Kr and Xe isotopic ratios on Venus with an accuracy in the neighborhood of 5%.

III. RADIOGENIC NOBLE GASES

The radiogenic noble gas ^{40}Ar, whose abundance is 34 ppm v/v, is less abundant by a factor of 4 on Venus than on Earth. Unfortunately, the abundance of radiogenic ^{4}He is uncertain by a factor of 20 (0.6 and 12 ppm), because it has been measured accurately only in the upper atmosphere and extrapolation to the mixed atmosphere is model dependent. There is between 175 to 3700 times as much ^{4}He in the atmosphere of Venus as there is in Earth's atmosphere. However, helium escapes from Earth in the polar wind at a globally average rate of $(3\pm1)\times 10^6$ cm^{2}s^{-1}. It is in a steady state in which escape and outgassing from the interior are balanced. The amount of ^{4}He outgassed on

Earth during its lifetime turns out to be larger than the amount in the atmosphere of Venus by a factor of at least 3 (perhaps 60). These nonradiogenic noble gas comparisons suggest that outgassing rates on Venus are less than on Earth by a factor of 3 to 4 (Krasnopolsky et al. 1994).

An accurate measurement of the ^{4}He abundance in the mixed atmosphere is required for an adequate understanding of the extent to which Venus has outgassed the radiogenic gases produced in its interior.

IV. WATER

Water is a prototypical volatile that should have been abundant on early Venus if the terrestrial planets were formed from well-mixed primordial nebular material, but is scarcely present at all today. Recently, a consensus has developed that most of the time and in most places the mixing ratio of water vapor below the clouds of Venus is only 30 ppm. The question that begs to be answered is whether this water is mostly the remnant of an early abundant supply of water, most of whose hydrogen has escaped to space, or, instead, consists to a large extent of water that has been introduced comparatively recently by comets or by volcanic outgassing. To answer this question it is necessary, in addition to knowing the present hydrogen abundance, at least to know the degree to which deuterium is enriched in the present water compared to primordial or exogenous water. This ratio (D/H) depends on the amount of hydrogen that has escaped compared to that originally present, because hydrogen escapes more readily from the atmosphere than deuterium does. It is also necessary to know the rate at which hydrogen and deuterium are escaping today. It would be good to be able to model also the rates at which they have escaped in the past. It would also be good to know the strength of cometary and volcanic sources. Although that is difficult, the evidence for volcanism and massive resurfacing events obtained by Magellan may provide some insights. Numerous measurements now firmly fix the deuterium:hydrogen ratio in water on Venus at 150 times that in terrestrial water. Water on Earth, which has scarcely suffered from fractionating escape of hydrogen, is presumably a good sample of primitive water on Venus. The escape fluxes of hydrogen from Venus have been modeled and measured. At most two processes are important. These are, first, escape of fast atoms, which were once fast ions but have exchanged electrons with slow atoms, and, second, escape of ions driven by charge separation electric fields. These processes vary strongly with solar activity. Averaged over the planet and over time, the H escape flux appears to lie between 7×10^6 cm^{-2}s^{-1} and 1.6×10^7 cm^{-2}s^{-1}. The fractionation factor, which measures the relative efficiency of deuterium and hydrogen escape, lies between 0.44 and 0.10. If sources are unimportant, this information is sufficient to determine that the original amount of water on Venus was 260 to 7700 times the amount present today. That is enough water to cover the planet with water 4 to 115 m deep, if it was liquefied. There is an interesting possibility that the water was originally much deeper—say as deep as a full terrestrial ocean—but was lost at such a catastrophic rate that deuterium was exhausted along with hydrogen. This would have been the consequence of vaporization of the water in a "moist" or "runaway" greenhouse atmosphere. Only when the water level reached the equivalent of a few meters would fractionation have begun. If a cometary source of deuterium-poor water is balancing hydrogen escape today, even more early water is required to explain the larger present D/H ratio—between 19.5 and 525 m. There are alternatives to this scenario. One is that hydrogen loss is balanced by volcanic injection of water with a D/H ratio 20 times terrestrial. Another is that a massive outgassing event injected water with a D/H ratio 1.5 times terrestrial within the past billion years. After the event, the atmosphere is supposed to have been 150 times wetter than it is today. Loss of hydrogen belonging to this water would have caused the present enhancement in the D/H ratio (chapter by Donahue et al.).

To provide a robust solution to the problem of the origin of the water on Venus, more information is needed about possible spatial and temporal variation in water abundance today and the strength of present potential endogenous and exogenous sources of water, along with their degree of fractionation.

V. SULFUR

Another important class of volatile substances on Venus are sulfur compounds. The clouds are 75% sulfuric acid, 25% water vapor. SO_2 is abundant—180 ± 70 ppm below the clouds in 1982 and also in 1992, according to near infrared spectral soundings. But above the clouds SO_2 seems to be quite variable, and to have decreased from 90 ppb to 3 ppb during the lifetime of the Pioneer Venus Orbiter (PVO) mission, according to observations made by the PVO ultraviolet spectrometer. Ultraviolet spectrophotometric observations with the Hubble Space Telescope indicate that its concentration has continued to decline since 1992. This behavior of SO_2 in the upper atmosphere has caused speculation that there was massive volcanic injection of SO_2 into the atmosphere shortly before PVO arrived. But there is a mystery: how can SO_2 vary above the clouds and remain constant below, as this suite of measurements suggests?

Another prominent sulfur species near the surface is carbonyl sulfide OCS. Its abundance decreases from 55 ppm near the surface to 0.5 ppm above 35 km. This variation is probably associated with the increase in CO from 12 ppm at 23 km to 23 ppm at 42 km (and perhaps 39 km ppm at 46 km) because of the removal of CO at the surface in a reaction with pyrite (FeS_2) that produces magnetite (Fe_3O_4) and carbonyl sulfide. H_2S has an abundance of 3 ppm near the surface and begins to disappear at 20 km. Other volatiles found in low concentration are HCl, HF and ethane (C_2H_6) (chapter by Fegley et al.).

One of the imperatives for exploration of Venus is the determination of the redox state of the lower atmosphere and understanding of the chemical interaction between the surface and the atmosphere. Gases that are volatilized near the surface in the lowlands may precipitate and form mineral layers which are conspicuous in radar images on highlands, such as Maxwell Montes. The reality of such processes needs to be established.

VI. THERMAL STRUCTURE

Greenhouse models (Fig. 3) give a reasonable explanation of the high surface temperature and the altitude profile. The atmosphere in these models is convective below 35 to 50 km. There are variations with latitude and local time significantly larger than expected theoretically before the PV and Venera probe missions. Where differences of the order of 0.1 K were expected, contrasts as large as 5 K were found below 10 km increasing to as much as 20 K near 60 km, with evidence for large oscillations. There is, however, no evidence for a *steady* variation in the deep atmosphere at low latitudes. Differences between high and low latitude temperatures are associated with maintenance of cyclostrophic balance of zonal winds at mid latitudes (chapter by Crisp et al.).

Determination of atmospheric structure parameters with high temporal and spatial resolution, horizontally and vertically, is highly desirable but technically all but unachievable because of the hostile environment at low altitudes.

VII. CLOUDS

Clouds cover Venus globally in three layers between 48 and 68 km. Above and below these layers is haze that extends as high as 90 km and as low as 32 (perhaps 10) km. There are three modes of cloud particles whose sizes and number densities are shown in the cartoon of Fig. 4. The large mode particles are found predominantly in the middle cloud layer. They are, perhaps, crystals. Those in the median mode have diameters of $2\,\mu$m. Small mode particles pervade the entire region from 30 to 70 km. Infrared soundings, especially from the Galileo spacecraft, have revealed that the cloud opacity evolves rapidly. It appears to vary rapidly in space and time. This behavior suggests high winds and a high level of meteorological activity near the tropopause of Venus. This is a state of affairs quite different than that expected from a dense, slowly moving lower troposphere.

The sulfuric acid which is the principal constituent of the cloud particles is probably produced as a result of the reaction of SO_2, rising from the troposphere, with dissociation products of water that produces SO_3. H_2O and SO_3 then react to make H_2SO_4. H_2SO_4 precipitating from the clouds will be thermally decomposed into SO_3, which is unstable against conversion to SO_2 at low altitudes. Another source of low altitude SO_2 is elemental sulfur, which is created by photolysis of SO_2 above the clouds and descends to levels at which it can be oxidized to SO_2 (chapter by Esposito et al.).

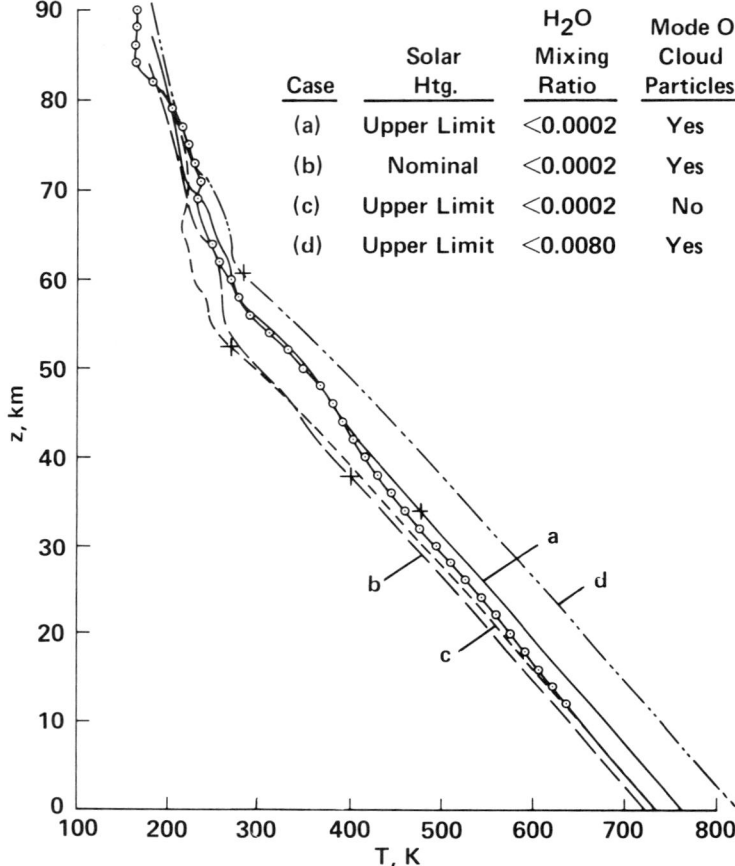

Figure 3. Theoretical greenhouse models. These are radiative, convective equilibrium models, adiabatic below the + symbols and superadiabatic above (figure from Pollack et al. 1980).

Discovering what makes the clouds yellow, confirmation of the cloud chemistry models, explanation of the large variability in cloud opacity and the nature of the mechanism that generates lightning, if it occurs in the clouds, should be the foci of investigations in the future.

VIII. WINDS: LOWER ATMOSPHERE

Zonal wind velocities increase with altitude (Fig. 5). High retrograde speeds of about 100 m s^{-1} are reached near 60 km. The velocity is small near the surface but grows to 10 m s^{-1} at 10 km. Near the subsolar point the atmosphere is neutrally stable from 20 to 30 km and from 50 to 55 km in the clouds. There is a stable layer between 30 and 50 km. This part of the atmosphere may support a regime of deep thermal convection. Recently,

CLOUDS

Layers	z km	μm	n cm^{-3}
Haze	90	<0.5	100
	68		
Upper		2.7-3.2	1-200
	57		
Middle		3.2-3.8	250
	51		
Lower		1.8-32	50
	48		
Haze	(32-10)	<0.5	>10^3

3(?) Particle Modes

$$H_2SO_4 + ?$$

Why Yellow??

$$Sulfur/FeCl_3 ?$$

UV markings

$$SO_2 + Cl_2 ?$$

Figure 4.   Venus cloud structure.

study of the distribution of ground streaks near impact craters in Magellan images has provided rather strong evidence for a Hadley circulation regime below 10 km (chapter by Gierasch et al.).

Our understanding of the dynamics of Venus' atmospheric circulation system is still much too primitive. Measurements with good temporal and spatial resolution and sufficient sensitivity to record very slow atmospheric motions at low altitude are required to repair this deficiency.

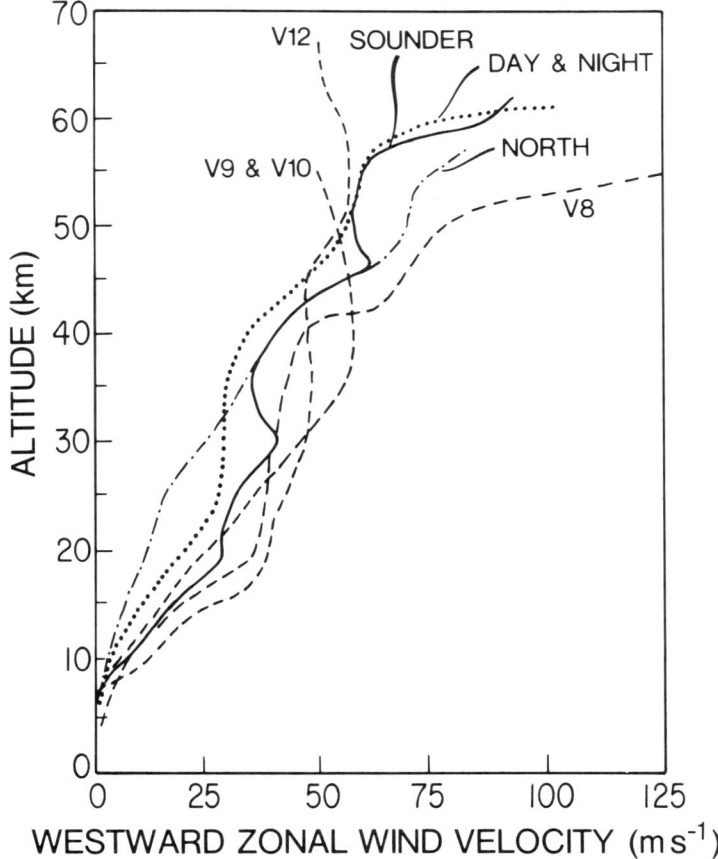

Figure 5. Vertical profiles of east-to-west wind speed from Doppler tracking of Veneras 8, 9, 10 and 12, and interferometric tracking of Pioneer Venus probes (figure from Schubert et al. 1980).

IX. WINDS: UPPER ATMOSPHERE

Gases in the upper atmosphere of Venus above 95 km (in the thermosphere and cryosphere) are driven in a strong flow from the dayside hemisphere to the nightside (Fig. 6). Here the gas descends and returns to the other hemisphere below 95 km. This circulation, which is different from that existing on Earth, and that believed to exist on Mars, is a consequence of the very slow retrograde rotation of Venus. Prolonged solar heating in one hemisphere and CO_2 in the other cause a great temperature contrast. Although radiative cooling keeps the temperature in the dayside thermosphere far below that of Earth, 300 K versus 1000 to 2000 K (Fig. 7), the night side has no thermosphere at all. Above 150 km the temperature drops below 100 K at night. This temperature contrast between the two hemispheres sets up a strong pressure gradient that drives the strong subsolar-antisolar circulation (chapter by Bougher et al.). Density, temperature and pressure change very rapidly near the terminator.

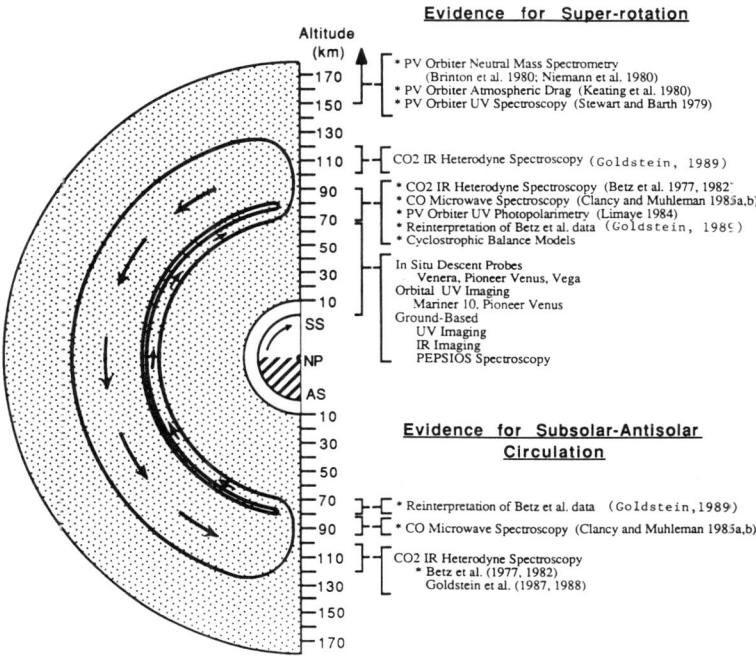

Figure 6. Dynamical regions of Venus' mesosphere and thermosphere (figure from Goldstein 1989).

The circulation is rather more complicated than a simple axially symmetric one. The temperature minimum on the night side is displaced toward the morning sector where there is also a maximum in the brightness of the airglow and a great bulge in the density of the light gases: hydrogen, deuterium and helium. This asymmetry may be due to a westward superrotation of the atmosphere with wind speeds greater than 50 m s^{-1} or the consequence of gravity wave breaking and production of drag which is larger at the morning terminator than at the evening terminator. The light species bulge on the night side is a consequence of the great scale height of these light gases which are dragged along by the dominant heavy gases. The circulation must be divergence free for the principal constituents, but such a circulation would lead to a net inflow of the light gases on the night side above 95 km, if not compensated for by a large asymmetry in density.

There is very little change in the temperature or density of the thermo-

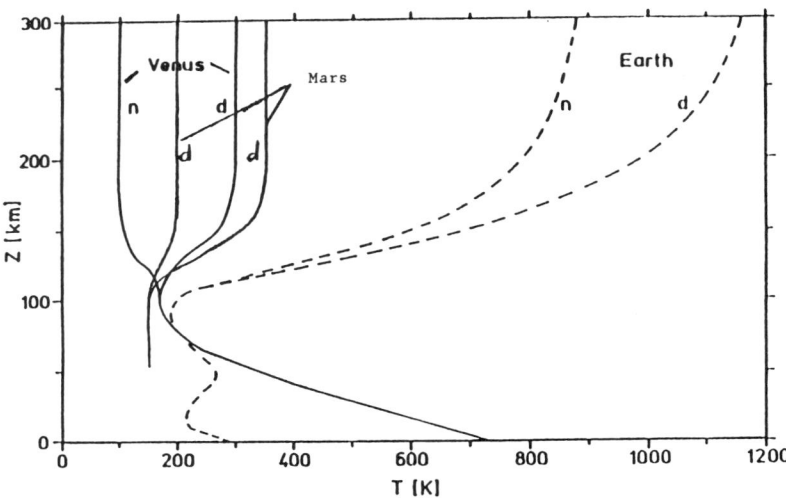

Figure 7. Temperatures of the neutral atmospheres of Earth, Venus and Mars (figure from Fox and Bougher 1991).

sphere and cryosphere with solar activity. This is less true for the short-term changes due, for example, to solar rotation and more so for the long term, solar cycle variations. The reason appears to be the balance between ultraviolet heating and CO_2 15-μm emission cooling (Bougher et al. 1994; chapter by Kasprzak et al.). The CO_2 emission is produced in a bending mode produced by collisions with atomic oxygen. As solar activity increases, so does the oxygen density, the collision rate and consequent CO_2 15-μm cooling (chapter by Bougher et al.).

X. THERMOSPHERIC COMPOSITION

At the homopause, where thorough mixing of different constituents weakens, and molecular diffusion begins to dominate, gases of different mass begin to assume their individual scale heights. The homopause is located somewhere between 130 and 145 km, depending on the gas species and the circulation model chosen. The number densities of the various species found in the thermosphere between 100 km and 200 km on the night side and day side are shown in Fig. 8. The atmosphere of Venus (at high and low altitude) is remarkable for the low concentration of the dissociation products of CO_2. O, CO, and a potential product of atomic oxygen recombination, CO_2 are very low in abundance. (Straightforward recombination of CO and O to reform CO_2 is a very slow process. Nevertheless, CO_2 dominates O_2 in the lower atmosphere.) This anomaly requires that O and CO be removed rapidly from the thermosphere and recombined efficiently in the middle atmosphere, presumably by a catalytic cycle. On Mars, which also has a CO_2-dominated atmosphere, the cycle involves products of water dissociation. Because of the low abundance of water on Venus, especially above the clouds, the same cycle will probably not work there. Instead, chlorine may be involved in chemistry of the CO_2 recombination process on Venus.

The upper atmosphere of Venus, like that of Earth and Mars, is the source of a great variety of airglow emissions. Solar resonance radiation excites hydrogen and oxygen. In the case of hydrogen, Lyman-α resonance radiation from the atmosphere has been observed by instruments carried on a number of spacecraft. Analysis of this spectral data provides the only source of information about atomic hydrogen above the escape level (200 km). Near the subsolar point the observations require a large upward flux of hydrogen in the lower thermosphere. This flow supports the inter-hemispherical thermospheric circulation pattern, already discussed. Strong emission bands of CO, O_2, CO^+, CO_2^+, O_2^+ and NO are seen in the nightglow, as well as emission lines of C, O, He, and their ions (chapter by Kasprzak et al.).

The mechanism of CO_2 recombination needs to be established. This requires composition measurements capable of identifying potential participants in catalytic recombination cycles in the middle atmosphere.

XI. IONOSPHERE

The ionosphere (Fig. 9) is produced primarily by solar EUV ionization of CO_2 on the day side of the planet. But CO_2^+, like its counterpart on Earth, N_2^-, is a minor species in the ionosphere. O_2^+ is the dominant ion up to about 200 km where O^+ becomes ascendant. The dominance of O_2^+ in the lower ionosphere is a consequence of the high reaction rate of CO_2^+ with O and O^+ with CO_2, both of which produce O_2^+.

The existence of a sometimes robust ionosphere at night is an anomaly. Given the slow rotation rate of the atmosphere, the night is long, and solar EUV absent long enough for the ions created in daylight to recombine. However, during solar maximum the dense ionosphere holds the solar wind well away from the planet. The daytime ionosphere is extensive (Fig. 9) and there is a large flux of O^+ ions across the terminator to the night side. These are the ions that maintain the night side ionosphere during times of high solar activity. During solar maximum, even minor ions such as H^+ and He^+ owe their existence at night (at least in large part) to charge exchange between O^+ and neutral H and He.

However, it is also clear that there are fluxes of energetic electrons on the night side large enough to create a significant amount of ionization under suitable conditions. Such conditions prevail during solar minimum when the ionopause moves close to the planet (Fig. 10). The spectacular compression of the ionosphere that is produced is illustrated by the electron density profiles obtained by the radio occultation observations at solar minimum in 1986. Transport of O^+ virtually terminates. The population of nighttime ions is

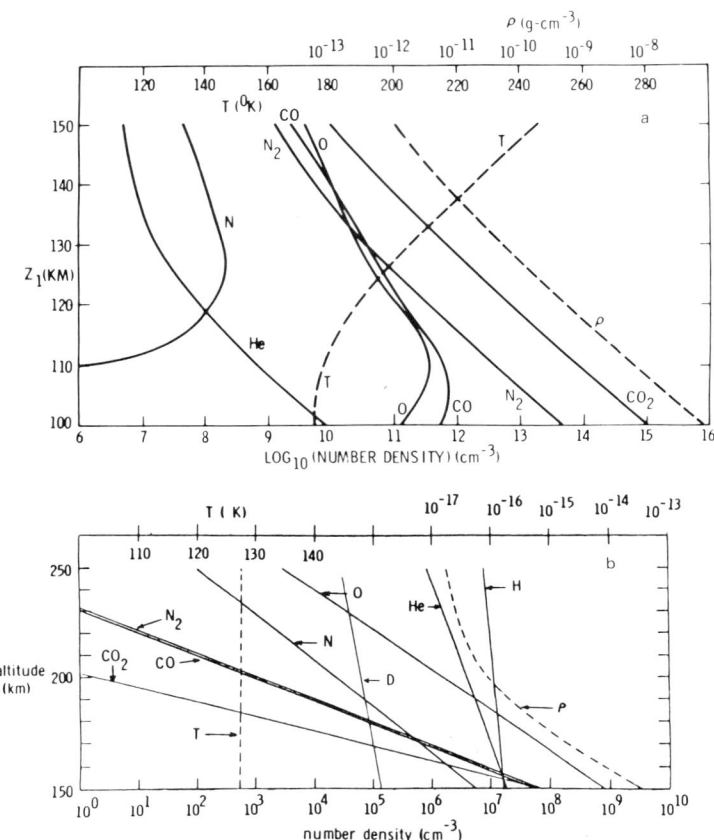

Figure 8. Neutral number densities from the VIRA model (a,b) day side; (c,d) night side (figure from Keating et al. 1985).

drastically reduced (Figs. 11, 12 and 13). Ionization by electrons with energies probably in the neighborhood of 1 keV then dominates at night.

The nightside ionosphere is by no means a region where ion densities vary slowly in space and time. The cartoon of Fig. 14 (Brace and Kliore 1991) illustrates the rich spectrum of temporal and spatial variations that occurs. Sometimes holes in the ionosphere are found extending to the lowest altitudes that have been explored. Filaments or rays extend far into regions called the ionotail and magnetotail. Detached clouds and streamers also populate these regions. All of these features, apparently, vary rapidly. Here some ions, such as O^+ have been observed to be traveling away from the planet with sufficient velocity to escape from the atmosphere into space (chapters by Fox et al. and

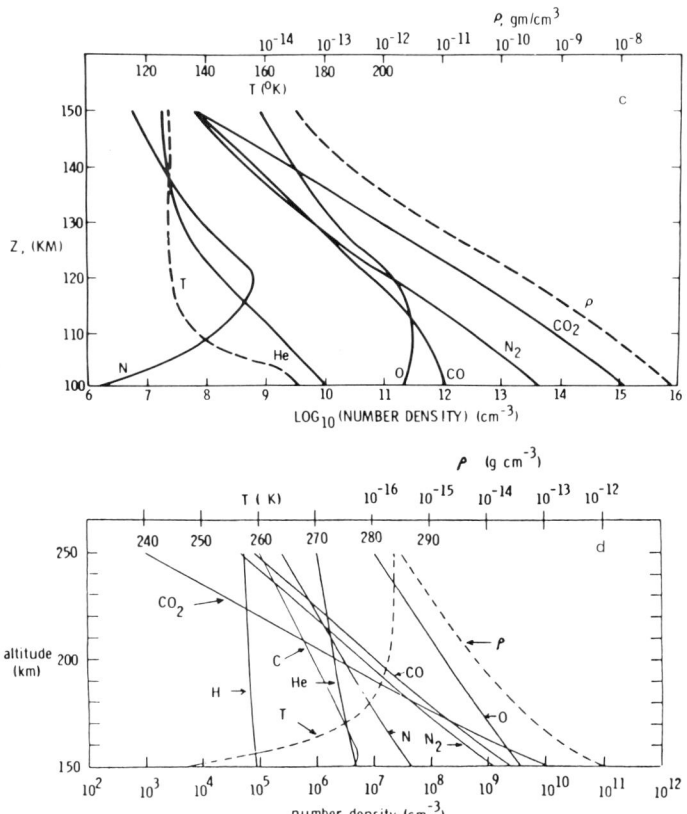

by Nagy et al.).

Despite the wealth of data acquired by the PVO mission, the difficult task of distinguishing spatial from temporal variations needs to be finished.

XII. ESCAPE OF HYDROGEN AND OXYGEN

It has already been mentioned that escape of hydrogen and deuterium from the planet occurs mainly from the light atom and light ion bulge in the dawn sector of the night side. Because of the large decrease in H^+ density there at solar minimum and because both dominant loss processes involve hydrogen ions, the escape flux decreases sharply as solar activity declines. This results in a significant increase in the population of hydrogen in the bulge at times of

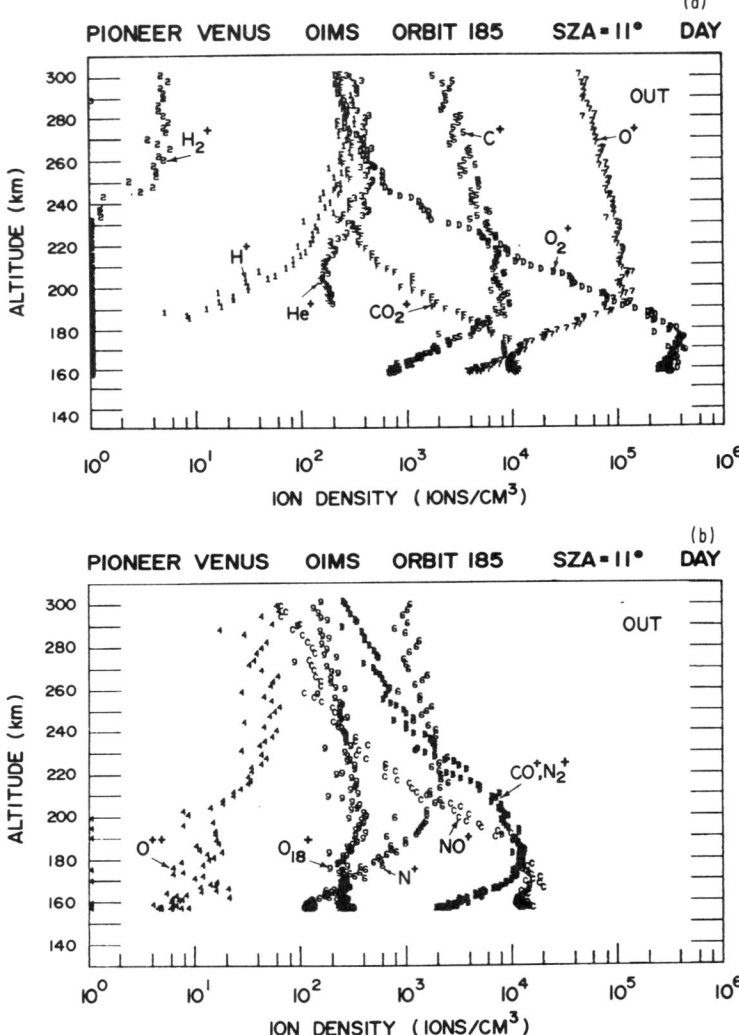

Figure 9. Ion composition in the subsolar region (figure from Brace and Kliore 1991).

low solar activity. Such an increase does not occur for the light gas helium, which does not escape in significant amounts (chapter by Donahue et al.).

Atomic oxygen is being lost by the planet at a rate comparable to that of the hydrogen loss rate. Approximately 90% of the O^+ ions created by photoionization above the ionopause are picked up by the solar wind. The gyroradius of their orbits in the solar wind is so small that, instead of flowing away freely, they re-impact the upper atmosphere. When they do so they

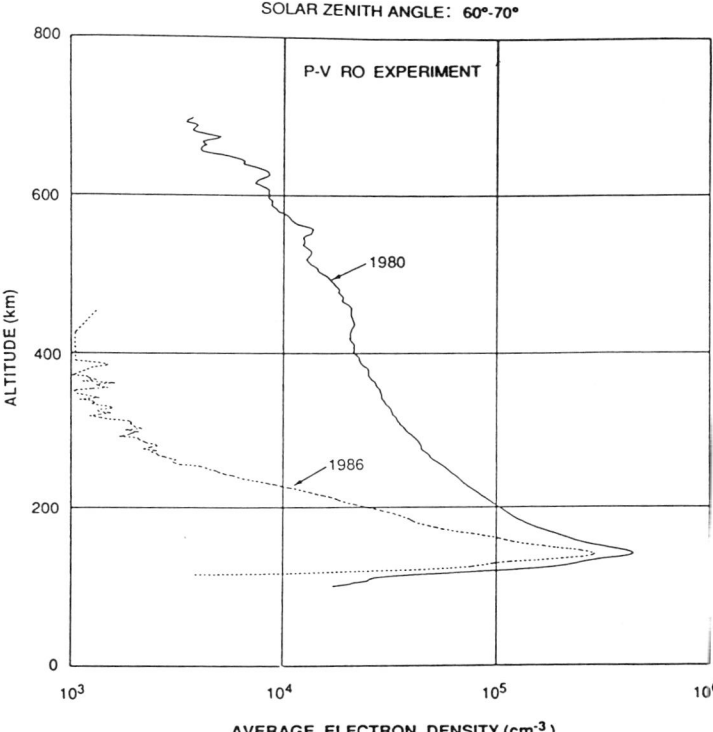

Figure 10. Electron density profiles under solar maximum and solar minimum conditions (figure from Brace and Kliore 1991).

collide with oxygen atoms, many of which attain escape velocity. This "sputtering" of oxygen produces a planet-wide escape flux of $6 \times 10^6 \text{ cm}^{-2}\text{s}^{-1}$. Thus, oxygen may be accompanying hydrogen away from the planet at a rate sufficient to prevent a change in its redox state, precluding the need to dispose of the oxygen by processes such as oxidation of the crust (see the chapter by Donahue et al.).

In our knowledge of the upper atmosphere, the most glaring gap is due to our comparative ignorance of the properties of the atmosphere between the cloud tops and 115 km. As on Earth, this is an ignorosphere. But it is a very important region because it is where the mixed undissociated atmosphere is coupled to the chemically active thermosphere above. It is also the region where the return flow of circulation in the upper atmosphere occurs.

XIII. LIGHTNING

In the terrestrial atmosphere, lightning, the common term for the electrical discharge of large potential differences in clouds, occurs in three quite distinct

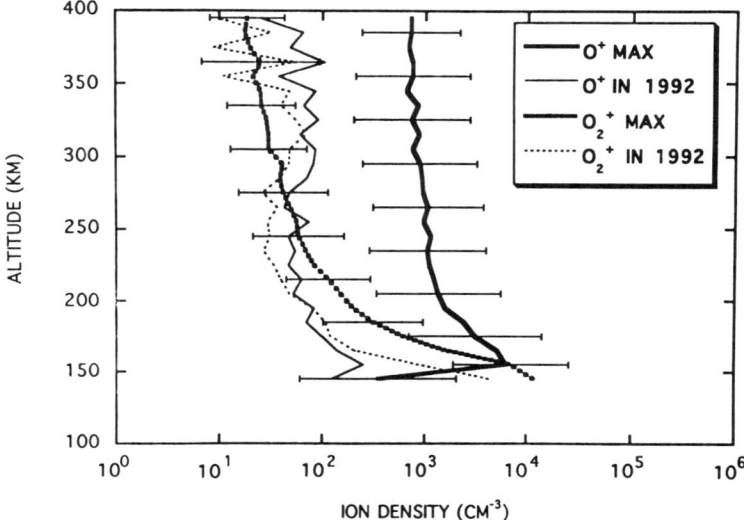

Figure 11. Solar cycle variation of major ions (figure from Kar et al. 1994).

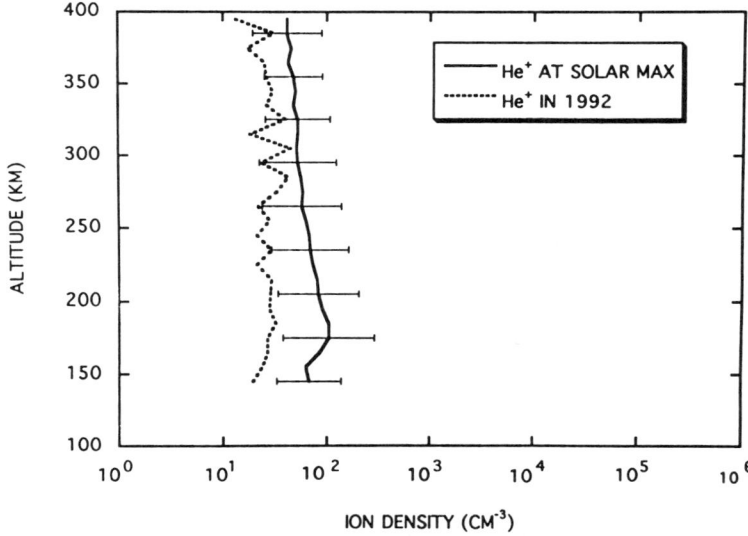

Figure 12. Solar cycle variation of ionized helium (figure from Kar et al. 1994).

ways. The most familiar discharge is from cloud to ground because it is the most readily discernible by the groundbased observer and the most dangerous to humans and machines. On Earth discharges within clouds (cloud-to-cloud) occur twice as often but are less well studied. A third type of discharge, extending upwards toward the ionosphere, has only been recognized recently and is only beginning to be explored. Of the three lightning discharge processes, we expect the latter two to predominate in ridding clouds on Venus of

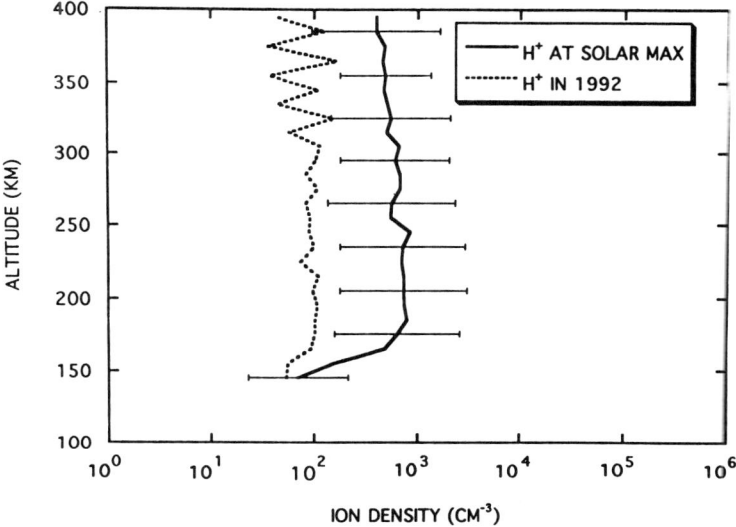

Figure 13. Solar cycle variation of ionized hydrogen (figure from Kar et al. 1994).

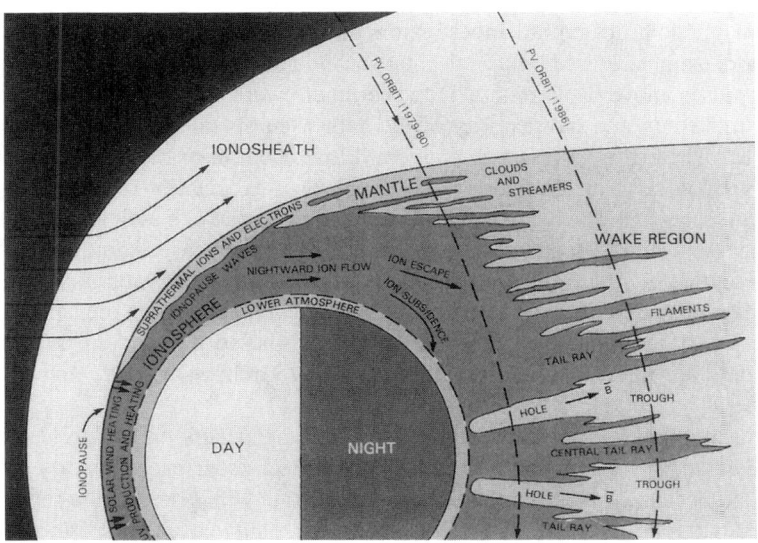

Figure 14. A cartoon representing the complex global configuration of the ionosphere of Venus. Radial scale expanded by a factor of two above the planet's surface (figure from Brace and Kliore 1991).

any build up of electrical potential because of the height of the clouds above the ground and the relative proximity of the ionosphere. The charging of terrestrial clouds is poorly understood even today but is generally believed to result from differential charging of small and large particles and the differential transport of these particles. Terrestrial dust clouds as well as water clouds are found to charge and abruptly discharge sometimes with catastrophic effects (such as in grain elevator explosions). Venus' clouds too have a range of particle sizes and are in a state of constant agitation.

Terrestrial cloud-to-ground lightning produces a column of superheated air that radiates in the visible and expands supersonically to produce the sound waves known as thunder. This column of superheated air also enables chemical reactions to occur that would not take place under standard temperature and pressure conditions. The importance of such processes in a planetary atmosphere depends on the rate of lightning occurrence (on Earth, about $100\ s^{-1}$ worldwide) and is a subject of much debate.

Lightning on Venus has been studied with a variety of techniques and on a variety of missions as illustrated in Fig. 15. The Venera 9 spectrometer apparently detected a lightning storm optically. The Venera 11 through 14 landers detected the impulsive electromagnetic signals in the VLF frequency range associated with terrestrial lightning. Pioneer Venus detected VLF signals in the ionosphere of Venus that could be propagating from the atmosphere below in the expected whistler mode. It also detected VLF signals with similar impulsive characteristics but that could not have reached PVO through electromagnetic propagation. Investigators on PVO further conducted an optical search using scattered light in the star sensor and found no evidence for optical pulses above background. However, there were not many data available for this study and the majority of the data were obtained over the morning sector where the VLF search suggested low occurrence rates for lightning. Nevertheless, some took this evidence, plus their expectation of low lightning occurrence rates due the low mass loading of the clouds, to indicate that lightning was a very infrequent phenomenon. This opinion eased somewhat when the Galileo plasma wave receiver observed electromagnetic radiation at about 1 MHz that could be due to lightning. It eased even more when optical signals similar to lightning were observed from Earth-based telescopes. Interested readers are referred to the review chapter by Grebowsky et al. and to the reviews by Russell (1991,1993).

Although it now seems clear that lightning occurs on Venus, detailed understanding is lacking. In particular, the rate of occurrence is poorly determined because it is difficult to determine the rate with present data sets over the daylit ionosphere. We also do not know the relative occurrence rates of the different possible discharge paths, and in what levels of the clouds lightning occurs. More radio frequency data at frequencies above 1 MHz are needed to measure the dayside rate.

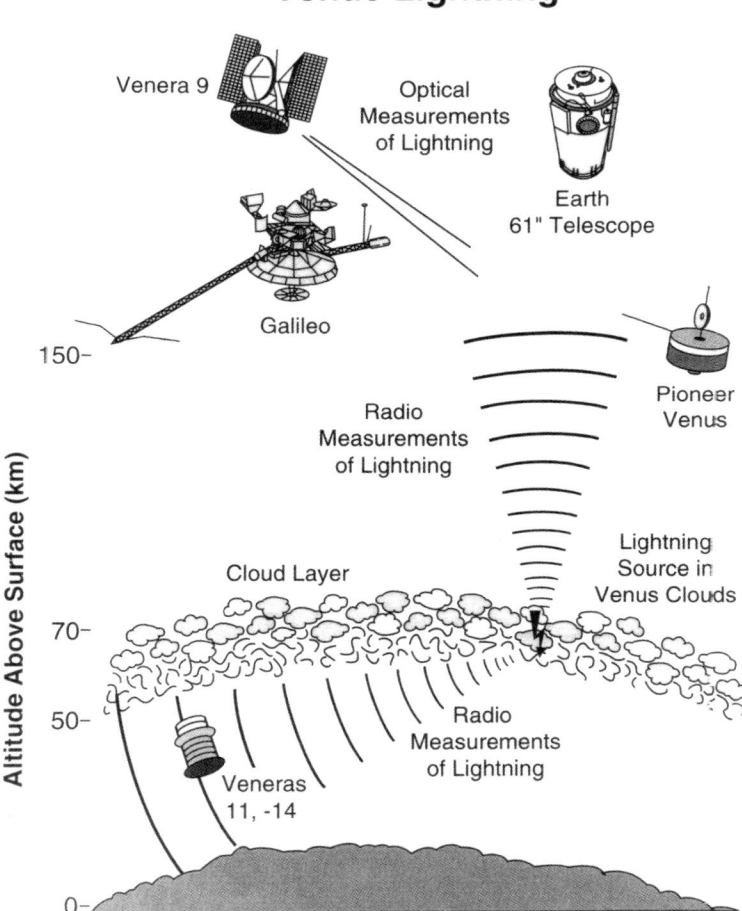

Figure 15. Schematic illustration of the observation of lightning on Venus with flyby, orbiter, and lander spacecraft as well as the 61-inch Earth-based telescope.

XIV. PLANETARY MAGNETIC FIELD

Venus is unique among the planets in that it has no detectable dynamo-driven or remanent magnetic field. (The Martian and lunar dynamos are no longer active, but evidence exists from meteorites and lunar samples that those two bodies have remanent magnetism, presumably induced by ancient dynamos.) Magnetic dynamos occur whenever the interior of a rotating planet has an electrically conducting region that is vigorously stirred by convection. Tiny Mercury has a dynamo believed to operate in a fluid shell on the outside of a relatively large solid inner core. The terrestrial dynamo operates in a fluid core with a small (by volume) inner core whose solidification is thought to provide much of the energy for generation of the magnetic field. Jupiter and Saturn have dynamos in their metallic hydrogen cores and Uranus and Neptune have dynamos thought to be generated in salt water layers at intermediate depths. Successive studies with Mariner 2 and 5, Venera 4, 9 and 10, and Pioneer Venus leave little doubt that the intrinsic field of the planet is far below that of these other planets (Russell et al. 1980; Phillips and Russell 1987). The lack of an intrinsic field at Venus suggests that the solid inner core expected to form in a terrestrial planet is not now condensing. The absence of any seismic data from the various landers that have reached the surface, leaves us almost completely in the dark about the interior of Venus.

The high surface temperatures approach or exceed the temperatures that allow most common planetary minerals to be magnetized. Thus, unlike the surfaces of Mars and the Moon, we do not expect the surface of Venus to be magnetized, and indeed there is no evidence for such magnetization at Venus.

The complete absence of a planetary magnetic field results in a much different interaction of the planet with the solar wind for all planets but Mars. As illustrated in Fig. 16, the magnetized solar wind flow is deflected by the planetary ionosphere (at Venus and also at Mars) at an altitude that permits direct interaction of the solar wind flow and the upper reaches of the planetary exosphere. Hence, a drizzle of atmosphere is continually lost as photoionization, charge exchange and impact ionization expose the exospheric particles to the solar wind electric field and acceleration to velocities far beyond that needed for escape. Sputtering of the atmosphere by these exospheric ions enhances the escape rate. However, the present day rates of escape related to the solar wind interaction are thought to be much less than those needed for the evolution of the atmosphere implied by the isotopic record.

To date no mission has addressed the chemical composition of the escaping ions nor do we have a reliable estimate of the total flux being lost and how this is controlled by the solar wind. This remains for future missions. We also underscore the need for long-term seismic measurements on the surface of Venus to probe the workings of the planetary interior.

XV. IONOSPHERIC MAGNETIZATION

The ionosphere of Venus can be found in two extreme magnetic states. At solar maximum when the solar EUV flux is strong, the ionosphere extends to high altitudes at times of low solar wind dynamic pressure. When such conditions occur the rate of classical diffusion of magnetic flux into the ionosphere is very low and to first order the ionosphere becomes nearly magnetic field-free. At the other extreme when the ionosphere is weak and the solar wind strong, a condition more frequently encountered at solar minimum, magnetic field rapidly diffuses into the ionosphere from the magnetic layer (magnetic barrier) on the inside of the magnetosheath. This magnetic flux is carried to low altitudes by the circulation of the ionosphere and deposited there. (Because

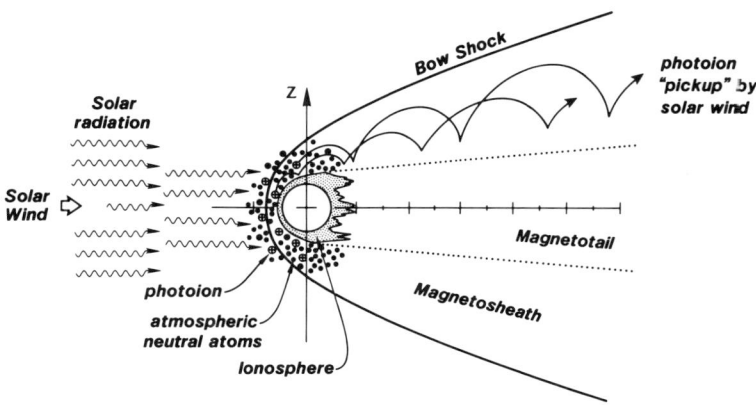

Figure 16. The formation of the ionosphere of Venus and the loss of ions by solar wind pickup (Luhmann 1995). The neutral atmosphere at low altitudes is ionized by the solar ultraviolet and extreme ultraviolet radiation. The neutral atmosphere that extends above the ionopause into the magnetosheath and solar wind is additionally ionized by charge exchange and impact ionization. The electric field of the solar wind accelerates these ions so that they spiral around the interplanetary magnetic field and drift away from the planet along cycloidal trajectories.

ionization occurs throughout the dayside ionosphere but recombination is most rapid at low altitudes, there is a net downward circulation in the subsolar ionosphere.) Ultimately this magnetic flux diffuses through the bottom of the ionosphere into the atmosphere maintaining a weak magnetic field in the ionospheric cavity. The day-to-night circulation of ionospheric plasma also carries this magnetic field into the tail region. The layered magnetic field structure on the day side becomes a more vertically structured magnetic field on the night side. Thus, the magnetic field appears to be draped over the planet on the day side but pulled down the wake in the antisolar direction at night. Because there must be pressure balance across these structures and because the pressure in the magnetic field and plasma are comparable, regions of strong magnetic field have low plasma densities and regions of high plasma density have low magnetic field strengths. Because the former regions are less common than the latter, the former regions are termed "ionospheric holes."

When the ionosphere is strong and the solar wind pressure is relatively weak, classical magnetic diffusion is not sufficient to bring magnetic flux into the ionosphere. Nonetheless, magnetic flux does appear in the ionosphere but in a form quite unlike that described above. Thin tubes of magnetic field, or flux ropes appear. This phenomenon is now realized to be quite prevalent in the magnetized plasmas of the solar system but is found in a novel form at Venus. High in the ionosphere the flux tube consists of fairly straight magnetic field lines and the pressure in the magnetic field is balanced by the pressure in the surrounding plasma. At low altitudes, as shown in Fig. 17, the magnetic field lines become tightly twisted around the axis of the flux tube so that the magnetic pressure forces in the tube balance themselves without the assistance of plasma pressure. Such structures are called force-free magnetic flux tubes. Force-free magnetic flux ropes are also found on the Sun but in a much different plasma regime in which the magnetic pressure forces dominate over the plasma pressure forces everywhere, not just in the rope. For more details on the magnetization of the ionosphere the interested reader is referred to the review by Luhmann and Cravens (1991) and the chapter by Cravens et al.

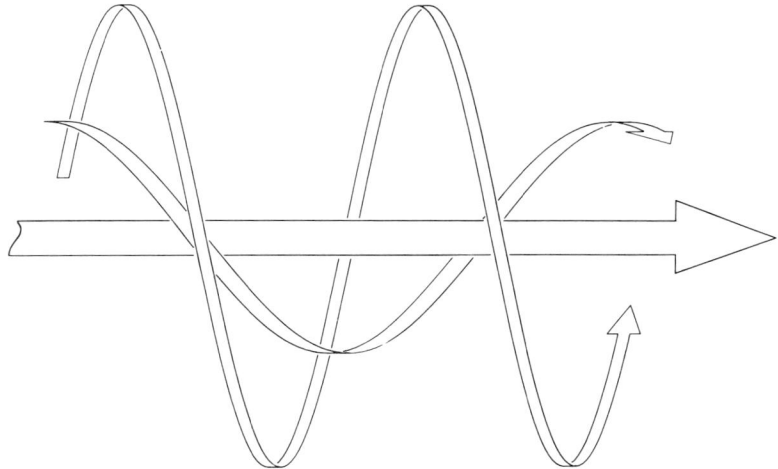

Flux Rope Magnetic Structure

Figure 17. The structure of a magnetic flux rope. The interior field points along the axis and is strong. The field weakens with distance from the center of the rope and twists around the axis with greater pitch. The outward pressure gradient associated with the decreasing magnetic field strength exactly balances the inward force due to the curvature of the magnetic field lines in a force-free magnetic rope (figure from Elphic and Russell 1983).

The study of the ionosphere of Venus has taught us much about the behavior of magnetized plasmas. However, we have gleaned little about how the magnetic flux ropes are formed, in part because of the lack of high temporal resolution plasma measurements in the structures. It is hoped that future missions may be designed to provide a complete set of plasma diagnostics at high temporal resolution to be able to understand more fully this intriguing and important plasma phenomenon.

XVI. SOLAR WIND INTERACTION

In some senses the solar wind interaction with Venus is the inverse of that with the Earth. The terrestrial magnetosphere is a region, in which the magnetic field forces exceed those of the plasma, confined by a flowing plasma, in which the plasma forces generally well exceed the magnetic. However, as shown in Fig. 18, at Venus the obstacle is essentially an unmagnetized plasma and the solar wind forces are transmitted to this plasma by a magnetic layer or barrier that forms between the solar wind and the ionosphere. The formation of the barrier depends on dynamical forces in the plasma that allows the plasma to be deflected along the magnetic field and around the planet. It can also be visualized in kinetic terms. As the plasma flows toward the ionosphere, its speed perpendicular to the magnetic field line decreases, but its thermal velocity remains nearly constant. Because of the three-dimensional geometry, hotter, faster particles can escape along the magnetic field line leaving the colder, slower particles behind. Thus as the field line gradually is pushed against the ionosphere, its plasma density decreases with time leaving particles with low velocities parallel to the magnetic field lines. The resulting magnetic layer only slowly diffuses into the ionosphere as discussed above, keeping the solar wind and ionospheric plasmas apart.

Because the solar wind plasma is almost completely deflected by the planet (at least at solar cycle maximum when the ionosphere is strong) and because the solar wind bulk velocity far exceeds its thermal velocity, the plasma must pass through a shock in order to be slowed, heated and deflected about the planet. This shock stands in the flow some distance in front of Venus like the bow wave of a ship. This distance is such that the plasma that passes through the shock can pass between the shock front and the planetary obstacle. The location of this bow shock varies with the solar cycle, being furthest from the planet at solar maximum and closest at solar minimum. Because the planet's size is constant during the solar cycle and the ionosphere is but a thin layer surrounding the planet, the explanation for this variation must lie in the absorption of the solar wind plasma at solar minimum when the ionosphere is weaker and less able to deflect the solar wind flow.

Behind Venus, as illustrated in Fig. 19, the solar wind wake stretches for at least 12 planetary radii and is largely magnetized as part of the solar wind interaction. The magnetic field is nearly parallel to the solar wind flow in two oppositely directed lobes, called the induced magnetotail. The magnetic structure evolves down the tail in a manner expected for a flow that is being accelerated back up to solar wind flow speeds by the magnetic forces in the tail. The two lobes are separated by a sheet of plasma apparently similar to the plasma sheet in the geomagnetic tail. More details on the solar wind interaction with Venus can be found in the reviews by Russell and Vaisberg (1983) and Luhmann (1986) and in the chapters by Luhmann et al. and by Huba and Strangeway.

The exact origin of the magnetotail and its plasmas still remain somewhat of a mystery because none of the plasma analyzers carried to Venus to date have had the sensitivities and the mass resolution to provide compositional constraints on the origin of the plasmas. It is hoped that future generations

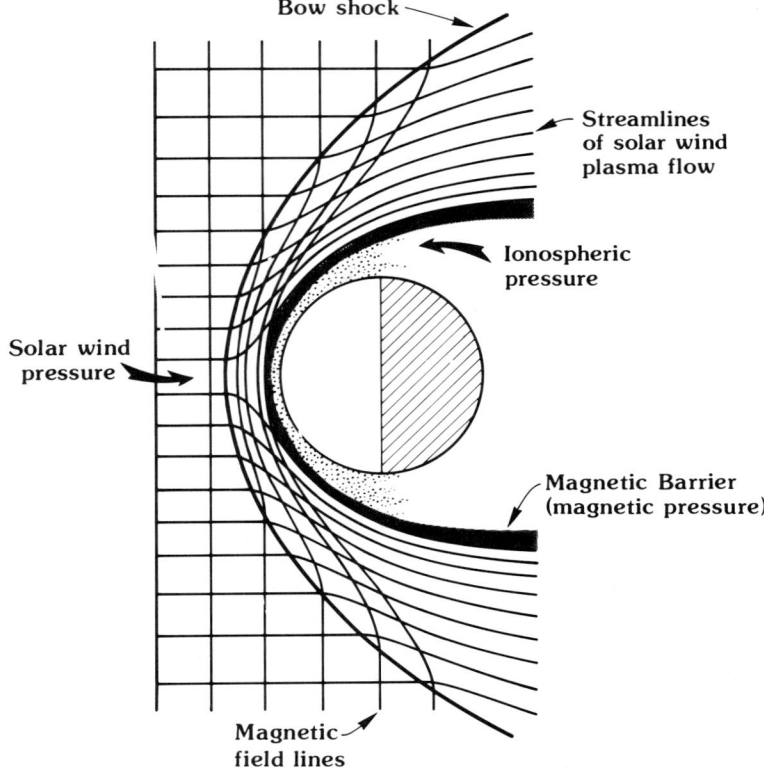

Figure 18. The solar wind interaction with Venus showing the streamlines of the flow, the magnetic field carried with the flow and the bow shock where the flow becomes subsonic and is deflected around the planet. The obstacle to the solar wind is the planetary ionosphere which the magnetized solar wind cannot penetrate. A magnetic layer builds up just above the ionosphere which itself forms part of the barrier to the solar wind flow (figure from Luhmann 1986).

of spacecraft will carry such instrumentation so that we will both unfold the processes that form the magnetotail of Venus and determine how much plasma is removed from Venus by tail acceleration processes.

XVII. CONCLUSIONS

Despite the great advances in our understanding of the atmosphere and plasma environment of Venus, achieved with the wealth of information gathered by the space science missions of the 1970s, serious gaps remain. These have been identified in a few sentences at the end of each section of this chapter. To summarize: Where atmospheric composition is concerned, the most important properties of the atmosphere that need to be more firmly established by future

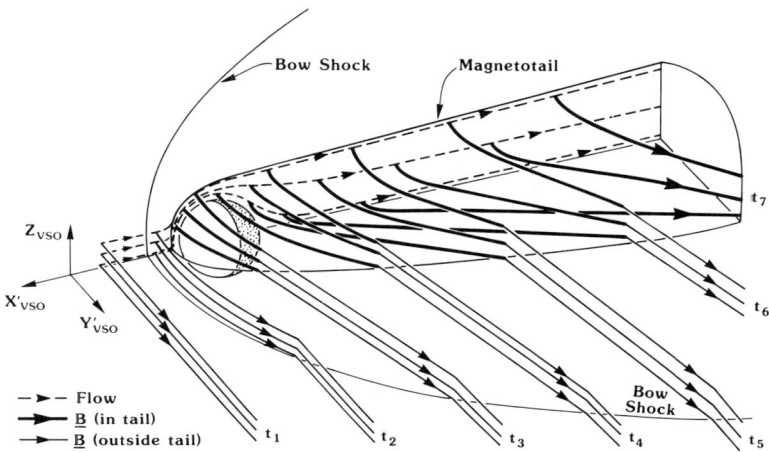

Figure 19. The formation of the magnetotail of Venus. Three magnetic field lines are followed as they flow over and around the planetary obstacle. The field lines closest to the planet move most slowly, become the most highly mass-loaded and are stretched most into a tail-like configuration. Magnetic pressure and curvature forces accelerate the plasma down the tail and the field lines gradually regain their original interplanetary directions (figure from Saunders and Russell 1986).

missions to Venus appear to be the isotopic ratios of krypton and xenon, the helium abundance, the sulfur chemistry and the redox state of the lower atmosphere. These are crucial for understanding the origin and evolution of the atmosphere and its interaction with the surface and interior. The rates at which hydrogen and deuterium are being introduced by outgassing and by cometary impact must be determined for comparison with the well-established escape rates and the deuterium to hydrogen ratio in atmospheric water vapor if the issue of the existence of a large early ocean is to be resolved. Determination of the abundance of CO, O_2 and potential catalysts, such as Cl, in the middle atmosphere is necessary before the chemical processes that so efficiently recombine CO_2 in the atmosphere can be identified.

Understanding the large-scale global atmospheric circulation system, which was a principal objective of the Pioneer Venus Multiprobe mission, has not been attained, partly because of a failure of atmospheric structure instruments on the probes below 10 km. Thus that goal, which is as important today as it was in the 1970s, remains to be attained.

At higher elevations the ubiquitous clouds are no longer impenetrable barriers to our sensors. Probes have cut through the cloud layers, balloon missions have floated in the clouds, radar measurements have covered the planetary surface, infrared sensors have detected spectroscopic emissions from all altitudes and even terrestrial telescopic data now can be obtained below the clouds. Yet the clouds still provide many mysteries, not the least of which is the mechanism for the generation of lightning and its rate of generation. While many sensors have provided evidence for lightning, controversy still abounds over the discharge rate. In particular we need a method to deduce the rate of lightning occurrence over the day side of the planet. This requires either new measurements, or possibly a study of the Magellan radiometer mode data for evidence of lightning discharges.

The ionosphere of Venus also was full of surprises. Prior to the Pioneer Venus mission the electrodynamics of the ionosphere was treated as a resistor, but soon it became evident that the ionosphere was more like a magnetohydrodynamic fluid with mass, momentum, magnetic forces, diffusion and dissipation. Nevertheless this more sophisticated treatment did not predict the formation of small magnetic flux bundles, or flux ropes, that at times permeate the entire dayside ionosphere. To study these structures requires a new generation of plasma sensors with much higher temporal resolution and perhaps even multiple probes.

The study of the solar wind interaction with Venus suffered from two deficiencies of the Pioneer Venus Orbiter mission. First, its ion spectrometer lacked mass resolution and, second, the orbit was not optimum for the study of the interaction. Thus, there are literally holes in our knowledge of the environment of Venus, and we do not know the present loss rates of the various species of ions. It is hoped that future missions to Venus can carry a more complete set of plasma diagnostics with high temporal resolution.

Early Earth was almost certainly much more similar to early Venus than to early Mars. To understand in what important ways these pristine planets differed and why Venus has evolved into a planet so dramatically different from Earth we must return to Venus.

Acknowledgment. While many missions from both the former Soviet Union and the United States have contributed to the study of Venus, the authors are particularly indebted to the many individuals who worked so hard on the Pioneer Venus mission. They made it a success and allowed scientists to reap the rewards of their efforts.

REFERENCES

Bougher, S. W., Hunten, D. M., and Roble, R. G. 1994. CO_2 cooling in terrestrial planet thermospheres. *J. Geophys. Res.* 99:14609–14622.

Brace, L. H., and Kliore, A. J. 1991. The structure of the Venus ionosphere. *Space Sci. Rev.* 55:81–163.

Donahue, T. M., and Pollack, J. B. 1983. Origin and evolution of the atmosphere of Venus. In *Venus*, eds. D. M. Hunten, L. Colin, T. M. Donahue and V. I. Moroz (Tucson: Univ. of Arizona Press), pp. 1003–1036.

Elphic, R. C., and Russell, C. T. 1983. Magnetic flux ropes in the Venus ionosphere: Observations and models. *J. Geophys. Res.* 88:58–72.

Fox, J. L., and Bougher, S. W. 1991. Structure, luminosity, and dynamics of the Venus thermosphere. *Space Sci. Rev.* 55:357–489.

Goldstein, J. 1989. Absolute Wind Speed Measurements in the Lower Thermosphere of Venus Using Infrared Heterodyne Spectroscopy. Ph.D. Thesis, Univ. of Pennsylvania, Philadelphia.

Jakosky, B. M., Pepin, R. O., Johnson, R. E., and Fox, J. L. 1995. Mars atmospheric loss and isotopic fractionation by solar-wind induced sputtering and photochemical escape. *Icarus* 111:271–288.

Kar, J., et al. 1994. Evidence of electron impact ionization on the nightside of Venus from PVO/IMS measurements near solar minimum. *J. Geophys. Res.* 99:11351–11355.

Keating, G. M., et al. 1985. Models of Venus neutral upper atmosphere: Structure and composition. In *The Venus International Reference Atmosphere*, eds. A. J. Kliore, V. I. Moroz and G. M. Keating (New York: Pergamon Press), pp. 117–171.

Krasnopolsky, V. A., et al. 1994. First measurement of helium on Mars: Implication for the problem of radiogenic gases on the terrestrial planets. *Icarus* 109:337–351.

Luhmann, J. G. 1986. The solar wind interaction with Venus. *Space Sci. Rev.* 44:241–306.

Luhmann, J. G. 1995. Plasma interactions with unmagnetized bodies. In *Introduction to Space Physics*, eds. M. G. Kivelson and C. T. Russell (Cambridge: Cambridge Univ. Press), pp. 203–226.

Luhmann, J. G., and Cravens, T. E. 1991. Magnetic fields in the ionosphere of Venus. *Space Sci. Rev.* 55:201–274.

Pepin, R. O. 1991. On the origin and early evolution of terrestrial planet atmospheres and meteoric volatiles. *Icarus* 92:2–79.

Pepin, R. O. 1995. Evolution of the Martian atmosphere. *Icarus* 111:289–304.

Phillips, J. L., and Russell, C. T. 1987. Upper limit on the intrinsic magnetic field of Venus. *J. Geophys. Res.* 92:2253–2263.

Pollack, J. B., Toon, O. B., and Boese, R. 1980. Greenhouse models of Venus' high surface temperature, as constrained by Pioneer Venus measurements. *J. Geophys. Res.* 85:8223–8231.

Russell, C. T. 1991. Venus lightning. *Space Sci. Rev.* 55:317–356.

Russell, C. T. 1993. Planetary lightning. *Ann. Rev. Earth Planet. Sci.* 21:43–87.

Russell, C. T., and Vaisberg, O. 1983. The interaction of the solar wind with Venus. In *Venus*, eds. D. M. Hunten, L. Colin, T. M. Donahue and V. I. Moroz (Tucson: Univ. of Arizona Press), pp. 873–940.

Russell, C. T., Elphic, R. C., Luhmann, J. G., and Slavin, J. A. 1980. On the search for an intrinsic magnetic field at Venus. *Proc. Lunar Planet. Sci. Conf.* 11:1897–1906.

Saunders, M. A., and Russell, C. T. 1986. Average dimension and magnetic structure of the distant Venus magnetotail. *J. Geophys. Res.* 91:5589–5604.

Schubert, G., et al. 1980. Structure and Circulation of the Venus atmosphere. *J. Geophys. Res.* 85:8007–8025.

GLOBAL MODELS OF THE SOLAR WIND INTERACTION WITH VENUS

J. G. LUHMANN
University of California at Berkeley

S. H. BRECHT
Berkeley Research Associates

J. R. SPREITER
Stanford University

S. S. STAHARA
RMA Associates

R. S. STEINOLFSON
Southwest Research Institute

and

A. F. NAGY
University of Michigan

In-depth observational analyses of the Venus-solar wind interaction over the last decade have both stimulated and responded to the synergistic development of global theoretical models. The latest global models include mass-loaded gas dynamic magnetosheath models. MHD models of the solar wind interaction with conducting spheres representing the planetary ionosphere, hybrid models in which both solar wind and planetary ion kinetics are included, and test particle models wherein ions are introduced into prescribed global field models. Each gives somewhat different but complementary insights into the physical consequences of the solar wind interaction with a planetary atmosphere. In particular, the gas dynamic magnetosheath and MHD models allow us to interpret the interplanetary magnetic field and solar wind flow disturbances created by the Venus obstacle. They also give us background fields to be used in test particle investigations of the characteristics of the planetary pick-up ion population. The global hybrid simulations uniquely combine self-consistent aspects of both, although they are constrained to lower spatial resolutions by current computational constraints. Several of these models are capable of evolving to include more realistic treatment of the "soft" solar wind-ionosphere interface, thus providing the possibility of a combined or coupled model of both the magnetotail/magnetosheath region and the ionopause/ionosphere. These models, and their future further-developed counterparts, will be of great value in the comparative study of the Venus and Mars solar wind interactions expected in the coming era of renewed Mars exploration.

I. INTRODUCTION

A. Background

Modeling the solar wind interaction with Venus presents a special challenge because it involves aspects of both magnetospheric and cometary solar wind interactions. As mentioned in the introductory chapter by Donahue and Russell, the solar wind induces currents in the planetary ionosphere that make it behave, in some respects, like an "impenetrable" closed magnetospheric obstacle. At the same time, heavy particles that form the planetary atmosphere can be ionized both in the surrounding solar wind and within the magnetosheath created by the obstacle. These heavy ions "mass-load" the plasma in these regions, thereby altering the flow and field properties. Moreover, the scale sizes of some of the pick-up ion gyroradii add kinetic influences to the solar wind interaction.

In this chapter we consider the various approaches that have been applied to the Venus-solar wind interaction to model its global characteristics. Many of these models were developed as either predictive tools or in response to measurements obtained by spacecraft. To maximize their realism, and in particular, to simulate the three-dimensional features of the solar wind interaction, they are necessarily numerical solutions. However, some have been adapted especially for comparisons with observations, with inputs and outputs designed to allow studies of how the interaction would appear along a spacecraft trajectory, or to accommodate changes for different interplanetary conditions. Most of the models are based on fluid approximations, wherein the solar wind (including in some cases the locally produced planetary ions) is treated by solving standard sets of fluid equations using finite difference methods on a spatial grid. However, a few treatments numerically solve the equations of motion for individual ions as a means of either determining the behavior of a minor (e.g., planetary) species, or of introducing ion kinetic effects.

In all cases it is assumed that an inner boundary exists at a highly conducting obstacle of either blunt body or spherical shape where the flow velocity and magnetic field become tangential to the surface. This presupposes that an effectively solid barrier is present to deflect the solar wind flow whether or not planetary ion pickup is significant in the interaction. Thus Venus models are unlike most purely cometary models which assume that only sources of heavy ions are present in the ambient flow. In some senses, the Venus models with planetary ion production can be considered as equivalent to cometary models with conducting obstacles at their centers. Of course the Venus ionosphere does not always behave so ideally in providing a sharp boundary between what can be considered solar wind (or magnetosheath) and the ionosphere proper. Here we address that lack of clear separation to only a minor extent in the discussion of ion kinetics. The details of the ionopause boundary behavior as well as the consequences of that boundary's sometimes diffusive nature are described in the chapter by Cravens et al.

As of this writing, most work on the global nature of the Venus–solar wind interaction involving models has made use of the gas dynamic model of Spreiter and Stahara (1980a,b) because of its relative simplicity, adaptability, and availability. For this reason, it is emphasized here. The descriptions of the recent, more physically comprehensive models are limited because they are generally still in the development and testing stage and have not yet been extensively probed or compared with observations.

B. Solar Wind at the Orbit of Venus

The solar wind flow parameters and planet size ultimately determine the appropriateness of the various simulation approaches. At the heliocentric distance of Venus, Luhmann et al. (1993) find the following median properties:

TABLE I
Median Properties at the Heliocentric Distance of Venus[a]

	Solar Maximum	Solar Minimum
Density	~ 18 cm^{-3}	~ 22 cm^{-3}
Velocity	~ 400 km s^{-1}	~ 410 km s^{-1}
Ion temperature	$\sim 1 \times 10^5$ K	$\sim 1 \times 10^5$ K
Magnetic field	~ 12 nT	~ 10 nT
Parker Spiral angle	~ 38 deg	~ 35 deg
Dynamic pressure	~ 5 nPa	~ 6 nPa
Sonic Mach	~ 6.8	~ 7.0
Alfvén Mach	~ 7.0	~ 9.0
Magnetosonic Mach	~ 4.5	~ 5.5
Beta[b]	~ 1.5	

[a] Based on PVO measurements.
[b] Requires estimates of electron temperature.

In general, the solar cycle changes are modest. Because of conversion of the much greater flow energy to thermal energy at the bow shock, the thermal solar wind proton gyroradius of ~ 15 km near Venus can increase to 100s of km in the post-shock magnetosheath. However, it is again reduced (by $\sim 50\%$) closer to the planet due to the compression of the field. Given the planetary radius of ~ 6050 km, this finite gyroradius motion causes only a small perturbation on the proton flow pattern around Venus. On the other hand, exospheric oxygen ions that are produced in the proton flow and accelerated by the solar wind convection electric field to solar wind speeds can have gyroradii that are a substantial fraction of a Venus radius.

II. GLOBAL MODELS

A. Gas Dynamic Models

1. Basic Model. The earliest and most extensively used global models of the Venus-solar wind interaction by Spreiter and coworkers (see, e.g.,

Spreiter et al. 1966,1970; Spreiter and Alksne 1969; Spreiter and Rizzi 1974; Spreiter and Stahara 1980a,b) considered the numerical solution, by finite difference methods, of the problem of hypersonic fluid or gas flow around an impenetrable blunt body. For this physical picture, the interaction is characterized by the formation of a bow shock that stands upstream of the obstacle where it serves to slow the flow to subsonic speeds prior to its divergence around the obstacle. Although these "magnetosheath" models do not address the physics of the wake region, they provide a means of analyzing the behavior of the dayside interaction in regions removed from the obstacle shadow. The equations that are solved:

continuity
$$\frac{\partial \rho}{\partial t} + \frac{\partial}{\partial x_k}(\rho v_k) = 0 \tag{1}$$

momentum
$$\rho \frac{\partial v_k}{\partial t} + \rho(\bar{v} \cdot \nabla)v_k + \frac{\partial p}{\partial x_k} = 0 \tag{2}$$

and energy
$$\frac{\partial \epsilon}{\partial t} + \frac{\partial}{\partial x_k}[(\epsilon + p)v_k] = 0 \tag{3}$$

where t is time, ρ is mass density, x_k a spatial coordinate, v_k the fluid velocity component along x_k, ϵ the total (internal plus kinetic) energy, and p the pressure given by the ideal gas law $p = \rho RT/m$ (related to the other variables by $p = (\gamma - 1)[\epsilon - \rho v^2/2]$ with γ the ratio of specific heats) are standards in fluid or gas dynamic theory (see, e.g., Spreiter et al. 1966). These have generally been solved for the Venus interaction case in conservative form using the cylindrical symmetry approximation with respect to the Venus–Sun line. The blunt-body shape of the obstacle is parameterized according to a value H/R_0, the atmosphere scale height (H) in obstacle nose radii (R_0). Shapes are obtained from the assumption of "Newtonian" pressure balance between the external solar wind ram pressure and an internal isothermal, spherically symmetric ionosphere with isotropic pressure obeying a hydrostatic law.

As mentioned above, the internal boundary condition of the numerical solution of these equations forces the normal component of the velocity to be zero on the blunt body. This is achieved by making the blunt body the inner boundary of the grid for the finite difference solutions. For the nose region portion of the solution, the outer boundary of the grid is located at an initial estimated position of the bow shock determined from an approximate formula. The appropriate jump conditions are applied at the bow shock discontinuity whose location is iteratively adjusted during the solution procedure until it converges to the actual shape. A sample grid is shown in Fig. 1. The spatial resolution in the inner cells is on the order of 0.01 R_0. An algorithm due to Beam and Warming (1978) allows the implicit solution of the unsteady Eqs. (1–3) in conservation form until a steady condition is achieved in the nose

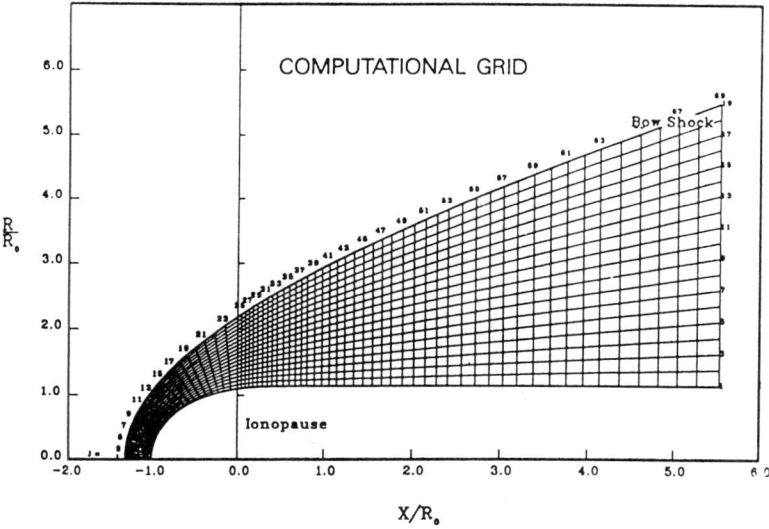

Figure 1. Computational grid used in Spreiter/Stahara gas dynamic models of the Venus magnetosheath (figure from Spreiter and Stahara 1992).

region. Downstream of this region, an explicit predictor-corrector scheme due to MacCormack (1969) is employed.

The only input parameters affecting the solution besides the obstacle shape are a (nominally sonic) Mach number and the ratio of specific heats γ. The resultant cylindrically symmetric, gas dynamic flow field variables (ρ, v, T), examples of which are given in Fig. 2, are normalized by the upstream values. As a consequence, they can be applied to a variety of upstream parameters following the initial solution as long as the assumed Mach number is taken into consideration. An interplanetary magnetic field of arbitrary strength and orientation can then be specified for the separate convected-field part of the gas dynamic model calculation.

The magnetic field $\bar{B}$ is assumed to be steady, frozen-in, and divergence-free, implying satisfaction of the two equations

$$\nabla \times (\bar{v} \times \bar{B}) = 0 \qquad (4)$$

$$\nabla \cdot \bar{B} = 0 \qquad (5)$$

but these are not solved using the usual finite difference methods. Instead, they are handled in a more computationally efficient way by making use of the known properties of the flow field and the magnetic field effectively to decompose the problem into the separate convection of the parallel (to the flow) and perpendicular contributions. The normalized parallel contribution

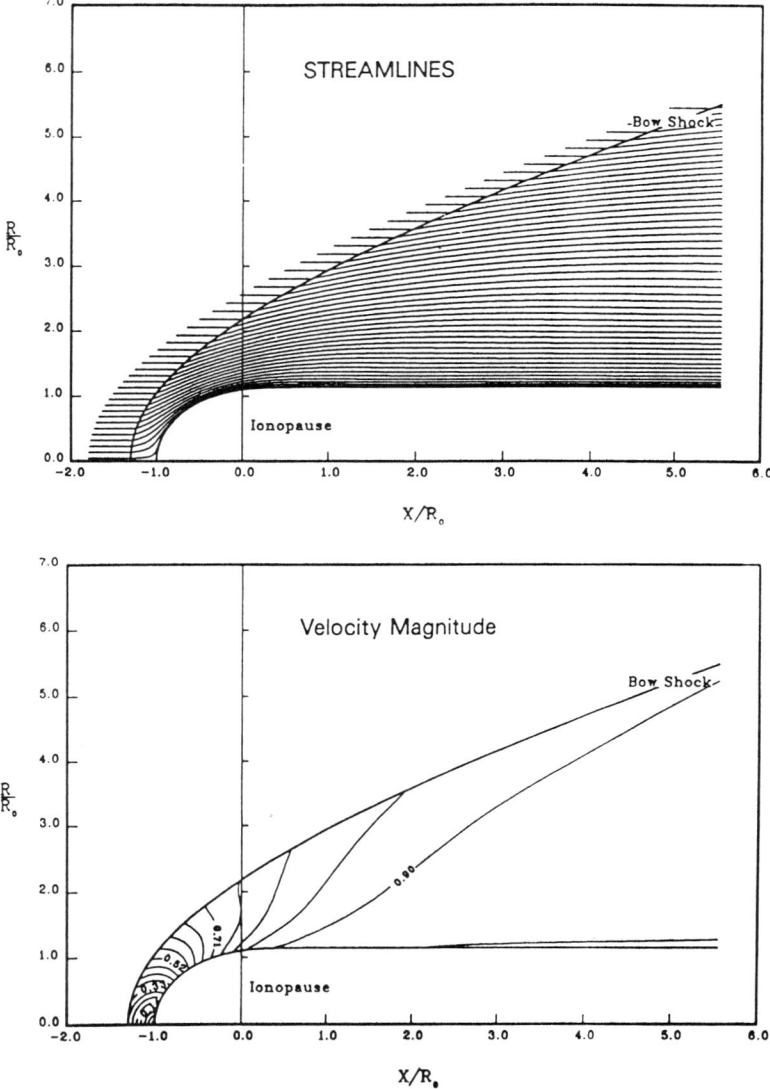

Figure 2. Samples of results from the (cylindrically symmetric) gas dynamic model (figure from Spreiter and Stahara 1992).

at any point follows the flow streamlines and has a magnitude given by

$$\left[\frac{B_{\text{local}}}{B_{\text{upstream}}}\right]_{\text{par}} = \frac{(\rho v)_{\text{local}}}{(\rho v)_{\text{upstream}}} \tag{6}$$

Similarly, the normalized contribution in the direction of the perpendicular interplanetary field component is given by

$$\left[\frac{B_{\text{local}}}{B_{\text{upstream}}}\right]_{\text{perp}} = \frac{(\rho \nabla l)_{\text{local}}}{(\rho \nabla l)_{\text{upstream}}} \qquad (7)$$

where ∇l is a "stretching factor" related to the change in the cross-stream tube distance between points originally aligned along the perpendicular direction upstream. The final orthogonal contribution (perpendicular to the plane of the upstream field) is

$$\left[\frac{B_{\text{local}}}{B_{\text{upstream}}}\right]_{\hat{n}} = \frac{(R\rho)_{\text{local}}}{(R\rho)_{\text{upstream}}}. \qquad (8)$$

Here R is the radial cylindrical coordinate of the streamline through the point of interest and $\hat{n}$ is the unit vector normal to the plane containing the upstream field. The local orientation of the parallel field component in the magnetosheath is the local streamline orientation, while the local orientation of the perpendicular component lies parallel to a vector determined from analysis of the distortion of a small element of stream tube as it propagates through the magnetosheath. These contributions add vectorially to give the total field.

Examples of the field solution are reproduced in Fig. 3. Spreiter and Stahara (1980b) built a diagnostic routine into their code to "fly through" the gas dynamic model along a spacecraft orbit. This allows direct comparison of the model gas or plasma and field parameters with the Venus magnetosheath (sometimes called ionosheath) observations. Examples of magnetic field comparisons obtained by Luhmann et al. (1986) using Pioneer Venus Orbiter (PVO) magnetic field data are shown in Fig. 4. Russell et al. (1988) found that the best fit of the gas dynamic model bow shock to the observations could be achieved by using $\gamma = 5/3$, together with the magnetosonic (rather than the sonic) Mach number. Similarly, results of statistical comparisons with magnetic field data in the magnetosheath by T. L. Zhang et al. (1993) demonstrated that an obstacle size larger than the ionopause was required to match the bow shock and outer magnetosheath measurements. Both of these studies indicate that neglect of the magnetic field influence on the plasma inherent in the gas dynamic convected field model has significant consequences. The importance of the magnetic field influence could also have been concluded from the observation of the "magnetic barrier" (chapter by Donahue and Russell) in the inner magnetosheath wherein magnetic field pressure dominates. However, these problems can be approximately corrected for by manipulation of the obstacle size and Mach number. Even time variations in interplanetary conditions can be accommodated to a satisfying degree in the gas dynamic model by employing sequential steady state models (see, e.g., Spreiter and Stahara 1992). The one observation that motivated a real modification of the rather successful gas dynamic models was that the bow shock position varied with the solar cycle.

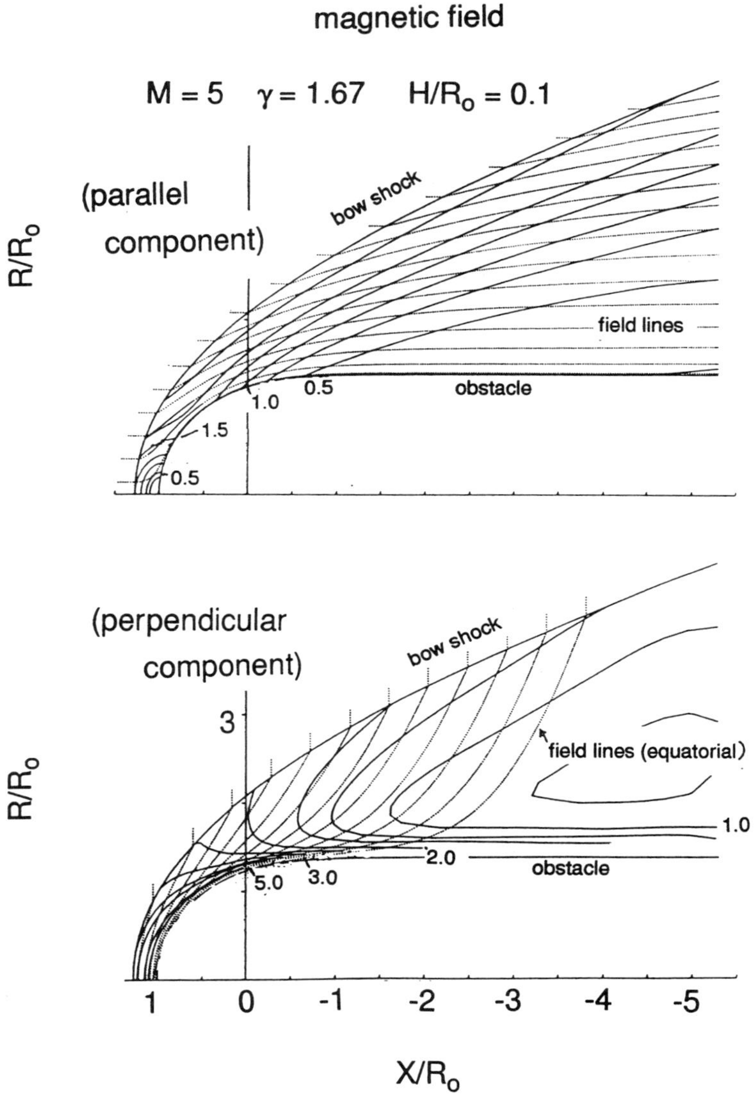

Figure 3. Magnetic field calculated by convecting a "frozen-in" interplanetary field through the gas dynamic model flow field in Fig. 2. The contours and field lines are for the equatorial plane.

2. Mass-Loaded Model. Russell et al. (1988) and Zhang et al. (1990) showed that the average terminator radius of the bow shock was $\sim$2.4 Venus radii (R_v) for solar maximum conditions, but only $\sim$2.1 R_v for solar minimum conditions. Average solar wind parameters do not change sufficiently during

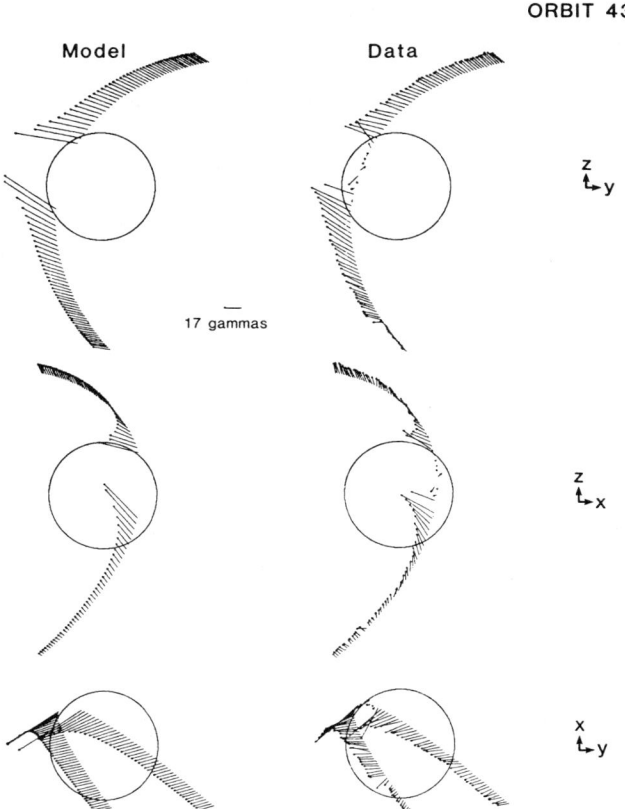

Figure 4. Comparison of magnetic field vectors computed from the gas dynamic model using the observed interplanetary magnetic field, with magnetometer measurements obtained along the orbit of the PVO. The views are (top) from the Sun, (center) in the noon-midnight meridian, and (bottom) as projected onto the plane of Venus' orbit (equivalent to the ecliptic) (figure from Luhmann et al. 1986).

the solar cycle to produce this variation, and so it was concluded that the shock movement must be related to a change in either the obstacle or in the rate of exospheric ion production in the magnetosheath and its effects. As mentioned in the overview chapter by Donahue and Russell, because the neutral atmosphere of Venus extends into the magnetosheath, substantial numbers of heavy (primarily atomic oxygen) ions are produced there by photoionization, impact ionization, or charge exchange (see, e.g., M. H. G. Zhang et al. 1993). The amount of photoionization increases at solar maximum as the solar EUV flux increases to ~ 2 to 4 times its solar minimum value. The EUV flux increase also enhances the upper atmosphere densities, and in particular affects the oxygen exosphere which depends on the ionospheric production rate for its source. To study a possible cause of this solar cycle variation of the bow shock position, two gas dynamic models which include a comet-like source term, describing ion production appropriate for the Venus oxygen exosphere, were developed independently by Belotserkovskii et al. (1987) and Spreiter and Stahara (1992). The object of this modification was to determine whether the anticipated production of heavy planetary ions in the magnetosheath, or "mass-loading" would inflate the bow shock as observed. For simplicity, this source is assumed due to photoionization only and is treated as part of the single fluid, with its mass density production rate reflecting the fact that the newly produced ions are heavier than the solar wind hydrogen.

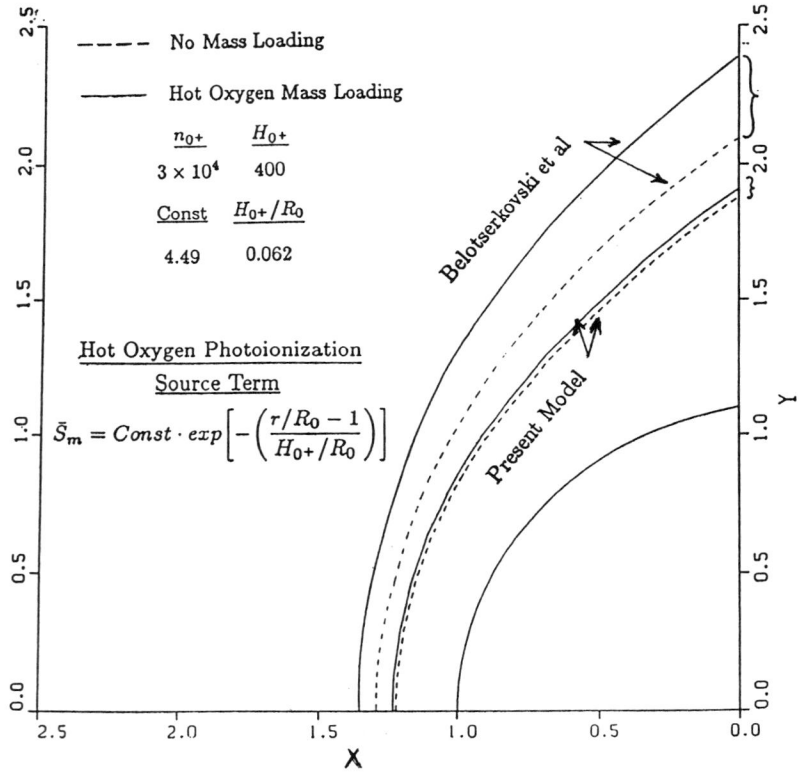

Figure 5. Comparison of gas dynamic model bow shocks with and without mass loading by oxygen exosphere ionization, as obtained by Belotserkovskii et al. (1987) and Spreiter and Stahara (1992). The form of the mass loading term is shown at left.

The atmospheric ion source term is added to the continuity equation (right-hand side of Eq. 1) only because the newly created photo-ions are presumed not to possess appreciable energy or momentum at the time of their creation (Spreiter and Stahara 1992; Belotserkovskii et al. 1987; see also

Breus et al. 1987). The form chosen:

$$S = S_0 \exp\left[-\frac{(r - R_0)}{H_0}\right] \quad (9)$$

assumes the atmospheric density profile is a simple exponential. Here, S_0 is a constant proportional to the number density of neutral oxygen atoms at the obstacle nose position $R_0 (\sim 3 \times 10^4$ cm^{-3} times their mass) divided by the characteristic time for photo-ionization of atomic oxygen ($\sim 1 \times 10^6$ s at Venus at solar maximum), r is the radial distance from the center of the planet, and H_0 is the scale height of the oxygen exosphere (~ 400 km). As illustrated by Fig. 5, the results from the two models developed were strikingly different. Belotserkovskii et al. (1987) and Breus et al. (1987,1989) concluded that the expected rate of oxygen ion production could inflate the Venus bow shock to its observed solar maximum position, while Spreiter and Stahara (1992) found the same rate insufficient to produce any movement approaching the observed inflation. Moore et al. (1991a) modified Spreiter and Stahara's treatment by introducing an ionospheric obstacle shape that converged on the night side. This approach allowed the simulation of a rudimentary induced "magnetotail." Based on both the amount of magnetic field draping in the modeled magnetotail and the bow shock position, Moore et al. (1991a) came to the same conclusion regarding the inability of these simple mass-loaded models to reproduce the shock position. Some possible origins of the contradictory results for similar approaches (albeit using different numerical methods to solve the gas dynamic equations) were discussed by Spreiter and Stahara (1992). We here adopt their conclusion that photo-ionization of the exosphere was insufficient to move the gas dynamic bow shock by the amount observed on PVO.

The apparent need for a stronger source term in the gas dynamic model can be interpreted in several ways. Provided that both the exospheric model (which has been verified with PVO ultraviolet spectrometer observations) and the photo-ionization rate are fairly accurate, possible additions to the apparent source could arise from: a thermospheric contribution with a smaller scale height at the lowest altitudes, or additional ionization mechanisms such as solar wind electron impact and charge exchange with solar wind protons (see, e.g., M. H. G. Zhang et al. 1993). Errors in the effective obstacle size or shape due to the neglected magnetic barrier, and magnetic field effects on the flow properties could also contribute to the inadequacy of the mass-loaded gas dynamic model predictions. The assumed "impenetrability" of the obstacle at solar minimum may also be violated. Finally, effects of the finite ion gyroradius, or the possibly inappropriate assumption that planetary photo-ions are adequately treated within the single fluid approximation, need to be considered.

The effects of the thermospheric contribution to the mass-loading, though not tested, were considered unlikely to provide the solution because these effects should be confined to regions near the inner magnetosheath boundary. On the other hand, additional ionization mechanisms can enhance the mass-loading effects at higher altitudes. M. H. G. Zhang et al. (1993) suggest that, even at solar maximum, the contribution of solar wind electron impact may be twice as important as photo-ionization. They further show that its distribution is such as to raise substantially the oxygen ion production rate in the subsolar middle magnetosheath. Because the mass-loading rates in the gas dynamic models have not yet been altered to reflect this additional source, the extent to which it may solve the bow shock location problem remains to be determined. However, more recent numerical simulations allow some assessment of the magnetic field effects, as well as of the influence of ion gyroradius-scale physics.

B. Magnetohydrodynamic Models

In general, numerical magnetohydrodynamic (MHD) models involve finite-difference solutions of the same set of fluid equations as the gas dynamic models, but with the addition of magnetic force terms:

continuity
$$\frac{\partial \rho}{\partial t} + \nabla \cdot (\rho v) = 0 \tag{10}$$

momentum
$$\rho \frac{\partial \bar{v}}{\partial t} + \rho (\bar{v} \cdot \nabla) \bar{v} + \nabla p = \frac{1}{\mu} (\nabla \times \bar{B}) \times \bar{B} \tag{11}$$

energy
$$\frac{\partial \epsilon}{\partial t} + \nabla \cdot ((\epsilon + p) \bar{v}) = \frac{-1}{\mu} (\nabla \times \bar{B}) \cdot (\bar{v} \times \bar{B}). \tag{12}$$

In addition, they include self-consistent solution of a subset of the Maxwell equations in the forms:

Ampere's Law
$$\mu \bar{J} = \nabla \times \bar{B} \tag{13}$$

Faraday's Law
$$\frac{\partial \bar{B}}{\partial t} + \nabla \times \bar{E} = 0 \tag{14}$$

Induction Equation (frozen-field)
$$\bar{E} + \bar{v} \times \bar{B} = 0 \tag{15}$$

where $\bar{E}$ is the electric field and $\bar{J}$ is the current density. (The induction equation is derivable from the generalized Ohm's Law.) The divergence-free field condition, described by $\nabla \cdot \bar{B} = 0$, is also enforced in these models by various means.

Although there have been several MHD modeling efforts specifically for magnetosheaths, in the spirit of the gas dynamic models described above (see, e.g., Wu 1992), we here look beyond these to consider others designed especially for the nonmagnetic planet-flow interactions including the wake region. Tanaka (1993), Cable and Steinolfson (1995), and DeZeeuw et al. (1996) independently solved the problem of hypermagnetosonic flow around a conducting sphere as a first approximation to the nonmagnetic planet interaction. Their computational methods differed significantly. DeZeeuw et al. used an adaptive grid, while Tanaka, and Cable and Steinolfson used a spherical grid with radial spacing increasing with distance from the planet. Moreover, Tanaka and DeZeeuw et al. used monotone upwind scalar-conservation law (MUSCL) algorithms (see Gombosi et al. [1994] for a further description and references), while Cable and Steinolfson used a form of Lax-Wendroff operator with a smoothing term to inhibit numerical instabilities. Their maximum spatial resolution also differs. Near the conducting sphere (of radius R_v), Tanaka worked with a grid spacing of .1 R_v, while Cable and Steinolfson imposed 0.017 R_v spacing. The near-planet resolution in the DeZeeuw et al. model is 0.063 R_v. The physical parameters of their external (solar wind) flows are, respectively, (magnetosonic) Mach numbers of $\sim$10, 3, 4 and plasma betas (thermal pressure/magnetic pressure) of 1.5, 0.6, 1.8 (or 6.0) for Cable and Steinolfson, DeZeeuw et al., and Tanaka. The interplanetary magnetic field orientation in the Tanaka model is perpendicular to the upstream velocity, in the Cable and Steinolfson model three different interplanetary field "cone" angles of 0°, 45°, and 90° are adopted, and in the DeZeeuw et al. model the upstream field is flow aligned. Because the results of each model are displayed somewhat differently, they are difficult to compare in any detail. However, some major features can be noted.

All of the models show upstream fast mode bow shocks with shock jumps in agreement with the Rankine Hugoniot relations. The terminator-to-nose radius ratios for the bow shocks are smaller by $\sim$0.1 than the observed value of $\sim$1.67 (Zhang et al. 1990) for the Tanaka (1993) and Cable and Steinolfson (1995) models, but larger for the DeZeeuw et al. (1996) model. The first two authors note that neglecting the the thickness of the ionosphere as part of the obstacle can cause their discrepancies. Cable and Steinolfson (1995) find that the quasi-parallel bow shocks are closer to the sphere than the quasi-perpendicular bow shocks. Pole–equator differences that have been observed in the bow shock radius in association with larger field cone angles (Russell et al. 1988) are found in the model results of both Tanaka (1993) and Cable and Steinolfson (1995). However, the degree of elongation in the polar directions (apparently produced by fast-mode wave speed differences in the polar and equatorial shock field and flow geometries) seems weaker than that derived from the measurements ($\sim$5% vs $\sim$10%).

All of the models also show a density depletion in the low altitude nightside wake. In the cases where a perpendicular component of the interplanetary magnetic field is present, a magnetic barrier forms against the dayside obstacle, and an induced magnetotail connected to this barrier appears in the wake to the lowest altitudes modeled. Figure 6 shows the global field line config-

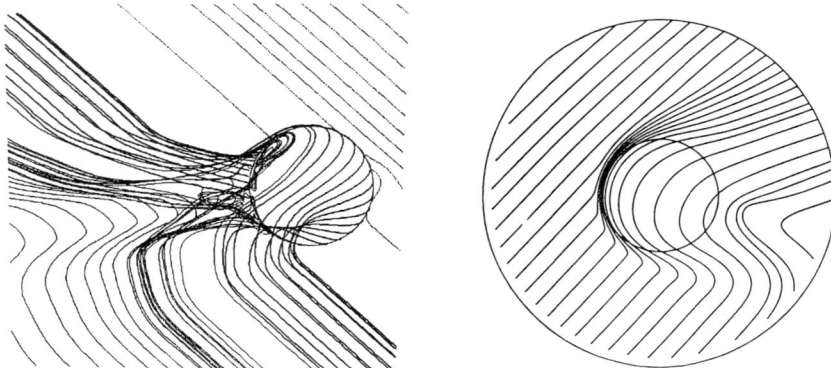

Figure 6. Draped magnetic field lines produced in the MHD simulations of the solar wind interaction with a conducting sphere carried out by Tanaka (1993) for perpendicular interplanetary magnetic field (left) and by Cable and Steinolfson (1995) for a 45° orientation (right). The latter is of selected field lines above the equatorial plane, projected down onto that plane.

uration from Tanaka (1993) for perpendicular upstream field and from Cable and Steinolfson (1995) for the 45° field. These configurations agree with general expectations based on the analysis of PVO magnetometer observations from solar maximum (see, e.g., the review by Luhmann 1986). A pressure maximum that occurs in the model between the magnetotail lobes is identified as a central plasma sheet. Both the Tanaka (1993) and Cable and Steinolfson (1995) results (for 45° and 90° cone angles) show accelerated "jets" of flow from the terminator regions where the magnetic field is sharply draped and undergoes a change of sign in its sunward–anti-sunward component. Indeed, the equatorial and meridional flow parameter contrasts, illustrated by Fig. 7, indicate that substantial asymmetries are present in the average interaction at Venus. Tanaka (1993) and Cable and Steinolfson (1995) also both found that the induced magnetotail lobes in their models appear weaker than in observations, possibly due to neglect of the atmosphere or ionosphere and/or mass-loading physics. Cable and Steinolfson moreover find reconnection of the low-altitude wake fields occurring in their model, although their numerical resistivity plays a role. Figures 8 and 9 show the magnetic field for the contrasting case of a precisely flow-aligned (radial) upstream field in which no traditional induced magnetotail is formed. The wake seems to contain weak vortices in both the flow and field (Fig. 8). This wake structure in fact shows good agreement with observations (Fig. 9), although the steady radial interplanetary field situation is rare. It is interesting that Tanaka (1993) also found indications of wake vortices in the magnetic equatorial region of their solution, but these did not manifest themselves so clearly in the presence of the induced magnetotail.

Overall, these models show considerable promise for probing the physics

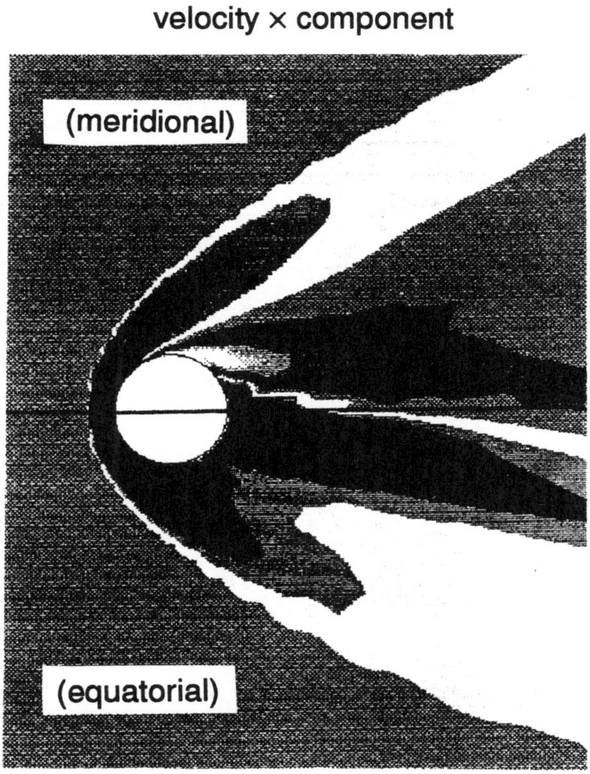

velocity × component

(meridional)

(equatorial)

Figure 7. Contours of fluid velocity in the meridional (top) and equatorial (bottom) planes from the Tanaka (1993) global MHD simulation.

of the induced magnetotail formation and the influence of both magnetic forces and mass-loading on the solar wind interaction with Venus. They have yet to be fully examined and exploited. At the same time, because of their isotropic "fluid" nature, they will not by themselves help us to understand details related to the effects of ion kinetics in the solar wind interaction with Venus.

C. Test Particle Models

Test particle methods, in which the equation of motion is solved for single particles (usually ions) in prescribed field configurations, have been applied to model the global characteristics of planetary pick-up ions around Venus. The large gyroradii expected for heavy (e.g., oxygen) pick-up ions relative to the subsolar interaction region and planetary radius scales were first noted by both Wallis (1972) and Cloutier et al. (1974) at an early stage in the investigation of the Venus–solar wind interaction. At that time, test particle approaches were used (see, e.g., Cloutier et al. 1974) to illustrate the asymmetry of the ion wake that would result from impact of exospheric pick-up ions on the

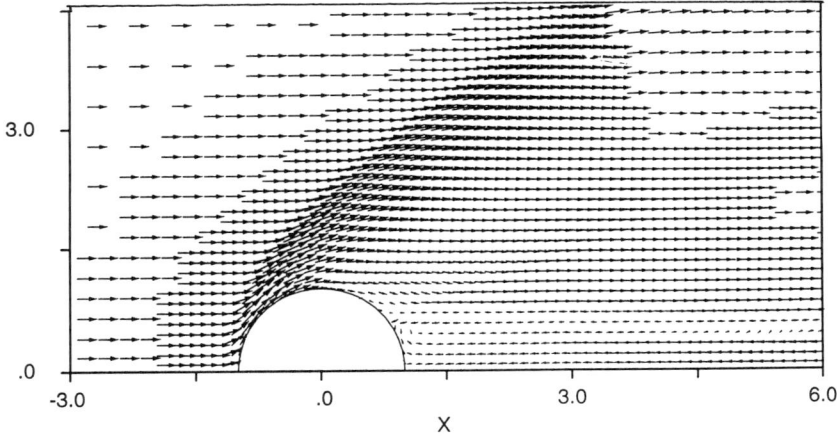

Figure 8. Magnetic field vectors in the wake of the flow-aligned field global MHD simulation by DeZeeuw et al. (1996). No classical induced magnetotail forms in this situation.

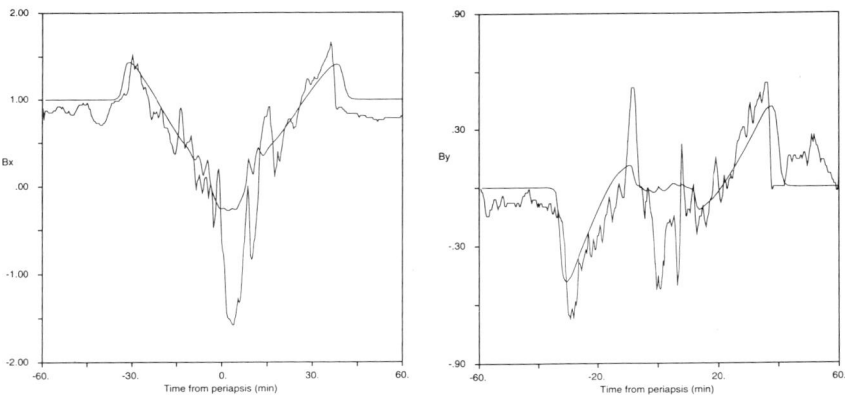

Figure 9. Comparison of some PVO magnetometer observations in the wake of Venus during a period of radial interplanetary field with the simulated field in the DeZeeuw et al. (1996) model.

dayside ionopause. The underlying field models in these early studies were highly simplified because of the limited computational capabilities available.

More sophisticated test particle models have since been constructed that make use of the results from the numerical models described above. These were in part motivated by the observations from the PVO (see, e.g., Phillips et al. 1987) which indicated first order agreement of the fluid model flows and fields with the actual situation, but required information outside of the framework of those models for interpretation of other data. For example,

Fig. 10 illustrates the general behavior of singly ionized oxygen (O^+) pick-up ions, which are "born" with negligible initial velocities in the dayside gas dynamic model of the Venus magnetosheath. The trajectories are computed by numerically solving the ion equation of motion,

$$\bar{a} = (q/m)(\bar{E} + \bar{V} \times \bar{B}) \qquad (16)$$

where $\bar{V}$ is the particle velocity, $\bar{a}$ is its acceleration, and q/m is the charge to mass ratio. The fluid models provide the description of $\bar{B}$ and the convection electric field $\bar{E} = -\bar{v} \times \bar{B}$ where $\bar{v}$ is the fluid bulk velocity. Phillips et al. (1987) used such trajectories to estimate the pick-up ion current perturbation of the magnetosheath field.

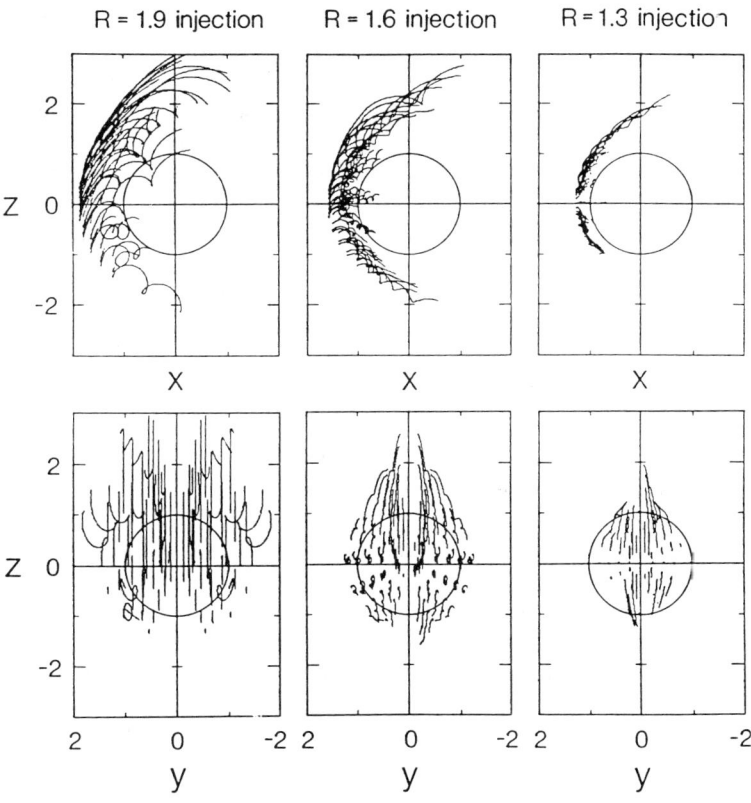

Figure 10. Picked up oxygen ion trajectories calculated using O^+ test particles in the Spreiter/Stahara gas dynamic magnetosheath model with the convected field.

Luhmann and Kozyra (1991) used a similar test particle model of the exospheric ion pickup to evaluate the flux of re-entering pick-up ions and to study their aeronomical effects; however, much of the interest in these ions relates to their role in induced magnetotail formation, a region not generally addressable with the blunt-body gas dynamic models. The modification introduced by Moore et al. (1991a), in which closure of the blunt-obstacle surface on the night side of the planet was assumed, provided one approximate solution. Moore et al. launched O^+ ions in the dayside magnetosheath, weighting them according to their altitude of origin using an exospheric model, to obtain spatial distributions and energy spectra in the wake region. Some of their ion trajectories are reproduced in Fig. 11. Based on this model, Moore et al. concluded that there must be a fluid-like mass addition at low altitudes that remains confined to the vicinity of the obstacle boundary, plus a high-energy, high-altitude component that exhibits an asymmetry controlled by the direction of the solar wind electric field $\bar{E} = -\bar{v} \times \bar{B}$. They further concluded that much of the lower energy, low-altitude-origin component was undetectable with the Pioneer Venus instruments.

Other test particle calculations by Luhmann (1993) used a comet tail model to represent the fields in the space occupied by the blunt body in the gas dynamic magnetosheath model. That construction, while implicitly suffering from a discontinuity in the flow and field at the blunt body boundary, is fairly consistent with PVO magnetometer measurements in the low-altitude wake. Equatorial field lines for this alternative magnetotail model are shown in Fig. 12. In this case, The self-consistent, anti-sunward flow underlying the comet-tail field lines determines the convection electric field in the pick-up ion equation of motion. Ionospheric particles launched tailward from the terminator within the space occupied by the comet tail model generally remain there to large distances. In particular, when the gravitational force is added to the equation of motion and the background velocity at the launch site (altitudes below ~ 1500 km) is assumed to be low (several tens of km s^{-1}), the pick-up ions obtain energies in the tens of eV range and form spatial patterns in the low-altitude wake reminiscent of the Venus ionospheric "tail rays" (Brace et al. 1987). Figures 13 and 14 show, respectively, examples of the low-altitude pick-up O^+ trajectories obtained by Luhmann (1993) and cross sections of the tail-ray-like structures. The behavior of these pick-up ions results from the combination of a low pick-up velocity, pickup where there is a substantial angle between the local magnetic field and underlying flow (e.g., "draping"), and the narrowness of the assumed low-altitude terminator source region. Recent work with analogous models to greater downstream distances further suggests that the observed, more energetic ($\sim$keV) ions observed at the PVO apoapsis of ~ 12 R_v (see, e.g., Moore et al. 1990) may be the distant wake counterparts of the ionospheric tail rays instead of the population of high-altitude exospheric origin. Indeed, the latter may produce fluxes too weak to be observed on PVO.

A great advantage of test particle models is their flexibility, which is

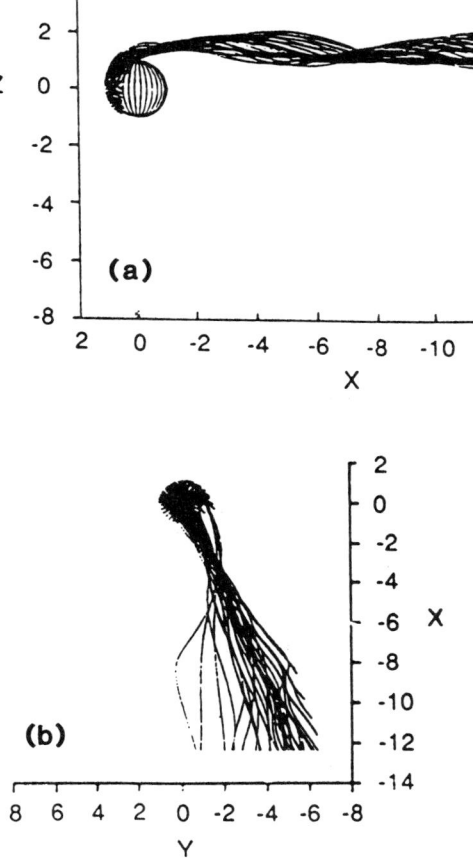

Figure 11. O^+ ion trajectories initiated on the day side of Venus and subjected to the modified gas dynamic model fields of Moore et al. (1990). The Parker spiral interplanetary field is assumed. (a) Meridional view. (b) Equatorial projection.

limited mainly by the availability of suitable background field and flow models. Provided that enough is known or can be assumed about the underlying "fluid" behavior, one can freely place particles of any mass, charge, and initial velocity, and characterize their subsequent behavior. However, these models can be criticized for their lack of physical self-consistency, which limits their usefulness for some applications.

D. Hybrid Models

The lack of self-consistency in test particle models largely vanishes in global hybrid models such as those developed by Brecht (1990), Moore et al. (1991b), and Brecht and Ferrante (1991), although adequate spatial resolution and ion statistics in these models comes at great computational cost. In typical cases,

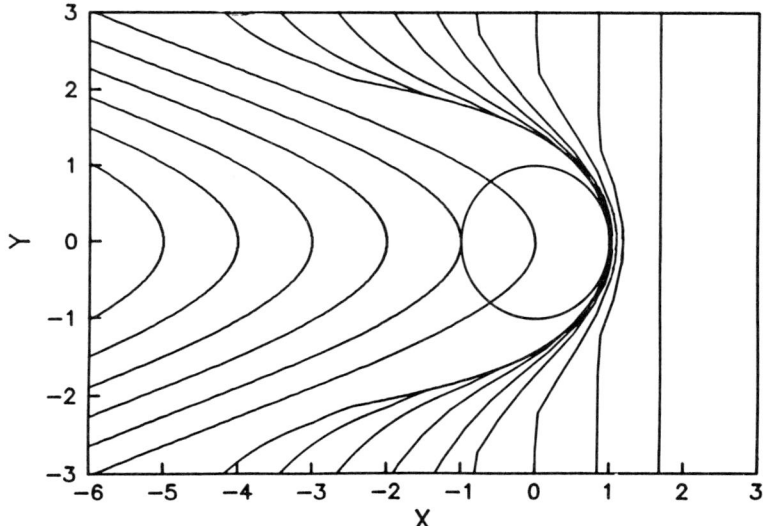

Figure 12. Modification introduced to the classical gas dynamic magnetosheath model by Luhmann (1993) to describe the fields in the wake.

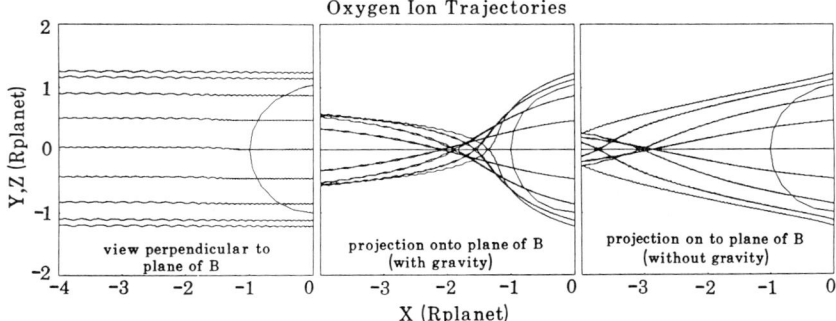

Figure 13. Low-energy O^+ ion trajectories obtained with a field of the form shown in Fig. 12. These ions were launched in a ring around the terminator at altitudes inside the gas dynamic magnetosheath model obstacle. Thus they move only in the comet-like "magnetotail" model.

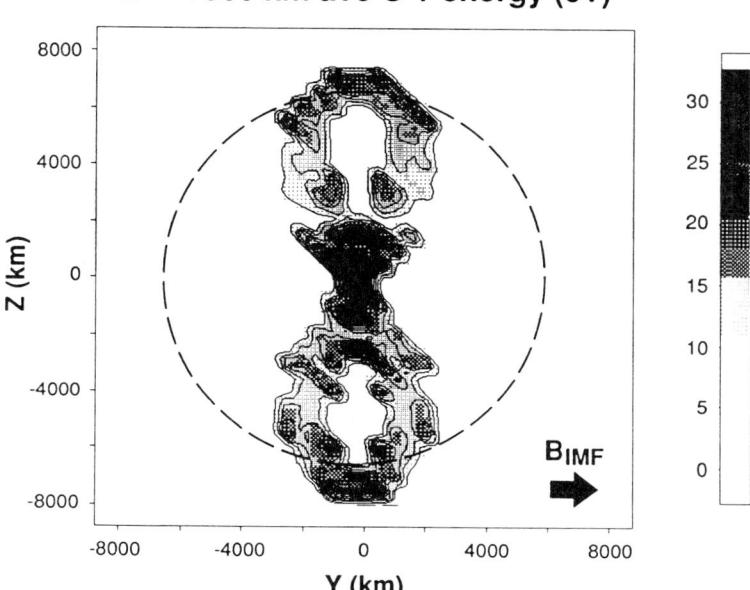

Figure 14. "Tail ray" structure resulting from an assumed low-altitude terminator source and the comet-like tail field model.

the ion equation of motion (Eq. 16) is solved for millions of particles to obtain the ion distribution function (and hence density and bulk velocity), while the electrons are treated as a massless fluid. Quasi-neutrality is assumed in the equations solved. The momentum equation for the electron fluid, with a resistive term $\eta \bar{J}$ added to $\bar{E}$ to allow some diffusion and dissipation of the magnetic fields, is solved for the electric field:

$$\bar{E} = \bar{v}_e \times \bar{B} - \frac{1}{qn}\nabla p_e + \eta \bar{J} \qquad (17)$$

where $\bar{v}_e$ is the electron fluid bulk velocity, n is the number density, η is the resistivity, and J is the current density. Faraday's Law (Eq. 14) is used to determine the self-consistent magnetic fields between steps of particle distribution advancement in time. Details concerning the numerical methods can be found in the paper by Brecht and Thomas (1988). High-frequency ($\gtrsim \omega_{ce}$, the electron gyrofrequency) electromagnetic waves are excluded by the assumption of zero electron mass, and quasi-neutrality. However, electromagnetic waves such as ion cyclotron, mirror modes, left-hand modes, and some electron whistlers, in addition to MHD waves are simulated, and can be driven physically by anisotropies that develop in the ion distribution functions.

Mass-loading is accommodated self-consistently in the hybrid models simply by introducing new ions with different masses heavier than protons at the appropriate spatial locations and rates to simulate their "birth" by ionization of the parent exospheric neutrals (Moore et al. 1991b). To date, these models have not included charge exchange or impact ionization collision processes. Moore et al. (1991b) considered the dayside interaction only, injecting O^+ ions at zero velocities at a rate that reflects the exospheric oxygen photo-production of 5×10^{24} s^{-1} for solar maximum. The assumed upstream solar wind density and velocity were 10 cm^{-3} and 500 km s^{-1}, respectively, and the interplanetary magnetic field of 12 nT was oriented 45° from the upstream flow direction. Solar wind ions were assumed cold, but solar wind electrons had 50 eV temperatures. The Alfvén, sonic, and fast Mach numbers were 8.5, 5.6, and 5.0. Because of computational limitations, Moore et al. used a spherical planetary obstacle about half the size of Venus, and then adopted an effective oxygen exosphere scale height and oxygen mass equal to half their actual values to obtain the appropriate relative scaling. The grid or cell size for the fluid solution, defining the model resolution, was $102 \times 360 \times 360$ km^3. Trajectories of some protons and oxygen ions from their simulation are reproduced in Fig. 15. As was also noted by Brecht (1990), asymmetries in apparent flow around the spherical obstacle in the hybrid model also occur for the solar wind protons, although they are exaggerated here due to the small obstacle assumed. Essentially all of the solar wind protons are diverted around the obstacle by a dayside magnetic barrier that forms as in the MHD models, except that in this case it is asymmetrical like the flow. Hall current associated magnetic gradient drifts in the barrier produce the asymmetric deflections. The oxygen ions, with their larger gyroradii, tend to impact the obstacle where they are removed from the simulation. Moore et al. (1991b) found that a realistic O^+ production rate did not significantly affect their hybrid model bow shock position. However, increasing the O^+ production rate by a factor of 5 caused an outward displacement on the order of what is observed between solar minimum and maximum on PVO. They also suggested that part of the explanation for the observed degree of ellipticity in the shock cross section may be that the quasi-parallel shock forms closer to the planet than the quasi-perpendicular shock by an amount in excess of that predicted from MHD alone.

Brecht and Ferrante (1991) produced a higher spatial resolution version of the three-dimensional hybrid model, ($\sim$50–100 km cell sizes), but were consequently restricted to just the nose region of the interaction. Like Moore et al. (1991b), they showed that the concept of a stagnation streamline does not apply in the hybrid model, and that the dayside magnetic barrier exhibits Hall-current induced asymmetries. They did not consider the effects of mass loading by O^+ ions, but experimented with the solar wind proton gyroradius to planet scaling and concluded that the spatial resolution in the Moore et al. (1991b) simulation was probably too coarse to reproduce the actual shock location. Of course, the position of their own subsolar bow shock was compromised by the limited part of the obstacle that was considered. These authors also tried hemispherical obstacles, which seemed to produce subso-

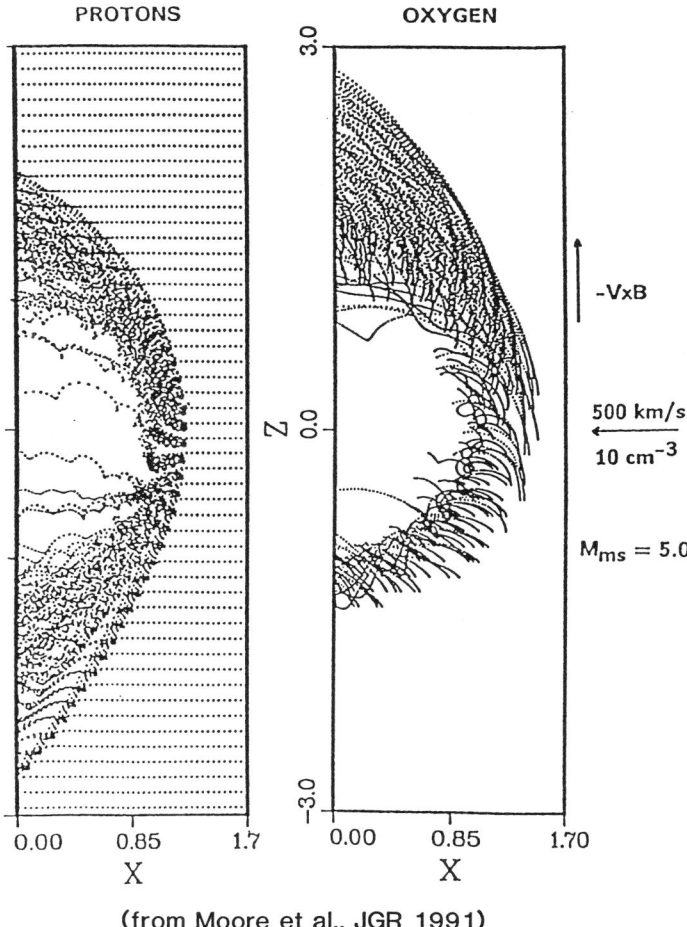

(from Moore et al., JGR 1991)

Figure 15. Dayside solar wind proton and exospheric O^+ ion trajectories from the hybrid model treatment of the solar wind interaction by Moore et al. (1991b).

lar bow shock positions in basic agreement with observations. They more recently extended their model, at lower spatial resolution, to a full sphere for the scaling of Mars in the solar wind (1995, unpublished manuscript). A fully global hybrid simulation has yet to be carried out for Venus. See the CD-ROM for animations CDP1C2M1 and CDP1C2M2.

In general, the hybrid models show considerable structure in their plasmas and fields, some of which may be noise introduced by the numerical methods and the finite numbers of ions, but some of which is physical. Examples of altitude profiles of dayside magnetosheath magnetic field and density from a hemispherical simulation of Brecht and Ferrante (1991) are shown in Fig. 16. Because this structure changes from time step to time step, such

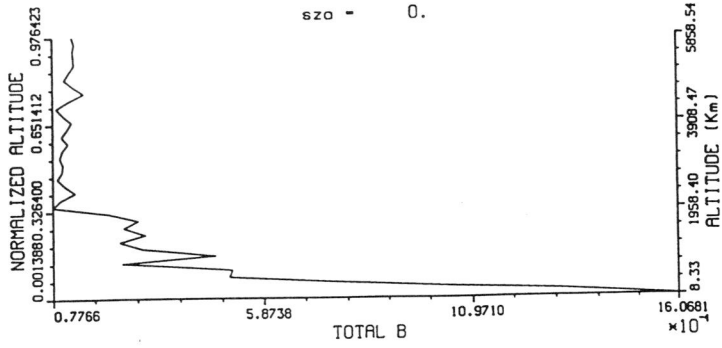

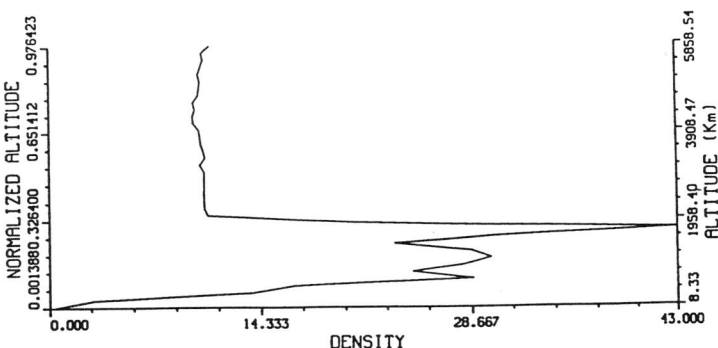

Figure 16. Simulated altitude profiles of the dayside magnetosheath magnetic field and proton density from the hybrid model of Brecht and Ferrante (1991). This model did not include O^+ ion effects.

displays must be regarded as "snapshots" from one instant. The amount of structure diminishes with averaging on the time scale of most observational data intervals.

Moore et al. (1991b) also experimented with the effect of finite obstacle conductivity. They found that the shock standoff distance is reduced if the magnetic fields are allowed to diffuse into the obstacle. Because this situation resembles, in some respects, what happens when the dayside ionosphere becomes magnetized (see the chapter by Cravens et al.), their simulations may have the potential of explaining the solar minimum and high solar wind dynamic pressure scenarios. Indeed, Zhang et al. (1990) inferred from bow shock observations and the gas dynamic model that the effective solar minimum subsolar obstacle would have to be located at or below the planet's surface. Instead, the explanation may be that the ionospheric obstacle becomes a less perfect conductor than at solar maximum. A further implication is that if the magnetic fields induced in the obstacle fail to set up an effective magnetic barrier, some of the solar wind particles will be absorbed. Clearly, much can be learned by experimenting with these state-of-the-art global self-consistent simulations including ion kinetics.

III. CONCLUSIONS

Each type of model described above makes unique contributions to the study of the solar wind interaction with Venus. The gas dynamic model provides a flexible tool for analyzing the dayside magnetosheath field configuration for a variety of interplanetary field orientations, and for carrying out first order studies of the bow shock position and the effects of planetary ion addition to the solar wind plasma (mass loading). The MHD models on average have poorer spatial resolution than the gas dynamic models, but they include the physics pertinent to the formation of the magnetic barrier and its extension into the wake. They also allow the study of flow interaction asymmetries introduced by the magnetic forces. Both types of fluid models provide background field and flow descriptions that can be used for test particle analyses of the behavior of the planetary pick-up ions. The most physically complete but computationally restrictive type of model is of course the global hybrid model, which includes both the MHD and ion kinetic effects as well as certain wave modes and Hall currents. This last model has the potential for explaining all of the features seen in the fluid and test particle models, including the bow shock, induced magnetotail, and pick-up ion attributes and consequences, with modifications (e.g., asymmetries) introduced by anisotropic ion distributions and finite ion gyroradii. However, it is not yet practical to "experiment" extensively with such models. While the available space does not permit discussion of all of the details of the various global model results, the above sampling illustrates the wealth of insights that can be gained from such simulations. Of course, the models' various advantages and shortcomings need to be considered before using them for the interpretation of observations or prediction of expected features.

As computational capabilities improve and become more accessible, it is easy to envision how global simulations may be utilized. Processes like charge exchange and impact ionization interactions between solar wind particles and exospheric neutrals can be incorporated into both fluid and hybrid models in order to test their individual effects. The boundary condition at the obstacle may be relaxed to allow not only diffusion of the magnetic field and absorption of solar wind plasma, but also active pressure-balance adjustment of a realistic ionosphere. This will allow better assessment of solar cycle and evolutionary time scale changes, and give indications of the induced magnetic fields at the planet surface. Time-dependent interplanetary conditions can be assumed in order to examine the responses of the simulations to various solar wind structures such as tangential or rotational discontinuities in the interplanetary field, and interplanetary shocks. Indeed, because no space physics missions to Venus are planned in the foreseeable future, much of our progress in further understanding the solar wind interaction and its consequences may be determined by these computational experiments. Comparisons with new observations from Mars, and applications of these models to that body, will also provide further understanding and discoveries relating to Venus. Thus with "bootstrapping" our knowledge should continue to grow in spite of the current outlook for *in-situ* measurements. Future generations who finally return to Venus exploration can benefit significantly from the on-going efforts described here, as our visions of near-Venus space and the global attributes and effects of the solar wind interaction with Venus' atmosphere become more complete.

REFERENCES

Beam, R. M., and Warming, R. F. 1978. An implicit factored scheme for the compressible Navier-Stokes equations. *AIAA J.* 16:393–402.

Belotserkovskii, O. M., et al. 1987. The effect of the hot oxygen corona on the interaction of the solar wind with Venus. *Geophys. Res. Lett.* 14:503–506.

Brace, L. H., et al. 1987. The ionotail of Venus: Its configuration and evidence for ion escape. *J. Geophys. Res.* 92:15–26.

Brecht, S. H. 1990. Magnetic asymmetries of unmagnetized planets. *Geophys. Res. Lett.* 17:1243–1246.

Brecht, S. H., and Ferrante, J. R. 1991. Global hybrid simulation of unmagnetized planets: Comparison of Venus and Mars. *J. Geophys. Res.* 96:11209–11220.

Brecht, S. H., and Thomas, V. A. 1988. Multidimensional simulations using hybrid particle codes. *Comput. Phys. Comm.* 48:135–143.

Breus, T. K., Krymskii, A. M., and Mitnitskii, V. Ya. 1987. Interaction of mass-loaded solar wind flow with a blunt body. *Planet. Space Sci.* 35:1221–1227.

Breus, T. K., Bauer, S. J., Krymskii, A. M., and Mitnitskii, V. Ya. 1989. Mass loading in the solar wind interaction with Venus and Mars. *J. Geophys. Res.* 94:2375–2382.

Cable, S., and Steinolfson, R. S. 1995. Three dimensional MHD simulations of the interaction between Venus and the solar wind. *J. Geophys. Res.* 100:21645–21658.

Cloutier, P. A., Daniell, R. E., and Butler, D. M. 1974. Atmospheric ion wakes of Venus and Mars in the solar wind. *Planet. Space Sci.* 22:967–990.

DeZeeuw, D. L., et al. 1996. A new axisymmetric MHD model of the interaction of the solar wind with Venus. *J. Geophys. Res.* 101:4547–4556.

Gombosi, T. I., Powell, K. G., and DeZeeuw, D. L. 1994. Axisymmetric modeling of cometary mass loading on an adaptive refined grid: MHD results. *J. Geophys. Res.* 99:21525–21540.

Luhmann, J. G. 1986. The solar wind interaction with Venus. *Space Sci. Rev.* 44:241–306.

Luhmann, J. G. 1993. A model of the ionospheric tail rays of Venus. *J. Geophys. Res.* 98:17615–17621.

Luhmann, J. G., and Kozyra, J. U. 1991. Dayside pickup oxygen ion precipitation at Venus and Mars: Spatial distributions, energy deposition and consequences. *J. Geophys. Res.* 96:5457–5467.

Luhmann, J. G., et al. 1986. A gas dynamic magnetosheath field model for unsteady interplanetary fields: Application to the solar wind interaction with Venus. *J. Geophys. Res.* 91:3001–3010.

Luhmann, J. G., et al. 1993. Solar cycle 21 effects on the interplanetary magnetic field and related parameters at 0.7 and 1.0 AU. *J. Geophys. Res.* 98:5559–5572.

MacCormack, R. W. 1969. The Effect of Viscosity in Hypervelocity Impact Cratering. AIAA Paper 69-354.

Moore, K. R., McComas, D. J., Russell, C. T., and Mihalov, J. D. 1990. A statistical study of ions and magnetic fields in the Venus magnetotail. *J. Geophys. Res.* 95:12005–12018.

Moore, K. R., McComas, D. J., Russell, C. T., Stahara, S. S., and Spreiter, J. R. 1991a. Gas dynamic modeling of the Venus magnetotail. *J. Geophys. Res.* 96:5667–5681.

Moore, K. R., Thomas, V. A., and McComas, D. J. 1991b. Global hybrid simulation of the solar wind interaction with the dayside of Venus. *J. Geophys. Res.* 96:7779–7791.

Phillips, J. L., Luhmann, J. G., Russell, C. T., and Moore, K. R. 1987. Finite Larmor radius effect on ion pickup at Venus. *J. Geophys. Res.* 92:9920–9930.

Russell, C. T., et al. 1988. Solar and interplanetary control of the location of the Venus bow shock. *J. Geophys. Res.* 93:5461–5469.

Spreiter, J. R., and Alksne, A. Y. 1969. Plasma flow around the magnetosphere. *Rev. Geophys.* 1:11–50.

Spreiter, J. R., and Rizzi, A. W. 1974. Aligned magnetohydrodynamic solution for solar wind flow past the Earth's magnetosphere. *Acta Astronaut.* 1:15–55.

Spreiter, J. R., and Stahara, S. S. 1980a. A new predictive model for determining solar wind-terrestrial planet interactions. *J. Geophys. Res.* 85:6769–6777.

Spreiter, J. R., and Stahara, S. S. 1980b. Solar wind flow past Venus: Theory and comparisons. *J. Geophys. Res.* 85:7715–7738.

Spreiter, J. R., and Stahara, S. S. 1992. Computer modeling of solar wind interaction with Venus and Mars. In *Venus and Mars: Atmospheres, Ionospheres and Solar Wind Interactions*, eds. J. G. Luhmann, M. Tatrallyay and R. O. Pepin (Washington, D. C.: American Geophysical Union), pp. 345–386.

Spreiter, J. R., Summers, A. L., and Alksne, A. Y. 1966. Hydromagnetic flow around the magnetosphere. *Planet. Space Sci.* 14:223–253.

Spreiter, J. R., Summers, A. L., and Rizzi, A. W. 1970. Solar wind flow past nonmagnetic planets—Venus and Mars. *Planet. Space Sci.* 18:1281–1299.

Tanaka, T. 1993. Configurations of the solar wind flow and magnetic field around the planets with no magnetic field: Calculation by a new MHD scheme. *J. Geophys. Res.* 98:17251–17262.

Wallis, M. K. 1972. Comet-like interaction of Venus with the solar wind I. *Cosmic Electrodyn.* 3:45–59.

Wu, C. C. 1992. MHD flow past an obstacle: Large scale flow in the magnetosheath. *Geophys. Res. Lett.* 19:87–90.

Zhang, M. H. G., et al. 1993. Oxygen ionization rates at Mars and Venus: Relative contributions of impact ionization and charge exchange. *J. Geophys. Res.* 98:3311–3318.

Zhang, T. L., Luhmann, J. G., and Russell, C. T. 1990. The solar cycle dependence of the location and shape of the Venus bow shock. *J. Geophys. Res.* 95:14961–14967.

Zhang, T. L., et al. 1993. On the spatial range of validity of the gas dynamic model in the magnetosheath of Venus. *Geophys. Res. Lett.* 20:751–754.

MAGNETOHYDRODYNAMIC PROCESSES: MAGNETIC FIELDS IN THE IONOSPHERE OF VENUS

THOMAS E. CRAVENS
University of Kansas

HIROYUKI SHINAGAWA
Nagoya University

and

JANET G. LUHMANN
University of California at Berkeley

Venus has virtually no intrinsic magnetic field which can divert the solar wind around that planet. However, Venus does possess a dense neutral atmosphere, and therefore it also has an ionosphere. The ionospheric plasma is a very good electrical conductor so that electrical currents can be induced that largely prevent the solar wind plasma and its associated interplanetary magnetic field from penetrating into the ionosphere—basically this is the diamagnetic effect. Hence, the ionosphere acts as an obstacle to the solar wind. Measurements made by the magnetometer and other instruments onboard the Pioneer Venus Orbiter (PVO) revealed that the ionosphere can exist in either of two states: (1) unmagnetized or (2) magnetized. The former state was observed when the solar wind dynamic pressure was less than the maximum ionospheric thermal pressure. In this case, large-scale magnetic fields are not present but small-scale structures (i.e., magnetic flux ropes) were observed by the PVO magnetometer. The ionosphere was observed to exist in a magnetized state when the solar wind dynamic pressure was high. The ionopause altitude varies with solar wind dynamic pressure and moves up as this pressure decreases. Magnetohydrodynamic (MHD) theory can be used to describe magnetized plasmas such as the ionospheric plasma at Venus. MHD provides a conceptual framework within which measurements can be interpreted. Quantitative calculations for the Venus ionosphere have been undertaken using numerical MHD models. The MHD theory used for Venus must be "non-ideal" and include collisional effects such as resistivity, ion-neutral collisions, and chemical production and loss terms. Both one- and two-dimensional MHD models have been developed for Venus, and both large-scale and small-scale properties of the ionosphere have been studied. Considerable progress has been made on understanding the large-scale structure of the dayside Venus ionosphere, but progress on the nightside magnetized ionosphere and on small-scale structure has been slower. For example, the small-scale structure of the ionopause current layer is not understood theoretically, although its thickness is known observationally to be only a couple of O^+ gyroradii. The formation and evolution of magnetic flux ropes also remain poorly understood although a fair amount of theoretical work has been undertaken.

The data gathered by the Mariner, Venera, and Pioneer Venus missions and the theoretical work motivated by these data have allowed our understanding of the ionosphere of Venus and its interaction with the solar wind to advance dramatically over the last twenty years. The *in-situ* measurements made in the ionosphere by the fields and particles instruments onboard the Pioneer Venus Orbiter (PVO) between 1978 and 1992 have been especially valuable. Many reviews on the ionosphere of Venus and its interaction with the solar wind have been published over the years (Russell and Vaisberg 1983; Brace et al. 1983; Nagy et al. 1983; Cloutier et al. 1983; Gringauz 1983; Luhmann 1986; Cravens 1989,1991; Mahajan and Kar 1988; Miller and Whitten 1991; Luhmann and Cravens 1991; Knudsen 1991). In particular, Luhmann and Cravens (1991) provide a good background for the current chapter, which will focus on magnetohydrodynamic processes in the ionosphere of Venus. The present chapter provides a general review of the ionospheric plasma and magnetic field, including observations, with particular emphasis on theoretical models discussed in the literature after about 1985. Both small-scale and large-scale structure will be discussed. Particular attention will be given to results published subsequent to the *Venus Aeronomy* book. Other chapters in this book complementary to this one include chapters by Luhmann et al., by Fox and Kliore, and by Nagy and Cravens.

I. SHORT TUTORIAL ON THE SOLAR WIND INTERACTION WITH VENUS

Venus has no measurable intrinsic magnetic field (cf., Russell and Vaisberg 1983; see the chapter by Donahue and Russell) that can act to divert the solar wind around the planet. But Venus does possess a substantial ionosphere that results from the ionization of its neutral atmosphere by solar radiation. The ionospheric plasma is an excellent electrical conductor so that currents are easily able to be induced in it by the solar wind. For the most part these currents act to exclude the solar wind plasma and its imbedded interplanetary magnetic field from the ionosphere. One consequence of this diamagnetic effect is that the ionosphere acts as an obstacle to the solar wind flow (see Fig. 1). A bow shock is present because the solar wind is both supersonic and super-Alfvénic and must become subsonic in order to flow around the obstacle. The shocked solar wind slows down as it approaches and flows around the planet. The magnetic field strength builds up as the flow slows, resulting in the formation of a magnetic barrier which is located outside the ionopause. The boundary between the cold ionospheric plasma and the hotter solar wind plasma and the magnetic barrier is called the ionopause. In the magnetic barrier the thermal pressure is much less than the magnetic pressure which thus accounts for almost all of the upstream solar wind dynamic pressure. In the tail region, the magnetic field assumes a draped geometry and the ionospheric plasma extends tailward in structures called tail rays (cf., Brace and Kliore 1991).

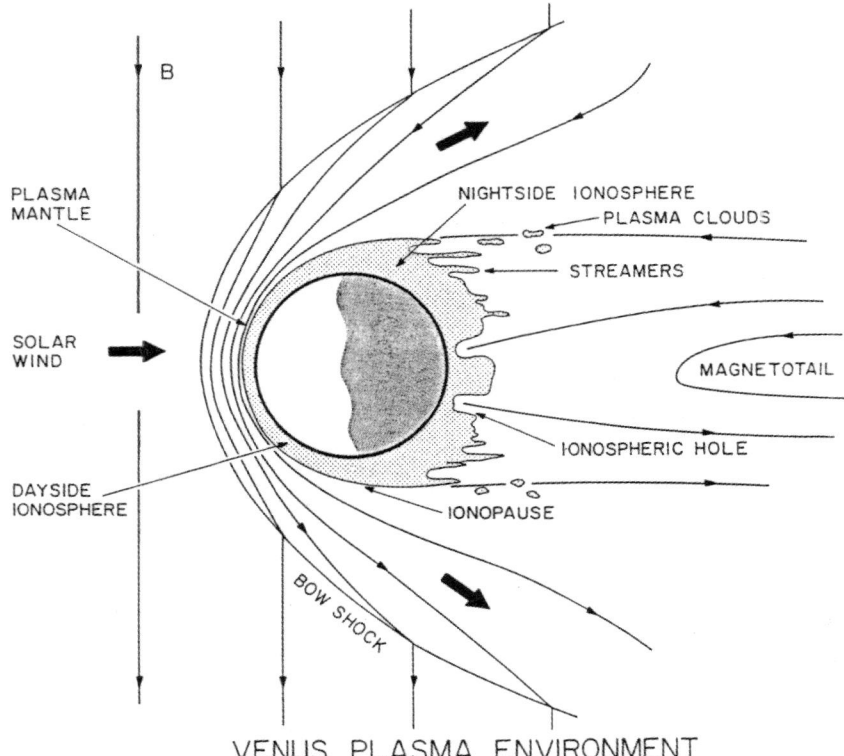

Figure 1. Schematic of the solar wind interaction with Venus. Interplanetary magnetic field lines are shown as are the bow shock and ionopause surfaces.

Measurements made by the magnetometer and other instruments onboard the Pioneer Venus Orbiter (PVO) revealed (cf., Luhmann and Cravens 1991; Russell and Vaisberg 1983) that the ionosphere can exist in either of two states: (1) unmagnetized, or (2) magnetized.

The unmagnetized state of the ionosphere has these characteristics:

1. It exists during periods when the solar wind dynamic pressure is less than the maximum ionospheric thermal pressure.
2. Large-scale magnetic fields are not present but small-scale structures (i.e., magnetic flux ropes) were observed by the PVO magnetometer. Flux ropes are narrow regions of enhanced magnetic field strength ($B \approx 30$ to 100 nT) which exist within an otherwise unmagnetized ionosphere.
3. The ionopause is located at high altitudes ($z > 300$ km) and is quite thin ($\Delta z \approx 25$ km).
4. A substantial ionosphere exists (electron densities, $n_e \approx 10^4$ cm^{-3} at the peak) throughout most of the nightside ionosphere, at least during solar

maximum conditions due to transport of plasma from the day to night (see the chapter by Fox et al.; see Brace et al. 1995).
5. Localized ionospheric "holes" exist deep on the night side. The ionospheric magnetic field in these holes is well-organized and roughly radial (i.e., vertical).

The magnetized state of the ionosphere is observed to have these characteristics:

1. It exists when the solar wind dynamic pressure is high.
2. Large-scale magnetic fields ($B \approx 50$ to 150 nT) are present throughout the dayside ionosphere, and the ionospheric magnetic field is mainly horizontal with a vertical structure characterized by a minimum near 200 km and a maximum near 170 km (cf., Luhmann et al. 1980).
3. The ionopause is located at lower altitudes ($z < 300$ km) and is thick ($\Delta z \approx 80$ km).
4. The night side ionosphere is weak or absent for altitudes greater than 150 km.

Figure 2 displays measured magnetic field strength and electron density as functions of height above the surface of Venus for three Pioneer Venus Orbiter (PVO) orbital passes. For Orbit 186 the solar wind "pressure" was low so that the magnetic field strength above the ionopause was low (this is the unmagnetized case). In this case, the magnetic field strength in the ionosphere itself was essentially zero, on the average, although narrow, small-scale, spikes of magnetic field were present (these are magnetic flux ropes; Russell and Elphic 1979). It has been shown that the magnetic pressure balances the thermal pressure of the ionosphere at the ionopause (cf., Phillips et al. 1985). Notice in Orbit 186 of Fig. 2 that the ionospheric density in the ionosphere above an altitude of 200 km varies with altitude as expected from hydrostatic equilibrium. On the other hand, the solar wind pressure was high for Orbits 176 and 177, as was the magnetic field strength in the magnetic barrier located above the ionopause (this is the magnetized ionosphere case). Notice the broad ionopause and also the characteristic vertical structure of the ionospheric magnetic field with its minimum near 200 km and maximum near 170 km. When the solar wind dynamic pressure exceeds, or is comparable to, the maximum ionospheric pressure, magnetic flux is pushed deep into the ionosphere and is convected downward to the lower ionosphere where it is subject to Ohmic dissipation. The characteristic shape of the magnetic field profile can be explained in terms of the vertical plasma flow velocity as a function of altitude (Cravens et al. 1984). Magnetohydrodynamic (MHD) theory can successfully explain this phenomenon.

Now we briefly introduce the ionosphere of Venus. The neutral thermosphere of Venus consists mainly of carbon dioxide (CO_2) at lower altitudes but at higher altitudes (altitude $z > 160$ km) atomic oxygen (O), from the dissociation of CO_2, becomes the major species (Niemann et al. 1980; cf., Fox

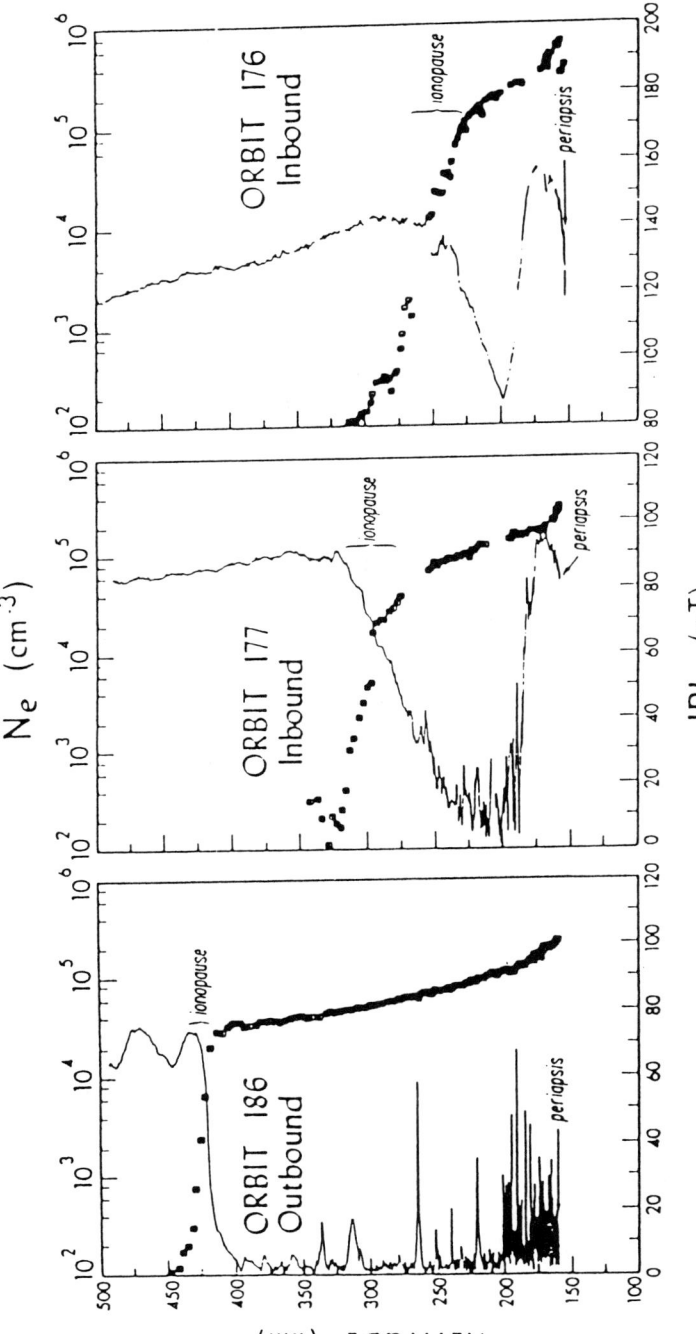

Figure 2. Measured magnetic field strength (Pioneer Venus Orbiter magnetometer) and measured electron density (Langmuir probe) in the ionosphere of Venus (figure from Russell and Vaisberg 1983).

and Bougher 1991). Other species are also present including CO, He, N_2, N, and H. These neutrals are ionized by solar extreme ultraviolet radiation producing many ion species (CO_2^+, O_2^+, O^+, H^+, ...). Ion-neutral chemical reactions and electron–ion recombination are both important processes in the lower ionosphere. CO_2^+ has the largest production rate at low altitudes, but O_2^+ is the major ion near the ionospheric peak due to the following rapid reaction: $CO_2^+ + O \rightarrow O_2^+ + CO$. The O_2^+ ions dissociatively recombine with electrons: $O_2^+ + e \rightarrow O + O$. However, at higher altitudes O^+ is the major species. This species is produced by photoionization of O or by dissociative photoionization of CO_2 and is chemically removed from the ionosphere by reaction with CO_2: $O^+ + CO_2 \rightarrow O_2^+ + CO$.

Transport is more important than chemistry in controlling the O^+ distribution for altitudes above about 200 km. O^+ ions are transported both vertically and horizontally in the ionosphere. The horizontal transport of ions from the day side to the night side in the unmagnetized ionosphere, with flow speeds that can reach 5 km s^{-1} near the terminator and for altitudes well above 200 km, is thought to be important for the maintenance of the nightside ionosphere (cf., Knudsen et al. 1980; chapter by Fox and Kliore). The O^+ vertical distribution can be described by diffusive equilibrium at higher altitudes in unmagnetized ionospheres. Significant abundances of H^+ have also been measured at higher altitudes, particularly on the night side. This species mainly comes from the charge transfer of O^+ ions with atomic hydrogen. A detailed discussion of the aeronomy of the ionosphere of Venus is outside the scope of this chapter and is discussed in other chapters of this book (e.g., Fox and Kliore) or in other reviews (Fox and Bougher 1991). A good general treatment of aeronomy can be found in Banks and Kockarts (1973).

In Sec. II of this chapter we will review MHD theory as it applies to the ionosphere of Venus. Plasma and magnetic field data will be reviewed in Sec. III. MHD models of large-scale structure will be discussed in Sec. IV and models of the small-scale structure in Sec. V. Section VI will briefly summarize our current state of knowledge and suggest future directions.

II. MHD THEORY

The set of equations needed to model the magnetized ionosphere of Venus are the multi-species magnetohydrodynamic equations, including all relevant aeronomical collision and production and loss terms. Models of the ionosphere of Venus and relevant theoretical equations have been presented in several review papers (Nagy et al. 1983; Cravens 1991; Luhmann and Cravens 1991). The fluid conservation equations are the velocity moments of the Boltzmann equation (Nicholson 1983). A relatively simple set of conservation equations is usually sufficient for the study of planetary ionospheres. The 5-moment equations consist of, for each charged particle species: a continuity equation, a momentum equation, and an energy equation for a scalar pressure. The scalar pressure of species s is equal to p_s, which can also be expressed in terms of density and temperature using the equation of state: $p_s = n_s k_B T_s$, where k_B is Boltzmann's constant and T_s is the temperature of species s. In this section we will discuss the continuity and momentum equations, as well as the magnetic induction equation. The energy equations for electrons and ions (that is, the energetics of the ionosphere of Venus) will be discussed in the chapter by Nagy and Cravens. Maxwell's equations must also be included in our set of equations.

A. Fluid Conservation Equations

The continuity equation for species s describes the time rate of change of the density n_s:

$$\frac{\partial n_s}{\partial t} + \nabla \cdot (n_s \mathbf{u}_s) = P_s - L_s \tag{1}$$

where P_s is the production rate of species s including primary production by photoionization (or by collisional ionization by energetic electrons or ions) as well as production due to chemical reactions. L_s is the loss rate of species s, due to chemistry of various sorts. The bulk flow velocity vector for species s is denoted $\mathbf{u}_s$.

The momentum equation describes the variation of the flow velocity $\mathbf{u}_s$ of species s:

$$n_s m_s \left[\frac{\partial \mathbf{u}_s}{\partial t} + \mathbf{u}_s \cdot \nabla \mathbf{u}_s \right] = -\nabla p_s + n_s e_s (\mathbf{E} + \mathbf{u}_s \times \mathbf{B})$$

$$+ n_s m_s \mathbf{g} - n_s m_s \sum_j \nu_{sj} (\mathbf{u}_s - \mathbf{u}_j) + P_s m_s (\mathbf{u}_s - \mathbf{u}_n) \tag{2}$$

where $\mathbf{g}$ is the acceleration due to gravity, m_s is the mass of species s, and ν_{sj} is the effective momentum transfer collision frequency between species s and j. The charge e_s on species s is equal to $\pm e$, depending on whether s denotes electrons or an ion species. $\mathbf{E}$ and $\mathbf{B}$ are the electric and magnetic field vectors, respectively. The terms on the right-hand side of this equation are, respectively: the pressure gradient force, the Lorentz force, the gravitational force, the frictional force due to collisions with other species, and the mass-loading term.

The electron momentum equation ($s = e$) can be approximated by neglecting the inertial terms (i.e., the left-hand side) and then rearranged so as to isolate the electric field:

$$\mathbf{E} = -\mathbf{u} \times \mathbf{B} + \frac{1}{n_e e} \mathbf{J} \times \mathbf{B} - \frac{1}{n_e e} \nabla p_e + \eta \mathbf{J} - \frac{m_e}{e} \nu_{en} (\mathbf{u} - \mathbf{u}_n). \tag{3}$$

$\mathbf{J}$ is the current density, p_e is the partial pressure due to electrons, $\mathbf{u}_n$ is the neutral flow speed, and η is the electrical resistivity. This is known as the generalized Ohm's law and it specifies the electric field needed to maintain electrical neutrality in the plasma (i.e., $n_e = n_i$, where n_i is the total ion density, summed over all ion species). The average (all species) plasma flow velocity is obtained by averaging over all the individual ion species (i.e., CO_2^+, O_2^+, O^+, etc., for Venus) and is denoted **u**. The terms on the right-hand side are called: the motional electric field, the Hall field, the ambipolar field, the Ohmic field, and another small collisional term. The motional electric field is usually the most important term in space physics applications, but in the Venus ionosphere the ambipolar and Ohmic terms are also important at some altitudes. This equation can be manipulated for different scenarios to obtain other "standard" electrical conductivity expressions (Cole 1990).

The momentum equations for all species can be combined into a single-fluid momentum equation which is more useful for some applications than the multi-species equations:

$$\rho \left(\frac{\partial \mathbf{u}}{\partial t} + \mathbf{u} \cdot \nabla \mathbf{u} \right) = \mathbf{J} \times \mathbf{B} - \nabla (p_e + p_i) + \rho \mathbf{g} - \rho v_{in} (\mathbf{u} - \mathbf{u}_n). \quad (4)$$

The mass density of the plasma (all species) is ρ and $\mathbf{J} \times \mathbf{B}$ is the force per unit volume due to an electrical current in a magnetic field (the "electric motor force"). Both the electron and ion thermal pressures (p_e and p_i) appear on the right-hand side and can be found using the energy equation for species s (chapter by Nagy and Cravens). The ion-neutral collision frequency is denoted v_{in}. The ambipolar diffusion equation, which has been so useful for ionospheric physics, can be derived from Eq. (4) by neglecting the left-hand side and solving for the plasma flow speed **u** that appears in the ion-neutral friction term.

B. Momentum Balance in the Ionosphere of Venus

The current density can be eliminated from the $\mathbf{J} \times \mathbf{B}$ force term by using Ampere's law, and the resulting expression can be manipulated with vector calculus to give:

$$\mathbf{J} \times \mathbf{B} = \frac{1}{\mu_0} (\nabla \times \mathbf{B}) \times \mathbf{B} = -\nabla \frac{B^2}{2\mu_0} + \frac{1}{\mu_0} (\mathbf{B} \cdot \nabla) \mathbf{B}. \quad (5)$$

The first term on the right-hand side is in the form of a gradient of a scalar "pressure" which is called magnetic pressure, $p_B = B^2/2\mu_0$. The second term on the right-hand side is the magnetic tension force which arises from curvature of field lines. The displacement current in Ampere's law was neglected in this expression. Using Eq. (5) without the tension force, the first two terms on the right-hand side of Eq. (4) appear together as: $-\nabla (p + p_B)$ where $p = p_e + p_i$ is the total thermal pressure. The plasma beta is defined as $\beta = p/p_B$. Thermal pressure effects control the plasma dynamics where the β is high, and magnetic forces dominate where the β is low.

Under some conditions, the plasma flow velocity is low enough that both the inertial terms (i.e., left-hand side) and frictional forces can be neglected in the momentum equation which then becomes a static force balance equation:

$$0 = \mathbf{J} \times \mathbf{B} - \nabla p + \rho \mathbf{g}. \quad (6)$$

In the low beta case, just the thermal pressure gradient and gravity terms remain, and expression (6) becomes the hydrostatic balance equation which can be solved to give an ionospheric pressure that varies exponentially with altitude (e.g., the electron density in the upper ionosphere for Orbit 186 in Fig. 2 has an exponential fall-off). Under other circumstances, like near the ionopause, the gravity term can be neglected and the magnetic tension part of Eq. (5) is not important so that Eq. (6) becomes $0 = -\nabla(p + p_B)$, which when integrated simply becomes $p + p_B$ = constant. This simple pressure balance relation applies across current sheets such as the ionopause. In the magnetic barrier just above the ionopause, the plasma β is low and magnetic pressure p_B dominates, whereas in the ionosphere just below the ionopause the plasma β is high and thermal pressure dominates. The resulting pressure balance at the ionopause can be represented:

$$\left(\frac{B^2}{2\mu_0}\right)_{\text{barrier}} = \{n_e k_B (T_e + T_i)\}_{\text{ionosphere}} \tag{7}$$

where the equation of state was used for the electron and ion pressures. T_e and T_i are the electron and ion temperatures, respectively. The magnetic pressure just above the ionopause and the thermal pressure just below the ionopause for Orbit 186 (Fig. 2) can be demonstrated to be almost equal to each other. The magnetic field strength is 70 nT just above the ionopause so that the magnetic pressure is $B^2/2\mu_0 = 2 \times 10^{-9}$ N m^{-2}. The electron density in the ionosphere, just below the ionopause, is $n_e = 3 \times 10^4$ cm^{-3}. The measured electron and ion temperatures are about T_e = 3800 K and T_i = 1800 K, respectively (Brace et al. 1983). The thermal pressure is thus $p = n_e k_B (T_e + T_i) = 2.2 \times 10^{-9}$ N m^{-2}, which is equal to the magnetic pressure to within 10%.

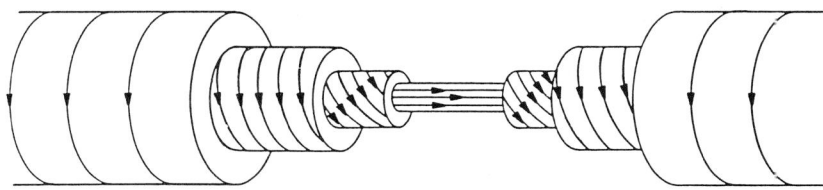

Figure 3. Schematic of a magnetic flux rope that is force-free (figure from Elphic and Russell 1983).

For some conditions, the momentum balance simply becomes $\mathbf{J} \times \mathbf{B} = 0$, and from Eq. (5) this implies that magnetic tension exactly balances the magnetic pressure gradient force. The resulting static balance is called force-free balance. Magnetic flux ropes in the ionosphere of Venus are mostly force-free (Fig. 3), and because they are cylindrically symmetric structures, the $\mathbf{J} \times \mathbf{B} = 0$ condition can be satisfied if $\mathbf{J}$ is parallel to $\mathbf{B}$: $\nabla \times \mathbf{B} = \alpha(r)\mathbf{B}$, where $\alpha(r)$ is a function of the distance from the axis of the flux rope.

Sometimes the plasma flow is fast enough that the inertial terms of Eq. (4) cannot be neglected. For simple one-dimensional geometries these terms can be written as the gradient of a dynamic pressure ρu^2. Hence, the dynamic pressure can be compared with the thermal or magnetic pressures in order to ascertain their relative importance.

C. Magnetic Convection–Diffusion Equation

The magnetic field appears in the MHD momentum equation and Faraday's law (or magnetic induction equation) can be used to determine this field: $\partial \mathbf{B}/\partial t = -\nabla \times \mathbf{E}$. Using the generalized Ohm's law, Eq. (3), minus some terms, for the electric field $\mathbf{E}$, the induction equation can be converted into the magnetic convection–diffusion equation:

$$\frac{\partial \mathbf{B}}{\partial t} = \nabla \times (\mathbf{u} \times \mathbf{B}) - \nabla \times \left(\frac{\eta}{\mu_0} \nabla \times \mathbf{B} \right). \tag{8}$$

The first term on the right-hand side of Eq. (8) is the magnetic convection term, and the second term is the magnetic diffusion term. Equation (8) plus the continuity, momentum, and energy equations are the magnetohydrodynamic (MHD) equations (also see Siscoe 1983). The one-dimensional version of Eq. (8) has been quite useful at Venus (Luhmann et al. 1984; Cravens et al. 1984). An alternate formulation of this equation appropriate for the ionosphere of Venus was provided by Cole (1990) and Cole and Hoegy (1994).

The resistivity divided by μ_0 is the magnetic diffusion coefficient $D_B = \eta/\mu_0$ where the resistivity η is proportional to the electron collision frequency with both neutrals and ions. The diffusion coefficient D_B increases with decreasing altitude (with values of 6×10^4 m^{2}s^{-1} at 150 km and 10^4 m^2 s^{-1} at 200 km) (see Fig. 32 in Luhmann and Cravens [1991]) due to the increasing neutral density and collision frequency. The magnetic Reynold's number is a dimensionless number that is the order of magnitude of the ratio of the convection term to the diffusion term: $R_m = Lu/D_B$ where L is a typical length scale. Where R_m is much greater than unity, the magnetic field is "frozen into" the plasma so that field lines stay "attached" to particular fluid parcels. On the other hand, where R_m is of the order of unity or less, magnetic diffusion dominates (i.e., Ohmic dissipation operates) and magnetic flux can "leak" out of a fluid parcel. Low magnetic Reynold's numbers can be found not only where the plasma resistivity is high but also in narrow current layers where the length scale is small. In the ionosphere of Venus, above about 150 km, R_m is large enough that the field can be considered to be frozen into the plasma, whereas below this altitude magnetic diffusion is important.

D. MHD Wave Modes and MHD Stability

Waves can propagate through a plasma, although the MHD equations only describe certain of these wave modes (i.e., MHD wave modes). Sound waves (i.e., ion acoustic) have propagation speed of $C_s = [\gamma p/\rho]^{1/2} =$

$[\gamma k_B (T_e + T_i)/m_i]^{1/2}$ where γ is the polytropic index ($\gamma = 5/3$ for an ideal gas although $\gamma = 1$ is often a better assumption for the electron gas) and m_i is the ion mass. Alfvén waves have speed $C_A = [B^2/(\mu_0 \rho)]^{1/2}$ and the magnetosonic mode has speed $C_{ms} = [C_s^2 + C_A^2]^{1/2}$. For the ionosphere of Venus, $C_s \approx 2$ km s^{-1}. For the magnetized ionosphere of Venus ($B \approx 100$ nT) the Alfvén speed C_A depends on altitude but at 200 km it is $C_A \approx 2$ km s^{-1}. The Mach number of a plasma flow is just the flow speed divided by the appropriate wave speed (e.g., sonic mach number $M = u/C_s$). Flows in the lower ionosphere of Venus are subsonic ($M \ll 1$) (unmagnetized case) and submagnetosonic (magnetized case), but in the upper ionosphere, especially near the terminator, supersonic horizontal ionospheric flow with $M > 1$ seems to be present (Knudsen et al. 1980).

Static equilibria in a plasma are also not always stable even if the static balance relation, Eq. (6), is satisfied. For example, if the velocity shear is too large the Kelvin-Helmholtz instability sets in. The Rayleigh-Taylor instability can operate if the thermal pressure gradient in a gravitational field has the "wrong" sign (i.e., a less dense fluid supporting a denser fluid) even though pressure balance is satisfied. The Venus ionopause is stable against the Rayleigh-Taylor instability, but some authors think that it might be unstable against the Kelvin-Helmholtz instability (see, e.g., Wolff et al. 1980). Magnetic flux ropes at lower altitudes are thought to be unstable against the MHD kink instability, in which the rope itself curls or "kinks" up (Elphic and Russell 1983).

Various forms of the MHD equations in this section have been solved numerically for the ionosphere of Venus and used to explain and interpret the observational data. Next, in Sec. III, we will briefly review the data relevant to MHD processes, and then in Secs. IV and V we will review MHD models (conceptual and numerical) of the Venus ionosphere.

III. REVIEW OF PLASMA AND FIELD DATA FOR THE IONOSPHERE OF VENUS

A. Large-Scale Global Structure on the Day Side

Many of the basic features of the dayside ionosphere were introduced earlier with Fig. 2. For example, it was indicated that a pressure balance exists at the ionopause. Solar wind dynamic pressure is converted to magnetic pressure in the barrier and this in turn is balanced by the ionospheric thermal pressure. This pressure balance has been demonstrated statistically using PVO data (see references and discussion in Brace and Kliore [1991]). One consequence of this is that the ionopause height adjusts to the solar wind pressure. As the solar pressure increases, a higher ionospheric pressure is needed to balance it (Eq. 7), and the ionopause moves to a lower altitude where the electron density (and thermal pressure) is higher. Another consequence of this is that the average ionopause height increases with increasing solar zenith angle

(SZA) because the effective (Newtonian) solar wind dynamic pressure varies with SZA as $\rho_{sw} u_{sw}^2 \cos^2$ (SZA). Statistical studies of the PVO plasma data indicate that on the average (for solar maximum conditions) the ionopause is located near 300 km for a SZA of 0 deg. (i.e., the subsolar point) and at 1000 km for SZA = 90 deg. (the terminator) (see Fig. 41 of Brace and Kliore [1991]). The ionopause altitude continues to increase on the night side but does not become infinite, although the nightside ionospheric structure is complex and includes tail rays, ionospheric holes, etc., so that it is often difficult to define an ionopause (see Brace and Kliore 1991).

B. Large-Scale Global Structure on the Night Side

First let us consider nightside ionosphere data from the main PVO mission (high solar activity) and low solar wind dynamic pressure conditions. For solar zenith angles less than 145 deg. the observed field was small ($B < 10$ nT) on the average (Luhmann et al. 1981). Thus, for these conditions the dayside and nightside ionospheres are similar in their magnetic field structure. But for solar zenith angles greater than 145 deg. the magnetic field was usually observed to be high ($B \approx 20$ to 40 nT) and to have a significant radial/vertical component in some localized regions (Luhmann et al. 1981; cf., Luhmann and Cravens 1991). The electron density was observed to be low in these regions of elevated field strength, which are called nightside ionospheric holes (Brace et al. 1982). Figure 4 is a plot of both the magnetic and the ionospheric thermal pressures, calculated from measured magnetic field and plasma parameters for a PVO passage through an ionospheric hole. The plasma β is low inside the hole and is high outside. Hence, the magnetic field must tightly constrain the plasma dynamics inside the hole. Furthermore, the constancy of the total pressure across the hole indicates that, at least horizontally, the hole boundary is in static MHD balance (see Eq. 6). Holes are often found in pairs with opposite magnetic polarity, and their statistical locations are consistent with a draped IMF pattern (Phillips and Russell 1987).

For the main PVO mission and high solar wind dynamic pressure conditions, the magnetic field tended to be high ($B \approx 10$ to 40 nT) throughout the entire (not just localized regions) nightside ionosphere and was oriented mainly in the horizontal direction (Luhmann et al. 1980). The plasma density was observed to be much lower than average for altitudes above 200 km ($n_e < 10^3$ cm^{-3} vs 10^4 cm^{-3}) and the plasma temperatures higher than average ($T_e > 10,000$ K vs 4000 K) for these conditions, which are given the name "disappearing ionospheres" (Cravens et al. 1982). The plasma β is of the order of unity for these conditions.

The nightside ionosphere during low solar activity conditions was observed during the PVO entry phase in September and early October 1992 (Strangeway 1993). The nightside plasma densities measured during this mission phase appear to be significantly lower than the average densities measured during the main mission (i.e., factor of ten reductions in n_e for altitudes above 200 km) (Theis and Brace 1993; Ho et al. 1993; Cloutier

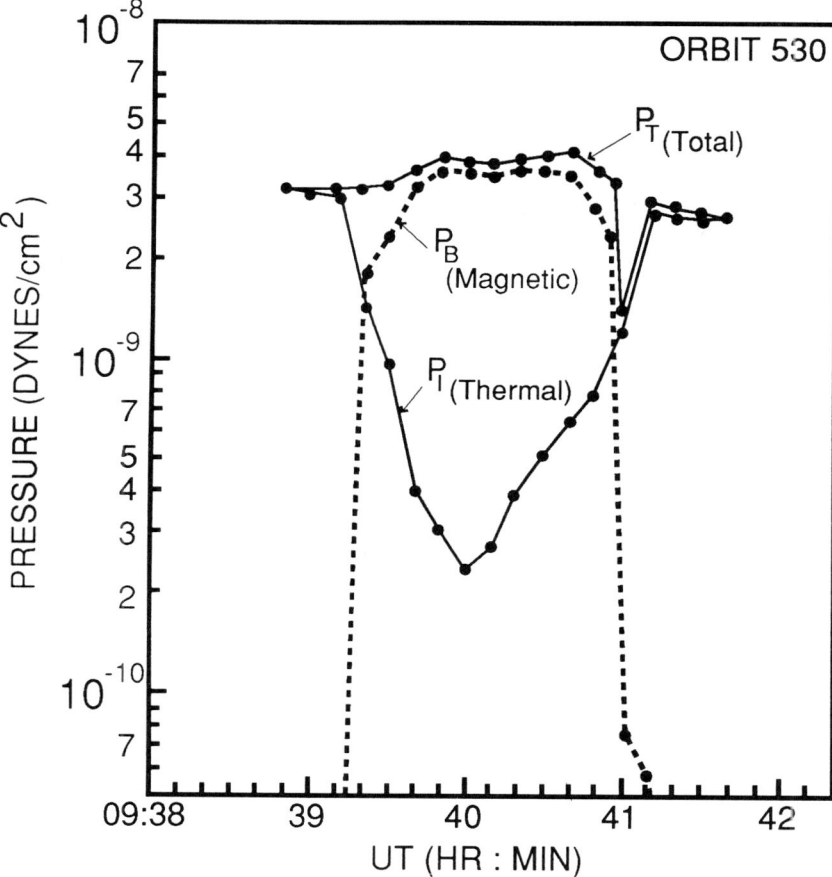

Figure 4. Magnetic pressure p_B and plasma pressure for a deep ionospheric hole. The total pressure, $P_T = P_B + P_i$, is also shown and is approximately constant across the hole. The hole is about 1000 km across (figure adapted from Brace et al. 1982; also see Luhmann and Cravens 1991).

et al. 1993). Ionospheric densities on the day side are also lower for lower solar activity conditions (Brace and Kliore 1991) because the ionizing solar extreme ultraviolet (EUV) flux is lower (chapter by Fox and Kliore). The average night side magnetic field strength for altitudes above 200 km was somewhat higher during the entry phase than during the main mission. Figure 5 displays both plasma pressure and magnetic pressure profiles on the night side calculated from measured parameters. The plasma β clearly exceeds unity for the main mission (on the average), whereas the β is about unity for the entry phase nightside ionosphere indicating an increased importance of the magnetic field in the ionospheric dynamics. The magnetic field strength

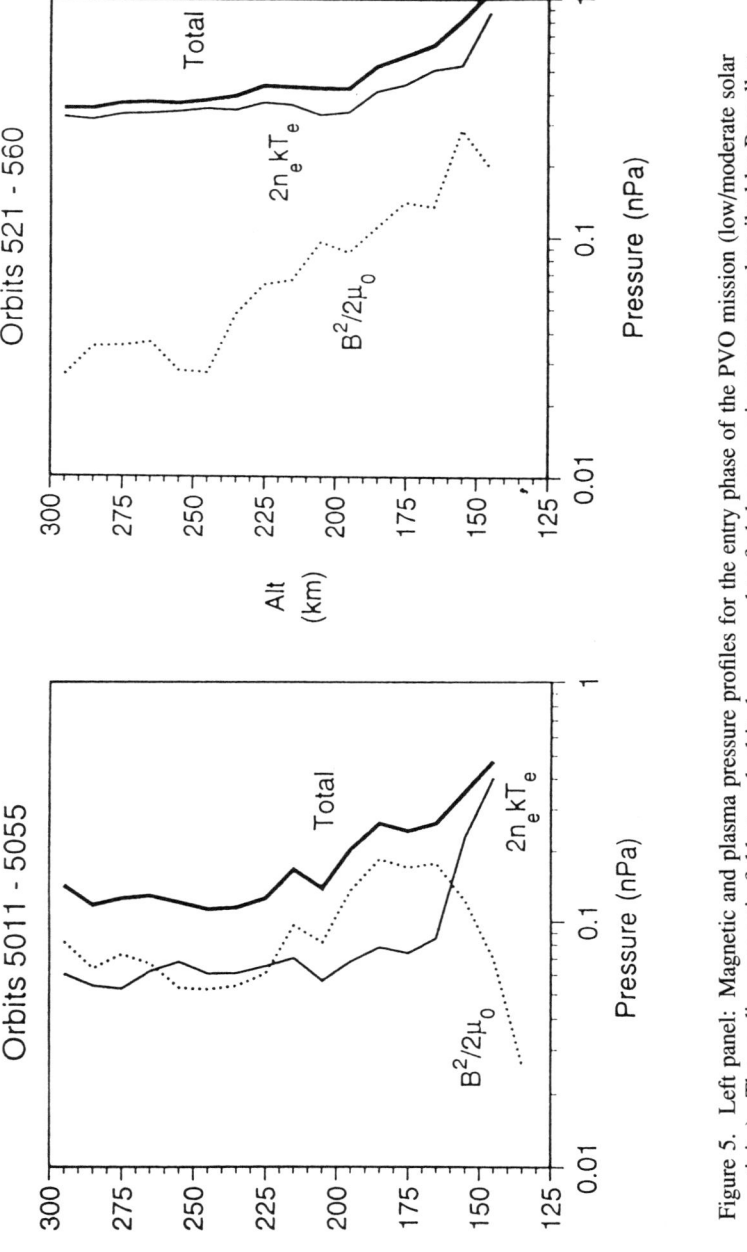

Figure 5. Left panel: Magnetic and plasma pressure profiles for the entry phase of the PVO mission (low/moderate solar activity). The median magnetic field at each altitude was used to find the magnetic pressure as described by Russell et al. (1993). The ionospheric pressure was obtained by using median values of the electron density and using main phase electron temperatures (figure from Russell et al. 1993). Right panel: Magnetic and plasma pressure profiles for the main phase of the PVO mission (high solar activity). The thermal plasma pressure was calculated assuming equal electron and ion temperatures (figure from Russell et al. 1993).

(Russell et al. 1993) also appears to be somewhat larger near 150 km than at other nightside altitudes, which is reminiscent of the day side low altitude magnetic layer during high solar wind dynamic pressure conditions. Overall, the nightside ionosphere at all times during the entry phase looks very much like the disappearing ionospheres of the main mission, which, however, occurred only during high solar wind dynamic pressure conditions. This is undoubtedly related to the lower ionospheric plasma densities (and pressures) during the lower solar activity period.

C. Small-Scale Structure Within the Ionosphere: Flux Ropes

Evident in the magnetic field profile of Orbit 186 (Fig. 2) are sharp spikes (called magnetic flux ropes) which were interpreted as force-free structures (Fig. 3). This interpretation is backed up in several ways (Elphic and Russell 1983). First, the electron density does not vary significantly across a flux rope. For an axial, non-force-free structure, with $B \approx 60$ nT, the plasma pressure (and electron density for roughly constant temperatures) would have to be reduced by about 30% to maintain force balance. At higher altitudes electron density reductions are often observed near ropes—but these are much less than 30%. The detailed structure of the magnetic field vectors during the passage of PVO through ropes also indicates a force-free structure. Figure 6 is a hodogram of the magnetic field vector during a passage of PVO though a flux rope; in this case PVO evidently passed near the center of the rope. The field is mainly axial near the rope's center and is azimuthally oriented near the outer edge (as illustrated in the schematic Fig. 3).

Studies have been made of the statistical properties of flux ropes (cf., Luhmann and Cravens 1991). At higher altitudes ropes appear to be about 15 km in diameter and to have some tendency to align horizontally, but below 250 km or so, the diameter is about 10 km and the alignment (i.e., orientation) of the ropes is random. Furthermore, the probability of finding a flux rope is highest in the 160 to 200 km altitude range and the maximum field strength in a rope statistically has a peak in the region below 200 km (Fig. 7). This region happens to be where the large-scale magnetic field has a maximum.

This flux rope discussion pertains to the dayside ionosphere in the general vicinity of the subsolar point. Magnetic structures have also been observed near the terminator, but they do not appear to be so cleanly force-free as flux ropes and they have a more axial structure with a much stronger tendency to align horizontally (Brace et al. 1983). The terminator waves, as they are called, also have significant (factors of 2) electron density and temperature variations associated with them, unlike flux ropes. Kilometer-sized waves in the electron density were also observed very deep in the nightside ionosphere during the entry phase of the PVO mission (Brace et al. 1993).

D. The Ionopause Current Layer and the Plasma Mantle

The ionopause current layer obviously has a finite thickness. Elphic et al. (1981) undertook a statistical study of this thickness and demonstrated that

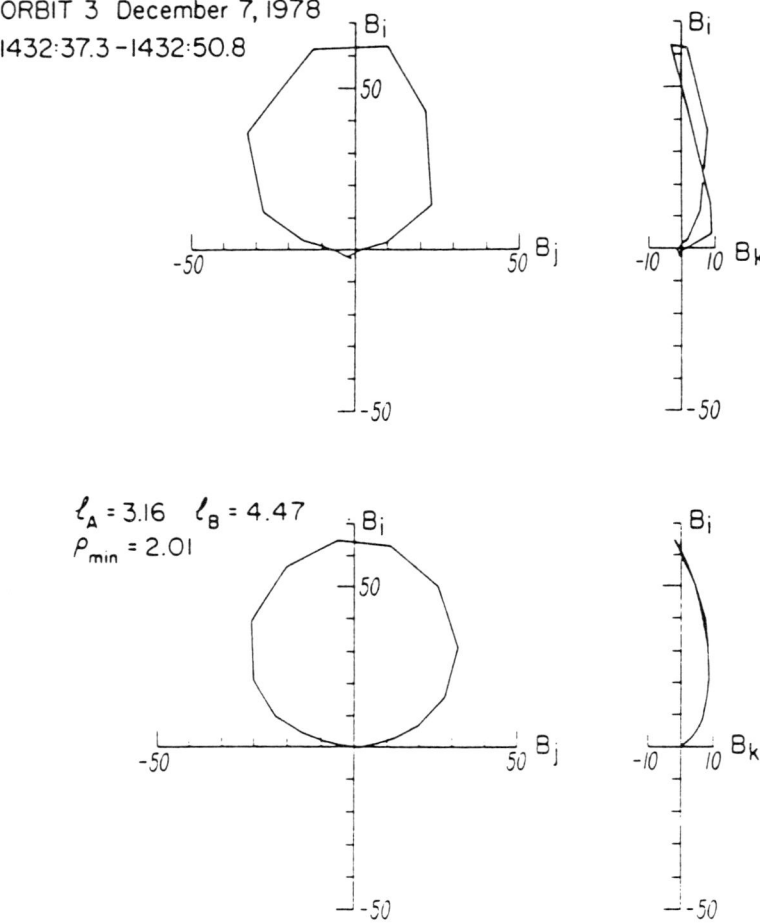

Figure 6. Hodogram in principal axis coordinates for a flux rope located at an altitude of 231 km for Orbit 185. The direction for B_k is chosen so that the field has the minimum variation (figure from Elphic and Russell 1983).

ionopauses located above about 300 km (i.e., low solar wind dynamic pressure conditions) have thicknesses of 20 to 30 km (that is, 2–10 ion gyroradii) whereas ionopauses located below 300 km (i.e., high solar wind dynamic pressures) have thicknesses of 20 to 90 km (see Fig. 11 of Russell and Vaisberg 1983). Furthermore, even for very high solar wind dynamic pressure and in the subsolar region, the ionopause has not been observed to go below 200 km.

The ionopause delineates the upper altitude limit of cold ionospheric plasma. The plasma in the magnetic barrier located just above the ionopause is much lower in density and higher in temperature than ionospheric plasma. The lower part of this magnetic barrier region has been denoted the "plasma

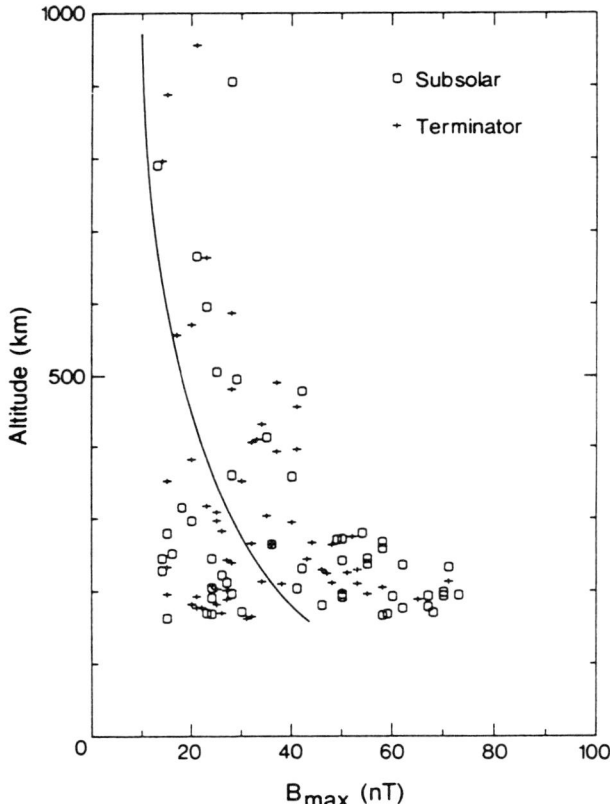

Figure 7. Statistical survey of the maximum magnetic field strength in flux ropes as a function of altitude (figure from Elphic and Russell 1983).

mantle" (Spenner et al. 1980; Vaisberg and Zeleny 1984) and contains relatively hot, or superthermal ions (Grebowsky et al. 1993; Kasprzak et al. 1982) and electrons which come from the ionization of exospheric neutrals and are "picked up" by the solar wind flow. However, due to the Zwan and Wolf effect (Zwan and Wolf 1976) and/or due to charge transfer with neutrals, the original solar wind protons seem to be highly depleted in this region. The plasma in this region is low β; magnetic pressure dominates the dynamics (cf., Russell and Vaisberg 1983; Cravens and Shinagawa 1991).

Wave activity is relatively high in the plasma mantle region (see Strangeway 1991), and observed wave intensities in the 100-Hz channel of the PVO electric field detector appear to be associated with the observation of superthermal ions either by the ion mass spectrometer (Grebowsky et al. 1993) or by the neutral mass spectrometer operating in its ion mode (Kasprzak et al. 1982). The 100-Hz wave activity also appears to be high in the upper portion of the ionopause layer itself but decreases rapidly as the electron density increases (Crawford et al. 1993). These waves have been interpreted as whistler-mode waves, and heating of the ionospheric electrons has been attributed to the Landau damping of these waves (cf., Strangeway 1991).

IV. MHD MODELS OF LARGE-SCALE STRUCTURE IN THE VENUS IONOSPHERE

A. Types of Theoretical Models

The numerous theoretical/numerical models published on the ionosphere of Venus can be classified according to what equations were solved in these models (see Cravens 1991). For example, in photochemical models only the production and loss terms of the continuity equation are solved (see, e.g., Kim et al. 1989). The energy equation by itself has often been used as a basis of models of ionospheric energetics (chapter by Nagy and Cravens). Sometimes the full set of equations, continuity, momentum, and energy are solved but for the unmagnetized ionosphere and without magnetic field terms (e.g., $\mathbf{J} \times \mathbf{B}$). This approach can be taken for a single-fluid (Eq. 4) or for many ion species (Eq. 3). This type of model has been constructed for the unmagnetized ionosphere of Venus (see, e.g., Nagy et al. 1980). Diffusion models are a special case of this type of model (the inertial terms are neglected in the momentum equation). Models of the unmagnetized ionosphere will be reviewed in the chapter by Fox and Kliore.

The ionosphere of Venus is often permeated by large-scale horizontal magnetic fields, and several approaches have been taken to numerically modeling this situation. The starting point, however, can be either the single-fluid or the multi-fluid MHD equations. For example, in the kinematic dynamo approach just the magnetic induction Eq. (8), by itself is solved. The plasma flow velocity appearing in the induction equation is not found self-consistently, but is merely adopted. Luhmann et al. (1984), Cravens et al. (1984), and Phillips et al. (1984) all presented the results of one-dimensional, time-dependent kinematic dynamo models. Cole (1990) also has worked with the one-dimensional magnetic induction equation. Luhmann (1988) constructed a three-dimensional model and Cravens et al. (1990) a two-dimensional model of the magnetic field in the ionosphere and magnetic barrier of Venus. These models have been important for understanding the magnetic field at Venus but will not be discussed further here (see Luhmann and Cravens 1991).

The continuity, momentum, and magnetic induction equations, and sometimes the energy equations, are solved for "full" MHD models. A time-dependent, one-dimensional, multi-species MHD model was constructed and used to study the magnetized ionosphere of Venus (Shinagawa et al. 1987; Shinagawa and Cravens 1988; Shinagawa 1993a). Shinagawa and Cravens (1989) used essentially the same multi-species, one-dimensional, MHD model to study the ionosphere of Mars. The Shinagawa et al. (1991) and Shinagawa

(1993a) models also included the electron and ion energy equations; the former was for solar maximum conditions and the latter was for solar minimum conditions. The inclusion of the energy equations did not significantly affect the results. Cloutier (1987) also started with the one-dimensional MHD equations but took a very different approach from the above MHD studies. The steady-state MHD equations, minus the magnetic induction equation, were integrated vertically and heat inputs were adjusted to give a magnetic field profile like the observed ones. However, the downward plasma velocity calculated for an altitude of 155 km was 600 m s^{-1}, which is much greater than values other models gave, and the calculated plasma density near this altitude was much less than observed values (see Luhmann and Cravens 1991). More recently, Shinagawa (1993b,1995) constructed a two-dimensional MHD model which will be discussed below.

B. Large-Scale Structure of the Dayside Ionosphere

Consider the Shinagawa and Cravens (1988) model (hereafter denoted SC). The equations solved numerically were essentially one-dimensional versions of the multispecies continuity and momentum equations presented in Sec. II (Eqs. 1, 2, 3, and 8) for the ion species: O_2^+, O^+, and H^+. The continuity equation alone was solved for CO_2^+. The magnetic field was assumed to be horizontal and was determined using the magnetic induction equation. Although the equations were one-dimensional (the only variable being altitude) the continuity equations included an effective loss term to take into account ion loss at high altitudes due to horizontal transport. Solutions were obtained for cases where the magnetic field strength in the barrier (top of the model) increases, decreases or remains steady over a time period of several hours. Magnetic field strength in the barrier is correlated with the solar wind dynamic pressure, as discussed earlier.

Figures 8 and 9 show density and magnetic field profiles at several times for the case of increasing solar wind dynamic pressure (i.e., transition from magnetized ionosphere to magnetized). Figure 10 shows vertical velocity profiles for O^+, O_2^+, and H^+ for another case although such profiles are similar for all cases. Note that the electron density is displayed and is the sum of the ion densities that were calculated. The ion composition resulting from the model is mainly O_2^+ at lower altitudes and O^+ at higher altitudes, in agreement with ion composition measured by PVO, but will not be discussed here (chapter by Fox and Kliore). The total ion vertical velocity which is defined as the sum over all species of the density-weighted individual ion velocity profiles (this is equal to the electron vertical velocity) is dominated by O_2^+ for altitudes below about 190 km and by O^+ at higher altitudes. The ion flow has a maximum downward value near 200 km. The magnetized ionosphere (early times) displays a low altitude (240 km) and broad ($\approx$40 to 50 km) ionopause both in the magnetic field profiles and in the electron density profiles. On the other hand, the ionopause shifts to higher altitudes (300 km) and is narrower ($\approx$20 km) at later times when the field strength in

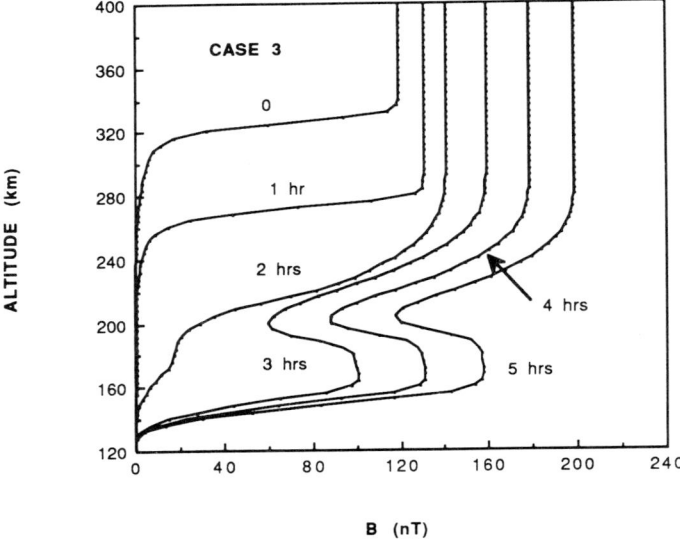

Figure 8. Magnetic field profiles from the SC multispecies MHD model—Case 3a (transition from unmagnetized to magnetized ionosphere) (figure from Shinagawa and Cravens 1988).

the barrier is weaker.

The magnetic field profile in the magnetized ionosphere exhibits the layer structure (minimum at 200 km and maximum near 160 km) present in the observed profiles (Fig. 2). Associated with this layer is a ledge in the electron density profile that has been observed (cf., Brace and Kliore 1991). The explanation for this structure is associated with the vertical velocity profiles (Cravens et al. 1984; Luhmann et al. 1984). The downward velocity profile (Fig. 10) has a maximum near an altitude of 200 km, and because the magnetic field is frozen into the flow above 150 km the field strength is lowest where the flow is fastest. Between 150 and 200 km, as the downward flow becomes slower, field lines pile up and the field strength increases. The field decreases again below about 150 km because magnetic diffusion (i.e., Ohmic dissipation) becomes important ($R_m < 1$) and the magnetic flux diffuses downward. The velocity is downward near 200 km primarily because O^+ ions produced at higher altitudes must be transported down to lower altitudes where the neutral density is higher and they are chemically removed by reaction with CO_2.

The electric field in the SC model is given by a simple version of the generalized Ohm's law and can be represented by: $E_y = -wB + D_B \partial B/\partial z$. The magnetic field is directed in the x-direction, in which case the electric field is in the y-direction. The vertical plasma velocity is denoted w. At higher altitudes where $R_m \gg 1$, $D_B \approx 0$, and only the motional electric field is

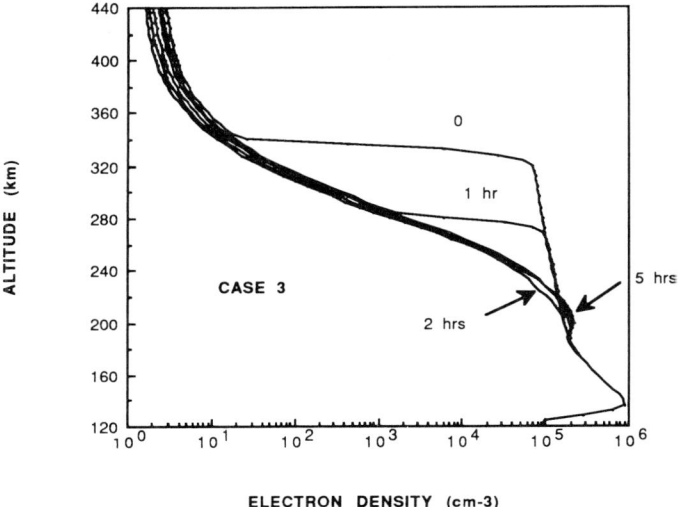

Figure 9. Electron density profiles from the SC multispecies MHD model—Case 3a (transition from unmagnetized to magnetized ionosphere) (figure from Shinagawa and Cravens 1988).

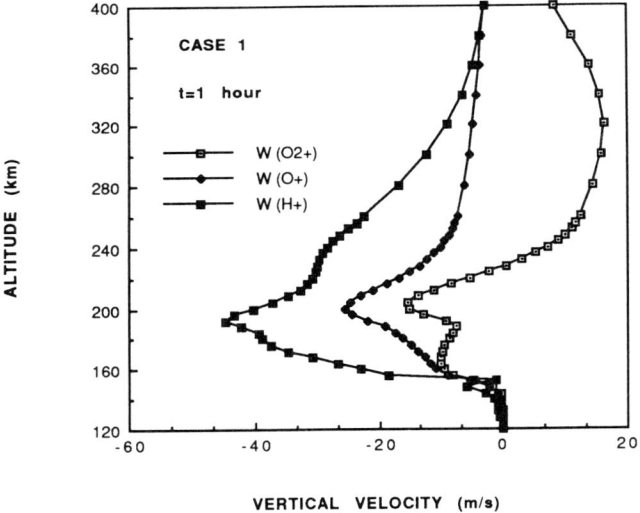

Figure 10. Vertical velocity profiles from the SC multispecies MHD model for 3 ion species—Case 2. Negative values mean downward velocity (figure from Shinagawa and Cravens 1988).

important ($E_y = -wB$), and at lower altitudes where $R_m \ll 1$, the Ohmic field dominates and $E_y \approx D_B \partial B/\partial z$. For steady state conditions, Faraday's law demands that E_y be constant, and the SC model for this case has values of $E_y \approx 1 - 2 \mu V/m$ depending on the upper boundary condition (i.e., magnetic field strength in the barrier).

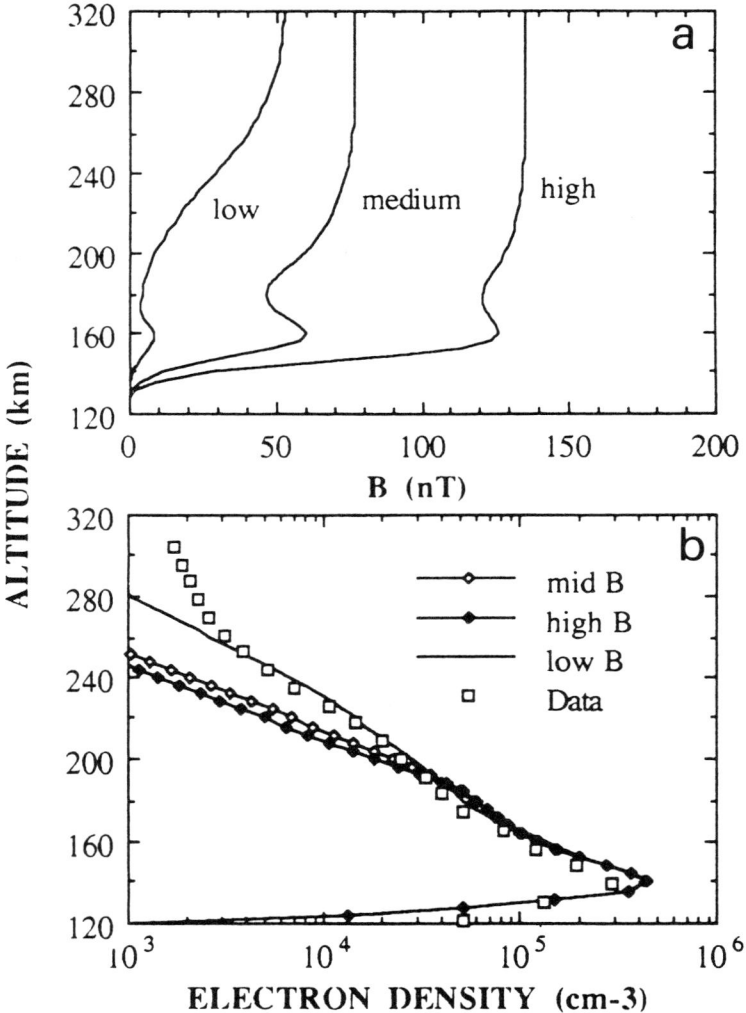

Figure 11. (a) Magnetic field profiles from one-dimensional MHD model for three cases for low solar activity; (b) electron density profiles from one-dimensional MHD model for three cases for low solar activity (figure from Shinagawa 1993a).

The SC model was for solar maximum conditions. The same model (actually the same as Shinagawa et al. [1991], with energy equations included) was applied by Shinagawa (1993a) to low/moderate solar activities conditions by adjusting the neutral atmosphere and lowering the photoionization rates. Again, the plasma velocity (not shown here) is directed downward, but the maximum speed is located near 180 km rather than the 200 km for higher activity conditions. The magnetic field and electron density profiles for three steadystate cases are reproduced in Fig. 11. There is still a general resemblance to the magnetized ionosphere of solar maximum conditions, even for relatively low high altitude magnetic fields. And the overall plasma density (and pressure) is lower when the solar activity is lower. The electron density profiles do not exhibit any obvious ionopause—there is a gradual density decrease as is evident in dayside electron density profiles measured by the radio occultation technique during solar minimum conditions (see review by Brace and Kliore 1991). Note that the magnetic layer is less pronounced than at solar maximum which is due to a less pronounced peak in the plasma velocity profile. Unfortunately, the PVO could not verify these dayside results because during its entry phase periapsis was on the night side. Both the density profiles and magnetic field profiles in this low solar activity model of the Venus ionosphere resemble the results from the Shinagawa and Cravens (1989) MHD model of the ionosphere of Mars.

C. Large-Scale Structure of the Dayside and Nightside Ionosphere: Two-Dimensional Model

A two-dimensional, single-fluid MHD model of the ionosphere of Venus was developed by Shinagawa (1993b,1995). The geometry is cylindrical with the magnetic field oriented along the axial direction. Figures 12 and 13 show magnetic field strength and electron density contours, respectively, from this model. The vertical flow (figure not included here) for SZA less than ≈ 40 deg. is downward at all altitudes, just as it was for the SC one-dimensional model, but for SZA between 40 deg. and the terminator the plasma flow is upwards for $z > 220$ km and in the antisolar direction. The figures indicate that the ionosphere for SZA < 40 deg. behaves very much like the one-dimensional SC model predicts it should. However, the ionosphere for larger solar zenith angles is field-free, although the ionopause is still broad. For solar zenith angles greater than about 140 deg. a weak magnetic field again appears in the ionosphere and the field strength has a maximum ($B \approx 20$ nT) at an altitude of about 160 km (see Fig. 12). A magnetic layer like this was observed on the night side by the PVO magnetometer (see Sec. II).

The electron density on the day side again appears similar to the results of the one-dimensional model for lower zenith angles. On the night side the density near 200 km is about 10^3 cm^{-3}, which appears to be intermediate between a "full" nightside ionosphere supported by day-to-night transport and a "disappearing ionosphere."

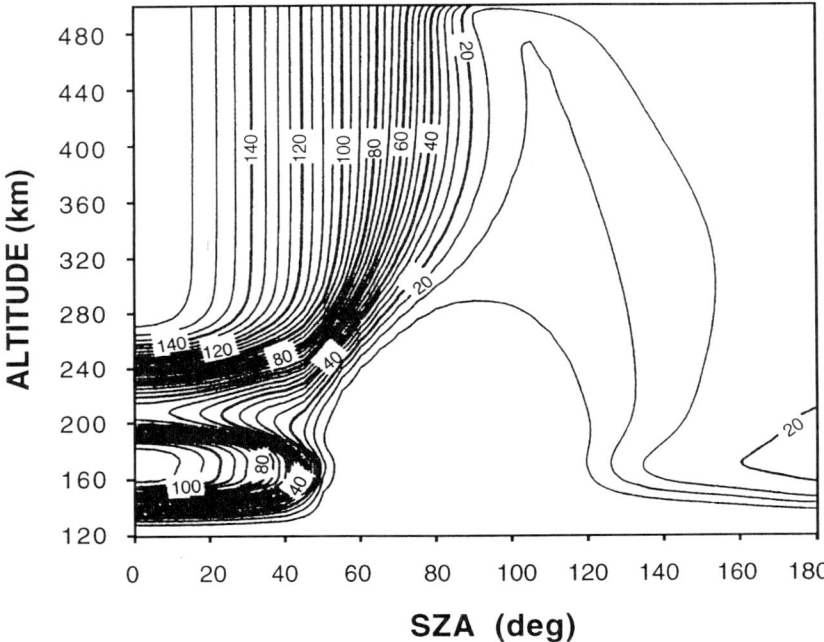

Figure 12. Magnetic field (units of nT) contours in altitude and solar zenith angle from the two-dimensional MHD model of Shinagawa (1995).

D. Interpretation of Global Magnetic Field

The plasma dynamics depends on the altitude and solar zenith angle, as well as on the state of magnetization of the ionosphere. Examining the results of the above MHD models and considering discussion in interpretation papers like Krymskii and Breus (1988), Cravens and Shinagawa (1991), and Luhmann (1992), we can reach some conclusions. For highly magnetized conditions the plasma flow probably has a significant downward component at all dayside solar zenith angles, but the plasma also rapidly flows from day to night, especially at high altitudes, driven by magnetic and thermal pressure gradients. The downward transport of ions and magnetic flux appears to be more important than horizontal transport over most of the day side in this case despite much larger horizontal flow speeds, because the relevant vertical length scale ($H \approx 20$ km) is much less than the horizontal length scale ($L \approx R_V = 6050$ km). However, during unmagnetized conditions (low solar wind pressure) the vertical extent of the ionosphere is much larger and although the plasma still flows downward below 220 km, at higher altitudes it flows upward on the day side and downward on the night side. The case modeled with two-dimensional MHD by Shinagawa (1995) appears to be intermediate between these two extremes (i.e., it is a moderately magnetized ionosphere).

Conceptual models of the three-dimensional nightside magnetic field

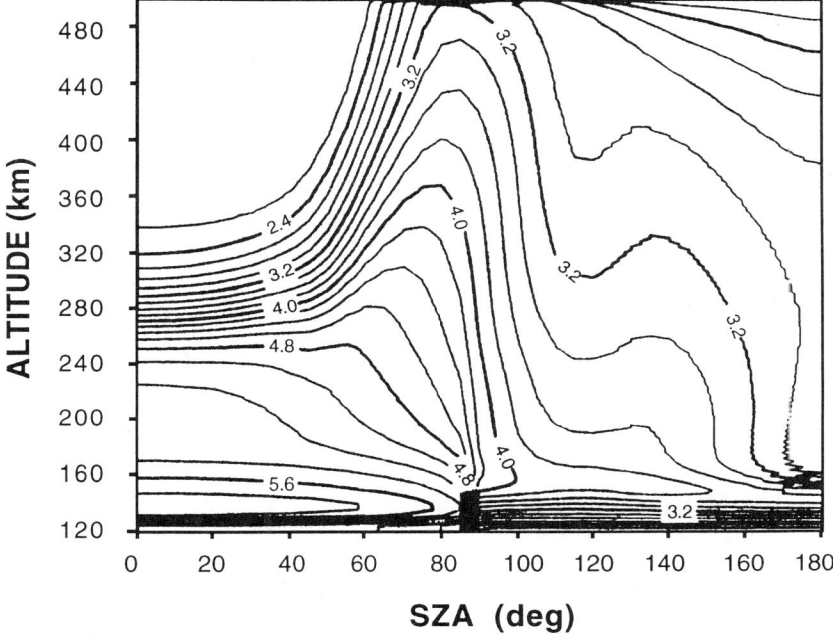

Figure 13. Electron density ($\log_{10}$ with units of cm^{-3}) contours in altitude and solar zenith angle from the two-dimensional MHD model of Shinagawa (1995).

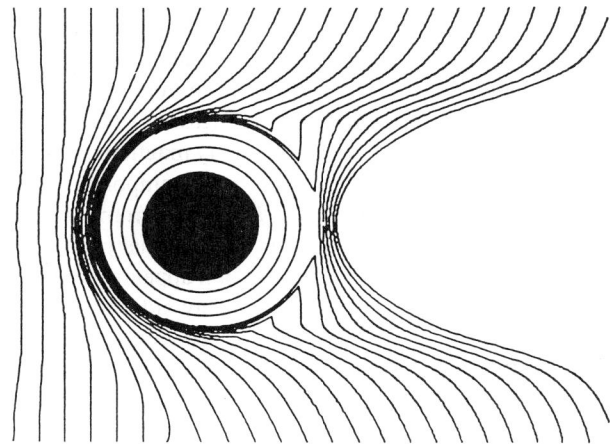

Figure 14. Field lines from Luhmann (1992) global conceptual model including ionospheric and magnetotail current systems, as well as leakage of magnetic flux through the dayside ionosphere and planetary crust and mantle. The black region is the planetary core.

were presented by both Brace et al. (1982) and Grebowsky and Curtis (1981). In these models the strong, almost radial fields found deep on the night side in ionospheric holes resulted from the convection of field lines from the day side plus draping. The Grebowsky and Curtis model also invoked reconnection in the near-Venus magnetotail which resulted in closed-off field lines around Venus at low altitudes. On the other hand, Knudsen et al. (1982) suggested that the magnetic fields in holes were the tail lobes of a weak intrinsic field "breaking through" the low-density nightside ionosphere. Luhmann (1992) suggested a conceptual model of the global electrical current pattern near Venus which included toroidal and cross-tail current systems, as well as some leakage from the day side to the night side of magnetic flux through the lower atmosphere and solid planet due to magnetic diffusion associated with high electrical resistivity (see Fig. 14). Some possible evidence in support of this leakage was given by Luhmann (1991), who presented measured dayside magnetic field profiles for very high solar wind dynamic pressure cases which suggested that the field might not go to zero at low altitudes.

V. MHD MODELS OF SMALL-SCALE STRUCTURE IN THE VENUS IONOSPHERE: FLUX ROPES AND THE IONOPAUSE

A. The Structure of the Ionopause/Plasma Mantle

Plasma dynamics in the vicinity of the ionopause and in the plasma mantle has been extensively discussed and some understanding has been obtained (see, e.g., Spenner et al. 1980; Vaisberg and Zeleny 1984; Elphic et al. 1980; Krymskii and Breus 1988; Mahajan et al. 1989; Gan et al. 1990; Cravens and Shinagawa 1991) although a fully satisfactory quantitative understanding has not yet been achieved. The main force on the plasma in the magnetic barrier, or in the magnetized ionosphere at higher altitudes, is the magnetic pressure gradient force. This force has a large vertical component but it also has a significant anti-solar horizontal component because the horizontal variation of the magnetic pressure is approximately Newtonian (i.e., $\approx \cos^2$ SZA). This force drives anti-solar plasma flow in the magnetic barrier and magnetized ionosphere. At lower altitudes ($z < 275$ km), the flow is limited by ion-neutral friction, and vertical transport is critical in determining the plasma distribution, but at higher altitudes inertia becomes important (i.e., the plasma is accelerated) and horizontal transport controls the plasma distribution. For this latter case, Cravens and Shinagawa (1991) derived simple analytical solutions to the single-fluid MHD continuity and momentum equations; the horizontal velocity u_h and the electron density n in the plasma mantle and upper magnetized ionosphere are:

$$u_h \approx \frac{p_{B0}}{mIn_n} \frac{y}{R_v^2} \approx \frac{(5 \text{km s}^{-1})}{[n_n (\text{cm}^{-3})/10^6]} \frac{y}{R_v}$$

$$n \approx \frac{mR_v^2 I^2 n_n^2}{p_{B0}} \approx 10^{-9} n_n^2 \qquad (9)$$

where $p_{B0} \approx 2 \times 10^{-8}$ dynes cm^{-2} is the magnetic pressure at the subsolar point, $m \approx 16$ amu is the ion mass (O$^+$), y is the horizontal distance from the subsolar point, R$_v$ is the radius of Venus, $I \approx 10^{-6}$ s^{-1} is the ionization frequency, and n_n is the neutral density. The flow speed is ≈ 5 km s^{-1} near 300 km and ≈ 50 km s^{-1} near 500 km according to Eq. (9). Such rapid ion flow could explain the measurements of superthermal ions mentioned earlier. The plasma density varies as the square of the neutral density giving density profiles that are in reasonable agreement with data, with the SC one-dimensional MHD model, and with the Shinagawa (1995) two-dimensional MHD model. In this scenario the shape of the (thick) ionopause for high solar wind dynamic pressure conditions (or for solar minimum ionospheres; Shinagawa 1993a) is controlled by ion-neutral friction and by vertical flow at lower altitudes and by horizontal flow at higher altitudes.

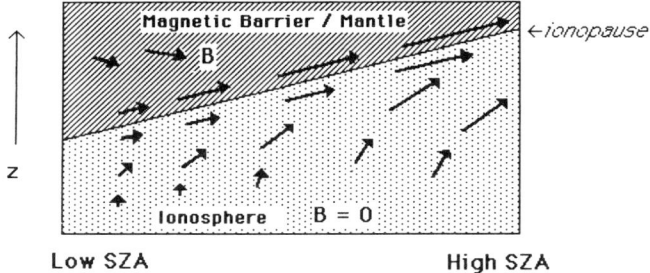

Figure 15. Conceptual plasma flow pattern in vicinity of a high altitude ionopause (from Cravens and Shinagawa 1991).

Why is the ionopause thinner for low solar wind dynamic pressure conditions? Cravens and Shinagawa (1991) suggested that convergence towards the ionopause of plasma flow (with frozen-in magnetic field lines) will naturally cause an ionopause layer to become thinner even if it started out thick. The plasma flow pattern shown in Fig. 15 was suggested, and two-dimensional MHD model (S95) results have verified the reasonableness of this pattern. The plasma flow is upward in the unmagnetized ionosphere as O$^+$ ions created in the dayside flow upward from 220 km (although downward at lower altitudes) and towards the night side in response to the thermal pressure gradient force. On the other hand, during high solar wind dynamic pressure conditions, the magnetic pressure gradient force associated with the bottom of the magnetic barrier necessitates downward plasma flow at all altitudes.

The question of the stability of the ionopause was reviewed by Brace and Kliore (1991). Clearly, the dayside ionopause is stable, as it is observed regularly on each PVO inbound and outbound pass. Johnson and Hanson (1979) suggested that the ionopause could be unstable to the flute instability and Dubinin et al. (1980) suggested that magnetic curvature would drive the interchange instability. Elphic and Ershkovich's (1984) analysis of the ionopause stability indicated that the Kelvin-Helmholtz (KH) instability could operate but that buoyancy of the plasma would stabilize the ionosphere against the flute instability. Brace et al. (1982) postulated that the breaking of ionopause waves near the terminator (perhaps caused by the KH instability) could explain the observed ionospheric clouds and streamers extending downstream from the ionosphere in the flanks.

B. Flux Ropes

Magnetic flux ropes and terminator waves are a three-dimensional MHD phenomenon and need to be studied with a high-resolution, time-dependent, three-dimensional MHD model. Unfortunately, such models have not yet been developed, and the formation and evolution of magnetic flux ropes (or terminator waves) remain poorly understood (cf., Russell 1990). Nonetheless, numerous conceptual models or more limited MHD models have appeared over the years and we will briefly review these.

Johnson and Hanson (1979) suggested that flux ropes form at the ionopause, and Wolff et al. (1980) suggested that the Kelvin-Helmholtz instability is responsible. Cloutier et al. (1983) suggested that KH operates within the ionosphere due to shear in internal plasma flow. Luhmann and Elphic (1985) and Luhmann (1990) also suggested that flux ropes form within the ionosphere, but from the redistribution of weak large-scale field by small-scale turbulent ionospheric plasma motions (assuming frozen-in magnetic field). This is a dynamo process. Kleeorin and Rogachevskii (1994) also suggested that a dynamo process plus turbulent plasma flow within the ionosphere can generate flux ropes. Cole (1994) also addressed the issue of the formation of flux ropes and their ionospheric implications. He suggested that the joule dissipation in flux ropes was so great that they must form near the location where they are observed, which thus implies that they cannot be generated by a Kelvin-Helmholtz instability near the ionopause. The evolution of flux ropes in the ionosphere, once formed, is not understood. Flux ropes if formed at the ionopause must be transported down to the lower ionosphere where they were most frequently observed by the PVO magnetometer (see Fig. 3). If a flux rope is strictly force-free then the plasma pressure (and plasma density to a lesser degree) inside the rope is the same as outside, and there is no buoyancy force on the rope. In this case, the rope will be inertly convected with the background plasma flow (neglecting the magnetic tension force on the rope as a whole). The plasma flow is downward in magnetized ionospheres, and force-free flux ropes created anywhere will be convected downward and towards the night side. Magnetic tension along the rope will tend to contribute to this downward motion. However, as Krymskii and Breus (1988) point out, if flux ropes are created at the high sharp ionopause found in unmagnetized ionospheres, then their transport down to the lower ionosphere would be opposed by the upward plasma convection and by the buoyancy force if the flux ropes are not perfectly force-free but have some density depletion. On the other hand, if flux ropes are created within the magnetized subsolar ionosphere, as Krymskii and Breus (1988) and Cloutier et al. (1983) suggested, then they would be convected downward and anti-sunward, in the same manner as the large-scale field in magnetized ionospheres. The question in this case is why the ropes would persist long after the large-scale field has disappeared. Shinagawa (unpublished manuscript) constructed a two-dimensional single-fluid MHD model of the evolution of a flux rope as it convects downward. The rope is assumed to start off force-free and the model shows that it remains force-free as it gradually is convected into the lower altitude region where magnetic diffusion gradually destroys the rope. The maximum field strength in the rope is qualitatively like that in the large-scale magnetic field and has also been seen in the statistical study of ropes (see Sec. II). None of the work carried out up to now has answered the question of why the ropes are force-free.

VI. SUMMARY AND FUTURE DIRECTIONS

Both one and two-dimensional MHD models have been developed for the ionosphere of Venus. Both large-scale and small-scale properties of the ionosphere have been studied with these models. Considerable progress has been made on understanding the large-scale structure of the dayside Venus ionosphere, but progress on the nightside magnetized ionosphere and on small-scale structure has been more limited. For example, the small-scale structure of the ionopause current layer is not understood theoretically, although its thickness is known observationally to be only a couple of O^+ gyroradii. The formation and evolution of magnetic flux ropes also remain poorly understood although a fair amount of theoretical work has been undertaken.

Future progress in understanding MHD processes in the ionosphere of Venus can be expected from:

1. Development of high-resolution, three-dimensional global MHD models of the ionosphere of Venus which properly incorporate aeronomical processes.
2. Ionospheric plasma and field data returned from the planned Japanese Planet-B mission and the Russian Mars-96 mission to Mars. The ionospheres of Mars and Venus are very similar, and the interaction of these two planets with the solar wind could also be very similar if the intrinsic magnetic field of Mars is as small as some recent studies suggest (see, e.g., Slavin and Holzer 1981; Luhmann et al. 1987; Shinagawa and Cravens 1989).
3. Possible future missions to Venus which focus on measurement of parameters relevant to MHD processes in the ionosphere. Perhaps the NASA Discovery program will provide such an opportunity.

Acknowledgments. Support from grants from NASA and the National Science Foundation is acknowledged.

REFERENCES

Banks, P. M., and Kockarts, G. 1973. *Aeronomy* (San Diego: Academic Press).
Brace, L. H. 1993. Kilometer-size waves in electron density in the Venusian nightside ionosphere. *Geophys. Res. Lett.* 20:2759.
Brace, L. H., and Kliore, A. J. 1991. The structure of the Venus ionosphere. *Space Sci. Rev.* 55:81–163.
Brace, L. H., et al. 1982. Holes in the nightside ionosphere of Venus. *J. Geophys. Res.* 87:199.
Brace, L. H., et al. 1983. The ionosphere of Venus: Observations and their interpretation. In *Venus*, eds. D. M. Hunten, L. Colin, T. M. Donahue and V. I. Moroz (Tucson: Univ. of Arizona Press), pp. 779–840.
Brace, L. H., Hartle, R. E., and Theis, R. F. 1995. The nightward ion flow scenario at Venus revisited. *Adv. Space Res.* 16(6):99.
Cloutier, P. A. 1987. Steady state flow/field model of solar wind interaction with Venus: Global implication of local effects. *J. Geophys. Res.* 92:7289.
Cloutier, P. A., et al. 1983. Physics of the interaction of the solar wind with the ionosphere of Venus: Flow/field models. In *Venus*, eds. D. M. Hunten, L. Colin, T. M. Donahue and V. I. Moroz (Tucson: Univ. of Arizona Press), pp. 941–979.
Cloutier, P. A., Kramer, L., and Taylor, H. A., Jr. 1993. Observations of the nightside Venus ionosphere: Final encounter of the Pioneer Venus orbiter ion mass spectrometer. *Geophys. Res. Lett.* 20:2731.
Cole, K. D. 1990. Electric currents in E-like planetary ionospheres. *Planet. Space Sci.* 38:1061.
Cole, K. D. 1994. Origin of flux ropes in Venus' ionosphere. *J. Geophys. Res.* 99:14951.
Cole, K. D., and Hoegy, W. R. 1994. Electric currents in the subsolar region of the Venus lower ionosphere. *J. Geophys. Res.* 99:8791.
Cravens, T. E. 1989. The solar wind interaction with non-magnetic bodies and the role of small-scale structures. In *Solar System Plasma Physics*, eds. J. H. Waite, Jr., et al. (Washington, D. C.: American Geophysical Union), p. 353.
Cravens, T. E. 1991. Ionospheric models for Venus and Mars. In *Venus and Mars: Atmospheres, Ionospheres, and Solar Wind Interaction*, ed. J. G. Luhmann (Washington, D. C.: American Geophysical Union), p. 277.
Cravens, T. E., and Shinagawa, H. 1991. The ionopause current layer at Venus. *J. Geophys. Res.* 96:11119.
Cravens, T. E., et al. 1982. Disappearing ionospheres on the nightside of Venus. *Icarus* 51:271–282.
Cravens, T. E., Shinagawa, H., and Nagy, A. F. 1984. The evolution of large-scale magnetic fields in the ionosphere of Venus. *Geophys. Res. Lett.* 11:267.
Cravens, T. E., Wu, D., and Shinagawa, H. 1990. A two-dimensional kinematic dynamo model of the ionospheric magnetic field at Venus. *Geophys. Res. Lett.* 17:2261.
Crawford, G. K., Strangeway, R. J., and Russell, C. T. 1993. VLF emissions at the Venus dayside ionopause. In *Plasma Environments of Non-Magnetic Planets*, vol. 4, ed. T. I. Gombosi (Oxford: Pergamon Press), p. 253.
Dubinin, E. M., Izrailevich, P. L., Shkolnikova, S. M., and Podgorny, I. M. 1980. Nature of flux ropes in the Venus ionosphere. *Pis'mu V A J.* 6:253.
Elphic, R. C., and Ershkovich, A. 1984. On the stability of the ionopause of Venus. *J. Geophys. Res.* 89:997.
Elphic, R. C., and Russell, C. T. 1983. Magnetic flux ropes in the Venus ionosphere: Observations and models. *J. Geophys. Res.* 88:58.

Elphic, R. C., Russell, C. T., Slavin, J. A., and Brace, L. H. 1980. Observations of the dayside ionopause and ionosphere of Venus. *J. Geophys. Res.* 85:7679.
Elphic, R. C., Russell, C. T., and Luhmann, J. G. 1981. The Venus ionopause current sheet: Thickness length scale and controlling factors. *J. Geophys. Res.* 86:11430.
Fox, J. L., and Bougher, S. W. 1991. Structure, luminosity, and dynamics of the Venus thermosphere. *Space Sci. Rev.* 55:357–489.
Gan, L., Cravens, T. E., and Horanyi, M. 1990. Electrons in the ionopause boundary layer of Venus. *J. Geophys. Res.* 95:19023.
Grebowsky, J. M., and Curtis, S. A. 1981. Venus nightside ionospheric holes—The signatures of parallel electric field acceleration regions. *Geophys. Res. Lett.* 12:1273.
Grebowsky, J. M., et al. 1993. Superthermal ions detected in Venus' dayside ionosheath, ionopause, and magnetic barrier regions. *J. Geophys. Res.* 98:9055.
Gringauz, K. I. 1983. The bow shock and the magnetosphere of Venus according to measurements from Venera 9 and 10 orbiters. In *Venus*, eds. D. M. Hunten, L. Colin, T. M. Donahue and V. I. Moroz (Tucson: Univ. of Arizona Press), pp. 980–993.
Ho, C. M., et al. 1993. The nightside ionosphere of Venus under varying levels of solar EUV flux. *Geophys. Res. Lett.* 20:2727.
Johnson, F. S., and Hanson, W. B. 1979. A new concept for the daytime magnetosphere of Venus. *Geophys. Res. Lett.* 6:581.
Kasprzak, W. T., Taylor, H. A., Brace, L. H., and Niemann, H. B. 1982. Observations of energetic ions near the Venus ionopause. *Planet. Space Sci.* 30:1107.
Kim, J., Nagy, A. F., Cravens, T. E., and Kliore, A. J. 1989. Solar cycle variations in the lower ionosphere of Venus. *J. Geophys. Res.* 94:11997.
Kleeorin, N., and Rogachevskii, I. 1994. A mechanism of magnetic flux rope formation in the ionosphere of Venus. *J. Geophys. Res.* 99:6475.
Knudsen, W. C. 1991. The Venus ionosphere from in situ measurements. In *Venus and Mars: Atmospheres, Ionospheres, and Solar Wind Interaction*, ed. J. G. Luhmann (Washington, D. C.: American Geophysical Union), p. 66.
Knudsen, W. C., Spenner, K., Miller, K. L., and Novak, V. 1980. Transport of ionospheric O+ ions across the Venus terminator and implications. *J. Geophys. Res.* 85:7803.
Knudsen, W. C., Banks, P. M., and Miller, K. L. 1982. A new concept of plasma motion and planetary magnetic field for Venus. *Geophys. Res. Lett.* 9:765.
Krymskii, A. M., and Breus, T. K. 1988. Magnetic fields in the Venus ionosphere: General features. *J. Geophys. Res.* 93:8459.
Luhmann, J. G. 1986. The solar wind interaction with Venus. *Space Sci. Rev.* 44:241.
Luhmann, J. G. 1988. A three-dimensional diffusion convection model of the large scale magnetic field in the Venus ionosphere. *J. Geophys. Res.* 93:5909.
Luhmann, J. G. 1990. "Wave" analysis of Venus ionospheric flux ropes. In *Physics of Magnetic Flux Ropes*, eds. C. T. Russell, E. R. Priest and L. C. Lee (Washington, D. C.: American Geophysical Union), p. 425.
Luhmann, J. G. 1991. Induced magnetic fields at the surface of Venus inferred from Pioneer Venus Orbiter near-periapsis measurements. *J. Geophys. Res.* 96:18831.
Luhmann, J. G. 1992. Pervasive large-scale magnetic fields in the Venus nightside ionosphere and their implications. *J. Geophys. Res.* 97:6103.
Luhmann, J. G., and Cravens, T. E. 1991. Magnetic fields in the ionosphere of Venus. *Space Sci. Rev.* 55:201.
Luhmann, J. G., and Elphic, R. C. 1985. On the dynamo generation of flux ropes in the Venus ionosphere. *J. Geophys. Res.* 90:12047.
Luhmann, J. G., et al. 1980. Observations of large scale steady magnetic fields in the dayside Venus ionosphere. *Geophys. Res. Lett.* 7:917.

Luhmann, J. G., et al. 1981. Observations of large scale steady magnetic fields in the nightside Venus ionosphere and near wake. *Geophys. Res. Lett.* 8:517.

Luhmann, J. G., Russell, C. T., and Elphic, R. C. 1984. Time scales for the decay of induced large-scale magnetic fields in the Venus ionosphere. *J. Geophys. Res.* 89:362.

Luhmann, J. G., et al. 1987. Characteristics of the Marslike limit of the Venus-solar wind interaction. *J. Geophys. Res.* 92:8545.

Mahajan, K. K., and Kar, J. 1988. Planetary ionospheres. *Space Sci. Rev.* 47:303.

Mahajan, K. K., Mayr, H. G., Brace, L. H., and Cloutier, P. A. 1989. On the lower altitude limit of the Venusian ionopause. *Geophys. Res. Lett.* 16:759.

Miller, K. L., and Whitten, R. C. 1991. Ion dynamics in the Venus ionosphere. *Space Sci. Rev.* 55:165.

Nagy, A. F., et al. 1980. Model calculations of the dayside ionosphere of Venus: Ionic composition. *J. Geophys. Res.* 85:7795.

Nagy, A. F., et al. 1983. Basic theory and model calculations of the Venus ionosphere. In *Venus*, eds. D. M. Hunten, L. Colin, T. M. Donahue and V. I. Moroz (Tucson: Univ. of Arizona Press), pp. 841–872.

Nicholson, P. R. 1983. *Introduction to Plasma Theory* (New York: J. Wiley and Sons).

Niemann, H. B., et al. 1980. Mass spectrometric measurements of the neutral gas composition of the thermosphere and exosphere of Venus. *J. Geophys. Res.* 85:7817–7827.

Phillips, J. L., and Russell, C. T. 1987. Upper limit on the intrinsic magnetic field of Venus. *J. Geophys. Res.* 92:2253.

Phillips, J. L., Luhmann, J. G., and Russell, C. T. 1984. Growth and maintenance of large-scale magnetic fields in the dayside Venus ionosphere. *J. Geophys. Res.* 89:10676.

Phillips, J. L., Luhmann, J. G., and Russell, C. T. 1985. Dependence of Venus ionopause altitude and ionospheric magnetic field on solar wind dynamic pressure. *Adv. Space Res.* 5:173.

Russell, C. T. 1990. Magnetic flux ropes in the ionosphere of Venus. In *Physics of Magnetic Flux Ropes* (Washington, D. C.: American Geophysical Union).

Russell, C. T., and Elphic, R. C. 1979. Observations of magnetic flux ropes in the Venus ionosphere. *Nature* 279:616.

Russell, C. T., and Vaisberg, O. 1983. The interaction of the solar wind with Venus. In *Venus*, eds. D. M. Hunten, L. Colin, T. M. Donahue and V. I. Moroz (Tucson: Univ. of Arizona Press), pp. 873–940.

Russell, C. T., Strangeway, R. J., Luhmann, J. G., and Brace, L. H. 1993. The magnetic state of the lower ionosphere during Pioneer Venus entry phase. *Geophys. Res. Lett.* 20:2723.

Shinagawa, H. 1993a. Model calculations of the dayside ionosphere of Venus at solar minimum. *Geophys. Res. Lett.* 20:2743.

Shinagawa, H. 1993b. A two-dimensional MHD model of the solar wind interaction with the Venus ionosphere. In *Plasma Environments of Non-Magnetic Planets*, vol. 4, ed. T. I. Gombosi (Oxford: Pergamon Press), p. 199.

Shinagawa, H. 1995. Two-dimensional model of the Venus ionosphere, 2: Magnetized ionosphere. *J. Geophys. Res.*, in press.

Shinagawa, H., and Cravens, T. E. 1988. A one-dimensional multispecies magnetohydrodynamical model of the dayside ionosphere of Venus. *J. Geophys. Res.* 93:11263.

Shinagawa, H., and Cravens, T. E. 1989. A one-dimensional multispecies magnetohydrodynamical model of the dayside ionosphere of Mars. *J. Geophys. Res.* 94:6506.

Shinagawa, H., Cravens, T. E., and Nagy, A. F. 1987. A one-dimensional time-dependent model of the magnetized ionosphere of Venus. *J. Geophys. Res.* 92:7317.

Shinagawa, H., Kim, J., Nagy, A. F., and Cravens, T. E. 1991. A comprehensive magnetohydrodynamic model of the Venus ionosphere. *J. Geophys. Res.* 96:11083.

Siscoe, G. L. 1983. Solar system magnetohydrodynamics. In *Solar Terrestrial Physics*, eds. R. L. Carovillano and J. M. Forbes (Dordrecht: D. Reidel), pp. 11–100.

Slavin, J. A., and Holzer, R. E., 1981. Solar wind flow about the terrestrial planets 1. Modeling bow shock position and shape. *J. Geophys. Res.* 86:11401.

Spenner, K., et al. 1980. Observation of the Venus mantle: The boundary region between solar wind and ionosphere. *J. Geophys. Res.* 85:7655–7662.

Strangeway, R. J. 1991. Plasma waves at Venus. *Space Sci. Rev.* 55:275.

Strangeway, R. J. 1993. The Pioneer Venus Orbiter entry phase. *Geophys. Res. Lett.* 20:2715–2722.

Theis, R. F., and Brace, L. H. 1993. Solar cycle variations of electron density and temperature in the Venusian nightside ionosphere. *Geophys. Res. Lett.* 20:2719.

Vaisberg, O. L., and Zeleny, I. M. 1984. Formation of the plasma mantle in the Venusian magnetosphere. *Icarus* 58:412.

Wolff, R. S., Goldstein, B. F., and Yeates, C. M. 1980. The onset and development of Kelvin-Helmholtz instability at the Venus ionopause. *J. Geophys. Res.* 85:7697.

Zwan, B. J., and Wolf, R. A. 1976. Depletion of solar wind plasma near a planetary boundary. *J. Geophys. Res.* 81:1636.

PLASMA WAVE PHENOMENA AT VENUS

J. D. HUBA
Naval Research Laboratory

and

R. J. STRANGEWAY
University of California at Los Angeles

A review of wave phenomena in the space environment surrounding Venus is presented. The emphasis of the review is on research carried out over the past decade using Pioneer Venus Orbiter (PVO) data. The types of wave behavior to be discussed are fluctuations in the electric field, the magnetic field, the electron density, and the electron temperature. Observational measurements from the Orbiter Electric Field Detector (OEFD), Orbiter Electron Temperature Probe (OETP), and Orbiter Magnetometer (OMAG) are presented to illustrate the wave activity. Theoretical analyses are described that have been proposed to explain the observations. The wave regions reviewed in detail are the foreshock, mantle, and ionosphere. Finally, a discussion of outstanding problems that remain is given.

I. INTRODUCTION

The Pioneer Venus Orbiter (PVO) mission has provided a wealth of information on the wave activity in the space environment surrounding Venus. Wave behavior is defined as fluctuations in the electric field, magnetic field, electron density, and/or electron temperature. In this chapter we primarily review observations and theoretical explanations that have received attention during the latter part of the PVO mission. Three major regions are discussed in detail: the foreshock, the mantle, and the nightside ionosphere. Additionally, a section is included on notable wave phenomena that do not yet have satisfactory theoretical explanations.

The primary instrument used to identify and study wave behavior during the PVO mission was the Orbiter Electric Field Detector (OEFD). This instrument was designed to measure electric field fluctuations at center frequencies 100 Hz, 730 Hz, 5.4 kHz, and 30 kHz with a 30% bandwidth; further details on the OEFD can be found in Scarf et al. (1980a). Early results of wave behavior discovered by the OEFD are reported in Scarf et al. (1980b); a more recent review of OEFD wave measurements is found in Strangeway (1991). It is important to emphasize that fluctuating magnetic field measurements were not made at the same frequencies as the OEFD. This has led to ambiguities over the identification of the electric field signals as to whether they are electrostatic or electromagnetic, and has contributed to the uncertainty surrounding the identification of 100 Hz signals in the nightside ionosphere as lightning-generated whistler waves (see the chapter by Grebowsky et al.).

In addition to the OEFD, other PVO instruments have provided information on wave activity near Venus; in particular, the Orbiter Electron Temperature Probe (OETP) and the Orbiter Magnetometer (OMAG). The OETP is a Langmuir probe which measures the current as the voltage on the collector is swept; the electron temperature T_e and electron density n_e can be obtained from the current-voltage characteristics. Further details on this instrument are given in Krehbiel et al. (1980). The OMAG instrument is a flux-gate magnetometer and provided simultaneous measurements of all three components of the magnetic field for most of the PVO mission; only one component of the field was measured during the final descent phase of the mission. Details of the instrument are provided in Russell et al. (1980).

II. FORESHOCK WAVES

The wave behavior observed in the Venus foreshock by the OEFD has been investigated in detail recently (Crawford et al. 1990,1993a; Strangeway and Crawford 1995). These studies are significant because they are the first comprehensive studies of wave behavior in the foreshock region of an unmagnetized body in the solar wind. Furthermore, this wave behavior can be compared to wave behavior in the Earth's foreshock; such a comparison can lead to valuable insights into the nature of the solar wind interaction with solar system bodies. One advantage to analyzing the Venus foreshock region over that of the Earth is the stability of the Venus shock wave. Zhang et al. (1990) has shown that the Venus bow shock essentially maintains a constant shape over the entire solar cycle; this allows the Venus foreshock to be modeled more accurately than the Earth's.

Figure 1 shows the configuration of the Venus foreshock region, and distinguishes the electron and ion foreshock regions. These regions are defined by the population of counterstreaming particles within them. Electrons and ions are energized by their interaction with the shock, and some can "escape" upstream of the shock along the magnetic field. Electrons typically have much higher velocities than ions and therefore can penetrate further upstream of the shock than the ions; additionally, both electrons and ions are convected downstream at the solar wind velocity. Thus, some distance away from the shock itself, the electron foreshock is slightly downstream of the tangent magnetic field line, and the ion foreshock is further downstream. The non-Maxwellian distributions associated with the energetic populations of electrons and ions in the foreshock region provide the source of "free energy" to drive plasma instabilities that generate the observed electric field fluctuations.

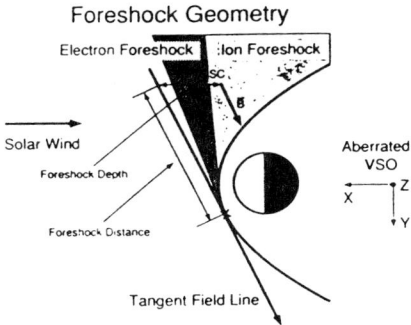

Figure 1. Schematic of the geometry of the Venus foreshock showing the electron and ion foreshock regions (figure from Crawford et al. 1993a).

A. Electron Foreshock Waves

The initial study of waves in the Venus electron foreshock is reported in Crawford et al. (1990). The dominant waves observed are in the 30 kHz band of the OEFD and are waves at the electron plasma frequency $f = f_{pe} = (n_e e^2/\pi m_e)^{1/2}$. A plasma density $n_e \simeq 10$ cm^{-3} corresponds to a plasma frequency $f \simeq 30$ kHz; the plasma density in the foreshock during the observational period was typically $n_e \simeq 5$ to 36 cm^{-3} as measured by the Orbiter Plasma Analyzer (OPA) (Intriligator et al. 1980).

An example of these wave observations is shown in Fig. 2 which plots electric field fluctuations in the 30 kHz channel, the position of the PVO, and magnetic field data during a period of time when the PVO was upstream of the bow shock crossing at 09:36. The salient point of the figure is the sudden onset of intense wave activity at time 08:53 when the magnetic field at the spacecraft becomes connected to the bow shock; this is indicated by the sign change of the depth parameter from negative to positive. Depth is defined as the distance along the solar wind flow direction from the tangent line to the spacecraft; it is positive for locations downstream of the tangent line. There is no significant wave activity in the lower OEFD frequency channels during this period.

Crawford et al. (1990) also found that the wave intensity is a maximum at the foreshock boundary, and it decreases as the depth into the foreshock increases. The peak amplitude of the waves was estimated to be 10 mV/m which is comparable to that of terrestrial electron foreshock waves (Filbert and Kellogg 1979). Finally, polarization studies indicate the fluctuating electric field is aligned with the ambient magnetic field. Further studies of electron foreshock wave behavior were carried out by Crawford et al. (1993c) and the results were similar to those of the initial study.

Recently, Strangeway and Crawford (1995) performed an extensive analysis of electron foreshock waves. They produced maps of VLF emissions at 30 kHz (and 5.4 kHz which will be discussed later) in the Venus space envi-

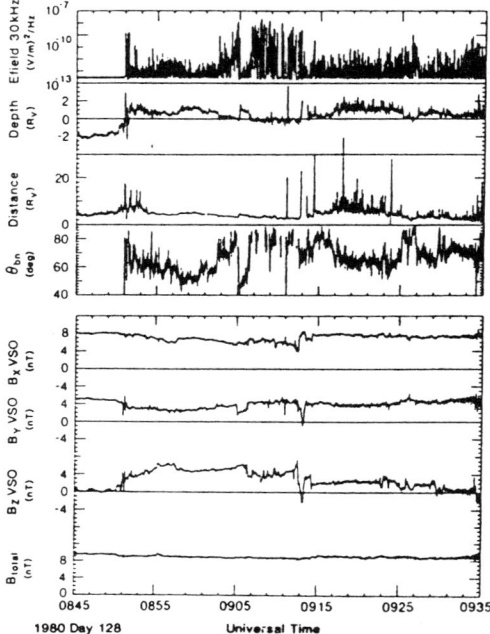

Figure 2. Wave data from the 30 kHz OEFD channel, the PVO position, and magnetic field data versus universal time from PVO orbit 519 (figure from Crawford et al. 1990). The sudden onset of intense electric field fluctuations occurs at 08:53 when the satellite crosses into the electron foreshock region.

ronment. An example of a VLF map is shown in Fig. 3. A grey-scale contour plot of the 9th decile of the 30 kHz wave intensity is shown as a function of spatial position in Venus radii (see Strangeway and Crawford [1995] for details on the spatial coordinates). The data is accumulated in 1×1 R_V bins. The number of samples per bin is also shown in Fig. 3, as well as the bow shock model and magnetic field tangent line.

The prominent features of Fig. 3 are the following. The 30 kHz waves are most intense just downstream of the magnetic field tangent line. Their maximum intensity only extends over a distance of several R_V from the tangent line. The intensity of the emissions falls off sharply for distances greater than ~ 15 R_V upstream of the tangency point. The emissions are presumed to be generated by an electron beam accelerated upstream from the Venus bow shock. The electrons are energized at the shock by fast Fermi acceleration (also known as shock drift acceleration) (Leroy and Mangeney 1984). The VLF emission maps suggest that the electron foreshock waves are controlled, in part, by the size of the Venus bow shock because of limitations on the fast Fermi energization process associated with shock curvature (Strangeway and Crawford 1995).

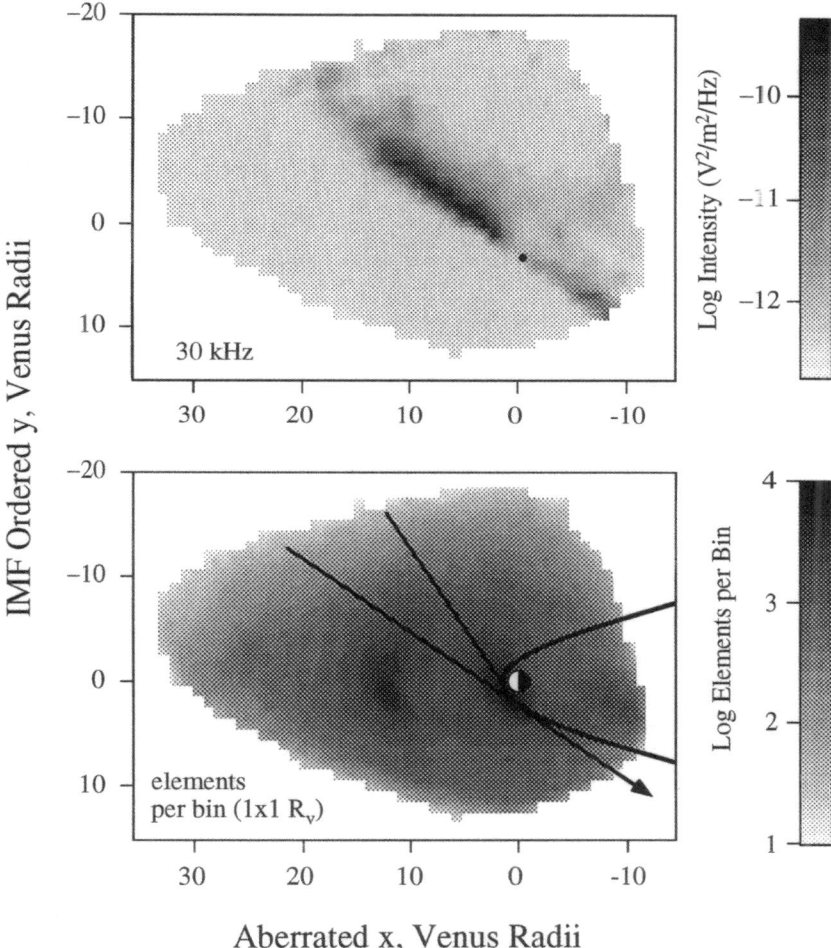

Figure 3. Maps of (a) 30 kHz intensity and (b) data elements per bin size 1×1 R_V vs spatial location (figure from Strangeway and Crawford 1995). Also shown in (b) are the positions of the bow shock model used in the analysis and the tangent magnetic field line.

Structured Langmuir waves (electron plasma oscillations) were also observed in the electron foreshock region by the Galileo satellite (Hospodarsky et al. 1994). The primary emission occurs around the electron plasma frequency (~43 kHz), but frequency shifts of up to 20 kHz above and below the plasma frequency are also observed. These shifts are correlated with the distance downstream from the tangent line; this suggests that they are associated with the electron beam velocity. Very fine scale structure, on time scales as short as 0.15 ms, is observed, as well as beat-type waveforms. This latter observation is interpreted to be caused by a parametric decay process.

As noted earlier, the cause of the observed electron oscillations is a plasma instability associated with the energetic, counterstreaming electron population reflected at the shock. This population is not observed because of instrument limitations on the PVO. However, because the Venus electron foreshock observations are very similar to those of the Earth (Filbert and Kellogg 1979), one is tempted to infer the cause of Venus foreshock oscillations from terrestrial data. For example, Fitzenreiter et al. (1984) made measurements of the electron velocity distribution in the Earth's electron foreshock region. "Bump-on-tail" distributions were observed in the velocity component parallel to the ambient magnetic; this type of distribution is unstable to electron plasma waves. More detailed theories have been proposed for the Earth's foreshock region that could be applied to the Venus foreshock (Onsager and Holzworth 1990).

Finally, we point out that electric field oscillations have been observed upstream of the Martian bowshock by Phobos 2 (Skalsky et al. 1992; Trotignon et al. 1992) and are similar to those observed at Venus and Earth. These waves are in the frequency range 5 to 40 kHz and have peak intensities at or near the local solar wind plasma frequency. The wave polarization is such that the electric field vector is parallel to the ambient magnetic field.

B. Ion Foreshock Waves

The Venus ion foreshock region is shown schematically in Fig. 1. This region is more difficult to identify than the electron foreshock region because the interaction of the solar wind ions with the shock is sensitive to the shock normal angle θ_{Bn}. Studies of the terrestrial ion foreshock region (Bonifazi and Moreno 1981a,b; Thomsen 1985) reveal that three distinct ion populations exist: field-aligned beams, intermediate distributions, and diffuse distributions. In general, field-aligned ion beams occur for $90° > \theta_{Bn} > 45°$ (quasi-perpendicular interaction), intermediate ion distributions for $70° > \theta_{Bn} > 25°$, and diffuse ion distributions for $45° > \theta_{Bn} > 0°$ (quasi-parallel interaction). In the Earth's ion foreshock both ULF and VLF waves are observed. The ULF waves are evident in magnetometer measurements (Thomsen 1985) and occur deep in the ion foreshock (i.e., far downstream of the ion foreshock line shown in Fig. 1) where the intermediate and diffuse ion populations dominate. The VLF waves are observed throughout the ion foreshock region and have frequencies below the electron plasma frequency (Gurnett 1985). It is generally believed that these waves are ion acoustic waves whose frequencies are Doppler-shifted by the solar wind to higher frequencies. Although the cause of these waves is not completely understood, Fuselier et al. (1987) presented evidence of ion beams occurring simultaneously with the VLF waves and argued an ion/ion acoustic instability generated the waves.

A problem with the OEFD data is that the antenna is sensitive to noise when the PVO is in sunlight because of the photoemission sheath enveloping the orbiter (Scarf et al. 1980b). The problem is not serious at the highest frequency channel (30 kHz), but becomes very severe at the lower frequency channels. Higuchi et al. (1992) developed a statistical method to remove most of the photoemission interference from OEFD data so that reliable wave measurements can be obtained. Recently, Crawford et al. (1993a) and Strangeway and Crawford (1995) used this method to study wave behavior in the Venus ion foreshock.

An example of waves observed in the Venus ion foreshock is shown in Fig. 4. Electric and magnetic field measurements, as well as the depth and shock normal angle, are shown as a function of time. Strong emissions in the frequency channels 5.4 kHz and 730 Hz are apparent at times 06:20 to 06:28 and 07:00 to 07:30. Statistical analysis of these waves indicate that they are parallel polarized ion acoustic emissions, and that the wave intensity is different in the upstream and downstream foreshocks (Crawford et al. 1993a). This latter point suggests that differences in the ion transmission and/or reflection processes are operative. Note that these emissions only occur for a quasi-parallel shock structure, i.e., $0° < \theta_{Bn} < 45°$. There are also strong ULF waves in all three magnetic field components associated with these waves; this is thought to be indicative of enhanced ion fluxes of bow shock origin. Finally, little or no emissions are seen at 100 Hz, possibly because of the high background noise level.

A very interesting comparison of Venus and terrestrial ion foreshock waves was made by Crawford et al. (1993a) using PVO data and terrestrial ion foreshock data from the AMPTE CCE spacecraft (Strangeway et al. 1988). This comparison is significant because the same plasma wave detectors are used on both spacecraft. Basically, the OEFD and CCE measurements are in very good agreement; this suggests that ion foreshock wave behavior is fundamentally the same at Venus as at the Earth, and one can posit similar mechanisms are responsible for the waves.

Finally, Strangeway and Crawford (1995) developed maps of the 5.4 kHz wave intensity and the magnetic field deviation σB_{xyz} (a surrogate for ULF wave activity) versus spatial location similar to those shown in Fig. 3 for 30 kHz waves. An example is shown in Fig. 5. Again, the bow shock and tangent magnetic field lines are shown; the straight line behind the tangent line is the ULF boundary based upon terrestrial studies. The ULF waves are most intense near the bow shock, and are generally located behind the expected ULF boundary. On the other hand, the 5.4 kHz waves tend to occur even farther behind the ULF boundary than the ULF waves. This suggests that perhaps different mechanisms are responsible for these waves. For example, Winske (1986) has proposed that large-amplitude hydromagnetic waves observed in the Venus magnetosheath downstream of the quasi-parallel shock are generated by the shock itself, and are not generated by an instability associated with the interaction of solar wind plasma with ionospheric plasma. Orlowski et al. (1994) compared ULF upstream wave characteristics (i.e., magnetic polarization and normalized magnetic compression ratio) with those predicted from both linear kinetic theory and fluid theory. The study did not address

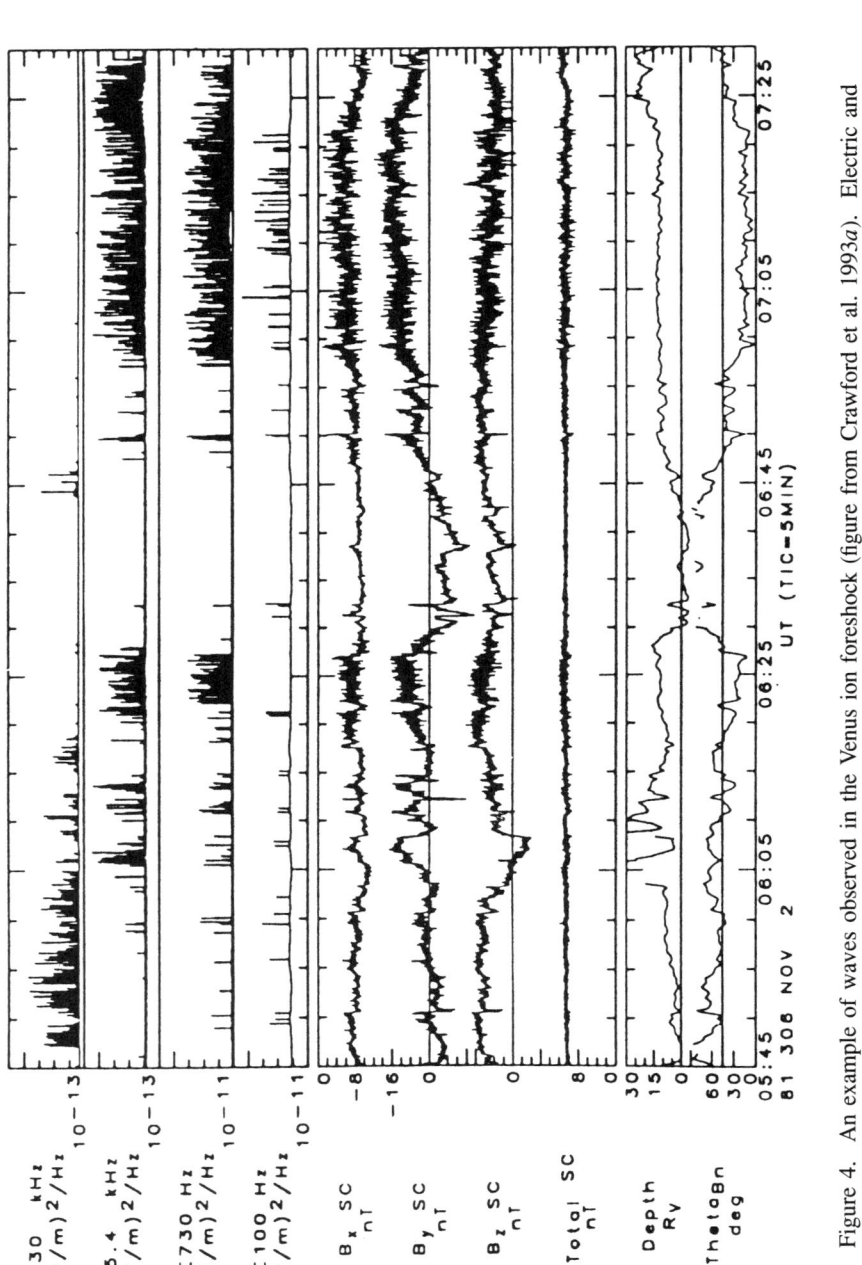

Figure 4. An example of waves observed in the Venus ion foreshock (figure from Crawford et al. 1993a). Electric and magnetic field measurements, as well as the depth and shock normal angle, are shown as a function of time. Strong emissions in the frequency channels 5.4 kHz and 730 Hz are apparent at times 06:20 to 06:28 and 07:00 to 07:30.

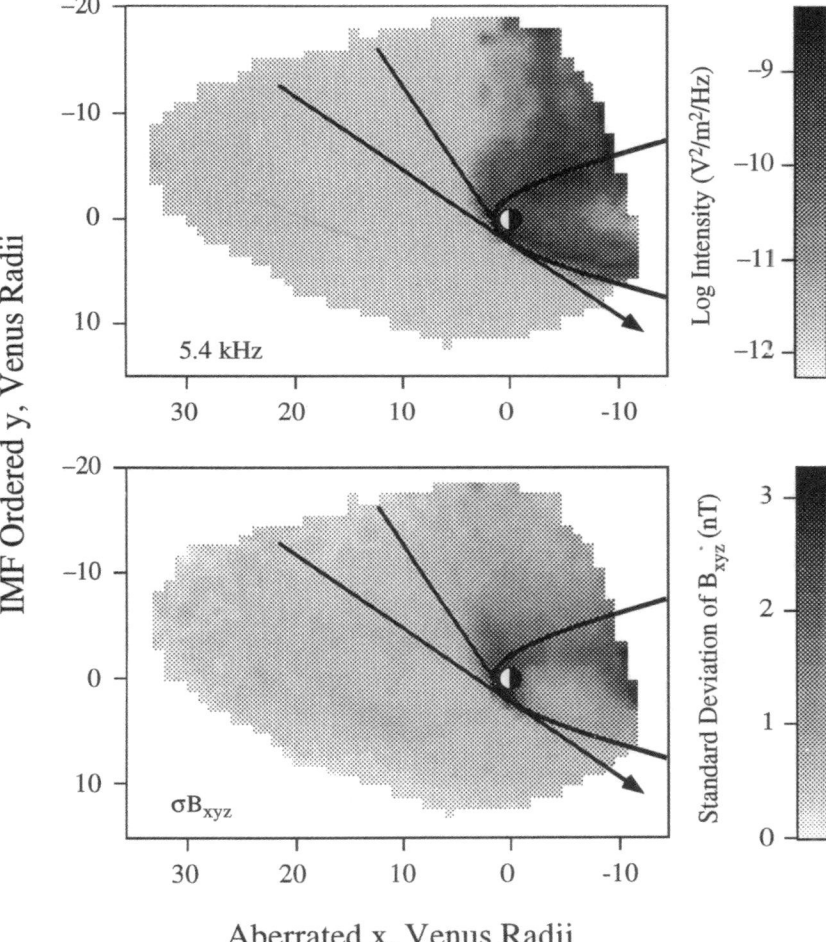

Figure 5. Maps of (a) 5.4 kHz intensity and (b) magnetic standard deviation σB_{xyz} (figure from Strangeway and Crawford 1995). Also shown are the positions of the bow shock model used in the analysis, the tangent magnetic field line, and the ULF boundary based upon terrestrial data.

the wave generation mechanism(s) but focused on the merits of the kinetic and fluid approaches. They found that the results based upon Vlasov theory were consistent with the data in the analyzed range $0.5 < \beta < 5.0$; results based upon Hall-MHD fluid theory were consistent with the data at low β for the fast mode, and at high β for the intermediate mode. And lastly, we mention that magnetic turbulence has also been observed upstream from the Mars subsolar bowshock when the magnetic field is connected to the bow shock and diffuse ions reach the Phobos 2 spacecraft (Russell et al. 1990a).

III. MANTLE WAVES

The Venus mantle is the transition layer between shocked solar wind and ionosphere, and, by definition, contains a mixture of solar wind plasma and ionospheric plasma (Spenner et al. 1980). A schematic showing the Venus boundary layers and mantle region is shown in Fig. 6. Scarf et al. (1979,1980b) reported wave activity in all channels of the OEFD; an example of their wave data is shown in Fig. 7. Scarf et al. (1980b) attributed the wave observations in the 730 Hz and 5.4 kHz channels to Doppler-shifted ion acoustic waves, and in the 100 Hz channel to shock-generated whistler waves. They further argued that the whistler waves propagated to the ionopause and were absorbed via electron Landau damping, thereby providing a mechanism to heat ionospheric electrons.

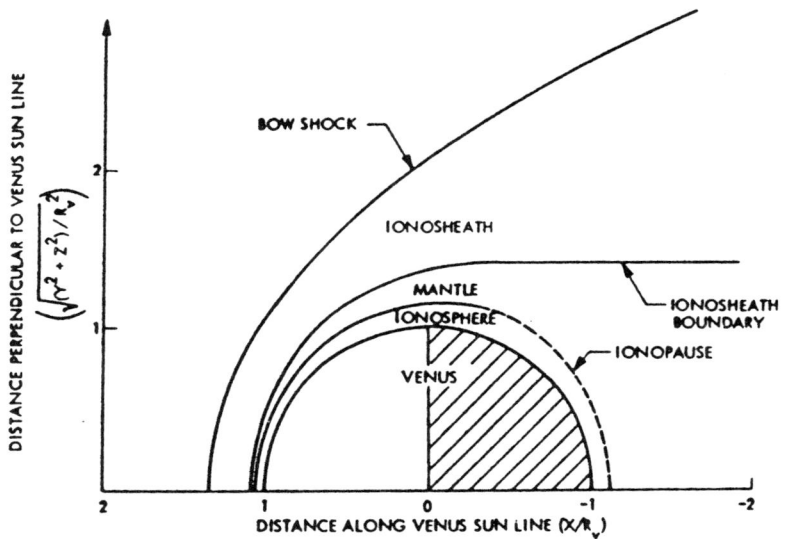

Figure 6. Schematic of the solar wind-Venus interaction showing the various regions and boundary layers (figure from Spenner et al. 1980).

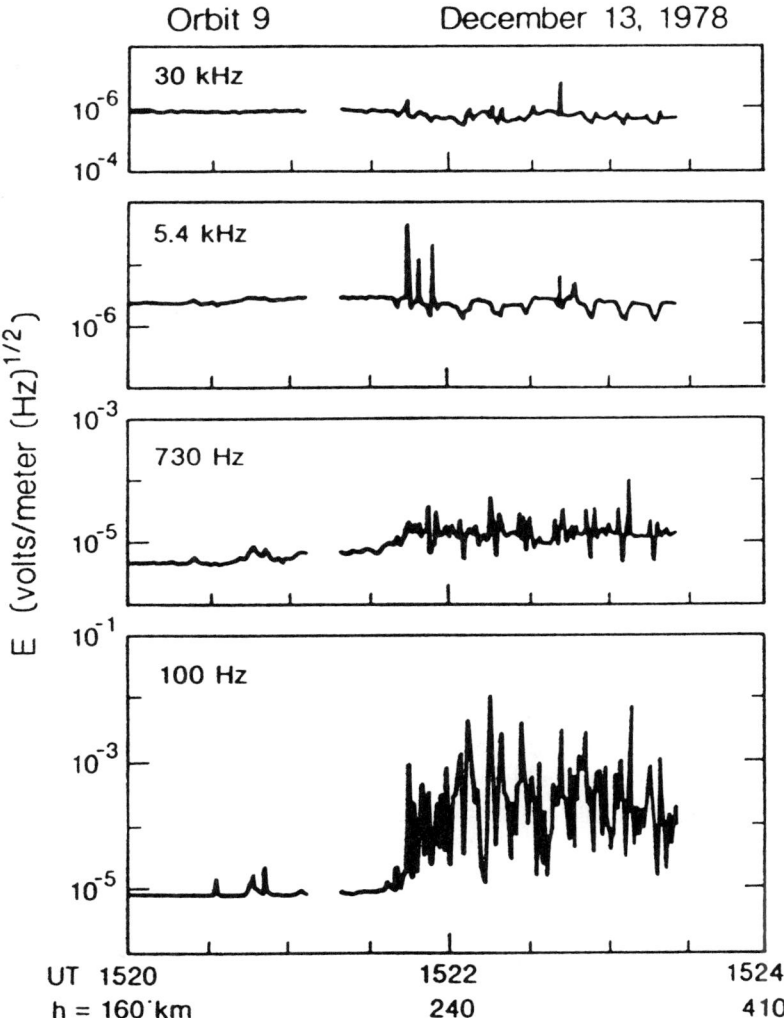

Figure 7. Example of OEFD wave measurements in the Venus mantle; the electric field fluctuations occur at altitudes above the ionopause (~230 km) (figure from Scarf et al. 1979).

Taylor et al. (1981) suggested that the 100 Hz waves are electrostatic and are locally generated by newly photoionized ions interacting with the flowing ionosheath plasma (i.e., shocked solar wind plasma). This is based upon Orbiter Ion Mass Spectrometer (OIMS) measurements of superthermal

O$^+$ ions (10–90 eV) above the ionosphere in conjunction with wave activity. Kasprzak et al. (1982) found a similar effect using data from the Orbiter Neutral Mass Spectrometer (ONMS).

Szegö et al. (1991) performed a more extensive data analysis of the mantle and examined 10 PVO orbits. They found that in six cases a well defined mantle region existed in which superthermal ions, 100 Hz waves, and unique electron signatures (i.e., low energy $\sim$5–60 eV) were observed. Based upon these findings, Szegö et al. (1991) performed a stability analysis of the mantle region in which they assumed the shocked solar wind encounters a cold population of planetary ions. The relative streaming velocity between these two components provides the free energy to drive the instability. The instability is the modified-two-stream instability and has the following wave characteristics: $\omega_r \simeq \omega_{lh}$ and $\gamma \lesssim \omega_{lh}$ where ω_r and γ are the real frequency and growth rate of the instability, and ω_{lh} is the lower hybrid frequency. The lower hybrid frequency is roughly $f_{lh} \simeq 40$ Hz for nominal magnetic field values in the mantle; this frequency is somewhat lower than the lowest frequency OEFD channel (100 Hz). The real frequency could be increased by Doppler-shifting and these waves could be detected by the OEFD. Szegö et al. (1991) further argue these waves propagate towards the ionopause heating the electrons, and then the ions, thereby creating the observed electron signatures and superthermal ions.

Strangeway and Crawford (1993) criticized the study of Szegö et al. (1991) for several reasons. First, the stability analysis of Szegö et al. (1991) assumed a cold plasma; however, in the mantle the electrons and ions have temperatures $\sim$10s eV. Temperature effects (e.g., finite electron Larmor radius effects, electron and ion resonances) are likely to reduce the growth rate of the instability. Second, although it may be possible to increase the wave frequency so that it is detectable in the lowest channel of the OEFD, it is highly unlikely that the modified-two-stream instability could generate observable waves at higher frequency OEFD channels (e.g., 730 Hz, 5.4 kHz, and 30 kHz). Finally, lower hybrid waves are quasi-electrostatic and the energy flux is carried primarily by the parallel Poynting flux. However, because the magnetic field is draped over the ionopause, the lower hybrid waves cannot transport significant wave energy to lower altitudes (i.e., transverse to the magnetic field); therefore, lower hybrid waves are not an important ionospheric heating mechanism.

A statistical study of VLF emissions in the Venus mantle was carried out by Crawford et al. (1993*b*). They investigated 52 ionopause crossings and found the following: 21% (11 crossings) contained only 100 Hz signals, 15% (8 crossings) contained only wideband signals (730 Hz, 5.4 kHz, and 30 kHz), and 64% (33 crossings) contained both 100 Hz and wideband signals. The waves are observed in or near current layers, indicated by strong rotations in the magnetic field, and are typically found above the ionopause proper. They conclude that the waves are some type of acoustic mode, and are not an important source of energy flux into the topside ionosphere.

Huba (1993a) proposed that the ion acoustic beam instability could generate the 100 Hz and 730 Hz waves in the Venus mantle. The physical model is similar to that used by Szegö et al. (1991). The plasma consists of newly ionized, cold O^+ ions, photoelectrons, and shocked solar wind protons and electrons. The O^+ ions appear as a beam relative to the solar wind; the relative drift between the ions and solar wind provides the free energy for the instability. The instability is driven by inverse electron Landau damping on an ion acoustic wave associated with the cold, ionospheric O^+ ions. The instability requires $T_i/T_e < 0.05$ to prevent O^+ Landau damping of the waves, as well as $0.1 < n_i/n_e < 0.9$. Following the wave generation period, the cold O^+ ions are "picked up" by the flowing plasma and attain the observed energies 10 to 90 eV. The instability can generate the observed 100 Hz waves directly, and the 730 Hz waves through the Doppler shift of the frequency caused by the satellite motion. However, the ion acoustic beam instability is unlikely to generate the higher frequency turbulence at 5.4 kHz and 30 kHz. A potential problem with this mechanism is that a significant population of cold ionospheric electrons can stabilize the instability via Landau damping (Shapiro et al. 1995); thus, strictly speaking, the model is valid at altitudes above the ionopause where the cold ionospheric electrons are a minor constituent of the plasma. Further work is needed in this area to quantify the effect of cold electrons on the ion beam acoustic mode.

Very recently, Shapiro et al. (1995) re-examined the physics of the Venus mantle. They presented a summary of the experimentally determined characteristics of the mantle, and a more detailed analysis of the microphysics than Szegö et al. (1991). Specifically, several of the shortcomings noted in Strangeway and Crawford (1993) were remedied (e.g., finite electron temperature effects were included in the analysis). Two branches of wave instability were shown to develop: a low-frequency, long-wavelength branch, and a high-frequency, short-wavelength branch. The former branch was identified as a hydrodynamic instability and corresponds to the conventional modified-two-stream instability; the latter branch was identified as a kinetic instability. In addition to linear analysis, Shapiro et al. (1995) investigated the nonlinear saturation and consequences of the unstable modes. The saturation mechanism is due to induced scattering of the waves on cold planetary ions. The consequences are an increased collision frequency between the solar wind protons and planetary ions, ion pickup and heating, electron acceleration parallel to the magnetic field, and ion heating.

It is important to remind the reader that the wave activity detected by the OEFD is limited: only electric field fluctuations are measured, and these are restricted to four frequency channels (100 Hz, 730 Hz, 5.4 kHz, and 30 kHz). This section has focussed on identifying plasma instabilities that can account for the OEFD wave observations, primarily in the 100 Hz channel. However, it is probable that plasma waves exist at lower frequencies (e.g., the lower hybrid frequency, the ion cyclotron frequency) and that these waves can affect the macroscopic interaction of the solar wind plasma and the ionospheric plasma.

Thus, the modified-two-stream instability is likely to be excited in the plasma mantle. Whether or not it generates the 100 Hz signals observed by the OEFD is debatable and is the subject of the aforementioned research. Although it has been argued that this instability is not responsible for the OEFD observations (Strangeway and Crawford 1993; Huba 1993a), this should not be construed as an argument that the instability does not exist in the Venus mantle.

Finally, we mention that a comparison between the structure and dynamics of the Venus mantle with that of the Mars planetosphere has also been made (Nagy et al. 1990; Szegö et al. 1992). There are a number of similarities between these two planetary space environments; this is not too surprising because both planets are essentially "nonmagnetic." Nagy et al. (1990) note that both environments have (1) a transition region between shocked solar wind plasma and cold, planetary ions; (2) a change in the electron population that is different from the solar wind electrons and photoelectrons; (3) a magnetic field signature (i.e., "pile-up") at the transition; and (4) low frequency electric field fluctuations within the transition region. An analysis of wave activity around Mars was made by Sagdeev et al. (1990) based upon Phobos 2 data. Intense wave activity was observed in the magnetosheath region ranging from a few hundred Hz to a few kHz. The high frequency waves were attributed to a current driven ion acoustic instability at the bow shock; the low frequency waves to an interaction between shocked solar wind plasma and cold, planetary ions, very similar to the mechanism(s) proposed in the Venus mantle.

IV. IONOSPHERIC WAVES

A comprehensive study of the spatial distribution of plasma wave activity in the nightside Venus ionosphere has been presented recently by Ho et al. (1994). In this work, 14 years of PVO OEFD data is used to characterize VLF wave activity in the nightside ionosphere. They identify four basic categories of VLF wave data. The first category is 100 Hz bursts only (i.e., no wave activity in the higher-frequency channels). These bursts are controlled by the orientation and strength of the magnetic field, and are weakly dependent upon the electron density. The occurrence rate decreases with altitude up to 600 km and is constant above 600 km. Ho et al. (1994) suggest that these signals are whistler waves generated by sub-ionospheric lightning. The second category is broadband turbulence: wave activity in all channels of the OEFD. This activity is observed at low altitudes (<300 km) and is also suggested to come from a sub-ionospheric source. The third category is strong, mid-frequency VLF signals. These signals are observed primarily in the 100 Hz, 730 Hz, and 5.4 kHz channels of the OEFD. They occur in the wake region of the nightside ionosphere and have durations of 1 to 2 min. Ho et al. (1994) suggest these waves are generated by a current-driven ion acoustic instability. The fourth category is narrowband waves in the 5.4 kHz or 30 kHz channels. They occur at high altitudes (>2000 km) in low-density regions of the Venus magnetotail and are believed to be Langmuir waves.

Probably the most controversial area of wave phenomena at Venus is the interpretation of the 100 Hz OEFD measurements in the nightside ionosphere. One interpretation is that these waves are lightning-generated whistlers propagating up from the lower atmosphere (Scarf et al. 1980c); the other interpretation is that they are produced locally by a plasma instability (Taylor et al. 1987). The lightning mechanism will not be explored in this chapter, but is discussed in the chapter on lightning by Grebowsky et al. The emphasis of this section will be on density fluctuations measured by the OETP but relevant OEFD measurements will also be discussed.

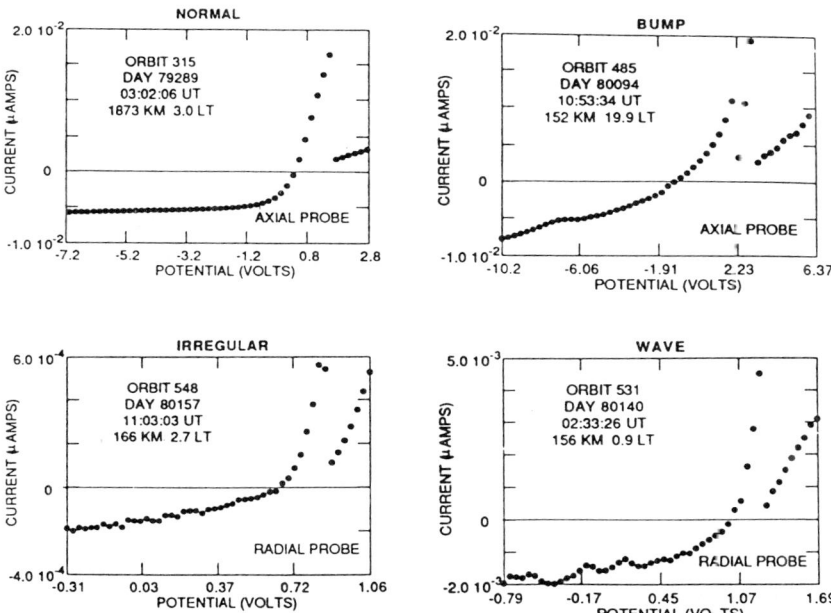

Figure 8. Examples of volt-ampere (I-V) curves which show the different types of profiles (figure from Grebowsky et al. 1991). The bottom plots show irregular and wave profiles in the ion collection region which are interpreted as electron density irregularities in the plasma.

Grebowsky et al. (1991) examined volt-ampere (I-V) curves from the OETP instrument and noted that wave-like irregularities often occurred in the nightside ionosphere. An example of the different I-V curves observed is shown in Fig. 8. Grebowsky et al. (1991) suggested these irregularities could be either stationary density structures with wavelengths $\lambda \simeq 0.1$ to 2 km, or high-speed waves in the frequency range $f \simeq 5$ to 100 Hz. They also showed that these irregularities are often associated with steep gradients in the electron density and the magnetic field as shown in Fig. 9. Finally, they showed that electric field signals in the 100 Hz channel of the OEFD were present during the majority (56%–73%, depending on the selection criteria used) of the density irregularity observations. This last finding suggests that, at least, some of the 100 Hz signals in the nightside ionosphere are associated with an *in-situ* plasma instability because whistler waves are electromagnetic and noncompressive (i.e., lightning-induced whistlers cannot produce the observed density waves).

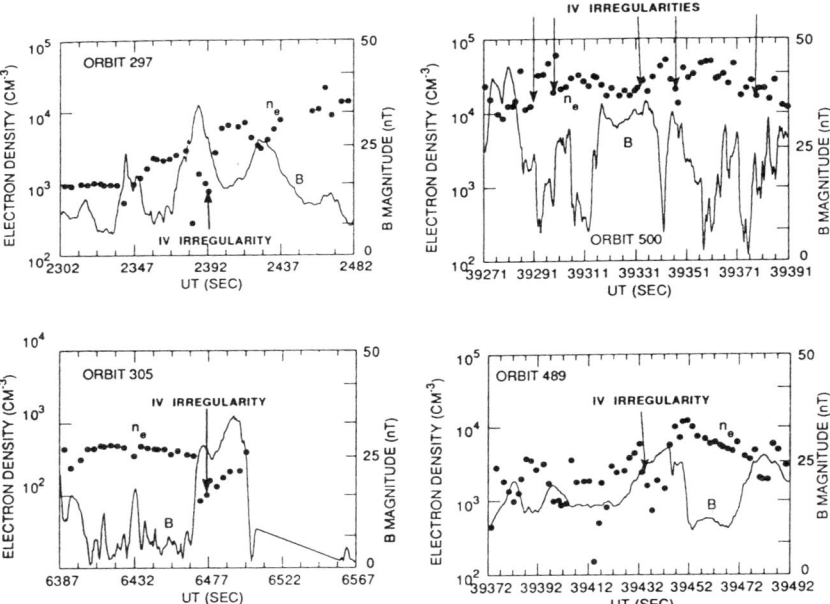

Figure 9. Plots of the electron density and magnetic field for orbital segments in which I-V irregularities were observed (figure from Grebowsky et al. 1991). The irregularities (denoted by arrows) occur in regions near density and field gradients.

A detailed statistical study of the relationship between OETP anomalies and OEFD 100 Hz burst observations has been performed by Strangeway (1995a). Strangeway argued that Grebowsky et al. (1991) did not test for random coincidence and used a burst rate threshold that was too low in some situations. He found that 20% of the Langmuir probe anomalies were statistically correlated with the 100 Hz bursts. Although this is less than that reported by Grebowsky et al. (1991), it is still a significant correlation.

Huba (1992) proposed the lower-hybrid-drift instability as a mechanism to produce the density and electric field fluctuations reported by Grebowsky et al. (1991). The instability is driven by the diamagnetic drift associated with the ambient density gradient (i.e., $V_d = (cT/eB)\partial \ln n/\partial x$). The real frequency and growth rate of the mode are typically $\omega_r \lesssim \omega_{lh}$ and $\gamma < \omega_r$ where ω_{lh} is the lower-hybrid frequency. The key features that favor the lower-hybrid-drift instability are the following. First, it is a flute mode and propagates orthogonal to the ambient magnetic field ($\mathbf{k} \cdot \mathbf{B} = 0$). Second, it is a relatively short-wavelength mode ($\lambda \sim \rho_e$ where ρ_e is the electron Larmor radius) and the Doppler shifted frequency can be $f \simeq 100$ Hz. (It is unlikely that the lower-hybrid-drift waves could be Doppler shifted to higher OEFD frequency channels.) Third, it can generate both density and electric field fluctuations. And fourth, it is most unstable in low β plasmas ($\beta = 8\pi n T / B^2 < 1$) so that it is likely to occur in the low-density, high magnetic field ionospheric holes. These features are consistent with the observational results of Taylor et al. (1987) and Grebowsky et al. (1991). In addition, Huba (1992) identified several approximate criteria for the generation of lower-hybrid-drift waves in the Venus ionosphere. Strangeway (1995a,b) analyzed 100 Hz observations as a function of these approximate criteria and found that the bursts are most likely to occur in regions of low electrons (i.e., $\beta_e = 8\pi n_e T_e / B^2 < 1$) where attenuation of whistler waves is small. A significant number of bursts also occur in the low-altitude, high-collision frequency regime in which the lower-hybrid-drift instability is stable. Strangeway (1995a,b) concluded that these results favor planetary lightning as the more likely source of VLF bursts.

Huba and Grebowsky (1993) performed a follow-up study on the lower-hybrid-drift instability in the nightside Venus ionosphere; they compared theoretical predictions based upon a linear stability analysis with observational data of small-scale density fluctuations and did not rely upon approximate criteria. The linear analysis used measured plasma and field parameters. Marginal stability boundaries ($\gamma = 0$) were calculated as a function of the magnetic field B and the electron density n_e; the occurrences of small-scale irregularities were also plotted in this parameter space. The results are shown in Fig. 10. The density fluctuations are separated into three altitude regimes: (1) altitudes <180 km are denoted by crosses, (2) 180 km < altitudes <800 km are denoted by circles, and (3) altitudes >800 km are denoted by plus signs. The marginal stability curves for the lower-hybrid-drift instability are plotted for $V_{di}/v_i = 1.0$ (diamonds) and $V_{di}/v_i = 2.0$ (squares) where V_{di} is the ion diamagnetic drift velocity and v_i is the ion thermal velocity; these values of the ion diamagnetic drift velocity correspond to density gradient scale lengths in the range $L_n \simeq 2$ to 10 km. The mode is unstable for values of B and n above the marginal stability curves, and stable for those below.

The important feature of Fig. 10 is that 80 to 85% of the observations of small-scale density irregularities lie in the unstable region of the lower-hybrid-drift instability; almost all of the observations between 180 km and 800 km lie in this region. The ambient conditions for the plasma density fluctuations observed in the stable region of the lower-hybrid-drift instability roughly fall into two categories: (1) low-altitude (<180 km) and high-density ($>2 \times 10^2$ cm^{-3}), and (2) high-altitude (> 800 km), low-density ($<2 \times 10^3$ cm^{-3}), and very low magnetic fields (<5 nT). This suggests that other

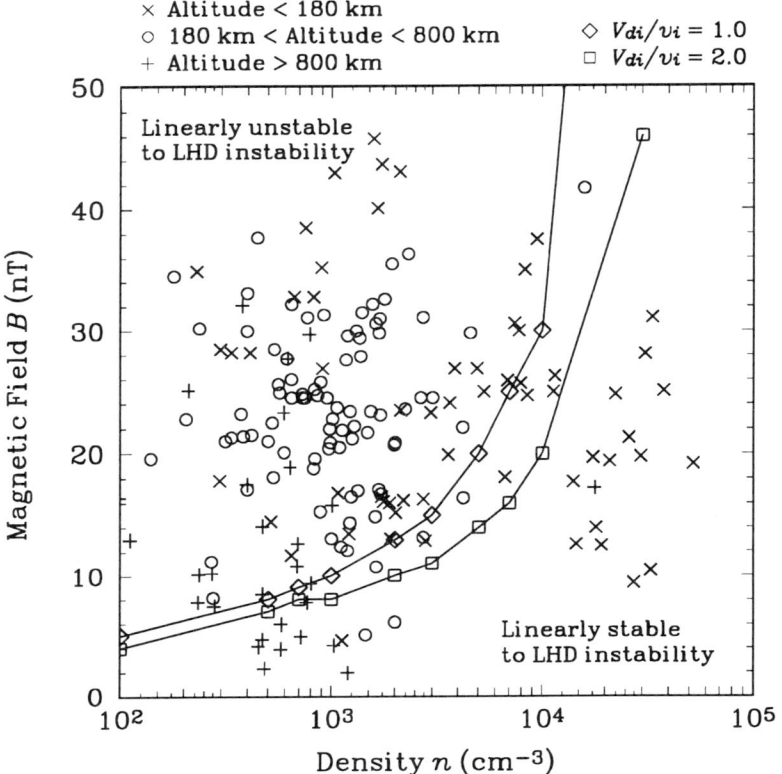

Figure 10. The occurrence of small-scale irregularities and the marginal stability boundaries for the lower-hybrid-drift instability as a function of the magnetic field B (nT) and density n (cm^{-3}) (figure from Huba and Grebowsky 1993). The data is restricted to the nightside ionosphere.

physical mechanisms are operative under these conditions to produce density fluctuations. For example, Scarf et al. (1985) noted that high-energy electrons in the tail region of Venus could generate such structure. Irregularities in the low-altitude, high-density regime will be discussed in more detail below.

Huba and Grebowsky (1993) also calculated the boundaries for the observation of 100 Hz waves as a function of magnetic field B and density n. The lower-hybrid-drift waves typically have real frequencies $f \lesssim f_{lh} \simeq 40$ Hz and must be Doppler shifted by the satellite motion to be detected in the 100 Hz OEFD channel. The results (see Fig. 7 of Huba and Grebowsky [1993]) suggest that a minimum magnetic field value $B \simeq 15$ nT is necessary for the observation of lower-hybrid-drift instability generated 100 Hz waves at low densities ($n < 10^3$ cm^{-3}). Interestingly, an analysis of selected PVO 100 Hz electric field data shows a cutoff at $B \simeq 20$ nT (Walker 1992); however, no similar cutoff was found by Strangeway (1995a,b).

A statistical study of the relationship between nightside plasma and field parameters, and VLF bursts has been presented by Ho et al. (1995). They found that 5.4 kHz burst activity is correlated with density depletions and electron temperature enhancements, but is relatively independent of magnetic field strength. These bursts have higher occurrence rates at steep density and temperature gradients. The 100 Hz bursts are correlated with density depletions and magnetic field enhancements; this corresponds to a low β plasma which favors both the propagation of whistler waves (Huba and Rowland 1993) and the lower-hybrid-drift instability. These bursts have a slightly higher burst rate for steep density and temperature gradients. Neither the 100 Hz nor 5.4 kHz waves are correlated with magnetic field gradients. Ho et al. (1995) conclude that "a gradient-driven instability cannot explain most of the 100 Hz narrow-band signals, but some 5.4 kHz signals may be associated with pressure gradients."

Ho et al. (1995) also found that most of the 100 Hz bursts ($\sim$95%) occurred in low-density gradient regions (>50 km); on the other hand, the theory of Huba (1992) and analysis of Huba and Grebowsky (1993) find that relatively short density gradient scale lengths (<20 km) are required to generate the lower-hybrid-drift instability such that it would be detected by the OEFD. One problem in trying to correlate lower-hybrid-drift wave activity with plasma or field gradients is that, for optimal wave growth, the wave vector is orthogonal to the ambient gradients. The maximum Doppler-shifted frequency occurs when the PVO direction is aligned with the wave vector; for this situation, the spacecraft could measure 100 Hz waves but would not measure any significant plasma or field gradients. Thus, 100 Hz waves generated by the lower-hybrid-drift instability need not be strongly correlated with measured density gradients. We conclude that a fraction ($\sim$20%) of the 100 Hz waves observed by the OEFD could be generated by the lower-hybrid-drift instability; it is difficult to reach a more definitive conclusion because of the limited data sets (i.e., specifically, the lack of magnetic field fluctuation data at frequencies similar to the OEFD).

At the end of the PVO mission, when the satellite descended to low periapsis, kilometer-size density waves were especially prominent in OETP I-V curves (Brace 1993). These waves were often quasi-sinusoidal with fluctuation levels $\delta n/n \sim$ 10 to 30%. They occurred primarily in the altitude range 140 to 160 km, and were located just above the peak in the electron density. The electron density gradient above the peak is very steep when the waves are observed ($L_n <$ few km). The electron temperature is inversely proportional to the electron density (Theis and Brace 1993) so that the hot electrons at higher altitudes (>160 km) maintain the steep density gradient.

Huba (1993b) proposed a mechanism to explain the observations of Brace (1993). The physical configuration and geometry considered by Huba (1993b) is shown in Fig. 11. In the region of interest, below $\sim$155 km, the ions are unmagnetized because $\nu_{in} > \Omega_i$ while the electrons are magnetized because $\nu_{en} < \Omega_e$ (here $\nu_{\alpha n}$ is the collision frequency between the α species and neutrals,

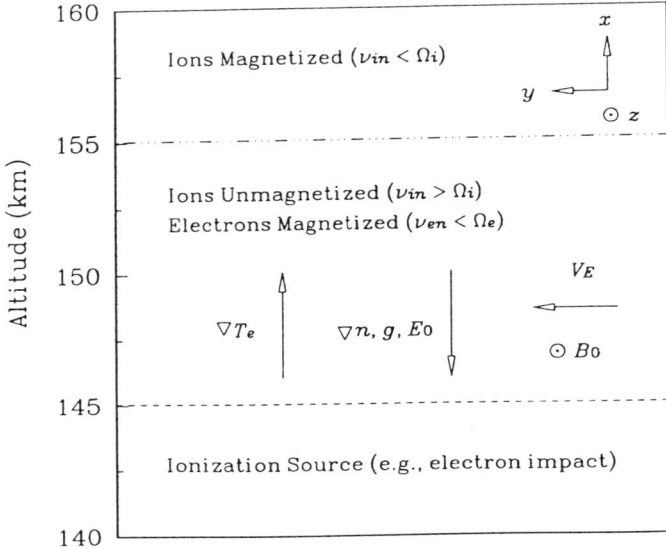

Figure 11. Schematic of the plasma configuration and geometry of the lower, nightside Venus ionosphere (figure from Huba 1993b).

and Ω_α is the cyclotron frequency of the α species). The magnetic field is taken to be horizontal in the z-direction, and gravity is in the $-x$-direction. The peak in the electron density is assumed to be in the range 140 to 145 km, consistent with observations (Kar et al. 1994).

The important feature of the model is the ambipolar electric field E_O. Under nominal conditions the polarization field is maintained by the downward gravitational acceleration of the ions; for this situation the ambipolar field is in the $+x$-direction. However, during periods of electron pressure enhancements, this ambipolar field reverses direction and can be enhanced as the plasma re-establishes equilibrium. (There is a large variability in the electron pressure from orbit to orbit which suggests that the nightside Venus ionosphere is very dynamic during solar minimum conditions [Theis and Brace 1993].) For example, if the electrons are suddenly compressed (i.e., pushed downward) by a temperature increase above 160 km, the ambipolar field will be directed downward; this field causes the ions to move with the electrons to maintain charge neutrality. This is shown schematically in Fig. 12. During this period the electrons undergo an $E_O \times B$ drift in the y-direction because of the ambipolar field; on the other hand, the ions do not $E_O \times B$ drift because they are effectively unmagnetized because $\nu_{in} > \Omega_i$.

Huba (1993b) performed a stability analysis for this configuration. He found that instability could occur when the ambipolar field E_O is in the same direction as the density gradient; this requirement is similar to that for the gradient drift instability which occurs in the Earth's equatorial electrojet (Rogister

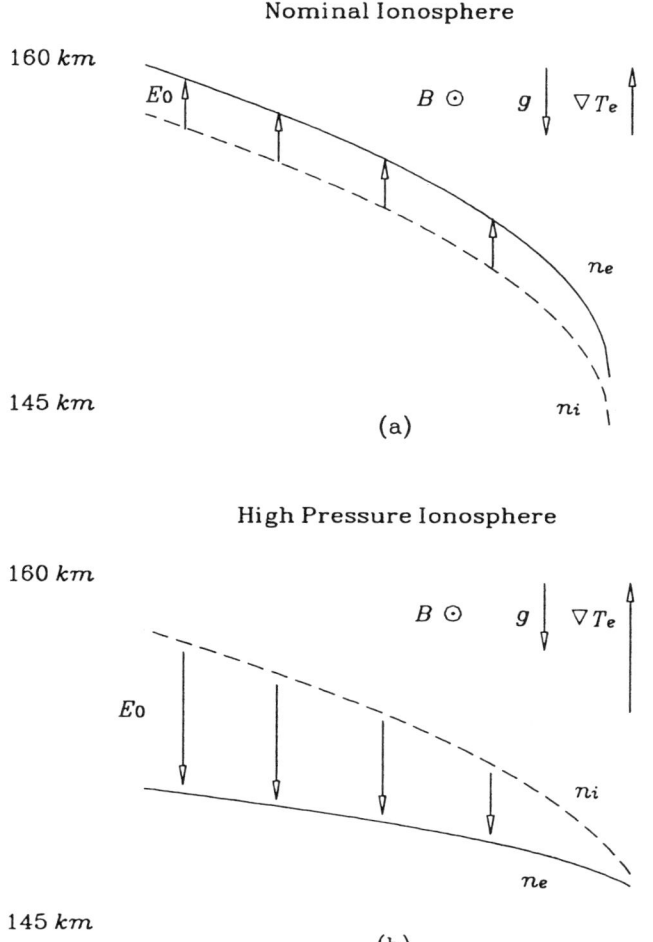

Figure 12. Schematic of the ambipolar electric field in Venus' nightside ionosphere for (a) nominal conditions, and (b) high-pressure conditions (figure from Huba 1993b).

and D'Angelo 1970). Thus, the nominal ambipolar field in the ionosphere (Fig. 12) is in the wrong direction to generate the instability. The direction of the ambipolar field must be reversed (e.g., by pressure enhancements) for instability to occur. The magnitude of the ambipolar field necessary for instability is roughly $E_O \sim 15\,\mu$V/m. Finally, the fastest growing waves have wavelengths $\lambda \sim 0.5$ to 1.0 km for parameters typical of the nightside Venus ionosphere; this finding is consistent with observations (Brace 1993). The key difference between this instability and the lower-hybrid-drift instability (Huba 1992) is that there is an additional source of free energy to drive the instability: the ambipolar electric field E_O.

V. OUTSTANDING ISSUES

In the previous sections we have presented wave observations and proposed theoretical explanations of these observations. However, there are several cases of interesting wave phenomena but no detailed model to explain them. Three significant areas are now discussed in some detail: post-terminator waves, wideband turbulence, and Langmuir waves.

A. Post-Terminator Waves

Interesting wave structure in the nightside ionosphere, downstream of the terminator, has been reported by Brace et al. (1983) using the OETP and OMAG instruments. The region where the waves have been observed is shown in Fig. 13; waves have also been observed in the dawn terminator. These waves have the following characteristics. Wave structure is seen in the electron density n_e, electron temperature T_e, and east-west component of the magnetic field B_E, mainly at altitudes below 200 km. The electron density and electron temperature fluctuations are 180° out of phase so that the plasma pressure is roughly constant across the wave train. The amplitude of these fluctuations is $\delta n_e/n_e$ and $\delta T_e/T_e \lesssim 2$ to 3. The magnetic field fluctuations are roughly 90° out of phase with the density and temperature fluctuations, and have amplitudes $\delta B \lesssim 30$ nT about a zero mean value. The wavelength in the north–south direction is ~ 150 km. The plasma pressure exceeds the magnetic field pressure by a factor ~ 4. Also, there did not appear to be any correlation with neutral atmosphere fluctuations. An example of this wave phenomenon is shown in Fig. 14. Brace et al. (1983) argue that these waves are significant because they represent an energy sink to the transterminator flow that is important to the maintenance of the nightside ionosphere.

Brace et al. (1983) suggest several mechanisms to generate these waves. The waves are first seen at the terminator density gradient; this suggests that the free energy to drive the instability is derived from the pressure gradient or the ion flow velocity generated by the gradient. The waves could then be advected downstream of the flow. Another source of free energy is the vertical shear in the horizontal ion flow velocity caused by ion-neutral drag at low altitudes. Alternatively, the instability may be generated at higher altitudes where the ion flow becomes supersonic again, and the effects are "mapped" into the ionosphere. Huba and Fedder (1993) explored the possibility that structured ion flows from high altitude could self-consistently generate the observed magnetic field structures in the lower ionosphere. Although this process is plausible, Huba and Fedder (1993) concluded that magnetic field fluctuations $B \sim 30$ nT could not be produced. Luhmann (1990) suggested that the interplanetary field, draped over the ionosphere, could generate the

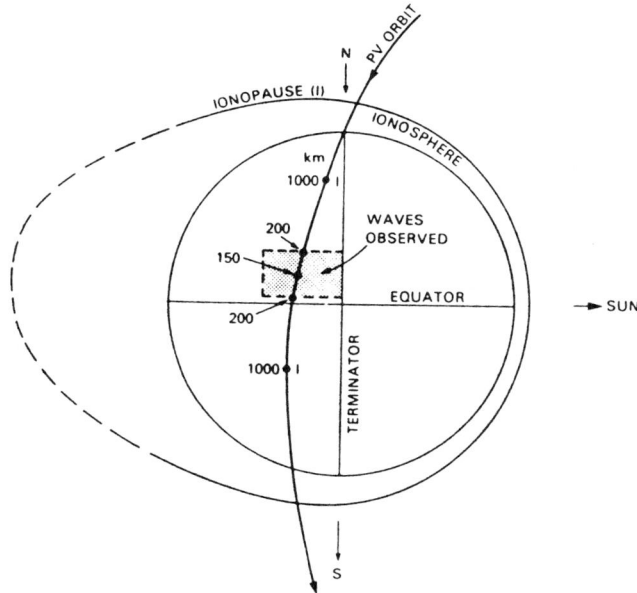

Figure 13. Spatial location of where post-terminator waves are observed (figure from Brace et al. 1983).

terminator waves via a subsolar turbulent dynamo process or nonexplosive magnetic reconnection. These concepts are appealing but detailed theoretical and numerical analyses are needed to determine if these processes can indeed generate terminator waves. No other models have been formulated to explain these waves.

Hoegy et al. (1990) studied the transition of the post-terminator waves into the anti-solar ionosphere. They found that the waves in the post-terminator region have wavelengths $\lambda \sim 40$ to 400 km (along the satellite track) and are coherent, similar to the results of Brace et al. (1983). However, as the solar zenith angle increases, the waves tend towards longer wavelengths and they become more irregular. In contrast to Brace et al. (1983) they find that the long wavelength modes are correlated with gravity waves, i.e., fluctuations in the neutral atmosphere.

B. Wideband Turbulence

The OEFD, as noted in Sec. I, measures electric field fluctuations in four center frequency channels, 100 Hz, 730 Hz, 5.4 kHz, and 30 kHz, with a bandwidth $\pm 15\%$. We define wideband turbulence as electric field signals in the upper three channels. The 100 Hz signals are typically below the electron cyclotron frequency, and therefore, have been interpreted as the consequence of lightning because they can propagate as whistlers out of the atmosphere

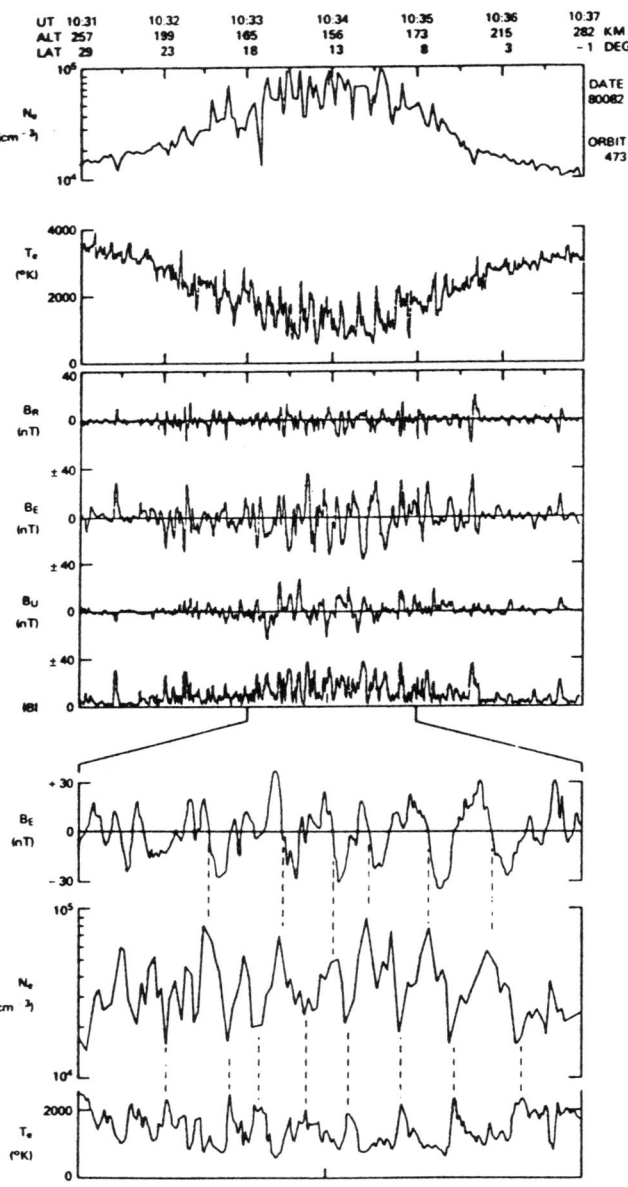

Figure 14. An example of post-terminator waves (figure from Brace et al. 1983). Shown are the electron density n_e, electron temperature T_e, and magnetic field components for a portion of orbit 473. An expanded view of n_e, T_e, and B_E (east-west component of the magnetic field) clearly shows the wave-like behavior of these quantities, and the dashed lines shows the phase relationships.

(Scarf et al. 1980c). For sufficiently strong magnetic fields, the 730 Hz signals can also propagate as whistlers (Huba and Rowland 1993). However, in general, the wideband signals have different characteristics than the 100 Hz signals and cannot be interpreted as whistlers (Russell et al. 1990b). For example, early in the PVO mission (during season III) the 100 Hz signals are prevalent at both low and high altitudes, but the wideband signals occur predominantly at low altitudes; this is shown in Fig. 15. Strangeway (1991) notes that the wideband turbulence could be a lightning-related transient process; this is based upon the observations of Kelley et al. (1985) who found anomalous transient signals associated with terrestrial lightning. We mention that in addition to wideband noise bursts, there is also a general enhancement of 730 Hz and 5.4 kHz wave activity at low altitudes in the nightside ionosphere. This enhancement is generally less impulsive than the wideband signals discussed above. Curtis et al. (1985) proposed that this enhancement is caused by an instability associated with impact ionization of CO_2 by the satellite. This mechanism appears to be plausible only at low altitudes, below 160 km.

Recent measurements during the final phase of the mission indicate a different situation (Strangeway et al. 1993a,b). Strangeway et al. (1993a) investigated intense wave bursts at very low altitudes during the final phase of the PVO mission. Although the emphasis of the paper was on 100 Hz emissions near 130 km, they also considered the relationship between the 5.4 kHz emissions and the CO_2 density as suggested by the theory of Curtis et al. (1985). They found that the relationship is consistent with impact ionization-driven waves above 137 km, but is not consistent below 137 km because the 5.4 kHz wave amplitude plateaus. They suggest that the instrument response to the impact ionization may be more complex than discussed by Curtis et al. (1985) and further research is needed in this area. Strangeway et al. (1993b) report that the wideband bursts tend to occur above 160 km, are a low magnetic field phenomenon, and no bursts occur in the 30 kHz channel. For these events they suggest that the wideband waves are ion acoustic waves, possibly excited by precipitating electrons from higher altitudes. Thus, although wideband turbulence is commonly observed in the nighttime Venus ionosphere, no satisfactory explanation for it has yet been put forth.

C. Langmuir Waves in the Magnetotail

Ho et al. (1993) reported evidence of Langmuir oscillations in the very low-density regions of the Venus magnetotail (altitudes $\gtrsim 2000$ km). The waves are field-aligned and are observed in both the 30 kHz and 5.4 kHz channels of the OEFD. The latter observation indicates that the electron density can be as low as 0.3 cm^{-3} in the magnetotail; this value is below the measurement range of the OETP. The waves primarily occur in the center of the two tail lobes on either side of the magnetotail current sheet; these lobes appear to be controlled by the orientation of the interplanetary magnetic field (IMF). Based upon the inferred electron density and measured magnetic field, and

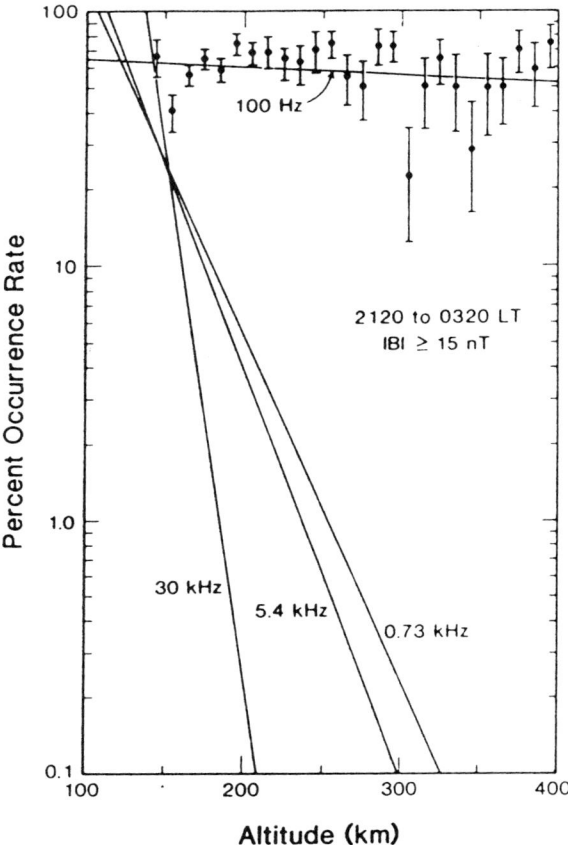

Figure 15. The percent occurrence rate as a function of altitude for the different frequencies measured by the OEFD (figure from Russell et al. 1990b).

assuming pressure balance, Ho et al. (1993) estimate the electron temperature to be ~ 13 eV in the low-density tail. They suggest that these waves are caused by the solar wind heat flux flowing through the cool ionosphere. However, no detailed theoretical calculations have been done to quantify this process, and further research is needed in this area to better understand the mechanism(s) that generates these Langmuir oscillations.

VI. SUMMARY

An overview of wave phenomena in the Venus space environment has been presented. The emphasis has been on studies made during the past decade, and has focused on wave behavior in three distinct regions: the foreshock, the mantle, and the nightside ionosphere. The types of waves considered are fluctuations in the electric field at frequencies $f = 100$ Hz, 730 Hz, 5.4 kHz, and 30 kHz as measured by the OEFD, fluctuations in electron density and temperature as measured by OETP, and fluctuations in the magnetic field as measured by OMAG. The plasma and field fluctuations are low frequency ($f < 100$ Hz), and have wavelengths ranging from $\lambda \simeq 1$ km to $\lambda \simeq 100$ km. Overall, we have shown that Venus wave phenomena is diverse and varied, resulting ultimately from the complex interaction between the solar wind and the Venus ionosphere. Although there have been satisfactory theoretical explanations for a number of the wave observations, there remain several unexplained phenomena that require further study (such as post-terminator waves, wideband turbulence, and Langmuir waves). Finally, an important aspect not discussed in detail but mentioned for the sake of completeness is the importance of comparing Venus wave data with terrestrial and Mars wave data. Such a comparison can help us understand the similarities/differences between the solar wind with magnetized and unmagnetized bodies. For example, there is much more terrestrial wave and space data than Venus data. If the wave processes are the same (or very similar) then we can infer what the Venus space environment is like based upon knowledge of the Earth's space environment.

Acknowledgments. We thank J. Grebowsky and K. Szegö for constructive comments on an early draft of this chapter. This work has been sponsored by the National Aeronautics and Space Administration, and the Office of Naval Research.

REFERENCES

Bonifazi, C., and Moreno, G. 1981*a*. Reflected and diffuse ions backstreaming from the earth's bow shock, 1, Basic properties. *J. Geophys. Res.*86:4397–4404.
Bonifazi, C., and Moreno, G. 1981*b*. Reflected and diffuse ions backstreaming from the earth's bow shock, 2, Origin. *J. Geophys. Res.* 86:4405–4414.
Brace, L. H. 1993. Kilometer-sized waves in electron density in the Venusian nightside ionosphere. *Geophys. Res. Lett.* 20:2759–2762.
Brace, L. H., Elphic, R. C., Curtis, S. A., and Russell, C. T. 1983. Wave structure in the Venus ionosphere downstream of the terminator. *Geophys. Res. Lett.* 10:1116–1119.
Crawford, G. K., Strangeway, R. J., and Russell, C. T. 1990. Electron plasma oscillations in the Venus foreshock. *Geophys. Res. Lett.* 17:1805–1808.
Crawford, G. K., Strangeway, R. J., and Russell, C. T. 1993*a*. VLF emissions in the Venus foreshock: Comparison with terrestrial observations. *J. Geophys. Res.* 98:15305–15317.
Crawford, G. K., Strangeway, R. J., and Russell, C. T. 1993*b*. VLF emissions at the Venus dayside ionopause. In *Plasma Environments of Non-Magnetic Planets*, ed.

T. I. Gombosi (New York: Pergamon Press), pp. 253–258.

Curtis, S. A., Brace, L. H., Niemann, H. B., and Scarf, F. L. 1985. CO_2 impact ionization-driven plasma instability observed by Pioneer Venus Orbiter at periapsis. *J. Geophys. Res.* 90:6631–6636.

Filbert, P. C., and Kellogg, P. J. 1979. Electrostatic noise at the plasma frequency beyond the earth's bow shock. *J. Geophys. Res.* 84:1369–1381.

Fitzenreiter, R. J., Klimas, A. J., and Scudder, J. D. 1984. Detection of bump-on-tail reduced electron velocity distributions at the electron foreshock boundary. *Geophys. Res. Lett.* 11:496–499.

Fuselier, S. A., et al. 1987. Ion beams and the ion/ion acoustic instability upstream from the earth's bow shock. *J. Geophys. Res.* 92:4740–4744.

Grebowsky, J. M., Curtis, S. A., and Brace, L. H. 1991. Small-scale plasma irregularities in the nightside Venus ionosphere. *J. Geophys. Res.* 96:21347–21359.

Gurnett, D. A. 1985. Plasma waves and instabilities. In *Collisionless Shocks in the Heliosphere: Reviews of Current Research*, eds. B. T. Tsurutani and R. G. Stone (Washington, D. C.: American Geophysical Union), pp. 207–224.

Higuchi, T., Crawford, G. K., Strangeway, R. J., and Russell, C. T. 1992. Separation of spin synchronous signals. *Res. Mem.* No. 430 (Tokyo: Inst. of Statistical Mathematics).

Ho, C. M., Strangeway, R. J., and Russell, C. T. 1993. Evidence for Langmuir oscillations and a low density cavity in the Venus magnetotail. *Geophys. Res. Lett.* 20:2775–2778.

Ho, C. M., Strangeway, R. J., and Russell, C. T. 1994. Spatial distribution of plasma wave activity in the nightside ionosphere of Venus. *Planet. Space Sci.* 42:813–823.

Ho, C. M., Strangeway, R. J., and Russell, C. T. 1995. Venus nightside ionospheric irregularities and their relationship to VLF bursts. *J. Geophys. Res.* 100:9697–9705.

Hoegy, W. R., Brace, L. H., Kasprzak, W. T., and Russell, C. T. 1990. Small-scale plasma, magnetic, and neutral density fluctuations in the nightside ionosphere. *J. Geophys. Res.* 95:4085–4102.

Hospodarsky, G. B., et al. 1994. Fine structure of Langmuir waves observed upstream of the bow shock at Venus. *J. Geophys. Res.* 99:13363–13371.

Huba, J. D. 1992. Theory of small-scale density and electric field fluctuations in the nightside Venus ionosphere. *J. Geophys. Res.* 97:43–50.

Huba, J. D. 1993*a*. Generation of waves in the Venus mantle by the ion acoustic beam instability. *Geophys. Res. Lett.* 20:1751–1754.

Huba, J. D. 1993*b*. Theory of kilometer-size density waves in the nightside Venus ionosphere. *Geophys. Res. Lett.* 20:2763–2766.

Huba, J. D., and Fedder, J. A. 1993. Self-generation of magnetic fields by sheared flows in weakly ionized plasmas. *Phys. Fluids B* 5:3779–3788.

Huba, J. D., and Grebowsky, J. M. 1993. Small-scale density irregularities in the nightside Venus ionosphere: Comparison of theory and observations. *J. Geophys. Res.* 98:3079–3086.

Huba, J. D., and Rowland, H. L. 1993. Propagation of electromagnetic waves parallel to the magnetic field in the nightside Venus ionosphere. *J. Geophys. Res.* 98:5291–5300.

Intriligator, D. S., Wolfe, J. H., and Mihalov, J. D. 1980. The Pioneer Venus orbiter plasma analyzer experiment. *IEEE Trans. Geosci. Remote Sensing* GE-18:39–43.

Kar, J., et al. 1994. Evidence of electron impact ionization on the nightside of Venus from Pioneer Venus Orbiter ion mass spectrometer measurements near solar minimum. *J. Geophys. Res.* 99:11351–11355.

Kasprzak, W. T., Taylor, H. A., Jr., Brace, L. H., and Niemann, H. B. 1982. Observa-

tions of energetic ions near the Venus ionopause. *Planet. Space Sci.* 30:1107–1115.
Kelley, M. C., et al. 1985. Electrical measurements in the atmosphere and the ionosphere over an active thunderstorm 1. Campaign overview and initial ionospheric results. *J. Geophys. Res.* 90:9815–9826.
Krehbiel, J. P., et al. 1980. Pioneer Venus orbiter electron temperature probe. *IEEE Trans. Geosci. Remote Sensing* GE-18:49–54.
Leroy, M. M., and Mangeney, A. 1984. A theory of energization of solar wind electrons by the earth's bow shock. *Annales Geophys.* 2:449–456.
Luhmann, J. G. 1990. "Wave" analysis of Venus ionospheric flux ropes. In *Physics of Magnetic Flux Ropes* (Washington, D. C.: American Geophysical Union), pp. 425–432.
Nagy, A. F., et al. 1990. Venus mantle–Mars planetosphere: What are the similarities and differences? *Geophys. Res. Lett.* 17:865–868.
Onsager, T. G., and Holzworth, R.H. 1990. Measurement of the electron beam mode in earth's foreshock. *J. Geophys. Res.* 80:19–26.
Orlowski, D. S., Russell, C. T., Krauss-Varban, D., and Omidi, N. 1994. A test of the Hall-MHD model: Application to low-frequency upstream waves at Venus. *J. Geophys. Res.* 99:169–178.
Rogister, A., and D'Angelo, N. 1970. Type II irregularities in the equatorial electrojet. *J. Geophys. Res.* 75:3879–3888.
Russell, C. T., Snare, R. C., Means, J. D., and Elphic, R. C. 1980. Pioneer Venus orbiter fluxgate magnetometer. *IEEE Trans. Geosci. Remote Sensing* GE-18:32–35.
Russell, C. T., et al. 1990a. Upstream waves at Mars: Phobos observations. *Geophys. Res. Lett.* 17:897–900.
Russell, C. T., von Dornum, M., and Scarf, F. L. 1990b. Impulsive signals in the nightside ionosphere of Venus: Comparison results obtained below the local electron gyrofrequency with those above. *Adv. Space. Res.* 10:37–40.
Sagdeev, R. Z., et al. 1990. Wave activity in the neighborhood of the bowshock of Mars. *Geophys. Res. Lett.* 17:893–896.
Scarf, F. L., Taylor, W. W. L., and Green, I. M. 1979. Plasma waves near Venus: Initial observations. *Science* 203:748–750.
Scarf, F. L., Taylor, W. W. L., and Virobik, P. F. 1980a. The Pioneer Venus orbiter plasma wave investigation. *IEEE Trans. Geosci. Remote Sensing* GE-18:36–39.
Scarf, F. L., Taylor, W. W. L., Russell, C. T., and Elphic, R. C. 1980b. Pioneer Venus plasma wave observations: The solar wind-Venus interaction. *J. Geophys. Res.* 85:7599–7612.
Scarf, F. L., Taylor, W. W. L., Russell, C. T., and Brace, L. H. 1980c. Lightning on Venus: Orbiter detection of whistler signals. *J. Geophys. Res.* 85:8158–8166.
Scarf, F. L., et al. 1985. Current-driven plasma instabilities and auroral-type particle acceleration. *Adv. Space Res.* 5:185–191.
Shapiro, V. D., et al. 1995. On the interaction between the shocked solar wind and the planetary ions on the dayside of Venus. *J. Geophys. Res.* 100:21289–21305.
Skalsky, A., et al. 1992. The Martian bow shock: Wave observations in the upstream region. *J. Geophys. Res.* 97:2927–2933.
Spenner, K., et al. 1980. Observation of the Venus mantle, the boundary region between the solar wind and ionosphere. *J. Geophys. Res.* 85:7655–7662.
Strangeway, R. J. 1991. Plasma waves at Venus. *Space Sci. Rev.* 55:275–316.
Strangeway, R. J. 1995a. Plasma wave evidence for lightning on Venus. *J. Atmos. Terres. Phys.* 57:537–556.
Strangeway, R. J. 1995b. An assessment of plasma instabilities or planetary lightning as a source for the VLF bursts detected at Venus. *Adv. Space Res.* 15:89–92.
Strangeway, R. J., and Crawford, G. K. 1993. On the instability and energy flux of

lower hybrid waves in the Venus plasma mantle. *Geophys. Res. Lett.* 20:1211–1214.
Strangeway, R. J., and Crawford, G. K. 1995. VLF waves in the foreshock. *Adv. Space Res.* 15:29–42.
Strangeway, R. J., Scarf, F. L., Zanetti, L. J., and Klumpar, D. M. 1988. AMPTE CCE plasma wave measurements during magnetospheric compressions. *J. Geophys. Res.* 93:14357–14368.
Strangeway, R. J., Russell, C. T., and Ho, C. M. 1993*a*. Observation of intense wave bursts at very low altitudes within the Venus nightside ionosphere. *Geophys. Res. Lett.* 20:2771–2774.
Strangeway, R. J., Russell, C. T., Ho, C. M., and Brace, L. H. 1993*b*. Plasma waves observed at low altitudes in the tenuous Venus nightside ionosphere. *Geophys. Res. Lett.* 20:2767–2770.
Szegö, K., et al. 1991. Physical processes in the plasma mantle of Venus. *Geophys. Res. Lett.* 18:2305–2308.
Szegö, K., Sagdeev, R.Z., Shapiro, V.D., and Shevchenko, V. I. 1992. On the dayside mantle region around those nonmagnetic solar system bodies which have ionosphere, *Adv. Space Res.* 12:291–298.
Taylor, H. A., Jr., et al. 1981. Dynamic variations observed in thermal and superthermal ion distributions in the dayside ionosphere of Venus. *Adv. Space Res.* 1:247–258.
Taylor, H. A., Jr., Cloutier, P. A., and Zheng, Z. 1987. Venus "lightning" signals reinterpreted as in situ plasma noise. *J. Geophys. Res.* 92:9907–9919.
Theis, R. F., and Brace, L. H. 1993. Solar cycle variations of electron density and temperature in the Venusian nightside ionosphere. *Geophys. Res. Lett.* 20:2719–2722.
Thomsen, M. F. 1985. Upstream suprathermal ions. In *Collisionless Shocks in the Heliosphere: Reviews of Current Research*, eds. B. T. Tsurutani and R. G. Stone (Washington, D. C.: American Geophysical Union), pp. 253–270.
Trotignon, J. G., et al. 1992. Electron density in the Martian foreshock as a by-product of the electron plasma oscillation observations, *J. Geophys. Res.* 97:10831–10840.
Walker, S. 1992. Plasma Instabilities in the Venus Nightside Ionosphere. Masters Thesis, Rice Univ.
Winske, D. 1986. Origin of large magnetic fluctuations in the magnetosheath of Venus. *J. Geophys. Res.* 91:11951–11957.
Zhang, T.-L., Luhmann, J. G., and Russell, C. T. 1990. The solar cycle dependence of the location and shape of the Venus bow shock. *J. Geophys. Res.* 95:14961–14967.

EVIDENCE FOR VENUS LIGHTNING

J. M. GREBOWSKY
NASA Goddard Space Flight Center

R. J. STRANGEWAY
University of California at Los Angeles

and

D. M. HUNTEN
University of Arizona

How strong is the evidence for lightning on Venus? Optical observations of the nightside Venus atmosphere from Venera 9 and terrestrial telescope observations have recorded flashes from the dusk-midnight sector. Optical results from Vega balloon lightning detectors in the cloud layer and from Pioneer Venus Orbiter (PVO) star sensor measurements were negative. Magnetic wave bursts recorded by the Venera landers have been attributed to dayside atmospheric discharges. A majority of the nightside 100-Hz electric field bursts measured on PVO have characteristics expected for electromagnetic waves propagating in the whistler mode from the atmosphere. These bursts were predominantly detected on the dawn side of midnight, where ionospheric conditions favor whistler propagation. Non-whistler signals observed on PVO near 2100-hr local time have been suggested to arise from lightning, but local nonlightning plasma generation processes cannot yet be ruled out. Several radio frequency pulses detected on Galileo's flyby of Venus could have a lightning origin, but there is uncertainty about whether conditions in Venus' ionosphere would have allowed them to escape. Each investigation found unique event characteristics, as though each sampled different discharge/storm distributions. The observational evidence supports the hypothesis that lightning occurs on Venus with intracloud lightning being the likely discharge process, but concrete evidence for the presence of conditions in the clouds needed to produce discharges is still missing. Volcanism plays no role.

I. INTRODUCTION

A. Background

The presence of conditions for the production of electric discharges were, and still are, conjectural because the Venus environment significantly differs from the only environment with which we have innate familiarity—Earth. From everything we know, there is no clear evidence that the Venus environment generates and separates electrical charge, two conditions that seem to be needed for the occurrence of an electrical discharge.

Optical and electromagnetic impulses detected by the Venera 9 optical spectrometer from orbit and by magnetic wave observations from the Venera 11 and 12 landers provided the initial evidence for the presence of Venus lightning discharges. Later, very low-frequency impulsive electric wave signals were observed by the electric field detector on the Pioneer Venus Orbiter (PVO) which had characteristics expected for signals produced by atmospheric electrical discharges. These observations were discussed in *Venus* (see Hunten et al. 1983).

The inference that lightning occurs in the Venusian atmosphere was not unreasonable because lightning signatures appear to be common to all the other planets in the solar system with dense atmospheres (see, e.g., Desch 1991). Strong pro-and-con positions were taken on the interpretation of the Venus lightning measurements, particularly the electric wave activity measured on PVO. A plethora of papers (see, e.g., the comprehensive bibliographies in the reviews of Russell [1991,1993]) have been published since the publication of *Venus* (Hunten et al. 1983) and it is not a simple matter for casual readers to extract the significant results from the often convoluted exchanges within the literature. More recently, new observations have been made, and intensive theoretical and statistical investigations have begun to improve upon early heuristic arguments and individual sample studies.

This chapter will attempt to document and put into perspective, the current catalog of information, experimental and theoretical, that exists regarding lightning on Venus. It is not our intent to provide a history delineating the often heated controversies that occurred over the years.

II. EFFECTS AND SOURCES

A. Significance of Lightning

For a demonstration of public interest in the topic, one only has to think of one's own response to lightning. Another example can be seen in the extensive public media coverage of recent observations of cloud to stratosphere-mesosphere lightning strokes at Earth. At the same time, fundamental understanding of processes that lead to atmospheric electrical breakdown expands with the exploration of discharge phenomena in new environments. The study of lightning on Venus is another step to furthering our understanding of similar phenomena in the terrestrial and outer planet atmospheres.

In regard to the Venus environment itself, the presence of lightning and its spatial/temporal distribution may provide clues to meteorological variations and cloud processes that have not been explored by the limited Venus spacecraft observations made to date (see Borucki et al. 1991, and references therein). For example, discharges will have an impact on the planet's fair weather electric field circuit as is the case at Earth, and would produce electromagnetic disturbances whose propagation into the ionosphere could confuse studies of disturbances resulting from *in-situ* plasma processes. The radiation may also be a significant source of ionospheric heating. Discharges in the

Venus atmosphere will locally produce minor atmospheric species at a rate which could be competitive with ambient photochemical processes. In order to study these possibilities, one must first unravel the available evidence for and against the presence of lightning, its properties, and source mechanisms.

B. Mechanisms

1. Terrestrial Processes. Because the only planet for which we have extensive understanding of the lightning process is Earth, it has been used as the model for inferring the presence of electrical discharges, and their sources, at other planets. Hence it is necessary to discuss some attributes of the terrestrial discharges as a lead-in to the discussion of Venus.

Lightning discharges on Earth have several established source mechanisms. Details of the discharge processes can be found in Uman (1987) and Levin et al. (1983). Basically, the discharge requires as its prelude a charging mechanism and a charge separation mechanism to separate opposite polarity charges against their electrical attraction until the developed electric potential difference exceeds the breakdown field of the atmosphere. Terrestrial lightning discharges occur, although infrequently, in volcanic plumes and within dust storms and even, although extremely rare, in clear air. Recently cloud to mesosphere/ionosphere discharges (Sentman and Wescott, 1993) have been the subject of excited interest, with source mechanisms currently being established. The most common lightning events are those produced within strongly convective cumulus clouds with intense precipitation. Thunderstorms are either isolated (i.e., the typical hot humid summer storms), or take place over broad horizontal weather fronts. The isolated storms are the result of localized pockets of buoyant warm air formed near the surface; the frontal storms develop when a cold front of air undercuts and lifts up warm moist air from near the surface. Lightning is most frequent at low latitudes. Over land, the highest frequency of occurrence is in the afternoon and early evening (e.g., Orville 1986). The frequency of occurrence is much lower over the oceans and does not exhibit a clear local time preference.

Terrestrial cloud discharges are usually linked to precipitation and vertical air motions. Charging typically results from collisions between different size particles with a net transfer of charge between them. The particle charging process depends on the electrochemical properties of water or ice and/or on the polarizability of water exposed to ambient background electric fields. Ion attachment and cosmic ray ionization can also contribute. The large E field generation in clouds that leads to electrical breakdown is produced by the separation of the charges via upward/downward drafts produced by convection and gravity. Differential, mass-dependent flows physically separate the charges of aerosol populations in which the sign of the charge on a particle depends on its size. Basically any build-up of strong E fields within clouds requires a phase change to produce liquid or solid aerosols, a population of different sized aerosols, production of charge with a sign dependent on particle size, and strong vertical, mass dependent transport. Phase changes, through the release of latent heat feed the process by an intensification of the vertical drifts.

2. *Venus Processes.* Evidence acquired early in the PVO mission indicated a sulfur dioxide enhancement that was attributed to intense volcanic activity (Esposito 1984). However, as Borucki (1982) early pointed out, something that other studies have yet to refute, the absence of aerosols near the surface is an indication that the explosive volcanism which lead to terrestrial discharges is not present at Venus. Also the dense atmosphere, composed of CO_2 with its large dielectric constant, would require electric fields much stronger than at Earth to produce breakdown of the gas near the surface. Hence volcanism is not a likely source of lightning. A similar conclusion can be made for dust-generated discharges; the requisite intense surface winds and dense aerosol distributions are not a characteristic of the lower Venus atmosphere. Cloud-ionosphere lightning events cannot be ruled out at Venus, particularly because the clouds on Venus are 30 to 40 km nearer to the base of the ionosphere than on Earth. The mechanisms behind the terrestrial phenomenon have yet to be resolved but the upward flashes have a frequency of occurrence small compared to the other discharges in storm clouds. Cloud-to-ground discharges would be extremely difficult to excite at Venus due to the large breakdown voltage of the dense CO_2 atmosphere and the extreme height of the cloud layers above the surface. On Earth, intracloud events are 2 to 6 times more frequent than cloud-to-ground discharges with the relative occurrence ratio increasing with altitude. Hence, the intracloud discharge is likely to be the most dominant possibility at Venus.

If lightning is present, it is probably the result of electric fields developed within the cloud region, i.e., from $\sim$48 km up to $\sim$68 km. Multi-sized aerosol constituents do exist in the clouds and the sulfuric acid droplets present have electrochemical properties that could lead to charge exchange between different-sized particles. However, the mass loading in the clouds of Venus is low compared to Earth (Borucki 1982) and there is no observational evidence of strong precipitation; the particle sizes are more consistent with a drizzle. Strong upward convection has not yet been detected in the cloud layers. Vega balloon observations (Sagdeev et al. 1986) in the middle cloud layer in the midnight-dawn quadrant recorded vertical motions with speeds typically less than 2 m s^{-1}, but did encounter a downdraft of $\sim$3.5 m s^{-1}. Maximum vertical speeds in terrestrial storms are larger and have been observed to exceed 20 m s^{-1} (see, e.g., Williams 1985). Young et al. (1987) interpreted the Vega measured drifts to be gravity waves generated along mountain ridges. Because solar energy absorption is greatest in the clouds in the subsolar region the presence of active convection might be anticipated on the day side of the planet. Levin et al. (1983) have offered a rigorous computational model demonstrating that electric fields can build up to discharge magnitudes if regions exist with larger size particles and greater mass loading than were observed by the probes sent to the Venus surface. Prominent convective cumulus clouds characteristic of terrestrial storm cells are not evident at

Venus where the clouds are stratiform. If strong convection is required, then the generation of the large breakdown electric field would likely take place in the middle cloud layer between 50 and 52 km which is characterized by an adiabatic lapse rate.

Under the assumption that lightning exists in the Venus clouds, there is currently not enough knowledge of atmosphere conditions to infer global behavior. Earth has a predominance of electrical storms in the afternoon and evening with occurrence frequencies being greatest at low latitudes (see, e.g., Orville 1986). The terrestrial lightning statistics are dominated by storms resulting from intense convection generated by local heating of moist air near the surface. This is not the case at Venus where hot spots on the surface are not produced. The near-surface behavior on Venus is more favorably compared to the terrestrial oceans; the lower atmosphere is rather uniform in the horizontal direction and the time scale for dynamic processes is much less than the radiative time scale. Over the oceans lightning storms are rather infrequent and the occurrence rates do not maximize in the afternoon/evening region. Further, terrestrial lightning storms generated by the interaction of weather fronts can occur anywhere in local time. One must be cautious in using the global distribution of terrestrial lightning as a framework for Venus.

III. OBSERVATIONS

A. Expectations

Although caution must be taken in using some characteristics of terrestrial lightning strokes to interpret observations at Venus, the geometry and electrical characteristics of the terrestrial discharges provide the only available standards of comparison. Charge generation conditions at the two planets may differ but the discharges likely have similar electrical and optical signatures for strokes originating in the clouds. The atmospheric pressures in the cloud regions are similar for both planets. It is difficult to see why the discharge process should differ significantly between the two planets.

A terrestrial intracloud discharge is viewed as a continually propagating leader with several visible return strokes. Cloud-to-ground flashes start with a leader which nearing the ground initiates the first return stroke and may be followed by subsequent leaders/strokes (see Uman [1987] for more details). The length of intracloud flashes range from <0.5 km to ~10 km. Typical optical flash intensities range from 10^5 to 10^6 J corresponding to 10^{-3} to 10^{-4} of the total dissipated power in the discharge. Most of the energy of the discharge is dissipated in heating of the air and production of ionization. The duration of one flash (comprised of several short duration strokes along a common discharge channel) ranges from 300 to 500 ms. The average global occurrence rate of flashes is ~100 flashes per second (2×10^{-7} km^{-2}s^{-1}) with a peak rate (from Bliokh et al. 1980) of ~2×10^{-6} km^{-2}s^{-1}. The number of electromagnetic wave bursts from a single flash ranges from 1 to 20 or more. The power spectrum of the electromagnetic radiation is dependent upon where it is measured relative to the source. The attenuation and transmission of electromagnetic waves depends upon frequency dependent propagation conditions imposed by the conducting ionosphere with its imbedded magnetic field, the atmosphere, and the conducting planetary surface. Given this background, subsequent sections will describe the Venus observations.

B. Optical Measurements

1. Venera 9 and 10. The first optical evidence for lightning in the nightside atmosphere of Venus (Krasnopolsky 1983*a,b*) was obtained from the orbiting Venera 9 spectrometer (300–800 nm wavelength band). Similar observations were made on Venera 10, but no nightside signals were detected. From ~2 months of observations, bursts were only detected for one 70-s period on 26 October 1975. They appeared on seven consecutive 10-s sweeps of the wavelength band. Figure 1 displays the data for one of the sweeps; the raw signals are depicted at the bottom of the figure and the deduced incident ambient spectral intensities on the top (a format originally used by Russell [1993]). The plots do not depict individual data points but are eyeballed averages of the 450 telemetered samples in the 10-s sweep. The burst amplitudes roughly follow the instrument response providing evidence that the observations are not due to bit errors. If the signal was internal to the instrument it is difficult to envision why only this short interval out of many hours of observations was affected.

The noisiness of the spectra plus the fact that laboratory arc discharges produce spectral lines on top of a very strong continuum spectrum, makes it improbable that spectral line information can be culled from the 10-s sweeps. A superposition of all seven spectra in which signals were recorded has not been published, but Krasnopolsky (1983*b*) says the cumulative spectrum is relatively flat. The brevity of individual lightning bursts coupled with anticipated stroke-to-stroke variability in the net light intensity penetrating the clouds would obscure details of the spectral structure. It was argued (Borucki 1982; Krasnopolsky 1983*a*) that the source of the emissions was in the clouds on the basis of the rapid falloff of signal strength below 4000 Å expected for ultraviolet absorption in the clouds. Emissions from lower near-surface altitudes would have been attenuated at wavelengths up to 5500 Å contrary to the data.

The one 70s event was observed in the vicinity of 1930 local time, 9°S latitude as the spectrometer field of view on the cloud layer ($\sim 160 \times 9$ km^2) moved horizontally 450 km. There were 10 to 20 flashes per 10-s scan with ~ 0.25 s peak widths—a duration comparable to that of terrestrial flashes. Borucki (1982) pointed out that there was no fine structure analogous to the short duration strokes comprising a terrestrial flash, making the observation somewhat puzzling. Due to the comparability of the instrument response time (5 ms) with the sampling time (20 ms), Krasnopolsky (1983*a,b*) suggested that the flashes consisted of 10 to 20 short-duration strokes which occurred at intervals shorter than 20 to 30 ms. Another alternative interpretation is that

EVIDENCE FOR VENUS LIGHTNING

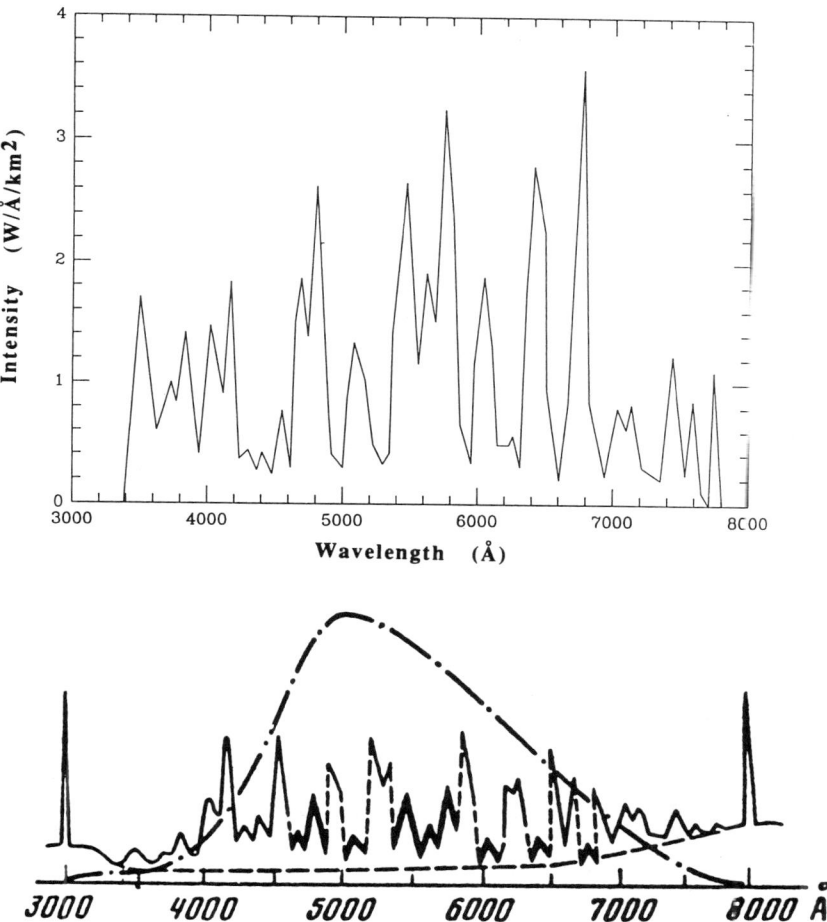

Figure 1. Venera 9 spectrum of a storm region. Bottom (from Krasnopolsky 1983b) shows measured signal with wavelength scale along the instrument zero. Dashed curve depicts the radiative background noise. The dash-dot curve is the response of the instrument to a 1 W/Å/km² source. Dark portion of spectra is for a scale less sensitive by a factor of 8. Top (provided by Krasnopolsky, personal communication) depicts the *in-situ* intensities deduced from the measurements, taking into account the instrument's response and gain states.

the flashes have a continuous optical emission. Terrestrial lightning flashes do occur which have continuous luminosity between the strokes.

Krasnopolsky (1983b) interpreted the 70-s event as evidence for a broad, frontlike thunderstorm region that was passed over by the spectrometer field of view. The cloud area in view of the spectrometer during the one storm event was $\sim 10^{-3}$ of the total area surveyed from Venera. The observed flash rate was ~ 1 s^{-1} with flashes detected over a $\sim 5 \times 10^4$ km² area as the view moved 450 km across the clouds. (In comparison, terrestrial frontal lightning storm sizes are $\sim 10^3$ km^2, 30 to 40 km width, with flash rates averaged over the storm duration of ~ 2 per min). The spatial/temporal uniqueness of this event makes any inference of global rates of occurrence pure speculation. Based on the energy of 3×10^7 J measured between 400 to 700 nm and the assumption of constant energy/wavelength between 400 to 1100 nm, 7×10^7 J was estimated for the optical energy emitted per flash assuming a long duration flash with no substrokes (Borucki 1982). Although this energy was more than 200 times greater than typical terrestrial levels, it was 30 times less than that of superbolts. On the other hand, Krasnopolsky estimated that if the flash consisted of 10 to 20 short strokes then the visible stroke energy would drop to $\sim 2 \times 10^6$ J, corresponding to an estimated $\sim 10^{10}$ J for the total energy dissipated in a single flash, which is similar to terrestrial flash energies.

2. *Pioneer Venus Orbiter.* The Pioneer Venus mission was the next opportunity to explore further details of the atmosphere of Venus. An optical spectrometer was not part of the payload, but Borucki et al. (1981) devised a creative use of the Pioneer Venus Star Tracker to search for optical bursts in the nightside ionosphere. The star sensor was mounted with its symmetry axis 56° from the spacecraft spin axis. During the spacecraft spin near periapsis, the sensor's field of view scanned across the cloud tops between sightings beyond the planet's limbs. The star sensor was turned on for 11.5° of each spin cycle when the planet's limb was visible only by scattering of light from the star sensor lens (due to nightside airglow the sensor saturated when it looked directly at the dark side of the Venus atmosphere). The rate of pulse detections was compared to the false-alarm rate (due to high-energy particles) measured outside of eclipse. Measurements from 36 darkside 1979 orbits, from near midnight to ~ 0500 local time were analyzed. The number of pulses detected was not different from the number of false alarms expected.

A later analysis (Borucki et al. 1991) noted that the field of view used in the earlier study was in error. In the revised study the actual response function (i.e., the dependence of the response on the angle between the optical axis and the light source) was measured for a lens identical to that employed in the Pioneer Venus star tracker. Scanner measurements near periapsis were studied for 53 orbits in 1988 and 55 orbits in 1990. The cloud coverage consisted of patches predominantly from 2230 local time to the dawn terminator but a patch near the dusk terminator was also viewed. As in the earlier study, event rates were those expected for false alarms. From the 1988 data an upper bound to the planetary flash rate was estimated to be 4×10^{-7} flashes km^{-2}s^{-1} for terrestrial-like short duration (few hundred microsecond) flashes at least 50% as bright as terrestrial flashes (i.e., median optical power of $\sim 2 \times 10^9$ W). Using the 1990 data an upper bound of 1×10^{-7} flashes km^{-2}s^{-1} was estimated for long duration flashes (i.e., the $\sim 1/4$ s bursts detected by the Venera spectrometer) that are at least 1.6% as bright as typical terrestrial flashes, or 33% as bright as the Venera 9 observed light pulses. Although this study surveyed a large area of the night side of Venus, the total time of search for the two-year 108 orbit study was only 83 s due to very stringent viewing requirements. Nevertheless, the analysis found no evidence of lightning.

3. Vega Balloons. Further searches for lightning at Venus were made by the two Vega balloon instrument packages inserted within the cloud layer of Venus on 11 and 15 June 1985, respectively (Sagdeev et al. 1986). Both were inserted near midnight local time—Vega 1 at 7°N latitude, Vega 2 at 7°S. They floated in the middle cloud layer at altitudes between ∼50 and ∼54 km. Each carried a silicon PIN diode light detector, sensitive to 400 to 1100 nm radiation, which looked downward with a field of view of ±60°. Each detector pointed directly at the top of the lower battery and nephelometer section which hung by straps from the section on which the detector was positioned. Nevertheless the field of view was wide enough to intercept light coming from below the vehicle. Surfaces were covered with a special white coating which provided a high surface albedo. Both balloons drifted with the wind for 30 hr from midnight through the dawn terminator. No lightning events were detected (Kremnev et al. 1986).

4. Mt. Bigelow, Arizona. A recent observation campaign by Hansell et al. (1995) used the 61" telescope on Mt. Bigelow to search for light flashes on the night side of Venus. Their study carefully employed coronagraphic optics, using 2 masks designed in accordance with the specific geometry for each individual night of viewing. An occulting mask was used in the imaging plane and a Lyot mask was used to block diffracted light by the edges and support structure of the secondary mirror. The CCD detector was operated at 18.8 frames s^{-1} for 30×30 pixel images of Venus. Each image pixel was a 2×2 summation of CCD pixels and each image had an approximate exposure time of 20 ms. Careful computer screenings of the images were made to eliminate non-lightning effects. A star was used to approximate a lightning image and to calibrate atmospheric effects as well as the threshold for detection. The observations were made for wavelengths at 777.4-nm (0.7 nm bandwidth) and 656.3 nm (2.0 nm bandwidth). Laser induced plasma spectra (Borucki et al. 1983) in a Venus-like atmosphere yielded strong emission associated with the excitation of neutral oxygen, with half of the emitted energy within the Mt. Bigelow experiment's 777.4-nm band. The 656.3-nm (H-alpha) band was selected as a control measurement—such emissions were not initially expected from lightning discharges on Venus.

Observations were made for 8 nights in February and March, 1993, midway between maximum elongation and inferior conjunction. The total viewing time was 3 hr at 777.4 nm and 45 min for the H-alpha. The dusk side of Venus was facing Earth. Seven events met the stringent criteria (including the requirement that an event must be seen on more that one pixel) that the experimenters used for isolating lightning flashes. Six events were detected in the 777.4-nm band. One of these is shown in Fig. 2a. The seventh occurred at the H-alpha wavelength. It was then realized that the spectrum of lightning includes a line at 658.0 nm which is within the rather broad passband of the filter. If on the other hand, the H-alpha burst is spurious, the spurious

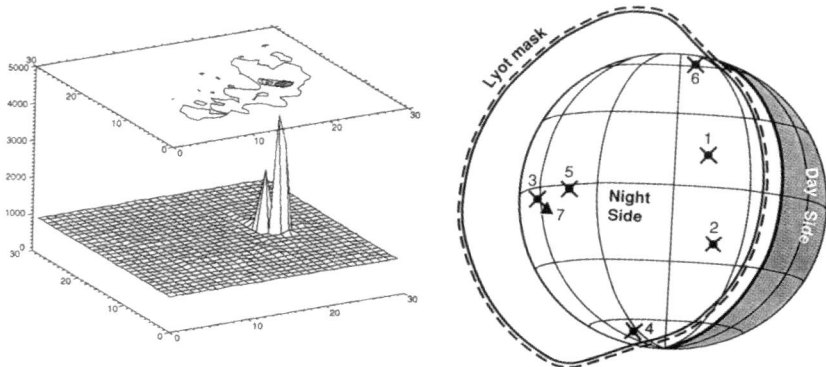

Figure 2. (a) Example of the CCD response for a flash detected at 777.4 nm from Mt. Bigelow, Arizona. (b) Locations of all 7 flash events. Event 7 is the lone burst detected at 658.0 nm and event 1 is the burst depicted in (a) (figure from Hansell et al. 1995).

detection rate would be 0.022 s^{-1}. The occurrence rate of the 777.4-nm events (0.033 s^{-1}) is only 50% greater, i.e., not significantly different to allow the absolute claim to be made that lightning is the only possible source for the events. However, the care with which the experiment was performed and the stringent tests performed to eliminate signals that could be confused with lightning bursts argues for accepting the bursts as light bursts from Venus.

The 777.4-nm bursts imply Venus lightning flashes with optical energies from $\sim 7 \times 10^7$ to $\sim 2 \times 10^9$ J (assuming a cloud transmission of 0.2 for red photon escape from the middle cloud layer). The calibrated threshold for detection was 5×10^6 J for 95% detection and 1.5×10^6 J for 50% detection. The computed flash rate/area was $\sim 3 \times 10^{-12}$ km^{-2}s^{-1}. The estimated flash optical energies exceed average terrestrial values by as much as 3 orders of magnitude and correspond to superbolt magnitudes. The lowest measured burst energy is similar to the flash energy deduced from the Venera 9 spectrometer measurements. Considering terrestrial flashes with optical energies exceeding the telescope experiment thresholds, Hansell et al. (1995) deduced that the observed Venus flash rate was ~ 1000 times smaller than the terrestrial rate.

There are other interesting features of the observations. First, the bursts did not persist from one readout of the CCD array to the next. Hence the durations are short, less than 20 ms, analogous to terrestrial stroke durations. On the other hand, the bursts are isolated in time with no evidence for the rapid sequences of flashes associated with terrestrial storms. Three events were detected within 10 min of one another on one day. On yet another night, no bursts were detected for the entire 75 min viewing period. There is no evidence for a preferred source location (see Fig. 2b) and no evidence of flashes persisting for hundreds of ms. Of course one cannot rule out the possibility that more frequent and longer duration flashes occurred at energy levels below the experiment threshold. The Venera deduced lightning characteristics are distinctly different from those inferred from the Mt. Bigelow data.

5. Optical Data Inferences. The optical measurements imply that Venus' nighttime flashes are sporadic and difficult to pin down unambiguously. The evidence is that lightning on the night side of Venus is not occurring continuously at one specific area. All positive sightings (i.e., the one Venera 9 event and the Mt. Bigelow 6 or 7 events) were made in the dusk–midnight sector of Venus. The Vega balloons' uneventful observations were in the midnight–dawn sector. The star sensor study survey was predominantly in the dawn sector, although patches of the clouds were viewed on the dusk side of midnight also. The isolated detection of only 70 s of activity from 2 months of Venera spectrometer observations and the Mt. Bigelow bursts indicate that the phenomenon is not persistent on the dusk side of midnight and may be absent for long periods of time at any location. The data are not yet sufficient to deduce confidently a local time dependence for the atmospheric events, but thus far only premidnight flashes have been recorded. An origin in a large storm-front region is implied by the Venera 9 observations. On the other hand, a discharge characterized by one intense stroke (and/or many weak ones?) that is locally and temporally rather singular is more consistent with the Mt. Bigelow results. The Venera bursts had durations of a few 100 ms, similar to those of terrestrial flashes. In contrast the bursts recorded in the telescope study lasted less than 20 ms, analogous to terrestrial stroke durations with no evidence of long-duration flashes composed of multiple strokes.

Overall, the optical observations provide, in the absence of any other known mechanism that could elicit the same instrumental responses, evidence for the presence of lightning on the night side of Venus. Electric/magnetic field wave measurements provide another means for studying electrical discharges in the Venus atmosphere. However, the interpretation of the latter observations, as will be seen, is complicated because electrostatic and electromagnetic wave sources exist apart from a lightning discharge origin.

C. Electromagnetic Wave Observations

1. Venera 11 and 12 Landers. The Venera 11 and 12 probes descended onto Venus on 21 and 25 December 1978, respectively, at similar midday, low latitude locations. Each lander carried a high-sensitivity loop antenna detector (Ksanfomality 1979) with four narrowband channels centered on 10, 18, 36, and 89 kHz, a wideband signal (8–90 kHz) detector, and an impulse counter. Measurements began at an altitude of ~60 km and continued during the ~1-hr descent and on the surface until contact was lost with the relay spacecraft.

During the descents bursts of impulsive radio noise were detected. The temporal burst characteristics differed significantly for Venera 11 and 12 even though they followed similar flight paths. No spacecraft discharges were measured on subsequent Venera 13 and 14 probes which carried coronal discharge detectors. Hence noise generated by spacecraft-atmosphere interactions was ruled out by Ksanfomality et al. (1983). These measurements were offered to be the first strong evidence for electromagnetic pulses generated by atmospheric electrical discharges. Although acoustic sounders were carried on the Venera 11 and 12 spacecraft, noise levels on descent overwhelmed any atmospheric sound that may have been present.

Five or six bursts of fine structured wave activity were detected on Venera 11 and two bursts of activity on Venera 12. The duration of the bursts ranged from several to more than 15 min. The highest intensities were typically observed in the lowest-frequency band centered on 10 kHz. On the surface Venera 12 recorded only one burst of activity 30 min after landing, while nothing was recorded from Venera 11. The bursts were composed of pulses occurring at rates ranging from ~ 10 s^{-1} to ~ 55 s^{-1}. The pulses seemed to be essentially continuous throughout a burst, with rates that at times exceeded those of radio frequency bursts generated in a terrestrial lightning flash which are typically $\lesssim 20$ s^{-1}.

If lightning discharges account for the observations, the source would likely be distant. For example, no thunder was heard by the instruments on the surface although a radio frequency burst was. Further, on descent, bursts did not appear until the probes reached ~ 30 km, far away from the most likely source region in the cloud layers. One Venera 11 burst sequence and part of another were modulated at the rotation rate of the spacecraft as expected if the antenna alternately faced toward and then away from a narrow source region with an angular dimension estimated at $\sim 5°$ (Ksanfomality et al. 1983). Hypotheses for the spatial location of several of the noise bursts have been given. These make the assumption that the signals recorded during descent are only functions of altitude and not time. Atmospheric refraction of waves (Croft and Price 1983), shadowing by the planetary surface, ionospheric absorption, and surface-ionosphere wave guide effects (Ksanfomality 1979) all have impacts on the wave propagation. Ksanfomality et al. (1983) estimated, on the basis of observed signal levels, that sources could be as far away as 1000 to 3000 km, and from the spin modulation of the signal estimated the presence of a broad horizontal storm region exceeding 100 km. Using a refraction model to analyze one burst, Croft and Price (1983) have suggested that the source may actually lay below the clouds. However, there are still many uncertainties in the propagation paths and unexplained features in the data. For example, the signal strength decreased very smoothly the last 9 km of the Venera 11 descent; bursts of narrowband activity sometimes were detected in only one of the high-frequency channels; the continuity of spikes in the bursts; and of course the separation of temporal from spatial variations.

2. Galileo. The Galileo spacecraft on 10 February 1990, flew by Venus on its way to Jupiter. During the flyby the plasma wave instrument was used (Gurnett et al. 1991) to search for Venus lightning signals in the 100-kHz to 5.6-MHz range. The data were acquired for 53 min at distances of 4 to 5 R_v on the dawn flank from the night side of Venus. All pulses were isolated. Observations from control periods separate from the Venus encounter determined the instrument background activity to eliminate false intrinsic noise events. During the Venus flyby, 9 pulses were identified and are shown in Fig. 3. An engineering model study demonstrated that radiation from midlatitude terrestrial lightning discharges would be detectable at the same distance. Although the Venus pulse amplitudes were low, it was concluded that lightning was the only source that appeared consistent with the bursts.

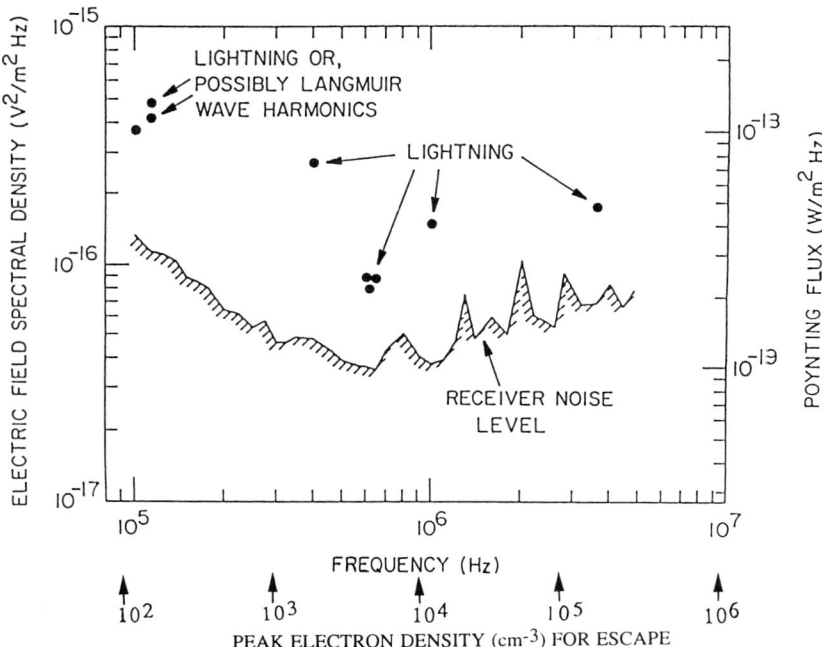

Figure 3. Nine impulses measured by the plasma wave instrument on Galileo's Venus flyby (reprinted with permission from Gurnett et al. [1991]; copyright 1991 American Association for the Advancement of Science). Signals are referenced to the measured noise background. A legend has been inserted below the published figure indicating the electron densities needed for the electron plasma frequencies to equal the measured frequencies.

However, from anticipated Venus ionosphere conditions under the solar maximum period of the Gallileo flyby, there is uncertainty about a lightning interpretation for the observations. The nightside ionosphere peak electron density for solar maximum conditions averages above 10^4 cm^{-3}, with individual peak values only rarely below several thousand cm^{-3} (see review by Brace and Kliore [1991]). Atmospheric generated radio frequency waves can propagate through the ionosphere only if their frequencies exceed the maximum ionospheric electron plasma frequency. As shown in Fig. 3, the three lowest frequency pulses are not consistent with an atmospheric source.

The four higher-frequency events below 1 MHz are borderline; somewhat exceptional ionosphere conditions had to prevail at the time of the observations for such signals to exit the atmosphere. Also, Hunten (1995) noted that Earth lightning is very weak at 5.6 MHz corresponding to the highest-frequency Galileo pulse (the terrestrial spectrum falls off as $\sim 1/f^2$). A very rough estimate for the flash rate is $\sim 10^{-11}$ flashes km^{-2}s^{-1} ($\sim$4 orders of magnitude less than that of Earth) assuming 6 valid flash signatures arising from anywhere on the night side.

3. Pioneer Venus Orbiter. PVO carried an electric field detector (OEFD) with four (30% bandwidth) frequency bands centered on 100 Hz, 730 Hz, 5.4 kHz and 30 kHz (Scarf et al. 1980). Of all the Venus spacecraft experiments with the potential for providing information on lightning discharges, the OEFD measurements comprised the largest data base. PVO's orbit early in the mission made three complete local time precessions about the planet (each complete local time cycle referred to as a season) when the periapsis altitude was below 200 km, extending as low as 138 km. Periapsis was then allowed to rise and ionospheric measurements were not made again until the final entry phase of the mission.

Because PVO was situated far above the cloud layers; because *in-situ* plasma processes and atmosphere interactions with the spacecraft can generate electric field waves; because propagation paths from the atmosphere into the ionosphere are not well defined; and because there are anomalous OEFD signals, the OEFD measurements in the lower nightside ionosphere have been the subject of much controversy in the literature for more than a decade. This has fortunately had the benefit of leading to significant insights and improved understanding of plasma wave processes firsthand in a planetary environment different from Earth. Unfortunately, it has also led to a labyrinth of pro-and-con lightning publications that has made it difficult for those looking from afar to understand the issues. Themes of published papers range from the position that all nonspurious OEFD low-altitude signals have a lightning origin (see, e.g., Russell 1991) to the opposite position of claiming that any inferred association of the OEFD waves with lightning is pure speculation (see, e.g., Taylor et al. 1995). A detailed historical analysis and a critique of individual studies within the limited allotted space would not help matters. The current state of understanding will be explored by focusing on an evaluation of the characteristics of the wave bursts highlighted in the most recent OEFD statistical studies which supplanted earlier coarser-resolution studies and individual case studies.

A framework for discussing how OEFD measurements might relate to lightning discharges is shown in Fig. 4, which portrays electromagnetic (e/m) wave propagation paths from sources in the clouds. Venus' ionosphere has a weak, embedded magnetic field. Low-frequency e/m waves with frequencies below the electron plasma frequency f_p at the peak of the ionosphere cannot penetrate the ionosphere unless the magnetic field lines extend through the ionosphere to the atmosphere. In such a case, only waves with frequency

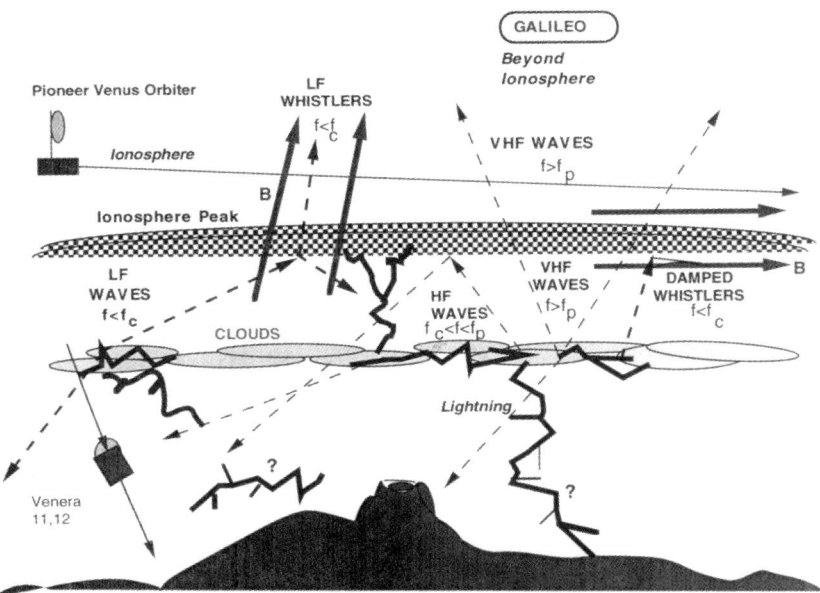

Figure 4. A diagram demonstrating the escape/nonescape of electromagnetic waves from lightning. The cross hatched band represents the ionosphere layer.

below the electron gyrofrequency f_g ($f_g < f_p$ in the Venus ionosphere) can, by propagating along B in the whistler mode, transit the ionosphere. (Note that the whistler mode exists only in the ionosphere; **B** plays no role in atmospheric propagation.)

On the day side of Venus the magnetic field is horizontal in the lower ionosphere and $f_{p\max}$ exceeds 1 MHz. Discharge radiation would not be detectable on the day side by OEFD. In the nighttime ionosphere **B** varies in space and it has configurations which suggest the possible escape of low-frequency signals from the atmosphere in the whistler mode. The most prominent of such regions are the ionospheric electron density troughs or "holes" (Brace et al. 1982), which are large, localized, depleted electron density regions in the antisolar region within which the magnetic field is intensified to 20 to 30 nT ($f_g \sim 560$–840 Hz) and tends to be aligned parallel to the ecliptic plane. Indeed it is only at night that OEFD recorded bursts of 100-Hz activity in the lower ionosphere.

If nightside 100-Hz bursts in the ionosphere are signatures of whistler mode activity, the local B magnitude must be large enough that $f_g > 100$ Hz (effectively, the required B threshold is higher because wave damping and dispersion of whistler mode signals reduce signal amplitudes for frequencies near f_g). A lightning origin requires that the magnetic field lines at the detection site extend to the base of the ionosphere, and that the E impulses should have the characteristically short lightning stroke durations. Another feature of whistler mode e/m waves is that they are transverse, i.e., their E vectors are orthogonal to the ambient **B**. All of these properties provide tests that the OEFD observations must pass to qualify as lightning messengers. And of course, it must be demonstrated that plasma waves generated in the ionosphere and signals generated in spacecraft-instrument-atmosphere-ionosphere interactions are not able to better explain the observations.

The extreme brevity of 100-Hz bursts (see example in Fig. 5) was established by W. W. L. Taylor et al. (1979) on the basis of OEFD response times. 100-Hz pulses were shorter in duration than the nominal telemetry sample time interval of 0.50 s. Ho et al. (1991) using nighttime data from the first three seasons of PVO demonstrated that the 100-Hz wave pulses (as well as those in the higher-frequency channels) were shorter than even the OEFD's highest-resolution sample time of 0.25 s. This is consistent with short lightning burst times. Although waves generated by plasma wave processes can have similarly short bursts (e.g., see the PVO examples in the plasma wave review of Strangeway [1991b]), no specific plasma source has been identified which could explain the shortness of the low-altitude impulses.

Given the burstiness of the phenomenon it was necessary to resort to statistical analysis to further explore the burst characteristics. This led to two basic approaches. The first of these was an occurrence rate (or activity rate) method (see, e.g., Russell et al. 1988a,b,1989), where the data were divided into 30-s intervals classified as either active or quiet. The second method employed counted individual bursts (see, e.g., Ho et al. 1991,1992), and determined burst rates. All the studies based on these methods have removed any telemetry interference and both result in normalized rates. This addresses issues raised over earlier studies by Taylor and Cloutier (1986,1988) and Taylor et al. (1987). In the following we will focus only on the more recent studies employing the burst rate approach.

The relation between the burst rate and magnetic field magnitude was studied by Ho et al. (1991) using counts of individual nightside bursts (see Fig. 6). The occurrence rates for all wave frequencies increased as the magnetic field exceeded 6 to 10 nT. The 100-Hz signals had the strongest dependence on B. It is peculiar to Venus' nightside ionosphere that the magnetic field is very variable when its magnitude is weak, and seems to lock in (i.e., becomes ordered) at ~ 10 nT. At low magnitudes of B, the field is too irregular to play a coherent role in generation/propagation of electric waves. Hence the correlation of burst rates of the 100-Hz signals with B fits the whistler requirement $f_g > f$.

A comparison of burst rates with the angle of the magnetic field from the horizontal was done by Ho et al. (1992). The distribution for narrowband 100-Hz signals (i.e., no coincident higher-frequency burst) as well as the broadband 100-Hz signals (coincident with a signal in one or more of the higher frequency channels) is shown in Fig. 7. Although most of the nightside bursts were detected in regions with horizontal **B** the highest rate of occurrence of the

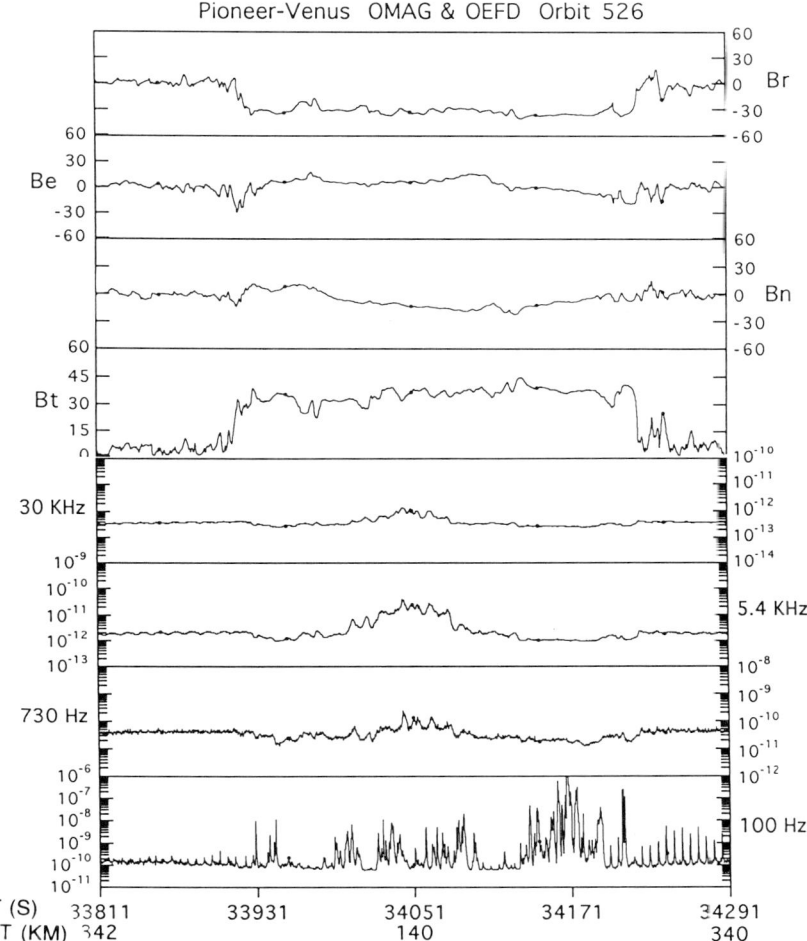

Figure 5. Narrowband 100-Hz bursts. E^2 is plotted in $(v/m)^2/Hz$. Signals were present when the local electron gyrofrequency, $f_g = 28B$ (in nT) Hz, exceeded 100 Hz and **B** (east, north and radial components are shown) was toward the vertical. The smooth bumps in the higher-frequency channels near periapsis are due to ionization produced by the impact of the spacecraft with the atmosphere (Curtis et al. 1985). The periodic spikes at the sides of the frame are, yet to be explained, interference signals.

narrowband 100-Hz is on nearly radial field lines. In contrast, wideband 100-Hz signals, rare compared to the narrowband bursts, favor horizontal fields, indicating that in the 100-Hz ensemble of bursts there are different populations with potentially different sources.

The inference from individual samples, such as in Fig. 5, that 100-Hz bursts needed vertical field lines led to early attempts to define precisely

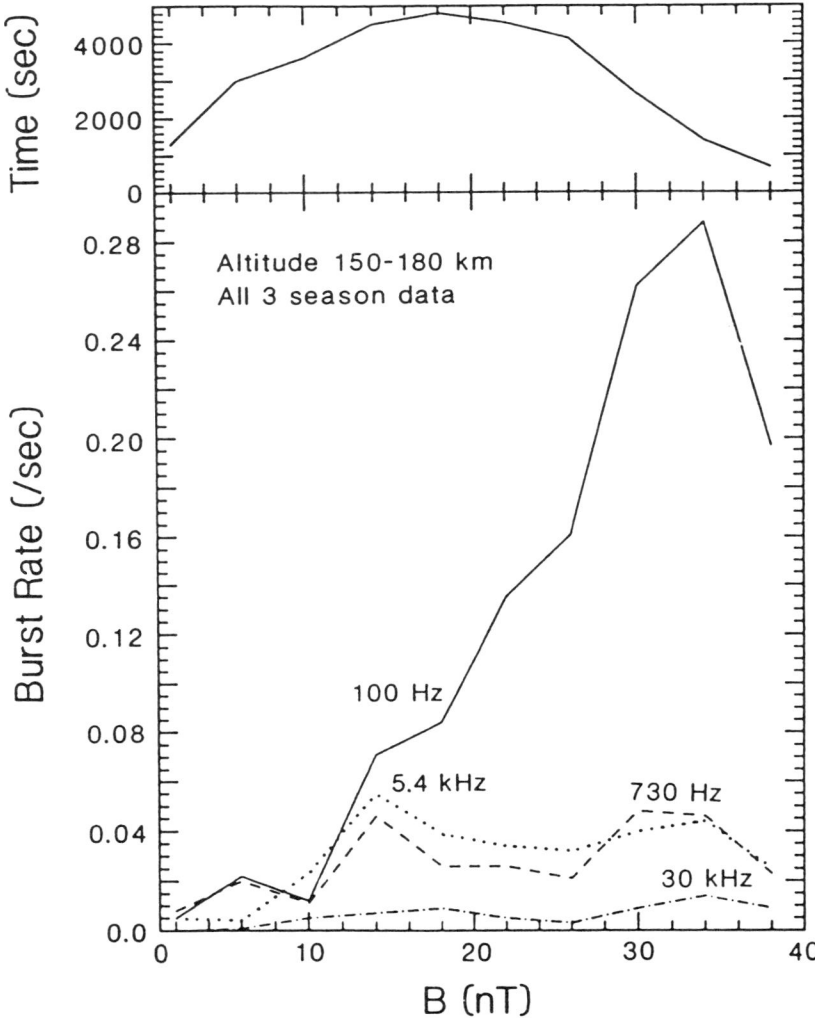

Figure 6. OEFD signal burst rate dependence on B in 4 nT bins. Burst rates (number of bursts/time spent in a window) are for all nightside measurements between 150 and 180 km from first three seasons (figure from Ho et al. 1991).

the source location. The **B** vectors at the locations of 100-Hz bursts were extrapolated down to the surface of Venus (Scarf and Russell 1983). Because there was a congregation of such extrapolations in the mountainous regions, the strong claim was made that Venus lightning existed because of volcanism. This conclusion was countered by Taylor et al. (1985,1987) who demonstrated that the preference of the bursts to occur over highlands was simply an artifact of PVO's orbital coverage, which oversampled the regions over the mountains.

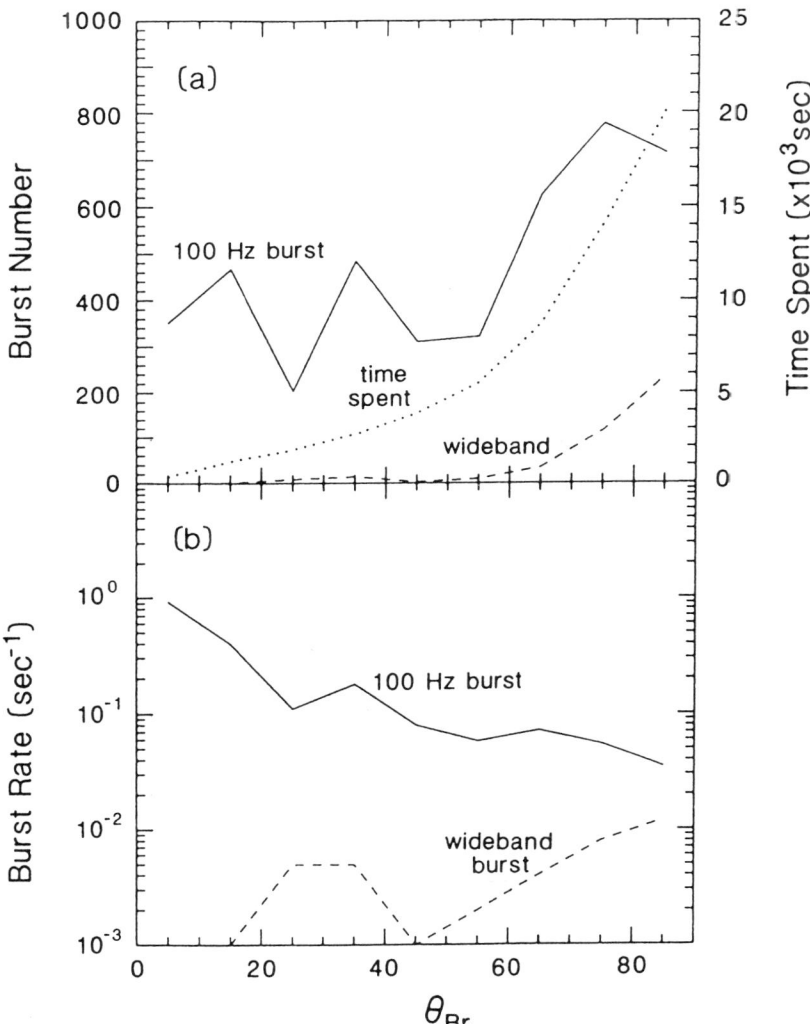

Figure 7. 100-Hz signals and the direction of **B**. Narrowband 100-Hz signals favor regions where **B** tends toward the vertical. The 100-Hz bursts appearing with higher frequency signals do not show the same behavior. (figure from Ho et al. 1992).

There was/is no evidence that volcanism on Venus would be of the explosive type needed for lightning. Further, it was not valid to estimate the propagation paths through the atmosphere to the surface by extrapolating the magnetic field lines to the surface because guidance of e/m waves by **B** only occurs in an ionized medium. The B field in the atmosphere below the ionosphere does not play a role in guiding atmospheric e/m atmospheric waves. In spite of an absence of any defensible evidence for volcanic lightning (see, e.g., Taylor et al. 1985,1987; Taylor and Cloutier 1986; Russell et al. 1989), it has been difficult to erase such a belief from the scientific community (Taylor and Cloutier [1992] document recent instances of this). Therefore it needs to be emphasized "there is no relation whatsoever between OEFD measured bursts and volcanism."

The partiality of the 100-Hz bursts for field lines inclined toward the vertical is consistent with propagation upwards from an atmospheric source. However, it is also consistent with propagation from above. A further test for the direction of propagation was suggested by Sonwalkar et al. (1991). Electromagnetic waves in the atmosphere propagate approximately in the free space mode with an index of refraction near unity. In the ionosphere the index for low-frequency whistler waves is very large (>1000). Low-frequency rays entering the ionosphere obliquely from below are refracted to the vertical. On the other hand, whistler mode propagation is allowed only at ray angles to the field line $<\cos^{-1}(f_g/f)$, the "resonance cone" angle. Wave normals outside the resonance cone are evanescent. Therefore a burst detected at PVO is consistent with whistler mode propagation from below if the radius vector from the planet to the spacecraft is within the resonance cone. Sonwalkar et al. (1991) applied their criterion to eleven OEFD segments of data. Of seven narrowband 100-Hz signal groupings explicitly singled out, only one failed to satisfy the resonance cone test. The four other samples were of broadband signal intervals for which the 100-Hz signals did not satisfy the whistler resonance cone test. Although it wasn't recognized in the paper, in the midst of the broadband examples were interwoven 100-Hz narrowband impulses, which also did not satisfy the resonance cone test (see Fig. 8). This paper thus demonstrated that not all narrowband 100-Hz bursts satisfy the resonant cone test and left it uncertain whether the bulk of the 100-Hz signals really satisfied the resonance cone condition for upward propagation. It should be noted that the resonance cone test will not discriminate upward from downward propagation when the field lines are nearly vertical. However, it is not obvious why an *in-situ* source would only occur for vertical fields.

Source location also might be inferred from the burst occurrence variation with altitude. Signal strengths and hence the number of detectable bursts should increase as the source is approached. The altitude dependence of the nightside burst rates for all four OEFD frequency ranges (Fig. 9) was sorted out by Ho et al. (1992). A resonance cone test was included. The burst rate of outside the resonant cone dropped off rapidly with altitude in contrast to the behavior of events inside the resonance cone, again demonstrating that not all 100-Hz bursts have common characteristics. The bulk of the 100-Hz bursts satisfied the resonance cone criteria, but had no significant variation in occurrence rate with altitude. The absence of a significant altitude dependence in occurrence rates was also noted earlier by Russell and Scarf (1990) for 100-Hz signals which were below half the electron gyrofrequency. This behavior indicates that whistler mode attenuation is small above 150 km, consistent

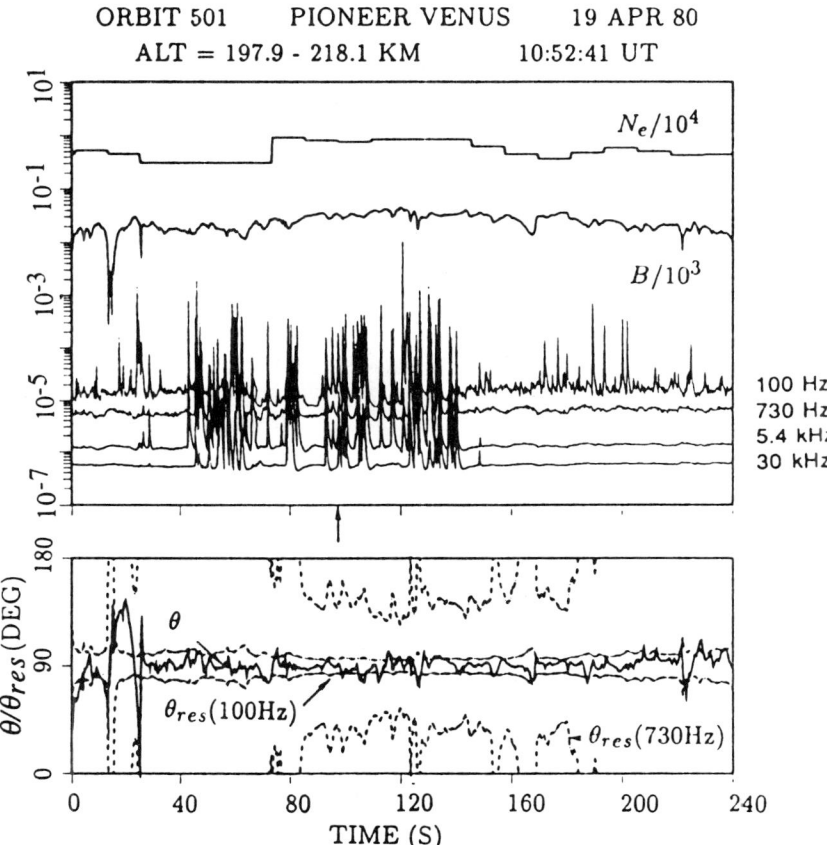

Figure 8. Signals outside resonance cone. Electron density (in cm^{-3}), B magnitude (nT), and the 4 OEFD frequency channel E measurements (units: v/m Hz$^{1/2}$) are shown on the top frame. The angle θ of the assumed vertical wave normal with respect to B and resonance cone angles θ_{res} for 100 and 730 Hz are shown on the bottom. Broadband and narrowband 100-Hz signals were detected outside the resonance cone (figure from Sonwalkar et al. 1991).

with modeled attenuation rate variations with altitude (see, e.g., Strangeway 1995). Results at altitudes below 150 km are uncertain owing to sparse orbital coverage and to uncertainty in the OEFD 100-Hz response that was singled out by Russell et al. (1989) to be a problem in dense plasmas with short Debye lengths.

A method for estimating the polarization of the OEFD 100-Hz signal was defined by Scarf and Russell (1988). They plotted the variation of the E signal (OEFD measured the E field between two separated wire cage probes on booms) with respect to B in the spacecraft spin plane over a PVO spin cycle. Strangeway (1991a) extended the analysis by analyzing the polariza-

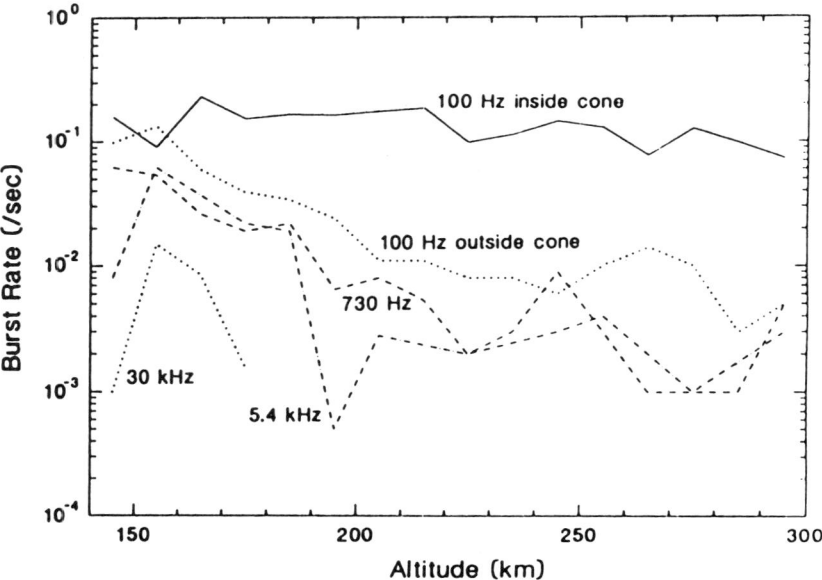

Figure 9. Altitude distribution of all OEFD bursts from 140 to 300 km in 10-km windows (figure from Ho et al. 1992).

tion statistics of the 100-Hz bursts for the periapsis segments of all the third season nightside PVO orbits. For those 100-Hz bursts that satisfied the resonance cone criterion, the average phase angle distribution (Fig. 10) showed polarization perpendicular to **B**. Thus another whistler mode requirement was satisfied.

The final two orbits of the mission provided nighttime measurements 10 km lower (to below 130 km) than was attained in the early mission ionosphere encounters. For such low altitudes, if discharges happened to be occurring frequently enough in the atmosphere, and the spacecraft lucky enough to be at the right place where **B** was intense and in a direction that allowed propagation from below, then one might see the wave amplitudes intensify near periapsis. Whistler mode attenuation lengths are shortest at the lower altitudes (Huba and Rowland 1993). Strangeway et al. (1993a) interpreted the final PVO OEFD measurements in this manner. The data from the last orbit are shown in Fig. 11. The 100-Hz burst amplitudes increased somewhat smoothly inbound towards periapsis and decreased again outbound. The asymmetry of the 100-Hz variation about periapsis was taken as evidence that spacecraft-atmosphere effects (which produce the bulges seen in the higher-frequency channels) was not causing the bursts. Although the two cited references were correct in concluding that wave attenuation would be greatest at the lowest PVO altitudes consistent with the last PVO periapsis measurements, the quantitative conclusions of the two cited publications need

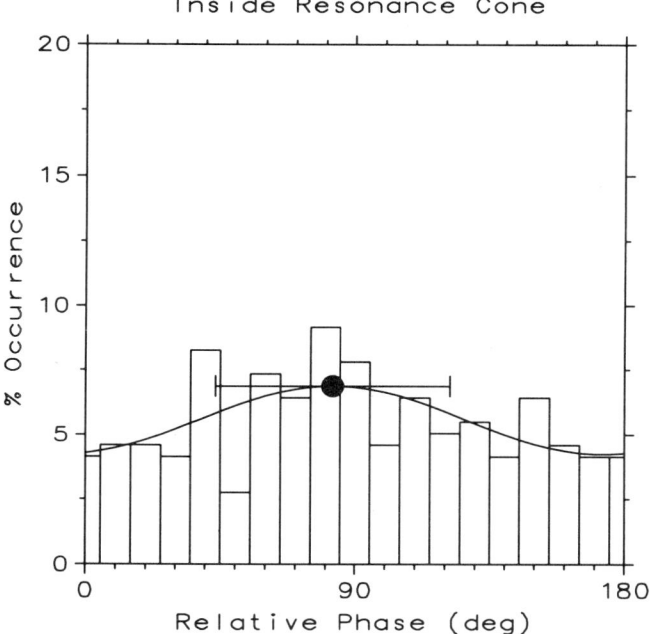

Figure 10. Statistics of 100-Hz E polarization within the resonance cone. The average phase of the E field signal maximum was perpendicular to **B** (30-s average value used). Extraneous signals not of external wave origin were filtered out (figure from Strangeway 1991b).

to be reconsidered. K. D. Cole and W. R. Hoegy (personal communication) pointed out that the electron-neutral collision frequency employed in both papers was in error by 2 orders of magnitude on the low side.

Strangeway (1996) did a full calculation of the propagation and ionospheric heating effects of 100-Hz E whistler signals with amplitudes consistent with the OEFD observations. He demonstrated that ionospheric heating at low altitudes would be significant (as originally suggested by Cole and Hoegy [1996]) but not inconsistent with ionospheric observations. This analysis included a recalculation of the vertical attenuation of upward propagating 100-Hz whistler signals that was originally modeled by Huba and Rowland (1993) but used the correct electron collision frequencies. The result showed that if the electron density in the lower ionosphere was very low, and the B field exceeded 10 to 20 nT, whistler waves from below could propagate to PVO's altitude. Although lower ionosphere conditions, in particular the existence of high electron densities below 135 km, may frequently lead to the

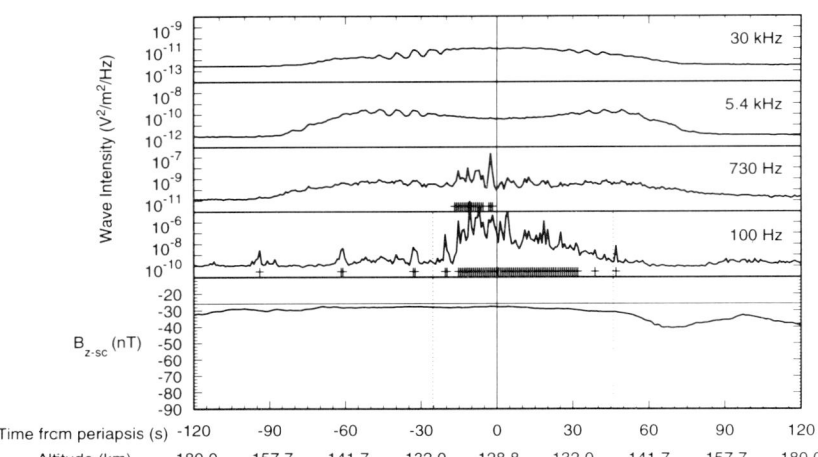

Figure 11. Electric wave activity and magnetic field from last Pioneer Venus orbit (figure from Strangeway et al. 1993a).

total attenuation of the waves, it is not implausible that local ionospheric conditions do prevail at times that will allow signals to escape. However, it must be cautioned that details of the complexly structured nightside-ionosphere morphology below 135 km are not clearly defined at the present time, making it impossible to predict quantitatively when or where the gate for propagation is open or closed.

The above discussion has demonstrated that a majority subset of the 100-Hz measurements in the lower ionosphere has characteristics of e/m whistler mode propagation as well as characteristics consistent with an atmospheric source. The next question is whether there are identifiable *in-situ* ionospheric plasma sources that could produce signals with identical signal properties, supplanting the lightning interpretation. Taylor et al. (1995) in fact argue that all events could be of a non-lightning origin. However, as will be seen, several studies not cited in that paper have already shown that established plasma source mechanisms cannot readily account for the bulk of the 100-Hz waves.

The tendency for narrowband 100-Hz bursts to appear in ionospheric hole regions which have steep electron-density-gradient boundaries led Taylor and Cloutier (1986) and Taylor et al. (1987) to suggest that the OEFD signals were the signatures of plasma-generated waves. Slow-propagating waves, such as ion-acoustic waves, were proposed for the 100-Hz bursts. The upper frequency of the ion-acoustic wave spectrum (>1 kHz in the nightside ionosphere) would have to be Doppler shifted in the spacecraft frame of reference so that only the 100-Hz channel would detect it. In defense of a whistler interpretation, Scarf and Russell (1988) showed that the Doppler shifts necessary to restrict measured activity solely to the 100-Hz channel were not consistent with the data. Also, ion-acoustic waves are longitudinally polarized, whereas the 100-Hz signals are on average transverse waves. Hence ion-acoustic waves cannot account for the bulk of the narrowband 100-Hz observations.

Another mechanism, one actually producing whistler radiation in the ionosphere, was proposed by Maeda and Grebowsky (1989). Very short terrestrial polar ionosphere whistler bursts called saucers have properties similar to the Venus narrowband 100-Hz signals. Using heuristic arguments based on current understanding of saucers, Landau resonance waves resulting from superthermal electron beams or waves generated by gyroresonant instability were suggested by Maeda and Grebowsky as possible explanation for the bursts at Venus. Strangeway (1992) investigated in detail the applicability of these mechanisms to Venus. He concluded that Landau resonant waves in the Venus ionosphere would be severely damped because the ionospheric electron thermal speeds were comparable to whistler mode wave speeds. He also looked at ionospheric conditions that would favor the generation of whistlers by gyroresonance. resulting from electron temperature anisotropies or superthermal electrons. He demonstrated that OEFD 100-Hz wave intensities were most intense for conditions under which the electron temperature anisotropy source was not operative, and through estimates of the growth rates (using ionospheric properties bounded by PVO measurements) showed superthermal electron beams could not produce the observed wave amplitudes. Thus whistler mode instabilities generated in the lower ionosphere of Venus are probably not significant wave sources. This does not rule out the possibility that whistlers are generated at much higher altitudes and propagate down to the lower ionosphere. There is at present no strong argument for this scenario.

The actual presence of *in-situ* plasma waves near 100 Hz was discovered using measurements of the OETP on PVO by Grebowsky et al. (1991). The waves were in the ionosphere electron density, with the most prominent events occurring with OEFD measured 100-Hz E bursts. Because whistler waves are noncompressive, this was the first substantial evidence for an ionospheric source of OEFD signals. Huba (1992) proposed that they were lower hybrid (LH) waves, generated by relative electron/ion drifts caused by density gradients perpendicular to **B**. Lower hybrid waves are polarized perpendicular to **B**. Because they are slow moving waves generated at frequencies below a few Hz they have to be Doppler shifted to appear in the 100-Hz band, but the maximum Doppler shift attainable at the 10 km s^{-1} spacecraft motion is not sufficient to shift them into the higher-frequency bands. Hence they would be detected as narrowband 100-Hz signals. They have all the properties that have been set up as criteria for whistlers. Although there is evidence for their presence, Strangeway (1995*a,b*) has provided a strong argument that a majority of the OEFD 100-Hz signals cannot be LH waves. He delineated the conditions needed for LH wave growth and showed that they were not present for the bulk of the 100-Hz pulses. Ho et al. (1995) later showed that only a small percentage of the ensemble of 100-Hz bursts were actually located on prominent density gradients. Hence, although *in-situ* plasma processes exist which generate 100-Hz E wave bursts, they have not yet been able to account for most of them. As seen, the consequences of an atmosphere discharge source are better defined and bracket the observed signal characteristics more easily than any other mechanism thus far proposed.

Signals appearing in the higher-frequency OEFD channels (in particular 30 kHz and 5.4 kHz) cannot propagate as e/m waves from an atmospheric electric discharge directly into the ionosphere. W. W. L. Taylor et al. (1979) originally interpreted all OEFD measured bursts, regardless of frequency, as due to lightning. Recognizing the difficulty of propagating the higher frequencies through the ionosphere, Scarf et al. (1980) restricted the analysis to only the 100-Hz emissions and it was those signals that were the prime focus of intensive lightning/plasma wave studies up to the present. Beginning with Singh and Russell (1986) attention was directed back to the broadband pulses as possible lightning signatures.

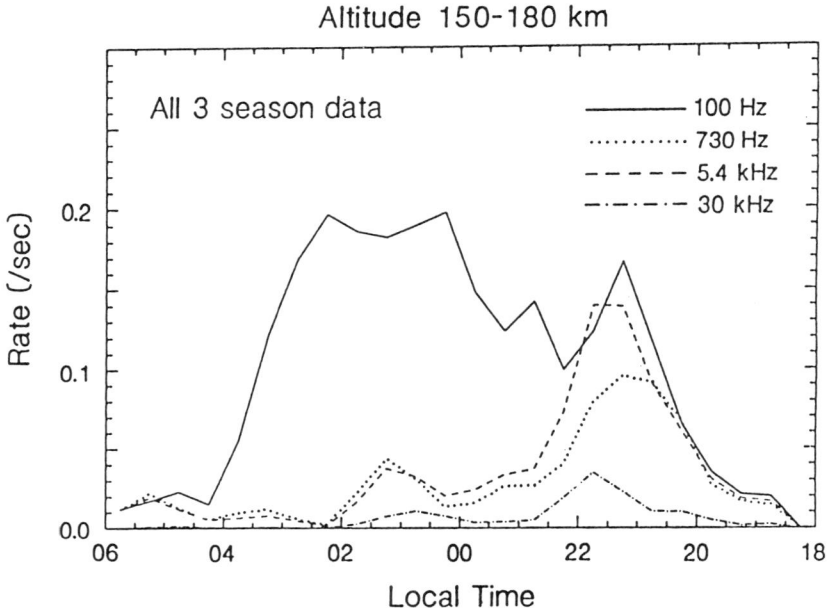

Figure 12. Local time distribution of burst rates in 1-hr local time bins (figure from Ho et al. 1991). Two peaks are prominent: a narrowband 100-Hz region on the dawn side of midnight and a broadband region centered near 2100 local time.

The broadband (i.e., with high-frequency signals) OEFD waves have distinctly different characteristics from the narrowband whistler-like 100-Hz signals. Russell et al. (1989) noted a rapid altitude dropoff of the high-frequency signals occurrence rates. As Fig. 9 shows, the non-whistler waves decrease rapidly in occurrence frequency with altitude above 160 km with a peak in the vicinity of 150 to 160 km. Unfortunately, the low altitude statistics are too poor to really be confident about the trends below 150 km. The local time distribution is shown in Fig. 12. The 100-Hz events are statistically more frequent in the 0000 to 0300 local time region than elsewhere. This is the local time region where ionosphere holes and their vertical, enhanced magnetic fields were most frequently encountered (Brace et al. 1982) and where PVO most frequently traversed ionospheric B fields exceeding 15 nT (Russell et al. 1988a). The 100-Hz signals in this local time zone are overwhelmingly narrowband. On the dusk side of midnight narrowband signals appear to be infrequent; the burst rates from all OEFD frequency channels peak near 2100 local time and the 5.4-kHz bursts are just as prevalent as the 100-Hz bursts.

The published observations suggest that the bulk of the 100-Hz bursts near 2100 local time, whether narrowband or part of broadband signals, are not whistler-like waves and hence their source mechanism is different from the 100-Hz burst region after midnight which have whistler mode properties. For example, the B field near 2100 local time when the broadband bursts were detected was mainly horizontal (Russell et al. 1988a). Further, Strangeway (1991a) showed that the 30-kHz bursts were polarized mainly parallel to the magnetic field. If this applies, as is plausible, for all the frequencies in the broadband signals, the signals comprising the post-dusk activity region are not e/m waves.

In-situ plasma wave sources, such as ion-acoustic waves, were suggested as explanations for the discrete bursts by Taylor et al. (1985,1987). Scarf et al. (1985) claimed that broadband bursts high in the nightside ionosphere were clear signals of ion-acoustic waves. Later, Strangeway et al. (1993b) came to the conclusion that the high-frequency signals observed in the disturbed lower ionosphere at the end of the PVO mission were electrostatic ion-acoustic waves. Both of these OEFD studies dealt with particularly disturbed ionosphere regimes compared to those statistically analyzed from the early mission period. Nevertheless they are concrete examples showing the presence of processes that may be applicable to the post-dusk broadband activity.

On the other hand, at Earth similar wave bursts have been detected in the ionosphere directly over lightning storms (Kelley et al. 1985), with a strong component of the wave electric field parallel to the ambient magnetic field. The source of the phenomenon is not yet firmly established for Earth. One theory (Kelley et al. 1990) viewed as attractive is the nonlinear interaction of the ionosphere with a packet of intense whistler mode wave fronts. Also Burke et al. (1992) have reported on an intense radio frequency pulse detected in the Earth's ionosphere over a hurricane's lightning storms, followed by a field aligned flow of keV electrons. But the terrestrial and Venus lower ionospheric wave propagation environments are different; e.g., the B field is horizontal where the Venus broadband bursts occur. It can be conjectured that the Venus bursts are the result of the interaction of discharge-related electron beams with the local ionosphere, the horizontal **B** providing a guiding path for energetic electrons produced via dayside afternoon discharge regions. At present there is no cogent theory of the coupling between lightning and the ionospheric pulses at Venus. Another possibility may be found from the studies of terrestrial phenomenon known as sprites (Sentman and Wescott 1993; Sentman et al. 1995). Recently, Inan et al. (1995) have shown that the ionospheric conductivity is modified above sprites, perturbing the great circle path of VLF signals in the Earth-ionosphere waveguide. Clearly the process responsible for sprites can perturb the ionosphere, and it seems reasonable that some wave signature of this process could be detected within the ionosphere. Whether or not this signature would correspond to the VLF bursts at Venus, or indeed whether sprite-like events exist there is not known. With such uncertainty, and the possible existence of actual ion acoustic wave signatures in the higher-altitude Venus ionosphere, it is premature to use the 2100 local time concentration of broadband bursts as evidence for a region of enhanced lightning activity as has been done (Russell 1991). What is needed is an in-depth study of ionospheric properties co-existent with the wideband bursts to determine whether conditions are more favorable for an *in-situ* or lightning generated wave interpretation.

In summary, the majority of the low-frequency 100-Hz bursts have characteristics of whistler mode e/m waves propagating from a lower altitude source region. Volcanism plays no role. There is uncertainty at the present time as to how readily 100-Hz e/m waves can propagate through the dissipating collisional region at the base of the ionosphere. With current understanding, ionospheric conditions would at times seem to allow the signals to transit this region. The measured 100-Hz burst rate can be used as a rough lower-limit estimate for the planetary flash rate; e/m radiation with frequencies below 100 Hz in the planet-surface-ionosphere waveguide will travel extremely long distances if the damping at the base of the ionosphere allows it. Atmosphere signals leaking out of the waveguide along the vertical field lines of an ionospheric hole as whistlers could come potentially from anyplace on the planet. On the other hand, the source of the broadband signals concentrated about 2100 local time is more undefined. If these impulses are related to lightning they might indicate a region of persistent lightning in the late afternoon or post dusk, but a generation mechanism appropriate to Venus' local ionosphere/magnetic field conditions has yet to be set up in even rudimentary form. *In-situ* ionosphere plasma processes are already established which could produce similar waves, but no study has yet considered the plausibility of, or the reason for, an ionospheric plasma-wave activity region in the post-dusk sector.

This complex deconvolution of the OEFD data leads to a rough estimate of planetary flash rates. The OEFD telemetry resolution of 0.25 to 0.5 s encompasses the duration of individual terrestrial flashes and the optically observed Venus bursts. Hence, each 100-Hz whistler-like burst can be taken as the signature of a flash discharge. In analogy to smoke coming out of the only open window of a house burning on the inside, the whistlers exiting along localized nightside clumps of vertical field lines could reflect atmospheric activity from around the planet. The observed concentration of 100-Hz bursts near 0000 to 0200 local time is due to the predominance of vertical magnetic field lines in this local time sector and not to an atmospheric source asymmetry. The peak nighttime 100-Hz rate of ~ 0.2 s^{-1}, if taken as a global rate, is much lower than the terrestrial rate of 100 s^{-1}. However this is clearly a lower limit because we have neglected any effects due to attenuation within the atmosphere, together with the attenuation as the waves propagate through the ionosphere. Ho et al. (1991) obtained a much higher global rate of ~ 250 s^{-1} under the hypothesis that the post-dusk OEFD 5.4 kHz bursts were signatures of lightning in the cloud region below the spacecraft. This estimate was made assuming that PVO could detect signals as far horizontally as the spacecraft is distant vertically from the clouds, and assuming that the lightning activity was restricted to within 30 degrees of the equator and localized to be the post-dusk local time region with an active area 1/8 of the planetary surface.

IV. CONCLUSIONS

With the exception of the experiments which found no evidence for electric discharges, the burst properties inferred from various experiments had little in common. The situation may be compared to making seven different types of measurements on Earth at separate local times, different years, different times of year, varying time intervals, diverse geographic locations and different fields of view. Lightning flash/burst characteristics would not likely be the same at all sites. Extended fronts of electrical storms and isolated convection cell storms which last for only a small part of the day and which are not uniformly distributed over the globe do not lend themselves to being easily abstracted from a limited set of measurements.

Of the seven separate experiments, three of the nighttime optical investigations (the 2 Vega lightning detectors and the Pioneer Venus star sensor experiment) detected nothing significant. Venera 9 spectrometer measurements detected only one prolonged 70-s event. Earth-based telescope observations from Mt. Bigelow recorded 6 to 7 spatially scattered nightside flashes with no evidence of a persistent stormlike pattern. The Galileo plasma wave experiment on the Venus flyby recorded radio frequency pulses that could be the product of nighttime atmospheric discharges, but there is uncertainty as to whether the required low nighttime ionospheric densities needed to allow the escape of the radiation were present. Low-frequency nighttime PVO measurements of 100-Hz E bursts grouped on the dawn side of midnight are consistent with escaping ionospheric whistler mode waves. The local time regime of their detection is where ionospheric conditions favor the propagation of whistler waves; it does not reflect any atmospheric source region preference. Due to the long distance propagation of such low-frequency waves in the atmosphere the 100-Hz whistler-like impulses could be signatures of discharges occurring practically anywhere about the planet. It is not yet clear how effectively wave damping in the lower ionosphere prevented their escape to PVO. The nighttime observations, with the potential exception of the one isolated storm-like period recorded from Venera 9, imply flash rates low in comparison to those at Earth. The magnetic wave experiments on all the Venera landers on descent detected noise bursts that have been interpreted as evidence for the persistent presence of dayside lightning frontal storms occurring within a few thousand km distance from the landing area.

Nighttime optical signatures have been detected only on the dusk side of midnight. The observations thus far have been too limited to deduce confidently that this reflects a planetary trend. Mt. Bigelow telescope observations had the widest instantaneous field of view of the planet, but thus far have only looked at the duskside nighttime crescent of Venus. The optical bursts recorded by the orbiting Venera spectrometer were in the dusk sector but were only seen for one 70-s period. All the investigations made on the dawn side of midnight (i.e., Vega and the PVO star sensor) have recorded nothing significant. The optical measurements provide no convincing statistical evidence yet that there is a distinctive region at night that is persistently flashing. The PVO OEFD broadband signals, concentrated near 2100 local time, have been hypothesized to be a consequence of lightning discharges but details of the underlying mechanism or alternate plasma-wave sources have yet to be considered. The several isolated sequences of radio frequency signals detected on each of the Venera landers indicate that dayside discharges could be a common occurrence. A puzzling difference between observations was in the measured flash durations. The Venera spectrometer event consisted of a 70-s sequence of $\sim 1/4$-s bursts, the Mt. Bigelow events were isolated <20 ms flashes, and the Venera lander field events were composed of apparently continuous rapid-fire noise bursts lasting many minutes.

There is no evidence whatsoever that volcanism plays a role in the origin of lightning. Lightning discharges most likely would occur within the cloud layer, but the actual presence of cloud aerosol properties or convection structures comparable to those associated with terrestrial discharges has yet to be substantiated. Further evidence for the presence of lightning has been found since the first *Venus* book (Hunten et al. 1983), but progress in understanding the source of the lightning phenomena has been essentially stagnant.

REFERENCES

Bliokh, P.V., Nicholaenko, A. P., Fillippov, Yu. F., and Llanwyn-Jones, D. 1980. *Schumann Resonances in the Earth-Ionosphere Wave Cavity* (New York: P.

Perigrinns Ltd.).
Borucki, W. J. 1982. Comparison of Venusian lightning observations. *Icarus* 52:354–364.
Borucki, W. J., Dyer, J., W., and Phillips, J. R. 1992. Pioneer Venus Orbiter search for Venus lightning. *J. Geophys. Res.* 96:11033–11043.
Borucki, W. J., et al. 1981. Optical search for lightning on Venus. *Geophys. Res. Lett.* 8:233–236.
Borucki, W. J., et al. 1983. Laboratory simulation of Venus lightning. *Geophys. Res. Lett.* 10:961–964.
Brace, L. H., and Kliore, A. J. 1991. The structure of the Venus ionosphere. *Space Sci. Rev.* 55:81–163.
Brace, L. H., et al. 1982. Holes in the nightside ionosphere of Venus. *J. Geophys. Res.* 87:199–211.
Burke, W. J., et al. 1992. Effects of lightning discharge detected by the DE-2 satellite over hurricane Debbie. *J. Geophys. Res.* 97:6359–6368.
Cole, K. D., and Hoegy, W. R. 1996. A major energy source for Venus' ionosphere. *J. Geophys. Res.* 101:2269–2278.
Croft, T. A., and Price, G. H. 1983. Evidence for a low-altitude origin of lightning on Venus. *Icarus* 53:548–551.
Curtis, S. A., Brace, L. H., Niemann, H. B., and Scarf, F. L. 1985. CO_2 impact ionization-driven plasma instability observed by Pioneer Venus Orbiter at periapsis. *J. Geophys. Res.* 90:6631–6636.
Desch, M. D. 1991. Lightning at planets in the outer solar system. In *Planetary Radio Emissions III Workshop* (Vienna: Verlag der Osterreichischen Akad. der Wissenshaft), pp. 371–390.
Esposito, L. W. 1984. Sulfur dioxide: Episodic injection shows evidence for active Venus volcanism. *Science* 223:1071–1074.
Grebowsky, J. M., Curtis, S. A., and Brace, L. H. 1991. Small scale plasma irregularities in the nightside Venus ionosphere. *J. Geophys. Res.* 96:21347–21359.
Gurnett, D. A., et al. 1991. Lightning and plasma wave observations from the Galileo flyby of Venus. *Science* 253:1522–1525.
Hansell, S., Wells, S. A., and Hunten, D. M. 1995. Optical detection of lightning on Venus. *Icarus* 117:345–351.
Ho, C.-M., Strangeway, R. J., and Russell, C. T. 1991. Occurrence characteristics of VLF bursts in the nightside ionosphere of Venus. *J. Geophys. Res.* 96:2361–2369.
Ho, C.-M., Strangeway, R. J., and Russell, C. T. 1992. Control of VLF burst activity in the nightside ionosphere of Venus by the magnetic field orientation. *J. Geophys. Res.* 97:11673–11680.
Ho, C.-M., Strangeway, R. J., and Russell, C. T. 1995. Venus nightside ionospheric irregularities and their relationship to VLF bursts. *J. Geophys. Res.* 100:9697–9705.
Huba, J. D. 1992. Theory of small scale density and electric field fluctuations in the nightside Venus ionosphere. *J. Geophys. Res.* 97:43–50.
Huba, J. D., and Rowland, H. L. 1993. Propagation of electromagnetic waves parallel to the magnetic field in the nightside Venus Ionosphere. *J. Geophys. Res.* 98:5291–5300.
Hunten, D. M. 1995. Venus lightning: Pros and cons. *Adv. Space Res.* 15(4):109–112.
Hunten, D. M., Colin, L., Donahue, T. M., and Moroz, V. I., eds. 1983. *Venus* (Tucson: Univ. of Arizona Press).
Inan, U. S., et al. 1995. VLF signatures of ionospheric disturbances associated with sprites. *Geophys. Res. Lett.* 22:1205–1208.
Kelley, M. C., et al. 1985. Electrical measurements in the atmosphere and the ionosphere over an active thunderstorm 1. Campaign overview and initial ionospheric

results. *J. Geophys. Res.* 90:9815–9823.
Kelley, M. C., Ding, J. G., and Holzworth, R. H. 1990. Intense ionospheric electric and magnetic field pulses generated by lightning. *Geophys. Res. Lett.* 17:2221–2224.
Krasnopolsky, V. A. 1983a. Lightnings and nitric oxide on Venus. *Planet. Space Sci.* 31:1363–1369.
Krasnopolsky, V. A. 1983b. Venus spectroscopy in the 3000–8000 Å region by Venera 9 and 10. In *Venus*, eds. D. M. Hunten, L. Colin, T. M. Donahue and V. I. Moroz (Tucson: Univ. of Arizona Press), pp. 459–483.
Kremnev, R. S., et al. 1986. VEGA Balloon system and instrumentation. *Science* 231:1408–1411.
Ksanfomality, L. V. 1979. Lightning in Venus' cloud layer. *Cosmic Res.* 17747–762.
Ksanfomality, L. V., Scarf, F. L., and Taylor, W. W. L. 1983. The electrical activity of the atmosphere of Venus. In *Venus*, eds. D. M. Hunten, L. Colin, T. M. Donahue and V. I. Moroz (Tucson: Univ. of Arizona Press), pp. 565–603.
Levin, Z., Borucki, W. J., and Toon, O. B. 1983. Lightning generation in planetary atmospheres. *Icarus* 56:80–115.
Maeda, K., and Grebowsky, J. M. 1989. VLF emission bursts in the terrestrial and Venusian nightside troughs. *Nature* 341:219–221.
Orville, R. E. 1986. Lightning phenomenology. In *The Earth's Electrical Environment* (Washington, D. C.: National Academy Press), pp. 23–29.
Rinnert, K. 1982. Lightning within planetary atmospheres, In *Handbook of Atmospheres*, ed. H. Volland, vol. 2 (Boca Raton, Fl.: CRC Press), pp. 1–22.
Russell, C. T. 1991. Venus lightning. *Space Sci. Rev.* 55:317–356.
Russell, C. T. 1993. Planetary lightning. *Ann. Rev. Earth Planet. Sci.* 21:43–87.
Russell, C. T., and Scarf, F. L. 1990. Evidence for lightning on Venus. *Adv. Space Res.* 10(5):125–136.
Russell, C. T., von Dornum, M., and Scarf, F. L. 1988a. VLF Bursts in the night ionosphere of Venus: Effects of the magnetic field. *Planet. Space Sci.* 36:1211–1218.
Russell, C. T., von Dornum, M., and Scarf, F. L. 1988b. The altitude distribution of impulsive signals in the night ionosphere of Venus. *J. Geophys. Res.* 93:5915–5921.
Russell, C. T., von Dornum, M., and Scarf, F. L. 1989. Source locations for impulsive electric signals seen in the night ionosphere of Venus. *Icarus* 80:390–415.
Sagdeev, R. V., et al. 1986. Overview of VEGA Balloon in situ meteorological measurements. *Science* 231:1411–1414.
Scarf, F. L., and Russell, C. T. 1983. Lightning measurements from the Pioneer Venus Orbiter. *Geophys. Res.* 10:1192–1195.
Scarf, F. L., and Russell, C. T. 1988. Evidence for lightning and volcanic activity on Venus: Pro and con. *Science* 240:222–224.
Scarf, F. L., Taylor, W. W. L., Russell, C. T., and Brace, L. H. 1980. Lightning on Venus: Orbiter detection of whistler signals. *J. Geophys. Res.* 85:8158–8166.
Scarf, F. L., et al. 1985. Current-driven plasma instabilities and auroral-type particle acceleration at Venus. *Adv. Space Res.* 5:185–191.
Sentman, D. D., and Wescott, E. M. 1993. Observations of upper atmosphere optical fluxes recorded from an aircraft. *Geophys. Res. Lett.* 20:2857–2860.
Singh, R. N., and Russell, C. T. 1986. Further evidence for lightning on Venus. *Geophys. Res. Lett.* 13:1051–1054.
Sentman, D. D., et al. 1995. Preliminary results from the Sprites94 aircraft campaign: 1. Red sprites. *Geophys. Res. Lett.* 22:3461–3464.
Sonwalkar, V. S., Carpenter, D. L., and Strangeway, R. J. 1991. Testing radio bursts observed on the nightside of Venus for evidence of whistler mode propagation from lightning. *J. Geophys. Res.* 96:17763–17778.

Strangeway, R. J.. 1991a. Polarization of the impulsive signals observed in the nightside of Venus. *J. Geophys. Res.* 96:22741–22752.

Strangeway, R. J. 1991b. Plasma waves at Venus. *Space Sci. Rev.* 55:275–316.

Strangeway, R. J. 1992. An assessment of lightning or in situ instabilities as a source for whistler mode waves in the night ionosphere of Venus. *J. Geophys. Res.* 97:12203–12215.

Strangeway, R. J. 1995a. An assessment of plasma instabilities or planetary lightning as a source for the VLF bursts detected at Venus. *Adv. Space Res.* 15(4):89–92.

Strangeway, R. J. 1995b. The plasma wave evidence for lightning on Venus. *J. Atmos. Terres. Phys.* 57:537–556.

Strangeway, R. J. 1996. Collisional Joule Dissipation in the ionosphere of Venus: the importance of electron heat conduction. *J. Geophys. Res.* 101:2279–2295.

Strangeway, R. J., Russell, C. T., and Ho, C. O. 1993a. Observation of intense wave bursts at very low altitudes within the Venus nightside ionosphere. *Geophys. Res. Lett.* 20:2771–2774.

Strangeway, R. J., Russell, C. T., Ho, C. O., and Brace, L. H. 1993b. Plasma waves observed at low altitudes in the tenuous Venus nightside ionosphere. *Geophys. Res. Lett.* 20:2767–2770.

Taylor, H. A., Jr., and Cloutier, P. A. 1986. Venus: Dead or alive. *Science* 234:1087–1093.

Taylor, H. A., Jr., and Cloutier, P. A. 1988. Telemetry interference incorrectly interpreted as evidence for lightning and present day volcanism at Venus. *Geophys. Res. Lett.* 15:729–732.

Taylor, H. A., Jr., and Cloutier, P. A. 1992. Non-evidence of lightning and associated volcanism at Venus. *Space Sci. Rev.* 61:387–391.

Taylor, H. A., Jr., Grebowsky, J. M., and Cloutier, P. A. 1985. Venus nightside ionospheric troughs: Implications for evidence of lightning and volcanism. *J. Geophys. Res.* 98:7415–7426.

Taylor, H. A., Jr., Cloutier, P. A., and Zheng, Z. 1987. Venus "lightning" signals reinterpreted as in situ plasma noise. *J. Geophys. Res.* 92:9907–9919.

Taylor, H. A., Jr., Kramer, L., Cloutier, P. A., and Walker, S. S. 1995. Signatures of solar wind interaction with the nightside ionosphere of Venus. *Earth, Moon Planets* 69:173–199.

Taylor, W. W. L., Scarf, F. L., Russell, C. T., and Brace, L. H. 1979. Evidence for lightning on Venus. *Nature* 279:614–616.

Uman, M. A. 1987. *The Lightning Discharge* (New York: Academic Press).

Williams, E. R. 1985. Large-scale charge separation in thunderclouds. *J. Geophys. Res.* 90:6013–6025.

Young, R. E., et al. 1987. Characteristics of gravity waves generated by surface topography on Venus: Comparison with the VEGA balloon results. *J. Atmos. Sci.* 44:2628–2639.

PART II
Upper Atmosphere
(≥100km)

IONOSPHERE: SOLAR CYCLE VARIATIONS

J. L. FOX
State University of New York at Stony Brook

and

A. J. KLIORE
Jet Propulsion Laboratory

We summarize here the current state of knowledge of solar cycle variations in the morphology of the Venus dayside and nightside ionospheres, and in the mechanisms for maintaining the nightside ionosphere. The solar cycle variability of the dayside ionosphere is well understood. The dayside peak ion or electron density varies approximately as $F_{10.7}^{0.35}$ for short-term or long-term solar variations. The non-Chapman behavior is mostly due to concomitant changes in the neutral atmosphere, and such changes also produce a large amplitude response to solar flux variations well above the peak. Similar behavior is observed on the night side, where the variation of the electron density is a factor of 2 near the peak, but is much larger at high altitudes. On the night side the ionotail extends thousands of kilometers behind the planet. The large response of the ionotail to solar flux variations and to solar wind dynamic pressure indicates that the ions found there originate on the day side. Pioneer Venus measurements have convincingly shown that the major source of nightside ionization at solar maximum is day-to-night plasma transport. Most estimates of the contribution of electron precipitation at high solar activity are in the range 20 to 30%, but the exact value is still not certain. Neither the solar cycle response of the nightside ionosphere, nor its behavior with solar zenith angle, nor its overall variability can be accounted for by electron precipitation as the major ion source.

I. INTRODUCTION

The ionosphere of Venus was first detected by the radio occultation experiment on Mariner 5 in 1967. The dayside electron density profile showed a peak of 5×10^5 cm^{-3} near 140 km, and a sharp cutoff at about 500 km (see, e.g., Kliore et al. 1967). The nightside ionosphere was found to extend to very high altitudes with a peak electron density of about 1×10^4 cm^{-3}. Electron density profiles were subsequently derived from similar measurements by U. S. and Soviet spacecraft over the period 1974 to 1992; a summary of these missions can be found in, for example, Kliore and Luhmann (1991). Most of our current knowledge of the Venus ionosphere, however, comes from measurements made by *in-situ* and remote sensing experiments on the

Pioneer Venus Orbiter (PVO) over the nearly 14-yr life of the spacecraft (December 1978–October 1992).

In addition to the orbiter radio occultation experiment (ORO), several instruments on the PVO returned *in-situ* data relevant to the ionosphere, including the Ion Mass Spectrometer (OIMS) (see, e.g., Taylor et al. 1980), the Retarding Potential Analyzer (ORPA) (see, e.g., Knudsen et al. 1980), and the Langmuir Probe (OETP), (see, e.g., Brace et al. 1980). Measurements made by the Orbiter Neutral Mass Spectrometer (ONMS) (see, e.g., Niemann et al. 1980), the Ultraviolet Spectrometer (OUVS) (see, e.g., Stewart et al. 1980), and the atmospheric drag experiment (OAD) (see, e.g., Keating et al. 1985) provided information about the background neutral atmosphere from which the ionosphere is formed.

Periapsis of the PVO was actively maintained near 150 km for about the first 600 orbits, which occurred once per Earth day. After that, however, the orbit was allowed to evolve, and periapsis rose to altitudes greater than 2000 km in 1986 before it began to descend to lower levels. *In-situ* measurements were therefore possible only during the first three Venus years (1978–1980), which occurred during a period of high solar activity, and during the re-entry period, which lasted about six weeks in the fall of 1992 before communication with the spacecraft was lost. The re-entry period was one of moderately low solar activity; the average value of $F_{10.7}$ was about 120.

In the course of the first 600 Pioneer Venus orbits, the spacecraft swept through all local times at least twice. Consequently, data returned from PVO instruments have revealed the basic chemical and dynamical structure and variability of the dayside and nightside ionospheres at high solar activity. During the re-entry period, measurements were confined to the post-midnight sector. Our knowledge of the low solar activity ionosphere is therefore limited largely to that gleaned from ORO measurements during the period 1983 to 1986, from *in-situ* measurements during the brief re-entry period in 1992, and from modeling.

Observations of the Venus ionosphere through 1982 were reviewed by Brace et al. (1983), and models and measurements of the ionospheric chemical structure and energetics by Nagy et al. (1983). Brace and Kliore (1991) have recently summarized our understanding of the structure and temporal behavior of the Venus ionosphere, with an emphasis on aspects of the subject not covered in earlier reviews. Measurements and theory related to ion dynamics and plasma motions are the subject of a review by Miller and Whitten (1991). Knudsen (1992) has presented a comprehensive discussion of the structure and solar cycle variability of the Venus ionosphere as revealed by the first 12 yr of PVO *in-situ* measurements. Cravens et al. (see their chapter) have discussed magnetohydrodynamic processes in the Venus ionosphere.

Our purpose here is to summarize our present understanding of, and data relevant to, the solar cycle variations of the dayside and nightside ionospheres. Our emphasis here is on the chemical structure and dynamics; energetics, including ion and electron temperatures, are the subject of a chapter by Nagy and Cravens (see their chapter). Although the relative contributions of the two major mechanisms for maintaining the nightside ionosphere, electron precipitation and plasma transport, have not been totally quantified, the extent of our knowledge is greater now and is more firmly grounded than even five years ago. We devote much of our discussion here to a summary of what is known and unknown about the sources of the nightside ionosphere and their solar cycle variability.

II. THE DAYSIDE IONOSPHERE

A. Chemical Structure

The first direct measurements of the chemical structure in the Venus ionosphere were carried out by the Pioneer Venus OIMS and ORPA. The OIMS measured the ion densities at discrete mass numbers, including 1 (H^+), 2 (D^+), 4(He^+), 8(O^{++}), 12 (C^+), 14 (N^+), 16 (O^+), 18 ($^{18}O^+$), 28 ($CO^+ + N_2^+$), 30 (NO^+), 32 (O_2^+), and 44 (CO_2^+), which were detected in measurable quantities on both the day and night sides. The OIMS was capable of measuring densities at mass positions 17, 24, 40, and 56, but these were not routinely detected. An example of the ion density profiles of the dayside ionosphere derived from OIMS data is shown in Fig. 1.

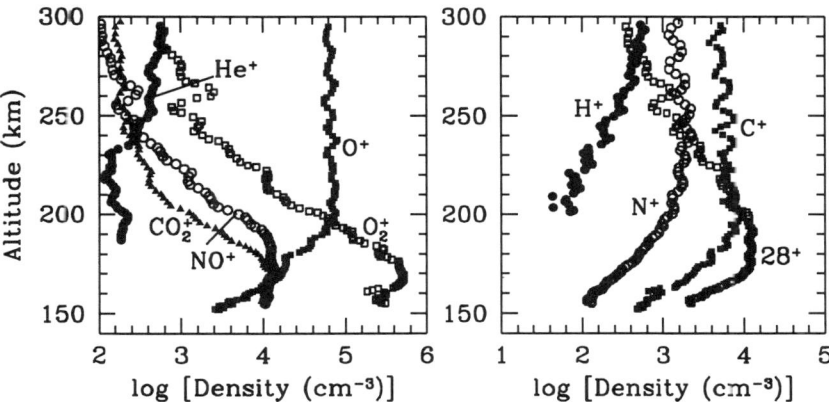

Figure 1. Smoothed ion density profiles measured by the OIMS on orbit 200. The solar zenith angle at periapsis was 25.7°, and the local time was about 13.4 hr. 28^+ denotes $CO^+ + N_2^+$.

In addition to ion temperatures and velocities, the ORPA measured total ion densities, as well as the densities of O^+, O_2^+, CO_2^+, and a pseudo-ion called $M29^+$, which represents the sum $NO^+ + CO^+ + N_2^+$ (see, e.g., Miller et al. 1984). The frequency and spatial resolution of the ORPA ion density measurements were generally too low for presentation as individual orbits, but statistical studies of the ORPA data have been carried out. Figure 2 shows smoothed profiles of median ion densities for the 60° to 70° solar zenith angle (SZA) interval as measured by the ORPA (Miller et al. 1984). These curves can be interpreted as typical profiles only insofar as the shape of the individual measured profiles does not exhibit significant temporal or spatial variability. This may be a valid approximation on the day side. PVO data show that the orbit-to-orbit variability of the low-altitude dayside neutral and ion densities are relatively small, and the altitude of the ion peak is fairly constant (see, e.g., Keating et al. 1985; Taylor et al. 1981). Kim et al. (1989) reported that the standard deviation of altitude of the dayside peak derived from 16 electron density profiles measured by the ORO is quite small: 0.22 km at solar maximum and 0.15 km at solar minimum.

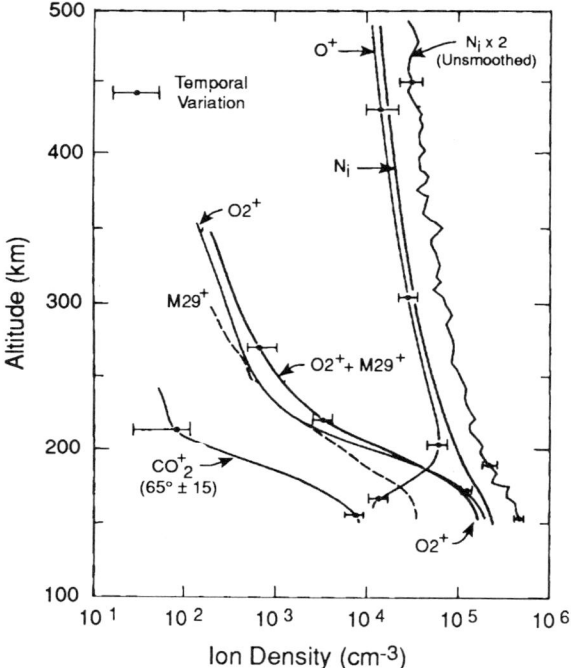

Figure 2. Smoothed median ion density profiles for CO_2^+, O_2^+, M29$^+$ (N_2^+ + CO$^+$ + NO$^+$) and the total ion density N_i as measured by the ORPA near 65° SZA. The curve labeled $N_i \times 2$ has not been smoothed (figure after Miller et al. 1984).

The chemistry of the major ions on the day side is fairly well understood and has been summarized previously (see, e.g., Nagy et al. 1980,1983; Fox 1982). Dalgarno and Fox (1994) have recently presented a comprehensive discussion of the chemistry of ions in atmospheric and astrophysical plasmas.

B. Solar Cycle Variability of the Dayside Ionosphere: Theory

For the case of a single major ion in the photochemical equilibrium region, the ion density n_i (and thus the electron density n_e) near the ion peak should be proportional to $\sqrt{P/\alpha}$, where P is the total ion production rate and α is the effective value of the recombination coefficient. Under certain conditions, the response of the major ion density profile to changes in solar flux or solar zenith angle can be predicted by applying Chapman (1931) theory. A Chapman model is a highly simplified model for the production rate and density profile of a major molecular ion produced by the interaction of monochromatic radiation with a single neutral species in an isothermal atmosphere. Chapman theory predicts that the ion production rate as a function of altitude z (relative to the altitude of maximum absorption for an overhead Sun) and solar zenith angle χ is given by:

$$q_i(z, \chi) = \frac{F^\infty \sigma_i}{H \sigma_a} \exp(-z/H - \sec \chi e^{-z/H}) \qquad (1)$$

where F^∞ is the solar ionizing flux outside the atmosphere, σ_i is the ionization cross section, σ_a is the absorption cross section, and H is the scale height, which is approximately constant for an isothermal atmosphere. If the molecular ion is assumed to be destroyed by dissociative recombination with a rate coefficient α, the ion density profile is given by

$$n_i(z, \chi) = \left(\frac{F^\infty \sigma_i}{H \sigma_a \alpha}\right)^{0.5} \exp[-z/(2H) - 1/2 \sec \chi e^{-z/H}]. \qquad (2)$$

To the extent that Chapman theory is applicable, the dayside ion densities should vary in a predictable way with changes in solar zenith angle and solar activity.

Ultraviolet solar fluxes, however, vary in a complicated way over the solar cycle. The $F_{10.7}$ index is often used as a proxy to represent the level of solar activity, but in fact the solar variability in the far and extreme ultraviolet is wavelength dependent, and may be larger or smaller than that of $F_{10.7}$. Figure 3a shows the solar photon flux as a function of wavelength from 18 to 2000 Å at low solar activity, and Fig. 3b the ratio of the solar flux at high solar activity ($F_{10.7} \approx 200$) to that at low solar activity ($F_{10.7} \approx 75$). The ratio is less than 1.2 for wavelengths near 2000 Å, but increases to factors of 2 to 4 in the range 500 to 1500 Å; below 500 Å, the ratio exhibits a great deal of structure and for some lines attains values close to 100. In addition, about 30% of the total ionization is caused by photoelectron impact at low solar activity. Because only photons with wavelengths less than 500 Å or so produce photoelectrons that are capable of further ionization, the importance of ionization by photoelectrons increases greatly with solar activity.

C. Solar Cycle Variability: Measurements and Models

Taylor et al. (1981) examined diurnal variations of O^+, O_2^+, C^+, and CO_2^+ densities at fixed altitudes between 180 and 220 km, and showed that the

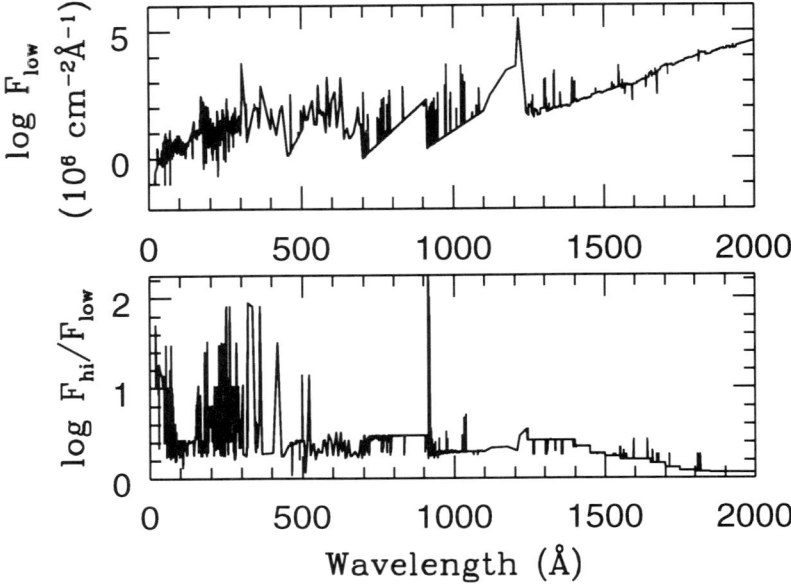

Figure 3. (Top) Solar photon flux as a function of wavelength from 18 to 2000 Å. The data is taken from the low solar activity SC#21REFW spectrum of Hinteregger (personal communication; see also Torr et al. 1979). (Bottom) The ratio of the high solar-activity fluxes to the low solar activity values. The high solar activity spectrum is the F79050N spectrum of Hinteregger.

densities exhibited evidence of a positive correlation with $F_{10.7}$, with peaks and troughs coinciding roughly with similar features in $F_{10.7}$. The first statistical study of the response of the dayside ionosphere to solar activity variations was, however, carried out by Elphic et al. (1984), who compared the electron densities and temperatures measured by the OETP over several 27-day solar cycles to a flux index V_{euv} derived from OETP data. V_{euv} is related to the production of solar photoelectrons on the radial Langmuir probe when it is outside the Venus atmosphere (see, e.g., Brace et al. 1988). Some of the results of the study of Elphic et al. (1984) are illustrated in Fig. 9 of the chapter by Nagy and Cravens.

Elphic et al. (1984) found that the electron densities correlated well with V_{euv}, and that n_e varied approximately as $V_{euv}^{0.33}$. The exponent of the solar flux index would be expected to be different from the value 0.5 predicted by Chapman theory for a number of reasons. First, V_{euv} reflects the work function of the Langmuir probe, and is not strictly proportional to the atmospheric ionizing flux F^∞ in Chapman theory. Slightly more than half of the photon flux represented by V_{euv} is due to Lyman alpha, which does not ionize the major species in the Venus thermosphere, and the rest to photons in the 300 to 1100 Å range. Second, the neutral thermospheric densities and temperatures also show a weak dependence on solar activity. Finally, the effective recombination rate coefficient is a function of the ion composition and the electron temperature, and thus may vary with solar activity. None of these effects are accounted for by Chapman theory.

Studies of the response of neutral thermospheric temperatures to changing solar activity show that the exospheric temperature on Venus varies from about 235 to 300 K from low to high solar activity (see, e.g., the chapter by Kasprzak et al.; Fox and Bougher 1991). From PVO and Magellan atmospheric drag data, Keating and Hsu (1993) reported that a 50 unit increase of $F_{10.7}$ produces an increase in the dayside temperature of about 30 K. As solar activity increases, the slightly higher temperatures and scale heights lead to a maximum in the solar EUV absorption at lower neutral densities, which partially compensates for the higher solar fluxes. The response of the ionization rate is thus damped compared to that predicted by Chapman theory.

Electron density profiles at all phases of the solar cycle are available from ORO measurements from the Pioneer Venus and Venera missions. Kliore and Mullen (1989) analyzed the response of the dayside peak to changes in solar activity, as indicated by 115 radio occultation profiles obtained during eight occultation seasons in solar cycle 21. From a statistical fit to their data, they derived an equation for the dayside peak electron density in units of cm^{-3} as a function of solar activity and solar zenith angle χ:

$$n_{e,\max}(F_{\mathrm{euv}}, \chi) = (5.92 \pm 0.03) \times 10^5 (F_{\mathrm{euv}}/150)^{0.376} (\cos \chi)^{0.511} \quad (3)$$

where F_{euv} is an estimated value of $F_{10.7}$ corrected to the orbital position of Venus. The variation of the peak density with solar zenith angle is in agreement with Chapman theory. Unlike a Chapman profile, however, for which the peak rises with increasing solar zenith angle, the altitude of the peak was found to remain near 140 km from the subsolar region to about 85° SZA. The invariance of the peak altitude with χ results from the compensating collapse of the neutral atmosphere as the solar zenith angle increases (Cravens et al. 1981). It is interesting to note that the exponent of the solar flux index for long-term solar cycle variations in Eq. (3), 0.376, is nearly the same as that for short-term solar rotation variations, 0.33, which was derived by Elphic et al. (1984).

The observed average shapes of the high and low solar activity electron density profiles for the SZA range 55 to 75°, as measured by the PVO radio occultation experiment, are illustrated in Fig. 4. In this figure, the average densities from 16 individual radio occultation height profiles from high solar activity (1979–1980) are compared to the average densities of 14 profiles from low solar activity (1984, 1986). The density near the peak increases by only about 50% from low to high solar activities, but the variation at high altitudes is much larger. The solar minimum electron densities at 300 km are about a factor of 10 smaller than those at solar maximum. Color contour plots of

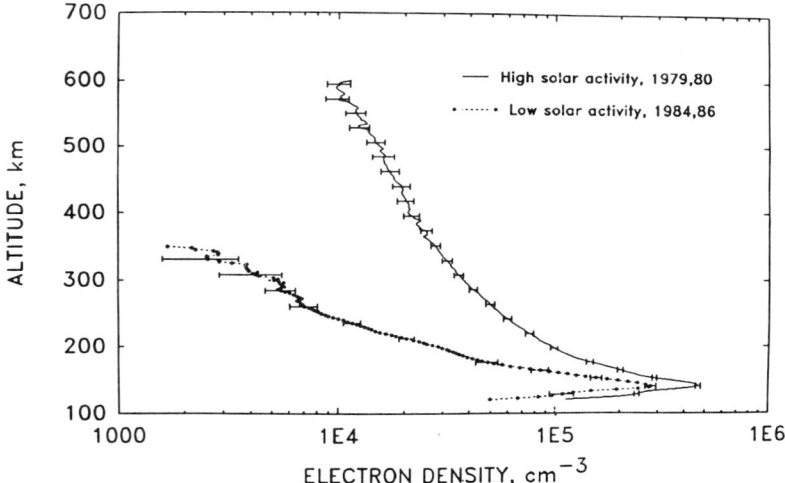

Figure 4. Average electron density profiles in the range 55 to 75° SZA from PV radio occultation data taken during solar maximum (1979–1980) and solar minimum (1984–1986) (figure from Kliore et al. 1991).

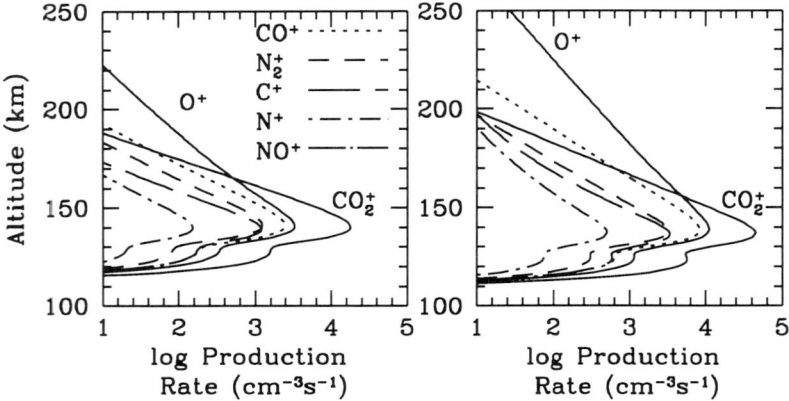

Figure 5. Computed altitude profiles of production rates of ions due to photoionization and photoelectron impact ionization (left) for the low solar-activity model, and (right) for the high solar-activity model.

the average dayside electron densities as a function of altitude and SZA from ORO measurements at low and high solar activities are shown in CDP2C1F1.

High and low solar-activity models similar to those presented for Mars by Fox et al. (1995) have been carried out. Computed production rates of the major ions by photoionization and electron-impact ionization at low and high solar activities are compared in Fig. 5. The peak production rate of CO_2^+ near 140 km in the high solar activity model is larger by a factor of about 2.5 than that for the low solar activity model. The production rate of O^+ increases by a larger factor of 3.3 near the ion peak and by a factor of approximately 9 at 200 km. The computed ion density profiles are compared in Fig. 6. The predicted O_2^+ profile exhibits a broad maximum in the 137 to 150 km range. The peak density is about 4.9×10^5 cm^{-3} at 143 km in the high solar activity model and about 3.4×10^5 cm^{-3} at 144 km in the low solar activity model. These values are in excellent agreement with those computed from the analytical expression (Eq. 3) derived by Kliore and Mullen (1989) of 4.7×10^5 and 3.2×10^5 cm^{-3}, for $F_{10.7} = 200$ and $F_{10.7} = 75$, respectively, at 45° SZA.

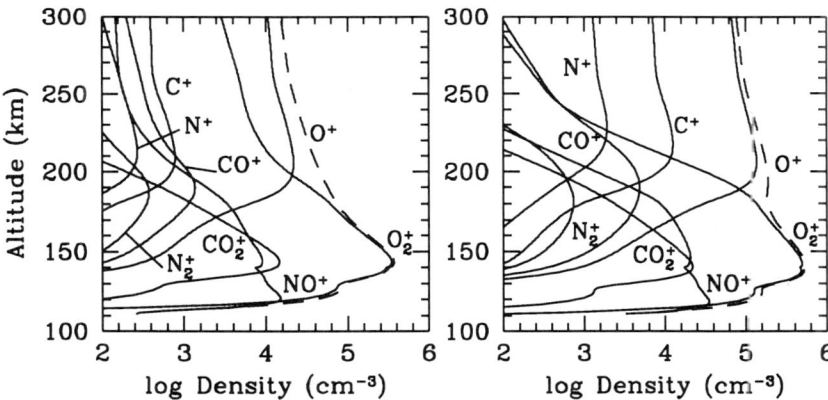

Figure 6. Computed altitude profiles for the major ions (left) for the low solar-activity model and (right) for the high solar-activity model.

Because the PVO penetrated below 145 km only rarely, the ion peak was not often sampled by *in-situ* instruments. The lowest altitude for which dayside median ORPA densities are available is 155 km, where the total ion densities are 2.2×10^5 and 3.1×10^5 cm^{-3}, and the the densities of O_2^+ + M29$^+$ are 1.9×10^5 and 2.7×10^5 cm^{-3}, for 65°±5° SZA and 25±5° SZA, respectively (Miller et al. 1984). Total electron density measurements by the OETP exhibited average values in the 150 to 155 km range of 3.14×10^5 and 2.48×10^5 for solar zenith angles of 25° and 55°, respectively, in the dawn sector (Theis et al. 1984). These values are in reasonable agreement with the high solar activity model above, which predicts an O_2^+ density of 3.5×10^5 cm^{-3} at 155 km for 45° SZA.

Taylor et al. (1981) reported OIMS measurements of density variations of individual ions as a function of local time over the first two diurnal cycles, when periapsis was at low northern latitudes. They showed that at 200 km, the O_2^+ density peaked at about 5×10^4 cm^{-3} at 12 hr local time and was about 3×10^4 cm^{-3} at local times of 9 and 15 hr. The high solar-activity model above agrees fairly well with these values; it predicts a density of 3.4×10^4 cm^{-3} at 200 km and 45° SZA.

The model illustrated in Fig. 6 also shows that, due to the peak in the O$^+$ density profile, the n_e profile exhibits a shoulder near 200 km at high solar activity. The appearance of such a shoulder in the electron density profile near that altitude has been cited as evidence that the dayside ionosphere is magnetized (see, e.g., Shinagawa et al. 1991). The shape of the model O$^+$ profile in Fig. 6 shows that caution should be exercised in drawing that conclusion in the absence of other evidence, especially at high solar activity.

Because O$^+$ is produced mostly by direct ionization of O and lost by reaction with CO$_2$, the peak O$^+$ density should be proportional to the product of the O density and the ionizing photon and suprathermal electron fluxes, and inversely proportional to the CO$_2$ density. Both the O density and the solar flux increase with solar activity, and the model calculations described above predict an increase in O$^+$ density of a factor of 6 at 300 km from low to high solar activity. This increase is slightly smaller than the factor of 10 found from ORO data, but because the low solar-activity neutral model used in the calculations (Hedin et al. 1983) was extrapolated from the high solar activity measurements of the PV ONMS and OAD, the agreement between measurements and model is acceptable.

Because the atomic ions C$^+$ and N$^+$ are produced by photoionization of atoms, and destroyed in ion-molecule reactions, their predicted variation with solar activity is similar to that of O$^+$. The NO$^+$ densities peak near 120 km in the models, with values of 1.5×10^4 and 3.5×10^4 cm^{-3} at low and high solar activities, respectively. The variability of NO$^+$ is larger than that of O$_2^+$ because the densities depend not only on the ionization and recombination rates, but on the abundance of atomic nitrogen, which reacts with O$_2^+$ to form NO$^+$. The abundance of N is predicted to increase with solar activity. There are, however, no *in-situ* measurements of ion composition at low solar activity with which to compare these predictions.

High and low solar-activity models were also constructed by Kim et al. (1989), who carried out photochemical equilibrium calculations of six ions in the lower ionosphere of Venus for a solar zenith angle of 65°. Their goal was to compare the models to averages of ORO measured electron density profiles in the region of the peak. The computed values for electron density peaks were about 6×10^5 cm^{-3} at 137 km and 4×10^5 cm^{-3} at 140 km for the high and low solar-activity models, respectively. The most notable difference between these models and those described above is in the altitude of the peak at high solar activity. The source of the discrepancy is not certain, but may relate to differences in the cross sections, neutral model, and electron temperature profile employed. The differences in the peak densities predicted by the two models, which are on the order of 20%, are not significant.

D. The Ionopause

The sharp drop in ion or electron density that occurs at the boundary between the thermal ionospheric plasma and the superthermal solar wind plasma on Venus is called the ionopause. Because Venus does not possess an intrinsic magnetic field, the ionosphere itself acts as a barrier to the solar wind, and the ionopause is found where the solar wind dynamic pressure (P_{sw}) is balanced by the plasma pressure of the ionosphere (see, e.g., Elphic et al. 1980; Luhmann 1986). The height of the ionopause is therefore determined by time-varying characteristics of both the solar wind and the ionosphere.

The exact location of the ionopause also depends on its definition, which is a function of the data set of interest. Elphic et al. (1980) defined the ionopause as the point where the plasma pressure is balanced by the component of the solar wind dynamic pressure that is perpendicular to the ionosphere. Brace et al. (1980) considered the ionopause to occur where the total electron density, as measured by the OETP, drops below a value of 100 cm^{-3} in a steep gradient. Knudsen et al. (1982a) adopted a similar definition for their study of the ionopause as indicated by ORPA measurements. For ORO data, the ionopause is defined as that altitude at which the electron density falls below 500 cm^{-3}, which is near the lower limit of sensitivity of the instrument (Kliore and Luhmann 1991).

Kliore and Luhmann (1991) discussed the altitude of the ORO ionopause as a function of solar zenith angle and solar activity. They showed that for solar zenith angles less than 40 to 45°, the ionopause is usually low, in the range 200 to 400 km. In this region, the ionopause heights determined from the different definitions are found to differ very little. At larger solar zenith angles, the observed ionopause altitudes are higher at intermediate and high solar activity than are those at low solar activity. The average ORO ionopause altitude at 85° SZA and high solar activity is approximately 500 km, although the value determined by the OETP or the ORPA is about 1000 km. The variability in the terminator ionopause height is large at high and intermediate solar activity, and a significant fraction of this variability can be ascribed to changes in P_{sw}. The terminator ionopause altitude decreases with increasing P_{sw} to a minimum near 200 to 300 km for values in excess of about 3×10^{-8} dyne cm^{-2} (Brace et al. 1980; Kliore and Luhmann 1991). At solar minimum, the dayside ionopause altitude is found to be uniformly low regardless of solar zenith angle or solar wind dynamic pressure. Further discussion of the ionopause can be found in the chapter by Cravens et al.

III. THE NIGHTSIDE IONOSPHERE

A. The Ionotail

Beyond the terminator, at high altitudes in the shadow of Venus, lies the ionotail, a region in which the ionosphere develops long comet-like tail rays that are aligned roughly in the anti-sunward direction, and in which the ion densities are about 50 to 500 cm^{-3} (Brace et al. 1987,1990). The tail rays are separated by regions of depleted plasma called troughs, in which the ion densities are 1 to 10 cm^{-3}. The horizontal dimensions of the rays are on the order of 1000 km, and they appear to extend thousands of kilometers behind the planet. Superimposed on the rays are smaller-scale structures called filaments; near the flanks of the tail rays, clouds of scavenged plasma appear (Brace et al. 1982). The ions in the tail rays have been observed to be mostly energetic O$^+$, with velocities exceeding the escape velocities. Brace et al. have suggested that substantial mass loss in the form of escaping ions occurs down the ionotail.

The ionotail is observed to change dramatically both from orbit to orbit and with solar activity. Brace et al. (1990) examined the response of the average electron density $\bar{n}_e$ measured by the OETP in the ionotail to solar flux and solar wind dynamic pressure. The orbit to orbit variations were found to be related to changes in P_{sw}. $\bar{n}_e$ was observed to decrease approximately linearly with increasing P_{sw}, with an increase of a factor of 10 in P_{sw} producing about a factor of 6 decrease in $\bar{n}_e$. The response of the ionotail to variations in V_{euv} was very large, with a 40% decrease in V_{euv} producing a factor of 15 depletion in $\bar{n}_e$. This amplification is evidently produced by strong solar control of both the dayside atomic ion densities and the terminator ionopause height, which regulate the day-to-night ion transport that is apparently responsible for populating the tail (see below).

B. The Nightside Lower Ionosphere

The nightside ionosphere near the peak has also been observed to be highly variable from orbit to orbit and with solar activity. Radio occultation profiles at solar maximum showed that the average value of the peak electron density for SZA>110° was about 2×10^4 cm^{-3}, but that the peak varied widely in magnitude, altitude, and width (Kliore et al. 1979; Zhang et al. 1990; Kliore et al. 1991). On many occasions a second peak appeared below the main peak at an altitude of about 124 km. Sometimes the nightside ionosphere was observed to be reduced to a few patches of scattered plasma (Cravens et al. 1982). The variability in the altitude of the peak is partly due to substantial changes in the nightside neutral densities, the standard deviation of which is about a factor of 2 (Keating et al. 1985), and partly to variations in the source of ionization.

Smoothed altitude profiles of median ion densities for the solar zenith angle range 120 to 180°, as measured by the ORPA are presented in Fig. 7. The median value for n_i in the region of the nightside peak (145±2.5 km) is about about 1.5×10^4 cm^{-3} (Knudsen 1987). Because the altitude of the nightside peak has been observed to vary substantially, this value is not strictly comparable to an average peak density. As the peak altitude changes, the high measured values will be averaged out, and the peak in the ORPA profile will be broader and flatter than a typical profile. In fact, the ORPA measured median density in the peak region is about 25% less than the mean peak value

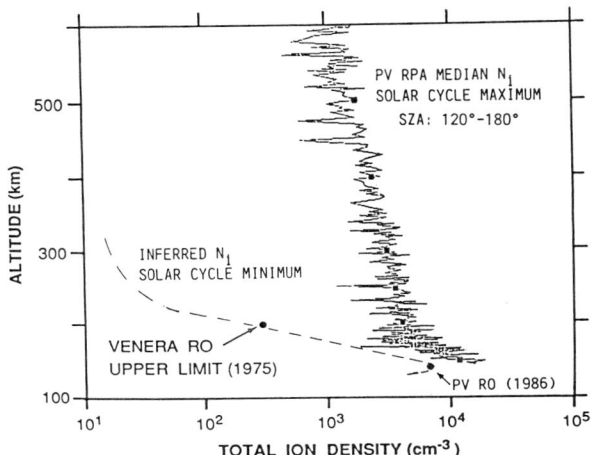

Figure 7. Vertical profile of medians of the total ion density N_i for the low nightside ionosphere from ORPA measurements. Also shown for comparison are the mean peak electron density from ORO measurements and the 200 km upper limit from the Venera 9 and 10 measurements (figure after Knudsen 1992).

of n_e reported from ORO profiles (Kliore et al. 1991).

Altitude profiles of median nightside electron densities derived from OETP data were also reported by Theis et al. (1984), and the values in the region of the peak are, however, significantly larger than those derived from ORPA measurements. The measured median densities at high solar activity are in the range $(2 \text{ to } 5) \times 10^4$ cm^{-3} in the lowest reported altitude bins, 145 to 150 km, for SZA$\geq 140°$ in both the dawn and dusk sectors. A model constructed by Theis and Brace by fitting the median OETP electron densities to an analytical functional form, however, exhibits values of n_e in the range $(1 \text{ to } 2) \times 10^4$ cm^{-3} in the same regions. The observed medians decrease by an order of magnitude from 150 km to about 450 km at high solar activity; the same relative decrease is seen in the ORPA data from 150 to about 600 km.

Density profiles measured by the OIMS during the initial high solar activity period showed that the composition of the nightside ionosphere was much like that of the dayside ionosphere, with densities of the heavy ions depleted relative to their dayside values by factors of 10 to 100 (Taylor et al. 1980,1981). Heavy ions such as O^+ and O_2^+ exhibited slightly larger densities in the dusk sector than near dawn. The light ions H^+, D^+ and He^+, on the other hand, were more abundant on the night side, maximizing in the pre-dawn sector near 4 a.m. local time. The variability in the measured densities at high solar activity showed a correlation with $F_{10.7}$, which Taylor et al. (1981) ascribed to solar control of the dayside ionization rate and of the

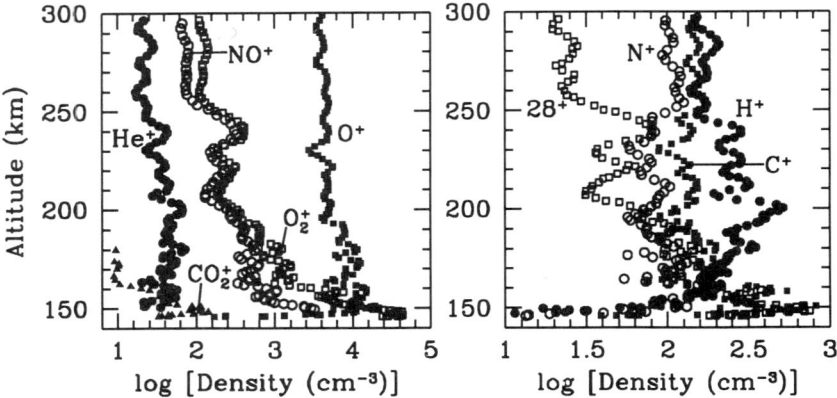

Figure 8. Smoothed OIMS ion density profiles from orbit 506. At periapsis, the SZA was 150.9° and the local time was 22.2 hr.

neutral circulation. Typical nightside density profiles from the OIMS data are shown in Fig. 8.

C. Sources of the Nightside Ionosphere

The appearance of a nightside ionosphere on Venus was surprising in view of the long period of the Venus night, which, due to the slow retrograde rotation of the planet, lasts for 58 Earth days. Of the many potential explanations that were advanced, only two are now considered significant: precipitation of suprathermal electrons of unknown origin that have been observed in the umbra, and transport of atomic ions from the day side. The relative importance of these sources appears to change from high to low solar activity, and even from orbit to orbit.

The first direct evidence for the electron precipitation source was obtained in November and December of 1975, when the plasma analyzers aboard the Venera 9 and 10 spacecraft measured integral suprathermal electron fluxes of (3 to 7) $\times 10^8$ cm^{-2} s^{-1} with energies up to 300 eV at high altitudes in the Venus umbra. Gringauz et al. (1977) estimated the ionization rate that would be caused by precipitation of these electrons into the nightside thermosphere. They showed that this rate was sufficient to produce a peak ion density of (5 to 8) $\times 10^3$ cm^{-3}. This value is roughly consistent with peak densities in the range (6 to 16) $\times 10^3$ cm^{-3}, which were inferred from the Venera radio occultation data at solar minimum. The PV ORPA also measured integral fluxes of ambient electrons from about 1 to 45 eV in the Venus wake; the derived differential fluxes were slightly larger than those inferred from the Venera measurement, but within the error bars of the two experiments (Knudsen and Miller 1985). Differential fluxes of suprathermal electrons derived from both the Venera and PV ORPA data are compared in Fig. 6 of the chapter by Nagy and Cravens. The integral fluxes were observed to decrease only slightly with solar zenith angle from the terminator toward the antisolar point (Knudsen and Miller 1985; Spenner et al. 1995b).

Because the suprathermal electrons were observed at high altitudes, the fraction of these electrons that reach the nightside thermosphere is uncertain. The appearance, however, of aurorally induced emissions of atomic oxygen at 1304 and 1356 Å in data from the PV orbiter ultraviolet spectrometer is evidence that at least some of the electrons precipitate into the atmosphere. Using a simplified model for electron transport, Fox and Stewart (1991) estimated the emission rate that would be produced by the measured fluxes, and showed that the observed average 1356 Å intensities of about 4 R could be produced by precipitation of 28% of the fluxes reported by Knudsen and Miller (1985). This estimate was later refined to 23% by Fox et al. (1992), from a model that employed a more accurate multi-stream electron transport code.

Soon after the PVO spacecraft was inserted into orbit around Venus, measurements carried out by the ORPA showed the presence of generally anti-sunward-flowing, mostly atomic ions from the subsolar region to about 150° SZA on the night side. The measured velocities were small near the subsolar point, but became supersonic, with velocities that approached 3 to 4 km s^{-1} at high altitudes in the terminator region (Knudsen et al. 1982b; Taylor et al. 1980). Because Venus does not possess an intrinsic magnetic field, the ions are free to respond to the day-to-night plasma pressure gradients that result from the difference between the large dayside ion source and the probably small *in-situ* nightside source. Models have in fact demonstrated that the plasma pressure gradient force is sufficient to accelerate the ions to the observed nightward flow velocities (Whitten et al. 1984; Elphic et al. 1984; Nagy et al. 1991). The single-species model of Nagy et al. (1991) showed that the converging flow produces downward ion velocities that approach 600 m s^{-1} for solar zenith angles of 90 to 150°.

Spenner et al. (1981) showed that downward fluxes of O$^+$ on the order of (1 to 2) $\times 10^8$ cm^{-2} s^{-1} were capable of producing the average nightside O$_2^+$ density peak through the reaction O$^+$ + CO$_2$ → O$_2^+$ + CO. Using the measured values of the terminator densities and velocities, Knudsen and Miller (1992) estimated the total flux of O$^+$ across the terminator as 5×10^{26} s^{-1}, which is equivalent to a uniform downward flux over the night side of 2.1×10^8 cm^{-2} s^{-1}.

Cravens et al. (1983) constructed a quasi-two-dimensional model that showed that the maintenance of the nightside ionosphere by day-to-night ion transport depends on an adequate supply of ions on the day side: by large ion densities and a high dayside ionopause. Figure 9 shows the peak electron density on the nightside as a function of solar wind dynamic pressure (P_{sw}), which has been demonstrated to be inversely correlated with near-terminator ionopause height. Although there appears to be little correlation at low-to-moderate solar activities, clearly at high solar activity there is some correlation, with very large values of P_{sw} associated with small peak electron densities. Evidently, large values of P_{sw} compress the dayside ionosphere; this

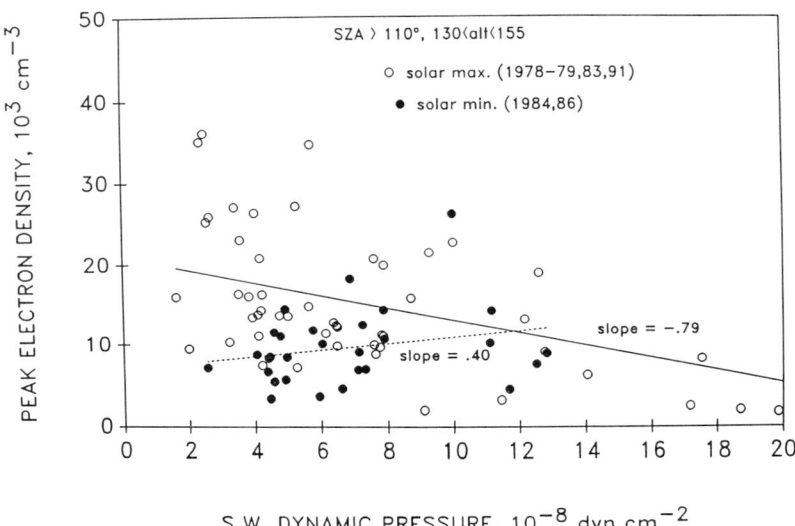

Figure 9. The ORO nightside main-peak electron density vs the solar wind dynamic pressure for data from solar maximum (open circles) and solar minimum (filled circles) (figure from Kliore and Luhmann 1991).

produces a low dayside ionopause and reduces the day-to-night ion transport.

As P_{sw} becomes smaller, the ionopause rises, increasing the anti-sunward fluxes of ions. This effect cannot continue indefinitely, as the day-to-night flux eventually becomes supply limited. Indeed, as Brace et al. (1995) have pointed out, the correlation between P_{sw} and nightside peak density (Fig. 9) appears poor at smaller values of P_{sw}. The maximum nightward flux is determined roughly by the production rate of O^+ and other atomic ions above the photochemical equilibrium region. Fox (1992) determined the maximum upward flux of O^+ that could be imposed at the upper boundary of a 45° SZA model of the Venus ionosphere as about 3.2×10^8 cm^{-2} s^{-1}. Brace et al. (1982) have called the maximum flux the "diffusion limit," and they derived a similar limiting value of 3×10^8 cm^{-2} s^{-1}. When O^+ fluxes close to the derived upper limits are imposed on a model of the nightside thermosphere, a peak ion density of about 4×10^4 cm^{-3} results (Fox 1992). Kliore et al. (1991) have in fact reported that the maximum nightside electron peak density observed at high solar activity is about 4×10^4 cm^{-3}.

The ionospheric velocity field as measured by the ORPA around the planet is shown in Fig. 10. To a first approximation the flow diverges in the subsolar region and converges on the night side. The flow is not, however, symmetrical around the Sun–Venus line, and at low altitudes seems to be influenced by the neutral thermospheric superrotation, with the regions of divergence and convergence displaced from noon and midnight, respectively, toward the dawn terminator. Near 150° SZA the flow becomes disordered,

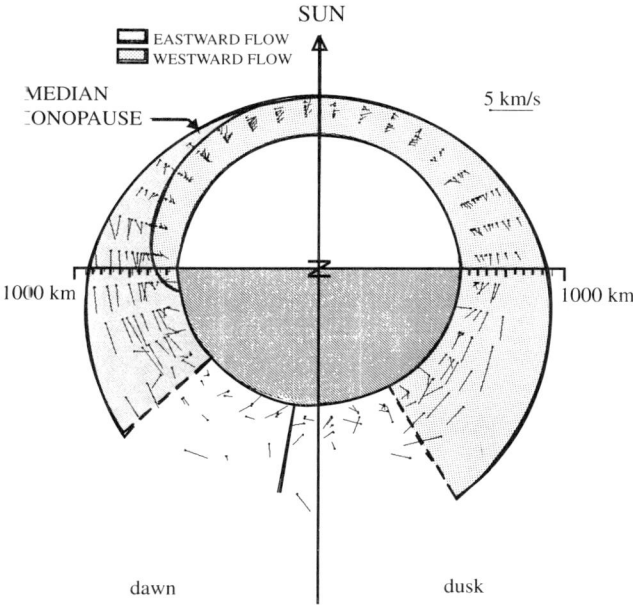

Figure 10. Average O^+ velocity field at 10° intervals of longitude. The altitude scale has been exaggerated by a factor of 4 relative to the planetary radius. The dashed radial lines represent the points where the velocities become chaotic. The light and dark shading represent regions of predominantly westward and eastward velocities (figure after Miller and Whitten 1991).

and the ion temperature exhibits a sudden large increase, which probably signifies the development of a shock (Knudsen et al. 1980). The appearance of such a shock wave, at a slightly smaller SZA, was demonstrated in the two-dimensional hydrodynamic ion-transport model of the Venus nightside ionosphere reported by Nagy et al. (1991).

Although the average velocity vectors in the anti-solar (unshaded) region in Fig. 10 appear smaller than those at solar zenith angles less than 150° (shaded region), this effect is mostly due to randomization in the direction of velocities, rather than to small ion speeds. At low altitudes (200–300 km) the antisunward velocities near the dusk terminator are larger than those at the dawn terminator, but at altitudes above about 350 km, the asymmetry is reversed (Miller and Knudsen 1987).

Information about the nightward transport of ions is also available from the OIMS and ORPA ion composition data. Model calculations showed that precipitation of electrons with fluxes similar to those measured by the Venera plasma analyzers or the ORPA produced an O_2^+ peak that was about half that observed, but the predicted O^+ profiles were an order of magnitude too small (Spenner et al. 1981; Cravens et al. 1983). From the ratio of model to observed peak densities, Spenner et al. (1981) concluded that electron precipitation could contribute up to one-fourth of the total ionization at high solar activity.

If the nightside O^+ is produced by transport from the day side, however, then the magnitude of the O^+ peak on the night side should be roughly proportional to the downward flux. Brannon and Fox (1994) constructed a map of the downward flux of O^+ over the night side by comparing O^+ peak densities measured by the OIMS to those of one-dimensional models in which downward fluxes of atomic ions were imposed at the upper boundaries of the models. Figure 11 shows the downward O^+ flux as a function of solar zenith angle, with evening and morning sectors shown separately. The error bars are standard deviations, and show that the inferred downward fluxes are highly variable. The downward flux is largest near the terminators, and decreases rapidly toward the antisolar point, with relative minima appearing near about 125° SZA and relative maxima near 155° SZA in both sectors. A small increase in the ion densities near 150° SZA was also observed in the two-dimensional model presented by Nagy et al. (1991).

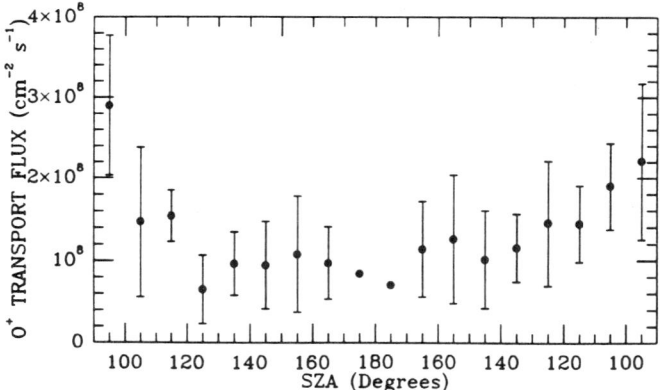

Figure 11. Downward O^+ flux as a function of solar zenith angle derived from OIMS O^+ data and modeling. The data were binned in 10° solar zenith angle intervals. The error bars are standard deviations of the data. No standard deviations were computed for the bins nearest the antisolar point, for which there were only two or three data points (figure from Brannon and Fox 1994).

Brannon and Fox (1994) found the average O^+ flux over the night side for the first 600 orbits to be about $1.5 \times 10^8 \, \text{cm}^{-2} \, \text{s}^{-1}$. The downward fluxes near the evening terminator were larger than those near the morning terminator; values of 2.9×10^8 and $2.2 \times 10^8 \, \text{cm}^{-2} \, \text{s}^{-1}$ were obtained for the 90 to 100° SZA range for the dusk and dawn sectors, respectively. These conclusions reflect the dawn-dusk asymmetries both in the OIMS measured O^+ densities (Taylor et al. 1980), and in the ORPA measured densities of n_i

and O^+ on the day side (Miller et al 1984). They are also consistent with the measured ion velocity field shown in Fig. 10, which indicates that at low altitudes the antisunward ion velocity is larger at the dusk terminator than at the dawn terminator.

From ORPA ion density profiles, Knudsen et al. (1980) estimated the nightside hemispheric molecular ion recombination rate as 2×10^8 cm^{-2} s^{-1}. This value is nearly equal to estimates of the transterminator ion fluxes of 2.1×10^8 and 2×10^8 cm^{-2} s^{-1} obtained from ORPA (Knudsen and Miller 1992) and OETP data (Brace et al. 1995), respectively. The near equality of the estimates for the nightward ion flux and nightside recombination rate would seem to indicate that electron precipitation does not play a major role in ion production. The uncertainty of these numbers, however, is of the order of 30%, so a contribution from electron precipitation of about that magnitude cannot be excluded. Knudsen and Miller (1992) placed an upper limit on the ion production rate by electron precipitation at high solar activity of 1×10^8 cm^{-2} s^{-1}. In addition, some of the nightward-transported ions may escape down the tail of the planet and thus do not contribute to the nightside ionization, although estimates of the atomic ion escape flux vary widely (see, e.g., Brace et al. 1987; McComas et al. 1986).

From OIMS O^+ density profiles, Brannon and Fox (1994) inferred a downward atomic ion flux over the night side of 1.65×10^8 cm^{-2} s^{-1}, about 20% less than the estimate of Knudsen et al. (1980) for the total hemispheric recombination rate. If accurate, these values imply that the ion production rate due to precipitation at high solar activity is about 27% of that due to ion transport. When the estimates for transterminator transport (Knudsen and Miller 1992) and downward ion flux are compared, it might be inferred that 20% of the nightward-flowing ions escape from the gravitational field of the planet. This estimate is similar to the ion escape fluxes inferred by Brace et al. (1987) and by Donahue and Hartle (1992). The uncertainty, however, in the downward ion flux derived by Brannon and Fox (1995) is probably on the order of 25%, and is comparable to the uncertainties in the estimates of the nightward fluxes or the total hemispheric recombination rate discussed above. Conclusions drawn by comparing these numbers should therefore be viewed with caution.

Estimating the contribution of electron impact relative to ion transport on any individual orbit is possible in theory but difficult in practice. Fox and Taylor (1990) pointed out that the mass-28 ion densities on different orbits of OIMS data from the early PVO mission varied by about 2 orders of magnitude, and they suggested that, because there are few chemical sources of N_2^+ and CO^+, high mass-28 ion densities in individual nightside orbits of OIMS data constitute a signature of electron precipitation. Although models have shown that the densities of mass-28 ions may in fact increase by up to an order of magnitude in the presence of electron precipitation, the extremely high mass-28 densities ($\sim 10^4$ cm^{-3}) that appeared in some of the orbits analyzed by Fox and Taylor (1990) and by Fox (1992) have been questioned by Fox and Grebowsky (manuscript in preparation) as being due to an instrumental effect. Densities of mass-28 ions on the order of a few times 10^3 cm^{-3} are possible, however, and should in fact indicate enhanced precipitation on an individual orbit.

Fox (1992) also pointed out that, if the O_2^+ is produced mostly by reaction of downward flowing O^+ with CO_2, then the ratio of the O_2^+ peak to that of O^+ should be characteristic of the O^+ flux. If the flux derived from the O_2^+/O^+ peak ratio is larger than that derived from the O^+ maximum density, production of O_2^+ by another source is indicated. Fox (1992) compared the O_2^+/O^+ peak ratio derived from OIMS data from the first 600 PVO orbits to the ratio that would be predicted for the transport-dominated ionosphere. The ratio for a substantial number of orbits lay above that predicted for the transport-dominated ionosphere, possibly indicating the presence of electron precipitation. This study was, however, hampered by the fact that the ion peak was not sampled very often, which would result in the O_2^+/O^+ ratio being underestimated. Nonetheless, Fox (1992) concluded that only three orbital segments of a total of 12 examined in detail showed evidence for significant electron precipitation.

D. Solar Cycle Changes in the Venus Nightside Ionosphere

The Venera 9 and 10 radio occultation measurements in 1975 and measurements made by instruments on PVO during the low solar activity period 1983 to 1986 provided information about the morphology of the low solar-activity nightside ionosphere, and solar cycle variability in the nightside maintenance mechanisms. Knudsen (1988) examined the high-altitude O^+ densities measured by the ORPA at high and low solar activities on the night side. The median densities from the two data sets extrapolated to values that differed by a factor of about 20 at 1800 km. Knudsen interpreted this as indicating that the nightside O^+ layer is absent at low solar activity, and the transport source is reduced by at least an order of magnitude.

Radio occultation measurements showed that the average electron density in the region of the nightside peak decreased by a factor of 2 from high to low solar activity, and the altitude of the peak decreased by about 6 km (Kliore et al. 1991). Averages of ORO electron density measurements at low, medium and high solar activities are presented in Fig. 12. The average density in the peak region at low solar activity was about 1×10^4 cm^{-3} at an altitude of 138 km. Substantial ionization appears above the peak during solar maximum that is absent at solar minimum.

Figure 9 shows that the anticorrelation between P_{sw} (and thus the height of the dayside ionopause) and the magnitude of the electron density peak that was demonstrated at high solar activity does not exist at moderate or low solar activity. Indeed, at low solar activity, the altitude of the ionopause is almost always low, like that at high solar activity for large values of P_{sw} (Kliore and Luhmann 1991). Kliore et al. (1991) interpreted the lack of correlation between the nightside peak and the dayside supply of ions as indicating that

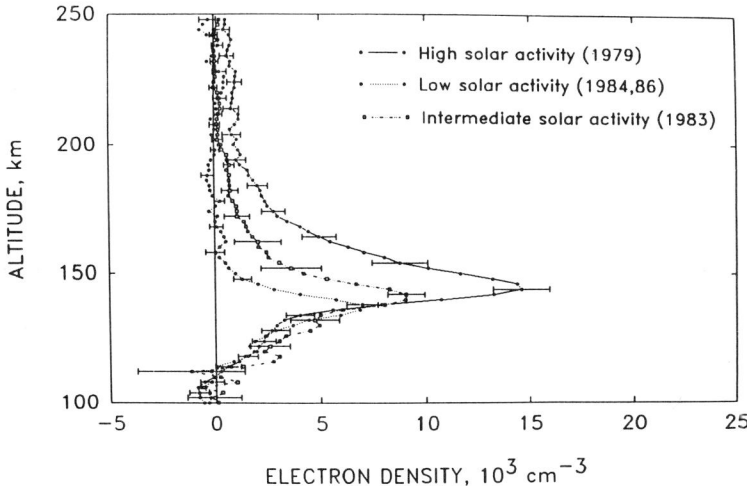

Figure 12. Averaged nightside electron density profiles derived from ORO measurements from high, intermediate, and low solar activity periods (figure from Kliore and Luhmann 1991).

the transterminator flow is shut off at solar minimum, leaving only the electron impact source to produce the nightside peak.

Contour plots of the nightside electron densities as a function of altitude and solar zenith angle, as measured by the ORO for high and low solar activities are displayed in CDP2C1F2. The solar minimum data show diminished electron density in the peak region, especially for solar zenith angles $\geq 130°$. The overall morphology of the nightside ionization structure is, however, similar at high and low solar activities. The reduction of the ion density above the peak from high to low solar activity merely reflects the behavior of the high-altitude dayside ionosphere. Both high and low solar activity nightside ionospheres show a maximum near the terminator in the ion densities, which then decline rapidly toward the antisolar point. Both exhibit a relative minimum in the region of the peak near 120 to 125° SZA, and a relative maximum near 140 to 150° SZA, although these features appear to be shifted slightly toward the terminator at low solar activity. The relative maximum corresponds to the region where the nightward flowing ions appear to converge in the velocity field shown in Fig. 10.

In-situ measurements were made by the PVO instruments at moderately low solar activity during the re-entry period. ORPA measurements of the high-altitude nightside suprathermal electron fluxes made during 15 orbits near the end of the mission showed that the median fluxes did not differ substantially from those at high solar activity, and their variability was about an order of magnitude from orbit to orbit (Spenner et al. 1995*a*). A small increase of the differential fluxes with altitude was also reported, but the fluxes were relatively uniformly distributed over the night side. Neither the orbit-to-orbit variability of the nightside ionosphere, nor the solar zenith angle dependence illustrated in CDP2C1F2, nor the response of the ionosphere to solar activity changes can be accounted for if the major source of the nightside ionosphere is precipitation of the observed suprathermal electron fluxes.

Theis and Brace (1993) reported that OETP measurements of n_e in the re-entry period showed that the electron densities decreased by a factor of about 7 at 200 km from high to moderately low solar activity, but the variability in the high-altitude densities was about 2 orders of magnitude. The densities in the peak regions of the models that were constructed to fit the median values of n_e, however, did not show much change. The latter conclusion, which is in obvious contradiction to the radio occultation measurements, may be an artifact of the function used to fit the altitude profile of medians.

Using a combination of modeling and analysis of OIMS data, Brannon et al. (1993) found that an auroral model could not reproduce the atomic ion density profiles measured during the re-entry period, and that ion transport must therefore still be important in maintaining the nightside ionosphere at moderately low solar activity. Brannon and Fox (1994) constructed a map of the downward O^+ flux over the night side using OIMS data from the re-entry period. Although the variability in the data was large, and there were no data near the terminator, the O^+ maxima indicated that the O^+ flux had decreased by a factor of at least 7 from high to moderately-low solar activity over the solar zenith angle range sampled.

Averaged density profiles for total ions, O^+, H^+, and O_2^+ derived from ORPA measurements during the re-entry period were presented by Spenner et al. (1995b). They showed that the total density was smaller at moderately low solar activity by factors of 4, 2, and 6 in the sectors at 90 to 120°, 120 to 150°, and 150 to 180° SZA, respectively, at altitudes from about 150 to 450 km. Spenner et al. (1995b) concluded that the transterminator fluxes were reduced by about 77%, but that the high-altitude flow that converges in the central night side was reduced by a larger factor due to the lower dayside ionopause at moderately low solar activity. They also pointed to the elevated ion temperatures in the antisolar region as evidence that plasma transport contributes to the ion production there at moderately low solar activity.

When all the ORPA total ion density data for the re-entry period were combined, the smoothed average profile could be compared to that derived from OETP data, and the results are shown in Fig. 13. The average ORPA ion densities in the 170 to 200 km region are similar to the OETP smoothed medians. The ORPA values in the region of the peak are, however, smaller, and above 200 km they are slightly larger than the OETP n_e medians. The latter difference may be a selection effect, because disturbances such as holes and disappearing ionospheres were not included in the ORPA averages, whereas such data were not excluded in the calculation of the OETP medians.

We end this section with a cautionary note. Several attempts have been made to estimate the relative importance of ion precipitation and plasma trans-

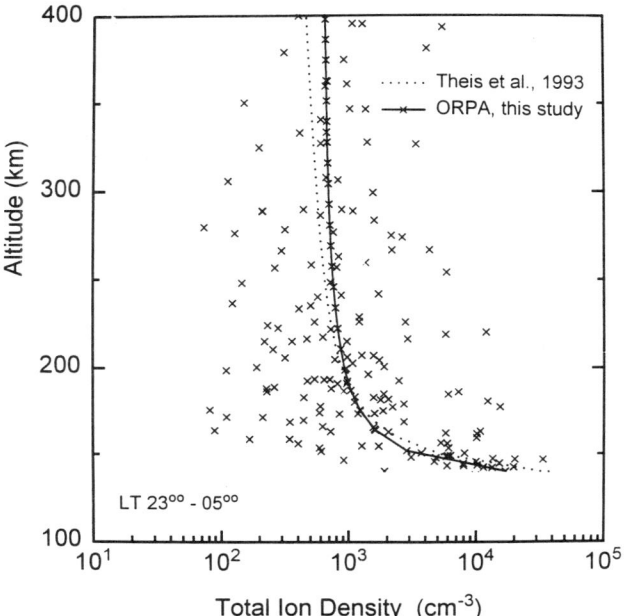

Figure 13. ORPA values of total ion density measured during the PV re-entry period. A smoothed average profile is also shown and is compared to the analytical function to which the OETP values of n_e were fitted by Theis and Brace (1993) (figure after Spenner et al. 1995a).

port in producing the nightside ionosphere by comparing the high and low solar-activity ion or electron densities at the peak, and those substantially above the peak. The relatively small change in the peak densities has been attributed to the presence of a constant source of ionization, electron precipitation, and the large changes at high altitudes to the highly variable day-to-night plasma transport. This is a dangerous line of reasoning for several reasons. First, the dayside ionosphere, for which there is a single source of ionization, exhibits just such a variability: the peak density shows a weak dependence on solar flux $\sim F_{10.7}^{0.35}$; the ion density at 200 or 300 km varies approximately as $\sim F_{10.7}^{2.5}$. The weak dependence of the peak region is due partly to the loss by dissociative recombination, which by itself would lead to a square-root dependence of the density on production rate, but a further weakening occurs due to compensating changes in the neutral atmosphere. The behavior of the nightside ionosphere, however, directly reflects the behavior of the dayside ionosphere. In general, the magnitude of an ionization source cannot be determined without knowledge of the neutral density profiles. Second, electron precipitation and plasma transport do not produce peaks at the same altitude, and therefore their effects are not strictly additive (cf., Brannon et al. 1993).

Third, for weak nightside ionospheres, chemical equilibrium may not be a good approximation, and diffusion may affect the altitude and shape of the peak (cf., Whitten et al. 1982; Fox et al. 1993). Consequently, the peak density may not reflect the magnitude of the source in the expected way, and the peak altitude may appear higher than the altitude of peak production. And finally, as we have pointed out previously, because the nightside peak is observed to vary in altitude substantially, averages or medians in the region of the peak do not necessarily correspond to average or median peak values, but will, in general, be smaller.

IV. SUMMARY

The chemical structure of the dayside ionosphere is well understood, and the variations with solar cycle are well reproduced by models. The electron densities near the peak vary approximately as (solar flux index)$^{0.35}$. The exponent is about the same for 27-day solar variations and for longer-term solar cycle variations. The depletion in n_e or n_i at high altitudes from high to low solar activity is much larger than the variability in the solar fluxes. This is due to a combination of reduced photon fluxes and reduced O densities at low solar activity, from which the dominant O^+ ions are formed.

Day-to-night ion transport is the most important source of the nightside ionosphere at high solar activity. Estimates of the importance of electron precipitation relative to ion transport are in the range 20 to 30%. Although reduced by factors that range from 4 to 7, day-to-night ion transport is still important at moderately low solar activity. Maps of the downward flux over the night side derived from OIMS O^+ data are consistent with contour plots of electron densities in the region of the peak derived from ORO data. The fluxes of suprathermal electrons in the umbra measured by the ORPA do not vary much with solar activity or solar zenith angle. Neither the solar cycle variations in the ion densities, nor the morphology of the ionosphere, as indicated by ORO electron density contours as a function of altitude and solar zenith angle, can be accounted for by the observed variations in the suprathermal electron fluxes.

Acknowledgments. This work was supported in part by NASA grants to the Research Foundation of the State University of New York at Stony Brook, and by NASA contracts at the Jet Propulsion Laboratory. J. L. F. thanks the Department of Physics at Wright State University for their hospitality during the writing of this chapter.

REFERENCES

Brace, L. H., and Kliore, A. J. 1991. The structure of the Venus ionosphere. *Space Sci. Rev.* 55:81–163.
Brace, L. H., et al. 1980. The dynamic behavior of the Venus ionosphere in response to solar wind interactions. *J. Geophys. Res.* 85:7663–7678.
Brace, L. H., Theis, R. F., and Hoegy, W. R. 1982. Plasma clouds above the ionopause of Venus and their implications. *Planet. Space Sci.* 30:29–37.
Brace, L. H., et al. 1983. The ionosphere of Venus: Observations and their interpretations. In *Venus*, eds. D. M. Hunten, L. Colin, T. M. Donahue and V. I. Moroz (Tucson: Univ. of Arizona Press), pp. 779–840.
Brace, L. H., et al. 1987. The ionotail of Venus: Its configuration and evidence for ion escape. *J. Geophys. Res.* 92:15–26.
Brace, L. H., Hoegy, W. R., and Theis, R. F. 1988. Solar EUV measurements at Venus based on photoelectron emission from the Pioneer Venus Langmuir probe. *J. Geophys. Res.* 93:7282–7296.
Brace, L. H., Theis, R. F., and Mihalov, J. D. 1990. Response of the nightside ionosphere and ionotail of Venus to variations in solar EUV and solar wind dynamic pressure. *J. Geophys. Res.* 95:4075–4084.
Brace, L. H., Hartle, R. E., and Theis, R. F. 1995. The nightward flow scenario at Venus revisited. *Adv. Space Res.* 16(6):99–112.
Brannon, J. F., and Fox, J. L. 1994. The downward flux of O^+ over the nightside of Venus. *Icarus* 112:396–404.
Brannon, J. F., Fox, J. L., and Porter, H. A. 1993. Evidence for day-to-night transport at low solar activity in the Venus pre-dawn ionosphere. *Geophys. Res. Lett.* 20:2739–2742.
Chapman, S. 1931. The absorption and dissociative or ionizing effects of monochromatic radiation in an atmosphere of a rotating earth. *Proc. Phys. Soc. London* 43:26–45.
Cravens, T. E., Kliore, A. J., Kozyra, J. U., and Nagy, A. F. 1981. The ionospheric peak on the Venus dayside. *J. Geophys. Res.* 86:11323–11329.
Cravens, T. E., et al. 1982. Disappearing ionospheres on the nightside of Venus. *Icarus* 51:271–282.
Cravens, T. E., Crawford, S. L., Nagy, A. F., and Gombosi, T. I. 1983. A two-dimensional model of the ionosphere of Venus. *J. Geophys. Res.* 88:5595–5606.
Dalgarno, A., and Fox, J. L. 1994. Ion chemistry in atmospheric and astrophysical plasmas. In *Unimolecular and Bimolecular Reaction Dynamics*, eds. C. Y. Ng, T. Baer and I. Powis (New York: Wiley), pp. 1–85.
Donahue, T. M., and Hartle, R. E. 1992. Solar cycle variations in H^+ and D^+ densities in the Venus ionosphere: Implications for escape. *Geophys. Res. Lett.* 19:2449–2452.
Elphic, R. C., Russell, C. T., Slavin, J. A., and Brace, L. H. 1980. Observations of the ionopause and ionosphere of Venus. *J. Geophys. Res.* 85:7679–7696.
Elphic, R. C., Brace, L. H., Theis, R. F., and Russell, C. T. 1984. Venus dayside ionospheric conditions: Effects of ionospheric magnetic fields and solar EUV flux. *Geophys. Res. Lett.* 11:124–127.
Fox, J. L. 1982. The chemistry of metastable species in the Venusian ionosphere. *Icarus* 51:248–260.
Fox, J. L. 1992. The chemistry of the nightside ionosphere of Venus. *Planet. Space Sci.* 40:1663–1681.
Fox, J. L., and Bougher, S. W. 1991. Structure, luminosity and dynamics of the Venus thermosphere. *Space Sci. Rev.* 55:357–489.

Fox, J. L., and Stewart, A. I. F. 1991. The Venus ultraviolet aurora: A soft electron source. *J. Geophys. Res.* 96:9821–9828

Fox, J. L., and Taylor, H. A., Jr. 1990. A signature of auroral precipitation in the nightside ionosphere of Venus. *Geophys. Res. Lett.* 17:1625–1628.

Fox, J. L., Brannon, J. F., Stewart, A. I. F., and Porter, H. S. 1992. Model calculations of the nightside ionosphere of Venus and Mars. *Bull. Amer. Astron. Soc.* 24:997.

Fox, J. L., Brannon, J. F., and Porter, H. S. 1993. Upper limits to the nightside ionosphere of Mars. *Geophys. Res. Lett.* 20:1339–1342.

Fox, J. L., Zhou, P., and Bougher, S. W. 1995. The thermosphere/ionosphere of Mars at high and low solar activities. *Adv. Space Res.* 17(11):203–218.

Gringauz, K. I., Verigin, M. I., Breus, T. K., and Gombosi, T. 1977. Electron currents measured in the optical shadow of Venus by the Venera-9 and Venera-10 satellites; the basic source of ionization in the nightside ionosphere of Venus. *Soviet Phys. Dokl.* 22:53–56.

Hedin, A. E., Niemann, H. B., Kasprzak, W. T., and Seiff, A. 1983. Global empirical model of the Venus thermosphere. *J. Geophys. Res.* 88:73–83.

Keating, G. M., et al. 1985. Models of the Venus neutral upper atmosphere: Structure and composition. *Adv. Space Res.* 5(11):117–171.

Keating, G. M., and Hsu, N. C. 1993. The Venus atmospheric response to solar cycle variations. *Geophys. Res. Lett.* 20:2751–2754.

Kim, J., Nagy, A. F., Cravens, T. E., and Kliore, A. J. 1989. Solar cycle variations of the electron densities near the ionospheric peak of Venus. *J. Geophys. Res.* 94:11997–12002.

Kliore, A. J., and Mullen, L. F. 1989. Long term behavior of the main peak of the dayside ionosphere of Venus during solar cycle 21 and its implications on the effect of the solar cycle upon the electron temperature in the main peak region. *J. Geophys. Res.* 94:13339–13351.

Kliore, A. J., and Luhmann, J. G. 1991. Solar cycle effects on the structure of the electron density profiles in the dayside ionosphere of Venus. *J. Geophys. Res.* 96:21281–21289.

Kliore, A. J., et al. Atmosphere and ionosphere of Venus from the Mariner 5 S-band radio occultation measurements. *Science* 205:99–102.

Kliore, A. J., Patel, I. R., Nagy, A. F., Cravens, T. E., and Gombosi, T. I. 1979. Initial nightside observations of the nightside ionosphere of Venus from Pioneer Venus orbiter radio occultation. *Science* 203:765–768.

Kliore, A. J., Luhmann, J. G., and Zhang, M. H. G. 1991. The effect of solar cycle on the maintenance of the nightside ionosphere of Venus. *J. Geophys. Res.* 96:11065–11072.

Knudsen, W. C. 1987. Frequency functions of Venus nightside ion densities. *J. Geophys. Res.* 92:7308–7316.

Knudsen, W. C. 1988. Solar cycle changes in the morphology of the Venus ionosphere. *J. Geophys. Res.* 93:8756–8762.

Knudsen, W. C. 1992. The Venus ionosphere from in situ measurements. In *Venus and Mars: Atmospheres, Ionospheres, and Solar Wind Interactions*, eds. J. G. Luhmann, M. Tatrallyay and R. Pepin (Washington, D. C.: American Geophys. Union), pp. 237–263.

Knudsen, W. C., and Miller, K. L. 1985. Pioneer Venus suprathermal electron flux measurements in the Venus umbra. *J. Geophys. Res.* 90:2697–2702.

Knudsen, W. C., and Miller, K. L. 1992. The Venus transterminator ion flux at solar maximum. *J. Geophys. Res.* 97:17165–17167.

Knudsen, W. C., Spenner, K., Miller, K. L., and Novak, V. 1980. Transport of ionospheric O^+ ions across the Venus terminator and implications. *J. Geophys. Res.* 85:7803–7810.

Knudsen, W. C., Miller, K. L., and Spenner, K. 1982a. Improved Venus ionopause altitude calculation and comparison with measurement. *J. Geophys. Res.* 87:2246–2254.
Knudsen, W. C., Banks, P. J., and Miller, K. L. 1982b. A model of plasma motions and planetary magnetic fields for Venus. *Geophys. Res. Lett.* 9:765–768.
Luhmann, J. G. 1986. The solar wind interaction with Venus. *Space Sci. Rev.* 44 241–306.
McComas, D. J., Spence, H. E., Russell, C. T., and Saunders, M. A. 1986. The average magnetic field draping and consistent plasma properties of the Venus magnetotail. *J. Geophys. Res.* 91:7939–7953.
Miller, K L., and Knudsen, W. C. 1987. Spatial and temporal variations of the ion velocity measured in the Venus ionosphere. *Adv. Space Res.* 7(12):107–110.
Miller, K. L., and Whitten, R. C. 1991. Ion dynamics in the Venus ionosphere. *Space Sci. Rev.* 55:165–199.
Miller, K L., Knudsen, W. C., and Spenner, K. 1984. The dayside Venus ionosphere. *Icarus* 57:386–409.
Nagy, A. F., et al. 1980. Model calculations of the dayside ionosphere of Venus: Ionic composition. *J. Geophys. Res.* 85:7795–7801.
Nagy, A. F., Cravens, T. E., and Gombosi, T. I. 1983. Basic theory and model calculations of the Venus ionosphere. In *Venus*, eds. D. M. Hunten, L. Colin, T. M. Donahue and V. I. Moroz (Tucson: Univ. of Arizona Press), pp. 841–872.
Nagy, A. F., Körösmezey, A., Kim, J., and Gombosi, T. I. 1991. A two-dimensional shock capturing, hydrodynamic model of the Venus ionosphere. *Geophys. Res. Lett.* 18:801–804.
Niemann, H. B., et al. 1980. Mass spectrometric measurements of the neutral gas composition of the thermosphere and exosphere of Venus. *J. Geophys. Res.* 85:7817–7828.
Shinagawa, H., Kim, J., Nagy, A. F., and Cravens, T. E. 1991. A comprehensive magnetohydrodynamic model of the Venus ionosphere. *J. Geophys. Res.* 96:11083–11095.
Spenner, K., et al. 1981. On the maintenance of the Venus nightside ionosphere: Electron precipitation and plasma transport. *J. Geophys. Res.* 86:9170–9178.
Spenner, K., Knudsen, W. C., and Lotze, W. 1995a. Suprathermal electron fluxes in the Venus nightside ionosphere at moderate and high solar activity. *J. Geophys. Res.* 101:4557–4564.
Spenner. K., Knudsen, W. C., and Lotze, W. 1995b. Ion density, temperature and composition of the Venus nightside ionosphere during a period of moderate solar activity: Implications for maintaining the central nightside. *J. Geophys. Res.* 100:14499–14506.
Stewart, A. I. F., Gerard, J. C., Rusch, D. W., and Bougher, S. W. 1980. Morphology of the Venus night airglow. *J. Geophys. Res.* 85:7861–7870.
Taylor, H. A., et al. 1980. Global observations of the composition and dynamics of the ionosphere of Venus: Implications for the solar wind interaction. *J. Geophys. Res.* 85:7765–7777.
Taylor, H. A., et al. 1981. Temporal and spatial variations observed in the ionospheric composition of Venus: implications for empirical modelling. *Adv. Space Res.* 1:37–51.
Theis, R. G., and Brace, L. H. 1993. Solar cycle variations of electron density and temperature in the Venusian nightside ionosphere. *Geophys. Res. Lett.* 20:2719–2722.
Theis, R. F., Brace, L. H., Elphic, R. C., and Mayr, H. G. 1984. New empirical models of the electron temperature and density in the Venus ionosphere with application to transterminator flow. *J. Geophys. Res.* 89:1477–1488.

Torr, D. G., Torr, M. R., Ong, R. A., and Hinteregger, H. A. 1979. Ionization frequencies for major thermospheric constituents as a function of solar cycle 21. *Geophys. Res. Lett.* 6:771–774.

Whitten, R. C., et al. 1982. The Venus ionosphere at grazing incidence of solar radiation: Transport of plasma to the night ionosphere. *Icarus* 51:261–270.

Whitten, R. C., et al. 1984. Dynamics of the Venus ionosphere: A two-dimensional model study. *Icarus* 60:317–326.

Zhang, M. H. G., Luhmann, J. G., and Kliore, A. J. 1990. An observational study of the nightside ionospheres of Mars and Venus with radio occultation. *J. Geophys. Res.* 95:17095–17102.

IONOSPHERE: ENERGETICS

ANDREW F. NAGY
University of Michigan

and

THOMAS E. CRAVENS
University of Kansas

A review of the theoretical understanding of the processes controlling the energy balance in the ionosphere is presented. The appropriate transport equations, along with expressions for the thermal conductivities and the collision terms are given. Photoelectron transport and heating rate calculations are also discussed. The large observational data base of electrons and ions is summarized and compared with results of model calculations. The theoretical model calculations of electron and ion temperatures have shown that using conventional EUV heating sources and classical thermal conductivities lead to temperature values significantly lower than the observed ones. Reasonably good agreement has been achieved between observed and calculated temperatures, when somewhat *ad hoc* topside heat inflows and/or reduced thermal conductivities are assumed. At this time it is not possible to separate the importance of these two assumed effects clearly or even establish unambiguously that they are truly responsible for the high observed temperatures. When the ionosphere is magnetized solar wind electrons are likely to have access to the upper ionosphere, and for that case, theoretical calculations agree with the data, without the need for further assumptions.

I. INTRODUCTION

The first measurements of an ionosphere, other than the terrestrial one, were the radio occultation observations made by the Mariner 5 spacecraft as it flew by Venus on 19 October 1967. A number of attempts were made to deduce upper atmospheric and ionospheric temperatures from the electron density profiles thus obtained. These efforts did not lead to the correct temperatures, because the wrong assumption was made about the major ion; at that time the major ion was thought to be CO_2^+, but we now know that it is O_2^+. The correct value of the thermospheric neutral temperature was not established until 1974, when Kumar and Hunten (1974), using a combination of airglow and electron density data, and the correct identification of the major ion species, O_2^+, came up with a value of about 300 to 350 K. The retarding potential analyzers, carried by the Viking 1 and 2 landers in 1976 (Hanson et al. 1977) provided information on the daytime ion temperatures at Mars, whose ionosphere is similar, in many respects, to the ionosphere of Venus.

It was soon demonstrated that the observed temperatures were higher than would be predicted using a direct and simple analogy from the terrestrial case (Chen et al. 1978; Johnson 1978; Rohrbaugh et al. 1979). This led to predictions that similar high ion and electron temperatures would be found by the Pioneer Venus Orbiter (PVO), when it began making measurements in the ionosphere of Venus, in December 1978 (Chen and Nagy 1978; Cravens et al. 1978). All our direct information of the ionospheric energetics at Venus comes from measurements made by instruments carried aboard PVO.

It is impossible to review and summarize ionospheric energetics, without considering the effects of various solar wind interaction and thermospheric processes. In order to avoid repetition we will not discuss in any detail these processes; the reader is referred to other chapters in this book, where they are reviewed in some detail. There have been numerous reviews published during the last twenty years, which touched on the topic of ionospheric energetics at Venus; however this topic has never had a specific and detailed review. Aspects, relevant to this topic, can be found in comprehensive reviews of the terrestrial F-region electron temperatures (Schunk and Nagy 1978), and the ionospheres of the terrestrial planets (Schunk and Nagy 1980). A number of related, Venus-specific reviews have also been published (Nagy et al. 1983a; Brace et al. 1983; Cravens 1991; Luhmann and Cravens 1991; Brace and Kliore 1991; Miller and Whitten 1991). In this chapter, we attempt to pull together and describe the basic processes controlling the electron and ion temperatures. We summarize the results from the *in-situ* PVO measurements and other related information, and discuss the various theoretical models which have been used to understand and interpret the results. Most of the discussion will concentrate on the average behavior of the dayside ionosphere, but some attention will also be given to the more structured, nightside conditions. Finally we indicate the status of our present understanding of the ionospheric energetics of Venus and also where the holes are in this understanding and what further information is needed to resolve the remaining questions.

II. THEORY

A. Overview

Solar extreme ultraviolet (EUV) radiation, shortward of about 100 nm, is the main source of ionization in the Venus ionosphere. This photoionization process produces photoelectrons, which lose their energy in elastic collisions with other electrons and ions and inelastic collisions with neutral atmospheric gas particles. Thus, these electrons undergo numerous collisions during their "lifetime" as they gradually keep losing energy until they eventually become low-energy, "thermal" electrons. In order to calculate the energy distribution of the electrons one needs to be able to calculate all of the production and loss processes and solve the appropriate Boltzmann equation. This has been such a formidable task that, except for one or two highly simplified cases (Ashihara and Takayanagi 1973; Krinberg 1973), all calculations have broken the energy distribution into two components: photoelectron and thermal electron components. Electron–electron collisions are inversely proportional to energy, thus at very low energies, below 1 to 2 eV, there are sufficient collisions among the electrons for them to be thermalized, and to be describable by a Maxwellian distribution, characterized by a single temperature. The bulk of the electron population is at these low, thermal energies and when we discuss ionospheric electron temperatures, we refer to this population. The electrons above the thermal energies are highly non-Maxwellian and are treated separately. Below about 180 to 200 km photoelectrons loose their energy locally, but at higher altitudes they can travel significant distances before becoming thermalized.

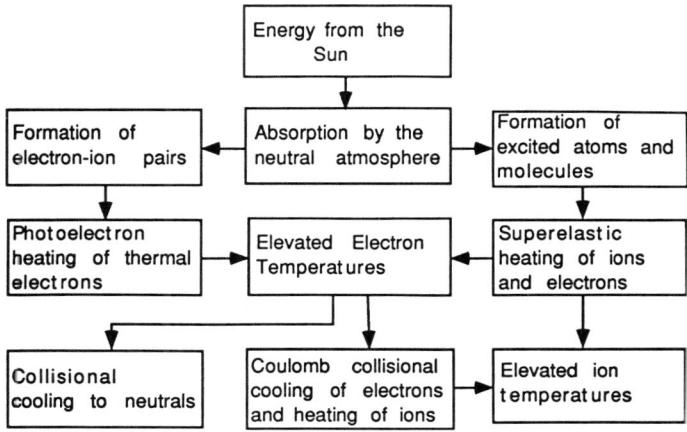

Figure 1.   Block diagram of the main energy flow processes in the ionosphere.

The ions are, in general, created with kinetic/thermal energies close to that of the parent neutral gas. These ions gain energy from the electrons, through Coulomb collisions and lose energy to the neutral gas via elastic and inelastic collisions. Superelastic collisions, through which an electron or ion can gain energy are not frequent, but are possible, e.g., a quenching collision with an excited atom or molecule. The overall energy flow is shown schematically in Fig. 1. It should be noted that only the most important processes are indicated in Fig. 1; at given times and altitude regions other processes, such as Joule, precipitating particle or chemical heating, can also play important roles. Furthermore, heat inflow from other regions or from outside the ionosphere may also play a crucial role in determining the temperature structure.

B. Transport Equations

The system of transport equations which has been used for ionospheric calculations of the thermal plasma behavior is based on the first few moments of the Boltzmann equation. These equations provide a valid description of the behavior of the bulk of the electron and ion population, in altitude regions, where there are sufficient number of collisions to lead to a near Maxwellian energy distribution. The most general system of equations are the so-called 20, 16 or 13 moment equations (cf., Schunk 1975,1977; Gombosi and Rasmussen 1991), which are based on the work of Grad (1949) and Burgers (1969). These formulations have the advantage of placing stress and heat flow on the same footing as density, flow velocity and temperature. These system of equations have been used in the last 10 to 20 years in a variety of models of the terrestrial ionosphere. However, in planetary ionospheric studies the equations that have been used in model calculations are a simplified form of the set given below (Schunk and Nagy 1980):

$$\frac{\partial n_s}{\partial t} + \nabla \cdot \{n_s \mathbf{u}_s\} = P_s - L_s n_s \tag{1}$$

$$n_s m_s \frac{D_s \mathbf{u}_s}{Dt} + \nabla p_s + \nabla \cdot \tau_s - n_s m_s \mathbf{G} - n_s e_s \{\mathbf{E} + \mathbf{u}_s \times \mathbf{B}\} = \frac{\delta \mathbf{M}_s}{\delta t} \tag{2}$$

$$\frac{D_s}{Dt}\left(\frac{3}{2} p_s\right) + \frac{5}{2} p_s (\nabla \cdot \mathbf{u}_s) + \nabla \cdot \mathbf{q}_s + \tau_s : \nabla \mathbf{u}_s = \frac{\delta E_s}{\delta t} + Q_s - \mathcal{L}_s \tag{3}$$

where $D_s/Dt = \partial/\partial t + \mathbf{u}_s \cdot \nabla$ is the convective derivative of species s, $p_s = n_s kT_s$ is the scalar partial pressure, n_s is the number density, m_s is the mass, e_s is the electric charge, T_s is the temperature, $\mathbf{u}_s$ is the drift velocity, $\mathbf{q}_s$ is the heat flow vector, τ_s is the stress tensor, P_s is the production/ionization rate, L_s is the loss/recombination rate, Q_s is the heating rate, $\mathcal{L}_s$ is the cooling rate, $\mathbf{G}$ is the acceleration due to gravity, $\mathbf{E}$ is the electric field, $\mathbf{B}$ is the magnetic field, $\partial/\partial t$ is the time derivative, ∇ is the spatial gradient and k is Boltzmann's constant. The double dot operator in Eq. (2) denotes the scalar product of two tensors. The quantities $\delta \mathbf{M}_s/\delta t$ and $\delta E_s/\delta t$ represent the rate of momentum and energy exchange due to collisions, respectively, between species s and the other species present.

Magnetohydrodynamic formulations and their application in the studies of the ionosphere of Venus are discussed in some detail in the chapter by Cravens et al.

C. Heat Flow and Thermal Conductivities

The continuity, momentum and energy equations, given above, are not a closed set; in planetary ionospheric studies it has been customary to neglect the stress tensor and use the following simple expression for the heat flow vector:

$$\mathbf{q}_s = -\kappa_s \nabla T_s \tag{4}$$

where κ_s is the thermal conductivity of species s. Equations (1) through (4) (with the stress tensor neglected) represent the eight moment set of transport equations. Banks (1966), using simple mean free path considerations, obtained a relatively simple expression for the electron thermal conductivity, which includes the effects of collisions with the background neutral gas particles. The expression given by Banks (1966), which assumed that no magnetic field and electric currents are present, had a numerical error in the original paper, that has propagated to subsequent publications; the correct expression is:

$$\kappa_s = \frac{7.7 \times 10^5 T_e^{5/2}}{\left[1 + 2.16 \times 10^4 \{\frac{T_e^2}{n_e}\} \sum_n n_n \langle q_D \rangle_n \right]} \text{ eV s}^{-1}\text{cm}^{-1}\text{ K}^{-1} \quad (5)$$

where n_e and n_n are the electron and neutral gas densities, respectively, $\langle q_D \rangle_n$ is the Maxwellian-averaged electron-neutral collision cross section and the summation is over the different neutral gas species. It is interesting to note that Schunk and Walker (1970) compared the thermal conductivity obtained by Banks (1966), with the more accurate expression, which used the formulation developed by Shkarovsky (1961) and found that the differences were less than 20%. However, this comparison used the expression given by Banks (1966), and the agreement becomes worse when the corrected Eq. (5) is used. Equation (5) can also be expressed in terms of the momentum transfer collision frequencies by recognizing the following relationship between the Maxwellian averaged momentum transfer cross section and collision frequency:

$$\nu_{es} = \left[\frac{8}{3\pi^{1/2}}\right] n_s v_T \langle q_D \rangle \quad (6)$$

where v_T is the most probable electron thermal speed, $(2kT_e/m_e)^{1/2}$. Equation (5) does reduce to the expression obtained by Spitzer and Harm (1953) for the electron thermal conductivity in a fully ionized plasma:

$$\kappa_e = 7.7 \times 10^5 T_e^{5/2} \text{ eV s}^{-1}\text{cm}^{-1}\text{ K}^{-1}. \quad (7)$$

At low altitudes electron-neutral collisions dominate and the conductivity approaches:

$$\kappa_e \approx 35.6 \frac{n_e T_e^{1/2}}{\sum_n n_n \langle q_D \rangle_n} \text{ eV s}^{-1}\text{ cm}^{-1}\text{ K}^{-1}. \quad (8)$$

Schunk (1975,1977) showed how to obtain an approximate expression for thermal conductivity, using the 13 moment set of equations and Burgers' "linear" collision terms (Burgers 1969), which are accurate enough for most ionospheric applications, after some simple corrections are applied. The general form of this expression for the thermal conductivity is:

$$\kappa_s = \frac{5k p_s}{2m_s} \left\{ \frac{4}{5}\nu_{ss} + \sum_{t \neq s} \nu_{st} \left[D_{st}^{(1)} + \frac{5 z_{st} \mu_{st}}{2m_s} \frac{T_s}{T_{st}} \right] \right\}^{-1} \quad (9)$$

where μ_{st} is the reduced mass, T_{st} is the reduced temperature, $(m_s T_t + m_t T_s)/(m_s + m_t)$, and z_{st} is a numerical constant which depends on the interaction potential (Schunk 1977) and all other symbols are as defined before. For Coulomb collisions z_{st} is 3/5 and $D_{st}^{(1)}$ is:

$$D_{st}^{(1)} = \frac{3m_s^2 + \frac{1}{10}m_s m_t - \frac{1}{5}m_t^2}{(m_s + m_t)^2}. \tag{10}$$

In the case of a single ion gas and no neutral gas background Eq. (9) becomes:

$$\kappa_i = \frac{75}{32\pi^{1/2}} \frac{k(kT_i)^{5/2}}{m_i^{1/2} e^4 \ln \Lambda} \tag{11}$$

where $\ln \Lambda$ is the Coulomb logarithm, k is Boltzmann's constant and m_i is the ion mass. Raitt et al. (1975) multiplied Eq. (11) with appropriate correction factors (about 1.3 for the major and 2.8 for the minor ion) to account for higher order approximations, when they gave numerical values for the ion thermal conductivities, which they used in their polar wind studies. Using their correction factors and the method they used to account for collisions with the neutral background gas, we obtain the following expressions for the major ion O^+, and the minor ion H^+, thermal conductivities, appropriate for Venus ionospheric applications:

$$\kappa(O^+) = \frac{7.75 \times 10^3 (n(O^+)/n_e) T(O^+)^{5/2}}{(1. + 1.33a + 1.46b + 1.54c + 2.88d)} \text{ eV s}^{-1} \text{ cm}^{-1} \text{ K}^{-1} \tag{12a}$$

where:

$$a = \frac{\nu(O^+ - CO_2)}{\nu(O^+ - O^+)}$$

$$b = \frac{\nu(O^+ - CO)}{\nu(O^+ - O^+)}$$

$$c = \frac{\nu(O^+ - O)}{\nu(O^+ - O^+)}$$

$$d = \frac{\nu(O^+ - H)}{\nu(O^+ - O^+)}$$

$$\kappa(H^+) = \frac{3.1 \times 10^4 (n(H^+)/n_e) T(H^+)^{5/2}}{(1. + 0.77e + 0.77f + 1.1g + 0.94h)} \text{ eV s}^{-1} \text{ cm}^{-1} \text{ K}^{-1} \tag{12b}$$

where:

$$e = \frac{\nu(H^+ - CO_2)}{\nu(H^+ - O^+)}$$

$$f = \frac{\nu(H^+ - CO)}{\nu(H^+ - O^+)}$$

$$g = \frac{\nu(H^+ - O)}{\nu(H^+ - O^+)}$$

$$h = \frac{\nu(H^+ - H)}{\nu(H^+ - O^+)} \tag{13}$$

$n(O^+)$, $n(H^+)$ and n_e are O^+, H^+ and the total electron number densities, respectively, and the ν's are the momentum transfer collision frequencies ($\nu(O^+ - O^+)$ is the self collision frequency). Expressions (12) and (13) are consistent with the expression given by Banks and Kockarts (1973) for the multi-ion thermal conductivity, neglecting neutral collisions, except for the numerical factor. Using the numerical values given by Raitt et al. (1975) the multi-ion expression, neglecting collisions with neutrals, becomes:

$$\kappa_i = \frac{3.1 \times 10^4}{n_e} \sum \frac{n_i T_i^{5/2}}{M_i^{1/2}} \text{ eV s}^{-1} \text{ cm}^{-1} \text{ K}^{-1} \tag{14}$$

where M_i is the weight of each ion in atomic mass units and the summation is over all ion species.

The above discussion of thermal conductivities assumed that magnetic fields are not present. The presence of a static magnetic field leads to an anisotropic thermal conductivity. The component of the conductivity parallel to the field lines is the same as that for the field free case, while the perpendicular component is reduced by the factor $(1 + w^2\tau^2)$, where w^2 is the cyclotron frequency and τ is the mean free time between collisions.

It has been found that, at times, spatially and temporally fluctuating magnetic fields are present in the Venus ionosphere (cf., Luhmann and Cravens 1991). Cravens et al. (1980) approximated the effect of such fields on the thermal conductivities by postulating a completely random (fluctuating) magnetic field, with a correlation length l. Given a charged particle mean free path λ, the thermal conductivity can be approximated as:

$$\kappa_m = \frac{3}{4} A_m n_m k \bar{v}_m \lambda \tag{15}$$

where n_m is the particle density, $\bar{v}_m$ is the mean thermal velocity and A_m is a correction factor of order unity. A typical gyroradius for an electron in the Venus ionosphere is about 200 m, in general shorter than the correlation length and, consequently, an electron will move along a field line. It was assumed that the field line performs a random walk with correlation length l; therefore, the effective mean free path for an electron is l. On the other hand, typical ion gyroradii are 20 km or greater, which is longer than the typical correlation length of a few km, therefore the mean free path is longer than the correlation length and the ions do not "feel" the presence of a fluctuating magnetic field.

In a more recent study Hoegy et al. (1990) noted that when heat sources and sinks can be neglected the heat flow is constant and the effect of the magnetic field fluctuations can be accounted for by averaging this term over the measured fluctuations of the dip angles. They noted that the heat flux in a constant magnetic field with a dip angle of I is:

$$q_z = -(K_1 \cos^2(I) + K_o \sin^2(I)) \frac{\partial T_e}{\partial z} \tag{16}$$

where K_o is the thermal conductivity parallel to the magnetic field and K_1 is the Pederson (or perpendicular) thermal conductivity. For a typical magnetic field of 10γ, $F = K_1/K_o$ is about 1.4×10^{-5}; even though this ratio is very small, it does lead to a significant reduction in the heat flow. Hoegy et al. (1990) defined an effective dip angle by the following relation:

$$\frac{1}{\left[\sin^2(I_{\text{eff}}) + F \cos^2(I_{\text{eff}})\right]} = \left\langle \frac{1}{\left[\sin^2(I) + F \cos^2(I)\right]} \right\rangle. \tag{17}$$

They carried out the averaging by using the measured spectrum of the magnetic field fluctuations.

D. Collision Terms

General expressions for the appropriate collision terms appearing on the right-hand side of Eqs. (2) and (3) do not exist. However, the expressions given below are appropriate for arbitrary interaction potentials and low relative drift speeds (Schunk 1977):

$$\frac{\delta \mathbf{M}_s}{\delta t} = \sum_t n_s m_s \nu_{st} \{\mathbf{u}_t - \mathbf{u}_s\} \tag{18}$$

$$\frac{\delta E_s}{\delta t} = \sum_t \frac{n_s m_s \nu_{st}}{m_s + m_t} (3k [T_t - T_s] + m_t [\mathbf{u}_s - \mathbf{u}_t]^2) \tag{19}$$

where ν_{st} is the momentum transfer collision frequency between species s and t (see Eq. 6). Expressions for the electron–ion, electron–neutral, ion–ion and ion–neutral collision frequencies have been presented by Schunk and Nagy (1980) and will not be repeated here. It should be remembered that the momentum transfer collision frequency is not symmetric with respect to change of indices, but satisfies the following relation:

$$n_s m_s \nu_{st} = n_t m_t \nu_{ts}. \tag{20}$$

E. Photoionization and Photoelectron Production Rates

Solar extreme ultraviolet radiation photoionizes the neutral constituents of the upper atmosphere, producing free electrons and ions. The peak in the daytime ionization rate occurs around 140 km and is due mainly to the absorption of radiation with wavelength less than 91 nm, the ionization threshold of atomic oxygen. The major initial ions produced are CO_2^+, as shown in Fig. 2. The energy of the ionizing photons exceeds, in general, the energy needed for ionization and the excess energy goes mainly to electron and ion kinetic energies and to ion excitation energies. The large mass difference between the electrons and ions means that, in effect, all the available kinetic energy goes to the electrons. The initial photoelectron energy depends not only on the energy of the ionizing photon and the ionization potential of the neutral gas, but also on the excitation state of the newly created photoion. The expression for the photoelectron production rate, $P_e(E, z)$ is:

$$P_e(E, z) = \sum_k \sum_l n_k(z) \int_0^{\lambda_k} d\lambda I_\infty(\lambda) \exp(-\tau(\lambda, z)) \sigma_k^i p_k(\lambda, E_l) \quad (21)$$

where the optical depth τ is:

$$\tau(\lambda, z) = \sum_k \sigma_k^a(\lambda) \int_z^\infty n_k(z') dz' \quad (22)$$

and n_k is the number density of the kth neutral constituent, $\sigma_k^i(\lambda)$ and $\sigma_k^a(\lambda)$ are the wavelength dependent total ionization and absorption cross sections, respectively, $p_k(\lambda, E_l)$ is the branching ratio for the excited ion state with an ionization energy of E_l, $E = E_\lambda - E_l$, E_λ is the energy corresponding to wavelength λ, λ_k is the ionization threshold wavelength for neutral species k and finally $I_\infty(\lambda)$ is the incident solar radiation flux at wavelength λ. Relevant solar flux information and wavelength averaged absorption and ionization cross sections, for a wide variety of atmospheric species, have been presented by Torr and Torr (1985), Solomon and Abreu (1989) and Gan (1991).

F. Photoelectron Transport and Thermal Electron Heating Rate Calculations

After its creation a photoelectron undergoes numerous inelastic and elastic collisions in the process of losing its energy. At higher energies, above about 40 eV, ionization and optically allowed excitation of the neutral constituent is the dominant process; as the energy decreases excitation of metastable states becomes increasingly more important. As the photoelectron energy drops below about 5 eV electron–electron collisions become very important, although vibrational and rotational excitation of the neutral molecules can also play a role. In order to calculate the energy received by the thermal electrons from the higher energy photoelectrons, one needs to calculate (or measure) the equilibrium photoelectron flux $\Phi(E, z)$. Given this flux, the thermal electron heating rate $Q_e(z)$, can be calculated from the following relation:

$$Q_e(z) = \int_{E_t}^\infty \Phi(E, z) \left[\frac{dE}{dx}\right]_e dE \quad (23)$$

where $(dE/dx)_e$ is the rate at which an electron of energy E loses energy to the ambient electrons in traveling a unit distance. The issue of the lower limit

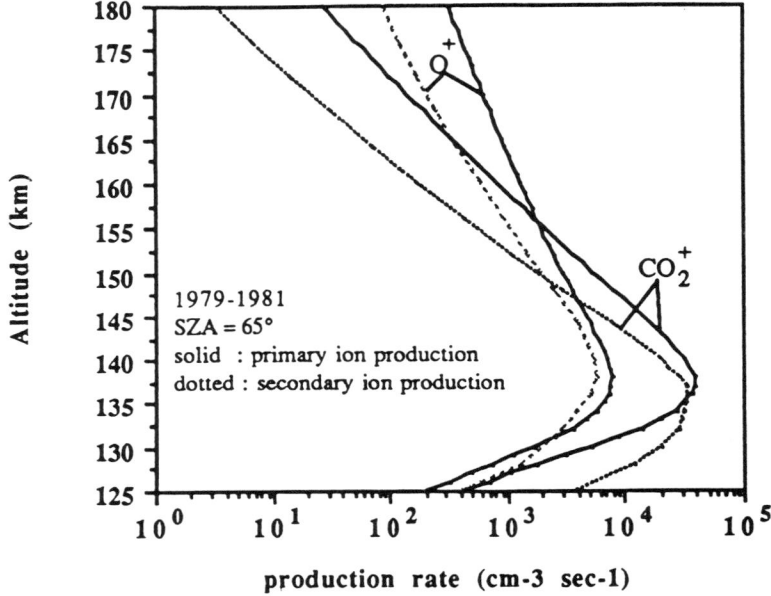

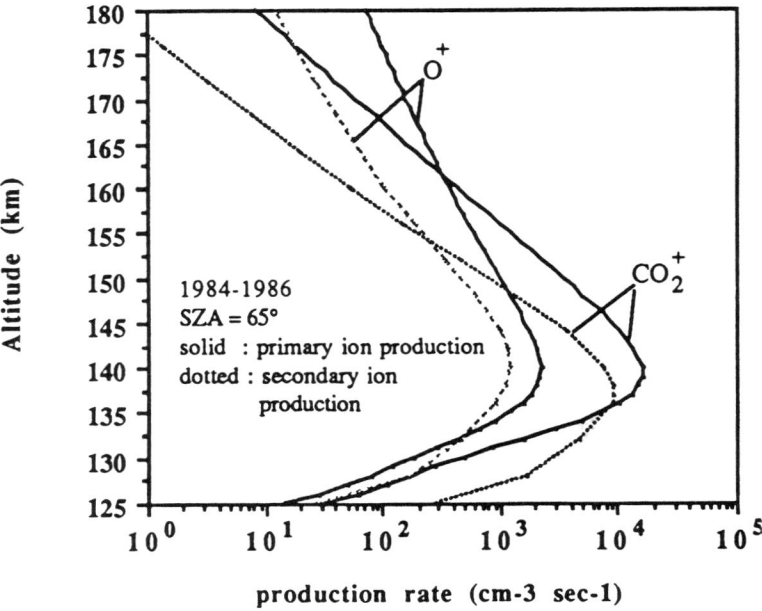

Figure 2. Calculated photo and secondary electron impact ionization rates for solar cycle maximum (1979–1989) and minimum (1984–1986) conditions (figure from Kim 1991).

on the above integral is a very important one and is the result of separating the electrons into a thermal and nonthermal (photo) electron component. Different authors have used different criteria over the years in selecting this energy value. Hoegy (1984) showed that the term given by Eq. (23) is the dominant of the three terms that contribute to the thermal electron heating rate. Given the many uncertainties associated with calculating the heating rates, no significant effort has gone into improving these calculations during the last decade or two.

The first formulation of the photoelectron- thermal electron energy transfer/loss rate was given by Butler and Buckingham (1962). About a decade later a number of authors pointed out that the expression obtained by Butler and Buckingham had a number of shortcomings (e.g., their use of an approximate collision operator) and they presented corrected expressions (Takayanagi and Itikawa 1970; Schunk and Hays 1971). Swartz et al. (1971) found the following, relatively simple, analytic approximation to the complex expression for this loss function:

$$L(E, E_e) = \frac{3.37 \times 10^{-12}}{E^{0.94} N_e^{0.03}} (\frac{E - E_e}{E - 0.53 E_e})^{2.36} \text{ eV cm}^{-3} \text{ s}^{-1} \quad (24)$$

where $E_e = 8.618 \times 10^{-5} T_e$, T_e being the temperature of the thermal electrons in degrees K.

At lower altitudes, below about 200 km, photoelectron transport can be neglected; that is, the electrons do not move a significant distance from where they are created. In this case, the problem becomes a "local" one and the production rate can be simply equated to the loss rate. Given this local assumption and considering that the energy losses take place in discrete steps the expression for the equilibrium photoelectron flux is:

$$\Phi(E, z) = \frac{(P_e(E, z) + \sum_k \sum_j n_k \sigma_{kj}(E + \delta E_{kj}) \Phi(E + \delta E_{kj}, z))}{(\sum_k \sum_j n_k \sigma_{kj}(E))} \quad (25)$$

where σ_{kj} the cross section for the inelastic collision j between an electron and neutral species k, which results in an energy loss of δE_{kj}. This expression omits the energy loss to the thermal electrons; however, it could be incorporated by using an effective cross section derived from Eq. (24).

The more general form of the transport equation for suprathermal electrons (including photoelectrons and possibly solar wind electrons) can be obtained from the Boltzmann equation. This equation, using the guiding center approximation is (Khazanov et al. 1992,1994):

$$\sqrt{\frac{m}{2E}} \frac{\partial \Phi}{\partial t} + \mu \frac{\partial \Phi}{\partial s} - \frac{1-\mu^2}{2} \left(-\frac{F}{E} + \frac{1}{B} \frac{\partial B}{\partial s} \right) \frac{\partial \Phi}{\partial \mu} + EF\mu \frac{\partial}{\partial E} \left[\frac{\Phi}{E} \right] = \sqrt{\frac{m}{2E}} \frac{\delta \Phi}{\delta t} \quad (26)$$

where t, s, μ, B and E are time, distance along a given field line, the cosine of the pitch angle, the magnetic field strength and energy, respectively; Φ is the flux of electrons defined as $\Phi = 2Ef/m^2$, where f is the distribution function, the force due to the longitudinal electric field is $F = -eE_\parallel$, and finally $\frac{\delta \Phi}{\delta t}$ is the net change in the flux due to the various collisional processes. Some analytic solutions of this equation are possible for highly simplified cases (Khazanov et al. 1992), but in general numerical solutions are necessary (see, e.g., Khazanov and Liemohn 1995). The significantly simplified, steady state, two-stream transport equation, introduced by Nagy and Banks (1970), have been used widely and practically uniquely for calculations involving the energetics of the Venus ionosphere. These two stream equations are:

$$\langle \mu \rangle \frac{\partial \Phi^+}{\partial s} = -\sum_k n_k \sigma_k^t \Phi^+ + \sum_k \frac{n_k \sigma_k^e}{2}(\Phi^+ + \Phi^-) + \frac{Q_e}{2} + Q^+ \quad (27)$$

$$-\langle \mu \rangle \frac{\partial \Phi^-}{\partial s} = -\sum_k n_k \sigma_k^t \Phi^- + \sum_k \frac{n_k \sigma_k^e}{2}(\Phi^+ + \Phi^-) + \frac{Q_e}{2} + Q^- \quad (28)$$

where Φ^+ and Φ^- are the up and down going hemispherical fluxes, defined as:

$$\Phi^+(E, s) \equiv \int_0^1 \int_0^{2\pi} \phi(E, s, \mu, \varphi) d\varphi d\mu \quad (29)$$

$$\Phi^-(E, s) \equiv \int_{-1}^0 \int_0^{2\pi} \phi(E, s, \mu, \varphi) d\varphi d\mu \quad (30)$$

and ϕ is the flux per solid angle, φ is the azimuthal angle, μ is the cosine of the electron pitch angle, σ_k^t and σ_k^e are the total and elastic scattering cross sections for species k, respectively, and n_k is the density of the scattering background species. A number of simplifying assumptions (e.g., flux and production rate are independent of φ, scattering probability is a constant) were used in order to obtain Eqs. (27) and (28). Q_e is the direct (photo) ionization rate, assumed to be isotropic, and Q^+ and Q^- are production rates due to cascading from higher energies. Typical daytime photoelectron fluxes for Venus, calculated by using this two-stream method, are shown in Fig. 3; the fluxes shown were calculated assuming (1) no magnetic field, denoted as "not inhibited," and (2) a horizontal magnetic field which "inhibits" any vertical transport. Thermal electron heating rates, corresponding to the photoelectron flux calculations shown in Fig. 3, are presented in Fig. 4.

There are suprathermal electrons present in the Venus ionosphere, which do not appear to be the result of photoionization. These electrons have been observed on the night side (Gringauz et al. 1979; Spenner et al. 1981) and are believed to be shocked solar wind electrons moving into the ionosphere from the tail region. Suprathermal electrons, believed to be of solar wind origin have also been seen in the plasma mantle region, just outside the dayside ionopause (Spenner et al. 1980). The transport calculations of these electrons can be handled in the same manner as those of the photoelectrons (Gan et al. 1990).

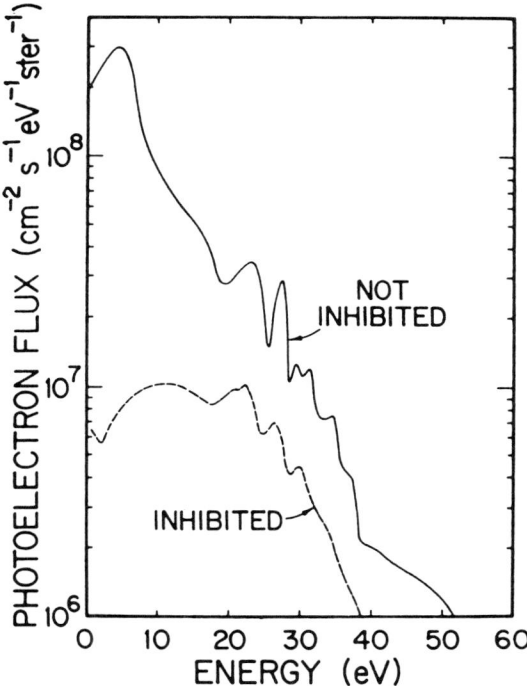

Figure 3. Calculated photoelectron flux at a given altitude (304 km) shown as a function of energy. The assumed solar zenith angle is 0°. Both the inhibited and uninhibited transport cases are shown (figure from Cravens et al. 1980).

III. OBSERVATIONS

A. Photo and Suprathermal Electrons

There were two instruments on the Pioneer Venus Orbiter which were potentially capable of measuring suprathermal electron fluxes, namely the retarding, potential analyzer (ORPA) (Knudsen et al. 1979a) and the solar wind plasma analyzer (OPA) (Intriligator et al. 1980). Microphonics and the long duty cycle of the latter meant that all meaningful ionospheric data, in this category, has come from the ORPA.

Retarding potential analyzers carried by Veneras 9 and 10 (Gringauz et al. 1977,1979) also measured the suprathermal electron fluxes on the night side of Venus during solar cycle minimum conditions. Figure 5 shows the suprathermal fluxes measured by the Veneras during solar cycle minimum and the ones measured by PVO during solar cycle maximum conditions (Knudsen and Miller 1985). Knudsen and Miller (1985) showed that there is great scatter (up to an order of magnitude) in the fluxes measured by both the PVO ORPA and the Venera 9 and 10 RPAs, which are believed to be due, at least

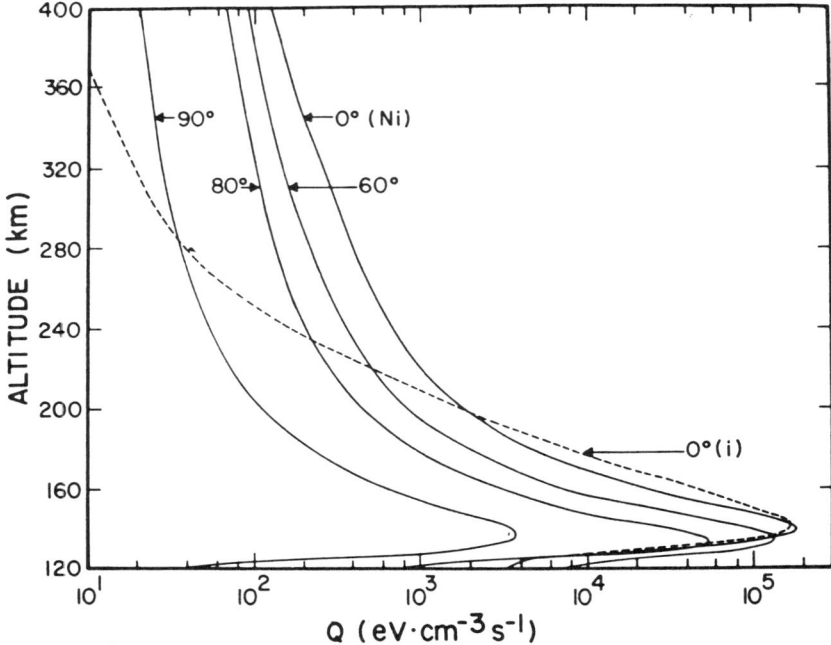

Figure 4. Calculated electron heating rates are shown for four different values of solar zenith angle for the case of uninhibited photoelectron transport. The heating rate for inhibited transport is only shown for the subsolar case and is labeled by the letter (i) (figure from Cravens et al. 1980).

partially, to day-to-day temporal variations. It appears that the solar cycle variations in these suprathermal electron fluxes is small; this conclusion is based on the comparison between the Venera and PVO values, as well as a very recent study carried out by Spenner et al. (1996), who compared the early solar cycle maximum PVO/ORPA results with data obtained at moderate solar cycle conditions, during the entry period of PVO in 1992.

B. Electron Temperatures

The only electron temperature measurements in the Venus ionosphere come from the Langmuir Probe (OETP) (Krehbiel et al. 1980) and the retarding potential analyzer (ORPA) (Knudsen et al. 1979a) carried aboard PVO. Theis et al. (1980,1984) combined all the electron temperature and density values obtained by the OETP instrument during the first four Venus years, while the periapsis of PVO remained in the ionosphere, and built an empirical model from this data base. This model provides electron densities and temperatures as a function of altitude and solar zenith angle (SZA); tabulated mean values of this electron temperature data base are shown in Table I. Theis and Brace (1993) also used the OETP results obtained during the entry phase of PVO

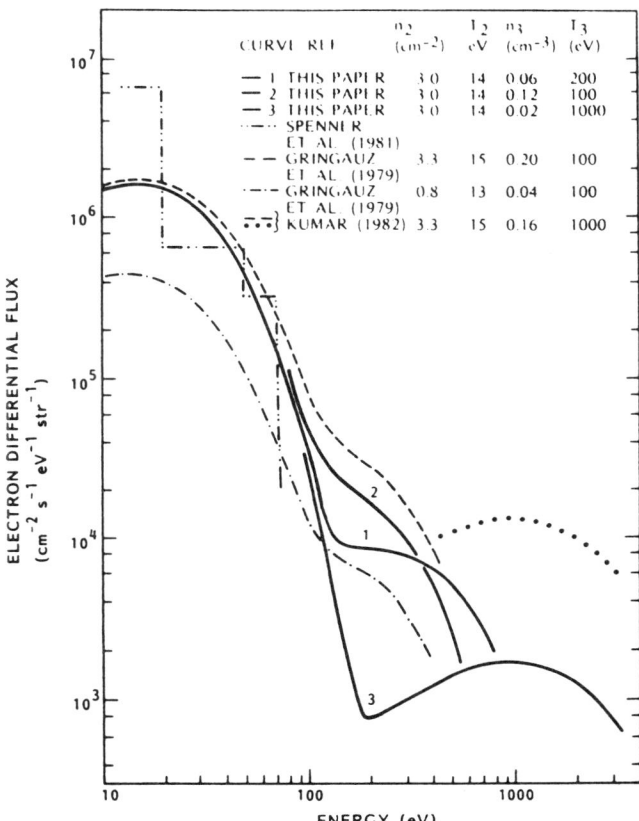

Figure 5. Nightside electron fluxes measured by the Venera 9 and 10 RPAs and the PVO/ORPA (figure from Knudsen and Miller 1985).

operations, corresponding to nighttime, moderate solar cycle conditions in 1992, in order to build an empirical model corresponding to that epoch (see Eqs. 31 and 32, and related discussion later in this section). Miller et al. (1980) summarized the early solar cycle maximum electron temperatures measured by the ORPA; these results are shown in Fig. 6. Both of these data sets indicate that on the average the electron temperatures are practically independent of the SZA.

The question of the energy sources responsible for the high observed ionospheric plasma temperatures has been debated for about twenty years and will be discussed in some detail in Sec. IV. Here we present some observational results bearing on this question. Elphic et al. (1984) examined the response of the lower (<200 km) dayside ionosphere to solar rotation, in order to establish the direct influence of solar EUV radiation. Figure 7 shows the ratios of individual orbit-averaged electron density and temperature values

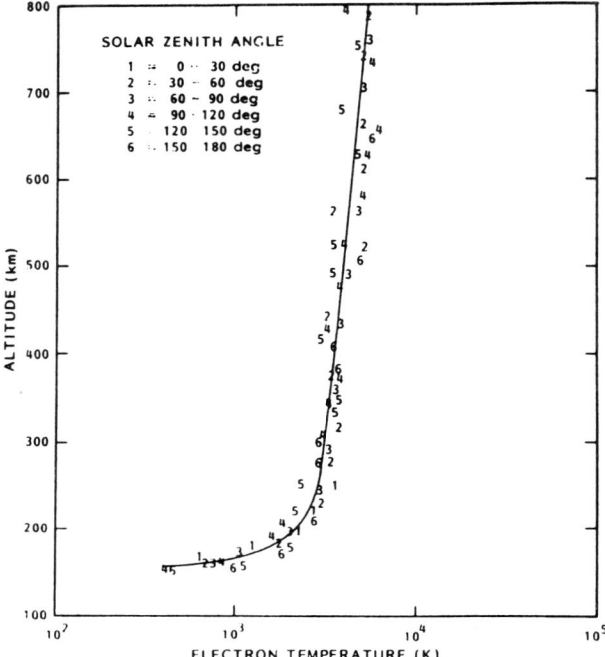

Figure 6. Measured electron temperatures as a function of SZAs obtained by the ORPA (figure from Miller et al. 1980).

to the Theis et al. (1984) model average, during four solar rotations. The bottom panel in this figure shows the solar EUV index, which is an indicator of the total EUV flux (Brace et al. 1988). This study clearly established the dependence of the electron density on solar EUV, but no such relation was noted for the electron temperature. Recently Dobe et al. (1993) used the PVO daytime electron temperature data base to see if they can establish any relations between the ionospheric electron temperatures and ionospheric magnetization, solar EUV intensities or 100 Hz wave intensities in the mantle region. They found that when the ionosphere is magnetized the electron temperatures are significantly higher than for the unmagnetized situation (see Fig. 8), but found no correlations between the electron temperatures and either the wave intensities or the EUV flux. Mahajan et al. (1994) examined the relation between the electron densities and temperatures in the ionopause region. They found a very strong inverse correlation, namely that decreasing densities corresponded to increasing temperature; this type of correlation was also noted in a general context by Knudsen et al. (1979b) much earlier.

The mean electron temperatures on the night side are not very different than the dayside values, as shown in Table I and Fig. 6. However, the night side is an extremely highly structured region. During the late entry period

TABLE I
Electron Temperature Values[a]

Altitude (km)	SZA (deg)					
	15	45	75	105	135	165
			Dawn			
147.5	—	—	941	722	—	726
152.5	1017	973	1003	845	—	939
157.5	1122	1099	1133	1197	1175	1194
162.5	1192	1242	1323	1438	1296	1449
167.5	1302	1417	1617	1623	1868	1606
175.0	1621	1849	1934	1873	1728	1809
185.0	2218	2263	2368	2208	2242	1755
195.0	2584	2575	2524	2503	2116	2598
210.0	2792	2694	2956	2844	2714	2519
235.0	3140	2964	2956	2844	2714	2519
275.0	3362	3241	3262	3214	3004	2306
325.0	3484	3516	3578	3539	3221	3061
375.0	3675	3761	3860	3732	3352	3226
450.0	4059	4048	4141	3999	3499	3401
600.0	4343	4449	4563	4494	4228	4015
850.0	—	4876	4906	5024	4572	4705
1250.0	—	5161	5093	5583	5265	5679
1750.0	—	—	—	5810	5561	6247
2500.0	—	—	—	5890	5807	6205
			Dusk			
147.5	—	—	—	—	626	752
152.5	1049	—	1126	1088	831	1024
157.5	1169	—	1252	1380	936	1147
162.5	1203	1368	1384	1448	1081	1446
167.5	1302	1365	1543	1546	1365	1932
175.0	1641	1681	2030	1761	2093	2207
185.0	2125	2224	2392	2077	2385	2259
195.0	2525	2524	2467	2177	2320	2584
210.0	2829	2772	2714	2404	2306	2842
235.0	3216	3037	2911	2699	2650	2625
275.0	3356	3243	3187	2967	3157	2983
275.0	3356	3243	3187	2967	3157	2893
325.0	3583	3502	3502	3280	3351	3284
375.0	3688	3704	3665	3565	3486	3392
450.0	4092	3924	3953	3864	3710	3893
600.0	4512	4299	4231	4286	4094	4215
850.0	—	4793	4632	5279	4923	4645
1250.0	—	—	4960	5712	5484	4770
1750.0	—	—	—	5592	5167	5114
2500.0	—	—	—	—	4909	5256

[a] Table from empirical model of Theis et al. (1984).

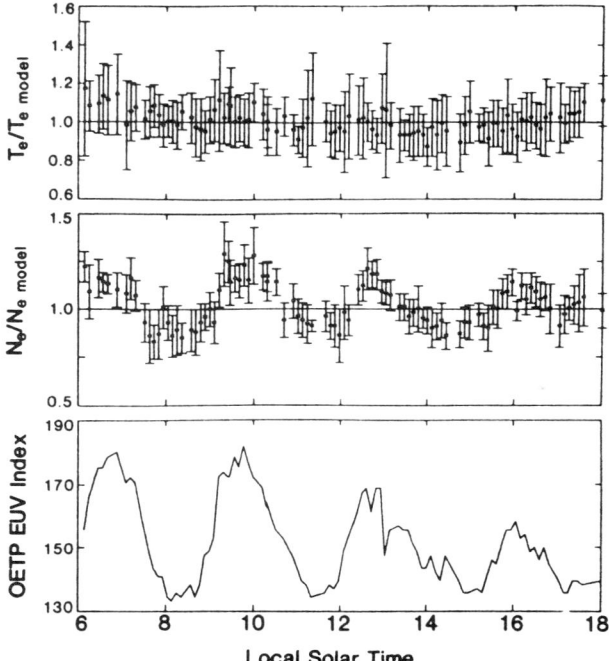

Figure 7. Ionospheric electron number density and temperature values, from below 200 km, normalized for altitude and SZA by the Theis et al. (1984) model. The OETP solar EUV index is also plotted. The error bars denote the standard deviations about the mean (figure from Elphic et al. 1984).

of PVO, July to October 1992, the periapsis of the spacecraft was located well within the ionosphere on the night side. During this time period the level of solar activity was relatively low ($F_{10.7} \sim 120$) compared to the early, prime mission phase, around 1979–1981 ($F_{10.7} \sim 200$). Theis and Brace (1993) presented the OETP electron density and temperature observations from the entry period. Figure 9 shows the comparison between the entry period and solar maximum empirical models for n_e and T_e. At the higher altitudes the electron densities are lower and the electron temperatures higher for the moderate solar activity period. For this latter period Theis and Brace (1993) provided the following relations for the electron densities and temperatures, covering the region above the ionospheric peak:

$$\log_{10}(n_e) = 2.902 + 1277.7/(h - 112.6)^2 - 6.2693 \times 10^{-4} h \qquad (31)$$

$$\log_{10}(T_e) = 3.471 - 1921.9/(h - 98.078)^2 + 8.5257 \times 10^{-4} h \qquad (32)$$

where h is the altitude in km and n_e and T_e are in cm^{-3} and degrees K, respectively.

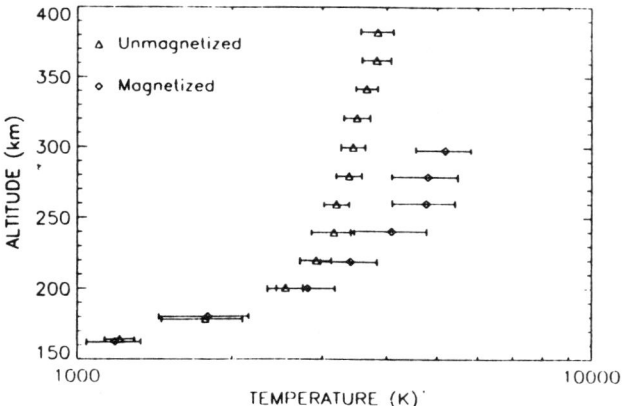

Figure 8. Measured electron temperature values obtained by the PVO/OETP from orbits with SZAs betwen 0° and 30°, grouped according to ionospheric magnetic field conditions. The horizontal bars indicate the standard deviations (figure from Dobe et al. 1993).

However, just as was the case during the prime mission, high solar activity period, the nightside electron temperatures were also quite variable during the entry period. For example, in the nightside ionospheric holes, where the electron density is low, the temperature of the "thermal" electron population is also low (Luhmann et al. 1982). On the other hand, in "disappearing ionospheres," found at times of high solar wind pressure conditions, the low electron densities are accompanied by quite high temperatures, exceeding 10^4K (Cravens et al. 1982). The electron temperatures can also vary on short time/spatial scales by about 10^3 K in "terminator waves," which, as the name implies are found near the terminator (Brace et al. 1983b). The changes in n_e and T_e were found to be anticorrelated in these waves. In fact, for almost all ionospheric structures, except for ionospheric holes, lower values of n_e are associated with higher values of T_e on both the day side and night side during both high and moderate solar activity periods.

C. Ion Temperatures

Ion temperatures were measured by the PVO/ORPA. A summary of these results is shown in Fig. 10. These data show that above about 300 km the ion temperatures are basically solar zenigth angle (SZA) independent, except in the region beyond 150°. Also below about 250 km the daytime ion temperatures are lower than nighttime values. Beyond about 150° SZA, the observed temperature rises significantly; this temperature increase is co-located with the region where the observed horizontal flow appears to slow down and randomize. Knudsen et al. (1980) suggested that these observed phenomena are an indication of a shock transition of the supersonic ion flow as it must slow down approaching the antisolar point. An examination of the

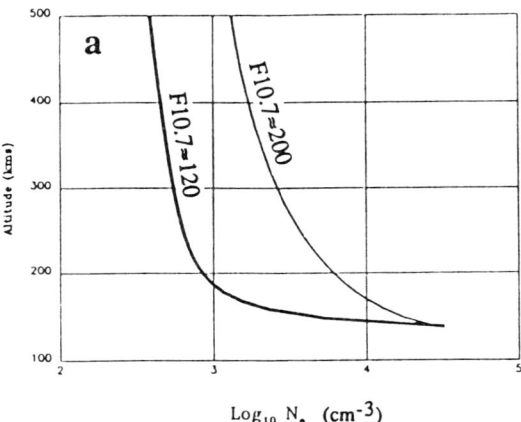

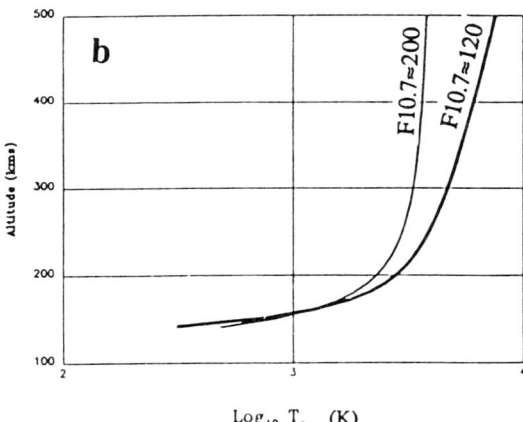

Figure 9. Comparison of measured nighttime electron densities and temperatures obtained by the PVO/ORPA during the entry period ($F_{10.7} \sim 120$) and solar maximum ($F_{10.7} \sim 200$) (figure from Theis and Brace 1993).

dayside ion temperature data base showed that the temperatures are basically independent of the magnetic conditions in the ionosphere (Nagy et al. 1997). Very recently Knudsen et al. (1996) presented results indicating that the H^+ temperatures are lower than the O^+ ones in the nightside hydrogen bulge region, during solar cycle maximum conditions (see Sec. IV.B).

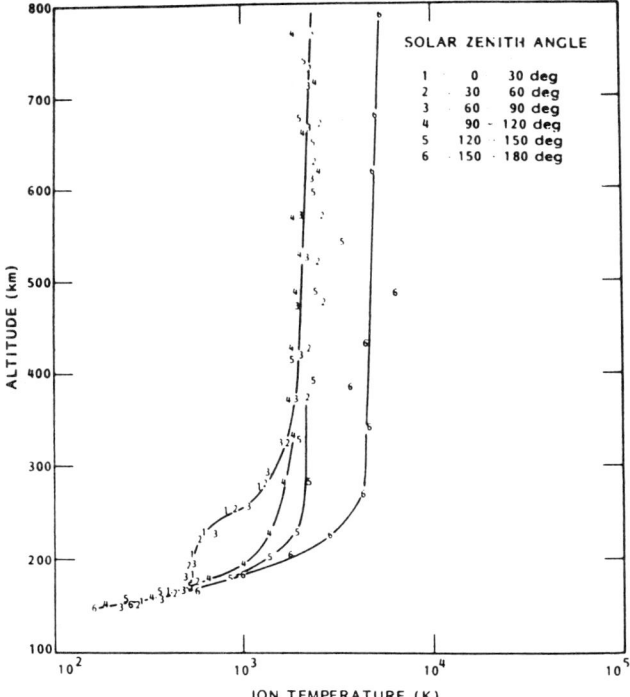

Figure 10. Median ion temperature values measured by the PVO/ORPA, grouped in 30° SZA increments. The solid curves represent fits to the data (figure from Miller et al. 1980).

IV. MODEL CALCULATIONS

A. One-Dimensional Models

Initial comparisons between the ion temperatures measured at Mars by the retarding potential analyzers carried by the Viking landers (Hanson et al. 1977), and model calculations indicated that, unlike the terrestrial mid-latitude ionosphere, conventional heat transport and solar EUV radiation alone cannot account for the observed values (Chen et al. 1978). The first self-consistent electron and ion temperature calculations for Venus were completed by Chen (1977), Chen and Nagy (1978) and Cravens et al. (1978) around the same time period. At that time, before the launch of Pioneer Venus, no direct information concerning the plasma temperatures in the Venus ionosphere were available, forcing modelers to make analogies with Mars. The Mars calculations indicated the need to invoke a reasonable, but *ad hoc*, topside heat inflow into the ionosphere in order to account for the observed temperatures.

Early results from the Pioneer Venus instruments, described in the previous section, allowed modelers to adjust their models in line with the ob-

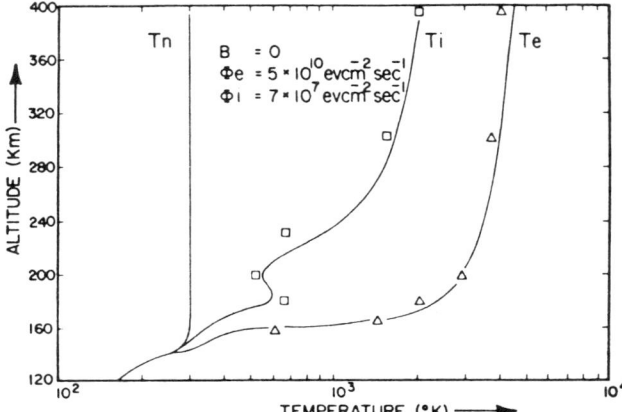

Figure 11. Calculated electron and ion temperatures obtained from a one-dimensional model. The calculations assumed that no magnetic field is present and that the heat fluxes at the top are as indicated. The squares and triangles indicate observed ion and electron temperature results from the ORPA and OETP, respectively (figure from Cravens et al. 1979).

servations (Cravens et al. 1979,1980; Knudsen et al. 1979b). These models established that using solar EUV as the only source of heating and classical thermal conduction results in temperatures significantly lower than the observed values. It was also found that, similar to the situation for Mars, invoking reasonable, but *ad hoc*, values of topside heat inflows ($\sim 5 \times 10^{10}$ and $\sim 7 \times 10^7$ eV cm^{-2} s^{-1} for electrons and ions, respectively) leads to dayside temperatures consistent with the observations. Typical results from such model calculations are shown in Fig. 11. Taylor et al. (1979) suggested that the 100 Hz waves that they observed in the mantle region are whistler mode waves and that they are absorbed in the ionopause region by Landau damping. They estimated that this mechanism can be responsible for a topside energy input into the ionosphere of about 3×10^{10} eV cm^{-2} s^{-1}, of just the right order of magnitude needed by the modelers. Cravens et al. (1980) also considered the potential role of magnetic fluctuations on the thermal conductivities (see Eq. 10). They found that the decreased thermal conductivities result in significant reduction in the downward heatflow and the eventual energy loss to the neutrals at the lower altitudes, which in turn means that considering only the classical solar EUV heating, solutions of the energy equation for daytime conditions give temperatures of the right magnitude. More recently, Hoegy et al. (1990) looked at the effect of magnetic field fluctuations on nightside electron temperatures. Using the approach mentioned at the end of Sec. II.C, they calculated an "effective" dip angle of 2.5° using data from 30 orbits (see Eq. 17); this value is considerably smaller than the average dip angle of 19.3° corresponding to the same 30 orbits. They found that this small effective dip angle reduced the necessary heat inflow by two orders of magnitude when compared to what would have been needed if the average dip angle would have been used. Figure 12 shows the results of their calculations in which they calculated electron temperature profiles using (i) their effective dip angles and (ii) a random distribution of dip angles based on the results of the observational data.

The models mentioned above were based, in general, on solving the steady-state coupled electron and ion energy equations, neglecting all bulk transport related terms (see Eq. 3) and using given plasma densities. It should also be mentioned that the quantitative heating rate calculations for the above mentioned models used the two stream approach (see Sec. II.F). One question, which received significant attention in these calculations, was the role of the weak observed magnetic fields on the photoelectron transport. Cravens et al. (1980) calculated photoelectron fluxes with and without considering the effect of weak, near horizontal magnetic fields (note that even a 10 nT field corresponds to a gyroradius of about 1.4 km for a 20 eV electron). The available observational data on photoelectron fluxes is very limited, but what is available seemed to indicate that the magnetic field does not inhibit photoelectron transport significantly. Furthermore, as indicated in Fig. 4, heating rates peak near 140 km, where transport is negligible anyway.

The neglect of the special transport terms prompted Schunk and St.-Maurice (1977,1981) to examine the potential importance of some of these terms. They showed that a relative drift between the ion species of only a few m s^{-1} results in an ion heat flow equivalent to a temperature gradient of 1 K km^{-1}. However, there is no information available at this time to indicate that such velocity differences are present. They also examined the issue of ion temperature anisotropy and heat flow that could result from high-velocity ion flows. Their parametric study showed that such anisotropies could become significant at altitudes below about 200 km, if the ion velocities are of the order of 1 km s^{-1}, but our limited data base on ion velocities (cf., Miller and Whitten 1991) suggests that the velocities are probably much smaller at these low altitudes.

B. Multi-Dimensional (or Multi-Species) Model Calculations: Unmagnetized Ionosphere

The first multi-dimensional electron temperature calculations were the ones by Hoegy et al. (1980) who solved a series of steady-state one-dimensional electron energy equations along magnetic field lines in order to try to reproduce the observed two-dimensional nightside electron temperature data base. Their energy calculations included solar EUV heating near the terminator, electron cooling to the ions and neutrals and adjustable topside heat inflows. Heat flow, which was assumed to be the dominant source of energy transport, was confined along magnetic field lines, therefore they used the magnetic field direction as a free parameter to obtain nightside electron temperatures that fitted the observed values. Figure 13 shows the magnetic field or heat

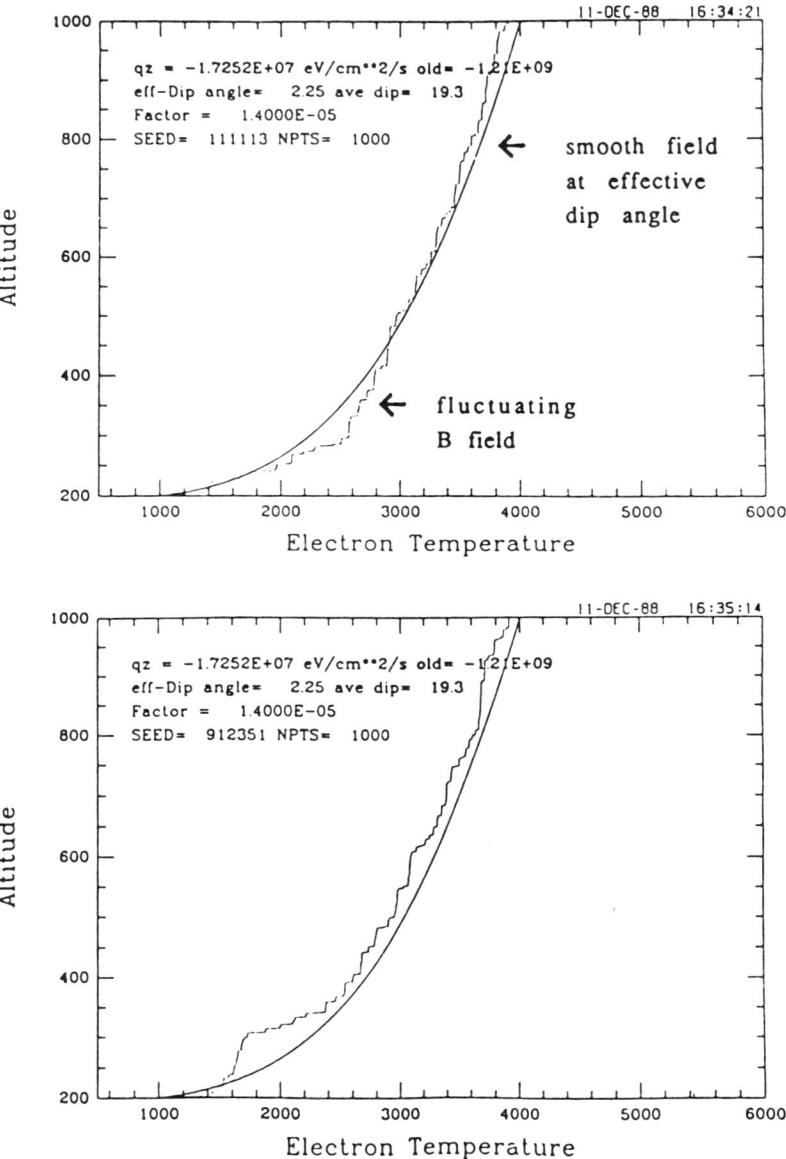

Figure 12. Electron temperature profiles calculated by solving the one-dimensional heat flow equation using (i) a constant "effective" dip angle (smooth line) and (ii) using a random distribution of dip angles based on observed magnetic field values (jagged line). The two altitude profiles were generated with two different random number seeds (figure from Hoegy et al. 1990).

flux direction that gave them their best fit to the data. Nightside ion temperatures were calculated, using a quasi-two-dimensional model by Bougher and Cravens (1984). They used an analytic fit to measured horizontal velocity values from the observations of Knudsen et al. (1980,1981) and solved the steady state energy equation, including both horizontal and vertical transport terms (e.g., adiabatic heating). They found that nightside ion temperatures are maintained at the high observed values out to about 150° SZA, by heat advection from the day side and adiabatic compressional heating due to the downward flow and the slowing of the horizontal flow. This model was based on subsonic transport and therefore is not expected to predict the correct temperatures beyond 150° SZA, if those elevated values are truly due to shock heating.

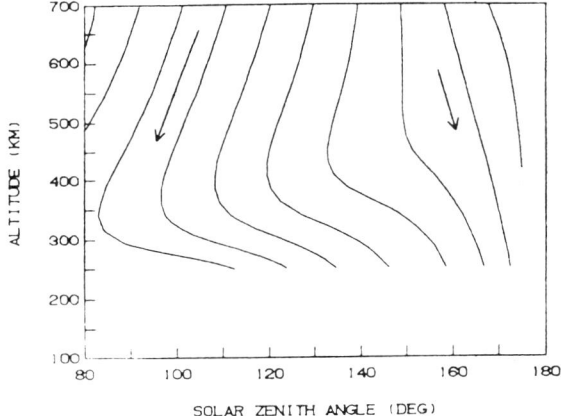

Figure 13. Heat flux lines or magnetic field lines that lead to electron temperatures consistent with the empirical model values, over the SZA range of 80 to 175° and over altitudes of 160 to 700 km (figure from Hoegy et al. 1980).

A somewhat similar two-dimensional model of the ion and electron temperatures was developed by Whitten et al. (1986). They used a spectral method to solve the heat conduction/advection equations. They assumed no topside heat inflow, but assumed the presence of an altitude distributed heat source. The plasma velocities used in the model were an eight term Legendre polynomial fit to the values calculated by Whitten et al. (1984). They found that in order to obtain electron and ion temperatures comparable to the observed ones they needed a nocturnal heat source about an order of magnitude smaller than the dayside heat source. Their estimate of the magnitude of this heat source depended very strongly on their assumption of the thermal conductivity; they assumed the presence of heat flux saturation (Merritt and Thompson 1980). They also found that adiabatic expansion followed by compression are responsible for the observed ion temperature dip just past the terminator

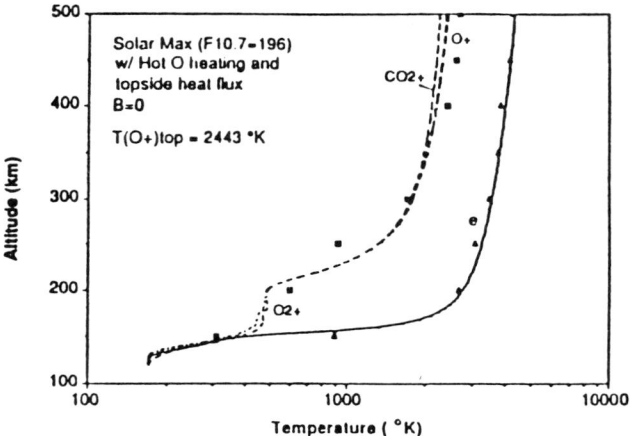

Figure 14. Calculated O^+, O_2^+ and CO_2^+ ion temperatures, assuming that no magnetic field is present. An electron heat inflow of 3×10^{10} eV cm^{-2} s^{-1} and a total ion heat inflow of 5×10^7 eV cm^{-2} s^{-1} was assumed. The squares and triangles are the corresponding VIRA ion and electron temperature values, respectively (figure from Kim et al. 1990).

and the following sharp rise, respectively.

Knudsen (1990) suggested that charge exchange between the hot neutral oxygen atoms, known to be present in the Venus upper atmosphere (Nagy and Cravens 1988), is an important and neglected source of ion heating. Kim et al. (1990) looked at that suggestion quantitatively by solving the one-dimensional, coupled electron and multi-species ion energy equations for daytime conditions and evaluating this proposed heating source in some greater detail. They found that while the proposed charge exchange does lead to some ion heating, it only makes a minor contribution to the necessary overall heating rate. They also found that the calculated temperature differences between the O^+, O_2^+ and CO_2^+ ions are very small, as shown in Fig. 14. Very recently Knudsen et al. (1996) found, by detailed examination of solar cycle maximum nighttime ORPA data, that the H^+ temperatures are significantly lower than the O^+ ones. These observational results were supported by solutions to the one-dimensional coupled electron and O^+ and H^+ ion energy equations, which also indicated the presence of such temperature differences, as shown in Fig. 15. The main reason for these significant temperature differences appears to be the difference in the thermal conductivities. Past calculations of nightside ion temperatures used either the fully ionized expression for thermal conductivity (Eq. 14) or some form of modified values (see, e.g., Whitten et al. 1986; Hoegy et al. 1990). Knudsen et al. (1996) were calculating ion temperatures corresponding to midnight conditions, a time at which the neutral hydrogen densities are nearly 2 orders of magnitude higher than during daytime. Ion collisions with neutral hydrogen are important under these conditions and reduce the thermal conductivity of O^+ more than that for H^+, leading to higher O^+ temperatures.

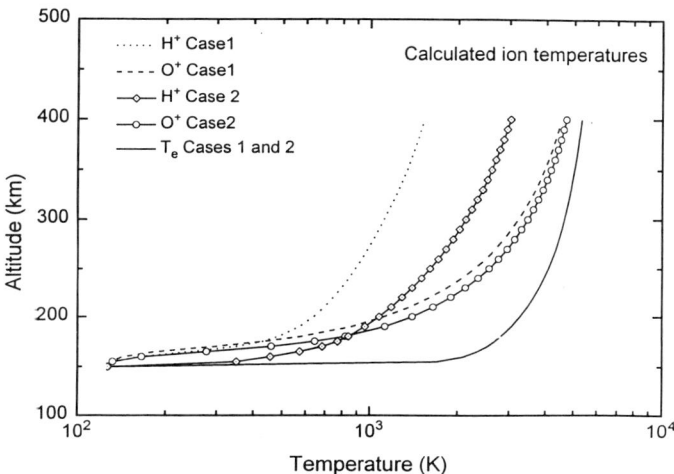

Figure 15. Calculated O^+ and H^+ ion and electron temperature values for solar cycle maximum midnight conditions. The curves denoted as Case 1 were obtained by assuming heat inflow values, at 400 km, of 3×10^{10} eV cm^{-2} s^{-1} and 1.5×10^8 eV cm^{-2} s^{-1} for electrons and the major ion O^+, respectively. The ratio of the O^+ and H^+ heat inflows were taken to be set by the ratio of the product of the thermal conductivity and density. The results denoted as Case 2 were obtained by setting the heat inflow value for both ions at 1.5×10^8 eV cm^{-2} s^{-1} (figure from Knudsen et al. 1996).

Nagy et al. (1991) used a shock-capturing numerical scheme (second-order Godunov scheme) to build a two-dimensional, single fluid, hydrodynamic model of the Venus ionosphere. The use of this numerical technique allowed for the first time an examination of possible shock formation in the nightside ionosphere, which was originally proposed by Knudsen et al. (1980) to explain the observed elevated ion temperatures near 150° (see Sec. III C). The calculated velocities reached values up to 3.5 km s^{-1}, which is in reasonably good agreement with the measurements. This flow has to slow down before it reaches the antisolar region and the calculations indicated the presence of a shock wave near 135°; compressional ion heating was observed to be present beyond the shock, raising the ion temperatures to over 5000°, consistent with the observed values. The model assumed topside heat inflows; dayside values of 5×10^9 eV cm s^{-1} and 2×10^7 eV cm^{-2} s^{-1} for the electrons and ions, respectively, were assumed and values equal to 30% of these were used for the night side. The calculations did show a drop in the ion temperatures just past the terminator, caused by the adiabatic expansion of the horizontal flow in this region.

C. Model Calculations for Magnetized Ionospheres and the Ionopause Region

The electron energetics of the plasma mantle region (i.e., the lower portion of the magnetic barrier) and upper ionosphere were modeled by Gan et al. (1990) for high solar wind dynamic pressure conditions, when the ionosphere is permeated by large scale magnetic fields. Transport of both suprathermal and thermal electrons is along magnetic field lines, which in this case is almost horizontal throughout most of the dayside ionosphere. Examples of the fluxes of suprathermal electrons, parallel and antiparallel to field lines, calculated by Gan et al. (1990), using the two stream method, are displayed in Fig. 16. In this study a solar wind electron spectrum was the input at the end of the field lines, which stretch from the low altitude subsolar region (~ 200 km) out into the solar wind in the tail region. At the lower altitudes photoelectrons are more important contributors to the total electron flux than the solar wind electrons, whose intensity is degraded by collisions with atmospheric neutrals and ionospheric thermal electrons. Gan et al. (1990) also solved the electron heat conduction equation (simplified form of the energy Eq. 3) along the same set of field lines as the suprathermal electron calculations. In the solution of this heat equation the heat input values were obtained from the two stream suprathermal electron calculations and the appropriate cooling rates were also included. Some of the results obtained from these calculations for the subsolar region are shown in Fig. 17, along with some of the measured OETP electron temperature values. The overall agreement between the calculated and measured temperatures indicates that "heat transport along magnetic field lines (via electron fluxes moving along the field and via heat conduction) can be an adequate heating source for the electrons both in the upper ionosphere, when magnetized, and in the mantle region" (Gan et al. 1990). Note that in this formulation the solar wind provides heat only via the suprathermal electron fluxes along the field lines. It should also be noted, as can be seen in Fig. 17, that the electron temperatures are higher at the higher altitudes, where the electron densities are lower; the ionopause was assumed to be near 300 km for this model.

Shinagawa et al. (1991) solved the time-dependent, one-dimensional coupled continuity, momentum equations for ions (O^+, O_2^+ and CO_2^+), the magnetic diffusion/convection equation for the magnetic field and the energy equations for ions and electrons in a self consistent manner. The main conclusions obtained from these comprehensive model calculations are that combining the energy calculations with the earlier MHD ones for densities, velocities and magnetic field values (Shinagawa and Cravens 1988) did not yield significantly different results from what was obtained when the temperatures were input values, based on observations. Conversely, the combined model did not provide any new insights into the energetics of the Venus iono-

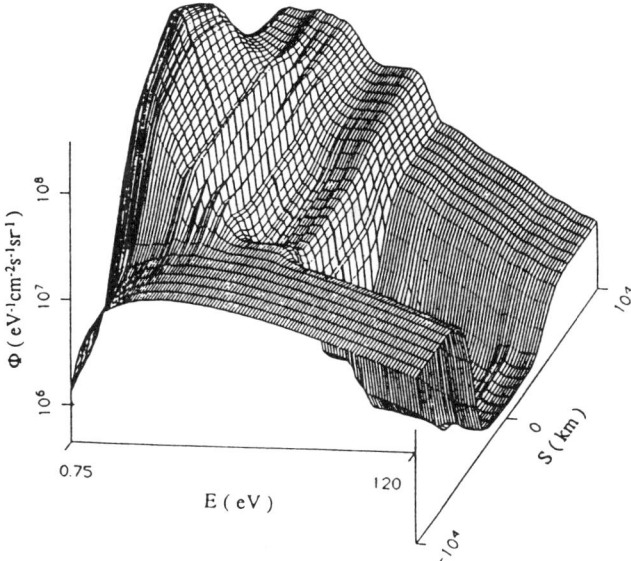

Figure 16. Calculated energy spectra of superthermal electron flux as a function of position along the draped magnetic field line. The subsolar altitude of the field line is 200 km. The electron flux at the beginning of the field line was assumed to have a Maxwellian distribution with $n = 28$ cm^{-3} and $T = 30$ eV. A transition region of about 3000 km separates the solar wind (or ionosheath) dominated region from the photoelectron dominated one (figure from Gan et al. 1990).

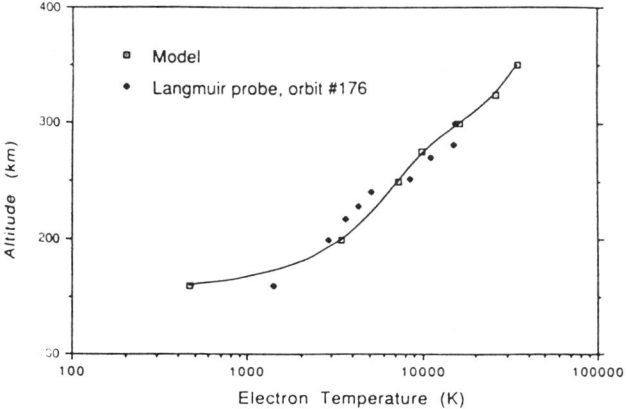

Figure 17. Comparison of electron temperatures calculated for the subsolar point with temperatures measured by PVO (figure from Gan et al. 1990).

sphere; this implies that solving the energy equations with appropriate density, velocity and magnetic field input parameters leads to reliable results.

V. SUMMARY

We have come a long way in advancing our understanding of the energetics of the Venus ionosphere during the last twenty years. As mentioned earlier, we know more about the ionosphere of Venus than that of any solar system body, except, of course, the terrestrial one. Most of the direct information has come from the PVO and a large variety of theoretical models. We now know that the Venus ionosphere is quite different from the terrestrial one. That is not a surprise because of many significant differences between the two planets. Venus has no intrinsic magnetic field of any significance, thus the solar wind interaction, as well as the ionospheric transport processes, are very different. Furthermore the major differences in atmospheric composition and thermospheric temperature also lead to important differences in the ionospheric behavior of the two planets. However, despite these major differences, as well as many other less important ones, many of the measurement and modeling tools are appropriate for both ionospheres.

Measurements by the instruments carried aboard PVO provided a good data base for day and nightside electron and ion temperatures during solar maximum conditions. During the brief entry period, corresponding to moderate solar cycle conditions, we also obtained nightside temperature information, but we have no comparable dayside data. All the data comes from a region near the equator, therefore we have no direct information on the latitude variation of the temperatures. We know that when the solar wind pressure is high, the ionopause is depressed and the ionosphere is magnetized. It has been found that at these times the electron temperatures are elevated, but that the ion temperatures do not change significantly.

A large number of theoretical models have been developed in order to study the energetics of the ionosphere. These studies have all demonstrated that assuming conventional EUV heating, combined with classical thermal conductivities, leads to temperature values significantly lower than the observed ones. In order to remedy this shortcoming, basically all the suggestions centered on reducing the applicable thermal conductivity and/or adding an *ad hoc* topside heat inflow. Both of these, individually or in combination, lead to temperature values in agreement with the observed ones. Unfortunately, there is not sufficient information available to establish the validity of, or distinguish between, these suggestions. There is considerable indication that both processes may play a role, but we cannot tell whether at times one or the other is dominant and/or whether other processes we have not considered are also important. As indicated in Sec. II.C, thermal conductivity, is likely to be less than the classical value, because of a number of factors, such as collisions with the neutral background atmosphere and magnetic field effects. There is also some observational indications that wave particle interaction processes in the mantle lead to energy flow into the ionosphere (see, e.g., Taylor et al. 1979; Shapiro et al. 1995).

We also know that there are important similarities between the ionospheres of Mars and Venus. During the past decade we have used our new PVO deduced information to advance our understanding of the ionosphere of Mars. There are no new missions planned to explore the ionosphere of Venus, but there are definite plans to launch two well-instrumented spacecraft to explore Mars. The Russian Mars/96 and the Japanese Planet-B mission should go a long way to help us elucidate the nature of the Mars ionosphere. The similarities between the ionospheres of the two planets may help us plug some of the holes remaining in our understanding of the Venus ionosphere.

Acknowledgments. The authors wish to thank R. W. Schunk for many, very helpful, discussions. The work involved in putting this chapter together was partially supported by two NASA Grants.

REFERENCES

Ashihara, O., and Takayanagi, K. 1973. Velocity distribution of ionospheric low-energy electrons. *Repts. Ionosp. Space Res. Japan* 77:65.
Banks, P. M. 1966. Charged particle temperatures and electron thermal conductivity in the upper atmosphere. *Ann. Geophysique* 22:577.
Banks, P. M., and Kockarts, G. 1973. *Aeronomy* (New York: Academic Press).
Bougher, S. W., and Cravens, T. E. 1984. A two-dimensional model of the nightside ionosphere of Venus: Ion energetics. *J. Geophys. Res.* 89:3837.
Brace, L. H., and Kliore, A. J. 1991. The structure of the Venus ionosphere. *Space Sci. Rev.* 55:81–163.
Brace, L. H., et al. 1983a. The ionosphere of Venus: Observations and their interpretation. In *Venus*, eds. D. M. Hunten, L. Colin, T. M. Donahue and V. I. Moroz (Tucson: Univ. of Arizona Press), pp. 779–840.
Brace, L. H., Elphic, R. C., Curtis, S. A., and Russell, C. T. 1983b. Wave structure in the Venus ionosphere downstream of the terminator. *Geophys. Res. Lett.* 10:1116–1119.
Brace, L. H., Hoegy, W. R., and Theis, R. F. 1988. Solar EUV measurements at Venus based on photoelectron emission from the Pioneer Venus Langmuir probe. *J. Geophys. Res.* 93:7282.
Burgers, J. M. 1969. *Flow Equations for Composite Gases* (New York: Academic Press).
Butler, S. T., and Buckingham, M. J. 1962. Energy loss of a fast ion in a plasma. *Phys. Rev.* 126:1.
Chen, R. H. 1977. The Ionosphere of Venus. Ph.D. Thesis, Univ. of Michigan.
Chen, R. H., Cravens, T. E., and Nagy, A. F. 1978. The Martian ionosphere in light of the Viking observations, *J. Geophys. Res.* 83:3871.
Chen, R. H., and Nagy, A. F. 1978. A comprehensive model of the Venus ionosphere. *J. Geophys. Res.* 83:1133.

Cravens, T. E. 1991. Ionospheric models for Venus and Mars. In *Venus and Mars: Atmospheres, Ionospheres and Solar Wind Interactions*, ed. J. G. Luhmann (Washington, D. C.: American Geophysical Union), p. 177.
Cravens, T. E., Nagy, A. F., Chen, R. H., and Stewart, A. I. 1978. The ionosphere and airglow of Venus: Prospects for Pioneer Venus. *Geophys. Res. Lett.* 5:613.
Cravens, T. E., et al. 1979. The energetics of the ionosphere of Venus: A preliminary model based on Pioneer Venus observations. *Geophys. Res. Lett.* 6:341.
Cravens, T. E., Gombosi, T. I., Kozyra, J. U., and Nagy, A. F. 1980. Model calculations of the dayside ionosphere of Venus: Energetics. *J. Geophys. Res.* 85:7778.
Cravens, T. E., et al. 1982. Disappearing ionospheres on the nightside of Venus. *Icarus* 51:271–282.
Dobe, Z., Nagy, A. F., Brace, L. H., Cravens T. E., and Luhmann, J. G. 1993. Energetics of the dayside ionosphere of Venus. *Geophys. Res. Lett.* 20:1523.
Elphic, R. C., Brace, L. H., Theis, R. F., and Russell, C. T. 1984. Venus dayside ionospheric conditions: Effects of ionospheric magnetic field and solar EUV flux. *Geophys. Res. Lett.* 11:124.
Gan, L. 1991. Electron Distributions and Solar Wind Interaction With Nonmagnetic Planets. Ph.D. Thesis, Univ. of Michigan.
Gan, L., Cravens, T. E., and Horanyi, M. 1990. Electrons in the iono-pause boundary layer of Venus. *J. Geophys. Res.* 95:19023.
Gombosi, T. I., and Rasmussen, C. R. 1991. Transport of gyration-domin-ated space plasmas of thermal origin. 1. Generalized transport equations. *J. Geophys. Res.* 96:7759.
Grad, H. 1949. On the kinetic theory of rarefied gases. *Comm. Pure Appl. Math.* 2:331.
Gringauz, K. I., Verigin, M. I., Breus, T. K., and Gombosi, T. I. 1977. The electron fluxes measured in the optical shadow of Venus on board Venera-9 and Venera-10 orbiters are the main ionization source of Venus' nighttime ionosphere. *Dokl. Akad. Naut. SSSR* 232:1039 (in Russian).
Gringauz, K. I., Verigin, M. I., Breus, T. K., and Gombosi, T. I. 1979. The interaction of electrons in the optical umbra of Venus with the planetary atmosphere-The origin of the nightside ionosphere. *J. Geophys. Res.* 84:2123.
Hanson, W. B., Sanatani, S., and Zuccaro, D. R. 1977. The martian ionosphere as observed by the Viking retarding potential analyzer. *J. Geophys. Res.* 82:4351.
Hoegy, W. R. 1984. Thermal electron heating rate: A derivation. *J. Geophys. Res.* 89:977.
Hoegy, W. R., Brace, L. H., Theis, R. F., and Mayr, H. G. 1980. Electron temperature and heat flow in the nightside Venus ionosphere. *J. Geophys. Res.* 85:7811.
Hoegy, W. R., Brace, L. H., Kasprzak, W. T., and Russell, C. T. 1990. Small scale plasma, magnetic, and neutral density fluctuations in the nightside Venus ionosphere. *J. Geophys. Res.* 95:4085–4102.
Intriligator, D. S., Wolfe, J. H., and Mihalov, J. D. 1980. The Pioneer Venus Orbiter plasma analyzer experiment, *IEEE Trans. Geosci. Remote Sensing* GE-18:39.
Johnson, R. E. 1978. Comment on ion and electron temperatures in the martian upper atmosphere. *Geophys. Res. Lett.* 5:989.
Khazanov, G. V., and Liemohn, M. W. 1995. Nonsteady state ionosphere-plasmasphere coupling of superthermal electrons. *J. Geophys. Res.* 100:9669.
Khazanov, G. V., Gombosi, T. I., Nagy, A. F., and Koen, M. A. 1992. Analysis of the ionosphere-plasmasphere transport of superthermal electrons: 1. Transport in the plasmasphere. *J. Geophys. Res.* 97:16887.
Khazanov, G. V., Neubert, T., and Gefan, G. D. 1994. A unified theory of ionosphere-plasmasphere transport of suprathermal electrons. *IEEE Trans. Plasma Sci.* 22:187.

Kim, J. 1991. Model Studies of the Ionosphere of Venus: Ion Composition, Energetics and Dynamics. Ph.D. Thesis, Univ. of Michigan.

Kim, J., Nagy, A. F., Cravens, T. E., and Shinagawa, H. 1990. Temperature of individual ion species and heating due to charge exchange in the ionosphere of Venus. *J. Geophys. Res.* 95:6569.

Knudsen, W. C. 1990. Role of hot oxygen in Venusian ionospheric ion energetics and supersonic antisunward flow. *J. Geophys. Res.* 95:1097–1101.

Knudsen, W. C., and Miller, K. L. 1985. Pioneer Venus superthermal electron flux measurements in the Venus umbra. *J. Geophys. Res.* 90:2695.

Knudsen, W. C., Bakke, J., Spenner, K., and Novak, V. 1979a. Retarding potential analyzer for the Pioneer-Venus Orbiter Mission. *Space Sci. Instrum.* 4:351.

Knudsen, W. C., et al. 1979b. Thermal structure and energy influx to the day and nightside Venus ionosphere. *Science* 205:105.

Knudsen, W. C., Spenner, K., Miller, K. L., and Novak, V. 1980. Transport of ionospheric O^+ ions across the Venus terminator and implications. *J. Geophys. Res.* 85:7803.

Knudsen, W. C., Spenner, K., and Miller, K. L. 1981. Anti-solar acceleration of ionospheric plasma across the Venus terminator. *Geophys. Res. Lett.* 8:241.

Knudsen, W. C., Nagy, A. F., and Spenner, K. 1996. Lack of thermal equilibrium between H^+ and O^+ temperatures in the Venus nightside ionosphere. *J. Geophys. Res.*, in press.

Krehbiel, J. P., et al. 1980. Pioneer Venus orbiter electron temperature probe. *IEEE Trans. Geosci. Remote Sensing* GE-18:49–54.

Krinberg, I. A. 1973. Description of the photoelectron interaction with ambient electrons in the ionosphere. *Planet. Space Sci.* 21:523.

Kumar, S., and Hunten, D. M. 1974. An ionospheric model with an exospheric temperature of 350 K. *J. Geophys. Res.* 79:2529.

Luhmann, J. G., and Cravens, T. E. 1991. Magnetic fields in the ionosphere of Venus. *Space Sci. Rev.* 55:201–274.

Luhmann, J. G., et al. 1982. Pioneer Venus observations of plasma and field structure in the near wake of Venus. *J. Geophys. Res.* 87:9205.

Mahajan, K. K., Gosh, S., Paul, R., and Hoegy, W. R. 1994. Variability of dayside electron temperature at Venus. *Geophys. Res. Lett.* 21:77.

Merritt, D., and Thompson, K. 1980. Thermal energy transport in the Venus ionosphere: Classical and saturated electron temperature profiles. *J. Geophys. Res.* 85:6778.

Miller, K. L., and Whitten, R. C. 1991. Ion dynamics in the Venus ionosphere. *Space Sci. Rev.* 55:165.

Miller, K. L., et al. 1980. Solar zenith angle dependence of ionospheric ion and electron temperatures and density on Venus. *J. Geophys. Res.* 85:7759.

Nagy, A. F., and Banks, P. M. 1970. Photoelectron fluxes in the ionosphere. *J. Geophys. Res.* 75:6260.

Nagy, A. F., and Cravens, T. E. 1988. Hot oxygen atoms in the upper atmospheres of Venus and Mars. *J. Geophys. Res.* 15:433.

Nagy, A. F., Cravens, T. E., and Gombosi, T. I. 1983. Basic theory and model calculations of the Venus ionosphere. In *Venus*, eds. D. M. Hunten, L. Colin, T. M. Donahue and V. I. Moroz (Tucson: Univ. of Arizona Press), pp. 841–872.

Nagy, A. F., Korosmezey, A., Kim, J., and Gombosi, T. I. 1991. A two dimensional shock capturing hydrodynamic model of the Venus ionosphere. *Geophys. Res. Lett.* 18:801.

Nagy, A. F., Sinkovics, A., Cravens, T. E., and Knudsen, W. C. 1997. The magnetic field control of the dayside ion temperatures in the ionosphere of Venus. *J. Geophys. Res.*, in press.

Raitt, W. J., Schunk, R. W., and Banks, P. M. 1975. A comparison of the temperature and density structure in high and low speed thermal proton flows. *Planet. Space Sci.* 23:1103.

Rohrbaugh, R. P., Nisbet, J. S., Bleuler, E., and Herman, J. R. 1979. The effect of energetically produced O_2^+ on the ion temperatures of the martian thermosphere. *J. Geophys. Res.* 84:3327.

Schunk, R. W. 1975. Transport equations for aeronomy. *Planet. Space Sci.* 23:437.

Schunk, R. W. 1977. Mathematical structure of transport equations for multispecies flows. *Rev. Geophys. Space Phys.* 15:429.

Schunk, R. W., and Hays, P. B. 1971. Photoelectron energy losses to thermal electrons. *Planet. Space Sci.* 19:113.

Schunk, R. W., and Nagy, A. F. 1978. Electron temperatures in the F region of the ionosphere: Theory and observations. *Rev. Geophys. Space Phys.* 16:355.

Schunk, R. W., and Nagy, A. F. 1980. Ionospheres of the terrestrial planets *Rev. Geophys. Space Phys.* 18:813.

Schunk, R. W., and St.-Maurice, J. P. 1977. Plasma transport in the topside Venus ionosphere. *Planet. Space Sci.* 25:921.

Schunk, R. W., and St.-Maurice, J. P. 1981. Ion temperature anisotropy and heat flow in the Venus lower ionosphere. *J. Geophys. Res.* 86:4823.

Schunk, R. W., and Walker, J. C. G. 1970. Transport properties of the ionospheric electron gas. *Planet. Space Sci.* 18:1535.

Shapiro, V. D., et al. 1995. On the interaction between the shocked solar wind and the planetary ions on the dayside of Venus. *J. Geophys. Res.* 100:21289–21305.

Shinagawa, H., and Cravens, T. E. 1988. A one-dimensional multispecies magnetohydrodynamic model of the dayside ionosphere of Venus. *J. Geophys. Res.* 93:11.

Shinagawa, H., Kim, J., Nagy, A. F., and Cravens, T. E. 1991. A comprehensive magnetohydrodynamic model of the Venus boundary region between solar wind and ionosphere. *J. Geophys. Res.* 96:11083.

Shkarofsky, I. P. 1961. Values of transport coefficients in a plasma for any degree of ionization based on a Maxwellian distribution. *Canadian J. Phys.* 39:1619.

Solomon, S. C., and Abreu, V. J. 1989. The 6300 nm dayglow. *J. Geophys. Res.* 94:6817.

Spenner, K., et al. 1980. Observation of the Venus mantle, the boundary region between solar wind and ionosphere. *J. Geophys. Res.* 85:7655–7662.

Spenner, K., et al. 1981. On the maintenance of the Venus nightside ionosphere: Electron precipitation and plasma transport. *J. Geophys. Res.* 86:9170.

Spenner, K., Knudsen, W. C., and Lotze, W. 1996. Suprathermal electron fluxes in the Venus nightside ionosphere at moderate and high solar activity. *J. Geophys. Res.* 101:4557–4564.

Spitzer, L., and Harm, R. 1953. Transport phenomena in a completely ionized gas. *Phys. Rev.* 89:977.

Swartz, W. E., Nisbet, J. S., and Green, A. E. S. 1971. Analytic expression for the energy transfer rate from photoelectrons to thermal electrons. *J. Geophys. Res.* 76:8425.

Takayanagi, K., and Itikawa, Y. 1970. Elementary processes involving electrons in the ionosphere. *Space Sci. Rev.* 11:380.

Taylor, W. W. L., Scarf, F. L., Russell, C. T., and Brace, L. H. 1979. Absorption of whistler mode waves in the ionosphere of Venus. *Science* 205:112.

Theis, R. F., and Brace, L. H. 1993. Solar cycle variations of electron density and temperature in the Venusian nightside ionosphere. *Geophys. Res. Lett.* 20:2719.

Theis, R. F., Brace, L. H., and Mayr, H. G. 1980. Empirical models of the electron temperature and density in the Venus ionosphere. *J. Geophys. Res.* 85:7787.

Theis, R. F., Brace, L. H., Elphic, R. C., and Mayr, H. G. 1984. New empirical models of the electron temperature of the Venus ionosphere with application to transterminator flow. *J. Geophys. Res.* 89:1477–1488.

Torr, M. R., and Torr, D. G. 1985. Ionization frequencies for solar cycle 21: Revised. *J. Geophys. Res.* 90:6675.

Whitten, R. C., et al. 1984. Dynamics of the Venus ionosphere: A two-dimensional model study. *Icarus* 60:317.

Whitten, R. C., Singhal, R. P., and Knudsen, W. C. 1986. Thermal structure of the Venus ionosphere: A two dimensional model study. *Geophys. Res. Lett.* 13:10.

SOLAR ACTIVITY BEHAVIOR OF THE THERMOSPHERE

W. T. KASPRZAK
NASA Goddard Space Flight Center

G. M. KEATING
George Washington University

N. C. HSU
Hughes STX Corporation

A. I. F. STEWART and W. B. COLWELL
University of Colorado

and

S. W. BOUGHER
University of Arizona

Venus neutral thermosphere and exosphere measurements above 100 km are reviewed with an emphasis on recently acquired data and further analysis since publication of the first Venus book (Hunten et al. 1983). Neutral thermosphere measurements above 100 km have been made over more than a solar cycle period by remote sensing, by *in-situ* and orbital drag from the Pioneer Venus Orbiter (PVO) spacecraft and by drag measurements of the Magellan Orbiter (MGN) spacecraft. Remote sensing of emissions due to CO, O and H from PVO provide a long-term monitoring of the thermosphere variations over the solar cycle. The PVO and MGN data suggest a relatively small change in the dayside and nightside thermosphere density and temperature with solar activity on both the solar rotation scale and the long-term (solar cycle) scale for all species except H and D. The weak dayside response is explained by the EUV and UV heating, and the CO_2 15-μm emission cooling which approximately balance each other in nearly the same altitude region. Atomic oxygen excites the bending mode of CO_2 into the strong 15-μm emission which is regulated by the relative O and CO_2 abundance. Eddy heat conduction plays a minor role. The nightside thermosphere, which is controlled by circulation and transport across the terminator, also shows very little variation with solar activity. In contrast, hydrogen and deuterium measurements in the bulge region show an increase in density with decreasing solar activity that is attributed to the reduction in exosphere escape of H and D with decreasing solar activity.

I. INTRODUCTION

This chapter will primarily review data and new analysis of results obtained on the neutral composition and temperature in the upper atmosphere above 100 km since 1982 when the first *Venus* book (Hunten et al. 1983) was published with reviews by von Zahn et al. (1983) and Seiff (1983). General reviews of this area since then have been given by Keating et al. (1985), Keating and Bougher (1987), Mahajan and Kar (1988,1990) and Luhmann (1991). Fox and Bougher (1991) reviewed the chemical and thermal structure, airglow and aurora emission, hot corona, dynamics of the thermosphere, and compared various empirical and chemistry models. Not included in that first *Venus* book were: (a) data obtained since 1982 by the PVO, which ceased operation in October of 1992; (b) data obtained by the MGN spacecraft on the neutral thermosphere from Fall 1992 to Fall 1994; (c) short and long-term solar activity effects; (d) temporal effects observed in the neutral temperature and density; and (e) advances made in understanding the roles of heating by EUV and ultraviolet radiation, and cooling by CO_2 radiation and eddy conduction.

The composition of the upper neutral atmosphere, as determined by *in-situ* measurements, consists of H, He, N, O, CO, N_2, and CO_2 as shown Fig. 1 (Niemann et al. 1980*b*; Brinton et al. 1980; Keating et al. 1980; see summaries by von Zahn et al. 1983 and Seiff 1983). At solar maximum, the inferred dayside exosphere temperatures are 275 to 300 K while inferred nightside exosphere temperatures are 100 to 110 K (Keating et al. 1980; von Zahn et al. 1980; Niemann et al. 1980*b*). The nightside thermosphere of Venus is also called the cryosphere (Shubert et al. 1980). The dominance of O in the upper dayside thermosphere above the "F2-ledge" was suggested by early models of the ionosphere based on Mariner 10 airglow observations (Bauer and Hartle 1974). Above the homopause the atmosphere is believed to be in diffusion equilibrium and at solar maximum the transition from CO_2 to O occurs at about 155 km on the day side and about 140 km on the night side; the O to He transition occurs above 250 km and near 180 km, respectively. The diurnal variation of all species except H, D and He, have maxima near noon with very steep gradients at the terminators leading to much lower densities on the night side than the day side (Niemann et al. 1980*b*; Keating et al. 1980; von Zahn et al. 1983; Hartle et al. 1996). The pre-dawn H, He and D bulges indicate that the upper thermosphere also superrotates (Mayr et al. 1980) and a recent model (Mengel et al. 1989) derives a period of about 6 days. This period is comparable to that observed at the cloud tops at lower altitudes (see, e.g., Schubert et al. 1980). Remote sensing observations have also identified NO (Feldman et al. 1979; Stewart and Barth 1979), O_2 (Krasnopolsky et al. 1976) and hot coronas due to nonthermal O (Nagy et al. 1981), C (Paxton 1985), N (Keating et al. 1985) and H (see summary of von Zahn et al. 1983). The exobase, where the atmospheric scale height is equal to the mean free path and particles can execute ballistic trajectories, is about 210 km on the day side, and about 154 km on the night side at high solar activity (Fox and

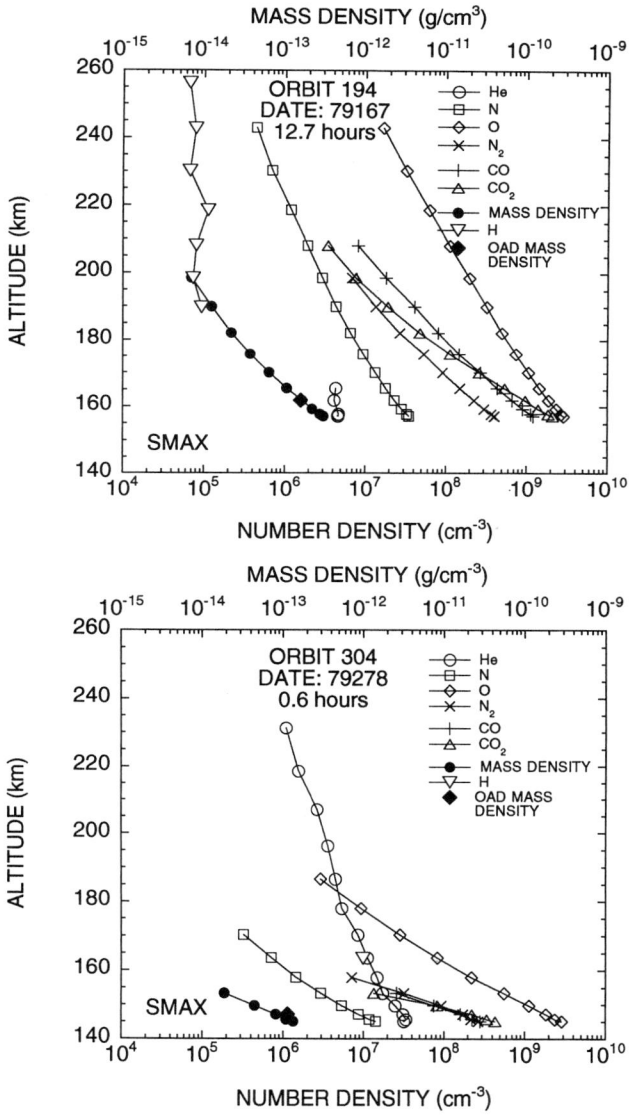

Figure 1. PVO neutral composition measured near noon (top) and near midnight (bottom) at solar maximum conditions. Number densities for He, N, O, N_2, CO and CO_2 and mass density from Niemann et al. (1980b), H from Brinton et al. (1980) and Grebowsky et al. (1995), and OAD mass density from Keating et al. (1980). The date 79167 is read as day 167 of year 1979. The PV ONMS densities displayed in this figure and elsewhere in the chapter have been raised by a factor of 1.63 as recommended by Hedin et al. (1983).

Bougher 1991) and is higher in altitude in the region of the dawn H bulge.

A. Spacecraft Missions

The Pioneer Venus (PV) Orbiter spacecraft was inserted into a highly elliptical, nearly polar, approximately 24-hr period, orbit on 4 December 1978 (Colin 1980) and operation continued (Brace and Colin 1984; Colin 1989) until loss of contact with the spacecraft after orbit 5055 on 7 October 1992 (Strangeway 1993). *In-situ* composition measurements were made by the Orbiter Neutral Mass Spectrometer (ONMS) (Niemann et al. 1980a), drag mass density measurements were made by the Orbiter Atmospheric Drag (OAD) experiment (Keating et al. 1979a; Shapiro et al. 1979), and main ionosphere measurements were made by the Orbiter Ion Mass Spectrometer (OIMS) (H. A. Taylor et al. 1980a) and Orbiter Electron Temperature Probe (OETP) (Krehbiel et al. 1980) over approximately 2.8 diurnal cycles during the first 630 orbits, when the periapsis altitude was below about 300 km and near 17° N latitude. Neutral density measurements were possible during the pre-entry phase beginning at orbit 4954 (29 June 1992) to orbit 5055, about 17.5 to 4.4 hr local solar time, near 10°S (Kasprzak et al. 1993a; Keating and Hsu 1993). During the intervening period the periapsis altitude was too high and neutral atmospheric densities were below the threshold for *in-situ* measurements. Remote sensing observations over the whole mission are available from the Orbiter Ultraviolet Spectrometer (OUVS) (Stewart et al. 1980a). Other neutral atmosphere measurements were also made by the Orbiter Infrared (OIR) instrument (Delderfield et al. 1980), Pioneer Venus Probes (LAS, SAS1, SAS2, SAS3) (Seiff et al. 1980) and the Bus Neutral Mass Spectrometer (BNMS) (Hoffman et al. 1980) during entry on 9 December 1978 (Hunten et al. 1983). All of the Pioneer Venus Mission instruments have been described in a special issue of IEEE Transactions on Geoscience and Remote Sensing (1980) except for the Cloud Photopolarimeter, which is described in Russell et al. (1977) and Travis (1979). A bibliography of Pioneer Venus and related publications up to 1994 is given in Lasher et al. (1994).

The primary mission of the Magellan Orbiter spacecraft (MGN) was radar mapping of the surface of Venus but atmosphere structure studies using spacecraft drag (Keating and Hsu 1993) and attitude control measurements were also conducted (Croom and Tolson 1994a,b). The orbital period of MGN ranged from 3 to 1.5 hr. Cycle 4 measurements began on 15 September 1992 and ended on 26 May 1993. During this period the orbit inclination was nearly polar, periapsis altitudes ranged from 165 to 185 km, in the atomic oxygen dominant regime, and periapsis latitude was near 11°N. After Cycle 4, aerobraking was performed placing the spacecraft into a near circular orbit. During aerobraking, drag measurements were obtained on the day side near 140 km in the CO_2 dominant regime (Keating and Hsu 1993). MGN has made measurements at higher latitudes up to 70°N during Cycles 5 and 6 after near circularization of the orbit (Espiritu and Tolson 1995).

On 10 February 1990 the Galileo spacecraft passed within 20,000 km of

Venus (Johnson et al. 1991). The Ultraviolet Spectrometer instrument (Hord et al. 1992) measured limb profiles of dayside HI 1216 Å and OI 1304 Å emissions (Hord et al. 1991).

B. Proxies for Solar Activity Variations

The thermosphere response to solar activity can be parameterized in terms of the $F_{10.7}$ flux, an Earth-based measurement of the 10.7-cm radio flux normalized to 1 AU and measured in units of 10^{-22} Wm^{-2} Hz^{-1}. Earth thermosphere density variations were found to be very similar to those observed in the solar decametric flux (Jacchia 1959). Overall, the $F_{10.7}$ flux correlates very well with the thermosphere densities and temperatures for short- and long-period time scales, and exhibits a high correlation with the EUV fluxes which are important in determining the exosphere temperature (Hedin 1984; Hedin and Mayr 1987). The $F_{10.7}$ flux is also used in the Venus thermosphere as a proxy for the EUV flux after adjusting for the orbital phase of the Earth and Venus (H. A. Taylor et al. 1982; Hedin et al. 1983; Keating and Bougher 1992a,b). Density and temperature variations are correlated with $F_{10.7}$: (a) over the solar rotation (nominally a 28-day period as observed on Venus); and (b) over the solar cycle (as reflected in a 3-solar rotation average) (Niemann et al. 1980b; Keating et al. 1980; Hedin et al. 1983; Keating et al. 1985). The equivalent $F_{10.7}$ index at Venus, for an average Venus–Sun distance of 0.72 AU, is a factor of 1.93 times larger than that measured at Earth. Orbital eccentricity effects in the thermosphere are not as important for Venus as they are for Earth and Mars. For future reference, the terms solar maximum (SMAX) ($F_{10.7}$ = 180–220), solar medium (SMED) ($F_{10.7}$ = 110–130), and solar minimum (SMIN) ($F_{10.7}$ = 68 − 80) will be used where the $F_{10.7}$ index refers to its value as measured on Earth at 1 AU.

The photoelectron current I_{pe} from the Orbiter Electron Temperature Probe (OETP) probe (Brace et al. 1988) is generated by the solar EUV flux for wavelengths in the approximate range 300 to 1300 Å with more than half of the spectral contribution for Ipe coming from Lyman-α at 1216 Å. Mahajan et al. (1990) showed that this EUV index was about as successful as $F_{10.7}$ in predicting exosphere scale height temperatures derived from the ONMS but no correlation was made with atmospheric densities. Short-term EUV variations (responsible for thermosphere heating) at times show distinctly different variations than the $F_{10.7}$ index (Hedin 1984). Paxton et al. (1988) have shown that the line center solar Lyman-α flux, measured by the PV OUVS instrument, is linearly correlated with the $F_{10.7}$ flux. Brace et al. (1988) give an approximate formula relating the I_{pe} and $F_{10.7}$ indices.

In Fig. 2 the average 3-solar rotation $F_{10.7}$ index as a function of year for the various satellite missions and Earth-based measurements is shown. Short-term solar activity variations at higher solar activity (SMAX) were observed in the PVO 1978–1980 *in-situ* data. Lower mean solar activity (SMED) *in-situ* measurements were simultaneously available in Fall 1992 from PVO and MGN. Remote sensing also covers the intervening years, when measurement

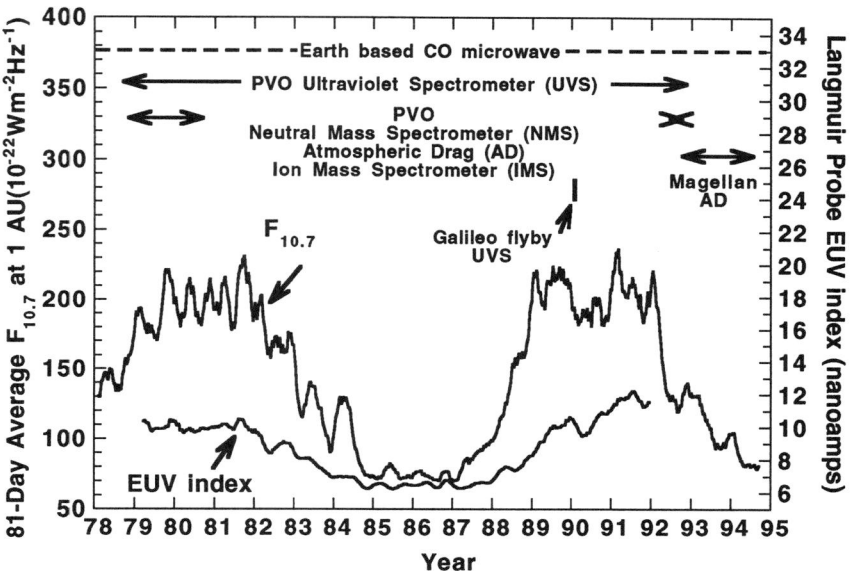

Figure 2. Solar activity as a function of year with date ranges for the various neutral density data sets obtained above 100 km. The 3-solar rotation 10.7-cm radio flux index, $F_{10.7}$, is measured at Earth and adjusted for the difference in phase of Earth and Venus. The solar EUV index, I_{pe} (Brace et al. 1988) is measured at Venus and has more than half of its contribution from Lyman-α.

conditions are suitable.

II. POST VENUS-I BOOK RESULTS

A. Data and Further Data Analysis

Figure 3 shows an example of the neutral atmosphere composition (He, N, O, N_2, CO and CO_2) measured by the PV ONMS instrument on orbit 5034 at 2.2 hr local solar time during the pre-entry period at SMED conditions (Kasprzak et al. 1993a). H is from Grebowsky et al. (1995). Mass density measured by the PV OAD experiment (Keating and Hsu 1993) is also shown along with ONMS derived mass density. CO_2 becomes dominant below about 140 km while O is dominant above this altitude.

Shown in Fig. 4 is a comparison of near equatorial orbiter drag measurements, normalized to 180 km, obtained at SMAX and at SMED conditions, around noon and midnight (Keating and Hsu 1993). The Venus International Reference atmosphere (VIRA) model (Keating et al. 1985) is calculated for SMAX conditions. PV OAD mass density measurements in the altitude range 165 to 180 km (day) and 160 to 180 km (night) were obtained during SMAX conditions (1978–1980). MGN data have been averaged using an 35 orbit (4 Earth day) running mean, to minimize the effect of day to day fluctuations,

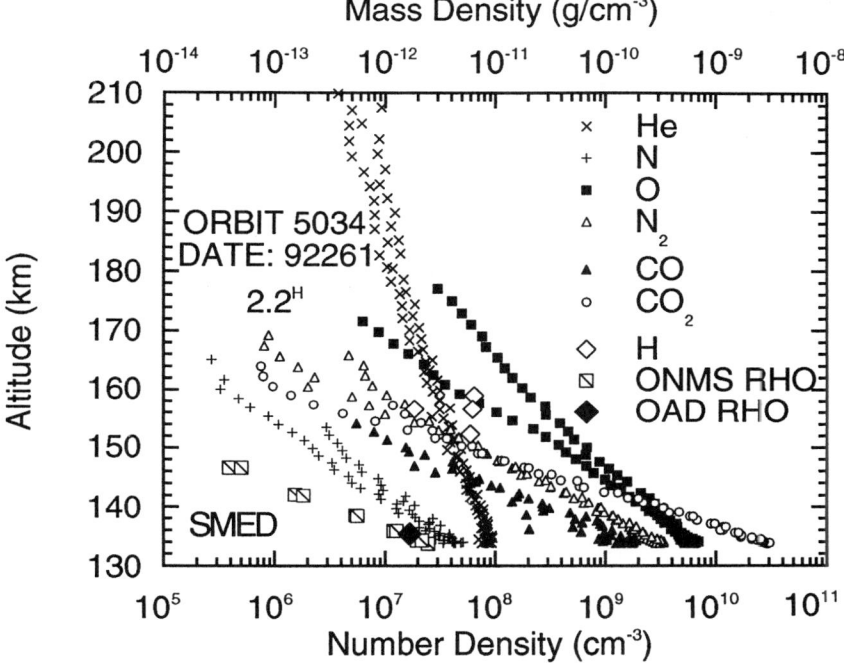

Figure 3. Composition data from the PV ONMS (He, N, O, N_2, CO, CO_2, RHO) (Kasprzak et al. 1993a; Niemann et al. 1980a), OIMS (H) (Grebowsky et al. 1995) and OAD RHO (drag determined mass density) (Keating and Hsu 1993) from pre-entry measurements at SMED conditions for orbit 5034 at 2.2 hr local solar time.

and were obtained during a period of SMED conditions (Fall 1992 to Spring 1993). Using an empirical model Keating and Hsu (1993) find that the density decrease from SMAX to SMED is about 20% for O at 150 km for both day and night, but smaller changes were detected near the terminators. The exospheric temperature decrease is about 30 K on the day side and about 10 K on the night side. Some experimental uncertainty may still exist in the absolute density comparison between the PV OAD and MGN measurements. MGN Cycles 5 and 6 show further nightside decreases as solar activity approaches SMIN (Keating and Hsu 1995). Latitude variations have also been observed in preliminary MGN density measurements for Cycles 5–6 (Espritu and Tolson 1995) that are not present in the low latitude VIRA model (Keating et al. 1985) (see the chapter by Bougher et al.).

An estimate has been made of the dayside exospheric temperature over the solar cycle (Fig. 5) by fitting 39 orbits of PV OUVS Lyman-α limb profiles with a model having three components: interplanetary background, a 1000 K "hot" component, and a "cold" component whose temperature was adjusted to provide a best fit (Stewart and Colwell 1995). The interplanetary

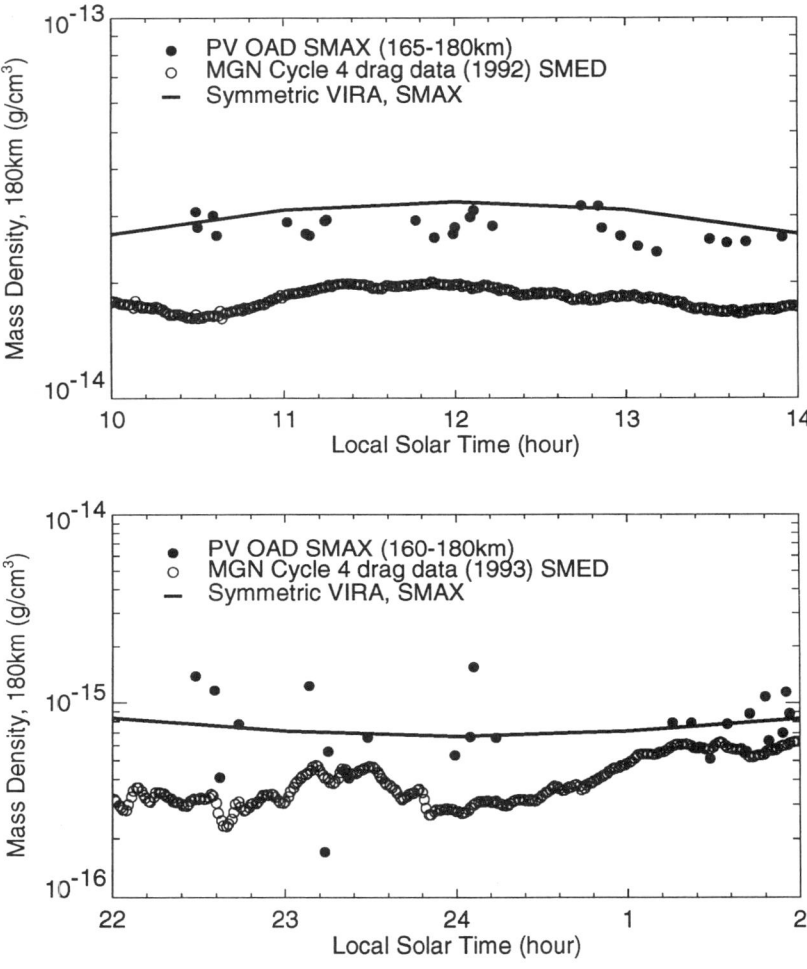

Figure 4. Mass density data (open circles) from Magellan cycle 4 drag measurements taken during SMED conditions, normalized to a constant altitude of 180 km, plotted as a function of local solar time (Keating and Hsu 1993) near noon (top) and near midnight (bottom). The data have been smoothed using a 35 orbit running mean. PV OAD data (Keating et al. 1980), from a selected altitude range and normalized to 180 km, are shown for SMAX conditions. The solid line is the VIRA model (Keating et al. 1985) prediction for SMAX conditions.

background is derived from OUVS measurements. The two planetary components were assumed to have temperatures independent of altitude. In this process, changes in the "scale height" of the cold component are interpreted as indicating changes in exosphere temperature. While the absolute values could be uncertain, as is evidenced somewhat by the data scatter, the trend

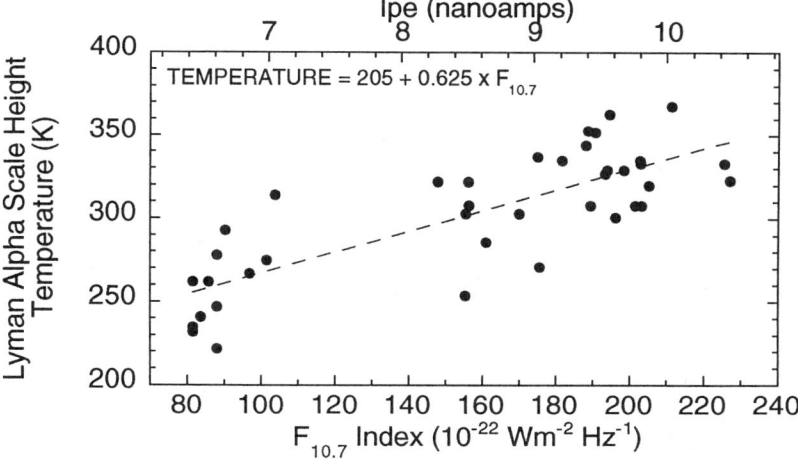

Figure 5. Exospheric temperature derived from Lyman-α limb scan data as a function of the solar activity as expressed by the $F_{10.7}$ and I_{pe} indices. The dashed line is a linear fit of the temperature to the $F_{10.7}$ index (Stewart and Colwell 1995).

of increasing temperature with increasing solar activity is clearly observed. From SMIN to SMAX the relative temperature change is small, only about 80 K for an $F_{10.7}$ change of 130 units.

Brinton et al. (1980) deduced atomic H concentrations using the charge exchange equilibrium reaction $O^+ + H \Leftrightarrow O + H^+$ and accounting for the loss of H^+ through a reaction with CO_2. McElroy et al. (1981) suggested that the OIMS mass 2 ion was D^+ rather than H_2^+. Hartle and Taylor (1983) confirmed this identification so that neutral deuterium concentrations can be deduced by a similar reaction scheme. The derived neutral H local time variation is very similar to that of the H^+ variation (H. A. Taylor et al. 1980b). Hydrogen data from the PV OUVS (Paxton et al. 1985) also show a diurnal variation similar to that derived by Brinton et al. (1980). The neutral H data were recalculated by Grebowsky et al. (1995) and D data by Hartle et al. (1996). The latest values of the rate constants were used and a criterion was used for determining the region of validity for H^+ equilibrium on an orbit-by-orbit basis. The H local time variation for PVO years 1 to 3 shows a predawn bulge similar to that measured for the He variation (see the chapter by Donahue et al.). Neutral D concentrations (Hartle et al. 1996) are about a factor of 100 lower than those for H. The maximum value of H in the bulge region is approximately 400 times that of the dayside value, compared to about a factor of 20 for He. The bulge H and D densities at SMAX are smaller in value than those at SMED so that the bulge density increases with decreasing solar activity. Using a spectral model Mayr et al. (1980,1985) and Mengel et al. (1989) have shown that the pre-dawn helium and hydrogen bulges are due to the effects of wind-induced diffusion, which would produce a bulge at midnight, combined with superrotation of the thermosphere, shifting the bulge to its position near the dawn terminator. Other evidence of superrotation comes from the NO airglow maximum around 2 a.m. (Stewart et al. 1980b; chapter by Bougher et al.) and diurnal temperature minimum near 2 a.m. (Mayr et al. 1980; Keating et al. 1980).

Measurements of the temperature at 100 km range from 160 K to 180 K over the years 1979 to 1992 (except for 205 K in 1985) as derived from CO microwave emission measurements on the night side (Clancy and Muhleman 1991; chapter by Lellouch et al.), from Magellan radio occultation measurements (Jenkins et al. 1994), from infrared measurements by the PV OIR instrument (F. W. Taylor et al. 1980) and from PV Probe measurements (Seiff and Kirk 1991; Kliore et al. 1985). Changes in the 100 km CO nightside temperatures are attributed to changes in the mesosphere circulation (chapter by Lellouch et al.). The nightside exospheric temperatures of nearly 100 K are clearly much less than the lower-altitude mesospheric temperatures of 160 K to 205 K. Microwave spectral observations of CO have also been used to determine CO mixing ratios (Clancy and Muhleman, 1991; chapter by Lellouch et al.).

The O_2/CO_2 ratio near 110 km has been estimated to be about 0.3% using CI emission measurements at 166- and 156-nm from the PV OUVS instrument (Paxton 1985). The peak atomic carbon density is 5×10^6 cm^{-3} at 155 km. Le Compte et al. (1989) analyzed 297 nm airglow data from the same instrument and concluded that atomic C could be the dominant source of that emission. Fox (1982) has estimated a mixing ratio for O_2/CO_2 of 10^{-4} or less based on C^+ densities obtained from the PV OIMS instrument above 210 km. These values are in contrast to the measurements of the integrated column density of molecular oxygen near the cloud tops which imply a much smaller mixing ratio of less than 3×10^{-7} (Trauger and Lunine 1983). Fox and Bougher (1991) state that the photochemistry for this topic needs to be revised and the subject revisited.

A re-analysis of the PV OUVS OI 1304 Å intensities (Meier et al. 1983) has been done by Paxton and Meier (1986) in the range 150 to 300 km altitude using revised estimates of the 1304 Å and 1356 Å electron impact cross sections. Atomic oxygen densities, determined from modeling of the OI emission, are now consistent with *in-situ* density measurements. Nagy and Cravens (1988) have revised their model of hot atomic oxygen using updated O_2^+ densities and there is now agreement with OUVS observations of the hot O corona (see Fig. 6).

Bertaux et al. (1985) have examined Lyman-α and He I (58.4 nm) emissions from Venera 11 and Venera 12 made in December 1978, at SMAX. Exosphere dayside temperatures and number densities of the thermal component at the exobase (250 km) are in the range (275–300)±25 K, and $[(4-6)\pm 2] \times 10^4$ cm^{-3}, respectively for Venera 11 and 12 with some indication of a local time or latitude effect. A nonthermal or hot component of H is observed up 40,000 km altitude which can be simulated by an exobase temperature of 10^3 K and an

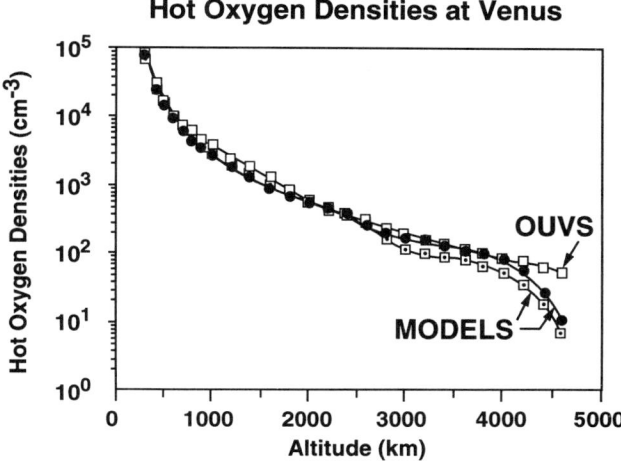

Figure 6. Hot O corona as observed by the PV OUVS instrument (Stewart et al. 1980a) on the day side and as predicted by several models (Nagy and Cravens 1988).

exobase density of 10^3 cm^{-3}. Measurements of He I during the same flyby (Chassefière et al. 1986; Bertaux et al. 1985) have been used to estimate the helium density at 150 km on the day side as 2×10^6 cm^{-3}, lower than *in-situ* measurements of 5×10^6 cm^{-3} from the Bus Neutral Mass Spectrometer and 7×10^6 cm^{-3} from the ONMS instrument (Hedin et al. 1983) but comparable to the Mariner 10 density (Kumar and Broadfoot 1975) of 2×10^6 cm^{-3} (145 km) at lower solar activity. There is an indication that the density increases at high latitudes above 60°.

B. Models

A comparison of some of the models discussed below is given in Fox and Bougher (1991). Niemann et al. (1980b) presented a semi-empirical model for the first diurnal cycle based on ONMS composition data for He, N, O, N_2, CO and CO_2. Temperatures were derived from atomic oxygen scale heights assuming diffusion equilibrium. A semi-empirical model of the PV OAD data mass density for the first PVO diurnal cycle with inferred temperature and composition (O, N_2+CO, CO_2) was presented by Keating et al. (1980). A one-dimensional chemistry model for the BNMS entry data was given in von Zahn et al. (1980) for a solar zenith angle of about 60°. The 1978–1980 PV ONMS data at SMAX were the basis of the "VTS3" semi-empirical model by Hedin et al. (1983) which was extended into the lower thermosphere, down to 100 km, based on additional data from the Pioneer Venus lower atmosphere probes, BNMS, and OIR. The lower atmosphere extension was constrained by hydrostatic equilibrium and the atomic nitrogen profile guided by the results of a dayside chemical one-dimensional model (Rusch and Cravens 1979).

The CO and O mixing ratios at 100 km were assumed to be equal based on the photochemical model of Yung and De More (1982). Species included He, N, O, CO, N_2, and CO_2. A factor of 1.6 was applied to the ONMS data for all species to match those from the BNMS and lower atmosphere probes, and a similar factor is needed to match approximately the OAD mass densities (Hedin et al. 1983).

Another model of temperature, pressure and mass density for solar zenith angles less than 50° and greater than 120° is given in Seiff (1983). The one-dimensional chemical model of Massie et al. (1983) was generated to bridge the gap between 140 and 100 km using available aeronomical data. Recent general circulation models have been more successful in this respect because they couple day side and night side through dynamic flow. A semi-empirical model has been presented by Keating and Hsu (1993) based on Cycle 4 Magellan, Magellan aerobraking, and Pioneer Venus mass density data derived from atmospheric drag.

The VIRA (Kliore et al. 1985) thermosphere model (Keating et al. 1985) above 150 km is based primarily on PV upper atmosphere composition from the ONMS instrument (Niemann et al. 1980b), H composition from Brinton et al. (1980) and mass density measurements from the OAD experiment (Keating et al. 1980). Tabulations were generated for medium solar activity conditions ($F_{10.7} = 150$), 16°N latitude, and 150 to 250 km altitude. The variation with solar activity is also presented. Included are species' densities of H, He, N, O, CO, N_2, CO_2, and temperature, mass density and pressure. The tabulations are symmetric in solar zenith angle so latitudinal variations could be easily estimated, but do not describe the asymmetries observed near the terminator nor the local time asymmetries of the He and H bulges which are displayed graphically. The photochemical region from 100 to 150 km is presented separately at noon and midnight and incorporates the mixing ratios given by the one-dimensional chemical model of Massie et al. (1983) near the homopause, except for increased CO based on a VTGCM calculation (Bougher et al. 1988). The nightside N, O and CO profiles are more realistic below 150 km than those of Hedin et al. (1983). The VIRA model also includes results from earlier VTGCM calculations, chemical models, and measurements and models of the hot exosphere from 250 to 3000 km. The Hedin et al. (1983) and VIRA (Keating et al. 1985) models were generated using data acquired in 1978–1980 for SMAX conditions and some caution should be exercised when extrapolating them to conditions for which there was no constraining data.

Gerard et al. (1988) using a photo-chemical diffusion model has been able to reproduce the dayside variation of N(^{4}S) measured by the PV ONMS instrument both in local time and altitude above 150 km. Computed ion species also compare favorably with PV OIMS measurements. The model computes N(^{4}S), N(^{4}D) and NO but only a comparison with N(^{4}S) above 150 km was given because *in-situ* NO measurements are not available. Computed NO profiles (Fox 1992) on the night side have mixing ratios at 131 km of about

3×10^{-6}, a factor of 10 less than that derived by Gerard et al. (1988) on the day side of 3×10^{-5} at 135 km.

The National Center for Atmospheric Research (NCAR) Venus Thermosphere General Circulation Model (VTGCM) (Bougher et al. 1988,1990; Fox and Bougher 1991; Bougher and Borucki 1994) is a three-dimensional model of the upper atmosphere which calculates the global distributions of CO_2, CO, N_2, and O (major species) and O_2, $N(^2D)$, and $N(^4S)$ (minor species) consistent with the observed day-night contrasts and the corresponding large-scale winds above 95 km. Superrotation is prescribed and added to the mean subsolar-to-antisolar circulation. A dayside photochemical ionosphere is calculated as a diagnostic of the dayside neutral structure. A general review of the VTGCM and similar models up to 1991 is given in Fox and Bougher (1991). The most recent VTGCM models are described and presented in the chapter by Bougher et al.

A spectral model of the thermosphere with super-rotation is described in Stevens-Rayburn et al. (1989) and Mengel et al. (1989) based on an earlier three-dimensional model of Mayr et al. (1985). This model is a nonlinear three dimensional, multi-constituent model with a spherical symmetric background atmosphere based on the Hedin et al. (1983) semi-empirical model. The perturbations due to atmospheric rotation, solar heating, cooling, thermal heat conduction and viscous dissipation are described in terms of spherical harmonic expansions about the mean atmosphere. Rayleigh friction (Bougher et al. 1986,1988) is added to the horizontal momentum equation to slow down the terminator winds, resulting in larger diurnal temperature and density amplitudes, and less heat transport to the night side. The model uses the heating rates of Fox (1988) with an efficiency close to 20%. The general circulation model reproduces the PV ONMS observed CO_2 and O diurnal thermosphere distributions and the He bulge maximum density near 5 a.m. with a best fit inferred superrotation rate of 6 days.

III. SOLAR ACTIVITY BEHAVIOR

Solar activity variations occur on two different time scales: (1) a variation with the solar rotation; and (2) with a period of about 11 years due to the solar cycle. The PVO data from 1978–1980 address the solar rotation variation as observed at SMAX ($F_{10.7} \approx 200$). PVO pre-entry data were primarily obtained on the night side at SMED ($F_{10.7} \approx 130$). Magellan Cycle 4 drag data during this same period covered both day side and night side at slightly lower solar activity ($F_{10.7} \approx 110$ to 130, still called SMED conditions) and Cycle 6 occurred near SMIN conditions. While the upper atmosphere changes slowly with the slowly changing solar activity, it is not totally quiescent. Wave-like structures and density perturbations have been observed in the ONMS neutral density measurements and in other measurements (Keating et al. 1979b; von Zahn et al. 1980; Niemann et al. 1980b; Kasprzak et al. 1988,1993; Mayr et al. 1988,1992; Seiff and Kirk 1991; Seiff et al. 1992; Keating and Hsu 1993) that

A. 28-day Solar Rotation Variation

The 27-day solar rotation period measured in the Earth-based $F_{10.7}$ index is observed as approximately a 28.3 day period on Venus. The 27-day variation has been correlated with changes in the atmospheric drag acceleration of Earth satellites (Jacchia 1964). A lag of about 1 day between the $F_{10.7}$ index and the inferred density changes was observed (Jacchia et al. 1973). The Venus Hedin et al. (1983) model uses a zero day delay. Density fluctuations were observed in PV ONMS data for all species, except He, which were correlated with the 28.3 day solar rotation period in the $F_{10.7}$ index. The fluctuations also increased with species mass and altitude implying an exospheric temperature variation. There was some ambiguity in assigning these density fluctuations to the variations in solar rotation flux because the PVO local solar time cycle at periapsis (225 days) was nearly commensurate with an integral number of 28.3 day solar rotation periods. In general, the model represents the variations fairly well except for CO_2, CO and N which have variations larger than predicted. No direct observation of either a density or a temperature effect correlated with the solar rotation could be detected in ONMS data near equatorial midnight because of the very large orbit-to-orbit density variations (Niemann et al. 1980b). The Hedin et al. (1983) model has no long- or short-term solar activity effect at midnight on the equator for this reason. Variations in dayside temperatures inferred from density measurements exhibit a solar rotation component. The top 3 panels of Fig. 7 show exospheric temperatures derived from ONMS atomic oxygen scale heights (Niemann et al. 1980b) correlated with the $F_{10.7}$ index and with the VEUV index which is linearly related to the EUV I_{pe} index (Brace et al. 1988). A similar correlation using PV OAD mass density measurements to infer temperature changes is shown in the bottom panel (Keating and Bougher 1992a,b). The correlation of the $F_{10.7}$ index with the Lyman-β and He II lines was noted early by H. A. Taylor et al. (1982) in a study of ionospheric response to solar activity. Keating and Bougher (1992a,b) have used the 28-day $F_{10.7}$ index as a proxy for the EUV He II line in a study of the neutral atmosphere response.

H. A. Taylor et al. (1982) found evidence for solar control of the H^+ bulge (proportional to the neutral H bulge) and the neutral He bulge in the second PVO diurnal cycle. A long-term (about 20 day) decrease in the solar flux is accompanied by a decrease in both the H^+ and He densities from 3 to 5 hr local solar time. The decrease reflects changes in solar heating and photoionization which occur in the near dawn sector on the day side and are transported into the bulge region. This observation points out the fact that the diurnal variations observed in the various neutral and ion species are not static but are part of an atmosphere in dynamic circulation. Species observed at noon on the day side are transported in two days to the night side above

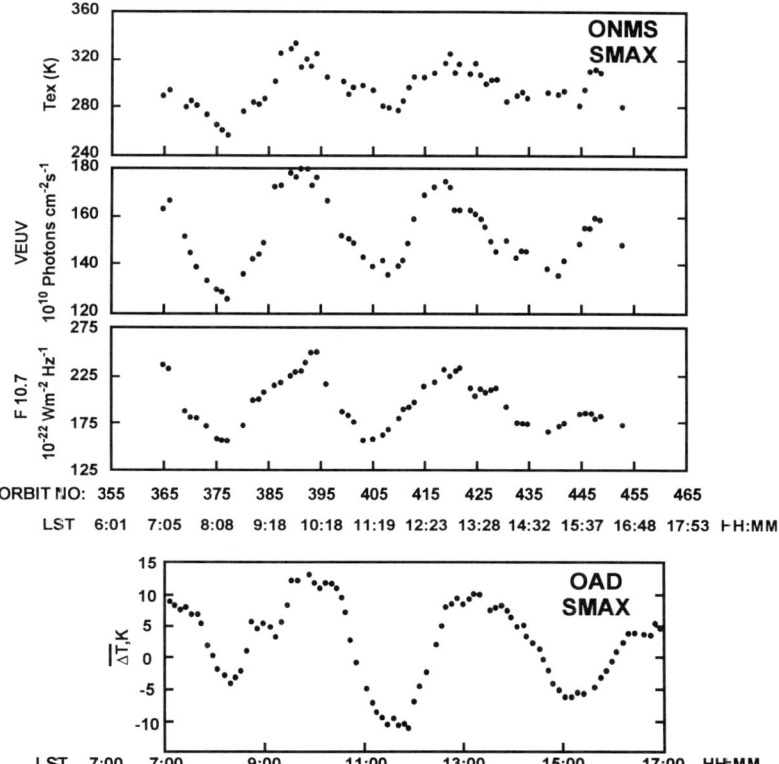

Figure 7. Scale-height temperatures, Tex, derived from PV ONMS dayside atomic oxygen measurements (Niemann et al. 1980b), the Langmuir Probe VEUV index (Brace et al. 1988), the daily $F_{10.7}$ index (corrected for the Earth-Venus phase difference) (Mahajan et al. 1990) and PV OAD exosphere temperature change derived from mass density measurements (Keating and Bougher 1992a,b) plotted as a function of local solar time.

110 km (Bougher et al. 1990). It was also speculated by H. A. Taylor et al. (1982) that day-to-day fluctuations in the observed H^+ and He bulges might be due to daily fluctuations in the day-to-night circulation as observed on the night side in the NO airglow (Stewart et al. 1980b).

Using PV ONMS scale-height temperature data for atomic oxygen at SMAX (Niemann et al. 1980b), H. A. Taylor et al. (1982) found that the dayside 28-day solar rotation temperature excursion was as much as 40 to 50 K per 80 units of $F_{10.7}$ index change (0.5–0.6 K/$F_{10.7}$ unit). The VIRA model shows a dayside temperature change of about 18 K per 50 units of $F_{10.7}$ (0.36 K/$F_{10.7}$ unit) at SMED (Keating et al. 1985, their Figs. 4–12). The global mean temperature excursion due to a solar rotation is 0.14 K/$F_{10.7}$ unit in the Hedin et al. (1983) model and 0.19 K/$F_{10.7}$ unit in the VIRA model

(Keating et al. 1985). A later study of Mahajan et al. (1990) using ONMS atomic oxygen scale-height temperatures found a value of about 0.17 K/$F_{10.7}$ unit. The Venus 28-day solar rotation response of 0.14 to 0.19 K/$F_{10.7}$ unit in the global mean temperature is about 1/10 that observed for Earth of about 1.7 K/$F_{10.7}$ unit (Hedin 1987). In the Hedin et al. (1983) model for Venus the temperature response to the solar rotation $F_{10.7}$ effect near the equator is approximately cancelled at night and doubled during the day.

Keating and Bougher (1992b) give a correlation between the EUV temperature response of OAD mass density over a solar rotation period and the corresponding $F_{10.7}$ index using VTGCM model results. The two dominant temperature contributions come from the He II line at 303.8 Å (a 23.4 K change in temperature for a 0.42% change in line intensity corresponding to a 1% change in $F_{10.7}$) and the 250 to 300 Å region (a 19.8 K change in temperature for a 0.54% change in band intensity corresponding to 1% change in $F_{10.7}$). Hedin (1984) also found a high correlation between Earth thermosphere N_2 densities (and temperatures), and 304 Å line and 255 to 300 Å band emissions.

B. Long-Term Solar Cycle Variation

Comparison of PVO data from 1978–1980 at solar maximum with PVO preentry data and Magellan data at solar medium conditions gives an indication of the thermosphere change due to long-term solar activity. The PVO pre-entry data occur only on the night side, from about 0 to 4 hr local solar time, and reflects changes in dayside heating and global circulation due to changes in the solar activity.

Figure 8 shows a comparison of the CO_2 and O from PV ONMS measurements at SMAX and SMED conditions (Kasprzak et al. 1993a; Niemann et al. 1980b), local time range 0.5 to 1.5 hr, where the data sets overlap significantly. Also shown are predictions from the semi-empirical model, VTS3, of Hedin et al. (1983) and the general circulation model, VTGCM, of Bougher and Borucki (1994). The density changes due to the solar activity change (i.e., a decrease of about 24% for O at 150 km and 1 a.m.) are small particularly in view of the highly scattered data and instrumental uncertainties. Model predictions also show small density changes (about 15% for the VTS3 model and 25% for the VTGCM model at 150 km and 1 a.m.) (Kasprzak et al. 1993a).

The H and D concentrations, in the bulge region, show an inverse correlation with the 3 solar rotation average $F_{10.7}$ index compared to He (Fig. 9). This effect in H and D has been attributed by Hartle et al. (1996) to a reduction in the escape rates of H and D during lower solar activity which are primarily due to the loss rates of H^+ and D^+ aided by polarization electric fields (Hartle and Grebowsky 1995). PV OUVS observations of the dayside exospheric Lyman-α indicate that the cold component decreases by a factor of 2, and the hot component increases by a factor of 2.5, from SMIN to SMAX. In the Earth's thermosphere global average cold H densities also increase with decreasing solar activity (Hedin and Mayr 1987). Kasprzak et al. (1993a)

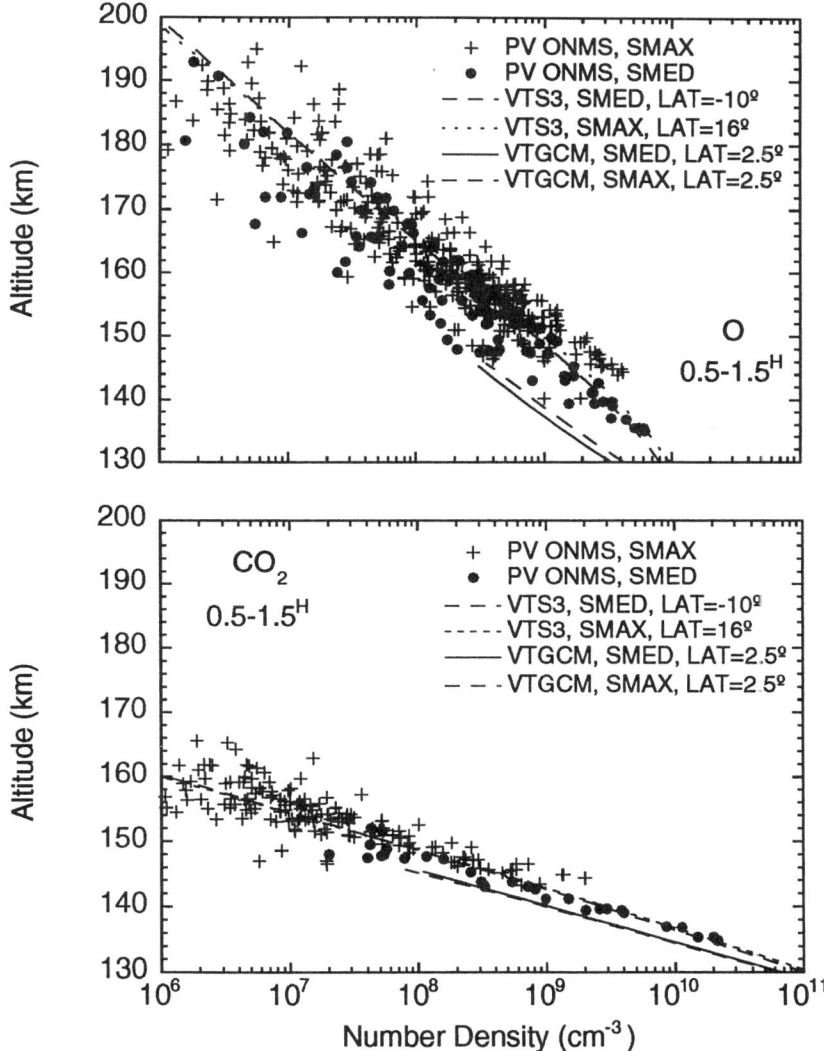

Figure 8. PV ONMS O and CO_2 number densities (Niemann et al. 1980b; Kasprzak et al. 1993a), and predictions from the VTS3 and the VTGCM models for SMAX conditions and SMED conditions plotted as a function of altitude. The data are restricted to a local time range of 0.5 to 1.5 hr for comparison.

have compared the pre-entry data for He with those of the first diurnal cycle in 1980 at solar maximum and find no observable change in the amplitude and phase. H. A. Taylor et al. (1984, 1985) found no significant interannual variation when the first 3 diurnal cycles of calculated H data from PVO were correlated with the He 304 Å EUV line flux.

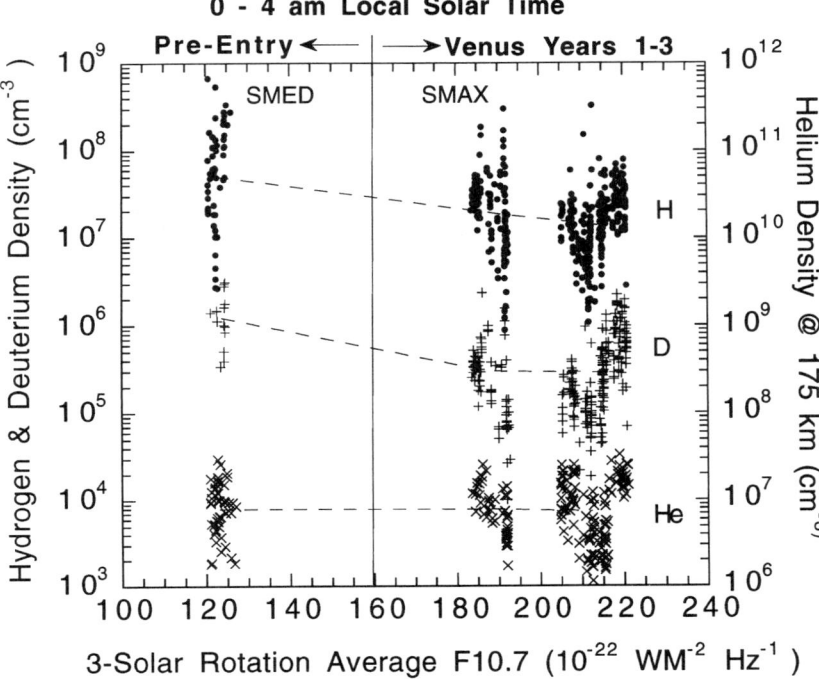

Figure 9. Variation of H, D, and He with solar activity for local solar time 0 to 4 hr. Dashed lines connect the logarithmic averages of the three major $F_{10.7}$ groupings around 120, 190 and 215, respectively (Grebowsky et al. 1995; Hartle and Grebowsky 1995; Niemann et al. 1980b; Kasprzak et al. 1993a).

Observations of the gamma and delta band nitric oxide nightglow (Stewart et al. 1980b) by the PV OUVS instrument have been extremely useful in constraining the day to night circulation (Massie et al. 1983; Bougher et al. 1990; Fox and Bougher 1991; chapter by Bougher et al.). Atomic nitrogen is created on the day side by photoelectron impact and photodissociation processes mainly in the 135 to 140 km region while atomic oxygen and carbon monoxide are largely produced near 110 km. Winds transport the radicals to the night side where they recombine around 115 km (Gerard et al. 1981) near 2 a.m. just south of the equator (Stewart et al. 1980b). Measurements from 1984–1986, at solar minimum, were not possible due to the high periapsis altitude of PVO. However, several measurements at apoapsis during this period indicate that the nightside NO airglow is approximately a factor of 3 smaller than at solar maximum in 1978–1980 (Fox and Bougher 1991; Colin 1989).

Other ultraviolet dayglow emissions show a marked dependence on solar activity (Stewart and Colwell 1995). The CO fourth positive band at 1392 Å is excited by the accidental resonance of the (14,0) band with solar Lyman-α.

The solar emission as measured by the Solar Mesosphere Explorer (SME) Earth satellite shows a direct proportionality to the PV OETP photoelectron current I_{pe} (Brace et al. 1988; T. N. Woods 1995, personal communication), and varies by a factor of 1.6 from SMIN to SMAX. The CO emission varies by a factor of 2.0 (Fig. 10, top), consistent with a 25% increase in thermospheric CO. The (0,1) CO band at 1597 Å increases by factor of 2.1 and the O resonance emission at 1304 Å increases by a factor of 2.2 (Fig. 10, bottom) (see also Alexander et al. 1993). However their excitation mechanisms involve both scattering of solar emissions and excitation by thermospheric processes such as photoelectron impact, photodissociation, and dissociative recombination, and a more careful interpretation is required.

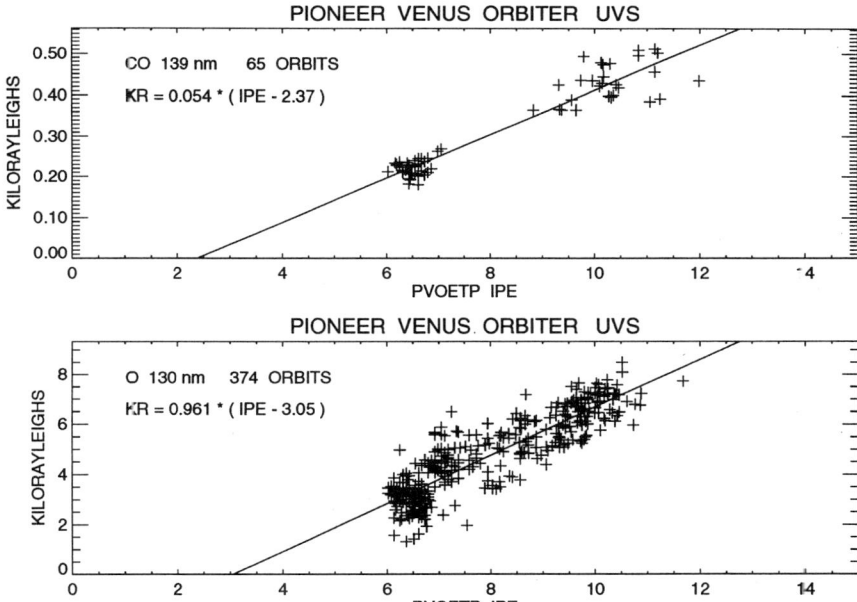

Figure 10. PV OUVS (Stewart 1980a) measurements of CO (top) and O (bottom) emissions as function of year and solar activity (Stewart and Colwell 1995). The variation of the CO emission, due to scattered Lyman-α radiation, shows a stronger dependence on solar activity than does the solar Lyman-α (see text) implying that the CO column emission varies with solar activity. The atomic oxygen emission is also well correlated with solar activity.

Keating and Hsu (1993) found that the exospheric temperature decrease inferred from an empirical model after including Magellan cycle 4 and aerobraking data was about 30 to 38 K on the day side and less than 15 K on the night side for an $F_{10.7}$ decrease of 180 to 130. Over the same $F_{10.7}$ range the dayside exospheric temperature change measured by the PV OUVS experiment (Fig. 5) is about 31 K. The exosphere temperature change from SMAX to SMIN had been previously estimated to be about 70 K (Fox and Bougher 1991) comparable to PV OUVS measurements of about 81 K. The small temperature change inferred from the PVO Neutral Mass Spectrometer (Kasprzak et al. 1993a) on the night side between the solar maximum value of 110 K and the pre-entry data of 105 to 120 K is consistent with that reported by Keating and Hsu (1993).

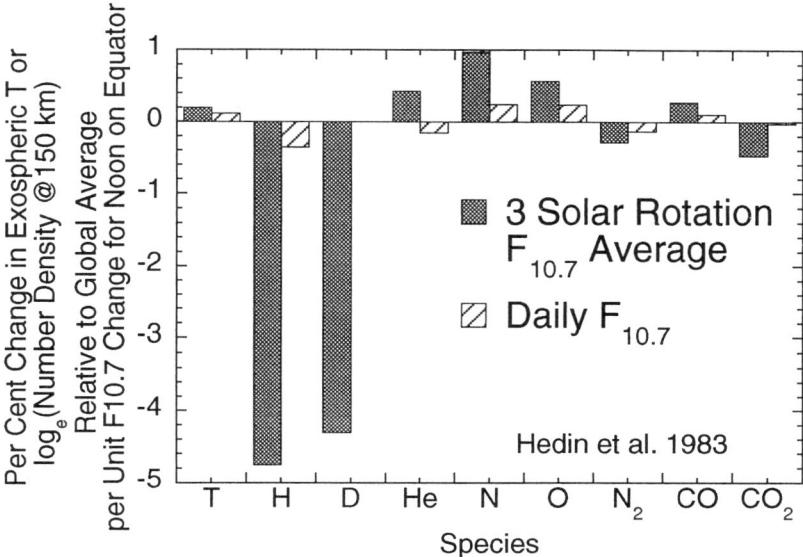

Figure 11. Response of the exospheric temperature T and natural logarithm of the species number density N at 150 km relative to the average for the 28-day and 3-solar rotation solar activity effects at noon on the equator (from the model of Hedin et al. [1983a] based on data from Niemann et al. [1980b], Grebowsky et al. [1995] and Hartle and Grebowsky [1995].

The Hedin et al. (1983) model, based on data obtained near SMAX, predicts about 0.23 K/$F_{10.7}$ unit change for the global long-term effect which is roughly twice the short-term response. For Earth the long-term response is roughly three times the short-term response (Hedin 1987). Figure 11 summarizes the composition changes at 150 km and the exospheric temperature response to both long and short term solar activity of the Hedin et al. (1983) model, with added H and D. The short-term response is less than the long term. H and D have large long-term effects compared to the other species with increasing densities for decreasing solar activity. N_2 and CO_2 decrease with increasing solar activity compared to N, O and CO as noted by Hedin et al. (1983).

Figure 12 (top) shows the long-term solar activity response in the dayside

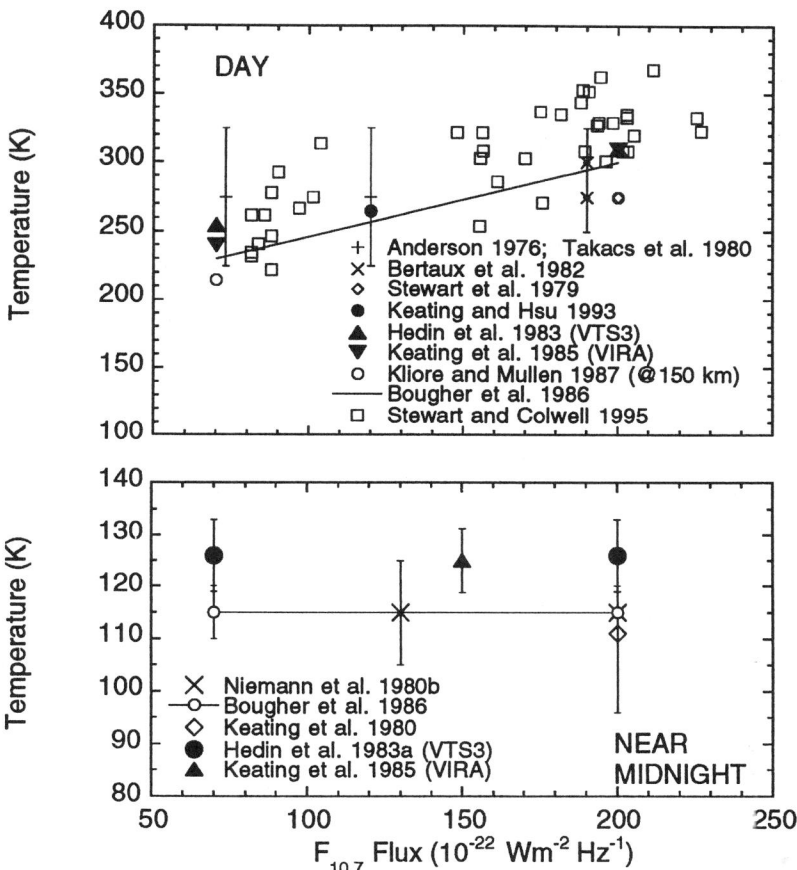

Figure 12. Dayside and near midnight exospheric temperatures measured over the solar cycle and predicted by several models for Venus.

exospheric temperature from several data sources and models as a function of the 3-solar rotation $F_{10.7}$ index. Fox and Bougher (1991) summarize the change as 50 to 70 K from solar minimum ($F_{10.7} = 70$) to solar maximum ($F_{10.7} = 200$) based on some of the same data. Over this same solar activity range, the Earth's exospheric temperature changes from about 725 K to 1200 K (Hedin et al. 1987). The relative temperature changes, however, are within a factor of 2. Nightside temperatures (Fig. 12, bottom) are relatively constant with solar activity. Given that Venus receives 1.9 times more radiation than does the Earth due to its closer proximity to the Sun, it is surprising that the dayside exospheric temperature is so low. This points to a very efficient cooling mechanism for the Venus thermosphere that is not as effective as comparable altitudes in the Earth's thermosphere.

NCAR two-dimensional model and VTGCM results (Bougher et al.

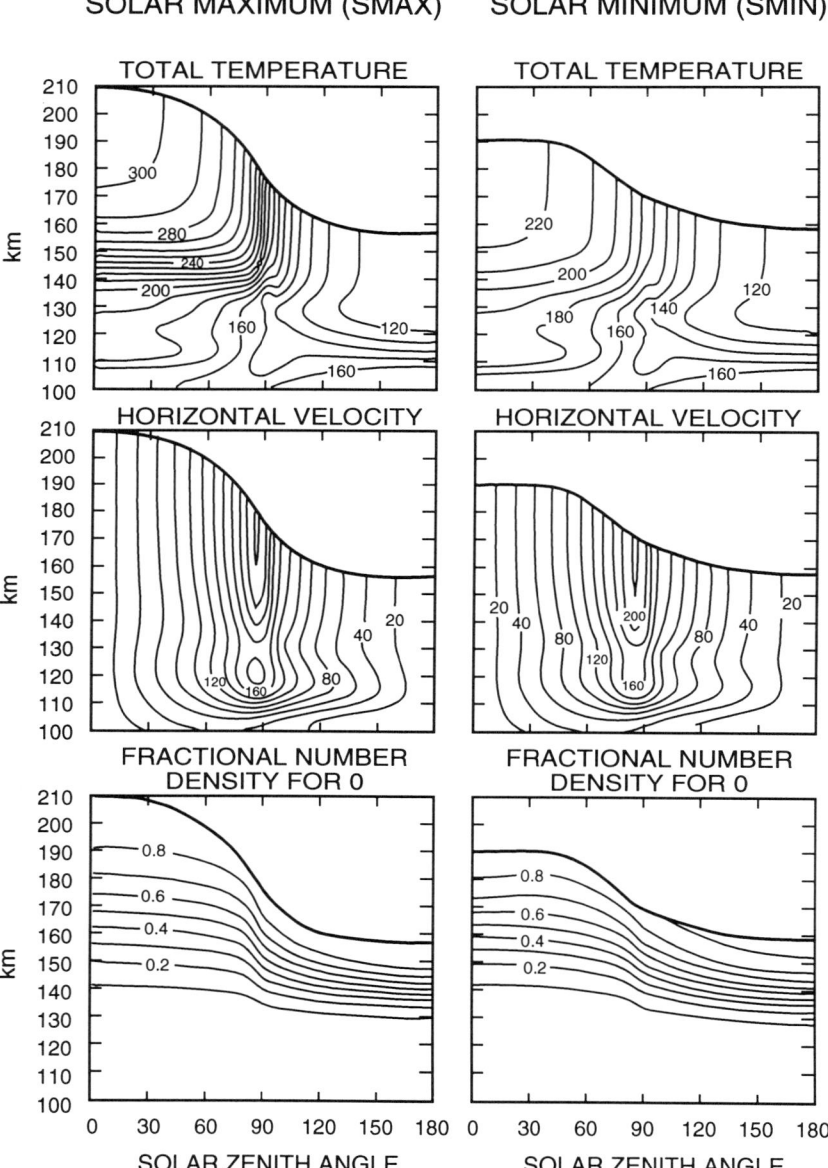

Figure 13. NCAR two-dimensional model predictions for SMAX and SMIN conditions showing the temperature, horizontal wind speed and fractional number density for O (figure from Bougher et al. 1986).

1986,1988,1990; Bougher and Borucki 1994) both show that from solar activity levels SMAX to SMIN EUV inputs for wavelengths less than 105 nm are reduced by about a factor of 3; O is reduced; there is reduced CO_2 cooling; wind speeds are lower by 20 to 30 m s^{-1}; the NO airglow is reduced by about a factor of 3; the nightside temperature remains about the same with a minimum near 2 a.m.; and the dayside temperature decreases by about 80 K (Fig. 13). The small response in temperature for Venus is due to the strong nonlinear temperature dependent CO_2 15-μm cooling which is regulated by the changing dayside atomic oxygen (Bougher and Roble 1991; Keating and Bougher 1992a,b; Bougher et al. 1994; Keating and Hsu 1993). This thermostat effectively buffers the response of the entire thermosphere to changes in the EUV activity which is the main heating source. In the VTGCM, the insensitivity to solar cycle change at night is due to the relative isolation of the dayside and nightside thermosphere, maintained by Rayleigh friction, which slows down the pressure-driven winds (chapter by Bougher et al.). Apparently, the small change in the dayside thermosphere does not greatly affect the nightside composition or temperature near 1 a.m. (Kasprzak et al. 1993a). The nightside MGN mass density decrease from SMAX to SMED is not inconsistent with that observed by the ONMS for atomic oxygen. At other local solar times MGN and PV OAD nightside mass density data (Fig. 4) (Keating and Hsu 1993) would seem to imply a somewhat larger density difference. This suggests that the nightside atmosphere may be more sensitive to solar activity changes than predicted by either the Hedin et al. (1983) or VIRA (Keating et al. 1985) empirical models which were generated using data from SMAX. Keating and Hsu (1993) interpret the MGN dayside density decrease at 180 km with decreasing solar activity (Fig. 4) to be the result of decreased O production combined with cooling of the thermosphere and on the nightside an additional decrease in the day-to-night transport is invoked due to the decrease of densities on the day side.

C. Possible Connections with Solar Wind and Ionosphere

Niemann et al. (1980b) have presented a series of 5 orbits that are an example of a disappearing atmosphere. The nightside orbits at local time approximately 20 to 21 hr are correlated with the plasma density (Luhmann 1986). A normal appearing atmosphere becomes drastically eroded on successive orbits with nearly an order of magnitude reduction in density values and finally relaxes back to nearly quiet conditions. During this time the plasma density also shows an irregular depletion typically associated with a disappearing ionosphere (Luhmann 1986; Cravens et al. 1982). Waves are observed in the neutral density profiles. Moderate correlation of neutral and plasma density waves, in general, has been found by Hoegy et al. (1990). Other examples have been observed of disturbed neutral atmospheres, ionospheres and superthermal ions (Kasprzak et al. 1982). In analysis of wave-like features in PV ONMS neutral composition, Kasprzak et al. (1988) noted that narrow pulse-like perturbations were observed that could be produced by an unidentified excitation source in the lower thermosphere (Mayr et al. 1988).

Keating et al. (1980) examined the correlation of the first diurnal cycle of PVO mass density with solar wind speed and concluded that there was no detectable effect. Luhmann (1986) pointed out that solar wind heating of the upper atmosphere should occur during times of high solar wind dynamic pressure. Exosphere temperatures inferred from ONMS density measurements were analyzed by Mahajan and Kar (1988) and Kar et al. (1991) for a possible dependence on the solar wind dynamic pressure. Only one orbit was found where a high solar wind pressure was accompanied by an apparent enhancement in the exosphere temperature.

IV. UPPER ATMOSPHERE HEAT BUDGET

A. Dayside and Global Mean Atmosphere

As Bougher et al. (1994) point out, resolving the heat budget of Venus has been a problem for almost 20 years (see review by Fox and Bougher 1991). Venus has a dayside exosphere temperature less than 300 K, is cooler than Earth and yet closer to the Sun. In order to cool the thermosphere large infrared or eddy thermal cooling must be invoked to balance the EUV heating. EUV heating efficiencies have been reviewed in Fox (1988). Some fraction of the heat produced by photoionization and photodissociation in the range from about 10 Å to about 2000 Å appears as heat, with the ratio of the heat produced to the energy absorbed being the heating efficiency. Other sources of heat input are ion-molecule reactions. The major source of energy above 130 km on the day side is O_2^+ dissociative recombination with quenching of metastable species and photodissociation being the main source below 130 km as shown in Fig. 14, left. The resulting heating efficiencies (Fig. 14, right, curves A and B) are in the range of 16 to 25% over the altitude range of 115 km to 200 km. The re-examination of the heating efficiencies was prompted by early models such as Dickinson and Bougher (1986) which used lower heating efficiencies in order to reproduce the low solar maximum dayside temperatures of 275 to 300 K (Keating et al. 1980; Niemann et al. 1980b; von Zahn et al. 1980). Bougher et al. (1986) stated that there seemed to be no way of reconciling heating efficiencies greater than about 10% with the current estimated magnitude of the cooling mechanisms used at that time. Strong eddy cooling was inconsistent with the eddy diffusion needed in hydrodynamic models. Circulation was not an effective thermostat in regulating the dayside temperature. Modest cooling by CO_2 at 15 μm was given by the standard non-LTE calculation that included enhancement by CO_2-CO_2 collisions, but relatively little enhancement by O-CO_2 collisions (Bougher et al. 1986,1988).

O-CO_2 collisions are now thought to be especially effective in exciting CO_2 vibrational states (e.g., $\nu_2 = 1$ bending mode), resulting in enhanced CO_2 15-μm emissions and strong cooling at thermosphere heights where non-LTE conditions prevail. The effectiveness of this enhancement process depends

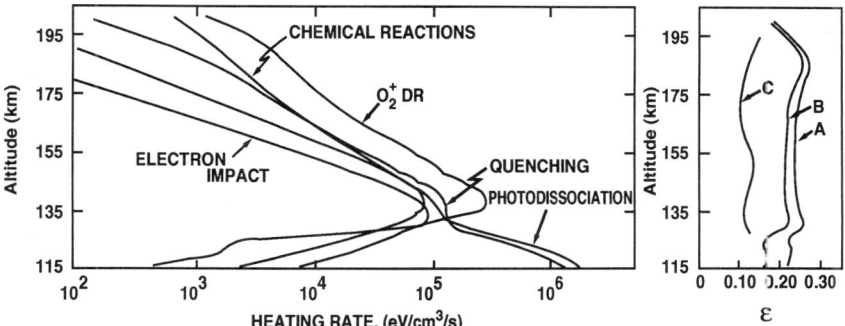

Figure 14. (Left) Heating rates due to various sources as a function of altitude for the dayside atmosphere (Fox 1988). (Right) Heating efficiencies estimated for a standard model (curve A) using a "best guess" value for the fraction of energy that appears as vibrational excitation in the quenching of metastable atomic oxygen and for a lower limit ("not unreasonable" value) model (curve B) (Fox 1988). Curve C is from Hollenbach et al. (1985). Preferred values are 16 to 25% (curves A, B) in the range of 115 to 200 km (Fox 1988).

upon the relative O densities at a given pressure level plus the collisional energy transfer rate coefficient for $O-CO_2$. In general, the measured quantity that is commonly quoted is the rate coefficient for quenching (relaxation), the reverse of excitation. Even though the importance of this $O-CO_2$ mechanism for enhancing 15-μm cooling in terrestrial planet upper atmospheres has long been recognized (Crutzen 1970; Bougher et al. 1994), the magnitude of the $O-CO_2$ relaxation rate has been the subject of vigorous debate. Only recently has significant progress been made that refines the estimates of this rate coefficient and improves the understanding of the role of CO_2 cooling in the terrestrial planet thermospheres. The corresponding $O-CO_2$ relaxation rate has been measured recently in laboratories at room temperature for the first time ($k = (1-2) \times 10^{-12}$ $cm^3 s^{-1}$) (Shved et al. 1991; Pollock et al. 1993). Also, derived values of this relaxation rate are becoming better constrained ($(1-6) \times 10^{-12}$ $cm^3 s^{-1}$) and are based primarily upon analyses of Earth CO_2 radiance and absorption measurements (Bougher et al. 1994). Certainly, these recently measured and derived values for the $O-CO_2$ rate are quite different. However, this rate was further constrained by incorporating its value into new terrestrial planet heat budget calculations.

Bougher and Roble (1991) have calculated the global mean temperatures for Venus, Earth and Mars. The computed temperature variation from solar maximum ($F_{10.7} = 240$) to solar minimum ($F_{10.7} = 70$) for Venus is 76 K, for Mars 110 K and for Earth 518 K. The reason for the small variation of Venus relative to Mars is that for Venus EUV heating and 15-μm CO_2 cooling occur in approximately the same altitude range and the O/CO_2 ratio is larger which increases the efficiency of the radiation. Larger EUV heating efficiencies were used, but less than those recommended by Fox (1988), and the collisional relaxation rate of 1×10^{-12} cm^{3}s^{-1} was used.

A study by Keating and Bougher (1992a,b) significantly advanced the understanding of relative magnitudes of the various cooling processes. A Venus one-dimensional model was used to examine the exosphere temperature variations due to the 28-day solar rotation at Venus. The temperature variations, inferred from mass density drag observations at solar maximum, have a peak-to-peak temperature amplitude of about 25 K (Fig. 7, bottom panel). The small temperature amplitudes combined with the necessity of maintaining a 300 K dayside temperature could be explained only by strong 15-μm cooling rather than eddy thermal conduction. A parameter study was conducted in order to quantify the values of the eddy diffusion coefficient, EUV heating efficiency and O-CO$_2$ relaxation coefficient which are consistent with observed 28-day exosphere temperature variations (Keating and Bougher 1992a,b). While the eddy diffusion coefficient might have long-term changes over a solar cycle, values should be essentially constant over a solar rotation period making it possible to isolate the relative significance of the eddy diffusion coefficient and the O-CO$_2$ relaxation coefficient in cooling the Venus thermosphere. In order to reproduce the observed weak solar rotation response in the exosphere temperature, the values of the VTGCM eddy diffusion coefficient K_t must be less than 2×10^7 cm^{2}s^{-1}. EUV efficiencies of 15 to 26% requires an O-CO$_2$ relaxation rate of $(1-4) \times 10^{-12}$ cm^{3}s^{-1}. A smaller range of EUV heating efficiencies (18–26%) require an O-CO$_2$ relaxation rate of $(2-4) \times 10^{-12}$ cm^{3}s^{-1}. The strong CO$_2$ cooling acts as a thermostat reducing the temperature sensitivity to EUV variations. Although the study dealt with only the 28-day variation there is an implication that the mean variation over the solar cycle should also be small. In addition, as solar activity increases, Venus O production increases along with CO$_2$ cooling and thus the thermosphere temperature response is further damped (Fox and Bougher 1991; Keating and Hsu 1993; Bougher et al. 1994). Keating and Hsu (1993) find, by combining in an empirical model the PVO and MGN data. that the ratio of O/CO$_2$ near 150 km at noon increases by 50% as solar activity increases from SMED to SMAX. This increases the O-CO$_2$ radiative cooling resulting in the changing O/CO$_2$ ratio acting as a natural thermostat (Bougher and Roble 1991).

Bougher et al. (1994) reviewed the current state of CO$_2$ cooling in the terrestrial planetary atmospheres (Venus, Mars, Earth), examining the values of the rate coefficient, temperature dependence and impact on the heat budgets. Values of the O-CO$_2$ relaxation rate of $(2-4) \times 10^{-12}$ cm^{3}s^{-1} at 300 K are suggested for future studies. This comparative approach provides the broadest range of conditions under which a common O-CO$_2$ relaxation rate should provide consistent results. Figure 15 shows the global mean heating rates for the Venus thermosphere (Bougher et al. 1994). The EUV heating and CO$_2$-O collisional cooling occur at about the same altitude resulting in a small response to solar activity changes. Eddy conduction plays a minor role in heat loss.

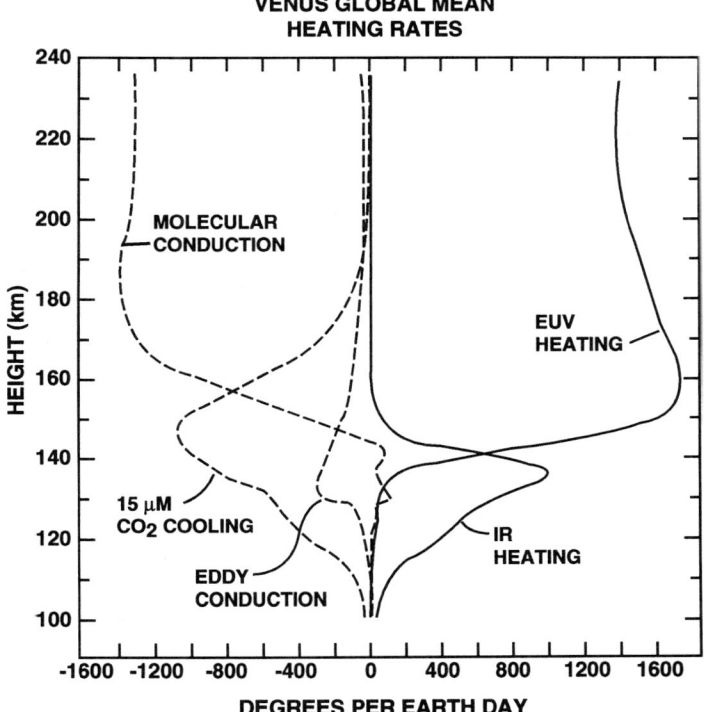

Figure 15 Global mean heating rates (figure from Bougher et al. 1994).

B. Nightside Upper Atmosphere

Nightside temperatures observed by PVO instruments, and PVO and MGN drag measurements, are very cold, of the order of 100 K. Isolation of the day side and night side is accomplished in the VTGCM by adding Rayleigh friction, due to wave drag, to the horizontal momentum equation (Bougher et al. 1988; Mengel et al. 1989). The effect of this addition is to slow down the flow from the day side to the night side, enhancing the day-to-night temperature and density contrasts, and further isolating the day and night thermosphere. The main source of heating is adiabatic compressional heating and that is lessened by reducing the cross-terminator winds (see review by Fox and Bougher [1991]; chapter by Bougher et al.). The source of the Rayleigh friction is assigned to upward propagating gravity waves from the lower atmosphere which break or saturate in the lower thermosphere and modify the horizontal flow. Alexander (1992) has presented a realistic gravity-wave braking model which supports this mechanism and suggests that gravity waves are also responsible for coupling the lower atmosphere superrotation observed at the cloud tops to the superrotation observed in the thermosphere by *in-situ* measurements. Upward propagating gravity waves are filtered by the lower thermosphere and upper mesosphere winds. The resulting wave drag is asymmetrical in local time, greater at the dawn terminator than the dusk terminator, yielding a substantial zonal wind component at thermosphere altitudes while simultaneously slowing down the subsolar-to-antisolar winds. The corresponding eddy diffusion produces asymmetries in the composition, especially O, with local solar time (Alexander et al. 1993).

Editor's Note

The CD-ROM contains a pointer to the VTS3 empirical model code.

REFERENCES

Alexander, M. J. 1992. A mechanism for the Venus thermospheric superrotation. *Geophys. Res. Lett.* 19:2207–2210.

Alexander, M. J., Stewart, A. I. F., Solomon, S. C., and Bougher, S. W. 1993. Local-time Asymmetries in the Venus thermosphere. *J. Geophys. Res.* 98:10849–10871.

Bauer, S. J., and Hartle, R. E. 1974. Venus ionosphere: An interpretation of Mariner 10 observations. *Geophys. Res. Lett.* 1:7–9.

Bertaux, J. L., Chassefière, E., and Kurt, V. G. 1985. Venus EUV measurements of hydrogen and helium from Venera 11 and Venera 12. *Adv. Space Res.* 5:119–124.

Bougher, S. W., and Borucki, W. J. 1994. Venus O_2 visible and IR nightglow: Implications for lower thermosphere dynamics and chemistry. *J. Geophys. Res.* 99:3759–3776.

Bougher, S. W., and Roble, R. G. 1991. Comparative terrestrial planet thermospheres: 1. Solar cycle variation of global mean temperatures. *J. Geophys. Res.* 96:11045–11055.

Bougher, S. W., et al. 1986. Venus mesosphere and thermosphere: II. Global circulation, temperature and density variations. *Icarus* 68:284–312.

Bougher, S. W., Dickinson, R. E., Ridley, E. C., and Roble, R. G. 1988. Venus mesosphere and thermosphere: III. Three-dimensional general circulation with coupled dynamics and composition. *Icarus* 73:545–573.

Bougher, S. W., Gerard, J. C., Stewart, A. I. F., and Fesen, C. G. 1990. The Venus nitric oxide night airglow: Model calculations based on the Venus thermospheric general circulation model. *J. Geophys. Res.* 95:6271–6284.

Bougher, S. W., Hunten, D. M., and Roble, R. G. 1994. CO_2 cooling in terrestrial planet thermospheres. *J. Geophys. Res.* 99:14609–14622.

Brace, L. H., and Colin, L. 1984. Pioneer Venus: Evolving coverage of the near Venus environment. *Eos: Trans. AGU* 65:401–402.

Brace, L. H., Hoegy, W. R., and Theis, R. F. 1988. Solar EUV measurements at Venus based on photoelectron emission from the Pioneer Venus Langmuir probe. *J. Geophys. Res.* 93:7282–7296.

Brace, L. H., et al. 1980. The dynamic behavior at the Venus ionosphere in response to solar wind interactions. *J. Geophys. Res.* 35:7663–7678.

Brinton, H. C., et al. 1980. Venus nighttime hydrogen bulge. *Geophys. Res. Lett.* 7:865–868.

Chassefière, E., Bertraux, J. L., Kurt, V. G., and Smirnov, A. S. 1986. Venus EUV measurements of helium at 58.4 nm from Venera 11 and Venera 12 and implications for the outgassing history. *Planet. Space Sci.* 34:585–602.

Clancy, R. T., and Muhleman, D. O. 1991. Long-term (1979–1990) changes in the thermal, dynamical, and compositional structure of the Venus mesosphere as inferred from microwave spectral line observations of ^{12}CO, ^{13}CO, and $C^{18}O$. *Icarus* 89:129–146.

Colin, L. 1980. The Pioneer Venus program. *J. Geophys. Res.* 85:7575–7598.

Colin, L., ed. 1989. *Pioneer Venus Orbiter—Ten Years of Discovery*, ed. L. Colin (Moffett Field, Calif.: NASA Ames Research Center).

Cravens, T. E., et al. 1982. Disappearing ionospheres on the nightside of Venus. *Icarus* 51:271–282.

Croom, C. A., and Tolson, R. H. 1994a. *Venusian Atmospheric and Magellan Properties from Attitude Control Data*, NASA CR-4619 (Hampton, Va.: Langley Research Center).

Croom, C. A., and Tolson, R. H. 1994b. Using Magellan attitude control data to study the Venusian atmosphere and various spacecraft properties. *Adv. Astronaut. Sci.* 87:451–467.

Crutzen, P. J. 1970. Comments on "Absorption and emission by carbon dioxide in the mesosphere," by J. T. Houghton *Q. J. Roy. Meteorol. Soc.* 96:769.

Delderfield, J., Schoefield, J. T., and Taylor, F. W. 1980. Radiometer for the Pioneer Venus Orbiter. *IEEE Trans. Geosci. Remote Sensing* GE-18:70–76.

Dickinson, R. E., and Bougher, S. W. 1986. Venus mesosphere and thermosphere: I. Heat budget and thermal structure. *J. Geophys. Res.* 91:70–80.

Espiritu, R. C., and Tolson, R. H. 1995. Determining Venusian upper atmosphere characteristics using Magellan attitude control data. In *AAS/AIAA Spaceflight Mechanics Meeting*, paper AAS 95-152 (San Diego: AAS Publications), pp. 1–17.

Feldman, P. D., Moos, H. W., Clarke, J. T., and Lane, A. L. 1979. Identification of the UV nightglow from Venus. *Nature* 279:221–222.

Fox, J. L. 1982. Atomic carbon in the atmosphere of Venus. *J. Geophys. Res.* 87:9211–9216.

Fox, J. L. 1988. Heating efficiencies in the thermosphere of Venus reconsidered. *Planet. Space Sci.* 36:37–46.

Fox, J. L. 1992. Chemistry of the nightside ionosphere of Venus. *Planet. Space Sci.* 40:1663–1681.

Fox, J. L., and Bougher, S. W. 1991. Luminosity and dynamics of the Venus thermosphere. *Space Sci. Rev.* 55:357–489.

Gerard, J. C., Stewart, A. I. F., and Bougher, S. W. 1981. The altitude distribution of the Venus ultraviolet nightglow and implications on vertical transport. *Geophys. Res. Lett.* 8:633–636.

Gerard, J. C., Deney, E. J., and Lerho, H. 1988. Sources and distribution of odd nitrogen in the Venus daytime thermosphere. *Icarus* 75:171–184.

Grebowsky, J. M., Kasprzak, W. T., Hartle, R. E., and Donahue, T. M. 1995. A new look at Venus' thermosphere H distribution. *Adv. Space Res.* 17:(11):191–195.

Hartle, R. E., and Grebowsky, J. M. 1995. Planetary loss from light ion escape on Venus. *Adv. Space Res.* 15(4):117–122.

Hartle, R. E., and Taylor, H. A. 1983. Identification of deuterium ions in the ionosphere of Venus. *Geophys. Res. Lett.* 10:965–968.

Hartle, R. E., Donahue, T. M., Grebowsky, J. M., and Mayr, H. G. 1996. Hydrogen and deuterium in the thermosphere of Venus: Solar cycle variation and escape. *J. Geophys. Res.* 101:4525–4538.

Hedin, A. E., Niemann, H. B., Kasprzak, W. T., and Seiff, A. 1983a. Global empirical model of the Venus thermosphere. *J. Geophys. Res.* 88:73–83; correction 88:6352.

Hedin, A. E. 1984. Correlations between thermospheric density and temperature, solar EUV flux, and 10.7-cm flux variations. *J. Geophys. Res.* 89:9828–9834.

Hedin, A. E. 1987. MSIS-86 thermospheric model. *J. Geophys. Res.* 92:4649–4662.

Hedin, A. E., and Mayr, H. G. 1987. Solar EUV induced variations in the thermosphere. *J. Geophys. Res.* 92:869–875.

Hoegy, W. R., Brace, L. H., Kasprzak, W. T., and Russell, C. T. 1990. Small-scale plasma, magnetic, and neutral density fluctuations in the nightside venus ionosphere. *J. Geophys. Res.* 95:4085–4102.

Hoffman, H. J., et al. 1980. The Pioneer Venus Bus Neutral Mass Spectrometer. *IEEE Trans. Geosci. Remote Sens.* GE-18:122–126.

Hollenbach, D. J., Prasad, S. S., and Whitten, R. C. 1985. The thermal structure of the dayside upper atmosphere of Venus above 125 km. *Icarus* 64:205–220.

Hord, C. W., et al. 1991. Galileo Ultraviolet Spectrometer experiment: Initial Venus and interplanetary cruise results. *Science* 253:1548–1560.

Hord, C. W., et al. 1992. Galileo Ultraviolet Spectrometer experiment. *Space Sci. Rev.* 60:503–530.

Hunten, D. M., Colin, L., Donahue, T. M., and Moroz, V. I., eds. 1983. *Venus* (Tucson: Univ. of Arizona Press).

IEEE. 1980. Special issue on instrumentation for the Pioneer Venus spacecraft. *IEEE Trans. Geosci. Remote Sens.* GE-18:1–130.

Jacchia, L. G. 1959. Two thermospheric effects in the orbital acceleration of artificial satellites. *Nature* 183:526–527.

Jacchia, L. G. 1964. Influence of solar activity on the Earth's upper atmosphere. *Planet. Space Sci.* 12:355–378.

Jacchia, L. G., Slowley, J., and Campbell, I. G. 1973. An analysis of the solar-activity effects in the upper atmosphere. *Planet. Space Sci.* 21:1835–1842.

Jenkins, J. M., et al. 1994. Radio occultation studies of the Venus atmosphere with the Magellan spacecraft. *Icarus* 110:79–94.

Johnson, T. V., Yeates, C. M., Young, R., and Dunne, J. 1991. The Galileo Venus encounter. *Science* 253:1516–1518.

Kar, J., et al. 1991. On the response of exospheric temperature on Venus to solar wind conditions. *J. Geophys. Res.* 96:7901–7904.

Kasprzak, W. T., et al. 1982. Observations of energetic ions near the Venus ionopause. *Planet. Space Sci.* 30:1107–1115.

Kasprzak, W. T., Hedin, A. E., Mayr, H. G., and Niemann, H. B. 1988. Wavelike perturbations in the neutral thermosphere of Venus. *J. Geophys. Res.* 93:11237–11245.

Kasprzak, W. T., et al. 1993a. Neutral composition measurements by the Pioneer Venus Neutral Mass Spectrometer during orbiter re-entry. *Geophys. Res. Lett.* 20:2747–2750.

Kasprzak, W. T., Niemann, H. B., Hedin, A. E., and Bougher, S. W. 1993b. Wavelike perturbations observed at low altitudes by the Pioneer Venus Orbiter Neutral Mass Spectrometer during orbiter entry. *Geophys. Res. Lett.* 20:2755–2758.

Keating, G. M., and Bougher, S. W. 1987. Neutral upper atmospheres of Venus and Mars. *Adv. Space Res.* 7:(12):57–71.

Keating, G. M., and Bougher, S. W. 1992a. Venus thermospheric response to short-term solar variations. *Adv. Space Res.* 12(9):111–128.

Keating, G. M., and Bougher, S. W. 1992b. Isolation of major Venus thermospheric cooling mechanism and implications for Earth and Mars. *J. Geophys. Res.* 97:4189–4197.

Keating, G. M., and Hsu, N. C. 1993. The Venus atmospheric response to solar cycle

variations. *Geophys. Res. Lett.* 20:2751–2754.
Keating, G. M., and Hsu, N. C. 1995. Venus thermosphere response to solar cycle variations. Division for Planetary Sciences Meeting, Oct. 9–13, Mauna, Hi., p. 22 (abstract).
Keating, G. M., Tolson, R. H., and Hinson, E. W. 1979a. Venus thermosphere and exosphere: First satellite drag measurements of an extraterrestrial atmosphere. *Science* 203:772–774.
Keating, G. M., Taylor, F. W., Nicholson, J. Y., and Hinson, E. W. 1979b. Short-term cyclic variations and diurnal variations of the Venus upper atmosphere. *Science* 205:62–64.
Keating, G. M., Nicholson, J. Y., III, and Lake, L. R. 1980. Venus upper atmosphere structure. *J. Geophys. Res./* 85:7941–7956.
Keating, G. M., et al. 1985. Models of the Venus neutral upper atmosphere: Structure and composition. *Adv. Space Res.* 5:117–171.
Kliore, A. J., and Mullen, L. 1987. Solar cycle influence on the topside plasma scale height of the Venus dayside ionosphere. *Bull. Amer. Astron. Soc.* 19:846.
Kliore, A. J., Moroz, V. I., and Keating, G. M. 1985. The Venus international reference atmosphere. *Adv. Space Res.* 5(11):1–305.
Krasnopolsky, V. A., Krysko, A. A., Rogachev, V. N., and Parashev, V. A. 1976. Spectroscopy of the Venus night airglow from the Venera 9, 10 orbiters. *Cosmic Res.* 14:789–795.
Krehbiel, J. P., et al. 1980. Pioneer Venus Orbiter electron temperature probe. *IEEE Trans. Geosci. Remote Sens.* GE-18:49–54.
Kumar, S., and Broadfoot, A. L. 1975. He 584 Å emission from Venus: Mariner 10 observations. *Geophys. Res. Lett.* 2:357–360.
Lasher, L. E., Wirth, F., and Colin, L. 1994. *Pioneer Venus Bibliography* (Moffett Field, Calif.: NASA Ames Research Center).
LeCompte, M. A., Paxton, L. J., and Stewart, A. I. F. 1989. Analysis and interpretation of observations of airglow at 297 nm in the Venus thermosphere. *J. Geophys. Res.* 94:208–216.
Luhmann, J. G. 1986. The solar wind interaction with Venus. *Space Sci. Rev.* 44:241–306.
Luhmann, J. G. 1991. Space plasma physics research progress 1987–1990: Mars, Venus, and Mercury. *Rev. Geophys. Suppl.* April, pp. 965–975.
Mahajan, K. K., and Kar, J. 1988. Planetary ionospheres. *Space Sci. Rev.* 47:303–397.
Mahajan, K. K., and Kar, J. 1990. A comparative study of Venus and Mars: Upper atmosphere, ionospheres and solar wind interactions. *Indian J. Radio Space Phys.* 19:444–465.
Mahajan, K. K., et al. 1990. Response of Venus exospheric temperature measured by neutral mass spectrometer to solar EUV flux measured by Langmuir probe on the Pioneer Venus Orbiter *J. Geophys. Res.* 95:1091–1095.
Massie, S. T., Hunten, D. M., and Sowell, D. R. 1983. Day and night models of the Venus thermosphere. *J. Geophys. Res.* 88:3955–3969.
Mayr, H. G., et al. 1980. Dynamic properties of the thermosphere infrared from Pioneer Venus mass spectrometer measurements. *J. Geophys. Res.* 85:7841–7847.
Mayr, H. G., et al. 1985. On the diurnal variations in the temperature and composition: A three-dimensional model with superrotation. *Adv. Space Res.* 5:109–112.
Mayr, H. G., et al. 1988. Gravity waves in the upper atmosphere of Venus. *J. Geophys. Res.* 93:11247–11262.
Mayr, H. G., Harris, I., and Pesnell, W. D. 1992. Properties of thermospheric gravity waves on Earth, Venus and Mars. In *Venus and Mars: Atmosphere, Ionospheres and Solar Wind Interactions* (Washington, D. C.: American Geophys. Union), pp. 91–111.

McElroy, M. B., Prather, M. J., and Rodriguez, J. M. 1981. Escape of hydrogen from Venus. *Science* 215:1614–1615.

Meier, R. R., Anderson, D. E., Jr., and Stewart, A. I. F. 1983. Atomic oxygen emissions observed from Pioneer Venus. *Geophys. Res. Lett.* 10:214–217.

Mengel, J. G., Stevens-Rayburn, D. R., Mayr, H. G. and Harris, I. 1989. Nonlinear three dimensional spectral model of the Venusian thermosphere with superrotation—II. Temperature, composition and winds. *Planet. Space Sci.* 37:707–722.

Nagy, A. F., and Cravens, T. E. 1988. Hot oxygen atoms in the upper atmospheres of Venus and Mars. *Geophy. Res. Lett.* 15:433–435.

Nagy, A. F., Cravens, T. E., Yee, J. H., and Stewart, A. I. F. 1981. Hot oxygen atoms in the upper atmosphere of Venus. *Geophys. Res. Lett.* 8:629–632.

Niemann, H. B., et al. 1980a. Pioneer Venus Orbiter neutral gas mass spectrometer experiment. *IEEE Trans. Geosci. Remote Sens.* GE-18:60–65.

Niemann, H. B., et al. 1980b. Mass spectrometric measurements of the neutral gas composition of the thermosphere and exosphere of Venus. *J. Geophys. Res.* 85:7817–7827.

Paxton, L. J. 1985. Pioneer Venus Orbiter ultraviolet spectrometer limb observations: Analysis and interpretation of the 166- and 156-nm data. *J. Geophys. Res.* 90:5089–5096.

Paxton, L. J., and Meier, R. R. 1986. Reanalysis of Pioneer Venus Orbiter ultraviolet spectrometer data: OI 1304 intensities and atomic oxygen densities. *Geophys. Res. Lett.* 13:229–232.

Paxton, L. J., Anderson, D. E., and Stewart, A. I. F. 1985. The Pioneer Venus Orbiter ultraviolet spectrometer experiment—analysis of hydrogen Lyman alpha data. *Adv. Space Res.* 5:129–132.

Paxton, L. J., Anderson, D. E., Jr., and Stewart, A. I. F. 1988. Analysis of the Pioneer Venus Orbiter ultraviolet spectrometer Lyman-alpha data from near the subsolar region. *J. Geophys. Res.* 93:1766–1772; correction 93:11551.

Pollock, D. S., Scott, G. B. I., and Phillips, L. F. 1993. Rate constant for the quenching of $CO_2(010)$ by atomic oxygen. *Geophys. Res. Lett.* 20:727–729.

Rusch, D. W., and Cravens, T. E. 1979. A model of the neutral and ion nitrogen chemistry in the daytime thermosphere of Venus. *Geophys. Res. Lett.* 6:791–794.

Russell, E. E., Watts, L. A., Pellicori, S. F., and Coffeen, D. L. 1977. Orbiter cloud photopolarimeter for the Pioneer Venus mission. *Proc. Soc. Photo-Optical Instrum. Eng.* 112:28–44.

Schubert, G., et al. 1980. Structure and circulation of the Venus atmosphere. *J. Geophys. Res.* 85:8007–8025.

Seiff, A. 1983. Thermal structure of the atmosphere of Venus. In *Venus*, eds. D. M. Hunten, L. Colin, T. M. Donahue and V. I. Moroz (Tuscon: Univ. of Arizona Press), pp. 215–279.

Seiff, A., and Kirk, D. B. 1991. Waves in Venus's middle and upper atmosphere: Implications of Pioneer Venus probe data above the clouds. *J. Geophys. Res.* 96:11021–11032.

Seiff, A., Juergens, D. W., and Lepetich, J. E. 1980. Atmosphere structure instruments on the four Pioneer Venus entry probes. *IEEE Trans. Geosci. Remote Sens.* GE-18:105–111.

Seiff, A., Young, R. E., Haberle, R., and Houben, H. 1992. The evidence of waves in the atmosphere of Venus and Mars. In *Venus and Mars: Atmosphere, Ionospheres and Solar Wind Interactions* (Washington, D. C.: American Geophys. Union), pp. 73–89.

Shapiro, I. I., et al. 1979. Venus: Density of upper atmosphere from measurements of drag on Pioneer Orbiter. *Science* 203:775–777.

Shved, G. M., et al. 1991. Measurement of the quenching rate constant for collisions $CO_2(010)$-O: The importance of the rate constant magnitude for the thermal regime and radiation of the lower thermosphere. *Atmos. Ocean. Phys.* 27:431–437.

Stevens-Rayburn, D. R., Mengel, J. G., Harris, I., and Mayr, H. G. 1989. Nonlinear three dimensional spectral model of the Venusian thermosphere with superrotation—I. Formulation and numerical technique. *Planet. Space Sci.* 37:701–705.

Stewart, A. I. F. 1980a. Design and operation of the Pioneer Venus Orbiter ultraviolet spectrometer. *IEEE Trans. Geosci. Remote Sens.* GE-18:65–70.

Stewart, A. I., and Barth, C. A. 1979. Ultraviolet night airglow of Venus. *Science* 205:59–62.

Stewart, A. I. F., and Colwell, W. B. 1995. Variations in Venus' thermospheric composition over solar cycle 21 deduced from Pioneer Venus Orbiter ultraviolet spectrometer observations of emissions from H, O and CO. Venus II, Jan. 4–7, Tucson, Ariz., p. 97, Abstract book.

Stewart, A. I. F., Gerard, J. C., Rusch, D. W., and Bougher, S. W. 1980b. Morphology of the Venus ultraviolet night airglow. *J. Geophys. Res.* 85:7861–7870.

Strangeway, R. J. 1993. The Pioneer Venus Orbiter entry phase. *Geophys. Res. Lett.* 20:2715–2717.

Taylor, F. W., et al. 1980. Structure and meteorology of the middle atmosphere of Venus: Infrared remote sensing from the Pioneer Orbiter. *J. Geophys. Res.* 85:7963–8006.

Taylor, H. A., Jr., et al. 1980a. Bennett ion mass spectrometers on the Pioneer Venus bus and orbiter. *IEEE Trans. Geosci. Remote Sensing* GE-18:44–49.

Taylor, H. A., Jr., et al. 1980b. Global observations of the composition and dynamics of the ionosphere of Venus: Implications for the solar wind interaction. *J. Geophys. Res.* 85:7765–7777.

Taylor, H. A., Jr., et al. 1982. Variations in ion and neutral composition at Venus: Evidence of solar control of the formation of the predawn bulges in H^+ and He. *Icarus* 52:211–220.

Taylor, H. A., Jr., et al. 1984. Interannual and short term variations of the Venus nighttime hydrogen bulge. *J. Geophys. Res.* 89:10669–10675.

Taylor, H. A., Jr., et al. 1985. In situ results on the variation of neutral atmospheric hydrogen at Venus. *Adv. Space Res.* 5:125–128.

Trauger, J. T., and Lunine, J. I. 1983. Spectroscopy of molecular oxygen in the atmospheres of Venus and Mars. *Icarus* 55:272–281.

Travis, L. D. 1979. Imaging and polarimetry with the Pioneer Venus Orbiter cloud photopolarimeter. *Proc. Soc. Photo-Optical Instrum. Eng.* 183:299–304.

von Zahn, U., et al. 1980. The upper atmosphere of Venus during morning conditions. *J. Geophy. Res.* 85:7829–7840.

von Zahn, U., Kumar, S., Niemann, H., and Prinn, R. 1983. Composition of the Venus atmosphere. In *Venus*, eds. D. M. Hunten, L. Colin, T. M. Donahue and V. I. Moroz (Tuscon: Univ. of Arizona Press), pp. 299–430.

Yung, Y. L., and Demore, W. B. 1982. Photochemistry of the stratosphere of Venus: Implications for atmospheric evolution. *Icarus* 51:199–247.

UPPER ATMOSPHERE DYNAMICS: GLOBAL CIRCULATION AND GRAVITY WAVES

S. W. BOUGHER
University of Arizona

M. J. ALEXANDER
University of Washington

and

H. G. MAYR
NASA Goddard Space Flight Center

The global circulation of the Venus upper atmosphere is much different from that of Earth, and can be decomposed into two distinct flow patterns: (1) a stable subsolar-to-antisolar (SS-AS) circulation cell driven by solar EUV heating, and (2) a retrograde superrotating zonal (RSZ) flow that seems to vary greatly over time. Wave-like perturbations have also been observed in the Venus thermosphere. These gravity waves are thought to be launched from the Venus cloud region (50–70 km), to propagate vertically, and to dissipate above 120 to 140 km due to saturation. This "wave breaking" process provides significant momentum and energy forcing to the thermosphere, thereby modifying the large-scale circulation and structure from that otherwise expected from *in-situ* solar forcing. Various multi-dimensional modeling tools are being used to interpret the available thermospheric data, thereby unfolding zonal and SS-AS wind magnitudes and gravity wave forcing parameters consistent with the observed thermospheric structure. This chapter reviews these observations and theoretical studies in order to examine the processes that maintain the wind structure of the Venus upper atmosphere.

I. INTRODUCTION AND OVERVIEW

A consistent picture of the dynamics of the Venus upper atmosphere over ~100 to 200 km is beginning to appear, in spite of the lack of *in-situ* wind measurements. This information is gleaned from an examination of Pioneer Venus Orbiter (PVO) neutral density distributions above 130 to 145 km, PVO ultraviolet night airglow distributions, visible and infrared O_2 nightglow distributions from Venera 9 and 10, Galileo, PVO, and the ground, and groundbased observations of lower thermospheric winds, CO densities, and temperatures. Many of these Pioneer Venus data were reviewed recently by Fox and Bougher (1991). Also, chapters by Kasprzak et al. and Lellouch et al. describe solar cycle and upper mesospheric (80–110 km) observations that support the dynamical characteristics of the upper atmosphere that are presented in this chapter. The global-scale dynamics of the Venus atmosphere can be separated into two distinct regimes in which very different flow patterns dominate: (1) a retrograde, superrotating, zonal (RSZ) flow from the surface up to $\sim$70 km, and (2) a mean subsolar- to-antisolar (SS-AS) symmetric cell above 120 km. In between (70–120 km), a transition region occurs in which zonal winds are highly variable yet generally decreasing with height while thermospheric-type SS-AS winds are increasing. The RSZ circulation appears to persist throughout the atmosphere, although it does not typically dominate SS-AS flow above $\sim$100 km (Goldstein et al. 1991). However, distinct periods are observed during which strong zonal winds, approaching 100 m s^{-1}, do persist up to $\sim$100 km or higher (Shah et al. 1991; Lellouch et al. 1994).

Gravity waves are believed to play a role in Venus thermospheric dynamics. Large values of parameterized wave drag and eddy diffusion are key ingredients in reproducing the observed day-to-night temperature contrasts and density distributions in general circulation models (see Fox and Bougher 1991). Wavelike density perturbations with horizontal scales of 100 to 600 km have also been observed in the Venus thermosphere by the PVO Neutral Mass Spectrometer (Kasprzak et al. 1988,1993b). These perturbations have been shown to be consistent with vertically propagating gravity waves from a source region well below thermospheric altitudes at $\sim$80 km (Mayr et al. 1988). In fact, these thermospheric gravity waves are thought to be launched from the Venus cloud region (50–70 km), to propagate vertically, and to break in the thermosphere providing a significant momentum and energy source that can give rise to the zonal asymmetry described as the thermospheric superrotation (Alexander 1992).

This chapter is organized with an historical perspective that parallels our increasing understanding of important dynamical processes affecting the upper atmosphere. Simple SS-AS flow is described first, and shown to be only one component of the global circulation pattern. The chapter ends with evidence and a discussion of how breaking gravity waves likely impact the global dynamics and small-scale mixing of the Venus thermosphere. Finally, we summarize the unknown issues that continue to challenge current model perceptions and beg for new Venus thermospheric measurements.

II. SYMMETRIC AND ZONAL THERMOSPHERIC FLOW

A. Neutral Species and Temperature Constraints

The most prominent feature of the thermosphere that suggests a strong SS-AS wind system is the large contrast in temperatures and densities between day and night. Temperatures on the day side of Venus increase from $\sim$170 to 180 K at the "mesopause" ($\sim$100 km) level (Taylor et al. 1980; Seiff et al. 1980) to near 300 to 310 K above 150 km for solar maximum conditions (von Zahn et al. 1979,1980; Niemann et al. 1979,1980; Keating et al. 1980; Bertaux et al. 1985). The rise of dayside temperatures above the mesopause is similar to that of the Earth's thermosphere, although the temperatures are much lower (see Fox and Bougher 1991). Venus nightside temperatures are shown to decrease from 170 K at the mesopause to values as low as 100 to 130 K above 150 km (Keating et al. 1979,1980; Niemann et al. 1979,1980; Seiff et al. 1980; Seiff 1982). This vertical structure is remarkably unlike the terrestrial thermosphere, in particular on the night side which has been termed a "cryosphere" (Schubert et al. 1980). In addition, changes in temperature across the terminator are very abrupt, with the minimum value observed just beyond midnight at local time (LT) = 0200 (Niemann et al. 1980; Keating et al. 1980; Hedin et al. 1983).

The kinetic temperatures in the Venus thermosphere above 150 km have been inferred primarily from the observed scale heights of the measured neutral densities. The diurnal variation of the heaviest species (CO_2, CO, N_2, and O) show a near symmetry about the SS-AS axis at a given altitude, with a density maximum at local noon and minimum at midnight (von Zahn et al. 1983). To first order, the dayside solar EUV heating results in this expansion of the dayside thermosphere; i.e., the height of a given pressure surface rises (falls) with increasing (decreasing) temperatures. The large diurnal density variation suggests a significant contraction of the nightside thermosphere with constant pressure surfaces decreasing abruptly in altitude across the terminators. Helium and hydrogen, the two lightest species, show a diurnal density reversal, with densities higher at night than during the day (Niemann et al. 1980; Brinton et al. 1980). This signature is consistent with their small atomic weights and correspondingly large-scale heights that lead to increased transport efficiencies by the large-scale winds, just as in the atmosphere of the Earth (Mayr et al. 1980). This transport is known as "wind-induced diffusion." The diurnal variation of helium at 165 km reveals a distinct (lasting) nightside bulge, with a maximum at LT = 0500 roughly 30 to 40 times that on the day side (von Zahn et al. 1983). Results obtained by Brinton et al. (1980) suggest a similar night–day bulge of hydrogen peaking at LT = 0500, with a diurnal ratio of nearly 400:1 at 165 km.

This temperature difference between day and night gives rise to horizontal pressure gradients, which to first order should drive a strong SS-AS circulation cell in Venus' upper atmosphere (Dickinson and Ridley 1977; Seiff 1982). A simplified two-dimensional flow pattern would be symmetric about the Sun-Venus line, with dayside upwelling centered on the subsolar point, strong cross-terminator flow, and subsidence centered on the antisolar point (Dickinson and Ridley 1977; Schubert 1983). Return flow, possibly between 70 to 90 km (Goldstein et al. 1991), would complete the circuit. This flow is driven primarily by solar EUV heating. Streamlines for transported species approximately follow constant pressure surfaces, which descend in altitude from day to night (Seiff 1982). Supersonic velocities are estimated across the terminators for laminar flow conditions; turbulent viscosity is likely important in modifying the pressure-driven wind speeds (Niemann et al. 1980; Seiff 1982; Bougher et al. 1986). This basic day–night flow pattern was hypothesized and studied in a series of numerical models by Dickinson and Ridley (1972,1975,1977) prior to Pioneer Venus measurements, and by Bougher et al. (1986,1988,1990), Bougher and Borucki (1994), Mayr et al. (1980), and Mengel et al. (1989) afterwards.

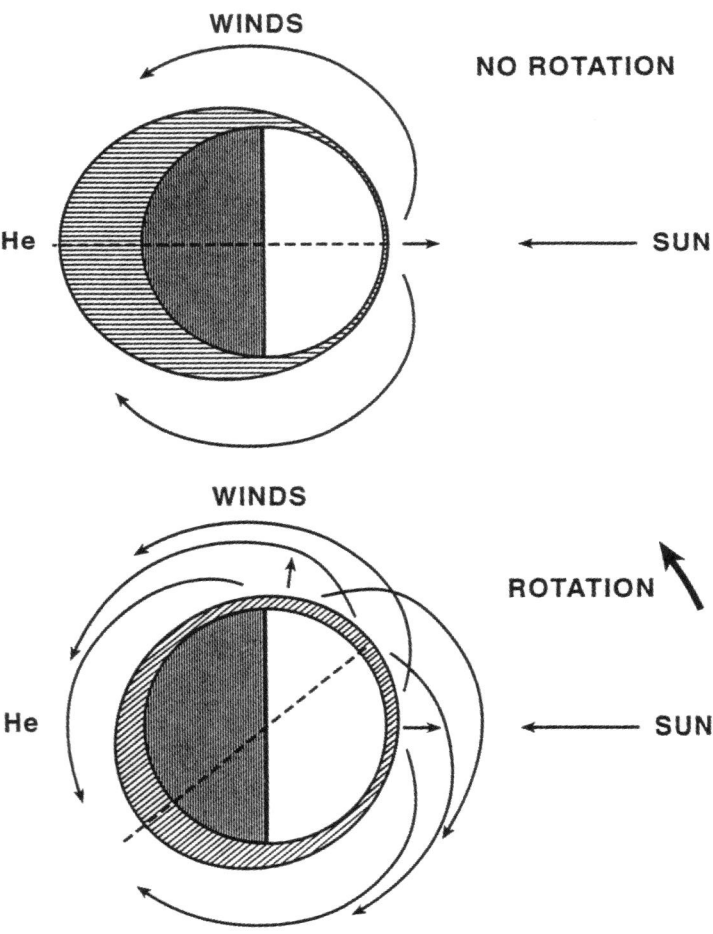

Figure 1. Cartoon of the SS-AS flow versus RSZ flow and its effects on the local time distribution of helium (figure adapted from Mayr et al. 1985).

This single symmetric cell, driven by *in-situ* solar heating, certainly exists. However, evidence suggests that convergence of the nightside flow does not occur precisely at midnight as expected from a purely symmetric SS-AS wind pattern. The actual motions in Venus' thermosphere are likely a superposition of the SS-AS flow and a RSZ wind component (see Fig. 1). Postmidnight maxima in the helium and hydrogen densities (Niemann et al. 1979; Brinton et al. 1980; Taylor et al. 1984), displacement of the minimum diurnal temperature to LT = 0200 (Niemann et al. 1980; Keating et al. 1980), and the NO night airglow maxima at LT = 0200 (Stewart et al. 1980) all suggest a westward superrotation of the upper atmosphere (≥ 150 km) with RSZ wind speeds of 50 to 100 m s^{-1}. This corresponds to a rotational period of 4 to 8 Earth days. This modified flow pattern now includes upwelling (divergence) near the subsolar point, and subsidence (convergence) offset from midnight, yielding a nightside flow which stagnates near LT = 0200 to 0300 (Niemann et al. 1980). H and He density distributions are particularly sensitive to this asymmetric flow (see Fig. 1). The RSZ winds add to the SS-AS flow across the evening terminator, giving stronger dusk winds than at dawn. This is consistent with convergence of the net global wind system after midnight, with corresponding light-specie density bulges near LT = 0300 to 0500.

This large day–night contrast of the thermospheric structure was also examined by monitoring the orbital changes of Venus spacecraft over the solar cycle. Drag measurements by Keating et al. (1980) during the PVO early mission period are quite consistent with those obtained at PVO entry and during the Magellan (MGN) Cycle 4 observational period (Keating and Hsu 1993) (see Fig. 4 in the chapter by Kasprzak et al.). The large day-to-night mass density variation is similar to that observed earlier, suggesting that the SS-AS flow persists over the solar cycle. Nightside neutral densities measured by the PVO Neutral Mass Spectrometer (ONMS) (Kasprzak et al. 1993*a*) also show little variation over the solar cycle. Thus, a relatively unchanged dayside and nightside thermosphere structure suggests that the SS-AS flow persists and is stable throughout the solar cycle.

Latitudinal variations of the thermospheric structure can also be used to constrain the net wind system. Such variations were unknown until MGN drag data became available during Cycle 5 and 6 observations. MGN reaction wheel measurements were obtained over September 6–13, 1994, yielding mass density variations between 10 and 70°N near LT = 1300. Data projected to a constant altitude of 165 km (Espiritu and Tolson 1995) show mass densities that drop by $\sim 60\%$ over this 60° latitude region. This behavior is consistent with global circulation model calculations (see Sec. III below) that combine SS-AS and RSZ wind effects (see, e.g., Bougher and Borucki 1994). Clearly, the latitudinal variation of mass density is not characterized precisely by a solar zenith angle (SZA) dependence. This behavior reflects the fact that horizontal density variations are driven by a combination of the thermal structure (heavy species) and the global wind system (lighter species). Also, upward propagating gravity waves may be affecting the MGN mass density variations observed from orbit to orbit.

Groundbased measurements of CO distributions can be used to monitor the large-scale circulation pattern between 90 and 110 km near the mesopause; i.e., CO densities are dynamically controlled in this altitude region. Lellouch et al. (see their chapter) discuss these observations in detail; we summarize here the relevant dynamical characteristics of the observed CO distributions. The Clancy and Muhleman (1991) measurements indicate that for most years the CO distribution over 90 to 110 km increased as a function of local time from the day side toward a peak at midnight. The day-to-night increase at 110 km was about a factor of 2 to 4. This behavior is consistent with weak RSZ winds over the 90 to 110 km altitude level; SS-AS symmetric winds are instead primarily responsible for the buildup and location of the observed nightside CO enhancement near 2400 LT. However, CO distributions have also been observed over 90 to 110 km showing a bulge centered near 0300 local time. Sometimes this a.m. bulge is large (10–100 night-to-day enhancement) (Clancy and Muhleman 1991; Gurwell et al. 1995); sometimes it is quite weak (uniform over the night side) (see, e.g., Lellouch et al. 1994). The position of this a.m. bulge implies that quite strong zonal winds do blow at the ~100 km altitude level. These observations confirm the changing nature of the lower thermosphere RSZ winds and their variation with altitude (see Sec. II.C).

B. Airglow Constraints

Several night airglow (nightglow) measurements were made by various spacecraft and groundbased observers, yielding additional constraints on the global thermospheric circulation throughout the solar cycle (see Fox and Bougher 1991; Bougher and Borucki 1994). Each of these nightglow emissions (NO ultraviolet, O_2 visible, and O_2 infrared) can be utilized as a remote tracer of the Venus upper atmosphere wind system over a specific altitude region. The airglow data approximately probe the (1) mid-to-upper thermosphere (115–150 km) via NO emissions, and (2) the lower thermosphere and upper mesosphere (100–130 km) via O_2 visible and infrared emissions. Dynamically important dayglow observations of atomic H and O were also made by the PVO Ultraviolet Spectrometer (OUVS) (see Fox and Bougher 1991; Paxton et al. 1988a,b; Alexander et al. 1993). These remote measurements provide a unique opportunity to characterize the Venus thermospheric structure as a function of latitude, unlike most *in-situ* measurements. In addition, variations as a function of solar activity signal changes in production mechanisms and the global circulation. An $F_{10.7}$ index is typically quoted as a proxy for solar EUV flux activity over the solar cycle period (see the chapter by Kasprzak et al.).

1. Nitric Oxide Nightglow. The ultraviolet NO night airglow was first mapped by the OUVS during the first diurnal period of the PVO mission (Stewart and Barth 1979; Stewart et al. 1980; Gerard et al. 1981). This nightglow was identified as the gamma and delta bands of nitric oxide (Feldman et al. 1979; Stewart and Barth 1979) as excited by the radiative recombination of N and O atoms. It was proposed that the airglow was the result of these atoms being transported from their source on the day side to the night side where recombination occurs. This mechanism was later validated using detailed three-dimensional modeling calculations (Bougher et al. 1990; see Sec. III below), implying that the NO airglow is a sensitive tracer of the Venus thermospheric circulation. Planet-wide observations by the OUVS at solar maximum (see Fig. 2A) over 35 orbits were statistically averaged and show that the NO(0,1) δ-band nightglow typically exhibits a bright spot reaching 1.9±0.6 kilo-Rayleigh (kR) near LT = 0200 just south of the equator (Bougher et al. 1990; Stewart et al. 1980). The NO airglow layer was found to exhibit a peak emission at 115±2 km (Gerard et al. 1981), in concert with the expected atomic-N profile.

The NO nightglow is particularly valuable in constraining the Venus thermospheric circulation over 115 to 150 km in the absence of *in-situ* measured winds (Stewart et al. 1980; Bougher et al. 1990). The average brightness distribution map of Fig. 2A can be viewed as an approximate map of the downward flux of N-atoms being supplied from the day side. The nightside downward flux depends on the dayside net column N-production plus the strength of the day-to-night global winds. Model simulations must reproduce typically observed dayside N densities above 150 km plus average nightside NO nightglow intensities in order to estimate the responsible horizontal wind speeds. In addition, the local time position of the average peak emission ("bright patch") constrains the zonal thermospheric wind speeds in this altitude region (Stewart et al. 1980; Bougher et al. 1990). The 2-hr offset of the average nightside peak toward the morning terminator implies RSZ winds over 115 to 150 km of roughly 40 to 60 m s^{-1}. Slightly faster RSZ winds having a period of 6 to 8 days are inferred from measured helium densities above 150 km (Mayr et al. 1980; Mengel et al. 1989). It is important to remember that individual orbit nightglow patches (see Fig. 2B) reveal a large spatial and temporal variability in the intensity distribution; i.e., time scales of an Earth day or less are implied. This suggests a highly variable thermospheric circulation from day to day.

2. O_2 Herzberg II Nightglow. A similar constraint on the Venus solar minimum thermospheric structure and circulation is given by examining the O_2 Herzberg II nightglow observed by Venera 9 and 10 instruments at visible wavelengths (400–800 nm) (Krasnopolsky et al. 1976; Krasnopolsky and Tomashova 1980). A recombination mechanism is again proposed in which day side created atomic-O is transported to the night side where it undergoes 3-body recombination at low altitude. The airglow layer is estimated to occur over ~90 to 115 km, giving an average integrated vertical intensity of the night side at solar minimum ($F_{10.7}$ = 75) of 2.7 kR; the airglow layer is typically 15 km thick. The mean nightglow distribution is not well defined due to limited observations. Nevertheless, an average of 25 individual brightness scans across the dark disk shows a shallow maximum at LT = 0030, just past midnight, and at 8 to 28° south of the equator. A rather weak zonal

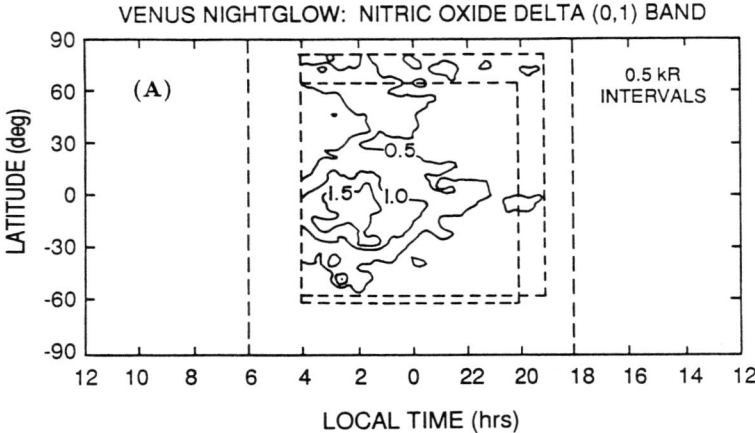

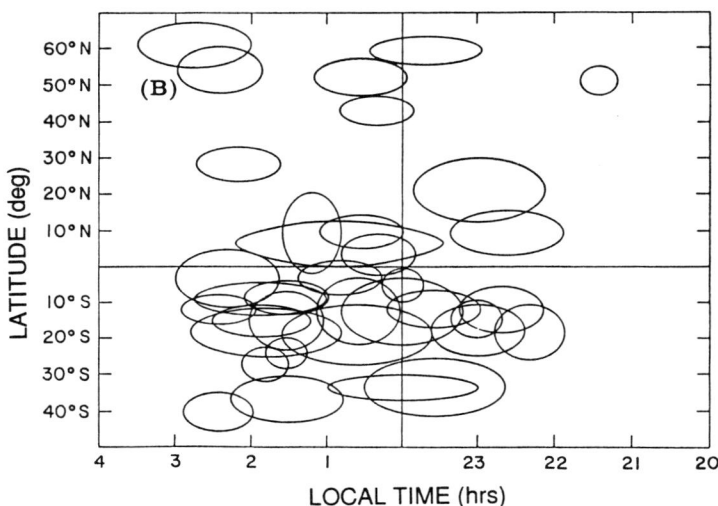

Figure 2. (a) The OUVS NO(0,1) δ-band vertical intensity distribution. The latitude and local time grid is filled with 198 nm data as obtained and revised in absolute intensity according to Bougher et al. (1990). This statistical map of the airglow is obtained over 35 orbits early in the PVO mission, and exhibits a dark-disk average intensity (120–180° SZA) of 460±120 R. (b) Spatial variability of the observed OUVS 198 nm individual bright airglow patches included in the statistical map of (a). The half maximum intensity spatial extent of each individual NO patch is plotted as a function of local time and latitude. The aspect ratio, defined as the local time to latitude ratio, of each equatorial to mid-latitude patch is on the order of 2:1. This value is largely independent of latitude, and it implies strong zonal winds (figure from Bougher et al. 1990).

asymmetric wind system near 100 km is implied by this data; i.e., RSZ wind speeds of less than 15 m s^{-1} are estimated. The lack of significant local time variation of this airglow across the night side also suggests that the time scale for day-to-night transport and nightside chemical losses of atomic-O are nearly equal (about 1–1.5 Earth day) at ~100 km (von Zahn et al. 1983; Bougher and Borucki 1994).

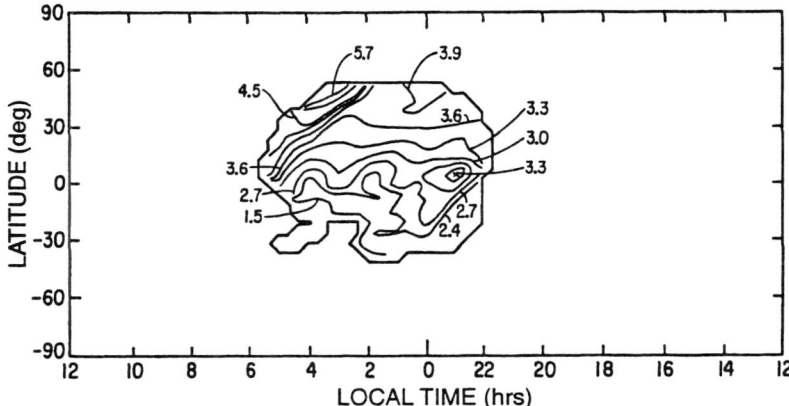

Figure 3. PVO star tracker O_2 visible airglow data from Season 16 (April–June 1988). This local time vs latitude map shows the coverage of the star tracker scans across the planet (40°S–60°N), and the magnitude of the integrated vertical intensity observed. In the equatorial region, a shallow maximum patch (3.3 kR) is located near midnight, indicating the presence of weak zonal winds (figure from Bougher and Borucki 1994).

More detailed (latitude-local time) maps of O_2 visible nightglow were later obtained by using off-axis visible light from the PVO Star Tracker (Bougher and Borucki 1994). The O_2 maps are built up and averaged over many orbits, thereby washing out much of the short-term variability and intensity contrasts. These averaged maps roughly approximate time-averaged conditions, and can be used to extract mean zonal winds. Figure 3 illustrates a sample map for PVO season 16, during which solar moderate conditions prevailed ($F_{10.7}$ = 150). There appears to be a systematic increase of the airglow from south to north across the entire night side during this season. A weak midnight intensity maximum of about 3.3 kR appears near the equator, indicating the presence of rather weak zonal winds. The responsible O-atoms follow streamlines that originate over ~100 to 130 km on the day side, near the CO_2 photolysis peak, and descend along constant pressure surfaces to near 100 km on the night side. Hence, the O_2 visible nightglow is a particularly good tracer of the zonal wind system over ~100 to 130 km.

3. *O_2 Infrared Nightglow.* On all three terrestrial planets, the infrared atmospheric band of O_2 at 1.27-μm is by far the strongest of all oxygen emis-

sions (see the detailed review in the chapter by Lellouch et al.). Groundbased point spectrometer observations of Connes et al. (1979) (day and night side) and Crisp et al. (1991) (night side) were made of this infrared emission. However, neither were adequate to resolve airglow features that could provide a tracer of the net wind system. Allen et al. (1992) reported additional images of the O_2 1.27-μm night glow taken with the infrared camera/spectrograph IRIS on the 3.9-m Anglo-Australian Telescope (AAT). Spatially resolved images were taken before inferior conjunction, in July 1991, and again in September and October of the same year when solar activity was near a maximum ($F_{10.7} = 200$). The resulting $O_2(^1\Delta)$ intensity distribution is not uniform, but instead shows intense discrete patches. The images reveal localized, often elongated regions of intense emission usually distributed around the anti-solar point, but sometimes seen at high latitudes or towards the dawn terminator. These features strongly resemble those seen in the daily NO nightglow maps (see Fig. 2B). Brightest regions usually are found at low- to mid-latitudes between midnight and 0300 LT. Spatial variations in intensity can exceed a factor of 10. The AAT technique also allows variability to be seen on a time scale of hours, revealing bright patches that disappear in less than one day (Crisp et al. 1996). These images seem to be recording the rapidly changing nightside downwelling of atomic O transported from the sunlit hemisphere by the global circulation system. Absolute intensities, once corrected for cloud reflection, appear to be of the same order as measured earlier by Connes et al. (1979) for the night side ($\sim$1 MR).

These infrared images suggest that the day-to-night circulation at the 100 to 130 km level is highly variable, on time scales of hours. Strong zonal winds will act to smooth out local time gradients in the downwelling oxygen (and the emission) while simultaneously moving the peak emission toward the dawn terminator. Weak zonal winds ($\leq$25 m s^{-1}) will permit maximum downwelling to occur near the anti-solar point, with a drop-off of the intensity toward both terminators. The sudden disappearance of "bright patches" on very short time scales is still a mystery (Crisp et al. 1996; see the chapter by Lellouch et al.). However, it is clear that the changing spatial distribution of the 1.27-μm discrete bright patches appears to correspond to changing zonal winds at the same altitude and of the same magnitude as implied by the CO data of Clancy and Muhleman (1991), Lellouch et al. (1994), and Gurwell et al. (1995).

4. Atomic Oxygen Airglow. The atomic oxygen resonance transition between the $^3S^0$ and the ground state produces a triplet of bright (O$\sim$10 kR) emission lines. Fox and Bougher (1991) summarize the history of the detection and interpretation, as well as the excitation mechanisms responsible for this bright dayglow emission from the Venus thermosphere. Alexander et al. (1993) describe the physics important to the interpretation of global images at this wavelength acquired by the Pioneer Venus OUVS experiment. The emission brightness is sensitive to the O/CO_2 ratio in the 130 to 250-km region in the thermosphere; O is the emitter, and CO_2 the absorber in the radiative transfer problem. CO_2 efficiently absorbs the emission at altitudes below 125 km. Line center scattering optical depths near the excitation peak at 130 km are quite large, so the contribution function to the emergent intensity peaks much higher, near 155 km. The 130-nm brightness is rather insensitive to temperature changes, even as large as 100 K. Alexander et al. (1993) analyzed a set of 98 images of Venus of the OUVS 130-nm images spanning the time period 1980 to 1990. These images showed solar-locked emission patterns on the day side, interpreted as global scale patterns in the O mixing ratio, that could not be explained by current global thermospheric models. These features were stationary in a solar latitude/local solar time frame of reference and were persistent; i.e., present in every image analyzed over more than a decade. This anomalous pattern is characterized by the averaged map shown in Fig. 4 which shows a strong gradient in brightness as a function of local time at all latitudes poleward of $\sim 30°$. This gradient suggests oxygen mixing ratios are up to a factor of 2.5 higher at the evening than the morning terminator, while global circulation models predict only a 10% variation from the asymmetric day/night advection of the zonal thermospheric superrotation (see Sec. III). An explanation for this local time gradient is proposed in Alexander (1992), as a consequence of asymmetric turbulent mixing due to breaking waves that originate at cloud levels. This is discussed further in Sec. V.

Patterns very similar to those in Fig. 4 also appear in OUVS observations of CO emissions from the 4th Positive band system (Stewart et al. 1993). Emissions from the (14,4) and (14,3) bands are optically thin, and show features in the CO mixing ratio from a fairly narrow layer near 130 km altitude. The appearance of the same solar-locked features in both the O and CO emissions suggest the same asymmetric turbulence mechanism is active at altitudes down to at least 130 km.

5. *Lyman-α Airglow.* The atomic-H distribution is a very sensitive tracer of the thermospheric circulation and the exospheric wind system (see Fox and Bougher 1991). Paxton et al. (1988*a,b*) describe OUVS images of the H-Lyman-α corona that can be used to further validate the dynamical picture first introduced by Brinton et al. (1980) and Taylor et al. (1984). Recall that the Brinton et al. ion-neutral chemical technique uses charge exchange equilibrium between O^+-O and H^+-H to estimate neutral H densities; this is appropriate below 200 km on the day side and 175 km at night. Measurements of *in-situ* neutral and ion species could be made only over 1978–80 and again in 1992 when PVO was in the main ionosphere and in a region where charge exchange equilibrium is valid. However, the OUVS was able to make H-Lyman-α observations at regular intervals over the entire PVO mission, which spanned the solar cycle. The local time variation in the observed Lyman-α emission (Paxton et al. 1985,1988*a,b*) seems consistent with the variation of atomic-H derived by Brinton et al. (1980); i.e., a strong H-bulge appears in the dawn sector just before the morning terminator. In addition, PVO observations of *in-situ* H show that part of the bulge is indeed present over 0 to 4 hours local time during medium solar activity (see the chapter by Kasprzak et al.). The H

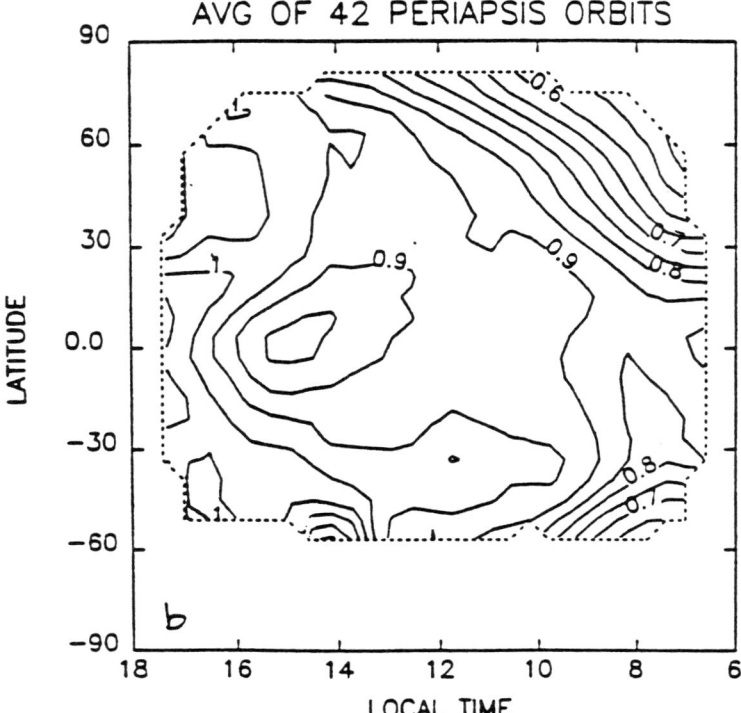

Figure 4. Contours of OUVS brightness ratio mapped into local solar time and solar latitude. The OUVS data has been ratioed to a standard model, which removes the known brightness dependences on solar zenith angle and view angle. The standard model includes no local time asymmetries in the O/CO_2 mixing ratio. "Periapsis" data are shown (figure from Alexander et al. 1993).

bulge also appears to increase in magnitude as solar activity decreases owing to a reduction in exospheric escape rates (see the chapter by Donahue et al.). In general, it is clear that H-atoms are being transported efficiently from the day side to the night side of the planet by the thermospheric wind system, with an asymmetry that is produced by superimposed RSZ winds above 150 km. An exospheric return flow, resulting from ballistic closed trajectories of H-atoms above the exobase, will also modify the H-distribution from that otherwise produced solely by the thermospheric SS-AS and RSZ flow. A detailed coupled thermosphere-exosphere model capable of distinguishing the relative roles of these exospheric and thermospheric wind systems is not yet available.

C. Groundbased Measurements of Winds

Only recently have the first direct measurements of the Venus thermospheric winds over 125 to 145 km been obtained (Maillard et al. 1995) using groundbased techniques. However, Venus' winds have been repeatedly measured near the base of the thermosphere, over approximately 95 to 120 km (see details in the chapter by Lellouch et al.). The best observing geometry for deconvolving the SS-AS and RSZ flow occurs near Venus inferior conjunction, during which time the RSZ and SS-AS components are expected to add on the astronomical west limb of the planet and to subtract on the east limb. Such dual-limb observations provide a straightforward mechanism for extracting unique SS-AS and RSZ component wind speeds.

The first reliable wind measurements were obtained in 1985, 1986, and 1987 from CO_2 10-μm infrared heterodyne spectroscopic observations (Goldstein et al. 1991). The December 1985 and March 1987 observations provided absolute wind velocities centered at 110 ± 10 km. A dominant SS-AS circulation (120 ± 30 m s^{-1}) was observed with a superimposed small (25 ± 15 m s^{-1}) RSZ wind component. Subsequent CO (1–0) millimeter measurements were used to probe RSZ and SS-AS winds at slightly lower altitudes ($\sim$99 km) during late April and early May 1988 (Shah et al. 1991). Very strong RSZ wind speeds of $\sim130\pm10$ m s^{-1} dominated the circulation at these times, with an uncertain but rather small SS-AS component still visible. Lellouch et al. (1994) later used several CO millimeter lines on Venus' night side during 1991 to measure simultaneously the absolute wind velocity in two layers (centered at 94.5 and 105 km), along with the vertical temperature profile and CO mixing ratios. The Venus circulation for this time period consisted of the superposition of an RSZ flow and an SS-AS flow of roughly equal magnitudes. These various groundbased measurements suggest an important temporal variability of the upper mesospheric circulation.

Most recently, emission lines at 4.7 μm, belonging to the (1–0) and (2–1) bands of CO, were used to probe thermospheric winds over 100 to 145 km (Maillard et al. 1995). These lines were discovered in 1987 on Venus' day side from Fourier transform observations at the Canada France Hawaii Telescope (CFHT) (de Bergh et al. 1988). Two competing nonthermal mechanisms can account for the observed band intensities: (1) fluorescence of CO excited by the solar infrared radiation, and (2) relaxation of vibrationally excited CO produced from CO_2 photolysis by solar ultraviolet radiation. The emissions originate essentially from layers around 100 to 110 km and 125 to 145 km for the (2–1) and (1–0) bands, respectively. Improved observations were obtained at the CFHT on September 22–23, 1994 (Maillard et al. 1995). The emissions were mapped on four points on Venus' lit crescent, and average rotational temperatures of 191 ± 9 K and 252 ± 20 K were measured from the (2–1) and (1–0) bands. In addition, the emission lines appeared to show Doppler shifts that allow absolute wind measurements. A constraint on the sum of the RSZ and SS-AS wind components is possible from these data, suggesting: (1) the presence of significant zonal winds at 100 to 110 km, and (2) the increase of the SS-AS winds with increasing altitude. These results are consistent with two-dimensional model predictions (see, e.g., Bougher et al. 1986), which yield SS-AS winds of $\sim$180 m s^{-1} at 135 km (see Sec. III).

A summary of both the direct and indirect estimates of RSZ and SS-AS flow speeds above ~80 km is given in Table I in the chapter by Lellouch et al. These estimates suggest that the net wind system in the Venus upper mesosphere/lower thermosphere (90–120 km) can be observationally characterized by a substantial stability of the SS-AS component over time, and a strong variability (on both short-term and long-term scales) of the RSZ component. Distinct periods of high zonal winds (~ 100 m s^{-1}) and relatively weak zonal winds (≤ 15 m s^{-1}) have both been observed near 100 to 110 km. Strong zonal winds are associated with warmer equatorial temperatures than found at mid-latitudes. This condition is the reverse of the PVO latitudinal temperature gradient observed in 1979. As a result, the cyclostrophic flow, resulting from a balance of the equatorward component of the centrifugal force and the poleward pressure gradient force, is maintained up to about 100 km. Weak zonal winds are connected to relatively cold equatorial temperatures in the upper mesosphere, leading to a breakdown of cyclostrophic balance above the cloud tops. Based on our present limited data, it is impossible to classify certain zonal wind periods as "anomalous" or as "nominal." However, the strong variability of the upper atmosphere RSZ flow is generally assumed to be related to wave activity (see Secs. IV and V; also see the chapter by Crisp and Titov).

III. MODELS OF THE THERMOSPHERIC CIRCULATION

Models of the Venus thermospheric circulation have been under development for nearly 25 years. A systematic Venus modeling strategy was employed, following a progression from one-dimensional to two-dimensional codes, that provided the necessary foundation for the development of detailed three-dimensional modeling schemes. One-dimensional composition and heat budget studies and their implications are described in the reviews by Fox and Bougher (1991) and by Kasprzak et al. (see their chapter). In this section we discuss those circulation models that have contributed most to understanding the large-scale wind structure of the Venus thermosphere.

A. Early Spectral and Two-Dimensional Models

Prior to Pioneer Venus, only the SS-AS circulation pattern was generally thought to be operating in the Venus thermosphere, due to the slow planetary rotation period and the small Venus obliquity. The Dickinson and Ridley (1972,1975,1977) two-dimensional model (DRM) of the temperatures, winds, and composition of the Venus thermosphere, correctly predicted the gross characteristics of this SS-AS mean circulation. Axial symmetry was assumed in choosing altitude and solar zenith angle (SZA) as the two model coordinates, in accord with the dominant SS-AS flow. Later versions of this model were fully self-consistent and nonlinear, and were exercised to calculate simultaneously thermal structure, winds, and CO_2, CO, and O densities above 100 km. Detailed results of the model, however, differed considerably from the Pioneer Venus observations (Schubert 1983; Dickinson and Bougher

1986). Dickinson-Ridley temperatures were somewhat too warm on the day side and much too warm, by as much as 100 K, on the night side. Their calculations failed to predict the nightside cryosphere entirely. Model atomic-O and CO bulges on the night side were also predicted by the DRM, that were not observed. Correspondingly, the dayside O-mixing ratios at the ionospheric F1-peak (140 km) were calculated to be 2.5 to 5.6%, about a factor of 4 to 6 smaller than actually observed (17–20%) (von Zahn et al. 1983).

Simplified spectral model calculations were conducted prior to Pioneer Venus that considered the potential importance of superrotation in the Venus thermosphere (Mayr et al. 1978; Hartle et al. 1978; Kumar et al. 1978). It was generally concluded that thermospheric winds would yield H and He bulges on the night side. The amplitudes of these bulges were found to be significantly different for cases with and without superrotation. However, like the DRM, the simulated nightside temperatures were too large and atomic-O was incorrectly predicted to peak on the night side.

An initial analysis of the Pioneer Venus ONMS measurements (Niemann et al. 1980) was carried out by Mayr et al. (1980) with the use a simplified three-dimensional spectral model. By independently varying the rates of thermospheric superrotation and eddy diffusion (at the turbopause), it was found that the inferred day-night temperature contrast in the thermosphere of Venus is likely associated with winds of about 200 m s^{-1} (to be superimposed on the superrotation velocity) which transport light species (e.g., O, He and H) towards the night side. An eddy diffusion coefficient of about 3.0×10^3 m^{2}s^{-1} is also required to buffer the horizontal mass transport so as to reproduce the observed daytime maximum in O and the day-to-night increase in He. Finally, the observed time response and relatively small day-night density variations, especially in He, require a thermospheric superrotation period between 5 and 10 days.

Several explanations were forwarded to account for the DRM model discrepancies in light of the initial ONMS studies of Mayr et al. (1980). Bougher et al. (1986) proposed that a global circulation system weaker than the DRM originally predicted would enable increased dayside O and CO densities to be maintained, while simultaneously reducing the buildup on the night side. Furthermore, because nightside heating is maintained primarily from the circulation (i.e., adiabatic compressional heating), slower winds would also result in cooler nightside temperatures more indicative of a cryosphere. Suggestions along this line were also made by von Zahn et al. (1983) who noted that dayside O/CO$_2$ ratios increased for the series of Dickinson and Ridley, (1972,1975,1977) models, due to progressively weaker day-to-night winds. Schubert (1983) also noted that dynamical processes may be required to explain the very low nightside thermospheric temperatures.

The Bougher et al. (1986) weakened circulation hypothesis seems to give the most self-consistent explanation for all the available Pioneer Venus composition, temperature, and airglow data. An updated version of the DRM model was constructed by Bougher et al. (1986) to examine how SS-AS winds, eddy diffusion/conduction, and strong CO_2 15-μm cooling affect the day-night contrasts in Venus densities and temperatures. It was learned that the basic subsolar-antisolar features of the Venus thermosphere (Hedin et al. 1983) could be simulated largely by adding a wave-drag mechanism, resulting from turbulent dissipation effects, which weakened the day-to-night flow. Eddy viscosity and Rayleigh friction were parameterized within the horizontal momentum equation to mimic the wave-mean-flow interaction required (see Secs. IV and V). Maximum terminator winds of $\leq$240 m s^{-1} were found to be consistent with observed global composition and temperature fields. Such a reduction of winds by nearly a factor of 2 (from those of DRM) permitted less O and CO to be carried to the night side; the dayside ionospheric-peak O-mixing ratios were calculated to be $\sim$15%, and observed dayside bulges of O and CO on a constant height surface were reproduced.

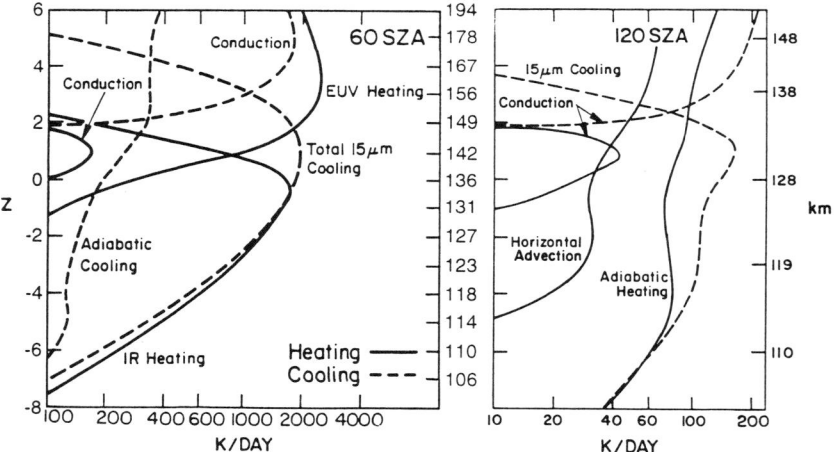

Figure 5. NCAR two-dimensional solar maximum model heating and cooling rates for 60 and 120° SZA. Heat balance for mean dayside and nightside conditions is presented in units of degrees K per Earth day (figure adapted from Bougher et al. 1986).

Subsequently, dayside temperatures were shown to be largely maintained by a balance of EUV/infrared heating and strong CO_2 15-μm cooling, with little influence by dayside adiabatic cooling due to upwelling flow (see the left-hand plot of Fig. 5). The increased isolation of the day and nightside thermospheres also serves to enhance the diurnal temperature contrast, giving midnight exospheric temperatures of about 115 K, characteristic of the observed cryosphere. This occurs because slower winds support less adiabatic compressional heating, the dominant source of nightside heating. The right-hand plot of Fig. 5 shows that adiabatic compressional heating from nightside subsiding winds largely balances CO_2 15-μm cooling below 140 km, with molecular conduction becoming important above. The magnitude of this adiabatic heating is reduced by about a factor of 3 from the pre-Pioneer Venus model of Dickinson and Ridley (1977), permitting the observed cold nightside temperatures to be reproduced. Early ideas that invoked nightside eddy heat conduction to give cold nightside temperatures (see, e.g., Niemann et al. 1980; Schubert 1983; Gordiets and Kulikov 1985) appear to be inappropriate. The Bougher et al. model simulations suggest that slower winds are sufficient to maintain cold nightside temperatures, and are simultaneously consistent with the observed diurnal density contrasts.

Figure 13 in the chapter by Kasprzak et al. illustrates a sampling of contour plots from two-dimensional model simulations (Bougher et al. 1986) for solar maximum (SMAX) and solar minimum (SMIN) conditions ($F_{10.7} = 200$ and 74, respectively). The SMAX terminator horizontal wind speeds approach 240 m s^{-1}; corresponding vertical winds peak at +0.7 m s^{-1} (day side) and -0.6 m s^{-1} (night side). The comparable set of SMIN simulations shows cooler dayside temperatures, which serve to collapse the dayside pressure surfaces. By contrast, the nightside temperatures and densities are virtually unchanged. This prediction of relatively stable mean nightside temperatures and densities over the solar cycle was verified by PVO entry measurements made in 1992 (Kasprzak et al. 1993a). Similarly, the dayside exospheric temperatures were predicted to vary by less than 70 K (near noon) over the solar cycle. The MGN drag measurements made during Cycle 4 at $F_{10.7} = 130$ show a temperature variation over half the solar cycle ($\sim$30 K) that is consistent with this two-dimensional model prediction (Keating and Hsu 1993). This confirms the important role played by very strong CO_2 15-μm cooling in the Venus dayside heat budget (Bougher et al. 1994; see the chapter by Kasprzak et al.).

B. Detailed Three-Dimensional Modeling

RSZ wind effects on thermospheric densities and temperatures can only be examined using three-dimensional coupled chemical-dynamical models. Such a self-consistent code coupling large-scale and sub-grid scale effects, including wave-drag and superrotation, is very difficult to formulate. Two research groups emerged in the 1980s with three-dimensional models that begin to address the feedbacks that exist among the Venus' thermospheric temperature, density and wind fields (see review by Fox and Bougher 1991).

A finite-difference model was first developed by Bougher et al. (1988) for Venus, based on the National Center for Atmospheric Research (NCAR) terrestrial thermospheric general circulation model (TGCM). The general framework of this terrestrial TGCM code (Dickinson et al. 1984) was adapted to Venus, in concert with new inputs and physical parameterizations, to address Venus' thermospheric processes (Bougher et al. 1988,1990; Bougher and Borucki 1994). The current benchmark Venus TGCM (VTGCM) code calculates global distributions of CO_2, CO, O, and N_2 (major species) which are consistent with the three-dimensional day–night temperature contrasts and the corresponding 3-component winds. Minor species (O_2, N(^{4}S), N(^{2}D), NO and He) can also be simulated, but have no impact on the wind system. The VTGCM model covers a 5° by 5° latitude-longitude grid, with 33 evenly spaced log-pressure levels in the vertical, extending from approximately 94 to 210 km at local noon. Dayside O and CO sources arise primarily from CO_2 net dissociation; catalytic CLO_x and HO_x reactions are also employed to specifically improve the chemical sources and sinks for O and CO below 120 km (Massie et al. 1983; Bougher and Borucki 1994). The VTGCM can be used to examine Venus' thermospheric structure and winds for solar maximum (SMAX), solar moderate (SMED), and solar minimum (SMIN) EUV-UV flux conditions; i.e., these cases correspond to terrestrial $F_{10.7}$ indices of 200, 110 to 130, and 68 to 80, respectively. The CD-ROM supplement accompanying this book gives color plots of benchmark VTGCM fields for these three cases: SMAX (CDP2C4F1-21), SMED (CDP2C4F22-42), and SMIN (CDP2C4P43-63).

CO_2 15-μm cooling, wave drag, and eddy diffusion formulations are incorporated into the VTGCM (Bougher et al. 1988). CO_2 15-μm emission is known to be enhanced by collisions with O-atoms, providing increased cooling in non-LTE (NLTE) regions of the atmosphere (see the chapter by Kasprzak et al.) . The corresponding collisional relaxation rate adopted for VTGCM simulations is 1 to 2×10^{-12} cm^{3}s^{-1}. This provides strong CO_2 cooling that is consistent with the use of EUV heating efficiencies of 16 to 20% (Keating and Bougher 1992; Bougher et al. 1994). Sub-grid scale processes (i.e., eddy diffusion, viscosity, conduction, and wave drag) are not currently self-consistently formulated in the VTGCM, but rather parameterized using standard aeronomical formulations. In particular, Rayleigh friction was used to mimic wave-drag effects on the mean flow (see Sec. III.C).

VTGCM simulations for SMAX conditions (early PVO mission) proved very useful in the interpretation of PVO density, temperature, airglow, and ionospheric data (see review by Fox and Bougher 1991). For example, the observed nightside "cryosphere" as well as the large day–night density contrasts are generally consistent with a global wind system that is mechanically slowed by turbulence induced friction. Average terminator horizontal winds of $\sim$180 to 200 m s^{-1} are required to maintain the basic day–night contrasts of measured temperatures and densities. The VTGCM also confirms previous NCAR two-dimensional model results which show the dayside heat budget is largely unaffected by the large-scale winds. Thermospheric temperatures are regulated instead by a strong CO_2 15-μm thermostat that is moderated by changing atomic O densities over the solar cycle (Bougher et al. 1994; see the chapter by Kasprzak et al.). VTGCM simulations were also conducted for SMED conditions corresponding to PVO entry in 1992 and the MGN Cycle 4 period in 1992–1993 (Bougher and Borucki 1994). A 25 to 30 K variation of SMAX to SMED dayside mean exospheric temperatures is simulated, in

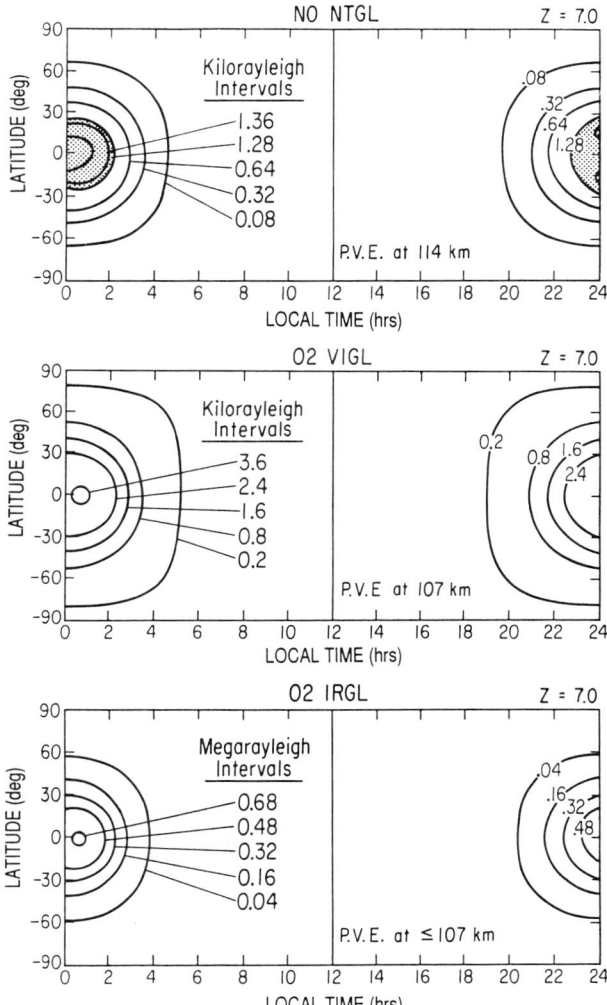

Figure 6. The VTGCM SMAX model. A set of night airglow distributions corresponding to the wind field is given for: (top) the NO(0,1) δ-band emission at 198 nm (NO NTGL); (middle) the O_2 Herzberg II visible nightglow over 400 to 800 nm (O2 VIGL); and (bottom) the O_2 1.27-μm infrared nightglow (O2 IRGL). The peak volume emission (PVE) rate altitude is indicated for each nightglow distribution.

accord with MGN drag measurements (Keating and Hsu 1993).

For the current benchmark VTGCM SMAX simulation, zonal winds are prescribed to be rather weak in the lower thermosphere (≤ 25 m s^{-1}, consistent with Goldstein et al. [1991] winds) and strong above 150 km (75 m s^{-1}, in accord with zonal winds inferred from hydrogen and helium diurnal distributions). The net VTGCM horizontal wind system consists of dusk winds of 240 m s^{-1} and dawn winds of 160 m s^{-1} near 150 km. The corresponding night airglow distributions of NO ultraviolet, O_2 visible, and O_2 infrared integrated vertical intensities are presented in Fig. 6. The resulting NO emission shows a peak at LT = 0200, consistent with strong zonal winds at higher altitudes, while both O_2 nightglow distributions show a peak near LT = 0100, corresponding to weaker zonal winds over 100 to 130 km. Strong zonal winds were later prescribed in the VTGCM lower thermosphere, in accord with the measurements obtained by Lellouch et al. (1994). The resulting O_2 infrared intensity distribution appears to be horizontally smoothed and its peak is shifted strongly toward the morning terminator. The corresponding CO mixing ratio distribution is rather smooth across the night side, closely matching the Lellouch et al. observations over 95 to 105 km. These VTGCM modeling exercises confirm that this combination of nightglow emissions provides a good method for constraining the average RSZ winds over 100 to 175 km (see Sec. II.B). However, the benchmark VTGCM is unable to reproduce the large temporal and spatial variability observed in the NO and O_2 infrared nightglow intensity distributions (see Sec. V).

The three-dimensional Goddard Space Flight Center (GSFC) spectral model of Stevens-Rayburn et al. (1989) and Mengel et al. (1989), referred to as SRM, also succeeds in describing the salient features of the Pioneer Venus thermospheric composition and temperature measurements. Unlike the model of Bougher et al. (1988), which is self consistent, the SRM model computes the departures from a height dependent background atmosphere which describes empirically the spherically symmetric components of temperature and composition inferred from Pioneer Venus (Hedin et al. 1983). In an expansion of spherical and Fourier harmonics, the nonlinear equations for mass, energy and momentum are solved for CO_2, O and He from 80 km up to 200 km, the exobase. Thermospheric superrotation is accounted for by assuming a height independent rotation period (4–8 days), which is treated as a free parameter chosen to reproduce the observations. In an earlier version of the nonlinear SRM model, Mayr et al. (1985) acheived such agreement by postulating a relatively large heating efficiency of 30%, which sufficiently increased the temperature contrast between day and night and produced SS-AS winds of about 200 m s^{-1} at the terminators. Distinct from that, a more realistic heating efficiency of about 20% (Fox 1988) is adopted in the current SRM model. Also, following Bougher et al. (1986), Rayleigh friction is introduced instead to reduce the SS-AS winds such as to increase the temperature contrast between day and night. A drag coefficient of 2×10^{-4}s^{-1} is adopted at 150 km varying with height at a rate inversely proportional to the square-root of the globally averaged ambient density. Such a height dependence is also adopted for the eddy diffusion rate, which is assumed to reach its maximum value of 3×10^3 m^{2}s^{-1} at 120 km. The SRM model solves the continuity and momentum equations of the major species CO_2 and O, allowing for the differences in the vertical and horizontal transport velocities and for the momentum exchange between the two constituents. Given the resulting temperature and wind fields, the variations in He and its vertical as well as horizontal transport velocities are then computed separately.

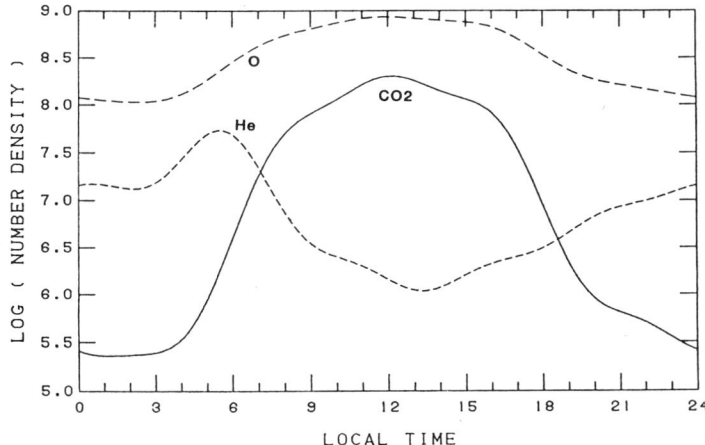

Figure 7. SMAX equatorial output fields vs local time at an altitude of 170 km for the standard GSFC spectral model. Number densities for CO_2, O, and He are shown (figure from Mengel et al. 1989).

Mengel et al. (1989) dayside and nightside temperature profiles are in rough agreement with ONMS derived temperatures. They concluded that the composition measurements from Pioneer Venus cannot be reproduced with a model EUV/UV heating efficiency significantly less than 20%. Superrotation was found to have relatively little effect on the temperature contrast between day and night; however, it did produce a dawn–dusk asymmetry in the flow that enabled the nightside helium bulge to peak near LT = 0500 (see Fig. 7). The diurnal variation in He is found to depend sensitively on the adopted thermospheric superrotation rate; i.e., the longer the superrotation period (i.e., the weaker the zonal wind component), the larger is the day to night buildup in the helium density, and the shorter is the time delay in the peak density after midnight (see Fig. 1). The helium density distribution is also strongly dependent on the eddy diffusion coefficient. Mengel et al. suggest the data are best fitted by a SS-AS circulation having terminator winds of $\sim$100 m s^{-1}, a moderate height-dependent global mean eddy diffusion ($K \leq 3 \times 10^3$ m^{2}s^{-1}), and a superrotation period of 6 Earth days. Exospheric transport, a density and temperature driven ballistic flow above the thermosphere, may also serve to weaken the magnitude of the nightside helium bulge, thereby permitting stronger SS-AS winds to be accommodated within the thermospheric wind system. Thus, use of helium as a reliable tracer of Venus thermospheric winds first requires that the role of exospheric wind transport be quantified.

It is apparent that differences in the mean state of the Venus thermospheric circulation (and small-scale processes) still exist in the most recent model simulations (Bougher et al. 1988,1990; Bougher and Borucki 1994; Mengel et al. 1989). However, the major tunable parameters have been identified: (1) the zonal superrotation period, (2) the magnitude of the maximum global mean eddy diffusion coefficient, and (3) the magnitude of the peak Rayleigh friction coefficient which determines SS-AS wind speeds. Section V will discuss how these "independent" parameters can be linked through gravity wave breaking theory.

C. Implications of Dynamical Processes

It is apparent from the observed Venus day–night density and temperature contrasts (Sec. II.A), the corresponding nightglow intensity distributions (Sec. II.B), and the requirement for strong CO_2 15-μm cooling (and significant O-atom abundance) on the day side (Sec. III.B), that some type of deceleration mechanism is necessary to slow the otherwise pressure-driven thermospheric winds (Seiff 1982; Bougher et al. 1986). Furthermore, this deceleration may not be symmetric in local time, because net zonal winds appear stronger at the dusk than the dawn terminator. Nonlaminar flow conditions are possible from: (1) *in-situ* instability of the mean flow, particularly near the terminators, and/or (2) the influence of upward propagating waves from the lower atmosphere.

The circulation models to date have elected to parameterize this "momentum drag" in the form of Rayleigh friction. The goal has been to quantify a time scale for the deceleration process so that self-consistent turbulence or wave breaking schemes, providing momentum stress and eddy diffusion, can be subsequently developed. A height-dependent Rayleigh friction profile is a first choice when little is known about wave parameters and their sources. The deceleration of the SS-AS winds was proposed by Bougher et al. (1986) to become important at a level where wave-breaking maximizes; i.e., also just above the dayside homopause (135–150 km). A profile for Rayleigh friction drag was prescribed which increases with height up to this region, consistent with a vertical propagating spectrum of waves each of which breaks at a slightly different level. A drag coefficient of ≤ 1 to 1.5×10^{-4} s^{-1} was chosen. This leaves SS-AS winds below 110 km largely unaffected, while those near 135 to 150 km are reduced from pressure-driven values by a factor of two. Adjustment of this drag profile was made enabling the NCAR two-dimensional and VTGCM models to reproduce correctly the observed day–night density and temperature contrasts. This global modeling procedure provides an estimate of the region where gravity wave breaking and corresponding momentum deposition should be most important, in substantial agreement with the 1978–80 and 1992 ONMS wave perturbations observed (Kasprzak et al. 1993*b*). The VTGCM studies argue strongly for a wave-breaking scheme that provides peak drag over 135 to 150 km of about 0.01 to 0.02 m s^{-2} (see Zhang et al. 1996). Wave drag will be discussed further in light of detailed gravity-wave breaking theory (see Secs. IV and V).

Fox and Bougher (1991) recently revisited the concept of the homopause level; i.e., where the eddy and molecular diffusion coefficients are equal. The Venus circulation model studies of Bougher et al. (1988,1990), Bougher and Borucki (1994), and Mengel et al. (1989) suggest that vertical small-scale eddy diffusion is only a moderate contributor to the maintenance of observed dayside thermospheric densities of the major species. Eddy coefficients required are about 2 to 3 times smaller than values used by previous one-dimensional continuity-diffusion models (see, e.g., von Zahn et al. 1980; Massie et al. 1983). This implies that the vertical eddy coefficient derived by these one-dimensional models is not solely a signature of small-scale vertical mixing. Rather it provides a reasonable description of compositional variations brought about by a combination of variable large-scale horizontal and vertical winds plus small-scale mixing.

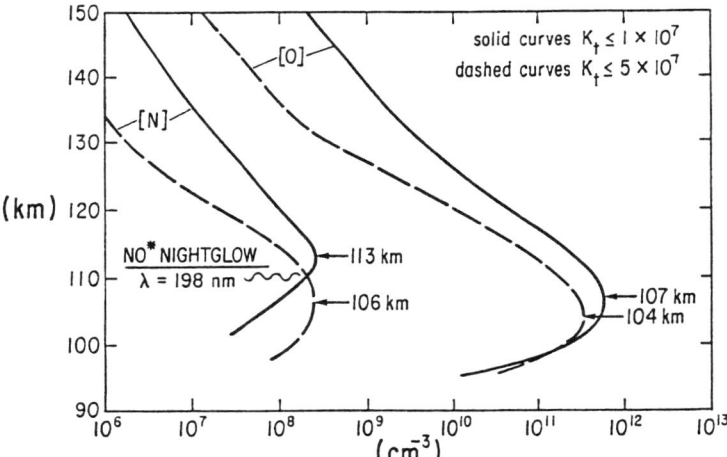

Figure 8. VTGCM nightside density profiles for N and O atoms are simulated for the period of PVO entry (LT = 2400). Nightside variations in eddy diffusion of a factor of 5 are predicted to have a large impact on the resulting N and O density profiles and the associated night airglow (NO and O_2) layers (figure from Bougher and Borucki 1994).

Recent Venus three-dimensional modeling efforts (Bougher and Borucki 1994; Mengel et al. 1989) still disagree as to the magnitude of this residual eddy diffusion required to supplement the global circulation in maintaining observed vertical and diurnal density distributions. The VTGCM cannot tolerate global eddy diffusion exceeding 2×10^3 m^{2}s^{-1}, in the context of $\sim$200 m s^{-1} SS-AS terminator winds, and still yield atomic-N and O densities that reasonably match Hedin et al. (1983) empirical model values. Furthermore, the altitude of the NO nightglow peak volume emission rate (115±2 km) is a very sensitive indicator of the magnitude of nightside eddy diffusion within this global circulation model. Figure 8 illustrates the sensitivity of the peak NO nightglow layer to a factor of 5 change in the peak eddy coefficient. Conversely, the Goddard SRM spectral model (Mengel et al. 1989) uses eddy diffusion reaching 3×10^3 m^2s^{-1}, in concert with maximum horizontal winds of ~ 100 ms^{-1}, to obtain observed diurnal helium distributions. The discrepancies in eddy diffusion and wave drag (wind) parameters in these three-dimensional models reflect the difficulty in obtaining a unique solution without further observations, namely simultaneous measurements of thermospheric temperatures, densities, and winds above 120 km.

IV. EVIDENCE FOR GRAVITY WAVES

The most profound evidence for gravity-wave saturation processes in the thermosphere lies in the apparent stability of Venus' primarily CO_2 atmosphere. Photochemical considerations alone suggest that CO_2 should be photolyzed rapidly by solar radiation in the ultraviolet ($\lambda \leq 169$-nm) to produce large quantities of the dissociation products, O and CO. Direct recombination (O + CO + M) is negligibly slow, while oxygen three-body recombination (O + O + M) is many orders of magnitude faster, leading to the false conclusion that CO and O_2 should dominate the composition. The solution to this photochemical puzzle is part chemical, and part dynamical (Chamberlain and Hunten 1987). Chlorine compounds in the Venus clouds catalyze the CO_2 recombination path; however strong vertical transport must be invoked to bring the O and CO photodissociation products down to the altitudes where this chemistry is efficient. The eddy diffusion coefficients prescribed in one-dimensional models (von Zahn et al. 1980; Massie et al. 1983) are 2 orders of magnitude larger than corresponding terrestrial values. Some of the vertical transport required in one-dimensional eddy diffusion calculations is accomplished by the SS-AS circulation in the thermosphere (see Sec. III.C). Nevertheless, the residual eddy mixing currently invoked in Venus circulation models to reproduce observations is still large. This eddy diffusion is presumably a result of gravity-wave breaking in the thermosphere.

Global thermospheric circulation models like the VTGCM also rely on gravity-wave saturation processes to reproduce the very cold observed nightside temperatures and the day-night density contrasts (see Sec. III.C). Without this wave drag scheme, as formulated by Rayleigh friction, the VTGCM would predict nightside bulges in O and CO and warm nightside temperatures, much like those simulated earlier by Dickinson and Ridley (1977) prior to Pioneer Venus observations.

In addition to evidence of gravity-wave saturation processes in the thermosphere, there is direct evidence of wave propagation at these altitudes. Periodic density fluctuations were observed by the Pioneer Venus ONMS during 1978–1980 and 1992 when the orbiter was dipping into the thermosphere at periapsis. In analyses by Kasprzak et al. (1988,1993*b*), periodic density fluctuations with wavelengths between 100 and 600 km were isolated. Longer wavelengths were filtered out in their analysis because the orbit of the spacecraft only allowed measurements along a track about 1500 km long. Shorter wavelengths were obscured by fluctuations associated with the spacecraft spin period. Amplitudes and relative phases of the fluctuations in He, N, O, N_2, and CO_2 were found to vary with the molecular weight of the species, consistent with the theoretical understanding of wave propagation in the heterosphere (Dudis and Reber 1976; Del Genio and Schubert 1979). This effect was modeled successfully in a companion paper by Mayr et al. (1988). Collisional momentum transfer and diffusion effects will tend to force O and CO_2 to vary in phase and to have similar amplitudes, while perturbations in the lightest constituent, helium, will be out of phase as observed.

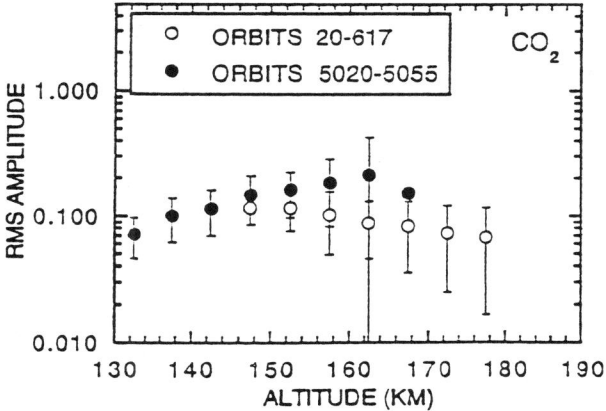

Figure 9. The rms wave amplitudes for CO_2 plotted as a function of altitude for the 1978–1980 ONMS data (Niemann et al. 1980) (open circles) and the 1992 data (Kasprzak et al. 1993a) (filled circles). Averages over a 5-km interval are shown with corresponding data scatter as error bars (figure from Kasprzak et al. 1993b).

Mayr et al. (1988) described these kinds of perturbations as gravity waves with periods of 30 minutes to an hour propagating from a localized source region in the middle atmosphere ($\sim$80 km). They found that a source at 130 km altitude was rapidly dissipated on a scale similar to the horizontal wavelength of the wave, ruling out a thermospheric source for the observed wave trains. Kasprzak et al. (1988) also observed larger amplitudes on the night side than the day side at a given altitude. Mayr et al. (1988) showed these observed amplitude variations were consistent with day/night temperature contrasts assuming the waves have a uniform energy per unit mass in the source region.

In Kasprzak et al. (1993b), rms wave amplitudes were observed to grow with height between 130 and 160 km (Fig. 9) with an effective energy growth scale height of about 30 km. Linear theory predicts conservative growth of gravity-wave amplitudes should vary with a scale height equal to twice the density scale height, which would be smaller than 10 km for these nightside observations. The observed 30 km scale height then implies gravity-wave amplitude growth is being limited by some nonlinear process like saturation or radiative damping in the thermosphere. The amplitudes appear to actually decrease in height above 160 km. Observations from the earlier PVO period (Kasprzak et al. 1988) did not sample much below 150 km, but showed a decrease in amplitude above that height. Indirect evidence of gravity wave effects in the thermosphere also appears in the airglow observations described in Sec. II.B. For example, the PVO Star Tracker was used to observe O_2 visible airglow along the limb, clearly indicating gravity-wave signatures on individual spacecraft orbits (Bougher and Borucki 1994).

Other evidence of waves in the thermosphere appeared in entry profiles from the Pioneer Venus (PV) probes and bus (Seiff et al. 1980; Seiff and Kirk 1991; von Zahn et al. 1980) and in the PV and Magellan orbiter drag data (Keating et al. 1979,1993). However, because these observations were not able to characterize the horizontal wavelengths or intrinsic periods of the perturbations, they may likely be associated with planetary scale waves rather than gravity waves.

There is abundant evidence for waves at Venus cloud levels. Schubert (1983) gives a review of early Pioneer Venus, Mariner 10 and Venera mission observations of waves in the Venus atmosphere. Gierasch et al. (see their chapter) address the most recent evidence provided by the Vega balloons. Albedo features observed at cloud levels in ultraviolet photographs of Venus show many structures which have been interpreted as waves with horizontal scales ranging from about 100 km to planetary scales. For example, ultraviolet features interpreted as convective cells (Del Genio and Rossow 1982) with horizontal scales of hundreds of km have been observed. The Vega balloons in 1985 (Linkin et al. 1986) probed the unstable layer at about 50 to 55 km altitude, and found vertical velocities of ~ 1 m s^{-1} associated with convective motions there. Hence, the ~ 50 to 80 km region within the clouds is a likely source level for vertically propagating waves. Small-scale cloud features have been observed to move with speeds close to the 4-day rotation of the mean atmosphere at cloud tops. Vertical propagation into the thermosphere of gravity waves with such large zonal phase speeds could lead to deposition of energy and momentum that is zonally asymmetric in local time according to Lindzen's (1981) theory of gravity-wave saturation. This idea was developed in Alexander (1992) and Alexander et al. (1993) to explain the thermospheric superrotation and local time asymmetries observed in O 130-nm images, and is described in the next section.

V. THEORY OF GRAVITY-WAVE BREAKING

The direct and indirect evidence for gravity waves in the Venus thermosphere points to gravity-wave forcing as an important process in explaining both the observed variability as well as some aspects of the general circulation. Research in this area has generally followed studies of gravity-wave effects in the Earth's upper atmosphere. Since Lindzen (1981) developed a parameterization for gravity-wave dissipation and momentum flux convergence, there has been a growing acceptance of the fundamental control these processes have on the composition and circulation in the Earth's mesosphere and lower thermosphere (see, e.g., Holton 1982, 1983; Dunkerton 1982; Garcia and Solomon 1985). Lindzen's theory used the concept of gravity-wave "saturation," a term defined in Holton (1982) as the condition where the local temperature lapse rate, produced by the sum of the wave and mean state, is dry adiabatic. A good review of the theory and consequences of gravity-wave saturation processes appeared in Fritts (1989).

A. Gravity-Wave Saturation Theory

Waves propagating vertically in a conservative fashion through a horizontally homogeneous background state will grow in amplitude conserving the momentum flux they carry. Saturation assumes that convective instability limits this conservative growth above the "breaking level," defined as the level where the local lapse rate equals the adiabatic lapse rate. Holton (1982) modified Lindzen's theory to make it more suitable for application in numerical models, and it is this basic form, with a discrete set of monochromatic source waves, that has been used successfully ever since.

In Lindzen's parameterization, between the breaking level and the critical level, the wave transfers energy and momentum to the background atmosphere. The wave induced force per unit mass is such that the flow is always accelerated towards the value of the phase speed of the wave. Lindzen's parameterization was a great advance over the Rayleigh friction method parameterizing gravity-wave effects. In particular, Rayleigh drag depends only linearly on the mean zonal wind, and so greatly underestimates the effects in a region of wind shear. Furthermore, Rayleigh friction can only slow wind magnitudes, whereas gravity wave forcing may be either an accelerative or decelerative force depending on the wave properties and the atmospheric conditions. Lindzen's parameterization also seems to capture the essential physics of gravity-wave breaking effects, particularly the relationship between wave energy dissipation and the drag force. Applications of Lindzen's theory to the Venus thermosphere are described in the next section.

More recently, Fritts and Lu (1993) have described a spectral parameterization based on empirical observations of gravity waves in the Earth's atmosphere and saturation theory. Their model overcomes many of the shortcomings of using a discrete set of monochromatic waves in numerical simulations. This parameterization has been applied in a model of the Earth's thermosphere (Roble and Ridley 1994), and is currently being modified for application to the Venus thermosphere in the VTGCM (Zhang et al. 1996).

B. Gravity-Wave Saturation Theory Applied to Venus

All of the essential physics captured in the Lindzen-Holton parameterization should also apply on Venus. The difference from those applications to the Earth's atmosphere will be primarily in the description of the wave sources. As discussed in Sec. IV, the upper cloud level (~50–80 km) is a likely source region for upper atmosphere gravity waves on Venus. Westward propagating waves generated near the surface tend to be absorbed at critical levels in the lower atmosphere, while those with eastward phase speeds tend to be reflected or trapped in unstable layers near 30 km and at 50 to 55 km (Schubert and Walterscheid 1984). Topographically forced stationary waves with horizontal wavelengths ≥100 km can propagate through these unstable regions, but are likely to be strongly damped at altitudes just above the clouds (see the chapter by Gierasch et al.).

Alexander (1992) hypothesized that the primary source region for waves that break in the thermosphere is near the cloud tops. Because strong westward zonal winds with speeds $\bar{u} \sim 100$ m s^{-1} prevail at these altitudes, gravity waves forced in this region will likely have large westward phase speeds, $c \sim 100$ m s^{-1}, similar to the mean winds at this level. This simple hypothesis along with the Lindzen-Holton gravity wave parameterization predict strong asymmetries in both wave drag and eddy diffusion in the thermosphere as a function of local time. The local time asymmetries arise from the interaction of the gravity waves with the SS-AS flow in the thermosphere. Lindzen's theory predicts the gravity-wave drag and diffusion to be quite sensitive to the magnitude of the intrinsic phase speed, $(\bar{u} - c)$. Qualitatively, then, it is expected that for $c \sim 100$ m s^{-1}, the drag and diffusion will be much larger at morning local times than at corresponding afternoon times, because the dawn and dusk SS-AS winds are roughly -200 m s^{-1} and $+200$ m s^{-1}, respectively. (Positive winds here are westward, in the same direction as the planet's rotation.) A strong asymmetry in eddy diffusion in the middle thermosphere at latitudes poleward of 30° was inferred from images of the atomic oxygen emission shown in Fig. 4. An asymmetry in drag on the SS-AS flow in this sense has also been observed indirectly as the thermospheric RSZ flow. Thus, gravity-wave momentum flux convergence in the thermosphere provides an *in-situ* forcing mechanism for the thermospheric superrotation if the sources for these waves are presumed to be at altitudes in the vicinity of the cloud tops.

Quantitative estimates of the thermospheric drag force per unit mass and eddy diffusion coefficient were described in Alexander (1992) and Alexander et al. (1993). Ratios of the drag and diffusion at dawn to dusk terminators were estimated to be on the order of 10:1 using some plausible scenarios for gravity-wave sources. The wave forcing derived in this work was in good agreement with the Rayleigh friction prescribed to reproduce the constraints on the thermospheric superrotation. The eddy diffusion asymmetry is of the correct magnitude to explain the asymmetries in the atomic oxygen images at high latitudes, although no such asymmetry was observed at low latitudes (Fig. 4). In short, this wave-breaking scheme provides a promising mechanism to explain the Venus thermosphere superrotation and the local time variation in eddy diffusion inferred from airglow observations.

VI. UNRESOLVED ISSUES

The processes responsible for maintaining the SS-AS and RSZ winds in the Venus upper atmosphere are still not well understood or quantified. It is apparent from available spacecraft and groundbased observations that some type of deceleration mechanism is necessary to slow the otherwise pressure-driven upper atmospheric winds. Furthermore, this deceleration is not symmetric in local time, because net zonal winds appear stronger at the dusk than the dawn terminator. The mechanism responsible for this deceleration and asymmetry is thought to be gravity-wave breaking and subsequent momentum and energy deposition in the thermosphere. The large variability of the zonal winds over 90 to 110 km is also attributed to the changing nature of this wave breaking. Further three-dimensional modeling, using a self-consistent gravity-wave breaking formulation, is needed to quantify the role of gravity waves in controlling the Venus thermospheric wind system. In this regard, the recent Fritts and Lu (1993) terrestrial wave breaking model, based upon the theory of wave saturation using a full spectrum of gravity waves, shows promise for adaptation to Venus (Zhang et al. 1996).

Finally, the three-dimensional modeling tools presently being used suffer from an inability to reproduce observed diurnal density and temperature variations with a unique set of wind fields and eddy/wave-drag parameters. This may reflect missing physical processes; i.e., exospheric transport affects the diurnal variations of helium and hydrogen. Nevertheless, the modeling task would be much improved if simultaneous wind, temperature, and density measurements could be made above 100 km. Such measurements may not be forthcoming for Venus, but should be considered for a future Mars upper atmosphere mission. Systematic monitoring of Venus upper mesospheric and lower thermospheric temperature and wind fields by groundbased observers is possible, and would serve to unfold the nature of cyclostrophic balance and the highly variable zonal winds above the cloud tops.

Acknowledgments. This work has been supported in part by the NASA Venus Data Analysis Program.

Editor's Note

The authors of this chapter have included an animation on the accompanying CD-ROM. Please see CDP2C4M1.

REFERENCES

Alexander, M. J. 1992. A mechanism for the Venus thermospheric superrotation. *Geophys. Res. Lett.* 19:2207–2210.

Alexander, M. J., Stewart, A. I. F., Solomon, S. C., and Bougher, S. W. 1993. Local time asymmetries in the Venus thermosphere. *J. Geophys. Res.* 98:10849–10871.

Allen, D., Crisp, D., and Meadows, V. 1992. Variable oxygen airglow on Venus as a probe of atmospheric dynamics. *Nature* 359:516–518.

Bertaux, J. L., Chassefiere, E., and Kurt, V. G. 1985. Venus EUV measurements of hydrogen and helium from Venera 11 and Venera 12. *Adv. Space Res.* 5(9):119.

Bougher, S. W., and Borucki, W. J. 1994. Venus O_2 visible and IR nightglow: Implications for lower thermosphere dynamics and chemistry. *J. Geophys. Res.* 99:3759–3776.

Bougher, S. W., et al. 1986. Venus mesosphere and thermosphere: II. Global circulation, temperature, and density variations. *Icarus* 68:284–312.

Bougher, S. W., Dickinson, R. E., Ridley, E. C., and Roble, R. G. 1988. Venus mesosphere and thermosphere III: Three-dimensional general circulation with coupled dynamics and composition. *Icarus* 73:545–573.

Bougher, S. W., Gerard, J. C., Stewart, A. I. F., and Fesen, C. G. 1990. The Venus nitric oxide night airglow: Calculations based on the Venus thermospheric general circulation model. *J. Geophys. Res.* 95:6271–6284.

Bougher, S. W., Hunten, D. M., and Roble, R. G. 1994. CO_2 cooling in terrestrial planet thermospheres. *J. Geophys. Res.* 99:14609–14622.

Brinton, H. C., et al. 1980. Venus nighttime hydrogen bulge. *Geophys. Res. Lett.* 7:865.

Chamberlain, J. W., and Hunten, D. M. 1987. *Theory of Planetary Atmospheres: An Introduction to Their Physics and Chemistry*, 2nd ed. (New York: Academic Press).

Clancy, R. T., and Muhleman, D. O. 1991. Long-term (1979–1990) changes in the thermal, dynamical, and compositional structure of the Venus mesosphere as inferred from microwave spectral line observations of ^{12}CO, ^{13}CO, and $C^{18}O$. *Icarus* 89:129–146.

Connes, P., Noxon, J. F., Traub, W. A., and Carleton, N. P. 1979. $O_2(^1\Delta)$ emission in the day and night airglow of Venus. *Astrophys. J. Lett.* 233:29–32.

Crisp, D., Allen, D. A., Grinspoon, D. H., and Pollack, J. B. 1991. The dark side of Venus: Near-infrared images and spectra from the Anglo-Australian Observatory. *Science* 253:1263–1266.

Crisp, D., et al. 1996. Ground-based near-infrared observations of the Venus nightside: 1.27-μm $O_2(^1\Delta)$ airglow from the upper atmosphere. *J. Geophys. Res.* 101:4577–4594.

de Bergh, C., Crovisier, J., Lutz, B. L., and Maillard, J. P. 1988. Detection of CO infrared emission lines in spectra of Venus. *Bull. Amer. Astron. Soc.* 20:831.

Del Genio, A. D., and Rossow, W. B. 1982. Temporal variability of ultraviolet cloud features in the Venus stratosphere. *Icarus* 51:391–415.

Del Genio, A. D., and Schubert, G. 1979. Characteristics of acoustic-gravity waves in a diffusively separated atmosphere. *J. Geophys. Res.* 84:1865–1879.

Dickinson, R. E., and Bougher, S. W. 1986. Venus mesosphere and thermosphere. I. Heat budget and thermal structure. *J. Geophys. Res.* 91:70–80.

Dickinson, R. E., and Ridley, E. C. 1972. A numerical solution for the composition of a steady state subsolar-to-antisolar circulation with application to Venus. *J. Atmos. Sci.* 29:1557.

Dickinson, R. E., and Ridley, E. C. 1975. A numerical model for the dynamics and

composition of the Venusian thermosphere. *J. Atmos. Sci.* 32:1219–1231.
Dickinson, R. E., and Ridley, E. C. 1977. Venus mesosphere and thermosphere temperature structure: II. Day–night variations. *Icarus* 30:163–178.
Dickinson, R. E., Ridley, E. C., and Roble, R. G. 1984. Thermospheric general circulation with coupled dynamics and composition. *J. Atmos. Sci.* 41:205–219.
Dudis, J. J., and Reber, C. A. 1976. Composition effects in thermospheric gravity waves. *Geophys. Res. Lett.* 3:727–730.
Dunkerton, T. J. 1982. Theory of the mesopause semiannual oscillation. *J. Atmos. Sci.* 39:2681–2690.
Espiritu, R. C., and Tolson, R. H. 1995. Determining Venusian upper atmospheric characteristics using Magellan attitude control data. *AAS/AIAA Spaceflight Mechanics Meeting*, Publ. AAS-95-152.
Feldman, P. D., Moos, H. W., Clarke, J. T., and Lane, A. L. 1979. Identification of the UV night airglow from Venus. *Nature* 279:221.
Fox, J. L. 1988. Heating efficiencies in the thermosphere of Venus reconsidered. *Planet. Space Sci.* 36:37–46.
Fox, J. L., and Bougher, S. W. 1991. Structure, luminosity, and dynamics of the Venus thermosphere. *Space Science Rev.* 55:357–489.
Fritts, D. C. 1989. A review of gravity wave saturation processes, effects, and variability in the middle atmosphere. *Pure Appl. Geophys.* 130:343–371.
Fritts, D. C., and Lu, W. 1993. Spectral estimates of gravity wave energy and momentum fluxes. Part II: Parameterization of wave forcing and variability. *J. Atmos. Sci.* 50:3695–3713.
Garcia, R. R., and Solomon, S. 1985. The effect of breaking gravity waves on the dynamics and chemical composition of the mesosphere and lower thermosphere. *J. Geophys. Res.* 90:3850–3868.
Gordiets, B. F., and Kulikov, Y. N. 1985. On the mechanisms of cooling of the nightside thermosphere of Venus. *Adv. Space Res.* 5(9):113–117.
Gerard, J. C., Stewart, A. I. F., and Bougher, S. W. 1981. The altitude distribution of the Venus ultraviolet nightglow and implications on vertical transport. *Geophys. Res. Lett.* 8:633–636.
Goldstein, J. J., Mumma, M. J., Kostiuk, T., and Espenak, F. 1991. A self-consistent picture of circulation in Venus' atmosphere from 70 to 200 km altitude. *Icarus* 94:45–63.
Gurwell, M. A., et al. 1995. Observations of the CO bulge on Venus and implications for mesospheric winds. *Icarus* 115:141–158.
Hartle, R. E., Mayr, H. G., and Bauer, S. J. 1978. Global circulation and distribution of hydrogen in the thermosphere of Venus. *Geophys. Res. Lett.* 5:719–722.
Hedin, A. E., Niemann, H. B., Kasprzak, W. T., and Seiff, A. 1983. Global empirical model of the Venus thermosphere. *J. Geophys. Res.* 88:73-83.
Holton, J. R. 1982. The role of gravity wave induced drag and diffusion in the momentum budget of the mesosphere. *J. Atmos. Sci.* 39:791–799.
Holton, J. R. 1983. The influence of wave breaking on the general circulation of the middle atmosphere. *J. Atmos. Sci.* 40:2497–2507.
Kasprzak, W. T., Hedin, A. E., Mayr, H. G., and Niemann, H. B. 1988. Wavelike perturbations observed in the neutral thermosphere of Venus. *J. Geophys. Res.* 93:11237-11246.
Kasprzak, W. T., et al. 1993a. Neutral composition measurements by the Pioneer Venus Orbiter Neutral Mass Spectrometer during orbiter re-entry. *Geophys. Res. Lett.* 20:2747–2750.
Kasprzak, W. T., Niemann, H. B., Hedin, A. E., and Bougher, S. W. 1993b. Wavelike perturbations observed at low altitudes by the Pioneer Venus Orbiter Neutral Mass Spectrometer during Orbiter entry. *Geophys. Res. Lett.* 20:2755–2758.

Keating, G. M., and Bougher, S. W. 1992. Isolation of major Venus cooling mechanism and implications for Earth and Mars, *J. Geophys. Res.* 97:4189–4197.

Keating, G. M., and Hsu, N. C. 1993. The Venus atmospheric response to solar cycle variations. *Geophys. Res. Lett.* 20:2751–2754.

Keating, G. M., Taylor, F. W., Nicholson, J. Y., III, and Hinson, E. W. 1979. Short-term cyclic variations and diurnal variations of the Venus upper atmosphere. *Science* 205:62.

Keating, G. M., Nicholson, J. Y., III, and Lake, L. R. 1980. Venus upper atmosphere structure. *J. Geophys. Res.* 85:7941–7956.

Keating, G. M., et al. 1985. Models of Venus neutral upper atmosphere: Structure and composition. In *Advances in Space Research: The Venus International Reference Atmosphere*, eds. A. J. Kliore, V. I. Moroz and G. M. Keating (New York: Pergamon Press), pp. 117–171.

Keating, G. M., Hsu, N. C., Ryne, M., and Wong, S. K. 1993. Discovery of 4-day oscillation of Venus thermosphere and low-altitude characteristics of cryosphere during low solar activity. *Eos: Trans. AGU* 74:187.

Krasnopolsky, V. A., and Tomashova, G. V. 1980. Venus nightglow variations. *Cosmic Res.* 18:766.

Krasnopolsky, V. A., Krysko, A. A., Rogachev, V. N., and Parshev, V. A. 1976. Spectroscopy of the Venus night airglow from the Venera 9 and 10 orbiters. *Cosmic Res.* 14:789–795.

Kumar, S., Hunten, D. M., and Broadfoot, A. L. 1978. Non-thermal hydrogen in the Venus exosphere: The ionospheric source and the hydrogen budget. *Planet. Space Sci.* 26:1063.

Lellouch, E., et al. 1994. Global circulation, thermal structure, and carbon monoxide distribution in Venus' mesosphere in 1991. *Icarus* 110:315–339.

Lindzen, R. S. 1981. Turbulence and stress owing to gravity wave and tidal breakdown. *J. Geophys. Res.* 86:9707–9714.

Linkin, V. M., et al. 1986. VEGA balloon dynamics and vertical winds in the Venus middle cloud region. *Science* 231:1417–1419.

Maillard, J.-P., et al. 1995. Carbon monoxide 4.7 μm emission: a new dynamical probe of Venus' thermosphere. *Bull. Amer. Astron. Soc.* 27:1080.

Massie, S. T., Hunten, D. M., and Sowell, D. T. 1983. Day and night models of the Venus thermosphere. *J. Geophys. Res.* 88:3955–3969.

Mayr, H. G., Harris, I., Hartle, R. E., and Hoegy, W. R. 1978. Diffusion model for the upper atmosphere of Venus. *J. Geophys. Res.* 83:4411.

Mayr, H. G., et al. 1980. Dynamic properties of the thermosphere inferred from Pioneer Venus mass spectrometer measurements. *J. Geophys. Res.* 85:7841–7847.

Mayr, H. G., et al. 1985. On the diurnal variations in the temperature and composition: A three-dimensional model with superrotation. *Adv. Space Res.* 5(9):109.

Mayr, H. G., et al. 1988. Gravity waves in the upper atmosphere of Venus. *J. Geophys. Res.* 93:11247–11262.

Mengel, J. G., Mayr, H. G., Harris, I., and Stevens-Rayburn, D. R. 1989. Non-linear three-dimensional spectral model of the Venusian thermosphere with superrotation: II. Temperature, composition, and winds. *Planet. Space Sci.* 37(6):707–722.

Niemann, H. B., et al. 1979. Venus upper atmosphere neutral gas composition: First observations of the diurnal variations. *Science* 205:54–56.

Niemann, H. B., et al. 1980. Mass spectrometer measurements of the neutral gas composition of the thermosphere and exosphere of Venus. *J. Geophys. Res.* 85:7817–7827.

Paxton, L. J., Anderson, D. E., and Stewart, A. I. F. 1985. The Pioneer Venus Orbiter Ultraviolet Spectrometer experiment: Analysis of H-Lyman-α data. *Adv. Space*

Res. 5(9):129–132.
Paxton, L. J., Anderson, D. E., and Stewart, A. I. F. 1988*a*. Analysis of the Pioneer Venus Ultraviolet Spectrometer Lyman-α data from near the subsolar region. *J. Geophys. Res.* 93:1766–1772.
Paxton, L. J., Anderson, D. E., and Stewart, A. I. F. 1988*b*. Correction to: Analysis of the Pioneer Venus Ultraviolet Spectrometer Lyman-α data from near the subsolar region. *J. Geophys. Res.* 93:11551.
Roble, R. G., and Ridley, E. C. 1994. A thermosphere ionosphere mesosphere electrodynamics general circulation model (TIME-GCM) : Equinox solar cycle minimum simulations (30–500 km). *Geophys. Res. Lett.* 21:417–420.
Schubert. G. 1983. General circulation and the dynamical state of the Venus atmosphere. In *Venus*, eds. D. M. Hunten, L. Colin, T. M. Donahue and V. I. Moroz (Tucson: Univ. of Arizona Press), pp. 681–765.
Schubert. G., and Walterscheid, R. L. 1984. Propagation of small-scale acoustic-gravity waves in the Venus atmosphere. *J. Atmos. Sci.* 41:1202–1213.
Schubert. G., et al. 1980. Structure and circulation of the Venus atmosphere. *J. Geophys. Res.* 85:8007–8025.
Seiff, A. 1982. Dynamical implications of the observed thermal contrasts in Venus's upper atmosphere. *Icarus* 51:574–592.
Seiff, A.. and Kirk, D. B. 1991. Waves in Venus's middle and upper atmosphere: Implications of Pioneer Venus Probe data above the clouds. *J. Geophys. Res.* 96:11021–11032.
Seiff, A., et al. 1980. Measurements of thermal structure and thermal contrasts in the atmosphere of Venus and related dynamical observations: Results from the four Pioneer Venus probes. *J. Geophys. Res.* 85:7903–7933.
Shah, K. P., Muhleman, D. O., and Berge, G. L. 1991. Measurement of winds in Venus' upper mesosphere based on Doppler shifts of the 2.6-mm ^{12}CO line. *Icarus* 93:96–121.
Stevens-Rayburn, D. R., Mengel, J. G., Harris, I., and Mayr, H. G. 1989. Non-linear three dimensional spectral model of the Venusian thermosphere with superrotation: I. Formulation and numerical technique. *Planet. Space Sci.* 37(6):701–706.
Stewart, A. I., and Barth, C. A. 1979. Ultraviolet night airglow of Venus. *Science* 205:59.
Stewart, A. I. F., Gerard, J. C., Rusch, D. W., and Bougher, S. W. 1980. Morphology of the Venus ultraviolet night airglow. *J. Geophys. Res.* 85:7861–7870.
Stewart, A. I. F., Alexander, M. J., and Pryor, W. R. 1993. The morphology of far-UV emissions from carbon monoxide in the Venus thermosphere. *Bull. Amer. Astro. Soc* 25:1095.
Taylor, F. W., et al. 1980. Structure and meteorology of the middle atmosphere of Venus: Infrared remote sensing from the Pioneer Orbiter. *J. Geophys. Res.* 85:7963–8006.
Taylor, H. A., Jr., et al. 1984. Interannual and short-term variations of the Venus nighttime hydrogen bulge. *J. Geophys. Res.* 89:10669–10675.
von Zahn, U., et al. 1979. Venus thermosphere: In-situ composition measurements, the temperature profile, and the homopause altitude. *Science* 203:768.
von Zahn, U., et al. 1980. The upper atmosphere of Venus during morning conditions. *J. Geophys. Res.* 85:7829–7840.
von Zahn, U., Kumar, S., Niemann, H. B., and Prinn, R. 1983. Composition of the Venus atmosphere. In *Venus*, eds. D. M. Hunten, L. Colin, T. M. Donahue and V. I. Moroz (Tucson: Univ. of Arizona Press), pp. 299–430.
Zhang, S., Bougher, S. W., and Alexander, M. J. 1996. The impact of gravity waves on the Venus thermosphere and O_2 IR nightglow, *J. Geophys. Res.* 101:23195–23205.

PART III
Lower Atmosphere
(≤100 km)

MONITORING OF MESOSPHERIC STRUCTURE AND DYNAMICS

E. LELLOUCH
Observatoire de Paris

T. CLANCY
Space Science Institute, Boulder

D. CRISP and A. J. KLIORE
Jet Propulsion Laboratory

D. TITOV
Space Research Institute, Moscow

and

S. W. BOUGHER
University of Arizona

The structure and dynamics of Venus' mesosphere can be investigated by a number of means, including (i) temperature measurements from spectroscopy and radio-occultations; (ii) wind measurements from tracking of cloud features and from Doppler shifts; (iii) spatial distribution of minor species (CO and O_2 emissions). The thermal structure in the lower mesosphere (70–85 km) suggests a decrease of the zonal winds with height, but direct measurements in this region are missing. Wind measurements in the upper mesosphere (85–110 km) indicate the co-existence of highly variable zonal winds with a more stable subsolar-to-anti-solar flow. The spatial distribution of CO and O_2 is a tracer of the global circulation, but the detailed structure is still largely not understood. The mesosphere appears to be a dynamically intermediate region characterized by large variability.

I. INTRODUCTION

Dynamical characterization naturally divides Venus' atmosphere in three different regions. From the surface to the upper cloud, the troposphere is dominated by an intense retrograde superrotation with maximal velocities in excess of 100 m s^{-1} (chapter by Gierasch et al.). In the thermosphere, above 110 km altitude, day/night insolation contrasts drive a predominantly axisymmetric circulation with the gas flowing from the subsolar point to the anti-solar point (chapters by Kasprzak et al. and Bougher et al.). The mesosphere separates the two regions and acts as a dynamical transition region, where the tropospheric and thermospheric regimes co-exist. The three regions can also be distinguished by the physical processes that control their thermal structure. Heat is transported primarily by convection and radiation in the troposphere. At higher levels, radiation (in the mesosphere) and thermal conduction (in the thermosphere) are the most important processes.

The first picture of the mesospheric structure and dynamics was derived from early Pioneer Venus observations, which revealed an inverted equator-to-pole temperature gradient at 70 to 100 km, associated with the decay of the cloud-top superrotation. The intrusion of subsolar-to-antisolar thermospheric flow into the mesosphere was first suggested by early studies of carbon monoxide diurnal variability and by theoretical models. However, the description of the mesosphere was still incomplete when *Venus* (Hunten et al. 1983) was published, because essential observational information was missing, in particular absolute wind measurements and a systematic monitoring of the thermal field and the minor species. Many new observations have been acquired in the last 15 yr, providing a much more detailed view of Venus' mesosphere. After an overview of the dynamical indicators and means of investigation (Sec. II), this chapter presents a comprehensive review of these observations: (i) temperature and wind measurements derived from spacecraft observations (Sec. III); (ii) absolute wind measurements from CO_2 spectroscopy (Sec. IV); (iii) CO millimeter observations (Sec. V); (iv) O_2 airglows at infrared and visible wavelengths (Sec. VI). Finally, Sec. VII attempts to propose a coherent picture of Venus' mesosphere, and summarizes some important unresolved issues. The Chapter emphasizes measurements obtained after the nominal Pioneer Venus mission (December 1978 to June 1979) or not mentioned in *Venus*. Earlier results are only mentioned when necessary. Measurements of minor species are addressed only if they provide constraints on the mesospheric dynamics (CO, O_2). Other compositional measurements above the clouds (SO_2, SO, H_2O) are described in the chapter by Esposito et al.

There are no strict altitude or pressure bounds for Venus' mesosphere. The usual criteria, based on the sign of the vertical temperature gradient, cannot be used at the lower boundary, because, except at high latitudes (∼50–80°), Venus has no stratosphere. A tropopause can still be conventionally defined as the point of sudden change in the vertical lapse rate. Tropopause altitudes vary with latitude, from ∼55 to 62 km (Kliore 1985; chapter by Crisp and Titov). For our purpose, we here take the lower limit of the mesosphere at 60 km (250 mbar), i.e., within the upper cloud. At the upper end, the average dayside temperature gradient becomes positive near 100 km ($p = 3 \times 10^{-5}$ bar). This is also the altitude where day and night temperature profiles start to diverge. Here, however, the discussion will be extended to ∼115 km, because some of the measurements we will describe naturally provide information up to the lower thermosphere.

II. DYNAMICAL TRACERS AND METHODS OF INVESTIGATION

A. Temperature Measurements

The most extensive information regarding the mesospheric structure and dynamics comes from remote or *in-situ* measurements of the temperature as a function of pressure (altitude), location on the planet, and time. *In-situ* temperature measurements at these low pressures result from thermal sensors or from density profiles locally obtained by accelerometers. Most data were obtained by the Venera 8–12 and Pioneer Venus descent probes. Results were given in the first Venus book (Seiff 1983; Avduevsky et al. 1983), except for a recent analysis of wave structures from Pioneer Venus Probe Data (see Sec. III.C).

Remote observations, which afford much better spatial and temporal coverage, include spectroscopic (groundbased or spaceborne) and radio-occultation measurements. Spectroscopic methods generally rely on inversion of molecular band or line profiles in the thermal infrared. This technique allows vertical discrimination because the radiation measured in a given frequency interval $\Delta \nu$ originates in a well-defined altitude region determined by the weighting function $WF = dT_r(p)/d\log(p)$, where T_r is the transmission (averaged over $\Delta \nu$) from pressure level p to space. The width of the weighting functions defines the vertical resolution. This method has been applied to the 15 and 4.3 μm bands of CO_2, observed from space (Pioneer Venus, Venera 15, Galileo), and to the rotational lines of CO measured from the ground. In the latter case, the absorber vertical profile must be known. In practice, several CO lines with different relative sensitivities to the CO and $T(p)$ profiles are observed, permitting a simultaneous retrieval of the two parameters (see Sec. V). Figure 1 shows examples of weighting functions, calculated for the CO_2 15 μm band (at 7.5 cm^{-1} spectral resolution and including aerosol contribution, adequate for Venera 15 observations) and for the CO(2–1) rotational line. The sounded pressure ranges are 0.03 to 200 mbar (60–100 km) and 0.0003 to 20 mbar (75–115 km) respectively, and the vertical resolution is between 5 and 10 km.

Temperature information can also be obtained from measurements of linewidths or from relative strengths of rotational transitions in airglow emission bands. Rotational temperatures have been derived from the O_2 nightglow emission at 1.27 μm and the CO dayside emission at 5 μm (Sec. VI). They represent physical temperatures, because even if the emissions are produced by non-LTE mechanisms, the rotational populations are thermalized. However, assigning altitude levels to the measurements requires a detailed understanding of these mechanisms. Kinetic temperature measurements were obtained from CO_2 10 μm emissions (Sec. IV) and from H_2O/HDO rotational lines (Sec. V).

Radio occultations constitute the other main source of temperature information. Abel transformation of radio occultation data (primarily the ray bending angle and impact parameter vs time) provides refractivity profiles,

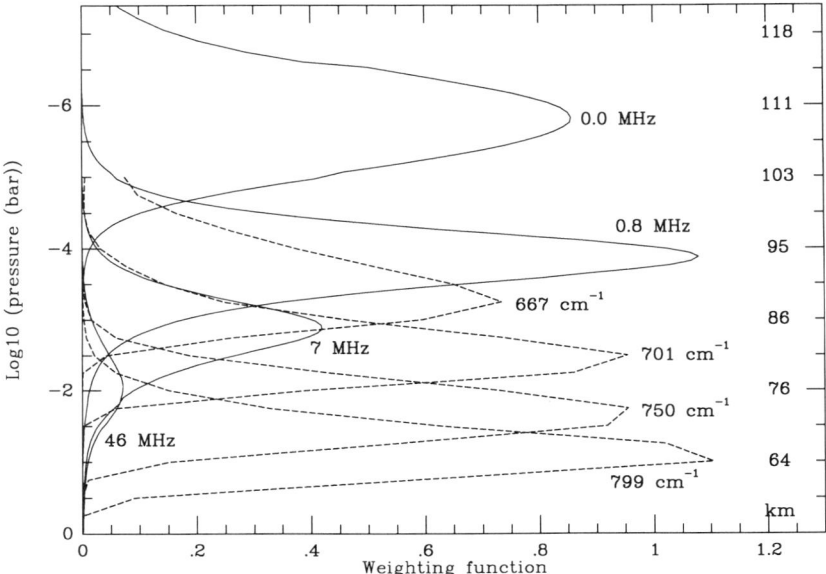

Figure 1. Example of temperature weighting functions for the ^{12}CO(2-1) rotational line at 230.538 GHz (solid lines, from top to bottom: 0.0, 0.8, 7 and 46 MHz from line center) and for the CO_2 15-μm band (dashed lines, from top to bottom: 667, 701, 750, and 799 cm^{-1}). The altitude scale is approximate.

which can be translated into atmospheric parameters (number density, pressure, and temperature as a function of altitude) if the composition is known. In Venus' case, the probed range extends from about 40 to 90 km. The advantage of the method over spectroscopic techniques is the better vertical resolution, diffraction-limited to $\sim$1 km. Information on even smaller scales can be extracted from intensity scintillations (see, e.g., Hinson and Jenkins 1995).

B. Wind Determinations

1. Thermal Winds. The zonal (east–west) winds in Venus' upper troposphere and lower mesosphere are approximately in cyclostrophic balance, such that the equatorward component of the centrifugal force on a zonally rotating parcel is equilibrated by the poleward pressure gradient force (Schubert 1983; chapter by Gierasch et al.). Mathematically the vertical shear in the zonal wind $u(p, \phi)$ and the zonal mean temperature $T(p, \phi)$ at pressure level p and latitude ϕ are related by:

$$2u \frac{\partial u}{\partial z} = -\frac{R}{\tan \phi} \frac{\partial T}{\partial \phi} \qquad (1)$$

where $z = \ln(p_0/p)$. This provides the only wind information in regions where direct measurements are not available (in particular the lower mesosphere at 70–90 km), but with important limitations: (i) only the zonal component of the wind field can be inferred; (ii) the equation cannot be applied in equatorial regions; (iii) a boundary $u(p_0, \phi)$ condition is required; and (iv) the spatial distribution of temperatures must be known with sufficient accuracy and resolution. More complex descriptions of the momentum balance have been proposed. Early models (Elson 1979; Taylor et al. 1980) employed a simple eddy diffusion formulation to include the effects of small scale disturbances. They provided estimates of the meridional circulation as well as the zonal winds, but they required unrealistically large eddy diffusion coefficients to produce a stable circulation. In addition, they predicted rapid upward flow over the poles with associated adiabatic cooling rates as large as 150 K day^{-1}. Mechanisms capable of maintaining the high mesospheric polar temperatures in the presence of this rapid upwelling have never been identified (Crisp 1989). More recent models (Baker and Leovy 1987; Newman and Leovy 1992) explicitly include the momentum forcing of the mesosphere by vertically propagating atmospheric tides (see Sec. VII.A).

2. Absolute Wind Measurements Above the Clouds. At the cloud tops (near 65 km), winds have been inferred from feature tracking in ultraviolet images taken by Mariner 10, Venera 9, Pioneer Venus (see reviews by Schubert [1983], Kerzhanovitch and Limaye [1985], and Limaye [1990]), and recently by Galileo (Belton et al. 1991). Above the clouds, in the absence of diagnostic features, the only method relies on the analysis of Doppler shifts on selected molecular lines.

Winds of 10 to 100 m s^{-1} in planetary atmospheres induce Doppler shifts of 3×10^{-8} to 3×10^{-7} on molecular lines. Measuring such shifts requires a resolving power of at least 10^5 (high dispersion photographic or interferometric spectroscopy in the visible and infrared) and preferably 10^6 to 10^7 (heterodyne spectroscopy in the mm range and at 10μm). Observing a molecular line and positioning the line center at several points on a planet permits, in principle, to measure line-of-sight projected winds at these points, allowing to reconstruct the global wind field. For this method to work in practice, the line must be strong, isolated and quasi-symmetric. In the case of Venus, visible and 10-μm lines of CO_2, and millimeter and 4.7 μm lines of CO have been used. Spatial resolution and coverage is important for model discrimination. Seeing-limited resolution of 1 to 2" ($\sim$500–1000 km on Venus at quadrature) can be obtained in the infrared and visible, while the diffraction-limited resolution in the millimeter range is only $\sim$10" for single dish observations and $\sim$2 to 3" for interferometers (but at the expense of signal-to-noise).

Obtaining spectra suitable for wind measurements requires cautious handling of systematic effects (e.g., the absolute calibration of the frequency scale, the topocentric velocity variations of the source, etc., see, e.g., Goldstein 1989). Doppler shifts are then measured by cross-correlating the observed spectra with a reference profile within a specified frequency interval. Several methods are possible to estimate error bars (see, e.g., Goldstein et al. 1991), but, generally, the accuracy is of order $W/[S/N]$, where W is the line width (natural or instrumental, in velocity units) and S/N is the signal-to-noise on the line contrast. S/N-limited uncertainties as low as 2 to 5 m s^{-1} can be reached (Fig. 2).

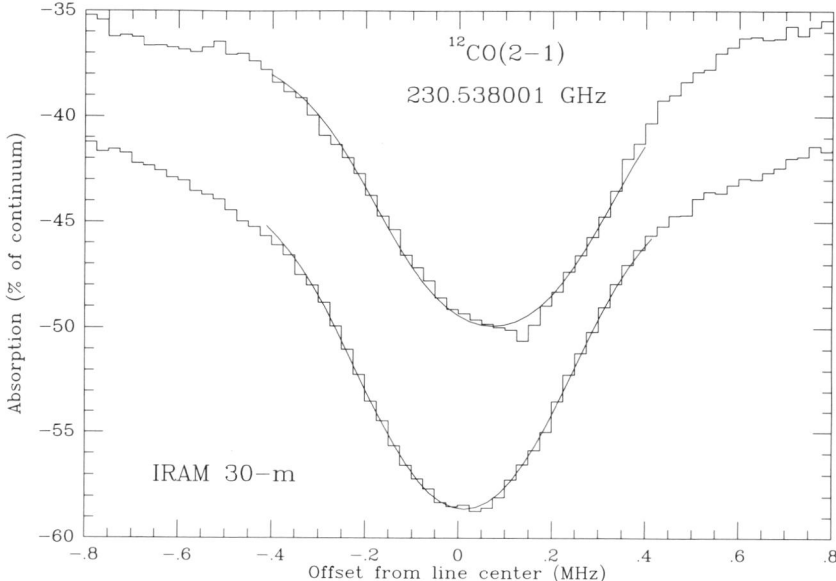

Figure 2. Two examples of the ^{12}CO (2–1) line core, observed at 25-kHz resolution on two positions on Venus' disk, along with two model fits (solid lines), allowing to measure the absolute wind velocity. The 0-MHz offset corresponds to the rest frequency. Winds of -16 ± 2 m s^{-1} (lower curve) and -95 ± 4 m s^{-1} are inferred (figure adapted from Lellouch et al. 1994).

Line-of-sight wind velocities reflect dynamical contributions from all regions of the planet intercepted by the telescope beam, and their interpretation requires detailed geometrical modeling. For Venus, most data have been modeled by a combination of zonal and subsolar-to-antisolar (SSAS) flows. Specific descriptions (as a function of latitude and solar zenith angle) are assumed, with a small number of free parameters (the equatorial velocity V_{eq} for the zonal flow, the cross terminator velocity V_{ter} for the SSAS flow), which are determined by comparison to the data. For some observations (Lellouch et al. 1994), a meridional component was included, and the latitude dependence of the zonal flow could be extracted. Assigning an altitude level for the inferred wind field relies on modeling the physical process responsible for producing the observed transition. When the lines are fully resolved, this altitude depends on the frequency interval chosen to measure the Doppler shifts, and can be estimated by averaging the temperature weighting functions within this interval with appropriate weights (see, e.g., Lellouch et al. 1994).

C. Minor Species Distribution

Trace species in Venus' upper atmosphere are subject to spatial redistribution by the global wind system. In the thermosphere (≥ 110 km), the diurnal variations of O, N, He and H are the primary tracers of the circulation (chapter by Bougher et al.). In the mesosphere, this role is played by CO and atomic oxygen. Photodissociation of CO_2 at mesospheric/thermospheric levels produces CO and O on Venus' day side. Chemical lifetimes of O and CO are of order 1 and 10 days, respectively, comparable to horizontal transport time scales. Transport to the night side and subsidence results in the recombination of O into radiatively excited molecular O_2, producing visible (Herzberg II) and infrared ($^1\Delta$) emissions (Sec. VI). The spatial distribution of the O_2 emissions and of CO therefore partly reflects the global mesospheric wind field, although the information that can be extracted is still semi-quantitative. Gross features of these distributions (local time of maximum, magnitude of diurnal contrasts) can be interpreted in terms of relative zonal/SSAS velocities, but the detailed structures remain poorly understood (Sec. VII).

III. TEMPERATURE MEASUREMENTS AND WIND INFERENCES FROM SPACECRAFT OBSERVATIONS

Spacecraft observations have contributed a lot to our knowledge of Venus' mesospheric structure, through extensive spectroscopic (CO_2), radio occultations, and probe entry temperature measurements, and from repeated absolute wind measurements at cloud tops.

A. Thermal Field Measurements from CO_2 Observations

1. Pioneer Venus Orbiter Infrared Radiometer. The Pioneer Venus Orbiter Infrared Radiometer (OIR) instrument observed Venus' northern hemisphere between December 1978 and June 1979 (Taylor et al. 1980). Results of this experiment (and of earlier Venera 9 and 10 radiometric measurements) were described in detail in *Venus* (Taylor et al. 1983). The temperature field exhibits longitudinal and latitudinal variations. Diurnal variations occur with remarkable tidal structure. Amplitudes are small at cloud tops (a few degrees), with the co-existence of wave numbers 1 and 2. Above 70 km, diurnal variations increase to reach ~ 5 K at 90 km, as the semi-diurnal tide (wave number 2) becomes progressively dominant. Above 95 km, the regime switches back to a wavenumber 1 mode, with increasing (≥ 10 K) day/night contrasts. Latitudinal temperature variations are small below 70 km (with temperatures mildly decreasing poleward), except in the region of the "cold collar" at 60 to 75° latitude. There, temperatures near 67 km are up to 30 K colder than at the equator and a large (15 K) vertical temperature inversion develops. Northward of 75°, the $T(p)$ profile is quasi-isothermal at 65 to 75 km. Above 75 km ($p \leq 0.02$ bar) and up to about 100 km, the temperature increases with latitude. Applying the thermal wind equation to this temperature field shows that a general decrease of the retrograde superrotation must occur between

70 and 90 km, at which altitude cyclostrophic balance tends to break down. However, depending on the lower boundary assumed for these calculations, significant zonal winds can be maintained up to 100 km or more, particularly at mid-latitudes (Schubert 1983).

2. *Venera 15 Infrared Fourier Spectrometer Experiment.* About 2000 spectra of Venus in the range 250 to 1600 cm^{-1} were recorded at a resolution of 5 to 7.5 cm^{-1} by the Venera 15 Infrared Fourier Spectrometer Experiment (IFSE) instrument from October 12 to December 14, 1983 (Oertel et al. 1985). Most spectra sample the northern hemisphere in two ranges of local time (morning: 4–10 am and evening: 4–10 pm). They provide simultaneous information on the cloud top composition (H_2O and SO_2) and the upper cloud structure (see the chapter by Esposito et al.), and on the temperature field at 58 to 100 km (3×10^{-5}–0.3 bar) with a vertical resolution of 3 to 5 km and an estimated accuracy of 2 K in most of the sounded range.

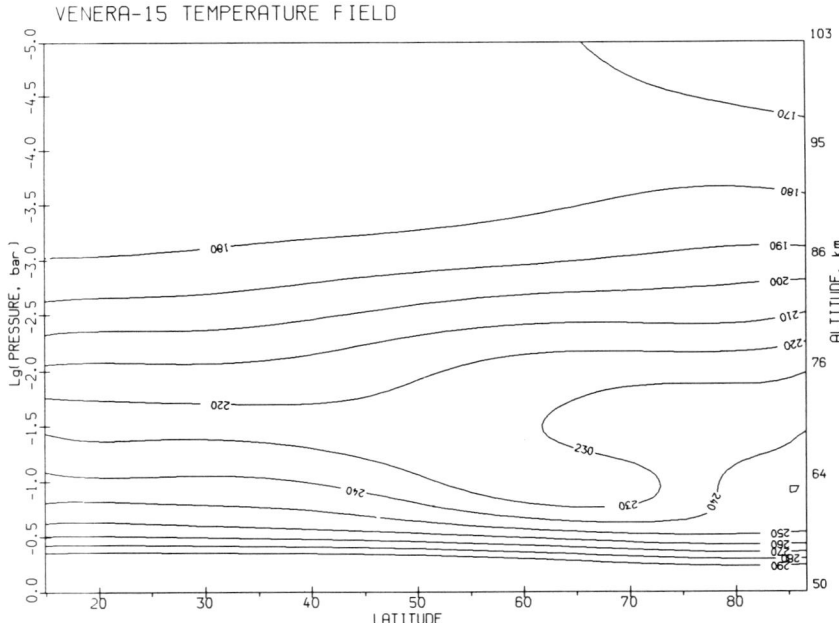

Figure 3. Temperature contours derived from Venera-15 IFSE measurements in the morning sector. Retrieval procedure is similar to Schäfer et al. 1990, but with improved regard of aerosol scattering. The altitude scale is approximate.

Figure 3 shows an example of the retrieved temperature field. The "cold collar" is located at $\sim$65°N. The associated vertical temperature inversion is deeper at the morning terminator ($\sim$15 K) than at the evening (0–5 K). The rapid decrease of the temperature at constant pressure level from 45° to 65°

results in a mid-latitude cyclostrophic jet centered at ~53° and 68 km altitude. Its peak velocity is higher at the morning terminator (140 m s^{-1}) than at the evening (120 m s^{-1}). Temporal variations in the jet position and strength were observed. From 75 to 90 km, the increase of the temperature with latitude is clearly visible, causing the overall decay of the zonal winds (falling to zero within ~10 km from cloud tops at latitudes ≥65°). Above 90 km, a poleward temperature decrease was tentatively identified.

3. *Galileo Near-Infrared Mapping Spectrometer.* The thermal field at 70 to 90 km was also inferred from spectral images of Venus' night side recorded by the Galileo Near Infrared Mapping Spectrometer (NIMS) in the region of the CO_2 4.3 μm band (Roos-Serote et al. 1995). Observations were limited to a single day (10 February 1990), but, unlike previous data sets, they probed both hemispheres (between ~60°S and 65°N). The main feature of the mesospheric structure, the poleward increase of temperatures above 75 km, was clearly present, and no significant differences between the two hemispheres were seen. Temperatures were consistent with the Pioneer Venus based VIRA reference profiles (Seiff et al. 1985) at 75 to 83 km and somewhat warmer (by 5–10 K) at 90 km. Thermal winds, computed with a boundary condition taken from Galileo/SSI wind measurements near 50 mbar (67 km), confirmed the general decrease of the zonal flow with height, with again, the possible presence of mid-latitude winds of order 50 m s^{-1}.

B. Temperature Measurements from Radio Occultations

1. Pioneer Venus. Between December 1978 and February 1991, 246 temperature/pressure measurements of Venus' atmosphere at 40 to 90 km were obtained from dual frequency (S- and X-bands) radio-occultations with the Pioneer Venus Orbiter radio-occultation (ORO) experiment. These measurements sampled two periods of solar maximum (1978–1983 and 1989–1991) and one period of solar minimum (1984–1986). Thermal and wind fields were obtained from the individual temperature profiles using a smoothing and polynominal fitting method (see, e.g., Newman et al. 1984). For all three solar activity seasons, the temperature vs latitude field showed a persistent pattern, essentially consistent with that inferred from the CO_2 measurements. However, the magnitude of the inversions near 70° differed between the north and south hemispheres for each season and between like hemispheres for different seasons. In addition, on some occasions (1984–1985), the temperature increase toward the poles above 70 km was weaker than on average.

Thermal zonal wind contours for the 1978 to 1983 season indicate a considerable north/south asymmetry of the strength of the mid-latitude jet near 70 km (Fig. 4 and CDP3C1F1). For the other two seasons, zonal winds could not be computed in the southern hemisphere because of insufficient data, but, interestingly, the northern hemisphere jet shows significant variability with time.

2. Veneras. The Venera 15 and 16 spacecrafts recorded 176 radio occultation profiles between October 1983 and September 1984, in a study similar to

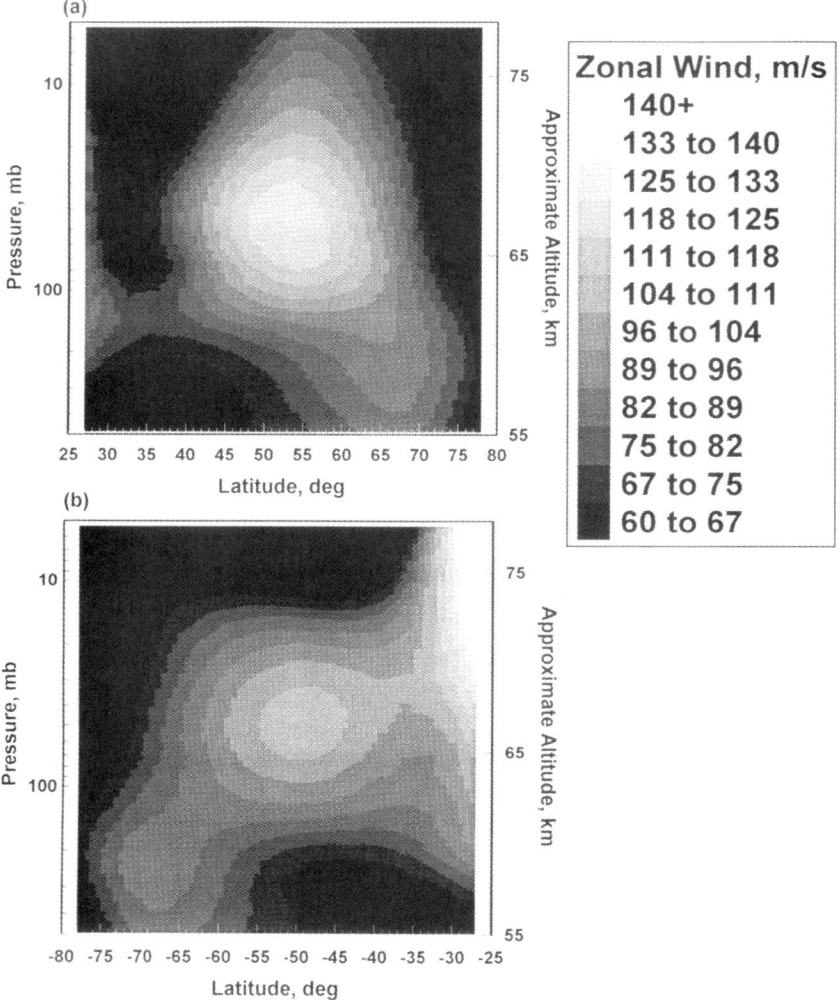

Figure 4. Thermal zonal wind field derived from Pioneer Venus radio occultation temperature measurements in 1978–1983 (high solar activity): (a) northern hemisphere (b) southern hemisphere.

earlier investigations with Venera 9 and 10 in 1975 to 1976 (see Seiff 1983). The emphasis of the analysis was on high-latitude regions and north/south comparisons (Yakovlev et al. 1991). The "circumpolar" atmosphere (60–80° latitude) was found to show a global N/S symmetry, whereas the polar atmosphere ($\geq 80°$) was 10 to 20 K warmer in the north than in the south at 57 to 80 km altitude.

3. Magellan. Radio occultation experiments were conducted with

Magellan in October 1991, December 1992 and June–August 1994. Only the first sequence has been analyzed, providing high S/N temperature profiles (34–98 km) at 67°N latitude. These profiles were similar to the Pioneer Venus results, although the inversion associated with the cold collar was much weaker (Jenkins et al. 1994). Atmospheric waves with small (2–4 km) and large (5–10 km) vertical wavelengths were observed. The small-scale oscillations have an amplitude of $\sim$4 K at 65 km, and are well described by a spectrum of vertically propagating gravity waves with aspect ratio (horizontal to vertical wavelength) of $\sim$100. The oscillations with the largest vertical wavelengths are not identified unambiguously, but could be thermal tides (Hinson and Jenkins 1995).

C. Temperature Measurements from Pioneer Venus Probe Data

A recent re-assessment of the Pioneer Venus Probe data provided further insight into the wave structure at 70 to 138 km (Seiff and Kirk 1991). Data recorded by the four Probes at a vertical resolution of 1.5 to 3 km show large temperature oscillations, $\sim$10 K at 70 to 80 km for the night probe, growing to 35 K at 130 km (roughly consistent with a $\rho^{-1/2}$ dependence). Wave profiles for the two morning Probes (Large and Day, separated by 1 hour in local time) are remarkably similar, both in waveform and amplitude (slightly larger than for the Night probe). Although not taken simultaneously, Pioneer Venus Orbiter Atmospheric Drag (OAD) measurements at 140 to 160 km (Keating et al. 1980) appear to extend the Night probe profile taken at the same local time, not only the mean temperature, but also the wave structure, suggesting a solar-fixed oscillation.

D. Wind Measurements at the Cloud Tops

Absolute wind measurements at cloud tops ($\sim$65 km) are obtained from tracking ultraviolet features from spacecraft. Results from Mariner 10 in 1974 and the nominal Pioneer Venus mission in 1979 were discussed by Schubert (1983) and Kerzhanovitch and Limaye (1985). Additional Pioneer Venus/OCPP (1980, 1982) and similar Galileo/SSI 418 nm (1990) observations confirmed the gross features of the circulation near 65 km, with a strong zonal retrograde flow (90–100 m s^{-1} at equator) and a significant poleward meridional component, increasing with latitude to 10 m s^{-1} at 45°N and 45°S (Limaye et al. 1988; Limaye 1990; Rossow et al. 1990; Belton et al. 1991; chapter by Gierasch et al.). However, the detailed structure of the zonal winds differed from case to case, with essentially solid-body rotation in 1979, almost uniform velocity in 1980, and mid-latitude jets (10–15 m s^{-1} extra velocity) in 1974, 1982, and 1990. In addition, a solar-locked mode is observed in the cloud top circulation; it has a zonal amplitude of $\sim$10 m s^{-1}, with minimum winds seen at $\sim$10:40 am local time, and a meridional amplitude of $\sim$1 m s^{-1}, with maximum poleward flows seen near noon (Limaye 1988; Del Genio and Rossow 1990). The presence of planetary-scale waves of similar amplitude and of 4-day period, identified as Kelvin waves, has also been inferred in 1979 to 1980 and 1990, but this mode was absent in 1982 to 1983 (Del Genio and Rossow 1990; Belton et al. 1991).

IV. WIND MEASUREMENTS FROM CO_2 SPECTROSCOPY

A. Visible Observations

Observations of Doppler shifts in Venus CO_2 lines and solar Fraunhofer lines at visible wavelengths have provided the only direct wind measurements just above the cloud tops (65–85 km). These measurements generated a great deal of controversy in the late 1970s (when the existence of the cloud-top 4-day winds was still being debated) because observations acquired by different observers and techniques resulted in a wide range of wind velocities. Some measurements indicated very large (≥ 100 m s^{-1}) retrograde zonal winds, supporting the existence of the cloud-top super-rotation, while others suggested a much weaker circulation (see references of measurements in Young et al. 1979; Traub and Carleton 1979). Young (1975) showed that much of this variance can be attributed to errors introduced by (1) the neglect of the finite angular size of the Sun, and its rapid equatorial rotation; and (2) apparent shifts due to the blending of telluric and Venus lines. Once these effects are taken into account, most of the data indicate wind velocities closer to zero than 100 m s^{-1} (Young et al. 1979), consistent with a rapid decrease of the the zonal super-rotation above the cloud tops.

B. 10-μm Observations

1. Early Measurements. In 1975, from heterodyne observations of Venus' day side, Betz et al. (1976) detected emission lines near 950 cm^{-1}, belonging to the 00^01–$[10^00, 02^00]$ bands of CO_2. The line half-widths indicated a gas temperature of ~ 200 K, but the intensities were much too strong for thermal emission at this temperature. The most detailed study of the emission mechanism was performed by Deming and Mumma (1983). Solar flux is absorbed in the near infrared, mostly in the 2.7 μm bands. Collisional transfer populates the 00^01 state, which then relaxes. Deming and Mumma (1983) found that the 10 μm emissions are produced in a ~ 20 km wide layer centered at 109 km.

Betz et al. (1976) used one of these lines to measure absolute winds from Doppler shifts. Line-of-sight velocities were in the range 0 to 35 m s^{-1} (with considerable short-term variability), ruling out a strong zonal flow at 120 km altitude. While this conclusion was reasonable, the data indicated suspiciously comparable horizontal and vertical velocities (measured at disk center). Betz et al. (1977) repeated their measurements in 1976 and 1977 and reported the discovery of a 130 m s^{-1} (cross-terminator) SSAS circulation near 110 km. However, implausible vertical velocities of ~ 30 m s^{-1} were also seen and line-of-sight winds in polar regions were anomalous with very large (~ 100 m s^{-1}) north–south asymmetries and a similar day-to-day variability. In April/May 1977, measurements were also performed on a thermal

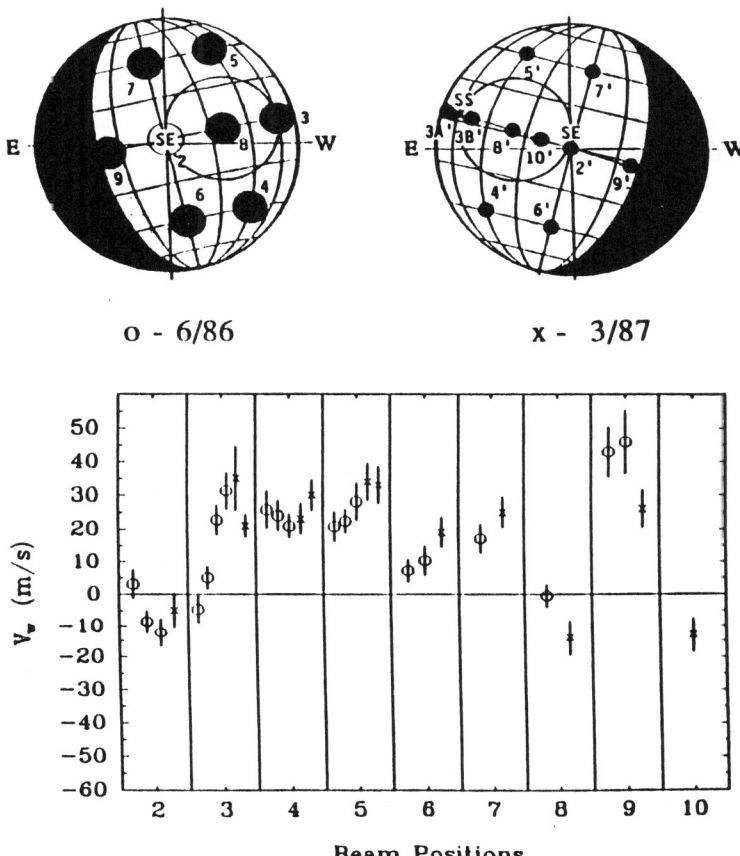

Figure 5. IRHS/GSFC 10-μm wind measurements near 109 km altitude. Planetary aspect, beam positions, and line-of-sight wind velocities (positive for recession) for 15–27 June 1986 (McMath Solar Telescope) and March 21–23, 1987 (NASA/IRTF). Associated primed and unprimed beam positions must exhibit identical line-of-sight SSAS components and opposed zonal components, to first order. Measured wind velocities reveal the dominance of the SSAS flow, along with a smaller contribution of the zonal retrograde flow (figure from Goldstein et al. 1991).

absorption line of $^{13}CO_2$ probing the 70 to 80 km region. This yielded a plausible 90 m s^{-1} zonal retrograde flow, but once again, disk center velocities were large (−45 m s^{-1}) and the polar winds showed very erratic behavior, casting serious doubts on the overall results. In spite of these problems, these pioneering measurements opened a new avenue for studies of Venus' mesospheric/thermospheric dynamics.

2. The Goddard Space Flight Center Infrared Heterodyne Spectrometer Measurements. The first unambiguous results were obtained with the IRHS (Infrared Heterodyne Spectrometer) instrument developed at GSFC. In 1980,

Deming et al. (1983) inferred an average kinetic temperature of 204±10 K at 109 km from the CO_2 R(8) (10.33 μm) linewidth. The instrument was later frequency stabilized to an absolute accuracy better than 0.1 MHz (equivalent to 1 m s^{-1}) for the purpose of wind measurements. Observations of the 10.33 μm line (Goldstein et al. 1991) were conducted at the 1.5-m McMath telescope (December 1985 and June 1986) and at the 3-m NASA/IRTF (October 1986 and June 1987). The June 1986 and March 1987 observing geometries, in particular, allowed easy decoupling of the SSAS and zonal components and revealed the dominance of the SSAS flow (Fig. 5). Error bars on the line-of-sight winds were 3 to 6 m s^{-1}. Quantitative modeling of the measurements was accomplished independently for the four observing periods, after correction for the systematic effects and along the method outlined in Sec. II.B.2, assuming a solid-body zonal component, and solving for V_{ter} and V_{eq}. Results were: V_{ter} = 105±15 m s^{-1}, V_{eq} = −30±15 m s^{-1} (retrograde) for Dec. 1985; V_{ter} = 130±20 m s^{-1}, V_{eq} = −20±5 m s^{-1} for March 1987. For October 1986, the geometry did not allow full decoupling of the components, but the data constrained $V_{ter} - V_{eq}$ = 140 to 155 m s^{-1}. For June 1986, the best fit was obtained for V_{ter} = 130±20 m s^{-1}, V_{eq} = +5 m s^{-1} (prograde), but had a higher χ^2. Remarkably, all four observing periods could be fit with V_{ter} = 120±30 m s^{-1}, V_{eq} = −25±15 m s^{-1}.

3. Re-analysis of Early Measurements. Goldstein (1989) re-assessed the Betz et al. (1977) measurements, in particular the $^{13}CO_2$ data. He assumed that they were suffering from a miscalibration of the frequency scale, causing the large apparent winds at disk center, and affecting all measurements in the same way. After applying an appropriate correction to all line-of-sight winds (so as to get zero wind at disk center), quantitative modeling indicated the superposition of a retrograde flow with V_{eq} = −94±6 m s^{-1} (consistent with the Betz et al. (1977) analysis) and an *anti-solar-to-solar* component with cross terminator velocity 35±6 m s^{-1}. The data may thus represent the first detection of the return branch of the thermospheric flow at mesospheric levels, but this conclusion is tentative as it *a priori* assumes instrumental miscalibration (Goldstein 1989).

V. MILLIMETER WAVE CO OBSERVATIONS

A. Disk-Averaged Observations

Venus' CO millimeter-wave lines have been routinely observed since the detection of the $J = 1-0$ ^{12}CO line at 115 GHz in August 1975 (Kakar et al. 1975). These lines appear as strong, primarily collisionally broadened absorptions, formed in the cool mesosphere (75–115 km) against the hotter (∼300 K) CO_2 continuum formed at 0.4 to 1 bar. Kakar et al. (1975) showed from the line shape that the CO mixing ratio increased with altitude at 85 to 100 km (∼2–0.02 mbar), in a way reasonably consistent with predictions from Venus photochemical and early dynamical-photochemical models (see, e.g., Dickinson and Ridley 1975).

This initial measurement and all pre-1986 observations had angular resolutions poorer than 30", primarily providing disk-averaged information. Nevertheless, observations at different Venus phases have allowed the study of the diurnal variations of mesospheric CO. Early observations (Gulkis et al. 1977, Schloerb et al. 1980) revealed variations in the ^{12}CO(1–0) spectrum with the Venus phase, with the line deeper and narrower when the observations emphasized the night side. This implied that the CO mixing ratio is larger on the night side than on the day side above $\sim$100 km, qualitatively consistent with the existence of a subsolar-to-antisolar (SSAS) Venus thermospheric flow, as modeled by Dickinson and Ridley (1975,1977) (see a review of the thermospheric models in the chapter by Bougher et al.). In addition, the data indicated a surprising factor of 10 enhancement in the CO dayside vs nightside abundance at $\sim$80 to 85 km altitude. In the most comprehensive study to date, Clancy and Muhleman (1985a) re-analyzed the ^{12}CO(1–0) 1977–1979 measurements, along with new spectra (^{12}CO(1–0) and (2–1, 230 GHz)) recorded in 1979–1982, and obtained further insight into the CO diurnal behavior. The CO peak above 95 km was observed to occur at 0.6±0.6 a.m. local time, whereas the dayside maximum below 90 km was found to be shifted to 8.5±1 a.m. In both cases, the diurnal range of CO abundance spanned a factor 2 to 4. Clancy and Muhleman (1985b) interpreted these variations in terms of dynamical redistribution of the CO photochemically produced on the day side. The near midnight peak above 95 km indicated that the SSAS thermospheric flow extends to below 100 km and that retrograde zonal winds are substantially reduced at this altitude. The morning peak below 90 km is more challenging. Clancy and Muhleman (1985b), employing the photochemical model of Yung and Demore (1982), did not find photochemical forcing an adequate explanation. They instead proposed a rather *ad hoc* explanation of the morning peak, involving an extension of the SSAS flow down to $\sim$80 km and significant retrograde zonal winds ($\sim$50 m s^{-1}) at this level. This explanation has been criticized by Goldstein (1989) who also questioned the observational reality of the CO maximum at 8 to 9 a.m. (see Sec. VII.B).

Increasing sensitivities of microwave spectrometers have facilitated measurements of the weak CO isotopic lines. These measurements have led to conflicting results (see discussion in Clancy and Muhleman 1991), but it is now clear that if allowance is made for variations in the thermal profile, all spectra are consistent with terrestrial isotopic ratios in CO. As regards the mesospheric structure and its variability, the most important implication is that combined analyses of several CO lines of different opacities are necessary to fully separate between CO abundance and temperature effects. In this spirit, Clancy and Muhleman (1991) analyzed ^{12}CO(1–0), ^{12}CO(2–1) and ^{13}CO(2–1) Venus' nightside spectra obtained at Kitt Peak NRAO in 1982, 1985, 1986, 1988, and 1990. These observations pointed to very large interannual temperature and CO variations. Nightside mesospheric (85–105 km) temperatures in 1982 were only $\sim$10 K cooler than the Pioneer Venus reference, but they were 40 K warmer in 1985 (Fig. 6). This effect, comparable to that observed by Venera 10 in 1975, is probably associated with the changes in the latitudinal structure seen in late 1984 by Pioneer Venus ORO measurements (Sec. III.B.). The nightside-averaged CO vertical distribution was also found to significantly vary, with, for example, a factor of 4 increase in the CO column density above 86 km from 1982 to 1985, and a similar change in the CO mixing ratio at 80 km between 1985 and 1986. All observations were consistent with CO spatial variations across the night side well below a factor of 10, except the December 1986 measurements, which were characterized by unusually deep $^{13}CO(2-1)$ and $C^{18}O(2-1)$ lines, yet normal $^{12}CO(1-0)$ and $^{12}CO(2-1)$ lines, suggesting the presence of a dramatically enhanced nightside CO "bulge." This phenomenon was more directly observed by interferometric mapping at OVRO (see below). The time scales, periodic vs aperiodic nature, and physical mechanisms associated with these thermal and compositional changes in 1985–1986 remain unanswered questions. Similar variability has been observed at upper cloud level with the decay of upper cloud level SO_2 abundances (Esposito et al. 1988; chapter by Esposito et al.).

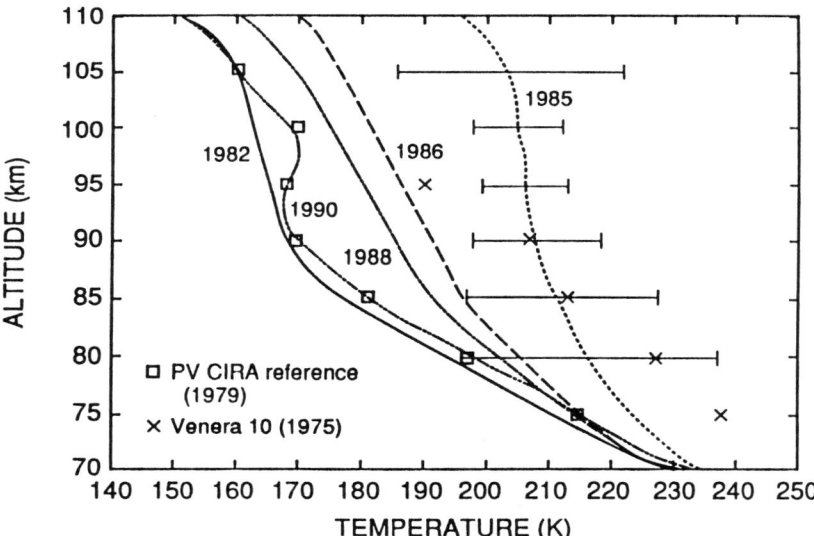

Figure 6. Variability of the (disk-averaged) mesospheric temperature profile, as inferred from CO millimeter-wave observations (figure from Clancy and Muhleman 1991).

Similarly to CO, disk-averaged temperature measurements in the upper mesosphere were obtained from millimeter-wave observations of H_2O (183 GHz) and HDO (226 GHz) conducted at the Institut de Radio Astronomie Millimetrique (IRAM) 30-m and the Caltech Submillimeter Observatory (CSO) 10.4-m telescopes. Mean dayside temperatures at 95 km were inferred to be

150±5 K in August 1990, 140±10 K in January 1991 and 155±10 K in July 1993, indicating overall cold conditions (Encrenaz et al. 1995).

B. Disk-Resolved Observations

Large radio telescopes and interferometers can now resolve Venus' disk at millimeter wavelengths. The first published spatially resolved CO observations were obtained with the 3×6 m Berkeley-Illinois-Maryland Array (BIMA) in January 1987 (de Pater et al. 1991). The shape of the spectra (^{12}CO(1–0) only) varied with local time in a way essentially consistent with inferences from the disk-averaged observations. A factor of 3 decrease in CO from night side to day side was estimated above 90 km, although the quality of the data did not permit a full recovery of the CO profile. Searches for latitudinal variability of mesospheric CO were also attempted for the first time, and no clear variations were seen.

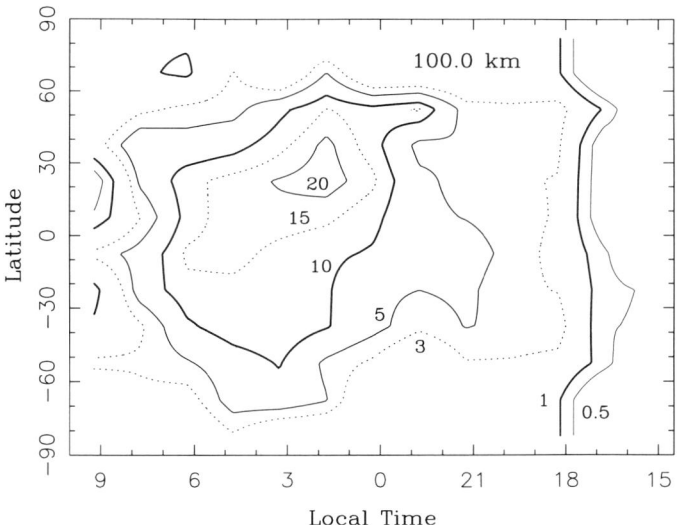

Figure 7. The CO "bulge" at 100 km observed in December 1986. Isocontours indicate the CO mixing ratio, in units of 10^{-4} (figure adapted from Gurwell et al. 1995).

Similar measurements, again restricted to the ^{12}CO(1–0) line, were conducted in December 1986 and April–May 1988 with the 3×10.4 m Owens Valley Radio Observatory (OVRO) interferometer. The sampled local time intervals on Venus were 10 p.m. to 10 a.m. and 1.20 p.m. to 1.20 a.m., respectively. In both cases, the CO diurnal distribution showed a factor of several gradient over a few hours across the (morning or evening) terminator (Gurwell et al. 1995). Very different CO vertical profiles were found for the day and night sides. Dayside mixing ratios were $\sim 10^{-4}$ at all altitudes, while nightside values increased from several 10^{-5} at 70 km to $\sim 10^{-3}$ at 105 km. The 1986 observations also indicated a well-marked nightside "bulge" at 90 to 100 km, i.e., a relatively confined region of large CO enhancement. The shape, spatial extent and contrast of the bulge and the latitude/local time position of its center varied with altitude; at 100 km for example, the bulge was stretched in an east–west direction and centered at 25°N latitude and local time = 2 a.m. (Fig. 7); the peak bulge-to-dayside ratio exceeded a factor of 20 and may have been as large as 50 to 100. Finally, the 1988 observations included high resolution spectra dedicated to wind measurements (Shah et al. 1991). They indicated a strong retrograde zonal flow with velocity 132±10 m s^{-1} in a 12-km layer centered near 99 km and a maximum 40 m s^{-1} SSAS flow. (This upper limit may have been too restrictive though [Lellouch et al. 1994] and is difficult to understand in the framework of the thermospheric models [chapter by Bougher et al.].)

Repeated multi-line (^{12}CO(2–1), ^{13}CO(2–1) and ^{12}CO(1–0)) observations with the IRAM 30-m telescope have permitted a global approach to the CO/dynamics problem from simultaneous wind/CO/temperature retrievals on Venus' night side near inferior conjunction (Lellouch et al. 1994; Rosenqvist et al. 1995). In August 1991, a zonal retrograde and a SSAS flow of approximately equal velocity coexisted in the mesosphere ($V_{eq} \sim V_{ter}$, increasing from 40±15 m s^{-1} at 95 km to 90±15 m s^{-1} at 105 km), with no large (≥ 20 m s^{-1}) day-to-day variations. At 105 km, a cos(latitude) dependence of the zonal flow was inferred and a poleward meridional component (35±30 m s^{-1}) was marginally detected. The high-latitude warming was seen, but mid-latitudes were colder than the Equator, allowing the maintenance of the cyclostrophic balance up to $\sim$105 km. The horizontal distribution of CO on Venus' night side was essentially uniform in latitude and local time. This unusual situation is consistent with the measured wind field because the strong zonal winds tend to smooth out diurnal contrasts (see Sec. VII.B). In May–June 1993, no wind velocities were obtained but the CO distribution showed a maximum around 3 a.m. above 100 km and a factor of 2 to 3 contrast over the disk. In November 1994, provisional wind measurements indicate comparable zonal and SSAS winds at 95 km, but a stronger SSAS flow at 105 km (see Table I). These data have not yet been inverted in terms of CO/temperature structure but almost simultaneous OVRO measurements (Gurwell 1996) show a CO peak significantly shifted past midnight (2 a.m. on Nov. 8, 4–4.30 a.m. on Dec. 9). Interannual variability of the mesospheric CO diurnal distribution and of the zonal flow is therefore suggested.

Finally, from similar ^{12}CO (2–1) Kitt Peak NRAO observations, Buhl et al. (1994) reported the presence, near 100 km altitude, of a puzzling return antisolar-to-solar flow in December 1989 (120 m s^{-1}) and March 1993 (60 m s^{-1}), reversing to a direct SSAS flow of 140 m s^{-1} in April 1993.

TABLE I
Mesospheric and Thermospheric Wind Constraints

Z (km)	Time Period	Wind Velocity (m s^{-1}) Zonal[a]	SSAS[b]	Method	Reference[d]
~70–90	1–6/79	weak, poss. jet		IR temp., PV OIR	1
~70–90	10–12/83	weak		IR temp., Venera 15	2
~70–90	2/91	weak, poss. jet		IR temp., Galileo	3
~70–90	1978–1991	weak, variable		PV occ. temp.	4
~70–80	4–5/77		−35±6[c]	10 µm heterodyne	5
~100	2–4/77, 1978	94±6	present	CO mm, CO distr.	6, 7
~90–105	1977–1982	weak	present	CO mm, CO distr.	8
~90–105	3/85, 5/88, 2/90	weak	present	CO mm, CO distr.	9
~90–105	12/86, 4–5/88	strong		CO mm, CO distr.	9, 10
99±6	4–5/88	132±10	≤40	CO mm, winds	11
94.5±6	8/91	35±15	45±15	CO mm, winds	12
105±9	8/91	95±10	90±15	CO mm, winds	12
~90–105	8/91	equal magnitude		CO mm, CO distr.	12
~90–105	5–6/93	SSAS moderately stronger		CO mm, CO distr.	13
94.5±6	11/94	45±30	50±35	CO mm, winds	13
105±9	11/94	75±20	110±20	CO mm, winds	13
109±10	12/85, 10/86, 03/87	25±15	120±20	10 µm CO$_2$ heterodyne	14
~95–110	7–10/91, 3–5/93	variable		O$_2$ IR nightglow	15
~100–130	1975	weak		O$_2$ vis., Venera	16
~100–130	1988–1990	weak		O$_2$ vis., PVO	17
~100–110	9/94	$V_{eq} + 0.7 \times V_{ter} = 140 \pm 45$		CO 5 µm, winds	18
~125–145	9/94	$V_{eq} + 0.7 \times V_{ter} = 200 \pm 50$		CO 5 µm, winds	18
115–150	1979–1980	40–60		NO UV nightglow, PVO	19, 20, 21
≥150	1979–1980	45–90		He density, PVO	22, 23, 24
≥150	1979–80, 1992	50–100	~200	nightside temps	22, 23, 25, 26

[a] Zonal retrograde flow, equatorial velocity.
[b] Subsolar-to-anti-solar flow, cross terminator velocity.
[c] Anti-solar-to-solar. Tentative re-assessment of Betz et al. (1977) data.
[d] Reference: 1. Taylor et al. 1980; 2. Schäfer et al. 1990; 3. Roos-Serote et al. 1995; 4. Kliore 1985; 5. Goldstein 1989; 6. Gulkis et al. 1977; 7. Schloerb et al. 1980; 8. Clancy et al. 1985a; 9. Clancy and Muhleman 1991; Gurwell et al. 1995; 11. Shah et al. 1991; 12. Lellouch et al. 1994; 13. Rosenqvist et al. 1995; 14. Goldstein et al. 1991; 15. Crisp et al. 1996; 16. Krasnopolsky 1983; 17. Bougher and Borucki 1994; 18. Maillard et al. 1995; 19. Steward et al. 1980; 20. Bougher et al. 1990; 21. Gérard et al. 1981; 22. Niemann et al. 1980; 23. Mayr et al. 1980; 24. Mengel et al. 1989; 25. Keating et al. 1980; 26. Bougher et al. 1986.

VI. THE O_2 INFRARED AND VISIBLE NIGHTGLOWS

1. Discovery. An airglow in the 400 to 800 nm region with total intensity of $\sim$3 kR was discovered in October 1975 on Venus' night side by the Venera 9 and 10 instruments and quickly identified as the Herzberg II band system of O_2 ($c^1\Sigma_u^- - X^3\Sigma_g^-$) (Krasnopolsky 1983 and references therein). Its spatial distribution was not well defined due to limited observations, but found to be generally uniform over the night side, with a shallow maximum just past midnight (0030 LT) at 8 to 28°S latitude. A month later, a similar, but much stronger, emission was discovered at 1.27 μm from groundbased Fourier Transform Spectrometer (FTS) observations, due to the O_2 ($a^1\Delta_g - X^3\Sigma_g^-$) band (Connes et al. 1979). Its total intensity was comparable on Venus' day and night sides (1.5 and 1.2 MegaRayleigh (MR) respectively, after correction for backscatter by the underlying clouds), but the spatial resolution was not sufficient to search for smaller-scale horizontal variations. Individual lines were spectrally separated, providing a measurement of the rotational temperature on the night side (185±15 K).

2. Production Mechanisms. The production of Venus' O_2 airglows is primarily a consequence of the recombination of atomic oxygen. Atomic O, produced on Venus' day side from CO_2 photolysis, mostly at 100 to 120 km, has a lifetime of $\sim$1 day. It is transported by the global circulation to the night side, where downwelling and three-body and catalytic recombination reactions occur. These exothermic reactions produce O_2 molecules in a variety of excited states, including $a^1\Delta_g$ and $c^1\Sigma_u^-$, which radiatively decay (within 1 to 2 hours for the first one) to the $X^3\Sigma_g^-$ ground level.

The intensity of the infrared airglow implies a globally integrated O_2(a) production rate comparable to the CO_2 photodissociation rate ($\sim$6 × 10^{12} cm^{-2}s^{-1}; Yung and Demore 1982). Direct three-body recombination and several proposed catalytic reactions involving chlorine or hydrogen radicals (Leu and Yung 1987) have quantum yields of only a few percent. However, production of O_2 (a) can occur through collisional quenching of other excited states. Bougher and Borucki (1994), considering the excitation of the $O_2(^5\Pi)$ state, obtained effective quantum yields for the O_2(a) and O_2(c) states of 27% and 37% respectively, and simulated maximum O_2 infrared intensities $\sim$70% of the hemispheric average value of Connes et al. (1979). Based on recent laboratory work, Crisp et al. (1996) estimated that 60 to 75% of the O_2 formed by recombination should eventually be converted to O_2 (a), presumably satisfying the observed emission rates. Bougher and Borucki (1994) further modeled the O vertical distribution and obtained (for benchmark conditions) a broad dayside peak of about 5 × 10^{10} O atoms cm^{-3} at 100 to 120 km, and a 10 times stronger nightside peak near 107 km. They concluded that the O_2 visible and infrared nightglows result from transport along streamlines descending from $\geq$110 km on the day side to $\sim$107 km on the night side. Probed altitude ranges are nominally 100 to 130 km (thermosphere) for the visible airglow and 95 to 110 km (upper mesosphere/lower thermosphere) for the infrared airglow, although these altitudes may be affected by unexplored catalytic reactions and additional collisional production processes below 100 km. Further discussion about recent observations of the O_2 visible airglow and other related airglows and their production mechanisms (notably the NO airglow) is given in the chapter by Bougher et al.

3. Recent Observations and Structure of the O_2 Infrared Nightglow. New spectroscopic and imaging data acquired since 1990 have provided insight into the structure of the O_2 infrared nightglow. Galileo/NIMS spectral images recorded on February 10, 1990, showed that Venus' night side was fairly uniform at 1.27 μm, except for one small ($\sim$100 km) area at 39.3°S latitude where the O_2 infrared emission was enhanced by a factor of 2 (Drossart et al. 1993). The (corrected) peak emission rate in this region was $\sim$4 MR. A bright airglow was also detected in concomitant observations at the Anglo-Australian Telescope (AAT). High-resolution FTS observations of selected regions of Venus were obtained at the Canada-France-Hawaii Telescope (CFHT) in June, July and October 1991. These spectra (Fig. 8), in which individual O_2 emissions lines near 7880 cm^{-1} were separated and clearly distinguished from the deep atmosphere thermal emission peak (near 7830 cm^{-1}), revealed the existence of large spatial (factor of 4 between 10°N and 5°S latitudes on June 27, 1991) and temporal (factor of 6 between June 27 and July 1 at 15°S) variations in the emission rates. Crisp et al. (1996) derived a rotational temperature of 186±6 K, similar to Connes et al. (1979).

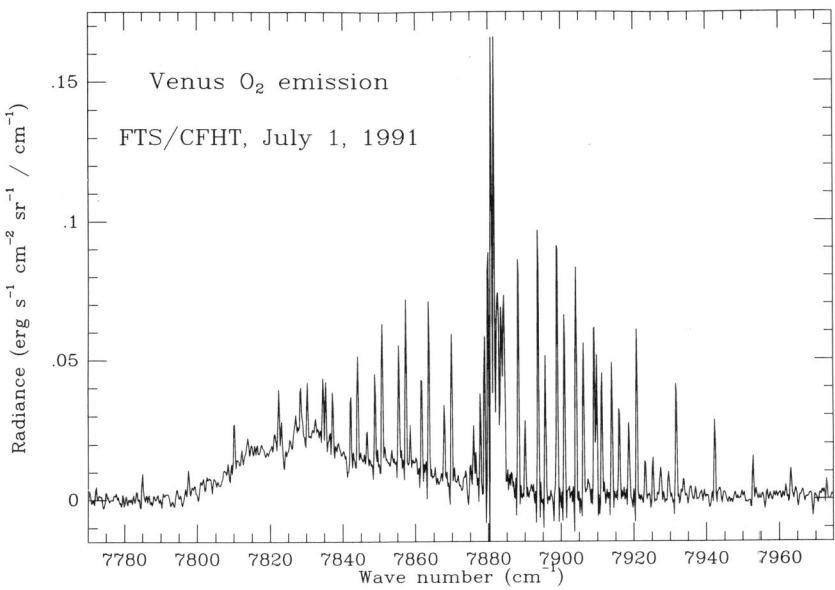

Figure 8. High-resolution (0.4 cm^{-1}) spectrum of the O_2 nightglow, observed with CFHT/FTS on 1 July 1991 (latitude = 15°S).

Detailed images of the $O_2(^1\Delta)$ nightglow were obtained at the AAT in July, September, and October 1991, and in April–May 1993 (Allen et al. 1992; Crisp et al. 1996). These images show localized, often elongated regions of intense emission (up to 6 MR, corrected), broadly distributed around or past the antisolar point (1 a.m.–3 a.m.), but sometimes at high latitudes near the limb, with largest characteristic dimensions of 1000 km. This structure is reminiscent of that of the NO ultraviolet and OI 130 nm emissions produced at higher altitudes (see the review in the chapter by Bougher et al.). Very dark regions are often juxtaposed with the brightest features and spatial contrasts in intensity can be as large as 10:1. Intensities in the bright regions can vary by as much as 20% over one hour and these regions can disappear entirely within 1 day (Fig. 9 and CDP3C1F2). The O_2 emission distribution on one day is sometimes anti-correlated with that seen on the following day. Comparable structure was seen in data obtained in October 1991 and March 1993 with the CFHT/FTS in spectro-imaging mode. On October 18, 1991, the O_2 image showed three discrete bright spots (the brightest being at 80°S), surrounding a dark region centered at the equator. In March 1993, the emission was subdued and relatively uniform, with a rotational temperature of 190±10 K (Maillard et al. 1993).

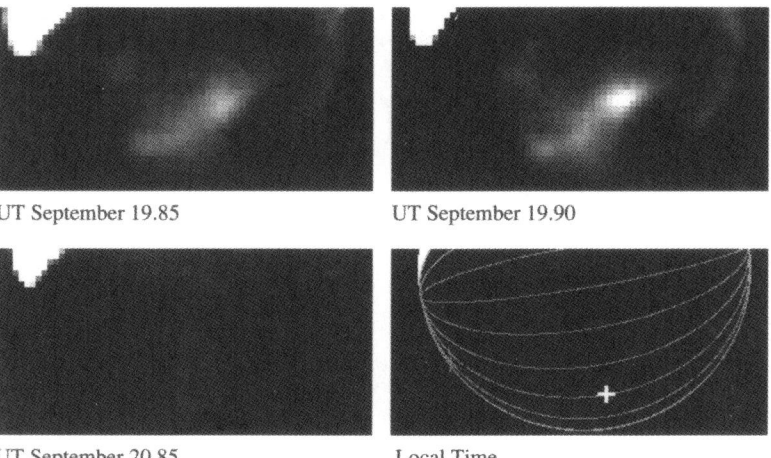

Figure 9. AAT/IRIS images of O_2 nightglow emission taken on 19–20 September 1991. The last panel shows the orientation of Venus, with the cross at the antisolar point (figure adapted from Allen et al. 1992).

A new airglow emission from Venus, due to the CO (1–0) and (2–1) bands at 4.7 μm, has been discovered from CFHT/FTS observations obtained in 1987 (de Bergh et al. 1988). Recent improved observations have allowed rotational temperatures and absolute wind velocities to be measured from individual lines of the two bands (Maillard et al. 1995; Table I). These emissions, however, probe Venus' thermosphere rather than the mesosphere (125–145 km for the 1–0 band and 100–110 km for the 2–1 band), so they are described more extensively in the chapter by Bougher et al.

VII. A GENERAL PICTURE OF VENUS' MESOSPHERE

A. Thermal Profiles, Winds, and Their Variations

Temperature and wind measurements show that the mesophere is dynamically an intermediate region characterized by significant time variability. (An overview of all wind direct measurements and inferences at mesospheric/thermospheric levels is given in Table I). The magnitude of the variations, however, is different for the lower (60–85 km) and upper (85–110 km) mesosphere. In the lower mesosphere, the main features of the thermal field (the cold collar at 65–70 km, and the inverted equator-to-pole gradient above 75 km) have been repeatedly observed for almost two decades, in spite of some noticeable amplitude variations. Cloud tops ($\sim$65 km) zonal winds are relatively stable with equatorial velocities between 90 and 100 m s^{-1}, but their meridional dependence varies with time. The reversed pole-to-equator temperature gradients observed at mesospheric levels by the Pioneer Venus and Venera 15 orbiters indicate that the zonal wind velocities generally decrease between the cloud tops and 80 km. Because the amplitude of this wind shear is strongly coupled to the horizontal temperature gradient through the cyclostrophic balance relation, the zonal winds at these levels may have varied substantially during the 1980s as the mesospheric temperature structure changed, although most observations document changes in the mean thermal structure rather than changes in the latitudinal temperature gradient. Cyclostophic balance models also indicate the presence of a mid-latitude jet at $\sim$70 km. Interestingly, the calculated jet velocity is $\sim$40% larger than velocities measured from ultraviolet feature tracking (130–140 vs 90–100 m s^{-1}), and its core is located at a slightly higher latitude (45–55° vs 35–45°). These differences suggest that at these latitudes, zonal winds actually increase between 65 and 70 km altitudes.

Variations in the upper mesosphere (85–110 km) are more dramatic. Interannual disk-average temperature changes by tens of K can occur (see Sec. V.A), with, in some occasions, modifications of the latitudinal gradient, maintaining cyclotrophic balance up to $\sim$105 km. Direct wind measurements consistently find a significant zonal velocity at 95 to 110 km (suggesting that there exists a region of minimum zonal flow at 85–90 km), but with a very large (several tens of m s^{-1}) interannual variability (Lellouch et al. 1994). Large short-term variability is also indicated, from the variable O$_2$ infrared nightglow. On the other hand, wind data (except the Buhl et al. [1994] observations) suggest a substantial stability of the SSAS flow, and measurements of its vertical profile at 90 to 110 km (Goldstein et al. 1991; Lellouch et al. 1994) match thermospheric model predictions (Bougher et al. 1986).

Many types of waves, including small-scale gravity waves and the diurnal and semidiurnal thermal tides, are superimposed on the mean thermal and dynamical state. The tidal model of Pechman and Ingersoll (1984) successfully matches the tidal structure (amplitude and phase of the two modes) seen in Pioneer Venus/OIR data, but provides only qualitative agreement with the PV Probe data. Waves have a significant impact on the overall circulation (see, e.g., Elson 1983; chapter by Gierasch et al.). Recent models (Baker and Leovy 1987; Crisp 1989; Newman and Leovy 1992) show that semidiurnal tides transport and deposit westward momentum at levels (≥ 70 km) where they are radiatively damped, exerting a drag on the zonal flow (hence causing the decay on the superrotation) and inducing an equator-to-pole meridional circulation, with rising motion in the tropics, poleward flow above the cloud tops, and subsidence at high latitudes. The adiabatic heating associated with the polar subsidence maintains the high temperatures observed in the polar mesosphere. Gravity waves can also induce a significant drag (Hinson and Jenkins 1995) and their activity may control the variability of the zonal flow over different timescales. Upward-propagating gravity waves generated at cloud tops grow in amplitude until they become unstable and break, depositing momentum and inducing vertical eddy mixing (Alexander 1992). Explicit incorporation of gravity wave breaking in the Venus thermospheric general circulation model (VTGCM) simulations is however required for further understanding (chapter by Bougher et al.).

B. The Minor Species Distribution in the Upper Mesosphere and their Relation with Dynamics

The 20-yr old discovery of a CO nightside peak in Venus' upper mesosphere has demonstrated that the CO distribution is linked to the global circulation. Subsequent monitoring, along with circulation models, has led to the following semiquantitative paradigm. For pure SSAS circulation, the SS and AS points are "stagnation points," i.e., points where the horizontal velocity is zero. In the presence of an additional, weaker, retrograde flow (V_z), both stagnation points are shifted toward the morning terminator. Light species like CO and O flow to the post-midnight stagnation point where they pile up. The SSAS velocity varies almost linearly with solar zenith angle (Bougher et al. 1986), allowing the approximate location of the CO maximum as: local time (max CO) = $V_z/V_{ter} \times 6$ h. The magnitude of the diurnal contrasts depends on the relative zonal and SSAS flows, because the latter tends to concentrate the CO at the AS point, while the former redistributes it over all local times. If $V_z \geq V_{ter}$, the stagnation points disappear and no pronounced diurnal variations are expected. With this picture, variations in the CO distribution reflect temporal variations of the wind field, and more specifically of the zonal flow. For example, with the Goldstein et al. (1991) 1985 to 1987 wind measurements at 110 km, a CO maximum is expected at (1.25 ± 0.8) a.m., in agreement with the 1978–1982 observations analyzed by Clancy et al. (1985a). For 1991, Lellouch et al. (1994) find $V_{eq} \sim V_{ter}$ at 95 to 105 km, consistent with the absence of strong diurnal CO variations on the night side.

The distribution of the O_2 visible and infrared emissions is also clearly linked to the mesospheric/thermospheric circulation. VTGCM simulations (Bougher and Borucki 1994) show that the O_2 intensities and spatial variations are sensitive to (i) the prescribed zonal wind profile; (ii) the Rayleigh friction controlling SSAS winds; and (iii) the nightside eddy diffusion coefficient. Models generally indicate oval O_2 distributions (elongated in local time), symmetric in latitude, with maximum at the equator and shifted from midnight (1 a.m. to 5 a.m. local time), and variations over the night side by factors 2 to 10. Increasing the zonal wind further offsets the maximum from midnight and smoothes out spatial variations. Comparison with the data suggests variable zonal winds (10–50 m s^{-1}) on short (hour) time scales at 90 to 110 km. Application of these models to CO shows similar sensitivity to the same factors and supports the above paradigm (Lellouch et al. 1994). For example, prescribing a zonal flow consistent with the Goldstein et al. (1991) measurements leads to a factor of 2.9 variation of CO over the night side at 105 km, consistent with Clancy et al. (1985a). This factor is reduced to $\sim$1.7 if the Lellouch et al. (1994) 1991 wind field is used.

While the existence of a dynamical control of CO and O_2 seems well established, the detailed situation may be complex. In the case of CO, results from OVRO (Sec. V.B) do not match the standard picture. The presence of an extreme CO "bulge" at 2 to 3.30 a.m. in 1986 is a challenge to interpretation, because such a large shift from midnight implies strong zonal winds, therefore reduced diurnal contrasts. The position of the bulge at 100 km, displaced to the northern hemisphere, is also not understood. In 1988, a factor of $\geq$5 CO increase over a few hours interval across the evening terminator is paradoxical in presence of a dominant 132$\pm$10 m s^{-1} zonal retrograde flow (VTGCM simulations in these conditions predict CO variations $\leq$25–40% over the night side). These results, if confirmed, clearly show the limitations of the paradigm. Gurwell et al. (1995) recognize, however, that ignorance of the temperature structure on a local scale may affect their inferred CO distribution. The issue of the CO variations in the lower mesosphere (below 90 km) also remains unclear. From the 1978 to 1982 data, Clancy and Muhleman (1985a) find a CO maximum at 8.5$\pm$1 a.m. Such a post-terminator maximum, however, cannot be explained by the "stagnation points" model. Goldstein (1989) has argued that the data rather suggest a CO minimum at 8 to 9 p.m. Such a minimum would result from the combination of a return anti-solar-to-subsolar flow and a weaker zonal retrograde flow in the lower mesosphere, qualitatively confirming his re-interpretation of the Betz et al. (1977) wind data.

In the case of the O_2 nightglows, VTGCM models currently do not explain the detailed structure of the emissions, in particular (i) the latitudinal distribution, not hemispherically symmetric; (ii) the multiplicity of local maxima; and (iii) the striking variability on a $\sim$1-hr time scale and the occasional disappearance of the bright spots over less than 1 day. These features suggest the existence of local chemical or dynamical processes (Crisp et al. 1996). The brightest patches may be regions of strong subsidence where the O-atom supply and recombination rates are largest. Conversely, the darkest areas might indicate regions of upwelling. If this is the case, bright regions should be warmer than dark areas because strong downdrafts (≥ 10 cm s^{-1}) produce significant adiabatic (compressional) heating, that possibly cannot be compensated for by radiative cooling (Crisp 1989; chapter by Crisp and Titov). The temperature increase may be large enough to inhibit further subsidence, and produce a dark region over a ~ 1 day radiative time scale. If this view is correct, spectral maps of the infrared O_2 emission should show a positive correlation between brightness and rotational temperature. Current datasets are still inconclusive on this issue, but we note that the temperature measured from an infrared O_2 bright patch in July 1991 (186±6 K) was higher than that obtained from the millimeter observations in August 1991 (~ 168 K near equator at 95 km), consistent with this hypothesized correlation.

C. Outstanding Issues and Unresolved Questions

In spite of the significant progress accomplished since the Pioneer Venus mission, a number of important issues remain unresolved. As a matter of conclusion, we list here some key questions that we feel should serve as directions for future research.

1. What causes the variations of the upper mesosphere thermal and dynamical state, in particular the spectacular variations of the zonal retrograde flow? What is the role of wave activity?
2. What is the reason for the occasional existence of a CO "bulge" and of large CO variations across the terminator? How can these be maintained in the presence of a strong zonal flow? What is the latitudinal CO distribution? Why is the structure of the O_2 infrared nightglow so complex and changing with time?
3. Where and how strong is the return branch of the thermospheric flow? The evidence so far for an AS-SS flow is either indirect or confusing. The best prospect to detect it is from wind measurements from 10 μm $^{13}CO_2$ lines, probing the 70 to 80 km range (Betz et al. 1977).
4. How strong is the meridional flow as a function of altitude? How does it vary with varying zonal flow? Does it impact on the (unknown) meridional distribution of minor species? Is cyclostrophic balance a satisfactory description of the mesospheric momentum balance?
5. What is the origin of the "cold collar" inversions? What mechanisms control their variability?

Finally, as is natural for time variable phenomena, a statistical evaluation of the different situations is needed. It has been customary to consider the 1978–1982 configuration (with vanishing zonal winds above 80 km) as "normal," and the intrusion of zonal winds into the upper mesosphere as a deviation from this normal state, but several direct wind measurements at 90 to 105 km suggest that this characterization may not be correct. Many more observations are required to establish a realistic climatology of Venus' mesosphere.

REFERENCES

Alexander, M. J. 1992. A mechanism for the Venus thermospheric superrotation. *Geophys. Res. Lett.* 19:2207–2210.

Allen, D. A., Crisp, D., and Meadows, V. S. 1992. Variable oxygen airglow on Venus as a probe of atmospheric dynamics. *Nature* 359:516–518.

Avduesky, V. S., et al. 1983. Structure and parameters of the Venus atmosphere according to Venera probe data. In *Venus*, eds. D. M. Hunten, L. Colin, T. M. Donahue and V. I. Moroz (Tucson: Univ. of Arizona Press), pp. 280–298.

Baker, N. L., and Leovy, C. B. 1987. Zonal winds near Venus' cloud top level: A model study of the interactions between the zonal mean circulation and the semidiurnal tide. *Icarus* 69:202–220.

Belton, M. J. S., et al. 1991. Images from Galileo of the Venus cloud deck. *Science* 253:1531–1536.

Betz, A. L., Johnson, M. A., McLaren, R.A., and Sutton, E. C. 1976. Heterodyne detection of CO_2 emission lines and wind velocities in the atmosphere of Venus. *Astrophys. Lett. J.* 208:41–44.

Betz, A. L., Sutton, E. C., McLaren, R. A. and McAlary, C. W. 1977. Laser heterodyne spectroscopy. In *Proc. of the Symposium on Planetary Atmospheres*, ed. A. V. Jones (Ottawa: Royal Soc. of Canada), pp. 29–33.

Bougher, S. W., and Borucki, W. J. 1994. Venus O_2 visible and IR nightglow: Implications for lower thermosphere dynamics and chemistry. *J. Geophys. Res.* 99:3759–3776.

Bougher, S. W., et al. 1986. Venus mesosphere and thermosphere. II. Global circulation, temperature and density variations. *Icarus* 68:284–312.

Bougher, S. W., Gérard, J. C., Stewart, A. I. F., and Fesen, C. G. 1990. The Venus nitric oxide night airglow: Calculations based on the Venus thermospheric general circulation model. *J. Geophys. Res.* 95:6271–6284.

Buhl, D., Goldstein, J. J., and Chin, G. 1994. Observations of variations in the mesospheric winds of Venus. *Bull. Amer. Astron. Soc.* 26:1144.

Clancy, R. T., and Muhleman, D. O. 1985a. Diurnal CO variations in the Venus mesosphere from CO microwave spectra. *Icarus* 64:157–182.

Clancy, R. T., and Muhleman, D. O. 1985b. Chemical dynamical models of the Venus mesosphere based upon diurnal microwave CO variations. *Icarus* 64:183–204.

Clancy, R. T., and Muhleman, D. O. 1991. Long-term (1979–1990) changes in the thermal, dynamical, and compositional structure of the Venus mesosphere as inferred from microwave spectral lines observations of ^{12}CO, ^{13}CO, and $C^{18}O$. *Icarus* 89:129–146.

Connes, P., Noxon, J. F., Traub, W. A., and Carleton, N. P. 1979. $O_2(^1\Delta)$ emission in the day and night airglow of Venus. *Astrophys. Lett. J.* 233:29–32.

Crisp, D. 1989. Radiative forcing of the Venus mesosphere. II. Thermal fluxes, cooling rates, and radiative equilibrium temperatures. *Icarus* 77:391–413.

Crisp, D., et al. 1996. Ground-based near-infrared observations of the Venus night side: Near-infrared $O_2(^1\Delta)$ airglow from the upper atmosphere. *J. Geophys. Res.* 101:4577–4594.

de Bergh, C., Crovisier, J., Lutz, B. L., and Maillard, J.-P. 1988. Detection of CO infrared emission lines in spectra of Venus. *Bull. Amer. Astron. Soc.* 20:831.

Del Genio, A. D., and Rossow, W. B. 1990. Planetary-scale waves and the cyclic nature of cloud top dynamics on Venus. *J. Atmos. Sci.* 47:293–318.

Deming, D., et al. 1983. D. Observations of the 10-μm natural laser emission from the mesospheres of Mars and Venus. *Icarus* 55:347–355.

Deming, D., and Mumma, M. J. 1983. Modeling of the 10-μm natural laser emission from the mesospheres of Mars and Venus. *Icarus* 55:356–368.

de Pater, I., Schloerb, F. P., and Rudolph, A. 1991. Venus images with the Hat Creek interferometer in the J = 1–0 CO line. *Icarus* 90:282–298.

Dickinson, R. E., and Ridley, E. C. 1975. A numerical model for the dynamics and composition of the Venusian thermosphere. *J. Atmos. Sci.* 32:1219–1231.

Dickinson, R. E., and Ridley E. C. 1977. Venus mesosphere and thermosphere temperature structure. II. Day–night variations. *Icarus* 30:163–178.

Drossart, P., et al. 1993. Search for spatial variations of the H_2O abundance in the lower atmosphere of Venus from NIMS-Galileo. *Planet. Space Sci.* 41:495–505.

Encrenaz, T., et al. 1995. The thermal profile and water abundance in the Venus mesosphere from H_2O and HDO millimeter wave observations. *Icarus* 117:162-172.

Elson, L. S. 1979. Preliminary results from the Pioneer Venus Orbiter infrared radiometer: Temperature and dynamics in the upper atmosphere. *Geophys. Res. Lett.* 6:720–722.

Elson, L. S. 1983. Solar related waves in the Venusian atmosphere from the cloud tops to 100 km. *J. Atmos. Sci.* 40:1535–1551.

Esposito, L. W., et al. 1988. Sulfur dioxide at the Venus cloud tops, 1978–1986. *J. Geophys. Res.* 93:5267–5276.

Gérard, J. C., Stewart, A. I. F., and Bougher, S. W. 1981. The altitude distribution of the Venus ultraviolet nightglow and implications on vertical transport. *Geophys. Res. Lett.* 8:633–636.

Goldstein, J. J. 1989. Absolute Wind Velocities in the Lower Thermosphere of Venus Using Infrared Heterodyne Spectroscopy. Ph.D. Thesis, Univ. of Pennsylvania, Philadelphia.

Goldstein, J. J., et al. 1991. Absolute wind velocities in the lower thermosphere of Venus using infrared heterodyne spectroscopy. *Icarus* 94:45–63.

Gulkis, S., Kakar, R. K., Klein, M. J., and Olson, E. T. 1977. Venus: Detection of variations in stratospheric carbon monoxide. In *Proc. of the Symposium on Planetary Atmospheres*, ed. A. V. Jones (Ottawa: Royal Soc. of Canada), pp. 61–65.

Gurwell, M. A. 1996. Planetary Atmospheres: Probing Structure Through Millimeter Wave Observations of Carbon Monoxide. Ph.D. Thesis, California Inst. of Technology.

Gurwell, M. A., et al. 1995. Observations of the CO bulge on Venus and implications for mesospheric winds. *Icarus* 115:141–158.

Hinson, D. P., and Jenkins, J. M. 1995. Magellan radio occultation measurements of atmospheric waves on Venus. *Icarus* 114:310–327.

Jenkins, J. M., Steffes, P. G., Hinson, D. P., Twicken, J. D., and Tyler, G. L. 1994. Radio occultation studies of the Venus atmosphere with the Magellan spacecraft. 2. Results from the October 1991 experiment. *Icarus* 110:79–94.

Kakar, R. K., Waters, J. W., and Wilson, W. J. 1975. Venus: microwave detection of carbon monoxide. *Science* 191:379–380.

Keating, G. M., Nicholson, J. Y., III, and Lake, L. R. 1980. Venus upper atmosphere structure. *J. Geophys. Res.* 85:7941–7956

Kerzhanovitch, V. V., and Limaye, S. S. 1985. Circulation of the atmosphere from the

surface to 100 km. *Adv. Space Res.* 5:59–83.
Kliore, A. J. 1985. Recent results on the Venus atmosphere from Pioneer Venus radio-occultations. *Adv. Space Res.* 5:41–49.
Krasnopolsky, V. A. 1983. Venus spectroscopy in the 3000–8000 Å region by Veneras 9 and 10. In *Venus*, eds. D. M. Hunten, L. Colin, T. M. Donahue and V. I. Moroz (Tucson: Univ. of Arizona Press), pp. 681–765.
Lellouch, E., et al. 1994. Global circulation, thermal structure and carbon monoxide distribution in Venus' mesosphere in 1991. *Icarus* 110:315–339.
Leu, M. T., and Yung, Y. L. 1987. Determination of $O_2(^1\Delta)$ and $O_2(^1\Sigma)$ yields in the reaction $O + ClO = Cl + O_2$: Implications for photochemistry in the atmosphere of Venus. *Geophys. Res. Lett.* 14:949–952.
Limaye, S. S. 1988. Cloud level circulation during 1982 as determined from Pioneer Cloud Photopolarimeter images. II. Solar longitude dependent circulation. *Icarus* 73:212–226.
Limaye, S. S. 1990. Venus atmospheric circulation: Known and unknown. *Adv. Space Res.* 10(5):91–101.
Limaye, S. S., Grassotti, C., and Kuetemeyer, M. J. 1988. Venus: Cloud level circulation during 1982 as determined from Pioneer cloud photopolarimeter images. I. Time and zonally averaged circulation. *Icarus* 73:193–211.
Maillard, J.-P., et al. 1993. Spectro-imaging of the dark side of Venus in the 1.27 µm O_2 emission with an imaging FTS. *Bull. Amer. Astron. Soc.* 25:1095.
Maillard, J.-P., et al. 1995. Carbon monoxide 4.7 µm emission: A new dynamical probe of Venus' thermosphere. *Bull. Amer. Astron. Soc.* 27:26.
Mayr, H. G., et al. 1980. Dynamic properties of the thermosphere inferred from Pioneer Venus mass spectrometer measurements. *J. Geophys. Res.* 85:7841–7847.
Mengel, J. G., Mayr, H. G., Harris, I. and Stevens-Rayburn, D. R. 1989. Non-linear three-dimensional spectral model of the Venusian thermosphere with superrotation: II. Temperature, composition, and winds. *Planet. Space Sci.* 37(6):707–722.
Niemann, H. B., et al. 1980. Mass spectrometer measurements of the neutral gas composition of the thermosphere and exosphere of Venus *J. Geophys. Res.* 85:7817–7827.
Newman, M. G., and Leovy, C. B. 1992. Maintenance of strong rotational winds in Venus' middle atmosphere by thermal tides. *Science* 257:647–650.
Newman, M. G., Schubert, G., Kliore, A. J., and Patel. I. R. 1984. Zonal winds in the middle atmosphere of Venus from Pioneer Venus radio occultation data. *J. Atmos. Sci.* 41:1901–1913.
Oertel, D., et al. 1985. Infrared spectrometry of Venus from "Venera-15" and "Venera-16." *Adv. Space Res.* 5(9):25–36.
Pechmann, J. B., and Ingersoll, A. P. 1984. Thermal tides in the atmosphere of Venus: Comparison of model results with observations. *J. Atmos. Sci.* 41:3290–3313.
Roos-Serote, M., et al. 1995. The thermal structure and dynamics of the atmosphere of Venus between 70 and 90 km from the Galileo-NIMS spectra. *Icarus* 114:300–309.
Rosenqvist, J., Lellouch, E., Encrenaz, T., and Paubert G. 1995. Global circulation in Venus' mesosphere from IRAM CO observations (1991–1994): A tribute to Jan Rosenqvist. *Bull. Amer. Astron. Soc.* 27:26.
Rossow, W. B., Del Genio, A. D., and Eichler, T. 1990. Cloud tracked winds from Pioneer Venus OCPP images. *J. Atmos. Sci.* 47:2053–2084.
Schäfer, K., et al. 1990. Infrared Fourier-Spectrometer experiment from Venera-15. *Adv. Space Res.* 10(5):57–66.
Schubert, G. 1983. General circulation and the dynamical state of the Venus atmosphere. In *Venus*, eds. D. M. Hunten, L. Colin, T. M. Donahue and V. I. Moroz

(Tucson: Univ. of Arizona Press), pp. 681–765.

Schloerb, F. P., Robinson, S. E., and Irvine, W. M. 1980. Observations of CO in the stratosphere of Venus via its J = 0–1 rotational transition. *Icarus* 43:121–127.

Seiff, A. 1983. Thermal structure of the atmosphere of Venus. In *Venus*, eds. D. M. Hunten, L. Colin, T. M. Donahue and V. I. Moroz (Tucson: Univ. of Arizona Press), pp. 215–279.

Seiff, A., and Kirk, D. B. 1991. Waves in Venus's middle and upper atmosphere: Implications of Pioneer Venus Probe data above the clouds. *J. Geophys. Res.* 96:11021–11032.

Seiff, A., et al. 1985. Models of the structure of the atmosphere of Venus from the surface to 100 km altitude. *Adv. Space Res.* 5(11):3–58.

Shah, K. P., Muhleman, D. O., and Berge, G. L. 1991. Measurements of winds in Venus' upper mesosphere based of Doppler shifts of the 2.6 mm ^{12}CO line. *Icarus* 93:96–121.

Stewart, A. I. F., Gérard, J. C., Rusch, D. W., and Bougher, S. W. 1980. Morphology of the Venus ultraviolet night airglow. *J. Geophys. Res.* 85:7861–7870.

Taylor, F. W., et al. 1980. Structure and meteorology of the middle atmosphere of Venus: Infrared remote sensing from the Pioneer Orbiter. *J. Geophys. Res.* 85:7963–8006.

Taylor, F. W., Hunten, D. M., and Ksanfomaliti, L. V. 1983. The thermal balance of the middle and upper atmosphere of Venus. In *Venus*, eds. D. M. Hunten, L. Colin, T. M. Donahue and V. I. Moroz (Tucson: Univ. of Arizona Press), pp. 650–680.

Traub, W. A., and Carleton, N. P. 1979. Retrograde winds on Venus: Possible periodic variations. *Astrophys. J.* 227:329–333.

Yakovlev, O. I., Matyugov, and Gubenko, V. N. 1991. Venera-15 and -16 middle atmosphere from radio occultations: Polar and near-polar atmosphere of Venus. *Icarus* 94:493–510.

Young, A. T. 1975. Is the four-day "rotation of Venus" illusory? *Icarus* 24:1–10.

Young, A. T., et al. 1979. Spectroscopic observations of winds on Venus. I. Technique and data reduction. *Icarus* 38:435–450.

Yung, Y. L., and Demore, W. B. 1982. Photochemistry of the stratosphere of Venus: Implications for atmospheric evolution. *Icarus* 51:199–247.

NEAR-INFRARED SOUNDING OF THE LOWER ATMOSPHERE OF VENUS

F. W. TAYLOR
University of Oxford

D. CRISP
Jet Propulsion Laboratory

and

B. BÉZARD
Observatoire de Paris

The recent discovery of near-infrared emission from the night side of Venus provides a powerful new technique for studying its lower atmosphere and surface. This emission is most intense within spectral windows between strong molecular absorption bands in the 0.9 to 2.5 μm wavelength region. The resulting emission from the planet is most easily detected on the nightside, where it is not overwhelmed by the more intense solar flux reflected from the clouds. Near-infrared imaging and spectroscopic observations of this emission by groundbased and spacecraft instruments have been used to investigate cloud particle sizes and optical thickness, the winds within the middle and lower cloud decks, and the abundances of several important trace gases, including water vapor, halides, carbon monoxide, sulfur dioxide, and carbonyl sulfide. They have provided new information about the near-surface temperature lapse rate, and the deuterium-to-hydrogen ratio. This chapter describes the physical basis for near-infrared sounding of Venus' lower atmosphere, reviews the results obtained to date, and looks briefly at the prospects for further progress in exploring an important and largely mysterious planetary regime which previously was accessible only to microwave sounding or entry probes.

I. THE LOWER ATMOSPHERE OF VENUS

This section provides a brief summary of our knowledge of the lower atmosphere of Venus before the discovery of the near-infrared windows in 1984. At that time, only microwave observations, most significantly of deep atmosphere temperatures and water vapor amount, were thought to penetrate the thick cloud cover. Knowledge of the structure and composition of the region, in so far as it was known at all, had been obtained mainly by direct-sensing instruments on the various probes. For example, *in-situ* spectroscopy was carried out by instruments on some of the Russian entry probes, beginning with

Venera 11 in 1978. The U. S. Pioneer Venus (PV) entry probes of December 1978 obtained simultaneous temperature-pressure data at four sites, giving the first horizontal gradients in the deep atmosphere, and compositional and cloud microphysics profiles from a single large probe. The results of these pre-1983 investigations have been extensively published, and most of them are summarized in the *Venus* book (Hunten et al. 1983).

A. Composition

The lower atmosphere of Venus is composed mainly of CO_2 (96.5%) and N_2 (3.5%), with small amounts of the noble gases, principally argon, and of chemically active species including H_2O, CO, OCS, SO_2, HCl, and HF. Variability in space and time is to be expected of reactive species, but the scale of the variability is virtually unknown in every case, and even the mean values are uncertain. For the best studied example, water vapor, values ranging from 20 to 25,000 ppmv have been published over the last 30 yr. The highest of these came from direct sampling instruments on entry probes, and were in direct conflict with groundbased microwave observations. Measurements at frequencies near 1 cm wavelength placed upper limits on the water vapor abundance of 2 to 300 ppmv, assuming uniform mixing below the clouds (Janssen et al. 1973; Janssen and Klein 1981). The Venera 11,12,13 and 14 probes carried spectrophotometers which scanned the range from 0.45 to 1.2 μm with moderate spectral resolution (0.02 μm). Again using water vapor as the most important example of the results obtained from these early experiments, the probes found (Moroz 1983; Young et al. 1984) mixing ratios close to the modern value in the lower troposphere, with larger and more ambiguous values at higher levels near the cloud decks (see Sec. IV below).

B. Clouds

The Venusian clouds totally cover the planet at heights between about 45 and 70 km, with thinner hazes above and below. Instruments on the Venera 9 and 10 landers and the four PV entry probes provided simultaneous *in-situ* measurements of the vertical structure and particle size distributions in the clouds (see *Venus* [Hunten et al. 1983] and the chapter by Esposito). They found that the main cloud consists of three distinct layers, separated by relatively clear regions. These layers, known as the upper ($\sim$57 to 70 km altitude), middle ($\sim$49 to 57 km), and lower ($\sim$47 to 49 km) clouds, are bounded above and below by more diffuse haze layers. Sulfuric acid (H_2SO_4) aerosols are the principal constituent of all three layers, but the particle size distributions differ from layer to layer. The PV Orbiter Cloud Photopolarimeter measurements showed that the upper haze layer is composed primarily of very small "mode 1" particles, which have modal radii between 0.25 and 0.4 μm. The PV Large Probe Cloud Particle Spectrometer showed that the lower haze was also composed primarily of these particles. The upper cloud consists primarily of a second particle type, called "mode 2," which has a $\sim$1 μm equivalent radius, but also includes significant numbers of mode 1 particles. A third particle type, "mode 3," is the principle component of the middle and lower clouds. This consists of relatively large particles, in a distribution with a mode radius of 3.85 μm and some particles as large as 35 μm equivalent radius. Mode 3 may have a crystalline component of uncertain composition. The smaller modes are found mixed in with mode 3 at all levels where the latter occurs.

The Vega-2 lander also carried a particle size spectrometer, only the second to be deployed on Venus to date. The Vega measurements revealed very few large particles at altitudes below 55 km, in marked contrast to the PV Large Probe, which of course entered at a different place and time. The measurements of cloud thickness by the nephelometers on the four PV probes differed by less than ~20% in the middle cloud (50–57 km), but revealed much larger differences in the particle densities within the lower cloud (47–50 km). Similar results were obtained by the nephelometers aboard Veneras 9, 10, and 11. It is now clear that the cloud opacity, the relative abundances of size modes, and possibly the modes themselves, vary considerably across the planet, and evolve in time. Studies of cloud variability using near-infrared mapping are discussed in detail in Sec. III below.

C. Dynamics

Before near-infrared sounding became a reality, the only data on winds in the troposphere were ten profiles obtained by tracking the movements of entry probes during their descent, and surface measurements by the Venera soft landers. They show a steady decline from the high values near the cloud tops to low values near the surface (~1 m s^{-1} according to the Venera 9 and 10 landers; Avduevsky et al. 1983). Interesting vertical structure and substantial differences between locations are apparent in the profiles, but are difficult to interpret due to the small sample size.

Feature tracking in near-infrared maps provides global-scale wind fields in the upper troposphere, mainly around 40 to 50 km where the thickest cloud layers occur. The cloud-tracked winds are generally consistent with the entry probe data in that they represent rapid zonal motion (about 50 m s^{-1} at this level, compared to nearer 100 m s^{-1} in the ultraviolet features near 70 km altitude) coupled with slower equator-to-pole drifts of a few m s^{-1}. These mean motions take place against a background of considerable meteorological activity, which is seen in the high contrast and very variable cloud morphology, as well as the winds themselves.

Tracking of entry probes provided little information about the wind profile in the lowest scale-height (~16 km) because, compared to the accuracy of the experiments, the winds there are small. Neither are there any cloud features to track at these heights, so far as we know. The situation has changed little since Schubert (1982) wrote in *Venus* (Hunten et al. 1983): "The actual circulation in the lowest scale height of Venus' atmosphere is completely unknown, and the dynamics of this region, which contains 65% of the atmospheric mass, is probably crucial to the build-up of the large zonal winds at greater heights because it is likely that the atmospheric angular momentum must be transported upwards from the surface." He calls for a network of long-lived meteorological stations on the surface of Venus, an attractive idea but one which, bearing in mind the conditions there, is well beyond current technology.

II. THE NEAR-INFRARED WINDOWS

A. Discovery

The first spatially resolved near-infrared images and spectra of the night side of Venus, revealing high-contrast markings at wavelengths near 1.74 and 2.3 μm, were obtained by Allen and co-workers (Allen and Crawford 1984) in 1983. While at first the observers attributed the origin of these features to scattered sunlight from the day side, their further work indicated, and radiative transfer modeling studies later confirmed, that they are produced when thermal radiation from the hot lower atmosphere (25–40 km) escapes through differing optical depths in the planet-wide cloud layers (Fig. 1). It was clear from the outset that an analysis of the spectral properties of this radiation offered a new way to probe the temperature, composition and cloud properties of the deep atmosphere.

It is surprising, perhaps, that such important behavior had not been predicted or observed earlier. In fact, the connection had not been made between three factors that contribute to the transparency of the Venus atmosphere at these wavelengths. First, concentrated sulfuric acid cloud droplets scatter almost conservatively in the near infrared. Second, there are a number of spectral windows between strong CO_2 and H_2O absorption bands, where the opacity is dominated by the far wings of strong spectral lines. Finally, the CO_2 absorption in these wings is substantially less than traditional pressure-broadened line shapes predict. Thus, upwelling radiation in the near-infrared windows is scattered very conservatively when it passes through the planet-wide cloud cover, and a significant fraction escapes to space. At wavelengths beyond 2.5 μm the refractive index of H_2SO_4 changes, and the cloud droplets begin to absorb strongly, blocking any further windows in the spectrum of the gaseous part of the atmosphere.

B. Summary of Observations

Allen's pioneering observations were obtained at the Anglo-Australian Telescope (AAT) during June and July of 1983. The original detection of nightside emission from Venus was made on 15 June, 1983, as part of a near-infrared survey of the solar system (Allen and Crawford 1984). Much better imaging was possible later that year when the planet presented only a narrow crescent but was still far enough from the Sun to be observed, a condition which occurs for only about two weeks every 19 months. Time-resolved images taken during 18–23 September 1983 showed that the near-infrared features moved from east to west, in the direction of the cloud-top superrotation, but with a period of 5.4±0.1 days. Images and spectra taken at wavelengths between 3

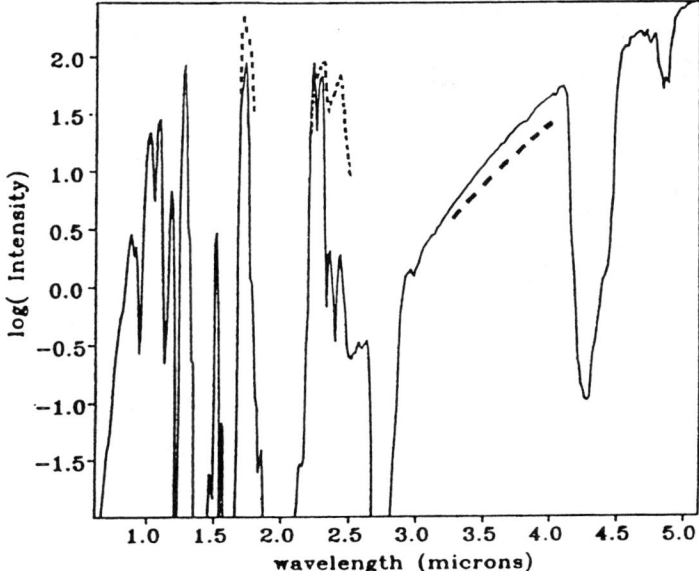

Figure 1. Early observed (dashed lines from Allen and Crawford [1984]) and modeled (solid line from Kamp et al. [1988]) spectra for the night side of Venus.

and 4.8 μm showed only the faint, nearly spatially uniform thermal emission from the cloud tops at ∼240K.

In 1990, Venus was observed from a global network of groundbased telescopes, timed to support the Galileo spacecraft, which flew past Venus on 10 February 1990. The spaceborne and groundbased observations revealed further spectral windows at 1.0, 1.10, 1.18, 1.27, and 1.31 μm, as predicted by modeling studies (Sec. II.C). The first three of these windows sound the Venusian surface and the lowest scale height of the atmosphere, while the last two probe just above this level. The spectra at these wavelengths reveal emission contrasts associated with surface temperature variations, and have been used to investigate the thermal structure in the lowest 6 km of the atmosphere (Sec. V).

The motions of the near-infrared features due to clouds have been measured by tracking them in groundbased and Galileo Near Infrared Mapping Spectrometer (NIMS) images (Fig. 2). These data allowed wind vectors, and the approximate level of formation of the features, to be inferred. They also revealed spatial variations in the spectral properties of the cloud particles, most likely due to variable size distributions (Sec. III).

Water vapor bands are detectable in the 1.10, 1.18, 1.74 and 2.3 μm windows. Near-infrared spectra also provide information on the mixing ratio, horizontal distribution, and variability, of CO, HF, HCL, OCS and SO_2 below the clouds (Sec. IV). Representative (high spatial, low spectral resolution)

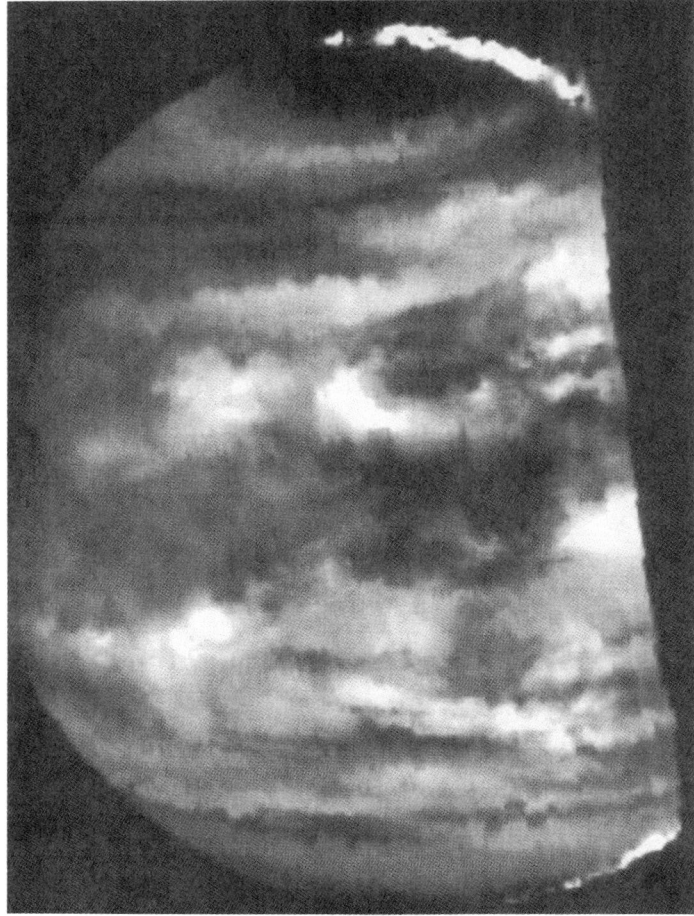

Figure 2. The dark side of Venus, mapped at a wavelength of 2.3 μm in one of the most prominent of the near-infrared windows. It shows the highly variable nature of the opacity in the main cloud deck, illuminated from behind by thermal emission from the lower atmosphere. This view was obtained by NIMS during the Galileo flyby of Venus on 10 February 1990 (Carlson et al. 1993). (A version of this image is available on the accompanying CD-ROM, figure CDP3C2F1.)

NIMS and (low spatial, high spectral resolution) groundbased spectra are presented in Figs. 3 and 4, respectively.

C. Interpretation and Models

Allen and Crawford (1984) speculated on the origin of the nightside near-infrared emission and the associated bright and dark features, but were not able to discriminate between a deep-atmosphere thermal source and a solar source. They noted that the observed high brightness temperatures at 1.74 μm

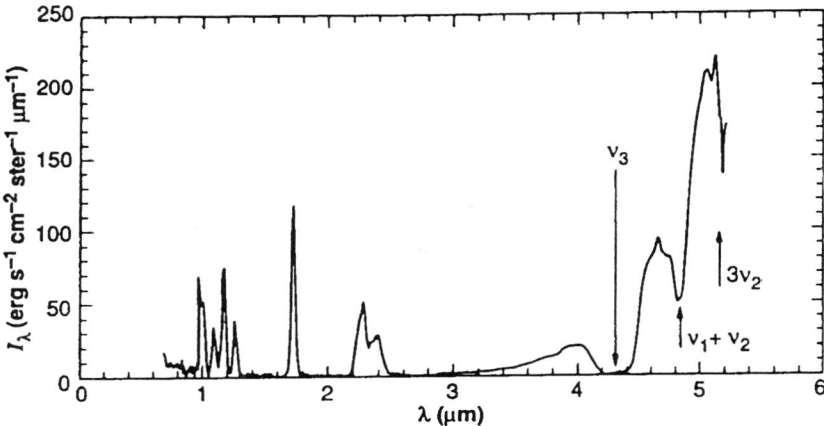

Figure 3. A representative spectrum of the dark side of Venus obtained by the NIMS instrument during the 1990 Galileo flyby (Carlson et al. 1991). The positions of some of the strongest CO_2 bands, which appear in absorption at the longer wavelengths, are indicated.

(~450 K) and the large absorber pathlengths (200 km atm) for the 2.33 μm CO band were consistent with a deep-atmosphere source. However, these observations alone did not completely preclude scattered sunlight as a source of the observed emission and contrasts. They also noted that even though the bright and dark near-infrared features superficially resembled the ultraviolet features seen at the cloud tops, their rotation periods were about 1 day longer than that of the ultraviolet features, and their contrast was much greater. These observations suggested a different level of formation for the ultraviolet and near-infrared features, but did not identify that level. Subsequent observations of the Venus night side taken during May 1985 confirmed the persistence of the near-infrared features and provided additional constraints on their origin (Allen 1987). In particular, Allen concluded that these observations clearly precluded a solar source because (1) the brightness and contrast of the near-infrared features did not decrease with distance from the terminator, (2) the equivalent width of the 2.33 μm CO band did not increase with distance from the terminator, and (3) there was no detectable polarization in the nightside near-infrared emission. It followed that the near-infrared radiation must originate as thermal emission from the deep atmosphere. Similarities between spectra of bright and dark regions on the night side also suggested that the spatial contrasts were produced as thermal radiation from the deep atmosphere escapes to space through regions of the clouds that have different optical thicknesses.

These discoveries generated great interest in the planetary atmosphere community because they provided new opportunities to study the deep atmosphere of Venus. Kamp et al. (1988) were the first to use a sophisticated

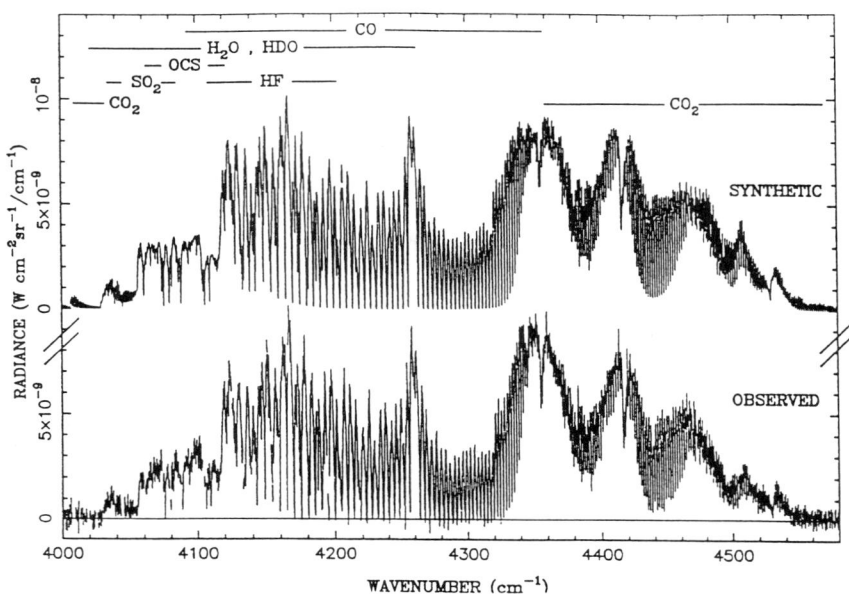

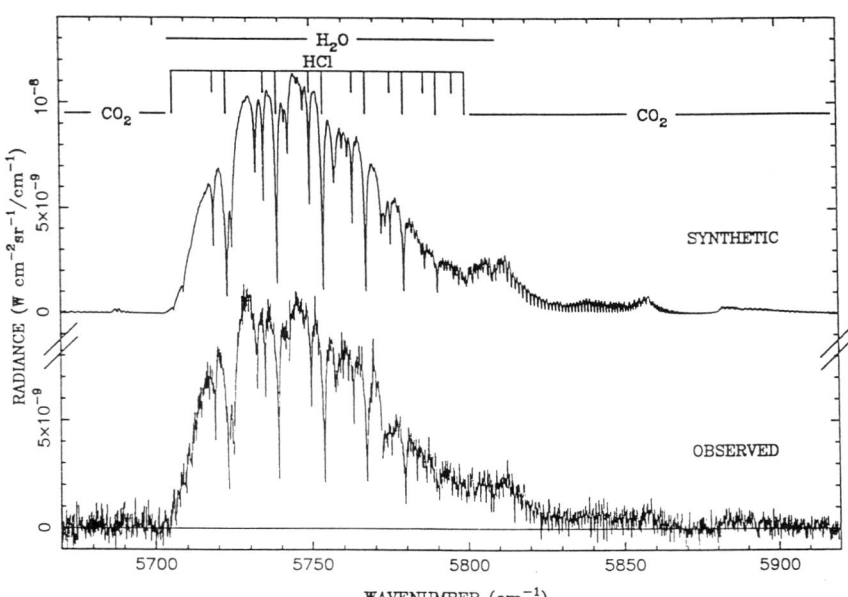

Figure 4. High-resolution groundbased spectra of the night side of Venus in the 2.3 μm and 1.7 μm windows, with the main absorption features identified (Bézard 1994).

radiative transfer model to analyze the 1983 and 1985 AAT observations. Their modeling study confirmed that the near-infrared radiation could originate as thermal emission from pressure-broadened CO_2 and H_2O lines at altitudes between 25 and 55 km, and showed that ~10% horizontal variations in the cloud optical depths could produce the observed contrasts. It was found that even an approximate fit to the observations required the absorption in the far wings of CO_2 spectral lines to be much less than that predicted by the standard pressure-broadened line profiles. In fact, the near-infrared windows on Venus would vanish altogether if Lorentzian profiles applied. Subsequent modeling work (Crisp et al. 1989; Kamp and Taylor 1990; Bézard et al. 1990) gradually improved the fits to the measured spectra, in particular by using better spectral parameters to represent CO_2 at high temperatures and pressures. The modeling results show which regions are sounded in the various windows; a summary is given in Table I. It can be seen that in general the windows probe deeper as the wavelength decreases, essentially because the molecular bands get weaker and the very strong CO_2 bands get farther apart, so that their far wings absorb less.

TABLE I
Principal Near-Infrared Windows Observed on Venus, and Estimates of the Approximate Height Ranges Probed[a]

Window Region (μm)	Approximate Depth Probed
1.01	Surface (>90%)
1.10	Surface (~60%) plus 0 to 15 km
1.18	Surface (~40%) plus 0 to 15 km
1.27	15 to 30 km
1.31	30 to 50 km
1.74	15 to 30 km
2.3	26 to 45 km

[a] These are fairly uncertain, especially the contribution from the atmosphere in the three shortest-wavelength windows.

The radiative transfer models have been used to interpret groundbased and spacecraft observations of near-infrared spectra. The key features which need to be represented are first, the very high single-scattering albedo of sulphuric acid droplets at wavelengths short of 2.5 μm, and second, the strongly sub-Lorentzian nature of the spectral line shapes for CO_2 under the conditions found on Venus (see, e.g., Brodbeck et al. 1991). The former ensures that, although photons emitted from gases in the deep atmosphere must encounter many particles in their path through the cloud layers to space, these encounters predominantly take the form of conservative scattering events and there is relatively little absorption. The fact that the CO_2 lines do not have strong wings means that the continuum absorption between the numerous bands is much weaker than a calculation based on the Lorentz pressure-broadened line shape would predict. Indeed, no complete theory for the line shape exists which predicts from first principles the small absorption in the far wings that is observed, and it is difficult to obtain the combination of high pressures and long path lengths required to study it in the laboratory. The results from Venus therefore contribute to our understanding of the forces present during the collisions of individual molecules under conditions of high temperature and pressure.

D. The Leaking Greenhouse

The near-infrared windows radiate energy from the surface and lower atmosphere directly to space, and are therefore a new factor in the debate about how the high surface temperature is maintained. It was confirmed at the time of Pioneer Venus (Tomasko et al. 1980) that the lower atmosphere of Venus is heated primarily by solar radiation propagating downwards through the clouds. The energy balance of the region depends on the subsequent removal of the energy so deposited, eventually by re-emission to space in the infrared. The high surface temperature is a conseqence of the fact that the loss process is quite inefficient overall, due to the high optical thickness of the overlying atmosphere at most wavelengths. The energy deposited in the lower atmosphere, mostly at low and mid-latitudes, is transported horizontally and vertically by advection and eventually radiated from the middle atmosphere, mostly at levels near the cloud tops (see the chapter by Crisp and Titov, for a full discussion of the energy budget of Venus' atmosphere).

This convective heat engine is bypassed to some extent by the flux emitted directly to space in the near-infrared windows. However, the windows occupy only a small fraction of the total spectrum, and fall at wavelengths where the Planck function has a small value compared to the mid-infrared, even when the relatively high temperatures of the emitting layers are taken into account. The net effect, integrated over wavelength, is that the atmosphere below about 60 km radiates very little energy directly to space. It can be estimated from the information in Fig. 1, for example, that the integrated flux escaping from Venus via the near-infrared windows is only about 0.05% of the total, enough to lower the equilibrium temperature of the surface of Venus by less than 1 K.

III. HIGH SPATIAL RESOLUTION NEAR-INFRARED MAPPING OF THE CLOUD STRUCTURE

Remote sensing in the near-infrared windows probes the properties of the lower cloud layers on Venus. Ultraviolet and thermal infrared wavelengths do not penetrate this region, so except for isolated sampling by entry probes, there was very little data on the main cloud deck and its variability until the near-infrared approach was developed. (The PV Orbiter Infrared Radiometer had a 0.4 to 4.0 μm channel, for measuring the reflected solar flux, and a narrow channel centered at 2.0 μm in a strong CO_2 band, for the detection of high clouds, but, although the former collected the flux from the near-infrared windows, this is such a small part of the total that no spatially-variable features were detected). In fact, at any given point in time the intensity emanating from Venus at wavelengths within the spectral windows varies strongly with location on the planet. The net optical thickness of the cloud deck is therefore also highly variable in space and time. This implies that the lower atmosphere is very active meteorologically, one of the most exciting, and in some ways perplexing, consequences of the discovery of the properties of the near-infrared windows, and one which promises much interesting exploitation in the future.

Most of the structure seen in the groundbased and NIMS images is believed to be in the main cloud deck at around 50 km altitude, which contains most of the cloud mass, i.e., well above the atmospheric layers which emit most of the near-infrared flux. In other words, we are seeing the cloud deck as a cold screen, back-lit by the glow from the hot atmosphere below. The remarkable variability of the cloud opacity in a single map, and the evolution with time of the features seen in successive maps, suggests high winds and vigorous convective and wave activity below the tropopause on Venus. Various features sometimes described as bars, bands, and ovals have been seen in the ground-based data, and there is a suggestion of a linear feature in the NIMS data at 45°N (see Fig. 5 of Carlson et al. 1991). However, in general the cloud morphology is sufficiently complex (cf., Fig. 2), and its origins sufficiently obscure, to discourage attempts to describe the meteorological regime at these levels on Venus in more specific terms at present.

The best spatial resolution obtained in near-infrared mapping of Venus was by the Galileo NIMS instrument, which achieved about 25 km during closest approach on 10 February 1990. These high-resolution maps showed intensity variations of about a factor 20 between the brightest and the darkest features in the 2.3 μm window. Carlson et al. (1991) examined the wavelength dependence of the cloud opacity in different regions of the disk, and found an asymmetry in the cloud types seen in the two hemispheres. Carlson et al. (1993) refined the analysis, and found 5 distinct cloud types altogether, which they interpreted as regions of distinct mixes of mode 2 and 3 particles.

Grinspoon et al. (1993) analyzed the global contrasts between NIMS observations in the 1.7, 2.3 and 3.75 μm windows, and provided more stringent constraints on the amplitude and vertical distribution of the optical depth anomalies. They found that these variations were largely confined to altitudes between 48 and 50 km, in the lower cloud, and were mainly attributable to mode 3 particles with an effective mean radius (r_{eff}) of 3.65 μm. In the upper cloud they found a best fit with ~85% of mode 2 (r_{eff} = 1 μm) and ~15% of mode 1 (r_{eff} = 0.3 μm) particles.

As the Galileo encounter with Venus lasted only one or two days, the NIMS observations give no information about the longer-term evolution of global structure in the middle and lower clouds. They have been augmented, therefore, using observations of the night side in the near-infrared windows.

Groundbased observations taken between 1988 and 1994 revealed a long-lived (>6 weeks), large-scale (zonal wavenumber 1), opaque, cloud feature at low latitudes (<40°), that extended about half-way around the planet (Crisp et al. 1989,1991b). This feature had a rotation period of 5.5±0.15 days, comparable to that of the middle cloud region. Its maximum optical thickness could not be determined precisely, because little or no near-infrared radiation escapes through it. Groundbased observations taken in January and February of 1990 show that the most transparent part of this cloud covered the night side on 10 February, when the Galileo NIMS observations were taken.

Mid-latitudes (40–60°) are usually occupied by bright zonal bands that appear to rotate with the large-scale, low-latitude feature in the middle clouds. These long-lived, partial clearings in the middle cloud might explain the greater infrared cooling detected near the cloud base by the PV North Probe Net Flux Radiometer (Revercomb et al. 1985). This cooling was originally attributed to meridional water vapor gradients, which now seem precluded by the near-infrared observations (Crisp et al. 1991a; Drossart et al. 1993).

The highest observable latitudes (>60°) are almost always dark and featureless in the near-infrared, indicating relatively high cloud opacity. These dark clouds are at the same latitudes as the cold collar detected by the Pioneer Venus OIR (Taylor et al. 1980) and groundbased mid-infrared observers (Diner et al. 1976).

IV. COMPOSITION MEASUREMENTS

A. Water Vapor and D/H Ratio

Water vapor measurements acquired by entry probes at altitudes below the clouds of Venus have produced conflicting results. The gas chromatographs aboard the Pioneer and Venera 14 probes indicated concentrations larger than 1000 ppmv near 40 km, while the PV mass spectrometer first reported a constant water concentration around 100 ppmv between 10 and 25 km, decreasing down to ~20 ppmv at the surface (Hunten et al. 1989). In a more recent reanalysis of these data, Donahue and Hodges (1992) derived an H_2O abundance close to 70 ppmv in the 10 to 25 km range, still sharply decreasing at lower altitudes. The scanning spectrophotometers on board the Venera 11 and 12 probes also indicated a strong decrease of the water abundance from ~150 ppmv at 42 km to ~20 ppmv near the surface. These results were based mainly on the analysis of the 0.94 μm H_2O band. However, Young et al. (1984) pointed out that the data from the 1.13 μm water vapor band could be interpreted with a constant H_2O mixing ratio of 20 to 30 ppmv below the clouds, while the measurements at 0.94 μm could be affected by an additional absorber.

More recent spectroscopic investigations of the nightside near-infrared emission have contributed significantly to this issue. Water vapor exhibits significant absorption in the windows centered at 2.3, 1.74, 1.18 and 1.10 μm. Because these windows probe different atmospheric regions in Venus' deep atmosphere, vertically resolved information can be retrieved on the H_2O concentration profile. The observed emission originates from deeper layers as wavelength decreases; the 2.3 to 2.5 μm window senses a region extending roughly from 26 to 45 km; the 1.7 μm emission originates between 15 and 30 km, while the 1.18 μm originates from the 5 to 15 km range. Information on the horizontal variability of the water vapor profile is also available, from spectral imaging of the night side from groundbased telescopes or, with much higher spatial resolution, from the Galileo spacecraft.

Kamp et al. (1988) and Kamp and Taylor (1990) analyzed the low-resolution ($\lambda/\Delta\lambda = 100$) 2.3 and 1.7 μm spectra of Allen (1987) using radiative transfer models. The best model fits using a sub-Lorentzian line shape, in which the line shape factor was a free parameter, suggested H_2O mixing ratios about 4 times less than the PV results, or about 25 ppmv. However, even the best calculated spectra did not match the observed spectra everywhere, mainly because the HITRAN data base they used for CO_2 was tested under terrestrial conditions, and proved inadequate to model the high temperature and pressure conditions on Venus.

In 1990, Bézard et al. published the first high-resolution ($\lambda/\Delta\lambda$ up to 20,000) spectra of the 2.3 and 1.7 μm windows recorded at the Canada-France-Hawaii Telescope (CFHT). Figure 4 shows an example covering the 2.3 μm window, in which features due to CO, COS, SO_2, CO_2, H_2O, HDO, and HF are clearly seen as distinct spectral features. Bézard et al. obtained a satisfactory fit to the observations by using a more complete description of the CO_2 spectrum than before, and by including an additional continuum opacity, likely to be due to far wings of strong CO_2 bands lying outside the windows. A sub-cloud H_2O abundance around 40 ppmv was derived by fitting the H_2O absorption lines. More recently, de Bergh et al. (1995) presented high-resolution spectra of the 2.3, 1.74, and 1.18 μm windows recorded in June 1991 at the 1991 at the CFHT over a region centered at 15°S latitude on the night side. All three windows were best fitted with a 30±10 ppmv (±15 ppmv at 1.18 μm) mixing ratio, suggesting that the H_2O profile is essentially constant from the cloud decks to the lowest scale height of the atmosphere. No measureable horizontal variation in the H_2O abundance could be found in a set of 2.3 μm spectra recorded at various locations on the night side, from 15°S to 30°N latitude.

de Bergh et al. (1991) took advantage of the presence of both H_2O and HDO lines in the 2.3 μm window to obtain the relative abundances of both species beneath the clouds. They obtained a value for the D/H ratio equal to 120±40 times the telluric value (Fig. 5). Such a large proportion of deuterium in Venus relative to Earth was first proposed as a result of the analysis of PV orbiter and probe data (discussed in the chapter by Donahue et al.), and more recently confirmed by airborne observations of H_2O and HDO lines near 2.6 μm probing the Venus cloud tops (Bjoraker et al. 1992). The enrichment probably results from photodissociation of normal and heavy water molecules in the high atmosphere, and subsequent mass-selective hydrogen escape. Whether it constitutes evidence for a lost primordial ocean, or just the steady state between loss by escape and supply by comets or volcanic outgassing, remains uncertain due to limitations in our knowledge of the escape flux of hydrogen, the cometary influx of water, and the outgassing rate of water from the surface.

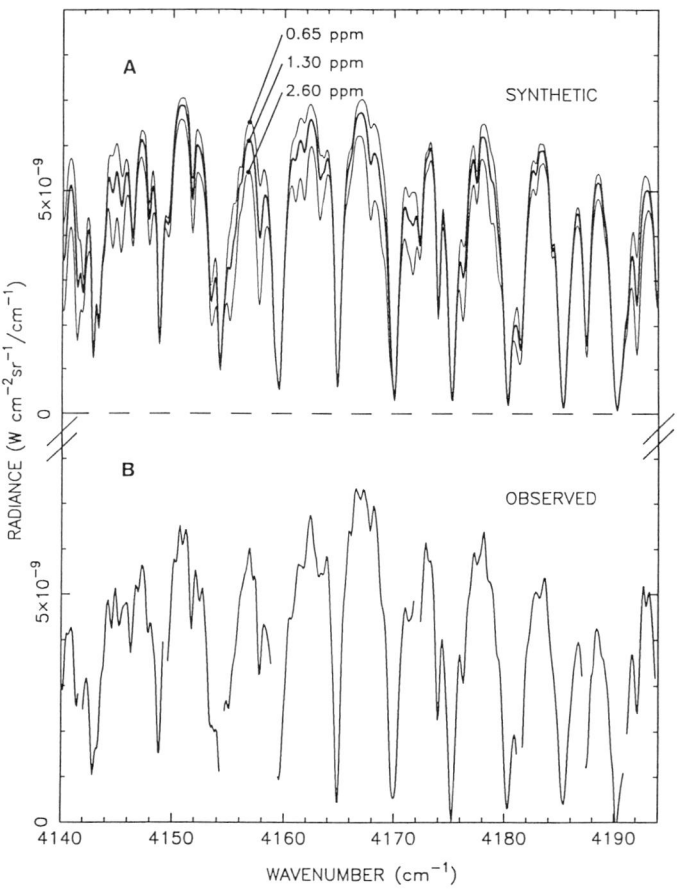

Figure 5. Groundbased spectra of Venus showing the sensitivity to the mixing ratio of HDO. The best fit between computed and measured spectra is for 1.30 ppmv of HDO and 34 ppmv of H_2O (figure after de Bergh et al. 1991).

On 29 January 1990, Bell et al. (1991) used the Cooled Grating Array Spectrograph at the NASA Infrared Telescope Facility to record moderate-resolution ($\lambda/\Delta\lambda = 1500$) spectra of dark and bright near-infrared features on the Venusian night side. These spectra indicated very different H_2O abundances according to the brightness of the region (Fig. 6). Mixing ratios near 40 ppmv were derived from the dark spot spectra, while the spectra of an anomalously bright spot indicated H_2O mixing ratios as high as 200 ppmv. The experimenters suggested that the bright near-infrared emission and the factor of 5 increase in H_2O mixing ratios might be due to the subsidence and evaporation of a significant faction of the H_2SO_4 droplets in the middle and lower cloud decks, with subsequent thermal dissociation of the H_2SO_4 vapor into H_2O and SO_3. About 10 days after these observations, on 9 February 1990, narrowband filter images taken within the 2.3 μm window with the Palomar 5-m prime focus near-infrared camera observed the bright spot seen by Bell et al., but, like all subsequent measurements, found no significant horizontal variation in H_2O.

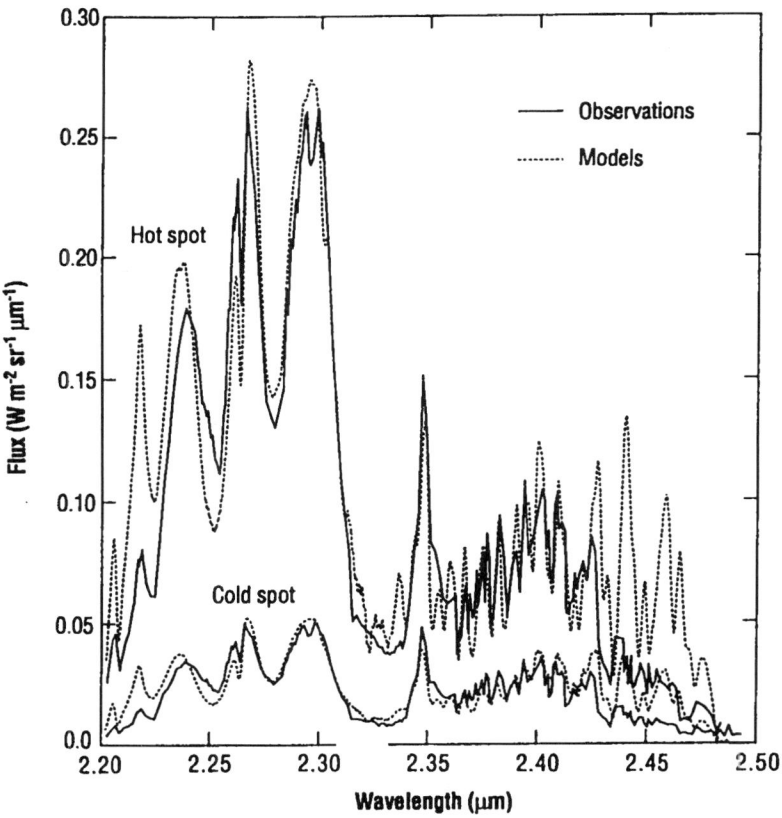

Figure 6. Spectra of hot and cold spots on Venus, with model fits (Bell et al. 1991). The amount of water vapor required to achieve a good fit was substantially different for the two regions.

An analysis of 2.3 and 1.27 μm spectra with resolutions up to ~2000 recorded in 1990 at the Anglo-Australian Telescope (AAT) indicated H_2O mixing ratios of 40±20 ppm (Crisp et al. 1991a). Images of the night side using a set of narrow filters within the 2.3 μm window indicated no significant horizontal variation in H_2O. More recently, Pollack et al. (1993) presented a more complete analysis of AAT spectra including all windows from 1.1 to 2.3 μm. They also made use of Bézard et al.'s (1990) CFHT spectra at higher spectral resolution to better constrain the atmospheric model. They deduced that the H_2O mixing ratio has a constant value of 30±10 ppmv in the altitude range from 10 to 40 km.

A preliminary analysis of the Galileo NIMS 2.3 and 1.7 μm window spectra (Carlson et al. 1991) yielded H_2O mixing ratios around 25 and 50 ppmv respectively, but with a larger uncertainty than the groundbased data due to the limited spectral resolution (0.026 μm). The data at 1.18 μm were later analyzed to search for possible horizontal variations of water vapor in the 5 to 15 km range (Drossart et al. 1993). A best fit to the spectrum was obtained with a water vapor abundance similar to that inferred from groundbased observations (30±15 ppmv), and no horizontal variation in excess of 20% was found in the dataset, which covered a band of longitudes between latitudes of 60°S and 64°N at a spatial resolution up to 25 km.

Measurements acquired by the PV Large Probe Neutral Mass Spectrometer (LNMS) indicate that the water mixing ratios decrease substantially between 10 km and the surface (Donahue and Hodges 1992), a result which has important implications for the presence of sources or sinks of water near the surface. The vertical distribution of H_2O in this region of the atmosphere cannot be retrieved from individual near-infrared spectra because the contribution function is too broad (~20 km, Pollack et al. 1993). However, it can be derived from spatially resolved spectra, because the column-integrated H_2O optical depths in adjacent high and low surface elevation regions can be subtracted to yield the temperature lapse rate and the H_2O column abundance between the two levels. This experiment was attempted using spatially-resolved low spectral resolution ($\lambda/\Delta\lambda \sim 400$) spectra of the night side collected with the Infrared Imaging Spectrometer on the AAT in July 1991 (Meadows and Crisp 1996). The best fit to the data was obtained with a water vapor mixing ratio of ~20 ppmv at the cloud base, increasing to 45(+10,−15) ppmv at the 10 bar level then remaining constant at 45 ppmv down to the surface. These results appear to preclude any large decrease in the H_2O mixing ratios near the surface, contrary to the PV LNMS observations.

The lack of evidence of any significant concentration gradient below the 10 bar level (~30 km) also contrasts with the predictions of thermochemical calculations (Lewis and Grinspoon 1990). In their model, H_2O reacted with CO_2 in the lower atmosphere to form carbonic acid gas (H_2CO_3), yielding an H_2O vertical profile roughly consistent with the Venera spectrophotometric measurements at 0.94 μm. However, the water profiles from nightside spectroscopic observations indicate that this conversion to H_2CO_3 probably does not occur. The low abundance of H_2O also has some important implications for the stability of hydrated silicates on the surface of Venus, as discussed in the chapter by Fegley et al.

In summary, the net evidence based on remote sensing observations suggests that the atmosphere of Venus is extremely dry, on the average at least, much drier than indicated by most of the *in-situ* probe measurements. The water vapor mixing ratio appears to be approximately constant both horizontally and vertically, in the lowest 30 km at least, with a value between 20 and 50 ppmv.

B. Carbon Monoxide

The (2-0) band of CO, visible in the early observations of Allen and Crawford (1984), is a prominent feature located around 2.35 μm near the center of the 2.3-μm window. First analyses of these low resolution observations (Kamp et al. 1988; Kamp and Taylor 1990) seemed to indicate CO abundances considerably lower than those measured by the PV gas chromatograph (which were 30 ppmv at 52 and 42 km; 20 ppmv at 22 km, according to Oyama et al. [1980]). However, as for water vapor, these values had uncertainties originating in the poor state of the spectral line data base on weak CO_2 bands under Venusian conditions. Using an improved representation of the CO_2 opacity, Bézard et al. (1990) fitted their high-resolution CFHT spectra with a CO profile around 1.5 time the PV profile. A more recent attempt to reproduce these observations using a still more sophisticated data base for CO_2 led to CO abundances similar to the PV measurements (Bézard 1994).

Pollack et al. (1993) analyzed the AAT spectra of Crisp et al. (1991a), deriving a CO mixing ratio equal to 23±5 ppmv at 36 km. They were also able to infer the CO vertical gradient around this level by simultaneously fitting the P- and R-branches of the CO band. They found that the concentration declines by a factor of 1.5 between 32 and 40 km, a trend in agreement with the PV profile. Further implications of these findings are discussed in Sec. IV.D.

A preliminary modeling of the CO absorption feature in the Galileo NIMS spectra of the 2.3 μm window estimated a mixing ratio around 50 ppmv near 30 km (Carlson et al. 1991). A more comprehensive analysis of the NIMS spectra by Collard et al. (1993) reduced this to around 30 ppmv, and detected a definite latitudinal trend (Fig. 7). Investigating the whole set of ~500 spectra spanning latitudes between 60°S and 70°N, they found that the depth of the CO absorption band was significantly larger northward of ~47°N. This corresponded to a CO mixing ratio which was 20 to 50% higher in the northern polar region than the value of 1.17±0.1 times the PV profile found at lower latitudes in either hemisphere. It was not possible to say whether a similar enhancement existed in the southern polar region, because the trajectory of Galileo favored observations of the northern hemisphere of Venus and high-latitude data was not obtained in the south. High latitudes are also inaccessible to Earth-based observers due to the small inclination of Venus' axis of rotation. Only new observations from future high-inclination spacecraft can confirm the existence of the northern polar CO enhancement, see if the equivalent is present in the southern hemisphere, and observe how they vary with time. It should then be possible to say how the anomaly arises. The various possibilities include localized volcanic activity, or (more probably) transport of CO from the upper atmosphere in the polar vortices, with major implications for the general circulation of the whole atmosphere (Taylor 1995).

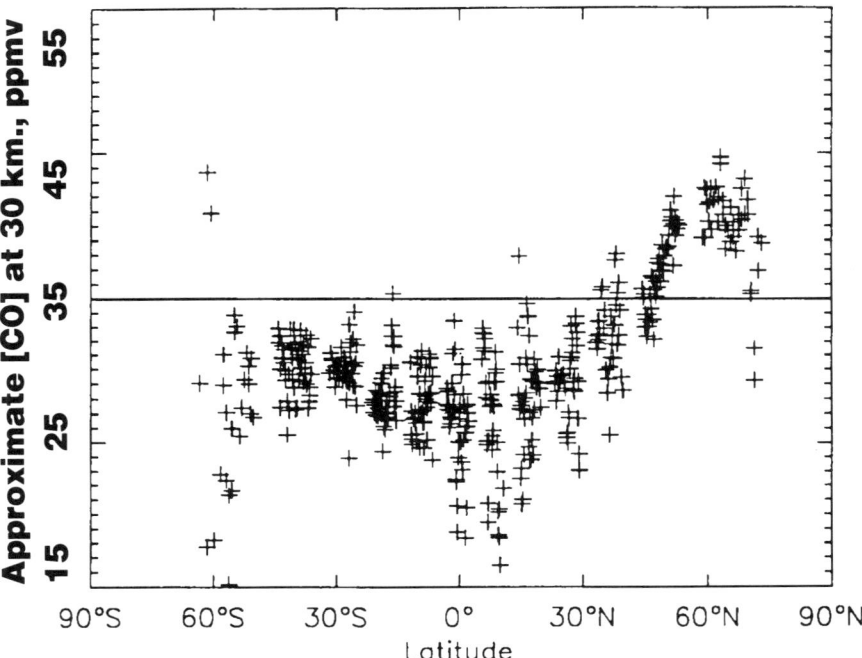

Figure 7. The distribution of carbon monoxide with latitude, inferred from the 0-2 band observed near 2.3 μm by Galileo-NIMS (after Collard et al. [1993] and Taylor [1995]). The abundance scale refers to the mean mixing ratio in a broad layer about 20 km thick centered on about 40 km altitude.

C. Hydrogen Halides

HCl and HF were detected originally at the cloud tops of Venus from high-resolution near-infrared spectroscopy of reflected light from the day side. Connes et al. (1967) derived mixing ratios of 0.6±0.1 ppmv for HCl and $5.0^{+5.0}_{-2.5}$ ppbv for HF. More recent observations near 3.6 μm by de Bergh et al. (1989) indicated ~0.4 ppmv HCl near 70 km altitude. These abundances were below the sensitivity levels of the instruments on the Pioneer and Venera probes.

Spectroscopic analyses of the nightside near-infrared emission provided the first measurements of the HCl and HF mixing ratios below the clouds. Bézard et al. (1990) detected a few HF lines in their high-resolution CFHT spectra at 2.3 μm, and several lines from the ^{35}Cl and ^{37}Cl isotopes of HCl in the 1.7 μm window. They derived mixing ratios of ~0.5 ppmv for HCl (pertaining to the 15–30 km altitude range), and ~4.5 ppbv for HF in the 30 to 40 km region. Pollack et al. (1993) inferred concentrations of 0.48±0.12 HCl, and between 1 and 5 ppbv for HF, using lower resolution AAT spectra.

The HCl and HF mixing ratios below the clouds thus appear similar to those measured at the cloud tops 20 years before. This constancy suggests that the observed abundances are equilibrium values rather than controlled by episodic volcanism. Fegley and Treiman (1992) emphasised that these reactive gases are likely to be in equilibrium with the surface mineralogy, and described different possible assemblages, involving alkaline rocks, capable of buffering their abundances (see the chapter by Fegley et al.).

D. Sulfur-Bearing Compounds

Carbonyl sulfide (OCS) was first unambiguously detected by Bézard et al. (1990) in their 2.3 μm CFHT spectra of the night side of Venus. They derived a mixing ratio of about 0.25 ppmv under the assumption that it was uniformly mixed below the clouds. Pollack et al. (1993) showed that it is possible to infer the vertical gradient of the OCS concentration from the shape of the absorption feature at the OCS band head near 2.43 μm (Fig. 8). Fitting Crisp et al.'s (1991a) AAT spectra, they found that the OCS mixing ratio increases sharply with decreasing altitude. A mixing ratio equal to 4.4±1 ppmv was derived for the 33-km level, increasing to some 20 to 40 ppmv at 28 km. A re-analysis of the high-resolution CFHT spectra, allowing for a vertical gradient in the OCS profile, gave consistent results: a mixing ratio of 0.35±0.10 ppmv at 38 km from the absorption at the band head at 2.43 μm, and an increase to 10 to 20 ppmv at and below 30 km to reproduce the low-wavelength wing of the band (Bézard 1994).

As noted by Pollack et al. (1993), there is some evidence that the decrease in OCS mixing ratio by about 10 ppmv between 30 and 38 km is matched by an increase in CO of about the same amount. The retrieved CO and OCS profiles are in good agreement with recent chemical kinetic modeling by Krasnopolsky and Pollack (1994) in which OCS is destroyed by reaction with SO_3 to yield CO above ~33 km. If the OCS mixing ratio is constant between ~28 km and the surface, this would be consistent with the equilibrium abundance calculated from pyrite chemical weathering, suggesting that iron sulfides are stable on Venus' surface (Fegley and Treiman 1992).

The SO_2 abundance beneath the cloud decks was first measured by the gas chromatographs carried by the PV and Venera 11 and 12 probes in 1978. The PV data yielded 185±43 ppmv at 22 km altitude, while a mixing ratio of 130±35 ppmv was derived below 42 km from the Venera measurements (von Zahn et al. 1983). More recently, an analysis of the ultraviolet spectra recorded by the Vega 1 and 2 probes in June 1985 yielded a SO_2 profile strongly decreasing downwards below the cloud decks and reaching 20 to 25 ppmv at 12 km (Bertaux et al. 1996; chapter by Fegley et al.).

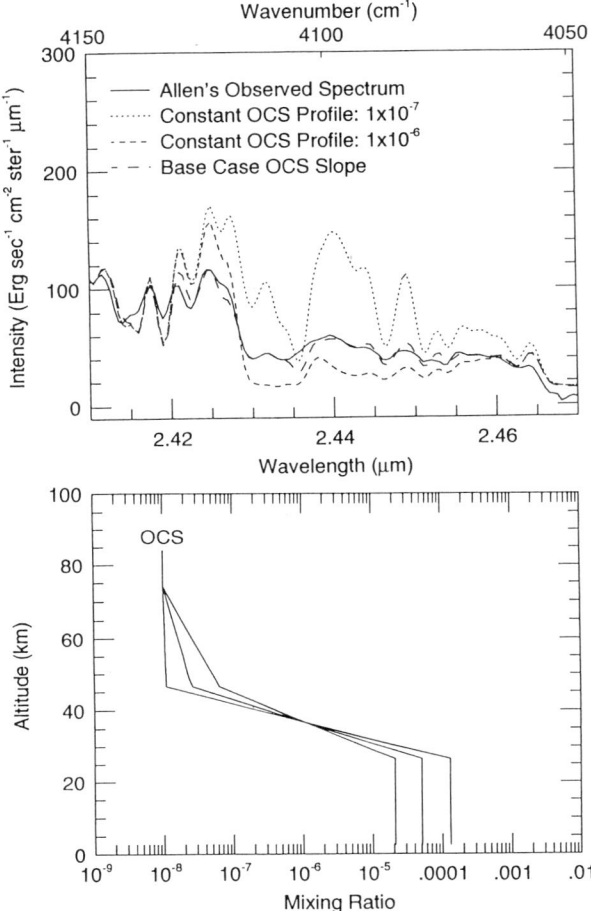

Figure 8. (Top) Observed and measured spectra in the 2.3 μm window showing features due to carbonyl sulfide (OCS). (Bottom) The range of OCS vertical profiles which provide a possible fit to the spectra (figure from Pollack et al. 1993).

Absorption by the weak $3\nu_3$ band of SO_2 near 2.45 μm was detected in CFHT spectra of the night side by Bézard et al. (1993). They derived an average mixing ratio of 130±40 ppmv over the 35 to 45 km region (Fig. 9). No spatial variation was detected over a set of 4 spectra recorded in 1989 and 1991 at different latitudes between 15°S and 30°N. Consistent with this, Pollack et al. (1993) derived a SO_2 mixing ratio of 180±70 ppmv using AAT spectra recorded in February 1990.

Thus, within the uncertainties, the deep-atmosphere SO_2 mixing ratios derived from nightside spectra are the same as those measured by the Venera, PV and probes from 1978 to 1985. This stability contrasts with the apparent

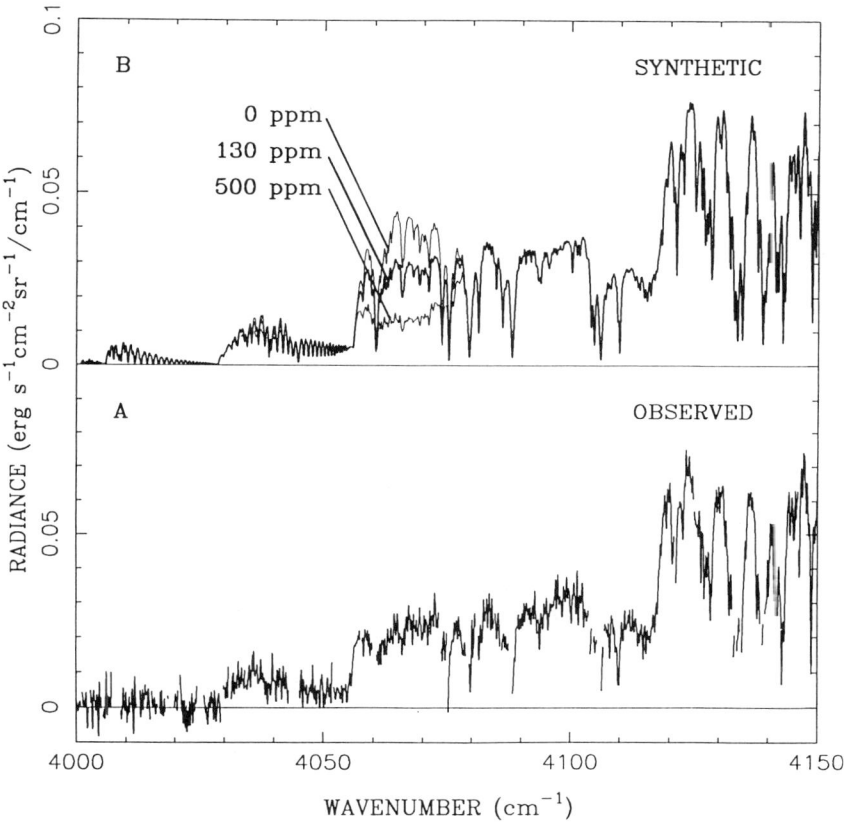

Figure 9. Observed and synthetic spectra for a region near 2.45 μm containing the $3\nu_3$ band of SO_2, showing the best fit value of 130 ppmv and the effect on the spectrum of varying this from 0 to 500 ppmv (figure from Bézard et al. 1993).

massive decrease observed at the cloud tops from ultraviolet spectroscopy over the period 1978–1988 (chapter by Esposito et al.). Because it is not matched by changes in the lower atmosphere, this decrease may indicate long-term changes in the circulation of the upper atmosphere, affecting the composition there, rather than a manifestation of a major volcanic eruption as originally proposed by Esposito (1984). The constant values observed below the cloud decks are consistent with the extremely low reaction rates of SO_2 with calcium-bearing minerals at the surface, which imply a removal time constant larger than 10^6 yr (Fegley and Treiman 1992).

E. Summary of Current Knowledge of the Composition

Table II provides current best estimates, with realistic uncertainties, of the mean mixing ratios of those gases which have been identified to date in nightside spectra of Venus.

V. DYNAMICS

At levels within the clouds (47–70 km), the Venusian atmosphere rotates westward 50 to 60 times faster than the solid body of the planet, driven by mechanisms which remain mysterious after more than two decades of detailed study (chapter by Gierasch et al.). A comprehensive description of the winds at the cloud-top levels where this super-rotation is most intense has been derived by tracking features in Mariner 10, PV and Galileo ultraviolet images, but they contribute little insight into the processes that must transport momentum from the planet's surface to these altitudes.

The winds below the cloud tops were obtained by tracking the PV, Venera, and Vega entry probes as they descended through the atmosphere. These measurements show that the winds blow from east to west at all levels, with amplitudes that decrease almost monotonically with altitude. The descent probe measurements are again too limited in spatial and temporal sampling to discriminate between atmospheric waves and the time- and space-averaged motions. They also lack the accuracy needed (± 1 m s^{-1}) to describe the much smaller zonal winds in the lowest scale height (~ 16 km), or the meridional winds at any level, well enough to understand the role which momentum transport by the meridional circulation and vertically propagating waves may play in the maintenance of the super-rotation.

The deep-atmosphere probe afforded by near-infrared spectroscopy offers the possibility of a more comprehensive global description of dynamics in the middle and lower cloud regions to complement the probe and balloon data. Winds derived from near-infrared feature rotation periods have been acquired during the each of the Venus inferior conjunctions since the discovery of the features in 1983. Allen and Crawford (1984) found that the night-side near-infrared features at wavelengths near 1.74 and 2.3 μm moved in the same direction as the cloud-top super-rotation, but with a somewhat longer period (~ 5.5 days). Subsequently, simulations with radiative transfer models showed that these features are associated with optical depth variations within the middle (49–57 km) and lower (47–49 km) clouds. During the observing campaign conducted to support the Galileo flyby before and after the January 1990 inferior conjunction, the markings were tracked to produce a global, two-dimensional description of horizontal winds in both of these cloud decks (Crisp et al. 1991b). The middle cloud was dominated by a long-lived wavenumber 1 feature, with a rotation period of 5.5$\pm$0.15 days, indicating equatorial velocities near 80 m s^{-1}. The lower cloud was characterized by short-lived small-scale features that had rotation periods of 7.4$\pm$1 days. The meridional structure of the zonal motions of both cloud decks was indistinguishable from solid-body rotation at latitudes equatorward of 60 deg. No systematic north–south motions were observed in the 6-week-long groundbased feature tracking experiments, indicating an upper limit of about 0.5 m s^{-1} on mean meridional winds in the middle cloud, and about 7 m s^{-1} for those in the lower cloud.

TABLE II
Recommended Values for Mixing Ratio of Minor Gases Measured from Spectroscopic Investigations of the Nightside Emission

Gas	Mixing Ratio (ppmv)	Altitude (km)	Reference[a]
H_2O	30±10	26–45	(1,2)
	30±10	15–30	(1,2)
	30±15	0–15	(1,3)
D/H	120±40⊕	26–45	(4)
CO	23±7	30	(2,5)
	29±7	40	(2,5)
HF	0.005±0.002	30–40	(5)
HCl	0.5±0.15	15–30	(5)
OCS	14±6	30	(2,5)
	0.35±0.1	38	(2,5)
SO_2	130±40	35–45	(6)

[a] The references are: (1) de Bergh et al. (1995); (2): Pollack et al. (1993); (3) Meadows and Crisp (1996); (4) de Bergh et al. (1991); (5) Bézard (1994); (6) Bézard et al. (1993). The symbol ⊕ refers to the terrestrial value.

The two Galileo NIMS images acquired during the 10 February 1990 flyby have also been used to infer wind velocities in the northern hemisphere (Carlson et al. 1991). These observations provide much more stringent constraints on small near-infrared feature displacements because their resolution is about a factor of 10 better than that of the best groundbased results. In contrast to the (almost contemporaneous) groundbased observations, which indicate solid-body rotation, the NIMS results appear to show an increase in the westward wind velocity with latitude. Crisp et al. (1991b) proposed that this apparent increase may be a consequence of the meridional variations in the altitude of near-infrared feature formation, rather than an actual meridional wind gradient. At the time of the NIMS observations, the lower latitudes were dominated by features in the more slowly moving lower cloud layer, while the high latitude regions were dominated by features in the faster-moving upper cloud region. The NIMS observations also revealed slow poleward motions in both hemispheres, of the order of a few m s^{-1}, but with error bars of the same magnitude.

VI. THE SURFACE AND THE NEAR-SURFACE ATMOSPHERE

In the very short wavelength windows, at 1.0, 1.1, and 1.18 μm, the total column opacity of the atmosphere is low enough that thermal emission from the surface can be detected. The surface contributes ~60% of the emission in the 1.18-μm and ~40% in the 1.1-μm windows; and more than 95% within the 1.0-μm window (Meadows and Crisp 1996). Maps made at these wavelengths reveal the topography with fairly high contrast, because the higher features are cooler and the rate of change of the Planck function with temperature is large at 1 μm and 730 K. For example, the highest feature, Maxwell Montes, which was also the first detected in this way (Carlson et al. 1991), is about 10 km higher, and therefore about 80 K cooler, than its surroundings. The radiance contrast at 1 μm, assuming the same emissivity, is more than an order of magnitude. Surface features such as Beta Regio, Phoebe Regio, and Ulfrum Mons are 20 to 50% darker than the surrounding plains because they are up to 45 K cooler. Contrasts much smaller than this are detectable in NIMS images of Venus, and the cooler features observed in the near-visible infrared windows correlate well with regions of elevated topography found in maps made by the PV and Magellan radars.

Surface features have since been detected in groundbased near-infrared observations with a maximum spatial resolution of about 100 km horizontally, but as small as 1 km vertically (Meadows et al. 1992; Lecacheux et al. 1993; Meadows 1994; Meadows and Crisp 1996). Meadows (1994) and Meadows and Crisp (1996) used AAT/IRIS image cubes taken in July 1991 to provide new constraints on the atmospheric temperature lapse rates near the surface. Their analysis used simultaneous observations taken within the 1.31 and 1.18-μm windows to correct for intensity differences associated with horizontal variations in the cloud optical depths. Synthetic spectra were calculated for a range of topographic elevations and combined with PV altimetry data to produce spatially resolved maps of the near-infrared thermal emission from the night side. Comparisons between these synthetic radiance maps and the IRIS observations indicated temperature lapse rates between -7 and -7.5 K km^{-1} in the lowest $\sim$6 km of the atmosphere, significantly smaller than the values of -8 to -8.5 K km^{-1} inferred from earlier measurements and greenhouse models.

VII. MAJOR OPEN QUESTIONS AND FUTURE WORK

An obvious goal is for future spacecraft missions to Venus is to exploit the remarkable properties of the near-infrared windows to study the meteorology of the deep atmosphere at much higher spatial and time resolution than is possible from the Earth. Instruments on high-inclination orbiting spacecraft could provide good coverage of the polar regions, where the atmospheric behavior is known to be markedly different from the lower-latitude regions accessible to Earth-based observers, and seems to have a key role in the atmospheric general circulation on Venus. In particular, it is fascinating to imagine how the polar vortices might appear in movies constructed from near-infrared images. Investigations of the deep structure around the poles may shed light on the complex dipole and cold polar collar features discovered by Pioneer Venus (Taylor et al. 1980).

Insight into the zonal, meridional and vertical motions at various levels, and meteorological activity on Venus in general, can be obtained from studies of the character and morphology of the features at different near-infrared window wavelengths, particularly if high spatial resolution and long-term coverage are available. The behavior of the major features seen in ground-based images suggests planetary-scale waves, while the general appearance of the clouds and the large opacity contrasts indicate large-scale convective cells at many locations (cf., Fig. 2). The middle latitudes of both hemispheres are often occupied by bright, quasi-zonal bands (Crisp et al. 1991b) that may be evidence for large-scale subsidence. If this is the case, they may reveal under closer study from spacecraft the poleward extent of the low-latitude Hadley cell, with important implications for models of the general circulation and processes affecting the maintenance of the cloud-level super-rotation.

In principle, it should be possible to derive a more global description of the near-surface winds from the temperature and pressure fields, but this is not currently possible because, despite various attempts, these fields are as poorly constrained as the winds themselves. The Magellan orbiter obtained passive microwave observations of the surface, but these could not yield a global description of the temperature field because of uncertainties in the absolute radiometric calibration of its antenna (Pettengill et al. 1992). Both approaches offer the possibility, in the future, for temperature mapping of the surface and lower troposphere with sufficient precision to shed light on the circulation at these depths.

High-resolution near-infrared spectroscopy can be used to obtain information on the species known or suspected to be variable in space and/or time, for example carbon monoxide, sulfur dioxide, and water vapor. A detailed study of these will lead to a better understanding of the coupling of dynamics and chemistry in the cloud layers and near the surface. Monitoring trace gases like OCS, SO_2, H_2O, or CH_4 at 2.3 μm is also a way of looking for volcanic activity. Again, these measurements would be most effectively conducted from an orbiting spacecraft.

REFERENCES

Allen, D. 1987. The dark side of Venus. *Icarus* 69:221–229.
Allen, D., and Crawford, J. 1984. Discovery of cloud structure on the dark side of Venus. *Nature* 307:222–224.
Avduevsky, V. S., et al. 1983. Structure and parameters of the Venus atmosphere according to Venera Probe data. In *Venus*, eds D. M. Hunten, L. Colin, T. M. Donahue and V. I. Moroz (Tucson: Univ. of Arizona Press), pp. 280–298.
Bell, J., III, et al. 1991. Spectroscopic observations of bright and dark emission features on the night side of Venus. *Science* 252:1293–1296.
Bertaux, J. L., et al. 1996. Vega-1 and Vega-2 entry probes: an investigation of local UV absorption in the atmosphere of Venus. *J. Geophys. Res.* 101:12709–12745.
Bézard, B. 1994. The deep atmosphere of Venus probed by near-infrared spectroscopy of the night side. Communication at the 30th COSPAR Scientific Assembly (Session C3.1), 11–21 July, Hamburg, Germany.

Bézard, B., de Bergh, C., Crisp. D., and Maillard, J. P. 1990. The deep atmosphere of Venus revealed by high-resolution nightside spectra. *Nature* 345:508–511.

Bézard, B., et al. 1993. The abundance of sulphur dioxide below the clouds of Venus. *Geophys. Res. Lett.* 20:1587–1590.

Bjoraker, G. L., et al. 1992. Airborne observations of the gas composition of Venus above the cloud tops: Measurements of H_2O, HDO, HF and the D/H and $^{18}O/^{16}O$ isotopic ratios. *Bull. Amer. Astron. Soc.* 24:995 (abstract).

Brodbeck, C., et al. 1991. Measurements of pure CO_2 absorption at high densities near 2.3 μm. *J. Geophys. Res.* 96:17497–17500.

Carlson, R. W., and Taylor, F. W. 1993. The Galileo Encounter with Venus: Results from the Near Infrared Mapping Spectrometer. *Planet. Space Sci.* 41(7):475–476.

Carlson, R. W., et al. 1991. Galileo infrared imaging spectroscopy measurements at Venus. *Science* 253:1541–1548.

Carlson, R. W., et al. 1993. Distinct Venus cloud types as observed by the Galileo Near Infrared Mapping Spectrometer. *Planet. Space Sci.* 41(7):477–486.

Collard, A. D., et al. 1993. Latitudinal distribution of carbon monoxide in the deep atmosphere of Venus. *Planet. Space Sci.* 41(7):487–494.

Connes, P., Connes, J., Benedict, W. S., and Kaplan, L. 1967. Traces of HCl and HF in the atmosphere of Venus. *Astrophys. J.* 147:1230–1237.

Crisp, D., et al. 1989. The nature of the features on the Venus night side. *Science* 246:506–509.

Crisp, D., Allen, D. A., Grinspoon, D., and Pollack, J. B. 1991*a*. The dark side of Venus: Near infrared images and spectra from the Anglo-Australian Observatory. *Science* 253:1263–1266.

Crisp, D., et al. 1991*b*. Ground-based near-infrared imaging observations of Venus during the Galileo encounter. *Science* 253:1538–1541.

de Bergh, C., et al. 1989. Ground-based high resolution spectroscopy of Venus near 3.6 microns. *Bull. Amer. Astron. Soc.* 21:926 (abstract).

de Bergh, C., et al. 1991. Deuterium on Venus: Observations from Earth. *Science* 251:547–549.

de Bergh, C., et al. 1995. Water in the deep atmosphere of Venus from high-resolution spectra of the night side. *Adv. Space Res.* 15(4):79–88.

Diner, D. J., Westphal, J. A., and Schloerb F. P. 1976. Infrared imaging of Venus: 8–14 micrometers *Icarus* 27:191–195.

Donahue, T. M., Hoffman, J. H., Hodges, R. R., Jr., and Watson, A. J. 1982. Venus was wet: A measurement of the ratio of deuterium to hydrogen. *Science* 216:630–633.

Drossart, P., et al. 1993. Search for spatial variations of the H_2O abundance in the lower atmosphere of Venus from NIMS-Galileo. *Planet. Space Sci.* 41(7):495–504.

Esposito, L. W. 1984. Sulfur dioxide: episodic injection shows evidence for active Venus volcanism. *Science* 223:1072–1074.

Fegley, B., Jr., and Treiman, A. H. 1992. Chemistry of atmosphere-surface interactions on Venus and Mars. In *Venus and Mars: Atmospheres, Ionospheres, and Solar Wind Interaction*, eds. J. Luhmann, M. Tatrallyay and and R. O. Pepin, AGU Geophysical Mono. 66, pp. 7–72.

Gierasch, P. J. 1987. Waves in the atmosphere of Venus. *Nature* 328:510–512.

Grinspoon, D. H., et al. 1993. Probing Venus's cloud structure with Galileo NIMS. *Planet. Space Sci.* 41(7):515–542.

Hunten, D. M., Colin, L., Donahue, T. M., and Moroz, V. I., eds. 1983. *Venus* (Tucson: Univ. of Arizona Press).

Hunten, D. M., Donahue, T. M., Walker, J. C. G., and Kasting, J. F. 1989. Escape of atmospheres and loss of water. In *Origin and Evolution of Planetary and Satellite Atmospheres*, eds. S. K. Atreya, J. B. Pollack and M. S. Matthews (Tucson: Univ.

of Arizona Press), pp. 386–422.
Janssen, M. A., and Klein, M. J. 1981. Constraints on the composition of the Venus atmosphere from microwave measurements near 1.35 cm wavelength. *Icarus* 46:58–69.
Janssen, M. A., Hills, R. E., Thornton, D. D., and Welch, W. J. 1973. Venus: New microwave measurements show no atmospheric water vapor. *Science* 179:994–997.
Kamp, L. W., and Taylor, F. W. 1990. Radiative transfer models of the night side of Venus. *Icarus* 86:510–529.
Kamp, L. W., Taylor, F. W., and Calcutt, S. B. 1988. Structure of Venus' atmosphere from modelling of night side infrared spectra. *Nature* 336:360–362.
Krasnopolsky, V. A., and Pollack, J. B. 1994. H_2O-H_2SO_4 system in Venus' clouds and OCS, CO, and H_2SO_4 profiles in Venus' troposphere. *Icarus* 109:58–78.
Lecacheux, J., et al. 1993. Detection of the surface of Venus at 1.0 μm from ground-based observations. *Planet. Space Sci.* 41:543–549.
Lewis, J., and Grinspoon, D. H. 1990. Vertical distribution of water in the atmosphere of Venus: A simple thermochemical explanation. *Science* 249:1273–1275.
Meadows, V. S. 1994. Infrared Observations with IRIS: From Planets to Galaxies. Ph.D. Thesis, Univ. of Sydney.
Meadows, V., and Crisp, D. 1996. Ground-based near-infrared observations of the Venus night side: The thermal structure and water abundance near the surface. *J. Geophys. Res.* 101:4595–4622.
Meadows, V., Crisp, D., and Allen, D. A. 1992. Ground-based near-IR observations of the surface of Venus. In *Intl. Colloquium on Venus*, LPI Contribution No. 789, pp. 70–71.
Moroz, V. I. 1983. Summary of the preliminary results of the Venera 13 and 14 missions. In *Venus*, eds. D. M. Hunten, L. Colin, T. M. Donahue and V. I. Moroz (Tucson: Univ. of Arizona Press), pp. 45–68.
Oyama, V. I., et al. 1980. Pioneer Venus gas chromatography of the lower atmosphere of Venus. *J. Geophys. Res.* 85:7891–7902.
Pettengill, G. G., Ford, P. G., and Wilt, R. J. 1992. Venus surface radiothermal emission as observed by Magellan. *J. Geophys. Res.* 97:13091–13102.
Pollack, J. B., Toon, O. B., and Boese, R. 1980. Greenhouse models of Venus' high surface temperature, as constrained by Pioneer Venus measurements. *J.Geophys. Res.* 85:8223–8231.
Pollack, J. B., et al. 1993. Near infrared light from Venus' nightside: a spectroscopic analysis. *Icarus* 103:1–42.
Revercomb, H. E., Sromovsky, L. A., Suomi, V. E., and Boese, R. 1985. Net thermal radiation in the atmosphere of Venus. *Icarus* 61:521–538.
Taylor, F. W. 1995. Carbon monoxide in the deep atmosphere of Venus. *Adv. Space Res.* 16(6):81–88.
Taylor, F. W., et al. 1980. Structure and meteorology of the middle atmosphere of Venus: Infrared remote sounding from the Pioneer Orbiter. *J. Geophys. Res* 85:7963–8006.
Tomasko, M. G., et al. 1980. The thermal balance of Venus in the light of the Pioneer Venus mission. *J. Geophys. Res.* 85:8187–8199.
von Zahn, U., Kumar, S., Niemann, H., and Prinn, R. 1983. Composition of the Venus atmosphere. In *Venus*, eds. D. M. Hunten, L. Colin, T. M. Donahue and V. I. Moroz (Tucson: Univ. of Arizona Press), pp. 299–430.
Young, L. D. G., Young, A. T., and Zasova, L. V. 1984. A new interpretation of the Venera 11 spectra of Venus. *Icarus* 60:138–151.

THE THERMAL BALANCE OF THE VENUS ATMOSPHERE

DAVID CRISP
Jet Propulsion Laboratory

and

DMITRI TITOV
Space Research Institute, Moscow

Below the Venus cloud tops (~65 km), the atmospheric thermal structure is characterized by high surface temperatures (735 K), moderately-stable vertical temperature lapse rates (-7.7 K km^{-1}), and small, negative, equator-to-pole temperature gradients. Above the clouds, in the mesosphere (60–100 km), the vertical temperature gradients are much more stable, and the equator-to-pole temperature gradients reverse sign such that polar temperatures are up to 20 K warmer than those at low latitudes. The physical processes responsible for these features of the atmospheric thermal structure were poorly understood in the early 1980s. This chapter reviews new constraints on the thermal balance provided by recent spacecraft and ground-based observations. It also describes recent advances in modeling methods that have addressed some of the most perplexing thermal balance issues, including (i) the need for unidentified thermal opacity sources to explain the high surface temperatures and atmospheric thermal flux distribution, (ii) the effects of reduced water vapor and cloud abundances on the greenhouse, and (iii) the processes that maintain the anomalously warm polar mesosphere.

I. INTRODUCTION

Microwave observations collected during the 1950s provided the first evidence for the high surface temperatures and hot lower atmosphere of Venus (Mayer et al. 1958; Drake 1962; Kuz'min and Salomonovich 1962). The high (>600 K) brightness temperatures inferred from these measurements were initially surprising because they were not consistent with the available groundbased infrared observations, which indicated cloud-top temperatures between 225 and 240 K (Pettit and Nicholson 1955; Sinton and Strong 1960). They also appeared to be inconsistent with measurements of the albedo of Venus, which showed that its clouds reflect almost 80% of the incident sunlight. Venus therefore absorbs only about 65% as much solar radiation as the Earth. With this solar insolation, simple thermal balance models suggest that Venus should have radiative equilibrium brightness temperatures more like those inferred from the infrared observations (Moroz 1983). The origin of these anomalously high microwave brightness temperatures, and their implications for the thermal balance of the Venus atmosphere were a major focus of both the U. S. and Soviet space programs during the first two decades of the space age. Direct measurements by the Venera 4 through 14 and Pioneer Venus (PV) entry probes confirmed that the atmospheric temperatures increase from $\sim$225 K at the cloud tops ($\sim$65 km) to values near 735 K, at the surface (Fig. 1; cf. Seiff 1983). These observations also constrained the composition and optical properties of the massive (90 bar), predominantly CO_2 atmosphere. During the 1970s, this information was incorporated into numerical radiative transfer models, which showed that the high surface temperatures could be maintained by an efficient, solar-driven, atmospheric greenhouse mechanism (Pollack et al. 1980).

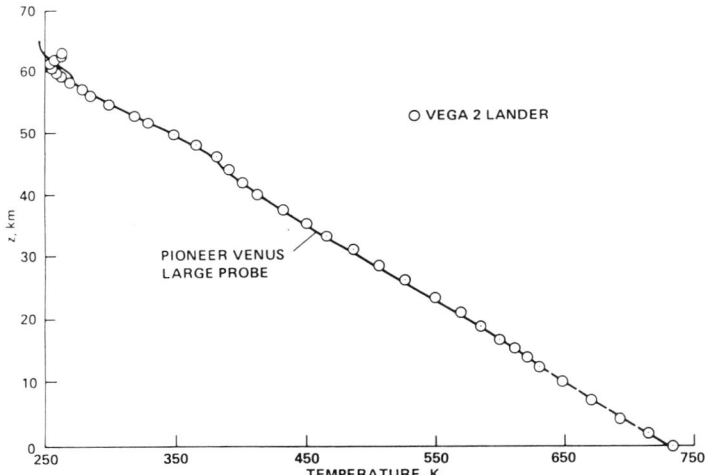

Figure 1. Temperature structure of the Venus lower atmosphere from the Pioneer Venus sounder probe (solid) and the Vega-2 lander (⊙). The Pioneer Venus probes transmitted no data between the surface and 12 km (figure from Seiff et al. 1987).

Several details of the deep-atmosphere thermal balance were still poorly understood in the late 1970s and early 1980s, however. In particular, observations and theoretical models available at that time indicated that the thermal opacity provided by known infrared-active gases (CO_2, H_2O, SO_2, HF, HCl) and H_2SO_4 aerosols was not adequate to maintain temperatures as high as those observed near the surface. To address the opacity deficit in these models, large numbers of undetectable sub-micron particles were added to the clouds, almost doubling their mass (Suomi et al. 1980). The moderately stable temperature lapse rates observed throughout the deep atmosphere (Fig. 2) were also poorly understood within the context of these greenhouse models, which required near-neutral stability (i.e., adiabatic lapse rates) at these levels. More recent observations, laboratory investigations, and modeling studies have provided improved constraints on the thermal opacity of the deep atmosphere, and contributed additional support for the greenhouse hypothesis. However, a complete, self-consistent model of the Venus deep atmosphere thermal structure has not yet been developed. The thermal balance of the deep atmosphere is reviewed in greater detail in Sec. III.

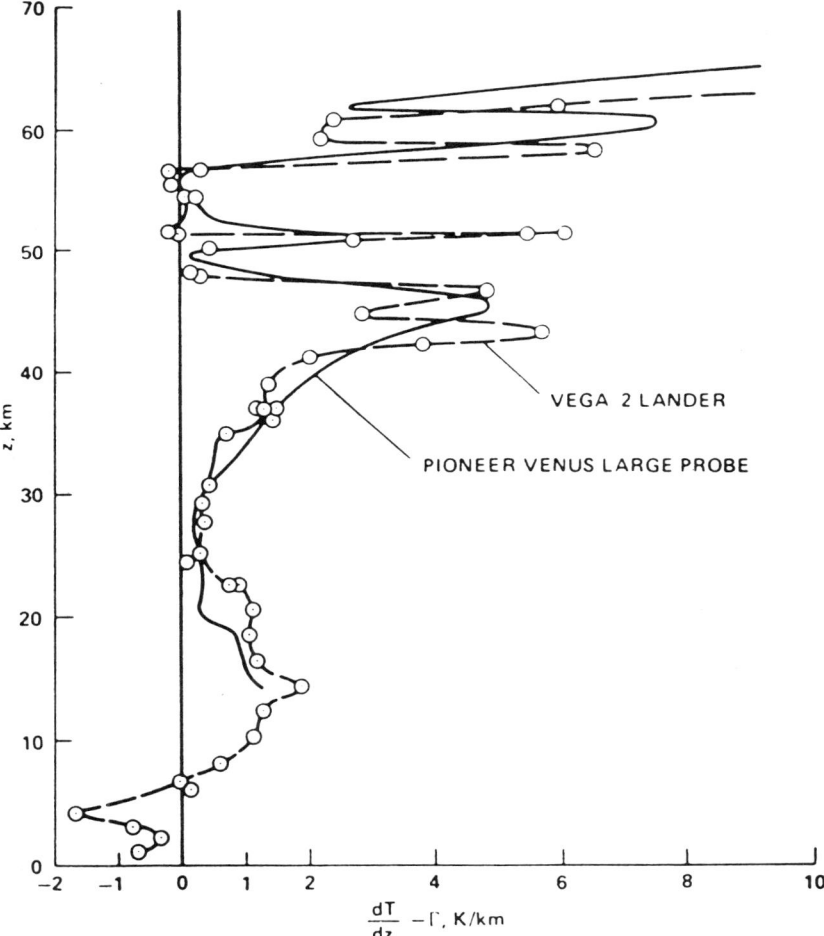

Figure 2. Static stability $(dT/dz - \Gamma)$ of the Venus lower atmosphere from the Pioneer Venus sounder probe (solid) and the Vega-2 lander ($\odot$) (figure from Seiff et al. 1987).

Pioneer Venus Orbiter (PVO) observations of the Venus mesosphere also provided the first direct evidence for the enigmatic thermal structure at altitudes between 70 and 90 km. At these levels, the equator-to-pole temperature gradients reverse, such that polar regions are up to 20 K warmer than the equator (Taylor et al. 1979,1980; Kliore and Patel 1982). The persistence of this anomalous thermal structure was confirmed by routine PVO and Magellan radio occultation observations taken during the 1980s and early 1990s, and by infrared observations obtained by the Venera 15 Infrared Fourier Spectrometer Experiment (IFSE; Zasova and Moroz 1992) and the Galileo Near Infrared Mapping Spectrometer (Carlson et al. 1991). The radiative and dynamical processes responsible for maintaining the mesospheric temperature gradients were not well understood in the early 1980s when Taylor et al. (1983) reviewed this topic. Modeling studies conducted since that time have provided additional insight into these processes. They show that the observed thermal structure may be a consequence of interactions between the vertically-propagating atmospheric thermal tides and the zonal superrotation (Baker and Leovy 1987; Newman and Leovy 1992). The mesospheric thermal balance is described in Sec. IV.

II. TEMPERATURES AND OPTICAL PROPERTIES

Thermal balance studies require a detailed description of the thermal structure and optical properties of the Venus atmosphere. These properties have been derived from a broad range of spacecraft and groundbased observations.

A. The Venus Thermal Structure

The first direct, *in-situ* measurements of temperatures below the Venus clouds were acquired in 1967 by the Venera 4 probe (Avduevsky et al. 1968). These measurements confirmed that the temperatures increase monotonically from values near 300 K within the clouds to values as high as 544 K at ~25 bar, but communications ceased before this probe reached the surface. Venera 7 successfully landed on the Venus surface in December 1970 and recorded temperatures near 747±20 K (Avduevsky et al. 1983). Temperature measurements by the Venera 8 through 14 Landers and the PV probes (Seiff et al. 1980) provided a refined description of the thermal structure below the cloud tops (Fig. 1). They showed that the temperature lapse rates are statically stable below the cloud tops, with mean values near -7.7 K km^{-1}. The region just below the cloud base (36–48 km) is much more stable (-6.6 K km^{-1}), while layers centered within the middle cloud (49–55 km) and just below 20 km, are only marginally stable (Fig. 2; cf. Seiff 1983; chapter by Gierasch et al.). At mesospheric levels (60–100 km), the static stability is much greater than in the deep atmosphere. The stability is greatest near the cloud tops at latitudes near 70°, in the cold collar, where deep temperature inversions are usually seen. At more poleward latitudes, the mesospheric temperatures are almost isothermal between 60 and 80 km.

The meridional (pole-to-equator) temperature gradients at altitudes below ~40 km are poorly constrained by *in-situ* observations because only one probe (the PV north probe) entered the Venus atmosphere as far north as 60°. All other probes entered at latitudes equatorward of 32°. However, if the

THERMAL BALANCE OF THE VENUS ATMOSPHERE

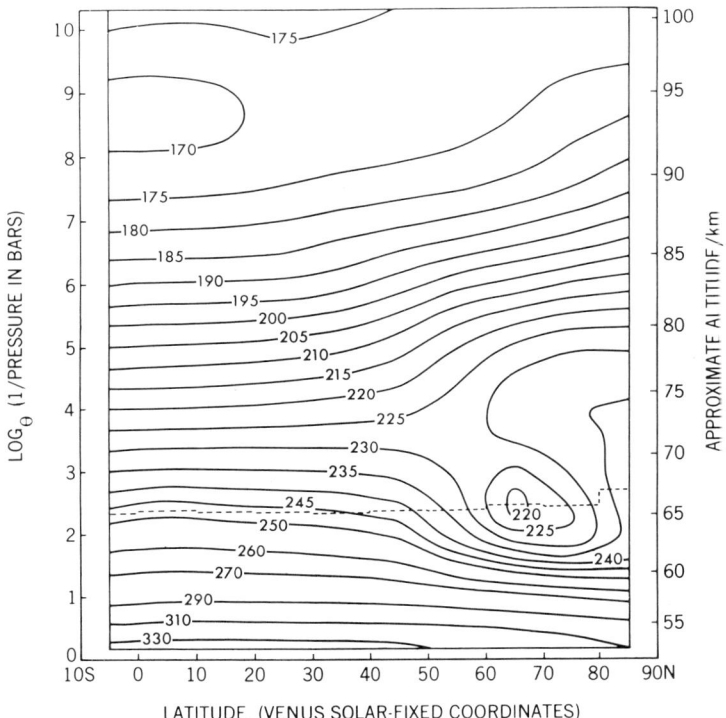

Figure 3. The zonally averaged temperature field as a function of latitude from PV OIR observations. The dashed line shows the level of cloud optical depth unity (figure from Schofield and Taylor 1983).

atmosphere is roughly in cyclostrophic thermal wind balance at these levels (cf. Schubert et al. 1980), the available temperature and wind measurements suggest that the meridional gradients are small (<1 K) near the surface. Radio occultation measurements provide a more complete description of the temperature distribution at altitudes between 40 and 100 km (Kliore and Patel 1982). Within the main cloud deck (47–70 km) these observations indicate that the meridional temperature gradients are negative, such that the polar regions (70–90°) are 25 to 30 K cooler than mid and low latitudes. At mesospheric levels between 65 and 90 km, PVO observations reveal reversed meridional temperature gradients, with polar regions up to 20 K warmer than the tropics (Fig. 3) (Taylor et al. 1979,1980; Schofield and Taylor 1983; Kliore and Patel 1982). Mesospheric temperatures derived from the Venera 15 Orbiter IFSE observations taken between October and December 1983 confirmed the persistence of this anomalous mesospheric thermal structure. They also show that the meridional gradients reverse once again above 95 km, yielding negative gradients between 95 and 105 km (Zasova and Moroz 1992; Schäfer et al. 1990).

B. Atmospheric Optical Properties

Both gases and aerosols contribute to the opacity of the Venus atmosphere. The principal absorbing gases include CO_2, H_2O, SO_2, CO, OCS, HCl, and HF. Their abundances and distributions are reviewed by van Zahn et al. (1983) and by Esposito et al. (see their chapter). Figure 4 shows the wavelength-dependent infrared transmission by each of these gases for optical paths extending over the entire atmospheric column, and for paths extending between the cloud base (~50 km) and the top of the atmosphere. CO_2 is the most important absorbing gas at most far-ultraviolet and infrared wavelengths. It also contributes significantly to scattering in the deep atmosphere, where the Rayleigh scattering optical depths exceed unity at wavelengths as long as 1 μm. CO_2 absorption effectively blocks thermal emission from the surface at wavelengths longer than 1.2 μm (<8300 cm^{-1}). At altitudes between the surface and the cloud base, several spectral windows open between the strongest CO_2 bands, allowing some radiation to escape. Most of the windows in the CO_2 spectrum at wavelengths >2.5 μm (<4000 cm^{-1}) are effectively blocked by the H_2SO_4 clouds and (to a lesser extent) by the other gases. At shorter wavelengths, where the clouds are only weakly absorbing, some thermal emission escapes from the surface and lower atmosphere (Allen and Crawford 1984; chapter by Taylor et al.). Studies of this emission have provided new constraints on the cloud optical depths and the abundances of several radiatively active trace gases in the lower atmosphere (Sec. III.E).

The planet-wide clouds play a major role in the thermal balance of the Venus atmosphere. These clouds reflect ~76% of the incident sunlight and absorb about half of the solar energy deposited on Venus. This absorption is contributed by the unknown cloud-top ultraviolet absorber and the H_2SO_4 cloud particles, which absorb strongly at wavelengths longer than 3 μm (Crisp 1986). The clouds act as greenhouse agents by trapping the upwelling thermal radiation from the surface and lower atmosphere that is emitted at wavelengths between strong gas absorption bands (cf. Pollack et al. 1980).

The composition and vertical distribution of the Venus clouds is reviewed in the chapter by Esposito et al. Most thermal balance studies have assumed that all cloud particles (modes 1, 2, 2' and 3) are composed of concentrated (75–85%) liquid H_2SO_4 droplets. This assumption is supported by most existing spectroscopic measurements, and by microwave observations of H_2SO_4 vapor pressures near saturation within and below the clouds (Jenkins et al. 1994). It is not entirely consistent with one interpretation of the PV LCPS observations, which suggests that the large mode 3 may be high-aspect crystals, rather than spherical liquid droplets (Knollenberg and Hunten 1980), but that interpretation has been challenged by Toon et al. (1982).

The wavelength-dependent, single-scattering optical properties of these particles have been derived from imaginary index data (Palmer and Williams

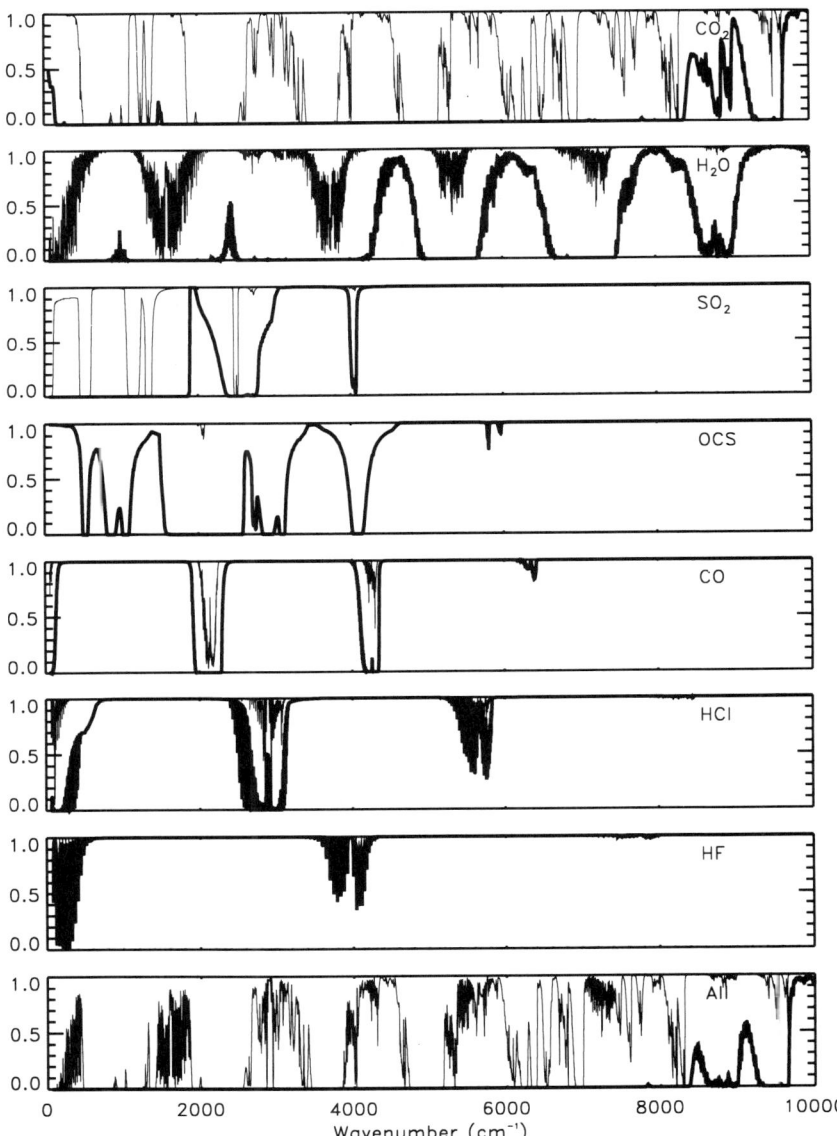

Figure 4. The atmospheric transmission is shown for the entire column (thick line) and for a path between the 50 km level and the top of the atmosphere (thin line) for CO_2, H_2O, SO_2, OCS, CO, and HCl, and for all gases. These results were generated with the line-by-line model and nominal atmospheric mixing ratios described in Meadows and Crisp (1996). The spectral resolution is 1 cm^{-1}.

1975; Pinkley and Williams 1976) with the aid of Mie Scattering models (cf. Pollack et al. 1980; Tomasko et al. 1985; Crisp 1986,1989).

The composition, distribution, and wavelength-dependent optical properties of the cloud-top ultraviolet absorber are not yet adequately constrained by existing observations, but PV Large Probe Solar Flux Radiometer (LSFR) measurements, and Venera 14 ultraviolet photometer observations (Ekonomov et al. 1984) indicate that this absorber is confined to altitudes above 58 km. In most thermal balance studies, the ultraviolet absorber is included as an impurity in the mode 1 (haze) particles at levels within the upper cloud. This is accomplished by reducing the single scattering albedos of these particles at near-ultraviolet wavelengths to simulate the observed spectrally dependent geometric albedo (cf. Crisp 1986).

III. THERMAL BALANCE OF THE DEEP ATMOSPHERE

Tomasko (1983) reviewed much of the early work on the thermal balance of the deep atmosphere in *Venus* (Hunten et al. 1983). We have therefore only summarized that work to set the stage for the more recent observational and modeling studies.

A. Historical Development

Three classes of models were initially proposed to explain the origin of the high microwave brightness temperatures reported by Mayer et al. (1958). The greenhouse (Sagan 1961) and aelosphere (Öpik 1961) models attributed these high temperatures to the Venus surface, while the ionospheric model (Sagan et al. 1961) ascribed them to a very hot, dense, ionosphere. Sagan (1960) proposed that a conventional, solar-driven atmospheric greenhouse mechanism could produce high surface temperatures if the atmosphere were sufficiently transparent at visible wavelengths to allow some sunlight to penetrate to the surface, and virtually opaque at thermal wavelengths, so that little of this radiative energy can escape to space. These criteria could not be evaluated directly during the early 1960s because the total pressure, composition, dynamics, and optical properties of the Venus atmosphere were not known.

The ionospheric model was the first to be eliminated. Sagan et al. (1961) proposed that the intense microwave emission could originate at high altitudes if the Venus ionosphere was both very hot, and about 100 times denser than its terrestrial counterpart. Unlike thermal emission from the deep atmosphere, however, this high-altitude emission should produce a limb-brightened disk. This constraint could not be tested directly from the ground because the microwave instruments available in the early 1960s did not have the spatial resolution needed to resolve the Venus disk. The Mariner 2 flyby of Venus in December of 1962 provided the first opportunity to test this hypothesis. Its high spatial resolution microwave observations showed that the Venus disk was limb darkened at wavelengths near 1.9 cm, largely precluding an ionospheric origin for the high brightness temperatures.

Both observations and theoretical models of the Venus atmosphere had evolved substantially by 1967, when Venera 4 entered the atmosphere and Mariner 5 flew past the planet. Their measurements confirmed that the atmospheric temperatures increase monotonically with depth beneath the cloud tops, and that CO_2 was the principal atmospheric constituent at all atmospheric levels (Avduevsky et al. 1968; Kliore et al. 1969). Pollack (1969b) incorporated these constraints into a non-grey radiative-convective-equilibrium (RCE) model for studies of the thermal balance of the deep atmosphere. This model represented the state-of-the-art at that time. At solar wavelengths outside of strong near-infrared gas absorption bands, it employed a multi-level version of the Schuster-Schwarzschild two-stream approximation (Sagan and Pollack 1967; Pollack 1969a) to account for absorption and multiple scattering by clouds and Rayleigh scattering by gases. Within near-infrared CO_2 and H_2O bands, gas absorption was estimated from empirical band models derived from laboratory measurements (Howard et al. 1956; Burch et al. 1969). Multiple scattering was not explicitly included at these wavelengths, but its effects were approximated by increasing the effective absorber path length in the broadband gas absorption calculations. Empirical band models were also used to estimate the broadband CO_2 and H_2O opacity at thermal wavelengths (2 to 100 μm). Multiple scattering was ignored at these wavelengths, but clouds were included as opaque blackbodies (unit emissivity, infinite absorption optical depth). The principal atmospheric constituents were assumed to be CO_2 (~85%), N_2 (~10%), and H_2O (< 0.5%). With these constraints, Pollack (1969b) showed that surface temperatures as high as those observed could be simulated if the clouds were composed of liquid water droplets and had visible optical depths between 20 and 40. His calculations also showed that the atmosphere should have an adiabatic temperature lapse rate between the surface and the cloud base. In this model, water vapor was found to be the most important greenhouse gas at most thermal infrared wavelengths.

These results were not universally accepted because the water vapor and liquid water abundances required to produce the greenhouse warming were much larger than those inferred from some groundbased observations (Kuiper et al. 1967; Belton and Hunten 1966). Venera 4 observations provided a tentative detection of water vapor below the clouds, but existing groundbased infrared spectroscopic observations appeared to preclude water as a major constituent of the clouds. If the clouds were not composed of water, they might not be sufficiently transparent at visible wavelengths, or sufficiently opaque at thermal wavelengths to produce the required greenhouse warming.

Observations acquired during the early 1970s provided new information about the thermal balance of the Venus atmosphere. Groundbased spectroscopy (Young 1973; Pollack et al. 1974) and polarimetry (Hansen and Hovenier 1974) showed that the clouds were composed of concentrated H_2SO_4 droplets, with modal radii near 1 μm. Venera 8 solar flux measurements confirmed that some of the incident sunlight penetrated to the surface, and that the clouds extended down to an altitude of about 35 km (cf. Avduevsky et al. 1983). Pollack and Young (1975) incorporated these results into their one-dimensional RCE model, and adopted reduced water vapor mixing ratios. Within the clouds, the water vapor pressure was determined by the saturation vapor pressure of H_2O over H_2SO_4 aerosols. Below the clouds, the water vapor mixing ratio was set to 0.3%. The atmospheric thermal opacity in this model was not adequate to maintain the observed surface temperatures when 1.0 μm radius H_2SO_4 cloud particles were used, but these temperatures could be produced if the clouds were composed primarily of 3 μm particles.

Measurements acquired by the PV and Venera 9 through 14 entry probes provided much better constraints on: (i) the thermal structure of the atmosphere; (ii) the composition, vertical distribution, and particle size distributions in the clouds (Knollenberg and Hunten 1980; Knollenberg et al. 1980; Ragent and Blamont 1980); (iii) the abundances of radiatively active trace gases (H_2O, SO_2) (Hoffman et al. 1980; Oyama et al. 1980); and (iv) the upward and downward solar and thermal fluxes in lower atmosphere (Tomasko et al. 1980; Suomi et al. 1980). With this information, Pollack et al. (1980) developed an updated one-dimensional RCE model of the Venus greenhouse. This model explicitly included thermal opacity by CO_2, H_2O, SO_2, CO, HCl, and HF, and four discrete H_2SO_4 aerosol particle modes (modes 1, 2, 2' and 3). An improved treatment of CO_2 pressure-induced absorption also enhanced the infrared opacity in this model. Instead of solving for the solar flux distribution, they simply adopted values derived from PV LSFR measurements (Tomasko et al. 1980). In these simulations, the principal sources of thermal opacity were CO_2, H_2O, the H_2SO_4 clouds, and SO_2, in order of decreasing importance. The PV and Venera entry probes measured between 3 and 10 times less water vapor below the clouds (100–1000 ppm) than that assumed in earlier greenhouse models ($\geq$3000 ppmv), but the reduced water opacity was largely compensated by CO_2 pressure-induced bands and the strong SO_2 vibration-rotation fundamentals at mid infrared wavelengths, which were neglected in the earlier models. Because the observed cloud particle populations did not provide enough thermal opacity to maintain the observed temperature structure, Pollack et al. (1980) postulated the presence of large numbers ($>10^5$ particles cm^{-3}) of tiny ($r_{\text{eff}} \leq 0.07 \mu$m) "mode 0" particles in the upper cloud. With this opacity, they were able to simulate the high surface temperatures, and many other features of the observed thermal structure including the stable region near the base of the clouds, and the neutrally stable region within the middle cloud.

B. Unresolved Issues for the Deep Atmosphere Thermal Balance

These investigations added significantly to our understanding of the thermal balance of the deep atmosphere, but they also raised a number of new concerns. For example, these models required additional sources of thermal opacity (large numbers of undetectable mode 0 particles) to reproduce the observed thermal structure (Tomasko 1983). Even though these particles cannot be precluded by the PV Large Probe Cloud Particle Spectrometer (LCPS) measurements, the apparent need for any additional opacity sources compromises our understanding of the thermal balance. The reduced H_2O mixing ratios inferred from recent near-infrared observations of the Venus night side have also raised concerns about the effectiveness of this gas as a greenhouse agent (Bézard et al. 1990; Drossart et al. 1993; de Bergh et al. 1995; Meadows and Crisp 1996). Pollack et al. (1980) adopted the Venera 11 and 12 spectrophotometer profile in their nominal model, and found that H_2O was the second most important greenhouse constituent (after CO_2). These mixing ratios decrease from 200 ppm at the cloud base to 20 ppm at the surface. The recent near-infrared observations indicate mixing ratios closer to 30 ppm throughout the lower atmosphere. The effects of these much drier conditions on the thermal balance have not yet been investigated in detail, but some preliminary results are presented in Sec. III.F.

The validity of the thermal balance calculations reported by Pollack et al. (1980) was further compromised by approximations adopted in their radiative transfer models. The empirical band absorptance models used to compute the absorption by CO_2, H_2O, and SO_2 at infrared wavelengths can introduce $\sim$100% errors for pressures and absorber path lengths beyond those used to fit their empirical coefficients (cf. Crisp et al. 1986). Other errors were introduced by shortcomings in the knowledge of the absorption by gases at high temperatures and pressures, like those encountered in the deep atmosphere of Venus. The neglect of multiple scattering at all thermal infrared wavelengths introduces another potential source of error. These errors are investigated in Sec. III.F.

C. Re-Analysis of PV Solar and Thermal Flux Measurements

Tomasko et al. (1985) combined observations collected by the PV LSFR, LCPS, OCPP, and OIR with Venera 11 spectrophotometer results to provide more stringent constraints on the cloud properties and the solar flux distribution in the Venus atmosphere. Revised cloud optical depths were used in their doubling-adding model to estimate the solar fluxes and heating rates as a function of solar zenith angle. They found globally averaged heating rates that decreased monotonically from values near 6 K day^{-1} (one day = 86,400 s) at the cloud tops ($\sim$68 km), to 1 K day^{-1} at the base of the upper cloud. The heating at these levels is associated primarily with the absorption of sunlight by the unknown cloud-top ultraviolet absorber. Within the middle and lower clouds, the globally averaged heating rates decrease from 0.19 K day^{-1} at 55.3 km, to 0.052 K day^{-1} at 48.9 km. Below the clouds, the globally averaged heating rates continue to decrease monotonically to values near 0.008 K day^{-1} at 36.1 km and 0.001 K day^{-1} at the surface. The computed globally averaged solar flux at the surface is 17 W m^{-2}, or about 2.6% of the incident solar flux.

The PV Small Probe Net Flux Radiometers (SNFR) and Large Probe Infrared Radiometer (LIR) were designed to provide detailed vertical profiles of the net thermal fluxes below the cloud tops, but their preliminary results included large errors that seriously compromised their value for thermal balance studies (Suomi et al. 1980; Ingersoll and Pechmann 1980). Revercomb et al. (1982,1985) re-analyzed the SNFR and LIR measurements, identified plausible sources for the measurement errors, and derived corrected fluxes. They then used a radiative transfer model (based on Pollack 1969a) to derive new constraints on the cloud particle populations, the global distribution of water vapor, and the thermal cooling rates in the deep atmosphere. They confirmed earlier SNFR results (Suomi et al. 1980), which suggested that an additional source of thermal opacity was needed in the upper cloud (58–71 km), and attributed this opacity to either the sub-micron mode 0 particles, or enhancements in the mode 2 number density (by up to a factor of 6). This re-analysis also confirmed that the net thermal flux increased with latitude. To explain these variations, they proposed that the water vapor mixing ratios in the lower atmosphere decreased from > 500 ppmv at the sounder probe site (4°N) to 20–50 ppmv at the north probe entry site (60°N). The recent near-infrared observations (Crisp et al. 1991a,b; Drossart et al. 1993) appear to preclude these large meridional water vapor gradients. Instead, they suggest that the large meridional net thermal flux gradients are associated with meridional variations in the optical depths of the middle and lower clouds. Because the clouds act as greenhouse agents that trap thermal radiation emitted by the deep atmosphere, these variations in the cloud opacity can affect net thermal fluxes and cooling rates within and below the clouds. The thinner mid-latitude clouds should trap less of this upwelling thermal emission, enhancing the cooling rates near the cloud base.

Regardless of their origin, these meridional gradients in net thermal fluxes have direct implications for the thermal balance and dynamics of the deep atmosphere (Revercomb et al. 1985). Because the net solar fluxes are expected to decrease with latitude, the observed net thermal flux increase with latitude indicates that the deep atmosphere is not in local radiative-convective equilibrium. Large-scale dynamical processes must therefore transport a significant amount of heat between low and high latitudes at these levels. A conventional, equator-to-pole Hadley cell might provide the meridional heat transport needed to balance the observed variations in the net radiative cooling (Gierasch 1975; Schubert 1983; Hou and Goody 1985). This cell would have rising motion at low latitudes, poleward flow near the top of the domain (cloud tops?), descending motion at high latitudes, and equatorward flow at depth. Because the temperature lapse rates are moderately stable throughout much of the lower atmosphere, the rising motion at low latitudes will be associated with adiabatic expansional cooling. Similarly, adiabatic compressional heating will occur in the descending, high-latitude branch of this cell. The net dynamical heating produced by this cell could therefore compensate for the horizontal variations in net radiative heating.

D. Vega Balloon and Lander Measurements

On 11 and 15 June 1985, the Soviet Vega mission deployed surface landers and meteorological balloons into the Venus atmosphere. The two Vega balloons entered the atmosphere at similar latitudes (7°11'N and 6°28'S, respectively), but they sampled air masses separated by about 135° of longitude as they floated in the middle cloud layer (50–55 km). Both balloons observed temperature lapse rates that were near adiabatic (stability ranging from 0 to 2 K km^{-1}), but the two temperature profiles were offset by 6.5 K (Seiff et al. 1987). The amplitude of this temperature difference was surprising because it was comparable to the pole-to-equator gradient at these altitudes. In addition, both balloons encountered vertical winds with amplitudes sometimes exceeding 3 m s^{-1}. Comparisons of observed temperatures and winds revealed upward convective heat fluxes between 0 and 360 W m^{-2}, but the mean value for both balloons was ~40 W m^{-2}. This is comparable to the globally averaged downward solar flux at these levels (Ingersoll et al. 1987). These data confirm that convection is responsible for the majority of the vertical heat transport through the middle cloud.

The Vega-2 Lander provided the first high-resolution measurements of the atmospheric temperatures near the Venus surface (Figs. 1 and 2). This probe entered the atmosphere at 6.6°S, 180.7°E (near Dali Chasma on Aphrodite Terra). Its measurements are similar to those made by the PV Large Probe at altitudes above ~12 km (where the external sensors on all of the PV probes failed). Layers with enhanced stability are centered near 45 and 14 km, but the stability decreases dramatically toward the surface. Superadiabatic lapse rates were measured at altitudes below 6.5 km at this site (Seiff et al. 1987). Within the lowest 1.5 km, the lapse rates vary from -1.6 K km^{-1} to -10 K km^{-1}, with a mean value near -8 K km^{-1}. These measurements have not yet been verified, but if they are valid, the mechanisms that maintain the alternating stable and unstable layers are not yet known. The implications of these measurements for the deep atmosphere thermal balance are also ambiguous because they provide no constraints on the spatial or temporal extent of the conditions observed at this site.

E. Near-Infrared Observations of the Venus Night Side

Recent observations of near-infrared (0.8–2.5 μm) emission from the night side of Venus have added significantly to our understanding of the thermal balance (Allen and Crawford 1984; Allen 1987; Kamp and Taylor 1990; Bézard et al. 1990; Crisp et al. 1991a,b; Pollack et al. 1993; chapter by Taylor et al.). This near-infrared emission accounts for less than 0.1% of the total emission from the top of the atmosphere, but it has provided valuable constraints on the atmospheric thermal structure and optical properties. For example, near-infrared spectra of the night side have provided improved estimates of the mixing ratios of H_2O and several other important greenhouse gases at altitudes below the clouds. NIR images of the night side also provided information about spatial and temporal variations in the optical properties of the clouds (Crisp et al. 1991b; Grinspoon et al. 1993). More recently, spatially resolved spectra taken at wavelengths between 1.0 and 1.18 μm have been used to constrain the temperature lapse rates near the Venus surface. Meadows and Crisp (1996) find night-side averaged temperature lapse rates of -7 to -7.5 K km^{-1} in the lowest 6 km (Meadows and Crisp 1996). These lapse rates indicate much greater static stability than those inferred from entry probe measurements and greenhouse models (-8 to -8.5 K km^{-1}) (cf. Seiff 1983; Seiff et al. 1987). If they are confirmed by subsequent observations, these results might indicate the presence of significant radiative losses from the surface during the Venus night.

F. Advances in Radiative Transfer Modeling Methods

Additional insight into the thermal balance of the deep atmosphere has been provided through the development of more sophisticated and reliable numerical radiative transfer models. In the late 1970s, when the last comprehensive investigations of the deep atmosphere were conducted, the approximations and simplifications required for computational efficiency (empirical band models, two-stream solutions to the equation of transfer, neglect of scattering processes at thermal wavelengths) substantially limited the reliability and range of validity of the available models. Their accuracy was further compromised by uncertainties in the optical properties of gases at high temperatures and pressures. Recent advances in our understanding of these properties, along with improved constraints on the opacity sources in the deep atmosphere, advanced radiative transfer modeling methods, and faster computers offer new opportunities to address many of these issues.

The most sophisticated methods available today employ multi-level, multi-stream, multiple scattering algorithms and line-by-line descriptions of gas absorption coefficients. Because these models are still computationally expensive, they are primarily used to simulate high-resolution spectra in narrow spectral regions, and to verify calculations performed with more efficient methods. Two approaches have been used to increase the computational speed of these methods for broadband radiance and thermal balance calculations. The first approach employs a monochromatic description of the atmospheric optical properties, but replaces the multi-stream multiple scattering algorithm with a two-stream flux algorithm (cf. Bézard et al. 1990; Crisp et al. 1991a,b; Pollack et al. 1993). The second approach retains the accurate multi-stream multiple scattering algorithm, but uses correlated-k or spectral mapping methods to reduce the number of monochromatic multiple scattering calculations required to resolve the spectral variations in the atmospheric optical properties. For example, the Spectral Mapping Atmospheric Radiative Transfer (SMART) model (Meadows and Crisp 1996) combines a multi-level, multi-stream, discrete ordinate method (Stamnes et al. 1988), and a high-pressure line-by-line model, with high-resolution spectral mapping methods to produce a comprehensive description of the radiances and heating rates throughout the Venus atmosphere. Here, we used SMART to address several features of the Venus thermal balance that could not be studied with earlier models. In particular, it has been used to estimate : (i) the errors introduced by the neglect of multiple scattering at infrared wavelengths; (ii) the effects of the reduced deep-atmosphere water vapor mixing ratios inferred from near-infrared spectra of the Venus night side; and (iii) the radiance and heating rate differences associated with the cloud optical depth variations inferred from near-infrared images of the night side.

Radiative transfer models of the Venus deep atmosphere have also benefited from recent advances in our understanding of the optical properties of gases at high pressures and temperatures. The Direct Numerical Diagonalization (DND) approach for computing the positions and strengths of molecular absorption lines has provided one of the most significant advances in this area (Wattson and Rothman 1992). The DND method has been used to generate the HITEMP data bases for CO_2 and H_2O (Pollack et al. 1993). These data bases include large numbers of hot and isotopic bands that were omitted from earlier spectral line data bases. They also provide a much more comprehensive description of ground-state transitions. In addition, recent improvements in the spectral line data bases for sulfur gases, including SO_2 (Lafferty et al. 1992), OCS (Fayt et al. 1986), and H_2S (Bykov et al. 1994) have contributed to our understanding of their role in the deep atmosphere thermal balance. Finally, ongoing investigations of infrared continuum absorption by gases promise significant improvements in our understanding of their effects. These processes include the finite duration of collisions (Ma and Tipping 1992; Clough et al. 1989), collision-induced transitions (Manzanares et al. 1984; Brodbeck et al. 1991), and collisional line mixing (cf. Lévy et al. 1992).

These new tools were first used to determine the effects of multiple scattering by the gases and aerosols at thermal infrared wavelengths. In the nominal case, both the absorption and scattering cross sections for gases and aerosols were included. In the nonscattering case, gas and aerosol absorption were included at thermal wavelengths, but their scattering cross sections were set to zero (cf. Pollack 1969a; Pollack et al. 1980; Crisp 1989). The nominal temperatures, gas mixing ratios, and aerosol distributions specified in Meadows and Crisp (1996) were used in both cases. The computed fluxes at the top of the Venus atmosphere are compared in Fig. 5. The largest differences occur at 300 to 500 cm^{-1} (20 to 33 μm) and 800 to 1000 cm^{-1} (10 to 12.5 μm), where the clouds provide a significant fraction of the thermal opacity. At these wavelengths, multiple scattering reduces the fluxes emitted at the top of the atmosphere by 6 to 10%. Much larger fractional errors are seen at near-infrared wavelengths, where the H_2SO_4 cloud particles have single scattering albedos near unity. If the thermal fluxes for each case are integrated over the entire thermal spectrum (50 to 10^4 cm^{-1}), we find that multiple scattering reduces the upwelling thermal flux by $\sim$10% from 179 to 162 W m^{-2}. These flux reductions are remarkably similar to those obtained by adding mode 0 particles to the upper cloud (Pollack et al. 1980; Suomi et al. 1980). The neglect of multiple scattering at thermal wavelengths also produces cooling rate errors as large as 10% at some levels of the deep atmosphere (Fig. 6).

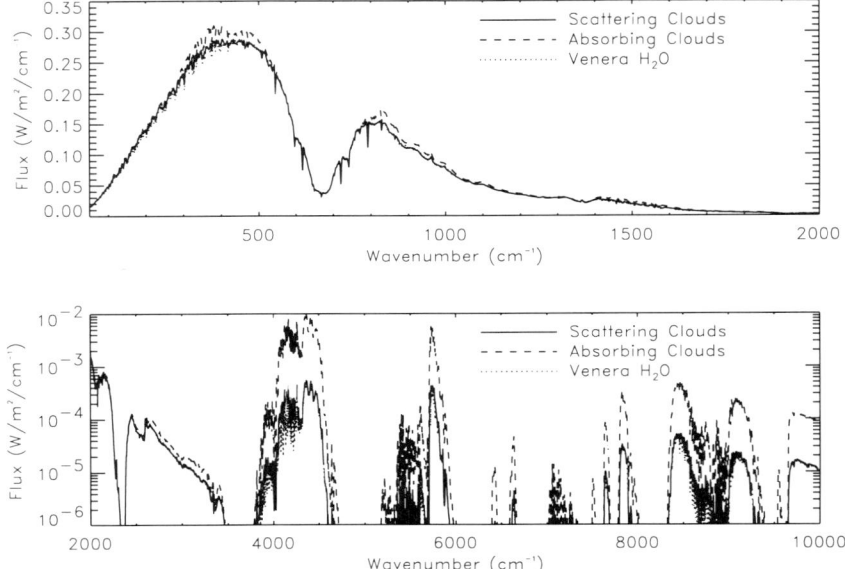

Figure 5. Synthetic spectra of the thermal flux emitted by the Venus atmosphere for (i) nominal gas and aerosol optical properties, (ii) nominal gas optical properties, with aerosol absorption, but no aerosol scattering, and (iii) nominal aerosol optical properties, but H_2O abundances from Venera 11 and 12 spectrophotometer results (20–200 ppmv).

The nominal atmospheric H_2O profile used in these calculations has mixing ratios that vary from 20 ppmv at the cloud base, to 45 ppmv at 10 bar (~30 km) and then remain constant between that level and the surface (Meadows and Crisp 1996). The alternate "wet" model employs the Venera 11 and 12 spectrophotometer profile (Moroz et al. 1978), which Pollack et al. (1980) adopted as their nominal profile. The largest flux differences are seen in the strong H_2O rotation band, (<300 cm^{-1}), and within the strong water vibration-rotation bands centered near 1580 and 4000 cm^{-1} (Figs. 5 and 6). Surprisingly, the net spectrally integrated difference between the fluxes at the top of the atmosphere for the nominal and "wet" cases is less than 1%. Water vapor therefore plays a much smaller role in the global thermal balance than was assumed in earlier models. Much of the thermal opacity originally attributed to H_2O is provided by CO_2 hot bands and pressure-induced bands, and absorption by SO_2 and OCS. In spite of its small contribution to the fluxes at the top of the atmosphere, the H_2O abundance and vertical distribution can have a significant impact on the thermal balance, by reducing the cooling rates at the cloud base, and increasing them at the top of the middle cloud (Fig. 6).

Spatial variations in the optical depths of the middle and lower clouds, like those associated with the bright and dark contrasts seen on the night side

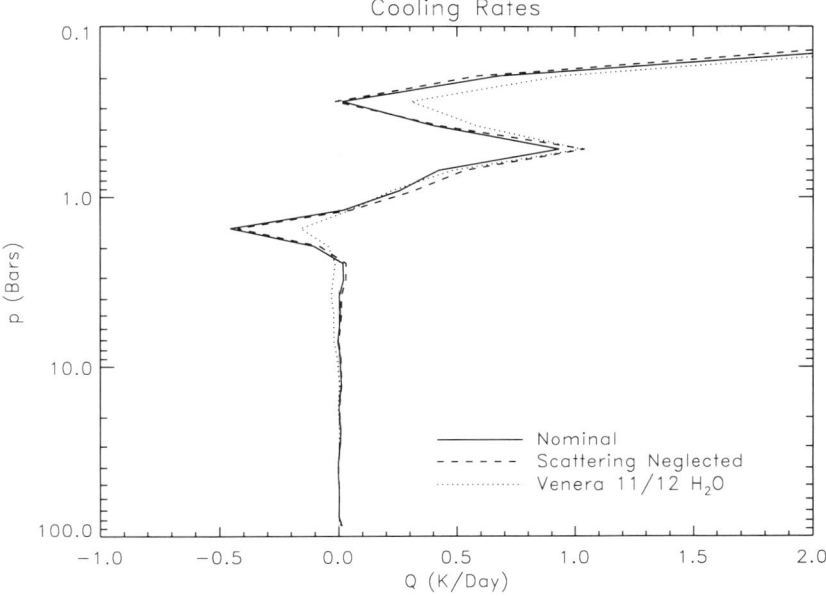

Figure 6. Radiative cooling rates for (i) nominal gas and aerosol optical properties, (ii) nominal gas optical properties with aerosol absorption, but no aerosol scattering, and (iii) nominal aerosol optical properties, but H_2O abundances from Venera 11 and 12 spectrophotometer results (20–200 ppmv).

of Venus (Allen and Crawford 1984), might also affect the thermal balance of the Venus atmosphere. The long-lived, wavenumber-1 opacity variation in the middle cloud described by Crisp et al. (1991b) is of particular interest because its lifetime exceeds the 1 to 2 week radiative time scale at those altitudes (Crisp 1989; Crisp et al. 1991b). To estimate the radiative forcing by such features, heating and cooling rates were derived for cloud models with nominal (Crisp 1986), enhanced ($\times 1.6$), and depleted ($\times 0.5$) particle populations within the middle and lower clouds (Fig. 7). Because the clouds are optically thick, reductions in their particle number densities produced larger changes in the heating and cooling rates than the increases. Decreases in the middle and lower cloud densities reduced both the solar heating rates and the thermal cooling rates at levels in the upper cloud (58–71 km, 0.028–0.5 bar) and in the upper regions of the middle cloud, and increases in these quantities at lower levels. Increases in the middle and lower cloud optical depths produced somewhat smaller heating and cooling rate changes of the opposite sign. Because the thermal cooling rates were almost twice as sensitive to these cloud optical depth changes, the net effect of a 50% decrease in the cloud particle density is to increase the net radiative heating throughout the upper cloud (0.6–0.028 bar) by ~ 0.4 to 1.2 K day^{-1}, and increase the net radiative

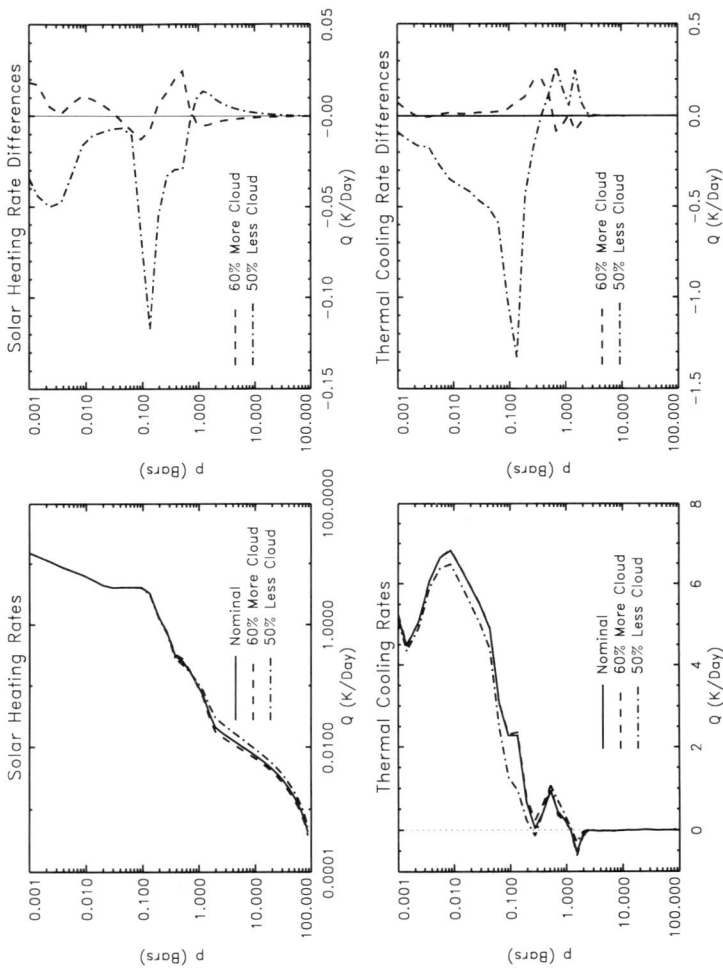

Figure 7. Global-average solar heating rates and thermal cooling rates for (i) nominal gas and aerosol optical properties, (ii) 50% reductions in the middle and lower cloud aerosol optical depths, (iii) 60% increase in the middle and lower cloud aerosol optical depths: (a) globally averaged solar heating rates, (b) heating rate differences, (c) thermal cooling rates, (d) cooling rate differences.

cooling in the middle and lower clouds by about 0.2 K day^{-1}. These results suggest that long-lived variations in the cloud particle number densities, like those inferred from groundbased near-infrared observations of the night side, could produce temperature variations within the middle and lower clouds as large as those observed by the two Vega balloons (6.5 K). If this were the case, the warmer temperatures measured by Vega balloon 1 suggests that it may have floated in an air mass where the clouds were more dense, while the cooler Vega balloon 2 profile suggests that it floated in a less dense region of the cloud. This conclusion is supported by the available cloud density observations obtained by that mission. The nephelometer carried by Vega balloon 1 recorded some of the highest particle densities ever seen within the Venus clouds (Ragent et al. 1987). The nephelometer did not work on the second Vega balloon, but the Vega-2 lander cloud particle experiment detected few large particles in the middle and lower clouds (Moshkin et al. 1986).

IV. THERMAL BALANCE OF THE MESOSPHERE

A pioneering study of the mesospheric thermal balance was conducted by Dickinson (1972). He derived the globally averaged radiative equilibrium temperature structure at altitudes between 66 and 130 km for a pure CO_2 atmosphere. The radiative transfer model included the effects of nonlocal thermodynamic equilibrium, but did not include a rigorous treatment of the overlap between spectral lines. The derived radiative equilibrium temperatures decreased from about 250 K at 66 km to 158 K at 88 km, and then increased to 190 K at 113 km. For this thermal structure, the solar heating and thermal cooling rates increased from $\sim$1.5 K day^{-1} at 65 km to $\sim$50 K day^{-1} at 100 km. Dickinson also determined the contribution to the solar heating and thermal cooling by each of the CO_2 fundamental, hot, isotopic, combination, and ultraviolet absorption bands. Thermal cooling rates were dominated by cooling to space by the strong 15 μm fundamental. The associated Newtonian cooling times decreased from about 30 days at 65 km to 0.05 day at 120 km.

Ramanathan and Cess (1974) largely confirmed Dickinson's results and estimated the diurnally averaged, meridional temperature gradients at mesospheric levels as well. In their experiments, the temperatures were fixed at 58 km, and allowed to evolve at higher altitudes. They derived temperatures at 80 km that decreased from $\sim$170 K at the equator, to about 160 K at 60°, and then to about 140 K at 85° latitude. These estimated meridional gradients suggested that the super-rotating cloud-top zonal winds would accelerate to speeds as high as 250 m s^{-1} at 85 km (Ramanathan and Cess 1975). It was impossible to verify these results before the PV mission because the only available constraints on the thermal structure above the cloud tops were provided by the Mariner 5 and 10 radio-occultation results, which provided limited meridional coverage. The only wind velocity measurements available at these altitudes were provided by groundbased spectroscopic observations of Doppler shifts in spectral lines, but these measurements were confusing because some indicated strong retrograde zonal winds (Traub and Carleton 1975), while others showed little or no wind (Young et al. 1979).

The PV OIR experiment provided the first spatially resolved description of the thermal structure of the Venus mesosphere (Fig. 3). Unlike the results predicted by Ramanathan and Cess (1974), these observations showed that the meridional temperature gradients change sign at altitudes near 70 km, such that the polar regions are warmer than the equator at most mesospheric levels (Taylor et al. 1979,1980). These results were surprising for two reasons. First, most of the solar energy is absorbed at mesospheric levels by the predominately CO_2 atmosphere, cloud-top ultraviolet absorber, and H_2SO_4 cloud particles. This solar heating should be most intense at low latitudes because Venus has an obliquity of $<4°$. Second, the radiative cooling at mesospheric levels was thought to be dominated by cooling to space in the strong CO_2 15 μm band (Dickinson 1972; Ramanathan and Cess 1974). The efficiency of this process is proportional to the local temperature. The enhanced solar heating at low latitudes, combined with enhanced thermal cooling from the warmer polar regions should therefore act to warm the equator and cool the poles, destroying the observed thermal structure.

The radiative and dynamical processes responsible for maintaining the anomalous mesospheric thermal structure were not yet identified in the early 1980s when Taylor et al. (1983) reviewed this topic in *Venus* (Hunten et al. 1983). In their review, they proposed that the high polar mesospheric temperatures might be produced by: (i) compressional heating in the descending branch of an axially-symmetric, equator-to-pole Hadley cell; (ii) enhanced high-latitude solar heating associated with increases in the atmospheric path lengths and abundances of (optically-thin) polar mesospheric aerosols; (iii) the absorption of the intense upwelling thermal radiation emitted by the polar hot spots by CO_2 and aerosols at mesospheric levels, or some combination of these processes. Subsequent modeling studies have shown that none of these mechanisms can account for the observed thermal structure (Crisp 1983,1986,1989). Instead, they indicate that the reversed mesospheric thermal structure is a consequence of the interactions between the vertically propagating atmospheric thermal tides and the zonal superrotation (Fels and Lindzen 1974; Baker and Leovy 1987; Newman and Leovy 1992). These studies are described in the next two subsections (Secs. IV.A and B).

A. Thermal Balance Results from PV Observations

Crisp (1983,1986,1989) developed a two-dimensional (latitude vs altitude) RCE model to study the radiative forcing of the Venus mesosphere. Unlike earlier models used for mesospheric thermal balance studies, this model included all radiative processes known to be important in the Venus atmosphere, including absorption, emission, and multiple scattering by gases (CO_2, H_2O, SO_2), the H_2SO_4 aerosols, and the unidentified ultraviolet absorber. The vertical and horizontal distributions of these constituents were derived from a broad range of spacecraft and groundbased observations. At solar wavelengths, the model used a multi-level δ-Eddington/adding model to evaluate the equation of transfer in the presence of absorption and multiple scattering. It also employed a novel physical band model, rather than empirical broad-band absorptance formulas, that facilitated a more rigorous treatment of multiple scattering in the presence of non-grey gas absorption. Like other contemporary thermal balance models, however, the thermal flux algorithm included absorption and emission by gases and aerosols, but ignored multiple scattering.

Global-average RCE experiments were performed to determine the accuracy of the radiative transfer modeling methods, the radiative forcing produced by each atmospheric constituent, and the sensitivity of the computed thermal structure to uncertainties in the mesospheric optical properties. For the nominal global-average gas, aerosol, and ultraviolet absorber distributions used in this study, the model produced spectrally dependent fluxes that are in good agreement with available spacecraft and groundbased observations (Tomasko et al. 1985). The RCE temperatures obtained for this nominal model atmosphere also agreed with globally averaged PV OIR results (within ± 7 K) at altitudes between 60 and 95 km. Finally, global-average RCE temperatures were determined for a range of aerosol optical depths, H_2O and SO_2 mixing ratios, and ultraviolet absorber distributions. These tests rarely produced RCE temperature differences larger than 5 K, indicating that the global-average thermal balance is relatively insensitive to existing uncertainties in the mesospheric optical properties (Crisp 1983,1989).

To establish the effects of meridional variations in the radiative forcing, temperatures were fixed at 55 km, and the diurnally averaged RCE thermal structure was derived at 10 latitudes between the equator and the pole. Experiments employing the nominal, low-latitude model atmosphere at all latitudes produced high-latitude temperatures that were $\sim$40 K cooler than equatorial temperatures at most levels above the cloud tops (Figs. 8 and 9). The nominal polar aerosol distribution, with a low, warm, polar cloud top, and enhanced mesospheric haze, actually produced somewhat cooler polar temperatures ($\sim$5 K) throughout the mesosphere, contradicting earlier predictions (cf. Schubert et al. 1980; Taylor et al. 1983). These latitude-dependent RCE experiments indicate that if the mesospheric optical properties are adequately constrained by existing observations, radiative processes will act to destroy the observed thermal structure, producing equatorial temperatures 5 to 15 K warmer than those observed, and polar temperatures up to 60 K cooler than those observed.

The observed thermal structure can be maintained in the presence of this radiative drive if there exists a mechanism for heating the polar mesosphere by >12 K day^{-1}, and cooling the low-latitude mesosphere by $\sim$8 K day^{-1} (Fig. 10). Crisp (1983,1989) proposed that this thermal forcing might be provided by an equator-to-pole meridional circulation cell, with rising motion at low latitudes, poleward flow within the upper mesosphere, subsidence over the pole, and return flow beneath the cloud tops. Like the Hadley cell proposed

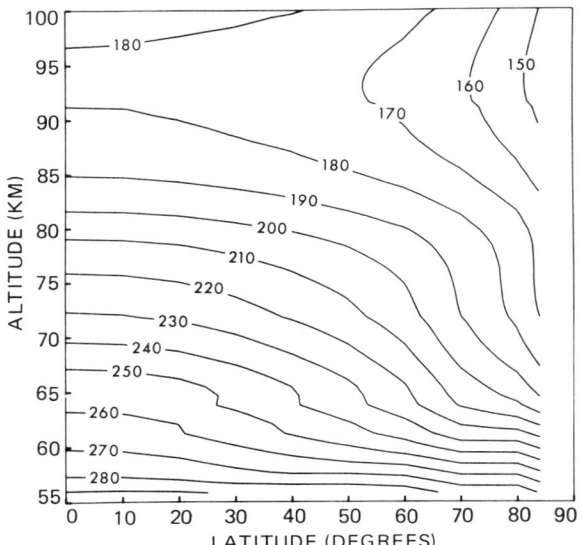

Figure 8. The zonally averaged radiative-equilibrium thermal structure of the Venus mesosphere as a function of latitude and altitude (figure after Crisp 1989).

by Taylor et al. (1983), this meridional cell produces adiabatic expansional cooling in its rising branch at low latitudes, and adiabatic compressional heating in its descending polar branch. Unlike a thermodynamically direct Hadley cell, however, this cell must transport heat against the thermal gradient to produce polar temperatures that are warmer than those at the equator. Crisp (1983) developed a diagnostic dynamical model to illustrate this dynamical heating mechanism, and found that the observed thermal structure could be maintained in the presence of the nominal radiative forcing by a meridional cell with upward vertical velocities <1 cm s^{-1} at low latitudes, poleward velocities of <10 m s^{-1} at altitudes near the mesopause, downward vertical velocities of <2.5 cm s^{-1} at high latitudes ($>85°$), and a weak (<1 m s^{-1}) equatorward flow at altitudes between 60 and 64 km (Fig. 11).

Because this meridional circulation cell is thermodynamically indirect, it must be driven by an external momentum source. Crisp (1983, 1989) proposed that this forcing could be provided by the interaction between the vertically propagating atmospheric thermal tides and the mean zonal superrotation (cf. Fels and Lindzen 1974). In these models, the tides generated within the upper cloud deck (57 to 71 km) propagate to higher mesospheric levels (80 to 100 km) where they are strongly damped by radiative cooling. This radiative damping causes the tides to transfer their momentum to the zonal flow at these levels. Because the tides have phase speeds near 4 m s^{-1} (prograde), they act as source of drag on the strong (-100 m s^{-1}) retrograde zonal super-rotation at their damping altitudes. This tidal drag decelerates the super-rotation in the

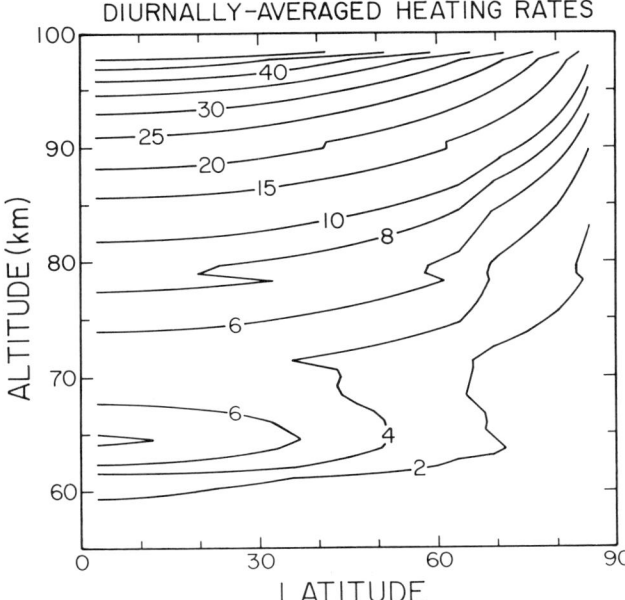

Figure 9. The zonally averaged solar heating rates in the Venus mesosphere as a function of latitude and altitude (figure after Crisp 1986).

upper mesosphere and forces a thermodynamically indirect, equator-to-pole, meridional circulation. This tidal momentum forcing mechanism has since been verified by numerical experiments (Fels 1986; Baker and Leovy 1987; Newman and Leovy 1992; cf. chapter by Lellouch et al.).

B. Thermal Balance Results from the Venera 15 Spectrometer

Thermal radiance spectra of the Venus mesosphere were taken with the Venera 15 Orbiter Infrared Fourier Spectrometer Experiment (IFSE) between 12 October and 14 December 1983. These spectra provide additional constraints on the thermal structure and optical properties of the mesosphere. The IFSE collected more than 1500 moderate-resolution (5 to 7 cm^{-1}) thermal infrared (270–1650 cm^{-1}) spectra at latitudes between 60°S and 87°N, for morning (4 hr to 10 hr) and evening (16 hr to 22 hr) sectors. These spectra provided a self-consistent description of the thermal structure, aerosol distribution, and mixing ratios of H_2O and SO_2 at altitudes between 60 and 105 km (Schäfer et al. 1990; Zasova and Moroz 1992).

Schäfer et al. (1990) were the first to use IFSE temperatures and aerosol distributions to derive solar heating rates and thermal cooling rates at mesospheric levels. Their solar radiative transfer model incorporated the modified Eddington approximation (Meador and Weaver 1980) and the two-stream adding method (Crisp 1986) to find solar fluxes and heating rates in the

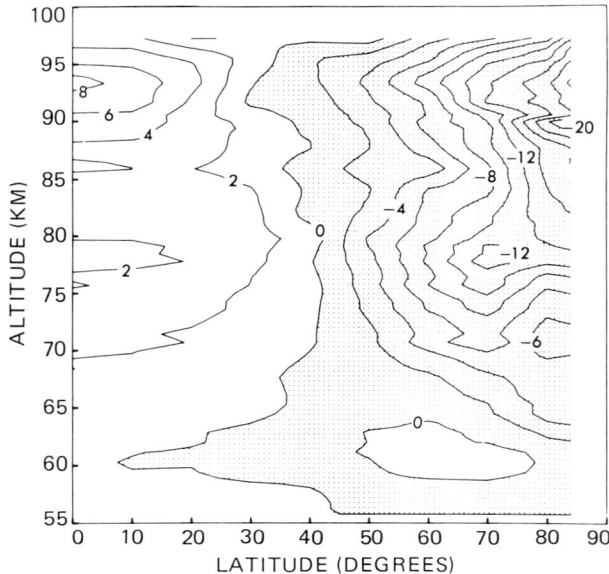

Figure 10. The net radiative heating rates (K day^{-1}) in the Venus mesosphere as a function of latitude and altitude for the PV OIR thermal structure (figure after Crisp 1989).

presence of gas and aerosol absorption and scattering. A line-by-line model was used to derive the monochromatic optical properties of gases. Unlike Crisp (1983, 1989), Schäfer et al. (1990) derived thermal fluxes from a model that included multiple scattering as well as absorption by gases and aerosols. These thermal balance experiments, and those conducted subsequently by Haus and Goering (1990) confirmed that the solar heating rates in the lower mesosphere (60–70 km) are very sensitive to the assumed distribution of the H_2SO_4 aerosols and the ultraviolet absorber near the cloud top, but these properties had not yet been derived from the IFSE data. The nominal aerosol model adopted by Haus and Goering (1990) included larger H_2SO_4 aerosol and ultraviolet absorber optical depths at levels above 65 km than those inferred from PV observations (Tomasko et al. 1980, 1985; Crisp 1983, 1986). Because of this, Haus and Goering find solar heating rates at levels between 65 and 73 km that are 2 to 3 K day^{-1} larger than those derived in the earlier studies. At altitudes above the cloud tops, Haus and Goering (1990) find solar heating rates that are significantly lower than those derived in the earlier studies. The heating rate differences at these levels are probably caused by the use of different CO_2 absorption line data bases. While Haus and Goering used line parameters from the 1976 version of the Air Force Geophysics Laboratory (AFGL) data base (Rothman and McClatchey 1976), Crisp (1983, 1986) used values from the 1980 version of this data base (Rothman 1981), which in-

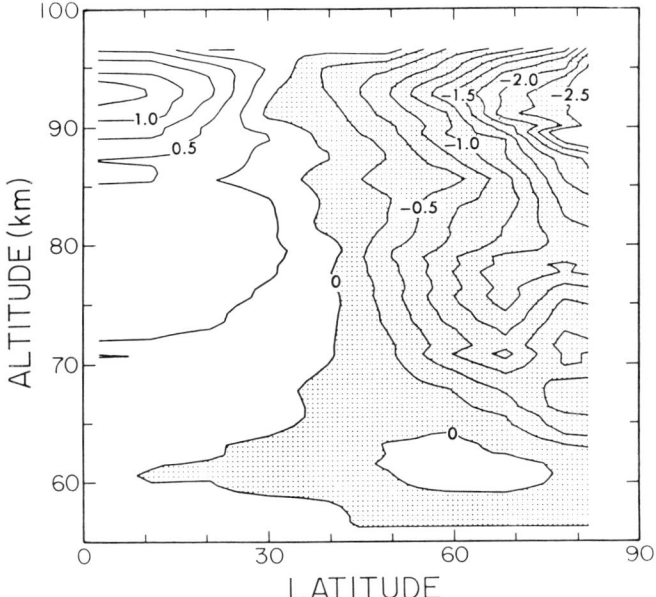

Figure 11. Vertical velocity distribution needed to produce adiabatic heating and cooling rates large enough to balance the net radiative drive for the PV OIR thermal structure (figure after Crisp 1983).

cluded a more complete description of the weak near-infrared overtone bands at wavelengths less that 2.15 μm. The absorption of sunlight in these bands dominates the near-infrared heating rates at most mesospheric levels.

Haus and Goering (1990) confirmed that the thermal cooling rates at mesospheric levels are not very sensitive to multiple scattering, or the abundance of H_2O, but depend strongly on the distribution of aerosols in the upper cloud. Even though the latitude-dependent solar heating and thermal cooling distributions derived by Haus and Goering (1990) are qualitatively similar to those obtained from PV observations (Crisp 1983,1986,1989), the net radiative heating rates are significantly different. Crisp found net radiative heating at latitudes equatorward of $\sim40°$, and net radiative cooling at higher latitudes at most mesospheric levels. This net heating distribution indicates that the mesosphere is roughly in global radiative equilibrium at each level. In contrast, Haus and Goering (1990) find net heating at all latitudes at levels within the upper cloud, and net cooling at all latitudes at altitudes between 70 and 80 km. The net cooling at altitudes above the clouds (~73 km) might be related to underestimates of the CO_2 solar heating rates at these levels, while the net heating at lower altitudes may be a consequence of different H_2SO_4 aerosol and ultraviolet absorber distributions.

Once self-consistent, latitude-dependent temperature, aerosol, and SO_2

distributions had been retrieved from the Venera 15 IFSE observations, Titov et al. (1992) repeated these thermal balance experiments. This investigation and their subsequent work (Titov 1995) largely confirmed the principal results reported by Haus and Goering (1990), but indicated that cooling prevailed throughout the mesosphere at the latitude of the *cold collar* (65–75°N). These differences resulted primarily from the use of a more accurate high-latitude aerosol distribution by Titov (1995). This result is somewhat more consistent with the results derived from PV observations (Crisp 1983,1989).

C. Temporal Variations in the Mesospheric Temperatures

During the mid-1980s, groundbased microwave observations revealed dramatic changes in the thermal structure at mesospheric levels. The largest changes were seen during 1985 and 1986, when whole-disk observations of CO emission indicated 20 to 40 K increases in the mid- and low-latitude temperatures at altitudes between 80 and 100 km (Clancy and Muhleman 1991; chapter by Lellouch et al.). These large temperature increases suggest a reversal of the anomalous temperature gradients revealed by PV and Venera 15 Orbiter observations. The night-time CO bulge also appeared to intensify and shift from near midnight to about 3 a.m. local time during this period, suggesting that the zonal super-rotation extended throughout the mesosphere. By 1988, these changes had been reversed once again, yielding a mesospheric thermal structure and circulation more like that seen during the late 1970s and early 1980s by the PV and Venera 15 Orbiters.

These dramatic variations in the mesospheric thermal structure are not yet understood, but they may be related to changes in the radiative and dynamical forcing at these levels of the Venus atmosphere. For example, several thermal balance studies (Crisp 1983,1986; Tomasko et al. 1985; Haus and Goering 1990) have shown that both the vertical extent and amplitude of the solar heating peak within the upper cloud are very sensitive to the vertical distribution of the H_2SO_4 aerosols and the ultraviolet absorber. Furthermore, the vertical distribution of the heating can have a significant impact on the phase and amplitude of the atmospheric thermal tides generated at these levels (Fels and Lindzen 1974). Changes in the distribution of these absorbing constituents could therefore significantly alter the tidal momentum forcing mechanism that is thought to be responsible for maintaining the anomalous, reversed thermal structure (Baker and Leovy 1987; Newman and Leovy 1992). There is currently no direct evidence for such changes in the cloud properties, but variations of this kind might have been associated with the factor of 10 decrease in mesospheric SO_2 mixing ratios that has been observed since the late 1970s. More systematic observations of the thermal structure, albedo, and cloud distribution are needed to test this hypothesis.

V. SUMMARY AND CONCLUSIONS

The principal features of the Venus thermal structure, including the high surface temperatures, hot lower atmosphere, and reversed pole-to-equator temperature gradients at mesospheric levels are now reasonably well understood. Groundbased and spacecraft observations acquired since the late 1970s, combined with improved modeling methods have largely confirmed that the thermal balance of the deep atmosphere is maintained by an efficient atmospheric greenhouse mechanism. The warm polar mesosphere has been attributed to the presence of a thermodynamically indirect meridional circulation cell that is driven by interactions between atmospheric thermal tides and the zonal super-rotation. However, there are several other aspects of the Venus thermal structure that are not yet well understood. For example, we still know very little about the thermal structure in the lowest atmospheric scale height, and we have no direct measurements of temperatures poleward of 60°. The radiative and dynamical processes associated with the large-scale variations in the thermal structure and the cloud particle number densities in the middle and lower cloud are largely unknown. The origin of the cloud-top temperature inversions in the vicinity of the cold collar and the mechanisms that produce the polar dipole structures are not well understood. The mechanisms responsible for the dramatic temporal variations in the SO_2 abundance, thermal structure, and dynamics of the mesosphere are inadequately constrained by theory and observations. Finally, the interactions between the radiative forcing and the zonal retrograde super-rotation are not well understood. Some of these issues can only be addressed by additional entry probe or orbiter measurements, but others can be resolved by carefully designed groundbased astronomical observations, and by more comprehensive theoretical models.

Acknowledgments. This chapter is dedicated to the memory or Dr. James B. Pollack, whose pioneering studies contributed greatly to our understanding of the thermal balance of the Venus atmosphere. Part of this work was conducted at the Jet Propulsion Laboratory of the California Institute of Technology, under contract to the National Aeronautics and Space Administration. DC acknowledges generous support from the NASA Planetary Atmospheres Program and the Venus Data Analysis Program.

REFERENCES

Allen, D. A. 1987. The dark side of Venus. *Icarus* 69:221–229.
Allen, D. A., and Crawford, J. W. 1984. Cloud structure on the dark side of Venus. *Nature* 307:222–224.
Avduevsky, V. S., Marov, M. Ya., and Rozhdestvensky, M. K. 1968 Model of the planet Venus based on results of the measurements made by the Soviet automatic interplanetary station Venera-4. *J. Atmos. Sci.* 25:537–545.

Avduevsky, V. S., et al. 1983. Structure and parameters of the Venus atmosphere according to Venera probe data. In *Venus*, eds. D. M. Hunten, L. Colin, T. M. Donahue and V. I. Moroz (Tucson: Univ. Arizona Press), pp. 280–298.

Baker, N. L., and Leovy, C. B. 1987. Zonal winds near Venus' cloud top level: A model study of the interactions between the zonal mean circulation and the semidiurnal tide. *Icarus* 69:202–220.

Belton, M. J. S., and Hunten, D. M. 1966. Water vapor in the atmosphere of Venus. *Astrophys. J.* 146:307–308.

Bézard, B., de Bergh, C., Crisp, D., and Maillard, J. P. 1990. The deep atmosphere of Venus revealed by high-resolution nightside spectra. *Nature* 345:508–511.

Brodbeck, C., et al. 1991. Measurements of pure CO_2 absorption at high densities near 2.3 μ. *J. Geophys. Res.* 96:17497–17500.

Burch, D. E., Gryvnak, D. A., Patty, R. R., and Bartky, C. E. 1969. Absorption of infrared radiant energy by CO_2 and H_2O. IV. Shapes of collision broadened lines. *J. Opt. Soc. Amer.* 59:267–280.

Bykov, A. D., et al. 1994. The infrared spectrum of H_2S from 1 to 5 μm. *Canadian J. Phys.* 72:989–1000.

Carlson, R. W., et al. 1991. Galileo infrared imaging spectroscopy measurements at Venus. *Science* 253:1541–1548.

Clancy, R. T., and Muhleman, D. O. 1991. Long-term (1979–1990) changes in the thermal, dynamical, and compositional structure of the Venus mesosphere as inferred from microwave spectral line observations of ^{12}CO, ^{13}CO, and $C^{18}O$. *Icarus* 89:129–146.

Clough, S. A., Kneizys, F. X., and Davies, R. W. 1989. Line shape and the water vapor continuum. *Atmos. Res.* 23:229–241.

Crisp, D. 1983. Radiative Forcing of the Venus Mesosphere. Ph.D. Thesis, Princeton University.

Crisp, D. 1986. Radiative forcing of the Venus mesosphere. I. Solar fluxes and heating rates. *Icarus* 67:484–514.

Crisp, D. 1989. Radiative forcing of the Venus mesosphere. II. Thermal fluxes, cooling rates, and radiative equilibrium temperatures. *Icarus* 77:391–413.

Crisp, D. S., Fels, B., and Schwarzkopf, M. D. 1986. Approximate methods of finding CO_2 15-μm band transmission in planetary atmospheres. *J. Geophys. Res.* 91:11851–11866.

Crisp, D., Ingersoll, A. P., Hildebrand, C. E., and Preston, R. A. 1990. VEGA balloon meteorological measurements. *Adv. Space Res.* 10:109–124.

Crisp, D., Allen, D. A., Grinspoon, D. H., and Pollack, J. B. 1991*a*. The dark side of Venus: Near-infrared images and spectra from the Anglo-Australian Observatory. *Science* 253:1263–1266.

Crisp, D., et al. 1991*b*. Ground-based near-infrared imaging observations of Venus during the Galileo encounter. *Science* 253:1538–1541.

de Bergh, C., et al. 1995. Water in the deep atmosphere of Venus from high-resolution spectra of the night side. *Adv. Space Res.* 15:79–88.

Dickinson, R. E. 1972. Infrared radiative heating and cooling in the Venusian mesosphere. I. Global mean radiative equilibrium. *J. Atmos. Sci.* 29:1531–1556.

Drake, F. D. 1962. 10-cm observations of Venus near superior conjunction. *Nature* 195:894.

Drossart, P., et al. 1993. Search for spatial variations of the H_2O abundance in the lower atmosphere of Venus from NIMS-Galileo. *Planet. Space Sci.* 41:495–504.

Ekonomov, A. P., et al. 1984. Scattered UV solar radiation within the clouds of Venus. *Nature* 307:345–346.

Fels, S. B. 1986. An approximate analytical method for calculating tides in the atmosphere of Venus. *J. Atmos. Sci.* 43:2757.

Fels, S. B., and Lindzen, R. S. 1974. The interaction of thermally excited gravity waves with mean flows. *Geophys. Fluid Dyn.* 6:149–192.

Fayt, A., Vandenhaute, R., and LaHaye, J. G. 1986. Global rovibrational analysis of carbonyl sulfide. *J. Molec. Spectrosc.* 119:233–266.

Gierasch, P. J. 1975. Meridional circulation and the maintenance of the Venus atmospheric circulation. *J. Atmos. Sci.* 32:1038–1044.

Grinspoon, D. H., et al. 1993. Probing Venus's cloud structure with Galileo NIMS. *Planet. Space Sci.* 41:515–542.

Hansen, J. E., and Hovenier, J. W. 1974. Interpretation of the polarization of Venus. *J. Atmos. Sci.* 31:1137–1160.

Haus, R., and Goering, H. 1990. Radiative energy balance of the Venus mesosphere. *Icarus* 84:62–82.

Hoffman, J. H., Hodges R. R., Donahue T. M., and McElroy M. B. 1980. Composition of the Venus lower atmosphere from the Pioneer Venus Mass Spectrometer. *J. Geophys. Res.* 85:7882–7891.

Hou, A. Y., and Goody, R. 1985. Diagnostic requirements on the superrotation on Venus. *J. Atmos. Sci.* 42:413–432.

Howard, T. N., Burch, D. E., and Williams, D. 1956. Infrared transmission of synthetic atmospheres. *J. Opt. Soc. Amer.* 46:186–190, 237–245, 334–338, 452–455.

Hunten, D. M., Colin, L., Donahue, T. M., and Moroz, V. I., eds. 1983. *Venus* (Tucson: Univ. of Arizona Press).

Ingersoll, A. P., and Pechmann, J. B. 1980. Venus lower atmosphere heat balance. *J. Geophys. Res.* 85:8219–8222.

Ingersoll, A. P., Crisp, D., Grossman, A. W., and the VEGA Balloon Science Team. 1987. Estimates of convective heat fluxes and gravity wave amplitudes in the Venus middle cloud layer from VEGA Balloon measurements. *Adv. Space Res.* 7:343–349.

Jenkins, J. M., et al. 1994 Radio occultation studies of the Venus atmosphere with the Magellan spacecraft, 2. Results from the October 1991 experiments. *Icarus* 110:79–94.

Kamp, L. W., and Taylor, F. W. 1990. Radiative-transfer models of the night side of Venus. *Icarus* 86:510–529.

Kliore, A. J., et al. 1969. Atmosphere and ionosphere of Venus from the Mariner 5 S-band radio occultation measurement. In *The Venus Atmosphere*, eds. R. Jastrow and S. I. Rasool (New York: Gordon and Breach), pp. 105–127.

Kliore, A. J., and Patel, I. R. 1982. Thermal structure of the atmosphere of Venus from Pioneer Venus radio occultations. *Icarus* 52:320–334.

Knollenberg, R., et al. 1980. The clouds of Venus: A synthesis report. *J. Geophys. Res.* 85:8059–8081.

Knollenberg, R. G., and Hunten, D. M. 1980. The microphysics of the clouds of Venus: Results of the Pioneer Venus particle size spectrometer experiment. *J. Geophys. Res.* 85:8038–8058.

Kuiper, G. P., Forbes F. F., and Johnson J. L. 1967. Program of astronomical spectroscopy from aircraft. *Commun. Lunar Planet. Lab.* 6:155–170.

Kuz'min, A. D., and Salomonovich A. E. 1962. Observations of the radio emission of Venus and Jupiter at wavelength 8 cm. *Astron. J.* 39:660–668.

Lafferty, W. J., et al. 1992. The 3-ν_3 band of $^{32}S^{16}O_2$: Line positions and intensities. *J. Molec. Spectrosc.* 154:51–60.

Lévy, A., Lacome, N., and Chakerian, C., Jr. 1992. Collisional line mixing. In *Spectroscopy of the Earth's Atmosphere and Interstellar Medium*, eds. K. N. Rao and A. Weber (Boston: Academic Press), pp. 261–337.

Ma, Q., and Tipping, R. H. 1992. A far wing line shape theory and its application to the water continuum absorption in the infrared region (I). *J. Chem. Phys.*

95:6290–6301.

Manzanares, C., Muñoz, A., and Hidalgo, D. 1984. Collision-induced absorption of infrared radiation by N_2, O_2, and CO_2. *Chem. Phys.* 87:363–371.

Mayer, C. H., McCullough, T. P., and Sloanaker, R. M. 1958. Observations of Venus at 10.2-cm wavelength. *Astrophys. J.* 127:1–10.

Meador, W. E., and Weaver, W. R., 1980. Two stream approximations to radiative transfer in planetary atmospheres: A unified description of existing methods and a new improvement. *J. Atmos. Sci.* 37:630–643.

Meadows, V. S., and Crisp, D. 1996. Ground-based near-infrared observations of the Venus nightside: The thermal structure and water abundance near the surface. *J. Geophys. Res.* 101:4595–4622.

Moroz, V. I. 1983. Summary of preliminary results of the Venera 13 and Venera 14 missions. In *Venus*, eds. D. M. Hunten, L. Colin, T. M. Donahue and V. I. Moroz (Tucson: Univ. of Arizona Press), pp. 45–68.

Moroz, V. I., et al. 1978. *Spectrophotometric Experiment On Board the Venera-11, -12 Descenders: Some Results of the Analysis of the Venus Day-Sky Spectrum*, Space Res. Inst. Acad. Sci. Leningrad Publ. 270.

Moshkin, B. E., et al. 1986. Vega 1, 2 optical spectrometry of Venus atmospheric aerosols at the 60–30 km levels: preliminary results. *Soviet Astron. Lett.* 12:36–39.

Newman, M., and Leovy, C. B. 1992. Maintenance of strong rotational winds in Venus' middle atmosphere by thermal tides. *Science* 257:647–650.

Öpik, E. J. 1961. The aeolosphere and atmosphere of Venus. *J. Geophys. Res.* 66:2807–2819.

Oyama, V. I., et al. 1980. Pioneer Venus gas chromatography of the lower atmosphere of Venus. *J. Geophys. Res.* 85:7891–7902.

Palmer, K. F., and Williams, D. 1975. Optical constants of sulfuric acid: Application to the clouds of Venus? *Appl. Opt.* 14:208–219.

Pettit, E., and Nicholson, S. B. 1955. Temperatures on the bright and dark sides of Venus. *Publ. Astron. Soc. Pacific* 67:293–303.

Pinkley, L. W., and Williams, D. 1976. Infrared optical properties of sulfuric acid at 250 K. *J. Opt. Soc. Amer.* 66:122–124.

Pollack, J. B. 1969*a*. Temperature structure of nongray planetary atmospheres. *Icarus* 10:301–313.

Pollack, J. B. 1969*b*. A non-gray CO_2-H_2O greenhouse model of Venus. *Icarus* 10:314–341.

Pollack, J. B., and Young, R. 1975. Calculations of the radiative and dynamical state of the Venus atmosphere. *J. Atmos. Sci.* 32:1025–1037.

Pollack, J. B., et al. 1974. Aircraft observations of Venus' near-infrared reflection spectrum: Implications for cloud composition. *Icarus* 23:8–26.

Pollack, J. B., Toon, O. B., and Boese, R. 1980. Greenhouse models of Venus' high surface temperature, as constrained by Pioneer Venus measurements. *J. Geophys. Res.* 85:8223–8231.

Pollack, J. B., et al. 1993. Near-infrared light from Venus' nightside: A spectroscopic analysis. *Icarus* 103:1–42.

Ragent, B., and Blamont, J. 1980. The structure of the clouds of Venus: Results of the Pioneer Venus nephelometer experiment. *J. Geophys. Res.* 85:8089–8106.

Ragent, B., and the VEGA Balloon Science Team. 1987. Results from the VEGA I Balloon nephelometer experiment. *Adv. Space. Res.* 7:315–322.

Ramanathan, V., and Cess, R. D. 1974. Radiative transfer within the mesospheres of Venus and Mars. *Astrophys. J.* 188:407–416.

Ramanathan, V., and Cess, R. D. 1975. An analysis of the strong zonal circulation within the stratosphere of Venus. *Icarus* 25:89–103.

Revercomb, H. E., Sromovsky, L. A., and Suomi, V. E. 1982. Reassessment of net-radiation measurements in the atmosphere of Venus. *Icarus* 52:279–300.
Revercomb, H. E., Sromovsky, L. A., Suomi, V. E., and Boese, R. W. 1985. Net thermal radiation in the atmosphere of Venus. *Icarus* 61:521–538.
Rothman, L. S. 1981. AFGL atmospheric absorption line parameters compilation: 1980 version. *Appl. Opt.* 20:791–795.
Rothman, L. S., and McClatchey, R. A. 1976. Updating of the AFCRL atmospheric absorption line parameters compilation. *Appl. Opt.* 15:2616–2617.
Sagan, C. 1960. *The Radiation Balance of Venus*, JPL Tech. Report 32–34.
Sagan, C. 1961. The planet Venus. *Science* 133:849–858.
Sagan, C., Siegel, K. M., and Jones, D. E. 1961. On the origin of the Venus microwave emission. *Astron. J.* 66:52.
Sagan, C., and Pollack, J. B. 1967. Anisotropic, nonconservative scattering and the clouds of Venus. *J. Geophys. Res.* 72:469–477.
Schäfer, K., et al. 1990. Infrared Fourier spectrometer experiment from Venera-15. *Adv. Space Res.* 10:57–66.
Schofield, J. T., and Taylor, F. W. 1983. Measurements of the mean, solar-fixed temperature and cloud structure of the middle atmosphere of Venus. *Quart. J. Roy. Meteorol. Soc.* 109:57–80.
Schubert, G. 1983. General circulation and dynamical state of the Venus atmosphere. In *Venus*, eds. D. M. Hunten, L. Colin, T. M. Donahue and V. I. Moroz (Tucson: Univ. of Arizona Press), pp. 681–765.
Schubert, G., et al. 1980. Structure and circulation of the Venus atmosphere. *J. Geophys. Res.* 85:8007–8025.
Seiff, A. 1983. Thermal structure of the atmosphere of Venus. In *Venus*, eds. D. M. Hunten, L. Colin, T. M. Donahue and V. I. Moroz (Tucson: Univ. of Arizona Press), pp. 215–279.
Seiff, A., et al. 1980. Measurements of thermal structure and thermal contrasts in the atmosphere of Venus and related dynamical observations: Results from the four Pioneer Venus probes. *J. Geophys. Res.* 85:7903–7933.
Seiff, A., and the VEGA Balloon Science Team. 1987. Further information on the structure of the atmosphere of Venus derived from the VEGA Venus Balloon and Lander Mission. *Adv. Space Res.* 7:323–328.
Sinton, W. M., and Strong, J. 1960. Radiometric observations of Venus. *Astrophys. J.* 131:470–490.
Stamnes, K., Tsay, S. C., Wiscombe, W., and Jayaweera, K. 1988. Numerically stable algorithm for discrete-ordinate-method radiative transfer in multiple scattering and emitting layered media. *Appl. Opt.* 27:2502–2509.
Suomi, V. E., Sromovsky, L. A., and Revercomb, H. E. 1980. Net radiation in the atmosphere of Venus: Measurements and interpretation. *J. Geophys. Res.* 85:8200–8218.
Taylor, F. W., et al. 1979. Temperature, cloud structure and dynamics of Venus middle atmosphere by infrared remote sensing from Pioneer orbiter. *Science* 205:65–67.
Taylor, F. W., et al. 1980. Structure and meteorology of the middle atmosphere of Venus: Infrared remote sensing from the Pioneer orbiter. *J. Geophys. Res.* 85:7963–8006.
Taylor, F. W., Hunten, D. M., and Ksanfomality, L. V. 1983. The thermal balance of the middle and upper atmosphere of Venus. In *Venus*, eds. D. M. Hunten, L. Colin, T. M. Donahue and V. I. Moroz (Tucson: Univ. Arizona Press), pp. 650–680.
Titov, D. V. 1995. Radiative balance in the mesosphere of Venus from the Venera-15 infrared spectrometer results. *Adv. Space Res.* 15:73–77.
Titov, D. V., Haus, R., and Schäfer, K. 1992. Thermal fluxes and cooling rates in the

Venus atmosphere from Venera-15 infrared spectrometer data. *Adv. Space Res.* 12:73–77.

Tomasko, M. G. 1983. The thermal balance of the lower atmosphere of Venus. In *Venus*, eds. D. M. Hunten, L. Colin, T. M. Donahue and V. I. Moroz (Tucson: Univ. of Arizona Press), pp. 604–631.

Tomasko, M. G., Doose, L. R., Smith, P. H., and Odell, A. P., 1980. Measurements of the flux of sunlight in the atmosphere of Venus. *J. Geophys. Res.* 85:8167–8186.

Tomasko, M. G., Doose, L. R., and Smith, P. H. 1985. The absorption of solar energy and the heating rate in the atmosphere of Venus. *Adv. Space Res.* 5:71–79.

Toon, O. B., et al. *Icarus* 57:143–160.

Traub, W. A., and Carleton, N. P. 1975. Spectroscopic observations of winds on Venus. *J. Atmos. Sci.* 32:1045–1059.

von Zahn, U., Kumar, S., Niemann, H., and Prinn, R. 1983. Composition of the Venus atmosphere. In *Venus*, eds. D. M. Hunten, L. Colin, T. M. Donahue and V. I. Moroz (Tucson: Univ. of Arizona Press), pp. 299–430.

Wattson, R. B., and Rothman, L. S. 1992. Direct numerical diagnolization: wave of the future. *J. Quant. Spectrosc. Rad. Transfer* 48:763–780.

Young, A. T. 1973. Are the clouds of Venus sulfuric acid? *Icarus* 18:564–582.

Young, A. T., Schorn, R. A., Young, L. D. G., and Crisp, D. 1979. Spectroscopic observations of winds on Venus; I. Technique and data reduction. *Icarus* 38:435–450.

Zasova, L. V., and Moroz, V. I. 1992. Latitude structure of upper clouds on Venus. *Adv. Space Res.* 12:79–90.

ION/NEUTRAL ESCAPE OF HYDROGEN AND DEUTERIUM: EVOLUTION OF WATER

T. M. DONAHUE
University of Michigan

D. H. GRINSPOON
University of Colorado

R. E. HARTLE
NASA Goddard Space Flight Center

and

R. R. HODGES, Jr.
University of Texas at Dallas

Infrared remote soundings and *in-situ* mass spectrometric measurements place the present water vapor abundance in the lower atmosphere of Venus at 30±10 ppm and there is some evidence of spatial and temporal variability. These measurements also agree that the D/H ratio in this water vapor is 150±30 times that of terrestrial water. No other hydrogen compounds are present with mixing ratios as large as 5 ppm v/v. The significance of these measurements for the evolution of water in the planet depends on the history of escape of hydrogen and deuterium and their replenishment by exogenous and endogenous sources, such as comets and volcanoes. A charge separation electric-field-driven upward flow from the nightside bulge region produces escape fluxes averaged over the entire planet of $(1.5 \pm 0.2) \times 10^7$ cm^{-2}s^{-1} in H$^+$ and $(5.6 \pm 1) \times 10^4$ cm^{-2}s^{-1} in D$^+$ during solar maximum (fractionation factor = 0.15). Very serious depletion of H$^+$ and D$^+$ in the bulge during periods of low solar activity reduces the escape to 0.15 of these rates. The solar cycle average loss rates are $(9 \pm 1.0) \times 10^6$ cm^{-2}s^{-1} for H and $(3.2 \pm 0.6) \times 10^4$ cm^{-2}s^{-1} for D. Until charge exchange loss rates are calculated for atmospheric models that conform to Pioneer Venus observations they can only be estimated. The best estimate is that H escapes at rates between 9 and 7×10^6 cm^{-2}s^{-1} during the solar cycle so that the solar cycle average is 8×10^6 cm^{-2}s^{-1}. The fractionation factor f should be small, about 0.02. The total flux could be as large as 1.7×10^7 cm^{-2}s^{-1} and as small as 9×10^6 cm^{-2}s^{-1}, depending on the actual charge exchange contribution. The loss rates can also be calculated from the solar cycle modulation of bulge densities. For hydrogen, a solar cycle planet-wide average of (6 to 7.5) $\times 10^6$ cm^{-2}s^{-1} is deduced. f is very large (0.44). This result suggests that the electric-field-driven flow in the thermosphere is responsible for the entire escape flux of about 8×10^6 cm^{-2}s^{-1}, $0.15 \leq f \leq 0.44$. If today's water vapor is predominantly a remnant of an ancient supply of water and the fractionation factor has not changed, Venus has lost the equivalent of at least 4 m of liquid water spread uniformly over its surface, 0.12% of a full terrestrial ocean. This lower limit would be 30 times larger if f is 0.44. If there is a source of water with $D/H = 1.6 \times 10^{-4}$, the amount of early water required to account for today's large D/H ratio increases to something between 0.6 and 16% of a full terrestrial ocean. It is also possible that the water present today has been supplied by outgassing of highly fractionated water from the mantle, perhaps during a massive resurfacing event.

I. INTRODUCTION

Today, the surface and atmosphere of Venus are very dry. How dry is a question that this chapter will address. How dry the interior may be is a question much more difficult to answer and thus one this chapter will consider only in passing. The planet is losing water because hydrogen escapes into space after water molecules have been dissociated in the upper atmosphere. Observations made by instruments on the Pioneer Venus Orbiter (PVO) have greatly improved our understanding of escape processes and their magnitude. This chapter will examine these findings in considerable detail. The planet's atmosphere may also be acquiring water at a rate comparable to the loss rate because of encounters with comets and contemporaneous outgassing of the interior. Thus, in principle, an appreciable component of the water vapor in the atmosphere today may have been contributed by one or both of these sources. On the other hand, much of the water vapor, and particularly the deuterated water vapor, may be the residue of water Venus acquired as it was being created. It is also possible that the provenance of much of the deuterated water vapor (HDO) is very different than that of ordinary vapor (H_2O) because the light isotope of hydrogen might be in a virtual steady state, balancing escape and injection of deuterium- poor water, while the heavy isotope is mostly a still decaying remnant of the primordial planetary water inventory. In such a case most of the light isotope in the atmosphere today could have arrived there relatively recently. This chapter will devote much of its attention to a discussion of the degree to which the source of today's water can be determined. In this regard explaining the remarkably high value of the deuterium to hydrogen ratio in the water vapor will be the principal objective of the chapter. This review will elaborate on developments since the last major reviews on this topic were published (Hunten et al. 1989; Kasting and Toon 1989).

II. PLANETOGENESIS: INITIAL VOLATILE INVENTORY

There is almost a consensus today that the terrestrial planets were formed by collisions among planetesimals that grew by accreting gases and dust from the solar nebula during several tens of million years after the Sun itself developed. Models of the latter stages of this process, after 10^{25} to 10^{26} g embryos had been formed, place the maturing planetesimals on very eccentric orbits that would have thoroughly mixed the material out of which the four planets were made (Wetherill 1991). Some or most of the volatile inventory of the planets, including water, may have been acquired during this phase of accretion. In this case, Venus and Earth, in particular, should have started with about the same endowment of volatiles, although they may have had very divergent experiences with outgassing and release of volatiles from their interiors after they reached their present size. Another very important source of volatiles for these planets may have been a heavy bombardment by volatile-rich comet-like objects arriving from the outer solar system toward the end of the accretionary period. The crater record clearly shows that such a bombardment occurred. How rich the intruders were in volatiles is less evident, but it is plausible to postulate that they were an important source of volatiles. Again, it is difficult to understand how they would have contributed very different amounts of volatiles to Earth and Venus. On the other hand, singular events such as giant impacts by Mars-sized objects that caused large-scale atmospheric erosion may well have affected one planet and not another. A recent attempt to explain the great difference in the rare gas abundances in the atmospheres of Earth and Venus (Pepin 1991) invokes such a giant impact to power an atmospheric blow off event on Earth that did not occur on Venus, destroying an initial balance in the two rare gas planetary abundances. This model also invokes early blow off of a dense hydrogen atmosphere driven by strong extreme ultraviolet radiation from the young Sun to explain the elemental and isotopic rare gas fractionation patterns.

An alternative to fractionation by hydrodynamic escape to explain volatile abundances is that the terrestrial planets received their argon, krypton and xenon, already fractionated, from impacting icy planetesimals (Owen et al. 1992). To account for the large abundance of argon on Venus, a late impact by a massive comet would be invoked. This proposal has been criticized on the ground that the gases would be eroded in the very impact that is meant to supply them (Kaula 1995). A criterion is needed to determine whether impactors are to be sources or eroders of planetary volatiles.

Further support for the "twin planet" model is the near equality of the nitrogen and oxidized carbon inventory. This balance is attained by comparing the CO_2 converted to carbonate on Earth with atmospheric CO_2 on Venus. The conversion on Earth depends on water for the weathering of silicate rocks and on an ocean in which the carbonate can precipitate. Thus, the absence of liquid water on Venus is crucial for the difference in presentation of carbon but begs the question of whether there was ever abundant water on that planet. Among the volatiles whose abundances have been assessed water is the odd one out.

On the other hand, Lewis (1972) has shown that, during the time when dust and gas were condensing to form the primitive solar system condensates, the temperature in the planetary disk at 0.7 AU was so high that minerals containing water of hydration would not have been formed there. If the terrestrial planets accreted most of their components from material that condensed near their present orbits, primitive Earth might have contained much more water than primitive Venus did. Indeed, Lewis (1972) argues that the density differences among the terrestrial planets can be explained if this is the case, whereas it is difficult to understand the apparently systematic density variation otherwise. Prinn and Fegley (1989), however, have pointed out that gas-solid reactions are kinetically inhibited at typical nebula temperatures, so that neither Venus nor Earth or Mars could have obtained their water this way.

If, on the other hand, Earth received an appreciable contribution of water from a bombardment of volatile-rich objects at the end of accretion, it is hard to understand why Venus would not. However, volatiles acquired as a late accreting veneer and those stored in the interior of the planet may well have undergone very different evolutionary paths (Pepin 1996). In particular, the late arriving volatiles may have been severely eroded by planetesimal impact and enhanced euv heating not long after their arrival, while those in the interior had to wait for their release by outgassing processes. Thus, if for some reason the orbits of large planetesimals did not commonly cross each other in the later stages of accretion, the volatile endowment of Earth and Venus might have been very different.

III. EVOLUTION OF ESCAPING D AND H

The measurement long considered diagnostic to the determination of whether or not Venus once had abundant water interacting with its atmosphere is that of the ratio of deuterium to hydrogen (D/H ratio) in its water today. Most processes by which hydrogen isotopes escape from a planet discriminate strongly against loss of deuterium, leading to enrichment of deuterium. If the relative efficiency of deuterium and hydrogen loss—the fractionation factor—is known, the ratio of the hydrogen content of a reservoir supplying the escaping hydrogen at two arbitrary times can be determined from the ratio of the D to H abundances. This is irrespective of the absolute magnitudes of the content of the reservoirs or the escape flux (unless the *net* loss of hydrogen has a zero rate). Thus, if the abundance of hydrogen in the atmosphere of Venus today (probably in the form of H_2O and perhaps some H_2) can also be determined, the present D to H ratio measured and the D to H ratio in primitive planetary water measured or plausibly estimated, the size of the early reservoir can be fixed.

If the amount of atomic hydrogen in a vertical column of hydrogen compounds is [H] and that of deuterium [D], the respective escape fluxes φ_1 and φ_2, and source strengths P_1 and P_2 are

$$\frac{d[H]}{dt} = P_1 - \varphi_1 \tag{1}$$

$$\frac{d[D]}{dt} = P_2 - \varphi_2. \tag{2}$$

The fractionation factor is normally defined as

$$f = \frac{\varphi_2/[D]}{\varphi_1/[H]} \tag{3}$$

$$= \frac{\varphi_2}{\varphi_1}\left(\frac{1}{R}\right) \tag{4}$$

where

$$R = [D]/[H] \tag{5a}$$

and, in case water is the dominant hydrogen component,

$$R = \frac{[HDO]}{2[H_2O]}. \tag{5b}$$

Integration of Eq. (4) gives the Rayleigh distillation relationship for the ratio of the size of the reservoir at some time t_2 to a later time t_1,

$$r(t_1, t_2) = \frac{[H(t_2)]}{[H(t_1)]} = \left(\frac{R(t_1)}{R(t_2)}\right)^{\frac{1}{1-f}} \tag{6}$$

independent of φ_1 and P_1, if f is constant. If, at t_1, [H] is indistinguishable experimentally from the hydrogen steady state [H_{SS}], that is if $\varphi_1 = P_1$, f would be undefined. Equation (6) then could not be used to relate [H] at a very early time, t_2, to [H] at t_1.

If the effective D to H ratio in the sources is R_S, Eq. (2) becomes

$$\frac{d[D]}{dt} = R_S P_1 - Rf\varphi_1. \tag{7}$$

Equations (1) and (7) can be integrated, numerically if necessary, provided there is a way to get a handle on the escape rates and source strengths and how they evolve. An analytical solution can be found if the hydrogen loss rate is a linear function of the total amount of hydrogen in the system,

$$\varphi_1 = K[H] \tag{8}$$

where K and both P_1 and P_2 are time independent. In that case,

$$[H] = [H_{SS}] + [H_O - H_{SS}]\,e^{-Kt} \tag{9}$$

$$[D] = [D_{SS}] + [D_O - D_{SS}]\,e^{-fKt} \tag{10}$$

where

$$[H_{SS}] = P_1/K = \varphi_{1SS}/K \tag{11}$$

with a similar definition for [D_{SS}]. [H_O] and [D_O] are the initial abundances. The ratio of Eqs. (10) and (9) (Gurwell 1995) gives

$$R(t) = \frac{R_O([H_O]/[H_{SS}])\,e^{-ft/\tau_H} + (R_S/f)(1 - e^{-ft/t_H})}{([H_O]/[H_{SS}])\,e^{-t/\tau_H} + (1 - e^{-t/\tau_H})} \quad (12)$$

where

$$\tau_H = [H_{SS}]/\varphi_{SS}. \quad (13)$$

In a typical case a hydrogen inventory of 8×10^{22} atoms per cm^2 is being evacuated at a rate of 10^7 atoms per cm^2 s^{-1} so that τ_H is about 0.25 Gyr. $R(t)$ is a function which increases with time to a maximum if R_S is small. It then decreases with time to a limit

$$R_{SS} = R_S/f. \quad (14)$$

This will be the approximate value of R in the extreme or "mature" steady state where neither [H] nor [D] is any longer varying appreciably with time.

[H], of course, reaches an approximate steady state long before [D] does. During the time when [H] is virtually in a steady state there is a useful approximation to Eq. (12), obtained by setting the denominator in Eq. (12) equal to unity, or setting

$$P = K[H] \quad (15)$$

$$\frac{d[H]}{dt} = 0 \quad (16)$$

$$[H] = [H_{SS}] \quad (17)$$

in Eq. (9). Then Eq. (12) may be represented by the approximation

$$R_l(t) = R_S/f + [r_O R_O - (R_S/f)]e^{-ft/\tau_H} \quad (18)$$

where

$$r_0 - r(0, \infty) = [H_0]/[H_{SS}]. \quad (19)$$

This approximation has appeared several times in the literature with [H_{SS}] set equal to [H_O]. Sometimes this has not been appropriate, particularly in the late stages of Rayleigh fractionation of an early reservoir [H_O], where the hydrogen escape flux and the hydrogen source strengths are approximately in balance and R_S very different from $R_l(t)$.

The behavior of $R_l(t)$ in the approximation Eq. (18) is strongly dependent on r_0. Long after [H] has reached a virtual steady state, [D] can be much larger than [D_{SS}] and R_t very large compared to R_S/f, its steady state limit. Thus, long after hydrogen has lost all memory of the original deuterium may retain it. The expression (18) seems to have appeared in the literature first in a paper dealing with oxygen fractionation on Mars (McElroy and Yung 1976). The conditions specified in that paper were that loss of oxygen was balanced with its production from water at all times so that from the very beginning oxygen was in the steady state. Enrichment of ^{18}O occurred because of a fractionation factor effectively determined by diffusive separation of ^{16}O and ^{18}O above the homopause. The counterpart of Eq. (18) was appropriate to that situation as it is to the application of Eq. (18) to hydrogen when water is taken always to be in a steady state between cometary impacts and escape of hydrogen (Grinspoon 1987; Grinspoon and Lewis 1988). But it was not to others in which its hydrogen analog with [H_O] set equal to [H_{SS}] has been applied to Martian and Cytherean hydrogen.

In Eq. (12) two characteristic times appear. One is the time constant associated with establishment of a steady state for hydrogen τ_H, the other, for the establishment of a steady state for deuterium and, thus, for R is

$$\tau_D = \tau_H/f \tag{20}$$

or 1.7 Gyr if f is 0.15 and τ_H is 0.25 Gyr. In Fig. 1 $R(t/\tau_H)$ is plotted for several values of r_O and f, with $R_S = R_0 = 1.6 \times 10^{-4}$. The approximation Eq. (18) fits the curves in Fig. 1 for $R(t/\tau_H)$ well if t/τ_H is greater than 15. In the case of Venus, with a hydrogen escape flux of 10^7 cm^{-2}s^{-1} and H_{SS} equal to 8.3×10^{22} atoms cm^{-2} (corresponding to 30 ppm of H_2O), t/τ would be 15 or larger for all times after 4 Gyr. In Fig. 1 curves are also plotted for the case $r_0 e^{-t/\tau_H} \gg (1 - e^{-t/\tau_H})$ corresponding to Rayleigh fractionation of a primitive ocean and $r_0 = 1$, which is a pure steady state. Practical application of these relationships to the atmosphere of Venus will be made in Sec. VI.

A useful approximation for r_0 can be obtained by rearranging Eq. (12), setting $R_S = R_0$ and $\rho = R(t)/R_0$, noting that in all practical cases

$$t/\tau_H > \frac{5.05}{1-f} \tag{21}$$

$$e^{-(t/\tau_H)} \ll 1 \tag{22}$$

and

$$\rho e^{-(t/\tau_H)} \ll e^{-f(t/\tau_H)} \tag{23}$$

for

$$t/\tau_H \geq 8 \tag{24}$$

and any reasonable value of f. Thus,

$$r_0 \cong \rho e^{ft/\tau_H} - (e^{ft/\tau_H} - 1)/f. \tag{25}$$

In most cases,

$$r_0 \cong (\rho - f^{-1}) e^{ft/\tau_H} \tag{26}$$

because usually

$$e^{ft/\tau_H} \gg 1. \tag{27}$$

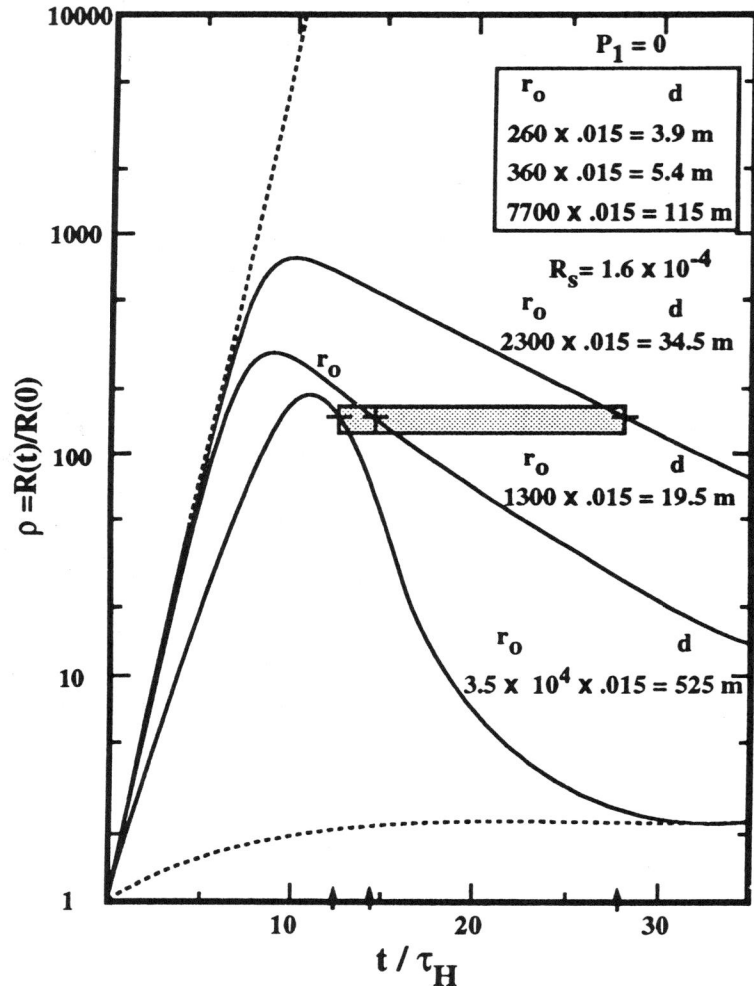

Figure 1. Evolution of D/H for pure Rayleigh fractionation, extreme steady state and composite models. The bar represents the possible range for current Venus atmosphere age and D/H ratio.

r_0 is plotted in Fig. 2 along with the approximation (Eq. 26) for a large range of fractionation factors and for $\rho = 150$. Such a large fractionation cannot be achieved at times

$$t < 5.05\tau_H/(1-f) \qquad (28)$$

because Eq. (25) becomes singular at this time. But as t increases above the limit set by Eq. (16) the value of r_0 required to achieve ρ equal to 150 in the presence of a source with $R_S = 1.6 \times 10^{-4}$ decreases rapidly to a minimum.

It then increases exponentially for values of t/τ_H larger than 6 to 8 depending on f. For times much greater than τ_H, large values of r_0 are required if an enhancement of D/H as large as 150 is still realized after a long period of quasi-steady state between escape and a source of low R_S. It is most interesting that, for $f \geq 0.02$, an enhancement R of 150 cannot be obtained at any time unless r_0 is larger than about 180. The signature of a massive loss of hydrogen will have persisted, no matter what the hydrogen steady state lifetime, if R is large.

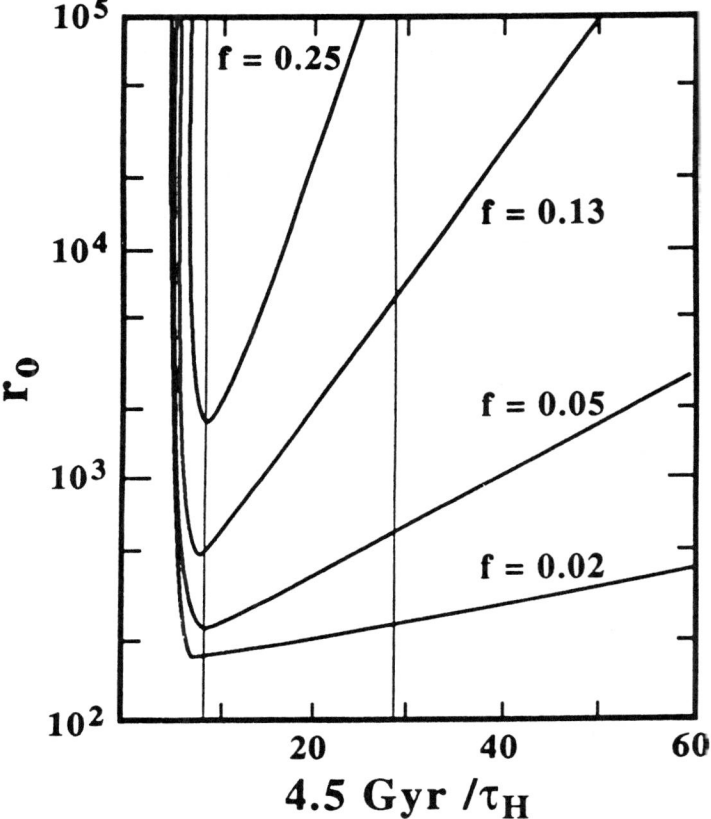

Figure 2. $r_O = [H_O]/[H_{SS}]$ as functions of time measured in units of τ_H for various fractionation factors and a D/H enrichment, ρ of 150.

On the other hand, if the planet never had a large amount of water exposed to its atmosphere and the sources always have been balanced with escaping hydrogen, so that r_0 is identically unity, $R(t)$ would have risen from R_S to R_S/f monotonically. An enrichment of 150 times r_0 would have been attainable with a source R_S of 1.6×10^{-4} only if f is less than 6.4×10^{-3}. τ_{SS}

would then be 50 Gyr. If f is as large as 0.15, R_S would have to be 3.75×10^{-3}, 22 times r_0, to produce a steady state of 150 times R_0 if $R_0 = 1.6 \times 10^{-4}$.

Unfortunately, it is not clear that the ideal case in which φ_1 varies linearly with [H] is often attained on any planet including Venus. One practical problem is that the tropospheric hydrogen may be mostly bound in a condensible, such as water vapor, which is prevented by a cold trap from passing freely through the tropopause. This is certainly the case on Earth, where stratospheric CH_4 and H_2 are the sources of much of the hydrogen above the tropopause. Only if the CH_4 and H_2 abundances can be shown to vary linearly with the water vapor content of the troposphere will φ_1 be proportional to [H]. This discussion neglects the water in liquid form, which provides a source for atmospheric hydrogen that surely will balance escape until the oceans run dry.

On Venus the clouds constitute a barrier between the lower atmosphere and the upper atmosphere. The hydrogen mixing ratio is much less above them than it is below them. It remains to be determined how the mixing ratio in the upper atmosphere, determining the escape flux, would be affected if the amount of water vapor in the lower atmosphere were very much—say orders of magnitude—larger than it is today.

In the case of Mars, H_2O is optically thick to the solar ultraviolet radiation which photolyzes it. Increasing the H_2O mixing ratio in the troposphere merely increases the altitude at which H is produced from H_2O. The consequence is that stratospheric H and H_2, which are generated by a complex of chemical reactions involving odd hydrogen and odd oxygen, change only slightly, even when H_2O changes by very large amounts. Consequently, the hydrogen escape flux changes nonlinearly with the total amount of hydrogen below the diffusion bottleneck near the homopause.

IV. DIAGNOSTICS: REQUIRED

The properties of the atmosphere to be determined if the problem of the evolution of water is to be addressed effectively include (1) the D/H ratio in water or some suitable substance such as atomic hydrogen, molecular hydrogen, H_2S or their ions; (2) the abundance of water and other hydrogen constituents of the mixed atmosphere, with their temporal and spatial variations, if any, understood; and (3) the escape fluxes of hydrogen and deuterium suitably averaged globally, diurnally, and over the solar cycle. Also required, but more difficult to obtain, are the average rates for injection of H and D by comets and by outgassing of the interior. Thanks mostly to the Pioneer Venus mission and remote spectrographic sensing from Earth and the Galileo spacecraft, most of the items on the first of these shopping lists by this time have been acquired. It is also possible to set reasonable bounds to the rate at which comets inject water, and the D/H ratio for water on one comet, P/Halley, is now known with good precision (Balsinger et al. 1995; Eberhardt et al. 1995). Magellan has provided enough information about volcanism on Venus to allow speculation about the outgassing history of the planet to be at least informed speculation.

V. DIAGNOSTICS: MEASUREMENTS

A. D/H Ratio

By this time the D/H ratio is one of the most thoroughly measured properties of the planet. These measurements, all concordant, fall into several classes.

1. Measurement of the HDO/H_2O ratio in the clouds by mass spectroscopy;
2. Measurement of the HDO/H_2O ratio in the mixed atmosphere spectrographically from Earth;
3. Measurements of the 1 amu and 2 amu ions in the ionosphere and the interpretation of the measurements in terms of ion chemistry;
4. Calculation of the H and D densities in the upper atmosphere from measurements of H^+, D^+, O^+ and O densities at altitudes where these ions and neutrals are in a chemical steady state.

The first of these measurements was made by the neutral mass spectrometer (LNMS) on the Large Pioneer Venus Probe as the probe descended (Donahue et al. 1982). In the clouds, between about 50 km and 25 km, the inlet leaks to the LNMS were sealed by sulfuric acid droplets. Very large signals in several mass channels near 18 amu and 19 amu appeared at this time. The spectrometer was then sampling H_2O and HDO from the sulfuric acid drops. There were also greatly enhanced signals from CH_3D and HD, apparently created by transfer of deuterium from Cytherean HDO to artifact CH_4 and H_2 in the LNMS (Donahue and Hodges 1992). Taking account of all the D and H represented in HDO, CH_3D, HD and H_2O gave a D/H ratio of $(2.5 \pm 0.5) \times 10^{-2}$ for Venus water, 157 ± 30 times the D/H ratio in standard mean ocean water (SMOW).

McElroy et al. (1982) first pointed out that the ion of mass 2 in the ionosphere could not be H_2^+ and must, therefore, be D^+. They concluded that the measurements implied a hundredfold enrichment of deuterium in the atmosphere of Venus. Measurement of the densities of ions at 2 amu and 1 amu by the PVO ion mass spectrometer were analyzed with the help of data from the neutral mass spectrometer. The assumption was made on the one hand that the 2 amu species was H_2^+ and on the other hand that it was D^+. The reactions removing H_2^+ and D^+ are quite different, the first being an ion molecule reaction with CO_2 forming CO_2H^+ and H, and the second a charge exchange reaction with O forming O^+ and D. The alternatives lead to very different height profiles for the 2 amu ions. The D^+ option was clearly favored. One set of analyses by Hartle and Taylor (1983) gave $(2.2 \pm 0.6) \times 10^{-2}$ for the D/H ratio after extrapolation to the homopause. The other set by Kumar and Taylor (1984) gave 1.4×10^{-2} and 2.5×10^{-2} for the ratio when corrected for diffusive separation.

Neutral mass spectrometers have serious problems in measuring the density of neutral atomic hydrogen because of its extreme reactivity. Fortunately, the atomic hydrogen and deuterium densities in the lower thermosphere may be calculated from measurements of the densities of O^+, H^+, D^+ and O with ion and neutral mass spectrometers in the region where the lifetimes of these species for ion molecule reactions are short compared to diffusion times. Because the resonant reactions

$$H^+ + O \leftrightarrow H + O^+ \tag{29}$$

and the H^+ removal reaction

$$H^+ + CO_2 \rightarrow COH^+ + O \tag{30}$$

determine the densities of H^+, O and H in this region, the H density may be calculated from the relationship

$$n(H) = \{n(H^+)/n(O^+)\}\{k_{29}n(O)(T_i/T)^{\frac{1}{2}} + k_{30}n(CO_2)\} \tag{31}$$

where n are number densities, $k_{29,30}$ are the reaction rate coefficient for reactions (29) and (30) and T_i and T are the ion and neutral temperatures (Brinton et al. 1980). An analogous expression holds for $n(D)$. The measurements of these quantities will be discussed in detail in Sec. V.F. H and D densities were obtained in the night time bulge region of the thermosphere during the first three Cytherean years that PVO was in orbit and during its reentry during 1992. When extrapolated to the homopause the ratio of deuterium to hydrogen densities obtained was 2.17×10^{-2}. Considering the uncertainty introduced by the need to extrapolate the thermospheric ratios, the agreement among all these determinations must be considered excellent.

Near infrared spectral bands of H_2O and HDO emitted deep in the atmosphere of Venus in a window near 2.3 μm have been analyzed to determine the mixing ratios of H_2O and HDO by two groups. One working from Mauna Kea (de Bergh et al. 1991) analyzed their observations to derive a D/H ratio of $(1.9 \pm 0.6) \times 10^{-2}$ (120±40 times SMOW) in the mixed atmosphere. The other, observing from the Kuiper Observatory Aircraft (Bjoraker et al. 1992) obtained $(2.5 \pm 0.3) \times 10^{-2}$ (157±18 times SMOW).

All of these measurements are in good agreement with each other. The ionospheric measurement tends to be a little lower than those effectively obtained by observing the mixed atmosphere. This may be because of the need to extrapolate measured densities to the homopause in the case of the measurements made in the upper atmosphere. Because of the possibility of error in this procedure, we shall prefer the "tropospheric" measurements and use a value of $(2.4 \pm 0.5) \times 10^{-2}$ (150 ± 30 × SMOW) in this chapter.

B. Water Vapor Abundance

There is no shortage of measurements of the water vapor mixing ratio in the atmosphere of Venus. Unfortunately, they lack the concordance of the D/H measurement. The measurements reported before April 1992 are recorded in Fig. 3. Those obtained by gas chromatographs and humidity sensors on

EVOLUTION OF WATER 397

Venera 13 and 14 and the Pioneer Venus Large Probe are orders of magnitude higher than the rest. Given the difficulties with contamination confronting *in-situ* measurements such as these and their gross disagreement with other measurements free of these problems we regard these data as unreliable. Analysis of near infrared spectra of the day sky obtained by the Venera 11 and 12 spectrophotometer yields a mixing ratio that decreases monotonically from 200 ppm at 50 km to 20 ppm at the surface. This result is perplexing because such a profile has been obtained by no other instrument, including other infrared soundings on the night side. Most of the near infrared soundings in the 1.7 and 2.3 μm H_2O bands radiated on the night side of Venus and reported before 1992 gave results that clustered around 40 ppm from 50 km to low altitudes. There is one exception reported by Bell et al. (1991) where one observation gave a mixing ratio of 200 ppm among others much lower. A large latitude dependence in water vapor was invoked to account for the variation of infrared net fluxes measured by the four Pioneer Venus entry probes (Revercomb et al. 1985). These refer to heights above 40 km. The values shown in Fig. 3 decrease with latitude. The lowest are near 60°, the next near 30° and the very high measurement near the equator.

The LNMS data at 18 and 19 amu have been analyzed by Donahue and Hodges (1992,1993). Their results are plotted in Fig. 4. The mixing ratio obtained by fitting the H_2O counting rate profile to the counting rate profile for ^{36}Ar was $28\pm^{18}_{5}$ ppm, between 10 and 26 km, $6.3\pm^{4}_{3}$ ppm at 51.3 km, $4.2\pm^{3}_{2}$ ppm at 55.3 km, and an apparent decrease below 10 km to 7 ± 3 ppm at 0.9 km. These mixing ratios are normalized to the ^{36}Ar abundances of 30 ppm obtained by the LNMS. The decrease in H_2O mixing ratio below 10 km may only be apparent. It is a consequence of a failure of the HDO counting rate, which is due mostly to Cytherean water, to rise below 10 km as rapidly as the ^{36}Ar counting rate. The H_2O signal is dominated by the terrestrial contaminant and could easily mask a variety of profiles for Venus H_2O. It is quite possible that deuterium atoms are being transferred to the very abundant deuterium-poor methane in the mass spectrometer. If there is really a decrease in the H_2O mixing ratio, there should be another hydrogen compound created to take the place of water vapor. There is no evidence for a hydrogenic compound in the LNMS spectrum into which the H_2O has been transformed by a gas phase reaction. At no mass channel belonging to a candidate reaction product can there be found evidence for a count rate sufficiently large at 0.9 km to make up the deficiency in H_2O there and which then decreases to zero at 10 km.

Since 1992 the near-infrared spectra of emissions from the dark side of Venus, obtained by Crisp et al. (1991a,b) and Bézard et al. (1990), have been simulated by Pollack et al. (1993). These data are fit best by a water vapor mixing ratio that is constant at 30 ± 10 ppm from 40 km to about 10 km, but may decrease between 10 km and the surface. The case for the decrease below 10 km is by no means robust. The profile obtained by the LNMS

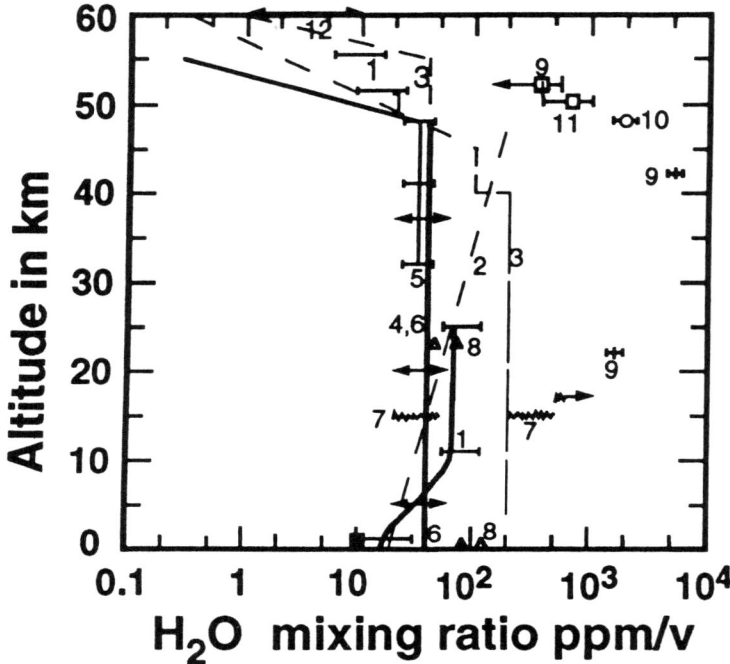

Figure 3. Water-vapor mixing ratios. Sources are (1) Pioneer Venus large-probe mass spectrometer (Donahue and Hodges 1992); (2) Venera 11 and 12 spectrophotometer (Moroz et al. 1982); (3) Bell et al. (1991); (4) Bell et al. (1991); (5) de Bergh et al. (1991); (6) Crisp et al. (1991a,b); (7) Revercomb et al. (1985); (8) Venera 11 and 12 mass spectrometers; (9) Pioneer Venus sounder probe gas chromatograph, Oyama et al. (1980); (10) Venera 13/14 humidity sensors; (11) Venera 14 gas chromatograph; (12) terrestrial spectroscopy; and (13) Galileo, Carlson et al. (1991).

(Fig. 4) well represents their reconstruction. High-resolution spectra obtained by Bézard et al. (1991) and de Bergh et al. (1992) in the 1.18 μm, 1.7 μm and 2.3 μm windows are also well fit by a constant mixing ratio profile of 30±15 ppm. Galileo NIMS measurements also gave 30±15 ppm with no horizontal variations exceeding 20% over a wide latitude range (Drossart et al. 1993). Water vapor measurements are also discussed in the chapter by Esposito et al.

C. Methane and H_2

The PVO LNMS registered very large signals at 2, 3, 15, 16 and 17 amu (Donahue and Hodges 1993). All of the 2 and 3 amu signal detected can be accounted for by terrestrial H_2 emanating from instrument surfaces, probably in the getter pumps and, except when the leaks were plugged, from H_3^+ produced in the ion source by electron bombardment of CH_4. A contribution of H_2 from the Venus atmosphere with an appropriately high HD/H_2 ratio associated with more than about 2 ppm of H_2 can be excluded, and the data

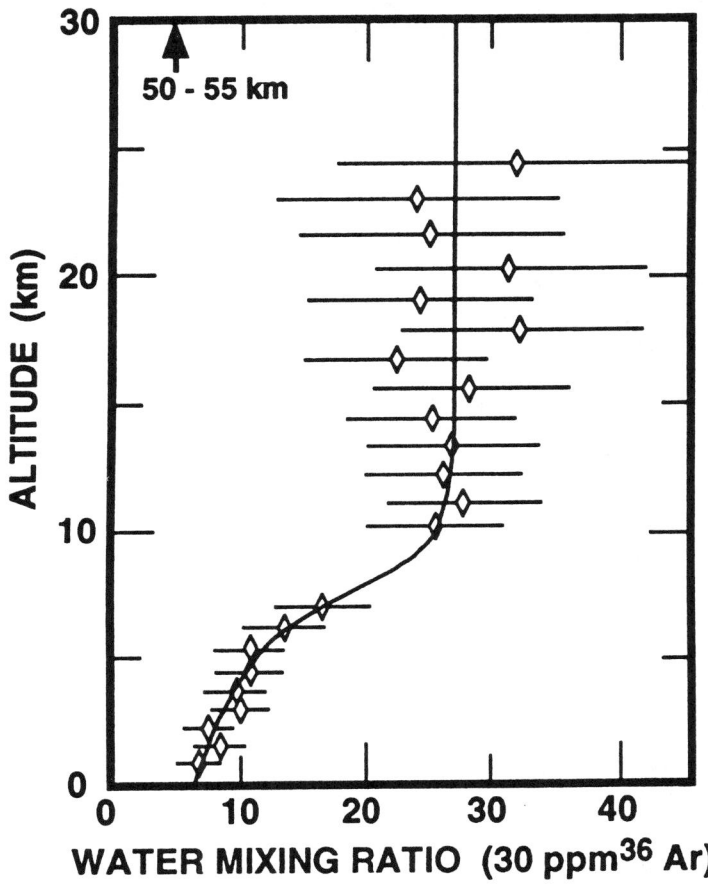

Figure 4. Water-vapor mixing ratios assuming a ^{36}Ar mixing ratio of 30 ppm (3σ error bar).

are consistent with no H_2 at all. The CH_4, in addition to that introduced into the LNMS deliberately to help in mass peak deconvolution, can be accounted for by methanation, probably occurring on getter surfaces and involving instrumental H_2 and CO from the atmosphere. Thus, this CH_4 component mimics an atmospheric gas (Donahue and Hodges 1993). The feature at 17 amu, except when the leaks were plugged, and below 10 km, appears to consist of a mixture of $^{13}CH_4$, and CH_3^+ generated in the ion source by an ion-molecule reaction involving CH_4^+ and CH_4.

D. Hydrogen in the Thermosphere: Lyman-α Scattering

Ultraviolet spectrometer (OUVS) Lyman-α spectra from 20 orbits spanning the first three Venus years have been analyzed by Paxton et al. (1988). By fitting emission rates to theoretical models for resonantly scattered solar Lyman-

α they have determined the atomic hydrogen number density and vertical flux at the base of the exosphere (200 km). Both remained remarkably constant at $(6.0 \pm 1.5) \times 10^4$ cm^{-3} and $(7.5 \pm 1.5) \times 10^7$ cm^{-2}s^{-1} in the subsolar region as solar activity varied. The integrated vertical column of hydrogen above 110 km was found to be $(3.6 \pm 1) \times 10^{13}$ cm^{-2}. Recently, re-analysis of these measurements by Hartle et al. (1995) has revised these values to $(7.5 \pm 2) \times 10^4$ cm^{-3} at 200 km and $(2.9 \pm 0.6) \times 10^7$ cm^{-2}s^{-1}. The large flux in a region where the escape flux is much smaller than 10^7 cm^{-2}s^{-1}, supplies lateral flow of hydrogen to the night side, where the hydrogen eventually escapes to space or descends below 100 km and returns to the day side. Light atoms participate in the planetary-wide circulation of winds to the night side in the thermosphere and back to the day side in the middle atmosphere. Consequently, the densities of the light species H, D and H_2 display a prominent bulge in the post-midnight sector of the nighttime hemisphere. The circulation of the high mass major atmospheric species is divergence free. Because of their much greater scale height, this results in a net loss of light gases on the day side in the thermosphere and a net gain on the night side. The consequent nightside bulge in abundance is shifted into the dawn sector by the superrotation of the atmosphere. Further discussion of the effects of superrotation in the upper atmosphere is contained in the chapter by Bougher et al.

E. Escape of Hydrogen and Deuterium

Until recently, two processes were thought to make an important contribution to hydrogen escape. Exospheric temperatures are only 300 K on the day side and 123 K on the night side. At such temperatures, the planet-wide average Jeans escape flux is about 5×10^6 cm^{-2}s^{-1}, utterly negligible compared to nonthermal loss processes. Collision of energetic oxygen atoms produced by dissociative recombination of O_2^+ ions with thermal hydrogen was once considered a major source, contributing a planet-wide average flux of 1.2×10^6 cm^{-2}s^{-1}, with zero fractionation factor (Rodriguez et al. 1984) or 3.5×10^7 cm^{-2}s^{-1} with a large fractionation factor of 0.31 according to Gurwell and Yung (1992). However, Hodges (1993) has pointed out that these calculations neglected the effect of collisions by the initially hot atoms with thermal oxygen that degrade their velocities before they can collide with hydrogen in the exosphere. After the effect of these collisions has been accounted for, the hydrogen loss rate following such collisions is reduced to insignificance, compared to the dominant processes involving hydrogen ions. The first of these is charge exchange in the exosphere between H^+ ions with greater escape velocity in the tail of the hot ion distribution and thermal hydrogen atoms. Two calculations of the loss rate due to these collisions gave discordant results. Most of the loss occurs in the 20% of the exosphere comprising the bulge on the night side. The Monte Carlo model of Hodges and Tinsley (1986) predicted a planet-wide average loss rate of 2.8×10^7 cm^{-2}s^{-1} for this process. A treatment by Rodriguez et al. (1984) gave 3.2×10^6 cm^{-2}s^{-1}. Donahue and Hartle (1992) have criticized both models for using unrealistic

H and H^+ distributions. They also pointed out that the calculations consider only solar maximum conditions whereas H^+ ion densities are reduced to at most 0.15 their solar maximum values during solar minimum (Fig. 5). When solar activity is high the ionopause is far from the planet. O^+ ions created in the daytime ionosphere are transported in great numbers to the night side, where they create H^+ ions in the bulge region by charge transfer collisions with neutral hydrogen. As solar activity decreases the density of ions in the ionosphere decreases and the ionopause closes in. Transport of O^+ to the night side is reduced and production of H^+ with the bulge consequently decreases considerably (see the chapter by Fox and Kliore). Another theoretical effort to recalculate the charge exchange loss rate using the densities measured by PVO for a complete range of solar activity is required. In the meanwhile, Donahue and Hartle (1992) recommend a rate of 9×10^6 $cm^{-2}s^{-1}$ for the planet-wide average escape flux during solar maximum. This recommendation is based on an estimate of the effect of adjusting the properties of the two exospheric models in question to realistic ones. Deuterium also is lost by charge exchange. The fractionation factor for the process has been estimated to be 0.02 (Krasnopolsky 1985).

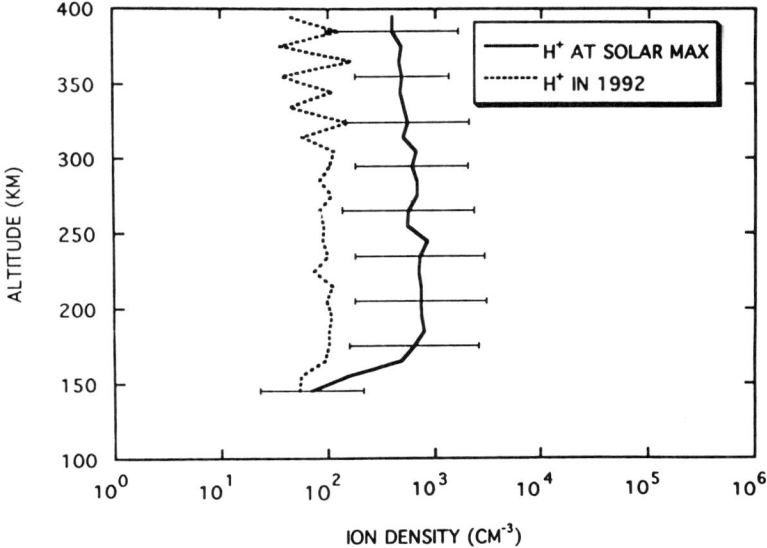

Figure 5. Venus hydrogen ion densities measured by PVO IMS.

Recently, another important escape mechanism for light ions has been identified. It is the analog of the terrestrial polar wind in which light ions are accelerated to escape velocity in the charge separation electric field that maintains the major ions and electrons of the ionosphere in a stable vertical distribution despite their mass difference. Hartle and Grebowsky (1993,1995) have shown that hydrogen and deuterium ions are flowing upward in these fields with more than enough velocity to escape. The outward flux of H^+ in the bulge during solar maximum is 7.5×10^7 cm^{-2}s^{-1}, for this mechanism (E) giving a planet-wide average flux of H^+ during solar maximum of 1.5×10^7 cm^{-2}s^{-1}. The D^+ planet-wide rate was measured to be 5.6×10^4 cm^{-2}s^{-1}. The fractionation factor for the process is thus 0.15.

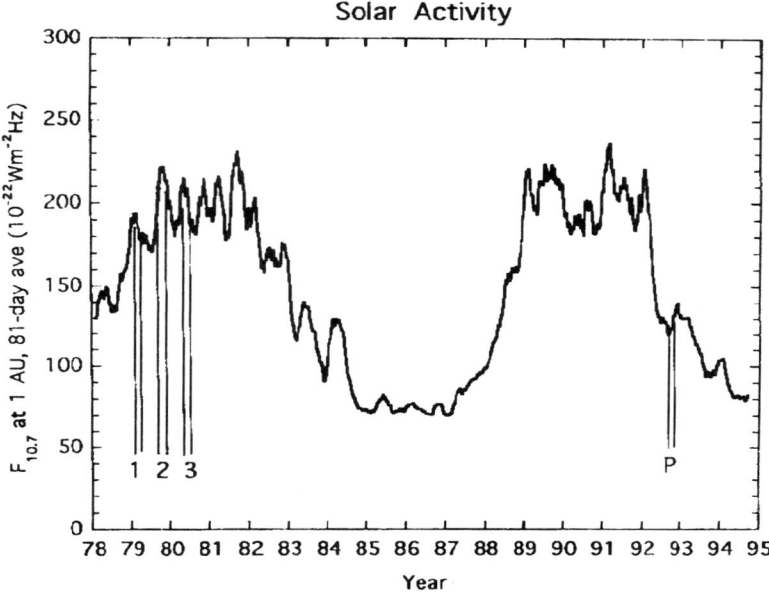

Figure 6. Solar activity from 1978 to 1995.

These escape fluxes will vary with solar activity. During the re-entry phase of PVO in the fall of 1992 the 10.7 cm radio flux was at a level of 125×10^{-22} Wm^{-2}Hz^{-1}. In 1978 and 1979 when the orbiter was exploring during solar maximum, the 10.7 cm flux was between 200 and 250×10^{-22} Wm^{-2}Hz^{-1} (Fig. 6). Because the H^+ ion densities in 1992 were reduced by a factor of 0.15 and the electric field loss is proportional to the ion densities, the planet-wide average (E) flux of H^+ was reduced to 2.3×10^6 cm^{-2}s^{-1}. f remained at 0.15. Thus, the solar cycle average electric field driven flux ϕ_{1E} is 9×10^6 cm^{-2}s^{-1} with a fractionation factor of 0.15. Because the charge exchange loss rate (CE) is proportional to the product of the ion and neutral densities it should be reduced from 9×10^6 cm^{-2}s^{-1} to only 7×10^6 cm^{-2}s^{-1} in 1992. If the electric field escape flux and the charge exchange flux are additive and independent, the solar maximum planet-wide hydrogen escape flux ($E + CE$) would be 2.4×10^7 cm^{-2}s^{-1} with a fractionation factor of 0.10. The low solar activity flux ($E + CE$) in 1992 would be 8.9×10^6 cm^{-2}s^{-1} with f equal to 0.06. The solar cycle average escape flux would be 0.9 and

0.8×10^7 cm^{-2}s^{-1} for the (E) and (CE), respectively, making the solar cycle average flux $(EC + CE)$, 1.7×10^7 cm^{-2}s^{-1} and 0.1 the effective fractionation factor. These measured and estimated fluxes are listed in Table I.

TABLE I
Planetwide $\langle \Phi_1 \rangle$ in 10^7 cm^{-2}s^{-1}

Process	ϵ	Eq. (33)	Sec. V.E	$\langle \bar{\varphi}_{1e} \rangle$
E	0.15	6.3–5.7	7.5	0.9
CE	0.73	7–6.3	4.5	0.8
$E + CE$	0.37	6.6–5.9	12	1.7

F. Transition from Solar Maximum to Solar Minimum: Flux Calculations

The PVO entered the atmosphere in the autumn of 1992. As it did so its instruments, including the neutral and ion mass spectrometer, determined ion and neutral densities in the nighttime ionosphere. As solar activity declined the hydrogen escape flow out of the bulge diminished, but the horizontal inflow from the dayside hemisphere continued undiminished. Consequently, the hydrogen and deuterium content of the bulge grew until the downward flux had increased sufficiently to balance the inflow. The densities of hydrogen and deuterium during the first three years of the mission and in 1992 have been calculated from the orbiter data following the technique discussed in Hartle et al. (1996) and in Sec. V.A. The results are shown in Figs. 7 and 8. The average hydrogen density in the bulge increased by a factor of 6.7 and that of deuterium by a factor of 4. This increase is due to the reduced escape flux and not to a change in the circulation from the night to daytime hemispheres as suggested by the smaller deuterium effect and the absence of any modulation for helium (Fig. 8). (In Fig. 9 we show the thermospheric D/H ratios during the four years for which data are available. These are the basis for the discussion in Sec. V.A.)

The number of hydrogen atoms in the thermosphere and exosphere Ψ is governed by the simple relationship

$$\frac{d\Psi}{dt} = \varphi_S A/2 - (\varphi_D + \varphi_e) A_B \qquad (32)$$

where φ_S is the upward flux into the thermosphere on the day side (area $A/2$), φ_D is the return flux into the mesosphere at the base of the bulge (area A_B, which is about 20% of the planetary area A), and φ_e is the escape flux from the bulge. It is possible to calculate the escape flux Φ_e during solar maximum and ϕ_e from the bulge during solar minimum in the steady state when $d\Psi/dt \cong 0$ in Eq. (32).

$$1.5 \Phi_S = \Phi_D + \Phi_1 \qquad (33)$$

$$2.5 \phi_S = \phi_D + \phi_1. \qquad (34)$$

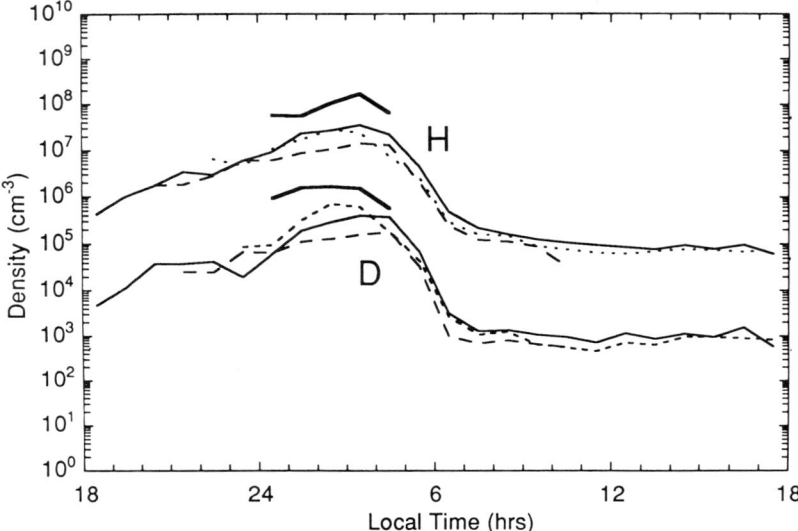

Figure 7. Hydrogen and deuterium PVO densities near 220 km, day time, and 160 km, night time. Heavy solid curve, 1992; light solid curve, Venus year 1; dotted curve, year 2; dark curve, year 3.

Φ_D and ϕ_D are proportional to the density in the bulge during solar maximum and solar minimum respectively, therefore

$$\Phi_D = 0.15\phi_D. \tag{35}$$

As was noted in Sec. V.A, Φ_S is known from PVUVS data (Paxton et al. 1988), as revisited by Hartle et al. (1996), to be 2.3×10^7 cm^{-2}s^{-1}. Unfortunately, the same is not true of ϕ_S, but there is no apparent reason that atmospheric circulation which is responsible for the flow of hydrogen should depend on solar activity. Thus, it is probably safe to assume that ϕ_S and Φ_S are the same. In any event diffusion limitation prevents ϕ_S from exceeding 1.5 Φ_s and so the change in escape flux between solar maximum and solar minimum should lie between 2.5 Φ_s and 1.9 Φ_s with the first value more likely. In terms of the solar maximum escape flux this means that

$$6.2 \times 10^7 \geq (1 - 0.15\epsilon)\Phi_1 \geq 5.6 \times 10^7 \text{ cm}^{-2} \text{ s}^{-1} \tag{36}$$

where ϵ is the appropriate ratio ϕ_1/Φ_1. In Table I the escape fluxes called for by this relationship are contained with the measured electric field flux and estimated charge exchange flux discussed in Sec. V.E. Even paying due respect to all the uncertainties in the measurements on which these estimates are based the electric field escape process alone appears to give reasonable agreement between the measured observed value and the one that explains the solar cycle variations in bulge densities. It is difficult to reconcile a very large

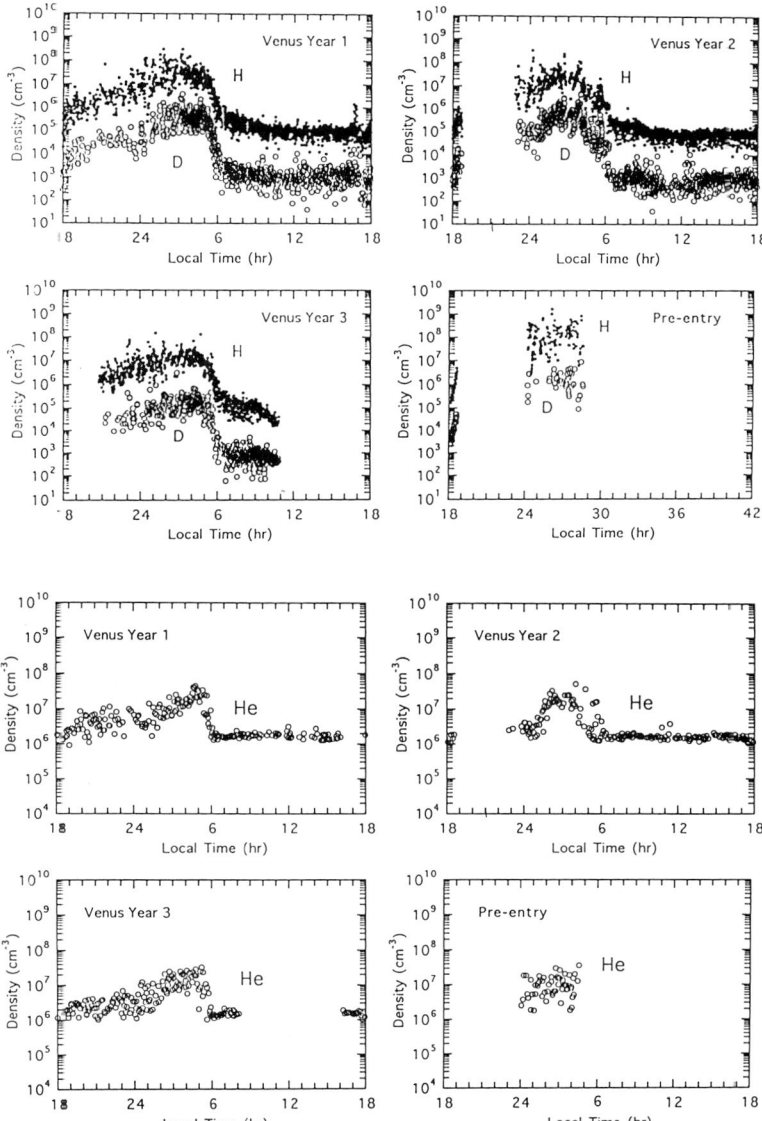

Figure 8. All H, D and He PVO density measurements.

contribution from H^+, H charge exchange with these results. In any event a very small planetary average flux of about 1.5×10^7 cm^{-2}s^{-1} during solar maximum and 2×10^6 cm^{-2}s^{-1} during times of low solar activity is indicated.

The effective H and D solar cycle average planet wide escape fluxes $\langle \bar{\varphi} \rangle$ resulting from this analysis of the bulge are also tabulated in Table II. The

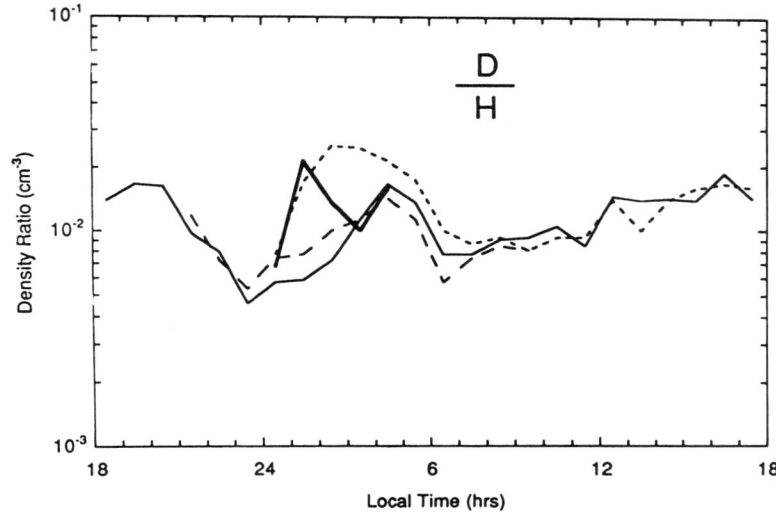

Figure 9. PVO ratio of D to H density. Heavy curve, 1992; light solid curve, year 1; dotted curve, year 2; dashed line, plot of year 3.

TABLE II
$\langle \bar{\varphi}_1 \rangle$ Hydrogen in 10^7 cm^{-2}s^{-1}

	E	CE	E + CE	Bulge	Synthesis
ϵ	0.15	0.73	0.37	0.15	0.15
$\langle \bar{\varphi}_1 \rangle$	0.9±0.1	0.8	1.6	0.6–0.75	0.8±0.2
f	0.15	0.02	0.10	0.44	0.15–0.44

$\langle \bar{\varphi}_2 \rangle$ Deuterium in 10^4 cm^{-2}s^{-1}

$\langle \bar{\varphi}_2 \rangle$	3.2±0.6	0.5	3.7	8–10	NA

"synthesis" flux in the table captures both the measured E flux and the flux determined from bulge properties in the case of hydrogen. ϵ is assumed to be 0.15. No such synthesis is possible for deuterium.

A similar analysis for deuterium calls for a solar maximum steady state flux, of about 1.4 to 1.7×10^5 cm^{-2}s^{-1} planet wide, (8 to 10×10^4 cm^{-2}s^{-1} solar cycle average) with only electric field acceleration important. Combined, these fluxes result in a very large fractionation factor of 0.44. It is difficult to conceive of a loss mechanism so highly fractionating in the modern atmosphere of Venus. The "observed" loss rate for deuterium ions (Hartle and Grebowsky 1993) was only one third as large, 5.6×10^4 cm^{-2}s^{-1}. This discrepancy could be removed if φ_S for deuterium were reduced by a factor of about 3, which can be accomplished only by reducing D/H to about 1.0×10^{-2}, given the observed density at 220 km during solar maximum years 1, 2 and 3. This would be an unacceptably large correction in the HDO/H$_2$O ratio.

VI. EVOLUTION OF WATER

We shall first consider two cases based on the fluxes discussed in Sec. V.E. In one the measured charge exchange flux will be added to the electric field flux. In the other it will not. In the first case, the average escape flux is 1.7×10^7 cm^{-2}s^{-1} with a fractionation factor of 0.10. In the second, it is $(9 \pm 1.0) \times 10^6$ cm^{-2}s^{-1} and f is 0.15. If the present hydrogen abundance [H] is set by the 30 ppm of water vapor in the lower atmosphere, the characteristic time required to establish a steady state in hydrogen is

$$\tau_H = (160 - 300) \text{ Myr.} \qquad (37)$$

τ_D, to establish a steady state in deuterium and in R, becomes

$$\tau_D = (1.6 - 2.1) \text{ Gyr.} \qquad (38)$$

Finally, the relationship of R_{SS} to R_S is

$$R_{SS} = (6.7 - 10) R_S. \qquad (39)$$

Such large values of f do not allow a highly differentiated "mature" steady state if the source is not rich in deuterium.

One end member scenario for evolution of water on Venus has the hydrogen and deuterium today as remnants of an early larger supply of water. In the case of Rayleigh fractionation, the factor r in Eq. (6) for $[H_O]/[H]$ is either 262 or 363, depending on choice of f. Originally the water on Venus would have been between 3.9 and 5.4 m deep if spread uniformly over the planet's surface as a liquid. This is between 0.12 and 0.17% of a full terrestrial ocean. The mixing ratio of this water in the vapor phase would have been between 1.0 and 1.4% if Venus then had the same dense atmosphere it has now. According to Watson et al. (1981), vigorous hydrodynamic escape or "blowoff" of hydrogen, driven by solar extreme ultraviolet radiation heating, would have occurred if the mixing ratio of H_2 in the upper atmosphere exceeded 2%. Because extreme ultraviolet radiation from the early Sun was almost surely enhanced, this coincidence led Donahue et al. (1982) to suggest that Venus may have started with the equivalent of a full terrestrial ocean, either in steam or liquid form. Blowoff, for which the fractionation factor was close to zero, would have exhausted all but a trace of this water in 280 Myr, without changing the D/H ratio appreciably. This concept was elaborated by Kasting and Pollack (1983), Kasting et al. (1984) and Kasting (1988) who calculated the blowoff escape rate that would have been reached in a runaway greenhouse induced when Venus developed an atmosphere in which an ocean had been converted to steam. They assumed a radiatively equilibrated atmosphere in which there were 90 bar of CO_2 and showed that before the Sun reached its present luminosity, surface temperatures on Venus would become so high that all the water in an ocean would have been converted to steam with a surface pressure of 270 bar. Blowoff would then follow. This model was superseded by one in which the effects of convection and condensation were taken into account. Initial conditions called for a surface pressure of one bar of an N_2, O_2 mixture, one bar of H_2O and an ocean, liquid because the moist convection left the surface temperature below 100° C. CO_2 was presumed to be converted to carbonate as on Earth. The stratosphere was wet in this "moist greenhouse." Hydrogen was rapidly lost as in the "runaway greenhouse." A cold trap developed only after nearly all of the water was lost, because of the low pressure of the noncondensible gases. Thus, a serious defect of the previous model, in which the development of a cold trap in the presence of 90 bar of CO_2 stopped blowoff with 20 bar of water left was avoided. In both models, atomic oxygen was presumed to have been lost, swept along in the flood of rapidly flowing hydrogen, and also oxidizing the crust. After the blowoff stopped, the remaining oxygen could have been reduced to its present insignificant level by reaction with CO. This gas should have accompanied outgassing CO_2 in a much larger relative abundance than prevails today. After the water had gone, no medium in which carbonate could form would have remained, and CO_2 would have accumulated in the atmosphere to its present level.

As was discussed in Sec. II, the simple Rayleigh fractionation relationship Eq. (6) cannot be used to determine the initial hydrogen inventory if hydrogen is in or near a steady state, that is, if $P_1 \cong \varphi_1$ and $d[H]/dt \cong 0$. It is not the amount of hydrogen remaining in the atmosphere [H] that is decisive in establishing this situation. It is the size of φ_1 compared to P_1. The fact that φ_1 is so small makes the possibility that it is balanced by a source of hydrogen, particularly from comets, very great (Grinspoon 1987; Grinspoon and Lewis 1988). The Venus D/H ratio is much larger than it would be if there were a mature steady state between escape and a source of incoming cometary water. The D/H ratio in Comet P/Halley is $3.16 \pm 0.34 \times 10^{-4}$ (Balsinger et al. 1995; Eberhardt et al. 1995). So, if f is 0.15, the steady state D/H ratio on Venus when cometary water having such a D/H ratio is injected, becomes 2×10^{-3}—less than a tenth of its present value. As Gurwell has pointed out, if there is a source with R_S equal to 1.6×10^{-4}, balancing escape of H, a ratio ρ of 150 today, when t/τ_H is between 14.5 and 28 can be attained only if r_O is between 1300 and 2300 (Figs. 1 and 2). Thus, even more early water would be needed than if the cometary source were absent. 19.5 to 34.5 m of water, 5 to 6.3 times as much as needed in the case of simple Rayleigh fractionation, would be required. On the other hand, an outgassing source, such as suggested by Grinspoon (1993) involving a very depleted mantle, would be sufficient to balance the current escape flux. The challenge is to demonstrate how it can be sufficiently fractionated.

A similar analysis based on the fluxes deduced in Sec. V.F—$\langle \bar{\varphi} \rangle \cong 7.5 \times 10^6$ cm^{-2}s^{-1}, $f = 0.44$—calls for the ratio of the primeval reservoir to the present one, r in Eq. (6), to be very large, namely 7700, if there are no sources. This would require 115 m of water, 3.5% of a terrestrial ocean.

In the case of a hydrogen steady state based on Eq. (25), with τ/τ_H equal to 12.5, r becomes 3.5×10^4. The required early reservoir would grow to 520 m of water, 16% of a full terrestrial ocean. τ_{SS} would be only 830 Myr, but the R_S required for a steady state with $R = 2.4 \times 10^{-2}$ would grow to 1.1×10^{-3}. (Some quantities determined here differ slightly from those quoted by Hartle et al. [1996] because ρ is assumed to be 150 in this chapter but 157 in that paper.)

In fact, the influx of cometary water on Venus probably fails to balance the loss of hydrogen to space by about an order of magnitude. Shoemaker and colleagues (Shoemaker and Wolfe 1982; Shoemaker et al. 1990) estimate the fluence of active cometary mass on Earth, and consequently also Venus, to 5×10^{17} g yr^{-1} for the past 100 Myr and 1.6×10^{17} g yr^{-1} for the past Gyr. Shoemaker et al. (1994) also argue that the flux of extinct periodic comets may be comparable to the flux of active comets. If so and if 30% of the cometary mass is water, the effective or average source of cometary water P_1 on Venus is presently between 0.7 and 1.4×10^6. The average value of P_1 over the past billion years would have been a factor of 3 larger. If the hydrogen source from comets is that small the estimates of H_O/H from Eq. (12), based on steady state assumptions, are too large. Those based on simple Rayleigh fractionation (Eq. 6) should be valid unless they are important endogenous sources of highly fractionated water, such as those Grinspoon (1993) has discussed. A constant ratio K between the escape flux φ_1 and the hydrogen burden of the atmosphere H, now in the neighborhood of 0.3 Gyr^{-1}, is obviously incompatible with an amplification factor H_O/H of approximately 400 given by Eq. (6). To obtain such a "small" loss of hydrogen, τ_H must be about 0.75 Gyr. Alternatively, the present state of the atmosphere would need to be very close to a steady state

$$[H_{SS}] = 0.9996 \, [H]. \quad (40)$$

This would require that

$$P_1 = 0.9996 \varphi_1. \quad (41)$$

It is much easier to accomodate a constant value of τ_H to obtain the amplification ratio H_O/H of 7700 called for by the Rayleigh fractionation relationship (Eq. 6) if f is as large as 0.44:

$$\left(1 - \frac{P_1}{\varphi_1}\right) = 7700^{\frac{-\varphi_1 t}{H}} \quad (42)$$

This condition would be satisfied if φ_1 is 7.2×10^6, as given by the bulge study, P_1 is 1.4×10^6 cm^{-2}s^{-1} and the mixing ratio of water vapor is 40 ppm instead of 30 ppm. Alternatively, for example, it can be satisfied with 30 ppm of water vapor, $\varphi_1 = 5.5 \times 10^6$ cm^{-2}s^{-1} and $P_1 = 1.5 \times 10^6$ cm^{-2}s^{-1}.

Proposals for rather different scenarios have been made by Grinspoon (1987, 1993) and Grinspoon and Lewis (1988). These generally have involved comets or outgassing of the interior, as a consequence of volcanism or massive

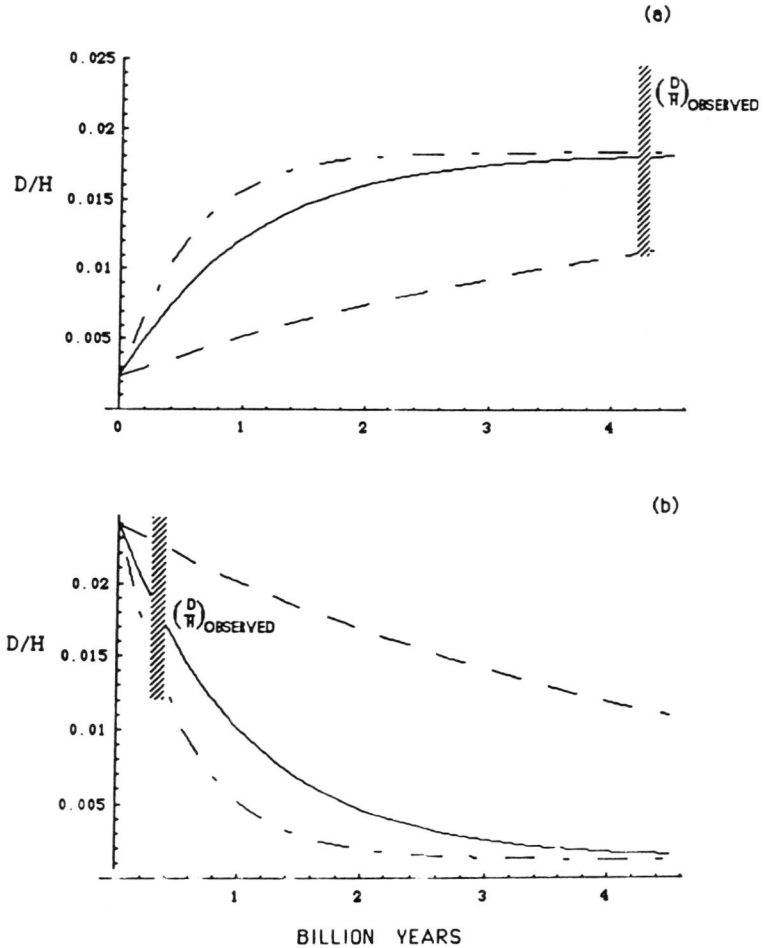

Figure 10. Evolution of the deuterium-to-hydrogen ratio from Eq. (18) using $f = 0.13$. Solid curve is for an H escape flux of 2×10^7 cm^{-2}s^{-1}, dashed and dot-dash curves are for fluxes of 0.4 and 3.7×10^7 cm^{-2}s^{-1}, respectively (de Bergh et al. 1991). Observed D/H shown is 120 ± 40 times the terrestrial value. (a) is for the case of $(D/H)_0 = 15$ times terrestrial. In (b), R_S = terrestrial, and $(D/H)_0 = 150$ times terrestrial.

resurfacing events, as sources of hydrogen. Given the large fractionation factor that now appears to characterize escape of hydrogen and deuterium, it is not possible for planetary escape to approach a mature steady state in which R is as large as it is today if cometary sources of water are as poor in deuterium as comet Halley or terrestrial water. The required value of R_S

is too large. Grinspoon (1993) has proposed as steady state candidates two examples of possible mantle sources. In one case, the outgassing water would be fractionated to within an order of magnitude of the present atmospheric value of R, the outgassing rates and hydrogen loss rates would be balanced and R close to its steady state value. Grinspoon proposed that the mantle of Venus may have attained its enhanced R_S because of fractionation in the creation of a severely depleted mantle, or because of retention of the large D/H signature in water frozen into the mantle after the planet had undergone an early period of highly fractionating loss of hydrogen. Alternatively, Grinspoon suggests that a massive resurfacing event within the past billion years released water from the interior. The atmosphere was then supposed to have experienced rapid loss of hydrogen in which R was enhanced to a value even higher than it has today. Subsequent loss of hydrogen would then have caused R to decrease to its present value as it declines toward a steady state in both D and H. The suggested evolution of R for these examples is shown in Fig. 10 (Grinspoon 1993). Note that the D/H range in this figure is based on the measurement of de Bergh et al. (1991), which is centered at 120 times terrestrial, rather than 150 times terrestrial.

VII. OUTSTANDING PROBLEMS

The present hydrogen inventory of Venus and the D/H ratio seem to be well understood, although there are some observations which make the atmosphere out to be much wetter than it usually seems to be. Recalculation of charge exchange loss rates, which are based on our present understanding of the temporal history of Venus neutral and ionized hydrogen densities and their variation with solar activity, are needed. The type of calculation once performed by Kumar et al. (1983) to determine the evolution of hydrogen and deuterium loss mechanisms over the 4.5 Gyr history of the planet but based on our present understanding of loss mechanisms, such as the electric field process, needs to be repeated. In view of the long time required for deuterium to reach a steady state, it would be instructive to investigate the effect on the present D/H ratio and water abundance of including the contribution of the remnants of a primitive "ocean" 10 m deep on "steady-state" degassing models, such as those proposed by Grinspoon (1993). Models of catastrophic mantle outgassing as sources of today's highly fractionated water vapor need to be bolstered with more robust and detailed foundations than those with which they are currently provided. It would be useful to have a study of the Venus thermosphere and mesosphere that continuously spans a complete solar cycle.

REFERENCES

Balsinger, H., Altwegg, K., and Geiss, J. 1995. D/H and $^{18}O/^{16}O$ ratio in the hydronium ion and in neutral water from in situ ion measurements in comet Halley. *J. Geophys. Res.* 100:5827–5834.

Bell, J. F., III, et al. 1991. Spectroscopic observations of bright and dark emission features on the night side of Venus. *Science* 252:1293–1296.

Bézard, B., de Bergh, C., Crisp, D., and Maillard, J.-P. 1992. The deep atmosphere of Venus revealed by high-resolution night-side spectra. *Nature* 345:508–511.

Bézard, B., et al. 1991. High resolution spectroscopy of Venus' nightside in the 2.3, 1.7, and 1.1–1.3 μm windows. *Bull. Amer. Astron. Soc.* 23:1192 (abstract).

Bjoraker, G. L., et al. 1992. Airborne observations of the gas composition of Venus above the cloud tops: Measurements of H_2O, HDO, HF and the D/H and $^{18}O/^{16}O$ isotopic ratios. *Bull. Amer. Astron. Soc.* 24:995 (abstract).

Brinton, H. C., et al. 1980. Venus nighttime hydrogen bulge. *Geophys. Res. Lett.* 7:865.

Carlson, R. W., et al. 1991. Galileo infrared imaging spectroscopy measurements at Venus. *Science* 253:1541–1548.

Crisp, D., Allen, D. A., Grinspoon, D. H, and Pollack, J. B. 1991a. The dark side of Venus: Near-infrared images and spectra from the Anglo-Australian Observatory. *Science* 253:1263–1266.

Crisp, D., et al. 1991b. Ground based near-infrared observations of Venus during the Galileo encounter. *Science* 253:1538–1541.

de Bergh, C., et al. 1991. Deuterium on Venus: Observations from Earth. *Science* 251:547–549.

de Bergh, C., et al. 1992. The H_2O abundance in the deep atmosphere of Venus from near-infrared spectroscopy of the night side. *World Space Congress*, Aug. 28–Sept. 5, Washington, D. C.

Donahue, T. M., and Hartle, R. E. 1992. Solar cycle variation in Venus H^+ and D^+ densities in the Venus ionosphere: Implications for escape. *Geophys. Res. Lett.* 19:2449–2452.

Donahue, T. M., and Hodges, R. R., Jr. 1992. Past and present water budget of Venus. *J. Geophys. Res.* 97:6083–6091.

Donahue, T. M., and Hodges, R. R., Jr. 1993. Venus methane and water. *Geophys. Res. Lett.* 20:591–594.

Donahue, T. M., Hoffman, J. H., Hodges, R. R., Jr., and Watson, A. J. 1982. Venus was wet: A measurement of the ratio of deuterium to hydrogen. *Science* 216:630–633.

Drossart, P., et al. 1993. Search for spatial variations in the H_2O abundance in the lower atmosphere of Venus from NIMS-Galileo. *Planet. Space Sci.* 41:495–504.

Eberhardt, P., Reber, M., Krankowsky, D., and Hodges, R. R., Jr. 1995. The D/H and $^{18}O/^{16}O$ ratios in water from comet P/Halley. *Astron. Astrophys.* 302:301–316.

Grinspoon, D. H. 1987. Was Venus wet? Deuterium reconsidered. *Science* 238:1702–1704.

Grinspoon, D. H. 1993. Evolutionary implications of a steady state water abundance on Venus. *Nature* 363:1702–1704.

Grinspoon, D. H., and Lewis, J. S. 1988. Cometary water on Venus: Implication of stochastic impacts. *Icarus* 74:21–35.

Gurwell, M. A. 1995. Evolution of deuterium on Venus. *Nature* 378:22–23.

Gurwell, M. A., and Yung, Y. L. 1992. Fractionation of hydrogen and deuterium on Venus due to collisional ejection. *Planet. Space Sci.* 40:1620–1628.

Hartle, R. E., and Grebowsky, J. M. 1993. Light ion flow in the nightside ionosphere of Venus. *J. Geophys. Res.* 98:7437–7445.

Hartle, R. E., and Grebowsky, J. M. 1995. Planetary loss from light ion escape on Venus. *Adv. Space Res.* 15(4):117–122.

Hartle, R. E., and Taylor, H. A., Jr. 1983. Identification of deuterium ions in the ionosphere of Venus. *Geophys. Res. Lett.* 10:965–968.

Hartle, R. E., Donahue, T. M., Grebowsky, J. M., and Mayr, H. G. 1996. Hydrogen and deuterium in the thermosphere of Venus: Solar cycle variations and escape. *J. Geophys. Res.* 101:4525–4538.

Hodges, R. R., Jr. 1993. Isotopic fractionation of hydrogen in planetary exospheres due to ionosphere-exosphere coupling: Implications for Venus. *J. Geophys. Res.* 98:10833–10838.

Hodges, R. R., Jr., and Tinsley, B. A. 1986. The influence of charge escape on the velocity distribution of hydrogen in the Venus exosphere. *J. Geophys. Res.* 91:649–658.

Hunten, D. M., Donahue, T. M., Walker, J. C. G., and Kasting, J. F. 1989. Escape of atmospheres and loss of water. In *Origin and Evolution of Planetary and Satellite Atmospheres*, eds. S. K. Atreya, J. B. Pollack and M. S. Matthews (Tucson: Univ. of Arizona Press), pp. 386–422.

Kasting, J. F. 1988. Runaway and moist greenhouse atmosphere and evolution of Earth and Venus. *Icarus* 74:472–494.

Kasting, J. F., and Pollack, J. B. 1983. Loss of water from Venus, I, Hydrodynamic escape of hydrogen. *Icarus* 53:479–508.

Kasting, J. F., and Toon, O. B. 1989. Climate evolution on the terrestrial planets. In *Origin and Evolution of Planetary and Satellite Atmospheres*, eds. S.K. Atreya, J. B. Pollack and M. S. Matthews (Tucson: Univ. of Arizona Press), pp. 423–449.

Kasting, J. F., Pollack, J. B., and Ackerman, T. P. 1984. Response of Earth's surface temperature to increases in solar flux and implications for loss of water from Venus. *Icarus* 57:335–355.

Kaula, W. 1995. In *Volatiles in the Earth and Solar System*, ed. K. A. Farley (New York: American Inst. of Physics), pp. 139–142.

Krasnopolsky, V. A. 1985. Total injection of water vapor into the Venus atmosphere. *Icarus* 62:221–229.

Kumar, S., and Taylor, H. A., Jr. 1984. Deuterium on Venus: Model comparisons with Pioneer Venus observations of the predawn bulge ionosphere. *Icarus* 62:494–504.

Kumar, S., Hunten, D. M., and Pollack, J. B. 1983. Non-thermal escape of hydrogen and deuterium from Venus and implications for loss of water. *Icarus* 55:369–375.

Lewis, J. S. 1972. Low temperature condensation from the solar nebula. *Icarus* 16:241–252.

McElroy, M. B., and Yung, Y. L. 1976. Oxygen isotopes in the martian atmosphere: Implications for the evolution of volatiles. *Planet. Space Sci.* 24: 1107–1113.

McElroy, M. B., Prather, M. J., and Rodriguez, J. M. 1982. Escape of hydrogen from Venus. *Science* 215:1614–1615.

Moroz, V. I., et al. 1982. Spectrophotometrical experiment on Venera 13 and Venera 14. *Pisma Astron. Zh.* 8:4044–4100.

Owen, T., Bar-Nun, A., and Kleinfeld, I. 1992. Possible cometary origin of heavy noble gases in the atmospheres of Venus, Earth and Mars. *Nature* 358:43–45.

Oyama, V. I., et al. 1980. Pioneer Venus gas chromatography of the lower atmosphere of Venus. *J. Geophys. Res.* 85:7891–7902.

Paxton, L. J., Anderson, D. E., and Stewart, A. I. F. 1988. Analysis of Pioneer Venus Orbiter Ultraviolet Spectrometer Lyman-α data from near the subsolar region. *J. Geophys. Res.* 93:1766–1772.

Pepin, R. O. 1991. On the origin and early evolution of terrestrial planet atmospheres and meteoric volatiles. *Icarus* 92:2–79.

Pepin, R. O. 1996. Evolution of Earth's noble gases: Consequences of assuming hydrodynamic loss driven by giant impact. In preparation.

Pollack, J. B., et al. 1993. Near-infrared light from Venus' nightside: A spectroscopic analysis. *Icarus* 103:1–42.

Prinn, R. G., and Fegley, B., Jr. 1989. Solar nebula chemistry: Origin of planetary, satellite and cometary volatiles. In *Origin and Evolution of Planetary and Satellite Atmospheres*, eds. S. K. Atreya, J. B. Pollack and M. S. Matthews (Tucson: Univ. of Arizona Press), pp. 78–136.

Revercomb, H. E., Sromovsky, L. A., Suomi, V. E., and Boese, R. W. 1985. Net thermal radiation in the atmosphere of Venus. *Icarus* 61:521–538.

Rodriguez, J. M., Prather, M. J., and McElroy, M. B. 1984. Hydrogen on Venus: Exospheric distribution and escape. *Planet. Space Sci.* 32:1235–1355.

Shoemaker, E., and Wolfe, R. 1982. Cratering time scales for the Galilean satellites of Jupiter. In *The Satellites of Jupiter*, ed. D. Morrison (Tucson: Univ. of Arizona Press), pp. 277–339.

Shoemaker, E., Wolfe, R., and Shoemaker, C. 1990. Asteroid and comet flux in neighborhood of Earth. In *Global Catastrophes in Earth History*, eds. V. L. Sharpton and P. D. Ward, GSA SP-247 (Boulder: Geological Soc. of America), pp. 155–170.

Shoemaker, E., Weissman, P., and Shoemaker, C. 1994. The flux of periodic comets near Earth. In *Hazards Due to Comets and Asteroids*, ed. T. Gehrels (Tucson: Univ. of Ariona Press), pp. 313–335.

Watson, A. J., Donahue, T. M., and Walker, J. C. G. 1981. The dynamics of a rapidly escaping atmosphere: Applications to the evolution of Earth and Venus. *Icarus* 48:150–166.

Wetherill, G. W. 1991. Formation of the terrestrial planets from planetesimals. In *Planetary Sciences: American and Soviet Research*, ed. T. M. Donahue (Washington, D. C.: National Academy Press), pp. 98–115.

CHEMISTRY OF LOWER ATMOSPHERE AND CLOUDS

LARRY W. ESPOSITO
University of Colorado

JEAN-LOUP BERTAUX
Service d'Aeronomie CNES

VLADIMIR KRASNOPOLSKY
Goddard Space Flight Center

and

V. I. MOROZ and L. V. ZASOVA
Space Research Institute, Russian Academy of Sciences

Venus is totally covered by clouds, up to 60 km in vertical extent. The photochemistry that creates the cloud aerosols grades into thermal chemistry that is dominant below the upper clouds. We review selected measurements from Venera, Vega, Hubble Space Telescope (HST) and groundbased telescopes that constrain the chemistry of the lower atmosphere and clouds. The *in-situ* Vega measurements of SO_2 abundance in the deep atmosphere are at variance with earlier gas chromatograph measurements and with present chemical models. Russian and American measurements of SO_2 in the upper cloud have been reconciled. SO_2 continues its long decline seen since 1978 above the cloud tops. We review chemical models of the atmosphere above the clouds and recent progress in modeling of the cloud layer and sub-cloud atmosphere. We still do not know the species responsible for Venus' blue absorption. The existence of large and/or crystalline size modes of cloud particles is still open. Advances in understanding Venus atmospheric chemistry will require future measurements of sulfur and chlorine compounds.

The effort to determine the composition and chemistry of the Venus atmosphere had been an objective of numerous space missions to Venus and extensive ground-based telescopic observations. In this chapter, we review some of the recent data bearing on the chemistry of the lower atmosphere and cloud region (with particular emphasis on data from Venera 15, Vega 1 and 2, and ultraviolet spectroscopy). We discuss the chemical models that have been developed to explain the observations in this region, and draw some critical comparisons among them.

Some of the most important new information of the last decade has come from near-infrared measurements of the Venus night side, which probe the deeper Venus atmosphere through windows in the CO_2 absorption spectrum.

The nightside emissions at 1.7 and 2.3 μm discovered by Allen and Crawford (1984) were readily identified (see, e.g., Krasnopolsky 1986, p. 181) as spectral windows to the lower atmosphere. Later, windows at 1.31, 1.27, 1.18, and 1.01 μm were found, and lines of CO_2, H_2O, HDO, SO_2, CO, OCS, HCl, and HF have been identified. These data are discussed in more detail in the chapters by Taylor et al. and by Crisp and Titov. In this chapter, we restrict our attention to the Venus atmosphere below 90 km; the chemistry above this altitude is discussed in the chapters by Fox et al. and by Bougher et al. The mesosphere is discussed in the chapter by Lellouch et al.

The Venus clouds and hazes have enormous vertical extent, with a lower haze down to ∼30 km and an upper thick haze up to 90 km altitude; the entire system covers a vertical depth of ∼60 km, with the average visibility in the Venus clouds better than several km. The main cloud deck extends from ∼70 km (the level of unit optical depth in the ultraviolet) down to altitudes between 45 to 50 km.

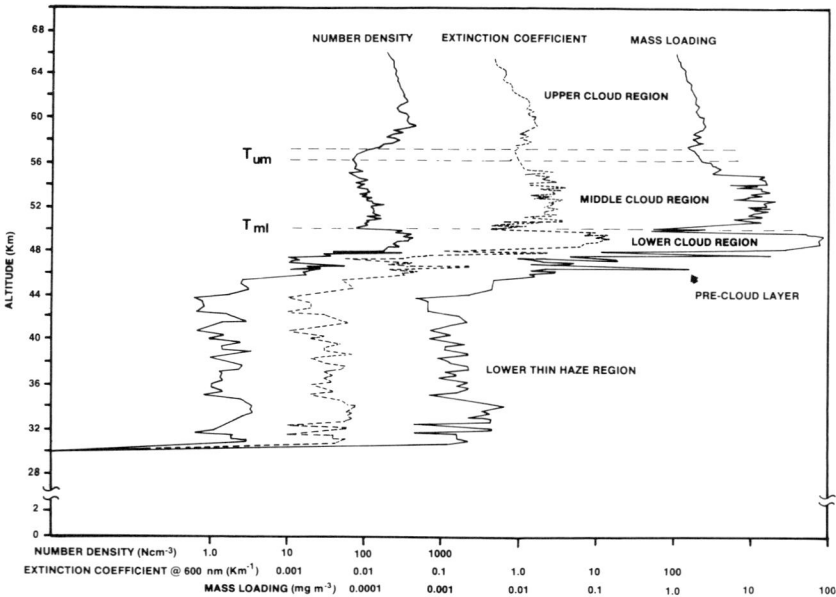

Figure 1. Vertical structure of the Venus atmosphere from direct computations based on the Pioneer Venus LCPS data (Knollenberg and Hunten 1980). The boundaries between the upper and middle cloud, T_{um} (about 1 km) and the middle and lower T_{ml} (several hundred meters) are shown. The mass loading is calculated for a density of 2 g cm^{-3}.

Spacecraft *in-situ* measurements allow us to divide the cloud system into upper, middle and lower clouds (see Fig. 1 and Esposito et al. 1983). Considering all the Pioneer Venus (PV) and Venera nephelometer results and the LCPS (Pioneer Venus Cloud Particle Size Spectrometer) measurements, it appears that the middle and upper cloud structure are planetwide features. In all cases the opacity is higher in the middle than upper cloud, typically by a factor of 2. The lower cloud is well defined and highly variable from location to location. Sharp layers are evident and these have the highest opacity at the Pioneer Large and Night probe sites.

The clouds within the main deck would all be thin stratiform in terrestrial classification. Instabilities are slight and latent convection potential is negligible (see Knollenberg et al. 1980). Only the middle cloud region appears to have any potential for convective overturning.

The chemistry of the lower Venus atmosphere divides naturally into three regions. In the main cloud layers above 60 km, photon-driven processes are important, and the term "photochemistry" applies well here. Below the clouds, thermal processes and vertical transport are key, and this region is dominated by "thermal chemistry." At the lower altitudes, interactions with the surface may dominate.

The recent reviews of Venus atmospheric chemistry by Prinn (1985) and Krasnopolsky (1986) are not outdated; the reader is referred to these for more detail on a number of chemical processes discussed in this chapter. The material in this chapter is divided into the following three sections: selected recent observations; the atmospheric chemistry; and open questions.

I. SELECTED RECENT OBSERVATIONS

A. Results from Venera 15 and Vega 1 and 2

1. Introduction. Venera 15 and 16 were launched in 1983. Both spacecraft were inserted on elliptical orbits around Venus. The main goal of this twin mission was mapping the Venus surface in the northern hemisphere with synthetic aperture radar (SAR). However, two atmospheric experiments were also included: observations of the thermal radiation spectra and radio-occultation measurements. Spectrometric observations were successful only on Venera 15. The instrument failed after two months on orbit, but considerable information about the temperatures, aerosols, sulfur dioxide and water vapor in the upper clouds and above was obtained in this short period of time.

Two years later (1985) the Vega 1 and Vega 2 spacecraft flew by Venus on their way to Halley's comet. As they passed Venus, each of them delivered two descent and two balloon probes. A set of *in-situ* atmospheric studies that included balloon experiments gave the first horizontal profiles of the Venusian atmosphere in the cloud region.

A short description of the Vega balloons and their preliminary results were presented by Sagdeev et al. (1986), and for Vega descent probes by Deriugin et al. (1987) and Moroz (1987). These spacecraft provided the last point in the long history of the studies of Venus by Soviet missions: 13 successful entry vehicles (including 10 with soft landings) beginning in 1967, when Venera 4 provided the first *in-situ* studies of another planet.

The following short review of the results of atmospheric studies on Venera 15 and Vega 1, 2 missions emphasizes the new findings useful to update the descriptions of the chemical composition of gas, clouds and hazes published more than 10 years ago in the *Venus* book (Hunten et al. 1983) by von Zahn et al. (1983) and Esposito et al. (1983).

2. *Infrared Spectra Measured by Venera 15: Observational Results and Interpretation.* The infrared Fourier spectrometer (FS) on board the Venera 15 orbiter (Moroz et al. 1986; Oertel et al. 1987) covered a spectral range from 270 to 1600 cm^{-1} (6–35 μm). The field of view of the instrument corresponded to an area about 100×100 km in the polar region. The goal of this experiment was to study the planet's thermal radiation spectra with higher spectral resolution (5–7 cm^{-1}) and continuous spectral coverage compared to the previously available thermal infrared spectra of Venus from the Earth (see, e.g., Kunde et al. 1977). Terrestrial measurements were not disk resolved and were corrupted by strong terrestrial absorptions; those of Venus were from filter spectroscopy of the PV OIR experiment (Taylor et al. 1980).

Venera 15 observed mainly the northern hemisphere, but a few tens of spectra were obtained in the equatorial region and south mid-latitudes. The total number of available spectra exceeds 1500. Some representative examples are shown in Fig. 2. Absorption bands of three atmospheric gases (CO_2, H_2O and SO_2) and also H_2SO_4 aerosols are clearly visible. The very strong 15 μm CO_2 band has different morphology at different locations, mostly owing to variability of the temperature profiles. Band center emission arises from atmospheric layers at about 90 km altitude. Differences are clearly visible in the continuum between the bands. These reflect variability in vertical structure of the upper clouds and haze.

Interpretation of FS data appears in a set of works from the last decade (Moroz et al. 1985,1986,1990; Spankuch et al. 1985,1990; Zasova et al. 1985,1989,1993; Schaefer et al. 1987,1990; Linkin et al. 1985). Their general approach included the following:

(1) Upper cloud and upper haze particles are presumed to consist of sulfuric acid water solutions. This critical assumption was confirmed by the good agreement of synthetic and observed spectra.

(2) The spectral dependence of the extinction coefficient is computed from the assumption that the particle size distribution corresponds to the measured "mode 2" from the PV LCPS, with parameters as proposed by Pollack et al. (1980) (log-normal distribution with mean cross-section weighted size $r = 1.05 \mu$ and variance $\sigma = 1.21$).

(3) CO_2 pure gas transmission functions for a set of channels within the 15 μm band are combined with computed aerosol transmission functions. Then an iteration procedure (e.g., using relaxation as by Zasova et al. [1989]) is adopted for the simultaneous retrieval of the temperature and aerosols profiles.

(4) Temperature and aerosol profiles retrieved by step (3) can be used to compute synthetic spectra for the SO_2 and H_2O bands, which allow us to derive their abundances (see below Sec. I.A.4).

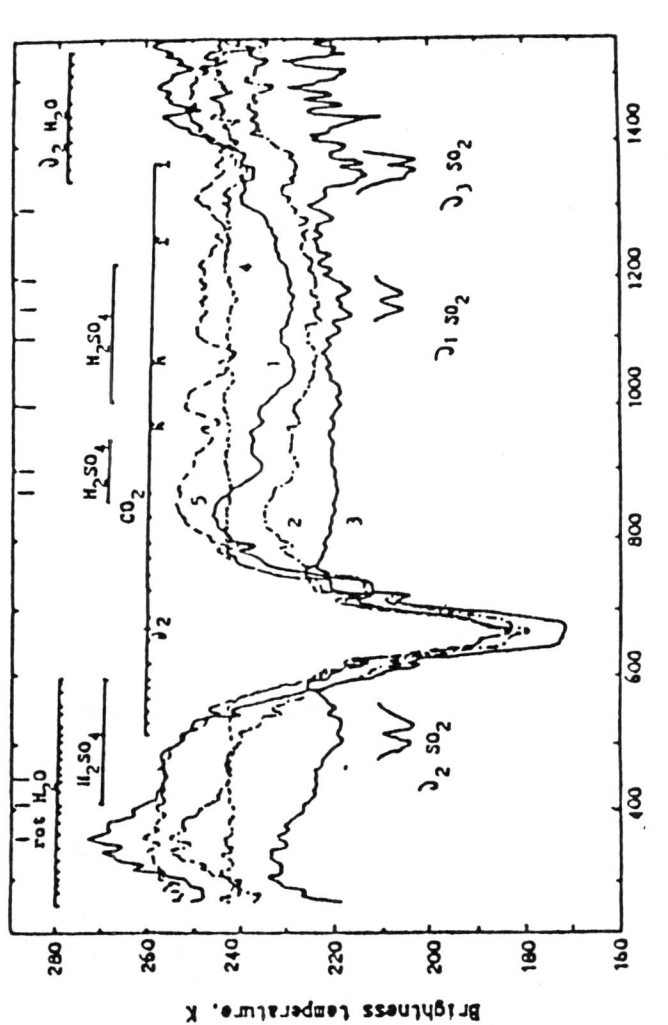

Figure 2. Venera 15 FS (infrared spectrometer experiment) brightness temperature vs wavenumber for five characteristic groups of spectra, averaged within latitude zones; 1 is for equatorial zone; 2 for 30 to 50°, 3 for "cold collar" on 60 to 80°, 4 is typical for near polar region; 5 for hot high-latitude regions (dipole?). From 5 to 12 spectra were averaged in each latitude bin. Some of the spectra observed between 50 and 80°N but outside of the cold collar can be explained as a linear combination of type (2) and (4) (Zasova et al. 1993). This means that the horizontal structure of clouds consists of small unresolvable spots having different vertical cloud structure.

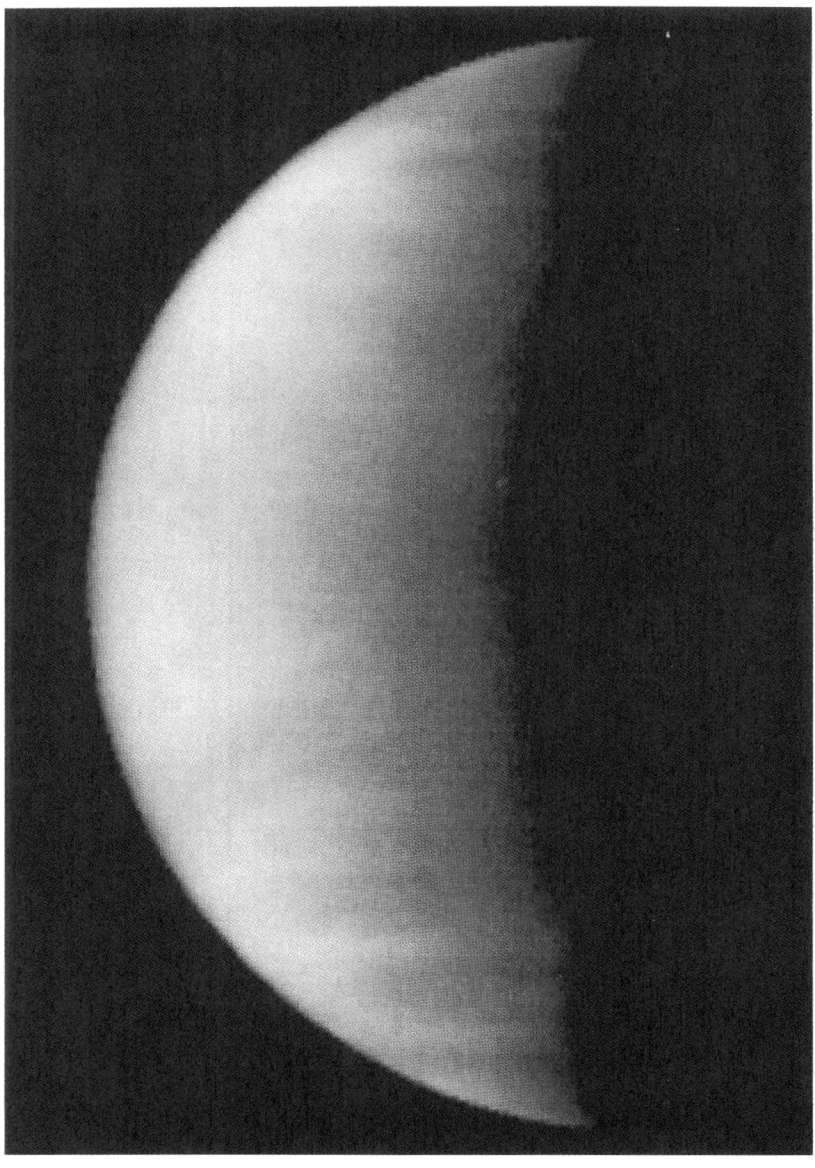

Figure 3. Hubble image of Venus (WFPC2 at 255 nm) from 25 January 1995. Note the muted contrasts and similarity to previous ultraviolet images (see, e.g., Rossow et al. 1980) (figure from Na and Esposito 1996).

3. *Vega Experiments Concerning Clouds, Hazes and Atmospheric Chemistry.* Much new information about winds and turbulence was obtained from ground radiotracking of the balloons and *in-situ* measurements. Those parts of the balloon results are outside of the scope of this chapter. However, the balloons provided the first *in-situ* nephelometric measurements for long horizontal paths (Sagdeev et al. 1986) and this was an important achievement in Venusian clouds studies.

Most of the atmospheric experiments on the descent probes targeted aerosols studies, including: (a) combined particle size spectrometer/nephelometer ISAV-A (Gnedyk et al. 1987*a,b*); (b) particle size spectrometers LSA, prepared by another team (Zhulanov et al. 1987); (c) gas-chromatograph (SIGMA-3) with aerosol collector (Porshnev et al. 1987); (d) mass-spectrometer (MALAHIT) with aerosol collector (Surkov et al. 1987); and (e) X-ray fluorescent analyzer IFP, also with an aerosol collector (Andreichikov et al. 1987).

All three instruments for chemical analysis of particle composition employed aerosol filters on their intakes to trap small particles during the descent of the probe through the atmosphere. Heating converted the collected material into the gas phase which were input to SIGMA-3 and MALAHIT. Both of these instruments were targeted mainly toward the *in-situ* detection of sulfuric acid, and this was actually accomplished (see below). IFP provided an analysis of the elemental composition of cloud particles. SIGMA-3 and IFP worked successfully on both probes, MALAHIT only on Vega 1.

All instruments mentioned above ceased their measurements between the altitudes of 35 and 45 km because they were mounted in a special section of the probe with only limited pressure and temperature protection.

The ultraviolet spectrometer ISAV-1 and 2 were used for sulfur dioxide measurements (Bertaux et al. 1987; Widemann et al. 1993; Bertaux et al. 1996) and humidity sensors for water vapor measurements (Surkov et al. 1987). Two different type sensors ("thermoelectrical" and "electrometrical") were used for water vapor measurements on both probes.

The landing sites of the Vega 1, 2 descent probes were located on the night side of the planet, in contrast to all previous Soviet missions to Venus beginning with Venera 9 and 10.

4. *Sulfur Dioxide and Water Vapor Measurements.* SO_2 was first detected in the atmosphere of Venus by Barker (1979) with groundbased observations, and it was subsequently confirmed by Stewart et al. (1979) and Conway et al. (1979). The abundances of SO_2 in 1978–1979 were larger than the previously established upper limits (Owen and Sagan 1972) by orders of magnitude. Continuous observations by Pioneer Venus from 1978 to 1986 showed a steady decline in the cloud top SO_2 abundance toward values consistent with previous upper limits (Esposito et al. 1988). This decline has been confirmed by IUE observations (Na et al. 1990) and by HST (Na and Esposito 1996); see Figs. 3, 4 and 5. The preliminary analysis of SO_2 abundance from HST is less than 25 ppb at the cloud tops (Na and Esposito 1996). Figure 5 is a compilation

of SO_2 cloud-top measurements. Explanations that have been advanced for the likely rapid increase and observed slow decline of SO_2 include active volcanism (Esposito 1984) and changes in atmospheric dynamics (Clancy and Muhleman 1991). The volcano hypothesis uses the volcanic eruption as a source of bouyancy that allows the abundant SO_2 below the Venus clouds to break through the stable upper cloud layer. This entrained SO_2 is then visible remotely to ultraviolet observations at the cloud top. The SO_2 abundance below the clouds varies much more slowly, related to the amount of volcanic activity over geologic time scales of millions of years (see the chapter by Fegley et al.). It is therefore puzzling why a similar reduction is seen between the Pioneer Venus and Venera 12 *in-situ* measurements of 1978 and the Vega results in 1986.

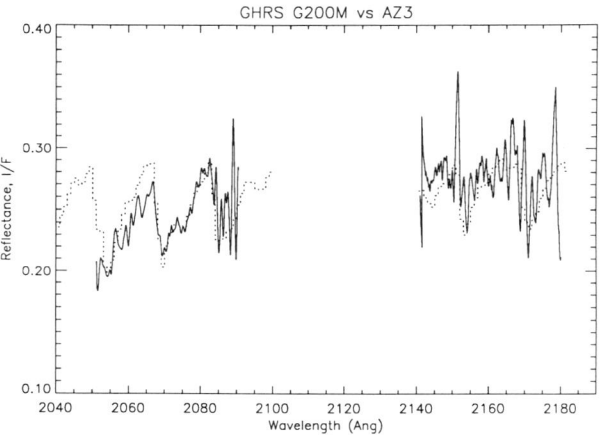

Figure 4. Hubble spectrum of Venus (GHRS G200M at 207 nm and 216 nm) from 1995 observations. Overplotted is a model spectrum for 12.5 ppb SO_2 at 40 mb with an SO_2 scale height of 3 km (figure from Na and Esposito 1996).

The changes in SO_2 above and within the clouds of Venus may have a significant effect on the photochemistry of the clouds of Venus. Pioneer Venus observations have shown that the clouds of Venus are created by the photochemical processes that oxidize upwelling SO_2 (Winick and Stewart 1980; Yung and DeMore 1982; see Sec. II). Thus, any significant changes in SO_2 may have an effect on the chemistry and dynamics of the clouds.

Venera 15 FS vs Pioneer Ultraviolet Spectroscopy. Three new sulfur dioxide absorption bands were found in the spectra observed by FS: about 519, 1150 and 1360 cm^{-1} (the strongest). All the ultraviolet observations have shown that, for the cloud-top level of 40 mb (height about 69 km), sulfur dioxide is more abundant at low latitudes, and is much less observable at latitudes more than 50° (see, e.g., Esposito et al. 1979).

Venera 15 FS observations gave the opposite result; the infrared sulfur

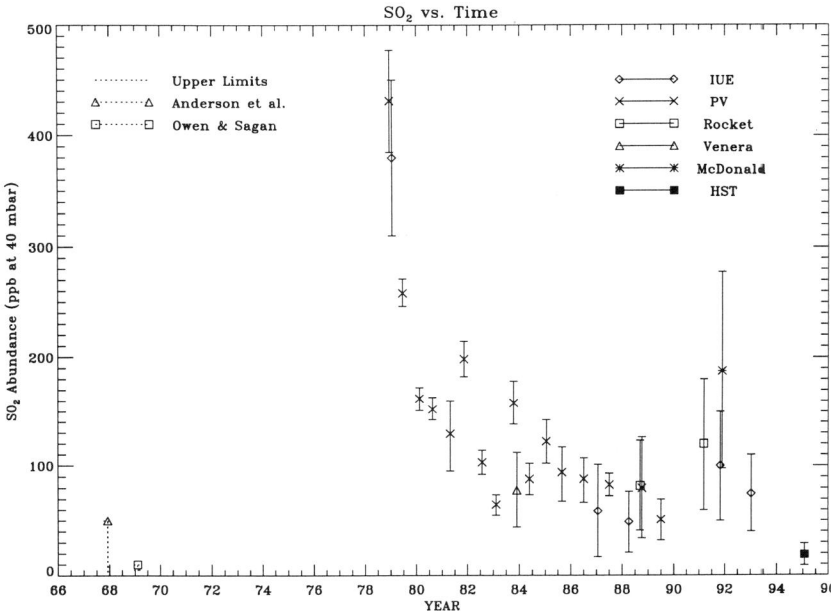

Figure 5. SO_2 abundance at 40 mb from 1968 to 1995, including groundbased, spacecraft and rocket measurements. Only upper limits were given by measurements prior to 1978 (left hand side of figure) (figure from Na and Esposito 1996).

dioxide bands were much more pronounced at high latitudes than at low. The shape and equivalent width of a band in thermal emission depends not only on the abundance of the corresponding gas but also on the temperature and aerosol extinction profiles. Fortunately, all of these can be obtained from the same spectrum in which the SO_2 bands are seen (see Fig. 6). The first attempts to retrieve the SO_2 abundances from these spectra did not show any pronounced latitude dependence for the heights 60 to 62 km (Moroz et al. 1990).

A joint analysis of both sets of data to resolve the apparent disagreement between the Venera 15 and PV OUVS SO_2 distribution was carried out by a small IKI/University of Colorado working group, and included the most recent observations from U. S. sounding rockets (McClintock et al. 1994; Na et al. 1994). The final conclusion (Zasova et al. 1993) is that all of the observations are consistent with the following latitude distribution. Around 1983, at the 40 mb (69 km) level, the SO_2 mixing ratio is a few tens of ppb at the low and middle latitudes (below 45°, approximately), 1 to 10 ppb in the cold collar and 100 to 200 ppb in the near polar hot regions ("dipole"), and in some places it reaches about 1000 ppb (Fig. 7).

The same analysis for the level 150 mb (62 km) yields the mixing ratio of 0.3 to 0.5 ppm at low and middle latitudes and increases to 1 to 2 ppm in

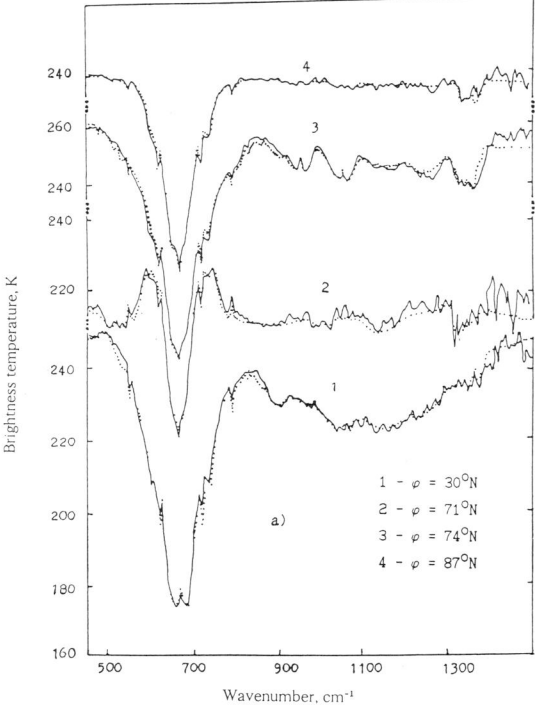

Figure 6. (a) Four individual Venera 15 FS spectra and (b) temperature and aerosol profiles derived from them. Dotted lines in part (a) are the derived synthetic spectra.

the polar region. The scale height of SO_2 is 1.5 to 2.5 km at low latitudes. The mixing ratio and scale height derived for middle latitudes are in good agreement with photochemically predicted values (Yung and DeMore 1982).

These two altitude reference levels (near 62 and 69 km) were selected because the first one is close to the lower boundary where the infrared measurements are sensitive to the vertical SO_2 profile and the second one is close to the upper boundary. The upper reference level is also convenient for comparison with ultraviolet observations (Fig. 8). At latitudes above 50 degrees the scale height varies from 1 to 6 km showing some pronounced correlation with the SO_2 mixing ratio.

The slant geometry typical for the high-latitude PV OUVS observations can explain the smaller amount of ultraviolet SO_2 absorption seen there. So, at the moment we have removed the controversy in understanding of the sulfur dioxide horizontal distribution at the level of the cloud tops. However, some of the discrepancies between ultraviolet and infrared remain. The SO_2 scale height from infrared data at high latitudes is 3 to 5 km, from ultraviolet it is about 1 km. At low latitudes, a rather good ultraviolet-infrared agreement was established for both the abundances, and for SO_2 scale height: here it is about 1.5 to 2.5 km. Different observational geometry and vertical variation of scale height (it could decrease above 69 km) can probably explain the remaining disagreement. Temporal variations are also very possible, because the Venera

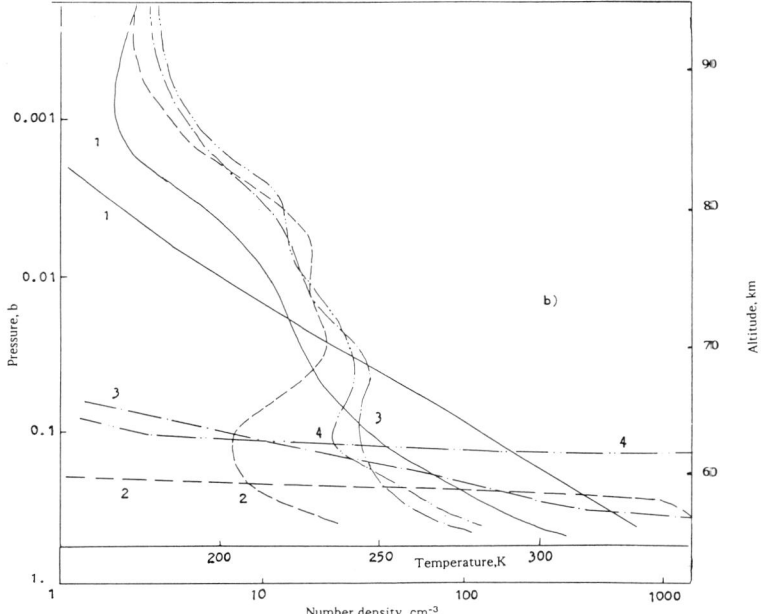

15 FS data are from a single two-month period in 1983.

Water Vapor Estimations from Venera 15 Infrared Spectra. Water vapor features are visible on all Venera 15 FS spectral records at wavenumbers below 400 cm^{-1} (wavelengths greater than 25 μm). These features show the short wavelength edge of the H$_2$O pure rotational band. The depth of these features varies, but a systematic effort to separate mixing ratio of H$_2$O from cloud variations (as was done for SO$_2$ bands) has still not been carried out. Tentative analysis (Moroz et al. 1990) showed that H$_2$O abundance at altitudes 58 to 60 km is 20±10 ppm and the scale height is between 2 and 3 km at altitude 58 km. Local mixing ratio variations up to a factor of 5 are possible. There is no clear conclusion about latitude variations. There is definitely no diurnal variation of the water vapor abundance.

The estimated mixing ratio 20±10 ppm at heights 58 to 60 km is in approximate agreement with the results of the analysis of PV OIR data (Schofield et al. 1982). However, there is no confirmation of any substantial increases of H$_2$O abundance (to 40 ppm) at night, as derived from OIR data. Moreover, it is not clear how water vapor and cloud absorptions could have been unambiguously separated using only the broadband filters of the OIR instrument.

The fundamental rotation-vibration water absorption band near 6.3 μm is also visible on Venera 15 FS spectra but the signal-to-noise ratio is here much poorer than for the pure rotational band.

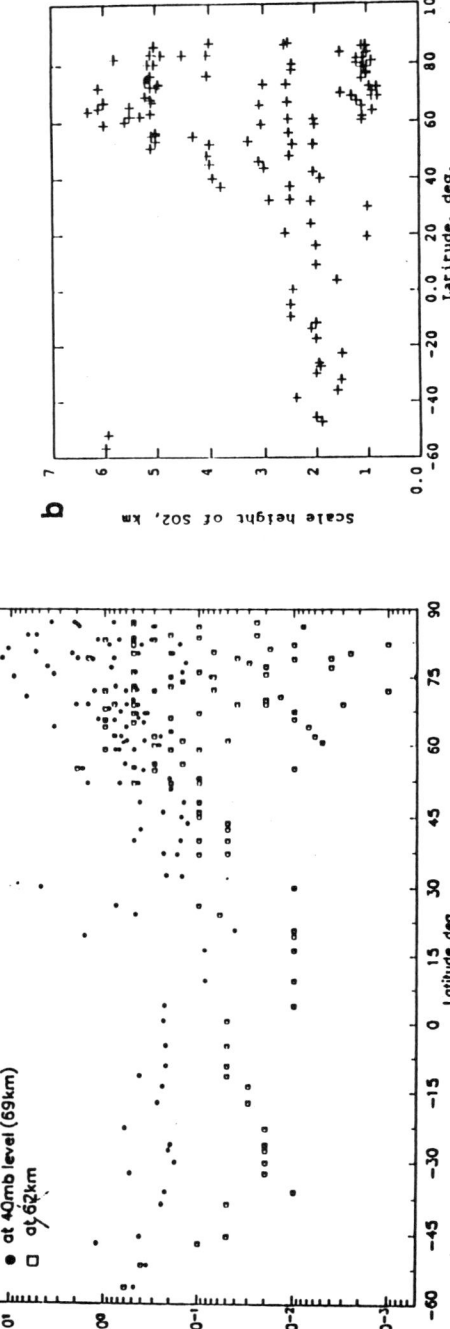

Figure 7. (a) The sulfur dioxide mixing ratio at 69 km (40 mb) vs latitude is given by filled circles, and the same at 62 km (150 mb) by squares. (b) scale height of the sulfur dioxide vs latitude (figure from Zasova et al. 1993).

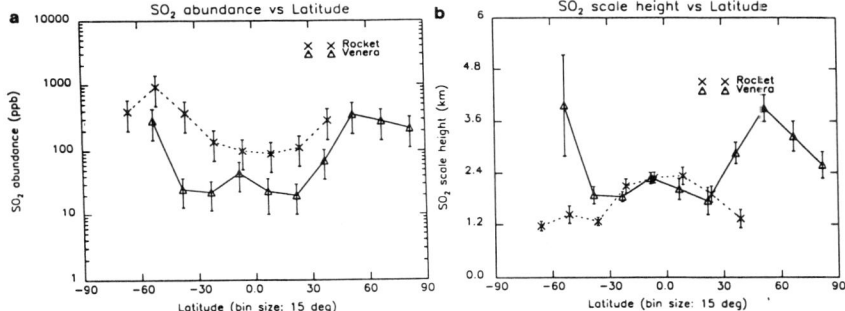

Figure 8. Comparison of the infrared Venera 15 FS and ultraviolet rocket data averaged over 15 degree latitude bins: (a) mixing ratio, (b) scale height (figure from Zasova et al. 1993).

Water Vapor from Vega Humidity Sensors. New *in-situ* H_2O measurements were also made by Vega 1 and 2 sounding/landing probes (Surkov et al. 1987). Surprisingly large abundances (about 1000 ppm with ±50% error bar) were found between 52 and 60 km. One of the proposed explanations was the absence of photochemical reactions at night. However, Venera 15 FS data do not show any daily variations, as was mentioned in the previous section. Maybe these very large abundances are really artifacts created by other constituents (cloud particles), but this was never checked by any additional analysis.

At heights 30 to 45 km the same humidity sensors have measured abundances of about 150 to 200 pm on both probes. These values are compatible with those obtained in previous missions (see reviews in Hunten [1983]) but are too high in comparison with those inferred from the near-infrared night glow spectra (see below).

5. Clouds and Hazes: Aerosol Profiles Inferred from Infrared Venera 15 FS Spectra. Brightness temperature (Tb) spectra of the thermal emission of Venus obtained in Venera 15 infrared experiment at low and high latitudes show significant differences in the characteristics of the continuum, visibility of hot CO_2 bands, and so forth (Fig. 2). This diversity can be partly explained by differences in the corresponding vertical temperature profiles. Aerosol vertical profiles would need to be different there also. The main properties of upper clouds and haze derived from infrared measurements (Zasova et al. 1993) are given in Table I.

Vega Descent Probes Nephelometer and Particle Spectrometer Results. The particle size distribution was measured by one of two particle size spectrometers of the Vega probes (ISAV-A; Gnedykh et al. 1987) at heights 48.5 to 52 km ("lower cloud"). This distribution can be matched by a superposition of two different size spectra: mode 1, with the relation between diameter D

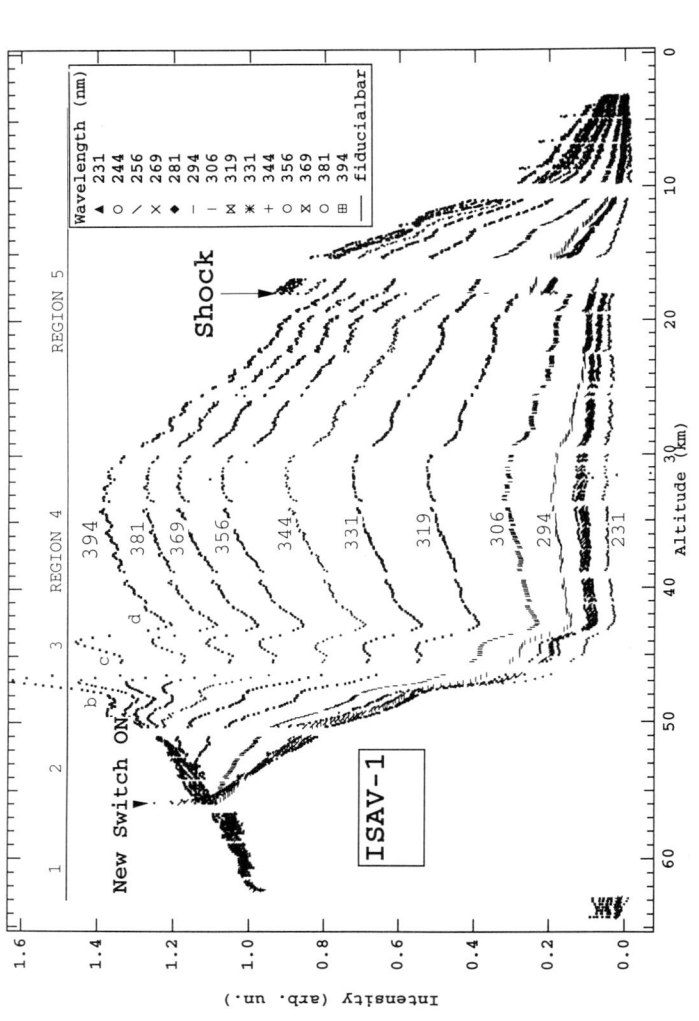

Figure 9. Vega ISAV-1 data as a function of altitude in the Venus atmosphere. Lightcurves for different wavelengths have been normalized to unity at altitude $z = 62.5$ km. In the region $20 < z < 40$ km, the lightcurves are in decreasing order of wavelength, with $\lambda = 394$ nm at the top. The event at $z = 18$ km where all lightcurves simultaneously increase is interpreted as a mechanical shock that released Venus aerosols collected on the ISAV optics at cloud altitudes. At both $z = 56$ km and $z = 50$ km, several wavelengths show a sharp increase, which disappears in about 20 s. This is interpreted as a spurious instrument effect occurring just after a new switch-on of the xenon flash. Five different regions (see Table II) are identified from the lightcurve variation. Narrow absorbing layers in the $\lambda = 394$ nm curve are labeled b, c, d (figure from Bertaux et al. 1996).

TABLE I
Aerosol Scale Height and Unit Optical Depth from Venera 15 FS

Latitude	Aerosol Scale Height (km)	Altitude of Unit Optical Depth for Two Wavenumbers (km)	
		365 cm^{-1}	1152 cm^{-1}
Less than 55°	3.5–4	57–59	67–69
55 to 75°			
Cold collar	<1	58–60	60–62
Inhomogeneous regions	<4	56–60	70–72
75 to 85°			
Hot dipole	1–1.5	56–58	59–64
Outside of dipole	<1	61–63	63–64
Greater than 80°	0.5	62–64	62–64

and number density N given by

$$dN(D)/dD = \text{const}^* D^n \tag{1}$$

with $n = 5\pm1$ for Vega 1 and 4 ± 1 for Vega 2. Mode 2 is as described by Pollack et al. 1980, although this gives some excess of particles with sizes between 1.5 and 4 μm.

This distribution differs from results of a similar experiment on the Pioneer Large probe (LCPS, Knollenberg and Hunten 1980). The LCPS identified a "mode 3," with sizes more than 5 μm as most important for the cross section and mass loading in lower clouds. There is a continuing controversy about the reality of mode 3 as a separate peak in the size distribution (the other possibility is that they can be described as a tail of some smooth size spectrum; see below, Sec. III.B), but there is no doubt of the presence of some large particles and their role in the integrated properties of Venusian clouds. Large particles also were not identified in the data of the second particle spectrometer (LSA; Zhulanov et al. 1986). There is a suspicion that most of large particles were simply missed in both of these experiments because they did not have a special channel to measure them as did the LCPS. The integrated number densities ($D>0.8\,\mu$m) are nearly the same as measured by LCPS at heights above 48.4 km, but are 30 times larger below 48 km. These experiments detect no sharp boundary at the lower cloud bottom, as has been identified in many previous missions beginning from Venera 9 and 10. Nephelometric data also do not show this boundary.

So the absence of the sharp lower boundary of the main cloud deck is probably a real feature of the aerosol vertical structure above the Vega 1 and 2 landing sites. Previous data provide only one instance where the lower cloud boundary was not found, in the Venera 8 mission. Perhaps this peculiarity is connected with the local time; both Vega landings were done in late night and Venera 8 was near the terminator. However, the Night Probe nephelometer on Pioneer Venus did not show any difference in cloud structure compared with those obtained during daytime.

ISAV-A provided simultaneous measurement of the light scattered by individual particles in four directions: forward, back and two with 90° scattering angle and 90° azimuthal difference. Comparison of signals in the two last channels provided a possibility to discriminate between spherical (liquid) and nonspherical particles. It was found that majority of particles with $D>0.8$ μm are spherical.

Comparison of signals corresponding to three different scattering angles provided estimations of the refractive index m. All particles with $D>2$ μm had $m = 1.4\pm05$. Particles with diameters $0.8<D<2$ μm can be separated in two groups: (1) with $m = 1.4\pm0.1$ (about 80% of registered events), and (2) with $m = 1.7\pm0.1$ (about 20%). No height dependence was found for these properties, but most of the particles were measured in the lower and middle clouds.

Vega Balloon Nephelometer Results. Simple nephelometers were included in the Vega balloon instrumental package. A backscattering coefficient about 1.5×10^{-4} m^{-1} sr^{-1} was measured at the height the balloons floated (54.5 km), in approximate agreement with nephelometers on the Vega landers. These measurements showed that middle clouds are horizontally very homogeneous near the equatorial region of the planet.

Chemical Composition of Particles: Sulfuric Acid. Venera 15 FS spectra provided a strong confirmation of an aqueous solution (75–85%) of sulfuric acid as the particulate material in the upper clouds. Otherwise, such good coincidence of synthetic and observed spectra would be impossible (see Fig. 6a).

The first in situ detection of sulfuric acid was made in experiments SIGMA-3 and MALAHIT. The analysis of SIGMA-3 results yields the average mass density of sulfuric acid about 1 mg m^{-3}, estimated for heights 48 to 54 km and "much lower" at about 54 km (Porshnev et al. 1987). Conclusions from MALAHIT analysis are even less definite and not in a good agreement with SIGMA-3 results; they give an average mass density between 2 and 10 mg m^{-3} (Surkov et al. 1987). Neither experiment can say anything about the upper clouds.

Chemical Composition of Particles: Other Constituents. Very interesting results (but quite difficult for quantitative interpretation) were obtained by elemental X-ray analysis of the thin layers collected by the IPF experiment (Andreichikov et al. 1987). Three elements were identified: sulfur, chlorine and phosphorus. Sulfur and chlorine in cloud particles had been detected previously by similar measurements, but not phosphorus. It is clear that some P-bearing substance can be important as a particulate in lower clouds. Phosphoric acid H_3PO_4 is a likely candidate for this substance and phosphorous anhydride P_4O_6 may be the gas responsible for its production (Andreichikov

1987). A critical review of these data was given by Krasnopolsky (1989). The key conclusion in the paper is that the lower subcloud boundary at the level about 33 km registered on Venera 8 and later on some other missions can be explained by phosphoric acid particles dominating in the subclouds.

The refractive index 1.7 estimated for some of the particles registered in ISAV-A experiment can be understood if they consist of free condensed sulfur. Some evidence for the presence of free sulfur in clouds were obtained from the analysis of the SIGMA-3 and IPF results.

B. ISAV Measurements of Ultraviolet Absorption

1. Introduction. The ISAV spectrometer (220–400 nm) was a part of the optical instrumentation flown on board the two Vega descent probes, together with the ISAV-A equipment dedicated to the aerosol cloud particles, for which results have been reported elsewhere (Moshkin et al. 1986; Gnedykh et al. 1987; see above).

The fly-by geometry at Venus was constrained by the later encounter with comet Halley, and as a result, the atmospheric probes were released on the night side of Venus. This was an ideal situation to conduct an "active spectroscopy" experiment, in which the ultraviolet spectrum of an artificial light source was continuously measured after traversing a gas-cell of 1.70 m path length, within which was circulated the ambient atmosphere all along the descent.

Useful measurements were made from 62.5 km of altitude down to the ground, with a vertical resolution ranging from 40 to 180 m, thanks to the relatively high sampling rate (one spectrum every 4 s). This was the first time such an instrument was flown in another planet's atmosphere, and we do not expect to have new opportunities to repeat it in the foreseeable future.

The results presented by Bertaux et al. (1996) rely on a number of assumptions made about the behavior of the instrument and the descent probe during its course to the ground. Bertaux et al. present both their primary data and their interpretation of the data along with a detailed discussion of the assumptions involved, in such a way that the reader may be able to establish his own opinion about the data and their conclusions. Because of the uniqueness of the observations, this interpretation deserves serious attention.

2. Discussion of the Light Curves. A total of 584 "short" spectra were acquired for ISAV-1 and 594 for ISAV-2 during the flights, at a constant rate of one every 4 seconds (except for a small interruption every 200 s for 20 s). All normalized lightcurves $I(\lambda_j, z)$ are shown on Fig. 9 for ISAV-1 and for λ_j varying from 231 to 394 nm, as a function of altitude above the landing point. The spread of the individual points along a lightcurve is representative of the total instrumental noise and is in most cases quite small.

Bertaux et al. (1996) identify several features in the lightcurves as purely instrumental. The rest of the intensity variations are interpreted as arising from gaseous absorption, along with the effect of a deposit of cloud particles on the instrument mirrors. This deposit builds up in the range of 54 to 44 km, corresponding to the middle and lower cloud layers (Fig. 1). The authors obtain the spectrum for the absorption due to the deposit. The SO_2 profile is determined via differential absorption with respect to 384 nm. Four thin absorbing layers were found by ISAV-2, three having exact counterparts seen with ISAV-1.

A list of the vertical features seen in these profiles is given in Table II.

3. *Comparison with Previous SO_2 Measurements*. In Table III we show a summary of previous measurements of SO_2 below 60 km. At an altitude of $z = 22$ km, the PV Large probe gas chromatograph and the Venera 12 lander gas chromatograph measured a concentration above 130 ppm: (130 ± 35) ppm for Venera 12 (Gel'man et al. 1979) and (185 ± 43) ppm for Pioneer Venus (Oyama et al. 1980, Table 2). At 22 km, the ISAV experiment obtains $n_{SO_2}(z) = 38$ ppm both for ISAV-1 and ISAV-2.

Table III also includes more recent measurements obtained from Earth through high-resolution spectroscopy of the infrared light emerging on the night side through the clouds of Venus (Bézard et al. 1993). They have interpreted their superb spectra to give a mixing ratio of 130 ± 40 ppmv in the altitude range of 35 to 45 km, quite compatible with the ISAV-1 value of 125 ppmv at the peak of 42 km. The ISAV-1 average over the range 35 to 45 km is rather 90 to 100 ppmv, a value still compatible with Bézard et al. estimate.

The fact that ISAV SO_2 data are in good agreement with groundbased spectroscopy in the region 35 to 45 km gives some confidence in the SO_2 retrieval algorithm and also the optical thickness measurement in the whole range 10 to 60 km. Although Bertaux et al. consider that the low values measured in the lower atmosphere of Venus are real, the many assumptions in the analysis and thus the somewhat uncertain performance of the experiment in these most extreme Venus conditions may justify some skepticism.

The concentration of SO_2 (in cm^{-3}) is slowly increasing with decreasing altitudes, but the mixing ratio of SO_2 is decreasing. If this decline of SO_2 mixing ratio at lower altitudes is real, as advocated by Bertaux et al. (1996), it implies a large transport of SO_2 downward, with a flux of $\cong 4 \times 10^{13}$ mol $cm^{-2}s^{-1}$ at 40 km and $\cong 16 \times 10^{13}$ mol $cm^{-2}s^{-1}$ at 10 km. The measurements stop at 6 km altitude, but a reasonable extrapolation down to the surface shows that this downward flux might vanish on the surface. Therefore, it is not in contradiction with a reaction rate of SO_2 with calcite at the surface of the order of 5×10^{10} mol $cm^{-2}s^{-1}$ (see the chapter by Fegley et al.). On the other hand, the decrease of the downward flux of SO_2 with decreasing altitude would imply the chemical transformation of SO_2 to another sulfur bearing chemical species, whose production in the lower atmosphere (10–40 km) would provoke an upward flux equivalent to the SO_2 downward flux, which is so far undetected. Further, our present understanding of Venus thermal chemistry does not reproduce such a decline of SO_2 at low altitudes (see below, Sec. II).

TABLE II
Thickness and Boundary Altitudes of Various Regions from ISAV Observations[a]

	ISAV-1			ISAV-2		
	Upper Boundary	Lower Boundary	Thickness	Upper Boundary	Lower Boundary	Thickness
Region 1[b]	62.5	56.6	6.0	62.5	56.5	6.0
Upper cloud[c]	66	56	10			
Region 2	56.5	49.6	6.9	56.5	51	5.5
Middle cloud	56	50	6			
Region 3	49.6	41.3	8.3	51	43.1	7.9
Lower cloud	50	47	3			
Layer a				51.1	49.8	1.3
Layer b	49.5	47.0	1.5	49.8	47.8	2.0
Layer c	47.0	43.9	3.1	47.8	44.7	3.1
Layer d	43.8	41.0	2.8	44.7	43.1	1.6
Region 4	41.3	31.3	10	42	32.3	9.7
Event[d]	18.0	18.0	NA	NA	NA	NA
Region 5	31.3	0	31.3	32.3	0	32.3

[a] All altitudes and thicknesses are in km.
[b] Boundary altitudes of ISAV regions and absorbing layers are visually defined on the plotted lightcurves (Fig. 9).
[c] Boundaries of Venus clouds are quoted from the night probe of Pioneer Venus (Ragent and Blamont 1980).
[d] Altitude of mechanical event in ISAV-1 lightcurve.

TABLE III
Summary of SO$_2$ Measurements

Instrument, observations	12 km	22 km	42 km	52 km	Reference
					In-situ
PV Large probe gas chromatograph		185±43	>176	<600	Oyama et al. (1980)
PV Large probe mass spectrometer		<300		<10	Hoffman et al. (1980)
Venera 12 lander gas chromatograph		130±35			Gel'man et al. (1979)
ISAV active spectroscopy	25±2	38	125	150 (ISAV-1)	Bertaux et al. (1996)
					Earth-based
Infrared night spectra			130±40 (35–45 km)		Bézard et al. (1993)

II. CHEMISTRY OF VENUS' ATMOSPHERE

Venus' atmosphere can be divided into several regions with different chemical conditions. The lower atmosphere extends up to 60 km, and only solar radiation longer than the ultraviolet can reach this region. The middle atmosphere is located between 60 and 110 km. This region is under control of photochemistry. The upper atmosphere is above 110 km, and dissociation, ionization, and ionospheric reactions occur there. Neutral reactions are scarce in this upper region, and a variety of transport phenomena are of interest in chemical-dynamical modeling. This division is close to, though it does not exactly coincide with the division into troposphere, strato-mesosphere, and thermosphere. In what follows, we restrict our discussion to the two lower regions. The Venus thermosphere is discussed in the chapters by Fox et al. and by Bougher et al.

A. Lower Atmosphere (0–60 km)

1. Introduction. High temperature and pressure and the absence of effective photolysis processes are typical of this region. The atmosphere consists of stable components, and chemical reactions are very slow despite the high temperature and pressure. At first, we consider some tools to be used to study this region.

Thermochemical Equilibrium. The first studies for the lower atmospheric composition were based on the assumption of thermochemical equilibrium at each height (see, e.g., Florensky et al. 1978). This assumption is valid if the time to reach equilibrium (which may vary from minutes to millions of years) is less than the mixing time, which is of the order of a few years in this region. Krasnopolsky and Parshev (1979) argued that thermochemical equilibrium may be valid for most of species near the surface due to possible catalytic action of surface rocks, while constant mixing ratios better approximate the vertical density profiles at higher altitudes. Therefore, we recommend the equilibrium assumption either for processes which are known to be fast (e.g., $H_2SO_4 = H_2O + SO_3$) or for conditions near the surface of Venus. If measurements show a mixing ratio varying with height, then the local thermochemical equilibrium calculation may help to understand this variation.

Constancy of Element Mixing Ratio. A chemical element is conserved in chemical reactions, therefore its flux is constant throughout an atmosphere. Without condensation and below the homopause this flux is defined by

$$\phi_i = -K\left(\frac{dn}{dz} + \frac{n_i}{H_a}\right) = -Kn\frac{df_i}{dz}. \qquad (2)$$

Here subscript i and a refer to the element and all atmospheric species, respectively, n is the number density, H is the scale height, f is the mixing ratio (fraction), K is the eddy diffusion coefficient. The flux may be neglected ($\phi_i \approx 0$), if $\phi_i \ll Kn_i/H_a = Kn_a f_i/H$. Therefore, f_i = constant if $f_i \gg \phi_i H_a/(Kn_a)$, the element mixing ratio is constant with height. As an example, this rule is valid for elemental hydrogen below 45 km if its mixing ratio exceeds 2×10^{-11}, given the measured hydrogen escape flux of 7×10^6 cm^{-2}s^{-1} (Rodriguez et al. 1984) and $K \approx 10^4$ cm^{2}s^{-1}. Therefore, to explain the water vapor mixing ratio decreasing from 200 ppm at 45 km to 20 ppm near 20 km, as proposed by Moroz et al. (1983), one should suggest another hydrogen-bearing component with mixing ratio 360/k ppm at 20 km. Here k is the number of hydrogen atoms in the molecule.

This rule is applicable to all elements on Venus below the cloud layer. However, it is exact only in a steady-state one-dimensional formulation. Advection and convection may break this rule, but typically not by very much. For example, to explain a wet location with $f_{H_2O} > 200$ ppm below 20 km, a strong upward flow must be transported horizontally to form a downward flow at the observed site. This would presume that the measured profile of Moroz et al. (1993), is not typical, and the mean profile has f_{H_2O} constant with abundance 20 to 200 ppm below 20 km.

Local Instability. As discussed in Sec. II.A.1, local thermochemical equilibrium does not necessarily hold in the atmosphere, therefore thermochemically unstable species may exist. The best example is sulfuric acid which is thermochemically unstable on Venus. Its existence is explained because sulfuric acid is formed photochemically. Species formed photochemically or thermochemically may exist outside the region of formation if the transport time (mixing and precipitation) is shorter or comparable with the chemical destruction time. For example, H_2SO_4 is formed above 60 km and disappears below 40 km because loss processes are slower than transport between 60 and 40 km. Furthermore, the lifetime of a solution of FeCl$_3$ in concentrated sulfuric acid is close to one week at room temperature, but much longer between $-20°$ and $0°C$. Compared with the precipitation time of the mode 2 particles (one month at 60–57 km), this helps to explain a coloration of the cloud layer at these heights by $FeCl_3$ (Zasova et al. 1981; Krasnopolsky 1985,1986), even though $FeCl_3$ is unstable in the presence of sulfuric acid.

2. Indirect Identifications of Aerosol Composition. Extensive experimental data on vertical profiles of aerosol particles, their multimodal distribution, the 300 to 500 nm absorption, its spectrum and location have resulted in many speculations about the aerosol chemical composition. Only sulfur has thermochemical plausibility as an aerosol material. Therefore, we come to the conclusion that the clouds must consist mostly of species of photochemical origin (e.g., H_2SO_4, Sn). In addition to H_2SO_4 which is indisputable, three candidates for the mode 3 (largest) particles have been considered: $HClO_4$, $AlCl_3$, and H_3PO_4 (see Krasnopolsky 1986,1989).

3. Sulfur Chemistry. This is clearly very important in the lower atmosphere of Venus. The chemical scheme proposed by Prinn (1975,1978,1979) is based on a prediction by Lewis (1970) of sulfur species with mixing ratios of 60 ppm for OCS, 6 ppm for H_2S, and 0.3 ppm for SO_2. In the 1970s, sulfuric acid was clearly identified in the clouds, while a search for gaseous sulfur components was not successful until 1979 (see Esposito et al. 1983). Prinn (1975) suggested a scheme of photochemical formation of sulfuric acid from carbonyl sulfide OCS and later (Prinn 1978) proposed the inverse processes leading to OCS and elemental sulfur from H_2SO_4:

$$SO_x + CO \longrightarrow SO_{x-1} + CO_2$$

$$\frac{S + CO + M \longrightarrow OCS + M}{\text{net } H_2SO_4 + 4CO \longrightarrow H2O + 3CO_2 + OCS.} \quad (3)$$

The predicted SO_2 mixing ratio was of a few ppm above 30 km and much larger than that near the surface. Prinn (1979) supposed that dissociation of S_3 and S_4 by the near ultraviolet ($\lambda \approx 350$ nm) might produce hot sulfur atoms which could drive the chemistry.

Pioneer Venus, the Veneras, and groundbased observations showed that the main sulfur species is SO_2 with a mixing ratio close to 150 ppm, and that elemental sulfur in the vapor phase is an important absorber below 25 km. At these heights photons with $\lambda > 450$ to 500 nm are the only ones available, and the energy of sulfur atoms released by photolysis of S_3 and S_4 is too low to drive effectively the exothermic reactions. Later Prinn (1985; see also von Zahn et al. 1983) summarized the sulfur chemistry on Venus considering three sulfur cycles, one geological and two atmospheric (see Fig. 6 in the chapter by Fegley et al.).

Slow Atmospheric Cycle of Sulfur: Recent Results. The measured abundances and particularly the gradients of OCS and CO should be directly connected with the slow atmospheric cycle of sulfur and have been analyzed by Krasnopolsky and Pollack (1994). Their scheme is opposite to a certain extent to that of Prinn (1978); Prinn argued that SO_3 forms OCS, gradually losing oxygen atoms in the reactions with CO, while Krasnopolsky and Pollack shcwed that SO_3 depletes OCS and CO below the clouds. Krasnopolsky and Pollack emphasize that this is only a partial solution to the problem of explaining the processes which govern profiles of OCS, CO, H_2SO_4, and SO_3, and their study does not cover some aspects of sulfur chemistry below 25 km, e.g., the reactions stimulated by the formation of SO. The main parameter of the model is the sulfuric acid column production rate $\phi_{H_2SO_4}$. Then, the H_2O mixing ratio below 30 km, f_{H_2O} (30 km), is chosen to fit the observed f_{H_2O} = 1 to 10 ppm at 62 to 65 km. Results of two models are given in Figs. 10 and 11 and in Table IV. The difference between the two models is that $\phi_{H_2SO_4}$ is equal to 2.2×10^{12} and 6.4×10^{12} cm^{-2}s^{-1} in models 1 and 2, respectively. Then f_{H_2O} (30 km) = 30 ppm and 90 ppm. Therefore, model 1 fits the data of the nightside near-infrared spectroscopy, while model 2 is closer to the earlier interpretation of Venera 11–14 data for water vapor (Moroz et al. 1983). The required fluxes of CO are 1.7×10^{12} and 4.2×10^{12} cm^{-2}s^{-1}, respectively. As discussed in Sec. II.A.1, fluxes of sulfuric acid and CO are determined by the net processes of photochemistry above 58 km:

$$CO_2 + SO_2 + h\nu \longrightarrow CO + SO_3$$

$$3SO_2 + 2h\nu \longrightarrow 2SO_3 + S. \qquad (4)$$

Therefore, it is easy to determine a flux of elemental sulfur, if fluxes of H_2SO_4 and CO are known. This gives the sulfur to sulfuric acid mass flux ratios of 1:27 and 1:18 in models 1 and 2, respectively. The predicted mixing ratios of sulfuric acid vapor are equal to 5 ppm and 10 ppm near the lower cloud boundary. The latter fits better to the radio occultation studies with the Magellan spacecraft (Jenkins et al. 1994). Both models require the reaction rate coefficients

$$k_1 = 10^{-11} \exp(-(13100 \pm 1000)/T) \text{ cm}^3 \text{ s}^{-1}$$

$$k_2 = 10^{-11} \exp(-(8900 \pm 500)/T) \text{ cm}^3 \text{ s}^{-1} \qquad (5)$$

and the OCS mixing ratio near the surface f_{OCS} (0 km) = 28±1 ppm. The derived rate coefficients are in reasonable agreement with some analogs of the considered reactions. The OCS mixing ratio is close to 20 ppm and 5 to 30 ppm given by Krasnopolsky and Parshev (1979) and Fegley and Treiman (1992), respectively.

This approach does not cover all aspects of sulfur chemistry below Venus' cloud layer and the problem needs further study.

4. Processes in the Clouds: Position of the Sulfuric Acid Lower Cloud Boundary. Analysis of processes in Venus' clouds requires saturated water and sulfuric acid vapor pressures p_1 and p_2 as functions of temperature and concentration of the acid solution (Krasnopolsky and Pollack 1994). Using these relationships, Krasnopolsky and Pollack (1994) developed a method

TABLE IV
Comparison of Two Chemical Models for the Lower and Middle Atmosphere of Venus

Parameter	Model 1	Model 2
Photochemical analog	Yung and DeMore (1982) Model C	Krasnopolsky and Parshev (1981)
k(CICO cycle)	Supported by kinetic data	100 K (Model 1)
(O_2 abundance)/ (upper limit)	13	2.5
Production of H_2SO_4	$\approx 2 \times 10^{12} cm^{-2} s^{-1}$	$\approx 6 \times 10^{12} cm^{-2} s^{-1}$
S/H_2SO_4 in aerosol	(0–4)%	(6–15)%
f_{H2O} (30 km)	30 ppm (agrees with spectroscopy of nightside (Pollack et al. 1993)	90 ppm (close to Venera results (Moroz et al. 1983)
Lower boundary of H_2SO_4 aerosol	48.4 km (agrees with *in-situ* data)	46.5 km (agrees with radio occultation)
f_{H2SO4} (45–47 km)	5 ppm	10 ppm (close to radio occultation data (Jenkins et al. 1994)
H_2SO_4-Mode 3 extinction in the middle cloud layer	0.3 km^{-1}	0.9 km^{-1}
Mode 3 LCPS data corrected for high aspect crystals	≈ 1	Contradicts this hypothesis

to calculate the lower cloud boundary (henceforth LCB), for a given flux of H_2SO_4. LCB is at 48.4 km in model 1 (see above) and at 46.6 km in model 2. These values may be compared with measurements: nephelometers on four PV probes give 48.4±0.75 km (Ragent and Blamon 1980); nephelometers on five Venera probes 49.3±0.57 km (Marov et al. 1983); and photometric data of four Venera probes 48.7±0.45 km. The measurements favor model 1. However, the PV radio occultation observations (Cimino 1982) show the LCB varying from 47 to 48 km at the low and middle latitudes to 47 to 43 km at high and subpolar latitudes.

Variations of the Lower Cloud Layer. The LCB varies due to variations of the H_2SO_4 vapor mixing ratio, the water vapor mixing ratio below the clouds, and temperature and pressure. Krasnopolsky and Pollack (1994) found that variations of LCB are produced mostly by variations of the sulfuric acid abundance and temperature. Variations of density of the lower cloud layer reflect variations of gaseous sulfuric acid, because water contributes only

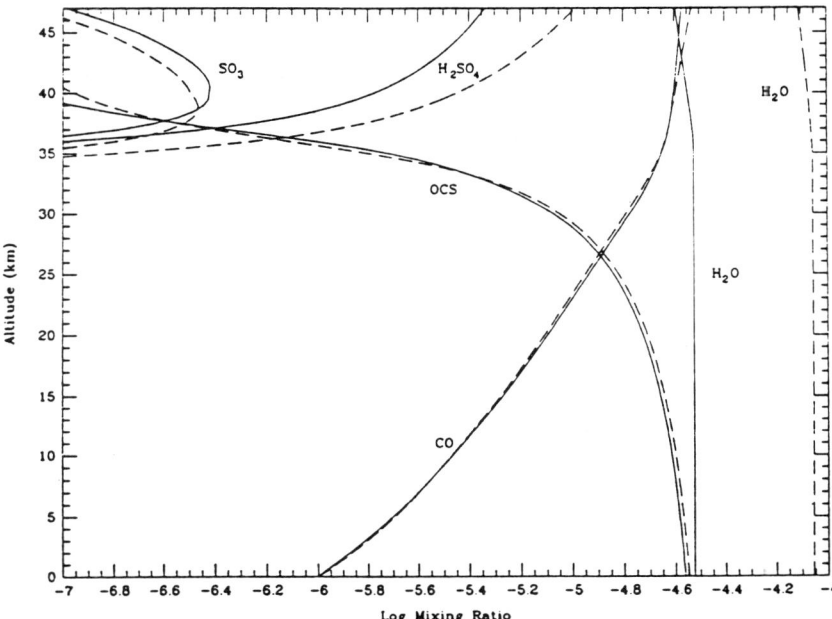

Figure 10. Mixing ratios for OCS, CO, SO_3, H_2SO_4 and H_2O in the lower atmosphere for models 1 (solid lines) and 2 (dashed line). For details, see text and Table IV (figure from Krasnopolsky and Pollack 1994).

slightly to the sulfuric acid aerosol in the lower cloud layer. Due to dynamics of the atmosphere at 60 to 70 km, more or less sulfur dioxide is exposed to solar radiation $\lambda < 219$ nm which transforms SO_2 to sulfuric acid, and more or less oxygen which affects this transformation may be available at different locations. Temperature variations are not very important in variations of density of the lower cloud layer; a temperature decrease of 6 K which lowers the LCB by 1 km, results in an increase of the cloud density by 13%.

Structure of the Cloud Layer. As discussed in the next section, photochemical models by Krasnopolsky and Parshev (1981b,1983) and Yung and DeMore (1982) show that sulfuric acid is produced mostly in a thin layer of 2 km depth centered at 62 km. Therefore, we model this layer by a Gaussian of this width, while the value of the integrated production, which is equal to the downward flux of H_2SO_4 below 60 km, $\phi_{H_2SO_4}$, is the main parameter for modeling the cloud layer and chemistry of the atmosphere below the clouds (see above).

Krasnopolsky and Pollack (1994) proved that the flux and concentration of liquid sulfuric acid and mixing ratios of H_2O and H_2SO_4 vapors are

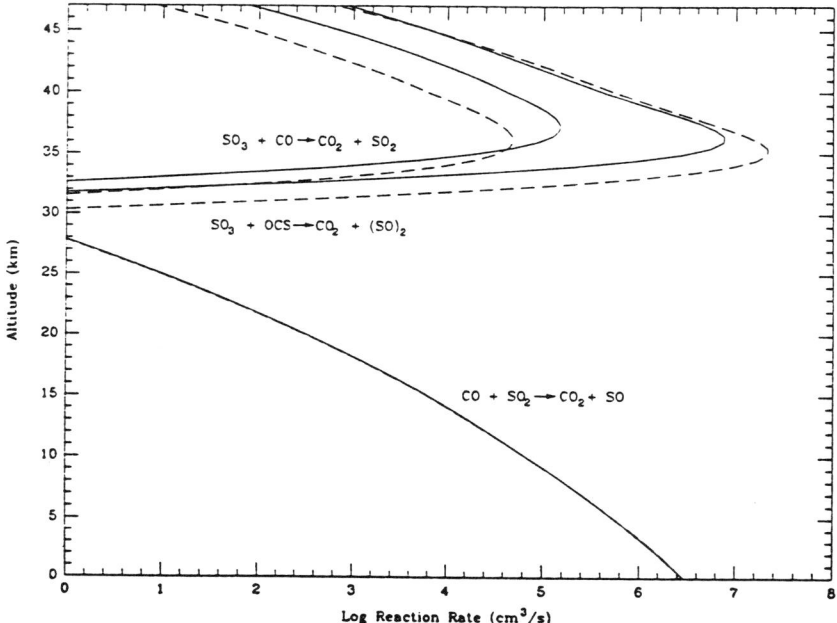

Figure 11. Reaction rates for model 1 (solid lines) and model 2 (dashed lines) from Krasnopolsky and Pollack (1994).

connected by equation

$$\phi_1 = \frac{K[CO_2]}{1+m}\left(\frac{df}{dz} - m\frac{df_1}{dz}\right) \quad (6)$$

for the single unknown $m(z) = H_2O/H_2SO_4$ in liquid sulfuric acid. This value is equivalent to concentration and more convenient, because chemical potentials μ_1 and μ_2 are given in Giauque et al. (1960) as function of m. If $m(z)$ is known, then $f_1(z)$ and $f_2(z)$ are also known, therefore $m(z)$ is really the only unknown in this equation (which is valid only above the LCB).

To solve this equation, one needs the sulfuric acid flux (ϕ_1 in this formulation), its mixing ratio at LCB, and the H_2O mixing ratio at LCB. The H_2SO_4 mixing ratio as function of ϕ_1 was discussed in Sec. II.A.3, and is determined by chemical processes in the atmosphere below the clouds. A choice of the H_2O mixing ratio below the clouds is determined by spectroscopic observations (Fink et al. 1972; Barker 1975) which give $f_{H_2O} = 1$ to 10 ppm at 62 to 65 km. The calculated structure of the cloud layer for models 1 and 2 ($\phi_{H_2SO_4} = 2.2 \times 10^{12}$ and 6.4×10^{12} cm^{-2}s^{-1}, respectively) is shown in Fig. 12.

Considering the flux of sulfuric acid droplets, it is possible to divide the clouds into three layers. Due to the photochemical formation of sulfuric acid,

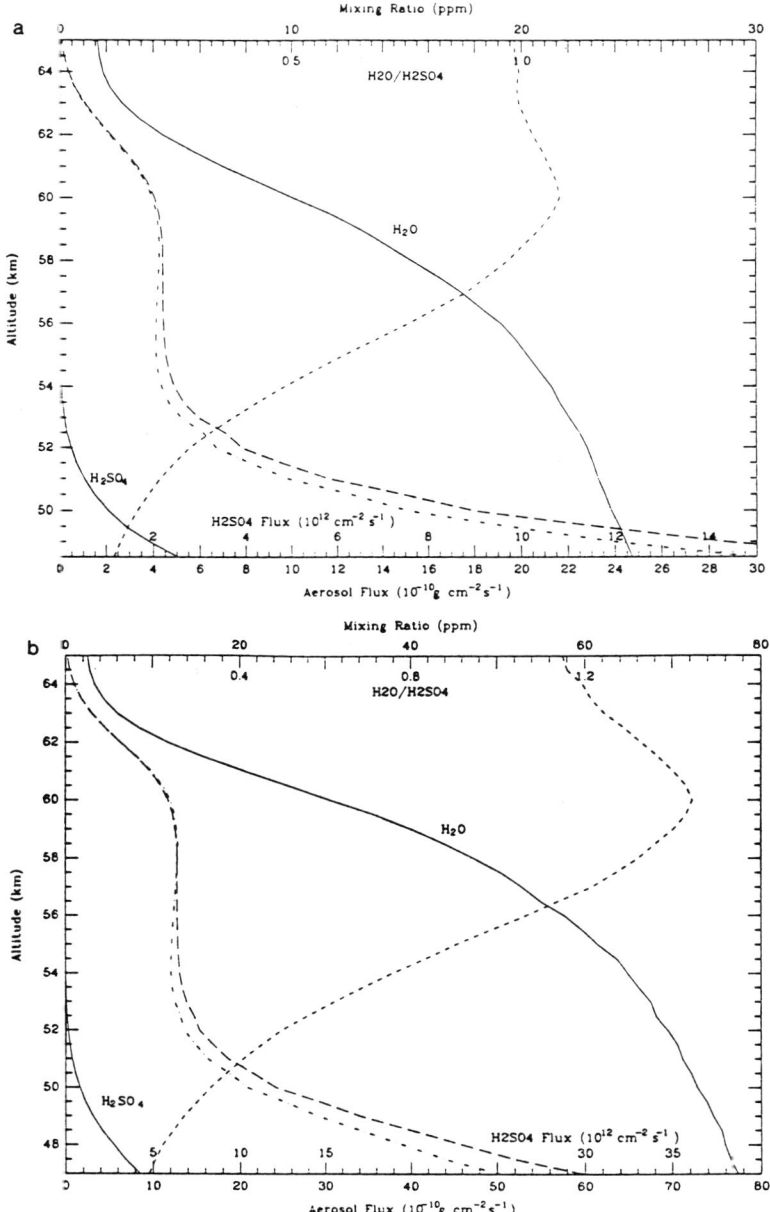

Figure 12. Calculated results for vertical distribution of H_2O and H_2SO_4 in the cloud layer from Krasnopolsky and Pollack (1994). Solid lines: mixing ratios. Short-dashed curves: H_2O/H_2SO_4 ratio in the cloud particles. Long-dashed curve: downward flux of liquid phase H_2SO_4. Dash-dotted curve: downward flux of sulfuric acid droplets. Panel a: model 1; panel b: model 2. See Table IV and text.

its flux increases steeply in the upper cloud layer which ends near 59 km (the measurements give 57 km). This increase correlates with the increasing H_2O mixing ratio while m and concentration of sulfuric acid is relatively constant at 84% and 81% in models 1 and 2, respectively. Measurements show 75 to 85% (Pollack et al. 1978; Reed et al. 1978) and agree with these models.

The flux of liquid sulfuric acid is constant in the middle cloud layer at 59 to 52 km (57–50 km according to the measurements). The increase of f_{H_2O} is only by a factor of 1.5, while m decreases from the top to the bottom of the middle cloud layer by a factor of 3 to 4. This variation changes the concentration from 81 to 84% to 93 to 94%.

In the lower cloud layer at 52 to 48 km, the flux of liquid sulfuric acid exceeds that in the middle cloud layer by a factor of 4 to 7. The H_2O mixing ratio is rather constant, and m continues to decrease by a factor of 3 until the LCB is reached. This corresponds to the concentration increasing to 97 to 98% at the LCB. The important feature of the lower cloud layer is the decrease of sulfuric acid vapor from 5 to 10 ppm at LCB to ≈ 0.5 ppm at the top of the lower cloud layer. A strong gradient of gaseous sulfuric acid drives this upward flux which condenses and forms a strong downward flux of liquid sulfuric acid with the sum of both fluxes being constant in the lower and middle cloud layers. This is the mechanism of formation of the lower cloud layer.

Comparison of the Models with Measurements. For the middle cloud layer, the nephelometric measurements and the photometric measurements show a spread of 20% between the data of the various probes. This shows it is stable around the planet; the conclusion is confirmed by the small horizontal variations seen by the Vega balloons. Therefore, Krasnopolsky and Parshev compare their models to the PV LCPS data (see Fig. 1).

The aerosol flux precipitating at the Stokes velocity is given by

$$\phi = \frac{8\pi}{27} \frac{g\rho^2 r^5}{\eta} n. \tag{7}$$

Coupling the values of the flux with the relationships for the flux and extinction coefficient, one can find the mode 3 extinction coefficient is 0.3 km^{-1} and 0.9 km^{-1} for models 1 and 2, respectively. Then model 1 means that either H_2SO_4 is the only species of the Mode 3 particles (if the interpretation in Esposito et al. [1993] is preferable) or it constitutes a third of large particle material (if the initial LCPS data are adopted), the rest may be $AlCl_3$ or other species (see Sec. II.A.2). Model 2 favors the initial LCPS data and sulfuric acid as a main component of the middle cloud layer. We prefer model 1 because it corresponds to the recent measurements of 30 ppm of H_2O below 30 km, though this model fits worse than model 2 to the sulfuric acid vapor mixing ratio at LCB.

B. Photochemistry from 60 to 110 km

1. Early Models. Actually, the period of intense studies of Venus' photochemistry was rather short and lasted from 1971 to 1982. Papers published before 1971 are of comparatively low interest, and there are no original publications on the subject in the last thirteen years. Another interesting and important aspect of the problem (mentioned by Prinn 1985) is that studies of chlorine and sulfur chemistry in and above Venus' clouds preceded a discovery of the importance of the similar processes in the Earth's atmosphere. This case shows a direct impact from planetary atmospheres studies for understanding phenomena in our own atmosphere.

   2. Post-Pioneer Venus Models. Among the many new atmospheric results from Pioneer Venus and the Venera 11 and 12 probes, the most important facts for photochemistry were (1) SO_2 is the main sulfur-bearing species with the mixing ratio of 150 ppm in the lower atmosphere; (2) H_2O has a mixing ratio of 200 ppm near the lower cloud boundary (which is now thought to be 25 ppm—see model 1 in Sec. II.A.3); (3) a steep decrease of the SO_2 mixing ratio with altitude to 100 ppb at the 40 mbar level (69 km) where the SO_2 scale height was equal to 1 km (later results showed a continuous decline with time of SO_2 at this level from 100 ppb to 20 ppb and scale height in the range 2 to 3 km; more recently the abundances were corrected to be larger by a factor of 4; the currently accepted values are 100 ppb and 3 km for the period of 1982 to 1992; see Fig. 5); and (4) data for O, CO, H, and N in the upper atmosphere. Photochemical papers published soon after (Winick and Stewart 1980; Krasnopolsky and Parshev 1980,1981a,b,1983) reflected these findings.

   Winick and Stewart (1980) modeled photochemistry of HO_x, ClO_x, SO_x, and O_x between 58 and 96 km, thus combining chlorine and sulfur chemistries. Their kinetic data were substantially improved compared with Prinn (1975) and Sze and McElroy (1975) due to five years of progress in the field. For example, the reactions of ClO and ClOO with CO were known as extremely slow and unimportant at that time. Formation of CO_2 via the reaction CO + OH is not effective, and the total production of sulfuric acid is smaller by a factor of 30 than the column CO_2 photolysis in that model. Almost all CO and O_2 formed by the CO_2 photolysis does not recombine to CO_2, and they are thus transported downward to the lower atmosphere. The spectroscopic limit of $f_{O_2} < 0.3$ ppm or the O_2 column abundance $N(O_2) < 1.5 \times 10^{18}$ cm^{-2} at 62 km (Trauger and Lunine 1983) is exceeded by a factor of 150 in this model. However, at that time contradictory data on O_2 from the PV gas chromatograph prevented the proper understanding of the O_2 problem. The H_2O mixing ratio was fixed at 1 ppm throughout the atmosphere, and the SO_2 abundance of 4 ppm was chosen at 58 km as a parameter to fit the measured SO_2 mixing ratio at 70 km.

   Krasnopolsky and Parshev (1981a,b,1983) developed a model for the altitude range from 50 to 200 km. To a certain extent this is the most complete model of Venus' photochemistry, because it considers together the problems of formation of sulfuric acid, its concentration, and vertical profiles of H_2O and SO_2; further, this is without the simplifying assumptions of $f_{H_2O} = 1$ ppm throughout the atmosphere and of SO_2 at the lower boundary as a fitting parameter. However, the kinetic data used in the model are from compilations by Kondratiev (1971) and Baulch et al. (1976) and are poor compared to those used in Winick and Stewart (1980) and especially in Yung and DeMore (1982). A very important cycle which is currently thought to be responsible for recombination of CO and O_2 to CO_2 on Venus was suggested in that paper:

$$\begin{array}{ll} CO + Cl + M \longrightarrow ClCO + M & k_1 \\ ClCO + M \longrightarrow CO + Cl + M & k_2 \\ \underline{ClCO + O_2 \longrightarrow CO_2 + ClO} & k_3 \\ net\ CO + O_2 + Cl \longrightarrow CO_2 + ClO. & \end{array} \quad (8)$$

ClO reacts with either O to form O_2 or SO to form SO_2. The collisional decomposition of ClCO was underestimated by Krasnopolsky and Parshev by 2 orders of magnitude due to a misprint in Kondratiev (1971). Therefore the effective rate coefficient of the cycle $k = k_1 k_3 / k_2$ appears to be overestimated by two orders of magnitude. Another cycle suggested is

$$\begin{array}{l} CO + Cl + M \longrightarrow ClCO + M \\ \underline{ClCO + O \longrightarrow CO_2 + Cl} \\ net\ CO + O \longrightarrow CO_2. \end{array} \quad (9)$$

Due to the fact that the model of Yung and DeMore overestimates the O_2 abundance using the more realistic cycle efficiency by a factor of 10, the Krasnopolsky and Parshev model which gives the extreme case is of some interest, because it overpredicts oxygen by only a factor of 2.5. In the model, all atomic oxygen produced by photolysis of CO_2 above this height recombines to form O_2. Below 87 km, a strong sink for both O_2 molecules moving downward and those newly formed at these heights is due to the ClCO cycle. The O_2 column density is under control of eddy diffusion and is equal to 3.7×10^{18} cm^{-2} which was less than the upper limit of 5×10^{18} cm^{-2} that existed at that time (Traub and Carleton 1974). To fit the current upper limit of 1.5×10^{18} cm^{-2}, the eddy diffusion coefficient at 80 to 90 km should be equal to 10^6 cm^{2}s^{-1}, i.e., larger by a factor of 3 than that used in the model. Oxygen is so scarce at 60 to 65 km, that oxidation of SO_2 to SO_3 with further formation of sulfuric acid must take oxygen also from SO_2 to form sulfur aerosols. The number of photons in the range of 202 to 219 nm which dissociate SO_2 is equal to 1.5×10^{13} cm^{-2}s^{-1} for global-mean conditions; therefore, the full production of sulfuric acid is close to 5×10^{12} cm^{-2}s^{-1}. Sulfuric acid and sulfur aerosols are formed with a mass ratio 7.3:1. Photolysis of SO_2 occurs in a thin layer centered at the altitude where the slant optical depth for SO_2 absorption is unity.

This corresponds to 64 km for the measured mixing ratio of 10^{-7} at 69 km using current SO_2 data (Zasova et al. 1993; Na et al. 1994) which differ from those used by Krasnopolsky and Parshev. The SO_2 flux is constant up to the thin layer centered at 64 km and is equal to $\phi_0 = 7.5 \times 10^{12}$ cm^{-2}s^{-1} (see equation in Sec. I.2).

This gives $f_{SO_2} = 80$ ppm at 50 km which may be compared with 50 ppm at this height from the photometry at the Venera 14 descent probe (Ekonomov et al. 1983). Each SO_3 molecule captures two H_2O molecules forming 85% sulfuric acid. Therefore the fact that the H_2O mixing ratio at 64 km is much smaller than that at 50 km, requires f_{H_2O} (50 km) $\approx f_{SO_2}$ (50 km) $\times 4/3 = 110$ ppm. We see that this model correlates with model 2 considered in Sec. II.A where the H_2SO_4 flux was equal to 6.4×10^{12} cm^{-2}s^{-1} and $f_{H_2O} = 75$ ppm at 50 km.

3. *Models by Yung and DeMore (1982)*. These models give the most comprehensive study of the photochemistry in the altitude region 58 to 110 km. Their important advantage is a very healthy kinetic data set. Though the models are thirteen years old now, we have not found that later studies of chemical processes relevant to these models significantly affect their results. Similar to Winick and Stewart (1980), Yung and DeMore do not consider the water vapor profile (assuming constant $f_{H_2O} = 1$ ppm) and concentration of sulfuric acid aerosol, and use the SO_2 mixing ratio at the lower boundary of 58 km as a fitting parameter to match their models with the SO_2 observations available at that time.

As discussed above, the central problem of Venus' photochemistry is the low abundance of O_2 above the clouds. Hydrogen, chlorine, and sulfur chemistries are each involved in the fate of the CO_2 photolysis products. The main source of active hydrogen and chlorine species (henceforth H* and Cl*) is photolysis of HCl. Though H* and Cl* production rates are equal and their losses are equal as well, due to the reactions

$$Cl + HO_2 \longrightarrow HCl + O_2$$

$$H + Cl_2 \longrightarrow HCl + Cl \tag{10}$$

their densities may be quite different because of the reactions (Krasnopolsky 1986)

$$OH + HCl \longrightarrow H_2O + Cl$$

$$H + HCl \longrightarrow H_2 + Cl$$

$$Cl + H_2 \longrightarrow HCl + H. \tag{11}$$

The first two reactions transform H* to Cl*, and the third reaction vise versa. Their column rates are balanced, and thus we have

$$\frac{[H^*]}{[Cl^*]} \approx 0.1 \frac{[H_2]}{[HCl]}. \tag{12}$$

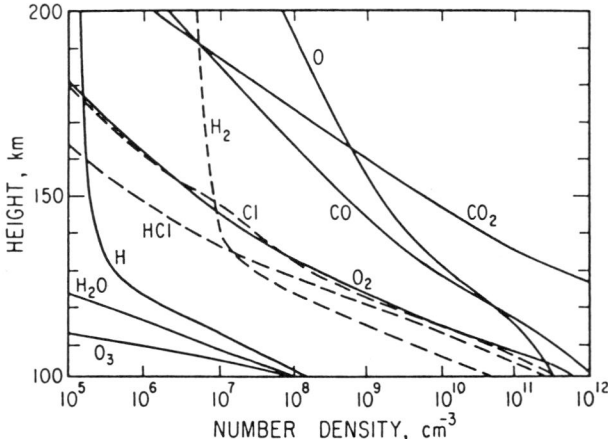

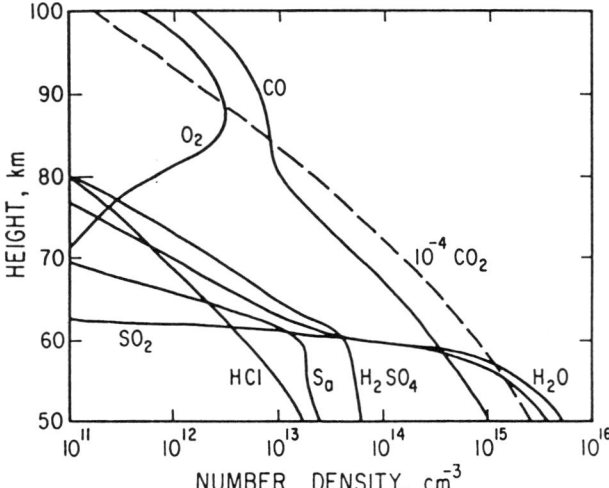

Figure 13. Chemical composition of Venus atmosphere from Yung and DeMore (1982) model C.

Here 0.1 is the ratio of the rate coefficients for a mean temperature of 230 K. Therefore relative importance of the H* and Cl* cycles depends strongly on the H_2 abundance which was rather uncertain.

Model A is based on the suggestion by Kumar et al. (1991) that ions of a mass number 2 measured by the PV ion mass spectrometer were H_2^+. We do not consider this model viable, because the mass 2 ions are currently recognized to be D^+ (McElroy et al. 1982) which requires the D/H ratio of

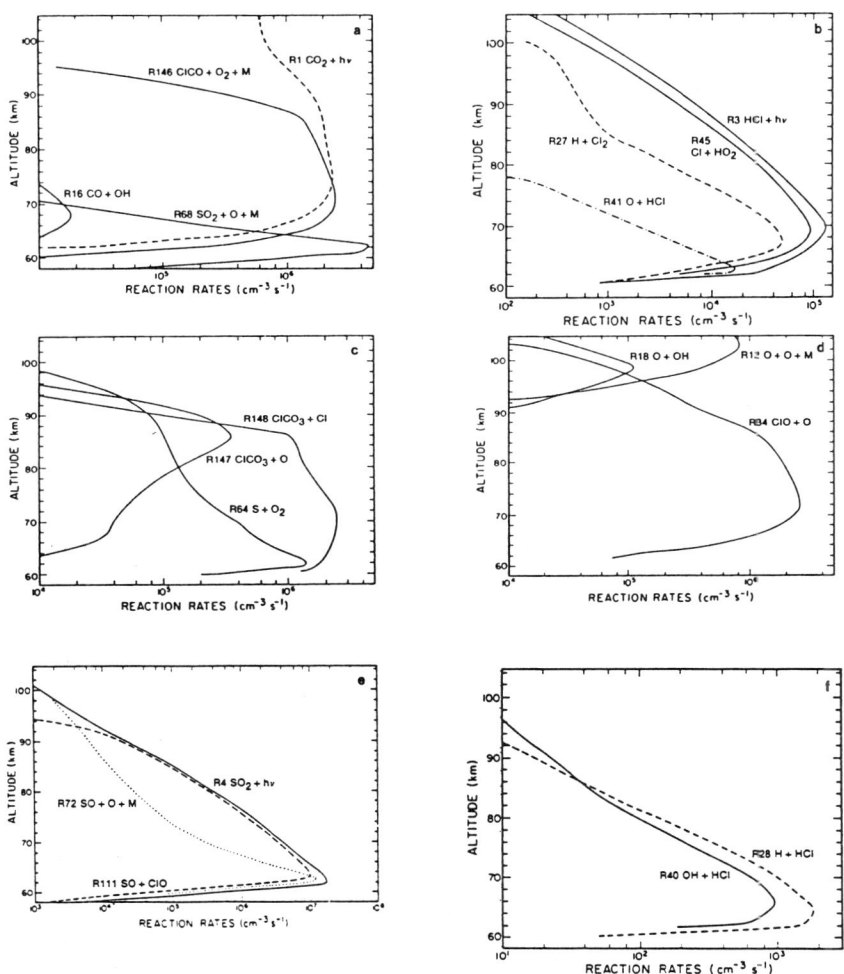

Figure 14. Results from Yung and Demore (1982). Model C: (a) sources and sinks of free oxygen; (b) HO_x and ClO_x; (c,d) formation of O_2; (e) equilibrium between SO_2 and SO; (f) production rates for H_2 and H_2O.

10^{-2} (see the chapter by Donahue et al.). Later, an upper limit of $f_{H_2} < 0.1$ ppm was established from those measurements (Kumar and Taylor 1985).

Model B adopts $f_{H_2} = 0.5$ ppm which is still higher than the upper limit. This model is likewise not viable. Model C assumes that all H_2 molecules in the atmosphere above the lower boundary at 58 km are produced photochemically. Except for the CO_2 and H_2O profiles, the model includes conditions at the lower boundary: $f_{CO} = 45$ ppm, $f_{SO_2} = 4$ ppm (free parameter), $f_{HCl} = 0.8$ ppm (twice the measured value); velocities of O_2, Cl_2, H_2, and O;

$v = -0.6$ K/H $= -0.02$ cm s^{-1}; and photochemical equilibrium for other species. Downward fluxes of 10^{12} cm^{-2}s^{-1} (which are equal to the column photolysis rate of CO$_2$ above 110 km) are taken for CO and O at 110 km. Fluxes of other species are assumed to be zero at the upper boundary. The results of calculations are shown in Figs. 13 and 14. The main cycles which determine balance of CO, O$_2$, and O are those with ClCO (see the previous section). Rates of the key reactions of these cycles are shown in Fig. 14. The O+O+M and O+OH reactions add to production of O$_2$, and the reaction S+O$_2$ reduces slightly its abundance which is equal to 1.8×10^{19} cm^{-2}, and thus exceeds the upper limit by an order of magnitude. The balance of H* and Cl* and formation of H$_2$O and H$_2$ are also shown in Fig. 14. Photolysis of CO$_2$ having the column rate of 7.6×10^{12} cm^{-2}s^{-1} is not completely balanced by production of CO$_2$, and therefore some fraction of the total production of CO equal to 1.4×10^{12} cm^{-2}s^{-1} moves down to the lower atmosphere. The SO$_2$ mixing ratio is equal to 32 ppb at 70 km, and the SO$_2$ scale height is 2.5 km. Both values agree with the measurements. SO$_2$/SO $= 25$ at 70 km which is slightly larger than 10 from the measurements (Na 1992). Production and loss of SO are shown in Fig. 14e. Formation of sulfuric acid and its downward flux is equal to that of CO, and the net result of photochemistry is

$$CO_2 + SO_2 \longrightarrow CO + SO_3. \qquad (13)$$

The downward flux of Cl$_2$ is smaller than those for CO and sulfuric acid by three orders of magnitude and equals 3×10^9 cm^{-2}s^{-1}. It is formed by the net process

$$2HCl + CO_2 \longrightarrow H_2O + CO + Cl_2. \qquad (14)$$

The calculated Cl$_2$ mixing ratio is equal to 25 ppb near 58 km which is much smaller than 1 ppm of Cl$_2$ required to produce the 320 to 500 nm absorption (Pollack et al. 1980).

One of the photochemical problems is to explain a very intense airglow of O$_2$($^1\Delta$) at 1.27 μm with intensities of 1.2 MR and 1.5 MR on the night and day sides, respectively (Connes et al. 1979; see the chapter by Lellouch et al.).

Chlorine-SO$_2$ Interaction. This possibility is discussed by DeMore et al. (1985) and has two important aspects. The first one is the formation of sulfuryl chloride SO$_2$Cl$_2$ via

$$\begin{array}{c} 2[Cl + SO_2 + M \longrightarrow ClSO_2 + M] \\ \underline{2ClSO_2 \longrightarrow SO_2Cl_2 + SO_2} \\ \text{net } 2Cl + SO_2 \longrightarrow SO_2Cl_2. \end{array} \qquad (15)$$

DeMore et al. claimed that SO$_2$Cl$_2$ should be the most abundant chlorine species in the mesosphere with the expected mixing ratio of 4 ppm. The second one is enhanced removal of O$_2$ and formation of sulfuric acid:

$$Cl + SO_2 + M \longrightarrow ClSO_2 + M$$

$$ClSO_2 + O_2 + M \longrightarrow ClSO_4 + M$$
$$ClSO_4 + Cl \longrightarrow + SO_3 + ClO$$
$$SO_2 + h\nu \longrightarrow SO + O$$
$$ClO + SO \longrightarrow SO_2 + Cl$$
$$\underline{SO_2 + O + M \longrightarrow SO_3 + M}$$
$$\text{net } O_2 + 2SO_2 \longrightarrow 2SO_3. \tag{16}$$

They suggest that a model with 4 ppm of SO_2Cl_2 and 0.4 ppm of HCl at the lower boundary might fit much better the experimental results than the models of Yung and DeMore (1982). However, this model has not been further developed. On the other hand, Pollack et al. (1993) concluded from the observed constancy of the HCl mixing ratio from the cloud tops to the surface that other chlorine species may be present in the atmosphere only with mixing ratios smaller than ≈ 0.2 ppm, therefore precluding 4 ppm of SO_2Cl_2. The chlorine-sulfur interaction is a promising way to explain the very low abundance of molecular oxygen and should be considered carefully in further improvements of Venus' photochemistry modeling.

Summary of the Models. There are some parallels between model 1 (low H_2SO_4) and model 2 (higher H_2SO_4 flux) for the lower atmosphere (Sec. II.A.) and the photochemical models of Yung and DeMore (1982) and Krasnopolsky and Parshev (1981b), respectively. Models 1 and 2 are determined by the column production rate (i.e., the downward flux) of H_2SO_4 which is equal to 2.2×10^{12} and 6.4×10^{12} $cm^{-2}s^{-1}$, respectively. The photochemical models respectively produce values of H_2SO_4 production close to these inputs. Comparison of the models is shown in Table IV.

We prefer model 1 because it is based on good kinetic data and agrees with the recent spectroscopic measurements of water vapor and with the position of the lower boundary of sulfuric acid in the clouds. Advantages of model 2 are: (1) the much smaller abundance of O_2; (2) presence of sulfur aerosol which is supported by measurements; (3) support by radio occultation measurements of the lower cloud layer (Cimino 1982), although Jenkins (1992) reprocessed these data and found no detection of cloud material; (4) sulfuric acid vapor mixing ratio (Jenkins et al. 1994); and (5) by water vapor spectroscopic measurements from the Venera probes. Further experimental and photochemical studies are needed. Clearly, chlorine-sulfur interactions will be a promising field for further development of Venus' photochemistry.

III. OPEN QUESTIONS AND FUTURE WORK

A. Unidentified Ultraviolet and Blue Absorption

Currently, the only positively identified absorbing species in the visible atmosphere are SO_2 and SO. However, they do not absorb longward of 3200 Å. Thus, other absorbers must explain the absorption in the Venus spectrum and dark markings that extend to 5000 Å (Esposito 1980; Pollack et al. 1980).

In addition, these other absorbers must explain the phase angle dependence of the ultraviolet dark markings (Barker et al. 1975) as well as their short lifetime above the clouds (from hours to days; see Esposito et al. 1983). This must also be consistent with the solar flux observations of Tomasko et al. (1980) which show absorption at 58 to 62 km, and little absorption below. Similar solar flux absorption results from Venera 14 (Ekonomov et al. 1983,1984) provide an additional constraint. Esposito and Travis (1982) noted the correlation between dark markings seen longward of 3200 Å and SO_2 enhancements seen at 2070 Å. Beyond the absorption spectrum, a good candidate must also match the vertical distribution, lifetime and correlation with SO_2 enhancement. This last correlation could be either chemical or dynamical, because the SO_2 visible in the far ultraviolet is likely the result of local upwelling (Esposito and Travis 1982). We briefly review the suggested candidates.

S_8. Hapke and Nelson (1975) and Young (1977,1983) proposed S_8 as the second absorber because it absorbs strongly in the ultraviolet. However, Pollack et al. (1980) showed that the spectral characteristics of S_8 were inconsistent with that of the second absorber. Another shortcoming of S_8 as the second absorber is its vertical profile. S_8 is not expected to disappear rapidly below the upper cloud layer because it precipitates as a solid, and the idea of these particles hiding inside sulfuric acid aerosols has been discounted by Young (1983). Thus, the vertical profile of S_8 does not match that of the second absorber. Furthermore, the chemical lifetime of S_8 above the clouds is much longer than the time scale of the dark markings, thus it is difficult to explain the rapid disappearance (lifetime <3 hr) of small scale dark markings (see, e.g., Rossow et al. 1980).

S_3 and S_4. Toon et al. (1982) suggested metastable sulfur allotropes, S_3 and S_4, as the most likely candidate for the second absorber. The absorption cross sections of S_3 peak around 400 nm and S_4, around 520 nm. The combination of these two sulfur gases with SO_2 provides a very close match to the albedo of Venus. The peak in the absorption cross sections of S_3 around 400 nm lines up with a kink in the albedo spectrum of Venus. S_3 and S_4 are metastable, and once produced they quickly relax to S_8 which exists as particulates. S_8 particles could then become incorporated into the sulfuric acid aerosols and fall out of the cloud region. This scenario thus explains the short lifetime of the dark features and the absence of the second absorber below the upper clouds. Furthermore, these sulfur allotropes can account for the high real refractive index of the upper cloud material, and the bimodal size distribution observed in the Venus clouds.

The production of sulfur allotropes may happen in oxygen poor areas in the atmosphere of Venus. With little oxygen in the atmosphere, sulfur allotropes are produced from SO_2 photolysis instead of sulfuric acid (Prinn 1975,1985).

Elemental sulfur can form photochemically from SO_2 if the efficiency of photochemical cycles which restore CO_2 from CO and O_2 is high. Then formation of sulfuric acid may result from

$$3SO_2 + 2h\nu \longrightarrow 2SO_3 + S \qquad (17)$$

and a mass ratio of productions of $H_2SO_4 \cdot H_2O$ to S is seven. However, if the efficiency is low, then sulfuric acid forms via

$$CO_2 + SO_2 + h\nu \longrightarrow CO + SO_3 \qquad (18)$$

without elemental sulfur production (see Sec. II.B). S_3 and S_4 may be produced in areas where sulfur vapor is enriched. However, there have been no positive detection of S_3 or S_4 to date. Further, these allotropes are rapidly photodissociated, giving lifetimes close to 1 s in the upper cloud.

One major problem with the above scenario is that sulfur particles would still absorb ultraviolet photons below the upper clouds.

S_2O. Hapke and Graham (1985,1989) proposed that the disulfur monoxide (S_2O) and polysulfur oxides may be responsible for the ultraviolet markings in the clouds of Venus. They measured the relative reflectance of S_2O frost at 77 K, and found that it has low reflectivity in the wavelength region from 200 to 500 nm. Na and Esposito (1995) estimate the the chemical lifetime and vertical distribution of S_2O, both of which match the second absorber. Its obvious chemical connection with SO_2 could explain the correlations of the dark markings with SO_2 enhancements. Unfortunately, we do not have a good spectrum measured for the gas phase.

$FeCl_3$. Krasnopolsky (1985,1986) showed that many properties of the clouds can be explained if condensation of Fe_2Cl_6 occurs at 47.5 km at the PV sounder probe site. This means that this species mixing ratio is equal to 15 ppbv below 47.5 km. The calculated profile of the $FeCl_3$ condensate coincides with that of the mode 1 particles in the lower and middle cloud layer. The mode 1 $FeCl_3$ particles are transported by eddy diffusion to the upper cloud layer where they serve as condensation centers for the mode 2 H_2SO_4 particles. These particles are liquid below 62 to 63 km, and the $FeCl_3$ flux to the H_2SO_4 production rate ratio corresponds to a solution with concentration of $FeCl_3$ close to 1%. It is this solution which can explain the 320 to 500 nm absorption (Zasova et al. 1981). The reaction between $FeCl_3$ and concentrated H_2SO_4 is rather slow at temperatures 250 to 280 K at 62 to 58 km, and the lifetime of the solution is close to the precipitation time of one month. Colorless ferric sulfate replaces $FeCl_3$ near 58 km. Thermochemical equilibrium is

$$Fe_s(SO_4)_3 + 6HCl + 3CO = Fe_2Cl_6 + 3H_2O + 3SO_2 + 3CO_2. \qquad (19)$$

Thus, elemental sulfur, S_2O, and ferric chloride solution in sulfuric acid are good candidate species responsible for the observed absorption at 320 to 500 nm, perhaps even in combination.

B. The Mode 3 Controversy

The LCPS measurement of mode 3 particles has provided a controversy that is still unresolved (see Esposito et al. 1983). The starting point for the mode 3 controversy comes from direct evidence for asymmetric (possible crystalline) particles provided by Knollenberg and Hunten (1980). Knollenberg et al. (1980) further state that only such crystals of high aspect ratio could satisfy the Pioneer Venus LCPS, LSFR, and LN results simultaneously. However, because the largest amount of mass (~80% according to Knollenberg and Hunten [1980]) is within the mode 3 particles, it is extremely important to verify their existence and determine their composition.

The LCPS undoubtedly detected large particles, but the evidence for solid particles is indirect. There were internal inconsistencies in the LCPS measurements as well as inconsistencies between the LCPS measurements and the measurements made by other instruments. Some of these inconsistencies were:

1. Calculations employing LCPS size distributions do not give the backscatter observed by the PV nephelometer in the lower clouds if reasonable refractive indices are used.
2. The LCPS size distributions do not yield the optical depths derived by the LSFR, assuming spherical particles.
3. Overlapping size ranges of the LCPS give conflicting measurements in the lower clouds.

In addition, independent Venera results show some oddities at the same altitudes: (1) Venera nephelometer phase function measurements are inconsistent with spherical particles having reasonable refractive indices in the lower cloud; (2) X-ray fluorescence measurements (Surkov et al. 1979) show about ten times as much chlorine as sulfur in the Venus clouds. The various inconsistencies can be explained by the simple hypothesis that mode 3 is composed of solid, nonspherical particles. However, this explanation requires an abundant gas-phase chemical in the clouds as the source for these particles. No such gas has yet been discovered.

Toon et al. (1982) reexamined the evidence that solid particles form a distinctive size mode. They find that mode 3 is defined by a discontinuity located between two size ranges of the LCPS. Although this could be real, it could also be the result of a small calibration shift of the PV instrument. A shift in the calibration removes the discontinuity, along with the internal inconsistency of the LCPS. The revised size spectrum is consistent with the Venera and Pioneer optical data in the lower clouds; all the modes can be composed of sulfuric acid droplets without any solid particles. The only unexplained data are those showing large amounts of chlorine compared to sulfur in the clouds. We note, though, that the more recent Soviet measurements from Veneras 13 and 14 show a large sulfur to chlorine ratio, the opposite of findings by Surkov et al. (1981). The Vega landers detected no large particles (see Sec. I.A.).

From the data in hand, it seems impossible to disprove the existence of mode 3. Two self-consistent, alternative interpretations of the data exist. Accepting the spacecraft observation at face value, we are led to the existence of a mode of large solid particles whose composition is unknown and whose source vapor has escaped detection. On the other hand, we may conclude that the large particle mode is merely the (mis-measured) tail end of the Mode 2 sulfuric acid droplets. This allows a simple understanding of the source of all the cloud particles, but at the cost of disbelieving some of the measurements.

C. Future Work

To advance our understanding of Venus chemistry, we require:

1. Better measurements of composition, including vertical and horizontal variation. Key species include SO_2, SO, H_2S, Cl_2, COS, and H_2O.
2. Better measurements of the cloud particle properties and their variation. The size distribution, shape and composition of the majority of the aerosol mass are still open, despite our assurance that "mode 2" (the aerosols visible at the cloud tops) are spherical droplets of concentrated sulfuric acid.
3. Determination of the rate and importance of surface–atmosphere interactions (see the chapter by Fegley et al.).
4. Detailed comparison of *in-situ* and remote determinations of water abundance in the deep atmosphere (see chapter by Taylor et al.).

This new information could justify more detailed models of the chlorine and sulfur cycles important in the atmosphere, and studies of the many interactions between radiation, clouds, chemistry and dynamics.

Acknowlegments. Esposito and Krasnopolsky acknowlege support from the Venus Data Analysis Program (VDAP). Moroz and Zasov acknowlege grants from the International Science Foundation Grants and the Russian Fundamental Science Foundation. We are grateful for careful reading of the manuscript by C. Na and P. Steffes.

REFERENCES

Allen, D. A., and Crawford, J. W. 1984. Cloud structure on the dark side of Venus. *Nature* 307:222–224.

Allen, D. A., Crisp, D., and Meadows, V. 1992. Variable oxygen airglow on Venus as a probe of atmospheric dynamics. *Nature* 359:516–519.

Andreychikov, B. M. 1987. Chemical composition and structure of the clouds of Venus inferred from the results of X-ray fluorescent analysis on descent probes VEGA 1 and 2. *Kosmich. Issled.* 25:737–743 (in Russian).

Andreychikov, B. M., et al. 1987. X-ray radiometric analysis of the cloud aerosol of Venus by the Vega 1 and 2 probes. *Kosmich. Issled.* 25:721–736 (in Russian).
Barin, I. 1989. *Thermochemical Data of Pure Substances* (Weinheim: VCH).
Barker, E. S. 1979. Detection of SO_2 in the UV spectrum of Venus. *Geophys. Res. Lett.* 6:117–120.
Barker, E. S., et al. 1975. Relative spectrophotometry of Venus from 3067 to 5960 Å. *J. Atmos. Sci.* 32:1205.
Bauer, S. H., Jeffers, P., Lifshitz, A., and Yavada, B. 1971. Reaction between CO and SO_2 at elevated temperatures: A shock-tube investigation. In *Proc. 13th Symp. on Internal Combustion*, pp. 417–425.
Baulch, D. L., Drysdale, D. D., Duxbury, J., and Grant, S. 1976. *Evaluated Kinetic Data for High Temperature Reactions* (London: Butterworths).
Bertaux, J. L., et al. 1987. Investigation of UV absorption in the atmosphere of Venus at VEGA 1,2 descent probes. *Cosmic Res.* 25(5):691–706.
Bertaux, J. L., et al. 1996. VEGA-1 and VEGA-2 entry probes: An investigation of local UV absorption (220–400 nm) in the atmosphere of Venus (SO_2, aerosols, cloud structure). *J. Geophys. Res.* 101:12709–12745.
Bézard, B., et al. 1993. The abundance of sulfur dioxide below the clouds of Venus. *Geophys. Res. Lett.* 20:1587–1590.
Carlson, R. W., et al. 1991. Galileo infrared imaging spectroscopy measurements at Venus. *Science* 253:1541–1548.
Cimino, J. 1982. The composition and vertical structure of the lower cloud deck on Venus. *Icarus* 51:334–357.
Clancy, R. T., and Muhleman, D. O. 1991. Long-term (1979–1990) changes in the thermal, dynamical, and compositional structure of the Venus mesosphere as inferred from microwave spectral observations of ^{12}CO, ^{13}CO and $C^{18}O$. *Icarus* 89:129–146.
Connes, P., Noxon, J. F., Traub, W. A., and Carleton, N. P. 1979 $O_2(^1\Delta)$ emission in the day and night airglow of Venus. *Astrophys. J. Lett.* 233:29–32.
Conway, R. R., McCoy, R. P., Barth, C. A., and Lane, A. L. 1979. IUE detection of sulfur dioxide in the atmosphere of Venus. *Geophys. Res. Lett.* 6:629–631.
DeMore, W. B., Leu, M. T., Smith, R. H., and Yung, Y. L. 1985 Laboratory studies on the reactions between chlorine, sulfur dioxide, and oxygen: Implications for the Venus stratosphere. *Icarus* 63:347–353.
Deriugin, V. A., et al. 1987. Automatic stations VEGA-1 and VEGA-2. Functioning of descent probes in the atmospheres of Venus. *Cosmic Res.* 25(5):494.
Ekonomov, A. P., et al. 1983. UV photometry at the Venera 13 and 14 landing probes. *Cosmic Res.* 21:194–206.
Esposito, L. W. 1980. Ultraviolet contrasts and the absorbers near the Venus cloud tops. *J. Geophys. Res.* 85:8151–8157.
Esposito, L. W. 1984. Sulfur dioxide shows evidence for Venus volcanism. *Science* 223:1072.
Esposito, L. W., and Travis, L. D. 1982. Polarization studies of the Venus UV contrasts. *Icarus* 51:374–390.
Esposito, L. W., Winick, J. R., and Stewart, A. I. 1979. Sulfur dioxide in the Venus atmosphere: Distribution and implications. *Geophys. Res. Lett.* 6:601–604.
Esposito, L. W., et al. 1983. The clouds and hazes on Venus. In *Venus*, eds. D. M. Hunten, L. Colin, T. M. Donahue and V. I. Moroz (Tucson: Univ. of Arizona Press), 484–564.
Esposito, L. W., et al. 1988. Sulfur dioxide at the Venus cloud tops 1978–1986. *J. Geophys. Res.* 93:5267–5276.
Fegley, B., Jr., and Treiman, A. H. 1992. Chemistry of atmosphere-surface interactions on Venus and Mars. In *Venus and Mars: Atmospheres, Ionsopheres, and Solar*

Wind Interactions, AGU Geophysical Mono. 66, pp. 7–71.
Fink, U., Larson, H. P., Kuiper, G. P., and Poppen, R. F. 1972. Water vapor in the atmosphere of Venus. *Icarus* 17:617–631.
Florensky, C. P., Volkov, V., and Nikolaeva, O. 1978. A geochemical model of the Venus troposphere. *Icarus* 33:537–553.
Gel'man, B. G., et al. 1979. Gas chromatograph analysis of the chemical composition of the Venus atmosphere. *Space Res.* 20:219.
Gnedykh, V. I., et al. 1987a. Vertical structure of the Venus cloud layer at the VEGA-1 and VEGA-2 landing points. *Kosmich. Issled.* 25:707–714.
Gnedykh, V. I., et al. 1987b. Vertical structure of cloud layer above landing sites of VEGA 1 and VEGA 2. *Kosmich. Issled.* 25:707–714 (in Russian).
Giauque, W. F., Horning, E. W., Kunzler, J. E., and Rubin, T. R. 1960. The thermodynamic properties of aqueous sulfuric acid solutions and hydrates from 15 to 300 K. *J. Amer. Chem. Soc.* 82:62–67.
Golovin, Yu. M., Moshkin, B. E., and Ekonomov, A. P. 1981. Aerosol component properties as measured by the Venera 11 and 12 spectrophotometer. *Cosmic Res.* 19:295–302.
Hapke, B., and Graham, F. 1985. Disulfur Monoxide and the spectra of Io and Venus. *Lunar Planet. Sci.* XV:316–317 (abstract).
Hapke, B., and Graham, F. 1989. Spectral properties of condensed phases of disulfur monoxide, polysulfur oxide, and irradiated sulfur. *Icarus* 79:47–55.
Hapke, B. W., and Nelson, R. M. 1975. Evidence for an elemental sulfur component of the clouds from Venus spectroscopy. *J. Atmos. Sci.* 32:1212–1218.
Hartley, K. K., Wolf, A. R., and Travis, L. D. 1989. Croconic acid: An absorber in the Venus clouds? *Icarus* 77:382–390.
Hoffman, J. H., Hodges, R. R., Jr., Donahue, T. M., and McElroy, M. B. 1980. Composition of the Venus lower atmosphere from the Pioneer Venus mass spectrometer. *J. Geophys. Res.* 85:7882–7890.
Hunten, D. M., Colin, L., Donahue, T. M., and Moroz, V. I., eds. 1983. *Venus* (Tucson: Univ. of Arizona Press).
Jenkins, J. M. 1992. Variations in the 13-cm Opacity Below the Main Cloud Layer in the Atmosphere of Venus Inferred from the Pioneer-Venus Radio Occultation Studies 1978–1992. Ph.D. Thesis, Georgia Inst. of Technology.
Jenkins, J. M., et al. 1994. Radio occultation studies of the Venus atmosphere with the Magellan spacecraft. *Icarus* 110:79–94.
Kamp, L. W., and Taylor, F. W. 1990. Radiative-transfer models of the night side of Venus. *Icarus* 86:510–529.
Knollenberg, R. G., and Hunten, D. M. 1980. Results of the Pioneer Venus particles size spectrometer experiment. *J. Geophys. Res.* 85:8039–8058.
Knollenberg, R. G., et al. 1980. The clouds of Venus: A synthesis report. *J. Geophys. Res.* 85:8059–8081.
Kondrat.ev, V. N. 1971. *Rate Coefficients of Gas Phase Reactions* (Moscow: Nauka Press).
Krasnopolsky, V. A. 1985. Chemical composition of Venus' clouds. *Planet. Space Sci.* 33:109–117.
Krasnopolsky, V. A. 1986. *Photochemistry of the Atmospheres of Mars and Venus* (New York: Springer-Verlag).
Krasnopolsky, V. A. 1989. Vega mission results and chemical composition of Venusian clouds. *Icarus* 80:202–210.
Krasnopolsky, V. A., and Parshev, V. A. 1979. On the chemical composition of the troposphere and cloud layer of Venus based on the Venera 11 and 12 and Pioneer Venus measurements. *Cosmic Res.* 17:763–770.
Krasnopolsky, V. A., and Parshev, V. A. 1981a. Initial data of calculation of chemical

composition of the Venus atmosphere down to 50 km. *Cosmic Res.* 19:87–109.
Krasnopolsky, V. A., and Parshev, V. A. 1981b. Photochemistry of the Venus atmosphere down to 50 km: Results of calculations. *Cosmic Res.* 19(2):61–280.
Krasnopolsky, V. A., and Parshev, V. A. 1983. Photochemistry of the Venus atmosphere. In *Venus*, eds. D. M. Hunten, L. Colin, T. M. Donahue and V. I. Moroz (Tucson: Univ. Arizona Press), pp. 431–458.
Krasnopolsky, V. A., and Pollack, J. B. 1994. H_2O-H_2SO_4 system in Venus' clouds and OCS, CO, and H_2SO_4 profiles in Venus' troposphere. *Icarus* 109:58–78.
Kumar, S., and Taylor, H. A. 1985. Deuterium on Venus: Model comparisons with Pioneer Venus observations of the predawn bulge ionosphere. *Icarus* 62:494–504.
Kumar, S., Hunten, D. M., and Taylor, H. A. 1981. H_2 abundance in the atmosphere of Venus. *Geophys. Res. Lett.* 8:237–239.
Kunde, V. G., Hanel, R. A., and Herath, L. W. 1977. High spectral resolution ground-based observations of Venus in the 450–1250 cm^{-1} region. *Icarus* 32:210–224.
Levine, J. S., et al. 1982. Production of nitric oxide by lightning on Venus. *Geophys. Res. Lett.* 9:893–896.
Lewis, J. S. 1970. Venus: atmospheric and lithospheric composition. *Earth Planet. Sci. Lett.* 10:73–80.
Linkin, V. M., et al. 1985. VENERA 15 and VENERA 16 infrared experiment. 5. Preliminary results of analysis of brightness temperature and thermal flux fields. *Cosmic Res.* 23(2):212–221.
Marov, M. Ya., et al. 1983. Study of Venus cloud structure by the Venera 13 and 14 nephelometers. *Cosmic Res.* 21:207–215.
McClintock, W. E., Barth, C. A., and Kohnert, R. A. 1994. Sulfur dioxide in the atmosphere of Venus: Sounding rocket observations. *Icarus* 112:381–388.
McElroy, M., Prather, M., and Rodriguez, J. 1982. Escape of hydrogen from Venus. *Science* 215:1614–1615.
Moroz, V. I. 1987. Scientific results of the VEGA mission. *Kosmich. Issled.* 25:659–672 (in Russian).
Moroz, V. I., et al. 1983. The Venera 13 and 14 spectrophotometric experiment. II. Preliminary analysis of H_2O absorption bands in spectra. *Cosmic Res.* 21:187–194.
Moroz, V. I., et al. 1985. VENERA 15 and VENERA 16 infrared experiment. 4. Preliminary results of spectral analyses in the region of H_2O and SO_2 absorption bands. *Cosmic Res.* 23(2):202–211.
Moroz, V. I., D. et al. 1986. Venus spacecraft infrared radiance spectra. Some aspects of their interpretation. *Applied Opt.* 25(10).
Moroz, V. I., et al. 1990. Water vapor and sulfur dioxide abundances at the Venus cloud tops from the VENERA 15 infrared spectrometry data. *Adv. Space Res.* 10(5):77.
Moshkin, B. E., et al. 1983. Veneras 13 and 14 spectrophotometric experiment. I. Methodics, results, and preliminary analysis of the measurements. *Cosmic Res.* 21:177–186.
Na, C. Y. 1992. Sulfur Oxides in the Middle Atmosphere of Venus. Ph.D Thesis, Univ. of Colorado.
Na, C. Y., and Esposito, L. W. 1996. UV observations of Venus with HST. In preparation.
Na, C. Y., Esposito, L. W., and Skinner, T. E. 1990. International Ultraviolet Explorer observation of Venus SO_2 and SO. *J. Geophys. Res.* 95:7485.
Na, C. Y., Esposito, L. W., McClintock, W. E., and Barth, C. A. 1994. Sulfur dioxide in the atmosphere of Venus: Modeling results. *Icarus* 112:389–395.
Oertel, D., et al. 1985. VENERA-15 and VENERA-16 infrared experiment. 1. Technique and first results. *Cosmic Res.* 23(2):162–175.

Oertel, D., et al. 1987. Infrared spectrometry from VENERA-15 and VENERA-16. *Adv. Space Res.* 5(9):25.
Owen, T., and Sagan, C. 1972. Minor constituents in planetary atmospheres: Ultraviolet spectroscopy from the Orbiting Astronomical Observatory. *Icarus* 16:557–568.
Oyama, V. I., et al. 1980. Pioneer Venus gas chromatography at the lower atmosphere of Venus. *J. Geophys. Res.* 85:7891–7902.
Petryanov, I. V., et al. 1981. Iron in the Venus clouds. *Dokl. AN SSSR* 260:834.
Pollack, J. B., et al. 1978. Properties of the clouds of Venus as inferred from airborne observations of its near infrared reflectivity spectrum. *Icarus* 34:28–45.
Pollack, J. B., et al. 1980. Distribution and source of the UV absorption in Venus atmosphere. *J. Geophys. Res.* 85:8141–8150.
Pollack, J. B., et al. 1993. Near-infrared light from Venus' nightside: A spectroscopic analysis. *Icarus* 103:1–42.
Porshnev, N. V., et al. 1987. Gas chromatographic analysis of products of thermal reactions of the cloud aerosol of Venus by the Vega 1 and 2 probes. *Cosmic Res.* 25:715.
Prinn, R. G. 1975. Venus: chemical and dynamical processes in the stratosphere and mesosphere. *J. Atmos. Sci.* 32:1237–1247.
Prinn, R. G. 1978. Venus: chemistry of the lower atmosphere prior to the Pioneer Venus mission. *Geophys. Res. Lett.* 5:973–976.
Prinn, R. G. 1979. On the possible role of gaseous sulfur and sulfanes in the atmosphere of Venus. *Geophys. Res. Lett.* 6:807–810.
Prinn, R. G. 1985. The photochemistry of the atmosphere of Venus. In *The Photochemistry of Atmospheres,* ed. J. S. Levine (Orlando: Academic Press), pp. 281–336.
Ragent, B., and Blamont, J. E. 1980. The structure of the clouds of Venus: Results of the Pioneer Venus nephelometric experiment. *J. Geophys. Res.* 85:8089–8105.
Reed, R. A., Forrest, W. J., Houck, J. R., and Pollack, J. B. 1978. Venus: The 17 to 38 micron spectrum. *Icarus* 33:554–557.
Rodriguez, J. M., Prather, M. J., and McElroy, M. B. 1984. Hydrogen on Venus: Exospheric distribution and escape. *Planet. Space Sci.* 32:1235–1251.
Rossow, W. B., et al. 1980. Cloud Morphology and Motions from Pioneer Venus images. *J. Geophys. Res.* 85:8107–8128.
Sagdeev, R. Z., et al. 1986. Overview of VEGA Venus balloon in situ meteorological measurements. *Science* 231:1411–1413.
Schaefer, K., et al. 1987. Structure of the middle atmosphere of Venus from analyses of Fourier-spectrometer measurements aboard VENERA-15. *Adv. Space Res.* 7(12):17.
Schaefer, K., et al. 1990. Infrared Fourier Spectrometer experiment from VENERA-15. *Adv. Space Res.* 10(5):57.
Schofield, J. T., Taylor, F. W., and McCleese, D. J. 1982. The global distribution of water vapor in the middle atmosphere of Venus. *Icarus* 52:263–278.
Sill, G. T. 1983. The clouds of Venus: sulfuric acid by the lead chamber process. *Icarus* 53:10.
Spankuch, D., et al. 1985. VENERA-15 and VENERA-16 infrared experiment. 2. Preliminary results of temperature profile retrieval. *Cosmic Res.* 23(2):176–188.
Spankuch, D., Matsygorin, I. A., Dubois, R., and Zasova, L. V. 1990. Venus middle-atmosphere temperature from VENERA-15. *Adv. Space Res.* 10(5):67.
Stewart, A. I. F., Anderson, D. E., Esposito, L. W., and Barth, C. A. 1979. Ultraviolet spectroscopy of Venus: Initial results from the Pioneer Venus Orbiter. *Science* 203:777–778.
Surkov, Yu. A., et al. 1981. A study of the Venus cloud aerosol by Venera 12

(preliminary data). *Cosmic Res.* 20:435.
Surkov, Yu. A., Ivanova, V. F., Pudov, A. N., and Caramel, D. 1987. Determination of the aerosols chemical composition in the Venusian clouds by means of the mass-spectrometer MALAHIT on the VEGA-1 probe. *Kosmich. Issled.* 15:744–750 (in Russian).
Sze, N. D., and McElroy, M. B. 1975. Some problems in Venus' aeronomy. *Planet. Space Sci.* 23:763–786.
Taylor, F. W., et al. 1980. Structure from the Pioneer Venus Orbiter. *J. Geophys. Res.* 85:7963–8006.
Titov, D. V. 1983. On the possibility of aerosol formation by the reaction between SO_2 and NH_3 in Venus' atmosphere. *Cosmic Res.* 21:401.
Tomasko, M. G., Doose, L. R., Smith, P. H., and Odell, A. P. 1980. Measurements of the flux of sunlight in the atmosphere of Venus. *J. Geophys. Res.* 85:8167–8186.
Toon, O. B., Turco, R. P., and Pollack, J. B. 1982. The ultraviolet absorber on Venus: Amorphous sulfur. *Icarus* 51:358.
Traub, W. A., and Carleton, N. P. 1974. A search for H_2O and O_2 on Venus. In *Exploration of Planetary Atmospheres*, eds. A. Woszcyk and C. Iwanishewska (Dordrecht: D. Reidel), p. 223.
Trauger, J. T., and Lunine, J. I. 1983. Spectroscopy of molecular oxygen in the atmospheres of Venus and Mars. *Icarus* 55:272.
Volkov, V. P., Sidorov, Y. I., Khodakovsky, I. L., and Barsukov, V. L. 1982. On the possible condensates in the Venus main cloud layer. *Geokhimia* 3.
von Zahn, U., Kumar, S., Niemann, H., and Prinn, R. G. 1983. Composition of the Venus atmosphere. In *Venus*, eds. D. M. Hunten, L. Colin, T. M. Donahue and V. I. Moroz (Tucson: Univ. of Arizona Press), pp. 299–430.
Watson, A. J., et al. 1979. Oxides of nitrogen and the clouds of Venus. *Geophys. Res. Lett.* 6:743–746.
Widemann, T., Bertaux, J. L., Moroz, V. I., and Ekonomov, A. P. 1993. VEGA-1 and VEGA-2 descent modules: In-situ measurements of ultraviolet absorption and relationship with present active volcanism on Venus. *Bull. Amer. Astron. Soc.* 25:1094 (abstract).
Winick, J. R., and Stewart, A. I. 1980. Photochemistry of SO_2 in Venus' upper cloud layers. *J. Geophys. Res.* 85:7849–7860.
Young, A. T. 1977. An improved Venus cloud model. *Icarus* 32:1–26.
Young, A. T. 1983. Venus cloud microphysics. *Icarus* 56:568.
Yung, Y. L., and DeMore, W. B. 1982. Photochemistry of the stratosphere of Venus: Implications for atmospheric evolution. *Icarus* 51:199–247.
Zasova, L. V., Krasnopolsky, V. A., and Moroz, V. I. 1981. Vertical distribution of SO_2 in the upper cloud layer of Venus and origin of the UV absorption. *Adv. Space Res.* 113–16.
Zasova, L. V., et al. 1985. VENERA-15 and VENERA-16 infrared experiment. 3. Some spectral analyses results on the cloud structure. *Cosmic Res.* 23(2):189.
Zasova, L. V., et al. 1989. Venusian clouds from VENERA-15 data. *Veroffenlichungen Forschungsbereichs Geo-Kosmoswissenschaften*, 18 (Berlin: Academie-Verlag).
Zasova, L. V., Moroz, V. I., Esposito, L. W., and Na, C. Y. 1993. SO_2 in the middle atmosphere of Venus: IR measurements from VENERA-15 and comparison to UV data. *Icarus* 105:92–109.
Zhulanov, Yu. V., Mukhin, L. M., and Nenarokov, D. F. 1986. Preliminary results of particles number densities measurements on clouds of Venus on heights 47–63 km on board VEGA-1 and VEGA-2 landing probes. *Pisma Astron. Z.* 12:97–130.

THE GENERAL CIRCULATION OF THE VENUS ATMOSPHERE: AN ASSESSMENT

P. J. GIERASCH
Cornell University

R. M. GOODY
Harvard University

R. E. YOUNG
NASA Ames Research Center

D. CRISP, C. EDWARDS, R. KAHN, D. McCLEESE and D. RIDER
Jet Propulsion Laboratory

A. DEL GENIO
Goddard Institute for Space Studies

R. GREELEY
Arizona State University

A. HOU
NASA Goddard Space Flight Center

C. B. LEOVY
University of Washington

and

M. NEWMAN
University of Colorado

The overall spin or "superrotation" of the Venus atmosphere is a striking phenomenon. In the 15 years since the NASA Pioneer Venus mission, a first-order understanding has been reached of the dynamics of the atmospheric region near and just above the Venus cloud tops. Tidal motions induced by solar heating produce a traveling disturbance whose vertical momentum transports are balanced by mean flow advection. The balance explains the shear of the mean flow above the clouds, and partially explains the strength of the mean flow at the cloud level where the strongest superrotation of the atmosphere occurs. But the fundamental cause of the global superrotation remains a mystery in spite of data from Earth-based observatories, from Pioneer Venus, from several Russian probes, from a Russian/French balloon experiment, and from the NASA Galileo flyby. The key missing knowledge is of momentum transfer processes in the deep atmosphere, between the surface and the cloud deck. Neither the forcing nor the drag and dissipation mechanisms are known. The existing data are reviewed here, and theoretical suggestions are listed. It is concluded that further measurements, in conjunction with numerical modeling, will be required to resolve this puzzling and challenging question. New data must improve by an order of magnitude on the accuracies achieved by the Pioneer Venus probes. Velocities in the deep atmosphere must be measured to better than 0.1 m s^{-1}, and relative temperatures to better than 0.1 K near the surface.

I. INTRODUCTION

The rotation of the Venus atmosphere is one of the most intriguing unexplained phenomena in planetary science. At cloud-top level, where the pressure is about 50 mb, the mean rotation period is about 4 terrestrial days, in contrast to the 243 day rotational period of the solid planet. The atmosphere is primarily CO_2, and the surface pressure is about 90 bar. The total atmospheric mass is roughly equivalent to 1 km of water, and as a consequence the thermal time constant (evaluated in Sec. III) is very long, approximately one century. Through most of its depth the atmosphere is stably stratified and it is likely that frictional dissipation is weak. In comparison, the relevant flow time constants based on advection are a few days. Thus the Venus atmosphere is weakly damped and weakly forced, and the mean flow may be only loosely coupled to the fundamental drives. The rapid mean flow may be the end state of kinetic energy transformations that are several steps removed from the solar heating that drives the system. It is an interesting system on general dynamical grounds, as well as presenting a puzzle specifically to the atmospheric and planetary sciences.

Our purpose is an assessment of the understanding of the Venus circulation rather than a detailed review of work that has been done, although we attempt to be complete in describing the observational basis. A pre-Pioneer Venus assessment was presented by Hunten and Goody (1969). Detailed post-Pioneer Venus reviews are by Moroz (1981), covering general Venus atmospheric science, and Schubert (1983) covering dynamics.

In Secs. II through VII below the presently available data is surveyed, and in Secs. VIII through XVI theoretical ideas are discussed.

II. THERMAL STRUCTURE OBSERVATIONS

We discuss in this section only those features of the thermal structure of the Venus atmosphere that are particularly relevant to understanding the circulation. For additional information, the reader is referred to an accompanying chapter by Crisp and Titov, which presents a detailed summary of what is known concerning thermal structure and energy balance.

Pioneer Venus entry probes determined temperature profiles at four widely separated locations on the planet, with an uncertainty of ±1 K (Seiff et al. 1980). The Sounder Probe entered approximately on the equator near the morning terminator at local time 7:38 a.m. The North Probe entered at 3:35 a.m. at about 60 deg latitude. The Day and Night probes entered at about 30 deg south latitude, at local times of 6:46 a.m. and 0:07 a.m. Potential temperature profiles from all four probes are displayed in Fig. 1, calculated from the temperature, pressure tables given by Seiff et al. (1980). These profiles expand the temperature scale, display the stability, and permit comparison of temperatures. They show horizontal temperature contrasts less than 10 deg, and they show that the stratification is similar at all probe locations. There have also been about a dozen Russian Venera probes into the Venus atmosphere, and all show temperature profiles very similar to those of Fig. 1 (Avduevskiy et al. 1983). Figure 1 does not display Venera data because the accuracy of these measurements (5 K; Avduevsky et al. 1983) is not sufficient to give a reliable stratification.

The Pioneer Venus Orbiter (PVO) carried an infrared radiometer with bands selected to permit retrieval of temperature profiles within the middle atmosphere (above the clouds). Figure 2 shows a latitude, height cross section and a longitude, height cross section in solar-fixed coordinates (Schofield and Taylor 1983). There is very little longitudinal temperature variation near the 1 bar pressure level but at higher altitudes a diurnal variation occurs, with an amplitude of about 5 K at the 10 mb level. The semi-diurnal component is the largest. These temperatures are evidence of solar-induced tides, and are extremely valuable for constraining the modeling of the tides on Venus (see Sec. XI below). Infrared spectra of Venus were obtained from the Venera 15 orbiter in 1983 (Schäfer et al. 1987), and thermal structure conclusions are consistent with those from the 1979 PVO.

The Russian/French Vega balloon experiment returned a large amount of data about temperatures within the cloud region from 50 to 55 km elevation, confirming that the stability is extremely small (Crisp et al. 1990). These results are described in Sec. VI below.

The thermal structure of the Venus atmosphere consists of a stable middle atmosphere, above the 0.2 bar, 60 km level, overlying a deep, less stable troposphere. At elevations above 50 km, where pressures are less than 1 bar, latitude-height temperature cross sections can be constructed from radio occultation data and thermal emission data, and the nature of the cyclostrophic balance can be examined (see Sec. III below). Longitudinal variation of temperature at these levels shows evidence of tides, but better vertical resolution would be useful. In the deep atmosphere there is an important gap in information within the bottom scale height because all the Pioneer Venus probes experienced electrical anomalies which affected the thermal sensors. Above 12 km the Pioneer probes determined the stability of the atmosphere. Figure 1

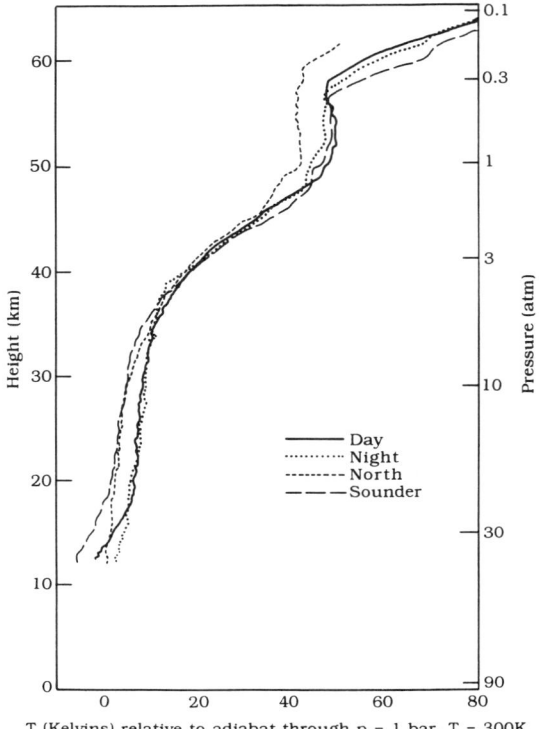

Figure 1. Temperatures from the Pioneer Venus probes, relative to an adiabat. This "potential temperature" would be uniform in a neutrally stable layer. Thermodynamic properties of the gases were taken from Hilsenrath et al. (1960).

shows the four temperature profiles relative to a reference adiabat. There are two layers with low stability, one in the cloud deck between about 0.5 and 1 bar, and one between the 8 and 30 bar level. Between 1 and 8 bars and in the upper atmosphere the thermal structure is stable. In Sec. VII we shall examine the correlation of the stability structure with the wind field at the Pioneer Venus probe entry sites.

III. RADIATION BALANCE OBSERVATIONS

The height profiles of upward and downward solar fluxes were measured by the Pioneer Venus Sounder Probe from the surface to about 64 km elevation (Tomasko et al. 1980; see also the chapter by Crisp and Titov). From this data one can calculate the net flux and the profile of energy deposition in the atmosphere. Approximately 17 W m^{-2} are absorbed at the surface. This is 11% of the net global average solar absorption $\overline{F} = 157$ W m^{-2}, and 2.6% of the global average insolation, 655 W m^{-2}. More than half of the solar

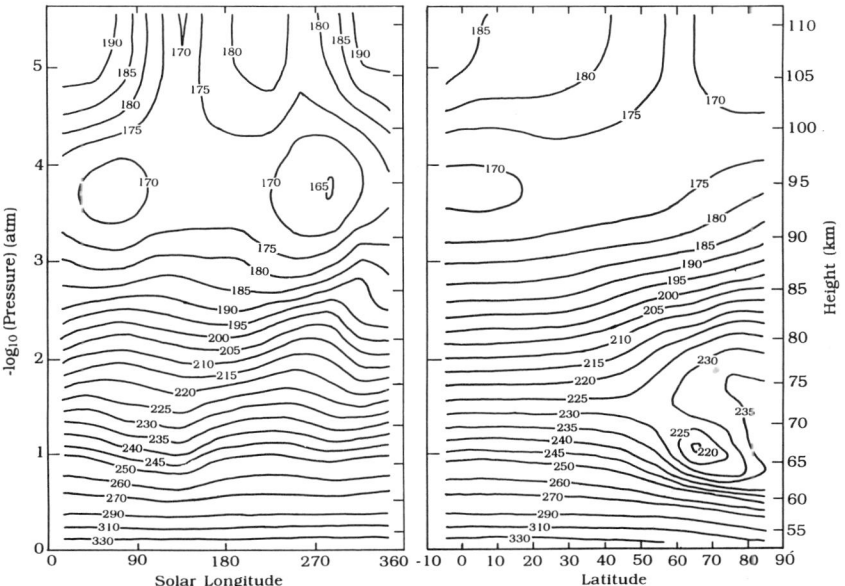

Figure 2. Temperature cross sections for the middle atmosphere, from Schofield and Taylor (1983). Local noon is at 0 deg longitude. The height–longitude cross section is above the equator.

energy is absorbed in the upper cloud layer between 55 and 70 km elevation. Of this, approximately half is absorbed by an unknown ultraviolet absorber in the cloud particles, and most of the rest is absorbed by H_2SO_4 aerosols (Crisp 1986).

The net global solar absorption is balanced by thermal emission to space. A radiative time constant based on the average thermal cooling rate and the entire mass of the atmosphere is $t_R = c_p p \overline{T}/(g \overline{F}) \sim 100$ yr, for $c_p = 800$ J kg^{-1}, a surface pressure of 90 bar, a mean temperature of 600 K, and an acceleration of gravity of 8.7 m s^{-2}. On the other hand, a radiative time constant based on the atmospheric mass within and above the cloud deck is only a few days (Crisp 1989), consistent with the observations indicating the existence of solar-induced tides.

In spite of uncertainties in the details, the fundamental radiative drive for large-scale motions in the deep atmosphere is known. Horizontal temperature gradients are small, and therefore the chief contribution to horizontal variations of net radiative heating is the variation of the insolation with distance away from the subsolar point, which has been established by the Pioneer probe measurements reported by Tomasko et al. (1985).

IV. CLOUDS

The Venus clouds are discussed in the chapters by Esposito et al. and by Crisp and Titov. Here we treat only topics related to dynamics.

Is the Venus cloud deck coupled actively to the atmospheric motion field? A major portion of the solar energy absorption is in the cloud, but if the cloud position is determined by microphysical and chemical processes that are independent of the atmospheric dynamics, then the answer could be negative. For the purposes of dynamical investigations, the cloud would then be a prescribed location of certain radiative forcings. The answer to the question is not yet fully understood, but several remarks can be made.

Nowhere in the Venus clouds did the Pioneer probes indicate a mass loading exceeding 100 mg m^{-3} (Knollenberg and Hunten 1980), corresponding to a mass mixing ratio of less than 10^{-3}, comparable to a fairly thick cirrus cloud on Earth. The latent heat of evaporation is an order of magnitude larger than the heat content of the condensate, but because the mixing ratio is small, a phase change can produce a temperature perturbation of at most a degree or two. In contrast, the large-scale latitudinal contrasts near cloud level are on the order of 10 K (Fig. 2). Latent heating is therefore small in the Venus clouds, and coupling to motions by this mechanism is not likely to be important except possibly for effects on local convection.

Esposito et al. (1983) discuss the microphysics of the Venus clouds, and show that droplet formation times near the cloud tops are probably on the same order as fallout times (based on a scale height). Both are on the order of several weeks. Large-scale vertical velocities have been self consistently estimated by Newman and Leovy (1992) in a tidal computation that includes adjustment of the mean state, and the mean flow advection time (vertical displacement equal to a scale height) is also a few weeks. Therefore it appears that the large-scale motion field should cause a major perturbation to the cloud position, but probably would not be the controlling factor. Indeed there is a latitudinal gradient of cloud-top properties, including a "polar collar" of bright material at near 60° latitude and a polar region of high infrared emission at still higher latitudes (Taylor et al. 1980) which corresponds to a depression of the cloud-top elevation. Thus the large-scale motion field does appear to alter the cloud-top position but only as a perturbing effect.

Near the cloud base at about 50 km elevation the situation is more complicated. Thermal emission at near-infrared wavelengths escapes to space through gaps in the CO_2 spectrum and permits determination of the total cloud optical depth. Groundbased observers of the night side of the planet have been able to map opacity variations (Crisp et al. 1991), and the Jupiter-bound Galileo spacecraft obtained infrared images with spatial resolution better than 100 km (Carlson et al. 1991). The opacity variations, which almost certainly arise in the lowest few km of the cloud, are approximately 25%. Radiative fluxes are altered by opacity variations, and there is the possibility of feedback between dynamics and the cloud structure. The pattern of cloud opacity variations is streaky at high latitudes and more patchy at low latitudes, similar to the ultraviolet features. This region of the atmosphere may contain important dynamical activity and is not well understood.

V. FLOW OBSERVATIONS: LARGE SCALE-FEATURES

Here we discuss flows at cloud level and below. For discussion of middle atmosphere dynamics, see the chapter by Lellouch et al. Figure 3 displays profiles of the zonal and meridional wind with latitude deduced from tracking small scale features in the blue or ultraviolet seen in spacecraft images. Figure CDP3C6F1 displays an image of Venus that shows a typical pattern of features. They are probably formed just below the level where the ultraviolet optical depth is unity, which is about 40 mb pressure and 70 km elevation (Kawabata et al. 1980). The depth may vary by a few km with latitude and with time, but probably not by much more than this because the gradient of optical depth is large; the particle scale height is approximately the same as that of the gas, or about 4.8 km (Kawabata et al. 1980). The observed velocity profiles, which are each averages over several days of data, vary from epoch to epoch (Rossow et al. 1990). These profiles are not necessarily representative of a zonal mean, because only the sunlit side of the planet is observed and there may be a solar-fixed tide with zonal or meridional amplitude on the order of 10 m s^{-1} (Del Genio and Rossow 1990; Newman and Leovy 1992).

The meridional velocities of Fig. 3 are of particular interest because they might give an indication of the strength of the Hadley circulation at cloud-top level. But unfortunately the uncertainty about the zonal mean is particularly important in this case because the tide and the mean may well be of the same order. The calculations of Newman and Leovy predict that the zonal mean is considerably smaller than the velocities shown in Fig. 3. If this is correct, the nightside velocities are small or even equatorward. On the other hand Schinder et al. (1990) have argued that the global spiral shape of the ultraviolet cloud features is consistent with poleward drift with zonal mean meridional velocities approximately equal to the observed dayside values. Further work with the cloud patterns may yield information on the important question of the relative strength of tidal and Hadley circulations. But there is no doubt that a tidal component exists in the cloud-tracking wind velocities. Figure 4 displays the latitude and longitude variation of the time-averaged zonal and meridional wind from Pioneer data (Del Genio and Rossow 1990), and compares it with the predictions of Newman and Leovy. Other observation periods give similar results (Limaye 1987; Del Genio and Rossow 1990).

Another strong flow component at cloud top is the traveling four-day wave that produces the global pattern shaped like a "Y" or a "Ψ" rotated to the left to lie on its side. Early observers noted the motion of this feature (Boyer and Camichel 1961; Boyer and Guérin 1969). The structure of the albedo pattern became clear in Mariner 10 images (Belton et al. 1976*a*) With Pioneer Venus and Galileo data it became possible to measure perturbations

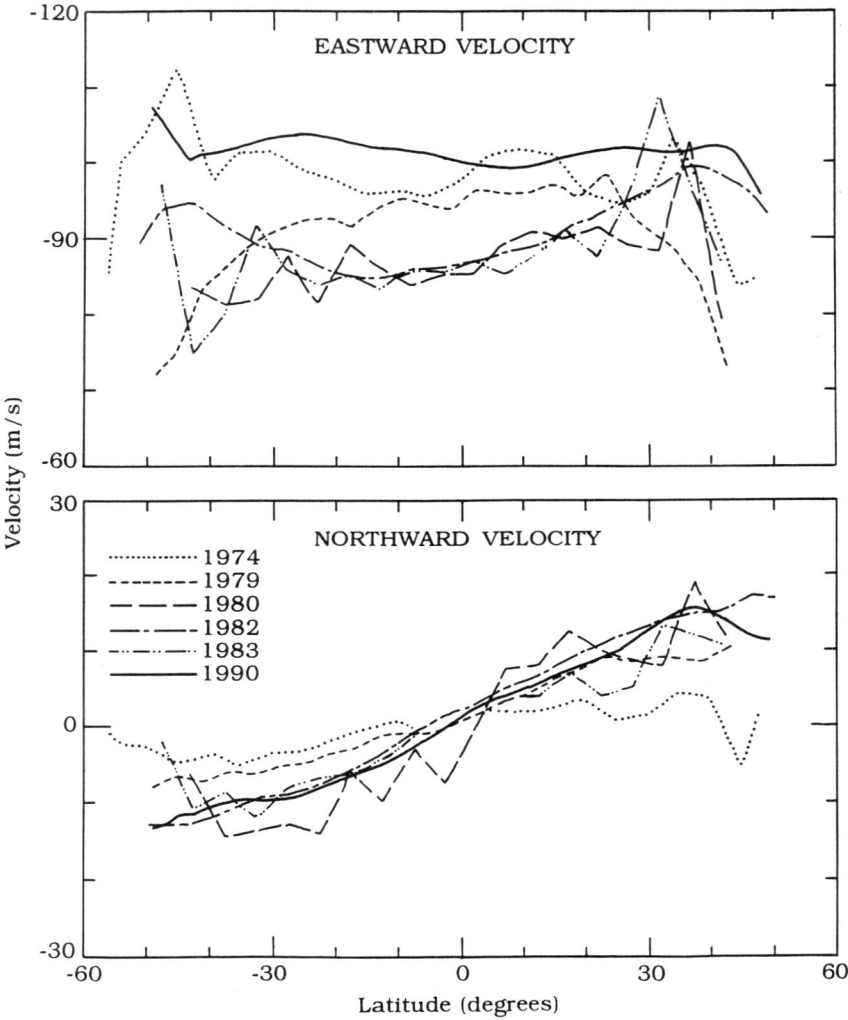

Figure 3. Velocities of cloud features. These are averages over the illuminated portion of the disk of the eastward (top frame) and northward velocities. Profiles from 1974 are from Mariner 10, measured by Limaye and Suomi (1981), profiles from 1979, 1982, 1980 and 1983 are from Pioneer Venus, measured by Limaye et al. (1982,1988) and Belton et al. (1991), and profiles from 1990 are from Galileo, measured by Belton et al. (1991).

associated with the feature (Del Genio and Rossow 1990; Rossow et al. 1990; Smith et al. 1993). The zonal wind oscillation at cloud top level is variable in magnitude but reaches 10 m s^{-1}. Meridional winds are smaller. This is consistent with early suggestions that the feature is a wave of Kelvin type

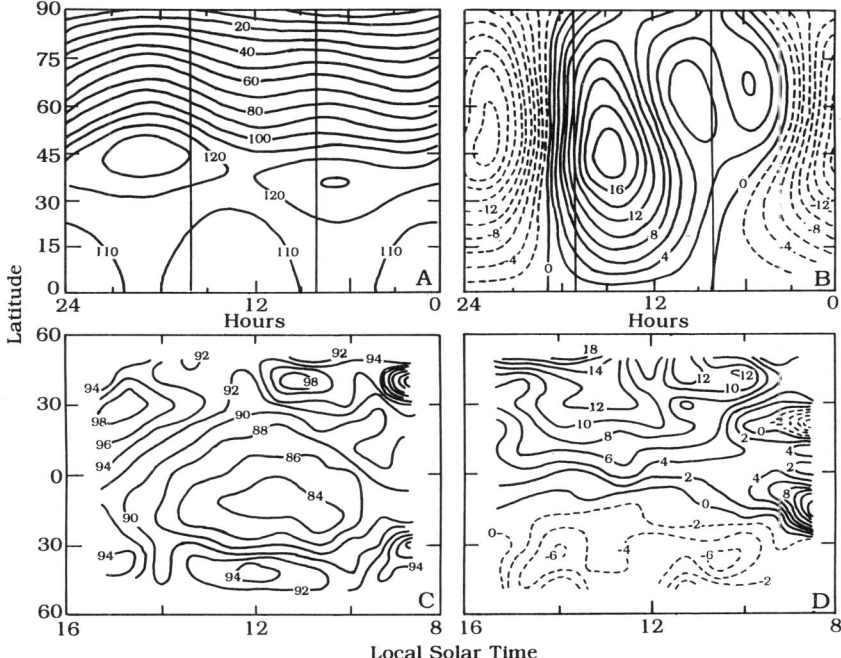

Figure 4. Time averaged velocity fields in a solar-fixed reference frame, at cloud top. Zonal wind is on the left and poleward wind is on the right. Panels (A) and (B) display calculated results from Newman and Leovy (1992), and panels (C) and (D) give the zonal and meridional components from measurements by Del Genio and Rossow (1990).

(see, e.g., Belton et al. 1976b). Theoretical calculation of wave structure was carried out by Covey and Schubert (1982), who showed that more than one free mode exists with the observed frequency. Smith et al. (1993), show that the Kelvin mode alone can produce the Y pattern and the flows associated with the waves as observed during the Galileo flyby. They also speculate that the wave is generated by a radiative-dynamical feedback at cloud base. It is important to learn the forcing and dissipation mechanisms for the four-day wave, because its momentum transfers depend on them and it appears to be a major dynamical phenomenon. Del Genio and Rossow and Rossow et al. also detected a disturbance with a five-day period, and identify it as a Rossby mode.

The solar tide, the four-day wave and the mean flow are the major components of the Venus large-scale flow near the cloud top. In fact, there is remarkably little else at this level. After removing the solar-fixed component and the four-day wave from the 1990 Galileo imaging data, Toigo et al. (1994) find an upper limit of about 4 m s^{-1} on the residuals, which would include all synoptic and mesoscale time-dependent eddy and wave activity. Del Genio and Rossow (1990), analyzing PVO images, find transient waves at some epochs but also with amplitudes of only a few meters per second.

Flow at deeper levels has been obtained by tracking entry probes. Unfortunately the Pioneer and Venera probes, because of their aerodynamic properties, acquired the horizontal velocity of the atmosphere to within a few m s^{-1} only beneath the 60 km level, and therefore there is a gap of about a scale height between the top of the vertical profile from the probes and the ultraviolet feature tracking level. At altitudes less than 55 km the probes give the horizontal wind to within about 1 m s^{-1}, with an altitude resolution of 1 km deep in the atmosphere and a few km near 55 km elevation (Counselman et al. 1980). Figure 5 displays the four Pioneer Venus entry probe profiles of zonal velocity, from Counselman et al., and also the locations of the entry points. These profiles show that the spin of the atmosphere is a global phenomenon, and that the spin begins to build up in the top half of the bottom scale height and continues to amplify right through the low and middle portions of the cloud deck.

Meridional velocities are also displayed in Fig. 5. There is no clear evidence for a Hadley circulation, but we shall see in Sec. VII below that the expected amplitude beneath the cloud is so small that none should be apparent in this data even if it exists as an important dynamical flow component.

Magalhães and Seiff (1995) have carefully looked for eddy structures in measurements of temperature from the atmospheric structure experiment on each of the Pioneer Venus entry probes using a spatial filtering technique. The filter separates structure with vertical scales of ≤ 10 km (eddy structures) from larger-scale structures (background structures). Between approximately 42 km to 60 km altitude, all the probes show eddy structure with amplitudes of about 1 K, well above the uncertainty level of the measurements, with a well-defined vertical wavelength of about 6 km. This structure is quite well correlated between the three small probes and some correlation with the fourth Large Probe is evident as well. Correlating the temperature fluctuations with velocity fluctuations is more difficult and has not yet been done, because the velocity fluctuations, of order 2 m s^{-1}, are the same order as the velocity measurement errors, about 1 m s^{-1} (Counselman et al. 1980). The correlation in temperature between the probes, which were well separated in horizontal spatial position, suggests a dynamical phenomenon with a large horizontal scale, such as a planetary wave or atmospheric tide.

The two Vega balloons sampled the large-scale flow in 1985 at about 53 km elevation and at latitudes of about ±7 deg (see description in Sec. VI). The east-west velocities were about 65 m s^{-1}, confirming measurements from descent probes (Crisp et al. 1990). The north–south velocity showed a poleward drift by the southern probe with an average velocity of about 2.5 m s^{-1}, and a much smaller northward velocity for the other. There is also a large scale acceleration pattern that may be due to a solar-fixed or topographically induced flow component.

It has recently been demonstrated that tracking of the near infrared hot

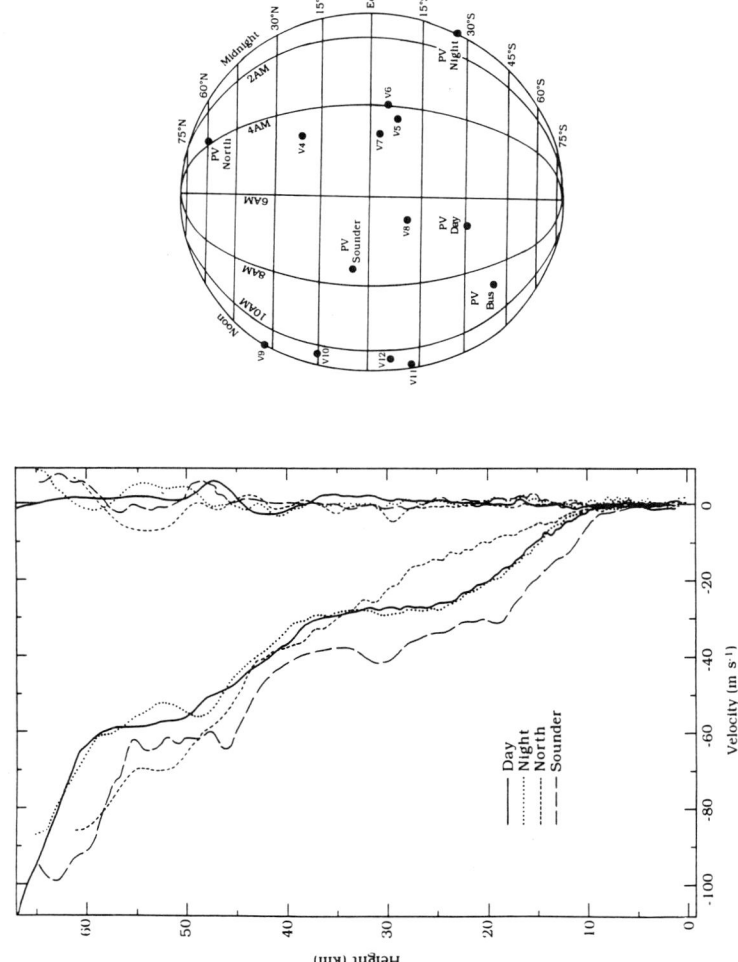

Figure 5. (Left) Zonal and meridional velocity profiles from Pioneer Venus probe tracking (Counselman et al. 1980). (Right) Locations of entry probes. "Day" and "Night" are designations for the two probes that entered in the southern hemisphere. Also indicated are the descent locations of Russian Venera probes.

spots, described in Sec. IV, gives velocities in the lower cloud region (Crisp et al. 1991; Carlson et al. 1991). In addition, Galileo imaging at 1 μm wavelength showed features that moved with about the same velocity (Belton et al. 1991). Thus it appears that near-infrared remote sensing can provide dynamical data from deep cloud layers, supplementing the traditional ultraviolet dynamical data from the cloud tops. At present the deeper data is sparse and not of high accuracy, but it offers promise of statistically useful results from a region of the Venus atmosphere that is not easily accessible by other means (see the chapter by Taylor et al.).

Newman et al. (1984) and Walterscheid et al. (1985) have used Pioneer Venus radio occultation data to estimate the latitude and height dependence of the zonal mean thermal structure for pressures between about 2 mb and 1.4 bar, and then integrated the cyclostrophic balance Eq. (4) to determine winds. They assumed a certain latitudinal profile of velocity at the 1.4 bar level. The dominant feature in their deduced zonal wind cross sections is a jet about one scale height above the ultraviolet cloud level with a maximum of about 120 m s^{-1} at about 50 deg latitude. At higher levels the wind decreases. There is some evidence that the position and strength of the jet vary in time.

In the deep atmosphere, probe data suggests that the latitudinal velocity profile is not far from solid body rotation at each height (Schubert [1983] examines Pioneer data; Venera probes are consistent), although the sampling is sparse. Thus the evidence suggests that a transition takes place near 40 km elevation, above which the latitudinal profile of angular velocity develops a jet-like structure that is more pronounced with increasing height, until the core of the jet is reached just above the ultraviolet cloud-top level.

Another dynamical region of interest is at high latitudes, in either hemisphere. Pioneer Venus infrared measurements show peculiar double hot spots at approximately 80 deg northern latitude, located 180 deg apart in longitude (Taylor et al. 1980). These authors attribute the structure to depressions in the cloud top elevation. Ultraviolet images of high latitudes appear symmetric in longitude and do not show the double structure. It is possible that there is a disturbance in the middle cloud region, too deep to be detected in the ultraviolet images. Stability studies of the circumpolar jet show that it may be unstable to wavenumber 2 (Elson 1982; Young et al. 1984; Michelangeli et al. 1987).

VI. FLOW OBSERVATIONS: SMALL SCALES

Ultraviolet images of the Venus cloud tops show blotches that are roughly circular, streaks that spiral toward the poles, and occasionally long linear features (Belton et al. 1976a; Rossow et al. 1980). Small features are typically between 200 and 1000 km in size. Interpretation is hampered because the ultraviolet absorber is unknown, and the process that causes darkening or lightening is not known. The albedo modification probably involves vertical displacement or vertical velocity. Del Genio and Rossow (1990) argue from the physics of Kelvin waves that positive vertical velocity corresponds to the dark region of the large scale "Y," but the exact phasing is uncertain, and small-scale features may behave differently from large ones. Schinder et al. (1990) showed that the pattern of streaks spiraling toward the pole can be explained by poleward advection and shearing of cloud patches formed at low latitudes, using the observed large-scale pattern of zonal and meridional velocity. Toigo et al. (1994), using high-resolution Galileo feature tracking data, show that small features have a lifetime of about two days. The Galileo data sequence was designed to search for small-scale high-frequency wave activity (Belton et al. 1991) and none was detected. It has been speculated (Belton et al. 1976b; Baker and Schubert 1992) that the isotropic blotches in the cloud top are manifestations of convection, but this is uncertain. Toigo et al. 1994 argue that the features are mesoscale dynamical cells driven by local radiative heating inhomogeneities caused by variable amounts of the mysterious ultraviolet absorber, with dynamical feedback reinforcing the absorber inhomogeneities. Contrasts in the ultraviolet are on the order of a few percent, enough to produce dynamically important heating.

Two balloons with tethered instrument packages were placed in the Venus atmosphere in 1985 by the Russian-French Vega mission. Meteorological measurements are reported by Crisp et al. (1990). The balloons were inserted four days apart and each drifted for approximately two days, covering more than 100 deg of longitude, at an elevation between 50 and 55 km. One was about 7 deg north of the equator and the other the same distance south. During their vertical excursions of 2 or 3 km they measured temperature gradients very close to adiabatic, consistent with the low stability zone apparent in Fig. 1 near the 600 mb level. Remarkably, the absolute temperatures measured by the two differed by 6.5 K, in spite of their small difference in latitude. If correct, this measurement indicates surprisingly large temperature variations. Vertical motions inferred from pressure variations are a few meters per second, and correlations of velocity with temperature imply a vertical heat flux on the order of 30 W m^{-2} with one burst an order of magnitude larger. These results are consistent with thermally driven convection due to the high opacity of the cloud to infrared radiation. There is a suggestion that activity is correlated with the position of topographic relief on the planet's surface. Young et al. (1987,1994) show that topographically excited stationary waves can penetrate to this height from the surface.

Schinder et al. (1990) show that the regions just above and just below the convecting layer within the cloud can act as ducts for horizontally propagating gravity waves. The strong wind shear above the low stability zone and the high static stability below it produce the ducts. Leroy and Ingersoll (1995a) show that waves do not propagate a large distance before being absorbed, however. Radio occultations show scintillations both above and below the convective layer that may be due to turbulence or trapped gravity waves (Woo et al. 1980; Leroy and Ingersoll 1995b).

The best observational evidence for the existence of small scale internal waves comes from two recent studies, one involving radio occultations obtained by the Magellan spacecraft (Hinson and Jenkins 1995), and the other an analysis of Pioneer Venus entry probe data (Magalhães and Seiff 1995). Hinson and Jenkins show that near and above the middle clouds, small-scale oscillations in retrieved temperature profiles and scintillations in received signal intensity are consistent with a spectrum of vertically propagating radiatively damped gravity waves. There appears to be one wave that predominates, having a vertical wavelength between 2 to 3 km and wave amplitude of about 4 K at 65 km altitude. The wave is nearly stationary with respect to the surface or the Sun. Computed wave attenuation due to radiative damping suggests that the wave contributes to wave drag on the mean zonal winds with a magnitude of about 0.4 m s^{-1} per day. Convective activity in the neutrally stable layer of the middle cloud deck would be unlikely to produce a wave stationary with respect to either the surface or Sun. Possible wave sources might be a stationary patch of turbulence in the middle cloud deck due to flow over topography (Gierasch 1987), bow waves set up by a subsolar disturbance (Belton et al. 1976), or topographically generated waves (Young et al. 1987,1994).

Below 40 to 42 km altitude, the Large, North, and Day probes show small-scale structures with amplitudes of about 0.2 to 0.5 K and a well-defined vertical wavelength of 4 to 8 km (Magalhães and Seiff 1995). No correlation between the probes is evident for the structures at these altitudes, implying that the fluctuations are probably due to small-scale or mesoscale activity. Magalhães and Seiff (1995) find that the phase and amplitude variation of the temperature fluctuations observed below 42 km altitude are well accounted for by the WKB solution for a single linear internal gravity wave mode propagating through the observed background structure. The large variation in static stability and zonal wind with altitude over the range of the observations leads to a strong sensitivity of the phase variation of the internal gravity wave to the assumed zonal phase speed and horizontal aspect ratio of the wave (defined as the ratio of the meridional wavenumber to zonal wavenumber), thus providing a constraint on these properties. A three-dimensional internal gravity wave with a zonal phase speed of approximately 30 m s^{-1} and horizontal aspect ratio of about 2 accounts best for the Large Probe observations. The North Probe observations are well reproduced by a wave with zonal phase speed 0 to 15 m s^{-1} and aspect ratio near 5. Waves generated near the surface either by convection or topography would be expected to have phase speeds near zero. However, from geometry, large zonal phase speeds can be associated with waves propagating mostly in the meridional direction even though the phase speed in the direction of propagation along the total wave vector is small. As will be discussed later in connection with convectively driven waves, such waves would not be expected to have significant zonal momentum transport. A wave generated higher in the atmosphere would likely propagate slowly with respect to the mean wind at the altitude of wave generation, and could thereby have a relatively large phase speed with respect to the surface.

Another indication of small or mesoscale activity on Venus is the diverse directionality of wind streaks observed by the Magellan radar. These are features typically a few km in size that are probably formed in the planetary boundary layer (Greeley et al. 1992,1995). The north–south component of the direction is somewhat more often equatorward than poleward (Greeley et al. 1992,1994), suggesting that a Hadley circulation exists, but a large-scale pattern is weak, if it exists at all. This is to be contrasted with the large (500 km) parabolic dark markings that occur (Campbell et al. 1992) and invariably point toward the east. These features are probably due to large impacts, with the open end of the parabola representing the downwind direction in the upper atmosphere. They suggest that the atmospheric rotation existed in the past in the same direction as at present.

VII. FLOW OBSERVATIONS: DISCUSSION

Figure 7 displays profiles of the squared Brunt frequency $N^2 = g/T[(dT/dz) - (dT/dz_{adiabatic}]$ and the squared shear, $S^2 = (\partial u/\partial z)^2 + (\partial v/\partial z)^2$, calculated from the Pioneer Venus probe profiles. Here g is the acceleration of gravity, T is temperature, z is height, and u and v are the eastward and northward velocity components. There appears to be a positive correlation between the thermal stratification and the shear from 20 km to 60 km elevation. There are two layers, one from about 50 to 55 km and one from 20 to about 35 km, where both the stability and the shear are relatively small. Between 10 and 20 km elevation there is a maximum in the shear, and there is some indication that it is reflected in the stratification, but unfortunately the absence of Pioneer probe temperature data below 12 km makes it impossible to evaluate the stability (which requires a gradient) below about 15 km.

The Richardson number, $Ri = N^2/S^2$, can be evaluated from the ratio of the stability and the shear and is displayed in Fig. 7. It is the ratio of two noisy quantities and is displayed as a scatter plot to help the reader assess its significance. The stability and shear profiles were smoothed again before forming the ratio. Figure 7 shows that there are two layers where Ri is of order unity, each about a scale height thick, one centered near 24 km and one near 54 km. Ri is on the order of 10 in a deep layer between 30 and 50 km. There is some indication that Ri increases below 20 km, but the lack of thermal data makes this very uncertain. There is also the possibility that near the base of the atmosphere there is another layer with small stability. Radiative heating usually destabilizes the boundary layer of an atmosphere when any solar heating reaches the surface (Goody and Yung 1989). Thus there are at least two, and possibly three, layers in the Venus atmosphere with small Ri (of order unity or less) sandwiched between more stable layers. The Richardson number is an important stability parameter, and in addition, Allison et al. (1994) have shown that it can be used to place constraints on the latitudinal shear at low latitudes. More extensive and accurate determinations would be valuable.

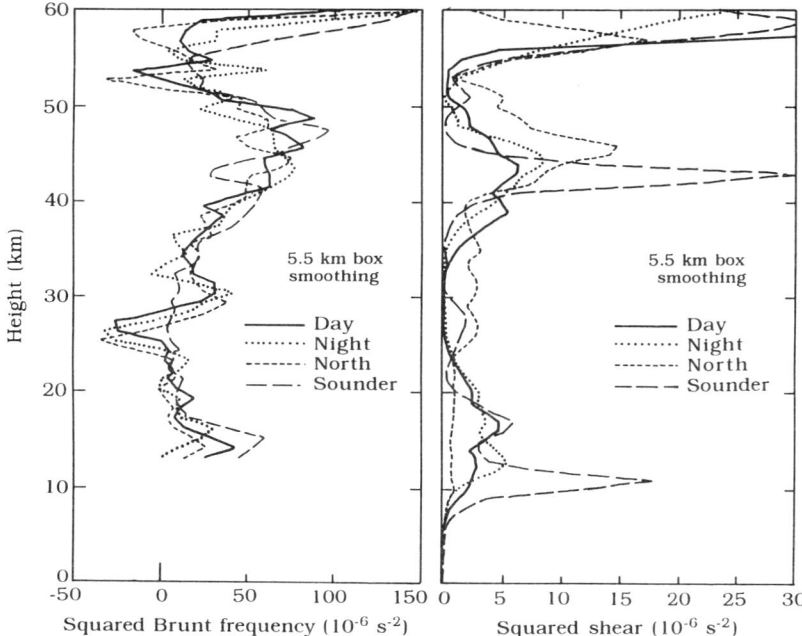

Figure 6. Shear and static stability profiles from the Pioneer Venus probes. The data was smoothed with a 5.5 km sliding box average before taking derivatives.

In Fig. 8 we have plotted the Pioneer Venus meridional velocities compared with two amplitude envelopes. The kinetic energy density is $\rho V^2/2 \sim$ const., where V is the magnitude of the full three-dimensional velocity and ρ is the gas density. Vertically propagating waves tend to conserve their kinetic energy density and therefore display velocity components that are approximately proportional to $\rho^{-1/2}$. Another characteristic behavior would arise if turbulence existed at all levels and dissipated energy per unit volume at approximately equal rates at all heights. In isotropic turbulence the kinetic energy dissipation rate is proportional to ρV^3, leading to a velocity perturbation that would vary as $\rho^{-1/3}$. Figure 8 shows no preference for one of these particular behaviors over the other, but it convincingly shows a height dependence of the velocity amplitude that is inversely related to the density. The velocity profiles do not show any obvious correlations between different probes, nor do they show vertical coherence that would indicate long trains of vertically propagating waves. The profiles, especially beneath 30 km elevation, must be viewed with caution because of ± 1 m s^{-1} uncertainties in Pioneer Venus velocities. Improved data of this kind is needed to answer questions about waves and turbulence, and samplings at more closely spaced horizontal stations are needed to establish the horizontal coherence scale of the motions.

It was first pointed out by Leovy (1973) that the large-scale latitudinal

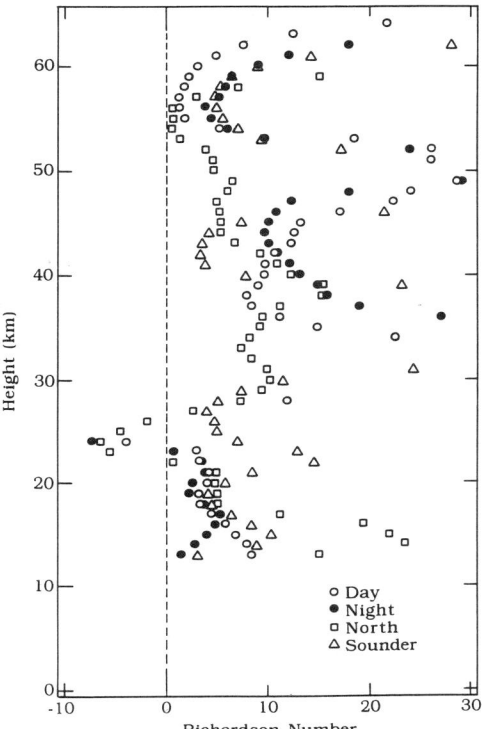

Figure 7. Richardson numbers calculated from Pioneer Venus probe data. A 5.5 km sliding box average was applied to the data twice before differentiating.

force balance in the Venus atmosphere is cyclostrophic. This balance is the analog of the geostrophic balance that obtains on rapidly rotating planets, with the centrifugal acceleration replacing the Coriolis acceleration (Holton 1992). The vertical force balance on large scales is hydrostatic. Let u_a be the absolute zonal flow velocity relative to a nonrotating frame, a be the planetary radius, ϕ be latitude and z be height. The cyclostrophic analog of the *thermal wind equation* of meteorology (Holton 1992) is

$$H\frac{\partial}{\partial z}u_a^2 = -\cot\phi\left[\frac{\partial}{\partial \phi}(RT)\right]_{p\,\text{constant}} \quad (1)$$

where the derivative with respect to latitude is at constant pressure, and H is the scale height, $H = RT/g$, with R the gas constant. In the deep atmosphere Eq. (1) can be used to estimate horizontal temperature gradients from observed wind shears. Near and above cloud top, it has been used to infer wind shear from measured temperatures, as discussed in Sec. V under the topic of radio occultation data.

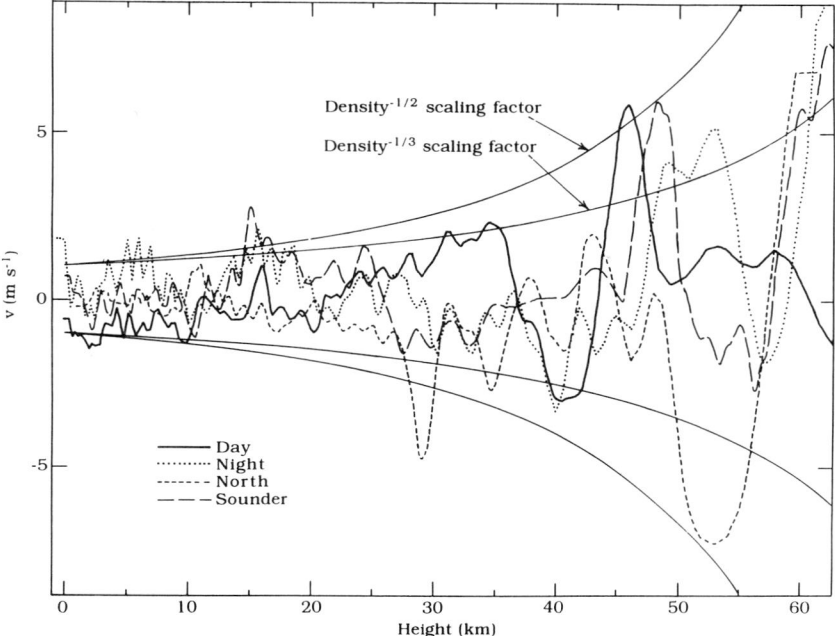

Figure 8. Meridional velocity profiles for the Pioneer Venus probes compared against two envelopes.

It is interesting to compare the Pioneer Venus probe measurements with simple dynamical scaling estimates based on various extreme assumptions. One case is the meridional circulation that just balances the latitudinal gradient of solar heating. The vertical velocity is obtained by assuming that adiabatic cooling balances the solar heating at low latitudes (Holton 1992):

$$\frac{N^2}{g} w = \frac{Q_R}{\rho c_p}. \tag{2}$$

The radiative heating Q_R can be evaluated by using the solar flux, because it is the latitudinal gradient of the solar flux that gives the equator to pole radiative heating differential, to a first approximation. Solar fluxes from Tomasko et al. (1980) were used. Figure 9 displays a smoothed profile of the Brunt frequency that was adopted. The horizontal velocity is estimated from the vertical velocity by multiplying by the ratio of the planet's radius to the local scale height.

An exception occurs where the Brunt frequency is very small because adiabatic cooling cannot balance the solar heating. In that case, the estimate

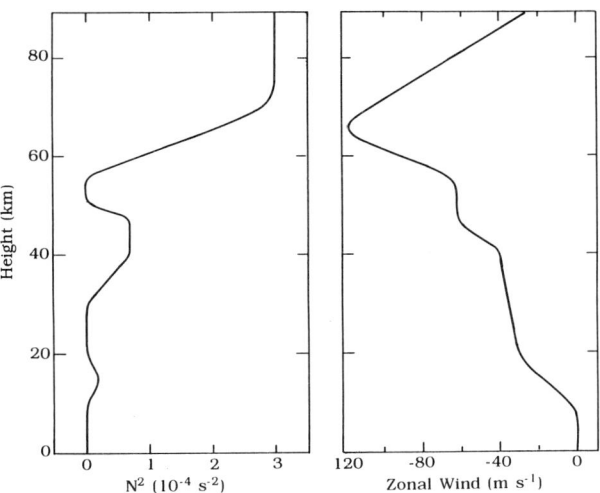

Figure 9. Smoothed profiles of the Brunt frequency and the zonal wind used in analytical estimates.

is replaced by a nonlinear one based on the assumptions (Gierasch et al. 1970)

$$\frac{v \Delta T}{a} = \frac{Q_R}{\rho c_p}$$

$$v^2 = H g \frac{\Delta T}{T}.$$

(3)

The horizontal temperature contrast ΔT can be estimated from the cyclostrophic balance result Eq. (1). Equations (1), (2) and (3) give

$$v = \frac{ag}{HN^2} \frac{Q_R}{\rho c_p T} \qquad N^2 > \left(\frac{ag}{H^2} \frac{Q_R}{\rho c_p T}\right)^{2/3}$$
$$v^3 = aHg \frac{Q_R}{\rho c_p T} \qquad \text{for} \qquad N^2 < \left(\frac{ag}{H^2} \frac{Q_R}{\rho c_p T}\right)^{2/3}.$$

(4)

The meridional velocity is displayed in Fig. 10. Near 15 and 40 km, where the assumed stratification is large, the velocity is on the order of a few cm s^{-1}. At sub-cloud levels it is only 2 or 3 m s^{-1} even at those heights where the assumed stratification is very small. Furthermore, these estimates assume that the entire equator to pole solar heating imbalance is carried by a Hadley cell, and therefore they are upper limits. At cloud top levels the estimates give about 10 m s^{-1}, which is comparable to calculated (and observed) tidal amplitudes.

Another interesting index for comparison is the free convection velocity of mixing length theory, for a mixing length equal to a scale height. This velocity is based on the assumptions (Priestley 1959)

$$WT' = \frac{F}{\rho c_p} \qquad \frac{W^2}{H} = g \frac{T'}{T}$$

(5)

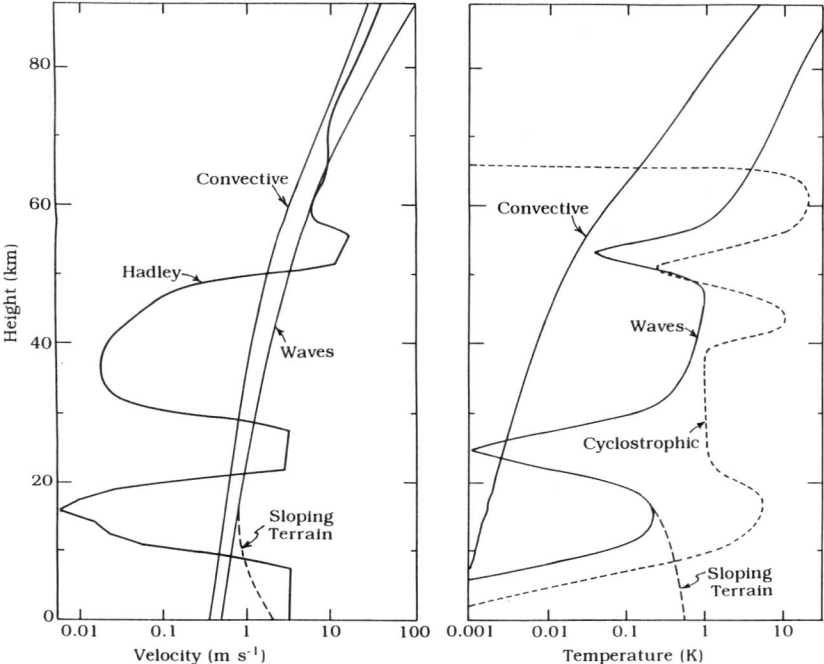

Figure 10. Scaling estimates of the amplitudes of velocity and temperature contrasts under different simple assumptions. Convective velocities are from mixing length theory, based on a mixing length of a scale height and the energy flux profile from Fig. 4. Hadley velocities are based on the same flux profile and the Brunt frequency profile from Fig. 9. The Hadley curve is in segments because two different kinds of estimates are made, depending on the stratification at different heights, as discussed in the text. Wave amplitudes are scaled inversely by the square root of the density, and arbitrarily set to 10 m s^{-1} at 60 km elevation. The cyclostrophic temperature contrast is of global scale, and is based on Eq. (1) and the zonal wind profile from Fig. 9. The other temperature amplitude estimates are discussed in the text. The dashed curves show theoretical boundary layer slope wind estimates (Sec. XIV).

where F is the forcing heat flux, in this case due to solar radiation, W the velocity, H is the mixing length, taken as the local scale height, and T' the temperature fluctuation amplitude. Solving for W and T' gives

$$W^3 = \frac{R}{c_p}\frac{HgF}{p} \quad \frac{T'}{T} = \frac{W^2}{Hg} \qquad (6)$$

after using $H = RT/g$. All three velocity components are assumed to be the same magnitude in this turbulent convection scaling estimate. The profile of W with height is displayed in Fig. 10 as the "convective" curve, with F

evaluated from the estimated profile of global mean solar flux (Tomasko et al. 1980). This velocity scale might be relevant near 55 km, near 25 km, and possibly near the surface, where the stratification is weak or negative.

Figure 10 also displays a velocity envelope proportional to $\rho^{-1/2}$ like that indicated in Fig. 8, which might be a result of vertical wave propagation. The surface amplitude was set at 0.5 m s^{-1}, which gives fairly good agreement with meridional velocity fluctuations of about 10 m s^{-1} at 60 km elevation in the data (Fig. 8). Figure 10 also indicates wave amplitude predictions (dashed line) for slope-induced waves or diurnal slope winds, which would exist only over elevated or sloping terrain (Sec. XV).

Another kind of motion that might well exist in the stratified regions of the Venus lower atmosphere is two dimensional eddy activity. The source could be instability of the horizontal shear. These motions would not propagate vertically. Their amplitude would be given by the strength of the horizontal shear in the zonal mean wind, which is unknown because the sampling density in the lower atmosphere is so small. Layers of quasi-two-dimensional turbulence can exist if the Richardson number is greater than unity (Herring and Métais 1989).

Figure 10 also presents temperature amplitude estimates for various flow components. The latitudinal temperature contrast associated with cyclostrophic balance is calculated from Eq. (4), using the smoothed zonal velocity profile shown in Fig. 10 and assuming a solid body rotational profile at each elevation (constant pressure surface). The mean temperature profile from the Pioneer Venus Sounder probe (Seiff et al. 1980) is used to evaluate the scale height and the temperature where they occur as coefficients. The heat equation, continuity equation and dispersion relation for gravity waves (Holton 1992) give an estimate of the temperature oscillation associated with internal waves

$$\frac{T'}{T} = V \frac{N}{g} \qquad (7)$$

where V is the horizontal velocity perturbation associated with the wave. Finally, the convective temperature fluctuation, from Eq. (6), is also displayed. The latter is too small to be responsible for the measured probe temperature fluctuations. The wave estimate, although a bit small, is consistent with the observations. Also consistent with the observed fluctuations would be two-dimensional eddy activity with horizontal advection of the latitudinal temperature contrasts associated with cyclostrophic balance.

VIII. MAINTENANCE OF THE CIRCULATION: GENERAL DISCUSSION

The angular momentum per unit mass is

$$M = \Omega r^2 \cos^2 \phi + ur \cos \phi \qquad (8)$$

where r is the spherical radius, ϕ is latitude, and u is the zonal (tangential) velocity. The equations of motion give

$$\frac{\partial}{\partial t}(\rho M) + \vec{\nabla} \cdot (\rho \mathbf{v} M) + \frac{\partial p}{\partial \lambda} = \vec{\nabla} \cdot (\tau \cdot \hat{\mathbf{z}} \times \mathbf{r}) \qquad (9)$$

where λ is longitude, $\hat{\mathbf{z}}$ is the unit vector in the pole direction, and τ is the viscous stress tensor. If Eq. (9) is averaged in longitude the pressure gradient term drops out. The remaining terms express the effects of advection and of friction on the longitudinally averaged angular momentum distribution. The continuity equation is

$$\frac{\partial \rho}{\partial t} + \nabla \cdot (\rho \underline{v}) = 0. \qquad (10)$$

For axisymmetric flow with negligible friction Eqs. (9) and (10) give

$$\frac{\partial M}{\partial t} + \mathbf{v} \cdot \nabla M = 0. \qquad (11)$$

Thus under the influence of axisymmetric advection, fluid rings conserve angular momentum as they move in height and latitude, and there is never an extremum of angular momentum along a particle trajectory. The highest angular momentum that a fluid ring can acquire from the planet's surface is that at the equator, and this is the maximum that axisymmetric advection of fluid rings can produce in the atmosphere (Hide 1969). Mid-latitude jets can be formed by poleward drift of material, because the angular velocity and the linear velocity will increase as the fluid moves closer to the rotation axis.

On Venus the angular momentum per unit mass of the atmosphere has an internal maximum at low latitudes and at an elevation of approximately 70 km. The angular velocity at this location is approximately 50 times that of the solid planet. Hide's theorem shows that nonaxisymmetric motions must be at play, and must pump momentum toward this location. Specifically, Eq. (9) shows that there must be convergence of the angular momentum flux $\mathbf{F}_M = \rho \mathbf{v} M$ toward the location of the maximum of the angular momentum per unit mass. The key terms in producing these are the velocity correlation, or *Reynolds' stress* terms

$$\overline{\mathbf{F}}_M = \rho r \cos \phi \left(\overline{u'v'}, \overline{u'w'} \right) \qquad (12)$$

where v' and w' are the meridional and vertical components of the eddy (nonaxisymmetric) portion of the velocity field.

For conceptual simplicity, the hypotheses about the maintenance of the rotation of the Venus atmosphere can be divided into two classes, depending on whether the vertical or meridional Reynolds' stress is most important. Under the first category a vertical flux of angular momentum by eddies at low latitudes drives the circulation. The key balance would then be vertical pumping of angular momentum by eddies, balanced by dissipative processes that might involve other eddies, turbulence or small-scale friction. The eddies that produce the vertical pumping might be caused by solar heating (tides), by convectively driven gravity waves, or other processes. Under the second class of hypotheses the meridional flux of eddy angular momentum is most important. In this case angular momentum is pumped horizontally into low latitudes from high latitudes. At high latitudes angular velocity is generated by axisymmetric poleward drift in a Hadley circulation. The vertical balance of angular momentum does not involve eddies under this hypothesis. A Hadley circulation generally pumps angular momentum upward because the angular momentum per unit mass is larger at low latitudes (where the Hadley circulation is rising) than at high latitudes, and the vertical balance can be maintained by an upward angular momentum transport by the Hadley cell, offsetting downward momentum flux by dissipative loss, friction or turbulence.

These angular momentum flux requirements can be stated quantitatively by integrating Eq. (9) over different portions of the atmosphere. The rate of change of the net angular momentum of a layer of atmosphere from radius r_1 upward to infinity is given by

$$\frac{\partial}{\partial t} \int_{r_1}^{\infty} \int_{-\pi/2}^{\pi/2} \int_{0}^{2\pi} \rho M r^2 \cos\phi \, d\lambda \, d\phi \, dr$$

$$= \left[\int_{-\pi/2}^{\pi/2} \int_{0}^{2\pi} \rho w M r^2 \cos\phi \, d\lambda \, d\phi - \int_{-\pi/2}^{\pi/2} \int_{0}^{2\pi} \tau_{r\lambda} r^3 \cos^2\phi \, d\lambda \, d\phi \right]_{r=r_1} \quad (13)$$

where $\tau_{r\lambda}$ is the r, λ component of the viscous stress tensor. For a steady state the radial flux of angular momentum by the flow must offset viscous drag. The radial flux of angular momentum can arise either by a large-scale axisymmetric Hadley circulation, because M is largest at low latitudes where w is positive, or it can arise because of a correlation of M and w associated with eddies. One or the other (or both) must exist.

In reality it is probably not viscous dissipation that must be overcome by organized angular momentum fluxes, but some combination of small-scale turbulence and wave drag. We shall elaborate on this point in Sec. IX. For simplicity, all "dissipative" mechanisms can be incorporated into τ for our purposes in this section.

Integrating over the entire depth of the atmosphere and over a latitude range from $-\phi_1$ to ϕ_1 gives another important statement,

$$\frac{\partial}{\partial t} \int_{r_{0+}}^{\infty} \int_{-\phi_1}^{\phi_1} \int_{0}^{2\pi} \rho M r^2 \cos\phi \, d\lambda \, d\phi \, dr$$

$$= \left[\int_{-\phi_1}^{\phi_1} \int_{0}^{2\pi} (\rho w M - \tau_{r\lambda} r \cos\phi) r^2 \cos\phi \, d\lambda \, d\phi \right]_{r=r_{0+}} \quad (14)$$

$$\left[\int_{r_{0+}}^{\infty} \int_{0}^{2\pi} (\rho v M - \tau_{\phi\lambda} r \cos\phi) r \cos\phi \, d\lambda \, dr \right]_{\phi=-\phi_1}^{\phi=\phi_1}.$$

There can be drag at the irregular bottom surface. In writing Eq. (14) we account for this by integrating only from r_{0+} to ∞, where r_{0+} is high enough above the mean surface position r_0 not to intercept the topography. The first term on the right-hand side of Eq. (14) would normally be expected to act as a drag (although it has been suggested that wave propagation from near-surface layers might carry net momentum upward; see Sec. XII below). In the second term on the right latitudinal fluxes into the volume are evaluated. The viscous stress term here is expected to be small, leaving the transport of momentum by latitudinal motions as the dominant term. This can be decomposed into the transport by the mean Hadley circulation and the transport by eddies. One expects the Hadley circulation to transport angular momentum poleward, because the flow is poleward at high levels where the spin rate is greatest. Therefore the Hadley circulation will act as a net drag on low latitude flow, and a balance will require that eddies transport angular momentum equatorward. The only alternative is that the near-surface term at r_{0+} provides sufficient acceleration to offset the Hadley circulation.

Note that the vertical flux of angular momentum can be either by eddies or by the mean Hadley circulation, but that in the latitudinal direction the Hadley circulation is always expected to move angular momentum out of low latitude regions. Thus the Hadley circulation can act to assist the equatorial superrotation by producing an *upward* momentum flux, but is always expected to hinder it *horizontally* by moving angular momentum out of low latitudes.

The required magnitude of the acceleration by eddy momentum flux convergence can be estimated by evaluating the Hadley circulation accelerations, using Eq. (4) to estimate the Hadley circulation strength. In fact Eq. (4) gives an upper limit, because eddies can transport part of the equator–pole heat flux. For example, Newman and Leovy (1992) find that the Hadley circulation is very much suppressed at cloud-top level at low latitudes by the effects of tidal waves. Nevertheless, estimates based on Eq. (4) are probably correct in order of magnitude, and useful if viewed with caution. The acceleration produced by poleward flow is

$$\frac{\partial u}{\partial t} = -\frac{v}{r\cos\phi}\frac{\partial}{\partial \phi}(u\cos\phi) \sim \frac{vu}{r} \tag{15}$$

where it is assumed that the zonal flow, on a global mean, is not far from uniform rotation at each height. Values of the meridional velocity can be read from Fig. 10, and the zonal wind from Fig. 5. For example, at heights of 40 km and 70 km, Eq. (15) gives estimates of $1\text{ cm s}^{-1}\text{ day}^{-1}$ and $10\text{ m s}^{-1}\text{ day}^{-1}$, respectively. Near the equator it would be more appropriate to use $w\,du/dz$ to make this estimate, and the result would be approximately the same because w and v in the Hadley circulation are related by continuity. Eddy strengths can then be estimated. For example, if horizontal momentum transports are to be important

$$\frac{v'u'}{r} \sim \frac{vu}{r}. \tag{16}$$

Assuming that the two eddy components are the same magnitude gives v' equal to the geometric mean of the Hadley meridional velocity and the zonal flow, or about 1 m s^{-1} at 40 km and 20 m s^{-1} at 70 km. Similar estimates can be made for vertical eddy requirements.

IX. FRICTION AND WAVE DRAG

The spin of the Venus atmosphere is remarkable and puzzling because intuition tells us that in the absence of forcing, friction will bring the atmosphere into corotation with the solid planet. But the nature of "friction" is also unclear, as is its magnitude. It may be that the Venus atmospheric rotation is controlled by balances between different large-scale flow components, and that friction, in the sense of molecular viscosity or small-scale turbulent dissipation, is unimportant. This is one of the key questions to be resolved by future observations. In this section we shall make a few remarks about candidates for frictional mechanisms.

The kinematic molecular viscosity of CO_2 is 0.07 cm^{-2} s^{-1} at STP. It is approximately proportional to $T^{1/2}\rho^{-1}$, but even at high altitudes above the Venus clouds where the density is reduced by a factor of 100 from STP the time scale H^2/ν is about 1000 yr. Molecular viscosity is therefore too small to have a direct influence on motions whose vertical scale of variation is a scale height.

Turbulence can act to produce an effective viscosity. At scales small enough so that anisotropies caused by stratification or mean shear become unimportant, turbulence is isotropic and kinetic energy migrates to small scales by a nonlinear cascade. This is the inertial, or Kolmogorov, regime. In this regime kinetic energy is removed by eventual viscous dissipation at very small scales. The fluid is brought toward its lowest energy state, which is rigid body rotation. Thus small-scale turbulence acts analogously to molecular viscosity and tends to erode angular velocity gradients. Prandtl's mixing length concepts can be used to estimate the effective viscosity if the eddy scales and velocities are known at the outer scale of the isotropic regime. The difficulty is that on Venus (and, incidentally, on Earth) the details of the velocity field and the stratification are not known well enough to describe the distribution of turbulent dissipation. Turbulence at larger scales can be anisotropic and does not necessarily act analogously to molecular viscosity. In fact, two dimensional turbulence is a candidate mechanism for producing equatorward momentum transports that help sustain the Venus atmospheric rotation (see Sec. XIII).

Wave drag is important in the Earth's atmosphere and is very likely the major deceleration mechanism that needs to be overcome on Venus. Waves of all scales are ubiquitous in atmospheres and are generated by a variety of sources, including convection, internal turbulent patches, flow over topography, and instabilities. The behavior of the Earth's atmosphere shows that the net effect of wave absorption is usually to brake the atmosphere (see, e.g.,

Holton 1982 or Andrews et al. 1987). Although many different sources exist with varied phase speeds, the presence of stationary irregularities at the bottom surface of an atmosphere produces a peak in the wave spectrum at zero horizontal velocity. Because absorption of waves accelerates the atmosphere toward the wave velocity, the net effect is to drag the atmosphere toward zero velocity.

There are notable exceptions when waves of one particular direction are favored and absorption tends to produce a net acceleration. The quasi-biennial oscillation in the Earth's equatorial stratosphere is an example (Holton and Lindzen 1972), and the solar tides on Venus are another (see Sec. XI). But these are exceptions, and drag is the general rule. To evaluate the magnitude of wave drag on Venus we must determine the spectrum of waves, particularly in the lower atmosphere, and we must determine the absorption mechanisms. In the Earth's mesosphere the dominant absorption mechanism is breaking of waves (Lindzen 1981; Holton 1982). At lower elevations radiative damping may play an important role in wave absorption (see, e.g., Holton and Lindzen 1972).

X. DIAGNOSTIC STUDIES

Hou (1984) and Hou and Goody (1985,1989) have deduced the momentum sources needed to maintain the Venus circulation for an assumed latitude-height distribution of zonal winds, temperatures and radiative heating. They assume that the vertical divergence of the eddy heat flux is smaller than the effect of vertical heat advection. Under these conditions the equations of motion can be inverted to give the momentum fluxes. These include the effects of both eddies and friction, but through most of the atmosphere friction is probably negligible relative to the eddies.

Two important conclusions about eddy Reynolds' stresses are probably valid in spite of any uncertainties about the mean flow and the neglect of eddy heat fluxes. One point is that over the height range between about 15 and 50 km, positive eddy accelerations (in the direction of the atmospheric spin) are necessary at low latitudes. This is the fundamental requirement for maintenance of the spin of the lower atmosphere. It is needed to offset the deceleration due to low latitude rising motion in a positive vertical shear. The other point is that from 50 to 80 km, also at low latitudes, an approximately offsetting pair of requirements exists, with *deceleration* at the highest levels, peaking between 70 and 80 km, and acceleration peaking near 65 km. This offsetting pair of requirements can be met by a *downward* flux of eddy momentum that is confined to the region near and just above the cloud tops. It is needed to balance the accelerations due to the Hadley circulation penetrating through the mean zonal flow vertical shear, that is positive just below the zonal flow maximum (at about 70 km) and negative above this.

This result suggests that two different eddy momentum transfer processes exist, and perhaps are accomplished by two different kinds of eddies. The distribution of zonal flow near and above the cloud tops requires a downward momentum flux, which is *opposite* to the global requirement to maintain the atmospheric spin. Several scale heights deeper, the maintenance of the basic spin requires convergence of a momentum flux that is either upward from the surface or equatorward from high latitudes. It now seems likely that the solar tides accomplish the cloud-top requirements, as we shall discuss in the next section. The deep requirements remain unexplained.

XI. THE TIDES

Recent work has demonstrated convincingly that the solar tides produce a downward angular momentum flux just above the Venus clouds, which decelerates the stratosphere and accelerates the upper part of the cloud layer. These tides are global-scale motions forced by solar heating. They are internal gravity waves modified by rotational effects. The internal gravity wave vertical wavenumber is approximately $k_V = (\omega/N)k_H$, where ω is the frequency in the frame moving with mean flow and k_H is the (imposed) horizontal wavenumber. It turns out that for the semi-diurnal tidal component, k_V matches well the vertical half-width of the solar heating distribution (Crisp 1986), and as a result there is effective driving. Leovy (1987) argues that this is not a coincidence, as we discuss below.

Soon after the rotation of the Venus atmosphere was discovered, Schubert and Whitehead (1969) showed that a moving heat source can produce a mean flow in a laboratory annulus, and suggested that the flow on Venus is produced by the motion of the subsolar point. Theoretical work showed that phase shifts across the laboratory fluid produce Reynolds' stresses that drive the mean flow in the laboratory situation (see, e.g., Malkus 1970). Fels and Lindzen (1974) pointed out that thermally excited gravity waves also produce Reynolds' stresses. Wave phase shifts arise because of propagation, while the laboratory annulus develops phase lags because of viscous and conductive effects. Waves carry momentum away from the excitation region, causing a reaction on the excitation region that accelerates it in the opposite direction to that of the transmitted wave. Figure 11 illustrates the concept. In the case of Venus the atmospheric rotation carries the fluid past the subsolar point and a wave is established that is stationary with respect to the Sun, and moves "upstream" in the fluid, against the direction of the mean flow. The wave is excited by the absorption of sunlight primarily between 60 and 70 km elevation. Vertically, it propagates upward and downward from the principal excitation region. The upward component is absorbed by radiative damping a few scale heights higher, and produces a deceleration, which is balanced by acceleration of the cloud tops. The downward component is weaker and is not as important.

Pechman and Ingersoll (1984) calculated the structure of both the diurnal (zonal wavenumber 1) and semidiurnal tidal waves (wavenumber 2) and showed that the semidiurnal tidal component agreed well with Pioneer Venus observations of the stratospheric temperature (see Fig. 2). A series of increas-

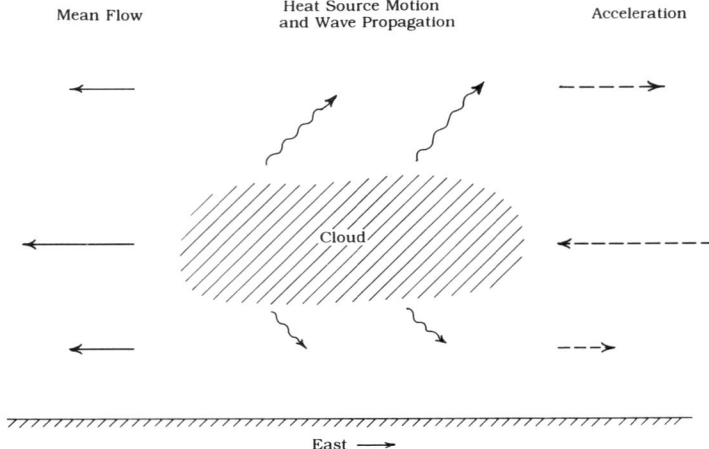

Figure 11. Schematic of tidal waves and accelerations. Solar motion is eastward but slow and most of the velocity of the heat source relative to the fluid is due to the westward motion of the fluid. Momentum is deposited where waves are absorbed, and a reaction force is produced where waves are generated. In practice, the details of dissipation, heating and wave propagation characteristics are all important in determining the three dimensional distribution of accelerations. On Venus, the downward momentum flux by tidal waves is smaller than the upward flux.

ingly detailed theoretical treatments of the tides followed (Fels 1986; Leovy and Baker 1987; Hou et al. 1990; Newman and Leovy 1992). Newman and Leovy allow the mean flow to evolve, including the Hadley circulation, under the influence of both the diurnal and semidiurnal tidal components. At the base of the computational model, near 40 km elevation, they impose a mean flow as observed. They find excellent agreement with the latitude and height dependence of the mean zonal wind above the cloud tops that was deduced from radio occultations and with the latitude and longitude dependence of cloud tracked winds (Fig. 4). Leovy (1987) points out that the tidal forcing will tend to establish a value of the mean flow near $u = Nh$, where N is the Brunt frequency and h is the thickness of the heated forcing layer (approximately two scale heights), because this mean flow speed produces a good match of the tidal vertical wavelength with the scale of the forcing. The vertical wavelength of a low frequency internal gravity wave is smaller than the horizontal scale by the factor ω/N, where ω is the Doppler shifted frequency in the frame moving with the gas. In the case of the tides, the horizontal scale is the radius of the planet a and the Doppler shifted frequency is u/a, so that the vertical scale of the wave is u/N. Putting this equal to h gives $u = Nh$. He shows that for any forcing strength above a certain threshold value, the coupled system, mean flow plus tide, will settle on a mean wind speed near Nh and quite insensitive to the forcing strength.

The vertical momentum transfer in the tidal calculations by Newman and

Leovy is primarily between the level near 65 km, where the wave is driven, and approximately the 90 km level, where it is absorbed. The results nicely explain the dynamical structure of the top portion of the Venus atmosphere, at heights above about 50 km. The principal momentum balance is between vertical fluxes by the tides and by the mean flow. To an order of magnitude,

$$\overline{u'w'} \sim \overline{uw}. \qquad (17)$$

Note that the tide and the mean Hadley circulation are both driven by solar heating, so the order of magnitude of w' and $\overline{w}$ are the same. Thus under this balance $\overline{u} = O(u')$. In reality Newman and Leovy find that numerical factors work in favor of the mean flow, and the mean flows that they calculate are several times larger than u'. Nevertheless this argument shows why vertical momentum fluxes by the thermally driven tides are unlikely to explain the mean flow at deep levels in the atmosphere. An estimate of the tidal velocity is given by Eq. (4) and is displayed in Fig. 10 (it is the same as the Hadley circulation estimate). At heights less than 50 km the velocities are a few m s^{-1} or less.

Horizontal momentum fluxes by the tides are under a similar restriction. The balance Eq. (14) states the requirement for redistribution of momentum horizontally to maintain an angular momentum gradient that increases toward the equator, and we see that if $v' \sim \overline{v}$, it follows that $\overline{u}$ cannot greatly exceed u'.

The Newman and Leovy simulation evolves in time and is not restricted to periodic components. It produces an equatorward transport of momentum near the cloud base by free eddies, which are not restricted in amplitude in the same manner as the tides. They feel that the free eddies are probably not accurately simulated in this particular model, however, because the atmospheric stratification cannot be allowed to take on small values. The appearance of these eddies may be significant (see Sec. XV).

XII. EDDY HORIZONTAL MOMENTUM TRANSFER MECHANISMS

In a stratified atmosphere, eddies can transport momentum out of a midlatitude jet and transfer it to lower latitudes. This has the effect of pumping angular momentum outward, toward the equator. Rossby (1947) speculated that "vorticity transport" would accelerate equatorial regions in a predominantly two-dimensional flow. One of the possible applications he had in mind was the Sun, which displays an equatorial angular velocity maximum. Gierasch (1975) speculated that this mechanism might be a key one on Venus, with the vertical transport by a Hadley cell being the other. The effect of eddy momentum transport in a spherical shell geometry was first quantitatively studied by Rossow and Williams (1979) in a numerical simulation. They found an inverse kinetic energy cascade, with a pumping of energy into two large-scale modes, rigid body rotation and a pair of eddies that formed a wavenumber one disturbance. The phenomenon has since been documented in numerous three-dimensional computations (see, e.g., Del Genio et al. 1993; Suarez and Duffy 1992).

Although it may seem unphysical that eddies can transfer angular momentum equatorward against its gradient, it is not. Ordinary viscosity does this. The lowest energy state of a rotating disk of fluid, consistent with a given angular momentum, is rigid body rotation, with a maximum angular momentum density at the outer edge. The perplexing question is why momentum transfer occurs horizontally over large scales, rather than dissipating a midlatitude jet locally, via either vertical eddies or small scale horizontal eddies. One key lies in the general properties of two dimensional flow. Fjørtoft (1953) proved that kinetic energy cannot cascade to small scales in two-dimensional flow, but must undergo an inverse cascade toward larger scales. Kraichnan (1967) described the physics of an inverse cascade in terms analogous to the well-known three-dimensional inertial cascade regime of Kolmogorov.

Numerical experiments by Herring and Métais (1989) show that quasi-two-dimensional motions develop, with an inverse kinetic energy cascade, when the Richardson number based on the local fluid properties exceeds unity. This condition is met within the three major stable layers of the Venus atmosphere, which are centered near 15 km and 40 km, and within the stratosphere. The numerical experiments show horizontal flow confined to layers whose thickness is given by u'/N, where u' is the fluctuation velocity. Within the Venus stable layers a typical $u' \approx 3$ m s^{-1} and $N \approx 0.003$ s^{-1}, leading to a thickness estimate of about 1 km. This is consistent with the fluctuations displayed in Fig. 5a, although both the vertical and the velocity resolution of the Pioneer probe data is marginal for this conclusion. Possible origins of eddies on Venus are important questions. Baroclinic or barotropic instabilities are likely to exist near the mid latitude jet above the clouds (Young et al. 1984; Michelangeli et al. 1987). At deeper levels the mean structure is not well enough known to diagnose stability conditions. It is also possible that eddies are triggered by vertically propagating waves.

XIII. BOUNDARY LAYER AND CONVECTIVELY DRIVEN WAVES

Hou and Farrell (1987) propose that gravity waves are excited by convection within the lowest scale height of the Venus atmosphere, that the waves propagate vertically, and that those with horizontal phase velocity in the same direction as the mean flow are absorbed at critical levels, thereby accelerating the mean flow and contributing to its maintenance. One usually expects waves generated in the boundary layer to have small horizontal phase velocities and to contribute to drag rather than acceleration. This is the implication from the study of the convective region of the middle cloud deck by Leroy and Ingersoll (1995a), a region analogous to a surface boundary layer in the sense of being convectively active. Their study indicates that, as expected, convectively driven gravity waves typically have small phase speeds relative to the mean zonal wind where they are generated. Because of vertical shear in the mean wind, waves carrying momentum that accelerates the mean zonal wind are absorbed within a few km of the convective zone, and therefore such waves cannot sustain the mean zonal wind beyond several km from the region of generation. Gravity waves which propagate obliquely to the zonal direction may have larger zonal phase speeds, but their contribution to the vertical flux of zonal momentum is small, and hence they also are not a significant source of momentum for the mean zonal wind. Nonlinear effects can considerably alter wave-mean flow interactions and change the location of critical levels (Fritts and Dunkerton 1984), but whether nonlinear processes would substantially enhance the role of convectively driven gravity waves in maintaining the superrotation is a subject for future study.

In the Earth's atmosphere there are examples of important exceptions to the general rule that boundary layer driven waves contribute more to wave drag than acceleration, as discussed above in Sec. IX. However, the wave phase speeds in the terrestrial case are comparable to the magnitude of the mean wind which they are supposed to drive (the quasi-biennial oscillation) because they have planetary-scale horizontal wavelengths. At tropical latitudes on Earth the effects of moisture can act to organize convection on a large scale and cause a coupling of convective activity to large scale waves (see, e.g., Holton 1992). Longitudinal variations in convective intensity may also contribute to forcing of long waves (Hitchman and Leovy 1988). On Venus it is not clear how large-scale organization might arise. Longitudinal variations of topography are a possible cause.

In addition to convection, boundary layer slope winds (Dobrovolskis 1993) or nonlinear eddies produced by the interaction of the mean flow with topography can potentially drive gravity waves into the upper atmosphere. Nonlinear calculations by Young et al. (1994) show that stationary waves induced near the surface can increase in amplitude with height, and may deposit momentum at higher levels. They also showed that nonlinear cascade of energy can generate dynamical structures with smaller scales than the originally excited waves. However, because the waves are stationary with respect to the planet surface, they act as a drag on the superrotation wherever they deposit momentum.

Consider a horizontal average of the angular momentum balance Eq (14), in order to focus on the vertical momentum exchanges. The data suggests that the averaged rotation of the Venus atmosphere is predominantly, and perhaps exclusively, in one direction at all heights. If one makes the assumption that dissipative processes, such as wave drag, exist throughout the atmosphere, then the offsetting momentum flux from near-surface layers must be predominantly of one sense. In order to satisfy this requirement, forcing by westward waves would need to dominate over eastward if wave momentum transfer is to explain the superrotation. This is the requirement that the right-hand side of Eq. (14) be equal to zero on a time average. Thus, under the assumption that the average drag is oppositely directed to the average rotation, there must be an anisotropy in the wave generation process if waves originating near the surface are to provide the forcing for the superrotation. This can come about, for example, because of feedback of waves or mean flow onto the convection field or because of special wave ducting properties of the atmosphere. None of these questions has been addressed in detail. To further examine the hypothesis by Hou and Farrell, the stratification, mean shear, and eddy properties in the lowest atmospheric scale height need to be better measured in order to determine wave ducting properties and the depth and intensity of convection.

XIV. BOUNDARY LAYER FLOWS

The solar flux reaching the surface is maximum at mid-day on the equator and is zero in the night hemisphere. Tomasko (1983), using Pioneer probe flux measurements and estimated optical properties of the atmosphere, presents an analytical approximate expression for the flux reaching the surface in the sunlit hemisphere as

$$F \approx 80 \, \text{W m}^{-2} (\cos \zeta)^{1/4} \tag{18}$$

where ζ is the solar zenith angle. The diurnal harmonic component on the equator is roughly 40 W m^{-2}. The time dependent thermal forcing of the boundary layer can drive global tidal motions, global or local slope winds, and varying convection layer depth and strength that might in turn drive waves upward. The unknown infrared opacity and thermal structure of the deep atmosphere creates a major uncertainty in estimates of these boundary layer effects, because it is not known whether the mean profile is strongly stratified or not. If it is close to adiabatic, the diurnal thermal oscillation will be relatively deep and of small amplitude. If it is strongly stratified, the oscillation will be shallow and of larger amplitude.

The known amplitude of the forcing permits evaluation of the total buoyancy forcing, integrated through the diurnal boundary layer depth. In the case with heat advection unimportant, the heat equation reduces to

$$\rho c_p \frac{\partial T}{\partial z} = -\frac{\partial F}{\partial z} \tag{19}$$

where F is the upward heat flux. The heat storage in the surface is negligible (Gierasch and Goody 1970) and the forcing heat flux is the full surface insolation. For a harmonic forcing flux amplitude ΔF with frequency $\omega = 2\pi/(\text{solar day}) \approx 2\pi/(117 \, \text{days})$, the vertical integral of Eq. (19) gives

$$\int T \, dz \equiv \Delta T \, \Delta z = \frac{R}{c_p} \frac{T \, \Delta F}{\omega p} \approx 1.2 \, \text{K km} \tag{20}$$

where p is the surface pressure and the boundary layer has been assumed to be thin enough so that the ambient density and pressure are approximately constant through it. The depth Δz is meant to be a measure of the boundary layer thickness. Thus the forcing ΔF, which is quite accurately known, fixes the product of the boundary layer depth and thermal oscillation amplitude, but neither the amplitude nor depth separately.

A lower limit on the boundary layer depth is given by the purely radiative case. For this limit to obtain, the boundary layer must be stratified with the Richardson number greater than 1/4, so that turbulence is suppressed and all heat transfer is by radiation. The properties of a radiative boundary layer can be estimated for the case of a simple gray infrared transfer model. Hunten and Goody (1969) pointed out that the thermal optical depth in such a model can be estimated if the solar flux is known, by requiring that the greenhouse effect give the correct surface temperature. Using a solar flux at the surface of 20 W m^{-2} we estimate an optical depth $\tau_R \approx 1600$. The opacity is $\kappa = \tau_R/H$, where H is the scale height near the surface, and the heat Eq. (19) becomes

$$\rho c_p \frac{\partial T}{\partial t} = -\frac{\partial R}{\partial z} = \frac{\partial}{\partial z}\left(\frac{4\sigma T^3}{\kappa}\frac{\partial T}{\partial z}\right) \qquad (21)$$

in the opaque limit (Goody and Yung 1989). Linearizing to examine the diurnal temperature perturbation for a thin boundary layer, one finds a radiative diffusivity

$$K_R = \frac{4\sigma T^3}{\rho c_p \kappa} \approx 0.02 \, \text{m}^2 \, \text{s}^{-1}. \qquad (22)$$

Using $\kappa = \tau_R/H$ with an optical depth of 1600, a scale height of 15 km, a surface pressure and temperature of 90 bar and 730 K gives a diurnal boundary layer depth estimate of about 170 m, which implies from Eq. (20) an amplitude of about 7 deg. This estimate is probably a minimum depth and maximum amplitude, because most effects omitted from this model will act to increase the diffusivity. Nongray radiation would introduce smaller opacities over portions of the spectrum, and turbulence would enhance the heat transfer.

The boundary layer has potentially important effects on large-scale dynamics. It has been mentioned that turbulent convection in the boundary layer can be an important source of gravity waves. In addition, the semidiurnal boundary layer volume change is a forcing for an atmospheric tidal response that produces a quadrupole moment of the mass distribution and hence a net torque by the solar gravitational field (Dobrovolskis 1983,1993; Dobrovolskis and Ingersoll 1980). The torque is in the correct sense to accelerate the observed atmospheric spin and was suggested by Gold and Soter (1971) as a possible cause of the superrotation. But the detailed calculation by Dobrovolskis and Ingersoll (1980) gives a net torque of about 1.8×10^{16} J, corresponding to a surface stress of about $\tau_S = 2 \times 10^{-4}$ dyne cm^{-2}. As a rough estimate of a boundary layer wind necessary to create an offsetting stress, one may use the drag formula

$$\tau_S = C_D \rho U^2. \qquad (23)$$

The density of the Venus atmosphere at the surface is about 0.06 g cm^{-3}. Using $C_D = 0.003$, appropriate for a neutral boundary layer a few meters above the surface (Priestley 1959) gives a flow speed estimate of $U \approx 1$ cm s^{-1}. Although this appears to be a small velocity, the tidal torque might possibly resolve an indeterminacy and define the sense of the mean flow, and it should not be ignored while our understanding is still incomplete.

Finally, Dobrovolskis (1993) shows that the diurnal boundary layer thermal disturbance will create up-slope and down-slope buoyancy forces that drive oscillating winds over topographic relief. Surface elevations and slopes are known from Pioneer Venus and Magellan radar data (Ford and Pettengill 1992). He calculates the winds for the case of a weakly stratified boundary layer, self-consistently estimates the turbulent diffusivity from mixing length theory, and then computes the associated boundary layer thickness and temperature amplitude. These come out to be approximately 5 km and 0.2 K. The surface stresses are on the order of 20 dyne cm^{-2}, which should be large enough to initiate saltation and soil movement (Iversen and White 1982). Observations, however, do not show correlation of aeolian features with surface slope direction (Greeley et al. 1994).

XV. GENERAL CIRCULATION EXPERIMENTS

Realistic numerical modeling of the Venus atmosphere is a challenging task. The range of time scales that must be described is very large. Dynamical activity, such as the tides, involves time scales less than a day. Thermal adjustment of the mean atmospheric structure involves time scales of decades. The range of length scales is also large. The tides and the mean flow are global in scale. Currently the minimum horizontal scale that must be resolved is not known because the key processes are not yet identified. In the vertical direction the resolution should be better than a scale height, and the circulation is more than eight scale heights deep.

Early three-dimensional numerical simulations of Venus-like atmosphere either did not produce rapid rotation of the atmosphere, or employed formulations of diffusion that rendered the results difficult to interpret (Kálnay de Rivas 1975; Young and Pollack 1977; Rossow 1983). Recent work has focused on parametric studies based on terrestrial models. A few parameters or processes are altered, but the model is otherwise fixed in the terrestrial configuration. This approach avoids the problem of the extremely slow Venus thermal adjustment, and also permits easy comparison with the well documented and better-understood terrestrial case. Del Genio and Suozzo (1987) varied the rotation rate in a model driven only by a latitudinal heating gradient (no tides) and found that mid-latitude jets are prominent features of even slowly rotating regimes, but the model did not achieve strong equatorial superrotations. Del Genio et al. (1993) modified the radiative heating in the same model so that friction due to convective activity was reduced, and found that both equatorial and mid-latitude superrotation developed. The mechanism is upward angular momentum transport by the Hadley circulation and equatorward angular momentum transport by eddies generated by mid latitude instabilities. In further work, Del Genio and Zhou (1996) found that in terrestrial models having Venus' rotation rate the computed superrotation was sensitive to numerical precision because of the smallness of individual terms in the momentum balance. They showed that in general, the strength of superrotating winds increases with planetary rotation rate, but the efficiency of superrotation, i.e., the ratio of total angular momentum in the superrotating state to an atmosphere corotating with the solid planet, increases with decreasing planetary rotation rate.

Important results have also been obtained by workers who are primarily interested in the stability of the terrestrial circulation. Suarez and Duffy (1992) found that a very simple two layer numerical model can be induced into a superrotating state by imposing a longitudinally varying (but stationary) low latitude heating. There is a hysteresis effect, and the superrotating state can be maintained even when the amplitude of the triggering heating is reduced. Again the mechanism for superrotation is equatorward momentum transport by eddies and vertical transport by the Hadley circulation. These studies raise the interesting possibility that the Venus circulation pattern (and the Earth's) may not be steady and may have undergone historical variability, switching between a superrotating state and a "normal" state.

XVI. SUMMARY AND ASSESSMENT

The success story of the past two decades is the understanding that has been reached of the dynamics of the middle atmosphere region, from about 60 km to 90 km elevation. The PVO, in 1979, gave a first order characterization of the thermal and dynamical structure here. Fels and Lindzen (1974) had suggested that tidal momentum transports control the mean flow. Increasingly detailed theoretical work followed, by Pechman and Ingersoll (1984), Fels (1986), Leovy (1987), Baker and Leovy (1987) and Hou et al. (1990). Newman and Leovy (1992) calculated a balanced solution, showing tidal momentum fluxes balanced by mean flow advection. The predicted velocities and temperature perturbations are in good agreement with observations. We may conclude that to first order the dynamics of this height region is understood, at least at low and mid-latitudes where data is most complete. The high latitude polar vortex is not as well documented, and may contain surprises.

But the middle atmosphere dynamics turns out to have little to do with maintaining the overall spin of the Venus atmosphere. In fact, a major element of the tidal solutions is a downward eddy flux of momentum into the velocity maximum near 65 km elevation. This eddy flux acts to accelerate the cloud-top flow, but at the expense of higher regions, and is in fact in the opposite direction to the eddy fluxes that would be needed to pump momentum from the surface of the planet upward into the atmosphere. Fluxes of the latter type are also part of the tidal solutions, as pointed out by Fels and Lindzen, but the detailed calculations have shown that tidal amplitudes below the clouds, where fluxes would be upward, are small. Tides cannot yet be ruled out as important beneath the clouds, but it seems unlikely.

The key to unlocking the mystery of the Venus atmospheric superrotation is in the momentum transports in the deep atmosphere. The crucial location is the region where the atmospheric spin increases from near corotation with the solid surface, within the lowest few km, to the height where a strong mean flow of about 50 m s^{-1} exists, at about 45 km elevation. In order to understand the flow, the balance between forcing and drag must be diagnosed and explained. Based on experience with the terrestrial atmosphere, wave drag probably operates on Venus and is likely to be the dominant drag to be overcome. To test this hypothesis, the nature of waves and turbulence needs to be determined. Two different classes of hypotheses for the acceleration mechanisms have been advanced. Under one, eddies or waves accomplish an upward momentum transport that drives the flow, and no other flow components are essential. Under the other, eddies or waves accomplish an equatorward transport and the Hadley circulation accomplishes the upward transport, and both components are essential. The two hypotheses are not mutually exclusive and a mix may be present.

Experience with geophysical fluid systems shows that theoretical modeling will not be conclusive except in conjunction with observations. We now know that the eddies or waves that are important in producing drag and in producing the momentum transports that offset it probably have amplitudes in the deep atmosphere between a few cm s^{-1} and a few m s^{-1} and are therefore beneath the accuracy threshold of the Pioneer Venus probe data. The thermal stratification is also crucial to establish, especially within the lowest 15 km where very little data exist. Specific goals for observation, based on the foregoing discussions, especially Sec. VII and Fig. 11, are:

1. Stratification of the bottom scale height at several locations on the planet, with an accuracy of at least 0.1 K and a measurement at least every 0.1 km in height. Processes constrained by this observation are boundary layer convection, waves generated by boundary layer convection, near-surface waves and wave propagation characteristics, and slope winds.
2. Three velocity components in the bottom scale height at several locations on the planet with an accuracy of at least 0.1 m s^{-1}, profiled vertically with spatial resolution of 0.1 km. Two locations as near as possible to dawn and sunset are desirable to look for changes in convection intensity. Processes addressed are the same as for (1).
3. Vertical scale (velocity coherence scale) and vertical structure (periodic or random) of eddies in the lower atmosphere. Three velocity components are desirable, with accuracy of 0.1 m s^{-1}, sampled at least every 1 km in height. Temperatures are desirable at the same spatial resolution and with accuracy of 1 K. Processes addressed are waves, turbulence, and momentum transfers. Gravity waves would be identified by partic-

ular phase correlations between velocities, and between velocities and temperatures, and also by spatially periodic structure.
4. Latitudinal and vertical structure of the zonal mean flow and the temperature. These are needed to determine the Richardson number and the horizontal shear, in order to define the stability properties of the mean flow. To determine the horizontal shear there should be at least three or four latitudinal points sampled in a hemisphere (north or south).
5. Relationship of eddy intensity to underlying topography or convection. Measurements (1)–(3) should be made at locations over both strong and weak topography. Processes addressed are wave drag, or wave generation in general, due to mean flow over topography or topographically induced slope winds.
6. Horizontal scale of eddies in the lower atmosphere. Measurements (2) and (3) should be made at different horizontal separations between about 200 km and 2000 km to determine the horizontal scale of dynamical activity.

For the purpose of understanding the superrotation, improved measurements of cloud microphysical properties and of radiative fluxes are not of high priority. A detailed understanding of the Venus greenhouse effect will require better information, especially on the thermal radiation field, but the heating imbalances that drive dynamics are quite well determined by the Pioneer Venus solar flux measurements and by the observed thermal field. Similarly, the Venus clouds present interesting and important questions of local cloud dynamics, microphysics, and chemistry, but these are probably not crucial issues for the superrotation.

Theoretical general circulation modeling, on the other hand, will be essential to assimilation of future data. Probe measurements can define the flow along only a few profiles, and only at certain instants of time. The statistical properties of the flow must be established by theoretical arguments that are constrained by these measurements. The tools will certainly include numerical general circulation computations. A hierarchy of these will probably be necessary, stressing different aspects of the flow. Boundary layer slope wind simulations, for example, may be necessary to isolate and treat with a single model, whose results are then abstracted and included in parameterized form in a global computation. Particular questions of importance that can only be answered by computations are the nature of the Hadley circulation, which is probably too weak at deep levels to be directly measured, and the global distribution of eddy momentum fluxes.

Acknowledgments. This work has been supported in part by the NASA Planetary Atmospheres Program.

REFERENCES

Allison, M., Del Genio, A. D., and Zhou, W. 1994. Zero-potential vorticity envelopes for the zonal-mean velocity of the Venus/Titan atmospheres. *J. Atmos. Sci.* 51:694–702.

Andrews, D. G., Holton, J. R., and Leovy, C. B. 1987. *Middle Atmosphere Dynamics* (Orlando: Academic Press).

Avduevsky, V. S., et al. 1983. Structure and parameters of the Venus atmosphere according to Venera probe data. In *Venus*, eds. D. M. Hunten, L. Colin, T. M. Donahue and V. I. Moroz (Tucson: Univ. of Arizona Press), pp. 280–298.

Baker, N. L., and Leovy, C. B. 1987. Zonal winds near Venus' cloud top level: A model study of the interaction between the zonal mean circulation and the semidiurnal tide. *Icarus* 69:202–220.

Baker, R. D., II, and Schubert, G. 1992. Cellular convection in the atmosphere of Venus. *Nature* 355:710–712.

Belton, M. J. S., et al. 1976a. Space-time relationships in the UV markings on Venus. *J. Atmos. Sci.* 33:1383–1393.

Belton, M. J. S., Smith, G. R., Schubert, G., and Del Genio, A. D. 1976b. Cloud patterns, waves and convection in the Venus atmosphere. *J. Atmos. Sci.* 33:1394–1417.

Belton, M. J. S., et al. 1991. Images from Galileo of the Venus cloud deck. *Science* 253:1531–1536.

Boyer, Ch., and Camichel, H. 1961. Observations photographiques de la planète Vénus. *Ann. Astrophys.* 24:531–535.

Boyer, Ch., and Guérin, P. 1969. Etude del la rotation rétrograde, en 4 jours, de la couche extérieure nuageuse de Vénus. *Icarus* 11:338–355.

Campbell, D. B., et al. 1992. Magellan observations of extended impact crater related features on the surface of Venus. *J. Geophys. Res.* 97:16249–16257.

Carlson, R. W., et al. 1991. Galileo infrared imaging spectroscopy measurements at Venus. *Science* 253:1541–1548.

Counselman, C. C., III, et al. 1980. Zonal and meridional circulation of the lower atmosphere of Venus determined by radio interferometry. *J. Geophys. Res.* 85:8026–8030.

Covey, C., and Schubert, G. 1982. Planetary-scale waves in the Venus atmosphere. *J. Atmos. Sci.* 39:2397–2413.

Crisp, D. 1986. Radiative forcing of the Venus mesosphere. I. Solar fluxes and heating rates. *Icarus* 67:484–514.

Crisp, D. 1989. Radiative forcing of the Venus mesosphere. II. Thermal fluxes, cooling rates, and radiative equilibrium temperatures. *Icarus* 77:391–413.

Crisp, D., Ingersoll, A. P., Hildebrand, C. E., and Preston, R. A. 1990. Vega balloon meteorological measurements. *Adv. Space Res.* 10(5):109–124.

Crisp, D., et al. 1991. The nature of the near-infrared features on the Venus night side. *Science* 246:506–509.

Del Genio, A. D., and Rossow, W. B. 1990. Planetary-scale waves and the cyclic nature of cloud top dynamics on Venus. *J. Atmos. Sci.* 47:293–318.

Del Genio, A. D., and Suozzo, R. J. 1987. A comparative study of rapidly and slowly rotating dynamical regimes in a terrestrial general circulation model. *J. Atmos. Sci.* 44:973–986.

Del Genio, A. D., and Zhou, W. 1996. Simulations of superrotation on slowly rotating planets: Sensitivity to rotation and initial conditions. *Icarus* 120:332–343.

Del Genio, A. D., Zhou, W., and Eichler, T. P. 1993. Equatorial superrotation in a slowly rotating GCM: Implications for Titan and Venus. *Icarus* 101:1–17.

Dobrovolskis, A. R. 1983. Atmospheric tides on Venus. III. The planetary boundary layer. *Icarus* 56:165–175.
Dobrovolskis, A. R. 1993. Atmospheric tides on Venus. IV. Topographic winds and sediment transport. *Icarus* 103:276–289.
Dobrovolskis, A. R., and Ingersoll, A. P. 1980. Atmospheric tides and the rotation of Venus. I. Tidal theory and the balance of torques. *Icarus* 41:1–17.
Elson, L. S. 1982. Wave instability in the polar region of Venus. *J. Atmos. Sci.* 39:2356–2362.
Esposito, L. W., et al. 1983. The clouds and hazes of Venus. In *Venus*, eds. D. M. Hunten, L. Colin, T. M. Donahue and V. I. Moroz (Tucson: Univ. of Arizona Press), pp. 484–564.
Fels, S. B. 1986. An approximate analytical method for calculating tides in the atmosphere of Venus. *J. Atmos. Sci.* 43:2757–2772.
Fels, S. B., and Lindzen, R. S. 1974. The interaction of thermally excited gravity waves with mean flows. *Geophys. Fluid Dyn.* 6:149–192.
Fjørtoft, R. 1953. On the changes in the spectral distribution of kinetic energy for 2-dimensional, non-divergent flow. *Tellus* 5:225–230.
Ford, P. G., and Pettengill, G. H. 1992. Venus topography and kilometer-scale slopes. *J. Geophys. Res.* 97:13103–13114.
Fritts, D. C., and Dunkerton, T. J. 1984. A quasi-linear study of gravity wave saturation and self acceleration. *J. Atmos. Sci.* 41:3272–3289.
Gierasch, P. J. 1975. Meridional circulation and the maintenance of the Venus atmospheric circulation. *J. Atmos. Sci.* 32:1038–1044.
Gierasch, P. J. 1987. Waves in the atmosphere of Venus. *Nature* 328:510–512.
Gierasch, P. J., and Goody, R. M. 1970. Models of the Venus clouds. *J. Atmos. Sci.* 27:224–245.
Gierasch, P. J., Goody, R. M., and Stone, P. H. 1970. The energy balance of planetary atmospheres. *Geophys. Fluid Dyn.* 1:1–18.
Gold, T., and Soter, S. 1971. Atmospheric tides and the 4-day circulation on Venus. *Icarus* 14:16–20.
Goody, R. M., and Yung, Y. L. 1989. *Atmospheric Radiation* (New York: Oxford Univ. Press).
Greeley, R., et al. 1992. Aeolian features on Venus: Preliminary Magellan results. *J. Geophys. Res.* 97:13319–13345.
Greeley, R., et al. 1994. Wind streaks on Venus: Clues to atmospheric circulation. *Science* 263:358–361.
Greeley, R., et al. 1995. Wind-related features and processes on Venus: Summary of Magellan results. *Icarus* 115:399–420.
Herring, J., and Métais, O. 1989. Numerical experiments in forced stably stratified turbulence. *J. Fluid Mech.* 202:97–115.
Hide, R. 1969. Dynamics of the atmospheres of the major planets with an appendix on the viscous boundary layer at the rigid bounding surface of an electrically-conducting rotating fluid in the presence of a magnetic field. *J. Atmos. Sci.* 26:841–853.
Hilsenrath, J. C., et al. 1960. *Tables of Thermodynamic and Transport Properties of Air, Argon, Carbon Dioxide, Nitrogen, Oxygen and Steam* (New York: Pergamon Press).
Hinson, D. P., and Jenkins, J. M. 1995. Magellan radio occultation measurements of atmospheric waves on Venus. *Icarus* 114:310–327.
Hitchman, M. H., and Leovy, C. B. 1988. Estimation of the Kelvin wave contribution to the semiannual oscillation. *J. Atmos. Sci.* 45:1462–1475.
Holton, J. R. 1982. The role of gravity wave induced drag and diffusion in the momentum budget of the mesosphere. *J. Atmos. Sci.* 39:791–799.

Holton, J. R. 1992. *An Introduction to Dynamic Meteorology* (San Diego: Academic Press).
Holton, J. R., and Lindzen, R. S. 1972. An updated theory for the quasi-biennial cycle of the tropical stratosphere. *J. Atmos. Sci.* 29:1076–1080.
Hou, A. Y. 1984. Axisymmetric circulations forced by heat and momentum sources: A simple model applicable to the Venus atmosphere. *J. Atmos. Sci.* 41:3437–3455.
Hou, A. Y., and Farrell, B. F. 1987. Superrotation induced by critical-level absorption of gravity waves on Venus: An assessment. *J. Atmos. Sci.* 44:1049–1061.
Hou, A. Y., and Goody, R. M. 1985. Diagnostic requirements for the superrotation on Venus. *J. Atmos. Sci.* 42:413–432.
Hou, A. Y., and Goody, R. M. 1989. Further studies of the circulation of the Venus atmosphere. *J. Atmos. Sci.* 46:991–1001.
Hou, A. Y., Fels, S. B., and Goody, R. M. 1990. Zonal superrotation above Venus' cloud base induced by the semidiurnal tide and the mean meridional circulation. *J. Atmos. Sci.* 47:1894–1901.
Hunten, D. M., and Goody, R. M. 1969. Venus: The next phase of planetary exploration. *Science* 165:1317–1323.
Iversen, J. D., and White, B. R. 1982. Saltation threshold on Earth, Venus and Mars. *Sedimentology* 29:111–119.
Kálnay de Rivas, E. 1975. Further numerical calculations of the circulation of the atmosphere of Venus. *J. Atmos. Sci.* 32:1017–1024.
Kawabata, K. D., et al. 1980. Cloud and haze properties from Pioneer Venus polarimetry. *J. Geophys. Res.* 85:8129–8140.
Knollenberg, R. G., and Hunten, D. M. 1980. The microphysics of the clouds of Venus: Results of the Pioneer Venus particle size spectrometer experiment. *J. Geophys. Res.* 85:8039–8058.
Kraichnan, R. H. 1967. Inertial ranges in two-dimensional turbulence. *Phys. Fluids* 10:1417–1423.
Leovy, C. B. 1973. Rotation of the upper atmosphere of Venus. *J. Atmos. Sci.* 30:1218–1220.
Leovy, C. B. 1987. Zonal winds near Venus' cloud top level: An analytic model of the equatorial wind speed. *Icarus* 69:193–201.
Leovy, C. B., and Baker, N. L. 1987. Zonal winds near Venus' cloud top level: A model study of the interaction between the zonal mean circulation and the semidiurnal tide. *Icarus* 69:202–220.
Leroy, S. S., and Ingersoll, A. P. 1995a. Convective generation of gravity waves in Venus' atmosphere: Gravity wave spectrum and momentum transport. *J. Atmos. Sci.* 52:3717–3737.
Leroy, S. S., and Ingersoll, A. P. 1995b. Radio scintillations in Venus' atmosphere: Application of a theory of gravity wave generation. *J. Atmos. Sci.*, submitted.
Limaye, S. S. 1987. Atmospheric dynamics on Venus and Mars. *Adv. Space Res.* 7(12):39–53.
Limaye, S. S., and Suomi, V. E. 1981. Cloud motions on Venus: Global structure and organization. *J. Atmos. Sci.* 38:1220–1235.
Limaye, S. S., Grund, C. J., and Burre, S. P. 1982. Zonal mean circulation at the cloud level on Venus: Spring and fall 1979 OCPP observations. *Icarus* 51:416–439.
Limaye, S. S., Grassotti, C., and Kuetemeyer, M. J. 1988. Venus: Cloud level circulation during 1982 as determined from Pioneer cloud photopolarimeter images. I. Time and zonally averaged circulation. *Icarus* 73:193–211.
Lindzen, R. S. 1981. Turbulence and stress due to gravity wave and tidal breakdown. *J. Geophys. Res.* 86:9707–9714.
Magalhães, J. A., and Seiff, A. 1995. Atmospheric wave activity in Pioneer Venus entry probe data acquired within and below the clouds. *Icarus*, submitted.

Malkus, W. V. R. 1970. Hadley-Halley circulation on Venus. *J. Atmos. Sci.* 27:529–535.

Michelangeli, D. V., Zurek, R. W., and Elson, L. S. 1987. Barotropic instability of midlatitude zonal jets on Mars, Earth and Venus. *J. Atmos. Sci.* 44:2031–2041.

Moroz, V. I. 1981. The atmosphere of Venus. *Space Sci. Rev.* 29:3–127.

Newman, M., and Leovy, C. B. 1992. Maintenance of strong rotational winds in Venus' middle atmosphere by thermal tides. *Science* 257:647–650.

Newman, M., Schubert, G., Kliore, A. J., and Patel, I. R. 1984. Zonal winds in the middle atmosphere of Venus from Pioneer Venus radio occultation data. *J. Atmos. Sci.* 41:1901–1913.

Pechman, J. B., and Ingersoll, A. P. 1984. Thermal tides in the atmosphere of Venus: Comparison of model results with observations. J. Atmos. Sci. 41:3290–3313.

Priestley, C. H. B. 1959. *Turbulent Transfer in the Lower Atmosphere* (Chicago: Univ. of Chicago Press).

Rossby, C.-G. 1947. On the distribution of angular velocity in gaseous envelopes under the influence of large-scale horizontal mixing processes. *Bull. Amer. Meteorol. Soc.* 28:53–68.

Rossow, W. B. 1983. A general circulation model of a Venus-like atmosphere. *J. Atmos. Sci.* 40:273–302.

Rossow, W. B., and Williams, G. P. 1979. Large-scale motion in the Venus stratosphere. *J. Atmos. Sci.* 36:377–389.

Rossow, W. B., Del Genio, A. D., Limaye, S. S., and Travis, L. D. 1980. Cloud morphology and motions from Pioneer Venus images. *J. Geophys. Res.* 85:8107–8128.

Rossow, W. B., Del Genio, A. D., and Eichler, T. 1990. Cloud-tracked winds from Pioneer Venus OCPP images. *J. Atmos. Sci.* 47:2053–2084.

Schäfer, K., et al. 1987. Structure of the middle atmosphere of Venus from analysis of Fourier spectrometer measurements aboard Venera 15. *Adv. Space Res.* 7(12):17–24.

Schinder, P. J., Gierasch, P. J., Leroy, S. S., and Smith, M. D. 1990. Waves, advection and cloud patterns on Venus. *J. Atmos. Sci.* 47:2037–2052.

Schofield, J. T., and Taylor, F. W. 1983. Measurements of the mean, solar-fixed temperature and cloud structure of the middle atmosphere of Venus. *Quart. J. Roy. Meteorol. Soc.* 109:57–80.

Schubert, G. 1983. General circulation and the dynamical state of the Venus atmosphere. In *Venus*, eds. D. M. Hunten, L. Colin, T. M. Donahue and V. I. Moroz (Tucson: Univ. of Arizona Press), pp. 681–765.

Schubert, G., and Whitehead, J. 1969. The moving flame experiment with liquid mercury: Possible implications for the Venus atmosphere. *Science* 163:71–72.

Seiff, A., et al. 1980. Measurements of thermal structure and thermal contrasts in the atmosphere of Venus and related dynamical observations: Results from the four Pioneer Venus probes. *J. Geophys. Res.* 85:7903–7933.

Smith, M. D., Gierasch, P. J., and Schinder, P. J. 1993. Global-scale waves in the Venus atmosphere. *J. Atmos. Sci.* 50:4080–4096.

Suarez, M. J., and Duffy, D. G. 1992. Terrestrial superrotation: A bifurcation of the general circulation. *J. Atmos. Sci.* 49:1541–1554.

Taylor, F. W., et al. 1980. Structure and meteorology of the middle atmosphere of Venus: Infrared remote sensing from the Pioneer Orbiter. *J. Geophys. Res.* 85:7963–8006.

Toigo, A., Gierasch, P. J., and Smith, M. D. 1994. High resolution feature tracking on Venus by Galileo. *Icarus*, 109:318–336.

Tomasko, M. G. 1983. The thermal balance of the lower atmosphere of Venus. In *Venus*, eds. D. M. Hunten, L. Colin, T. M. Donahue and V. I. Moroz (Tucson:

Univ. of Arizona Press), pp. 604–631.

Tomasko, M. G., Doose, L. R., Smith, P. H., and Odell, A. P. 1980. Measurements of the flux of sunlight in the atmosphere of Venus. *J. Geophys. Res.* 85:8167–8186.

Tomasko, M. G., Doose, L. R., and Smith, P. H. 1985. The absorption of solar energy and the heating rate in the atmosphere of Venus. *Adv. Space Res.* 5:71–79.

Travis, L. D., et al. 1979. Orbiter cloud photopolarimeter investigation. *Science* 203:781–785.

Waltersheid, R. L., et al. 1985. Zonal winds and the angular momentum balance of Venus' atmosphere within and above the clouds. *J. Atmos. Sci.* 42:1982–1990.

Woo, R., Armstrong, J. W., and Ishimaru, A. 1980. Radio occultation measurements of turbulence in the Venus atmosphere by Pioneer Venus. *J. Geophys. Res.* 85:8031–8038.

Young, R. E., and Pollack, J. B. 1977. A three-dimensional model of dynamical processes in the Venus atmosphere. *J. Atmos. Sci.* 34:1315–1351.

Young, R. E., Houben, H., and Pfister, L. 1984. Baroclinic instability in the Venus atmosphere. *J. Atmos. Sci.* 41:2310–2333.

Young, R. E., et al. 1987. Characteristics of gravity waves generated by surface topography on Venus: Comparison with the VEGA balloon results. *J. Atmos. Sci.* 44:2628–2639.

Young, R. E., et al. 1994. Characteristics of finite amplitude stationary gravity waves in the atmosphere of Venus. *J. Atmos. Sci.* 51:1857–1875.

PART IV
Surface Processes

REMOTE SENSING OF SURFACE PROCESSES

BRUCE A. CAMPBELL
Smithsonian Institution

RAYMOND E. ARVIDSON
Washington University

MICHAEL K. SHEPARD
Bloomsburg University

and

ROBERT A. BRACKETT
Washington University

Surface weathering processes on Venus occur at rates several orders of magnitude slower than those on Earth. This is due in large part to the absence of water, minimal diurnal temperature changes, and lack of strong winds near the surface. Erosion rates are thus similar to those on the Moon, but the dense atmosphere and relatively young age of the surface have prevented the formation of a thick regolith. Much of the terrain on Venus is comprised of smooth bedrock and closely packed fragmental debris, of basaltic composition. Mechanical and chemical erosion reduce the meter-scale roughness of rocky surfaces over time scales of millions of years, but a major source of fine-grained material which can be moved by the wind is likely impact crater formation. Young craters have associated parabolic haloes of fine material, which are dispersed on time scales of a few tens of millions of years. The close association of the few identified dune fields with impact craters further argues for these events as the primary source of fine sediment and transient high winds. Enhanced radar backscatter and low emissivity in the highlands may be caused by loading of rocks with conductive or ferroelectric minerals which are stable only at higher elevations, or by cold trapping of such materials transported by the atmosphere. Questions as to the specific mineralogy of the highlands, the distribution of sediments, and the role of chemical weathering at the surface remain to be addressed by future missions.

I. INTRODUCTION

Our knowledge of the surface properties of Venus comes from observations by Venera, Vega, and Pioneer Venus landers, radar maps of the planet produced by Pioneer Venus Orbiter, Venera 15 and 16, and Magellan, and from Earth-based radar mapping by the Arecibo and Goldstone antennas. Geochemical data from the landers imply a predominantly basaltic composition for the surface, with one site indicating possibly more alkalic rock (Surkov 1983; Basilevksy et al. 1992; Weitz and Basilevsky 1993). Radar images reveal a surface dominated by relatively smooth plains, interspersed with gently sloping shield volcanoes and upland plateaus of highly deformed terrain. The bedrock geology is overprinted by several processes: (1) comminution of surface rocks by chemical or mechanical (including impact) weathering; (2) redistribution of fine-grained soil by mass movements or winds (due either to global circulation or impact cratering); and (3) "highlands" weathering and/or modification related to surface-atmosphere interactions. We use the term "soil" here to indicate layers of debris with variable, though small, grain sizes. Impact cratering in particular produces fine material, scours large areas of the surface, overturns rock layers, and creates transient high winds. Each of these surface processes is strongly affected by the unique environment in terms of the high surface and atmospheric temperature, the high atmospheric pressure, the lack of diurnal temperature excursions, the trace element chemistry of the atmosphere, and the lack of water (cf., McGill et al. [1983] for a pre-Magellan review of surface processes on Venus).

In this chapter we first provide an overview of surface processes which operate in the Venus lowlands, and the remote sensing methods used to characterize them. We then focus on the unusual microwave properties of the Venus highlands and models which may explain these observations. We conclude with a brief summary of important issues which remain to be addressed by future Earth-based and spacecraft measurements.

II. OVERVIEW OF SURFACE PROCESSES OPERATING IN THE LOWLANDS

A. Mechanical and Chemical Weathering

The mineralogy of the surface of Venus has never been directly measured, because the lander experiments carried by the Venera and Vega spacecraft permitted derivation of only bulk elemental abundances. Interpretation of these data varies, but some form of basaltic composition is generally agreed to fit the majority of the surface measurements. The initial appearance of Venus from Magellan radar images is of a relatively fresh surface; impact craters have sharply defined rims, wrinkle ridges contrast strongly with the plains, and even very steep slopes do not exhibit large talus deposits. The subdued structures of older lunar or Martian craters and highlands are not evident.

The relative youth of the surface is confirmed by the impact crater population. While the specific details of crustal recycling and overturn on Venus remain contentious, a concensus has emerged that the planet was largely resurfaced, and the crater population reset, on the order of 300 to 500 Myr ago (Schaber et al. 1992; Phillips et al. 1992). Based on this average age, Arvidson et al. (1992) estimated that the rate of physical weathering in the lowlands is no higher than $\sim 10^{-3} \mu m\ yr^{-1}$ (or ~ 0.5 m over 500 Myr), comparable to regolith formation rates on the Moon (Horz et al. 1991).

Chemical weathering on Venus can be studied by assuming that the surface and atmosphere are in thermodynamic equilibrium, and calculating the stability of various mineral assemblages on the surface based on the measured composition of the Venusian lower atmosphere (see the chapter by Fegley et al.). Mueller (1963) first proposed that the high surface pressure and temperature of Venus made it similar to metamorphic environments on the Earth, and as a result, (1) reaction kinetics might be expected to be geologically rapid, and (2) thermodynamic equilibrium between the atmosphere and the surface should be attained. Laboratory experiments on the gas-solid kinetics of anhydrite formation, pyrite decomposition, and oxidation of basalts on the surface of Venus have demonstrated that chemical reactions proceed geologically quickly. For example, the rate of chemical alteration of calcite to anhydrite is $\sim 1\ \mu m\ yr^{-1}$ (Fegley and Prinn 1989) and the rate of chemical alteration of pyrite to iron oxide is $>1\ cm\ yr^{-1}$ (chapter by Fegley et al.), or more than 7 orders of magnitude more rapid than the expected physical weathering. These estimates of weathering rate are only theoretical upper bounds, and do not necessarily imply a rapid rate of chemical erosion on the surface. The specific interaction between the surface rocks and the atmosphere has not been modeled, and a variety of possible mechanisms may occur on Venus.

In the broadest sense, two types of chemical reactions will occur at the surface on Venus: (1) volume-increasing and (2) volume-decreasing. Volume-increasing reactions, which include the formation of calcite from calcium oxide, reduce porosity and lead to sintering and cementing of the reaction products. These processes may effectively compete against physical weathering as they increase the grain size of materials, and it has been proposed that such reactions may be partly responsible for the relatively pristine condition of the Venus surface (Burke et al. 1994). Volume-decreasing reactions result in an increase in porosity and may speed physical weathering by causing the mechanical disintegration of surface rocks. Decomposition of pyrite, for example, results in a decreased volume of product (relative to reactant) and increased porosity (chapter by Fegley et al.). The extent to which individual chemical processes contribute to changes in the texture of the surface and produce soil is unknown, but we will address the observational evidence for sediment accumulation below.

B. Sediment Production and Redistribution

The surface of Venus lacks a deep regolith, primarily due to the relatively young age of the surface (less than ~ 500 Myr), the protection from meteorite bombardment afforded by the thick atmosphere, and the lack of major erosional action by wind or water. The Venera lander images demonstrate that fine debris (down to at least the cm-scale) does exist on Venus, and that the quantity and grain size of the material varies from site to site (Garvin et al. 1984). Weathering of the surface rocks by mechanical or chemical processes appears to produce material that is shallow, coarse, and relatively immobile. As noted above, the pristine state of many geologic features argues for a weathering rate of $<10^{-3}$ μm yr^{-1}, comparable to those found for the lunar surface. Such *in-situ* weathering should be most effective in the highlands, where slopes are relatively steep and we could expect a constant exposure of fresh surface rock. Analysis of mass movements within the highlands demonstrates, however, that most such deposits are dominated by bedrock collapses; no evidence for sediment slumps, flows, or falls has been found (Malin 1992). The Venera 9 site on the flank of Theia Mons contains a substantial population of surface rocks, but no apparent concurrent development of a major soil component. The global average dielectric constant is $\sim$4.5, consistent with moderate-low density terrestrial basalts or more dense material with a very thin surficial soil layer (Pettengill et al. 1992; Ulaby et al. 1988).

Another possible source of fine material is impact crater formation, which produces ejecta deposits with a broad range of grain sizes and radial distances from the cavity (cf., Oberbeck 1975; McGetchin 1973). Young craters on Venus are characterized by radar-bright floors and parabolic patterns of low-backscatter, low-emissivity material. These parabolas extend for hundreds of km (typically to the west of the crater), and were evidently emplaced as airfall deposits from ejecta clouds thrown high into the atmosphere (Campbell et al. 1992; Schultz 1992). Grain sizes of materials in the parabolas are likely no greater than 1 cm, based on both the ability of the atmosphere to support the loaded cloud of hot debris and the surface roughness of the emplaced material. Izenberg et al. (1994) estimated that these features have a lifetime of a few tens of Myr based on a global mean cratering rate, which supports an average dispersion rate of $\sim 10^{-3}$ μm yr^{-1}. Many impact craters also have concentric, near-circular low-backscatter areas (Phillips et al. 1991). These regions may be covered with a shallow layer of fine material produced by shock comminution of the surface rock by the atmospheric blast associated with a bolide's entry (Schultz 1992; Ivanov et al. 1986). Areas of higher backscatter surrounding some craters may be regions where fine material has been swept from the surface or where coarse comminuted material is produced by the blast wave.

If the majority of craters form parabolas, concentric low-return haloes, or some combination of these features, then impact processes may contribute the bulk of fine-grained soil on the surface. This hypothesis is supported by the strong tendency for wind streaks and dunes to occur in relationship to craters (e.g., Aglaonice and Fortuna-Meshkenet dune fields), indicating that only this material is fine enough to be moved easily by the wind (Greeley et al. 1992). Similar wind streaks are not observed randomly about the plains, so any soil formed by *in-situ* weathering is probably more granular and immobile. Garvin (1990) calculated a possible sediment budget for Venus based on crater counts from the Venera 15 and 16 radar images, and suggested that very fine material (<30 μm) would not likely be produced in adequate amounts to supply a global sedimentary layer. Large-scale coverage of the surface to depths observed at the Venera lander sites would require transport of particles up to 1 cm in size over large regions. While the Magellan data have confirmed the paucity of small craters shown by the Venera images, other mechanisms of impact-derived sediment formation have also been identified. Specifically, the large low-backscatter haloes and the parabolas represent styles of sediment generation and transport not included in typical crater ejecta models (see, e.g., Melosh and Schaller 1996). The issue of whether impact-produced sediments can dominate the global budget of such materials thus remains open.

A final possible source of fine material is explosive volcanism. The high Venus atmospheric presure inhibits the exsolution of volatiles from rising magma bodies, and thus limits pyroclastic eruptions to cases of high volatile content (and preferably higher elevation) (Garvin et al. 1982). While evidence for such eruptions is present (i.e., Maat Mons [Klose et al., 1992]), they are likely to be too limited to contribute a significant fraction of the global sediment budget.

Redistribution of sediment on Venus can occur by only two processes: mass movement (slumping) or wind transport. For the lowlands, topographic slopes are typically so small (on the order of 1° or less) that gravitational movement of fine soil is negligible. Steeper slopes in the highlands may create local sediment-collecting troughs, but these have not been directly observed. Greeley et al. (1992) found that the global pattern of wind streaks indicates neither a zonal nor a meridional circulation pattern. This suggests that many of the surface streaks may reflect the local winds at the time of crater formation, which is consistent with the arcuate wind features found concentric to many large impact craters (Schultz 1992). More recent work (chapter by Greeley et al.) demonstrates that the streaks do represent two populations: one indicative of Hadley cell circulation, the other linked to impact-produced transient winds.

In summary, the surface of Venus is likely comprised of relatively pristine bedrock, that undergoes a slow process of mechanical rubbling due to surface-atmosphere interactions. Such *in-situ* weathering may reduce blocky surface textures to more granular deposits, though again at a very slow rate. Fine soil accumulates predominantly as a result of impact crater formation, and this material can be moved by the wind, at least during subsequent nearby cratering events. High backscatter areas found surrounding some craters may thus be regions of coarse comminuted debris or areas stripped of all past mantling by the strong winds, while low backscatter concentric deposits may be regions where fine soil has been produced by atmospheric blast waves. Erosion in the highlands is still poorly understood, because topography and chemical processes in these areas are significantly different from those in the plains.

III. DERIVATION OF LOWLAND SURFACE PROPERTIES FROM MICROWAVE DATA

With the exception of the Venera and Vega lander sites, all estimates of the texture, density, and composition of the Venus surface are based on analysis of microwave data for backscatter and emission. A thorough characterization of the terrain requires analysis of both surface roughness and dielectric constant variations. In this section we discuss tools for such analysis and the results obtained from global mapping.

A. Surface Roughness

Radar backscatter is a function of both roughness at a variety of scales and the bulk dielectric properties of the target material. Roughness on scales that are large with respect to the wavelength can produce facet-like (quasi-specular) echoes, while roughness on scales close to that of the radar wavelength tends to produce diffuse returns (Hagfors 1970). Magellan used a 12.6-cm radar system to map the surface primarily in a horizontal-transmit, horizontal-receive (HH) polarization sense. Incidence angles varied from $\sim 20°$ to $45°$ with latitude and radar viewing configuration. The measured power was converted to values of specific radar backscatter cross section.

Terrestrial radar data sets, such as those produced by the NASA/JPL AIRSAR system, provide a good reference point for understanding radar backscatter from natural surfaces and for comparison with Magellan data. A synthesized S-band (12.6-cm) HH radar image of Kilauea Volcano, Hawaii, is shown in Fig. 1, and the scattering behaviors for lava flows of varying roughness are presented in Fig. 2. There is a good separation in echo strength due to changes in lava texture for angles $>30°$, which makes it possible to directly infer Venus surface roughness from Magellan data at similar viewing geometries (Campbell et al. 1993; Campbell and Campbell 1992; Gaddis 1992; Arvidson et al. 1992; Plaut 1991). At angles less than about $30°$, the backscatter data suffer from an ambiguity between diffuse and quasi-specular scattering mechanisms which can effectively mask large differences in wavelength-scale surface structure (Campbell and Campbell 1992). Echoes from relatively smooth surfaces have angular scattering functions that drop rapidly from a very high nadir value, while rough surfaces exhibit a moderate radar return and a gentle decline in power with angle. For rocky terrains, these two types of angular scattering function cross one another near $30°$, resulting in a diminished sensitivity to small-scale surface roughness.

The inference of relative surface roughness by comparison to terrestrial analogs requires a few caveats: (1) the surface dielectric constant is assumed to be relatively uniform among geologic units and is comparable for the Earth and Venus; (2) the Venusian surface is not covered to a large degree by fine soil; and (3) the response of the radar to roughness reflects only the statistical nature of the height fluctuations rather than some particular scatterer shape. The first assumption is verified by global mapping of bulk dielectric constant, which ranges from about 4 to 6 for most Venus lowland lava flows and plains, demonstrating that rock or closely packed debris dominates the surface (Pettengill et al. 1992; Campbell 1994). The second issue appears to be important only in areas of impact crater debris or volcanic mantling materials; porous material on a surface will tend to reduce the backscatter coefficient by lowering the dielectric contrast between the rock layer and free space and

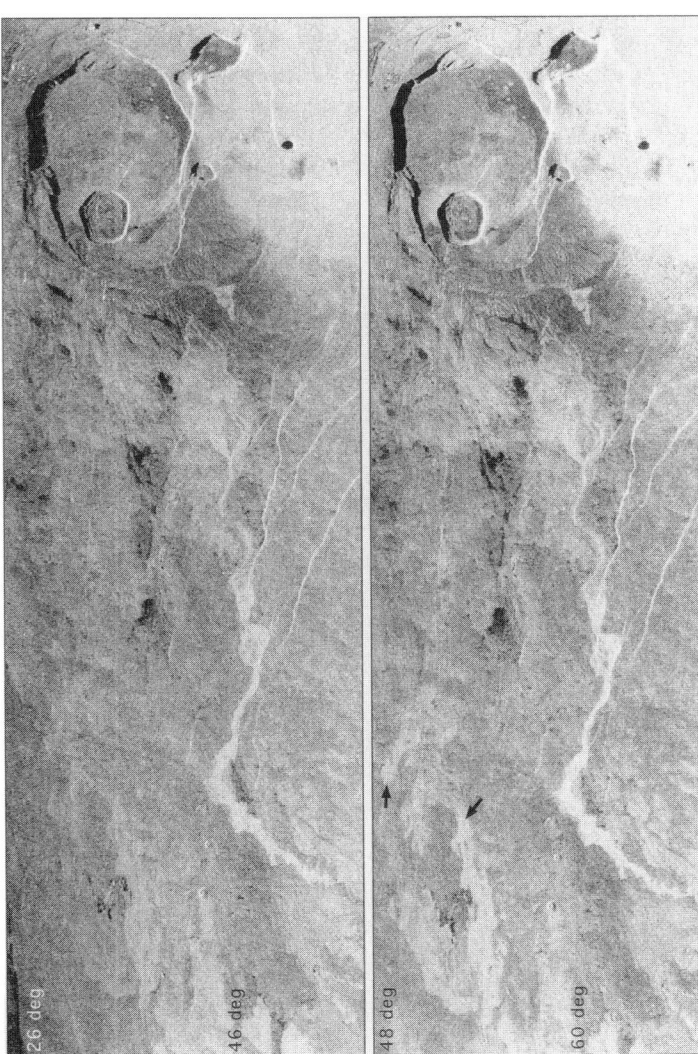

Figure 1. Radar images of the Kilauea Volcano summit area and Kau desert region in Hawaii, collected on two parallel flightlines which provide complementary incidence angle coverage. Data are synthesized S-band (12.6-cm) HH polarization, inferred from AIRSAR measurements at 5.7 and 24 cm. Image resolution 10 m/pixel; image width 21 km. Logarithmic scaling, with data normalized to the Muhleman scattering law applied to Magellan data for Venus. The range of incidence angles is marked along the side of each scene; radar flight direction is across the top of the image. Note that the rough a'a lava flows from the Mauna Iki complex, shown by black arrows in the lower scene, are indistinguishable from their smoother surroundings at a 26° incidence angle in the upper scene.

by attenuation within the fine-grained deposit. The third assumption merely states that natural surfaces tend to have random distributions of scattering centers (i.e., fields of rocks or cracks).

The terrestrial analog data can be used to discern levels of roughness among Venus surface units, and we use an example area to illustrate this methodology. A Magellan image for the area surrounding Sapas Mons (viewed at an incidence angle of $\sim 45°$) is shown in Fig. 3. The majority of plains regions surrounding Sapas Mons have backscatter values comparable to smooth pahoehoe flows in Hawaii (-20 dB$<\sigma_o<-9$ dB), consistent with their probable origin as large fissure-fed basaltic eruptions. Slow mechanical weathering may also contribute to the low roughness of these regions. Some areas do have considerably lower backscatter values, indicating either extremely smooth textures or the presence of fine-grained mantling material, which attenuates the incident energy. Several small lava flows on southeast Sapas Mons display this signature, as do a number of plains patches to the south of the edifice. The radar-dark flows on Sapas are younger than the surrounding plains, so it is likely that their smooth surfaces are indicative of the originally emplaced morphology rather than fine soil formed by *in-situ* weathering.

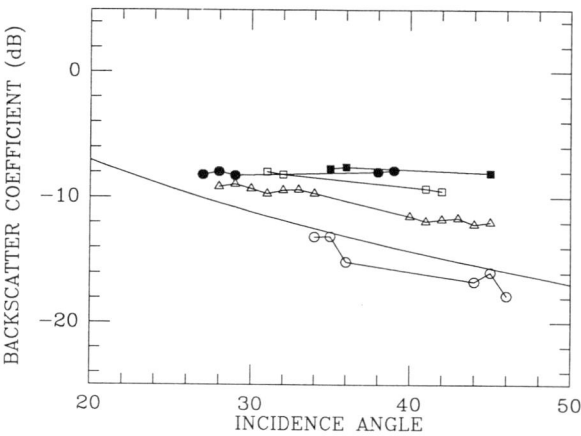

Figure 2. Plot of backscatter versus incidence angle for Hawaiian lava flows. HH polarization cross section values for 12.6-cm wavelength, interpolated from 5.7- and 24-cm AIRSAR measurements. Solid symbols are two a'a lava flows made up of jagged or platy textures; remaining three plots are for progressively smoother pahoehoe units. Solid line is average backscatter behavior for Venus (Muhleman law). Note the drop in overall power at high angles and the steeper decline with angle for smoother surfaces.

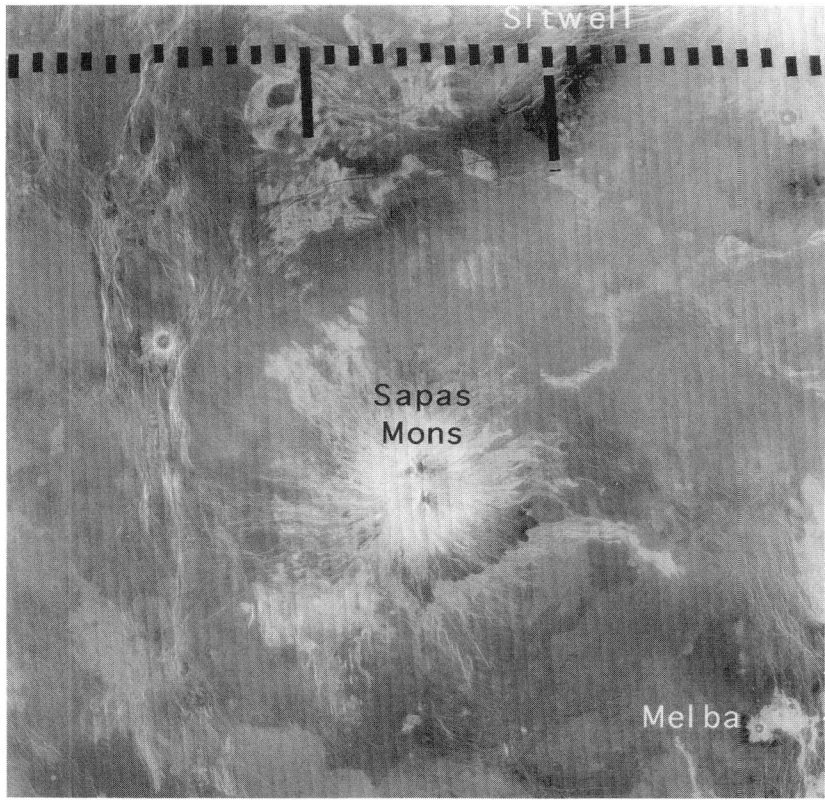

Figure 3. Magellan image of region surrounding Sapas Mons (3.5–16.5°N, 181.4–194.6°E). The high backscatter values for the summit area of Sapas Mons are due to increased reflectivity rather than to ubiquitously high roughness. Note the low radar backscatter from areas covered by fine ejecta from Sitwell crater.

Analysis of global roughness behaviors illustrates that much of Venus is very smooth, relative to young terrestrial basaltic flows, at both the meter and tens of meter scales (Ford and Pettengill 1992; Campbell and Campbell 1992). Many large Venus plains regions have scattering functions which drop more rapidly with incidence angle than even smooth terrestrial lava flows. Given the lack of deep soil layers at the Venera sites, it is most likely that the low backscatter values and steep angular scattering functions reflect a lack of significant blocky or hummocky textures in the plains. Arvidson et al. (1992) observed that some plains regions have subdued lobate flows which appear to be stratigraphically older than flows with similar shapes but higher radar backscatter. This indicates that weathering processes reduce the meter-scale roughness, to the extent that it originally exists, of plains basalt flows over long periods of time.

An arcuate band of very low radar return occurs north of Sapas Mons. Examination of the image data shows that the reduced echo from these plains is due to a covering of fine-grained impact ejecta from the crater Sitwell (16.68°N, 190.35°E, 34.5 km diameter). As noted by several authors (Schaber et al. 1992; Schultz 1992; Campbell et al. 1992), these dark haloes typically surround a relatively unmodified or radar-bright central area closer to the parent crater, where the increased brightness is due to either formation of rough comminuted material or the removal of pre-existing fine mantling debris. The atmospheric blast wave associated with bolide entry plays a major role in forming these concentric deposits, and the resulting distribution of distal impact material on Venus differs considerably from that for an airless body.

Areas of higher backscatter occur in the near-rim ejecta blankets of impact craters, but also in some crater outflow deposits (note the 24-km crater Melba in the southeast corner of the image). Numerous lava flows from Sapas Mons have quite high surface roughness, suggesting either high effusion rates or viscous magma. A survey of other large edifices shows that such rough flows are uncommon on Venus, with most shield volcanoes built up of more moderate- to smooth-surfaced flows consistent with low-volume eruption rates or low magma viscosity. The highest σ_o values in the example region occur on Sapas Mons and along the ridged terrain in the northeast corner at elevations above ~6054 km radius. As discussed below, these regions have enhanced Fresnel reflectivities, probably due to the formation of a high-dielectric (low emissivity) surficial coating. Their elevated backscatter is thus not a direct indicator of high roughness, and additional work is required to define textural changes in highland areas.

B. Dielectric Constant

The complex dielectric constant (permittivity) of the surface determines its reflectivity and the degree of attenuation experienced by radar energy passing through the material. Several approaches can be used to derive estimates of the dielectric constant from Magellan data. The Magellan altimeter system collected backscatter data for the surface at small angles of incidence ($<10°$) and these echoes were modeled using the Hagfors expression to derive estimates of reflectivity ρ and rms slope (Ford and Pettengill 1992; Tyler et al. 1992). A correction for diffuse scattering based on the SAR echo at large incidence angles was added to refine the estimates of ρ.

The Magellan radar receiver was also used to measure the microwave thermal emission from Venus. Emissivity is defined as the ratio between the observed emitted radiation and that expected for a Planck blackbody at the same kinetic temperature (determined for Venus from a standard model of temperature lapse with elevation) (Pettengill et al. 1992). The microwave emission from a surface is controlled in part by the dielectric constant, with higher values of directional hemispherical reflectivity leading to lower values of emissivity at the same viewing geometry. For a perfectly smooth surface, the emissivity E at emission angle ϕ and real dielectric constant ϵ is given by the Fresnel transmission coefficients (Stratton 1947):

$$E_h = T_h = \frac{\sin 2\phi \sin 2\theta}{\sin^2(\theta + \phi)} \tag{1}$$

$$E_v = T_v = \frac{\sin 2\phi \sin 2\theta}{\sin^2(\theta + \phi) \cos^2(\phi - \theta)} \tag{2}$$

$$\theta = \sin^{-1}\left(\frac{\sin \phi}{\sqrt{\epsilon}}\right) \tag{3}$$

where the h and v subscripts refer to horizontal and vertical polarization states. The emissivity from a rough surface has not been rigorously modeled, so an approximate behavior is often used (Hagfors 1970; England 1975; Ulaby et al. 1982). As the surface becomes rougher at the scale of the radar wavelength, the emissivity is assumed to move toward the average of the plane-surface transmission coefficients for the two polarizations (Ulaby et al. 1982; Campbell 1994). At lower dielectric constants or smaller values of ϕ the range of emissivity between smooth and rough surfaces decreases (chapter by Pettengill et al., Fig. 1). For emission angles of less than ∼30°, the difference in emissivity between a perfectly smooth and rough surface is negligible, so we can use the Fresnel expressions given above to solve for the dielectric constant (the "plane surface" assumption) from E_h or E_v. Volume scattering and emission processes are neglected in this approach.

A method for refining the dielectric constant estimate in rough areas was proposed by Campbell (1994) from analysis of the global correlation between SAR HH backscatter and H-polarized emissivity. For areas below ∼6054 km in radius, these parameters are systematically correlated, implying (1) a random distribution of dielectric constant with texture, and (2) a mutual increase in backscatter and emissivity with wavelength-scale roughness. By assuming that the emissivity varied between the smooth- and rough-surface values as the backscatter increased, an empirical model for roughness and dielectric constant was derived. This technique appears to suppress roughness-related changes in emissivity across the tesserae and other rough areas, but the results are dependent upon the model assumptions.

Another method for dielectric constant estimation involves use of dual-polarization emissivity data. Magellan collected global emission measurements for Venus in the horizontally polarized sense, but vertically polarized values for emissivity are also available for a few hundred orbits. The average emissivity can be shown to be less sensitive to roughness than either individual component; and Arvidson et al. (1994) used this behavior to map dielectric variations across a swath of Ovda Regio. Nearly all approaches to dielectric constant estimation assume a single-dielectric interface and a simple separation of the surface into plane and rough components. Where mixed dielectric values occur, such as for the case of partial mantling by fine soil, the derived dielectric constant will reflect an average of the two components

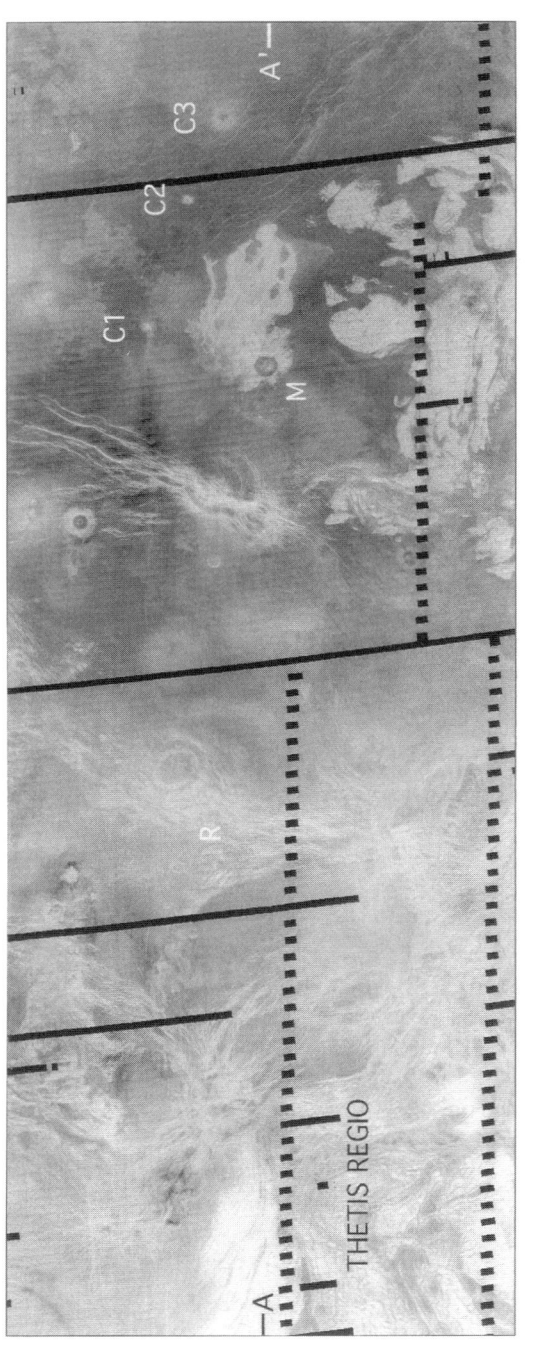

Figure 4a. Radar backscatter image for the region between Thetis Regio and Markham crater (indicated by an M). Area covered is 10°S–2°N, 132–164°E. The three smaller impact features discussed in the text are noted by C1, C2, and C3 designations. Note the difference in radar brightness between tessera regions at lower elevations, such as those south of Markham, and plateaus in Thetis Regio which exceed 6054 km in radius.

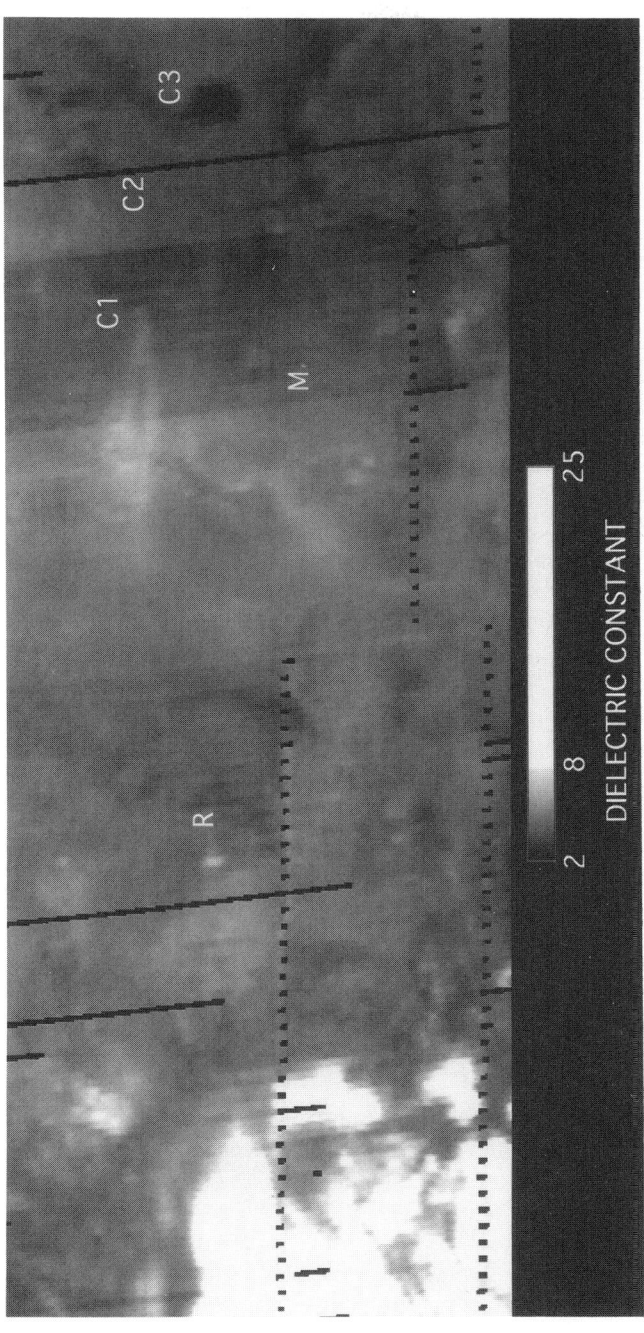

Figure 4b. Dielectric constant map for the region between Thetis Region and Markham crater (indicated by an M).

weighted by their areal distribution. The effect of such partial mantling on the radar backscatter and emission will be considerably different, so care is required when making roughness estimates for regions of potential soil cover.

Despite these caveats, the methods decribed above can yield reasonable estimates of the dielectric properties of the Venus surface. To illustrate the range of dielectric properties found on Venus, we selected an area that includes eastern Thetis Regio and Markham crater (Fig. 4). Dielectric constants were derived from the global map of Campbell (1994). The topography, radar backscatter, emissivity, and model-derived dielectric constant values for an east–west profile (A-A′) through Markham are shown on Fig. 5.

The plains, typified by the area between a small ridge belt (R on Fig. 4) and Markham crater, are characterized by gently undulating topography, moderate to low backscatter, and a narrow range of dielectric values ($\epsilon = 3$–5). Across Venus, dielectric constants for the plains tend to stay within the range from 3 to 6, consistent with values found for low-moderate density terrestrial basalts (Ulaby et al. 1988). Variations in dielectric constant among plains-forming units may be due to minor variations in oxide mineral content (e.g., ilmenite) or to differing amounts of surficial fine material. Changes in the proportions of minerals such as ilmenite have been shown to have a significant effect on the dielectric constant of lunar basalts (partly through the increased bulk density of the rock) (Carrier et al. 1991), while fine superposed material will attenuate the incident energy in a manner which varies with the loss properties of the soil and its depth. Lava flows from volcanic edifices display a slightly larger range in dielectric constant, with values reaching 8 to 9. The highest dielectric flows typically appear to be younger than other deposits based on superposition, again suggesting that over time these flows experience chemical weathering (i.e., oxidation of metallic minerals) or soil emplacement due to mechanical weathering or impact ejecta deposition. The ridge belts and tessera terrains within the lowlands (<6054 km radius) have no strong dielectric differences from the plains, indicating that these areas are probably deformed basaltic material. The large radar echoes from the ridge belts along the profile arise from a combination of folding (which lowers the local incidence angle) and meter-scale roughening due to this deformation.

Distal impact crater ejecta deposits on Venus have a marked dichotomy in their dielectric properties. Some fine-grained layers have ϵ values of 2 to 3, consistent with low-density rock powders. These deposits mantle the underlying terrain and subdue the backscattered return. In contrast, the parabolic ejecta from apparently young craters (Campbell et al. 1992) is typically too thin to have a significant effect on the radar echo, but can have dielectric constant values of 6 to 8 (early observations of these properties were made by Jurgens et al. 1988). This implies that the fine debris has a significant component of metallic minerals, possibly formed during the impact melting process. Conversely, many of the large circular radar-dark haloes around craters have no associated emissivity anomaly.

A number of impact-related features occur in the area shown in Fig. 4,

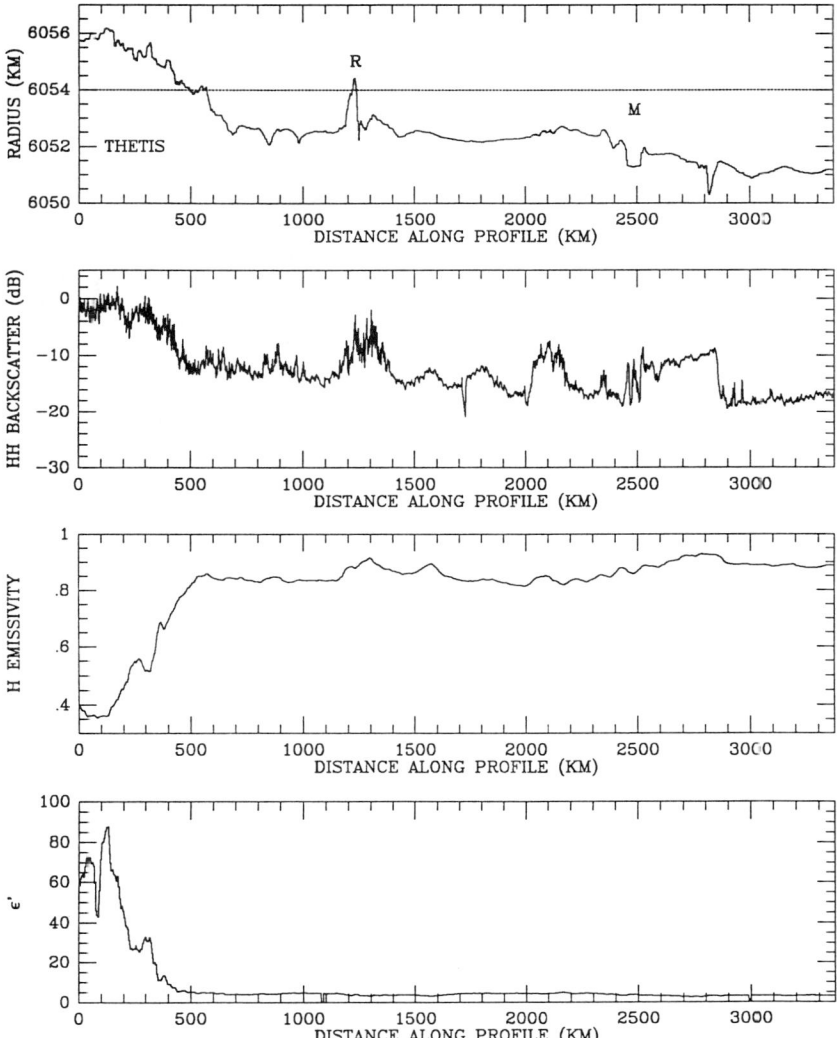

Figure 5. Average values for topography, backscatter cross section, H-polarized emissivity, and model-derived dielectric constant (from Campbell 1994) for a 25-km wide profile (A-A' in Fig. 4) through Thetis Regio and Markham crater (noted by an M). A small ridge belt west of Markham is indicated by an R. Note the rapid rise in dielectric constant (reduction in emissivity) and backscatter cross section for areas above 6054 km in radius (shown by dotted line). The abrupt drop in topography at the distal margin of the outflow may be a spurious value, based on poor near-nadir echoes from the very rough surface.

and demonstrate the degree to which such deposits can vary in their properties. Figure 6 presents a closer view of some of these impacts deposits. Markham (indicated by an M on Figs. 4 and 6) is a 60-km diameter crater with a partial peak ring complex, low-return floor materials, and an east-trending outflow deposit which formed contemporaneously with the ejecta blanket (Johnson and Baker 1994; Schultz 1992). None of the units associated with Markham can be distiguished clearly from the plains on the basis of their dielectric constant, indicating that any mineralogical or porosity changes (i.e., fine-grained mantling deposits) related to the impact event have been eliminated by aeolian or chemical weathering. The high radar returns from the Markham crater outflow deposit thus indicate a rough surface texture.

Figure 6 also illustrates flow-like features northeast of Markham which have relatively subdued radar signatures relative to the younger crater outflow deposit. In many plains areas, a similar pattern can be observed, with stratigraphically older units exhibiting lower or less varied radar backscatter than apparently younger lobate deposits. Such behaviors may imply that: (1) plains areas with few discernible lobate features are older and weathering has reduced the meter-scale roughness of the surface, or (2) plains may form by a variety of eruption styles, some of which produce discernible lobate forms and some which lead to relatively featureless smooth surfaces. Use of the degree of lobate texture in the plains to infer relative age is thus ambiguous unless supported by clear stratigraphic relationships.

North of Markham is a 4-km crater (C1 on Fig. 4) with a westward-pointing parabolic deposit whose bulk dielectric constant reaches values of ~ 7. While such values can be satisfied by relatively dense terrestrial basalts, the parabolic deposits are assumed to be fine-grained material that thinly mantles the plains (Campbell et al. 1992; Schultz 1992). Such porous deposits would typically have lower permittivity than the solid parent rock, and the increase in dielectric constant may be due to reduction of metallic minerals such as iron oxides during the impact melting and recondensation process (Campbell 1994; Brackett 1995). Reduced materials might have a limited lifespan at the Venus surface, but their inclusion within glasses may slow the re-oxidation process. The radar backscatter of the parabola is lower only in the distal reaches of the deposit, indicating a possible thickening of the material (or greater smoothing due to finer grain sizes) to the west. An 11-km crater (marked C2 on Fig. 4) northeast of Markham lacks any emissivity signature. This emphasizes the ephemeral nature of the fine-grained parabola-forming material, which has evidently been removed from this crater by aeolian processes (Izenberg et al. 1994). Another impact feature east of Markham (C3 on Fig. 4) has a circular annulus of apparently low dielectric material ($\epsilon = 2$), but this is an artifact of the solar specular reflection observed during the Magellan mapping Cycle 1 (Pettengill et al. 1992).

IV. HIGHLAND SURFACES

Areas at elevations above ~ 6054 km radius on Venus typically have high backscatter cross section, high microwave Fresnel reflectivities, and low emis-

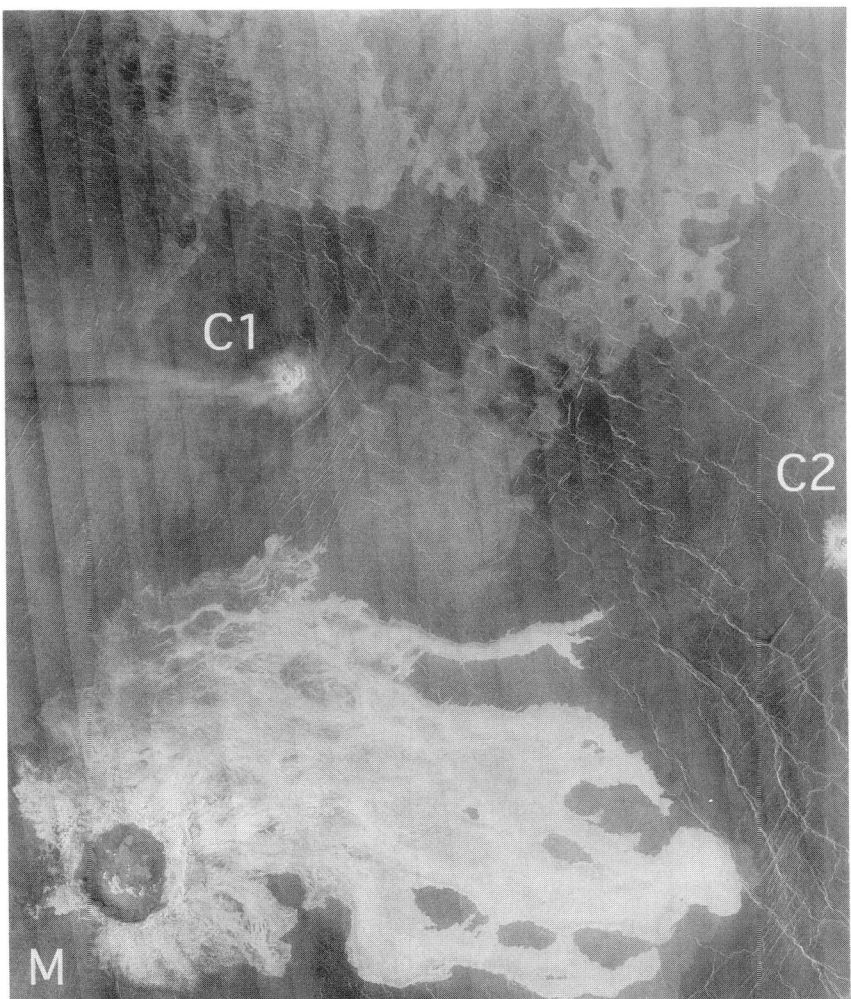

Figure 5. Closeup view of Markham crater (60-km diameter) and nearby impact features discussed in text. Designations correspond to those on Fig. 4.

sivity (Pettengill et al. 1982, 1988, 1992). At the highest elevations (greater than about 6058 km radius), emissivity shifts back toward values consistent with bare rock (Klose et al. 1992; Arvidson et al. 1994). The progressive decrease in emissivity (increasing dielectric constant) is directly correlated with increasing elevation, and this behavior is seen across the planet, with dielectric values reaching 250 or more in the Maxwell Montes (chapter by Pettengill et al.). The dotted line in Fig. 5 indicates the approximate elevation at which the dielectric enhancement occurs for Thetis Regio; note that model-derived values for ϵ of up to 90 occur in this region. The estimation of dielectric constant for highland areas based on the models discussed above is still tentative, because the effect of enhanced multiple scattering on backscatter and emission from such highly reflective terrains is poorly understood.

Some areas above the nominal elevation for dielectric enhancement (~6054 km) exhibit little or no such change in their backscatter and emission properties. Most notable is the summit of Maat Mons (1.5°N, 194°E), which reaches a maximum height of 9 km above the datum but has a model-derived dielectric constant of 6 to 8 (Klose et al. 1992; Campbell 1994). The lack of a dielectric enhancement for the summit of Maat has been interpreted to indicate a young age for the surface lava flows, based on the assumption that the highland chemical changes occur ubiquitously and on a short time scale (Klose et al. 1992; Robinson and Wood 1993; Robinson et al. 1995). While a youthful surface near the summit of Maat Mons and in other low-ϵ highland areas is plausible, the current lack of a candidate mineralogy and reaction sequence for the weathering process argues for caution. Alternative hypotheses include the inability of the loading phases or cold-trapped materials to occur on lavas of certain chemistries, or the presence of fine-grained ash layers at the surface which depress the apparent dielectric constant.

The enhanced backscatter cross section in the highlands has been interpreted as due to the increasing Fresnel reflectivity of the surface materials, but surface roughness also plays a role, with the tessera areas generally having a higher degree of rocky texture than the plains. Global analyses show that, to first order, the value of the dielectric constant has little relation to the type of landform (smooth intra-tessera plains, tessera, or ridges), indicating that the emissivity change occurs due to a ubiquitous surficial coating or mineralogic change (Klose et al. 1992; Arvidson et al. 1992,1994; Campbell 1994). Whether this chemical process influences the surface roughness remains an open issue.

A variety of possible scattering mechanisms and dielectric constant variations could account for the observed highland properties, but only a few are consistent with the geologic evidence and the analysis of bistatic radar echoes (chapter by Pettengill et al.). Tryka and Muhleman (1992) proposed that high dielectric scatterers imbedded in a low-loss soil could permit efficient multiple scattering, while Pettengill et al. (1992) proposed that wavelength scale voids in a basalt matrix could also account for the observations. Recent bistatic observations of select highland regions (chapter by Pettengill et al.) indicate the existence of a Brewster angle, i.e., a bistatic angle of observation at which vertically polarized scattered radiation disappears. This phenomenon is associated exclusively with surface scattering. Furthermore, the value of the Brewster angle is indicative of the dielectric constant of the surface, which is inferred to be on the order of 100 or more, consistent with values extracted from emissivity observations if simple Fresnel emission is assumed. Volume scattering models also require a background medium (i.e., the soil) which is nearly transparent to the radar, whose ubiquitous presence in the Venus highlands would be problematic (Wilt 1993). All current evidence thus points to a surface scattering mechanism for the highlands.

There have been several surface-scattering models proposed to explain the high radar reflectivity of the Venus uplands. The earliest was the loaded dielectric model, which proposed that iron sulfide or oxide comes into thermochemical equilibrium only in the highlands because of the strong altitude-dependence of pressure and temperature that exist on Venus (Pettengill et al. 1982; Klose et al. 1992). The presence of a conducting phase interspersed within a rock-like matrix would give rise to a very high effective dielectric constant and therefore a low observed emissivity. Recent experimental kinetic work suggests that pyrite (FeS_2) is not a viable candidate for a loaded dielectric phase because it is thermodynamically unstable over the entire range of Venusian surface conditions (Fegley et al. 1993a,b,1995; Klingelhofer et al. 1994). Any candidate phase must be (1) highly conductive, (2) present in significant quantities (>10% by volume), and (3) come into thermochemical equilibrium between elevations of ~6054 and 6058 km.

Fegley et al. (1992) suggested that perovskite ($CaTiO_3$), a thermodynamically unstable mineral phase under all Venusian surface conditions, might be a viable candidate material for the loading phase. Perovskite is known to occur in terrestrial alkaline basalts, and dielectric constant measurements indicate values of 1100 to 2200 at frequencies of 10 kHz (Timco 1977). The decomposition of perovskite is likely to be governed by an Arrhenius relationship, with the rate of decomposition proportional to temperature. The gradual decrease in emissivity with elevation can thus be explained by the temperature dependence of decomposition kinetics. The return to normal emissivities at the highest elevations is attributed to an abrupt phase transition to rutile (TiO_2) and flourite (CaF_2) caused by reaction with atmospheric HF (Fegley et al. 1992). Further experimental work is needed to assess the thermodynamic stability of perovskite, the kinetics of perovskite decomposition, and the dielectric constants of perovskite under Venusian conditions.

Shepard et al. (1994) proposed that the high dielectric constants observed in the highlands are due to the presence of ferroelectric minerals (many minerals in the perovskite family are ferroelectric). Ferroelectric minerals have unusual dielectric properties; below a compositionally dependent Curie temperature, the dielectric constants are in the range of 10 to 100. However, at the Curie temperature, the dielectric constants dramatically increase, often to values as high as 100,000 (Burfoot 1967). Mixing models for two-component mixtures of dielectrics indicate that as little as 0.1% by volume of a ferroelectric mineral phase can explain the lowest emissivities observed on Venus (Shepard et al. 1994). This model explains the return to normal emissivities as the altitude at which the surface temperature drops below the Curie point. The primary weakness of the ferroelectric model is that no single phase has yet been identified as a strong potential candidate; in fact, most ferroelectric phases are considered exotic by terrestrial standards.

Brackett et al. (1995) proposed that high-dielectric metal halides and chalcogenides are deposited in the highlands and are responsible for the observed low emissivities and high backscatter cross sections. They proposed that the high vapor pressures of metal halides and chalcogenides on Venus, coupled with the altitude dependent temperature gradient, lead to the diffusive transport of highly volatile metal phases from the hot lowlands to the (relatively) cold highlands. Deposition rates are highly variable (up to 1 μm yr^{-1}) and depend primarily upon the volatility of the phase. A sufficiently thick layer of these materials (i.e. mm to cm) deposited on existing surfaces could produce emissivities and backscatter cross sections similar to those observed. The bistatic observations of Pettengill et al. (see their chapter) indicate that the material responsible for the high radar reflectivities may be confined to a thin surface layer, and thus strongly favor this model. The increasing abundance of metal phases with decreasing temperature (i.e., increasing elevation) is consistent with the observed altitude dependent behavior of emissivity and backscatter cross section. However, the trend back toward normal emissivity at the highest elevations is more difficult to explain, and a specific phase has yet to be identified as a strong potential candidate.

To summarize, the observations to date strongly suggest that the low emissivities and high radar backscatter cross sections in the highlands are caused by a surface scattering and emission phenomenon. Of the four models currently proposed, each has strengths and weaknesses. The loaded dielectric model suffers due to the lack of a proposed thermodynamically stable conductive phase and the large quantity of conductive material required to produce the observed emissivities. The perovskite and ferroelectric models are related and may be considered together. Ferroelectric phases explain the observed trends in emissivity with elevation, and the return to normal emissivities at the highest elevations, very simply. However, the geochemistry of appropriate ferroelectric and/or perovskite phases has yet to be adequately explained. Perhaps the strongest of the hypotheses at this point is the metal phase deposition model, which is plausible given the current understanding of Venusian atmospheric chemistry. At least one set of observations from bistatic experiments seems to favor this model over the others. However, a specific phase has yet to be identified as a strong candidate and much geochemical work remains before this can occur. Inference of possible relative ages for geologic units in the highlands based solely on their emissivity is speculative at present, because the role of original rock chemistry in the development of loaded dielectric phases, ferroelectric minerals, or the growth of a metal phase by cold trapping is unknown.

V. SUMMARY AND FUTURE DIRECTIONS

The lowland plains of Venus, which make up more than 80% of the surface area of the planet, are characterized by primary volcanic, impact crater, and tectonic landforms which are only very slowly modified by in situ weathering. Weathering rates on Venus are orders of magnitude slower than those on Earth, and are comparable to erosional processes on the lunar surface. If Venus was resurfaced on the order of 500 Myr ago, then the various weathering effects must operate at less than $\sim 10^{-3}$ μm yr^{-1} to retain the pristine appearance of so many landforms. This estimate is supported by the distribution of fine-grained parabolic deposits from impact craters, which survive for several tens of Myr despite a very shallow depth. Resurfacing of the planet within the last 500 Myr and slow weathering processes have prevented the development of a lunar-like regolith or the fine dust which dominates the Martian surface.

In contrast to the other terrestrial planets, Venus presents a relatively straightforward view of how different features were created, and successive periods of tectonic or volcanic overprinting can often be identified (see, e.g., Solomon et al. 1992; Senske et al. 1992; Campbell and Rogers 1994). Lowland surface dielectric constants of 3 to 8, the absence of widespread wind-related features not associated with craters, and the lack of observed sediment mass movements, argue that fine soil does not develop rapidly under Venus conditions, and that the surface is composed to a much larger degree by bare rock and coarse fragmental debris than that of the Moon, Mercury, or Mars. Fine material is created during impact cratering (as part of the crater excavation and as a consequence of atmospheric blast waves), and is redistributed across the planet by upper-level winds during the ejection process and by global or crater-related local winds after deposition. This source likely constitutes the bulk of the very fine soil component on the lowlands surface. The radar properties of the highlands can be explained by surface scattering from a high-dielectric interface, but identification of a specific phase which produces this effect within the required narrow band of elevation remains a problem.

The Magellan mission provided a broad framework for understanding geologic processes on Venus. A number of important issues remain, however, which cannot be addressed solely with current data: (1) What is the total budget of sediment on the surface of Venus? (2) What is the role of chemical weathering, in terms of cementing and lithification, in the preservation of Venusian landforms? (3) What is the specific material which causes the high surface dielectric constant in the uplands? Does the formation of this material alter the surface texture in any way? (4) What is the nature of the dielectric enhancement associated with parabolic crater ejecta deposits and some crater floors? (5) How does the presence of the dense atmosphere affect the impact process, and to what extent can atmospheric blast waves comminute the upper surface? (6) What information, if any, on the surface today records the past climate and activity of the planet? Future Earth-based observations of Venus may answer some of these questions, but further understanding of most problems will come only from new lander (or near-surface) missions that focus on the geology, chemistry, and mineralogy of the surface materials.

Acknowledgments. The authors thank an anonymous reviewer for helpful comments. This work was supported in part by a grant from the NASA Venus Data Analysis Program.

REFERENCES

Arvidson, R. E., et al. 1992. Surface modification of Venus as inferred from Magellan observations of plains. *J. Geophys. Res.* 97:13303–13318.

Arvidson, R. E., et al. 1994. Microwave signatures and surface properties of Ovda Regio and surroundings, Venus. *Icarus* 112:171–196.

Basilevsky, A. T., Nikolaeva, O. V., and Weitz, C. M. 1992. Geology of the Venera 8 landing site region from Magellan data: Morphological and geochemical considerations. *J. Geophys. Res.* 97:16315–16336.

Brackett, R. A., Fegley, B., Jr., and Arvidson, R. E. 1995. Volatile transport on Venus and implications for surface geochemistry and geology. *J. Geophys. Res.* 100:1553–1563.

Burfoot, J. C. 1967. *Ferroelectrics: An Introduction to the Physical Principles* (London: Van Nostrand).

Burke, K., Fegley, B., and Sharpton, V. 1994. Are steep slopes on Venus preserved as a result of chemical cementation of pore-space in surface rocks? *Lunar Planet. Sci. Conf.* XXV:201–202 (abstract).

Campbell, B. A. 1994. Merging Magellan emissivity and SAR data for analysis of Venus surface dielectric properties. *Icarus* 112:187–203.

Campbell, B. A., and Campbell, D. B. 1992. Analysis of volcanic surface morphology on Venus from comparison of Arecibo, Magellan, and terrestrial airborne radar data. *J. Geophys. Res.* 97:16293–16314.

Campbell, B. A., and Rogers, P. G. 1994. Bell Regio, Venus: Integration of remote sensing data and terrestrial analogs for geologic analysis. *J. Geophys. Res.* 99:21153–21171.

Campbell, B. A., Arvidson, R. E., and Shepard, M. K. 1993. Radar polarization properties of volcanic and playa surfaces: Applications to terrestrial remote sensing and Venus data interpretation. *J. Geophys. Res.* 98:17099–17113.

Campbell, D. B., et al. 1992. Magellan observations of extended impact crater related features on the surface of Venus. *J. Geophys. Res.* 97:16249–16278.

Carrier, W. D., Olhoeft, G. R., and Mendell, W. 1991. Physical properties of the lunar surface. In *Lunar Sourcebook: A User's Guide to the Moon*, eds. G. Heiken, D. Vaniman and B. M. French (Cambridge: Cambridge Univ. Press).

England, A. W. 1975. Thermal microwave emission from a scattering layer. *J. Geophys. Res.* 80:4484–4496.

Fegley, B., Jr., and Prinn, R. G. 1989. Estimation of the rate of volcanism on Venus from reaction rate measurements. *Nature* 337:55–58.

Fegley, B., Jr., and Treiman, A. H. 1992. Chemistry of atmosphere-surface interactions on Venus and Mars. In *Comparative Studies of Venus and Mars: Atmospheres, Ionospheres, and Solar Wind Interactions* (Washington, D. C.: American Geophysical Union), pp. 7–71.

Fegley, B., Jr., Lodders, K., and Klingelhofer, G. 1993*a*. Kinetics and mechanism of pyrite decomposition on the surface of Venus. *Bull. Amer. Astron. Soc.* 25:1094 (abstract).

Fegley, B., Jr., Treiman, A. H., and Sharpton, V. L. 1993b. Venus surface mineralogy: Observational. and theoretical constraints. *Proc. Lunar Planet. Sci. Conf.* 22:3–19.
Fegley, B., Jr., Lodders, K., Treiman, A. H., and Klingelhöfer, G. 1995. The rate of pyrite decomposition on the surface of Venus. *Icarus* 115:159–180.
Ford, P. G., and Pettengill, G. H. 1992. Venus topography and kilometer-scale slopes. *J. Geophys. Res.* 97:13102–13114.
Gaddis, L. R. 1992. Lava flow characterization at Pisgah volcanic field, California, with multi-parameter imaging radar. *Geol. Soc. Amer. Bull.* 104:695–703.
Garvin, J. B. 1990. The global budget of impact-derived sediments on Venus. *Earth, Moon, Planets* 50/51:175–190.
Garvin, J. B., Head, J. W., and Wilson, L. 1982. Magma vesiculation and pyroclastic volcanism on Venus. *Icarus* 52:365–372.
Garvin, J. B, Head, J. W., Zuber, M. T., and Helfenstein, P. 1984. Venus: The nature of the surface from Venera panoramas. *J. Geophys. Res.* 89:3381–3399.
Greeley, R., et al. 1992. Aeolian features on Venus: Preliminary Magellan results. *J. Geophys. Res.* 97:13319–13345.
Hagfors, T. 1970. Remote probing of the moon by infrared and microwave emissions and by radar. *Radio Sci.* 5:189–227.
Horz, F., et al. 1991. Lunar surface processes. In *Lunar Sourcebook: A User's Guide to the Moon*, eds. G. Heiken, D. Vaniman and B. M. French (Cambridge: Cambridge Univ. Press).
Ivanov, B. A., Basilevsky, A. T., Kryuchkov, V. P., and Chernaya, I. M. 1986. Impact craters of Venus: Analysis of Venera 15/16 data. *Proc. Lunar Planet. Sci. Conf.* 16, *J. Geophys. Res. Supp.* 91:413–430.
Izenberg, N. R., Arvidson, R. E., and Phillips, R. J. 1994. Impact crater degradation on the venusian plains. *Geophys. Res. Lett.* 21:289–292.
Johnson, J. R., and Baker, V. R. 1994. Surface property variations in venusian fluidized ejecta blanket craters. *Icarus* 110:33–70.
Jurgens, R. F., Slade, M. A., and Saunders, R. S. 1988. Evidence for highly reflecting material on the surface and subsurface of Venus. *Science* 240:1021–1023.
Klingelhofer, G., Fegley, B., and Lodders, K. 1994. ^{57}Fe Mossbauer studies of the kinetics of pyrite decomposition on the surface of Venus. *Lunar Planet. Sci. Conf.* XXV:707–708 (abstract).
Klose, K. B., Wood, J. A., and Hashimoto, A. 1992. Mineral equilibria and the high radar reflectivity of Venus mountaintops. *J. Geophys. Res.* 97:16353–15369.
Malin. M. C. 1992. Mass movements on Venus: Preliminary results from Magellan cycle 1 observations. *J. Geophys. Res.* 97:16337–16352.
McGetchin, T. R., Settle, M., and Head, J. W. 1973. Radial thickness variation in impact crater ejecta: Implications for lunar basin deposits. *Earth Planet. Sci. Lett.* 20:226–236.
McGill, G. E., et al. 1983. Topography, surface properties, and tectonic evolution. In *Venus*, eds. D. M. Hunten, L. Colin, T. M. Donahue and V. I. Moroz (Tucson: Univ. of Arizona Press), pp. 69–130.
Melosh, H. J., and Schaller, C. J. 1996 The abundance of fine-grained impact ejecta on Venus. *Lunar Planet. Sci. Conf.* XXVII:861–862 (abstract).
Mueller, R. F. 1963. Chemistry and petrology of Venus: Preliminary deductions. *Science* 141:1046–1047.
Oberbeck, V. R. 1975. The role of ballistic erosion and sedimentation in lunar stratigraphy. *Revs. Geophys.* 13:337–362.
Pettengill, G. H., Ford, P. G., and Nozette, S. 1982. Venus: Global surface radar reflectivity. *Science* 217:640–642.
Pettengill, G. H., Ford, P. G., and Chapman, B. D. 1988. Venus: Surface electromag-

netic properties. *J. Geophys. Res.* 93:14881–14892.
Pettengill, G. H., Ford, P. G., and Wilt, R. J. 1992. Venus surface radiothermal emission as observed by Magellan. *J. Geophys. Res.* 97:13091–13102.
Phillips, R. J., et al. 1991. Impact craters on Venus: Initial results from Magellan. *Science* 252:299–297.
Phillips, R. J., et al. 1992. Impact craters and resurfacing history. *J. Geophys. Res.* 97:15923–15948.
Plaut, J. J. 1991. Radar Scattering as a Source of Geological Information. Ph.D. Thesis, Washington University.
Robinson, C. R., and Wood, J. A. 1993. Recent volcanic activity on Venus: Evidence from radiothermal emissivity measurements. *Icarus* 102:26–39
Robinson, C. R., Thornhill, G. D., and Parfill, E. A. 1995. Large-scale volcanic activity at Maat Mons: Can this explain fluctuations in atmospheric chemistry observed by Pioneer-Venus? *J. Geophys. Res.* 100:11755–11764
Schaber, G. G., et al. 1992. Geology and distribution of impact craters on Venus: What are they telling us? *J. Geophys. Res.* 97:13257–13302.
Schultz, P. H. 1992. Atmospheric effects of ejecta emplacement and crater formation on Venus from Magellan. *J. Geophys. Res.* 97:16183–16248.
Senske, D. A., Schaber, G. G., and Stofan, E. R. 1992. Regional topographic rises on Venus: Geology of Western Eistla Regio and comparison to Beta Regio and Atla Regio. *J. Geophys. Res.* 97:13395–13420.
Shepard, M. K., Arvidson, R. E., Fegley, B., Jr., and Brackett, R. A. 1994. A ferroelectric model for the low emissivity of highlands on Venus. *Geophys. Res. Lett.* 21:469–472.
Solomon, S. C., et al. 1992. Venus tectonics: An overview of Magellan observations. *J. Geophys. Res.* 97:131399–13255.
Stratton, J. A. 1947. *Electromagnetic Theory* (New York: Wiley).
Surkov, Y. A. 1983. Studies of Venus rocks by Veneras 8, 9, and 10. In *Venus*, eds. D. M. Hunten, L. Colin, T. M. Donahue and V. I. Moroz (Tucson: Univ. of Arizona Press), pp. 154–158.
Timco, G. W. 1977. High Pressure Dielectric Properties of Perovskite Ferroelectrics. Ph.D. Thesis, Univ. of Western Ontario.
Tryka, K. A., and Muhleman, D. O. 1992. Reflection and emission properties on Venus: Alpha Regio. *J. Geophys. Res.* 97:13379–13394.
Tyler, G. L., Simpson, R. A., Maurer, M. J., and Holmann, E. 1992. Scattering properties of the Venusian surface: Preliminary results from Magellan. *J. Geophys. Res.* 97:13115–13139.
Ulaby, F. T., Moore, R. K., and Fung, A. K. 1982. *Microwave Remote Sensing: Active and Passive* (Reading, Mass.: Addison-Wesley).
Ulaby, F. T., et al. 1988. Microwave Dielectric Spectrum of Rocks. Rept. 23817-1-T (Ann Arbor: Univ. of Michigan Radiation Lab.).
Weitz, C. M., and Basilevsky, A. T. 1993. Magellan observations of the Venera and Vega landing site regions. *J. Geophys. Res.* 98:17069–17098.
Wilt, R. J. 1993. A Study of Low Radiothermal Emissivity on Venus. Ph.D. Thesis, Massachusetts Inst. of Technology.

SURFACE SCATTERING AND DIELECTRIC PROPERTIES

GORDON H. PETTENGILL
Massachusetts Institute of Technology

BRUCE A. CAMPBELL
National Air and Space Museum

DONALD B. CAMPBELL
Cornell University

and

RICHARD A. SIMPSON
Stanford University

Radio and radar observations of Venus, made over several decades, have determined that the typical surface at elevations below about 6054-km radius has a bulk dielectric constant lying between 4.0 and 4.5. This value is consistent with dry compacted rocks of various compositions, presumably similar to those found on Earth and seen on the other terrestrial planets. Occasional lower values probably arise from an overlying layer of dusty or granular material. The typical surface is quite smooth at radio wavelengths, exhibiting 10 to 20% depolarization in its total scattered energy, and a root-mean-square average surface slope of 2° to 4° on scales of decimeters to meters. Regions of much higher roughness are found, however. Terrain lying above a planetary radius of about 6054 km often displays surface electrical properties vastly different from those found at lower elevations. A recent bistatic radar observation over the Maxwell Montes region finds a complex dielectric constant of $-i\,100$, that can be interpreted as a surface conductivity of 13 mhos m^{-1}. Candidates for surface materials include ferro-electrics, a plating of magnetite or a thin layer of tellurium frost.

I. INTRODUCTION

The massive, visually opaque, Venus atmosphere limits global remote sensing of a planet's surface to radio wavelengths that can penetrate the atmosphere with acceptable transmission loss (wavelengths of about 3 cm or longer). Both "active" (radar) observations and "passive" (thermal radioemission) measurements yield the discontinuity in index of refraction at the surface-atmosphere interface, a discontinuity which is primarily controlled by the bulk dielectric permittivity of the surface. While admittedly not the optimum way to discriminate among different types of minerals, measurements of the electromagnetic properties of the Venus surface nevertheless place useful constraints on the types and density of surface constituents. They presently represent the only surface properties accessible to direct measurement from above the thick Venus atmosphere, and join a small number of *in-situ* lander results.

There are several ways that radio waves can give information on the electromagnetic properties of a surface: (1) through the intensity of radar echoes; (2) through the polarization changes introduced by the scattering; (3) through the thermal emission intensity from a surface at a known physical temperature; and (4) through the polarization of that emission. The first, second and fourth of these methods require a surface having sufficient coherence (smoothness) for the Fresnel formulas for reflection and emission to hold, although corrections can often be made if the surface is not entirely rough. The third method is the least sensitive to surface roughness, requiring corrections for roughness only in second order, if averaged over polarization. Because each of the methods has a different dependence on surface structure, and generally suffers from very different systematic errors, more credible results are usually obtained when two or more of the methods are applied to the same terrain, and the results compared.

The first measurements of the radar reflectivity of the (unresolved) disk of Venus took place in 1961, from Earth (Victor and Stevens 1961; Pettengill et al. 1962). Since then, mostly at times of inferior conjunction when the planet is closest to Earth and most easily observed, measurements have been made at wavelengths covering the interval from 7.84 m to 3 cm (see Pettengill [1978] for a summary). Because of limitations imposed on the Earth-based observing geometry by the small obliquity of Venus, echoes at normal incidence could be obtained only near the Venus equator. At wavelengths longer than 15 cm, a value of about 0.15 for the disk-averaged reflectivity was obtained. Beginning in 1978, and extending to March, 1981, the high-inclination Pioneer Venus Orbiter's radar altimeter carried out vertical incidence measurements of the planet's surface reflectivity (Pettengill et al. 1982,1988), yielding a near global distribution of results that were free of the geometric limitations imposed on Earth-based data.

Most observations of the brightness temperature of Venus prior to the arrival of Pioneer Venus were limited to whole-disk averages and were made near inferior conjunction, where the angular size of the disk is largest (see Pettengill et al. [1988] for a summary); the most accurate disk-averaged result for surface brightness temperature emerging from these observations was 636 ± 25 K. Assuming a surface physical temperature of 735 ± 5 K (Marov 1978) leads to a corresponding disk-averaged emissivity of 0.87 ± 0.04, a value in excellent agreement with a mean observed disk-averaged radar reflection coefficient of 0.13 ± 0.03 (Pettengill 1978).

Pettengill et al. (1982) were the first to report (based on Pioneer Venus data) the existence of regions on the surface of Venus that exhibited unexpectedly high values of radar reflectivity (up to 0.4). Most of these regions were located at relatively high altitudes. Shortly afterwards, Ford and Pettengill

(1983) reported low values of surface emissivity (as low as 0.54), also obtained from the Pioneer Venus radar instrument, which were associated with the same regions that exhibited the anomalously enhanced radar reflectivities reported earlier. Additional observations of Venus, in which the distribution of brightness temperature could be resolved over surface areas as small as 200 km were made in 1983 from Earth, using the Very Large Array (VLA) in New Mexico (Pettengill et al. 1988). The latter data confirmed the existence of regions having radio emissivities as low as 0.58; it was the discovery and confirmation of the unusual behavior of these regions that largely justified the incorporation of a radiometer in the Magellan radar mission sent to Venus in 1989.

The Magellan mission carried a very competent radar system (Pettengill et al. 1991) that allowed detailed, near-global measurements of both the radar reflectivity and radiothermal emission of the surface. It is these measurements that provide our best current knowledge of the electrical properties of Venus, and that form the bulk of the data discussed in this review. The major shortcoming of the Magellan radar system was its inability to determine the full polarization state of the received radar echo or radiothermal emission; Earth-based observations have been used where available to supply this (and other) useful additional information.

II. RADIO WAVES AND SURFACES

Backscattering from a single interface is often modeled as a combination of quasi-specular reflection, involving coherent processes on an undulating surface (Hagfors 1970) that may be described by the Fresnel equations (Stratton 1941), and a diffusely scattering component arising from rough surface structure of the order of a wavelength in size (Pettengill and Thompson 1968). The quasi-specular component is frequently approximated by the Hagfors scattering function, given by

$$\sigma_o^{QS}(\theta) = \frac{R_o C}{2(\cos^4 \theta + C \sin^2 \theta)^{3/2}} \quad (1a)$$

where $\sigma_o(\theta)$ is the specific (dimensionless) radar cross section observed at incidence angle, θ, $C = \alpha^{-2}$, α is the root mean square (rms) average value of the undulating surface slope, in radians, and R_o is the Fresnel power reflection coefficient at normal incidence. The diffuse component is often modeled as

$$\sigma_o^D(\theta) = B \cos^n \theta \quad (1b)$$

where B is a constant chosen to fit data at large incidence angles (where the diffuse component predominates), and n is usually a small number in the range: $1 \leq n \leq 3$. The combination of undulating surface and wavelength-sized roughness enables this parameterized model to be fitted to almost any observed angular behavior when the echo polarization has the expected sense for a quasi-specular surface. However, neither the Hagfors, nor any other quasi-specular, model can account for depolarized scattered power. Viewed in emission, this type of surface will emit preferentially in the vertical linearly polarized mode, as shown by Eq. (3).

The Fresnel equations (Stratton 1941) controlling the power reflectivity, $R = |\rho|^2$, of a smooth surface yield the complex reflection coefficient ρ as

$$\rho_h = \frac{a-b}{a+b}, \quad \text{and} \quad \rho_v = \frac{\varepsilon a - b}{\varepsilon a + b} \qquad (2)$$

where $a = \cos\theta$, and $b = \sqrt{\varepsilon - \sin^2\theta}$.

The corresponding radiothermal emission efficiency, usually called the emissivity, $e = 1 - R$ (for a smooth surface), may be written as

$$e_h = \frac{4a\Re(b)}{|a+b|^2}, \quad \text{and} \quad e_v = \left|\frac{4\varepsilon a\Re(b)}{(\varepsilon a + b)^2}\right| \qquad (3)$$

where $\Re(b)$ is the real part of b.

In these equations, θ is the angle of incidence or emergence and ε is the complex ratio of the relative dielectric permittivities (often called dielectric constants) characterizing the two sides of the interface. Contributions from a discontinuity in the magnetic permeability at the interface are assumed to be negligible. The subscripts h and v refer to radiation polarized perpendicularly ("horizontal" to the surface) or parallel ("vertical" to the surface), respectively, to the plane of scattering or emission. For reference, the relative dielectric permittivity of the carbon dioxide atmosphere just above the planet's surface ($P = 92$ atm; $T = 735$ K) is 1.034.

According to the quasi-specular model, radar backscatter is dominated by reflection from locally smooth facets, oriented at normal incidence to the illuminating radar beam (at any particular angle to the mean surface, of course, these facets normally comprise only a small fraction of the total illuminated area). For normal incidence

$$R_h = R_v = \left|\frac{1-\sqrt{\varepsilon}}{1+\sqrt{\varepsilon}}\right|^2 \qquad (4)$$

and the polarization of the specularly reflected component is preserved (for incident elliptically polarized radiation, the rotational sense is reversed by inversion of the direction of propagation).

The situation for emissivity is more complicated, because we are usually viewing at oblique angles to the surface. In general, emissivity depends on the angle of emission, as shown by Eq. (3), although for the range of values of ε encountered here that dependence is not significant at angles below 45°, when averaged over polarization (Fig. 1a). The emissivity of a rough surface is harder to calculate, although measurements reported by Ulaby et

SURFACE ELECTRICAL PROPERTIES

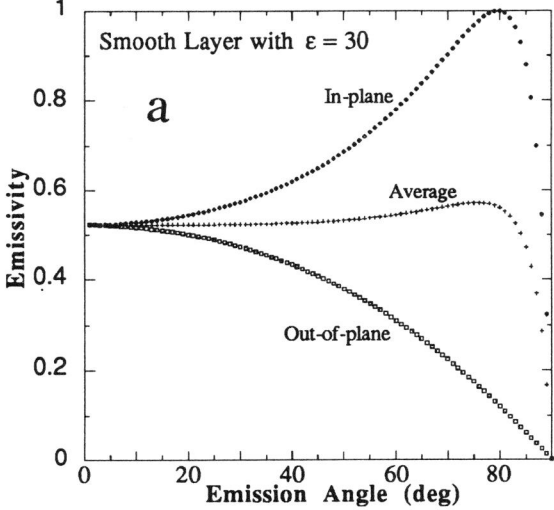

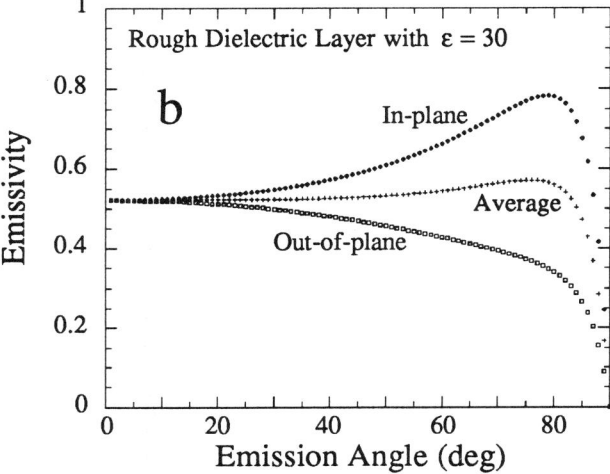

Figure 1. (a). Variation of emissivity with emission angle, for a smooth (Fresnel) interface having a dielectric discontinuity ratio of 30. The term "in-plane" refers to the linearly polarized (vertical) component lying in the scattering plane, normal to the interface; "out-of-plane" refers to the linearly polarized (horizontal) component normal to that plane. Note how little variation with angle is displayed by the average of these two components, at least below 70°. (b) Variation of emissivity for a partially roughened interface, which is otherwise similar to that shown in (a).

al. (1982) suggest that, for a given value of surface dielectric constant, the total polarization-averaged emissivity increases slowly as the surface becomes rough at scales smaller than the wavelength, still tending to remain constant over angles of emission less than about 45°. The degree of linear polarization is reduced by the roughening, of course, as the coherence of the surface is destroyed (Fig. 1b).

Detailed thermodynamic balance requires that, at a given wavelength and angle of emergence, a surface's emissivity be equal to its absorptivity for radiation incident at that angle. The absorptivity, in turn, must be the unit complement of the reflectivity (summed over all scattering directions) for power incident at that angle, in order to conserve energy. The observed emissivity can therefore be considered the complement of the reflectivity measured in a totally uncollimated outgoing geometry. This implies that the observed brightness temperature for Venus is lower than the physical surface temperature, because of a contribution from reflected cold sky integrated over the upper hemisphere and weighted by the appropriate bistatic surface scattering function. For a smooth, or smoothly undulating (quasi-specular) surface, most of the contribution will come from the direction corresponding to specular reflection. For a highly diffuse Lambert-law surface, the sky's contribution will derive from the entire upper hemisphere, weighted only by the cosine of the zenith angle.

Volume scattering results from successive interactions with one or more interfaces or inhomogeneities internal to the planetary surface, and is known to play an important role in radar scattering from the icy surfaces of the Galilean satellites, as well as from the ice deposits at the poles of Mars and Mercury. It is not thought to play a significant role on Venus, where ice is absent.

III. OBSERVATIONS

A. Monostatic Intensity Observations

The Magellan radar system provided three single-antenna (monostatic) data sets that can be used to extract information on surface roughness and dielectric constant; these were obtained by the synthetic aperture radar (SAR), the radar altimeter, and the microwave emission radiometer (Pettengill et al. 1991). All three were taken at 12.6-cm wavelength, with the SAR and radiometer looking to one side in a common polarization and viewing geometry. The altimeter collected vertical-incidence measurements of backscatter near the spacecraft's nadir. Each data set has particular sensitivities to roughness and dielectric properties, which can be exploited to create a more complete picture of the Venus surface.

The backscattered radar intensity and state of polarization vary with both the dielectric properties of the surface-atmosphere interface and the detailed structure of the surface. At the angles viewed by Magellan (20° to 45°), the SAR echoes depend primarily on diffuse scattering, typically associated with surface structure on the scale of the radar wavelength, but are also affected strongly by surface tilts on the scale of the image resolution. The Magellan SAR data are augmented for portions of the surface by Earth-based radar images collected at the Goldstone and Arecibo Observatories; these images contain data for both rotational senses of circular echo polarization and cover a range of incidence angles that nicely complement the Magellan observations.

The radar altimeter system has a much poorer linear spatial resolution (about 10 km) than that achieved either with the Magellan SAR (typically 120 m) or at Arecibo (1 to 2 km), but can collect measurements of the radar cross section within about 15° of normal incidence. As discussed above, we expect that large quasi-specular facets will dominate the radar return at such low angles of incidence. The scattered energy is modeled assuming the Hagfors behavior given in Eq. (1a) (Ford and Pettengill 1992). For gently undulating terrain, the Hagfors model also provides a useful estimate of the rms surface slope at scales of decimeters to meters (but see Sec. IV.A). This model also estimates the normal-incidence power reflection coefficient of the surface, as shown in Fig. 2, but only to the extent that the area within the footprint satisfies the quasi-specular assumptions. Diffuse roughness will tend to spread the incident energy over a wide range of scattering angles and thus reduce the inferred reflectivity. A correction for this "lost" power is included in altimeter processing algorithms, using the SAR data to estimate the amount of roughness, but its reliability in areas of very high roughness is uncertain (Pettengill et al. 1988).

In contrast to the strong correlation between radar backscatter and wavelength-scale roughness, when viewed at oblique angles, microwave thermal emission is primarily controlled by changes in the bulk dielectric constant of the surface. As noted earlier, surface roughness has a significant effect on the polarization state of the emitted energy (Fig. 1b). For those limited areas on the surface where dual-polarization SAR and emission data have been collected, the polarization-averaged emissivity can provide a robust estimate of the dielectric constant largely independent of roughness. In most areas, however, we must estimate the roughness and dielectric constant using only the horizontally polarized components of radar backscatter cross section σ_o, and emissivity e. For regions below 6054 km in altitude, the two parameters tend to be highly anticorrelated, suggesting a relatively narrow variation in dielectric constant over the plains. A model has been presented by Campbell (1994), which attempts to use values of σ_o and e to estimate both a diffusely scattering surface fraction and the dielectric constant.

For any given location on Venus, we usually have only a single observation of radar backscatter at an incidence angle defined by a latitude-dependent viewing geometry profile (Saunders et al. 1992). In its initial observations, Magellan viewed the surface at incidence angles between 25° and 45°. However, later surveys covered most of the surface at a different angle. Furthermore, Arecibo data are available that augment our knowledge of the angular scattering behavior, especially for areas seen from Earth at high incidence angles, e.g., Maxwell Montes, Beta Regio and Mylitta Fluctus.

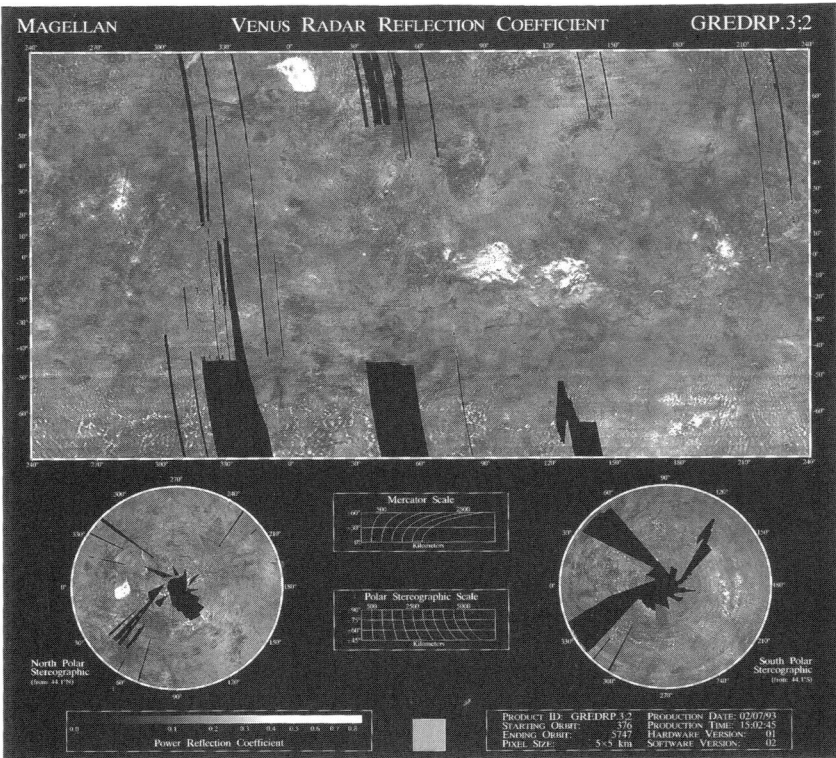

Figure 2. Mercator and polar stereographic projections of the distribution of power reflection coefficient over the Venus surface, as determined by the Magellan radar altimeter. Values are rendered by a gray scale, shown in the inset. Black areas represent missing data. Data south of approximately 60°S are noisy, because of lower system sensitivity at the high spacecraft altitudes in this portion of the elliptical orbit.

The low-emissivity highlands have backscatter cross sections much larger than those of any dry, rough terrestrial surface. Examination of Arecibo data shows that these areas are characterized by high circular polarization ratios (see Sec. III.B) and unusual angular backscattering functions. Backscatter from Maxwell Montes, the Beta Regio mountains, and Ovda Regio tends not only to be very strong, but also varies only slowly with incidence angle over the range 0° to 25°; however, it drops sharply as the angle increases toward 45° and beyond (Fig. 3). This behavior is not seen in scattering from rocky terrains on Earth.

B. Monostatic Polarization Observations

Single-antenna, i.e., monostatic, measurements of the polarization state of the backscattered radar return have been used to estimate the degree of wavelength-scale surface roughness (Pettengill and Thompson 1968). How-

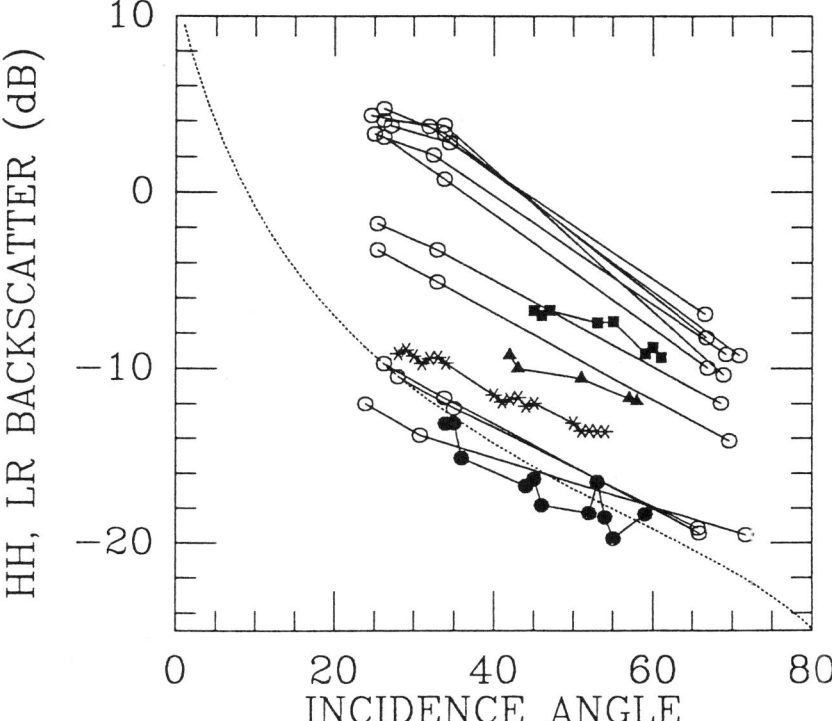

Figure 3. Specific radar cross section (in decibels) of the quasi-specular component of scattering for planetary surfaces as a function of the angle of incidence. Open circles detail the behavior of areas within Maxwell Montes, drawing on both Magellan and Arecibo data taken at 13-cm wavelength; filled symbols refer to terrestrial data for Hawaiian lava flows of varying surface roughness. The dotted curve represents the Venus mean scattering function (Muhleman law), as adopted by the Magellan Project.

ever, if subsurface or volume scattering contributes significantly to the echo, the polarization state may be affected by the polarization-dependent transmission of the wave both into and out of the surface. Hagfors et al. (1965) suggested that scattering from low-porosity powdered surfaces like the upper lunar regolith represents an example of this process, and used such a model to estimate the dielectric constant of the lunar surface, on the assumption that the backscattered echo at high incidence angles originated entirely from diffuse subsurface scatter. Hagfors and Campbell (1974), using a circularly polarized transmitted signal, failed to find any corresponding linearly polarized component in Arecibo 70-cm wavelength radar returns from areas on Venus viewed at high incidence angles. This result suggests that, for Venus, the majority of the radar echo comes from scattering at the atmosphere-surface boundary, with little or no subsurface contribution.

While the electromagnetic properties of the Venus surface cannot be directly deduced from the polarization state of the backscattered radar return, the polarization can be used to help discriminate among the various models invoked to explain the low radiothermal emission observed above 6054 km. The most extensive studies of the polarization properties of the backscattered Venus radar echo have been made at Arecibo, using its 13-cm wavelength radar system (the same wavelength used by Magellan). The Arecibo system normally transmits left circularly polarized radiation, and receives both the left (LL) and right (LR) circularly polarized echo components. The LL (depolarized sense) and LR (expected sense) cross sections are used to calculate the LL/LR polarization ratio (Campbell and Campbell 1992). Tryka and Muhleman (1992) mapped the RR cross section of Alpha Regio at 3.5-cm wavelength, using the 70-m Goldstone Deep Space Network antenna for transmission, and the VLA antenna array to synthesize an angularly resolved image of the depolarized echoes. In this experiment, the RL polarized echoes were sufficiently strong to saturate the receiving system, and were lost. The large, two-way atmospheric absorption for Venus at a wavelength of 3.5 cm forced this system to limit its RR imaging to those regions which could be viewed at relatively low angles of incidence, principally Alpha Regio. The depolarized observations of Alpha Regio, a region that does not exhibit unusual values of emissivity, showed several small areas of localized RR enhancement.

Typical values for the LL/LR polarization ratio of echoes from the Venus lowlands and rolling plains, viewed at incidence angles above 40° from Arecibo, fall near 0.2; in contrast, the high-altitude, high-reflectivity (and low-emissivity) regions show ratios that often exceed unity. In the Maxwell Montes, the polarization ratio is everywhere close to unity, with maximum values rising to 1.25 along its northern and southern edges (Fig. 4). For Beta Regio, values of the ratio slightly lower than unity are found near the summit of Theia Mons, although there are a few areas where the ratio is close to one. High ratios correlate well with areas of large LR cross section, but the correlation is not perfect. We note also, that regions with a large polarization ratio tend to correspond to regions having very low values of emissivity in the Magellan data, although the latter have a poorer spatial resolution than do the corresponding radar images. Figure 5 shows an LL (depolarized) image of Maxwell Montes obtained at Arecibo, indicating the track along which the ratios given in Fig. 3 were obtained. (This track is also the path of the bistatic observations discussed in Sec. III.C).

C. Bistatic Observations

Bistatic radar experiments, where the reflecting surface element is illuminated and viewed from different directions, allow a wider variety of useful measurements, particularly of the polarization dependence, of the scattering process than is available in the more usual monostatic geometry, where the two directions are retro-aligned (Simpson 1993). The Magellan spacecraft has been used in this mode to study Venus by illuminating the planet's surface with its

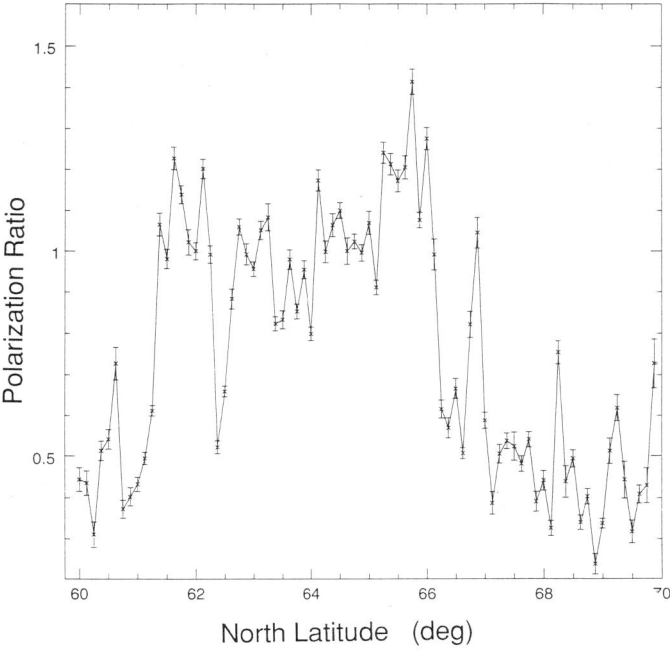

Figure 4. The LL/LR polarization ratio of 13-cm echoes from Maxwell Montes, as obtained at the Arecibo Observatory, taken along the track outlined in Fig. 5.

13-cm-wavelength telemetry transmitter, in a geometry chosen to satisfy the requirements for specular reflection as viewed from Earth (Pettengill et al. 1996). The echo signals were received using dual phase-coherent, orthogonal, circularly polarized receiving channels, which enabled recovery of a full Stokes-vector description of the received energy.

The 13-cm-wavelength continuous wave (CW) downlink radiation was linearly polarized at transmission. By adjusting the orientation of the spacecraft, the polarization angle of the incident polarization was maintained at 45° to the scattering plane. Thus, the power striking the surface in both the in-plane (vertical) and out-of-plane (horizontal) linearly polarized components was equal and in-phase. The polarization angle of the reflected signal was, therefore, determined simply by the inverse tangent of the ratio of the amplitude reflection coefficients given in Eqs. (2). For rough surfaces, we expect weak echoes, since the scattered energy is diffused over a wide range of emerging solid angle. Where the surface is reasonably smooth, the signal is concentrated in the specular direction and remains highly polarized. An important advantage of this observing technique is that the ratio of the in-plane and out-of-plane reflected components depends only on the dielectric constant of the surface, thus eliminating the need for accurate calibration of the received power. Equally useful is that only a small portion of the scattering

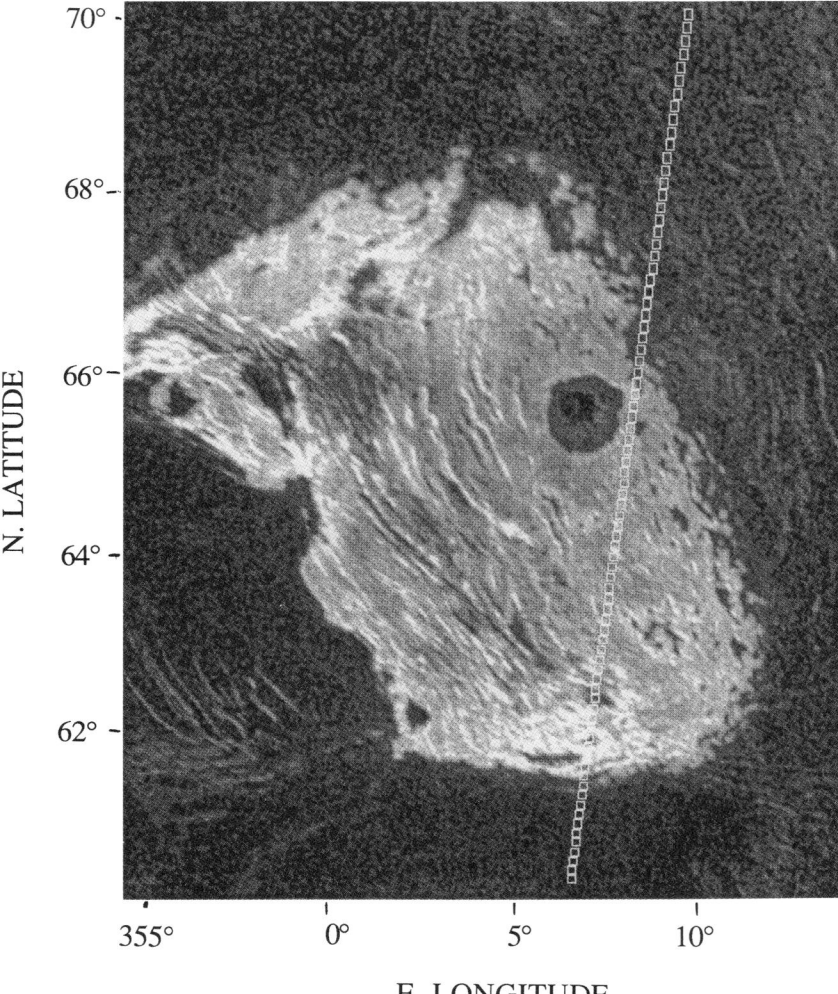

Figure 5. Arecibo radar image of Maxwell Montes, using the depolarized (LL) sense of received circular polarization. The data for this image, which has a resolution of approximately 1.5 km, were obtained during the June 1988, inferior conjunction of Venus. The track starting northeast of the 100-km diameter crater Cleopatra, and extending to the south–southwest is the track of the specular reflection point for the 5 June 1994, Magellan bistatic experiment described in the text.

surface need be smooth to yield a measurement of the rotation angle of the incoming polarization, and thus the ratio of components described above. In this way, even an extremely rough surface can yield credible estimates of its dielectric properties, as long as a few smooth horizontal facets are present.

In bistatic measurements carried out 5 June 1994 (Pettengill et al. 1996),

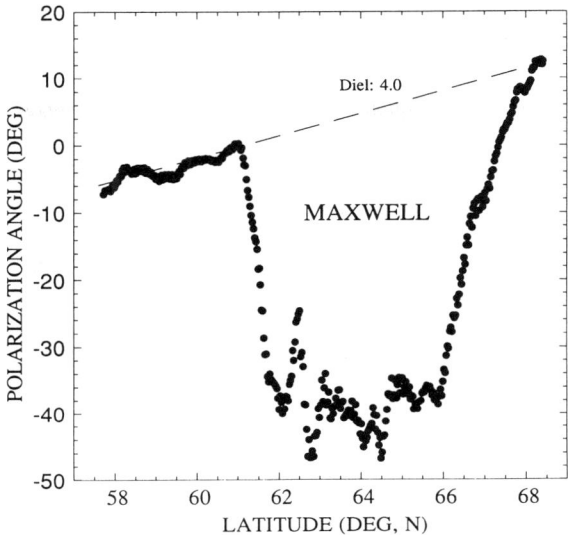

Figure 6. The polarization position angle observed during the bistatic experiment carried out by the Magellan spacecraft on 5 June 1994, as it flew over the Maxwell Montes. A polarization angle of zero degrees corresponds to reflection at the Brewster angle, resulting in a polarization direction parallel to the Venus surface. The dashed line shows the theoretical behavior (varying with incidence angle) expected for a smooth surface having a dielectric constant of 4.0.

the observing geometry permitted the Magellan spacecraft to illuminate a surface swath that traversed much lowland terrain, but also passed over the high-altitude low-emissivity regions in the southeast of Maxwell Mons (Fig. 5). Near Maxwell Mons, because of its high latitude, the angle of incidence (67°) was close to the Brewster angle (64°) for material having a dielectric constant of about 4.5; thus, where the illuminated footprint comprised "typical" surface material, very little in-plane echo was generated and the polarization angle of the reflected ray was nearly perpendicular to the scattering plane. In fact, for typical lowlands regions, the observed polarization angle rotation was consistent with a dielectric constant of 4.0±0.5. As the footprint moved into the low-emissivity regions in Maxwell, however, the plane of received polarization suddenly rotated through nearly 45° (Fig. 6). Analysis shows a polarization angle of 36.9°±2° in the Maxwell Montes region at the edge of the crater Cleopatra, implying a surface dielectric constant there of order 100. A particularly surprising result was the appearance over Maxwell of a component of right circularly polarized power corresponding to about 10% of the total reflected signal. The amount of this component (presumably arising from a finite phase difference between the complex amplitudes given in Eqs. (2), as a result of electrical loss in the surface), the observed polarization angle, and the previously measured emissivity of about 0.33 place constraints on the electrical properties of the reflecting surface of the Maxwell Montes. The effects seen may conceivably result from a lossy dielectric, but a more likely assumption of a semiconducting layer having an imaginary dielectric constant of $-i100 \pm i50$ appears to work quite well (Pettengill et al. 1996). The consequences of this unusual finding are explored in the next section.

IV. SURFACE STRUCTURE AND COMPOSITION

A. The Typical Surface

Venus backscatter values have been interpreted by a number of workers through comparison with similar radar measurements of putative terrestrial analogs (Gaddis 1992; Arvidson et al. 1992; Campbell and Campbell 1992; Plaut 1991). With the exception of some crater-related deposits, variations in the surface dielectric constant for most areas on Venus below 6054 km radius are small enough to be neglected in a broad-brush survey.

The plains that make up more than half the planet's surface area vary widely in their backscatter characteristics, depending on the density of superposed tectonic ridges and fractures, but the underlying material was clearly very smooth when emplaced, a property that is consistent with its assumed origin as fluid lava erupting from linear vents or fissures. The inference of smoothness is based on the relatively low backscatter coefficient of the Venus plains relative to terrestrial lava flows, and on the rapid decrease in backscattered power as the incidence angle increases. A survey of volcanic edifices on Venus suggests that, in general, the magma issuing from many of the large shield volcanoes appears similar to smooth terrestrial basaltic flows (Campbell and Campbell 1992). The rough textures often seen in terrestrial flows emplaced at high eruption rates, or resulting from lavas with high silica content, are rare on Venus (Moore et al. 1992). Some radar-dark flows, typically those appearing stratigraphically recent, are smooth down to the centimeter level.

Dielectric constants vary only slightly (from 3 to 6) over the plains and edifice lava flows; the changes seen probably arise from relatively small differences in mineral content or magma density. At altitudes below a planetary radius of about 6054 km, the mean surface has a bulk dielectric constant of between 4.0 and 4.5 (Pettengill et al. 1992,1996). This value is consistent with dry compacted rocks of a variety of compositions, as seen on Earth and other planets (Campbell and Ulrichs 1969); the occasional surface areas having lower values are consistent with layers of overlying, less compact, dusty or granular material (Gold et al. 1970; Carrier et al. 1991).

The dielectric values derived from the altimeter and the emissivity data, respectively, agree well for these surfaces, presumably because corrections for small-scale roughness are small. Areas possessing a higher dielectric constant (up to 8), usually associated with parabolic crater ejecta deposits and radar-bright crater floors, have been identified in the plains (Campbell et al. 1992; Plaut and Arvidson 1992). The crater parabolas probably result from mantling by fine material produced in relatively recent impact events; if not recent, they would likely have been dispersed by the wind over a few tens of Myr (Izenberg et al. 1994; Arvidson et al. 1992; Greeley et al. 1992). The emissivity data suggest a dielectric enhancement in the fine-grained material, that may be related to the impact melting process. In any event, the enhancement evidently survives in the lowland environment for a significant period of time. Low-dielectric areas around impact craters and some volcanoes are attributed to fine-grained material which mantles the underlying rocky surface, but which lacks enhancement in its constituent dielectric constant. Based on the reflectivity and emission data, the amount of porous material on the surface seems relatively low, and is consistent with slow erosional mechanisms and limited aeolian transport (Greeley et al. 1992).

The coherent radar echoing properties of most of the Venus surface appear to be well represented by models assuming quasi-specular scattering from a single interface. Because the probability density function for surface slopes is proportional to the specific radar cross section under certain conditions (Barrick 1968), probability functions $p(\beta)$ (where β is the tilt of a reflecting element with respect to the normal to the mean spherical surface) can be extracted from the Magellan altimeter echoes and used to characterize surface morphology in a statistical sense over areas as small as a few thousand square km. Tyler et al. (1992) have shown that the Hagfors function given in Eq. (1a) provides good agreement with the data, but that an exponential form for $\sigma_o(\theta)$ yields lower residuals in most areas. The typical surface is quite smooth at radio wavelengths, undulating with values of rms slope between $2°$ and $4°$ (Ford and Pettengill 1992; Tyler et al. 1992), and exhibiting only 10 to 20% depolarization in its scattered energy (Carpenter 1966).

Hagfors (1970) assumed that the quasi-specular scattering surface had a Gaussian height distribution and an exponential lateral autocorrelation function, but argued that high frequencies in the surface structure are effectively filtered out by the choice of radar wavelength. Surfaces with "sawtooth" profiles are thus permissible as long as we allow the sharp corners to be rounded during the scattering process. The corners themselves may contribute to the diffuse component of the echo, which is distributed widely in angle compared with the quasi-specular component. The fact that the Magellan data often match an exponential better than the Hagfors scattering function suggests that not only can much of the Venus surface be modeled as a set of flat, tilted plates, but that there are more plates with zero tilt than we would expect if the tilts were randomly distributed. At least in selected locations on the broad Venus plains, Venera lander images confirm that the surface is composed of larger and flatter facets than we see on Mars and the Moon, and that there are fewer discrete, rounded rocks exposed (Garvin et al. 1981,1984).

The average tilt of reflecting facets may not, in fact, always be zero. A surprising result of the analysis carried out by Tyler et al. (1992) showed that Magellan altimetry echoes are often asymmetrically distributed in frequency. In some regions, echoes seem to arise from surface elements that are preferentially inclined for backscatter ahead of the spacecraft (positive Doppler shifts), while in others the echoes seem to come preferentially from behind (negative Doppler shifts). Although large-scale topographic tilt can account for some of this offset, in many areas, the observed effects are an order of magnitude larger than could be explained by large-scale topography. One possibility is that the reflecting facets are oriented to have long, gradual rises in one direction, followed by sharp drops. In this way more of the surface could tilt toward the north (for example) while the net topographic change along the track would be zero. Similar Doppler offsets have since been found in Venera altimetry data (A. Zakharov, personal communication; Maurer et al. 1993). East-west asymmetries in the distribution of backscattered energy have been reported from analysis of Magellan SAR data for a few areas linked to aeolian features on the surface (Plaut et al. 1992; Weitz et al. 1994).

In mountainous areas on Venus, as compared with the lowlands, the surface roughness inferred from nadir radar sounding is generally higher, and the specific radar cross section tends to be better represented by an expression using a lateral auto-correlation of Gaussian form, rather than either the truncated exponential function used in the Hagfors formulation (Eq. 1a) or a simple exponential (Tyler et al. 1992). In most such areas, the primary scattering still appears to be quasi-specular, suggesting that the surface is likely to be more angular, and have fewer flat, plate-like components, than in the lowlands. The panoramic view from Venera 9 (Garvin et al. 1984) is consistent with such a more angular surface at higher elevations. For a small region in Alpha Regio, Tryka and Muhleman (1992) suggest that the polarization properties are consistent with volume scattering. Volume scattering of any type implies very low loss tangents (negligible amounts of absorbing material, such as iron-bearing minerals), which would likely require a nonbasaltic composition (Ulaby et al. 1982; Carrier et al. 1991; Wilt 1992). The emission associated with the Alpha Regio areas, although locally low, is still much greater than that seen in most highland regions, where values of horizontally polarized emissivity (at $45°$ viewing angles) below 0.33 are sometimes found.

B. The Low-Emissivity Surface

In contrast to data for the lowlands, observations of the Venus highlands often yield dielectric constants of 50 or more (Pettengill et al. 1992). In general, dielectric values extracted from emissivity observations tend to be significantly larger than those obtained from altimeter data; however, the accuracy of the latter may suffer from the large diffuse scattering corrections needed for the rough highland surfaces. The transition from "normal" dielectric values typical of the lowlands to the large values found at high altitudes occurs rapidly above a critical elevation; the altitude of the onset appears to vary slightly with location on the planet (Arvidson et al. 1992,1994; Klose et al. 1992; Wilt 1992), moving from just below 6054 km planetary radius near the equator, to just above 6055 at high northern latitudes. While roughness in areas above these levels tends to be high, this roughness appears to stem from the formational history of the terrain, i.e., producing a preponderance of tesserae, rather than from a weathering of the high-dielectric material (Arvidson et al. 1994; Campbell 1994). There is relatively little correlation between the visible structure of any given terrain and its dielectric enhancement.

Several possible explanations for the low emissivity of the Venus highlands have been proposed since the discovery of the phenomenon. The first of these, postulating a surface loaded electrically with small conductors (iron sulfides of various forms) was put forward by Pettengill et al. (1982), Ford and Pettengill (1983), and Pettengill et al. (1988). As the required value of surface dielectric permittivity has been pushed higher and higher by the observations (currently it is around 100), materials with ferroelectric properties have been proposed (Shepard et al. 1994), which also allow for a rapid variation in electromagnetic reactivity with small changes in ambient temperature. But the ferroelectric mineral with the precise properties required by the observations has yet to be identified. With the recent results obtained from the bistatic experiment described in Sec. III.C, however, it is clear that a semiconducting surface layer will also suffice to meet the electrical requirements of the observations. Pettengill et al. (1996) derive a value for the surface electrical conductivity of 13 mhos m^{-1}. Of the chemical elements, only germanium and tellurium have values of conductivity that might accommodate this value; tellurium appears particularly intriguing, because it has a freezing point temperature of 723 K, almost precisely that of the limiting highlands elevation above which the anomalous effect is seen. In fact, the small (1 km) variation in onset altitude may reflect corresponding differences in atmospheric temperature (8 K) with latitude. Brackett et al. (1995) have investigated the problem of volatile transport from the lowlands to higher and cooler locations on Venus, although specific calculations for tellurium (or tellurides) have not yet been carried out. It is interesting to note that, if the semiconducting material is tellurium, a thin layer (less than 3 mm, and perhaps only several μm thick) may be able to support the results seen; it is estimated that only between 10^{-3} to 10^{-5} of the tellurium contained in the crust and mantle of Venus need be outgassed to support this hypothesis (Pettengill et al. 1996). Another possibility involves the slow deposition of magnetite above the critical altitudes (see the chapter by Wood).

In addition to understanding the dielectric properties of the low emissivity surface, it is necessary to explain the large depolarization ratios seen in these areas (Fig. 4). An intriguing possibility, enhanced by the efficient reflectivity of a semiconducting surface, is that whiskers or other highly inclined dendritic structures on the surface might allow the multiple (primarily dual) reflections needed to give rise to the the coherent backscatter opposition effect (Hapke 1990).

V. CONCLUSIONS

The surface of Venus is dominated by plains comprised largely of flat plates of basaltic material. These plates exhibit complex regional tilt patterns, which have not as yet been plausibly explained. The typical plains region has a surface with an rms slope between 2° and 4°, at lateral scales greater than a few meters; a 5 to 10% component of small-scale (several cm) roughness; and a nominal dielectric constant falling between 4.0 and 4.5. Variations in the dielectric constant across the plains likely arise from minor changes in chemical composition, density, or post-emplacement soil formation. The large volcanic edifices are often characterized by relatively smooth-surface lava flows, suggesting low magma viscosity or slow effusion rates. Occasional high surface roughness in these areas probably results from a viscous magma or large eruption rates.

Fine-grained material is found mostly surrounding impact craters, and has widely varying dielectric properties. Some low-dielectric deposits mantle the underlying terrain to depths of 10 to 50 cm, attenuating the backscattered echo. Parabolic crater deposits often have low-intensity backscatter signatures, as a result of their relative smoothness, but their relatively low values of emissivity imply a moderately high dielectric constant (about 8). Porous layers should have lower dielectric constants than their parent rock, so a moderately higher intrinsic dielectric constant is implied for these deposits. Because similar emissivity values are found in some crater floors, the enhancement may arise from impact melt processes.

The radar-bright highlands appear to be underlain by normal volcanic or tectonically deformed terrain, covered by a lossy conducting material yielding an imaginary dielectric permittivity of about 100. A possible candidate for this layer is a composite of basaltic rock and magnetite, or a frost of tellurium. It is likely that a full understanding of the low-emissivity surface composition and structure must await *in-situ* studies. The highland regions of Venus obviously form high-priority targets for future Venus landers.

Acknowledgments The authors acknowledge support under several grants from NASA. D. Campbell acknowledges support from the National Astronomy and Ionosphere Center, operated by Cornell University under cooperative agreement with the National Science Foundation, with supplemental funding from NASA.

REFERENCES

Arvidson, R. E., et al. 1992. Surface modification of Venus as inferred from Magellan observations of plains. *J. Geophys. Res.* 97:13303–13318.

Arvidson, R. E., et al. 1994. Microwave signatures and surface properties of Ovda Regio and surroundings, Venus. *Icarus* 112:171–186.
Barrick. D. E. 1968. Rough-surface scattering based on the specular point theory. *IEEE Trans.* AP–16:449–454.
Brackett, R. A., Fegley, B., Jr., and Arvidson, R. E. 1995. Volatile transport on Venus and implications for surface geochemistry and geology. *J. Geophys. Res.* 100:1553–1563.
Campbell, B. A. 1994. Merging Magellan emissivity and SAR data for analysis of Venus surface dielectric properties. *Icarus* 112:187–203.
Campbell, B. A., and Campbell, D. B. 1992. Analysis of volcanic surface morphology on Venus from comparison of Arecibo, Magellan, and terrestrial airborne radar data. *J. Geophys. Res.* 97:16293–16314.
Campbell, D. B., et al. 1992. Magellan observations of extended impact-crater related features on the surface of Venus. *J. Geophys. Res.* 97:16249–16278.
Campbell, M. J., and Ulrichs, J. 1969. Electrical properties of rocks and their significance for lunar observations. *J. Geophys. Res.* 74:5867–5881.
Carpenter, R. L. 1966. Study of Venus by CW radar—1964 results. *Astron. J.* 71:142–152.
Carrier, W. D., Olhoeft, G. R., and Mendell, W. 1991. Physical properties of the lunar surface. In *Lunar Sourcebook* (Cambridge: Cambridge Univ. Press).
Ford, P. G., and Pettengill, G. H. 1983. Venus: global surface radio emissivity. *Science* 220:1379–1381.
Ford, P. G., and Pettengill, G. H. 1992. Venus topography and kilometer-scale slopes. *J. Geophys. Res.* 97:13103–13114.
Gaddis, L. R. 1992. Lava flow characterization at Pisgah volcanic field, California, with multi-parameter imaging radar. *Geol. Soc. Amer. Bull.* 104:695–703.
Garvin, J. B., Mouginis-Mark, P. J., and Head, J. W. 1981. Characterization of rock populations on planetary surfaces: Techniques and a preliminary analysis of Mars and Venus. *Moon and Planets* 24:355–387.
Garvin, J. B., Head, J. W., Zuber, M. T., and Helfenstein, P. 1984. Venus: The nature of the surface from Venera panoramas. *J. Geophys. Res.* 89:3381–3399.
Gold, T., Campbell, M. J., and O'Leary, B.T. 1970. Optical and high-frequency electrical properties of the lunar sample. *Science* 167:707–709.
Greeley, R., et al. 1992. Aeolian features on Venus: Preliminary Magellan results. *J. Geophys. Res.* 97:13319–13345.
Hagfors, T., et al. 1965. Tenuous surface layer on the moon: evidence derived from radar observations. *Science* 150:1153–1156.
Hagfors, T. 1970. Remote probing of the moon by infrared and microwave emissions and by radar. *Radio Sci.* 5:189–227.
Hagfors, T., and Campbell, D. B. 1974. Radar backscattering from Venus at oblique incidence at a wavelength of 70 cm. *Astron. J.* 79:493–502.
Hapke, B. 1990. Coherent backscatter and the radar characteristics of outer planet satellites. *Icarus* 88:407–417.
Izenberg, N. R., Arvidson, R. E., and Phillips, R. J. 1994. Impact crater degradation in the Venus plains. *Geophys. Res. Lett.* 21:289–292.
Klose, K. B., Wood, J. A., and Hashimoto, A. 1992. Mineral equilibria and the high radar reflectivity of Venus mountaintops. *J. Geophys. Res.* 97:16353–16370.
Marov, M. Ya. 1978. Results of Venus missions. *Ann. Rev. Astron. Astrophys.* 16:141–169.
Maurer, M. J., Tyler, G. L., and Simpson, R. A. 1993. Comparison of Magellan Venus differential altimetry with observed doppler anomalies. Paper presented at the AAS-DPS Meeting, Boulder, Co.
Moore, H. J., Plaut, J. J., Schenck, P. M., and Head, J.W. 1992. An unusual volcano

on Venus. *J. Geophys. Res.* 97:13479–13494.
Pettengill, G. H. 1978. Physical properties of the planets and satellites from radar observations. *Ann. Rev. Astron. Astrophys.* 16:265–292.
Pettengill, G. H., and Thompson, T. W. 1968. A radar study of the lunar crater Tycho at 3.8-cm and 70-cm wavelengths. *Icarus* 8:457–471.
Pettengill, G. H., et al. 1962. A radar investigation of Venus. *Astron. J.* 67:181–471.
Pettengill, G. H., Ford, P. G., and Nozette, S. 1982. Venus: Global surface radar reflectivity. *Science* 217:640–642.
Pettengill, G. H., Ford, P. G., and Chapman, B. D. 1988. Venus: Surface electromagnetic properties. *J. Geophys. Res.* 93:14881–14892.
Pettengill, G. H., et al. 1991. Magellan: radar performance and data products. *Science* 252:260–265.
Pettengill, G. H., Ford, P. G., and Wilt, R. J. 1992. Venus surface radiothermal emission as observed by Magellan. *J. Geophys. Res.* 97:13091–13102.
Pettengill, G. H., Ford, P. G., and Simpson, R. A. 1996. Electrical properties of the Venus surface from bistatic radar observations. *Science* 272:1628–1631.
Plaut, J. J. 1991. Radar Scattering as a Source of Geological Information on Venus and Earth. Ph.D Thesis, Washington University.
Plaut, J. J., and Arvidson, R. E. 1992. Comparison of Goldstone and Magellan radar data in the equatorial plains of Venus. *J. Geophys. Res.* 97:16279–16291.
Plaut, J. J., et al. 1992. Anomalous scattering behavior of selected impact "parabola" features: Magellan cycle-to-cycle comparisons. In *Intl. Colloquium on Venus*, LPI Contrib. No. 789, pp. 92–93.
Saunders, R. S., et al. 1992. Magellan mission summary. *J. Geophys. Res.* 97:13067–13090.
Shepard, M. K., Arvidson, R. E., Brackett, R. A., and Fegley, B., Jr. 1994. A ferroelectric model for the low emissivity highlands on Venus. *Geophys. Res. Lett.* 21:469–472.
Simpson, R. A. 1993. Spacecraft studies of planetary surfaces using bistatic radar. *IEEE Trans. Geosci. Remote Sensing* 31:465–482.
Stratton, J. A. 1941. *Electromagnetic Theory* (New York: McGraw-Hill).
Tryka, K. A., and Muhleman, D. O. 1992. Reflection and emission properties on Venus: Alpha Regio. *J. Geophys. Res.* 97:13379–13394.
Tyler, G. L., Simpson, R. A., Maurer, M. J., and Holmann, E. 1992. Scattering properties of the Venus surface: Preliminary results from Magellan. *J. Geophys. Res.* 97:13115–13139.
Ulaby, F. T., Moore, R. K., and Fung, A. K. 1982. *Microwave Remote Sensing*, vol. 2 (Reading, Mass.: Addison Wesley).
Victor, W. K., and Stevens, R. 1961. Exploration of Venus by Radar. *Science* 134:46–48.
Weitz, C. M., Plaut, J. J., Greeley, R., and Saunders, R.S. 1994. Dunes and microdunes on Venus: Why were so few found in the Magellan data? *Icarus* 112:282–295.
Wilt, R. J. 1992. A Study of Areas of Low Radiothermal Emissivity on Venus. Ph.D. Thesis, Massachusetts Inst. of Technology.

AEOLIAN PROCESSES AND FEATURES ON VENUS

RONALD GREELEY and KELLY C. BENDER
Arizona State University

R. STEPHEN SAUNDERS
Jet Propulsion Laboratory

GERALD SCHUBERT
University of California at Los Angeles

and

CATHERINE M. WEITZ
Brown University

Aeolian features on Venus include dune fields, eroded hills (yardangs), wind streaks, and possibly microdunes (miniature dunes of 10 to 30 cm wavelength). Although repetitive imaging by Magellan did show changes in the appearance of the surface, these changes are attributed to radar artifacts as a consequence of look direction rather than to physical changes of the surface. Nonetheless, measurements of wind speeds near the surface of Venus and wind tunnel simulations suggest that aeolian processes could be currently active on Venus. Study of radar images of terrestrial analogs shows that radar wavelength, polarization, and viewing geometry, including look direction and incidence angle, all influence the detection of dunes, yardangs, and wind streaks. For best detection, dune crests and yardangs should be oriented perpendicular to look direction. Longer wavelength systems can penetrate sand sheets a meter or more thick, rendering them invisible, especially in arid regions. For wind streaks to be visible, there must be a contrast in surface properties between the streak and the background on which it occurs. Nonetheless, more than 6000 aeolian features have been found on Magellan images of Venus, the most common of which are various wind streaks. Mapping wind streak orientations enables near-surface wind patterns to be inferred for the time of their formation. Type P streaks are associated with parabolic ejecta crater deposits and are considered to have formed in association with the impact event. Most Type P streaks are oriented westward, indicative of the upper altitude superrotation winds of Venus. Non Type P streaks have occurrences and orientations consistent with Hadley circulation. Some streaks in the southern hemisphere are oriented to the northeast, suggesting a Coriolis effect.

I. INTRODUCTION

Prior to the space age, there was considerable speculation on the surface environment of Venus. Ideas ranged from steamy tropical climates to hot barren deserts (Colin 1983). With measurements of atmospheric composition, atmospheric density, and surface temperature, models emerged suggesting that aeolian processes could exist on Venus (Ronca and Green 1970). This possibility was further explored by Sagan (1975), Hess (1975), and Iversen et al. (1976).

Images returned by the Venera 8 and 9 landers and later by Veneras 13 and 14 revealed a surface littered with fine-grained particles amenable to entrainment by wind (Florensky et al. 1977a,b; Basilevsky et al. 1985). Near-surface (<2 m) wind speeds were measured by the Venera landers (Ksanfomality et al. 1983) and extrapolated to the surface by the Pioneer Venus atmospheric probe (Counselman et al. 1979) at 0.5 to 1 m s^{-1}. These values were well within the range predicted to move sand and dust. The Venusian Wind Tunnel (VWT) (Greeley et al. 1984) was put into operation in 1982 and enabled simulation of the Venusian aeolian environment. Results from the VWT confirmed that particles a few hundred microns in diameter could be entrained easily by winds measured on Venus. Thus, as reviewed by Kuzmin (1989), there were several lines of evidence to suggest that aeolian processes could be active on Venus.

A. Relevance of Aeolian Processes and Features

Aeolian processes link the atmosphere to the surface, while aeolian features provide direct clues of that link. Consequently, features such as wind streaks (Fig. 1) can indicate characteristics of the atmosphere, including wind direction at the time of streak formation. Mapping the patterns formed by wind streaks on local, regional, and global scales enables near-surface atmospheric circulation patterns to be deduced. Aeolian features also provide information on the nature of the surface. For example, sand dunes signal the presence of loose, fine-grained material that might otherwise be difficult to detect by remote sensing.

Windblown sand and dust could influence the composition of the atmosphere. Fine grains in suspension or in saltation expose mineral surfaces to the atmosphere, enhancing chemical reactions. For example, Nozette and Lewis (1982) assessed potential weathering reactions on Venus and suggested that windblown particles could react chemically with the atmosphere. They outlined several models for common minerals and atmospheric compositions. Fegley et al. (see their chapter) review these models and report on the results of laboratory experiments on possible weathering processes. In most cases, rates of reactions would be enhanced by the presence of windblown material because fine-grained materials in the atmosphere expose larger surface areas than a static surface.

Aeolian features on Earth often represent the most recent geologic processes and provide clues to the current and recent surface environment. By analogy with Earth, aeolian processes on Venus may play a role in the evolution and modification of the surface.

Thus, aeolian processes and features hold the potential for providing insight into aspects of the atmosphere, characteristics of the surface, and models of surface evolution. This potential was reviewed prior to the Magellan mission, and led to the formation of the following fundamental questions regarding aeolian processes (Greeley and Arvidson 1990) and atmospheric circulation (Schubert 1983) on Venus:

1. What is the nature of particle movement by wind on Venus? What is the rate of particle transport? Are the velocities and trajectories of entrained particles sufficiently energetic to cause abrasion of rocks and concomitant particle attrition and comminution (the accretion of minute amounts of particulate material due to collision with the surface)?
2. What is the sedimentary "budget" on Venus if materials are transported by wind? Do certain regions (e.g., highlands) act predominantly as source areas while other regions (e.g., plains) are deposition sites? How does the global aeolian system influence weathering and erosion processes, and the general "resurfacing" of the planet?
3. Is the transported material in sufficient quantity on a regional basis to cause significant chemical changes of surface deposits and to affect the atmospheric composition by subsequent gas-solid reactions?
4. Does the transport of material give rise to familiar aeolian features such as ripples and dunes? If so, what are the characteristics of these sedimentary features? Are there sedimentary deposits such as dust layers? How thick are they and do they suggest that current conditions have prevailed through geologic time?
5. What is the nature of near-surface atmospheric circulation?

About 65% of the Venus atmosphere by mass lies within 16 km of the surface (the lowest scale height of the atmosphere). Prior to Magellan very little was known about the circulation of this region. Doppler tracking of Venera (Kerzhanovich and Marov 1983) and Pioneer Venus (Counselman et al. 1980; Schubert et al. 1980) entry probes provided vertical profiles of zonal and meridional wind velocities at a few locations and times; in addition, near-surface wind speeds were recorded by Veneras 9 and 10 (Avduevskii et al. 1976). Although a westward circulation of the atmosphere prevailed about 10 km above the surface in all these measurements, the zonal circulation in the lowest several km and the meridional circulation of the lowest scale height of the atmosphere could not be inferred from these observations (Schubert 1983). Theoretical considerations (Schubert 1983) suggest a near-surface Hadley circulation which latitudinally redistributes solar energy absorbed in the lower atmosphere and at the ground by upflow at the equator, poleward winds at altitude, and equatorward surface return flow in both hemispheres.

In this chapter, we review current knowledge of aeolian features and processes on Venus gained from laboratory simulations of Venus, considerations of terrestrial analogs, and results from the Magellan mission. The

Figure 1. Wind streaks form in response to near-surface winds and are found on Earth, Mars, and Venus. As shown here in these Venus images, wind streaks are visible as differences in radar backscatter cross section; wind is inferred to blow from top of images toward the bottom. (a) Radar "bright" wind streaks northwest of Hestia Rupes at about 9°N, 67°E, showing streaks oriented to the south (bottom) and associated with small hills (A), ridges (B), and on open plains (C); area shown is about 95 × 105 km (F-MIDR 10N65). (b) Radar "dark" streaks east of Danu Montes at about 58.3°N, 344°E showing streaks associated with linear fractures and ridges; area shown is about 50 × 50 km (F-MIDR 60N344).

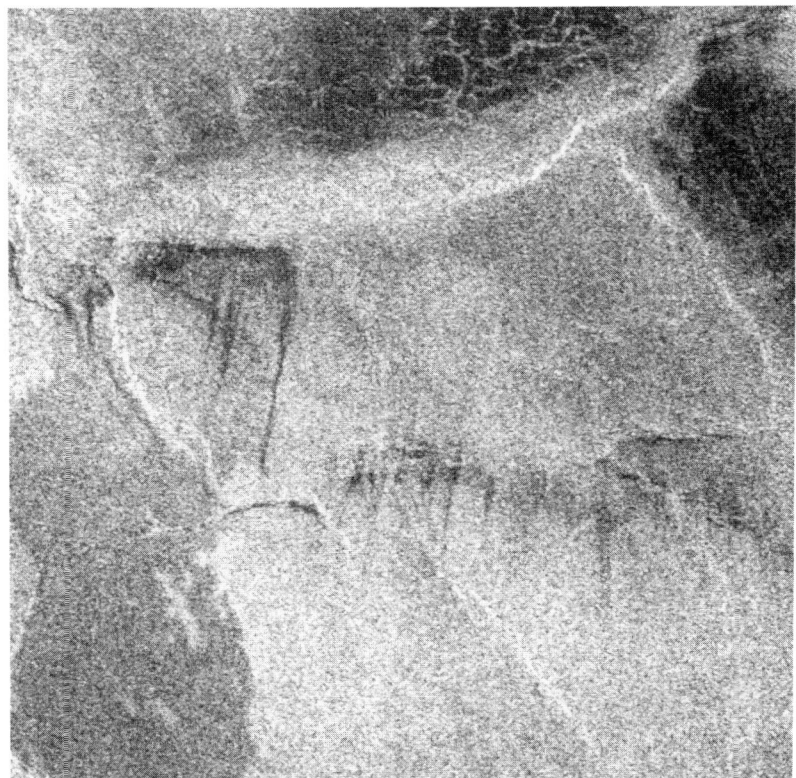

implications of aeolian features for near-surface atmospheric circulation are also considered.

II. LABORATORY STUDIES

The entrainment of sand and dust on Earth was studied by Bagnold (1941) using a wind tunnel. His approach was adapted to study aeolian processes for Venus, using the VWT (Greeley et al. 1984). In this section we review the requirements to move particles by wind on Venus (i.e., threshold conditions), infer the flux and speed of windblown grains, consider wind abrasion, and explore possible small bedforms such as ripples and microdunes that may form under Venusian conditions.

A. Particle Threshold

Bagnold (1941) described three modes of aeolian transport on Earth: surface creep, saltation, and suspension. Typically, surface creep involves very coarse grains (2000 to 4000 μm), saltation involves sand grains ($\sim$40 to 2000 μm), and suspension involves fine particles ($\lesssim 40\ \mu m$). On Earth, both surface creep and suspension result primarily from the impact of saltating grains.

Saltation threshold wind speed (designated u_{*_t}, the minimum wind friction speed necessary to initiate saltation) is the fundamental parameter governing most aeolian processes. Saltation threshold has been analyzed for Earth and Mars (reviewed by Greeley et al. 1984). Theoretical predictions of Venusian threshold were made by several investigators (Sagan 1975; Hess 1975; Iversen et al. 1976; Iversen and White 1982), who showed that threshold is a function of grain size and density, gravitational acceleration, and atmospheric temperature and density. Experiments were run in the VWT to study particle threshold under simulated Venusian conditions (Greeley et al. 1984). In contrast to predictions of Hess (1975) and others who suggested that the particle size most easily moved would be 32 to 34 μm in diameter, experiments showed that the most easily moved particles are $\sim$75 μm in diameter, similar to Earth and Mars.

Experiments also were conducted to assess modes of aeolian transport, such as saltation, on Venus. Greeley and Marshall (1985) noted three types of particle motion: (a) wobbling, (b) rolling, and (c) fully developed saltation (Fig. 2). It was inferred that suspension of fine grains and impact creep of very coarse grains would also occur on Venus. Continuous saltation corresponds to the single threshold curves predicted for Venus by Iversen and White (1982) and earlier workers. However, when the rolling mode of wind transport is taken into account, minimum wind speeds on Venus for aeolian activity may be 30% less than previous estimates which were based solely on saltation threshold. This result suggested that aeolian activity could be more frequent than previously suspected.

The rolling mode of transport observed in the VWT did not lead to a cascading effect, which on Earth ordinarily transforms the surface of a particle bed into a saltation cloud. However, observations showed that the rolling mode can be maintained in the VWT for an indefinite period. Thus, Venusian aeolian transport differs significantly from that on Earth. As noted by Iversen et al. (1987), the movement of particles resembles that of water-driven particles on Earth in some respects, and may lead to the formation of unusual bedforms, which are discussed below.

B. Flux of Aeolian Particles

An expression for saltation flux on Earth was derived from theory and experiments by Kawamura (1951). This expression was applied to planetary environments by White (1979) and refined by Blumberg and Greeley (1993) as

$$q = 2.61 u_*^2 \rho_a (1 - u_{*_t}/u_*)(1 + u_{*_t}/u_*)^2/g \qquad (1)$$

in which q = saltation flux (in g cm^{-1} s^{-1}), ρ_a = atmospheric density, u_* = friction velocity, u_{*_t} = threshold friction velocity, and g = acceleration due to gravity. The constant, 2.61, was determined in wind tunnel tests for

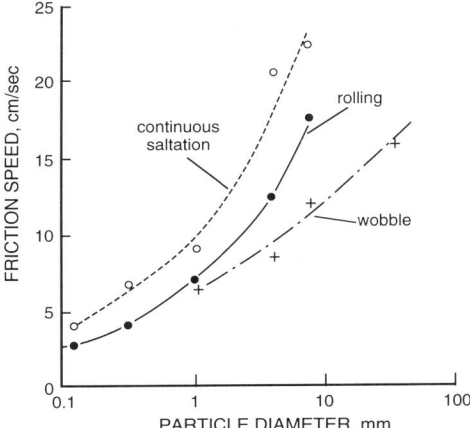

Figure 2. Wind threshold friction velocities for particle motion as a function of particle size under Venusian conditions simulated in the Venus Wind Tunnel. Wobble refers to particle motion without downwind displacement; continuous saltation refers to cascading grains similar to motion observed on Earth; rolling refers to continuous downwind movement without grains becoming airborne for appreciable periods. These experiments demonstrate that grains can move under winds lower than for saltation.

subrounded quartz sand 100 μm in diameter. However, a different constant may be required for different materials, as q is known to vary with particle shape, size distribution (Williams 1964), and density (Greeley et al. 1980). To assess this expression for Venus, saltation flux was determined using the VWT (Greeley et al. 1984; Williams and Greeley 1985) in which a given mass of sand was spread over the test plate and the tunnel was run at a given wind speed and duration. The amount of sand remaining on the test plate was recovered and the mass was determined. The total flux Q was then derived as the amount of sand removed from the plate divided by the plate width and wind duration. Results agree closely with the theoretical prediction for fine (75–90 μm) quartz particles. However, there was a systematic discrepancy between experimental results and predictions for coarse particles (500–600 μm). This was explained by Williams and Greeley (1987,1994) who showed that the high fluid density of the Venusian atmosphere can entrain large masses of particles, but probably reaches a critical stage in which the flux is effectively "choked" as the fluid becomes saturated with grains. If a steady supply of sand-size particles is assumed, the flux of particles on Venus could be as high as 2.5×10^{-5} g cm^{-1}s^{-1}.

C. Particle Speed

In order to assess potential abrasion of rocks and erosion of landforms it is essential that particle speeds be known as a function of wind speed and height above the surface. Experiments were run in the VWT to obtain appropriate data (Greeley et al. 1983; White 1986). Figure 3 shows results for 500 to 600 μm quartz particles under conditions for Earth, Mars, and Venus. Particle speeds are given as a percentage of wind speed. In general, it was found that particle speed increased with height. Under Venusian conditions, particles are accelerated to nearly the same speed as the wind, probably because the higher atmospheric density enhances particle movement.

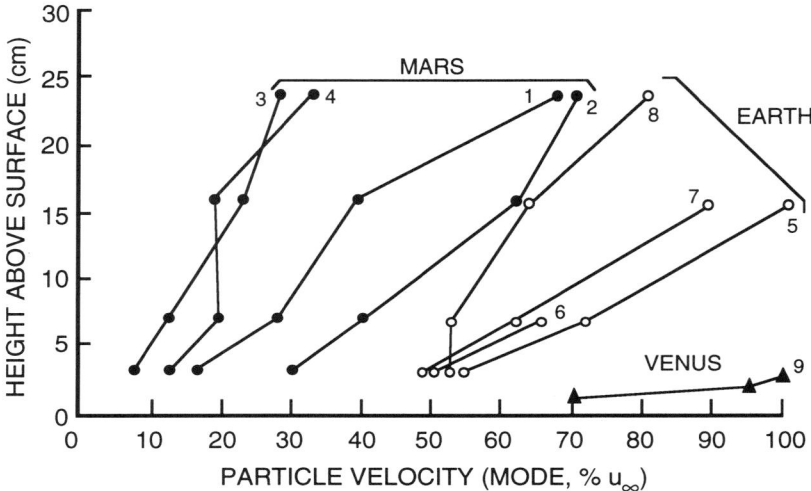

Figure 3. Particle velocities expressed as a percentage of free-stream wind speed for Mars, Earth, and Venus at different heights above the surface. Because of the high density of the atmosphere on Venus, there is a more efficient coupling with particles, accelerating the grains to the same, or nearly the same speed as the wind.

D. Wind Abrasion

Based on results for particle speeds, experiments were then run in an apparatus capable of duplicating the full temperature and pressure environment of Venus to simulate the abrasion of rocky materials. Tests were conducted at six atmospheric pressures (12, 22, 42, 65, 80, 95 bar) with the temperature held constant at 737 K, appropriate for Venus. In addition, tests were conducted at 47 bar, 660 K and 105 bar, 750 K (conditions representative of the highest and lowest elevations on Venus), respectively (Greeley et al. 1987; Marshall et al. 1988).

In initial experiments, target rocks were subjected to 10^5 impacts by particles at velocities of 0.5 to 0.7 m s^{-1}, appropriate for simulating the low winds measured near the surface on Venus. In later experiments, the number of impacts was reduced to 2×10^4. Most tests involved basalt targets and basalt particles, consistent with the probable basaltic compositions on Venus (Surkov et al. 1984,1987). Figure 4 shows results from the experiments. Despite the very low impact velocities, the edges of the colliding grains became rounded. Material was removed by chipping to produce pits, and by fine-scale abrasion that led to smoothing of areas between the pits. Average attrition was estimated from analysis of scanning electron microscope images as ≤1% volume reduction of the grains for the 14-hr experiments. Experiments showed that the rate of abrasion was apparently unaffected by temperature or pressure for the ranges tested.

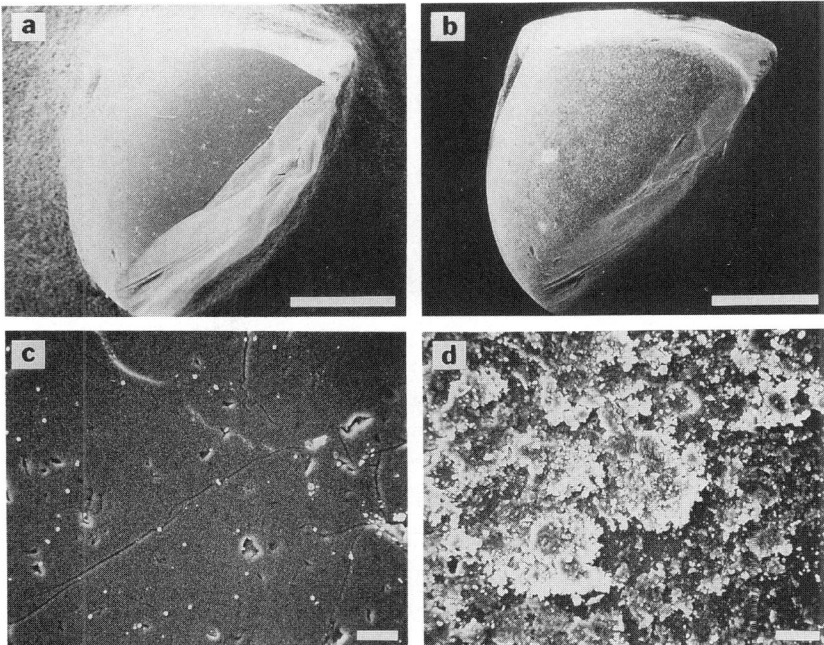

Figure 4. Results of wind abrasion experiments run in the Venus Simulator. Frames a and b show the windblown grains before (a) and after (b) abrasion. Frames c and d show the basalt target before (c) and after (d) the experiment; scale bars for grain = 10 μm, scale bars for target = 1 mm.

Profiles of the rock targets impacted by grains before and after each test showed an increase in thickness of the weathered surface of several micrometers. Both the surface appearance and the increase in thickness (determined in cross section) are attributed to accretion of comminuted material derived from the incident grains. Although the experiments did not show abrasion of the target surface by the incident grains, the placement of the rock target in the chamber exposed only a flat surface to impact. It is likely that abrasion of target edges would occur at a rate similar to the abrasion of edges on the impacting grains, if they were exposed. Thus, angular projections on rocks and bedrock surfaces on Venus would probably be abraded, even in the gentle winds recorded on Venus, as suggested by Garvin et al. (1984).

Wind-sculpted rocks may have a different appearance on Venus in comparison to ventifacts on Earth. On Earth, rocks are commonly undercut at a height of 15 to 20 cm above the surface. This height represents the maximum kinetic energy of impacting windblown grains considering the increase in particle velocity with height above the surface and the decrease in flux with height. Because the saltation trajectories are lower on Venus (White 1981), the height of maximum abrasion may be only a centimeter above the surface.

E. Bedforms

Potential aeolian bedforms on Venus have received relatively little attention, although the Soviets predicted miniature bedforms termed microdunes (Florensky et al. 1983). In addition, White (1981) suggested that ripples on Venus may have relatively short (a few cm) wavelengths, owing to the relatively short path lengths of saltating particles.

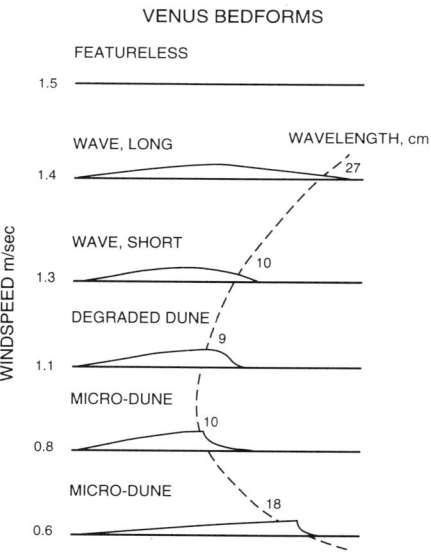

Figure 5. Cross section of bedforms produced in fine ($\sim 100\,\mu$m) quartz sand in the Venus Wind Tunnel at various wind speeds. Microdunes develop only at speeds between slightly above saltation threshold and 1.1 m s^{-1}.

Bedforms were produced in the VWT under Venus conditions by Greeley et al. (1984), Bougan and Greeley (1985), and Marshall and Greeley (1992) as a function of particle size and wind speeds (Fig. 5). From these experiments, three flow regimes were identified based on threshold wind speeds for rolling/intermittent saltation, continuous saltation, and suspension. Regime 1 is characterized by longitudinal bedforms, Regime 2 has transverse bedforms, and Regime 3 has flat (featureless) beds. Bedforms produced in Regime 1 were generally parallel to the wind and were predominantly longitudinal furrows and ridges, but also included "chaotic topography" and chevron-shaped features. These bedforms were typically restricted to particles <100 μm, resulted mostly from erosion of loose sediments, and occurred at sub-saltation wind speeds. Bedforms in Regime 2 were transverse to the wind, regularly spaced, and occurred at wind speeds of 0.7 to 1.8 m s^{-1}. Their wavelengths were 10 to 15 cm and their heights were <1 cm. At the lowest wind speeds, asymmetric "microdunes" with well-defined slip faces formed (Fig. 6). Symmetrical waves with no slip faces formed at the higher wind speeds of this regime. Ridges transitional between microdunes and waves occurred between wind speeds of 1.1 and 1.4 m s^{-1}. Ridges and waves also formed in sand ~250 μm in diameter, the upper particle-size limit for microdune development. Ridges graded imperceptibly into both bedform types, suggesting that all transverse bedforms may be part of a morphological continuum. In Regime 3 the bed was generally flat. The transition from Regime 2 to Regime 3 occurred over a relatively narrow range of wind speeds (within ~0.1 m s^{-1}) and was characterized by degradation of microdunes into ridges, then waves, with an overall increase in wavelength. The size of the microdunes and the length of the longitudinal grooves may increase with time and sand supply on Venus. Although such growth cannot be tested in the VWT because of the limited size of the chamber, there is nothing inherent in the processes of formation to retard their size.

Laboratory simulations provide insight into the physics of particle entrainment by wind on Venus. They also enable study of possible small-scale bedforms and other aeolian features. Until long-term, highly capable spacecraft land on the surface of Venus, these and similar experiments provide the only means to study the effects of aeolian processes on the Venusian surface for extrapolation to data obtained from orbit.

III. TERRESTRIAL ANALOGS

Aeolian features on Earth imaged by radar provide the opportunity to study the factors that influence their appearance on radar images and enable better interpretation of the Magellan data for Venus. In this section, we discuss the appearance of terrestrial dunes, yardangs, and streaks in synthetic aperture radar (SAR).

A. Dunes

Sand dunes on Earth have been imaged by many radar systems, including AIRSAR, Seasat, and SIR-A, -B, and -C (Blom and Elachi 1981,1987; Blom 1988; Lancaster et al. 1992). There are several factors that influence the identification of sand dunes in radar images, including presence and amount of vegetation and moisture, characteristics of intradunal areas, and radar parameters such as wavelength, look direction, incidence angle, and polarization.

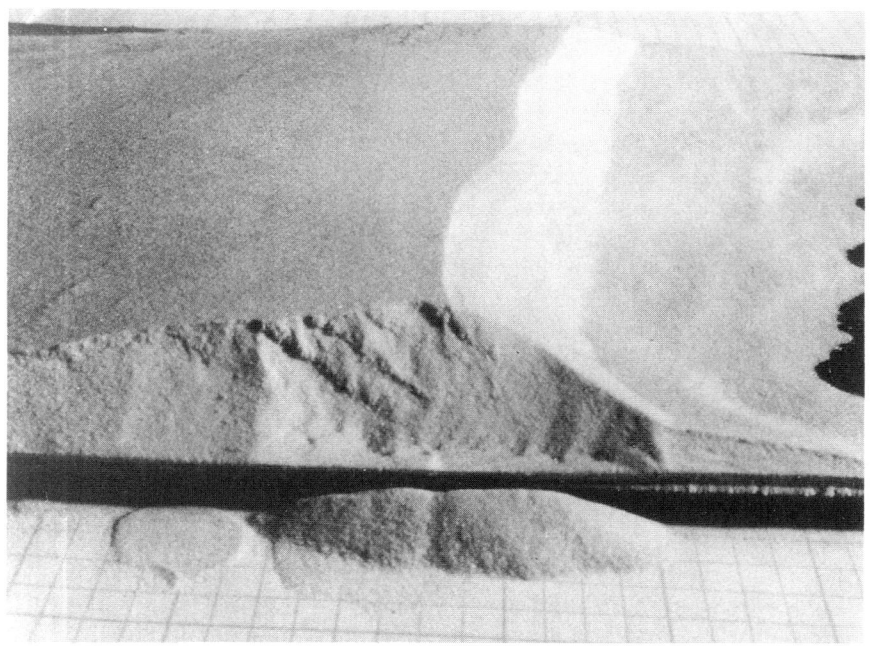

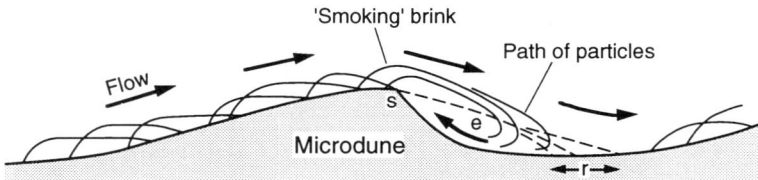

Figure 6. Cross sections of microdunes (~10 cm wavelength) developed in the Venus Wind Tunnel [(a) photograph; (b) diagram] showing typical internal bedding and particle motions characteristic of full-size transverse dunes on Earth.

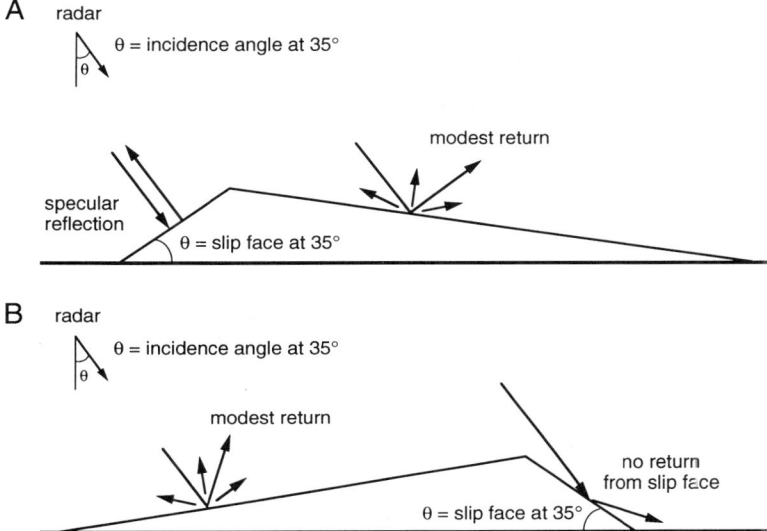

Figure 7. Visibility of dunes as a function of radar incidence angle and dune geometry. (a) Dunes with the slip face (downwind side) oriented toward the radar system would produce a specular return and appear bright; (b) Dunes oriented with the slip face on the opposite would return little energy to the radar system and might not be detected.

The orientation of dunes with respect to the radar viewing geometry is particularly critical. Most dunes are asymmetric, with steep slip faces having slopes ~35° because of the angle of repose for sand and gentle windward faces typically ≲15°. Vegetation covering dunes can affect the radar scattering from the dune surface (Blom and Elachi 1981,1987). Incidence angles ≲35° can produce a strong radar return if viewed normal to slip faces (Fig. 7). Moreover, the axis of the dune is important. If linear dunes or the crests of transverse dunes are parallel to look direction, they may not be seen on radar images. For example, the Cadiz dune field in California was imaged by Seasat on two passes (Fig. 8). The dunes are about 28 km long, 4 km wide, and 5 to 10 m high (Blom and Elachi 1981). On one pass, the radar look direction was oriented perpendicular to the dune crests and the dunes are clearly visible from specular reflections. However, in the other pass, the look direction was parallel to the dunes and the dunes are not visible. In addition, Blom and Elachi (1981) found that some dunes could be seen in Seasat images because of reflections from sub-pixel sized dune faces (only a few wavelengths in size in each dimension) oriented perpendicular to the radar look direction.

Where radar incidence angle and/or look direction are not conducive for imaging, two other factors can allow dune identification—vegetation and intradune surfaces (Blom and Elachi 1987). For example, active dunes are typically free of vegetation while inactive areas may support vegetation. Be-

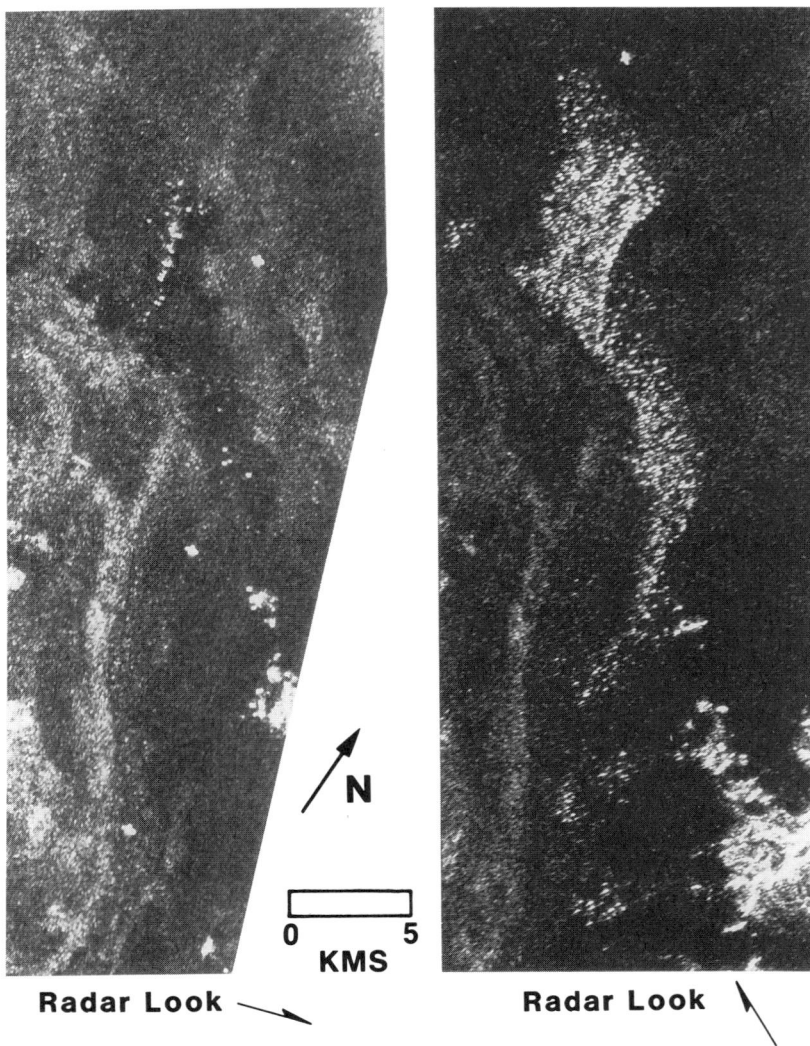

Figure 8. Seasat radar image of the Cadiz dunes in California. Dunes are not visible in the ascending pass (left image) because their trend (N70°E) is nearly parallel to the radar look direction. They are visible as bright points in the descending pass (right image) because of specular returns from dune faces.

cause vegetation tends to increase the backscatter energy, the dunes may be visible as relatively radar-dark zones set on a brighter background. Similarly, some intradunal areas consist of rough surfaces or gravel deposits which would also appear bright. The radar dark dunes then stand out in contrast against the bright intradunal areas.

Lack of vegetation and moisture on Venus results in incidence angle, dune orientation, and intradunal surface characteristics being the primary factors influencing the identification of dunes. In the Magellan data, incidence angles <35° were acquired at latitudes greater than 45°N and 25°S for cycle 1 data and for all cycle 2 data. Right and left look directions were used along the generally north-south flight track. Thus, the detection of dunes would tend to be limited to those having crests oriented north–south at all latitudes, and with a bias for dunes having slip faces on the west side for cycle 1 data and the east side for cycle 2.

B. Yardangs

Yardangs are elongate hills and ridges typically composed of relatively soft or friable materials which are eroded by the wind. Yardangs generally have the appearance of an inverted boat hull and can have a keel-like ridge along the crest (McCauley et al. 1977). Terrestrial yardangs were subdivided by Cooke et al. (1993) into three size classes, with megayardangs referring to those a km or more in length. This class is typified by yardangs of the African Tibesti massif which average about 200 m high, several hundred meters wide by many km long, and are spaced 0.5 to 2 km apart.

Terrestrial yardangs are found in desert regions where vegetation and soils are minimal and where strong unidirectional winds occur during most of the year. Relatively few yardangs have been imaged by radar and data for comparison with Magellan images are limited. The Borkow region of Chad contains yardangs and was imaged by SIR-B and SIR-C (Fig. 9). This region is characterized by winds which blow from the north–northwest 8 months of the year with average velocities of 3 m s^{-1} during summer and 22 m s^{-1} during winter which have sculpted the sandstone bedrock into yardangs (McCauley et al. 1977). In the SIR-B radar image, the bare bedrock surface of the yardang produces a bright radar return. Sands deposited in the troughs between individual yardangs yield a dark tone on the image. As with all linear structures, the features are best seen when they trend perpendicular to the radar look direction. The overall appearance of the Borkow yardangs is of long, linear, closely spaced features, giving a "combed" appearance to the surface.

C. Wind Streaks

Wind streaks on Earth typically form by the interaction of local topography, sediments, and prevailing winds. They are visible on optical images because of differences in albedo and spectral properties of bedrock exposures and mantling sediments, and differences in vegetation. They are also visible in SAR images because some of these factors influence radar backscatter cross sections.

Terrestrial wind streaks have been imaged by radar in several areas, including Bolivia, Afghanistan, Saudi Arabia, and China. In the Altiplano of Bolivia, wind streaks were identified in SIR-A radar images (Fig. 10).

Figure 9. Shuttle Imaging Radar (SIR-B) view of wind-eroded features in Borkow Province, northern Chad. The eastern (right) part of the image shows prominent yardangs eroded from Devonian sandstones; the dark streaks to the west (left) represent streaks and sand-filled depressions eroded by the wind. Area shown is about 35×85 km (SIR-B data-take 05-107.20, scene 006) (figure from Ford et al. 1986).

Figure 10. Shuttle Imaging Radar (SIR A) view of part of the Altiplano in Bolivia showing various radar-dark streaks (A,B) associated with small hills (C) and a ridge (D). Sand-size particles in the streaks are derived partly from volcanic ash erupted from the volcano, Cerro Quemado (E). Prevailing winds are from the west (left side); north is to upper right. The area shown is 39 × 65 km (SIR-A data-take 31, ASU photograph 2461-H).

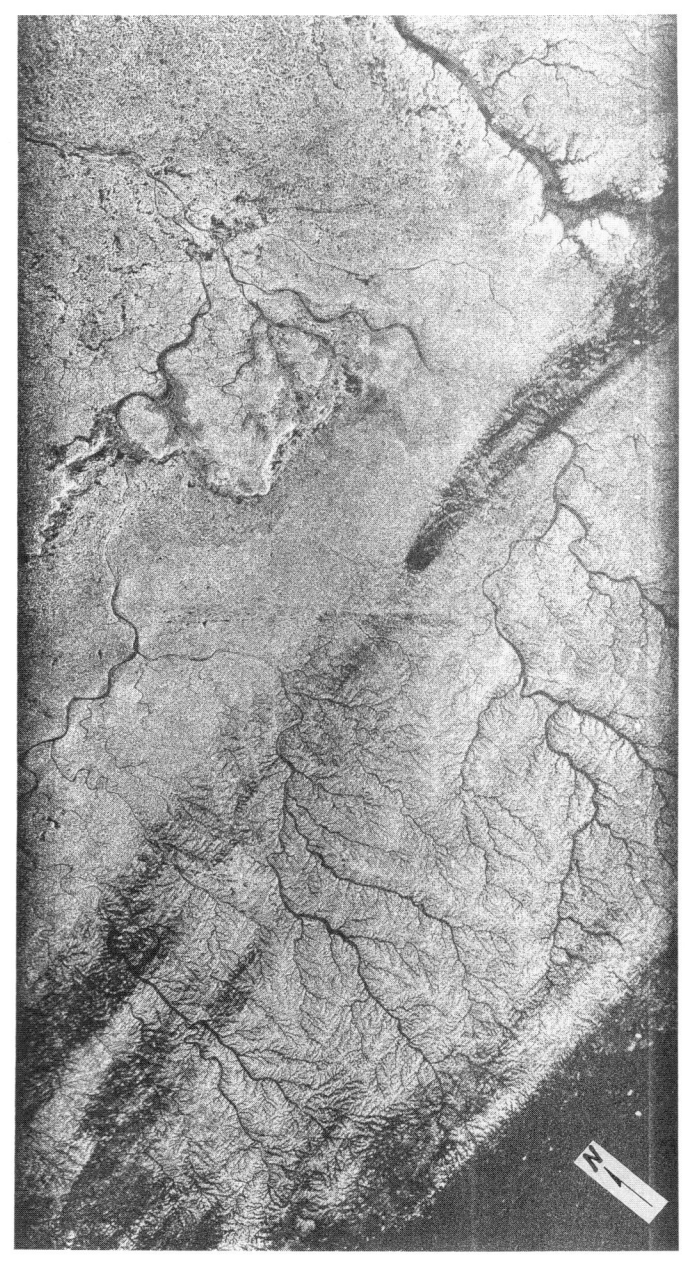

Figure 11. Shuttle Imaging Radar (SIR-A) image of the eastern edge of An Nafud, Saudi Arabia, showing wind streaks formed from sand. Prevailing winds are from the west (upper left). The area shown is about 65×50 km (SIR-A data-take 28) (figure from Breed et al. 1982).

Field work showed that streaks were visible as areas of active sand movement set against a vegetated background (Greeley et al. 1989). Other streaks in the area consist of volcanic ash deposits that were funneled through narrow gaps in ridges. The deposits are radar dark and can be tracked several km downwind from the ridge. AIRSAR data for this area show that the appearance of the streaks varies with radar wavelength and polarization (Blumberg and Greeley 1995). Wavelength-derived variations may be due to interaction with the vegetation and/or the different depths of sand penetration as a function of wavelength. Like-polarized data show more of the wind streaks than cross-polarized data for the shorter wavelengths, with HH showing a slight enhancement of the streaks in comparison to VV.

Figure 11 shows a prominent radar dark wind streak imaged by SIR-A in Saudi Arabia. This feature consists of a sand deposit more than 20 km long and numerous smaller deposits which form streaks (Breed et al. 1982). Other streaks imaged during SIR-A include multiple linear features in Afghanistan. These streaks consist of sand and gravel deposits of contrasting radar backscatter. Wind streaks at Roter Kamm, Namibia were imaged by SRL-1 in 1994. It was found that streak visibility is dependent on several instrument parameters. In X-band (3-cm wavelength), the streaks are indistinct at an incidence angle of 25°, but are prominent at an incidence angle of 53° (Greeley and Blumberg 1995).

IV. MAGELLAN RESULTS

A. General

The Magellan spacecraft, launched from the space shuttle Atlantis in 1989, mapped 98% of the surface at SAR spatial resolutions of $\sim$100 m per pixel. The Magellan mission provided the first comprehensive view of the geologic diversity of Venus. Aeolian features were clearly seen in the initial analyses of the data (Arvidson et al. 1991). More detailed studies of aeolian features were reported by Greeley et al. (1992,1994,1995). Among the features identified are dunes, yardangs, and wind streaks. To facilitate discussion, these features are described by their various radar backscatter cross sections as radar-bright, radar-dark, radar-mottled, etc., as they appear in the SAR images.

B. Dunes and Other Bedforms

Two distinctive dune fields and several areas of possible microdunes have been identified in Magellan data. As outlined by Weitz et al. (1994), other dunes and dune fields may be present on Venus, but the available data may be inadequate for their detection.

Aglaonice Dune Field (Menat Undae). The Aglaonice dune field (Fig. 12), known formally as Menat Undae, is located at 25°S,340°E, approximately 100 km north of the impact crater Aglaonice (Greeley et al. 1992; Weitz et al. 1994). The dune field consists of hundreds of 200 m and

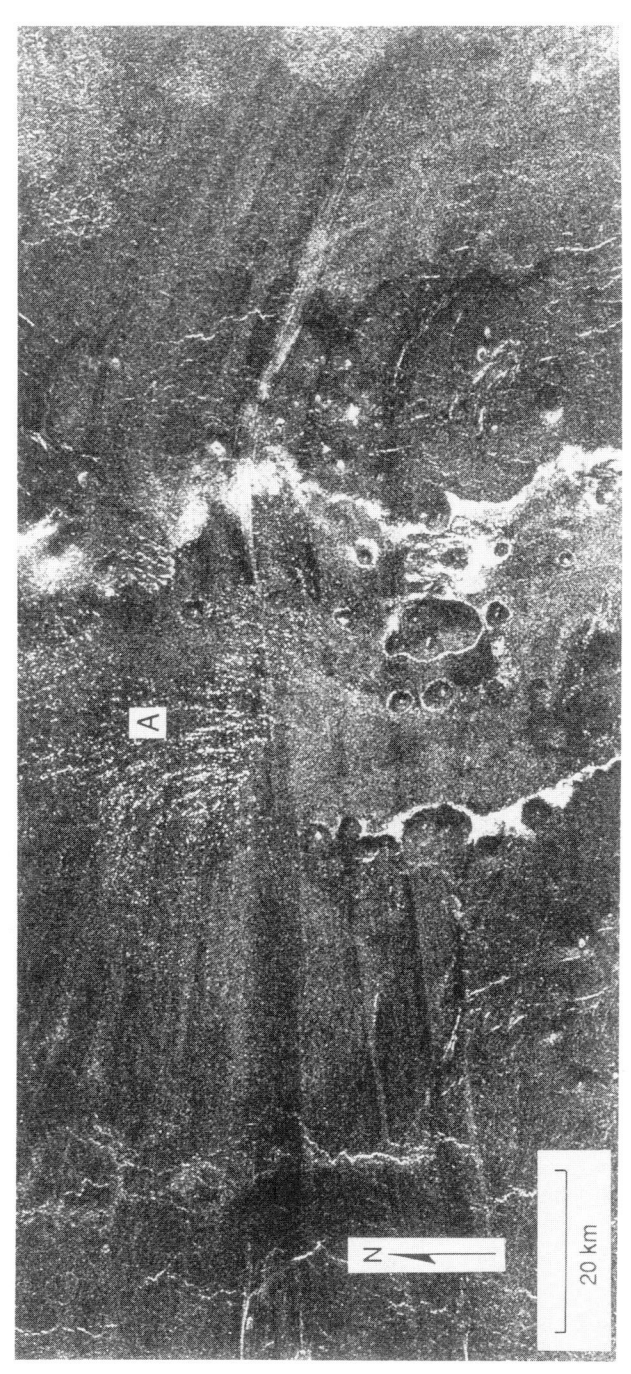

Figure 12. Aglaonice dune field (Menat Undae), centered at 24.8°S; area shown is 78×180 km. This dune field, indicated by the specular pattern at A, is located within an outflow associated with the Aglaonice impact crater. Radar-dark linear streaks sweep across the area, suggesting winds from the east (right) toward the west (left). If this wind orientation is correct, the proposed dunes would be transverse forms (Magellan MRPS 34032).

larger dunes that have crests oriented north–south. Their speckle-like appearance results from near-normal reflections at the Magellan 35° incidence angle. The dune field covers an area of about 1300 km² and is found inside an ejecta outflow deposit from Aglaonice crater. Radar-bright and -dark wind streaks associated with these dunes suggest winds from the east, indicating that the dunes are transverse to the wind flow. The dune deposits apparently were derived from the fine-grained ejecta generated from the Aglaonice impact, which were subsequently reworked by the wind.

Fortuna-Meshkenet Dune Field (Al-Uzza Undae). The Fortuna-Meshkenet dune field is centered at 67°N,91°E in a valley between Ishtar Terra and Meshkenet Tessera (Fig. 13). Within the dune field, about 40 radar-bright wind streaks have been identified in association with radar-bright hills (Weitz et al. 1994). Both the dunes and wind streaks have orientations which indicate a southeast to northwest wind flow that shifts to a westward direction in the northern part of the field. As shown in Fig. 13, a cycle 1 image taken at 25° incidence angle from a left viewing geometry (Fig. 13a) is compared to a cycle 2 (Fig. 13b) image taken at 25° incidence angle from a right viewing geometry. Comparison of the two images taken 8 months apart shows no significant differences; however, this should not be surprising considering that even the largest changes expected in terrestrial dunes over the same time span would not be resolved with Magellan data.

Figure 14 shows part of the dune field for cycles 1 and 2. Radar-bright and -dark features range from 0.5 to 10 km in length and are 0.2 to 0.5 km wide with an average spacing of 0.5 km. The dunes are transverse to the wind streaks and do not cross through the streaks. Although the dunes appear to be transverse forms with bright backscatter from slip faces oriented near-normal to the radar illumination, there are several observations which suggest that this explanation is incorrect. Both images were acquired at 25° incidence angle but from opposite directions. If the dunes are asymmetric transverse dunes with slip faces greater than 25°, then the backscatter cross sections and apparent positions of the dunes should change between cycles 1 and 2 because of the different viewing geometry. However, because the shape and backscatter of the dunes vary only slightly, the dunes probably do not represent radar-bright returns from near-normal slip faces. Moreover, many of the dunes in the southern part of the field appear to be parallel to the radar illumination. In order for these dunes to be radar-bright, there must be some other scattering effect, such as change in the dielectric constant or surface roughness. If grains with a high dielectric constant were concentrated near the crests of symmetrical dunes, then this could explain the bright backscatter in both images.

The most likely explanation for the appearance of the dunes is that they represent radar-dark features superposed on a radar-bright basement (Weitz et al. 1994). The bright basement could have been produced by the 13-km diameter crater, Jadwiga, either by scour from the impact shock wave or by lack of deposition of fine particulate ejecta. Jadwiga or the impact crater to the east,

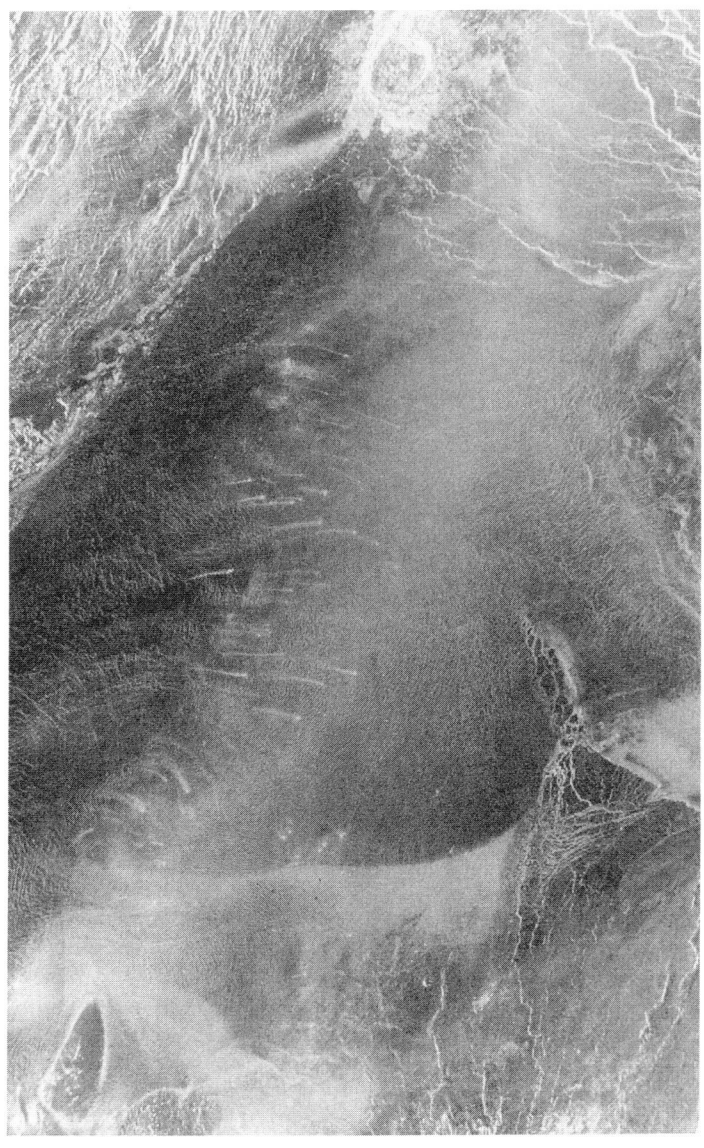

Figure 13a. The Fortuna-Meshkenet dune field (Al-Uzza Undae) centered at 67°N,91°E. About 40 radar-bright wind streaks are located in the field and they indicate a wind flow to the northwest at the bottom of the field and a westward flow in the rest of the field. The impact crater Jadwiga is located at the top of the image. Radar illumination is from the left at 25° incidence angle. Area shown is 280×170 km (Magellan MIRPS 39824).

AEOLIAN PROCESSES AND FEATURES 569

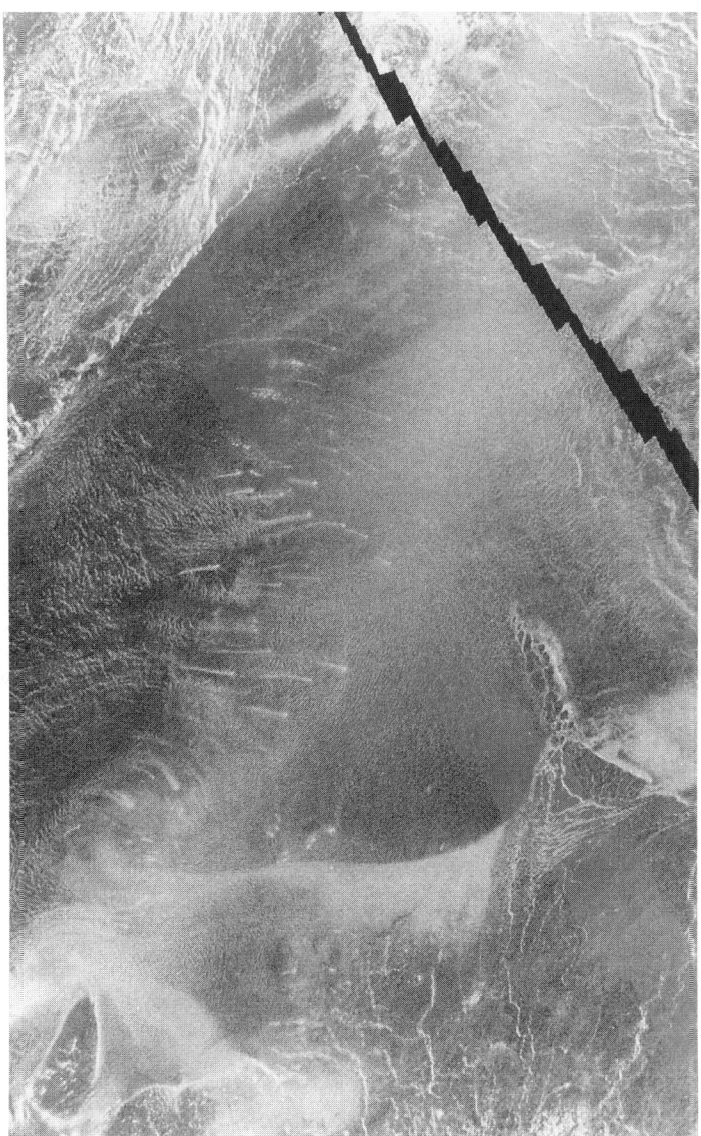

Figure 13b. Image showing the same area covered in Fig. 13a except the radar illumination is from the right at 25° incidence angle. No movement of the dunes or wind streaks can be identified over the 8-month interval between the cycle 1 and 2 acquisition.

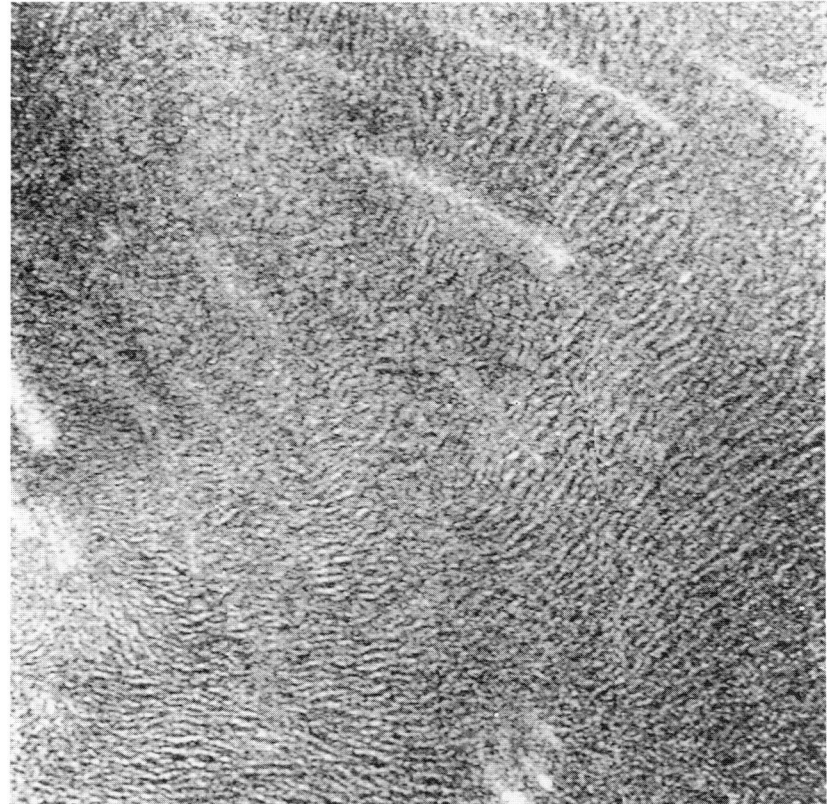

Figure 14a. Enlargement of center of part of Fig. 13a. The dunes are oriented transverse to the radar-bright wind streaks, even as the orientation of the streaks changes across the field, supporting an aeolian origin for the linear features. Area shown is 55 by 55 km.

La Fayette, could have generated sand-size debris in the Fortuna-Meshkenet depression, with subsequent winds modifying the deposits to produce the dunes. As long as the underlying terrain has a greater radar backscatter cross section than the dunes viewed at 25° incidence angle, the dunes will appear dark on a brighter background, regardless of illumination direction.

Microdunes. Several potential microdune fields were identified in the southern hemisphere of Venus (Weitz et al. 1994). The microdunes are inferred from variations in radar brightness observed on different Magellan mapping cycles. One region near the crater Stowe (43.2°S,233.0°E) is shown in Fig. 15. The cycle 1 and 3 images were acquired with radar illumination from the left at an incidence angle of 25° while the cycle 2 image was taken from the right at 25°. A change in the incidence angle or a change in surface properties

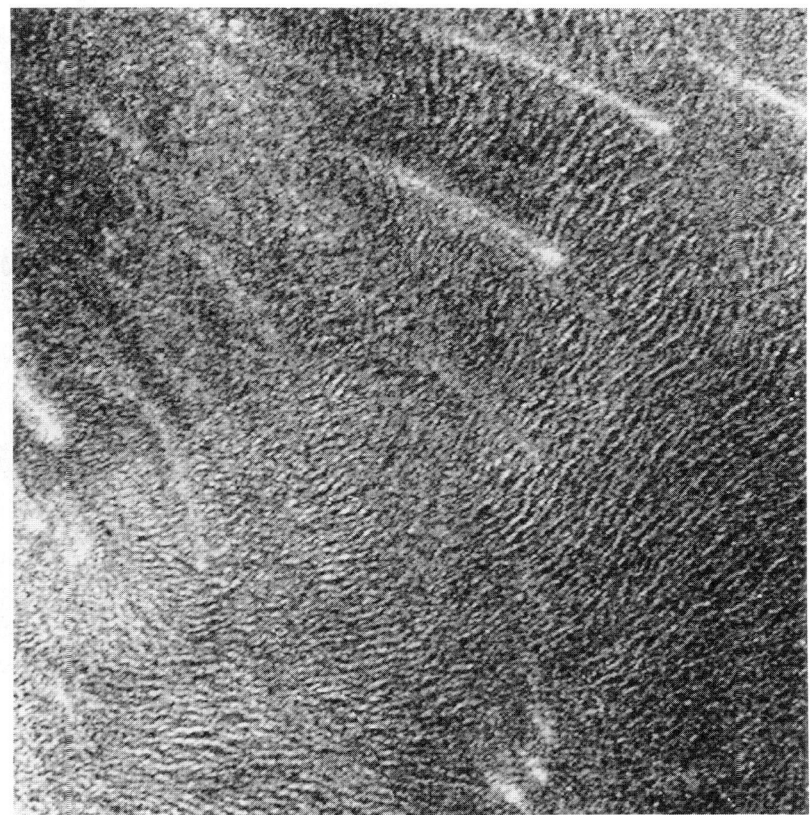

Figure 14b. Enlargement of the cycle 2 image (Fig. 13b). Even though the radar illumination is now from the opposite side as that shown in Fig. 14a, the shape and backscatter cross sections do not change for the dunes under the different viewing geometries. If the Fortuna-Meshkenet field consists of smooth, radar-dark dunes on a rough, radar-bright basement, then this would explain the cycle 1 and 2 images.

between the cycles cannot explain the observed increase in brightness in the cycle 2 image. Microdunes and/or ripples with asymmetry in the east–west direction may explain the radar brightness variations (Plaut et al. 1992; Weitz et al. 1994).

Microdunes were also proposed for the area near the craters Guan Daosheng (61.1°S,181.8°E) and Eudocia (59.1°S,201.9°E). Guan Daosheng has a radar-dark parabolic deposit surrounded by a radar-bright parabolic deposit observed in the cycle 1 image, taken at 20° incidence angle from the left (Fig. 15a). The cycle 2 right-looking image (Fig. 16b), taken at 23° incidence angle, shows a radar-dark plain surrounding the crater but no bright parabolic deposits. Cycle 1 and 2 images of Eudocia show a similar disappearance of the parabolic deposits in the cycle 2 image (Fig. 17).

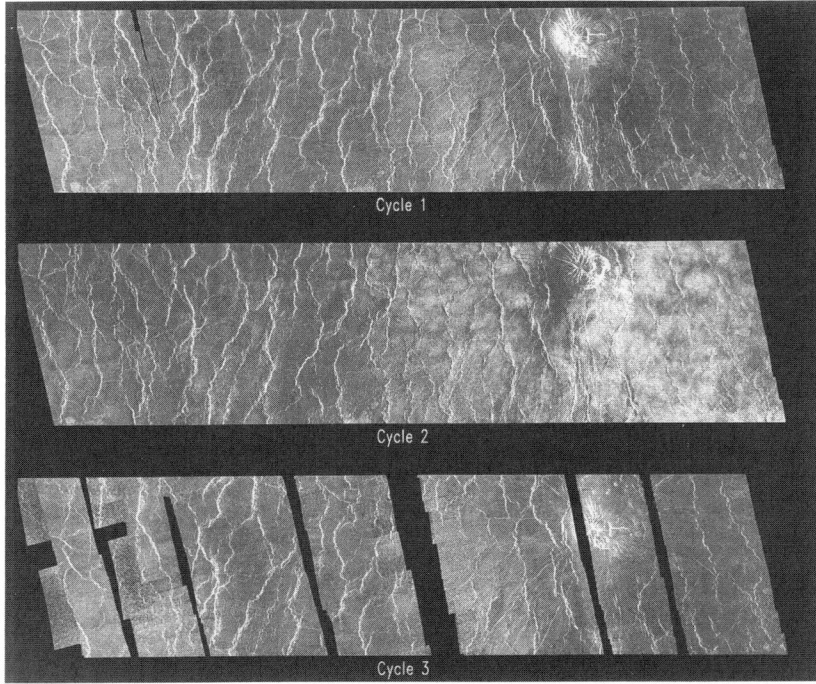

Figure 15. Set of three images showing the same area centered at 47.5°S,226°E in all three Magellan cycles. The cycle 1 and 3 images were taken from the left at 25° incidence angle while the cycle 2 image was acquired at 25° incidence angle from the right. The bright patches seen in the cycle 2 image may result from microdunes that have an asymmetry in the east–west direction. The impact crater Stowe, located 500 km to the east, created fine-grained material during the impact event that may have settled out to form the unusual surface deposits in this region. Each image covers 100×400 km (Magellan P-40691).

To account for the change in backscatter, proposed microdunes at Guan Daosheng and Eudocia may have slopes of 20 to 21° on the west and much different angles on the east. Bragg scattering from the microdunes is possible if the surface undulations have a horizontal wavelength spacing that allows the reflections to be constructive. Weitz et al. (1994) calculated that the microdunes in the Stowe region would need to be spaced 15 cm apart for Bragg scattering, which is comparable to the 20-cm wavelength spacing for the microdunes produced in laboratory experiments (Greeley et al. 1984). In addition, subpixel reflections from near-normal slopes on the microdunes in only one direction could explain the bright backscatter in only one look direction.

The location of microdune fields, all associated with craters having parabolic deposits, are shown in Fig. 18. The microdunes could have formed during the emplacement of the parabolic deposits; perhaps they were generated when convection cells from the impact interacted with the westward

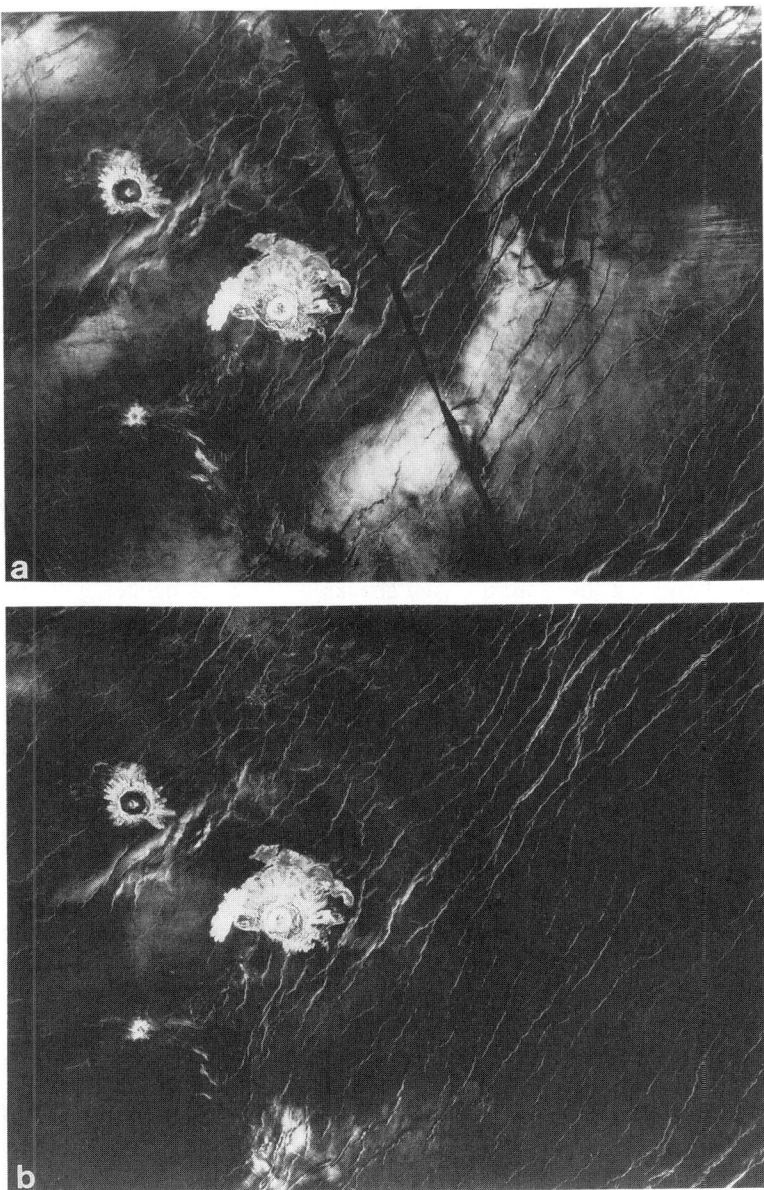

Figure 16. (a) The radar-bright crater Guan Daosheng with an associated parabolic radar-dark deposit surrounding a radar-bright deposit in this cycle 1 image taken at 20° incidence angle from the left. Area shown is 860×700 km. (b) Cycle 2 image at 23° incidence angle from the right of the same area. The radar-bright parabola and many of the wind streaks seen in the cycle 1 image are not visible in this image. The bright patches in the cycle 1 image most likely result from reflections from small dunes formed during or after the Guan Daosheng impact event (Magellan P-40157).

zonal winds in the upper atmosphere at the time of the impact (Greeley et al. 1994).

Two dune fields and several inferred fields of microdunes have been identified on Venus. The apparent scarcity of dunes could be an artifact of the Magellan data. To be identified in the radar data, dunes must be larger than several pixels across (>500 m); moreover, unless smaller dunes are oriented perpendicular to the radar look direction to produce sub-pixel reflections, they will not be discernible. As discussed above, many terrestrial dunes in radar images appear as radar-dark lineations at larger (>33°) look angles (Blom and Elachi 1987). Hence, some dunes on Venus may be interpreted incorrectly in the Magellan data as radar-dark wind streaks.

A lack of sand-size particles and appropriate wind speeds may also contribute to the apparent paucity of dunes on Venus (Weitz et al. 1994). On Earth and Mars, most sand-size particles result from weathering and erosion involving water. In the absence of running water on Venus, sand-size particles probably form by mechanical weathering associated with processes such as volcanism, mass wasting, and impact cratering, perhaps enhanced by chemical weathering. Mass movements have been identified on steep terrain in the tessera (Malin 1992) and on the sides of small volcanic domes (Guest et al. 1992) but there are no wind streaks or dunes associated with debris. Schultz (1992) identified several craters with associated radar-bright zones that he interprets as microdunes produced by uprange wake blasts from the impact process. However, because most impact craters have no identifiable dunes associated with them, either the impacts do not produce sufficient sand-size particles (consistent with the estimates of Garvin 1990) or local winds generated by impact craters are inadequate to produce dunes. Impact-related winds apparently generated some wind streaks (Schultz 1992; Greeley et al. 1994), but the winds may be too strong or too short in duration to produce dunes. Simulations in the VWT show that microdunes are destroyed at wind speeds above ~ 1.5 m s^{-1}, supporting the hypothesis that wind speeds may be too strong during the impact process to form dunes. The combination of a globally starved sediment supply of sand-size particles, the lack of winds capable of moving enough of these particles to form dunes, and the difficulty in achieving the proper viewing geometry required for identifying dunes may all explain why only a few dunes were identified in the Magellan images.

C. Yardangs

Features on Venus having the appearance of yardangs were found southeast of Mead crater (Fig. 19) (Greeley et al. 1992). However, the definitive identification of these features as yardangs is lacking because of limited image resolution. For example, it is not possible to determine confidently if the observed features are yardangs or wind streaks. They differ from streaks, however, in that the margins are sharply defined and have a combed appearance whereas streak margins tend to be diffuse. In addition, the planimetric form and spacing are comparable to the megayardangs described by Cooke et

AEOLIAN PROCESSES AND FEATURES 575

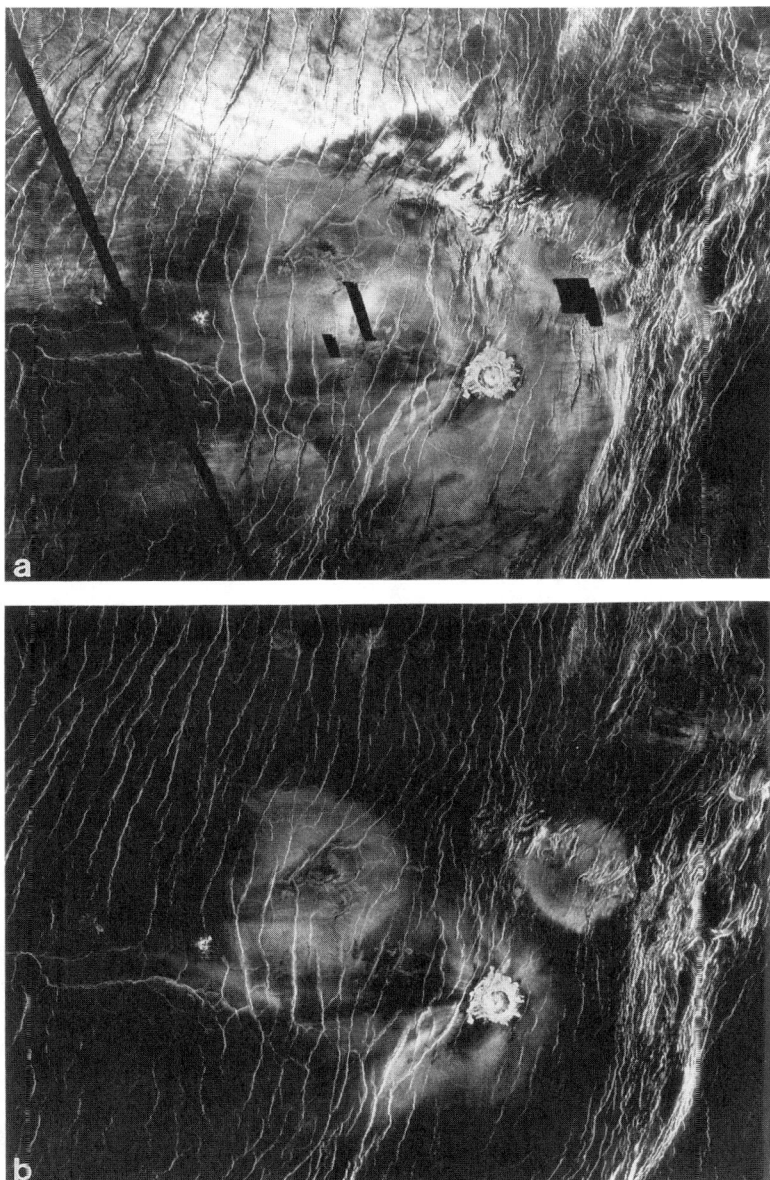

Figure 17. (a) The crater Eudocia with a radar-bright parabola surrounded by a radar-dark parabola and numerous wind streaks associated with the parabolas in this cycle 1 image taken at 21° incidence angle from the left. Area shown is 860 × 700 km. (b) Cycle 2 image at 24° incidence angle from the right of the same area. Although several bright patches are still visible, the radar-bright parabola and wind streaks seen in the cycle 1 image have disappeared. The changes between cycle 1 and 2 may be due to microdunes that return strong reflections only in the left-looking images (Magellan P-40156).

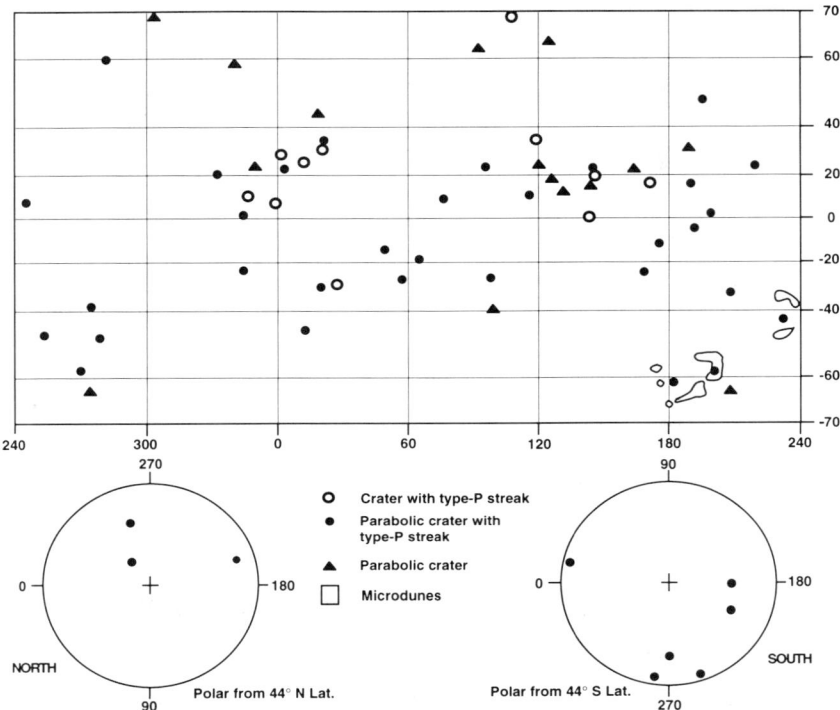

Figure 18. All of the microdune fields (shaded areas) on Venus associated with craters having parabolic deposits (filled symbols). In addition the three craters with microdunes also have associated Type P wind streaks.

al. (1993) on Earth.

The Venusian yardang field covers some 40,000 km² and lies about halfway between Mead Crater and Aphrodite Terra, west of Hestia Rupes. The center of the field is 480 km southeast of the center of Mead Crater. An elevated belt of topography trending northeast is found between the yardang field and Mead Crater. Reticulate plains (volcanic plains modified by wrinkle ridges) lie to the north and west of the field. Individual yardangs average 25 km long by 0.5 km wide and are spaced 0.5 to 2 km apart (Fig. 19). The yardangs occur in curvilinear sets which strike N33°E to N54°E. These orientations do not correlate with the linear patterns of the reticulate plains in the area, suggesting that their formation is not structurally controlled, at least not by structural features visible in the near vicinity.

In order to gain insight into the possible origin of the yardangs, wind streaks were analyzed to assess local and regional wind patterns (Greeley et al. 1995). The general area of the yardang field and Mead Crater has the greatest concentration of wind streaks on Venus. As noted below, many

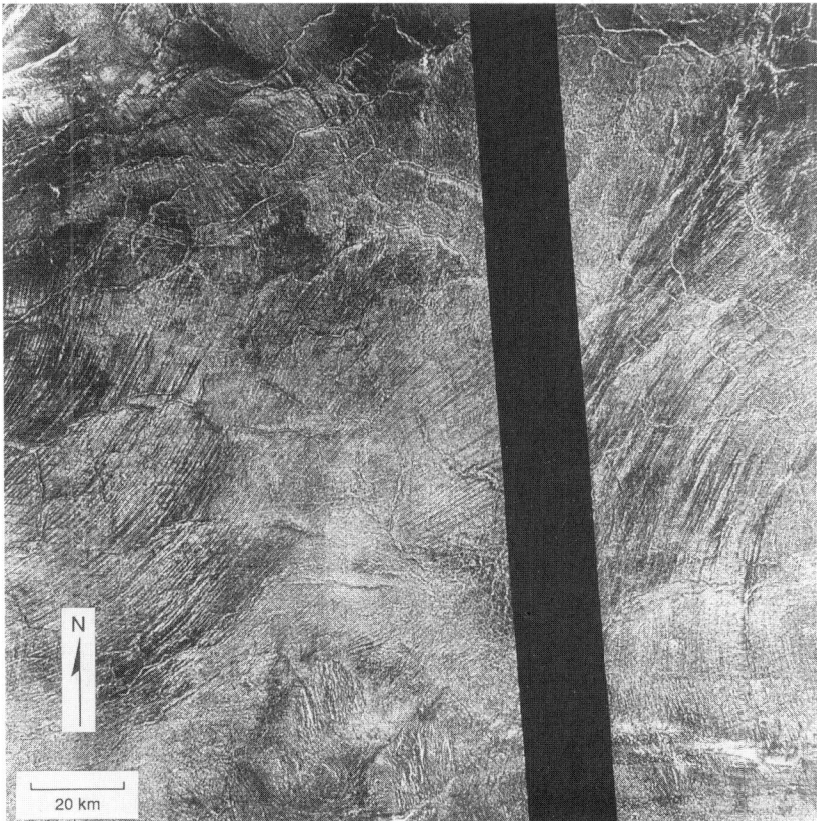

Figure 19. Venus yardangs, centered at 9°N, 60.7°E. Area shown is ~200×200 km (Magellan MRPS 37879).

streaks are found in association with impact craters, and it is presumed that the presence of fine-grained impact ejecta enables streaks to develop. Mead Crater, northeast of the yardang field, and Adivar Crater, 1000 km east of the field, probably generated sufficient ejecta deposits which could account for the abundance of the streaks. Many of these streaks, however, are thought to have resulted from transient winds produced by impact processes and might not reflect long-term near-surface wind patterns of the sort required for the formation of yardangs. Consequently, these streaks were not used in the analysis of wind directions for comparison with the yardangs. Rather, orientations of fan-type streaks were studied; this category of streaks is considered to reflect local, long-term winds (Greeley et al. 1994). Fan-type streaks occur north and northeast of the yardang field and are oriented south toward the equator and consistent with a Hadley circulation. Examination of regional topography (Ford and Pettengill 1992) shows that the yardang field is located in a shallow, elongate depression between Aphrodite Terra to the southeast and a ridge belt to the northwest. Greeley et al. (1995) proposed that the generally southward-flowing winds (indicated by the fan-type wind streaks) are deflected southwestward by the high-standing Aphrodite massif and are funneled through the linear depression. This proposed wind flow is parallel with the strike of the yardangs. In this model, the yardangs are inferred to result from erosion of plains materials by long-term winds that are accelerated locally through this area.

D. Wind Streaks

A global search of all Magellan BIDR (Basic Image Data Record) and MIDR (Mosaicked Image Data Record) prints yielded 5736 wind streaks on Venus (Greeley et al. 1995). A data base for each streak was compiled which includes: (1) location key by image number; (2) latitude; (3) longitude; (4) length; (5) width; (6) azimuth; (7) streak type (see explanation below); (8) reflectivity (see explanation below); (9) topographic feature with which streak is associated (such as hill or ridge); (10) terrain type in which streak occurs (such as smooth plains); (11) local slope azimuth; (12) local slope magnitude; (13) slope gamma (the angle between the downslope azimuth and the downwind streak azimuth); (14) elevation at streak origin; (15) incidence angle; (16) multi-cycle image identifier—BIDR number and comments for streaks imaged in multiple cycles (see below); and (17) general notes/comments. (Complete listing of the data set can be found on the Venus II CD-ROM [CDP4C3T1].)

Wind streaks on Venus occur in different shapes and radar backscatter cross sections, which are the primary factors for streak classification. Streak types (Fig. 20) include (a) linear (length to width ratio $>20:1$, occurs as single or multiple features); (b) fan (length to width ratio $<20:1$); (c) transverse; and (d) wispy. Streak reflectivities include bright, dark, mixed, and "zebra." Bright reflectivity is interpreted to indicate sediment removal by the wind, thus exposing a rougher (radar brighter) surface; dark reflectivity is interpreted to represent sediment deposition, causing a relatively smooth surface. Zebra streaks occur in multiple sets, and are either dark streaks on a bright background or bright streaks on a dark background. An additional qualifier is identified for dark or zebra linear streaks which occur in association with parabolic ejecta deposits—these are denoted as "Type P" streaks in the data set (see Fig. 18 for location). Type P streaks account for 1856 of the total 5736 streaks identified. The number of streaks by type and reflectivity are listed in Table I. No streaks were seen to change during the Magellan mission, although repetitive data that can be compared are limited.

Analysis of the data base shows the global distribution of streaks and has been used to assess the various characteristics of wind streaks on Venus (Table I). Most streaks (79%) are linear forms of dark and zebra reflectivities. These streaks are also the longest, with one measuring >700 km. Wind streaks have been identified as short as 1 km long, although the identification

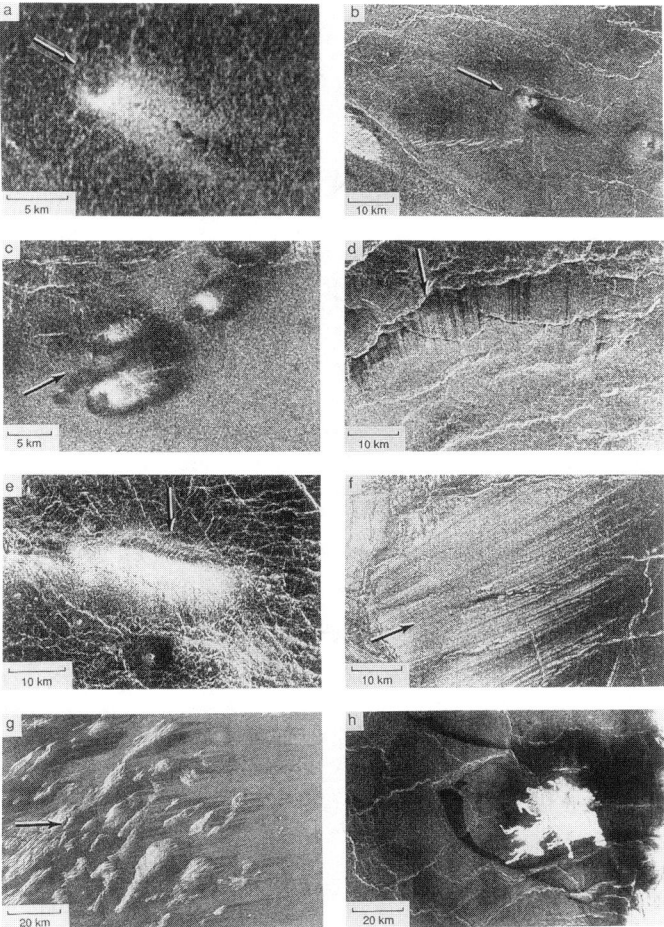

Figure 20. Venus wind streaks (arrows indicate inferred downwind direction). (a) Radar-bright fan-shaped wind streak 10.5 km long associated with a small hill in eastern Niobe Planitia, centered at 36.5°N, 174.6°E (Magellan F-BIDR 1194). (b) Radar-dark fan-shaped wind streak about 10 km long associated with a small hill centered at 29.4°N,57°E (Magellan MRPS 40983). (c) Radar-bright and -dark (mixed) fan-shaped wind streak in the Carson crater area, centered at 23°S,344.9°E. Area shown is about 25×36 km (Magellan F-MIDR 23S345). (d) Multiple linear radar-dark streaks associated with a ridge centered at 37.5°N,65.5°E. Area shown is about 44×64 km (Magellan MRPS 3883). (e) Transverse wind streak (radar-bright) associated with a ridge in Guinevere Planitia, centered at 26.2°N,331.4°E. Area shown is about 39×57 km (Magellan F-MIDR 25N333). (f) "Zebra" streaks in the vicinity of Mead crater, centered at 15°N,65°E. Area shown is about 44×64 km (Magellan MRPS 37877). (g) Multiple linear streaks (radar-dark) in western Aphrodite, centered at 0.9°S,7.1°E. Area shown is 82×120 km (Magellan F-MIDR 00N070). (h) Radar-dark wispy streak in eastern Sedna Planitia, centered at 37°N,2°E. Area shown is about 87×128 km (Magellan C1-MIDR 30N009).

TABLE I
Numbers of Wind Streaks By Type and Radar Reflectivity Characteristics

	Bright	Dark	Mixed	Zebra	Totals
Linear	427	1476	33	3075	5011
Fans	399	76	76	0	551
Transverse	56	4	0	0	60
Wispy	0	114	0	0	114
Totals	882	1670	109	3075	5736

of streak lengths is limited by the spatial resolution of the Magellan data and it is likely that shorter wind streaks exist.

Histograms of streaks by latitude and longitude (Fig. 21) show that their distribution is not uniform. Most streaks are found on plains (Fig. 22). However, this may be due in part to the discernibility of streaks on the uniform SAR-return produced by the plains. Areas with abundant structures (fractures and ridges) produce complex SAR returns and make streak identification difficult, if not impossible. As systematic geologic mapping becomes available, the wind streak data base will be expanded to include the geologic unit on which streaks occur and enable possible correlations. Streaks also tend to occur in association with: (1) parabolic ejecta deposits; (2) fields of small volcanoes which may provide particulate material such as ash; and (3) the area around the 280-km-diameter Mead Crater, the largest impact crater preserved on Venus.

One of the primary goals for analyzing wind streaks is to assess near-surface wind patterns. The distributions of wind streak azimuths (the downwind direction) are shown for the northern and southern hemispheres of Venus in Fig. 23. The bimodal distributions in both hemispheres results from the inclusion of parabolic ejecta deposit (Type P) streaks. When these streaks are deleted, distribution becomes somewhat unimodal (Fig. 24). Azimuths are generally equatorward in the northern hemisphere and to the NNE in the southern hemisphere. Azimuths of the Type P streaks are westward for both hemispheres.

Type P streaks comprise 32% of the data base. They are classified by their association with parabolic ejecta deposits of craters and probably formed in response to the impact event. The westward azimuth of this class of streak is most likely related to wind patterns other than the general near-surface winds. Type P streaks are dark- or zebra-reflectivity linear types which "drape" over the topography, and are seemingly unaffected by ridges, hills, or large volcanic features. They generally exceed 30 km in length and occur in groups numbering from tens to hundreds of streaks. Type P streaks are found in association with 44 of the 60 parabolic ejecta deposits identified on Venus by Campbell et al. (1992).

E. Atmospheric Circulation

The Magellan wind streak data set can be interpreted to reveal different aspects

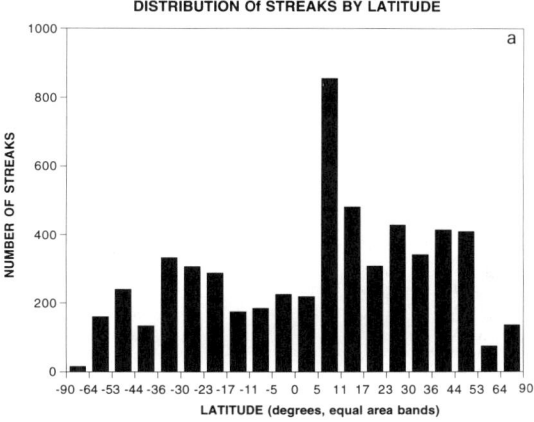

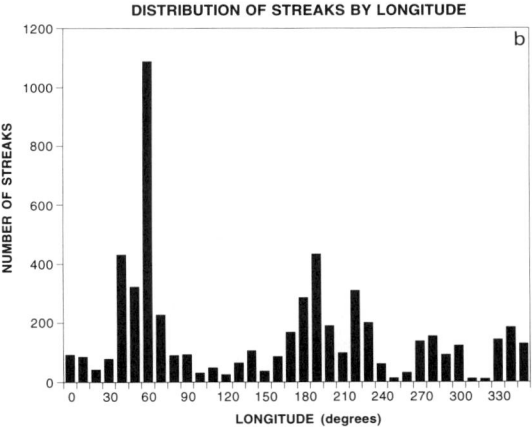

Figure 21. Distribution of Venus wind streaks by equal-area bands of latitude (a) and by longitude (b).

of the Venus atmospheric circulation. As noted above, Type P streaks are associated with parabolic ejecta deposits of craters and probably formed as a direct consequence of the impact event. However, the winds generated directly by the impact were probably not responsible for producing the Type P streaks because of the great lengths of these streaks and their alignment with the westward direction of the high altitude retrograde atmospheric superrotation. Instead, the Type P streaks were likely created by a process similar to the one involved in the emplacement of the parabolic ejecta deposits (Arvidson et al. 1991; Vervack and Melosh 1992; Campbell et al. 1992). As suggested by W. I. Newman and described in Greeley et al. (1994), Type P streaks may form by the deposition of impact ejecta raised to great heights and transported downwind (westward) by the high altitude westward superrotation. The zebra-pattern of Type P streaks might be a consequence of a modulation

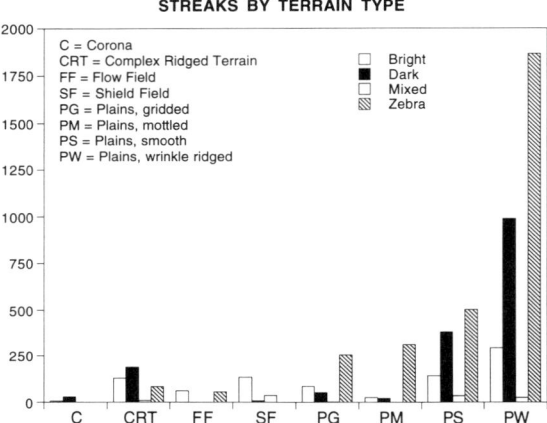

Figure 22. Distribution of streaks by terrain types on which they occur.

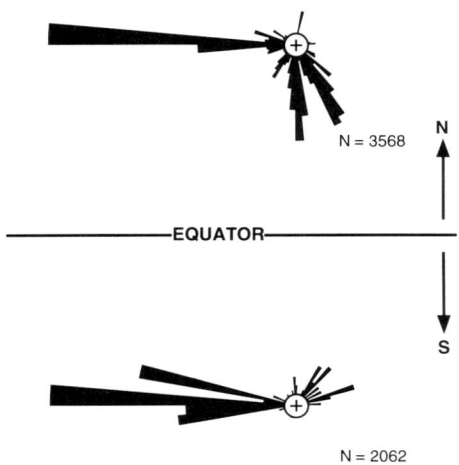

Figure 23. Azimuths of all wind streaks on Venus given in downwind direction.

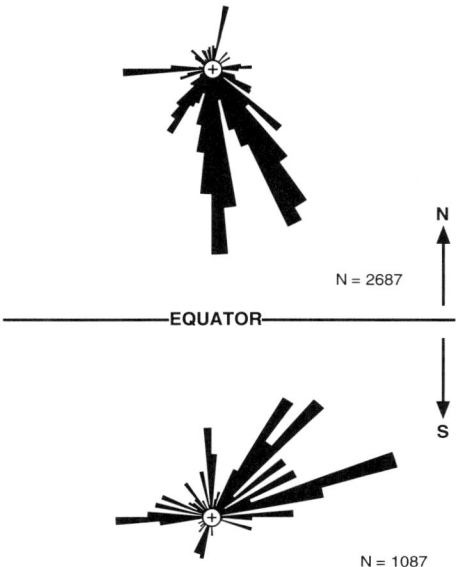

Figure 24. Azimuths of Venus wind streaks excluding Type P streaks, given in downwind direction.

of the deposition pattern of the ejecta by near-surface roll convection cells generated by impact heating of the surface and atmosphere and oriented with their axes east–west by the high altitude westward winds (Fig. 25). This model for the formation of Type P streaks is supported by the observations of a decrease in streak length down-range from some craters (such as Stowe crater) and of a lack of Type P streaks in association with craters less than 18 km in diameter, suggesting insufficient impact-generated heat to induce near-surface atmospheric convection and/or insufficient energy to eject material to the height of the westward winds. In this model, Type P streaks reflect not the present high altitude westward superrotation of the atmosphere but a similar atmospheric circulation in the past at the times of the impact events. Although it is not possible to assign precise ages to the Type P streaks and the inferred atmospheric circulation, they span a geologically long period represented by the ages of the impact craters. This argues against the suggestion (chapter by Gierasch et al.) that the Venus high altitude circulation pattern may not be steady, but might switch between a "normal" and a superrotating state.

Streaks other than Type P features reveal different aspects of the atmospheric circulation and, as noted above, the non-P streaks are generally oriented equatorward in the northern hemisphere and to the NNE in the southern hemisphere (Fig. 24). The predominant equatorward azimuth of the non-P streaks was previously interpreted as evidence in support of a lower atmosphere Hadley circulation (Greeley et al. 1992,1994,1995). This is the only

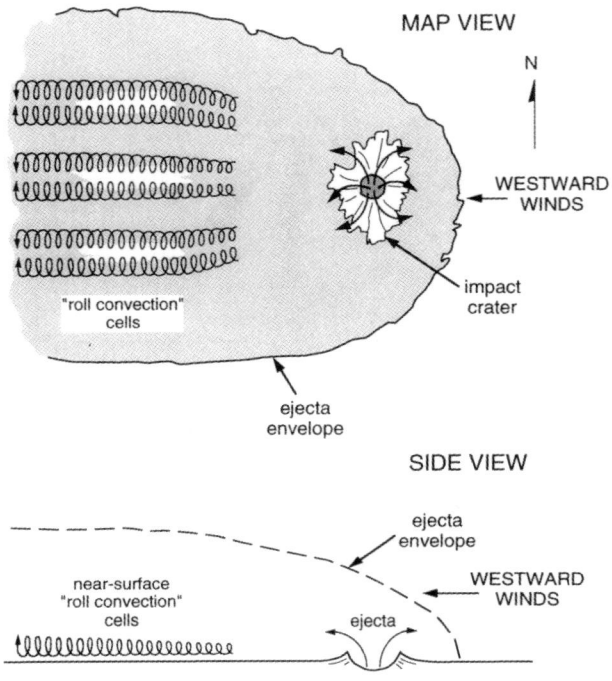

Figure 25. Schematic diagram of possible mechanism for generating the zebra pattern of Type P wind streaks. Impact heating sets up near-surface roll convection cells oriented with their long axes parallel to westward winds aloft. The cells modulate the deposition of ejecta to produce the zebra pattern.

observational evidence to constrain lower atmosphere circulation because Pioneer Venus and Venera probe measurements of wind speeds in the deep atmosphere are neither accurate enough nor sufficiently numerous to infer anything about the lower atmosphere circulation. The inferred Hadley circulation represents an average of the lower atmospheric wind patterns over the unknown time period recorded by the wind streaks. At any instant, the circulation of the lower atmosphere may be different from the Hadley pattern and the broad distribution of wind streak azimuths (Fig. 23) does suggest temporal and/or spatial variability in lower atmospheric winds. The inferred Hadley pattern also represents an overall global organization of a circulation that might look quite different on a smaller scale, thereby accounting for some of the directional variability in wind streak azimuths. Equatorward oriented streaks occur at high latitudes in both hemispheres suggesting that the inferred Hadley circulation might extend to the poles.

The differences in wind streak azimuth distributions between the northern and southern hemispheres suggest differences in the lower atmosphere circulation regimes of the two hemispheres, hemispheric differences in the supply or transportability of small particles, or as-yet unrecognized geologic or topographic hemispheric influences on wind streaks. Aphrodite Terra lies mostly south of the equator and Ishtar Terra lies poleward of 60°N; these major highland regions might have an influence on the lower atmosphere circulation system. As noted previously, in the discussion of the yardang field, the Aphrodite Massif might deflect generally southward flowing winds southwestward through a linear elongate depression and parallel to the strike of the yardangs.

The NNE trend in the distribution of wind streak azimuths in the southern hemisphere might have an explanation in terms of the deflection of northward winds by the Coriolis force. However, the effect of the Coriolis force should be small in the Venus atmosphere because the Rossby number, which measures the ratio of the inertial force to the Coriolis force, is estimated to be large compared with unity (Schubert 1983). It seems more reasonable to interpret the eastward orientation of many southern hemisphere wind streaks as indicative of an eastward component to the lower atmosphere winds in this hemisphere. Although there is no explanation for this sense of lower atmospheric circulation, the suggestion in the wind streak azimuths for an eastward near-surface atmospheric circulation may be significant because it has been assumed previously that the zonal atmospheric winds are westward throughout the atmosphere. The existence of eastward winds on a global scale in the lower atmosphere of Venus would have important implications for our understanding of Venus' atmospheric circulation, its angular momentum balance, and the drive for the higher altitude superrotation.

Diurnal thermal forcing will drive global tidal winds in the lower atmosphere of Venus (Dobrovolskis and Ingersoll 1980; Dobrovolskis 1983,1993; Covey et al. 1986) and also drive oscillatory winds up and down topographic slopes at speeds up to meters per second (Dobrovolskis 1993). Globally, there is no significant correlation of wind streak azimuth with topographic slope (Greeley et al. 1994), but some wind streaks are oriented parallel to the topographic gradient and these may be due to diurnal forcing. Further investigations should be undertaken to search for the signature of tidally forced surface winds in the wind streak data base.

V. SUMMARY

Prior to the Magellan mission, fundamental questions regarding aeolian processes and features were posed, as reviewed in the introduction to this chapter. Some have been answered or at least partly addressed, while others await information from future missions or research.

Particle Movement By Wind. As reviewed by Garvin et al. (1984), although there are good indications for current aeolian activity on Venus, repetitive SAR imaging by Magellan did not reveal any changes on the surface that could be attributed to aeolian processes. However, repetitive coverage is very limited and may be of too coarse a scale to reveal such changes. Details of the physics of particle movement and rates of rock abrasion on Venus must await future missions to the surface. Meanwhile, laboratory simulations afford a means to gain insight into surface processes in the Venusian environment.

Windblown Sediments. The presence of dunes and wind streaks signals the presence of sand and dust on Venus in many regions, and potentially winds of sufficient strength to set them into motion. The thicknesses of the sediments, however, are not known. More detailed mapping of aeolian features and correlations with local geology and physical characteristics of the surface derived from Magellan radar data may enable some source regions for windblown materials to be determined, along with transport pathways and sites of deposition.

It is possible that in addition to the observed dune fields, extensive sedimentary deposits are present on Venus, but are not currently recognized in Magellan data because they have been lithified. For example, exploratory experiments by Marshall et al. (1991) suggest a type of "cold welding" process on Venus which could alter the lithologic properties of sediments and give them the appearance of solid rocks. Additional study of this process and assessment of the resulting radar characteristics are required to test this hypothesis.

Aeolian Features. More than 6000 wind-related features have been identified on Venus, including dunes, yardangs, and wind streaks. They occur at all latitudes and longitudes, but are most common on Venusian plains and in association with impact crater ejecta deposits. Given the difficulty of identifying terrestrial aeolian features on radar, it is likely that many more aeolian features are present on Venus.

Near-Surface Wind Circulation. Orientations of wind streaks by type shed light on near-surface wind patterns. On a global scale, a Hadley circulation appears to extend to the surface. On regional scales, there is a suggestion of topographic influence on circulation patterns. Further analysis of wind streak data will enable assessment of topographic influences and possible Coriolis effects.

Acknowledgments. This work was supported by the National Aeronautics and Space Administration Planetary Geoscience Program. We are grateful to the entire Magellan Project for completing a highly successful mission and in obtaining the data to allow the analysis of aeolian features on Venus. This chapter benefited from reviews by J. Garvin and R. Blom.

REFERENCES

Arvidson, R. E., et al. 1991. Magellan: Initial analysis of Venus surface modification.

Science 252:270–275.
Avduevskii, V. S., et al. 1976. Measurement of wind velocity on the surface of Venus during the operation of stations Venera 9 and Venera 10. *Cosmic Res.* 14:622–625.
Bagnold, R. A. 1941. *The Physics of Blown Sand and Desert Dunes* (London: Methuen).
Basilevsky, A. T., et al. 1985. The surface of Venus as revealed by the Venera landings: Part II. *Geol. Soc. Amer. Bull.* 96:137–144.
Blom, R. G. 1988. Effects of variation in look angle and wavelength in radar images of volcanic and aeolian terrains, or now you see it, now you don't. *Intl. J. Remote Sensing* 9:945–965.
Blom, R. G., and Elachi, C. 1981. Spaceborne and airborne imaging radar observations of sand dunes. *J. Geophys. Res.* 86:3061–3073.
Blom, R. G., and Elachi, C. 1987. Multifrequency and multipolarization radar scatterometry of sand dunes and comparison with spaceborne and airborne radar images. *J. Geophys. Res.* 92:7877–7889.
Blumberg, D. G., and Greeley, R. 1993. Field studies of aerodynamic roughness length. *J. Arid Environ.* 25:39–48.
Blumberg, D. G., and Greeley, R. 1995. New observations of Bolivian wind streaks by JPL airborne SAR; preliminary results. Submitted to the JPL Airborne Geoscience Workshop.
Bougan, S., and Greeley, R. 1985. Microdunes and other aeolian bedforms in high-density atmospheres. *Inst. Theoret. Stat. Mem.* (Aarhus, Denmark) 8:369–376.
Breed, C. S., et al. 1982. Dunes on SIR-A images. In *Cimino*, eds. J. B. Cimino and C. Elachi, JPL Publ. 82-77, pp. 4-52–4-87.
Campbell, D. B., et al. 1992. Magellan observations of extended impact-crater related features on the surface of Venus. *J. Geophys. Res.* 97:16249–16278.
Colin, L. 1983. Basic facts about Venus. In *Venus*, eds. D. M. Hunten, L. Colin, T. M. Donahue and V. I. Moroz (Tucson: Univ. of Arizona Press), pp. 10–26.
Cooke, R., Warren, A., and Goudie, A. 1993. *Desert Geomorphology* (London: Univ. College London Press).
Counselman, C. C., et al. 1979. Venus winds are zonal and retrograde below the clouds. *Science* 205:85–87.
Counselman, C. C., et al. 1980. Zonal interferometry. *J. Geophys. Res.* 85:8026–8030.
Covey, C., Walterscheid, R. L., and Schubert, G. 1986. Dissipative tides: Application to Venus' lower atmosphere. *J. Atmos. Sci.* 43:3273–3278.
Dobrovolskis, A. R. 1983. Atmospheric tides on Venus. III. The planetary boundary layer. *Icarus* 56:165–175.
Dobrovolskis, A. R. 1993. Atmospheric tides on Venus. IV. Topographic winds and sediment transport. *Icarus* 103:276–289.
Dobrovolskis, A. R., and Ingersoll, A. P. 1980. Atmospheric tides and the rotation of Venus. I. Tidal theory and the balance of torque's. *Icarus* 41:1–7.
Florensky, C. P., Ronca, L. B., and Basilevsky, A. T. 1977a. Geomorphic degradations on the surface of Venus: An analysis of Venera 9 and Venera 10 data. *Science* 196:869–871.
Florensky, C. P., et al. 1977b. The surface of Venus as revealed by Soviet Venera 9 and 10. *Geol. Soc. Amer. Bull.* 88:1537–1545.
Florensky, C. P., et al. 1983. Venera 13 and Venera 14: Sedimentary rocks on Venus? *Science* 221:57–59.
Ford, P. G., and Pettengill, G. H. 1992. Venus topography and kilometer-scale slopes. *J. Geophys. Res.* 97:13103–13114.
Ford, J. P., Cimino, J. B., Holt, B., and Ruzek, M. R. 1986. *Shuttle Imaging Radar Views the Earth from Challenger: The SIR-B Experiment*, JPL Publ. 86-10.

Garvin, J. B. 1990. The global budget of impact-derived sediments on Venus. *Earth, Moon, Planets* 50/51:175–190.
Garvin, J. B., Head, J. W., Zuber, M. T., and Helfenstein, P. 1984. Venus: The nature of the surface from Venera panoramas. *J. Geophys. Res.* 89:3381–3399.
Greeley, R., and Arvidson, R. 1990. Aeolian processes on Venus. *Earth, Moon, Planets* 50/51:127–157.
Greeley, R., and Blumberg, D. G. 1995. Preliminary analysis of shuttle radar laboratory (SRL-1) data to study aeolian features and processes. *IEEE Trans. Geosci. Remote Sensing*, submitted.
Greeley, R., and Marshall, J. R. 1985. Transport of Venusian rolling 'stones' by wind? *Nature* 313:771–773.
Greeley, R., et al. 1980. Threshold wind speeds for sand on Mars: Wind tunnel simulations. *Geophys. Res. Lett.* 7:121–124.
Greeley, R., Williams, S. H., and Marshall, J. R. 1983. Velocities of windblown particles in saltation: Preliminary laboratory and field measurements. In *Eolian Sediments and Processes*, eds. M. E. Brookfield and T. S. Ahlbrandt (Amsterdam: Elsevier), pp. 133–148.
Greeley, R., et al. 1984. Windblown sand on Venus: Preliminary results of laboratory simulations. *Icarus* 57:112–124.
Greeley, R., Marshall, J. R., and Pollack, J. B. 1987. Physical and chemical modification of the surface of Venus by windblown particles. *Nature* 327:313–331.
Greeley, R., Christensen, P., and Carrasco, R. 1989. Shuttle radar images of wind streaks in the Altiplano, Bolivia. *Geology* 17:665–668.
Greeley, R., et al. 1992. Aeolian features on Venus: Preliminary Magellan results. *J. Geophys. Res.* 97:13319–13345.
Greeley, R., et al. 1994. Wind streaks on Venus: Clues to atmospheric circulation. *Science* 263:358–361.
Greeley, R., et al. 1995. Wind related features and processes on Venus: Summary of Magellan results. *Icarus* 115:399–420.
Guest, J. E., et al. 1992. Small volcanic edifices and volcanism in the plains of Venus: *J. Geophys. Res.* 97:15949–15966.
Hess, S. C. 1975. Dust on Venus. *J. Atmos. Sci.* 32:1076–1078.
Iversen, J. D., and White, B. R. 1982. Saltation threshold on Earth, Mars, and Venus. *Sedimentology* 29:111–119.
Iversen, J. D., Greeley, R., Marshall, J. R., and Pollack, J. B. 1987. Aeolian saltation threshold: The effect of density ratio. *Sedimentology* 34:699–706.
Iversen, J. D., Greeley, R., and Pollack, J. B. 1976. Windblown dust on Earth, Mars, and Venus. *J. Atmos. Sci.* 33:2425–2429.
Kawamura, R. 1951. Study on sand movement by wind. *Inst. Sci. Tech. Rept.* (Tokyo) 5:95–112.
Kerzhonovich, V. V., and Marov, M. Y. A. 1983. The atmospheric dynamics of Venus according to Doppler measurements by the Venera entry probes. In *Venus*, eds. D. M. Hunten, L. Colin, T. M. Donahue and V. I. Moroz (Tucson: Univ. of Arizona Press), pp. 766–778.
Ksanfomality, L. V., Gorshkova, N. V., and Khondyrev, V. K. 1983. Wind velocity near the surface of Venus from acoustic measurements. *Cosmic Res.* 21:161–167.
Kuzmin, R. O. 1989. Aeolian processes. In *Planet Venus*, III.7.3 (Moscow: Nauka Press), pp. 278–291.
Lancaster, N., Gaddis, L., and Greeley, R. 1992. New airborne imaging radar observations of sand dunes: Kelso dunes, California. *Remote Sensing Environ.* 39:233–238.
Malin, M. C. 1992. Mass movements on Venus: Preliminary results from Magellan cycle 1 observations. *J. Geophys. Res.* 97:16337–16352.

Marshall, J. R., and Greeley, R. 1992. An experimental study of aeolian structures on Venus. *J. Geophys. Res.* 97:1007–1016.

Marshall, J. R., Greeley, R., Tucker, D. W., and Pollack, J. B. 1988. Aeolian weathering of Venusian surface materials: Preliminary results from laboratory simulations. *Icarus* 74:495–515.

Marshall, J. R., et al. 1991. Adhesion and abrasion of surface materials in the venusian aeolian environment. *J. Geophys. Res.* 96:1931–1947.

McCauley, J. F., Grolier, M. J., and Breed, C. S. 1977. Yardangs. In *Geomorphology in Arid Regions*, ed. D. O. Doehring (New York: SUNY at Binghamton), pp. 233–269.

Nozette, S., and Lewis, J. S. 1982. Venus: Chemical weathering of igneous rocks and buffering of atmospheric composition. *Science* 216:181–183.

Plaut, J. J., et al. 1992. Anomalous scattering behavior of selected impact "parabola" featurescomparisons. In *Intl. Colloquium on Venus*, LPI Contrib. No. 789, pp. 92–93.

Ronca. L. B., and Green, R. R. 1970. Aeolian regime of the surface of Venus. *Astrophys. Space Sci.* 8:59–65.

Sagan, C. 1975. Windblown dust on Venus. *J. Atmos. Sci.* 32:1079–1083.

Schubert, G. 1983. General circulation and the dynamical state of the Venus atmosphere. In *Venus*, eds. D. M. Hunten, L. Colin, T. M. Donahue and V. I. Moroz (Tucson: Univ. of Arizona Press), pp. 681–765.

Schubert, G., et al. 1980. Structure and circulation on the Venus atmosphere. *J. Geophys. Res.* 85:8007–8025.

Schultz, P. H. 1992. Atmospheric effects of ejecta emplacement and crater formation on Venus from Magellan. *J. Geophys. Res.* 97:16183–16248.

Surkov, Yu. A., et al. 1984. New data on the composition, structure, and properties of Venus rock obtained by Venera 13 and Venera 14. *Proc. Lunar Planet. Sci. Conf.* 14, *J. Geophys. Res. Suppl.* 89:393–402.

Surkov, Yu. A., et al. 1987. Uranium, thorium, and potassium in the venusian rocks at the landing sites of Vega 1 and 2. *Proc. Lunar Planet. Sci. Conf.* 17, *J. Geophys. Res. Suppl.* 92:537–540.

Vervack, R., and Melosh, H. J. 1992. Wind interaction with falling ejecta: Origin of the parabolic features on Venus. *Geophys. Res. Lett.* 19:525–528.

Weitz, C. M., Plaut, J. J., Greeley, R., and Saunders, R.S. 1994. Dunes and microdunes on Venus: Why were so few found in the Magellan data? *Icarus* 112:282–295.

White B. R. 1979. Soil transport by wind on Mars. *J. Geophys. Res.* 84:4643–4651.

White B. R. 1981. Venusian saltation. *Icarus* 46:226–232.

White B. R. 1986. Particle transport by atmospheric winds on Venus: An experimental wind tunnel study. In *Aeolian Geomorphology*, ed. W. G. Nicklin (Boston: Allen and Unwin), pp. 57–73.

Williams, G. P. 1964. Some aspects of the eolian saltation load. *Sedimentology* 3:257–287.

Williams, S. H., and Greeley, R. 1985. Aeolian activity on Venus: The effect of atmospheric density on saltation flux. *Lunar Planet. Sci. Conf.* XVI:908–909 (abstract).

Williams, S. H., and Greeley, R. 1987. Particle speed and concentration in the saltation cloud: Full saltation development and choking. *Lunar Planet. Sci. Conf.* XVIII:1088–1089 (abstract).

Williams, S. H., and Greeley, R. 1994. Windblown sand on Venus: The effect of high atmospheric density. *Geophys. Res. Lett.* 21:2825–2828.

GEOCHEMISTRY OF SURFACE–ATMOSPHERE INTERACTIONS ON VENUS

B. FEGLEY, Jr.
Washington University

G. KLINGELHÖFER
Institut für Kernphysik, Darmstadt

K. LODDERS
Washington University

and

T. WIDEMANN
Service d'Aéronomie du CNRS

We review the consequences of atmosphere-surface interactions for the geochemistry of Venus. Earth-based, Earth-orbital, and spacecraft observations of the atmosphere and surface are combined with experimental studies of reaction rate kinetics and theoretical models of thermochemical equilibria to discuss mineral buffering of CO_2, HCl, and HF in the atmosphere; the evidence for carbonatites and related alkaline igneous rocks on the surface; the oxidation state of the surface of Venus, the sulfur geochemical cycle; and the origin of low radar emissivity regions on the surface of Venus. Although available experiments, observations, and theory provide important constraints on atmosphere-surface interactions, some of the key issues that remain unresolved include (1) the mineralogy of the surface of Venus, (2) the chemical composition of the lower 22 km of the atmosphere of Venus, (3) the oxidation state of the surface of Venus, and (4) the identity of the high dielectric phase(s) present in low radar emissivity regions. These issues can be best addressed by *in-situ* measurements of atmospheric composition and surface mineralogy on future spacecraft missions.

I. INTRODUCTION

Venus is a natural laboratory for studying thermochemical interactions between a planetary surface and the overlying atmosphere (Fegley and Treiman 1992). The temperature is $\sim$740 K and the global surface pressure is about 95.6 bar (at the modal planetary radius of 6051.4 km (Ford and Pettengill 1992), which is taken as 0 km altitude in this chapter). Chemically reactive and highly corrosive gases such as HF, HCl, H_2SO_4, SO_2, and supercritical CO_2 are present in Venus' atmosphere at abundances orders of magnitude greater than in the Earth's atmosphere and therefore indicate the lower atmosphere of Venus is probably reacting with the surface. Only a few percent of the incident solar flux penetrates to below the clouds and the short-wavelength ultraviolet sunlight that photolyzes CO_2 is stopped at high altitudes above the clouds. Thus, photochemically driven atmosphere-surface reactions are less important than thermochemically driven atmosphere–surface reactions.

In the 1960s the microwave radiometer on the Mariner 2 mission (Barath et al. 1964) confirmed the earlier Earth-based microwave observations (Mayer et al. 1958) of high brightness temperatures for Venus. The Mariner 2 observations led Mueller (1963) to realize that the deduced surface temperature of ∼700 K "corresponds with those [temperatures] attained during moderately high degrees of metamorphism on Earth. It is therefore possible that large parts of the atmosphere of Venus are partially equilibrated with the surface rocks. From this assumption, it follows that the composition of the atmosphere should reflect the mineralogical character of the rocks."

Mueller's hypothesis of chemical equilibrium formed the basis for theoretical work by many investigators who explored the consequences for the geochemistry and mineralogy of the surface and the composition of the atmosphere (see, e.g., the review by Fegley and Treiman 1992). Despite these theoretical studies and the great advances in Earth-based, Earth-orbital, and spacecraft observations of Venus many important questions about the nature of the surface remain unanswered.

In this chapter we use constraints derived from observations of the surface and atmosphere of Venus; from chemical equilibrium models of atmosphere-lithosphere chemical reactions; and from laboratory studies of the kinetics of some important gas-solid reactions to discuss three fundamental issues about the geochemistry and mineralogy of the surface of Venus:

1. What are the major minerals, and rock types that are present on the surface of Venus? Related questions involve the possible hypsometric control of mineralogy and whether or not chemical reactions between the atmosphere and surface are kinetically or thermodynamically controlled.
2. What is the oxidation state of the surface of Venus? Is magnetite (Fe_3O_4) stable on the surface of Venus, or is hematite (Fe_2O_3) stable on the surface of Venus, or are both oxides stable? Related questions involve whether or not the gases in the near-surface atmosphere of Venus are in thermochemical equilibrium with each other and with minerals present on the surface.
3. What plausible phases are responsible for the low radar emissivity regions on the surface? Related questions are whether or not the low radar emissivity regions reflect altitude-dependent chemical weathering reactions (see point 1 above), whether or not pyrite (FeS_2) is stable on the surface of Venus, or if volatile metal compounds analogous to those commonly observed around volcanic vents and fumaroles on the Earth are found on Venus.

After discussing these issues we conclude by identifying key questions about the geochemistry and mineralogy of the surface of Venus and suggest Earth-based observations, spacecraft and laboratory experiments, and theoretical studies that can improve our knowledge.

II. MINERALOGY OF THE SURFACE OF VENUS

We begin by considering constraints on the presence of carbonates on the surface of Venus. Because we have no direct information on the mineralogy of the surface, the question of whether or not carbonates are present on Venus is a long-standing and controversial issue (see, e.g., Barsukov et al. 1980,1982; Fegley and Treiman 1992; Fegley et al. 1992; Klose et al. 1992; Lewis 1968,1970,1971; Mueller 1963,1964; Orville 1974; Vinogradov and Volkov 1971; Volkov et al. 1986). We then discuss halogen-bearing minerals, which can buffer HCl and HF in the atmosphere. We conclude this section by discussing possible high dielectric constant phases present in low radar emissivity areas on the surface.

A. Thermochemical Calculations of Carbonate Stability on Venus

As Table I shows, CO_2 is the major gas in the atmosphere of Venus, and is a supercritical fluid under Venus' surface conditions. Carbon dioxide is involved in terrestrial metamorphic reactions which take place at temperatures of several hundred degrees, and it would be somewhat surprising if CO_2 were not also involved in chemical reactions with reactive minerals on the hot surface of Venus. Although unsuitable for providing kinetic data relevant to Venus (see, e.g., Fegley and Treiman 1992), laboratory experiments show that (even under anhydrous conditions) $CO_2(g)$, quartz (SiO_2), calcite ($CaCO_3$), and wollastonite ($CaSiO_3$) react within hours at 830 to 900°C and CO_2 pressures of 1 to 68 bars (Kridelbaugh 1973).

Urey (1952) suggested that carbonate-silicate reactions such as

$$MgCO_3 + SiO_2 = MgSiO_3 + CO_2 \ (g) \qquad (1)$$

$$CaCO_3 + SiO_2 = CaSiO_3 + CO_2 \ (g) \qquad (2)$$

could regulate or buffer the CO_2 pressure in a planetary atmosphere. But when Urey (1952) proposed this concept, the low surface temperature accepted at that time for Venus led him to dismiss reactions (1) and (2) as irrelevant to Venus because the equilibrium CO_2 pressures at low temperatures are much smaller than the CO_2 pressure in the atmosphere of Venus.

However, the situation changed dramatically after the Mariner 2 microwave radiometer observations confirmed Venus' high surface temperature. Mueller (1963,1964) and others (see, e.g., Lewis 1968,1970) revived Urey's concept that carbonate-silicate equilibria are CO_2 buffers. Apparent discrepancies between observations by the American Mariner 5 flyby and Soviet

TABLE I
Abundances of Chemically Reactive Gases in the Lower Atmosphere of Venus[a,b]

Gas	Abundance	References and Notes
CO_2	96.5±0.8%	Uncertainty due to uncertainty in N_2; von Zahn et al. (1983)
N_2	3.5±0.8%	Recommended value from von Zahn et al. (1983)
SO_2	150 ppm	52 km; Bertaux et al. (1996) Vega 1 UV spectroscopy
	65 ppm	52 km; Bertaux et al. (1996) Vega 2 UV spectroscopy
	125 ppm	42 km; Bertaux et al. (1996) Vega 1 UV spectroscopy
	200 ppm	42 km; Bertaux et al. (1996) Vega 2 UV spectroscopy
	130±35 ppm	≤42 km; Gel'man et al. (1979) Venera 11/12 GC data
	130±40 ppm	35–45 km; Bézard et al. (1993) Earth-based IR spectra
	180±70 ppm	42 km; Pollack et al. (1993) Earth-based IR spectra
	185±43 ppm	22 km; Oyama et al. (1980) Pioneer Venus GC data
	38 ppm	22 km; Bertaux et al. (1996) Vega 1 UV spectroscopy
	25±12 ppm	12 km; Bertaux et al. (1996) Vega 1 UV spectroscopy
H_2O	150 ppm	42 km; Moroz et al. (1979) Venera 11/12 Vis-IR spectrophotometry[c]
	60 ppm	22 km; Moroz et al. (1979) Venera 11/12 Vis-IR spectrophotometry[c]
	40 ppm	35–45 km; Bézard et al. (1990) Earth-based IR spectra
	30±10 ppm	10–40 km; Pollack et al. (1993) Earth-based IR spectra
	30±10 ppm	15–25 km; de Bergh et al. (1995) Earth-based IR spectra
	30±15 ppm	0–15 km; de Bergh et al. (1995) Earth-based IR spectra
	20 ppm	0 km; Moroz et al. (1979) Venera 11/12 Vis-IR spectrophotometry[c]
CO	45±10 ppm	Cloud top; Connes et al. (1968) Earth-based IR spectra
	51 ppm	Cloud top; Young (1972) re-analysis of Connes et al. (1968) data
	45 ppm	35–45 km; Bézard et al. (1990) Earth-based IR spectra
	23±5 ppm	36 km; Pollack et al. (1993) Earth-based IR spectra
	30±18 ppm	42 km; Oyama et al. (1980) Pioneer Venus GC data
	28±7 ppm	36–42 km; Gel'man et al. (1979) Venera 11/12 GC data
	20±3 ppm	22 km; Oyama et al. (1980) Pioneer Venus GC data
	17±1 ppm	12 km; Marov et al. (1989) Venera 11/12 GC data
H_2S	3±2 ppm	<20 km; Hoffman et al. (1980a,b) Pioneer Venus MS data[d]
HCl	0.6±0.12 ppm	Cloud top; Connes et al. (1967) Earth-based IR spectra
	0.4 ppm	Cloud top; Young (1972) re-analysis of Connes et al. (1967) data
	0.4 ppm	Cloud top; de Bergh et al. (1989) Earth-based IR spectra
	0.5 ppm	35–45 km; Bézard et al. (1990) Earth-based IR spectra
	0.48±0.12 ppm	23.5 km; Pollack et al. (1993) Earth-based IR spectra
OCS	4.4±1.0 ppm	33 km; Pollack et al. (1993) Earth-based IR spectra
	40±20 ppm	29–37 km; Mukhin et al. (1983) Venera 13/14 GC data
SO	20±10 ppb	Cloud top; Na et al. (1990)
HF	$5^{+5}_{-2.5}$ ppb	Cloud top; Connes et al. (1967)
	10 ppb	Cloud top; Young (1972) re-analysis of Connes et al. (1967) data
	4.5 ppb	35–45 km; Bézard et al. (1990) Earth-based IR spectra
	1–5 ppb	33.5 km; Pollack et al. (1993) Earth-based IR spectra

[a] Modified from Fegley et al. (1992). [b] Excluding the noble gases and isotopically substituted species. [c] Re-analysis of these data by Young et al. (1984) gives a constant H_2O abundance of 20 to 30 ppm if only the 1.13 μm band is considered. [d] Reported by Hoffman et al. (1980a,b) as a preliminary value. Oyama et al. (1980) reported an upper limit of 2 ppm at ≤22 km altitude.

Venera 4–6 entry missions caused the temperature and pressure at the surface of Venus to remain a matter of debate until the early 1970s (Lewis 1968, 1969, 1970). However, it was generally accepted that the CO_2 pressure from reaction (2) at ~700 to 740 K was comparable to the CO_2 pressure at the surface of Venus.

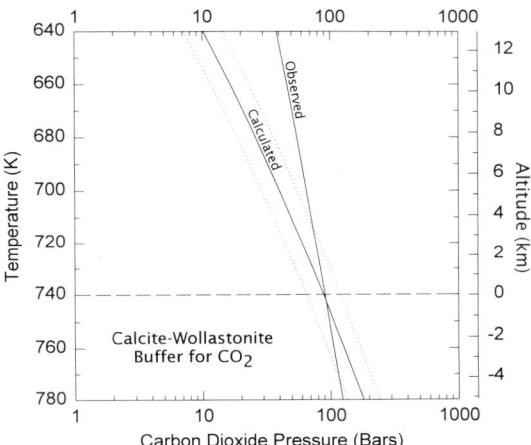

Figure 1. A comparison of the observed and calculated CO_2 pressures on Venus. The observed CO_2 pressure is 0.965 times the total pressure from Seiff (1983). The calculated CO_2 pressure is for the calcite-quartz-wollastonite buffer. The dashed lines are for 2 sigma uncertainties in the thermodynamic data. The observed and calculated curves intersect at 740±30 K (0±4 km altitude) (based on calculations presented in Fegley and Treiman 1992).

This point is illustrated by Fig. 1 which shows that the equilibrium CO_2 pressure buffered by the coexistence of quartz, calcite, and wollastonite intersects the observed CO_2 pressure in the lower atmosphere of Venus at 740±30 K (0±4 km altitude). This agreement is better than that for any other carbonate-silicate buffer (Fegley and Treiman 1992), and for over two decades has been regarded as evidence that reaction (2) is buffering the CO_2 pressure on Venus (Lewis 1970). (The details of thermodynamic calculations for this buffer, and other gas-solid reactions are discussed by Fegley and Treiman [1992].)

Equilibria similar to reactions (1) and (2) have been used to predict the thermodynamic stability of other carbonates on the surface of Venus. Lewis (1970) considered reactions involving dolomite ($CaMg(CO_3)_2$), magnesite ($MgCO_3$), and siderite ($FeCO_3$). He showed that only calcite was stable at 740 K. Fegley and Treiman (1992) calculated the hypsometric control of carbonate stability on Venus initially proposed by Florensky et al. (1977a). They showed that dolomite and magnesite could be stable at lower temperatures and higher altitudes where the equilibrium CO_2 pressures are equal to or less than the CO_2 partial pressure in the atmosphere of Venus. Their results are summarized in Table II.

TABLE II
Carbonate Stability on the Surface of Venus[a]

Reaction[a]	Stability Field[b] T_{max}(K) Z_{min}(km)
$CaCO_3 + SiO_2 = CaSiO_3 + CO_2$ (g)	740±30 0±4
$MgCO_3 + MgSiO_3 = Mg_2SiO_4 + CO_2$ (g)	700±30 5±4
$CaMg(CO_3)_2 + 4MgSiO_3 = 2Mg_2SiO_4 + CaMgSi_2O_6 + 2CO_2$ (g)	660±70 10±9
$2CaMg(CO_3)_2 + SiO_2 = 2CaCO_3 + Mg_2SiO_4 + 2CO_2$ (g)	645±25 12±3

[a] Thermodynamic data sources are given in Fegley and Treiman (1992).
[b] The maximum temperature and minimum altitude at which the carbonate-bearing mineral assemblage is stable.

Fegley and Treiman (1992) and Fegley et al. (1992) also calculated the stability of siderite, rhodochrosite ($MnCO_3$), and cerussite ($PbCO_3$) as a function of altitude because these minerals were potential candidates for high dielectric constant phases on Venus (Garvin and Head 1985; Pettengill et al. 1988). However, the pure carbonates are predicted to be unstable on Venus. Other potential carbonate-forming elements, such as Ba, Sr, and Zn are present at very low abundance in Venusian rocks. Their pure carbonates would not be important, even if stable.

However, the relevance of carbonate-silicate equilibria to Venus has been questioned by several authors (Orville 1974; Klose et al. 1992), who argued that calcite is absent on the surface of Venus. Because of the importance of carbonates for atmosphere-surface reactions on Venus, and in order to guide the design of geochemical experiments on future spacecraft missions to Venus, it is worth examining these arguments in some detail.

B. Geochemical Arguments For and Against Carbonates on Venus

1. Calcite Paragenesis. On the Earth calcite is generally, but not exclusively, found in sedimentary and metamorphic rocks (Deer et al. 1963). As noted by Fegley and Treiman (1992) and discussed in more detail by Fegley et al. (1992) and Kargel et al. (1994), calcite also typically is found at the level of a few percent in many alkaline igneous rocks such as nepheline-syenites, melilite-nephelinites, okaites, ijolites, and melteigites (Johannsen 1937). Calcite is also a major phase in kimberlites (Mitchell 1986) and some carbonatites (Woolley and Kempe 1989). Wollastonite occurs with calcite in many of the same alkaline igneous rocks. Kargel et al. (1994) pointed out that wollastonite occurs with quartz in carbonatites (Deer et al. 1963; Sorensen 1974; Hogarth 1989). As discussed in Sec. II.C below, the XRF and γ-ray analyses of the surface of Venus by several Venera and Vega spacecraft provide indirect evidence for the presence of calcite-bearing rocks on Venus. The XRF data and geomorphological evidence also suggest that carbonatites and related alkaline igneous rocks, while rare on Earth, are apparently more common on Venus (Kargel et al. 1993,1994). Several other arguments also suggest the presence of calcite on Venus. We mention these points here, and discuss some of them in more detail with hydrogen halide buffers in Sec. II.D.

2. Geomorphology. Kargel et al. (1991,1994) argued that carbonatite magmas, which normally have water-like rheological properties, could be responsible for the canali, outflow channels, and associated volcanic deposits seen in Magellan radar images. Kargel and colleagues point out that these features on Venus more closely resemble fluvial landforms than common volcanic features made by basaltic magmas on the Earth and Mars.

3. Hydrogen Halide Buffers. These geological and geophysical arguments are consistent with independent geochemical arguments that carbonatites and related types of alkaline rocks provide the necessary mineral assemblages to buffer the observed HCl and HF abundances in the atmosphere of Venus (Fegley et al. 1992; Kargel et al. 1994). As discussed later in Sec. II.D, mineral assemblages containing wollastonite + sodalite ($Na_4[AlSiO_4]_3Cl$) or nepheline ($NaAlSiO_4$) give the best agreement between the calculated and observed HCl pressure in the atmosphere of Venus. Nepheline is widespread and abundant in alkaline igneous rocks and carbonatites. Wollastonite and sodalite also occur in these rock types but typically are not as widespread and abundant as nepheline. The best agreement between calculated and observed HF pressures in the atmosphere of Venus is provided by assemblages involving feldspathoids and fluoramphiboles or feldspar and fluorphlogopite. These mineral assemblages are also found in alkaline igneous rocks and carbonatites.

4. Dielectric and Ferroelectric Minerals. Carbonatites and associated alkaline igneous rocks also contain perovskites and pyrochlores, which are high dielectric phases, that may explain the low radar emissivity on Venus (Pettengill et al. 1982,1988; chapter by Pettengill). The observed abundances of perovskites and pyrochlores in terrestrial alkaline rocks and carbonatites are consistent with the amounts of ferroelectrics required to increase the bulk dielectric constant of rocks on Venus. Perovskites are common minor and trace phases, found at abundances of a few percent, in terrestrial carbonatites and igneous alkaline rocks (Hogarth 1989; Semenov 1974). For example, $\sim$2 to 9% (by volume) perovskite is found in melilite-rich alkaline rocks from the Kola peninsula, the Oka hills in Quebec, and in South Africa and 5 to 9% (by volume) perovskite is found in leucite-rich rocks from Uganda and in nepheline-rich rocks from Colorado (Johannsen 1937; Deer et al. 1963; Sorensen 1974). Pyrochlores ($A_{16-x}B_{16}O_{48}(O,OH,F)_{8-y}$), with a wide variety of compositions, are also common minor and trace minerals present at percent levels in carbonatites and associated alkaline igneous rocks (Hogarth 1989; Semenov 1974).

Synthetic perovskite ($CaTiO_3$) has dielectric constants of 165 to 170 at radar frequencies (Von Hippel 1954) and some natural perovskites have dielectric constants of 1100 to 2200 at 10 kHz (Timco 1977). Fegley et al. (1992) calculated the stability of pure $CaTiO_3$ against chemical weathering by CO_2, SO_2, HCl, and HF on Venus. At the highest elevations (e.g., 10 km at the summit of Maxwell Montes), $CaTiO_3$ is converted to fluorite+rutile at a rate that exceeds the production of fresh, perovskite-bearing rock. Thus, the top of Maxwell displays normal radar emissivity because it is perovskite poor. Along the flanks of Maxwell, at slightly lower elevations, fluorite formation is thermodynamically unfavorable and the rate of $CaTiO_3$ weathering by the less reactive gases CO_2 and SO_2 is slow enough that perovskite can persist on a geologic time scale. These regions display low radar emissivity because they are perovskite rich. At the base of Maxwell, the temperatures are sufficiently high that chemical weathering of $CaTiO_3$ by CO_2 and SO_2 proceeds fast enough to destroy perovskite in a geologically short time. As a result, the low altitude regions also display normal radar emissivity.

Kargel et al. (1994) expanded on this model by noting that the variations in the critical elevation (where low emissivity regions begin) are plausibly due to variations in the compositions, and hence, reactivities, of natural perovskites. They also pointed out that some regions within the perovskite metastability zone (at intermediate altitudes) that show normal emissivity, may simply not contain any perovskite-bearing rocks.

Other perovskites, in particular Nb-rich ones, are ferroelectric phases with even higher dielectric constants (Lines and Glass 1977). Many synthetic pyrochlores studied in the laboratory are also ferroelectric phases. Figure 2 in Shepard et al. (1994) dramatically illustrates that the variation of dielectric constant with temperature for a typical ferroelectric compound almost exactly reproduces the altitude dependent emissivity variations that are observed at Ovda Regio. Arvidson et al. (1994) also showed that the ferroelectric model can reproduce the emissivity vs altitude curves for 13 areas in geologically different regions on the surface of Venus.

Using theoretical models for the bulk dielectric constant of a binary mixture, Shepard et al. (1994) showed that only 0.09% (by volume) of ferroelectric phases with dielectric constants of 100,000 can raise the bulk dielectric constant of Venus surface rock from values of ~4 to 5 to a dielectric constant of 65. Larger amounts of ferroelectrics would yield even larger bulk dielectric constants. Shepard et al. (1994) also showed that ferroelectrics of slightly different composition can plausibly explain the observed trend of emissivity vs altitude and the observed variations in the critical elevation from region to region on Venus.

Although the phases studied in the laboratory are generally not the same minerals found in carbonatites and alkaline igneous rocks, there are important similarities which suggest that it is important to measure electrical properties of appropriate minerals and rocks to test the model before it can be uncritically accepted or arbitrarily dismissed.

5. Ancient Sedimentation or High Altitude Weathering. Calcite could also be produced by other processes that operated on Venus in the past or operate on Venus today. If water were present on Venus in the past, as the high atmospheric D/H ratio suggests (see, e.g., Donahue et al. 1982), then sedimentary calcite formation may have taken place. Alternatively, as proposed by Florensky et al. (1977a) and Nozette and Lewis (1982), chemical weathering of rocks at high elevations could produce calcite (and other carbonates) that are then abraded and blown into the hotter lowlands below where these fine-grained carbonate weathering products react with quartz to form wollastonite and establish the calcite-quartz-wollastonite buffer.

None of these arguments for or against the presence of calcite on Venus can be proven or disproven at the present time. However, alkaline igneous rocks, with or without associated carbonatites, provide the minerals involved in the calcite-quartz-wollastonite buffer, several petrologically plausible buffers for HCl and HF, plausible high dielectric constant phases to explain the origin of low radar emissivity regions, and also plausibly explain the observed variations of emissivity with altitude and location. As discussed below, elemental analyses of the surface of Venus are compatible with the presence of alkaline igneous rocks on Venus.

C. Evidence for Carbonates from X-ray Fluorescence and Gamma Ray Spectroscopy on Venera and Vega Landers

Table III lists the seven Venera and Vega missions which made geochemical analyses of the surface of Venus. Descriptions of these experiments are given in the Appendix on the CD-ROM (CDP4C4A1). Magellan radar imagery, interpretations of the γ-ray spectroscopy data for K, U, and Th (Table IV), and interpretations of the X-ray fluorescence (XRF) major element data (Tables V and VI) indicate that several different rock types were sampled by these geochemical analyses (see, e.g., Basilevsky et al. 1992; Barsukov et al. 1986; Kargel et al. 1993; Volkov et al. 1986).

For example, Kargel et al. (1993) interpreted the Venera 9, 10, 14, and Vega 1 and 2 landing sites as tholeiitic basalts, the Venera 13 site as a weathered olivine leucitite, and the Venera 8 site as a leucitite, rhyolite, or monzonite. These interpretations are not unique but the other groups cited above have come to similar conclusions. Of particular interest here are the interpretations (Basilevsky et al. 1992; Barsukov et al. 1986; Volkov et al. 1986) of the Venera 8 γ-ray and the Venera 13 XRF analyses indicating that alkaline igneous rocks on Venus (Tables III and V), may be more abundant than on the Earth (see, e.g., Kargel et al. 1993).

The mass deficits in the XRF elemental analyses (Table V) were interpreted by Fegley et al. (1992) and Kargel et al. (1994) as evidence for carbonates on Venus. The Venera and Vega XRF instruments (Surkov et al. 1983,1984,1986), like those flown on the Viking landers (Clark et al. 1977), were sensitive only to Mg and heavier elements, and could not detect C and other light elements with atomic numbers $Z<12$. The Venera 13, 14, and Vega 2 XRF elemental analyses, show mass deficits of $\sim$4 to 7% (Table V), before the addition of $\sim$2 to 2.5% Na_2O, that was estimated on the basis of petrochemical trends (Barsukov et al. 1986; Surkov et al. 1984). Similar mass deficits of $\sim$7 to 10% occur in the Viking XRF analyses (see, e.g., Clark et al.

TABLE III
Geochemical Analyses and Imaging on the Surface of Venus[a]

Probe	Latitude[b] (deg)	Longitude[b] (deg)	Altitude[c] (km)	Location and Suggested Rock Types[f]	Experiment	Reference(s)
Venera 8	−10.7	335.25	0.4±0.2	Mottled volcanic plains east of Navka Planitia; leucitite? lamprophyres? rhyolite?	γ-ray	Vinogradov et al. 1973
Venera 9	31.0	291.64	1.2±0.5	NE slope of Beta Regio, lander site has 15 to 20° slope with decimeter-size rock fragments with soil between them, MORB-like basaltic tholeiite?	γ-ray TV image[e] Photometry[d]	Surkov 1977 Florensky et al. 1977a,b Ekonomov et al. 1980
Venera 10	15.42	291.51	0.7±0.6	Lowlands near SE edge of Beta Regio, lander site has soil between 10 to 15 cm high outcrops of bedrock, MORB-like basaltic tholeiite?	γ-ray TV image[e] Photometry[d]	Surkov 1977 Florensky et al. 1977a,b Ekonomov et al. 1980
Venera 13	−7.55	303.69	0.7±0.3	Navka Planitia at east end of Phoebe Regio rise, landscape similar to Venera 10 site, mafic alkaline rocks such as weathered olivine leucitite?, nephelinite?	XRF Redox experiment TV imaging[d,e]	Surkov et al. 1984 Florensky et al. 1983a Selivanov et al. 1983
Venera 14	−13.05	310.19	0.8±0.3	Southern Navka Planitia on flank of a volcano, landing site is a plain dominated by layered bedrock and minor amount of soil, weathered MORB-like tholeiite?	XRF Redox experiment TV imaging[d,e]	Surkov et al. 1984 Florensky et al. 1983a Selivanov et al. 1983
Vega 1	8.10	175.85	−0.2±0.1	Rusalka Planitia, north of Aphrodite Terra, no TV panoramas, MORB-like tholeiite?	γ-ray	Surkov et al. 1987a,b
Vega 2	−7.14	177.67	1.1±0.2	Transitional zone between Rusalka Planitia and E. edge of Aphrodite Terra rise, no TV panoramas, MORB-like tholeiite?	γ-ray XRF	Surkov et al. 1987a,b Surkov et al. 1986

[a] Positional and elevation data are from Basilevsky et al. (1992); R. A. Brackett (personal communication, 1995).
[b] Typical uncertainties on latitude and longitude are ±1.5° (Basilevsky et al. 1992).
[c] Relative to a mean radius of 6051.5 km with one standard deviation uncertainties.
[d] Also see Florensky et al. (1983b); Garvin et al. (1984) and Pieters et al. (1986) for interpretations of photometric data and imaging.
[e] Black and white TV image for Venera 9 and 10; Red-green-blue TV imaging for Venera 13 and 14.
[f] Suggested rock types are taken from the literature and are discussed in more detail in the text.

TABLE IV
Gamma Ray Analyses of the Surface of Venus[a]

Lander	K (wt%)	U (ppm)	Th (ppm)
Venera 8	4.0±1.2	2.2±0.7	6.5±2.2
Venera 9	0.47±0.08	0.60±0.16	3.65±0.42
Venera 10	0.30±0.16	0.46±0.26	0.70±0.34
Vega 1	0.45±0.22	0.64±0.47	1.5±1.2
Vega 2	0.40±0.20[b]	0.68±0.38	2.0±1.0

[a] Table from Surkov et al. (1987a,b).
[b] Compare to 0.1±0.08 wt% from XRF analysis on Vega 2.

TABLE V
Major Element Composition of the Surface of Venus and of Some Terrestrial Rocks

	Mass Percent (±1σ)					
Oxide	Venera 13[a]	Venera 14[a]	Vega 2[b,e]	N-MORB[f]	Leucitite[g]	Lamprophyre[h]
SiO_2	45.1±3.0	48.7±3.6	45.6±3.2	48.77	46.2	46.3
TiO_2	1.59±0.45	1.25±0.41	0.2±0.1	1.15	1.2	2.6
Al_2O_3	15.8±3.0	17.9±2.6	16±1.8	15.90	14.4	13.5
FeO[c]	9.3±2.2	8.8±1.8	7.7±1.1	9.82	8.09	11.0
MnO	0.2±0.1	0.16±0.08	0.14±0.12	0.17	0.0	0.21
MgO	11.4±6.2	8.1±3.3	11.5±3.7	9.67	7.0	9.1
CaO	7.1±0.96	10.3±1.2	7.5±0.7	11.16	13.2	10.7
Na_2O[d]	2±0.5	2.4±0.4	2	2.43	1.6	3.1
K_2O	4.0±0.63	0.2±0.07	0.1±0.08	0.08	6.4	2.9
SO_3	1.62±1.0	0.88±0.77	4.7±1.5			
Cl	<0.3	<0.4	<0.3			
Total	98.1	98.7	95.4	99.15[f]	98.09	99.41

[a] Surkov et al. (1984).
[b] Surkov et al. (1986).
[c] All Fe reported as FeO for all analyses.
[d] Calculated by Surkov et al. (1984,1986).
[e] In addition to Cl, Surkov et al. (1986) also report the following upper limits (in mass %): Cu, Pb<0.3; Zn<0.2; Sr, Y, Zr, Nb, Mo<0.1; As, Se, Br<0.08.
[f] N-type, or normal MORB (Wilson 1989). Also contains 0.09% P_2O_5 and 0.30% H_2O.
[g] Leucitite, an alkaline basalt (Philpotts 1990). Also contains 0.4% P_2O_5.
[h] Lamprophyre, which is an ultrapotassic rock (Wilson 1989). Also contains 0.9% P_2O_5, 2.6% H_2O, and 2.5% CO_2.

1977,1982; Toulmin et al. 1977). The Na content of Martian soil was either estimated by assuming that enough Na was present to balance the measured Cl and S (Toulmin et al. 1977) or from analyses of the SNC meteorites (Banin et al. 1992).

As discussed by Toulmin et al. (1977), the mass deficits in the Viking

TABLE VI

Normative Mineralogy of the Venera and Vega Landing Sites[a]

Mineral		CIPW Norm (Mass %)		
		Venera 13	Venera 14	Vega 2
Orthopyroxene	En	20.8	0.0	14.2
	Fs	9.2	0.0	10.1
Clinopyroxene	Wo	0.8	9.4	5
	En	0.5	5.8	2.8
	Fs	0.2	3.0	2.0
Olivine	Fo	7.4	16.2	2.6
	Fa	3.6	9.3	2.1
Plagioclase	Ab	17	0.0	17.0
	An	39.3	13.1	40.6
K-feldspar	Or	0.7	10.2	1.2
Feldspathoids	Lc	0.0	11.2	0.0
	Ne	0.0	18.8	0.0
Oxides	Il	0.4	3.0	2.4
Total		99.9	100.0	100.0

[a] Calculated by Kargel et al. (1993) on a volatile-free basis. En: enstatite; Fs: ferrosilite; Wo: wollastonite; Fo: forsterite; Fa: fayalite; Ab: albite; An: anorthite; Lc: leucite; Ne: nepheline; Il: ilmenite.

XRF analyses, and by analogy in the Venera and Vega XRF analyses, must be attributed to some or all of the following factors: (1) analytical error, (2) the presence of low Z elements, and (3) the sum of smaller amounts of heavier elements that are individually below their respective detection limits. Toulmin et al. (1977) discuss why it is unlikely that analytical errors account for the observed mass deficits in the Viking XRF analyses. Based on descriptions of the Venera and Vega XRF experiments, the data processing procedures and algorithms (Surkov et al. 1983,1984,1985,1986,1987c), and the similar uncertainties in individual elemental abundances in the Viking, and Venera/Vega XRF analyses, it is unlikely that analytical errors can account for all of the observed mass deficits in the Venera 13,14 and Vega 2 XRF elemental analyses. The amounts of heavier elements that are individually below their respective detection limits are also probably insufficient to account for the remaining mass deficit. As emphasized by Toulmin et al. (1977) for the Viking XRF data, the cosmochemical abundances of other high Z elements (Anders and Grevesse 1989) and geochemical considerations suggest that the other undetected, high Z elements are not present in large enough amounts to contribute significantly to the mass deficit. The possible exceptions are Ni and Mn (Toulmin et al. 1977). However, Mn is already included in the Venera and Vega XRF analyses (Table III), and by analogy with the Earth's crust, the average Ni content of the surface of Venus would be $\sim$0.003 to 0.02%.

As with the Viking XRF analyses, the most plausible explanation for the mass deficits in the Venera and Vega XRF analyses is the presence of compounds formed by low Z elements. Some low Z elements (Li, Be, B) and their compounds can be ruled out as major contributors to the mass deficits because of their low cosmochemical and geochemical abundances. Oxygen and Na, which are not directly measured, are already considered in the XRF analyses. Nitrogen, in the form of nitrides and nitrates can be ruled out because these compounds are not stable on the surface of Venus (Fegley and Treiman 1992). Dissolved N in minerals is probably only a few ppm by mass. Fegley and Treiman (1992) concluded that hydrated silicates are unstable on the surface of Venus. Recent Earth-based infrared observations of H_2O in Venus' near-surface atmosphere and more recent thermodynamic data for hydrated minerals strengthen this conclusion. Thus, the mass deficits are most plausibly due to compounds formed by fluorine and carbon. Although F-bearing minerals are probably present on the surface, their abundance is unlikely to account for the entire mass deficit (as also argued by Toulmin et al. [1977] for the Viking XRF analyses). Assuming that the mass deficit is indeed due to carbonates, Kargel et al. (1994) calculated that calcite made up $\sim$4% (by mass) of the Venera 13 sample, $\sim$3% of the Venera 14 sample, and $\sim$10% of the Vega 2 sample. In principle, these inferred carbonate abundances can be tested by future Venus landers equipped with low Z sensitive analytical experiments such as an APX (α-proton-X-ray) analyzer or with infrared reflection spectrometers.

D. HCl and HF Mineral Buffers

Table I shows that HCl and HF are only trace constituents of the Venusian atmosphere. However, they are still many times more abundant on Venus than on Earth. The HCl mixing ratio of $\sim$0.5 ppm corresponds to a column density of $\sim 7 \times 10^{20}$ HCl molecules cm^{-2}. On Earth, the total Cl mixing ratio (CH_3Cl + $2CF_2Cl_2$ + $3CFCl_3$ + $3CH_3CCl_3$ + $4CCl_4$ + HCl) in the nonurban terrestrial troposphere is $\sim$2.7 ppb (Fegley 1995). This is equal to a total Cl column density of $\sim 6 \times 10^{16}$ Cl atoms cm^{-2}, or about 0.009% of the Cl column density in the atmosphere of Venus. The Cl content of the terrestrial troposphere is dominated by man-made compounds such as chlorofluorocarbons, while volcanic HCl, sea spray, and some biogenic halocarbons comprise a smaller percentage of the chlorine in the Earth's troposphere. Thus, the Cl content of the terrestrial troposphere would be much smaller if the anthropogenic, biogenic, and oceanic components were removed.

Likewise, the 1 to 5 ppb of HF in the atmosphere of Venus corresponds to an HF column density of $\sim(1-7) \times 10^{18}$ cm^{-2}. The total F mixing ratio ($2CF_2Cl_2$ + $CFCl_3$ + $4CF_4$ + $2CHClF_2$ + $3C_2Cl_3F_3$) in the nonurban terrestrial troposphere is $\sim$1.3 ppb (Fegley 1995). The corresponding total fluorine column density of $\sim 3 \times 10^{16}$ F atoms cm^{-2} is about 0.4 to 3% of the HF column density on Venus. Again, total atmospheric fluorine on Earth is dominantly due to man-made compounds, and the natural fluorine column density is significantly lower.

The much higher abundances of HCl and HF on Venus are directly due to the high surface temperature. The corrosive nature of HCl and HF implies that these acids are not inert components of the Venusian atmosphere, but are instead chemically reacting with minerals and rocks on the surface of the planet.

Lewis (1968,1970) and Mueller (1968,1969) proposed that HCl and HF in Venus' atmosphere are buffered by gas-solid reactions. The essentially identical HCl and HF abundances observed above the clouds (Connes et al. 1967) and below the clouds (Pollack et al. 1993) support this concept. Both HCl and HF are found in volcanic exhalations on the Earth (e.g., at Kilauea in Hawaii) and are also likely to be found in volcanic gases on Venus. Although HCl and HF in the atmosphere of Venus plausibly originated from volcanoes, their constant abundances during the 30 years since their discovery in 1967 imply that HCl and HF are probably being buffered by chemical reactions with minerals on the surface of Venus.

Thermochemical models of nebular condensation chemistry predict that Venus and Earth have similar Cl and F inventories (Fegley and Lewis 1980). The Cl/C and F/C ratios at the surface of the Earth (atmosphere + oceans + sediments + crust) are $\sim$0.12 and 0.09, respectively (Smith 1981; Ronov and Yaroshevsky 1976), and the Cl/F ratio is approximately unity. In contrast, the atmospheric Cl/C and F/C ratios on Venus (essentially the HCl and HF mixing ratios) are orders of magnitude smaller, and the Cl/F ratio is $\sim$100 to 500. Thus, Cl- and F-bearing phases are plausibly present in the interior and on the surface of Venus today.

Fegley and Treiman (1992) reconsidered the HCl and HF buffers suggested previously (Lewis 1968,1970; Mueller 1968,1969) using more recent thermodynamic data. They found that many of the reactions originally suggested by Lewis and Mueller as HCl and HF buffers actually did not provide a good match to the observed abundances. The discrepancies are due to changes in thermodynamic data for some minerals (e.g., for sodalite) and to increased knowledge of conditions, in particular the temperature and water vapor abundance, at the surface of Venus. Furthermore, as originally noted by Orville (1974), the buffers that did seem to match the observed hydrogen halide abundances involved mineral assemblages that were petrologically unreasonable. For example, jadeite, which is unstable below $\sim$10 kilobar pressure, or the assemblage nepheline + quartz, which reacts to form albite, were used in buffer reactions.

Fegley and Treiman (1992) proposed new buffers for HCl and HF on Venus. They did so using terrestrial petrology and experimental phase equilibria as guides. Their preferred buffers, that most closely match the HCl abundance in the atmosphere of Venus, are the wollastonite-sodalite-halite buffer

$$12HCl\ (g) + 6CaSiO_3 + 5Na_4[AlSiO_4]_3Cl =$$

$$17NaCl + 6CaAl_2Si_2O_8 + 3NaAlSi_3O_8 + 6H_2O\ (g) \quad (3)$$

and the nepheline-albite-sodalite buffer

$$2HCl\ (g) + 9NaAlSiO_4 = Al_2O_3 +$$

$$NaAlSi_3O_8 + 2Na_4[AlSiO_4]_3Cl + H_2O\ (g) \qquad (4)$$

Both buffers provide the observed amount of HCl within the uncertainties of the thermodynamic data and spectroscopic observations. Reaction (4), involving nepheline+albite+sodalite, does not match the observed HCl abundance as closely as reaction (3), but the sodalite+nepheline+albite assemblage is common in terrestrial nepheline-syenites and by analogy would also be expected to be common in alkaline rocks on Venus. Reaction (3) involves the assemblage of wollastonite+sodalite which, as discussed earlier in Sec. II.B, is also found in alkaline igneous rocks.

The preferred HF buffers of Fegley and Treiman (1992) include the fluorphlogopite buffer first proposed by Lewis (1970):

$$2HF\ (g) + KAlSi_3O_8 + 3MgSiO_3 =$$

$$KMg_3AlSi_3O_{10}F_2 + 3SiO_2 + H_2O\ (g) \qquad (5)$$

and two buffers involving fluoredenite

$$2HF\ (g) + NaAlSiO_4 + 2CaMgSi_2O_6 + 3MgSiO_3 =$$

$$NaCa_2Mg_5Si_7AlO_{22}F_2 + SiO_2 + H_2O\ (g) \qquad (6)$$

$$2HF\ (g) + NaAlSiO_4 + 2CaMgSi_2O_6 + Mg_2SiO_4 +$$

$$MgSiO_3 = NaCa_2Mg_5Si_7AlO_{22}F_2 + H_2O\ (g). \qquad (7)$$

The HF buffers, like the HCl buffers and the calcite-quartz-wollastonite CO_2 buffer are generally compatible with the mineralogy of terrestrial alkaline igneous rocks and carbonatites (Kargel et al. 1994). However, pure enstatite is not found, or at least is not at all common, in terrestrial carbonatites (Hogarth 1989), but augites in alkaline igneous rocks do contain an enstatite component which can take part in buffering reactions. Kargel et al. (1994) also noted that fluoramphiboles are involved in HF buffers and that amphiboles containing up to 3 weight percent F are found in terrestrial carbonatites (Hogarth 1989).

III. OXIDATION STATE OF THE LOWER ATMOSPHERE AND SURFACE OF VENUS

A. CO Abundance and Oxygen Fugacity of the Near-Surface Atmosphere

Carbon monoxide is a minor gas in the atmosphere of Venus (Table I) and it is generally believed (Lewis 1970; Lewis and Kreimendahl 1980; Prinn 1985; Volkov et al. 1986; Fegley and Treiman 1992; Zolotov 1994) that the oxidation state of the lower atmosphere of Venus is controlled by the net thermochemical reaction:

$$2CO\ (g) + O_2\ (g) = 2CO_2\ (g). \tag{8}$$

The oxygen fugacity as a function of temperature for reaction (8) is

$$\log_{10} f_{O_2} = 2\log_{10}\left(\frac{X_{CO}}{X_{CO_2}}\right) + 9.17 - \left(\frac{29607}{T}\right) \tag{9}$$

A combination of Earth-based and *in-situ* measurements of CO in the lower atmosphere of Venus indicates that the CO concentration decreases with decreasing altitude toward the surface of Venus (Table I). This trend was discussed by von Zahn et al. (1983), and the data in Florensky et al. (1983*a,c*), Marov et al. (1989), and Pollack et al. (1993) provide more evidence for a decreasing CO mixing ratio toward the surface of Venus. However, at present there are no direct measurements of the CO concentration at altitudes below 12 km on Venus, so the CO mixing ratio, and by inference, the oxygen fugacity and redox state at the surface are open questions.

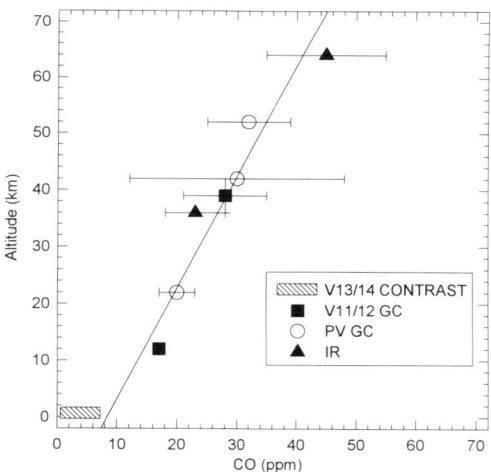

Figure 2. Observations of the CO abundance in the atmosphere of Venus (∼0–70 km altitude) from the Venera 11/12 gas chromatograph (black squares), the Pioneer Venus (PV) gas chromatograph (open circles), the Venera 13/14 CONTRAST experiment (striped box), and from Earth-based infrared spectroscopy (black triangles). Table I lists the data sources. The CONTRAST experiment gave lower limits on the CO abundance (see text). Due to uncertainties in the thermodynamic data for reactions (14) and (15), these lower limits vary from 0.6 to 7 ppm CO.

The Venera 11 and 12 gas chromatograph experiment observed 17 ppm CO at 12 km altitude. Assuming no further decrease in the CO concentration with decreasing altitude and using a temperature of 740 K, Eq. (9) yields an oxygen fugacity of $10^{-21.3}$ bar at 0 km altitude. Zolotov (1996) independently calculated an oxygen fugacity of $10^{-21.3\pm0.2}$ bar at the same temperature from his thermochemical models. Alternatively, we can assume that the CO concentration continues to decrease toward the surface of Venus at altitudes below 12 km. A linear least squares fit to the CO mixing ratios at 12, 22, 36, 39, 42, 52, and 64 km determined by the Pioneer Venus and Venera 11 and 12 gas chromatographs and Earth-based infrared spectra of Venus (Connes et al. 1968; Oyama et al. 1980; Gel'man et al. 1979; Marov et al. 1989; Pollack et al. 1993) is shown in Fig. 2 and gives the equation

$$CO(ppm) = 8.4(\pm 3.0) + 0.51(\pm 0.07) Z(km) \qquad (10)$$

where the uncertainties are $\pm 1\sigma$ errors. The extrapolated CO mixing ratio at 0 km on Venus is then 8.4 ± 3.0 ppm and the corresponding oxygen fugacity is $10^{-20.7\pm 0.4}$ bar. For comparison, Pollack et al. (1993) deduced a CO gradient of 1.20 ± 0.45 ppm km^{-1} around 36 km altitude from their analysis of Earth-based infrared spectra. This gradient is ~ 2.4 times larger than that in Eq. (10) above. However, the CO concentration gradient derived by Pollack et al. (1993) cannot be constant from 0 to 36 km because the derived CO mixing ratio at 0 km on Venus is then less than zero, even when the stated uncertainty is considered.

The calculations discussed above indicate that the CO mixing ratio at the surface of Venus is in the range of ~ 5 to 17 ppm depending on whether or not the trend observed from 12 to 64 km continues downward to the surface. The corresponding oxygen fugacities at 740 K range from $10^{-20.3}$ bar (5 ppm CO) to $10^{-21.3}$ bar (17 ppm CO). As discussed in the next sections, the CO mixing ratio and oxygen fugacity in the near-surface atmosphere of Venus are very important factors that influence the oxidation state of the surface and the relative abundance of reduced (e.g., OCS, H_2S) and oxidized (e.g., SO_2) sulfur gases. Thus *in-situ* measurements of the CO mixing ratio and of the oxygen fugacity below 12 km altitude on Venus are highly desirable.

B. Thermochemical Equilibrium Models of the Oxidation State of the Surface of Venus

Going back to Mueller (1963,1964), theoreticians (Lewis 1970; Lewis and Kreimendahl 1980; Prinn 1985; Volkov et al. 1986; Fegley and Treiman 1992; Zolotov 1987,1994) assumed that the CO/CO_2 ratio in the hot, near-surface atmosphere of Venus controls the oxidation state of the surface of Venus via net thermochemical reactions such as

$$CO_2\,(g) + 2Fe_3O_4 = CO\,(g) + 3Fe_2O_3 \qquad (11)$$

$$CO_2 (g) + 2FeSiO_3 =$$
$$Fe_2O_3 + 2SiO_2 + CO (g) \qquad (12)$$
$$CO_2 (g) + Fe_2SiO_4 = Fe_2O_3 + SiO_2 + CO (g). \qquad (13)$$

The equilibrium CO/CO_2 ratios, and the equilibrium oxygen fugacities for reactions (11–13), are independent of the total pressure because reactions (11–13) have the same number of gas molecules on each side of the equations. The fugacity and partial pressure of CO_2 are identical to $\sim 1\%$ under Venus surface temperatures and pressures, so deviations from ideal gas behavior are negligible. Thus, total pressure cancels out of the equilibrium constant expressions.

Thermochemical models can be used to calculate the stable Fe-bearing minerals on the surface of Venus for a given temperature, pressure, and assumed CO/CO_2 ratio in the near-surface atmosphere. This was done by several authors who concluded that the CO/CO_2 ratio in the near-surface atmosphere of Venus, while too oxidizing for either metallic Fe or wüstite ($Fe_{0.947}O$), is sufficiently reducing to fall within the magnetite stability field (see, e.g., Mueller 1964; Lewis 1970; Lewis and Kreimendahl 1980; Prinn 1985; Fegley and Treiman 1992). In contrast, Zolotov (1987) predicted that hematite is thermodynamically stable at high elevations on the surface of Venus. Furthermore, the recent work of O'Neill (1988) and Hemingway (1990) indicates that hematite is more stable than indicated by thermodynamic data in compilations used by the theorists cited above. Thus, we recalculated stabilities of magnetite and hematite on the surface of Venus.

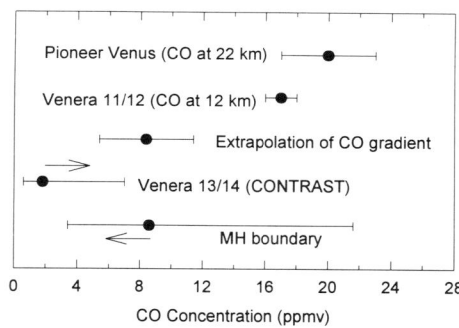

Figure 3. A comparison of observed (Table I) and predicted CO concentrations in the lower atmosphere of Venus. The extrapolated point is from Eq. (10). The uncertainty in the CO concentration at the magnetite-hematite (MH) phase boundary (Eq. 11) is from the 2 sigma uncertainties in the thermodynamic data (Robie and Hemingway 1995). The lower limit from the Venera 13 and 14 CONTRAST experiment is described in the text. The calculated CO concentrations are for 735 K.

In Fig. 3, we compare observations of CO in the lower atmosphere of Venus to the magnetite-hematite phase boundary. The figure shows that within the uncertainties of the observations and calculations, either hematite, or magnetite, or both oxides are stable at $\sim$0 km altitude on Venus. Unfortunately, there are no direct measurements (e.g., by X-ray diffraction or Mössbauer spectroscopy) of the mineralogy of the surface to test these predictions.

C. Venera Spacecraft Observations

Two indirect observations of the oxidation state of the surface and lower atmosphere of Venus have been made. Initially, it appeared that these two measurements contradicted one another (Florensky et al. 1983a,c; Pieters et al. 1986). However, it now appears that the results of the two experiments are consistent and indicate that hematite is present on the surface of Venus (Zolotov 1996; Fegley et al. 1996).

One indirect observation of the redox conditions at the surface of Venus comes from the CONTRAST experiment on the Venera 13 and 14 landers (Florensky et al. 1983a,c). This experiment qualitatively measured the CO content from the color change from white to dark blue for the reaction

$$Na_4V_2O_7 + CO_2 \text{ (g)} + CO \text{ (g)} = V_2O_4 + 2Na_2CO_3 \tag{14}$$

or from white to black for the reaction

$$Na_4V_2O_7 + 2CO(g) = V_2O_3 + 2Na_2CO_3 \tag{15}$$

for asbestos paper impregnated with sodium pyrovanadate ($Na_4V_2O_7$). Florensky et al. (1983a,c) reported that the observed darkening of the asbestos paper gave a lower limit of $\geq$10 ppm for the CO mixing ratio and, assuming chemical equilibrium between CO and CO_2, an upper limit of $\leq 10^{-21}$ bar for the oxygen fugacity. This oxygen fugacity falls inside the magnetite stability field at Venus surface temperatures.

However, Zolotov (1996) recalculated the thermodynamics of reaction (14) and derived a lower limit of $\geq$2.5 ppm CO and an upper limit of $\leq 10^{-19.7}$ bar for the oxygen fugacity (at 740 K). This oxygen fugacity implies that hematite, and not magnetite, is stable on the surface of Venus. Zolotov's (1996) results are different than those reported earlier because the thermodynamic data compilation (Naumov et al. 1971) used by Florensky et al. (1983a,c) did not take into account high temperature phase transitions in sodium carbonate, sodium pyrovanadate, and the vanadium oxides.

Further analysis (Fegley et al. 1996) of the thermodynamics of reactions (14) and (15) confirms Zolotov's (1996) conclusion. Fegley et al. (1996) calculate lower limits of 1.8 ppm CO for reaction (14) and 4.0 ppm CO for reaction (15) at the Venera 13 and 14 landing sites (0.7–0.8 km altitude, and 733–734 K). Because of the uncertainties in the thermodynamic data, the lower limits on the CO abundance from CONTRAST ranges from 0.6 to 0.7 ppm (Fegley et al. 1996).

As Fig. 3 shows, either hematite or magnetite may be consistent with the revised limits on CO. The uncertainties in the thermodynamic data for the magnetite-hematite boundary and for the CONTRAST experiment preclude any stronger statements. Although CONTRAST was a qualitative experiment, Fig. 2 in Florensky et al. (1983a) shows that the soil-free and soil-covered portions of the indicator could be distinguished. Recent image processing by Gektin (1996) shows that the low albedos of the soil-free areas indicates vanadium oxide formation by either reaction (14) or (15) or both. Thus, we must regard the results as suggestive, but not conclusive.

The second indirect observation of the oxidation state of the surface of Venus was made by the imaging and spectrophotometry experiments on the Venera 9, 10, 13, and 14 spacecraft (Golovin et al. 1983; Selivanov et al. 1983). Pieters et al. (1986) used a combination of the Venera 9 and 10 wide angle photometer measurements and the Venera 13 and 14 color images to derive the spectral reflectance of the Venusian surface in the 0.54 to 1.0μm range. They found that at visible wavelengths the spectral reflectance of the Venusian surface was not diagnostic about the Fe oxidation state (with either Fe^{2+}- or Fe^{3+}-bearing minerals being acceptable), but that a high reflectance in the near infrared region apparently required Fe^{3+}-bearing phases such as hematite.

Initially, the inferred presence of hematite disagreed with the results of the CONTRAST experiment on the Venera 13 and 14 spacecraft. However, as discussed above, either magnetite or hematite may be consistent with the revised CO lower limits derived for the CONTRAST experiment Thus, at present, the imaging and spectrophotometry data (Pieters et al. 1986) and the revised interpretation of the CONTRAST experiment (Zolotov 1996; Fegley et al. 1996) are both consistent with, but do not prove, the presence of hematite on the surface of Venus.

D. Kinetic Problems with CO-CO_2 Gas Phase Equilibration

The thermochemical models for iron oxide stability on the surface of Venus rest upon the assumption that CO and CO_2 in the near-surface atmosphere reach chemical equilibrium and thus control the oxygen fugacity. However, this assumption may not be valid at the relatively low temperatures ($\sim$740 K at 0 km to $\sim$660 K at 10 km) over the surface of Venus. For example, Huebner (1975) measured the oxygen fugacity of CO-CO_2 gas mixtures as a function of temperature, composition, and gas flow rate. He found that chemical equilibrium was reached at high temperatures (above $\sim$1373 K), but not at lower temperatures. To good first approximation, Huebner's results were independent of the CO/CO_2 ratio and gas flow rate.

More recent experiments by Fegley et al. (1995b) tested whether or not CO-CO_2 gas mixtures reached thermochemical equilibrium at temperatures of 665 to 873 K. These experiments were done at temperatures below 973 K, which is the lower temperature limit for zirconia oxygen sensors. Oxidation and reduction reactions of metal oxides were used to constrain the oxygen fugacities of the gas mixtures.

One set of experiments was done by heating CuO powder in different CO-CO_2 gas mixtures at different temperatures. After heating, the CuO had quantitatively converted to Cu metal. The quantitative reduction of CuO to Cu metal proves that the f_{O_2} values of the CO-CO_2 gas mixtures were less than the oxygen fugacity of the CuO-Cu phase boundary because no CuO was left in the sample after heating. These experiments give upper limits to the oxygen fugacity of the CO-CO_2 gas mixtures.

Another set of experiments was done by heating weighed aliquots of magnetite in different CO-CO_2 gas mixtures at different temperatures. After heating the magnetite had partially converted to hematite. The conversion was indicated by the color change, the weight gain, the X-ray diffraction pattern, and the Mössbauer (MB) spectrum.

The latter method is especially sensitive to the amount of magnetite and hematite in the experimental samples. For example, Fig. 4 and Table VII show the MB spectra and parameters for the magnetite starting material and the almost pure hematite product from one experiment. The MB spectra "before" and "after" heating are clearly different.

TABLE VII

Mössbauer Parameters for Magnetite and Hematite (Sample 118)

Component	Isomer Shift (mm s^{-1})	HF^a (Tesla)	ΔEQ^a (mm s^{-1})
Magnetite A	0.25	49.2	0.0
Magnetite B	0.64	46.2	0.0
Hematite	0.38	51.9	−0.17

^{a} HF = hyperfine splitting; ΔEQ = electrical quadrupole splitting.

The quantitative oxidation of magnetite to hematite shows that the oxygen fugacity of the gas mixture was above the magnetite-hematite boundary. This experiment gives a lower limit to the oxygen fugacity of the CO-CO_2 gas mixture.

In experiments done with the same CO-CO_2 gas mixture (CO_2/CO ratio = 98.0) at the same temperature (688 K, which corresponds to $\sim$7 km altitude on Venus) CuO was quantitatively reduced to Cu metal while magnetite was partially oxidized to hematite. The combination of these two experiments shows that the oxygen fugacity of the CO-CO_2 gas mixture was somewhere between $10^{-24.9}$ bar (MH boundary) and $10^{-14.1}$ bar (Cu-CuO boundary). However, Eq. (9) predicts that this CO-CO_2 gas mixture should have an even lower oxygen fugacity of $10^{-29.9}$ bar at 688 K. This is $\sim$5 orders of magnitude below the lower limit obtained from the magnetite oxidation experiment and $\sim$15 orders of magnitude below the upper f_{O_2} limit obtained from the CuO reduction experiment. Similar results were obtained for other CO-CO_2 gas mixtures (Fegley et al. 1995b). The metal oxide redox experiments show that

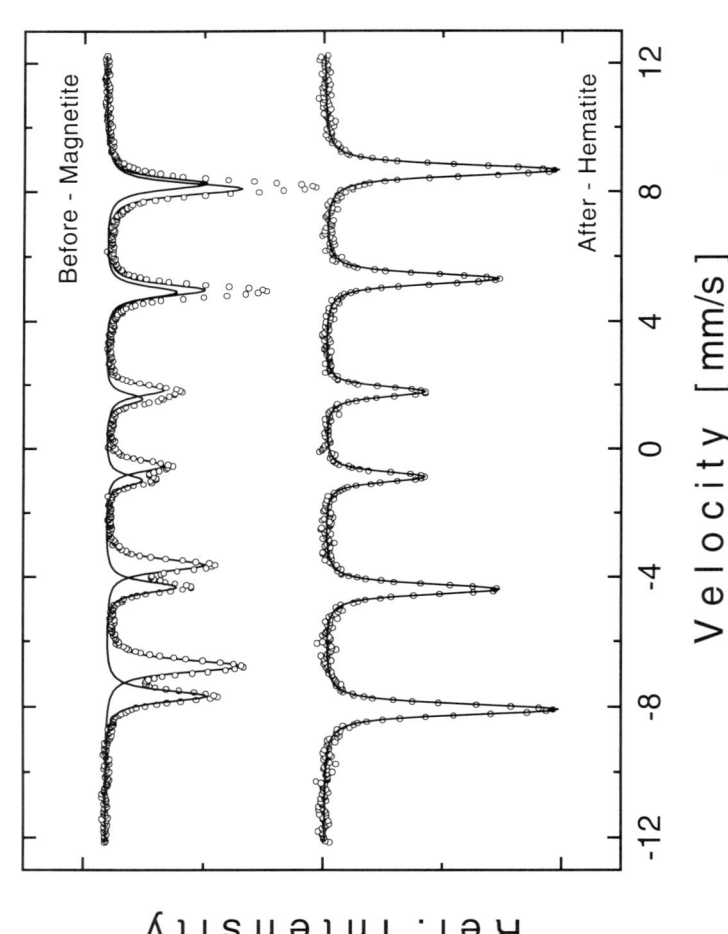

Figure 4. Mössbauer spectra "before" and "after" heating synthetic magnetite (experiment R118) for 3.8 days at 690 K in a CO-CO_2 gas mixture containing 100 ppm CO. Even though the CO/CO_2 ratio of the gas is in the magnetite stability field, almost all of the magnetite oxidized to hematite. The MB parameters are listed in Table VII.

at low temperatures CO-CO_2 gas mixtures, while extremely reducing, are still several orders of magnitude more oxidizing than predicted by thermochemical equilibrium calculations. As a result, magnetite which is heated in "reducing" CO-CO_2 gas mixtures is either partially or totally converted to hematite. A similar situation may be present on Venus.

E. Hematite Formation from Basalt in CO-CO_2 Gas Mixtures

Fegley et al. (1995b) also heated basalt samples in CO-CO_2 gas mixtures with CO_2/CO ratios inside the magnetite stability field. The CO molecular number density at the surface of Venus is ~0.5 (5 ppm CO) to ~2 (20 ppm CO) times that in the experiments. The mineralogy of the basalt starting material and the heated samples was determined by several analytical methods including MB spectroscopy. The results of the MB analyses are summarized in Table VIII and the MB spectra are illustrated in Fig. 5. Details of the experiments and analytical methods are described by Fegley et al. (1995b).

The MB spectrum of the unheated basalt powder shows that most of the Fe is present as Fe^{2+} in pyroxene, olivine, and ilmenite. The pyroxene also contains some Fe^{3+}, but no hematite or magnetite were detected by MB spectroscopy, or other methods.

As shown in Table VIII and Fig. 5, hematite and magnetite are present in the MB spectra of the heated samples. In addition, the heated samples contain smaller amounts of Fe^{2+} in olivine and pyroxene, and larger amounts of Fe^{3+} in pyroxene than the starting material. The MB data for sample ZT1, heated at 505°C for 7 days, show that magnetite, which is nearly, but not exactly stoichiometric, and hematite have formed and comprise about 23.5% of the total Fe atoms in the heated basalt. The Fe^{3+} in pyroxene has also increased from ~28% of the total Fe atoms in the unheated basalt to ~34% of total Fe. Sample BBB1, heated at 803°C for 10 days, contains more hematite (~37%) and Fe^{3+} in pyroxene is 41%.

Although hematite was formed by heating the basalt in a CO-CO_2 mixture containing 1000 ppm CO, the predicted f_{O_2} of the gas mixture is inside the magnetite stability field and thus significantly lower than the actual f_{O_2}, which led to hematite formation. In the case of sample ZT1, thermochemical equilibrium calculations predict an oxygen fugacity of $10^{-22.9}$ bar while the presence of magnetite and hematite shows that the actual f_{O_2} was $\leq 10^{-19.0}$ bar. For sample BBB1, thermochemical equilibrium calculations predict an f_{O_2} of $10^{-12.3}$ bar and the presence of only hematite shows that the actual value is $>10^{-9.4}$ bar. These results suggest that the red color observed by Pieters et al. (1986) at several Venera landing sites is hematite formed by sub-aerial oxidation of basalt on the surface of Venus.

F. Hematite Formation from Pyrite and Magnetite in CO-CO_2-SO_2 Gas Mixtures

Hematite formation was also observed in the pyrite decomposition experiments done in CO-CO_2, and CO-CO_2-SO_2 gas mixtures by Fegley et al.

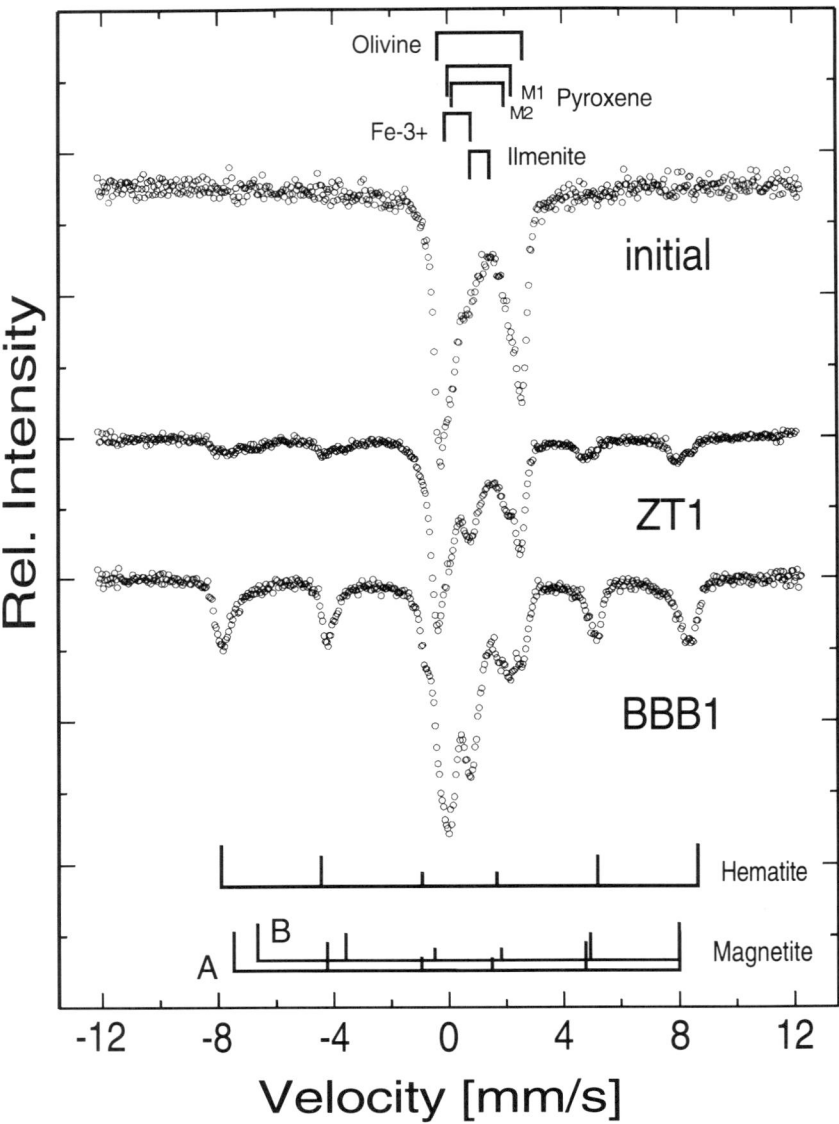

Figure 5. A comparison of the Mössbauer spectra of the unreacted basalt and two oxidized samples. The stick diagrams show the positions and relative intensities of the peaks for Fe^{2+} in olivine, Fe^{2+} in the M1 and M2 sites of clinopyroxene, Fe^{3+} in clinopyroxene, Fe^{2+} in ilmenite, Fe^{3+} in hematite, and for the A and B sites of magnetite. Sample ZT1 was heated for 7 days at 505° C and sample BBB1 was heated for 10 days at 803° C. Both samples were heated in the same CO-CO_2 gas mixture, containing 1000 ppm CO. Table VIII lists the MB parameters for the spectra, which are discussed in Fegley et al. (1995b).

TABLE VIII
Mössbauer Parameters for Unreacted and Oxidized Basalts

Component	Isomer Shift (mm s^{-1})	QS^a (mm s^{-1}) HF (Tesla)	Relative Intensity (%)^b	Relative Mass (%)
Unreacted Basalt				
Olivine Fe^{2+}	1.15	2.94	23.5	10.5
Cpx Fe^{3+}	0.37	0.88	28.2	31.0
Cpx Fe^{2+} (M1)	1.11	2.25	25.0	33.0
Cpx Fe^{2+} (M2)	1.05	1.80	22.0	25.0
Ilmenite Fe^{2+}	1.11	0.71	1.3	0.5
Oxidized Basalt ZT1				
Olivine Fe^{2+}	1.12	2.94	22.0	14.0
Cpx Fe^{3+}	0.37	0.95	34.0	51.0
Cpx Fe^{2+} (M1)	1.04	2.35	10.6	23.0
Cpx Fe^{2+} (M2)	1.12	1.85	4.9	8.0
Ilmenite Fe^{2+}	1.11	0.68	5.0	1.1
Hematite	0.37	51.6 T	3.5	0.5
Fe$_3$O$_4$ A site	0.26	48.3 T	7.0	0.8
Fe$_3$O$_4$ B site	0.63	45.7 T	13.0	1.6
Oxidized Basalt BBB1				
Olivine Fe^{2+}	1.12	2.95	6.3	4.2
Cpx Fe^{3+}	0.36	0.87	41.0	63.3
Cpx Fe^{2+} (M1)	1.14	2.16	12.4	23.3
Cpx Fe^{2+} (M2)	1.13	1.90	3.0	5
Ilmenite Fe^{2+}	1.08	0.71	0.4	0.2
Hematite	0.36	49.76 T	10.8	1.2
	0.378	51.43 T	11.3	1.3
	0.37	43.64 T	8.5	1.0
	0.38	47.52 T	6.3	0.8

^{a} QS = quadrupole shift (mm s^{-1}) and HF = hyperfine splitting in Tesla (T). Hyperfine splittings are listed for hematite and magnetite and quadrupole splittings are listed for the other components.

^{b} The apparent variation in the ilmenite content gives an indication of the typical uncertainty in the relative intensities of strongly overlapping components. Typical uncertainties in the relative intensities of clearly resolved components, i.e., magnetite and hematite, are significantly smaller.

(1995a). As discussed below in Sec. IV.B, upon heating pyrite decomposed to pyrrhotite, which then underwent oxidation to hematite via magnetite and maghemite formation. In some cases, hematite occurred without detectable magnetite. In all cases, the CO/CO$_2$ ratios of the binary and ternary gas mixtures fall inside the magnetite stability field.

Of most interest here are the experiments done with CO-CO$_2$-SO$_2$ gas mixtures having CO molecular number densities ~10 (20 ppm CO) to ~40 (5 ppm CO) times that at the surface of Venus, and SO_2 molecular number densities from ~1.2 times (150 ppm SO_2) to ~7.6 times (25 ppm SO_2) of that at the surface of Venus. The experiments indicate that the presence of SO_2 does not catalyze the equilibration of CO and CO_2, at least on laboratory time scales, and that the actual oxygen fugacity of the ternary mixture is still greater than that calculated assuming thermochemical equilibrium.

G. Mechanisms for Equilibrating CO and CO_2 at the Surface of Venus

The CO_2/CO ratios of the gas mixtures used in the experiments ranged from ~26 to ~10,000. In comparison, the CO_2/CO ratios at the surface of Venus range from ~50,000 (assuming 20 ppm CO) to ~200,000 (assuming 5 ppm CO). The large CO_2/CO ratios at the surface of Venus qualitatively suggest that equilibration of CO and CO_2 will also be kinetically inhibited there. Here we explore the kinetic problems involved in the equilibration of CO and CO_2 and discuss the mechanisms by which these two gases are converted.

The simplest way to convert CO to CO_2 is via the elementary reaction

$$CO + O + M \longrightarrow CO_2 + M \qquad (16)$$

where M is any third body. However, the rate constant for reaction (16) is very slow (Baulch et al. 1976) because it is spin forbidden. As a consequence reaction (16) is ineffective for converting CO back to CO_2. Instead the elementary reaction

$$CO + OH \longrightarrow CO_2 + H \qquad (17)$$

with a rate constant of (Baulch et al. 1995)

$$k_{17} = 1.05 \times 10^{-17} T^{1.5} \exp\left(\frac{250}{T}\right) cm^3 \ s^{-1} \qquad (18)$$

is an important component of catalytic cycles that reconvert CO to CO_2 in the atmosphere of Mars (McElroy and Donahue 1972; Nair et al. 1994). Reaction (17) also plays a major role for CO oxidation in flames and combustion processes (Warnatz 1984) but is probably less effective in the stratosphere of Venus than it is on Mars because of the extremely low abundance of hydrogen compounds above the clouds of Venus. Other catalytic cycles involving Cl and NO_x chemistry may be necessary for transforming CO back to CO_2 in the Venusian stratosphere (Yung and DeMore 1982; chapter by Esposito et al.).

Fegley et al. (1995b) suggested that reaction (17) is also of limited importance at the surface of Venus because of the low water vapor abundance. Fegley and Lodders (1995) and Fegley et al. (1996) compared the chemical lifetime for CO oxidation via reaction (17)

$$t_{chem}(CO) \sim \frac{1}{(k_{17}[OH])} \qquad (19)$$

to the vertical mixing time in the lower atmosphere of Venus

$$t_{mix} \sim \frac{H^2}{K_{eddy}} \qquad (20)$$

The [OH] number densities were taken from thermochemical equilibrium calculations and K_{eddy}, the vertical eddy diffusion coefficient was taken as $\sim 10^4$ cm^{2}s^{-1}. Higher and lower K_{eddy} values were also investigated. The pressure scale height H (=$RT/\mu g$) is ~ 16 km at the surface of Venus. If $t_{chem} < t_{mix}$, then CO will be oxidized via reaction (17), while if $t_{chem} > t_{mix}$, then CO will not be oxidized. At 740 K, Eq. (19) yields $t_{chem}(CO) \sim 66$ yr vs $t_{mix} \sim 8$ yr from Eq. (20). For the nominal K_{eddy} value of 10^4 cm^{2}s^{-1}, reaction (17) quenched at all temperatures below ~ 775 K (i.e., at 4.5 km below the modal planetary radius). Fegley et al. (1996) also studied other reactions that may be the slow step in the conversion of CO to CO$_2$ at the surface of Venus. They found that the fastest reaction

$$CO + SO_2 \rightarrow CO_2 + SO \qquad (21)$$

quenched at 748±13 K (−1.0±1.7 km below the modal radius). They concluded that CO and CO$_2$ only equilibrate, if at all, in the lowest levels of Venus' atmosphere below about 0.7 km altitude. A disequilibrium region, that is more oxidizing than predicted by equilibrium calculations, exists at higher elevations. Their conclusions are consistent with the inferred presence of hematite at several Venera landing sites (Pieters et al. 1986) and the experiments described above.

IV. VENUS SULFUR CYCLE

A. Background

Sulfur chemistry on Venus is a complex and controversial topic that has been discussed for the past three decades because of its importance for the global sulfuric acid cloud cover, the energy budget and greenhouse effect in the lower atmosphere, volcanism, and chemical weathering of the surface. The complexity arises because sulfur in the atmosphere is either known to, or strongly suspected to, exist as aqueous sulfuric acid cloud droplets, sulfuric acid vapor, $SO_3(g)$, $SO_2(g)$, $SO(g)$, $(SO)_2(g)$, $OCS(g)$, $H_2S(g)$, elemental sulfur vapor S_{1-8}, and solid sulfur particles (von Zahn et al. 1983; Prinn 1985; Krasnopolsky and Pollack 1994; chapter by Esposito et al.). Sulfur has been reported on the surface of Venus (Table V), and has been suggested to exist as anhydrite ($CaSO_4$), pyrite (FeS_2), and pyrrhotite (Fe_7S_8) (see Lewis and Kreimendahl 1980; Pettengill et al. 1982). Controversy arises because the available measurements of sulfur gas abundances in the atmosphere do not always agree (Table I), because we do not know the composition of the lower 22 km of the atmosphere, and because we have no direct information (from MB spectroscopy or X-ray diffraction) on the mineralogy of the surface.

Over a decade ago, data from the Pioneer Venus mission, the Venera 11 and 12 missions, and Earth-based remote sensing were used to propose the Venus sulfur cycle (von Zahn et al. 1983; Prinn 1985). An expanded and revised version of the original Venus sulfur cycle is illustrated in Fig. 6. This cycle schematically shows the gas phase, gas-aerosol, and gas-rock reactions which continuously cycle sulfur between the atmosphere, the clouds, and rocks on the surface of Venus.

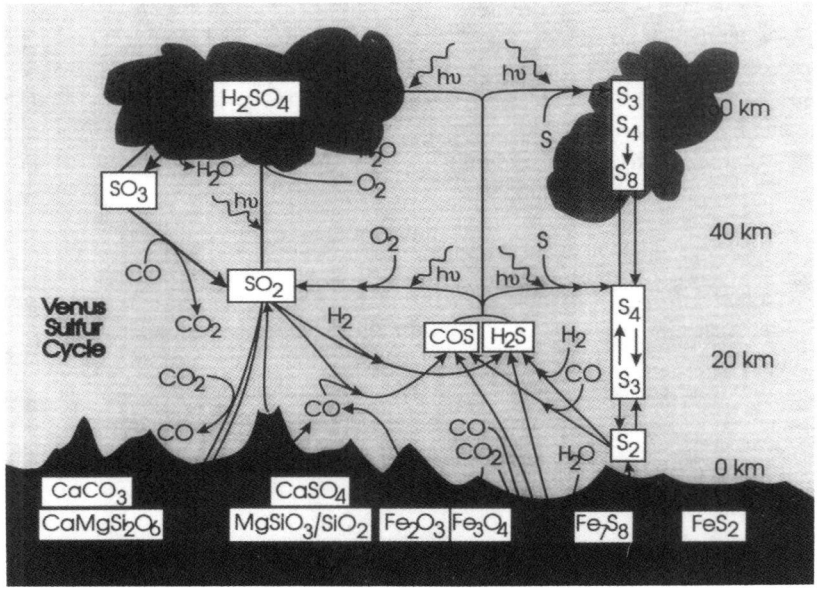

Figure 6. An expanded and revised version of the Venus sulfur cycle proposed by von Zahn et al. (1983) and Prinn (1985). Experimental results show that anhydrite ($CaSO_4$) formation is a sink for atmospheric SO_2 (Fegley 1988, 1990; Fegley and Prinn 1989), that pyrite decomposition is a source for atmospheric sulfur vapor (Fegley et al. 1995a), that pyrrhotite oxidation is a source for atmospheric OCS and H_2S (Treiman and Fegley 1991; Fegley et al. 1995a), and that magnetite oxidizes to hematite (Fegley et al. 1995b).

Below we consider in some detail the gas-rock reactions suggested for the Venus sulfur cycle and also briefly discuss some observations of sulfur gases in the lower atmosphere. The gas phase and gas-aerosol reactions involved in the sulfur cycle, and the observations of sulfur gases above the clouds of Venus are discussed in more detail in the chapter by Esposito et al.

B. Anhydrite Formation on Venus

Sulfur dioxide was the most abundant sulfur gas detected by the Pioneer Venus and Venera 11 and 12 spacecraft (Table I). The Venera 11 and 12 gas chromatography experiment reported 130 ± 35 ppm at 42 km altitude and below

(Gel'man et al. 1979) and the Pioneer Venus gas chromatography experiment reported 185±43 ppm at 22 km (Oyama et al. 1980). These observations were surprising at the time because thermochemical equilibrium models predicted that OCS and H_2S were the dominant sulfur gases, and that SO_2 was a minor sulfur gas in the lower atmosphere of Venus (Lewis 1968,1969,1970; Lewis and Kreimendahl 1980). The thermochemical equilibrium models also predicted that the SO_2 abundance was ~100 times smaller than observed and was regulated by the reaction:

$$CaCO_3 + SO_2 (g) = CaSO_4 + CO (g) \qquad (22)$$

However, the Pioneer Venus and Venera 11 and 12 measurements indicate that the sulfur gas chemistry in the lower atmosphere of Venus is controlled by kinetic factors (von Zahn et al. 1983; Prinn 1985).

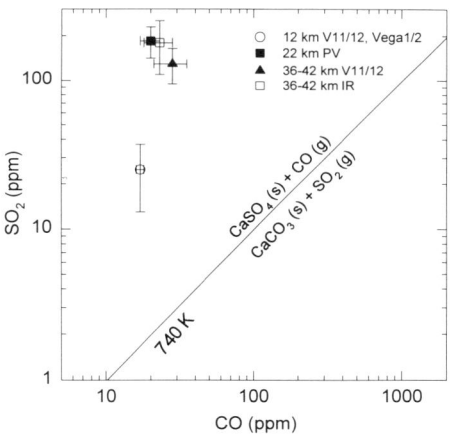

Figure 7. A comparison of the observed and calculated SO_2 abundances in the lower atmosphere of Venus. Observations of SO_2 and CO from the Pioneer Venus gas chromatograph (Oyama et al. 1980) the Venera 11 and 12 gas chromatograph (Gel'man et al. 1979; Marov et al. 1989), Earth-based infrared observations of the night side of Venus (Bézard et al. 1990,1993; Pollack et al. 1993), and from the Vega 1 and 2 ultraviolet spectroscopy experiments (Bertaux et al. 1996) are higher than the abundances of SO_2 and CO in thermochemical equilibrium with co-existing calcite and anhydrite (diagonal line). The data graphed are listed in Table I. Note that the pairs of SO_2 and CO observations refer to the same altitude or altitude range (e.g., 22 km for the Pioneer Venus gas chromatograph). Only the 12 km SO_2 data from Vega 1 and 2 are shown because the higher altitude data fall within the range of the Pioneer Venus, Venera 11 and 12, and infrared data points.

Figure 7 displays a comparison of the SO_2 abundances observed by the Pioneer Venus and Venera 11 and 12 spacecraft and the thermochemical equilibrium SO_2 abundance predicted by reaction (12) at the global mean temperature of 740 K on Venus. Also shown are subsequent measurements of SO_2 from Earth-based infrared spectra of the lower atmosphere of Venus (Bézard et al. 1993; Pollack et al. 1993), and by ultraviolet spectroscopy on the Vega 1 and 2 probes (Bertaux et al. 1996). In each case the corresponding CO abundance measured at the same altitude is also plotted. All the data show that SO_2 in the lower atmosphere of Venus is present at abundances significantly higher than predicted by the thermochemical equilibrium models (diagonal line).

von Zahn et al. (1983) and Prinn (1985) suggested that the most plausible explanation for the overabundance of SO_2 is that the reaction of SO_2 with Ca-bearing minerals on Venus is slow and does not reach chemical equilibrium. von Zahn et al. (1983) also noted that "Laboratory experiments on the rates of various gas and gas-mineral reactions involving sulfur compounds are needed to further clarify the sulfur cycle on Venus." Experimental studies of anhydrite formation kinetics via reaction (22) (Fegley 1988,1990; Fegley and Prinn 1989; Fegley and Treiman 1992) provided the required gas-solid kinetic data.

The rate at which SO_2 reacts with calcite to form anhydrite was measured from 873 to 1123 K by isothermally heating high purity calcite crystals in flowing CO_2-SO_2 gas mixtures at ambient atmospheric pressure. Data at Venus surface temperatures were not obtained because of the extremely slow reaction rates. The gas mixtures either had about the same SO_2 number density as at the surface of Venus (if the Pioneer Venus, Venera 11 and 12, or infrared abundance data are assumed to hold at 0 km altitude), or several times higher SO_2 number densities (if the Vega 1 SO_2 abundance at 12 km altitude is taken at face value). The experimental procedures and analytical methods are described in Fegley and Prinn (1989).

The experimental rate equation (Fegley and Prinn 1989) for anhydrite formation via reaction (22) is

$$\frac{d[CaSO_4]}{dt} = 10^{19.64(\pm 0.28)} \exp\left(\frac{-15250(\pm 2970)}{T}\right) \text{mol cm}^{-2}\text{s}^{-1}. \quad (23)$$

The global mean rate of anhydrite formation of Venus was derived from Eq. (23) by convolving the surface area distribution as a function of altitude, the variation of temperature with altitude, the variation of the reaction rate with temperature, and the measured Ca content from the Venera 13,14, and Vega 2 XRF analyses (Table V).

Figure 8 shows that this global mean rate is about 4.6×10^{10} mol cm^{-2}s^{-1}, which corresponds to an anhydrite deposition rate of ~ 1 μm per year. Because the growing anhydrite layer becomes less porous with time, the reaction may slow down as the anhydrite layer becomes thicker. However, aeolian weathering (Marshall et al. 1988) may plausibly remove the thin anhydrite layers fairly rapidly, so that the reaction can easily proceed. As already emphasized by several authors (Prinn 1985; Zolotov and Khodakovsky 1989; Fegley and Treiman 1992), the S/Ca ratios measured by the Venera 13,14,

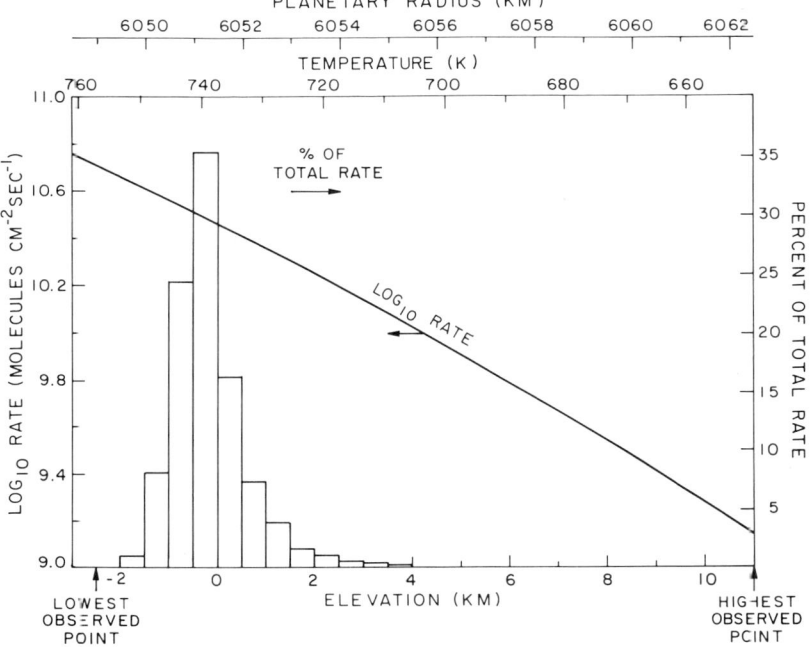

Figure 8. The rate of anhydrite formation as a function of elevation and temperature on the surface of Venus. The bar graph shows the percent of the total rate in each elevation bin. The solid line shows how the rate decreases with temperature and is an extrapolation of the rate Eq. (23). The calculations also take into account that the surface of Venus is not pure calcite. The weighted mean CaO content calculated from the Venera 13,14, and Vega 2 XRF analyses is 7.90($\pm$0.51) wt. %. The global mean rate of anhydrite formation is about 4.6×10^{10} mol cm^{-2}s^{-1}, and corresponds to an anhydrite deposition rate of ~ 1 μm per year (figure from Fegley 1990).

and Vega 2 spacecraft are less than that expected if all Ca were combined in CaSO$_4$. Therefore, at present the surface of Venus still contains reactive Ca which is a net sink for SO$_2$ loss via anhydrite formation.

However, the rate of SO$_2$ loss is very slow, and the chemical lifetime of SO$_2$ in the atmosphere of Venus is given by

$$t_{\text{chem}}(SO_2) \sim \frac{\sigma_{SO_2}}{\frac{d[CaSO_4]}{dt}} \quad (24)$$

where σ_{SO_2} is the SO$_2$ column density on Venus (given by the product of the pressure scale height H and the [SO$_2$] number density) and d[CaSO$_4$]/dt is the global mean anhydrite formation rate from Fig. 8, which is the same as the SO$_2$ gas depletion rate. Fegley and Prinn (1989) calculated $t_{\text{chem}}(SO_2) \sim 1.9$ Myr based on 150 ppm SO$_2$ recommended by von Zahn et al. (1983). If the lower SO$_2$ mixing ratio of 25 ppm reported by the ultraviolet spectroscopy experiment on Vega 1 at 12 km altitude is taken at face value, $t_{chem}(SO_2)$ ~320,000 yr results. However, in any case, the formation of anhydrite will remove almost all (~99%) of the SO_2 from the atmosphere of Venus within geologic time scales unless the SO_2 is replenished in some fashion.

Fegley and Treiman (1992) calculated that reactions of SO_2 with other calcium minerals such as diopside ($CaMgSi_2O_6$), anorthite ($CaAl_2Si_2O_8$), and wollastonite ($CaSiO_3$) are also thermodynamically favorable under Venus surface conditions. However, our unpublished laboratory experiments show that these reactions are apparently an order of magnitude or more slower than that of SO_2 with calcite. Thus, as mentioned earlier by Fegley and Treiman (1992) and Bézard et al. (1993), chemical reactions of SO_2 with diopside, anorthite, and wollastonite cannot buffer SO_2 in the lower atmosphere of Venus because of the extremely slow rate of these reactions.

C. Rate of Volcanism on Venus

Fegley and Prinn (1989) argued that maintenance of the global sulfuric acid cloud cover, which is produced by photooxidation of SO_2, requires a source to replenish the SO_2 currently being lost by reaction with calcite on the surface of Venus. If the SO_2 were not replenished, the global sulfuric acid clouds would eventually disappear within a geologically short time.

Fegley and Prinn (1989) argued that volcanism is the most plausible source for replacing atmospheric sulfur. The required volcanism rate is a function of the sulfur content of the erupted material (magma + gas) and of the degassing efficiency of the erupted material. The degassing efficiency is close to unity as indicated by pyrite and pyrrhotite decomposition experiments (Fegley et al. 1995a; Treiman and Fegley 1991). Different geochemical models for the sulfur content of the erupted material led to estimated volcanism rates ranging from 0.4 to 11 km^3 yr^{-1}. The preferred rate of ~1 km^3 yr^{-1} was calculated by assuming that material erupted onto the surface of Venus had the same composition as that determined by XRF analyses of the surface at the Venera 13,14, and Vega 2 landing sites. This geochemically estimated rate is comparable to the rate of sub-aerial volcanism on the Earth (Ivanov and Freney 1983) and is about 5% of the plate creation rate of ~20 km^3 yr^{-1} on the Earth (Parsons 1981). The geochemical rate of ~1 km^3 per year is also supported by independent geophysical estimates of volcanism rates on Venus (Grimm and Solomon 1987; Spohn 1991).

D. Pyrite Decomposition on the Surface of Venus

As illustrated in Fig. 6, an important component of the Venus sulfur cycle is pyrite decomposition, which was predicted to be a source of reduced sulfur gases at the surface of Venus (see, e.g., von Zahn et al. 1983; Prinn 1985) via the net thermochemical reactions

$$FeS_2 + 2CO_2\ (g) \longrightarrow FeO + 2COS\ (g) + \frac{1}{2}O_2\ (g) \qquad (25)$$

$$\text{FeS}_2 + \text{CO}_2 \text{ (g)} + \text{CO (g)} \longrightarrow \text{FeO} + 2\text{COS (g)} \quad (26)$$

$$\text{FeS}_2 + 2\text{H}_2\text{O (g)} + \text{CO (g)} \longrightarrow \text{FeO} + 2\text{H}_2\text{S (g)} + \text{CO}_2 \text{ (g)}. \quad (27)$$

However, when the sulfur cycle was first proposed, no kinetic data were available to model the rates of these suggested pyrite decomposition reactions.

Fegley et al. (1995a) experimentally studied pyrite decomposition because of its importance for the hypothesized Venus sulfur cycle and showed that pyrite rapidly decomposes over the entire surface of Venus. They also showed that pyrite decomposition in CO_2 and CO_2 gas mixtures is a multistage process which is more complex than originally proposed. This work is relevant to models of atmosphere-surface interactions, for understanding the origin of the low emissivity regions seen in radar imaging, to studies of the origin and evolution of the present surface conditions on Venus, for the design of future spacecraft missions to Venus, and for the rapidly evolving Earth-based infrared observations of the lower atmosphere of Venus.

The pyrite decomposition experiments were done by isothermally heating pyrite slices for different time periods in CO_2 or CO_2 gas mixtures. This is the same general technique used earlier in the anhydrite formation experiments (Fegley and Prinn 1989), and is the same way in which many gas-solid kinetic experiments are conducted in materials science. The gas mixtures used were chosen to study the effects of the CO_2 and CO number densities on the kinetics and to study the kinetics in a ternary gas mixture with CO and SO_2 number densities similar to those at the surface of Venus. The proposed decomposition of pyrite by water vapor, reaction (27), is being studied in a subsequent series of experiments. The experiments were done along five isotherms from $\sim$660 to $\sim$800 K, which span the temperatures over the surface of Venus.

X-ray diffraction and MB spectroscopy of the reacted samples showed that pyrite decomposed to pyrrhotite (Fe_7S_8), which on continued heating lost more sulfur leaving behind more Fe-rich pyrrhotites. The pyrrhotite forms a layer covering the unreacted pyrite. With increasing time, more and more pyrite decomposes to pyrrhotite, until pyrrhotite is the only sulfide left in the sample. During this process, the pyrrhotites were also being oxidized to magnetite. As discussed earlier in Sec. III.F, the magnetite was oxidized to hematite via maghemite as an intermediate. These, and other data led Fegley et al. (1995a) to conclude that pyrite decomposition occurs via the net thermochemical reaction

$$7\text{FeS}_2 \longrightarrow \text{Fe}_7\text{S}_8 + 3\text{S}_2 \text{ (g)}. \quad (28)$$

They also suggested that pyrrhotite oxidation and magnetite oxidation to hematite occurs by net thermochemical reactions exemplified by

$$3\text{Fe}_7\text{S}_8 + 28\text{CO}_2 \text{ (g)} \longrightarrow 7\text{Fe}_3\text{O}_4 + 12\text{S}_2 \text{ (g)} + 28\text{CO (g)} \quad (29)$$

$$\text{Fe}_9\text{S}_{10} + 12\text{CO}_2 \text{ (g)} \longrightarrow 3\text{Fe}_3\text{O}_4 + 5\text{S}_2 \text{ (g)} + 12\text{CO (g)} \quad (30)$$

$$2Fe_3O_4 + CO_2 \text{ (g)} \longrightarrow 3\gamma - Fe_2O_3 + CO \text{ (g)} \tag{31}$$

$$\gamma - Fe_2O_3 = \alpha - Fe_2O_3. \tag{32}$$

These proposed reactions for pyrite decomposition, pyrrhotite oxidation, and magnetite oxidation are schematically illustrated in Fig. 6. Alternative reactions for pyrrhotite oxidation, which involve the direct production of OCS, may occur in concert with or in preference to reactions suggested by Fegley et al. (1995a). The experimental data confirm the hypothesized importance of pyrite chemical weathering in the Venus sulfur cycle (von Zahn et al. 1983; Prinn 1985), although pyrite decomposition does not occur as a single step via reactions (25–26).

The experimental data show that pyrite decomposition follows zero-order kinetics and is independent of the amount of pyrite present. The data also show that within the experimental uncertainties, the rate of pyrite decomposition and the activation energy ($\sim$150 kJ mole^{-1}) for pyrite decomposition are independent of the gas compositions used. This suggests that the rate determining step is sulfur loss from pyrite (Fegley et al. 1995a).

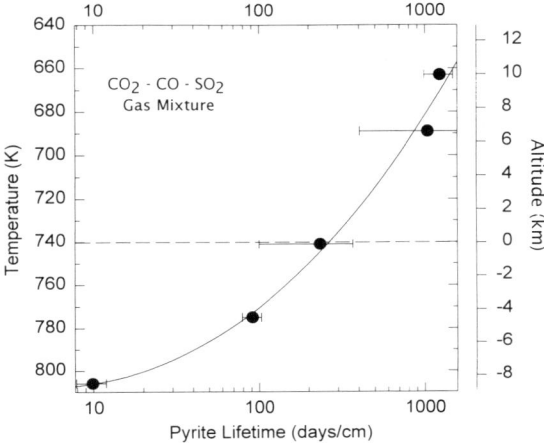

Figure 9. Pyrite decomposition lifetimes (days/cm) as a function of temperature and altitude on the surface of Venus. The error bars show the one sigma uncertainties in the pyrite lifetimes. As discussed by Fegley et al. (1995a), pyrite decomposition follows zero-order kinetics and is independent of the amount of pyrite present. Pyrite decomposition has the same rate, within experimental uncertainties, in CO_2 and in CO_2 gas mixtures. The pyrite lifetimes shown are for a CO(1.9%)-CO_2(96.3%)-SO_2(1.8%) gas mixture with CO and SO_2 number densities close to those at the surface of Venus. Pyrite decomposition kinetics were also independent of the CO_2 number density, within experimental uncertainties.

Figure 9 shows the rate of pyrite decomposition in a CO-CO_2-SO_2 gas mixture with CO and SO_2 number densities close to those at the surface of Venus. The decomposition rate in days/cm varies from ~ 10 days at ~ 800 K (-7 km) to ~ 230 days at 740 K (0 km) to ~ 1200 days at ~ 660 K (~ 10 km). These lifetimes are very short and show that pyrite cannot exist on Venus for any significant time.

The experiments also show that the rate of pyrrhotite oxidation is significantly slower than the rate of pyrite thermal decomposition. The fine-grained pyrrhotite produced by pyrite decomposition is oxidized to magnetite during the course of the experiments. Pyrrhotite oxidation is a diffusion-controlled process that depends on the particle size (Asaki et al. 1983; Treiman and Fegley 1991). The preliminary rate data (Treiman and Fegley 1991) predict that millimeter-sized pyrrhotite grains oxidize completely in hundreds of years, while decameter-sized masses of pyrrhotite, such as occur on Earth in magmatic sulfide deposits, have lifetimes of millions of years on the surface of Venus.

The presence of pyrrhotite, which is a source of OCS via oxidation by atmospheric CO_2 and CO (Fegley and Treiman 1992; Fegley et al. 1992), is supported by the observed increase of OCS with decreasing altitude on Venus (Pollack et al. 1993). In fact, Pollack et al. (1993) state that "these results are in remarkable agreement with calculations of buffering by surface minerals (Fegley and Treiman 1992) and thermodynamic equilibrium of gases close to Venus' surface (Krasnopolsky and Parshev 1979), both of which predict a sharply increasing OCS mixing ratio toward the surface, with a surface value of several 10s of ppm. In both cases, CO acts as one of the source gases for making OCS." Krasnopolsky and Pollack (1994) modeled sulfur chemistry in the lower atmosphere of Venus and commented that "the OCS mixing ratio near the surface, 28 ppm, is close to the value of 20 ppm and 5–30 ppm given by Krasnopolsky and Parshev (1979) and Fegley and Treiman (1992), respectively. Pyrite decomposition to pyrrhotite and pyrrhotite oxidation to iron oxides on the Venus surface (Fegley and Treiman 1992; Fegley et al. 1993) may be a source of OCS." Thus, there appears to be a convergence of experiment, observation, and theory suggesting that iron sulfide chemical weathering is a source of reduced sulfur gases in the near-surface atmosphere of Venus.

E. Sulfur Chemistry in the Near-Surface Atmosphere

Finally, we briefly consider sulfur chemistry, in particular oxidized (e.g., SO_2) and reduced (e.g., OCS, H_2S, and S_{1-8}) sulfur gases. The recent reinterpretation by Bertaux et al. (1996) of the ultraviolet spectrometer data from the Vega 1 and 2 spacecraft, suggests that the abundance of SO_2 drops to ~ 25 ppm at 12-km altitude. Taking these observations at face value we need to ask where the remaining ~ 125 ppm or so, of sulfur originally contained in SO_2 has gone?

One possibility is that SO_2 was removed by reaction with crustal minerals.

Another possibility is that sulfur has been redistributed into other gases. The reactions of SO_2 with calcium-bearing minerals discussed above, are very slow. The fastest reaction with calcite takes ~ 1.9 Myr to reduce SO_2 to its equilibrium abundance. One could imagine that SO_2 would be removed faster if fine-grained calcite were present on the surface of Venus. However, laboratory studies of anhydrite formation from SO_2 and synthetic calcium carbonate powder show that after a short period of time the fine-grained material has turned into coarse-grained lumps that are chemically cemented together. An example of this behavior is illustrated in Burke et al. (1994). Another problem is the source of the fine-grained calcite required for the rapid reaction. This is closer to beach sand than to the igneous carbonate probably present in carbonatites and alkaline rocks on the surface of Venus. Also, as discussed earlier, the XRF data show that only a few percent of the surface can be calcite. Thus, we consider this possibility unlikely.

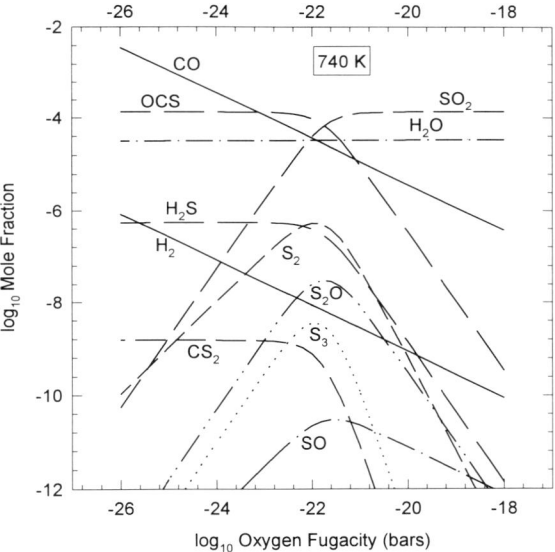

Figure 10. Thermochemical equilibrium abundances of important C-O-S-H gases as a function of the oxygen fugacity in the near-surface atmosphere of Venus (740 K, 0 km). The calculation assumes gas phase thermochemical equilibrium and neglects all gas-solid reactions. SO_2 remains the dominant sulfur-bearing gas, and OCS and H_2S are less abundant at the surface of Venus for all plausible oxygen fugacities (i.e., f_{O2} values $> 10^{-21.3}$ bar). The thermodynamic data are taken from Fegley et al. (1996).

The second possibility, that sulfur has been redistributed into other gases, is considered in Fig. 10. This shows the results of thermochemical equilibrium calculations of gas abundances as a function of the oxygen fugacity in the near-surface atmosphere of Venus. These calculations neglect any reactions with the surface. The results are in good agreement with the prior gas phase thermochemical equilibrium calculations of Krasnopolsky and Parshev (1979). Sulfur dioxide remains the dominant sulfur gas and OCS and H_2S are less abundant species for all oxygen fugacities $>10^{-21.8}$ bar. As discussed earlier, plausible CO mole fractions at the surface range from 5 to 17 ppm with the corresponding oxygen fugacities ranging from $10^{-20.3}$ bar to $10^{-21.3}$ bar. Figure 10 shows that SO_2 remains the dominant sulfur gas under these conditions.

von Zahn et al. (1983) proposed that the SO_2/H_2S, the SO_2/S_2, and by implication the SO_2/OCS and SO_2/SO_3 ratios in the lower atmosphere of Venus are maintained at approximately the gas phase equilibrium values when reactions with the surface are negligible. Thus, we do not feel that the bulk of the sulfur originally in SO_2 is redistributed into reducing sulfur gases near the surface of Venus. The recent kinetic models of Krasnopolsky and Pollack (1994) predict an OCS abundance of 28 ppm near the surface of Venus. This is similar to results in Fig. 10 at plausible oxygen fugacities. More modeling of sulfur chemistry in the near-surface atmosphere of Venus along with further analysis of the Vega 1 and 2 data for OCS, H_2S, and elemental sulfur vapor is needed to understand the Vega results.

V. GEOCHEMISTRY OF VOLATILE ELEMENTS

Interest in the volatile element geochemistry of Venus dates back to Mueller (1964) who speculated about chemical mass transport of volatile phases on Venus. Lewis (1968) also considered this concept and proposed that volatiles may be transported toward the polar regions on Venus. However, we now know that the maximum temperature range with altitude is ~80 K, while the maximum temperature range with latitude is probably 10 to 20 K (Seiff 1983).

The concept of volatile transport on Venus was recently revived by Brackett et al. (1995) who showed how it could explain the low radar emissivity of the highlands of Venus, the low altitude hazes observed by two Pioneer Venus entry probes (Ragent and Blamont 1980), and possibly the Pioneer Venus entry probe anomalies at 12.5 km.

Briefly, their model, which is schematically illustrated in Fig. 11, is as follows. Many volatile metals (e.g., Cu, Zn, Sn, Pb, As, Sb, Bi) form halides and chalcogenides that have high vapor pressures. Compounds of these metals are typically found around terrestrial volcanic vents and fumaroles, are enriched in volcanic aerosols, and are observed in volcanic gases. For example, CuCl(g) has been observed in volcanic gases at Kilauea in Hawaii and at Nyiragongo in Zaire (Brackett et al. 1995). Thermochemical equilibrium calculations predict that many volatile metals are transported as halides and sulfides, even in steam-rich terrestrial volcanic gases (Symonds et al. 1987). Volatile metal compounds emitted from volcanoes on Venus may form aerosol hazes in the near-surface atmosphere of Venus, or they may immediately snow out and condense out over the entire surface of the planet, and/or if sufficiently

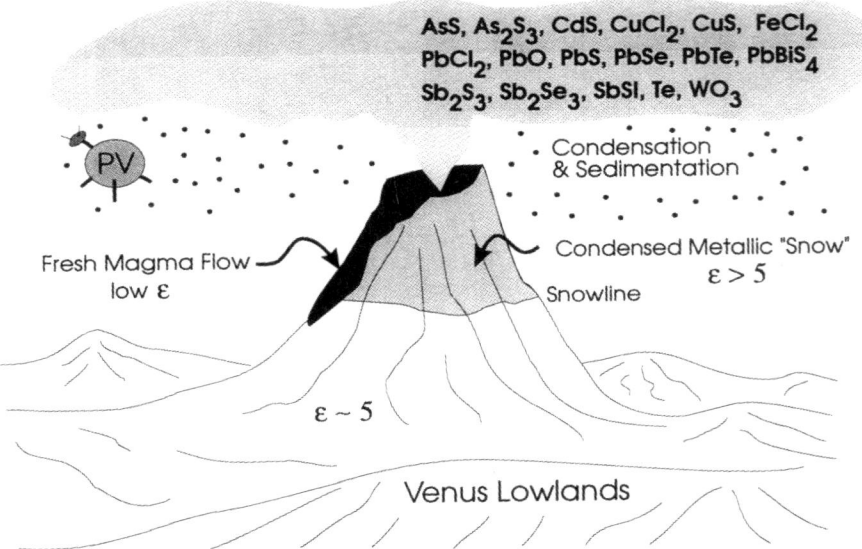

Figure 11. A cartoon schematically illustrating the metallic frost model of Brackett et al. (1995). Volatile metal halides and chalcogenides (i.e., sulfides, selenides, and tellurides), are erupted into the atmosphere of Venus. Some vapor may immediately form aerosol hazes in the near-surface atmosphere of Venus, which may be the cause of the low-altitude hazes observed by two Pioneer Venus entry probes (Ragent and Blamont 1980). Some vapor may immediately snow out and condense out over the entire surface of the planet, and other more volatile species may remain in the near-surface atmosphere. See text for details.

volatile, such as many Hg compounds (Lewis 1968,1969), they may remain in the near-surface atmosphere.

The higher vapor pressures of volatile metal condensates in the hotter plains of Venus, coupled with the altitude-dependent temperature gradient over the surface of Venus eventually leads to the vapor transport of the volatile metal compounds to the cooler highland regions of Venus. Brackett et al. (1995) modeled this transport process and concluded that, depending on the vapor pressure of the particular species, significant amounts of material could be transported in geologically short time scales. Many metal halides and chalcogenides have high dielectric constants (see, e.g., Young and Frederikse 1973) and many are ferroelectric (Lines and Glass 1977), leading to the deposition of high dielectric constant "frost" at high elevations.

Because the emission of volatile metal compounds is an ongoing process on the Earth, and because of the presence of S, Cl, and F on Venus, it is plausible that volcanic emissions on Venus contain volatile metal compounds. What is unknown, and what cannot be determined without remote sensing or *in-situ* analyses, are the exact metals and compounds present.

VI. SUMMARY

We conclude by identifying some of the key questions about the geochemistry and mineralogy of the surface of Venus and by suggesting observational studies, spacecraft experiments, laboratory experiments, and theoretical studies that can improve our knowledge of these important issues.

Perhaps the single most fundamental question facing us today is the mineralogy of the surface of Venus. von Zahn et al. (1983) emphasized this point by stating: "If we are ignorant of some of the details of atmospheric composition at the venusian surface, we are infinitely more ignorant of the nature of surface minerals." In discussing our lack of knowledge about the mineralogy of the surfaces of Venus and Mars, Fegley and Treiman (1992) emphasized that: "This information is crucial to understanding the chemical interactions between the atmospheres and surfaces of these two planets and needs to be determined by spacecraft missions. We recommend development of a X-ray diffraction experiment suitable for spacecraft operations in order to address this issue. In this regard, we note that Vaniman et al. (1991) have recently discussed development of a combined X-ray diffraction and X-ray fluorescence spacecraft instrument. Many of the questions currently facing us about gas-solid reactions on these two planets can be solved with a knowledge of the presence and abundance of the major rock-forming minerals." Here we can only second these recommendations and once again emphasize the overriding importance of developing spacecraft instruments, using methods such as X-ray diffraction (see, e.g., Blake et al. 1994), Mössbauer spectroscopy (see, e.g., Klingelhöfer et al. 1995), and nuclear magnetic resonance, to directly determine the mineralogy of the surface of Venus. It is important to determine whether or not the minerals which are apparently buffering CO_2, HCl, and HF are present. These techniques are also needed to determine the nature of the high dielectric phases in lower radar emissivity regions.

The second key question is the chemical composition of the lower 22 km ($\sim$80% by mass) of Venus' atmosphere. The remarkable advances during the past few years in Earth-based infrared spectroscopy (Bézard et al. 1990,1993; Pollack et al. 1993) have provided a wealth of new information about the composition of the sub-cloud atmosphere of Venus. Additional advances will undoubtedly occur. However, even this method has its limitations. For example, the positions of the near-infrared "windows" that allow observations of the near-surface atmosphere do not match the positions of absorption bands for all gases of interest. Thus, there is still a pressing need for a deep atmospheric spacecraft to measure the chemical composition of the lower 22 km of the atmosphere. The gases of most interest are H_2O, CO, SO_2, OCS, S_{1-8}, and H_2S and the method of choice for measuring them is probably infrared spectroscopy, although other techniques such as mass spectroscopy and gas chromatography should also be considered.

The third key issue that needs to be addressed is the oxidation state of the surface of Venus. The work of Pieters et al. (1986) suggesting the presence of hematite on the surface of Venus, the revised interpretation (Zolotov 1996; Fegley et al. 1996) of the CONTRAST experiment which now also suggests the presence of hematite, and the recent laboratory studies (Fegley et al. 1995a,b) showing that hematite is formed by basalt oxidation and magnetite oxidation in gases with CO/CO_2 ratios inside the magnetite stability field, suggests that the surface of Venus is more oxidized than previously believed by many workers. In order to determine the oxidation state of the surface and near-surface atmosphere it will be necessary to measure the CO abundance and oxygen fugacity below 12 km, to determine the mineralogy of iron-bearing minerals on the surface, for example by MB spectroscopy or visible/infrared reflection spectroscopy, and to determine the relative abundances of oxidized and reduced sulfur gases.

Acknowledgments. Work at Washington University was supported by a grant from NASA. The collaborative work between Washington University and the Technische Hochschule, Darmstadt was supported by a NATO Collaborative Research grant. We thank J. S. Kargel, J. S. Lewis, and M. Yu. Zolotov for their constructive reviews and comments. We also thank E. Kankeleit and P. Held for their support and assistance.

REFERENCES

Anders, E., and Grevesse, N. 1989. Abundances of the elements: Meteoritic and solar. *Geochim. Cosmochim. Acta* 53:197–214.

Arvidson, R. E., et al. 1994. Microwave signatures and surface properties of Ovda Regio and surroundings, Venus. *Icarus* 112:171–186.

Asaki, Z., Matsumoto, K., Tanabe, T., and Kondo, Y. 1983. Oxidation of dense iron sulfide. *Metall. Trans. B* 14:109–116.

Banin, A., Clark, B. C., and Wänke, H. 1992. Surface chemistry and mineralogy. In *Mars*, eds. H. H. Kieffer, B. M. Jakosky, C. W. Snyder and M. S. Matthews (Tucson: Univ. of Arizona Press), pp. 594–625.

Barath, T. T., et al. 1964. Mariner 2 radiometer experiment results. *Astron. J.* 69:49–58.

Barsukov, V. L., Volkov, V. P., and Khodakovsky, I. L. 1980. The mineral composition of Venus surface rocks: A preliminary prediction. *Proc. Lunar Planet. Sci. Conf.* 11:765–773.

Barsukov, V. L., Volkov, V. P., and Khodakovsky, I. L. 1982. The crust of Venus: Theoretical models of chemical and mineral composition. *Proc. Lunar Planet. Sci. Conf. 13, J. Geophys. Res. Suppl.* 87:3–9.

Barsukov, V. L., Surkov, Yu. A., Dmitriyev, L. V., and Khodakovsky, I. L. 1986. Geochemical studies on Venus with the landers from the Vega 1 and Vega 2 probes. *Geochem. Intl.* 23:53–65.

Basilevsky, A. T., Nikolaeva, O. V., and Weitz, C. M. 1992. Geology of the Venera 8 landing site region from Magellan data: Morphological and geochemical

considerations. *J. Geophys. Res.* 97:16315–16335.
Baulch, D. L., Drysdale, D. D., Duxbury, J., and Grant, S. J. 1976. Evaluated kinetic data for high temperature reactions. In *Homogeneous Gas Phase Reactions of the O_2-O_3 System, the CO_2-O_2-H_2 System and of Sulphur-Containing Species* (London: Butterworths).
Baulch, D. L., et al. 1995. Evaluated kinetic data for combustion modelling supplement I. *J. Phys. Chem. Ref. Data* 23:847–1033.
Bertaux, J. L., et al. 1996. Vega-1 and Vega-2 entry probes: An investigation of local UV absorption (220-400 nm) in the atmosphere of Venus (SO_2, aerosols, cloud structure). *J. Geophys. Res.* 101:12709–12745.
Bézard, B., de Bergh, C., Crisp, D., and Maillard, J. P. 1990. The deep atmosphere of Venus revealed by high-resolution nightside spectra. *Nature* 345:508–511.
Bézard, B., et al. 1993. The abundance of sulfur dioxide below the clouds of Venus. *Geophys. Res Lett.* 20:1587–1590.
Blake, D. F., Vaniman, D. T., and Bish, D. L. 1994. A mineralogical instrument for planetary applications. *Lunar Planet. Sci.* XXV:121–122 (abstract).
Brackett, R. A., Fegley, B., Jr., and Arvidson, R. E. 1995. Volatile transport on Venus and implications for surface geochemistry and geology. *J. Geophys. Res.* 100:1553–1563.
Burke, K., Fegley, B., Jr., and Sharpton, V. L. 1994. Are steep slopes on Venus preserved as a result of chemical cementation of pore-spaces in surface rocks? *Lunar Planet. Sci. Conf.* XXV:201–202 (abstract).
Chase, M. W., Jr., et al. 1985. *JANAF Thermochemical Tables* (New York: American Chemical Soc. and American Inst. of Physics).
Clark, B. C., et al. 1977. The Viking X ray fluorescence experiment: Analytical methods and early results. *J. Geophys. Res.* 82:4577–4594.
Clark, B. C., et al. 1982. Chemical composition of Martian fines. *J. Geophys. Res.* 87:10059–10067.
Connes, P., Connes, J., Benedict, W. S., and Kaplan, L. D. 1967. Traces of HCl and HF in the atmosphere of Venus. *Astrophys. J.* 147:1230–1237.
Connes, P., Connes, J., Kaplan, L. D., and Benedict, W. S. 1968. Carbon monoxide in the Venus atmosphere. *Astrophys. J.* 152:731–743.
de Bergh, C., et al. 1989. Ground-based high resolution spectroscopy of Venus near 3.6 microns. *Bull. Amer. Astron. Soc.* 21:926 (abstract).
de Bergh, C., et al. 1995. Water in the deep atmosphere of Venus from high-resolution spectra of the night side. *Adv. Space Res.* 15:79–88.
Deer, W. A., Howie, R. A., and Zussman, J. 1963. *Rock-Forming Minerals*, vols. 1–5 (London: Longmans Green).
Donahue, T. M., Hoffman, J. H., Hodges, R. R., Jr., and Watson, A. J. 1982. Venus was wet: A measurement of the ratio of deuterium to hydrogen. *Science* 216:630–633.
Ekonomov, A. P., Golovin, Yu. M., and Moshkin, B. E. 1980. Visible radiation observed near the surface of Venus: Results and their interpretation. *Icarus* 41:65–75.
Fegley, B., Jr. 1988. Thermochemical kinetics of SO_2 reactions with possible Venus crustal minerals: First data for calcite. *Lunar Planet. Sci. Conf.* XIX:315–316 (abstract).
Fegley, B., Jr. 1990. The applications of chemical thermodynamics and chemical kinetics to planetary atmospheres research. In *First International Conference on Laboratory Research for Planetary Atmospheres*, eds. K. Fox, J. E. Allen, Jr., L. J. Stief and D. Quillen, NASA CP-3077, pp. 267–302.
Fegley, B., Jr. 1995. Properties and composition of the terrestrial oceans and of the atmospheres of the Earth and other planets. In *Global Earth Physics A Handbook of Physical Constants*, ed. T. Ahrens (Washington, D. C.: American Geophysical

Union), pp. 320–345.

Fegley, B., Jr., and Lewis, J. S. 1980. Volatile element chemistry in the solar nebula: Na, K, F, Cl, Br, and P. *Icarus* 41:439–455.

Fegley, B., Jr., and Lodders, K. 1995. Chemical models of the lower atmosphere of Venus. *Bull. Amer. Astron. Soc.* 27:1072 (abstract).

Fegley, B., Jr., and Prinn, R. G. 1989. Estimation of the rate of volcanism on Venus from reaction rate measurements. *Nature* 337:55–58.

Fegley, B., Jr., and Treiman, A. H. 1992. Chemistry of atmosphere-surface interactions on Venus and Mars. In *Venus and Mars: Atmospheres, Ionospheres, and Solar Wind Interactions*, eds. J. G. Luhmann, M. Tatrallyay and R. O. Pepin (Washington, D. C.: American Geophysical Union), pp. 7–71.

Fegley, B., Jr., Treiman, A. H., and Sharpton, V. L. 1992. Venus surface mineralogy: Observational and theoretical constraints. *Proc. Lunar Planet. Sci. Conf.* 22:3–20.

Fegley, B., Jr., Lodders, K., and Klingelhöfer, G. 1993. Kinetics and mechanism of pyrite decomposition on the surface of Venus. *Bull. Amer. Astron. Soc.* 25:1094 (abstract).

Fegley, B., Jr., Lodders, K., Treiman, A. H., and Klingelhöfer, G. 1995a. The rate of pyrite decomposition on the surface of Venus. *Icarus* 115:159–180.

Fegley, B., Jr., Klingelhöfer, G., Brackett, R. A., Izenberg, N., Kremser, D. T., and Lodders, K. 1995b. Basalt oxidation and hematite formation on the surface of Venus. *Icarus* 118:373–383.

Fegley, B., Jr., Zolotov, M. Yu., and Lodders, K. 1996. The oxidation state of the lower atmosphere and surface of Venus. *Icarus*, in press.

Florensky, C. P., et al. 1977a. First panoramas of the venusian surface. *Proc. Lunar Sci. Conf.* 8:2655–2664.

Florensky, C. P., et al. 1977b. The surface of Venus as revealed by Soviet Venera 9 and 10. *Geol. Soc. Amer. Bull.* 88:1537–1545.

Florensky, C. P., et al. 1983a. Redox indicator "CONTRAST" on the surface of Venus. *Lunar Planet. Sci. Conf.* XIV:203–204 (abstract).

Florensky, C. P., et al. 1983b. Venera 13 and Venera 14: Sedimentary rocks on Venus? *Science* 221:57–59.

Florensky, C. P., et al. 1983c. The oxidizing-reducing conditions on the surface of Venus according to the data of the "KONTRAST" geochemical indicator on the Venera 13 and Venera 14 spacecraft. *Cosmic Res.* 21:278–281.

Ford, P. G., and Pettengill, G. H. 1992. Venus topography and kilometer-scale slopes. *J. Geophys. Res.* 97:13103–13114.

Garvin, J. B., and Head, J. W. 1985. High dielectric surfaces on the terrestrial planets. *Lunar Planet. Sci. Conf.* XVI:264–265 (abstract).

Gektin, Yu. M. 1996. Personal communications to M. Yu. Zolotov.

Gel'man, B. G., et al. 1979. Analysis of chemical composition of Venus atmosphere by gas chromatography on Venera 12. *Cosmic Res.* 17:585–589.

Golovin, Yu. M., Moshkin, B. Ye., and Ekonomov, A. P. 1983. Some optical properties of the Venus surface. In *Venus*, eds. D. M. Hunten, L. Colin, T. M. Donahue and V. I. Moroz (Tucson: Univ. of Arizona Press), pp. 131–136.

Grimm, R. E., and Solomon, S. C. 1987. Limits on modes of lithospheric heat transport on Venus from impact crater density. *Geophys. Res. Lett.* 14:538–541.

Hemingway, B. S. 1990. Thermodynamic properties for bunsenite, NiO, magnetite, Fe_3O_4, and hematite, Fe_2O_3, with comments on selected oxygen buffer reactions. *Amer. Mineral.* 75:781–790.

Hoffman, J. H., Hodges, R. R., Donahue, T. M., and McElroy, M. B. 1980a. Composition of the Venus lower atmosphere from the Pioneer Venus mass spectrometer. *J. Geophys. Res.* 85:7882–7890.

Hoffman, J. H., Oyama, V. I., and von Zahn, U. 1980b. Measurements of the Venus lower atmosphere composition: A comparison of results. *J. Geophys. Res.* 85:7871–7881.

Hogarth, D. D. 1989. Pyrochlore, apatite, and amphibole: Distinctive minerals in carbonatite. In *Carbonatites: Genesis and Evolution*, ed. K. Bell (London: Unwin Hyman), pp. 105–148.

Huebner, J. S. 1975. Oxygen fugacity values of furnace gas mixtures. *Amer. Mineral.* 60:815–823.

Ivanov, M. V., and Freney, J. R., eds. 1983. *The Global Biogeochemical Sulphur Cycle* (New York: J. Wiley and Sons), pp. 25–127.

Johannsen, A. 1937. *A Descriptive Petrography of the Igneous Rocks*, vol. IV, (Chicago: Univ. of Chicago Press).

Kargel, J. S., et al. 1991. Compositional constraints on outflow channel-forming lavas on Venus. *Lunar Planet. Sci. Conf.* XXII:685–686 (abstract).

Kargel, J. S., Komatsu, G., Baker, V. R., and Strom, R. G. 1993. The volcanology of Venera and VEGA landing sites and the geochemistry of Venus. *Icarus* 103:253–275.

Kargel, J. S., Kirk, R. L., Fegley, B., Jr., and Treiman, A. H. 1994. Carbonate-sulfate volcanism on Venus? *Icarus* 112:219–252.

Klingelhöfer, G., et al. 1995. Mössbauer spectroscopy in space. *Hyp. Int.* 95:305–339.

Klose, K. B., Wood, J. A., and Hashimoto, A. 1992. Mineral equilibria and the high radar reflectivity of Venus mountaintops. *J. Geophys. Res.* 97:16353–16369.

Krasnopolsky, V. A., and Parshev, V. A. 1979. Chemical composition of Venus' troposphere and cloud layer based on Venera 11 and 12 and Pioneer Venus measurements. *Cosmic. Res.* 17:630–637.

Krasnopolsky, V. A., and Pollack, J. B. 1994. H_2O-H_2SO_4 system in Venus' clouds and OCS, CO, and H_2SO_4 profiles in Venus' troposphere. *Icarus* 109:58–78.

Kridelbaugh, S. J. 1973. The kinetics of the reaction calcite + quartz = wollastonite + carbon dioxide at elevated temperatures and pressures. *Amer. J. Sci.* 273:757–777.

Lewis, J. S. 1968. An estimate of the surface conditions of Venus. *Icarus* 8 434–456.

Lewis, J. S. 1969. Geochemistry of the volatile elements on Venus. *Icarus* 11:367–385.

Lewis, J. S. 1970. Venus: Atmospheric and lithospheric composition. *Earth Planet. Sci. Lett.* 10:73–80.

Lewis, J. S. 1971. Venus: Surface temperature variations. *J. Atmos. Sci.* 28:1084–1086.

Lewis, J. S., and Kreimendahl, F. A. 1980. Oxidation state of the atmosphere and crust of Venus from Pioneer Venus results. *Icarus* 42:330–337.

Lines, M. E., and Glass, A. M. 1977. *Principles and Applications of Ferroelectrics and Related Materials* (Oxford: Clarendon Press).

Marov, M. Ya., et al. 1989. Lower atmosphere. In *The Planet Venus: Atmosphere, Surface, Interior Structure*, eds. V. L. Barsukov and V. P. Volkov (Moscow: Nauka Press), pp. 25–67.

Marshall, J. R., Greeley, R., and Tucker, D. W. 1988. Aeolian weathering of Venus surface materials: Preliminary results from laboratory simulations. *Icarus* 74:495–515.

Mayer, C. H., McCullough, T. P., and Sloanaker, R. M. 1958. Observations of Venus at 3.15 cm wavelength. *Astrophys. J.* 127:1–10.

McElroy, M. B., and Donahue, T. M. 1972. Stability of the martian atmosphere. *Science* 177:986–988.

Mitchell, R. H. 1986. *Kimberlites: Mineralogy, Geochemistry, and Petrology* (New York: Plenum Press).

Moroz, V. I., Parfent'ev, N. A., and San'ko, N. F. 1979. Spectrophotometric experiment on the Venera 11 and 12 descent modules. 2. Analysis of Venera 11 spectra by layer-addition method. *Cosmic. Res.* 17:601–614.

Mueller, R. F. 1963. Chemistry and petrology of Venus: Preliminary deductions. *Science* 141:1046–1047.

Mueller, R. F. 1964. A chemical model for the lower atmosphere of Venus. *Icarus* 3:285–298.

Mueller, R. F. 1968. Sources of HCl and HF in the atmosphere of Venus. *Nature* 220:55–57.

Mueller, R. F. 1969. Planetary probe: Origin of atmosphere of Venus. *Science* 163:1322–1324.

Mukhin, L. M., et al. 1983. Gas chromatograph analysis of the chemical composition of the atmosphere of Venus by the landers of the Venera 13 and Venera 14 spacecraft. *Cosmic. Res.* 21:168–172.

Na, C. Y., Esposito, L. W., and Skinner, T. E. 1990. International ultraviolet explorer observations of Venus SO_2 and SO. *J. Geophys. Res.* 95:7485–7491.

Nair, H., Allen, M., Anbar, A. D., and Yung, Y. L. 1994. A photochemical model of the Martian atmosphere. *Icarus* 111:124–150.

Naumov, G. B., Ryzhenko, B. N., and Khodakovsky, I. L. 1971. *Handbook of Thermodynamic Data* (Moscow: Atomizdat).

Nozette, S., and Lewis, J. S. 1982. Venus: Chemical weathering of igneous rocks and buffering of atmospheric composition. *Science* 216:181–183.

O'Neill, H. St. C. 1988. Systems Fe-O and Cu-O: Thermodynamic data for the equilibria Fe-"FeO", Fe-Fe_3O_4, "FeO"-Fe_3O_4, Fe_3O_4-Fe_2O_3, Cu-Cu_2O and Cu_2O-CuO from emf measurements. *Amer. Mineral.* 73:470–486.

Orville, P. 1974. Crust-atmosphere interactions. In *The Atmosphere of Venus*, ed. J. E. Hansen (New York: Goddard Institute for Space Studies), pp. 190–195.

Oyama, V. I., et al. 1980. Pioneer Venus gas chromatograph of the lower atmosphere of Venus. *J. Geophys. Res.* 85:7891–7902.

Parsons, B. 1981. The rate of plate consumption and creation. *Geophys. J. Roy. Astron. Soc.* 67:437–448.

Pettengill, G. H., Ford, P. G., and Nozette, S. 1982. Venus: Global surface radar reflectivity. *Science* 217:640–642.

Pettengill, G. H., Ford, P. G., and Chapman, B. D. 1988. Venus: Surface electromagnetic properties. *J. Geophys. Res.* 93:14881–14892.

Philpotts, A. R. 1990. *Principles of Igneous and Metamorphic Petrology* (Englewood Cliffs, N. J.: Prentice-Hall).

Pieters, C. M., et al. 1986. The color of the surface of Venus. *Science* 234:1379–1383.

Pollack, J. B., et al. 1993. Near infrared light from Venus' nightside: A spectroscopic analysis. *Icarus* 103:1–42.

Prinn, R. G. 1985. The photochemistry of the atmosphere of Venus. In *The Photochemistry of Atmospheres*, ed. J. S. Levine (New York: Academic Press), pp. 281–336.

Ragent, B., and Blamont, J. 1980. The structure of the clouds of Venus: Results of the Pioneer Venus Nephelometer Experiment. *J. Geophys. Res.* 85:8089–8105.

Robie, R. A., and Hemingway, B. S. 1995. *Thermodynamic Properties of Minerals and Related Substances at 298.15 K and 1 Bar (10^5 Pascals) Pressure and at Higher Temperatures*, U.S.G.S Bull. No. 2131.

Ronov, A. B., and Yaroshevsky, A. A. 1976. A new model for the chemical structure of the Earth's crust. *Geochem. Intl.* 13:89–121.

Seiff, A. 1983. Thermal structure of the atmosphere of Venus. In *Venus*, eds. D. M. Hunten, L. Colin, T. M. Donahue and V. I. Moroz (Tucson: Univ. Arizona Press), pp. 215—279.

Selivanov, A. S., et al. 1983. The first color panoramas of the surface of Venus, sent by Venera 13 and 14. *Cosmic Res.* 21:129–136.
Semenov, E. I. 1974. Economic mineralogy of alkaline rocks. In *The Alkaline Rocks*, ed. H. Sorensen (London: Wiley), pp. 543–552.
Shepard, M. K., Arvidson, R. E., Brackett, R. A., and Fegley, B., Jr. 1994. A ferroelectric model for the low emissivity highlands on Venus. *Geophys. Res. Lett.* 21:469–472.
Smith, J. V. 1981. Halogen and phosphorus storage in the Earth. *Nature* 289:762–765.
Sorensen, H. 1974. Alkali syenites, feldspathoidal syenites and related lavas. In *The Alkaline Rocks*, ed. H. Sorensen (London: Wiley), pp. 22–52.
Spohn T. 1991. Mantle differentiation and thermal evolution of Mars, Mercury, and Venus. *Icarus* 90:222–236.
Surkov, Yu. A. 1977. Geochemical studies of Venus by Venera 9 and 10 automatic interplanetary stations. *Proc. Lunar Sci. Conf.* 8:2665–2689.
Surkov, Yu. A., et al. 1983. Determination of the elemental composition of rocks on Venus by Venera 13 and Venera 14 (preliminary results). *Proc. Lunar Planet. Sci. Conf.* 13, *J. Geophys. Res. Suppl.* 88:481–493.
Surkov, Yu. A., et al. 1984. New data on the composition, structure, and properties of Venus rock obtained by Venera 13 and Venera 14. *Proc. Lunar Planet. Sci. Conf.* 14, *J. Geophys. Res. Suppl.* 89:393–402.
Surkov, Yu. A., et al. 1985. Method, instruments, and results of the determination of elements contained in Venusian rock by the Vega-2 interplanetary probe. *Astron. Vestn.* 19:177–186.
Surkov, Yu. A., et al. 1986. Venus rock composition at the Vega 2 landing site. *Proc. Lunar Planet. Sci. Conf.* 17, *J. Geophys. Res. Suppl.* 91:215–218.
Surkov, Yu. A., et al. 1987a. Uranium, thorium, and potassium in the venusian rocks at the landing sites of Vega 1 and 2. *Proc. Lunar Planet. Sci. Conf.* 17, (*J. Geophys. Res. Suppl.* 92:537–540.
Surkov, Yu. A., et al. 1987b. Abundances of natural radioactive elements in the rocks of Venus from data from the Vega-1 and Vega-2 stations. *Cosmic Res.* 25:590–594.
Surkov, Yu. A., et al. 1987c. Element makeup of the rocks of Venus in the northeastern part of Aphrodite Terra (from Vega-2 lander data). *Cosmic Res.* 25:751–761.
Symonds, R. B., et al. 1987. Volatilization, transport and sublimation of metallic and non-metallic elements in high temperature gases at Merapi Volcano, Indonesia. *Geochim. Cosmochim. Acta* 51:2083–2101.
Timco, G. W. 1977. High-Pressure Dielectric Properties of Perovskite Ferroelectrics. Ph.D. Thesis, Univ. of Western Ontario.
Toulmin, P., III, et al. 1977. Geochemical and mineralogical identification of the Viking inorganic chemical results. *J. Geophys. Res.* 82:4625–4634.
Treiman, A. H., and Fegley, B., Jr. 1991. Venus: The chemical weathering of pyrrhotite, $Fe_{1-x}S$. *Lunar Planet. Sci. Conf.* XXII:1409–1410 (abstract).
Urey, H. C. 1952. *The Planets* (New Haven: Yale Univ. Press).
Vaniman, D. T., Bish, D. L., and Chipera, S. J. 1991. In-situ planetary surface analyses: The potential of X-ray diffraction with simultaneous X-ray fluorescence. *Lunar Planet. Sci.* XXII:1429–1430 (abstract).
Vinogradov, A. P., and Volkov, V. P. 1971. On the wollastonite equilibrium as a mechanism determining Venus' atmospheric composition. *Geochem. Int'l.* 8:463–467.
Vinogradov, A. P., Surkov, Yu. A., and Kirnozov, F. F. 1973. The contents of uranium, thorium, and potassium in the rocks of Venus as measured by Venera 8. *Icarus* 20:253–259.
Volkov, V. P., Zolotov, M. Yu., and Khodakovsky, I. L. 1986. Lithospheric-atmospheric

interactions on Venus. In *Chemistry and Physics of Terrestrial Planets*, ed. S. K. Saxena (New York: Springer-Verlag), pp. 136–187.

Von Hippel, A. 1954. *Dielectric Materials and Applications* (Cambridge: MIT Press).

von Zahn, U., Kumar, S., Niemann, H., and Prinn, R. 1983. Composition of the Venus atmosphere. In *Venus*, eds. D. M. Hunten, L. Colin, T. M. Donahue and V. I. Moroz (Tucson: Univ. Arizona Press), pp. 299–430.

Warnatz, J. 1984. Rate coefficients in the C/H/O system. In *Combustion Chemistry*, ed. W. C. Gardiner, Jr. (New York: Springer-Verlag), pp. 197-360.

Wilson, M. 1989. *Igneous Petrogenesis* (Boston: Unwin Hyman).

Woolley, A. R., and Kempe, D. R. C. 1989. Carbonatites: Nomenclature, average chemical compositions, and element distribution. In *Carbonatites Genesis and Evolution*, ed. K. Bell (London: Unwin Hyman), pp. 1–14.

Young, K. F., and Frederikse, H. P. R. 1973. Compilation of the static dielectric constants of inorganic solids. *J. Phys. Chem. Ref. Data* 2:313–409.

Young, L. D. G. 1972. High resolution spectra of Venus–A review. *Icarus* 17:632–658.

Young, L. D. G., Young, A. T., and Zasova, L. V. 1984. A new interpretation of the Venera 11 spectra of Venus. *Icarus* 60:138–151.

Yung, Y. L., and DeMore, W. B. 1982. Photochemistry of the stratosphere of Venus: Implications for atmospheric evolution. *Icarus* 51:199–247.

Zolotov, M. Yu. 1987. Redox conditions on Venus surface. *Lunar Planet. Sci.* XVIII:1134–1135 (abstract).

Zolotov, M. Yu. 1994. Near-surface atmosphere of Venus: New estimations of redox conditions based on new data. *Lunar Planet. Sci. Conf.* XXV:1569–1570 (abstract).

Zolotov, M. Yu. 1996. A model of thermochemical equilibrium in the near-surface atmosphere of Venus. *Geochem. Intl.* 33:80–100.

Zolotov, M. Yu., and Khodakovsky, I. L. 1989. Exogenic processes. In *The Planet Venus: Atmosphere, Surface, Interior Structure*, eds. V. L. Barsukov and V. P. Volkov (Moscow: Nauka Press), pp. 262—290.

ROCK WEATHERING ON THE SURFACE OF VENUS

JOHN A. WOOD
Harvard-Smithsonian Center for Astrophysics

The basaltic surface rock of Venus is to some degree weathered to a secondary mineral assemblage, probably consisting chiefly of Mg pyroxene, Na plagioclase, anhydrite, andalusite, K feldspar, quartz, an Fe mineral (magnetite, pyrite, or hematite), and rutile. The identity of the Fe mineral depends upon the altitude and the redox state of the atmosphere; the latter has not been fixed. The rate of weathering is unknown. The S content of surface materials, surface morphologies, and physical properties of surface materials at the Soviet lander sites point to substantial weathering. The thickness of surface material the atmosphere is capable of weathering is small, $\lesssim 1$ m. The degree to which reaction with surface minerals buffers the atmospheric composition is unknown; the evidence is that the CO_2 pressure of the atmosphere is not buffered in this way. An extreme manifestation of weathering, or something akin to it, appears to have endowed the Venus mountaintops with a continuous coating of a semi-conducting mineral several mm thick. Such a layer could be created by the reaction of Fe minerals at plains altitudes with HCl in the atmosphere to form $FeCl_2$ vapor, which convects upward and plates out as pyrite or magnetite on the lower-temperature mountaintops. Anomalous material having modestly elevated dielectric constants also occurs at plains altitudes, especially as crater ejecta. The nature of this material and its origin are probably quite different from that on mountaintops.

I. INTRODUCTION

Basalt is the most abundant type of surface rock in the solar system. It covers most of the Earth's surface (i.e., the ocean floors beneath a thin layer of sediment). From the dominance of lava flood and shield volcano landforms on Venus, and the Venera 13 and 14 and Vega 2 chemical analyses of Venus surface materials, it is clear that basalt is abundant on that planet too. The processes that have created (or exposed) other rock types on the Earth and Moon are not applicable to Venus; these are stream transport and marine deposition of weathered material, which create chemically diverse sedimentary rock types on our planet; deep erosion of uplifted mountain masses, which exposes a range of igneous rock types created deep in the Earth; and preservation of primordial crustal material, which in the case of the Moon is chemically distinctive anorthosite formed by crystal fractionation in an initial magma ocean. In the absence of these sources of alternative rock types on Venus, it is likely that the primary surface rock type almost everywhere is basalt.

Basalt on Earth consists of the collection of minerals that was thermodynamically stable in the temperature range of crystallization (~1350–1100°C) and in the chemical environment that the hot lava carried with it, which was not much different from that deep inside the Earth. Once cooled, of course, the rock is exposed to quite a different environment, one determined by the temperature and chemistry of Earth's atmosphere or hydrosphere. Another set of minerals is thermodynamically stable under these new conditions, and chemical reactions between the rock and its new environment set about *weathering* the primary minerals (pyroxenes, feldspar, oxides of Fe, Ti, Cr) into these secondary minerals (typically clay minerals and oxides—the inorganic components of terrestrial soil). The weathering reactions proceed slowly, requiring millions of years to turn rock into soil. Where a surface is not subject to mechanical erosion, a layer of residual soil many meters deep can accumulate above unweathered rock.

It is important to draw a distinction between *chemical weathering*, discussed above, and *mechanical erosion and transport*, as by running water, wind, or glacial ice. Chemical weathering can occur without erosion, as in the case of the residual soil layers cited. Erosion and transport also can occur without chemical weathering, as by the plucking and scouring of fresh bedrock by glacial ice. More often, however, both processes work in concert on Earth.

There is no reason to think the set of primary basaltic minerals that crystallizes on Venus is significantly different from that which forms on Earth. And though the surface environment on Venus is very dissimilar to Earth's, it is again the case that the primary basaltic minerals would not be thermodynamically stable once the lava cooled and the Venus atmosphere permeated it. It is important to ask what collection of minerals basalt would weather to under these alternative conditions, how fast it would happen, and how profoundly altered various elements of the Venus surface are. This information bears importantly on questions of the electrical properties and radar response of the surface material, the rate of degradation of visible landforms, and the supply of particulate matter for aeolian transport.

II. WEATHERING ON VENUS

A. The Equilibrium Minerals produced by Weathering

Thermodynamic principles can be used to deduce the mineral assemblage at equilibrium on the Venus surface. One of the reactions that might be important in weathering is

$$CaMg(SiO_3)_2 + SO_2(g) + CO_2(g) \rightleftharpoons$$
$$CaSO_4 + MgSiO_3 + SiO_2 + CO(g) \qquad (1)$$

wherein SO_2 in the atmosphere reacts with diopside, $CaMg(SiO_3)_2$, an abundant primary mineral component in basalt, to form the secondary mineral anhydrite (CaSO$_4$) and other products. The Gibbs free energies of formation of these and other compounds, as a function of temperature, can be found in the chemical literature. If the free energies of the reaction products, at the temperature of the Venus surface, are summed and those of the reactants subtracted, the difference (ΔG^0) can be related to the gas pressures required for diopside and anhydrite to coexist in equilibrium (see, e.g., Wood and Fraser 1976):

$$\Delta G^0 = -RT \ln \left[\frac{p(CO)}{p(SO_2) \times p(CO_2)} \right] \quad (2)$$

(this assumes ideal behavior of gases, a good approximation under conditions on Venus). Here R is the universal gas constant, T is the temperature (K), and $p(x)$ is the partial pressure of gas species x in the system, normalized to whatever pressure was used when the values of Gibbs free energy used were measured (generally 1 bar). (Actually, each pressure should be shown raised to a power equal to the number of moles of the corresponding gas in reaction (1). Because only one mole of each gas appears in this particular reaction, they were omitted. For other reactions, however, this may not be the case.) Then, after transposing and exponentiating, if the ratio

$$\frac{p(CO)}{p(SO_2) \times p(CO_2)} > e^{\frac{\Delta G^0}{RT}} \quad (3)$$

it means the gas is not sulfurous enough to push reaction (1) to the right, and diopside is the stable Ca mineral. If the left side of Eq. (3) is smaller than the right, then the gas *is* sulfurous enough and diopside should weather to anhydrite plus enstatite (MgSiO$_3$). Because the concentrations and hence partial pressures of CO, SO$_2$, and CO$_2$ in the Venus atmosphere have been measured, and the surface temperature and ΔG^0 are known, it would seem straightforward to calculate whether basaltic diopside tends to weather to anhydrite or not on Venus, but this is not the case. One problem is that other competing reactions also can be imagined, such as

$$CaMg(SiO_3)_2 + CO_2(g) \rightleftharpoons CaCO_3 + MgSiO_3 + SiO_2 \quad (4)$$

which would tend to convert diopside to calcite (CaCO$_3$) rather than anhydrite. Which reaction has first call on the Ca? In fact, there is a very large number of possible mineral products of weathering on Venus, and an even larger number of interlocking reactions that can be written to express their hypothetical formation from various combinations of primary basaltic minerals and atmospheric gases. To recognize the few dominant reactions and use them to make broad statements about the equilibrium mineral assemblage created by weathering on Venus is a difficult task, requiring the exercise of expert chemical intuition and many trial calculations.

Klose et al. (1992) review the history of thermodynamic analyses of this sort, which have proceeded concurrently with improvements in our knowledge of the surface temperature and atmospheric chemistry of Venus. Most recently some workers (Khodakovsky et al. 1979; Barsukov et al. 1982a,1987; Zolotov 1991a; Klose et al. 1992) have adopted an approach that eliminates the need to consider individual weathering reactions. Tables of thermodynamic data for a large number of minerals and gas species are assembled and incorporated in a computer program; the success of the method depends upon including all minerals and gases that might be important in the Venus system. A temperature and total gas pressure are entered as input, as well as abundances of elements in the Venus atmosphere and abundances of cationic elements (Si, Mg, Ca, etc.) in the Venus surface material. Following an iterative procedure, the program searches for that combination of surface minerals which (a) adds up to the same relative abundances of cationic elements as were assumed, and (b) contains the lowest possible total amount of Gibbs free energy. These are the thermodynamically stable minerals under the conditions assumed. This method effectively considers all possible reactions and the ways they would interact with one another.

Using the energy-minimization approach and assuming a broad range of plausible atmospheric compositions (discussed below), Barsukov et al. (1982a,1987) and Klose et al. (1992) found similar assemblages of secondary minerals that should be stable at plains altitude on Venus. Barsukov et al. found (in order of decreasing abundance) plagioclase ($CaAlSi_2O_8$-$NaAlSi_3O_8$), clinoenstatite ($MgSiO_3$), pyrite (FeS_2) or magnetite (Fe_3O_4), anhydrite ($CaSO_4$) or diopside ($CaMg[SiO_3]_2$), microcline ($KAlSi_3O_8$), quartz (SiO_2), fluorapatite ($Ca_5[PO_4]_3F$), and in some cases sphene ($CaTiO_3$) and/or hercynite ($FeAl_2O_4$). Klose et al. found the same mineralogy with the addition of andalusite (Al_2SiO_5), the substitution of rutile (TiO_2) for sphene and sanidine ($KAlSi_3O_8$) for microcline, and the omission of fluorapatite and hercynite. The differences are largely attributable to somewhat different tables of thermodynamic data used by the minimization programs.

Klose et al. determined the equilibrium mineralogy over a range of conditions. They studied the effect of altitude on the weathering product, because the surface temperature and gas pressure vary with this parameter; and also the effect of the redox state of the atmospheric gas, because there is no consensus on the correct value of this important parameter (Sec. II.B). Results were presented in phase diagrams containing fields corresponding to various stable minerals or mineral groups, as a function of altitude and a parameter δ that describes the redox state of the atmosphere. (The Venus atmosphere contains very nearly enough O to oxidize all its C to CO_2, all its S to SO_2, and all its H to H_2O; it falls short of this by a tiny amount. δ expresses this *oxygen deficiency*; for example, $\delta = 10^{-5}$ means $1 - 10^{-5} = 0.99999$ of the O needed to completely oxidize C, S, and H is present in the system. δ is a straightforward expression of the amount of the element O in the gas phase. Other parameters that quantify the redox state (e.g., fO_2 or CO/CO_2) are derivative, and depend upon temperature, pressure, and degree of equilibration as well as the amount of O in the gas). Only Klose et al. (1992) have

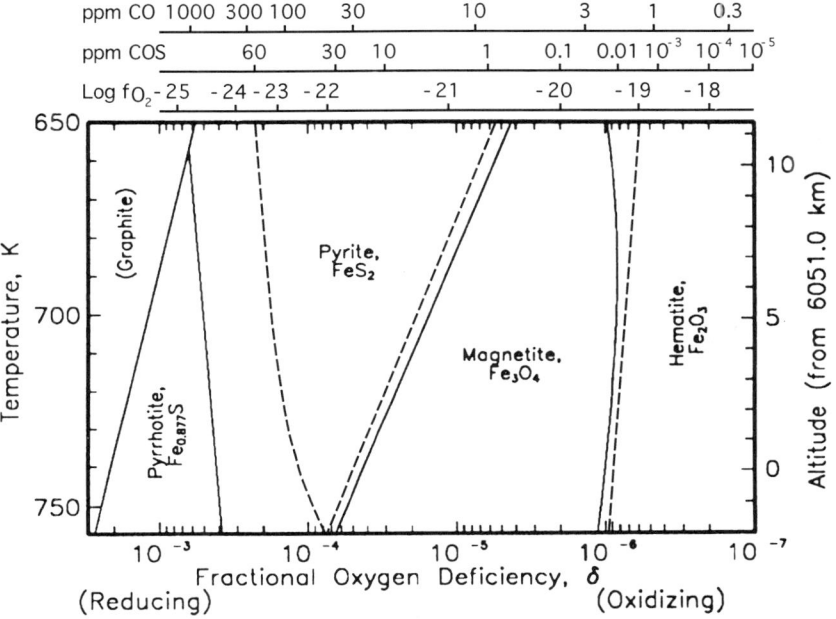

Figure 1. Fields showing which Fe phase is stable on the surface of Venus, as a function of altitude and δ. Solid lines are for 100 ppm H_2O and 185 ppm SO_2 in the atmosphere, dashed lines for 40 ppm H_2O and 60 ppm SO_2. The field labeled "Graphite" is disallowed; there conditions are so reducing that atmospheric CO_2 converts to graphite. Equilibrium atmospheric abundances of CO and COS, and fO_2, for the 100 ppm H_2O to 185 ppm SO_2 case at 0 altitude (6051.0 km planetary radius), are shown across the top. These parameters vary with altitude; δ is invariant (figure adapted from Klose et al. 1992).

published diagrams of this type, which are reproduced in Figs. 1 and 2.

To make the energy-minimization calculations that underlie these figures, values had to supplied for the H and S contents of the atmosphere. Because these elements appear mostly as H_2O and SO_2 in the Venus atmosphere, Klose et al. used best estimates for these gas abundances based on measurements made by probes that entered the Venus atmosphere: 100 ppm of H_2O, from von Zahn et al.'s (1983) assessment of the entry probe data, and 185 ppm of SO_2 from the Pioneer Venus gas chromatograph results of Oyama et al. (1980). However, some studies of the infrared and microwave absorption spectra of the Venus atmosphere measured at groundbased observatories have suggested lower concentrations of these atmospheric constituents, e.g., ~40 ppm of H_2O (Crisp et al. 1991) or less (Pollack et al. 1993), and ~60 ppm of SO_2 (Fahd and Steffes 1992). I have repeated the calculations of Klose et al. for 40 ppm H_2O and 60 ppm SO_2 and superimposed the alternative field boundaries on Figs. 1 and 2. Predictably, the lowered concentration of SO_2 shrinks the fields of stability of S-bearing minerals: that of the S-rich iron

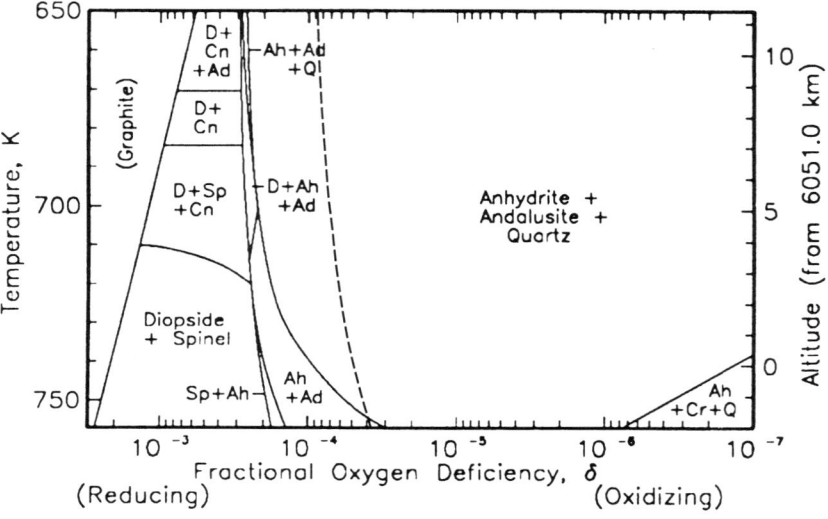

Figure 2. Fields of stability for other minerals on the surface of Venus, for 100 ppm H_2O and 185 ppm SO_2 (Klose et al. 1992). Ad: andalusite, Al_2SiO_5; Ah: anhydrite, $CaSO_4$; Cn: corundum, Al_2O_3; Cr: cordierite, $Mg_2Al_3(AlSi_5O_{18})$; D: diopside, $CaMg(SiO_3)_2$; Q: quartz, SiO_2; Sp: spinel, $MgAl_2O_4$. Formal results of the energy-minimization calculation are shown, but details of the complex region to the left cannot really be resolved because of uncertainties in the thermodynamic data. Only one phase boundary, that for the crucial diopside-anhydrite reaction (reaction 1), is shown as a dashed curve for the case of an atmosphere containing 40 ppm H_2O and 60 ppm SO_2.

mineral pyrite (FeS_2) in Fig. 1, and the field for anhydrite (formed by, e.g., reaction 1) in Fig. 2.

These calculations do not allow for the formation of oxide solid solutions. They could be made more realistic by including activity data for the hematite (Fe_2O_3)-ilmenite ($FeTiO_3$) and magnetite (Fe_3O_4)-ulvöspinel (Fe_2TiO_4) solid solution series, as was done by Zolotov (1994a).

Fegley et al. (1995) describe a series of kinetic experiments which has been widely represented as proving that the mineral pyrite is not, in fact, stable on the surface of Venus. The experiments were based on the premise that "pyrite is thermodynamically unstable on the surface of Venus," a conclusion reached in an earlier paper by Fegley and Treiman (1992). Their paper calculated that pyrite stability at, say, 700 K (representative of highland areas) requires an atmospheric concentration of COS $\gtrsim 12$ ppm and an H_2S concentration $\lesssim 0.07$ ppm. The authors then cited Bézard et al.'s (1990) estimate, from groundbased infrared spectral studies, of a COS concentration of ~ 0.25 ppm at altitudes <50 km, and Hoffman et al.'s (1980) Pioneer Venus neutral mass spectrometer detection of ~ 3 ppm H_2S (no altitude dependence stated) as violations of these constraints, and evidence that pyrite cannot be stable on the Venus surface.

However, the COS content of the Venus atmosphere has been revised upward by Pollack et al. (1993). These authors derive, from their modeling of infrared spectra, a concentration of COS of 4.4±1.0 ppm at 33 km altitude and a concentration gradient of 1.58±0.30 ppm km^{-1} increasing downward. They (speculatively) extrapolate the COS profile to a concentration of ~120 ppm at ground level. Krasnopolsky and Pollack (1994) refine this value to 28±1 ppm COS, which falls within the range calculated by Fegley and Treiman (1992) for pyrite *stability*.

The H_2S number of Hoffman et al. (1980) is not solidly founded; these authors note that the 34 amu spectral peak attributed to H_2S might be H_2O_2 instead, but "for now" they assume it is H_2S. Von Zahn et al. (1983) state that the H_2S detection "is not yet regarded as definite." Moreover, an atmospheric H_2S concentration *greater* than the range calculated by Fegley and Treiman for equilibrium with pyrite and magnetite is not inconsistent with pyrite stability.

Thus the conclusion reached by Fegley and Treiman (1992) is unfounded. Pyrite *may be* unstable on the Venus surface; Fig. 1 shows a range of δ where this would be the case. However, the calculations of Fegley and Treiman (1992) do not establish that it is unstable. The kinetic experiments of Fegley et al. (1995) also do nothing to establish the instability of pyrite, because the experiments did not adequately reproduce conditions on Venus (Wood and Brett 1997). In particular, the gas mixtures Fegley and Treiman flowed over their charges contained no S_2. Pyrite evolves S_2 gas when it is heated, but if the S_2 partial pressure in the coexisting gas phase is high enough pyrite continues to exist because it regains as much S_2 from the gas as it loses to it. The experiments of Fegley and Treiman blew S_2 away from their pyrite charges as fast as it was evolved, and gave the pyrite no chance to replace its losses. Under these circumstances destruction of their pyrite charges was inevitable.

B. Redox State at the Surface of Venus

Only one value of δ actually applies to the surface of Venus, and opinion varies as to what it is. Three items of evidence must be considered: measurements of the composition of the Venus atmosphere; spectral evidence for the presence of hematite on the Venus surface; and the *Kontrast* experiment carried by Soviet landers.

1. Atmospheric Composition. Concentrations of the most abundant gas species in the Venus atmosphere, which define its redox state, have been measured by entry probes and by groundbased studies of infrared and microwave absorption spectra. The values reported are for altitudes ≥ 20 km, not ground level. Best estimates of the species concentrations a few years ago indicated that the atmosphere was out of equilibrium (Zolotov 1991*b*; Klose et al. 1992); it contained too little COS to be in equilibrium with its CO/CO_2 ratio at any altitude. The disequilibrium mixture was thought to have been circulated downward from high altitudes, where it was produced by photochemical reactions. However, more recent work (Pollack et al. 1993; Krasnopolsky and

Pollack 1994) has increased the estimate of COS concentration in the atmosphere to a level consistent (to within uncertainties) with equilibrium, which would make the concentrations of CO and other measured species reliable indicators of the redox state (Zolotov 1994b). Most estimates of the redox state made from the composition of the Venus atmosphere fall near the centers of Figs. 1 and 2: Barsukov et al. (1982a) at $\delta \sim 3 \times 10^{-5}$; Klose et al. (1992), 2.6×10^{-5}; Fegley et al. (1992), 2×10^{-5}; Zolotov (1995), $\sim 4 \times 10^{-5}$. For any of these redox states the Klose et al. figures indicate that the stable Fe mineral at plains altitude is magnetite, Fe_3O_4, and primary diopside probably should weather to anhydrite, $CaSO_4$ (the thermodynamic uncertainty associated with the diopside to anhydrite reaction is particularly large; Sec. II.D).

2. Visible and Near-Infrared Spectra. The Venera 9 and 10 landers made photometric measurements of surrounding surface material in five wavelength bands with peaks from 0.53 to 0.87 μm (Ekonomov et al. 1980). Reflectivity was found to increase with wavelength in a manner characteristic of Fe^{3+} absorption. Pieters et al. (1986) showed that the Venus spectrum resembles the spectra of hematite and hematite-bearing weathered basalt at 500°C, rather than that of magnetite, which has an almost flat reflectance spectrum. A relatively oxidizing environment is required for hematite stability, $\delta \lesssim 10^{-6}$ according to Fig. 1. The stable Ca mineral in this case is again probably anhydrite rather than diopside.

However, hematite is not the only possible explanation for a Venus spectrum that displays Fe^{3+} absorption, as Pieters et al. (1986) point out. For example, 2 out of 3 of the Fe atoms in magnetite are also in the Fe^{3+} oxidation state, but normally magnetite absorbs light incident on it so strongly that its spectral character is suppressed. However, if weathering of ferrous pyroxenes on Venus produced extremely small magnetite inclusions, small enough to permit strong selective absorption of light passing through them without causing complete attenuation, the weathered material would show Fe^{3+} absorption (Wood 1996). Ramdohr (1980) notes that: "In extremely thin films a few hundred molecules thick, as, e.g., in inclusions occurring in some micas, [magnetite] is transparent brownish-grey...."

3. Kontrast Experiment. The Venera 13 and 14 landers carried a very simple experiment designed specifically to measure the redox state of the Venus environment. The landers carried small samples of asbestos paper impregnated with sodium pyrovanadate. This compound is white (albedo $\sim 80\%$) under oxidizing conditions, but a reducing CO_2-rich environment ($fO_2 < \sim 10^{-20}$ at the Venus surface temperature; $\delta \lesssim 7 \times 10^{-5}$) will react with it to produce blue V_2O_4 and/or black V_2O_3 (Florensky et al. 1983a,b). The *Kontrast* test samples were uncovered when the spacecraft had descended to altitudes of ~ 12 km. The samples were exposed to the hot atmosphere during the ~ 20 min needed to reach the Venus surface from 12 km, and for another 5 to 10 min before they were photographed by the Venera panoramic cameras (O. V. Nikolaeva, personal communication). They were found to have darkened (albedos $\sim 2\%$ for Venera 13, $<30\%$ for Venera 14). If the darkening was due to reaction of sodium pyrovanadate with the atmosphere, Fig. 1 indicates that the stable Fe mineral on the Venus surface at plains level is not hematite but magnetite, or even pyrite or pyrrhotite.

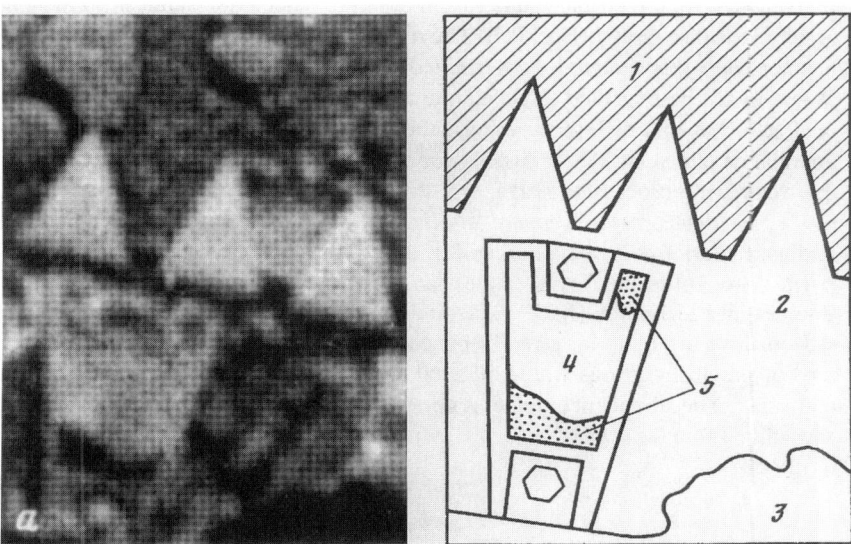

Figure 3. The *Kontrast* experiment on Venera 13. 1: Venus surface; 2: base of Venera lander; 3: shadow; 4: exposed test surface; 5: deposits of soil (figure from Florensky 1983*a*).

A problem is that soil from the Venus surface was thrown onto the sample surfaces during or after landing of the spacecraft. Deposits of soil can be seen in the images of the *Kontrast* samples (Fig. 3). In addition to these discrete deposits, a thin even film of soil may have covered the whole test surface in both cases, and this could be responsible for the lowered albedos seen by the panoramic cameras. Members of the *Kontrast* team are confident at the 90% level that the low albedos detected were caused by reacted sodium pyrovanadate (O. V. Nikolaeva, personal communication), but this other possibility cannot be ruled out.

4. An Attempt to Accommodate All the Evidence Bearing on Redox State.
A recent re-evaluation of chemical equilibria on Venus has defined conditions that may satisfy all three redox criteria (Zolotov 1995). The conditions are: 17 ppm CO, 30 ppm H_2O, 130 ppm SO_2, and 28 ppm COS in the near-surface atmosphere ($\log fO_2 = -21.3$); the use of thermodynamic data of Hemingway (1990) for magnetite and hematite; and the assumption that magnetite and hematite occur in solid solution with other oxide minerals rather than as pure minerals. Under these conditions, to within the uncertainties of the data, reaction of the *Kontrast* experiment to a dark oxide is consistent with the stability of magnetite *and* hematite in the surface material, and the

hematite straightforwardly explains the reflection spectrum of the latter.

C. Degree to which Primary Basalt has been Weathered to Secondary Minerals

It takes time for the primary minerals in basalt to react to the stable secondary assemblage discussed in Sec. II.B. The reaction rate in the Venus environment has not been experimentally determined. The higher surface temperature on Venus might be expected to accelerate reaction rates relative to their values on Earth, but this difference is probably more than offset by the very low abundance of H_2O in the Venus environment. Water is intimately involved in the weathering process on Earth.

1. Evidence from Chemical Analyses of Surface Material. Weathered surface material should contain sulfur, in the form of anhydrite and possibly pyrite. Several of the Soviet landers analyzed surface material by the X-ray fluorescence technique (XRF). The landers carried rock drills that penetrated ~3 cm into the material directly beneath the spacecraft. Approximately 1 cm^3 of the drill cuttings were collected and brought into each spacecraft for analysis. The abundances of S reported for the three lander missions are shown in Table I.

TABLE I[a]
Abundances of Sulfur in Venus Surface Material

	Venera 13	Venera 14	Vega 2
Latitude	7.6°S	13.2°S	6.45°S
Longitude	303.5°E	310.1°E	181.1°E
Plan. radius, km	6052.3	6052.3	6052.7
Temperature, K	730.6	730.6	727.4
Pressure, bar	90.4	90.4	88.3
SO_2 = 185 ppm H_2O = 100 ppm			
$\delta = 10^{-6}$	S = 3.8%	5.2	4.2
$\delta = 10^{-5}$	3.7	5.0	4.0
$\delta = 10^{-4}$	10.6	10.4	9.3
SO_2 = 60 ppm H_2O = 40 ppm			
$\delta = 10^{-6}$	3.8	5.1	4.1
$\delta = 10^{-5}$	3.5	3.7	2.9
$\delta = 10^{-4}$	7.9	6.9	6.8
Reported by XRF analysis	0.64±0.40	0.36±0.32	1.9±0.60

[a] Comparison of the amounts of sulfur (wt. %) that should be present in the equilibrium mineral assemblage, for various assumptions (above), with the amounts actually detected by the XRF surface analysis experiment (below), at the three Soviet lander sites.

For comparison, I have recalculated (as in Sec. II.A) the fully weathered (equilibrium) mineral assemblages at the three sites, using the proportions of cationic elements found in each analysis and the atmospheric temperature and pressure appropriate to the altitude of its site (Garvin and Head 1986). Calculations were made for 185 ppm SO_2/100 ppm H_2O and 60 ppm SO_2/40 ppm H_2O, and δ of 10^{-4}, 10^{-5}, and 10^{-6}. These results are also shown in Table I. Sulfur contents are low where small δ is assumed; in this case, anhydrite is the only S mineral, and part of the Ca that might form anhydrite occurs as anorthite instead. Sulfur contents are highest when SO_2 = 185 ppm and $\delta = 10^{-4}$ are assumed. In this case, more anhydrite forms and a second S mineral, pyrite, is also stable.

A mass-balance calculation suggests how deeply the Venus surface might be weathered in this fashion; 185 ppm of SO_2 would correspond to a total atmospheric S content of $\sim$14 g cm^{-2} of planetary surface. If this S were all used to weather surface material to the extent shown in Table I, $\sim$1 m of material would be transformed. (More likely, weathering reactions would only partly affect the surface material, but by penetrating through cracks and pore space it might reach greater depths than 1 m.)

The concentration of S found at the Vega 2 site is consistent with a substantial degree ($\sim$50%) of weathering of the surface material, but smaller amounts of S than this were reported and lower degrees of weathering ($\sim$7–20%) are inferred at the two Venera sites. It is difficult to know how much confidence to attach to the Venera S numbers, as no report has ever been published of how the S data were reduced for the Venera missions. The early papers in which the Venera XRF experiment was described (Surkov et al. 1982; Barsukov et al. 1982b) did not discuss S or even include it in the surface compositions they reported. Only later, in company with the Vega 2 analysis, were low concentrations of S reported for the Venera missions.

2. Evidence from the Morphology and Mechanical Properties of the Venus Surface. The surface of Venus seen by the panoramic cameras of Venera 10, 13, and 14 (Fig. 4) consists of slabs of material a few centimeters thick and the order of a meter wide (Florensky et al. 1977,1983c; Basilevsky et al. 1985). Objects on the surface at the Venera 9 site are somewhat thicker and blockier, and can less accurately be described as "slabs." Varying amounts of unconsolidated soil can also be seen in the panoramas, between the slabs and partly covering them. The slabs have split loose along planes parallel to the planetary surface at vertical spacings of a few cm, and fractured laterally along irregular boundaries at spacings of $\sim$1 m. In the absence of water, ice, strong winds, growing plants, diurnal and seasonal temperature cycling, and micrometeorite bombardment, this degree and style of fragmentation of the surfaces of lava flows is surprising. It most resembles the exfoliation that can occur during weathering of terrestrial igneous rock, when expansion of the weathered zone beneath a free face splits it away from deeper unweathered rock as a discrete layer. A volume increase of several percent would accompany the weathering of primary basalt to the stable secondary assemblage on

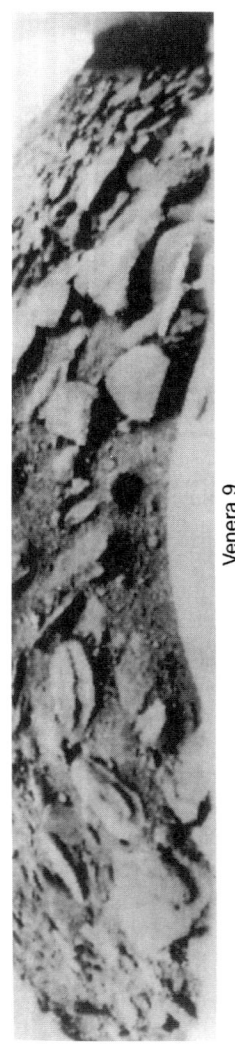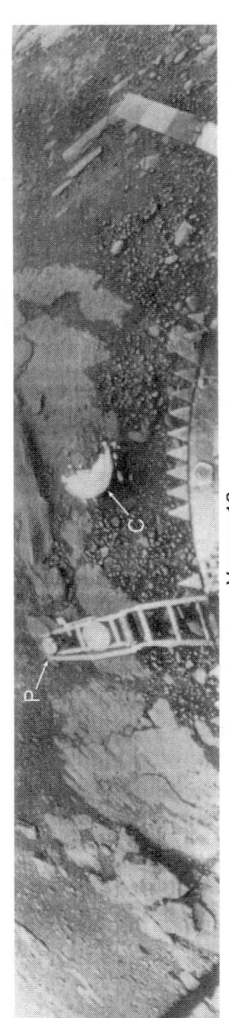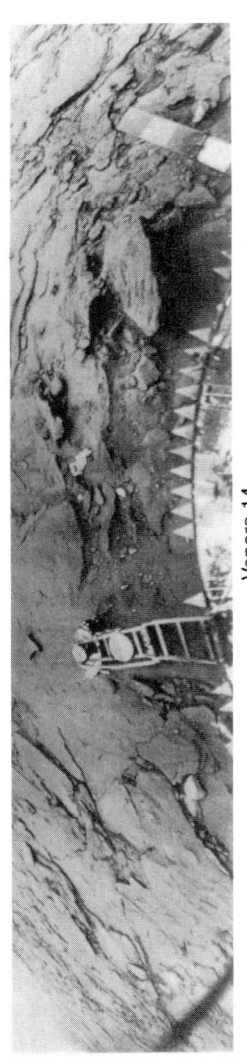

Figure 4. Panoramic photographs taken by the Venera 9, 13, and 14 landers. P: the Venera 13 penetrometer arm; C: discarded cover for camera port. Surface material at the Venera 10 site, not shown, has the same slabby character as that at Veneras 13 and 14.

Venus (Klose et al. 1992), so this may be what has split slabs from the surface of the Venus flood basalts. Basilevsky et al. (1985) summarize the evidence that Venus surface material is chemically altered.

Components of the particulate soil seen in lander panoramas may be impact ejecta and volcanic pyroclastics, but a more copious likely source is the wasting away of slabs and boulders of surface material as continued weathering reduces their strength and the differential expansion of newly formed minerals at their surfaces pries bits loose from them. Losses of this sort from the surfaces of angular blocks tend to increase their roundness, which could account for the crudely rounded forms of some of the objects in the Venera 9 panorama. Windblown mineral grains are not capable of mechanically eroding basalt in the Venus surface environment (Marshall et al. 1988), but they might help disintegrate weaker chemically weathered material.

Mechanical properties of the Venus surface material were inferred from a comparison of the stress-strain profiles generated when the Soviet spacecraft landed, with profiles measured when a spacecraft mockup was dropped on various known substances on Earth (summarized by Basilevsky et al. 1985). The landing dynamics of Venera 13 were found to be consistent with impact on weak, porous material (bearing strength, 4 to 5 kg cm^{-2}; density 1.4 to 1.5 g cm^{-3}, implying $\sim$50% porosity). At the Venera 14 site, the stress-strain profile is consistent with similar surface material covered by $\sim$10 cm of weaker porous material. For comparison, fresh massive basalt has a bearing strength of $\sim$2000 kg cm^{-3}.

The Venera 13 and 14 landers also carried penetrometers designed to determine the physical strength of the Venus surface material (Kermurdzian et al. 1983). These can be seen, deployed, in Fig. 4. A spring-loaded arm unfolded outward and downward, driving a pointed metal cone and blade against the surface at known velocity. Another spring then attempted to rotate the cone and blade inside the depression it had made. The depth of penetration and the angle of rotation achieved could be read by the panoramic camera. These data were compared with the behavior of the device on various types of material on Earth. Unfortunately, Fig. 4 shows that the Venera 14 penetrometer landed on (or at least was interfered with by) the discarded cover for the camera port. From the Venera 13 penetrometer data a bearing strength of 2.6 to 10 kg cm^{-2} was inferred, consistent with the result from landing dynamics. One other estimate of the mechanical properties of surface material came from the resistance it offered to drilling, which was likened to that of "compacted ash material of a volcanic tuff type" at both sites (Surkov et al. 1984). This would be consistent with the properties of weathered basalt.

The landers also measured the electrical resistivity of surface soil and found values of 89 and 73 ohm-meters at the Venera 13 and 14 sites, respectively (Kermurdzian et al. 1983). These are surprisingly low values, in the semi-conductor range. Coauthor V. V. Gromov (personal communication) ascribes this low resistivity to the presence of a thin film of electrically conductive material on the soil particles.

D. Weathering as a Buffer of the Atmospheric Composition

Weathering involves the movement of atoms of anionic elements (O, C, S, H) between the atmosphere and the solid planetary surface. If weathering affects only a microscopic surface layer of planetary material, very small in mass compared to the atmosphere, then the composition of that layer can be altered dramatically without significantly affecting the composition of the atmosphere. The composition of the mineral system in the thin weathered layer is said to be "open" to the gaseous components of the atmosphere, because the partial pressures of these components remain effectively constant in the atmosphere in spite of withdrawals of some of the components to react with surface minerals. The mineral systems treated by Klose et al. (1992) were assumed to be open to atmospheres of specified and invariant composition.

If a sufficiently large volume of solid planetary material is affected by weathering, on the other hand, this no longer holds and the planetary minerals can influence the composition of the atmosphere. The various weathering reactions do this in a particular way. Consider a layer of partly weathered basalt that contains some secondary anhydrite, but also some remaining primary diopside; i.e., minerals on both sides of reaction (1). This configuration is thermodynamically stable only if equality (2) holds, meaning that the partial pressures of CO, SO_2, and CO_2 must be such as to produce exactly the right quotient on the right hand side of Eq. (2). If $p(SO_2)$ is too high and/or $p(CO)$ is too low for the amount of CO_2 in the atmosphere, Eqs. (2) and (3) are unbalanced and the reaction is driven to the right, converting more diopside to anhydrite. This has the effect of removing SO_2 (and CO_2) from the atmosphere and adding CO to it, which pushes it closer to the composition needed for the "exactly right quotient" just referred to. With continued weathering and adjustment of the atmospheric composition, this quotient will eventually be attained, and at that point there will be no further impetus for diopside to react to anhydrite. Weathering, or at least this particular weathering reaction, will cease.

This works in the other direction, too. If the atmosphere has too little SO_2 and/or too much CO for its CO_2 pressure, reaction (1) is driven to the left: anhydrite, enstatite and quartz (if all are available) will react to form diopside, absorbing CO from the atmosphere and releasing SO_2 and CO_2, until the "exactly right quotient" is achieved. Thus the surface mineralogy acts from either direction to drive the abundances of gas species in the atmosphere to a particular set of values and hold them there. The surface minerals are said to *buffer* the composition of the atmosphere. Clearly certain conditions need to be met for buffering to be effective; all the minerals needed for the buffering reaction must be present and effectively in contact with one another, and the rate of the reaction must be fast enough for it to predominate over other processes that perturb the atmospheric composition, such as the addition of gases from volcanoes, or volatiles from meteoroids.

The Gibbs free energy change (left-hand side of Eq. 2) for the reaction by which diopside is weathered to anhydrite and other compounds (reaction 1), from data in Robie et al. (1979), equals $39{,}192 \pm 10{,}210$ J/mol at 749 K, the temperature at plains level on Venus (taken to be 6051.0 km radius). If the right-hand side of Eq. (2) is evaluated for $p(CO_2) = 96.4\% \times 103.1$ bar (appropriate to 6051.0 km radius) and $p(CO) = 20$ ppm $\times 103.1$ bar, $p(SO_2) = 180$ ppm $\times 103.1$ bar (values favored by Pollack et al. 1993), a value of 42,324 J/mol is obtained. The nominal inequality in Eq. (2), if correct, would mean that diopside and anhydrite are not both stable in contact with the atmosphere; additional reaction of diopside into anhydrite is needed to bring the atmospheric concentrations of SO_2 and CO to the buffered values. However, the uncertainty in ΔG^0 is large enough to embrace the value calculated for the right hand side of Eq. (2). Thus it may be that Eq. (2) *is* an equality, diopside and anhydrite coexist stably on the plains of Venus, and reaction (1) is buffering the SO_2 content of the atmosphere; but because of the large uncertainties attached to the comparison this cannot be proven. It is even possible that $\Delta G^0 >$ the right-hand side of Eq. (2), which would mean anhydrite is not a stable weathering phase at all.

The concept of buffering is an engaging one, and some atmospheric chemists feel impelled to search for mineral buffers to account for the concentrations of many or all of the key constituents of atmospheres. The idea that the pressure of the principal component of the Venus atmosphere, CO_2, is buffered by a reaction similar to reaction (4) has long been propounded (see, e.g., Lewis 1970; Lewis and Kreimendahl 1980; Fegley et al. 1992). (Typically these arguments speak of wollastonite, $CaSiO_3$, as the Ca silicate mineral rather than diopside. However, as a practical matter wollastonite is a rare mineral on Earth, formed by the metamorphism of limestone, and it probably does not exist on Venus. It is more realistic geochemically to think in terms of the $CaSiO_3$ component of pyroxene, i.e., diopside, as the Ca silicate mineral [Walker 1975].) An analysis similar to that given above for the diopside-anhydrite reaction, carried out for reaction (4), does show that diopside and calcite would coexist stably at the temperature and for the atmospheric partial pressure of CO_2 at the surface of Venus (to within the considerable uncertainty of the calculation). However, this buffering reaction comes into conflict with the diopside-anhydrite reaction that potentially buffers the partial pressure of SO_2 in the Venus atmosphere, because both depend upon diopside as one of the reactants. Which atmospheric gas, CO_2 or SO_2, has first claim on the Ca in diopside? For both to share the Ca it would be necessary for calcite and anhydrite to coexist stably in the Venus environment. A reaction relating the two minerals is

$$CaCO_3 + SO_2(g) \rightleftharpoons CaSO_4 + CO(g). \qquad (5)$$

Thermodynamic analysis shows that for calcite and anhydrite to coexist stably the ratio of partial pressures of CO to SO_2 would have to be 14.7 ± 7.8. For a CO concentration of 20 ppm and 180 ppm of SO_2, the ratio is only ~ 0.1. If a concentration of 40 ppm is used for SO_2, the ratio only rises to 0.5. Thus the two minerals cannot coexist; reaction of diopside with the Venus atmosphere should produce only anhydrite, not calcite (Barsukov et al. 1982a; Klose et al. 1992). A buffer system should be capable of absorbing any extra CO_2 that joins the atmosphere, in excess of the equilibrium partial pressure, or of supplying CO_2 if the partial pressure of that element for some reason falls below the equilibrium value. The Venus mineral assemblage cannot absorb CO_2, as just shown, and there is no reason to think it can supply it, as there is no plausible source for significant amounts of carbonate minerals in the surface rock and they could not be stored near the surface if there were. It seems likely that all of Venus' near-surface inventory of C is in the atmosphere, and the atmospheric CO_2 pressure is not buffered by surface mineral reactions.

III. ANOMALOUS SURFACE MATERIAL ON MOUNTAIN TOPS

A. Observations

The radar altimeter on the Pioneer Venus Orbiter spacecraft discovered, and the Venera 15 and 16 and Magellan radar altimeters confirmed, that the highest elevations on Venus ($\gtrsim 4$ km above the plains) are areas of greatly enhanced radar reflectivity (Pettengill et al. 1982; Garvin et al. 1985; Ford and Pettengill 1992). This enhanced reflectivity manifests itself as extreme brightness (high radar backscatter cross section) in the Magellan SAR images. Pioneer Venus also found and Magellan confirmed that these same mountainous areas display anomalously low radiothermal emissivity (Ford and Pettengill 1983; Pettengill et al. 1992), consistent with the approximate complementarity that is expected to relate these two parameters ($e \cong 1 - r$, where e is emissivity and r the power reflection coefficient). Studies of the anomalous areas have tended to focus on the emissivity rather than the reflectivity data set, in spite of the far superior spatial resolution of the reflectivity data, because the latter are more strongly affected by variations in surface roughness than are the emissivity data. The effects of roughness cannot yet be subtracted from the reflectivity data with confidence.

Klose et al. (1992) used scatterplots of emissivity vs altitude (a) for points representing individual emissivity footprints to display the lowered emissivity at high altitudes within defined mountainous areas. Examples of two dissimilar types of a/e relationship are shown in Fig. 5. At Maxwell Montes, the highest peak on Venus, e is almost constant with a value, unexceptional for Venus, of ~0.85 at altitudes less than ~4.8 km. Above that altitude, the a/e trend abruptly turns to much lower values of emissivity. (Klose et al. [1992] referred to the altitude at which this inflection occurs as the "critical altitude.") At $e \sim 0.4$ and $a \sim 6$ km, the trend again turns upward (no further systematic change in e with a). The trend at Ovda Regio is much more blurred and the turnover to lower values of e is more gradual and begins at a lower altitude. The trend never turns vertical again as it does at Maxwell, but this is probably because Ovda is not high enough to display such an effect. Comparison of the

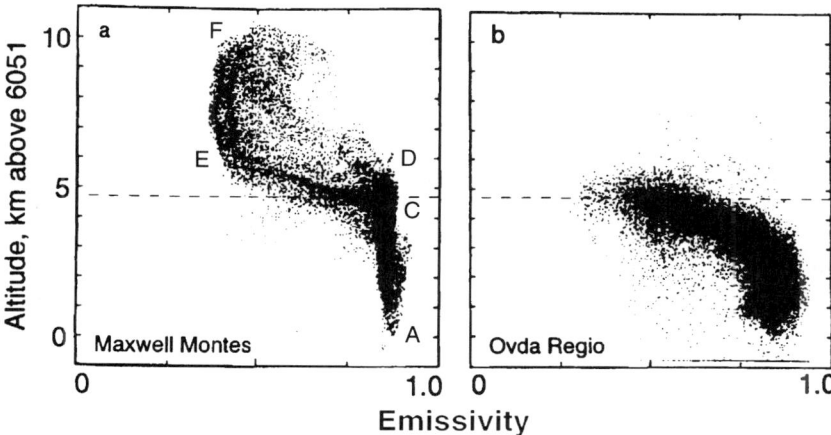

Figure 5. Scatterplots of emissivity vs altitude for two mountainous regions on Venus; points represent individual Magellan radiometer footprints. The fiducial dashed line records the critical altitude for Maxwell Montes. Letters A–F in the Maxwell Montes plot are discussed in the text (figure adapted from Klose et al. 1992).

Maxwell and Ovda a/e trends also demonstrates that critical altitudes vary from one mountainous region to another; Klose et al. found a range of 2.5 to 4.75 km for this parameter.

The radar reflectivity coefficient of a material is controlled by its dielectric constant (permittivity). The high reflectivity (low emissivity) displayed by mountainous regions on Venus was initially interpreted to mean that surface materials in those areas have a relatively high real value of relative dielectric permittivity, mostly in the range 11 to 20 but sometimes greater, compared with a value of ~4 over most of the Venus surface (Pettengill et al. 1982). A permittivity of ~4 would be normal for a variety of dry rock materials with some pore space, but 11 to 20 is outside the range of all but the most exotic types of rock (Campbell and Ulrichs 1969).

A final experiment carried out by the Magellan spacecraft, two years after it ceased its radar mapping program, has redefined the character of the anomalous surface material at high altitudes (Pettengill et al. 1996). On 5 June 1994 the Magellan S-band telemetry transmitter and high-gain antenna were used to reflect linearly polarized radio signals (at incidence angle 66.8–67.2°) off the Maxwell Montes region, toward Earth. Received on Earth, the plane of polarization of the signals was found to have been rotated ~37°, and a component (~10%) of circularly polarized power was also present. Dielectric permittivity (ϵ) has a real (κ) and an imaginary component

$$\epsilon = \kappa + i \times [L + \sigma/(\epsilon_0 \omega)] \tag{6}$$

where L is the frequency-independent dielectric loss, σ the volume conductivity, ϵ_0 the permittivity of free space, and ω the angular frequency of the incident radiation. The large angle through which the plane of polarization was rotated requires that the surface off which it reflected has a high (either real or imaginary) value of relative dielectric permittivity. The creation during reflection of a measurable component of circular polarization requires that the surface material is electrically lossy, and means there is a significant imaginary term in Eq. (6). Pettengill et al. (1996) find that a value of $\epsilon = -i \times (100 \pm 50)$ produces the best agreement with observations. Here the required large dielectric permittivity resides in the imaginary term, not in the real term, as (for example) Pettengill et al. (1982) had assumed. Pettengill et al. (1996) consider that a continuous skin of semi-conducting material, coating the surface of Venus at high altitudes, provides the most likely explanation for a large imaginary component of the dielectric permittivity. They calculate from Eq. (6) and the expression for ϵ just given that σ, the conductivity of the skin, is ~ 13 s m^{-1}. Its thickness is ≥ 3 mm, the electrical skin depth for attenuation of electromagnetic radiation at S-band wavelengths.

B. Interpretations of Anomalous Surface Material

1. Models Employing Materials with Large Real Dielectric Permittivities. Prior to the 1994 bistatic radar experiment, efforts to understand the electrically anomalous material on Venus mountain tops invoked the presence of surface materials with unusually large real dielectric permittivities. These included the pyrite-bearing loaded dielectric models of Pettengill et al. (1982) and Klose et al. (1992), the perovskite ($CaTiO_3$)-rich surface material of Fegley et al. (1992), and the ferroelectric mineral model of Shepard et al. (1994). None of these models accounts for the large imaginary component of dielectric permittivity required for surface material to account for the outcome of the bistatic radar experiment of Pettengill et al. (1996). In addition, the models have other failings. The loaded dielectric models do not provide even real permittivities that are as great as those required in some areas, and they do not explain the observation that the critical altitude above which surface material has anomalous electrical properties is not constant for all mountain tops. The models of Fegley et al. (1992) and Shepard et al. (1994) are geochemically implausible, in calling for unprecedented concentrations in high-altitude surface materials of a known mineral (perovskite), or the presence of minerals not known to occur in nature (the ferroelectric minerals named by Shepard et al.).

2. Models Involving the Vapor Transport of Volatile Elements or Compounds. Brackett et al. (1995) cite the abundance of volatile metal halides and sulfides (of, e.g., Cu, Zn, Sn, Pb, As, Sb, Bi) emitted as vapors by terrestrial volcanos, and postulate that compounds of this sort on Venus make their way from low altitudes, where they can evaporate to the extent of their saturation vapor pressures, to high altitudes, where they plate out on the surface because there their saturation vapor pressures are lower. Some of these minerals have high dielectric constants, so the deposited layer is capable of lowering the emissivity at high altitudes. Pettengill et al. (1996) point out that a surface skin of elemental Ge or Te would have approximately the right conductivity (as do some of the deposited minerals Brackett et al. picture) to be consistent with the lossiness and the consequent large imaginary dielectric permittivity that high-altitude surface material appears to have on Venus. Pettergill et al. postulate that Te, which is relatively volatile, evaporates at plains altitudes and is circulated to higher altitudes where it is deposited on the "cooler" mountain tops.

The concept of volatile transport of minerals from the hot plains to the "cool" mountain tops is probably the key to understanding the anomalous surface electrical properties of those elevated regions, but it is not necessary to invoke the presence of exotic volcanic sublimates or the concentration of a rare element like Te to establish these properties. Thermodynamic calculations like those of Sec. II.D show that the reaction

$$Fe_3O_4 + 6HCl(g) + CO(g) \rightleftharpoons 3FeCl_2(g) + 3H_2O(g) + CO_2(g) \quad (7)$$

would establish an equilibrium partial pressure of $FeCl_2(g)$ of $\sim 3.2 \times 10^{-10}$ bar at plains level on Venus. There is nothing exotic or speculative about this vapor species: it contains one of the most abundant elements in the crust of Venus, Fe, and Cl consistent with the concentration of HCl (0.5 ppm) found by groundbased studies of the Venus atmospheric absorption spectrum (Pollack et al. 1993). Barsukov et al. (1981) proposed that $FeCl_2$, vaporized in this fashion, condensed into microparticles at high altitudes to form a component of the Venus cloud layer. (The equilibrium vapor pressures of FeCl and $FeCl_3$ at the Venus surface are less than that of $FeCl_2$.)

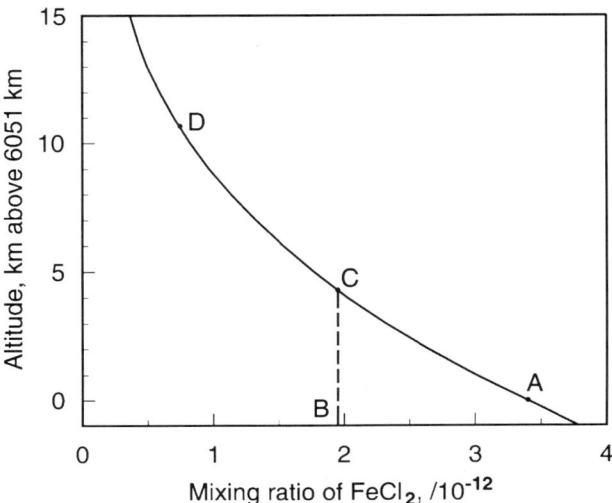

Figure 6. Equilibrium mixing ratio of $FeCl_2$ in the Venus atmosphere vs altitude (curve). Letters and dashed line are discussed in Sec. III.B.2.

This partial pressure of $FeCl_2$ corresponds to a concentration (mixing ratio) of 3.4×10^{-12} (A in Fig. 6). Figure 6 shows the decrease of the equilibrium mixing ratio of $FeCl_2$ with altitude. In practice, a parcel of Venus atmosphere at plains level would not have time to become saturated with $FeCl_2$ before it was convected upward. The concentration of $FeCl_2$ in the gas might reach B in Fig. 6, for example, after which convection would move it up line BC. At C it would reach the saturation mixing ratio curve, and if the gas were flowing over a mountain slope the supersaturated gas would precipitate an iron mineral on the planetary surface. This would not be solid $FeCl_2$, but whatever Fe mineral is stable at that altitude for the value of δ that holds for the Venus atmosphere (Fig. 1). It might be magnetite, in which case reaction (7) would simply run to the left, depositing magnetite on the planetary surface; or for a lower value of δ and/or a higher altitude, SO_2, H_2O, and CO could react the $FeCl_2$ to pyrite, HCl, and CO_2 (in this case a slightly different equilibrium mixing ratio curve would be applicable). This mechanism predicts a sharp lower boundary to the altitude range where magnetite (or pyrite) begins to plate out; immediately below C in Fig. 6 magnetite (or pyrite) is still capable of being destroyed by reaction to $FeCl_2$.

If gas rising up a slope is able to lose its $FeCl_2$ fast enough by this reaction mechanism, the mixing ratio of that species declines along CD. If not, and for gas not near mountain slopes when it convects upward, $FeCl_2$ becomes supersaturated above altitude C and may eventually precipitate out as a fine dust of magnetite or pyrite.

The DC conductivity of magnetite at 300 K is $\sim 10^4$ s m^{-1}, that of pyrite is $\sim 10^3$ s m^{-1} (Olhoeft 1989). These minerals are in the same semi-conducting category as the elemental Te cited by Pettengill et al. (1996). A deposited 3-mm skin of magnetite or pyrite could have the electrical properties dictated by the bistatic radar experiment for the Venus highland surface material. It is difficult to estimate the rate at which such a buildup could occur, taking into account the kinetics of the reactions that form $FeCl_2$ and magnetite or pyrite and rates of transfer from the plains to mountain tops, and assess whether it is realistic for a suitable semi-conducting skin to form in, say, the time since the ~ 0.3 Gyr resurfacing event. Certainly there is no problem with availability of the Fe, Cl, and O or S needed for the process.

The vapor-deposition mechanism provides a natural explanation for the nonuniformity of critical altitudes on Venus. The position of B in Fig. 6, i.e., the amount by which gas falls short of the saturation concentration of $FeCl_2$ when it begins to convect upward, is controlled by kinetics and depends upon several things: the altitude and temperature of the plains; the age and state of weathering of (and hence the accessibility of Fe in) plains material; the wind field; and the proximity of sinks for atmospheric $FeCl_2$, in the form of mountain peaks on which it can react out. All these factors vary to some degree from one mountainous locale to another on Venus, so the position of B must vary and therefore the altitude of C, which is the critical altitude.

Venus air is expected to contain already a substantial amount of $FeCl_2$ left over from previous convective cycles when it returns to the plains for a new cycle, so the last variable named, the proximity of a mountain system that can deplete this store of $FeCl_2$, may be particularly important. It is reasonable that Maxwell Montes, the highest and "coldest" of Venus mountains, would be the most effective at depleting atmospheric $FeCl_2$ in its vicinity. This would place B (Fig. 6) farthest to the left for Maxwell Montes and the critical altitude at the highest level on Venus, which is the observed state of affairs.

The model described, with its sharp transition from normal weathered surface material to high-dielectric-constant material at higher altitudes, can account straightforwardly for most of the Maxwell Montes a/e plot. Points between A and C in Fig. 5a are normally weathered surface material; E–F is high-altitude material covered with a plated-out semi-conducting mineral; C–D consists of radiometer footprints that overlap both types of surface. The turnback toward higher emissivities at F may result from an insufficiently thick skin of semi-conducting mineral at highest altitudes, where growth was stunted by a decrease in the equilibrium partial pressure of $FeCl_2$ and the precipitation rate of magnetite or pyrite with altitude.

Then how are we to explain the less angular shape of the Ovda Regio a/e plot (Fig. 5b)? Many other Venus mountain areas have a/e plots similar to Ovda Regio, e.g., Thetis Regio, Atla Regio, Beta Regio (Klose et al. 1992). In fact, Maxwell Montes is unique in the angularity of its a/e plot. Each of the other a/e plots named embraces data from numerous mountain slopes within a broadly defined area. The a/e trends on individual slopes vary because of differences in critical altitude, roughness, overall slope, and the relationship between slope angle and incidence angle. The relationship between topographic slope, incidence angle, and the a/e trend is shown in Fig. 7. When many dissimilar a/e trends are overlain, a thick blurred scatterplot like that of Ovda Regio results. This compositing completely washes out the angles in trends that correspond to critical altitudes.

Pathare (1992) divided the Maxwell Montes area represented by the scatterplot in Fig. 5a, 1.28×10^6 km^2, into an 8×8 grid and studied the a/e plot of each subarea separately in order to understand the origin of the various features seen in Fig. 5a. He found that most of the coherent line of data points that define the distinctive CE segment of the trend were generated in an area east of Cleopatra Patera, where lava that welled out of Cleopatra flowed east into Fortuna Tessera and filled the tessera valleys. Fortuitously, the critical altitude for Maxwell Montes crosses this relatively smooth, shallow-sloping lava surface. Its trace is plainly marked by areas of low and high radar backscatter coefficient (i.e., dark and bright areas) in the SAR image of the area (Fig. 8). It is largely because of the unique circumstance that a portion of Maxwell's critical altitude contour lies on a shallow, eastward-sloping smooth surface (see Fig. 7) that the a/e plot for Maxwell Montes displays the sharp angle ACE.

Though the model of this subsection explains the low-emissivity properties of elevated regions very differently than Klose et al.'s (1992) model did,

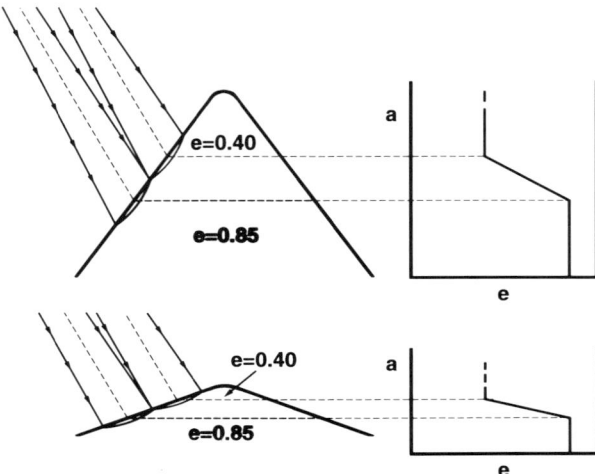

Figure 7. Relationship between topographic steepness, and steepness of the trend of points representing footprints that overlap low- and high-emissivity terrains in an a/e plot (schematic). (Truncated) cones shown at left, representing emissivity measurements made on west-facing slopes, have the same incidence angles and angles of convergence. In both cases the two cones shown define the range of footprint positions that overlap the contact between low- and high-emissivity surface materials. Dashed lines are centerlines that define altitudes of those footprints. All footprints between these two altitudes sample both low- and high-emissivity surface materials. The spread in altitudes is greater for steep mountain slopes (above) than for shallow ones (below). It can be seen that the sloping segment of the a/e trend would be even steeper if the footprints seen by an east-looking spacecraft were on the eastern rather than the western slopes of mountains.

the chronometric principle these authors applied still holds; if the time scale for alteration of the surface material is similar to the time scales of tectonism and volcanism on Venus, then the presence or absence of an emissivity anomaly for a mountaintop provides a crude measure of the time when the area was uplifted or resurfaced. The absence of an emissivity anomaly on Maat Mons can still be interpreted in terms of relatively recent volcanic activity on that mountain. Pathare (1992) found that most of the points in bulge D of Fig. 5a were generated by a high-altitude, high-emissivity link in the mountain chain between Maxwell and Freyja Montes (66–67°N, 349–353°E), and speculated that this area was uplifted relatively recently.

IV. LOW-EMISSIVITY SURFACE MATERIAL AT LOW ALTITUDES

The most dramatic examples of low-emissivity surface material, requiring a special mineralogical explanation, occur on the Venus mountaintops. However, anomalously-low-emissivity materials have also been observed on the plains of Venus and at intermediate altitudes (Jurgens et al. 1988; Weitz 1992;

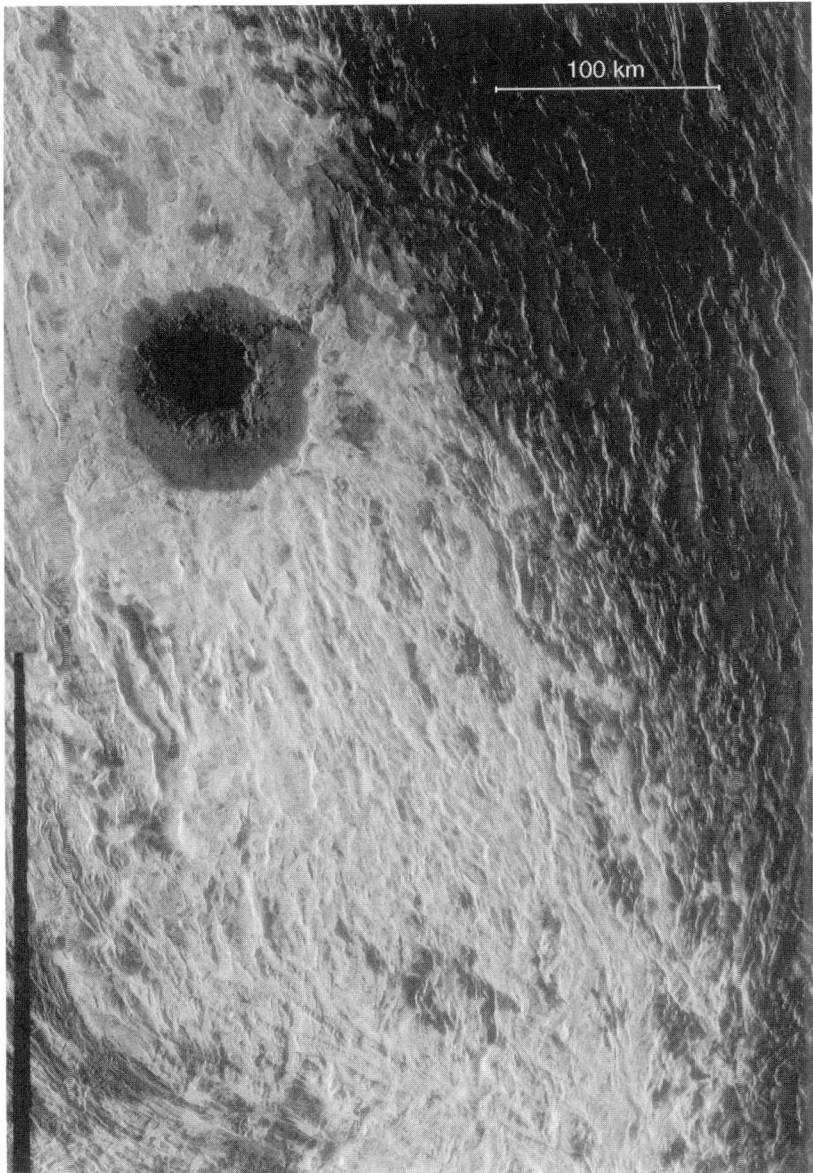

Figure 8. SAR image of a portion of the elevated Maxwell Montes region, showing Cleopatra Patera. Bright (high-radar-backscatter) areas to the left in the figure are above the critical altitude; dark areas to the right, below it (image from Magellan F-MIDR 65N006).

Pettengill et al. 1992). The minimum emissivity found in these cases is ~0.6. It is likely that this more modest anomaly arises from some cause altogether different from the mechanism that produces lower emissivities on mountain tops. For example, an emissivity of 0.6 can be explained by the loaded dielectric model of Pettengill et al. (1982); this emissivity corresponds to a dielectric constant (presumably real) of ~20, which would require ~9 volume percent of semi-conducting inclusions in the surface material (Pettengill et al. 1988). Semi-conducting inclusions (pyrite) in this abundance can be obtained by the complete weathering of basalt in the Venus environment (Klose et al. 1992).

The craters Boleyn, Stanton, Stuart, and Mead display emissivities between 0.61 and 0.70 (Pettengill et al. 1992). The widths of ejecta blankets for these craters are small compared to crater diameters and emissivity footprint sizes, so it is not possible to separate the contributions of crater floors and ejecta blankets. Eight Venus craters have low-emissivity (0.6–0.8) parabolas associated with them (Campbell et al. 1992), analogous to the much more abundant radar-dark parabolas visible in SAR images. The emissivity parabolas, like the SAR parabolas, presumably are deposits of fine crater debris that was ejected high into the atmosphere and blown westward as it rained out. Only one crater, Carson, has both types of parabolas (the emissivity parabola nests inside the SAR parabola). In other cases the SAR parabolas are regions of slightly lowered emissivity, and the emissivity parabolas are regions of slightly lowered radar backscatter coefficient.

It is hard to escape the conclusion that the Venus crustal rock excavated by craters has a higher dielectric constant than material exposed at the surface, at least in some locations. This material, as crater ejecta, displays anomalously low emissivities for some period of time, then presumably weathers to the same mineralogy and dielectric constant as other surface material. Weitz (1992) suggested that subsurface material is a loaded dielectric with a large enough content of conducting inclusions to produce the observed emissivities, but did not attempt to explain why it differed in this way from surface material or unweathered primary basalt (which is incapable of accounting for the low emissivities). Wood (1994) speculated that the high-dielectric-constant subsurface material consists of basalt flows that were weathered during an earlier (pre-resurfacing) epoch, when the Venus environment and weathering products were different than they are now. However, this model is untenable because of the very limited thickness (<1 m) of weathered material the atmosphere is capable of creating (Sec. II.C.1). This material would be too highly mixed and diluted by unweathered debris during a cratering event to retain a detectable low-emissivity signature in the ejecta deposits.

It may be that the high dielectric constant of (at least some) buried crustal material on Venus, relative to surface material, is a product of metamorphic mineralization that occurs in basalt once it is emplaced and buried beneath later flows. Gases effusing outward from the interior of Venus have a composition different from that of the atmosphere, and at some modest depth the hot crustal rock is altered in the presence of this endogenic environment rather than the atmospheric environment that controls the weathering of surface rock. It is possible that the altered mineralogy imparts a higher dielectric permittivity and a lower emissivity to the rock than those of unaltered basalt or basalt weathered at plains level in contact with the Venus atmosphere.

Acknowledgments. I am very grateful to M. I. Petaev for extensive literature research, translations, and advice concerning the essential Russian component of Venus research. Also helpful and patient were A. T. Basilevsky, S. V. W. Beckwith, O. V. Nikolaeva, G. H. Pettengill, C. Phillips, and M. Yu. Zolotov.

EDITOR'S NOTE

The reader/student should examine this chapter and that of Fegley et al. in order to assimilate arguments on both sides of a debate regarding issues of rock weathering on Venus' surface.

REFERENCES

Barsukov, V. L., et al. 1981. The metal chloride and elemental sulfur condensates in the Venusian troposphere—Is it possible? *Proc. Lunar Planet. Sci. Conf.* 12:43–45.

Barsukov, V. L., Volkov, P., and Khodakovsky, I. L. 1982a. The crust of Venus: Theoretical models of chemical and mineral composition. *Proc. Lunar Planet. Sci. Conf.* 13:3–9.

Barsukov, V. L., Surkov, Yu. A., and Moskaleva, L. P. 1982b. Geochemical explorations of the surface of Venus by the spacecraft Venera 13 and Venera 14. *Geokhimiya* 7:899–919 (in Russian).

Barsukov, V. L., et al. 1987. Mineral composition of Venusian regolith at the Venera-13, Venera-14 and Vega-2 landing sites, as estimated from thermodynamic calculations. *Dokl. Akad. Nauk SSSR* 287:154–156 (in Russian).

Basilevsky, A. T., et al. 1985. The surface of Venus as revealed by the Venera landings: Part II. *Geol. Soc. Amer. Bull.* 96:137–144.

Bézard, B., de Bergh, C., Crisp, D., and Maillard, J.-P. 1990. The deep atmosphere of Venus revealed by high-resolution nightside spectra. *Nature* 345:508–511.

Brackett, R. A., Fegley, B., Jr., and Arvidson, R. E. 1995. Volatile transport on Venus and implications for surface geochemistry and geology. *J. Geophys. Res.* 100:1553–1563.

Campbell, D. B., et al. 1992. Magellan observations of extended impact crater related features on the surface of Venus. *J. Geophys. Res.* 97:16249–16277.

Campbell, M. J., and Ulrichs, J. 1969. Electrical properties of rocks and their significance for lunar radar observations. *J. Geophys. Res.* 74:5867–5881.

Crisp, D., Allen, D. A., Grinspoon, D. H., and Pollack, J. B. 1991. The dark side of Venus: Near-infrared images and spectra from the Anglo-Australian Observatory. *Science* 253:1263–1266.

Ekonomov, A. P., Golovin, Yu. M., and Moshkin, B. E. 1980. Visible radiation observed near the surface of Venus: Results and their interpretation. *Icarus* 41:65–75.

Fahd, A. K., and Steffes, P. G. 1992. Laboratory measurements of the microwave and millimeter-wave opacity of gaseous sulfur dioxide (SO_2) under simulated conditions for the Venus atmosphere. *Icarus* 97:200–210.

Fegley, B., Jr., and Treiman, A. H. 1992. Chemistry of atmosphere-surface interactions on Venus and Mars. In *Venus and Mars: Atmospheres, Ionospheres, and Solar Wind Interactions*, eds. J. G. Luhmann, M. Tatrallyay and R. O. Pepin (Washington, D. C.: American Geophysical Union), pp. 7–72.

Fegley, B., Jr., Treiman, A. H., and Sharpton, V. I. 1992. Venus surface mineralogy: Observational and theoretical constraints. *Proc. Lunar Planet. Sci. Conf.* 22:3–19.

Fegley, B., Jr., Lodders, K., Treiman, A. H., and Klingelhöfer, G. 1995. The rate of pyrite decomposition on the surface of Venus. *Icarus* 115:159–180.

Florensky, C. P., et al. 1977. The surface of Venus as revealed by Soviet Venera 9 and 10. *Geol. Soc. Amer. Bull.* 88:1537–1545.

Florensky, C. P., et al. 1983a. The redox state of the Venus surface, based on the "Kontrast" geochemical indicator on the Venera-13 and Venera-14 spacecraft. *Kosmich. Issled.* 21:351–354 (in Russian).

Florensky, C. P., et al. 1983b. Redox indicator "Contrast" on the surface of Venus. *Lunar Planet. Sci. Conf.* XIV:203–204 (abstract).

Florensky, C. P., et al. 1983c. Venera 13 and Venera 14: Sedimentary rocks on Venus? *Science* 221:57–59.

Ford, P. G., and Pettengill, G. H. 1983. Venus: Global surface radio emissivity. *Science* 220:1379–1381.

Ford, P. G., and Pettengill, G. H. 1992. Venus topography and kilometer-scale slopes. *J. Geophys. Res.* 97:13103–13114.

Garvin, J. B., and Head, J. W. 1986. Characteristics of the Venera and Vega landing sites from Pioneer Venus radar data. *Lunar Planet. Sci. Conf.* 7:253–254.

Garvin, J. B., Head, J. W., Pettengill, G. H., and Zisk, S. H. 1985. Venus global radar reflectivity and correlations with elevation. *J. Geophys. Res.* 90:6859–6871.

Hemingway, B. S. 1990. Thermodynamic properties for bunsenite, NiO, magnetite, Fe_3O_4, and hematite, Fe_2O_3, with comments on selected oxygen buffer reactions. *Amer. Mineral.* 75:781–790.

Hoffman, J. H., Hodges, R. R., Donahue, T. M., and McElroy, M. B. 1980. Composition of the Venus lower atmosphere from the Pioneer Venus mass spectrometer. *J. Geophys. Res.* 85:7882–7890.

Jurgens, R. F., Slade, M. A., and Saunders, R. S. 1988. Evidence for highly reflecting materials on the surface and subsurface of Venus. *Science* 240:1021–1023.

Kermurdzian, A. L., et al. 1983. Preliminary results of the measurement of physicomechanical properties of soil by the Soviet interplanetary automatic stations Venera 13 and Venera 14. *Kosmich. Issled.* 21:323–330 (in Russian).

Khodakovsky, I. L., et al. 1979. Venus: Preliminary prediction of the mineral composition of surface rocks. *Icarus* 39:352–363.

Klose, K. B., Wood, J. A., and Hashimoto, A. 1992. Mineral equilibria and the high radar reflectivity of Venus mountaintops. *J. Geophys. Res.* 97:16353–16369.

Krasnopolsky, V. A., and Pollack, J. B. 1994. H_2O-H_2SO_4 system in Venus' clouds and OCS, CO, and H_2SO_4 profiles in Venus' troposphere. *Icarus* 109:58–78.

Lewis, J. S. 1970. Venus: Atmospheric and lithospheric composition. *Earth Planet. Sci. Lett.* 10:73–80.

Lewis, J. S., and Kreimendahl, F. A. 1980. Oxidation state of the atmosphere and crust of Venus from Pioneer Venus results. *Icarus* 42:330–337.

Marshall, J. R., Greeley, R., Tucker, D. W., and Pollack, J. B. 1988. Aeolian weathering of Venusian surface materials: Preliminary results from laboratory simulations. *Icarus* 74:495–515.

Olhoeft, G. R. 1989. Electrical properties of rocks. In *Physical Properties of Rocks*, eds. Y. S. Touloukian, W. R. Judd and R. F. Roy (New York: Hemisphere Publishing), pp. 257–329.

Oyama, V. I., et al. 1980. Pioneer Venus gas chromatography of the lower atmosphere of Venus. *J. Geophys. Res.* 85:7891–7902.

Pathare, A. V. 1992. Reflections on Venus: A Study of the Emissivity-Altitude Relationship of Maxwell Montes Using Data from Magellan. B.A. Honors Thesis, Harvard University.

Pettengill, G. H., Ford, P. G., and Nozette, S. 1982. Venus: Global surface radar reflectivity. *Science* 217:640–642.

Pettengill, G. H., Ford, P. G., and Chapman, B. D. 1988. Venus: Surface electromagnetic properties. *J. Geophys. Res.* 93:14881–14892.

Pettengill, G. H., Ford, P. G., and Wilt, R. J. 1992. Venus surface radiothermal emission as observed by Magellan. *J. Geophys. Res.* 97:13091–13102.

Pettengill, G. H., Ford, P. G., and Simpson, R. A. 1996. Electrical properties of the Venus surface from bistatic radar observations. *Science*, 272:1628–1631.

Pieters, C. M., et al. 1986. The color of the surface of Venus. *Science* 234:1379–1383.

Pollack, J. B., et al. 1993. Near-infrared light from Venus' nightside: A spectroscopic analysis. *Icarus* 103:1–42.

Ramdohr, P. 1980. *The Ore Minerals and their Intergrowths* (Berlin: Academie-Verlag).

Robie R. A., Hemingway, B. S., and Fisher, J. R. 1979. *Thermodynamic Properties of Minerals and Related Substances at 298.15 K and 1 Bar (10^5 Pascals) Pressure and at Higher Temperatures*, Geol. Survey Bull. 1452 (Washington, D. C.: U. S. Govt. Printing Office).

Shepard, M. K., Arvidson, R. E., Brackett, R. A., and Fegley, B., Jr. 1994. A ferroelectric model for the low emissivity highlands on Venus. *Geophys. Res. Lett.* 21:469–472.

Surkov, Yu. A., et al. 1982. Investigation of Venus with Venera-13 and -14 probes: First data on composition of rocks. *Astron. Vest.* 16:139–152 (in Russian).

Surkov, Yu. A., Barsukov, V. L., Moskalyeva, L. P., and Kharyukova, V. P. 1984. New data on the composition, structure, and properties of Venus rock. *Proc. Lunar Planet. Sci. Conf.* 14:393–402.

von Zahn, U., Kumar, S., Niemann, H., and Prinn, R. 1983. Composition of the Venus atmosphere. In *Venus*, eds. D. M. Hunten, L. Colin, T. M. Donahue and and V. I. Moroz (Tucson: Univ. Arizona Press), pp. 299–430.

Walker, J. C. 1975. Evolution of the atmosphere of Venus. *J. Atmos. Sci.* 32:1248–1256.

Weitz, C. M. 1992. Low-emissivity impact craters on Venus. *Lunar Planet. Sci. Conf.* XXIII:1513–1514 (abstract).

Wood, B. J., and Fraser, D. G. 1976. *Elementary Thermodynamics for Geologists* (Oxford: Oxford Univ. Press).

Wood, J. A. 1994. Occurrences of low-emissivity surface material at low altitudes on Venus: A window to the past. *Lunar Planet. Sci. Conf.* 25:1509–1510.

Wood, J. A. 1996. Must the Venus surface material contain hematite? *Lunar Planet. Sci. Conf.* XXVII:1451–1452 (abstract).

Wood, J. A., and Brett, R. 1997. Comment on "The rate of pyrite decomposition on the surface of Venus." *Icarus*, in press.

Zolotov, M. Yu. 1991*a*. Chemical weathering of olivines and ferromagnesian pyroxenes on the surface of Venus. *Lunar Planet. Sci.* 22:1567–1568.

Zolotov, M. Yu. 1991*b*. Redox conditions of the nearsurface atmosphere of Venus. I. Some reevaluations. *Lunar Planet. Sci.* 22:1571–1572.

Zolotov, M. Yu. 1994*a*. Phase relations in the Fe-Ti-Mg-O oxide system and hematite stability at the condition of Venus' surface. *Lunar Planet. Sci. Conf.* XXV:1571–1572 (abstract).

Zolotov, M. Yu. 1994*b*. Near-surface atmosphere of Venus: New estimations of redox conditions based on new data. *Lunar Planet. Sci. Conf.* XXV:1569–1570 (abstract).

Zolotov, M. Yu. 1995. A model of thermochemical equilibrium in the near-surface atmosphere of Venus. *Geokhimiya* 11:1551–1569 (in Russian).

PART V
Geologic Structure

PHYSIOGRAPHY, GEOMORPHIC/GEOLOGIC MAPPING, AND STRATIGRAPHY OF VENUS

KENNETH L. TANAKA
U. S. Geological Survey

DAVID A. SENSKE
Jet Propulsion Laboratory

MARIBETH PRICE
South Dakota School of Mines and Technology

and

RANDOLPH L. KIRK
U. S. Geological Survey

The physiography of Venus has been unveiled by radar imaging and altimetry measurements from Pioneer Venus, Arecibo, Venera 15 and 16, and particularly Magellan. These data reveal a geologically complex surface and a unimodal hypsometry in which the majority of the elevations lie within a km of the mean planetary radius. Important geographic features seen in Magellan images now have names, but some refinement in the planet's nomenclature is yet needed. Most of the planet (70%) is made up of lowland and intermediate-elevation plains, whereas much of the highlands comprise two continental-size regions (Ishtar and Aphrodite Terrae) made up mostly of tesserae. The topography, geomorphology, geology, and radar characteristics of plains, tesserae, volcanic rises, volcanoes, coronae, rift valleys, ridge belts, fracture belts, mountain belts, lava channels, craters, and dunes and yardangs have been characterized, which assists in their interpretation. Geomorphic/geologic mapping at intermediate resolution has further distinguished many of the rock materials, terrains, and structures of Venus. Superposition relations among geologic materials and features determined from mapping and from crater densities have been used to reconstruct the planet's recent geologic history. The geologic record appears to be spread across a few hundred million years, and follows the general sequence: (1) tesserae, (2) lowland smooth plains (channeled in places), (3) ridge and fracture belts, coronae and coronalike features, volcanic flow fields and large edifices, and mountain belts, and (4) surficial aeolian materials. Much uncertainty remains among the details of this history, because relative age relations are not yet fully documented and crater data can only offer general indications. Ongoing, systematic geologic mapping at the full Magellan resolution should greatly improve our present knowledge of the geologic history of Venus.

I. INTRODUCTION

Beginning with the altimetry and radar imagery provided by Pioneer Venus and Venera 15 and 16 spacecraft (supplemented by Earth-based Arecibo radar mapping), cartographers and scientists have been mapping the terrains and landforms of Venus for the purposes of physiographic characterization and geologic interpretation (see, e.g., Masursky et al. 1980; Schaber 1982; Barsukov et al. 1986; Senske 1990). Ongoing geologic mapping from Magellan image mosaics and associated crater-density studies define the stratigraphy at local and regional scales and the volcanic and tectonic evolution of Venus. In this chapter we first describe the physiography (including nomenclature and hypsometry) of Venus; second, we review the findings of mapping studies completed thus far; and finally we discuss stratigraphic inferences produced by geologic mapping and crater-density data. Although the physiography of Venus is relatively well known, geologic mapping studies are still in their early stages, resulting in only a broad-scale understanding of planetwide stratigraphy at present.

II. PHYSIOGRAPHY

Physiography, in its broadest sense, is the descriptive study of a planet's surface. In this section, we deal mainly with the history and current status of the physical description of Venusian surface features, including their geographic nomenclature, broad-scale topography, and small-scale geomorphology. Other aspects of the altimetry data and genetic interpretation of the landforms are discussed in other chapters on surface processes, volcanism, tectonism, impact cratering, and geodynamics.

A. Nomenclature

On the basis of the Goldstone radar images and Pioneer Venus topographic maps, Venus geographic nomenclature began with the discrimination of mostly broad-scale topographic and reflectivity features (Masursky et al. 1980; U. S. Geological Survey 1981,1984). More identification and naming of features was extended to the part of the northern hemisphere imaged by Venera (Kotelnikov et al. 1989; U. S. Geological Survey 1989) and then for the remainder of the planet based on Magellan data (Russell 1994). Owing to this variability in data coverage, Venusian nomenclature inadvertently has acquired some inconsistencies (which are discussed in the section on terrains and landforms). Moreover, systematic naming of features using the Magellan data set is still in progress; equatorial and southern latitudes particularly need further work. Finally, the distinction between geographic and geologic/geomorphic terminology needs to be remembered, for in notable cases the two approaches do not correspond well (see Table I).

The International Astronomical Union governs the naming of Venusian features, including defining rules, setting policy, and approving proposed

TABLE I
Feature Types Approved for Use on Venus[a]

Features[b]	Number[c]	Description (common geologic interpretation)	Source of Name
Chasma, chasmata	29	Canyon (rift or corona chain)	Goddesses of hunt; Moon goddesses
Collis, colles	5	Small hills or knobs (small volcanoes, etc.)	Miscellaneous goddesses
Corona, coronae	157	Raised, ovoid feature (tectonic structure)	Fertility and Earth goddesses
Crater, craters	535	Bowl-shaped depression (impact)	Famous women; <20 km, female first names
Dorsum, dorsa	38	Ridge	Sky goddesses
Farrum, farra	8	Pancake-like structure (volcano)	Water goddesses
Fluctus, fluctūs	9	Flow terrain (lava flow)	Goddesses, miscellaneous
Fossa, fossae	14	Long, narrow, shallow depression (graben)	Goddesses of war
Linea, lineae	11	Elongate marking (graben)	Goddesses of war
Mons, montes	45	Mountain (large volcano, mountain belt)	Goddesses, miscellaneous
Patera, paterae	44	Shallow crater; scalloped, complex edge (caldera, large volcano)	Famous women
Planitia, planitiae	20	Low plain (lava plain)	Mythological heroines
Planum, plana	1	Plateau or high plain (lava plain)	Goddesses of prosperity
Regio, regiones	17	Region (tessera, volcanic rise, large volcano)	Giantesses, Titanesses
Rupes, rupēs	6	Scarp	Goddesses of hearth and home
Terra, terrae	3	Extensive land mass (highland, upland plain)	Goddesses of love
Tessera, tesserae	17	Tile; polygonal ground	Goddesses of fate or fortune
Tholus, tholi	10	Small domical mountain or hill (volcano)	Goddesses, miscellaneous
Undae	3	Dunes (dunes, yardangs)	Desert goddesses
Vallis, valles	16	Valley (lava channel)	Word for planet Venus in various languages; river goddesses for smaller valles

[a] Table modified from Russell (1994).
[b] Singular and plural forms given.
[c] Formally approved by the International Astronomical Union as of November 1995.

names. The most obvious rule is the feminine theme (with few exceptions). Depending on feature type, names are taken from well-known yet relatively noncontroversial deceased individuals or mythical figures as well as common feminine first names. Planetologists can propose names for various features as needed (for guidelines, see Russell [1994], or Tanaka [1994, Appendix C]). Twenty feature types have been approved for Venus, and 988 features thus far have been named (Table I). These include regional-scale features such as planitiae, plana, regiones, terrae, and tesserae (Fig. 1) as well as many local features (particularly craters and coronae). Their characteristics as they relate to geomorphologic features are described in more detail below.

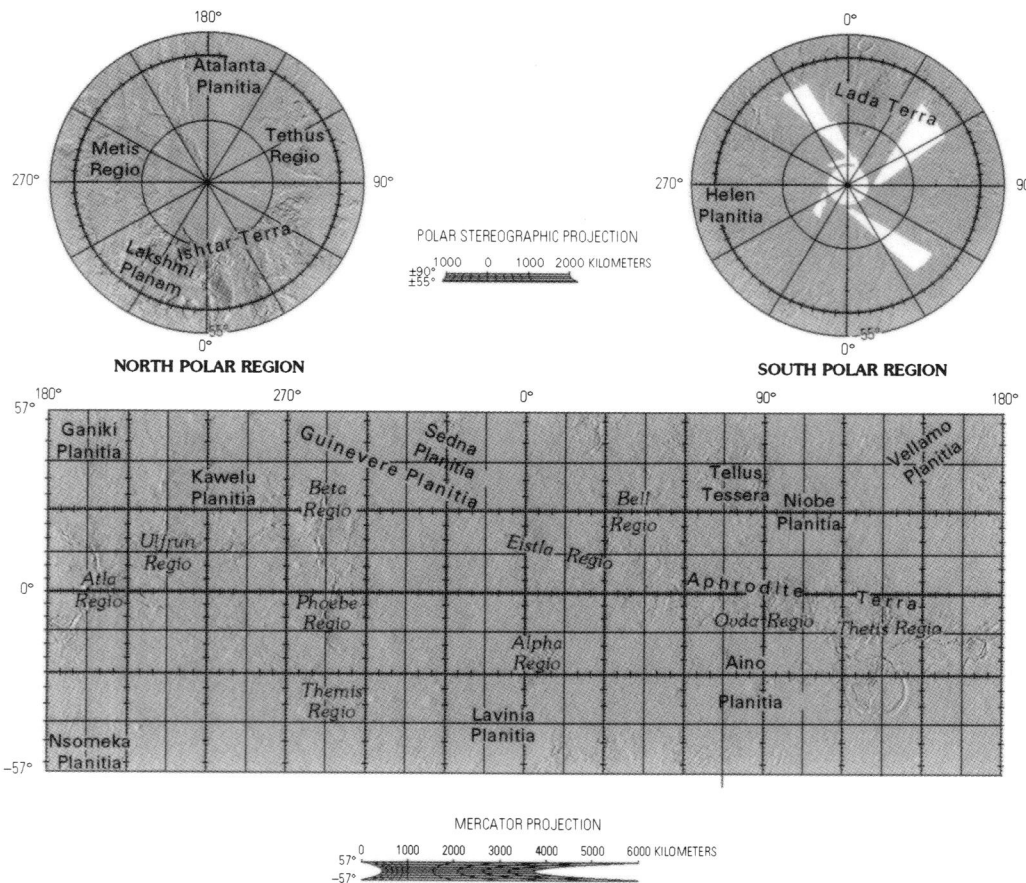

Figure 1. Global digital shaded relief map showing nomenclature of major geographic features; shown in Mercator and polar projections; scale 1:300,000,000.

B. Altimetry and Hypsometry

Pioneer Venus provided a global topographic perspective of the planet at a few tens of km spatial resolution and 200 m vertical accuracy between 74°N and 63°S latitudes (Masursky et al. 1980; Pettengill et al. 1980). On the basis of the values of mean, median, and modal planetary radii determined by Pioneer Venus, 6051.0 km was chosen arbitrarily as the cartographic datum used on topographic maps of Venus (Masursky et al. 1980). Next, Venera 15 and 16 altimetry filled in elevation information for the north polar region at spatial resolution and vertical accuracy similar to those of Pioneer Venus (Kotelnikov et al. 1989). Magellan altimetry provided the latest, most refined topography of Venus (vertical resolution obtained is ~100 m and horizontal resolution varies but at best is about 10 km) (Pettengill et al. 1991; Ford and Pettengill 1992). The resolution yields fine-scale topography, including steep scarps with slopes that exceed 30° and low-relief landforms (such as steep-sided domes, grabens, valleys, and impact craters). In addition, Magellan more precisely defined the broad topography of the south polar region (which includes the southern uplands known as Lada Terra) and the mean planetary radius at 6051.84 km.

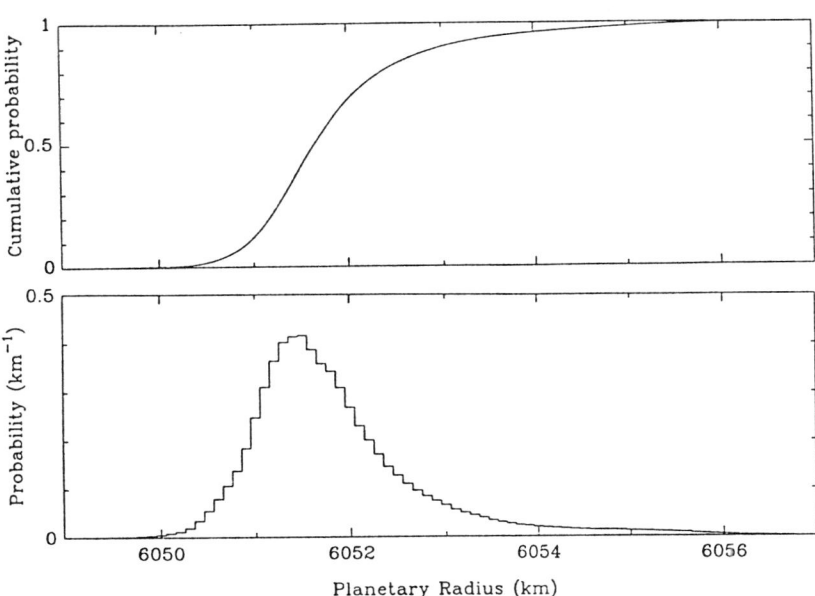

Figure 2.   Hypsometric histogram of Venus (modified from Ford and Pettengill 1992).

Unlike the bimodal hypsometry of the Earth, Venus has a unimodal elevation distribution (Fig. 2); >90% of the surface has an elevation between −1.0 and 2.5 km (Ford and Pettengill 1992). The surface can be divided into lowlands (27% of the surface; 0 to 2 km below the planetary datum of 6051.0 km), upland rolling plains (65%; 0 to 2 km elevation), and highlands (8%; 2 to 11 km elevation) (Masursky et al. 1980); other topographic divisions have been used as well (see, e.g., McGill et al. 1983). The lowland plains appear generally smooth at radar wavelengths (i.e., low radar backscatter), whereas the upland plains have generally intermediate and diverse backscatter caused by somewhat rougher surfaces at radar wavelengths and a plethora of small-scale landforms (scarps, ridges, troughs, hills, etc.). Rugged highland features distinctly rise above surrounding plains and include the continental-size Ishtar and Aphrodite Terrae and Alpha and Beta Regiones (Fig. 1). Maxwell Montes in Ishtar Terra, the highest area on the planet, rises to 12 km over planetary datum. The lowest known area, Diana Chasma, is only about 2 km below datum.

C. Visualization

Advanced scientific visualization techniques are an invaluable aid in describing the complex interplay between hypsometry, local relief, surface textures, and physical properties that characterize and define landforms (see, e.g., Kirk et al. 1992). Magellan data are of sufficiently high resolution and quality for useful application of such techniques. The most important technique is perspective rendering, in which an image of a planetary surface and a co-registered topographic model are used to create an artificial view of the surface from an arbitrary position (and usually with exaggerated relief). The interpretability of perspective views can be greatly increased by incorporating and exaggerating as many of the following natural depth cues as possible:

1. motion (i.e., viewing sequential images as a movie);
2. stereopsis (by rendering image pairs from nearby vantage points);
3. texture gradients (by enhancing fine image details);
4. shading (by adding exaggerated relief shading calculated from the topography);
5. perspective convergence (defined by the geometry of the edges of the image).

Of these, the first two are by far the most powerful but of course the least applicable to the printed page.

Independent of the rendering of perspective views, color may be used to encode surface properties (such as elevation or reflectivity), and this color can then be merged with the monochrome image. As Kirk et al. (1992) point out, this is a "natural" use of the visual faculties, in that we are accustomed to associating color with surface properties. In addition, the low resolution of Magellan physical-properties data (5 km/pixel) compared to that of the image (generally 75 or 225 m/pixel) is appropriate to the lower acuity of color (as opposed to intensity) discrimination. Using color to encode relief information is considerably less natural, and does not lead to the immediate apprehension of surface shape as with perspective views. It can nevertheless be useful because the relations between elevation and surface details can be presented on a global scale. Our images included in the accompanying CD-ROM show Magellan (and pre-Magellan) radar image and altimetry data of Venus that have been merged in this way. Note that the color scale involves hue variations only; the spectrum has been adjusted in accord with the hypsometric curve for Venus so that all colors are equally represented spatially and their discriminability is maximized. For more detailed information on visualization and Magellan-based image products, consult the CD-ROM.

D. Geomorphology

The Magellan synthetic aperture radar (SAR) data, as well as previous radar measurements, have been used to characterize the geomorphology of Venus. In addition, radar brightness, emissivity, and other properties aid in distinguishing variations in surface texture and composition related to elevation and terrain type. The following discussion describes the physical characteristics of major terrains (plains, highlands, and tesserae) and landforms on Venus. Because this book focuses on scientific results, we present this discussion in terms of types of geomorphic features (related types of geographic features used in the nomenclature system, where different from the commonly used geomorphic term, are noted in italics). Also, see other chapters for Magellan images of these features.

1. Plains. Plains form broad, relatively smooth, flat surfaces that cover about 70% of the planet. They typically lie within 1 to 3 km of planetary datum and span as much as several thousand km (Guinevere Planitia extends some 7,500 km). Plains below the datum are generally radar dark (smooth at the 12.6 cm wavelength) and are referred to as lowland plains or *planitiae*. Parts of Atalanta Planitia, the deepest plain, dip below -1.5 km elevation. Most lowland plains have low roughness at centimeter to meter scales consistent with the smooth, slabby surfaces seen in Venera lander images (Campbell and Campbell 1992; Guest et al. 1992; Basilevsky et al. 1985). Some of the individual plains adjoin one another to form larger, composite plains (U. S. Geological Survey 1984).

Upland or rolling plains, whose elevations are above datum, also extend for great distances and lie above the lowland plains and below the highlands, tesserae, and volcanic rises. These plains tend to have somewhat higher radar backscatter and root mean square (rms) slopes than their lowland counterparts (see, e.g., Ford and Pettengill 1992). Lada Terra forms a more pronounced upland plain 0.5 to 1.5 km above datum and about 6,500 km across, and it includes extensive coronae and fracture belts (Baer et al. 1994). The unique highland plateau Lakshmi Planum occurs at 3 to 4 km elevation within western Ishtar Terra (discussed further below).

Mottled plains appear to be made up of lobate flows of varying radar backscatter (Arvidson et al. 1992); rarely, discernible scarps demarcate flow boundaries (e.g., Ovda Fluctus). (Some of the more pronounced flow fields have been named *fluctūs*.) Commonly, the more distinctive plains-forming lava flows have recognizable source areas (see, e.g., Roberts et al. 1992), whereas in other areas sources cannot be identified and even a volcanic origin for the plains-forming material, while generally assumed, cannot be firmly established (alternatively, some plains material could be sedimentary). In addition, the Venusian plains, though primarily characterized by their flat surfaces, are marked by various relatively small geomorphic features, including channels, irregular networks of valleys, domes, small shields, fractures, grabens, and wrinkle ridges (see, e.g., Baker et al. 1992; Banerdt and Sammis 1992; Guest et al. 1992; Head et al. 1992; Johnson and Sandwell 1992) and are interrupted by larger features, such as coronae, large volcanoes, ridge belts, and rift valleys (see below).

2. Tessera Highlands. Ishtar and Aphrodite Terrae form continental-size highlands ranging from 1 to >5 km above datum. These highlands comprise mostly tesserae (described below) and some volcanic features and are complexly deformed by many tectonic structures.

Ishtar Terra, about 5600 km across from east to west and extending as much as 2000 km from north to south, includes the highest mountain belts on the planet. Western Ishtar Terra consists of Lakshmi Planum surrounded by high-standing Akna, Freyja, Danu, and Maxwell Montes (4 to 11 km elevation), which in turn are surrounded by lower tesserae and plains. Lakshmi Planum hosts two flat-floored, irregular volcanic depressions, Colette (80×120 km) and Sacajawea (140×280 km) Paterae. In contrast, eastern Ishtar Terra is made up of 3- to 5-km-high Fortuna Tessera and is transected by several rift valleys.

Aphrodite Terra, 1 to 5 km high, is strung out across one-third of the planet's equatorial region, or about 10,000 km. The overall shape of Aphrodite resembles a scorpion (Masursky et al. 1980). The western part consists of high ridges (or "claws") that bound a broad, circular area of low-lying tessera that is about 2000 km in diameter. Two large central plateaus, Ovda and Thetis Regiones (about 4000 and 3000 km across, respectively, and 3 to 4 km high), consist mainly of tessera. The eastern part of Aphrodite reaches an elevation of only 1 to 1.5 km and includes a long string of rift valleys, fracture belts, and coronae that joins the large Beta/Atla/Themis triangle of fracture belts and large volcanoes.

3. Other Tesserae. The term "tessera" was first applied to patterned ridged terrain revealed by Venera images in the northern high latitudes of Venus (see, e.g., Basilevsky et al. 1986; Sukhanov et al. 1986,1989). Because in many places tessera does not have a tile-like appearance, the term "complex ridged terrain" (or CRT) has also become popular (Bindschadler et al. 1992*a,b*; Solomon et al. 1992). Meanwhile, the usage of the term tessera has also been expanded to include the same variety of terrain; for consistency, we will restrict ourselves to that term. Bindschadler and Head (1991) identified three morphologic types of tessera: (1) linear ridged terrain (subparallel ridges and troughs and cross-strike lineations), (2) trough and ridge terrain (subparallel troughs and orthogonal ridges and troughs), and (3) disrupted terrain (chaotic patterns of relatively short structures). Of these, disrupted terrain is most common and trough and ridge terrain is rare. Magellan images show the complex structural fabric; specifically, the presence of tightly packed, overlapping and crosscutting broad ($\sim$10–20 km wide) linear and arcuate ridges; narrow (<3 km wide) ridges; and linear scarps, troughs, and grabens (as much as several km wide) (Bindschadler et al. 1992a). Recently, Hansen and Willis (1996) have described even more varieties of tessera. Tesserae are also rough at radar wavelength to meter scales, resulting in high radar backscatter and high rms slope values (see, e.g., Masursky et al. 1980; Bindschadler et al. 1990; Ford and Pettengill 1992).

Tesserae, dispersed planetwide, protrude from surrounding plains as dozens of large areas hundreds to a couple thousands of km across (including the highland regions of Ishtar and Aphrodite Terrae described above) as well as hundreds of smaller patches (see Ivanov and Head 1996). In equatorial and southern latitudes, large areas of tesserae are labeled *regiones* (e.g., Bell, Ovda, Thetis, and Phoebe) because of their strong radar brightness as observed in early radar datasets, whereas high northern latitude regions are shown as *tessera* (e.g., Fortuna, Laima, and Tellus) because of their geomorphology as revealed by Venera 15 and 16 images. Tesserae cover about 10% of the northern quarter of Venus and about 8 to 9% of the entire planet (Bindschadler and Head 1989; Price and Suppe 1994; Ivanov and Head 1996).

4. Volcanic Rises. Domical highs, known as volcanic rises, in equatorial regions of Venus extend more than a thousand km across and rise 1 to 3 km above surrounding plains (Stofan et al. 1989; Bindschadler et al. 1992a,b; Senske et al. 1992; Campbell and Rogers 1994; McGill 1994). The rises include the most prominent volcanoes on Venus, which sit astride or on the flanks of rift valleys, fracture belts, and numerous coronae. The volcanic edifices typically have diameters in the range of 100 to 200 km and rise to elevations of 1 to 9 km above planetary datum (e.g., Theia Mons in Beta Regio, Maat Mons in Atla Regio, and Sif Mons in western Eistla Regio; Schaber 1991). Although the volcanic rises are morphologically similar, they vary in their detailed geologic characteristics. A general classification shows that they can be distinguished as rift-dominated (Beta and Atla Regiones), volcano-dominated (Dione, Bell, western Eistla, and Imdr Regiones), and corona-dominated (eastern and central Eistla Regiones) (Stofan et al. 1995). In addition, stratigraphic relations indicate that radial fracturing often predated large-scale lava emplacement, suggesting that volcanism was associated with the uplift that formed the regional topographic rises (Senske et al. 1992).

5. Volcanoes. Thousands of volcanoes, from kilometer-size (smallest resolvable) vents to broad shields hundreds of km across, have been identified on Venus. Their morphologies, distribution, and geologic associations are described in greater detail in the chapters on volcanism (see also Head et al. 1992; Crumpler et al. 1993).

Thus far, 167 large volcanoes (>100 km across; a few exceed 700 km) on Venus have been cataloged (chapter by Crumpler et al.). (Many of the most striking and largest volcanoes are named *mons* and a few, *regio* or *patera*.) Typically, digitate lava flows radiate from their edifices and form broad circular to elliptical aprons, except where surrounding topography constrains flow direction. The flows range from radar bright to dark, indicating the presence of various surface textures (i.e., rough to smooth at radar wavelength). The summits may be cut by calderas and dotted by parasitic vents, and summits and flanks commonly are cut by fractures where they overlie a rift zone or fracture belt. Other large volcanoes form in close association with fracture centers and coronae.

Intermediate-size volcanoes (generally 20 to 100 km in diameter) include both smaller (or partly buried) versions of the large volcanoes and some distinctive steep-sided types having domical or pancake-like forms. (Some of the domical volcanoes are named *tholus*, and the pancake ones are named *farrum*.) Some steep-sided volcanoes have scalloped edges. Together, nearly 300 intermediate volcanoes have been cataloged planetwide.

The smallest volcanoes mostly form in groups of moderate to dense concentrations (tens to hundreds of volcanoes total and several volcanoes per thousand square km) known commonly as shield fields. In detail, they vary in shape and size, and many have discernible summit pits. Most have a shield profile, but others are domical, conical, or flat-topped. Shield fields may be surrounded by their own flows, associated with large volcanoes or tectonic features, or isolated as patches in plains material. More than 600 shield fields have been identified (in only a few cases, they have been named *colles*).

6. *Coronae and Related Features.* The term "corona" covers a diverse group of features that have been clearly recognized only on Venus, although similar structures may also be present on Mars and Earth (Watters and Janes 1995). Coronae were initially characterized using Pioneer Venus, Arecibo, Goldstone, and Venera 15 and 16 data (Barsukov et al. 1986; Basilevsky et al. 1986; Stofan and Head 1990), but Magellan provided the opportunity to compile a detailed global inventory (Squyres et al. 1992a; Stofan et al. 1992). Contributions from all of these studies are included in the following summary.

Commonly, coronae are circular to elliptical features characterized by an annulus of tectonic deformation, a domical or plateau-type topography with a surrounding moat, and moderate to extensive volcanism. Topographic signatures of coronae are highly variable, however, and include both plateaus with interior or rim depressions and those with simple depressions. The annulus is usually composed of concentric or en echelon fractures, but compressional folds (ridges) also occur. Asymmetric or angular coronae are common, and multiple coronae also can form as two or more linked structures having a common annulus. Other types of volcanic features, such as calderas with a central depression and concentric faulting, may exhibit characteristics of coronae; establishing a clear-cut demarcation between them is often difficult. Radially fractured domes occur both within and near coronae and possibly represent an early stage of corona formation.

Coronae range in diameter from 60 to 1060 km, with a median value of ~200 km (Stofan et al. 1992). These statistics exclude the anomalously large Artemis Corona, a 2600-km-diameter structure with a surrounding chasma and ridge belt that is one of the most prominent features on the planet (and may have a different origin from other coronae). Many coronae and related features are intimately associated with rift zones, forming linear chains connected by throughgoing extensional faults and graben. In other places, coronae occur individually or in clusters in the plains. The hypsometric distribution of coronae is unimodal but peaks at higher elevations than does the global hypsometry, with a small shoulder on the upper side due to Artemis.

7. Rift Valleys (Chasmata). Rifts on Venus, interpreted to be sites of moderate crustal extension, lie almost exclusively within the equatorial region and southern latitudes of the planet; they extend eastward from Aphrodite Terra to Beta Regio and southeast from Atla Regio to Themis Regio (Fig. 1; Masursky et al. 1980; Schaber 1982). On the basis of map patterns and topographic characteristics, two principal classes of rifts are identified: simple rifts and corona chains (Senske and Head 1992). Simple rifts are made up of single, continuous, linear troughs having lengths of thousands of km and widths of 60 to 200 km. These depressions are bounded by steep scarps and reach depths of 1 to 4 km; faulting is restricted primarily within the trough. The outer flanks of the rifts are typically elevated and may reach as much as 2 km above the surrounding plains. Simple rifts associated with regional volcanic rises (Beta, Atla, western Eistla, and Imdr Regiones) commonly radiate from large edifices (e.g., Theia Mons in Beta Regio and Ozza Mons in Atla Regio). Volcanism is concentrated at the edifices and at isolated locations on the rift flanks, with the troughs being generally free of infilling.

Corona chains (Dali, Hecate, and Parga Chasmata) range from 500 to more than 1000 km wide, are made up of coronae (diameters of 200 to 600 km) connected by zones of fracturing and faulting, and are surrounded by interconnected arcuate troughs bounded by major scarps. In numerous areas, volcanic deposits are localized around the rim of the coronae (lava flow lengths of 50–100 km) and, as with simple rifts, these deposits do not provide significant infilling of the troughs (Roberts et al. 1992). Along the rim of several coronae are arch-like ridges interpreted to be compressional. Corona chains are found exclusively within the plains where they link the volcanic rises to each other and connect the large region of tessera at Aphrodite to the volcanic center at Atla Regio (Senske et al. 1992).

8. Ridge Belts (Dorsa). Ridges form prominent radar-bright belts that lie within the Venusian plains and along some margins of the larger tessera blocks. From Venera images, Kryuchkov (1990) identified three types of ridge belts, and Frank and Head (1990) defined five ridge morphologies. The first type of ridge belt consists of broad, arcuate to sinuous ridges. In the northern hemisphere, these ridge belts are 200 to 800 km long, 20 to 40 km across, and as much as 1.5 km high; the broad ridges are superposed by narrow ridges (e.g., Lukelong Dorsa). Magellan confirmed the presence of prominent areas with ridge belts of this type in the southern hemisphere, Helen and Lavinia Planitiae (Campbell et al. 1991; Senske et al. 1991b). Here, the ridge belts are smaller (a few km wide) and commonly bifurcate and rejoin; an estimate of their relief is 200 to 300 m (Squyres et al. 1992b). The second type of ridge belt consists of long (mean length 43 km), closely spaced narrow ridges (mean spacing 8.5 km). Overall the belts extend from 500 to several thousand km in length and 70 to 300 km in width and have as much as 1.0 to 1.5 km of relief (U. S. Geological Survey 1989; Squyres et al. 1992b). Commonly, the ridges trend around elliptical depressions within these belts. The longest belts extend for thousands of km toward the north pole across Vellamo, Vinmara, and Snegurochka Planitiae. The third type consists of narrow (mostly <1 km wide), low sinuous ridges having lengths of a few tens of km and spacing of a few to about 20 km or more apart. Magellan images demonstrate that these ridges resemble the wrinkle ridges common on the lunar mare and the high plains of Mars (see, e.g., Watters 1988; Squyres et al. 1992b).

9. Fracture Belts (Lineae). Fracture belts occur throughout much of the plains regions (e.g., in Lavinia Planitia) and in eastern Aphrodite Terra. Fracture belts generally consist of fractures and grabens having common orientations and patterns; linear, sinuous, and arcuate patterns are common. In comparison with ridge belts, fracture belts have similar lengths but commonly have larger widths. Magellan altimetry revealed that most fracture belts specifically form on elevated terrain; belt intersections and bends commonly form the highest parts (Solomon et al. 1992; Squyres et al. 1992b). On a broader scale, fracture belts in many places connect and entwine other fracture belts as well as ridge belts, coronae, rift valleys, large scarps, and large volcanoes (Solomon et al. 1992). Long, narrow fractures, grabens, and troughs that show no distinctive associated topographic signature occur in parallel sets across many broad areas of plains; some of these sets have been named "fossae."

10. Mountain Belts. Within Ishtar Terra, Lakshmi Planum forms an elevated volcanic plateau that is surrounded by Akna, Freyja, Danu, and Maxwell Montes. These unique mountain belts are made up of parallel sets of ridges and troughs arrayed around the margin of the central part of the Ishtar highland. The belts range in width from 50 km at Danu Montes to over 400 km at Maxwell; sets of ridges typically spaced between 3 km and 15 km form broad arches and anastomosing patterns. The highest elevations and some of the steepest slopes on the planet are found at the mountain belts—Freyja Montes rises more than 6 km above planetary datum, whereas Maxwell sits at an elevation of 12 km. Slopes are found to be as steep as 35° over distances of several km.

11. Channels (Valles). Channels comprise a relatively rare yet striking volcanic feature occurring mostly in plains settings (see Baker et al. 1992).

More than 200 large channels and channel segments have been documented to have simple (sinuous rilles and long canali), complex (anastomosing, braided, and branching), and integrated (perhaps formed by sapping) and compound patterns. Channel lengths are remarkable; Baltis Vallis, the longest canali, extends 6800 km and Kallisto Vallis, an outflow-channel complex, reaches 1200 km. The outflow channels include fluvial erosional landforms indicative of exotic lava compositions. The longest channels mostly occur in the lowland plains.

12. Craters. Impact craters on Venus share similarities with their counterparts on other bodies (e.g., circular form, elevated rim, central depression, and surrounding ejecta blanket). Nevertheless, the ensemble of Venusian craters is unusual in many respects. Here we describe the attributes of the craters themselves. Major papers focusing on various issues with regard to cratering on Venus include Phillips et al. (1991,1992), Schaber et al. (1992). Herrick and Phillips (1994), Herrick et al. (1995), and Strom et al. (1994). For discussion of diffuse radar-dark and radar-bright textural (i.e., non-physiographic) features that appear related to impact phenomena, see chapters on impact cratering and surficial processes.

Fewer than 1000 craters have been identified in the Magellan images, which cover about 98% of Venus. The size-frequency distribution for large craters is similar to that on other inner solar-system bodies, though with a low total density, but smaller craters (diameter $D<35$ km) are increasingly deficient as a consequence of atmospheric shielding. (The implications of the distribution of craters for the planet's resurfacing history is discussed in a later section.) The largest crater on Venus, Mead, has a diameter of 270 km; the smallest craters are ~ 1.5 km in diameter.

Morphologies of larger craters are generally similar to those on other silicate bodies: multi-ring basins for $D>100$ km, double-ring (peak-ring) craters at intermediate diameters, and craters with central peaks in the 20 to 50 km diameter range. The floors of all these craters, exclusive of the ring or peak structures, are generally flat and may be radar bright (in $\sim 1/3$ of larger craters) or dark. Smaller craters lack central peaks or rings but do not exhibit the bowl shape common on other bodies. Craters with structureless floors dominate the 10- to 20-km size range but also include many of the smallest craters. Irregularly shaped craters and clusters of multiple, overlapping small craters are increasingly common at smaller diameters, again as a consequence of atmospheric disruption of the impactors. Depth-diameter relations have been assessed by a combination of altimetry (only the largest craters were fully resolved by the Magellan altimeter) and parallax measurements (Schaber et al. 1992). Venusian crater depths follow a trend similar to that for other silicate bodies but are generally intermediate between those for the Earth and those for the Moon and Mars. Craters with radar-dark floors appear shallower than "fresh" craters with bright floors, either as a result of volcanic flooding within the crater or simultaneous isostatic relaxation and weathering of the bright deposit.

Ejecta blankets (and the floors of small craters) are radar bright due to the presence of wavelength scale and larger facets, and typically show textural variations, particularly in the proximal zone around larger craters. Ejecta margins range from generally diffuse in smaller craters to lobate in larger ones. Many craters have ejecta blankets with a marked bilateral ("butterfly") symmetry, including ejecta-free sectors, that results from oblique impact. Radar-bright outflows of shock-melted target rock are associated with many of the largest craters, particularly those with butterfly ejecta. These flows are extremely thin but reach lengths of 600 km. Flow morphology is highly variable and includes both erosional and depositional features (see Asimow and Wood 1992; Schultz 1992; Chadwick and Schaber 1993; Johnson and Baker 1994).

13. Dunes (Undae) and Yardangs. Greeley et al. (1992) have identified two large dune fields (>100 km across) on Venus—Al-Uzza Undae (between Ishtar Terra and Meshkenet Tessera) and Menat Undae (on the outflow deposit of Aglaonice crater)—and one possible yardang field (225 km across and 300 km southeast of Mead crater). Dunes range in length from 0.5 to 10 km and have widths and spacings of a few hundred meters. The yardanglike features are about 25 km long, 0.5 km wide, and 0.5 to 2 km apart and may occur in deposits of Mead crater. The steep surfaces of the dunes and yardangs appear bright in Magellan SAR images where favorably oriented for quasi-specular radar returns.

III. GEOMORPHIC/GEOLOGIC MAPPING

Global-scale imaging, at various resolutions, has been acquired for several of the terrestrial planets and for the outer planet satellites from a variety of mainly space-based platforms. The construction of geologic maps from these data, when interpreted in conjunction with crater statistical data, provide a basis to assess the processes and sequence of events by which the surfaces of these bodies formed and evolved. For Venus, the progressive improvement in data quality and interpretation is paralleled by mapping that has progressed from primarily geomorphic (landforms) to largely geologic (rock units).

A. Venera 15 and 16, Pioneer Venus, and Arecibo-Based Mapping

1. Soviet Maps (1986–1989). The first geomorphic/geologic mapping of Venus began with the 27 Venera radar mosaics of the northern quarter of Venus at 1:5,000,000 scale, supplemented by Pioneer Venus and Arecibo radar data (Kotelnikov et al. 1989). Of the 27 quadrangles in this series, all or parts of 16 quadrangles were published as figures in English-translated journal articles (for references, see Sukhanov et al. 1989). This work included definition, mapping, and description of various terrain types (e.g., smooth and hilly plains and tessera) and rock units (e.g., lava flows), structures (e.g., coronae and calderas), and topographic features (scarps and ridges). In addition, geologic relations (including superposition) were noted among many of the major features and units, leading to interpretations of geologic histories.

These quadrangle maps, along with unpublished maps, were used to compile a synoptic map of the northern part of Venus covered by Venera radar mosaics at 1:15,000,000 scale in polar stereographic projection (Sukhanov et al. 1989). The summary map consists of 13 terrain units divided into the categories of volcanic constructs, plains units, and rough-terrain units (including tesserae, coronae, and ridge belts); 9 topographic and geomorphic structure types (e.g., ridges, arachnoids, and impact craters); and outlines of radar-bright areas. Map units were distinguished purely on the basis of physiographic expression, and their relative elevations were shown.

2. Schaber and Kozak (1990). Another 1:15,000,000-scale, geomorphic/geologic map of the same northern region of Venus was produced by Schaber and Kozak (1990), using the same data as the Soviet mappers. Their mapping is notable for the level of detail, including 34 units divided into plains units, tesserae, coronae, domed terrains, marginal belts, lineated and fractured terrains, landforms interpreted as volcanic, ridge terrains, and other miscellaneous units. Furthermore, they mapped nine types of structures and geomorphic features. In a follow-up report (Schaber 1990), additional data were provided on each unit's area, elevation, rms slope, reflectivity, and impact-crater density, as well as detailed statistics of crater distribution as a function of longitude, latitude, and elevation (however, the identification of many craters was uncertain because of resolution).

3. Senske et al. (1991a,b). During the inferior conjunction of 1988, radar observations of Venus were performed from the Arecibo Observatory (Campbell et al. 1989,1991). These data, obtained at a resolution comparable to that achieved from orbit by the Venera 15 and 16 spacecraft, extended medium- to high-resolution coverage (0.75 to 1.0 km) into the equatorial region and southern high latitudes. Eleven geomorphic/geologic units were identified on the basis of radar textural properties (Senske et al. 1991*a,b*). In many places the Arecibo data facilitated the identification of local stratigraphic relations; however, age relations between many global-scale units were typically difficult to determine. Additional analyses to quantify the backscatter of individual units were performed using Pioneer Venus SAR data (Ford and Senske 1990), which provided a basis to characterize better geologic units mapped from radar data.

B. Magellan-Based Mapping

1. Feature-Based Mapping. An alternate mapping approach delineates types of tectonic and volcanic features, which in many instances is similar to the geologic/geomorphic approach. Magellan-based catalogs of individual volcanic features (Head et al. 1992), coronae and corona-like features (Stofan et al. 1992), and impact craters (Herrick and Phillips 1994; Schaber et al. 1992,1995) served as the basis for later global maps that outlined those features (Namiki and Solomon 1994; Price and Suppe 1994,1995). The Magellan data have also contributed to improved global maps of structural and tectonic features. Schaber (1982) presented one of the earliest maps of the equatorial rift system, and Crumpler et al. (1993) showed the distribution of fracture belts. Suppe and Connors (1992) compiled a listing of fold belts on Venus and a preliminary fold belt map which was incorporated into the global map of Price and Suppe (1995). Other publications showing the distribution of features include maps of wrinkle ridges (Bilotti et al. 1993), mafic dike swarms (Grosfils and Head 1994), and tessera terrain (Ivanov et al. 1992; Price and Suppe 1995; Ivanov and Head 1996).

2. Geomorphic/Geologic Unit Mapping. At the conclusion of the Magellan mission's three imaging cycles (one cycle is equal to 243 days or one complete rotation of Venus on its axis), more than 98% of the surface had been viewed at much higher resolution ($\sim$75 m/pixel) than previously seen. These data confirmed that the global geology is complex and that the overall density of impact craters is low (Saunders et al. 1991; Phillips et al. 1991; Schaber et al. 1992; Herrick and Phillips 1994; Strom et al. 1994). Geomorphic/geologic mapping was carried out to identify 16 global-scale units and assess their stratigraphy (Senske et al. 1996). As in previous studies, map units were defined on the basis of variations in radar backscatter, surface textures, topography, and morphology (at this scale, map units do not precisely define rock strata; see summary of unit properties in Table II). Additionally, superposition and cross-cutting relations provide relative age information (see Sec. IV on stratigraphy).

3. GIS Mapping. One useful development in Venus mapping has been the use of a geographic information system (GIS) to map and analyze the distribution of features such as volcanoes, coronae, and impact craters. Price and Suppe (1995) have created an extensive GIS map and data base for Venus that includes impact craters, volcanoes, coronae, corona-like features, rifts, tesserae, shield fields, fold belts, wrinkle ridge trends, and four plains units based on the distinctiveness of flow morphology (Fig. 3). The GIS data were compiled by mapping directly on digital Magellan C1-MIDRs (225 m/pixel), incorporating tabulations of geologic features, and digitizing existing maps. Maps of each feature type are stored digitally in separate layers using a common coordinate system, and corresponding information about individual structures, such as area, name, morphologic type, etc., are maintained in a data base linked to the map.

The conventional geomorphic/geologic map of Senske et al. (1996), which also is based on C1-MIDRs, has also been digitized in a raster GIS format (see CD-ROM). Digital mapping techniques hold great promise for fully exploiting the extensive data sets generated on planetary mapping missions such as Magellan. GIS offers both mapping and quantitative spatial analysis tools (e.g., area calculations and relations among mapped features and units and other datasets), which so far have been used: (1) to investigate and model the distribution of impacts within different map features and the hypsometric characteristics of features and craters (Price and Suppe 1995);

TABLE II
Areas, Elevations, Properties, and Crater Statistics of Venusian Geologic/Geomorphic Units[a]

Unit	Area (10⁶ km²)	% of Map Area	Mean Elevation (km)	Mean rms slope (deg)	Mean σ^0 (db)[b]	No. of Craters	Relative Crater Density	Standard Error of Density
Volcanic edifices	18.05	4.03	1.28	5.91	2.23	29.2	0.78	0.28
Shield fields	3.49	0.78	0.68	5.39	1.22	3.2	0.44	0.49
Σ volcanics	21.54	4.81				32.4	0.72	0.25
Ridge belts	11.31	2.53	0.62	7.12	2.57	21.2	0.90	0.39
Mountain belts	0.76	0.17	6.05	9.35	9.31	4.0	2.53	2.53
Σ belts	12.07	2.70				25.2	1.00	0.39
Fracture belts	42.90	9.59	1.43	7.23	5.55	65.8	0.74	0.18
Bright plains	9.48	2.12	0.37	5.89	2.01	28.9	1.47	0.54
Digitate plains	23.39	5.23	1.02	5.42	2.00	27.2	0.56	0.21
Dark plains	0.34	0.08	0.42	3.55	−1.50	1.0	1.40	2.80
Mottled plains	49.34	11.02	0.62	5.37	2.19	103.8	1.01	0.19
Reticulate plains	229.70	51.03	0.56	4.93	0.51	531.0	1.11	0.06
Lineated plains	13.48	3.01	0.69	5.63	1.88	30.6	1.09	0.39
Σ plains	325.71	72.79				722.5	1.07	0.04
Complex ridged terrain	39.09	8.74	2.02	8.30	5.59	71.8	0.88	0.20
Ridged/fractured terrain	6.18	1.38	0.95	7.03	3.71	13.3	1.03	0.56
Σ ridged terrain	45.27	10.12				85.1	0.90	0.19

[a] This table is based on the global geologic map of Venus compiled by Senske et al. (1996) at 1:50,000,000 scale from maps generated by the Magellan Science Team (see accompanying CD-ROM). The map covers a total of 447.5×10⁶ km², or slightly over 97% of Venus. It was digitized by Kirk and Schaber (1996) at a resolution of 2.5 km/pixel, and areas of crater material and impact-related diffused deposits were re-mapped base on C1-MIDRs to show the geologic units underlying the impact deposits. Elevation, rms slope, and backscatter σ^0 averages for each unit were calculated from the Magellan GTDR, GSDR, and MIDR data sets, respectively. Elevation is with respect to the 6051-km datum; global mean is 0.84 km (Ford and Pettengill 1992). Backscatter is in decibels relative to Muhleman's (1964) function describing globally averaged backscatter properties. Crater counts and densities relative to the global mean ($\sim 2.0 \times 10^{-6}$ km^{-2}) are also shown for each unit and group of units.

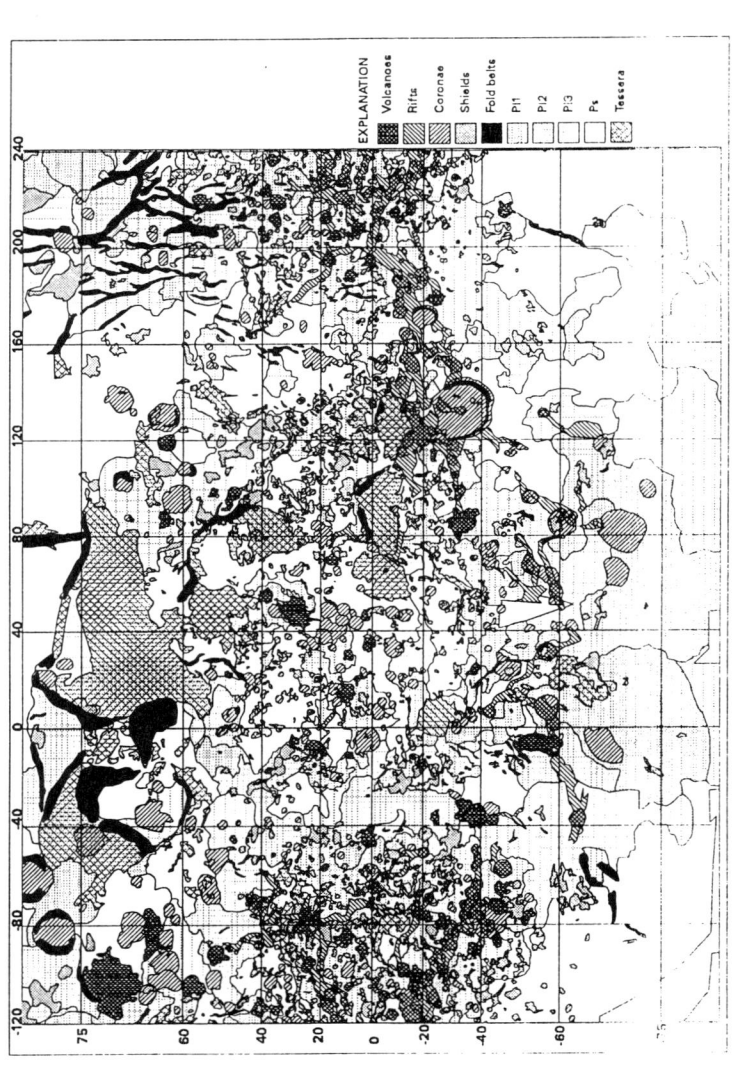

Figure 3. Feature-based geomorphic/geologic map of Venus (from Price 1995a). The features were mapped at C1-scale from digital cycle 1 to 3 Magellan images using a GIS. The features include large volcanoes, coronae and corona-like features, rift zones, fold belts, fields of small shield volcanoes, tesserae, and four plains units (highly lobate plains [Pl], moderately lobate plains [Pl2], slightly lobate plains [Pl3], and non-lobate plains). The legend lists the units in order of increasing mean age, based on crater density, although the actual stratigraphic reltationships are far more complex. Contributors include M. Price, P. Coburn, F. Bilotti, C. Connors, and E. Gibbons.

(2) to determine impact-crater densities and crater-based ages for different map units (Price and Suppe 1994; Kirk and Schaber 1996); and (3) to study lineament trends in Fortuna Tessera (Riley et al. 1995).

4. 1:5,000,000-Scale Geologic Mapping Program. Beginning in 1993, systematic geologic mapping of 1:5,000,000-scale quadrangles of Venus began (see Tanaka 1994). Unlike previous Magellan mapping, which largely relied on lower-resolution image mosaics, this new program takes advantage of the full-resolution images mosaics at 75 m/pixel (the FMAP series). The mapping philosophy includes portraying mainly rock units (bodies of material) and thus avoiding, where possible, geomorphic units (areas characterized by distinctive surface features); also, radar physical properties of map units will be documented in a consistent manner (Campbell 1995).

However, even at this scale, several significant uncertainties in the mapping should be noted, including: (1) the thickness of rock units is generally unknown; (2) transitions in morphology (or other properties), even where abrupt, do not necessarily indicate different rock units; and (3) overlap and embayment relations, which may clearly define superposition at a given locality, may not necessarily apply to contacts between the same units many km away.

IV. STRATIGRAPHY

Stratigraphy deals with the study of rock strata, particularly their characteristics, distribution, and sequence. As such, stratigraphy provides the primary information for the interpretation of rock formational processes and geologic history. In this section, we focus on the emerging geologic history of Venus (as limited by the extant rock record) as documented primarily by three types of approaches. First, local superposition relations produce reliable geologic sequences that collectively can be used to infer global stratigraphy. Second, types of volcanic and tectonic features can be mapped and their mean ages determined by crater density. Third, broad geomorphic/geologic unit mapping can elucidate general superposition relations and provide another basis for crater-density determinations. A summary and intercomparison of the most comprehensive study to date of each approach is shown in Fig. 4.

A. Local Stratigraphies

The assessment of the stratigraphy of Venus by Basilevsky and Head (1995) was compiled from mapping 36 sites covering 1000-km-wide squares ($\sim$8% of the planet; the sites were chosen based on the presence of young impact craters having dark parabolic haloes) distributed across the planet. At these sites, they mapped and documented the relative ages of 16 major units, which include various rock and surficial materials as well as tectonic features and terrains. For any given site, 5 to 11 units were present (tessera and plains units being most common). Basilevsky and Head found that the sequences of formation of these units were sufficiently consistent to propose a global

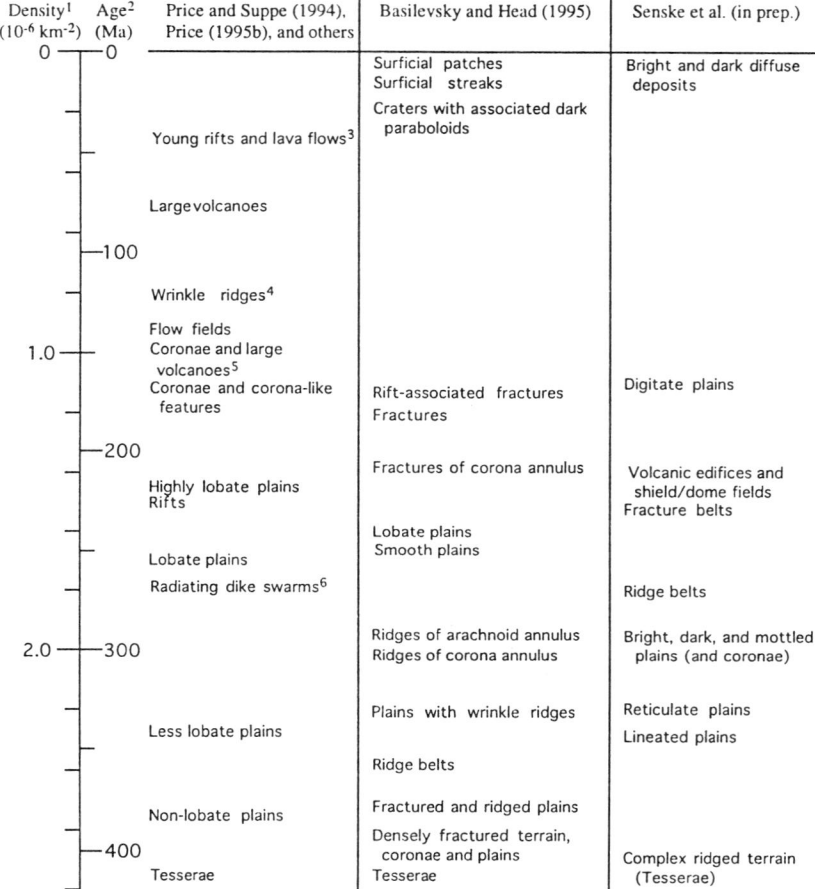

Figure 4. Relative and absolute ages for Venusian features based on crater densities and superposition relations.

stratigraphy; their sequence is shown in Fig. 4. As noted by Basilevsky and Head (1995), the detailed relative ages of many intermediate-age structures, including younger tessera structure, coronae, ridge belts, mountain belts, and rifts, are still highly uncertain and require more detailed investigation.

Other studies (many based on ongoing geologic mapping at 1:5,000,000 scale) provide some examples of other local stratigraphic relations and their complexity. Moore et al. (1992) inferred relative ages of the volcanic flows of Mahuea Tholus and the surrounding plains and crater materials using superposition relations. McGill (1993) showed that older sets of wrinkle ridges are more continuous than cross-cutting younger sets; he also found that, in central Eistla Regio, coronae formed soon after widespread emplacement of plains material, followed by the large volcanoes Sappho and Anala Paterae (McGill 1994). Baer et al. (1994) documented the close spatial and temporal relations of coronae, fracture belts, and flow fields of northern Lada Terra. Near Phoebe Regio, Chapman (1995) documented an age progression of arachnoids. Aubele (1995) discovered that the oldest plains unit in the Vellamo Planitia region is marked by shield fields. Conversely, in nearby Vinmara Planitia, shield fields are interpreted to be relatively young, and ridge-belt formation is coeval with plains emplacement (Rosenberg 1995).

B. Crater Densities

Impact-crater densities have been routinely used to define relative ages of geologic units on the Moon and Mars (see, e.g., Tanaka 1986; Wilhelms 1987). For Venus, the low number of craters ($\sim$938; G. G. Schaber, personal communication) limits the application of this technique (Herrick and Phillips 1994; Schaber et al. 1992,1995; Strom et al. 1994,1995; Herrick et al. 1995). Although Venusian craters are spatially uniform on a global scale, both Monte Carlo simulations and arguments based on approximate but well-studied statistical distributions show that significant density and, therefore, age differences among geologic units on Venus can exist (Price and Suppe 1994; Price 1995b; Price et al. 1996). The scarcity of craters requires that the units to be dated must contain at least 8 craters (Price et al. 1996). Most individual features on Venus, such as large volcanoes, coronae, and rifts, are too small to contain enough craters, so they must be grouped into global-scale map units to acquire reliable dates.

With these caveats in mind, we present the following general history of Venus based on impact-crater densities of major types of geologic features. The impact-based ages agree with the general sequence of events determined by local and regional stratigraphic studies (McGill 1994; Basilevsky and Head 1995). They also provide information on the duration and rate of resurfacing by different features, because crater densities can be used to estimate absolute age based on theoretical impact rates. All unit ages reported in the remainder of this section are ratios with respect to the global crater density of $T = 1.98 \times 10^{-6}$ craters km^{-2}, corresponding to a mean surface age for which the best current estimate is 288 ($-98,+311$) Myr (Strom et al. 1994).

Crater counts appear to delineate four main periods, as follows: (1) The tesserae are the oldest materials with an age of 1.93 to 1.01 T (560–290 Myr; Ivanov and Basilevsky 1993), once the density has been corrected for an observed deficit of small impact craters. (2) The plains have a mean age of 1.1 ± 0.1 T (320 Myr), but subunits within the plains, discriminated by the degree of preservation of flow morphology, indicate a protracted emplacement spanning 1.3 ± 0.2 to 0.7 ± 0.3 T (370–200 Myr) (Price 1995b). The oldest plains are preserved in the planitiae and in regions where tesserae are concentrated, whereas giant radiating graben swarms (interpreted to be formed in association with intrusive dikes) have a similar age to younger plains units (0.9–0.8 T or 260–230 Myr; Grosfils and Head 1995). (3) Other tectonic and volcanic features superimposed on the plains have crater ages substantially younger than the plains—about 0.4±0.3 T (120 Myr) for the rifts and coronae and 0.3±0.2 T (90 Myr) for the large volcanoes. Namiki and Solomon (1994) report ages of 0.5 T (140 Myr) for the coronae and large volcanoes using slightly different populations of features. These features also have higher proportions of faulted and embayed craters (20–50%) than the plains (∼6%). Wrinkle ridges surrounding Aphrodite Terra have a similar age (0.4±0.2 T or 120 Myr; Bilotti and Suppe 1994). (4) Impacts with parabolic haloes, considered to be the youngest 10% (0.1 T or 30 Myr) of the crater population (Campbell et al. 1992), have been used to establish that the age of rifting and volcanism in several regions may be less than 50 Myr (Basilevsky 1993; Ghail 1995). Such young ages suggest that Venus is still geologically active.

On the basis of the calculated mean ages, the resurfacing history of Venus shows a substantial decrease in area resurfacing rates (Fig. 5), from about 4 km^2 yr^{-1} during plains emplacement to about 0.5 km^2 yr^{-1} over the past 1.0 T (290 Myr). This reduction has been accompanied by a shift in volcanic style from flood-type volcanism in the plains to more localized development of volcanoes, coronae, and rifts. This shift may have lead to greater resurfacing in the uplands, but the lowland plains have seen little tectonic or volcanic activity (although a crater deficit exists for the highlands; Kirk et al. [1995] noted that it consists of only about six missing craters smaller than 8 km). The total area resurfaced by volcanism or tectonism since plains emplacement has been variously estimated at 20% (14% volcanic, 6% tectonic) (Price and Suppe 1995), 22% volcanic (Namiki and Solomon 1994), 30% volcanic and tectonic (Herrick 1994), and 3 to 19% volcanic (Strom et al. 1994,1995).

C. Global Mapping

General age relations have been identified among the global map units of Senske et al. (1996), but in cases where units do not touch each other, the stratigraphy is not always clear. Stratigraphic relations indicate that the earliest rock exposures on Venus consist of highly deformed regions of tessera (complex ridged terrain). Next, widely distributed and isolated outcrops of lineated plains were emplaced. The distinct truncation of lineaments at the contact with reticulate plains provides evidence that a second, later episode of regional plains formation occurred. Owing to the absence of identifiable individual flows, these later reticulate plains are thought to be associated with widespread lava flooding. More recent magmatic activity associated with bright, digitate, dark, and mottled plains is linked to individual vents, fractures, and coronae (Guest et al. 1992). Other sites of recent volcanism correspond to large edifices and shield/dome fields. In a number of areas, ridge belts and fracture belts deform older plains, corresponding to some of

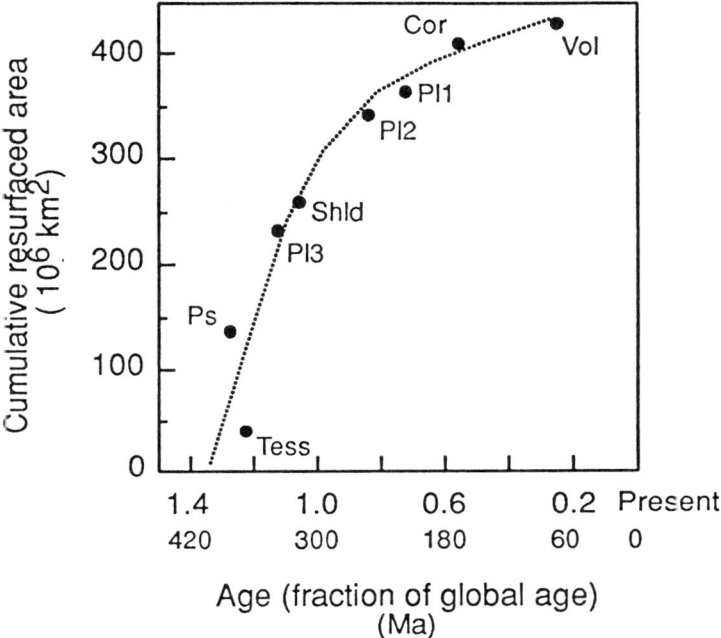

Figure 5. Resurfacing history of Venus (based on the mapped areas and crater–density dates for each map unit of Price and Suppe [1995], and Price [1995b]). The cumulative volcanically resurfaced area is plotted against age. On the basis of this graph, the area resurfacing rate has decreased from 4.1 km^{2}yr^{-1} during plains emplacement to 0.5 km^{2}yr^{-1} in the more recent past. The units include: Tess = tessera, Ps = nonlobate plains, Pl3 = subtly lobate plains, Shld = shield fields, Pl2 = lobate plains, Pl1 = strongly lobate plains, Cor = coronae and corona-like features, and Vol = large volcanoes.

the most recent tectonic activity. Locally, rifting and volcanism at large edifices in Beta and Atla Regiones appear to be contemporaneous with volcanic deposits both filling the rift and cut by faulting. The timing between regional topographic uplift and rifting is typically less clear. Finally, surficial units make up some of the youngest materials.

Crater densities determined for the map units by Kirk and Schaber (1996) are summarized in Table II and Fig. 3. This work differs from that of Price and others, because it relies on an independent crater data base (Schaber et al. [1992,1995], rather than Herrick and Phillips [1994]), geologic map (Senske et al. 1996), and method of allocating craters by automatically counting pixels of the map inside each crater. Two-sigma error bars were derived as in Price and Suppe (1994). The only significant discrepancy from the results of Price and Suppe is for volcanic edifices—the relative crater density is 0.78 T (225 Myr; Kirk and Schaber) vs 0.25 T (72 Myr; Price and Suppe) and may be due in part to the exclusion of embayed craters by Price and Suppe

Schaber emphasize that, because of the rather large error bars, it is important not to over interpret subtle differences in crater density). (The crater density on coronae could not be calculated in this study because they are mapped as structures rather than areal units.)

Acknowledgments. The authors wish to thank A. T. Basilevsky (Vernadsky Institute), L. S. Crumpler (Brown University), and H. J. Moore and G. G. Schaber (U. S. Geological Survey) for their thoughtful and constructive reviews. J. Blue (U. S. Geological Survey) assisted in and checked for accuracy our presentation of Venusian nomenclature.

Editor's Note

Tanaka et al. have provided an animation on the accompanying CD-ROM. Please see CDP5C1M1.

REFERENCES

Arvidson, R. E., et al. 1992. Surface modification of Venus as inferred from Magellan observations of plains. *J. Geophys. Res.* 97:13303–13318.

Asimow, P. D., and Wood, J. A. 1992. Fluid outflows from Venus impact craters: Analysis from Magellan data. *J. Geophys. Res.* 97:13643–13666.

Aubele, J. C. 1995. Stratigraphy of small volcanoes and plains terrain in Vellamo Planitia-Shimti Tessera region, Venus. *Lunar Planet. Sci. Conf.* XXVI:59–60 (abstract).

Baer, G., Schubert, G., Bindschadler, D. L., and Stofan, E. R. 1994. Spatial and temporal relations between coronae and extensional belts, northern Lada Terra, Venus. *J. Geophys. Res.* 99:8355–8369.

Baker, V. R., et al. 1992. Channels and valleys on Venus: Preliminary analysis of Magellan data. *J. Geophys. Res.* 97:13421–13444.

Banerdt, W. B., and Sammis, C. G. 1992. Small-scale fracture patterns on the volcanic plains of Venus. *J. Geophys. Res.* 97:16149–16166.

Barsukov, V. L., et al. 1986. The geology and geomorphology of the Venus surface as revealed by the radar images obtained by Veneras 15 and 16. *J. Geophys. Res.* 91:378–398.

Basilevsky, A. T. 1993. Age of rifting and associated volcanism in Atla Regio, Venus. *Geophys. Res. Lett.* 20:883–886.

Basilevsky, A. T., and Head, J. W. 1995. Global stratigraphy of Venus: Analysis of a random sample of thirty-six test areas. *Earth, Moon, Planets* 66:285–336.

Basilevsky, A. T., et al. 1985. The surface of Venus as revealed by the Venera landings. *Geol. Soc. Amer. Bull.* 96:137–144.

Basilevsky, A. T., et al. 1986. Styles of tectonic deformation on Venus: Analysis of Venera 15 and Venera 16 data. *Proc. Lunar Planet. Sci. Conf.* 16, *J. Geophys. Res. Suppl.* 91:D399–D411.

Bilotti, F., and Suppe, J. 1994. Determining ages of wrinkle ridge deformation on Venus using the relative dating of craters and faults. *GSA Abst/w Programs*

26:A-264 (abstract).

Bilotti, F., Connors, C., and Suppe, J. 1993. Global distribution of wrinkle ridges on Venus: Relationship to long-wavelength topography and gravity. *Eos: Trans. AGU* 74:191.

Bindschadler, D. L., and Head, J. W. 1989. Characterization of Venera 15/16 geologic units from Pioneer Venus reflectivity and roughness data. *Icarus* 77:3–20.

Bindschadler, D. L., and Head, J. W. 1991. Tessera terrain, Venus: Characterization and models for origin and evolution. *J. Geophys. Res.* 96:5889–5907.

Bindschadler, D. L., et al. 1990. Distribution of tessera terrain on Venus: Prediction for Magellan. *Geophys. Res. Lett.* 17:171–174.

Bindschadler, D. L., et al. 1992a. Magellan observations of Alpha Regio: Implications for formation of complex ridged terrains on Venus. *J. Geophys. Res.* 97:13563–13577.

Bindschadler, D., Schubert, G., and Kaula, W. 1992b. Coldspots and hotspots: Global tectonics and mantle dynamics of Venus. *J. Geophys. Res.* 97:13495–13532.

Campbell, B. A. 1995. *Use and Presentation of Magellan Quantitative Data in Venus Mapping*, U. S. G. S. Open-File Rept. 95-519.

Campbell, B. A., and Campbell, D. B. 1992. Analysis of volcanic surface morphology on Venus from comparison of Arecibo, Magellan, and terrestrial airborne radar data. *J. Geophys. Res.* 97:16293–16314.

Campbell, B. A., and Rogers, P. G. 1994. Venus: Integration of remote sensing data and terrestrial analogs for geologic analysis. *J. Geophys. Res.* 99:21153–21171.

Campbell, D. B., et al. 1989. Styles of volcanism on Venus: New Arecibo high resolution radar data. *Science* 246:373–377.

Campbell, D. B., et al. 1991. Venus southern hemisphere: Geologic characteristics and age of major terrains in the Themis-Alpha-Lada region. *Science* 251:180–183.

Campbell, D. B., et al. 1992. Magellan observations of extended impact crater related features on the surface of Venus. *J. Geophys. Res.* 97:16249–16278.

Chadwick, D. J., and Schaber, G. G. 1993. Impact crater outflows on Venus: Morphology and emplacement mechanisms. *J. Geophys. Res.* 98:20891–20902.

Chapman M. G. 1995. Relations between plate movement and Pheobe Regio tessera? *Lunar Planet. Sci. Conf.* XXVI:233 (abstract).

Crumpler, L. S., Head, J. W., and Aubele, J. C. 1993. Relation of major volcanic center concentration on Venus to global tectonic patterns. *Science* 261:591–595.

Ford, P. G., and Pettengill, G. H. 1992. Venus topography and kilometer-scale slopes. *J. Geophys. Res.* 97:13103–13114.

Ford, P. G., and Senske, D. A. 1990. The radar scattering characteristics of Venus landforms. *Geophys. Res. Lett.* 17:1361–1364.

Frank, S. L., and Head, J. W. 1990. Ridge belts on Venus: Morphology and origin. *Earth, Moon, Planets* 50/51:421–470.

Ghail, R. C. 1995. Dali Vinculum: Rifting, coronae, and subduction. *Lunar Planet. Sci. Conf.* XXVI:457–458 (abstract).

Greeley, R., et al. 1992. Aeolian features on Venus: Preliminary Magellan results: *J. Geophys. Res.* 97:13319–13346.

Grosfils, E. B., and Head, J. W. 1994. The global distribution of giant radiating dike swarms on Venus: Implications for the global stress state. *Geophys. Res. Lett.* 21:701–704.

Grosfils, E. B., and Head, J. W. 1995. Giant radiating dike swarms on Venus: Stratigraphic constraints upon their time of emplacement. *Lunar Planet. Sci. Conf.* XXVI:523–524 (abstract).

Guest, J. E., et al. 1992. Small volcanic edifices and volcanism in the plains of Venus: *J. Geophys. Res.* 97:15949–15966.

Hansen, V. L., and Willis, J. J. 1996. Structural analysis of a sampling of tesserae:

Implications for Venus geodynamics. *Icarus*, in press.
Head, J. W., et al. 1992. Venus volcanism: Classification of volcanic features and structures, associations, and global distribution from Magellan data. *J. Geophys. Res.* 97:13153–13197.
Herrick, R. R. 1994. Resurfacing history of Venus. *Geology* 22:703–706.
Herrick, R. R., and Phillips, R. J. 1994. Implications of a global survey of Venusian impact craters: *Icarus* 111:387–416.
Herrick, R. R., Izenberg, N., and Phillips, R. J. 1995. Comment on "The global resurfacing of Venus" by R. G. Strom, G. G. Schaber, and D. D. Dawson. *J. Geophys. Res.* 100:23355–23359.
Ivanov, M. A., and Basilevsky, A. T. 1993. Density and morphology and impact craters on tessera terrain, Venus. *Geophys. Res. Lett.* 20:2579–2582.
Ivanov, M. A., and Head, J. W. 1996. Tessera terrain on Venus: A survey of the global distribution, characteristics, and relation to surrounding units from Magellan data. *J. Geophys. Res.* 101:14861–14908.
Ivanov, M. A., Tormanen, T. and Head, J. W. 1992. Global distribution of tesserae: Analysis of Magellan data. *Lunar Planet. Sci. Conf.* XXIII:581–582 (abstract).
Johnson, C. L., and Sandwell, D. T. 1992. Joints in Venusian lava flows. *J. Geophys. Res.* 97:13601–13610.
Johnson, J. R., and Baker, V. 1994. Surface property variations in Venusian fluidized ejecta blanket craters. *Icarus* 110:33–70.
Kirk, R. L., and Schaber, G. G. 1996. Spatial distribution, physical properties, and geologic context of Venusian impact craters. In preparation.
Kirk, R. L., Soderblom, L. A., and Lee, E. M., 1992. Enhanced visualization for the interpretation of Magellan radar data: Supplement to the Magellan special issue. *J. Geophys. Res.* 97:16371–16380.
Kirk, R. L., Schaber, G. G., and Strom, R. G. 1995. New statistical results on the spatial distribution and physical properties of impact craters on Venus: *Lunar Planet. Sci. Conf.* XXVI:757–758 (abstract).
Kotelnikov, V. A., Yashchenko, V. R., Zolotov, A. F., and 14 others. 1989. *Atlas of Venus Surface* (Moscow: Main Dept. of Geodesy and Cartography of the U.S.S.R. Council of Ministers), in Russian.
Kryuchkov, V. P. 1990. Ridge belts: Are they compressional or extensiona structures? *Earth, Moon, Planets* 50/51:471–491.
Masursky, H., et al. 1980. Pioneer Venus radar results: Geology from images and altimetry. *J. Geophys. Res.* 85:8232–8260.
McGill, G. E. 1993. Wrinkle ridges, stress domains, and kinematics of Venusian plains. *Geophys. Res. Lett.* 20:2407–2410.
McGill, G. E. 1994. Hotspot evolution and Venusian tectonic style. *J. Geophys. Res.* 99:23149–23162.
McGill, G. E., et al. 1983. Topography, surface properties, and tectonic evolution. In *Venus* eds. D. M. Hunten, L. Colin, T. M. Donahue and V. I. Moroz (Tucson: Univ. of Arizona Press), pp. 69–130.
Moore, H. J., Plaut, J. J., Schenk, P. M., and Head, J. W. 1992. An unusual volcano on Venus. *J. Geophys. Res.* 97:13479–13493.
Muhleman, D. O. 1964. Radar scattering from Venus and the Moon. *Astron. J.* 69:34–41.
Namiki, N., and Solomon, S. C. 1994. Impact crater densities on volcanoes and coronae on Venus: Implications for volcanic resurfacing. *Science* 265:929–933.
Pettengill, G. H., et al. 1980. Pioneer Venus radar results: Altimetry and surface properties. *J. Geophys. Res.* 85:8261–8270.
Pettengill, G. H., et al. 1991. Magellan: Radar performance and data products. *Science* 252:260–265.

Phillips, R. J., et al. 1991. Impact craters on Venus: Initial analysis from Magellan. *Science* 252:288–297.
Phillips, R. J., et al. 1992. Impact craters and Venus resurfacing history. *J. Geophys. Res.* 97:15923–15948.
Price, M. 1995*a*. Dating Resurfacing on Venus Using Impact Crater Densities from GIS-based Global Mapping. Ph.D. Thesis, Princeton University.
Price, M. 1995*b*. Resurfacing history of the Venusian plains based on the distribution of impact craters. *Lunar Planet. Sci. Conf.* XXVI:1143–1144 (abstract).
Price, M., and Suppe, J. 1994. Mean age of rifting and volcanism on Venus deduced from impact crater densities. *Nature* 372:756–759.
Price, M., and Suppe, J. 1995. Constraints on the resurfacing history of Venus from the hypsometry and distribution of volcanism, tectonism, and impact craters. *Earth, Moon, Planets* 71:99–145.
Price, M. H., Watson G. S., Suppe, J., and Brankman, C. 1996. Dating volcanism and rifting on Venus using impact crater densities. *J. Geophys. Res.* 101:4657–4672.
Riley, K. M., Anderson, R. C., and Peer, B. J. 1995. Lineament analysis of Fortuna Tessera, Venus: Results from an ongoing study. *Lunar Planet. Sci. Conf.* XXVI:1171–1172 (abstract).
Roberts, K. M., Guest, J. E., Head, J. W., and Lancaster, M. G. 1992. Mylitta Fluctūs, Venus: Rift-related, centralized volcanism and the emplacement of large-volume flow units. *J. Geophys. Res.* 97:15991–16016.
Rosenberg, L. 1995. Areal geology of the Pandrosos Dorsa quadrangle (V5), Venus. *Lunar Planet. Sci. Conf.* XXVI:1185–1186 (abstract).
Russell, J. F. 1994. *Gazetteer of Venusian Nomenclature*, U. S. G. S. Open-File Rept. 94-235.
Saunders, R. S., et al. 1991. An overview of Venus geology. *Science* 252:249–252.
Schaber, G. G. 1982. Venus: Limited extension and volcanism along zones of lithospheric weakness: *Geophys. Res. Lett.* 9:499–502.
Schaber, G. G. 1990. *Venus: Quantitative Analyses of Terrain Units Identified from Venera 15/16 Data and Described in Open-File Report 90-24*, U. S. G. S. Open-File Rept. 90-468.
Schaber, G. G. 1991. Volcanism on Venus as inferred from the morphometry of large shields. *Proc. Lunar Planet. Sci.* 21:3–11.
Schaber, G. G., and Kozak, R. C. 1990. *Geologic/Geomorphic and Structure Maps of the Northern Quarter of Venus*, U.S.G.S. Open-File Rept. 90-24, 1:15 000,000 scale.
Schaber, G. G., Kirk, R. L., and Strom, R. G., 1995. Update on the USGS impact crater database for Venus: *Lunar Planet. Sci. Conf.* XXVI:1227–1228 (abstract).
Schaber, G. G., et al. 1992. Geology and distribution of impact craters on Venus: What are they telling us? *J. Geophys. Res.* 97:13257–13301.
Schultz, P. H. 1992. Atmospheric effects on ejecta emplacement and crater formation on Venus from Magellan. *J. Geophys. Res.* 97:116183–16248.
Senske, D. A. 1990. Geology of the Venus equatorial region from Pioneer Venus radar imaging. *Earth, Moon, Planets* 50/51:305–327.
Senske, D. A., and Head, J. W. 1992. Zones of extension and rifting on Venus: Characteristics and distribution. *Lunar Planet. Sci. Conf.* XXIII:1269–1270 (abstract).
Senske, D. A., et al. 1991*a*. Geology and tectonics of Beta Regio, Guinevere Planitia, Sedna Planitia, and Western Eistla Regio, Venus: Results from Arecibo image data. *Earth, Moon, Planets* 55:163–214.
Senske, D. A., et al. 1991*b*. Geology and tectonics of the Themis Regio-Lavinia Planitia-Alpha Regio-Lada Terra area, Venus: Results from Arecibo image data. *Earth, Moon, Planets* 55:97–161.

Senske, D. A., Schaber, G. G., and Stofan, E. R. 1992. Regional topographic rises on Venus: Geology of western Eistla Regio and comparison to Beta Regio and Atla Regio. *J. Geophys. Res.* 97:13395–13420.

Senske, D. A., Saunders, R. S., and Stofan, E. R., and Members of the Magellan Science Team. 1996. The global geology of Venus: Characterization of units and geologic history. In preparation.

Solomon, S. C., et al. 1992. Venus tectonics: An overview of Magellan observations. *J. Geophys. Res.* 97:13199–13256.

Squyres, S. W., et al. 1992a. The morphology and evolution of coronae on Venus. *J. Geophys. Res.* 97:13611–13634.

Squyres, S. W., et al. 1992b. Plains tectonism on Venus: The deformation belts of Lavinia Planitia. *J. Geophys. Res.* 97:13359–13599.

Stofan, E. R., and Head, J. W. 1990. Coronae of Mnemosyne Regio, Venus: Morphology and origin. *Icarus* 83:216–243.

Stofan, E. R., et al. 1989. Geology of a rift zone on Venus: Beta Regio and Devana Chasma. *Geol. Soc. Amer. Bull.* 101:143–156.

Stofan, E. R., et al. 1992. Global distribution and characteristics of coronae and related features on Venus. Implications for origin and relation to mantle processes. *J. Geophys. Res.* 97:13347–13378.

Stofan, E. R., et al. 1995. Large topographic rises on Venus: Implications for mantle upwelling. *J. Geophys. Res.* 100:23317–23327.

Strom, R. G., Schaber, G. G., and Dawson, D. D. 1994. The global resurfacing of Venus. *J. Geophys. Res.* 99:10899–10926.

Strom, R. G., Schaber, G. G., Dawson, D. D., and Kirk, R. L. 1995. Reply (to Comment on "The global resurfacing of Venus" by R. G. Strom, G. G. Schaber, and D. D. Dawson). *J. Geophys. Res.* 100:361–365.

Sukhanov, A. L., et al. 1986. A geologic and morphologic description of Ishtar Terra (a photographic map of the surface of Venus, plate V-5). *Solar System Res.* 20:64–71.

Sukhanov, A. L., et al. 1989. *Geomorphic/Geologic Map of Part of the Northern Hemisphere of Venus*, U.S.G.S. Misc. Inv. Series Map I-2059, scale 1:15,000,000.

Suppe, J., and Connors, C. 1992. Critical taper wedge mechanics of fold-and-thrust belts on Venus. *J. Geophys. Res.* 97:13545–13561.

Tanaka, K. L. 1986. The stratigraphy of Mars. *Proc. Lunar Planet. Sci. Conf.*, 17, *J. Geophys. Res. Suppl.* 91:E139–E158.

Tanaka, K. L., 1994. *The Venus Geologic Mappers' Handbook*, U. S. G. S. Open-File Rept. 94-438.

U. S. Geological Survey. 1981. *Altimetric and Shaded Relief Map of Venus*, U.S.G.S. Map I-1364, scale 1:50,000,000.

U. S. Geological Survey. 1984. *Topographic and Shaded Relief Maps of Venus*, U.S.G.S. Map I-1562, scale 1:50,000,000.

U. S. Geological Survey. 1989. *Maps of Part of the Northern Hemisphere of Venus*, U.S.G.S. Map I-2041, scale 1:15,000,000.

Watters, T. R. 1988. Wrinkle ridge assemblages on the terrestrial planets. *J. Geophys. Res.* 93:10236–10254.

Watters, T. R., and Janes, D. M. 1995, Coronae on Venus and Mars: Implications for similar structures on Earth. *Geology* 23:200–204.

Wilhelms, D. E. 1987. *The Geologic History of the Moon*, U.S.G.S. Prof. Paper 1348 (Washington, D. C.: U. S. Government Printing Office).

PART VI
Volcanism

VOLCANOES AND CENTERS OF VOLCANISM ON VENUS

LARRY S. CRUMPLER and JAYNE C. AUBELE
Brown University

DAVID A. SENSKE
Jet Propulsion Laboratory

SUSAN T. KEDDIE
Science Applications International Corporation

KARI P. MAGEE
Jet Propulsion Laboratory

and

JAMES W. HEAD
Brown University

Magmatic centers on Venus include those characterized dominantly by surface volcanism and those that are characterized dominantly by structural features associated with the emplacement and transport of magmas at depth. Those characterized by surface volcanism are referred to here as "centers of volcanism" and include constructional volcanic edifices, clusters of small volcanic edifices, centers from which significant volumes of magma have erupted (flow fields), and calderas. There are 1194 volcanic centers larger than 20 km, or 2.6 volcanic centers per million square km, globally. More than one-half of the population of centers occur in <30% of the global area (Beta-Atla-Themis region). Broad mantle upwelling in the Beta-Atla-Themis region and enhanced shallow magma reservoir formation at upland elevations may together account for this hemispheric concentration of volcanoes. Globally, 167 large volcanoes (volcanic edifices >100 km in diameter) have been identified and mapped and can be classified into 9 fundamental types based on their geologic characteristics. Two contrasting styles of volcanism are recognized: individual large centers (volcanic edifices and calderas) and fields of small volcanoes. The difference between a single large edifice or voluminous eruption and a group of small edifices reduces to the conditions necessary for shallow magma reservoir formation. Thermal conditions necessary for formation of a stable shallow magma reservoir on Venus predict that a field of volcanoes is characterized by magma replenishment rates of approximately 10^{-4} $km^3 yr^{-1}$. Rates higher than this are predicted to result in shallow reservoirs and large volcanoes, in agreement with estimated global eruption rates of approximately 1.7×10^{-2} $km^3 yr^{-1}$ associated with the production of large volcanoes. Although eruption rates are probably high for large flow fields and lava floods, durations are probably low and residence time in the crust is insufficient for shallow reservoir formation.

I. INTRODUCTION

Knowledge of volcanism on Venus has increased due to the successful Magellan mission to Venus (Saunders et al. 1992; Ford and Pettengill 1992; Cattermole 1994). In this chapter we expand on the observations made in the initial mission reports (Saunders and Pettengill 1991; Saunders et al. 1992) with additional data, closure of the remaining gaps in global SAR coverage, and on the basis of continued mapping and data analysis in general. The volume of data, the variety of data types, and the necessity to insure careful evaluation, correlation, and interpretation of the data has challenged efforts to assimilate this knowledge into a coherent understanding of volcanic centers and volcanic processes on Venus. Documentation and classification of the volcanological diversity are far from complete.

The great range in scales and morphology of volcanic centers on Venus is evidence that central volcanism and volcanic effusion on Venus are associated with a large range of volume scales of magma transport, storage, and emplacement in the lithosphere. Determining the origin of this diversity on Venus, relative to the smaller terrestrial planets, is essential to the task of interpreting the geologic history of the surface and current state of the interior of the planet. Many of these fundamental geologic questions cannot be addressed because we first need to better understand mantle, lithospheric, crustal, thermal, and rate-related influences and processes, and the effects of the tectonic characteristics of Venus, on the process of magma transport and emplacement at individual volcanic centers and within volcanic complexes. This chapter is necessarily abbreviated, but will attempt to synthesize what has been learned regarding the geologic characteristics of centers of volcanism.

The term "centers of volcanism" includes large constructional volcanic edifices, clusters of small volcanic edifices, centers from which significant volumes of magma have erupted (flow fields), and volcanic structures associated with very shallow magma reservoirs (calderas). Throughout the following discussions we adopt a definition in which "volcanic" refers to surface products of magma emplacement and "magmatic" refers to both surface volcanic products and intrusive magmatism.

II. BACKGROUND

Prior to the Magellan mission, synthetic aperture radar images from Venera 15 and 16 of the northern 25% of the surface of Venus (Barsukov et al. 1986) initially demonstrated the morphologic diversity of volcanism on Venus. Concurrent and subsequent Earth-based radar image data from Arecibo confirmed the presence of similar characteristics of volcanism in low latitudes and in the southern hemisphere (Campbell et al. 1991; Senske et al. 1991a). Venera 15 and 16 also enabled identification of individual large volcanoes for the first time, each characterized by patterns of flows radiating for tens to hundreds of km (Barsukov et al. 1986; Slyuta 1990), circular structural patterns up to several hundred km across with associated volcanism (named "coronae" and "arachnoids") (Pronin and Stofan 1990; Ivanov and Basilevsky 1990; Stofan et al. 1991; Stofan and Head 1990), patterns of lava flow fields, extensive low-lying regions of lava plains, and low relief domical edifices (Barsukov et al. 1986), >20 km in diameter, that were interpreted to represent small, globally abundant, shield volcanoes (Aubele and Slyuta 1990; Garvin and Williams 1990).

Observations up to the time of the initial results from Magellan confirmed that volcanism on Venus was morphologically diverse and that volcanic features appeared to represent a wide range in scales of magmatic, tectonic, and mantle convective processes. However, the incomplete global coverage and moderate resolution limited the assessment of the types of processes, styles and scales of volcanism, and precluded a synthesis of the global characteristics of volcanism.

Initial surveys and analysis of Magellan data expanded on these earlier results both through increased resolution (Pettengill et al. 1991) and greater areal coverage. These data identified numerous additional morphological types of volcanic centers and edifices from volcanoes <1 km to >1000 km in diameter (Guest et al. 1992a; Head et al. 1991 1992a), widespread plains-type volcanism (Guest et al. 1992a), digitate lava flows and fields less than several km long to vast flow fields, lava floods, and lava channels in excess of 1000 km long (Guest et al. 1992a; Baker et al. 1992; Gregg and Greeley 1993; Komatsu et al. 1993; Lancaster et al. 1992; Parker 1992; Roberts Magee et al. 1992) and enigmatic ridged flows (Moore et al. 1992). The great abundance of apparently youthful volcanic features and the relatively low global crater abundance over the surface (Schaber et al. 1992; Phillips et al. 1992) imply that global rates of resurfacing due to volcanism on Venus over late solar system history have been high relative to that on the smaller terrestrial planets.

Assessment of the spatial distribution of volcanic centers, magmatic centers, coronae, and related features (Head et al. 1992a; Crumpler and Aubele 1992; Crumpler et al. 1992,1993a; Squyres et al. 1992b; Stofan et al. 1992) initially indicated that a significant concentration of volcanic centers occurs in the Beta, Atla, and Themis (BAT) Regiones (Head et al. 1992a; Crumpler and Aubele 1992; Crumpler et al. 1992). Several large volcanic rises and many smaller concentrations of volcanic centers appear to be associated with large rifts (Senske et al. 1992; Grimm and Phillips 1992; Jurdy and Stefanick 1993; Hamilton and Stofan 1994; Crumpler et al. 1992; 1993a) and complex global-scale patterns of extension and rifting (Crumpler et al. 1993a,c). Because features believed by many researchers to be representative of localized, shallow mantle upwelling, such as coronae (Stofan et al. 1992), are also common in these clusters, large concentrations of magmatic centers have been interpreted to represent global mantle anomalies resulting from patterns of convection and regional mantle upwelling (Crumpler et al. 1992,1993a,c). The overall arrangement of volcanic centers and coronae on

Venus appears to be more analogous to the distribution of hot spots related to mantle plumes on Earth than volcanism associated with plate boundaries (Stofan and Saunders 1990; Crumpler 1994; Phillips et al. 1991); however Schubert and Sandwell (1995) have interpreted some trench-like structures in coronae as features similar to terrestrial subduction zones and postulate that rifting, due to hotspots, and subduction may occur in close proximity.

Early recognition that the surface environment is characterized by extremely high temperature and high pressure prompted studies of the influence of the environment on volcanic eruption characteristics (Wood 1979; Garvin et al. 1982; Head and Wilson 1986; Thornhill 1993). As a general conclusion, it has been argued that pyroclastic eruptions capable of dispersing magma as ash are unlikely unless unusually high magma volatile contents (Head and Wilson 1986) or minor transient conditions are considered (Fagents and Wilson 1995). (This is in contrast to the results of studies of volcanism on Mars, where the low atmospheric pressure raises the problem of how to produce the effusive eruptions and lava flows responsible for the late shield volcanoes [Wilson and Head 1994]. On Venus, the problem is whether it is possible to produce ash at all.) Therefore, significant injection of volcanic material, including gases, into the atmosphere of Venus is not anticipated. On the other hand, although magmas that are relatively gas charged may not be dispersed, they may upon eruption be excessively vesiculated and result in a frothy form of mafic lava flow with unusually high apparent bulk viscosity (Head and Wilson 1992). Lava flows have been shown to be relatively little affected by the surface environment. Although the surface temperature is high, it is well below the solidus of most silicates. Together with the higher convective heat transfer from lava flows resulting from the dense CO_2 atmosphere, the corresponding radiative and convective cooling of lava flows is predicted to be equivalent to that in subaqueous environments on Earth (Head and Wilson 1986; Aubele and Slyuta 1990).

III. CHARACTERISTICS OF CENTERS OF VOLCANISM

All magmatic centers on Venus can be classified into those characterized by surface volcanism and those that are characterized by structural features associated with the emplacement and transport of magmas at depth (Head et al. 1992a) (Fig. 1). The first type, which we call volcanic centers, includes volcanic edifices resulting from accumulation of erupted material. These can be further subdivided into individual vents or clusters of vents. Calderas are frequently associated with extrusive materials and edifices and thus are included with volcanic centers for the purposes of this discussion. The second type, which we call magmatic centers, are generally defined by structural characteristics and are interpreted by many Venus researchers to arise from the emplacement and evolution of magmas in the subsurface, either as reservoirs or as diapirs. We consider this second type to be fundamentally intrusive. This class includes coronae and arachnoids (originally named by Barsukov et al. 1986 to designate two different morphological classes of magmatic center), and stellate fracture centers (originally named "nova" by Schubert et al. 1991). These features are discussed elsewhere (chapter by Stofan et al.).

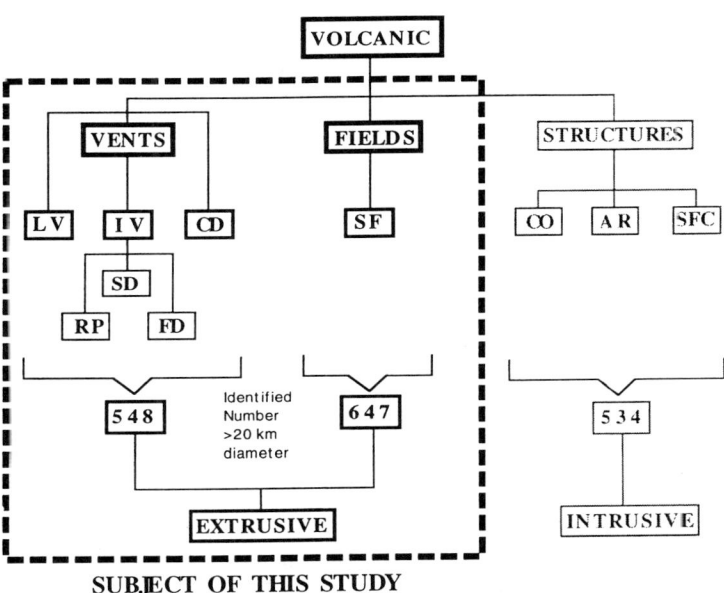

LV = Large Volcano
IV = Intermediate Volcano
IV/ RF = radial flows
IV/ SD = steep-sided dome
IV/ FD = fluted dome
CD = Caldera
SF = Shield Field
CO = Corona
AR = Arachnoid
SFC = Stellate Fracture Center

Figure 1. Volcanic center classification tree. Over two thirds of the population of magmatic centers are fundamentally extrusive. Two primary classes of volcanic center are identified: single volcanoes, or edifices, and "fields" of volcanoes. The latter accounts for most of the small volcanoes (<20 km) and comprises the most common single type of volcanic center.

We follow the original scheme outlined on the basis of Venera 15 and 16 results (Barsukov et al. 1986; Slyuta and Kreslavsky 1990) in which volcanic edifices are classified broadly on the basis of size as "large," "intermediate" and "small." Large volcanoes include those volcanoes equal to or larger than 100 km; intermediate volcanoes, those ≥20 km and <100 km; and small volcanoes, those <20 km in diameter. These specific size divisions are in part arbitrarily defined, but also derive from the overall cumulative size frequency distribution (Fig. 2) which bins approximately one-half of the measured edifices into the size range between 20 km and 100 km. No particular geologic significance is attached to the division of volcanoes into these three particular size categories and many recurring characteristics occur throughout the three size categories, although some characteristics ("steep-sided" or "pancake" domes, see later discussions) appear to occur only at diameters <80 km. The 20 km upper size limit for small volcanoes originated in part because it was convenient during early phases of the Magellan mission because 20 km is approximately the width of individual SAR image strips acquired during each orbital pass. Relatively few volcanoes on Earth exceed 20 km in diameter, so the term "small" is used here only in the relative sense.

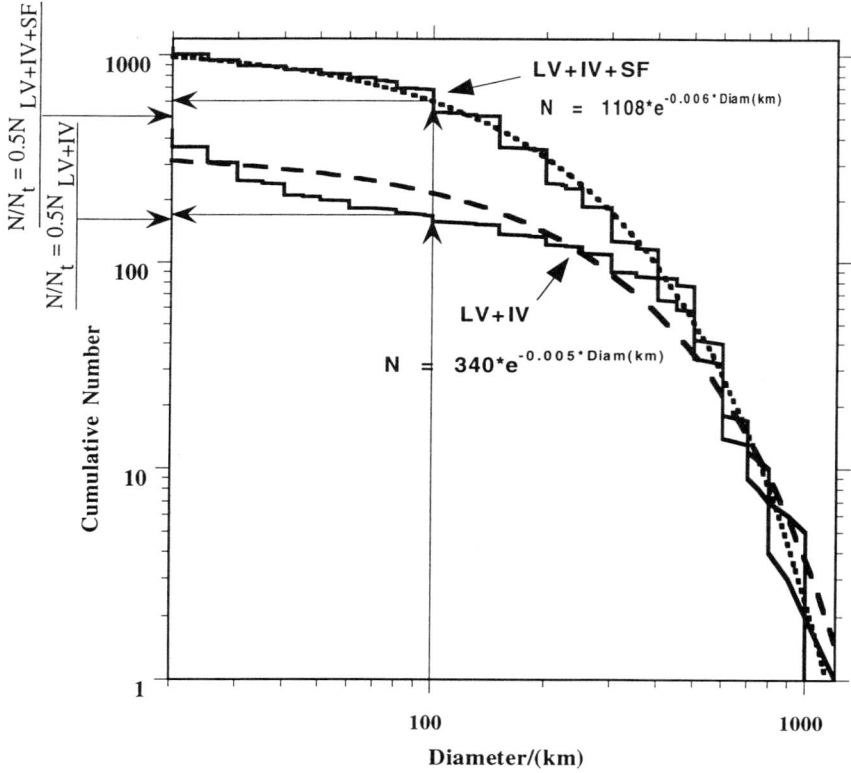

Figure 2. Cumulative size distribution of volcanic edifices >20 km. The distribution of sizes follows an exponential distribution. The upper curve includes shield fields with intermediate volcanoes and large volcanoes; the lower curve includes large volcanoes and intermediate volcanoes only. In both cases approximately half of the population occurs at diameters >100 km. There is a net deficiency in intermediate volcanoes and an excess of large volcanoes relative to a normal exponential distribution. Competition between size production rates and size obliteration rates could account for departures from a uniformly exponential distribution law.

In the following we have occasion to refer to many centers of volcanism individually. Yet because of the great number of volcanic centers, many are unnamed. For this reason, throughout the following discussions we will identify each center, where necessary, with a Magellan Volcano Catalog (MVC) number which consists of the latitude, longitude, and abbreviation of the volcano type. (Example: MVC 19/80LV = large volcano listed in the catalog and located at 19°N and 80°E.) A preliminary version of the MVC (Crumpler et al. 1993d;1996b) is accessible at www.planetary.brown.edu/planetary/databases/venus_cat.html. It was created following astronomical sky survey catalog numbering systems and the numbering system often used in terrestrial volcanic field mapping by the U.S. Geological Survey. Coordinates are recorded to the nearest 0.5° and generally refer to the geometric center (which may differ from the altimetric summit) of each volcanic center.

A. Large Volcanoes ($D \geq 100$ km)

Large volcanoes are defined as centers of eruption >100 km in diameter, identified principally by lava flows centered on a region of current or former positive topography as indicated by multiple lava flows radially distributed over 360° of azimuth. Those of very low relief may be obscured topographically amid complex regional structural patterns such that relief determined through the altimetry measurements alone is an unreliable indicator of the presence of a volcano. It is for this reason that we prefer to define them based on radial flow patterns indicative of quaquaversal slopes during the time of construction.

Globally, 167 large volcanoes with these characteristics are identified (Table I, CDP6C1T1). The lateral dimensions of large volcanoes are measured out to the average distal ends of the associated digitate lava flows comprising the circular part of the edifice. Unusual flow fields that run out from a central volcano over great distances in one direction are not specifically included in the dimensional estimate. Using these criteria for dimensions, the sum of the apparent areas of edifices associated with all identified large volcanoes accounts for 2.4×10^7 km^2 or about 5% of the surface area of Venus. This figure does not account for any plains volcanism that may be associated with an individual volcano. The significance of large volcanoes does not lie in their contribution to the global surface area, but in the fact that they represent areas where volcanism is concentrated such that some of the large-scale processes of magmatism on Venus are repeatedly recorded as discreet stratigraphic units.

The cumulative size-frequency distribution of large volcanoes (Fig. 2) based on these criteria is characterized by a flat distribution between 100 and 700 km diameter, with only a few volcanoes in excess of 700 km. However, lava flows associated with some edifices may be more extensive and account for areas of regional plains. The component of regional plains erupted from central volcanoes remains to be determined in most cases, and the characteristics reported here are based solely on apparent radial flows.

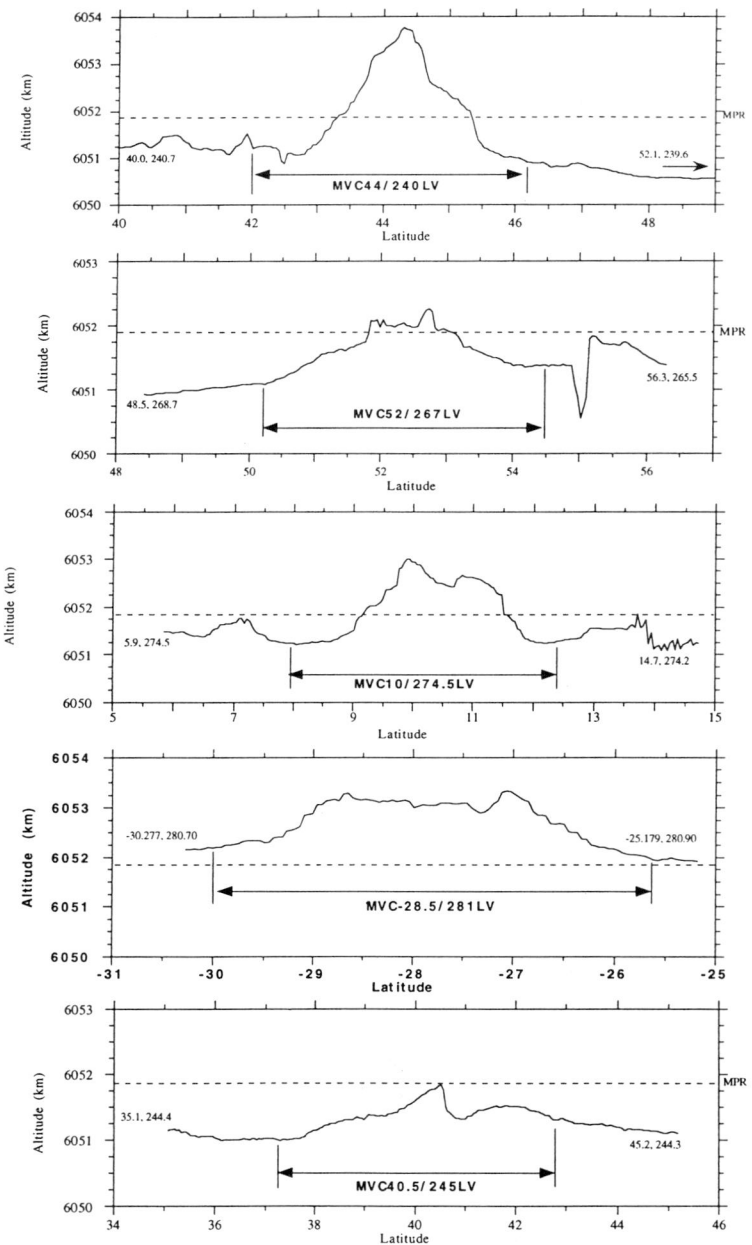

Figure 3. Altimetric profiles of a representative sample of large volcanoes illustrating the five basic topographic shapes. Horizontal axes are segments of a great circle; $1° \approx 105.7$ km.

1. Morphometry of Large Volcanoes. Altimetric profiles of selected examples of some prominent large volcanoes are shown in Fig. 3. Slopes rarely exceed a few degrees over a distance of several hundred km. Although their lateral dimensions are frequently equivalent to that of the largest Martian shield volcanoes, median summit relief relative to their bases is 1.43 ± 1.1 km. Large volcanoes on Venus are therefore an order of magnitude lower in edifice height, and their volumes are correspondingly much smaller, than Martian volcanoes.

The primary variations in slope occur at the summit where superposed circular and radial fractures, frequent parasitic smaller edifices, and complex calderas or caldera-like features may occur. As a result, three fundamental types of altimetric profiles are distinguished (Fig. 3): 1. straight-sloped cone or shield; 2. straight-sloped cone or shield with (i) truncated, (ii) shallow upper flank slope, or (iii) depressed summit area; and 3. irregular, asymmetric, or domical. This sequence is interpreted to reflect increasing complexity, and perhaps evolution, of the edifices from simple lava accumulations to volcanoes with differing degrees of structural modification.

2. Morphologic Classification. The morphometry and the geologic characteristics of large volcanoes suggest a classification from simple to more evolved. Geologic mapping of individual examples identifies those characterized geologically as basic constructional edifices, resulting from lava accumulation, and those characterized as constructional edifices with superposed structural deformation (such as summit calderas, radial rifts, fractures, and corona-like interiors). In the latter case, many large volcanoes exhibit characteristics transitional to other large magmatic centers such as coronae and stellate fracture centers. The structural diversity implies that there are variable levels of volcanologic complexity and geologic development recorded in the observed population of large volcanoes. Of interest to the understanding of magmatic processes on Venus in general is the question of whether these differences reflect characteristic evolutionary patterns for large magmatic systems on Venus, or whether the differences are random variations in complexity of the population at large.

Nine fundamental types of large volcanoes were identified from initial analysis of Magellan data (Fig. 4A,B) (Head et al. 1992*b*; Crumpler et al. 1993*b*). These divisions emphasize individual features and associations, but are not mutually exclusive; that is, individual large volcanoes are in many cases composites of several fundamental geologic styles.

Class I (Simple). Simple large edifices characterized by a relatively symmetrical outline and distribution of radial flows extending away from a summit region. Calderas and associated flanking structures are absent or not prominent. *Example:* Tuulikki Mons (MVC 10/275LV).

Class II (Caldera). Edifice surmounted by central caldera(s). *Example:* MVC 19/80LV.

Class III (Flanking Rift Zones). Edifice with one or more flanking rift zones arrayed generally radially to the edifice, similar to the flanking rift zones

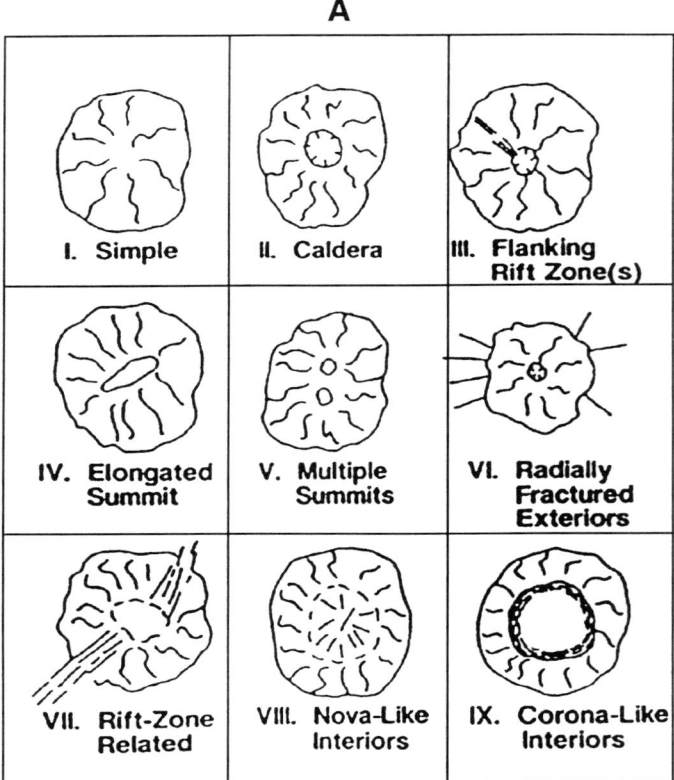

Figure 4. (a) Summary of the nine identified classes of large volcanoes according to summit structure, interaction with linear tectonic elements, and superposition of other magmatic center characteristics.

seen on terrestrial volcanoes such as Kilauea. (Note that this class differs from Class VII below.) *Example:* MVC 30/48.5LV.

Class IV (Elongated Summit). Edifice with elongated summit, often with multiple caldera-like features. *Example:* Gula Mons (MVC 22/358.5 LV).

Class V (Multiple and Steep Summits). Edifice with multiple or steep topographic summits. This class also includes volcano summits bearing multiple edifices of diverse morphologies that may include parasitic intermediate volcanoes (Fig. 5C). *Example:* Sapas Mons (MVC 09/188LV).

Class VI (Radially Fractured Exterior). Edifice surrounded by an exterior set of fractures that generally appear to be radial to the volcanic edifice and commonly predate many of the flow units making up the central part of the edifice. *Example:* Sekmet Mons (MVC 44.5/240.5LV).

Class VII (Rift-related). Edifice arranged along the axis of a rift trend.

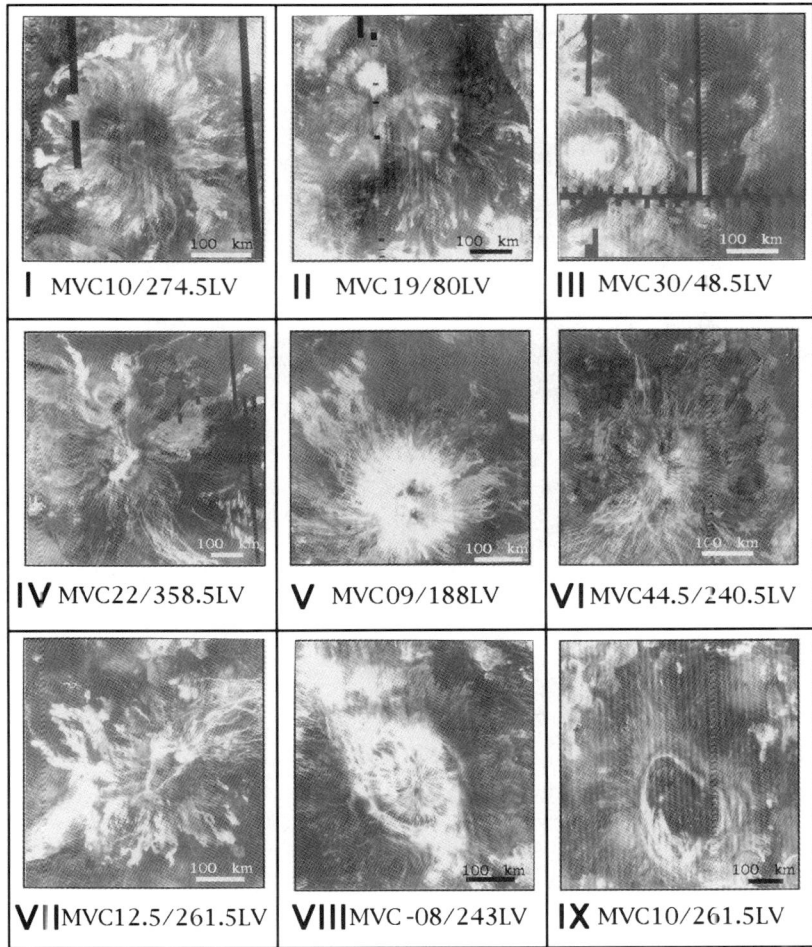

Figure 4. (b) SAR images of the nine type areas of large volcanoes. The degree of apparent complexity and interaction with exterior features increases from upper left to lower right. Bar scales are approximately 200 km.

In contrast to Class III volcanoes in which the rifts are confined to the flanks of the edifice, these are distinguished by association with large regional rift zones (illustrated by Gula Mons and Guor Linea) or through-going rift zones (well illustrated by Theia Mons and Devana Chasma). *Example:* MVC −12/261.5LV.

Class VIII (Radially Fractured Interior). Radial fractures occur at the center or topographic summit of some volcanoes, frequently with a very high density. The pattern of fractures is similar to the plexus of fractures classified as "stellate fracture centers" (originally named "nova" by Schubert et al.

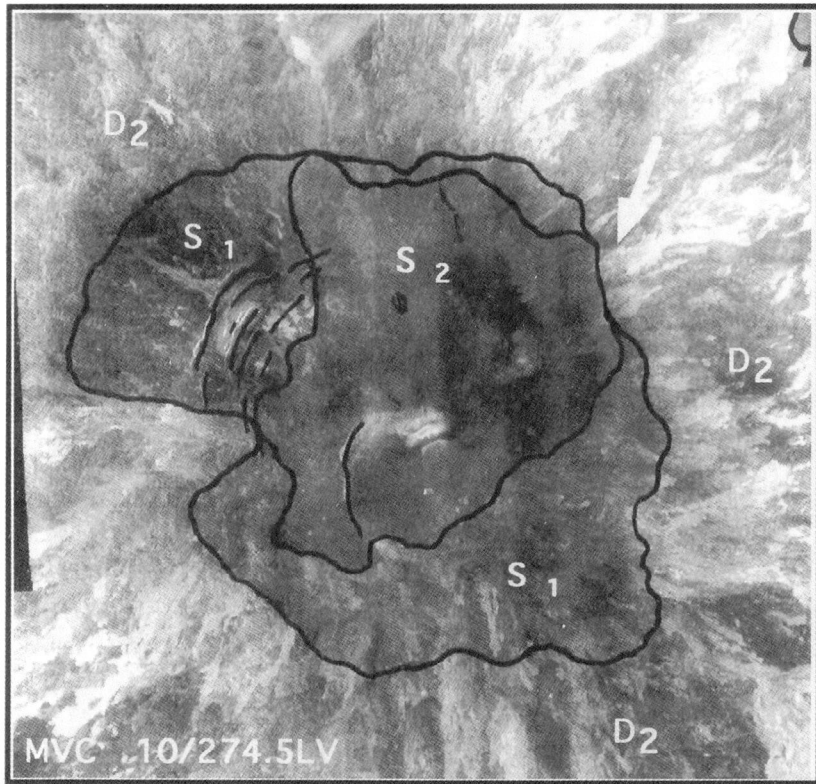

Figure 5. Images illustrating geologic details for several type examples of large volcanoes: (A) Obscuring deposits on the summit of type example of class I (Tuulikki Mons). Digitate flows are increasingly obscured by units S1 and S2; D2 indicates area of unaffected digitate flows on lower flanks. Arrow indicates an example where a single bright digitate flow appears gradationally from a region of summit darkening. Similar dark deposits occur on the summits of several large volcanoes. Image width, 260 km.

1991). In some cases, flanking flows clearly emerge from the radial fractures. The high concentration of radial fractures in the inner zone distinguishes these volcanoes from class III and VII. *Example:* MVC −8/243LV.

Class IX (Corona-like Interior). Large shield volcanoes with corona-like interiors, where the corona-like structure comprises 50% or more of the volcano diameter, but does not obscure the distinctive radial flow-like features that define large shield volcano edifices. There is clearly a morphologic transition between some of the large volcanoes and coronae; but it is not clear that one is always a preliminary stage and the other is always the final stage of development. *Example:* MVC 10/261.5LV (associated with Aruru Corona).

3. *Summary of Geologic Characteristics.* Geologic mapping of examples of the nine classes of large volcanoes illustrates that both lava flow

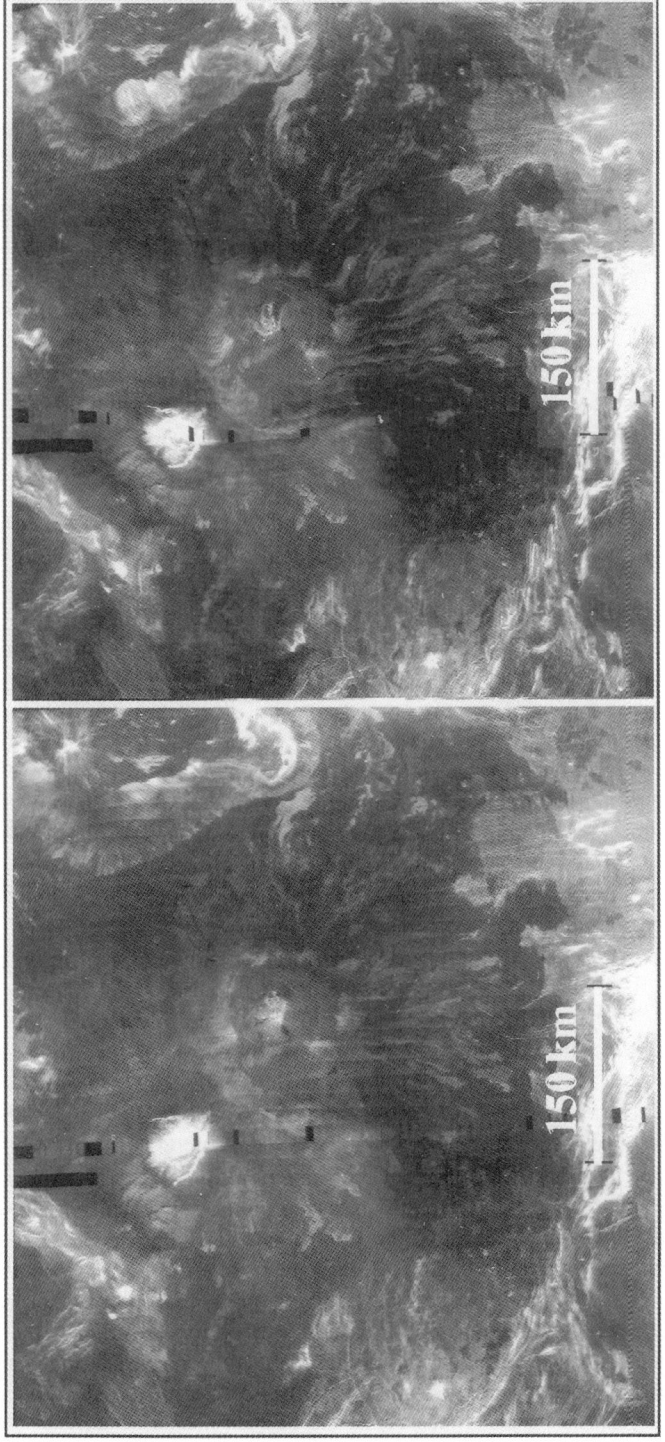

Figure 5. (B) Vertically exaggerated synthetic stereo image (merged altimetry and SAR data) showing details of summit caldera crest, shape of flank, and relationship to digitate flows (MVC 19/80LV).

accumulation and structural deformation have been important in large volcano evolution. At least two of the classes (VIII and IX) appear transitional in morphology to other types of magmatic features and illustrate the close relation that exists between these types of magmatic centers and large volcanoes.

The eruptive sequence of large volcanoes frequently follows a pattern of early, voluminous lava flows to late, radially patterned, digitate flows. This is in turn frequently followed by some type of central structural deformation. However, in many examples the volcano has developed at a site of preexisting structural characteristics, including both stellate fracture centers and coronae. A simple sequence in which large magmatic centers evolved from one feature type to another (for example, large volcano to corona, or vice versa) is not universal, and considerable diversity in evolution of each center is indicated. It is concluded that generally random conditions are present in the development of reservoirs and the eruptive behavior evolves with time, frequently in nonlinear fashion.

Large volcanoes are notable because they tend to be relatively high in the regional and local stratigraphic section, and thus appear to overlie regional plains and pre-date only the youngest extensional tectonism. This may be a simplification, and many examples are less clear, but large volcanoes as a class, if not individually, appear to be a relatively late occurrence in the global volcanic record of Venus. Tests of these stratigraphic observations have been attempted with crater age determinations and the results (Namiki and Solomon 1994; Price 1995; Grosfils and Head 1995) all suggest that large volcanoes, as a group, tend to be among the younger types of surface on Venus, are similar in age to the most recent geologic processes such as rifting (Basilevsky and Head 1995), and are generally younger than the extensive areas of plains. If these results are correct, they support the conclusion that the resurfacing of large areas on Venus occurred in a relatively short time interval geologically (Schaber et al. 1992; Strom et al. 1994) and that volumetric eruption rates have been less since that time.

B. "Volcanic Rises" and Major Volcanic Center Provinces

Large volcanic rises, as a class, demonstrate the interaction between volcanism and tectonism at large scales and warrant discussion in the context of edifice evolution. We distinguish large volcanoes from very large volcanically complex topographic swells exceeding several thousand kilometers in diameter, such as the volcanic "rises" of Western Eistla, Atla and Dione Regiones (Senske et al. 1992; Keddie and Head 1995). These are frequently associated with broad tectonic junctions (Senske 1990) and some or much of the observed topography may be nonconstructional (uplift, "dynamic swells," etc.). Nonetheless, the distinction between constructional and tectonic relief is unclear in many cases; and it is appropriate to consider that at least a significant component of the radially patterned lava flows within these rises may be accumulative.

Regional volcanic rises form a major part of the highlands in the equatorial region of Venus. These broad domical uplands, 1000 to 3000 km across, contain centers of volcanism forming large edifices, and are associated with extension and rifting. In particular, most of the rises are the sites of large gravity anomalies and may be evidence for dynamic mantle phenomena associated with volcanism and tectonism.

The morphology and geology of several topographic rises have been discussed by Senske et al. (1992) and Keddie and Head (1995). Two classes of rises are observed: (1) those that are dominated by tectonism acting as major centers for converging rifts, such as Beta Regio (Campbell et al. 1984) and Atla Regio, and are termed tectonic junctions (Senske 1990); and (2) those forming uplands characterized primarily by large-scale volcanism forming edifices, Western Eistla Regio and Bell Regio, where zones of extension and rifting are less clearly developed. Within this second class of features, the edifices are typically found at the end of a single rift or are associated with a linear belt of deformation (Senske 1990; Senske et al. 1991*b*).

C. Intermediate Volcanoes ($D<100$ km)

Intermediate-size volcanoes are characterized by a range of morphologies that are similar to those seen in large volcanoes but, in addition, include a variety of steep-sided and modified morphologies. The number of intermediate volcanoes identified in the Magellan volcano catalog totals 289; they can be divided into four basic morphologies that might be indicative of differing styles of eruption or unusual magma characteristics (Fig. 6). As with terrestrial volcanoes, it is likely that no single interpretation of the significance of the different classes should be assumed other than that there is a range of conditions for the ascent and emplacement of magmas that are perhaps more clearly manifested in the size range of intermediate volcanoes.

1. Four Classes of Intermediate Volcanoes. The simplest class of intermediate-size volcano is circular in planform with no discernible radial flows, characterized by a radar bright/dark dichotomy indicative of shallow flank slopes, and typically surmounted by a small caldera, volcanic crater, or pit. These appear to be essentially larger-scale examples of the class of edifice that frequently form the more abundant small ($D<20$ km) volcanoes. In several cases the circularity appears to arise from burial of the lower flanks by surrounding plains lavas. As a result, some of the intermediate volcanoes may be large-size volcanoes that are partially buried such that only the upper flanks are evident.

A distinctive second class of intermediate volcanoes is characterized by radially patterned flows (described informally as "anemone-type" in Head et al. 1992*a*). These appear to be smaller-scale versions of the morphologically similar radially patterned large volcanoes (Class 1 "simple" large volcanoes; Sec. III.A.3 above). Morphologically similar features also occur among the smallest ($D<20$ km) volcanic edifices (see discussion of small volcanoes below). These are interpreted to be shield volcanoes constructed from the accumulation of multiple lava flows of more or less uniform width and length

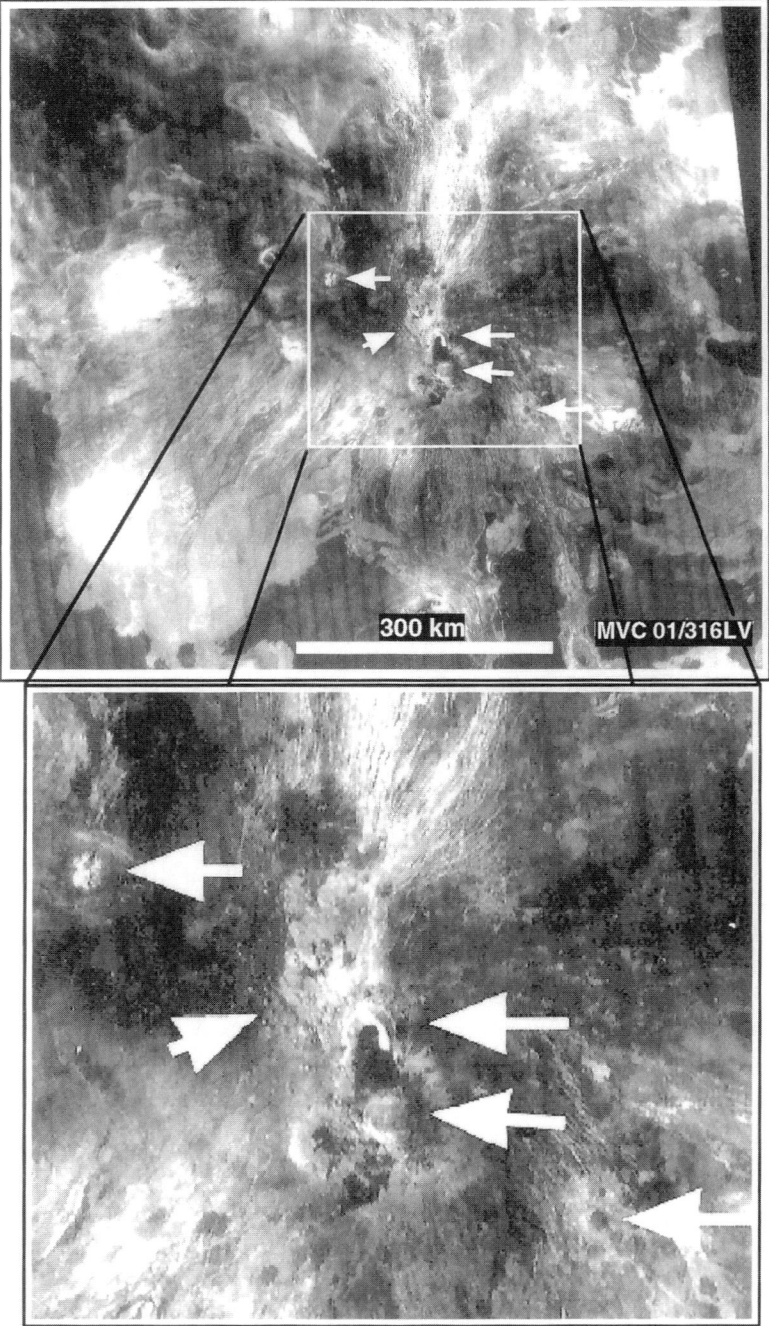

Figure 5. (C) Multiple small and intermediate volcanoes on summit of a class V large volcano (MVC 01/316 LV).

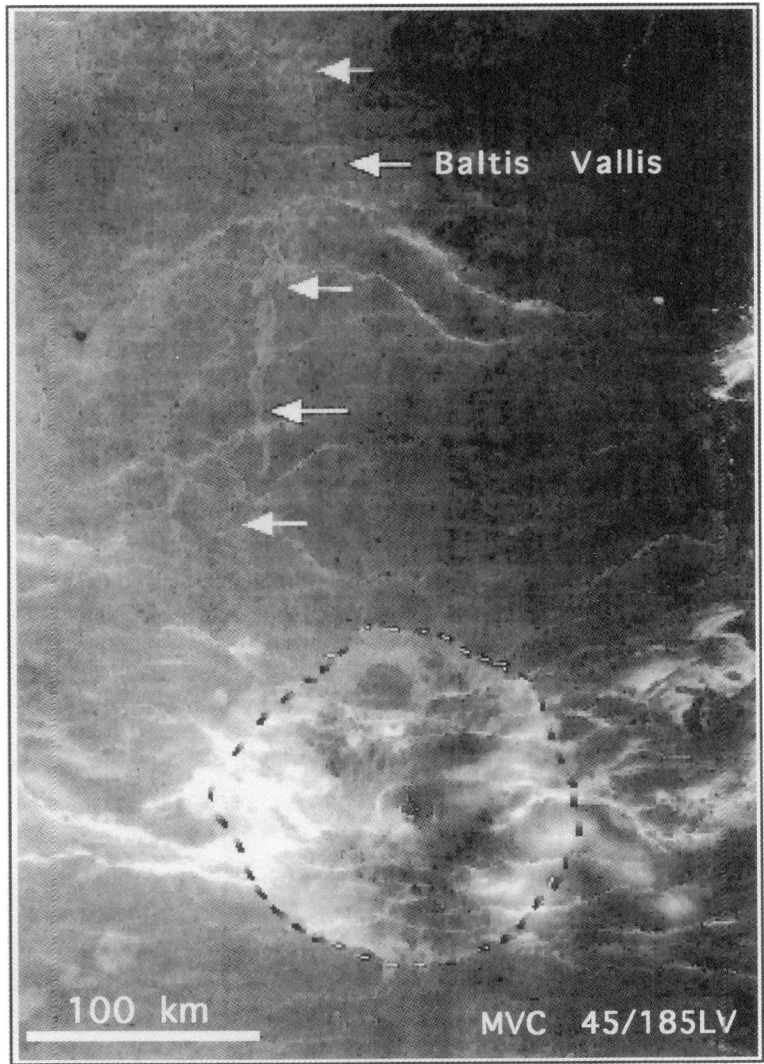

Figure 5. (D) Simple (class 1) large volcano at the apparent head of Baltis Vallis (arrows), the longest known lava channel.

erupted from a central, circular or elongate, vent.

The third class of intermediate volcano, identified previously as "pancake" domes (McKenzie et al. 1992) and "steep-sided" domes (Pavri et al. 1992; Head et al. 1992a), are characterized by relatively steep sides, circular planforms, relatively flat or upwardly convex profiles and variable fractures and pits (Pavri et al. 1992). Although the average flank slopes, even on the

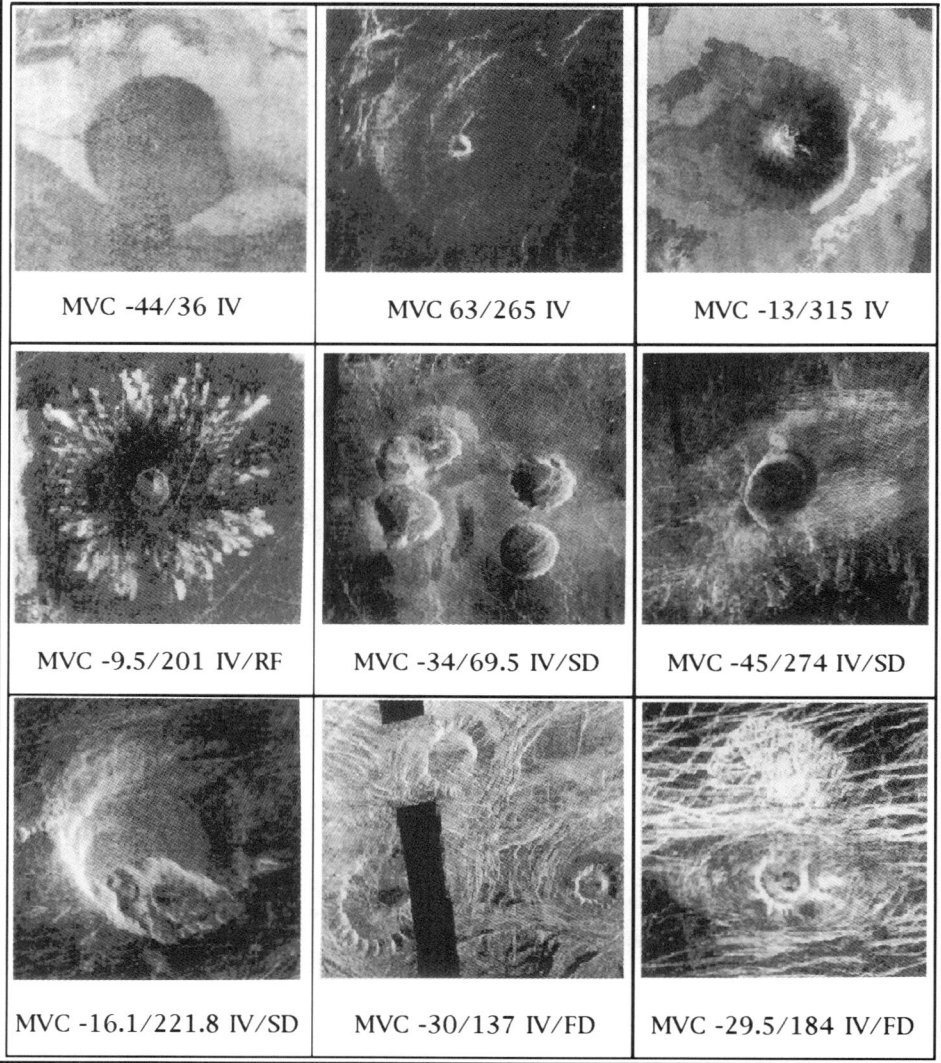

Figure 6. Selected examples of the four principal classes of intermediate volcano ($20 < D < 100$ km). These include simple circular edifices, edifices with radially patterned digitate lava flows, steep-sided domes, and modified and fluted steep-sided domes. Note that MVC −34/69.5 IV/SD is from right-looking cycle 2 SAR data. All others shown here are left-looking SAR images. ("Look" direction is specified as the radar ilumination direction if facing south along the north to south orbital direction of the Magellan spacecraft.)

more domical of the steep-sided domes, are generally quite low by terrestrial standards, the "steep-sided" designation refers to the fact that their flanks are relatively steep, and characterized by the effects of radar "lay over" in many cases, as compared with the other classes of intermediate volcanoes on Venus. The distribution, classification, morphology, and models of origin for these have been discussed by Bridges (1995); Bulmer (1994); Anderson et al. (1994); Plaut et al. (1994); Pavri et al. (1992); McKenzie et al. (1992); Guest et al. (1992a,b); and Bulmer et al. (1991), many of whom proposed or discussed composition-dependent models comparing the Venus steep-sided domes to terrestrial silicic domes.

The fourth class of intermediate volcano includes the steep-sided edifices of irregular shape variously referred to as "ticks," "scalloped-margin" domes, "fluted" domes, or "modified" domes. "Modified domes" appears to be a more encompassing term, although it has genetic implications in that it presupposes that the morphology is due to some alteration of a primary constructional morphology. Scalloped-margin domes and fluted domes are useful descriptive terms because some edifices appear more characterized by arcuate "scallops" and others by groove-like "flutes." Regardless of terminology, this class of volcano, occurs throughout the small and intermediate size ranges. The characteristic radial "fluting" was first described by Guest et al. (1991b;1992a,b). A hypothesis suggested by Head et al. (1992a) is that some of the fluting may represent rift zones linked to radial dike intrusion. Guest et al. (1991b,1992a,b); Bulmer and Guest (1994,1995); Bulmer (1994); and Bulmer et al. (1992a,1993) have interpreted them as due to collapse of the margins of initially circular steep-sided domes.

2. Significance of Domical and "Modified" Domical Volcanoes. As a class, "steep-sided" or "pancake" domes are of interest because they appear to be evidence for either distinctive magma chemistry or unusual characteristics of mafic rheology. At least three models have been proposed for the apparent viscous rheology exhibited by steep-sided domes: (1) chemical evolution of shallow magma bodies and derivation of relatively viscous intermediate and high-silica magma compositions (Pavri et al. 1992; Moore et al. 1992; McKenzie et al. 1992; Fink et al. 1993), (2) unusually crystal-rich magmas (Sakimoto and Zuber 1995), or (3) unusually gas-charged and vesiculated mafic magmas, or "basaltic foams" (Head and Wilson 1992; Head et al. 1992a).

Given the potential for abundant shallow magma reservoirs and corresponding chemical and thermal evolution of isolated shallow magma batches, variations in magma chemistry, temperature, or volatile content are potential candidates for the unusual characteristics. Petrologic evolution of shallow magma bodies is likely on Venus, but evolved magmas need not be rhyolites only. In addition to truly silicic magmas such as rhyolites, (discussed by Pavri et al. 1992; Moore et al. 1992; McKenzie et al. 1992; Fink et al. 1993), we note that intermediate-basaltic compositions representing more extreme differentiates of silica-undersaturated alkalic magmas, such as mugearite, benmoreite, and trachyte, may also be characterized by high viscosity. Endogenic domes of trachyte and alkalic intermediate variants are relatively rare, but known terrestrial examples are structurally similar to silicic domes and are characterized in some cases by summit craters surrounded by pyroclastic materials of the same composition indicative of strombolian venting in association with endogenic dome formation (Crumpler 1980). The Venera 8 site is near an identified pancake dome and the similarity between the alkali element content of the rocks at that site (Nikolayeva 1990) with major element content of trachyte (or the intrusive equivalent, syenite) may be significant, as pointed out by Moore et al. (1992) and Kargel et al. (1993).

In the crystal-rich model, the higher viscosity is a result of relatively cooler magma temperatures (Sakimoto and Zuber 1995) and correspondingly greater magma crystallinity. In support of the crystal-rich model, we note from our own experience with the trachyte domes discussed above that crystallinity is the primary difference between the more fluid strombolian late vent activity on their summits and the lavas comprising the domes. The latter are chemically indistinguishable but are characterized microscopically by masses of very small, lath-like anorthoclase crystals arrayed in classic "trachytic" textures. The presence of abundant crystals in a silicate melt may enhance viscous behavior through a polymerizing effect (Shaw 1969). For this reason high viscosity may also appear in more mafic lavas. There are many examples of basaltic lava flows with characteristics of more silicic lavas in large scale that differ from other basaltic lava flows of identical chemistry only in that they possess unusual (>10%) phenocrysts of pyroxene and feldspar (see, e.g., Crumpler et al. 1994).

The basaltic foam model has not been the subject of a detailed assessment as yet, but is essentially a variation on the crystal-rich model in which bubbles are the polymerizing constituent rather than crystals. Evidence for generally weak and unconsolidated behavior of the flanks, and the morphological gradation between steep-sided domes and modified domes (discussed below) support a model in which steep-sided domes are comprised of unusually low density, vesiculated mafic lavas. In addition, their great circularity and lack of complex flank characteristics are not consistent with the variations and asymmetries that would arise from eruptions of individual pulses of magma responsible for the lobes of individual flows. In this respect steep-sided domes are generally dissimilar to terrestrial domes resulting from eruption of silicic magmas (Anderson et al. 1994; Plaut et al. 1994), in which breakouts and asymmetric lateral growth is common. The conditions on Venus are such that if a magma is unusually gas charged, formation of a foam-like high volume magma eruption that collapses as it erupts is favored rather than a dispersed ash eruption or ash flow. The characteristic circularity of the domes supports a foam model; rapid eruption and limited cooling between pulses of extruded basaltic foam would enhance an endogenic behavior and contribute to azimuthally uniform characteristics in contrast to lobate flows and other asymmetries that might develop and propagate quickly during extrusion of silicic or crystal-rich magmas. The viability of the foam model as it applies to the Venus environment remains to be tested, however, and the crystal-rich model might better explain more lobate- or "festoon-"shaped lavas (Moore et al. 1992).

Many examples of scalloped-margin domes, as well as steep-sided domes, occur near the summits of large volcanoes. The change in eruption style which they imply, from digitate lavas to more localized eruptive forms, is evidence for either evolved basaltic geochemistry (i.e., mugearite to trachyte) or gas-rich late-stage magma eruptions associated with the shallow magma reservoirs underlying large central volcanic edifices. Frequent occurrence of steep-sided domes in association with coronae and coronae-like structures (Pavri et al. 1992) further suggests that long-term evolution of a magma reservoir or magma chamber, either petrochemically, in terms of volatiles, or rheologically, may be responsible for their unusual characteristics.

Another unusual characteristic of scalloped margin domes suggests a possible association with shallow magma reservoirs in some cases. Several examples (for example, Fig. 6 bottom center and bottom right) lie within enclosing fractures characterized by oval or augen-shaped patterns (Head et al. 1992a). The "augen"-shaped areas are frequently topographic depressions similar to calderas. Sag-like depressions of this type with superposed volcanic features of large volume can arise from rapid evacuation of a shallow reservoir, particularly in ash-flow and magma-foam eruptions where the magmas may erupt rapidly, and the substrate overlying the reservoir subsequently sags along the enclosing fractures. While this agrees in principle with the model of steep-sided dome formation by frothy magma eruption, the association might also arise from interaction with regional stress patterns with any shallow magma, so this type of feature is not diagnostic. In any case, we agree with the essential conclusions of Sakimoto and Zuber (1995) that the large-scale characteristics of steep-sided and modified domes are not diagnostic indicators of magma composition, and that additional factors, unrelated to composition, may be equally important in governing their morphology.

D. Small Volcanoes, Small Shields with $D<20$ km, and Shield Fields

Small volcanoes (Fig. 7) were initially identified in Venera 15 and 16 and Arecibo data (Barsukov et al. 1986; Campbell et al. 1989; Sinilo and Slyuta 1989). During analysis of Venera 15 and 16 data they were called small "domes," after the lunar usage of the term, but interpreted to be predominantly shield-type volcanoes by Aubele and Slyuta (1990) and Garvin and Williams (1990); an interpretation generally confirmed by Magellan data (Head et al. 1991,1992a; Guest et al. 1991a,1992a).

The basal diameter of a measured subset of 1000 randomly selected small volcanoes in the Magellan images ranged from <1 to 16 km with a mode of 4 km (Guest et al. 1992a). They follow an exponential distribution similar to the size frequency distribution of seamounts as measured from GLORIA sonar images (Aubele and Slyuta 1990; EEZ-SCAN 1986). A dearth of volcanoes <1 km in diameter may simply reflect the inability to identify

Small Volcanoes (D < 20 km)

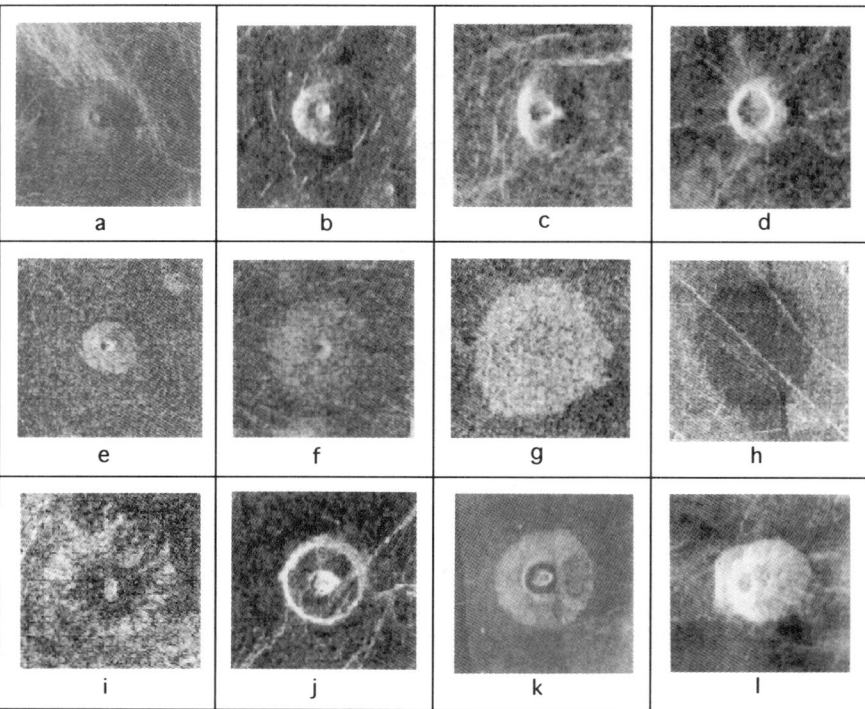

Figure 7. Selected examples of small volcanoes ($D<20$ km) on Venus showing range in morphology. (a) illustrates the presence of a much broader apron of lava flows (visibly embaying local fractures) surrounding the topographic edifice. Many small volcanoes may have similar extensive associated lavas that are not apparent due to lack of contrast with surrounding plains. (b,c,d) Examples of small volcanoes with more domical or conical edifice flank slopes. (e,f,g,h) Examples of small volcanoes ranging from apparent topographic edifices with radar backscatter contrast with surroundings (e and f, the latter is arguably the most common appearance of small volcanoes on Venus) to those identified by backscatter contrast only (g and h). (i) Small volcano with visible radial flows, similar to class I (simple) large volcanoes; also compare with MVC $-9.5/201$IV/RF in previous figure. (j and k) Small volcanoes characterized by unusual flat-topped relief and scarp-like margins. (l) Unusual small domical volcano with radial striations on the flanks. Volcanoes location and size are as follows: (a) 45.4°N/ 117.4, 16 km including apron; (b) 44.2°N/115.6, 8 km; (c) 46°N/19, 6 km; (d) 41.5°N/18.6, 6 km; (e) 39.1°N/334.7, 5 km; (f) 34.7°N/331.2, 5 km; (g) 11.3°N/175.2, 8 km; (h) 5.7°S/211.4, 12 km; (i) 10.1°N/188.6, 7 km; (j) 7.6°S/200.6, 8 km; (k) 2°N/228.5, 20 km to edge of bright striations; (l) 48°N/192.5, 18 km.

very small volcanic edifices in the Magellan data. Alternately, a diameter of 1 to 2 km may represent a significant minimum diameter resulting from the intrinsic processes of edifice formation. Conditions leading to recurring size characteristics of small shield volcanoes are thought to prevail on Earth and relate to controls on the dimensions of small and shallow chambers, the details of their connection with the source reservoirs and the elastic properties of the country rock in which the reservoirs reside (Gudmundsson 1986,1987). Summit pits do not occur in all of the small volcanoes, but when present, their diameters generally range from 0.2 to 1.8 km. Neither the occurrence of summit pits nor their diameters have a direct relationship to basal edifice diameter; however, pits large in relation to the diameter of the edifice occur more commonly in more domical or conical shaped edifices. A different pit diameter to basal edifice diameter ratio appears to apply to a class of flat-topped small edifices (Slyuta et al. 1992) described in more detail below.

1. Classification of Small Edifices. Guest et al. (1992a) originally identified three fundamental edifice shapes (shields, cones, and domes) with fourteen sub-classes based on a combination of summit pit size and shape, flank characteristics, and radar backscatter characteristics. Aubele (1993) subsequently suggested a different classification scheme, assigning all fourteen sub-classes to eight major classes, four of which were defined by apparent edifice profile and four of which were defined solely by radar backscatter. These two classification schemes reflect different underlying assumptions One possibility, reflected by the Guest et al. classification scheme, is that the small volcanoes are produced by variations in lava composition (mafic to silicic) and that differences in their flank slopes and shapes represent these fundamentally different types of volcanism. In this case, there would be definite differences between shields, cones and domes, and it would be important to delineate the percentage of each morphological type. The other possibility, reflected by the Aubele classification, is that all of the small volcanoes are basaltic eruptions and that differences in their morphology are nonuniquely produced by minor variations in eruption parameters and are of interest in understanding basaltic eruption processes but do not imply fundamental differences in composition. In this case, shield, dome and cone shapes represent minor gradational variations in flank slope and it is the range of morphologies that is of key importance rather than the absolute classification and percentage of each.

2. Characteristics of Small Volcanoes. Regardless of classification scheme, the most common small volcano edifice appears to be shield shaped with variable flank slopes and a single summit pit (Fig. 7a,e,f). The assumption can be made that these volcanoes represent terrestrial style shield volcanoes. Although individual flow units cannot be identified on most of the small edifices at Magellan resolution, the small volcanoes can be modeled by multiple, centralized small-volume lava flows with a contribution by localized pyroclastic deposition. Small volcanoes that appear to be mesa-like or "flat-topped" (described by Guest et al. [1992a] as type S2 and by Aubele [1993] as one of the four main edifice profile types) are similar to steep-sided domes in the intermediate volcanoes but have quite distinctive characteristics (see Fig. 7j,k). They differ from "pancake" or steep-sided domes of the intermediate size volcanoes in that they consist of a radar-dark, flat-appearing edifice with a small central crater. Edifice margins appear sheer and scarplike, and sometimes have radially striated, radar-bright aprons around their bases. These small edifices are almost identical to some seafloor volcanoes (seamounts) imaged by GLORIA (Aubele and Slyuta 1990; Bridges 1995; Bridges and Fink 1994; Sakimoto 1994). Although they differ from the intermediate sized steep-sided domes, they could be variations on the same mechanism of formation as influenced by scale. The flat-top profile is difficult to model by construction due to multiple centralized small-volume lava flows alone, but could result from perched and flooded lava ponds (Guest et al. 1992a) in addition to the variety of mechanisms discussed above for pancake domes. Slyuta et al. (1992) have also suggested steeper central areas in some small volcanoes could arise from the presence of a "cinder," or pyroclastic, superstructure due to late-stage explosive eruptions, which is typical of many small terrestrial shield volcanoes in the ~2 to 5 km size range.

Many small volcanoes are visible only as variations in radar backscatter and are interpreted to have extremely low topographic slopes (see Fig. 7g,h,i). A radial flow pattern type (classified as type S6 by Guest et al. 1992a) appears to be very low in profile although individual flows are clearly visible and well defined. This type of small volcano is also similar to an intermediate-size volcano type, the radially patterned type ("anemone") first described early in the Magellan mission (Head et al. 1992a). These volcanoes appear to be produced by long linear flows extruded from a circular to elongate fissure-type vent and are frequently aligned along local structural trends. Radar bright (or less commonly dark) aureoles or haloes (Aubele and Slyuta 1990; Klose 1990) are sometimes associated with small but otherwise typical topographically defined edifices and may be evidence that the total volume of extrusive material associated with each vent is larger than the edifice alone. Although haloed edifices are somewhat rare, they commonly occur in fields rather than as a single example. This implies that the origin of the halo effect may be a regional phenomenon. The halo is either constructional and emplaced on pre-existing plains in association with the volcanic edifice or it may represent the exposed radar-bright lower slopes or surrounding apron of flows of the edifice encircled by later plains material at a constant contour level. Circular radar bright spots with no apparent topographic edifice may represent very low slope shield volcanoes. Radar-dark circular or irregular "spots" (classified as type S3 by Guest et al. [1992a]) frequently occur associated with graben and may represent individual flows and small edifices erupted from a point source on a fissure, similar to small localized Hawaiian rift eruptions.

3. Fields of Small Volcanoes. Enhanced concentrations of small volcanoes have been called "dome fields" (Aubele and Slyuta 1990; Senske et al. 1991a,b) or "shield fields" (Aubele and Crumpler 1992; Aubele et al. 1992; Head et al. 1992a) following terrestrial volcanological usage of the term "volcanic field" to describe any prominent cluster of volcanic vents. These fields of small volcanoes can be considered to be a distinct type of volcanic center. A typical Venus shield field consists of volcanoes which are predominantly shield-type edifices, numbering in the tens or hundreds of volcanoes and ranging in density from 4 to 10 edifices per 10^3 km² within an area $\geq 10^4$ km². Typical shield fields are roughly equant in outline with diameters from 50 to 350 km, with a mode from 100 to 150 km. The cumulative size distribution of shield fields appears to follow the trend of coronae/arachnoids and stellate fracture centers more closely than it does the trend of large and intermediate-sized volcanoes (Fig. 8). This similarity to features assumed to be dominantly indicative of subsurface magma or diapir emplacement may reflect melting source dimensions and provides a clue to the geologic significance of the shield fields. A possible significance of this is discussed in Sec. V.C.2.

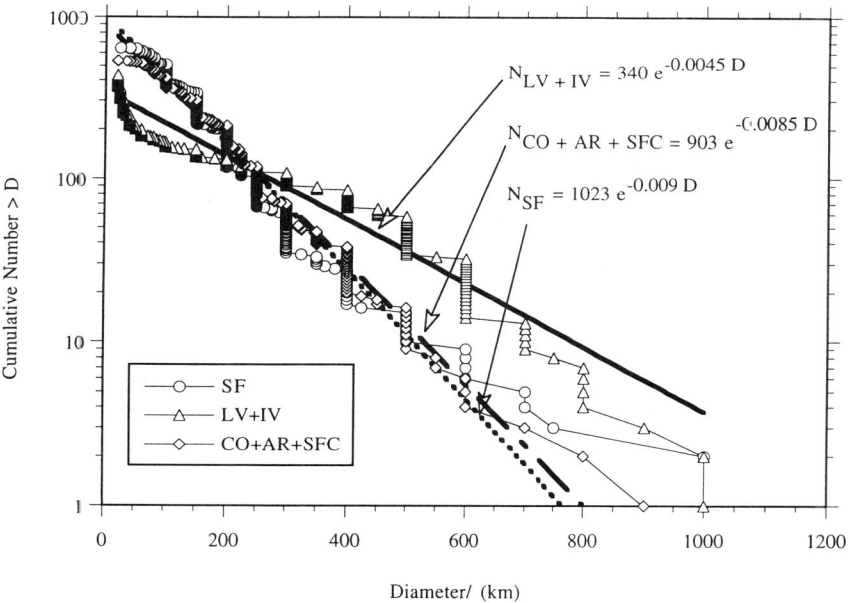

Figure 8. Size distribution of shield fields in relation to other large magmatic centers. The diameter distribution of shield fields more closely follows that of coronae, arachnoids, and stellate fracture centers than large and intermediate volcanoes. The distribution of vents that define shield fields, like the circular, annular, and radial structures of coronae, arachnoids, and stellate fracture centers, may be indicative of the lateral extent of deep magma reservoirs or magma source regions rather than the physical limits on lateral effusion of erupted lavas that typifies volcanic edifices.

Radar-dark or radar-bright material occasionally surrounds the edifices within a shield field and is interpreted to represent associated volcanic material, probably thin lava flow units, although minor amounts of ash or cinder may produce a thin local veneer in some areas (Guest et al. 1992a). If the visible flow fields associated with some shield fields are typical of average size, then the area of resurfacing associated with a shield field may to be comparable to that of the area of a single large volcano on Venus. A few edifices within a few shield fields may be aligned along dominant structural trends, and summit pits occasionally occur along structural trends; however, the majority of shield fields appear to contain edifices that are randomly scattered.

Based on the 647 identified shield fields, we have divided fields of small volcanoes into four basic classes (Fig. 9). A "simple shield field" appears as a cluster of randomly scattered edifices on a plains unit. There is no apparent association between the plains unit and the edifices, and the stratigraphic relationship between the edifices and the plain unit is frequently obscure. "Apron shield fields" show clusters of small volcanoes that are spatially associated with patterns of radar-bright or radar-dark material interpreted to be volcanic flows. "Companion shield fields" are spatially associated with another type of volcanic center. This class includes fields close to a large or intermediate-size volcano or one or more coronae, and fields that are located on the summits or flanks of large volcanoes or within the ring structure of coronae or arachnoids. Instead of a well-defined cluster of small volcanoes within a relatively small area, the fourth class consists of plains units, with associated abundant small shield volcanoes, that have consistent stratigraphic relationships with other mappable units. One such region in Akkruva Colles, originally recognized as an area with a high density of small volcanoes by Schaber (1988) and Barsukov et al. (1986), has been defined as a local stratigraphic unit through geologic mapping and informally called "shield plains" (Aubele 1994,1995).

4. Total Number of Small Volcanoes. A reliable estimate of the number and size distribution of the total population of volcanoes and calderas can be used to understand global rates of resurfacing and magma production (see Sec. V.B). However, the abundance of small volcanoes on Venus, and the difficulty of identifying all of the less obvious examples, precludes easily identifying, measuring and counting them. Two separate projects (one at JPL and one in the U. K.) are attempting to use artificial intelligence pattern recognition to produce ultimately a global data set of the small volcanoes (Aubele et al. 1995; Burl et al. 1993; Smyth et al. 1992; Wiles and Foreshaw 1992). Until these projects are concluded, total number must be estimated. Pre-Magellan analysis of the 25% of the planet imaged by Venera 15 and 16 (Aubele and Slyuta 1990) estimated the total number as 10^5 to 10^6 planetwide using an exponential distribution while Garvin and Williams (1990) estimated 10^7 using a power law distribution. Using current Magellan data we may estimate the total population in several ways. The total number may be estimated by extrapolating from a known cumulative frequency distribution. Using an exponential distribution determined from measured small volcanoes within a known area, and given a minimum size of 1 km in diameter, we can then estimate the total number over the area of the planet. This number is 931,000. The planetwide population may also be estimated if we use the 647 shield fields that have been identified and assume an average number of volcanoes per shield field. This total is an underestimate, however, because

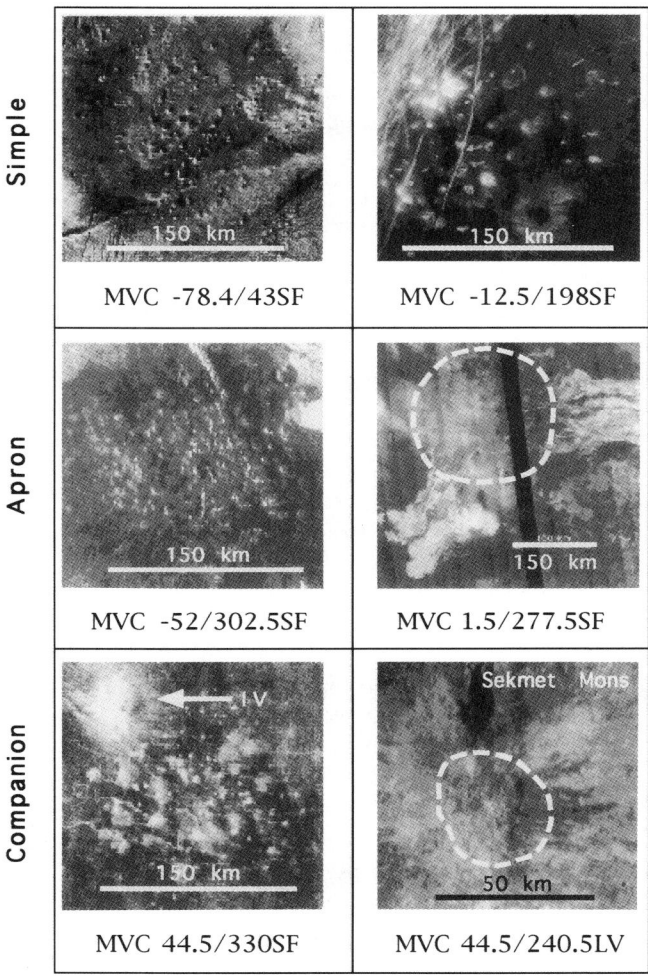

Shield Fields

Figure 9. Selected SAR image examples of three types of shield field occurrences. These include simple clusters of small edifices that include either morphologically uniform and similar, or morphologically diverse edifices; shield fields with an apron of surrounding materials, either forming local plains or radial, digitate lava flows, presumably erupted from vents within the field; or fields associated with companion intermediate edifices or near the summit of large edifices. The example in the lower left is analogous to many terrestrial cinder cone type volcanic fields in that a single intermediate edifice occurs in association with a field of small volcanoes. The example in the lower right is of unusually small edifices clustered in the central vent region of Sekmet Mons (MVC 44.5/240.5LV).

not all small volcanoes occur in fields. The differences in these results indicate that any estimate is dependent on the assumed uniformity of the characteristics measured from sample populations and illustrate the difficulty in estimating the total population based on the disparate characteristics of size, spacing, and regional abundance of small shield volcanoes and shield fields. A direct counting method will be required to obtain a reliable number.

E. Calderas

A variety of circular, volcanic depressions occur on Venus (Fig. 10) ranging from simple volcanic depressions to the more complex class of centers identified as arachnoids (Barsukov et al. 1986; Head et al. 1992*a*; Dawson and Crumpler 1993) and coronae (Barsukov et al. 1986; Stofan et al. 1992; Squyres et al. 1992*a*; Janes et al. 1992; chapter by Stofan et al.). In contrast to coronae and arachnoids, the calderas on Venus are similar to relatively simple depressions associated with extrusive volcanic activity common on the other terrestrial planets.

1. Definition of Caldera. Because the suitability of terms for various features on Venus, both volcanic and tectonic, is still under debate, a short discussion of volcanic depression terminology is warranted. Confusion about the choice of terms to be used for volcanic depressions on Venus is understandable because the terminology and dimensional limits for volcanic depressions originating through a wide range of volcanic deformation processes on Earth is not universally agreed upon. This is in part because the terminology of volcanic depressions on Earth is still evolving as our understanding of the actual process of volcanic depression formation continues to evolve (McBirney 1990). "Volcanic depressions" is an encompassing term that refers to pits, volcanic craters, calderas (of which there are several defined types), cauldrons, and volcano-tectonic depressions. The generally accepted definition of "caldera" is that provided by Williams and McBirney (1979) who originally defined a caldera as "a large volcanic collapse depression, more or less circular, or cirque-like in form, the diameter of which is many times greater than that of any included vents." The current model of caldera formation is catastrophic collapse correlated with a major explosive eruption, rapid effusive eruption, or rapid lateral intrusion. Walker (1984) has pointed out that some calderas on Earth have detailed characteristics that do not conform to a single catastrophic formation, but instead appear to have formed incrementally such that the subsidence is distributed over a broad area by downsagging. A downsag caldera, therefore, is a genetic term, based on proposed mechanism of formation of the caldera. The type example, described by Walker (1984) is Taupo in New Zealand. Arsia Mons on Mars, although not directly a counterpart, may be comparable to a downsag caldera in some morphologic respects and in terms of probable long-term accumulation of downward displacements (Crumpler et al. 1996*a*).

No lower size for calderas is universally accepted, although dimensions between 1 km (Wood 1984; Francis 1976) and 5 km (Ollier 1969) have

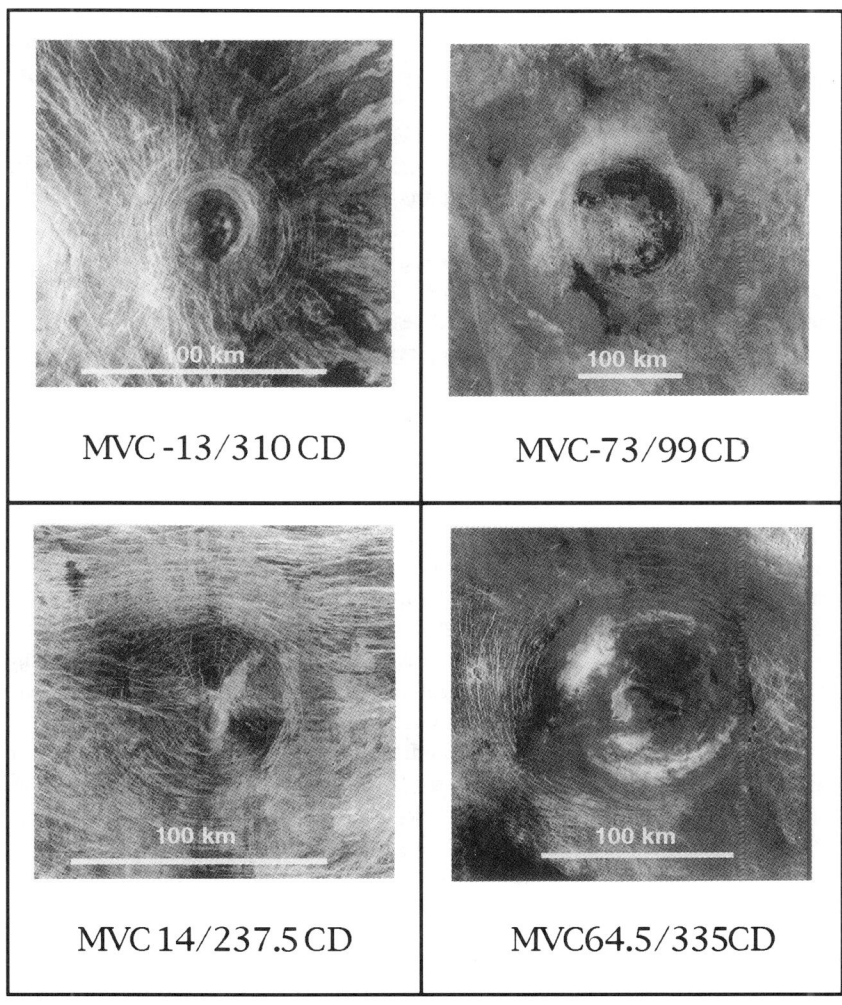

Figure 10. Selected SAR image examples of calderas. Calderas on Venus are frequently circular, deep, and encircled by fractures indicative of deformation within the brittle crust overlying shallow magma reservoirs. Overlapping calderas of the type common on Earth and Mars are unusual on Venus at diameters >20 km.

been cited and used in the past. Williams and McBirney (1979) note that "subsidence features less than a kilometer across are usually called 'collapse pits'." Many pits on Venus, including those exceeding 1 km in diameter that are associated with "small" ($D < 20$ km) volcanoes, are likely to be "calderas" by terrestrial definition. In discussing small volcanoes we have included summit depressions up to 1.5 km in diameter in the class of collapse pits even though that is slightly larger than the upper limit of collapse pits on Earth.

The reason for this expanded upper limited is the inability to distinguish collapse pits from volcanic craters at diameters much smaller than 1 km using Magellan data; a volcanic crater is generally much smaller and genetically different (Williams and McBirney 1979) from both collapse pits and calderas but overlaps the lower diameter range of calderas.

An upper size limit for features identified as calderas is less clear and is convoluted with various distinctions between calderas, cauldrons, and volcano-tectonic depressions. Williams and McBirney (1979) use a genetic definition which defines cauldrons as those depressions resulting from "passive collapse," meaning collapse of the type that may follow accumulation of extensional stresses resulting from growth and expansion of a shallow reservoir (Gudmundsson 1988), and defines calderas as the result of actual withdrawal of magma from a shallow reservoir. Smith and Baily (1968) and Elston (1984) and support a definition of caldera as a topographic feature associated with eruption at the surface, and a nongenetic definition of cauldron as a structural term for any large volcanic subsidence structure "regardless of shape or size or depth of erosion or connection with surface volcanism." Their definition of cauldron would thus encompass "coronae" and "arachnoids" as these are identified largely on the basis of their morphology and structure regardless of connection with surface volcanism and independently of the underlying geologic units. Features referred to as cauldrons generally occur at the upper size range of circular volcanic depressions on Earth. For a variety of reasons relating to the evolved composition of stable continental crust where most cauldrons occur (Walker 1984) and other factors exclusive to Earth (Shaw 1985), the largest magmatic systems are relatively silicic, and in practice the term "cauldron" has become identified with silicic magmatism on Earth; however, composition is not a defining characteristic in principle and the term could certainly be used for any large volcanic subsidence structure defined solely on the basis of morphology and structure.

From the above discussion it should be gathered that no particular significance should be assumed about a volcanic depression based on terms used without attribution to a reference because the existing definitions have not become universally used.

2. General Characteristics of Calderas. Clearly identifiable calderas (as defined by Smith and Bailey [1968] and Elston [1984]) occur at the summits of a few of the large volcanoes; for example, Sif Mons (Senske et al. 1992) and Tepev Mons (Campbell and Rogers 1994). But unmistakable calderas, as defined above, are less frequent when one examines the total population of large volcanoes. More frequently there are depressions or circular structures of undetermined topography at the summits of many volcanoes that may or may not be true calderas. Because of this frequent uncertainty, we have chosen not to include the few calderas on summits of large volcanoes in our compilation of calderas as a class. The discussion in the following is also confined to volcanic depressions larger than 20 km in diameter in order to insure that the characteristics and size distribution of the complete population within the size range is observed. Small volcanic depressions, that may actually be calderas, volcanic craters, or volcanic collapse depressions, associated with the abundant intermediate to small-size volcanoes of Venus, are far too abundant to examine every example. And, at smaller dimensions, detailed characteristics of calderas are obscured by complex background geology such that the characteristics of the total population are underdetermined. This is not to imply that calderas do not occur at smaller sizes. Bulmer et al. (1992b, 1995) have identified approximately 100 calderas from a range of caldera sizes, from those on small volcanic edifices to those associated with large volcanoes to those not associated with volcanic edifices, and have classified them by type of associated edifice and caldera size.

In this discussion, however, we are concerned with the size distribution and characteristics of a complete population of one type of Venus caldera; large calderas, >20 km in diameter, that occur independently of volcanic edifices and therefore represent a different style of volcanic center than that associated with the formation of a volcanic edifice. Following the topographic definition of calderas (subsidence in association with eruption of volcanic materials at the surface) we have identified 96 calderas that meet these criteria (Table II, CDP6C1T2). These calderas occur in a variety of regional settings. Several of these calderas lie at the center of radiating patterns of lava flows, others appear to be the source of great flow fields distributed in a restricted azimuth from the caldera. It is possible that some of these radiating flow patterns are indicative of the former presence of constructional edifices, but no associated topographic edifice exists currently (for example, Fig. 11). A few calderas on Venus appear to be the source of extensive flow fields. Those that are not, frequently occur in the presence of extensive regional fractures. This latter observation suggests that drawdown in these cases might be aided by lateral propagation of dikes from the reservoir along patterns of regional extension rather than eruption of the magmas at the surface. A salient characteristic of many of these 96 calderas in the observed size range is a complex concentric pattern of fractures, frequently so dense in spacing that the character of lava flows becomes obscured near the margins of the calderas. It is on the basis of this observation that Bulmer et al. (1992b) proposed all large Venus calderas not associated with edifices to be the counterparts of terrestrial down-sag calderas. As is the case with Arsia Mons, some of the calderas on Venus may certainly be comparable to down-sag calderas; however the detailed criteria by which downsag calderas are recognized on Earth is difficult to ascertain on a planetary image and the topography and depth of some of the calderas on Venus are not diagnostic of a down-sag caldera.

3. Topographic Relief and Plan Shape. Topographic details of the caldera margins are variable and some are characterized by apparent raised rims, whereas others are simple depressions with either straight, steeply sloped walls or with walls that gradually steepen with depth to an apparently smooth, and approximately flat, floor. Many of the calderas are relatively deep by terrestrial standards and attain depths up to several km (Fig. 12). An addi-

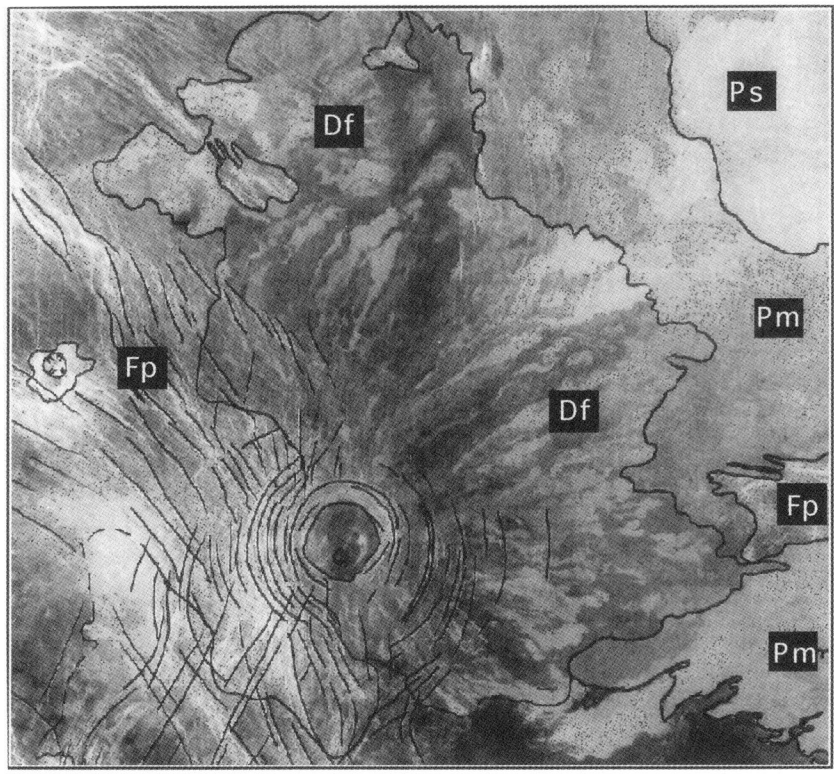

Figure 11. Example of a large flow field (#130, Table III, CDP6C1T3) associated with a caldera (MVC −13/310 CD). Explanation of mapped unit abbreviations: Df, digitate flow field; Fp, fractured plains; Pm, mottled plains; Ps, smooth plains; image width, 300 km.

tional distinguishing characteristic is their great circularity with respect to the frequently irregular, often overlapping, and generally flat-floored morphology of typical calderas on Earth or Mars. In this respect, and in terms of their large dimensions, many belong to a class of feature comparable to the great Arsia-type calderas of Mars (Crumpler et al. 1995,1996a). However, the only comparable caldera in terms of depth is the summit caldera of Pavonis Mons on Mars which is 5 km deep (Bibring et al. 1990) and interpreted to have resulted from drawdown of the underlying reservoir during voluminous eruptions from the flanks. The greater depth and circularity of calderas on Venus could arise from the relatively greater predicted depth of reservoirs.

F. Centers of Large Flow Field Eruption

A preliminary description of lava flood-type flow fields on Venus (Head et al. 1992a) identified lava floods not associated with other volcanic centers and

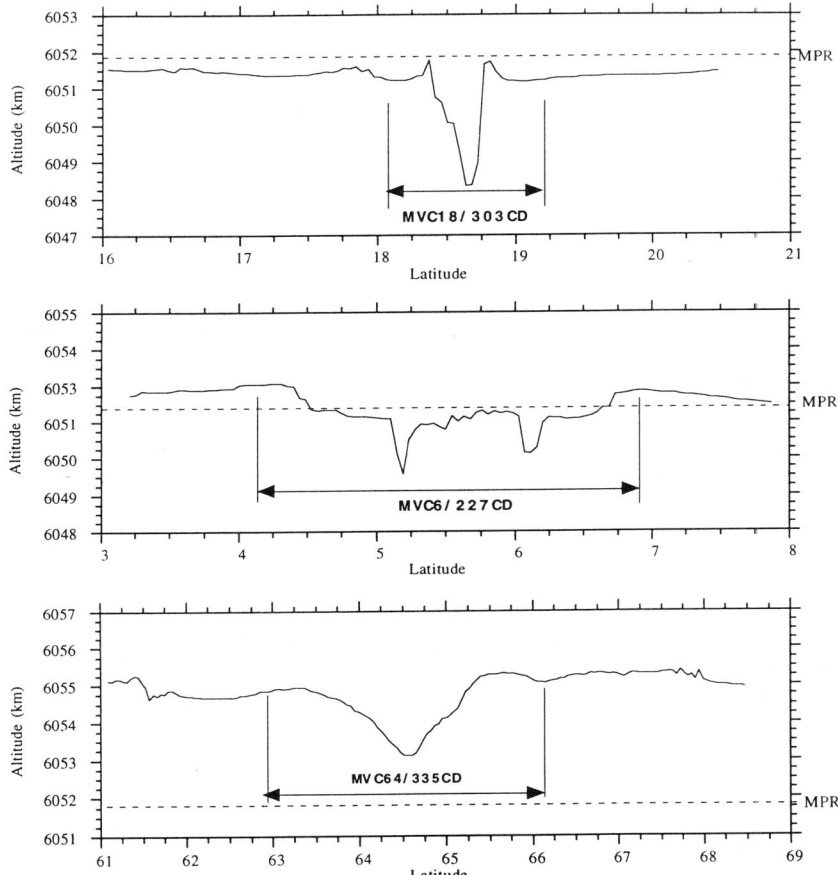

Figure 12. Representative topographic profiles across several calderas. Many calderas are relatively deep in comparison with those on Earth and Mars, and attain depths of several km. Horizontal axes are segments of a great circle; $1° \approx 105.7$ km.

informally described the following sub-types (1) dark streaked flows (nicknamed "amoeboid" after their amoeba-like shapes) frequently emanating from fractures; (2) "fluctus," large flow fields frequently flowing in one direction from a source area; and (3) "festoon," radar-bright flows that show organized patterns of internal ridges and flow bands. An example of the "festoon"-type has been described in detail by Moore et al. (1992). For the purposes of this chapter, we are concerned with large flow fields (exemplified by the "fluctus"-type) as another specific and unique type of volcanic center. Detailed work by Lancaster et al. (1993a,b,1992); Roberts Magee et al. (1992); Roberts Magee and Head (1993a,b,1994) has focused on the global population and analysis of these large flow fields. Lancaster et al. 1993a, has defined a "flow field" as a collection of individual lava flows, erupted from the same source or source area, that forms a spatially contiguous field. The presence of large flow fields on Venus similar in scale to terrestrial flood basalts was first determined from the analysis of Mylitta Fluctus, a flow field that covers approximately 300,000 km^2 in southern Lavinia Planitia (Campbell et al. 1991; Roberts Magee et al. 1992).

Two hundred and eight flow fields larger than $\approx$50,000 km^2 have been identified and an additional four flow fields are only marginally smaller than 50,000 km^2 (Table III, CDP6C1T3). Average values for individual flow lengths (measured in the along-flow direction) for the population of flow fields range from 60 to 570 km, with a mean value of $\sim$210 km.

Large flow fields have been previously classified by morphology (Lancaster et al. 1992,1993a; Roberts Magee et al. 1992). No correlation is observed between flow field morphology and total area. Sheet-like flow fields are typically large expanses of uniform radar backscatter. They comprise 9 to 18% of the total population of flow fields. Transitional flow fields are similar to sheet-like flow fields but contain one or more broad flow lobes and comprise $\sim$1% of the flow field population. Digitate flow fields have widely varying backscatter properties interpreted as multiple, overlapping radar-bright and -dark flows, and are divided into three sub-categories on the basis of variations in the amount of downstream divergence. Flows that are generally radially symmetric about a central source (e.g., a volcanic shield or corona) are termed "aprons" and characterize 50 to 52% of all flow fields on Venus. Fan-shaped flow fields widen substantially in their distal regions and represent $\sim$1% of all flow fields on Venus. Sub-parallel flow fields, such as Mylitta and Kaiwan Fluctus, are not radially symmetric about a central source and do not diverge substantially downstream. They comprise about 16 to 18% of all flow fields on Venus. Approximately 9% of all flow fields are apron/sub-parallel combinations. Typically the sub-parallel flow field is the most lengthy. Together the apron and sub-parallel categories represent the majority (74–79%) of large flow fields on Venus.

The most common source features for large flow fields are coronae, large volcanoes (>100 km in diameter), calderas (e.g., Fig. 11), fissures and fractures within rifts, fracture belts or other linear deformation belts, clusters of small volcanoes or "shield fields," and stellate fracture centers ("novae"). Coronae, large volcanoes, and fractures within rifts and fracture belts are associated with 82% of the large flow fields; and their associated flow fields are, on average, greater than the flow fields associated with the other source types. The smallest mean flow field areas are associated with stellate fracture centers ("novae").

The tectonic environments of large flow fields on Venus are dominated by zones of extension. Approximately 54% of all large flow fields are located within the BAT zone. 85% of the large flow fields associated with rifts and fracture belts post-date the onset of extension but may have been deformed by subsequent extension along the same deformation belt. Less than 2%

appear to predate rifting (or at least to have been substantially deformed by continued extension).

IV. SPATIAL DISTRIBUTION OF VOLCANIC CENTERS

Definitions of Venus volcanic features are currently under debate and different catalogs use different definitions and different size ranges. The Magellan Volcano Catalog (Crumpler et al. 1993d,1996b) was begun during the mission in an attempt to provide a basic data set, using consistent criteria, of all volcanic centers larger than 20 km on the surface of Venus. With completion of analysis of the cycle 2 and 3 image data, this data set now consists of 1194 volcanic edifices (intermediate and large size), shield fields, and calderas (not associated with edifices), all larger than 20 km (Fig. 13a,b). If distributed uniformly, these features would represent 2.6 volcanic centers/10^6 km^2 globally. However, volcanic centers are not uniformly distributed over the surface. Several areas of significant concentration are identified within the global population on the basis of density distribution data. The largest and most obvious concentration occurs in the region between longitudes 180°E and 340°E and between latitudes 40°N and 50°S and is is bounded by Atla Regio on the west, Beta Regio on the northeast and Themis Regio on the south. A somewhat smaller concentration occurs between longitudes 0°E and 50°E and latitudes 40°N and the equator and bounded by Eistla Regio on the west and Bell Regio on the northeast. Consideration of global densities of volcanic centers, independent of map projection, shows that these areas are centered about low latitudes and may be characterized broadly as equatorial. Conversely, few volcanic centers occur in the polar areas. We note that the orientation of the principal axes of the moment of inertia determined for Venus (Bills et al. 1987) may in this respect be coupled somewhat with the present distribution of volcanic centers. An analogous correlation of the near-equatorial arrangement of the Tharsis region (Ward et al. 1979) and hot spots (Jurdy 1983) occurs relative to the principal axes of the moments of inertia on Mars and Earth, respectively.

The areal density of large volcanic centers within the ~30% of the global surface area encompassed by the Beta-Atla-Themis concentration is three times the areal concentration in the remaining 70% of the global surface, and twice the global mean concentration (Fig. 14). In this respect, the global distribution of volcanic centers is similar to that of hot spots on Earth and volcanoes on Mars (Crumpler 1994); on both planets most of the global occurrences of anomalous hot spots and volcanic centers are also concentrated in a small fraction of the total surface area. The observed areal abundances of volcanic centers over significant fractions of the surface of Venus are thus as high as 4.3 volcanic centers/10^6 km^2 (1,990/global area). Considering the differing abundance of large volcanoes over several equal areas, and employing a χ^2 test similar to that used previously in analysis of hot spot distributions on Earth (Stefanik and Jurdy 1984), the probability that the

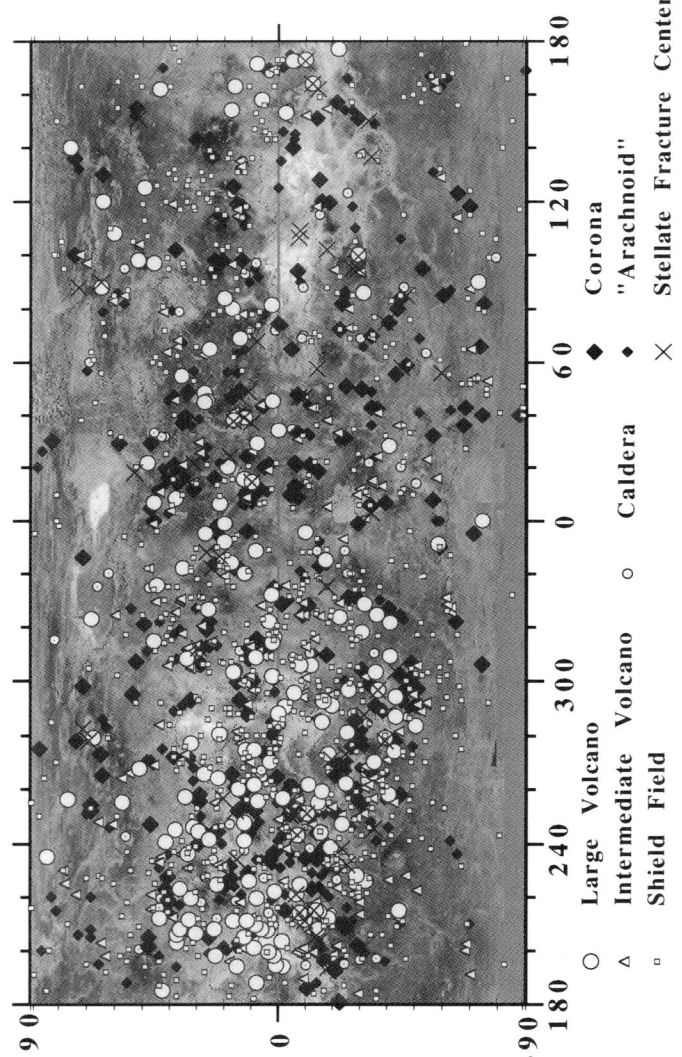

Figure 13. (A) Map showing the global distribution of large and intermediate volcanic edifices, shield fields (white symbols), and calderas (open circle with dot), in relation to arachnoids, coronae, and stellate fracture centers ("novas") (black symbols). Base map is a global SAR image mosaic; modified Miller cylindrical projection.

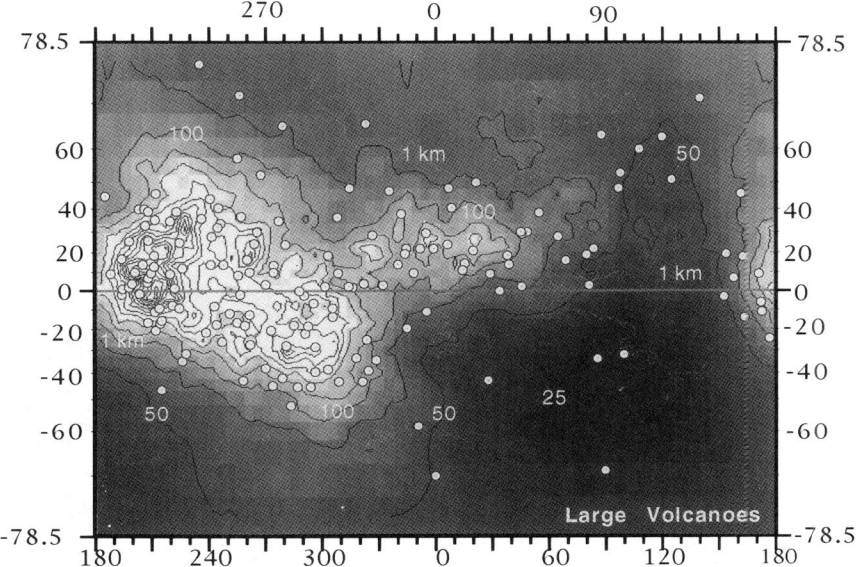

Figure 13. (B) Contours of areal density of large volcanoes in a mercator projection. The trace of the 1 km global relief contour (white lines) is shown for reference. Significant concentrations of volcanic centers occur in the western hemisphere. The global density of all volcanic edifices differs from the density of large volcanoes primarily in magnitude and in precise location of the highs, but preserves the same overall apparent distribution pattern. (Following hot spot density map convention for Earth, the density is expressed as the number that would occur in the global area based on the local density. Map was produced by contouring the density determined on a 106° grid; the density at each grid point was obtained by determining the circular area containing the nearest *nn* large volcanoes. The value of *nn* was selected to be equal to some meaningful indication of regional cluster tendency, in this case the number of centers [~50] in the average regional cluster of large volcanoes.)

observed concentrated distribution of large volcanoes results from chance alone is less than 0.001 (Table IV, CDP6C1T4). The apparent nonuniformity of observed distribution is therefore unlikely to be a random occurrence and represents a significant global anomaly in the distribution of volcanism and in the processes responsible for enhanced regional volcanism. If large volcanoes and coronae are plotted together for the purpose of comparison, this hemispheric enhancement of magmatic center abundance is impressive (inset, Fig. 14).

A. Physiographic Associations with Variation in Density Distribution

The nonuniform distribution of all of the volcanic centers is exemplified by the areal density of large volcanoes alone (Fig. 13b). Large volcanoes preferentially occur in the broad "rises" and tectonic junctions of intermediate elevation (Senske 1990) on the Equatorial Highlands and cluster in Atla Regio, Beta-Atla-Themis Regiones, and Eistla Regio. When corrected for

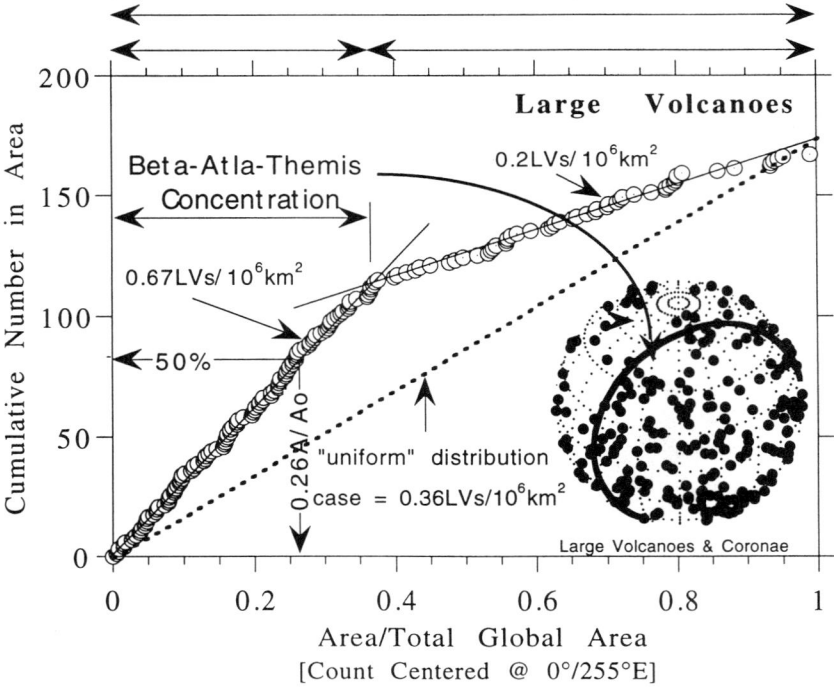

Figure 14. Cumulative distribution of large volcanoes about the center of apparent concentration in the western hemisphere (0°N, 255°E). One half of the observed population occurs within 26% of the global area. The significantly enhanced concentration is confined to a circle encompassing approximately 40% of the surface area. If the population were uniformly distributed, the trend would form a diagonal; plotted on a similar diagram; the observed crater population does not depart significantly from the diagonal. Graph illustrates large volcanoes, inset hemispherical plot illustrates large volcanoes and coronae.

hypsometric distribution, large volcanoes occur with greatest frequency in intermediate to upland altitudes (Keddie and Head 1994b; Crumpler et al. 1993a), and are sparse in areas of low and high altitudes.

In addition to variations in abundance of volcanic and magmatic centers, local differences in the most common type of magmatic center occur. For example, the topographic rise of Themis Regio is dominated by coronae (Hamilton and Stofan 1994) whereas Atla Regio is characterized by large edifices. The more prominent topographic rises, whether comprised largely of coronae or volcanic edifices tend to lie along strong tectonic junctions where many rifts or fracture patterns appear to converge (Senske et al. 1992).

The areas of low concentration represent all of the primary areas of lowland plains on Venus, eastern Atalanta, Vinmara, Sedna, Aino, Helen, and Lavinia Planitiae. Although *volcanic materials* are present as the dominant surface unit in lowland plains and numerous fissure vents may be located in the lowlands, there are few obvious volcanic centers. On the basis of the preceeding discussion, this distinction may be important. Identified sources for the plains-forming lava flows lie within the uplands areas of concentration. For example, the source for Baltis Vallis, the longest known lava channel and an apparent source for one of the widespread plains-forming lavas of Rusalka Planitia (Basilevsky and Head 1996), originates at a large volcano (MVC 45/185LV) on the northwest flank of the Beta-Atla-Themis concentration (see Fig. 5D). Perhaps significantly, many of these areas are peripheral to the largest concentration between Beta, Atla, and Themis Regio.

Volcanic centers are infrequent in the tesserae, which include most of the major highland areas, Ishtar and Aphrodite Terrae. With a few exceptions, notably central Beta Regio and central Lakshmi Planum, the highland areas represent "holes" in the global occurrence of volcanic centers (Fig. 15). Due to their location within the major Beta-Atla-Themis concentration, the relative dearth of centers within Beta and Phoebe Regiones emphasizes the obvious "gaps" in the distribution of volcanic centers in areas of tessera. However, there are volcanic centers present in Beta and Phoebe and these are associated with prominent rifts. This underscores the tendency for volcanic centers to occur in association with rifts, even within terranes where they are otherwise generally infrequent.

Magmatic centers that we characterize as being of dominantly intrusive origin, such as calderas, stellate fracture centers ("nova"), arachnoids, and coronae, may occur more frequently within tessera than large and intermediate volcanoes, suggesting that deep magma reservoirs are more easily developed than shallow reservoirs in tessera settings. That tesserae are not an impenetrable barrier to magma ascent is indicated by small volcanoes that occur within some tessera surrounded by small areas of intra-tessera lava plains. Instead, shallow magma chambers appear to be suppressed within tessera. Accordingly, in the absence of shallow reservoirs the long-term eruption and accumulation of many lava flows from a central source has not occurred within tessera and no large central volcanoes have developed. An obvious exception is Colette Patera in Lakshmi Planum (and there is no correlation with a through-going rift as there is in Beta Regio). The presence of major centers of eruption and lava accumulation in this highland, and not others, is in keeping with the variety of anomalous geophysical characteristics of Lakshmi Planum (Kaula et al. 1992). It also implies that lithospheric physical characteristics are not the sole determining factor in sustaining magma chambers; and that additional factors related to regional magmatic productivity and deep-seated mantle thermal and dynamic characteristics are equally important.

B. Geologic and Tectonic Associations with Variation in Density Distribution

Correlation with elevation in rises is assumed to result from primary formation factors, thermal uplift associated with mantle plumes (Bindschadler et

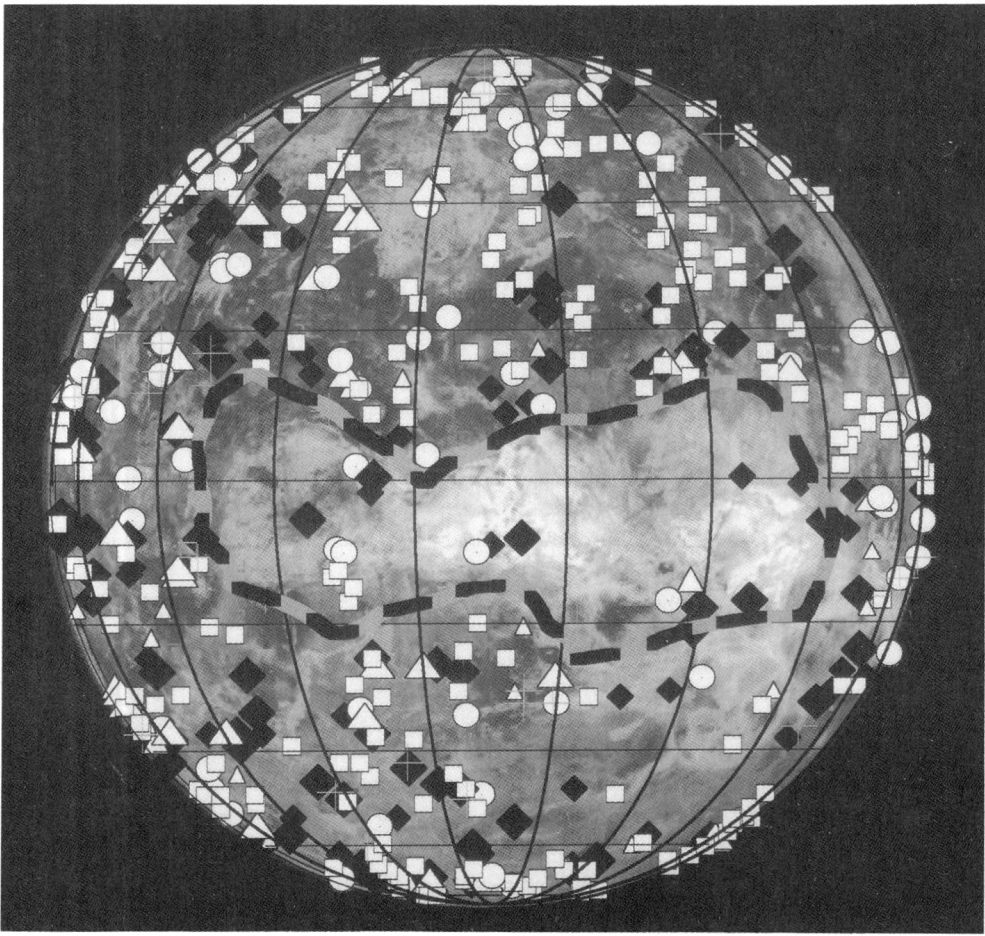

Figure 15. Hemisphere centered at 0°N, 90°E showing the distribution of magmatic centers in relation to the major tessera and highland area of Aphrodite Terra. Few centers occur within the area of Aphrodite Terra (dashed outline). Tessera such as Aphrodite have been shown to be stratigraphically lower than the surrounding plains and uplands on which many of the volcanoes occur. The absence of volcanoes within the highlands of Aphrodite is not a result of late resurfacing relative to the age of the observed volcano population; the occurrence of volcanic edifices appears to be suppressed within tessera highlands. Symbols are the same as those in Fig. 13a. Orthographic projection on global SAR image mosaic base.

al. 1992), or altitude-related effects on neutral buoyancy zones and edifice growth (Head and Wilson 1992). Although factors relating to preservation (post-formation subsidence and covering by plains-forming lava flows) could also play a part in causing an apparent concentration of volcanic centers in intermediate to high elevations, detailed geologic mapping thus far has shown little evidence for the array of partially buried volcanic centers on the lower flanks of the rises as might be expected if they represented preserved "islands" in lava-flooded plains.

The apparent global correlation of volcanic centers with intermediate elevations is a consequence of the fact that over one-half of the global population occurs within the most extensive region of intermediate (uplands) elevations, the Beta-Atla-Themis region. This correlation in turn derives from the observation that volcanic centers are frequently associated with rifting and, therefore, with the regional rises of intermediate (upland) elevation associated with rifting. Because the largest concentration of rifting and associated topographic rises occurs in the Beta-Atla-Themis region, volcanic centers and uplands tend accordingly to be associated, and this association is reflected in the global population

At a global scale, areas of low volcanic center concentration are correlated with areas of apparent crustal shortening (Crumpler et al. 1993*a,c*). Areas of crustal shortening of the type associated with ridge belts and ridged plains are contiguous and occur in interconnected regions of lowlands (Bindschadler et al. 1992; Solomon et al. 1992) many of which are peripheral to the Beta-Atla-Themis concentration (Crumpler et al. 1993*a,c*). This arrangement might be interpreted to represent the presence of broad regions of peripheral mantle downwelling surrounding the broad mantle upwelling in the Beta-Atla-Themis region (Crumpler et al. 1993*a,c*). In this model the characteristics of the Beta-Atla-Themis region are largely a result of associated higher heat flow and corresponding mantle melting as well as uplift, extensional, and regional rifting. Return mantle flow from this broad region of upwelling is accommodated by downwelling in the surrounding lowlands, and, as stressed by Bindschadler et al. (1992), with corresponding low heat flow, decreased mantle melt production and associated compressional deformation. A similar arrangement has been suggested to occur at regional scale for Eistla Regio (Grimm and Phillips 1992) and some of the highlands (Bindschadler et al. 1992). Thus we interpret the characteristics of the hemisphere centered on the Beta-Atla-Themis concentration as a large-scale example of the types of processes previously proposed for many of the large volcanic rises. These models are consistent in principle with variable scales of mantle dynamic phenomena, but, if so, require additional observation and testing to determine to what extent they are relics of past or more recent events. Observation of the inferred stress pattern and ages of radial or stellate fracture centers (Grosfils and Head 1994) suggest that the events responsible for this arrangement may be closer in timing with the estimated age of global resurfacing events than to the present.

V. GENERAL INTERPRETATION AND IMPLICATIONS

A. Size Distribution of Volcanoes and Calderas

An exponential distribution generally fits the observed population of volcanoes, but at diameters less than 200 km the observed population is notably less than that expected from larger diameters. Volcanoes appear distinctly depleted below 100 km (the upper range of intermediate volcanoes), and cumulative numbers significantly rise again below 50 km (Figs. 2 and 8). Thus the cumulative size distribution is notably nonlinear and the slope of the exponential curve varies smoothly in small diameter ranges.

Several factors may be operating to generate this departure from the general exponential form of the size distribution: (1) the cumulative production is discontinuous, the exponent is a function of size range, the exponential fit varies in slope at different size intervals, and certain diameter ranges follow a different distribution law, and (2) the cumulative production is continuous, a single exponential curve defines the production, and post-emplacement obliteration selectively removes portions of the population over certain size intervals.

The dimensions of many geologic phenomena resulting from nonproportional growth follow an exponential distribution law, and considering the nearly linear exponential size distribution-dependence of many different types of volcanic features on other terrestrial planets, hypothesis (1) seems unlikely, although it cannot be rejected. The implications of hypothesis (2) are potentially significant for other reasons. Considering that the observed population is the result of accumulation over geologically significant time, a component of the cumulative population statistics is potentially not a production population and could arise from partial destruction of the population. Those portions of the distribution for volcanoes that appear under-represented (50 km $< D <$ 200 km) might result from loss of edifices that are relatively infrequently formed and subject to global processes of resurfacing and at least partial burial. Those portions of the population that appear greater than the global average distribution curve ($D \leq 50$ km), may represent those edifices that, although relatively small, are more frequently produced and a correspondingly larger number have been accumulated in late geologic time either subsequent to most of the resurfacing events or at a rate exceeding obliteration rates. In any case, a detailed accounting of the sizes of volcanoes and their stratigraphic relationship to regional geologic units is needed as it may have some bearing on the question of magma production over the age of the preserved surface and since the time of proposed global resurfacing.

Similar overall arguments may be applied to the size-distribution of calderas. Comparison of the size distribution of calderas larger than 20 km in diameter on Venus with the calderas of Mars and Earth (Fig. 16) reveal information about the calderas on Earth as well as those on Venus. The cumulative distribution of caldera diameters for all three planets is exponential in form, which appears to characterize the populations of volcanic centers of all types. For both the calderas of Mars and Venus, the slopes of the cumulative distribution curves are similar and differ only in magnitude. Because calderas reflect strains at the surface associated with volume changes in an underlying magma reservoir, the similarity in distribution for Venus and Mars predicts that reservoir development and caldera formation on both planets fol-

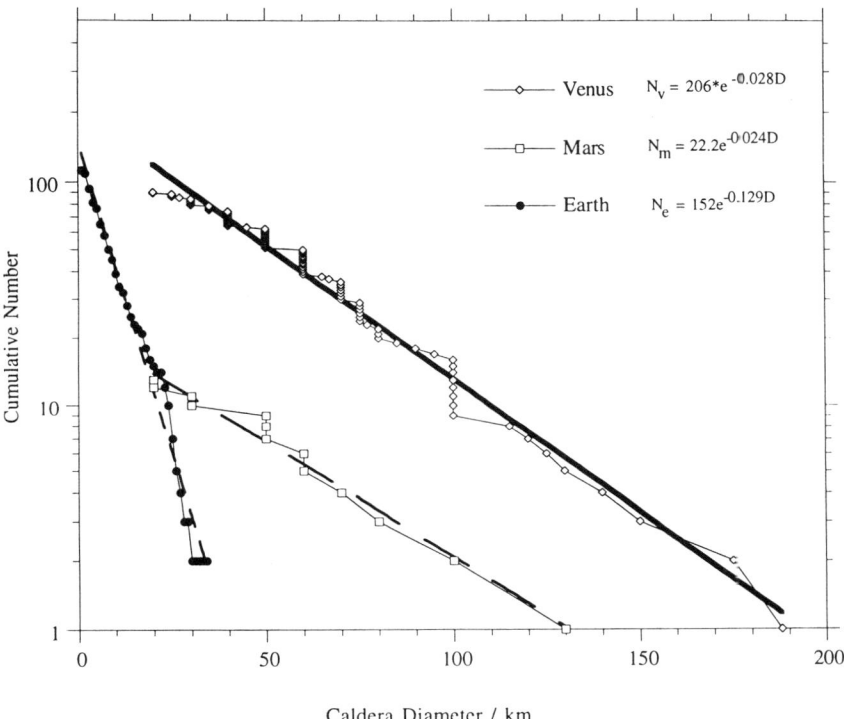

Figure 16. Comparison of the cumulative distribution of caldera dimensions on Earth, Mars, and Venus. All three planets show exponentially distributed cumulative size distributions. The similarity in slopes of the curves for Mars and Venus imply that the process of reservoir growth is similar in silicate planets with stable lithospheres. The steep slope of the cumulative curve for calderas on Earth may reflect the exclusion of large calderas from the recent geologic record due primarily to the significantly lower relative rates of formation of large calderas and corresponding lower accumulated number within recent geologic time; alternately, large calderas may be excluded from forming on Earth due to dynamic plate motions and the corresponding inability to grow large magma reservoirs. Data for Earth are Quaternary calderas (from Walker 1984); data for Mars are from Crumpler et al. (1996a).

low similar fundamental processes that are common to silicate lithospheres in general and result from thermal and mechanical conditions governing growth, stability, and longevity of shallow magma reservoirs.

The total number of calderas (1 km diameter and greater) currently known on Earth is similar to the number (20 km diameter and greater) estimated for Venus, assuming the exponential form of the distribution continues at small diameters. This assumption is important in the comparison because the smallest calderas in our sample for Venus approach the dimensions of the larger calderas on Earth, and thus the curves do not converge. The distribution for Earth also differs in that it is markedly steeper, even provided that only calderas >20 km are considered, and denotes a dearth of large calderas relative to Mars and Venus. There are several possible origins of this difference.

First, the caldera population used for Earth is of Quaternary age, only because accurate determination of caldera diameters or identification of calderas much older than this is difficult due to erosion and removal by tectonic destruction. Cauldrons (large structural volcanic subsidence features) and volcano-tectonic depressions might be included in the compilation, but they are neither all identified nor preserved, and the resulting statistics would inaccurately reflect the population. The Quaternary calderas of Earth are therefore reliable as an indicator of the slope of the production curve, if not the magnitude of the production. In contrast, calderas of Venus and Mars reflect the accumulated record of caldera formation throughout the observable age of their surfaces. In this case, the total number of calderas N_c accumulated on the surface over a given period of time will depend on the instantaneous production and destruction rate $dN_c/dt = P_c - vN_c$. Both P_c, the rate of production, and v_c, the fractional rate of destruction, are probably exponential functions of diameter for most processes of formation, obscuration, and obliteration, such that many small calderas form and are obliterated in the time interval over which only a few large ones are formed and are obliterated. If the rate of production is high and an inverse exponential function of diameter, but destruction is an inverse exponential function with a lower slope (as appears to be the case for the efficient process of erosion and for tectonic removal), then the relative size distribution accumulated within a short time interval, such as the Quaternary of Earth, will be depleted in large diameter calderas relative to the production curve. Significant changes in the overall rate balance of destruction at different diameters for an extended period of time could suffice to impart different slopes to different size ranges of the overall cumulative distribution. If a detailed counting of small calderas between 1 and 20 km was shown to be characterized by a significantly greater cumulative slope, then it may be evidence for significant resurfacing over geologically large intervals of time. Provided that the rate of caldera formation could be reliably estimated, then it would be possible to estimate the magnitude of the time interval. In the comparison of Fig. 16, the difference in diameter distributions for Earth could easily reflect the selective preservation, and counting, of smaller calderas.

A second hypothesis considers the assumed inherent stability of the surfaces of Mars and Venus relative to Earth. On Mars and presumably on Venus, the lateral dimensions of reservoirs are limited only by the longevity, rate of supply of magmas, and mechanical characteristics of the crust capable of sustaining an active reservoir. On Earth, reservoirs larger than a certain size may be infrequent due to lithospheric motion of plate tectonics; reservoirs may move away from the source before growth to large sizes can occur.

Finally, the assumed continuity of the exponential distribution may be incorrect, and calderas between 1 and 20 km in diameter may be more abundant than predicted by extrapolation from large diameters. This is possible because calderas of different scale on Earth may form through a variety of processes which differ in detail depending on the host rock, stress environment, and the manner in which the ascent and eruption progresses. Differences in the size distributions relative to Earth could arise because there are as yet unconsidered conditions related to magma chemistry (temperature, density, dike width, or viscosity), that we do not see on Earth, such as more primitive mafic magmas that are able to accumulate larger reservoirs. Because calderas imply the presence of shallow reservoirs, any variation in the conditions favoring shallow reservoir formation, particularly as regards size and depth, will influence the observed size distribution of calderas. If so, then the slope of the exponential curve representing the size distribution could steepen at small diameters implying, for instance, that smaller reservoirs are more easily formed.

B. Significance of Large Volcano Dimensions and Volumes

Although large in diameter, the volume of individual large volcanoes is low (Keddie and Head 1994*b*) due to their relatively low heights. The relatively small overall volume of the largest volcanoes on Venus in relation to volcanoes of similar size on Mars and Earth, despite the apparently stable configuration of the lithosphere with respect to probable sub-lithosphere sources on Venus, may be significant in deducing characteristics of the magma supply and total volume of the source regions. Models for the formation of large centers that rely on the presence of plumes may require plumes of considerably smaller volume than those used to account for similar volcanic provinces on Earth or Mars. Alternately, the extrusive component of volcanic centers may be lower on Venus for the same overall volume of the magmatic system and the intrusive component might be a larger proportion of the overall magma output. The great abundance of coronae and other features arising largely from fundamentally intrusive subsurface effects of magma emplacement may be consistent, in principle, with a generally greater ratio of intrusive to extrusive magmatism relative to Earth and Mars.

The total global volume of the observed population is 8.7×10^6 km^3 if the average dimension of large volcanoes is assumed for each of the 167 observed examples. The estimated volume rate of magma erupted in order to accumulate the observed population of volcanoes over the 500 to 800 Myr since a global resurfacing event is 1×10^{-2} to 1.7×10^{-2} km^{3}yr^{-1}. This estimate is for the rate of large volcano production alone and does not take into account the intrusive component of magmatism or any other types of volcanism that occurred during the same time interval. Volume rates of this magnitude, together with the low magma replenishment rates we estimate as necessary to yield shallow reservoirs (discussed below), predict that evidence for shallow magma reservoirs should be common on Venus where environmental conditions are favorable for their initiation. Furthermore, given the low necessary rates, it is also clear that increased rate regimes at individual centers of volcanism may be accompanied by extensive lateral growth of reservoirs by lateral dike injection. This agrees in principle with the contention (Grosfils and Head 1994) that many of the radial fracture patterns observed in association with magmatic and volcanic centers may be evidence for the propagation of dikes from shallow reservoirs. The growth of unusually large reservoirs is consistent with formation of the large scale magmatic centers such as coronae and arachnoids.

C. Origin and Emplacement of Shield Fields

Several fundamental questions are raised by the presence of abundant small volcanoes on Venus: (1) what type of volcanic activity do they represent? (2) what is the volume contribution of numerous but small volcanic source vents to the global volcanic budget? (3) what is the volcanological significance of abundant volcanoes occurring in "fields"; and (4) what might shield fields imply for the resurfacing history of Venus?

1. Structure and Stratigraphy of Small Shield Volcano Fields. An assessment of these questions requires an understanding of their emplacement. Three possible models for the origin and stratigraphic relationship of fields of small volcanoes have been discussed (Aubele and Crumpler 1992): (1) a field represents an "island" of higher topography subsequently surrounded by later plains materials; (2) a field represents the area of a region of anomalous melting; or (3) a field represents the area of a magma reservoir. Model 1 would imply that the fields represent portions of a stratigraphic "layer" of small edifices produced globally in an earlier period of greater small shield productivity and that there has been a change in eruption style with plains formation occurring predominantly after the production of the small edifices. This model may explain the "shield plains"; however, local stratigraphic relationships of the 647 shield fields seem to indicate a possible range of some shield field ages in relation to the surrounding regional plains units and associated larger volcanic features, implying that, although enhanced small volcano formation may have occurred at some time or in some regions, all small volcano formation did not occur planetwide as a single event. Models 2 and 3 imply that the fields represent areas of melting anomalies, either a region of melting or the actual extent of a magma reservoir, and might explain the generally equant shape and average diameter of many shield fields.

Another important question is the relationship of shield fields to the extensive ridged plains of Venus. Three models, some of which are also related to the models of origin described above, have been proposed (Aubele and Crumpler 1992): (1) the edifices are all formed contemporaneously with, or may be the source of, the ridged plains; (2) the edifices all predate the plains; or (3) the edifices all postdate the plains. These are simplified hypotheses of a potentially more complex situation, and they have differing implications for the volume contribution from each edifice, both extrusive and intrusive, and for possible changes in eruption style and thermal evolution of Venus with time. Preliminary results of the geologic mapping of Venus indicate that model 3 may represent the correct stratigraphic relationship (Aubele 1995).

2. Origin of Shield Fields. The origin of shield fields may be significant for understanding rates of magmatism on Venus. Large volcanoes and calderas are evidence for the presence of shallow magma reservoirs. Calderas are the most obvious manifestation of shallow reservoirs and are important as a means of inferring the presence of shallow magma bodies. The concentric patterns of fractures and central depression of calderas are physical evidence for the strain and brittle fracture of the surface material resulting from volume changes at shallow depths. Radially distributed lava flows erupted from a single vent or vent region that have constructed an edifice are a less obvious, but equally important, indicator of shallow magma reservoirs. In contrast, areal concentrations of many small volcanoes are evidence for the repeated ascent of small batches of magma without long-term residence in a shallow reservoir. The difference between a single large edifice and a group of small edifices therefore reduces to the conditions necessary for shallow magma reservoir formation.

Variations in scales and long-term rates of magma transport, storage, and eruption are correlated with differences in large-scale eruption characteristics, magma crystallinity, morphology of volcanic centers, and the presence or absence of shallow magma reservoirs on Earth (Smith 1979; Fedotov 1981; Hardee 1982; Crisp 1984; Shaw 1985; Hardee 1986; Tsukui et al. 1986). The largest and shallowest magma reservoirs and the most voluminous eruptions are associated frequently with moderate rates of magma emplacement in mid-crustal levels and corresponding development of large magma reservoirs (Shaw 1985). On Earth this moderate rate is equivalent to the rate of magma intrusion and eruption at the Hawaiian hot spot, averaged over the past 70 Myr. Rates of emplacement lower or higher than this moderate rate tend not to yield well-developed, isolated, long-lived magma reservoirs that evolve chemically and structurally in association with single volcanic edifices. Rates of emplacement lower than this value on Earth frequently result in numerous short-lived volcanic centers and are characteristic of "fields" of small volcanoes (see, e.g., Settle 1979; Hasenaka and Carmichael 1985; Condit et al. 1989; Connor et al. 1992; Crumpler et al. 1994). Rates greater than the moderate rate regime frequently result in massive flow fields, incompletely developed volcanic centers and magma reservoirs, and relatively minor associated amounts of evolved magma compositions (Fedotov 1981; Shaw 1985; Hardee 1986).

The rates necessary to sustain and grow a reservoir in the Venus environment can be estimated. Using the numerical models and methods outlined in terrestrial studies (Fedotov 1981; Shaw 1985; Hardee 1986), we have calculated the influence of the higher thermal environment on magma transport and thermal losses within intrusions in the shallow crust of Venus. The results predict that, for a given magma replenishment rate into a shallow reservoir, relatively lower rates of magma supply are necessary to support stable magma reservoirs on Venus relative to Earth, but nonetheless are comparable in magnitude to that associated with small volcanic fields on Earth. On Venus the calculated rate is 5×10^{-4} km^{3}yr^{-1} and on Earth it is 10^{-3} km^{3}yr^{-1}. The difference between a single large edifice and a group of small edifices are predicted to be a result of the long-term rates of magma emplacement. Fields of small shield volcanoes on Venus can be interpreted to represent a low and steady supply of magma to the surface from a melting anomaly for a brief interval of geologic time, but at rates insufficient to form shallow magma reservoirs. The coincidence of large magmatic centers and shield fields need not refute this assertion. The occurrence of shield fields within coronae and at the summits of many large volcanic centers may reflect late stage waning of magma supply rates analogous to that possibly responsible for summit volcanic fields of Mauna Kea and other terrestrial centers that are otherwise characteristic of larger replenishment rates.

Thus, two contrasting styles of volcanic center are identified on Venus, those resulting from eruption of magmas from melt source regions productive enough to support shallow reservoirs and those resulting from eruption of magmas from melt source regions at rates too low to support shallow reservoirs. The ability to assess the rates of magma supply necessary for these two fundamentally different styles of volcanic center may be of significance in estimating global magma rates through the observable history of Venus.

D. Significance of Large Flow Fields

Because of the similarity in scale, large flow fields are considered to be Venus analogs to terrestrial flood basalts. The origin of flood basalts on Earth has been linked to large-scale mantle upwellings or plume heads and lithospheric extension, although the relative importance of these processes is a matter of controversy (see, e.g., McKenzie and Bickle 1988; Richards et al. 1989; White and McKenzie 1989; Campbell and Griffiths 1990; Hooper 1990; Hill 1991; Anderson et al. 1992; White 1992). Observations indicate that the formation of flood basalts on Venus, under current conditions, generally requires the presence of thinned and rifted lithosphere to enhance decompression melting in upwelling mantle material. In addition, an association of large flow fields on Venus with possible mantle plumes is indicated by their presence in regions of broad, domical uplifts, such as Atla and Beta Regiones, and clusters of volcanic features that may represent upwellings formed as secondary instabilities from large plume heads at depth.

Several large flow fields originate at large calderas where fractures encircle a depression indicative of a reservoir. In general, large volume flood eruptions are likely to be derived from eruptions directly from the source region; a shallow reservoir is unlikely to attain volumes large enough to feed lava floods of the observed dimensions. However, if a shallow magma trap exists, there is nothing to exclude a shallow reservoir from forming in association with a large volume eruption.

The large flow fields on Venus, i.e., those we discuss here, are stratigraphically younger than the plains on which they occur; however, they are not all necessarily the youngest unit or event in each individual region and there is no evidence that they all formed at the same time globally or occur at identical stratigraphic positions. Flood-type eruptions may have been responsible for the formation of both the population of large flow fields and the background plains on Venus. However, the population of large flow fields discussed here clearly post-dates the formation of extensive ridged plains and is restricted in large part to zones of rifted and presumably thinned lithosphere. The presence of flood-scale volcanic flow fields younger than the widespread lowland plains of Venus indicates that a minimum of 9% of the planet has been resurfaced, by lava flow fields, in relatively recent geologic history.

E. Significance of Regions of Enhanced Volcanic Center Concentration

The importance, and underlying premise, of determining global volcano distribution is that variations in distribution in general, and concentrations in particular, indicate areas of anomalous melting in the mantle. Once the distribution is known, attempts can be made to reconcile these variations with observable regional geologic characteristics. These characteristics are useful as they can provide some insight into the nature and potential origins of the melting anomaly. Given that some of the large volcanoes and volcanic rises are associated with significant gravity anomalies (Sjogren et al. 1983) in excess of that attributable to topography alone, an additional premise is that the melting anomaly may be related to dynamic mantle phenomena, diapirs, or plumes (Smrekar 1994; Grimm and Phillips 1992; Bindschadler et al. 1992; Senske et al. 1992).

The principal examples of enhanced concentration include the Beta-Atla-Themis region, at the large end of the scale spectrum, and the individual volcanic rises on the lower end of the size spectrum. Both examples occur at elevations higher than the mean planetary radius. Most of the areas of low volcanic center concentration are areas that lie at or below mean planetary radius and are characterized by the presence of extensive ridged plains consisting of vast sheets of relatively smooth flood-type lavas. Taken together with the proposed single age of global resurfacing (Schaber et al. 1992; Strom et al. 1994) this distribution might be interpreted to reflect the relative preservation of volcanic edifices in high elevations and their burial in low elevations during a near-global volcanic resurfacing event. However, there appear to be few, if any, partially buried edifices other than small volcanoes on the lower reaches of these elevated regions of edifice concentration. Several studies have argued also that, collectively, the large volcanic and magmatic centers, including large volcanoes (Namiki and Solomon 1994; Price 1995), stellate fracture centers (Grosfils and Head 1995), and coronae (Namiki and Solomon 1994), have significantly lower crater abundances. Based on these studies, as a class, the large volcanic and magmatic centers appear to have accumulated chiefly in post global-resurfacing time, but for reasons outlined in the discussion of large volcanoes, the detailed stratigraphy of the total population of large centers must be determined before this conclusion can be validated.

Global concentrations appear to represent sites where the process of large volcanic and magmatic center formation has actually been concentrated. Two models appear to satisfy this arrangement. They arise from consideration of either large-scale mantle dynamic phenomena or from controls on the process of shallow reservoir formation exerted principally by elevation. In the first model, the concentrations of large magmatic centers are the surface manifestation of anomalous melting at depth. The origin of the melting anomaly may itself be fundamental, that is, arising from long-lived physical and chemical heterogeneities of the deep interior, or possibly residual from the process (Arkani-Hamed 1994; Parmentier and Hess 1992; Turcotte 1993) which initiated large-scale global resurfacing. The frequent association with areas of enhanced rifting and of significant gravity anomalies in this model would imply the response of the shallow mantle to warm, upwelling deeper mantle material, presumably as diapirs and plumes. In this interpretation, the areas of low concentration may be the corresponding sites of downwelling, relatively cool mantle, and correspondingly limited opportunities for melting and volcanism. The presence of many of the more prominent examples of contractional deformation (ridges, ridge belts, and mountain belts) in the regions of low concentration, where converging strain is predicted in association with converging mantle flow (Bindschadler et al. 1992), support this scheme (Crumpler et al. 1993a,c). This large-scale model assumes that the extensive ridged lava plains, which may be more voluminous than all of the volcanic centers combined, either have sources within the areas of upwelling, and have flowed down into and filled the lowlands, or predate the large-scale upwelling anomalies.

The second model derives from the prediction (Head and Wilson 1992) that the high atmospheric pressure on Venus will reduce volatile exsolution in magmas as they ascend to the surface relative to that on Earth, leading to a relatively high-density crust that inhibits formation of neutral buoyancy levels ultimately necessary for shallow reservoir formation. As a result, the formation of large reservoirs and the corresponding large centers of volcanism are inhibited in low elevations. The details of the model predict that formation of shallow reservoirs at high elevations will be possible in certain instances. Those magmas that ascend in low elevations will ascend directly from deep source regions and will be characterized by high total volumes and possibly high effusion rates. In this model, the primary difference in central volcano abundances arises from existing topographic variations, perhaps enhanced in areas of rifting and local uplift at the sites of regional upwelling by relatively easier access to the surface.

Currently both models appear applicable, and both may operate in combination to generate the observed distribution of magmatic centers. The first model may be tested by determining to what extent the observed deformation of lowland plains predates the Beta-Atla-Themis concentration. The latter model is testable if it can be shown that the vents for the plains-forming lavas are located within the lowlands. In general, more work needs to be done in determining the relative stratigraphic relationships of individual volcanic centers at global scales to ascertain the extent of differences in age of the different styles of eruption, the sources of plains lavas, and the stratigraphic position of current and potential past sites of mantle upwelling and downwelling. Many of these issues may be resolved with further global geologic mapping and modeling of the Magellan gravity results. Of key importance will be the continued identification of characteristics from each of these models, and additional models, that may be tested with geologic and geophysical observations. Currently many of the identified tests are unconstrained.

VI. CONCLUSIONS

There are 1194 volcanic edifices, shield fields, and isolated calderas larger than 20 km identified on Venus from the results of Venera, Magellan, and Earth-based observations. Smaller volcanic centers probably encompass many times this number, and need to be identified, classified, and cataloged before a complete assessment can be made of their significance. Together with tectonism, volcanic centers and volcanic deposits account for most of the geologic characteristics of the surface, in contrast to all other smaller terrestrial planets where the geologic characteristics of the surfaces are profoundly influenced by impact cratering. It is also in contrast to Earth where atmospheric effects, chiefly water erosion, have obscured many of the volcanic and tectonic events that have occurred throughout geologic history. The ultimate significance of Venus from the volcanological perspective is that the processes of volcanism have been largely preserved. Many of the fundamental processes of volcanism may be better understood because of the level of detail preserved, evidence of the evolution of each center, and the great surface area and number of examples of each type of volcanic center available for inspection. On the basis of the existing survey we may make some general observations about volcanic centers in the size range studied.

Volcanic centers on Venus follow a general exponential size distribution. Details of the size distribution at diameters <20 km remain to be determined. Based on the existing data, some departures from linear cumulative size frequency appear likely. Significant departures of a cumulative size distribution from an exponential form may reflect differences in preservation of different diameter portions of the cumulative populations. Either differences in the production rate of volcanic centers in different size ranges has occurred, or significant obliteration has affected only a part of the population. Steeper slopes of cumulative size-frequency curves at smaller diameters, if observed in future studies, may affect estimation of the rate and scales of resurfacing over the observed surface age.

If distributed uniformly, the observed population of volcanic centers would represent 2.6 volcanic centers per million square km globally. Volcanic centers on Venus are not uniformly distributed and more than one-half of the population of centers and large flow fields occur in <30% of the global area (which primarily includes the Beta-Atla-Themis region). In general, large volcanoes occur with the greatest frequency in intermediate to upland altitudes where rifting is common; they are infrequent in areas of lowland plains where characteristics of crustal shortening are common, and are infrequent at high altitudes (areas that also are dominated chiefly by tesserae). Broad mantle upwelling in the Beta-Atla-Themis region and enhanced shallow magma reservoir formation at upland elevations may together account for this hemispheric concentration of volcanoes.

Two distinct styles of volcanic center are identified: (1) those that are characterized by voluminous flows generally emanating from a single well-defined point or vent region over the course of many eruptions; and (2) regions on the order of 100 to 200 km in diameter characterized by numerous small edifices and evidence for relatively limited associated volumes of lava flows. The first type of center includes large and intermediate-size volcanic edifices, flow fields and calderas. The second type includes clusters of small size edifices. The difference between a single large edifice or voluminous eruption and a group of small edifices reduces to the conditions necessary for shallow magma reservoir formation.

There is much evidence from Earth and the terrestrial planets that the long-term rate of magma eruption, emplacement, and replenishment (the magma rate regime) may be the most significant influence on the presence of shallow magma reservoirs, long-lived volcanic centers, and the overall morphology of volcanic centers. The formation of shallow magma reservoirs requires both trapping, or stalling, of magma at shallow depths, as well as specific rates of magma delivery to shallow depths in order for a reservoir to form and grow. The development of shallow magma reservoirs is necessary for the formation of centers of repetitive volcanic eruption (i.e., large volcanoes). The primary difference in the characteristics of large volcanic centers and fields of edifices, therefore, is not petrologic or tectonic association, but differences in the magma ascent and emplacement rates and the presence or absence of stable shallow magma reservoirs during the period over which accumulation of the center of volcanism takes place at the surface. Other factors such as the geochemical character of magmas, surface environment of eruption, and buoyancy considerations, traditionally viewed as important, may be important in controlling the character of eruptions in individual volcanic centers; but the long-term style of the eruptions appear to be correlated with magma rate regime.

Our calculations indicate that Venus shield fields form where magma replenishment rates are approximately 10^{-4} km^{3}yr^{-1} or lower. Rates higher than this are predicted to result in shallow reservoirs and large volcanoes. Estimated global eruption rates for the population of large volcanoes based on inferred surface age and volume estimates are approximately 1.7×10^{-2} km^{3}yr^{-1} and support this result. The predicted magma rates associated with fields of small volcanoes should be testable in a similar manner. A global estimate of eruption rates for the population of shield fields will require a detailed global accounting of their numbers and associated lava volumes as well as their stratigraphic positions. The combined available theoretical and observational estimates imply that there has been a spectrum of long-term magma replenishment rates operating throughout the observed history of Venus from shield fields, on the low end, through individual large volcanoes and calderas on the higher end. Higher short-lived rates in which the local tectonic environment favors direct magma ascent without substantial residence in the shallow crust are responsible for brief periods of flood-like lava eruption. The large range of magma rates, together with the differing environments and magma ascent conditions among volcanic centers, have resulted in a great morphological diversity. And the great abundance of shield fields in relation to other types of magmatic features on Venus is a predicted result of a relatively more frequent occurrence of low magma ascent rates from relatively deep reservoirs or source layers. A major question that remains to be addressed, and that has significance for the long-term thermal evolution of Venus, is whether or not the relative frequency of occurrences of low and high magma rate events has changed with time. These estimates together with estimates of magma production from stratigraphic studies and other means will ultimately be important in testing models for the global rate of magma production over the observable geologic history of Venus.

Acknowledgments. Support for initial cataloging was provided by NASA grants. Helpful reviews by E. R. Stofan and K. Tanaka and comments from M. Bulmer enabled clarification of many issues discussed in this chapter and are gratefully acknowledged. The discussion on volcanic depressions benefited from two decades of off and on discussion with those, notably including W. E. Elston, who have debated this very issue.

REFERENCES

Anderson, D. L., Zhang, Y.-S., and Tanimoto, T. 1992. Plume heads, continental lithosphere, flood basalts and tomography. In *Magmatism and the Causes of Continental Break-up*, eds. B. C. Storey, T. Alabaster and R. J. Pankhurst, GSA SP-68 (Boulder: Geological Soc. of America), p. 404.

Anderson, S. W., Crown, D. A., Plaut, J. J., and Stofan, E. R. 1994. Surface characteristics of steep-sided domes on Venus and terrestrial silicic domes: A comparison. *Lunar Planet. Sci. Conf.* XXV:33–34 (abstract).

Arkani-Hamed, J. 1994. On the thermal evolution of Venus. *J. Geophys. Res.* 99:2019–2033.

Aubele, J. C. 1993. Venus small volcano classification and description. *Lunar Planet. Sci. Conf.* XXIV:47–48 (abstract).

Aubele, J. C. 1994. Stratigraphy of small volcanoes and plains terrain in Vellamo Planitia, Venus. *Lunar Planet. Sci. Conf.* XXV:45–46 (abstract).

Aubele, J. C. 1995. Stratigraphy of small volcanoes and plains terrain in Vellamo Planitia-Shimti Tessera region, Venus. *Lunar Planet. Sci. Conf.* XXVI:59–60 (abstract).

Aubele, J. C., and Crumpler, L. S. 1992. Shield fields: Concentrations of small volcanoes on Venus. In *International Colloquim on Venus*, LPI Contrib. No. 789, pp. 7–8.

Aubele, J. C., and Slyuta, E. N. 1990. Small domes on Venus: characteristics and origin. *Earth Moon Planets* 50/51:493–532.

Aubele, J. C., Head, J. W., and Crumpler, L. S. 1992. Fields of small volcanoes on Venus (shield fields): Characteristics and implications. *Lunar Planet. Sci. Conf.* XXIII:47–48 (abstract).

Aubele, J. C., et al. 1995. Locating small volcanoes on Venus using a scientist-trainable analysis system. *Lunar Planet. Sci. Conf.* XXVI:61–62 (abstract).

Baker, V. R., et al. 1992. Channels and valleys on Venus: Preliminary analysis of Magellan data. *J. Geophys. Res.* 97:13421–13444.

Barsukov, V. L., et al. 1986. The geology and geomorphology of the Venus surface as revealed by the radar images obtained by Veneras 15 and 16. *Proc. Lunar Planet. Sci. Conf.* 17, *J. Geophys. Res.* 91:378–398.

Basilevsky, A. T., and Head, J. W. 1995. Global stratigraphy of Venus: analysis of a random sample of thirty-six test areas. *Earth Moon Planets* 66:285–336.

Basilevsky, A. T., and Head, J. W. 1996. Stratigraphic studies in the Baltis Vallis region, Venus: Implications for areal extent and timing of volcanic resurfacing events. *Geophys. Res. Lett.*, in press.

Bibring, J.-P., Combes, M., Langevin, Y. 1990, ISM observations of Mars and Phobos; first results. *Proc. Lunar Planet. Sci. Conf.* 20:461–471.

Bills, B. G., Kiefer, W. S., and Jones, R. L. 1987. Venus gravity: a harmonic analysis. *J. Geophys. Res.* 92:10335–10351.

Bindschadler, D. L., Schubert, G., and Kaula, W. M. 1992. Coldspots and hotspots: Global tectonics and mantle dynamics of Venus. *J. Geophys. Res.* 97:13495–13532.

Bridges, N. T. 1995. Submarine analogs to Venusian pancake domes. *Geophys. Res. Lett.* 22:2781–2784.

Bridges, N. T., and Fink, J. H. 1994. Aspect ratios of lava domes on Earth, Moon, and Venus. *Lunar Planet. Sci. Conf.* XXIII:159–160 (abstract).

Bulmer, M. H. 1994. An Examination of Small Volcanoes in the Plains of Venus, With Particular Reference to the Evolution of Domes. Ph.D. Thesis, Univ. of London.

Bulmer, M. H., and Guest, J. E. 1994. Modified lava domes on Venus. *Lunar Planet. Sci. Conf.* XXV:193–194 (abstract).

Bulmer, M. H., and Guest, J. E. 1995. Modified volcanic domes and associated debris aprons on Venus. *Special Publ. Geol. Soc. London*, in press.

Bulmer, M. H., Guest, J. E., and Wiles, C. R. 1991. Morphological characteristics of small monogenetic volcanoes in southern Guinevere Planitia: Implications for eruption conditions. *Lunar Planet. Sci. Conf.* XXII:505 (abstract).

Bulmer, M. H., et al. 1992a. Debris avalanches and slumps on the margins of volcanic domes on Venus; characteristics of deposits. In *International Colloquium on Venus*, LPI Contrib. No. 789, pp. 14–15.

Bulmer, M. H., Guest, J. E., and Stofan, E. R. 1992b. Calderas on Venus. *Lunar Planet. Sci.* XXIII:177–178 (abstract).

Bulmer, M. H., Guest, J. E., Michaels, G., and Saunders, S. 1993. Scalloped margin domes: what are the processes responsible and how do they operate? *Lunar Planet. Sci. Conf.* XXIV:215–216 (abstract).

Bulmer, M. H., Stofan, E. R., and Guest, J. E. 1995. Phenomenology of calderas on Venus. *Lunar Planet. Sci. Conf.* XXVI:187–188 (abstract).

Burl, M. C., et al. 1993. A pattern recognition system for locating small volcanoes in Magellan SAR images of Venus. *Lunar Planet. Sci. Conf.* XXIV:227–228 (abstract).

Campbell, B. A., and Rogers, P. G. 1994. Bell Region, Venus: Integration of remote sensing data and terrestrial analogs for geologic analysis. *J. Geophys. Res.* 99:21153–21171.

Campbell, D. B., Head, J. W., Harmon, J. K., and Hine, A. A. 1984. Venus volcanism and rift formation in Beta Regio. *Science* 226:167–170.

Campbell, D. B., et al. 1989. Styles of volcanism on Venus: New Arecibo high resolution radar data. *Science* 246:373–377.

Campbell, D. B., et al. 1991. Venus southern hemisphere: Character and age of terrains in the Themis-Alpha-Lada region. *Science* 251:180–183.

Campbell, I. H., and Griffiths, R. W. 1990. Implications of mantle plume structure for the evolution of flood basalts. *Earth Planet. Sci. Lett.* 99:79–83.

Cattermole, P. 1994. *Venus: The Geological Story* (Baltimore: Johns Hopkins Univ. Press).

Condit, C. D., Crumpler, L. S., Aubele, J. C., and Elston, W. E. 1989. Patterns of volcanism along the southern margin of the Colorado Plateau: The Springerville field. *J. Geophys. Res.* 94:7975–7986.

Connor, C. B., Condit, C. D., Crumpler, L. S., and Aubele, J. C. 1992. Evidence of regional structural controls on vent distribution: Springerville volcanic field, Arizona. *J. Geophys. Res.* 97:12349–12359.

Crisp, J. A. 1984. Rates of magma emplacement and volcanic output. *J. Volc. Geotherm. Res.* 20:177–211.

Crumpler, L. S. 1980. Alkali basalt through trachyte suite and volcanism, Mesa Chivato, Mount Taylor volcanic field, New Mexico. *Geol. Soc. America Bull.* 91:253–255, 1293–1313.

Crumpler, L. S. 1994. The distribution of hot spots and its relation to global geology: Venus, Mars, and Earth. *Lunar Planet. Sci. Conf.* XXV303–304 (abstract).

Crumpler, L. S., and Aubele, J. C. 1992. Two global concentrations of volcanism on Venus: Geologic associations and implications for global pattern of upwelling and downwelling. *Lunar Planet. Sci. Conf.* XXIII:275–276 (abstract).

Crumpler, L. S., Head, J. W., Aubele, J. C., Guest, J., and Saunders, R. S. 1992. Venus volcanism: global distribution and classification from Magellen data. *Lunar Planet. Sci. Conf.* XXIII:277–278 (abstract).

Crumpler, L. S., Head, J. W., and Aubele, J. C. 1993a. Relation of major volcanic center concentration on Venus to global tectonic patterns. *Science* 261:591–595.

Crumpler, L. S., Head, J. W., and Aubele, J. C. 1993b. Large volcanoes on Venus: Examples of geologic and structural characteristics from different classes. *Lunar Planet. Sci. Conf.* XXIV:365–366 (abstract).

Crumpler, L. S., Aubele, J. C., and Head, J. W. 1993c. Synthesis of global thematic mapping, Venus: Geologic correlations/questions for the Magellan gravity mission. *Lunar Planet. Sci. Conf.* XXIV:363–364 (abstract).

Crumpler, L. S., Aubele, J. C., and Head, J. W. 1993d. The Magellan volcanic and magmatic feature catalog. *Lunar Planet. Sci. Conf.* XXIV:361–362 (abstract).

Crumpler, L. S., Aubele, J. C., and Condit, C. D. 1994. Volcanoes and neotectonic characteristics of the Springerville volcanic field, Arizona. *New Mexico Geol. Soc. 45th Field Conf. Guidebook, Mogollon Slope*, pp. 147–164.

Crumpler, L. S., Head, J. W., and Aubele, J. C. 1995. Magma chambers associated with calderas on Mars: Significance of long-term magma replenishment rates. *Lunar Planet. Sci. Conf.* XXVI:305–306 (abstract).

Crumpler, L. S., Head, J. W., and Aubele, J. C. 1996a. Calderas on Mars: Characteristics, structure, and associated flank deformation. *J. Geol. Soc. London, Spec. Pub.* 110:307–347.

Crumpler, L. S., Aubele, J. C., and Head, J. W. 1996b. Volcanic and magmatic features on Venus: A global survey. *Geol. Soc. Amer. Special Paper*, in preparation.

Dawson, C. B., and Crumpler, L. S. 1993. Characteristics of arachnoids from Magellan data. *Lunar Planet. Sci. Conf.* XXIV:383–384 (abstract).
EEZ-SCAN Scientific Staff. 1986. *Atlas of the Exclusive Economic Zone*, U. S. G. S. Misc. Invest. Series I-1792.
Elston, W. E. 1984. Mid-Tertiary ash flow tuff cauldrons, southwestern New Mexico. *J. Geophys. Res.* 89:8733–8750.
Fagents, S. A., and Wilson, L. 1995. Explosive volcanism on Venus: transient volcanic explosions as a mechanism for localized pyroclastic dispersal. *J. Geophys. Res.* 100:26327–26338.
Fedotov, S. A. 1981. Magma rates in feeding conduits of different volcanic centers. *J. Volc. Geotherm. Res.* 9:379–394.
Fink, J. H., Bridges, N. T., and Grimm, R. E. 1993. Shapes of venusian "pancake" domes imply episodic emplacement and silicic composition. *Geophys. Res. Lett.* 20:261–264.
Ford, P. G., and Pettengill, G. H. 1992. Venus topography and kilometer-scale slopes. *J. Geophys. Res.* 97:13103–13114.
Francis, P. 1976. *Volcanoes* (New York: Penguin Press).
Garvin, J. B., and Williams, R. S. 1990. Small domes on Venus: Probable analogs of Icelandic lava shields. *Geophys. Res. Lett.* 17:1381–1384.
Garvin, J. B., Head, J. W., and Wilson, L. 1982. Magma vesiculation and pyroclastic volcanism on Venus. *Icarus* 52:365–372.
Gregg, T. K. P., and Greeley, R. 1993. Formation of Venusian canali: Considerations of lava types and their thermal behaviors. *J. Geophys. Res.* 98:10873–10882.
Grimm, R. E., and Phillips, R. J. 1992. Anatomy of a Venusian hot spot: Geology, gravity, and mantle dynamics of Eistla Regio. *J. Geophys. Res.* 97:16035–16054.
Grosfils, E. B., and Head, J. W. 1994. The global distribution of giant radiating dike swarms on Venus: implications for the global stress state. *Geophys. Res. Lett.* 21:701–704.
Grosfils, E. B., and Head, J. W. 1995. The timing of giant radiating dike swarm emplacement on Venus: Implications for resurfacing of the planet and its subsequent evolution. *J. Geophys. Res.*, in press.
Gudmundsson, A. 1986. Mechanical aspects of post-glacial volcanism and tectonics of the Reyklanes Peninsula, Southwest Iceland. *J. Geophys. Res.* 91:12711–12721.
Gudmundsson, A. 1987. Formation and mechanics of magma reservoirs in Iceland. *Geophys. J. Roy. Astron. Soc.* 91:27–41.
Gudmundsson, A. 1988. Formation of collapse calderas. *Geology* 16:808–810.
Guest, J. E., Head, J. W., Bulmer, M. H., and Wiles, C. R. 1991a. Volcanic plains and small edifices. *Lunar Planet. Sci. Conf.* XXII:503 (abstract).
Guest, J. E., et al. 1991b. Slope failure on the margins of volcanic domes on Venus. *Eos: Trans. Amer. Geophys. Union* 72:44, 278.
Guest, J. E., et al. 1992a. Small volcanic edifices and volcanism in the plains of Venus. *J. Geophys. Res.* 97:15949–15966.
Guest, J. E., et al. 1992b. Gravitational collapse on the margins of volcanic domes on Venus. *Lunar Planet. Sci. Conf.* XXIII:461–462 (abstract).
Hamilton, V. E., and Stofan, E. R. 1994. The geology and evolution of Hecate Chasma, Venus. *Lunar Planet. Sci. Conf.* XXV:501–502 (abstract).
Hardee, H. C. 1982. Incipient magma chamber formation as a result of repetitive intrusions. *Bull. Volc.* 45:41–49.
Hardee, H. C. 1986. Replenishment rates of crustal magma and their bearing on potential sources of thermal energy. *J. Volc. Geotherm. Res.* 28:275–296.
Hasenaka, T., and Carmichael, I. S. E. 1985. The cinder cones of the Michoacan-Guanajuato, central Mexico, their age, volume, distribution, and magma discharge rate. *J. Volc. Geotherm. Res.* 25:105–204.

Head, J. W., and Wilson, L. 1986. Volcanic processes and landforms on Venus: Theory, predictions, and observations. *J. Geophys. Res.* 91:9407–9446.

Head, J. W., and Wilson, L. 1992. Magma reservoirs and neutral buoyancy zones on Venus: Implications for the formation and evolution of volcanic landforms. *J. Geophys. Res.* 97:3877–3903.

Head, J. W., et al. 1991. Venus volcanism: Initial analysis from Magellan data. *Science* 252:276–288.

Head, J. W., et al. 1992a. Venus volcanism: Classifications of volcanic features and structures, associations and global distribution from Magellan data. *J. Geophys. Res.* 97:13153–13198.

Head, J. W., Crumpler, L. S., and Aubele, J. C. 1992b. Large shield volcanoes on Venus: distribution and classification. *Lunar Planet. Sci. Conf.* XXIII:513–514 (abstract).

Hill, R. I. 1991. Starting plumes and continental break-up. *Earth Planet. Sci. Lett.* 104:398–416.

Hooper, P. R. 1990. The timing of crustal extension and the eruption of continental flood basalts. *Nature* 345:246–249.

Ivanov, M. A., and Basilevsky, A. T. 1990. Coronae and major shields on Venus: Comparison of their areas, basal altitudes, and areal distribution. *Earth Moon Planets* 50/51:409–420.

Janes, D. M., et al. 1992. Geophysical models for the formation and evolution of coronae on Venus. *J. Geophys. Res.* 97:16055–16067.

Jurdy, D. M. 1983. Early Tertiary subduction zones and hot spots. *J. Geophys. Res.* 88:6395–6402.

Jurdy, D. M., and Stefanick, M. 1993. Distribution of coronae. *Eos: Trans. Amer. Geophys. Union* 74:16, 189.

Kargel, J. S., Komatsu, G., Baker, V. R., and Strom, R. G. 1993. The volcanology of Venera and Vega landing sites and the geochemistry of Venus. *Icarus* 103:253–275.

Kaula, W. M., et al. 1992. Styles of deformation in Ishtar Terra and their implications. *J. Geophys. Res.* 97:16085–16120.

Keddie, S. T., and Head, J. W. 1994a. Sapas Mons, Venus: Evolution of a large shield volcano. *Earth Moon Planets* 65:129–190.

Keddie, S. T., and Head, J. W. 1994b. Height and altitude distribution of large volcanoes on Venus. *Planet. Space Sci.* 42:455–462.

Keddie, S. T., and Head, J. W. 1995. Formation and evolution of volcanic edifices on the Dione Regio rise, Venus. *J. Geophys. Res.*, 100:11729–11754.

Klose, K. B. 1990. Radar-bright haloes of Loukha Planitia: Implications for Venus volcanic style. *Lunar Planet. Sci. Conf.* XXI:639–640 (abstract).

Komatsu, G., Baker, V. R., and Gulick, V. C. 1993. Venusian channels and valleys: Distribution and volcanological implications. *Icarus* 102:1–25.

Lancaster, M. G., Guest, J. E., Roberts Magee, K., and Head, J. W. 1992. "Great" Lava Fields on Venus. *Lunar Planet. Sci. Conf.* XXIII:753–754 (abstract).

Lancaster, M. G., Guest, J. E., and Magee, K. 1993a. Great lava flow fields on Venus. *Icarus*, submitted.

Lancaster, M. G., Guest, J. E., and Roberts Magee, K. 1993b. Sheet flows on Venus. *Lunar Planet. Sci. Conf.* XXIV:843–844 (abstract).

McBirney, A. R. 1990. An historical note on the origin of calderas. *J. Volc. Geotherm. Res.* 42:303–306.

McKenzie, D., and Bickle, M. J. 1988. The volume and composition of melt generated by extension of the lithosphere. *J. Petrol.* 29:625–679.

McKenzie, D., Ford, P. G., Liu, F., and Pettengill, G. H. 1992. Pancake-like domes on Venus. *J. Geophys. Res.* 97:15967–15976.

Moore, H. J., Plaut, J. J., Schenk, P. M., and Head, J. W. 1992. An unusual volcano on Venus. *J. Geophys. Res.* 97:13479–13494.
Morgan, P., and Phillips, R. J. 1983. Hot spot heat transfer: Its application to Venus and implications to Venus and Earth. *J. Geophys. Res.* 88::8305–8317.
Namiki, N., and Solomon, S. C. 1994. The impact crater density on volcanoes and coronae on Venus: Implications for volcanism and global resurfacing. *Science* 265:926–933.
Nikolayeva, O. V. 1990. Geochemistry of the Venera 8 material demonstrates the presence of continental crust on Venus. *Earth Moon Planets* 50/51:329–341.
Ollier, C. 1969. *Volcanoes* (Cambridge, Mass.: MIT Press).
Parker, T. J. 1992. Application of left- and right-looking SAR stereo to depth measurements of the Ammavaru outflow channel, Lada Terra, Venus. In *International Colloquium on Venus*, LPI Contrib. No. 789, pp. 84–85.
Parmentier, E. M., and Hess, P. C. 1992. Chemical differentiation of a convecting planetary interior: Consequences for a one-plate planet such as Venus. *Geophys. Res. Lett.* 19:2015–2018.
Pavri, B., Head, J. W., Klose, K. B., and Wilson, L. 1992. Steep-sided domes on Venus: Characteristics, geologic setting, and eruption conditions from Magellan data. *J. Geophys. Res.* 97:13445–13478.
Pettengill, G. E., et al. 1991. Magellan: Radar performance and data products. *Science* 252:260–265.
Phillips. R. J., Grimm, R. E., and Malin, M. C. 1991. Hot-spot evolution and the global tectonics of Venus. *Science* 252:651–658.
Phillips, R. J., et al. 1992. Impact craters and Venus resurfacing history. *J. Geophys. Res.* 97:15923–15948.
Plaut, J. J., Stofan, E. R., Crown, D. A., and Anderson, S. W. 1994. Topographic and surface roughness properties of steep-sided domes on Venus and Earth from radar remote sensing and field measurements. *Lunar Planet. Sci. Conf.* XXV:1091–1092.
Price, M. 1995. Resurfacing history of the Venusian plains based on distribution of impact craters. *Lunar Planet. Sci. Conf.* XXVI:1143–1144 (abstract).
Pronin, A. A., and Stofan, E. R. 1990. Coronae on Venus: Morphology, classification, and distribution. *Icarus* 87:452–474.
Richards, M. A., Duncan, R. A., and Courtillot, V. E. 1989. Flood basalts and hot-spot tracks: Plume heads and tails. *Science* 246:103–107.
Roberts Magee, K. P., and Head, J. W. 1993a. Large-scale volcanism associated with coronae on Venus: Implications for formation and evolution. *Geophys. Res. Lett.* 20:1111–1114.
Roberts Magee, K. P., and Head, J. W. 1993b. The role of rifting in the generation of melt: Implications for the origin and evolution of the Lada Terra-Lavinia Planitia region of Venus. *J. Geophys. Res.*, submitted.
Roberts Magee, K. P., and Head, J. W. 1994. Venus: Morphology and morphometry of volcanism in rifting environments. *Lunar Planet. Sci. Conf.* XXV:823–824 (abstract).
Roberts Magee, K. P., Guest, J. E., Head, J. W., and Lancaster, M. G. 1992. Mylitta Fluctus, Venus: rift-related, centralized volcanism and the emplacement of large-volume flow units. *J. Geophys. Res.* 97:15991–16015.
Robinson, C. A., Thornhill, G. D., and Parfitt, E. A. 1995. Large scale volcanic activity at Maat Mons: can this explain fluctuations in atmospheric chemistry observed by Pioneer Venus? *J. Geophys. Res.* 100:11755–11763.
Sakimoto, S. E. H. 1994. Terrestrial basaltic counterparts for the Venus steep-sided or pancake domes. *Lunar Planet. Sci. Conf.* XXV:1189–1190 (abstract).
Sakimoto, S. E. H., and Zuber, M. T. 1995. The spreading of variable-viscosity ax-

isymmetric radial gravity currents: Applications to the emplacement of Venusian "pancake" domes. *J. Fluid Mech.* 301:65–77.
Saunders, R. S., and Pettengill, G. E. 1991. Magellan: Mission summary. *Science* 187:247–248.
Saunders, R. S., et al. 1992. Magellan mission summary. *J. Geophys. Res.* 97:13067–13090.
Schaber, G. G. 1988. Elevations of Venusian shields as indicators of lithospheric thickness. *Lunar Planet. Sci. Conf.* XIX:1023–1024 (abstract).
Schaber, G. G., et al. 1992. Geology and distribution of impact craters on Venus: What are they telling us? *J. Geophys. Res.* 97:13257–13302.
Schubert, G., and Sandwell, D. T. 1995. A global survey of possible subduction sites on Venus. *Icarus* 117:173–196.
Schubert, G., et al. 1991. Magellan observations of Venusian coronae: Geology, topography, and distribution. *Eos: Trans. AGU* 72:175.
Senske, D. A. 1990. Geology of the Venus equatorial region from Pioneer Venus radar imaging. *Earth Moon Planets* 50/51:305–327.
Senske, D. A., and Head, J. W. 1992. Zones of extension and rifting on Venus: Characteristics and distribution. *Lunar Planet. Sci. Conf.* XXIII:1269–1270 (abstract).
Senske, D. A., et al. 1991*a*. Geology and tectonics of the Themis Regio-Lavinia Planitia-Alpha Regio-Lada Terra Area, Venus: Results from Arecibo image data. *Earth Moon Planets* 55:97–161.
Senske, D. A., et al. 1991*b*. Geology and tectonics of Beta Regio, Guinevere Planitia, Sedna Planitia, and Western Eistla Regio, Venus: results from Arecibo image data. *Earth Moon Planets* 55:163–214.
Senske, D. A., Schaber, G. G., and Stofan, E. R. 1992. Regional topographic rises on Venus: Geology of western Eistla Regio and comparison to Beta Regio and Atla Regio. *J. Geophys. Res.* 97:13395-13420.
Settle, M. 1979, The structure and emplacement of cinder cone fields. *Amer. J. Sci.* 279:1089–1107.
Shaw, H. R. 1969. The rheology of basalt in the melting range. *J. Petrol.* 10:510–535.
Shaw, H. R. 1985. Links between magma-tectonic rate balances, plutonism, and volcanism. *J. Geophys. Res.* 90:11275–11288.
Sinilo, V. P., and Slyuta, E. N. 1989. Radarclinometry: Implications for the morphology of small dome-like hills on Venus. *Lunar Planet. Sci. Conf.* XX:1016–1017 (abstract).
Sjogren, W. L., et al. 1983. Venus gravity anomalies and their correlations with topography. *J. Geophys. Res.* 88:1119–1128.
Slyuta, E. N. 1990. Large shield volcanoes (>100 km in diameter) on Venus: Morphologic types. *Lunar Planet. Sci. Conf.* XXI:1172–1173 (abstract).
Slyuta, E. N., and Kreslavsky, M. A. 1990. Intermediate (20–100 km) sized volcanic edifices on Venus. *Lunar Planet. Sci. Conf.* XXI:1174–1175 (abstract).
Slyuta, E. N., Shalimov, I. V., and Nikishin, A. M. 1992. Different types of small volcanoes on Venus. In *International Coll. on Venus*, LPI Contrib. No. 789, pp. 112–114.
Smith, R. L. 1979. Ash-flow magmatism. In *Ash Flow Tuffs*, ed. C. E. Chapin, GSA SP-180 (Boulder: Geological Soc. of America), pp. 5–28.
Smith, R. L., and Bailey, R. A. 1968. Resurgent cauldrons. *Geol. Soc. Amer. Memoir* 116:613–662.
Smrekar, S. 1994. Evidence for active hot spots on Venus from ananlysis of Magellan gravity data. *Icarus* 112:2–26.
Smyth, P., Anderson, C. H., Aubele, J. C., and Crumpler, L. S. 1992. Multi-resolution pattern recognition of small volcanoes in Magellan data. In *International Collo-*

quium on Venus, LPI Contrib. No. 789, pp. 116–117.
Solomon, S. C., et al. 1992. Venus tectonics: An overview of Magellan observations. *J. Geophys. Res.* 97:13199–13256.
Squyres, S. W., et al. 1992a. The morphology and evolution of coronae on Venus. *J. Geophys. Res.* 97:13611–13634.
Squyres, S. W., et al. 1992b. The spatial distribution of coronae on Venus. In *International Colloquium on Venus*, LPI Contrib. No. 789, pp. 119–120.
Stefanick, M., and Jurdy, D. M. 1984. The distribution of hot spots. *J. Geophys. Res.* 89:9919–9925.
Stofan, E. R., and Head, J. W. 1990. Coronae of Mnemosyne Regio: Morphology and origin. *Icarus* 83:216–243.
Stofan, E. R., and Saunders, R. S. 1990. Geologic evidence of hotspot activity on Venus: Predictions for Magellan. *Geophys. Res. Lett.* 17:1377–1380.
Stofan, E. R., Bindschadler, D. L., Head, J. W., and Parmentier, E. M. 1991. Coronae structure on Venus: Models of origin. *J. Geophys. Res.* 96:20933–20946.
Stofan, E. R., et al. 1992. Global distribution and characteristics of coronae and related features on Venus: Implications for origin and relation to mantle processes. *J. Geophys. Res.* 97:13347–13378.
Stofan, E. R., Smrekar, S. E., Bindschadler, D. L., and Senske, D. A. 1996. Large topographic rises on Venus: Implications for mantle upwelling. *J Geophys. Res.*, in press.
Strom, R. G., Schaber, G. G., and Dawson, D. 1994. The global resurfacing of Venus. *J. Geophys. Res.* 99:10899–10926.
Takada, A. 1989. Magma transport and reservoir formation by a system of propagating cracks. *Bull. Volc.* 52:118–126.
Thornhill, G. D. 1993. Theoretical modeling of eruption plumes on Venus. *J. Geophys. Res.* 98:9107–9111.
Tsukui, M., Sakuyama, M., Koyaguchi, T., and Ozawa, K. 1986 Long-term eruptiona rates and dimensions of magma reservoirs beneath quaternary polygenetic volcanoes in Japan. *J. Volc. Geotherm. Res.* 29:189–202.
Turcotte, D. L. 1993. An episodic hypothesis for Venusian tectonics. *J. Geophys. Res.* 98:17061–17068.
Walker, G. P. L. 1984. Downsag calderas, ring faults, caldera sizes, and incremental caldera growth. *J. Geophys. Res.* 84:8407–8415.
Ward, W. R., Burns, J. A., and Toon, O. B. 1979. Past obliquity oscillations of Mars: The role of the Tharsis uplift. *J. Geophys. Res.* 84:243–259.
White, R. S. 1992. Magmatism during and after continental break-up. In *Magmatism and the Causes of Continental Break-up*, eds. B. C. Storey, T. Alabaster and R. J. Pankhurst, GSA SP-68 (Boulder: Geological Soc. of America), p. 404.
White, R. S., and McKenzie, D. 1989. Magmatism at rift zones: The generation of volcanic continental margins and flood basalts. *J. Geophys. Res.* 94:7685–7729.
Wiles, C. R., and Foreshaw, M. R. B. 1992. Automated detection and measurements of small volcanoes on Venus. *Lunar Planet. Sci. Conf.* XXIII:1527–1528 (abstract).
Williams, H., and McBirney, A. R. 1979. *Volcanology* (San Francisco: Freeman Cooper).
Wilson, L., and Head, J. W. 1994. Mars: Review and analysis of volcanic eruption theory and relationships to observed landforms. *Rev. Geophys.* 32(3):221–264.
Wood, C. A. 1979. *Venusian Volcanism: Environmental Effects on Style and Landforms*, NASA TM-80339, pp. 244–246.
Wood, C. A. 1984. Calderas: A planetary perspective. *J. Geophys. Res.* 89:8391–8406.

CHANNELS AND VALLEYS

VICTOR R. BAKER and GORO KOMATSU
University of Arizona

VIRGINIA C. GULICK
NASA Ames Research Center

and

TIMOTHY J. PARKER
Jet Propulsion Laboratory

More than 200 channels and valleys have been identified on the Magellan images of Venus. These are classified, on the basis of morphology, as simple channels (including sinuous rilles, simple channels with flow margins, and canali), complex channels (with or without flow margins), compound channels, and valley networks (including rectangular, labyrinthic, and pitted or irregular networks). Sinuous rilles closely resemble their lunar counterparts. Canali are exceptional for their remarkably constant width along very extenuated flow paths, exceeding 500 km. One of the compound channels, the outflow complex of Kallistos Vallis, extends over 1200 km and is up to 30 km wide. Venusian channels are globally distributed, but each class has a preferential topographic association. The canali are developed on the volcanic plains, while sinuous rilles occur at higher elevations, associated with volcanic complexes and coronae. Both canali and sinuous rilles have been deformed by post-emplacement tectonism. Highly fluid lavas, erupted at sustained, high discharges seem best to explain many of the channel features, particularly for the canali and the compound channels. Explanation of the canali morphologies may also involve unusual low-viscosity lavas, perhaps of exotic composition.

I. INTRODUCTION

Because liquid water was known to be unstable on the surface of Venus at current ambient conditions, the number and diversity of channels revealed on the surface of Venus by Magellan radar mapping was unexpected. The very first channel identified in Magellan imagery, later recognized as a canali-type channel, was sinuous, with constant width, no tributaries, no clear association with flow units, and extraordinary length (>500 km) (Head et al. 1991). This unique combination of characteristics has not been observed in terrestrial and Martian fluvial channels, nor in lava channels on the terrestrial planets. This anomaly attracted the attention of researchers who hypothesized diverse mechanisms for the channel formation on Venus (Baker et al. 1992; Komatsu et al. 1992b; Gregg and Greeley 1993; Komatsu et al. 1993; Kargel et al. 1994; Bussey et al. 1995). As the mission progressed, researchers discovered a wide range of channel morphologies, including those more typical for lava channels on other terrestrial planets (Komatsu et al. 1993).

There are a number of Venusian channels similar in morphology to typical lava channels on the terrestrial planets. They have clear associations with flow units (levees/lateral flow deposits, or explained as flow margins in the classification of Komatsu et al. [1993]), and they tend to narrow and shallow downstream. Although these Venusian lava channels are commonly as much as an order of magnitude or two larger in scale than their counterparts on Earth, their mechanisms of formation can be considered analogous to those of terrestrial lava channels. Similar reasoning has been applied to the very large lava channels of Mars (Mouginis-Mark et al. 1992). The following discussion, therefore, focuses on the more unusual channels and valleys of Venus, such as canali, sinuous rilles, compound channels, and valley networks.

II. CLASSIFICATION AND MORPHOLOGY

A. Channels

More than 200 channels and valleys have been identified on the Magellan images of Venus. Of these there are approximately a dozen valley networks and more than 200 channels exhibiting a wide variety of morphological characteristics. Preliminary classifications were presented by Gulick et al. (1991,1992a), and Baker et al. (1992). These were subsequently updated by Komatsu et al. (1993). The morphological characteristics of the classification scheme are illustrated in Fig. 1. Channels will be discussed first, according to three types: (1) simple, (2) complex, and (3) compound.

Simple Channels. These are generally composed of a single main conduit which lacks complex branching and anastomosing (see Fig. 2 in Baker et al. 1992). Simple channels are further subdivided into sinuous rilles, simple channels with flow margins, and canali. Sinuous rilles are similar in morphology and size to those observed on the Moon (see Fig. 3a,b in Baker et al. 1992). They emanate from distinct, circular or elongated regions of collapse (generally several km in diameter); they are approximately 1 to 2 km wide and several tens of km long; and they become narrower and shallower distally. As on the Moon, most sinuous rilles on Venus are not associated with detectable lava flow margins. The morphological similarity between Venusian and lunar sinuous rilles suggests an origin by the thermal erosion of flowing lava (Hulme 1973,1982). However, the exact origin of these channels could have been vastly different given that Venusian sinuous rilles tend to cluster on or near geologic features that are not found on the Moon, such as coronae, corona-like features, and arachnoids.

Some simple channels are located on well-defined flow deposits or flow fields (see Fig. 5 in Komatsu et al. 1993). These channels are somewhat

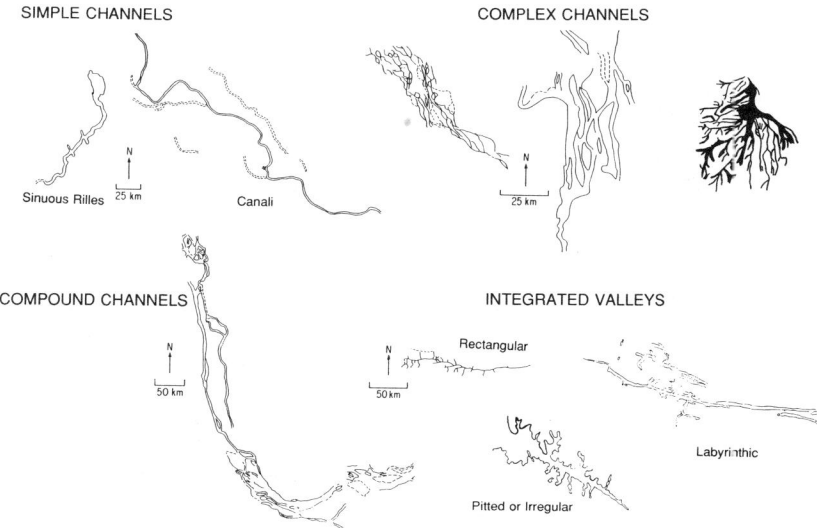

Figure 1. Morphological classification scheme of channels and valleys on Venus (figure updated from Gulick et al. 1992*b*).

similar to channels formed on terrestrial lava flows except that levees are usually not apparent on the Magellan radar imagery. Because these channels have formed entirely on lava flows and do not deeply incise surrounding terrain, they appear to be constructional in origin, similar to their terrestrial counterparts. In general, these simple channels exhibit indistinct source and terminal regions, in contrast to the Venusian sinuous rilles. Simple channels with flow margins commonly feed extensive lava flows associated with coronae, shield volcanoes, rift and fracture zones (Komatsu et al. 1993).

Canali are channels having a constant width and depth (Fig. 2). They are typically as wide as 3 km and are up to 500 km in length, although a few canali are as wide as 10 km, and the length reaches 6800 km in one case (see Fig. 6a in Komatsu et al. 1993). Canali may locally exhibit abandoned channel segments, cut-off meander bends, levees (see Fig. 5b in Baker et al. 1992) and radar dark terminal deposits (Gulick et al. 1991,1992*a*). Sources and termini are generally indistinct. Canali are generally located in plains regions, particularly in Guinevere Planitia, Helen Planitia, as well as in the plains southeast of Artemis Planitia and north of Rusalka Planitia (Komatsu et al. 1993). Canali morphology suggests their probable formation by, and conveyance of, large discharges of low viscosity lava to distant regions over prolonged periods (Komatsu et al. 1993).

Complex Channels. These form anastomosing, braided or distributary patterns that are usually, although not everywhere, on flow deposits (see Fig. 9 in Komatsu et al. 1993). Individual channel widths range from approximately

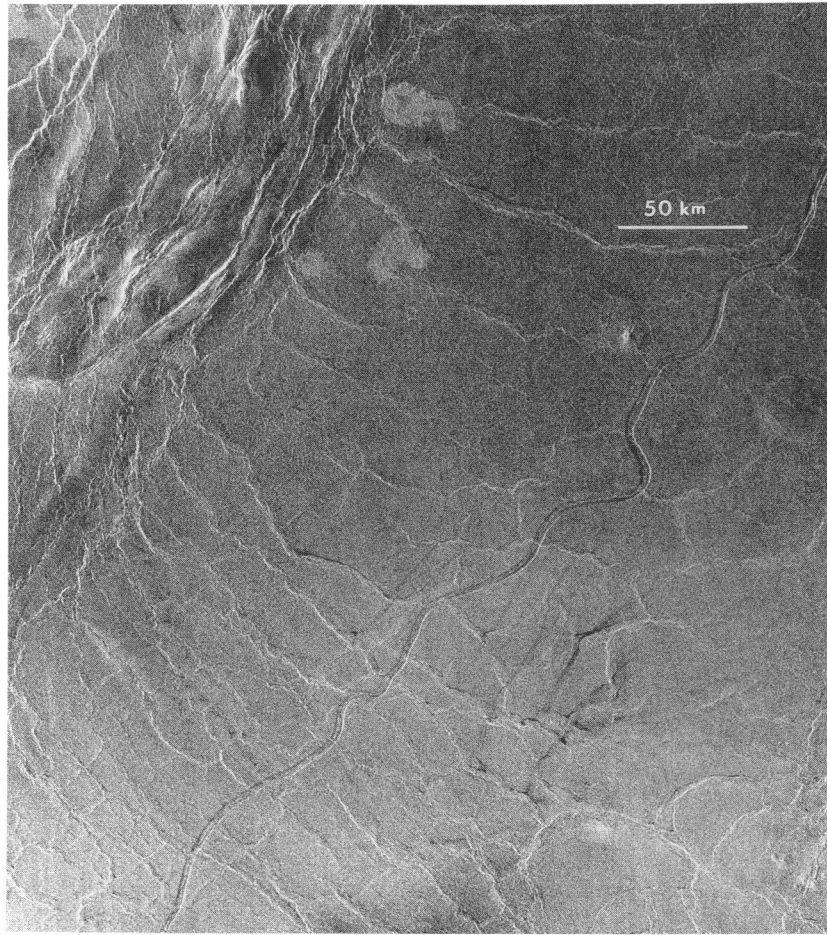

Figure 2. Radar image of Baltis Vallis, a canali-type channel located at 49 to 51°N, 165 to 168°E (50N163; left-looking radar illumination). Channel width is approximately 1 to 4 km; scale bar equals 50 km. Baltis Vallis is the longest channel yet discovered in the solar system with a total length of ~6800 km.

3 km down to the limit of resolution, whereas the channel system may be 20 to 30 km wide and up to hundreds of km in length. Complex channels without associated flow deposits are commonly observed as segments of other channel types, such as the middle reach of a compound channel or as a subsidiary anastomosing, branching segment in a canali type channel. The lack of related flow deposits suggests that these channels may be eroded into surrounding terrain. This particular subclass of complex channels is known simply as "complex channels without flow margins" (Komatsu et al. 1993)

(also called "complex erosional channels" in Gulick et al. 1992a).

Most complex channels are located on flow deposits and are classified as complex channels with flow margins (Komatsu et al. 1993) (also called "complex constructional channels" in Gulick et al. 1992a). Channels are commonly separated by "islands" of radar-bright (or radar-dark in some cases) material (see Fig. 10b in Komatsu et al. 1993). Channel margins are sometimes lined with radar-bright material. Complex channels with flow margins are common on fluidized ejecta blankets (see Fig. 7a in Baker et al. 1992).

Complex channels are also located along with simple channels on lava flow deposits. For example, both complex and simple channels occur on the enormous flow deposits of Mylitta Fluctus (Roberts 1992) in Lavinia Planitia (54°S, 354.4°E). Complex channels form delta-like distributary patterns along terminal reaches of flow deposits in Aphrodite Terra (see Fig. 7c,d in Baker et al. 1992).

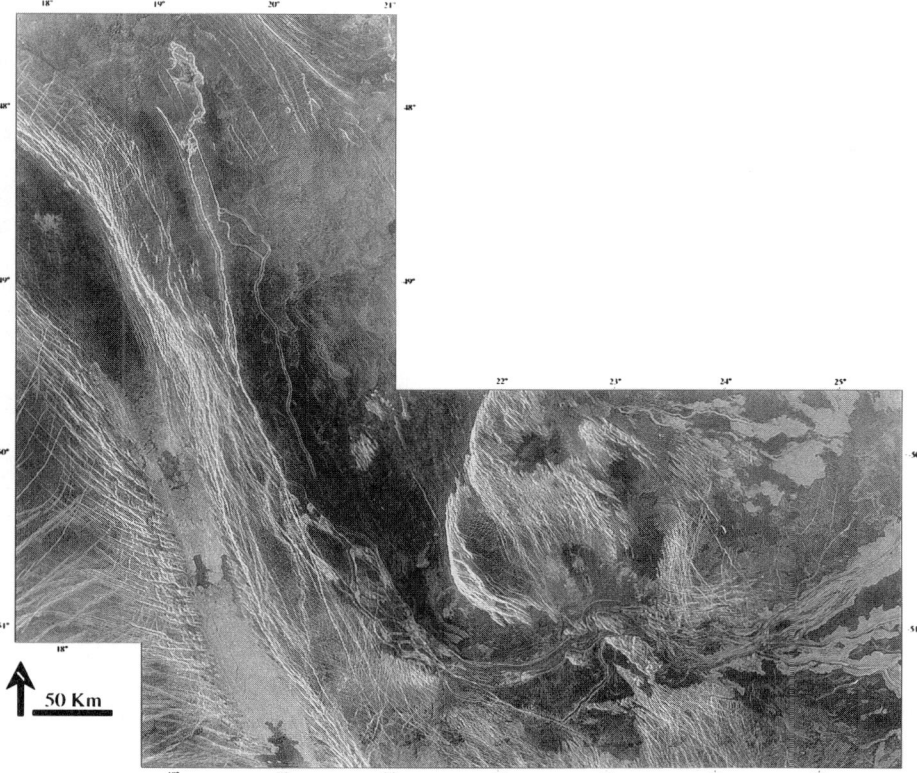

Figure 3. Portion of FMIDR 50S021 (left-looking, Cycle 1 data) coverage of Kallistos Vallis, from source collapse terrain (at top) through upper "distributary reach" (at right).

Compound Channels. These display both simple and complex segments (see Fig. 8 in Baker et al. 1992). Channels vary greatly in size, with widths ranging between several tens of km in complex regions down to the limit of resolution in simple reaches. Lengths of compound channels can range from 75 km to thousands of km. The upper segment of a compound channel commonly, although not in every case, resembles the morphology of a sinuous rille in that a single conduit emanates from a distinct collapsed region (Fig. 3). Instead of becoming narrower and shallowing distally, however, a compound channel bifurcates and anastomoses, resulting in stream-lined islands (Fig. 3). The lower reach may terminate in a flow deposit which can be quite extensive in some cases (Fig. 3). A detailed description of a representative compound channel, Kallistos Vallis, is provided in Sec. III.

B. Valley Networks

The valley networks are distinct from the Venusian channels in that they lack bedforms that are direct indicators of fluid flow. These integrated valley systems are somewhat similar in morphology to those produced by groundwater sapping processes on Earth and Mars. Like their terrestrial and Martian counterparts, Venusian valley networks probably did not form by fluids being conveyed through conduits at bank-full stage. Instead, these landforms appear to have formed by the undermining and subsequent collapse of surface material by outflow of low-viscosity subsurface fluids, augmented by surface flow. The Venusian valley networks are divided into three morphological classes: (1) rectangular, (2) labyrinthic, and (3) pitted or irregular (Gulick et al. 1992b). Some networks display a morphology that is transitional between rectangular and labyrinthic. See Table I.

TABLE I
Classification and Location of Valley Networks on Venus

	Location	Type	Magellan Image	CD-ROM
1.	2°N, 70.5°E	Rectangular	F-00N070	MG009
2.	0°N, 69.5°E	Rectangular	F-00N070	MG009
3.	57.5°S, 165°E	Rectangular to labyrinthic	C1-60S153	MG0031
			F-60S164	MG0063
4.	8.5°S, 87°E	Labyrinthic	F-10S087	MG0010
5.	34.5°N, 261.5°E	Labyrinthic	C1-30N261	MG0040
6.	33°N, 268.5°E	Labyrinthic	C1-30N261	MG0040
7.	19°S, 196°E	Rectangular to labyrinthic	C1-15S197	MG0033
8.	68.5°S, 1°E	Labyrinthic	C1-75S023	MG0054
9.	15°S, 86°E	Labyrinthic	C1-15S077	MG0019
10.	56°N, 204.5°E	Labyrinthic	F-55N208	MG0028
11.	16°S, 57°E	Labyrinthic	C1-15S060	MG0016
12.	5°N, 304°E	Pitted or irregular	C1-00N300	MG0047
			F-05N307	MG0053

Rectangular Networks. These form the most integrated valley systems on Venus. The networks are approximately 100 km long and less than a km wide. Rectangular valley networks generally occur in smooth, radar dark plains regions.

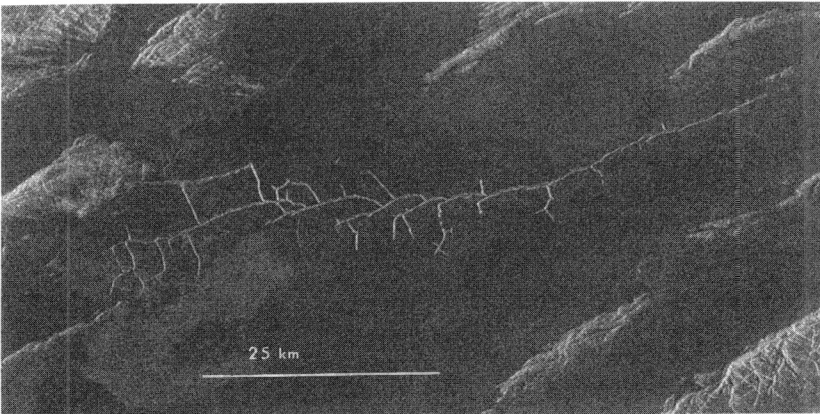

Figure 4. Rectangular valley network located at 2°N, 70.5°E. Image width is approximately 100 km (left-looking radar illumination).

An example is located at 2.1°N, 70.4°E within a narrow 15 km wide radar dark plains region (Fig. 4). This network is over 50 km long, less than a km wide and contains at least 15 first-order and 5 second-order tributaries. The right-angled bends of the tributaries and the unusually straight valley segments suggest that the morphological pattern is structurally controlled, although fractures are noticeably absent in the surrounding terrains. Pre-existing (probably subsurface) fractures may have been exploited by subsurface fluid flow (probably of low-viscosity lavas) to create the rectangular networks.

Labyrinthic Valley Systems. These are the most common type of valley networks on Venus (see Fig. 14 in Komatsu et al. 1993). Valleys are typically several km wide and a couple hundred of km long, and they appear to have been modified through wall collapse and removal of material along the valley floor. Tributaries have stubby, theater-shaped heads, and the overall pattern is less integrated when compared with rectangular systems. Labyrinthic valley networks are observed within or near tectonically deformed terrains, or near volcanic landforms (such as coronae). The complexity of tectonically deformed terrains precludes clear identification of the relative orientations of valley segments and tectonic structures, although some networks appear to merge with scalloped troughs, which are interpreted to be graben (Solomon et al. 1992). Sinuous rilles have also been observed to connect with labyrinthic valley systems (Komatsu et al. 1993). These morphologic characteristics are similar to those of terrestrial and Martian valleys which were formed or modified by groundwater sapping processes. Lack of significant labyrinthic valley enlargement suggests that headward erosion along fractures took place at a rate which was several orders of magnitude faster than that at which valley sidewalls could erode. This may have been caused by resistant underlying rock layers, by the permeability of the subsurface layers which contained the eroding fluid, or by the viscosity of the fluid itself.

Pitted or Irregular Valley Systems. These are observed in areas of relatively smooth plains containing linear ridges (see Fig. 15 in Komatsu et al. 1993). Pitted or irregular valleys have irregular, scalloped margins which appear to have formed from a series of coalesced pits or scalloped depressions having diameters up to several km. The widest main valley reach is surrounded by a series of larger, loosely organized pits with diameters of several km. Narrower sections of these valleys are bordered by smaller, coalesced pits. These systems may be as long as 300 km and as wide as several tens of km; main valley reaches are typically 50 to 75 km wide. This particular subclass of Venusian valley networks is morphologically similar to areas of thermokarst on Mars (see Figs. 3b and 4b in Gulick et al. 1992*b*) and to alas valleys in permafrost regions on Earth, suggesting that pitted or irregular valley systems on Venus may have formed by an analogous mechanism. The necessary subsurface erosion would have to involve removal of a fluid (lava) or sublimation of a solid. Fields of collapse depressions on Earth are morphologically similar to pitted or irregular valley systems, although these terrestrial pits do not typically coalesce to form valleys (Heacock et al. 1966).

All types of Venusian valley networks are integrated systems of smaller channels (or valleys) whose arrangement appears to have been structurally controlled. They tend to occur in or near tectonically deformed terrains, and they are commonly aligned with fractures or connected to scalloped troughs (graben). The high degree of tributary integration strongly suggests that a fluid (most likely a low-viscosity lava), probably moving in near-surface regions and locally intersecting the surface, formed or modified these networks. Where subsurface fluids encounter pre-existing fractures, surface and/or subsurface erosion is enhanced because fluid flow becomes concentrated along these zones of weakness.

Although the morphology of Venusian valley networks appears to have been structurally controlled, the integrated nature of individual valleys within the systems suggests that fluid flow was also involved in their formation and evolution. Venusian valley networks do not approach the degree of complexity or diversity typically attained by fluvial (water) valleys on either Earth or Mars (Gulick et al. 1992*b*), but have a higher degree of valley integration than that observed in lunar sinuous rilles. Thus, we propose that the fluid that formed the valleys on Venus had properties which are intermediate between those of water and lunar basaltic lava.

C. Meander Properties

The range of sinuosities for the measured Venusian channels (Fig. 5) are concentrated between 1.0 and 1.3; published sinuosities of terrestrial rivers range from 1.0 to 2.3 (Schumm et al. 1972). The relationship between wavelength and width for Venusian channels lies along the same trend as that for terrestrial river meanders (Fig. 6). Wavelength-to-width ratios for Venusian channels vary from 1.99 to 51.68. The similarities between the measured Venusian channels and terrestrial river channels suggests that the channel-forming fluid—which was water on Earth, and (presumably) lava on Venus—behaved in a comparable manner on both planets. Venusian channels with flow margins and canali-type channels tend to have slightly higher L/W ratios (4.56–26.99 and 3.38–51.68, respectively) for the measured samples (Fig. 6). Although data are limited, terrestrial lava channels-collapsed tubes exhibit a similar trend.

Venusian sinuous rilles tend to be wider for a given wavelength (L/W ratios in the range 1.99–16.59; Fig. 6) than other channel types on Venus, but have similar meander properties (Fig. 6) to those of lunar sinuous rilles. Experimental lava channels, simulated with hot water flowing on polyethylene glycol (Huppert and Sparks 1985), indicate that meander sinuosity is primarily determined by initial flow discharge (the higher the discharge, the lower the sinuosity). Because the hot water thermally widens and deepens the channel while retaining the original meander sinuosity (Huppert and Sparks 1985), L/W ratios decrease with time. Because both Venusian and lunar sinuous rilles display unusually low L/W ratios (Fig. 6), flowing lava may have eroded (through thermal and mechanical processes) both the floors and walls of these channels (Hulme 1973; Komatsu and Baker 1994a).

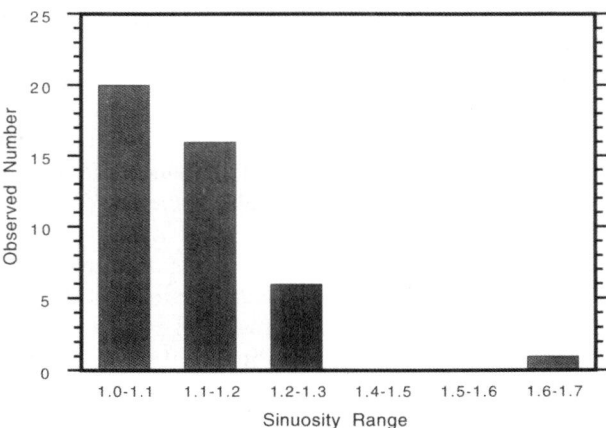

Figure 5. Sinuosity ranges for a representative sample of Venusian channels, illustrating the various classes. Sinuosity is defined as the ratio of the channel length to the length of the meander belt axis (Brice 1964).

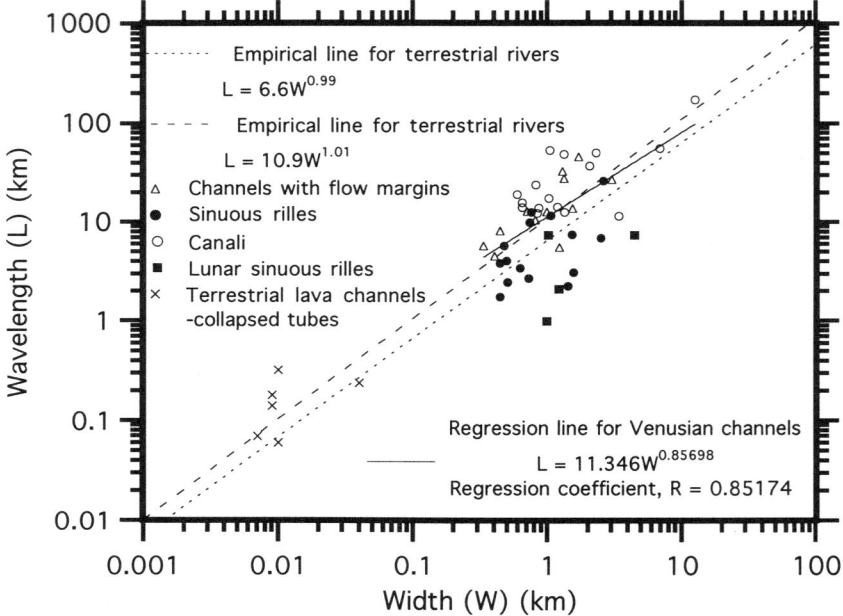

Figure 6. The relation between meander wavelength (L) and channel width (W) for various Venusian channel types, lunar sinuous rilles, terrestrial lava channels, and collapsed terrestrial lava tubes. Regression line (solid line) is derived for Venusian channels. Two empirical lines, $L = 6.6W^{0.99}$ and $L = 10.9W^{1.01}$ (Leopold and Wolman 1960), derived from terrestrial rivers, are shown for comparison.

Studies of terrestrial rivers (Leopold and Wolman 1957; Schumm et al. 1972) and flume experiments (Schumm and Khan 1972) show that meandering occurs within specific ranges of slope and discharge. Theoretical considerations, using energy minimization principles (Chang 1988), also suggest that channel sinuosity is a function of discharge. The morphologic similarity between Venusian channels and terrestrial fluvial channels (Komatsu et al. 1993) suggests that meanders observed in Venusian channels also developed within a limited range of discharge and slope.

On Earth flowing water generates meanders in alluvial sediments by erosion and deposition over some prolonged period ($\sim$1 to 10^2 yr). For bedrock meanders (or valley meanders), erosion takes even longer. In general, terrestrial lava channels and tubes are constructional and, as a result, meanders develop where channel-forming lava retains its fluidity (see, e.g., Greeley 1974). In contrast to fluvial systems, meander properties of lava channels are determined by the flow characteristics (discharge and slope) during a relatively short-term eruption. Venusian channels presumably generated by flowing lava have been proposed to have formed via construction, erosion, or by a combination of the two (Komatsu et al. 1992a,1993; Baker et al. 1992;

Gregg and Greeley 1993). Because terrestrial lava flows remain active for less time than do river channels, the relatively short period of formation may explain why the meanders tend to be less sinuous than in terrestrial rivers.

Specific Venusian channels have multiple wavelength meanders. This phenomenon is observed mostly on sinuous rilles, and it may be caused by either changes in discharge rates or by structural control. At least one canali-type channel (at 33°S, 157.5 to 158.5°E) shows an exceptionally high sinuosity (>1.5). This may indicate a relatively long time scale of flow duration, an attribute consistent with other morphological characteristics of canali, such as their long length and local cut-off bends (Baker et al. 1992).

III. KALLISTOS VALLIS

While it is not possible to provide a detailed description of the many Venusian channels, we provide here a description of a prominent compound channel, Kallistos Vallis, which illustrates the evidence leading to inferences concerning channel-forming lavas. Kallistos Vallis is a catastrophic lava flood channel, morphologically similar to the "outflow" channels on Mars (Mars Channel Working Group 1983), though smaller. The channel is over 1200 km long and up to 30 km or more in maximum width. Kallistos Vallis is centered at 50°S, 21°E (Fig. 3). The channel source is a collapse feature on the southwest flank of a broad volcanic complex with gently sloping flanks near the center of Derceto Corona (47°S,19°E). Derceto Corona is a large, elongate ovoid structure that is bounded by a prominent arcuate ridge belt on its northeast and east sides and a rift valley complex on its west side. This volcanic complex is an elongate feature with broad topographic swells to the northeast and southwest, and a quasi-circular collapsed caldera at its northwest end. The ridge belt is typical of many found at the periphery of several Venusian coronae.

We used detailed geologic mapping to characterize the geologic relations and relative timing of the events which produced Kallistos Vallis. Radar geomorphic units were identified based on morphology, texture, radar brightness, and degree of fracturing using stereo pairs of F-MIDR 50S021 (75 m/pixel, centered at 50°S, 21°E). Both left-looking and right-looking SAR mosaics exist for the Kallistos Vallis region; Cycle 3 SAR (obtained with lower incidence angles than Cycle 1 SAR specifically so that the two sets could be used for stereo geologic mapping) is not available. Despite the opposite illumination geometry in the Cycle 1 and 2 data, stereo viewing is possible for much of the region, because most of the topography is rather gentle. Furthermore, the most interesting reaches of Kallistos Vallis trend east–west, parallel to the look direction, and the incidence angles for the Cycle 1 and 2 images are similar (23° and 25°, respectively) so that the radar-backscatter of nearly horizontal surfaces is comparable. An added benefit from the opposite illumination geometry is that channel walls trending perpendicular to the look direction, if not visible in one view, are usually strong reflectors in the other

(Fig. 7). An advantage to "left-right" stereo over "left-left" stereo available for parts of the planet is that topographic resolution is much better because the parallax angle is much larger. At the latitude of Kallistos Vallis, a single pixel of parallax displacement corresponds to approximately 17 m vertical precision. These stereo data coupled with the lower-resolution topography from the radar altimeter make it possible to construct high-precision longitudinal and cross profiles of the channel (Figs. 7, 8, and 9).

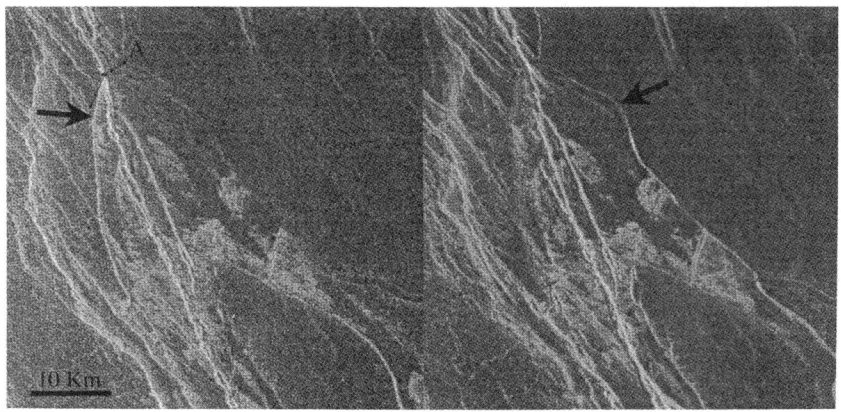

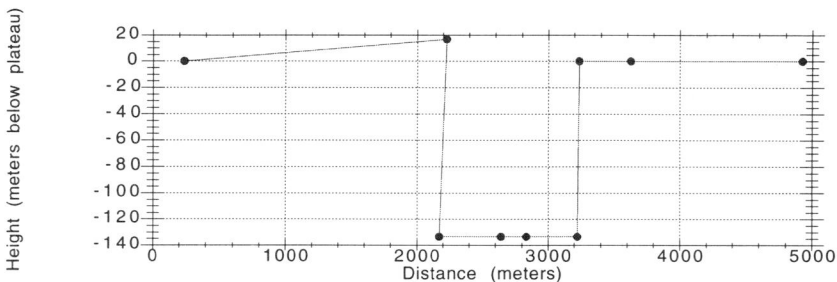

Figure 7. (a) SAR stereo showing where Kallistos Vallis (Fig. 3) spills out of trough reach and begins anastomosing reach. Channel walls are most visible where oriented facing incident radar (arrowed left and right-looking scenes). Cycle 2 (right-looking) image on left, Cycle 1 (left-looking) image on right in this and subsequent figures. (b) Cross-sectional channel profile taken just beyond where the fluid within the trough spilled out. Depth of trough is approximately 135 m.

Parker et al. (1991) divided Kallistos Vallis into three reaches: (1) a collapsed terrain source/trough reach characterized by a deep collapse pit at the head of the system and a deep, structurally controlled trough that contained much of the fluid flow for the first 275 km; (2) a 200-km anastomosing reach exhibiting numerous streamlined islands; and (3) a distributary reach where

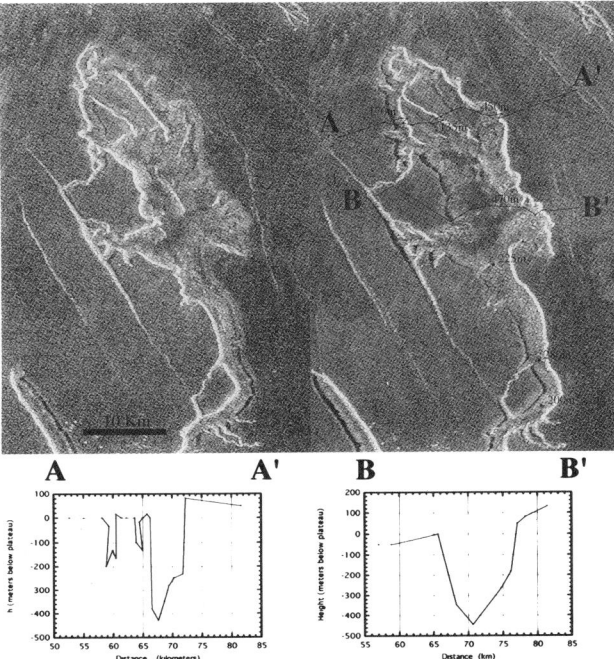

Figure 8. Stereo SAR of collapsed terrain source region for Kallistos Vallis. Depths plotted on right image based on stereo parallax measurements. Note that greatest depth measurement is associated with central, radar-dark floor of structure. Shallower values in conduit to south may be due to infilling of valley by subsequent wall collapse (radar-bright debris aprons from walls).

the flow began to spread across the existing plains to collect finally into a vast, radar-bright lobate deposit over 100,000 km^2 in area. Kallistos Vallis trends south-southeast from the caldera complex for about 400 km, at which point it encounters an elevated unit of fractured terrain that deflected the flow eastward.

The source of Kallistos Vallis is a collapsed region 17.5 km wide by 33 km elongate in the northwest direction, parallel to the local structural grain, and over 400 m deep (Baker et al. 1992). This source region is morphologically similar to, though less complex and smaller, than those associated with the well-known Martian outflow channels (Baker 1982; Mars Channel Working Group 1983). Like many of the Martian chaotic terrains, the source of Kallistos Vallis exhibits isolated mesa remnants and foundered blocks of the surrounding plains in its interior, a smooth central floor, and prominent aprons of debris at the bases of the cliffs. This trough appears to have been enlarged and deepened where it joins the larger structure, suggesting withdrawal of a subterranean fluid (Fig. 8).

The collapsed terrain source is linked to the trough reach by a 45 km long, incised gorge which appears to have been controlled by local structure

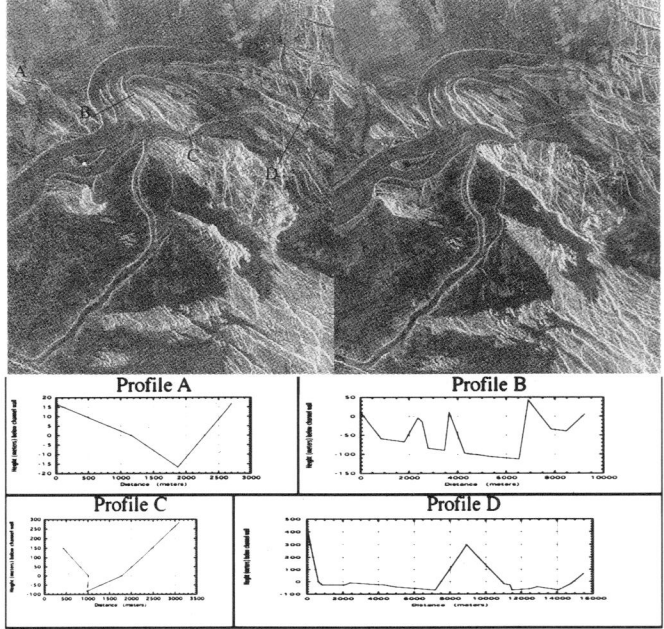

Figure 9. Stereo SAR of spillover from ponded region of Kallistos Vallis at lower end of anastomosing reach. Fractured terrain was breached at two locations: northwest-trending fractured terrain at location of profiles A–C; and north-trending fractured terrain at profile D. Cross-profiles at these locations are indicated at bottom.

and topography. Where the collapse region connects with the trough reach, additional mesa remnants indicate local enlargement of the conduit and trough through collapse. A smaller, relatively shallow collapse trough contained a portion of the flow east of the main trough. There, the fluid spilled out across the plains surface in the form of a shallow, 1.5 km wide sinuous canali (Baker et al. 1992) approximately 175 km long.

The trough reach of Kallistos Vallis appears to be structurally controlled. It is 380 km long, as wide as 4 km, and has a maximum depth of more than 600 m. At its southern end, the trough reach enters the fractured terrain adjacent to the rift valley. Because the fractured terrain is topographically higher than the adjacent plains, the southern end of the trough reach is higher relative to its proximal end.

The anastomosing reach of Kallistos Vallis begins where the trough attains its lowest elevation before crossing onto the fractured terrain. At this point, the trough reach narrows to 1.5 km and shallows to less than 150 m. The channel expands across the plains on either side of the trough to a maximum width of 18 km (Fig. 7). Sixty-five km downstream from this point, the flow leaves the trough completely, having been deflected to the southeast by the elevated fractured terrain. Streamlined islands, consisting largely of plains material, are most abundant in this reach, the largest being 15 km wide and

45 km long. Here, the floor of Kallistos Vallis exhibits a range of radar brightnesses, from bright to dark, probably reflecting variations in roughness on the channel floors on Earth (Fig. 10). At the downstream end of the 300-km long, anastomosing reach, the channel encounters a broken, north-south trending ridge cutting across its path (Fig. 9). Faint channels to either side of the main channel appear to be the remnants of lateral spreading of the fluid flow as this ridge temporarily dammed the flow. The main channel within this ponded region was probably formed after the ridge was breached and the lava lake surface dropped. Beyond the ridge barrier, the fluid spreads out into a series of distributary channels with radar-bright lobate deposits spreading laterally on either side. The main channel continues for another several hundred km on the depositional plain to the east.

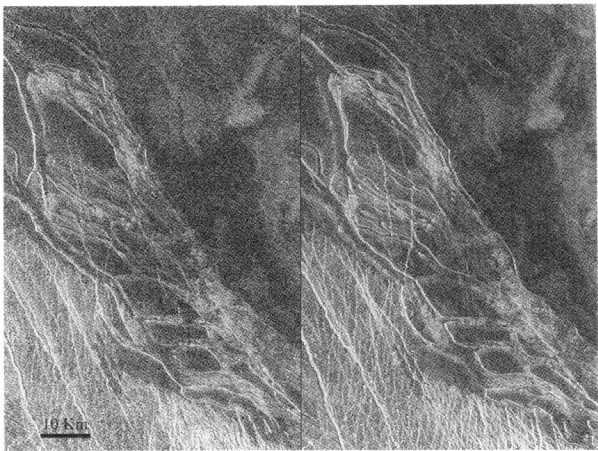

Figure 10.   Stereo SAR of upper anastomosing reach of Kallistos Vallis. Post-channel faulting has occurred, as evidenced by numerous north-trending lineaments left of center and northwest-trending lineament at lower right.

Kallistos Vallis and its terminal deposits are relatively pristine, suggesting that its formation was one of the latest events in the mapped area. Few faults cut the channel, although north-south trending faults cross-cut both the channel and its midstream islands, and in the upper anastomosing reach, as well as where the trough cross-cuts the breached ridge. A prominent 135-km-long fault trends northwest, subparallel to the anastomosing reach (Fig. 8). Although this fault appears to cross-cut the channel, it also clearly contained part of the flow through a southern branch for about 60 km, where that branch veered northeastward back toward the main channel. This fault may have been reactivated after the Kallistos Vallis fluid flow event.

IV. GLOBAL DISTRIBUTION AND GEOLOGIC SETTINGS

Venusian channels are widely distributed, but their distribution is not uniform.

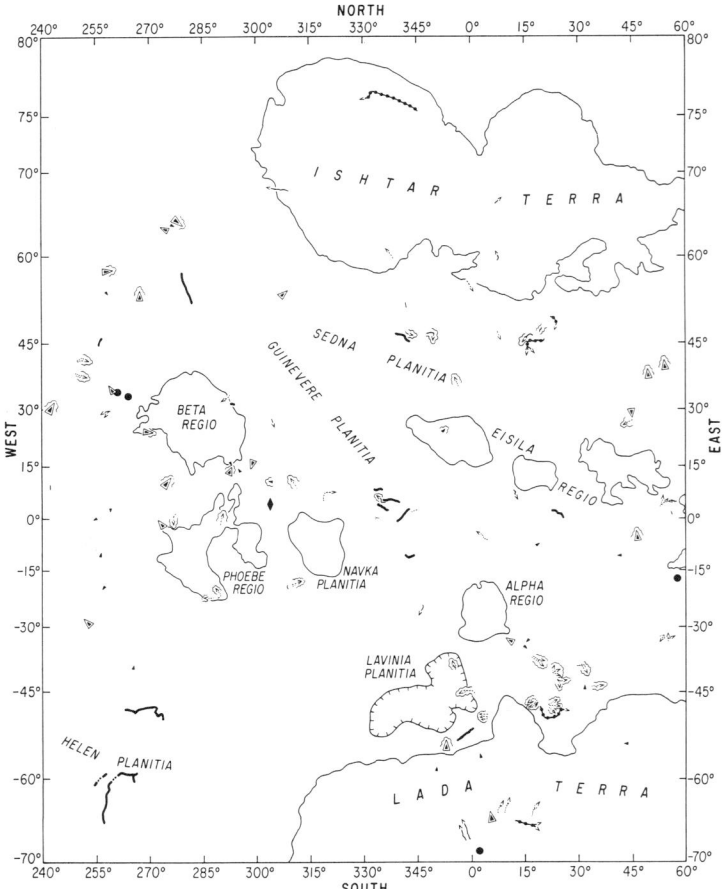

Figure 11. A global distribution of Venusian channels and valleys. The areas north of 80°N and south of 70°S are not shown because of the limited numbers of channels in these small regions (where no valleys are observed). Many simple channels, except for canali, are shorter than 100 km (distinguished by different symbols), while complex channels are generally longer than 100 km (some are clustered, but represented by the longest channels in the map). This may be an observational bias because of image resolution.

Equatorial regions seem to have unusually high channel densities. These areas include highlands, rift and fracture zones associated with large shield volcanoes, coronae, and other volcanic features. These regions contain areas (e.g., the Atla-Beta region) with the highest densities of various volcanic features on the planet (Head et al. 1992). However, as discussed below, there are other regions of low volcanic feature density that have high concentrations of canali-type channels.

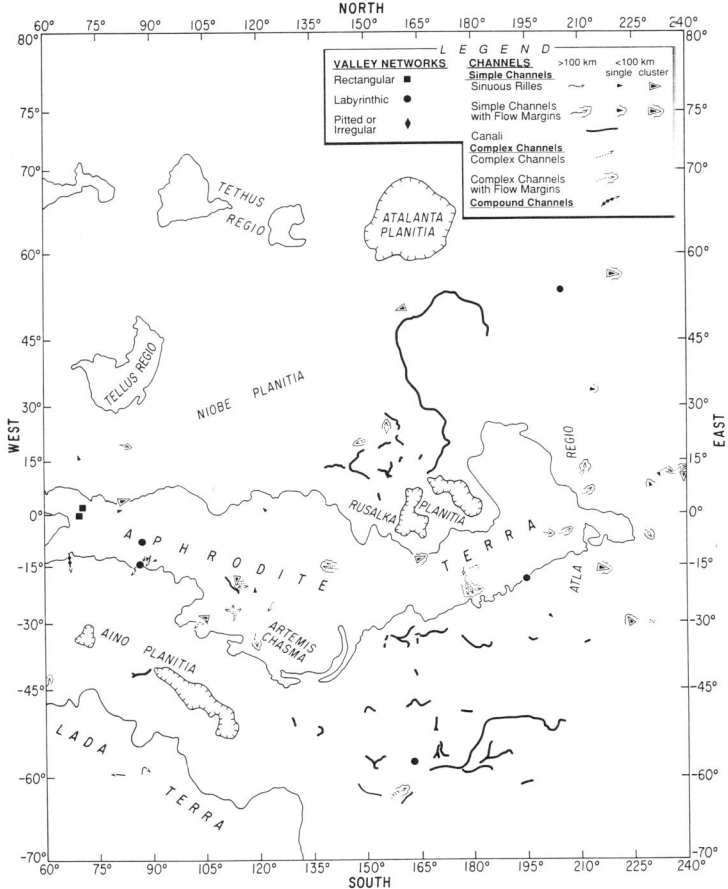

Sinuous rilles are widely distributed, but there are several regions of especially high concentration. More than half (39 out of 59, in which clusters of very small sinuous rilles were counted as one) occur on or near coronae, corona-like features or arachnoids, suggesting that channel-forming processes may be closely related to coronae evolution. Flow directions are generally consistent with the regional slope if the rilles occur on highlands.

Some simple channels occur on discernible flow deposits or flow fields with well-defined flow margins. We classify these as simple channels with flow margins. These channels are generally shallow and do not appear to incise into the surrounding terrain. This is consistent with a constructional lava-flow origin (Komatsu and Baker 1992b). Such channels are sinuous, and narrow toward their termini. However, they commonly lack source depressions and do not show the distal shallowing typical of sinuous rilles. We find that simple channels with flow margins are widely distributed (Fig. 11) and that they commonly feed extensive lava flows associated with various volcanic edifices (e.g., coronae, shield volcanoes, rift and fracture zones).

Canali are concentrated in several plains regions, including southern Guinevere Planitia, Helen Planitia, eastern Aino Planitia (southeast of Artemis Chasma), and the plains north of Rusalka Planitia. The channels trend in somewhat random directions, and occur on relatively smooth plains that have a low density of tectonic and volcanic features. Few canali have clearly defined sources or termini. All regions of canali concentration are topographically smooth, leading to a potential observational bias, whereby we might more easily recognize canali in plains than in other geologic units. For other geologic units, the channels might be destroyed by subsequent geologic processes, or they might be more difficult to discern. However, not all plains have a high concentration of canali. Small-scale canali may be found in future analysis of increased-resolution images. Nevertheless, the fact that canali larger than certain scale are observed only in some specific plains suggests that those plains favor their formation. All canali are more or less modified by subsequent tectonism and/or volcanism, indicating that their formation was not geologically recent.

Complex channels with flow margins are widely distributed (Fig. 11). These channels feed extensive lava flows that are associated with various volcanic edifices (e.g., coronae, shield volcanoes, rift and fracture zones).

Networks are rare. They are observed in the highlands, particularly in Aphrodite Terra. They are sometimes closely associated with corona, corona-like features or arachnoids.

V. GENETIC PROCESSES

A. Canali

Canali-type channels are typically longer than 500 km, and the longest ones are as long as 6800 km (Komatsu et al. 1993). These channels are unique for their nearly constant width along their entire length. Cross-sections of possible canali structures are summarized in Fig. 12. Unlike typical lava channels on Earth, most canali lack obvious levees or lateral flow deposits, although these are observed locally (Komatsu et al. 1992*a*; Gregg and Greeley 1993). Some canali have features which resemble those of terrestrial fluvial channels, such as oxbows (see Fig. 5b in Baker et al. 1992), point-bars, lateral migration of meanders (see Figs. 1 and 2 in Kargel et al. 1994), highly sinuous reaches (see Fig. 6b in Komatsu et al. 1993), fan-shaped terminal deposits (see Fig. 5a in Baker et al. 1992), and deltas (see Fig. 6 in Kargel et al. 1994). The analogy between canali and fluvial channels implies erosional processes for the formation of canali. However, Komatsu et al. (1992*a*) and Gregg and Greeley (1993) presented some arguments against erosional processes by low viscosity silicate lavas, including mafic alkaline lavas and ultramafic lavas, as a primary canali formation mechanism. It is unlikely that silicate lavas (such as basalt or komatiite) would be able to maintain turbulence along the observed canali lengths—they would cool and become too viscous—and turbulence greatly enhances the processes of thermal and mechanical erosion. Thus, the ability of a lava flow to erode decreases with increasing distance from the vent, implying that erosional channels should shallow distally. This distal shallowing is not observed in canali.

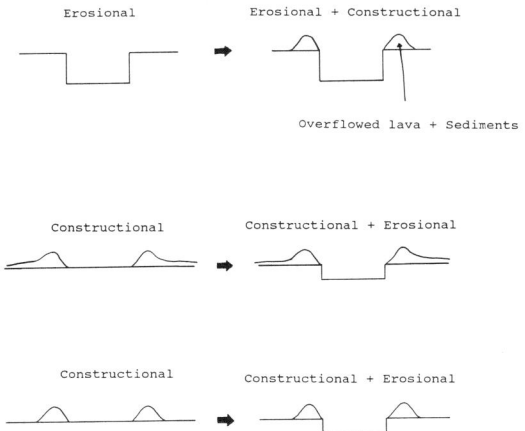

Figure 12. Cross-sections of hypothetical canali structures based on whether they were formed primarily by constructional or by erosion.

The constructional origin of canali has been also proposed and discussed (Komatsu et al. 1992a; Gregg and Greeley 1993). Mafic, including mafic alkaline, and ultramafic, lavas can flow the required distances beneath an insulating crust (Komatsu et al. 1992a). The thermophysical properties of the Venusian atmosphere cause it to cool efficiently the surface of a mafic lava flow via convection, enhancing the growth of a lava crust and promoting longer flows (Gregg and Greeley 1993). Gregg and Greeley (1993) suggested that thermal erosion on Venus is less efficient due to the smaller temperature differences between the lava and the ground rock on Venus than on Earth. Hence, they concluded that the canali are most likely the product of mechanical or constructional processes, but this does not preclude the formation of fluvial-like morphologies by the lava while it is still fluid. Alternatively, canali may be collapsed lava tubes; roofing over of the channel would thermally insulate the lava and decrease the cooling range. Assuming that the roof of a lava tube can be approximated as an unsupported beam, Oberbeck et al. (1969) proposed that the maximum width of uncollapsed, unsupported terrestrial basaltic lava tubes is $\sim$30 m. Maximum lava tube width on Venus would be similar, because gravitational acceleration is only slightly lower. Without lava supporting the solid crust, channels greater than 30 m in width would be unable to sustain a crustal roof, and typical canali are much wider than 30 m. Thus, if canali are collapsed lava tubes, their floors should be littered with collapsed roof segments, and there is no clear evidence for this. However, it is also possible that some canali may have roofed reaches (Bussey and Guest 1992). The general absence of levees or lateral flow deposits also argues against a constructional origin for canali (Komatsu et al. 1992a,1993). It is possible that the levees or lateral flow deposits are not observed because the channel-forming lava had a very low yield strength, resulting in a very narrow levee. Alternatively, a fluid lava may generate levees or lateral deposits which are too smooth to be distinguished from the background plains on Magellan imagery. Finally, degradational processes may have obliterated the levees or lateral flow deposits. It is possible that canali may have been formed by lavas which are exotic in composition (by terrestrial standards), and both sulfur and carbonatite flows have been proposed (Komatsu et al. 1992a; Baker et al. 1992; Kargel et al. 1994; Treiman 1994; Gregg and Greeley 1994). Sulfur's melting temperature is below the surface temperature of Venus at ambient conditions, and thus, would never solidify (Fig. 13). The viscosity of sulfur under Venus surface conditions is comparable to that of liquid water on Earth. Thus, although sulfur cannot thermally erode silicate ground rock, sulfur flows should be able to travel great distances even at relatively low discharge rates. Sulfur flows could possibly construct channel banks by deposition of eroded materials, similar to terrestrial rivers. However, the low humidity of sulfur in the Venusian atmosphere causes liquid sulfur to evaporate rapidly upon exposure to the surface. Alkali-rich carbonatites have melting points comparable to or slightly greater than the Venus ambient temperatures, potentially allowing considerable travel distance and mechanical erosional capacity (Fig. 13). Furthermore, carbonatites have melting points lower than those of silicate lavas, and viscosities which are orders of magnitude lower. Like sulfur flows, carbonatite lavas could not thermally erode silicate materials, but should be able to travel great distances before solidifying, even with low flow rates. Because carbonatite lavas should eventually solidify at Venusian surface conditions, they may form a channel constructionally. However, the low viscosity of carbonatite should promote mechanical erosion. Although the CO_2 pressure of the Venusian atmosphere is near carbonate saturation, there is some evidence that solid carbonatite is unstable at Venusian surface conditions, and may eventually sublime.

Even though sulfur and carbonatite are rheologically satisfactory, these lavas are only generated in small volumes on Earth today, and it is difficult to explain how they could be produced in such quantities on Venus. The dimensions of the longest canali-type channel (6800 km long by 1 km wide by 20 m deep) indicates that the total amount of sulfur lava required to scour the channel is about 10^{12} m^3, assuming the ratio of lava to eroded material is 10. Because carbonatite lavas are capable of forming constructional channels on Venus, they may be able to form a similar channel with less lava; however, the required volumes will still be much larger than observed carbonate eruptive volumes on Earth. No known terrestrial examples of sulfur and carbonatite volcanism match this scale. If canali were formed by such exotic lavas, the erupted volumes would indeed be unique among the terrestrial planets.

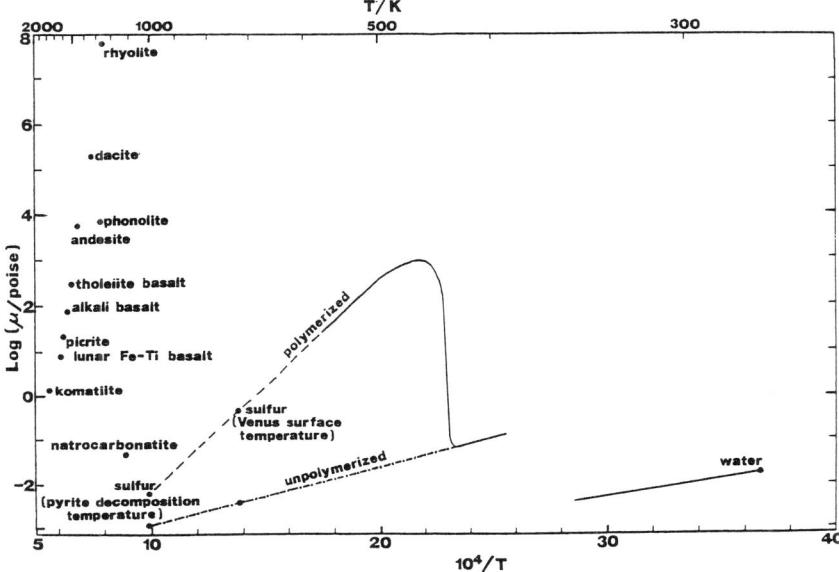

Figure 13. Viscosities and eruption temperatures of possible Venusian lavas.

B. Sinuous Rilles

Sinuous rilles received their name because of their close resemblance to lunar sinuous rilles. Like the latter, Venusian sinuous rilles narrow and shallow downstream, originate from collapse pits, have sections as deep as several hundreds meters, and do not normally have observable levees or lateral flow deposits. Various origins have been proposed for the lunar sinuous rilles, including lava drainage channels (Strom 1965), collapsed lava tubes (Oberbeck et al. 1969; Greeley 1971) and thermally eroded lava drainage channels (Hulme 1973,1982; Carr 1974).

Venusian sinuous rilles were most likely formed as a high-temperature, low-viscosity lava thermally eroded the underlying material during a high-discharge-rate, long-duration eruption. Sinuous rille depths, obtained from foreshortening relations and stereo images, may be too great to have been formed by simple lava drainage or construction processes (Komatsu et al. 1993), and thus may be the result of thermal erosion processes. The meander characteristics of Venusian and lunar sinuous rilles are distinct from those of canali and typical terrestrial lava channels (Fig. 6) (Komatsu and Baker 1994a; Kargel et al. 1994). These differences may be due to the involvement of erosional processes, particularly thermal erosion, in the genesis of sinuous rilles. Because more than half of the Venusian sinuous rilles are associated with coronae, corona volcanism may have contributed to the conditions for the channel formation (Komatsu et al. 1993).

The lunar sinuous rilles which are most similar in morphology to Venusian sinuous rilles are those that most probably formed by lava with high Fe-Ti content in comparison with terrestrial tholeiitic basalt (Murase and McBirney 1970). High Fe-Ti basalt has a higher melting temperature (and presumably higher eruption temperature) and lower viscosity than typical tholeiite (Murase and McBirney 1970). The one-dimensional thermal erosion rate E is expressed by the equation (Huppert and Sparks 1985)

$$E = h(T_l - T_{mg})/\rho_g[C_g(T_{mg} - T_o) + L_g] \tag{1}$$

in which h is the empirical heat transfer coefficient, T_l is lava temperature, T_{mg} is ground melting temperature, ρ_g is a ground density, C_g is ground heat capacity, T_o is initial ground temperature, and L_g is ground latent heat of fusion. The empirically determined heat transfer coefficient h (for Earth) is (Huppert and Sparks 1985)

$$h = 0.02k/d \, Pr^{0.4} \, Re^{0.8} \tag{2}$$

in which k is a thermal conductivity; Pr is the Prandtl number as a function of dynamic viscosity η; specific heat C and thermal conductivity of the lava (given below); Re is the Reynolds number determined by two-dimensional discharge rate (Q_{2D}) and kinematic viscosity of the lava ν; and d is the thickness of the flow and a function of discharge rate. These variables are related as follows:

$$Pr = \eta C/k \tag{3}$$

$$Re = Q_{2D}/\nu. \tag{4}$$

As shown in Eqs. (1), (2), (3), and (4), the thermal erosion rate is controlled by lava temperature and viscosity when other conditions are similar. Figure 14 illustrates erosion rate estimates for various lava types incising tholeiitic ground material. For tholeiitic lava to erode tholeiitic ground efficiently, the lava must be superheated or the flow duration must be very prolonged. Superheating can happen if the magma ascends through the crust without significant heat loss, or if the basaltic crust is heated from below by hotter ultramafic magma from a mantle plume. Unless superheated, tholeiitic lava might construct sinuous rilles, but it could not thermally deepen and widen them to a major extent. Lunar basalt, high-MgO lavas, such as komatiite and picrite, and mafic alkaline lava all have more capability for efficient thermal erosion than does tholeiitic basalt, because of high eruption temperatures and low viscosities. These factors will lead to high thermal erosion rates, particularly when the lava flows across ground materials of relatively low melting temperature (Huppert and Sparks 1985).

We conclude that the origin of Venusian sinuous rilles is a problem that can be resolved by direct analogy to the origin of lunar sinuous rilles. Highly prolonged and/or very hot lavas are required to achieve considerable thermal

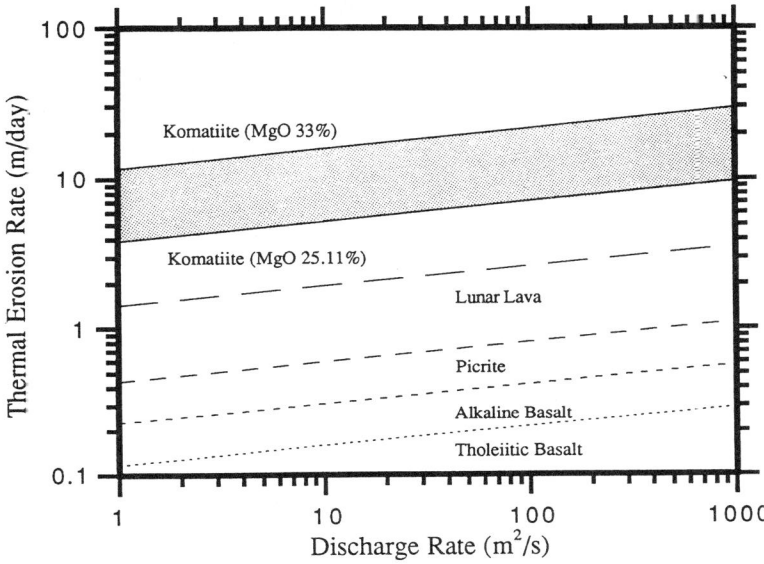

Figure 14. Thermal erosion rates of tholeiitic ground rock on Venus by various lava compositions with various flow rates. The Chezy equation was used to relate flow thickness and discharge rate (Komatsu et al. 1992a). Lavas were assumed to erupt at their respective temperatures, and the ground rock was assumed to have a melting temperature of 1270°C. For simplicity, thermal conductivity and specific heat were assumed to be 1 W m^{-1}K^{-1} and 730 J kg^{-1}K^{-1}, respectively, for all rock types. Other parameters, liquidus temperature, viscosity, density and latent heat of fusion for the ground material were derived from Murase and McBirney (1970,1973), Ryerson et al. (1988), and Hyppert and Sparks (1985). For the komatiite, a liquidus temperature range of 1500 to 1650°C (MgO content 25.11–33%) was chosen (erosion rate range is shown by the dotted area). Lunar lava is represented by a mare basalt collected during the Apollo 11 mission. Picrite is represented by a sample collected at the Kilauea Iki Lava Lake. Alkaline basalt is represented by Galapagos Island Basalt. Tholeiite is represented by Columbia River Plateau Basalt within an assumed melting temperature of 1300°C. For simplicity, all flows are assumed to be turbulent.

erosion. The canali, in contrast, have morphologies that are unique for lava channels. Their genesis requires highly fluid lavas, erupted at sustained, high discharges. Unusually low viscosities are required to explain fluvial-like features. Only the most fluid of silicate lavas would be candidates. Carbonatite and sulfur lavas are also possibilities, though their abundance would be highly unusual for terrestrial planetary geochemistry.

C. Genesis of Kallistos Vallis

The compound channel Kallistos Vallis deserves special attention because its morphology indicates massive outpourings of low-viscosity fluid (Baker et al. 1992, Parker et al. 1995). This channel is wide, originates from a collapsed pit, and has braided reaches. These characteristics are typical of Martian outflow channels, which implies the channels origin by high discharge flows. Kallistos Vallis forms radar bright levees/lateral flow deposits at its terminus, and this indicates that the channel was formed by lava.

The channel seems to have been formed by highly turbulent flow as suggested by the shapes of streamlined, probably erosional landforms (Fig. 3). For highly turbulent flow, the velocity can be approximated by a Chezy equation of the form

$$v = C(d\alpha)^{1/2} \qquad (5)$$

where C is the Chezy coefficient ($C = 1/n(d)^{1/6}$; n is the Manning coefficient), d is flow thickness, and slope $\alpha = 0.0005$. Because C is proportional to $g^{1/2}$ (where Venus gravity g_V is 8.87 m s^{-2}), the Manning coefficient for Venus (n_V) can be related to the empirical terrestrial Manning coefficient (n_E), by the equation $n_V = n_E(g_E/g_V)^{1/2} = 1.05 n_E$. This correction is very small compared with the range of possible Manning coefficients. Note that Eq. (5) ignores the viscosity factors that usually dominate in lava-flow mechanics. However, we cannot conceive of how flows with appreciable viscous effects could generate the landforms observed for Kallistos, so we are presuming highly turbulent flows in the calculations to follow.

Using Eq. (5) we calculate the velocities for a range of flow thicknesses ($d = 1, 10, 100$ m). The range of discharge rates (three-dimensional) is estimated as 5×10^2 to 5×10^7 m^{3}s^{-1} ($d = 1$–100 m and $n_V = 0.015$–0.150), because the range of width is about 3 to 15 km. A range of $d = 1$ to 100 m was chosen, assuming that the flow thickness was less than the depth of the channel. In the case of terrestrial catastrophic flood channels, the high water marks come close to the edge of the banks. If the morphology of the Kallistos Vallis is a result of flow processes similar to those of catastrophic flood channels on Earth, we can assume that the maximum flow thickness was equal to the depth of the channel, and this gives the peak discharge rates.

If lava solidified on the channel floor, the depth may not really represent the flow thickness. Indeed, the depth/width ratio of the channel is much smaller than is typical for terrestrial and Martian outflow channels. Moreover, the slope may have been significantly different at the time of formation, because the draining of a large quantity of lava could have caused a subsidence of the region. Our estimate is subject to the above limitations. The estimated discharge or effusion rate is comparable to those for some maximum-scale volcanic and fluvial processes in the solar system (Table II).

We also estimated the range of power per unit area W of the flows. The flow power is given by the formula

$$W = \gamma Q\alpha/w \qquad (6)$$

where $\gamma = \rho g$ is the specific weight of the fluid (23949 N m^{-3} for silicate lava) and w is the width of the channel. For a flow 10 m thick, and slope $\alpha = 0.0005$, the range of W is 83 W m^{-2} (at $n_V = 0.15$) to 830 W m^{-2} (at $n_V = 0.015$). This range of W is equivalent to powers of some terrestrial floods (Baker and

TABLE II
Discharge Rates of Major Volcanic and Fluvial Events in the Solar System

	Two-dimensional	Three-dimensional	References
Volcanic process			
Lunar channel forming event		10^4 to 10^6 m^3 s^{-1}	Hulme and Fielder 1977
		10^7 to 6×10^8 kg s^{-1}	Head and Wilson 1980
Lunar flood basalt event		10^4 to 10^5 m^3 s^{-1}	Head and Wilson 1986
		8×10^4 m^3 s^{-1}	Hulme 1974
Terrestrial flood basalt event	1 km^3 day^{-1} km^{-1} (11.5 m^2 s^{-1})		Swanson et al. 1975
Terrestrial komatiite flow	1 to 10^2 m^2 s^{-1}		Huppert et al. 1984
Fluvial Process			
Martian outflow channels		10^5 to 10^8 m^3 s^{-1}	Baker 1982
Mississippi River		3×10^4 m^3 s^{-1}	Baker 1982
Missoula Floods		10^5 to 10^7 m^3 s^{-1}	Baker 1982

Costa 1987). This large power is strongly suggestive of mechanical erosion, which may have played a very important role in the formation of this channel.

D. Valley Networks

Because the Venusian valley networks are morphologically similar to terrestrial and Martian valleys modified through ground-water sapping processes (Gulick et al. 1991; Komatsu et al. 1996), a lava-sapping process was proposed for their origin (Komatsu et al. 1992*b*,1996). Although lava sapping is not known on Earth, the process is thought to be somewhat analogous to ground-water sapping with low-viscosity lava as the eroding fluid. In this hypothetical scenario, outflow of low viscosity lava from the subsurface would remove substrate, resulting in collapse of overlying material and subsequent enlargement of existing valleys in a manner similar to ground-water sapping on Earth and Mars (Baker et al. 1990). Low viscosity lava with properties which approach that of water would flow preferentially along buried fracture systems. High-temperature, low-viscosity silicate lavas (such as basalt or komatiite) would warm the fracture walls, possibly resulting in melting or mechanical plucking of wall material and its subsequent incorporation into the flowing lava. This process of thermal erosion (see, e.g., Hulme 1973; Carr 1974) would create a self-enhancing relation; thermal expansion of the fractures allows more lava to flow through, which in turn increases the rate of thermal expansion. The process would cease either when subsurface outflow of lava was insufficient to maintain sapping or when lava viscosity increased due to cooling.

Carbonatite and sulfur lavas are attractive candidates for the sapping fluid, because of their low viscosities and melting temperatures (see discussion in Sec. V). Although, these fluids cannot thermally erode silicate host rocks, they can physically remove unconsolidated material. Exotic lavas are especially favorable for forming the pitted or irregular valley networks where sulfur may have evaporated in a manner similar to sublimating ice in thermokarst landforms on Earth and Mars. Solidified carbonate may sublime for present Venus surface conditions (Kargel et al. 1994). Carbonatite and sulfur flows have appropriate rheologies to have formed the Venusian sapping valleys. However, as with the Venusian channels, the necessary volumes of these magmas are problematic. In terrestrial sapping processes, the amount of water required to erode valleys in the Colorado Plateau would be on the order of 10^2 (if the ground is unconsolidated sediments) to 10^5 (if the ground rock is consolidated rocks) times more than the amount of eroded material (Howard et al. 1988). Similarly, Gulick (1993) and Gulick and Baker (1993) concluded that the ratios of the amount of water required to erode valleys in the Hawaiian volcanic terrains is approximately 3000:1. Assuming that low-viscosity lavas on Venus can mechanically and chemically erode their host rock in a manner similar to that of terrestrial groundwater, the volumes of magma required to form the Venusian valley networks can be estimated. For example, using a volume of $\sim 10^{10}$ to 10^{11} m^3 for the valley networks at 57.5°S, 166°E, the required lava volume would be $\sim 10^{12}$ to 10^{16} m^3. As discussed in the section on channel genesis, the production of such large volumes of exotic lava has not been documented on Earth. Thus, it is difficult to understand the generation of these amounts of carbonatite or sulfur flows on Venus.

VI. DEFORMATION OF LONGITUDINAL PROFILES

The longest canali-type channel, Baltis Vallis (previously named Hildr Fossa), is located west of Atla Regio (Fig. 15). The channel longitudinal profile (Fig. 16) is highly undulatory. The undulations in the profile occur at multiple superimposed scales implying that the deformation pattern is hierarchical (Komatsu and Baker 1994*b*). Because the channels are sinuous, scales of the profile undulations do not precisely correlate exactly with the linear scales of tectonic deformation observed in the surrounding plains. Thus, listed scales of deformation are approximate and relative to the sinuous trend of the datum provided by the channel path. In Fig. 16 at least two scales are observed: a smaller one at a few hundred km and a larger one at a few thousand km.

We hypothesize that these two wavelengths represent the characteristic modes of tectonic deformation in the plains regions (Figs. 16 and 17). The long scales ($\sim 10^3$ km) correspond to the characteristic scales of basin structures, and the short scales ($\sim 10^2$ km) correspond most closely to belts of compressional ridges visible on the SAR imagery (Fig. 15). Note that on a three-dimensional perspective of the topography (Fig. 17), a complex of quasi-circular domes and basins occurs at $\sim 10^2$-km scale, which corresponds to the ridge belts and intervening plains.

Not appearing on the regional-scale SAR scene (Fig. 15) are numerous wrinkle ridges (average spacing 20–40 km) with multiple orientations which are nearly ubiquitous on the Venusian plains (McGill 1993). These are not crucial in the topographic profiles (Fig. 16) because of the large spacing of altimetric sampling and the large area of the altimetric footprint. Wrinkle ridge formation postdates the emplacement of Baltis Vallis in this region, although the canali-type channel is superposed by more recent lavas. Highlands probably already existed at the time of channel formation because the channel seems to avoid the northern arms of Atla Regio (Fig. 15). This channel is postdated by lava flows, impact cratering and various episodes of tectonic deformation, as indicated by the disruptions in the channel reaches (Fig. 15).

The second longest Venusian channel is located in eastern Aino Planitia (Fig. 18). Undulations in this channel's longitudinal profile also occur at multiple scales (Fig. 19). The long scale ($\sim 10^3$ km) does not obviously correspond with any geologic feature in the SAR image. However, at a more regional scale, this concave undulation can be related to an elongate basin oriented transverse to the channel's flow direction (approximately east two-third of the profile is traversing the basin). The short-scale undulations ($\sim 10^2$ km) can be associated with belts of compressional ridges that parallel the long axes of the basin. The north-trending lineaments in Fig. 18 are wrinkle

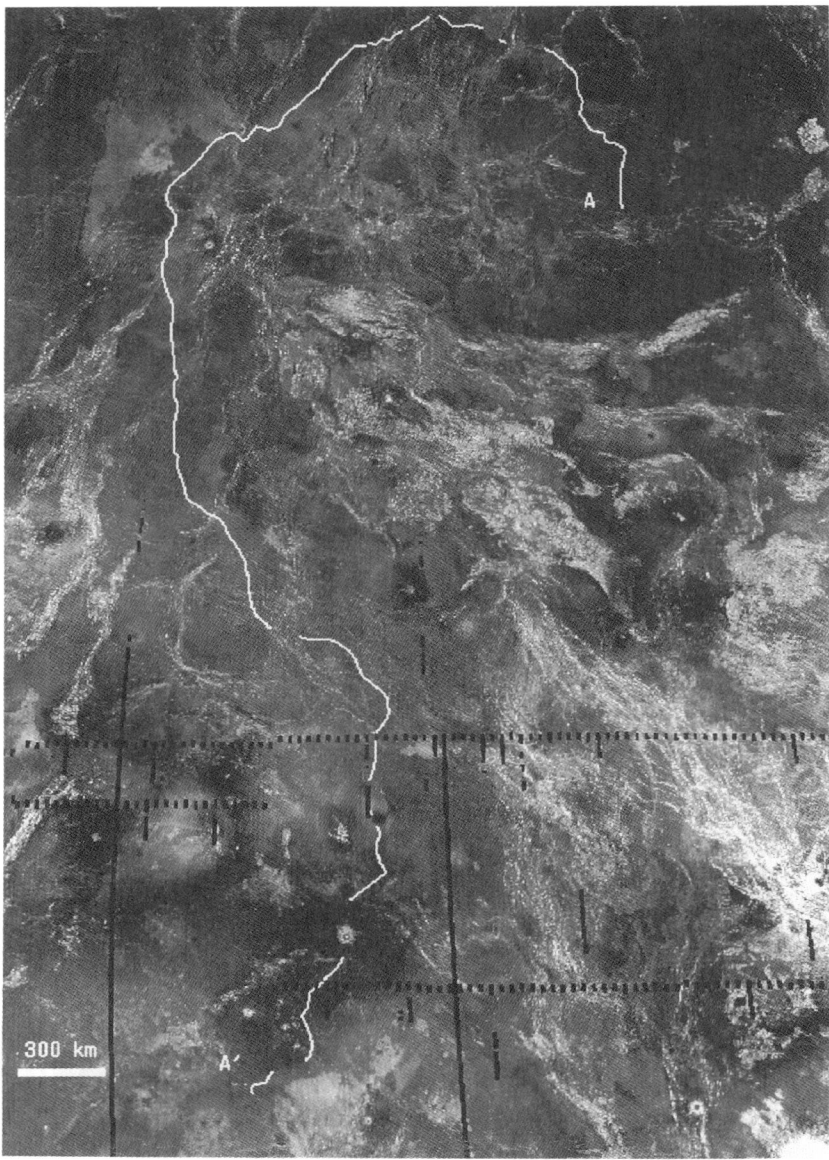

Figure 15. Atla Regio, plains and Baltis Vallis (highlighted). The image covers latitude 9 to 53°N and longitude 150 to 200°E. Longitudinal topographic profile is derived between A and A''. The northern reaches of the channel avoid the northern arm of Atla Regio, indicating that channel formation postdates this highland region. The channel is disrupted by lava flows, impact cratering and various episodes of tectonic deformation.

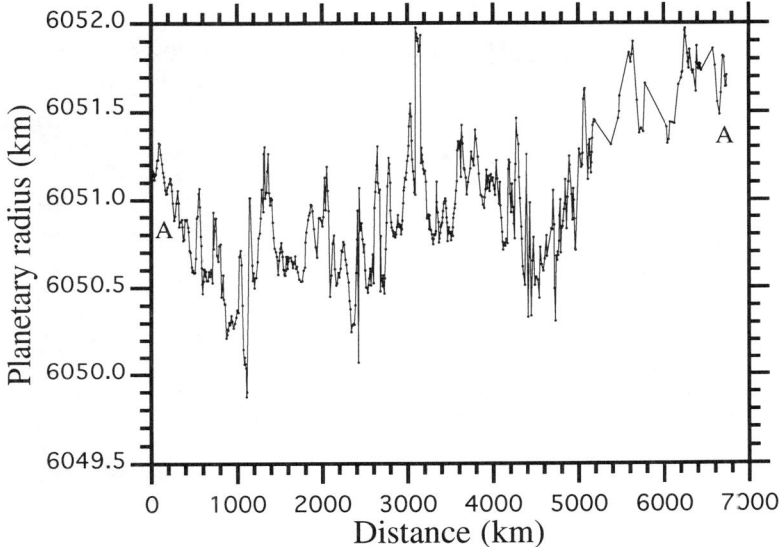

Figure 16. Longitudinal profile of Baltis Vallis.

ridges with an average spacing of 20 to 40 km. As mentioned previously, wrinkle ridges do not appear in the topographic profile. Some wrinkle ridges may have formed concurrently with canali, as suggested by locally diverted channel directions. The tectonic explanation for profile undulations seems to be post-channel downwarping of the basin concurrently with formation of compressive ridge belts and wrinkle ridges.

Sinuous rilles also show signs of deformation in their longitudinal profiles. Figure 20 illustrates a sinuous rille which trends downhill. The channel may have been deformed, but this deformation was not extensive enough to reverse the trend. Figure 21 illustrates an example of upwardly deformed sinuous rille. Among the sinuous rilles investigated, all of them are less than 500 km and most of them are less than 150 km (Fig. 22), indicating that the deformation scales recorded on sinuous rilles are much smaller than for canali.

On Earth, where a river crosses a region of slow, continuous deformation, the river adjusts its bed to the deformation to maintain a continuous graded profile (Mackin 1948). The profile is maintained through time by deposition at the downwarps and by incision at upwarps. The classic example is the Colorado River, which maintained its grade across the rising domal uplift of the Colorado Plateau (Powell 1875; Hunt 1969). On Venus, canali probably formed in a relatively short time interval at the end of the major phase of volcanic plains emplacement (Komatsu et al. 1993). The original profiles, while

Figure 17. Three-dimensional perspective of Atla Regio and plains and Baltis Vallis (highlighted).

not horizontal, probably reflected a rather uniform gradient of topography that was followed by gravity-driven, sustained flows of lava (Komatsu et al. 1992a,1993). Thus, all cumulative post-channel, long-acting tectonic warping is reflected in the profiles. On Earth, active rivers would have maintained grade across such warpings, and erosion of uplands or filling of basins would have masked their surficial expression.

On Venus, the multiple deformational scales recorded in canali longitudinal profiles fall into clearly distinct size classes. This scaling hierarchy suggests that each deformational scale may be associated with a different tectonic process of the Venusian interior. Solomon et al. (1992) recognized multiple scales of tectonic features and pointed out that 10- to 30-km-scale structures can be attributed to the response of a strong upper crustal layer (Zuber and Parmentier 1990). In contrast, the deformation of a strong mantle layer can account for tectonic features with characteristic scales of a few hundred to a few thousand km (Zuber and Parmentier 1990), particularly where these scales can be identified in the long-wavelength gravity as well as the topography (Kiefer et al. 1986). This large scale is hypothesized to be dominated by

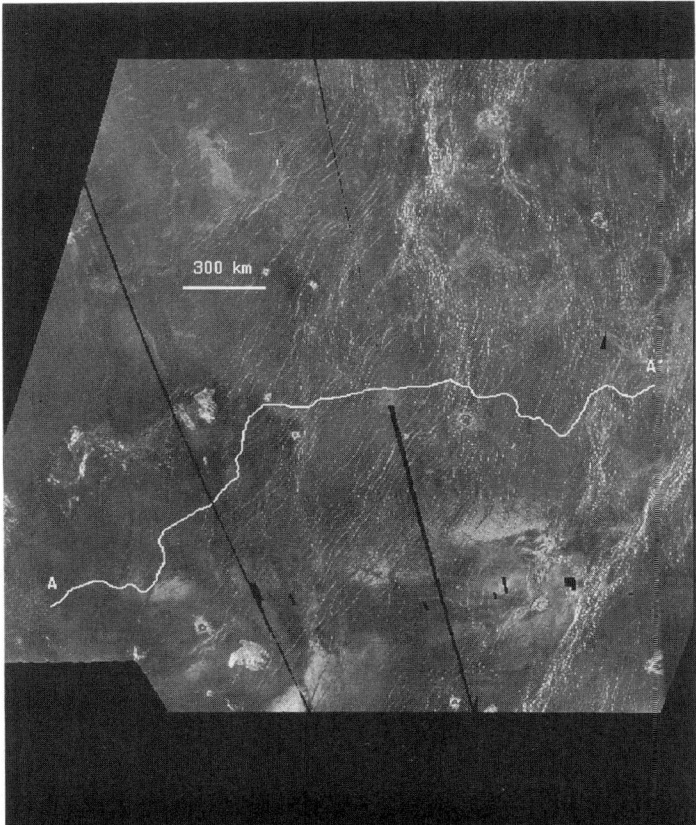

Figure 18. A canali-type channel in the eastern Aino Planitia (highlighted). The image covers latitude 40 to 63°S and longitude 175 to 208°E. Longitudinal topographic profile was derived between A and A'.

mantle convection and its associated dynamic stresses and heat transport. The thousand-km-scale features include broad rises and basins in plains regions. It has been suggested that plains are sites of mantle downwelling (Zuber 1990; Phillips et al. 1991). Bindschadler et al. (1992) identified Atalanta Planitia and Lavinia Planitia as surface expressions of young mantle coldspots or regions of mantle downwelling. Longitudinal profiles, particularly for the longest and second longest channels (Figs. 16 and 19) clearly indicate formation of basins and associated compressional ridge belts after channel formation. The scale of a few km and less involves either internal deformation of the upper crust or tectonic disruption of a thin surficial layer decoupled thermally or mechanically from the remainder of the otherwise strong upper layer (Solomon et al. 1992).

The time sequence derived from the stratigraphic relationships indicates

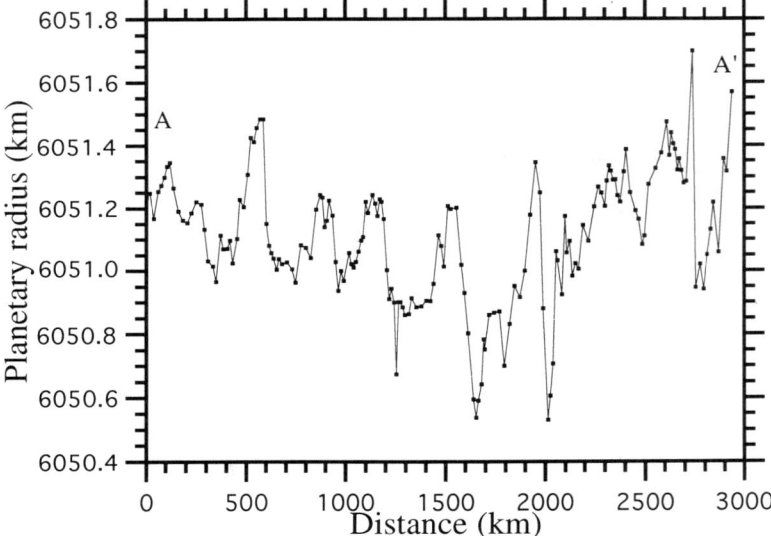

Figure 19. Longitudinal profile of canali-type channel in eastern Aino Planitia (Fig. 18).

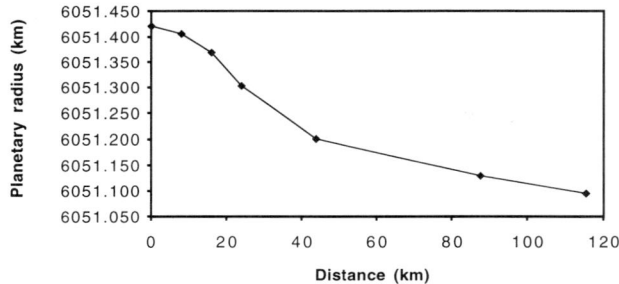

Figure 20. Longitudinal profile of a sinuous rille at 0°N,255°E. This channel trends downhill. Any deformation that may have occurred did not deform the profile to trend uphill.

that canali are among the younger landforms on the plains, and that their formation is closely related to the last phases of extensive plains volcanism, possibly induced by the hypothesized global resurfacing event (Schaber et al. 1992). The multiple scales of plains deformation recorded in canali profile undulations therefore reflect the various deformation processes of the Venusian interior that operated after the major plains resurfacing epoch (Komatsu and Baker 1994b). This event is estimated to have occurred at approximately 190 to 600 Myr (Strom et al. 1994).

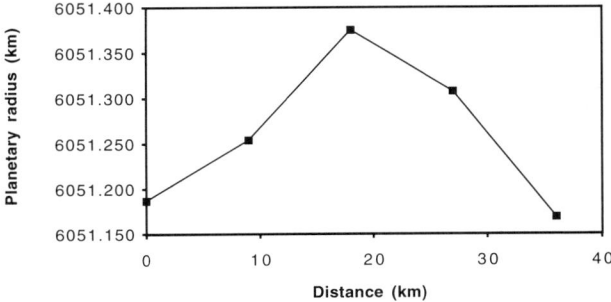

Figure 21. Longitudinal profile of a sinuous rille at 1°N,259°E. This channel has an upwardly deformed profile. This deformation may be caused by tectonic activity of nearby volcanic dome.

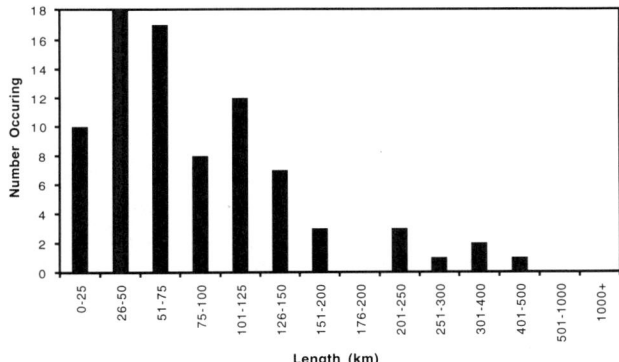

Figure 22. Length distribution of sinuous rilles.

VII. CONCLUSIONS

More than 200 Venusian channels and valleys are recognized in a global context. Some channels with associated flow deposits are similar in appearance to terrestrial lava channels. Sinuous rilles, similar in morphology to lunar sinuous rilles, most likely formed through erosion processes in high-effusion-rate, long-lived lava flows. Valley networks were also probably developed by the action of low-viscosity lavas, possibly by an unusual lava sapping process.

Canali-type channels are unique because of their great lengths (up to 6800 km) and nearly constant width along their paths. These dimensions require unusual eruptive conditions, most probably including an extended eruption of a large volume of channel-forming lava. If canali formed mainly by constructional processes, laminar tholeiitic flows of relatively high, sustained discharge rates might travel the observed distances, but the absence of levees or lateral flow deposits would need to be explained. An exotic low-temperature, low-viscosity lava, such as carbonatite or sulfur, seems to be required for the erosional genesis of canali.

Undulatory topography imposed on canali and sinuous rille profiles seem to be the result of tectonic activity imposed since channel formation. This implies that some of the youngest plains materials, which may have been emplaced coincident with the channels, experienced significant postemplacement tectonic deformation.

Acknowledgments. We benefitted from discussions with J. Kargel, J. Johnson, and R. Strom. J. Aubele and T. K. P. Gregg provided unusually thorough and helpful reviews of the initial version of this chapter. J. Langdon assisted with the preparation of Figs. 20–22. Our research was supported by a NASA Planetary Geology and Geophysics grant.

Editor's Note

Baker et al. have provided a Venus flyover animation on the accompanying CD-ROM. Please see CDP6C2M1.

REFERENCES

Baker, V. R. 1982. *The Channels of Mars* (Austin: Univ. of Texas Press).
Baker, V. R., and Costa, J. E. 1987. Flood power. In *Catastrophic Flooding*, eds. L. Mayer and D. Nash (Boston: Allen and Unwin), pp. 1–21.
Baker, V. R., Kochel, R. C., Laity, J. E., and Howard, A. D. 1990. Spring sapping and valley development. In *Groundwater Geomorphology*, eds. C. G. Higgins and D. R. Coates, GSA SP-252 (Bouder: Geological Soc. of America), pp. 235–265.
Baker, V. R., et al. 1992. Channels and valleys on Venus: Preliminary analysis of Magellan data. *J. Geophys. Res.* 97:13421–13444.
Bindschadler, D. L., Schubert, G., and Kaula, W. M. 1992. Coldspots and hotspots: Global tectonics and mantle dynamics of Venus. *J. Geophys. Res.* 97:13495–13532.
Brice, J. C. 1964. *Channel Patterns and Terraces of the Loup Rivers in Nebraska*, U. S. G. S. Prof. Paper 422-D, D1-D41.
Bussey, D. B. J., and Guest, J. E.1992. Erosion vs. construction: The origin of Venusian channels. In *International Colloquium on Venus*, LPI Contrib. No. 789, pp. 18–19 (abstract).
Bussey, D. B. J., Sorenson, S.-A., and Guest, J. E. 1995. Factors influencing the capability of lava to erode its substrate: Application to Venus. *J. Geophys. Res.* 100:16941–16948.
Carr, M. H. 1974. The role of lava erosion in the formation of lunar rilles and Martian channels. *Icarus* 22:32–43.
Chang, H. H. 1988. *Fluvial Processes in River Engineering* (New York: J. Wiley & Sons).
Greeley, R. 1971. Lunar Hadley Rille: Consideration of its origin. *Science* 172:722–725.

Greeley, R., ed. 1974. *Geologic Guide to the Island of Hawaii* (Washington, D. C.: NASA).
Gregg, T. K. P., and Greeley, R. 1993. Formation of canali: Considerations of lava types and their thermal behaviors. *J. Geophys. Res.* 98:10873–10882.
Gregg, T. K. P., and Greeley, R. 1994. Reply. *J. Geophys. Res.* 99:17165–17167.
Gulick, V. C. 1993. Magmatic Intrusions and Hydrothermal Systems: Implications for the Formation of Martian Fluvial Valleys. Ph.D. Thesis, Univ. of Arizona.
Gulick, V. C., and Baker, V. T. 1993. Fluvial erosion on Mars: Implications for paleoclimatic change. *Lunar Planet. Sci. Conf.* XXIV:587–588 (abstract).
Gulick, V. C., et al. 1991. Channels on Venus: A preliminary morphological assessment and classification. *Lunar Planet. Sci. Conf.* XXII:507–508 (abstract).
Gulick, V. C., Baker, V. R., and Komatsu, G. 1992a. Channel and valley morphology on Venus: An updated classification. *Lunar Planet. Sci. Conf.* XXIII:465–466 (abstract).
Gulick, V. C., Komatsu, G., and Baker, V. R. 1992b. Integrated valley systems on Venus: A comparative morphological study. *Lunar Planet. Sci. Conf.* XXIII: 467–468 (abstract).
Heacock, R. L., et al. 1966. *Ranger VIII and IX, Part II. Experimenters' Analyses and Interpretations*, NASA TR-32-800.
Head, J. W., and Wilson, L. 1980. Lunar sinuous rille formation by thermal erosion: Eruption conditions, rates and durations. *Lunar Planet. Sci. Conf.* XI:427–429 (abstract).
Head, J. W., and Wilson, L. 1986. Volcanic processes and landforms on Venus: Theory, predictions, and observations. *J. Geophys. Res.* 91:9407–9446.
Head, J. W., et al. 1991. Venus volcanism: Initial analysis from Magellan data. *Science* 252:276–299.
Head, J. W., et al. 1992. Venus volcanism: Classification of volcanic features and structures, associations, and global distribution from Magellan data. *J. Geophys. Res.* 97:13153–13197.
Howard, A. D., Kochel, R. C., and Holt, H. E. 1988. *Sapping Features of the Colorado Plateau*, NASA SP-491.
Hulme, G. 1973. Turbulent lava flow and the formation of lunar sinuous rilles. *Mod. Geology* 4:107–117.
Hulme, G. 1974. The interpretation of lava flow morphology. *Geophys. J. Roy. Astron. Soc.* 39:361–383.
Hulme, G. 1982. A review of lava flow processes related to the formation of lunar sinuous rilles. *Geophys. Surveys* 5:245–279.
Hulme, G., and Fielder, G. 1977. Effusion rates and rheology of lunar lavas. *Phil. Trans. Roy. Soc. London A* 285:227–234.
Hunt, C. B. 1969. *Geologic History of the Colorado River*, U. S. G. S. Prof. Paper 669-C, 59-130.
Huppert, H. E., and Sparks, R. S. J. 1985. Komatiites I: Eruption and flow. *J. Petrology* 26:694–725.
Huppert, H. E., Sparks, R. S. J., Turner, J. S., and Arndt, N. T. 1984. Emplacement and cooling of komatiite lavas. *Nature* 309:19–22.
Kargel, J. S., Fegley, B., Jr., Treiman, A., and Kirk, R. L. 1994. Carbonatite-sulfate volcanism on Venus. *Icarus* 112:219–252.
Kiefer, W. S., Richards, M. A., Hager, B. H., and Bills, B. G. 1986. A dynamic model of Venus' gravity field. *Geophys. Res. Lett.* 13:14–17.
Komatsu, G., and Baker, V. R. 1992a. Venusian sinuous rilles. In *International Colloquium on Venus*, LPI Contrib. No. 789, pp. 60–61.
Komatsu, G., and Baker, V. R. 1992b. Formation of Venusian channels and valleys, and styles of volcanism. *Lunar Planet. Sci. Conf.* XXIII:715–716 (abstract).

Komatsu, G., and Baker, V. R. 1994a. Meander properties of Venusian channels. *Geology* 22:67–70.
Komatsu, G., and Baker, V. R. 1994b. Plains tectonism on Venus: Inference from canali longitudinal profiles. *Icarus* 110:275–286.
Komatsu, G., Kargel, J. S., and Baker, V. R. 1992a. Canali-type channels on Venus: Some genetic constraints. *Geophys. Res. Lett.* 19:1415–1418.
Komatsu, G., Gulick, V. C., Kargel, J. S., and Baker, V. R. 1992b. Venus lava sapping valleys. *Lunar Planet. Sci. Conf.* XXIII:719–720 (abstract).
Komatsu, G., Baker, V. R., Gulick, V. C., and Parker, T. J. 1993. Venusian channels and valleys: Distribution and volcanological implications. *Icarus* 102:1–25.
Komatsu, G., Gulick, V. C., and Baker, V. R. 1996. Valley networks on Venus. *J. Geomorph.*, submitted.
Leopold, L. B., and Wolman, M. G. 1957. *River Channel Patterns: Braided, Meandering, and Straight*, U. S. G. S. Prof. Paper 282-B, pp. 39–85.
Leopold, L. B., and Wolman, M. G. 1960. River meanders. *Geol. Soc. America Bull.* 71:769–794.
Mackin, J. H. 1948. The concept of the graded river. *Geol. Soc. America Bull.* 59:463–512.
Mars Channel Working Group. 1983. Channels and valleys on Mars. *Geol. Soc. America Bull.* 94:1035–1054.
McGill, G. E. 1993. Wrinkle ridges, stress domains, and kinematics of Venusian plains. *Geophys. Res. Lett.* 20:2407–2410.
Mouginis-Mark, P. J., Wilson, L., and Zuber, M. T. 1992. The physical volcanology of Mars. In *Mars*, eds. H. H. Kieffer, B. M. Jakosky, C. W. Snyder and M. S. Matthews (Tucson: Univ. of Arizona Press), pp. 424–452.
Murase, T., and McBirney, A. R. 1970. Viscosity of lunar lavas. *Science* 167:1491–1493.
Murase, T., and McBirney, A. R. 1973. Properties of some common igneous rocks and their melts at high temperatures. *Geol. Soc. America Bull.* 84:3563–3592.
Oberbeck, V. R., Quaide, W. L., and Greeley, R. 1969. On the origin of lunar sinuous rilles. *Mod. Geology* 1:75–80.
Parker, T. J., et al. 1991. An outflow channel in Lada Terra, Venus. *Lunar Planet. Sci. Conf.* XXII:1035–1036 (abstract).
Parker, T. J., Komatsu, G., and Baker, V. R. 1995. Kallistos Vallis, Venus: Geology of a catastrophic flood channel. In preparation.
Phillips, R. J., Grimm, R. E., and Malin, M. C. 1991. Hot-spot evolution and the global tectonics of Venus. *Science* 252:651–658.
Powell, J. W. 1875. *Exploration of the Colorado River of the West and Its Tributaries* (Washington, D. C.: U. S. Government Printing Office).
Roberts, M., Guest, J. E., Guest, J. W., and Lancaster, M. G. 1992. Mylitta Fluctus, Venus: Rift-related, centralized volcanism and the emplacement of large-volume flow units. *J. Geophys. Res.* 97:15991–16015.
Ryerson, F. J., Weed, H. C., and Piwinskii, A. J. 1988. Rheology of subliquidus magmas. 1. Picritic compositions. *J. Geophys. Res.* 93:3421–3436.
Schaber, G. G., 1992. Geology and distribution of impact craters on Venus: What are they telling us? *J. Geophys. Res.* 97:13257–13302.
Schumm, S. A., and Khan, H. R. 1972. Experimental study of channel patterns. *Geol. Soc. America Bull.* 88:1755–1770.
Schumm, S. A., Khan, H. R., Winkley, B. R., and Robbins, L. G. 1972. Variability of river patterns. *Nature* 237:75–76.
Senske, D. A., Saunders, R. S., Stofan, E. R., and Members of the Magellan Science Team. 1994. The global geology of Venus: Classification of landforms and geologic history. *Lunar Planet. Sci. Conf.* XXV:1245–1246 (abstract).

Solomon, S. C., et al. 1992. Venus tectonics: An overview of Magellan observations. *J. Geophys. Res.* 97:13119–13255.

Strom, R. G. 1965. *Interpretations of Ranger VII Records*, JPL Tech. Rept. 32, chp. III.

Strom. R. G., Schaber, G. G., and Dawson, D. D. 1994. The global resurfacing of Venus. *J. Geophys. Res.* 99:10899–10926.

Swanson, D. A., Wright, T. L., and Helz, R. T. 1975. Linear vent systems and estimated rates of magma production and eruption for the Yakima Basalt on the Columbia Plateau. *Amer. J. Sci.* 275:877–905.

Treiman, A. H. 1994. Comment on "Formation of Venusian canali: Considerations of lava types and their thermal behaviors" by T. K. P. Gregg and R. Greeley. *J. Geophys. Res.* 99:17163–17164.

Zuber, M. T. 1990. Ridge belts: Evidence for regional- and local-scale deformation on the surface of Venus. *Geophys. Res. Lett.* 17:1369–1372.

Zuber, M. T., and Parmentier, E. M. 1990. On the relationship between isostatic elevation and the wavelengths of tectonic surface features on Venus. *Icarus* 85:290–308.

PART VII
Tectonism

TECTONIC OVERVIEW AND SYNTHESIS

VICKI L. HANSEN and JAMES J. WILLIS
Southern Methodist University

and

W. BRUCE BANERDT
Jet Propulsion Laboratory

Geomorphically and geologically Venus is divisible into plains, tesserae, coronae and chasmata, volcanic rises, crustal plateaus, and Ishtar Terra. In this chapter we discuss work in progress for each of these geologic provinces. The tectonism responsible for the surface geology, as reflected in Magellan data, comprises dominantly vertical motions with a lithosphere that remains relatively fixed in place horizontally, and a mantle convection system that changes below it through time. Venus tectonic environments can be divided into mantle downwelling, mantle upwelling, plumes, and Ishtar Terra. The plains are regions of broad mantle downwelling as supported by long-wavelength gravity data and contractional ridge belts indicative of compressive stress. Diffuse mantle upwellings are marked by regions of lithospheric extension and coronae-chasmata. Crustal plateaus may represent the lithospheric signature of ancient deep thermal plumes (hotspots), and volcanic rises represent younger thermal plumes. The difference between plateaus and rises may reflect the interaction of the plumes with thin lithospheres (plateaus) and thick lithosphere (rises). The lithosphere might have been thinner earlier in Venus' history, thermally evolving to thicker lithosphere more recently. Ishtar Terra represents a unique region of ponded mantle residuum, thickened lower crust and folded upper crust. Mantle upwellings and downwellings change location through time, and plains provinces may become "intruded" by coronae and cut by chasmata, whereas corona-chasma regions may founder and become plains as the upwellings beneath them decay. Some tessera inliers may represent flooded crust originally deformed within coronae-chasmata and crustal plateaus. Thus Venus hosts a vertical, rather than horizontal, tectonic cycle, and Venus' lithosphere thickened with time.

I. INTRODUCTION

Venus is similar to the Earth in mass and radius and appears to cool itself in a different manner. NASA's Magellan satellite returned breathtaking radar imagery, and altimetry and gravity data covering nearly the entire surface of Venus. From a tectonics standpoint, the most striking aspect of the planet, revealed by these near globally complete data sets, is the lack of a global-scale system of linear features such as the ocean spreading centers, volcanic arcs, and deep-sea trenches present on Earth (Fig. 1). The most obvious interpretation of these observations is that plate tectonic processes are not presently (nor in the recent past) operative on Venus (Solomon et al. 1991,1992). In addition, Venus exhibits unimodal hypsometry, unlike Earth's bimodal distribution. Most of the planet lies at mean planetary radius (MPR) with local deep troughs (2–7 km below MPR) and highland regions (2–11 km above MPR). If Venus does not have plate tectonics how does heat escape to the surface? What is the nature of Venus' global tectonism? Through synthesis of radar images, and altimetry and gravity data, the following sections address these questions aimed at understanding the tectonic processes that led to the formation of individual geologic provinces.

Geomorphically and geologically, the surface of Venus is divisible into seven types of features or provinces (Fig. 1). Plains, or planitiae, lie at or below MPR and comprise ~80% of the planet. Volcanic rises marked by domical topographic profiles (e.g., Atla, Beta, Eistla, Themis), crustal plateaus marked by steep-sided flat-topped profiles (e.g., Alpha, Ovda, Phoebe, Tellus, and Thetis), and Ishtar Terra all reside within the highlands. Coronae, circular features commonly topographically higher than the surrounding terrain, and chasmata, fractured linear troughs (e.g., Dali, Diana, Hecate, and Parga), are found together in the equatorial region, mostly within eastern Aphrodite Terra, but also extending farther east. Locally coronae and chasmata are spatially associated with volcanic rises. Tesserae, terrain characterized by seemingly complex deformation, occur (1) within the highland regions, (2) as islands within the plains, and (3) in association with volcanic rises and some corona chains. The largest tracts of tesserae lie within the crustal plateaus.

In this chapter we briefly review the types of Magellan data sets and the types of constraints imposed by these data. We then review each of the geologic provinces, and finally attempt to relate these features spatially and temporally within a global framework. Our understanding of each province, and the global relations between provinces, is very much work in progress. It will take many years of dedicated research to sift through the clues afforded by the integrated Magellan data sets. This chapter should be viewed as a report of investigations to date, and directions for future research.

A. Magellan Data Sets

In order to understand how individual types of structural features relate to one another and to tectonic processes, we turn to the Magellan data sets. These include synthetic-aperture radar (SAR) images, altimetry, emissivity, and high-resolution gravity. Each data set can be used to help constrain different aspects of Venus tectonics.

SAR Imagery. Nearly 98% of Venus is imaged by SAR and is available as compressed or full (~100–250 m) resolution images. This vast data set is vital for understanding surface deformation and volcanism. It is particularly useful because Venus lacks an asthenosphere, and thus the surface is more directly "tied" to mantle flow fields than on Earth (Phillips 1986,1990). In

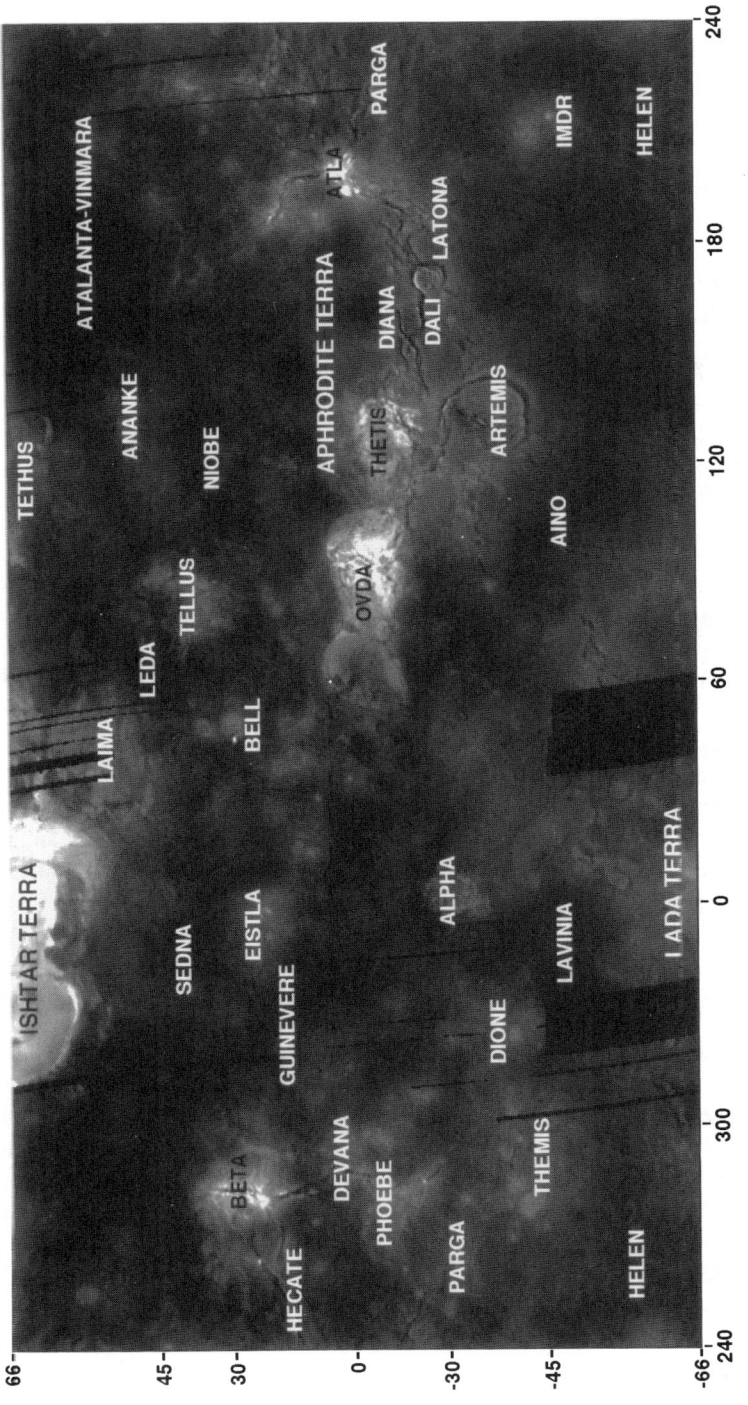

Figure 1. Mercator projection of Venus topography with locations of features discussed in the text.

addition, the absence of water inhibits erosion (Barsukov et al. 1986; Kaula 1990)—thus the surface records a virtually unmodified response to tectonic processes. The high resolution afforded by SAR allows one to differentiate fractures, folds, and graben with relative confidence (Ford et al. 1993). Spatial and temporal relations between structures or suites of structures can commonly be determined. SAR imagery together with roughness, reflectivity, and emissivity data yield constraints on individual volcanic flow sources, extent of flows, flow direction, and relative temporal relations between flows and structures. Geologic mapping and analysis of SAR images results in constraints on the geologic and structural history of the surface, which constrains, in turn, crustal thickening processes.

Impact Craters. About 930 impact craters decorate the surface in a spatial distribution indistinguishable from a random population (Schaber et al. 1992; Phillips et al. 1992; Strom et al. 1994). Although most of the craters appear pristine and unmodified by deformation or volcanism, Herrick and Sharpton (1996) proposed that the number of embayed craters may be seriously underestimated. The crater distribution, like the topography, signals that Venus' cooling mechanism and tectonic cycle (if indeed it has a cycle) are different from that of the Earth. Some workers have used the random crater distribution to argue for catastrophic resurfacing of Venus at $\sim$500 to 300 Myr—the average age of the surface based on crater density (Schaber et al. 1992; Strom et al. 1994)—and hence catastrophic tectonics (Turcotte 1993). However, others have demonstrated that fractured and/or embayed craters have an affinity for low-density crater regions, which also correlate with increased radar backscatter—regions of high topography and high roughness, reflecting relative geologic youth. Thus, tectonically-disturbed areas may be younger than nontectonic regions (Phillips et al. 1992), and the apparent random crater population records a range of ages, achieved through either cyclic or evolutionary processes (Phillips 1993).

Although catastrophic and noncatastrophic processes might be differentiated by dating individual surface areas of the planet, several problems are inherent to this type of analysis. Problems arise from the limited number of craters, the paucity of small craters, and the nature of the crater distribution. In order to date a surface (crater density dates only the surface, not necessarily topography, gravity signature, or even deformation), a region must be $>10^7 \text{km}^2$ to ensure that anomalous crater densities are statistically important, and do not result merely by chance (Phillips et al. 1992). Thus the large required area limits the usefulness of crater dating. Crater densities cannot constrain temporal relations of specific tectonic elements or processes, though they may prove useful for global-scale modeling.

Study of individual craters is useful both for constraining local relative temporal relations and as local quantitative strain indicators (Senske et al. 1992). For example, fractures that cross-cut crater floors must postdate the impact. Impact melt-derived flows that flood adjacent fabrics indicate relative youth of the flows. Deformed craters may provide local estimates of strain, although the estimates may not reflect total accumulated local or regional strains (Solomon et al. 1992).

Although most craters appear pristine, recognition of degradation stages may allow for the development of a relative age hierarchy within the crater population (Izenberg et al. 1994). A crater's relative youth could indicate the relative youth of flow surfaces or structures that postdate young craters (Phillips and Izenberg 1995).

Altimetry. The altimeter instrument had an average footprint of 10 to 30 km with centers of successive footprints spaced ~20 km. Resultant data have been processed and interpolated to ~5-km resolution. Though features on the order of tens of km can be resolved, individual structures, such as folds and faults, cannot be resolved using this "averaged" data base. However, higher-resolution altimetry information is available by several sources and/or methods (Ford and Pettengill 1992; Plaut 1993a,b). Doppler-resolved altimetry can be obtained using the full along-track resolution. In this case, the data have not been averaged, resulting in increased resolution, but this method is limited to narrow regions along individual tracks. Altimetry information can also be obtained from the SAR data set, using patterns of SAR return echoes (i.e., using the SAR instrument as an altimeter) and/or parallax measurements between objects imaged from different incidence angles (i.e., images from two or more cycles). Digital elevation models constructed using these photogrammetric methods can resolve features approaching image resolution (~100–250 m) (Plaut 1993a). Altimetry data are also used in combination with the comparatively low-resolution gravity data (below), although the altimetry data must be filtered so that the effective resolution is commensurate with that of the gravity data.

Emissivity. Emissivity data can be used in certain circumstances to identify recent volcanism or recent tectonic uplift (Robinson and Wood 1993; Klose et al. 1992). The emissivity increases with surface elevation up to ~6055 km; above this altitude, a sharp decrease occurs. This pattern in emissivity is interpreted to be the result of weathering and/or precipitation of surface minerals, as a function of temperature (see, e.g., Klose et al. 1992; Brackett et al. 1994; Pettengill 1995, AGU Whipple Lecture). Deviations from the regional emissivity trend may indicate recent or near-recent activity.

Gravity. Any deviation from a uniform, spherically symmetric distribution of density will cause an anomaly in the gravity field which may be detected through its effect on an orbiting satellite's trajectory. Such lateral density variations may consist of actual differences in the composition or temperature, or in local radial (i.e., vertical) deflections of a horizontal density interface. Because these variations also imply forces (through additional weight or buoyant uplift), gravity can be an extremely useful tool in understanding subsurface processes. The nature of the correlation of a planet's gravity with its topography reflects much of its internal density structure. The calculated gravity signature due to the topography is virtually always much less than the observed gravity anomaly, because processes that support topographic load tend to generate gravity anomalies that partially cancel the effect of the surface relief. Topography can be supported isostatically by the buoyancy of low-density material at depth or by the downward deflection of the low-density crust, displacing higher-density mantle. The elastic strength of the mechanical lithosphere may contribute to its support, as may stresses generated by motions in the underlying region which themselves may be driven by density variations. Horizontal and depth variations in composition, temperature, and dynamic stresses each can contribute to how topography is compensated, making it difficult in most cases to delineate clearly the processes involved. However other constraints, combined with differing dependencies on spatial dimensions for the various processes, can be used to determine the structure of the planet's interior.

Two quantities that geophysicists use to interpret gravity anomalies are gravity-to-topography ratios (GTRs) and apparent depths of compensation (ADCs). Large GTRs indicate that topography is either only partially compensated, or more likely, that the compensating mass is located at great depth. With some assumptions, GTRs can be used to help distinguish between isostatic support and dynamic support. ADCs are derived from GTRs assuming a simple, single-level isostatic compensation model. They provide a qualitative indication of the depth at which the topographic features are supported. Caution must be exercised in interpreting the ADC, however, as other support mechanisms (such as elastic flexure) and multiple levels of isostatic support (such as crustal thickening over a depleted, higher-density mantle) can distort the ADC from the actual depths. Generally, however, shallow ADCs are taken to indicate crustal compensation, whereas deep ADCs are interpreted as reflecting dynamic support by thermal anomalies or convection within the mantle.

Flow Laws. Material flow laws, derived from experimental and theoretical studies, constrain the rheologic or strength properties of mineral phases at given pressure-temperature conditions. These constraints can impose strong biases on the way we view deformation. For example, prior to 1994, essentially all models for the crustal portion of the mechanical lithosphere used a flow law from diabase that was not completely dry (Caristan 1982). This flow law, when combined with the much better-understood flow law for dunite (Avé Lallemant 1978) (representing mantle rocks), resulted in a very thin, weak elastic layer in the upper crust, separated from the strong elastic zone in the upper mantle by a ductile lower crust (the so-called "jelly sandwich structure") (Zuber 1987; Banerdt and Golombek 1988). New flow laws determined from extremely dry Columbia diabase indicate much greater strength (Mackwell et al. 1995) and may provide a better analog for Venusian crust given that the outermost portions of Venus lack water (Kaula 1990) which weakens silicate minerals. This flow law predicts a crust that is almost as strong as the dunite mantle portion of the lithosphere, allowing a three-layer rheological structure only over a very limited temperature range, and with considerably less strength contrast. However, both flow laws might be applicable at different stages of Venus' tectonic evolution—that is, Venus might have had more mineralogical water in the past and thus hosted weaker lithosphere that has become drier and stronger with time.

Relations Between Data Sets. It is important to keep in mind the assumptions inherent in each of the data sets and the analyses that employ these data sets, and the spatial and temporal constraints imposed by each model. For example, gravity data might constrain the structure of present *subsurface* support; SAR images record the response of the *surface* to *past* geologic events. The manner in which one relates separate interpretations is nonunique and model dependent. Therefore, the strongest or most robust models will address the constraints imposed by each of these independent data sets.

Throughout the rest of this chapter we review each of the geologic provinces, we highlight the major conclusions drawn from study of each of these types of features, and we point out future directions of research. We discuss tessera terrain first because it occurs across topographic and geologic divisions. From there our discussion is organized by topography beginning with the lowland plains, to coronae and chasmata, to features within the highlands. Tesserae and crustal plateaus are not specifically covered elsewhere within this volume; therefore, we treat them in greater detail in the following discussion. We end with an attempt to relate these features spatially and temporally within a global framework.

II. TESSERA TERRAIN

Tessera terrain is characterized by at least two intersecting sets of structural elements, high relief compared to the surrounding volcanic plains, and unusually high surface roughness at centimeter- to meter-scale (Barsukov et al. 1985,1986; Basilevsky et al. 1986; Sukhanov 1986,1987; Bindschadler et al. 1990*a*). Tesserae constitutes about 8 to 10% of the Venusian surface (Ivanov and Basilevsky 1993; Price and Suppe 1994) and occurs as a dominant tectonic terrain of crustal plateaus, Ishtar Terra, and the plains.

When and how tesserae formed is a challenge in deciphering the history of Venus. Many authors consider that tesserae record complex deformation histories (see, e.g., Barsukov et al. 1985,1986; Bindschadler and Head 1991; Solomon et al. 1991,1992; Bindschadler et al. 1992*a,b*) and represent some of the oldest, preserved crust on Venus (Bindschadler and Head 1989; Kaula et al. 1992; Senske et al. 1992; Squyres et al. 1992*b*; Ivanov and Basilevsky 1993; Basilevsky and Head 1995*a,b*). The latter interpretation led to the conclusion that globally tesserae are of similar age, and can therefore be used as a global time-stratigraphic marker (Solomon 1993*a,b*; Grimm 1994*a*; Ivanov and Head 1995; Tanaka et al. 1995; Basilevsky and Head 1995*a,b*). Tesserae as a global stratigraphic unit is appealing because, if true, regional correlation becomes less challenging. In addition, accepting that Venus experienced an early period of global tesserization is appealing because it simplifies the delineation of the stages of Venusian evolution. A stage of tesserization might be used to assume a time of globally weak lithosphere which can be dynamically modeled (Solomon 1993a; Grimm 1994b). However appealing global synchroneity of tessera formation might be we must find robust support for this currently unproven theory.

The conclusion that tesserae record complex deformation stems from the first description of tesserae supplied by Soviet Venera 15 and 16. These early images have coarse resolution compared to that of some tessera deformation fabrics. In many cases the improved resolution provided by Magellan SAR allows for the delineation of individual linears and for the differentiation between folds, fractures, and graben with much higher confidence. In some cases, temporal relations between families of structures can be interpreted.

A. Types of Tesserae

Using Magellan SAR imagery, Hansen and Willis (1996) examined tessera terrain in Ishtar Terra, crustal plateaus, and as islands within the plains; they described several types of tesserae that are found in specific tectonic environments (Fig. CDP7C1F1). The following discussion of tesserae stems primarily from this work.

Approximately seven types of tessera have already been described, and several more might be identified. Tessera types are recognized on the basis of the structural fabrics. The structurally simplest of the tessera types, fold terrain, is found only in the Ishtar deformed belts, tesserae and montes, and contractional ridge belts, and it consists of a well-defined fold fabric comprised of elongate (>100 km) ridges and valleys with near constant wavelength resulting from simple horizontal unidirectional contraction. The "lava flow terrain" is composed of curvilinear fold ridges and exhibits remarkable similarities to pahoehoe lava flows, as originally noted by Sukhanov (1986), hence its name. Lava flow terrain may have formed when the upper crust was displaced and deformed differentially by movement of material beneath the deformed surface layer.

S-C terrain, found only in western Itzpapalotl Tessera, Ishtar Terra, appears complex, but records a simple geologic history involving distributed crustal-scale noncoaxial shear (Hansen 1992).

Extended fold terrain characterizes the boundaries of many crustal plateaus and large tessera inliers. The fold ridges are cut by wide, lens-shaped graben. Fold axes and graben are typically coaxial; they commonly represent the same overall bulk strain regime (no rotation of principal strain axes). Temporal relations indicate surface contraction predates surface extension. An additional extensional fabric, common throughout crustal plateaus and large tessera inliers, locally parallels the lens-shaped graben, but typically exhibits variable orientation with respect to folds and graben. This "ribbon terrain" is comprised of thin (2–7 km), long (50–150 km), ribbon-like ridges (Fig. CDP7C1F2). Ribbon terrain formed prior to fold fabrics represents an early stage extension.

The interior regions of crustal plateaus host arcuate ridges and troughs, referred to as basin-and-dome terrain. These fabrics variably formed by consecutive polyphase folding (Solomon et al. 1992; Bindschadler et al. 1992a; Phillips and Hansen 1994) followed by late extension, and/or polyphase extension (Hansen and Willis 1996). Early ribbon terrain development is evident in many cases of basin-and-dome terrain.

Star terrain, comprised of extensional fractures and graben, dominates central Phoebe Regio: linears form a "star" pattern, possibly as the result of uplift and dilation of a previously fractured crust.

B. Implications

The different structural histories of tesserae (1) demonstrate that tesserae are not necessarily formed by complex geologic histories, (2) illustrate that not all tesserae were created by the same mechanism, and (3) bring into question the treatment of tesserae as a single global map unit.

Arguments that tesserae are globally synchronous because they typically represent the oldest local unit are potentially severely flawed. The range of deformation histories recorded by various tessera types indicates that tesserae were not created by the same mechanism, and so should not be considered a single map unit. For example, terrestrial crystalline metamorphic rocks are commonly the oldest rocks at separate localities; however, absolute dating techniques indicate that gneissic rocks range in age from greater than three billion years to just a few million years old. Synchroneity of formation of spatially discreet and morphologically and structurally distinct tesserae must be independently justified. Such justification requires a global time-frame. However, the size of the largest individual patches of tesserae is less than the minimum patch size needed to record anomalous crater density. Even if tesserae could be shown to be "on average" statistically older than non-tessera terrains, this condition does not imply that *all* tesserae are older than *all* other terrains (Phillips and Hansen 1994).

Because deformation style (e.g., brittle vs ductile) is related to rheology, the character of tessera deformation may provide clues regarding the crustal rheology at the time of deformation. For example, the progressive increases in the structural wavelength of ribbons, folds and late graben in the same region record rheological changes within shallow crustal layers through time (Hansen et al. 1996). The observation that fold terrain tesserae show little or no obvious strain gradient for hundreds of km perpendicular to their structural trends suggests that these tesserae deformed due to stress transmitted from below (shear traction), rather than due to edge, or in-plane, forces (see, e.g., Hansen and Phillips 1995). Following the three-layer strong-weak-strong model of Zuber (1987), a middle layer must be weak enough to allow the surface layer to decouple from the lower layer, yet strong enough to transmit stress from the lower layer to the surface layer. Thus, the presence of dominantly contractional tesserae might place constraints (in time and/or space) on the rheological layering of the mechanical lithosphere, but it does not robustly constrain the strength or rheology of the overall mechanical lithosphere.

Accepting that different types of tesserae form in different geologic environments, we can make some preliminary proposals about tesserae. Fold and S-C terrain are found uniquely in western Ishtar Terra, and thus we consider their style of formation to be separate from other tessera types. Ribbon, lava flow, and basin-and-dome terrains reside within the interior of crustal plateaus, whereas extended folded terrains (with or without early ribbon fabrics) dominantly lie within the margins of plateaus. Large tessera inliers also host these types of tesserae, supporting a model in which large tessera inliers represent flooded crustal plateaus (Phillips and Hansen 1994). Small inliers marked by densely fractured terrain display evidence of minor extension. The fracture patterns of these small inliers are similar in size, style, variation, and changes in orientation to the fracture patterns associated with coronae and chasmata (Stofan et al. 1992; Hamilton and Stofan 1996). It is possible that these tessera inliers represent flooded (deflated or "sunken") ancient corona-chasma.

Tesserae, which are nothing more than deformed surface layers, could form in several types of tectonic environments, including (1) by subsurface flow in western Ishtar Terra, (2) as variable sequences of surface-layer extension and contraction in crustal plateaus, recording changes in crustal rheology through time, (3) as extended and fractured, previously deformed crustal plateaus which have collapsed, become flooded, and preserved as large plains inliers, and (4) as surface layers that were densely fractured during corona and chasma formation which have since become variably flooded and thus preserved as isolated, scattered, highly fractured inliers. Thus tesserae would *not* form a global onion skin; they would not represent a globally synchronous unit; and they would *not* record a single period of deformation in Venus' tectonic history. In the same sense that deformed rocks on Earth form in a wide variety of tectonic environments, tesserae record a range of spatially and temporally discreet tectonic processes.

III. LOWLAND PLAINS

Plains make up the bulk of Venus, covering ~80% of its surface. Thus, although the volcanically and tectonically complex highlands and mesolands tend to attract much more scientific attention, it is the tectonic style of the plains that, in a sense, characterizes the planet. Plains are generally surfaced by flat, radar-smooth units believed to be due to flood volcanism. Virtually all of these areas, except for isolated patches of smooth terrain in the highlands, lie at or below the 1.5 km above MPR. Although the plains appear generally smooth overall, even a cursory inspection reveals that almost no area is free from some sort of fractures or faults. This is in sharp contrast to the other terrestrial planets, which have large areas which do not display any recognizable tectonic deformation. Detailed geologic study of the plains is still in its infancy. The full value of the information awaits the completion of studies of plains regions, as well as studies of the highlands (which localy affects lowland deformation observed in the lowlands). There are several observations that can begin to cast some light onto the processes, structure and history of the planet's surface.

Plains tectonic features can be classified into three categories: distributed deformation, concentrated deformation, and local fracture patterns, based on the intensity and areal extent of deformation. Distributed deformation consists of sets of features which individually have a small, but discernible, amount of deformation which over an extended region can comprise a significant amount of contraction or extension. Concentrated deformation is manifested in quasi-linear deformation belts (ridge belts and fracture belts), separated by areas of relatively undisturbed terrain. Local fracture patterns are sets of features with limited areal extent, and whose individual features have widths at or below the resolution of the Magellan imaging system ($\sim$100–250 m). Thus, this type of deformation does not contribute significantly to the larger-scale strain of the crust. In the following sections we summarize the salient characteristics of these categories of tectonism, following the more complete treatment in the chapter by Banerdt et al.

A. Distributed Deformation

Virtually all plains are characterized by abundant wrinkle ridges. Wrinkle ridges are long ($\leq$300–400 km), narrow (<1 km), sinuous features, and are inferred to be due to compression (McGill 1993). They tend to occur in sets of approximately evenly spaced, parallel ridges. It is common for two or three of these sets to occur in the same area. Wrinkle ridges are younger than the plains flows they deform. Also, the abundance of wrinkle ridges commonly decreases with decreasing age of specific plains units, suggesting a progressive contractional deformation of crustal rocks during emplacement of plains materials (Solomon et al. 1992; Squyres et al. 1992*b*). For example, young lobate and digitate lava flows, such as those typically associated with large shield volcanoes, are nearly devoid of wrinkle ridges.

Most plains regions are characterized by sets of wrinkle ridges that maintain a roughly uniform trend or uniform curvature of trend over hundreds to thousands of km. These sets appear to define large-scale stress domains that are relatively constant over large areas, although these stress domains could be confined to the shallow crust. In some areas the dominant wrinkle-ridge set appears related to a major tectonic or topographic feature (Bilotti et al. 1993), but in many areas the dominant trend does not have a clear relation to nearby features (e.g., Lancaster and Guest 1994). In many areas there is a second (or even a third) set of wrinkle ridges in addition to the dominant set. In areas so far studied, the additional sets are younger than the dominant set. In many places, the younger sets appear to be more closely related to local tectonics and topography.

Distributed extension is represented by narrow (<2 km) graben and radar-bright linears, which are interpreted as joint zones, dikes or scarps. These can be extremely long, sometimes reaching lengths of several thousand km. Unlike wrinkle ridges, these extensional features do not generally define structural domains over wide areas; rather, most are associated (at one end, at least) with volcanic constructs or coronae, although their subsequent distal trajectories appear to be influenced by regional stress fields (McKenzie et al. 1992b; Grosfils and Head 1994; Ernst et al. 1995).

B. Deformation Belts

Some plains regions contain sets of belts of relatively intense deformation. The dimensions of these belts vary widely; widths range up to 300 km, and lengths range from <100 to >2000 km. Most of these belts stand a few hundred meters to >1 km above the surrounding plains. Deformation belts are generally "ridge belts" characterized by subparallel ridges, or "fracture belts" characterized by subparallel scarps and graben, although composite belts exist locally (Solomon et al. 1992; Squyres et al. 1992b).

Ridge belts are commonly composed of an array of individual ridges that may exhibit diverse morphologies. The most thoroughly studied examples are in Atalanta, Vinmara, and Lavinia planitiae. They are generally interpreted as resulting from global-scale compressive stresses oriented normal to their trends (Basilevsky and Head 1988; Basilevsky et al. 1986). The arch-like morphology of some individual ridges support the interpretation of these features as anticlinal folds (Solomon et al. 1992; Squyres et al. 1992b). Given the broad, spatially coherent style of deformation of some ridge belts, the source of stress that produced these features is consistent with formation by convergent flow in the mantle associated with downwelling (Zuber 1990; Phillips et al. 1991; Squyres et al. 1992b).

Fracture belts are broad, elongate swells or arches composed of numerous, roughly belt-parallel linears, scarps, and graben that commonly occur as sets that intersect at low angles. The best-studied fracture belts, in Guinevere and Lavinia planitiae, range in size up to about 200 km wide and 1000 km long, and stand a few hundred meters to more than a km above adjacent plains. However, it is often difficult in other areas to identify fracture belts clearly, because similar structures grade along trend into corona chains or ridge belts. The formation of the scarps and graben making up fracture belts can most easily be explained as due to tension normal to the belt trends. This is consistent with the observation in Lavinia Planitia that ridge belts (which are presumably contractional) and fracture belts that occur together tend to have orthogonal trends. On the other hand, the broad rises characteristic of fracture belts suggest crustal thickening due to compression, as for ridge belts (Squyres et al. 1992b), and the presence of fault-bounded rhomboidal depressions on some fracture belts suggests that transtension may be important (Solomon et al. 1991).

C. Local Deformation

Local deformation is manifested in patterns of fine-scale structures developed on the km- to sub-km scale. Unlike the deformation styles discussed above, these features do not generally represent a significant amount of strain. However, they are potentially valuable for the insight they can provide into the structure of the upper crust.

Local deformation patterns are classified as parallel sets, irregular patterns and polygonal patterns. Parallel sets are composed of parallel, thin (a single pixel), straight lineations. The patterns typically cover areas with dimensions of hundreds of km. Average lineation spacing ranges between 1 and 2.5 km. They have been interpreted as tension fractures in the brittle upper layers of the volcanic plains material (Banerdt and Sammis 1992). The very close, regular spacing has been explained using a shear-lag model, in which a relatively thin surface layer is partially decoupled from similar material below by a frictional contact. A key implication of this model is that the tectonic episode responsible for the parallel fracture set must have occurred relatively soon after the emplacement of the plains material, while it was still relatively intact, predating other strong deformation events recorded on that surface.

Irregular patterns, such as those in Guinevere Planitia (Solomon et al. 1991; Bowman and Sammis 1995), are composed of sub-parallel, sinuous extensional features composed of curvilinear en echelon segments. The length distribution of these structures shows a distinct break in slope at 80 km. In the model of Bowman and Sammis (1995), this break corresponds to the change from the two-dimensional growth of a semi-circular crack in an half-space to simple one-dimensional horizontal elongation after it has penetrated to the bottom of the brittle lithosphere, inferred to be about 40 km.

Polygonal patterns of bright lineations are another common feature on the plains. These lineations are 10 km long down to SAR resolution limit and 1 to 2 km across. Their isotropic orientations and apparent tensional nature led to the interpretation of these features as the result of thermal cooling stresses (Johnson and Sandwell 1994). Thus their characteristics may preserve something of the conditions extant during the emplacement of the plains.

D. Implications for Lithosphere Structure

Two observations of the pattern of plains tectonism have been used to infer the structure of the Venusian lithosphere. The first is the spacing of features such as wrinkle ridges (dominant wavelength ~20 km) and deformation belts (characteristic spacing of 200–300 km); the second is the persistence over large distances of relatively mild tectonic deformation with a consistent orientation. Until recently, these observations appeared to support a self-consistent story, in which a thin, brittle upper crustal layer was separated from an elastic upper mantle by a ductile lower crust. This type of structure leads naturally to two distinct length scales of deformation (Zuber 1987; Banerdt and Golombek 1988), and provides a natural mechanism for transmitting stresses from the interior to the surface layer through shear coupling at the base, which could lead to uniformly distributed deformation over large areas corresponding to the length scales of mantle deformation. This scenario was supported by preliminary laboratory results on rock rheology, that predicted appropriate transitions between brittle and ductile behavior for temperatures thought to be realistic for the shallow portions of the planet. However, measurements of the ductile properties of very dry diabase (Mackwell et al. 1995), which may be more analogous to Venusian crust, suggest that it is much stronger at high temperatures than previously thought. These results indicate that the crustal lithosphere may be nearly as strong as the upper mantle, and that a three-layer structure would have much reduced strength contrasts and could exist only over a very limited range of thermal gradients. The rheological results are consistent with recent gravity analyses that indicate a thick, strong lithosphere. This has caused a general re-evaluation of the geophysical models of the lithosphere, which is still ongoing.

Global models of the internal density distribution from inversion of gravity and topography data (Banerdt 1986; Herrick and Phillips 1992) show evidence for isolated upwellings (confined beneath highlands) among an interconnected network of downwellings. Major regions of downwelling are all associated with plains regions. Power spectral ratios of gravity and topography also indicate a large-scale pattern of downwelling flow beneath the plains lithosphere. These results are all consistent with a generally compressive regime in the plains, as evidenced by the presence of ridge belts and the ubiquity of wrinkle ridges. Gravity/topography relationships of plains regions indicate a thickness of the mechanical lithosphere of ∼50 to 100 km (see, e.g., Bindschadler 1994; Simons et al. 1994; Banerdt et al. 1994). This strong lithosphere, which would be able to maintain its integrity even when subjected to large stresses, could explain the persistence of limited deformation over great distances, but must be reconciled with the presence of both long and short length scales of tectonic features in the plains.

IV. CORONAE AND CHASMATA

Coronae are circular to quasi-circular features that display radial fractures and/or concentric fractures and locally folds (Stofan et al. 1992; Squyres et al. 1992*a*). There were initially thought to be unique to Venus, but they may also be present on Mars and Earth, and the Uranus moon Miranda (Janes and Squyres 1993; Pappalardo et al. 1994; Watters and Janes 1995). Coronae occur as isolated features, in clusters associated with volcanic rises, and in chains associated with chasmata (chapter by Stofan et al.). They range in size from ∼100 km to 1000 km in diameter with a strong mode at 200 to 300 km. Artemis "Corona" is unusual—at 2600 km diameter, over twice the size of the next largest corona, Artemis shares some similarities to coronae, and was therefore classified as a corona (Stofan et al. 1992); however, recent work (Brown and Grimm 1995) reveals that Artemis development does not follow the standard evolutionary model of coronae. Thus, Artemis' origin becomes important—if it is a corona, albeit atypical, then observed relations should be incorporated into models of corona formation; however, if it is not a corona, then incorporating Artemis constraints into model formulation may be detrimental. In this short discussion, we do not consider Artemis a corona, although we do address its formation with regard to corona-related topics.

A. Characteristics

Coronae display characteristic structural features, although their topographic expression is highly variable (chapter by Stofan et al.). Coronae commonly display a central region higher than the surroundings, a raised rim, and a peripheral moat. Concentric fractures parallel the rim-moat transition. Radial fractures may be preserved only within the coronae, or may extend beyond the outermost concentric fracture set that marks the typically narrow (<150 km) tectonic annuli, that parallel corona rims just inboard of the moat. Extensional faults typically mark corona annuli, although fold belts and small-scale strike-slip structures are present locally along parts of individual coronae, having formed in response to local stresses (Cyr and Melosh 1993; Willis and Hansen 1996). Moderate to large amounts of lava, escaping through radial and concentric fractures, accompany corona formation. Double-ring concentric coronae show a sequence of inner concentric faulting followed by volcanism, followed in turn by faulting and volcanism within an outer ring, indicating that coronae can grow outward by deformation associated with magmatic pulses (Hansen and Phillips 1993b).

Coronae cluster about MPR, and show no correlation with either strong positive or strong negative geoid anomalies (Herrick and Phillips 1992), indicating a complex relation between coronae and mantle convection. Comparison of deg-60 Magellan gravity with short-wavelength topography (~600 km) shows little resolution of most coronae because they are smaller than the observational wavelength; resolvable large coronae yield ADCs ranging from 75 to 150 km—Artemis yields an ADC of 200 km (Schubert et al. 1994).

Coronae are commonly associated with chasmata, topographic troughs with trough-parallel fractures and faults. Chasmata comprise two groups, symmetric and asymmetric, depending on topographic profiles (Hansen 1993). Symmetric chasmata are associated with corona chains (e.g., Parga and Hecate), and commonly trend outward from volcanic rises defining a broad zone of short trough segments that host syntectonic volcanism, and hence are interpreted as rifts marked by minor extension (McGill et al. 1981; Schaber 1982; Senske et al. 1992). Asymmetric chasmata, the deepest chasmata, are bounded on one side by a ridge that is as high as the trough is deep. The trough-to-ridge differential can be as much as 7 km, over a 30-km distance. Asymmetric chasmata typically overlap coronal troughs of large coronae (diameter ≥ 500 km).

B. Formation

There is widespread agreement that coronae result from buoyant diapirs impinging on the base of the lithosphere (Basilevsky et al. 1986; Schubert et al. 1989,1990; Pronin and Stofan 1990; Stofan and Head 1990; Stofan et al. 1991,1992; Squyres et al. 1992a; Janes et al. 1992; Hansen and Phillips

1993b). Radial fractures, typically the oldest tectonic features of coronae, result from impingement of the diapir on the base of the elastic lithosphere, and concentric fractures and interior sag result from lateral expansion, and cooling and relaxation of the diapir.

There was early disagreement as to whether coronae-forming diapirs were thermally driven plumes arising from the core-mantle boundary—genetically related to volcanic rises (Stofan et al. 1992)—or if they were compositional diapirs resulting from melt instabilities within the upper mantle (Tackley and Stevenson 1991,1993; Hansen and Phillips 1993b; Phillips and Hansen 1994). However, the small size and large number of coronae relative to the large size and limited number of volcanic rises indicate that corona diapirs probably do not originate at the core-mantle boundary (Phillips and Hansen 1994; Hamilton and Stofan 1996). Tackley and Stevenson (1991,1993) discussed a mechanism for spontaneous and self-perpetuating magmatism that could account for coronae. Regions of partial melt exist in the upper mantle where the geotherm approaches the solidus temperature. A Rayleigh-Taylor-like instability develops as a result of an infinitesimal upward velocity perturbation resulting in pressure-release partial melting and formation of buoyant melt. Melt buoyancy increases the upward velocity leading to more partial melting forming a compositional plume. The instability could result from lithospheric extension, and the correlation of many coronae with extensional tectonics (e.g., Parga Chasma) supports this view. Hamilton and Stofan (1996) used the spacing between coronae in Hecate Chasma to constrain the depth of melt generation; calculations yield depths of ~150 to 200 km.

Thus, there seems to be agreement that diapirs originate in the upper mantle, although debate remains whether the diapirs are compositional (Tackley and Stevenson 1991,1993; Hansen and Phillips 1993b) or thermal (Roberts and Head 1993; Hamilton and Stofan 1996) and may be dependant on the local tectonic environment (Phillips and Hansen 1994). Corona clusters associated with volcanic rises may be dominantly thermal resulting from the break-up of a deep thermal plume (Stofan et al. 1995), whereas coronae within extensional environments (Baer et al. 1994) may be more compositionally derived. The intimate association of magmatism with coronae, including multiple volcanic phases with apparently different composition, favors a compositional origin for at least some coronae (Hansen and Phillips 1993b), and the compositional nature of the diapir may allow some coronae to retain their elevated interior following cooling. Roberts and Head (1993) interpreted the correlation of coronae with large volcanic flow fields with regions of lithospheric extension to result from increased partial melting of thermal plumes that reached shallower depths than normal because of lithospheric stretching. However, the Tackley-Stevenson model also predicts a positive correlation between lithospheric stretching and melt volume; in addition, Venus' high surface temperature allows even small, slow magma bodies to reach the surface in a wide range of thermal environments (Sakimoto and Zuber 1995) so the presence of a thermal plume is not required.

Coronae within the equatorial highlands gain their elevation from thickened crust and localized thermal buoyancy (while young and warm; hence local isostatic and thermal buoyancy) and from their present locations above recent mantle upwellings, which contribute ~1 km of regional thermal elevation above MPR (Phillips and Malin 1984). If and when the large-scale thermal perturbation decays, the corona chain will sink and may become progressively flooded by volcanism. The observation that densely fractured and embayed terrain within Hecate and Parga chasma regions resembles corona and chasma fracture patterns (Hamilton and Stofan 1996) supports this interpretation.

Is Subduction Important in Corona Evolution? The final stage of evolution of some coronae is the topic of some debate. Sandwell and Schubert (1992) and Schubert and Sandwell (1995) argued that some large coronae have experienced subduction along their margins. This argument stems from the topographic similarity of several asymmetric chasmata with terrestrial subduction zones (McKenzie et al. 1992*a*; Sandwell and Schubert 1992). Sandwell and Schubert (1992) proposed retrograde subduction in which the "trench" migrates toward its convex side (away from the corona interior), increasing the arc radius of curvature; the corona interior expands as the trench migrates.

Sandwell and Schubert (1992) and Schubert and Sandwell (1995) submit that Latona Corona is perhaps the best example of subduction. Hansen and Phillips (1993*b*) discussed kinematic problems with this interpretation. They showed that radial fractures can be traced across the southwestern part of the southern trough, and that the spacing of the fractures on either side of the trough does not change. Because the fractures are radial with respect to the corona interior, and because they predate trough formation (Hansen and Phillips 1993*b*; Schubert and Sandwell 1995), there should be an obvious change in fracture spacing across the proposed trench if subduction had occurred.

Deformation associated with Artemis Chasma appears more boundary-like than many other deformation zones on Venus; strain increases across the annulus from trough to interior (Brown and Grimm 1995). However, the subduction model predicts radial roll-back subduction outboard of the trough and radial expansion of the interior; evidence for neither of these strain regimes is observed. Folds parallel the southeastern part of Artemis Chasma, but the folds feather into chasma-oblique structures indicative of left-lateral and right-lateral strike-slip translation along the east/northeastern and southern boundaries, respectively, indicating overall southeast-directed overthrusting across a quasi-circular boundary, but not indicative of radial roll-back subduction (Brown and Grimm 1995). In addition, although extensional features occur within the interior of Artemis (Stofan et al. 1992), the extension is not kinematically or temporally consistent with the subduction model. Retrograde subduction predicts interior radial extension synchronous with trough formation, whereas Artemis's interior records southeast–northwest contraction *and* extension, both of which predate formation of structures related to Artemis Chasma (Brown and Grimm 1995). Thus, the interior of Artemis may simply represent "captured real-estate," with the interior structural fabric having formed prior to the formation of Artemis Chasma.

Schubert and Sandwell (1995) postulated that retrograde subduction zones form in regions of lithospheric extension. In their model the lithosphere is broken along rifts formed above diffuse mantle upwellings. Although Earth's spreading centers mark breaks through the lithosphere, the ability of these zones to form may depend, in large part, on the recycling process in subduction zones thousands of km away—which allows, within the global budget, for high elongational strain to focus at rift forming boundaries. Breaking the lithosphere on a one plate planet may be much more difficult than on Earth. In order to break, lithosphere must experience high elongational strain—the amount depends on the thickness and the rheological structure of the lithosphere. Even if Venus' lithosphere is broken, an upwelling environment is not conducive to whole lithospheric subduction, a process driven by negative density contrast. Hot lithosphere is buoyant and thinned lithosphere is buoyant; thus, lithosphere ruptured within a thermal upwelling environment is not likely to subduct. Density-driven subduction requires thickened crust in order to drive the high density-eclogite transition, but such thickness is at odds with fracturing the lithosphere as a result of thinning within a rift environment. In addition, the subduction proposed by Schubert and Sandwell (1995) is *along strike* of rift zones, a proposal which is counterintuitive to *extensional* rifting, and therefore requires detailed kinematic justification.

Schubert and Sandwell (1995) stated that if subduction is a viable process on Venus, it may contribute significantly to mantle cooling. Despite the problems outlined above, if one accepts that $\sim$9000 km of subduction zones exist on Venus, equivalent to $\sim$25% of Earth's length of subduction zones, these zones can account for significant cooling only if the entire lithosphere is recycled to the mantle at relatively fast rates (Schubert and Sandwell 1995). In order for cooling to be even $\sim$25% of Earth's annual cooling accommodated by subduction, Venus must subduct lithosphere at a rate similar to the Earth. This would require 5000 to 10,000 km of lithosphere to have been subducted across each of the proposed Venusian subduction zones over the past 100 Myr. If this much lithosphere had been subducted, surface strain patterns at the planform "ends" of the laterally discontinuous subduction zones should display evidence indicating how displacement transfer was accommodated. SAR images reveal no structural/kinematic evidence for transfer zones, nor do they reveal regions where significant lithosphere is created to balance loss by subduction. One might be able to justify a "rip" in the lithosphere where one side is downwarped relative to the other; but such a tear would not allow for subduction and recycling of the lithosphere to the mantle and thus would not contribute to significant mantle cooling. The Venusian lithosphere shortens along low-angle fault zones (Suppe and Connors 1992) resulting in local imbrication; however, local imbrication does not equate to subduction.

Therefore, although the topography of large corona troughs may be consistent with slab flexure within a retrograde subduction model, this interpretation is nonunique and few other lines of evidence support it.

Recent modeling of coronae by Smrekar and Stofan (1996) using a finite difference scheme to model temperature and chemistry variations and a penalty function finite element formulation to solve buoyant viscous flow equations may be able to accommodate both the geologic observations against a subduction hypothesis and the topographic profiles attributed to subduction by some workers. In their model, cold lithosphere at the edge of a plume head is sucked downward—explaining the subduction-like troughs—however, delamination within the lithosphere allows the lower lithosphere to sink to great depths, while the surface crust does not. Thus, surface fractures and flows might continue across deep troughs. By this model corona trenches may represent delamination and sinking of cold lithosphere. Sinking is initially driven by viscous flow at the edge of a plume head and is sustained by the density difference between the lithosphere and the mantle and the difference in the thickness of the lithosphere above the plume and in the cooling lithosphere.

In summary, coronae result from buoyant diapirs that originate in the upper mantle as melt instabilities within localized or diffuse regions of mantle upwelling beneath lithospheric extension. The diapirs blister the local lithosphere resulting in radial and concentric deformation, and magmatism. As mantle upwellings decay coronal regions sink and become embayed, or repainted, by flood lava flows. Much remains to be understood, however, about the nature of the diapirs, whether compositional or thermal or both, about the mechanisms leading to pulses of magmatism and deformation, about the details of the mechanism(s) responsible for corona topography, and about the influence of the local lithospheric structure on the character of individual coronae, corona clusters, and corona chains.

V. VOLCANIC RISES

Volcanic rises, crustal plateaus, and Ishtar Terra comprise the three classes of morphologic highland provinces. Although volcanic rises vary in morphology, topography, and gravity, they are generally distinguished by their overall broad, gentle, dome-like topography, and by their gravity signature, which implies relatively deep apparent depths of compensation. Nine volcanic rises have been identified (Atla, Beta, Bell, Dione, Imdr, Themis, and western, central, and eastern Eistla regiones) (Fig. 1). They range in diameter from 1000 to 2500 km, and in height from about 1 to 2.5 km. They exhibit volcanism in the form of large-to-intermediate volcanic edifices and extensive flows, and more than half have coronae on their crest or flanks.

Volcanic rises have been the focus of a great deal of geophysical attention, due both to their obvious importance in the thermal and tectonic history of Venus, and to the fact that the magnitudes and spatial extents of the gravity signatures are well within the capabilities of Magellan tracking data to resolve, allowing the application of many powerful analysis techniques developed for terrestrial problems. A general consensus has been reached on their identification as analogs to terrestrial hotspot swells (see the chapter by Smrekar et al.). Our current understanding of these features is strongly shaped by this conceptual framework, although there is considerable debate as to the degree to which specific processes operate on either planet and at specific hotspots. Volcanic rises are interpreted as the surface manifestation of convective mantle upwelling. The broad swell is due to a combination of uplift due to lithospheric thinning due to heating from below, and crustal thickening by volcanic activity, both intrusive and extrusive. Volcanic activity is driven by pressure-release melting in the rising plume, and associated coronae are thought to be manifestations of the breakup of this plume or secondary convection.

Volcanic rises have been classified into three types; rift-, volcano- and corona-dominated (Stofan et al. 1995). Rift-dominated rises are characterized by a major rift valley that extends beyond the topographic rise itself. The alignment of the valley is apparently controlled by the large-scale linear trend defined by adjoining highlands. These rifts extend for thousands of km, and are associated with intense deformation. The two rift-dominated rises (Atla and Beta regiones) are the highest volcanic rises on Venus, with crests over 2 km above the surrounding plains. They are the site of some of the largest volcanic edifices on the planet, but they host few coronae. Volcano- and corona-dominated rises have volcanism manifested in edifices and/or extensive flows, respectively, but no large-scale rifts. Evidence of extension is present to some degree in these rises, ranging from rifting (at a smaller scale than in the rift-dominated rises), to concentrated bands of extension, to isolated graben formation. Topographic relief varies widely among these rises, ranging from about 1 to 2 km. The classification into rift-, volcano- and corona-dominated rises is somewhat arbitrary, as many have rifts, over half have coronae, and all show evidence of volcanic activity of some sort. There are some notable characteristics that are common among volcanic rises. All exhibit abundant volcanism, consistent with the presumed origin by upwelling in the mantle. Most have wrinkle ridges, and all exhibit some degree of extensional tectonics.

Volcanic rises have been shaped almost exclusively by vertical processes. But did they form by extrusive volcanism, piling up flow upon flow on the surface? Was there massive intrusion, thickening the crust and isostatically raising the surface? Is lithospheric uplift by thermal or dynamic forces principally responsible for the broad domes? Or (more likely) is a combination of these processes in play? The answers to these questions have a direct bearing on the thermal state of the Venusian lithosphere, and the processes that are responsible for the evolution of the rises have implications on the thickness of the crust and mechanical lithosphere.

Several observations can be used to estimate the relative contribution of various processes to the present surface elevation. The total amount of extrusive volcanism is difficult to estimate, as young flows cover old flows. However, most rise surfaces (away from major volcanic edifices) are covered with flows that are indistinguishable from the surrounding plains. These units have a similar crater density to the plains (although there are indications that densities on some rises may be lower than average; Phillips and Izenberg 1995), and contain wrinkle ridges in similar numbers and with similar orientations to the adjoining plains. This indicates that either the pre-existing plains surface was lifted to its present elevation, or that both plains and rises have been resurfaced since rise formation. Several paleoslope indicators, consisting of flows and a lava channel with apparent uphill gradients (Stofan et al. 1995) and a tilted pond of impact melt (Connors 1992), have been identified that support the former conclusion. These observations argue against extrusive construction as the major mode of rise formation.

The relative contributions of intrusion ("inflation" of the crust) and lithospheric uplift are somewhat more difficult to determine. Both result in the vertical displacement of a surface layer, either a thin layer of crust or the entire lithosphere, respectively. An indirect estimate of the volume of intrusive volcanism can be derived from the observed volume of extrusives. Using typical terrestrial ratios of extrusive to intrusive volume of 1:2 to 1:10, Smrekar and Parmentier (1996) derived a lower bound on intrusive volume of 5 to 50% of the total rise volume, based on the volume of extrusives comprising major edifices (Stofan et al. 1995). However, if the thickening is isostatically compensated, even the high end of this estimate will only account for ~ 100 m of elevation (Smrekar and Parmentier 1996).

The tectonics of the rises also support an origin by primarily uplift rather than construction. Contractional wrinkle ridges are present on volcanic rises, but they appear at least in some cases to predate uplift (McGill 1993; Stofan et al. 1995) and thus do not bear directly on rise formation. This is especially true for the rift-dominated rises, but also holds for the corona- and volcano-dominated rises. The timing of extension is typically contemporaneous with volcanism and uplift. Extrusive construction loads the top of the lithosphere, causing downwarping of the underlying lithosphere. This results in compressional stress in the upper lithosphere within the high terrain; extension is confined to the periphery (e.g., Tharsis on Mars; Banerdt et al. 1992). Uplift, on the other hand, is effective in creating tensile stress within the raised topography. If there is an elongation of the rise, or if it is in close proximity to another rise, the maximum tensile stresses will be oriented in such a fashion as to create extensional structures such as rifts with a preferential orientation along the axis of the topographic feature (e.g., Fig. 9 of Banerdt 1986). The presence of major rifts within some rises suggests a strong component of uplift, as extension of a mechanical layer whose thickness is a significant fraction of the rift width is thought to be necessary for the formation of such structures. On the other hand, there is also considerable evidence for large amounts of intrusion. Numerous extensional features interpreted as dikes are found on volcanic rises, many associated with volcanic edifices. They commonly occur in bands that are similar in width to rifts, but that lack normal faults. The injection of magma into a zone of extension induced by tensile stress, would tend to suppress the isostatic subsidence of a central block and the concomitant uplift of its flanking blocks, and at the same time relieve the horizontal tension within the lithosphere.

VI. CRUSTAL PLATEAUS

Crustal plateaus define a major tectonic province of Venus. Although crustal plateaus exhibit individual peculiarities, all share the following general characteristics: they comprise major plateau-shaped highland regions, up to ~4 km above MPR, with steep outer slopes; they exhibit gravity signatures typified by low GTR and ADC values; and they reveal complex deformation patterns, including complexly deformed interiors, with local marginal fold belts. Crustal plateau evolution is controversial, and many questions remain, including questions regarding the temporal and spatial development of tectonic fabrics, the relations of crustal plateaus to mantle dynamics, and whether crustal plateaus are still forming today or represent remnants of an earlier geologic epoch. We outline what characterizes crustal plateaus, discuss models of their origin, and conclude with suggested directions of future research.

A. Characteristics of Crustal Plateaus

The crustal plateaus range in size from ~3000 by 2000 km to ~1700 by 1000 km, and include from largest to smallest, Ovda, Thetis, Tellus, Alpha, and Phoebe regions (Fig. 1). Although some authors classify western Ishtar Terra as a crustal plateau (Bindschadler et al. 1992b), western Ishtar differs in its topography, deformation gravity signature, and its relations between deformation and volcanism. Eastern Ishtar Terra might exhibit plateau-like characteristics. We discuss plateau characteristics in order from topography and gravity to surface geologic patterns, corresponding to the order in which these data sets were obtained.

Pioneer Venus topography and gravity data formed the basis for distinguishing plateaus from rises. Venera 15 and 16, Arecibo, and especially Magellan provided detailed radar images of the surface. Plateaus are elevated 1 to 4 km above adjacent plains, with steep outer slopes and margins locally elevated ~1 to 2 km above the interior regions (Masursky et al. 1980; Pettengill et al. 1980; Smrekar and Phillips 1991; Bindschadler et al. 1992a,b; Phillips and Hansen 1994). Although interior regions display rough topography laterally at km to tens of km scale, the overall geometry defines steep-sided plateau-shaped highlands (Fig. 2).

Plateaus also exhibit distinct gravity signatures, specifically small gravity anomalies, low GTRs, and thus shallow ADCs (Herrick et al. 1989; Smrekar and Phillips 1991; Bindschadler et al. 1992b; Simons et al. 1994; Grimm 1994b; Herrick 1994). Most workers agree that crustal plateaus are isostatically compensated by thickened crust (35–45 km; Grimm 1994b), with little

TECTONIC OVERVIEW AND SYNTHESIS 819

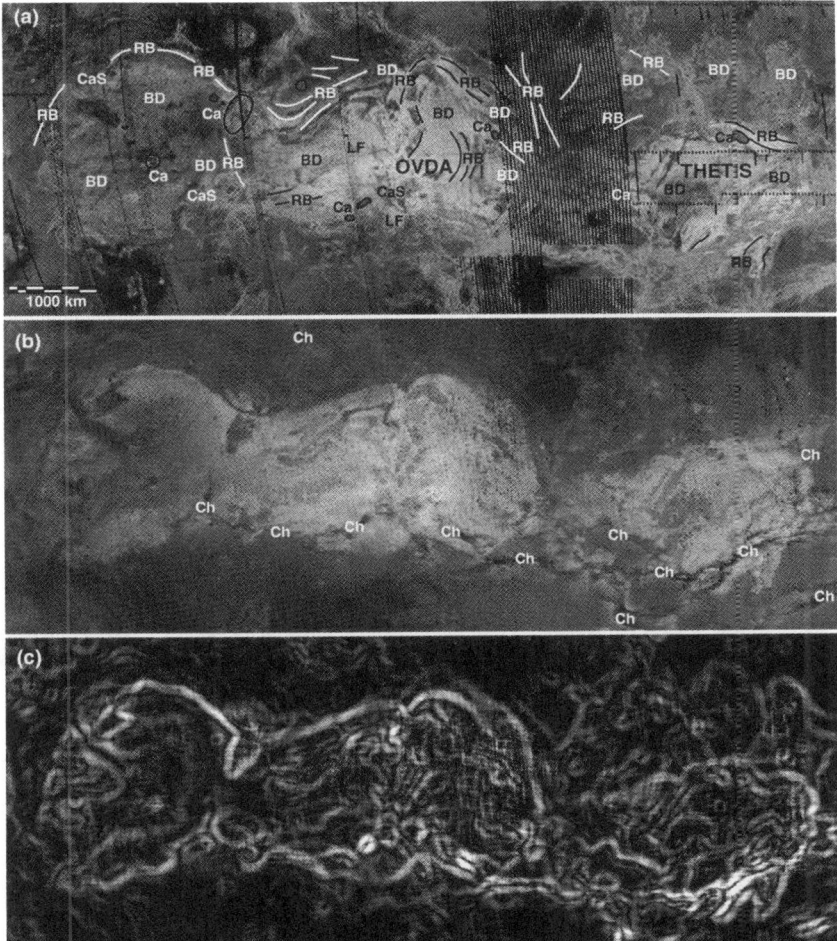

Figure 2. (a) SAR image of Ovda and Thetis regiones, western Aphrodite Terra. Light tones correspond to greater radar brightness. Black stripes represent data gaps. BD: basin-and-dome tesserae, undifferentiated between contraction- and extension-dominated terrains. LF: lava flow tesserae. RB: well-defined ridge belts (typically of extended fold tessera type); lines indicate general fold orientation (white versus black used for visual discrimination purposes only). Ca: major intratessera calderas; derived volcanic flows flood adjacent tesserae, yielding local radar-dark patches. CaS: complex suite of calderas. (b) Topographic image of western Aphrocite Terra. Light tones indicate greater elevation. Ch: chasmata. Calderas appear as circular to elongate depressions. (c) Surface slope image of western Aphrodite Terra. Surface slopes derived from a nonlinear edge-enhancement operator. Light tones indicate steeper slopes. Chasmata form paired steep slopes, and calderas exhibit steep-sided margins.

correlation to long-wavelength geoid (upper mantle) signatures. Thetis Regio, an exception, may require significant mantle contribution to its compensation (Grimm 1994b; Herrick 1994).

Complex deformation patterns typify the plateaus. Venera and Arecibo data revealed that tesserae dominate Tellus and Alpha regions, respectively (Barsukov et al. 1985,1986; Basilevsky et al. 1986; Campbell et al. 1989). Using Pioneer reflectivity and rms slope (roughness) data, Bindschadler et al. (1990a) predicted that Ovda and Thetis regions host primarily tesserae, which Magellan data confirmed (Solomon et al. 1992; Bindschadler et al. 1992b). In general, plateau interiors are characterized by dominantly basin-and-dome and lava-flow tessera types (Fig. 2), whereas the margins locally exhibit elevated ridge belts cut by perpendicular graben (extended fold terrain) (Bindschadler et al. 1992b; Phillips and Hansen 1994; Törmänen 1993,1995; Hansen and Willis 1996). Phoebe Region exhibits star terrain. The complexity of interior deformation fabrics, both contractional and extensional, contribute to observed rough topography.

Some basin-and-dome type tesserae record polyphase contraction and interference folding, although a dominant fold trend may be present (Bindschadler et al. 1992a; Solomon et al. 1992; Phillips and Hansen 1994; Hansen and Willis 1996). Reconnaissance mapping reveals that trends of fold ridges within crustal plateau interiors are widely variable, but in places describe broadly radial patterns, revealing axisymmetry, at least along partial arcs. For example, the internal basin-and-dome fabric at Alpha Regio forms radial or "spoke-like" pattern (see Fig. 2b of Bindschadler et al. 1992a), rather than the margin-parallel pattern suggested by Bindschadler.

Marginal ridge belts, typically with higher elevation than plateau interiors, do not entirely encircle plateaus, and in several examples they trend into surrounding plains or into plateau interior regions at high angles (Bindschadler et al. 1992a,b; Törmänen 1993; Phillips and Hansen 1994). For example, the northern margin of Ovda Regio is dominated by an extensive ridge belt system, but well-defined ridges are rare along the southern margin. Ridge belts locally occur within crustal plateau interiors, and separate regions of basin-and-dome terrain.

Extensional structures are common in crustal plateau tesserae. Many workers recognized contraction (ridges or folds) postdated by extension (graben) (Solomon et al. 1991,1992; Bindschadler and Head 1991; Bindschadler et al. 1992a,b; Phillips and Hansen 1994; Basilevsky and Head 1995a). Recently recognized ribbon terrain records early widespread extension, in contrast to previous interpretations. Ribbon terrain also imposes strict constraints on the depth to the brittle-ductile transition in the crust ribbon fabric formation; and the increased spacing of the younger folds, and still younger graben, requires a progressive increase in the depth to the brittle-ductile transition with time. Thus, plateaus record initial crustal extension, then limited contraction, and late extension, and a progressive increase in the depth to the crustal brittle-ductile transition (Hansen et al. 1996).

Intratessera plains are common within plateaus (Bindschadler and Head 1991; Bindschadler et al. 1992*a*; Hansen and Phillips 1994; Gilmore and Head 1994; Head 1995). Many represent structural basins that host deformation fabrics that cross-cut and therefore postdate tesserae. However, others seem to represent undisturbed regions between deformation belts reflecting pre-existing plains that were isolated or "captured" by deformation fabrics formed during crustal plateau evolution (Phillips and Hansen 1994; Head 1995). Both types may be structurally controlled, the former by collapse (discussed below), and the latter by isolation due to adjacent tessera deformation.

Collapse-related intratessera plains inhabit basins, which are typically 50 to 200 km with depth ranging from hundreds of meters to over 2 km (Fig. 2). Circumferential fractures cut adjacent tessera fabrics. Wrinkle ridges commonly deform the basin interior. Volcanic flows derived from these basins cover thousands of square km of adjacent tesserae. Intratessera basins commonly occur within fracture belts that transect plateau fabrics, although isolated basins without obvious relations to fracture belts or rift-like systems are also present. For example, a north–northeast trending fracture belt that subdivides Ovda Regio from westernmost Aphrodite Terra hosts several large (up to $\sim 550 \times 300$ km) elongate basins (Fig. 2). Geologic relations at some intratessera basins are typical of calderas, although their origin is questionable, perhaps forming by partial melting due to lithospheric extension with subsequent collapse due to depletion of magma reservoirs, and/or as "eclogite sinkholes" (negative eclogite diapirs) associated with crustal roots extending below the basalt-eclogite transition (Head 1995).

Approximately 40 impact craters pepper crustal plateau surfaces. Crater densities of plateaus and of tesserae in general have been used to infer an old age relative to other geologic surfaces of Venus, and to date individual plateaus (Ivanov and Basilevsky 1993; Basilevsky and Head 1995*a*). However, the areas of individual crustal plateaus are too small to date by crater density. Although the crater record alone cannot constrain ages of individual plateaus, other methods of dating surfaces, for example by detailed mapping and regional correlations, may constrain the relative ages and may support the apparent antiquity of *mean* plateau ages determined from crater studies. Gilmore et al. (1995) demonstrated that tectonized craters within plateaus show only late graben extension. Thus the majority of the plateau tectonism occurred prior to the mean surface age of the plateaus, as defined by craters.

B. Models of Crustal Plateau Formation

Two general end-member models have been developed for plateau formation: "hotspot," or mantle upwelling (Phillips and Malin 1984; Head and Crumpler 1987, 1990; Herrick and Phillips 1990; Phillips et al. 1991; Phillips and Hansen 1994), and "coldspot," or mantle downwelling (Bindschadler and Parmentier 1990; Bindschadler et al. 1990*b*, 1992*b*; Lenardic et al. 1991). We compare the two models.

The upwelling model evolved from Pioneer Venus topography and gravity studies at western Aphrodite Terra. Two classes of models developed, those invoking crustal spreading or not. Head and Crumpler (1987,1990) proposed an origin comparable to terrestrial mid-oceanic ridges and sea-floor spreading. Subsequent Magellan imagery did not show evidence of structural features typical of terrestrial spreading ridges, including transform faults and fracture zones, thus discounting a mid-ocean ridge model (Solomon et al. 1992; Bindschadler et al. 1992*b*). Herrick and Phillips (1990) and Phillips et al. (1991) proposed that areas within western Aphrodite Terra and Beta Regio, record evolutionary stages of interaction between ascending mantle plume heads and the lithosphere. According to their model, plateaus evolved from rises due to partial melting and lateral spreading of the plume. The following evolutionary sequence was proposed, based on dynamic modeling of a cylindrical plume. Early dynamic uplift and associated circumferential extension due to the rising plume produce broadly radial fabrics, and minor volcanism. Massive partial melting and crustal thickening by magmatism occur as the plume head impinges on the base of the lithosphere and spreads laterally. Topography transforms from a domal swell to a plateau morphology—marginal contractional deformation results due to ductile crustal detachments down the slope of dynamic topography, but as topography loses dynamic support, membrane compressional stresses yield radially oriented contraction. At this stage, a marginal rim of thickened crust is left, presumably standing topographically highest, as compensation is ultimately controlled by crustal thickness variations. Finally, extensional and contractional fabrics result from relaxation and spreading of the crustal plateau. The upwelling model explained available topographic and gravity data, and made testable predictions for the Magellan mission.

Specific objections to the upwelling model were addressed following acquisition of Magellan data (Bindschadler et al. 1992*b*; Phillips and Hansen 1994), and include: (1) the lack of major volcanic constructs, such as large shield volcanoes (although large calderas are present, they appear to postdate plateau deformation fabrics); (2) deformation fabrics which were interpreted to record initial contraction postdated by extension; and (3) the absence of transitional forms between rises and plateaus, as documented by geologic observations and the statistical clustering of GTRs into two distinct families. Additional concerns of the upwelling model include the apparent lack of radial and concentric patterns (implying a lack of axisymmetry) for some plateaus, the lack of a circumferential extension belt as predicted during final stages of collapse, and flooding of deformation fabrics at rises by associated volcanism, opposite to the predicted age progression (Bindschadler et al. 1992*b*). We return to these concerns in our summary of plateau models.

In the downwelling model, subsolidus flow of lower crustal material due to viscous coupling with a mantle coldspot results in the observed crustal thickening at plateau highlands (Bindschadler and Parmentier 1990; Bindschadler et al. 1990*b*,1992*b*). Modeling of mantle downflow predicts initial downward deflection of the surface, yielding a circular to linear lowland (dependent on cylindrical or sheet-like downwelling, respectively) and formation of margin-parallel contractional fabrics. As crust thickens over the downflow, an elevated plateau morphology evolves. As topography builds, a zone of margin-normal contractional fabrics forms in the interior region; with increased topography this zone of hoop compression migrates toward the margin, and is replaced by interior strike-slip faulting, followed in turn by hoop extension. As downwelling wanes and ends, thereby removing dynamic support of the crustal plateau, gravitational collapse occurs, decreasing elevation and resulting in extensional deformation. Bindschadler and coworkers favored a downwelling origin for plateaus because (a) overall topographic patterns were suggested to be more consistent with horizontal strains and associated thickening, rather than dominantly vertical tectonics of hotspots; (b) low GTR and ADC values indicate compensation by crustal thickening, in contrast with high GTRs and ADCs of rises attributed to dynamic support; (c) of the presence of margin-parallel contractional fabrics along elevated margins, and contractional fabrics within plateau interiors; and (d) the sequence of deformation, as interpreted at the time, from initial contraction to later extension corresponds to predicted surface strain patterns predicted for mantle downflow.

Points (a) and (b) of the downwelling model are not exclusive of an upwelling origin. Downwelling and upwelling models satisfy observed topography and gravity data. For example, the hotspot model predicts steep outward slopes, elevated margins, and subsided interior observed at crustal plateaus (Herrick and Phillips 1990; Grimm and Phillips 1991). Downwelling and upwelling models result in regions of thickened crust, but differ in the mode of thickening, whether by subsolidus (coldspot model) or supersolidus (hotspot model) flow (see Phillips and Hansen 1994). The mode of thickening cannot be constrained by gravity or topography but might be constrained by surface structural and kinematic patterns that record the history of the surface response to thickening processes.

The amount of contraction within plateau interiors may be less than previously considered, and the previously interpreted deformational sequence of contraction postdated by extension may be incomplete. Although some basin-and-dome terrains are contraction-dominated, others are dominated by extensional fabrics (Hansen and Willis 1996), thereby "reducing" total contraction. Newly recognized ribbon terrain fabrics, common throughout crustal plateaus, indicate that extension predates contraction. This evidence disfavors a downwelling model, which does not predict initial surface extension.

Additional drawbacks of the downwelling model are based on the large amounts of crustal material required to flow and thereby thicken the crust (Phillips and Hansen 1994). Specific shortcomings include (1) the extensive distances that lower crustal material must be transported—with no surface features identified to create or transport the necessary crust, (2) the absence of negative GTRs, (3) gravitational evidence against a crustal asthenosphere, and (4) the exceedingly long time scales required to thicken the crust. The latter is especially problematical. Phillips and Kidder (1995) modeled downwelling by combining convection models with viscoelastic non-Newtonian finite element code. They stated that "for a range of parameters describing current Venus rheology, temperature gradient, crustal thickness and convective vigor, the crust is thickened an insignificant amount in one billion years," and that "plateau heights in excess of one kilometer are possible within the lifetime of a mantle downwelling only for a weak crustal flow law." It therefore seems improbable for the plateaus to have formed by subsolidus crustal thickening above downwellings, unless the crustal strength was considerably weaker than that suggested by Mackwell et al. (1995) for dry diabase.

Thus, the shortcomings of the downwelling model for building plateaus are serious, yet where does this leave us with respect to an upwelling origin? The previous lack of observed initial extension may be remedied by the recognition of ribbon terrain. Broadly radial patterns of contractional fabrics within plateau interiors may result from membrane compressive stresses as predicted by the upwelling model. The progressive deformation of many plateau tesserae from ribbons to folds to graben record temporal changes in near-surface crustal rheology. Early-stage ribbons require a shallow brittle-ductile transition, whereas late-stage large graben require deeper depths to the brittle-ductile transition. The relatively shallow brittle-ductile transition during ribbon formation, and the downward migration of this transition with time are most easily envisioned by an initial thermal source—a plume—which raises the ductility limit to shallow levels, and that cools with time resulting in an increase in the depth to the brittle-ductile transition (Hansen et al. 1996).

If plateaus are formed by mantle upwellings, the lack of transitional forms between plateaus and rises becomes a concern, although temporal and/or spatial differences in lithospheric rheological structure and resultant interaction with an ascending plume may contribute to the formation of plateaus versus rises (Phillips and Hansen 1994). The apparent absence of early-stage large volcanic constructs during plateau formation has been suggested as a possible drawback of an upwelling origin (Bindschadler et al. 1992b; Phillips and Hansen 1994), yet the *necessity* of such constructs is unclear. For example, lithospheric conditions during early stages of plateau formation may inhibit the formation of early-stage large volcanic constructs, and furthermore crustal thickening at crustal plateaus is likely dominated by subsurface magmatism rather than extrusional surface flow (Herrick and Phillips 1990; Phillips et al. 1991; Phillips and Hansen 1994). Plateaus may represent the surface signature of a plume interacting with relatively thin lithosphere, and rises may represent the surface signature of similar plumes interacting with thicker and stronger lithosphere (Hansen et al. 1996). Phoebe Regio, which lacks ribbon terrain yet which is dominated by extensional fabrics, may represent a transition between early formed plateaus and younger rises, marking a time of rheological transition in Venus' lithosphere. Thus, plateaus and rises may have formed by a similar hotspot mechanism but they may represent evolutionary changes in Venus' lithosphere and rheology, rather than stages of hotspot evolution.

Cylindrical plumes would likely result in azimuthally symmetric topography and deformation patterns; plateaus typically host areas that describe regional axisymmetric patterns (at least for partial arcs), although some areas exhibit considerable variability. For example, western Aphrodite Terra displays widely variable patterns overall, but within that framework distinct regions with radial and concentric patterns can be identified. Phillips and Hansen (1994) recognized that "most of the outward convective heat transfer in the Venusian mantle is caused by internal heating and thus takes the form of *diffuse* upwellings...Lithospheric stretching following buoyant uplift allows diffuse upwellings to rise to relatively shallow depths. The resulting partial melting adds to the thickness of the crust, and the accompanying edifice stresses maintain the lithosphere in tension, which continues the upwelling and melting processes." Thus, single plumes versus plumes within an overall diffuse upwelling system may explain variable topographic and deformation patterns in general. Alpha Regio, isolated from other plateaus, may record products from a single large-scale plume, whereas western Aphrodite Terra may record products resulting from multiple plumes, each comparable to an Alpha-type plume, but within an overall system of diffuse upwelling. Different areas may be affected at different times, as individual plumes rise from the mantle, perhaps breaking into smaller plumes in their ascent. For example, tesserae at Thetis Regio covers a vast area and hosts deformation fabrics typical of crustal plateaus, yet only the triangular central block is topographically elevated with steep sides (i.e., a present-day crustal plateau).

Overall, observations appear to favor upwelling over downwelling as the causal mechanism of plateau formation, yet the debate is far from over. Some aspects concerning crustal plateaus, however, are independent of preferred model of mantle dynamics. For example, both models allow for gravitational collapse of thickened and elevated crustal blocks following loss of dynamic support. Thus, smaller plateaus may evolve from larger plateaus (Phillips and Hansen 1994). Large inliers, including Laima and Ananke tesserae, commonly host tessera fabrics typical of plateaus, including basin-and-dome, extended fold, and folded ribbon terrains; other large inliers locally describe arcuate boundaries that host margin-parallel contractional and margin-normal extensional fabrics (e.g., Dekla Tessera, and unnamed tessera west of Beta Regio). Volcanic flooding models by progressive "drowning" of plateau highlands reveal many characteristics typical of large tessera inliers, including arcuate marginal traces, and highly-irregular boundaries for flooded interior regions (Head and Ivanov 1993). Thus large inliers and some small inliers may evolve by continued collapse and loss of topographic elevation of former plateaus, and flooding by plains volcanism—inliers may represent the "death rattle" of crustal plateaus (Phillips and Hansen 1994). Not all plains inliers, however, represent collapsed plateaus, as noted earlier; many small inliers exhibit structural fabrics more typical of smaller-scale coronae and chasmata, and likely represent the demise of those tectonic features.

Finally, plateaus may be deformed by strain supplied by sources outside the plateau environment (Solomon et al. 1991,1992; Grimm and Phillips 1990; Phillips and Hansen 1994). Phillips and Hansen (1994) stated: "Once a crustal plateau is created...it may represent a relatively weaker portion of the lithosphere because it has a thicker than average crust....Because of their inherent strength boundaries, crustal plateau may deform significantly in response to large-scale regional strain fields that may wax and wane on Venus over geological time." Thus, a crustal plateau, may serve as "strain magnet" (Grimm and Phillips 1990), recording applied regional compressive or tensile stresses. Marginal deformation that extends into adjacent plains (Phillips and Hansen 1994; Saunders 1995) may reflect such behavior. Thus, spatial and temporal patterns of plateau deformation may shed light on not only plateau origin, but also on regional strain conditions

C. Directions of Future Research

The mode of crustal thickening, and thus how crustal plateaus are built, cannot be constrained by topography and gravity, because both sets record present conditions and any relation to past activity is model dependent. Structural and kinematic patters, however, record the history of the surface response to thickening processes, and provide a key constraint for models. Several important issues remain, including plateau developmental history, plateau relations to driving mantle mechanisms, and how plateaus subsequently degrade following loss of dynamic support. Analysis should particularly include detailed documentation of deformation relations in the plateaus themselves, plateau-like inliers, and their smaller counterparts. Integration of all available data sets, and comparison to model predictions, will ultimately yield the best model for crustal plateau evolution.

VII. WESTERN ISHTAR TERRA

The origin of western Ishtar Terra is fundamental to understanding Venusian tectonism. Although the definitive model is yet to be proposed, mounting evidence restricts the possible models. In this section we briefly present the implications of various types of data for models of western Ishtar Terra, and discuss models in light of these constraints. Western Ishtar Terra straddles a broad topographic welt; it includes Lakshmi Planum (4 km above MPR) and surrounding deformed belts including interior mountain belts Maxwell, Danu, Akna, and Freyja montes (3.5–11 km above MPR), and outlying tesserae western Fortuna, Clotho, Atropos, and Itzpapalotl (1–4.5 km above MPR) (Fig. 3). Montes slope steeply inward toward Lakshmi and slope gently outward toward their respective tesserae resulting in a distinctive, and globally unique, topographic profile. Uorsar Rupes marks the steep slope along the northern boundary of Itzpapalotl, and Vesta Rupes, Danu, and South Scarp and Basin bounds Lakshmi to the south. North Basin (Kaula et al. 1992) is a topographic deep between Maxwell and Freyja montes. Ishtar Terra is notably free of coronae, but a northwest-trending belt of calderas crosses Ishtar from the boundary between Itzpapalotl and Atropos to the South Scarp and Basin area (Willis and Hansen 1995). Ishtar crater density indicates a *mean* surface retention age of several hundred Myr (Phillips et al. 1992; Schaber et al. 1992).

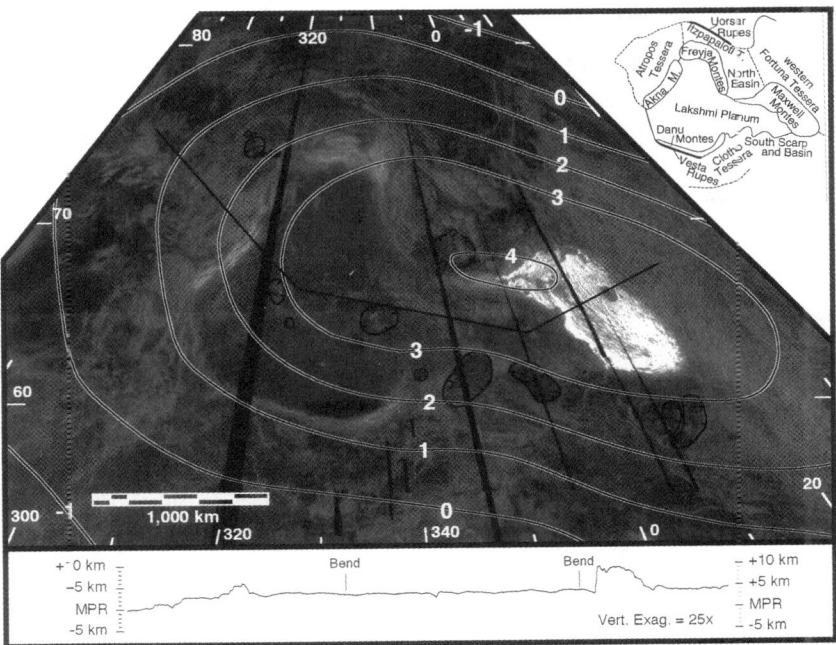

Figure 3. SAR image of Ishtar with long-wavelength topography in km (gray on white), calderas (black; hachures indicate downdropped block), and topographic profile (figure modified after Hansen and Phillips 1995; Willis and Hansen 1995). MPR is mean planetary radius.

A. Observations

Montes and tesserae each have length to width ratios of <4, with the exception of Danu Montes. In comparison, most Earth mountain belts have length/width ratios >10. Interpreted most simply, these relations suggest that Ishtar deformation is not likely related to plate boundary processes.

Magellan SAR images hold important constraints on the surface structural evolution of Ishtar Terra, which in turn reflects subsurface processes (Kaula et al. 1992, Smrekar and Solomon 1992; Keep and Hansen 1994; Phillips and Hansen 1994; Hansen and Phillips 1995). Lakshmi Planum, which is relatively structureless except for minor contractional wrinkle ridges chiefly along its periphery, hosts calderas and flooded tesserae—comprised of dominantly fracture fabrics and exposed at the highest elevations of Lakshmi (Roberts and Head 1990*a,b*; Kaula et al. 1992). Contractional fold ridges and valleys (Campbell et al. 1983; Barsukov et al. 1986; Crumpler et al. 1986; Kaula et al. 1992; Solomon et al. 1991; Suppe and Connors 1992), which are continuous along strike for hundreds of km, dominate the tectonic fabric of the montes and tesserae. In Maxwell-western Fortuna and in Akna-Atropos, parallelism of folds defines a coherent tectonic fabric that extends from Lakshmi wrinkle ridges across the mountain belts and into adjacent tesserae (Keep and Hansen 1994; Hansen and Phillips 1995). Fold spacing and orientation remain unchanged despite several km elevation change from montes to tesserae. These observations are important because (1) they indicate that topography—not tectonic fabric—distinguishes montes from tesserae, and (2) the apparent lack of strain gradient across hundreds of km requires a mechanism to deform huge expanses of crust without developing strong strain gradients and regardless of changes in topography (Hansen and Phillips 1993*a*; Keep and Hansen 1994). If deformation resulted from stress transmitted horizontally through the crust across a (sub)vertical mechanical boundary, strong strain gradients might be expected, with the highest strain being closest to the mechanical discontinuity. No obvious strain gradient is observed, and thus strain probably resulted from shear stresses transmitted upward across the expanse of deformed crusts. Basilevsky et al. (1986) and Phillips (1986) suggested that large regions of deformation on the surface of Venus might result from sublithospheric flow.

Structural history differentiates western Ishtar tesserae from other tessera terrains globally. Ribbon terrain is not observed in western Ishtar Terra. Atropos and western Fortuna tesserae display simple fold structures and perpendicular extension fractures. Itzpapalotl Tessera represents a family of structures that formed as a result of progressive noncoaxial (sinistral) deformation distributed across its 250 to 300 km width (Hansen 1992; Kaula et al. 1992). Thus Atropos, Itzpapalotl, and parts of western Fortuna Tessera should not be considered *a priori* as globally correlative with other tesserae.

Approximately ten calderas (or caldera complexes) form a band parallel to the long axis of Ishtar's long-wavelength topographic bulge (Fig. 3) (Barsukov et al. 1986; Gaddis and Greeley 1990; Roberts and Head 1990*a,b*; Head et al. 1992; Willis and Hansen 1995). The calderas range in size from 40×20 km to 480×220 km, and in depth from <1 km to ~ 3.5 km. South Scarp and Basin apparently result from caldera collapse, accounting for the serrated nature of this southern boundary, and North Basin exhibits caldera-related characteristics (Willis and Hansen 1995). Reconnaissance analysis indicates that volcanism is late synkinematic with local deformation. Extensive volcanic flooding resurfaced much of Lakshmi Planum, the Ishtar deformed belts, and adjacent plains.

Magellan's high-resolution gravity provides important constraints for the nature of Ishtar's roots. Pioneer Venus gravity allowed comparison of long-wavelength topography and gravity (>2500 km), and indicated that only 4 km of elevation could be supported dynamically (Grimm and Phillips 1991).

Magellan gravity data can resolve features with wavelength ≤360 km (chapter by Kaula et al.) and therefore these data can constrain the compensation depths of individual montes and tesserae. Degree-75 Magellan gravity data show a strong correlation of short-wavelength topography and gravity, particularly for regions higher than 4.5 km above MPR; short-wavelength (500 to 2500 km) features are compensated at depths ranging from 25 to 75 km, and the long-wavelength (>2500 km) bulge is compensated at ~130 km. The gravity data indicate that short- and long-wavelength features must be compensated by different mechanisms; short-wavelength features could be supported by variations in crustal thickness, and the long-wavelength topographic welt could be supported buoyantly, thermally, or in some combination (Hansen and Phillips 1995). Individual short-wavelength features have variable depths; of note, Lakshmi is compensated at the same depth as the surrounding plains (~30 km), Maxwell Montes at 35 to 45 km, Freyja at 30 to 40 km, Akna at 60 to 70 km, and North Basin at 15 to 25 km.

Given Venus' high surface temperature and low erosion rate it becomes important to consider how a high-standing feature can survive, and not gravitationally collapse with time. Early flow laws for Venusian crust did not allow for Ishtar to be preserved for more than 10 Myr (Smrekar and Solomon 1992), however, the new flow law for super-dry diabase (Mackwell et al. 1995) allows Ishtar enough strength to survive for at least 500 Myr without gravitational collapse (Freed and Melosh 1995).

B. Ishtar Formation

The key observations that models for Ishtar Terra must satisfy include:

A. Ishtar's topographic profile—a broad topographic dome, with superimposed mountain ranges standing up to 11 km above MPR, and extending up to 700 km wide, with steep slopes oriented inward toward Lakshmi and gentle slopes transitional to adjacent tesserae.
B. Structural fabrics, dominated by folds parallel to the orientation of their respective host montes with Lakshmi, reflect widespread crustal contraction around Lakshmi; Lakshmi itself is generally undeformed at the surface.
C. Coherent fold patterns across hundreds of kilometers normal to strike record no obvious strain gradient despite several km change in elevation.
D. Lack of evidence for extensional collapse at the highest elevations.
E. Numerous caldera complexes with late-syntectonic volcanism track the long axis of Ishtar's long-wavelength topographic bulge.
F. Mean surface retention age of several hundred Myr.
G. Different compensation mechanisms for short- and long-wavelength features—short-wavelength (500 to 2500 km) features have variable depths of compensation, ranging from 15 to 75 km, whereas long-wavelength (>2500 km) features are compensated at ~130 km.

Pre-Magellan convergent plate-tectonic models called for subduction all around Lakshmi, or at least along Uorsar Rupes and Vesta Rupes (Crumpler et al. 1986; Head 1990; Roberts and Head 1990*a*,*b*). Southward subduction along Uorsar Rupes would require that Akna and Maxwell are left-lateral and right-lateral shear zones, respectively, whereas northward subduction along Vesta Rupes would require the opposite shear sense in each belt. Although right-lateral translation within Maxwell was proposed based on Venera data (Vorder Bruegge et al. 1990; Vorder Bruegge and Head 1991), evidence for these kinematics is not observed in Magellan SAR imagery (Keep and Hansen 1994; Vorder Bruegge 1994). Subduction along the eastern, northern, and western margins (to accommodate observed contractional structures) is kinematically difficult; subduction would require radial contraction in the subducting lower plate, or radial extension in the upper plate, or both; evidence for neither is observed.

Mantle upwelling models (Pronin 1986; Basilevsky et al. 1986; Grimm and Phillips 1991), in which Ishtar Terra comprises the surface expression of a hotspot that produces topography dynamically and by volcanic construction, and the deformed belts result from incipient mantle return flow that has encountered a lateral strength heterogeneity, could account for two distinct mechanisms of compensation (crustal thickening beneath mountain belts and deep dynamic support of long-wavelength topography), and might allow for late-syntectonic caldera-related volcanism. However, these models do not explain the mountain belt-scale topography nor the surface strain patterns (A–D above). Upwelling predicts early radial extension, but no evidence for this is observed. Although deceleration of outward plume flow at the plateau margin might result in accumulation of the crust to form mountain belts, the hotspot model would predict steep slopes outward from the mountain belts contrary to the observed steep inner slopes (Phillips and Hansen 1994).

Ishtar crustal shortening and long-wavelength topography have been interpreted by many as the result of mantle downwelling and the resultant thickening of basaltic crust (Bindschadler and Parmentier 1990; Kiefer and Hager 1991; Lenardic et al. 1991; Vorder Bruegge and Head 1991; Bindschadler et al. 1990*b*,1992*b*; Arkani-Hamed 1993; Namiki and Solomon 1993). Mantle downwelling (or coldspot) models assert that thickened *crust* supports the long-wavelength welt of Ishtar Terra together with a sinking mantle diapir, which pulls *crustal* material inward and downward beneath Lakshmi. Initially the surface is deflected downward until the crust thickens, resulting in high surface elevation. When downwelling ceases, the thickened crust spreads gravitationally decreasing in elevation. Downwelling may provide the necessary environment for contractional strain. However, flaws with this model lie in the role required of the crust (see Phillips and Hansen 1994; Hansen and Phillips 1995). (1) The downwelling model calls on huge volumes of crust to thicken Ishtar Terra, yet it does not identify evidence of structures that produce or transport large volumes of crust. (2) Dynamic support of high-standing topography driven by inward flow of the lower crust places the surface and upper crust in tension (Kiefer and Hager 1991); therefore, this model predicts surface extension postdating early contraction. (3) The basalt-eclogite phase transition limits the elevation and lifetime of mountains formed as a result of thickened basaltic crust (Namiki and Solomon 1993; Jull and Arkani-Hamed 1995). In order to achieve elevations of 11 km, the Ishtar mountains must be <50 Myr old, in contrast with crater density data. (4) If the crust participates in compensation of both the long- and short-wavelength topography as required by the model, topographic and gravity constraints cannot be met (Grimm and Phillips 1991). (5) This model does not explain the surface strain patterns—although it predicts overall contractional strain, it does not justify the lack of strain gradient across topography from montes to tesserae, nor does it justify the lack the strain at Lakshmi Planum.

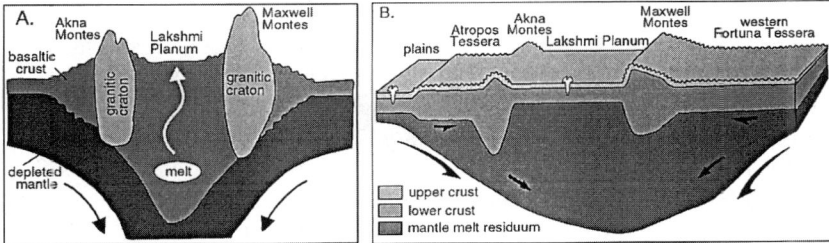

Figure 4. Schematic diagram of the formation of Ishtar Terra (a) from the convergence of granitic cratons over a mantle downwelling (figure after Jull and Arkani-Hamed 1995), and (b) as the result of ponded residuum (figure after Hansen and Phillips 1995).

In light of concerns (1), (3) and (4) above, Jull and Arkani-Hamed (1995) argued that wavelength-dependent compensation depths cannot be explained by simple thickening of a uniform density crust, nor can thickened basaltic crust result in elevations up to 11 km. They proposed a model in which Maxwell and Akna result from lower-density "granitic cratons" which converge over a mantle downwelling (Fig. 4a). In their model Lakshmi and caldera volcanism resulted from melting at the base of a thickened basaltic crust. This model does not specifically address surface strain patterns except that contraction is broadly predicted. The similarity in structural style from tesserae to montes may create a problem for this model. Granitic cratons might be expected to deform quite differently than layered basalt, contrary to structural patters (above). Granite is weaker than basalt given its lower melting temperature. In addition, OH-bearing minerals in the granitic cratons would likely break down at high temperatures within Ishtar Terra. The release of mineralogical water would probably result in ductile flow of the granite at depth, causing collapse of the granitic cratons. granitic cratons. Thus, granitic cratons might not have sufficient strength to form the montes root zones. Similarly, as alluded to by the authors of this model, granitic cratons might be difficult to form in the Venus environment. Once formed, granitic cratons would have to collect in western Ishtar Terra.

Zuber and Parmentier (1995) proposed that Ishtar topography is, in part, stress supported. In their model, Ishtar's long-wavelength topography results from horizontal shortening of a laterally heterogeneous lithosphere, and Ishtar's mountain-scale topography results from deformation of the mechanically strong outer layer. This model provides another means to compensate partially Ishtar's mountain-scale topography. A potential problem lies in the required relative youth of the mountains. Because these workers are basically matching the mountain belt-scale shape as a result of rheology, total strain, and time, the model imposes very stringent constraints. For example, the present cross-sectional shape of Akna Montes can be achieved using the new flow law (Mackwell et al. 1995), a thermal gradient of 18 K km^{-1}, a convergence rate of 10 to 15 s^{-1}, and a strain of 0.1, over 3 to 4 Myr (Zuber and Parmentier 1995). If strain is allowed to accumulate for longer than 4 Myr, the profile changes. Thus the relative youth of Ishtar mountain belts inferred by this model may contrast with Ishtar's apparent antiquity. Further proposals building on this general theme must demonstrate that the model can meet the constraints imposed by high-resolution gravity data, surface strain patterns, and calderas. This model, however, is important because it indicates that dynamic stress support may be an important component to be considered for Venusian mountain ranges. In short, this work illustrates that consideration of strictly isostatic support of montes root zones may not be valid.

The most comprehensive model to date attempts to address all of the constraints outlined above, except (E) (Hansen and Phillips 1995). This model evolved out of constraints determined from structural analysis of SAR images (Kaula et al. 1992; Hansen and Phillips 1993a; Keep and Hansen 1994; Phillips and Hansen 1994), and was modified as a result of Magellan degree-75 gravity data. The distribution and parallelism of fold structures across montes and adjacent tesserae led these workers to conclude that Ishtar deformation resulted from subsurface displacements, rather than large-scale translation of the surface layer(s). In addition, the apparent lack of correlation of structural style with topography from montes to tesserae indicates that mountain-scale topography cannot be solely and directly related to increased crustal thickness. If this were the case then the mountains should show evidence of higher strain gradients given that they sit several km above adjacent tesserae. Thus a model emerged in which surface deformation and mountain-scale topography result from displacement and thickening of the lower crust that occurred dominantly beneath a surface "crustal rug." This model had the advantage that it did not require the translation of vast regions of crust toward Ishtar to accommodate thickening of the entire Ishtar welt—mechanisms for which there is no evidence in SAR images (Phillips and Hansen 1994). In an early formulation of this model (Phillips and Hansen 1994), mantle partial melt residuum thickened at depth around an ancient crustal plateau (Lakshmi) comprised of thickened crust, which acted like a buttress at depth; thickened stable residuum (upper layer) accommodated mountain-scale topography, and unstable residuum (lower layer) supported the broad Ishtar welt.

The acquisition of degree 75 gravity data did not change the surface constraints, but changed identity of subsurface layers in the model. In the revised model (Hansen and Philips 1995), modified in light of the variable compensation depths imposed by gravity data, mantle downflow resulted in ponding and thickening of partial melt residuum that compensates the long-wavelength bulge of Ishtar; variably thickened lower crust, resulting from shear forces associated with the structurally deeper residuum, isostatically compensates short-wavelength topography; and relative displacement of the lower crust beneath the upper crust—which deforms in rug-like fashion—results in the surface strain patterns (Fig. 4b). In the revised model, Lakshmi is underlain by "normal" plains crust rather than a crustal plateau buttress. The local topography within Ishtar is related to the thickness of the lower crust and the thickness of the ponded residuum layer. Ishtar's long-wavelength topography reflects an inverse of the residuum pond at depth (Fig. 3).

Residuum is a natural byproduct of partial melting of the mantle. On Earth, residuum is probably swept from the oceanic upper mantle into the continents by plate motion (Kaula 1990), where it forms a mantle "keel" that extends beneath the continents (Jordan 1975, 1981). Venus lacks large-scale plate motions (Solomon et al. 1992), so residuum should be more widely distributed, but subject to disruption due to remixing associated with instabilities "dripping" from the cold upper thermal boundary layer of mantle convection. The lower part of the residuum, being less viscous and negatively buoyant, is likely the only part that would participate in the thermal boundary-layer mechanics (Parmentier and Hess 1992). The upper part of the residuum would pond and form a bouyant keel. The upper part of the residuum flows in response to shear forces associated with flow instability developed in the lower residuum. The shear forces are transmitted, in turn, to the crust and are at a maximum in those regions where the velocity gradients in the residuum are greatest—at intermediate distances from the geographical center of the residuum, where inward-directed flow turns downward. These regions mark the locations of maximum shear thickening of the crust and formation of mountain belts. Relative westward, southward, and eastward displacements of lower crust toward Lakshmi are responsible for Maxwell, Freyja, and Akna montes and their outboard tesserae, respectively. The narrow character of Danu Montes implies little relative northward displacement of the lower crust, as might be expected given the location of this belt relative to the shape of the residuum at depth as reflected by the long-wavelength topography (Fig. 3). The surface strain patterns of Ishtar (A–D above) resulted from differential translation of the lower crust beneath the upper crust. The upper crust is partially decoupled from the lower crust (Zuber 1987) such that shear stresses are transmitted from the lower crust to the upper crust, across a structure analogous to a "roof thrust"; the upper crust crumples and folds (or faults) like a rug. Translation of the lower crust inward toward Lakshmi, in response to upper residuum shear traction, could result in a broadly synchronously strained surface over thousands of km. The surface would deform wherever the lower crust was displaced at depth relative to the upper crust. Gravitational instabilities might occur along steep slopes or in regions of enhanced crustal thickening due to polyphase deformation, such as along the steep north and south slopes of Maxwell Montes (Kaula et al. 1992; Smrekar and Solomon 1992; Keep and Hansen 1994).

Hansen and Phillips's (1995) revised model does not directly address the band of caldera complexes (E, above). These calderas lie along a trend parallel to the long axis of Ishtar's long-wavelength topography, and appear to "avoid" regions of thickened crust (Fig. 3). Head (1995) proposed that three of these features—the examples at Atropos/Itzpapalotl junction, North Basin, and south of Maxwell Montes—formed as the result of negative eclogite diapirs at the base of thickened crust. Alternatively, given that the calderas avoid thickened crust and lie along the area of maximum interpreted residuum it might be easier to justify derivation of melts from thickened residuum (Willis and Hansen 1995), but such a proposal must be supported by petrologic modeling.

Although the presence of Ishtar calderas may tend to favor a hotspot-type model for Ishtar formation, too many other data sets seem to fly in the face of models requiring divergence. Hotspot models must specifically address mechanisms to account for the detail of Ishtar's mountain-scale topographic profile and the detail of the surface strain patterns. There seems to be a growing consensus that successful models for Ishtar Terra will include lateral/vertical variations in the thermal/mechanical structure of the lithosphere and global-scale convergent forces. Mounting data and a list of seven observations (A–G above) representing a variety of data sets impose strong constraints on models for Ishtar formation.

VIII. GLOBAL SYNTHESIS

The global tectonic model we envision for Venus follows that of Phillips and Hansen (1994) with some important modifications. We propose that Venus' tectonic evolution encompasses a few major types of tectonic processes that are either cyclic or evolutionary, and which are temporally and spatially diachronous. The tectonism responsible for the surface geology comprises dominantly vertical motions with a lithosphere that remains relatively fixed horizontally, and a mantle convection system that changes below it through time (Fig. 5). Locally, deep thermal plumes rise to encounter the lithosphere. The evolution of the lithosphere through time results in differences in the surface response above mantle upwellings. Tectonic and magmatic resurfacing occurs at regional and local scales. Venus' high surface temperature allows even small, slow magma bodies to reach the surface in a wide range of thermal environments (Sakimoto and Zuber 1995).

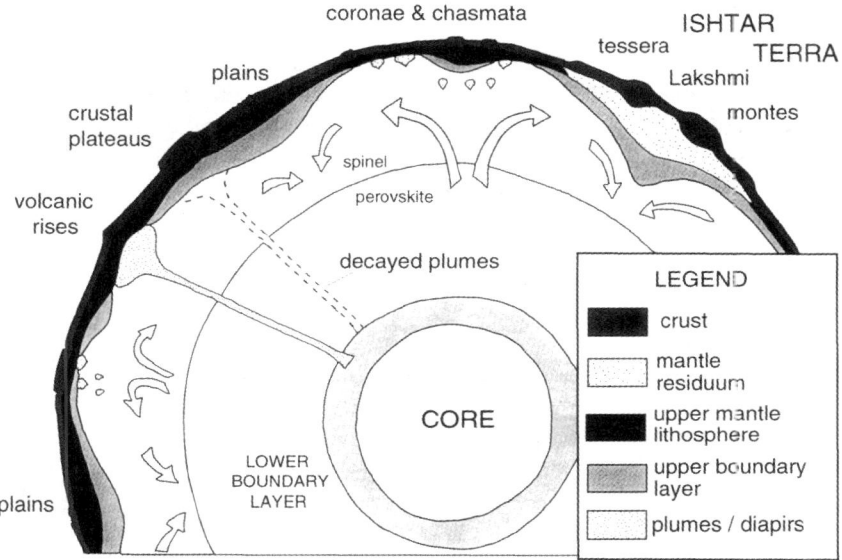

Figure 5. Global synthesis of Venus (figure modified from Phillips and Hansen 1994).

Venus' tectonic environments can be divided into mantle downwelling, mantle upwelling, plumes, and Ishtar Terra. The plains are regions of broad mantle downwelling as supported by long-wavelength gravity data (Banerdt 1986; Herrick and Phillips 1992), and the contractional ridge belts indicative of compressive stress (Zuber 1987). Diffuse mantle upwellings are marked by regions of lithospheric extension and coronae-chasmata. Crustal plateaus may represent the lithospheric signature of ancient deep thermal plumes (hotspots), and volcanic rises represent young thermal plumes. The difference between plateaus and rises may reflect the interaction of the plumes with thin lithosphere (plateaus) and thick lithosphere (rises). The lithosphere might have been thin early in Venus' history and has thickened with time. Ishtar Terra represents a unique region of ponded residuum, thickened lower crust and folded upper crust. Mantle upwellings and downwellings change location through time, and plains provinces may become "intruded" by coronae and cut by chasmata, whereas corona-chasma regions may founder and become plains as the upwellings beneath them decay. Tessera inliers may represent flooded crust originally deformed within coronae-chasmata and crustal plateaus. Thus Venus hosts a vertical, rather than horizontal, tectonic cycle, and Venus' lithosphere thickens with time.

The equatorial highlands comprise a region of diffuse mantle upwelling and lithospheric extension, an environment conducive to formation of coronae chasmata. Coronae result from melt instabilities within the upper mantle. Chasmata form synchronously with coronae. Coronae gain their elevations (above MPR) from thickened crust and localized thermal buoyancy (while young and warm; hence local isostatic and thermal buoyancy) and from their location above present mantle upwellings. Isolated coronae may mark minor mantle upwellings, not yet (or perhaps never to be) organized into sheet-like upwellings, as beneath corona chains. These smaller upwellings—perhaps more point-source-like—could decay in the same manner as their counterparts in the equatorial highlands.

Crustal plateaus may have resulted from massive partial melting over a thermal plume at a time when Venus' lithosphere was relatively thin and weak. Volcanic rises formed over similar thermal plumes as the Venusian lithosphere thickened with time. Phoebe Regio may represent a transitional stage between relatively thin plateau-forming lithosphere and thicker lithosphere which could support rises. If this model is correct, crustal plateaus are not currently forming. Some early formed crustal plateaus may have collapsed and become variably flooded and embayed; collapse could result in a variety of ways, whether the result of changes in global temperature, or related to local tectonics. Collapsed plateaus might form the large, typically arcuate, tracts of plains tessera (e.g., Ananke and Laima tesserae in the northern hemisphere).

Venus is continually vertically resurfacing herself. Crust thickens as a result of flood lava and corona and early crustal plateau formation. Venus is repainted in piecemeal fashion mostly in the topographically low plains— the ultimate base-level for low-viscosity flows. But the plains of today may have been elevated regions in the past marked by coronae-chasmata and crustal plateaus.

Deformed belts, or ridge belts, may reflect large-scale lithospheric inhomogeneities that form a linear anastomosing pattern in Atalanta-Vinmara planitiae and an orthogonal pattern in Lavinia Planitia. What could cause these lithospheric weaknesses? They might represent ancient chasmata or rift systems. As the region sank due to thermal decay, regions of thinned lithosphere—rifts forming topographic lows—would become flooded with lava. When a downwelling formed beneath this region the ancient rift systems might concentrate strain and become tectonically inverted, similar to inverted rift basins on Earth (Cooper and Williams 1989).

Volcanic rises represent hotspot plumes. The plumes may have finite lifetimes. The three different types of volcanic rises, rift-dominated, volcano-dominated, and corona-dominated, may result from differences in the local tectonic environment or lithospheric structure. Rift-dominated and corona-dominated rises form where hotspot plumes overlap with diffuse mantle upwelling, such as within the equatorial highlands (e.g., Atla and Beta). Volcano-dominated rises form above isolated hotspot plumes. Volcanic rises formed only in the relatively recent past after Venus' lithosphere thickened enough to support these structures. Thus we do not expect there to be extremely old volcanic rises; ancient plumes are recorded by crustal plateaus.

Ishtar Terra, unique on Venus, comprises a region in which mantle residuum collects at depth above a mantle downwelling. The lower crust thickens as the result of shear forces associated with the structurally deeper residuum. Displacement of the lower crust results, in turn, in translation and folding of the upper crust which deforms in rug-like fashion becoming detached from the lower crust. Thickened mantle residuum supports the broad topographic welt of Ishtar Terra, whereas thickened lower crust supports mountain-scale topography. Lakshmi Planum, within the interior of Ishtar Terra, represents captured plains crust elevated due to its location above deep ponded residuum. The late syntectonic formation of Ishtar's many calderas may signal another stage in Ishtar's long evolution.

The present is an exciting time in Venus tectonic thought—we have a wealth of data to analyze and assimilate, we have the technology to employ these immense data sets, and we have the recognition that Venus tectonics is very different from Earth plate tectonics. Undoubtedly further understanding of Venus tectonics will lead to a broader understanding of terrestrial planet evolution, terrestrial tectonic styles, and geodynamics; and it will likely prove to be a fertile environment for a fresh look at Earth's tectonic evolution.

Acknowledgments. Partial funding for this work was provided by a grant from the National Aeronautics and Space Administration (awarded to VLH) as part of the Planetary Geology and Geophysics program. B. Banerdt's work was performed at the Jet Propulsion Laboratory of the California Institute of Technology, under contract with the National Aeronautics and Space Administration. Critical reviews by G. McGill, K. Hansen, G. Bixler, D. Oliver, and J. Walther helped clarify many aspects of this manuscript.

REFERENCES

Arkani-Hamed, J. 1993. On the tectonics of Venus. *Phys. Earth Planet. Int.* 76:75–96.
Avé Lallemant, H. G. 1978. Experimental deformation of diopside and websterite. *Tectonophys.* 48:1–27.
Baer, G., Schubert, G., Bindschadler, D. L., and Stofan, E. R. 1994. Spatial and temporal relations between coronae and extensional belts, northern Lada Terra, Venus. *J. Geophys. Res.* 99:8355–8369.
Banerdt, W. B. 1986. Support of long wavelength loads on Venus and implications for internal structure. *J. Geophys. Res.* 91:403–419.
Banerdt, W. B., and Golombek, M. P. 1988. Deformation models of rifting and folding on Venus. *J. Geophys. Res.* 93:4759–4772.
Banerdt, W. B., and Sammis, C. G. 1992. Small-scale fracture patterns on the volcanic plains of Venus. *J. Geophys. Res.* 97:16149–16166.
Banerdt, W. B., Golombek, M. P., and Tanaka, K. L. 1992. Stress and tectonics on Mars. In *Mars*, eds. H. H. Kieffer, B. M. Jakosky, C. W. Snyder and M. S. Matthews (Tucson: Univ. of Arizona Press), pp. 249–297.
Banerdt, W. B., et al. 1994. The isostatic state of Mead crater. *Icarus* 112:117–129.

Barsukov, V. L., et al. 1986. The geology and geomorphology of the Venus surface as revealed by the radar images obtained by Veneras 15 and 16. *J. Geophys. Res.* 91:378–398.

Barsukov, V. L., et al. 1985. Main types of structures of the northern hemisphere of Venus. *Solar System Res.* 19:1–9.

Basilevsky, A. T., and Head, J. W. 1988. The geology of Venus. *Ann. Rev. Earth Planet. Sci.* 16:295–317.

Basilevsky, A. T., and Head, J. W. 1995*a*. Global stratigraphy of Venus: Analysis of a random sample of thirty-six test areas. *Earth, Moon, Planets* 66:285–336.

Basilevsky, A. T., and Head, J. W. 1995*b*. Regional and global stratigraphy of Venus: A preliminary assessment and implications for the geological history of Venus. *Planet. Space Sci.* 43:1523–1553.

Basilevsky, A. T., et al. 1986. Styles of tectonic deformations on Venus: Analysis of Venera 15 and 16 data. *J. Geophys. Res.* 91:399–411.

Bilotti, F., Connors, C., and Suppe, J. 1993. Global organization of tectonic deformation on Venus. *Lunar Planet. Sci. Conf.* XXIV:107–108 (abstract).

Bindschadler, D. L. 1994. Magellan LOS gravity of Venus plains regions: Lithospheric properties and implications for global tectonics. *Lunar Planet. Sci. Conf.* XXV:113–114 (abstract).

Bindschadler, D. L., and Head, J. W. 1989. Characterization of Venera 15/16 geologic units from Pioneer reflectivity and roughness data. *Icarus* 77:3–20.

Bindschadler, D. L., and Head, J. W. 1991. Tessera terrain, Venus: Characterization and models for origin and evolution. *J. Geophys. Res.* 96:5889–5907.

Bindschadler, D. L., and Parmentier, E. M. 1990. Mantle flow tectonics: The influence of a ductile lower crust and implications for the formation of topographic uplands on Venus. *J. Geophys. Res.* 95:21329–21344.

Bindschadler, D. L., et al. 1990*a*. Distribution of tessera terrain on Venus: Prediction for Magellan. *Geophys. Res. Lett.* 17:171–174.

Bindschadler, D. L., Schubert, G., and Kaula, W. M. 1990*b*. Mantle flow tectonics and the origin of Ishtar Terra, Venus. *Geophys. Res. Lett.* 17:1345–1348.

Bindschadler, D. L., de Charon, A., Beratan, K. K., and Head, J. W. 1992*a*. Magellan observations of Alpha Regio: Implications for formation of complex ridged terrains on Venus *J. Geophys. Res.* 97:13563–13577.

Bindschadler, D. L., Schubert, G., and Kaula, W. M. 1992*b*. Coldspots and hotspots: Global tectonics and mantle dynamics of Venus. *J. Geophys. Res.* 97:13495–13532.

Bowman, D. D., and Sammis, C. G. 1995. Implications of small-scale fracturing on Guinevere Planitia, Venus. *Lunar Planet. Sci. Conf.* XXVI:155–156 (abstract).

Brackett, R. A., Fegley, B., and Arvidson, R. E. 1994. Vapor transport, weathering, and the highlands of Venus. *Lunar Planet. Sci. Conf.* XXV:157–158 (abstract).

Brown, D. D., and Grimm, R. E. 1995. Tectonics of Artemis Chasma: A venusian "plate" boundary. *Icarus* 117:219–249.

Campbell, D. B., Head, J. W., Harmon, J. K., and Hine, A. A. 1983. Venus: Identification of banded terrain in the mountains of Ishtar Terra. *Science* 221:644–647.

Campbell, D. B., et al. 1989. New Arecibo high-resolution radar images of Venus: Preliminary interpretation. *Lunar Planet. Sci. Conf.* XX:142–143 (abstract).

Caristan, Y. 1982. The transition from high temperature creep to fracture in Maryland diabase. *J. Geophys. Res.* 87:6781–6790.

Connors, M. 1992. Crater floor slope and tectonic deformation on Venus. *Eos: Trans. AGU* 73:331 (abstract).

Cooper, M. A., and Williams, G. D. 1989. *Inversion Tectonics* (Boston: Blackwell Scientific).

Crumpler, L. S., Head, J. W., and Campbell, D. B. 1986. Orogenic belts on Venus.

Geology 14:1031–1034.
Cyr, K. E., and Melosh, H. J. 1993. Tectonic patterns and regional stresses near Venusian coronae. *Icarus* 102:175–184.
Ernst, R. E., et al. 1995. Giant radiating dyke swarms of Earth and Venus. *Earth Sc. Rev.* 39:1–58.
Ford, P. G., and Pettengill, G. H. 1992. Venus topography and kilometer-scale slopes. *J. Geophys. Res.* 97:13103–13114.
Ford, J. P., et al. 1993. *Guide to Magellan Image Interpretation*, JPL Publ. 89-41.
Freed, A. M., and Melosh, H. J. 1995. Long term survival of the topography of Ishtar Terra, Venus. *Lunar Planet. Sci. Conf.* XXVI:421–422 (abstract).
Gaddis, L. R., and Greeley, R. 1990. Volcanism in northwest Ishtar Terra, Venus. *Icarus* 87:327–338.
Gilmore, M. S., and Head, J. W. 1994. Intratessera volcanism of Alpha and Tellus tesserae on Venus. *Lunar Planet. Sci. Conf.* XXV:425–426 (abstract).
Gilmore, M. S., Ivanov, M. I., Head, J. W., and Basilevsky, A. T. 1995. Deformation of craters on tessera terrain, Venus. *Eos: Trans. AGU* 76(46):341.
Grimm, R. E. 1994*a*. Recent deformation rates on Venus. *J. Geophys. Res.* 99:23163–23171.
Grimm, R. E. 1994*b*. The deep structure of Venusian plateau highlands *Icarus* 112:89–103.
Grimm, R. E., and Phillips, R. J. 1990. Tectonics of Lakshmi Planum, Venus: Tests for Magellan. *Geophys. Res. Lett.* 17:1349–1352.
Grimm, R. E., and Phillips, R. J. 1991. Gravity anomalies, compensation mechanisms, and the geodynamics of western Ishtar Terra, Venus. *J. Geophys. Res.* 96:8305–8324.
Grosfils, E. B., and Head, J. W. 1994. The global distribution of giant radiating dike swarms on Venus: Implications for the global stress state. *Geophys. Res. Lett.* 21:701–704.
Hamilton, V. E., and Stofan, E. R. 1996. The geomorphology and evolution of Hecate Chasma, Venus. *Icarus*, 121:171–194.
Hansen, V. L. 1992. Regional non-coaxial deformation on Venus: Evidence from western Itzpapalotl Tessera. *Lunar Planet. Sci. Conf.* XXIII:478–479 (abstract).
Hansen, V. L. 1993. Asymmetric Venusian rifts: Arguments against subduction. *Eos: Trans. AGU* 74(16):377 (abstract).
Hansen, V. L., and Phillips, R. J. 1993*a*. Ishtar deformed belts: Evidence for deformation from below? *Lunar Planet. Sci. Conf.* XXIV:603–604 (abstract).
Hansen, V. L., and Phillips, R. J. 1993*b*. Tectonics and volcanism on eastern Aphrodite Terra, Venus: No subduction, no spreading. *Science* 260:526–530.
Hansen, V. L., and Phillips, R. J. 1995. Formation of Ishtar Terra, Venus: Surface and gravity constraints. *Geology* 23:292–296.
Hansen, V. L., and Willis, J. J. 1996. Structural analysis of a sampling of tesserae: Implications for Venus geodynamics. *Icarus* 123:296–312.
Hansen, V. L., Willis, J. J., and Phillips, R. J. 1996. Structural evolution of crustal plateaus, Venus. *Eos: Trans. AGU* 77(22):83 (abstract).
Head, J. W. 1990. Formation of mountain belts on Venus: Evidence for large-scale convergence, underthrusting, and crustal imbrication in Freyja Montes, Ishtar Terra. *Geology* 18:99–102.
Head, J. W. 1995. Processes of crustal and depleted mantle layer loss on Venus: Evidence from basins in tesserae, uplands, and plains. *Lunar Planet. Sci. Conf.* XXVI:577–578 (abstract).
Head, J. W., and Crumpler, L. S. 1987. Evidence for divergent plate boundary characteristics and crustal spreading on Venus. *Science* 238:1380–1385.
Head, J. W., and Crumpler, L. S. 1990. Venus geology and tectonics: Hotspot

and crustal spreading models and questions for the Magellan mission. *Nature* 346:525–533.

Head, J. W., et al. 1992. Venus volcanism: Classification of volcanic features and structures, associations, and global distribution from Magellan data. *J. Geophys. Res.* 97:13153–13197.

Head, J. W., and Ivanov, M. 1993. Tessera terrain on Venus: Implications of tessera flooding models and boundary characteristics for global distribution and mode of formation. *Lunar Planet. Sci. Conf.* XXIV:619–620 (abstract).

Herrick, R. R. 1994. Resurfacing history of Venus. *Geology* 22:703–706.

Herrick, R. R., and Phillips, R. J. 1990. Blob tectonics: A prediction for western Aphrodite Terra, Venus. *Geophys. Res. Lett.* 17:2129–2132.

Herrick, R. R., and Phillips, R. J. 1992. Geological correlations with the interior density structure of Venus. *J. Geophys. Res.* 97:16017–16034.

Herrick, R. R., and Sharpton, V. L. 1996. Geologic history of the Mead impact basin, Venus. *Geology* 24:11–14.

Herrick, R. R., Bills, B. G., and Hall, S. A. 1989. Variations in effective compensation depth across Aphrodite Terra, Venus. *Geophys. Res. Lett.* 16:543–546.

Ivanov, M. A., and Basilevsky, A. T. 1993. Density and morphology of impact craters on tessera terrain, Venus. *Geophys. Res. Lett.* 20:2579–2582.

Ivanov, M. A., and Head, J. W. 1995. Tessera terrain on Venus: Global distribution and characteristics and implications for the geologic history of Venus. *Venus II: Geology, Geophysics, Atmosphere, and Solar Wind Environment*, Jan. 4–7, Tucson, Ariz., Abstract Booklet, p. 68.

Izenberg, N. R., Arvidson, R. E., and Phillips, R. J. 1994. Impact crater degradation on Venusian plains. *Geophys. Res. Lett.* 21:289–292.

Janes, D. M., and Squyres, S. W. 1993. Radially fractured domes: A comparison of Venus and the Earth. *Geophys. Res. Lett.* 20:2961–2964.

Janes, D. M., et al. 1992. Geophysical models for the formation and evolution of coronae on Venus. *J. Geophys. Res.* 97:16055–16068.

Johnson, C. L., and Sandwell, D. T. 1994. Lithospheric flexure on Venus. *Geophys. J. Int.* 119:627–647.

Jordan, T. H. 1975. The continental tectosphere. *Geophys. Space Phys.* 13:1–12.

Jordan, T. H. 1981. Continents as a chemical boundary layer. *Trans. Roy. Phil. Soc. London A* 301:359–373.

Jull, M. G., and Arkani-Hamed, J. 1995. The implications of basalt in the formation and evolution of mountains on Venus. *Phys. Earth Planet. Int.* 89:163–175.

Kaula, W. M. 1990. Mantle convection and crustal evolution on Venus. *Geophys. Res. Lett.* 17:1401–1403.

Kaula, W. M., et al. 1992. Styles of deformation in Ishtar Terra and their implications. *J. Geophys. Res.* 97:16085–16120.

Keep, M., and Hansen, V. L. 1994. Structural evolution of Maxwell Montes, Venus: Implications for Venusian mountain belt formation. *J. Geophys. Res.* 99:26015–26028.

Kiefer, W. S., and Hager, B. H. 1991. Mantle downwelling and crustal convergence: A model for Ishtar Terra, Venus. *J. Geophys. Res.* 96:20967–20980.

Klose, K. B., Wood, J. A., and Hashimoto, A. 1992. Mineral equilibria and the high radar reflectivity of Venus mountaintops. *J. Geophys. Res.* 97:16353–16370.

Lancaster, M. G., and Guest, J. E. 1994. Volcanism and tectonism in Rusalka Planitia and Atla Regio, Venus. *Lunar Planet. Sci. Conf.* XXV:767–768.

Lenardic, A., Kaula, W. M., and Bindschadler, D. L. 1991. The tectonic evolution of western Ishtar Terra, Venus. *Geophys. Res. Lett.* 18:2209–2212.

Mackwell, S. J., Zimmerman, M. E., Kohlstedt, D. L., and Scherber, D. S. 1995. Experimental deformation of dry Columbia diabase: Implications for tectonics on Venus. In *Rock Mechanics: Proc. 35th U. S. Symposium*, eds. J. J. K. Daemen and R. A. Schultz (Brookfield, Vt.: A. A. Balkema), pp. 207–214.

Masursky, H., et al. 1980. Pioneer Venus radar results: Geology from images and altimetry. *J. Geophys. Res.* 85:8232–8260.

McGill, G. E. 1993. Wrinkle ridges, stress domains, and kinematics of Venusian plains. *Geophys. Res. Lett.* 20:2407–2410.

McGill, G. E. 1994. Evolution of a hot spot, central Eistla Region, Venus. *Lunar Planet. Sci. Conf.* XXV:877–878 (abstract).

McGill, G. E., Steenstrup, S. J., Barton, C., and Ford, P. G. 1981. Continental rifting and the origin of Beta Regio, Venus. *Geophys. Res. Lett.* 8:737–740.

McKenzie, D., et al. 1992*a*. Features on Venus generated by plate boundary processes. *J. Geophys. Res.* 97:13533–13544.

McKenzie, D., McKenzie, J. M., and Saunders, R. S. 1992*b*. Dike emplacement on Venus and on Earth. *J. Geophys. Res.* 97:15977–15990.

Namiki, N., and Solomon, S. C. 1993. The gabbro-eclogite phase transition and the elevation of mountain belts on Venus. *J. Geophys. Res.* 98:15025–15031.

Namiki, N., and Solomon, S. C. 1994. Impact crater densities on volcanoes and coronae on Venus: Implications for volcanic resurfacing. *Science* 265:929–933.

Pappalardo, R., Greeley, R., and Reynolds, S. J. 1994. Extensional tectonics of Arden Corona, Miranda: Evidence for an upwelling origin of coronae. *Lunar Planet. Sci. Conf.* XXV:1047–1048 (abstract).

Parmentier, E. M., and Hess, P. C. 1992. Chemical differentiation of a convecting planetary interior: Consequences for a one plate planet such as Venus. *Geophys. Res. Lett.* 19:2015–2018.

Pettengill, G. H., et al. 1980. Pioneer Venus radar results: Altimetry and surface properties. *J. Geophys. Res.* 85:8261–8270.

Phillips, R. J. 1986. A mechanism for tectonic deformation on Venus. *Geophys. Res. Lett.* 13:1141–1144.

Phillips, R. J. 1990. Convection-driven tectonics on Venus. *J. Geophys. Res.* 95:1301–1316.

Phillips, R. J. 1993. The age spectrum of the Venusian surface. *Eos: Trans. AGU* 74(16):187 (abstract).

Phillips, R. J., and Hansen, V. L. 1994. Tectonic and magmatic evolution of Venus. *Ann. Rev. Earth Planet. Sci.* 22:597–654.

Phillips, R. J., and Izenberg, N. 1995. Ejecta correlations with spatial crater density and Venus resurfacing history. *Geophys. Res. Lett.* 22:1517–1520.

Phillips, R. J., and Kidder, J. G. 1995. Subsolidus thickening of Venusian crust. *Eos: Trans. AGU* 76(46):341 (abstract).

Phillips, R. J., and Malin, M. C. 1983. The interior of Venus and tectonic implications. In *Venus*, eds. D. M. Hunten, L. Colin, T. M. Donahue and V. I. Moroz (Tucson: Univ. of Arizona Press), pp. 159–214.

Phillips, R. J., and Malin, M. C. 1984. Tectonics of Venus. *Ann. Rev. Earth Planet. Sci.* 12:411–443.

Phillips, R. J., Kaula, W. M., McGill, G. E., and Malin, M. C. 1981. Tectonics and evolution of Venus. *Science* 212:879–887.

Phillips, R. J., Grimm, R. E., and Malin, M. C. 1991. Hotspot evolution and the global tectonics of Venus. *Science* 252:651–658.

Phillips, R. J., et al. 1992. Impact crater distribution and the resurfacing history of Venus. *J. Geophys. Res.* 97:15923–15948.

Plaut, J. J. 1993*a*. *Stereo Imaging*, JPL Publ. 93(24):33–41.

Plaut, J. J. 1993*b*. *The Non-SAR Experiments*, JPL Publ. 93(24):19–31.

Price, M., and Suppe, J. 1994. Mean age of rifting and volcanism on Venus deduced from impact crater densities. *Nature* 372:756–759.

Pronin, A. A. 1986. The structure of Lakshmi Planum, an indication of horizontal asthenospheric flows of Venus. *Geotectonics* 20:271–281.

Pronin, A. A., and Stofan, E. R. 1990. Coronae on Venus: Morphology and distribution. *Icarus* 87:452–474.

Roberts, K. M., and Head, J. W. 1990*a*. Lakshmi Planum, Venus: Characteristics and models of origin. *Earth, Moon, Planets* 50/51:193–249.

Roberts, K. M., and Head, J. W. 1990*b*. Western Ishtar Terra and Lakshmi Planum, Venus: Models of formation and evolution. *Geophys. Res. Lett.* 17:1341–1344.

Roberts, K. M., and Head, J. W. 1993. Large-scale volcanism associated with coronae on Venus: Implications for formation and evolution. *Geophys. Res. Lett.* 20:1111–1114.

Robinson, C. A., and Wood, J. A. 1993. Recent volcanic activity on Venus: Evidence from radiothermal emissivity measurements. *Icarus* 102:26–39.

Sakimoto, S. E. H., and Zuber, M. T. 1995. Effects of planetary thermal structure on the ascent of cooling of magma on Venus. *J. Volcan. Geothermal Res.* 64:53–60.

Sandwell, D. T., and Schubert, G. 1992. Flexural ridges, trenches and outer rises around Venus coronae. *J. Geophys. Res.* 97:15923–15948.

Saunders, R. S. 1995. Ovda margin tectonic relationships. *Eos: Trans. AGU* 76(46):342 (abstract).

Schaber, G. G. 1982. Venus: Limited extension and volcanism along zones of lithospheric weakness. *Geophys. Res. Lett.* 9:499–502.

Schaber, G. G., et al. 1992. Geology and distribution of impact craters on Venus: What are they telling us? *J. Geophys. Res.* 97:13257–13301.

Schubert, G., and Sandwell, D. T. 1995. A global survey of possible subduction sites on Venus. *Icarus* 117:173–196.

Schubert, G., Bercovici, G. D., Thomas, P. J., and Campbell, D. B. 1989. Venus coronae: Formation by mantle plumes. *Lunar Planet. Sci. Conf.* XX:968–969 (abstract).

Schubert, G., Bercovici, G. D., and Glatzmaier, G. A. 1990. Mantle dynamics in Mars and Venus: Influence of an immobile lithosphere on three-dimensional mantle convection. *J. Geophys. Res.* 95:14105–14129.

Schubert, G., Moore, W. B., and Sandwell, D. T. 1994. Gravity over coronae and chasmata on Venus. *Icarus* 112:130–146.

Senske, D. A., Schaber, G. G., and Stofan, E. R. 1992. Regional topographic rises on Venus: Geology of western Eistla Regio and comparison to Beta Regio and Atla Regio. *J. Geophys. Res.* 97:13395–13420.

Simons, M., Hager, B. H., and Solomon, S. C. 1994. Global variations in the geoid/topography admittance of Venus. *Science* 264:798–803.

Smrekar, S. E., and Parmentier, E. M. 1996. The interaction of mantle plumes with surface thermal and chemical boundary layers: Applications to hotspots on Venus. *J. Geophys. Res.* 101:5397–5410.

Smrekar, S. E., and Phillips, R. J. 1991. Venusian highlands: Geoid to topography ratios and their implications. *Earth Planet. Sci. Lett.* 107:582–597.

Smrekar, S. E., and Solomon, S. C. 1992. Gravitational spreading of high terrain in Ishtar Terra, Venus. *J. Geophys. Res.* 97:16121–16148.

Smrekar, S. E., and Stofan, E. R. 1996. Towards a comprehensive model of corona formation on Venus. *Lunar Planet. Sci. Conf.* XXVII:1227–1228 (abstract).

Solomon, S. C. 1993*a*. A tectonic resurfacing model for Venus. *Lunar Planet. Sci. Conf.* XXIV:1331–1332 (abstract).

Solomon, S. C. 1993*b*. The geophysics of Venus. *Physics Today* 46:48–55.

Solomon, S. C., et al. 1991. Venus tectonics: Initial analysis from Magellan. *Science* 252:297–312.

Solomon, S. C., et al. 1992. Venus tectonics: An overview of Magellan observations.

J. Geophys. Res. 97:13199–13255.
Squyres, S. W., et al. 1992*a*. The morphology and evolution of coronae on Venus. *J. Geophys. Res.* 97:13611–13634.
Squyres, S. W., et al. 1992*b*. Plains tectonism on Venus: The deformation belts of Lavinia Planitia. *J. Geophys. Res.* 97:13579–13599.
Stofan, E. R., and Head, J. W. 1990. Coronae of Mnemosyne Regio, Venus: Morphology and origin. *Icarus* 83:216–243.
Stofan, E. R., Bindschadler, D. L., Head, J. W., and Parmentier, E. M. 1991. Corona structures on Venus: Models of origin. *J. Geophys. Res.* 96:20933–20946.
Stofan, E. R., et al. 1992. Global distribution and characteristics of coronae and related features on Venus: Implications for origin and relation to mantle processes. *J. Geophys. Res.* 97:13347–13378.
Stofan, E. R., Smrekar, S. E., Bindschadler, D. L., and Senske, D. A. 1995. Large topographic rises on Venus: Implications for mantle upwelling. *J. Geophys. Res.* 100:317–327.
Strom, R. G., Schaber, G. G., and Dawson, D. D. 1994. The global resurfacing of Venus. *J. Geophys. Res.* 99:10899–10926.
Sukhanov, A. L. 1986. Parquet: Regions of areal plastic dislocations. *Geotectonics* 20:294–305.
Sukhanov, A. L. 1987. Parquet on Venus: Areas of regional deformations. *Lunar Planet. Sci. Conf.* XVIII:972–973 (abstract).
Suppe, J., and Connors, C. 1992. Critical tape-wedge mechanics of fold-and-thrust belts on Venus: Initial results from Magellan. *J. Geophys. Res.* 97:13545–13561.
Tackley, P. J., and Stevenson, D. J. 1991. The production of small Venusian coronae by Rayleigh-Taylor instabilities in the uppermost mantle. *Eos: Trans. AGU* 72(44):287 (abstract).
Tackley, P. J., and Stevenson, D. J. 1993. Volcanism without plumes: Melt-driven instabilities, buoyant residuum and global implications. *Eos: Trans. AGU* 74(16):188 (abstract).
Tanaka, K. L., Senske, D. A., Schaber, G. G., and Price, M. 1995. The geologic framework of Venus. *Venus II: Geology, Geophysics, Atmosphere, and Solar Wind Environment*, Jan. 4–7, Tucson, Ariz., Abstract Booklet, p. 47.
Törmänen, T. 1993. Complex ridged terrain-related ridge belts on Venus: Global distribution and classification. *Lunar Planet. Sci. Conf.* XXIV:1439–1440 (abstract).
Törmänen, T. 1995. Topographic and stratigraphic characteristics of ridge belts associated with tessera boundaries on Venus: Examples from northern Ovda Regio margin and Kutue Tessera. *Lunar Planet. Sci. Conf.* XXVI:1415–1416.
Turcotte, D. L. 1993. An episodic hypothesis for Venusian tectonics. *J. Geophys. Res.* 98:17061–17068.
Vorder Bruegge, R. W. 1994. Variation in compressional structures across Maxwell Montes: Evidence for a sequence of events in a Venusian orogeny. *Lunar Planet. Sci. Conf.* XXV:1449–1450 (abstract).
Vorder Bruegge, R. W., and Head, J. W. 1991. Processes of formation and evolution of mountain belts on Venus. *Geology* 19:885–888.
Vorder Bruegge, R. W., Head, J. W., and Campbell, D. B. 1990. Orogeny and large-scale strike-slip faulting on Venus: Tectonic evolution of Maxwell Montes. *J. Geophys. Res.* 95:8357–8381.
Watters, T. R., and Janes, D. M. 1995. Coronae on Venus and Mars: Implications for similar structures on Earth. *Geology* 23:200–204.
Willis, J. J., and Hansen, V. L. 1995. Caldera-related volcanism and collapse at Ishtar Terra, Venus. *Eos: Trans. AGU* 76(46):341.
Willis, J. J., and Hansen, V. L. 1996. Conjugate shear fractures at "Ki Corona,"

southeast Parga Chasma, Venus. *Lunar Planet. Sci. Conf.* XXVII:1443–1444.

Zuber, M. T. 1987. Constraints on the lithospheric structure of Venus from mechanical models and tectonic surface features. *J. Geophys. Res.* 92:541–551.

Zuber, M. T. 1990. Ridge belts: Evidence for regional- and local-scale deformation on the surface of Venus. *Geophys. Res. Lett.* 17:1369–1372.

Zuber, M. T., and Parmentier, E. M. 1995. Formation of fold-and-thrust belts on Venus by thick-skinned deformation. *Nature* 377:704–707.

LARGE VOLCANIC RISES ON VENUS

SUZANNE E. SMREKAR
Jet Propulsion Laboratory

WALTER S. KIEFER
Lunar and Planetary Institute

and

ELLEN R. STOFAN
Jet Propulsion Laboratory

Large volcanic rises on Venus have been interpreted as hotspots, or the surface manifestation of mantle upwelling, on the basis of their broad topographic rises, abundant volcanism, and large positive gravity anomalies. Hotspots offer an important opportunity to study the behavior of the lithosphere in response to mantle forces. In addition to the four previously known hotspots, Atla, Bell, Beta, and western Eistla Regions, five new probable hotspots, Dione, central Eistla, eastern Eistla, Imdr, and Themis, have been identified in the Magellan radar, gravity and topography data. These nine regions exhibit a wider range of volcano-tectonic characteristics than previously recognized for Venusian hotspots, and have been classified as rift-dominated (Atla, Beta), coronae-dominated (central and eastern Eistla, Themis), or volcano-dominated (Bell, Dione, western Eistla, Imdr). The apparent depths of compensation for these regions ranges from 65 to 260 km. New estimates of the elastic thickness, using the 90 deg and order spherical harmonic field, are 15 to 40 km at Bell Regio, and 25 km at western Eistla Regio. Phillips et al. (see their chapter) find a value of 30 km at Atla Regio. Numerous models of lithospheric and mantle behavior have been proposed to interpret the gravity and topography signature of the hotspots, with most studies focusing on Atla or Beta Regiones. Convective models with Earth-like parameters result in estimates of the thickness of the thermal lithosphere of approximately 100 km. Models of stagnant lid convection or thermal thinning infer the thickness of the thermal lithosphere to be 300 km or more. Without additional constraints, any of the model fits are equally valid. The thinner thermal lithosphere estimates are most consistent with the volcanic and tectonic characteristics of the hotspots. Estimates of the thermal gradient based on estimates of the elastic thickness also support a relatively thin lithosphere (chapter by Phillips et al.). The advantage of larger estimates of the thermal lithospheric thickness is that they provide an explanation for the apparently modest levels of geologic activity on Venus over the last half billion years.

I. INTRODUCTION

Broad regional topographic rises, over 1000 km in diameter, with large volcanic edifices were identified in Earth-based radar data (Saunders and Malin 1977; Campbell et al. 1984) and in Pioneer Venus data (Masursky et al. 1980; Schaber 1982; McGill et al. 1981), and were found generally to be concentrated in the equatorial region. McGill et al. (1981) argued that Beta Regio, the largest volcano on Venus, was analogous to terrestrial hotspots because of evidence for a broad topographic swell, rifting, and volcanism. These characteristics define hotspots, or the surface manifestation of mantle upwellings. The large positive gravity anomalies over such features as Beta Regio identified in Pioneer Venus gravity data, which were interpreted to indicate deep thermal anomalies at the base of the lithosphere, furthered the mantle upwelling hypothesis for large volcanic rises (Phillips and Malin 1983; Smrekar and Phillips 1991).

Major topographic highs (see Fig. 1) interpreted to be hotspots based on their topographic shape, associated volcanism, and apparent depth of compensation, include: Atla Regio (Phillips and Malin 1983,1984; Senske et al. 1992), Beta Regio (McGill et al. 1981; Campbell et al. 1984), Bell Regio (Basilevsky and Janle 1987; Janle et al. 1987), Dione Regio (Keddie and Head 1995), eastern, central, and western Eistla Regio (Senske et al. 1992; Grimm and Phillips 1992; McGill 1994), Imdr Regio (Stofan et al. 1995), and Themis Regio (Stofan et al. 1992). Other areas exhibit some but not all of the characteristics of hotspots. Ulfrun Regio is an elongated rise with a series of volcanic peaks to the northeast of Atla Regio interpreted as a hotspot by Basilevsky and Head (1988), but it has a relatively shallow compensation depth (Smrekar and Phillips 1991). Asteria Regio, to the west of Beta Regio, has volcanic edifices and a large depth of compensation, but it is a topographic plateau rather than a dome (Smrekar and Phillips 1991).

Rises on Venus exhibit great variations in their morphologic and topographic characteristics (Stofan et al. 1989; Bindschadler et al. 1992; Senske et al. 1992), as well as their gravity signatures (Smrekar and Phillips 1991; Smrekar 1994; Simons et al. 1994; Stofan et al. 1995). Many topographic rises exhibit abundant volcanism, generally in the form of large shield volcanoes and lava flood plains (Head et al. 1992). Other rises, such as western Eistla, also contain small (<50 km diameter) and intermediate scale edifices. Over half of the volcanic rises have coronae, circular tectonic features which are interpreted as small scale mantle upwellings (see, e.g., the chapter by Stofan et al.) on or near their flanks (Senske et al. 1992; Stofan et al. 1992). Evidence of extensional deformation is present at nearly all volcanic rises, ranging from small grabens to belts of fractures and troughs to major rift systems. In some regions, Magellan gravity data is sufficiently high resolution to estimate the thickness of the elastic lithosphere (Phillips 1994; Smrekar 1994).

The interpretation of Magellan data, in particular the impact cratering record (Phillips et al. 1992; Schaber et al. 1992), have also provided a foundation for a new interpretation of the tectonic history of Venus. The impact crater statistics give a large average crater retention age for the surface of approximately 300 to 500 Myr. Two puzzling observations are that the dis-

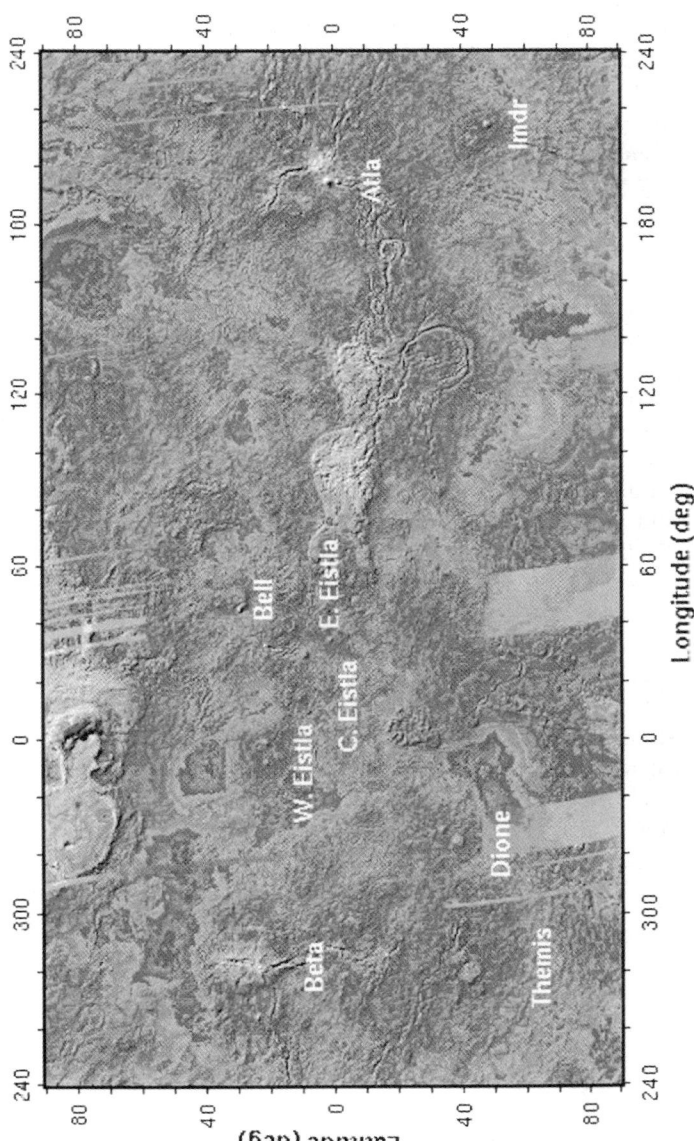

Figure 1. Location of nine probable hotspots on a base map of Magellan topography in Mercator projection, where the highest regions are brightest and the lowest regions are darkest. Regional names are centered beneath each topographic rise. The total range is approximately 12 km.

tribution can not be distinguished from a random one, and that there are few craters modified by volcanism or tectonism. One interpretation is that a global resurfacing event is required to reset the age of the entire surface (Schaber et al. 1992). Another end member case is one in which resurfacing occurs at a steady rate in small patches (Phillips et al. 1992). Both end member models are unlikely (Phillips et al. 1992). More detailed analyses indicates that there are regions of the planet with significantly different ages (Namiki and Solomon 1994; Price and Suppe 1994; Phillips and Izenberg 1995). Some hotspots are among the regions with younger ages (Phillips and Izenberg 1995). New theories proposed to explain the tectonic evolution of Venus include catastrophic resurfacing though episodic lithospheric overturn (Parmentier and Hess 1992; Turcotte 1995), chaotic mantle convection (Arkani-Hamed 1993), a transition to stagnant lid convection (Solomatov and Moresi 1996), and gradual lithospheric cooling (Solomon 1993; Grimm 1994a).

Attempts to explain the cratering record and the thermal history of the planet have fostered a debate over whether the lithosphere on Venus is essentially Earth-like in thickness ($\sim$100 km) or several times thicker (see the chapter by Phillips et al. for more discussion). Studies of hotspots have played a key role in this debate. Many studies that argue for either a thick or a thin lithosphere fit models of either mantle upwelling or thermal thinning of the lithosphere to the gravity and topography data of one or more hotspot regions. As we will discuss, models of the structure of the lithosphere and mantle do not provide a unique solution. However, the observed volcanism and rifting, as well as estimates of elastic thickness, are more consistent with a fairly thin lithosphere.

The wealth of information about terrestrial hotspots is the starting point for understanding the processes that form Venusian hotspots. For this reason, we review here what is known and hypothesized about terrestrial hotspots. Although much of what we know about hotspots is based on in-depth studies of a few regions, such as Hawaii for the Earth, and Beta and Atla Regions for Venus, it is clear that hotspots on both planets exhibit a wide range of characteristics. The ongoing study of terrestrial hotspots continues to illuminate the process of hotspot formation, yet there remains considerable debate over the extent to which processes such as dynamic uplift, thermal thinning of the lithosphere, and flexural versus crustal compensation of volcanoes operate at individual hotspots. This uncertainty is even greater for Venus.

Models of mantle convection are becoming increasingly sophisticated and offer new insights on mantle upwellings. Models that include temperature-dependent viscosity predict considerable energy at short wavelengths that can complicate the estimation of elastic thickness (Moresi and Parsons 1995; Kiefer 1995). Strongly temperature-dependent viscosity models can lead to new convective modes, such as the stagnant lid convection models of Solomatov and Moresi (1996). Models that include pressure-release melting are able to use estimates of volcanic volumes as an additional constraint on mantle upwelling models (Smrekar and Parmentier 1996).

The purpose of this chapter is to examine the role of hotspots in the tectonic evolution of Venus. First, we review the geophysics of terrestrial hotspots, the geologic characteristics of Venusian hotspots, models of hotspot formation and constraints provided by gravity, topography, and radar data. We then discuss the implications of these studies for the present-day level of hotspot activity and the thickness of the Venusian lithosphere, as well the relationship between Venusian hotspots, coronae, and highland plateaus. Here we favor the interpretation that some hotspots may be still active and that the lithosphere is relatively thin, on the order of 100 to 150 km.

II. TERRESTRIAL HOTSPOTS

The analogy to terrestrial hotspots is an important tool for better understanding Venusian hotspots. Roughly 40 hotspots are recognized on Earth (Crough 1983). However, there is a considerable variability in hotspot characteristics (see, e.g., Sleep 1990), and much of the understanding of this class of tectonic features is based on studies of Hawaii and several of the other major hotspots. Most occur on oceanic plates and can be easily recognized by their "hotspot tracks," or the chain of topographic swells and associated volcanoes left behind as the tectonic plate moves over the relatively stationary plume. The observation that plumes appear to have a fixed position with respect to the mobile plates is one line of evidence suggesting that they originate from deep in the interior, most likely at the core–mantle boundary. The heights of topographic swells associated with the plumes are in the range of 0.4 to 2.1 km and radii vary from 900 to 1800 km (Monnereau and Cazenave 1990). These swell heights are typically measured below water, and the isostatic loading effect of sea water must be removed from the measurement for comparison with Venus. The equivalent sub-aerial swell heights, based on simple mass balance arguments, would be 30% less.

Numerous data sets provide constraints on the structure of the crust, lithosphere, and upper mantle beneath hotspots. In a survey of 23 oceanic hotspots, all have geoid anomalies of less than 10 m (Monnereau and Cazenave 1990). These geoid, or equivalently gravity, anomalies are all far less than would be expected given the large topographic swells and evidence for mantle plumes at depth. These small geoid anomalies have been interpreted to indicate the presence of a low viscosity zone in the upper mantle that acts to decouple the topography from the plume buoyancy forces at depth (Robinson and Parsons 1988). An increase in the topography and the geoid-to-topography ratio at hotspots on older oceanic plates provides evidence that the low viscosity zone thins as the thermal lithosphere thickens (Cazenave et al. 1988). Gravity data also help constrain the thermal profile beneath hotspots by yielding estimates of the elastic thickness of the lithosphere. Typical values are 30 to 55 km (McNutt and Shure 1986; McNutt 1988; Sheehan and McNutt 1989; Ebinger et al. 1989). However, these elastic thickness estimates may be somewhat biased by a short wavelength convective signature, as discussed below.

The volcanic edifices visible at the surface are likely to be only a fraction of the total contribution of volcanism to the swell topography. Studies of gravity and seismic data at several oceanic hotspots indicate a large amount of volcanic material accumulates in the lower crust (White 1993; Wolfe et al. 1994). Given the large uncertainties inherent in these approaches, the bounds on the likely ratio of extrusive to intrusive volcanism for terrestrial hotspots, and by analogy Venusian hotspots, is 1:2 to 1:10 (see Smrekar and Parmentier [1996] for more discussion). In addition to the intrusion and extrusion of volcanic melt, pressure-release melting leaves behind low density residuum beneath the thermal lithosphere that may isostatically compensate up to 25% of the broad swell topography (Phipps Morgan et al. 1995).

Seismic imaging of mantle plumes to date is consistent with a plume head that extends up to several hundred km in depth (Nataf and VanDecar 1993; Bjarnason et al. 1996). Because plate motions can be used to determine the time period that a plume head has been in contact with the plate, the topographic swell can be used to estimate the buoyancy flux for hotspots. Somewhat different techniques yield a total value of ~ 50 Mg s^{-1} for all terrestrial hotspots (Davies 1988; Sleep 1990), or approximately 10% of the Earth's total heat flux (Davies 1988). These approaches, as well as petrologic evidence (Watson and McKenzie 1991) indicate that the near-surface plume-mantle temperature difference is approximately 200°C. Because plate tectonics continually destroys plates, the duration of mantle plumes is poorly constrained. The oldest active hotspot may be Crozet, which is believed to be the source of the 200 Myr Karoo flood basalts (White and McKenzie 1989).

Numerous processes contribute to the characteristic broad topographic swell and edifices found at hotspots. There is evidence for compensation of the swell topography by low density mantle residuum, intrusive volcanism, and thermal buoyancy of the plume, as well as flexural compensation of volcanoes. Debate continues about the relative contributions of the various processes, which in turn affects the accuracy of estimates of parameters such as lithospheric thickness and plume strength. With respect to the swell topography, there is debate about the relative contributions of the thermal and chemical density anomalies (Phipps Morgan et al. 1995), the extent to which the thermal lithosphere is thinned by the plume (see, e.g., McNutt 1988), and the amount of dynamic uplift, where dynamic uplift refers to a rising thermal plume as opposed to a static thermal anomaly (see, e.g., McNutt and Judge 1990). As discussed above, the amount of intrusive volcanism is also poorly constrained.

III. VOLCANO-TECTONIC CHARACTERISTICS

Volcanic rises on Venus exhibit a wide range of size and geologic characteristics. Swell heights and diameters are given in Table I. Locations of the nine hotspots are shown in Fig. 1. Volcanic volumes quoted below are from Stofan et al. (1995). Volcanic rises on Venus are clearly separable on the basis of the relative importance of rifts, major volcanic edifices, and coronae (Stofan et al. 1995). Below we describe the geologic characteristics of individual hotspots and the reasons for assigning each hotspot to a specific category.

A. Characteristics of Individual Hotspots

Atla Regio contains some of the largest Venusian volcanoes. The volcanic edifices Ozza and Maat Montes have minimum volcanic volumes of 3.0×10^5 km^3, and 2.1×10^5 km^3, respectively. Maat Mons is the tallest volcano on Venus and has a topographic rise of over 5 km. Another large volcanic edifice, Sapas Mons, lies to the west of the swell. The topographic rise lies at the junction of four major rift systems: Dali, Parga, Hecate and Ganis Chasmata. Senske et al. (1992) found that rifting and volcanism overlapped in time at Atla Regio. Atla is also associated with several coronae and some fragments of complexly deformed (or tessera) terrain.

Beta Regio is volumetrically the largest rise, and is cut by the major rift, Devana Chasma (Fig. 2). Beta Regio has one major volcanic edifice, Theia Mons, which is superposed on the rift, but has also been cut by subsequent rifting (Campbell et al. 1984; Stofan et al. 1989; Senske et al. 1991). Theia Mons has an approximate volume of 1.6×10^5 km^3. At the northern end of Beta, a large region of uplifted tessera terrain is also cut by Devana Chasma, indicating that the rise may have formed in a region of pre-existing tessera terrain (Senske et al. 1992).

Bell Regio includes several large volcanic centers, including Tepev Mons, which has a minimum volume of 3.2×10^4 km^3, as well as several coronae (Fig. 3). No evidence of rifting is seen at Bell Regio. Campbell and Rogers (1994) found that volcanism at Bell Regio has changed over time, from production of low relief volcanic centers to steep-sided edifices such as Tepev Mons.

Dione Regio has three major volcanic edifices, Ushas, Innini and Hathor Montes. Hathor is the largest edifice, with a minimum volume of 1.6×10^5 km^3 (Stofan et al. 1995). Dione has a very poorly defined topographic swell approximately 1000 km in diameter and less than 1.0 km high (Stofan et al. 1995). Some evidence for rifting is present at Dione Regio, although not as well expressed as at Atla, Beta or western Eistla Regiones. Keddie and Head (1995) interpret the presence of large volcanoes and a poorly defined topographic rise to indicate either secondary upwellings from a single large plume or near-contemporaneous upwelling of several smaller plumes. An alternative explanation is that the poorly defined topographic swell indicates a very late stage hotspot with little remaining thermal anomaly at depth.

Eistla Regio is composed of three separate highlands: western, central and eastern Eistla Regio. Western Eistla Regio is the largest of the three segments and has two large volcanic edifices, Sif and Gula Montes, with volumes of 1.6×10^4 km^3 and 2.3×10^4 km^3, respectively. The topographic rise is cut by a rift, Guor Linea, that is interpreted to be coeval with volcanism at Gula Montes (Senske et al. 1992). Sif and Gula Montes appear to be

TABLE I
Characteristics of Volcanic Rises on Venus

Volcanic Rise	Volume of Volcanics × 10^6 (km^3)	Minimum–Maximum Diameter (km)	Apparent Depth of Compensation (km)	Swell Height (km)
Rift dominated				
Beta Regio	160.0	1900–2500	225[a]	2.1
Atla Regio	514.0	1200–1600	175[a]	2.5
Volcano dominated				
Imdr Regio	48.0	1200–1400	260	1.6
W. Eistla Regio	39.1	2000–2400	200[a]	1.8
Dione Regio	~200	?	130	0.5
Bell Regio	32.6	1100–1400	125[a]	1.2
Corona dominated				
Themis Regio	?	1650–2300	100	1.5
C. Eistla Regio	?	1000–1400	120[b]	1.0
E. Eistla Regio	?	1600–1800	65	1.0

[a] Smrekar (1994).
[b] Grimm and Phillips (1992).

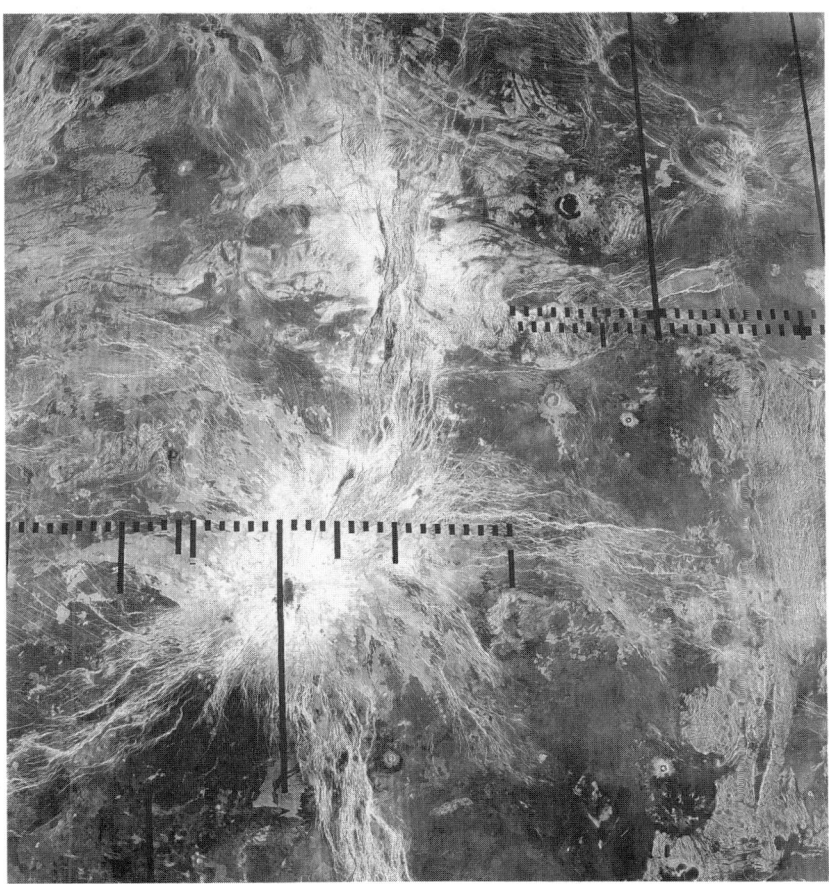

Figure 2. Magellan radar image of Beta Regio. This and other images are in sinusoidal equal area projection, with north at the top, and are mosaics of right-looking radar observations. Black strips are gaps in the data at the time when the mosaic was made. Bright areas are either rough on the cm scale or are topographic surfaces facing the radar antenna. Most bright areas are either rough lava flows or tectonically fractured regions. Dark areas are smooth surfaces, such as smooth lava flows. The image is approximately 2400 km across. The bright region just to the southwest of the center is Theia Mons. The bright lineations extending to the north and south of Theia Mons are Devana Chasma, a major rift. Another rift, Hecate Chasma begins at Theia Mons and continues to the southwest. Additional, unnamed fractures run east–west. A corona is seen in northeast corner. Most of the bright-to-gray, finely lineated regions are areas of pre-existing complex ridged terrain. Eight impact craters are also visible.

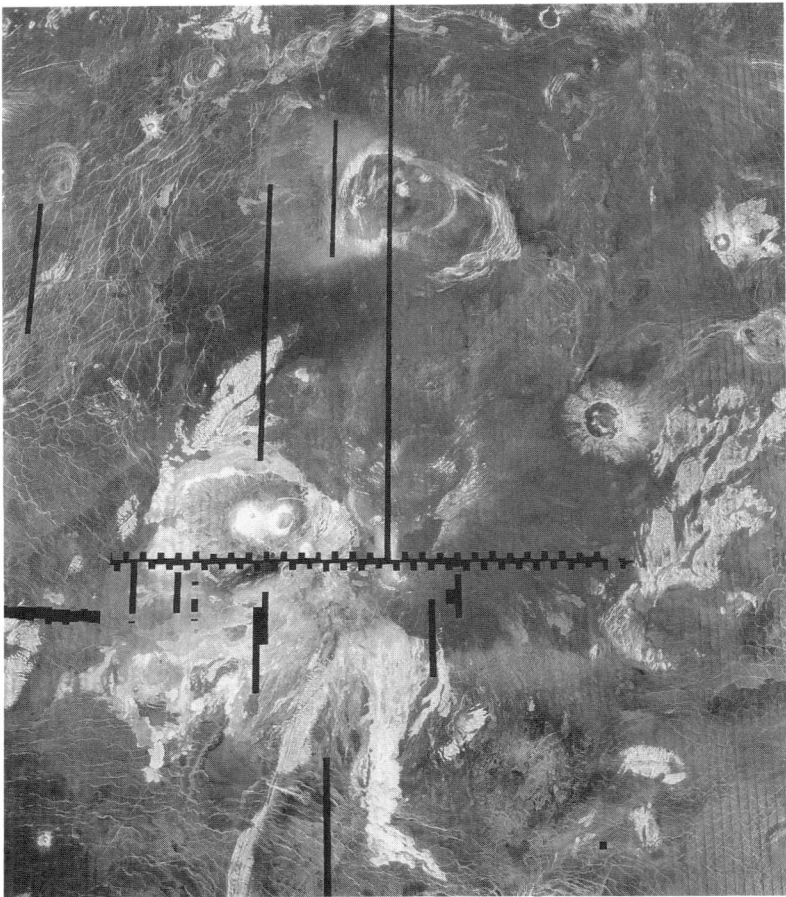

Figure 3. Magellan radar image (see Fig. 2 for discussion) of Bell Regio. The image is approximately 1800 km across. The bright region to the right of center is Tepev Mons, a 5 km high shield volcano, with associated flows. The dark region to the west is another volcanic center. The bright annulus near the top of the image is Nefertiti Corona.

coeval (Senske et al. 1992), although some flows from Gula overlie Sif flow units. Two coronae on the northern flank of western Eistla have formed after Gula Mons.

Central Eistla Regio has several large volcanoes and coronae. McGill (1994) interpreted the volcanoes to predate corona formation at central Eistla Regio. Evidence for uplift and volcanic construction has also been identified in central Eistla Regio (McGill 1994). No major rifts are present within this highland. Plains surrounding central Eistla are interpreted to have formed generally after formation of the highland, and to have been emplaced in a relatively short time period (McGill 1994). Grimm and Phillips (1992) analyzed the Pioneer Venus gravity data for Western and Central Eistla Regiones. They found that the tectonic fracture pattern was generally consistent with the stress pattern inferred from a two layer model of crustal and deep compensation used to fit the gravity and topography data. They interpreted Western Eistla to be underlain by an active plume and Central Eistla to be underlain by a waning plume.

Eastern Eistla Regio is a cluster of five coronae with diameters ranging from 300 to 600 km sitting on an irregular topographic rise. Despite their small size, local gravity highs are centered over four of the five coronae (Schubert et al. 1994). Each of the coronae are characterized by a raised rim, and are associated with large flow deposits. No major extensional deformation is seen at eastern Eistla Regio.

Imdr Regio has one major unnamed volcanic edifice, with an approximate volume of 4.8×10^4 km^3. Wrinkle ridge patterns indicate that the plains were uplifted a minimum of 200 m to form the topographic rise (Stofan et al. 1995). The rise is cut by a minor rift structure. Imdr has the least complex surface morphology of all Venusian volcanic rises, with no associated coronae and the smallest amount of associated volcanism.

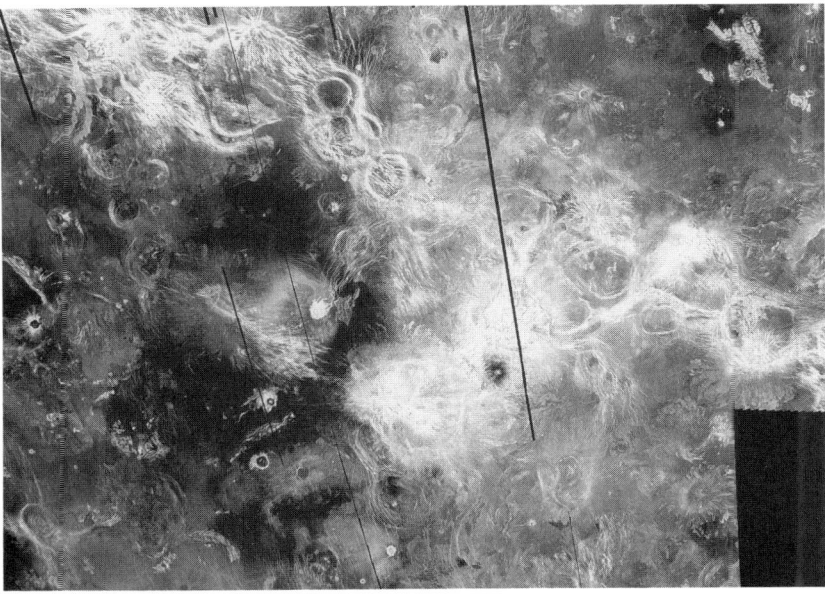

Figure 4. Magellan radar image (see Fig. 2 for discussion) of Themis Regio. The image is approximately 2800 km across. Several of the larger coronae are cut by radial fractures. Themis Regio is at the southeast end of Parga Chasma.

Themis Regio is dominated by five major coronae (Stofan et al. 1992). The coronae at Themis Regio are typically 200 to over 500 km across, and vary in spacing (Fig. 4). The Themis Regio coronae exhibit abundant volcanism, including extensive radiating flows and small to intermediate scale edifices. Themis Regio lies at the termination of Parga Chasma. Extensional deformation at Themis Regio is manifested by a graben lying along the axis of the highest topography within the largest corona. Minor extensional features also cut across the swell, continuing from Parga Chasma. Fractures and graben are much less common than along the rest of Parga Chasma, and are embayed by corona-related flows in places. Some fractures and graben bend around the coronae, indicating that corona formation occurred prior to some extensional deformation.

B. Classification of Hotspots

Atla and Beta Regiones are classified as rift-dominated. Each is cut by a major axial rift valley that is 50 to 100 km wide, up to 2 km deep (Solomon et al. 1992) and that extends for thousands of km, continuing far beyond the topographic swell (see Fig. 2). Some other rises are associated with minor rift structures or belts of grabens, but these zones of extension extend for hundreds rather than thousands of km and are less intensively deformed. Atla and Beta also contain large shield volcanoes, and have few associated coronae. Volcanism and extension overlapped in time at both Atla and Beta Regiones (see, e.g., Stofan et al. 1989; Senske et al. 1992). The rift-dominated rises are also the topographically highest of volcanic rises on Venus and have large apparent depths of compensation (see Table I).

Imdr, western Eistla, Bell, and Dione Regiones, classified as volcano-dominated, contain one or more large-scale (>300 km diameter) volcanic edifices. The edifices at each rise are about 1 to 2 km high, and are surrounded by extensive flows. At each volcano-dominated rise, only minor extension is present. An evolutionary sequence can be determined at several of the volcano-dominated rises, based primarily on interpretation of the gravity data (Smrekar 1994; Smrekar and Parmentier 1996). Imdr and western Eistla Regiones are likely to be an intermediate stage of evolution, with an active plume. Bell and Dione Regiones are likely to be in a late stage of evolution, based on their significantly smaller apparent depths of compensation (see below for more discussion of gravity signatures).

The corona-dominated rises are Themis, central Eistla, and eastern Eistla Regiones. The coronae at each rise are typically 200 to over 500 km across and exhibit abundant volcanism. Themis Regio is the only one of the three corona-dominated rises that contains any significant extensional deformation, probably because of its association with Parga Chasma. The volcano- and corona-dominated swells all lie 1.0 to 1.8 km above the surrounding plains, closer to the typical heights of terrestrial swells.

The three morphologic classes of hotspots were interpreted by Stofan et al. (1995) to be the result of variations in plume or lithospheric properties rather than different stages of evolution. Both of the rift-dominated rises occur at tectonic junctions along large-scale chasmata systems (Schaber 1982), where Stofan et al. (1995) interpret the large-scale extensional stress state, which is responsible for the regional chasmata system to control the rift-dominated morphology of each rise. Corona-dominated rises are interpreted to reflect breakup of a plume or secondary convection. Volcano-dominated rises appear to be the simplest manifestation of a mantle plume and are the most similar to terrestrial oceanic hotspots.

IV. GRAVITY STUDIES

Gravity data furnish clues about the structure of the lithosphere and mantle beneath hotspots on Venus. Density variations in the subsurface such as low density crustal roots or thermal anomalies support the load of the surface topography. One of the simplest approaches to interpreting gravity data is to calculate an apparent depth of compensation (ADC). The term apparent is used because compensation is assumed to be constrained to a single depth and density contrast. The ADC gives some first-order insights into whether compensation occurs in the crust at shallow depths, or is related to lithospheric and/or mantle processes, that are deeper. A comparable approach is to estimate the geoid-to-topography ratio (Smrekar and Phillips 1991; Simons et al. 1994). The spectral admittance technique is used to calculate the ratio of the gravity (or geoid) to topography as a function of wavelength (see, e.g., Dorman and Lewis 1970; Forsyth 1985). This approach affords more insight into which compensation mechanisms may be operating and is best used if the data resolution is adequate to decompose spectrally the observed gravity field.

The large ADCs at Beta and Atla Regiones were first recognized in the Pioneer Venus gravity data (Phillips and Malin 1983,1984) and interpreted to indicate the presence of mantle plumes. Later studies showed that Bell and western Eistla Regio also had large depths of compensation and were also likely hotspots (Smrekar and Phillips 1991; Grimm and Phillips 1992). Additionally, gravity studies indicated that, in contrast to Earth, the upper mantle could not contain a significant low viscosity region or asthenosphere (Phillips 1986,1990; Kiefer and Hager 1991; Smrekar and Phillips 1991).

Doppler tracking of the Magellan spacecraft provided a significant improvement in the resolution of the gravity field, which is now modeled to spherical harmonic degree 90, or wavelengths longer than 420 km, over much of the planet (chapter by Sjogren et al.). Because the effects of lithospheric flexure are most pronounced at short wavelengths, the resolution of the Magellan gravity field has allowed the gravity and topography data to be used as a constraint on the thickness of the elastic lithosphere, as discussed more fully in the next section. It is now also possible to estimate ADCs for all nine probable hotspots (see Table I). The ADCs range from 65 to 260 km.

It is worth noting that initial studies of the Magellan data may not have been based on the highest resolution data ultimately obtained for a region.

Additionally, errors in the data reduction software lead to minor errors in the gravity field which have been corrected in the 90 deg and order spherical harmonic field (chapter by Sjogren et al.). The general result of the improved resolution and corrections is to decrease estimates of elastic thickness somewhat, as shown below for western Eistla and Bell Regiones and in the chapter by Phillips et al. for Atla Regio. Estimates of the long wavelength ADCs are generally unaffected by the improvements in the new fields. The interpretation of gravity and topography data in the context of proposed hotspot models is discussed in the next section.

V. HOTSPOT MODELS

In this section we discuss the range of models that have been proposed to explain the present-day subsurface structure beneath Venusian hotspots by fitting model predictions to topography and gravity data. A wide range of models can fit the data. In the discussion section (Sec. VI), we will offer some suggestions for discriminating between models based on both the fit to the gravity and topography data and additional information.

The most common approach is to model a mantle upwelling using numerical or analytic methods (Kiefer and Hager 1991; Smrekar and Phillips 1991; Koch 1994; McKenzie 1994; Moresi and Parsons 1995; Smrekar and Parmentier 1996; Solomatov and Moresi 1996). Another approach is to assume that the topography is supported isostatically by thermal thinning of the lithosphere (Morgan and Phillips 1983; Kucinskas and Turcotte 1994; Moore and Schubert 1995). Alternatively one can assume that a density contrast occurs both at the crust-mantle interface and at a deeper interface such as the base of the lithosphere (Banerdt 1986; Williams and Gaddis 1991; Bills and Fischer 1992; Grimm and Phillips 1992; Herrick and Phillips 1992). A third approach is to include the flexure of the elastic lithosphere in the predictions of gravity and topography (Phillips 1994; Smrekar 1994; chapter by Phillips et al.).

A. Mantle Upwelling Models

One example of a numerical simulation of axisymmetric mantle upwelling is shown in Fig. 5. This figure compares a model with Rayleigh number of 10^7 with gravity and topography data across Beta Regio. The Rayleigh number, Ra, describes the strength of the convection, with higher values indicating more vigorous convection (see, e.g., the chapter by Schubert et al.). A more complete discussion of the model and parameters are given in Kiefer and Hager (1991), and the high Rayleigh number calculations shown in Figs. 5 and 11 are discussed in Kiefer (1994). The observed profiles are taken on an east–west profile at 29°N, with the profiles centered at 282°E, roughly midway between Theia and Rhea Montes, in order to minimize the influence of volcanic construction on the topography. The model geoid is an excellent fit to both the shape and amplitude of the observations. The topography model is a good match to the overall shape and amplitude of the observed topography. In detail, however, there are a number of disagreements between the model and observed topography. For example, the very large topographic depression near the center of the profile is Devana Chasma. The high-frequency variations along the observed topography profile most likely represent variations in crustal thickness. Based on gravity modeling in other regions, tessera units are thought to be regions of thickened crust (Smrekar and Phillips 1991; Grimm 1994) and hence elevated topography. Most of the region between −1400 and −2200 km on this profile (258°–267°E) consists of tessera (see Fig. 2), which probably explains why the observed topography is consistently above the model. Similarly, the narrow topographic peak near +1080 km (294°E) is associated with tessera.

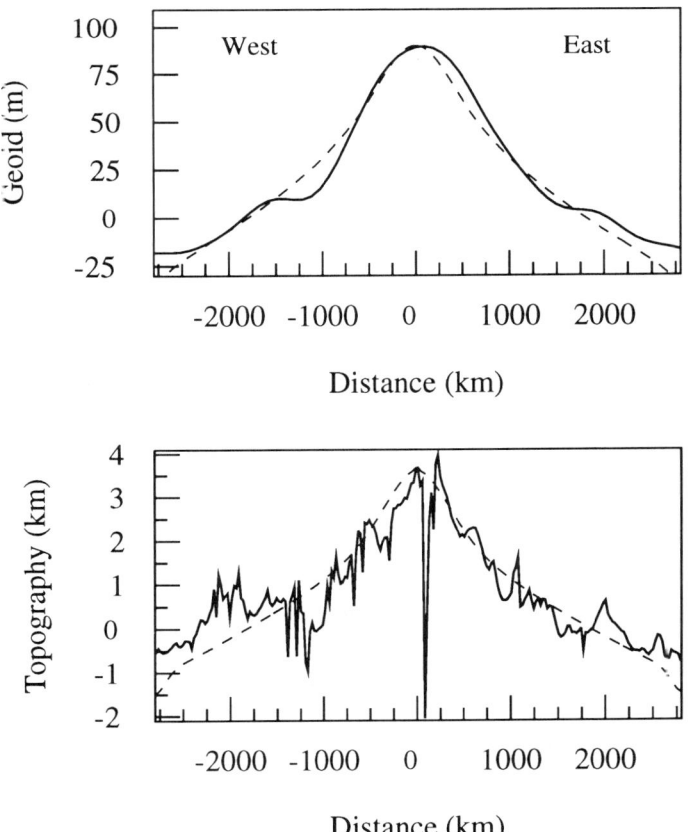

Figure 5. Comparison between Magellan observations for Beta Regio (solid lines) and the Rayleigh number of 10^7 plume model (dashed lines). The observed values are taken on an east–west profile along 29°N, centered at 282°E. (a) Geoid anomalies. (b) Topography.

The overall fit to the broad-scale geoid and topography indicates that much of the topography of Beta Regio might be supported dynamically. The model topography in Fig. 5b has a peak amplitude of 3.7 km. In contrast, Stofan et al. (1995) estimated that the swell topography without the effects of volcanic constructs at Beta Regio has a maximum amplitude of only 2.1 km. Their estimate was based on topographic profiles that were several hundred km north and south of the profile shown in Fig. 5. Both estimates have tried to minimize the influence of volcanism, but have arrived at very different estimates of swell topography. If only 2.1 km of topography is plume related at Beta Regio, then the plume-related geoid is probably no more than 50 m. An additional 40 m could come from other sources, such as flexurally compensated volcanoes (Smrekar and Parmentier 1996) or low density residuum from pressure-release melting (Moore and Schubert 1995; Smrekar and Parmentier 1996).

Although the model presented above provides a good fit to the data, it is not unique. A number of parameters influence the geoid and topography produced by mantle plumes. Two important parameters are the variation of viscosity with depth and the Rayleigh number. An equally good fit of the Pioneer Venus Orbiter observations of Beta Regio were done using $Ra = 10^6$ models by Kiefer and Hager (1991). Increasing the Rayleigh number decreases the thickness of the boundary layers and the widths of the upwellings and downwellings, which has the effect of reducing both the geoid and the dynamic topography produced by the convection (Kiefer and Hager 1992). With suitable adjustments in model parameters, one can produce several distinctly different models that nevertheless have nearly identical geoid and dynamic topography (Kiefer 1994).

Other convection models applied to Venus differ in several respects, particularly in the choices of convective layer thickness and the temperature contrast between the plume and the average mantle. In the model discussed above, the convective layer thickness is 2800 km, which is assumed to be the mantle thickness on Venus by analogy to the thickness of Earth's mantle. The temperature contrast between the plume and the average mantle is about 300 K, which is consistent with a variety of geophysical and petrological observations of terrestrial hotspots (Sleep 1990).

Solomatov and Moresi (1996) use a temperature-dependent viscosity model of mantle upwelling with a Rayleigh number and a temperature contrast that are similar to those described above, a convecting layer thickness of 1600 km, and a Cartesian geometry. Their model also fits observations (Fig. 6) across Beta Regio, but with a near-surface high viscosity layer about 400 km thick. They propose that Venus is currently experiencing stagnant lid convection in which there are no mobile plates and very little surface deformation. For other hotspots with smaller gravity and topography amplitudes, their model predicts a convective layer thickness of 600 to 900 km, a stagnant lid thickness of 200 to 400 km, and a viscosity of 10^{20} to 10^{21} Pa s. McKenzie (1994) also examined hotspot regions on Venus and argued for a convecting layer of only 1000 km, although he did not provide a specific comparison between model and observations. As emphasized by both Solomatov and Moresi (1996) and McKenzie (1994), the physical meaning of such small convective layer depths is unclear.

Koch (1994) used a boundary integral method to describe a rising viscous drop beneath a free-fluid surface as a means of modeling plume evolution. The predicted topography evolved from a dome to a plateau and finally decays while the GTR starts out very large and gradually approaches zero. Based on the GTRs for Venusian hotspots, Koch inferred that many of the hotspots are in an early stage of evolution and highland plateaus are in a late stage of evolution, in agreement with a hypothesis put forward by Phillips et al. (1991). No lithosphere is explicitly included in Koch's models.

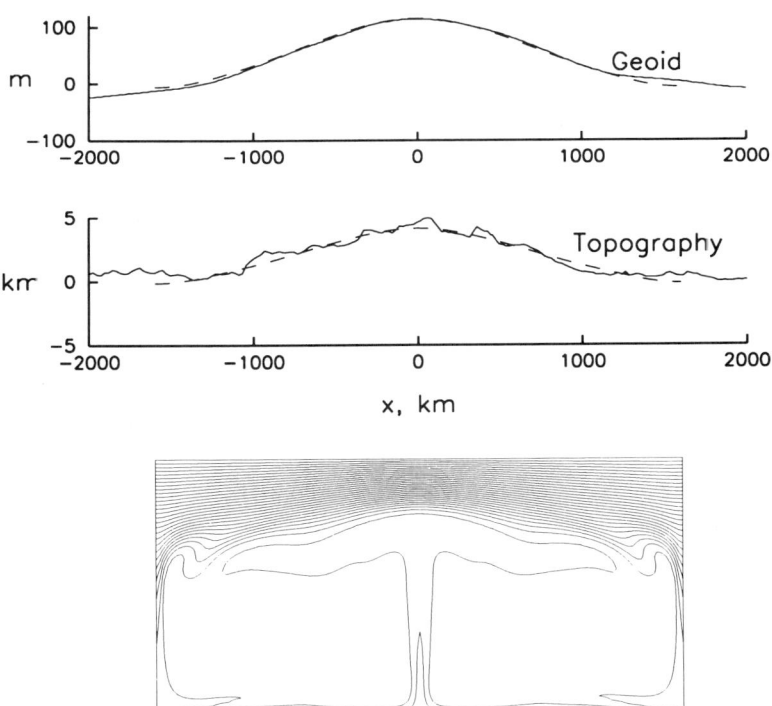

Figure 6. Comparison of the stagnant lid model of Solomatov and Moresi (1996) to profiles of geoid and topography at Beta Regio. The profiles are taken from $-86.62°E$, $11.77°N$ to $-53.39°E$, $37.13°N$. Temperature contours at $50°C$ intervals are shown for the convection model at the bottom. The model is for convection in a square box with a height of 1600 km, a Rayleigh number of 3×10^7 and a viscosity contrast of 10^6.

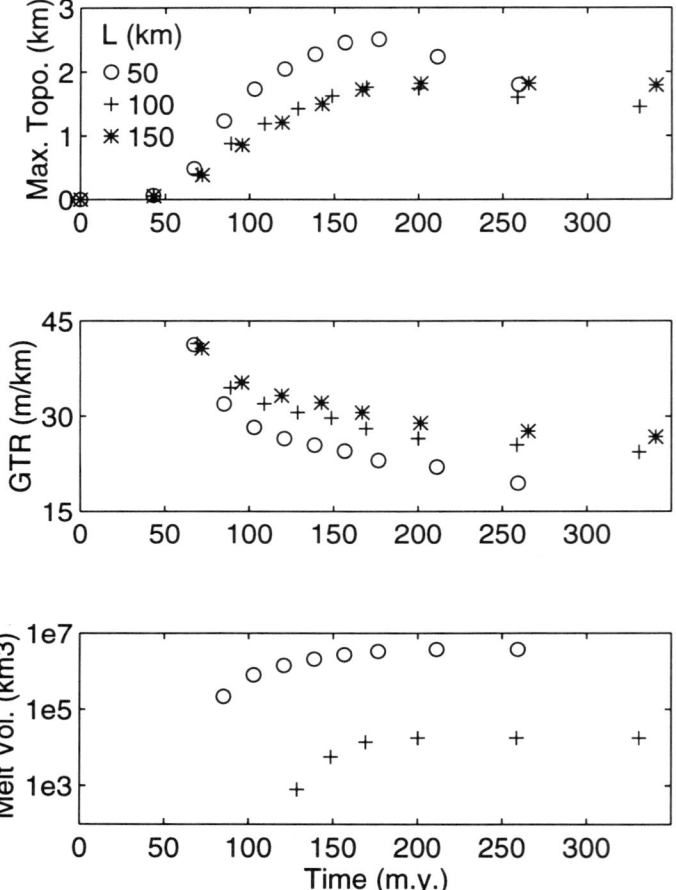

Figure 7. Topography, geoid-to-topography ratio, and the cumulative volume of pressure release melting predicted by an axisymmetric model of mantle upwelling with temperature-dependent viscosity in which the plume reaches the lithosphere after ~70 Myr, and the plume begins to die out at ~150 Myr (see Smrekar and Parmentier [1996] for details of model). Symbols show cases with thermal lithospheric thicknesses of 50, 100 and 150 km. The geoid-to-topography ratio range shown approximately corresponds to an ADC range of 100 to 300 km.

Smrekar and Parmentier (1996) created a suite of numerical convection models with temperature-dependent viscosity and pressure-release melting intended to fit the overall range of estimated volcanic volumes, and gravity and topography signatures found for Venusian hotspots. Earth-like values of lithospheric thickness (~100–150 km) and plume-mantle temperature differences (~200 K) can fit the observed range of gravity, topography, and estimated volcanic volumes (see Fig. 7) for Venusian hotspots. Like Koch (1994) this study emphasized the time evolution of the plume. Smrekar and Parmentier's study differed in that it did not include the development of plateaus, predicted much longer time scales of evolution, and resulted in a different interpretation of the evolutionary stage.

Smrekar and Parmentier (1996) showed that the predicted range of gravity and topography over the lifetime of the plume can account for much of the variability of the data for the nine likely hotspots on Venus (Fig. 7). In the earliest stage of evolution, the ADC is very large, the swell topography is minimal, and there is no volcanism. This stage is relatively short-lived, lasting on the order of 10 Myr. No hotspots were found to be in this stage on Venus (Stofan et al. 1995; Smrekar and Parmentier 1996). Intermediate and late stage hotspots are somewhat difficult to distinguish, as volcanism that occurs in the intermediate stage will remain at the surface, and the topography and ADC decrease slowly from the intermediate to the late stage. All nine of the possible hotspots on Venus are interpreted to be in an intermediate to late stage of evolution (Stofan et al. 1995; Smrekar and Parmentier 1996).

B. Isostatic Compensation Models

Morgan and Phillips (1983) originally proposed that much of the topography on Venus could be explained by variations in the thickness of the thermal lithosphere. Rosenblatt et al. (1994) also confirmed this result using Magellan data. Both studies found that topography at higher elevations, including some of the hotspots, do not fit the model of thermal thinning of the lithosphere, based on an assumed reference lithospheric thickness of approximately 100 km. Using Magellan gravity data for Beta Regio, Moore and Schubert (1995) fit a thermal compensation model with a reference thermal boundary layer thickness of 270 km, thinning to 100 km, and a temperature contrast of 800 to 1000 K between the hot mantle and the lithosphere. As pointed out by Moore and Schubert (1995), such a large temperature contrast at depths of 100 to 150 km would produce more melting than appears to be consistent with surface volcanism, leading them to suggest that other mechanisms must also be operating. Kucinskas and Turcotte (1994) also used a thermal thinning model to fit the data for Atla and Beta Regiones. They estimated a thermal lithospheric thickness of 350 km, with thinning of the lithosphere to 113 km beneath Beta and 88 km beneath Atla.

The large ADCs for hotspots indicate that they can not be crustally compensated. However, crustal compensation could partially support the topography along with a thermal anomaly at depth. Grimm and Phillips (1992) used a model with a crustal thickness of 20 km and an average ADC of 230 km, presumed to be the approximate location of a mantle upwelling, to fit the gravity and topography for Eistla Regio. Using a spatial fit to the data they found an ADC of 210 km for western Eistla Regio, and a value of 120 km for central Eistla Regio. Bills and Fischer (1992) also used a two layer mass model with density variations at 50 and 500 km. Both of the above models assumed Stokes flow on the lower boundary. The results of Bills and Fischer

(1992) supported the interpretation of Atla and Beta Regiones as upwellings.

Another model of hotspot compensation is one in which a low-density residuum root, produced by pressure-release melting as the plume head rises, supports the topographic swell after the thermal anomaly decays. The density contrast in the residuum root may be as high as 50 kg m^{-3}, equivalent to 20% melting for a basaltic crust (see Smrekar and Parmentier 1996). Using an isostatic model of compensation and assuming a mantle density of 3300 kg m^{-3}, the thickness of a residuum root required to support 1 km of topography is 66 km. The volume of residuum that is required to support the entire swell is probably inconsistent with the estimated volume of volcanics. Phipps Morgan et al. (1995) estimate that 25% of the compensation at terrestrial hotspots comes from a residuum layer. This estimate is probably applicable to Venus as well because of the similarity in volcanic and swell volume estimates (Stofan et al. 1995). In contrast, Phillips et al. (1990) estimated that 8 km of relief in the Tharsis volcanic region of Mars could be supported largely by a residuum root. Their approach differs in that they try to match the observed topography rather than constrain the amount of pressure-release melting with a combination of upwelling models and estimated volume of volcanics, as was done in Smrekar and Parmentier (1996). They thus allow a larger volume of material to undergo much larger degrees of partial melting, as well as permit a greater fraction of intrusive volcanism.

C. Flexural Models and Admittance Studies

Studies of admittance spectra are the most useful approach to distinguishing between various compensation models, and are particularly useful for determining flexural contributions to topography. Because the flexural signature occurs at short wavelengths (~1000 km or less), it is important to use the highest resolution data available. Comparisons between admittance spectra calculated from the 60 vs 90 deg and order gravity fields along with new model fits are shown for Bell and western Eistla Regiones in Fig. 8. A similar comparison for Atla Regio is found in the chapter by Phillips et al. Data are shown from three different sources. The asterisk shows data from a local inversion of the line-of-sight data (see Smrekar [1994] for details), the circles show the data for the 60 deg and order spherical harmonic field (Konopliv and Sjogren 1994), and the pluses show data for the 90 deg and order spherical harmonic field (chapter by Sjogren et al.). There is general agreement between the three gravity fields. The 90 deg and order field should be considered the best representation of the gravity field, because prior data sets included small data reduction errors. Further, it has the smallest error estimates (see Smrekar [1994] for error formula) and produces the smoothest admittance curve.

In addition, the accuracy of the model fit also depends on estimating the wavelength at which the data resolution decays. One estimate of the appropriate wavelength cut off comes from the degree strength estimate of the spherical harmonic gravity field (chapter by Sjogren et al.). For Bell and western Eistla Regiones, the cut offs are approximately 475 and 500 km,

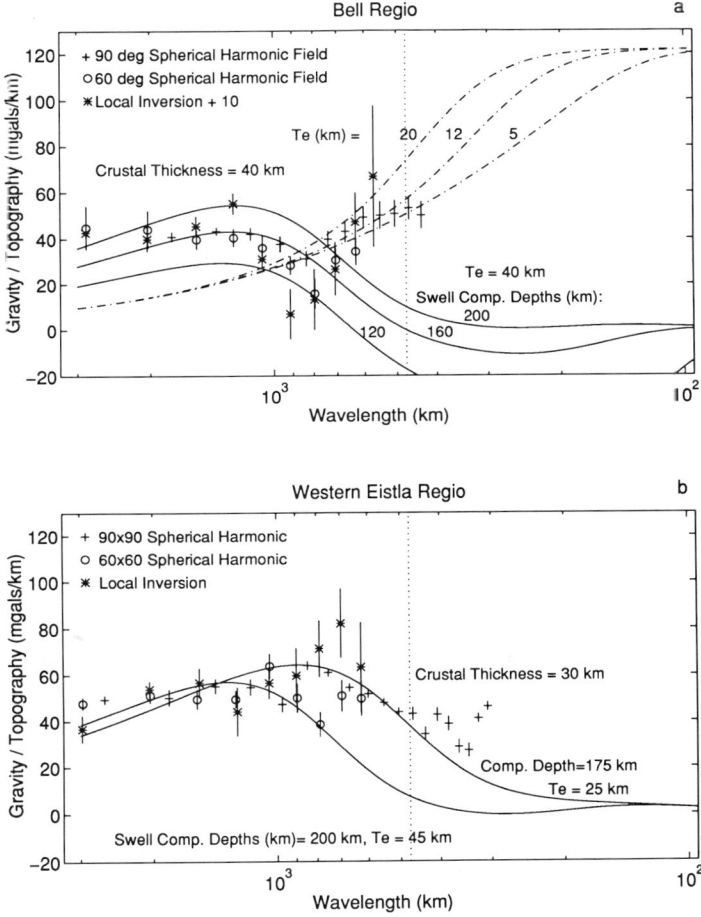

Figure 8. Admittance spectra for (a) Bell Regio and (b) western Eistla Regio. Spectra are shown using three different gravity fields. Solid model curves include a crustal thickness and an elastic lithosphere loaded from below by a deep density anomaly at depth. The three solid curves indicate different depths for the deep density anomaly. The dashed-dot curves include a crust and top loading of the elastic plate, assuming three different crustal thicknesses. The vertical solid line is at the degree strength cut off wavelength. See the text for more details and Smrekar (1994) for a discussion of the effect of various model parameters.

respectively. These estimates are consistent with the point at which the admittance spectra begin to flatten out (Fig. 8), suggesting that there is no power in the gravity spectra.

The modeling approach taken to fitting the admittance spectra illustrated in Fig. 8 is to make the explicit assumption that top loading will be important at short wavelengths and bottom loading will dominate at long wavelengths

(McNutt and Shure 1986). This type of model fits the admittance data at Bell Regio and Atla Regio very well (Smrekar 1994). Purely isostatic models do not provide a good fit to the data at any of the hotspots to which spectral admittance techniques have been applied, including Bell, Beta, Atla, or western Eistla Regiones (Smrekar 1994). The admittance spectrum for Bell Regio is shown with two sets of model curves (Fig. 8a). The solid lines are predicted by bottom loading of the elastic lithosphere due to a deep density contrast, assumed to be the thermal anomaly in the plume head. The dashed lines are models of loading of the elastic plate from above, presumably by a volcano, and crustal compensation. The elastic thickness estimate at short wavelengths is 15 ± 15 km, and 40 ± 15 km at long wavelength. These values are lower than the elastic thickness estimates of 30 ± 5 km at short wavelengths and 50 ± 5 km found using earlier versions of the Magellan gravity data (Smrekar 1994). Modeling of the 90 deg and order field also gives a larger estimate of the ADC than that found using the 60 deg and order field. However, this difference of 160 km vs 125 km is a result of using a different crustal thicknesses, as required to fit the slope of the data at short wavelengths and does not represent a significant change in the data at long wavelength. Because the uncertainty in the crustal thickness suggests an error of ±25 km in the estimate of the ADC (Smrekar 1994), these two estimate are within the error bars.

One interpretation of the difference in the elastic thickness as a function of wavelength is that the elastic lithosphere is locally thinned under the volcanoes. Sheehan and McNutt (1989) interpreted the admittance data for the Bermuda Rise, which resembles the admittance spectrum for Bell Regio, in this way. An alternative interpretation is that the short-wavelength value may be a "remnant thickness" that represents the value of the elastic thickness when the volcanoes were formed (see Smrekar [1994] for more discussion).

Western Eistla Regio has only the signature of loading from below (Fig. 8b). The absence of a top loading signature is surprising given the presence of the volcanic edifices, Sif and Gula Montes. However, the estimated volumes for Sif and Gula Montes are substantially smaller than for Tepev Mons at Bell Regio, and an order of magnitude less than the volcanoes at Atla Regio (see Table I). There is again general agreement between admittance curves produced using different gravity fields. If one tries to fit a single model curve to the admittance data calculated with the 90 deg and order field, a model curve with an elastic thickness of 25 km produces the best overall fit. A model curve for an elastic thickness of 45 km is also shown, as it provides a better fit to one or two of the long wavelength admittance points. However, the case for different elastic thicknesses at different wavelengths is not strong.

In the absence of additional constraints, there are several caveats to consider when interpreting both apparent depths of compensation and estimates of elastic thickness. If the elastic thickness varies over the study region, the estimate will be biased towards those regions with the most power in the gravity and topography, essentially the regions with the highest topography (Forsyth 1985). Additionally, in going from the 60 deg and order spherical harmonic gravity field to the 90 deg and order field, the estimates of elastic thickness decreased (see also the chapter by Phillips et al.). Apparently the lower resolution of the earlier field lead to an overestimate of the elastic thickness.

Convective processes also introduce ambiguity into interpretations of the elastic thickness. Recent numerical simulations using depth-dependent or temperature-dependent rheology have shown that mantle upwellings have a significant structure at short wavelengths (Smrekar 1994b; Moresi and Parsons 1995). An example shown is in Fig. 9, which shows results from the $Ra = 10^7$ plume previously illustrated in Fig. 5. The dashed line is the complete free-air gravity anomaly at all wavelengths from 5600 km (the diameter of the model domain) to 420 km. The solid line is the free-air gravity filtered to include only wavelengths between 1100 km and 420 km, corresponding to spherical harmonics degrees between 35 and 90. The free-air gravity anomaly in this waveband has an amplitude of 47 mgal at the center of the upwelling, which indicates that high Rayleigh number mantle plumes can be an important contributor to short-wavelength gravity anomalies. The topography for this model is shown in Fig. 8b; the filtered topography has an amplitude of 0.52 km at the plume center. For comparison, a model with $Ra = 10^6$ has a short-wavelength gravity anomaly of only 19 mgal. The diameter of the plume decreases with increasing Rayleigh number, which implies an increasing amount of short-wavelength power with increasing Ra. This explains the large difference in the short-wavelength gravity anomalies for the two models. Moresi and Parsons (1995) have shown that temperature-dependent rheology increases the short-wavelength admittance produced by convection. Temperature-dependent rheology also increases the amplitude of the short-wavelength convective gravity anomaly (Kiefer 1995).

As an illustration of how the short wavelength convective signature might be mistakenly interpreted as a flexural signature, the short-wavelength admittances for the $Ra = 10^7$ model are shown in Fig. 9c. Two different elastic flexure model fits to this admittance spectrum are also shown. These flexure models were calculated by R. Phillips (personal communication, 1995) for a four-parameter top and bottom loading model (Forsyth 1985; Phillips 1994). The results of a Monte Carlo inversion for the best fitting model parameters are illustrated with the short-dashed line. This model has a crustal thickness of 192 km, which is implausibly large. Accordingly, the long-dashed model shows the result of a parameter inversion when the crustal thickness is constrained to be 50 km; the results are nearly as good as for the thicker crust. The remaining three parameters in the inversion are the ratio of bottom-to-top loading, the depth of bottom loading, and the thickness of the elastic lithosphere. These three parameters are nearly the same in both model fits. The ratio of bottom loading to top loading is infinite, as expected for a mantle plume and no surface load. The bottom loading depth is 227 km for the flexure fit, and should be interpreted as an effective average loading depth. The elastic lithosphere in this model fit has a thickness of 88 km, despite the fact that the convection model used to calculate the admittance actually has no elastic layer in it.

The uncertainties in interpreting the elastic thickness estimates suggest that larger estimates of the uncertainty are warranted than those required by the fit to the data alone. An uncertainty of ± 10 km in the elastic thickness estimates is perhaps an appropriate value. We suggest this value based on comparisons to terrestrial studies where the interpretation of admittance curves are subject to most of the same uncertainties. However, additional constraints on the elastic thickness estimates, such as the age of the plate, seismic data, and heat flow data indicate that the estimates obtained from admittance studies are reasonable. The similarity between admittance curves for terrestrial hotspots and Bell and Atla Regiones, the Venusian hotspots with the best behaved admittance spectra, suggests both that the estimates of elastic thickness are well enough constrained to be of value and that similar processes are operating.

As shown in this section, the relationship between gravity and topography does not allow a unique determination of the viscosity structure of the lithosphere or the position of the plume density anomaly. For hotspots, one problem is that none of the models include all of the processes believed to be operating at hotspots due to the difficulty of modeling elastic and viscous processes simultaneously. The effect of applying a model that does not include all of the processes acting at a hotspot is to bias the estimate of the ADC and the associated estimate of thermal lithospheric thickness. The elastic thickness values can be used to estimate the thickness of the thermal lithosphere, but still include the uncertainty of the amount of heat flow generated by the plume relative to the background heat flux (chapter by Phillips et al.). In the next section, we discuss additional constraints on lithospheric thickness.

VI. DISCUSSION

A. Are Venusian Hotspots Active?

As there is no plate motion to constrain the duration of Venusian hotspots, the only information on hotspot age is indirect. Although large volcanic rises as a class cannot be dated with any accuracy, the crater density for large volcanoes and coronae together is less than the average density of the planet (Namiki and Solomon 1994). Because large volcanoes and coronae form concurrently with most large volcanic rises, the relative youth of volcanoes and coronae overall is consistent with hotspots having an age younger than the average age of the planet. Additionally, some of the large volcanic rises appear to have a younger than average age based on analyses of extended ejected deposits (Phillips and Izenberg 1995).

Gravity and topography data also provide some clues about hotspot activity. Comparison of ADCs (or GTRs) of hotspots to models of hotspot evolution indicate that all Venusian hotspots are in an intermediate to late stage of evolution (Smrekar and Parmentier 1996). The bottom-loading signature observed in the gravity and topography for Atla, Bell, and W. Eistla

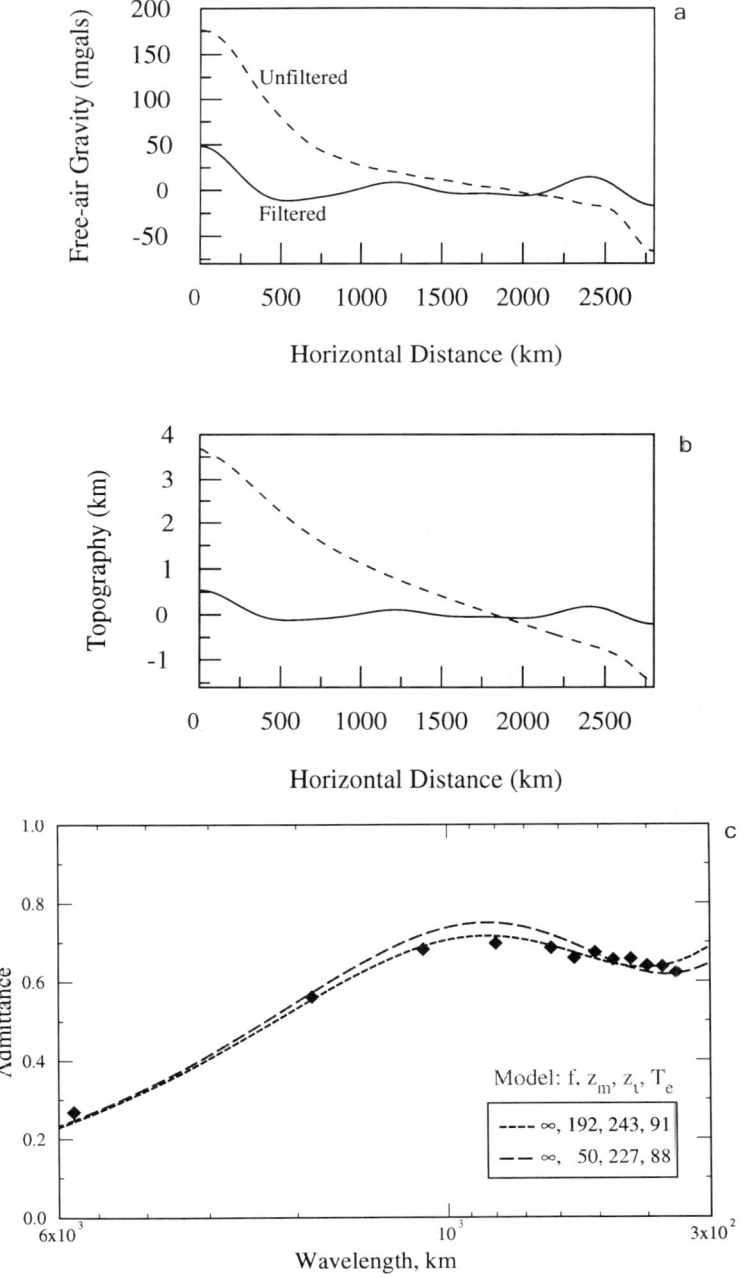

Figure 9. (a) The free-air gravity anomaly for the Ra = 10^7 mantle plume model. The dashed line is the unfiltered gravity anomaly and the solid line is the gravity anomaly filtered to include only wavelengths between 420 and 1100 km. (b) The topography for this plume model, using the same line conventions as in panel a. (c) The non-dimensional admittance spectrum for this plume model. The two dashed lines are elastic flexure model fits. The model parameters are shown in the box using the nomenclature of Phillips (1994). See text for details.

Regiones suggests the presence of a thermal anomaly at depth (Smrekar 1994). Smrekar and Parmentier (1996) estimate that the thermal anomalies under Venusian hotspots require at least 100 to 200 Myr to cool, which gives a lower bound on the age of hotspots displaying a bottom loading signature. An alternate interpretation is that the bottom-loading signature may be caused by a low-density mantle residuum root. However, the observed volcanism at Venusian hotspots is inconsistent with at least the larger volcanic rises being entirely compensated by residuum. This is based on the assumption that the extrusive-to-intrusive ratio is no less than 1:10, by analogy to terrestrial hotspots (see Smrekar and Parmentier [1996] for more discussion). Dione Regio could be an exception, in that it has a very large volume of volcanics with little swell topography. The evidence at this point suggests that at least a few hotspots are active.

If we make the assumption that all nine Venusian hotspots are active, we can estimate an upper bound on their buoyancy flux. The duration of the plume is necessary to estimate accurately plume buoyancy, which is, of course, unknown. The average crater retention age, 300 to 500 Myr, is probably a reasonable upper bound. For comparison, the oldest hotspot that is still active on Earth is probably Crozet, with an age of 200 Myr. Smrekar and Parmentier (1996) showed that the buoyancy flux for a typical individual Venusian hotspot is comparable to that of a typical terrestrial hotspot, assuming an active plume for 150 Myr. Longer-lived plumes would have lower buoyancy fluxes. Venus has at most nine active hotspots, in comparison to the approximately 40 hotspots on Earth (Davies 1988; Sleep 1990). Even if highland plateaus and large coronae are counted, there would still be only about half the number identified on Earth. Thus the total heat flux coming from all Venusian hotspots is likely to be a fraction of that coming from terrestrial hotspots. The total flux from terrestrial hotspots is estimated to be approximately 10% of Earth's total heat flux (Davies 1988).

B. Implications for Lithospheric Thickness

The thickness of the lithosphere beneath hotspots cannot be constrained solely by fitting convection or thermal models to the gravity and topography data. Models that predict a thick lithosphere ($\sim$300 km) include thermal thinning of the lithosphere and stagnant lid convection. However, mantle upwelling models with Earth-like parameters, including a thin lithosphere ($\sim$100 km), also fit the data. In this section we will argue that additional constraints and considerations are most consistent with a thin lithosphere.

One such constraint is the presence of large volcanic edifices and extension found at most of the large volcanic rises. Beta Regio, the hotspot used to constrain the majority of models discussed in the previous section, has very large edifice volumes (see Table I) and prominent rifts. As discussed by Solomatov and Moresi (1996), the stagnant lid model predicts a transition to a regime with no pressure-release melting and little if any surface deformation. For the thermal thinning model, the opposite problem applies, in that the extreme thinning of the lithosphere and very large temperature contrasts predict too large a volume of melt (Moore and Schubert 1995). The similarity in the volume of volcanics on Earth and Venus is consistent with similar lithospheric thicknesses (Smrekar and Parmentier 1996). The mantle and/or plume temperature can be increased to allow for a lithosphere thicker than that of the Earth, but one must then explain very high mantle temperatures combined with a thick thermal lithosphere.

The estimates of elastic thickness for Venusian hotspots based on the 90 deg and order gravity field range from 15 to 40 km (see also the chapter by Phillips et al.), with 25 km being the best average. Phillips et al. discuss the range of elastic thickness estimates for different regions of Venus, the results of various convection models, and the implications for the thickness of the lithosphere. They conclude that the thickness of the lithosphere based on translating elastic thickness to mechanical thickness and thermal gradient and on estimates of heat flow from the interior is not well constrained and allows for thin to intermediate lithospheric thicknesses (perhaps 100–200 km).

Finally, both thick and thin lithosphere models have intrinsic problems. With the stagnant lid model (Solomatov and Moresi 1996), a convective layer 600 to 1600 km in thickness is required to fit the range of GTRs. The interpretation of such a shallow convecting layer thickness, not to mention the variations required to fit the observed range of hotspot parameters, is unclear. On the other hand, if the lithosphere is essentially Earth-like, than why is the resurfacing history so different and the level of volcanic and tectonic history so low? Available information to date does not permit a conclusive answer to this controversy (see the chapter by Phillips et al. for more discussion). However, for the reasons discussed above, we favor the thin lithosphere interpretation and continue to search for answers as to why Venus and Earth are so different.

C. Hotspots and Coronae

Coronae are believed to be caused by small-scale upwellings that are likely to originate at shallower depths than the upwellings causing mantle plumes (Janes et al. 1992; Stofan et al. 1992). Although coronae are typically an order of magnitude smaller in diameter than hotspots, there are a few large coronae that overlap in size with the smaller hotspots (see the chapter by Stofan et al. for a comprehensive reference list and discussion). The volcanic, topographic, and tectonic signatures of coronae are very different from hotspots. Despite the wide variation in the topography of coronae, the morphology is most consistent with an upwelling diapir origin. A survey of ten coronae using Magellan data (Schubert et al. 1994) show that their ADCs overlap with the full range found for hotspots (see the chapter by Stofan et al., their Fig. 17).

The primary difference between coronae and hotspots of similar size is likely to be the lithospheric structure. Models of coronae indicate that the elastic thickness must be fairly small (Janes et al. 1992; Cyr and Melosh 1993). The absence of large volcanic edifices suggests that the lithosphere may be too weak to support large loads. However, this simple interpretation is in conflict with estimates of elastic thickness at large coronae, which are in the range of 30 to 45 km (Sandwell and Schubert 1992; Smrekar and Yu 1996). As discussed by Stofan et al. (see their chapter), models of coronae formation are still incomplete. Outstanding questions are why coronae are found only on Venus, where the diapirs originate, and why some upwellings develop into hotspots while others develop into coronae.

D. Hotspots and Highland Plateaus

There is an ongoing debate about whether highland (or crustal) plateaus are the result of mantle upwellings or downwellings (Phillips and Hansen 1994). In the downwelling scenario, the plateaus form as the crust is thickened in response to traction associated with mantle downwelling (Bindschadler and Parmentier 1990; Bindschadler et al. 1990). In the upwelling model, the plateau is a crustal block formed by massive pressure-release melting above a mantle upwelling (Herrick and Phillips 1990; Phillips et al. 1991). This mechanism forms terrestrial flood basalts. The block is susceptible to subsequent deformation because the more differentiated crust is weaker than the surrounding plains (Grimm and Phillips 1991). In either case, the plateaus are likely to be remnants from an earlier, more active stage in the history of Venus. Stratigraphy shows that plateaus are locally the oldest features. The downwelling model requires a thin, weak lithosphere. In the upwelling model, extensive pressure-release melting is facilitated by a thin thermal lithosphere and hot mantle, which is more likely to be present earlier in the planet's history (Smrekar and Parmentier 1996).

Both models of plateau formation present problems. Recent re-evaluations of the downwelling model using a new, strong flow law for dry diabase to describe the behavior of the crust (Mackwell et al. 1995) suggest that the time scales for downwelling are prohibitively slow (Phillips and Kidder 1995; Kaula and Lenardic 1995). New evidence in favor of the upwelling model comes from detailed geologic analysis of highland plateaus that indicates an early extensional phase of deformation (Hansen and Willis 1996). Compression in the upwelling model has not been fully modeled. Possible explanations include external tectonic forces, possibly aided by a thinner lithosphere under the plateau (Zuber and Parmentier 1995), traction supplied by the plume (Basilevsky 1986; Grimm and Phillips 1991), or delamination resulting directly from the upwelling (Smrekar and Stofan 1996).

VII. SUMMARY

The global high resolution picture of Venus afforded by Magellan has shown a greater variability in the geology of hotspots than was previously known. Imdr, Dione, Themis, central Eistla, and eastern Eistla Regions have been interpreted to be hotspots on the basis of their topographic morphology, evidence for volcanism, and deep ADCs (Stofan et al. 1995), in addition to the four regions previously identified as likely hotspots, Atla, Bell, Beta, and western Eistla Regiones. The nine hotspots have been classified as rift-dominated (Atla, Beta Regiones), volcano-dominated (Bell, Dione, western Eistla, and Imdr Regiones), and corona-dominated (central Eistla, eastern Eistla, and Themis Regiones). Rift-dominated rises cut by major rifts that continue beyond the topographic rise and appear to be shaped by the influence of regional extension (Schaber 1982; Stofan et al. 1995). Coronae-dominated rises are clusters of coronae that occur on a broad topographic swell and may indicate break-up of a plume head. This variation in hotspot tectonic signature may reflect differences in the properties of the plume, lithosphere, or both.

Terrestrial and Venusian hotspots share many characteristics. For example, the ranges of swell diameter and height, and volcanic edifice volume are similar (Stofan et al. 1995). The volcano-dominated class of Venusian hotspots most closely resembles typical terrestrial hotspots. Rift-dominated hotspots have some terrestrial analogs, such as the East African Rise (McGill 1981) and possibly the Baikal Rift. Coronae-dominated rises are unique to Venus, as are coronae. Hotspots on the two planets both exhibit a wide variation in surface expression. Gravity signatures of hotspots on the two planets vary significantly due to the presence of a low viscosity zone on Earth.

The range of ADCs for the five additional hotspots overlaps with that of the four previously known hotspots, giving a total range of 65 to 260 km. Factors that are likely to contribute to the range of ADCs include variations in the thickness of a residuum layer, the thickness of the lithosphere, plume strength, the evolutionary stage, and the amount of flexural compensation of volcanoes. Some hotspots may be active or recently active, based on their bottom-loading signature (Smrekar 1994) and on the overall fit of ADCs to models of evolutionary sequence (Smrekar and Parmentier 1996). The resolution in the gravity data also permits the estimation of the elastic thickness of the lithosphere in some regions. The value at Atla Regio is 25 km (chapter by Phillips et al.). At Bell Regio, the elastic thickness is estimated to be 15 km at short wavelengths and 40 km at long wavelengths. At western Eistla Regio, the best estimate is 25 km. These values have errors of approximately ± 10 km, based on uncertainties in the admittance technique and on the contribution of convection to the short wavelength signature.

A wide range of models have been proposed to explain the gravity and topography of hotspots, with most of the models applied specifically to Atla or Beta Regiones. The central issue in these studies is the estimation of the thickness of the thermal lithosphere and the strength of any thermal anomaly beneath the hotspot. This question is directly tied to understanding the tectonic evolution of Venus and resurfacing processes. Models of axisymmetric upwellings with Earth-like parameters or extrapolations of the elastic thickness estimates yield Earth-like values of the thickness of the thermal lithosphere, or approximately 100 km (Kiefer and Hager 1991; Smrekar and Phillips 1991; McKenzie 1994; Moresi and Parsons 1995; chapter by Phillips et al.; Smrekar and Parmentier 1996). Models of stagnant lid convection (Solomatov and Moresi 1996) or thermal thinning of the lithosphere (Kucinskas and Turcotte 1994; Moore and Schubert 1995) predict that the thickness of the thermal lithosphere is on the order of 300 km.

Without additional constraints, any of the above compensation models provide a reasonable fit to the gravity and topography profiles across Beta and Atla Regiones. Models with a thin thermal lithosphere are more consistent with observed extension and the estimates of volcanic volumes. Models with a thick thermal lithosphere are inconsistent with these observations, but do offer the advantage of providing one explanation for the apparently low present-day levels of geologic activity on Venus. Here we favor the thin lithosphere interpretation because it consistent with all available data for hotspots.

Additional outstanding questions for Venusian hotspots are their relationship to coronae and to highland plateaus. Further insights on the thickness of the lithosphere, the interaction of mantle upwellings with the lithosphere, and the role of mantle upwellings in the resurfacing history of Venus are likely to come from solving these questions through improved modeling approaches, more detailed geologic studies, and possibly additional gravity studies. The collection of seismic, heat flow, and additional geochemical data could also greatly improve our understanding of why Venus has evolved along a different path than the Earth.

Acknowledgments. We thank R. Phillips for his review, which greatly improved the manuscript. This work was carried out in part at the Jet Propulsion Laboratory, California Institute of Technology, sponsored by the National Aeronautics and Space Administration. The authors gratefully acknowledge the support from four NASA grants.

REFERENCES

Arkani-Hamed, J. 1993. On the tectonics of Venus. *Phys. Earth Planet. Int.* 76:75–96.
Banerdt, W. B. 1986. Support of long-wavelength loads on Venus and implications for internal structure. *J. Geophys. Res.* 91:403–419.
Basilevsky, A. T. 1986. Structure of central and eastern areas of Ishtar Terra and some problems of venusian tectonics. *Geotect.* 20:282–288.
Basilevsky, A. T., and Head, J. W. 1988. The geology of Venus. *Ann. Rev. Earth Planet. Sci.* 16:295–317.
Basilevsky, A. T., and Janle, P. 1987. Geological-morphological and gravimetric characteristics of the Bell Region on Venus. *Astron. Vestnik* 21:109–121.
Bjarnason, I. T., Wolfe, C. J., Solomon, S. C., and Gudmunson, G. 1996. Initial results from the ICEMELT experiment: Body-wave delay times and shear-wave splitting across Iceland. *Geophys. Res. Lett.* 23:459–462.

Bills, B. G., and Fischer, M. A. 1992. A spatial domain Stokes flow model for the gravity of the middle latitudes of Venus. *J. Geophys. Res.* 97:18285–18294.

Bindschadler, D. L., and Parmentier, E. M. 1990. Mantle flow tectonics: The influence of a ductile lower crust and implications for the formation of topographic uplands on Venus. *J. Geophys. Res.* 95:21329–21344.

Bindschadler, D. L., Schubert, G., and Kaula, W. M. 1990. Mantle flow tectonics and the origin of Ishtar Terra. *Geophys. Res. Lett.* 16:1345–1348.

Bindschadler, D. L., Schubert, G., and Kaula, W. M. 1992. Coldspots and hotspots: Global tectonics and mantle dynamics of Venus. *J. Geophys. Res.* 97:13495–13532.

Campbell, B. A., and Rogers, P. G. 1994. Bell Region, Venus: Integration of remote sensing data and terrestrial analogs for geological analysis. *J. Geophys. Res.* 99:21153–21171.

Campbell, D. B., Head, J. W., Harmon, J. K., and Hine, A. A. 1984. Venus: volcanism and rift formation in Beta Regio. *Science* 226:167–170.

Cazenave, A. K., Dominh, K., Rabinowicz, M., and Ceuleneer, G. 1988. Geoid and depth anomalies over ocean swells and troughs: Evidence of an increasing trend of the geoid to depth ratio with age of the plate. *J. Geophys. Res.* 93:8064–8077.

Crough, S. T. 1983. Hotspot swells. *Ann. Rev. Earth Planet. Sci.* 11:165–193.

Cyr, K. E., and Melosh, H. J. 1993. Tectonic patterns and regional stresses near venusian coronae. *Icarus* 102:175–184.

Davies, G. F. 1988. Ocean bathymetry and mantle convection, 1: Large-scale flow and hotspots. *J. Geophys. Res.* 93:10467–10480.

Dorman, L. M., and Lewis, B. T. R. 1970. Experimental isostasy, I: Theory of the determination of the Earth's isostatic response to a concentrated load. *J. Geophys. Res.* 75:357–3365.

Ebinger, C. J., Bechtel, T. D., Forsyth, D. W., and Bowin, C. O. 1989. Effective elastic plate thickness beneath the East African and Afar plateaus and dynamic compensation of the uplifts. *J. Geophys. Res.* 94:2883–2901.

Forsyth, D. W. 1985. Subsurface Loading and estimates of the flexural rigidity of continental lithosphere. *J. Geophys. Res.* 90:12623–12632.

Grimm, R. E. 1994. Recent deformation rates on Venus. *J. Geophys. Res.* 99:23163–23171.

Grimm, R. E. 1994. The deep structure of Venusian plateau highlands. *Icarus* 112:89–103.

Grimm, R. E., and Phillips, R. J. 1991. Gravity anomalies, compensation mechanisms, and the geodynamics of western Ishtar Terra, Venus. *J. Geophys. Res.* 96:8305–8324.

Grimm, R. E., and Phillips, R. J. 1992. Anatomy of a Venusian hot spot: Geology, gravity, and mantle dynamics of Eistla Regio. *J. Geophys. Res.* 97:16035–16054.

Hansen, V. L., and Willis, J. J. 1996. Structural analysis and geodynamic implications of tessera terrain, Venus. *Lunar Planet. Sci. Conf.* XXVIII:489–490 (abstract).

Head, J. W., et al. 1992. Venus volcanism: Classification of volcanic features and structures, associations, and global distribution from Magellan data. *J. Geophys. Res.* 97:13153–131197.

Herrick, R. R., and Phillips, R. J. 1990. Blob tectonics: a prediction for western Aphrodite Terra, Venus. *Geophys. Res. Lett.* 17:2129–2132.

Herrick, R. R., Izenberg, N., and Phillips, R. J. 1995. Comment on "The global resurfacing of Venus" by R. G. Strom, G. G. Schaber, and D. D. Dawson. *J. Geophys. Res.* 100:23355–23359.

Janes, D. M., et al. 1992. Geophysical models for the formation and evolution of coronae on Venus. *J. Geophys. Res.* 97:2961–2964.

Janle, P., Jannsen, D., and Basilevsky, A. T. 1987. Morphologic and gravimetric investigations of Bell and Eistla Regiones on Venus. *Earth Moon Planets* 39:251–273.

Kaula, W. M., and Lenardic, A. 1995. Ishtar Terra. *Eos: Trans. AGU* 76:341.

Keddie, S. T., and Head, J. W. 1995. Formation and evolution of volcanic edifices on the Dione Regio rise, Venus. *J. Geophys. Res.* 100:11729–11754.

Kiefer, W. S. 1994. Mantle plumes on Venus: New high Rayleigh number models and applications to Magellan observations. *Lunar Planet. Sci. Conf.* XXV:699–700 (abstract).

Kiefer, W. S. 1995. Mantle plumes with temperature-dependent rheology and implications for the origin of volcanic rises on Venus. *Eos: Trans. AGU* 76:F342 (abstract).

Kiefer, W. S., and Hager, B. H. 1991. A mantle plume model for the equatorial highlands of Venus. *J. Geophys. Res.* 96:20947–20966.

Kiefer W. S., and Hager, B. H. 1992. Geoid anomalies and dynamic topography from convection in cylindrical geometry: applications to mantle plumes on Earth and Venus. *Geophys. J. Intl.* 108:198–214.

Koch, D. M. 1994. A spreading drop model for plumes on Venus. *J. Geophys. Res.* 99:2035–2052.

Konopliv, A. S., and Sjogren, W. L. 1994. Venus spherical harmonic gravity model to degree and order 60. *Icarus* 112:42–54.

Kucinskas, A. B., and Turcotte, D. L. 1994. Isostatic compensation of equatorial highlands on Venus. *Icarus* 112:104–116.

Mackwell, S. J., et al. 1995. Experimental deformation of dry Columbia diabase: implications for tectonics on Venus. In *Rock Mechanics: Proc. 35th U. S. Symposium*, eds. J. J. K. Daeman and R. A. Schultz (Brookfield, Vt.: A. A. Balkema), pp. 207–214.

Masursky, H., et al. 1980. Pioneer Venus radar results: Geology from images and altimetry. *J. Geophys. Res.* 85:8232–8260.

McGill, G. E. 1994. Hotspot evolution and Venusian tectonic style. *J. Geophys. Res.* 99:23149–23161.

McGill, G. E., Steenstrup, S. J., Barton, C., and Ford, P. G. 1981. Continental rifting and the origin of Beta Regio, Venus. *Geophys. Res. Lett.* 8:737–740.

McKenzie, D. 1994. The relationship between topography and gravity on Earth and Venus. *Icarus* 112:55–88.

McNutt, M. 1988. Thermal and mechanical properties of the Cape Verde Rise. *J. Geophys. Res.* 93:2784–2794.

McNutt, M. K., and Judge, A. V. 1990. The superswell and mantle dynamics beneath the South Pacific. *Science* 248:969–975.

McNutt, M., and Shure, L. 1986. Estimating the compensation depth of the Hawaiian swell with linear filters. *J. Geophys. Res.* 91:13915–13923.

Monnereau, M., and Cazenave, A. 1990. Depth and geoid anomalies over oceanic hotspot swells: A global survey. *J. Geophys. Res.* 95:15429–15438.

Moore, W. B., and Schubert, G. 1995. Lithospheric thickness and mantle/lithosphere density contrast beneath Beta Regio, Venus. *Geophys. Res. Lett.* 22:429–432.

Moresi, L., and Parsons, B. 1995. Interpreting gravity, geoid, and topography for convection with temperature dependent viscosity: Application to surface features on Venus, *J. Geophys. Res.* 100:21155–21171.

Morgan, P., and Phillips, R. J. 1983. Hot spot heat transfer: Its application to Venus and implications to Venus and Earth. *J. Geophys. Res.* 88:8305–8317.

Morgan, W. J. 1972. Plate motions and deep mantle convection. *Geol. Soc. Amer. Memoir* 132:7–22.

Namiki, N., and Solomon, S. C. 1994. Impact crater densities on volcanoes and

coronae on Venus: Implications for volcanic resurfacing. *Science* 265:929–933.
Nataf, H. C., and VanDecar, J. 1993. Seismological detection of a mantle plume? *Nature* 364:115–120.
Parmentier, E. M., and Hess, P. C. 1992. Chemical differentiation of a convecting planetary interior: Consequences for a one plate planet such as Venus. *Geophys. Res. Lett.* 19:2015–2018.
Phillips, R. J. 1986. A mechanism for tectonic deformation on Venus. *Geophys. Res. Lett.* 13:1141–1144.
Phillips, R. J. 1990. Convection-driven tectonics on Venus. *J. Geophys. Res.* 92:7403–7418.
Phillips, R. J. 1994. Estimating Lithospheric Properties at Atla Regio, Venus. *Icarus* 112:147–170.
Phillips, R. J., and Hansen, V. L. 1994. Tectonic and magmatic evolution of Venus. *Ann. Rev. Earth Planet. Sci.* 22:597–654.
Phillips, R. J., and Izenberg, N. R. 1995. Ejecta correlations with spatial crater density. *Geophys. Res. Lett.* 22:1517–1520.
Phillips, R. J., and Kidder, J. G. 1995. Subsolidus thickening of venusian crust. *Eos: Trans. AGU* 76:341 (abstract).
Phillips, R. J., and Malin, M. C. 1983. The interior of Venus and tectonic implications. In *Venus*, eds. D. M. Hunten, L. Colin, T. M. Donahue and V. I. Moroz (Tucson: Univ. of Arizona Press), pp. 159–214.
Phillips, R. J., and Malin, M. C. 1984. Tectonics of Venus. *Ann. Rev. Earth Planet. Sci.* 12:411–443.
Phillips, R. J., Sleep, N. H., and Banerdt, W. B. 1990. Permanent uplift in magmatic systems with application to the Tharsis Region of Mars. *J. Geophys. Res.* 95:5089–5100.
Phillips, R. J., Grimm, R. E., and Malin, M. C. 1991. Hot-spot evolution and the global tectonics of Venus. *Science* 252:651–658.
Phillips, R. J., et al. 1992. Impact crater distribution on Venus: Implications for planetary resurfacing. *J. Geophys. Res.* 97:15923–15948.
Phipps Morgan, J., Morgan, W. J., and Price, E. 1995. Hotspot melting generates both hotspot volcanism and a hotspot swell? *J. Geophys. Res.* 100:8045–8062.
Price, M., and Suppe, J. 1994. Mean age of rifting and volcanism on Venus deduced from impact crater densities. *Nature* 372:756–759.
Robinson, E. M., and Parsons, B. 1988. Effect of a shallow low-viscosity zone on the formation of midplate swells. *J. Geophys. Res.* 93:3144–3156.
Rosenblatt, P., Pinet, P. C., and Thouvenot, E. 1994. Comparative hypsometric analysis of Earth and Venus. *Geophys. Res. Lett.* 21:465–468.
Saunders, R. S., and Malin, M. C. 1977. Geologic interpretation of new observations of the surface of Venus. *Geophys. Res. Lett.* 4:547–550.
Schaber, G. G. 1982. Venus: Limited extension and volcanism along zones of lithospheric weakness. *Geophys. Res. Lett.* 9:499–502.
Schaber, G. G., et al. 1992. Geology and distribution of impact craters on Venus: What are they telling us? *J. Geophys. Res.* 97:13257–13420.
Schubert, G., Moore, W. B., and Sandwell, D. T. 1994. Gravity over coronae and chasmata on Venus. *Icarus* 112:130–146.
Senske, D. A., et al. 1991. Geology and structure of Beta Regio: Results from Arecibo radar imaging. *Geophys. Res. Lett.* 118:1159–1162.
Senske, D. A., Schaber, G. G., and Stofan, E. R. 1992. Regional topographic rises on Venus: Geology of western Eistla Regio and comparison to Beta Regio and Atla Regio. *J. Geophys. Res.* 7:13395–13420.
Sheehan, A. F., and McNutt, M. K. 1989. Constraints on thermal and mechanical structure of the oceanic lithosphere at the Bermuda Rise from geoid height and

depth anomalies. *Earth Planet. Sci. Lett.* 93:377–391.
Simons, M., Hager, B. H., and Solomon, S. C. 1994. Global variations in the geoid/topography admittance of Venus. *Science* 264:798–803.
Sleep, N. H. 1990. Hotspots and mantle plumes: Some phenomenology. *J. Geophys. Res.* 95:6715–6736.
Sleep, N. H. 1992. Hotspot volcanism and mantle plumes. *Ann. Rev. Earth Planet. Sci.* 20:19–43.
Smrekar, S. E. 1994. Evidence for active hotspots on Venus from analysis of Magellan gravity data. *Icarus* 112:2–26.
Smrekar, S. E., and Parmentier, E. M. 1996. The interaction of mantle plumes with surface thermal and chemical boundary layers: Applications to hotspots on Venus. *J. Geophys. Res.* 101:5397–5410.
Smrekar, S. E., and Phillips, R. J. 1991. Venusian highlands: Geoid to topography ratios and their implications. *Earth Planet. Sci. Lett.* 107:582–597.
Smrekar, S. E., and Stofan, E. R. 1996. Towards a comprehensive model of coronae formation on Venus. *Lunar Planet. Sci. Conf.* XXVII:1227–1228, (abstract).
Smrekar, S. E., and Yu, J. 1996. Admittance spectra and elastic thickness estimates for venusian coronae. *Lunar Planet. Sci. Conf.* XXVII:1229–1230 (abstract).
Solomatov, V. S., and Moresi, L. N. 1996. Stagnant lid convection on Venus. *J. Geophys. Res.* 101:4737–4754.
Solomon, S. C. 1993. The geophysics of Venus. *Phys. Today* 46:48–55.
Solomon, S. C., et al. 1992. Venus tectonics: An overview of Magellan observations. *J. Geophys. Res.* 97:13199–13256.
Stofan, E. R., et al. 1989. Geology of a rift zone on Venus: Beta Regio and Devana Chasma. *Geol. Soc. Amer. Bull.* 101:143–156.
Stofan, E. R., et al. 1992. Global distribution and characteristics of coronae and related features on Venus: Implications for origin and relation to mantle processes. *J. Geophys. Res.* 97:13347–13378.
Stofan, E. R., et al. 1995. Large topographic rises on Venus: implications for mantle upwelling. *J. Geophys. Res.* 100:23317–23327.
Turcotte, D. L. 1995. How does Venus lose heat? *J. Geophys. Res.* 100:16931–16940.
Watson, S., and McKenzie, D. 1991. Melt generation by plumes: A study of Hawaiian volcanism. *J. Petrol.* 32:501–537.
White, R. S. 1993. Melt production rates in mantle plumes. *Phil. Trans. Roy. Soc. London A* 342:137–153.
White, R., and McKenzie, D. P. 1989. Magmatism at rift zones: The generation of volcanic continental margins and flood basalts. *J. Geophys. Res.* 94:7685–7729.
Williams, D. R., and Gaddis, L. 1991. Stress analysis of Tellus Regio, Venus, based on gravity and topography: Comparisons with Venera 15/16 radar imags. *J. Geophys. Res.* 96:18841–18859.
Wolfe, C. J., McNutt, M. K., and Detrick, R. S. 1994. The Marquesas archipelagic apron: Seismic stratigraphy and implications for volcano growth, mass wasting and crustal underplating. *J. Geophys. Res.* 99:13591–13608.
Zuber, M. T., and Parmentier, E. M. 1995. Formation of fold and thrust belts on Venus by thick-skinned deformation. *Nature* 377:290–308.

ISHTAR TERRA

W. M. KAULA, A. LENARDIC and D. L. BINDSCHADLER
University of California at Los Angeles

and

J. ARKANI-HAMED
McGill University

Ishtar Terra has the highest topography on Venus and the third greatest geoid height. However, it is not as readily interpretable as the other two major peaks, Beta Regio and Atla Regio, because it has a great variety of features of conflicting indications, such as tessera, volcanic plains, and ridge systems, varying in character on length scales of a few 100 km. Most of the margins between geologic provinces are abrupt, making a comprehensive stratigraphy difficult. The steep slopes of the mountain belts, particularly Maxwell Montes, suggest either recent tectonic activity or high crustal strength. The crater density, two (million km)$^{-2}$, and some sparse stratigraphic relationships with the surrounding plains indicate that Ishtar Terra's most recent resurfacing was earlier than 300 Myr. The pristine character of craters, preservation of narrow ridge belts, and other features indicate negligible later resurfacing, tectonic or volcanic. The extraordinarily low radio emissivities at altitudes above 4.0 km in the mountain belts are consistent with a very old surface. Gravity data from the circularized Magellan orbit have a resolution of about 180 km (half wavelength), and sharply localize the highs over mountain belts of more than 4.5 km altitude: Maxwell, Freyja, and Akna Montes. Apparent depths-of-compensation (ADCs), based on gravity: topography admittance ratios, vary from about 200 km for wavelengths greater than 3000 km to 15 km for 400 km. New experiments that determined the rheology of dry diabase find it to have a viscosity comparable to that of dry olivine. Hence hypotheses of static support for Ishtar Terra and features within it are viable. The long wavelength ADC requires support of the entire plateau in the mantle. A keel of high Mg:Fe ratio is consistent with the long time scale for Ishtar Terra, but support from sluggish-lid convection cannot be ruled out. Static mechanisms are more strongly indicated for support of the mountain belts. However, the basalt:eclogite transition limits the thickness of a gabbroic crust to about 50 km. Hence support of the highest topography appears to require variations in crustal density. Use of the new diabase viscosity in dynamic models lengthens the time scale of crustal tectonics to several 100 Myr, making it possible that the mountain belts are older than 500 Myr. Hypotheses for the lateral variation in crustal density require secondary differentiations in the crust either in previous orogenic episodes, which could be well over 10^9 yr ago, or in the process of forming the belts. The latter hypothesis implies positive sinkers hundreds of km below the belt, entailing stress differences of a few 100 MPa. Also requiring differentiations are the extensive lava flows that surfaced Lakshmi Planum and the smooth scarps. The great extent of these flows indicates basic composition, and thus magma sources most likely in the upper mantle. This volcanism was plausibly contemporaneous with the tectonism raising the mountain belts.

I. INTRODUCTION

Ishtar Terra is one of the most marked features on Venus, but also one of the most enigmatic; it is not clearly either a divergent feature, like Beta Regio or Atla Regio, nor a convergent feature, like Ovda Regio or Thetis Regio. Ishtar Terra does not fit easily into systematic classification, such as "...Venus can be divided into three basic types of regions: tesserae, plains, and rift-volcano-coronae zones" (Herrick 1994). While Ishtar Terra east of 15°E is dominated by tessera, this is its least remarkable part. West of 15°E there are several patches of tessera, but they are submerged in a variety of other geomorphic types. Within distances of a few hundred km the character of its terrain changes markedly; e.g., marked ridge belts at 7 km elevation in Maxwell Montes are only 300 km from smooth volcanic plain at 4 km elevation in Lakshmi Planum. While Ishtar Terra is atypical, because of its extreme characteristics it should be a severe tester of hypotheses of the evolution of Venus. Hence after summarizing the data on Ishtar Terra, we discuss general considerations of Venusian evolution as a context for interpretation and models of Ishtar Terra.

II. DATA

A. Imagery

The synthetic aperture radar (SAR) of the Magellan spacecraft greatly improved knowledge of Ishtar's surface. Space limitations prevent showing much of this imagery in this chapter, and reference should be made to the papers referenced in this section, particularly Kaula et al. (1992) and Smrekar and Solomon (1992). The discussion by Kaula et al. (1992) of Magellan radar imagery available mid-1991 divided central Ishtar Terra into seven physiographic provinces, each named after its principal feature: Lakshmi, Danu, Freyja, North Scarp and Basin, South Scarp and Basin, Maxwell, and Western Fortuna. These provinces have markedly different character. But within each there are complex patterns suggestive of regional evolutions that are interleavings of tectonism and volcanism, difficult to relate to any grand pattern characteristic of Ishtar as a whole (see Figs. 1 and 2). Kaula et al. (1992) synthesized the geologic provinces into five types. These types appear applicable to the Magellan images obtained since mid-1991, which cover western Lakshmi Planum and Akna Montes, but only a minor part of central and eastern Fortuna. The discussion below incorporates results of analyses by Smrekar and Solomon (1992), Keep and Hansen (1994a,b), Vorder Bruegge (1994a,b), Basilevsky (1995), and Willis and Hansen (1995).

Upland Plains. Lakshmi Planum is a remarkably smooth and extensive lava-flooded plain, at an elevation of about 3.5 km above sea level. It is central to the other regional types. The smoothness is interrupted by two large caldera, Sacajewea and Colette. Willis and Hansen (1995) identify eight more caldera, and suggest others may have been covered with lava

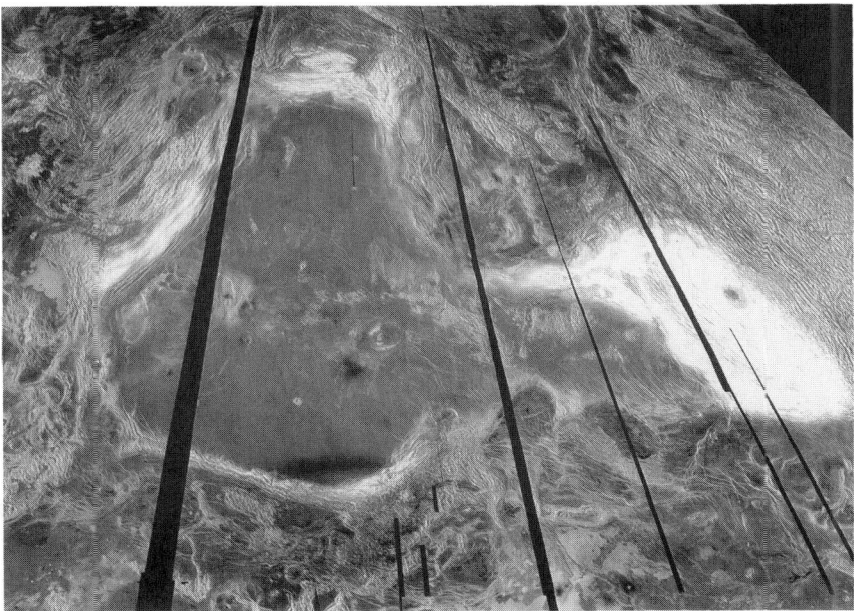

Figure 1. Western Ishtar Terra. Latitudes 55° to 80°N, longitudes 60°W to 30°E (figure from C2-MIDRP 60N333E).

flows. They infer that these caldera are the sources for extensive lava flows. In the northeast, at higher elevations, there is some tessera (ridged terrain) partially lava-flooded. Also in the east there are some grabens, apparently lava channels. Nowhere do there appear any major rifts. Kaula et al. (1992) did not identify any embayments of the ridge belts east of 30°W, but Basilevsky (1995) suggests some at the southwestern edge of Maxwell Montes. Akna Montes, 60°N x 50°W, has appreciable embayments, and distorts lobate flows from Colette, indicating near contemporaneity of tectonism and volcanism (Willis and Hansen 1995).

Mountain Belts. Maxwell, Freyja, and Danu Montes are similar in being concave to Lakshmi Planum; Akna Montes, which forms the western margin, is virtually straight. The sparsity of embayments indicates that these ranges have been thrust over the plains since termination of the bulk of the lava flow. All the mountain belts are characterized by series of ridges with spacings of 3 to 10 km. In many places, the ridges are intersected by differently oriented trends of fine lineaments, which appear to reflect earlier kinematic patterns (Smrekar and Solomon 1992; Keep and Hansen 1994a). Trenches on the Lakshmi side are less than 1 km deep, and indications of volcanism are confined to the outer slopes. Gravitational slumping is confined to steep marginal slopes, except in Danu Montes, which has appreciably greater indications

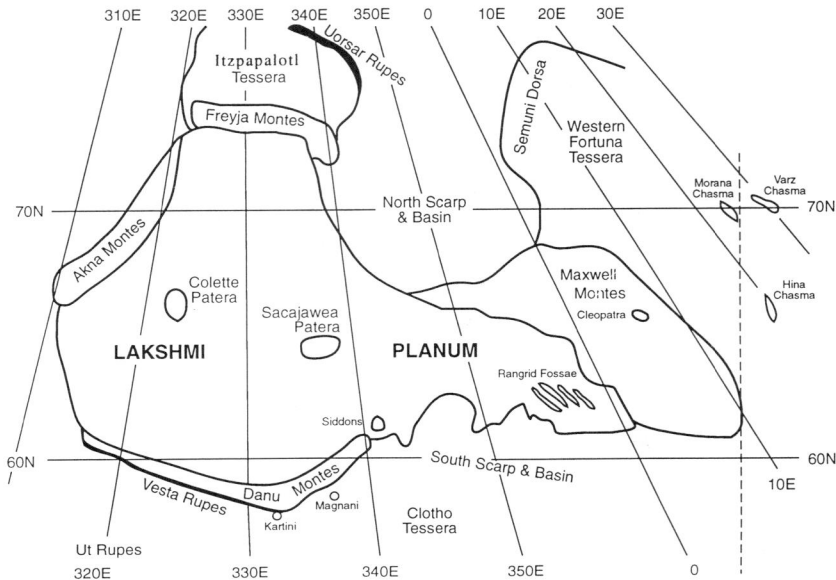

Figure 2. Sketch map of western Ishtar Terra. A degree of latitude is 105.6 km; at the latitudes of Ishtar Terra, a degree of longitude varies from 61 km to 22 km. The dashed line indicates the eastern limit of Fig. 1.

of volcanism, extensional tectonics, and gravitational modification than the other belts. This character extends to the adjacent Clotho Tessera (Smrekar and Solomon 1992; Keep and Hansen 1994b). Maxwell Montes, the highest and broadest mountain belt, is remarkable in having little indication of gravitational modification, except on its steep northern and southern slopes.

Kaula et al. (1992) inferred Maxwell to be the most recent feature. However, later analyses infer modifications since its ridge belts were emplaced, such as depositional units (volcanism or debris) on the slope toward Lakshmi (Vorder Bruegge 1994a), and embayment by the volcanic plains of the foot of the west slope (Basilevsky 1995). The 3 to 10 km scale ridge structure seems inconsistent with the magnitude of the topography, and hence reflects an earlier tectonic regime (Keep and Hansen 1994a; Vorder Bruegge 1994b). Considerable emphasis is placed on a sinuous lineament near 0° longitude. This lineament starts in the plains unit south of Maxwell Montes, coincides with its western boundary over latitudes 61° to 65°N, cuts across the western arm of Maxwell Montes, 65° to 68°N (see Fig. 3), runs along the northwest facing boundary, and extends into the lowlands at least as far as 72°N. Basilevsky (1995), applying stereoscopy, infers that the lineament is a scarp, an upthrust fault with a sinistral strike-slip component. From the relationship of the lineament to features in the plains, he infers that it formed at the time of

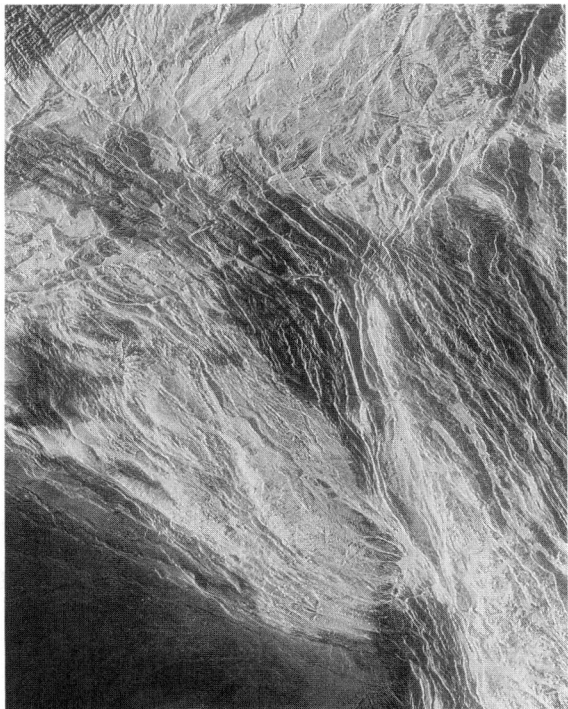

Figure 3. SAR image of northwest Maxwell Montes, 64.7°N to 67.5°N, 356°to 1°E = 295×228 km. The 0°lineament runs from the northeast corner down the right half of the image (figure from F-MIDRP.65N354).

ridging of the plains, 300 to 500 Myr ago (Basilevsky and Head 1995). Hence the ridge system of Maxwell must be even older. The existence of the great crater Cleopatra (66°N, 7°E) is thus less improbable. Cleopatra is remarkably deep, 2.6 km, for its diameter, 108 km. Around Cleopatra is an extensive plains unit of melt or finely divided ejecta, blanketing Maxwell Montes and western Fortuna Tessera out to 250 km. Akna Montes appears to have more embayment by volcanism than the other mountain belts, and is intermediate between Danu and Freyja Montes in the extent of gravitational modification, evidenced by grabens on the outer slope (Smrekar and Solomon 1992) (see Fig. 4).

Outboard Plateaus. On the side of the mountain belts away from Lakshmi Planum are regions 150 to 1000 km wide characterized by tessera: highly deformed ridged terrain, including ridged terrain, with a main pattern parallel to the mountain belts, but intersected by lineaments at angles to these trends, extending into the plains. In Itzpapalotl Tessera, to the north of Freyja Montes, the ridge complexes are interrupted by a smooth belt 30 km wide. The most

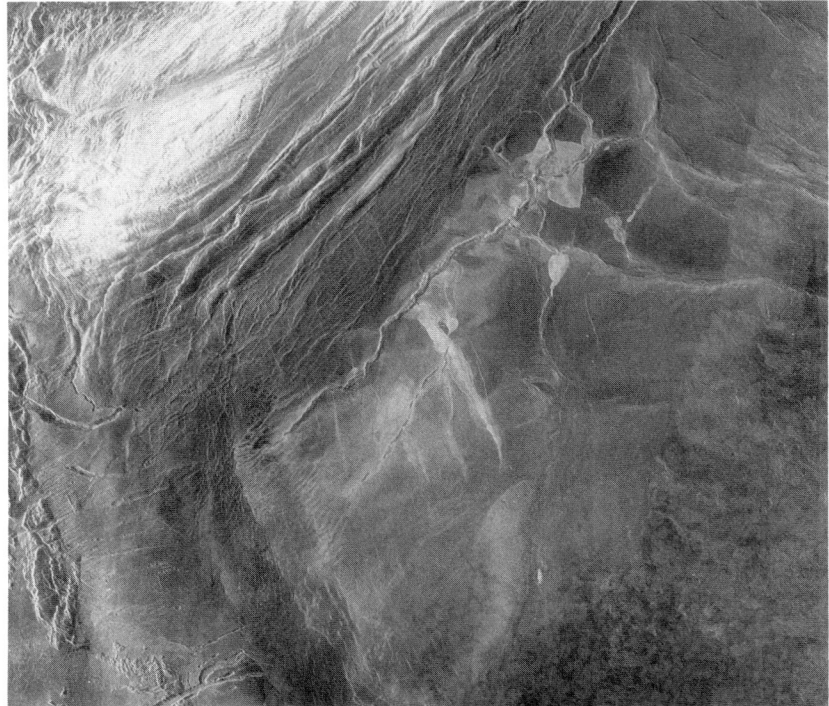

Figure 4. SAR image of southern Akna Montes, smooth scarp of western Vesta Rupes, and adjacent Lakshmi Planum; 64.0° to 67.3°N, 40° to 48°W = 349×348 km (figure from F-MIDRP.65N318).

extensive outboard plateau is Fortuna Tessera, east of Maxwell Montes. The most intense interlacing of ridges is the region between Maxwell and the gap in imagery, 30° to 53°E. Figure 5 shows longitudes 53° through 70°E. It still is mainly tessera, but there are more patches of smooth areas.

Smooth Scarps. On both the north and south side of Lakshmi Planum, to the west of Maxwell Montes at about 20°W to 0° longitude, there are relatively smooth drops of 2 to 3 km within 40 to 70 km horizontally. These scarps are convex toward Lakshmi Planum. They are cut by many rilles, which appear to be of volcanic origin. Another smooth scarp is Vesta Rupes, between Akna Montes and Danu Montes (Fig. 4). The rilles and the smoothness of the scarps indicate lava flows, but of remarkable volume and extent (Willis and Hansen 1995).

Basins. Adjacent to the smooth scarps are regions of 2 to 3 km elevation, generally similar in morphology to the plains that dominate most of the planet: a variety of subtler patterns, suggestive of a long tectonic his-

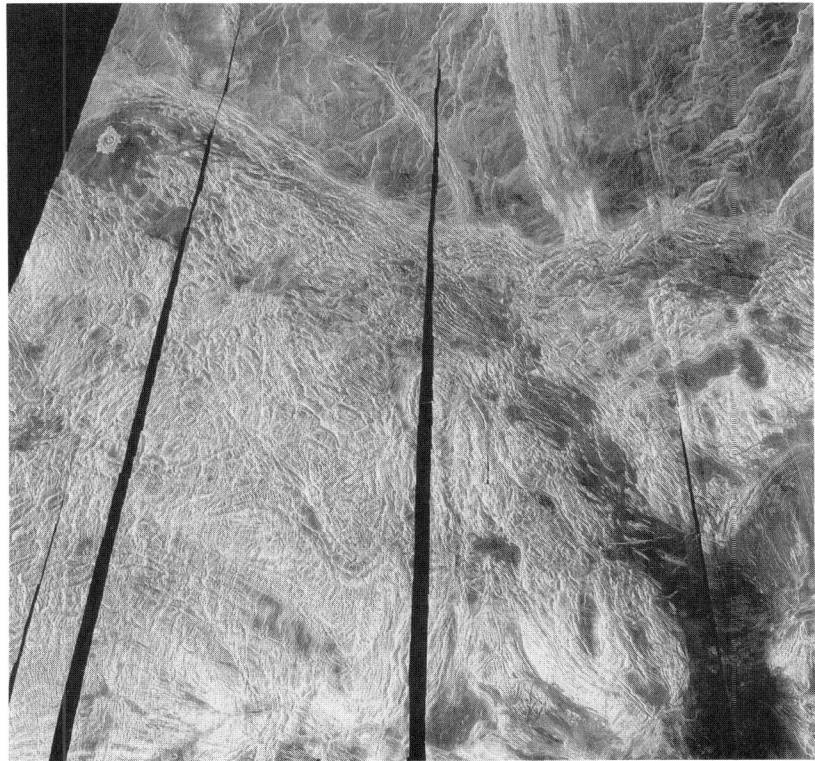

Figure 5. SAR image of eastern Fortuna Tessera; 67.2° to 78.0°N, 55° to 96°E = 1140 × 1220 km (figure from C1-MIDRP.75N074).

tory. Stratigraphies for each province are given by Kaula et al. (1992); for Maxwell Montes, by Keep and Hansen (1994a). Relationships among these provinces are at best tentative. Most striking is the inference by Basilevsky (1995) that the 0° lineament indicates the ridge belts in Maxwell Montes to be quite ancient.

B. Gravity

Analysis of line-of-sight (LOS) residuals of the X-band (8.4258 GHz) tracking by the JPL Deep Space Network (DSN) of the Magellan spacecraft cycles 5 and 6, in orbits at 180 to 500 km altitude, obtains a resolution of 180 km (half wavelength) at best over Ishtar, despite using more than 200,000 data points. The main limitation on resolution at high latitudes appears to be the electromagnetic environment of Venus out to several hundred km (Kaula 1996).

Figure 6 is a gravity map of western Ishtar Terra, and Fig. 7 is the complementary topographic map. Figures 8 and 9 are equivalent maps of eastern

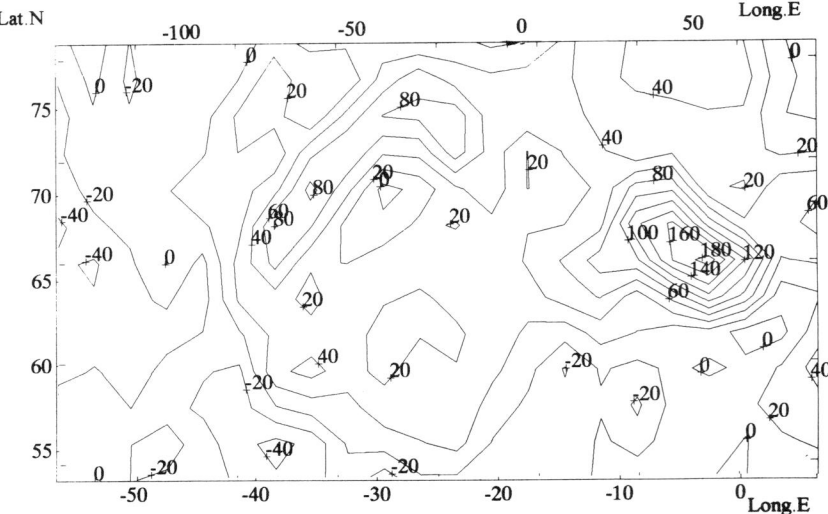

Figure 6. Free air gravity anomalies, in milligals, over western Ishtar Terra; 180 km resolution, centered on 66°N, 25°W. From a 40th degree global harmonic solution (Konopliv and Sjogren 1994) plus analysis of 251,025 LOS acceleration residuals (in groups of 5) from 1239 orbits (Kaula 1996). Figures 5 through 8 are sinusoidal projections, but referred to great circles through the center points, so as to preserve nearly uniform spacing in both directions, as well as preserving equal area.

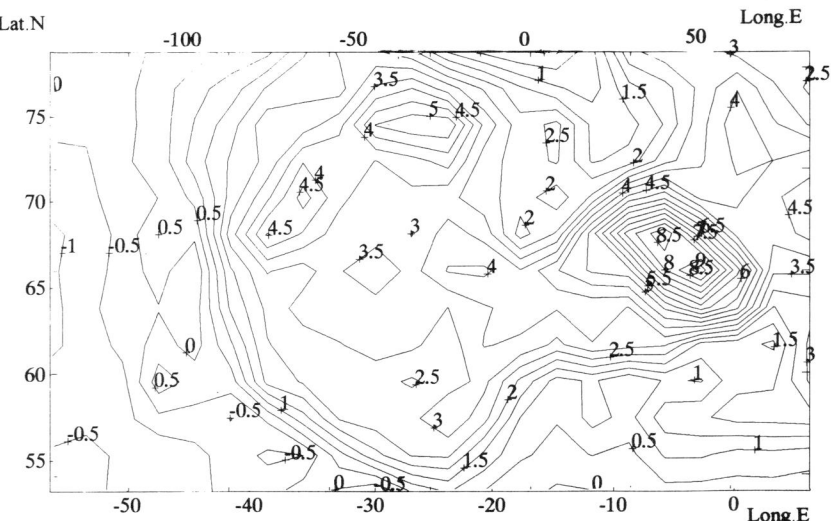

Figure 7. Topographic elevations, in km, over western Ishtar Terra, 180 km resolution, centered on 66°N, 25°W. From degrees 2 to 108 of a harmonic representation by Rappaport and Plaut (1994).

TABLE I
Apparent Depths of Compensation (ADC) by
Bands of Spherical Harmonic Degree ℓ

Location	ADC	
	ℓ = 17 to 75 km	ℓ = 76 to 100 km
Maxwell Montes	40–45	15–20
Freyja Montes	25–30	<10
Akna Montes	60–70	20–25
Metis Regio	15–25	10–15
Western Fortuna Tessera	25	10–15
Lakshmi Planum	40–50	20

Ishtar Terra. The main result of the gravity data from the cycle 5–6 Magellan orbits is to obtain an even stronger correlation of gravity and topography at the shorter wavelengths. This correlation is particularly marked for topography of more than 4.0 km altitude. On the other hand, the gravitational signal is mild for the steep topographic drop extending around the south side of western Ishtar Montes from Akna Montes to Maxwell Montes: less than 40 mgal for a 2.5 km topographic drop. The gravity variations east of 10°E are markedly milder than those in the west, the most prominent feature being a north–south trough of −40 milligals at 40°E, not correlated with topography. Grimm and Phillips (1991) initially estimated that Ishtar had an apparent depth of compensation (ADC) of 100 to 180 km—less than Atla or Beta Regio. This value applies to the entire plateau. Subsequent estimates of ADC obtain shallower values for shorter wavelengths (Herrick and Phillips 1992; McKenzie 1994; Simons et al. 1994; Hansen and Phillips 1995; Arkani-Hamed 1996). Table I is an extension of Table 1 in Hansen and Phillips (1995), using analyses of LOS residuals to get wavelengths shorter than those from the JPL solutions (Kaula 1995b). The depth of compensation continues to decrease with spherical harmonic degree above 75, but the resolution is questionable above degree 90 for features north of latitude 70°.

III. GENERAL CONSIDERATIONS

The radar imagery of the Magellan project shows a crater abundance equivalent to about 400 Ma (±100 Myr) and a distribution close to random (Phillips et al. 1992; Schaber et al. 1992). The main debate is whether its decline is more monotonic (see, e.g., Arkani-Hamed and Toksoz 1984; Solomon 1993; Herrick 1994), or episodic, with enhanced tectonism and volcanism yet to come (Turcotte 1993; Herrick and Parmentier 1994). Arkani-Hamed (1994) computes a transition from oscillatory behavior to monotonic decline. Since 400 Ma, the low level of activity inferred from the abundance of craters favors the monotonic hypothesis. The low atmospheric abundance of radiogenic argon, implying retention of heat sources at depth, favors the episodic (Kaula 1995a). A major problem, common to all hypotheses of Venus's evolution,

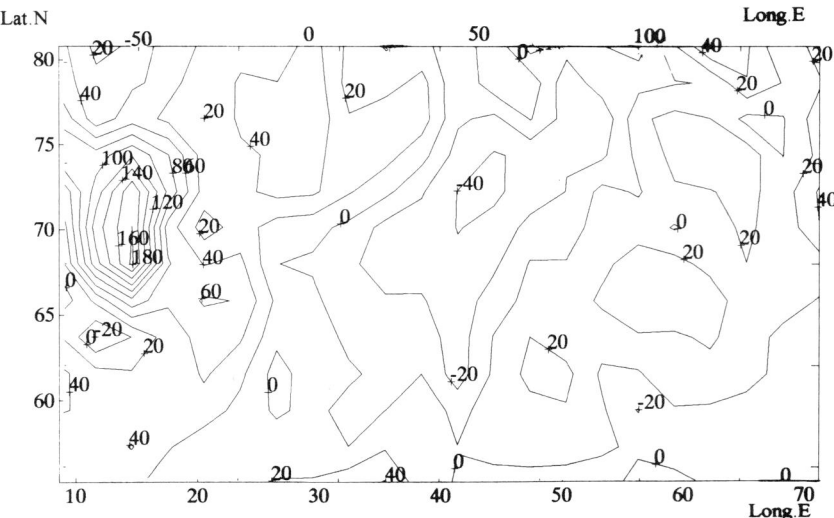

Figure 8. Free air gravity anomalies, in milligals, over eastern Ishtar Terra, 180 km resolution, centered on 68°N, 40°E. From a 40th degree global harmonic solution (Konopliv and Sjogren 1994) plus analysis of 280,920 LOS acceleration residuals (in groups of 5) from 1512 orbits (Kaula 1996).

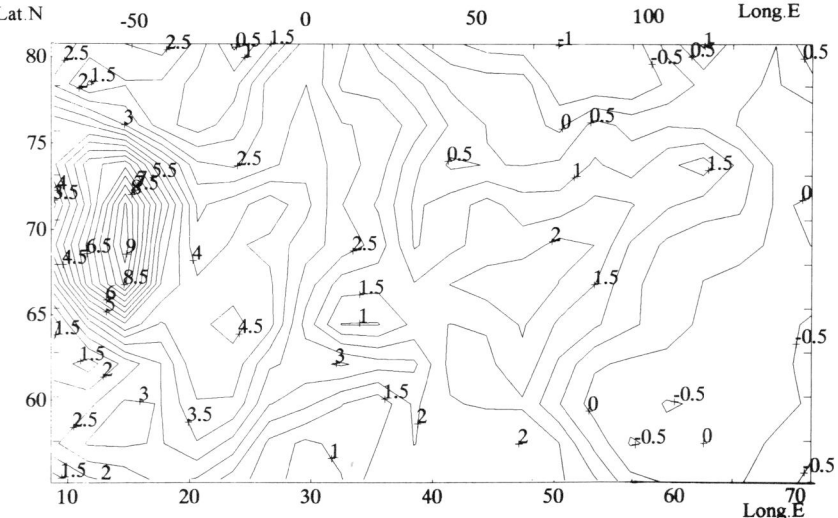

Figure 9. Topographic elevations, in km, over eastern Ishtar Terra, 180 km resolution, centered on 68°N, 40°E. From degrees 2 to 108 of the harmonic representation of Rappaport and Plaut (1994).

is an apparently rapid decline of thermal and tectonic activity some 400 Myr ago (Strom et al. 1994): the nature of the instabilities causing it, and their propagation. These factors are discussed by Kaula (1995a), who conjectures that the dry rheology inhibited plate tectonics, because of the lack of weak margins, but caused much larger yield stresses, and thus catastrophic release thereof. However, removal upward of heat sources seems important to the quiesence of the last 400 Myr. An important recent development, solving some major problems, is the new rheology for dry diabase by Mackwell et al. (1995), who find it to have a viscosity comparable to that of dry olivine. This removes the constraint of crater depth:diameter ratios on crustal thickness (Grimm and Solomon 1988), thus making more plausible thick crust, and thence the upward concentration of radioactive heat sources.

It also becomes plausible for mountains as high and steep as Maxwell to support themselves on the time scale of 400 Ma. However, such a stiff rheology leads to a very slow evolution of crust: mantle interaction (Lenardic et al. 1995), difficult to reconcile with a rapid transition around 400 Myr. Ishtar Terra has an average crater density west of 10°E, two per million km^2, with no significant variation. Hence this part must have shared in the transition from active resurfacing to quiescence. But the stretched time scale allowed by the dry rheology may help explain the seeming incongruity of features a few hundred km apart. A long time scale is consistent with the marked difference in emissivity of the higher altitudes in Maxwell and Freyja Montes from that of Lakshmi Planum, regardless of mechanism (Klose et al. 1992; Bracken et al. 1995; Pettengill et al. 1996; chapter by Fegley et al.), because the elements that might cause the reduction in emissivity all have low abundance. Maxwell Montes differs significantly in this respect from the other two highest topographic features, Atla Regio and Beta Regio (Robinson and Wood 1993). Ishtar Terra is archetypical of the enduring conflict between geologists and geophysicists; the latter's models still fall far short of the former's data. This problem is compounded on Venus by the lack of seismological and dating constraints (including the absence of erosion to reveal stratigraphy), and the absence, to date, of a global paradigm comparable to plate tectonics. Furthermore, the absence of water plausibly has a significant effect on the physics and chemistry underlying tectonic and magmatic processes.

IV. MODELS

A. Static

The poor resolution of the gravity data at high latitudes makes flexural rigidity analysis (Forsyth 1985) somewhat frustrating. Some inferences could be made from the topography of the trenches west of Maxwell Montes, but these are not strongly constraining on the flexural rigidity, being less than one km deep. Hence the main question is the nature of the compensation indicated by the ADCs. The regional ADCs can be accommodated by Airy isostasy (variations in thickness of a gabbroic crust) for all regions except the three mountain belts with high topography and gravity, where such compensation would extend below the depth at which the gabbro eclogite phase transition stabilizes: 50 to 70 km (Arkani-Hamed 1993; Namiki and Solomon 1993; Herzog et al. 1995; chapter by Grimm and Hess). A kinetic delay in the transition requires high strain rates, about 10^{-15} s^{-1} (Namiki and Solomon 1993), much higher than inferred by Grimm (1994) from craters less pristine than those in Ishtar. Hence other mechanisms must be invoked: either lateral variations of density within the crust, arising from intracrustal differentiation (Jull and Arkani-Hamed 1995; Kaula 1995a), or within the mantle, arising from concentration of residuum with high Mg/Fe ratio (Phillips and Hansen 1994). Probably both mechanisms are significant, because it is likely that Ishtar Terra has had a long and complex evolution. The lower ADCs for harmonic degrees above 75 are consistent with intracrustal differentiation. The strong indications of convergence creating Maxwell Montes suggests a further complication: a positive buried load, maintaining its convergent character (Grimm and Phillips 1991; Kaula 1995a). While minimum stress models forbid such a load, solutions adding it to the standard model of Forsyth (1985) obtain compensation by lower crustal density, entailing stresses of a few 100 MPa (see Fig. 10). A similar result is obtained by Arkani-Hamed (1996).

B. Dynamic

There are three scales of problems. First, there is the global, of wavelength more than 3000 km, associated with support of the entire plateau, and having an ADC more than 100 km (Grimm and Phillips 1991; Herrick and Phillips 1992; Simons et al. 1994; Hansen and Phillips 1995); Second, there is a regional scale, at shorter wavelengths, associated with support of the mountain belts, most notably Maxwell Montes, because it is the greatest load. Third, there is a local scale: features in the imagery which must depend on shallow rheology and tectonics for their creation, such as the ridge belts, or their maintenance, such as the sparse evidence of gravitational relaxation at the highest elevations. Although it was suggested that Ishtar may be a Venusian continent (Phillips et al. 1981), most pre-Magellan models operated under the assumption that Ishtar was more akin to a terrestrial crustal plateau. Extreme models of plateau formation are (1) upwelling: magmatic addition and thermal effects within, or just below the lithosphere; and (2) downwelling: shortening of crust and mantle lithosphere (by thrust stacking, folding, ductile thickening, or underthrusting). An example of (1) is Iceland; of (2), Tibet. The majority of terrestrial plateaus are considered to be closer to extreme (1), attributed to mantle plumes, which are dynamic upwelling limbs within the convecting mantle. Plumes provide a heat source for the magmatism needed to form crust and can also contribute to topography, either dynamically through thermal buoyancy or through added chemical buoyancy associated with depleted mantle residuum, or through some combination thereof. The earliest upwelling models of Ishtar's formation responded

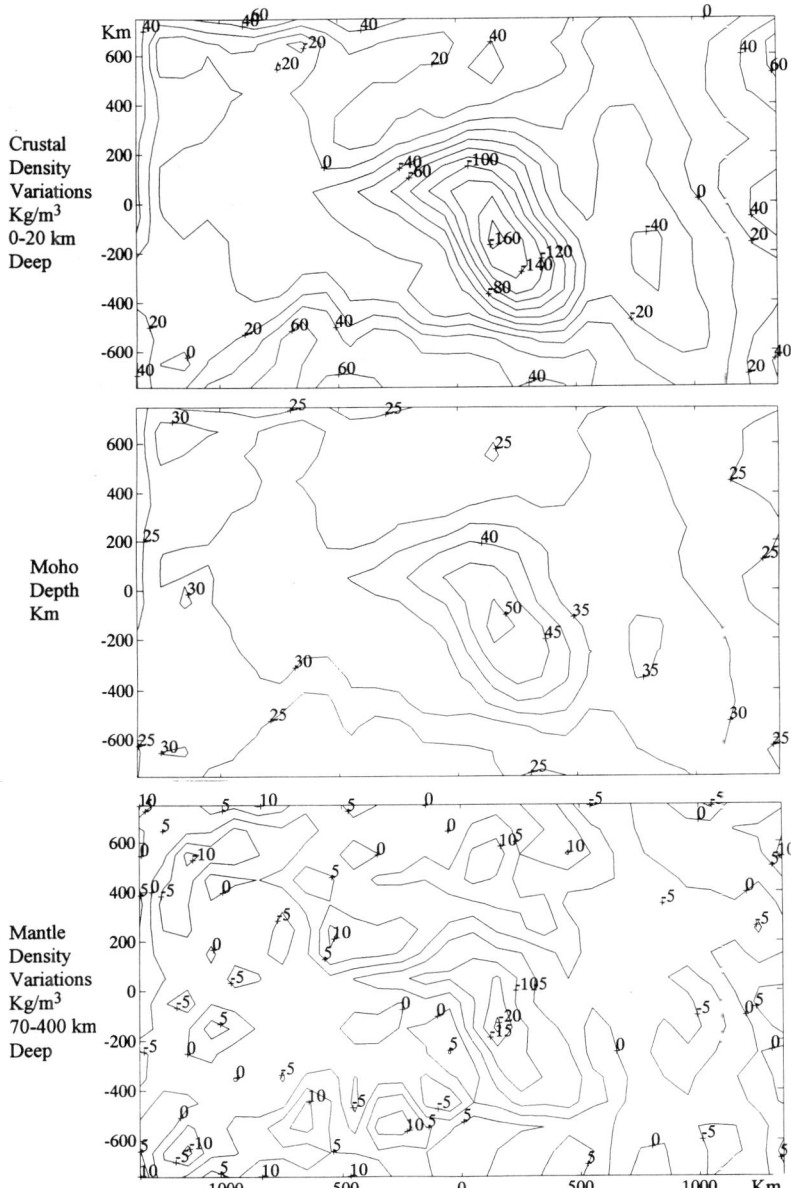

Figure 10. Simultaneous solution for surface, buried, and deep loads from gravity and topography over Maxwell Montes. Solved as a least stress adjustment to an *a priori* model assuming a deep sinker proportionate to topographic elevation. The solution included error in the acceleration data, leading to smoothing of features below the resolution level of the gravity field (180 km). Coordinates are with respect to a center at 65° latitude, 0° longitude.

primarily to radar imagery (Pronin 1986; Basilevsky 1986) but later models (Grimm and Phillips 1990,1991; Phillips et al. 1991) used observed gravity and altimetry. The long wavelength gravity and topography are well predicted, with minimal assumptions, by a mantle upwelling below Ishtar. For a downwelling model, more complicated models are required to match the ADCs. The major criticism of upwelling models was that they could not directly account for the tectonic fabric of Ishtar, particularly the peripheral mountain belts which required compressive stress. This criticism was met in two ways. One suggested that secondary downflow associated with a mantle plume centered below Ishtar occurred below the peripheral mountain belts (Pronin 1986; Basilevsky 1986). The other suggested that Ishtar was bordered by strong lithosphere, its own lithosphere being rather weak, and that stress due to spreading plume material below the lithosphere was concentrated at the rheologic discontinuity that framed the highland. This stress, it was argued, could provide the compression the mountains required (Grimm and Phillips 1990).

Downwelling models that attributed Ishtar to crustal shortening were stimulated mainly by the compressive stress apparently required for the highlands. This hypothesis divides into two as to the cause of the compressive stress; either from convective downwelling in the mantle below Ishtar (Bindschadler and Parmentier 1990; Bindschadler et al. 1990; Kiefer and Hager 1991; Lenardic et al. 1991) or from horizontal normal forces at the edges of Ishtar (Roberts and Head 1990; Head et al. 1990). Convective mantle downwelling provides a natural source of compression and the shear stresses associated with it could couple into the base of the crust causing it to thicken over a downflow thereby forming a high-standing plateau. This scenario was explored quantitatively and advocated as a means of forming Ishtar by Bindschadler et al. (1990,1992). It predicted long wave-length topography and tectonic deformation patterns, but fell short of accounting for much detail inferred from the imagery. It could also account for the gravity, but, as noted, entailing appreciably higher stress differences. These models are consistent with close coupling of mantle convection to the lithosphere, as inferred by Phillips (1990). A second class of shortening models invoked regional compression, by stresses normal to the edge of the plateau, as a means of creating Ishtar (Roberts and Head 1990; Head et al. 1990; Vorder Bruegge and Head 1990). These models usually entailed underthrusting and imbrication, resulting from the regional stress state directed toward the center, as the means of forming peripheral mountain belts. Roberts and Head (1990) hypothesized that Lakshmi was a strong block of ancient tessera that served as a buttress against which peripheral mountains formed. These models are the most Earthlike, in that they depend on a lithosphere essentially detached from the interior to transmit horizontal forces.

After receipt of the Magellan imagery there was no further development of purely upwelling models. A model that relied on both upwelling and downwelling was put forward based primarily on analysis of Magellan imagery

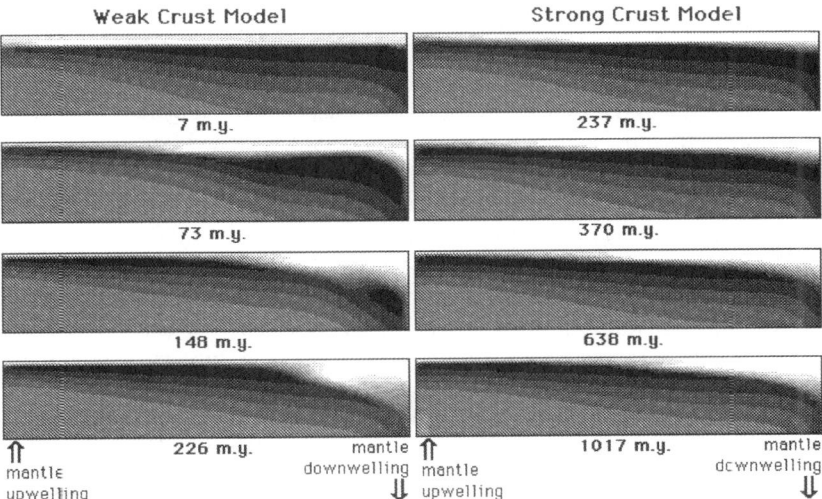

Figure 11. Evolution of coupled mantle convection–crustal deformation models. Shown are density image frames from the upper 140 km of 700×700 km upper mantle convection cells. White represents crust and black represents dense mantle lithosphere. Flow is left to right and the crust thickens above a thermal mantle downwelling. Parameters are the same for both models except that the strong crust model assumes a temperature dependent viscosity for the crust that is equal to that used for the mantle while the weak crust model assumes a crustal viscosity that is about an order of magnitude lower for the crust than for the mantle at equal temperatures. Further model details are in Lenardic et al. (1993,1995).

(Kaula et al. 1992). This compromise model was somewhat reminiscent of Pronin (1986) in that it invoked upflow below Lakshmi and downflow below peripheral mountain belts (a configuration that could plausibly result from crustal thickening above a downflow followed by local mantle insulation, caused by the thickened crustal plateau, forcing mantle downflow to migrate to the plateau periphery [Lenardic et al. 1993]). The compromise model was more geared toward explaining regional tectonic patterns and left the problem of long wavelength support as a residual of regional processes (akin to earlier regional compression models [see, e.g., Roberts and Head 1990]), or due to poorly defined mantle effects. The rock deformation experiments of Mackwell et al. (1995) that showed the strength of dry basalt to be comparable to that of olivine compelled a rethinking of downwelling models (see, e.g., Bindschadler et al. 1990,1992), because they greatly lengthened the time scale of crustal thickening to be much longer than experienced with Earth. Numerical experiments that inferred a formation time of about 100 Myr with the low crustal viscosity (Lenardic et al. 1991,1993) obtained 1 Gyr with the new value (Lenardic et al. 1995) (see Fig. 11). While the time scale of near-surface tectonics on Venus may be very long due to dryness, the same does not necessarily apply to the mantle, where temperature dependence of viscosity might lead to an adjustment proportionate to the heat to be removed (Turcotte 1993). Computer modelings of thermally convecting systems of high Rayleigh number indicate that it would be unlikely for a mantle downwelling pattern to remain fixed in space for 1 Gyr (chapter by Schubert et al.). Furthermore, a long thickening time scale seems inconsistent with a widespread resurfacing event occurring within 100 Ma at about 400 Myr (Strom 1994). While the 100 Myr resurfacing entails some significantly different factors than the long term (10^9 yr) removal of heat from the interior, a correlation of the two is intuitively expected. The essential problem is inferring the nature of the tectonic/convective instabilities that occurred in the epoch of the resurfacing (Kaula 1995a). This problem is keenest for Ishtar Terra, because, unlike the other prominent features in Venus gravity and topography, it does not have indications of resurfacing occurring in more recent times. The aforedescribed dynamic models do not take into account differentiation of the mantle, as is necessary for static support of Ishtar by a keel, discussed in the previous section. Modified downwelling models to form and support the long-wavelength topography of Ishtar have been advanced by Phillips and Hansen (1994), Keep and Hansen (1994a), and Hansen and Phillips (1995). This modified downwelling model was driven partly by deformational constraints derived from SAR imagery and partly by the range of ADCs inferred for Ishtar (Hansen and Phillips 1995). They attributed the long wavelength support to thickened residuum, with a mantle downwelling serving as the agent of thickening, and short-wavelength support to crust that had thickened against the residuum, which came to serve as a buttress. Whether forward models of residuum thickening can account for long-wavelength support, which requires explaining the long-term stabilization of thickened residuum against convective remixing into the mantle, remains to be tested, as numerical codes of mantle convection are just beginning to cope with multi-component flows (chapter by Schubert et al.). Different ADCs for different wavelengths of Ishtar topography was also used by Jull and Arkani-Hamed (1995), and Arkani-Hamed (1996) to argue that support of short wavelength topography could not be accomplished simply by thickening crust of a uniform chemical composition. They hypothesize that blocks of differentiated crust support the peripheral mountain belts of Ishtar, i.e., a component of Pratt compensation is required. Their preferred model invoked blocks of light crust (a secondary differentiate akin to continental crust of Earth or the anorthositic highlands of the Moon) being swept toward Ishtar by mantle downflow. This light crust, being of high LIL content, provided the energy source for the subsequent volcanism forming Lakshmi Planum. An alternative model (Kaula 1995a), is that low density crust was formed in place as a consequence of crustal convergence above a convective mantle downflow, followed by radioactive heating of the thickened crust. Partial melting from this heating would have caused upward concentration of heat sources with a denser crustal component sinking below the gabbro:eclogite transition. This sinker could be a driver for continuing convergence as Venus evolved toward its present quiescence and the vigor of thermal mantle downflow below Ishtar waned. This would provide a means of maintaining compression locally under the mountain belts and tension in the adjacent areas, as tectonic quiescence prevailed. This scenario also needs to be tested by forward modeling.

Short-wavelength details in the mountain belts strongly indicate a different thermal and tectonic regime in the past. The spacing of ridge belts is consistent with a competent layer only 20 km thick (Zuber 1987). Thus there must have been a phasing from dynamic to passive support. The extensive plains unit around Cleopatra Crater also is suggestive of high temperatures at shallow depths. However, Grieve and Cintala (1995) infer from the high depth and diameter ratio of the crater that this unit could be entirely drained impact melt. The finite element computations of Smrekar and Solomon (1992) with weak crustal rheology indicated that the steep slopes of Maxwell, Freyja, and Akna Montes could be maintained without slumping for a time scale of only 10 Myr. However, the computer experiments of Freed and Melosh (1995), while primarily addressing the global and regional scale problems, had a finite element grid fine enough near the region of steep surface slope to infer a time scale of 400 Myr for the local problem of maintaining slopes. The occurrence of extensive volcanism to form the surface of Lakshmi Planum is consistent with the idea that rifting, leading to volcanism, occurs adjacent to convergence (Magee and Head 1995). The flows creating Lakshmi Planum must be extensive enough to conceal most evidence of rifting. The imagery of the margins between Lakshmi Planum and the mountain belts forbids the lava flows from being mostly later than uplift of the mountain belts, but they could be roughly contemporaneous (Willis and Hansen 1995).

V. CONCLUSIONS

The evolution of ideas about Ishtar Terra since acquisition of the Magellan imagery and the drastically revised rheology of dry diabase has been toward a much greater age and longer evolution. Ishtar shares the general resurfacing that occurred at 300 to 500 Ma, but it lacks the fresh coronae, rifts, and volcanic structures indicative of recent activity in the equatorial belt (Namiki and Solomon 1994; Herrick and Phillips 1994; Price and Suppe 1994; Phillips and Izenberg 1995), thus it is probably older than Atla Regio or Beta Regio. Hence it is plausible that Ishtar's broad support is now by a continental keel that is a residuum of high Mg/Fe ratio (see Fig. 12) (Hansen and Phillips 1995; Jull and Arkani-Hamed 1995). Contemporary convection could leave such a keel relatively undisturbed, as it apparently does for the terrestrial keels of Africa and Australia. But there is the counter argument that the higher viscosity of Venus would more closely couple any keel to mantle convection, tearing it apart. This question does not seem resolvable until we get either seismology on Venus or persuasive models of the conditions under which mantle convection can sweep aside a residuum layer.

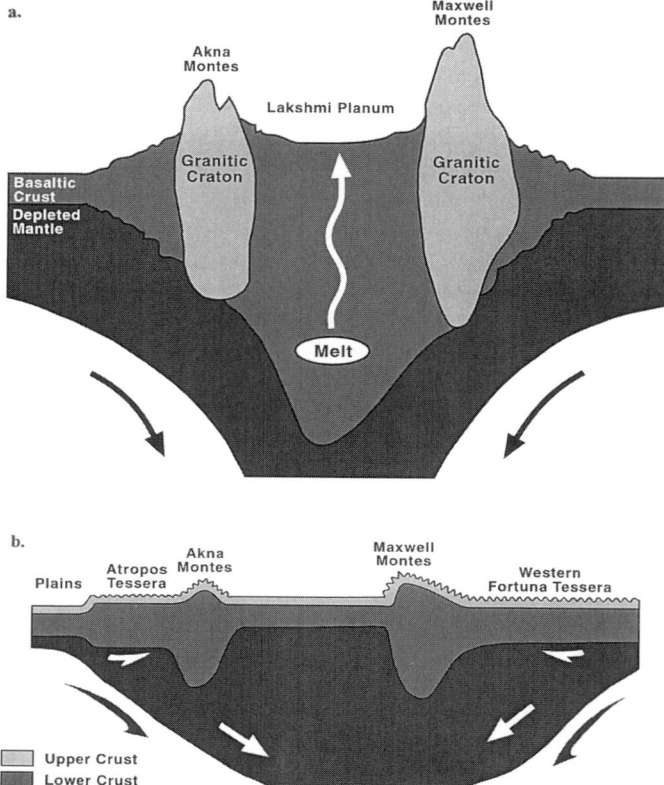

Figure 12. Hypotheses of Ishtar structure: (a) by Jull and Arkani-Hamed (1995); (b) by Hansen and Phillips (1995).

Other questions are (1) the source of the Lakshmi Planum volcanism: upper mantle, favoring low viscosity (Hansen and Phillips 1995), or the crust, favoring high heat sources (Jull and Arkani-Hamed 1995), and (2) the timing of intracrustal differentiation: in an earlier epoch (Jull and Arkani-Hamed 1995), or contemporaneous with the tectonism forming the mountain belts (Kaula 1995a). These differentiations could be more anorthositic than silicic (as on the Moon) (Kaula 1993). Quantitative models are still far from accounting for the wealth of detail in deformational patterns on Ishtar; even qualitative models (see, e.g., Hansen and Phillips 1995) are just beginning to do so. Improvements in forward modeling require not only bigger and faster computers, but also more comprehensive rheological experiments, taking into account a broader range of mineralogy, grain size, and strain rate, as well as higher pressures (Zuber 1994; Kaula 1995a). A persisting problem is how to get the uplands of Ishtar Terra to share in a global resurfacing over a few

10 Myr, if it is done by low viscosity lavas issuing from low-elevation fissures, as suggested by Strom et al. (1994) and Herrick (1994). Whereas Lakshmi seems effusive enough to cover the craters on the 3.5 km high plateau, to keep the crater counts down in the mountain belts and their outboard tesseral complexes seems to require a coincidence in timing of tectonic resurfacing with volcanic resurfacing. As for the overall problem of decline in the vigor of Venus' thermal and tectonic evolution, it still seems most plausible that it arises from upward concentration of heat sources, consequent upon the dryness of Venus inhibiting plate tectonics and crustal recycling.

Acknowledgments. This chapter has benefitted from thorough reviews by V. L. Hansen and R. E. Grimm. This work was partially supported by a grant from NASA.

REFERENCES

Arkani-Hamed, J. 1993. On the tectonics of Venus. *Phys. Earth Planet. Int.* 76:75–96.
Arkani-Hamed, J. 1994. On the thermal evolution of Venus. *J. Geophys. Res.* 99:2019–2033.
Arkani-Hamed, J. 1996. Analysis and interpretation of the high-resolution surface topography and gravity of Ishtar Terra, Venus. *J. Geophys. Res.* 101:4691–4710.
Arkani-Hamed, J., and Toksoz, M. N. 1984. Thermal evolution of Venus. *Phys. Earth Planet. Int.* 34:232–250.
Basilevsky, A. T. 1986. Structure of central and eastern areas of Ishtar terra and some problems of Venusian tectonics. *Geotektonika* 20:282–288.
Basilevsky, A. T. 1995. Compositional heterogeneity and late-stage deformation in Maxwell Montes, Venus. *Lunar Planet. Sci. Conf.* XXVI:79–80 (abstract).
Basilevsky, A. T., and Head, J. W. 1995. Global stratigraphy of Venus: Analysis of a random sample of thirty-six test areas. *Earth, Moon, Planets* 66:285–335.
Bindschadler, D. L., and Parmentier, E. M. 1990. Mantle flow tectonics: The influence of a ductile lower crust and implications for the formation of topographic uplands on Venus. *J. Geophys. Res.* 95:21329–21344.
Bindschadler, D. L., Schubert, G., and Kaula, W. M. 1990. Mantle flow tectonics and the origin of Ishtar Terra, Venus. *Geophys. Res. Lett.* 17:1345–1348.
Bindschadler, D. L., Schubert, G., and Kaula, W. M. 1992. Coldspots and hotspots: Global tectonics and mantle dynamics of Venus. *J. Geophys. Res.* 97:13495–13578.
Brackett, R. A., Fegley, B., and Arvidson, R. E. 1995. Volatile transport on Venus and implications for surface geochemistry and geology. *J. Geophys. Res.* 100:1553–1563.
Forsyth, D. W. 1985. Subsurface loading and estimates of flexural rigidity of continental lithosphere. *J. Geophys. Res.* 90:12623–12632.
Freed, A. M., and Melosh, H. J. 1995. Long term survival of the topography of Ishtar Terra, Venus. *Lunar Planet. Sci. Conf.* XXVI:421–422 (abstract).

Grieve, R. A. F., and Cintala, M. J. 1995. Impact melting on Venus: Some considerations for the nature of the cratering record. *Icarus* 114:68–79.

Grimm, R. E. 1994. Recent deformation rates on Venus. *J. Geophys. Res.* 99:23163–23171.

Grimm, R. E., and Phillips, R. J. 1990. Tectonics of Lakshmi Planum, Venus: Tests for Magellan. *Geophys. Res. Lett.* 17:1349–1352.

Grimm, R. E., and Phillips, R. J. 1991. Gravity anomalies, compensation mechanisms, and the geodynamics of western Ishtar Terra, Venus. *J. Geophys. Res.* 96:8305–8324.

Grimm, R. E., and Solomon, S. C. 1988. Viscous relaxation of impact crater relief on Venus: Constraints on crustal thickness and thermal gradient. *J. Geophys. Res.* 93:11911–11929.

Hansen, V. L., and Phillips, R. J. 1995. Formation of Ishtar Terra, Venus: Surface and gravity constraints. *Geology* 23:292–296.

Head, J. W., Vorder Bruegge, R. W., and Crumpler, L. S. 1990. Venus orogenic environments, architecture and origin. *Geophys. Res. Lett.* 17:1337–1340.

Herrick, R. R. 1994. Resurfacing history of Venus. *Geology* 22:703–706.

Herrick, D. L., and Parmentier, E. M. 1994. Episodic large-scale overturn of two-layer mantles in terrestrial planets. *J. Geophys. Res.* 99:2053–2062.

Herrick, R. R., and Phillips, R. J. 1992. Geological correlations with the interior density structure of Venus. *J. Geophys. Res.* 97:16017–16034.

Herrick, R. R., and Phillips, R. J. 1994. Implications of a global survey of Venusian impact craters. *Icarus* 111:387–416.

Herzog, S. G., Hess, P. C., and Parmentier, E. M. 1995. Constraints on the basalt to eclogite transition and crustal recycling on Venus. *Lunar Planet. Sci. Conf.* XXVI:591–592 (abstract).

Jull, M. G., and Arkani-Hamed, J. 1995. The implications of basalt in the formation and evolution of mountains on Venus. *Phys. Earth Planet. Int.* 89:163–175.

Kaula, W. M. 1990. Venus: A contrast in evolution to Earth. *Science* 247:1191–1196.

Kaula, W. M. 1993. Compositional evolution of Venus. In *Evolution of the Earth and Planets*, eds. E. Takahashi, R. Jeanloz and D. Rubie (Washington, D. C.: American Geophysical Union), pp. 27–40.

Kaula, W. M. 1995a. Venus reconsidered. *Science* 270:1460–1464.

Kaula, W. M. 1995b. A one-degree square mean, or 180th degree harmonic, solution for the gravity field of Venus. *Eos Suppl.* 76:F331.

Kaula, W. M. 1996. Regional gravity fields on Venus from Magellan LOS acceleration residuals. *J. Geophys. Res.* 101:4683–4690.

Kaula, W. M., et al. 1992. Styles of deformation in Ishtar Terra and their implications. *J. Geophys. Res.* 97:16085–16120.

Keep, M., and Hansen, V. L. 1994a. Structural history of Maxwell Montes, Venus: Implications for Venusian mountain belt formation. *J. Geophys. Res.* 99:26015–26028.

Keep, M., and Hansen, V. L. 1994b. Structural evolution of Danu Montes, Venus: Deformation around a curved boundary. *Lunar Planet. Sci. Conf.* XXV:681–682 (abstract).

Kiefer, W. S., and Hager, B. H. 1991. Mantle downwelling and crustal convergence: A model for Ishtar Terra, Venus. *J. Geophys. Res.* 96:20967–20980.

Klose, K. B., Wood, J. A., and Hashimoto, A. 1992. Mineral equilibria and the high radar reflectivity of Venus mountain tops. *J. Geophys. Res.* 97:16353–16369.

Konopliv, A. S., and Sjogren, W. L. 1994. Venus spherical harmonic gravity model to degree and order 60. *Icarus* 112:42–54.

Lenardic, A., Kaula, W. M., and Bindschadler, D. L 1991. The tectonic evolution of Western Ishtar Terra, Venus. *Geophys. Res. Lett.* 18:2209–2212.

Lenardic, A., Kaula, W. M., and Bindschadler, D. L. 1993. A mechanism for crustal recycling on Venus. *J. Geophys. Res.* 98:18697–18705.

Lenardic, A., Kaula, W. M., and Bindschadler, D. L. 1995. Some effects of a dry crustal flow law on numerical simulations of coupled crustal deformation and mantle convection on Venus. *J. Geophys. Res.* 100:16949–16957.

Mackwell, S. J., Zimmerman, M. E., Kohlstedt, D. L., and Scherber, D. S. 1995. Experimental deformation of dry Columbia diabase: Implications for tectonics on Venus. In *Rock Mechanics*, eds. J. J. K. Daemen and R. A. Schutz (Rotterdam: Balkema), pp. 207–214.

Magee, K. P., and Head, J. W. 1995. The role of rifting in the generation of melt; implications for the origin and evolution of the Lada Terra–Lavinia Planitia region of Venus. *J. Geophys. Res.* 100:1527–1552.

McKenzie, D. P. 1994. The relationship between topography and gravity on Earth and Venus. *Icarus* 112:55–88.

Namiki, N,. and Solomon, S. C. 1993. The gabbro-eclogite phase transition and the elevation of mountain belts on Venus. *J. Geophys. Res.* 98:15025–15031.

Namiki, N., and Solomon, S. C. 1994. Impact crater densities on volcanoes and coronae on Venus: Implications for volcanic resurfacing. *Science* 265:929–933.

Pettengill, G. H., Ford, P. G., and Simpson, R. A. 1996. Electrostatic properties of the Venus surface from bistatic radar. *Science*, in press.

Phillips, R. J. 1990. Convection-driven tectonics on Venus. *J. Geophys. Res.* 95:1301–1316.

Phillips, R. J., and Hansen, V. L. 1994. Tectonic and magmatic evolution of Venus. *Ann. Rev. Earth Planet. Sci.* 22:597–654.

Phillips, R. J., and Izenberg, N. R. 1995. Ejecta correlations with spatial crater density and Venus resurfacing history. *Geophys. Res. Lett.* 22:1517–1520.

Phillips, R. J., Kaula, W. M., McGill, G. E., and Malin, M. C. 1981. Tectonics and evolution of Venus. *Science* 212:879–887.

Phillips, R. J., Grimm, R. E., and Malin, M. C. 1991. Hot-spot evolution and the global tectonics of Venus. *Science* 252:651–658.

Price, M., and Suppe, J. 1994. Mean age of rifting and volcanism on Venus deduced from impact crater densities. *Nature* 372:756–759.

Pronin, A. A. 1986. The structure of Lakshmi Plateau, an indication of asthenosphere horizontal flows on Venus. *Geotectonika* 20:271–280.

Rappaport, N., and Plaut, J. J. 1994. A 360-degree and -order model of Venus topography. *Icarus* 112:27–33.

Roberts, K. M., and Head, J. W. 1990. Western Ishtar Terra and Lakshmi Planum, Venus: Models of formation and evolution. *Geophys. Res. Lett.* 17:1341–1344.

Robinson, C. A., and Wood J. A. 1993. Recent volcanic activity on Venus: Evidence from radio-thermal emissivity measurements. *Icarus* 102:26–39.

Simons, M., Hager, B. H., and Solomon, S. C. 1994. Global variations in the geoid/topography admittance of Venus. *Science* 264:798–803.

Smrekar, S. M., and Solomon, S. C. 1992. Gravitational spreading of high terrain in Ishtar Terra, Venus. *J. Geophys. Res.* 97:16121–1648.

Solomon, S. C. 1993. A tectonic resurfacing model for Venus. *Lunar Planet. Sci. Conf.* XXIV:1331–1332 (abstract).

Strom, R. G., Schaber, G. G. and Dawson, D. D. 1994. The global resurfacing of Venus. *J. Geophys. Res.* 99:10899–10926.

Turcotte, D. L. 1993. An episodic hypothesis for Venusian tectonics. *J. Geophys. Res.* 98:17061–17068.

Vorder Bruegge, R. W. 1994a. Depositional units in western Maxwell Montes: Implications for mountain building processes on Venus. *Lunar Planet. Sci. Conf.* XXV:1447–1448 (abstract).

Vorder Bruegge, R. A. 1994b. Variation in compressional structures across Maxwell Montes: Evidence for a sequence of events in a Venusian orogeny. *Lunar Planet. Sci. Conf.* XXV:1449–1450 (abstract).

Vorder Bruegge, R. W., and Head, J. W. 1990. Tectonic evolution of eastern Ishtar Terra, Venus. *Earth, Moon, Planets* 50/51:251–304.

Willis, J. J., and Hansen, V. L. 1995. Caldera-related volcanism and collapse at Ishtar Terra, Venus. *Eos Suppl.* 76:F341.

Zuber, M. T. 1987. Constraints on the lithospheric structure of Venus from mechanical models and tectonic surface features. *J. Geophys. Res.* 92:541–551.

Zuber, M. T. 1994. Rheology, tectonics, and the structure of the Venus lithosphere. *Lunar Planet. Sci. Conf.* XXV:1575–1576 (abstract).

PLAINS TECTONICS ON VENUS

W. BRUCE BANERDT
Jet Propulsion Laboratory

GEORGE E. McGILL
University of Massachusetts

and

MARIA T. ZUBER
Massachusetts Institute of Technology

Tectonic deformation in the plains of Venus is pervasive, with virtually every area of the planet showing evidence for faulting or fracturing. This deformation can be classified into three general categories, defined by the intensity and areal extent of the surface deformation: distributed deformation, concentrated deformation, and local fracture patterns. Each of these styles offers information about the tectonic history of the surface, as well as the physical properties and processes of the crust and upper mantle. Distributed deformation is manifested as individually narrow wrinkle ridges, troughs, and fractures which occur in subparallel sets that commonly extend over hundreds of km. The orientations of these sets tend to be generally constant over large regions, providing a valuable indicator of the regional stress and implying a basal (e.g., mantle convection) origin to this stress. Cross-cutting relationships with other geologic features (such as craters and volcanic landforms) and among different sets of fractures in the same region can provide relative timing information about the evolution of the surface. Concentrated deformation occurs in deformation belts, the nature and origin of which are problematic. The characteristics of ridge belts are generally consistent with an origin related to a regional compressive stress, although some evidence suggests that their formation is not simply related to a regional stress field. It is not evident how fracture belts formed, with both extensional features and elevated topography; the process by which deformation is concentrated into belts is also unclear. Local fracture patterns range in character from faint, parallel fracture sets to polygonal fracturing reminiscent of cooling or desiccation cracking (albeit on a much larger scale). The nature, spacings, and lengths of these features can offer insight into the mechanical properties of the uppermost layers of the crust and mechanical lithosphere. New flow laws available for dry diabase have important implications for modeling tectonic features and using them to infer the structure of the crust and mantle. In particular, these results call into question the importance, or even the existence of a weak lower crustal channel.

I. INTRODUCTION

Plains are the most widespread geologic province on Venus, making up over 80% of its surface. Originally, based on the relative flatness of the non-highland areas of Venus at the resolution of the Pioneer Venus altimeter, plains on Venus were defined in terms of elevation (Masursky et al. 1980), comprising all terrain below roughly the 1.5 km contour. With the higher resolution provided by Magellan, most of this area has indeed been found to be surfaced by flat, generally radar-smooth units believed to be due to flood volcanism, analogous to the volcanic plains found on the Moon, Mars, and (probably) Mercury. In this chapter we are concerned with the tectonic deformation of these plains, particularly that deformation that is not intimately associated with highlands, tessera, or volcanic features such as shield volcanoes and coronae.

One of the surprises to emerge from the Magellan images was the ubiquity of tectonic deformation evident in the plains. Almost no area is free from some sort of fracturing or faulting. This is in sharp contrast to the other terrestrial planets. On the Earth, tectonic deformation tends to be concentrated near plate boundaries. Similarly, albeit for different reasons, there are large areas on the Moon, Mars and Mercury which do not display any recognizable deformation at all.

The existence of deformational structures in a region provides a tool, in addition to such things as craters and volcanic flows, to unravel the temporal geologic sequence of events through superposition and cross-cutting relationships. Tectonic features also provide key information on the structure of the crust and lithosphere. This is particularly important for Venus, for which we have no subsurface seismic information or very high-resolution regional gravity data. As discussed later, the length scales of tectonic features, which presumably formed due to horizontal extension or contraction of lithospheric layers of varying thickness and mechanical competence, can be used in combination with geologic observations and experimental information on rock strength at varying pressure-temperature conditions to constrain the thicknesses and vertical strength distribution of the crust and lithosphere. Thus, the study of the tectonics of Venus' plains can offer important insights contributing to our understanding of the history and processes of its crust and upper mantle.

Primarily for organizational purposes we divide tectonic features observed on the surface of Venus into three categories: distributed deformation, concentrated deformation, and local fracture patterns. These divisions are based on the intensity and areal extent of deformation. Distributed deformation consists of sets of features which individually have a small, but discernible, amount of contraction, extension or shear, and which in aggregate over an extended region can comprise a significant strain. Concentrated deformation is manifested in quasi-linear zones of intense deformation, separated by areas of relatively undisturbed terrain. Local fracture patterns are sets of features with limited areal extent, and whose individual features have widths at or below the resolution of the Magellan imaging system (no better than ~100 m). Thus this type of deformation is not expected to contribute significantly to the larger-scale strain of the crust.

II. DISTRIBUTED DEFORMATION

A. Wrinkle Ridges

Wrinkle ridges are long, narrow, sinuous features ranging in width from the limit of SAR resolution to about 1 km, and with lengths up to several hundred km (Fig. 1). They generally are brighter on Magellan SAR images than the surfaces upon which they occur, and this brightness contrast is unrelated to the orientation of the ridges with respect to radar look direction (McGill 1993), implying that the enhanced radar return is more related to wavelength-scale surface roughness than it is to topography. Most Venusian wrinkle ridges occur in sets of approximately evenly spaced, parallel ridges. It is common for two, and sometimes three, of these sets to occur in the same area. Where more than one set occurs, it generally is clear that they are of different ages, based on stratigraphic relationships or on the intersection relationships among individual ridges of the several sets (McGill 1993).

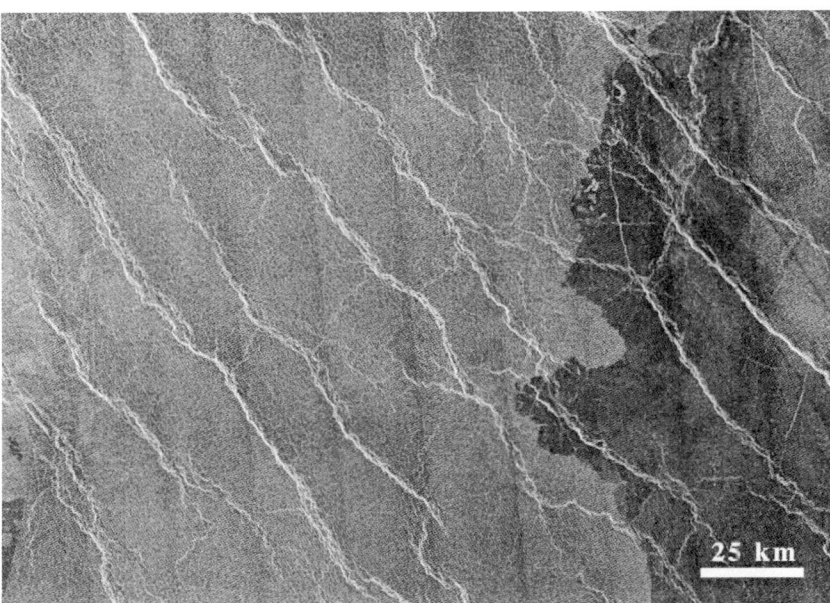

Figure 1. Typical wrinkle ridges in Rusalka Planitia (177°E, 2.5°N). Ridges are sinuous, less than 1 km wide, and exhibit an average spacing of about 20 km; part of C1–MIDR00N180, tile 19.

Distribution. Almost all plains areas on Venus are characterized by abundant wrinkle ridges. As plains constitute approximately 80% of the surface of Venus (Masursky et al. 1980), wrinkle ridges are extremely abundant on Venus; more so than on any other planet or moon. Wrinkle ridges are evidently not present (or not discernible) on tessera terrain, and are very rare or absent on lobate and digitate lava flows associated with relatively young shield volcanoes. They also appear to be less common in areas of intense post-plains deformation and within mountain belts.

Identification as Wrinkle Ridges. The primary criteria for the identification of wrinkle ridges on Venus are the long, narrow, sinuous morphology, and the occurrence in organized sets (Plescia and Golombek 1986; Watters 1991). Some of the larger wrinkle ridges exhibit the lateral brightness contrast on SAR images expected of ridges, but most features interpreted as wrinkle ridges on Venus have insufficient relief to cause resolvable shadows or back-slope darkening on SAR images. Some wrinkle ridges have ponded flows derived from impact, and others appear to define sharp contacts of local plains units (see, e.g., Figs. 1 and 2 of McGill 1993). These relationships imply positive relief, and thus it is inferred that other planimetrically similar sinuous features also are positive relief features. Wrinkle ridges on other planets commonly have a complex morphology characterized by a gentle arch surmounted by a much narrower, sinuous ridge that can be in the center of the arch or to either side (Strom 1972; Bryan 1973; Maxwell et al. 1975; Watters 1988). Some Martian and lunar wrinkle ridges involve an even broader, subtle rise (Lucchitta 1977; Plescia and Golombek 1986). Venusian wrinkle ridges either do not include broad rises and gentle arches, or these features are topographically too subdued to affect the radar return. Consequently, Venusian wrinkle ridges appear comparable to the simpler forms found on Mercury, Earth, the Moon and Mars. Wrinkle ridges generally are much smaller than the ridges or ridge belts (see below), and wrinkle ridge trends do not seem to exhibit any consistent relationship to those of ridge belts on a global scale; in places they are parallel, elsewhere they are not. Locally, however, features that appear to be wrinkle ridges grade into ridges of ridge belts by a gradual increase in width and apparent relief along trend. In addition, many coronae and corona chains include radial and concentric structures that appear to be morphologically similar to typical wrinkle ridges found on the plains. Our discussion of the structural and tectonic significance of wrinkle ridges is confined to the well-defined sets of parallel features found on the plains.

The origin of wrinkle ridges has been under debate for at least three decades. Both igneous and structural hypotheses have been proposed, as reviewed by Plescia and Golombek (1986). Recent research has narrowed the controversy to two main models: (1) buckling with or without some faulting (Watters 1991); and (2) thrust or reverse faulting with or without some folding (Plescia and Golombek 1986; Golombek et al. 1991). Both of these models imply that wrinkle ridges result from compressive stresses in the crust that are oriented approximately normal to the length of the ridges.

There is some independent evidence on Venus supporting this inference where wrinkle ridges interact with well-defined topographic features (see, e.g., Fig. 3 of McGill 1993).

Age Relations. Almost all Venusian wrinkle ridges are superposed on the materials that make up the plains, and thus the ridges must be younger than these plains. Within the plains, the abundance of wrinkle ridges commonly decreases with decreasing age of specific plains units, suggesting a progressive contractional deformation of crustal rocks during emplacement of the plains materials (Solomon et al. 1992; Squyres et al. 1992). In some places, wrinkle ridges appear to postdate at least the initial stages of corona formation (McGill 1993,1994), but this relationship has not been systematically tested globally. In contrast, young lobate and digitate lava flows, especially those associated with large shield volcanoes, are nearly devoid of wrinkle ridges. Although the relative ages of impact craters and wrinkle ridges are commonly ambiguous, almost all ridges that can be dated relative to craters are older than the craters. Thus wrinkle ridges appear to be temporally related to plains formation, with most ridge-forming deformation occurring relatively early, based on the limited crater data available. During the waning stages of ridge formation the trends of ridge sets maintained the same orientation in some places, e.g., parts of Lavinia and Guinevere Planitiae (Solomon et al. 1992; Squyres et al. 1992), whereas in other places trends changed significantly, e.g., Eistla Regio (Basilevsky 1994; McGill 1994).

Structural Domains. Most plains regions of Venus are characterized by a dominant, throughgoing set of wrinkle ridges that maintains a roughly uniform trend or uniform curvature of trend over hundreds to thousands of km. For example, in most of Lavinia Planitia the dominant trend is northeast (Squyres et al. 1992), over a very large area of Sedna and Niobe Planitiae the dominant trend is approximately east–west, in Rusalka Planitia the dominant trend is northwest (Lancaster and Guest 1994), and within Aino Planitia the dominant wrinkle ridge set defines a large arc concentric to Artemis Chasma (McGill 1992; Bilotti and Suppe 1992; Bilotti et al. 1993). These regions represent stress domains that are presumably confined to the shallow crust. Although studies are in progress (see Fig. 2 for a map of ridge directions in the Aphrodite/Eistla region), a global map of stress domains defined by wrinkle ridges has not yet been completed. In some areas, such as Aino Planitia, the dominant wrinkle ridge set appears related to a major tectonic or topographic feature (Bilotti et al. 1993). In other areas, such as Sedna and Rusalka Planitiae (Lancaster and Guest 1994), the dominant trend is clearly oblique to nearby tectonic and topographic features. In many, and perhaps most, plains areas there is a second (and, rarely, a third) set of wrinkle ridges in addition to the dominant set that defines the domain. In all areas so far studied in sufficient detail, these additional sets are younger than the dominant set. In many places, the younger sets are clearly related to local tectonic and topographic features. For example, on the plains adjacent to Eistla Regio younger wrinkle ridges define sets concentric to the large shield volcanoes (Basilevsky 1994; McGill

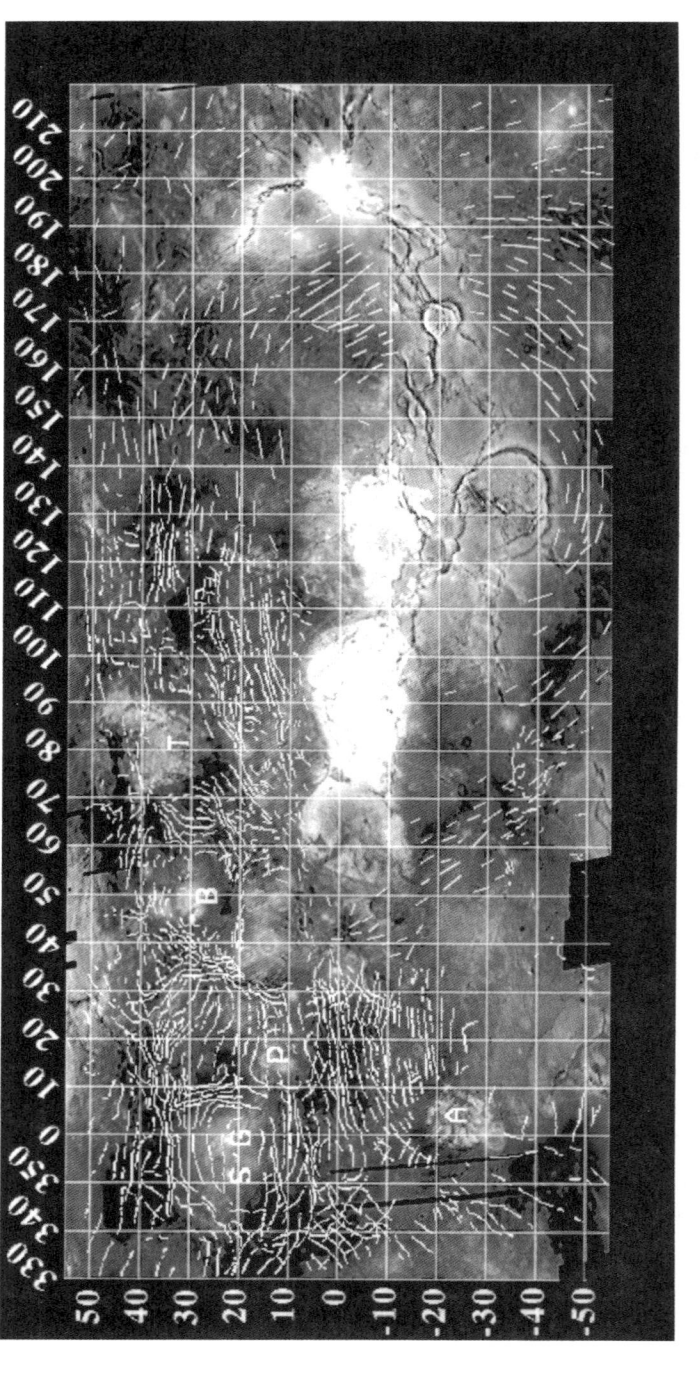

Figure 2. Trends of wrinkle ridges in plains surrounding Aphrodite Terra and Eistla Regio, plotted on a topographic base. Trends are both parallel and oblique to these major topographic features, but the dominant trends generally are oblique. Data from Bilotti et al. (1993), Basilevsky (1994), Lancaster and Guest (1994), McGill (1994), and McGill (unpublished). S = Sif Mons, G = Gula Mons, P = Sappho Regio, A = Alpha Regio, B = Bell Regio, T = Tellus Regio.

1994; Copp and Guest 1995).

The tens of km spacings typical of many wrinkle ridges are indicative of a depth penetration of regional compressive stresses on the order of the thickness of the strong upper crustal layer (i.e., a few to 10 km). The observed widths of 1 km or less, which are generally maintained along strike for tens to hundreds of km, imply small amounts ($\sim$1–5%) of upper crustal shortening.

B. Grabens, Linears, and Dikes

Linear features of extensional or inferred extensional origin are abundant on the Venusian plains. Many of these show the brightening or darkening on SAR images that one would expect for slopes facing towards or away from the radar; these are inferred to be fault scarps. Many of these scarps occur in facing pairs bounding troughs that are morphologically identical to grabens on Earth, Mars and the Moon. More common are radar-bright linears that are too narrow to define their boundaries because they are only one or two pixels across. This limited pixel width also means that any relief associated with these linears is not resolvable.

Very few areas underlain by the global plains are completely devoid of faults or linears. Plains surrounding volcanic constructs and coronae are especially likely to have abundant extensional structures, commonly with multiple trends. Large (width greater than two km or so) grabens seem to be more common adjacent to tessera terrain.

Scarps, especially those bounding troughs, are readily interpreted as extensional by direct analogy with similar features on Earth. The narrow linears are inferred to be extensional as well, based on less direct evidence. Because many of these features parallel larger structures resolvable as grabens or fault scarps, and because some increase in width along trend into scarps or grabens (Fig. 3), it is very likely that all of them represent extensional structures. In addition, the planimetric patterns defined by some of these linears suggest origins by extension, as will be discussed in more detail below. Thus they could be either fractures or small faults. Because the brightness of these features on SAR images seems unrelated to their orientation with respect to the radar look direction, this brightness is most likely due to enhanced roughness at radar wavelength scale rather than due to topography. This characteristic would seem to favor interpreting narrow linears as fractures (joint zones?) rather than as faults.

Many extensional structures occur as radial or concentric sets associated with coronae or volcanic centers. Individual linears of these sets extend hundreds or even thousands of km from the source feature. These radial sets very likely represent surface fractures or narrow grabens overlying dikes derived from the central feature, analogous to the well-studied examples in northwestern Scotland and the Spanish Peaks region of Colorado (Anderson 1951; Odé 1957; Muller and Pollard 1977). McKenzie et al. (1992) present a mechanical analysis of the Venusian radial and concentric linears based on models developed for these terrestrial examples. At large distances from the central source,

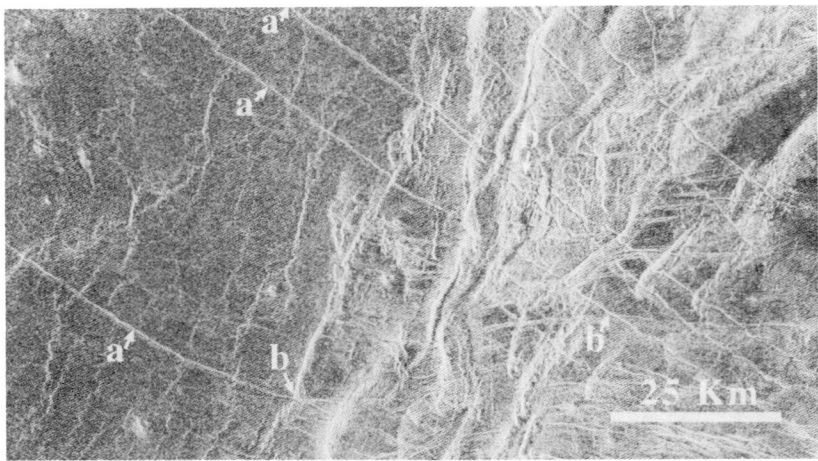

Figure 3. Examples of typical bright, narrow linears with unresolvable geometry (points "a") that widen along trend into resolvable grabens (points "b"). Thus these linears, and parallel members of the same set, are interpreted as due to extensional strain; parts of FMIDR45S350, tiles 9 and 10.

the radial sets commonly follow preferred trends that most likely define the regional trajectory of the maximum principal compression. These sets locally diverge around holes in the crust (McKenzie et al. 1992), a pattern consistent with the behavior of compressive stress trajectories around holes in elastic plates. In the vicinity of Aphrodite Terra, the inferred maximum compression is normal to the long wavelength topography (Grosfils and Head 1994), in general agreement with the results of Bilotti et al. (1993) based on trends of wrinkle ridges in Aino Planitia. Elsewhere, however, there is no globally consistent relationship between the inferred stress orientations and large-scale topography (Grosfils and Head 1994), a result that also is consistent with the wrinkle ridge data. Where available, stratigraphic relationships indicate that linears of these large radiating swarms formed after the regional plains but before almost all impact craters (Grosfils and Head 1995). Assuming that the determinable relative ages are characteristic of the global relationships, this implies that the dikes inferred lie at depth beneath the linears, and thus also the source centers, all formed very soon after emplacement of the global plains.

In addition to the large swarms of linears associated with well-defined central sources, the plains regions of Venus commonly are cut by much shorter linear features. These can occur as closely spaced, parallel linears generally scores to hundreds of km long, or as en-echelon lines a few km to a few tens of km long. Most commonly, there is no connection between these smaller linears and topographic, structural, or volcanic features. Locally, these smaller linears are so closely spaced that they define a fabric that is penetrative at the scale of the Magellan images (see, e.g., Squyres et al. 1992, Figs. 3 and 14). It seems unlikely that these smaller linears can be explained as the surface manifestations of dike emplacement. Some of these small linears are parallel to sets of larger extensional structures radial to coronae or volcanic centers, and thus they probably are simply mode I (purely extensional) cracks formed in response to the same regional stress field that is controlling the far-field orientations of the radial swarms. Other small linears are not parallel to larger structures, but their similarity in size and geometry to sets that are parallel to larger structures suggests that they, too, represent mode I cracks.

Most of the grabens occurring as members of linear sets are narrow, ranging in width from the limit of resolution to perhaps 2 km. Grabens with widths up to 15 km are common in tessera inliers within plains. Many of these are partially filled by plains materials, indicating that they are older than plains formation. However, there are local indications of continued motion on these structures after emplacement of plains (see, e.g., Fig. 33 of Solomon et al. 1992), suggesting that the stress regime responsible for these grabens still existed at the very beginning of plains formation.

C. Broad-Scale Vertical Deformation

Vertical deformation of the lithosphere can result in characteristic tectonic patterns. These patterns are formed by a combination of extensional and contractional structures that either follow the vertical deformation gradient or are orthogonal to it (e.g., radial or circumferential orientation for a circularly symmetric deformation), depending on the horizontal scale of the uplift or depression and the thickness of the elastic lithosphere (see Banerdt et al. 1992). Such patterns have been tentatively identified around a number of relatively small features, such as volcanic edifices and coronae. On a broader scale (say, comensurate with highland plateaus) these types of pattern have not been identified, and either they do not exist on Venus or the tectonic complexity obscures them. However, the existence of lava channels and digitate lava flows provide us with snapshots of local slope directions at the time of their emplacement that can be compared with present slopes to infer vertical deformation.

Both lava channels and digitate lava flows can provide direct evidence of paleoslopes if the initial flow direction is unambiguous. Even if the flow direction is ambiguous, these features can serve as reference surfaces because the direction of slope cannot have reversed during emplacement; any slope reversals now present must be due to deformation of the original slope at a scale smaller than the entire flow or channel length. Lava channels prove especially useful in this regard because many are very long and thus define an initially gentle slope. Published results suggest that many long lava channels have experienced significant post-emplacement deformation with as much as 2 km of relief between adjacent low and high points along the channels

(Parker et al. 1992; Komatsu and Baker 1994; McLeod and Phillips 1994). The relief defines two scales of deformation: one with a wavelength of a few thousand km, the other with a wavelength of a few hundred km. The large-scale undulations have been defined for only the two longest channels (Komatsu and Baker 1994); these authors infer that this deformation scale corresponds to that of large basins. The shorter scale corresponds closely with the characteristic spacing of ridge belts (Zuber 1986; Frank and Head 1990; Squyres et al. 1992).

Lava channels must, of course, be younger than the plains units they transect. Although a complete global evaluation of their ages has yet to be completed, channels commonly are disrupted by impact craters and cut by wrinkle ridges (McGill 1993; Komatsu and Baker 1994). This suggests that they formed either late in the global episode of plains formation, or very soon thereafter. Thus channels appear to be relatively old plains features.

III. CONCENTRATED DEFORMATION

Several of the plains regions of Venus are characterized by long, narrow belts of relatively intense deformation. The dimensions of these belts vary widely; widths range from narrow tips a few km across to broad zones as much as 300 km wide, and lengths range from <100 to 2000 km or more (Frank and Head 1990; Senske et al. 1991). Most of these belts stand a few hundred meters to more than a km above the surrounding plains, although a very few deformation belts occur within shallow depressions. Deformation belts tend to be bright on SAR images compared to adjacent plains, and this contrast is probably due both to topographic effects and to the greater roughness of the belts at radar wavelength scale. Initial studies of these features were based on Venera 15 and 16 images (Barsukov et al. 1986; Basilevsky et al. 1986; Kryuchkov 1988; Sukhanov and Pronin 1989; Sukhanov et al. 1989; Frank and Head 1990) and Arecibo images (Campbell et al. 1991; Senske et al. 1991). Because of the very low incidence angle of the Venera radar, and the kilometer-scale resolutions of both data sets, it was difficult to determine if the individual linear structures within these belts are scarps, graben-like grooves, or ridges (see, e.g., Senske et al. 1991). The better resolution and more favorable incidence angles of Magellan images allow a clear distinction in most instances between deformation belts characterized by scarps and grabens, referred to as "fracture belts," and deformation belts characterized by ridges, referred to as "ridge belts" (Solomon et al. 1992; Squyres et al. 1992). Although it is convenient and logical to discuss fracture belts and ridge belts separately, as we do, there are composite belts in places.

A. Ridge Belts

Description and Classification. Ridge belts are morphologically diverse. Some consist essentially of a single broad arch, with smaller ridges superposed in some places. More commonly, ridge belts include a complex array of individual ridges that also exhibit diverse morphologies. The most thorough descriptions of ridge belts, and the only classifications, are based on Venera 15 and 16 images (Kryuchkov 1988; Frank and Head 1990). Kryuchkov (1988) divided ridge belts into three broad classes: (I) belts consisting essentially of a single wide swell with a broad summit (Fig. 4); (II) belts made up of many closely spaced smaller ridges (Fig. 5); and (III) "spaced" ridges, with individual ridges very far apart compared to their widths. Classes I and II are easily recognizable on Magellan images, but it is not clear that there is a distinct group of belts corresponding to class III (see below).

Frank and Head (1990) developed a classification of individual ridges based on Kryuchkov's three classes of ridge belts. Class I ridge belts are termed "broad arches," and consist of single arches 20 to 40 km wide, commonly with smaller ridges superposed. Based on Venera images, these ridges were considered to be analogs of lunar, Mercurian and Martian wrinkle ridges (Frank and Head 1990), which commonly consist of a broad arch with a narrower superposed sinuous ridge. This analogy is less convincing on Magellan images because there generally are several superposed smaller ridges rather than one, and because these superposed smaller ridges are less sinuous and significantly broader than typical first-order wrinkle ridges on other planets (Watters 1988). It is clear on the Magellan images of Nephele Dorsa (Fig. 4), a class I ridge belt or broad arch (Frank and Head 1990), that the superposed small ridges most similar to first-order wrinkle ridges are, in fact, members of a wrinkle-ridge set on the adjacent plains that cross the ridge belt at a low oblique angle. The superposed small ridges confined to Nephele Dorsa are larger and straighter than typical wrinkle ridges.

Class II ridge belts were subdivided into three subclasses by Frank and Head (1990) according to whether the individual ridges within the belts are discontinuous, paired, or anastomosing. Most class II belts are dominated by anastomosing patterns of individual ridges, but discontinuous and paired ridges are generally interspersed among the anastomosing ridges. The overwhelming majority of all ridge belts are in class II. These also are the largest ridge belts, with widths as great as 300 km and lengths up to several thousand km locally. Individual ridges within class II belts are typically 5 to 15 km wide, with a slightly larger inter-ridge spacing (Frank and Head 1990; Squyres et al. 1992). In Vinmara and Atalanta Planitiae class II ridge belts are sufficiently abundant to determine the typical spacing between them. Spacing varies from 70 to 670 km, but most values lie between 325 and 425 km (Zuber 1986; see also Fig. 4b of Frank and Head 1990). The ridge belts of Lavinia Planitia exhibit similar spacings of ridges and belts (Squyres et al. 1992).

Frank and Head (1990) did not provide a detailed description of a class III ridge belt. The illustrated example is the same one used by Kryuchkov (1988), incorrectly located in Bezlea Dorsa. The area is actually in easternmost Sedna Planitia, about 25° to the west of Bezlea Dorsa. The ridges shown are widely spaced, sinuous, and <1 km wide; they are part of a completely typical set of

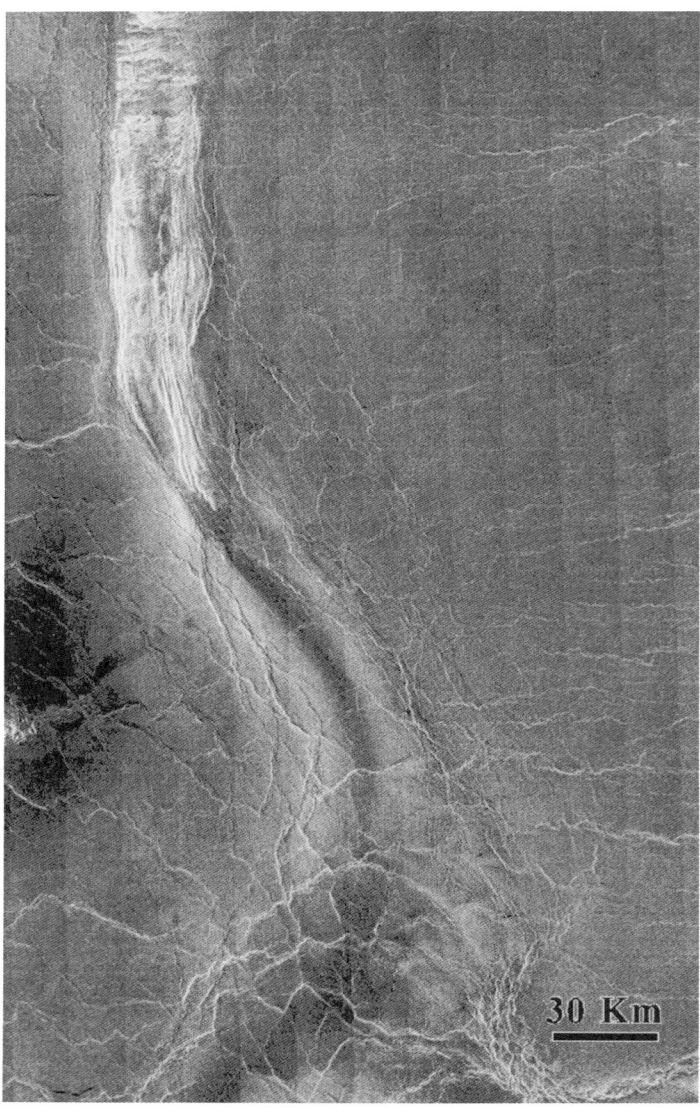

Figure 4. Northern part of Nephele Dorsa, a class I ridge belt located at 40°N, 140°E in northern Niobe Planitia. Nephele is dominated by a broad, gentle arch about 40 km wide. The narrow, sinuous ridges superposed on this arch in the southern half of the figure are wrinkle ridges associated with the surrounding plains, and are not related to Nephele. The ridges superposed on the brighter part of the arch in the northern half of the figure do appear to be related to Nephele. However, these are wider than typical wrinkle ridges and not sinuous; C1–MIDR45N138, tiles 37 and 45.

plains wrinkle ridges. Thus there is reason to doubt if this is a valid class of ridge belts.

Age and Origin. The most thoroughly studied areas of ridge belts are in Atalanta, Vinmara, and Lavinia Planitiae. In Lavinia Planitia, the ridge belts have deformed a radar-bright, "textured" plains formation that clearly is older than the materials characteristic of most of the global plains (Squyres et al. 1992, Figs. 3 and 14). The ridges and the textured plains are embayed by flood-type lavas. It thus seems clear that in this area the ridges formed relatively early in the evolution of the plains. In the northern hemisphere, however, relationships are more complex. In Vinmara Planitia, the relative ages of ridge-belt and plains materials are ambiguous for many, and perhaps most, of the class II ridge belts (Rosenberg 1995). Some ridge belts in the Atalanta-Vinmara area deform tessera terrain, as do belts south and east of Fortuna Tessera (Sukhanov and Pronin 1989), but these belts also extend across and are locally younger than adjacent plains. As discussed briefly in Solomon et al. (1992), it thus is not clear whether (a) ridge belts always formed early in plains development but plains in different areas are not the same age, or (b) ridge belts formed at diverse times in the development of coeval plains.

Ridge belts are generally interpreted as resulting from global-scale compressive stresses oriented normal to their trends (Basilevsky and Head 1988; Basilevsky et al. 1986). The arch-like morphology and topography of some individual ridges (Barsukov et al. 1986; Frank and Head 1990) supports the interpretation of these features as anticlinal folds. The common tendency for ridge belts to be elevated relative to their surroundings is also consistent with localized crustal thickening due to the same global-scale stress field (Solomon et al. 1992; Squyres et al. 1992). The elevated topography will induce gravity sliding stresses (Turcotte and Schubert 1982), and the importance of these stresses can be estimated from the predicted strength of the lithosphere and the elevation of the ridge belts over the surrounding plains. This can be as much as 2 km (Ford and Pettengill 1992; Frank and Head 1990), but is more typically hundreds of meters (Squyres et al. 1992). As discussed below, the depth-averaged strength of the Venus lithosphere can be assumed to be controlled by Byerlee's Law in the brittle regime and by flow laws for dry diabase (Mackwell et al. 1995) for the crust and dry olivine (Chen and Morgan 1990) for the mantle in the ductile regime. The maximum elevation of the ridge belts results in a maximum gravity sliding stress of $\sim$50 MPa, and, for the average strength indicated by laboratory experiments (see below), the stresses could produce failure in the upper several km of the Venus lithosphere. Gravity sliding stresses associated with the elevation of individual ridge belts may thus contribute to the development of shallow, small-scale deformation features within the belts. But given the greater strength of the Venus lithosphere at depth (which would inhibit longer wavelength deformation due to gravity sliding) and the fact that ridge belts occur in areas of average or low elevation relative to the global mean, broader-scale ridge belt deformation requires

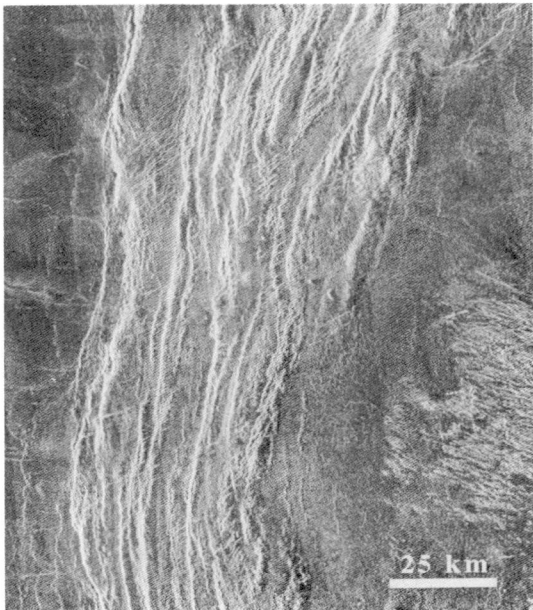

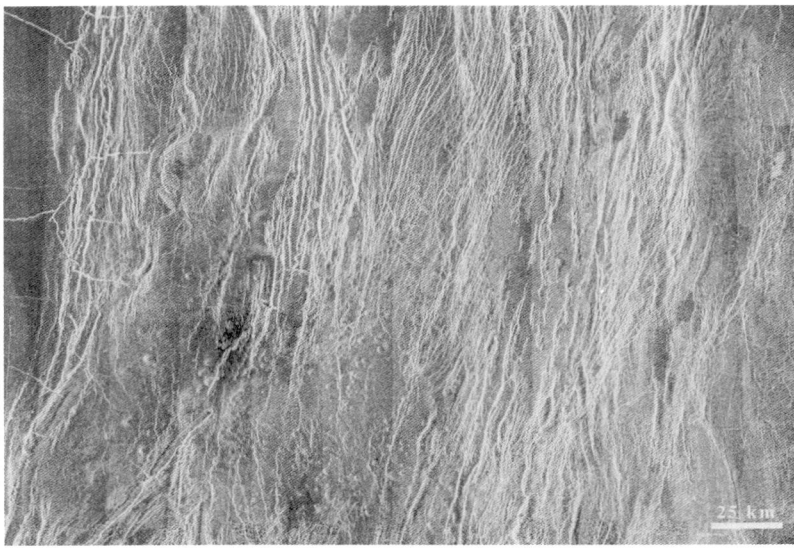

Figure 5. Pandrosos Dorsa, a ridge belt in Vinmara Planitia. (*Top*) Northern, relatively simple portion of the belt (214°E, 63.5°N). The belt here is 40 km wide; the individual ridges are about 1.5 km wide and spaced about 5 km apart; part of C1–MIDR60N208, tile 14. (*Bottom*) Southern, complex portion of the belt (207.5°E, 56°N). The width of the belt is 220 km; individual ridges are similar to, but more closely spaced than ridges in the northern portion. Abundant bright linears (fractures?) are present in addition to ridges, and the belt exhibits evidence for a prolonged history of continuous or alternating deposition and deformation; parts of C1–MIDR60N208, tiles 44 and 45.

an additional or alternative mechanism, such as mantle flow-related stresses (Phillips 1990). Given the broad, spatially coherent style of deformation of some ridge belts (Zuber 1990), the source of stress that produced these features is consistent with formation by mantle downwelling (Phillips et al. 1991; Squyres et al. 1992; Zuber 1990). The lack of a broad, regional uplift due to thickening of the crust that would be a consequence of the downwelling process (Bindschadler et al. 1992) indicates that the ridge belt fan in Atalanta Planitia could be a site of incipient downflow (Zuber 1990). It has also been proposed that ridge belts may mark the sites of former downwelling, with belts formed due to thrusting associated with rebound subsequent to the cessation of downward flow (Phillips et al. 1991).

An alternative explanation of at least some ridge belts as due to extension has been proposed (Kryuchkov 1990; Raitala and Tormanen 1990; Sukhanov and Pronin 1989) based on the recognition that features such as volcanic centers associated with some ridge belts are difficult to explain in a compressional stress environment. However, the defining characteristics of these belts as ridge-like arches would not be expected in a tensional stress regime. Tensional bending stresses at the crests of flexural folds produced due to remote compression may provide an explanation for at least some of the observed localized extension. A flexural (as opposed to penetrative deformation) response of the lithosphere to compressive stress would be consistent with the notion of significant mechanical competence of the lithosphere, as suggested from gravity modeling (see, e.g., the chapter by Phillips et al.).

B. Fracture Belts

Description. Fracture belts are broad, elongate swells or arches that are characterized by numerous roughly belt-parallel linears, scarps, and grabens. In detail, the fractures and faults of these belts commonly occur as sets that intersect at low angles. The most thoroughly studied examples occur in Guinevere and Lavinia Planitiae in the southern hemisphere (Solomon et al. 1991,1992; Squyres et al. 1992). Fracture belts in Guinevere and Lavinia Planitiae range in size up to about 200 km wide and 1000 km long, and stand a few hundred meters to more than a km above adjacent plains. Globally, it is difficult to clearly define fracture belts, because in places structures similar to those described in Lavinia Planitia grade along trend into corona chains (Fig. 6) or ridge belts (see, e.g., Figs. 29 and 30 of Solomon et al. 1992).

Age and Origin. Where mapped in Lavinia and Guinevere Planitiae, fracture belts clearly fault and elevate materials of the surrounding plains (Squyres et al. 1992; Solomon et al. 1992, Figs. 31 and 32). Because these plains embay ridge belts, the fracture belts of Lavinia Planitia are clearly younger than the ridge belts (Fig. 7). However, young digitate flows in Lavinia Planitia were diverted by pre-existing fracture belts and thus, like ridge belts, fracture belts appear to be older than the youngest flows on Venus.

The linears (fractures?), scarps, and grabens making up fracture belts are all consistent with formation by tension normal to the belt trends. The

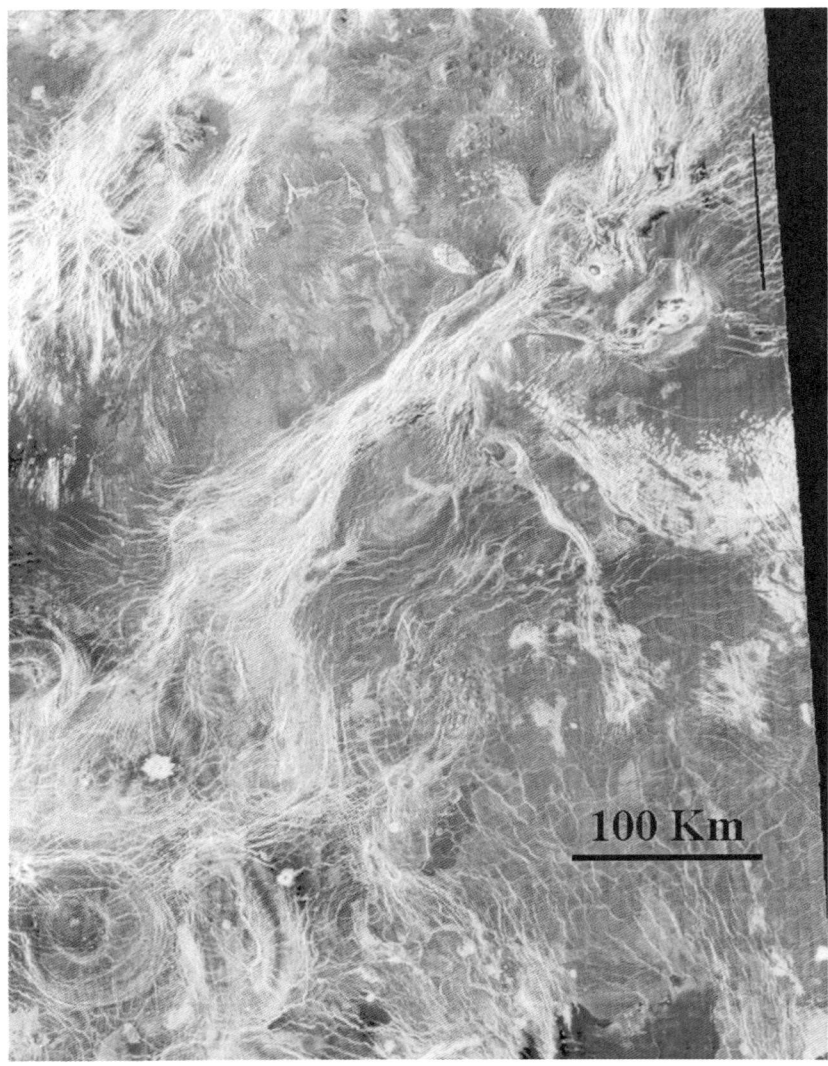

Figure 6. Fracture belt in northern Bereghinya Planitia, located at 50°N, 20°E. This belt, which is morphologically similar to those found in Lavinia Planitia, grades to a chain of coronae to the south; C2–MIDR30N026, tiles 4, 5, 12, and 13.

presence of rhomboidal depressions bounded by faults on some fracture belts (Solomon et al. 1991) suggests that many are actually due to transtension. On the other hand, the broad ridges characteristic of these belts suggests crustal thickening due to compression, as for ridge belts (Squyres et al. 1992). But the ridge and fracture belts of Lavinia Planitia are nearly orthogonal to each

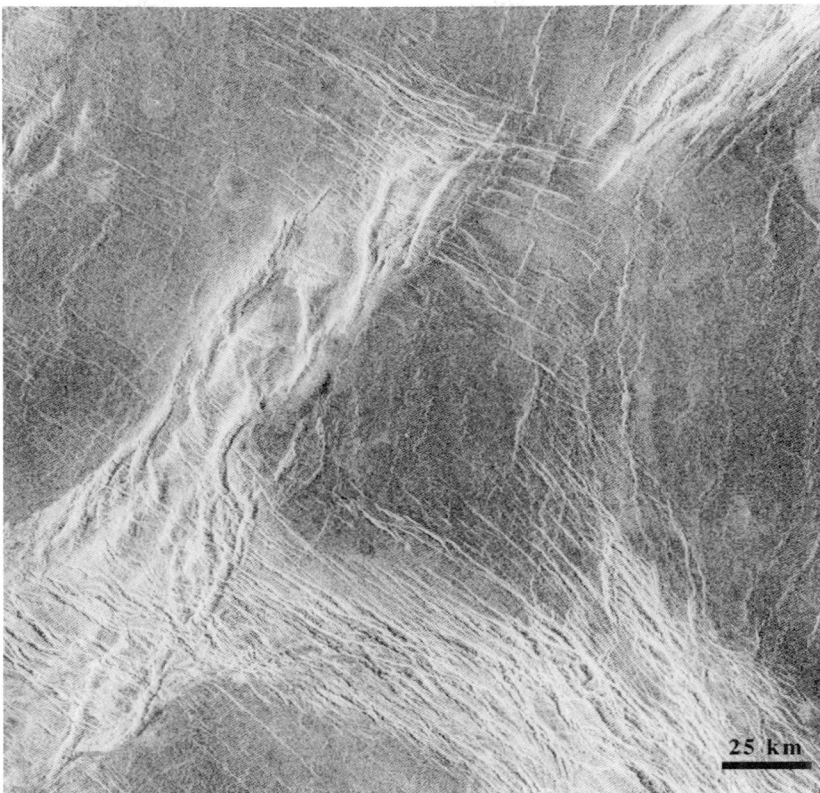

Figure 7. Ridge and fracture belts in Lavinia Planitia (348°E, 38°S). The ridge belt, trending NE, is 30 km wide and generally consists of 2 or 3 individual ridges 3.5 to 4 km wide. The ridges and the material they deform appear to be embayed by the surrounding plains material. The transecting fracture belt, trending NW, cuts the ridges and the surrounding plains, and thus is younger than the ridge belt. Note that the wrinkle ridges on the plains are generally not parallel to the ridges of the ridge belt; C1–MIDR45S350, tile 4.

other (e.g., Fig. 7). It is possible in Lavinia Planitia, at least, to account for this apparent paradox by assuming a nearly 90° reorientation of principal compression trajectories between formation of ridge belts and fracture belts. However, this appears inconsistent with the persistence of NNE trends of wrinkle ridges. If fracture belts are due to compression rather than tension, then the characteristic extensional structures must result from local tension due to uplift and bending (Solomon et al. 1991). However, for a typical belt width of 200 km and height of 1 km, simple sinusoidal bending would have induced a maximum strain of only about 1% (for a 40 km plate), insufficient to account for the abundant fractures and faults.

IV. LOCAL DEFORMATION

Local deformation as defined above is commonly manifested on the Venusian plains in patterns of fine-scale structures, which are developed on the kilometer to sub-kilometer scale. These patterns can be classified generally into parallel sets, polygonal patterns and irregular patterns.

A. Parallel Fracture Sets

Solomon et al. (1991) first noted remarkably linear features which are developed with a regular spacing of about a km in the "gridded plains" of Guinevere Planitia, where faint, regular lineations form the NE-trending component of the grid (see Fig. 3 of Solomon et al. 1991 and Fig. 1 of Banerdt and Sammis 1992). Although such a well-developed orthogonal grid is unique to this location, Banerdt and Sammis (1992) found similar sets of regularly spaced lineations in many locations on Venus, and concluded that such features are relatively common throughout the plains.

An example of such a set is shown in Fig. 8. They are composed of parallel, thin (a single pixel in Magellan images), straight lineations whose microwave reflectivity does not depend on radar illumination direction. The patterns typically cover areas with dimensions of hundreds of km. The average spacing of lineations is small, between 1 and 2.5 km, and the scatter in individual spacings is about $\pm 1/3$ the average. Based on these observations, especially the narrow, linear geometry and the azimuthal independence of radar reflectivity, Banerdt and Sammis (1992) concluded that they are tension fractures in the brittle upper layers of the volcanic plains material.

The very close spacing of these features is perplexing, as conventional geophysical models for regular spacing require an unreasonably thin lithospheric layer (<1 km; Solomon et al. 1991). One possible solution to this problem is suggested by the observation from structural geology that jointing within a sedimentary layer often occurs with a spacing roughly proportional to the layer thickness (see, e.g., Pollard and Aydin 1988). Calculations of the state of stress around a vertical crack (see, e.g., Lachenbruch 1961; Pollard and Segall 1987) show that the relief of horizontal tensile stress (the "stress shadow") occurs mostly within a distance comparable to the crack depth, inhibiting the development of subsequent cracks within that region. However, Banerdt and Sammis (1992) observed that these patterns appeared to have virtually the same spacing (1–2.5 km) everywhere they were observed. They proposed a shear-lag model in which a relatively thin surface layer is partially decoupled from similar material below by a frictional contact. This results in a spacing between features that is independent of the thickness of the layer, as both the frictional resistance and the layer strength scale similarly with thickness. An implicit requirement of this model is that the layer have a relatively large tensile strength (implying only a small amount of pre-existing fracturing). This implication and the observation that the parallel fracture sets do not appear to follow other superimposed structural trends suggest that the

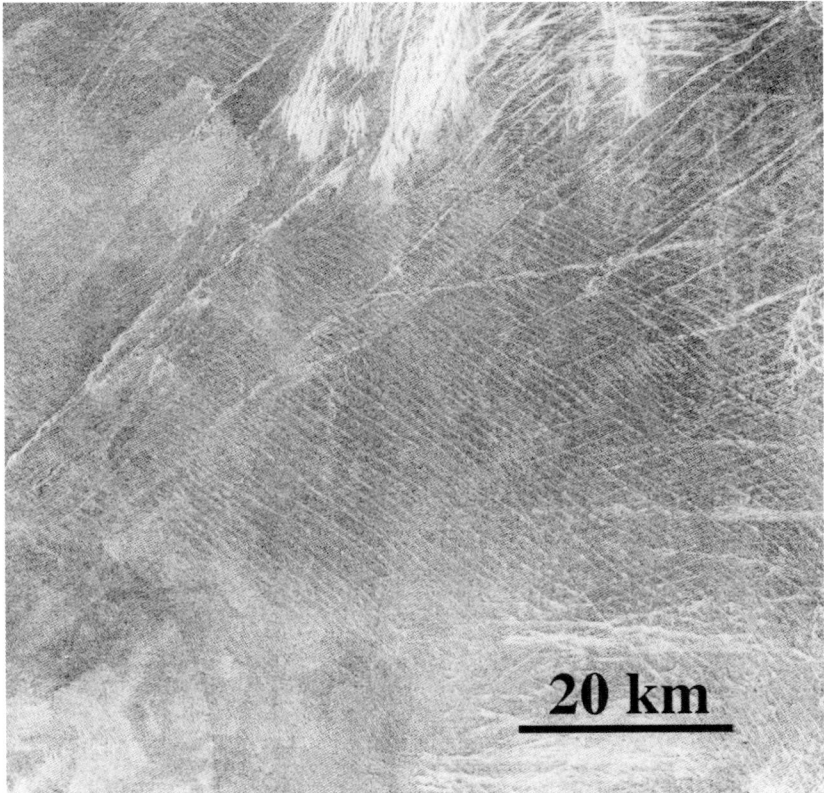

Figure 8. Typical parallel fracture pattern on plains units in Eistla Regio (15°N, 44°E); parts of FMIDR15N043, tiles 29, 30, 37, and 38.

tectonic episode responsible for the parallel fracture set must predate other strong deformation events recorded on that surface. Thus these structures may serve as a relative temporal marker for a deformation sequence, showing the orientation and sense (extensional) of the earliest tectonic event to which the plains unit was subjected.

B. Irregular Structures of the Gridded Plains

The set of features orthogonal to the thin parallel lineations in the gridded plains have been the subject of several studies which have attempted to use their morphology and length distribution to constrain lithospheric properties (Sammis and Banerdt 1991; Banerdt and Sammis 1992; Bowman and Sammis 1995). These features are primarily extensional in origin, because they grade into recognizable grabens to the north. However, their distinct curvilinear en-echelon morphology suggests a component of shear as well. Bowman and Sammis (1995) have inferred that these structures formed by the propagation of deeper ("basement") fractures through a surficial basalt layer to the surface. The length distribution of these structures is found to have a distinct break in slope at a length of about 80 km. In the model of Bowman and Sammis (1995), this break corresponds to the change from the two-dimensional growth of a semi-circular crack in an half-space to simple one-dimensional horizontal elongation after it has penetrated to the bottom of the brittle lithosphere. With this model the length distribution implies a thickness for the mechanical lithosphere of about 40 km, in agreement with independent determinations in other areas.

C. Polygonal Patterns

Polygonal patterns of bright lineations, broadly similar in appearance to cooling and dessication crack patterns (albeit at a much larger scale), are another common feature on the plains (Fig. 9). These appear at scales from 10 km down to the resolution of the Magellan radar, and are typically 1 to 2 km across. Because of their isotropic orientations and apparent tensional nature, Johnson and Sandwell (1992) interpreted these features to be due to thermal stresses. They investigated two scenarios for the generation of these stresses, cooling of an initially liquid lava flow and local heating of the lithosphere from below. Both process were found to generate sufficiently large stresses to produce the fracturing for reasonable parameters, but the reheating model was favored due to the difficulty in scaling the dimensions of the polygons from the meter-scale structures seen in terrestrial lava lakes to the kilometer-scale features observed on Venus.

There are also common occurrences of more complex and irregular patterns of radar-bright lineations with length scales of the order of a few km (Solomon et al. 1991; Johnson and Sandwell 1992). In some of these areas a background grid of parallel lineations is discernible (Banerdt and Sammis 1992). It is possible that processes similar to those responsible for the kilometer-scale spacing of the parallel fracture patterns also control the length scales of the more irregular patterns.

V. DISCUSSION

The likely lack of a low viscosity zone in the Venus mantle (Kiefer et al. 1986) implies that mantle convective stresses could be strongly coupled to the overlying lithosphere. Numerical experiments have demonstrated that mantle flow-related stresses transmitted to the lithosphere could attain magnitudes sufficient for tectonic deformation (Phillips 1990). It is thus prudent to compare the distribution of plains tectonism to the internal density structure implied from gravity and topography data. Global gravity fields (Reasenberg and Goldberg 1992; McNamee et al. 1993; Nerem et al. 1993; Konopliv et al. 1993; Konopliv and Sjogren 1994) and a regional high-resolution line-of-sight inversion (Barriot and Balmino 1994) show the plains of Venus to be relatively gravitationally featureless. This is not surprising, as the resolution of

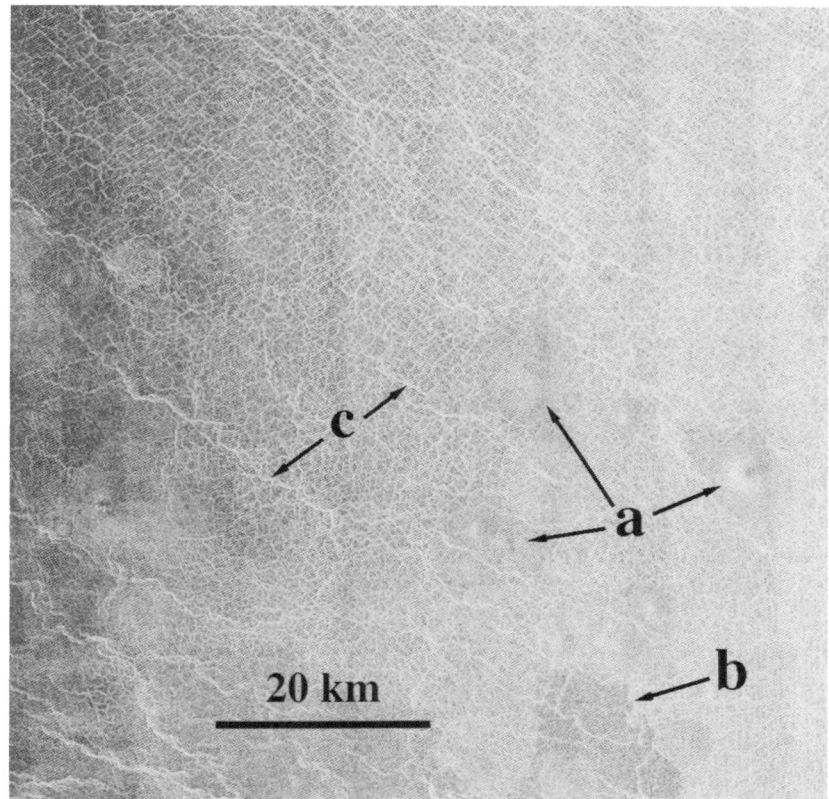

Figure 9. Polygonal patterns in Guinevere Planitia (21°N, 334°E). The lineations are locally obscured by presumably younger volcanic domes and associated deposits (e.g., "a"), and are truncated at the boundary between lighter and darker plains units ("b"). Brighter, more continuous lineations trending NE (e.g., "c") are parallel to, and in places grade into, wrinkle ridges in the adjoining plains. We interpret them to be younger than the polygonal pattern, as they appear to follow the polygonal grid; parts of FMIDR20N334, tiles 20, 21, 28, and 29.

the gravity is much less than the length scales of individual tectonic features. Topographically, the plains are, by definition, at or below the planetary mean (Ford and Pettengill 1992; Rappaport and Plaut 1994), which is generally consistent with the ubiquity of contractional tectonic structures.

Global models of the internal density distribution from inversion of gravity and topography data (Banerdt 1986; Herrick and Phillips 1992) show evidence for isolated upwellings among an interconnected network of downwellings. Major regions of downwelling are all associated with plains units, some of which contain ridge belts. Power spectral ratios of gravity and topography are also consistent with a large-scale pattern of downwelling flow beneath the plains lithosphere. An alternative interpretation is that the plains are regions of crustal thinning associated with mantle uplift (Buck 1992), but this hypothesis has not been tested in the context of the observed distribution of surface tectonics. Gravity/topography relations of plains regions indicate a thickness of the mechanical lithosphere of ∼50 to 100 km (Bindschadler 1994; Simons et al. 1994; Banerdt et al. 1994), which must be reconciled with the presence of both long and short length scales of tectonic features in the plains.

The pre-Magellan view that Venus exhibits a thin mechanical lithosphere was based on observations of the length scales of tectonic features defined by stretching and shortening instabilities (Zuber 1987; Zuber and Parmentier 1990) or characteristic elastic wavelengths (Banerdt and Golombek 1988), and on the depths of impact craters as compared to models of viscous relaxation of surface relief (Grimm and Solomon 1988). These models were all characterized by common assumptions concerning the composition of the Venus crust and mantle: (1) the Venus mantle is similar in composition to Earth's mantle and therefore the primary constituent is olivine, and (2) the crust in areas where observed impact and tectonic structures are located is similar to that determined for the Soviet Venera and Vega landers, and is appropriately described by the mechanical properties of diabase (Surkov et al. 1983,1984,1987). Results for both classes of models indicated a range of Venus crustal thicknesses of ∼10 to 30 km and associated thermal gradients of <25 K km^{-1}. However, the results are critically dependent on knowledge of the brittle and ductile deformational behavior of diabase and olivine.

Numerous experiments on terrestrial rocks indicate that the brittle strength of near-surface rocks is essentially independent of rock type, strain rate and grain size (Byerlee 1968); strength depends almost solely on pressure (depth) and is described in a simple linear relation by Byerlee's Law. The ductile strength of crustal and mantle materials is significantly more problematic, as it is sensitive to temperature, strain rate, composition, and modal mineralogy (Kohlstedt 1992). Because of its importance with regard to flow in the Earth's mantle, the ductile rheology of single crystal olivine is relatively well understood (Goetze 1978; Kirby and Kronenberg 1987; Chen and Morgan 1990; Kohlstedt et al. 1995), albeit for strain rates many orders of magnitude greater than characterize the mantle. Crustal rheologies are much less well characterized. A particularly important issue is that the data used in pre-Magellan studies of crustal rheology were derived from experiments in which the samples were not completely dried (Shelton 1981; Shelton and Tullis 1981; Caristan 1982). Water remaining in the samples during deformation led to low (by terrestrial comparison) flow strengths at the near-surface temperature of Venus, with the depth of transition from brittle deformation by shear failure to ductile flow by temperature-controlled dislocation creep occurring within a few km of the surface (Fig. 10a).

Given the lack of water on Venus, at least near the surface (Kaula 1990), experiments performed at exceptionally dry conditions are applicable. Recent

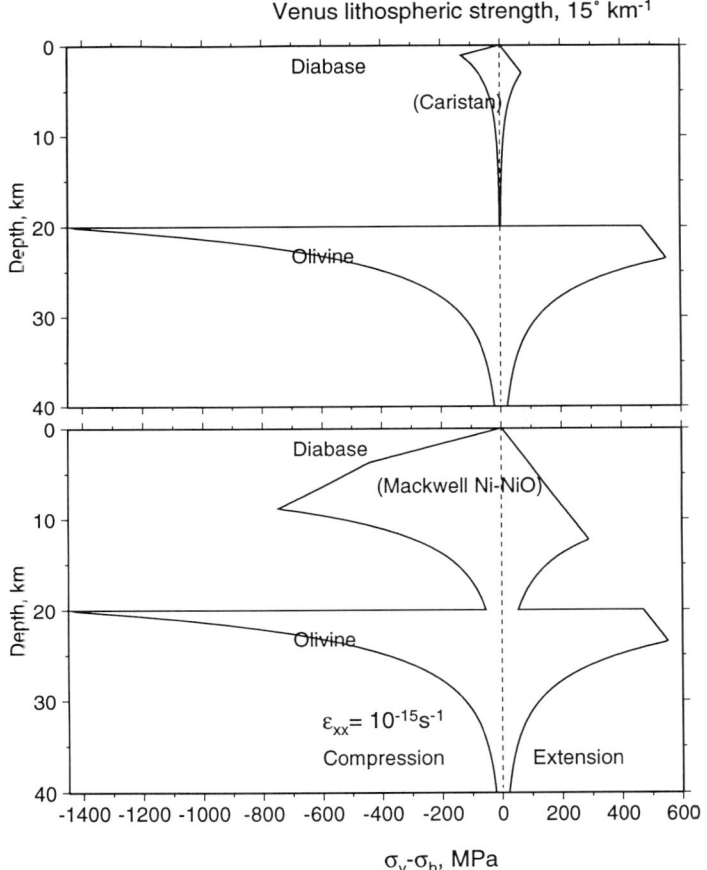

Figure 10. Strength profiles of the Venus lithosphere in uniform horizontal compression and extension assuming diabase crustal compositions using flow laws from (a) Caristan (1982) and (b) Mackwell et al. (1995). Both calculations assume a surface thermal gradient of 15 K km^{-1}, a crustal thickness of 20 km and an olivine mantle with the flow law of Chen and Morgan (1990).

experiments have been performed on thoroughly dried samples of Columbia diabase (Mackwell et al. 1995; see also the chapter by Phillips et al.), with this rock type chosen on the basis of gross similarity to chemical compositions determined at the Venera lander sites (Surkov et al. 1987). The yield envelope illustrated in Fig. 10b indicates that the absence of water results in a Venus crust that is much stiffer than previously thought. The much higher strength is consistent with Magellan findings from topography and gravity data of a thick present-day elastic lithosphere (Johnson and Sandwell 1992,1994; Moore et al. 1992; Sandwell and Schubert 1992a,b; Phillips 1994; Schubert et al. 1994; Banerdt et al. 1994; chapter by Phillips et al.). However, one should not discount the possibility that some of the structures we now see were formed under conditions that were significantly different (such as higher crustal temperature gradients and water concentrations) than those at present.

With the new experimental data on rock rheology, a systematic re-analysis to understand the structure of the crust from the length scales of plains structures is necessary. The first step in any such analysis is the identification and quantification of widths and periodic length scales, where they exist, from Magellan SAR imagery and altimetry. This is particularly important for areas that were either not imaged before Magellan or that contain features with short tectonic length scales that were imaged at lower resolution.

Future analyses should necessarily incorporate advances in modeling techniques for the development of tectonic features. For example, previous models that related wrinkle ridge and graben spacings to the thicknesses of lithospheric layers utilized continuum folding (for contraction) and necking (for extension) instability models that did not take into account the almost inevitable presence of faults associated with those features. Current numerical techniques have the ability to incorporate the presence of faulting, either by *a priori* inclusion (Melosh and Williams 1989) or by strain localization (Scholz 1990) techniques (see, e.g., Neumann and Zuber 1995). Such approaches are also relevant to models of highland deformation features such as rifts and mountain belts.

A major consequence of the new experimental rheological data is the absence or at least a significant decrease in size of a weak lower crustal channel. The crustal channel was thought to act as a decoupling zone between the strong upper crust and the upper mantle and was believed to enable the simultaneous development of multiple tectonic length scales within a given region (Zuber 1987; Banerdt and Golombek 1988). Certain classes of ridge and fracture belts have well-defined widths in association with an apparent regular development of kilometer-scale deformation. If the regularity of the small-scale fracturing can be established, the question arises how to develop multiple length scales of deformation in the absence of a lower crustal channel that separates strong lithospheric layers near the surface and at depth. Possible explanations include the existence of other compositional or rheological layers in the crust (such as individual lava flow units; Banerdt and Sammis 1992), strain weakening during deformation in areas of finite strain (Zuber 1994), locally high near-surface thermal gradients, and additional mechanisms for the development of tectonic length scales. Alternatively, the length scales (especially the longer scales) may reflect periodicities in the forces that formed them (perhaps related to convective processes) rather than the mechanical properties of the elastic lithosphere itself. Establishing whether long and short length scales of deformation developed concurrently would provide an important constraint on the origin of the features. While there is no current evidence contrary to the assertion that all length scales developed contemporaneously, such relationships are difficult to document from SAR images.

Another important scale observation is the persistence of tectonic trends over large areas, sometimes for thousands of km. This appears to require either a very strong mechanical lithosphere capable of transmitting stresses over large distances without undergoing large non-elastic strains, or else the processes that are causing the deformation over wide plains regions are dominated by tractions on the bottom of the lithosphere rather than edge forces.

Parallel fracture patterns on the Venusian plains contain various enigmatic elements. Interpreted by Banerdt and Sammis (1992) as tension fractures, these features have been explained using a shear-lag mechanism, in which the spacing of the lineations is controlled by a relation between the tensile strength of the brittle layer and a shear traction on its base. In this scenario there is no relationship between fracture spacing and the thickness of the brittle layer. Though shown to be plausible, models that incorporate layer thickness dependencies have yet to be tested. In such models it will be necessary to investigate whether the controlling layer is the brittle crust or other sub-layering. A possible alternative model also involves shear at the base of a surface brittle layer, but has the spacing of surface features dependent on the thickness and mechanical properties of the layer as well as the shear traction at the base of the layer. In either model the source of regional tension must be identified. Flexural uplift, lithospheric cooling and mantle flow have been suggested (Banerdt and Sammis 1992), but these possibilities have not been quantitatively tested.

VI. SYNTHESIS AND FUTURE DIRECTIONS

The full value of the information available from the tectonics of Venus' plains awaits the completion of detailed geologic studies of the various plains regions, as well as the highlands, because the processes which formed one have undoubtedly affected the other. This will allow the undertaking of the global syntheses necessary to put the bewildering array of tectonic features into a consistent framework, and allow the integrated tectonic history of the planet to be inferred.

In assessing the many unique aspects of plains deformation on Venus, the effect of the absence of water on deformational style provides a natural focus for future work. Experiments relevant to the ductile strength of the Venus lithosphere, both crust and mantle, should be performed for a broader range of modal mineralogy (for the crust) and grain size, as well as for larger strains. Theoretical models should incorporate general distributions of lithospheric strength, and also must adapt to deal with combined continuum and fault deformation, as well as time dependencies. The greatest challenge in future studies of plains tectonism will be to understand how to relate the complex time history of deformation to the post-resurfacing global stress state of Venus. Such analyses will be essential to understand the complex thermal evolution of Venus and its differences from Earth.

Acknowledgments. V. Hansen provided a thorough review of the original manuscript. We gratefully acknowledge the support of NASA's Planetary Geology and Geophysics Program. Portions of this work were carried out at the Jet Propulsion Laboratory of the California Institute of Technology, under contract to the National Aeronautics and Space Administration.

REFERENCES

Anderson, E. M. 1951. *The Dynamics of Faulting and Dyke Propagation with Applications to Britain* (Edinburgh: Oliver and Boyd).
Banerdt, W. B. 1986. Support of long wavelength loads on Venus and implications for internal structure. *J. Geophys. Res.* 91:403–419.
Banerdt, W. B., and Golombek, M. P. 1988. Deformational models of rifting and folding on Venus. *J. Geophys. Res.* 93:4759–4772.
Banerdt, W. B., and Sammis, C. G. 1992. Small-scale fracture patterns on the volcanic plains of Venus. *J. Geophys. Res.* 97:16149–16166.
Banerdt, W. B., Golombek, M. P., and Tanaka, K. L. 1992. Stress and tectonics on Mars. In *Mars*, eds. H. H. Kieffer, B. M. Jakosky, C. W. Snyder and M. S. Matthews (Tucson: Univ. of Arizona Press), pp. 249–297.
Banerdt, W. B., et al. 1994. The isostatic state of Mead crater. *Icarus* 112:117–129.
Barriot, J.-P., and Balmino, G. 1994. Analysis of the LOS gravity data set from Cycle 4 of the Magellan Probe around Venus. *Icarus* 112:34–41.
Barsukov, V. L., et al. 1986. The geology and geomorphology of the Venus surface as revealed by the radar images obtained by Veneras 15 and 16. *Proc. Lunar Planet. Sci. Conf.* 16, *J. Geophys. Res. Suppl./* 91:378–398.
Basilevsky, A. T. 1994. Concentric wrinkle ridge pattern around Sif and Gula. *Lunar Planet. Sci. Conf.* XXV:63–64 (abstract).
Basilevsky, A. T., and Head, J. W. 1988. The Geology of Venus. *Ann. Rev. Earth Planet. Sci.* 16:295–317.
Basilevsky, A. T., et al. 1986. Styles of tectonic deformations on Venus: Analysis of Venera 15 and 16 data. *Proc. Lunar Planet. Sci. Conf.* 16, *J. Geophys. Res. Suppl.* 91:399–411.
Bilotti, F., and Suppe, J. 1992. Planetary distribution and nature of compressional deformation around Artemis Corona, Venus. *Lunar Planet. Sci. Conf.* XXIII:101–102 (abstract).
Bilotti, F., Connors, C., and Suppe, J. 1993. Global organization of tectonic deformation on Venus. *Lunar Planet. Sci. Conf.* XXIV:107–108 (abstract).
Bindschadler, D. L. 1994. Magellan LOS gravity of Venus plains regions: Lithospheric properties and implications for global tectonics. *Lunar Planet. Sci. Conf.* XXV:113–114 (abstract).
Bindschadler, D. L., Schubert, G., and Kaula, W. M. 1992. Coldspots and hotspots: Global tectonics and mantle dynamics of Venus. *J. Geophys. Res.* 97:13495–13532.
Bowman, D. D., and Sammis, C. G. 1995. Implications of small-scale fracturing on Guinevere Planitia, Venus. *Lunar Planet. Sci. Conf.* XXVI:155–156.
Bryan, W. B. 1973. Wrinkle ridges as deformed surface crust on ponded mare lava. *Proc. Lunar Sci. Conf.* 4:93–106.

Buck, W. R. 1992. Global decoupling of crust and mantle: Implications for topography, geoid and mantle viscosity on Venus. *Geophys. Res. Lett.* 19:2111–2114.

Byerlee, J. D. 1968. Brittle-ductile transition in rocks. *J. Geophys. Res.* 73:4741–4750.

Campbell, D. B., et al. 1991. Venus southern hemisphere: Geologic character and age of terrains in the Themis-Alpha-Lada region. *Science* 251:180–183.

Caristan, Y. 1982. The transition from high temperature creep to fracture in Maryland diabase. *J. Geophys. Res.* 87:6781–6790.

Chen, Y., and Morgan, W. J. 1990. A nonlinear rheology for mid-ocean ridge axis topography. *J. Geophys. Res.* 95:17583–17604.

Copp, D. L., and Guest, J. E. 1995. Geology of the V31 Sif and Gula quadrangle of Venus. *Lunar Planet. Sci. Conf.* XXVI:283–284.

Ford, P. G., and Pettengill, G. H. 1992. Venus topography and kilometer-scale slopes. *J. Geophys. Res.* 97:13103–13114.

Frank, S. L., and Head, J. W. 1990. Ridge belts on Venus: Morphology and origin. *Earth Moon Planets* 50/51:421–470.

Goetze, C. 1978. The mechanisms of creep in olivine. *Phil. Trans. Roy. Soc. London A* 288:99–119.

Golombek, M. P., Plescia, J. B., and Franklin, B. J. 1991. Faulting and folding in the formation of planetary wrinkle ridges. *Proc. Lunar Planet. Sci. Conf.* 21:679–693.

Grimm, R. E., and Solomon, S. C. 1988. Viscous relaxation of impact crater relief on Venus: Constraints on crustal thickness and thermal gradient. *J. Geophys. Res.* 93:11911–11929.

Grosfils, E. B., and Head, J. W. 1994. The global distribution of giant radiating dike swarms on Venus: Implications for the global stress state. *Geophys. Res. Lett.* 21:701–704.

Grosfils, E. B., and Head, J. W. 1995. Giant radiating dike swarms on Venus: Stratigraphic constraints upon their time of emplacement. *Lunar Planet. Sci. Conf.* XXVI:523–524 (abstract).

Herrick, R. R., and Phillips, R. J. 1992. Geological correlations with the interior density structure of Venus. *J. Geophys. Res.* 97:16017–16034.

Johnson, C. L., and Sandwell, D. T. 1992. Variations in lithospheric thickness on Venus. *Intl. Colloquium on Venus*, LPI Contrib. No. 789, pp. 51–52.

Johnson, C. L., and Sandwell, D. T. 1994. Lithospheric flexure on Venus. *Geophys. J. Intl.* 119:627–647.

Kaula, W. M. 1990. Venus: A contrast in evolution to Earth. *Science* 247:1191–1196.

Kiefer, W. S., Richards, M. A., Hager, B. H., and Bills, B. G. 1986. A dynamic model of Venus's gravity field. *Geophys. Res. Lett.* 13:14–17.

Kirby, S. H., and Kronenberg, A. K. 1987. Rheology of the lithosphere: Selected topics. *Rev. Geophys.* 25:1219–1244.

Kohlstedt, D. L. 1992. Rheology of the crust and upper mantle of Venus: Constraints imposed by laboratory experiments. In *Workshop on Mountain Belts on Venus and Earth* (Houston: Lunar and Planetary Inst.).

Kohlstedt, D. L., Evans, B., and Mackwell, S. J. 1995. Strength of the lithosphere: Constraints imposed by laboratory experiments. *J. Geophys. Res.* 100:17587–17802.

Komatsu, G., and Baker, V. R. 1994. Plains tectonism on Venus: Inferences from canali longitudinal profiles. *Icarus* 110:275–286.

Konopliv, A. S., and Sjogren, W. L. 1994. Venus spherical harmonic gravity model to degree and order 60. *Icarus* 112:42–54.

Konopliv, A. S., et al. 1993. Venus gravity and topography: 60th degree and order model. *Geophys. Res. Lett.* 20:2403–2406.

Kryuchkov, V. P. 1988. Ridge belts on the plains of Venus. *Lunar Planet. Sci. Conf.* XIX:649–650 (abstract).

Kryuchkov, V. P. 1990. Ridge belts: Are they compressional or extensional structures? *Earth, Moon, Planets* 50/51:471–491.

Lachenbruch, A. H. 1961. Depth and spacing of tension cracks. *J. Geophys. Res.* 66:4273–4292.

Lancaster, M. G., and Guest, J. E. 1994. Volcanism and tectonism in Rusalka Planitia and Atla Regio, Venus. *Lunar Planet. Sci. Conf.* XXV:767–768 (abstract).

Lucchitta, B. K. 1977. Topography, structure, and mare ridges in southern Mare Imbrium and northern Oceanus Procellarum. *Proc. Lunar Sci. Conf.* 8:2691–2703.

Mackwell, S. J., Zimmerman, M. E., Kohlstedt, D. L., and Scherber, D. S. 1995. Experimental deformation of dry Columbia diabase: Implications for tectonics on Venus. In *Rock Mechanics: Proc. 35th U. S. Symposium*, eds. J. J. K. Daemen and R. A. Schultz (Brookfield, Vt.: A. A. Balkema), pp. 207–214.

Masursky, H., et al. 1980. Pioneer Venus radar results: Geology from images and altimetry. *J. Geophys. Res.* 85:8232–8260.

Maxwell, T. A., El Baz, F., and Ward, S. H. 1975. Distribution, morphology, and origin of ridges and arches in Mare Serenitatis. *Geol. Soc. Amer. Bull.* 86:1273–1278.

McGill, G. E. 1992. Wrinkle ridges on venusian plains: indicators of shallow crustal stress orientations at local and regional scales. In *Intl. Colloquium on Venus*, LPI Contrib. No. 789, pp. 67–68.

McGill, G. E. 1993. Wrinkle ridges, stress domains, and kinematics of venusian plains. *Geophys. Res. Lett.* 20:2407–2410.

McGill, G. E. 1994. Hotspot evolution and venusian tectonic style. *J. Geophys. Res.* 99:23149–23161.

McKenzie, D., McKenzie, J. M., and Saunders, R. S. 1992. Dike emplacement on Venus and on Earth. *J. Geophys. Res.* 97:15977–15990.

McLeod, L. C., and Phillips, R. J. 1994. Venusian channel gradients as a guide to vertical tectonics. *Lunar Planet. Sci. Conf.* XXV:885–886 (abstract).

McNamee, J. B., Borderies, N. J., and Sjogren, W. L. 1993. Venus: Global gravity and topography. *J. Geophys. Res.* 98:9113–9128.

Melosh, H. J., and Williams, C. A. 1989. Mechanics of graben formation in crustal rocks: A finite element analysis. *J. Geophys. Res.* 94:13961–13973.

Moore, W., Schubert, G., and Sandwell, D. T. 1992. Flexural models of trench/outer rise topography of coronae on Venus with axisymmetric spherical shell elastic plates. In *Intl. Colloquium on Venus*, LPI Contrib. No. 789, pp. 72–73.

Muller, O. H., and Pollard, D. D. 1977. The state of stress near Spanish Peaks, Colorado, determined from a dike pattern. *Pure Appl. Geophys.* 115:69–86.

Neumann, G. A., and Zuber, M. T. 1995. A continuum approach to the development of normal faults. In *Rock Mechanics: Proc. 35th U. S. Symposium*, eds. J. J. K. Daemen and R. A. Schultz (Brookfield, Vt.: A. A. Balkema), pp. 191–198.

Nerem, R. S., McNamee, J. B., and Bills, B. G. 1993. A high resolution gravity model for Venus: GVM-1. *Geophys. Res. Lett.* 20:599–602.

Odé, H. 1957. Mechanical analysis of the dike pattern of the Spanish Peaks area, Colorado. *Bull. Geol. Soc. Amer.* 68:567–576.

Parker, T. J., Komatsu, G., and Baker, V. R. 1992. Longitudinal topographic profiles of very long channels in Venusian plains regions. *Lunar Planet. Sci. Conf.* XXIII:1035–1036 (abstract).

Phillips, R. J. 1990. Convection-driven tectonics on Venus. *J. Geophys. Res.* 95:1301–1316.

Phillips, R. J. 1994. Estimating lithospheric properties at Atla Regio, Venus. *Icarus* 112:147–170.

Phillips, R. J., Grimm, R. E., and Malin, M. C. 1991. Hot-spot evolution and the global tectonics of Venus. *Science* 252:651–658.
Plescia, J. B., and Golombek, M. P. 1986. Origin of planetary wrinkle ridges based on study of terrestrial analogs. *Geol. Soc. Amer. Bull.* 97:1289–1299.
Pollard, D. D., and Aydin, A. 1988. Progress in understanding jointing over the past century. *Geol. Soc. Amer. Bull.* 100:1181–1204.
Pollard, D. D., and Segall, P. 1987. Theoretical displacements and stresses near fractures in rock: With applications to faults, joints, veins, dikes, and solution surfaces. In *Fracture Mechanics of Rock*, ed. B. K. Atkinson (San Diego: Academic Press), pp. 277–349.
Raitala, J., and Tormanen, T. 1990. Cytherean ridge belts connected with tessera areas: Tensional or compressional structures? *Earth Moon Planets* 49:57–83.
Rappaport, N., and Plaut, J. J. 1994. A 360-degree and -order model of Venus topography. *Icarus* 112:27–33.
Reasenberg, R. D., and Goldberg, Z. M. 1992. High-resolution gravity model of Venus. *J. Geophys. Res.* 97:14681–14690.
Rosenberg, E. 1995. Areal geology of the Pandrosos Dorsa quadrangle (V5), Venus. *Lunar Planet. Sci. Conf.* XXVI:1185–1186 (abstract).
Sammis, C. G., and Banerdt, W. B. 1991. Self organized critical faulting on Venus. *Lunar Planet. Sci. Conf.* XXII:1163–1164 (abstract).
Sandwell, D. T., and Schubert, G. 1992a. Evidence for retrograde lithospheric subduction on Venus. *Science* 257:766–770.
Sandwell, D. T., and Schubert, G. 1992b. Flexural ridges, trenches, and outer rises around coronae on Venus. *J. Geophys. Res.* 97:10069–10083.
Scholz, C. H. 1990. *The Mechanics of Earthquakes and Faulting* (New York: Cambridge Univ. Press).
Schubert, G., Moore, W. B., and Sandwell, D. T. 1994. Gravity over coronae and chasmata on Venus. *Icarus* 112:130–146.
Senske, D. A., et al. 1991. Geology and tectonics of the Themis Regio-Lavinia Planitia-Alpha Regio-Lada Terra area, Venus: Results from Arecibo image data. *Earth Moon Planets* 55:97–161.
Shelton, G. 1981. Experimental Deformation of Single Phase and Polyphase Crustal Rocks at High Pressures and Temperatures. Ph.D. Thesis, Brown University.
Shelton, G., and Tullis, J. 1981. Experimental flow laws for crustal rocks. *EOS Trans. Amer. Geophys. Union* 62:396.
Simons, M., Hager, B. H., and Solomon, S. C. 1994. Global variations in the geoid/topography admittance of Venus. *Science* 264:798–803.
Solomon, S. C., et al. 1991. Venus tectonics: Initial analysis from Magellan. *Science* 252:297–312.
Solomon, S. C., et al. 1992. Venus tectonics: An overview of Magellan observations. *J. Geophys. Res.* 97:13199–13255.
Squyres, S. W., et al. 1992. Plains tectonism on Venus: The deformation belts of Lavinia Planitia. *J. Geophys. Res.* 97:13579–13599.
Strom, R. G. 1972. Lunar mare ridges, rings and volcanic ring complexes. *Modern Geol.* 2:133–157.
Sukhanov, W. L., and Pronin, A. A. 1989. Ridge belts on Venus as extensional features. *Proc. Lunar Planet. Sci. Conf.* 19:335–348.
Sukhanov, W. L., et al. 1989. *Geomorphic/Geologic Map of Part of the Northern Hemisphere of Venus*, U. S. G. S. Misc. Invest. Map I–2059.
Surkov, Yu. A., et al. 1983. Determination of the elemental composition of rocks on Venus by Venera 13 and Venera 14 (preliminary results). *J. Geophys. Res.* 88:481–493.
Surkov, Yu. A., et al. 1984. New data on the composition, structure, and properties of

Venus rock obtained by Venera 13 and Venera 14. *J. Geophys. Res.* 89:393–402.
Surkov, Yu. A., et al. 1987. Uranium, thorium, and potassium in the venusian rocks at the landing sites of Vega 1 and 2. *J. Geophys. Res.* 92:537–540.
Turcotte, D. L., and Schubert G. 1982. *Geodynamics: Applications of Continuum Physics to Geological Problems* (New York: J. Wiley).
Watters, T. R. 1988. Wrinkle ridge assemblages on the terrestrial planets. *J. Geophys. Res.* 93:10236–10254.
Watters, T. R. 1991. Origin of periodically spaced wrinkle ridges on the Tharsis Plateau of Mars. *J. Geophys. Res.* 96:15599–15616.
Zuber, M. T. 1986. A dynamic model for ridge belts on Venus and constraints on lithospheric structure. *Lunar Planet. Sci. Conf.* XVII:979–980 (abstract).
Zuber, M. T. 1987. Constraints on the lithospheric structure of Venus from mechanical models and tectonic surface features. *J. Geophys. Res.* 92:541–551.
Zuber, M. T. 1990. Ridge belts: Evidence for regional- and local-scale deformation on the surface of Venus. *Geophys. Res. Lett.* 17:1369–1372.
Zuber, M. T. 1994. Rheology, tectonics, and the structure of the Venus lithosphere. *Lunar Planet. Sci. Conf.* XXV:1575–1578 (abstract).
Zuber, M. T., and Parmentier, E. M. 1990. On the relationship between isostatic elevation and the wavelengths of tectonic surface features on Venus. *Icarus* 85:290–308.

CORONAE ON VENUS: MORPHOLOGY AND ORIGIN

ELLEN R. STOFAN
Jet Propulsion Laboratory

VICTORIA E. HAMILTON
Arizona State University

DANIEL M. JANES
Cornell University

and

SUZANNE E. SMREKAR
Jet Propulsion Laboratory

Coronae are features first identified on Venus. They are surrounded by an annulus of concentric fractures and lineaments, are associated with large amounts of volcanism, and many have radially oriented fractures in their interiors. Most coronae exhibit partially raised topography, primarily high topography associated with fracturing along the corona annulus. Coronae form through a sequence of volcanism, uplift of topography, and tectonic deformation, followed by reduction of topographic relief and continued volcanism. However, many variations to this sequence have been identified. Coronae occur at volcanic rises, as isolated features in the plains, and most commonly, along chasmata systems. Most coronae form due to upwelling followed by gravitational relaxation, but current models do not adequately explain the tectonic, volcanic, and topographic variations observed at coronae. Coronae provide a direct link to interior processes on Venus as these features are the most likely to result from mantle/lithosphere interactions, and therefore can provide clues to mantle dynamics, as well as information on the interior structure of the planet.

I. INTRODUCTION

Coronae were first identified in Venera 15 and 16 data of Venus (Fig. 1) (Barsukov et al. 1984), although large circular structures of enigmatic origin had previously been identified in Pioneer Venus (Masursky et al. 1980) and Arecibo data (Campbell and Burns 1980). The term "corona" is a morphologic term, indicating any circular to elongate structure defined primarily by an annulus of concentric fractures and ridges (Barsukov et al. 1986; Basilevsky et al. 1986; Pronin and Stofan 1990; Stofan et al. 1992). Coronae (from the Latin word for "crown") are apparently unique to Venus, although similarities to structures such as Alba Patera on Mars have been noted (Janes and Squyres 1993; Watters and Janes 1995), and their abundance on the surface of Venus and their possible link to interior processes has resulted in intense study of their characteristics and possible modes of origin.

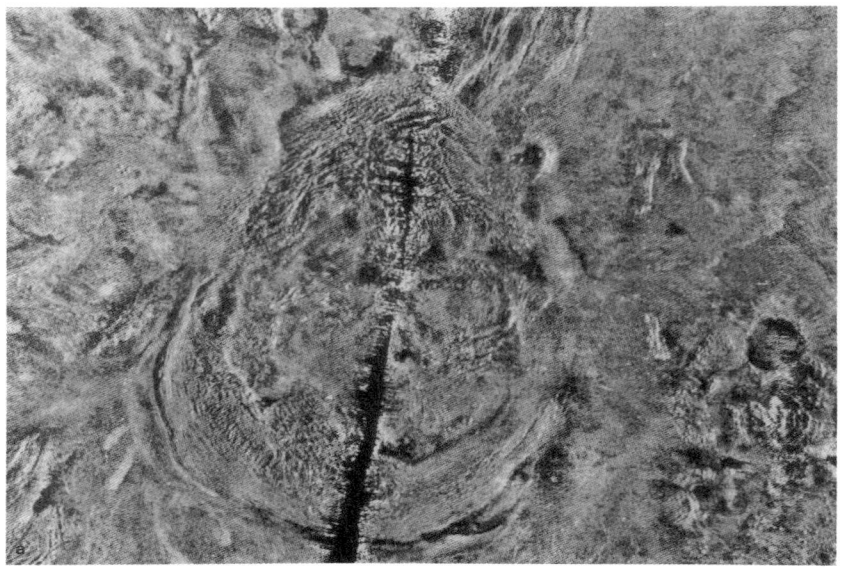

Figure 1. Venera 15/16 image of Nightingale Corona (63°N, 130°). Nightingale Corona is approximately 560×480 km across.

Magellan data revealed a population of approximately 360 coronae with diameters (defined by the extent of the annulus) of approximately 75 to over 2000 km (Stofan et al. 1992), with complex volcanic, tectonic, and topographic variations (Squyres et al. 1992; Stofan et al. 1992). Stofan et al. (1992) strictly applied the original definition of coronae (Barsukov et al. 1986; Basilevsky et al. 1986), with all features having an annulus of concentric tectonic features classified as coronae. The features, under International Astronomical Union convention, are named for goddesses of fertility and the harvest. Analysis of the basic characteristics of coronae has led many workers to favor mantle upwelling for the origin of coronae (Basilevsky et al. 1986; Stofan and Head 1990; Stofan et al. 1991,1992; Squyres et al. 1992; Janes et al. 1992; Koch 1994), although the extreme range in size and morphology of circular features which have an annulus of tectonic features may indicate that not all coronae may form due to mantle upwelling.

Early studies of coronae enabled workers to define a preliminary classification scheme for the 36 features identified in Venera 15 and 16 data (Pronin and Stofan 1990), which was later expanded and altered to classify the over 360 features identified in Magellan data (Stofan et al. 1992). This classification was developed in the initial phases of data return from the Magellan mission, and therefore is based primarily on planform shape and the tectonic characteristics of coronae. Coronae were classified as concentric, radial/concentric, concentric-double ring, asymmetric or multiple. Concentric coronae, the most common type of coronae, have well-defined, symmetric annuli (Fig. 2). Radial/concentric coronae are dominated by interior radial fractures, whereas concentric-double ring coronae are characterized by two distinct, concentric annuli. Asymmetric coronae are irregular to polygonal features exhibiting a marked asymmetry of form. Multiple coronae are made up of two to three linked structures with a continuous annulus (Fig. 3); two or more coronae for which a clear superposition relationship exists indicating a progression in age were not included in this class. Twenty-six corona-like features (features similar to coronae but lacking a well-defined annulus) were classified as radial and volcanic (Stofan et al. 1992). The radial class have extensive radiating fractures, whereas the volcanic class features have radiating flows with some degree of radial and concentric fracturing.

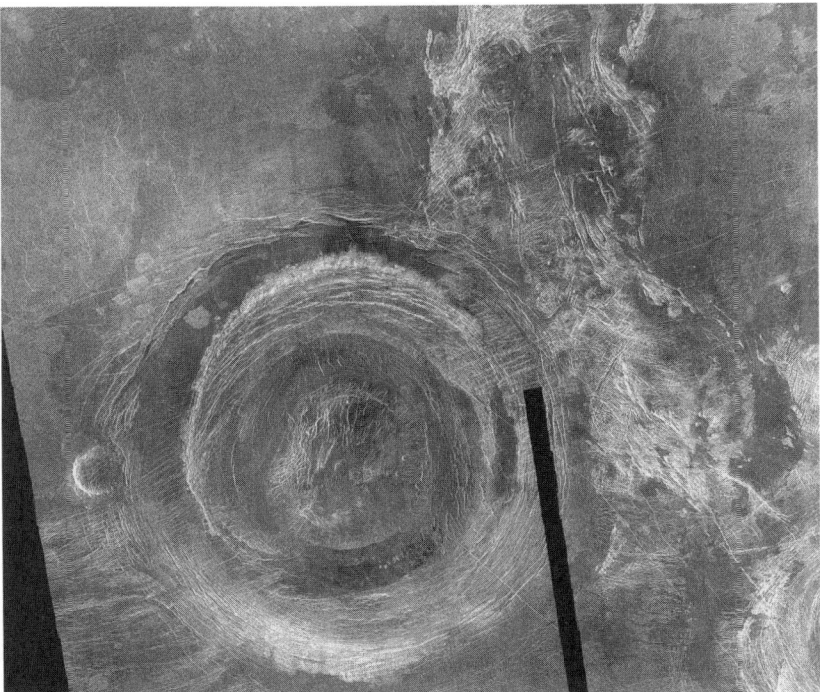

Figure 2. Aramaiti Corona (26°S, 170°) is a concentric corona, 375 km in diameter, and has an approximately 80 km wide annulus of fractures and ridges. The corona is surrounded by flows, and has a steep-sided dome superposed on the annulus.

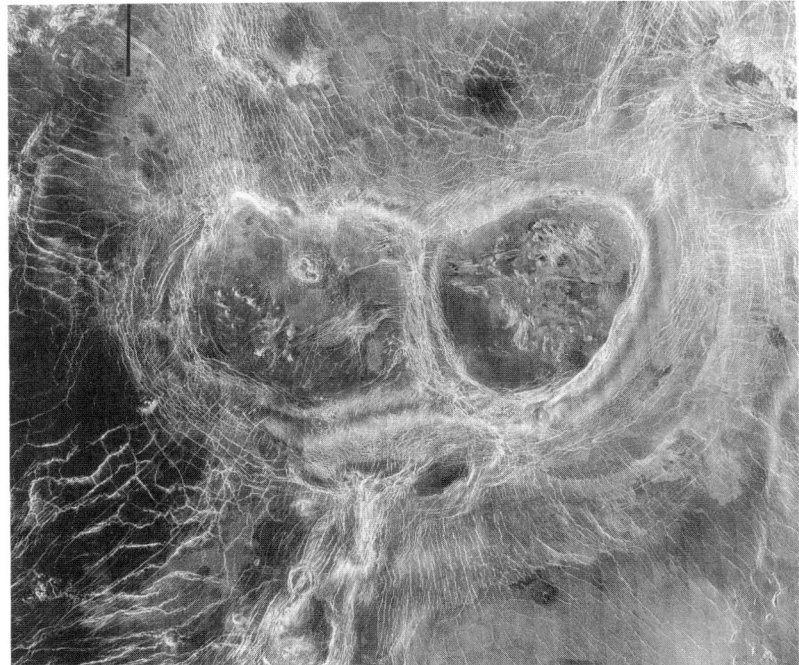

Figure 3. Neyterkob Corona (40°N, 203°) is a multiple corona, 225×210 km across, with two linked corona structures. No apparent superposition relations can be discerned. The structure has abundant interior and exterior volcanism.

The variations in morphology of coronae may result from variations in lithospheric structure and/or regional tectonic environment, as well as differences in mode of origin. In this chapter, we review the major characteristics of coronae that were described in more detail in Squyres et al. (1992) and Stofan et al. (1992), and discuss new results on volcanic and tectonic characteristics, topography, evolutionary sequence, and gravity data. We discuss the geologic settings and gravity signatures of coronae, with particular emphasis on chasmata coronae, as this is by far the most common geologic setting. We review the proposed models of origin for coronae, summarizing models that account for the previously identified phases of corona evolution (see, e.g., Stofan et al. 1991; Janes et al. 1992; Koch 1994). The majority of coronae are interpreted to result from upwelling mantle diapirs. The depth of origin of these diapirs is thought to be shallower than the mantle plumes thought to produce volcanic rises on Venus (Stofan et al. 1992). Existing models do not adequately explain the wide range in morphology and topography of coronae.

II. CORONA MORPHOLOGY

A. Volcanic Characteristics of Coronae

All coronae have some association with volcanism (Fig. 4). The interiors of most coronae contain small (<30 km diameter) cones and shields and smooth plains deposits interpreted to be volcanic in origin. A small number of coronae have intermediate (30–100 km) diameter edifices in their interiors, many of which are characterized by radiating fractures interpreted as the surface expression of dikes. Other volcanic features found within coronae include small calderas (2–50 km in diameter), collapse pits and chains of collapse pits, and sinuous rilles. Small cones, domes, and shields are also found within the annuli of many coronae. Coronae are frequently surrounded by volcanic deposits, some with distinct digitate and/or lobate patterns and others with more sheet-like characteristics. These deposits sometimes appear to originate from fractures within the annulus, but in other cases the source is not identifiable.

In an initial survey using C1-MIDR data, Stofan et al. (1992) found that 9% of coronae were associated with large amounts of volcanism (abundant flows and edifices), 68% with moderate amounts of volcanism, and 23% with little associated volcanism (few small edifices or distinct surrounding flow deposits). Roberts and Head (1993) found that 41% of large coronae had extensive flow deposits surrounding them.

B. Tectonic Characteristics of Coronae

Coronae, by definition, are surrounded by an annulus of concentric fractures. The width of this annulus varies from 10 to over 150 km across (Stofan et al. 1992), but tends to be less than half of the radius of the feature. Corona annuli are dominated by graben and other fractures interpreted to be extensional in origin. At many features, however, ridges interpreted to be the result of crustal shortening are present also. Numerous lineaments within the annuli are of indeterminate origin. Most corona annuli are partially embayed by flows that originated either from within the annuli or from within the corona interior.

Whereas coronae are defined primarily by their annulus of tectonic structures, interior deformation is also common (Squyres et al. 1992; Stofan et al. 1992). Radial, concentric and obliquely trending fractures have been identified, with radially oriented fractures the most ubiquitous (Fig. 5). Most interior deformation at coronae appears to predate annulus formation, although radial fractures at several coronae postdate annulus formation. Only approximately one third of coronae have interiors dominated by deformational features; most coronae are dominated by interior smooth plains. A small subset of coronae have small embayed patches of tessera terrain in their interior or in the annulus.

C. Topography of Coronae

Coronae exhibit a great range of topographic types. Using Magellan altimetry data, the topography of 360 corona and corona-like features was determined

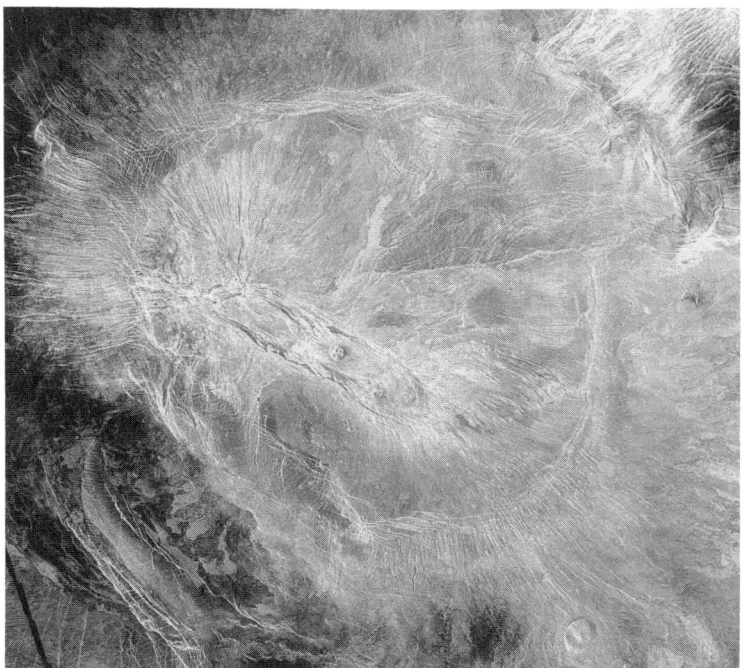

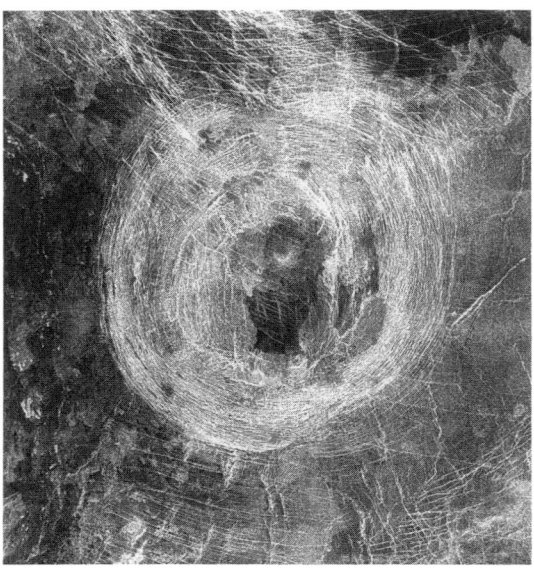

Figure 4. Examples of volcanism at coronae. (*Top*) Shiwanokia Corona 375 km diameter corona in Themis Regio (42°S, 280°) with a very high degree of associated volcanism. (*Bottom*) A concentric 170 km diameter corona (Sunrta Corona, 8.3°N, 11.7°) in western Eistla Region with a low to moderate amount of associated volcanism.

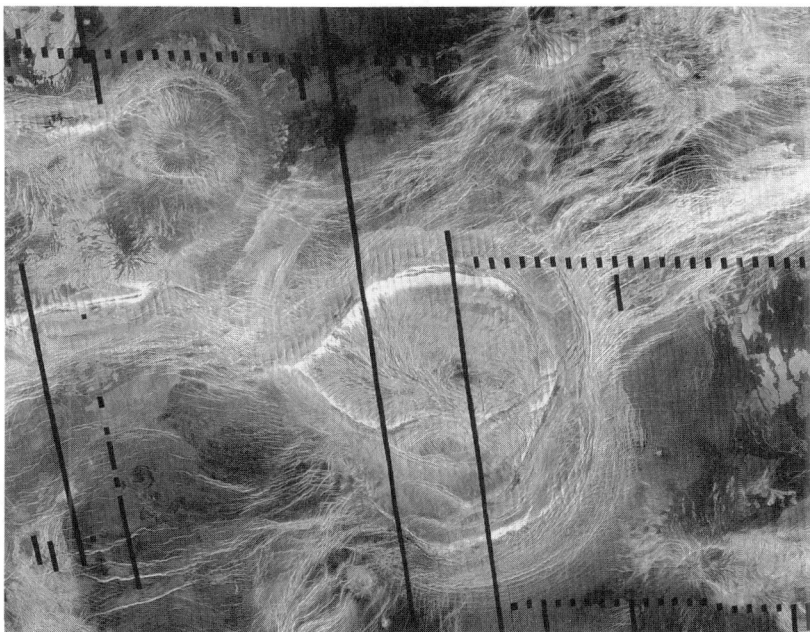

Figure 5. Latona Corona (20°S, 171°) illustrates the complex tectonics assocated with some coronae. The corona is 870×750 km across, and has an annulus of ridges and fractures encircled by a deep trough. The interior of the corona has radially, obliquely and concentrically oriented fractures.

(Stofan 1995). The topography was examined both in planform and with a number of representative profiles to determine the topographic type. The nine topographic groups identified include: (1) depressions; (2) rim surrounding depressions with or without an outer trough; (3) rim surrounding interior highs with or without an outer trough; (4) rim only; (5) outer rise, trough, interior high; (6) outer rise, trough, rim, interior depression; (7) plateaus with or without a surrounding trough; (8) domes with or without a surrounding trough; and (9) no discernible topographic signature (Fig. 6). Approximately half of coronae have interior high topography (domes, rims with interior highs, and plateaus), whereas one-third have interior depressions. Twenty-four percent of coronae are surrounded by exterior troughs, with outer rises occurring at 6% of coronae. Twenty-one percent of coronae have only a rim or no discernible topographic relief. No correlations exist between corona diameter and topographic groups. Even the most complex topographic groups such as groups 5 and 6 (features with troughs and outer rises) occur at very small and very large diameters.

The morphologic classes of coronae do not correspond well with the topo-

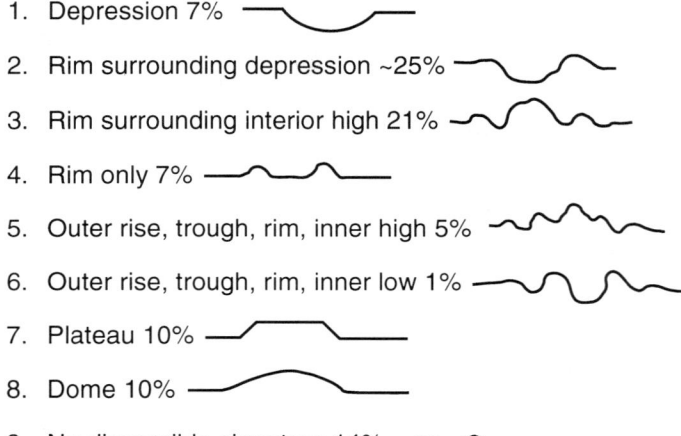

Figure 6. Nine types of topographic profiles associated with coronae: (1) depressions; (2) rim surrounding depressions with or without an outer trough; (3) rim surrounding interior highs with or without an outer trough; (4) rim only; (5) outer rise, trough, interior high; (6) outer rise, trough, rim interior depression; (7) plateaus with or without a surrounding trough; (8) domes with or without a surrounding trough; and (9) no discernible topographic signature.

graphic groups. An exception to this are groups 1 and 2, depressions without or with rims, which are almost exclusively concentric features. However, all other topographic groups also contain concentric coronae.

The relationship between the annuli of coronae and topography is complex. At most coronae, the annulus corresponds to relatively elevated terrain, although the annulus frequently extends down the slope of an annular moat, if present (Stofan and Bindschadler 1993). No contractional features have been identified within the moat. At a few coronae, the deformation extends into the moat and beyond, indicating that the deformational annuli at many coronae were originally more extensive and are partially buried by volcanic deposits. At other coronae, the tectonic features do not parallel the corona rim topography, indicating that each may preserve distinct events in corona formation.

D. Gravity Data of Coronae

The Magellan gravity data (chapter by Sjogren et al.) provides information on the deep structure of coronae. The gravity signature of coronae was first examined by Schubert et al. (1994), using the 60 degree and order spherical harmonic model field MGNP60FSAAP (Konopliv and Sjogren 1994) and gridded Magellan topography of Ford and Pettengill (1992). Schubert et al. (1994) found that most coronae had no resolvable signature. Most coronae are too small to be resolved as the resolution of the 60 deg and order field is at best ~600 km. Positive gravity anomalies were found at Artemis, Latona,

Fatua, and Heng-o Coronae, at three coronae along Parga Chasma, at one corona along Hecate Chasma, and at a cluster of four coronae in eastern Eistla Regio. This group of twelve coronae includes many of the largest coronae, such as Artemis and Heng-o. Some coronae several hundred km in diameter, such as Fatua and those in eastern Eistla Regio, were also resolved (at 1/2-wavelength resolution), perhaps because of their isolation from other major topographic features with large gravity anomalies.

TABLE I
Corona Gravity Data[a]

	Peak Anomaly	Peak Anomaly (Revised)	ADC
Artemis	42	74	200
Latona	49	93	150
Eastern Eistla Corona	45	66	75
Parga Corona	38	70	125
8°S, 225°	49	96	75
Hecate Corona	38	97	150
Heng-o	20	31	150
Fatua	22	31	75

[a] Data in columns 1 and 3 from Schubert et al. (1994).

For Artemis, Latona, the region containing four coronae in eastern Eistla, a corona centered at 16°S, 243.5° in Parga Chasma, the larger corona in western Parga, centered at 8°S, 225°, a corona in Hecate Chasma, Heng-o, and Fatua Coronae, Schubert et al. (1994) found apparent compensation depths ranging from 75 to 200 km (Table I, Fig. 7). These values were found using an Airy compensation model to find the compensation depth that best fit the peak anomaly for each corona. They used compensation depths at intervals of 25 km. In addition to the four corona in Parga and Hecate Chasmata, Latona and Artemis Coronae are associated with chasmata in Aphrodite Terra. Fatua and Heng-o Coronae are isolated features surrounded by plains. Stofan et al. (1995) studied the gravity signature for two additional clusters of coronae. Although these regions have numerous individual coronae, altogether they have the characteristics of large volcanic rises. Stofan et al. (1995) classified Themis, and eastern and central Eistla Regiones as large volcanic rises, which are interpreted to be analogous to terrestrial hotspots, because of their 1 to 1.5 km high broad topographic swells, the presence of volcanism, and their large positive gravity anomalies (Fig. 7). Apparent depths of compensation of 100 and 120 km were found for Themis Regio and central Eistla Regio, respectively, using single line-of-sight Magellan gravity data orbits. The apparent depth of compensation at eastern Eistla Regio was estimated to be approximately 65 km, in good agreement with the value of 75±25 km found by Schubert et al. (1994).

The gravity signature of coronae is better resolved in the 90 degree

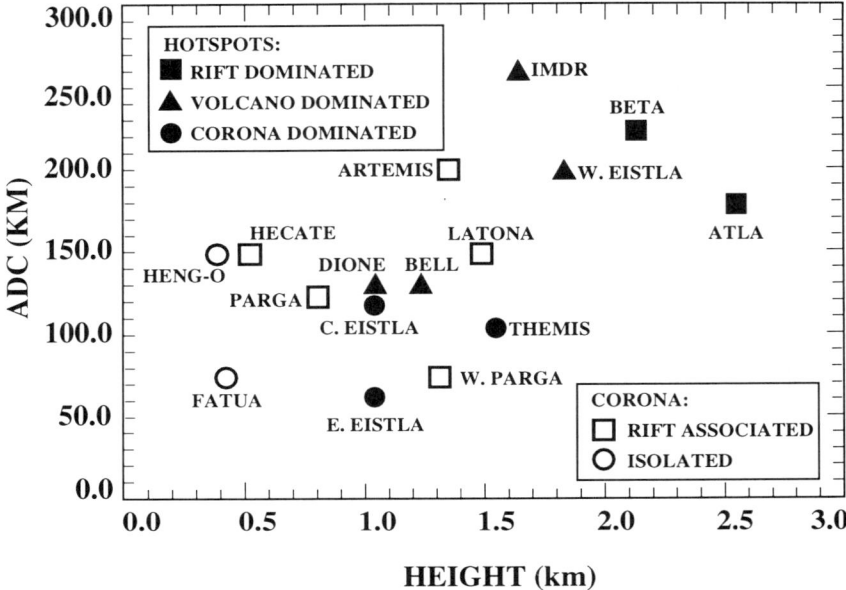

Figure 7. Graph showing the relationship between the apparent depth of compensation (ADC) as a function of the height of some coronae and volcanic rises on Venus (figure after Stofan et al. 1995).

and order field produced by Sjogren et al. (see their chapter), that has a resolution of up to ~400 km. The new field has much more power at short wavelengths, which has increased the peak gravity anomalies for the coronae discussed above by 35 to 155%. The largest increases in peak anomalies were primarily among coronae associated with chasmata. The new peak anomalies for Artemis, Latona, eastern Eistla, Parga, western Parga, Hecate, Heng-o and Fatua are given in Table I. Although detailed analysis has not yet been done, these increases in the peak anomalies will result in changes in the estimates of apparent depths of compensation in these regions.

E. Evolutionary Sequence Based on Morphology

Mapping of coronae in Venera 15 and 16 data led to the proposal of a sequence of events in corona formation (Stofan and Head 1990). Coronae were proposed to go through an initial stage of uplift/construction, accompanied by interior faulting and fracturing and volcanism (Fig. 8). This is followed by annulus and trough formation. In their final stage of evolution, coronae have continued volcanism and reduction of topographic relief (generally still remaining positive topographic features). This sequence of events was supported by mapping of coronae in Magellan data by Squyres et al. (1992). Squyres et al. (1992) identified a number of features having extensive radial faulting (dominantly the radial class of features) that they interpreted to be coronae in the early stages of formation. This stage had been predicted by modeling (Stofan et al. 1991), but not seen in Venera 15 and 16 data. In addition, over half of coronae are cross cut by tectonic structures that postdate and appear to be unrelated to corona formation.

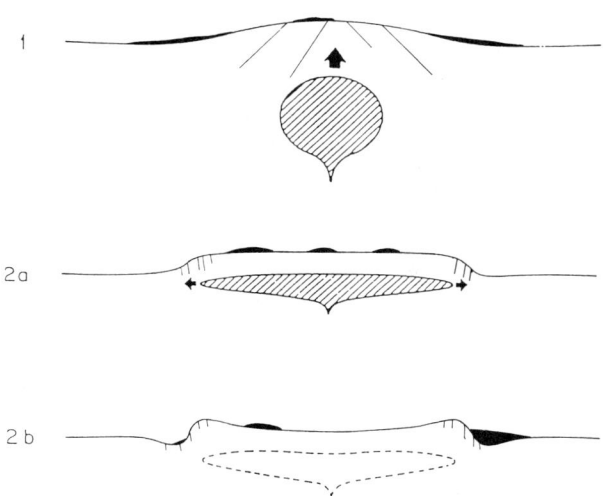

Figure 8. The major steps in the generalized conceptual model of formation of a Venusian corona (figure from Squyres et al. 1992). Step 1, a mantle diapir rises toward the surface, setting up a flow that forces domical uplift of the surface, radial fracturing, and volcanism. Step 2a, the diapir impinges on the underside of the lithosphere, spreading radially and flattening and transforming the shape of the uplift from domical to plateau-like. Volcanism continues. Step 2b, the diapir cools, removing thermal support of topography and allowing gravitational relaxation of the topography to form a moat, rim, and interior depression. Concentric fracturing takes place. Steps 2a and 2b may overlap in time.

While this evolutionary sequence broadly fits a number of coronae, many coronae do not appear to follow it. For example, some coronae are depressions or have depressed interiors. These features either never went through the uplift phase or represent an additional evolutionary stage. While a large number of coronae have radial fractures predating the annulus, at several coronae radial fractures postdate annulus formation. In addition, it appears that the nature of volcanism associated with coronae has varied through time. Many coronae initially produce sheet-like radial deposits that extend several corona radii beyond the annulus. In general, this stage appears to predate annulus formation. Later stage volcanism tends to produce more digitate flows, as well as more localized edifices. Detailed geologic mapping of Venus currently being conducted will undoubtedly provide important modifications to the evolutionary sequence for coronae, and a better understanding of the variations between features.

III. GEOLOGIC SETTINGS OF CORONAE

Based on Venera 15 and 16 data, coronae appeared to form preferentially in clusters, notably to the east and west of Ishtar Terra, as seen in their data (Pronin and Stofan 1990), and near volcanic rises, that form preferentially in the equatorial region (Schaber 1982). The global data set now available indicates that coronae occur over most of the planet's surface, although the distribution is nonrandom (Stofan et al. 1992; Squyres et al. 1993). The majority of coronae are clustered along chasmata in the so-called BAT (Beta-Atla-Themis) zone that also contains a concentration of other types of volcanic features (Crumpler et al. 1993). Analysis of the global distribution indicates that coronae occur in three distinct geologic environments: at volcanic rises (rise coronae); along chasmata (chasmata coronae); and as relatively isolated features in the plains (isolated coronae) (Stofan 1995). We review each of these settings briefly, then concentrate on two chains of coronae located along Parga and Hecate Chasma, as examples of the most typical setting of coronae, in order to better constrain the origin and evolution of coronae.

A. Rise Coronae

Volcanic rises with clusters of coronae include central and eastern Eistla, Themis, and Mnemosyne Regiones (see the chapter by Smrekar et al.). The surface morphology of most volcanic rises are not characterized primarily by coronae, but instead are dominated by rifts or large volcanoes (i.e., Beta, Atla, Dione Regiones) (Stofan et al. 1995). Central Eistla Regio has three coronae and several large volcanic constructs and calderas (Stofan et al. 1992; McGill 1994). Sappho is classified as a corona, as it possesses a distinct annulus of tectonic features; however, it has many similarities to a volcanic edifice (McGill 1994). There are four coronae in eastern Eistla Regio including the 525-km diameter corona Pavlova (Fig. 9). Themis Regio has five coronae, and is the only one of the three volcanic rises dominated by coronae that contains significant extensional deformation in the form of rifts and fractures. Themis Regio lies at the termination of Parga Chasma. Coronae appear to form both prior to, during and after extension at Themis Regio. At central Eistla Regio, McGill (1994) found that coronae formed prior to volcanic edifices. The coronae in each cluster at volcanic rises are typically 200 to over 500 km in diameter, and vary in spacing. At each rise, the coronae exhibit abundant volcanism, including extensive radiating flows and small to intermediate scale edifices.

Stofan et al. (1995) examined the differences between Venusian volcanic rises in order to better understand volcanic rise origin and evolution. They did not interpret the presence of coronae at volcanic rises to indicate a specific stage of rise evolution. The coronae are interpreted to indicate small-scale convection in the lithosphere induced by the rising plume interpreted to form the rise, or instabilities and break up of the plume head, creating small-scale diapiric upwellings that form coronae (Stofan et al. 1995).

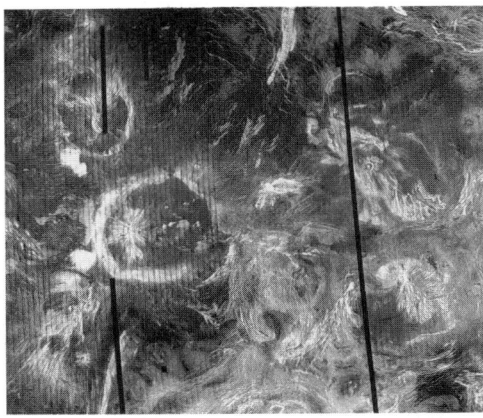

Figure 9. Magellan image (C1-MIDR15N043) of eastern Eistla Regio, approximately 1800 km across. Eastern Eistla, a corona-dominated volcanic rise, has four coronae with diameters of about 300 to 525 km.

B. Isolated Coronae

Eleven percent of coronae are classified as isolated coronae, including some very large coronae such as Heng-o Corona (Stofan 1995). Plains volcanism appears to embay most of the isolated coronae. The isolated coronae correspond predominantly to topographic groups 4 and 9, features with little or no topographic signature. These coronae are interpreted to be in the late stages of evolution, with topography consistent with gravitationally relaxed features. Given the predominance of coronae occurring along fracture belts and chasmata, and the degree to which many of these fracture belts are embayed, it is possible that the isolated coronae also formed along fracture belts that have been subsequently covered by volcanic flows.

C. Chasmata Coronae

The majority of coronae (68%) occur along chasmata or fracture belts, most in the BAT region as discussed above. Chasmata coronae occur in chains, unlike the rise coronae which occur in a cluster on a topographic high. Major corona chains include Parga and Hecate Chasmata, eastern Aphrodite, and the Alpha-Lada fracture zone. Not all chasmata coronae are associated with major rifts; some coronae lie along fracture belts that have a low density of fractures and are not associated with deep troughs, such as the zone of deformation between Beta and Eistla Regiones. Coronae located along some chasmata have large amounts of associated volcanism, possibly indicating that the intersection of upwellings with extensional zones is a controlling factor in the amount of melt produced at a corona (Roberts and Head 1992; Magee and Head 1995). Here, we concentrate on Parga and Hecate Chasma, as these are the two largest occurrences of chasmata coronae and contain variations typical of all corona chains.

The Hecate and Parga deformational zones were identified in Pioneer Venus and Arecibo data (Schaber 1982; Masursky et al. 1980) as two of several linear zones of deformation preferentially located at equatorial latitudes. In Magellan data, these and other fracture zones have numerous associated coronae and corona-like features (Stofan et al. 1992). Hecate and Parga Chasmata extend to the northeast and southeast, respectively, from Atla Regio. The fracture belt and corona chain of Hecate Chasma extends over 8000 km, and terminates at Beta Regio, whereas the corona chain and zone of deformation associated with Parga Chasma extends over 10,000 km to Themis Regio. Both the Beta and Themis Regiones highlands exhibit extensive volcanism and rifting (Stofan et al. 1995), and in the case of Themis Regio, numerous coronae. Along strike, the Hecate and Parga deformational zones have discontinuous, en-echelon troughs. Neither zone has a well-defined rift system such as the one at Devana Chasma in Beta Regio; instead, Hecate and Parga are complex networks of anastomosing fractures with multiple offsets of the rift and associated fractures along the trend of the chasmata.

Both Hecate and Parga Chasmata are discontinuous troughs 0.5 to 2.0 km deep, 20 to 100 km wide, interrupted in places by local topographic highs corresponding to coronae. Both systems extend for over 2000 km and lie along the axes of broad topographic rises. An 80 to 200 km wide band of radar-bright linear features overlie the trough. The fractures occur en-echelon or anastomosing along the chasma. We interpret the paired linear features to be normal faults defining graben in a zone of extension (Hamilton and Stofan 1996). The origin of these (and other) fracture systems and their associated coronae has been a matter of debate, with theories including formation by extension and diapiric upwelling (Stofan et al. 1992; Hansen and Phillips 1993; Hamilton and Stofan 1996; Stofan and Hamilton 1996) and subduction (Sandwell and Schubert 1992a,b,1993; Schubert and Sandwell 1995).

D. Hecate Chasma

The tectonic and volcanic characteristics of Hecate Chasma, as well as the associated corona chain, are described by Hamilton and Stofan (1996). Normal faults, graben, and fractures dominate Hecate Chasma, and are most densely concentrated immediately adjacent to or in the topographic troughs (Fig. 10). Individual linear features range in width from 2 to 50 km, and from less than 50 to over 600 km in length. Coronae along Hecate Chasma typically have radial sets of normal faults and graben.

The fracture system along Hecate Chasma exhibits two morphologic types, both associated with coronae: diffusely fractured regions with minor trough development, and a region of dense fracturing with a well-developed trough (Hamilton and Stofan 1996). Diffusely fractured regions, mostly in western Hecate Chasma, presumably accommodate tectonism in a broad zone rather than in narrow regions of localized deformation. The main trough, highly asymmetric in profile in the region of concentrated fracturing, crosses a 525-km diameter corona. To the west of the corona, the high topography

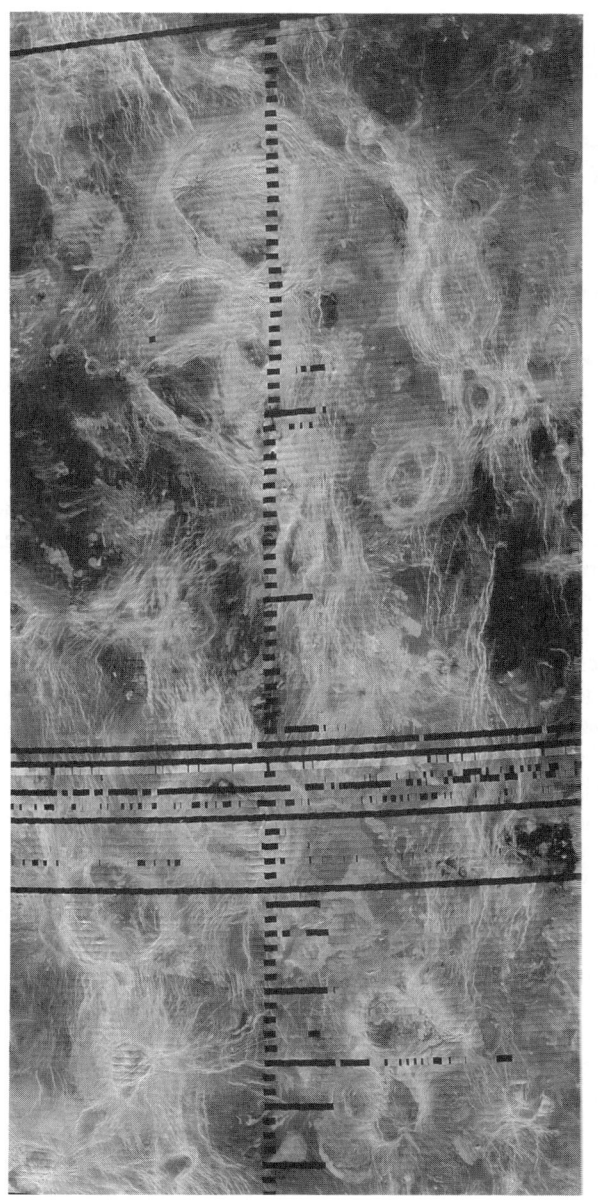

Figure 10. Central Hecate Chasma (image 3400 km across) showing the highly fractured terrain with trough development and the highest concentration of coronae. The two morphologic regions described in the text are shown.

is on the north side of the chasm, and to the east, the south side of the trough has the highest topography. Whether or not the corona postdates the trough remains unclear; lack of deformation suggests that the corona might be younger than the trough, but it is also possible that the corona predates the trough and supplies a structural barrier to the trough's propagation (Hamilton and Stofan 1996).

Coronae along Hecate Chasma display a wide variety of structural, volcanic, and topographic characteristics, fitting into nearly every category of the classification of Stofan et al. (1992). The coronae along the Hecate chain range in diameter from 85 to over 500 km, and are spaced 460 km apart on average, although they may be as close as 170 km or as far apart as 900 km. The coronae are not located at a single elevation nor are the coronae linked to any structural pattern, indicating that the spacing of coronae is not controlled structurally or topographically. Hamilton and Stofan (1996) classify 46 coronae in this chain, and find that the population represents all stages of corona evolution. Hecate Chasma coronae do not display a progression of evolutionary stages along trend. Thirty coronae are in the intermediate to advanced stages of development. The majority of coronae display evidence of volcanism, generally in the form of radial flows and/or interior cones and shields. Fractures defining the corona generally cut the volcanic flows, implying, consistent with models of corona formation, that volcanism is most active in the early stages of corona formation. Hamilton and Stofan (1996) classified the coronae along Hecate as either pre-tectonic, syntectonic, or post-tectonic, on the basis of their cross-cutting relationships with local tectonic and volcanic features. Along the Hecate corona chain, the population of coronae fall approximately evenly into each of the three classes, indicating that coronae have been forming throughout the evolution of the area.

The characteristics of many coronae along Hecate Chasma and the two morphologic types of fractured regions described earlier correlate. Regions of diffusely concentrated fractures are typically associated with coronae that have plateau-shaped profiles, dominantly exhibit sub-radial fracturing, and commonly have poorly developed annuli. Most of these coronae are classified as being in the early to intermediate stages of their development. Coronae in trough-dominated regions are more likely to be topographic depressions with elevated rims, usually lack radial fractures, and tend to have well-developed annular fractures. Hamilton and Stofan (1996) suggest that the high concentration of advanced-stage coronae in the trough dominated, or densely fractured, region, may be indicative of more advanced stage of rift development than in the diffusely fractured region, where most coronae are in early to intermediate stages of development.

Lava flows, small cones, and shield edifices are common along Hecate Chasma. Digitate lava flows are most commonly found emanating from the dominant linear fractures, particularly graben. These flows display flow directions down current topographic slopes, indicating that volcanism has postdated the latest regional uplift or that subsequent uplift has not altered local slope orientations. We find small cones and shields in the plains adjacent to the most concentrated fracturing, which are characterized by small circular or elongate summit pits, and generally do not display radial or circumferential fracturing. Cones and shields commonly occur in groups, and may also be found in the interiors of some coronae. We find the most massive volcanism to be restricted to the volcanic centers at Atla, Asteria, and Beta Regions. The two volcanoes in Asteria Regio occur in depressions along arms of the Hecate Chasma fracture zone. Both edifices display numerous volcanic flows and are dissected by normal faults and graben defining the fracture zone.

E. Parga Chasma

Stofan and Hamilton (1996) describe similar tectonic and volcanic characteristics along Parga Chasma: linear structures, a segmented trough, and coronae (Fig. 11). Linear features along Parga Chasma are dominantly paired sets of lineaments interpreted as graben. Normal faults and fractures of undetermined origin are the next most common linear features. All of these features form both broad, diffuse groups as well as locally concentrated, narrow zones of deformation. The narrower zones typically correlate with the deepest trough segments. Wrinkle ridges are generally found in the plains adjacent to the primary fracture zone. Only limited extension appears to have taken place along Parga Chasma, as there is no evidence for split coronae.

Trough segments forming Parga Chasma are much more discontinuous, generally shallower, have a lower fracture density, and less of a correlation of fracturing with the trough segments than their counterparts along Hecate Chasma. There are roughly seven trough segments defining Parga Chasma. Segments are commonly en-echelon, resulting in an overall northwest–southeast trend. The deepest parts of the troughs may reach 3.5 km below the mean planetary radius (MPR). Trough segments typically display raised rims reaching elevations of 1.5 to 2.2 km above MPR, possibly indicative of flexural support of the lithosphere. Similar to Hecate Chasma, the correlation between extensional fracturing and trough formation at Parga is suggestive of a region of broad extension.

Twenty seven coronae form a "chain" of features associated with the deformation at Parga Chasma (Stofan et al. 1993). Features range in diameter from 67 to 570 km. Numerous radial and arcuate features interpreted as remnant or mostly buried coronae (Stofan et al. 1993) are found along Parga Chasma. Coronae both superpose and cut graben and faults defining Parga Chasma and, in addition to remnant features, suggest that deformation and related corona formation has been ongoing. The Parga Chasma corona chain terminates at Themis Regio, a corona-dominated rise. Themis is unique among corona-dominated rises, as discussed above, due to the fact that it contains well-developed extensional deformation. This deformation lies along the same trend of the Parga deformation and appears to be synchronous. Extension at the southeastern end of the Parga system is less prevalent as there are fewer fractures. Many fractures curve around coronae; however, they

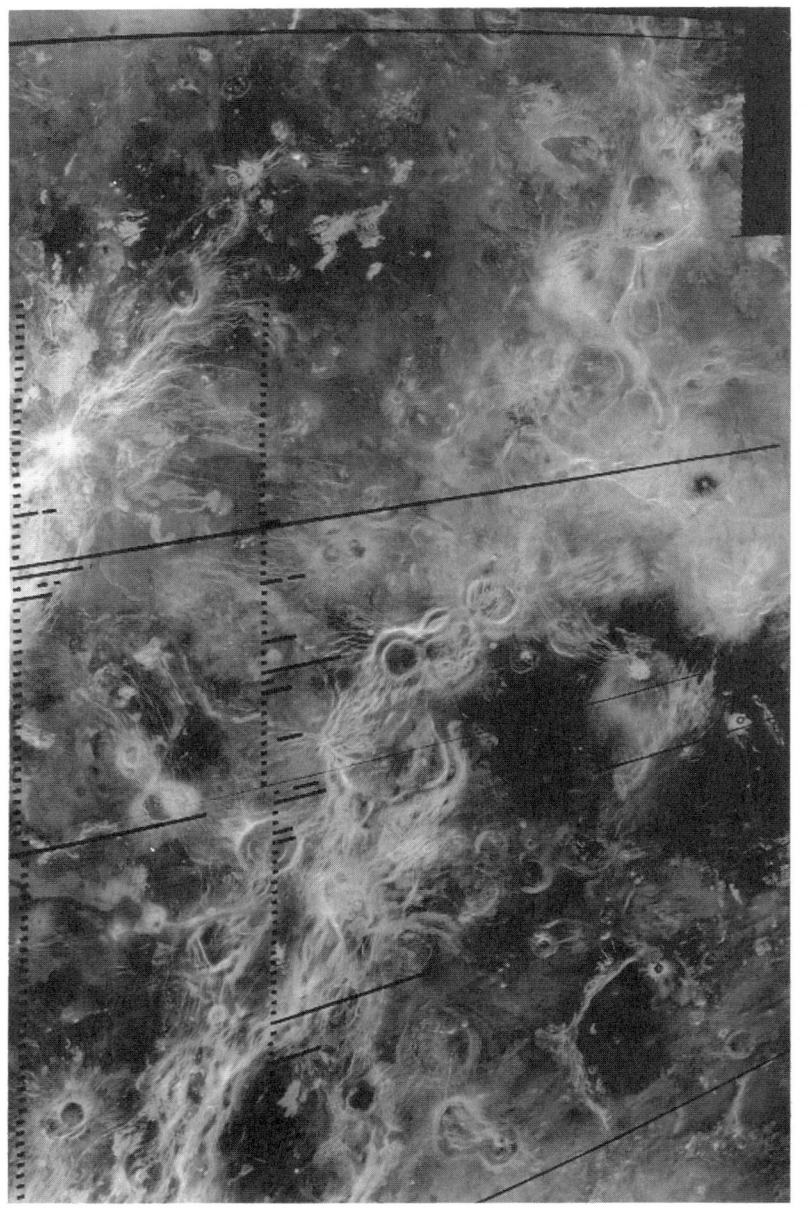

Figure 11. Eastern Parga Chasma, terminating at Themis Regio in the southeast. This Magellan radar image is approximately 4700 km across.

have gone through multiple episodes of volcanism and tectonism indicating a probable significant overlap in time with regional extension.

There is extensive volcanism associated with both the coronae and the linear fractures along Parga Chasma. Volcanic features include radar-dark and -bright flows, shields, cones, calderas, and domes. Edifices are commonly found adjacent to or within coronae, and the coronae themselves may be sources of numerous volcanic flows. Volcanism typically was the most recent event along most of Parga Chasma, with flows embaying coronae, infilling portions of the trough, and covering fractures. However, at some locations, flows have been fractured by subsequent tectonic activity. Volcanic flows emanating from coronae are also commonly fractured and/or uplifted. Volcanism is most abundant in the part of the trough system adjacent to Atla Regio, at the point where Parga, Hecate, and the Dali/Diane corona chain converge. Also, volcanism tends to be abundant at locations along the trough system where offsets and changes in trend are prominent.

The relative timing between the formation of the Hecate and Parga deformational zones is uncertain. Atla Regio is the only location where any cross-cutting relationship between the two fracture systems is evident. On the eastern side of Ozza Mons (a large volcanic edifice in Atla Regio), two sets of graben cut the flows of the volcano. One set, associated with Hecate Chasma, trends northeast, and the other set trends southeast along Parga Chasma. Nearly all of the northeast trending graben are either partially or completely flooded by lava flows from Ozza Mons, attesting to volcanism subsequent to the formation of these fractures. The graben from Parga Chasma cut across the northeast trending graben of Hecate Chasma, indicating that, at least in this region, deformation along the Parga zone is younger than deformation in the Hecate Chasma fracture belt. In addition, Parga Chasma has a higher concentration of coronae interpreted to be in the early stages of evolution (Stofan 1993), supporting the interpretation that it may have formed subsequent to Hecate Chasma.

IV. MODELS OF CORONA ORIGIN

The first Venera 15 and 16 images revealed coronae to be characterized by a pattern of concentric ridges that suggested formation by slumping and buckling along the flank of growing domes (Barsukov et al. 1984). The later addition of altimetric data showing that many coronae typically consist of an elevated, bowl-shaped interior surrounded by a raised rim and exterior moat, and the frequent association of coronae with volcanism reinforced the view that coronae formed as a result of gravitational modification to hotspot or diapir-induced uplift (Basilevsky et al. 1986). Alternative formative mechanisms for coronae have been proposed, including the rejuvenation of old craters (Nikolaeva et al. 1986), ring dike intrusion (Masursky 1987), cold mantle sinkers (Stofan et al. 1987), and lithospheric subduction (Sandwell and Schubert 1992a; McKenzie et al. 1992). The raised topography and narrow size range of coronae are inconsistent with an impact origin. Formation of coronae by ring dikes would require greater erosion rates than are evident on Venus to expose dikes at the surface (Stofan and Head 1990; see the chapter by Campbell et al.). Numerical modeling of corona formation to date has therefore addressed mantle flow and gravitational relaxation (see, e.g., Stofan et al. 1991; Janes et al. 1992; Koch 1994; Janes and Squyres 1995), and possible subsequent subduction and flexure of the lithosphere (Sandwell and Schubert 1992a,b). The greater resolution of the Magellan SAR and altimetric data sets have significantly aided these modeling efforts.

In broad outline, the prevailing theory of corona formation involves a three-stage process (Fig. 8), based on the evolutionary sequence determined from the observed geologic history of coronae (i.e., Stofan and Head 1990; Squyres et al. 1992). This interpreted geologic history based on mapping indicates that corona formation is consistent with initiation by a rising mantle diapir which raises the overlying lithosphere into a domical uplift. The diapir then flattens as it impinges against the underside of the mechanical lithosphere and the surface topography takes on a more plateau-like shape with concentric fractures forming on the rim of the plateau. Finally the diapir cools and loses buoyancy and the raised topography relaxes under the force of gravity resulting in an additional ring of concentric fractures, outside those formed on the plateau rim, as the lithosphere flexes under the now unsupported load. As an alternative to the thermal relaxation of the third stage outlined above, Sandwell and Schubert (1992a,b) suggest a final tectonic stage in which the diapiric magma fractures and breaks through the lithosphere, emplaces a load on its surface and initiates subduction of the broken lithosphere which flexes under the load. We will look at the modeling carried out on each of these stages. To date, no completely self-consistent model of this entire hypothesized sequence has been developed, but each of the individual stages has been studied in some detail.

A. Uplift

Stofan et al. (1991) employed a layered viscous model (Bindschadler and Parmentier 1990) to investigate the rise of diapirs through the mantle and the uplift and stress produced at the surface. Viscosities in the crust and mantle layers were linearized from experimentally determined flow laws for diabase (Shelton and Tullis 1981) and olivine (Goetze 1978), whereas the diapir was treated as an axisymmetric Gaussian density anomaly at depth. The diapir was initially placed at a depth of 300 km and allowed to rise to a depth of 100 km. The rising diapir produced a generally domical uplift that becomes higher and more narrowly peaked as the diapir rises. In this viscous model, thick crusts or high thermal gradients result in a lower viscosity value for the lower crust, allowing a portion of the stress to be dissipated as flow in the lower crust. A stiffer crustal rheology, such as that proposed by Mackwell et al. (1993a,b; see also the chapter by Phillips et al.), would ameliorate this effect and produce greater uplift for a given diapir rise. This uplift is accompanied by the development of a central area of extensional hoop and radial stresses at the surface. The hoop stress is greater than the radial stress so that the formation of radially oriented extensional fractures is favored. This central extensional area is surrounded by a region where the radial stress becomes compressional, predicting strike-slip or concentrically oriented folding or thrust faulting. However, these outer compressional features are not commonly observed. As with the topography, stresses become more pronounced as the diapir nears the surface. Similarly, stresses are enhanced by thin, strong near-surface layers.

Stofan et al. (1991), using their elastic layered viscous model, also investigated the possibility that doming of the surface could be due to an unstable, high-density anomaly forming at the base of the thermal lithosphere, detaching and sinking through the mantle. Assuming that such a sinking, negatively buoyant diapir would create a cavity in the lithosphere that would be filled by lower-density mantle material, they found that an initial downwarping of the surface during the detachment of the sinking diapir would be followed by isostatic uplift similar to that produced by a rising diapir. However, the lack of basins of the appropriate sizes on Venus led them to favor the rising diapir model for corona formation.

Janes et al. (1992) employed a somewhat different model that treated the rheologic structure of the lithosphere as an elastic spherical shell with basaltic elastic properties over a constant viscosity mantle using an analytical thick-shell model (Janes and Melosh 1988). The diapir was modeled as a swarm of buoyant point masses in a spherical region at depth in the mantle. They fit the topographic profiles across the radially fractured domes Makh and Mokos, finding that both could be well fit with diapirs approximately 90 km in radius located just below the elastic lithosphere (Fig. 12). The higher topography of Mokos required a greater density contrast between the diapir and the mantle as well as a thinner effective elastic lithospheric thickness. The inferred lithospheres were fairly thin, 8 km for Makh and 4 km for Mokos, and density contrasts were 18 and 33 kg m^{-3}, respectively. As with the viscous model, stresses in the center of the uplift were also predicted to be extensional with the hoop stress greater than the radial stress, leading to radial fracturing. This modeling also predicts that the radial stress becomes compressional as the distance from the center of the uplift increases. For both Makh and Mokos, the differential stress at the center of the uplift is large enough to initiate fracturing, and the observed fractures extend to distances where the model indicates that differential stress between the hoop and radial stresses falls below 50 MPa.

Fits to topography using these methodologies are nonunique because models have more free parameters (e.g., lithospheric or crustal thickness, size, depth, and density contrast of the mantle, thermal gradient, etc.) than observational constraints (e.g., height and width of the topography, extent of radial fracturing). However, Janes and Squyres (1993), using the elastic shell model, were able to show that, for reasonable density contrasts between the diapir and surrounding mantle, uplifted topography with height and width similar to that observed for radially fractured domes on Venus rather tightly

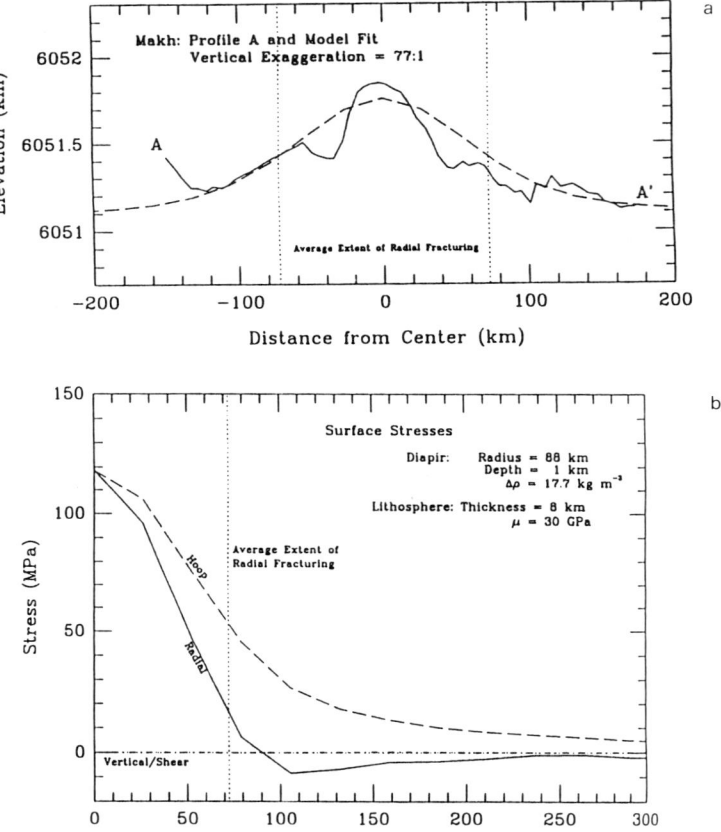

Figure 12. (a) Altimetric profile A-A′ (solid line) across Makh with the model fit (dashed line) superposed from Janes et al. (1992). The fit has a central height of 660 m and a full-width half-maximum of 145 km. The dotted vertical lines mark the average extent of radial fracturing around Makh. (b) Surface stresses for (a) from Janes et al. (1992). Surface stresses shown are due to a rising diapir 88 km in radius located 1 km below an 8-km thick lithosphere. The diapir is 17.7 kg m^{-3} less dense than the surrounding mantle. The dotted vertical lines mark the average extent of radial fracturing around Makh.

constrained the effective elastic lithospheric thickness of Venus to be less than 17 km and the diapir to be less than 50 km below the base of the lithosphere. The radius of the diapir was limited to be between 60 and 110 km. The presence of regional stresses during uplift can lead to significant alteration in the predicted radial pattern of extensional fractures; deflections of the radial fractures can be used to determine regional stresses (Cyr and Melosh 1993).

Hamilton and Stofan (1996) build upon the idea of Tackley and Stevenson (1991) and Tackley et al. (1992) in suggesting that small gravitational instabilities in the upper mantle, (Rayleigh-Taylor instabilities) may provide a possible mechanism for the formation of coronae, particularly those occurring along chasmata. These instabilities arise when a layer of low density is overlain by a layer of higher density. The boundary between the two layers develops a perturbation resulting in diapiric upwellings with a characteristic spacing that is dependent, in part, on the depth to the low density layer. Hamilton and Stofan (1996) estimate that if coronae in this region were formed by Rayleigh-Taylor instabilities, the depth to the boundary layer lies somewhere between 150 and 200 km. They conclude that the likelihood of a chemical boundary at this depth is low, and that the most plausible explanation for the unstable layer is a zone of partial melting which forms diapirs that become concentrated along a zone of thermal weakness in the lithosphere.

B. Flattening

The above models assume spherical diapirs or Gaussian driving forces. The profiles of some uplifted features, such as the corona Selu, is more broadly plateau-like (Squyres et al. 1992), implying that any diapir responsible for producing this topography has begun to flatten out against the base of the lithosphere. Janes et al. (1992) modeled Selu as an uplift over a flattened, elliptical diapir. Their best fit model had a 4-km thick elastic lithosphere over a diapir with a density contrast of 40 kg m^{-3}, radius of 224 km and thickness of 18 km. Although they could fit the topography reasonably well, the stresses derived from their model did not account for the circumferential graben observed on the rims of Selu and other plateau-like features.

Koch (1994) applied a dynamical flow model to follow the rise and deformation of a buoyant initially spherical diapir through constant viscosity mantle. In this fully viscous model, a spherical diapir is initially placed at some depth below a rigid surface and allowed to rise. As the diapir rises, a broad topographic swell develops with radial fracturing predicted as with the modeling discussed previously. For a 100 km radius diapir with a 10% density contrast with the mantle, it predicts a 1 km uplift. This dynamic model then allows the diapir to deform as it impinges on the underside of the lithosphere. The model predicts it will flatten into a disk shape producing a broad plateau at the surface. The radius of the plateau formed is nearly equal to the radius of the flattened diapir. The model predicts that at the edge of the plateau the radial stress will be the most tensional and the vertical stress the most compressional, consistent with the development of concentric extensional fracturing around the rim of the plateau, as stresses beyond the plateau predict concentric contractional features. However, the differential stresses predicted on the rim are relatively small, and probably cannot account for the observed concentric fracturing.

C. Relaxation

Eventually, the thermal buoyancy support of the diapir should diminish as the diapir cools, and the upraised topography would be left uncompensated by lower-density material at depth and subject to gravitational relaxation. This process was first modeled for coronae by Stofan et al. (1991) using the same layered viscous model they employed for diapiric uplift. They found that gravitational relaxation of an uncompensated plateau will produce a general profile shape associated with some coronae, namely a raised central bowl, annular rim, and surrounding moat (Fig. 13). For this general profile shape to form, it was necessary both that the initial topography be uncompensated in order to relax, and that it be plateau shaped to produce the central bowl and outer moat. During relaxation radial tensional stresses developed primarily in the deepening trough which formed at the edge of the initial plateau. This radial stress would account for the concentric radial fracture system around many coronae (Squyres et al. 1992).

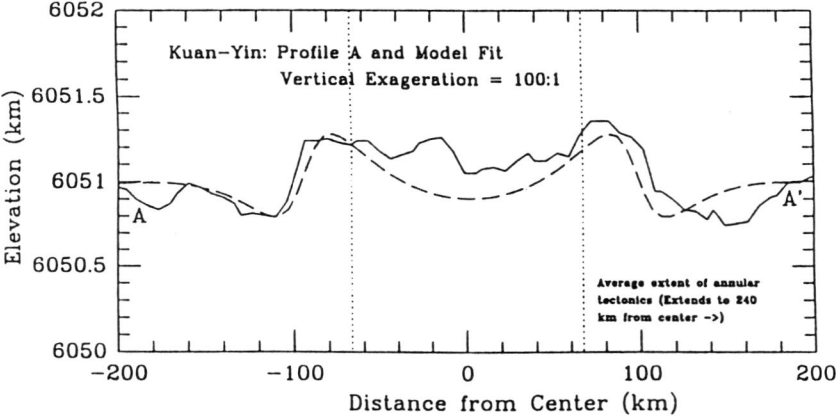

Figure 13. Fit to a corona profile from Janes et al. (1992). Altimetric profile A-A' (solid line) across Eve with the model fit (dashed line) superposed. The dotted vertical lines mark the average inner and outer limits of the annulus of tectonics surrounding Eve.

Janes and Squyres (1995) studied gravitational relaxation using a viscoelastic finite element model with layered power-law rheology (Goetze 1978; Shelton and Tullis 1981) and a failure envelope (Brace and Kohlstedt 1980) to determine the dominant response (elastic or viscous) at depth. They also found that the characteristic topography of coronae also required, as a starting condition, that the relaxing load be plateau shaped. Their model produced concentric extensional fracturing concentrated on the outer wall of the moat which develops around the plateau during relaxation. They showed that corona topography is more pronounced and tectonic fractures are better developed when the crust is thicker or the thermal gradient steeper, or when the crust is thinned and replaced by mantle material. Use of the more recent dry diabase flow law (Mackwell et al. 1993a,b) subdued the dependence of the final topography and tectonics on crustal thickness and thermal gradient but did not prevent the formation of corona through the relaxation process. In addition, they showed that relaxed corona topography will persist over geologic time scales, supported by flexure of the lithosphere and isostatic buoyancy.

D. Flexure

As a possible alternative to the formation of final corona topography by gravitational relaxation of an upraised lithosphere, Sandwell and Schubert (1992a,b) proposed a process they termed "retrograde subduction." Noting that corona rims and moats resemble terrestrial subduction zones in both planform (McKenzie et al. 1992) and profile, they suggested that the last stages of corona formation resulted when the upwelling mantle material breeched the lithosphere, emplaced a load on the surface, and began to roll back the lithosphere. They examined this idea using a thin plate flexure model to fit the topography of the moat and outer rise for four large coronae, finding good fits for lithospheric thicknesses ranging from 15 km at Eithinoha and Latona to 40 km or greater at Heng-o (Fig. 14). They also examined (Sandwell and Schubert 1992a) whether the rim of the corona supplied a sufficient load to account for the flexure observed. Finding that this was not the case for Latona led them to posit that the subducted lithosphere had become negatively buoyant, adding to the downward force on the flexed lithosphere. However, subducted lithosphere cannot reach the basalt-eclogite transition depth for coronae smaller than 200 km in diameter (Sandwell and Schubert 1992a). Johnson and Sandwell (1994) expanded the plane strain model used by Sandwell and Schubert to axisymmetry to examine the possibility that the moat and outer rise of smaller corona could also be due to flexure, finding that it could for coronae as small as 250 km in diameter for elastic thicknesses of 12 to 25 km.

E. Time Scales

Determinations of the time scales involved in corona formation and development are all model dependent, but are relatively consistent between models. Diapir rise times during the formation of radially fractured domes depend on the depth at which the diapirs form, their size and density contrast, and the viscosities of the diapir and surrounding mantle. Stofan et al. (1991) give the time to rise through a diapir radius, based on Stokes flow, of about 10 Myr for a 100 km diapir 10% less dense than the mantle which has a viscosity of 10^{21} Pa s, so that such a diapir should rise from a depth of 500 km to near the surface in about 25 Myr. Koch (1994) finds that for the same parameters, and for a diapir viscosity equal to the mantle viscosity, the diapir is expected to rise from a depth of 500 km to near the surface in about 15 Myr (note that the time constant in her Fig. 3 should be divided by 4/15, D. Koch, personal communication), and will take an additional 50 to as much as 200 Myr to achieve significant flattening after it hits the base of the lithosphere. These times assume that the density of the diapir remains unchanged during flattening.

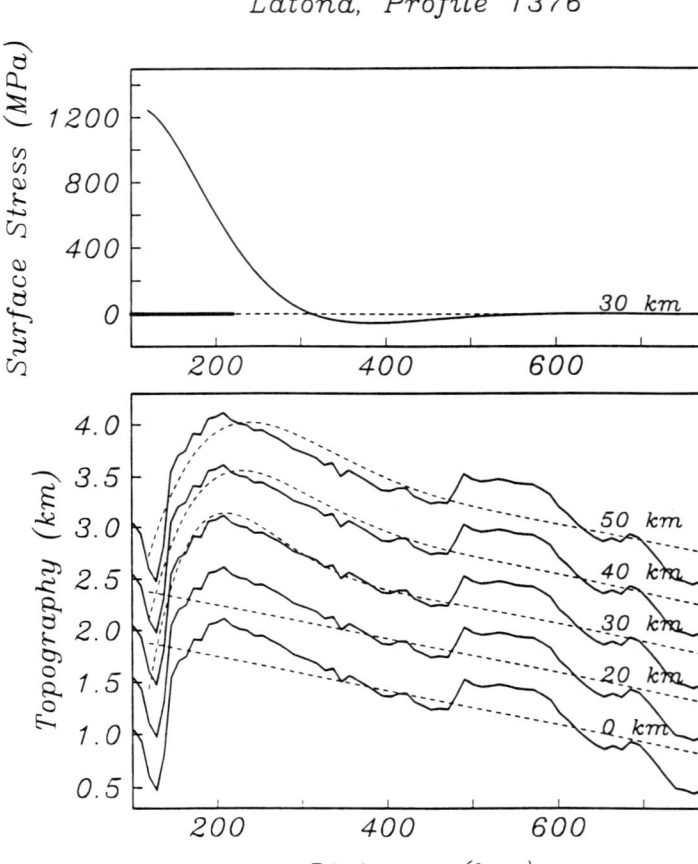

Figure 14. Flexural fits to corona from Sandwell and Schubert (1992b). The fits illustrated are of plate flexure models to profile 1376 across southern Latona (bottom). The 30-km thick elastic plate provides the minimum rms fit as well as the best visual fit. Model surface stress is tensile between the trench axis and ~300 km (top). Circumferential fractures (thick line, top) occur out to a distance of 230 km.

However, cooling times for a diapir in contact with the lithosphere, estimated from simple one dimensional models, are on the order of 50 to 85 Myr (Stofan et al. 1991; Janes and Squyres 1995). In contrast, model-derived gravitational relaxation times are fairly short, varying from 25,000 yr (Stofan et al. 1991) to 1 Myr (Janes and Squyres 1995) depending on the rheological structure, but in any case are significantly shorter than cooling times for the buoyant diapir. Therefore, relaxation should be able to proceed nearly instantaneously as the diapir cools. The total time for corona formation will therefore be controlled by the diapir rise, flattening and cooling times, with the total evolutionary time scale expected to be on the order of 100 to 250 Myr.

V. DISCUSSION

A. Origin of Chasmata Coronae

Approximately linear groups of volcanic features on Earth, such as the Hawaiian-Emperor seamount chain, can be attributed to hotspot volcanism, a process whereby the movement of a lithospheric plate over an active mantle upwelling produces a linear set of volcanic features parallel to the direction of plate motion. Volcanoes along such a chain display a progression in age with the youngest features overlying the active upwelling, and the oldest features at the distal end of the chain. Due to the lack of plate tectonics on Venus (Solomon et al. 1992), and the absence of any progression in the stages of evolution of coronae along corona chains (Stofan et al. 1992; Hamilton and Stofan 1996), a hotspot model with a fixed, continuous source and a moving lithosphere is not a likely origin for corona chains.

As discussed above, both upwelling (Basilevsky et al. 1986; Stofan et al. 1991) and delamination/subduction (Sandwell and Schubert 1992a,b) have been proposed to explain corona formation. In this section, we discuss evidence from chasmata-related coronae relevant to distinguishing between these models. Several observations suggest that the subduction model may not be applicable to the formation of trough systems with corona chains. Subduction zones on Venus have been identified based on four criteria: narrow, deep trenches elongate along strike/arcuate planform; ridge-trench-outer rise topography; large outer rise curvature; and fractures parallel to strike of trench on outer trench wall and outer rise with no cross-strike fractures across the trench (Schubert and Sandwell 1995). While these are characteristic of the subduction model, they are also consistent with the diapir model, and thus do not provide a unique interpretation. The ridge-trench-outer rise asymmetric profiles (noted by Sandwell and Schubert 1992a,b; Schubert and Sandwell 1995) at coronae along chasmata also tend to be present along the chasmata where there are no coronae, similar to asymmetries seen at terrestrial rifts. These troughs around coronae can also be explained by other models such as gravitational relaxation (i.e., Stofan et al. 1991), however, at some features high initial topography is required by the gravitational relaxation model to produce the observed topography (Janes et al. 1992).

Schubert and Sandwell (1995) suggest that chasmata coronae form by subduction, while those isolated from chasmata form by diapirism. All coronae along the Hecate and Parga chains do not exhibit morphologies like those of the coronae suggested to be sites of subduction (Schubert and Sandwell 1995), and many display fracture patterns that are inconsistent with adjacent subduction. Coronae also form all along the chasmata, not just in the lowland regions where Schubert and Sandwell (1995) predict corona formation/subduction. If the chasmata segments are formed by subduction, then they should terminate on transform faults, as suggested by McKenzie et al. (1992). However, there is no evidence of such features bounding the trough segments of Hecate and Parga, nor is there evidence for transform faults in the Dali/Diana area (Hansen and Phillips 1993). In general, the styles of deformation, circularity of coronae, and cross-cutting relationships argue against a subduction origin (Hansen and Phillips 1993; Stofan and Bindschadler 1993; Hansen 1993).

Features consistent with the regional extension with diapiric upwelling model (Stofan et al. 1992; Hansen and Phillips 1993; Hamilton and Stofan 1996) are extensional zones associated with volcanic flows, concentrated fracturing, and vertical displacement due to normal faulting. Fissures and faults that run parallel to the axis of such a zone are the result of tension. Coronae would be predicted to occur anywhere along the zone due to diapiric upwelling accompanying rifting. In profile, extension-formed trough segments have inward-dipping walls and may or may not appear symmetrical depending on the nature of the faulting and the response of the lithosphere to stress. The presence of a major trough with parallel normal faulting and associated outflows at Hecate Chasma is interpreted by Hamilton and Stofan (1996) to be consistent with other Venusian (Baer et al. 1994; Senske et al. 1991a,b,1992) and terrestrial (see, e.g., Vening Meinesz 1950) rift morphologies. The main portion of the Hecate trough system has an asymmetric profile marked by vertical variations in topography of several km that can be explained by uneven normal faulting (Senske et al. 1992). Troughs forming Parga Chasma have shallower profiles, but similarly exhibit lava flows and axis-parallel faults concentrated in and near the trough segments. These data are consistent with the modeling of Hamilton and Stofan (1996), who suggest that the chasmata coronae originate from a zone of partial melting along a zone of thermal weakness in the lithosphere, as discussed above. In addition, the diapir model easily explains the circularity of coronae, which must be explained by coincidences in the positions of oppositely subducting arc segments in the subduction model (Schubert and Sandwell 1995).

Does rifting drive the formation of coronae along chasmata, or do coronae cause localization of faulting in these regions? The heterogeneity of relationships between tectonism and corona formation along chasmata systems and corona chains indicates that tectonism and corona formation have been coeval. A study of northern Lada Terra by Baer et al. (1994) found geologic evidence similar to that along Hecate and Parga Chasmata that neither the influences exerted by coronae on the location of surface expressions of regional extensional stresses, or vice versa, appear to have dominated the development of the region. These relationships support the idea that neither corona formation nor extension alone can be said to drive the evolution of these regions. This is in contrast to the hypothesis of Schubert and Sandwell (1995), where corona formation and subduction follow rift propagation.

B. Implications of Gravity Data

Schubert et al. (1994) interpret the apparent depths of compensation between 75 and 200 km as due to either thermal compensation at the base of the lithosphere, or as shallow, crustal compensation together with deep compensation by a dense subducting slab. As with the large depth of compensation at large volcanic rises (chapter by Smrekar et al.), the depths are too large to be due to crustal compensation alone. As with the volcano-tectonic and topographic signature of coronae, the gravity data can be interpreted as supporting both the subduction and upwellings models of coronae formation.

Schubert et al. (1994) interpret the results of their gravity investigations as providing further evidence for the subduction hypothesis. In particular, they point out that the gravity highs along chasmata-associated coronae occur on the concave side of the trench, which would be consistent with the hypothesis that retrograde subduction occurs on the concave portions of chasmata. They suggest that the variation in the gravity anomaly observed along Venusian chasmata cannot be used to argue for or against the presence of subducted slabs, as there is a great deal of variation in the gravity signature along terrestrial subduction zones due to variations in lithospheric thickness and back-arc spreading.

A number of arguments can be made in favor of the upwelling hypothesis. The apparent compensation depth range for coronae, 65 to 200 km, is similar to the range of 100 to 260 km found for large volcanic rises on Venus (Fig. 7) (Stofan et al. 1995; chapter by Smrekar et al.). Coronae associated with rifting tend to have higher topography, but have the same range of apparent depths of compensation as isolated coronae. For large volcanic rises, rift-associated extension was interpreted to facilitate lithospheric thinning, greater pressure-release melting and uplift, and large apparent depths of compensation (Smrekar 1994; Stofan et al. 1995). The same conclusions about the influence of rifting on volcanism and topographic uplift in coronae were discussed above; no clear pattern in apparent depths of compensation can be determined from this limited sample. Although Schubert et al. (1994) argue that the presence of a gravity high on the concave side of the trench is consistent with subduction, the same position of the gravity high is predicted for an upwelling model. Although the magnitude of the gravity anomaly does vary a great deal along terrestrial subduction zones, there is a positive signature along the length of terrestrial subduction zones. On Venus, there is no pronounced gravity signature that follows the trend of the chasmata. The only prominent gravity highs are centered over coronae, which suggests that the primary deep changes in density, as predicted by the upwelling model, are localized under the coronae and not distributed along the chasmata.

C. Remaining Questions

Numerical modeling of the various processes thought to be involved in coronae formation and comparisons to the observed features has generally verified the original hypothesis that most are the result of diapiric uplift followed by gravitational relaxation of the raised topography. Even so, a number of questions remain unanswered. For example, the source of the driving diapirs remains a problem (Phillips and Hansen 1994). The size of the diapirs during the formation of the radially fractured domes indicate that they form at relatively shallow depths, in the upper mantle rather than at the core-mantle boundary. The uplifts observed at radially fractured domes and plateaus require relatively large density contrasts between the diapir and the mantle. If these density contrasts are thermally driven, they require localized large temperature contrasts. The density difference between the diapir and the mantle might be due to pressure-release melting (Squyres et al. 1992; Tackley et al. 1992; Janes and Squyres 1993; Hansen and Phillips 1993; Phillips and Hansen 1994). This scenario is supported by the association of coronae with areas of extension (Stofan and Head 1990; Stofan et al. 1992; Squyres et al. 1992; Baer et al. 1994; Magee and Head 1995). However, if the diapir responsible for uplift results from the segregation of low density minerals from the surrounding mantle, it is difficult to see how its buoyant support could be removed so that gravitational relaxation could act on the raised topography to produce a coronae. This seeming paradox may indicate that the diapirs derive their buoyancy from a combination of thermal and compositional effects.

Further modeling of the diapir/upwelling process is required as current diapir models do not predict some aspects of coronae on Venus. The width and the location of the annulus is not always consistent with model predictions. At some coronae, the annulus does not correspond with the topographic rim, indicating possible unmodeled topographic stages. Current models do not predict coronae that have topography below the surrounding plains, while a significant number of coronae correspond to depressions. Late-stage radial fracturing may indicate that some coronae go through multiple phases of uplift and relaxation. The complexities observed from continuing mapping of coronae indicate that corona formation may have more stages than previously recognized, and may take place over a significant period of time (Stofan and Smrekar 1996)

VI. CONCLUSIONS

Coronae represent a significant new type of geomorphologic feature, apparently unique to Venus. The majority of coronae are thought to be produced by upwelling (Stofan et al. 1991; Janes et al. 1992). The tectonic, volcanic and topographic characteristics of coronae show a great deal of variation. This variation may indicate that not all coronae follow the same evolutionary sequence. For example, in the population of approximately 360 coronae described by Stofan et al. (1992), Sacajawea and Colette Paterae were included because they fit the definition of a corona (possessing an annulus of concentric tectonic features), although they are widely interpreted to be volcanic calderas (Roberts and Head 1990; Head et al. 1991,1992). The general similarity of large-scale Venusian calderas to some coronae may indicate that late-stage collapse or downsagging is an important process in corona evolution. In addition, Artemis Chasmata/Corona has been interpreted as a corona (Stofan and Saunders 1990), a subduction zone (McKenzie et al. 1992; Sandwell and Schubert 1992*a,b*) or a location of underthrusting and convergence (Brown and Grimm 1995). Artemis is much larger than other coronae, and may not have formed by the same processes as other coronae. Many of the features classified as coronae are also transitional in morphology with large volcanic edifices, such as the volcano Mokosha Mons. Therefore, although we interpret most coronae to have formed by upwelling followed by gravitational relaxation, coronae should be regarded as a useful morphologic classification rather than a strict genetic classification.

No strong correlation exists between the morphologic classes of coronae described here and their geologic setting. The exception is the radial/concentric class, which has been interpreted to be coronae in relatively early stages of evolution (Stofan et al. 1992; Squyres et al. 1992; Stofan 1993). All of the radial/concentric coronae occur along chasmata, which may be the only location where the lithosphere is currently thin enough to permit corona formation. In all geologic settings, a wide range of topographic types can be seen. For example, along Parga Chasma, all topographic groups except for coronae with an outer rise, trough, rim, interior depression (topographic group 6) can be found. In general, the observed range in corona tectonic patterns and topography are not correlated, nor do they relate in a clear fashion to regional geologic setting. If one makes the assumption that different geologic settings are likely to have different lithospheric structure, such as that the plains regions have thicker lithosphere than chasmata, there does not appear to be any correlation between the topographic signature of coronae and lithospheric structure.

Models of corona formation generally support the hypothesis that coronae form over rising mantle diapirs with subsequent modification by gravitational relaxation and flexure of the lithosphere. These models predict effective elastic lithosphere thicknesses during initial upwelling of less than 10 km but perhaps as much as 30 or 40 km during relaxation and flexure. This range of elastic thicknesses is consistent with other studies of the Venusian lithosphere (see, e.g., Johnson and Sandwell 1995; Sandwell and Schubert 1994; Smrekar 1994; Phillips 1994; chapters by Phillips et al. and Smrekar et al.) and indicate that the lithosphere is thinned, either thermally or mechanically, during diapir rise. The diapirs are generally 100 km in radius for the majority of coronae which have diameters between 100 and 400 km and probably form at fairly shallow depths in the mantle. Diapir buoyancy is most likely due to a combination of thermal and compositional segregation, as discussed above, and are likely to form within a few hundred kilometers of the surface. The timespan between the formation of the diapir at depth and relaxation to the final corona form is likely to be on the order of 100 Myr. However current models of corona origin (Stofan et al. 1991; Janes et al. 1992; Koch 1994) do not predict the wide range of morphologies and complexities of deformation observed at coronae. It is quite possible that more than one mechanism is involved and particularly that mechanisms which may form the bulk of 200 to 400 km diameter coronae are not the same as those responsible for the few much larger features such as Artemis, Heng-o and others.

Two models, subduction/delamination and extension with upwelling, offer explanations of many features identified along corona chains and chasmata systems. Based on geologic observations, we favor an extensional regime for the formation of chasmata on Venus. Diapiric upwellings are consistent with tectonic extension at the surface, and suggest a region of increased lithospheric heat flow, which would tend to inhibit subduction. Further detailed mapping studies of coronae and chasmata regions should provide further constraints for more complex models of the origin and evolution of coronae.

Acknowledgment. This work was carried out in part at the Jet Propulsion Laboratory, California Institute of Technology, sponsored by grants from the National Aeronautics and Space Administration. A review by V. Hansen was extremely helpful. ERS gratefully acknowledges the support of J. Guest and University College London where some of this work was conducted, and SCGD for a timely arrival.

REFERENCES

Baer, G., Schubert, G., Bindschadler, D. L., and Stofan E. R. 1994. Spatial and temporal relations between coronae and extensional belts, northern Lada Terra, Venus. *J. Geophys. Res.* 99:8355–8369.

Barsukov, V. L., et al. 1984. Preliminary evidence on the geology of Venus from radar measurements by the Venera 15 and 16 Probes. *Geokhimia* 12:1811–1820.

Barsukov, V. L., et al. 1986. The geology and geomorphology of the Venus surface as revealed by the radar images obtained by Veneras 15 and 16. *J. Geophys. Res.* 91:378–398.

Basilevsky, A. T., et al. 1986. Styles of tectonic deformation on Venus: Analysis of Veneras 15 and 16 data. *J. Geophys. Res.* 91:399–411.

Bindschadler, D. L., and Parmentier, E. M. 1990. Mantle flow tectonics and a ductile lower crust: Implications for the formation of large-scale features on Venus. *J. Geophys. Res.* 95:21329–21344.

Brace, W. F., and Kohlstedt, D. L. 1980. Limits on lithospheric stress imposed by laboratory experiments. *J. Geophys. Res.* 85:6248–6252.

Brown, C. D., and Grimm, R. E. 1995. Tectonics of Artemis Chasma: A venusian "plate" boundary. *Icarus* 117:219–249.

Campbell, D. B., and Burns, B. A. 1980. Earth-based radar imagery of Venus. *J. Geophys. Res.* 85:8271–8281.

Crumpler, L. S., Head, J. W., and Aubele, J. C. 1993. Relation of major volcanic center concentration on Venus to global tectonic patterns. *Science* 261:591–595.

Cyr, K. E., and Melosh, H. J. 1993. Tectonic patterns and regional stresses near Venusian coronae. *Icarus* 102:175–184.

Ford, P. G., and Pettengill, G. H. 1992. Venus topography and kilometer-scale slopes. *J. Geophys. Res.* 97:13103–13114.

Goetze, C. 1978. The mechanisms of creep in olivine. *Phil. Trans. Roy. Soc. London* A 288:99–119.

Hamilton, V. E., and Stofan, E. R. 1996. The geomorphology and evolution of Hecate Chasma, Venus. *Icarus*, in press.

Hansen, V. L. 1993. Asymmetric venusian rifts: Arguments against subduction. *Eos: Trans. AGU* 377.

Hansen, V. L., and Phillips, R. J. 1993. Tectonics and volcanism of eastern Aphrodite Terra, Venus: No subduction, no spreading. *Science* 260:526–530.

Head, J. W., et al. 1991. Venus volcanism: Initial analysis from Magellan data. *Science* 252:276–288.

Head, J. W., et al. 1992. Venus volcanism: Classifications of volcanic features and structures, associations, and global distribution from Magellan data. *J. Geophys. Res.* 97:13153–13198.

Janes, D. M., and Melosh, H. J. 1988. Sinker tectonics: An approach to the surface of Miranda. *J. Geophys. Res.* 93:3127–3143.

Janes, D. M., and Squyres, S. W. 1993. Radially fractured domes: A comparison of Venus and the Earth. *Geophys. Res. Lett.* 20:2961–2964.

Janes, D. M., and Squyres, S. W. 1995. Viscoelastic relaxation of topographic highs on Venus to produce coronae. *J. Geophys. Res.* 100:21173–21187.

Janes, D. M., et al. 1992. Geophysical models for the formation and evolution of coronae on Venus. *J. Geophys. Res.* 92:16055–16068.

Johnson, C. L., and Sandwell, D. T. 1994. Lithospheric flexure on Venus. *Geophys. J. Int.* 119:627–647.

Koch, D. M. 1994. A spreading drop model for plumes on Venus. *J. Geophys. Res.* 99:2035–2052.

Konopliv, A. S., and Sjogren, W. L. 1994. Venus spherical harmonic gravity model to degree and order 60. *Icarus* 112:42–54.

Mackwell, S. J., Zimmerman, M. E., Kohlstedt, D. L., and Scherber, D. S. 1993a. Dry deformation of diabase: Implications for tectonics on Venus. *Lunar Planet. Sci. Conf.* XXV:817–818 (abstract).

Mackwell, S., Kohlstedt, D. L., Scherber, D. S., and Zimmerman, M. E. 1993b. High temperature deformation of diabase: Implications for tectonics on Venus. *Eos: Trans. AGU* 378.

Magee, K. P., and Head, J. W. 1995. The role of rifting in the generation of melt: Implications for the origin and evolution of the Lada Terra-Lavinia Planitia region of Venus. *J. Geophys. Res.* 100:1527–1552.

Masursky, H. 1987. Geological evolution of coronae. *Lunar Planet. Sci. Conf.* XVIII:598–599 (abstract).

Masursky, H., et al. 1980. Pioneer Venus radar results: Geology from images and altimetry. *J. Geophys. Res.* 85:8232–8260.

McGill, G. E. 1994. Hotspot evolution and Venusian tectonic style. *J. Geophys. Res.* 99:23149–23161.

McKenzie, D., et al. 1992. Features on Venus generated by plate boundary processes. *J. Geophys. Res.* 97:13533–13544.

Nikolaeva, O. V., Ronca, L. B., and Basilevsky, A. T. 1986. Circular structures on the plains of Venus as indicating geologic history. *Geokhimia* 5:579–589.

Phillips, R. J. 1994. Evaluation of venusian highland models. *Icarus* 112:27–33.

Phillips, R. J., and Hansen, V. L. 1994. Tectonic and magmatic evolution of Venus. *Ann. Rev. Earth Planet. Sci.* 22:597–654.

Pronin, A. A., and Stofan, E. R. 1990. Coronae on Venus: Morphology and distribution. *Icarus* 87:452–474.

Roberts, K. M., and Head, J. W. 1990. Western Ishtar Terra and Lakshmi Planum, Venus: Models of formation and evolution. *Geophys. Res. Lett.* 17:1341–1344.

Roberts, K. M., and Head, J. W. 1993. Large-scale volcanism associated with coronae on Venus: Implications for formation and evolution. *Geophys. Res. Lett.* 20:1111–1114.

Sandwell, D. T., and Schubert, G. 1992a. Evidence for retrograde lithospheric subduction on Venus. *Science* 257:766.

Sandwell, D. T., and Schubert, G. 1992b. Flexural ridges, trenches, and outer rises around coronae on Venus. *J. Geophys. Res.* 97:16069.

Sandwell, D. T., and Schubert, G. 1993. A global survey of possible sites of subduction on Venus. *Eos: Trans. AGU* 377.

Schaber, G. 1982. Limited extension and volcanism along zones of lithospheric weakness. *Geophys. Res. Lett.* 9:499–502.

Schubert, G., and Sandwell, D. T. 1995. A global survey of possible subduction sites on Venus. *Icarus* 117:173–196.

Schubert, G., Moore, W. B., and Sandwell, D. T. 1994. Gravity over coronae and chasmata on Venus. *Icarus* 112:130–146.

Senske, D. A., et al. 1991a. Geology and tectonics of Beta Regio, Guinevere Planitia, Sedna Planitia, and Western Eistla Regio, Venus: Results from Arecibo image data. *Earth, Moon, Planets* 55:163–214.

Senske, D. A., et al. 1991b. Geology and tectonics of the Themis Regio-Lavinia Planitia-Alpha Regio-Lada Terra Area, Venus: Results from Arecibo image data. *Earth, Moon, Planets* 55:97–161.

Senske, D. A., Schaber, G. G., and Stofan, E. R. 1992. Regional Topographic rises on Venus: Geology of western Eistla Regio and comparisons to Beta Regio and Atla Regio. *J. Geophys. Res.* 97:13395–13420.

Shelton, G., and Tullis, J. 1981. Experimental flow laws for crustal rocks. *Eos: Trans. AGU* 62:396.

Smrekar, S. E. 1994. Evidence for active hotspots on Venus from analysis of Magellan gravity data. *Icarus* 112:2–26.

Solomon, S. C., et al. 1992. Venus tectonics: An overview of Magellan observations. *J. Geophys. Res.* 97:13199–13256.

Squyres, S. W., et al. 1992. The morphology and evolution of Coronae on Venus. *J. Geophys. Res.* 97:13611–13634.

Squyres, S. W., et al. 1993. The spatial distribution of coronae and related features on Venus. *Geophys. Res. Lett.* 20:2965–2968.

Stofan, E. R. 1993. Coronae: Relative ages and stratigraphic significance. *Eos: Trans. AGU*, pp. 188–189.

Stofan, E. R. 1995. Coronae on Venus: Topographic variations and correlations between morphology and regional setting. *Lunar Planet. Sci. Conf.* XXVI:1361–1362 (abstract).

Stofan, E. R., and Bindschadler, D. L. 1993. Corona annuli: Characteristics and modes of origin. *Eos: Trans. AGU* 377.

Stofan, E. R., and Hamilton, V. E. 1996. The geology of Parga Chasma. In preparation.

Stofan, E. R., and Head, J. W. 1990. Coronae of Mnemosyne Regio, Venus: Morphology and origin. *Icarus* 83:216–243.

Stofan, E. R., and Smrekar, S. E. 1996. New insights into corona evolution on Venus: Implications for models of origin. *Lunar Planet. Sci. Conf.* XXVII:1279–1280 (abstract).

Stofan, E. R., Bindschadler, D. L., Head, J. W., and Parmentier, E. M. 1987. Corona structures on Venus: Models of origin. *Lunar Planet. Sci. Conf.* XVIII:954–955 (abstract).

Stofan, E. R., Bindschadler, D. L., Head, J. W., and Parmentier, E. M. 1991. Corona structures on Venus: Models of origin. *J. Geophys. Res.* 96:20933–20946.

Stofan, E. R., et al. 1992. Global distribution and characteristics of coronae and related features on Venus: Implications for origin and relation to mantle processes. *J. Geophys. Res.* 97:13347–13378.

Stofan, E. R., Hamilton, V. E., and Cotugno, K. 1993. Parga and Hecate Chasma, Venus: Structure, volcanism and models of formation. *Lunar Planet. Sci. Conf.* XXIV:1361–1362 (abstract).

Stofan, E. R., Smrekar, S. E., Bindschadler, D. L., and Senske, D. A. 1995. Large topographic rises on Venus: Implications for mantle upwelling. *J. Geophys. Res.* 100:23317–23327.

Tackley, P. J., and Stevenson, D. J. 1991. The production of small Venusian coronae by Rayleigh-Taylor instabilities in the uppermost mantle. *Eos: Trans. AGU* 72:287.

Tackley, P. J., Stevenson, D. J., and Scott, D. R. 1992. Volcanism by melt-driven Rayleigh-Taylor instabilities and possible consequences of melting for admittance ratios on Venus. In *International Colloquium on Venus*, LPI Contrib. No. 789, pp. 123–124.

Vening Meinesz, F. A. 1950. Les "graben" africains résultat de compression ou de tension dans la croéte terrestre? *K. Belg. Kol. Inst. Bull.* 21:539–552.

Watters, T. R., and Janes, D. M. 1995. Coronae on Venus and Mars: Implications for similar structures on Earth. *Geology* 23:200–204.

PART VIII
Impact Cratering

CRATERING ON VENUS: MODELS AND OBSERVATIONS

WILLIAM B. McKINNON
Washington University

KEVIN J. ZAHNLE
NASA Ames Research Center

BORIS A. IVANOV
Institute for Dynamics of Geospheres, Moscow

and

H. J. MELOSH
University of Arizona

Venus has the densest atmosphere of the terrestrial planets. High-resolution radar images demonstrate the important influence of this atmosphere on would-be impactors. Small multiple and slightly larger irregular craters ($\sim$1.5–15 km in diameter) are clear evidence for deceleration and breakup of impactors $\lesssim$1 km in diameter. The crater size-frequency distribution observed on Venus is distinctive. It is deficient in craters as large as $\sim$30-km diameter and probably larger and lacks smaller craters entirely. As craters on Venus are nearly all well preserved, the lack of small craters is logically caused by the failure of small bodies to strike the surface. This dearth is qualitatively and quantitatively explained here by a model of atmospheric intervention. The atmospheric screen is very effective, but careful attention to atmospheric deceleration and impactor flattening is necessary to constrain reliably the crater retention age of the surface. Our nominal models, which incorporate the flattening seen in most of the best-resolved numerical simulations, yield ages of $\sim$750 Myr. The crater size-frequency distribution is consistent with a constant atmospheric pressure over the period of observed crater accumulation. Objects that fail to reach the surface often effectively explode. Strong shock waves during cratering or from such airbursts can shatter and fragment surface rock and create powerful, high-speed surface winds that may do the same. The results are dark and bright halos and splotches, characteristic radar-albedo features that may form, Tunguska-like, even in the absence of a central crater. More than 50 fresh impact craters are associated with west-opening, radar-dark "parabolas." These are most plausibly explained as deposits of small particles that were initially ejected through the atmosphere with the expanding vapor plume above the impact site; the particles then coasted ballistically, re-entered the atmosphere, and drifted westward under the influence of the steady, high-altitude zonal winds before being deposited. Craters on Venus are morphologically complex, i.e., modified by gravity-driven restoring forces, yet they are deep, considering Venus' gravity, compared with similar craters on the Moon. This difference is more apparent than real, and can be explained by a modestly greater transient strength for cratered Venusian crust and/or Venusian craters being so flat that mechanical equilibrium is determined by their apparent, or below-ground-plane, depths. The abundance of impact melt in Venusian crater ejecta should be much higher than on the Moon, with the proportion increasing with increasing crater size and impact angle (to the zenith). This is probably one of the major causes of the extensive ejecta flows seen. The majority of large impacts are ringed, and well preserved, but measurements and interpretations of them are controversial.

I. INTRODUCTION

The most profound lessons Venus has for impact cratering as a physical process concern the interactions of impactors and atmospheres. The most profound lesson impact craters have for Venus is that the planet's surface appears both geologically young and of a relatively uniform age. As such, this review focuses on the effects of atmospheres on impactors (Sec. II) and the effects of impactors on atmospheres (Sec. III), taking as its major theme the question of the age of the Venusian surface (Sec. II.D). We review in detail the question of impactor flattening and deceleration, which is so important to the age question (Secs. II.C and II.E). We also discuss the role of the atmosphere in the formation of non-craterform but clearly impact-related radar-albedo features: dark and bright halos and splotches (Sec. II.F) and dark parabolas (Sec. III.B). Finally, we discuss aspects of Venusian crater morphology where a strong role for the atmosphere is not called for (Sec. IV), focusing on the depths of complex craters, the generation of impact-melt-rich ejecta as one cause of extensive, run-out ejecta flows, and the formation of ringed basins (more on morphology can be found in the chapter by Herrick et al.).

II. INFLUENCE OF THE ATMOSPHERE ON IMPACTORS

A. Small Multiple and Larger Irregular Craters

The question of cratering on Venus has tantalized researchers for years. Pioneer Venus altimetry could not reveal them, and it was long hoped that the characteristics of the craters, once seen, would reveal the influence of the atmosphere and indeed allow detection of the time evolution of the atmosphere due to the changing effects on the crater size-frequency distribution (see, e.g., Tauber and Kirk 1976; Kahn 1982). Arecibo, Venera 15 and 16, and ultimately Magellan radar images did not disappoint; Plate 2 in Phillips et al. (1992) shows that Venus is pitted with craters of a broad variety of sizes, and that smaller ones are absent.

Our conceptions of potential atmosphere-impactor interactions predate the radar imaging missions. Figure 1a outlines what might happen when a small cosmic body enters an atmosphere at hypervelocity. The pressure encountered causes the impactor to fracture, and possibly, ultimately to be crushed as its brittle strength is exceeded. The fragments can disperse as their bow shocks interact, and Passey and Melosh (1980) formulated a simple model for how far fragments might spread (these authors were interested in

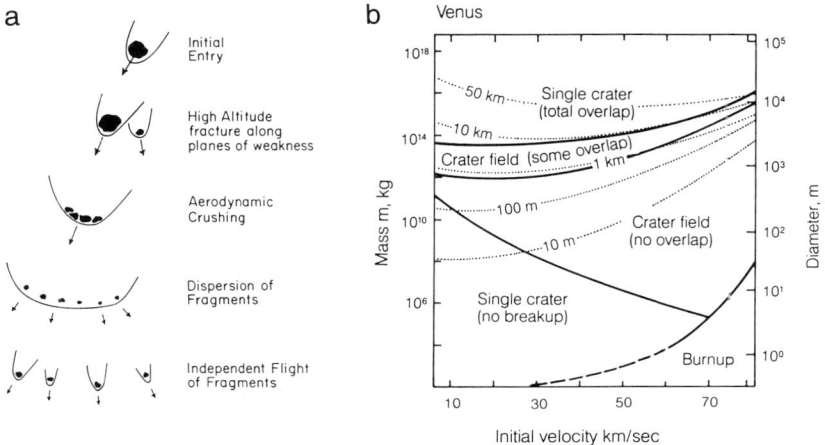

Figure 1. (a) Schematic illustration of the atmospheric entry and breakup of a large impactor on Venus. Modified from Melosh (1981). (b) The fate of iron impactors entering the atmosphere of Venus at an angle of 15 deg; a 500 MPa crushing strength is assumed. Different final outcomes are separated by solid lines, and the approximate size of the largest final crater (if there is more than one) is given by the dotted lines (figure modified from Passey 1980).

the formation of crater fields on the Earth, such as Henbury in Australia). While necessarily imprecise, the model predicts that there is a maximum cross-range spreading distance irrespective of impactor size (although down-range spreading can increase considerably for low-angle impactors). Hence if the craters formed are comparable or larger than the cross-range spread, then the individual fragment craters may all blend into one.

The predicted outcomes from the atmospheric breakup model are shown more quantitatively in Fig. 1b. The figure illustrates the outcomes for strong (iron-like) objects entering Venus' atmosphere as a function of impactor mass (on left), or diameter (on right), and impactor velocity. The impact angle assumed is 15°, which exaggerates the effects of spreading (45° is the average), but the comparisons are useful nevertheless. Small objects moving very fast burn up, but focusing on the median asteroidal velocity of ~ 20 km s^{-1} (Shoemaker et al. 1991) and going up in size, it can be seen that somewhat larger objects decelerate (aerobrake but do not break) and fall to the surface. Still larger objects break into fragments and make a crater field (or mound field if they fall slowly enough). At the largest scales the separate craters begin to overlap and finally effectively form a single crater. The crater sizes predicted are given by the dotted lines. We see that the transition from crater field to single crater occurs between ~ 1 km diameter and ~ 10 km diameter. This is certainly true in a general way; i.e., it is what is seen in Magellan radar images. Quantitative application of aspects of this model to Magellan observations can be found in Phillips et al. (1992) and Herrick and Phillips (1994a).

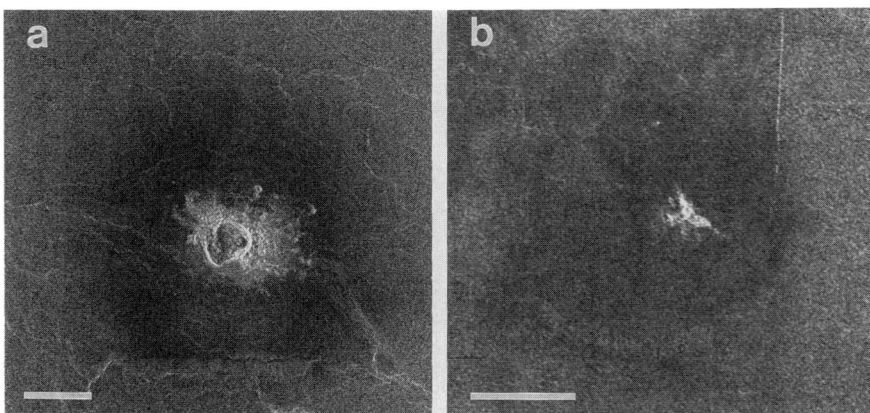

Figure 2. Dark-haloed impact features: (a) Irregular crater ~5 km in diameter formed on Sedna Planitia at 42.6°N, 349.5°E is surrounded by a low-radar-albedo zone ~50 km across; (b) bright feature, composed at least in part of several small pits, is centered within a ~30-km diameter dark splotch (at 47.2°N, 333° E). North is up in both images; scale bars are 10 km.

Examples of the ultimate fate of a dispersed impactor include the irregular crater in Fig. 2a. It is surrounded by what have been variously described as dark margins, dark halos, or dark shadows. Both the images in Fig. 2 are part of a continuum of irregular craterforms that dwindle to essentially nothing except the dark (or sometimes bright) halo. There are 350 of these dark splotch-like features on Venus (Strom et al. 1994), and they have generally been ascribed to interaction of the impactor shock wave with the surface (Phillips et al. 1991). In these almost-craters, the impactor does not quite reach the ground. The impactor dumps its energy in the air, and the blast wave and moving wake from what is effectively an explosion pummels the ground. Zahnle (1992) modeled this process and found that the impactor needs to decelerate in the bottom-most scale height if a strong shock is to be coupled into the ground. Essentially, then, dark halos and splotches are Tunguskas without the trees. Dark (and bright) halos will be returned to in Sec. II.F.

B. Atmospheric Shielding and Impactor Populations

The thick atmosphere of Venus prevents all but the largest impactors from cratering the surface. The number of small craters on Venus provides an interesting and statistically significant test of models for the disruption and deceleration of impacting bodies. The cratering record is also the only means presently available for dating the surface of Venus.

One of the key results of the Magellan mission is that Venusian impact craters are fresh-seeming almost without exception (Phillips et al. 1991; Strom et al. 1994). The craters are also sparsely and randomly distributed. Hence to first approximation, they *appear* to define a single surface of uniform age, clearly younger than the age of the solar system, yet once formed not subject to significant erosion or volcanic resurfacing. This fact has obvious implications for the history of the planet. We leave discussion of these implications to other chapters of this book.

For our present purposes, what is important is that the craters on Venus appear to represent a geologically recent production function rather than crater saturation. In particular, the relative lack of craters smaller than $\sim$30-km diameter, or as we will argue, smaller than $\sim$50-km diameter, is mainly caused by smaller objects not reaching the surface.

In this review we compare simulated crater distributions with the observed crater distribution on Venus. The simulation requires the mass and velocity distributions of the potential impactors, their incidence angles, and the rates at which Venus is struck by different types of objects; rules for predicting crater diameters given impact parameters; and a description of how the atmosphere interacts with incident bodies. These simulations are meant to supersede those presented by Zahnle (1992).

Incidence Angle. Incidence angles are taken to be isotropic, as in Zahnle (1992).

Impact Velocity. Velocity distributions are relatively less important than the mass and incidence angle distributions, because impactor masses span many orders of magnitude while impact velocities do not and because shallow entry so accentuates atmospheric interaction. For asteroids striking Venus we have fit the impact velocity distribution used by Herrick and Phillips (1994a), which was based on the Venus-crossing asteroid collision velocities and Öpik-derived probabilities in Shoemaker et al. (1991), to $v_A = 13(1 + 0.5x + 2x^3)$ km s^{-1}, where x is a random number between 0 and 1. We use the same distribution for short-period comets. For long-period comets we have assumed that impact velocities go as $v_{LP} = 15 + 75x^{0.6}$ km s^{-1}. This latter velocity distribution has a reasonable median impact velocity of 65 km s^{-1} and spans the full range of possible impact velocities; otherwise it is more convenient than realistic.

An arbitrary lower cutoff on the terminal velocity v_f of 300 m s^{-1} has been imposed on impact velocities producing craters. There must be some lower limit to hypervelocity cratering. For plausible choices, the cutoff on v_f affects production of only the smallest ($\lesssim$5 km) craters, which because of this, and impactor flattening (discussed at length below), are harder to simulate than larger craters. Our velocity limit is, however, consistent with secondary crater formation on the Moon (see, e.g., Vickery 1986).

Impactors. Following the work of Shoemaker and others (Shoemaker et al. 1990,1994), we consider six categories of impactors. Each is simulated separately, and the total number of simucraters is summed. We include S-type and C-type asteroids after Shoemaker et al. (1990), and supplement these by adding iron asteroids at 3% by mass. The S- and C-types are of comparable importance. The iron asteroids, although few, contribute a disproportionate share of the smallest craters. For lack of better information we treat the three types of asteroids as following the same mass and velocity distributions. This assumption is probably wrong, but as we shall see, there are other considerations against which the proportions of asteroids by type appear unimportant.

Shoemaker et al. (1994) split comets into three categories: long period (LP); short period, Jupiter family (SP); and short period, Halley-family. The mass fluxes in all three categories are very uncertain, mostly because it is difficult to measure the size of comet nuclei, but also because density, albedo, etc., are not settled for comets. The only reliable density is ≈ 0.5 to 0.6 g cm^{-3} for the now extinguished P/Shoemaker-Levy 9 (Asphaug and Benz 1994; Solem 1994), which may or may not be representative. Using parameters preferred by Shoemaker et al. (1990), the flux of LP comets striking the Earth is 10^{-7} yr^{-1} for masses $m > 8 \times 10^{12}$ kg. The impact flux of LP comets on Venus is ostensibly about 20% higher (Zimbelman 1984). The resulting flux is

$$\dot{N}_{\rm LP}(>m) = 5 \times 10^{-7} \left(\frac{10^{12} \text{ kg}}{m}\right)^{0.66} \text{yr}^{-1}. \qquad (1)$$

The (inverse) mass exponent here is that used by Shoemaker et al. (1990), $b = 0.66$. It is based on work by Roemer and colleagues in the 1950s and 1960s (see, e.g., Roemer et al. 1966), and is consistent with some other published estimates (Donnison 1986; Hughes 1988). It is not consistent with Everhart's (1967) distribution, in which large comets are much less abundant and follow a $b = 1.5$ distribution. By either count, over 1 Gyr Venus is struck 500 times by LP comets with masses greater than 10^{12} kg, very few of which we shall see reach the surface.

Active SP comets are of negligible importance, but inactive ("extinct") SP comets are expected to be much more numerous. Following Shoemaker et al. (1994), we deduce a flux of Jupiter-family extinct comets of 7×10^{-7} yr^{-1} for $m > 5 \times 10^{11}$ kg. We will assume that, as for Earth-crossing asteroids, the impact flux of SP comets on Venus is about 95% that on Earth (Shoemaker et al. 1991). This gives

$$\dot{N}_{\rm SP}(>m) = 4.3 \times 10^{-7} \left(\frac{10^{12} \text{ kg}}{m}\right)^{0.66} \text{yr}^{-1} \qquad (2)$$

where the power $b = 0.66$ has been assumed the same as it is for LP comets (Shoemaker et al. 1994). We assume comets have bulk densities of 1 g cm^{-3}.

Information about extinct Halley-family comets is not abundant (see Shoemaker et al. 1994), but it is unlikely that they constitute a large overlooked fraction of the total. A simple way to add the Halley-family comets would be to increase the number of LP comets by a few tens of percent, as they share similarly high impact velocities. For the present we will omit the Halley-family comets from the simulations.

Shoemaker et al. (1990) estimate that the flux of near-Earth asteroids (NEAs) striking Earth is $4.3 \pm 2.6 \times 10^{-6}$ yr^{-1} for objects with absolute V magnitudes (H) fainter than 17.7. By magnitude, 2/3 of these are S-type (stones), 1/3 C-type (carbonaceous). Because they are darker, at the same magnitude C-type objects are larger. Shoemaker et al. (1990) assume densities, allowing for bulking due to fragmentation, of $\rho_C = 1.7$ and $\rho_S = 2.4$ g cm^{-3}. At $H = 17.7$, a C-type object has a diameter of 1.7 km and an S-type has a diameter of 0.9 km.

According to Shoemaker et al. (1990), the present population of NEAs is ill-described by a single power law. The larger (brighter) asteroids are distinctly less numerous than would be expected. These authors fit the distribution with two power laws, with a change of slope at $H = 15.7$. Using an impact flux on Venus about 95% that on Earth (Shoemaker et al. 1991), which includes a factor for undiscovered Venus-crossers whose orbits lie entirely inside the orbit of the Earth, we deduce cumulative impact rates of

$$\dot{N}_S(>m) = 2.6 \times 10^{-6} \left(\frac{10^{12} \text{ kg}}{m}\right)^{0.67} \text{ yr}^{-1}; \quad m < 1.5 \times 10^{13} \text{ kg} \quad (3a)$$

$$\dot{N}_S(>m) = 4.2 \times 10^{-7} \left(\frac{1.5 \times 10^{13} \text{ kg}}{m}\right)^{1.4} \text{ yr}^{-1}; \quad m > 1.5 \times 10^{13} \text{ kg} \quad (3b)$$

and

$$\dot{N}_C(>m) = 3.8 \times 10^{-6} \left(\frac{10^{12} \text{ kg}}{m}\right)^{0.67} \text{ yr}^{-1}; \quad m < 7.3 \times 10^{13} \text{ kg} \quad (4a)$$

$$\dot{N}_C(>m) = 2.2 \times 10^{-7} \left(\frac{7.3 \times 10^{13} \text{ kg}}{m}\right)^{1.4} \text{ yr}^{-1}; \quad m > 7.3 \times 10^{13} \text{ kg}. \quad (4b)$$

These are literal fits to the Shoemaker et al. distribution. In the absence of better information, we implicitly assume that the proportions of C- and S-type objects remain the same at all magnitudes. Hence the break in slope among the darker C-type objects is placed at a much larger mass (Fig. 3).

Shoemaker et al. (1991) do not explicitly separate iron asteroids from the S-types. Irons are not especially numerous, but because they are strong and dense they are responsible for almost all the small impact craters on Earth (see, e.g., Grieve 1991). A few percent of the taxonomically classified NEAs are M-types (metallic) (Chapman et al. 1994). By mass, irons make up a few percent of the meteorites collected in Antarctica, which may be our least biased sample (although the prerequisite fragmentation of iron parent bodies

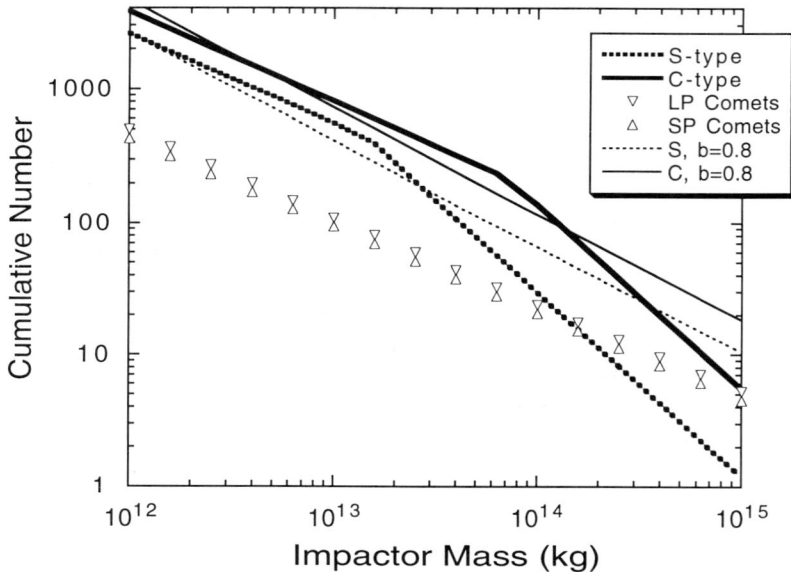

Figure 3. The Shoemaker et al. (1990,1994) asteroid populations (bold lines) and comet populations (open symbols) by mass (Eqs. 3–5); the cumulative numbers are relative at $m = 10^{12}$ kg, the C, S, and Fe-asteroid fluxes are proportionally 60:40:3 and the cometary mass flux is 18% of the asteroidal flux (the irons are not shown for clarity). Also shown are the master-power-law representations of the asteroid populations (thin lines, Eq. 6). In this case the C, S, and Fe-asteroid fluxes are proportionally 64:36:3.

is probably more difficult than for other meteorite types). We will assume that 3% of the asteroids are iron, that they have density $\rho_{Fe} = 8$ g cm^{-3}, and that they have the same albedos and orbital distributions as S-type asteroids. We deduce Venusian impact rates of

$$\dot{N}_{Fe}(> m) = 1.9 \times 10^{-7} \left(\frac{10^{12} \text{ kg}}{m} \right)^{0.67} \text{ yr}^{-1}; \quad m < 5 \times 10^{13} \text{ kg} \quad (5a)$$

$$\dot{N}_{Fe}(> m) = 1.4 \times 10^{-8} \left(\frac{5 \times 10^{13} \text{ kg}}{m} \right)^{1.4} \text{ yr}^{-1}; \quad m > 5 \times 10^{13} \text{ kg}. \quad (5b)$$

We will find, somewhat surprisingly, that the impact rates given by Eqs. (1) through (5) give a fine fit to the number-diameter distribution of craters on Venus. As the orbital evolution time scale of NEAs is measured in tens of Myr, there is no obvious reason to expect that the present distribution of impactors has been constant over the past 1000 Myr. As an alternative that might be more representative of the norm, we construct a *master power law* that is calibrated to the Shoemaker et al. (1990) distribution at $H = 17.7$:

$$\dot{N}_{ast}(> m) = A_{12} \left(\frac{10^{12} \text{ kg}}{m} \right)^{0.8} \text{ yr}^{-1} \quad (6)$$

where the power $b = 0.8$ is the slope of the power law predicted by a fragmentation cascade (Dohnanyi 1972; Safronov 1972), and $A_{12} = 2.6 \times 10^{-6}$, 4.6×10^{-6}, and 2.2×10^{-7} for S-, C-, and Fe-type asteroids, respectively. We will see that these master power laws give a less perfect, but nonetheless informative, fit to the Venusian data than do the present-day asteroidal distributions (Eqs. 3–5). The diverse distributions are compared in Fig. 3.

For a simple power law $\dot{N}(>m) \propto m^{-b}$, the impactor mass m is chosen randomly by $m = m_{\min} x'^{-1/b}$, where x' is another random number between 0 and 1 and $m_{\min}$ is the smallest object in the distribution. As a practical matter, for impacts on Venus $m_{\min} = 10^{11}$ kg is more than adequately small. More generally, the Monte Carlo masses are obtained from $m = \dot{N}^{-1}\{x'\dot{N}(m_{\min})\}$, where $\dot{N}^{-1}$ is the inverse of $\dot{N}$.

Crater Scaling. Crater scaling rules in the gravity regime have been established by laboratory experiments in strengthless targets. Based on Schmidt and Housen's (1987) and Holsapple's (1993) hard-rock scaling, which is derived from experiments in wet sand (a weak, non-porous target), we adopt for transient crater diameters D_{tr} measured rim-to-rim

$$D_{tr} = 1.27 g^{-0.22} \left(\frac{m_f v_f^2}{2\rho_t}\right)^{0.26} v_f^{-0.08} \left(\frac{\rho_i}{\rho_t}\right)^{0.07} \quad (7)$$

where g is surface gravity and ρ_i and ρ_t are impactor and surface (target) density, respectively. Specifically, we adopt their *volume* scaling law for hard rock in the gravity regime ($\pi_V = 0.2\pi_2^{-0.65}$, with the dimensionless π-groups defined as in Schmidt and Housen [1987]) as it is more physically fundamental, and assume a transient crater with a depth/diameter ratio of $1/2\sqrt{2}$ measured with respect to the ground plane (the apparent crater); the rim diameter is ≈ 1.15 times greater (these geometric factors are based on terrestrial craters; see Grieve and Garvin [1984]). Laboratory experiments indicate that oblique impacts produce smaller craters. To first approximation, it appears that only the normal component of the impact velocity contributes to the crater (see, e.g., Chapman and McKinnon 1986, p. 507). Hence we take m_f and v_f to refer to the mass and the *normal* component of the impact velocity when the impactor strikes the surface. Both can be considerably reduced by interaction with the atmosphere from their values at the top of the atmosphere. We use a surface density $\rho_t = 2.86$ g cm^{-3} (basalt) in all of the simulations.

The diameters given by Eq. (7) refer to so-called transient craters, which are presumed to be self-similar; i.e., the transient crater looks the same regardless of scale. Only small craters retain this shape. Large craters promptly relax, or slump, into wider, shallower bowls. The slumped crater is the observed crater. McKinnon et al. (1991) relate the diameter of the final crater to the diameter of the transient crater by

$$D \approx 1.17 D_{tr}^{1.13} / D_c^{0.13} \quad (8)$$

where D_c is the transition between simple and complex craters (and the numerical factors are calibrated to the Moon). For Venus this transition would occur at ∼3 to 4 km (see Sec. IV). For small craters, $D<D_c$, we assume that $D \approx 1.17 D_{tr}$, as even small craters (e.g., Meteor Crater) widen and form breccia lenses (Grieve and Garvin 1984; Melosh 1989).

Laboratory impact and explosion experiments performed on sand at elevated gravity and variable pressure imply that the effect of Venus' 90 bar atmosphere on cratering efficiency is probably small for impactors $\gtrsim 1$ km in diameter (Schmidt 1993; R. M. Schmidt, personal communication 1996), or $\sim 10^{12}$ kg for a stony composition. This is close enough to our minimum mass impactor that we ignore this pressure effect. An additional question concerns the possible effect of drag on outward moving ejecta, which may be important (Schultz 1992; Schmidt 1993). Theoretical modeling by Ivanov et al. (1986) and numerical code calculations of impacts at Venusian conditions by Ivanov et al. (1992) and Schmidt and Hassig (1995) show that after the initial atmospheric density increase caused by passage of the shock, the density of the shock-heated atmosphere over the crater rapidly declines by an order of magnitude compared with the unperturbed state for the duration of crater excavation. These results call into question the influence of atmospheric drag on the relatively great mass of ejecta launched by the Venusian craters we see.

C. Impactor Deceleration and Deformation

In this chapter we will use the semi-analytical model for the disruption and deceleration of large bodies striking deep planetary atmospheres described by Chyba et al. (1993) in the context of the famous Tunguska event of 1908 and by Zahnle and Mac Low (1994) and Mac Low and Zahnle (1994) in the context of the more recently famous Shoemaker-Levy 9 terminus of 1994. This model is based on a slightly more primitive model developed by Zahnle (1992), in the less famous context of craters and not-quite-craters on Venus. As this model is an important element of the calculations presented here, we will spend some space describing it.

The model is similar to models developed (over a span of years) by Grigoryan (1979), Passey and Melosh (1980), Melosh (1981), Ivanov et al. (1986), Hills and Goba (1993), Chevalier and Sarazin (1994), Field and Ferrara (1995), and Svetsov et al. (1995). In its simplest form, the basic idea is that the impactor is first pulverized by aerodynamic forces, then flattened into a pancake. This greatly increases the impactor's cross section, and thus speeds its deceleration. In detail, interpretations differ. The pancake has sometimes been taken too literally (see, e.g., Zahnle 1992), and this has led to criticism. The impactor does not really increase its radius by a factor of five or more while remaining a coherent body. Rather, material is shed by Rayleigh-Taylor instabilities, and the outermost pieces are swept off. This is seen in the numerical experiments. Yet the relevant quantity—the effective cross section of the body as a whole—does appear to be reasonably well-described by a pancake. For example, Svetsov et al. (1995) show that the growth of Rayleigh-Taylor instabilities is governed by essentially the same equation that describes the growth of the pancake, and this mechanism agrees with what is seen in many of the best-resolved numerical simulations (see, e.g., Boslough et al. 1994; Mac Low and Zahnle 1994). We return to this issue in greater detail in Sec. II.E.

There are three basic relations required to specify the interaction of a fragmenting bolide with the atmosphere. Two of these are the conventional drag and ablation equations (Bronshten 1983; see Zahnle 1992). The third describes how the impactor spreads in response to aerodynamic forces. To compute a crater diameter from Eqs. (7–8), we need to obtain the impact velocity at the surface as a function of impactor mass, velocity, and incidence angle. As we use it here, the semi-analytic model includes radiative ablation and a crude parameterization for strength. It neglects gravity, lift, and the curvature of the planet; these additional terms are usually small.

As discussed in Zahnle (1992), ablation is limited by the radiative transfer from the shocked and ionized gases concentrated in front of the impactor. For $v = 20$ km s^{-1} and a characteristic heat of vaporization (for ice) of 8×10^{10} erg g^{-1}, the maximum ablation rate is ~ 2 m s^{-1}. Thus ablation is of little importance for the $\gtrsim 1$-km-diameter objects that can approach the surface of Venus.

Impactors fragment as the result of aerodynamic stresses (Tauber and Kirk 1976; Melosh 1981; Bronshten 1983, p. 280ff; Ivanov et al. 1986; Melosh 1989, pp. 207–209). Fragmentation occurs when the aerodynamic pressure $\rho_a v^2$, where ρ_a is the atmospheric density, exceeds a characteristic strength of the material. On Venus a 60 km s^{-1} comet with the native strength of SL-9 (<1 kPa) would begin to break up at the 1-μbar level, while even a strong (chondritic) stone (breaking strength ~ 10 MPa) at 20 km s^{-1} would fail before it reached 100 mbar. For some observed falls, the meteor appears to have fragmented when subjected to aerodynamic stresses smaller than the compressive strengths of the recovered fragments by a factor of a few or more (Bronshten 1983, pp. 280–281). Impactors generally fragment high in Earth's atmosphere, typically above ~ 30 km for stuff as friable as carbonaceous chondrites. For example, Svetsov et al. (1995) note that the Sikhote-Alin meteorite, an iron, began to break up at 30 km. Fragmentation would occur much higher still on Venus (Ivanov et al. 1986). Near the surface of Venus, where $\rho_a = 0.066$ g cm^{-3}, an object falling at 20 km s^{-1} would see a ram pressure of 30 GPa, roughly 100 times the unconfined strength of iron, and 10 times the Hugoniot elastic limit (where the crystal structure of a material in shear gives way, and an effectively fluid condition obtains; see Melosh [1989] for a discussion of this limit). On Venus nearly any object that reaches the surface with cosmic velocity disaggregates scale heights above the ground.

Chyba et al. (1993) idealized the impacting body as a cylinder of radius r. The average aerodynamic pressure at the leading face is $1/2 C_D \rho_a v^2$, where C_D is the drag coefficient ($\simeq 0.9$ for a sphere). Pressures on the sides and rear are much smaller. When the aerodynamic stresses exceed the impactor's crushing strength, it breaks. If the fragments cannot relieve the building stress by moving out of the way, they themselves fragment: the result is a pulverized mass of rubble. Thereafter the rubble behaves like a fluid, in which pressure is isotropic, and which therefore flattens, or pancakes, in response to the aerodynamic pressure at the front. Chyba et al. took the pressure inside the cylinder to be the hydrostatic average $1/4 C_D \rho_a v^2$, and let this force act on the inertia of the whole cylinder (cf., Melosh 1981). The relatively small aerodynamic confining pressure is neglected. The radius of the cylinder spreads according to

$$\frac{d^2 r}{dt^2} = \frac{1}{2} C_D \frac{\rho_a v^2}{\rho_i}. \tag{9}$$

Written in terms of altitude z and flight angle from the zenith θ, this becomes

$$r \frac{d^2 r}{dz^2} + \frac{r}{v} \frac{dv}{dz} \frac{dr}{dz} = \frac{1}{2} C_D \frac{\rho_a \sec^2 \theta}{\rho_i}. \tag{10}$$

Equation (10) is solved numerically in concert with those for ablation and momentum loss using πr^2 as the cross section, subject to $\rho_a v^2 > Y$, where Y is the effective yield strength.

An implicit assumption behind Eqs. (9–10) is that the impactor deforms quasi-statically, i.e., that sound waves rebound inside the impactor quickly compared with the increase in $\rho_a v^2$ caused by descent. This assumption is reasonable for km-sized bodies with sound speeds of a few km s^{-1} (cf., Ivanov et al. 1992). Field and Ferrara (1995), in considering the crushing of a porous impactor with an effective sound speed of zero, argued that even under nonquasistatic deformation, Eq. (9) provides a reasonable approximation to the true rate of spreading. Further discussion of the physics of impactor deformation, supporting the simple model outlined here, can be found in Sec. II.E.

D. Simulations and the Age of the Venusian Surface

Figures 4 and 5 show simulated fits to the Venus cratering record. Both compare the differential crater counts for Venus (the actual number of craters on the surface) with simulations of the same based on the model described in the preceding two sections. Crater counts are summed in diameter bins that increase geometrically by factors of $\sqrt{2}$, and then plotted at the geometric mean diameter for the bin, with $\sqrt{N}$ errors (data from N. Izenberg, personal communication 1995). At the high diameter end we presumably have an expression of the production population, and moving to smaller sizes the distribution rolls over, the often predicted effect of atmospheric screening; at smaller and smaller scales the number of craters declines precipitously.

Figure 4 shows the fit using the detailed population distributions (Eqs. 1–5), based on a calibration to the total number of craters on Venus (940). The fit is quite good, even for the smallest craters. As the smallest craters are, for a variety of reasons, those least well modeled, we do not take a

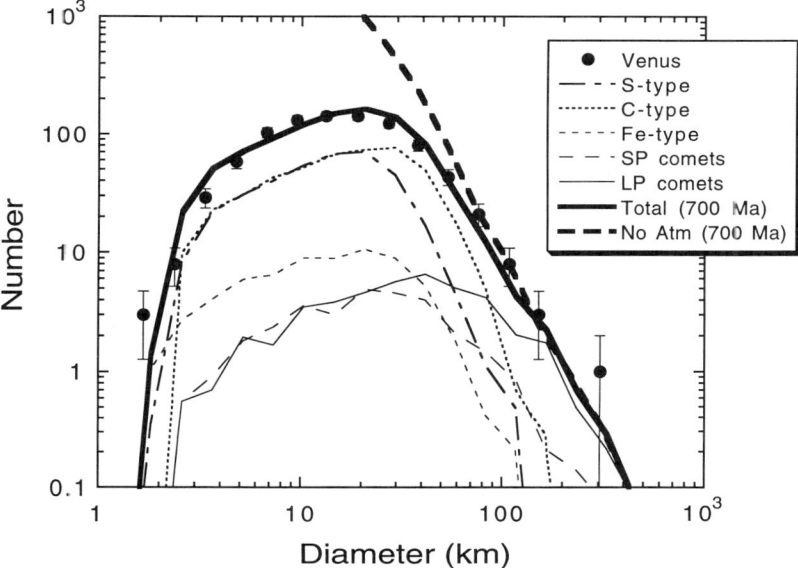

Figure 4. Simulated Venus craters, using the Shoemaker et al. (1990,1994) impactor populations (Eqs. 3–5). Calibrated to produce a cumulative 940 craters, the nominal age of the surface in the simulation is 700 Myr.

failure to reproduce exactly the craters in the three smallest bins to be a major shortcoming. Nevertheless, it is interesting to note that we have obtained a good fit throughout the diameter range without adjusting any parameter from its nominal value. This does not mean that the nominal values are the correct ones, but it is a hopeful result. For comparison, we show the crater production function that would be obtained in the absence of the atmosphere. Note also that the nominal model age of the surface is 700 Myr. This is a crater retention age, and is thus affected by resurfacing and tectonism. Strom et al. (1994) estimate that some 50 to 75 craters have been erased. Accounting for the lost craters raises the model age for remainder of the surface (however that is defined) to 750 Myr.

In Fig. 5, we show a simulation based on the simple, master power-law distributions for the asteroids (Eq. 6), supplemented by the comets in Eqs. (1) and (2), and again calibrated to the Venusian crater total. This simulation does an excellent job of reproducing the Venus data between ~5 and 80 km diameter, and as in Fig. 4, gives a decent match to the smaller craters considering how difficult they are to model. The nominal age of the surface in this model is 800 Myr.

The master power-law simulation (Fig. 5) overpredicts the number of large (100-km diameter) craters in comparison with Fig. 4. These excess large craters are a direct consequence of not having a dropoff in the number

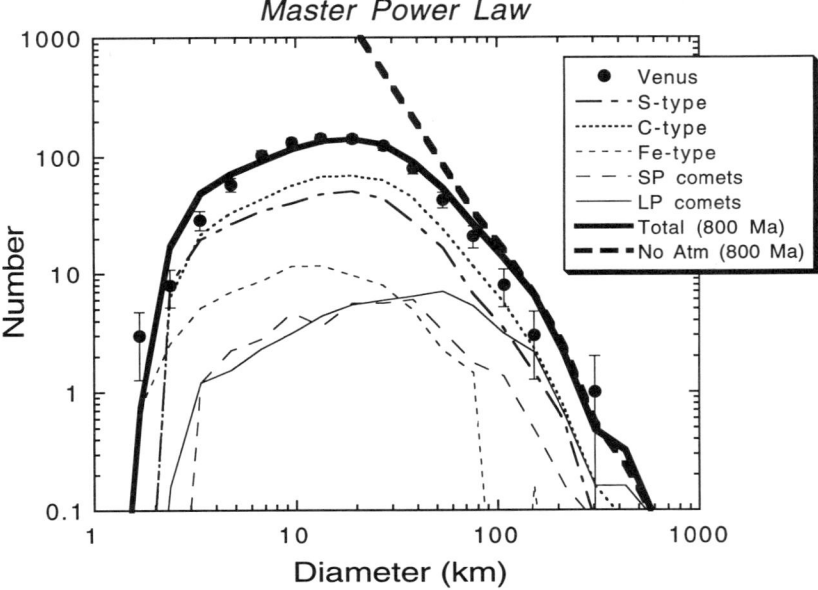

Figure 5. Simulated Venus craters, using the master power-law distributions for asteroids (Eq. 6) and the same cometary distributions as in Fig. 4. The fit is aesthetically pleasing between ∼5 and 80 km diameter. Calibrated to produce a cumulative 940 craters, the nominal surface age is 800 Myr.

of large (10-km diameter) asteroids, as in Eqs. (3–5). The breaks in these latter asteroid distributions correspond to the breaks near the peaks of the distributions of craters formed by C and S asteroids in Fig. 4. The sharpness of these breaks, which is somewhat artificial, causes the total simulation in Fig. 4 to be peaked near ∼20-km crater diameter. The actual incremental crater distribution is gently rounded in the ∼5 to 50-km diameter range, and the smoother asteroid distributions in Fig. 5 give a better fit in this range. Herrick and Phillips (1994a) argue that several different asteroid distributions with varying physical properties are necessary to account for the broad incremental maximum. We find that this is a natural consequence of our atmospheric interaction and cratering model (Fig. 5).

According to Shoemaker et al. (1990), the flux of Earth-crossing (and by extension, Venus-crossing) asteroids is coupled to collisions between mainbelt asteroids. Dynamical mechanisms are sufficient to deliver enough of the fragments of these collisions to account for more-or-less steady-state populations similar to those represented by Eqs. (3–5) (see, e.g., Wetherill 1988,1989). The generation of ∼10-km diameter asteroids, though, requires the disruption of larger mainbelt asteroids, which occurs much less frequently. In Shoemaker et al. (1990), these rarer events lead to an influx of larger asteroids into the terrestrial planet region every few hundred million yr, but there they dynamically evolve by collision and ejection (and thus die away) on time scales an order of magnitude shorter. Thus, averaged over several 100 Myr, Eqs. (3–5) underrepresent the number of larger asteroids, whereas Eq. (6) over-represents them. The true distribution(s) of asteroidal masses, averaged over many 100 Myr, would extend beyond the slope breaks in Fig. 3 and roll over more gently to a steeper power law. Equations (3–5) and Eq. (6) are, then, more akin to limiting cases, with reality somewhere in between. So rather than an overproduction of large craters (Fig. 5) or what can be taken to be a marginal underproduction (Fig. 4), we expect that a realistic representation of the long-term asteroid flux at Venus, with help from large comet nuclei, would with our atmospheric screening model give an excellent fit to the Venusian cratering record at all crater sizes $\gtrsim$5-km diameter.

To summarize so far, we have produced two acceptable models of the Venus cratering record, one using the present population of Venus-approaching asteroids and comets, and one using more generalized asteroidal populations. In both simulations approximately 26,000 impactors of $m > 10^{11}$ kg (our lower mass limit) enter the atmosphere to produce the $\sim$1000 craters (the simulations actually involve much larger numbers, for better statistics). *The atmospheric screen is very effective.* Examining the contribution due to comets in Figs. 4 and 5, we find that their contribution is reduced but not eliminated at intermediate sizes, and can in fact become dominant at the largest sizes (especially in Fig. 4). This dominance is not due to physics but to comets following slightly or manifestly shallower mass distributions than asteroids. It is entirely plausible that the largest craters on Venus, Isabella and Mead, were made by giant comets. The smallest craters are dominated by irons, as on the Earth.

Our models give nominal surface ages of $\sim$700 to 800 Myr. Such age estimates naturally should be viewed with caution. Several elements in the cratering simulation are less than precisely calibrated. The lead factor in the crater scaling (Eq. 7) is less well determined experimentally than is the power-law exponent. The simple-to-complex scaling is also uncertain; both Croft (1985) and Holsapple (1993) have proposed similar functional forms, but with different exponents (1.18 and 1.09, respectively). The interaction with the atmosphere is dependent on the exact value of the drag coefficient in Eq. (10), or alternately, on several factors of order unity in Eq. (10). And of course, estimates of the Venus-crossing asteroid population will continuously improve.

Nevertheless, our age estimates are somewhat larger than popular estimates of 300 to 500 Myr (see, e.g., Phillips et al. 1991,1992; Schaber et al. 1992; Strom et al. 1994). Shoemaker et al.'s (1991) estimate of 240 (+290, −85) Myr, revised in Strom et al. (1994) to 290 (+310, −100) Myr, is especially widely quoted. The difference between these latter earlier estimates and the $\sim$750 Myr derived here is substantial and deserves some discussion. Shoemaker et al. (1991) use their own crater scaling, for one thing, which gives somewhat larger craters than Eqs. (7–8). They also do not perform a full simulation of the cratering record. Their (asteroidal) cratering rate at 20-km diameter was simply adjusted to account for atmospheric drag (by a factor of 0.69). In our estimation, this is the key difference. The atmosphere may easily reduce the cratering rate by a factor of 2 at $D\sim20$ to 30 km (Ivanov 1990), and our estimates here of the crater production population in the absence of an atmosphere (Figs. 4 and 5) imply even more severe depletions, a total factor of ~3 at $D\sim30$ km and ~6 at 20 km.

Monte Carlo simulations *were* carried out by Phillips et al. (1992), and revised in Herrick and Phillips (1994a). Phillips et al. (1992) obtained a production age of 400 (+600, −100) Myr, plus a variety of other age estimates based on different interpretations of the fit at larger sizes ($D>32$ km), given that their overall fit to the shape of the incremental curve was not satisfactory. The results of Phillips et al. are difficult to compare directly with ours, because they (1) used a shallower impactor production function $dN/dm \propto m^{-0.5}$; (2) used a crater scaling law for dry sand, which is inappropriate (cf., Holsapple 1993); (3) used a fixed enlargement factor for crater slumping; (4) assumed comets cannot survive passage through the atmosphere; (5) adopted an atmospheric pressure correction to the cratering efficiency; and (6) modeled break-up but not flattening of the impactor during atmospheric passage. Herrick and Phillips (1994a) included short-period comets and flattening, but argued that flattening is limited (see next section). The relative proportion of irons, stones, and comets (and other parameters) were allowed by them to vary to obtain an improved fit to the incremental crater distribution, but this variation formally precludes a new age estimation for the surface.

To the extent that all the above estimates depend on a first principles understanding of atmospheric passage and crater growth, they are subject to revision. An end run around these uncertainties can be attempted by simply trying to calibrate the large-diameter end of the Venus crater distribution to the ostensibly reasonably well-known crater production rates for the Moon or Earth (see, e.g., Phillips et al. 1992). This only requires that relative factors be reasonably estimated (cf., Chapman and McKinnon 1986).

Ivanov and Basilevsky (1987) compared craters in the ~30 to 70-km diameter range, from Venera 15 and 16 images, with the cratering record in the North American and European-Russian cratons (Grieve and Dence 1979). Assuming that the Venusian atmosphere does not affect craters in this size range, which we argue above is *not* true, they estimated a retention age of ~500 Myr for Venus. In this regard, it is interesting to note that the cumulative areal density of large craters from Magellan images, covering 98% of Venus, is statistically indistinguishable from the terrestrial cratonic curve. Neukum and Ivanov (1994) calibrated the post-mare lunar cratering curve to Venusian conditions and obtained $\sim650\pm100$ Myr; lowering the impactor flux at Venus by a factor of 0.95, as done here, raises their estimate to ~700 Myr. Both Ivanov and Basilevsky (1987) and Neukum and Ivanov (1994) presumed a constant cratering rate over their respective calibration intervals, which is debated (cf., Grieve and Shoemaker 1994; Neukum and

Ivanov 1994). A higher recent cratering rate, or uncompensated effects of erosion on the terrestrial record (incompleteness) (Grieve 1984), would bring these age estimates down.

Our judgement is that the differences between all the estimates we have discussed are indicative of the overall uncertainty in the crater retention age. At this point, any age between ~300 Myr and 1 Gyr may be acceptable. Future work should involve a more detailed calibration of the Venusian crater curve to those of the Earth and the Moon.

E. Impactor Deformation Revisited

The modeling presented in the previous section depends on a reasonably accurate accounting of the effects of the atmosphere on the incoming impactor. Here we examine some of the detailed work being carried out to understand such interactions, included work motivated by the July 1994 Shoemaker-Levy 9 impacts with Jupiter. It also serves as preamble to the discussion of the effects of atmospheric deceleration and terminal breakup in Sec. II.F.

Fragmentation and Spreading. The initial stages of deformation of a 1.3-km diameter, strong stony impactor, entering the Venusian atmosphere at $45°$ with an initial velocity of 20 km s^{-1}, have been calculated by Ivanov and Melosh (1994; cf., Sec. 1 in Gyaznov et al. 1994). The SALE hydrocode as modified by Melosh et al. (1992) was used to compute the dynamic fragmentation of the impactor. The equation-of-state and Grady-Kipp fragmentation parameters for competent basalt were used, taken from Asphaug and Melosh (1993) (e.g., the impactor is not a rubble pile). The code was modified to incorporate Mohr-Coulomb behavior and damage accumulation in shear, and the ultimate or plastic strength (von Mises limit) was set at several GPa. A limitation of the model is that the atmospheric gas flow is not directly simulated by the hydrocode, but supplied by an independent gas dynamical numerical model; nor can the calculation be followed to arbitrary deformations. The calculation is, however, unique in that the strength of impactor is *not* neglected and the initial fragmentation can be modeled.

These calculations are illustrated in Fig. 6. The flight direction is "down." Failure in tension at any given stage is indicated by the black and grey areas, with black implying maximum tension in the azimuthal (hoop) direction and grey implying maximum tension within meridional planes; shear failure in compression is indicated by the cross-hatched regions. The calculation is initiated at sufficiently high altitude that deformation of the impactor is entirely elastic. The reverberation time $4a/c$, where a is the impactor radius and c is the bulk sound speed, is ≈ 0.5 s, about the same time it takes the impactor to traverse a scale height (at altitude), so equilibrium stress conditions within the body are fairly closely approached.

The external pressure on the "incoming" hemisphere creates compressive stresses that prevent tensile failure there. Failure begins at about 50 km altitude, when the stagnation pressure reaches ~300 MPa, but the first "cracks" appear in the rear (less pressurized) hemisphere, close to the impactor's sur-

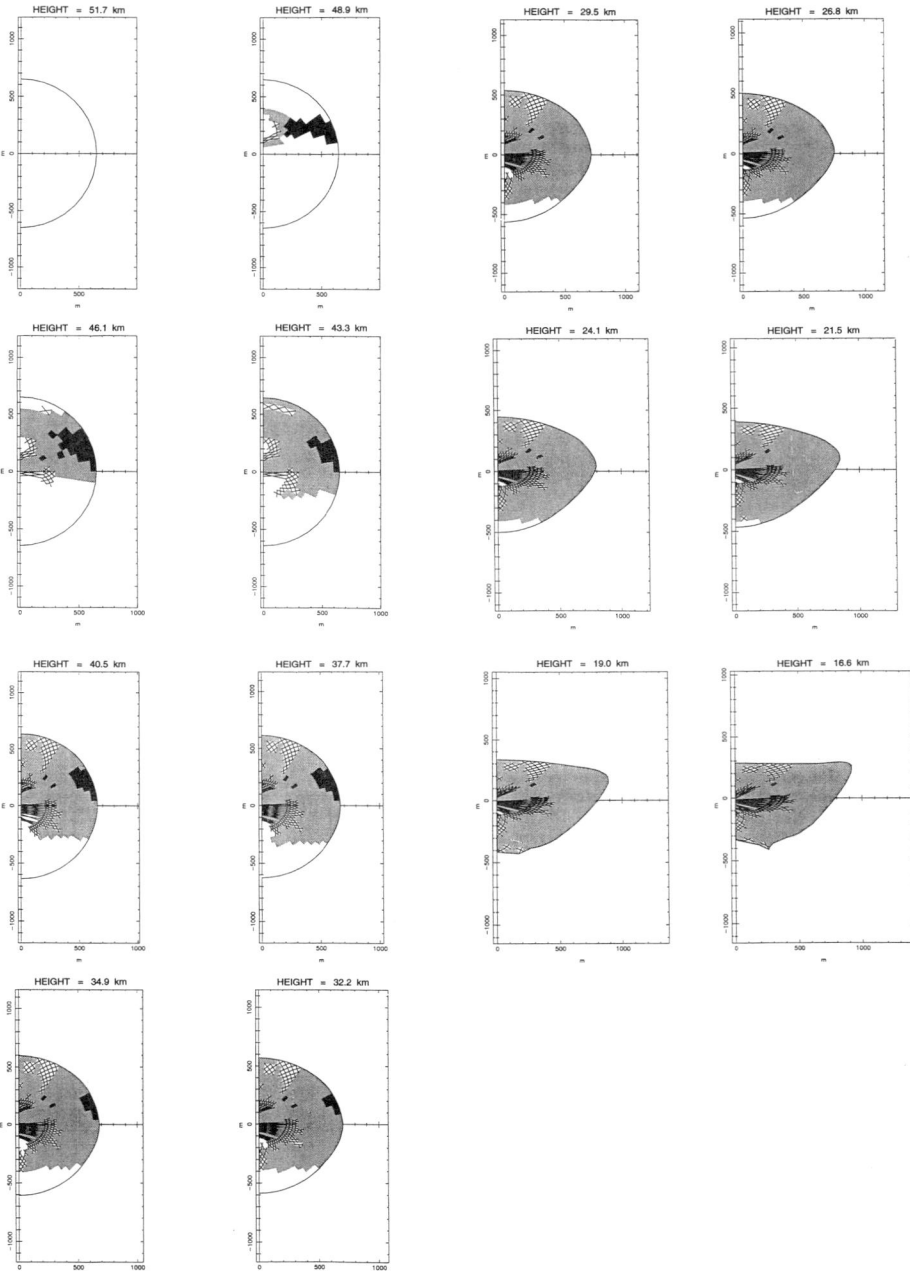

Figure 6. Fragmentation of a silicate (competent basalt) body of 1.3 km diameter falling to Venus at 45 deg. Black and grey shading show tensile damage, crossed zones are failing in shear. One can see that atmospheric breakup is a complex process, which begins 50 km above the surface. Complete failure is reached approximately 20 km above the surface. Growth of hydrodynamic instabilities controls the final breakup (see Fig. 7).

face, and then spread throughout that hemisphere. Failure is dominantly tensional; maximum tension initially occurs in the hoop direction, but such meridional fractures ("orange section" disruptions) are partly an artifact of the axisymmetry of the calculation (faults and other macroscopic heterogeneities are not incorporated either). As the object drops though the 40-km level the failure spreads farther and the impactor begins to distort. Dropping through a height of 25 km, one observes essentially full failure and significant flattening. Approaching the final scale height our impactor has an aspect ratio of $\sim$3, the relative internal motion is $\sim$0.5 km s^{-1}, and a high-pressure air pocket is beginning to develop along the flight axis. At this point the object is cohesionless and only friction controls the internal flow; the calculation is terminated due to large deformations of the computational cells.

Spreading and Breakup. What would logically follow from Fig. 6 can be seen in a different calculation (Fig. 7). This calculation, specifically, is for the entry of a strengthless comet into the atmosphere of Jupiter (from Mac Low and Zahnle 1994), using the general purpose (magneto)hydrodynamics code Zeus-3D (Stone and Norman 1992). The impactor is 1 km in diameter, and a stiffened gas equation of state appropriate for water (Melosh 1989, p. 231) is used to model the comet. The important point here is that the impactor is strengthless. The pressures indicated in Fig. 7 are not the ram or shock pressures but the atmospheric pressure at the altitude the impactor has reached, $P(z_o)$. Log density levels are indicated by the various shadings white to black.

As the ice impactor descends into the Jovian atmosphere it flattens, and a high density bow air pocket develops along the flight axis as in Fig. 6. Later the impactor spreads more and becomes very flattened; as this calculation is also in axial symmetry, the apparent plane of higher density is actually a disk, or pancake, with a lower density center. It is the high-pressure gas in the center of the disk that blows through to disrupt the object, via a Rayleigh-Taylor instability (Field and Ferrara 1995). This and other Rayleigh-Taylor instabilities (and because of the flow of shocked atmosphere past the impactor, Kelvin-Helmholtz instabilities) cause the impactor to disintegrate or effectively explode in flight (see, e.g., Ivanov 1988; O'Keefe et al. 1994; Crawford et al. 1995; Svetsov et al. 1995). For a 45° entry angle, this would occur within half a scale height (based on Mac Low and Zahnle 1994), $\lesssim$7 km on Venus.

Sufficient computational resolution is necessary to see a level of detail such as in Fig. 7 (the example shown has 50 computational zones in one impactor radius). We can also see in Fig. 7 that an overall effect during the terminal phase of an impactor's descent is a change in aspect ratio to about 8:1, which is essentially the same as argued for by Melosh (1981) in his simple spreading model. The amount of lateral spreading in Fig. 7 is, however, greater than the limit adopted by Melosh (1981), $\sim 8^{1/4} \approx 1.7$, because the impactor has distended as well as extended. The increase in impactor area in the bottom panel of Fig. 7 is a factor of $\sim$16. This degree of spreading is naturally incorporated in the Monte Carlo model used in the previous sections,

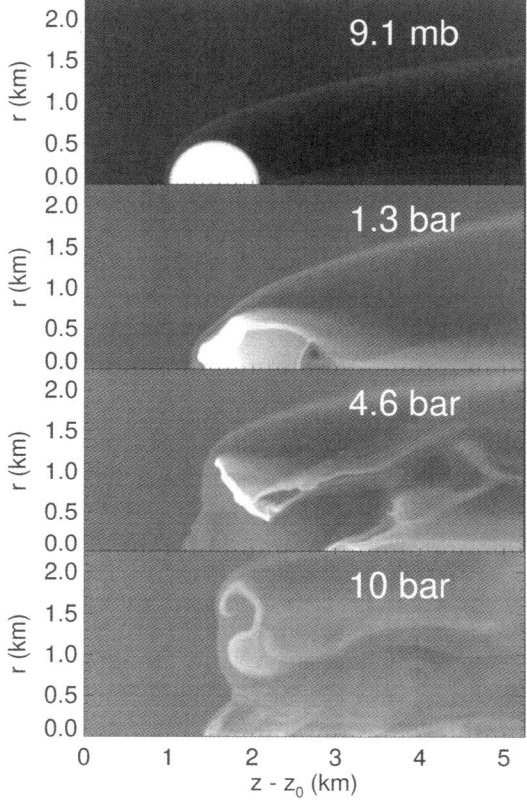

Figure 7. "Airburst" of a 1-km diameter comet entering Jupiter's atmosphere at 45 deg and 60 km s^{-1}. A nearly incompressible equation of state for water is used for the comet, with sufficiently high resolution to follow the development of the Rayleigh-Taylor and Kelvin-Helmholtz instabilities that dominate the disruption. The figure shows the log of density at times after the start of the calculation of 0.1, 2.5, 3.5, and 4.5 s; black corresponds to a density of 10^{-7} g cm^{-3}, and white to 1.3 g cm^{-3}. The altitudes above 1 bar for the four times are $z_0 = 96$, -6, -48, and -83 km, respectively, and the pressure at each height is shown.

but is not seen in all numerical models or if insufficient resolution is used in the computations (e.g., the hydrodynamic entry simulations in Ivanov et al. [1992]). If the bow shock is insufficiently numerically resolved, the computed bow shock penetrates the impactor itself, and thereby makes the surface of the impactor stable against Rayleigh-Taylor instabilities, and so limits spreading. (The effect of resolution on estimates of impactor spreading and energy deposition was in fact the major motivation for Mac Low and Zahnle [1994].) The amount of spreading also depends somewhat on the definition of the impactor/atmosphere boundary; the lateral extent may oscillate, spreading alternating with shedding, for example, and even material that is shed tends to follow along behind (Crawford et al. 1995). It goes without saying that any complex numerical calculation should not be taken at face value, but examined carefully in detail. Better yet, different codes should be benchmarked against each other.

The semi-analytic pancake spreading model has, in fact, been subject to several recent numerical tests. To date the numerical experiments remain somewhat controversial, in part because the different experiments are not quite comparable. The most directly comparable high resolution studies, however, are in reasonably good agreement with the predictions of the simple semi-analytic model. Mac Low and Zahnle (1994) and Svetsov et al. (1995) both compare numerical model results directly with the predictions of the semi-analytic model, and find essential agreement. Boslough et al. (1994) and Mac Low and Zahnle (1994) both address 1-km-diameter strengthless ice impactors, and get similar results, but because the two groups use different definitions of penetration and energy deposition, the fundamental similarity of their results was not immediately apparent. For example, the impactors in Crawford et al. (1995) also spread substantially (by about a factor of 5 in lateral extent), but Crawford et al. plot the energy deposited in the atmosphere *after* downward advection of the wake region. The moving wake can carry energy and momentum about a full scale height deeper (for normal impacts) than the altitude of the terminal deceleration (explosion), and is discussed in detail in Mac Low (1996).

The astute reader will note the relative stability of the impactor in Fig. 6 compared with that in Fig. 7. Part of this is due to resolution. Part is also due to the high plastic strength of competent basalt, which allows very high compression without shear failure at the impactor's leading edge until the impactor is deep in the atmosphere. This in turn depends on the perfect symmetry of a spherical impactor. Weaker or fractured rock, ice (see Gyaznov et al. 1994), and even metals (which have lower plastic strengths), as well as more irregular impactors, should fail at lower aerodynamic pressures. To the extent that strength effects can delay instability growth, once instabilities are initiated, their growth should be faster.

Restricted Spreading. Spreading *may* be limited in the dynamic loading regime, obtained for impactors large enough that the bulk-sound- or shock-crossing time is less than the ram-pressure e-building time (Svetsov et al. 1995), which defines an impactor radius limit

$$a > a_\mathrm{d} = \frac{Hc}{2u \sin \theta} \qquad (11)$$

where H is the scale height (≈ 15 km in the lower atmosphere of Venus). For an impact velocity $v \approx 20$ km s^{-1}, impact angle $\theta = 45°$, and bulk sound speeds c for water ice, basalt, and iron of $\sim 2, 4,$ and 5 km s^{-1}, respectively (cf., Brackett and McKinnon 1992, Table 1; Gaffney 1985, Fig. 4; Ivanov and Melosh 1994), the limiting radii for dynamic loading a_d are $\sim 1, 2,$

and 2.5 km, respectively. For impactors much larger than this, the initial deformation will propagate as a crushing shock front through the impactor, and spreading will initiate at the leading surface while the rear of the bolide remains undeformed. Kelvin-Helmholtz or other instabilities may erode or ablate the spreading material as it moves to the sides of the impactor, and this material may be swept into the wake region (Ivanov et al. 1992; Svetsov et al. 1995). Compression of the impactor may also play a role in promoting this aerodynamic erosion, but the latter remains to be better established, as does the role of resolution on the outcomes of the relevant calculations in Ivanov et al. (1992) and Svetsov et al. (1995). Given the range in a_d values above, however, it would appear that if limited impactor flattening is real, it is most applicable on Venus to incoming comets (Ivanov et al. 1992).

We conclude this section by noting that it is sometimes argued that the pancake model is simply wrong (Herrick and Phillips 1994a; cf., Crawford 1996; Mac Low 1996; Zahnle 1996). It is contended that spreading is limited, perhaps to no more than a factor of 2 in radius, and that hydrodynamic forces somehow shape the disintegrated impactor to make it more streamlined (see, e.g., Basilevsky et al. 1987; Ivanov et al. 1992). Consequently, to use a relevant example, the crater production model of Basilevsky et al. (1987) predicts smaller craters (for a given crater-scaling relation) and lower-energy airbursts (see next section) than Zahnle (1992) does. As discussed above, there are conditions in which aerodynamic forces do tend to work this way, but in general we find that placing arbitrary limits on the spreading pancake does not work. We have run a simulation similar to those in the previous section, but in which the diameter of the impactor is not allowed to spread by more than a factor 2. What we find is that the resulting crater distribution is bimodal, marked by a glut of small (2–4 km) craters. Herrick and Phillips (1994a) also impose such a restriction on lateral displacement, but because the simulation that results is also unacceptable, they then introduce an additional parameter that produces terminal explosions. By tuning this additional parameter they can force good fits, but this approach is, in our view, *ad hoc* and not otherwise recommended.

F. Radar-Dark and Radar-Bright Halos

In addition to the ≈940 impact craters on Venus, there are some ∼400 radar-dark and/or radar-bright "splotches," "margins," "halos," or "shadows" (Phillips et al. 1991; Schaber et al. 1992; Strom et al. 1994) (Figs. 2 and 8). Most (>90%) are radar-dark quasi-circular features that are more often than not (∼70% vs ∼30%) surrounded by radar-bright halos (Fig. 8). Because their radar albedo decreases with increasing radar incidence angle, the dark regions were assumed by Phillips et al. (1991) to be smooth areas of subdued surface roughness (on the scale of the Magellan radar wavelength of 12.6 cm). As introduced in Sec. II.A, a fraction (∼10%) of the dark splotches have central features plausibly caused by impact, and many small craters (and ∼15 to 50% of all craters, depending on who is doing the identifying) are surrounded by similar radar albedo features (Schaber et al. 1992; Herrick and Phillips 1994b; Izenberg et al. 1994). With respect to *crater-related* radar-albedo features, we follow Strom et al. (1994) and distinguish sharp-edged and sometimes flow-like dark margins from more diffuse dark halos. We concentrate here on the latter; the former may be related to the continuous ejecta (see discussion in the chapter by Herrick et al.). With regard to craterless splotches, we refer to the (often central) dark variety as umbrae and the the surrounding bright zones (if they exist) as penumbrae, in analogy to the nomenclature for sunspots.

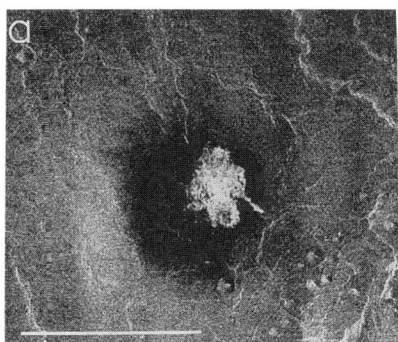
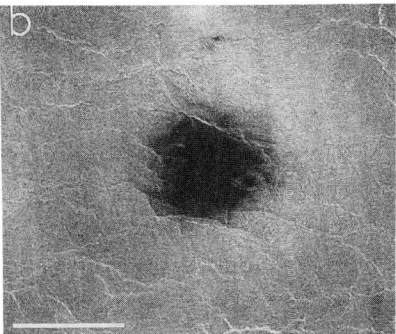

Figure 8. Magellan images of dark and bright halo and splotch features. (a) Unnamed crater, with mean diameter of ∼6 km at 16.5°N, 334.4°E; (b) Nested halo feature without crater (generically, a splotch) at 8.7°N, 333.5°S. Scale bars are 50 km.

A comparison of radar backscatter cross sections in Arvidson et al. (1992) and Takata et al. (1995) indicates that the dark halos are closer to lake playa in surface roughness than the weathered and degraded basalt flows (or fresh pahoehoe; Campbell and Campbell 1992) typical of the average Venusian surface; the bright halos presumably reflect a rougher surface, somewhere between mantled bedrock and fresh a'a lava in roughness. The bright halos were ostensibly seen by the Venera radar mappers, but all or nearly all of the bright halos identified in Ivanov et al. (1986) turn out, in retrospect, to be bright, rough ejecta blankets (see their Fig. 10). Regardless, because similar radar-bright and radar-dark albedo features surround many smaller impact craters, there is little doubt that the crater-less radar albedo markings are impact features of some kind.

In contrast to the impact craters, the ∼400 splotches are not randomly distributed. They are found preferentially in relatively flat lowlands (e.g., Sedna Planitia and Phoebe Regio); they avoid the rough, radar bright highlands (Strom et al. 1994). They would number ∼1100 were they uniformly distributed (Strom et al. 1994). The patchiness of their distribution must in part be traced to the terrain on which they formed; it also appears likely that splotches and halos, more so than craters, are subject to erosion (Izenberg et al. 1994; Phillips and Izenberg 1995). If so, then they would be fragile or superficial features. It would be useful to know whether the distribution of crater halos correlates with the distribution of craterless splotches.

We critically examine several formation models.

Shock Waves. Following discovery of the radar-dark features in Magellan images, Phillips et al. (1991) suggested that dark halos and splotches (umbrae) formed where shock waves generated by airbursting impactors slammed into the ground, pulverizing decimeter scale surface features, such as strewn rocks and rough lava. That shock waves from near-surface airbursts might be strong enough to shatter rocks on Venus had been pointed out earlier by Ivanov et al. (1986).

The shock damage hypothesis remains highly plausible. To show why, consider the force of a blast wave striking the surface of Venus. Atmospheric density at the surface of Venus is $\rho_a = 0.065$ g cm^{-3}. The density of a strongly shocked ideal gas is larger by the factor $(\gamma + 1)/(\gamma - 1)$, where γ is the ratio of specific heats. Taking $\gamma = 1.1$ to 1.3 (Brackett and McKinnon 1992; Takata et al. 1995), the strongly shocked gas would have a density of ≈ 0.5 to 1.5 g cm^{-3}, more like rock (or at least water) than thin air. This thick air would strike a heavy blow were it to hit the surface at several km s^{-1}, as it would if the airburst occurred near the ground. The shock pressure generated in the surface by shock wave transmission in principle can greatly exceed the overpressure of the airshock (a maximum of ~ 10 GPa vs ~ 1 GPa for shock speeds of a few km s^{-1}; Brackett and McKinnon 1992).

Dark halos and umbrae correspond to where the blast wave strikes the surface hard enough to reduce surface rocks to fine rubble (small cobbles, pebbles, and smaller). For reasonably strong rocks (yield strength ~ 200 MPa), Zahnle (1992) showed that with no central crater the umbrae would be expected to have diameters of some 15 to 30 km; with a central crater the dark halos could be more than 50 km across. If a weaker shock suffices to break stones, the umbrae and halos could be wider, with the diameter D_s of a region seeing a shock stronger than a pressure P going like $P^{1/4}$. If we modify the simulations in Sec. II.D to produce umbrae as well as craters, we find that to produce ~ 1100 dark splotches over 700 Myr (which we assume would be counted as ~ 400 after accounting for terrain effects and weathering), the shock pressure in the surface rock necessary to produce umbrae is $\gtrsim 150$ MPa.

Zahnle (1992) did not find cases where the blast wave delivers a shock that exceeds the Hugoniot elastic limit (roughly 4.5 GPa; Melosh 1989, p. 35) while not also forming a crater. Hence it is improbable that the surface of a halo or splotch is actually completely pulverized. Reaching the Hugoniot elastic limit is not necessary to shatter rocks, however; for explosive or percussive breakage of igneous rocks, high-pressure impulses of about 100 MPa are adequate (see, e.g., Grady and Lipkin 1980). On the other hand, the airshock may be too large compared to Venusian surface rocks to be considered impulsive. Still, compressive strengths of basalt are only slightly greater (a few 100 MPa; Table II in Mizutani et al. 1990), and interaction of the blast wave with the complex geometry of an actual surface may compensate as well, enhancing a given shock's destructive power (see, e.g., Ivanov et al. 1986; Head and Melosh 1995). We note that these strength levels are comparable to the limits derived in the numerical halo-formation simulations just discussed. Zahnle (1992) also found that for typical near-surface Venus airbursts strong shocks (0.1–1 GPa) may penetrate hundreds of meters or even several km into the crust. It is possible that the effects of the blast wave are also deep, which might make the radar-dark features longer lived.

Presumably the radar-bright halos and more distant penumbrae correspond to coarser, decimeter-scale rubble. One possibility is that the coarser fragments simply correspond to the impact of weaker blast waves, as would naturally result for higher or more distant explosions. Thus the gradient in surface radar properties would parallel the gradient in shock intensities: pulverized fields of gravel immediately beneath the airburst site (ground zero) grading into long stretches of disrupted lava flows, littered with scattered blocks and shattered rocks. The ground shock may also outrace the airshock, leading to a head-wave-like phenomenon and surface spallation.

Supersonic Winds. Another possibility is that bright halos and penumbrae correspond to the formation of the Mach stem (Zahnle 1992; cf., Takata et al. 1995). Here the full three-dimensional nature of the shock wave is considered; the Mach stem forms where the blast wave reflected off the surface overtakes and coalesces with the blast wave traveling directly from the explosion (the shock geometry is shown in Fig. 9 for the cases of vertical impact, oblique impact, and airblast, and a classic example of a Mach stem can be seen in Starrfield and Shore [1995]). Mach stem formation results in an approximate doubling of the strength of the blast wave, as the reflected shock reinforces the direct shock, and the blast wave travels parallel to the surface (Glasstone and Dolan 1977; Hornung 1986). Thus, once the Mach stem forms the surface ceases to be strongly shocked. Effectively, the blast wave is converted into a powerful (supersonic) wind blowing radially outward from the epicenter of the airburst. The wind would scour the region, leaving only a clean planetary surface and its larger rocks behind (see also Ivanov et al. 1986,1992; Provalov and Ivanov 1992). The result would be radar-bright.

Takata et al. (1995) stress the importance of the Mach stem, as well as regular reflection (no Mach stem) and diffusive reflection, given that all may lead to supersonic winds, and with a somewhat different twist. They conclude that they can explain the bright halos and penumbrae by the sweeping away of fine material in conjunction with the rupturing or overturning of surface rocks. Their explanation for dark halos and umbrae is that rubble shaken loose by the shock or by the winds (which may include boulders as large as ~ 1 m in diameter) is violently saltated and pulverized. This is different than simple shock fragmentation. Takata et al. (1995) emphasize the role of small impactors that reach the Venusian surface at nearly cosmic velocity. From the point of view advocated in this chapter (see previous section), splotches and haloed craters are more likely to be made by comparatively more massive impactors

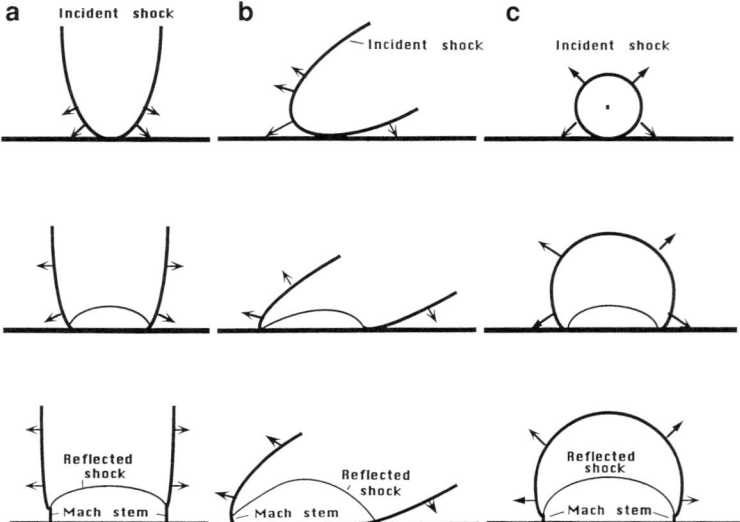

Figure 9. Cartoon illustrating possible Mach stem geometries. Propagation and interaction with the surface is shown for (a) a bow wave for impact at normal incidence, (b) a bow wave for an oblique impact, and (c) a spherical blast wave from an idealized atmospheric explosion (figure adapted from Takata et al. 1995).

that have spread laterally and decelerated substantially. Thus, unfortunately, we cannot easily make what we consider relevant, detailed comparisons of the predictions of the Takata et al. model with Venusian observations.

We note that the analyses in Zahnle (1992) and Takata et al. (1995) appeal primarily to air-to-ground shock effects and shock-generated winds. The role of the finite momentum in the wake region, as noted in the previous section, has yet to be incorporated in any splotch or halo model, but could prove important to halo formation.

Super-pressurized Pores. Another model appeals to "back venting" during the rarefaction stage of the shock wave passage (Ivanov et al. 1992; Provalov and Ivanov 1992; Schaber et al. 1992). Back venting supposes that during shock passage over a soil surface the high pressure induces a fast atmospheric gas filtration into the soil. The soil matrix can support relatively high compression loading, permitting gas flow into the porous space without total pore collapse. When compression ends and the rarefaction phase begins, the compressed gas under the surface vents back to the atmosphere. The soil matrix in tension has much lower strength than in compression, and the back venting ruptures the surface layer. Uplifted, crushed soil is deposited back onto the surface with a density lower than the original value, and thus of a lower radar albedo. Also, the crushed soil may be moved along the surface by atmospheric gas flows. Thus, a shallow quasi-circular zone of decreased density and/or smoothed surface may arise, appearing radar-dark.

Laboratory experiments demonstrate that back venting and rupture of a sand surface under shock wave action can be effective, but for this model to be applicable to Venus it must be established that sufficient soil exists on Venus, which is debatable (Arvidson et al. 1992; Greeley et al. 1992). As noted above, the average Venusian surface has radar properties matched by terrestrial weathered basalt or fresher pahoehoe, not sediments (Arvidson et al. 1992; Campbell and Campbell 1992); the panoramas returned by the Venera landers are consistent with this. Limited emissivity and backscatter measurements indicate that dark halos and umbrae may or may not possess lower dielectric constants (consistent with greater porosities) than their surroundings (Campbell et al. 1992, Campbell 1994; Brackett 1995). Furthermore, there is no obvious way for back venting to account for bright halos and penumbrae, unless it can be shown that hard rock can be pressurized and disrupted.

Impactor Debris. For completeness, we note that Schultz (1992) has proposed a "depositional" origin for dark halos and umbrae, the rainout of atmospherically crushed impactor material, acting in combination with shock-wave disruption of the surface. Rainout was also listed as a possibility by Phillips et al. (1991). Schultz (1992) also proposes that the bright halos and penumbrae are caused by fine material that has been worked by recovery winds (as opposed to the much stronger outward-directed winds generated by shock passage, described above) into small dunes that appear rough to the radar. The radar brightness of such roughly concentric mini-dune fields should vary with their alignment to the radar look direction, but no such azimuthal variation is seen. Strong winds are also not generally conducive to dune formation on Venus (see discussion in Weitz et al. 1994).

As a closing remark to this section, we judge that the picture of a strong blast wave crushing the surface to leave pans of gravel is compatible with the splotches avoiding highlands and rough, radar bright terrain. This kind of splotch requires a level surface. It would be much harder to bury a highland surface beneath shock-generated rubble. By the same token, scouring winds (as generated by a Mach stem) can do little to uncover a rough surface if the surface is already rough.

III. INFLUENCE OF IMPACTORS ON THE ATMOSPHERE

In this section we briefly consider the interactions of the hot, impactor-produced vapor cloud with the Venusian atmosphere, and some of the effects that may result.

A. Lessons from Shoemaker-Levy 9

It may seem unusual to bring up impacts on Jupiter, but the impact process on Venus is in some senses an intermediate case to that on the Earth and Jupiter. That is, Venus as a target is obviously similar to the Earth, but its deep 90-bar atmosphere resembles, in some ways, that of Jupiter. We take the 1994 Shoemaker-Levy 9 (SL-9) impacts as type examples. The SL-9 fragments apparently penetrated to depths in the 1 to 4 bar range, based on the carbon- and-sulfur-rich but oxygen-poor material blasted out into the Jovian stratosphere (Zahnle et al. 1995; Zahnle 1996) (an argument, however, that requires re-examination in light of the Galileo mission); these penetration depths are also supported by atmospheric disruption models (Mac Low 1996; Zahnle 1996) for fragments of the size and density favored by Asphaug and Benz (1994). If these same fragments had collided with Venus they would have penetrated to somewhat lower pressures (the mass intercepted goes as $P(z)/g$, and should be similar regardless of velocity due to momentum conservation; see, e.g., O'Keefe and Ahrens 1982). As spectacular as the SL-9 impacts appeared, such impacts on Venus would not make craters; they would not even make dark or bright halos. What would occur?

On Jupiter what were observed were fireballs that segued into plumes rising ~3000 km over the Jovian limb and then collapsing ballistically onto, and so heating, the Jovian stratosphere. These plumes are generated by the shock heating of gas along the impactor entry corridor and the near-explosive release of energy during the terminal airburst. For Venus, impact speeds are much lower on average, but the surface gravity is also lower by a factor of 2.6 compared with Jupiter, and the confining scale heights much smaller as well, so an expanding plume should have a greater size (see Zahnle and Mac Low 1994). Over heights of thousands of km, the radial variation of the Venusian gravity field is also important, so Venusian plumes should extend even farther. Hence, SL-9-like impacts could create plumes almost as big as the planet. These would wrap around the entire globe, ballistically re-entering the stratosphere, heating and polluting it. Observed from the Earth, they would be phenomenal.

SL-9-like impactors would penetrate into the Venusian cloud deck. Because of the strongly oxidizing nature of the Venus atmosphere, all or nearly all of the sulfur present as H_2SO_4 or in the impactor would be converted to and re-enter the stratosphere as SO_2 (along with original tropospheric SO_2).

Taking for the sake of argument an average cometary fragment mass of $\sim 10^{11}$ kg (Zahnle 1996) that penetrates to ~2 bar, a plume mass 100 times larger (a reasonable upper limit) could loft 2×10^9 kg of tropospheric SO_2 into the Venusian stratosphere (using a tropospheric SO_2 mixing ratio of 130 ppm from Bézard et al. [1993]). Moreover, impactor sulfur alone could contribute almost 10 times as much if the impactor were chondritic in composition (Anders and Grevesse 1989). This is the same order of magnitude of SO_2 as in the Venusian stratosphere today (2×10^{10} kg above the cloud tops at 40 mbar, based on Na et al. [1990]). As 10^{11} kg is also the minimum mass in our cratering simulation (Sec. II.D), an impact of this scale is an ~30,000 yr event for Venus. Further consideration of the effects of impacts on Venusian stratospheric chemistry, thermal structure, and stability may be warranted.

B. Parabolas

If the impactor survives to strike the ground at hypervelocity and make a crater, then the large vapor and entrained ejecta plume created may be able to leave a trace in the geologic record. Indeed, among the many interesting geologic features discovered by Magellan are more than 55 radar-dark, and in some cases low-emissivity, parabolic features associated with radar-bright young craters (Arvidson et al. 1991; Phillips et al. 1991; Campbell et al. 1992) (although the SAR signature of all but one emissivity parabola is faint [Campbell et al. 1992]). There are also about 10 extended, approximately circular low backscatter cross section features with craters at their centers (Campbell et al. 1992), and numerous other impact-related low-emissivity features may exist (Lawson and Plaut 1994). The radar-dark (low backscatter cross section) parabolas range from several hundred km to about two thousand km from the vicinity of their parent craters, nearly all opening toward the west (Fig. 10a). Arvidson et al. (1991) suggested that they formed as the distal ejecta of the craters interacted with the strong east–west high-altitude zonal winds already known to encircle the planet.

The radar backscatter data indicate that the parabolic deposits are indeed quite smooth, with an upper limit of 1 to 2 cm on the sizes of the particles of interest in the most radar-dark regions (Campbell et al. 1992). These relatively fine-grained ejecta (pebble-sized and less) probably fill up cracks and other small-scale roughness elements. Deposit thicknesses range between a few centimeters to 1 or 2 meters, based on their ability to obscure other features (Campbell et al. 1992). The hypothesis of Arvidson et al. has been incorporated in several more detailed models (Vervack and Melosh 1992; Campbell et al. 1992) that quantitatively describe the formation of these parabolas. Fits of the model of Vervack and Melosh (1992) to the observed parabolas show a consistent variation of model parameters with crater size and, as a bonus, yield information on the dependence of the mean ejecta fragment size on range from the crater for very fast ejecta (Schaller and Melosh 1994).

The basic idea of the models is that, for sufficiently large craters, the ejecta plume breaches the atmosphere around the impact site and hurls fine ejecta briefly into space before it re-enters ballistically some distance away (as in the previous section). The minimum crater size for this to occur is estimated at about 20 km diameter (Vervack and Melosh 1992). Smaller craters do not produce ejecta plumes capable of breaching the atmosphere, so their ejecta is caught up in a fireball that allows it to rain out in a simple streak downwind (westward) of the crater. Observations support this picture in a general way: the area of the parabolic feature declines toward zero for parent crater diameters between ~10 and 20 km diameter (see Fig. 5 in Campbell et al. [1992]).

The ballistic ejecta of craters larger than about 20 km fall into the upper atmosphere in an initially symmetric pattern centered around the crater. The mean particle size d at any range r is assumed to be an inverse power function

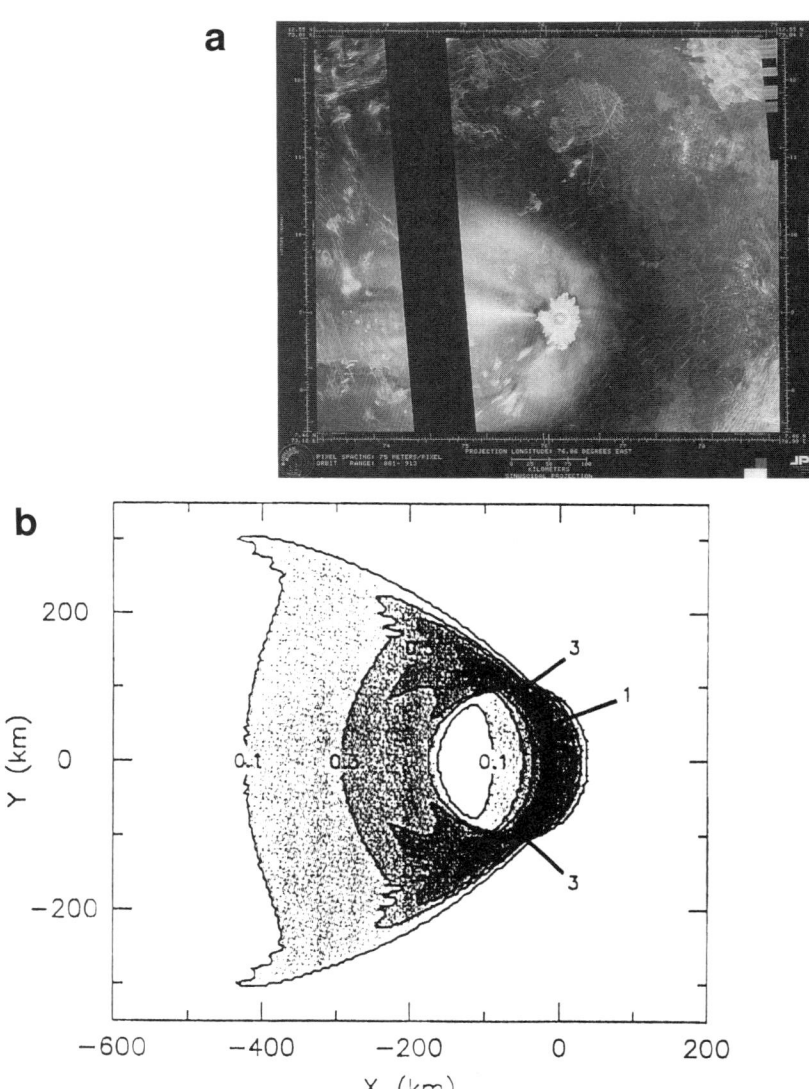

Figure 10. (a) Magellan cycle 1 SAR image of the 30-km diameter crater Adivar (at 8.9°N, 76.2°E) and its associated westward-opening dark parabola (radar incidence angle 46°). The origin of the inner bright zone is unclear. Possibilities include coarser ejecta, a surface scoured by turbulent surface winds generated by interaction of the zonal flow with the shock heated crater site (Greeley et al. 1994), or unresolved minidunes (cf., Fig. 7 in Weitz et al. 1994). (b) "Generic" contour plot of the thickness of the ejecta, in meters, about a 40-km diameter crater on Venus. (figure after Vervack and Melosh 1992).

of range

$$\bar{d} = \bar{d}_c \left(\frac{r_c}{r}\right)^\alpha \tag{12}$$

where $\bar{d}_c$ is the mean particle size extrapolated to the crater rim (at $\Omega = \Omega_c$), and α is the power-law dependence. These particles may be solid ejecta, impact melt droplets, or even recondensed impact vapor.

As the ejected particles fall through the atmosphere they are carried westward by the winds. The distance they are carried before they reach the ground depends on the wind velocity, altitude distribution of the winds and particle size. Because the wind velocity and distribution are relatively well known from entry probes (Seiff et al. 1985), the pattern of ejecta deposition on the ground is mainly controlled by the mean size of the ejecta particles (Fig. 10b). Thus, by fitting the predicted ejecta distribution patterns to the observed parabolas one can effectively measure α and $\bar{d}_c$ in Eq. (12) above. A fit (Schaller and Melosh 1994) of this type to some 17 Venusian parabolas around craters greater than 20 km in diameter yielded excellent results and a very consistent value of the parameter $\alpha = 2.65 \pm 0.03$, independent of crater size. Similarly, a relationship for $\bar{d}_c$ was found that depends on crater size, but again gave a very consistent relationship

$$\log(\bar{d}_c \text{ m}^{-1}) = (3.18 \pm 0.34) - (1.56 \pm 0.26) \log(r_c \text{ km}^{-1}). \tag{13}$$

In addition, the overall length of the parabolas could be fit only when the thickness of the ejecta at the edge of the radar-dark zone was about 1 cm, thus giving an idea of the thickness of a deposit of fine material necessary to blanket and smooth terrain at a radar wavelength of 12.6 cm. This thickness is consistent with those derived from the radar backscatter measurements summarized above. The Venusian data are also consistent with the ejecta size-distance relationship determined from the Chicxulub crater (Vervack and Melosh 1992).

Schultz (1992) has independently proposed a scenario in which the expanding fireball and plume "shoulder aside" the entire upper atmospheric wind field, and material only falls out of the turbulent plume at the plume boundary, resulting in a parabolic deposit. While densely imagined, this scenario is highly speculative. At this stage in our understanding, even exploratory numerical simulations of plume interactions would go far in improving our models of parabola formation.

Although a global analysis has not been attempted, regional and local studies indicate that parabolic deposits can have relatively high dielectric constants $\epsilon \sim 5$ to 8 (Plaut and Arvidson 1992; Campbell et al. 1992; Campbell 1994; Brackett 1995). The highest values have prompted discussion of loading parabola-bound ejecta with a conducting phase. The process of melting, vaporization and recondensation could drive a reduction of silicates toward simple oxides and metals, as occurs in tektites (Ganapathy and Larimer 1983; Schreiber et al. 1984), fulgarites (Essene and Fisher 1986), and apparently, in lunar soils (see, e.g., Morris 1980). Alternatively, or perhaps in addition, the emissivity parabolas may be due to the impact of iron or stony-iron asteroids (the simulations in Figs. 4 and 5 are consistent with this possibility). We caution, however, that the parabolic deposits, while smooth, are not ultra-fine debris or rock powders; such would be globally dispersed by the zonal winds. Rather, the parabolic deposits may represent more of a rain of glassy spherules, tektites and melt bombs; the resulting deposits may be more similar to Archean spherule beds (see, e.g., Lowe and Byerly 1986; Lowe et al. 1989), and not necessarily underdense in bulk (i.e., they would require less of a conducting phase to give higher ϵ values).

Overall, the results summarized here strongly support the ejecta-wind interaction model of parabola formation, and have incidentally yielded the first high-quality data on the mean size of ejecta fragments vs range (or, equivalently, velocity) for the distal ejecta of an impact crater. These results have implications (mainly still to be drawn) for the total abundance of impact-created fines on Venus, and the rate at which surface processes can remove thin deposits of fine grained material (see, e.g., Arvidson et al. 1992).

IV. CRATER MORPHOLOGY

In this section we concentrate on crater depth measurements and their interpretations, impact melt (and vapor) production on Venus and some of its possible consequences, and briefly, peak-ring craters and multiringed basins. The emphasis here is less on atmospheric interactions than it is with basic questions in comparative planetology that involve impact mechanics.

A. Depth-Diameter Relations

In contrast to some pre-Venera and pre-Magellan predictions of craters largely infilled by their own atmospherically decelerated ejecta (see, e.g., Settle 1980), Venusian craters are reasonably deep, considering the relatively high Venusian surface gravity. This has prompted discussion as to whether Venusian craters are deeper than would be predicted based on g^{-1} scaling.

There are two published sets of Venusian crater depth measurements. The first is based on the cross-track distortion, or offset, technique (Ivanov 1989; Schaber et al. 1992; Sharpton 1994). This technique exploits an interesting aspect of SAR image generation. Using SAR, one projects a synthesized pixel of an image at a given slant distance on the reference surface. The crater floor, situated below the surrounding surface, is therefore projected with a lateral offset with respect to the crater rim. The magnitude of this offset may be used to estimate the crater depth. Critical to the success of this technique are adequate image resolution and crater symmetry. For high-resolution Magellan data, only the second is a concern. Schaber et al. (1992) and Sharpton (1994) used the crater-rim outline to measure the offset of the flat-floor image, so their data correspond to rim-to-floor crater depths h. Application of this technique to earlier Venera 15 and 16 data by Ivanov (1989) suffered from relatively low resolution and in some cases from radar foldover, where the inner rimwall slope angle was larger than the radar incidence angle (see Leberl et al. 1992; Sharpton 1994). For whatever combination of reasons, the earlier, lower-resolution data set (Ivanov 1989) gives systematically shallower craters than the higher-resolution, higher-incidence angle Magellan data set (Schaber et al. 1992; Sharpton 1994).

The second set of crater depths is based on direct determination by altimetry. Unfortunately, in this case accuracy is constrained by the size of the footprint of the radar altimeter beam; radar altimetry tends to average over topographic details that are small compared with the footprint, and in regions of rapidly changing elevations such as crater rims, it may be difficult to identify the signature of the rim in the time-resolved altimeter signal. For Magellan, this technique allows one to measure depths reliably only for large craters (greater than 50 km diameter or so). Also, because of lowered sensitivity to the crater rim, only the so-called apparent depth, i.e., that measured with respect to the ground plane, h_a, may be reliably measured. The first such crater profiles were based on Venera 15 and 16 altimetry (Ivanov et al. 1986). Subsequent Magellan altimetry was used by Schaber et al. (1992) to measure the depths of 11 Venusian craters larger than 50 km diameter, and altimetric depths of all craters greater than 70 km diameter were reported in Ivanov and Ford (1993).

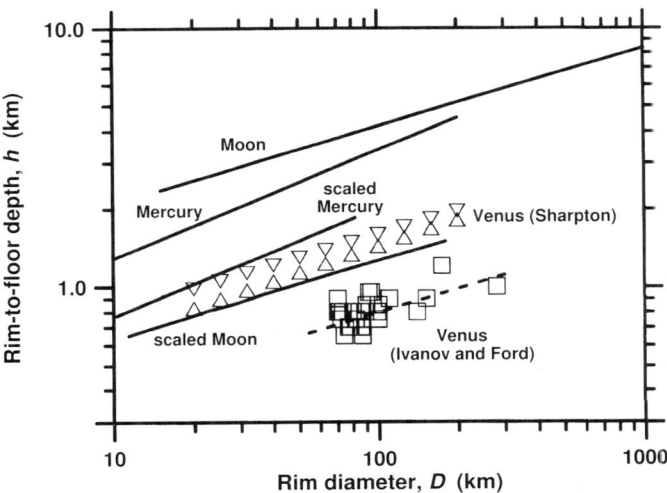

Figure 11. Depth-diameter relations for craters on the Moon (1977b), Mercury (Pike 1988), and Venus from Magellan altimetry (squares; measurements from Ivanov and Ford [1993]) and cross-track rim–floor offsets (upper and lower triangles; fits from Sharpton [1994]). Lunar and Mercurian fits scaled to Venusian gravity are shown for comparison. See text for discussion.

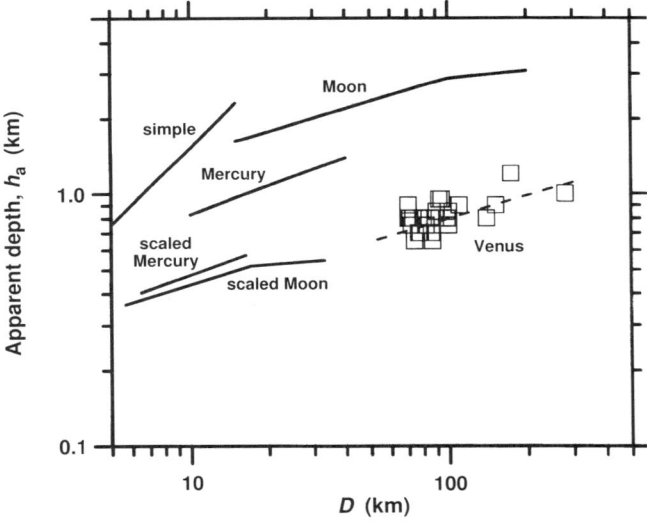

Figure 12. Apparent depth-diameter relations for craters on the Moon (Pike 1977a), Mercury (based on Pike 1988), and Venus from Magellan altimetry (squares; same as Fig. 11). Lunar and Mercurian fits scaled to Venusian gravity and apparent depths of simple terrestrial craters are shown for comparison. See text for discussion.

According to current understanding (e.g., Melosh 1989, ch. 5), small craters are similar to and form from a transient cavity, mostly by a modest amount of wall slumping and/or ejecta fallback. For larger transient craters, the effective strength of the surface material immediately surrounding the crater is insufficient to resist collapse, which occurs promptly not only by wall slumping, but also by central uplift. The onset of this process is controlled by the dimensionless ratio

$$R_c = \rho_t g h_{\text{tr}}/c^* \qquad (14)$$

where c^* is an effective, and transient, cohesion and h_{tr} is the transient cavity depth. When R_c rises above a certain value (around 5 to 7), the transient cavity begins to experience progressive collapse.

The exact physical meaning of the effective cohesion is not still well understood. Values of c^* as low as 1 to 3 MPa or so correspond to the observed onset of crater modification on the terrestrial planets (see, e.g., Melosh 1989; Leith and McKinnon 1991). This value is apparently lower for volatile-rich areas of the terrestrial and Martian crusts (Pike 1980).

Above the critical value of R_c, crater collapse continues until a final equilibrium is reached. Equation (14) can be generalized to this final depth, and ostensibly predicts a constant depth for the final crater. However, the effect of shape variation, a very small amount of internal friction (see discussion in Leith and McKinnon [1991]), results in a depth-diameter relationship where depth is proportional to a small power of the crater diameter ($h \propto D^{0.3-0.4}$).

While the exact meaning of such a power law is not understood, we can suppose that the depth/diameter ratio for complex craters is a function of a ratio D/D_c, or h/h_c, where D_c is the collapse onset diameter (e.g., Sec. II.B; Holsapple 1993), and h_c is the transient cavity depth when $D = D_c$. From such a hypothesis and Eq. (14), one may derive that depth-diameter curves for complex craters may be scaled in coordinates hg, Dg (i.e., all length scales multiplied by gravity). We use this hypothesis below.

We use data for the (volatile-free) crusts of the Moon and Mercury. Power-law depth-diameter relations for lunar craters are taken from Pike (1977b) for rim-to-floor depths h, and from Pike (1977a) for apparent depths h_a. Rim heights and rim-to-floor depths from Pike (1988) are used to estimate the apparent depth relation for craters on Mercury. We also gravity scale the lunar and Mercurian crater depth-diameter relations to Venusian gravity, and both the original and scaled fits are shown in Figs. 11 and 12.

For Venus we emphasize the most extensive set of measurements, Sharpton's (1994) cross-track (floor-offset) based determinations. In Fig. 11 we plot his power-law fit to the rim-to-floor depths of 94 Venusian craters (inverted triangles) and that for 22 fresh craters with parabolic deposits (normal triangles). For completeness we note that Schaber et al.'s measurements (not plotted) tend to be deeper and are more scattered. The Sharpton et al. fits lie comfortably within the scaled lunar and Mercurian fits.

Altimetry-based apparent depth measurements for Venus (from Ivanov and Ford 1993) plot below (i.e., are shallower than) the other data in Fig. 11, as expected (the altimetric depths from Schaber et al. [1992], not plotted, are [as above] somewhat deeper). For most of the measured craters above 70 km diameter, apparent depths vary from ~0.7 to 1.2 km. When we compare these depths with apparent crater depths on the Moon and Mercury (Fig. 12), we see that when gravity-scaled, the lunar, Mercurian and Venusian data all follow the same approximate trend, close to the power-law $h - D$ relationship usually seen for complex craters. We note that the depths of terrestrial complex craters (Grieve et al. 1981, not shown) are not dissimilar to those of Venusian craters (Fig. 12), but this comparison is complicated by the way complex crater depths on Earth are measured and corrected for erosion (Grieve et al. 1981; Sharpton 1994).

Because the depths of complex craters slowly increase with diameter, at *fixed diameter* the depths of complex craters on different planets with similar crustal lithologies do not actually scale as g^{-1}. Rather, they scale as a smaller power. For example, for $h \propto D^{0.3}$, as applies to the Moon and Venus (Pike 1977b; Sharpton 1994), h at fixed D scales as $g^{-0.7}$, yielding a depth ratio ≈ 3.3. This is close to the ratio of depths (at fixed D) between lunar complex craters and Sharpton's fit to fresh Venusian crater depths, ≈ 2.5. This is also consistent with the slight mismatch between the lunar and Venusian fits in Fig. 11, but more importantly, the divergence is easily absorbed into a modest variation in the transient strength (c^*) of cratered Venusian crust compared with that of the Moon (cf., Fig. 24 in Pike [1988]).

What should scale as g^{-1}, for similar crustal lithologies, are the simple-to-complex transition diameters D_c. For Venus this transition cannot be directly observed because of atmospheric interference with orderly cratering processes at small crater diameters. Extrapolating Sharpton's (1994) power-law depth-diameter fits to an assumed $h - D$ curve for simple craters yields $D_c \sim 3$ to 4 km, depending on the simple crater depth dependence adopted ($h \approx 0.2D$ is standard).

In summary, we conclude that there appears to be a fundamental similarity of depth-diameter relationships for complex craters on Venus, Mercury and the Moon. Furthermore, we conclude that Venusian impact craters are not deeper (or shallower) than those on the Moon or Mercury when the later are g^{-1} scaled. The depth-diameter relationships indicate that the dense atmosphere does not drastically change the main morphologic relations for complex craters. Further analyses of Magellan data and more sophisticated modeling should prove fruitful in improving our understanding of complex crater modification.

B. Impact Melt and Ejecta Flows

Yet another remarkable aspect of Venusian impact craters is the occurrence of very long run-out ejecta flows (Phillips et al. 1991; Schaber et al. 1992). Although characteristics of the flows vary, it is clear that many of them spread in a very fluid and lava-like manner, extending well beyond the limits of the continuous ejecta (often by several crater diameters). More than one basic mechanism may contribute to the development of these flows, but there seems little doubt that abundant impact melt is a major cause of this fluidity.

A good summary of flow morphology and research done on them can be found in the chapter by Herrick et al. The purpose of this subsection is simply to summarize the basic results of impact mechanics as related to melt (and vapor) production, which can then be applied to the interpretation of craters on Venus.

Several fundamental observations can be made concerning Venusian long, run-out ejecta flows. First, their occurrence is correlated with crater size, such that long, more extensive flows occur preferentially around larger craters (Schaber et al. 1992; Chadwick and Schaber 1993). Second, their occurrence is correlated with the angle of impact, such that more extensive flows occur preferentially around more oblique impacts (as determined by the asymmetry of their ejecta distributions); this is nicely illustrated in Fig. 39 of Schultz (1992) and statistically quantified in Schaber et al. (1992) and Chadwick and Schaber (1993). Third, the ejecta flows are generally radar-bright compared with the surrounding plains, and so have rougher surfaces.

The first two observations have been generally held as explainable by the following: (1) the higher surface temperature on Venus leads to greater impact melt production than on the Earth or Moon (Ivanov et al. 1986); (2) the higher gravity on Venus compared with the Moon leads to a greater proportion of melt (and vapor) for a given crater size (Phillips et al. 1991; Vickery and

Melosh 1991); (3) a greater relative proportion of impact melt is ejected in the down-range direction as the impact becomes more oblique (Schaber et al. 1991; Chadwick and Schaber 1993); (4) melt-rich ejecta flows start off hotter and thus more fluid than on the Earth or the Moon (Ivanov et al. 1986); (5) melt-rich ejecta flows take longer to equilibrate thermally with cooler, solid clasts, and so stay fluid longer than on the Earth or the Moon (Ivanov et al. 1986).

All of these factors are probably true to some extent, but some are much more important than others. The most detailed theoretical calculations for impact melt production on Venus, albeit preliminary ones, are those of Vickery and Melosh (1991). They combined an explicit (ANEOS) equation of state for dunite with simple geometric models of the shock pressure distribution (generally following Kieffer and Simonds 1980) and excavation flow (the Maxwell Z-model). The model allows estimation of the amount of ejected melt. Results were also determined for the Earth and Moon, and as a function of impact angle.

Interestingly, the model of Vickery and Melosh (1991) indicates only a very marginal enhancement in the amount of ejected melt for Venusian craters of a given size compared with terrestrial ones. This result can be compared to calculations by Grieve and Cintala (1992b), who predict an increase of $\sim 10\%$ in *total* melt production for (transient) craters in basalt, and much simpler estimates by Ivanov et al. (1992) of an ~ 20 to 40% increase. The comparable results of Vickery and Melosh (1991) and Grieve and Cintala (1992b) should not be seen as that surprising. Our geological experience with ejected melt is not based on terrestrial ejecta blankets (which have nearly all been eroded away or buried), but on the ejecta blankets of much smaller, lower-gravity worlds such as the Moon. What the model of Vickery and Melosh *does* show is a substantial enhancement in the proportion of ejected melt on Venus compared with the Moon, by a factor of several (Grieve and Cintala find a similar factor for total melt). In terms of absolute percentages, for an average impact angle of 45°, the melt fraction in Venusian ejecta increases with crater size and reaches $\sim 20\%$ for craters of ~ 100-km diameter.

The increasing proportion of ejected melt at larger sizes is easily understood from the point-of-view of scaling (and has been appreciated for some time; see, e.g., Chapman and McKinnon 1986; Melosh 1989; Grieve and Cintala 1992a; Cintala and Grieve 1994). Cratering efficiency, or mass excavated and displaced per impactor mass, goes as $(ga)^{-0.65}$ at large sizes (the gravity regime) and fixed velocity. Melting or fusion efficiency, in contrast, is constant at fixed velocity, so the ratio of melt to excavated and displaced mass goes as $(ga)^{0.65}$. This ratio is consistent with the results of Vickery and Melosh (1991) and Grieve and Cintala (1992b), keeping in mind that the ratio of melt produced to excavated and displaced mass (i.e., the apparent crater volume times the surface density) is not exactly the same as the melt fraction in the ejecta.

The effect of impact angle is more subtle, and much less well understood.

To make the same sized crater from an oblique impact requires a larger impactor if the velocity stays the same. From the crater scaling in Sec. II.B, $a^{2.35}(v\cos\theta)^{1.3}$ must remain constant; for fixed v, a should go as $(\cos\theta)^{-0.55}$. If the normal component of velocity is the controlling factor in impact melt generation as well, then the fusion efficiency should be reduced by an even greater degree. We do not yet know, however, how fusion efficiency scales with impact angle. If, as Vickery and Melosh (1991) assume, fusion efficiency is basically only a function of total kinetic energy, and thus independent of impact angle, then for the same crater size the ratio of melt produced to excavated mass goes as $\approx(\cos\theta)^{-1.66}$ at fixed v. In particular, Vickery and Melosh (1991) find that the melt fraction in Venusian ejecta may exceed $\sim$50% for large (100-km diameter) craters for impact angles $\gtrsim 70°$. Such loading of the ejecta with hot melt would surely profoundly affect the cooling times, viscosities, and flow and spreading behavior of the ejecta blanket.

The same general behavior, with regard to size and impact angle (whatever the latter is), should also apply to impact vaporization on Venus. An additional factor is the possibility of shock reverberation and multiplication in the compressed atmosphere trapped between impactor and surface at the beginning of the compression stage (Brackett and McKinnon 1992). Although only a theoretical model at this time, Brackett and McKinnon (1992) predict that, due to the increasing stiffness of impactor and surface under multiple shocks, the ultimate pressure enhancement over atmosphereless conditions may be several tens of percent (and much higher for the most compressible impactors, comets). The most significant effect is predicted to be a lowering of threshold velocities for melting and vaporization, for a given impactor-surface combination, which may be important for oblique impacts.

C. Peak-ring Craters and Multiringed Basins

Dozens of peak-ring basins, craters that are morphologically analogous to Schroedinger on the Moon, have been identified on Venus from Venera, Arecibo, and Magellan radar images (Basilevsky et al. 1987; Alexopoulos and McKinnon 1992,1994; Schaber et al. 1992; Herrick and Phillips 1994b). These peak-ring craters are all greater than $\sim$30 to 40 km diameter (Fig. 13), and comprise part of the characteristic scale-dependent morphological sequence that impact craters follow (see, e.g., Melosh 1989). There is, however, some disagreement over whether the crater–rim/peak–ring diameter ratio is $\sim$2 or a decreasing function of crater diameter. This is surprising, considering that identifying peak–ring elements and crater rims is easier on Venus than on any other planet, owing to the relatively undegraded morphologies of Venusian craters and the radar brightness of rims and peak–ring mountains compared with generally radar-dark crater floors and surrounding plains.

The measurements in Schaber et al. (1992) and Alexopoulos and McKinnon (1994) are in fundamental agreement, and show that crater–rim/peak–ring diameter ratios are relatively high, $\sim$4 to 5, near the transition from central–peak to peak–ring craters, and decline toward $\sim$2 at larger sizes (cf., Fig. 13

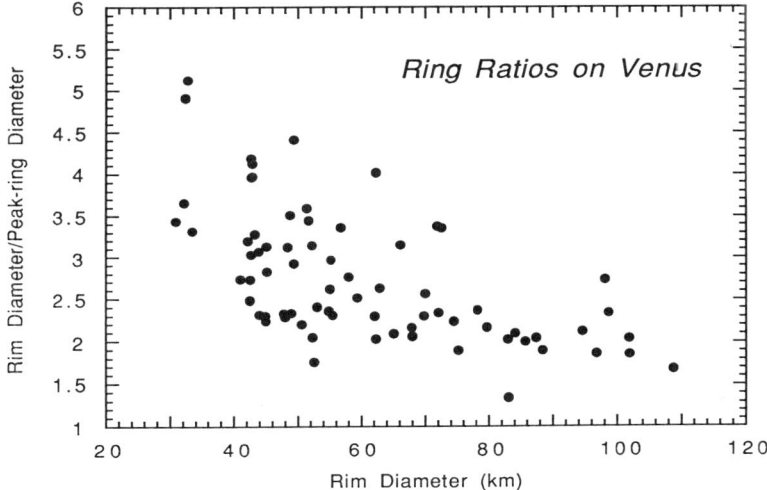

Figure 13. Ring ratios (crater–rim to peak–ring diameter) for 72 double-ringed craters or basins on Venus, as a function of rim diameter. The four largest multiringed basins ($D \gtrsim 150$ km) fall outside of the diameter range plotted here (for greater diameter resolution), but have still smaller adjacent ring ratios (three of the four have three rings) (figure modified from Alexopoulos and McKinnon 1994).

here, and Fig. 17 in Schaber et al. [1992]). There is also considerable scatter in the ring ratio at any given crater size. Herrick and Phillips (1994b), on the other hand, identify a significantly smaller number of peak–ring structures, preferring, as they state, to bin the others as multiple-peak craters (apparently, even if there is a distinct circular outline to the peaks), and to separate out peak–ring craters that have rim/ring ratios much larger than 2, terming them protobasins (see Alexopoulos and McKinnon [1994] for a discussion of protobasins on Venus and elsewhere). This selected sample of peak–ring craters naturally possesses ring ratios close to ≈ 2. Most of the excluded peak–ring-like craters would, we maintain, have been classified as peak–ring craters on any other planet, Mercury for example, because a much more forgiving set of identification criteria would have been used owing to contrast and resolution limitations and the severe effects of erosion. Nevertheless, it is not nomenclature that is important, it is the morphological characteristics of the impact structures. Taken together, peak–ring craters, peak–ring-like craters, and protobasins form an ensemble whose rim/ring ratios decrease with increasing crater size.

The peak–ring formation models proposed so far, hydrodynamic collapse of the central peak (Melosh 1982,1989; Alexopoulos and McKinnon 1992,1994), enhanced shock melting of the core of the central uplift (Grieve and Cintala 1992a,b; see Sec. IV.B), and the impactor depression/imprint hypothesis of Schultz (1992), all predict relative enlargement of the ring with increasing rim diameter. The hydrodynamic collapse model and the enhanced depth-of-melting model may not be mutually exclusive, especially as estimates of the crater diameter at which the depth-of-melting intersects the crater floor are in the 60 to 80 km range on Venus (Grieve and Cintala 1992b). Schultz (1992) claims that the rim/ring ratio decreases with increasing impact angle (his Fig. 49), which could be consistent with the greater melt production in more oblique impacts discussed in the last section. Schultz (1992) also estimates peak–ring to *impactor* diameter ratios of $\approx$5 to 6, so it is unclear how peaks rings might represent his proposed "uplift of the penetration zone." In contrast to the above, we note, and without prejudice, that no particular hypothesis has been offered to explain a fixed rim/ring ratio of $\approx$ 2 (which is itself a subset of the famous $\sqrt{2}$ hypothesis; see Melosh 1989), other than it is some sort of standing (gravity?) wave.

The four largest craters on Venus, Klenova, Meitner, Isabella, and Mead, and perhaps others, have been identified as multiringed basins, i.e., impact structures less similar to peak–ring craters and more similar to the Orientale Basin on the Moon (Alexopoulos and McKinnon 1992,1994; Schaber et al. 1992). What is remarkable are their small sizes compared with Orientale (up to $\sim$6.5 times less), an effect at least partially attributable to Venus' higher gravity ($\sim$5.5 times greater). If outer Cordilleran-style rings (named after the outer ring of Orientale) form by circumferential faulting and block rotation in a lithosphere, driven by inward asthenospheric flow (Melosh and McKinnon 1978), then the upper Venusian mantle may have to be near-solidus in temperature. This may conflict with other evidence for a thick lithosphere (see relevant chapters in this book). A crustal asthenosphere remains a possibility (Alexopoulos and McKinnon 1992) as the recent, stiff anhydrous basalt flow laws (Mackwell et al. 1995) are strongly non-Newtonian (power-law index $n\approx5$), so the crust is quite weak at high stress. As is all too obvious, much further work remains to be done on peak–ring and multiringed basin formation.

V. VENUS DEMENTIA

There are almost 1000 impact craters on Venus. Most are relatively morphologically pristine, giving them a scientific worth far beyond their simple numbers. As of early 1996, analysis of these craters, as with most Magellan data, has barely moved out of the preliminary stage. Consensus has not yet been reached on most of the matters discussed in this chapter. Much remains to be done. Unfortunately, budgetary and other pressures forced early termination of the data analysis program targeted at extracting Venus' secrets. Similar pressures advanced the date of the conference that this book is based on, ensuring that many views and conclusions represented in this and other chapters will likely undergo profound modification in the next few years. Space limitations have prevented us from even discussing many aspects of crater formation on Venus. We can only hope that we have stressed much of what is important, and leave to the future a fuller accounting of impact cratering and attendant atmospheric effects on the planet. The Magellan data set is both vast and rich. Those who choose to mine it will be rewarded.

Acknowledgments. We thank D. Grinspoon for discussions of Venusian atmospheric chemistry, M.-M. Mac Low for insight on numerical limitations, N. Izenberg for figure assistance, and the NASA Planetary Geology and Geophysics (WBM and HJM), Venus Data Analysis (WBM), and Exobiology (KZ) Programs for supporting this work. BAI thanks the International Science Foundation. WBM thanks FZ.

REFERENCES

Alexopoulos, J. S., and McKinnon, W. B. 1992. Multiringed impact craters on Venus: An overview from Arecibo and Venera images and initial Magellan data. *Icarus* 100:347–363; erratum, *Icarus* 103:161.
Alexopoulos, J. S., and McKinnon, W. B. 1994. Large impact craters and basins on Venus: Implications for ring mechanics on the terrestrial planets. In *Large Meteorite Impacts and Planetary Evolution*, eds. B. O. Dressler, R. A. F. Grieve and V. L. Sharpton, GSA SP-293 (Boulder: Geological Soc. of America), pp. 178–198.
Anders, E., and Grevesse, N. 1989. Abundances of the elements: Meteoritic and solar. *Geochim. Cosmochim. Acta* 53:197–214.
Arvidson, R. E., et al. 1991. Magellan: Initial analysis of Venus surface modification. *Science* 252:270–275.
Arvidson, R. E., et al. 1992. Surface modification of Venus as inferred from Magellan observations of plains. *J. Geophys. Res.* 97:13303–13317.
Asphaug, E., and Benz, W. 1994. Density of comet Shoemaker-Levy 9 deduced by modelling breakup of the parent "rubble pile." *Nature* 370:120–124.
Asphaug, E., and Melosh, H. J. 1993. The Stickney impact of Phobos: A dynamical model. *Icarus* 101:144–164.
Basilevsky, A. T., et al. 1987. Impact craters of Venus: A continuation of the analysis of data from the Venera 15 and 16 spacecraft. *J. Geophys. Res.* 92:12869–12901.
Bézard, B., et al. 1993. The abundance of sulfur dioxide below the clouds of Venus. *Geophys. Res. Lett.* 20:1587–1590.
Brackett. R. A. 1995. Dielectric properties of venusian impact craters and associated deposits. *Lunar Planet. Sci.* XXVI:157–158 (abstract).
Brackett, R. A., and McKinnon, W. B. 1992. Cratering mechanics on Venus: Pressure enhancement by the atmospheric "ocean." *Geophys. Res. Lett.* 19:2115–2118.
Boslough, M., Crawford, D., Robinson, A., and Trucano, T. 1994. Mass and penetration depth of Shoemaker-Levy 9 fragments from time-resolved photometry. *Geophys. Res. Lett.* 21:1555–1558.
Bronshten, V. A. 1983. *Physics of Meteoric Phenomena* (Dordrecht: D. Reidel).
Campbell, B. A. 1994. Merging Magellan emissivity and SAR data for analysis of Venus surface dielectric properties. *Icarus* 112:187–203.
Campbell, B. A., and Campbell, D. B. 1992. Analysis of volcanic surface morphology on Venus from comparison of Arecibo, Magellan, and terrestrial airborne radar

data. *J. Geophys. Res.* 97:16293–16314.

Campbell, D. B., et al. 1992. Magellan observations of extended impact crater related features on the surface of Venus. *J. Geophys. Res.* 97:16249–16277.

Chadwick, D. J., and Schaber, G. G. 1993. Impact crater outflows on Venus. *J. Geophys. Res.* 98:20891–20902.

Chapman, C. R., and McKinnon, W. B. 1986. Cratering of planetary satellites. In *Satellites*, eds. J. A. Burns and M. S. Matthews (Tucson: Univ. of Arizona Press), pp. 492–580.

Chapman, C. R., Harris, A. W., and Binzel, R. 1994. Physical properties of near-Earth asteroids: Implications for the hazard issue. In *Hazards Due to Comets and Asteroids*, ed. T. Gehrels (Tucson: Univ. of Arizona Press), pp. 537–549.

Chevalier, R., and Sarazin, C. 1994. Explosions of infalling comets in Jupiter's atmosphere. *Astrophys. J.* 429:863–875.

Chyba, C. F., Thomas, P. J., and Zahnle, K. J. 1993. The 1908 Tunguska explosion: Atmospheric disruption of a stony asteroid. *Nature* 361:40–44.

Cintala, M. J., and Grieve, R. A. F. 1994. The effects of differential scaling of impact melt and crater dimensions on lunar and terrestrial craters: Some brief examples. In *Large Meteorite Impacts and Planetary Evolution*, eds. B. O. Dressler, R. A. F. Grieve and V. L. Sharpton, GSA SP-293 (Boulder: Geological Soc. of America), pp. 51–59.

Crawford, D. 1996. Models of fragment penetration and fireball evolution. In *The Collision of Comet Shoemaker-Levy 9 with Jupiter*, ed. K. Noll (New York: Cambridge Univ. Press), in press.

Crawford, D. A., Boslough, M. B., Trucano, T. G., and Robinson, A. C. 1995. The impact of periodic comet Shoemaker-Levy 9 on Jupiter. *Intl. J. Impact Eng.* 17:253–262.

Croft, S. K. 1985. The scaling of complex craters. *Proc. Lunar Planet. Sci. Conf.* 15, *J. Geophys. Res. Suppl.* 90:828–842.

Dohnanyi, J. S. 1972. Interplanetary objects in review: Statistics of their masses and dynamics. *Icarus* 17:1–48.

Donnison, J. R. 1986. The distribution of cometary magnitudes. *Astron. Astrophys.* 167:359–363.

Essene, E. J., and Fisher, D. C. 1986. Lightning strike fusion: Extreme reduction and metal-silicate liquid immiscibility. *Science* 234:189–193.

Everhart, E. 1967. Intrinsic distributions of cometary perihelia and magnitudes. *Astron. J.* 72:1002–1011.

Field, G., and Ferrara, A. 1995. The behavior of fragments of comet Shoemaker-Levy 9. *Astrophys. J.* 438:957–967.

Gaffney, E. S. 1985. Hugoniot of water ice. In *Proc. NATO Workshop Ices in the Solar System*, eds. J. Klinger, D. Benest, A. Dollfus and R. Smoluchowski (Dordrecht: D. Reidel), pp. 119–148.

Ganapathy, R., and Larimer, J. W. 1983. Nickel-iron spherules in tektites: Non-meteoritic in origin. *Earth Planet. Sci. Lett.* 65:225–228.

Glasstone, S., and Dolan, P. J. 1977. *The Effects of Nuclear Weapons* (Washington: U. S. Dept. of Defense and U. S. Dept. of Energy).

Grady, D. E., and Lipkin, J. 1980. Criteria for impulsive rock fracture. *Geophys. Res. Lett.* 7:255–258.

Greeley, R., et al. 1992. Aeolian features on Venus: Preliminary Magellan results. *J. Geophys. Res.* 97:13319–13345.

Greeley, R., et al. 1994. Wind streaks on Venus: Clues to atmospheric circulation. *Science* 263:358–361.

Grieve, R. A. F. 1984. The impact cratering rate in recent time. *Proc. Lunar Planet. Sci. Conf.* 14, *J. Geophys. Res. Suppl.* 89:403–408.

Grieve, R. A. F. 1991. Terrestrial impact: The record in the rocks. *Meteoritics* 26:175–194.
Grieve, R. A. F., and Cintala, M. J. 1992a. An analysis of differential impact melt-crater scaling and implications for the terrestrial impact record. *Meteoritics* 27:526–538.
Grieve, R. A. F., and Cintala, M. J. 1992b. Venusian impact craters: Effects of differential scaling. *Lunar Planet. Sci.* XXIII:451–452 (abstract).
Grieve, R. A. F., and Dence, M. R. 1979. The terrestrial cratering record II. The crater production rate. *Icarus* 38:230–242.
Grieve, R. A. F., and Garvin, J. B. 1984. A geometric model for the excavation and modification of terrestrial simple impact craters. *J. Geophys. Res.* 89:11561–11572.
Grieve, R. A. F., Robertson, P. B., and Dence, M. A. 1981. Constraints of the formation of ring impact structures, based on terrestrial data. In *Multi-Ring Basins*, eds. P. H. Schultz and R. B. Merrill (New York: Pergamon Press), pp. 37–57.
Grieve, R. A. F., and Shoemaker, E. M. 1994. The record of past impacts on Earth. In *Hazards Due to Comets and Asteroids*, ed. T. Gehrels (Tucson: Univ. of Arizona Press), pp. 417–462.
Grigoryan, S. S. 1979. Motion and disintegration of meteorites in planetary atmospheres. *Cosmic Res.* 17:724–740.
Gryaznov, V. K., et al. 1994. Collision of the comet Shoemaker-Levy 9 with Jupiter: Interpretation of observed data. *Earth Moon Planets* 66:99–128.
Head, J. N., and Melosh, H. J. 1995. Airblast shattering of rocks on Venus: A numerical simulation. *EOS: Trans. AGU Suppl.* 76:F336 (abstract).
Herrick, R. R., and Phillips, R. J. 1994a. Effects of the Venusian atmosphere on incoming meteoroids and the impact crater population. *Icarus* 112:253–281.
Herrick, R. R., and Phillips, R. J. 1994b. Implications of a global survey of venusian impact craters. *Icarus* 111:387–416.
Hills, J. G., Goda, M. P. 1993. The fragmentation of small asteroids in the atmosphere. *Astron. J.* 105:1114–1144.
Holsapple, K. A. 1993. The scaling of impact processes in planetary sciences. *Ann. Rev. Earth Planet. Sci.* 21:333–373.
Hornung, H. 1986. Regular and Mach reflection of shock waves. *Ann. Rev. Fluid Mech.* 18:35–58.
Hughes, D. W. 1988. Cometary distribution and the ratio between the numbers of long- and short-period comets. *Icarus* 73:149–162.
Ivanov, B. A. 1988. Simple hydrodynamic model of atmospheric breakup of hypervelocity projectiles. *Lunar Planet. Sci.* XIX:535–536 (abstract).
Ivanov, B. A. 1989. Morphometry of impact craters on Venus. *Astron. Vestn.* 23:39–49 (in Russian).
Ivanov, B. A. 1990. Venusian impact craters on Magellan images: View from Venera 15/16. *Earth Moon Planet* 50/51:159–173.
Ivanov, B. A., and Basilevsky, A. T. 1987. A comparison of crater retention ages on the Earth and Venus. *Solar System Res.* 21:84–89.
Ivanov, B. A., and Ford, P. G. 1993. The depths of the largest impact craters on Venus. *Lunar Planet. Sci.* XXV:689–690 (abstract).
Ivanov, B. A., and Melosh, H. J. 1994. Dynamic fragmentation of a comet in the jovian atmosphere. *Lunar Planet. Sci.* XXV:597–598 (abstract).
Ivanov, B. A., Basilevsky, A. T., Kryuchkov, V. P., and Chernaya, I. M. 1986. Impact craters of Venus: Analysis of Venera 15 and 16 data. *Proc. Lunar Planet. Sci. Conf. 16, J. Geophys. Res. Suppl.* 91:413–430.
Ivanov, B. A., et al. 1992. Impact cratering on Venus: Physical and mechanical models. *J. Geophys. Res.* 97:16167–16181.

Izenberg, N. R., Arvidson, R. E., and Phillips, R. J. 1994. Impact crater degradation on venusian plains. *Geophys. Res. Lett.* 21:289–292.

Kahn, R. 1982. Deducing the age of the dense Venus atmosphere. *Icarus* 49:71–81.

Kieffer, S. W., and Simonds, C. H. 1980. The role of volatiles and lithology in the impact cratering process. *Rev. Geophys. Space Phys.* 18:143–181.

Lawson, S.L., and Plaut, J. J. 1994. Impact-related low-emissivity anomalies on Venus. *Lunar Planet. Sci.* XXV:781–782 (abstract).

Leberl, F. W., Thomas, J. K., and Maurice, K. E. 1992. Initial results from the Magellan stereo experiment. *J. Geophys. Res.* 97:13675–13689.

Leith, A. C., and McKinnon, W. B. 1991. Terrace width variations in complex Mercurian craters, and the transient strength of the cratered Mercurian and lunar crust. *J. Geophys. Res.* 96:20923–20931.

Lowe, D. R., and Byerly, G. R. 1986. Early Archean silicate spherules of probable impact origin, South Africa and Western Australia. *Geology* 14:83–86.

Lowe, D. R., Byerly, G. R., Asaro, F., and Kyte, F. J. 1989. Geological and geochemical record of 3400-million-year-old terrestrial meteorite impacts. *Science* 245:959–962.

Mackwell, S. J., Zimmerman, M. E., Kohlstedt, D. L., and Scherber, D. S. 1995. Experimental deformation of dry Columbia diabase: Implications for tectonics on Venus. In *Rock Mechanics: Proc. 35th U. S. Symposium*, eds. J. J. K. Daeman and R. A. Schultz (Brookfield, Vt.: A. A. Balkema), pp. 207–214.

McKinnon, W. B., Chapman, C. R., and Housen K. R. 1991. Cratering of the Uranian satellites. In *Uranus*, eds. J. T. Bergstralh, E. D. Miner and M. S. Matthews (Tucson: Univ. of Arizona Press), pp. 1177–1252.

Mac Low, M.-M. 1996. Entry and fireball models vs. observations: What have we learned? In *The Collision of Comet Shoemaker-Levy 9 with Jupiter*, ed. K. Noll (New York: Cambridge Univ. Press), in press.

Mac Low, M.-M., and Zahnle, K. 1994. Explosion of Comet Shoemaker-Levy 9 on entry into the Jovian atmosphere. *Astrophys. J. Lett.* 434:33–36.

Melosh, H. J. 1981. Atmospheric breakup of terrestrial impactors. In *Multi-Ring Basins*, eds. P. H. Schultz and R. B. Merrill (New York: Pergamon Press), pp. 29–35.

Melosh, H. J. 1982. A schematic model of crater modification by gravity. *J. Geophys. Res.* 87:371–380.

Melosh, H. J. 1989. *Impact Cratering: A Geological Process* (New York: Oxford Univ. Press).

Melosh, H. J., and McKinnon, W. B. 1978. The mechanics of ringed basin formation. *Geophys. Res. Lett.* 5:985–988.

Melosh, H. J., Ryan, E. V., and Asphaug, E. 1992. Dynamic fragmentation in impacts: Hydrocode simulation of laboratory impacts. *J. Geophys. Res.* 97:14735–14759.

Mizutani, H., Takagi, Y., and Kawakami, S. 1990. New scaling laws on impact fragmentation. *Icarus* 87:307–326.

Morris, R. V. 1980. Origins and size distribution of metallic iron particles in the lunar regolith. *Proc. Lunar Planet. Sci. Conf.* 11:1697–1712.

Na, C. Y., Esposito, L. W., and Skinner, T. E. 1990. International Ultraviolet Explorer observation of Venus SO_2 and SO. *J. Geophys. Res.* 95:7485–7491.

Neukum, G., and Ivanov, B. A. 1994. Crater size distributions and impact probabilities on Earth from lunar, terrestrial-planet, and asteroid cratering data. In *Hazards Due to Comets and Asteroids*, ed. T. Gehrels (Tucson: Univ. of Arizona Press), pp. 359–416.

O'Keefe, J. D., and Ahrens, T. J. 1982. The interaction of the Cretaceous/Tertiary extinction bolide with the atmosphere, ocean, and solid Earth. In *Geological Implications of Impacts of Large Asteroids and Comets on the Earth*, eds. L. T.

Silver and P. H. Schultz, GSA SP-190 (Boulder: Geological Soc. of America), pp. 103–120.
O'Keefe, J. D., Takata, T., and Ahrens, T. J. 1994. Penetration of large bolides into dense planetary atmospheres—Role of hydrodynamic instabilities. *Lunar Planet. Sci.* XXV:1023–1024 (abstract).
Passey, Q. R. 1980. Effects of atmospheric breakup on crater field formation. *Lunar Planet. Sci.* XI:863–864 (abstract).
Passey, Q. R., and Melosh, H. J. 1980. Effects of atmospheric breakup on crater field formation. *Icarus* 42:211–233.
Phillips, R. J., and Izenberg, N. R. 1995. Ejecta correlations with spatial crater density and Venus resurfacing history. *Geophys. Res. Lett.* 22:1517–1520.
Phillips, R. J., et al. 1991. Impact craters on Venus: Initial analysis from Magellan. *Science* 252:288–297.
Phillips, R. J., et al. 1992. Impact craters and Venus resurfacing history. *J. Geophys. Res.* 97:15923–15948.
Pike, R. J. 1977a. Apparent depth/apparent diameter relation for lunar craters. *Proc. Lunar Sci. Conf.* 8:3427–3436.
Pike, R. J. 1977b. Size-dependence in the shape of fresh impact craters on the Moon. In *Impact and Explosion Cratering*, eds. D. J. Roddy, R. O. Pepin and R. B. Merrill (New York: Pergamon Press), pp. 489–509.
Pike, R. J. 1980. Control of crater morphology by gravity and target type: Mars, Earth, Moon. *Proc. Lunar Planet. Sci. Conf.* 11:2159–2189.
Pike, R. J. 1988. Geomorphology of craters on Mercury. In *Mercury*, eds. F. Vilas, C. R. Chapman and M. S. Matthews (Tucson: Univ. of Arizona Press), pp. 165–273.
Plaut, J. J., and Arvidson, R. E. 1992. Comparison of Goldstone and Magellan radar data in the equatorial plains of Venus. *J. Geophys. Res.* 97:16279–16291.
Provalov, A. V., and Ivanov, B. A. 1992. Near surface soil-gas flow due to impact on Venus. *Lunar Planet. Sci.* XXIII:1115–1116 (abstract).
Roemer, E., Thomas, M., and Lloyd, R. E. 1966. Observations of comets, minor planets, and Jupiter VIII. *Astron. J.* 71:591–601.
Safronov, V. S. 1972. *Evolution of the Protoplanetary Cloud and Formation of the Earth and the Planets*, NASA Tech. Transl. F-677.
Schaber, G. G., et al 1992. Geology and distribution of impact craters on Venus: What are they telling us? *J. Geophys. Res.* 97:13257–13302.
Schaller, C. J., and Melosh, H. J. 1994. Venusian parabolic halos: Numerical model results. *Lunar Planet. Sci.* XXV:1199–1200 (abstract).
Schmidt, R. M. 1993. Pressure versus drag effects on crater size. *Lunar Planet. Sci.* XXIV:1253–1254 (abstract).
Schmidt, R. M., and Hassig, P. J. 1995. Asteroid entry into Venusian atmosphere: Pressure and density fields. *Lunar Planet. Sci.* XXVI:1239–1240 (abstract).
Schmidt, R. M., and Housen, K. R. 1987. Some recent advances in the scaling of impact and explosive cratering. *Intl. J. Impact Eng.* 5:543–560.
Schreiber, H. D., Minnix, L. M., and Balazs, G. B. 1984. The redox state of iron in tektites. *J. Non-Cryst. Solids.* 67:349–359.
Schultz, P. H. 1992. Atmospheric effects on ejecta emplacement and crater formation on Venus from Magellan. *J. Geophys. Res.* 97:16183–16248.
Seiff, A., et al. 1985. Models of the structure of the atmosphere of Venus from the surface to 100 kilometers altitude. *Adv. Space Res.* 5:3–58.
Settle, M. 1980. The role of fallback ejecta in the modification of impact craters. *Icarus* 42:1–19.
Sharpton, V. I. 1994. Evidence from Magellan for unexpectedly deep complex craters on Venus. In *Large Meteorite Impacts and Planetary Evolution*, eds. B. O.

Dressler, R. A. F. Grieve and V. L. Sharpton, GSA SP-293 (Boulder: Geological Soc. of America), pp. 19–27.

Shoemaker, E. M., Wolfe, R. F., and Shoemaker, C. S. 1990. Asteroid and comet flux in the neighborhood of Earth. In *Global Catastrophes in Earth History*, eds. V. L. Sharpton and P. D. Ward, GSA SP-247 (Boulder: Geological Soc. of America), pp. 155–170.

Shoemaker, E. M., Wolfe, R. F., and Shoemaker, C. S. 1991. Asteroid flux and impact cratering rate on Venus. *Lunar Planet. Sci.* XXII:1253–1254 (abstract).

Shoemaker, E. M., Weissman, P. R., and Shoemaker, C. S. 1994. The flux of periodic comets near Earth. In *Hazards Due to Comets and Asteroids*, ed. T. Gehrels (Tucson: Univ. of Arizona Press), pp. 313–335.

Solem, J. C. 1994. Density and size of comet Shoemaker-Levy 9 deduced from a tidal breakup model. *Nature* 370:349–351.

Starrfield, S., and Shore, S. N. 1995. The birth and death of Nova V1974 Cygni. *Sci. Amer.* 272:76–81.

Stone, J. M., and Norman, M. L. 1992. Zeus-3D. *Astrophys. J. Suppl.* 80:753–789.

Strom, R. G., Schaber, G. G., and Dawson, D. D. 1994. The global resurfacing of Venus. *J. Geophys. Res.* 99:10899–10926.

Svetsov, V. V., Nemtchinov, I. V., and Teterev, A. V. 1995. Disintegration of large meteoroids in the Earth's atmosphere: Theoretical models. *Icarus* 116:131–153.

Takata, T., Ahrens, T. J., and Phillips, R. J. 1995. Atmospheric effects on cratering on Venus. *J. Geophys. Res.* 100:23329–23348.

Tauber, M. E., and Kirk, D. B. 1976. Impact craters on Venus. *Icarus* 28:351–357.

Vervack, R., and Melosh, H. J. 1992. Wind interaction with falling ejecta: Origin of the parabolic features on Venus. *Geophys. Res. Lett.* 19:525–528.

Vickery, A. M. 1986. Size-velocity distributions of large ejecta fragments. *Icarus* 67:224–236.

Vickery, A. M., and Melosh, H. J. 1991. Production of impact melt in craters on Venus, Earth, and the Moon. *Lunar Planet. Sci.* XXII:1443–1444 (abstract).

Weitz, C. M., Plaut, J. J., Greeley, R., and Saunders, R. S. 1994. Dunes and microdunes on Venus: Why were so few found in the Magellan data? *Icarus* 112:282–295.

Wetherill, G. W. 1988. Where do the Apollo objects come from? *Icarus* 76:1–18.

Wetherill, G. W. 1989. Cratering of the terrestrial planets by Apollo objects. *Meteoritics* 24:15–22.

Zahnle, K. J. 1992. Airburst origin of dark shadows on Venus. *J. Geophys. Res.* 97:10243–10255.

Zahnle, K. J. 1996. Dynamics and chemistry of SL9 plumes. In *The Collision of Comet Shoemaker-Levy 9 with Jupiter*, ed. K. Noll (New York: Cambridge Univ. Press), in press.

Zahnle, K., and Mac Low, M.-M. 1994. The collision of Jupiter and Comet Shoemaker-Levy 9. *Icarus* 108:1–17.

Zahnle, K., Mac Low, M.-M., Lodders, K., and Fegley, B. 1995. Sulfur chemistry in the wake of Shoemaker-Levy 9. *Geophys. Res. Lett.* 22:1593–1596.

Zimbelman, J. R. 1984. Planetary impact probabilities for long-period comets. *Icarus* 57:48–54.

MORPHOLOGY AND MORPHOMETRY OF IMPACT CRATERS

ROBERT R. HERRICK and VIRGIL L. SHARPTON
Lunar and Planetary Institute

MICHAEL C. MALIN
Malin Space Science Systems

SUZANNE N. LYONS
Texas A and M University

and

KIMBERLY FEELY
Arizona State University

We present a qualitative and quantitative description of Venusian impact craters and crater-related features. The ~900 Venusian craters are non-overlapping and unaffected by erosion, and volcanism and/or tectonism has not seriously altered the appearance of most of the craters; however, a significant fraction of the craters have experienced some volcanic modification, and features such as a dark halo, a parabola, and a deep radar-bright floor may distinguish truly pristine craters. Qualitatively, the appearances of Venusian craters are similar to lunar craters overall, but significant differences occur, primarily because Venus has a dense atmosphere and higher surface gravity than the Moon. Comparison of Venusian crater depths and interior morphology with those on other planets suggest that gravity and target strength control interplanetary differences in complex crater shape, but no simple inverse-gravity dependence exists. The presence of a dense atmosphere has observable effects on the passage of the meteoroid through the atmosphere and the ejecta emplacement process: meteoroid disruption prevents small craters from forming and affects the number and appearance of craters up to ~30 km in diameter; most ejecta is emplaced nonballistically, sometimes as long ejecta flows; and fine-grained ejecta lofted into the prevailing easterlies produce west-facing parabolic features.

I. INTRODUCTION

Magellan data has revealed ~900 separate Venusian crater features of at least possible impact origin (Phillips et al. 1992; Schaber et al. 1992; Herrick and Phillips 1994*a*). Although this number is substantially less than the number observed on the other terrestrial bodies (excluding Earth), there are enough craters to provide a great deal of insight into several areas of study, including cratering mechanics, meteoroid behavior, age-dating of the surface, and stratigraphy. The Venusian craters are uneroded, nonoverlapping, and generally emplaced on smooth lava plains. Their relatively pristine appearance makes them well suited for studies that use crater morphology to infer the mechanics of crater formation (particularly in an atmosphere). The chapter by McKinnon et al. uses a model-based approach to address how the Venusian cratering record has influenced the understanding of cratering mechanics, and the chapter by Basilevsky et al. focuses on inferences about the global resurfacing history of the planet that can be made from the cratering record. In this chapter we characterize the Venusian craters both qualitatively and quantitatively, and we take a data-oriented approach to discussing the implications of this characterization.

II. WHAT DOES A VENUSIAN CRATER LOOK LIKE?

A. General Appearance

To first order, fresh impact craters on all of the rocky bodies of the solar system are similar in appearance, and Venus is no exception. Here we present a qualititative description of "typical" Venusian impact craters, using the familiar craters of the Moon as a reference frame for comparison (see Wilhelms 1987, Ch. 3 for a general discussion of lunar crater appearance). Figure 1 shows a size progression of typical craters on Venus, and Fig. 2 graphically shows the change in interior morphology with increasing size. Like the Moon, the larger craters have more "complex" interiors, with features such as central peaks, flat floors, terraces, peak rings, etc., appearing with increasing diameter (Herrick and Phillips 1994a). However, there are significant differences between craters on the two planets, primarily caused by differences in planetary gravity and the thick Venusian atmosphere.

There are no craters smaller than $\sim$2 km in diameter because the atmosphere shields the surface from meteoroids small enough to make these craters (Phillips et al. 1991). Many of the smaller craters are distinctly noncircular in appearance, almost certainly the result of the meteoroid fragmenting and hitting the surface as a tightly spaced cluster of material (Phillips et al. 1991). A consequence of these atmospheric effects is that only a handful of craters appear to have interiors similar to lunar "simple" craters, and it may be that higher resolution images would reveal these as not bowl-shaped. Under certain conditions a few fragments may separate enough to produce their own craters that are either separate from or overlap the main crater (Fig. 3; Phillips et al. 1991).

As crater size increases, the effects of meteoroid fragmentation become less important: fragment separation becomes negligible compared to final crater diameter. Most craters above $\sim$12 km are circular, and central peaks and wall terraces are the norm for craters in the 10 to 30 km size range. Craters above $\sim$18 km in diameter usually have a topographically flat floor that onlaps the terrace zone and any central structure. At larger diameters it is common for several distinct peaks to be observable, and most of the craters above 60 km have a well-defined ring of massifs in the interior. A few of the very largest craters have two or more of these rings (Schaber et al. 1992). Mead, the largest crater (268 km diameter), and perhaps a few other large craters have a ring that is an asymmetric scarp rather than a symmetric massif (Schaber et al. 1992). Lunar complex craters have interiors similar to complex crater interiors on Venus, but the corresponding features occur at larger crater diameters. This appearance at smaller diameters for Venusian craters is generally ascribed to the higher Venusian gravity causing the collapse (at smaller diameters) of a parabolic-shaped "transient cavity" (see, e.g., Herrick and Phillips 1994*a*). The ejecta deposits for Venusian craters are quite distinct in appearance from any of the other terrestrial planets. It is easier to define the extent of ejecta deposits for Venusian craters than for lunar craters, possibly because Venusian ejecta are deposited from a cloud rather than ballistically emplaced. Also, radar data emphasizes the blocky nature of ejecta, and Venusian craters are isolated rather than overlapping. For small craters the ejecta blanket has a ragged and somewhat diffuse boundary, and for the smallest craters it is typical for this blanket to encompass only a small portion of the crater's perimeter. Above $\sim$20 km diameter the peripheries of the ejecta blankets change, becoming sharply defined and lobate in nature. Secondary craters occur often as isolated rays about smaller craters, but are ubiquitous in and beyond ejecta blankets for craters greater than $\sim$50 km in diameter.

Beyond the radar-bright blocky material that is usually considered the boundary of the ejecta blanket there are, for many craters, a number of additional features clearly related to the impact. About a third of Venusian craters (Herrick and Phillips 1994*a*) are surrounded by a large (tens of km) radar-dark zone that can either have a diffuse dark halo or sharp dark margin boundary (e.g., Fig. 1a,c). Dark haloes and dark margins have been attributed to crushing of the surface into fine grain debris by the projectile-generated atmospheric shock wave (see, e.g., Phillips et al. 1991) and/or deposition of fine-grained ejecta (Schultz 1992). About 6% of the craters are associated with even larger (hundreds of km) radar-dark zones that are parabolic and open to the west (Campbell et al. 1992; Fig. 4). These are the result of deposition of fine-grained debris ejected high into the atmosphere and carried by the prevailing easterlies (Campbell et al. 1992; Vervack and Melosh 1992). Finally, many craters, particularly those at larger diameters, have ejecta flows extending from the ejecta blanket, sometimes for hundreds of km (e.g., Fig. 1g), that indicate relatively inviscid fluid flow (Asimow and Wood 1992).

B. What Is A Fresh Crater and How Does It Degrade?

Because the surface of Venus is young, and its atmosphere is dense, sluggish and dry, its craters do not show the conspicuous effects of degradation that are typical of those on other planets. Consequently, such conventional approaches as evaluating rim sharpness, crater infilling, and preservation of blocky ejecta

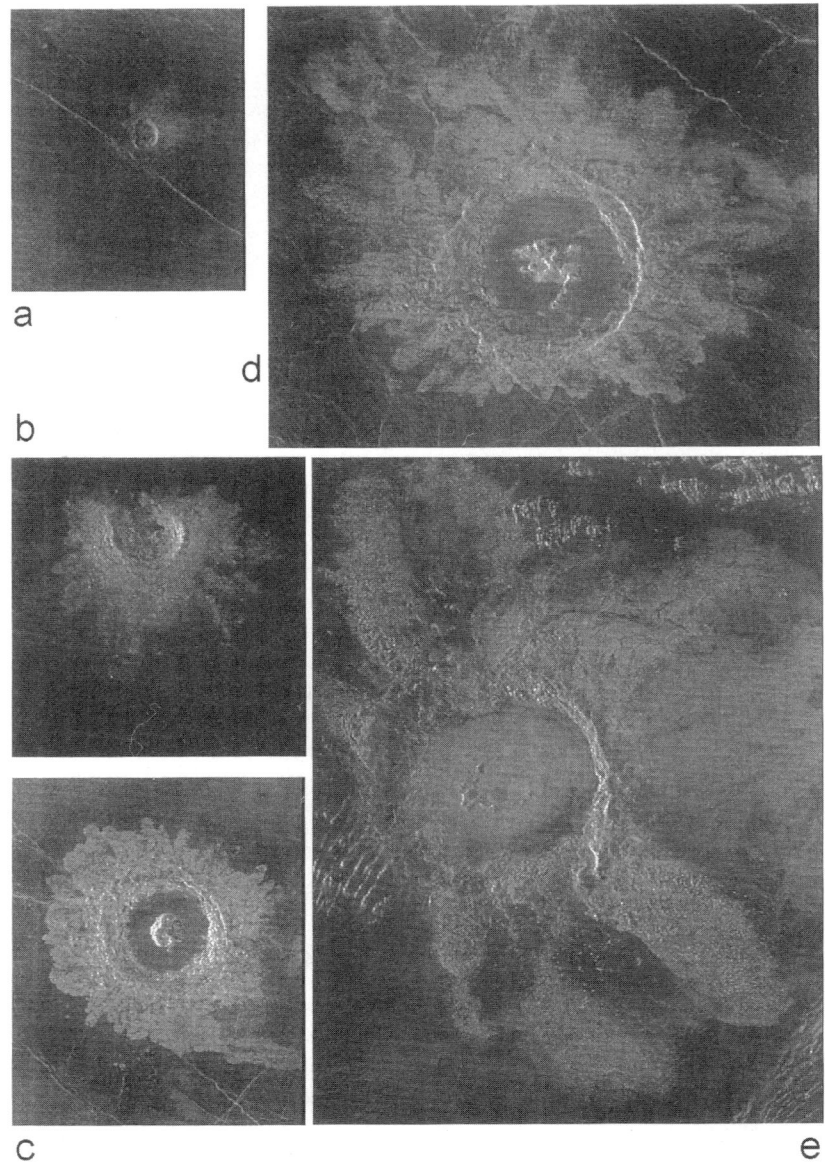

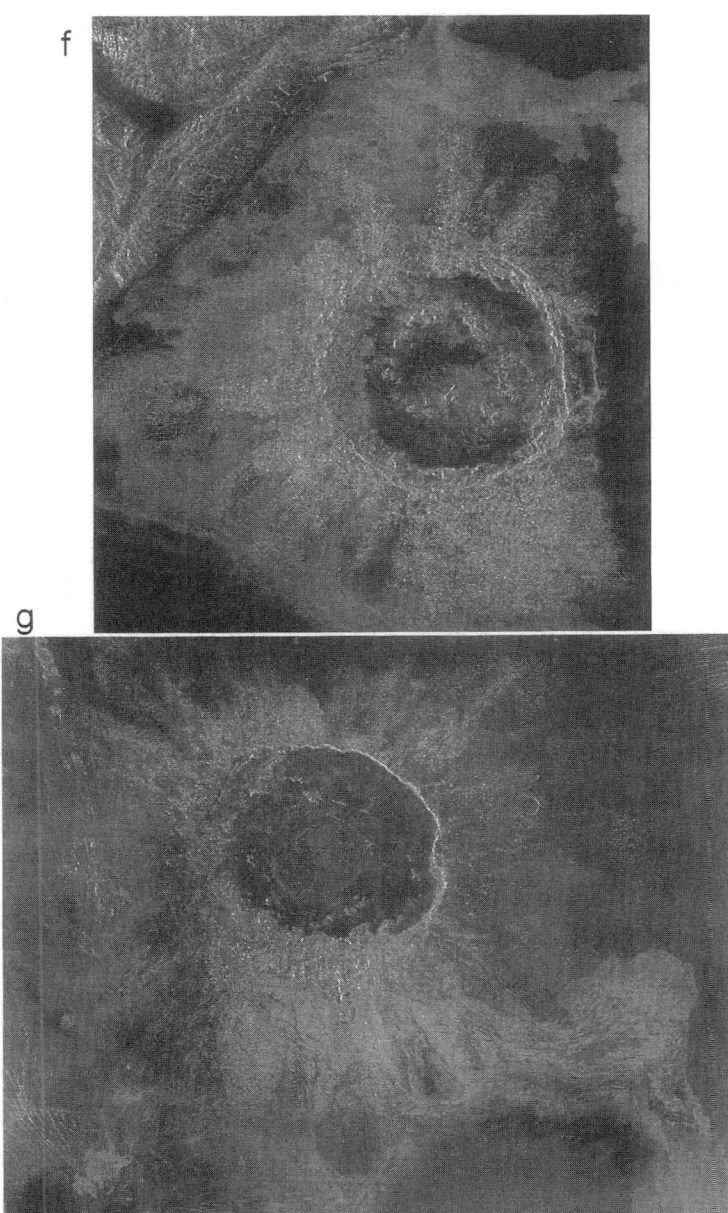

Figure 1. Size progression of typical impact craters. Diameters of craters are (a) 4.4 km (6.4°N, 272.2°E); (b) 13.6 km (60.0°N, 273.1°E, Margit); (c) 23.7 km (1.1°N, 284.3°E, Sikibu); (d) 37.5 km (17.4°N, 170.4°E); (e) 53.3 km (66.3°N, 125.7°E, Zhilova); (f) 98.1 km (51.9°N, 143.4°E, Cochran); and (g) 176 km (29.9°S, 204.2°E, Isabella).

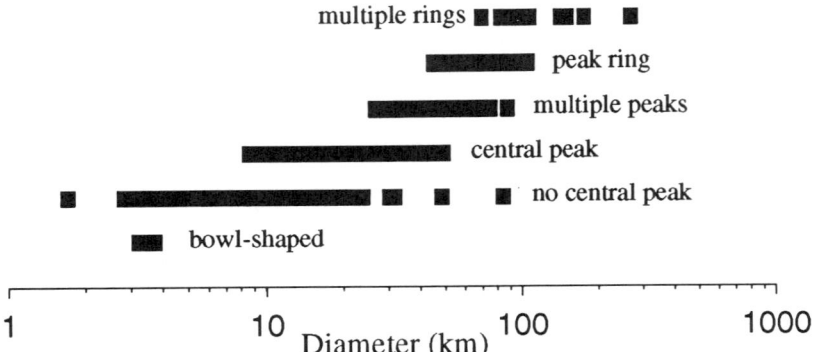

Figure 2. Morphologic type vs diameter for Venusian craters.

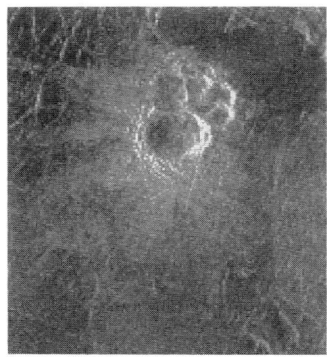

Figure 3. Clustered impact at 0°N, 142.6°E. Image is 35×40 km.

are not ideally suited for discriminating the youngest and freshest craters on Venus. However, distal impact units, i.e., those beyond the ejecta blanket such as distal ejecta and surface effects provide an expedient means of gauging the relative age and degradation state of impact craters on Venus.

As mentioned above and discussed later, ~6% ($n = 57$) of the total population of large craters on Venus retain dark (low radar backscatter) parabolic deposits of ejecta. The radar properties indicate that dark parabolas are thin deposits (several cm thick) composed of fine-grained particles (Campbell et al. 1992). Craters with these distal deposits seem to be randomly distributed about the planet and occur on all terrain types, suggesting that their presence is not due to some abnormality in target composition or structure. Because these dark parabolas do not appear to be extensively overlain by other geologic units, they seem to be among the youngest features on the surface of the planet (Campbell et al. 1992; cf. Izenberg et al. 1994). Craters with dark parabolas form a subset which appears to have a production slope (Campbell

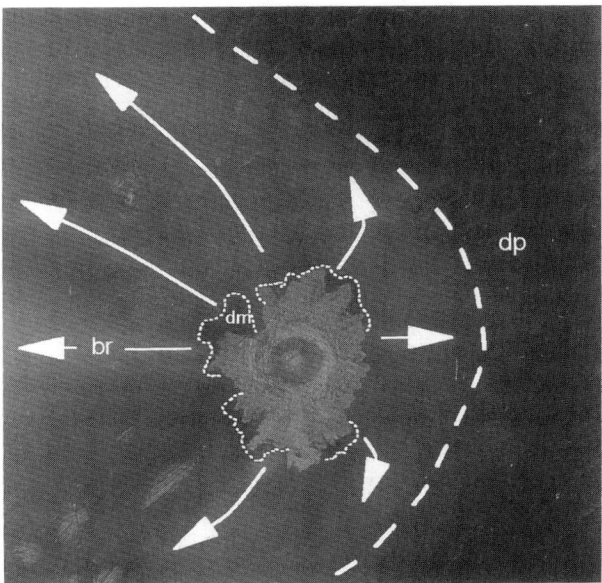

Figure 4. Sketch map of features around Adivar (8.9°N, 76.2°E, 29 km diameter).

et al. 1992; Herrick and Phillips 1994a). If all Venus craters on the production curve initially had parabolic deposits that were removed by a steady-state process, then these deposits persist over time scales of 10^7 to 10^8 yr, based on a global retention age of $\sim$500 Myr (Phillips et al. 1992; Schaber et al. 1992) indicating craters displaying parabolas are of this age or younger.

The 30-km diameter central peak crater Adivar (8.9°N, 76.2°E; Fig. 4) exhibits the most ornate assemblage of distal impact deposits observed on Venus. In addition to a well-developed dark parabola, it displays bright radial units, or streaks (br in Fig. 4) similar to the bright rays associated with the freshest craters on the Moon but not as extensive. The most prominent bright streak is directed due west of the crater; streaks extending in other directions appear to be shorter, with the eastward streaks being the shortest. This suggests that Adivar's bright streaks, like the parabolic deposits, are affected by the easterly zonal winds. (The 19-km diameter Cohn [33.3°S, 208.1°E] crater also displays distinct bright streaks.) If we make the same assumption as for the dark parabolas, namely that bright streak deposits are characteristic of all pristine craters but are removed by subsequent geologic processing, then Adivar would be about one million years old or less. The estimated crater production rate for Venus (Phillips et al. 1992) predicts only about one crater of this diameter or larger would be produced in this time frame. Consequently, Adivar is among the very youngest complex craters on Venus and is clearly the type example of a pristine Venusian impact crater.

Extending beyond the continuous ejecta blanket at Adivar crater, and many other craters on Venus, are radar dark deposits with sharp distal margins. These dark margin deposits have irregular or lobate shapes similar to the ejecta blanket and their outer contacts are sharp and cuspate or scalloped. Because of similarities in backscatter characteristics, these deposits are often combined with dark halo deposits (Phillips et al. 1991; Schaber et al. 1992),which themselves appear to be a varient of dark parabolas, or perhaps the surface effects of the impact shock wave (Ivanov et al. 1986; Phillips et al. 1991; Schaber et al. 1992; Zahnle 1992). However, dark margin deposits are considerably different in morphology from the diffusely bounded, roughly circular shaped dark halos, and these differences imply a different mode of formation. The lobate shapes and sharp margins suggest ground-hugging fluid or debris flow. Similarities between these deposits and the more proximal bright ejecta deposits suggests a genetic association. Schultz (1992) proposed that the lobate dark margin deposits could be extentions of the ejecta blanket due to flow separation created at the head of a basal debris flow. Craters with dark margin deposits and those with dark halos together constitute ~35% of the total crater population on Venus (Herrick and Phillips 1994a) suggesting that these deposits are easily removed with time. Unfortunately, no systematic study of Venusian craters has compiled information on the frequency of craters with just dark margin deposits and so it is not possible to quantify the average length of time these deposits last. Also, further study is required to determine whether dark margins can diffuse into dark haloes, or whether dark margins can be formed by embayment of dark haloes.

Crater Modification by Volcanism and Tectonic Deformation. With time, the pristine crater morphology is modified by subsequent geologic and atmospheric processes. Schaber et al. (1992) noted that ~38% of all craters on Venus have been modified to some extent by either tectonic deformation or volcanic flooding. Although the majority of these show evidence of faulting, tectonic deformation is not as important as volcanism is for planetary resurfacing. It is important to recognize that the ~4% of all craters recognized as volcanically modified represents only those that show conspicuous evidence of having been embayed by lavas.

Figure 5 shows the 37-km diameter crater Bashkirtseff (14.7°N, 194.1°E) that has been embayed by bright lavas from the north. It is the contrast in backscatter properties of the relatively young lava unit with the plains that clearly permits the superposition relationships between the volcanic and impact units to be discerned. To the north the lava encroached over the distal portions of the ejecta blanket but it ponded against the more high-standing proximal lobes and was deflected downslope and around the crater to the west and east. The result is that the volcanic embayments themselves mimic the shape of the ejecta blanket. It would be extremely difficult, if not impossible to recognize that this crater was volcanically modified if the post-crater lavas had a radar backscatter similar to the plains.

In fact, virtually every crater recognized to be volcanically modified

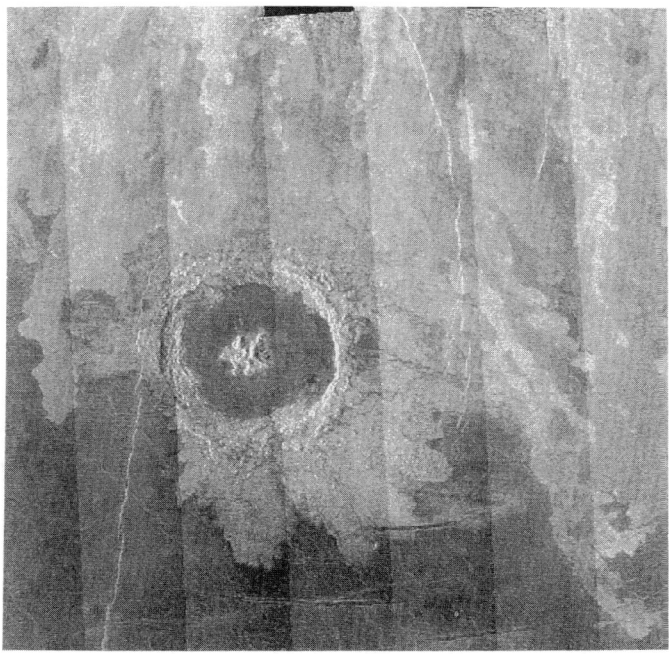

Figure 5. Crater Bashkirtseff (14.7°N, 194.1°E, 37-km diameter) showing post-impact lavas that embay the ejecta blanket without covering it. Had these lavas been radar-dark it would be difficult to discern that this crater has been embayed. Image is 120 km across.

shows post-crater units whose radar properties are distinct from those of the surrounding, older plains. But individual flow episodes cannot be discriminated for the vast majority of the Venusian plains units, either because of uniform original surface properties, or because their surface properties have been homogenized by prolonged weathering and aeolian activity (Arvidson et al. 1991,1992). There are many craters on Venus that have anomalously small deposits of continuous ejecta. We suggest that a significant number of these may be volcanically modified craters that have gone unrecognized. The craters that have been recognized as having been volcanically embayed to date, therefore, may represent only the youngest or most extensively modified cases.

Bright Crater Floors. The radar backscatter properties of crater floors vary greatly from extremely high (bright and blocky) to extremely low (dark and smooth) (Phillips et al. 1991). Grieve and Cintala (1995) have argued that dark floor deposits simply reflect the fact that more melt is produced in large Venusian impact events than in lunar impacts of comparable energies because of the enhanced gravity and surface temperature of Venus. They suggest that craters with bright floors are reflecting either higher viscosities or cooling rates possibly due to impacts into higher elevation terrains, or less melt production due to variations in impact conditions. However, none of the freshest 10% of the large craters on Venus, i.e., those retaining dark parabolas, have distinctly dark floor deposits. Instead, the floors of these craters are characteristically bright and blocky (Campbell et al. 1992; Izenberg et al. 1994). In fact, craters with bright floor deposits comprise ~20% of the craters large enough to have parabolas, and of these about half display dark parabolas (Herrick and Phillips 1994a). This suggests that bright crater floors are also a reliable indicator of crater youth, but that this trait persists over longer time scales than the dark parabolas.

The blocky, radar bright nature of pristine craters on Venus is consistent with greater impact melt production compared to craters on the Moon. Even massively thick melt sheets would be mantled by fall back ejecta and deformed by cooling features, so there is no *a priori* reason why any pristine crater floors should be smooth and radar dark. The largest crater on Venus, the 270-km Mead basin, which because of its great size should contain the thickest melt sheet, has a radar-bright interior.

Dark Floors as Evidence of Volcanic Modification. Because dark floor deposits are not observed within the freshest craters on Venus, it seems most reasonable to conclude that they are not produced during or even shortly after the time of impact. Therefore models of dark floor origins calling upon impact melting or impact-induced volcanism do not seem plausible. Campbell et al. (1992) noted that some process had to be darkening the crater floors either by infilling or by chemical weathering. Sharpton (1994) calculated floor depths for 21 bright floor craters with dark parabolas and 33 craters chosen because their floors were distinctly, completely, and uniformly dark. Although there is considerable scatter in the depth/diameter estimates, particularly within the dark floor crater subset, dark floor craters are on average 150 to 400 m shallower than bright floor craters. Consequently, dark floors seem to represent smooth deposits, probably lavas, which mantle and partially fill the original crater floor rather than chemical weathering which would not change floor depth appreciably. A volcanic origin for at least some dark floor deposits is supported by Aglaonice crater (26.5°S, 340.0°E), which exhibits a very dark floor and an ejecta blanket that appears to be embayed by lavas which originate from an east-west trending system of fractures just north of the crater (Sharpton 1994). These observations suggest that perhaps most large craters have been modified to some extent by post-impact lavas.

In summary, although only ~4% of the crater population shows conspicuous evidence of volcanic modification from sources outside the crater, it is likely that this percentage represents an extreme lower limit to the actual number of embayed craters.

III. MORPHOMETRY OF CRATER INTERIORS

The appearance of an impact crater, particularly a complex impact crater, is dependent on many target properties such as surface gravity, target density, the presence or absence of an atmosphere, and target composition. Unfortunately, in common geologic materials only simple craters can be created in the laboratory or even with large explosions. Only quantitative description, or morphometry, of impact craters on the solid bodies of the solar system can provide data on how various target properties affect complex crater formation. The recent addition of crater morphometric data for Venus (Herrick and Phillips 1994a; Sharpton 1994) and the icy satellites (Schenk 1989,1991) greatly extends the range of surface gravities and target compositions for which data exist, and Venus allows study of how a dense atmosphere affects the cratering process. Table I summarizes and updates morphometric data for Venusian craters presented in Herrick and Phillips (1994a) and Sharpton (1994). This section concentrates on the implications of measurements made on the crater interiors, and the next section focuses on ejecta deposits. The morphometry of Venusian craters must be compared in the context of similar measurements for the other planets. Figure 6 compares functions for Venus with other planets and Fig. 7 compares onset diameters. Within the error bars for the data, it is difficult to make a case that any target or impactor property has an effect on either the central peak diameter or the peak ring diameter of a crater relative to the crater diameter. Venusian peak rings, either as ring massifs or asymmetric scarps, fall close to the multiple of $\sqrt{2}$ spacing suggested for the other bodies in the solar system (see, e.g., Pike and Spudis 1987). It should be noted that although the functions listed in Table I do not show the dramatically increasing ratio of ring diameter vs crater diameter ratio as do Alexopoulos and McKinnon (1994), this difference is largely a question of semantics. Alexopoulos and McKinnon (1994) "connected the dots" of the massifs for many craters classified by Herrick and Phillips (1994a) as multiple-peak craters; measurements on craters that both papers classified as peak-ring craters are very similar. The presentation in Alexopoulos and McKinnon (1994) argues strongly that central peaks and peak-ring massifs form from fundamentally the same physical process; i.e., there is a gradual transition from central peaks to peak-rings with increasing diameter.

In contrast, the onset diameter of complex crater features and the vertical dimensions of crater interiors show the effects of target strength and surface gravity. Comparison of crater depths and the onset diameters for central peaks and peak rings for the Moon, Mercury, and Venus, indicate that increasing planetary gravity decreases the onset diameters and crater depths on planets with otherwise similar target properties. Similarly, comparisons of the same measurements on Mercury and Mars or the Moon and Ganymede indicate that decreasing target strength has a similar effect to increasing gravity. The presence of water ice is presumed to weaken the crusts of Mars and Ganymede. Unfortunately, the complex nature of the cratering process makes it unclear exactly what measure of crustal strength is actually being reflected by the observations.

Recently, we have applied stereo photogrammetric techniques to left-

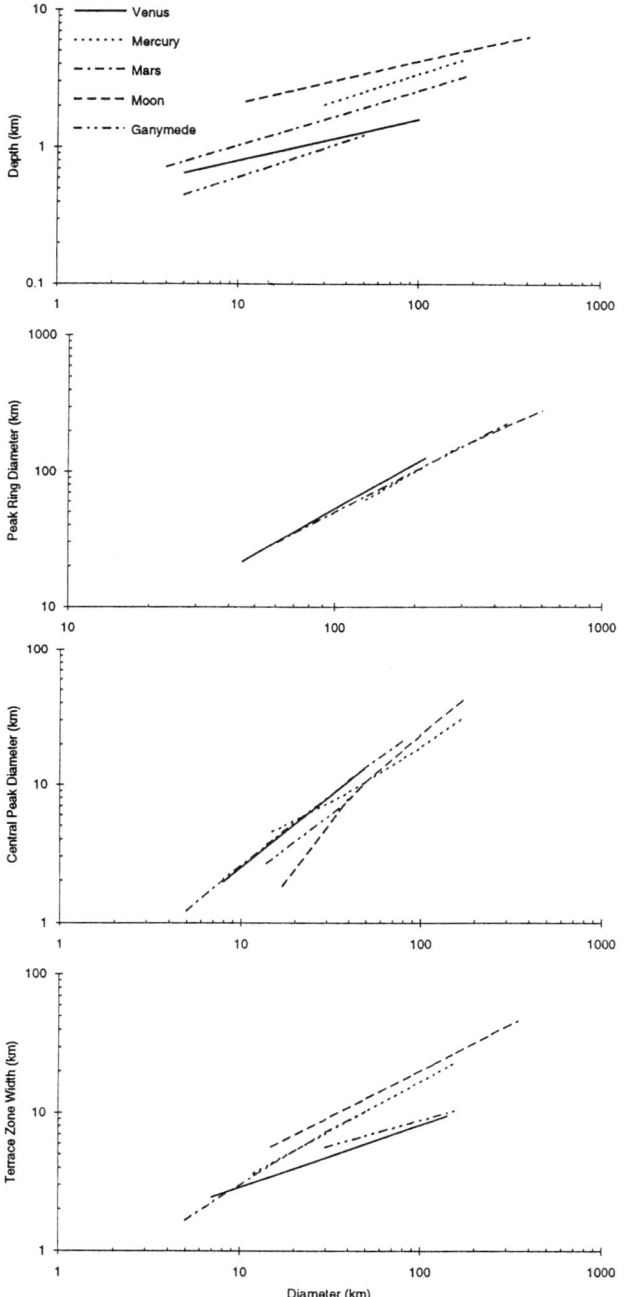

Figure 6. Comparison of several properties vs crater diameter on Venus and the other planets. Data are from Herrick and Phillips (1994a), Hale and Head (1979,1980, 1981), Pike and Spudis (1987), Pike (1977,1980a,1988), Schenk (1991), Melosh (1989), Malin and Dzurisin (1978) and Pike and Davis (1984); see Herrick and Phillips (1994a) for detailed listing of the functions.

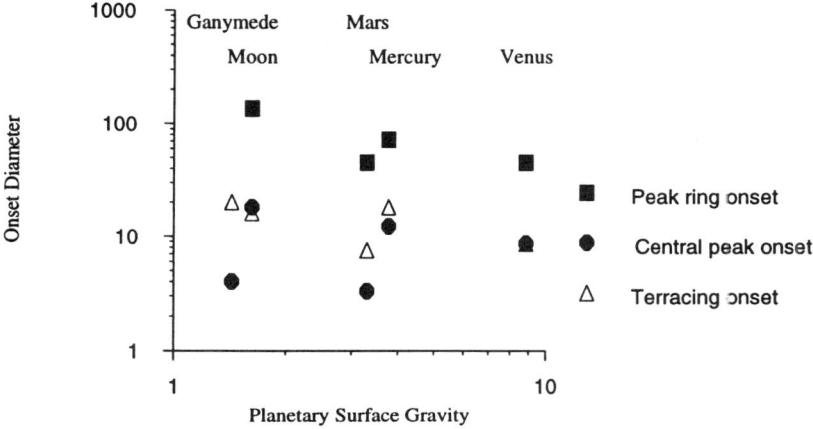

Figure 7. Comparison of crater diameter (km) vs surface gravity (m s^{-2}) for the onsets of terraces, central peaks, and peak rings. Data are from Herrick and Phillips (1994a), Wood and Andersson (1978), Gault et al. (1975), Pike (1980a), Pike and Spudis (1987), Wood (1980), Schenk (1991), Pike (1988), and Smith and Hartnell (1978); see Herrick and Phillips (1994a) for detailed listing of the functions. (Note that the gravity of Mars is intentionally plotted at a slightly lower value so the data points for Mars and Mercury can be distinguished.)

left looking stereo radar imagery to obtain high-resolution topography over several impact craters (Sharpton et al. 1994). For rim-floor depths these results generally coincide with the results from those using the floor-offset method in Sharpton (1994). Although the influence of target gravity is clearly demonstrated, comparison of depths with those on Mercury and the Moon do not follow a simple $1/g$ relationship: both Mercury and Venus have deeper craters than would be expected (Pike 1988; Sharpton 1994). Although we have a small data set and the scattering is large, it appears that this discrepancy cannot be attributed to unusually high rims on the Venusian craters, as the Venusian rim heights average 35 to 40% of the rim-floor depths, similar to the Moon and Mercury (Pike 1977, 1988). Central structure heights average about 300 m above the floor, and our current limited data suggest that there may be a slight decrease with increasing diameter. Although this is lower than the other terrestrial planets, for central peak craters it is actually substantially higher relative to the surrounding terrain than on the other terrestrial planets (Malin and Dzurisin 1978; Pike 1977; Pike and Davis 1984). For lunar craters the central peak is typically several hundred meters below the surrounding terrain, but for Venusian craters the central peak often rises to near the level of the surrounding terrain and occasionally exceeds it. Central peak craters on some of the smaller icy satellites are the only other bodies with proportionally high central peaks (Schenk 1991). Central peak heights are not the same for the different solar system bodies, but no clear trend with target properties is apparent.

TABLE I
Best-Fit Functions for Venusian Craters[a]

Measurement	Function[b] (km)	rms Log Error (log km)[c]	# Used	Size Range (km)[d]
Central peak diameter	$0.22D^{1.05}$	0.11	225	8–49
Peak ring diameter[e]	$0.13D^{1.33}$	0.08	26	45–269
inner ring	$0.32D^{1.11}$	0.01	24	45–176
outer ring	$0.42D^{1.11}$	0.02	6	70–269
Depth	$0.48D^{0.16}$			
fresh	$0.40D^{0.30}$			
Hummocky ejecta max extent	$0.52D^{0.82}$	0.14	355	3–270
Continuous ejecta max extent (all)	$3.09D^{0.60}$	0.15	529	3–270
$D>20$ km[f]	$16.9+(D-20)^{1.02}$	0.14		
$D<20$ km	$5.38D^{0.38}$			
Continuous ejecta average extent (all)	$1.57D^{0.61}$	0.18	201	3–67
$D>20$ km[f]	$8.77+(D-20)^{1.02}$	0.16		
$D<20$ km	$2.19D^{0.46}$			
Secondaries from rim	$4.73D^{0.53}$	0.19	150	6–270
Wall width	$1.02D^{0.45}$	0.11	279	7–142
Flow distance from rim	$2.93D^{0.90}$	0.20	176	4–176
Dark halo diameter	[g]$42+2.4D$	0.19	258	3–270
Peak onset diameter[h]	8.6	0.09		
Peak ring onset diameter[h]	44.9	0.02		
Terrace onset diameter[h]	8.6	0.05		

[a] Updated from Herrick and Phillips (1994a).
[b] Power-law function obtained from best-fit line (minimum rms error) on log–log plot.
[c] Root mean square difference between observations and best-fit function on log–log plot.
[d] Diameter range of observations used to determine best-fit functions.
[e] First line reflects equivalent diameter from area enclosed by central structure. The "inner" function is the diameter of the inner ring for a crater with two rings or the only ring for a single crater. The "outer" function is for two-ringed structures only (the rim is not counted as a ring). The "inner" and "outer" functions are calculated as in Pike and Spudis (1987) by fitting a circle to the tops of the ring massifs.
[f] Errors for combined data set of craters above and below 20 km diameter.
[g] Linear best fit.
[h] Error estimated by subtracting logarithm of diameters of smallest and fourth smallest craters with stated feature.

Similarly, a gravity or strength trend for the onset of terracing or terrace-zone width is not apparent. However, if terracing is a late-stage phenomenon that occurs after the final crater depth is determined (Pike 1980b), then the data support enhancement of terracing (lower onset diameter, wider terrace zones) by increased gravity and decreased crustal strength.

Whereas general comparisons provide insight into the cratering process, it is desirable to use the data to actually test physical models of the cratering process. The recent addition of data from Venus and the icy satellites now makes the inverse approach a feasible method for determining the factors controlling interplanetary differences in crater morphometry. Standard inversion techniques provide a structured framework for comparing models, incorporating data with errors, and determining ranges of acceptable parameters. As an example, here we consider two inversions where all interplanetary differences in crater morphometry are directly controlled by the ratio of crustal strength c to crustal density ρ multiplied by surface gravity g. This dependence can be approximated by an equation of the form

$$O_{ji} = A_i \left(\frac{c_j}{\rho_j g_j}\right)^b \tag{1}$$

where O_{ji} is an observation of type i for planet j, and A_i and b_i are constants. In log–log space

$$(\log O)_{ji} = (\log A)_i + b_i (\log c)_j - b_i (\log \rho g)_j. \tag{2}$$

TABLE II
Morphometry Inversion Results

Slopes in Varying b Model		Relative Crustal Strength		
			$b = 1$	b varies
Depth	0.76	Venus	21.9	20.4
Depth/diameter transition	1.07	Mercury	14.8	13.1
Central peak onset	1.17	Mars	10.7	8.9
Peak ring onset	1.09	Moon	10.0	10.0
Terrace onset	0.63	Ganymede	1.1	0.9
rms Error (log space)		Rhea	0.6	0.6
		Dione	0.5	0.4
Data	0.103			
$b = 1$	0.139	F test:	$F = 0.8$	
b varies	0.128			

Assuming ρ and g are known for each planet, then a set of nonlinear equations exists where the $(\log A)$s, $(\log c)$s, and bs are unknowns that can be solved for (only relative strengths can be determined). In the example presented here we have performed a damped, nonlinear, iterative, least-squares inversion (Menke 1989). Figure 8 and Table II show the measurements used, the results, and comparison of errors for two models using the same data. For the first model the b values were held constant at 1.0, making the problem an overconstrained linear problem with 29 equations and 12 unknowns. This model is equivalent to assuming that all complex crater features are controlled by constant thresholds in hydrostatic pressure vs crustal strength in a transient cavity of invariant shape. Allowing b to vary added 5 more unknowns and made the inversion nonlinear.

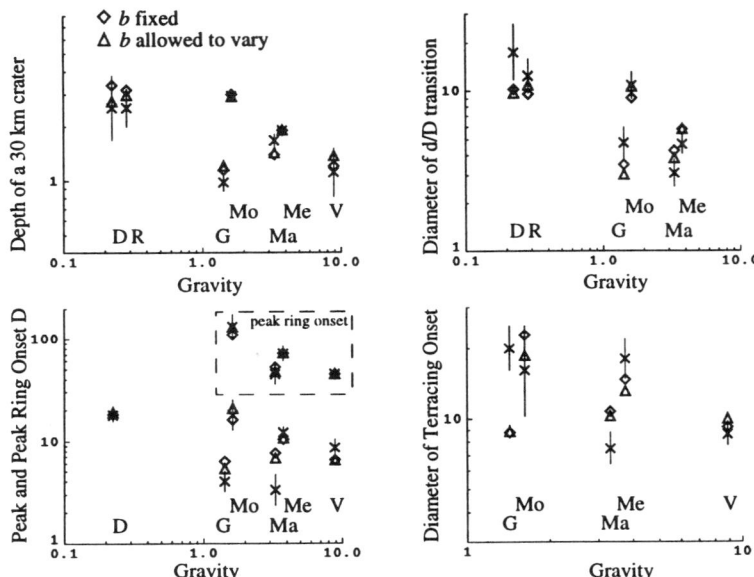

Figure 8. Comparison of inversion results with observed data plotted in terms of planetary gravity vs crater diameter of a particular feature. The "x" symbols represent the observations and the vertical lines are estimated 1 σ error bars. Open diamonds and triangles represent model results for linear and nonlinear models, respectively. Data from Dione (D), Rhea (R), Ganymede (G), the Moon (Mo), Mars (Ma), Mercury (Me), and Venus(V) were used. See figure captions for Figs. 6 and 7 for references for the data. (Note that the gravity of Mars is intentionally plotted at a slightly lower value so the data points for Mars and Mercury can be distinguished.)

To first order, reasonable solutions were obtained with the linear inversion, suggesting that the factor $c/(\rho g)$ has an important effect on interplanetary variations of crater morphometry. The results are fairly robust in showing an order of magnitude difference in strength between the rocky planets and the icy satellites. However, at least some of the measurements are influenced by factors other than $c/(\rho g)$, and neither model was able to fit the data overall with an error less than the rms error of the data. For example, no satisfactory value of c could be found to make the Mars data match the trends from other planets. Neither model fit the terracing onset data well, suggesting these data are controlled by at least one other factor besides $c/(\rho g)$, as suggested earlier. An F test (Menke 1989) comparing the two models indicates there is not a significant reduction in error when b is allowed to vary.

Overall, the planetary data seem to support a scenario advocated by Pike (1980b) for complex crater formation. Gravity and crustal strength are the dominant factors in determining crater depth, central peak onset, and peak ring onset. Terrace formation appears to be controlled by strength and gravity, but only after the crater has rebounded to nearly its final depth. If rim wall failure occurs after the depth is largely determined, then slumping and large-scale horizontal movement of material are not significant for determining crater depths or creating central structures. This interpretation, combined with planet-independent central peak and peak ring dimensions with respect to diameter, suggests that the shape of large craters is created primarily by vertical movement of material, with horizontal dimensions controlled by diameter of the initial transient cavity.

IV. EJECTA DEPOSITS

A. General Description

As discussed qualitatively above, Venusian impact craters have unique ejecta blankets whose nature changes with increasing diameter. Herrick and Phillips (1994a) showed that the maximum extent of the ejecta blanket vs crater diameter was not well fit by a single exponential function, but instead showed a marked change in slope at around 20 km (functions listed in Table I). For a randomly selected subset of 200 craters we have measured the planimetric area enclosed by the ejecta blanket, calculated an equivalent radius by treating the area as if it were circular, and subtracted the crater radius to calculate an estimate of the average extent of the ejecta vs crater diameter, and these results are plotted in Fig. 9. These data are also better fit by a two-part function, but the scatter in the data makes the choice of 20 km for the break point more arbitrary than for the maximum-extent data. The functions for average extent are nearly parallel to those for maximum extent, and to first order the average extent of the ejecta is about half the maximum extent.

Functions for the average extent have been calculated for the other terrestrial planets and are plotted with the Venusian data in Fig. 9. It should be noted that for the other planets the functions were calculated by estimating the average extent rather than calculating the area covered, and these estimations were made from photographs rather than radar images. Comparison of ejecta blankets for Mercury and the Moon show the expected trend that for ejecta blankets ballistically emplaced in a vacuum the extent of the ejecta is dependent on gravity, but both arms of the Venusian curve are above this trend (Phillips et al. 1991). However, the Venusian curve is well below that of Martian rampart craters which have ejecta blankets that appear to have been emplaced as mudflows (Carr et al. 1977). Herrick and Phillips (1994a) suggested that for craters <20 km in diameter the ejecta cloud could be treated as a block sliding outward from the rim, and craters >20 km in diameter could be

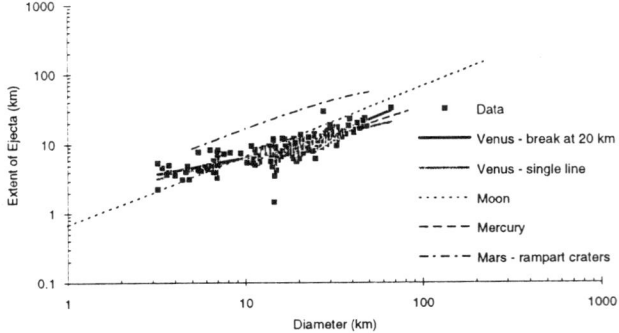

Figure 9. Comparison of extent of the continuous ejecta blanket for Venusian craters with functions for other terrestrial planets. Functions for other planets from Pike (1977), Gault et al. (1975), and Carr et al. (1977); see Herrick and Phillips (1994a) for detailed listing of the functions.

treated with approximations for "sturzstroms," large avalanches with low coefficients of friction. They calculated an expected run-out distance for the two mechanisms proportional to $D^{0.5}$ and $D^{0.9}$, respectively, in good agreement with the observations.

While the majority of the ejecta appears to be emplaced nonballistically, secondary craters do exist around Venusian craters. Table I shows a best-fit function for the maximum extent of secondaries. For craters >100 km in diameter, secondary craters exist in all directions around the crater and extend beyond the continuous ejecta. For craters 50 to 100 km in diameter secondary craters appear to exist within the continuous ejecta (see Fig. 1). Roughly 1/4 of the craters <100 km in diameter have isolated secondaries that extend beyond the continuous ejecta, and for craters <50 km in diameter they usually occur only as one or two rays of secondaries. For the larger craters where secondaries are ubiquitous they have traveled much farther than is possible in ambient Venusian atmosphere. Herrick and Phillips (1994a) suggested that for larger craters the largest ejecta blocks were emplaced ballistically but traveled part of the distance in rarefied atmosphere created by the outward blowing ejecta cloud, a mechanism suggested by Schultz and Gault (1979). For smaller craters Herrick and Phillips (1994a) suggested that perhaps an exotic process like spallation (Melosh 1985) may occasionally eject large, high-speed ejecta blocks to create the isolated rays of secondaries observed.

Beyond the main ejecta blankets are dark haloes/margins and parabolic features. To first order dark halo diameter increases linearly with crater diameter (Table I). Often dark haloes are surrounded by radar-bright areas (Phillips et al. 1991). While details of the process are model dependent, a general consensus is that dark haloes arise from interaction of the surface with the atmospheric shock wave produced by the downgoing meteoroid (Phillips et al. 1991; Zahnle 1992; Ivanov et al. 1992). However, fallout from fine-grained debris has been suggested as a cause of some dark haloes, and fine-grained ejecta flows have been suggested as causes for radar-dark margins around ejecta blankets (Schultz 1992).

The largest impact-related features are the radar-dark west-facing parabolas. Campbell et al. (1992) identified 57 of these features; their size ranges from several hundred to 2000 km in diameter and loosely correlates with crater diameter. Eight of these features are easily identifiable only in the emissivity data. Only four of the parabolas visible in the normal radar imagery are associated with craters <15 km in diameter, but five of the emissivity-only features are with craters <15 km (Campbell et al. 1992). The favored explanation of the parabolic features is that high-energy impacts inject small particles into the upper atmosphere where they are transported to the west by E–W zonal winds (Campbell et al. 1992; Vervack and Melosh 1992).

B. Ejecta Flows

As mentioned previously, many large craters have long flows reaching many crater radii from the continuous ejecta (Fig. 10). Although they greatly resemble lava flows in morphology, their intimate relationship to impact craters, with many characteristics nearly indistinguishable from the near-field crater ejecta, argues strongly for their origin through processes associated with impacts. These features have been called outflows, run-out flows, run-out ejecta, and fluidized ejecta, among other terms. We use the term *ejecta flows* here because it incorporates references both to the ejected materials and to the flow-like forms they create.

Phillips et al. (1991) first described these landforms as "...nonradial, flow-like ejecta indicative of flow of a low-viscosity material..." and noted several common morphologic attributes, including the apparent fluidity during emplacement (as indicated by the ability to flow around obstacles) and radar-rough lateral and distal margins (suggestive of rough surfaces). Many investigators of these flows have documented the strong correlation of ejecta flows with impact angle: the more oblique the impact, the greater the probability that one or more of the forms of ejecta flows will have been created. The location of ejecta flows circumferrentially around a crater also shows a high correlation with the apparent approach azimuth of the projectile; the preponderance of ejecta flows initiate in the downrange direction.

At least three varieties of ejecta flows have been described (Chadwick and Schaber 1993; Schultz 1992); proximal ejecta flows and flow fields (also called transitional ejecta by Johnson and Baker 1994) are mostly radar bright (implying rough surfaces), rough in local relief, broader than they are in length, with somewhat indistinct to well-defined lobate margins (Fig. 10a). Medial ejecta flows are areas of intermediate radar brightness with anastomosing channels, radar-bright margins, low relief, and digitate (sometimes referred to as "feathery") distal margins (Fig. 10b). Distal ejecta flows are radar bright, long, narrow, and lobate, with relatively low relief except near their margins (Fig. 10c). Chadwick and Schaber (1993) group the proximal and medial ejecta together, Johnson and Baker (1994) group the medial and distal ejecta

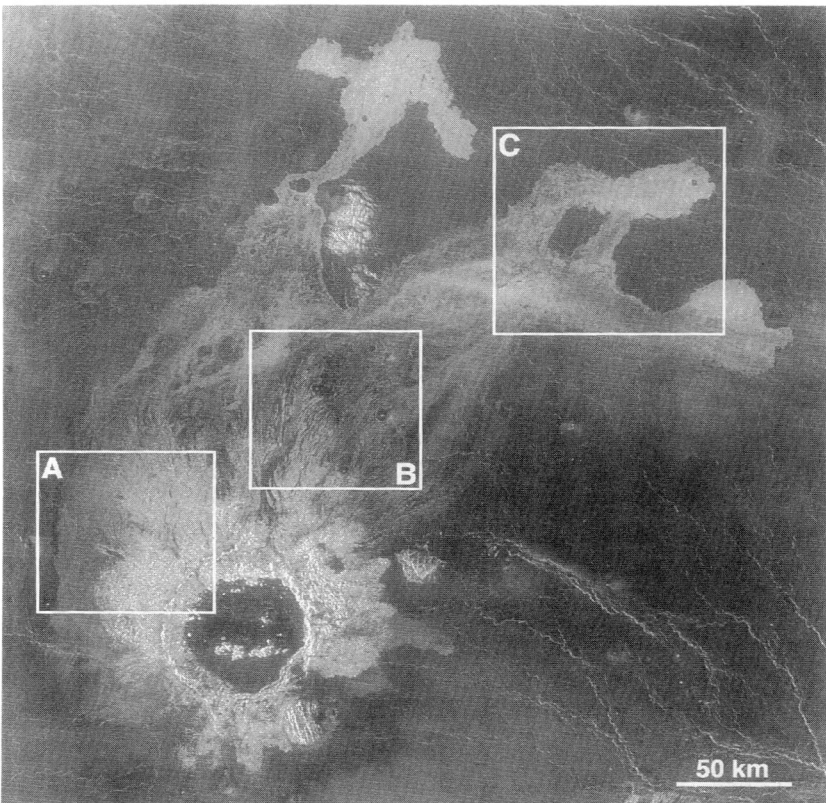

together, and Schultz (1992) considers all three forms transitional. In many instances, medial ejecta flow morphology grades longitudinally into the distal form, with broad areas of anastomosing channels transitioning to a single or small number of distributary, distal lobes. An observation noted by most investigators is that proximal ejecta overlies the distal forms (Fig. 11).

Two models for the origin of these ejecta flows were originally proposed by Phillips et al. (1991). One associated the flows with the fine-grained component of hot, turbulent, density clouds of impact melt and vapor generated by impacts, believed to behave much like volcanic pyroclastic flows. The second, simpler model, suggested the flows were impact melt, behaving like volcanic lava flows. Both of these models remain the keystones of contemporary interpretation. Much of the work of subsequent investigators has focused on one or more aspects of either or both of these models.

Schultz (1992) presents the most comprehensive interpretation of the formation of ejecta flows, attempting to account for a large range of morphologic, topographic, and textural relationships. His view suggests that flow morphology reflects primarily the fluid mechanical properties of the materials during, and at the cessation of, motion. This view is supported in part by the later anal-

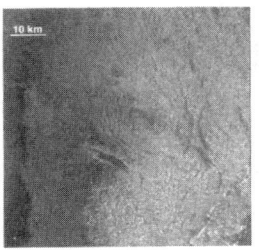
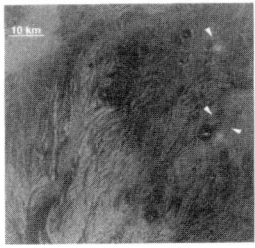
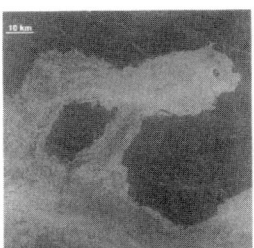

Figure 10. Crater Seymour (18.2°N, 326.5°E, 66 km diameter) showing (A) proximal, (B) medial, and (C) distal landforms of ejecta flows. (a) Proximal ejecta flow, showing characteristic high radar brightness (rough surface), hummocky relief, obscuration of underlying topography (implies moderate thickness), and subtle evidence of both radial and transverse forms. Grades to medial form. (b) Medial ejecta flow, showing characteristic channels and variagated radar brightness. Note depositional tails behind low-relief shield volcanoes (white arrows). Underlying relief is occassionally visible, suggesting variably thin deposits and relatively little erosion. Grades to distal form. (c) Distal ejecta flow, showing characteristic moderate to high radar brightness (rough surface), lobate perimeter, obscuration of underlying topography (implies moderate thickness), and marginal brightness variations suggestive of levees.

yses by Asimow and Wood (1992) and Johnson and Baker (1994), although they differ in many specific details. In Schultz's model, high-energy, inflated, turbulent motion of impact melt/vapor mixtures produces the medial flows, which collapse and condense to form laminar flows distally. Schultz argues that the impact melt/vapor ratio, and hence the nature of motion, reflects both compositional aspects of the impacting body (volatile-rich impactors produce

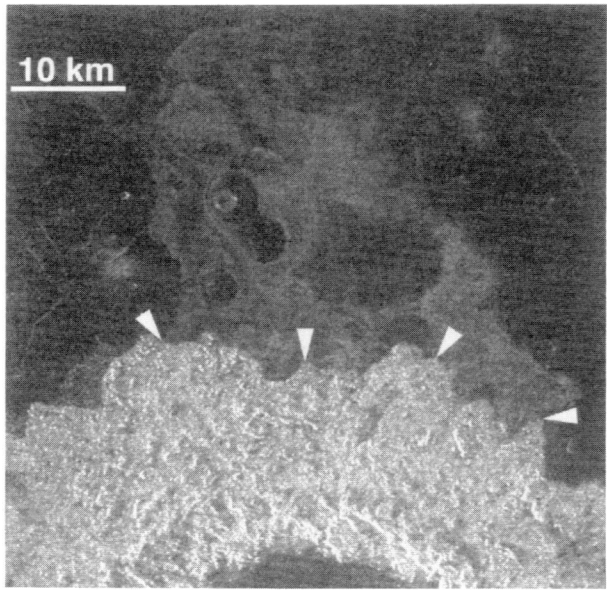

Figure 11. Crater Saskia (37 km in diameter), located at 28.6°S, 337.1°E, showing proximal ejecta material overlying medial/distal ejecta flow (white arrows).

turbulent flows, silicate- and metal-rich impactors form laminar flows) as well as the impact angle (vertical angles lead to laminar flow forms). The composition of the impacting body plays an important role, especially in oblique impacts, because of a "ricochet" effect wherein the projectile material interacts with the atmosphere prior to contacting the surface and is contained and displaced downrange. This "early-stage" process permits the fluidized materials to be emplaced prior to arrival of the ejecta from the crater proper.

Chadwick and Schaber (1993) present a substantially different view, in which the proximal materials are deposited from turbulent mixtures of impact melt and other primary ejecta, and the distal materials, formed of impact melt that has segregated from the proximal deposits, are deposited in a manner analogous to lava flows. In their model, the ejecta flows superpose the more traditional forms of crater ejecta (e.g., secondary craters and blocks, Fig. 12) and are hence "late-stage phenomena." They also believe that the channels seen in the proximal flows are "low energy" forms that would be destroyed by emplacement from inflated, turbulent flows. Chadwick and Schaber enunciate four factors that contribute to the formation of ejecta flows: the elevated temperature on Venus (1) makes melting surface materials easier and (2) allows melted ejecta to remain molten longer, permitting it to segregate and coalesce into flows, while projectile factors like (3) kinetic energy and (4) impact incidence angle dictate the melt/solid ratio and volume of melt to fall

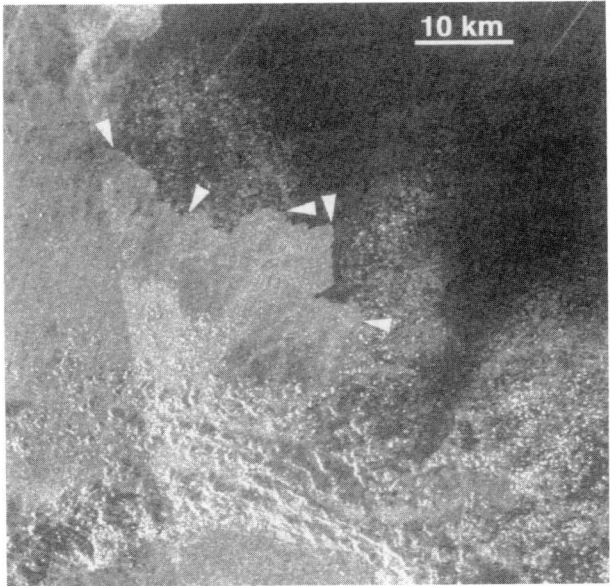

Figure 12. Crater Stuart (67 km in diameter), located at 30.8°S, 20.2°E, showing medial/distal ejecta flow overlying near-field crater ejecta (white arrows).

outside the transient crater, respectively.

From their analysis of morphology supplemented by quantitative measurements of backscatter, altimetry, and radiometry, Johnson and Baker (1994) conclude that the surface roughness of proximal deposits increases downflow, suggesting that they are formed by surface flow of impact melt, while distal deposits decrease in roughness downflow, which they argue more closely mimics gravity sediment flows (e.g., debris or volcanic pyroclastic flows). They note, however, that there are some occasions where the reverse may be the case, and where longitudinal variations appear to be result from changes in material properties (e.g., dielectric constant) rather than emplacement process (e.g., surface roughness). Given these variations, Johnson and Baker conclude that additional factors are at play, possibly including the effects of (1) topography on flow velocity and rheology, and (2) multiple flow paths and the effects of materials following these paths at contributary and distributary locations.

Given the diversity of forms, and the evidence for potentially diametrically reversed stratigraphic relationships, Asimow and Wood (1992) conclude that there must be at least two different processes forming ejecta flows. One, a late-stage process, follows ejecta deposition and modifies the continuous ejecta blanket as it moves distally away from the crater. They infer that the process involves segregation of impact melt from the ejecta and its subsequent flow, although they do not preclude ejecta-cloud debris flows or ballistically induced mass movements from participating in formation of the ejecta flows. The second, early-stage process, involves a vapor/melt/solid mixture of projectile and countryrock, moving as a relatively ground-hugging flow akin to a pyroclastic flow or dry avalanche. Factors favoring the first process are the higher production of impact melt on Venus owing to higher temperature and greater impact velocities, and the greater efficiency of melt segregation owing to higher temperature and pressure and larger particle size (lower surface area to volume ratios). Factors favoring the second process include the presence of a thick atmosphere to confine the impact vapor near the surface and to participate in the fluidization process downflow.

As one can see above, there are obvious points of agreement and disagreement between investigators as to flow formation mechanisms. Most of the disagreement appears to derive from simplifications each has made in order to formulate a tractable hypothesis. Some observations appear contradictory but may only reflect the emphasis of the investigators. For example, the photographic evidence clearly supports contentions that ejecta flows come both before (Fig. 11) and after (Fig. 12) emplacement of the continuous ejecta blanket, although both relationships do not appear to occur together at a single crater. Other observations are more debatable. Areas of greatest channeling in the medial ejecta flows show little evidence of full-fledged drainage integration, and it is difficult to determine the nature of radar brightness variations that define the channels. Contentions that the channels in the most channeled areas (e.g., Fig. 10b) are low-lying, smooth-floored, bounded by marginal levees, and forming dendritic patterns can be challenged in almost every instance, as the images would equally support the contention that the pattern is truly anastomosing, and formed of bars and scour of preferentially deposited coarse debris.

There is, however, clearly a consensus view in interpreting the general framework for the origin of ejecta flows. All models invoke at least two flow mechanisms, involving to a lesser or greater extent impact melt and vapor admixed with unmelted country rock fragments. They all acknowledge the importance of oblique impact angle on the generation of the melt and vapor phases, as well as the importance of kinetic energy in establishing the ratio of vapor, melt, and solid ejecta. They also argue as a group that the surface temperature and atmospheric pressure play substantive roles in creating conditions that promote melt and flow. Their differences, as noted above, lie in areas of specific timing (before or after emplacement of the near-crater ballistic ejecta) and sequence of phenomena (collapsing debris clouds creating surface flows, failing surface flows generating debris clouds, or ground-hugging mixtures of melt and rock fragments segregating with time/distance into flows of melt and lag deposits of rock). These differences are mostly minor when compared to the general agreement.

V. ASSORTED IMPACT ODDITIES

A. Phenomena Due to Meteoroid Disruption

There are several observable effects of the dense Venusian atmosphere on incoming meteoroids. Well before imaging data was available for Venus, Tauber and Kirk (1976) predicted that the minimum crater diameter on Venus would be 150 to 300 m, a value reasonably close to the observed value of 1.7 km. Small meteoroids are decelerated enough that they do not strike the surface with enough velocity to create an impact crater; and hence no small craters form on the surface. In some cases, the deceleration may be rapid enough that the incoming meteoroid can only release the kinetic energy through catastrophic disruption (the meteoroid explodes). It has been suggested that in certain cases this disruption has occurred close enough to the ground surface that a visible scar known as a *dark splotch* is formed (Fig. 13; Phillips et al. 1991). As with dark haloes the favored mechanism for forming a dark splotch is pulverization of the surface by an atmospheric shock wave (Phillips et al. 1991; Ivanov et al. 1992; Zahnle 1992). One group has identified ~400 splotches that they claim have formed as a result of meteoroid explosion (Schaber et al. 1992; Kirk and Chadwick 1994; Strom et al. 1994), and some researchers have estimated the meteoroid size responsible for these features to be a few kilometers in diameter (Zahnle 1992; Schultz 1992). It is important to consider, however, that the knowledge base for studying these features is far less than exists for impact craters, as the only well-studied non-Venusian example of a meteoroid exploding in the atmosphere and affecting the surface is the Tunguska event on Earth (Turco et al. 1982). Consequently, it is much more difficult to rule out nonimpact origin for at least some of the splotches, and estimates of the meteoroid size that produces a splotch must be based entirely on numerical models.

Meteoroids enter the atmosphere with an incremental angular probability distribution of $2 \sin \theta \cos \theta d\theta$, where θ is entry angle (Shoemaker 1962). Thus, the path length that meteoroids must travel through the atmosphere before striking the surface varies greatly, and consequently the minimum meteoroid size that can form a crater varies with incoming angle. For primarily this reason, the size-frequency distribution of craters vs diameter is not a straight-line exponential function with an abrupt cut-off at a minimum diameter, but instead exhibits a gradual roll over in the incremental size-frequency distribution towards smaller diameters. This roll over was first observed and modeled on Venus with data obtained from the Venera 15/16 mission (Ivanov et al. 1986; Basilevsky et al. 1987; Ivanov 1990). While a simple model that accounts for the effect of path length on survivability matches the observed size-frequency distribution to first order, in detail such a model produces a more abrupt roll over than is observed. As we discuss below, the more gradual roll over in the observations can be accounted for by differences in composition and density of the incoming meteoroids. Other factors which play a part, but do not appear to have a significant effect on the final

Figure 13. Example of a dark splotch (19.6°S, 338.8°E) in the Venusian plains. Image is 70 km across.

size-frequency distribution, are variations in incoming meteoroid velocity and target elevation.

There are 89 craters termed *multiple impacts* that are the result of an incoming meteoroid fragmenting and separating enough to form overlapping (multiple-floored craters) or completely separated (crater fields) impact craters. Formation of these features is also dependent on the properties of the incoming meteroid. Phillips et al. (1992) attempted to model the size-frequency distribution of these features along with the distribution of the entire crater population with the goal of determining the minimum crater diameter above which atmospheric effects on the size-frequency distribution are minimal (~32 km in diameter). Herrick and Phillips (1994b) extended this effort by using a more sophisticated model and attempting to model several properties of the multiple impacts also (e.g., impact angle, down-range and cross-range separation).

Herrick and Phillips (1994b) found that not only is a distribution of meteoroid densities required to explain the broad observed rollover in the size-frequency distribution, but other components of meteoroid composition must be important as well. Low-density meteoroids are inherently easier to decelerate due to larger drag effects, but to make the model match the size-frequency curve, low-density meteoroids must also be more likely to flatten or explode during atmospheric passage. The end result is that a disproportionately large percentage of small craters were probably formed by high-density iron meteoroids. Herrick and Phillips (1994b) also found that the separation of craters in multiple impacts was too great to be accounted for by meteoroid fragments merely drifting apart due to drag effects during atmospheric passage. The meteoroids that formed multiple impacts must have survived deep enough in the Venusian atmosphere for pressure effects to produce substantial transverse velocities upon breakup. Thus, most multiple impacts were produced by stronger, and presumably higher-density, meteoroids.

B. Multi-Asteroid Impacts

While compiling a global survey of Venusian craters, Herrick and Phillips (1994b) observed 20 sets of craters that they felt were too close together to be unrelated, but too large and too far apart to be from breakup of a single incoming meteoroid (Fig. 14). These sets of craters are inferred to be produced by simultaneous impact of multiple asteroids. Cook et al. (1995) carried out a specific search for multi-asteroid impact craters >10 km in diameter and separated by less than 150 km, and they classified 32 sets of craters as possible or probable doublets. While some of these may be chance occurrences of unrelated impacts, the results suggest that perhaps 1 to 5% of the asteroid population are multiple asteroids separated by a few tens of kilometers.

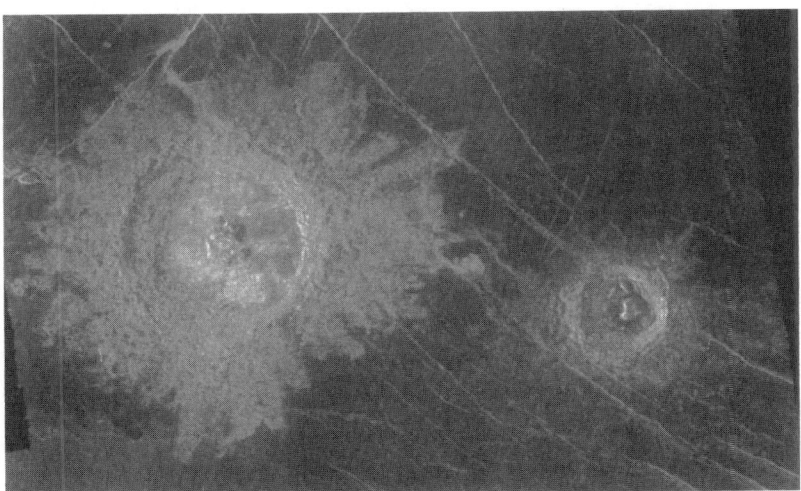

Figure 14. Example of a possible impact of a binary asteroid. Craters are Bathsheba (15.1°S, 49.4°E, 38 km diameter) and Gillian (15.2°S, 50.1°E, 16 km diameter). Image is 130×80 km.

C. Highly Oblique Impacts

Low impact-angle (<15 deg) craters created in the laboratory have distinctive features observed in impact craters of all the terrestrial planets (see, e.g., Gault and Wedekind 1978; Schultz 1992), and Venus is no exception. Very low impact angle craters are expected to have an elongate shape; the largest example of a suspected low-angle impact is the crater Graham (6°S, 6°E; Fig. 15). The ejecta blanket on Graham appears to have been emplaced as

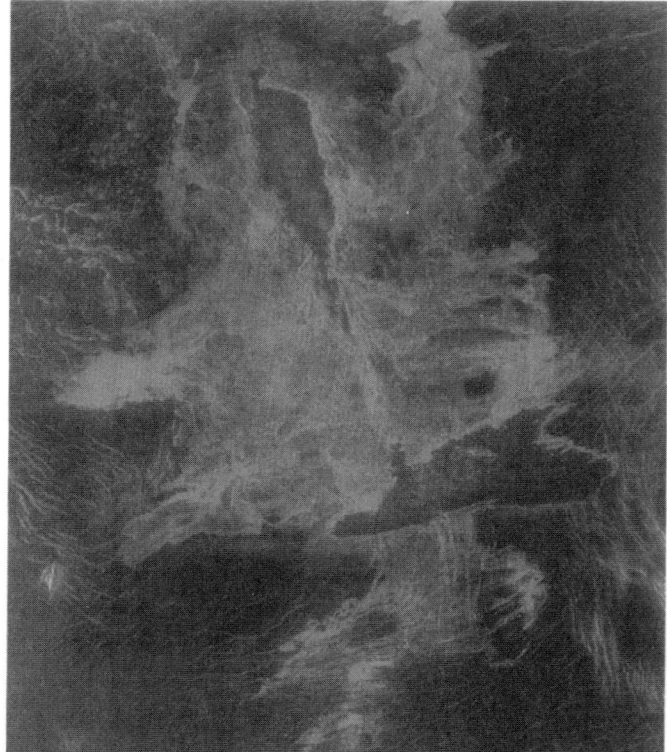

Figure 15. Crater Graham (6.2°S, 6.0°E, 190×50 km), an example of a very low-angle impact. Image is 375×400 km.

very fluid run-out flows, and this is typical for the other few elongate craters on Venus. Two craters on Venus have been identified as possible impacts where the impactor has ricocheted and formed a second crater (Fig. 16; Herrick and Phillips 1994*b*).

VI. SUMMARY: THE IMPACT PROCESS ON VENUS

Based on observations of the craters and crater-related features, we can begin to visualize the events that occur in the impact cratering process on Venus. The process starts with a meteoroid entering the atmosphere at a velocity of a few tens of kilometers per second. Small meteoroids (less than a few hundred meters in diameter) will either ablate away or be slowed to terminal velocity in the atmosphere. Slightly larger meteoroids may be slowed so rapidly that their kinetic energy release causes explosion of the meteoroid; if this occurs close to the ground the resulting atmospheric shock wave will pulverize and radar-darken the ground. Depending on composition, entry velocity, and entry

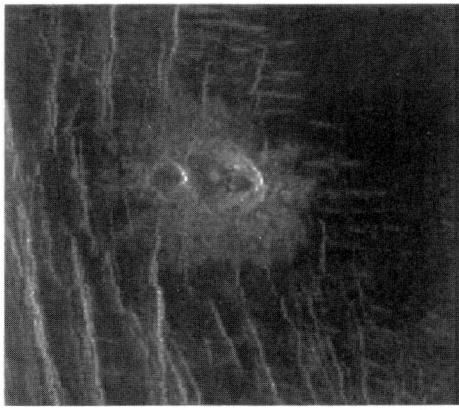

Figure 16. Example of a possible ricochet impact (26.7°N, 16.4°E). Projectile's direction of travel is right to left. Image is 43 × 40 km.

angle, incoming meteoroids a few km in diameter may or may not survive atmospheric passage with enough velocity to form an impact crater. Those meteoroids that survive atmospheric passage will be quickly trailed by an atmospheric shock wave capable of radar-smoothing the surrounding terrain for tens of km. In most cases fragments break off the incoming meteoroids. In some cases the fragments split far enough away from the main part of the meteoroid that separate or overlapping craters are formed. Typically though, the fragment separation is negligible and the crater forms as though a single meteoroid had struck.

The excavation flow probably proceeds in a manner similar to crater excavation on the other terrestrial planets: a parabolic cavity is formed as material is excavated in an outward growing cone. However, once the ejecta leaves the crater, it is caught up in a turbulent cloud moving outward. The cloud carries ejecta well beyond what ballistic emplacement would. Behavior of the cloud varies with crater diameter, and a major transition to more fluid-like behavior occurs for craters over 20 km in diameter. For larger craters the largest ejecta blocks are emplaced ballistically but travel at least some of the distance in a rarefied atmosphere. In some cases the amount of melt/vapor generation and the impact angle may have the right combination to produce lava-like flows of ejecta that travel several radii from the crater. Larger craters eject some fine material high up into the atmosphere to be carried west by upper level winds.

For all but the smallest craters, while the ejecta is being deposited the crater interior is collapsing from an initial parabolic shape. With increasing diameter this collapse forms central peaks and peak rings. The width of these features is controlled by the width of the precollapse transient cavity. If a large enough amount of melt is present, it ponds after collapse and forms a radar-bright flat floor. Collapse is enhanced by higher gravity and hindered by higher crustal strength. Terrace zones form very late in the impact process.

Some time after crater formation the appearance may be altered by volcanic or tectonic processes, but erosion will be negligible.

Acknowledgments. All authors except MCM were supported by a NASA grant to the Lunar and Planetary Institute. Part of this work was supported by a VDAP grant to VLS. MCM was supported by a NASA VDAP contract. We thank P. Schultz and an anonymous reviewer for comments. This is LPI Contribution 889.

Editor's Note

Herrick et al. have provided an animation on the accompanying CD-ROM. Please see CDP8C2M1. A data base of Venusian impact craters with their morphology and morphometry is also on the supplementary CD-ROM as well as digital images of the craters discussed in the text.

REFERENCES

Alexopoulos, J. S., and McKinnon, W. B. 1994. Large impact craters and basins on Venus, with implications for ring mechanics on the terrestrial planets. In *Large Meteorite Impacts and Planetetary Evolution*, eds. B. O. Dressler, R. A. F. Grieve and V. L. Sharpton, GSA SP-293 (Boulder: Geological Soc. of America), pp. 29–50.

Arvidson, R. E., et al. 1991. Magellan: Initial analysis of Venus surface modification. *Science* 252:270–275.

Arvidson, R. E., et al. 1992. Surface modifications of Venus as inferred from Magellan observations of plains. *J. Geophys. Res.* 97:13303–13318.

Asimow, P. D., and Wood, J. A. 1992. Fluid outflows from Venus impact craters: Analysis from Magellan data. *J. Geophys. Res.* 97:13643–13666.

Basilevsky, A. T., et al. 1987. Impact craters of Venus: A continuation of the analysis of data from the Venera 15 and 16 spacecraft. *J. Geophys. Res.* 92:12869–12901.

Campbell, D. B., et al. 1992. Magellan observations of extended impact crater related features on the surface of Venus. *J. Geophys. Res.* 97:16249–16278.

Carr, M. H., et al. 1977. Martian impact craters and emplacement of ejecta by surface flow. *J. Geophys. Res.* 82:4055–4065.

Chadwick, D. J., and Schaber, G. G. 1993. Impact crater outflows on Venus: Morphology and emplacement mechanisms. *J. Geophys. Res.* 98:20891–20902.

Cook, C. M., Melosh, H. J., and Bottke, W. J., Jr. 1995. Doublet craters on Venus. *Lunar Planet. Sci. Conf.* XXVI:275–276 (abstract).

Gault, D. E., and Wedekind, J. A. 1978. Experimental studies of oblique impact. *Proc. Lunar Planet. Sci. Conf.* 9:3843–3875.

Gault, D. E., et al. 1975. Some comparisons of impact craters on Mercury and the Moon. *J. Geophys. Res.* 80:2444–2460.

Grieve, R. A. F., and Cintala, M. J. 1995. Impact melting on Venus: Some considerations for the nature of the cratering record. *Icarus* 114:68–79.

Hale, W., and Head, J. W. 1979. Central peaks in lunar craters: Morphology and morphometry. *Proc. Lunar Planet. Sci. Conf.* 11:2623–2633.
Hale, W., and Head, J. W. 1980. Central peaks in Mercurian craters: Comparisons to the Moon. *Proc. Lunar Planet. Sci. Conf.* 11:2191–2205.
Hale, W., and Head, J. W. 1981. Central peaks in Martian craters: Comparisons to the Moon and Mercury. *Lunar Planet. Sci. Conf.* XII:386–388 (abstract).
Herrick, R. R., and Phillips, R. J. 1994a. Implications of a global survey of Venusian impact craters. *Icarus* 111:387–416.
Herrick, R. R., and Phillips, R. J. 1994b. Effects of the Venusian atmosphere on incoming meteoroids and the impact crater population. *Icarus* 112:253–281.
Ivanov, B. A. 1990. Venusian impact craters on Magellan images: View from Venera 15/16. *Earth Moon Planets* 50/51:159–173.
Ivanov, B. A., Basilevsky, A. T., Kryuchkov, V. P., and Chernaya, I. M. 1986 Impact craters of Venus: Analysis of Venera 15 and 16 data. *J. Geophys. Res.* 91:413–430.
Ivanov, B. A., et al. 1992. Impact cratering on Venus: Physical and mechanical models. *J. Geophys. Res.* 97:16167–16182.
Izenberg, N. R., Arvidson, R. E., and Phillips, R. J. 1994. Impact crater degradation on venusian plains. *Geophys. Res. Lett.* 21:289–292.
Johnson, J. R., and Baker, V. R. 1994. Surface property variations in Venusian fluidized ejecta blanket craters. *Icarus* 110:33–70.
Kirk, R. L., and Chadwick, D. J. 1994. Splotches on Venus: Distribution, properties, and classification. *Lunar Planet. Sci. Conf.* XXV:705–706 (abstract).
Malin, M. C., and Dzurisin, D. 1978. Modification of fresh crater landforms: Evidence from the moon and Mercury. *J. Geophys. Res.* 83:233–243.
Melosh, H. J. 1985. Ejection of rock fragments from planetary bodies. *Geology* 13:144–148.
Melosh, H. J. 1989. *Impact Cratering* (New York: Oxford Univ. Press).
Menke, W. 1989. *Geophysical Data Analysis: Discrete Inverse Theory* (San Diego: Academic Press).
Phillips, R. J., et al. 1991. Initial analysis of Venus impact craters with Magellan data. *Science* 252:288–297.
Phillips, R. J., et al. 1992. Impact crater distribution and the resurfacing history of Venus. *J. Geophys. Res.* 97:15923–15948.
Pike, R. J. 1977. Size dependence in shape of fresh impact craters on the moon. In *Impact and Explosion Cratering*, eds. D. J. Roddy, R. O. Pepin and R. B. Merrill (New York: Pergamon Press), pp. 489–509.
Pike, R. J. 1980a. Control of crater morphology by gravity and target type: Mars, Earth, Moon. *Proc. Lunar Planet. Sci. Conf.* 11:2159–2189.
Pike, R. J. 1980b. Formation of complex impact craters: Evidence from Mars and other planets. *Icarus* 43:1–19.
Pike, R. J. 1988. Geomorphology of impact craters on Mercury. In *Mercury*, eds. F. Vilas, C. R. Chapman and M. S. Matthews (Tucson: Univ. of Arizona Press), pp. 165–273.
Pike, R. J., and Davis, P. A. 1984. Towards a topographic model of Martian craters from photoclinometry. *Lunar Planet. Sci. Conf.* XV:645–646 (abstract).
Pike, R. J., and Spudis, P. D. 1987. Basin-ring spacing on the moon, Mercury, and Mars. *Earth Moon Planets* 39:129–194.
Schaber, G. G., et al. 1992. The geology and distribution of impact craters on Venus: What are they telling us? *J. Geophys. Res.* 97:13257–13302.
Schenk, P. M. 1989. Crater formation and modification on the icy satellites of Uranus and Saturn: Depth/diameter and central peak occurrence. *J. Geophys. Res.* 94:3813–3832.

Schenk, P. M. 1991. Ganymede and Callisto: Complex crater formation and planetary crusts. *J. Geophys. Res.* 96:15635–15664.

Schultz, P. H. 1992. Atmospheric effects on ejecta emplacement and crater formation on Venus from Magellan. *J. Geophys. Res.* 97:16183–16248.

Schultz, P. H., and Gault, D. E. 1979. Atmospheric effects on Martian ejecta emplacement. *J. Geophys. Res.* 84:7669–7687.

Sharpton, V. L. 1994. Evidence from Magellan for unexpectedly deep complex craters on Venus. In *Large Meteorite Impacts and Planetary Evolution*, eds. B. O. Dressler, R. A. F. Grieve and V. L. Sharpton, GSA SP-293 (Boulder: Geological Soc. of America), pp. 19–27.

Sharpton, V. L., Herrick, R. R., and Klaus, K. K. 1994. Topography of venusian craters and implications for cratering mechanics. *Eos: Trans. AGU* 75:413 (abstract).

Shoemaker, E. M. 1962. Interpretation of lunar craters. In *Physics and Astronomy of the Moon*, ed. Z. Kopal (New York: Academic Press), pp. 283–359.

Smith, E. I., and Hartnell, J. A. 1978. Crater size-shape profiles for the moon and Mercury: Terrain effects and interplanetary comparisons. *Moon and Planets* 19:479–511.

Strom, R. G., Schaber, G. G., and Dawson, D. D. 1994. The global resurfacing of Venus. *J. Geophys. Res.* 99:10899–10926.

Tauber, M. E., and Kirk, D. B. 1976. Impact Craters on Venus. *Icarus* 28:351–357.

Turco, R. P., et al. 1982. An analysis of the physical, chemical, optical, and historical impacts of the 1908 Tunguska meteor fall. *Icarus* 50:1–52.

Vervack, R. J., and Melosh, H. J. 1992. Wind interaction with falling ejecta: Origin of the parabolic features on Venus. *Geophys. Res. Lett.* 19:525–528.

Wilhelms, D. E. 1987. *The Geologic History of the Moon*, U. S. Geological Survey Prof. Paper 1348 (Washington, D. C.: U. S. Government Printing Office).

Wood, C. A., and Andersson, L. 1978. New morphometric data for fresh lunar craters. *Proc. Lunar Planet. Sci. Conf.* 9:3669–3689.

Wood, C. A. 1980. Martian double ring basins: New observations. *Proc. Lunar Planet. Sci. Conf.* 11:2221–2241.

Zahnle, K. J. 1992. Airburst origin of dark shadows. *J. Geophys. Res.* 97:10243–10256.

THE RESURFACING HISTORY OF VENUS

A. T. BASILEVSKY
Vernadsky Institute of Geochemistry and Analytical Chemistry, Moscow

J. W. HEAD
Brown University

G. G. SCHABER
U. S. Geological Survey, Flagstaff

and

R. G. STROM
University of Arizona

Roughly 90% of the Venus' history, from planetary formation until about 300 to 500 Myr ago, is not preserved in the surface geomorphological record that is distinguishable on the Magellan images. The beginning of this observable part of its history was characterized by intensive tectonic deformation of global or nearly global scale which resulted in the formation of tessera terrain. The duration of the deformational episode(s) is unknown but the rarity of on-tessera craters affected by the tessera-forming deformation shows that the termination of that deformation was rather sudden. The nature and composition of the tessera-forming material are unknown. After tessera formation, several stages of evidently basaltic, regional to global volcanism occurred forming what we see now as regional plains. Formation of tessera and emplacement of the regional plains together form what is often referred to as the global resurfacing event. Estimates of surface ages based on crater densities date this event as between about 300 and 500 Myr old but they cannot reliably discriminate the time of formation of tessera terrain from the time of formation of regional plains. Photogeologic observations show that following, and partly coincident with, the terminal stages of tessera formation, the early regional plains units formed. They were then heavily deformed by a very dense pattern of closely spaced extensional fractures and graben. This was followed by additional emplacement of volcanic plains, but deformation of the plains mostly switched from extensional to compressional, forming the broad ridges of the ridge belts and wrinkle ridges on the regional plains. The duration of these episodes of plains-forming activity has yet to be determined. Monte Carlo modeling shows that the global resurfacing event may have ended within a time interval of about 100 Myr or less. The average rate of emplacement of the regional volcanic plains is estimated to be about 3 to 10 km^3 yr^{-1}, about the same as ocean ridge volcanism on Earth in parallel with formation of regional plains. The formation and evolution of coronae occurred. Early stages in the formation of coronae were evidently contemporaneous with various early regional plains. Intermediate stages in the evolution of coronae were often contemporaneous with the emplacement of wrinkle-ridged plains while the final evolutionary stages of the coronae postdate the formation of regional plains. Finally Venus entered its present stage which was characterized by a significantly lower level of geologic activity distinguised by a predominance of rifting and localized rift- and corona-associated basaltic volcanism in the form of large shields and lava plains. When the present stage of lower geologic activity occurred depends on when emplacement of regional plains terminated. The average rate of Venus' volcanism at this time is estimated to be 0.1 to 0.2 km^3 yr^{-1}. This is comparable to the present rate of intraplate volcanism on Earth. In a few cases the rifting and the associated volcanism is interpreted by some to postdate emplacement of craters with dark parabolas; the latter are estimated to be 30 to 50 Myr old, thus supporting the idea that Venus is presently volcanically and tectonically active. Exogenic resurfacing on Venus is a complex total of aeolian deflation and accumulation with an unknown contribution of chemical weathering and on-slope mass wasting. Its average rate is estimated to be about 2 mm Myr^{-1}. This rate of exogenic resurfacing is orders of magnitude less than the rate of surface modification on Earth, or the rate of aeolian dispersal of fine-grained materials on Mars, and is comparable to the rate of formation of lunar regolith in post-mare time.

I. INTRODUCTION

The goal of this chapter is to suggest a scenario for the geologic history of Venus. Being constrained in this work by analysis of surface morphology, we see the history of the planet as a sequence of geologic events changing and renovating the planet's surface. Two major approaches are being used to understand the resurfacing history of Venus. One is stratigraphic analyses on local, regional, or even planetary scales to understand the resurfacing history. It establishes the relative time sequence of geologic formations without considering the question of absolute time.

Another approach has been an analysis of the time sequence of geologic formations and corresponding events, with direct or indirect consideration of absolute time. In the case of planetary bodies from which no samples have yet been returned, the absolute age dating is based solely on impact craters statistics.

The extensive Magellan data base provides us with the possibility to work out regional and then global stratigraphic models of Venus, to consider the general style of its geologic activity, to estimate the average absolute surface age of the whole planet and perhaps some stratigraphic units, and to synthesize all of this into a model for the geologic history of Venus. This will be done in the following sections of this chapter.

II. VENUS REGIONAL AND GLOBAL STRATIGRAPHY

We are presently at the stage of global-scale reconnaisance and characterization in the stratigraphic analysis of Venus (Saunders and Stofan 1992; Senske et al. 1992; Solomon et al. 1992; Stofan et al. 1992; Head et al. 1992; Phillips et al. 1992; Strom et al. 1994; Herrick and Phillips 1994) in addition to initial detailed local and regional studies (Senske et al. 1992; Kaula et al. 1992;

Bindschadler et al. 1992; Grimm and Phillips 1992; Aubele 1995; Copp and Guest 1995; Basilevsky 1993,1995). In this chapter we report the results of an initial stratigraphic analysis of the Magellan C1- and F-MIDRP mosaics for thirty-six 1000×1000 km randomly spaced areas, and several larger regions on Venus. A tentative model of Venus stratigraphy was suggested from this initial stratigraphic study (Basilevsky and Head 1995a,b,c). This model is one proposed model for the stratigraphy of Venus and we use it here as a convenient tool to describe the age relations among various geologic formations of this planet. Eventually it will be modified and perhaps replaced by the stratigraphic and geologic analyses that will result from the ongoing NASA Venus Geologic Mapping Program at 1:5 million scale (Tanaka 1994; chapter by Tanaka et al.).

A. Rock-Stratigraphic Units

There is virtually a consensus among those who study Venus geology that there are two major groups of terrains on the planet: (1) highly tectonized tessera terrain, and (2) various plains which embay and overlie tessera (Sukhanov et al. 1989; Sukhanov 1992; Gilmore and Head 1995; Ivanov and Head 1993,1995,1996; Strom et al. 1994; Head 1995a,b; Tormanen 1995). Considering analogies with Earth geology, this two-member situation resembles the continental platforms where the highly tectonized old basement is overlain by younger plains-forming sedimentary and volcanic formations. Of course, we did not assume that just because a terrain is heavily tectonized that it is old; these relationships are established on the basis of careful stratigraphic analysis at each location. Following the tradition giving stratigraphic units names of geographic areas where their typical representatives (stratotypes) occur (ACSN 1961) the first group was named by Basilevsky and Head (1995b,c) for Fortuna Tessera which is a large tessera massif containing several different varieties of this terrain. The second group consists in turn of two major subdivisions: (1) a sequence of plains-forming materials with varieties of morphologies composing the Guinevere Supergroup named after Guinevere Planitia; and (2) overlying radar-dark parabola-shaped mantles associated with some impact craters, and isolated streaks and patches of radar-dark debris, composing the Aurelia Group named after the crater Aurelia. Subdividing of the Fortuna Group into units of lower hierarchy is mostly a task for future studies. The Guinevere Supergroup is suggested to be subdivided into several groups, each of them, in turn, consisting of one or more units of lower hierarchy. Following is a description of the proposed preliminary stratigraphic sequence.

Fortuna Group. This group includes materials of all varieties of tessera terrain (Tt), which together form several large highlands and numerous small islands standing over the Venusian plains (Fig. 1). It covers about 8% of the surface of Venus (Ivanov and Head 1993,1996). This material is heavily modified by multiple deformation which determines the extremely rough ridge and valley morphology of tessera. As mentioned above, tessera is

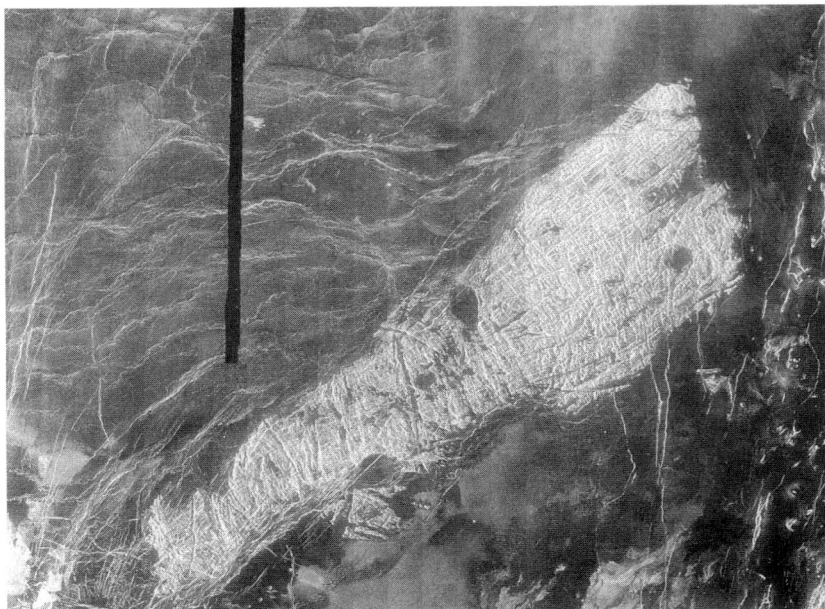

Figure 1. A remnant of tessera terrain embayed by plains with wrinkle ridges. Plains are darkened by the parabolic deposits associated with crater Yablochkina; area 200×300 km; part of F-MIDRP 45N195;1.

embayed and overlain by all plains units described below; thus establishing the pre-plains stratigraphic position of tessera (Basilevsky and Head 1995a,b). The true stratigraphic rank of tessera is not clear. Depending on whether its material was emplaced within certain relatively short time period or it is an assemblage of the materials formed at essentially different times the Fortuna Group may be equivalent in the stratigraphic rank to the majority of the overlaying groups or it may be a kind of analog of the Precambrian assemblage of the basement of some terrestrial continental platforms.

Also included in the Fortuna Group is the material of the mountain belts surrounding Lakshmi Planum. These belts (Maxwell, Freyja, Akna, and Danu) geomorphologically merge through gradual change of their patterns into adjacent tessera, forming together a large circum-Lakshmi concentric structure. The pre-plains age of the belts is directly established by their embayment by neighboring plains. These mountain belts are considered to be formed by compressional deformation (Crumpler et al. 1986; Pronin 1986; Basilevsky 1986; Vorder Bruegge et al. 1990; Keep and Hansen 1994) which might be contemporaneous and genetically related to the early compressional deformation of tessera mentioned above. Thus the circum-Lakshmi mountain belts may be structural facies of tessera.

The compositional nature of the material making up the Fortuna Group is unknown as none of the Venera/Vega landers made surface geochemical measurements on the Fortuna Group materials (Weitz and Basilevsky 1993). Recently some variations in the surface radio-reflectivity/emissivity in Maxwell mountain massif were interpreted as an indirect evidence of possible compositional inhomogeneity of the material of this component of the Fortuna Group (Basilevsky 1995; Basilevsky and Head 1995b,c).

Guinevere Supergroup. This is an assemblage of several types of plains-forming material units. It is subdivided into four groups separated one from another and from the underlying Fortunian and overlying Atlian (see below) materials by unconformities caused by episodes of tectonic deformation.

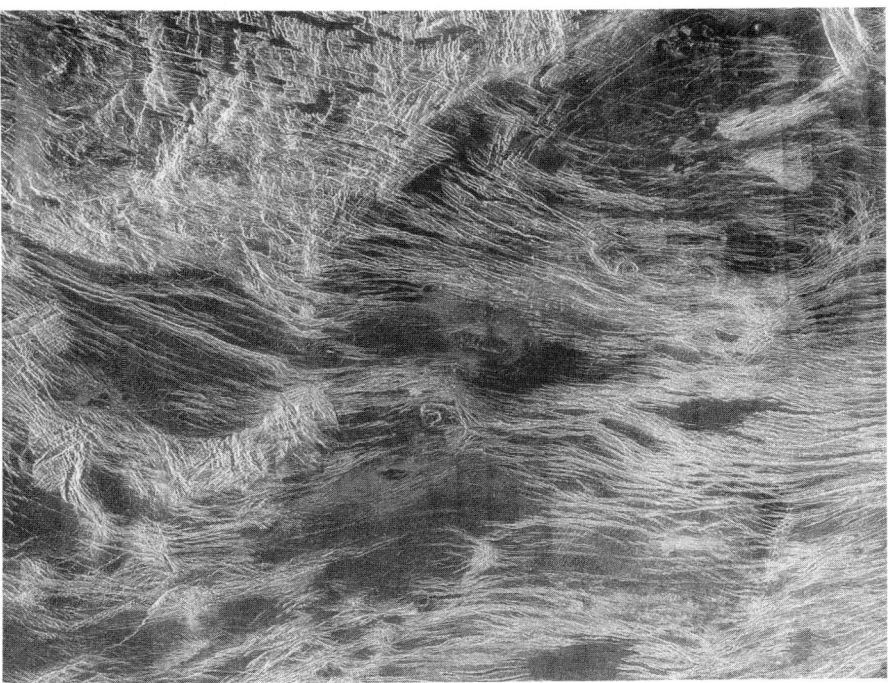

Figure 2. Densely fractured terrain of plains embaying SE part of Clotho Tessera and being locally embayed by plains with wrinkle ridges; area 270×310 km; part of F-MIDRP 55N337;201.

Sigrun Group. The densely fractured terrain units of plains (Pdf) and coronae (COdf) form this group named after Sigrun Fossae (Fig. 2). Sigrun Group materials are deformed by dense swarms of faults. But ignoring the faults, the terrains composed of this material appeared to be primarily plains. The materials of the Sigrun Group are observed on the surface as islands or kipukas embayed by different varieties of ridged and unridged plains described below. Their abundance on Venus is no more than 3 to 5% of the surface. The

Sigrun Group materials appear to be embaying tessera when the two terrains are in contact. There is no direct geochemical information for the materials of this group (Weitz and Basilevsky 1993) but the plain-forming character of them suggests that they are probably made of mafic volcanics.

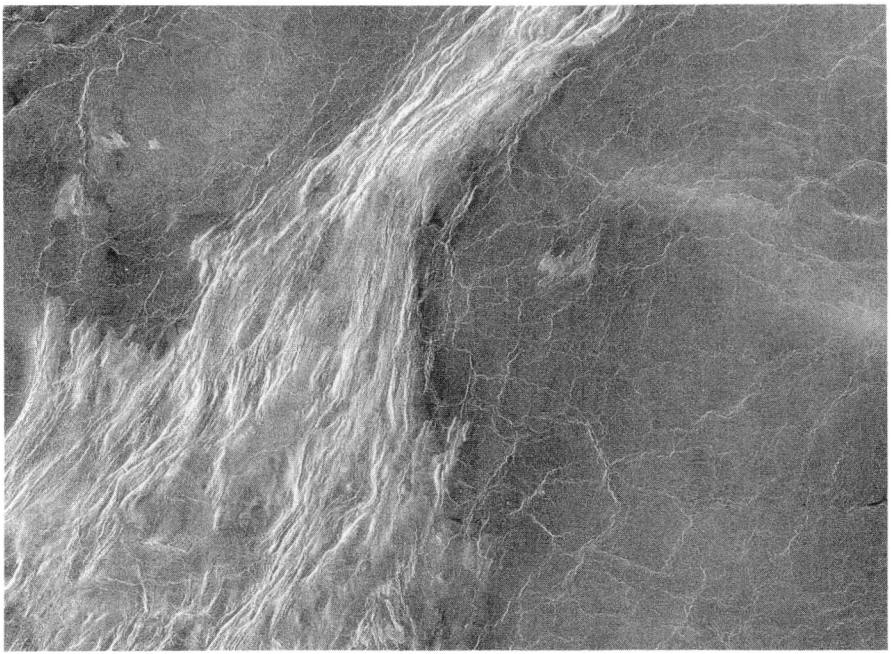

Figure 3. Ridge belt embayed by plains with wrinkle ridges; area 230×280 km; part of F-MIDRP 40N159;1.

Lavinia Group. This group (named after Lavinia Planitia) is composed of units of fractured/ridged plains (Pfr) and ridge belts (RB), which together occupy about 1 to 3% of Venus' surface (Basilevsky and Head 1995a,b). These units contain and embay remnants of the Sigrun Group and are embayed in turn, by plains with wrinkle ridges and smooth and lobate plains (Figs. 3 and 4). The materials making up the Lavinia Group are deformed by broad ridges and locally by wrinkle ridges and younger faults. Fractured/ridged plains and ridge belts are structural facies of the Lavinia Group materials which are distinguished based on the degree (higher for ridge belts) of their involvement in the formation of broad ridges. None of the Venera/Vega landers made surface geochemical measurements on the Lavinia Group materials. In the undeformed state, the Lavinian materials look very similar to the materials of the younger plains which have been characterized by Venera-Vega landers as being composed of basaltic lavas (Surkov 1986; Surkov et al. 1984,1986).

Rusalka Group. Units of this Group include a variety of subunits composing the plains with wrinkle ridges (Pwr) which occupy on total about 70%

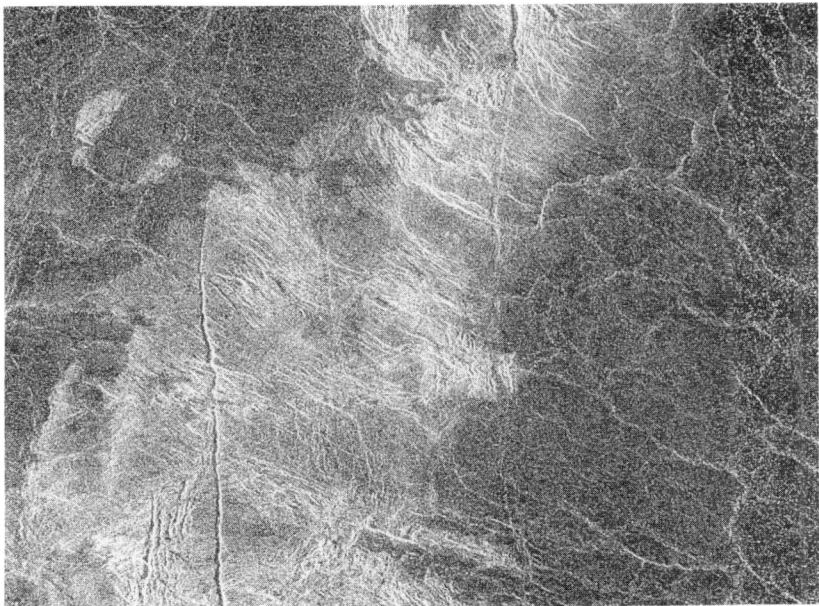

Figure 4. A remnant of fractured and ridged plains with inliers of densely fractured terrain embayed by plains with wrinkle ridges; area 50×70 km; part of F-MIDRP 35N161;1.

of the surface of Venus (Basilevsky and Head 1995*a,b,c*) (Figs. 3 and 4). The name Rusalka was given because Rusalka Planitia is mostly made of plains with wrinkle ridges and because geochemical measurements by Vega 1 and 2 landers in Rusalka Planitia provided a knowledge of chemical composition of these materials. They were found to be similar to tholeitic basalts (Surkov et al. 1986). Venera 9 and 10 landers, also sitting within the Rusalka Group plains, showed tholeitic composition too (Surkov et al. 1986). Rusalka Group materials embay materials of Fortuna, Sigrun and Lavinia Groups and are embayed by materials of smooth and lobate plains discussed below. Rusalka Group materials are deformed by wrinkle ridges forming regional to global networks (Bilotti 1992; McGill 1993) and by younger faults.

Small volcano plains (or shield plains, Psh) were identified by Aubele (1995,1996) in the Niobe Planitia region. It consists of clusters of 10 to 20 km in diameter, volcanic shields overlapping densely fractured terrain and fractured/ridged plains, ridge belt material embayed by the younger plains with wrinkle ridges (Fig. 5). Tentatively it has been suggested that this unit should be included into the model in between the materials of the Lavinia Group and Rusalka Group (Basilevsky 1996). However, in some other places, the position of the shield plains in the rock-stratigraphic sequence is not clear and new studies are necessary to clarify its stratigraphic significance.

Atla Group. Units of this group are the smooth (Ps) and lobate (Pl) plains

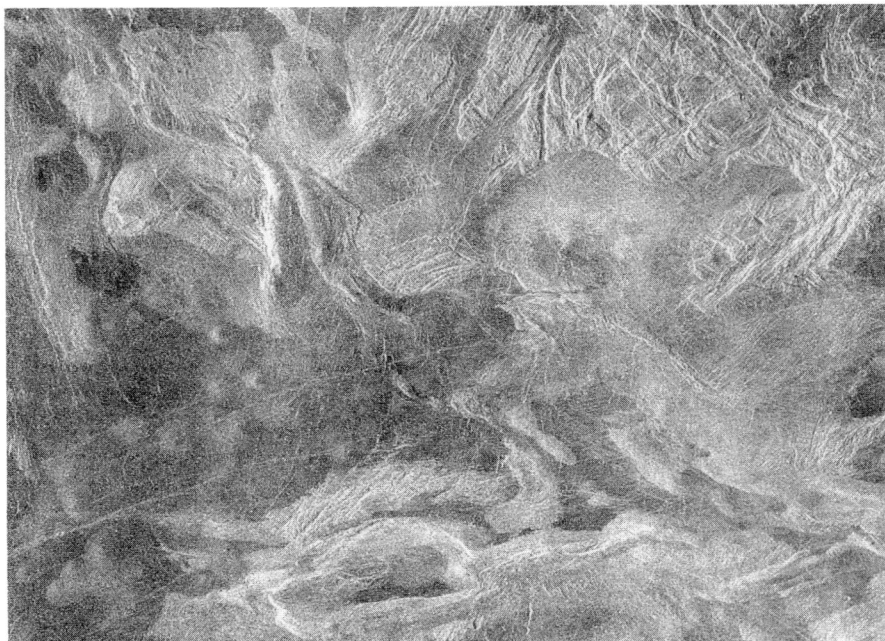

Figure 5. Shield volcano plains (left) overlapping fractured and ridged plains which, in turn, embay tessera (upper right); area 200×250 km; part of F-MIDRP 30N275;1.

with no wrinkle ridges and represent about 10% of the surface (Basilevsky and Head 1995a,b,c) (Fig. 6). The name Atla was suggested because smooth and lobate plains are very common in Atla Regio. Atla Group materials embay all the previously discussed units and are covered in some places by materials of dark-paraboloid features associated with some impact craters (Campbell et al. 1992). In a few cases Atla Group materials postdate materials of dark paraboloid features and are crossed by the youngest faults (Basilevsky 1993). Atla Group materials tend to be concentrated in association with the large-scale rift zones (Solomon et al. 1992; Senske et al. 1992) forming either areas of practically horizontal plains (Magee Roberts et al. 1992) or gentle-sloping mountains such as Sif, Gula, Maat or Ozza (Senske et al. 1992). They are also often observed in association with coronae (Fig. 7). Surface morphology and geochemical measurements (Venera 14; Surkov et al. 1984) show that these materials are mostly basalts.

Aurelia Group. Materials composing this Group are composed of parabolic radar-dark mantles typical of the youngest 10% or so of impact craters (Cdp) (Campbell et al. 1992; Basilevsky 1993; Strom 1993) as well as other crater materials (ejecta including outflow, materials of crater floors, walls, and central peaks) of this crater subpopulation (Fig. 8). Also included in this group are some debris accumulations that are evidently contemporaneous with Cdp, which form radar-dark patches (Sp) in local topographic lows against or

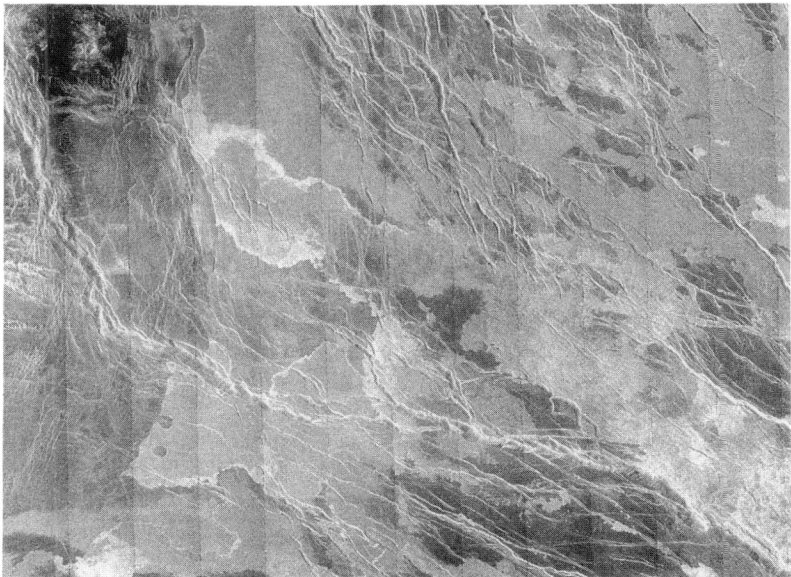

Figure 6. Lobate plains associated with Theia Mons volcano and fractures of the Devana Chasma rift zone; area 170×240 km; part of F-MIDRP 25N278;1.

behind positive topographic obstacles, as well as some radar-dark wind streaks (Ss) (Greeley et al. 1992) evidently contemporaneous with Cdp. The total area occupied by dark-paraboloid mantles is about 8% of the planet surface (Campbell et al. 1992). Aurelia Group materials overlay all stratigraphic units described above, except for the rare cases where volcanic materials belonging to the youngest part of Atla Group (e.g., some lavas of Maat Mons) postdate dark-paraboloid craters (Basilevsky 1993). Thus, the youngest materials of the Atla Group overlap in time with materials of the Aurelia Group. Some Cdp craters are criss-crossed by young, rift-associated fractures.

B. Structural Deformation

One of the characteristics of the identified stratigraphic units is the presence (or absence) of widely distributed deformation which differs from unit to unit in its pattern, abundance and spacing. The earliest and most intense deformation determines the observed morphology of tessera. Its complex pattern (Fig. 1) is believed to be a result of intensive multiple deformation probably changing in time (maybe during a very short period) from compression to extension (Bindschadler and Head 1991; Bindschadler et al. 1992; Ivanov and Head 1993,1995,1996; Chadwick and Schaber 1994) but some authors believe that it could be done simultaneously in one tectonic episode and emphasize the role of differences in tectonic environment which different varieties of tessera might form in (Willis and Hansen 1995; Hansen and Willis 1996). Subparallel ridging in the circum-Lakshmi mountain belts may be considered

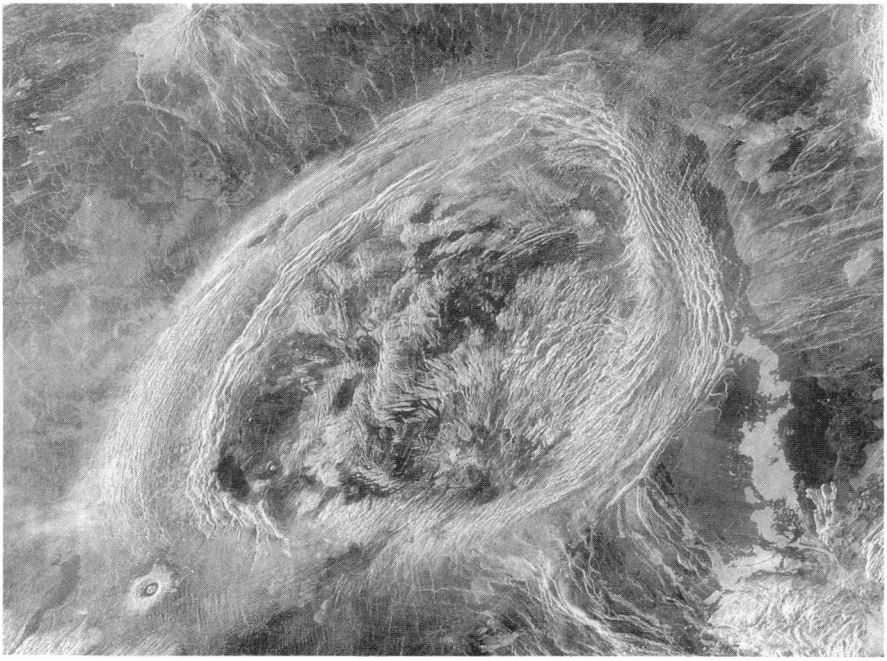

Figure 7. Anahit Corona (340×430 km) NW of Atropos Tessera; part of C1-MIDRP 75N299;1.

as a representative of the compressional stage almost not affected in this place by subsequent extensional deformation. We do not know the duration of the tessera-forming structural episode but think that its terminating part was concentrated in time because only a few of the impact craters superposed on tessera show evidence of significant deformation (Ivanov and Basilevsky 1993; Strom et al. 1994; Gilmore and Head 1995; Gilmore et al. 1996). Tessera islands are so widely distributed over the Venus globe that it leads one to think that the tessera-forming deformation was global-wide and that tessera terrain underlies the plains units over much of the planet and may be even all over the planet. Ivanov and Head (1995) estimated that tessera may underlie at least half of the Venusian plains. More work should be done to understand if tessera formed in one or several episodes of deformation and what mechanisms were responsible for that.

Following and maybe partly contemporaneous with the terminal stages of tessera deformation was extensional and perhaps shear faulting now visible in the remnants of the densely fractured material of the Sigrun Group that are widespread over Venus. In some places where the Sigrun Group units (mostly densely fractured terrain) are in contact with tessera (Fig. 2) one can observe a propagation of this dense fracturing into tessera. It is not clear if this global-wide dense fracturing did affect all the plains formed at the Sigrunian time. If not, some of the plains presently classified as Lavinian may actually represent a mixture of Lavinian and Sigrunian materials or Sigrunian materials only.

Next in time was a formation of the broad ridges characteristic of the ridge belts (Fig. 3), fragments of which are now seen as islands of the Lavinia Group material. This marks a transition from the dominance of extension to compression. This broad ridging also affected the stratigraphically lower units such as tessera and the densely fractured terrain of plains. Remnants of the ridge belts are currently observed in practically all regions of Venus, often forming structurally consistent systems up to several thousand km long (e.g., Lavinia and Atalanta Planitiae). Thus, the distribution of the broad ridging activity was evidently planet-wide.

Following in time was the formation of wrinkle ridges (Figs. 3 and 4), which are also considered as compressional features (Plescia and Golombek 1986; McGill 1993). They are dominantly superposed on primarily undeformed Rusalkian plains and sometimes on Lavinian plains already affected by broad ridges. However, wrinkle ridges are never observed on the Fortuna Group units and they are seen on the Sigrun Group densely fractured terrains only in cases where the latter are closely intermixed with younger volcanics. Obviously, highly fractured materials of the Fortuna and Sigrun Groups have physical properties that are not favorable for warping into wrinkle ridges.

Wrinkle ridges form regional to global network(s) (Bilotti 1992; McGill 1993) with local clusters around many coronae. It is not clear if these clusters were responses to regional or even global compressional deformation on local corona-related stress fields, and thus simultaneous to the regional wrinkle ridging, or they formed exclusively due to corona-related stresses independently from the regional/global stress field. In the latter case the corona-related wrinkle ridging may be or may not be contemporaneous to the emplacement of the regional network. Structural alignment of many corona-related wrinkle-ridge clusters with the ridges of the regional network favors the first option but additional studies are necessary to resolve this problem. In any case, the emplacement of all or at least the majority of wrinkle ridges evidently followed emplacement of the Pwr plains material very closely in time because among several hundred craters superposed on plains with wrinkle ridges only six are mentioned by Schaber et al. (1995) as deformed by wrinkle ridges. A special search for craters deformed by wrinkle ridges made by the first author of this chapter on the area of about 10% of Venus surface confirmed the data of Schaber et al. (1995); only one wrinkle-ridged crater was added to the Schaber et al. list (crater Deken, 47.1°N, 288.5°E, $D =$ 44 km).

Younger deformation is represented by faults that may (Figs. 6 and 8) or may not be associated with rifts. They are overprinted on practically all stratigraphic units. Only some Atlian volcanic plains and the majority of Aurelian radar-dark parabolic mantles overlay them. Practically all typical rift-associated faults are definitely younger than the Rusalka Group. One

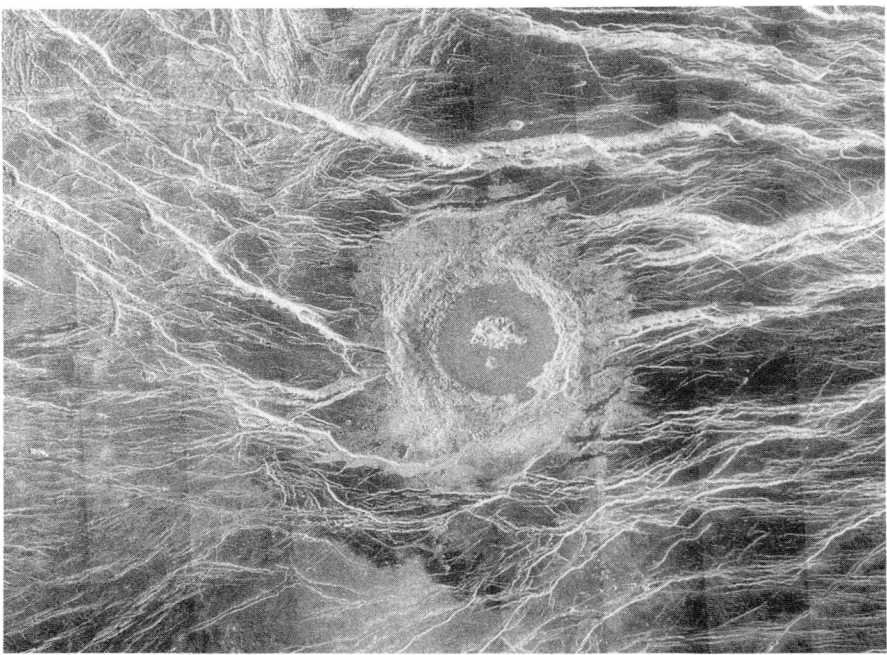

Figure 8. Crater Sitwell (16.6°N, 190.4°E, D = 33 km) superposed on Ganis Chasma rift zone; part of F-MIDRP 15N190;1.

possible example of older rifting is represented by a system of anastomosing fractures typical for rifts in Lavinia Planitia. These fractures criss-cross Lavinian ridge belts but do not extend into the surrounding plains with wrinkle ridges of Rusalkian age (see Fig. 10 in Basilevsky and Head 1995c). Thus, the question is whether the rifts started to form on Venus in the upper (post-Rusalkian) part of the studied sequence of geologic events or whether they formed also earlier, but extensive Rusalkian plain-forming volcanism buried them.

C. Coronae

The attributive for any corona is an annulus consisting in different proportions of densely fractured terrain of coronae (COdf) mentioned above, plains-forming materials embaying COdf and warped into concentric set of ridges (COar), and sets of concentric and/or radial fractures which criss-cross both COdf and COar-warped plains (Fig. 10) (Basilevsky and Head 1995a,b,c). The area inside the corona annulus is usually made of plains-forming materials not deformed, or deformed by corona-related or regional deformation. Lobate lava flows are often seen both inside the corona and radiating outside the corona annulus. The majority of them are not deformed by any structures, but some are deformed by wrinkle ridges and/or fractures.

Thus, each corona formed not in a single event but in a series of events or phases which may be correlated with geologic events represented by the regional rock-stratigraphic units. The oldest recognizable component of COdf is considered to be part of Sigrun Group. The plains-forming material, which embays COdf and is warped into ridges of corona annulus, seems to be Rusalkian and in some cases Lavinian. Wrinkle ridges of the regional network coming to coronae are often in a structural alignment with COar that probably indicates contemporaneity of their formation. Lobate plains associated with coronae are considered to be part of the Rusalka Group, if they are deformed by wrinkle-ridges, and part of the Atla Group, if they are not so deformed. Young radial and concentric fractures criss-crossing Rusalkian and Atlian plains may be considered as roughly contemporaneous to the young, regionally distributed faults (mentioned above) that are commonly (but not always) associated with rifts. All corona-related materials and structures were criss-crossed by the rift-associated faults in several cases where Basilevsky and Head (1995a,b,c) observed coronae in contact with rift-associated fractures.

This observation is in general agreement with observation of Magee and Head (1995) and Baer et al. (1994) who studied coronae along the extentional belts in Northern Lada and found that no corona along the belts was formed subsequent to the cessation of the regional extention. Jackson et al. (1995) studied eleven coronae in Diana and Dali Chasmata. They found that seven coronae began forming prior to rifting and continued to develop while early rifting was taking place. The remaining four coronae appear to have originated during early rifting, and did not develop further after the cessation of rifting. We restudied all these sites and found one circumstance not clearly stated in the text of their paper but vaguely present in the table; all eleven coronae are postdated by the late rift-associated fractures.

The information described above on the rock-stratigraphic units and structures can be tentatively summarized in the form of stratigraphic column (Fig. 9) and a diagram showing a time sequence of material (mostly volcanic) emplacements and deformational episodes (Fig. 10). The observed stratigraphic sequence can correspond to two different end-member options of geologic history of Venus: (1) the same stratigraphic units/structures and corresponding geologic events were roughly synchronous around the planet; or (2) this sequence is just a sequence of events which typically occurred in different places at different times. This dilemma is actually a variation of the current discussion on the character of geologic history of Venus which started to be well known to the community from the papers of Schaber et al. (1992) and Phillips et al. (1992). Now positions of these two research groups have become closer. However, significant difference in opinion still exists between these two groups so it is worthwhile, first, to present the early models as an

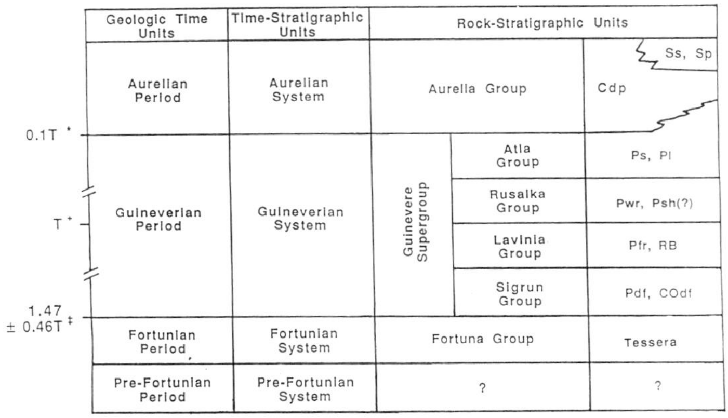

Geologic Time Units	Time-Stratigraphic Units	Rock-Stratigraphic Units			
Aurelian Period	Aurelian System	Aurelia Group		Cdp	Ss, Sp
Guineverian Period	Guineverian System	Guinevere Supergroup	Atla Group	Ps, Pl	
			Rusalka Group	Pwr, Psh(?)	
			Lavinia Group	Pfr, RB	
			Sigrun Group	Pdf, COdf	
Fortunian Period	Fortunian System	Fortuna Group		Tessera	
Pre-Fortunian Period	Pre-Fortunian System	?		?	

0.1T *

T +

1.47 ± 0.46T ++

Figure 9. Stratigraphic column suggested by Basilevsky and Head (1995a,b,c). (*) estimation of age of the dark parabola craters from Basilevsky (1993); (+) estimation of global average surface age from Phillips et al. (1992) and Strom et al. (1994); (++) estimation of tessera surface age from Ivanov and Basilevsky (1993).

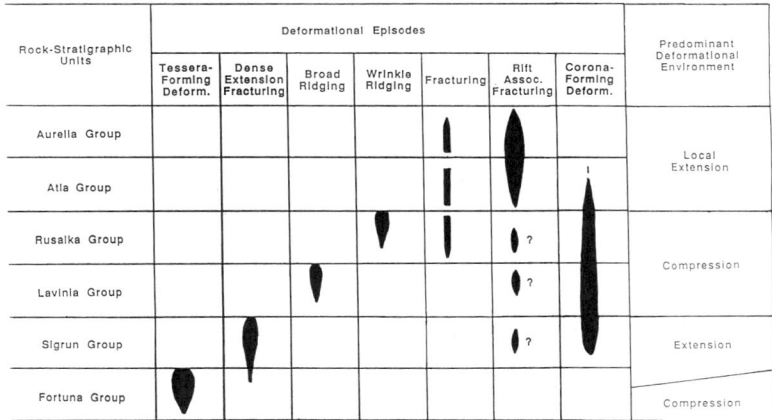

Figure 10. Correlation chart of stratigraphic units and structures illustrating deformational episodes and predominant deformational environments as a function of time.

introduction to the problem, even if none of the groups did follow them in the extreme form.

III. GENERAL CHARACTER OF GEOLOGIC HISTORY OF VENUS: END-MEMBER MODELS

Both of these end-member models are based on two observations on the Venus impact crater population: (1) most of the observed impact craters on Venus have a pristine appearance and do not look embayed by the surrounding volcanic plains but rather are superposed on them; and (2) the spatial distribution of craters around the planet is not distinguishable from a random one (Schaber et al. 1992; Phillips et al. 1992).

Schaber et al. (1992) considered the observed crater population as essentially a production one which accumulated on Venus since the time when the surface of this planet was flooded by extensive volcanic eruptions which destroyed the pre-existing crater population. In the later work of these authors this hypothetical global resurfacing event is considered as involving both volcanism and tectonism (Strom et al. 1994). Monte Carlo simulation of the interaction of the accumulating crater and crater-related feature population and the volcanic resurfacing phenomena led these authors to the conclusion that "the global resurfacing event ended within a time interval no greater than 10 Myr." This is why this hypothesis is often called the "catastrophic resurfacing model" (see, e.g., Phillips et al. 1992) although its authors prefer to call it "global resurfacing model" thus avoiding use of the term "catastrophe" (Schaber et al. 1992; Strom et al. 1994). (See note in proof at end of text.)

Following this global resurfacing event the volcanic and tectonic activity was minor and continued bombardment formed the observed crater population. This scenario is compatible with the first option of geologic history mentioned above: a sequence of almost synchronous geologic events all around the planet. In this case the Fortuna, Sigrun, Lavinia, and Rusalka Groups represent a sequence of events which composed together the global resurfacing, while the Atla Group represents the post-global-resurfacing limited endogenic activity. The Aurelia Group represents young exogenic events not related directly to endogenic life of Venus.

Here it is necessary to emphasize two important things which are often misunderstood. First, the estimate by Strom et al. (1994) "no greater than 10 Myr" is the time of terminating of the global resurfacing event, but not the time of the global resurfacing event itself. Strom et al. (1994) believe (but do not prove) that "the global resurfacing event probably lasted at least tens of millions of years." Second, the estimation "no greater than 10 Myr" was based on the assumption that all (33 at that time) of the lava embayed craters were embayed following the end of planet-wide global resurfacing. Unfortunately, this was ambiguously said in Strom et al. (1994), and was only made clear in their later publication (Strom et al. 1995). (See note in proof.)

Phillips et al. (1992) based their model on the same basic facts using the crater inventory collected at that time by R. R. Herrick. But unlike Schaber et al. (1992), Phillips et al. (1992) consider the crater population not as a production population but as an equilibrium one. They believe that the observed situation may be the result of endogenic activity, which was permanent on the global scale and which occurred in zones of limited size (no larger than about 400 km across) in different places at different times. This situation in the sense of the permanent character of the activity looks quite similar to what we see in the observed part of geologic history of Earth and for this reason looks more realistic than the global resurfacing model for people aquainted with the geology of Earth. In relation to the crater population this model considers that on average for the time when a new crater is formed, volcanic and tectonic events completely destroy the earlier formed one. This is why this hypothesis is often called the "equilibrium resurfacing model." This model corresponds to the second option of geologic history mentioned above: multiple repetition of a typical sequence of events in different places at different times. In this case, the sequence from the Fortuna Group to the Atla Group represents the local stratigraphic sequences which, for some reason, are very similar around the planet.

The polemics between supporters and opponents of each of these two models emphasized strong and weak sides of each of them. So Monte Carlo simulations by Strom et al. (1994) show that although the random distribution of impact craters around Venus is reached in the case of local resurfacing zones (<400 km across), the percentage of volcanically embayed craters should be much higher than what is actually observed. Responding to this Herrick et al. (1995) first of all expressed doubts about the Strom et al. (1994) list of impact craters and crater related features (splotches). They insist that the splotches should be taken out from the modeling because it is not proved yet that (1) all of them are formed by bolides, (2) that they are eliminated by processes other than those that eliminate craters. Herrick et al. (1995) also say that the population of 932 craters of Strom et al. (1994) is strongly polluted with features which are "almost certainly of volcanic origin" ($\sim$50 features) or their volcanic origin cannot be ruled out ($\sim$150 features). They assert that Strom et al. (1994) underestimate the percentage of volcanically embayed craters and that the simulation criteria used in the model for whether a crater is embayed or destroyed is unrealistic because the model is two-dimensional and does not consider whether a volcanic flow is thick enough to cover the ejecta blanket or breach the crater rim.

From the point of view of the first author of this chapter, the criticism of Herrick et al. (1995) on including splotches in the Monte Carlo modeling is reasonable. Even if all the splotches were formed by bolides, they, contrary to impact craters, are certainly sensitive to aeolian resurfacing which is not significantly affecting impact craters and is not considered by the modeling of Strom et al. (1994).

The pollution of the Strom et al. (1994) list of impact craters with volcanic features is theoretically possible but it is very unlikely in such large (50+150) amounts. The problem of age relations between craters and surrounding plains does exist and the underestimation of percentage of volcanically embayed craters by Strom et al. (1994), who found only 33 embayed craters, is also possible. Now after revision of the crater inventory Schaber et al. (1995) have in their list 46 craters volcanically embayed from the exterior (see note in proof). The major question in this case is how significant this underestimation may be. Phillips et al. (1992), Izenberg et al. (1994), and Herrick and Phillips (1994) report on about 60 such craters. In the new Herrick et al. (1995) crater data base 74 craters are listed as embayed from the exterior. However, if there are not 33 or 46 but 60 or even 100 volcanically embayed craters out of the more than 900 craters on Venus, it does not significantly change anything in the logic of Strom et al. (1994).

An important observation by Sharpton (1994) mentioned by Herrick et al. (1995) is that bright-floored craters are typically a few hundred meters deeper than similar-sized dark-floored ones. This observation, if confirmed, favors the idea that the dark-floored craters are partly infilled with volcanic lava. However, the total area represented by floors of all impact craters of Venus is only 0.12% of the planet surface (Schaber et al. 1995); that is about 2 orders of magnitude less than the area of Atlian rift-associated volcanism, so from the point of view of global geology of Venus this is a phenomenon of minor significance.

More convincing is the Herrick et al. (1995) criticism of the Strom et al. (1994) modeling of the interaction of the growing crater population and on-going volcanism. This modeling was indeed two-dimensional and significant progress in this area demands a large effort on determination of the thickness of volcanic floods on Venus which is unlikely based on the available data. Until this determination is done the results of this modeling in relation to percentage of volcanically embayed craters is vulnerable to criticism. But even this non-perfect modeling of Strom et al. (1994) is informative; if the lava flows in the center of the local resurfacing zone are not thick enough to completely destroy the craters (as it is assumed by the model), craters in this central part of the zone will be not destroyed but embayed. This should result in an increased amount of embayed craters which is not observed in reality (Strom et al. 1995).

Summarizing, a consideration of the end-member models in their "pristine" form showed that each of them was vulnerable to criticism and that the mutual criticism, highlighted in the Herrick et al. (1995) and Strom et al. (1995) polemics, clarified the general character of the problem. As a result scenarios of the geologic history of Venus worked out by different research groups are converging (Strom et al. 1994; Herrick 1994; Phillips 1993). The polemics were mostly concentrated on the characteristics of the impact crater population so it seems worthwhile to consider this problem involving data from other areas of observations. Serious constraints may be put on the general character of geologic history of Venus from consideration of age relations of Venusian formations.

IV. STRATIGRAPHIC OBSERVATIONS

The global resurfacing model agrees well with the consistency of stratigraphic relations in all sites and regions studied by Basilevsky and Head (1994,1995a,b,c,1996) without any additional requirements. The equilibrium resurfacing model does not contradict the consistency of stratigraphic relations inside the zones of local endogenic activity but it demands that on the boundaries between the neighboring zones the stratigraphic sequence of the younger zone should be superposed on the stratigraphic sequence of the older one. Observations of Basilevsky and Head (1995a,b,c) in thirty-six 1000×1000 km sites and several larger regions did not discover any cases of such superposition although the sites and regions are large enough to consist of several, or more, local activity zones. So the global resurfacing model does not contradict these observations while the equilibrium resurfacing model does. It is necessary to emphasize, however, that this conclusion is based on a limited number of observations. The current program of regional geologic mapping can test for superposition of local stratigraphies around the planet.

An example of such a test is the analysis of stratigraphic relations of units and structures along Baltis Vallis, the longest canali-type channel on Venus (Baker et al. 1992) evaluated by Basilevsky and Head (1995b,c). The channels of this type are excellent local stratigraphic markers (Baker et al. 1992; Parker et al. 1992) because the formation of each of them is geologically instantaneous (from 1 to 100 yr, according to Kargel et al. 1994). Observations of Basilevsky and Head (1995b,c) showed that along all of its 6800 km length and for the great distances from its margins, there are no violations of the stratigraphic sequence described above, no superpositions of one stratigraphic sequence over another. Over the whole of this region, which is equivalent in size to more than a hundred hypothetical local activity zones, Baltis Vallis is incised into Rusalka Group plains units which overlay units of the Lavinia, Sigrun and Fortuna Groups. Several late members of the Rusalka Group (radar-bright flows) were emplaced and partially flood Baltis prior to deformation of the plains into the characteristic wrinkle ridges. Earlier members of the Rusalka Group are clearly superposed on the Lavinia Group plains deformed in a very consistent NS system of broad ridges. This nonconforming broad ridging formed a distinct lower boundary of the Rusalka Group. The consistency of age relations between the Rusalka Group subunits, wrinkle ridges and Baltis Vallis as well as between the Rusalka and Lavinia Groups plains is evidence that the Rusalkian plains composing about 80% of this very large area is not a mosaic of volcanic floods emplaced at significantly different times. This contradicts the equilibrium resurfacing model and indirectly favors the global resurfacing model.

Another attempt to progress in this area using the stratigraphy approach was recently made by Collins et al. (1996); they compared the Venus crater data bases compiled by R. Herrick (see Herrick and Phillips 1994) and by Schaber et al. (1995). Collins et al. (1996) thoroughly studied the best of available images of craters, listed in these two data bases as embayed from the exterior, and made their own conclusions to whether they are embayed or not and, if embayed, what was the stratigraphic position of the embaying material. This study has shown, first of all, that each worker defines an embayed crater slightly differently and until the images with significantly higher resolution are available some ambiguity in such interpretation is inevitable.

Collins et al. (1996) found that only 58 craters of more than 900 craters of total crater population are embayed by lavas from the exterior. This number is in between and not very different from the numbers of Schaber (46) and

Herrick (74); 40 to 45 of those 58 craters are embayed by the Atlian smooth and lobate plains, which, as was discussed above, were formed after the global resurfacing event (assuming that it happened) while 13 to 18 of 58 craters are embayed by materials belonging to Rusalkian and older plains which were considered above as the products of the supposed global resurfacing event.

These numbers are very important for constraining of general character of the geologic history of Venus. Remember that the estimation by Strom et al. (1994) that the global resurfacing event may have ended very abruptly (<10 Myr) was based on the Monte Carlo modeling assuming that all (33 at that time) of the lava-embayed impact craters were embayed following the end of global resurfacing. But as we know now it is not the case. Now we can use the determined amount of craters embayed by lavas formed during the global resurfacing event for more reliable estimation of the time to terminate the global resurfacing. For this purpose we are using a plot from Strom et al. (1995) based on their latest Monte Carlo simulation of the crater formation vs volcanic floods interaction for the ending or tail of global resurfacing, during which the volcanic rate is assumed to be decreasing exponentially. This plot links the number of craters embayed by global resurfacing with the percent surface area resurfaced after the global resurfacing, and the time to terminate the global resurfacing. (See note in proof.)

If we rely on figures obtained by Collins et al. (1996) the diagram (case "craters only") shows that the percent surface area resurfaced after the global resurfacing event is between 8 and 10% and the time to terminate global resurfacing is estimated to be 140 to 190 Myr in the 500 Myr scale or 85 to 115 Myr in the 300 Myr scale. Basilevsky and Head (1995a,b,c) estimated the percent surface area occupied by Atlian post-global resurfacing volcanic plains as about 10% of the total surface of Venus. From measurements made by Price and Suppe (1994) the sum of areas of volcanoes and flows, which correspond evidently to the Atlian lobate and smooth plains of Basilevsky and Head (1995a,b,c), is about 28×10^6 km^2, that is about 6.4% of total area surveyed by them. Price and Suppe (1995) estimate about 14% of volcanic resurfacing is based on combined areas of volcanoes, flow fields and coronae. But it is necessary to keep in mind that only part of corona-related materials, namely corona-related Atlian lobate plains, corresponds to the post-global activity and in relation to the discussed topic their number is an overestimation. Therefore, the model percent surface area resurfaced after the global resurfacing event is reasonably close to the observations. This may be considered as an indirect evidence in favor of the correctness of this approach to the modeling of Venus' recent resurfacing.

Finally, one additional consideration of the nature of the geologic history of Venus can be found in the general character of the geology of the planet. Based on the stratigraphic sequence of material units and structures (described above), as well as on numerous publications on geology of various regions (Saunders et al. 1992; Solomon et al. 1992; Herrick 1994; Strom at al. 1994; Basilevsky and Head 1995a,b,c, and many others) the geology of the planet can be roughly summarized by three basic terrain types (from older to younger): (1) tessera terrain, (2) majority of plains, and (3) rifts and associated volcanics. This sequence shows the change in general style of endogenic activity of Venus with time and agrees well both with the global resurfacing model of Strom et al. (1994) and with the idea of Phillips (1993) that the surface of Venus is divisible into three major age groupings.

It is evident that serious progress in understanding what was the real character of the considered part of the geologic history of Venus could be achieved if we had a knowledge of the absolute time duration of the periods corresponding to the identified rock stratigraphic units. The only way to do it now is by a consideration of impact crater densities and this is the subject of the following section.

V. SURFACE AGE

A. Average Surface Age of the Planet

Analysis of Magellan imagery showed that about 1000 impact craters exist on the 460 million km^2 surface of Venus, yielding an average crater density of about 2 craters per million km^2. The main characteristics of the diameter-frequency distribution of the Venusian impact craters is a deficit of craters with $D<20$ km in comparison with the well-known power law of an increasing number of craters as the crater diameter decreases. This effect was predicted theoretically as a consequence of atmospheric shielding (see, e.g., Kahn 1982; Melosh 1981) and was confirmed by observations of Venera 15 and 16 (Ivanov et al. 1986; Basilevsky et al. 1987). The differential size-frequency distribution of craters with diameters above 35 km has a slope of about -3, typical for young production populations of airless planets (Schaber et al. 1992). If we assume that the atmosphere-free case for crater size-frequency distribution is described by the power law the models for atmospheric transit of meteoroids show that formation of craters >32 km in diameter is practically unaffected by the atmosphere (Ivanov et al. 1986; Phillips et al. 1992).

Shoemaker et al. (1991) and E. M. Shoemaker (personal communication, 1993) estimated that the crater production rate at Venus is $1.07\pm0.6 \times 10^{-15}$ km^{-2} yr^{-1} for craters >35 km in diameter and $1.29\pm0.6 \times 10^{-15}$ km^{-2} yr^{-1} for craters >32 km in diameter. Strom et al. (1994) have counted 147 craters >35 km in diameter and 162 craters >32 km in diameter on 446 million km^2 of the Venusian surface. Assuming that the crater population is a producing one, averaging the age estimates for the two diameter ranges gives an average surface age of 288+311/−98 Myr (190–600 Myr) (Strom et al. 1994). Similar numbers were obtained earlier based on Venera 15 and 16 images of the northern quarter of Venus: 500 to 1000 Myr by Basilevsky et al. (1985); about 1000 Myr by Basilevsky et al. (1987); and by Kryuchkov (1987); 100 to 500 Myr by Schaber et al. (1987); and 300 to 900 Myr by Ivanov and Basilevsky (1987) and Neukum (1988).

We should keep in mind, however, that the cratering rate is based on the estimate of the current collision rate of Venus-crossing asteroids, and includes the corrections for deceleration, impact speed and crater excavation efficiency. Extrapolation of the rate over several hundred Myr is an assumption. Moreover, Schultz (1992) warns that the crater excavation efficiency on Venus could be reduced due to the effect of very dense atmosphere. So the uncertainty in the average surface age of Venus may be even larger. For convenience the average surface age for Venus' surface is adopted by Strom et al. (1994) as about 300 Myr.

Phillips et al. (1992) used four methods for estimating the production age of the Venusian surface. Three involved the incoming flux of asteroids and one involved the crater distribution on the Moon. All four methods give global average production ages in the range 400 to 800 Myr. The authors realize the shortcomings of all these methods and believe that their estimates could be lower bounds but are probably not incorrect by more than a factor of 2. For the purpose of convenience the average surface age of Venus is adopted by these authors as about 500 Myr. They consider this estimate as a characteristic time over which Venus is globally resurfaced.

So the average surface age of Venus is estimated by different research groups from the impact crater record as about 300 or 500 Myr with possible variation of this estimate mostly due to the uncertainies of the crater production rate.

B. Surface Age of Terrains and Rock Stratigraphic Units

Determination of the absolute age of terrains and material units, if it is done with enough accuracy, could provide significant progress in resolving the problem of the general style of the geologic history of Venus. Strom et al. (1994, p. 10,903) made a statement that "because impact craters on Venus can not be distinguished from a random distribution, both spatially and hypsometrically. relative and absolute dating of local or regional terrains based solely on crater densities is statistically impossible." The sense of word "solely" should be clarified. It means that the crater density on Venus cannot be the only criterion for distinguishing areas with different relative or absolute age. But if some terrain or rock stratigraphic units are distinguished as being of different relative age, using other criteria, for example, a principle of superposition, then the absolute surface age of the distinguished area, terrain or unit can be determined solely on the crater density. This approach has been already used by several authors (see, e.g., Ivanov and Basilevsky 1993; Namiki and Solomon 1994; Price and Suppe 1994; Price 1995; Price et al. 1996). However, even if the terrains or units show a noticeable difference in crater density, the reason for the difference could be either a real difference in age or just a stochastic variation of the crater density. It is possible to estimate a probability that this difference is a stochastic variation and then the researcher using other criteria should decide what hypothesis (age difference or stochastic variation) looks preferable. Another feature of the ages of the terrains and material units determined through crater counts is a fact which is often ignored in the considerations. In the case of Venus, it is always the area-weighted mean of the actual distribution of ages in the unit and it alone puts no constraints on the range of ages of individual parts of this unit.

Tessera. Ivanov and Basilevsky (1993) found that the total density of craters (counting craters of all sizes together) on tessera is practically the same as the crater density on the plains outside of tessera; both nearly match the global average. However, the density of relatively small (<16 km in diameter) craters on tessera is less than the density outside of tessera while the density of large (>16 km and >32 km) craters on tessera is higher than the density outside of tessera. The lower density of small craters on tessera is evidently an observational effect caused by the difficulty of crater identification on the morphologically rough and radar-bright tessera surface, whereas the higher density of larger craters on tessera reflects its older (compared to the adjacent plains) surface age. Based on the density of relatively large craters, the tessera surface age was estimated as $1.47 \pm 0.46 \, T$, where T is global average of surface age of Venus and error bars are at the 1σ level. Large error bars, resulting from the rather small amount of craters used for dating, do not allow one to conclude how significant the age difference is between tessera and the plains.

Strom et al. (1994) confirmed the on-tessera crater statistics of Ivanov and Basilevsky (1993) but they considered the observed differences in crater densities for <16 km and >16 km size intervals on tessera and on the whole planet as a stochastic effect having no significance in terms of age differences. In their later work this research group accepted the idea that the small (<10 craters) deficit of small craters may be "an observational selection effect, caused by the difficulty of identifying very small craters on the rugged highland terrain" (Kirk and Schaber 1995), thus indirectly supporting the approach of Ivanov and Basilevsky (1993) in using only large craters for dating tessera.

In summary, the estimation of the surface age of tessera terrain through crater density showed that tessera terrain formation occurred prior to the majority of presently observed plains; however, the tessera/plains age difference may be either very small (a few Myr) or rather large (a few hundred Myr). Formation of the tessera-forming rock material preceded formation of the tessera terrain. The time gap between the tessera material emplacement and its deformation is of unknown duration.

Regional Plains. This consists mostly of Pwr plains with lesser amounts (a few percent of total) of densely fractured terrain/coronae, fractured and ridged plains, and ridge belts, as well as shield plains. Keeping in mind that the total crater density on tessera is practically same as the global average and that younger lobate and smooth plains (see below) occupy only about 10% of Venus' surface, the surface age of regional plains, which is essentially plains with wrinkle ridges, should be practically equal to the global average crater retention age of Venus (T). This conclusion is confirmed by Price and Suppe (1994) who found that the crater density on plains, which is evidently what we call regional plains, (2.23±0.18) is indistinguishable from the global average (2.01±0.14). This crater density corresponds to the age of about 300 Myr, which they called the reference age.

Price (1995) subdivided the Venus plains into four stratigraphic units and determined the surface age of them through crater density. Her approach is based on the conclusion of Arvidson et al. (1992) that morphological prominence of the plains-forming lava flows is degraded with time due to surface weathering and infill by aeolian deposits. The proposed plains units are (from younger to older): (1) Pl1, highly lobate plains such as Mylitta Fluctus, occupy about 5% of the total plains; (2) Pl2, plains with distinct but less dramatic flow morphology, about 25%; (3) Pl3, plains with subtle lobate appearance, about 30%; and (4) Ps, smooth plains with no discernable flow morhoplogy present, about 35%. Crater densities increase from Pl1 to Ps and, assuming that the planet's average global surface age is 300 Myr, the surface age of these plains units is estimated (Price 1995) to be 210±75 Myr for Pl1, 240±40 Myr for Pl2, 325±45 Myr for Pl3, and 370±45 Myr for Ps (2σ error bars). The percentage of embayed craters decreases along this sequence from about 17% for Pl1, to about 7% for Pl2, 4% for Pl3, and 3% for Ps. These results show that the plains appear to have been emplaced over a period of at least 100 Myr. Monte Carlo resurfacing simulation based on crater density variations in different map units show that this plains emplacement history is completely consistent with the random distribution of impact craters. These results led Price (1995) to the conclusion that "global resurfacing may be an overstatement."

A special work on comparison of Price (1995) map units, based on the prominence of lava flow morphology, with the units of the Kryuchkov (1996) map (which are actually the units of Basilevsky and Head (1995*a*,*b*,*c*), identified based on traditional principle of superposition) was done by Basilevsky et al. (1996). Their results show that at least in the area under study (40–80°N, 140–260°E) the plains units of Price (1995) are mixtures of units of Basilevsky and Head (1995*a*,*b*,*c*). In the sequence from Pl1 to Ps the role of younger plains units of Basilevsky and Head (1995*a*,*b*,*c*) systematically decreases while the role of older units increases. This evidently means that the approach of Price (1995) provides the generally correct ranges of large regions of the planet according to their area-weighted average age but it misses the key events of the geologic history of Venus such as an emplacement of separate units of plains with wrinkle ridges or lobate and smooth plains. Because the plains units of Price (1995) dated by crater counts are mixtures of some end-members it seems promising to apply mixing models to estimate the ages of those end-members.

Coronae. Trying to estimate absolute age of coronae we should not forget that they are products of multi-stage tectonic and volcanic activity. Namiki and Solomon (1994) used the corona data base of Stofan et al. (1992) and determined the crater density inside the outer boundary of the tectonic annulus of the coronae. They divided 358 coronae under study into five groups reflecting their evolutionary stage. For the three groups which include coronae with the predominance of features corresponding to the earlier stages of evolution, the crater density and therefore the surface age are found to be indistinguishable from the global average. For coronae of the so-called categories 2 and 3 having a significant amount of associated interior volcanism the crater density ($\sim 1 \pm 0.5$ craters per million km^2) is about one-half of the global average. As it follows from the observations of Basilevsky and Head (1995b,c), this corona-related volcanism is a mixture of volcanics belonging to the Rusalkian and Atlian groups. Averaging crater counts for all coronae Namiki and Solomon (1994) obtained a crater density of about 0.7 of the global average. Price and Suppe (1994) estimated the average crater density for 364 coronae as 1.12 ± 0.34 (2σ) craters per million km^2, thus generally confirming the results of Namiki and Solomon (1994). Making a correction for the possible presence of pre-corona craters in this crater subpopulation, Price and Suppe estimated the average age of the coronae as 0.28 ± 0.37 T where T is the average surface age of Venus. Because the most ancient component of coronae are stratigraphically as old as ancient components of the regional plains (Basilevsky and Head 1995a,b,c) this estimation of Price and Suppe (1994) is evidently appropriate for the late (Atlian) corona-related activity.

Large Volcanoes and Prominent Lava Flows. Namiki and Solomon (1994) estimated the crater density on 175 volcanoes of at least 50 km in diameter listed in the MIT unpublished volcano data base. This population of volcanoes is evidently dominated by large rift-associated volcanoes such as Theia, Sif, Gula, Ozza, and Maat, which are classified by Basilevsky and Head (1995b,c) as a part of the Atlian Ps/Pl volcanic plains; however, some volcanoes of the Rusalkian age may also be present. Only craters superposed on these volcanoes were taken into account while craters embayed by their lavas were not counted. The crater density on this volcano population (0.9 ± 0.2 craters per million km^2) is about half of the global average. Price and Suppe (1994) found an even lower crater density on 128 large volcanoes: 0.51 ± 0.32 (2σ) craters per million km^2. The difference with the results of Namiki and Solomon (1994) is probably due to slightly different volcano populations studied by these research groups and to the problem of the boundary craters considered in detail by Ivanov and Basilevsky (1993).

Price and Suppe (1994) also estimated the crater density on 48 flood-type lava flow fields. The majority of them correspond evidently to the Atlian lobate plains of Basilevsky and Head (1994). However, some prominent Rusalkian lobate flows deformed by wrinkle ridges, for example, the lava field at the north of Sedna Planitia, may be present among the studied lava fields. The average crater density on these lava fields is found to be 0.92 ± 0.65 (2σ) craters per million km^2 or about one-half of the global average.

Combining the results of Namiki and Solomon (1994) and Price and Suppe (1994), the average surface age of the Atlian volcanics associated both with rifts and coronae may be estimated as about half of the global average surface age or even younger. Basilevsky (1993) found that lavas of Maat volcano, which cover fractures crossing the dark-parabola crater Uvaisi, are probably younger than about 10% of global average surface age (0.1 T). This means that the youngest Atlian volcanoes are of this age. However, in order to make the average age of Atlian volcanics to be about a half of the global average some of the Atlian volcanics should be noticeably older than 0.5 T.

Rifts. Average crater density within the major rift zones was estimated by Price and Suppe (1994) as 1.48±0.40 (2σ) craters per million km^2 that is ~0.74 of the global average. Making corrections on stretching of the area as a result of rifting and on the presence of craters formed prior to the rifting, these authors estimate the average age of rifting as 0.27±0.39 T.

Radar-Dark Parabolas. These were distinguished by Campbell et al. (1992) as parabolic-shaped mantles of radar-dark debris mantles associated with the youngest 10% of Venus' crater population. Basilevsky and Head (1995b,c) consider them as the youngest stratigraphic unit of Venus, that is the Aurelia Group. No crater density counts were made for the Aurelia group and estimation of its absolute age was made based on another approach. Basilevsky (1993) and Strom (1993) confirmed the observation of Campbell et al. (1992) that radar-dark parabolas associate with the youngest 10% of population of Venus craters and concluded that the age of this youngest 10% of Venus' craters should not be more than 10% of age of Venus' crater population. The latter is estimated to be 300 to 500 Myr (Phillips et al. 1992; Schaber et al. 1992; Strom et al. 1994) thus the absolute age of the Aurelian impact craters and associated radar-dark parabolas should be not more than 30 to 50 Myr (Basilevsky 1993; Strom 1993).

So a review of attempts to determine the absolute age of Venus surface and some of its parts showed that the global average surface age is about 300 to 500 Myr, that is, Paleozoic in terms of the Earth geologic time units. The surface age estimates of various terrains show that stratigraphically upper ones have consistently younger ages than the lower units (Fig. 11), supporting to a first approximation the quasi-synchronous character of the stratigraphic units. The mentioned estimations of the surface age agree both with the model of global resurfacing (Strom et al. 1994) and with the idea of Phillips (1993) that the surface of Venus is divisible into three major age groupings.

VI. RATE OF RESURFACING

Both endogenic and exogenic resurfacing operate on Venus although their rates are very different. That endogenic resurfacing dominates is evident from the striking morphologic prominence of hundred-million old volcanic and tectonic landforms and terrains visible on the images. Endogenic resurfacing is the sum of tectonic and volcanic resurfacings. Their scales may be roughly estimated from morphologic observations. Their rates are strongly model dependent, becoming higher if the catastrophe-style models are involved and lower in the case of models which consider resurfacing as a long-term process.

Tectonic resurfacing with drastic renovation of the surface took place at least twice within the studied period of geologic history of Venus: (1) when tessera formed, and (2) when densely fractured terrain of plains and coronae were formed. In addition, a noticeable tectonic resurfacing was (and maybe still is) associated with the formation of the observed rift zones. During Fortunian time, the formation of tessera terrain probably affected from half to all the planet's surface. We do not know the duration of tessera formation, thus the rate of resurfacing in late Fortunian time is unknown. Remnants of densely fractured terrain of plains and coronae are distributed globally so one may guess that tectonic resurfacing in late Sigrunian time was global-wide or almost global-wide but this is, of course, an assumption. We do not know the duration of this tectonic episode(s) so the rate of the late-Sigrunian tectonic resurfacing is also unknown. Very dense faulting that obscures the precursor terrain is typical for significant areas of some rift zones (e.g., Dali, Diana, Artemis Chasmata), accounting for as much as 1 to 2% of Venus surface. As was shown above, this rifting occurred at Atlian time during the last half or quarter of the average global surface age of Venus. However, because it is not clear whether this rifting was distributed over all this time or concentrated in the shorter time periods, the rate of the Atlian tectonic resurfacing is also unknown. We may only conclude that its scale was about 2 orders of magnitude less than the scales of the late Fortunian and late Sigrunian tectonic resurfacings.

Volcanic resurfacing of the observable part of geologic history of Venus can be subdivided in two stages: (1) pre-Atlian resurfacing resulted in the formation of regional plains, and (2) Atlian resurfacing. Formation of regional plains led to burial of vast tessera areas a few km in elevation that demands a volcanic layer a few km thick (Ivanov and Head 1996). Assuming that the regional plains volcanics thickness is about 1 to 3 km, their total volume is estimated to be about 0.4 to 1.2×10^9 km^3. If according to Strom et al. (1994) this resurfacing stage "lasted at least tens of millions of years," the average rate of the volcanism formed regional plains was about 3 to 10 km^3 yr^{-1}, that is more by an order of magnitude higher than the estimated current rate of intraplate volcanism on Earth (0.33–0.5 km^3 yr^{-1}) and the same or by a factor of 3 larger compared to the Earth ocean ridge volcanism (Crisp 1984; Head et al. 1992).

Atlian volcanic resurfacing in the form of lava fields and large shield volcanoes affected about 10% of Venus' surface. Average thickness of the Atlian volcanics is probably comparable to 1 km so their total volume is estimated as about 4.6×10^7 km^3. If the regional plains emplacement lasted tens of Myr (Strom et al. 1994) or even about 100 Myr (Price 1995), the duration of Atlian time was about 200 to 250 Myr to 300 to 450 Myr (the case if the 500 Myr global average is more correct than 300 Myr). This means that the rate of Venus' volcanism during the Atlian time was about 0.1 to 0.2 km^3 yr^{-1} that, taking in mind the low accuracy of our estimations of time interval durations and volcanics volumes for Venus, is comparable to the current rate

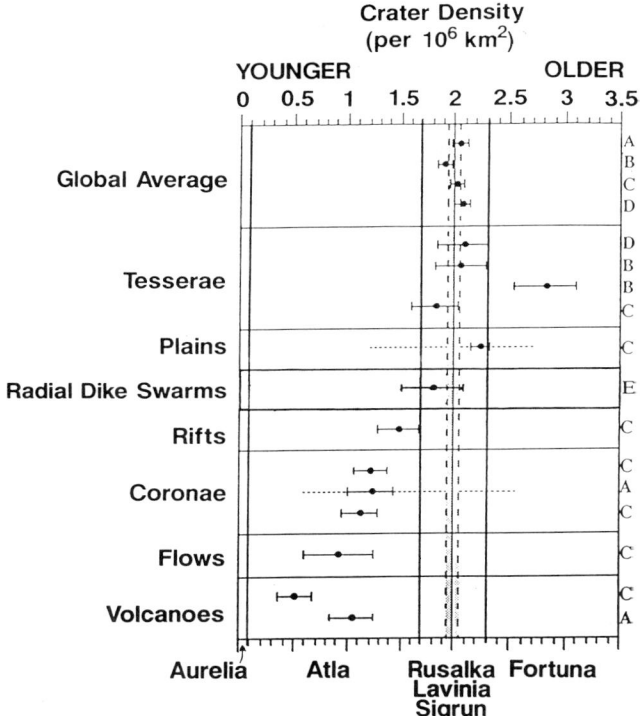

Figure 11. Comparison between the crater density reported for different geologic units and stratigraphic units of Basilevsky and Head (1995a,b). See text for details. Letter keys for the data sources are as follows: A = Namiki and Solomon (1994); B = Ivanov and Basilevsky (1993); C = Price and Suppe (1994); D = Strom et al. (1994); E = Grosfils and Head (1996).

of intraplate volcanism of Earth (Crisp 1984; Head et al. 1992).

Exogenic resurfacing on Venus is a complex total of aeolian deflation and accumulation with an unknown contribution of chemical weathering and on-slope mass wasting. Sources of sediment are predominantly impact cratering (Garvin 1990), as well as material contributed by tectonic deformation and chemical weathering. Estimate of average rate was first made based on the phenomenon of disappearance of radar-bright halos around impact craters that have undergone the process of morphological maturation interpreted from the Venera 15 and 16 images (Nikolaeva et al. 1986; Basilevsky et al. 1992). The halos are zones of impact-induced surface roughness of decimeter-decameter scale and they are present only around the youngest craters, totalling about one-fourth of the crater population. If the age of crater population is about 500 Myr, then these observations suggest that the decimeter-decameter roughness is smoothed out during a time period as long as 100 to 200 Myr. This leads to the resurfacing rate (in terms of thickness of the eroded or deposited layer) of 0.1 to 10 cm Myr^{-1}.

Arvidson et al. (1992), using the Magellan observations concluded that most plains surfaces have been modified in the upper meter and this led them to an estimation of the exogenic resurfacing rate as about 2 mm Myr^{-1} thus confirming the Venera-based estimates. This rate of exogenic resurfacing is orders of magnitude less than the rate of surface modification on Earth, or the rate of aeolian dispersal of friable materials on Mars and is comparable to the rate of formation of lunar regolith in the post-mare time (Arvidson et al. 1992; Nikolaeva et al. 1992; Basilevsky 1974).

Endogenic resurfacing on Venus is comparable in its scale and rate with those for Earth while exogenic resurfacing on Venus is orders of magnitude less efficient than on Earth. The latter is evidently a result of the stagnant environment of Venus' surface where the massive atmosphere is a giant buffer minimizing and even extinguishing possible environmental variations.

VII. A SCENARIO OF THE GEOLOGIC HISTORY OF VENUS

Based on the relations among various geologic formations described above, resulting from observations, modeling, assumptions and interpretations of many authors, a scenario of the geologic history of Venus is suggested (Fig. 12).

About 90% of the history, from planetary formation until about 300 to 500 Myr ago, is not preserved in the surface geomorphological record distinguishable on the Magellan images. The beginning of this observable part of the history was characterized by intensive tectonic deformation of global or semi-global scale which formed tessera terrain. Early stages of that deformation were evidently compressional and later (maybe within very short period of time) changed into extensional (Bindschadler and Head 1991; Ivanov and Head 1995,1996). The duration of the deformational episode(s) is unknown but the rarity of on-tessera craters affected by the tessera-forming deformation shows that termination of that deformation was rather fast, probably less than a few Myr for the compressional stages and less than about 30 Myr for the extensional stages (Ivanov and Basilevsky 1993; Gilmore et al. 1996).

The cause of the tessera-forming deformation is an area of many hypotheses (see summary in Ivanov and Head 1996). It might be a gravitational instability causing mantle overturn (Parmentier and Hess 1992; Head et al. 1994), oscillatory convective regime of the planets mantle (Arkani-Hamed and Toksoz 1984; Arkani-Hamed et al. 1993), episodic plate tectonics (Turcotte 1993), "catastrophic" convective episode caused by the phase transition(s) in the mantle (Steinbach and Yuen 1992; Weinstein 1993; Herrick 1994), or something else. This intensive tectonism could be accompanied by volcanic activity (Ivanov and Head 1996). The nature and composition of the tessera-forming material are unknown. The idea that it may be essentially feldspatic, being a counterpart of the continental crusts of other terrestrial planets (Nikolaeva et al. 1988,1992), is still neither proved nor rejected. Emplacement

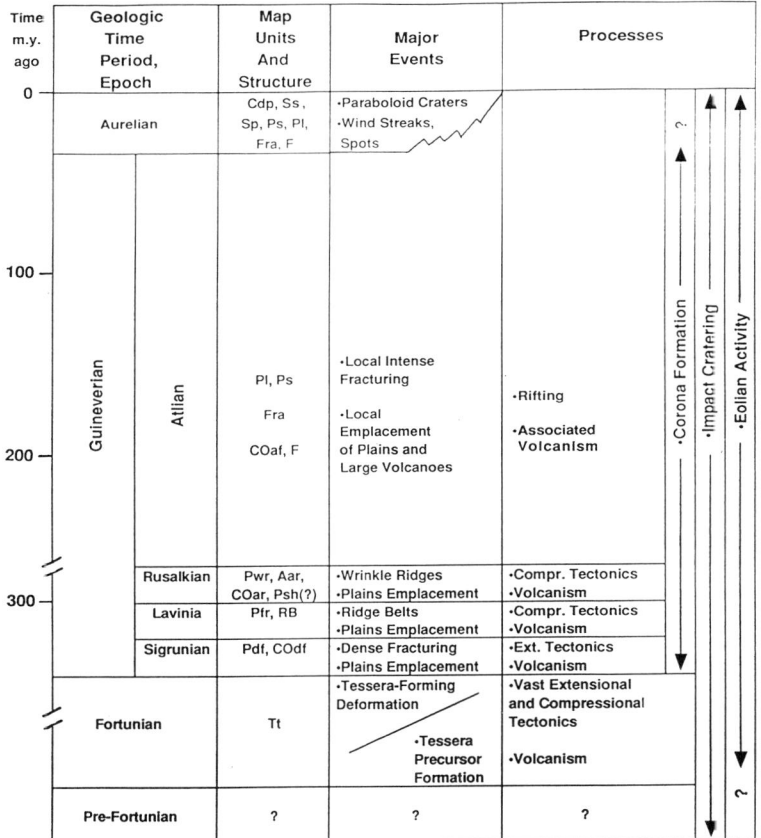

Figure 12. Summary of Venus global stratigraphy and geologic history. For convenience, the average crater retention age is considered to be 300 Myr.

of the tessera-forming material and its deformation in the tessera terrain are major geologic events of the Fortunian time.

After tessera formation, several stages of extensive evidently basaltic volcanism occurred burying vast areas of tessera and forming what we see now as regional plains. Following and partly coincident with the terminal stages of tessera formation, the early regional plains units formed (Sigrunian group) and then were deformed by a dense pattern of closely spaced extensional fractures and graben. Then the emplacement of volcanic plains continued, but deformation of the plains switched mostly from extensional to compressional, forming broad ridges of the Lavinian ridge belts and wrinkle ridges of the Rusalkian Pwr plains. The duration of these episodes of plains-forming activity is a problem to be resolved. Areas of tessera that survived this plains-forming activity bear structures formed in this later (Guineverian) period and overprinted on the earlier tessera-forming structures. Assessment of the relative role of different episodes and styles of deformation of what is now tessera terrain is a challenge for future studies.

Average rate of volcanism during the period of the regional plains emplacement was close to Earth oceanic ridge volcanism of the most recent geologic epoch. This, however, does not imply a similarity in styles of these two volcanic activities. The Venusian plains were emplaced in a tectonic environment very different from the terrestrial mid-oceanic ridges where almost continuous growth of the basaltic crust is realized on the surface in the form of individual extrusions with typical areal extent from tens-hundreds of meters to tens of km. Meanwhile regional plains of Venus were mostly emplaced in the form of a really vast (hundreds of km long and more) lava flows thus resembling the large-scale volcanism associating with episodic events ("superplumes") represented, for example, by the Decan or Columbia River flood basalts (Coffin and Eldholm 1994).

The formation and evolution of coronae occurred in parallel with the formation of regional plains. Early formative stages of the corona population were evidently contemporaneous with Sigrunian densely fractured plains and Lavinian fractured and ridged plains and ridge belts. Intermediate stages of evolution of the majority of coronae were evidently contemporaneous with the emplacement of the Rusalkian plains and their deformation by wrinkle ridges. The final stages of the life of the corona population extended into Atlian time.

On the basis of our data, an ideal corona would start its activity in the Sigrunian time and terminate it in the Atlian time. Real coronae show various options when (in the regional stratigraphy sense) the presently visible activity of each given corona started and terminated: early start-early end, early start-late end, intermediate start-late end and so on. An important characteristics of the corona evolution is that the early (in the regional stratigraphy sense) activity is usually in the form of Sigrunian type dense fracturing and Lavinian type broad ridging. The intermediate activity is usually in the form of volcanic emplacement and wrinkle ridging of the Rusalkian type. And the late activity is usually in the form of volcanic emplacement and fracturing of the Atlian type. This means that time, actually the regional tectonic environment at this given time period, determined somehow the style of the corona activity.

Finally, Venus entered its present stage with a predominance of rifting and localized rift-associated and corona-associated basaltic volcanism in the form of large shield volcanic constructions and volcanic plains-forming units. This is the Atlian time. The question of when it started is actually the same question as when the regional plains emplacement was terminated. The average rate of volcanism at this period was estimated to be comparable to the rate of intraplate volcanism of Earth. The tectonic environment of this stage of Venusian volcanism, associating with rifts and coronae, resembles the tectonic enviroment of the continental rifts and hot spots typical for the intraplate volcanism of Earth. In a few cases the rifting and associated volcanism postdated emplacement of craters with dark parabolas which are indicative for the Aurelian time. The estimated duration of the Aurelian time is about 30 to 50 Myr (Basilevsky 1993; Strom 1993).

The popular question is whether Venus is volcanically and tectonically active now. Comparison of images of Venus' surface taken by Magellan with 8- and 16-month time intervals did not reveal any changes for which current volcanism and tectonics might be responsible. However, it is evident that, if Magellan had surveyed not Venus but Earth, a certain luck would have to occur to have detected the recognizeable changes. The cases of rifting and associated volcanism postdating the dark parabola craters mentioned above show that Venus was still endogenically active in very recent geologic epoch and this is indirect evidence in favor of today's activity.

Parameters of the Venus atmosphere such as surface pressure and temperature for all relatively recent periods of geologic history of Venus were probably close to that now observed. This is evidenced by the fact that the impact crater population has a deficit of small craters which is the same as predicted by models of meteoroid passage through today's atmosphere (Ivanov 1988; Melosh 1981). This means that over the last 300 to 500 Myr exogenic processes on Venus' surface were approximately the same as now: low aeolian deflation and accumulation of debris mostly associated with impact craters and no processes involving liquid water. Thus, the epoch of significantly higher water content on Venus, supposed based on a very high D/H ratio measured in the Venusian atmosphere by the Pioneer Venus (Donahue et al. 1982), probably occurred (if ever) prior this 300 to 500 Myr period.

VIII. CONCLUSION

The Magellan mission strongly advanced our understanding of the geology of this planet and the character of its geologic evolution for the last 300 to 500 Myr. Analysis of the data is far from complete, and further analysis will provide new important discoveries. However, it is already clear that further significant progress in understanding of Venus geology, which is crucial for understanding the general principles of the geologic evolution of the terrestrial planets, including Earth, demands new data on geology, geochemistry and geophysics of this planet. Three directions of the studies seem to be of key importance and are listed below.

1. Imaging of significant areas of Venus' surface (about 0.1 to 1% of the planet surface) with resolution of a few meters. Analysis of such images could provide the possibility of seeing traces of pre-tessera events in Venus' history and of linking reliably the lander observations and measurements to regional stratigraphic units. It is necessary to study various options to make such imaging available: side-looking radar from orbiter, aerobots, and diving balloons.
2. The direct geochemical and mineralogical characterization of a number

of important elements of Venus geology such as tessera, steep-sided domes, floors and deltas of Venusian channels, various plains units and so on. This is the way to resolve the question if there is a continental geochemically evolved crust on Venus. This approach demands new landings on Venus. As a distant perspective, a design of a sample return mission from Venus should be considered as a tool to make very large progress in the problems of Venus' geochemistry and geochronology.
3. Seismic sounding of the Venusian interior and determination of the level of today's seismicity of Venus. This demands long-lived landers on Venus, and if it happens, it will provide a third dimension to Venus geology, and will answer unequivocally the question about today's endogenic activity of this planet.

The challenge for the planetary science community is to work out basic concepts, to initiate a general design of new missions to Venus and to put pressure on the national space agencies to fly these missions.

NOTE ADDED IN PROOF

A recent revision of the U.S.G.S. Venus impact crater inventory (contact http://www.flag.wr.usgs.gov) shows a total of 940 craters (743 craters appearing pristine, 107 slightly fractured, 27 heavily fractured, 26 embayed by lavas from external volcanic sources and slightly fractured, 6 embayed by lavas from external volcanic sources and heavily fractured, 8 compressed, 2 compressed and slightly fractured, and 5 craters embayed or mantled by the ejecta from other impact craters).

Recent two-dimensional Monte Carlo simulations by Strom, Dawson, and Schaber were undertaken to update their earlier results (see Fig. 2 and pp. 23,364–23,365 of Strom et al. [1995]) addressing (1) the percentage of Venus resurfaced by volcanism (alone) following the termination of the most recent global resurfacing event (about 300 Myr ago), and (2) the time estimated to end global resurfacing. The new computations for the first time include as simulation parameters the surface area covered by crater (usually radar-dark) halo deposits—in addition to impact crater ejecta and splotches. As earlier, the new simulations were run assuming that specific numbers of the observed volcanically-embayed craters were embayed either by the proposed global resurfacing, or subsequent to it (over the past 300 Myr or so). The results (in preparation) show that the inclusion of crater halos in the simulations change little the results reported earlier (Strom et al. 1995) that considered only craters and splotches. For example, when craters, halos and splotches are considered, and it is assumed that none, 10, and 20 (just under 50%) of the observed lava-embayed craters have been embayed during the global resurfacing event and the remainder subsequent to that event, then only 6.5%, 4.6% and 3.3%, respectively, of the planet could have been resurfaced by volcanism alone since the proposed global event ended. If splotches are not considered in the simulations, then the maximum percentage of the planet that could be resurfaced under the same assumptions would be 13.5%, 10.7%, and 6.8%, respectively. These are maximum values given that the two-dimensional simulations take into consideration the entire planetary surface, and overestimates the number of destroyed craters. Therefore, these values may in fact slightly overestimate the degree of volcanic (only) resurfacing over the past 300 Myr.

Given the same 0, 10, and 20 craters assumed to have been embayed during the global event, the time to end global resurfacing by volcanism was found by Strom, Dawson, and Schaber to be <10 Myr, 50 Myr, and 100 Myr, respectively, when considering craters, halos and splotches. When only craters and halos were considered in their simulations (splotches were excluded), the time to end global resurfacing is found to be <10 Myr, 60 Myr, and 160 Myr, respectively.

REFERENCES

ACSN. 1961. American code of stratigraphic nomenclature. *Amer. Assoc. Petrol. Geol. Bull.* 45:645–660.
Arkani-Hamed, J., and Toksoz, M. N. 1984. Thermal evolution of Venus. *Phys. Earth Planet. Int.* 34:232–250.
Arkani-Hamed, J., Schaber, G. G., and Strom, R.G. 1993. Constraints on the thermal evolution of Venus inferred from Magellan data. *J. Geophys. Res.* 97:5309–5315.
Arvidson R. E., et al. 1992. Surface modification of Venus as inferred from Magellan observations of plains. *J.Geophys. Res.* 97:13303–13317.
Aubele, J. C. 1995. Stratigraphy of small volcanoes and plain terrains in Vellamo Planitia-Shimti Tessera Region, Venus. *Lunar Planet. Sci. Conf.* XXVI:59–60 (abstract).
Aubele J. C. 1996. Akkruva small shield plains: Definition of a significant regional plains unit on Venus. *Lunar Planet. Sci. Conf.* XXVII:49–50 (abstract).
Baer, G., Schubert, G., Bindschadler, D. L., and Stofan, E. R. 1994. Spatial and temporal relations between coronae and extensional belts, northern Lada Terra. *J. Geophys. Res.* 99:8355–8369.
Baker, V. R., Komatsu, G., and Parker T. J. 1992. Channels and valleys on Venus: Preliminary analysis of Magellan data. *J.Geophys. Res.* 97:13421–13444.
Basilevsky, A. T. 1974. Estimation of thickness and degree of resurfacing cf lunar regolith from the crater density. *Kosmicheskie Issledovania* 12:606–609 (in Russian).
Basilevsky, A. T. 1986. The structure of the Ishtar Terra central and eastern parts and some tectonic problems of Venus. *Geotectonika* 4:42–54 (in Russian).
Basilevsky, A. T. 1993. Age of rifting and associated volcanism in Atla, Regio, Venus. *Geophys. Res. Lett.* 20:883–886.
Basilevsky, A. T. 1995. Compositional heterogeneity and late-stage deformation in Maxwell Montes, Venus. *Lunar Planet. Sci. Conf.* 26:798–780 (abstract).

Basilevsky, A. T. 1996. Geologic mapping of V-17 Beta Regio quadrangle: Preliminary results. *Lunar Planet. Sci. Conf.* XXVII:65–66 (abstract).

Basilevsky, A. T., and Head, J. W. 1994. Characteristics of the Geology of Thirty-Six Sites on Venus. (Providence, R. I.: Brown Univ. Library).

Basilevsky, A. T., and Head, J. W. 1995*a*. Global stratigraphy of Venus: Analysis of a random sample of thirty-six test areas. *Earth, Moon, Planets* 66:285–336.

Basilevsky, A. T., and Head, J. W. 1995*b*. Geologic history of Venus for the last 300–500 m.y. based on photogeologic analysis of the Magellan images. *Astron. Vestnik* 29(3):195–218 (in Russian).

Basilevsky, A. T., and Head, J. W. 1995*c*. Regional and global stratigraphy of Venus: A preliminary assessment and implications for the geologic history of Venus. *Planet. Space Sci.* 43(12):1523–1553.

Basilevsky, A. T., and Head, J. W. 1996. Evidence for rapid and widespread emplacement of volcanic plains on Venus: Stratigraphic studies in the Baltis Vallis region. *Geophys. Res. Lett.* 23:1497–1500.

Basilevsky, A. T., et al. 1985. Impact craters of Venus based on the Venera 15/16 radar images. *Dokl. Akad. Nauk SSSR* 282:671–674 (in Russian).

Basilevsky, A. T., et al. 1987. Impact craters of Venus: A continuation of the analysis of data from the Venera 15 and 16 spacecraft. *J. Geophys. Res.* 92:12869–12901.

Basilevsky, A. T., Nikolaeva, O. V., and Kuzmin, R. O. 1992. Resurfacing. In *Venus Geology, Geochemistry, and Geophysics*, eds. V. L. Barsukov, A. T. Basilevsky, V. P. Volkov and V. N. Zharkov (Tucson: Univ. of Arizona Press), pp. 153–160.

Basilevsky, A. T., Burba, G. A., and Kruchkov, V. P. 1996. Stratigraphy of venusian plains: Comparison of Price (1995) and Basilevsky and Head (1995) units. *Lunar Planet. Sci. Conf.* XXVII:73–74 (abstract).

Bilotti, F. 1992. Global organization of tectonic deformation on Venus. *Lunar Planet. Sci. Conf.* XXIV:107–108 (abstract).

Bindschadler, D. L., and Head, J. W. 1991. Tessera terrain, Venus: Characterization and models for origin and evolution. *J. Geophys. Res.* 96:5889–5907.

Bindschadler, D. L., Schubert, G., and Kaula, W. M. 1992. Coldspots and hotspots: Global tectonic and mantle dynamics of Venus. *J. Geophys. Res.* 97:16135–16335.

Campbell, D. B., et al. 1992. Magellan observations of extended impact crater related features on the surface of Venus. *J. Geophys. Res.* 97:16249–16277.

Chadwick, D. J., and Schaber, G. G. 1994. Evidence for episodic construction of Ovda Regio, Venus. *Lunar Planet. Sci. Conf.* XXV:229–230 (abstract).

Coffin, M. F., and Eldholm, O. 1994. Large igneous provinces: Crustal structure, dimensions and external consequences. *Rev. Geophys.* 32:1–36.

Collins, G. C., et al. 1996. Impact crater embayment on Venus and the termination of global resurfacing. *Lunar Planet. Sci. Conf.* XXVI:245–246 (abstract).

Copp, D. L., and Guest, J. E. 1995. Geology of the V31 Sif and Gula quadrangle of Venus. *Lunar Planet. Sci. Conf.* XXVI:283–284 (abstract).

Crisp, J. A. 1984. Rates of magma emplacement and volcanic output. *J. Volcanol. Geotherm. Res.* 20:177–211.

Crumpler, L. S., Head, J. W., and Campbell, D. B. 1986. Orogenic belts on Venus. *Geology* 14:1031–1034.

Donahue, T. M., Hoffman, J. H., Hodges, R. R., and Watson, A. J. 1982. Venus was wet: a measurement of the ratio of deuterium to hydrogen. *Science* 216:630–633.

Garvin, J. B. 1990. The global budget of impact derived sediment on Venus. *Earth Moon Planets* 50/51:175–190.

Gilmore, M. S., and Head, J. W. 1995. Formation of tessera terrain on Venus: A structural analysis of Tellus Regio. *Lunar Planet. Sci. Conf.* XXVI:461–462 (abstract).

Gilmore, M. S., Ivanov, M. A., Head, J. W., and Basilevsky, A. T. 1996. Deformation of craters on tessera terrain, Venus. *Lunar Planet. Sci. Conf.* XXVII:419–420 (abstract).
Greeley, R., et al. 1992. Aeolian features on Venus: Preliminary Magellan results. *J. Geophys. Res.* 97:13319–13345.
Grimm, R. E., and Phillips, R. J. 1992. Anatomy of venusian hot-spot: Geology, gravity, and mantle dynamics of Eistla Regio. *J. Geophys. Res.* 97:16035–16054.
Grosfils, E. B., and Head, J. W. 1996. The timing of giant radiating dike swarms emplacement on Venus: Implications for resurfacing of the planet and its subsequent evolution. *J. Geophys. Res.* 101:4645–4656.
Hansen, V. L., and Willis, J. J. 1996. Structural analysis and geodynamic implications of tessera terrain, Venus. *Lunar Planet. Sci. Conf.* XXVII:489–490 (abstract).
Head, J. W. 1995a. Formation and evolution of tessera terrain on Venus: Outstanding problems. *Lunar Planet. Sci. Conf.* XXVI:573–574 (abstract).
Head, J. W. 1995b. Tectonic facies in Venus tessera terrain: Classification and interpretation of sequence of deformation. *Lunar Planet. Sci. Conf.* XXVI:579–580 (abstract).
Head, J. W., et al. 1992. Venus volcanism: Classification of volcanic features and structures, associations, and global distribution from Magellan data. *J. Geophys. Res.* 97:13153–13197.
Head, J. W., Parmentier, E. M., and Hess, P. C. 1994. Venus: Vertical accretion of crust and depleted mantle and implications for geological history and processes. *Planet. Space Sci.* 42(10): 803–811.
Herrick, R. R. 1994. Resurfacing history of Venus. *Geology* 22:703–706.
Herrick, R. R., and Phillips, R. J. 1994. Implications of a global survey of venusian impact craters. *Icarus* 111:387–416.
Herrick, R. R., Izenberg, N., and Phillips, R. G. 1995. Comment on "The global resurfacing of Venus" by R. G. Strom, G. G. Schaber, and D. D. Dawson. *J.Geophys. Res.* 100:355–359.
Ivanov, B. A. 1988. On the breakup diameter of meteoroids in the venusian atmosphere. *Lunar Planet. Sci. Conf.* XIX:533–534 (abstract).
Ivanov, B. A., and Basilevsky, A. T. 1987. Comparison of crater retention age on Earth and Venus. *Astron. Vestnik.* 21:136–143 (in Russian).
Ivanov, M. A., and Basilevsky, A. T. 1993. Density and morphology of impact craters on tessera terrain, Venus. *Geophys. Res. Lett.* 20:2579–2582.
Ivanov, M. A., and Head, J. W. 1993. Tessera terrain on Venus: Global characterization from Magellan data. *Lunar Planet. Sci. Conf.* XXIV:691–692 (abstract).
Ivanov, M. A., and Head, J. W. 1995. Sequence of events in tessera formation, Northern Ovda Regio, Venus. *Lunar Planet. Sci. Conf.* XXVI:661–662 (abstract).
Ivanov, M. A., and Head, J. W. 1996. Tessera terrain on Venus: A summary of the global distribution, characteristics, and relation to surrounding units from Magellan data. *J. Geophys. Res.* 101:14861–14908.
Ivanov, B. A., Basilevsky, A. T., Kryuchkov, V. P., and Chernaya, I. M. 1986. Impact craters on Venus: Analysis of Venera 15/16 data. *Proc. Lunar Planet. Sci. Conf.* 16, *J. Geophys. Res. Suppl.* 91:423–430.
Izenberg, N. R., Arvidson, R. E., and Phillips, R. J. 1994. Impact crater degradation on venusian plains. *Geophys. Res. Lett.* 21:289–292.
Jackson, E., Brown, D., and Grimm, R. E. 1995. Rifting and corona formation in eastern Aphrodite Terra, Venus. *Lunar Planet. Sci. Conf.* XXVI:665–666 (abstract).
Kahn, R. 1982. Deducing the age of the dense Venus atmosphere. *Icarus* 49:71–85.
Kargel, J., Kirk, R. L., Fegley, B., and Treiman, A. H. 1994. Carbonate-sulfate

volcanism on Venus. *Icarus* 112:219–252.
Kaula, W. K., et al. 1992. Styles of deformation in Ishtar Terra and their implications. *J. Geophys. Res.* 97:16085–16120.
Keep, M., and Hansen, V. L. 1994. Structural history of Maxwell Montes, Venus: Implications for venusian mountain belt formation. *J. Geophys. Res.* 99:26015–26028.
Kirk, R. L., and Schaber, G. G. 1995. New statistical results on the spatial distribution and physical properties of impact craters on Venus. *Lunar Planet. Sci. Conf* XXVI:757–758 (abstract).
Kryuchkov, V. P. 1987. Analysis of population of venusian impact craters. *Izvestia U.S.S.R. Academy Sci. Geology* 6:75–83 (in Russian).
Kryuchkov, V. P. 1996. Problems of Venus stratigraphy revealed in the process of 1:10M photogeologic mapping in the area between Ananke Tessera and Lakshmi Planum, Venus. *Lunar Planet. Sci. Conf.* XXVII:713–714 (abstract).
Magee, K., and Head, J. W. 1995. The role of rifting in the generation of melt: Implications for the origin and evolution of the Lada Terra-Lavinia Planitia region of Venus. *Geophys. Res. Lett.* 100:1527–1552.
Magee Roberts, K. M., Guest, J. E., Head, J. W., and Lancaster, M. G. 1992. Mylitta Fluctus, Venus: Rift-related, centralized volcanism and the emplacement of large-volume flow units. *J. Geophys. Res.* 97:15991–16015.
McGill, G. E. 1993. Wrinkle ridges, stress domains, and kinematics of Venusian plains. *Geophys. Res. Lett.* 20:2407–2410.
Melosh, H. J. 1981. Atmospheric breakup of terrestrial impactors. In *Multi-Ring Basins*, eds. P. H. Schultz and R. B. Merrill (New York: Pergamon Press), pp. 29–35.
Namiki, N., and Solomon, S. C. 1994. Impact crater densities on volcanoes and coronae on Venus: Implications for volcanic resurfacing. *Science* 265:929–933.
Nikolaeva, O. V., Ronca, L. B., and Basilevsky, A. T. 1986. Circular fractures on the plains of Venus as an evidence of its geologic history. *Geochimia* 5:579–589 (in Russian).
Nikolaeva, O. V., et al. 1988. Are tesserae the outcrops of feldspatic crust on Venus? *Lunar Planet. Sci. Conf.* XIX:864–865 (abstract).
Nikolaeva, O. V., Basilevsky, A. T., and Kuzmin, R. O. 1992. Resurfacing. In *Venus Geology, Geochemistry, and Geophysics*, eds. V. L. Barsukov, A. T. Basilevsky, V. P. Volkov and V. N. Zharkov (Tucson: Univ. of Arizona Press), pp. 153–160.
Neukum, G. 1988. The cratering rate in the Earth-Moon system over the past 3 b.y. and in recent time. *Lunar Planet. Sci. Conf.* XIX:850–851 (abstract).
Parker, T. J., Komatsu, G., and Baker, V. R. 1992. Longitudinal topographic profiles of very long channels in venusian plains regions. *Lunar Planet. Sci. Conf.* XXIII:1035–1036 (abstract).
Parmentier, E. M., and Hess, P. C. 1992. Chemical differentiation of a convecting planetary interior: Consequences for a one plate planet such as Venus. *Geophys. Res. Letts.* 19:2015–2018.
Phillips, R. J. 1993. The age spectrum of the Venusian surface. *Eos: Trans. AGU* 74(16):187.
Phillips, R. J., et al. 1992. Impact craters and Venus resurfacing history. *J. Geophys. Res.* 97:15923–15948.
Plescia, J. B., and Golombek, M. P. 1986. Origin of planetary wrinkle ridges based on study of terrestrial analogs. *Geol. Soc. Amer. Bull.* 97:1289–1299.
Price, M. 1995. Resurfacing history of the venusian plains based on distribution of impact craters. *Lunar Planet. Sci. Conf.* XXVI:1143–1144 (abstract).
Price, M., and Suppe, J. 1994. Mean age of rifting and volcanism on Venus deduced from impact crater densities. *Nature* 372:756–759.

Price, M., and Suppe, J. 1995. Constraints on the resurfacing history of Venus from the hypsometry and distribution of volcanism, tectonism, and impact craters. *Earth, Moon, Planets* 71(1-2):99-145.

Price, M. H., Watson, G., Suppe, J., and Brankman, C. 1996. Dating volcanism and rifting on Venus using impact crater densities. *J. Geophys. Res.* 101:4657-4671.

Pronin. A. A. 1986. The Laksmi Plateau structure as an indicator of asthenospheric flow on Venus. *Geotektonika* 4:26-41 (in Russian).

Saunders, R. S., and Stofan, E. R. 1992. Science questions for the Magellan continuing mission. *Lunar Planet. Sci. Conf.* XXIII:1211-1212 (abstract).

Saunders, R. S., and the Magellan Science Team. 1992. Magellan mission summary. *J. Geophys. Res.* 97:13067-13090.

Schaber, G. G., Shoemaker, E. M., and Kozak, R. C. 1987. The surface age of Venus. *Solar System Res.* 21:89-94.

Schaber, G. G., et al. 1992. Geology and distribution of impact craters on Venus: What are they telling us? *J. Geophys. Res.* 97:13256-13301.

Schaber, G. G., Kirk, R. L., and Strom, R. G. 1995. *Data Base of Impact Craters on Venus Based on Analysis of Magellan Radar Images and Altimetry Data*, USGS Open-File Rept. 95-561.

Schultz, P. 1992. Atmospheric effects on ejecta emplacement and crater formation on Venus from Magellan. *J. Geophys. Res.* 97:183-248.

Senske, D. A., Schaber, G. G., and Stofan, E. R. 1992. Regional topographic rises on Venus: Geology of Western Eistla Regio and comparison to Beta Regio and Atla Regio. *J. Geophys. Res.* 97:13395-13420.

Sharpton, V. L. 1994. Evidence from Magellan for unexpectedly deep complex craters on Venus. In *Large Meteorite Impacts and Planetary Evolution*, eds. B. O. Dressler, R. A. F. Grieve and V. L. Sharpton, GSA SP-293 (Boulder: Geological Soc. of America), pp. 19-27.

Shoemaker, E. M., Wolfe, R. F., and Shoemaker, C. S. 1991. Asteroid flux and impact cratering rate on Venus. In *Reports to Planetary Geology and Geophysics Program-1990*, NASA TM-4300, pp. 389-390.

Solomon, S. C., et al. 1992. Venus tectonics: An overview of Magellan observations. *J. Geophys. Res.* 97:13199-13255.

Steinbach, V., and Yuen, D. A. 1992. The effects of multiple phase transitions on Venusian mantle convection. *Geophys. Res. Lett.* 19:2243-2246.

Stofan, E. R., et al. 1992. Global distribution and characteristics of coronae and related features on Venus: Implications for origin and relation to mantle processes. *J. Geophys. Res.* 97:13347-13378.

Strom, R. G. 1993. Parabolic features and the erosion rate on Venus. *Lunar Planet. Sci. Conf.* XXIV:1371-1372 (abstract).

Strom, R. G., Schaber, G. G., and Dawson, D. D. 1994. The global resurfacing of Venus. *J. Geophys. Res.* 99:10899-10926.

Strom, R. G., Schaber, G. G., Dawson, D. D., and Kirk, R. L. 1995. Reply. *J. Geophys. Res.* 100:23361-23365.

Sukhanov, A. L. 1992. Tesserae. In *Venus Geology, Geochemistry, and Geophysics*, eds. V. L. Barsukov, A. T. Basilevsky, V. P. Volkov and V. N. Zharkov (Tucson: Univ. of Arizona Press), pp. 82-95.

Sukhanov, A. L., et al. 1989. *Geomorphic/Geological Map of Part of the Northern Hemisphere of Venus*, 1:15,000,000 scale, I-2059 (Flagstaff, Ariz.: U. S. Geological Suvey).

Surkov, Yu. A. 1986. *Cosmochemical Studies of the Planets and Satellites* (Moscow: Nauka Press), in Russian.

Surkov, Yu. A., et al. 1984. New data on the composition, structure, and properties of Venus rock obtained by Venera 13 and Venera 14. *Proc. Lunar Planet. Sci. Conf.*

14, *J. Geophys. Res. Suppl.* 89:393–402.

Surkov, Yu. A., et al. 1986. Venus rock composition at the Vega 2 landing site. *Proc. Lunar Planet. Sci. Conf.* 16, *J. Geophys. Res. Suppl.* 91:212–218.

Tanaka, K. L. 1994. *Venus Geologic Mappers' Handbook*, 2nd ed., U. S. Geol. Survey Open File Rept. 94-438.

Tormanen, T. 1995. Topographic and stratigraphic characteristics of ridge belts associated with tessera boundaries on Venus: Examples from Northern Ovda Regio margin and Kutue tessera. *Lunar Planet. Sci. Conf.* XXVI:1415–1416 (abstract).

Turcotte, D. L. 1993. An episodic hypothesis for Venusian tectonics. *J. Geophys. Res.* 98:17061–17068.

Vorder Bruegge, R. W., Head, J. W., and Campbell, D. B. 1990. Orogeny and strike-slip faulting on Venus: Tectonic evolution of Maxwell Montes. *J. Geophys. Res.* 95:8357–8391.

Weinstein, S. A. 1993. Catastrophic overturns of the Earth's mantle driven by multiple phase changes and internal heat generation. *Geophys. Res. Lett.* 20:102–104.

Weitz, C. M., and Basilevsky, A. T. 1993. Magellan observations of the Venera and Vega landing site regions. *J. Geophys. Res.* 98:17069–17097.

Willis, J., and Hansen, V. 1995. Myths regarding tessera terrain on Venus. In *A Workshop on Venus Tessera Terain*, Brown Univ.-Vernadsky Inst. Microsymposium 21 (Houston: Lunar and Planetary Inst.).

PART IX
Geodynamics

VENUSIAN SPIN DYNAMICS

CHARLES F. YODER
Jet Propulsion Laboratory

Observations of Venus' figure orientation are presently limited to a mean rotation rate $\omega = 2\pi/243.0185 \pm 0.001$d, a mean obliquity ϵ of its spin axis relative to the orbit normal of 2.6° (detected by both groundbased radar and Magellan SAR imaging) and a Chandler-like wobble amplitude of 0.5° inferred from C_{21}/S_{21} gravity field coefficients and obtained from Doppler tracking of Magellan. The retrograde rotation is believed to be a steady state equilibrium which is maintained by a balance of the axial components of a solid tidal torque and a thermally driven atmospheric torque. The wobble is not entirely unexpected, although what internal excitation mechanism might support it is unclear. Active resurfacing is a plausible candidate. However, the residual obliquity has posed the most serious problem. The amplitude of the solar tide raised on Venus has been detected ($k_2 = 0.295 \pm 0.066$) and supports the hypothesis that Venus' core is liquid. Fluid friction at a core mantle boundary caused by differential precession of mantle and fluid core should damp it to a very small amplitude on a time scale as short as 10^6 yr if this boundary were spherical. Differential core mantle obliquity is significantly reduced for large core mantle boundary ellipticity e_c because the core torque tilting the spin axis is $\propto \sin^2 \varepsilon e_c^{-3}$. Further, the combined contribution from solid and atmospheric thermal tides generally tend to increase obliquity for most plausible models. Reasonable estimates to these torques lead to a constraint on core ellipticity $e_c \approx 4 \times 10^{-4}$ which although large is not physically unreasonable. A second possibility is that the influence of small, near resonant orbital variations may drive Venus' obliquity, and this mechanism may reduce the inferred e_c by a factor of perhaps 2. A variety of topics are reviewed here, including dynamical modeling and plausible tidal histories of Venus' obliquity.

I. INTRODUCTION

A. Background

The slow retrograde spin of Venus (Carpenter 1964; Goldstein 1964; Gold and Soter 1969), near alignment of its spin axis with the orbit normal (Ward and DeCampli 1979; Yoder 1995b) and nonprincipal axis rotation (Yoder and Ward 1979) are a unique combination of dynamical features among the planets and satellites of the solar system and suggest that an unusual combination of forces and structural features such as its deep atmosphere are at work and may intimate a strange history for this planet. Although the interplay of geophysics and dynamics has proved fruitful in the study of Earth (Lambeck 1980), the application of this type of study to Venus has been at best rudimentary until now, partly because of the scarcity of relevant data and partly due to the slow development of compelling and predictive models. The most significant observational results so far are (1) the retrograde rotation, (2) a 0.5° Chandler-like free wobble, (3) a 2.16° free obliquity, and (4) detection of the Venusian k_2 Love number. The objective shall be to review interpretations of these observations.

The earliest success was that of Gold and Soter (1969) who proposed that the slow retrograde rotation is near steady state and involves a balance between a solid tidal friction torque which drives Venus toward synchronous rotation and a thermally driven, atmospheric torque which drives it away from synchronous rotation. The spin stabilizes because the solid friction axial torque is believed to be insensitive to small changes in spin rate while the atmospheric torque is expected to be inversely proportional to the semidiurnal tidal frequency (Ingersoll and Dobrovolskis 1978). However, stability of pole orientation seems to require a different kind of frequency dependence for the other tidal constituents to keep Venus' spin axis from rolling over to a prograde orientation (Dobrovolskis 1978,1980) while maintaining a nonzero spin.

The successful explanation of Mercury's rotation rate in terms of a resonant spin-orbit couple with the Sun led Gold and Soter (1969) to propose a similar mechanism for Venus (also see Goldreich and Peale 1967,1970) involving Earth instead of the Sun as the external body controlling the spin rate. The predicted Venusian rotation rate ω necessary for this mechanism to maintain this spin-orbit lock is $2\pi/243.16$d. The results from geodetic analysis of overlapping Magellan SAR images (Davies et al. 1992; also see the chapter by Sjogren et al.) find an observed rate ω that is not close enough to the resonant rate to maintain this spin-orbit configuration. We revisit this hypothesis in Sec. IV and find that this interaction is weak and is an insignificant part of the predicted axial torque balance. However, the thermally driven atmospheric torque may be sensitive to changes in solar insolation or atmospheric albedo. Venus' spin may have randomly passed through this resonance several times in the past due to small, perhaps climate related changes in the atmosphere and hence in the spin rate for which torque balance is achieved. It turns out that capture into resonance is quite plausible if passage is smooth and slow. Therefore, the *absence* of a present-day lock says something about the size and rate of change of atmospheric angular momentum.

Ward and DeCampli (1979) proposed that turbulent fluid friction at a core-mantle boundary should drive the spin pole to a fully dampened obliquity state which ends with retrograde rotation. The only requirement is that Venus' orientation is retrograde at the time the core friction torque begins to act. The problem with this hypothesis (which they recognized) is that observed Venus' spin axis orientation is not fully dampened. In this context, "fully dampened" means that the rotation pole direction should be completely predictable from the known variations in Venus' orbit about the Sun. The predicted pole position depends on the size of the free precession rate (FMN) relative to the orbit motion and this free rate is proportional to the effective oblateness $e_\circ$ relative to the spin axis. If Venus' rotation axis vector is aligned with the axis defining the largest principal moment of inertia C ($C \geq B \geq A$), then this free rate is proportional to $e_\circ = (C - 1/2(A + B))/C$. Given a Venusian ellipticity of the same magnitude as Earth's nonhydrostatic ellipticity (see, e.g., see Yoder 1995a), one should expect that the predicted pole vector should be aligned to within $1°$ of the orbit normal vector. Instead, the obliquity is $2.6°$ and happens to be within $0.5°$ of the momentum axis of the solar system (also known as "invariable or Laplacian plane" normal). An obliquity end-state near the Laplacian normal is consistent with a small oblateness.

This last observation motivated Yoder and Ward (1979) to suggest that Venus' rotation axis is not aligned with the principal moment axis. This nonprincipal axis rotation results in a Chandler-like wobble of amplitude ϵ_W. It happens that the effective oblateness is reduced by a factor of $\sim(1 - 3/2 \sin^2 \epsilon_W)$. The figure spin axis for ϵ_W near $54°$ tends to be aligned with the invariable plane normal. This proposal was also motivated by the fact that Venus' spin pole would tidally evolve to this orientation by core friction while the wobble damping, which is entirely controlled by solid friction, damps much more slowly. This wobble damping rate is $7 \times 10^5 \, Q_W$ yr or order 10^7 to 10^8 yr for plausible dissipation factor Q_W in range: $\sim 10 \leq Q_W \leq 300$. Although the wobble amplitude might be large, the wobble period $\simeq 243$ d $C/(C_m[1 + \frac{3}{2}(n/\omega)^2]e_\circ) \simeq 1.5 \times 10^4$ yr is also quite long (C_m is mantle moment of inertia). Yoder and Ward used simple scaling arguments to show that if Earth-like excitation mechanisms were scaled to Venus, then climatic or tectonic activity might justify a wobble amplitude as large as $10°$ or more.

The simplest means for the detection of a wobble from spacecraft data involves solution for C_{21}/S_{21} gravity field coefficients (see, e.g., see Mottinger et al. 1985; Bills et al. 1987). These gravity coefficients are related to off-diagonal elements of the moment of inertia tensor I_{ij}

$$C_{21} = I_{13}/MR^2$$

$$S_{21} = I_{23}/MR^2 \qquad (1)$$

where the indices 1, 2 and 3 refer to the x, y and z figure axes, respectively. If the inertial tensor is obtained in a nonprincipal axis system (where the z-axis is aligned with the spin axis), then the off-diagonal components can be related to the principal moments C, B and A of the *diagonalized* tensor. The wobble traces out a path defined by: $(\widetilde{J_2} - 2\widetilde{C}_{22} \cos 2\phi_W) \sin^2 \epsilon_W = \text{const}$, with $\widetilde{J_2}MR^2 = [(C - \frac{1}{2}(A + B)]$ and $4\widetilde{C}_{22}MR^2 = (B - A)$. The phase ϕ_W is connected to the above off-diagonal moments by $\tan \phi_W = I_{23}/I_{13}$. The wobble amplitude is related to these parameters by

$$\sin^2 \epsilon_W \simeq \left(\frac{I_{13}}{C - B}\right)^2 + \left(\frac{I_{23}}{C - A}\right)^2 \qquad (2)$$

for small amplitude. Until recently, solutions for the Venus C_{21}/S_{21} coefficients have varied considerably (Nerem et al. 1993 [GVM-1]; McNamee et al. 1993 [VGM6A]; Konopliv et al. 1993; Konopliv and Sjogren 1994 [MGNP60],1996 [MGNP90L]) as illustrated in the scatter plot in CDP9C1F1. The latest solution for this wobble based on Magellan solutions for C_{21}/S_{21} finds

$$\epsilon_W = 0.52° \pm 0.03° \qquad (3)$$

and $\phi_W \simeq 26°$. This small wobble may support the proposition that Venus is significantly less active than Earth, where activity is due to climate or tectonics. Crater populations and estimates of the last major (and apparently nearly global) resurfacing event (Schaber et al. 1992) about 0.5±0.3 Gyr ago tend to support this interpretation. Arkani-Hamed (1993) argues that the lack of evidence of ongoing, large-scale resurfacing, the apparently small convergence rate ~ 0.1 cm yr^{-1} for some of the folded terrain, the possible dominance of hot spots for heat flow (Morgan and Phillips 1983) and the absence of a magnetic field support a cold Venus model (Arkani-Hamed and Toksöz 1984; Arkani-Hamed 1993), where the mean mantle temperature is ~ 100 K cooler than Earth's mantle at the same pressure. Numerical convection models constructed by Arkani-Hamed and Toksöz (1984) with active recycling of crust can result in a frozen core. On the other hand, the absence of a Venusian magnetic field has been linked to the absence of a solid inner core (Stevenson et al. 1983), partly based on the idea that core freezing provides the energy (through bottom layer buoyancy) necessary to drive fluid convection.

Wobble excitation by resurfacing is considered here (see Sec. V) and shown to be consistent with a resurfacing rate of ~ 0.05 km^3 yr^{-1} using a simple (perhaps simplistic) random walk model.

The major focus of this article is to review new models which account for the nonzero obliquity (Yoder 1995b). However, a short discussion of plausible mantle profiles based on small deviations from an Earth-like composition and thermal profile are given first. The objective here is to obtain a plausible range for structure parameters such as moments of inertia of mantle and core and Love numbers. Next, a brief derivation of the forced obliquity due to the motion of the orbit is obtained and the determination of the residual free component is given. Its purpose is to motivate the lengthy examination of various mechanisms which might account for the free obliquity.

Section II begins with a review of the important mechanisms affecting the orientation and stability of the spin axis. Brief descriptions of the solid friction tidal torque (Kaula 1964; Yoder 1995b), atmospheric thermal tidal torque (see, e.g., Dobrovolskis 1978) and core mantle friction torque (Yoder 1981,1995b) are first developed. The complex coupling of the fluid core and mantle through both the pressure exerted on an elliptical core mantle boundary (CMB) and frictional shear caused by differential motion are then presented as scaler versions of Euler equations. This model includes not only the tidal torques but the action of the Sun on Venus' figure. The effectiveness of core oblateness or dynamical ellipticity e_c in precessing the core spin vector is determined by the relative magnitude of the core free precession frequency $\sigma_c = -e_c\omega$ relative to whole body rate $\sigma_o = -3/2e_o n^2/\omega$. The core friction model is compared with the predictions of a laminar friction model and laboratory experiments (Vanyo 1991) which determine the onset of turbulence. Those who are uninterested can jump to Sec. II.E for a discussion of the torque balance argument.

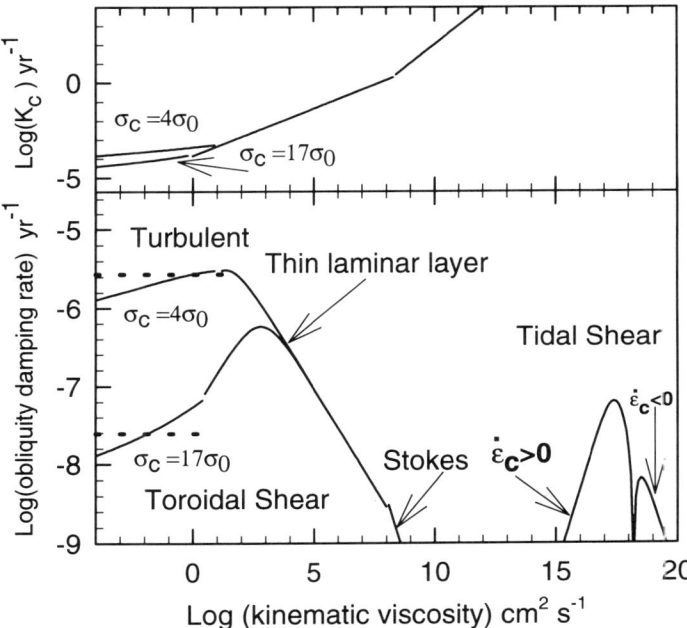

Figure 1. Venusian obliquity damping rate resulting from toroidal shear caused by either free precession or spheroidal tidal deformation of a uniformly viscous core. The dashed lines correspond to Vanyo's (1991; Eq. [13] and Fig. 5) result for the onset of turbulence assuming $\epsilon = 2.1°$, and are to be compared with the model calculation shown for two values of the ratio of mantle over core free precession frequencies: σ_0/σ_c. The Stokes limit applies when the viscosity ν satisfies $\nu/\omega R_c^2 \gg 1$. The equivalent core $K_c = 35\nu/R_c^2$. The obliquity growth or damping rate due to a solid tide in a viscous core is estimated using Eqs. (86, 87) where the angular flexing rates s_j are $2n - 2\omega$, $2n - \omega$ and ω, respectively. The effective Love number $k_s \simeq 0.4$ for the core is constrained by the elastic mantle deformation.

Figure 1 displays the inverse decay constant τ for obliquity damping as a function of core kinematic viscosity ν for four different types of frictional dissipation (which are more fully explained later) and two values of the ratio σ_c/σ_o and illustrates the dilemma we face. Model calculations of viscosity of liquid iron at Mbar pressures is estimated to be close to 0.01 cm^2 s^{-1} (Gans 1972; Poirier 1988) and therefore core damping rate is quite large unless the ratio $\sigma_c/\sigma_\circ$ is greater than about 15. There is a band of viscosity between 10^6 to 10^{16} cm^2 s^{-1} where core viscosity does not contribute to damping and yet would appear to be liquid if determined from the 58d semidiurnal tide on Venus. The size of the solar tide is scaled by the second harmonic potential Love number k_2 and determination of this parameter can determine the core state (solid or fluid) at this frequency. In Fig. 1, the Venusian core is effectively solid for viscosity $>10^{18}$ cm^{2}s^{-1}, as inferred by the tidal k_2. Magellan Doppler tracking data is sufficiently accurate that a useful constraint on k_2 can be obtained and is (Konopliv and Yoder 1996; chapter by Sjogren et al.)

$$k_2 = 0.295 \pm 0.066. \tag{4}$$

If core viscosity lies in the range 10^6 to 10^{16} cm^{2}s^{-1}, then the core follows any slow motion of the spin axis yet is fluid as inferred from a semidiurnal tidal flexing amplitude. Still, we shall adopt the position that because the theoretical estimates of fluid core viscosities are of order 0.01 cm^{2}s^{-1} while effective mantle viscosities near the solidus are $\gg 10^{16}$ cm^{2}s^{-1} (hence core viscosity near the solidus might also be of the same magnitude), the lesson from the Love number is that Venus' core is much like Earth's as far as fluidity is concerned.

Having demonstrated the central problem posed by Venus' free obliquity, we concentrate on dynamical modeling mentioned earlier. Once this task is accomplished, the model is used to determine the static balance of core frictional torque with the combined atmosphere and solid friction torque to determine core ellipticity. Next we consider how these mechanisms influence tidal evolution of Venus' obliquity and spin. The near resonant action of the slow orbit variations on part of the tidal history of Venus' obliquity are also explored and are shown to result in chaotic but limited variations of Venus' obliquity for some fraction of its tidal history (Laskar and Robutel 1993). This mechanism extends the range of initial obliquities with which Venus might have started, especially if its spin orientation were initially prograde. However, it turns out the most crucial feature is the magnitude of the semiannual thermal tide relative to the remaining tidal constituents. This particular contribution happens to produce a nonzero torque on the obliquity at 90° obliquity and can force primordial prograde orientation to evolve to retrograde orientation.

B. Structure Parameters

A suite of models have been constructed using a version of the parametric Earth model (PEM) (Dziewonski et al. 1975; Yoder 1995b) as a starting point. Mantle composition is varied by changing the molar fraction of Fe relative to Mg (f_{Fo} = Mg/(Mg + Fe)) using the properties of forsterite (Mg$_2$SO$_4$) and fayalite (Fe$_2$SO$_4$) as analogs. The density, rigidity and bulk modulus of this mixture is determined using composite solid theory (Hashim 1983). Scale factors for each of these parameters is obtained by dividing each of the estimated parameters with the corresponding estimated parameters of an Earth-like analog with molar fraction of 89%. These scale factors are then introduced into the PEM model to determine the variation of mantle properties with molar fraction. The advantage of this approach is that derivation from first principles is only accurate to 1 to 2% (Basaltic Volcanism Project 1981), and this procedure emphasizes the similarity of Venus with Earth.

The core density is varied from an Earth-like model by introducing a constant difference $\delta\rho_c$. Variation of a plausible f_{Fo} is limited to ±5% about an Earth-like mean of 89%. This constraint is supported by *in-situ* X-ray fluorescence spectra of surface rocks obtained by Venera landers which indicate an Earth-like basaltic composition (Moroz 1983). The corresponding fractional change in mantle density is ±0.03. Core density change $\delta\rho_c$ is limited to changes of less than ±0.5 g cm^{-3}. The upper bound is based on the idea that Earth's inner core density (which supposedly includes the effect of freezing and distillation of iron from its lighter component) is only 0.6 g cm^{-3} more dense than the fluid outer core at the inner core boundary and that maximum core density is very close to that of pure iron.

The corresponding constraints on total and core moments are

$$0.331 \leq \frac{C}{MR^2} \leq 0.341$$
$$0.041 \geq \frac{C_c}{MR^2} \geq 0.020. \tag{5}$$

Note that core moment size and total moment are anti-correlated.

The potential Love number k_2 for each model is also calculated because it is a factor in the tidal torque and might be a proxy for some facets of mantle internal structure. Figure 2 displays k_2 versus whole-body moment of inertia covering a wide range of mantle compositions and core size (see CDP9C1T1, Love numbers and moments). The most curious result is that k_2 is most strongly correlated with $\delta\rho_c$. A plausible range for k_2 is 0.23 to 0.29 from the theoretical constraints on mantle composition and core density. The Love number has also been calculated for cases where the core is completely solid and the results indicate a reduction in magnitude to k_2(solid) $\simeq$ 0.17. The k_2 solution from Magellan Doppler tracking (see Eq. 4 and Fig. 2) is sufficiently accurate to assure that Venus' core is liquid. However, at least a factor of 5 reduction in uncertainty is required for this parameter to provide a useful constraint on mantle properties.

C. Free Obliquity

Because it is central to my argument, we shall first derive the predicted pole direction based on a simplified model which momentarily omits the effects of core and tides. The solar torque acting on Venus' oblate figure is proportional to the difference $p - P$, where the whole-body obliquity variable p and orbit normal variable P are defined by (see Fig. 3)

$$p = \sin\epsilon\, e^{-i\phi}; \quad P = \sin I\, e^{-i\Omega}. \tag{6}$$

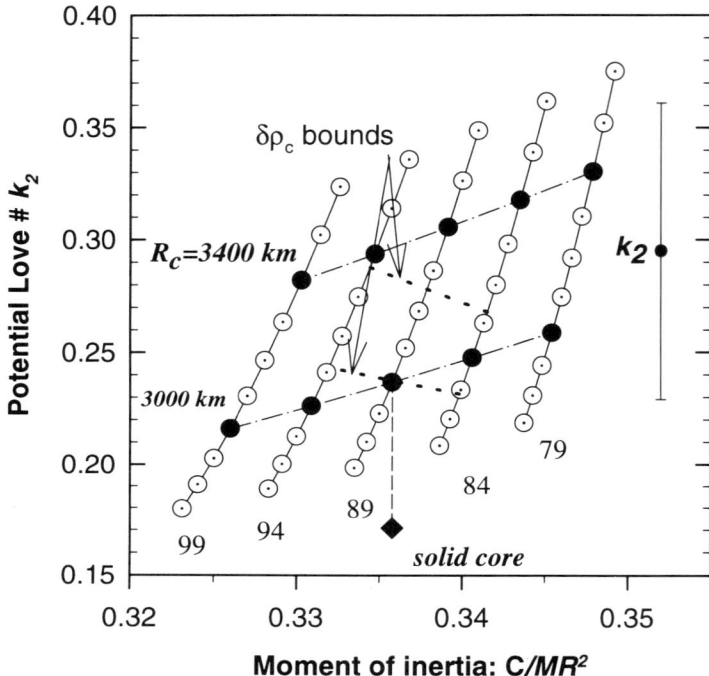

Figure 2. Love number k_2 vs moment of inertia with lines joining models with common mantle molar fraction $f_{Fo} = (Mg/(Fe+Mg))$. The solution parameters for $f_{Fo} = 0.89$ and $R_c = 3100$ km example are: $C = 0.3366MR^2$, $C_c = 0.0282MR^2$, mean core density $\bar{\rho}_c = 10.16$ g cm^3, $\delta\rho_c = 0.08$ g cm^3, and Love numbers: $k_2 = 0.252$ $h_2 = 0.5031$ and $l_2 = 0.077$.

The complex, Cartesian variable p ($p = p_x - ip_y$) describes variation of the obliquity ϵ and nodal orientation ϕ with respect to the invariable plane. The uppercase P variable is similarly related to the orbit inclination I and orbital nodal angle Ω. The changes in P arise from the orbit-averaged action of the other planets (Laskar 1988,1990) and can be expanded in a periodic series $P(t) = \sum_j \sin I_j e^{-i\Omega_j(t)}$. The frequencies $d\Omega_j/dt$ are nearly all negative. There are 8 primary terms corresponding to the linearized Laplace-Lagrange solution for orbit pole orientation (see CDP9C1T2) (Laskar 1990). The two largest $\sin I_j$ terms have amplitudes of 0.013 and 0.019 and rates $-5.62''$ yr^{-1} and $-18.85''$ yr^{-1}, respectively. A similar solution exists for orbit eccentricity, except that the 8 primary terms have positive frequency. This sparse frequency range is both densified and expanded in range to cover the positive frequency range out to about $60''$yr^{-1} from the influence of terms of order e^2 and I^2 smaller. Figure 4 displays the amplitudes of terms I_j in the appropriate range

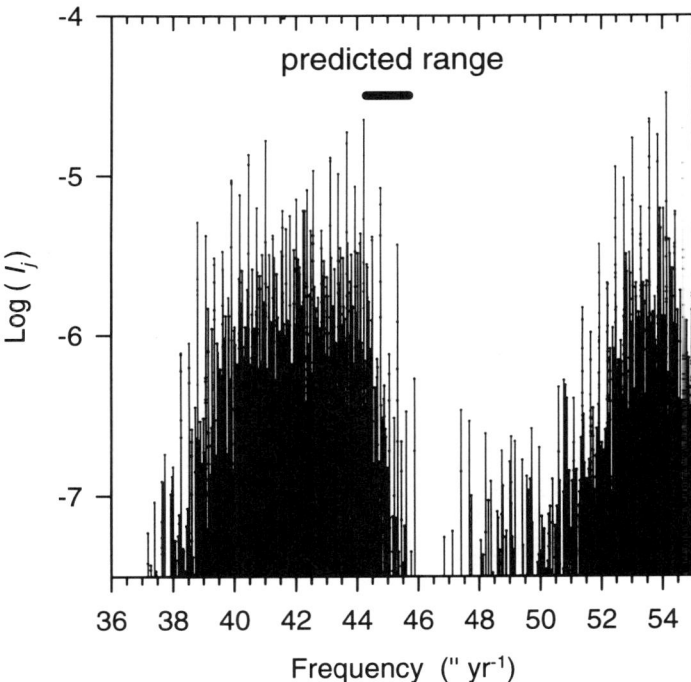

Figure 3. Amplitude spectra of terms in orbit inclination I near the estimated free precession rate for Venus.

affecting obliquity.

The linearized dynamical equation describing the variation in obliquity is

$$\frac{d}{dt}p + \left(i\sigma + \frac{1}{\tau}\right)(p - P) = 0 \qquad (7)$$

where τ is a decay time scale. The free precession rate σ (see, e.g., Smart 1961) is

$$\sigma = -\frac{3}{2}J_2\frac{n^2}{\omega}\frac{MR^2}{C}(1 - e^2)^{-3/2}\cos\epsilon \qquad (8)$$

with rotation rate $\omega = -2\pi/(243.0185 \pm 0.0003)$d (Davies et al. 1992), orbital mean motion $n = 2\pi/224.6954$d, mass $M = 48.685 \times 10^{23}$ kg and radius $R = 6051.8$ km are well determined quantities. Venus' present orbital eccentricity $e = 0.0068$; however, its long-term average is 0.034, while we shall argue that the present free obliquity is $2.1°$. These two factors change

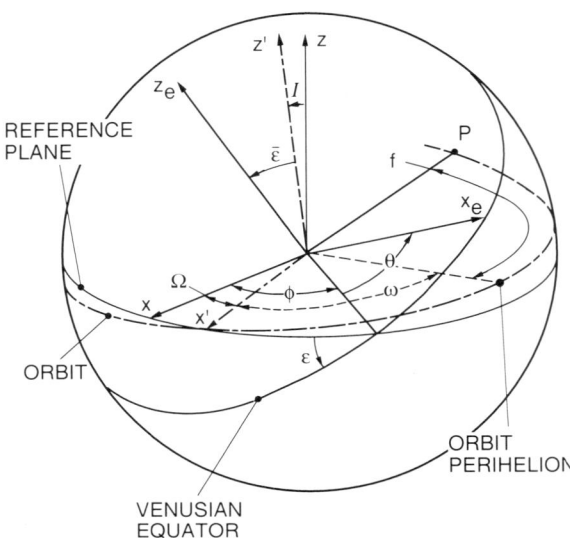

Figure 4. Angular coordinates of both the Venusian figure (x_e, y_e, z_e) and orbit (x, y, z) relative to a reference plane(x', y', z'). The figure orientation is defined by the obliquity ϵ, nodal angle ϕ and rotation angle θ. The orbit position at point P is defined by inclination I, node Ω, argument of pericenter ω and true anomaly f. The obliquity $\bar{\epsilon}$ relative to the orbit is also shown.

σ by only 0.1%. The most recent solution for the zonal harmonic gravity coefficient J_2 is

$$J_2(\text{gravity}) = 4.4098 \pm 0.0016 \times 10^{-6} \tag{9}$$

from a solution for a gravity field of degree and order 90. For principal axis rotation, J_2 is a linear combination of the principal moments: $J_2 = (C - \frac{1}{2}(A+B))/MR^2$. By the way, if both J_2(gravity) and the tidal k_2 Love number are obtained from analysis of Doppler data, then the total J_2 is

$$J_2 = J_2(\text{gravity}) + \frac{1}{2}k_2\frac{M_\odot}{M}\left(\frac{R}{a}\right)^3\left(1 - 3\sin^2\epsilon\right). \tag{10}$$

The latest solution for Venusian gravity (Konopliv and Sjogren 1996; Konopliv and Yoder 1996) does include k_2 as a solution parameter, with the result that the effective

$$J_2(\text{total}) = 4.4192 \pm 0.0016 \times 10^{-6}. \tag{11}$$

The polar moment is the most uncertain parameter and leads to the following estimate of σ:

$$44.3'' \text{yr}^{-1} \leq \sigma \leq 45.8'' \text{ yr}^{-1} \tag{12}$$

if the core is solid. The presence of a fluid core with polar moment C_c and core nutation frequency σ_c reduces σ by a factor $\simeq (1 - C_c \sigma / C \sigma_c)$ (see Eq. 70 below). The corresponding range for C is

$$44.1'' \text{ yr}^{-1} \leq \sigma \leq 45.6'' \text{yr}^{-1} \tag{13}$$

for a fluid core and mantle which satisfy constraints on σ_c and C_c.

If very small driving terms in P in the frequency band 43.7 to 47" yr^{-1} are excluded, then the predicted pole location is insensitive to uncertainties in σ. The purely forced value is

$$\begin{aligned} p_{x,\text{forced}} &= 0.01877 \text{ rad} \\ &= 1.0765° \\ p_{y,\text{forced}} &= 0.02204 \text{ rad} \\ &\simeq 1.263° \end{aligned} \tag{14}$$

and the Laplacian normal is the origin ($P_x = 0.0246$; $P_y = 0.0303$). The observed pole position is

$$\begin{aligned} p_{x,\text{obs}} &= -0.00287 \\ &= -0.164° \\ p_{y,\text{obs}} &= -0.00768 \\ &= -0.440°. \end{aligned} \tag{15}$$

The difference is the free obliquity

$$\begin{aligned} p_{x,\text{free}} &= -0.02163 \\ &= -1.239° \\ p_{y,\text{free}} &= -0.02972 \\ &= -1.703° \end{aligned} \tag{16}$$

or an amplitude of 2.1°. If one adopts a positive Venusian spin, then the corresponding obliquity is 177.9°.

II. DYNAMICAL MODEL

A. Solid Body Tides

The second harmonic component of the solar gravitational potential, i.e., $(GM_\odot R^2 / r^3) (3/2 \cos^2 S - 1/2)$, depends on the solar angle S, solar mass $M_\odot$ and orbit radius r and tidally distorts a nearly spherical body of radius R. The elastic change in its moment of inertia tensor is

$$I_{ij} = -k_2 \frac{M_\odot R^5}{r^3} \left(u_i u_j - \frac{1}{3} \delta_{ij} \right) \tag{17}$$

where k_2 is a potential Love number, δ_{ij} is the Kronecker delta function and $u_i = \mathbf{r} \cdot \mathbf{R}_i / rR$ are direction cosines of the Sun relative to the body-fixed coordinate frame (see, e.g., Lambeck 1980). Solid friction depends on knowledge of unknown material rheology. This dissipation may depend on tidal flexing frequency and perhaps, through plasticity distribution within the mantle, on tesseral order.

The general expression in terms of complex gravity coefficients C_{nm} of tesseral order m and degree n is

$$C_{nm}^t = k_n \frac{\overline{M'}}{M} \left(\frac{R}{\overline{r}}\right)^{n+1} M_{nm} N_{nm} Y_{nm}^*(\overline{\Theta}, \overline{\Phi}) \tag{18}$$

$$N_{nm}^2 = \frac{4\pi}{2n+1} \frac{(n+m)!}{(n-m)!}; \quad M_{nm} = (2-\delta_{m0})\frac{(n-m)!}{(n+m)!} \tag{19}$$

where $C_{nm} = C_{nm}^K + i S_{nm}^K$ (* is complex conjugate, the superscript K refers to the unnormalized, real coefficients defined by Kaula [1960]) and $C_{n0} = -J_n$. The normalized spherical harmonic function $Y_{nm}(\overline{\Theta}, \overline{\Phi})$ depends on the colatitude $\overline{\Theta}$ and longitude $\overline{\Phi}$ of the tide-raising body relative to the body-fixed frame. The convention adopted here is:

$$Y_{nm}(\overline{\Theta}, \overline{\Phi}) = N_{nm} P_{nm}(\cos \overline{\Theta}) \exp -i\overline{\Phi}$$

$$P_{nm}(x) = \frac{(-1)^m}{2^n n!}(1-x^2)^{m/2}\frac{d^{n+m}}{dx^{n+m}}(x^2-1)^n \tag{20}$$

for $m \geq 0$. The tide is raised by mass $\overline{M'}$ on mass M.

The tidal disturbing function is obtained by substituting the above into the spin-orbit disturbing function R after expanding the potential in the same manner. The expansion of the n, m components of the potential is

$$R_{nm}^t = \frac{GMM'R^n}{r'^{(n+1)}} C_{nm}^t N_{nm} Y_{nm}(\Theta, \Phi). \tag{21}$$

The next step is to relate the colatitude and longitude of both the tide raising body *and* the body acting on the figure in terms of the orbital variables and the figure orientation. Consider the rotational transformation of a spherical function (see, e.g., Rose 1957; Gottfried 1966; Levie 1971)

$$Y_{nm}(\Theta, \Phi) = \sum_{j=-n}^{n} d_{jm}^n(\beta) e^{-i(j\alpha + m\gamma)} Y_{nj}(\overline{\Theta}, \overline{\Phi}) \tag{22}$$

where α is a counterclockwise rotation about the z-axis, β is a rotation about the new y-axis and finally γ is a rotation about the new z-axis. The function $d_{jm}^n(\beta)$ is $O(\beta^{|j-m|})$ and is defined later. We can sequentially apply this rotation to each spherical harmonic function. First rotate from the body-fixed coordinate frame to a space-fixed frame, where the angle coordinates α, β, γ describe the orientation of the figure and the Θ, Φ angles describe the orientation of the perturbing body in this space frame. Second, rotate from the space-fixed frame to an orbit frame in which the final x- and z-axes are parallel to the orbit radius vector and orbit normal, respectively.

The first Euler rotation sequence corresponds to a rotation ϕ about the z-axis, a rotation ϵ about the new x-axis and a rotation θ about the new z-axis. Identify $\alpha = \phi - \pi/2$, $\beta = \epsilon$ and $\gamma = \theta + \pi/2$. The new angles $\Theta^\dagger$, $\Phi^\dagger$ are the spherical coordinates of the perturbing body in the space-fixed frame. Define the orbital angles, true anomaly f, argument of pericenter ω, node Ω and inclination I. For the second rotation, identify $\gamma = -\Omega + \pi/2$, $\beta = -I$ and $\alpha = -f - \pi/2$. The final spherical coordinates are: $[\Theta^{++}, \Phi^{++}] = [\pi/2, 0]$. The result is

$$Y_{nm}(\Theta, \Phi) = i^{k-m} d^n_{jm}(\epsilon) d^n_{kj}(-I) Y_{nk}\left(\frac{\pi}{2}, 0\right) \times \qquad (23)$$
$$\exp i(k(f + \omega) + j(\Omega - \phi) - m\theta).$$

The obliquity function $d^n_{jm}(\beta)$ for $j \geq m$ is

$$d^n_{jm}(\beta) = (-1)^{j-m} \left[\frac{(n+j)!(n-j)!}{(n+m)!(n-m)!}\right]^{1/2} \times \qquad (24)$$
$$\sum_s (-1)^s \binom{n+m}{s} \binom{n-m}{n-j-s} c_2^{2n-j+m-2s} s_2^{j-m+2s}$$

with $c_a = \cos\beta/h$ and $s_h = \sin\beta/h$. The sum over integer s is limited to only terms for which the binomial coefficients are defined: $\max(0, n - j)$ to $\min(n + m, n - m)$. These functions satisfy the symmetry relations $d^n_{jm}(\beta) = (-1)^{j-m} d^n_{-j,-m}(\beta) = d^n_{mj}(-\beta)$ and therefore, of the $(2n + 1)^2$ components of $d^n_{jm}(\beta)$, only $(n + 1)^2$ are unique. Also $d^n_{jm}(0) = \delta_{jm}$.

The $Y_{nk}(\pi/2, 0)$ function is (Abramowitz and Stegun 1965)

$$Y_{nk}\left(\frac{\pi}{2}, 0\right) = (-1)^k \frac{N_{nk}(n+k)!}{2^n \left(\frac{n-k}{2}\right)! \left(\frac{n+k}{2}\right)!} \cos(n-k)\frac{\pi}{2} \qquad (25)$$

and vanishes if $n - k$ is odd. Thus, replace k with $n - 2p$ in Eqs. (23) and (25). The inclination function $F_{nmp}(I)$ (Kaula 1960) is related to the above by ($t = 0$ for $n - m$ even and $t = 1$ for $n - m$ odd)

$$F_{njp}(I) = i^{n-m-2p+t} N_{nj} Y_{n,n-2p}\left(\frac{\pi}{2}, 0\right) d^n_{n-2p,j}(-I) \qquad (26)$$

and is similar to the expression of Allan (1967).

A general (and purely formal) expression can be derived from these operations (including an additional expansion in terms of orbit eccentricity e, mean anomaly ℓ and semimajor axis a) which explicitly takes into account both the influence of the orbital parameters and figure orientation (node ϕ, obliquity ϵ and rotation angle θ). The n' the harmonic component of the solid tidal potential is proportional to the potential Love number k_n and is (Yoder 1995b)

$$R_n^t = \frac{G\overline{M'}M'R^{2n+1}k_n}{\overline{a}^{n+1}a^{n+1}} \sum_{p,\overline{p}=0}^{2n} \sum_{m,j,\overline{j}=-n}^{n} \sum_{q,\overline{q}=-\infty}^{\infty}$$

$$(2-\delta_{m0})\sqrt{\frac{(n-j)!(n-\overline{j})!}{(n+j)!(n+\overline{j})!}} d_{jm}^n(\overline{\epsilon})d_{jm}^n(\epsilon)F_{n\overline{j}\overline{p}}(\overline{I})F_{njp}(I)$$

$$G_{n\overline{p},\overline{q}}(\overline{e})G_{npq}(e)\cos[\Psi_{nmpjq} - \overline{\Psi}_{nm\overline{p}\overline{j}\overline{q}} + \delta_{nm\overline{p}\overline{j}\overline{q}}(\dot{\overline{\Psi}})] \quad (27)$$

where the overbar on the summation indices denotes the tide-raising body and m is the tesseral order. This reduces to Kaula's (1964) tidal expansion in limit $\epsilon = \overline{\epsilon} = 0$.

There does not exist a simple expression for the eccentricity function $G_{npq}(e)$ (Kaula 1960; Jarnagin 1965) which is $O(e^{|q|})$ in magnitude; also, $G_{npq}(0) = \delta_{q0}$. As Venus' eccentricity is small, we shall neglect its contribution.

The usual approximation is that the variables a, e and I are nearly constant and that the angle variable

$$\Psi_{nmpjq} = (n-2p)(\ell+\omega) + q\ell + j(\Omega-\phi) - m\theta \quad (28)$$

varies linearly with time. Dissipation due to the periodic flexing of the solid is then modeled by introducing a phase lag $\delta_{nm\overline{p}\overline{j}\overline{q}}(\dot{\overline{\Psi}})$ that has the same sign as the rate $\dot{\overline{\Psi}}_{nmpjq}$.

Here, we are primarily interested in how Venus' figure responds to tidal friction because its effect on the orbit is order $C\omega/Mna^2 \sim 10^{-9}$ smaller. The tidal equations for obliquity and spin are obtained from partials of the unbarred variables

$$C\frac{d}{dt}\omega = \frac{\partial}{\partial\theta}R_n^t$$

$$C\omega\sin\epsilon\frac{d}{dt}\epsilon = -\frac{\partial}{\partial\phi}R_n^t + \cos\epsilon\frac{\partial}{\partial\theta}R_n^t. \quad (29)$$

If the tide-raising body and the body affecting the disturbing body are the same, then the terms contributing to the secular rates have $p = \overline{p}$, $j = \overline{j}$ and $q = \overline{q}$. The $\{nmpjq\}$ term in the obliquity rate is

$$C\omega\sin\epsilon\frac{d}{dt}\epsilon_{nmpjq} = (m\cos\epsilon - j)T_{nmpjq}\sin\delta_{nmpjq} \quad (30)$$

with

$$T_{nmpjq} = \frac{GM'^2R^{2n+1}k_n}{a^{2n+2}}M_{nj}G_{npq}^2F_{njp}(I)^2 d_{jm}^n(\epsilon)^2. \quad (31)$$

The special case where $e = I = 0$, has $n - 2p = m$ and $q = 0$. The obliquity rate reduces to

$$\frac{C\omega}{\sin \epsilon}\frac{d}{dt}\epsilon = -\frac{3}{4}n^2 MR^2 H_\odot k_2 \begin{bmatrix} c_2^6\delta(2n-2\omega) + s_2^6\delta(2n+2\omega) \\ +c_2^4(2-c_1)\delta(2n-\omega)+ \\ s_2^4(2+c_1)\delta(2n+\omega) - c_1^3\delta(-\omega) \\ -\frac{1}{2}s_1^2 c_1\delta(-2\omega) + \frac{3}{4}s_1^2\delta(2n) \end{bmatrix}. \quad (32)$$

For obliquity near 90° and $|\omega| \gg n$, all the terms either vanish or tend to cancel in pairs, except for the semiannual constituent with lag $\delta(2n)$. The above angular dependence of K_{tp} for phases $\delta(s)$ that are constant and equal in magnitude and with $|\omega| \gg n$ is

$$c_1(1 + \frac{3}{4}s_1^2)\delta(2n - 2\omega) + \frac{3}{4}s_1^2\delta(2n). \quad (33)$$

The solid tide semi-annual constituent ($\propto \delta(2n)$) always tends to drive the obliquity to the prograde state. However, we shall discover that the equivalent atmospheric term may dominate and can cause Venus to tidally evolve from a prograde to retrograde state if the initial obliquity is sufficiently large. In the limit of small obliquity, the rate reduces to

$$C\omega \sin \epsilon \frac{d}{dt}\epsilon = K_{tp} + K_{ap} \quad (34)$$

$$K_{tp} = -\frac{3k_2 n^2 MR^2 H_\odot}{4C\omega}[\delta_t(2n - 2\omega) + \delta_t(2n - \omega) - \delta_t(-\omega)]$$

and K_{ap} is the atmospheric contribution defined in the next section. Solid tides on Venus are set by the parameter $H_\odot$,

$$H_\odot = \frac{M_\odot}{M}\left(\frac{R}{a}\right)^3 = 7.15 \times 10^{-8}. \quad (35)$$

The $\{nmpjq\}$ component of the tidal spin rate change is

$$C\frac{d}{dt}\omega_{nmpj0} = mT_{nmpjq}\sin\delta_{nmpjq}. \quad (36)$$

Applying the same approximation to spin as was done for obliquity, we obtain

$$C\frac{d}{dt}\omega = \frac{3}{2}k_2 n^2 MR^2 H_\odot \begin{bmatrix} c_2^8\delta(2n-2\omega) - s_2^8\delta(2n+2\omega)+ \\ \frac{1}{2}(c_2^4 s_1^2\delta(2n-\omega) - s_2^4 s_1^2\delta(2n+\omega)) \\ +\frac{1}{4}s_1^4\delta(-2\omega) + \frac{1}{2}s_1^2 c_1^2\delta(-\omega) \end{bmatrix}. \quad (37)$$

The above angular dependence is

$$(5/8 + 3/4 c_1^2 - 3/8 c_1^4)\delta(2n - 2\omega) \quad (38)$$

for constant $\delta(s)$ and $|\omega| \gg n$. For small obliquity the spin rate change is

$$\frac{1}{\omega}\frac{d}{dt}\omega = K_{ts} + K_{as}$$

$$K_{ts} = \frac{3k_2 n^2 M R^2 H_\odot}{2C\omega}\delta_t(2n-2\omega). \tag{39}$$

Presently $\omega = -0.9246n$, where n is the orbital mean motion. Therefore, solid tidal friction tends to drive spin toward synchronous rotation where $\omega = n$. The change in the spin is a factor of 81 over 4.6×10^9 yr given a constant semidiurnal phase $\delta(2n-2\omega) = 1/Q = 1/50$, $k_2 = 0.25$, $C/MR^2 = 0.337$ and a small obliquity over this interval.

B. Atmospheric Thermal Tides

Simple intuition suggests that heating of the atmosphere by Sun during the course of a day should gradually expand the atmosphere. Therefore, atmospheric mass is greater in the morning than the afternoon, and we should expect that the gravitational attraction of Sun tends to pull on the atmosphere in the same sense as it is rotating, leading to an increase in angular momentum of atmosphere and planet. This torque is opposite to the effect of inelastic solid tides. Thermal tides generate air currents and can be sensitive to such feature as zonal wind and variations in energy absorption with height. An accurate estimate of the resultant torque requires detailed modeling of thermal, wind and density profiles (Dobrovolskis 1978,1980; Dobrovolskis and Ingersoll 1980). However, a simple model (Ingersoll and Dobrovolskis 1978) can be derived which ignores most of the dynamics of the atmosphere except its response to heating. First, consider the atmospheric solar heating rate $J_\odot(z, S)$ which is a function of atmospheric depth z and solar angle S. To lowest order, the expression is

$$J_\odot(z, S) = \begin{bmatrix} \left(\frac{a}{r}\right)^2 J_\odot(z) \cos S & -\frac{\pi}{2} \leq S \leq \frac{\pi}{2} \\ 0 & \frac{\pi}{2} \leq S \leq \frac{3\pi}{2} \end{bmatrix}. \tag{40}$$

This discontinuous function can be expanded in a periodic series in $\cos mS$, and the terms proportional to $\cos mS$ ($m = 1, 2$) are

$$\left(\frac{a}{r}\right)^2 J_\odot(z) \left(\frac{1}{2}\cos S + \frac{2}{3\pi}\cos 2S + \ldots\right). \tag{41}$$

The term proportional to $\cos S$ causes a displacement of center of mantle figure from center of the total mass. However, it does not lead to a torque on the figure of Venus. If this $\cos S$ term is dropped and a term independent of S is added, one can cast the above expression into a form nearly identical to the $n = 2$ part of the solar solid tidal potential

$$\frac{8}{9\pi}\left(\frac{a}{r}\right)^2 J_\odot(z)\left(\frac{3}{2}\cos^2 S - \frac{1}{2}\right) \tag{42}$$

except that factor $GM_\odot a/r$ is replaced by $8J(z)/9\pi$. We see that the orbital dependence of these two expressions differs only in its radial dependence, and for Venus this radial variation is unimportant. This vertically distributed heating can be replaced by the column integrated, absorbed flux F_a which if deposited near the surface is

$$F_a = J_\odot(0) \int dz \rho_a(z). \tag{43}$$

If one ignores advection and other small terms, the first law of thermodynamics relates internal energy e_a to pressure P_a and air density ρ_a: $\frac{d}{dt}e_a = -P\frac{d}{dt}\rho_a$. This internal energy rate is also related to the semidiurnal solar heating function in Eq. (41) by

$$\frac{d}{dt}e_a = \frac{2}{3\pi}J_\odot(z)\cos 2S. \tag{44}$$

Here, $\Re$ and c_p are the specific gas constant and heat capacity, respectively. Pressure can be related to density through the hydrostatic law: $dP_a/dt = -g\rho_a$ and temperature T_a through the ideal gas law: $P_a = \rho_a \Re T_a$. The next step is to connect the above heating to a change in column atmospheric density and surface pressure.

$$\int_0^\infty dz \frac{d}{dt}\delta\rho_a(z,S) = -g^{-1}\frac{d}{dt}\delta P_a(S) = -\frac{2}{3\pi}\frac{F_a}{c_p T(0)}\cos 2S. \tag{45}$$

Clearly, there must be some process which limits the amount of heat absorbed with time. We could replace d/dt with $d/dt + i/\tau_A$ in Eq. (45). The parameter τ_A might be identified with a Newtonian cooling time or a time scale associated with zonal wind circulation period. However, we shall assume that this time constant is large compared with Venus' rotation period.

From Eq. (45), the variation in air column mass depends on the ratio of heat absorbed during time $1/2\dot{S}$ to surface thermal energy per unit mass, $c_p T(0)$. Appropriate parameters for Venus are: $F_a \simeq 100$ W m^{-2} at the surface (Avduevskii et al. 1976) compared to a solar constant of 2600 W m^{-2}, surface temperature $T(0) \simeq 730$ K and atmospheric specific heat $c_p \simeq 950$ J kg^{-1} K^{-1}. We find $\delta P_a(2S) \simeq 22\sin 2S$ mbar. The fact that pressure decreases as heat is added supports the intuition that the maximum semidiurnal tide should occur in the morning and the minimum in the afternoon. For periodic heating, the pressure leads the heat flux by 90° and is inversely proportional to the frequency. The equivalent atmospheric tidal potential is obtained by integrating the above density variation over the sphere to obtain the perturbed moment of inertia tensor. The only complication is that the solar angle must be first transformed to a body-fixed system. Clearly, an equivalent atmospheric tidal potential must have the same form as the solid tide except for a difference in scale. This can be determined from evaluating, say, the I_{11} moment for the special case where the tide maximum is aligned with the x

axis (i.e., $u_1 = 1$ in Eq. [19]). We find $\dot{I}_{11}(\text{atm}) = -(64/135)R^4(F_a/c_pT(0))$ while $I_{11}(\text{solid}) = -2/3k_2(M_\odot R^2/r^3)$. The resulting potential is

$$R_2^t = \frac{GM_\odot R^4(1+k_2')}{a^3 g} \sum_{p,\overline{p}=0}^{4} \sum_{m,j,\overline{j}=-2}^{2} \sum_{q,\overline{q}=-\infty}^{\infty} \qquad (46)$$

$$(2-\delta_{m0})\sqrt{\frac{(2-j)!(2-\overline{j})!}{(2+j)!(2+\overline{j})!}} d_{jm}^2(\epsilon) d_{\overline{j}m}^2(\epsilon) F_{2\overline{j}\overline{p}}(\overline{I}) F_{2jp}(I)$$

$$G_{2\overline{p}\overline{q}}^{-1}(\overline{e}) G_{2pq}(e) \delta P_a(\dot{\Psi}_{2m\overline{p}\overline{j}\overline{q}}) \cos[\Psi_{2mpjq} - \overline{\Psi}_{2m\overline{p}\overline{j}\overline{q}} + \frac{\pi}{2}].$$

The factor $(1+k_2') = (1+k_2-h_2) \simeq 0.75$ compensates for elastic loading of the mantle by the atmosphere. The only functional change is the adoption of the generalized eccentricity function $G_{npq}^j(\overline{e})$ defined by the expansion (see, e.g., Plummer 1960)

$$\left(\frac{a}{r}\right)^{n+1+j} e^{i(n-2p)f} = \sum_q G_{npq}^j e^{i(n-2p+q)\ell} \qquad (47)$$

and the introduction of the pressure factor

$$\delta P_a(\dot{\Psi}_{2m\overline{p}\overline{j}\overline{q}}) = \frac{32}{45} \frac{g F_a}{c_p T_\circ \dot{\Psi}_{2m\overline{p}\overline{j}\overline{q}}}. \qquad (48)$$

One can identify an equivalent atmospheric phase shift $\delta_a(\dot{\Psi})$:

$$\delta_s(\dot{\Psi}) = -\frac{(1+k_2')}{k_2} \frac{\delta P_a R^2}{MgH_\odot}. \qquad (49)$$

The corresponding shift for the semidiurnal tide is $\simeq -0.026$.

Using this second formalism, the atmospheric contributions K_{as} and K_{ap} to the rate equations (see Eqs. 33 and 38) are

$$K_{as} = \frac{3k_2 n^2 M R^2 H_\odot}{2C\omega} \delta_a(2n-2\omega) \qquad (50)$$

$$K_{ap} = -\frac{3k_2 n^2 M R^2 H_\odot}{4C\omega} [\delta_a(2n-2\omega) + \delta_a(2n-\omega) - \delta_a(-\omega)]. \qquad (51)$$

Atmospheric tides drive the spin away from synchronous rotation and may also tend to increase obliquity. The spin rate achieves a stationary, stable state when $(K_{ts} + K_{as})$ vanishes. However, the factor controlling obliquity evolution $(K_{tp} + K_{ap}) \simeq -1.4 K_{ts} > 0$ unless one either chooses the tidal phase and pressure factors to have the same frequency dependence (in which case the spin balance is neutrally stable) or adopt a tuned atmosphere model which happens to introduce just the right counterbalance. Sophisticated atmospheric models that take into account winds and vertical structure and include strong bottom heating (Dobrovolskis 1978,1980; Pechmann and Ingersoll 1984; Shen and Zhang 1989,1990) obtain results similar to this simple model. Four models of Dobrovolskis (1978; also see Table IV in Yoder 1995b) which retain heating within the bottom 20 km, have $(K_{tp} + K_{ap})$ ranging from $-0.4K_{ts}$ to $-1.8K_{ts}$. This assumes that the solid phase lags are independent of frequency and the spin is at equilibrium.

C. Core Mantle Fluid Friction

The angle between the core and mantle spin vectors is $\Delta\epsilon$ while the core spin rate is $\cos\Delta\epsilon$ smaller (Busse 1968). This $\Delta\epsilon$ displacement of the spin vectors causes shear at a core mantle boundary where the typical velocity u is $u \sim |\omega| R_c \sin\Delta\epsilon$ and the typical lateral displacement d is u/ω. Fluid friction caused by shear at a core mantle boundary can be nonlaminar (and nonlinear) if this shear is sufficiently sharp as measured by the Reynolds number Re. Define a local $Re = ud/\nu$ where ν is kinematic viscosity. If Re is greater than 10^5 to 10^6 (the appropriate value determined by experiment) then fluid flow near the boundary is turbulent. A typical Re for precessional flow is therefore

$$Re = R_c^2 |\omega| \sin^2\Delta\epsilon/\nu \sim 3 \times 10^{10} \sin^2\Delta\epsilon (\text{cm}^2\text{s}^{-1}/\nu) \qquad (52)$$

for a Venusian core (core radius $R_c \simeq 1/2R$). The core Re number is so large that turbulence at the CMB is almost certain unless $\Delta\epsilon$ is exceptionally small. An estimate of the turbulent stress can be obtained using mixing length theory (Goldstein 1965) in which the laminar, viscous boundary layer is replaced with two layers: an interior, laminar viscous sublayer with thickness δ which is matched with an exterior turbulent boundary layer. Within each layer, the stress is assumed constant. The velocity profile in the laminar sublayer increases linearly with the distance ξ from the wall ($\mathbf{u} = \mathbf{u}_\delta \xi/\delta$) up to a layer thickness δ and velocity $\mathbf{u}_\delta$. The viscous stress is $\vec{\tau}_v = \rho\nu\mathbf{u}_\delta/\delta$. The laminar stress is matched with the turbulent stress at $\xi = \delta$ to determine δ in terms of the other parameters. The turbulent local stress $\vec{\tau}_t$ is governed by

$$\vec{\tau}_t \simeq \rho\kappa^2\xi^2 \left|\frac{d}{d\xi}\mathbf{u}\right| \frac{d}{d\xi}\mathbf{u} \qquad (53)$$

where $\kappa \simeq 0.40$ is the *Kármán* constant. The expected logarithmic velocity profile within the outer turbulent layer can be approximated by

$$\vec{\tau}_t \simeq \frac{\rho\kappa^2 |\mathbf{u}(\xi)| \mathbf{u}(\xi)}{\ln^2 |\xi/\delta + e - 1|} = \rho\kappa^2 |\mathbf{u}_\delta| \mathbf{u}_\delta \qquad (54)$$

which exhibits the appropriate behavior as $\xi \to \delta_+$. Matching the two stresses at the boundary $\xi = \delta$ determines δ: $\delta = \nu/\kappa^2 u_\delta$. The outer boundary at

$\xi \sim d$ where the velocity reaches its limiting value u_o leads to the following expression for the local stress:

$$\vec{\tau}_t \simeq \rho \kappa^2 \left(\frac{u_\delta}{u_o}\right)^2 |\mathbf{u}_o| \mathbf{u}_o \qquad (55)$$

$$\frac{u_\delta}{u_o} \simeq \frac{1}{\ln\left|\frac{u_\delta}{u_o}\kappa^2 Re\right|}. \qquad (56)$$

Now the velocity $\mathbf{u}_o = (\vec{\omega}_c - \vec{\omega}) \times \mathbf{R}_c$ and the local torque exerted by the local stress is $\mathbf{n} = \mathbf{R}_c \times \vec{\tau}$. Except near the polar circle with colatitude $\theta = \Delta\epsilon$, the shear velocity relative to the boundary location at θ and longitude β is periodic and has amplitude

$$u_o \simeq |\omega| R_c \sin \Delta\epsilon \sin \theta \sin(\omega t - \beta - \Delta\phi) \qquad (57)$$

where $\Delta\phi$ is the relative nodal intersection line. The CMB torque is obtained by integrating $\mathbf{R}_c \times \vec{\tau}$ over the sphere. Before performing this operation, replace $|\mathbf{u}_o|$ in Eq. (55) with its time averaged value $\left|\omega R_c \sin \Delta\epsilon \sin \theta /\sqrt{2}\right|$.

The resulting global torque $\mathbf{N}_t$ is,

$$\mathbf{N}_t = C_c K_c(\text{turb})(\vec{\omega}_c - \vec{\omega}). \qquad (58)$$

The core moment is $C_c \simeq \frac{2}{5} M_c R_c^2$ and the turbulent coupling parameter is

$$K_c(\text{turb}) \simeq \frac{45\pi}{32\sqrt{2}} \kappa^2 \left(\frac{u_\delta}{u_o}\right)^2 \sin \Delta\epsilon |\omega|. \qquad (59)$$

In this calculation, u_δ/u_o was held fixed and is obtained from Eq. (56) using the nominal value for the Reynold's number. The value $u_\delta/u_o \leq 1/10$ for the expected core parameters.

Vanyo's experiments on precessing liquid-filled spheres near the onset of fully developed turbulence (Vanyo 1973; Vanyo and Paltridge 1981; Vanyo 1991), predict $K_c(\text{turb}) \simeq 0.0035(\pm 50\%) |\omega| \sin \Delta\epsilon$. The two expressions for $K(\text{turb})$ are equivalent if $u_\delta/u_o \simeq 1/12$ and $Re \sim 1 \times 10^7$. Unfortunately, the range of Reynolds number explored by Vanyo is too restricted to see the predicted logarithmic dependence on Re.

Surface roughness is an additional complication which can increase the turbulent torque. An appropriate approximation for the local stress in this limit is (see, e.g., Lamb 1960)

$$\vec{\tau}_t \simeq \rho D_f |\mathbf{u}| \mathbf{u}. \qquad (60)$$

The coefficient $D_f \simeq 0.002$ while the coupling parameter is (Yoder 1981; Dickey et al. 1994)

$$K_c(\text{rough}) = \frac{45\pi}{32\sqrt{2}} D_f \omega \sin \Delta\epsilon. \quad (61)$$

This estimate is probably an upper bound for turbulent frictional coupling.

The equivalent parameter $K_c(\text{lam})$ expression for laminar boundary layer friction is (Busse 1968; Roberts and Stewartson 1965)

$$K_c(\text{lam}) = 2.6\sqrt{\nu |\omega|/R^2} = 2.6 \sin \Delta\epsilon \, |\omega| \, Re^{-1/2}. \quad (62)$$

$K_c(\text{lam})$ is smaller than $K_c(\text{turb})$ if $Re \geq 27(\ln|Re\kappa^2 u_\delta/u_\circ|)^4$ or if $\nu \leq 0.4$ cm^{2}s^{-1}. The laminar boundary layer thickness is $\sim R_c\sqrt{\nu/\omega R_c^2}$.

Poirier (1988) argues that at core pressures core viscosity $\eta = \rho\nu$ is close to the STP value for liquid iron near its melting point (0.06 P), although this is uncertain by at least an order of magnitude. Pinning the inner core boundary at the melting point, Poirier estimates that $\eta = 0.03$ P at the CMB. Gans' (1972) estimate for η based on Andrade's theory of liquids is similarly low (0.04 P $\leq \eta \leq 0.2$ P). In these calculations, I shall adopt $\nu = \eta/\rho = 0.01$cm^{2}s^{-1}, although changing this number by a factor of 10 changes the torque by only 10%.

The tidal equation for spin deceleration due to CMF is obtained from (Goldreich and Peale 1970)

$$\frac{d}{dt}\omega_{CMF} = \omega \tan \epsilon \frac{d}{dt}\epsilon_{CMF} \quad (63)$$

and is derived from the condition that the total component of spin momentum normal to the orbit $\cos \epsilon \, C\omega$ is conserved. The fact that $\cos \epsilon \, C\omega$ is nearly conserved once CMF begins to dominate implies that if the initial obliquity is large, the obliquity damping rate is initially proportional to ω^{-8} compared to a solid tide ω^{-1} dependence.

D. Dynamical Equations

The Euler equations describing the coupling of a fluid core, contained in an ellipsoidal cavity, to the forced, short-period nutational motion of the mantle are well developed for studying Earth's nutations. However, the general approach approximates the precessional motion as a zero frequency tilt-over mode in calculating the core nutations and its effect on the mantle. A more complete set of equations has been derived by comparing the linearized, rigid body development (Borderies and Yoder 1990; Yoder 1995b) and the core mantle equations derived by Sasao et al. (1977,1980).

The core equation in the inertial frame is ($m_c = (\omega_x^c - i\omega_y^c)e^{-i\omega t}$ and $\sin \Delta\epsilon = |m_c|\omega|$)

$$(D + i\sigma_c + K_c)m_c = -(D + i\omega)Dp \quad (64)$$

and $\sigma_c = -e_c\omega$ is the unperturbed, free core nutation (FCN) frequency, $e_c = (C_c - A_c)/C_c$ is core figure ellipticity and D is the time operator d/dt. The above form explicitly shows that for a slow precession (with frequency s) of the whole-body pole, the core response is of order s/σ_c smaller. Also, a nearly diurnal wobble with frequency $(-\omega + s)$ results in a core response which is s/ω smaller.

The whole-body dynamical equation is

$$((1 - e_\circ) D^2 + \omega (iD - \sigma_\circ - iK_{ta})) p - e_{22} (3n^2 q e^{-i2\omega t} - D[e^{-i2\omega t} Dq]) \\ = -\alpha Dm_c - \omega\sigma_\circ(P - \frac{e_{22}}{e_\circ} Qe^{-i2\omega t}). \quad (65)$$

Here,

$$K_{ta} = K_{tp} + K_{ap} \\ e_c = e_\circ = \frac{2C - A - B}{2C} \\ e_{22} = \frac{B - A}{2C} \quad (66)$$

with q and Q the complex conjugates of p and P, respectively. Examination of Eq. (65) for its free solutions in the limit $\alpha \to 0$ reveals two modes: a nearly diurnal mode with frequency

$$\omega \left(1 + \frac{3}{2} \left(\frac{n}{\omega}\right)^2\right) \sqrt{(e_o + e_{22})(e_o - e_{22})} \quad (67)$$

and a slow mode with frequency $\sigma_\circ$

$$\sigma_\circ = -\frac{3}{2} \frac{n^2}{\omega} e_\circ. \quad (68)$$

The former mode is the equivalent Venusian Chandler wobble and the latter mode (sometimes called the "tilt-over mode") is the unperturbed (by the core) free mantle nutation or precession (FMN). The eigenfunction equation for an unforced motion proportional to $\exp(-i\gamma t)$ is

$$(\gamma - \sigma_c + iK_c) \left(-\gamma^2(1 - e_\circ) + \omega(\gamma - \sigma_\circ - iK_{ta})\right) = \alpha\gamma^2(\omega - \gamma). \quad (69)$$

If $e_c \gg e_\circ$ and the K_c and K_{ta} factors are nonzero, then the two frequencies are

$$\gamma_+ \simeq \sigma_c (1 + \alpha) [1 - i(K_c - \alpha K_{ta})] \quad (70)$$

$$\gamma_- \simeq \sigma_\circ \left(1 - \alpha \frac{\sigma_\circ}{\sigma_c}\right) + i \left[K_{ta} - \alpha \left(\frac{\sigma_\circ^2}{(\sigma_c - \sigma_\circ)^2 + K_c^2}\right) K_c\right].$$

For $K_c \ll \sigma_c$, the FMN mode has a real frequency (neither damps or grows) if $K_{ta} - \alpha \left(\frac{\sigma_o}{\sigma_c}\right)^2 K_c = 0$. Now K_c is itself proportional to $\sin \Delta\epsilon$ and from Eq. (64)

$$\sin \Delta\epsilon \simeq \frac{\sigma_o}{\sqrt{(\sigma_c - \sigma_o)^2 + K_c^2}} \sin \epsilon. \tag{71}$$

E. Obliquity Torque Balance

A stationary end state is achieved when the ratio σ_o/σ_c satisfies Eqs. (39), (59), (66), (70) and (71):

$$\kappa^2 \alpha \left(\frac{\sigma_o}{\sigma_c}\right)^3 \left(\frac{u_\delta}{u_o}\right)^2 \simeq \frac{K_{ta}}{K_{ts}} \frac{3k_2}{2Q} \frac{n^2 M R^2}{\omega^2 C} \frac{M_\odot}{M} \left(\frac{R}{a}\right)^3 \frac{32\sqrt{2}}{45\pi} \frac{1}{\sin \epsilon}. \tag{72}$$

The left-hand side of Eq. (72) is reasonably well determined and independent of parameters on the right side. Plausible values for the input parameters are $C/MR^2 = 0.336$, $k_2 = 0.25$, $Q = 50$, $\sin \epsilon = 0.036$, the Kármán constant (see Eq. 53) $\kappa = 0.4$ and $K_{ta} \simeq -1.4 K_{ts}$. The resulting value for the right-side factor in Eq. (72) is

$$\alpha \left(\frac{\sigma_o}{\sigma_c}\right)^3 \left(\frac{u_\delta}{u_o}\right)^2 \simeq 1.5 \times 10^{-7}. \tag{73}$$

The constraints imposed on Venus' structure suggest that the moment ratio satisfies $0.070 \leq \alpha \leq 0.12$ as compared with $\alpha_\oplus \simeq 0.114$. Adopting $R_c = 3100$ km and $\alpha = 0.084$ and $\nu = 0.01$ cm^{2}s^{-1}, we obtain $Re = 2 \times 10^7$, $u_o/u_\delta \simeq 11$, $\sigma_c/\sigma_o \simeq 17$ and

$$e_c \simeq 30 e_o = 4 \times 10^{-4}. \tag{74}$$

Because $e_c = (a_c^2 - c_c^2)/(a_c^2 + c_c^2) \simeq (a_c - c_c)/a_c$, the difference in core equatorial and polar axes is 1.3 km. This is to be compared with Earth's non-hydrostatic core ellipticity, $\Delta e_c = 1.2 \times 10^{-4}$ (Herring et al. 1986), whole body $\Delta e_o = 3.15 \times 10^{-5}$ (Yoder 1995a) and ratio $\Delta e_c/\Delta e_o = 3.8$. Clearly, a firm determination of ϵ from observation can be combined with a modestly firm estimate of the tidal and core parameters to provide a useful estimate for the unknown e_c, because it depends on the cube root of these controlling factors.

III. OBLIQUITY RESONANCE

An alternative explanation for Venus' obliquity is that it may have recently passed through or is presently locked in an obliquity resonance arising from some very small orbital variations in inclination driven primarily by a fourth order ($e^2 I^2 \cos 2\omega$) interaction between Earth and Venus. Here ω is an orbit angle (argument of pericenter). The complex orbital perturbations in inclination have maximum amplitude $\sim 10^{-5}$, but can still produce significant obliquity perturbations as some terms have frequencies near the prograde FMN mode. Although such terms are formally included in modern theories of planetary quasi-secular perturbations, their amplitudes are significantly below usual cutoffs for published results. These small prograde terms have been reconstructed using Laskar's (1990) model as a starting point. CDP9C1T2 contains the 8 primary terms affecting orbit inclination and eccentricity. Each of these primary amplitudes can be expanded with secondary sidebands with frequencies which are largely (but not entirely) combinations of these 16 listed here. The $P(2\omega)$ expansion is

$$P(2\omega) = \sum I_{mnojkl} e^{-i\psi_{mnojkl}}. \tag{75}$$

The indices identify a particular frequency, and can be related to the fundamental frequencies in the planetary problem. Table I contains the four largest terms within the frequency band 44″ yr^{-1} to 47″ yr^{-1}.

TABLE I
Prograde Frequency Variations[a]

m	n	o	j	k	l	Frequency ″	Phase deg.	I_{mnojkl} ×10^{-8}
2	4	3	1	1	1	44.223	114.539	2966
2	4	3	1	1	4	44.502	213.753	787
2	4	3	1	3	1	44.775	327.003	582
2	3	3	1	5	1	45.326	359.336	443

[a] The above terms have been identified with the following combinations of the fundamental eigenfrequencies: $\frac{d}{dt}\phi_{243111} = g_2 + g_4 - s_3$, $\frac{d}{dt}\phi_{234114} = g_2 + g_4 - s_3 + \delta f_1$, $\frac{d}{dt}\phi_{243131} = g_2 + 2g_4 - g_3 - s_3$, $\frac{d}{dt}\phi_{233151} = g_2 + g_4 + s_4 - 2s_3$. Laskar (1990) found that he had to introduce new frequencies $\delta f_1 = 0.28″$ yr^{-1} and $\delta f_2 = 0.12″$ yr^{-1} to aid this identification; also $2\delta f_1 = g_3 - g_4$.

Numerical integration of the equations is the most straight forward method of assessing the importance of these terms, and factors which must be considered, besides initial conditions and proximity to a particular resonance are variations in the tidal torque models and a secularly changing J_2 gravity coefficient. The objective is to determine when the prograde rate terms are important and show what kind of conditions could account for the observed obliquity. There are two interesting cases: (1) the maximum CMF torque which can compete with resonance without totally overwhelming it; (2) the conditions where resonance excitation is due to chaotic wander of the driving frequencies rather than a changing J_2.

In order to compare numerical integrations with the present-day phase and amplitude of the free obliquity, the equations are integrated in the rotating frame of the near resonant driving frequency, $d\psi_k/dt$. Resonance passage occurs where $d\phi/dt - d\psi_k/dt$ reverses sign and can change the amplitude and phase of the free obliquity which is fixed relative to this frame. The p_x

and p_y variables in this frame are

$$p_x = \sin\epsilon \sin\left(\phi - \frac{d\psi_k}{dt}t\right)$$
$$p_y = \sin\epsilon \cos\left(\phi - \frac{d\psi_k}{dt}t\right). \quad (76)$$

The only limitation is that nonresonant terms will generate small oscillations whose phases do not match those of the present day, but one can visually adjust for their contribution. The following pair of equations omit the FCN mode as it damps so rapidly:

$$\frac{d}{dt}p_x = \left(\sigma_\circ \cos\epsilon - \frac{d}{dt}\psi_k\right)p_y + K_{ta}\left(1 - \frac{\sin\epsilon}{\sin\epsilon_\circ}\right)p_x \quad (77)$$
$$- \sigma_\circ \sum_j I_j \sin\left(\frac{d}{dt}\psi_k t - \psi_j\right)$$

$$\frac{d}{dt}p_y = -\left(\sigma_\circ \cos\epsilon - \frac{d}{dt}\psi_k\right)p_x + K_{ta}\left(1 - \frac{\sin\epsilon}{\sin\epsilon_\circ}\right)p_y \quad (78)$$
$$+ \sigma_\circ \sum_j I_j \cos\left(\frac{d}{dt}\psi_k t - \psi_j\right).$$

The core torque is modeled by the term $\propto \sin\epsilon/\sin\epsilon_\circ$.

First consider the case where the core friction equilibrium obliquity is actually smaller than 0.036 and the observed obliquity is potentially due to recent resonance passage. Figure 5 displays the results of an integration $\sin\epsilon_\circ = 0.012$. The system evolves from $\sigma_\circ = 43.8''\text{yr}^{-1}$ to $45.4''\text{yr}^{-1}$ after 750 Myr by introducing the constant rate $d\ln J_2/dt = 5 \times 10^{-11}\,\text{yr}^{-1}$. Passage through ψ_{243111} (see Fig. 5b) does excite a large amplitude obliquity of 0.06 rad, but it decays so rapidly that the curve fails to pass through the pole location. Figure 5c shows that the evolution due to ψ_{234114} does almost cover the pole location, partly due to the favorable phase of this term and nonresonant oscillations. Smaller $\epsilon_\circ$ fails to excite an obliquity that can cover the present pole location. Passage which is driven by a negative $d\ln J_2/dt = -5 \times 10^{-11}$ yr^{-1} results in even weaker excitations. Histories with a negative J_2 rate produces a weaker response, partly because capture into permanent resonance is physically impossible. We can safely conclude that resonant excitation can relax the constraint on core friction such that the equilibrium $\epsilon_\circ$ is reduced by *no more* than a factor of 2 to 3. This, in turn, reduces the required core ellipticity by a factor of only 1.25 to 1.5.

The positive rate chosen for J_2 allows for only a small fractional change of 0.03 in J_2 over the 750 Myr integration, which is longer than the 300 to 500 Myr estimate of the last resurfacing event. A significantly larger J_2 rate might push evolution so fast that the resonances have substantially less

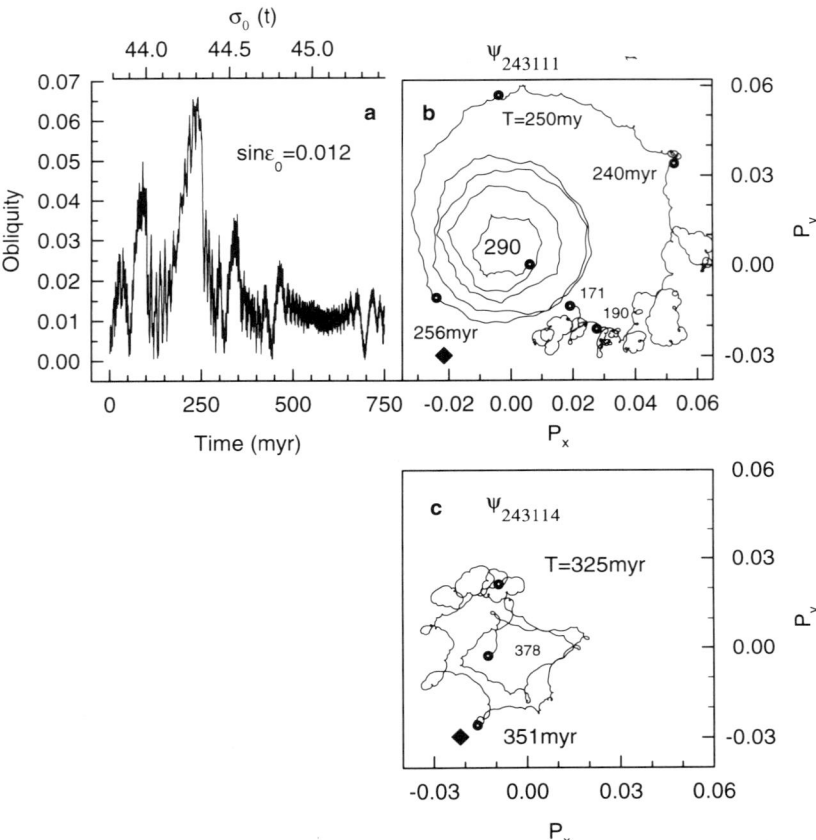

Figure 5. Obliquity history with positive rate $d \ln J_2/dt = 5 \times 10^{-11}$ yr^{-1} and $\sin \epsilon_0 = 0.012$ and initial $p_x = 0.002$ and $p_y = 0.000$. Panel (a) displays the (p_x, p_y) history for ψ_{243111} (171 Myr$\leq t \leq$290 Myr) while panel (b) shows ψ_{243114} (325 Myr$\leq t \leq$351 Myr). For each case, p_x and p_y are shown relative to rotating frame of each argument such that phase relative to the present is preserved. Observed pole location (diamond) is indicated, and origin is predicted pole position due to nonresonant, retrograde terms. These (p_x, p_y) panels cover only the interesting time interval of the pertinent variable.

effect. Figure 6 shows an integration in which $\sin \epsilon_\circ = 0.024$ and $d \ln J_2/dt$ is 4×10^{-10}. The largest resonance term dominates and helps to wash out the effect of nearby terms. The large, positive J_2 rate also helps to retain memory of the ψ_{243111} excitation to larger $\sigma_\circ$, which is within the model bound. This, in turn, suggests that $\epsilon_\circ$ might be reduced by a factor of 2 or less for some narrow range of $\sigma_\circ$ within the bound.

So far, we have ignored the fact that the driving frequencies and amplitudes vary slightly over time. Figure 7 shows the histories of 4 combinations

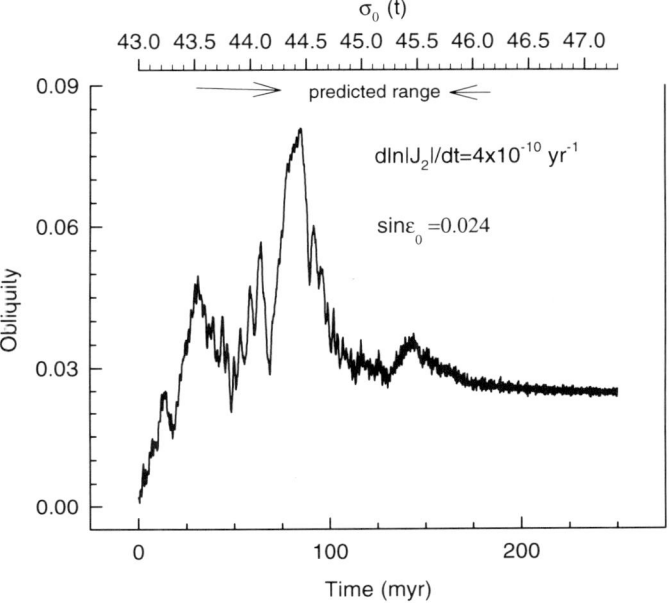

Figure 6. Obliquity history with $d \ln J_2/dt = 4 \times 10^{-10}$ yr^{-1} and $\sin \epsilon_0 = 0.024$ and initial $\sigma_\circ(J_2) = 43.5''$ yr^{-1}.

of the primary frequencies, based on Laskar's (1990) FFT analysis covering a 200 Myr integration into the past. We shall rely on these variations rather than a changing J_2 to excite obliquity, but this mechanism requires that $\sigma_\circ$ is close to one of the four frequencies. Consider the range of $\sigma_\circ(0)$ for which the excited obliquity amplitude is equal to or larger than the observed amplitude without regard to phase. From numerical studies, the range of $\sigma_\circ(0)$ which satisfies this constraint is

$$\begin{aligned} 43.80 &< \sigma_\circ(0) < 44.34 \quad '' \text{ yr}^{-1} \\ 44.42 &< \sigma_\circ(0) < 44.57 \\ 44.65 &< \sigma_\circ(0) < 44.77 \\ 45.23 &< \sigma_\circ(0) < 45.26. \end{aligned} \quad (79)$$

Given that the plausible range for $\sigma_\circ(0)$ is $44.1 < \sigma_\circ < 45.8''$ yr^{-1}, a large excitation is observed over about 30% of this range. Thus, chaotic excitation is a plausible explanation, but still not quite as compelling as any involving core friction.

IV. SPIN-ORBIT RESONANCE

Goldreich and Peale (1967) considered the possibility that Venus' axial spin is locked in a gravitational resonance with Earth involving the general angle

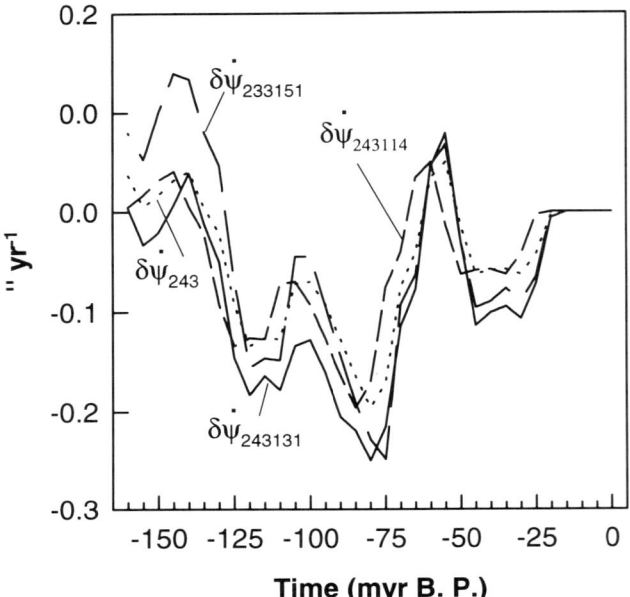

Figure 7. Frequency histories of four arguments (ψ_{243111}, ψ_{234114}, ψ_{243131}, ψ_{233151}) which cover the past 150 Myr (from Figs. 8 and 9 in Laskar 1990).

argument $\Psi_p = 2(\lambda_s - \lambda) - 2p(\lambda_\oplus - \lambda)$. The rate nearly vanishes for $p = -5$. The axial variation of the spin is described by

$$\frac{d^2}{dt^2}\lambda_s + \frac{3}{2} n_\oplus^2 \frac{M_\oplus}{M_\odot} \frac{B-A}{C_m} \widetilde{K}(-5) \cos^4 \frac{1}{2}\epsilon \sin \Psi_{-5} = N_3. \quad (80)$$

The coefficient $\widetilde{K}(-5) = 2.53$ for circular, coplanar orbits. The moment difference $B - A = (2.22 \pm 0.01) \times 10^{-6} M R^2$ (Konopliv and Sjogren 1996) and if Venus were trapped in this lock but had a large libration amplitude, then the maximum angular velocity produced by this torque is

$$\left|\frac{d}{dt}\Psi\right|_{max} \simeq n_\oplus \sqrt{3 \frac{M_\oplus}{M_\odot} \frac{B-A}{C_m} \widetilde{K}(-5)} \simeq 1.3 \times 10^{-5} n_\oplus. \quad (81)$$

Presently, the rate $\frac{d}{dt}\Psi_{[-5]} = -1.89 \times 10^{-3} n_\oplus$, and Venus' present-day spin rate is clearly too large for this spin-orbit torque to maintain a zero rate for the angle Ψ_{-5} or a "lock."

In the limit of small obliquity, the sum of torques N_3 acting on the spin is

$$N_3 = -\omega(K_{ts} + K_{as}) - \frac{C_c}{C} K_c \left[\left(\frac{\sigma_o}{\sigma_c}\right)^2 \omega \tan\epsilon \sin\epsilon + \frac{C}{C_m}\frac{d}{dt}\lambda_s\right]. \quad (82)$$

The constants K_{ts} (Eq. 39) and K_{as} (Eq. 50) arise from solid tidal friction within mantle and atmospheric thermal tide, respectively. A core fluid friction term $\propto K_c$ contains both a small constant part (which depends on obliquity and the ratio of mantle and core precession constants) and a periodic part arising from the differential motion of mantle and fluid core. The tidal coefficient is

$$\frac{\omega K_{ts} \simeq 3.2 \times 10^{-7} n^2 k_2}{Q}. \tag{83}$$

The core secular torque is $O(\sin^2 \epsilon)$ smaller than the solid tidal torque.

The ratio of resonant restoring force to solid tidal torque is

$$(B - A)\widetilde{K} \frac{M M_\oplus}{M_\odot^2} \frac{Ca^3}{C_m R^3} \frac{Q}{k_2} \simeq 2.6 \times 10^{-4} Q/k_2 \tag{84}$$

or about 1/20 using reasonable estimates for $k_2 = 0.25$ and $Q = 50$. Clearly, this spin-orbit torque can have only a minor effect on the spin balance even if this lock were established.

On the other hand, if solid and atmospheric tides are in exact balance at the present rate of rotation, then the secular torque at resonant spin ω_r is $\sim 0.5(\omega - \omega_r)K_{ts}$ and the above ratio of torques is increased from 1/20 to ~ 60. Thus, the spin-orbit torque should be able to stabilize the spin. Capture into resonance might be possible if Venus slowly evolved through this resonance during the tidal spin-down phase of planet history. A more interesting possibility is that small contemporary changes in the quasi-secular torque which might cause the spin to drift back and forth through this resonance in a random fashion. This variable torque might be caused by climatic changes in the planetary albedo (which should affect both the zonal wind and thermal tide) or dynamic changes in planetary obliquity and eccentricity.

Capture into resonance depends on the ratio of rotational energy dissipated by the term $\propto K_c \, d\lambda_s/dt$ to the energy change caused by the secular torque during one oscillation: $V = \frac{C_c}{C_m} K_c |\frac{d}{dt}\Psi|_{\max}/(\omega - \omega_s)K_{ts} \simeq 3$ for the adopted parameters. If this ratio is greater than unity then capture is automatic [more precisely if $P_c \geq 1$, where $P_c = 2V/(V + \pi/4)$] (Goldreich and Peale 1967; Yoder 1979). Because capture seems assured, one must question the implicit assumption that atmospheric changes are small, smooth and slow. Perhaps volcanic activity or asteroidal impacts loft enough dust to change the albedo and hence the absorbed radiation. This mechanism will change both the mean zonal wind and the thermal tidal torque since both depend on absorbed solar radiation. Another factor may be the long-term changes in insolation due to changes in orbital eccentricity, which varies by $0 \leq e \leq 0.065$ on a 10^5 time scale (Laskar 1990).

Successful detection and modeling of hypothetical variations in Venus' axial rotation at seasonal and perhaps 11 to 22 yr solar periods might justify these speculations concerning long-term changes in the axial torque balance

(Schubert 1983). The weighted super-rotation velocity is $\simeq 13$ m s^{-1} and the atmospheric mass $M_a = 1.0 \times 10^{-4}$. The excess atmospheric momentum is $1.4 \times 10^{-3} C_m \omega$ and if added to the mantle would *decrease* the rotation period by 0.3 d. A 1% fluctuation in atmospheric momentum changes ω by 1.4×10^{-3}%. The corresponding rotation angle amplitudes for 225 d and 11 yr periodic fluctuations are 2.1" and 48", respectively (or 80 m and 1.4 km at the equator). Analysis of Magellan SAR data for periodic variations in axial rotation certainly should result in at least useful upper bounds.

V. EULERIAN WOBBLE

The last aspect to discuss is the apparent wobble amplitude inferred from the gravity field C_{21}/S_{21} coefficients: $\epsilon_W \simeq \sqrt{C_{21}^2 + S_{21}^2}/J_2$. Excitation from a random walk source will grow until limited by the wobble damping due to mantle solid friction. The decay time τ_W is (Yoder and Ward 1979)

$$\tau_W^{-1} = \omega \frac{k_2}{Q_W} \frac{\omega^2 R}{3g} \frac{M R^2}{C_m} \left(1 + \frac{3}{2}\left(\frac{n}{\omega}\right)^2\right)^2 \frac{1+\sqrt{\alpha/\beta}}{\sqrt{\alpha/\beta}+\sqrt{\beta/\alpha}}. \quad (85)$$

Here, $\alpha = (C-B)/A$, $\beta = (C-A)/B$ and Q_W is the dissipation factor due to flexing at the wobble frequency σ_W. The decay time is $\tau_W = 1 \times 10^6 Q_W/k_2$ yr.

The wobble period $\simeq 15{,}000$ yr and the effective Love number might be closer to its fluid limit $k_s \simeq 1$, depending on the effective mantle viscosity. For a uniform viscous body (see, e.g., Lambeck 1980),

$$\frac{k_2}{Q_W} = k_s \frac{F}{1+F^2} \quad (86)$$

with

$$F = \frac{19}{2}\frac{\nu \sigma_W}{gR}. \quad (87)$$

The factor $F = 1$ for $\nu = 4 \times 10^{21}$ cm^{2}s^{-1} which is close to Earth's upper mantle viscosity (about 2×10^{22} cm^{2}s^{-1}).

Spada et al. (1996) propose that the free wobble is excited by steady, ongoing changes (due to mantle convection) in the off-diagonal moments of inertia or C_{21}/S_{21} coefficients. If we ignore wobble damping, the wobble amplitude is $p_W \simeq (MR^2/C)(\omega/\sigma_S^2) dC_{21}/dt$ for a constant dC_{21}/dt rate. The predicted C_{21}/dt which will account for a 0.5° wobble is $\simeq 3 \times 10^{-11}$ yr^{-1}. This rate is at least an order of magnitude larger than any estimates of the tectonic contribution to secular changes of Earth's gravity coefficients.

Consider random impulses which change C_{21} by ΔC_{21} every 2×10^4 yr, on average. This will grow by random walk by a factor of $\sqrt{\sigma_W t}$ until limited by solid friction. The mean amplitude is

$$C_{21} \simeq \sqrt{\tau_W \sigma_W} \Delta C_{21} \simeq 20\sqrt{Q_W/k_2} \Delta C_{21}. \quad (88)$$

A $\Delta C_{21} \simeq 4 \times 10^{-10}$ accounts for the observed wobble given a stiff mantle viscosity of 1×10^{23} cm^{2}s^{-1}. If the impulses are due to mass emplacements on the surface (e.g., volcanic activity) then each event amounts to about 1×10^3 km^3 of material. The equivalent resurfacing rate is about 0.05 km^3 yr^{-1} which is in reasonable agreement with independent estimates of its upper limit. However, the analysis of this process is not yet rigorous enough to choose this mechanism over others that might be constructed.

One can image long-period fluctuations in atmospheric surface pressure with $n = 2, m = 1$ symmetry might excite a wobble. The required variations in pressure are $P_a(M/M_a)4 \times 10^{-10} \simeq 0.4$ mbar. The unresolved issue here is how such pressure variations might be maintained and communicated to the surface. One avenue might be to examine the pressure field generated by a combination of the zonal wind and topography.

VI. TIDAL HISTORIES

How certain should we be that the obliquity has achieved its fully damped state? This question cannot be unequivocally resolved because both the initial rotation state and tidal model parameters are insufficiently well known to recover a unique tidal history. However, one can investigate a wide range of models and determine a probable history. The most significant difference between this study and previous investigations is that the CMF model is more accurate in that it includes core ellipticity and is nonlinear in obliquity. This torque displays a rapid increase in strength as Venus approaches its present spin state from larger initial spin and obliquity. In addition, Venus can evolve from prograde to retrograde rotation for a special suite of initial conditions and atmospheric tidal models. To avoid further confusion, we should again emphasize that we have adopted the convention that the spin is negative and the present obliquity is small. An equally valid convention is to choose spin to be positive and obliquity near 180°.

The total e_o is the sum of a quasi-rigid part due to internal strength and convection and a hydrostatic part due to rotation: $e_o = e_o(\text{rigid}) + e_o[\text{hyd}, \omega(t)]$. The changing hydrostatic contribution $e_o[\text{hyd}, \omega(t)]$ as a function of rotation is approximately (assuming $C/MR^2 = 1/3$)

$$e_o(\text{hyd}, \omega(t)) = \frac{39}{41} \frac{\omega^2 R}{g} = 6.7 \times 10^{-8} \left[\frac{\omega(t)}{\omega(\text{now})}\right]^2 \qquad (89)$$

and contributes only 0.5% to the total at the present. The core e_c similarly is made up of these two parts, and has $e_c(\text{hyd}) \simeq 3/4 e_o(\text{hyd})$ for an Earth-like core.

The ratio $\sigma_o/\sigma_c = 1.5(e_o/e_c)(n/\omega)^2$, and the hydrostatic contribution to oblateness due to rotation is small as long as the rotation period is >20 d. Because the contribution from CMF is proportional to $(\sigma_o/\sigma_c)^3 \omega$, the damping due to CMF was dramatically smaller when Venus' rotation period was smaller, although this reduction can be partially counterbalanced by a change in e_o/e_c to more Earth-like values $\sim 1/10$ during most of its history.

Two additional modifications of the linearized equations must be included before obtaining tidal histories. First, the FMN frequency must be modified for large obliquity and is $\sigma_o \cos \epsilon$. Second, we have attempted to account for both rotationally induced oblateness and a more Earth-like value for the ratio e_o/e_c during most of its history. Laminar CMF begins to dominate over turbulent CMF for Reynolds number $\sim 10^{10} \sin^2 \epsilon (\omega_o/\omega(t))^3 < 10^5$, but this occurs in a regime where CMF is already small compared to tides.

The relative rate of damping of obliquity and spin are shown in Fig. 8 for a suite of initial obliquities and an initial spin 30 times the present rate. A separatrix (dotted lines) delineates the two types of history: (1) evolution from prograde to either prograde or retrograde end state; and (2) evolution from prograde to prograde end state. The initial separatrix obliquity is $\sim 45°$ (prograde spin). This maximum can be reduced further by starting the system with faster spin.

If Venus ever happened to be prograde, then an additional complication is obliquity resonance. This time the retrograde orbital variations in inclination are of order 10^{-2} to 10^{-4} in amplitude (see CDP9C1T2). This effect is mostly chaotic because the spacing of 2 or more frequencies (including sideband frequencies) with comparable amplitudes is sometimes less than the obliquity libration width. The results shown here differ from Laskar and Robutel (1993) in that a quasi-constant nonhydrostatic contribution ΔJ_2 to the gravity J_2 coefficient is included.

$$J_2 = 0.0011 \left(\frac{\omega(t)}{\omega_\oplus}\right)^2 + \Delta J_2 \qquad (90)$$

where $\omega_\oplus = 2\pi/\text{day}$. During most of Venus' tidal history solid friction tides are the dominant torque changing the spin $\omega(t)$ with time t. CDP9C1F2 displays 4 typical histories with $\Delta J_2 = 0.44 \times 10^{-5}$, close to its present-day value. One of these curves starts with very small initial obliquity near $5°$ (red). As Venus' spin slows, it eventually enters a chaotic zone while the spin lies between $37n$ and $27n$. The obliquity upon exiting (residence time is probabilistic in character) is substantially larger and it thereafter evolves to a retrograde orientation, driven past $90°$ by the semi-annual tide. Core friction takes over once spin rate is less than $5n$ and rapidly damps obliquity. One significant point is that exit from the chaotic zone is only possible from higher obliquity over most of the history because tides tend to increase obliquity unless spin is very small. One unusual history is shown (black). After it exits the chaotic zone, the obliquity temporarily locks with a small, isolated perturbation with frequency $-0.69''\ \text{yr}^{-1}$ at an obliquity slightly less than $90°$.

CDP9C1F3 displays five histories with an Earth-like $\Delta J_2 = 1.0 \times 10^{-6}$. Curiously, this increase apparently broadens the chaotic zone. Actually, what is happening is that the range in ω for the chaotic zone has shifted toward

Venusian Obliquity Histories

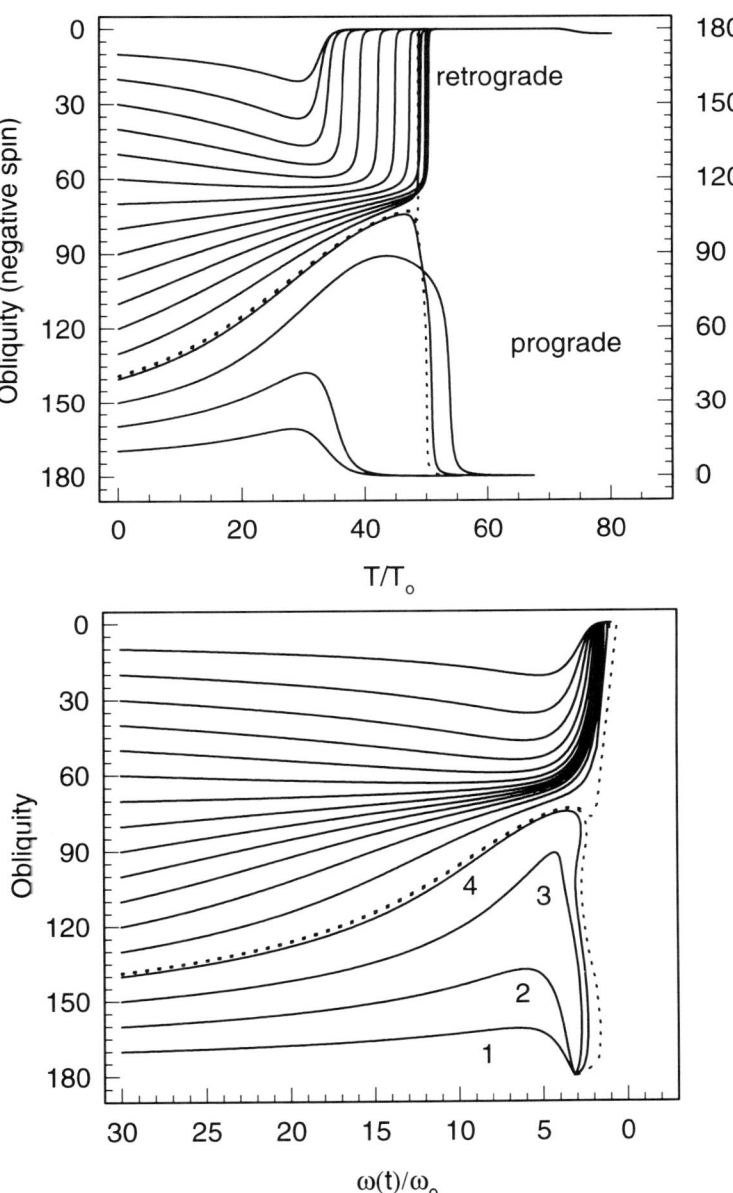

Figure 8. Tidal histories of obliquity $\epsilon(t)$ as a function of both time T/T_0 (a) and spin $\omega(T)/\omega(\text{now})$ (b). The time scale $T_0 = 1/K_{ts}(\text{now}) \sim 5 \times 10^7$ yr. The mantle nonhydrostatic e_o is set equal to $3e_o(\text{now})$ until $T = 70T_0$, after which $e_o = e_o\omega(T)/\omega(\text{now})$. This choice tends to speed up the onset of core friction. Here the separatrix starts at a prograde obliquity of $\simeq 45°$ (retrograde obliquity of 135°) and $\omega(0)/\omega(\text{now}) = 30$.

slower spins. One history (blue) does evolve to prograde orientation, partly because it remains in the chaotic zone long enough for core friction to become effective. Although evolution to retrograde orientation is not inevitable, the effect of chaos is to eliminate the connection between initial obliquity and the final obliquity state. Chaotic resonance excitation, when coupled with strong semidiurnal thermal tide, almost always leads to histories with retrograde orientation.

The potential dominance of the atmospheric semi-annual tide does open the possibility that Venus may have started with a prograde rotation. One can imagine a large impact on Venus which, unlike the proposed impact creating the Earth–Moon system (see, e.g., Newsom and Taylor 1989), dramatically tilted the spin axis to greater than $\sim 45°$ and perhaps slowed the planetary spin to, say, 5 to 10 d. A slow rotation guarantees that any debris disk is inside synchronous rotation orbit (for Venus, a 5d synchronous radius $\simeq 20R$) and will tidally evolve any primordial satellite onto Venus (Ward and Reid 1973).

VII. SUMMARY

The principal result here is that we now have two plausible explanations for Venus' nonzero free obliquity $\simeq 2.1°$. The first and least constrained hypothesis predicts that Venus' obliquity (like its spin) has achieved steady state due to a large CMB oblateness $e_c \simeq 4 \times 10^{-4}$, which reduces by 4 orders of magnitude the usually dominant CMB friction torque. A balance of CMB friction torque with the combined torque due to solid tides and atmospheric thermal tides is achieved at the present obliquity. The second option is that small, near-resonant excitations maintain the free obliquity. This mechanism eliminates the role played by the tides but not the constraint on the core oblateness. Either mechanism is qualitatively rigorous despite possible shortcomings related to the estimate of the turbulent core friction coupling constant, uncertainties in atmospheric thermal tide modeling and Venus' core size, even resonant excitation, all of which might conspire to reduce the inferred CMB oblateness by at most 1/2 to 1/3. The inferred Venusian core oblateness is equal to or greater than Earth's nonhydrostatic core oblateness, and is consistent with dynamic compensation within a D'' thermal boundary layer (Richards and Hager 1984; Hide et al. 1993). This result should provide an important constraint on Venus' mantle convection that must be accounted for in any detailed model (see, e.g., Turcotte 1993).

A more detailed understanding of Venusian internal structure from analysis of the various mechanisms affecting rotation require better data. The most important at this point is re-analysis of Magellan SAR images and Doppler tracking data for seasonal changes in axial rotation and perhaps a more accurate determination of the k_2 Love number.

The predicted precession rate $\dot{p}$ is $1.1'' \text{ yr}^{-1}$ or about 32 m yr^{-1} drift for a lander at Venus' north pole. Precise measurement of this parameter determines polar moment of inertia. The wobble rate in the body-fixed frame is $\dot{p}_W \simeq \sigma_W \sin \epsilon_W \simeq 0.8''$ yr^{-1}. Both wobble and seasonal rotation couple weakly to the fluid core (see Eqs. 64 and 82). Therefore the amplitude of each are inversely dependent on mantle moment and precise monitoring of these three signatures could contrain both mantle and core moments. However, inferring mantle moment from rotation also requires knowledge of the change in atmospheric momentum.

A future mission might include a mixture of surface landers, a swarm of small transponders which float at various pressure heights and latitudes within the atmosphere and orbiting spacecraft. The primary scientific theme of this mission might be Venusian atmospheric dynamics rather than just geophysical structure. Still, tracking accuracy, instrument sensitivity and spacecraft lifetime would be driven by the precision needed to measure these geophysical parameters to a useful accuracy.

Acknowledgments. The research described in this chapter was carried out by the Jet Propulsion Laboratory, California Institute of Technology, under a contract with the National Aeronautics and Space Administration.

REFERENCES

Abramowitz, M., and Stegun, I. I. 1965. *Handbook of Mathematical Tables* (New York: Dover).
Allan, R. R. 1967. Resonance effects due to the longitude dependence of the gravitational field of a rotating primary. *Planet. Space Sci.* 15:53–76.
Arkani-Hamed, J., and Toksöz, M. N. 1984. Thermal evolution of Venus. *Phys. Earth Planet. Int.* 34:232–250.
Arkani-Hamed, J. 1993. On the tectonics of Venus. *J. Phys. Earth Planet. Int.* 76:75–96.
Avduevskii, V. S., et al. 1976. Preliminary results of an investigation of the lighting conditions in the atmosphere and on the surface of Venus. *Cosmic Res.* 14:643–649.
Bills, B. G., Kieffer, W. S., and Jones, R. L. J. 1987. Venus gravity: a harmonic analysis. *J. Geophys. Res.* 92:10335–10351.
Borderies, N., and Yoder, C. F. 1990. Phobos' gravity field and its influence on its orbit and librations. *Astron. Astrophys.* 233:235–251.
Busse, F. H. 1968. Steady fluid flow in a precessing spherical shell. *J. Fluid Mech.* 33:739–751.
Carpenter, R. L. 1964. Study of Venus by CW radar. *Astron. J.* 69:2–11.
Davies, M. E., et al. 1992. The rotation period, direction of the north pole, and geodetic control network of Venus. *J. Geophys. Res.* 97:13141–13152.
Dickey, J. O., et al. 1994. Lunar laser ranging: a continuing legacy of the Apollo program. *Science* 265:482–490.
Dobrovolskis, A. R. 1978. The Rotation of Venus. Ph.D thesis, California Inst. of Technology).

Dobrovolskis, A. R. 1980. Atmospheric tides and the rotation of Venus. II. Spin evolution. *Icarus* 41:18–35.
Dobrovolskis, A. R., and Ingersoll, A. P. 1980. Atmospheric tides and the rotation of Venus. I. Tidal theory and the balance of torques. *Icarus* 41:1–17.
Dziewonski, A., Hales, A. L., and Lapwood, E. R. 1975. Parametrically simple Earth models consistent with geophysical data. *Phys. Earth Planet. Int.* 10:12.
Gans, R. F. 1972. Viscosity of the Earth's core. *J. Geophys. Res.* 77:360–366.
Gold, T., and Soter, S. 1969. Atmospheric tides and the resonant rotation of Venus. *Icarus* 11:356–366.
Goldreich, P., and Peale, S. J. 1967. Spin-orbit coupling in the solar system. II. The resonant rotation of Venus. *Astron. J.* 72:662–668.
Goldreich, P., and Peale, S. J. 1970. The obliquity of Venus. *Astron. J.* 75:275–284.
Goldstein, R. M. 1964. Venus characteristics by Earth-based radar. *Astron. J.* 69:12.
Goldstein, S. 1965. *Modern Developments in Fluid Mechanics* (New York: Dover).
Gottfried, K. 1966. *Quantum Mechanics*, vol. 1 (New York: Benjamin).
Hashim, Z. 1983. Analysis of composite solids—a survey. *J. Applied Mech.* 50:481–505.
Herring, T. A., Gwinn, C. W., and Shapiro, I. I. 1986. Geodesy by radio interferometry: Studies of the forced nutations of the Earth. I: Data analysis. *J. Geophys. Res.* 91:4755–4765.
Hide, R., Clayton, R. W., Hager, B. H., Speith, M. A., and Voorhies, C. V. 1993. Topographic core-mantle coupling and fluctuations in Earth's rotation. In *Relating Geophysical Structure and Processes*, AGU Mono. 76, pp. 107–120.
Ingersoll, A. P., and Dobrovolskis, A. R. 1978. Venus' rotation and atmospheric tides. *Nature* 275:37–38.
Jarnagan, M. P., Jr. 1965. Expansions in elliptic motions. *Astron. Papers Amer. Ephem. Naut. Almanac* XVIII:5.1–5.6.
Kaula, W. M. 1960. *Theory of Satellite Geodesy* (Waltham: Blaisdell).
Kaula, W. M. 1964. Tidal dissipation by solid body friction and the resulting orbital evolution. *Rev. Geophys.* 2:661–685.
Konopliv, A. S., and Sjogren, W. L. 1994. Venus Spherical harmonic gravity field to degree and order 60. *Icarus* 112:42–54.
Konopliv, A. S., and Sjogren, W. L. 1996. *Venus Gravity Handbook*, JPL Publ. 96-2.
Konopliv, A. S., and Yoder, C. F. 1996. Venusian k_2 tidal Love number from Magellan and PVO tracking data. *Geophys. Res. Lett.* 23:1857–1860.
Konopliv, A. S., et al. 1993. Venus gravity and topography: 60th degree and order model. *Geophys. Res. Lett.* 20:2403–2406.
Lamb, H. 1960. *Hydrodynamics* 6th ed. (New York: Dover).
Lambeck, K. 1980. *The Earth's Variable Rotation* (Cambridge: Cambridge Univ. Press).
Laskar, J. 1988. Secular evolution of the solar system over 10 million years. *Astron. Astrophys.* 198:341–362.
Laskar, J. 1990. The chaotic history of the solar system: A numerical estimate of the size of chaotic zones. *Icarus* 88:266–291.
Laskar, J., and Robutel, P. 1993. The chaotic obliquity of the planets. *Nature* 361:608–612.
Levie, S. L. 1971. Transformation of a potential function under coordinate rotations. *J. Astronaut. Sci.* 18:217–235.
McNamee, J. B., Borderies, N. J., and Sjogren, W. L. 1993. Venus: global gravity and topography. *J. Geophys. Res.* 98:9113–9128.
Nerem, R. S., Bills, B. G., and Mcnamee, J. B. 1993, A high resolution gravity model for Venus: GMV-1. *Geophys. Res. Lett.* 20:599–602.
Morgan, W. J., and Phillips, R. J. 1983. Hot spot heat transfer: Its application to Venus

and implications for Venus and Earth. *J. Geophys. Res.* 88:8305–8317.
Moroz, V. I. 1983. Summar of preliminary results of the Venera 13 and 14 missions. In *Venus*, eds. D. M. Hunten, L. Colin, T. M. Donahue and V. I. Moroz (Tucson: Univ. of Arizona Press), pp. 45–68.
Mottinger, N. A., Sjogren, W. L., and Bills, B. G. 1985. Venus gravity: A harmonic analysis and geophysical implications. *J. Geophys. Res.* 90:739–756.
Newsom, H. E., and Taylor, S. R. 1989. Geochemical implications of the formation of the Moon by a single giant impact. *Nature* 338:29–34.
Pechmann, J. B., and Ingersoll, A. P. 1984. Thermal tides in the atmosphere of Venus: Comparison of models results with observations. *J. Atmos. Sci.* 41:3290–3313.
Plummer, H. C. 1960. *An Introductory Treatise on Dynamical Astronomy* (New York: Dover).
Poirier, J. P. 1988. Transport properties of liquid metals and the viscosity of the Earth's core. *Geophys. J.* 92:99–105.
Richards, M. A., and Hager, B. H. 1984. Geoid anomalies in a dynamic Earth. *J. Geophys. Res.* 89:5987–6002.
Ringwood, A. E., and Anderson, D. L. 1977. Earth and Venus: a comparative study. *Icarus* 30:243–253.
Roberts P. H., and Stewartson, K. 1965, On the motion of a liquid in a spherical cavity of a precessing rigid body. II. *Proc. Cambridge Phil. Soc.* 61:279–288.
Rose, M. E. 1957. Elementary Theory of Angular Momenta (New York: Wiley).
Sasao, T., Okamota, I., and Sakai, S. 1977. Dissipative core-mantle coupling and nutational motion of the Earth. *Publ. Astron. Soc. Japan* 29:83–105.
Sasao, T., Okubo, S., and Saito, M. 1980. Linear theory on the dynamical effects of a stratified core upon the nutational motion of the Earth. In *Nutation and the Earth's Rotation*, eds. M. L. Smith and P. L. Bender (Dordrecht: D. Reidel), pp. 155–183.
Schaber, G. G., et al. 1992. Geology and distribution of impact craters on Venus: What are they telling us? *J. Geophys. Res.* 97:13257–13301.
Shen, M., and Zhang, C. K. 1989. Dynamical evolution of the rotation of Venus. *Earth, Moon, Planets* 43:275–287.
Shen, M., and Zhang, C. K. 1990. A numerical solution for thermal tides in the atmosphere of Venus. *Icarus* 85:129–144.
Schubert, G. 1983. General circulation and the dynamical state of the Venus atmosphere. In *Venus*, eds. D. M. Hunten, L. Colin, T. M. Donahue and V. I. Moroz (Tucson: Univ. of Arizona Press), pp. 681–766.
Smart, W. M. 1961. *Celestial Mechanics* (New York: J. Wiley).
Spada, G., Sabadini, R., and Boshi, E. 1996. The spin and inertia of Venus. *Geophys. Res. Lett.* 23:1997–2000.
Stevenson, D. J., Spohn, T., and Schubert, G. 1983. Magnetism and thermal evolution of the terrestrial planets. *Icarus* 54:466–489.
Turcotte, D. L. 1993. An episodic hypothesis for Venusian tectonics. *J. Geophys. Res. Planets* 98:17061–17069.
Vanyo, J. P. 1973. An energy assessment for liquids in filled precessing spherical cavities. *J. Applied. Mech.* 40:851–856.
Vanyo, J. P. 1991. A geodynamo powered by luni-solar precession. *Geophys. Astrophys. Fluid Dynamics* 59:209–234.
Vanyo, J. P., and Paltridge, G. W. 1981. A model for energy dissipation at the core mantle boundary. *Geophys. J. Roy. Astron. Soc.* 66:677–690.
Ward, W. R. 1975. Tidal friction and generalized Cassini's laws in the solar system. *Astron. J.* 80:64–70.
Ward, W. R., and Reid, M. J. 1973. Solar tidal friction and satellite loss. *Mon. Not. Roy. Astron. Soc.* 164:21–32.

Ward, W. R., and DeCampli, W. M. 1979. Comments on the Venus rotation pole. *Astrophys. J. Lett.* 230:17.

Yoder, C. F. 1979. Diagrammatic theory of transition of pendulum-like systems. *Celest. Mech.* 19:3–29.

Yoder, C. F. 1981. Free librations of a dissipative moon. *Phil. Trans. Roy. Soc. London A* 303:327–246.

Yoder, C. F. 1995a. Astrometric and geodetic properties of Earth and the solar system. In *Global Earth Physics: A Handbook of Geophysical Constants* (Washington, D. C.: American Geophysical Union), pp. 1–31.

Yoder, C. F. 1995b. Venus' free obliquity. *Icarus* 117:250–286.

Yoder, C. F., and Ward, W. R. 1979. Does Venus wobble? *Astrophys. J. Lett.* 233:33–37.

THE VENUS GRAVITY FIELD AND OTHER GEODETIC PARAMETERS

W. L. SJOGREN, W. B. BANERDT, P. W. CHODAS and A. S. KONOPLIV
Jet Propulsion Laboratory

G. BALMINO and J. P. BARRIOT
CNES/Groupe de Recherches de Géodesie Spatiale

J. ARKANI-HAMED
McGill University

and

T. R. COLVIN and M. E. DAVIES
RAND Corporation

The gravity field of Venus has been determined from Doppler radio tracking data of two Venus spacecraft orbiters, Magellan and Pioneer Venus Orbiter, which have provided nearly a complete global gravity data set. Together they have been used to solve for a 90th degree and order spherical harmonic gravity field of Venus as well as the GM of Venus (gravitational constant times the mass), tide or Love number (k_2) of Venus, pole and rotation rate of Venus. This data set has also been used to resolve and evaluate local or region gravity perturbations using line-of-sight acceleration profiles. The results are consistent with the spherical harmonics, but each has its own merits. The combining of gravity and topography results reveal the very high spatial correlation between the two surface models, more so than on any other planet. An improved Venus pole orientation and rotation period have also been determined from the synthetic aperture radar images and Doppler data.

I. INTRODUCTION

This chapter is divided into five main sections. The first section covers the history and data used to produce the results discussed in the following four sections. The next two sections discuss the determination of the Venusian gravity field using different reduction techniques. The first method fits classical spherical harmonics directly to the raw Doppler data whereas the other technique analyses line-of-sight (LOS) acceleration residuals relative to a fixed base gravity field model. The two methods produce very comparable results and each has its benefits and particular shortcomings. The fourth section incorporates the gravity results with the topography (see, e.g., the chapter by Phillips et al.) to infer the possible internal structure of Venus and its regional variations. The last section describes the determination of the Venusian spin pole location, the rotation rate of Venus and a geodetic control network. These geodetic parameters were extracted from detailed analysis of the Magellan Synthetic Aperture Radar (SAR) data in addition to Doppler radio tracking and altimeter data.

A. History

Prior to Magellan and Pioneer Venus Orbiter (PVO), the Mariner 2, 5, and 10 flybys of Venus provided mass estimates and an upper bound on the gravitation coefficient J_2 (Anderson and Efron 1969; Howard et al. 1974), showing that the oblateness for Venus was several orders of magnitude smaller than the Earth's oblateness. The Russian Venera 9 and 10 spacecraft provided a more accurate estimate of $J_2 = 4.0 \pm 1.5 \times 10^{-6}$ (Akim et al. 1978). The initial spherical harmonic determinations of the Venus gravity field from PVO were low degree and order solutions by Ananda et al. (1980) to degree and order 6 and Williams et al. (1983) to degree and order 7. Mottinger et al. (1985) extended the harmonic solution to degree and order 10 by using only high periapse altitude (about 1000 km) data from PVO, and Bills et al. (1987) solved for a degree and order 18 field by combining the high altitude data with low altitude data arcs. In addition to harmonic analyses, others such as Phillips et al. (1979), Sjogren et al. (1980,1983,1984), Reasenberg et al. (1981,1982), and more recently Reasenberg and Goldberg (1992) have solved for high resolution surface mass distributions either regionally or globally.

With the arrival of faster computers with more memory and disk space, the spherical harmonic solutions have shown a drastic increase in resolution. In support of the Magellan navigation effort, McNamee et al. (1992) reprocessed the low-altitude PVO data to produce a 21st degree and order model. Nerem (1991) with the additional high-altitude data set produced the Preliminary Goddard Venus Model 1, a 36th degree and order field, and ushered in dramatic increases in resolution for harmonic fields. Konopliv et al. (1992) followed with a 42nd degree and order field and Nerem et al. (1993) with a 50th degree and order field all based upon PVO low- and high-altitude data sets.

With Magellan in orbit, McNamee et al. (1993) produced another 21st degree and order field incorporating Magellan high altitude (periapse altitude of 250 km) data with PVO. After September 1992, Magellan began to be tracked during periapse (altitude of 170 km). Prior to this, the high gain antenna was pointed toward Venus to acquire SAR images and no Doppler tracking was obtained within 30 minutes of periapse. Konopliv et al. (1993) produced a 60th degree and order model by combining the PVO data with four months (or about one-half of longitude coverage) of Magellan data. After Magellan successfully aerobraked into a near-circular orbit in August of 1993, Konopliv and Sjogren (1994) produced another 60th degree and order model incorporating much of the near-circular orbit data, and Konopliv et al. (1994*a*) produced a 75th degree and order model with all the Magellan gravity data. This report presents the results for a 90th degree and order gravity solution (named MGNP90LSAAP). This model is the highest resolution spherical harmonic model for any planet (including the Earth) that is based solely upon spacecraft tracking data. The current solutions for the Earth gravity field extend to degree and order 70 for solutions based upon spacecraft tracking data only (Nerem et al. 1994) and to degree and order 360 with surface measurements included (Rapp et al. 1991). Nerem et al. (1995) summarize the history of gravity determinations for the Earth.

In addition to the spherical harmonic gravity field, LOS accelerations with respect to a spherical harmonic gravity field (degree and order 0, degree and order 75, JPL solution MGNP75ISAAP, for the pre-aerobraking data, and degree and order 40, JPL gravity solution MGN40E, for the post-aerobraking data) have been produced for the Magellan Doppler residuals. The procedure is to solve for the spacecraft state and other spacecraft specific dynamical variables using the base spherical harmonic model. The residuals then contain the remaining gravity signature in the data that is not incorporated into the harmonic field. Orbits are processed one at a time, and so this eliminates long-term modeling errors (and long-term gravity information). The pre-aerobraking LOS data was processed by Barriot and Balmino (1992,1994) to produce detailed gravity maps over Eistla Regio and the western portion of Aphrodite Terra and by Arkani-Hamed (1996*a*) to assess the high-resolution gravity of western Ishtar Terra. Kaula (1996*a*), Smrekar (1994), and others have produced regional gravity determinations from the LOS data as well.

B. Data

The gravity measurements used for the Venus gravity field determination are two-way coherent Doppler tracking of the PVO and Magellan spacecraft acquired at the Deep Space Network (DSN) complexes at Goldstone, California; Madrid, Spain; and Canberra, Australia. The PVO spacecraft operated at S-band with the DSN stations transmitting at 2.11 GHz. The transponder on board the spacecraft then multiplies the frequency by 240/221 to obtain a downlink frequency of 2.30 Ghz. For PVO, only two-way data were acquired, i.e., the receiving DSN station is the same as the transmitting station. The spacecraft transponder also provided a minimal amount of X-band downlink data (S-band and X-band were received simultaneously at the ground station), but these data were not processed. The Magellan spacecraft had an S-band transponder with an X to S-band uplink converter and an S to X-band downlink converter. The resulting system had either a X-band or S-band uplink and an S-band and/or X-band downlink. The X-band uplink and X-band downlink (8.43 GHz) provided the high resolution gravity data because of reduced charged particle effects on the X-band signal. All two-way Magellan data were processed beginning with the gravity data of Cycle 4. Three-way data (i.e., the receiving DSN complex is different from the transmitting complex) were also taken for generation of differenced Doppler data in support of navigation, but these data were not incorporated into the gravity solution.

PVO was inserted into orbit about Venus on 4 December 1978 and provided low periapse altitude data (150 to 170 km) until July of 1980 when no maneuvers were performed to maintain a low periapse altitude. By November of 1980, periapse altitude reached 400 km due to solar perturbations on the orbit. The low altitude data set is continuous from 9 December 1978 to 4 December 1980 except for a data gap from 12 July 1979 to 18 December 1979 during which superior conjunction occurred and periapse was occulted. Tracking data were acquired during this time, but these data have not been recovered from archive tapes and may or may not exist. Due to the superior resolution of the Magellan data, the PVO data were compressed substantially by averaging to 60 s for the interval within 30 min each side of periapse, 300 s for the next 2 hr around periapse, and 600 s near apoapse. The spacecraft velocity at periapse is 9 km s^{-1} and 60 s samples decrease the periapse resolution in the PVO data, but this is easily recovered with the Magellan data and the three to four longitudinal coverages of the PVO data. This compression still retains the long-term gravity, rotational, tidal, and ephemeris information and greatly reduces computer time required for filtering. Future solutions may re-introduce a higher number of samples at periapse. The original PVO data, as processed by Konopliv et al. (1993) and others has 5 s samples at periapse. The total number of Doppler observations processed for the low altitude PVO data is 170,000.

PVO had a highly eccentric ($e = 0.8$) orbit with a period of 24 hr and was nearly polar with an inclination of 106°. Because of the high eccentricity, the altitude climbs to over 1000 km for the high-latitude regions (>60°N, <30°S). The geometry of the orbit provides good gravity data because the plane-of-sky inclination (angle between LOS and orbit plane normal) rarely falls below 20° to near face-on geometries. Initially, periapse was occulted and became visible in March of 1979. Because the rotational period of Venus is 243 days, the low-altitude data provides several longitudinal coverages of Venus, with the high-resolution data contained within a narrow latitude band around periapse (15°N). The low-altitude data, with the exception of the first three months of data to March 1979, are the same data set as used by McNamee et al. (1993) and Nerem et al. (1993).

The second PVO data set extends from 6 November 1981 to 7 September 1982. During this time, periapse increased from 980 km to 1340 km due to the solar perturbation. At the beginning of this time span, periapse was just coming out of occultation, and near the end of the high-altitude data, periapse again entered occultation, in August of 1982. After 1982, periapse altitude increased unimpeded until its maximum of about 2500 km in 1986 and then began decreasing until 1992. At that time, periapse raise maneuvers were performed to keep PVO from burning up in the atmosphere. However, with propellant exhausted, PVO entered the atmosphere and the DSN lost signal on 8 October 1992. This data set, due to sparseness of tracking, is not included in this gravity solution. Again, future studies may include this tracking. The high-altitude PVO was tracked continuously for three days once a week for this part of the mission (Mottinger et al. 1985), and amounts to 34,000 observations with the same compression scheme as the low altitude PVO data. This data set is identical to the high-altitude data set used by Mottinger et al. (1985) and is included in the solution by Nerem et al. (1993).

The Magellan spacecraft was inserted into orbit about Venus on 10 August 1990. The first three cycles (one cycle is one Venus rotation period of 243 days) were dedicated to SAR imaging of the Venus surface. This required the high-gain antenna to be pointed to the Venus surface within about 30 min of periapse passage. Thus only high-altitude tracking (>2500 km) was obtained when the high-gain antenna was returned to point at Earth. These data still, however, have some long-term information on the gravity field and it was used by McNamee et al. (1993). The altitude of periapse for the first three cycles is 250 km and is higher than in Cycle 4. Except for the periapse altitude, the orbit shape for Cycle 4 is identical to the previous cycles and should contain all the gravity information that is in the previous cycles (semi-axis = 10425, eccentricity = 0.393, inclination = 85.5 to the Venus equator, periapsis altitude $\approx$280 km, apoapsis altitude $\approx$8500 km). In addition, for the first three cycles the modeling of the solar pressure force for a rotating antenna through periapse passage is complicated, and solar pressure is a significant force for the higher periapse due to diminished drag. For these reasons, the first three cycles of data were not included in the gravity solution.

Cycle 4 began on 15 September 1992 and continued to 24 May 1993. During the complete cycle, there were no periapse altitude adjustments and the altitude varied between 185 and 165 km. Magellan was tracked through periapse with a 2-s sample time, but for the spherical harmonic gravity solutions the sample time was compressed to 10 s. With a periapse velocity of 8.5 km s^{-1}, 10 s samples provide two or three samples per half wavelength of a 90th degree and order field. For higher solutions, we may need to increase the number of samples near periapse. There are 770,000 10-s observations (both X- and S-band) for Cycle 4. Due to smaller eccentricity (e = 0.4), the Magellan Cycle 4 data are much more sensitive at the higher latitudes than the PVO data, but still lacks the high resolution gravity information for these regions.

At the end of May 1993, Magellan periapse was lowered deep into the atmosphere to begin aerobraking. Over the next several months to early August, the atmospheric drag on the spacecraft changed the orbit to nearly circular to provide much lower altitude gravity tracking at the higher latitudes. From 6 August 1993 (17 August for beginning of X-band) to 10 October 1994, Magellan was tracked in this nearly circular orbit with apoapse altitude varying from 600 km to 350 km and periapse altitude from 155 km to 220 km. Post-aerobraking orbits include Cycle 5 and part of Cycle 6 until the Magellan spacecraft was "windmilled" into the atmosphere of Venus. Even if the spacecraft had not been deliberately terminated, Magellan would have been lost within several weeks due to degradation of the solar arrays and loss of power. It would have never been able to fill in the tracking data gaps that remain in the gravity field. The plane-of-sky inclination vs longitude is shown in Fig. 1, and clearly displays the coverage of the post-aerobraking data. Initially, periapse is occulted and apoapse is tracked from longitude 100°W to 60°E, then periapse becomes visible from 220°E to 90°E except for a gap due to superior conjunction. Apoapse tracking then resumes from 110°W to 130°E, and finally periapse is tracked from 60°W to 10°E at the conclusion of the mission. As a result, there is a gap between 140°E and 220°E where there is no direct low-altitude tracking for the high latitudes. This data gap especially shows up in the southern hemisphere because there is only coverage during Cycle 4 and periapsis is essentially constant at 10.7°N (Konopliv and Sjogren [1996], show altitudes). The same compression time of 10 s is used for the post-aerobraking data and amounts to 1,230,000 observations. The velocity of the Magellan spacecraft in the nearly circular orbit is about 7 km s^{-1}, providing a sample every 70 km.

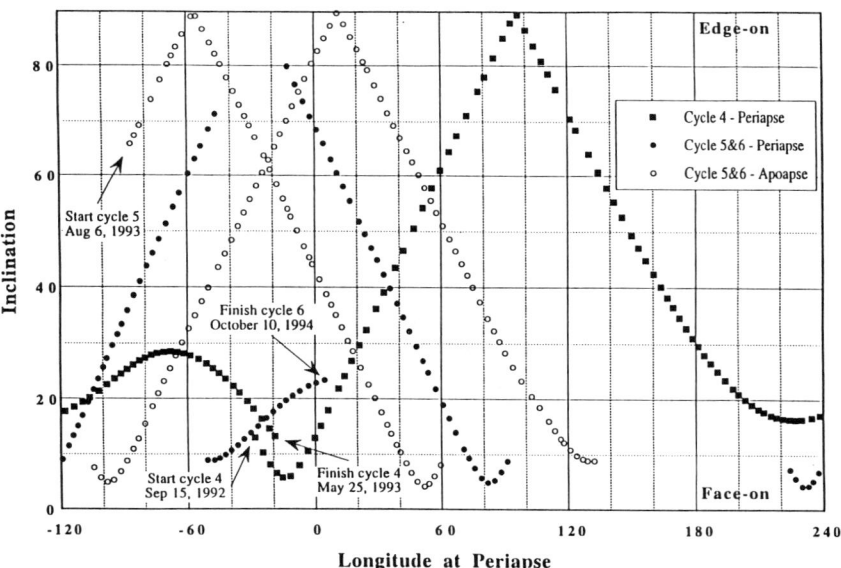

Figure 1. Plane of sky vs periapsis/apoapsis longitude.

The PVO and Magellan data were processed in many arcs (a data time span which is dynamically continuous) where for each arc the initial spacecraft state and other parameters are solved for. The PVO data arcs were chosen to be as long as practical given the imperfect knowledge of the spacecraft nongravitational accelerations. The arc lengths generally were shorter in regions where uncertainties in the nongravitational accelerations (primarily those due to atmospheric drag) were highest. In addition, the data arcs did not include any propulsive maneuvers. The arc lengths for the low-altitude PVO tracking data varied from a minimum of one day to a maximum of ten days, with six days as the typical length. The high-altitude PVO data arcs were three days in length to match the tracking schedule. For Magellan, the data arc lengths for pre- and post-aerobraking were generally one day in length with a maximum length of two days. The long term information in the gravity field can be enhanced by increasing the arc lengths if careful attention is given to the nongravitational forces acting on the spacecraft. This will be the focus of future work.

The Magellan orbits with X-band tracking provide the highest-resolution gravity information for Venus. The LOS accelerations for all X-band orbits, with respect to the gravity fields mentioned above, were delivered to the Planetary Data System (PDS) Geophysics Subnode of the Geosciences Node at Washington University in St. Louis. For the LOS data, 2-s samples were used where available. This amounted to delivery of 5,973,000 observations for 4600 orbits.

II. SPHERICAL HARMONICS

The geometrical models used in the reduction include Earth platform parameters such as station positions, precession, nutation, etc., and media calibrations for the observables due to the troposphere and ionosphere of the Earth. The dynamical models consist of all forces acting on the spacecraft and these are briefly described below. The gravitational attraction of the Sun and planets other than Venus are modeled as point masses with the positions given by the JPL DE403 ephemeris. The major contributor is the Sun with an acceleration of 1×10^{-9} km s^{-2}. The relativistic acceleration on the spacecraft (Moyer 1971) is largest due to Venus and is about 1×10^{-9} km s^{-2} at periapse with the Sun contribution 4 orders of magnitude smaller. The force is predominantly in the radial direction and inversely proportional to the distance squared (equivalent to a change in GM).

The gravitational potential of Venus is modeled by a spherical harmonic expansion with normalized coefficients ($\bar{C}_{nm}$, $\bar{S}_{nm}$) and is given by

$$U = \frac{GM}{r} + \frac{GM}{r} \sum_{n=2}^{\infty} \sum_{m=0}^{n} \left(\frac{a_e}{r}\right)^n \bar{P}_{nm}(\sin\phi)[\bar{C}_{nm}\cos m\lambda + \bar{S}_{nm}\sin m\lambda] \quad (1)$$

where n is the degree and m is the order, $\bar{P}_{nm}$ are the fully normalized associated Legendre polynomials, a_e is the reference radius of Venus (6051.0 km for our models), ϕ is the latitude, and λ is the longitude. The normalized coefficients are related to the unnormalized by (see Kaula 1966)

$$(\bar{C}_{nm} ; \bar{S}_{nm}) = \left[\frac{(n+m)}{(2-\delta_{0m})(2n+1)(n-m)}\right]^{1/2} (C_{nm} ; S_{nm}) \quad (2)$$

where δ_{0m} is the Kronecker delta function and $\bar{C}_{n0} = -\bar{J}_n$. The harmonic coefficients of degree one are fixed to zero because the origin of the coordinate system is chosen to be the center of mass of the body. The body-fixed coordinate system is nominally given by the 1991 IAU values (Davies et al. 1992a,b; see Sec. V) for Venus pole position and rotation rate. The pole and rate are fixed to the 1991 IAU values for the final gravity solution delivered to the scientific community.

The normalized gravity coefficients are estimated complete to degree and order 90 (8277 coefficients). However, the gravity analysis for Venus is not complete with a 90th degree and order solution. The Magellan Doppler tracking data contain information about the Venus gravity field to about harmonic 120 for much of the planetary surface with the equatorial band resolution perhaps to degree 180 or higher (Kaula 1996b). Determination of a higher degree and order gravity field for Venus will be a part of future work. A 120th degree and order field (15,000 parameters) is easily within reach on the JPL Cray T3D Supercomputer. The nominal gravity field for the 90th degree field presented here is MGNP90LSAAP.

Also modeled and estimated is the tidal effect on Venus due to the Sun. This force causes a time varying component of the second degree harmonics as a function of the body-fixed position of the Sun.

A major dynamic effect on the spacecraft is atmospheric drag with along-track accelerations for PVO reaching 1×10^{-6} km s^{-2} for the daytime atmosphere at 150 km altitude. The density profile for the Venus atmosphere is given by the Venus International Reference Atmosphere (VIRA) model. It is a multilayered exponential model with density values at 5-km intervals in altitude and profiles given at different local solar times (Keating et al. 1985). The local solar time (LST) is defined by the direction of Venus' rotation which is retrograde (the morning terminator = 6 a.m.), and is the angle between the longitude of periapse and the longitude of the Sun. The 23 atmospheric layers extend from 140 km to 250 km altitude and the VIRA model is symmetric about noon and midnight LST. The atmospheric drag on the orbit is estimated for every periapse passage for both Magellan and PVO. The exponential scale-height values for each layer are held fixed and the density at the lowest layer of 140 km is estimated (thus changing the density at each layer). For periapse altitudes above 250 km (including the 1000-km altitude PVO data), a single-layer atmosphere is used with scale-height values remaining a function of LST.

The solar radiation pressure acceleration for both PVO (1.3×10^{-10} km s^{-2}) and Magellan (2.7×10^{-10} km s^{-2}) is accounted for in each arc by estimating one coefficient in each of three orthogonal directions (Sun-spacecraft line, ecliptic north direction, and in the ecliptic normal to the Sun-spacecraft direction). The radiation force from Venus' albedo is a ring model (Knocke 1987) where a simple bus model is used for Magellan and a cylindrical model is used for PVO. The albedo force is basically undetermined for Magellan and the low-altitude PVO data due to the atmospheric drag. For the high-altitude PVO, the albedo force is significant. The mean albedo value is 0.76 (Taylor and Stowe 1984) and variations in albedo are allowed for by estimating a scale factor for the albedo for each data arc. The albedo results in an acceleration of approximately 1×10^{-10} km s^{-2}, and is about four times smaller than the effect of the tidal force (k_2).

The Magellan spacecraft attitude was maintained with the use of momentum wheels, and they were also used to change the orientation of the spacecraft. These changes did not impart any force on the spacecraft. However, due to atmospheric drag torques on the spacecraft, the momentum wheels had to be despun or desaturated every orbit (about 8 times per day for Cycle 4 and 15 times per day for Cycles 5 and 6). Prior to the gravity cycles, the desaturations occurred about three times per day. The desaturations imparted an incremental velocity to the spacecraft of about 1 mm s^{-1} and required estimation of the three components of the velocity vector. These desaturations greatly reduced the long-term gravity information in the Magellan data.

All the above forces on the spacecraft are included in the numerical integration of the spacecraft state in rectangular coordinates of the Earth mean equator of the $J2000$ coordinate system. The integrator is a multistep Adams type predictor-corrector that varies the order to obtain the largest possible step size. An absolute integration tolerance of 2×10^{-11} was used which results in numerical noise for the Doppler observable of less than 0.01 mm s^{-1}.

A. Estimation

The JPL gravity estimation software is based upon the Orbit Determination Program or ODP (see Moyer 1971), the software set used at JPL for navigation of all planetary spacecraft. The ODP was modified for use on the JPL Cray T3D supercomputer, a parallel computer with 256 Dec Alpha processors (for more specific details of the T3D software see Konopliv [1995]). The spacecraft state and other parameters were estimated using a weighted least-squares filter based upon the square root of the information matrix (see Lawson and Hanson 1995; Bierman 1977). The parameters that are estimated consist of arc-dependent or local variables (spacecraft state, atmospheric densities, etc.) that are determined separately for each data arc and global variables (harmonic coefficients, etc.) that are common to all the data arcs.

Initially, we converge the data arcs by estimating only the local variables using the nominal values for the global variables. The observations (raw Doppler, range rates) of each arc are weighted according to data root mean square (rms) within that arc for each tracking station pass and the rms includes corrections for the count times of the observations. The actual data weight used is the rms multiplied by a factor of 2 with an additional correction factor for the observation elevation. Because the PVO and Magellan orbits are nearly polar, the groundtracks converge near the pole and the observations become more dense. For this reason, the observation sigma is adjusted for latitude ϕ ($\sigma_{\text{new}} = \sigma_{\text{old}}^* \cos^{-1/2} \phi$).

Once the local variables are converged, the global parameters are determined with a technique described by Kaula (1966) and Ellis (1980) that merges only the global parameter portion of the square-root information (or SRIF) arrays from all the arcs, but is equivalent to solving for the global parameters plus local parameters of all arcs. For each data arc, the local variables estimated are the spacecraft state, three solar pressure coefficients, a factor for the Venus albedo, the base density for each periapse passage through the atmosphere, the lift-to-drag coefficient for the low altitude PVO orbits, velocity vector increments for the momentum wheel desaturations and star calibrations of Magellan, acceleration vectors for the "hide" manuevers of Magellan, and a UT1 bias for the PVO arcs. The *a priori* uncertainties for the spacecraft state are large (20 km). The *a priori* base density uncertainties for the PVO orbits are large but are more tightly constrained for Magellan $(1 \times 10^{-12} \pm 1 \times 10^{-12}$ g cm$^{-3})$. Future work will constrain the Magellan base densities more closely to the VIRA model. The *a priori* on the Magellan desaturations are 5×10^{-6} km s^{-1} and for the star calibrations are 3×10^{-7} km s^{-1}. The hide manuevers are constrained to 10^{-10} km s^{-2}.

Once all the global information is packed from all the data arcs, an *a priori* constraint is applied to the gravity field. The common method is to constrain each harmonic coefficient toward zero with an uncertainty given by the Kaula rule (1966) for that particular planet (used, for example, in Konopliv et al. 1993,1994b; Nerem et al. 1993; McNamee et al. 1993; Smith et al. 1993; Lemoine 1992; Lemoine et al. 1995). The second *a priori* constraint method is a spatial constraint and is provided in Konopliv et al. (1994b).

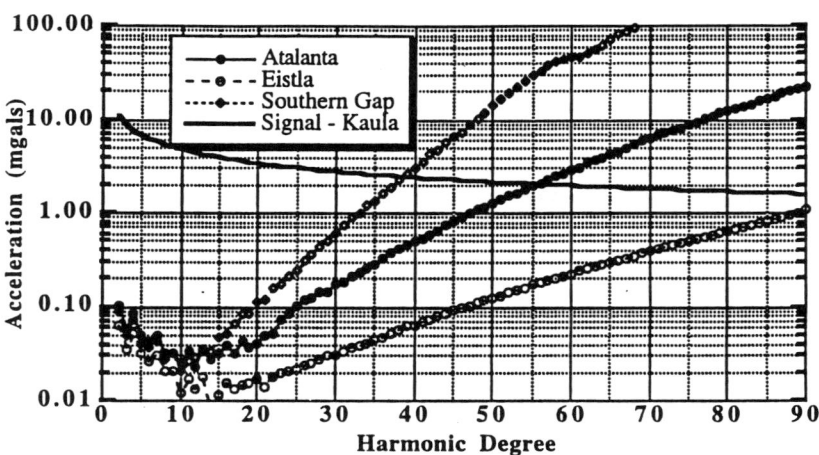

Figure 2. Acceleration profiles for Venus unconstrained surface gravity.

Figure 2 shows the expected acceleration profile from the Kaula rule and the unconstrained acceleration uncertainty profile (from the solution covariance matrix without *a priori* constraints on the gravity coefficients) for Atalanta, the periapse region for Magellan Cycle 4 (eastern Eistla Regio),

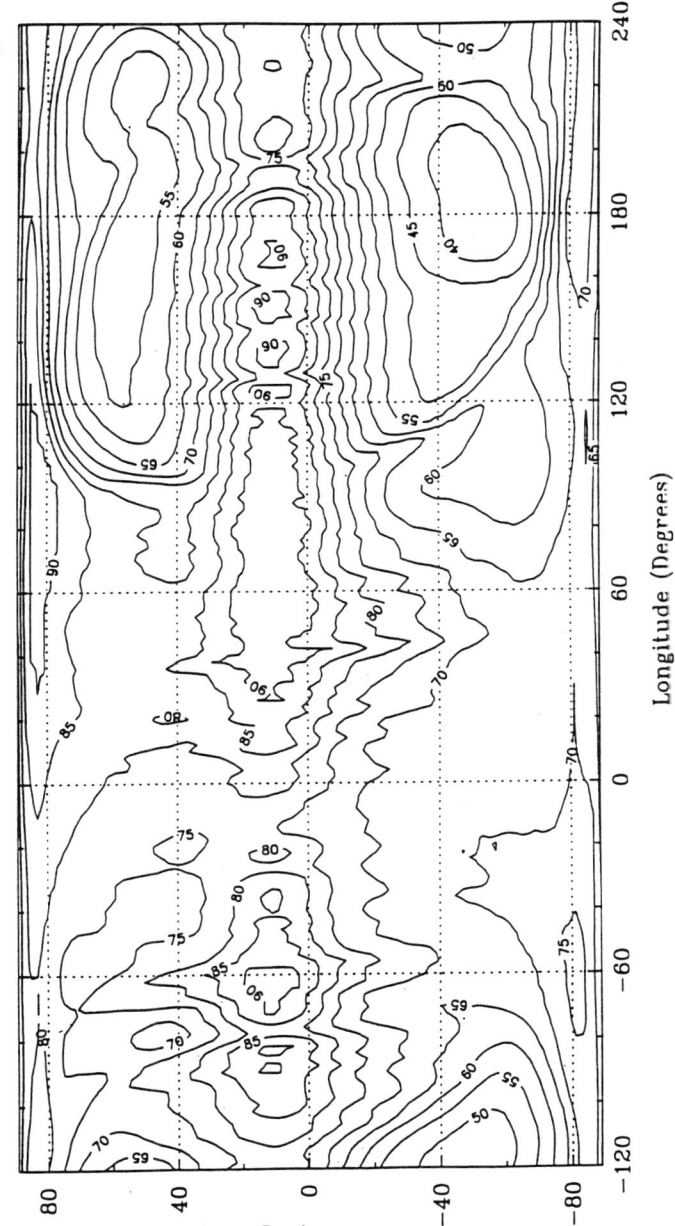

Figure 3. Degree strength for MGNP90LSAAP.

and the gap in Magellan Cycle 5 data in the southern hemisphere (160°E to 220°E, 30°S to 80°S). The crossing point of the Kaula signal with the acceleration uncertainty is called the degree strength of the gravity field for that particular latitude and longitude. For degrees greater than the degree strength, the "noise" in the data exceeds the "signal." Based upon the Kaula rule, the degree strengths for Atalanta, Eistla, and the southern gap are 55, 90, and 38, respectively. Figure 3 displays the spatial variation of degree strength. The maximum degree strength (greater than harmonic degree 90) occurs near the low-altitude periapse locations. For the highest resolution areas, the degree strengths from Fig. 3 are much lower than the values obtained from resolution analyses (Kaula 1995), and yet greater in the weakly determined locations. The Fig. 3 resolution is tied to the covariance from a complete dynamical fit to the observations. Because the data are processed in continuous multiple orbits, this dynamical information increases the resolution in the weakly determined areas. However, the Doppler data are weighted at their true accuracy for the longer wavelengths and are under-weighted for the short wavelengths. This is due to the non-Gaussian or red noise characteristic of the Doppler data (Woo 1975).

The basic idea of the gravity constraint method is to constrain the "noise" of the gravity field to zero with some uncertainty when the "noise" exceeds the "signal." The acceleration at the surface from all harmonic coefficients greater than or equal to the degree strength is constrained to zero with an uncertainty approximately equal to the expected signal at the degree strength. This amounts to generating *a priori* observations over the entire surface of the sphere based upon the degree strength at each latitude and longitude. The *a priori* observations should be spaced such that at least three are generated over the shortest harmonic wavelength. The weight used for an *a priori* observation is then proportional to the area between them and is approximately equal to the signal at the degree strength (10 to 20 milligals). The *a priori* observations are globally spaced on a rectangular grid of latitude and longitude with a spacing of two degrees. The relative weight between the *a priori* observations and the unconstrained gravity covariance is not explicitly determined, but is adjusted to provide the best correlation with topography and least amount of aliasing. To obtain truer peak values, there is no constraint near the peaks of Maxwell Montes and Maat Mons and Eistla and Beta Regio.

One of the main advantages of using this spatial constraint instead of a straight Kaula rule on the spectrum appears to be better determination of peak amplitudes. Because the well-determined degrees are not constrained directly (only somewhat through correlations), the amplitudes (and coefficients) for those degrees are not biased toward zero. It is also flexible in allowing relaxation of selected regions for any reason, such as incorrect data weighting or a region exhibiting greater signal than the power rule would predict. The amplitudes are reduced by about 5 to 10% when a Kaula power law is applied vs the spatial constraint (e.g., 10 milligals for Maxwell Montes and Gula Mons, and 35 milligals for Maat Mons). The correlation with topography for the two different constraint methods are very similar (especially through degree 25 where the strength of the raw data overide any *a priori*). The correlations from the Kaula constraint are generally slightly higher for the medium wavelength harmonics and slightly lower for the higher frequencies with the sum of correlations over the degrees in favor of the surface constraint. In general, there are only slight differences between the methods because Venus does not have strong local deviations from Kaula power spectrum. For Mars, the differences are more pronounced for the Tharsis region (Konopliv and Sjogren 1995).

B. Gravity Results

The normalized coefficients of the nominal gravity solution MGNP90LSAAP (SAAP = Surface Acceleration *a priori*) can be requested from the authors at ask@krait.jpl.nasa.gov or obtained from the Planetary Data System at Washington University in St. Louis, Missouri (http://pds.geophys/pds). A detailed discussion of Venus gravity analyses is available in Konopliv and Sjogren (1996).

GM solutions for Venus were made for different combinations of data. The best GM solution with a realistic error is 324858.601 ± 0.014 km^{3}s^{-2} (2 times formal uncertainty). This solution agrees with our previous solutions within about two formal uncertainties. The ionosphere calibrations play a major role in the determination of the Venus GM from the PVO data, and their neglect in previous estimates may be a reason for corrupted values.

The vertical (or radial) gravity at the Venusian surface for several areas is displayed in Fig. 4 with contours every 10/20 milligals. The radial gravity is given by the partial of the potential (Eq. 1) in the radial direction without the central mass term. In contrast to the Earth and Mars, a sphere is used as the reference surface for Venus because J_2 is comparable in size to the other spherical harmonics.

The uncertainties in the vertical gravity are given in Fig. 5. The uncertainties in the surface acceleration are given by the errors up to the resolution or degree strength of the data plus the error for omission of terms beyond the resolution. The error in the gravity field that the data can sense is globally uniform and equals about 4 to 5 milligals. After Magellan was aerobraked into a near circular orbit, there was no direct observation of the gravity field from about 160°E to 220°E and this gap is apparent in the uncertainty maps. The total errors are largest for the southern gap in the Magellan post-aerobraking data because there the resolution is poorest and we have the greatest error from omission of higher-degree terms. The largest vertical gravity error is 20 milligals. Also visible in the error map (and degree strength map Fig. 3), is the apoapse tracking in the Cycle 5 Magellan data from about 100°E to 160°E. The face-on orbit geometry of Magellan Cycle 4 data and the decrease in resolution is evident in Fig. 1 near longitudes 0° and 240°E.

The peak values of the vertical gravity for areas of interest are given in Table I for MGNP90LSAAP and the previous 75th degree and order solution

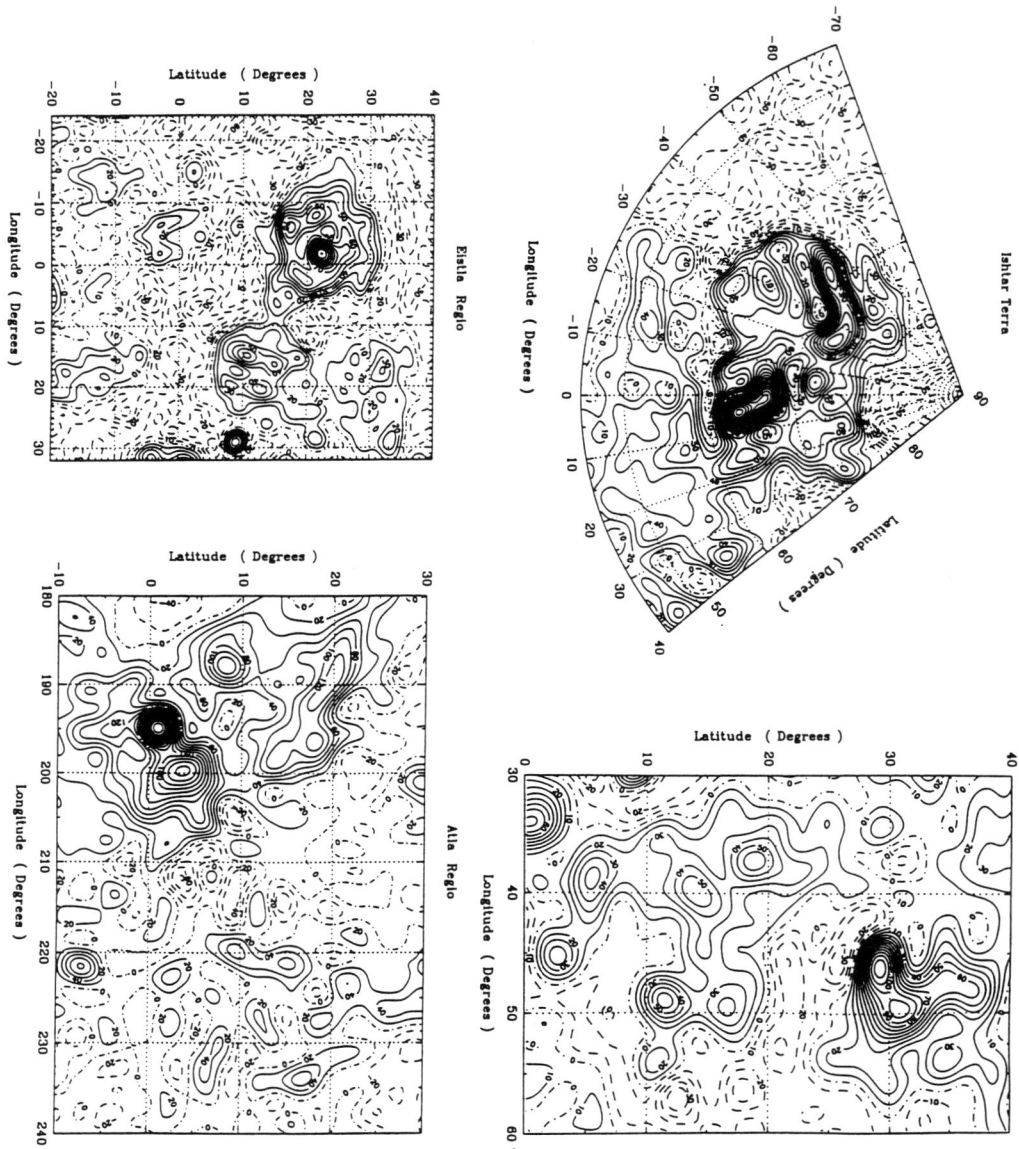

Figure 4. Vertical gravity over several areas of Venus at the surface (mgals).

(MGNP75ISAAP; Konopliv et al. 1994a,b) and 60th degree and order solution (MGNP60FSAAP; Konopliv and Sjogren 1994). The strongest gravity feature on Venus is Maat Mons, which should continue to increase in amplitude with

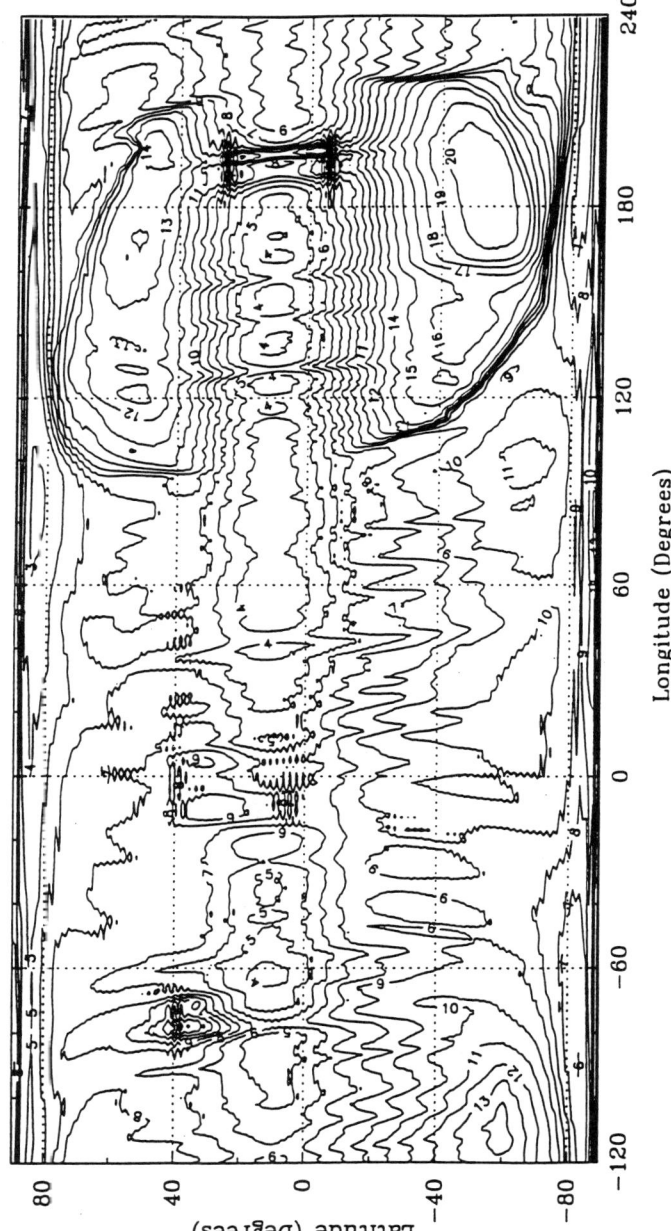

Figure 5. Vertical gravity uncertainty at the surface (mgals).

(MGNP75ISAAP; Konopliv et al. 1994a,b) and 60th degree and order solution (MGNP60FSAAP; Konopliv and Sjogren 1994). The strongest gravity feature on Venus is Maat Mons, which should continue to increase in amplitude with the increasing higher degree and order gravity solutions. The 90th degree postfit Doppler residuals still show substantial systematic trends from the gravity for Atla and Beta regions. The peaks of Bell Regio show noticable increase and also better alignment with the topographic highs.

TABLE I

Gravity Peaks for Venusian Features of Interest (milligals)

Feature	Lon	Lat	MGNP90LSAAP	MGNP75ISAAP	MGNP60FSAAP
Maxwell	4.5	63.5	244.68	220.65	184.30
Akna	−42.5	68.5	115.17	99.57	75.52
Freya	−23.5	73.5	126.34	123.98	105.51
Bell	46.0	29.0	126.25	116.40	102.88
Beta	−79.0	25.5	234.32	231.82	211.87
Gula	−2.0	22.0	138.27	121.77	99.90
Maat	195.0	1.0	356.41	308.46	228.63
Ozza	200.0	3.5	245.52	250.75	224.63
Nokomis	190.0	19.5	132.89	136.44	124.42
Sapas	188.0	8.5	157.54	135.76	126.92
Atalanta	164.5	62.5	−84.44	−83.96	−78.34
Mead	57.2	12.6	−49.67	−39.71	−29.85

A comparison of what other analysts obtained for some of these features is shown in Table II. They used the LOS accelerations derived from Doppler residuals and produced local estimates (except Esposito et al. [1982], who used the raw Doppler observations and surface mass disks to estimate the Beta gravity anomaly). The various estimates were obtained at different reference altitudes and therefore our harmonic estimates were evaluated at the same respective altitude to make comparisons valid. Except for the estimates of Kaula (1996a,b), all estimates are lower than the harmonic estimates by considerable amounts. This is rather surprising because the Doppler residuals should contain the very highest resolution of the data. On the other hand, the harmonics at degree 90 leave almost no systematic signature in their residuals. An explanation for this variance may be due to the model fitting to the LOS data. The experimenters must decide on optimum block sizes or mass distributions. The data are then smoothed to avoid distortions at the surface and may reduce the amplitudes. Also, there will be amplitude reductions as a result of a larger than needed spline interval for determination of the accelerations from the Doppler residuals.

For Venus, the most evident indicator of the plausibility of gravity variations at higher wave number has been the correlation with topography. As more tracking data were added to the solution and as modeling improved, the correlation with topography continued to show higher values. As mentioned

TABLE II
Comparisons of Spherical Harmonics with LOS Reductions (mgals)

Feature		Reference Altitude (km)			
Beta	Surface	187	200	250	Comments
Konopliv and Sjogren (1996)	234	131	128	114	
Kaula (1996a)	240				2% high
Kaula (1996b) Region	270			85	15% high
Kaula (1996b) G180A6	270	135	131	118	15% high
McKenzie (1995)		90			31% low
Sjogren et al. (1983)			73		43% low
Esposito (1982)			135		5% high
Smrekar (1994)				85	25% low
Maxwell	Surface	323			
Konopliv and Sjogren (1996)	245	68			
Kaula (1996b)	200				18% low
Kaula (1996b) Region	272				11% high
Kaula (1996b) G180A6	239	72			2% low
McKenzie (1995)		39			42% low
Maat	Surface	260			
Konopliv and Sjogren (1996)	356	106			
Kaula (1996b) Region	379				6% high
Kaula (1996b) G180A6	340	114			4% low
Sjogren et al. (1983)		64			40% low
Smrekar (1994)		75			29% low
Gula	Surface	180	202		
Konopliv and Sjogren (1996)	138	61	57		
Kaula (1996b) Region	161				17% high
Kaula (1996b) G180A6	174	64	59		26% high
Sjogren et al. (1983)		38			34% low
Barriot (1994)	110				20% low
McKenzie (1995)			40		30% low
Smrekar (1994)			50		12% low

in Konopliv et al. (1994a), the correlation uncertainties to about degree 30 are probably dominated by uncertainties in the topography, and for degree greater than 30 the errors are mostly from gravity. The dip in correlation at degree 15 (see Fig. 6) is real and the majority of it (80%) is due to the poor correlation in the zonal coefficient. The correlations for previous solutions are also presented in Fig. 6. Note the increase in correlation for the last 10 degrees (65–75) for MGNP90LSAAP vs MGNP75ISAAP due to the removal of aliasing for the higher degree solution.

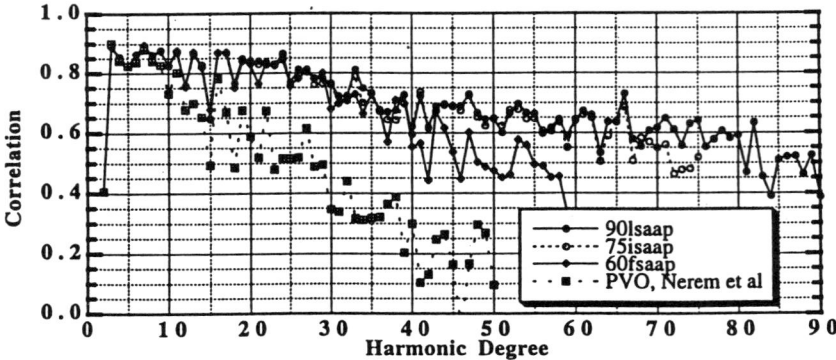

Figure 6. Comparison of correlation with topography for gravity solutions.

III. LINE-OF-SIGHT RESIDUAL ACCELERATION MODELING

As recalled in Sec. I.A, the Venusian gravity field has been extensively modeled in terms of a spherical harmonic representation. The derivation of such a model proceeds by iterating on initial trajectory states and physical parameters (separately or in ensemble, a process based on linearization) which usually converges to the extent that the data residuals are small with respect to a reference orbit. The modeled data are mainly the Doppler shifts affecting the radio signals issued by the spacecraft along the LOS, the line joining the observer on the Earth to the probe. These dynamic global methods are computationally expensive, and until recently could not easily extract from the Doppler data the full content of the gravity signal, especially in the case of a single spacecraft on an elliptical orbit. As a consequence, "residual" techniques have been developed (see, e.g., Phillips et al. 1978; Bowin et al. 1985; Von Frese et al. 1988) to extract from the Doppler residuals the remaining gravity signal. These methods analyze the time derivative of the Doppler residual signal, also known as the "LOS gravity data" (Muller and Sjogren 1968). Their computational advantage is that they can be locally applied, because the residual Doppler signal is strongly correlated with local gravity features. We present here two methods to determine from the LOS gravity data the residual radial gravity at the surface of a planet or at a constant altitude. The first one is well adapted to the Cycle 4 of Magellan LOS gravity data. The second one is better adapted to Cycles 5 and 6.

The LOS gravity data measurement $L(S)$ is obtained through a numerical differentiation of the residual Doppler shifts with respect to time (see Fig. 7). This quantity reflects not only the planet density function, but also the probe altitude and the viewing geometry. These two effects can be quite large and we have to remove them. Also, the LOS gravity value can be biased because this quantity is relative to spatial coordinates computed from an approximate orbit adjusted with a poor knowledge of the gravity field of the planet. These

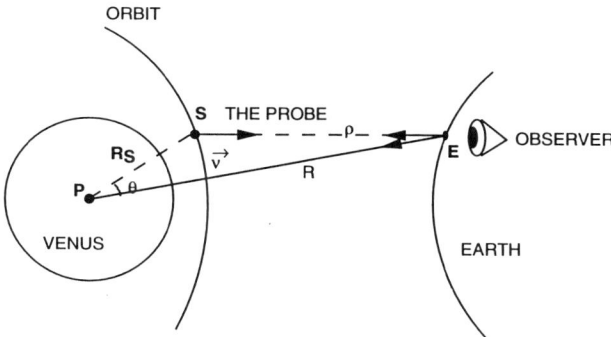

Figure 7. Geometry of a LOS gravity measurement. The point E represents the observer on the surface of the Earth; the point S represents a probe around a given planet. The vector ES is the LOS. We have $R \simeq \rho > R_S$. The quantity sensed is the projection, along the LOS, of the difference between the actual acceleration of the probe, and the acceleration computed using given theory and model.

coordinates can be affected by systematic errors, the main effect of which is along the radial direction, i.e., on altitude. This biasing effect is called the orbital acceleration error, and can show a strong correlation between successive orbits, with an amplitude that can be more than 10% of the total signal amplitude. Another strong effect is the electromagnetic environment of Venus reducing the resolution obtainable at higher latitudes.

The free air gravity anomalies used in Earth analyses are appropriate for surface measurements referred to heights above the geoid, but they are a superfluous complication on Venus, where the radial gravity perturbation suffices. In the planar approximation, perturbations of gravity and free air gravity anomalies reduce to the same quantity.

The mathematical link between the radial gravity perturbation $\delta g(M)$ at a point M on the planet surface (radius R_M) and the corresponding perturbing potential $\Delta U(S)$ at the satellite position S (altitude R_S) is

$$\Delta U(S) = \int_{\sigma_1} K(S, M) \delta g(M) d\sigma_M \tag{3}$$

with

$$K(S, M) = \frac{R_M}{4\pi} \left[\frac{2R_M}{d_{SM}} - \ln \frac{d_{SM} + R_M - R_S \cos \Psi_{SM}}{R_S(1 - \cos \Psi_{SM})} \right] \tag{4}$$

and

$$\delta g(M) = -\left[\frac{\partial}{\partial r} \Delta U \right]_{r=R_M} = -\left[\frac{\partial}{\partial r} (U - U_{\text{orb}}) \right]_{r=R_M} \tag{5}$$

where d_{SM} and Ψ_{SM} are respectively the rectilinear and angular distances between S and M; $K(S, M)$ is the kernel of the integral equation solution of the exterior Neumann's problem for the sphere; $\vec{\nabla}_S K(S, M)$ is the vector gradient of $K(S, M)$ in S; σ_1 is the unit sphere and $d\sigma_M$ the differential surface element of σ_1 relative to M, U is the true (unknown) gravity potential of the planet, U_{orb} is the estimation (known) of the gravity potential of the planet used to compute the approximate orbit of the spacecraft.

If $\vec{v}_S$ is the direction of the Earth as seen from the probe in S, we infer

$$L(S) = -\vec{v}_S \cdot \vec{\nabla}_S(\Delta U) = -\int_{\sigma_1} \vec{v}_s \cdot \vec{\nabla}_S K(S, M)\delta g(M) d\sigma_M. \quad (6)$$

By taking the Cartesian form of the gradient of the $K(S, M)$ kernel with respect to a reference frame to the planet's surface, we obtain the equivalent form

$$L(S) = \frac{\vec{v}_S \cdot \vec{r}_S}{4\pi} \int_{\sigma_1} \alpha(S, M)\delta g(M) d\sigma_M + \frac{1}{4\pi} \int_{\sigma_1} \beta(S, M)(\vec{v}_S \cdot \vec{r}_M)\delta g(M) d\sigma_M \quad (7)$$

where $\vec{r}_S = (X_S, Y_S, Z_S)$ are the vectors identifying the points S and M; and where $\vec{\nabla}_S = (\frac{\partial}{\partial X_S}, \frac{\partial}{\partial Y_S}, \frac{\partial}{\partial Z_S})$. We have, with $\chi_{SM} = d_{SM} + R_M - R_s \cos \Psi_{SM}$,

$$\alpha(S, M) = R_M \left(-\frac{2R_M}{d_{SM}^3} - \frac{1}{\chi_{SM} d_{SM}} + \frac{1}{R_S^2(1 - \cos \Psi_S M)} \right) \quad (8)$$

and

$$\beta(S, M) = R_M \left(\frac{2R_M}{d_{SM}^3} + \frac{1}{\chi_{SM} d_{SM}} + \frac{1}{\chi_{SM} R_M} - \frac{1}{R_M R_S(1 - \cos \Psi_{SM})} \right). \quad (9)$$

A. Dynamic

Our dynamical inversion procedure is a variation of the procedure described by Phillips et al. (1978). Basically in this approach, the U_{orb} term is limited to the central term of the gravity field of the planet. Because all known effects except the detailed lateral variation of the gravity field have been modeled (including conservative and nonconservative forces acting on the probe), the Doppler residuals only contain the high-frequency gravity field information. But, as already stated, because of the "central term only" model of the gravity field of the planet, the derived LOS information can contain significant biases. Nevertheless, when doing geophysical modeling, Phillips et al. (1978) argue that "this distortion problem can be circumvented, however, by integrating a spacecraft trajectory in the presence of the force field of a given mass model, using the initial state determined from the orbit determination program." Our dynamical method follows this strategy, but instead of modeling the remaining anomalous gravity by masses over a zone A on the planet surface

$$\Delta U(S) = GR_M^2 \int_A \frac{\rho(M)}{d_{SM}} d\sigma_M \approx GR_M^2 \sum_{M_i} \frac{\rho(M_i)}{d_{SM_i}} \Delta \sigma(M_i) = G \sum_{M_i} \frac{m_{M_i}}{d_{SM_i}} \quad (10)$$

where $\rho(M)$ is a surface density function and G is the gravitational constant, we use the representation of ΔU given by a discretization of Eq. (3). The great advantage of doing such a substitution is that we are then able to get the surface gravity in a one stage process. Conversely, Grimm and Phillips (1992) had first to evaluate the m_{Mi} coefficients, then to compute the radial gravity at some altitude from this discrete set, and finally to apply a downward filter to get the radial surface gravity. The differential equation giving the evolution with respect to time of the partial derivative of the probe position $\vec{r}_S$, relative to a particular equivalent $\mu_{M_i} = \delta g(M_i)\Delta\sigma(M_i)$ of a point mass at the planet surface is then, with zero initial values for the $\vec{\xi}^S_{M_i}$

$$\frac{d^2}{dt^2}(\vec{\xi}^S_{M_i}) = -\frac{GM}{R_S^5}\left[\vec{\xi}^S_{M_i} R_S^2 - 3\vec{r}_S(\vec{\xi}^S_{M_i}\cdot\vec{r}_S)\right] -$$

$$\frac{1}{4\pi}[\alpha(S, M_i)\vec{r}_S + \beta(S, M_i)\vec{r}_{M_i}] \qquad (11)$$

where $\vec{\xi}^S_{M_i} = (\frac{\partial X_S}{\partial \mu_{M_i}}, \frac{\partial Y_S}{\partial \mu_{M_i}}, \frac{\partial Z_S}{\partial \mu_{M_i}})$. The position $\vec{r}_S$ of the spacecraft in Eq. (11) follows a Keplerian law, with the initial state vector provided with each LOS gravity data arc. A truncated singular value decomposition of the design matrix formed by the partial derivatives $\vec{v}_S \cdot \frac{d^2}{dt^2}(\vec{\xi}^S_{M_i})$ then gives access to a least-squares estimate of the coefficients, where the truncation index realizes a compromise between the data fit and the solution smoothness.

As LOS gravity data from the Cycle 4 of the Magellan mission were computed with respect to a "central term only" method, they are particularly suited to our local dynamical inversion procedure. Figure 8 (left) shows the result of such a local inversion over the west part of Eistla Regio, and Fig. 8 (right) shows the associated topographic map.

B. Static

In the static approach, we assume that the LOS gravity data are directly the residual directional gravity sensed along the flight path of the spacecraft, i.e., with a sufficiently small associated orbital error. This implies that the maximum degree of the U_{orb} model is high enough (higher than degree 10 or 20) that orbital errors are not of significant concern. With a good knowledge of other forces, the altitude of the spacecraft is then computed with an accuracy better than a few tens of meters or less. Equation (6) is then directly applicable. It involves a linear noninvertible integral operator with a continuous kernel, that has to be solved with respect to δg.

A direct discretization of formula (6), to obtain a matrix equation, and then solving the resultant linear system is doomed to failure because adjacent rows of the matrix will become closer and closer as the discretization becomes finer and finer (Wahba 1990). We describe with considerable details in Barrict and Balmino (1992) a procedure to circumvent this problem. First, we discretize

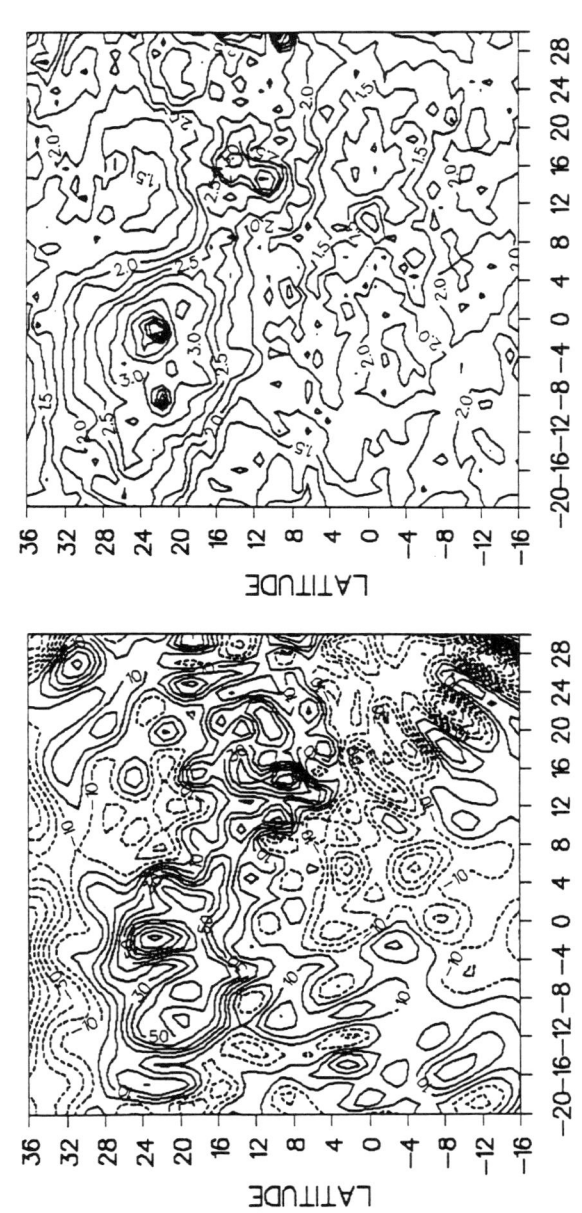

Figure 8. (*Left*) 1° × 1° radial perturbations of surface gravity over the topographic swell associated with Sif and Gula Montes volcanoes, in mgal. We estimated, through a dynamical inversion, a total of 3480 radial gravity values covering an area wider, on each edge, by four degrees than the area shown on the figure. We analyzed 75 regularly spaced orbits from number 5783 to 6080 (Cycle 4). We considered a total of 15,894 LOS gravity data points along these orbits. To build the regularized solution, we retained only the 384 largest singular values of the normal matrix, from the 3480 original values. Because of this low threshold, the associated formal variances are largely underestimated ($\approx$ 2 mgal) and meaningless. (*Right*) Topographic map associated with Fig. 8 (left) in km.

Eq. (6) as

$$L(S) \approx \sum_{\text{over} A} < -\vec{v}_S \cdot \vec{\nabla}_S K(S, M) > < \delta g(M) > \Delta\sigma_M \qquad (12)$$

where A is the region of interest of the planet, the brackets $<>$ denote the mean values of the kernel and of the gravity perturbations δg, for example over $1° \times 1°$ geographical degree areas. We do not use a cruder point-wise discretization as in the dynamical case, because the only utility of such a discretization is to allow an analytical form for the set of differential Eq. (11). The region A includes a peripheral buffer zone needed to obtain an optimal solution over a central region (see Barriot and Balmino 1992). The derivation of the set of equations of observations (12) is ten times faster (or more) than the derivation of the corresponding set for the dynamical modeling, because we do not have to integrate numerically the set of differential Eq. (11). To keep this speed advantage, the regularized inversion is not done through a singular value analysis of the normal matrix, but by considering an *a priori* covariance matrix taking into account the spectral properties (Kaula's rule) of the expected mean values of the perturbation of gravity $< \delta g >$. A direct Cholesky decomposition of the modified normal matrix then gives access to the regularized solution.

LOS gravity data from the Cycles 5 and 6 of the Magellan mission were computed with respect to a 40th degree and order model of the Venus gravity field and therefore do not present significant systematic biases. As the LOS gravity data from these cycles do not contain information for the wavelengths of the gravity signal corresponding to harmonic degrees lower than 40, these wavelengths are directly rebuilt into the final solution from the U_{orb} model. Figure 9 (left) shows the result of a static inversion of these data sets over a wide area from Guinevere Planitia to Eistla Regio. Figure 9 (right) shows an enlargement of Fig. 9 (left) over the same area as in Fig. 8 (left) for comparison purposes.

We have presented two different methods to reduce the LOS data to radial perturbations of the gravity over a zone on the Venus surface. The dynamical procedure is well suited to LOS data reduced with respect to a "central term only" reference field model, and is very popular among the community of planetary geophysicists. This method will become much more useful with the modification we propose. The second one stands on firmer geodetic grounds, but supposes that a sufficiently good model of the gravity field of the planet be first used for the modeling of the Doppler shifts. With a combination of global modeling in terms of spherical harmonic coefficients (long wavelengths) plus local modeling as described here (short wavelengths), we think that a global determination of the gravity field of Venus is feasible up to degree and order 180, which will be meaningful over selected areas with low-altitude data.

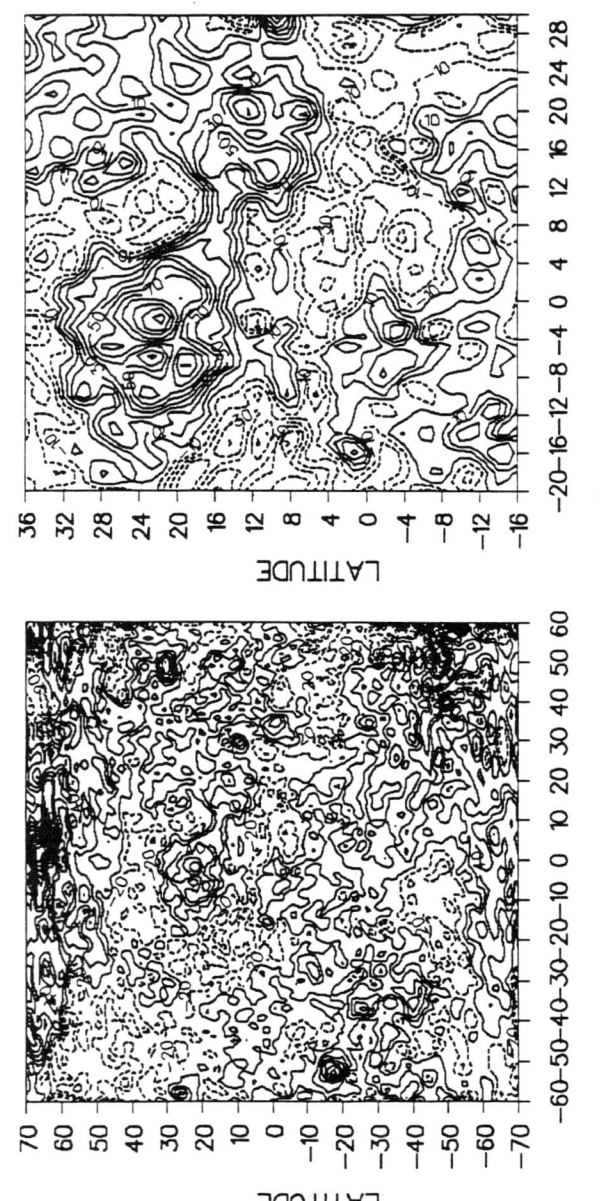

Figure 9. (*Left*) 1° × 1° radial perturbations of surface gravity over a wide area from Guinevere Planitia to Eistla Regio, in mgal. We estimated, through a static inversion, 16,800 radial gravity values covering an area from 70°S to 70°N, and from 60°W to 60°E. Two 10° wide strips (in latitude), at the north and south were discarded from the solution because they showed large artifacts related to the ground-track pattern. We used 112 regularly spaced orbits from Cycle 5 (8606–9968) and 190 similarly spaced orbits from Cycle 6 (12,250–13,933), for a total of 102,287 data points. The associated formal variances for the high-wavelength content of the solution (beyond $\ell = 40$, see text) range from 8 to 10 mgal. (*Right*) Enlargement of Fig. 9 (left) over the same area as in Fig. 8 (left), for comparison purposes.

IV. SOME GEOPHYSICAL IMPLICATIONS OF THE GRAVITY FIELD

The primary importance of the gravity field is in the information it can supply about the structure and processes of the interior of the planet. The new 90th degree and order field model offers a view of this field with an unprecedented level of clarity over the planet, with a half-wavelength resolution approaching 200 km over regions of the planet with the best tracking coverage.

Analyses of the gravity field can be performed in either the spatial domain or the spectral (wavenumber) domain. In the past, most work with harmonic gravity modeling has been done in the spectral domain, as the resolution available from the primary LOS tracking data far exceeded that obtained from the low-degree field models (see, e.g., Bills et al. 1987). However, with the advent of high degree and order models made possible by modern computational methods and resources, the spatial resolution can approach that of the tracking data, with the advantage that the full two-dimensionality of the field can be utilized in the analysis (see, e.g., Banerdt et al. 1994).

As an example of what can be done with such a high-resolution harmonic field, we will briefly consider Mead crater, a multi-ring basin with an outer ring diameter of 280 km. Mead was the subject of a previous investigation by Banerdt et al. (1994) with a 60th degree and order field. The absolute peak anomaly for that field was −30 mgal at the surface, with the anomaly offset from the crater by about a degree (∼100 km) to the east. Figure 10 shows an area centered on Mead crater, with gravity contours (at the surface) from the 90th degree and order field superimposed on the topography. In the 90th degree field the anomaly is almost perfectly centered on the crater, and the absolute amplitude of the anomaly has increased to about −50 mgal. Increasing the resolution of the field to degree and order 180 using regional techniques increases the peak magnitude of the anomaly by only about 4 mgal (Kaula 1996), so we are confident that we are utilizing most of the gravity signal available in the LOS data. Using the harmonic topography model of Rappaport and Plaut (1994), we derived a local digital topography model which included the full topography within 200 km of the center of the crater, then tapered to zero at a radius of 300 km. Using this DTM, we calculated the corresponding gravity anomalies from the isolated Mead topography at 60th and 90th degree resolution (assuming a surface density of 2800 kg m^{-3}), deriving −10 and −25 mgal, respectively. With the presence of other strong gravity features in the vicinity, it is difficult to unambiguously determine the relative magnitude of the Mead gravity anomaly with respect to the background field, but it appears to be very nearly −25 mgal, or perhaps greater. Isostatic calculations show that gravity signatures of −19 and −21 mgal would result from compensation at depths of 25 and 50 km, respectively. Thus results from this higher resolution field further substantiates the earlier conclusions of Banerdt et al. (1994) that Mead is essentially uncompensated (or partially compensated at depths in excess of

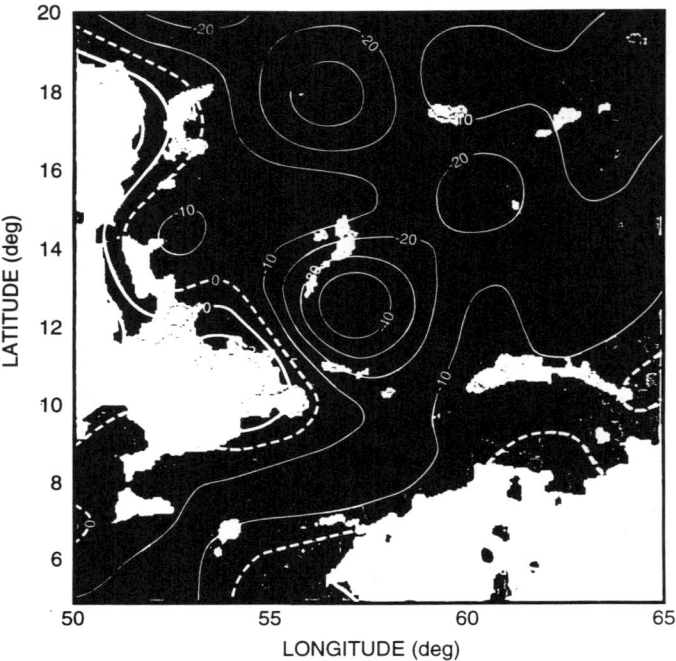

Figure 10. Topography and gravity of an area around Mead crater. The image is about 1600 km across. For topography, shading differences represent 250 m contours, with darker areas lower and lighter areas higher; the deepest portion of Mead's floor is at a radius of 6050.5 km. Data were taken from the Magellan GTDR.3;1. Contours represent vertical acceleration computed at the surface from the 90th degree and order gravity field MGNP90LSAAP. Solid heavy contours indicate positive values, the thin contour lines indicate negative values; the zero contour is denoted by a dashed line. The contour interval is 10 mgal.

50 km) and cannot have undergone significant flexural relaxation.

Several other chapters in this book describe gravity studies relative to specific processes and features (see, e.g., chapters by Kaula et al., by Smrekar et al., and by Phillips et al.), and we will not attempt a comprehensive review of that work. Instead, we will summarize some of the implications of the gravity spectrum for the global view of Venus (see also Arkani-Hamed 1996b). The inferences that can be made from the gravity spectrum alone are limited by the intrinsic non-uniqueness of the interpretation of potential fields. In addition, it should be born in mind that the inferences below are based on global characteristics of the gravitational potential, which will be areally weighted in a complex fashion. For example, admittance spectra tend to disproportionately reflect regions with higher-amplitude gravity signals. Still, these observations

POWER SPECTRA OF GRAVITATIONAL POTENTIALS

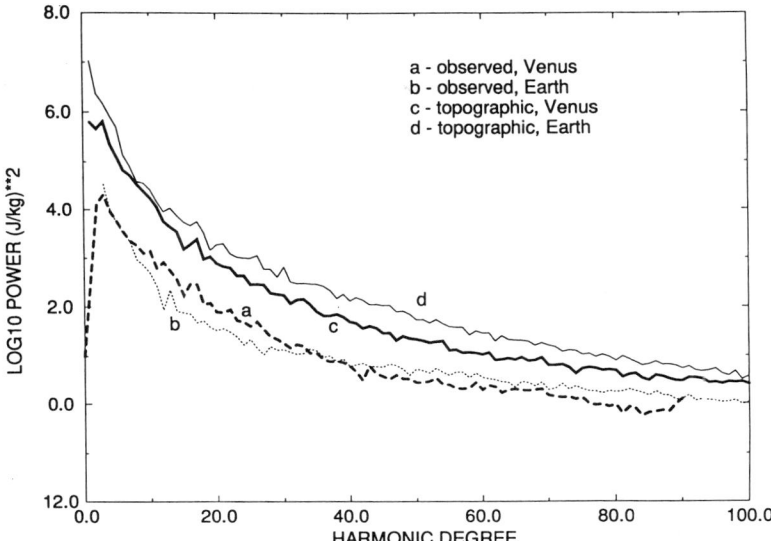

Figure 11. Power spectrum for gravity and topographic potential for Venus and the Earth. Venus coefficients are from MGNP90LSAAP (gravity) and Rappaport and Plaut (1994) (topography). Terrestrial coefficients were supplied by R. Rapp (personal communication).

represent, in some sense, average characteristics of the planet.

The Venusian power spectrum is shown in Fig. 11, along with that for the Earth. It can be seen that the gravity spectra are broadly similar for the two planets. At the longest wavelengths the two planets have nearly equal power, whereas Venus has somewhat more power in the middle wavelength band ($9 \lesssim n \lesssim 40$, about 1000 to 4000 km) and less at shorter wavelengths. Both spectra decrease as roughly n^{-3}.

As shown by Lambeck (1976), this spectral distribution can be produced by a random distribution of density anomalies throughout the mantle. However for Venus, as for the Earth, an examination of the relationship between the gravity and topography strongly suggests another interpretation, namely that the gravity spectrum is the result of the superposition of compensating density anomalies at different depths in the crust and mantle. Figure 12 shows the degree correlation between gravity and topography. The correlation is very high over virtually the entire spectrum (except degree 2), dropping to lower values only at the highest degrees. (As discussed above, these are the more poorly determined coefficients; historically each improvement in the gravity field solution has resulted in an increase in the short-wavelength correlation.) The clear implication of this correlation is that the gravity and topography are related, in that the same interior processes are responsible for both of them.

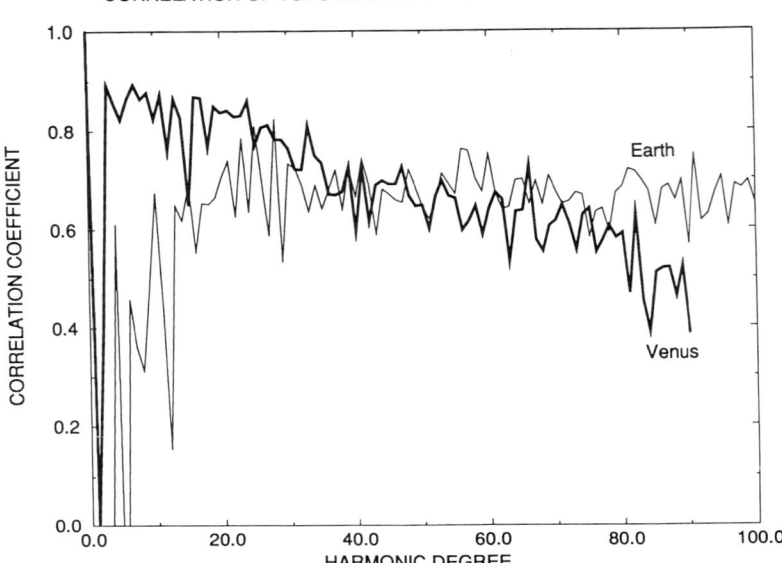

Figure 12. Correlation between gravity and topography for Venus and Earth.

This suggests that the topography is compensated, either buoyantly within the lithosphere (i.e., Airy or Pratt compensation, modified by elastic support) or through the dynamic compensation of topography by density variations associated with convective motions within the mantle.

Figure 11 includes the power spectra of the potential due to the topography of Venus (Rappaport and Plaut 1994) and the Earth (Rapp and Pavlis 1990), for comparison with the gravitational spectra. Surface rock densities of 2900 and 2700 kg m^{-3} were assumed for Venus and the Earth, respectively, for the potential calculation. For both planets the observed power is considerably lower than that expected from the topography, suggesting that a major portion of the topography is compensated. It can also be seen that the Earth's topography appears relatively more compensated than Venus', especially at the longer wavelengths, as the difference between the curves is greater. Note, however, that the fundamental assumption of compensation, that the gravity and topography are well correlated, is not valid on the Earth for degrees less than about 15 (Fig. 12). This does not require that topography is not compensated within this wavelength range, but rather that the situation is more complex than simple single-layer compensation, with significant uncorrelated (or perhaps even anti-correlated) density anomalies at depth.

A more quantitative look at the compensation of topography can be obtained by plotting the apparent depth of compensation (ADC). The ADC for Airy compensation is given by

$$D_n = a\left[1 - \left[1 - (2n+1)\frac{\bar{\rho}\sigma_n(T, V)}{3\rho_c\sigma_n(T, T)}\right]^{1/n}\right] \quad (13)$$

where a is the planet radius, $\bar{\rho}$ is the average planet density, ρ_c is the crustal density, and $\sigma_n(T, V)$ and $\sigma_n(T, T)$ are the cross covariance of the topography and gravity and the auto covariance of the topography, respectively (Phillips and Lambeck 1980). The degree ADCs from Eq. (13) are shown in Fig. 13. The compensation mechanisms are clearly different for the two planets. The compensation depth for the Earth is nearly constant at about 30 km for degrees above about 15 (i.e., those that are well correlated). Thus all but the longest wavelength topography is essentially compensated at or near the Moho. On Venus, however, the situation is clearly different. The ADC shows a sharp decrease from about 350 km at degree 3 to 150 km at degree 10, then a gradual transition to a nearly constant value of about 35 km above degree 40. The low-degree portion of the spectrum has been shown by Kiefer et al. (1986) to be consistent with dynamic compensation, with density variations distributed throughout the mantle. This results naturally in a decrease in ADC with degree, as the longer gravitational wavelengths sense deeper into the planet. The high-degree portion is likely compensated at shallow depths within the lithosphere, perhaps by Airy compensation at the base of a $\sim$35 km crust. It should be noted that the ADC represents an upper bound on the actual depth of Airy compensation in the event that topography is partially supported by the elastic strength of the lithosphere (although the very shortest wavelengths may underestimate the ADC if there is power in the gravity field that has not been absorbed into the harmonic model). Thus, if the spectral ratios of gravity and topography in Eq. (13) reflect a globally averaged crustal thickness, it may be some 5 to 10 km thinner than 35 km, especially if the mechanical lithosphere is relatively thick (50–100 km) as suggested by recent work (see, e.g., Phillips 1994; Simons et al. 1994; Smrekar 1994; Banerdt et al. 1994).

A by-product of the gravity field measurement is the determination of the time-varying part of the field. In particular for the tidal bulge due to the Sun, the only significant time-varying terms occur for C_{22} and S_{22}. With a solar "day" on Venus of 116.75 Earth days, the period of the variations in C_{22} and S_{22} is 58.4 days. The expected amplitude for a nominal Love number k_2 of 0.25 is 70×10^{-10} for the normalized coeficients ($<1\%$ of the C_{22} value and $<10\%$ of the S_{22} value). The standard deviation (formal) for these gravity coefficients from the MGNP90LSAAP gravity solution is 9×10^{-10} and indicates the Love number can be determined. The tidal effect must be separated from other forces that have dependence on the solar longitude. In particular, the tidal force on the spacecraft (about 2×10^{-10} km s^{-2}) is about four times the size of the force due to Venus albedo. By investigating the sensitivity of the Love number solution from a variety of solutions, Konopliv and Yoder (1996) have determined the Love number to

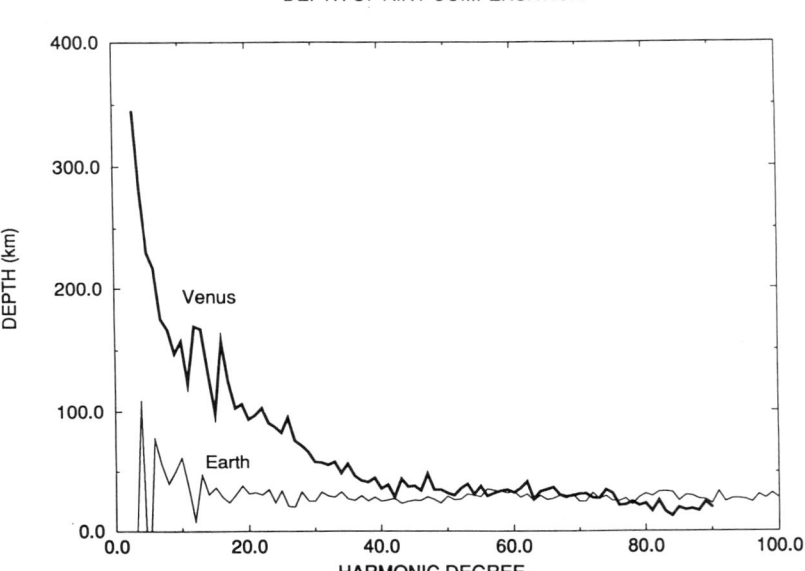

Figure 13. Apparent depth of compensation for Venus and Earth.

be 0.295±0.066 (2 times formal σ). Yoder (1995) has calculated that a solid core (Arkani-Hamed and Toksoz 1984) should yield a k_2 near 0.17, whereas for a liquid core (Stevenson et al. 1983) the Love number will range between 0.23 and 0.29.

The Love number at the very least indicates that the outer core is soft. There are two possible interpretations. First, as Yoder outlined in his chapter, the core viscosity, like the Earth's core, is less than 1 pa-s and the explanation for the free obliquity of 2.1 deg is that the liquid core friction at the core mantle boundary is reduced by large oblateness of the core mantle boundary in order to allow other mechanisms to account for this free obliquity. The second explanation is that the outer core viscosity is in the range of 10^6 to 10^{16} pa-s. This range of viscosities allows the fluid core to closely follow the precession of the figure of Venus and yet appear to be liquid at the 58-day tidal flexing period.

The pole offset to first order is given by the magnitude of the C_{21} and S_{21} gravity coefficients. The normalized second degree coefficients and formal uncertainties for a number of different data combinations have been computed. Using a factor of three times the formal error, the pole offset with a realistic error is 0.52±0.03 degrees. Hence, these results give some confidence that there is a pole offset and that there is a wobble for Venus (Yoder and Ward 1979).

V. GEODETIC PARAMETERS

A. Introduction

The rotation period and pole direction of Venus are important parameters in the modeling of the gravity field, as discussed in Sec. I. They are also fundamental in other investigations, such as studies of the rotational dynamics of Venus, from which the internal structure of the planet can be inferred (see the chapter by Yoder). Although they can be estimated as part of the gravity solution, these parameters are more accurately determined by tracking the motion of surface features over a series of radar images.

Before the Magellan mission, the best estimates of the Venus rotation period and pole direction were derived from Earth-based radar measurements. The 1988 IAU values for these parameters were adopted for use in processing the Magellan SAR data and positioning the images on a latitude/longitude grid. These values were: rotation period of 243.025 days retrograde, and north pole direction of right ascension 272.69°, and declination 67.17° in the $J2000$ coordinate system (Davies et al. 1989). However, it became evident after Magellan had observed Venus for a full rotation, that the parameters were in error, as the Venus-fixed coordinates of features seen for the second time were offset by up to a km from their first observed positions. The high resolution of the Magellan radar images provided an opportunity to improve our knowledge of the rotational parameters.

The Venus period and pole direction were estimated from Magellan data using a combination of two separate estimation procedures, which are described in the following two subsections. Although both methods used measurements of surface features from multiple SAR images, they took different approaches in dealing with spacecraft ephemeris errors, which were the main error source in determining the rotational parameters. In the first method, an improved spacecraft ephemeris was estimated along with the rotational parameters, by combining the surface feature measurements with Earth-based Doppler tracking measurements, and recomputing the ephemeris by integrating the equations of motion (Chodas et al. 1992). This approach was applied on small blocks of typically 5 to 8 Magellan orbits, although several blocks could be linked together via common landmarks. When blocks from different mapping cycles were linked in this fashion, a very accurate determination of the Venus rotation period was obtained (Chodas et al. 1994). In the second estimation procedure, features were selected and measured on several hundred orbits, but only from the polar region north of 80°, where there was much overlap in the images. Spacecraft ephemeris errors were corrected using the simpler approach of approximating the orbit by an ellipse and estimating corrections in two of the orbital elements on each orbit, although improved ephemerides from the first process were used when available. This second method yielded the best estimate for the direction of the spin vector of Venus.

In a collaboration with E. L. Akim, V. A. Stepanyantz, and Z. P. Vlasova of the Keldysh Institute of Applied Mathematics, and A. I. Zakharov of the Institute of Radio Engineering and Electronics, Moscow, an independent determination of the period and pole direction was made using features common to Venera 15/16 and Magellan images. Although the 7-yr time span between the acquisition of the Venera and Magellan images offered the potential for an accurate rotation period determination, the lower resolution and different radar look direction of the Venera images made precise matching of features very difficult; the results of this effort to date have been similar to, but less accurate than determinations based solely on Magellan data.

B. Ephemeris Improvement and Determination of the Rotation Period

The process used to improve the spacecraft ephemeris over selected blocks of orbits, and to estimate accurately the rotational period of Venus, will now be described. On each orbit, Magellan acquired a long narrow north–south swath of radar data, extending from the north polar region down to 70°S, and swaths from consecutive orbits overlapped at the edges along their entire lengths. Surface features were selected as landmarks along the full lengths of the swaths, and always in regions which overlapped with other swaths. Image correlation techniques were used to ensure that the point selected as a landmark was the same in all swaths in which it was observed. For each measurement, the radar burst closest to the landmark was identified, and the pixel coordinates of the point were converted to range and Doppler measurements using the resampling coefficients for the burst. Expected values of range and Doppler were computed using assumed values of the Venus-fixed coordinates of the landmark (latitude, longitude, and radial distance from the center of Venus), and the spacecraft position and velocity at the burst time, which were obtained from an initial spacecraft ephemeris. The radius values for landmarks located within regions covered by Magellan altimetry were obtained by interpolating within the altimetry data set. Residuals were formed by differencing the measured and computed range and Doppler coordinates.

Also computed were the partial derivatives of range and Doppler with respect to the parameters to be estimated. These included the initial position and velocity of the spacecraft, from which the improved spacecraft ephemeris was computed, latitudes and longitudes of landmarks, radii of landmarks for which altimetry was unavailable, and the rotation period of Venus. The landmark measurements were combined with Earth-based Doppler tracking measurements in a weighted least-squares process which minimized the sum of squares of the residuals (Bierman 1977), resulting in an improved ephemeris which optimally fit both the landmark measurements and the Earth-based Doppler tracking. An accurate solution for the rotation period of Venus was also obtained when this method was applied to overlapping blocks of orbits from different mapping cycles. One such solution, based on 52 landmarks measured on orbits 874 to 878 of Cycle 1 and orbits 4456 to 4458 of Cycle 3, yielded a period of 243.0185±0.0001 days (Davies et al. 1992b). This value was adopted for the 1991 IAU set of planetary constants (Davies et al. 1992a), and was used in the 90th degree gravity solutions described in Sec. I. More recently, a refinement to this solution has been made using 306 landmarks measured on orbits 376 to 384 at the beginning of Cycle 1, and on orbits 5741 to 5747, three planetary rotations later. The new rotation period solution was 243.01842±0.00006 days, consistent with the earlier value.

C. Determination of the Pole Direction and the North-Polar Control Network

The process used to estimate accurately the pole direction of Venus was as follows. Control points, usually craters, were selected to form a control network in the area north of 80° latitude. This region was covered by SAR swaths on even-numbered orbits from Cycle 1, orbits 376 to 2170, with gaps during superior conjunction (orbits 678–788) and apoapsis occultation (1046–1346). As before, the radar burst in which the point was most nearly centered was identified. The spacecraft position and velocity at the burst time were obtained either from the standard Magellan ephemeris, derived from Earth-based Doppler tracking measurements, or from the improved ephemeris described earlier, if available for that orbit. Improved ephemerides were used for about 50 orbits in total. Assumed Venus-fixed coordinates for each control point (latitude, longitude and radius from the center of Venus) were combined with the spacecraft position and velocity to compute the range and Doppler coordinates for each measurement. These were then converted to pixel coordinate values using the range and Doppler resampling coefficients, and residuals were formed by differencing these with the measured pixel coordinates.

Partial derivatives of the pixel coordinates with respect to selected free parameters were also computed. The residuals and partials for all control point measurements were combined in a least squares algorithm that minimized the sum of squares of the residuals. Free parameters included the latitude, longitude, and radii of the control points for which altimetry was unavailable, the right ascension and declination of the north pole and, for those orbits without improved ephemerides, the argument of periapsis and the orbital inclinations. These selected orbital elements were allowed to vary to account for errors in the standard groundbased ephemeris. Although this permits the orbits to slide in the along-track and cross-track directions, the relative positions of the control points were maintained. Orbits with improved ephemerides computed by the first procedure were used directly in the estimation, and were not allowed to vary.

The control network computation was based on 3942 measurements of 575 control points on 570 orbits, and yielded a north pole direction of right ascension 272.76°±0.02° and declination 67.16°±0.016° with respect to the $J2000$ frame (Davies et al. 1992b). These values were adopted for the 1991 IAU set of planetary constants (Davies et al. 1992a). A more recent refinement to this solution has been made using the improved rotation period described earlier, and a larger data set: 4421 measurements of 654 control points on 682 orbits. The resulting solution for the direction of the north pole was: right ascension $272.758° \pm 0.0206°$, and declination $67.157° \pm 0.012°$ in the $J2000$ system, again consistent with the earlier values.

Acknowledgment. We thank C. Yoder for his input on the Love number determination, M. Standish for work with the Venus ephemeris, N. Rappaport for the Venus topography model, C. Dang and H. Royden for their special efforts in delivering the ionospheric and tropospheric calibrations used in this work, Fred Krogh for help with the numerical integration, and A. Kucinskas for discussions on the geophysics of Venus. Special thanks are due C. Lawson and J. Giorgini for their efforts in developing the gravity software for the Cray. The Cray Supercomputer funding came from the NASA offices of Mission to Planet Earth, Aeronautics, and Space Science. Part of the research described in this chapter was carried out by the Jet Propulsion Laboratory, California Institute of Technology, under a contract with the National Aeronautics and Space Administration. Another part of the research was supported by the French Space Agency (CNES) and the Centre National de la Recherche Scientifique, while a third part of the research was supported by the RAND Corporation under a contract with the National Aeronautics and Space Administration.

NOTE ADDED IN PROOF

At the time this chapter went to press, a 120th degree and order spherical harmonic estimate and LOS accelerations on this field for Cycles 4, 5 and 6 had been obtained by A. S. Konopliv, but time did not permit their incorporation. One can obtain these results by contacting James Alexopoulos at the Geophysics Planetary Data Node, Washington University, St. Louis, Mo.

REFERENCES

Akim, E. L., Vlasova, Z. P., and Chuiko, I. V. 1978. Determination of the dynamical flattening of Venus from measurements of the trajectories of its first artificial satellites, Venera 9 and 10. *Soviet Phys. Dokl.* 23:313–315.

Ananda, M. P., et al. 1980. A low-order global gravity field of Venus and dynamical implications. *J. Geophys. Res.* 85:8303–8318.

Anderson, J. D., and Efron, L. 1969. The mass and dynamical oblateness of Venus. *Bull. Amer. Astron. Soc.* 1:231.

Arkani-Hamed, J. 1994. On the thermal evolution of Venus. *J. Geophys. Res.* 99:2019–2033.

Arkani-Hamed, J. 1996a. Analysis and interpretation of high-resolution topography and gravity of Ishtar Terra, Venus. *J. Geophys. Res.* 101:4691–4710.

Arkani-Hamed, J. 1996b. Analysis and interpretation of the surface topography and gravitational potential of Venus. *J. Geophys. Res.* 101:4711–4724.

Arkani-Hamed, J., and Toksoz, M. N. 1984. Thermal evolution of Venus. *Phys. Earth Planet. Int.* 34:232–250.

Banerdt, W. B., et al. 1994. The isostatic state of Mead crater. *Icarus* 112:117–129.

Barrio, J. P., and Balmino, G. 1992. Estimation of local planetary gravity fields using line-of-sight gravity data and integral operator. *Icarus* 99:202–224.

Barriot, J. P., and Balmino, G. 1994. An Analysis of LOS gravity data set from cycle 4 of the Magellan probe around Venus. *Icarus* 112:34–41.

Bierman, G. J. 1977. *Factorization Methods for Discrete Sequential Estimation* (New York: Academic Press).

Bills, B. G., Kiefer, W. S., and Jones, R. L. 1987. Venus gravity: A harmonic analysis. *J. Geophys. Res.* 92:10335–10351.

Bowin, C., Abers, G., and Shure, L. 1985. Gravity field of Venus at constant altitude and comparison with Earth. *Proc. Lunar Planet. Sci. Conf.* 15, *J. Geophys. Res. Suppl.* 90:757–770.

Chodas, P. W., Wang, T.-C., Sjogren, W. L., and Ekelund, J. E. 1992. Magellan ephemeris improvement using synthetic aperture radar landmark measurements. In *Astrodynamics 1991: Advances in the Astronautical Sciences*, vol. 76 (San Diego: Univelt), pp. 875–889.

Chodas, P. W., Lewicki, S. A., Hensley, S., and Masters, W. C. 1994. High-precision Magellan orbit determination for stereo image processing. In *Astrodynamics 1993: Advances in the Astronautical Sciences*, vol. 85 (San Diego: Univelt), pp. 279–296.

Davies, M. E., et al. 1989. Report of the IAU/IAG/COSPAR Working Group on Cartographic Coordinates and Rotational Elements of the Planets and Satellites: 1988. *Celest. Mech.* 46:187–204.

Davies, M. E., et al. 1992a. Report of the IAU/IAG/COSPAR Working Group on Cartographic Coordinates and Rotational Elements of the Planets and Satellites: 1991. *Celest. Mech.* 53:377–397.

Davies, M. E., et al. 1992b. The rotation period, direction of the north pole, and geodetic control network of Venus. *J. Geophys. Res.* 97:13141–13151.

Ellis, J. 1980. Large scale state estimation algorithms for DSN tracking station location determination. *J. Astronaut. Sci.* 28:15–30.

Esposito, P. B., et al. 1982. Venus gravity: Analysis of Beta Regio. *Icarus* 51:448–459.

Grimm, R. E., and Phillips, R. J. 1992. Anatomy of a Venusian hot spot: Geology, gravity, and mantle dynamics of Eistla Regio. *J. Geophys. Res.* 97:16035–16054.

Howard, H. T., et al. 1974. Venus: Mass, gravity field, atmosphere and ionosphere as measured by Mariner 10 dual frequency radio system. *Science* 183:1297–1301.

Kaula, W. M. 1966. *Theory of Satellite Geodesy* (Waltham, Mass.: Blaisdell).

Kaula, W. M. 1996a. Regional gravity fields on Venus from tracking of Magellan cycles 5 and 6. *J. Geophys. Res.* 101:4683–4690.

Kaula, W. M. 1996b. A one-degree square mean or 180th degree harmonic solution for the gravity field of Venus. *Geophys. Res. Lett.*, in preparation.

Keating, G. M., et al. 1985. Models of Venus neutral upper atmosphere: Structure and composition. *Adv. Space Res.* 5:117–171.

Kiefer, W. S., Richards, M. A., Hager, B. H., and Bills, B. G. 1986. A dynamical model of Venus's gravity field. *Geophys. Res. Lett.* 13:14–17.

Knocke, P., and Ries, J. 1987. Earth Radiation Pressure Effects on Satellites. Center for Space Research Tech. Memorandum (Austin: Univ. of Texas).

Konopliv, A. S. 1995. DDF Final Report. JPL Int. Doc. IOM 312.D-95-103.

Konopliv, A. S., and Sjogren, W. L. 1994. Venus spherical harmonic gravity model to degree and order 60. *Icarus* 112:42–54.
Konopliv, A. S., and Sjogren, W. L. 1995. *The JPL Mars Gravity Field, Mars50c, Based Upon Viking and Mariner 9 Doppler Tracking Data*, JPL Publ. 95-5.
Konopliv, A. S., and Sjogren, W. L. 1996. *Venus Gravity Handbook*, JPL Publ. 96-2.
Konopliv, A. S., and Yoder, C. F. 1996. Venusian k_2 tidal love number from Magellan and PVO tracking data. *Geophys. Res. Lett.*, in press.
Konopliv, A. S., Williams, B. G., and Christensen, E. J. 1992. A Venus gravity solution to degree and order 42 from PVO data only. Paper presented at AGU Spring Meeting, May 12–16, Montreal, Canada.
Konopliv, A. S., et al. 1993. Venus gravity and topography: 60th degree and order model. *Geophys. Res. Lett.* 20:2403–2406.
Konopliv, A. S., Sjogren, W. L., Graat, E., and Arkani-Hamed, J. 1994a. Venus gravity data reduction. Presentation at Fall 1994 AGU Meeting, Dec. 5–9, San Francisco, Calif.
Konopliv, A. S., et al. 1994b. A high resolution lunar gravity field and predicted orbit behavior. In *Proceedings of the AAS/AIAA Astrodynamics Specialist Conf.* (San Diego: Univelt), pp. 1275–1294.
Lambeck, K. 1976. Lateral density anomalies in the upper mantle. *J. Geophys. Res.* 81:6333–6340.
Lawson, C. L., and Hanson, R. J. 1995. *Solving Least Squares Problems* (Philadelphia: Soc. for Industrial and Applied Math.).
Lemoine, F. G. 1992. Mars: The Dynamics of Orbiting Satellites and Gravity Model Development. Ph.D. Thesis, Univ. of Colorado, Boulder.
Lemoine, F. G., Smith, D. E., Zuber, M. T., and Neumann, G. A. 1995. High degree and order spherical harmonic models for the Moon and historic S-band doppler data. Presented at the Int. Union of Geology and Geophysics, XXI General Assembly, July 2–14, Boulder, Colo.
McKenzie, D., and Nimmo, F. 1995. Elastic thickness estimates for Venus from line-of-sight accelerations. *Icarus*, in press.
McNamee, J. B., Kronschnabl, G. R., Wong, S. K., and Ekelund, J. E. 1992. A gravity field to support Magellan navigation and science at Venus. *J. Astron. Sci.* 40:107–134.
McNamee, J. B., Borderies, N. J., and Sjogren, W. L. 1993. Venus: Global gravity and topography. *J. Geophys. Res.* 98:9113–9128.
Mottinger, N. A., Sjogren, W. L., and Bills, B. G. 1985. Venus gravity: A harmonic analysis and geophysical implications. *J. Geophys. Res.* 90:739–756.
Moyer, T. D. 1971. *Mathematical Formulation of the Double-Precision Orbit Determination Program (DPODP)*, JPL Tech. Rept. 32-1527.
Muller, P. M., and Sjogren, W. L. 1968. Lunar mass concentrations. *Science* 161:680–684.
Nerem, R. S. 1991. An improved gravity model for Venus using tracking data from Pioneer Venus Orbiter. *Eos: Trans. AGU* 72(17):174–175.
Nerem, R. S., Bills, B. G., and McNamee, J. B. 1993. A high resolution gravity model for Venus: GVM-1. *Geophys. Res. Lett.* 20:599–602.
Nerem, R. S., et al. 1994. Gravity model development for TOPEX/Poseidon: Joint gravity models 1 and 2. *J. Geophys. Res.* 99:24421–24447.
Nerem, R. S., Jekeli, C., and Kaula, W. M. 1995. Gravity field determination and characteristics: Retrospective and prospective. *J. Geophys. Res.* 100:15053–15074.
Phillips, R. J. 1994. Estimating lithospheric properties at Atla Regio, Venus. *Icarus* 112:147–170.
Phillips, R. J., and Lambeck, K. 1980. Gravity fields of the terrestrial planets: Long-

wavelength anomalies and tectonics. *Rev. Geophys. Space Phys.* 18:27–76.
Phillips, R. J., et al. 1979. Gravity field of Venus: A preliminary analysis. *Science* 205:93–96.
Phillips, R. J., Sjogren, W. L., Abbott, E. A., and Zisk, S. H. 1978. Simulation gravity modeling to spacecraft-tracking data: Analysis and application. *J. Geophys. Res.* 83:5455–5464.
Rapp, R. H., and Pavlis, N. K. 1990. The development and analysis of geopotential coefficient models to spherical harmonic degree 360. *J. Geophys. Res.* 95:21885–21911.
Rapp, R. H., Wang, Y. M., and Pavlis, N. K. The Ohio State 1991 Geopotential and Sea Surface Topography Harmonic Coefficient Models. Rept. 410, Dept. of Geodetic Sci. and Surveying (Columbus: Ohio State Univ.).
Rappaport, N. J., and Plaut, J. J. 1994. A 360-degree and -order model of Venus topography. *Icarus* 112:27–33.
Reasenberg, R. D., and Goldberg, Z. M. 1992. High-resolution gravity model of Venus. *J. Geophys. Res.* 97:14681–14690.
Reasenberg, R. D., Goldberg, Z. M., MacNeil, P. E., and Shapiro, I. I. 1981. Venus gravity: A high resolution map. *J. Geophys. Res.* 86:7173–7179.
Reasenberg, R. D., Goldberg, Z. M., and Shapiro, I. I. 1982. Venus: Comparison of gravity and topography in the vicinity of Beta Regio. *Geophys. Res. Lett.* 9:637–640.
Simons, M., Hager, B. H., and Solomon, S. C. 1994. Global Variations in the geoid topography admittance of Venus. *Science* 264:798–803.
Sjogren, W. L., Phillips, R. J., Birkeland, P. W., and Wimberly, R. N. 1980. Gravity anomalies on Venus. *J. Geophys. Res.* 85:8295–8302.
Sjogren, W. L., et al. 1983. Venus gravity anomalies and their correlation with topography. *J. Geophys. Res.* 88:1119–1128.
Sjogren, W. L., Bills, B. G., and Mottinger, N. A. 1984. Venus: Ishtar gravity anomaly. *Geophys. Res. Lett.* 11:489–491.
Smith, D. E., et al. 1993. An improved gravity model for Mars: Goddard Mars model 1. *J. Geophys. Res.* 98:20871–20889.
Smrekar, S. E. 1994. Evidence for active hotspots on Venus from analysis of Magellan gravity data. *Icarus* 112:2–26.
Stevenson, D. J., Spohn, T., and Schubert, G. 1983. Magnetism and thermal evolution of the terrestrial planets. *Icarus* 54:466–489.
Taylor, V. R., and Stowe, L. L. 1984. Reflectance characteristics of uniform Earth and cloud surfaces derived from NIMBUS-7 ERB. *J. Geophys. Res.* 89:4987–4996.
Von Frese, R. R. B., Ravat, D. N., Hinze, W. J., and Macgue, C. A. 1988. Improved inversion of geopotential field anomalies for lithospheric investigations. *Geophysics* 53(3):375–385.
Wahba, G. 1990. *Spline Models for Observational Data—CBMS-NF* (Philadelphia: Soc. for Industrial and Applied Math.).
Williams, B. G., Mottinger, N. A., and Panagiotacopulos, N. D. 1983. Venus gravity field: Pioneer Venus Orbiter navigation results. *J. Geophys. Res.* 84:2381–2387.
Woo, R. 1975. Multifrequency techniques for studying interplanetary scintillations. *Astrophys. J.* 201:238–248.
Yoder, C. F. 1995. Venus' free obliquity. *Icarus* 117:250–286.
Yoder, C. F., and Ward, W. R. 1979. Does Venus wobble? *Astrophys. J.* 233:33–37.

LITHOSPHERIC MECHANICS AND DYNAMICS OF VENUS

ROGER J. PHILLIPS
Washington University

CATHERINE L. JOHNSON
Carnegie Institution of Washington

STEPHEN J. MACKWELL
Penn State University

PAUL MORGAN
Northern Arizona University

DAVID T. SANDWELL
Scripps Institution of Oceanography

and

MARIA T. ZUBER
Massachusetts Institute of Technology

Lithospheric mechanical and dynamical properties are constrained on Venus by both flexural modeling and convection modeling. Estimates of effective elastic thickness T_e on Venus are obtained by fitting flexural expressions to topographic profiles and by matching gravity/topography relationships, as a function of wavelength, to flexural models. T_e values from profile matching for coronae and other tectonic features lie in the general range of 10 to 40 km. Most estimates from gravity/topography modeling lie between 20 and 40 km. An earlier analysis for Atla Regio is updated with a degree 90 gravity model, and T_e is estimated to be about 25 km. In order to use T_e estimates to constrain lithospheric temperature gradients or heat flow, an understanding of lithospheric rheology is required. We update estimates of the steady state creep properties of diabase, and conclude that the crust of Venus could be nearly as strong as the mantle. Observed multiple scales of deformation in the absence of a weak lower crust may require new models of lithospheric behavior, including the possibilities of strain or velocity weakening. Both moment-matching methods and inelastic flexural modeling have been used to estimate surface heat flow $\hat{q}$. Moment-matching methods yield heat flow estimates for coronae and other flexural topography in the range 45 to 100 mW m^{-2}. High heat flow estimates at smaller coronae may reflect the thermal processes of coronae origin. Inelastic flexural modeling at Artemis Chasma gives $\hat{q} < 12$ mW m^{-2}, and this may reflect thermally old lithosphere. For the three hotspots Atla, Bell, and Western Eistla Regiones, $\hat{q}$ lies in the range 36 to 65 mW m^{-2}. If an estimated 0 to 10 mW m^{-2} of hotspot excess heat flow is subtracted, then the background flux is consistent with a planet that lies between the somewhat sluggish convective style predicted for the terrestrial planets in the absence of plate tectonics and a more Earth-like behavior. A plausible model is that there is a difference in heat flow between the two planets because Venus operates with less thermal efficiency than Earth due to a lack of significant subduction and to convective stratification.

I. INTRODUCTION

Building on a foundation of earlier missions, imaging and altimetry data from the Magellan mission (Saunders 1992), gravity data collected from Magellan's elliptical and circular orbits (chapter by Sjogren et al.), laboratory measurements of rock strength under Venusian conditions (Mackwell et al. 1995,1996), and improved numerical modeling all provide insight into the present and past mechanical and dynamical state of the Venusian lithosphere. Models for the mechanical state of the lithosphere are constrained by the following observables:

1. Gravity—both spherical harmonic models and inversion or forward modeling of line-of-sight (LOS) accelerations;
2. Topography—used alone or in conjunction with gravity to provide direct boundary conditions on interior mechanical/density models and used to constrain spatial instability modes of both extensional and compressional lithospheric deformation;
3. Images—mechanical models predict stress fields, which can be compared with normal faults and rifts, reverse faults and folds, and strike-slip faults observed (or inferred) in SAR images. Images also reveal dominant wavelengths of folding and faulting, which can be related to strength stratification of the lithosphere;
4. Rock strength—laboratory measurements of the ductile strength of rocks are essential to models of lithospheric mechanical response, particularly the time scales involved. Creep properties are used directly to determine thermomechanical properties from estimates of elastic lithospheric thickness;
5. Radiogenic heat production—as determined through geochemical measurements by landed spacecraft.

This chapter focuses on the ways in which elastic thickness can be estimated from topography and gravity data, the results of such analyses, and how such estimates can constrain the thermal structure of the Venusian lithosphere.

Consideration of the state of the lithosphere cannot be decoupled from the state of mantle convection beneath (chapter by Schubert et al.). Solomatov and his colleagues in a series of papers (Solomatov 1993,1995; Solomatov and Moresi 1996; Moresi and Solomatov 1995) argue that because of the temperature dependence of viscosity, most terrestrial planets should presently be in a state of "stagnant lid convection," wherein the thermal lithosphere is quite thick (hundreds of km) and is essentially immobile except very near its bottom. A planet can avoid this seemingly boring fate by having a plate tectonic regime wherein the entire lithospheric column can be recycled because brittle faults extend through the strong part of the lithosphere. A key question is whether or not the interpretation of the mechanical and dynamical properties of the present Venusian lithosphere are consistent with a thick stagnant lid, a more Earth-like behavior, or something in between. This is at present a point of considerable controversy. Convection models with wildly disparate surface heat fluxes are able to match geoid and topography data at volcanic rises on Venus. Elastic thickness estimates provide constraints on surface heat flux, and can help resolve this controversy.

The purpose of this chapter, then, is to review what has been inferred about the mechanical and dynamical state of the Venusian lithosphere and to indicate where there are differences of opinion and to explain, if possible, why such differences exist. A basic theme is to use the results of elastic and inelastic modeling to estimate the thermal state of the planet. We commence with two sections on estimating effective elastic thickness of the Venusian lithosphere from topographic profiles and from gravity/topography relationships. A review and update of crustal rheology and lithospheric strength then sets the stage to use elastic thickness estimates to infer mechanical lithospheric thicknesses and temperature gradients. The corresponding heat fluxes are interpreted in the context of mantle convection models, and the issue of lithospheric state (thick vs thin) is summarized. We also discuss the implications of the surface evidence for significant nonelastic deformation, this in light of the recent experimental measurements that imply the crust has high ductile strength.

II. FLEXURAL MODELING OF TOPOGRAPHY

A. Introduction

Lithospheric flexure can result from static or dynamic processes and provides constraints on spatial and/or temporal variations in lithospheric thickness and strength. Lithospheric thickness may be determined solely from modeling topographic flexure or by combining gravity and topography data. In this section we discuss flexural modeling of Magellan altimetry data (Ford and Pettengill 1992); Sec. III will review inferences of lithospheric thickness on Venus using both gravity and topography data. Lithospheric flexure on Venus was first inferred from Venera 15 and 16 altimetry data over Freyja Montes (Solomon and Head 1990). Magellan altimetry has revealed additional sites of possible flexural signatures; these are associated with coronae (Sandwell and Schubert 1992a; Moore et al. 1992; Johnson and Sandwell 1994; Brown and Grimm 1996; see the chapter by Stofan et al. for a description of coronae), chasmata (McKenzie et al. 1992) and rifts (Evans et al. 1992).

On Earth, flexural topography is observed predominantly at seamounts and subduction zones. If the flexure has persisted on geologic time scales, models involving the bending of a thin elastic or elastic-plastic plate may be appropriate and can provide an estimate of the effective elastic plate thickness. A purely elastic flexure model also assumes that the lithosphere can sustain infinite stresses; however, laboratory studies suggest that the strength of the upper lithosphere is limited by pressure-dependent brittle failure (Byerlee 1978) and the strength of the lower lithosphere is limited by temperature and strain-rate dependent ductile flow (Goetze and Evans 1979; Brace and Kohlstedt 1980). We are interested in that part of the lithosphere that can support stresses over geologic time scales, i.e., the thickness of the *mechanical lithosphere*. This is defined as the depth at which the lithosphere has little strength (less than tens of MPa, usually) for an assumed strain rate. The mechanical lithosphere is thicker than the elastic lithosphere, which is simply a convenient mathematical construct. Methods for using inferred elastic plate parameters and lithospheric rheological models to estimate mechanical thicknesses are discussed in Sec. V of this chapter.

Simple two-dimensional Cartesian elastic plate models (Solomon and Head 1990; Johnson and Sandwell 1994; Evans et al. 1992; Sandwell and Schubert 1992*a*; Brown and Grimm 1996) and two-dimensional axisymmetric models (Moore et al. 1992) have been used by several authors to model potential topographic flexural signatures on Venus. A full inelastic model has also been applied to the topography at Artemis Chasma (Brown and Grimm 1996; see Sec. V). Here we present a global synthesis of these results. The reader is referred elsewhere for details of the modeling techniques. Flexure of a viscous lithosphere is discussed, with reference to both a simple model developed for terrestrial flexure (DeBremaecker 1977), and viscous/viscoelastic relaxation mechanisms.

B. Data

Locations of Venusian features for which flexural modeling of topography has been attempted are shown in Fig. 1 (Ford and Pettengill 1992). Most of these features were studied as part of a global survey of lithospheric flexure using Magellan data (Johnson and Sandwell 1994) that incorporated a previous study of flexure at four large coronae (Sandwell and Schubert 1992*a*). Preliminary results from flexural analyses at rifts (Evans et al. 1992) and at small coronae (Moore et al. 1992) are also available. In most studies, altimetry orbit profiles across a given feature were modeled, either individually or as a combined data set. A few studies used topographic profiles extracted from the global gridded altimetry data (the GTDR; Ford and Pettengill 1992). Inferred flexural signatures exhibit a topographic low (generally referred to in flexure literature as a moat or trench) adjacent to a lower amplitude topographic high (see Fig. 2). In practice, however, topographic noise can mask low amplitude outer rises. Another complication is that faulting on the outer trench wall can lead to very rough topographic profiles, which are difficult to model. Magellan synthetic aperture radar (SAR) data is a useful secondary source of data in flexure modeling, as tectonic deformation observed in the radar images can be compared with model-predicted surface stresses.

C. Thin Elastic Plate Models

Thin plate models are based on the assumption that the plate thickness is small compared with the flexural wavelength. The general differential equation for plate flexure can be solved in either polar or Cartesian coordinates. Flexural features on Venus have been modeled using either two-dimensional Cartesian models in which the topography is assumed to be continuous along-strike, or two-dimensional axisymmetric models (ring or disk loads). In Cartesian coordinates the general differential equation for two-dimensional plate flexure (assuming no shear tractions on the base of the plate) is

$$\frac{d^2 M(x)}{dx^2} + \frac{d}{dx}\left[N(x)\frac{dw(x)}{dx}\right] + \Delta\rho g_0 w(x) = q(x) \quad . \tag{1}$$

In Eq. (1), $M(x)$ is the bending moment, $N(x)$ is the horizontal force per unit length (or "in-plane load"), $q(x)$ is the loading force, $\Delta\rho$ is the density contrast, g_0 is mean planetary gravitational acceleration, and $w(x)$ is the plate deflection. Equation (1) can be solved for either an inelastic rheology (moment-curvature relationship is nonlinear and a numerical solution is required; see Sec. V) or an elastic plate (linear moment-curvature relationship and analytical solutions possible). For the elastic case, the bending moment can be expressed in terms of the flexural rigidity, which in turn is a function of the elastic plate thickness T_e (see Sec. V).

In this section we focus on elastic plate solutions to Eq. (1). Most studies have assumed zero in-plane force, $N(x)$, and in fact it can be shown that $N(x)$ is in general poorly constrained in elastic plate models (Mueller and Phillips 1995). On Venus, where many features exhibiting potential flexural signatures are coronae, it is important to know whether an axisymmetric model is required or whether the planform geometry of the feature being modeled can be approximated by a two-dimensional Cartesian geometry. This has been discussed in some detail in Johnson and Sandwell (1994): numerical simulations show that a two-dimensional Cartesian model provides an accurate representation of the elastic plate thickness or equivalently the flexural parameter α (Turcotte and Schubert 1982; Eq. (2), Sandwell and Schubert 1992a) as long as the planform radius of the feature under study is several times the flexural parameter. However, the simulations show that even when the flexural parameter/elastic thickness is well determined using the Cartesian approximation, an axisymmetric model must be used to obtain a reliable estimate of the load/bending moment.

Details of modeling procedures vary from study to study. The more common approach is to minimize the rms misfit between the predicted elastic plate deflections and topographic profile(s) across a given feature. This provides an estimate of the elastic plate thickness, and allows the computation of surface

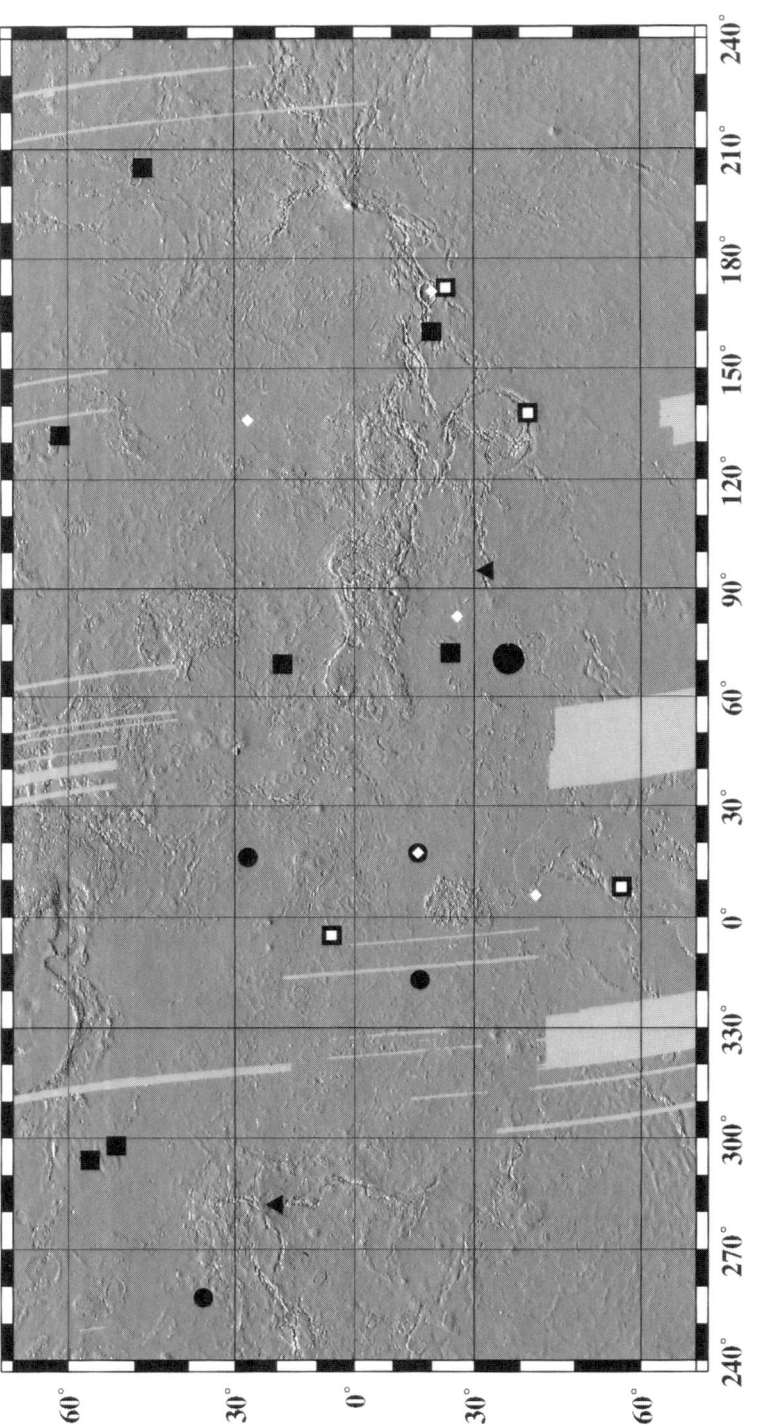

Figure 1. Mercator projection topography map of Venus illuminated from north (Ford and Pettengill 1992). Large black squares are features modeled using a two-dimensional Cartesian model and described in Johnson and Sandwell (1994); smaller white squares are features modeled using a two-dimensional Cartesian model and described in Sandwell and Schubert (1992a) (one area, Freyja Montes, is off the map at 78°N, 335°E). Black circles represent smaller coronae modeled in Johnson and Sandwell (1994), smaller white diamonds are coronae modeled with an axisymmetric model in Moore et al. (1992). Black triangles are rifts modeled using a two-dimensional Cartesian model and described in Evans et al. (1992).

stresses, bending moment and curvatures anywhere along a profile. An example of the application of this procedure to one corona on Venus (Nightingale corona) is shown in Fig. 2. The elastic plate thickness and plate curvatures can be used later, along with an assumed rheology, to estimate mechanical plate thickness and average lithospheric thermal gradients. In some cases only gross properties of the topographic profiles are modeled; e.g., the trench or outer-rise distance provides an estimate, albeit crude, of the elastic plate thickness in cases where modeling of the whole topographic profile is not possible (Johnson and Sandwell 1994). A summary of the results obtained from the application of elastic plate models to Venus is given in Table I. It can be seen that average or best-fit elastic thickness estimates fall in the range 5 to 56 km. Although elastic thickness does not provide direct information on lithospheric structure, it provides a lower bound on the mechanical lithospheric thickness and hence an upper bound on average thermal gradients. Estimates of mechanical thickness (or equivalently thermal gradients and heat flow) and implications for Venusian thermal and tectonic evolution are discussed in Secs. V and VII.

D. Flexure of a Viscous Plate

The elastic flexure model assumes that trench/outer rise features are statically maintained by large fibre stresses within a thin elastic lithosphere. Flexural features may also result from dynamical processes operating on a viscous lithosphere (DeBremaecker 1977; Melosh 1978). DeBremaecker (1977) derived a model for a hydrostatically supported viscous lithosphere, loaded at the trench and undergoing horizontal strain. The topography predicted by this model is indistinguishable from that predicted by an elastic plate model. In the viscous model, the parameter analogous to the flexural parameter of elastic plate models now incorporates both the horizontal strain rate and the thickness of the viscous lithosphere. Strain rate and viscous plate thickness are inversely proportional in these models. We have no evidence for large present-day strain rates on Venus (Grimm 1994); thus, as for elastic and inelastic flexure models, these viscous/viscoelastic plate models imply large viscous plate thicknesses and low thermal gradients. Another alternative is that the flexural signatures around, for example, coronae are the result of gravitational relaxation of topography produced earlier in the corona's evolution (Janes et al. 1992; Janes and Squyres 1995). New crustal rheologies for dry diabase (Mackwell et al. 1995, 1996; also discussed in this chapter) suggest extremely long relaxation time scales, and hence it may not be surprising that we see so few coronae with associated flexural outer rise signatures (chapter by Stofan et al.). If gravitational relaxation is the mechanism whereby flexural signatures are produced, then the small number of observed flexural features at coronae suggest that most coronae on Venus are relatively young, consistent with evidence from impact crater densities, which imply that, on average, coronae are young relative to the mean surface age of the planet (Namiki and Solomon 1994; Price and Suppe 1994). We note, however, that

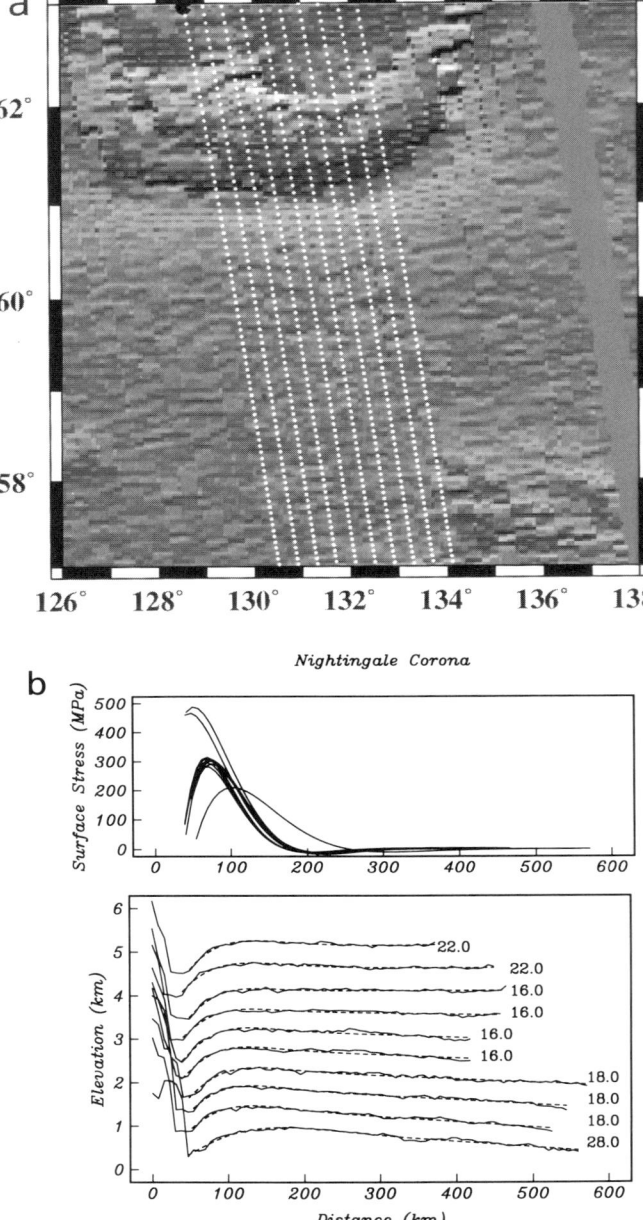

Figure 2. Results of Cartesian flexure modeling at Nightingale corona. (a) Shaded relief, with altimetry orbit tracks marked. (b) Lower plot shows altimetry orbits modeled (solid line), with the best-fit Cartesian elastic model (dashed line). Distance is calculated relative to the highest point of the topography inboard of the flexural moat. Elastic thickness corresponding to the best-fit model is given at the end of each profile. The upper figure shows the surface stresses predicted by the best fit models. The three anomalous surface stress profiles correspond to the upper 2 and lowermost altimetry profiles in the lower figure (those fit by a thicker plate) (figure from Johnson and Sandwell 1994).

TABLE I
Elastic Thickness Estimates from Topographic Profiles

Feature Name	Location		T_e (km)		Reference
	lat (°)	lon (°)	best/ave	Range	
Nishigri Corona	−24.5	72.0	12	10–16	JS[a]
Neyterkob Corona	48.5	205.0	14	12–18	JS
Ridge	18.6	69.0	18	14–22	JS
Nightingale Corona	61.0	131.0	22	16–30	JS
S. Demeter Corona	53.0	298.0	22	16–32	JS
N. Demeter Corona	57.0	294.0	24	18–28	JS
W. Dali Chasma	−20.0	160.0	34	26–40	JS
Artemis Corona	−41.5	138.0	37	30–45	SS[b]
″			56		BG[c]
Latona Corona	−23.5	172.0	35	30–45	SS
″	−20.0	171.0	30		Me[d]
Eithiroha Corona	−57.3	8.2	18	12–24	SS
Heng-O	6.0	355.0	38	35–40	SS
Freyja Montes	78.0	335.0	16	15–25	SS
″				11–18	SH[e]
Indrani Corona	−37.5	70.5		4–7	JS
Bhumidevi Corona	−17.0	343.0		16–29	JS
Unnamed Corona	37.0	257.0		8–15	JS
Beyla Corona	27.0	16.0		8–13	JS
Fatua Corona	−16.5	17.2		5–10	JS
″	−17.0	17.0	15		Me
Selu Corona	−43.0	6.0	10		Me
Aramaiti	−26.0	82.0	10		Me
Boann	27.0	136.0	5		Me
Juno Dorsum	−33.0	92.0		8–20	Ee[f]

[a] JS = Johnson and Sandwell (1994); [b] SS = Sandwell and Schubert (1992a); [c] BG = Brown and Grimm (1996); [d] Me = Moore et al. (1992); [e] SH = Solomon and Head (1990); [f] Ee = Evans et al. (1992).

the theoretical relaxation profiles of coronae, while exhibiting a moat, display only very subtle outer rise flexural bulges, which occur late in their evolution (Janes and Squyres 1995).

III. FLEXURAL MODELING USING GRAVITY DATA

A. Introduction

In addition to flexural modeling using topographic profiles alone, flexural studies have been carried out using relationships between gravity and topography, as a function of wavelength. This "transfer function" approach is well established for the Earth with both isostatic and flexural studies (Dorman and Lewis 1970; Lewis and Dorman 1970; McKenzie and Bowin 1976; Forsyth

1985; McNutt and Shure 1986; Bechtel et al. 1987; McNutt 1988; Ebinger et al. 1989; Zuber et al. 1989). Difficulties in transferring the technique to Venus arise because the short wavelengths that are often necessary to infer lithospheric properties are not uniformly resolved in the gravity data.

Phillips (1994), Smrekar (1994), Smrekar et al. (see their chapter), and McKenzie and Nimmo (1996) have all determined lithospheric parameters using models matched to estimates of the admittance and/or coherence spectra. Most of these analyses were carried out at large volcanic rises, interpreted to be hotspots (see the chapter by Smrekar et al.). One straightforward and useful model involves an effective elastic thickness T_e (or flexural rigidity D), a crustal thickness z_m, a depth to bottom or subsurface load z_l, and the ratio f of the weight of a subsurface load to the weight of a surface load (Forsyth 1985). The physical basis of the model is in the Fourier transform of the three-dimensional version of Eq. (1). The model can be generalized for multiple density interfaces and multiple subsurface loads.

B. Modeling Ambiguity

In the absence of specific knowledge, we might assume only two dominant density contrasts within the Venusian lithosphere—at the surface and at the crust-mantle boundary. On Venus, one might anticipate a lithosphere loaded from above by volcanic constructs, for example, and from below by the buoyancy associated with thermal thinning of the lithosphere and with convection in the mantle. Clearly, the challenge is to account for all of these effects, and assumptions are usually required to make any progress. One approach is to assume that different processes operate in different portions of the wavelength spectrum. This is the basis of the approach of McNutt (1988), who applied the linear filters of McNutt and Shure (1986) to isolate longer wavelength hotspot swells (representing loading from below) from volcanic top loading at shorter wavelengths.

But Moresi and Parsons (1995) demonstrated with temperature-dependent viscosity convection calculations that coupling between long-wavelength temperatures and short wavelengths in the viscosity field produces a flat free-air admittance spectrum at short wavelengths, which is characteristic of uncompensated topography. Inversion of a short wavelength admittance spectrum for the flexural properties of the lithosphere might tend to overestimate elastic thickness if the spectrum is attributed solely to a finite flexural rigidity. However, the flat admittance spectrum of uncompensated topography is quite different than the sloped spectrum of flexurally compensated topography, so *a priori* there would seem to be little chance for confusion if one or the other process dominates. Unfortunately, not all convection models predict a flat spectrum (see, e.g., Fig. 11c in the chapter by Smrekar et al.), so a general statement about the ability to use spectral shape to separate convection effects from flexural effects is not possible at this time.

It has been suggested that spectral methods that assume that the top and bottom loads are statistically independent (see, e.g., Forsyth 1985) will resolve the two contributions (short wavelength convective effects, finite flexural rigidity) to the behavior of the gravity and topography fields (Moresi and Parsons 1995). Additionally, it is reasonable to subdivide the spectrum in such an analysis, as the ratio of bottom to top loading f is not expected to be constant across the spectrum (McNutt and Shure 1986; Phillips 1994), and in general we should consider that f is a function of wavenumber k.

C. A Synthetic Inversion

The idea that a Forsyth-type model can separate convective loading from surface loading can be tested by inverting synthetic admittance spectra. Using a four-parameter flexural model (Forsyth 1985), we inverted a synthetic admittance spectrum generated by Kiefer (1995) from a plume convection model with a Rayleigh number of 10^7 and a 65-km-thick high viscosity lid (see Fig. 11c in the chapter by Smrekar et al.). Even without temperature-dependent viscosity, such convection models produce significant spectral energy over the wavelength band one might use for flexural studies (~ 1000 km down to the 420 km cutoff of a deg 90 gravity model [chapter by Sjogren et al.]) (see Fig. 11a,b in the chapter by Smrekar et al.). Clearly, this renders suspect flexural rigidity (and elastic lithospheric thickness) estimates derived from gravity data, particularly over volcanic rises, whose most likely origin is still thought to be mantle plumes (Phillips and Malin 1983). The same holds true, of course, for flexural estimates in terrestrial hotspot regions (see, e.g., McNutt 1988; Ebinger et al. 1989).

However, the inversion of the synthetic admittance spectrum yielded a ratio of bottom to top loading f of infinity, which is exactly consistent with the convection model used to generate the gravity and topography "data." Secondly, the depth to bottom loading z_l found (225–250 km) is consistent with the thermal boundary layer of the plume model. The elastic thickness of 90 km obtained from the inversion may or may not be a surrogate for the high viscosity lid. Until complex models involving both temperature-dependent convective loading and flexural loading are examined, perhaps nothing can be said confidently about lithospheric properties derived from Venusian gravity data. However, it is equally clear that some insight can be obtained from the inversion itself, particularly if methods such as that of Forsyth (1985) can really separate convective bottom loading from flexural top loading (Moresi and Parsons 1995), and from examining the geologic setting of the region in question. First, it is necessary to establish with reasonable certainty the statistical independence of top and bottom loading, which is an inherent assumption of the approach. If a short-wavelength inversion yields a bottom loading depth of a few hundred kilometers or more and a large value of f, then one ought to suspect a significant component of convective loading and extreme caution should be used in elastic thickness interpretations. Conversely, if both z_l and f are small, then it may be reasonable to conclude that convective loading is unimportant. If in this case the terrain being studied is dominated at short wavelengths by gravity anomalies associated

D. Atla Regio Revisited with a Degree 90 Gravity Model

With these caveats in mind, we revisited the estimate for effective elastic thickness obtained by Phillips (1994) for the Atla Regio area—an analysis of a $\sim 3500 \times 3500$ km region based on gravity data from a spherical harmonic model of degree $\ell = 60$ (Konopliv and Sjogren 1994). In 1993, the Magellan spacecraft was placed in a near circular orbit and higher degree spherical harmonic gravity solutions became possible. An $\ell = 90$ solution was available when this was written (MGNP90LSAAP; see the chapter by Sjogren et al.). Although tracking of the near-circular orbit did not take place over the Atla region, the inherent resolution in the tracking data from previous elliptical-orbit tracking exceeds that of $\ell = 60$ models at Atla.

Phillips (1994) argued that the loading ratio $f(k)$ at Atla Regio could be approximated as a piecewise constant function in two spectral bands with a common boundary at a wavelength λ of approximately 1000 km. Figure 3 plots admittance estimates for deg 60 and deg 90 gravity models, along with the mean Monte Carlo lithospheric model from Phillips (1994), in the short wavelength band. Good spectral estimates are possible down to $\lambda = \sim 700$ km and ~ 500 km for $\ell = 60$ and 90, respectively. Using the same annuli spacing for wavenumber averaging, the first three spectral estimates of $\ell = 90$ (circles) are close to $\ell = 60$ (diamonds). However, there is a significant scatter in the $\ell = 90$ estimates, which can be smoothed considerably by increasing the averaging interval slightly (squares). The smaller slope of this smoothed estimate presages an elastic thickness smaller than the 45-km value found by Phillips (1994).

Figure 4 shows an application of the model-fitting techniques of Phillips (1994) to the smoothed $\ell = 90$ admittance and Bouguer coherence spectra. The solid curves indicate the result of Monte Carlo modeling simultaneously of the admittance and coherence, while the medium- and short-dashed lines are the results, respectively, of Marquardt inversions (Press et al. 1992) of the admittance and coherence separately. There were no Monte Carlo results that produced solutions within the estimated errors. Allowing solutions to fit to within 6% produced 74 solutions in 10^7 runs. Note that only slightly different coherence curves produce significantly different elastic thickness estimates (10 km, Monte Carlo; 17 km, Marquardt), indicating that shorter wavelengths, where coherence spectra should have steeper slopes, are required to resolve the elastic thickness with a coherence model. This suggests that fitting the admittance only might be more useful, and a Marquardt inversion yields an elastic thickness of 27 km and is a better fit than the Monte Carlo result obtained jointly on admittance and coherence. For the Marquardt inversion on admittance, the crustal thickness z_m was set to 25 km and the solution

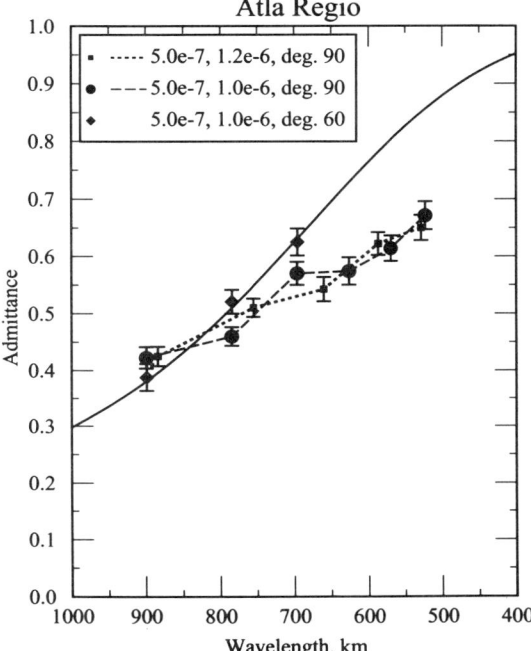

Figure 3. Dimensionless admittance spectrum in short wavelength band defined in Phillips (1994). Three spectral estimates are obtained in the $\ell = 60$ gravity model, and five in the $\ell = 90$ model. The first column in the legend refers to the wavenumber (m^{-1}) of the radius of the first annulus in the spectral averaging and the second column refers to the subsequent annuli spacing increment. The solid line shows the admittance of the mean Monte Carlo model from Phillips (1994): $f = 0.10$, $z_m = 30$ km, $z_l = 7$ km, and $T_e = 45$ km.

set is $f = 0.18$, $z_l = 107$ km, and $T_e = 27$ km. Setting z_m to 10 km yields $f = 0.19$, $z_l = 112$ km, and $T_e = 30$ km. (A slightly more conservative philosophy excludes the shortest-wavelength admittance estimate because at wavelengths shortward of ~600 km there is a sharp increase in the phase of the gravity/topography cross spectrum and a sharp decrease in the power in the two individual spectra. However, a new inversion ($z_m = 25$ km) increases T_e by only 2 km.)

To gain insight into parameter ranges and correlations, we carried out 10^8 Monte Carlo runs on admittance only; 3220 solutions were produced that generated model admittance values within the estimated errors. The search ranges were 0 to 1, 0 to 50 km, 0 to 400 km, and 0 to 50 km, respectively, for f, z_m, z_l, and T_e. A crustal thickness of 50 km is a sensible upper bound (see the chapter by Grimm and Hess). The solution properties are $f = 0.17 \pm 0.08$, $z_m = 38 \pm 9$ km, $z_l = 131 \pm 57$ km, and $T_e = 25 \pm 3$ km. The errors are the

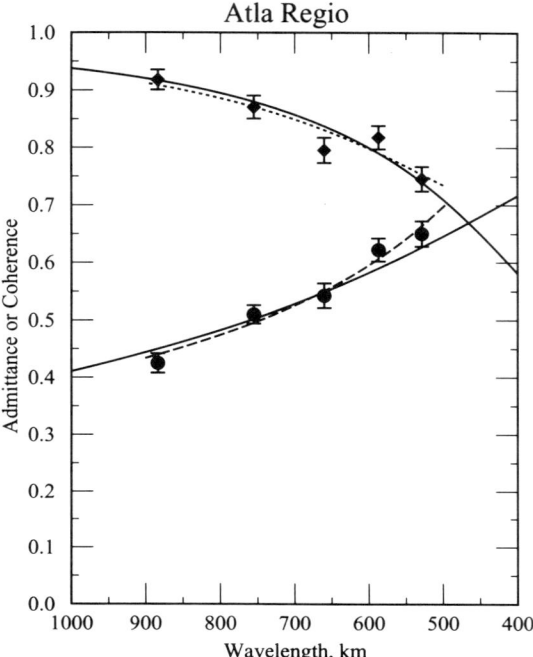

Figure 4. Results of Forsyth (1985) model fitting to smoothed $\ell = 90$ dimensionless free-air admittance (circles) and Bouguer coherence (diamonds) spectra. Solid curves are the results of Monte Carlo modeling, allowing solutions that matched all admittance and coherence estimates within 6%; in 10^7 runs, 74 solutions were obtained. The best fit is shown; it has solution parameters $f = 0.35$, $z_m = 4$ km, $z_l = 88$ km, and $T_e = 10$ km. The short-dashed line is the result of a Marquardt inversion on the Bouguer coherence and has solution parameters: $f = 0.37$, $z_m = 5$ km, $z_l = 95$ km, and $T_e = 17$ km. This suggests that in the Monte Carlo solutions, the first three parameters are controlled predominantly by the coherence. The medium-dashed line is the result of a Marquardt inversion on the admittance with z_m set to 25 km and has solution parameters: $f = 0.15$, $z_l = 108$ km, and $T_e = 27$ km.

estimated parameter standard deviations of the 3220 solutions. Parameter correlations were examined in solution scatter plots. Elastic thickness, T_e, was well bounded and poorly correlated with load depth z_l. There was also a poor correlation of z_l to loading ratio f. The results did indicate that top-loading dominates in this wavelength band, but either $f \approx 0.1$, and z_l is not well constrained, or $z_l \approx 80$ km, and f lies between 0.1 and 0.5. Crustal thickness was poorly constrained in these solutions and showed a strong negative correlation with the lower bound on elastic thickness. However, there was a minimum in elastic thickness at $z_m = 80$ km, and if the crustal search range was extended to an implausible upper limit of 100 km, then $T_e = 19 \pm 4$ km (and 24 ± 6 km for an absurd upper limit of 400 km).

E. Summary

A tabulation of elastic thickness estimates by gravity methods for different regions of Venus is given in Table II. Results depending on either the degree 90 spherical harmonic gravity model or LOS gravity data to extend spatial resolution to better than ~600 to 700 km seem to settle on a value for the elastic thickness at Atla Regio of 30±5 km. It is important to note that the Atla estimates have been obtained by distinct methods of data analysis: (*i*) localization of the spherical harmonic fields to the Atla area followed by a Cartesian spectral analysis (Phillips 1994); (*ii*) localized inversion of LOS gravity data to a surface density representation followed also by a Cartesian spectral analysis of the corresponding gravity field (Smrekar 1994; Phillips 1994); and (*iii*) Fourier transforms of localized LOS gravity data to estimate directly the admittance (McKenzie and Nimmo 1996).

All of the results in Table II are based on spectral methods; they use relationships between gravity (or geoid) and topography as a function of wavelength, in conjunction with simple models, to estimate elastic thickness. Phillips (1994) and the new results for Atla given above use a combination of coherence and admittance estimates to invert for model parameters. McKenzie and Nimmo (1996) differentiated LOS Doppler velocity residuals to obtain an estimate of LOS gravity that contained higher spatial resolution than the traditional version of this product, where the use of cubic splines can result in oversmoothing. They used a top-loading flexural model ($f \equiv 0$) to obtain estimates of z_m, T_e, and ρ_c over that part of the wavelength spectrum they argue is dominated by flexure.

Smrekar (1994) and Smrekar et al. (see their chapter) also used admittance and coherence estimates to estimate elastic thickness. In addition, they used the linear filter approach of McNutt and Shure (1986) to separate bottom-loading from top-loading effects. This technique represents a means of specifying the spectral dependence of the loading ratio f. Finally, we note that Simons (1996) used a spatio-spectral localization technique directly on the spherical harmonic models to examine local admittances. The advantages of this technique are that it works directly with the spherical harmonic fields to localize the fields and quantify the spectral behavior (e.g., shortest resolvable wavelength) in any given region. The latter step is accomplished in Cartesian spectral analysis using information from the power, phase, and coherence spectra. In Simons (1996, Fig. 3.17), the admittance spectra at Bell and Atla Regiones are relatively flat over most of the spherical harmonic spectrum but have a sharp upturn at about 900 to 1000 km wavelength. The flat long-wavelength spectra are consistent with certain classes of convection models and the upturn can be interpreted in terms of flexure. This same behavior is seen, for example, in Cartesian spectra (Phillips 1994, Fig. 4; Smrekar 1994, Figs. 10 and 11). Simons employs a top-loading flexural model, and his results (Fig. 3.17) for Bell and Atla are consistent with the information given in Table II.

TABLE II
Elastic Thickness Estimates from Gravity-Topography Spectral Analyses

Region	Location			T_e (km)		Reference
	Lat (°)	Lon (°)		Best/Avg.	Range	
Atla Regio	−10 to 25	180 to 215		45	37–52	RP[a]
				25	22–28	TC[b]
				30	20–40	SS[c]
				35		MN[d]
Bell Regio	20 to 40	40 to 60			25–35*	SS
					45–55+	
				15*		SK[e]
				40+		
Beta Regio	16 to 39	272 to 295		30		MN
W. Eistla Regio	10 to 33	343 to 10		30	30–50	SS
				25*		MN
				45+		SK
Ovda Regio	−15 to 7	70 to 110		≲20		MN
Ishtar Terra	55 to 80	305 to 80		≲15		MN
Alpha Regio	−35 to −18	355 to 10			~25–30	MN

[a] RP = Phillips (1994); [b] TC = this chapter; [c] SS = Smrekar (1994); [d] MN = McKenzie and Nimmo (1996); [e] SK = chapter by Smrekar et al.; * = short-wavelength result; + = long-wavelength result.

IV. RHEOLOGY AND LITHOSPHERIC STRENGTH

A. Introduction

An understanding of the tectonic behavior of the lithosphere and underlying mantle on Venus requires knowledge of the mechanical properties of the materials that comprise the Venusian crust and mantle. The elastic thickness of a lithosphere, as described in the previous two sections, is a surrogate for the complex rheology that comprises the real mechanical lithosphere. This does not diminish the importance of obtaining elastic thickness estimates, for it is a relatively straightforward parameter to obtain using topography and gravity data. In turn, one can often estimate the equivalent mechanical lithospheric thickness, or, more precisely, an equivalent inelastic bending moment. However, to make any use of this equivalent bending moment, or to solve the problem directly for loading of an inelastic lithosphere, we need to know the lithosphere's rheological properties. Assuming that the mineralogy of the mantle on Venus is essentially the same as on Earth, we can use experimental measurements of creep behavior for olivine aggregates (see, e.g., Karato et al. 1986; Hirth and Kohlstedt 1995) to constrain mantle viscosities for Venus. The major distinction between the conditions likely to prevail in the interior of Venus versus those of Earth is that continued volcanic activity on Venus in the presence of high surface temperatures has probably resulted in significant devolatilization of the crust and upper mantle on that planet (Kaula 1990,1995), whereas subduction beneath the oceans on Earth provides a return path for water into the interior. Thus, as the mechanical behavior of most silicate rocks is known to be affected by the presence of water, dry rheologies are likely to be the most appropriate for application to Venus.

Although detailed experimental studies have been performed on the mechanical behavior of upper mantle minerals and rocks (see, e.g., Kohlstedt et al. 1995), there have been few studies on crustal rocks with compositions similar to basalts. From measurements made on the composition of rocks for various parts of the surface of Venus by the Venera landers (Surkov et al. 1983,1984,1986), we infer that much of the surface of Venus is composed of basalts. The radar images of the surface of Venus by Magellan also indicate that there has been abundant large-scale deformation of the lithosphere, although there are few features that argue for plate tectonics as observed on Earth (see, e.g., Solomon et al. 1992; Phillips and Hansen 1994; Kaula 1995). Questions about the ability of the crust to dynamically support the topography, despite the high surface temperatures, and about the coupling of mantle dynamics to crustal deformation, require an assessment of the strength of the materials that comprise the crust on Venus. As noted above, we also wish to use the elastic thickness estimates of the previous two sections to assess lithospheric mechanical thickness and corresponding temperature gradient in various regions of Venus. This also requires a knowledge of crustal strength.

B. Previous Experimental Studies

There have been two published studies on the experimental rheologies of rocks of basaltic composition at conditions where plastic behavior will be dominant (Shelton and Tullis 1981; Caristan 1982). Both studies were performed on samples of a diabase (a rock of basaltic composition) from near Frederick, Maryland, that were either untreated or heated only on a hot plate prior to deformation. As this diabase contains several percent of hydrous minerals that do not dehydrate until temperatures above 500°C, it is believed that the hydrous minerals dehydrated during the deformation experiments, resulting in water-weakening of the minerals in the rock. Thus, these previous studies measured the mechanical behavior of diabase under undried conditions, which are not appropriate to the dry conditions expected in the crust on Venus.

C. Recent Experiments: Dry Diabase Measurements

Mackwell and his colleagues (Mackwell et al. 1995,1996) have performed creep experiments on dry diabase samples from near Frederick, Maryland (referred to as "Maryland diabase") and Columbia, South Carolina (referred to as "Columbia diabase"). Mineralogical properties of the samples and experimental procedures are described in the two references. Oxygen fugacity was controlled by using a Fe/FeO or Ni/NiO buffer.

The results of several of the deformation experiments on Columbia and Maryland diabase are illustrated in Figs. 5a,b. When the creep rate was measured at multiple temperatures or stresses in a single experiment, the initial temperature or stress was revisited later to test for irreversible changes in the sample during the deformation (such as grain growth, microfracturing, or partial melting). Given the sample-to-sample variability in modal composition and homogeneity, the reproducibility in creep rate between different samples of the same material deformed at the same conditions was remarkably good.

When all of the stress-strain rate data at a fixed temperature for each sample were fit using a linear least squares regression, values for the stress exponent n were determined. From these measurements, a value of $n = 4.8 \pm 0.6$ was calculated for the Columbia diabase, and $n = 4.7 \pm 1.0$ for the Maryland diabase, indicative of dislocation-controlled creep. Given the similarity of the stress exponents for the two diabase rocks, the stress exponents from all experiments were averaged to get $n = 4.7 \pm 0.8$ for dry diabase deformation. The creep activation energy Q for each diabase in the experiments was determined when the creep behavior was measured at a range of temperatures: $Q = 510 \pm 30$ kJ mol^{-1} for both the Columbia and Maryland diabases. The temperature ranges for each diabase were limited due to partial melting, the high activation energies for creep, and the limited range of stresses and strain rates available. Subsequently, flow laws for each rock were obtained.

For Columbia diabase,

$$\dot{\epsilon} = 4000\sigma^{4.7} \exp(-510/RT)$$

for the Fe/FeO buffered case;

$$\dot{\epsilon} = 1550\sigma^{4.7}\exp(-510/RT)$$

for the Ni/NiO buffered case; and for Maryland diabase,

$$\dot{\epsilon} = 77\sigma^{4.7}\exp(-510/RT)$$

for the Fe/FeO buffered case;

$$\dot{\epsilon} = 49\sigma^{4.7}\exp(-510/RT)$$

for the Ni/NiO buffered case, where the stress is in MPa and the activation energy is in kJ mol^{-1}. These flow laws are shown as dotted lines in Figs. 5a,b.

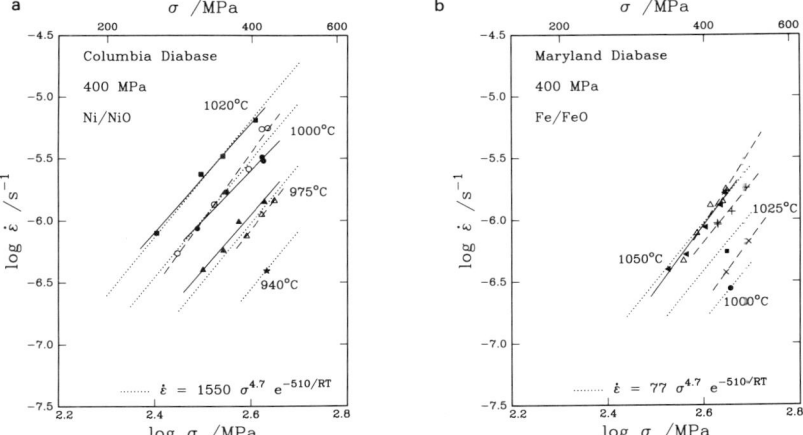

Figure 5. (a) Plot of strain rate versus stress for samples of Columbia diabase deformed at the Ni/NiO buffer at a confining pressure of 400 MPa. The solid symbols represent one experiment, the open symbols a second. The solid and dashed lines are linear least squares fits to the data at each temperature. Although the data from the two experiments at each temperature are in excellent agreement, two points at 1000°C appear to show higher strain rates than expected; we believe that these data probably include a component of microfracturing in the strain due to the high ratio of differential stress to confining pressure. The dotted lines represent the fit of the flow law at the bottom of the figure to the data. (b) Plot of strain rate versus stress for samples of Maryland diabase deformed at the Fe/FeO buffer at a confining pressure of 400 MPa. The solid symbols represent one experiment, the open symbols a second, and the + and × symbols a third; the results of the third experiment are somewhat stronger than the others, perhaps reflecting sample-to-sample variability in mineral composition. The solid and dashed lines are linear least squares fits to the data at each temperature. The dotted lines represent the fit of the flow law at the bottom of the figure to the data.

The ratio of the creep rates for the diabases from the two sources is about a factor of 30 to 50, with the Maryland diabase having the higher strength. By comparison, we observed little difference in the creep rate for the two diabase rocks resulting from the change in oxygen fugacity from Fe/FeO to Ni/NiO. The results of previous experimental studies of the deformation behavior of albite and anorthite (Shelton and Tullis 1981) and diopsidite (Kirby and Kronenberg 1984) bracket in strength the deformation behavior of the two diabases.

Optical microscope and scanning electron microscope investigations of the heat treated and deformed samples show little change in texture from the starting material. Transmission electron microscope investigations (J. C. White, personal communication) indicate that, while the bulk of the deformation in the Columbia diabase is localized within the plagioclase grains, the deformation in the Maryland diabase is distributed between the plagioclase and pyroxene grains. These observations are in general agreement with the observations of Kronenberg and Shelton (1980) on the undried Maryland diabase samples of Shelton and Tullis (1981) that show deformation predominantly in the plagioclase grains.

D. Comparison to Previous Flow Laws

The results of this study clearly illustrate the distinction between the strength of diabase under dry and wet conditions. Figure 6 is a lithospheric strength envelope (or yield strength envelope, YSE) plot of the strength of rock as a function of depth for a fixed strain rate of 10^{-15} s^{-1}; the relationship between temperature and depth is assumed to follow a simple linear form, and the surface temperature is taken as the mean surface temperature of Venus (470°C). At shallow depths (and low temperatures) the rock strength is defined by the brittle failure of the rock, which increases with increasing overburden pressure (depth); in this plot we use Byerlee's law (see, e.g., Goetze and Evans 1979; Brace and Kohlstedt 1980; Kohlstedt et al. 1995) to describe rock strength when the temperature is insufficient for plastic mechanisms to accommodate the stress. At a certain depth, depending on rock type, rates of deformation, and chemical environment, plastic processes within individual mineral grains become sufficiently active that the rock deforms predominantly by plastic mechanisms (the brittle-plastic transition). Below this depth, the rock strength follows the appropriate plastic flow law for the rock type appropriate for the depth range. In common geophysical application, the plastic flow law adopted is that of steady state creep.

In Fig. 6, the flow laws determined in this study for Columbia diabase and Maryland diabase (at the Fe/FeO buffer only) are plotted for representative Venus conditions, as well as flow laws for olivine aggregates (from Karato et al. 1986), and the previously published flow laws for Maryland diabase (MD, Caristan 1982; FD[ST], Shelton and Tullis 1981). As mentioned above, these latter studies on diabase were performed under conditions where the samples were wet, in that pretreatments only removed superficial water but

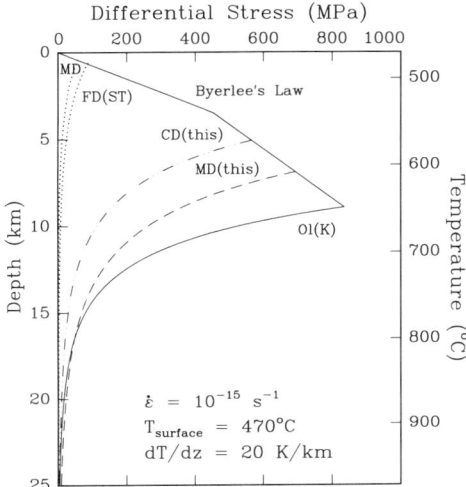

Figure 6. YSE plot of differential stress (essentially rock strength) vs depth for typical Venus conditions from the results of this study, (CD[this] and MD[this]), as well as the previous results on diabase from Shelton and Tullis (1981) (FD[ST]) and Caristan (1982) (MD), and olivine polycrystals from Karato et al. (1986) Ol(K). The relationship between rock strength and depth at shallow depths is assumed to follow Byerlee's law for frictional sliding (Kohlstedt et al. 1995).

not water trapped in hydrous minerals that dehydrated under run conditions. Thus, the new data for diabase deformation under dry conditions show it to be significantly stronger than the previous results, predicting much greater crustal strength and a deeper brittle-plastic transition.

E. Applications to Venus Rheology

Several general observations may be made based on application of these flow laws to Venus using relatively simple boundary conditions. One issue regards the existence of a strong-upper-crust/weak-lower crust/strong-upper-mantle lithospheric rheology (an "SWS" rheology, otherwise known as a "jelly-sandwich"). Uniform horizontal tension or compression applied to a rheologically stratified lithosphere leads, by instability growth, to creation of one or more dominant wavelengths of deformation depending on the stratification structure (Zuber 1987; see also Zuber and Parmentier 1990). Rift zones and ridge belts on Venus often have two dominant wavelength of deformation, indicative of an SWS rheology. The strong crust indicated by these new measurements of the ductile behavior of diabase provides an opportunity and a challenge to produce lithospheric rheology models that satisfy both the observational *and* experimental constraints (Zuber 1994). Section VI presents several hypotheses on how tectonic deformation exhibiting dominant wavelengths could exist in the face of a strong crustal rheology. Figure 7 shows YSEs for what might be considered "weak" and "strong" rheologies, having linear temperature gradients, strain rates, surface temperatures, and compositions (with fugacities) of 10 and 5 K km^{-1}, 10^{-17} and 10^{-15} s^{-1}, 470 and 447°C, and Columbia diabase (Fe/FeO) and Maryland diabase (Ni/NiO), respectively. Also shown are YSEs for weak and strong mantles using the olivine aggregate results of Karato et al. (1986). An apparent SWS rheology results from a 15-km-thick crust under weak conditions, whereas the crust shows no ductile behavior at all under strong conditions. On the other hand, a weak rheology with a 35-km-thick crust leads to both a weak lower crust and a weak upper mantle, and a strong rheology leads to a mild SWS. Finite deformation studies (see, e.g., Zuber and Parmentier 1996) in conjunction with these flow laws should provide important constraints on lithospheric structure.

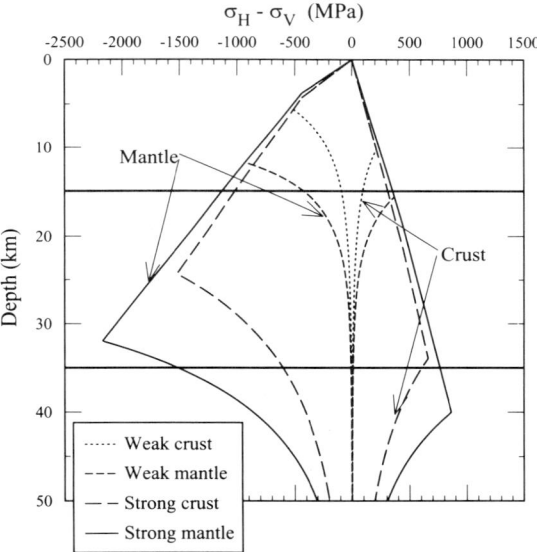

Figure 7. YSE plots for diabase (this work) and olivine aggregates under weak (short-dash for crust, medium-dash for mantle) and strong (long-dash for crust, solid for mantle) conditions. Weak: $dT/dz = 10$ K km^{-1}, $\dot{\epsilon} = 10^{-17}$ s^{-1}, surface temperature = 470°C, Columbia diabase (Fe/FeO buffer). Strong: $dT/dz = 5$ K km^{-1}, $\dot{\epsilon} = 10^{-15}$ s^{-1}, surface temperature = 447°C, Maryland diabase (Ni/NiO buffer). Horizontal lines mark crustal thicknesses discussed in text.

As the integrated area behind these strength versus depth plots is a measure of the lithospheric strength, we can make a number of general comments about predicted tectonic patterns on Venus. Although it is somewhat surprising given the high surface temperatures on Venus, the overall lithospheric strength is very high, and may be sufficiently high to permit isostatic (as well as flexural) support of topography over extended periods of geologic time. This observation may be sufficient to provide an explanation for the high correlation between gravity and topography in some regions, and limited relaxation of the impact craters. Another notable feature of these plots is that the contrast in strength of the rocks at the crust-mantle interface can be quite low under some conditions (Fig. 7). Such low contrast in rock strength permits strong dynamic coupling between the crust and mantle. Coupled with the probable lack of an asthenosphere or low viscosity channel on Venus (due to the absence of water in the Venusian uppermost mantle, and the large water-weakening effect in olivine, but also a reasonable inference from the deep apparent depths of isostatic compensation [Phillips 1986]), there may be a strong shear coupling of the convective motion in the interior of Venus to the upper portions of the lithosphere (Phillips 1986,1990). In the absence of the plate tectonic processes that occur on Earth, such coupling may explain the abundant tectonism and mountain building observed on Venus.

V. INELASTIC FLEXURE

A. Methodologies

With this review of lithospheric rheology, we can now reinterpret elastic thickness estimates presented in previous sections in terms of properties of the Venusian mechanical lithosphere. The elastic assumption inherent in the modeling in Secs. II and III is that bending moment $M(x)$ is linearly related, through the flexural rigidity D, to the curvature $C(x)$ of an elastic plate representing the lithosphere:

$$M(x) = -DC(x) \equiv -D\frac{d^2w(x)}{dx^2} \qquad (2)$$

where $D = Eh^3/[12(1-v^2)]$ (and E = Young's modulus, v = Poisson's ratio, and h = plate thickness). It follows that the bending, or fiber, stress in the plate is linear but unbounded in z. But we have argued in the previous section that stress in the lithosphere is limited by a YSE. Only when stresses are less than the yield value can they be supported elastically (the "elastic core"); otherwise, they do not exceed the yield stress. That is, the YSE describes a lithosphere that is *elastic, perfectly plastic* and the flexure is *inelastic*. It follows that a nonlinear moment-curvature relationship results (Fig. 8). As the curvature increases in magnitude, the slope of the stress vs depth in the elastic core approaches zero and no further increase in moment is possible. This is known as *moment saturation* or a *plastic hinge*. Figure 8 serves to illustrate that the elastic thickness estimates obtained by geophysical methods are mathematical constructs that may act as surrogates for more realistic lithospheric rheologies.

In purely elastic problems, Eq. (2) is substituted into Eq. (1) to yield a fourth-order differential equation in $w(x)$, which is subject to analytic solution. In the elastic–plastic case this relationship is decidedly nonlinear and numerical methods must be employed for solution (Phillips 1990; Mueller and Phillips 1995; Brown and Grimm 1996). The bending moment at any horizontal position on a thin plate is independent of rheology and is given, in the absence of in-plane forces, by (McNutt 1984)

$$M(x) = \int_x^\infty \Delta \rho g_0 w(x')(x' - x) \mathrm{d}x' \qquad (3)$$

and must be balanced by the bending moment found by integrating the stress moment within the plate, which therefore is also independent of rheology. Thus, for any specific position on a deformed plate, corresponding to a specific curvature $C(x)$, elastic–plastic moment, $M_{EP}(x)$, and elastic moment must be equivalent (McNutt 1984)

$$M_{EP}[C(x), T(z), \Re] = \frac{ET_e^3}{12(1-v^2)} C(x) \qquad (4)$$

where $T(z)$ is the temperature profile (often parameterized into a simple temperature gradient $\mathrm{d}T/\mathrm{d}z$) and $\Re$ is the collection of additional parameters needed to specify the YSE. If the appropriate curvature is specified, then the moments of the elastic thicknesses reviewed in Secs. II and III may be converted to their equivalent elastic–plastic moments and a temperature gradient can be estimated. Implicit in this approach is that the lithosphere deforms geometrically the same way for both the elastic and elastic–plastic rheologies (i.e., the curvatures match everywhere), but this is not always true.

Mueller and Phillips (1995) examined the conditions, on both Earth and Venus, under which a specific choice of curvature could accurately recover the mechanical thickness of the lithosphere, thus allowing reliable estimates of temperature gradient. Methods based on determining curvature from actual topographic profiles would seem to degenerate as the bending moment exceeds half of its saturation value. A method based on using the maximum curvature of the best fitting elastic profile to the observed topography seemed to do well up to at least 90% of moment saturation. However, small amounts of topographic noise can lead to large uncertainties in elastic thickness estimates and corresponding curvatures. In general, the moment-matching method can be expected to break down as saturation is approached (Brown and Grimm 1996).

B. Saturation Moments

Given the rheological models described in Sec. IV and estimates of saturation bending moment on Venus, one can estimate the temperature gradient within the mechanical portion of the lithosphere. Schubert and Sandwell (1995) categorized arcuate trenches to assess the possibility and significance of subduction on Venus. These arcuate features were chosen, in part, because they displayed downward flexures with curvatures exceeding about 2×10^{-7} m^{-1}. As shown in Fig. 8, such high curvatures indicate the plate is flexed beyond its elastic limit and is approaching a state of moment saturation, as discussed above. Saturation bending moments were estimated at 15 sites using the methods described in Sec. II. While these areas were mostly poor candidates for estimating elastic and mechanical thickness because of their high outer rise curvatures and high topographic noise, they are adequate candidates for estimating saturation moments. In all 15 areas, between 4 and 14 altimeter profiles were modeled using the standard elastic methods. While elastic thickness estimates for these areas are unreliable on an individual profile basis, the moment estimates at the first zero crossing outboard of the trench will accurately reflect the moment due to the outer rise topography (Geotze and Evans 1979). (Table I "best" values [JS, SS] are based on the value of elastic thickness that gives the minimum RMS misfit to *all* profiles.) The flexure model simply provides a smooth curve for integrating the topography times the moment arm (Eq. 3); any smooth curve will serve the same purpose.

These estimates of saturation moment can be equated to the maximum moment that a lithosphere of prescribed rheology, strain rate, and temperature gradient can sustain. These results are given in Table III for a dry olivine rheology (Karato et al. 1986), a surface temperature of 717 K, and strain rates of 10^{-16} and 10^{-17} s^{-1}. Insofar as the estimated moments are not in complete saturation, then temperature gradients estimated will be upper bounds. Low thermal gradients of 4.8 to 4.1 K km^{-1} are needed to maintain the large observed moment at Artemis Corona. Brown and Grimm (1996) find much higher bending moment magnitudes along the eastern margin of Artemis ($\sim 70 \times 10^{16}$N), which require even lower thermal gradients (an upper bound of 4 K km^{-1}). This results, in part, because of the large in-plane force required to match the topographic profiles when inelastic flexural modeling is used (see Sec. V.D). Most of the other large flexures suggest temperature gradients of 6 to 10 K km^{-1} while the smaller features indicate higher temperature gradients (10–30 K km^{-1}). The flexures displaying small moments and thus high thermal gradients may reflect flexure of the upper crust or a decoupling between the crust and the mantle due to the presence of a jelly sandwich rheology. In these cases, our estimates of the temperature gradient may be much too high.

C. Moment-Matching Results

When good estimates of elastic thickness obtained from topographic profile matching are available, then direct moment matching (Eq. 4) is possible. Solomon and Head (1990) were the first to apply the McNutt (1984) moment-matching technique to Venus. They inverted an elastic thickness estimate based on a flexural interpretation of the topographic profile of the Freyja Montes foredeep. For an elastic thickness range of 11 to 18 km, they obtained a temperature gradient range of 23 to 14 K km^{-1}, using a strain rate of 10^{-16} s^{-1} and curvatures in the range of (1 to 2) $\times 10^{-7}$ m^{-1}. The corresponding mechanical thickness range is about 13 to 20 km.

Johnson and Sandwell (1994; see Sandwell and Schubert 1992a) also converted elastic thickness estimates (Sec. II) to mechanical thickness estimates

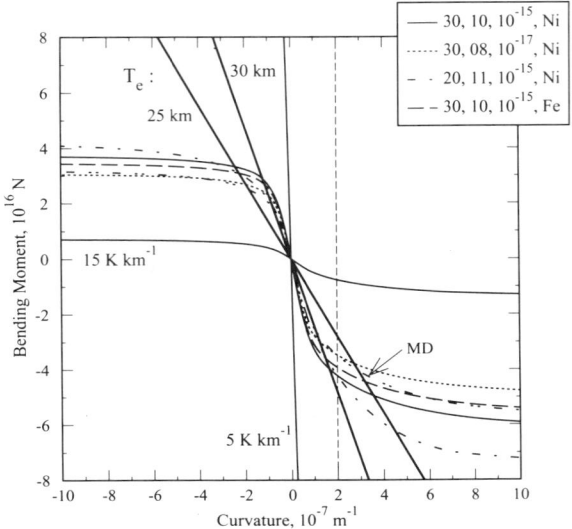

Figure 8. Bending moment vs curvature for 25- and 30-km-thick elastic plates and for elastic-perfectly plastic lithospheres whose stresses are limited by a YSE. Surface temperature is 720 K. The entries in the legend refer to crustal thickness, temperature gradient, strain rate, and oxygen fugacity conditions (see Sec. IV). The mantle has a dry olivine rheology (Goetze 1978). All curves are for Columbia diabase, except the curve labeled "MD," which is Maryland diabase. This curve has the same parameters as the bottom entry in the legend ($\equiv$ REF), except for a temperature gradient of 11 K km^{-1}. Using REF and invoking the dry olivine aggregate of Karato et al. (1986) produces a curve (not shown) that lies very nearly on the MD curve. Also shown are results with the same parameters as REF, except that temperature gradients are 5 and 15 K km^{-1}. Other parameter variations changed the curves very little. This included changing the surface temperature to 743 K (470°C) and changing the temperature profile from a linear to an error function form (Zuber and Parmentier 1990). The effective elastic lithosphere can be interpreted in terms of an elastic-plastic lithosphere where the two curves match for a specific curvature, in this case taken to be 2×10^{-7} m^{-1}.

h_m using Eq. (4) and the moment of the best fitting elastic profile at the point of maximum curvature and at the first zero crossing (Fig. 9). Temperature gradients are easily derived from mechanical thickness estimates in this work because the base of the mechanical lithosphere corresponds approximately to the 1030 K isotherm (inferred from Fig. 4 in Solomon and Head 1990) and the temperature gradient is constant. This yields $dT/dz \approx 290/h_m$, where h_m is in km and dT/dz is in K km^{-1}. Of the features modeled in Johnson and Sandwell (1994), five were considered sufficiently reliable to convert elastic thickness to mechanical thickness. These yielded for h_m a range of 21 to 37 km, and a corresponding temperature gradient range of 14 to 8 K km^{-1}.

Figure 8 invokes the moment-matching method to assess the temperature gradient beneath Atla Regio. The objective is to vary lithospheric rheology

TABLE III[a]
Saturation Moments and Maximum Temperature Gradients for Selected Features

Feature Name	Location		Moment	Curvature	dT/dz (K km^{-1}) $\dot{\epsilon} =$	
	lat (°)	lon (°)	10^{16} N	10^{-7} m^{-1}	10^{-16} s^{-1}	10^{-17} s^{-1}
Neyterkob Corona	48.5	205.0	0.8	5.9	19.7	17.6
Nightingale Corona	61.0	131.0	0.7	1.6	20.6	18.5
S. Demeter Corona	53.0	298.0	2.0	3.2	13.6	11.9
N. Demeter Corona	57.0	294.0	3.1	4.6	11.3	9.9
W. Dali Chasma	−20.0	160.0	12.9	5.7	6.4	5.6
Artemis	−41.5	138.0	25.4	10.3	4.8	4.1
S. Latona Corona	−23.5	172.0	6.4	3.1	8.5	7.5
N. Latona Corona	−20.0	171.0	4.7	34.5	9.5	8.4
Eithinoha Corona	−57.3	8.2	1.0	3.3	18.1	16.1
Derceto Plateau	−47.6	19.7	0.2	209.4	27.2	25.6
E. Diana Chasma	−15.6	157.2	12.4	19.9	6.5	5.7
Hecate Chasma	16.5	248.4	2.4	3.7	12.6	11.0
Parga Chasma	−15.4	245.4	1.6	17.2	15.0	13.2
Uorsar Rupes	77.8	331.2	0.28	33.2	26.1	24.0
Quetzalpetlatl Corona	−66.6	350.0	0.85	9.8	19.2	17.2

[a] Table after Schubert and Sandwell (1995). Estimates of temperature gradient based on dry olivine rheology (Karato et al. 1986) and surface temperature of 717 K. dT/dz = temperature gradient; $\dot{\epsilon}$ = strain rate.

parameters to find the range of solutions that will match the moment range of 25- to 30-km-thick elastic lithospheres at the estimated maximum curvature of 2×10^{-7} m^{-1} (Phillips 1994). Crustal thickness, temperature gradient, strain rate, diabase composition, and oxygen fugacity were varied. It can be seen that matching the elastic moment is determined almost entirely by the temperature gradient; other parameters play a secondary role within reasonable ranges of plausibility. Acceptable solutions are within the range 8 to 11 K km^{-1}, and temperature gradients of 5 and 15 K km^{-1} are easily excluded. Results are given for strain rates of 10^{-15} and 10^{-17} s^{-1}. We note that the Johnson and Sandwell (1994) results are based on a strain rate of 10^{-16} s^{-1}. Grimm (1994) has used the degree of faulting of impact craters to infer that the near-surface strain rate on Venus has an upper bound of 10^{-17} s^{-1} over the span of time represented by the crater population. This strain rate gives the lowest temperature gradient, 8 K km^{-1}.

D. Inelastic Flexure Solutions

The actual deformation and stress state of an elastic–plastic rheology can be found by solving numerically the flexure equation (Eq. 1) with the nonlinear moment curvature relationship. Phillips (1990) obtained inelastic flexure solutions for model Venus lithospheres. He used the gravity field to estimate

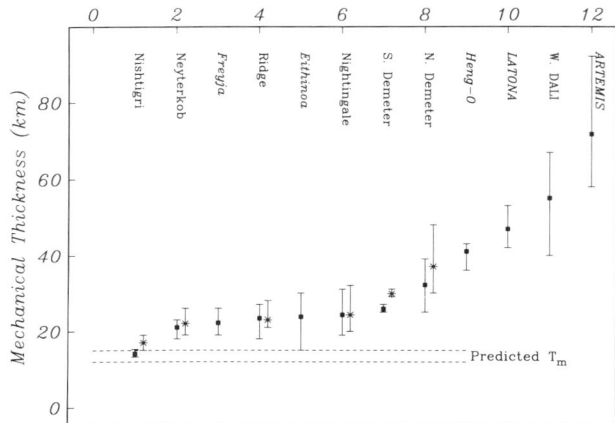

Figure 9. Mechanical thickness estimates derived from elastic thickness solutions for two-dimensional Cartesian flexure solutions of Johnson and Sandwell (1994), and from Sandwell and Schubert (1992a; italics in figure). See also Table I. Capitalized names are features that do not yield reliable estimates because they are thought to be moment saturated. Dashed lines indicate mechanical thickness based on Earth-scaled heat flow (Solomon and Head 1982). Square symbols and error bars are based on bending moment and curvature of first zero crossing of best fitting elastic profile (McNutt 1984); the asterisks and error bars correspond to the point of maximum curvature of the best-fitting elastic profile (Mueller and Phillips 1995) (figure from Johnson and Sandwell 1994).

in-plane force magnitudes under the assumption that the long wavelength gravity field is directly related to mantle convection that couples both shear and normal forces into the lithosphere. Solutions to the inelastic flexure equation were obtained for Venusian conditions using the dry websterite flow law of Avé Lallemant (1978). In response to the same surface load, deflection and bending moment of the lithosphere were significantly different under elastic rheology, elastic–plastic rheology, and elastic-plastic rheology with an estimated in-plane force of 3×10^{12} N m^{-1}.

Brown and Grimm (1996) carried out a detailed analysis of inelastic flexure at southeastern Artemis Chasma. Figure 10 is an "interaction diagram" showing inelastic solution properties as a function of applied bending moment and in-plane force. The solutions are bounded by a curve representing moment saturation as a function of the in-plane force. Contours to the right of this boundary are for the width of the outer rise at half of the maximum amplitude (FWHM). Also indicated in the figure is the domain of model solutions that satisfy the observed ranges of FWHMs and outer rise maximum amplitudes.

Given the large bending moment estimates, it is curious that normal faulting is not observed on the slope of the outer trench (chasma side of the outer rise) of southeast Artemis Chasma. If a compressional in-plane force is applied first, then application of a moment will cause unbending to

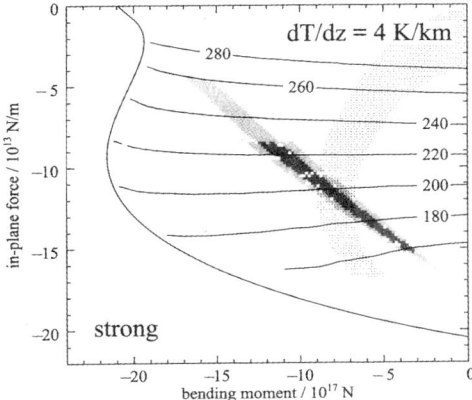

Figure 10. Interaction diagram for inelastic flexure of elastic–plastic plate with temperature gradient of 4 K km^{-1}. Axes are applied bending moment and applied in-plane load. Sinuous curve is the saturation bending moment resulting from the in-plane force. Contours are the outer rise width (km) at half of the maximum rise amplitude (FWHM). Diagonal shading indicates agreement (to within two standard deviations) with the observed FWHMs and outer rise maximum amplitudes at southeastern Artemis Chasma. The stippled region indicates those conditions for which no deformational failure (e.g, faulting) occurs at the surface, assuming that the surface has a finite strength of 50 MPa. A "strong" rheology was assumed to maximize temperature gradient (figure from Brown and Grimm 1996).

occur, leading to tensional failure at the surface. If, on the other hand, the in-plane force and bending moment are applied simultaneously, then Brown and Grimm (1996) propose that unloading can be neglected, leading to some solutions with surface stresses less than the assumed surface cohesion of 50 MPa. In this case, the interpretation is that the in-plane force does not arise from an external source (e.g., convective coupling; Phillips 1990), but from congestion at the trench itself. The stippled region in Fig. 10 indicates the solutions that have no surface faulting, and thus the intersection of this region with the acceptable FWHM/outer rise amplitude region indicates acceptable solutions overall. Brown and Grimm have devised a strong rheology model so as to maximize the temperature gradient (see their paper for details) and a temperature gradient of <4 K km^{-1} satisfies all three constraints over the spread of outer rise widths observed at southeast Artemis Chasma. A similarly low temperature gradient is obtained by the moment-matching technique (Sandwell and Schubert 1992a), but the elastic modeling results are insensitive to the magnitude of the in-plane force.

The in-plane force required for the Artemis inelastic model solutions, in the range -2 to -20×10^{13} N m^{-1} for the mean outer rise width, is large but similar to values inferred from modeling deformation in the central Indian Ocean lithosphere (McAdoo and Sandwell 1985; Karner and Weissel

1990). Trench congestion caused by regional impedance to subduction (the Indian-Asian collision impeding subduction of the Indo-Australian plate) can create forces this large. But the ultimate force driving the Indo-Australian plate is the subduction along a portion of its boundary. Mature subduction zones appear to be the only places where forces in excess of 10^{13} N m^{-1} in magnitude can be generated (Mueller and Phillips 1991), and in this case they appear to be transmitted across the plate, leading to intraplate deformation (see Cloetingh and Wortel 1986). Given the lack of water in the Venusian lithosphere and the low temperature gradients inferred at Artemis, in-plane forces in the upper end of the solution range are probably required to promote motion on throughgoing lithospheric faults in this region. Where such forces arise within the lithosphere and convecting mantle is a major question. The origin of the large bending moments required is also problematical.

We note that both the moment-matching and full inelastic flexural results have assumed that the crust and mantle are mechanically coupled. If they are not, due to a SWS rheology (see Sec. IV), then the bending moment decreases because the sum of separate crust and mantle bending moments is always less than the bending moment of the lithosphere taken as a single entity (see Burov and Diament 1995). A temperature gradient estimated from a coupled model will be an overestimate if the crust and mantle in the real lithosphere are, in fact, decoupled. Finally, if a lithospheric plate is progressively cooling, it will develop large thermal stresses. At regions of negative plate curvature, such as a trench, thermal stresses will increase the bending moment that can be supported. A plate will appear stronger (lower temperature gradient estimated) than it actually is if this effect is neglected (Wessel 1992). This mechanism, which can also delay normal faulting on the surface outboard of a trench, could be an important factor on Venus.

VI. TECTONICS AND STRENGTH STRATIFICATION

A. Introduction

Many areas of the Venus surface exhibit spatially extensive deformation features of both extensional and compressional origin that have been interpreted as rifts and mountain belts (Masursky et al. 1980; Campbell et al. 1983, 1984; Solomon and Head 1990; Solomon et al. 1992). The most prominent example of extensional deformation is Beta Regio, a rift zone associated with a major volcanic rise (Masursky et al. 1980; Campbell et al. 1984; Stofan et al. 1989; Senske et al. 1991). Fold and thrust structures are most commonly found in Ishtar Terra, particularly surrounding the Lakshmi Planum Plateau, though some have also been identified in the western part of Aphrodite Terra (Campbell et al. 1983; Kaula et al. 1992; Solomon et al. 1992; Suppe and Conners 1992; Keep and Hansen 1994; Williams et al. 1994). Extensional and compressional lineations are also observed over broad regions of the Venusian plains (Solomon et al. 1992; Squyres et al. 1992). Spatially extensive areas of more complex, and probably multiphase, deformation, termed tessera (or complex ridged terrain), represent an additional deformation type that does not have a direct terrestrial analog.

B. Determination of Lithospheric Structure

Analyses based on Pioneer Venus (Masursky et al. 1980) and Venera 15 and 16 (Basilevsky et al. 1986) radar imaging revealed that a number of prominent structural assemblages exhibited generally regular feature spacings or widths, usually in the range 10 to 20 km. The spacings have been interpreted as dominant wavelengths of folding or stretching of lithospheric layers (see, e.g., Solomon and Head 1984). When structural deformation occurs at a characteristic wavelength, the wavelength can be used to constrain the thickness of the lithospheric layer. Before Magellan, application of flexural elastic (Solomon and Head 1984; Banerdt and Golombek 1988) and viscous and plastic instability (Zuber 1987) models yielded an elastic thickness or depth to the brittle–plastic transition for Venus generally in the range of 1 to 6 km, which is much less than comparable values observed for Earth. This range of values was viewed to be consistent with the idea that Venus ought to have a much thinner lithosphere than Earth owing to its much higher surface temperature ($\sim$700 K).

The recognition that many areas of Venus that display the short tectonic length scales also exhibit longer length scales of deformation (100–300 km) led to the suggestion that at least parts of the lithosphere may have a strong-weak-strong or jelly sandwich strength structure corresponding to a strong upper crust, weak lower crust and strong upper mantle (Zuber 1987; Banerdt and Golombek 1988). The multiple wavelengths led to constraints on crustal thickness ($\sim$10–30 km) and thermal gradient ($dT/dz < 25$ K km^{-1}) (Zuber 1987; Banerdt and Golombek 1988; Zuber and Parmentier 1990) that were in agreement with independent estimates derived from viscous relaxation models constrained by crater depths (Grimm and Solomon 1988). However, both the dominant wavelength and viscous relaxation models were based on rheological data for diabase crustal material that was, as discussed earlier, not fully dried (Shelton and Tullis 1981; Caristan 1982). Experimental results for dry diabase (Mackwell et al. 1995,1996), discussed in Sec. IV, indicate that a weak, lower crustal channel may in fact not be present on Venus, or may be sufficiently narrow that the upper and lower crust are not decoupled (see Sec. IV.E). The rheological results combined with the evidence from Magellan gravity and topography for a relatively thick elastic lithosphere on Venus (Johnson and Sandwell 1994; Moore et al. 1992; Sandwell and Schubert 1992a,b; Schubert et al. 1994; Phillips 1994; Smrekar 1994; Secs. II and III) may be in conflict with the idea of a very shallow brittle–plastic transition.

A conundrum thus arises as to how to explain the regions of deformation with small length scale tectonic fabrics. Magellan images have revealed that these features contain even finer-scale structure than previously recognized. The styles of deformation in many areas appear to be indicative of a lithosphere that is deforming as a plastic or viscous material without a significantly strong elastic core. In fact, most models for fine scale compressional tectonics have invoked stacking or thrusting of layers, perhaps above a viscous or plastic detachment (Smrekar and Phillips 1988; Head 1990; Vorder Bruegge and Fletcher 1990; Suppe and Conners 1992; Williams et al. 1994). One possible explanation for this paradox is that much of the small-scale deformation dates from a time when the Venus lithosphere was much younger and hotter than present. If this is the case, it will ultimately be necessary to explain how the lithosphere thickness changed over time. Another possibility is that smallscale structure is controlled by intra-crustal layering, for example, by many superposed near surface volcanic flows. A third hypothesis is that mechanisms such as strain weakening (Zuber 1994) or velocity weakening (Neumann and Zuber 1995) could cause regional weakening of the lithosphere in areas of concentrated deformation. Yet another possibility is that the deformation is not controlled by a layer of a given thickness, but perhaps by subsurface shear. This has been suggested for lineations in some Venusian plains (Banerdt and Sammis 1992). It should be noted that all of these possibilities assume that the deformation indeed contains characteristic wavelengths. Magellan data have shown most areas to have more complex structures than previously appreciated, and length scales should be re-analyzed in a global sense.

While the small-scale aspect of the deformation is likely indicative of shallow structure, the broad spatial patterns of deformation provide evidence that compressional and extensional stresses that produced the deformation are a consequence of mantle dynamics (Phillips and Hansen 1994). Reconciling the spatial and temporal characteristics of extensional and compressional deformational structures into a self-consistent model for the thermal evolution is the major challenge in future study of these structures.

VII. SYNTHESIS

A. Introduction

Estimates of the effective elastic thickness remain, in the absence of seismic information and the direct measurements of heat flow, perhaps the only way geophysically in which to infer lithospheric thermal structure, providing such estimates can be converted to mechanical thicknesses. In this concluding section, we use such information to construct thermal models of the lithosphere. To proceed, we first review temperature gradients and heat fluxes inferred from elastic thickness estimates for various regions of the planet, and then come full circle to questions posed at the start of this chapter, comparing these results to convection models.

B. Inferences from Hotspot Results

1. Estimates of Temperature Gradient and Heat Flux. What do the hotspots tell us about temperature gradients on Venus, and how does this information relate to the heat flow at terrestrial hotspots? If, for consistency, we use just the results from the $\ell = 90$ gravity solution (chapter by Sjogren et al.), then elastic thickness estimates at three hotspots (Atla, Bell, Western Eistla Regiones) range from 15 to 25 km. This range includes only the short wavelength results in the chapter by Smrekar et al. (see Table II here). Using a reference lithosphere ("REF", see Fig. 8) and the moment matching technique, linear temperature gradient is well fit by a relationship

$$dT/dz = 9.54\left(\frac{T_e}{30}\right)^{-0.817} \tag{5}$$

over a range of T_e from 10 to 40 km. The interval 15 to 25 km corresponds to a range in dT/dz of 17 K km^{-1} to 11 K km^{-1}. This temperature gradient must include the "mean" or background temperature gradient plus the contribution of the hotspot. Hotspot plumes probably come from the core-mantle boundary (see, e.g., Sleep 1990) and in a sense are superimposed on the dominant internally heated convection of the mantle. It is this latter phenomenon we wish to constrain.

The surface heat flux $\hat{q}$ corresponding to a given temperature gradient depends on the thermal conductivity K, which will have a significant radiative component in the Venusian lithosphere, given the high surface temperature. Because the temperature gradients estimated are in effect average values through the lithospheric column, we also use average thermal conductivities. Using expressions by Schatz and Simmons (1972), the average values of K for temperature gradients of 11 and 17 K km^{-1} are 3.3 and 3.8 W m^{-1} K^{-1}, respectively. This leads to a $\hat{q}$ range of 36 to 65 mW m^{-2} for these three hotspot regions.

2. Estimating the Plume Background or "Mean" Heat Flow. The estimated hotspot range can be used to infer a number more representative of the mean global heat flux only if the excess heat flow of the hotspot over that of the surrounding "ambient" mantle can be estimated. On Earth, this has been a difficult prospect. The excess heat flow at Hawaii, the "type" hotspot, may be at most 10 mW m^{-2} and is arguably consistent with zero (Von Herzen et al. 1989).

But models for the entrainment of hot plume material beneath the lithosphere (see, e.g., Sleep 1994) are intimately involved with the competition between the time scales of plate motion and diffusive thinning of the lithosphere. On Venus, with its presumably stationary lithosphere, heat delivered by plumes to the base of the lithosphere must eventually thin this region and elevate the surface heat flow unless heated portions of the lower lithosphere can be entrained in return flow to the deeper mantle. On Earth, excess heat flow associated with hotspots on slow moving plates appears to be in the range 10 to 20 mW m^{-2} (see, e.g., Sleep 1990; Phillips 1994) and this may be the closest analog to Venus. That is, the terrestrial lithosphere may be close to thermal equilibrium for these features and thus comparable to the zero-plate-velocity Venus.

On the other hand, model calculations by Smrekar and Parmentier (1996) show that the buoyancy flux (thermally anomalous mass per second delivered by a plume) that produces topographic swells typical of Venusian hotspots is small by terrestrial standards. Typical model values are ~ 0.1 Mg s^{-1}. On the slow-moving Atlantic plate, the buoyancy fluxes estimated at the Cape Verde and Bermuda hotspots are 1.6 Mg s^{-1} and 1.1 Mg s^{-1}, respectively (Sleep 1990). The corresponding estimates of plume heat flux are 20 mW m^{-2} (Courtney and White 1986) and 11 mW m^{-2} (Detrick et al. 1986). Because the heat delivered by a plume is directly proportional to its buoyancy flux, these model results suggest that the excess heat flux delivered by Venusian mantle plumes may not be large.

Considering both types of arguments (terrestrial analogy, models), a reasonable estimate for the excess heat flux of Venusian plumes is 0 to 10 mW m^{-2}. The background, or "mean," heat flux would then lie between 36 and 55 mW m^{-2} if we link the two ranges (i.e., the upper end of the excess heat flux range is matched to the upper end of the Venusian range).

3. Comparison to Convection Modeling. This range of hotspot background or "mean" heat flux can be compared to estimates from parameterized convection models, from fully convective plume models matched to hotspot gravity/geoid and topography, and from Earth-scaled heat flow (Table IV).

TABLE IV
Estimates of Surface Heat Flux $\hat{q}$

Source (model or feature type)	$\hat{q}$, mW m^{-2}
Hotspots	
Atla, Bell, W. Eistla Regiones (see Table II)	36–65
"Background" from hotspots $\approx$ global mean	**36–55**[a]
Parameterized Convection Solutions (Global Mean)	
Non-stagnant lid (Phillips and Malin 1983 & updated)	~ 50
Stagnant lid (Solomatov and Moresi 1996)	~ 15
Phillips and Malin (1983) updated (stagnant lid, core cooling, modulated diffusion control of lithosphere)	~ 35
Earth-scaled Global Mean	
Solomon and Head (1982)	**74**
Turcotte (1995)	**63**
Artemis Chasma (Brown and Grimm 1996)	<12
Coronae, chasmata (see Table I)	45–100

[a] Numbers in bold are assessments of present-day "mean" heat flow on Venus.

A parameterized convection thermal model with no core heat component and with a chondritic heat source abundance produces a $\hat{q}$ of ~ 50 mW m^{-2} with a Nusselt number–Rayleigh number power exponent of 1/3 (see Phillips and Malin 1983). A similar value for heat flux was obtained by Solomatov and Moresi (1996) in parameterized convection calculations for Venus under approximately constant viscosity conditions (no stagnant lid).

Solomatov and Moresi (1996) obtain present-day estimates of stagnant lid ($\approx$ thermal boundary layer) thickness and $\hat{q}$ of $\sim$230 km and $\sim$15 mW m^{-2}, respectively, for a parameterized convection model in which a constant viscosity regime is switched to a stagnant lid regime 0.6 Gyr ago. After the switch, the heat flux and lithospheric thickening are controlled purely by diffusion cooling of the lithosphere. These results are consistent with an "average" Venus scaled from stagnant lid convection solutions (interior Rayleigh number Ra = 3×10^6 and vertical viscosity contrast $\Delta\eta = 10^6$) adjusted to match the geoid, gravity field, and topography at Beta Regio.

We have updated the code of Phillips and Malin (1983) by introducing core cooling, by equating the thermal lithosphere explicitly with the upper thermal boundary layer, and by allowing for stagnant lid convection to set in at a specified time in the thermal history run. When the stagnant lid regime sets in, the thermal boundary layer thickness is controlled by diffusion cooling (Solomatov and Moresi 1996), but in our case the lithospheric thickness is also modulated by convective heat flux from beneath. In a run with chondritic heat sources and a switch to the stagnant lid regime 2 Gyr ago, the present value of $\hat{q}$ is about 35 mW m^{-2}, and the thickness of the thermal lithosphere is 120 km. Further work is required to elucidate the range of solutions possible with this approach. We note that this code in a nonstagnant lid regime (and with the usual determination of boundary layer thickness from the reciprocal of the Nusselt number) produces a present-day $\hat{q}$ of 53 mW m^{-2}. Table IV compares the estimated "mean" heat flow to parameterized convection solutions, as well as to the Earth-scaled estimates of 74 mW m^{-2} (Solomon and Head 1982) and 63 mW m^{-2} (Turcotte 1995).

The matching of plume convection models to the gravity/geoid and topography of hotspots is another potential method for determining lithospheric properties (Kiefer and Hager 1991,1992; Solomatov and Moresi 1996; Moresi and Parsons 1995; Smrekar and Parmentier 1996). Models that depend on bottom heating to create both a plume *and* an upper thermal boundary layer may be unrealistic. We expect that the upper thermal boundary is maintained by internally heated convection; it is disturbed by plumes initiated from the lower thermal boundary layer, which is maintained by modest amounts of heat released by the core. Such bottom-heated models are able to produce a spectrum of lithospheric thicknesses and heat fluxes, with the relative contributions of the plume and the thinned lithosphere to the gravity and topography fields varying amongst the models. Thus such models do not help distinguish between thick and thin lithospheres. Smrekar and Parmentier (1996), however, came close in spirit to modeling plume behavior superimposed on internally heating convection by launching plumes into a prespecified thermal structure that mimicked this condition. They showed that models with Earth-like mantle, lithosphere, and plume parameters can fit the range of geoid-to-topography ratios (GTRs) found for all likely Venusian hotspots. As noted above, the corresponding plume excess heat flux is quite small.

4. Summary. Elastic thickness determinations at hotspots lead to a range of estimated background $\hat{q}$ values, which, if equated with global average $\hat{q}$, significantly exceed values predicted by geoid/topography scaled stagnant lid parametrized convection results. Nonstagnant lid parameterized convection solutions fall within this range, but simple Earth-scaled values fall above. The modulated stagnant lid parameterized convection solution produces an Earth-like lithospheric thickness (120 km), and a $\hat{q}$ at the low end of the estimated range. Thus our estimate of Venus average heat flow seems to lie between a somewhat sluggish convective style predicted for the terrestrial planets in the absence of plate tectonics and a more Earth-like behavior, but with Venus mantle convection operating with less thermal efficiency than its terrestrial counterpart. This may be due, in part, to differences between the two planets in the mass of lithosphere subducted, as on Earth mantle cooling is strongly influenced by the heating of subducted slabs. In addition, the absence of plate tectonics increases the tendency for stratified convection, which leads to a lower flux (chapter by Schubert et al.). Another piece of evidence for Earth-like behavior, or at least not a thick stagnant lid, is that pressure-release partial melting may require a thin lithosphere (or very high plume and mantle temperatures) (Smrekar and Parmentier 1996). Additionally, the very low stress in the upper parts of a thick stagnant lid would severely inhibit rifting, which is contrary to observation.

C. Inferences from Corona/Chasma Results

Artemis Chasma is the most thoroughly studied of the Corona/Chasma features in terms of inelastic flexural modeling. As reviewed above, the temperature gradient is estimated to be less than 4 K km^{-1} (Brown and Grimm 1996); this corresponds to a heat flow of <12 mW m^{-2}. This result is not necessarily at odds with the conclusions drawn above, and might be expected if the underthrusting observed at Artemis Chasma (Brown and Grimm 1995) is associated with cold mantle downwelling.

Topography on Venus about 1 km on either side of the planetary mean radius follows a linear relationship between elevation and the normalized square root of cumulative area (Morgan and Phillips 1983; Rosenblatt et al. 1994). This same relationship is true generally for Earth's ocean floor, and this is to be expected from the depth dependence on the square root of plate age (for ages $\lesssim$100 Myr) resulting from diffusion cooling of the lithosphere. A simple model for Venus, with its immobile lithosphere, is a stochastic one; random impingement of mantle plumes at the base of the lithosphere, locally resetting its thermal age, should lead to the same result. Regional differences in heat flow of tens of mW m^{-2} are easily realizable in this scenario. As a lithosphere ages thermally, its degree of negative buoyancy increases, and, all other factors being equal, it may be more likely to initiate subduction (but see Cloetingh et al. 1989; Mueller and Phillips 1991). The proposed nascent subduction at Artemis Chasma (Sandwell and Schubert 1992*a,b*; Brown and Grimm 1995) and the low temperature gradient may both be pointing to a lithosphere that is thermally old in this region.

Flexure at Dali Chasma and Latona Corona may be moment saturated (see Fig. 9) so temperature gradient estimates by the moment-matching techniques are probably unreliable (Johnson and Sandwell 1994), although saturation moments (Table III) provide more dependable upper bounds. The remainder of the coronae, chasmata (and one ridge) have a T_e range, with one exception, that falls in the interval 10 to 20 km (Table I), corresponding to a heat flow range of 45 to 100 mW m^{-2}. This corresponds to a range from approximately the nonstagnant parameterized convection results to a value about 20% in excess of Earth's mean heat flux. Can these generally higher heat flows be reconciled with conclusions above? One possibility is that these (mostly) corona features represent an earlier, hotter time in Venus' history, but this seems unlikely given the relatively young average age of these features (Namiki and Solomon 1994; Price and Suppe 1994). A second possibility is that the flexure results are consistent with the heightened thermal environment of the proposed magmatic origin of most coronae (chapter by Stofan et al.). A third possibility is, of course, that the corona results are more representative of Venus and that Earth-scaled heat flux is the best model for Venus.

D. Concluding Remarks

We have attempted to summarize what is understood early in 1996 about the mechanical and dynamic state of the lithosphere of Venus. Our emphasis has been on using properties of the lithosphere to constrain the thermal state of the planet, for this is one of the most fundamentally interesting goals of lithospheric studies on planets where seismic and direct heat flow information are not available.

Further progress can be expected from (i) combined modeling of lithospheric flexure and mantle convection, (ii) an improved understanding of regional age differences, (iii) improvements in gravity field representations—higher spherical harmonic degree expansions and improved local modeling, (iv) new dynamical models of the lithosphere to explain multiple scales of deformation, and (v) continuing experiments on rock deformation under Venusian conditions.

Acknowledgments. S. Smrekar and R. Grimm provided useful, constructive reviews of this chapter. M. Zimmerman and D. Kohlstedt worked with SJM in performing the deformation experiments on diabase, and J. White is currently working with SJM analyzing the deformation microstructures in the experimentally deformed samples. Many of us were supported by NASA grants.

REFERENCES

Avé Lallemant, H. G. 1978. Experimental deformation of diopside and websterite. *Tectonophysics* 48:1–27.
Banerdt, W. B., and Golombek, M. P. 1988. Deformational models of rifting and folding on Venus. *J. Geophys. Res.* 93:4759–4772.
Banerdt, W. B., and Sammis, C. G. 1992. Small-scale fracture patterns on the volcanic plains of Venus. *J. Geophys. Res.* 97:16149–16166.
Basilevsky, A. T., et al. 1986. Styles of tectonic deformations on Venus: Analysis of Veneras 15 and 16 data. *J. Geophys. Res.* 91:D399–D412.
Bechtel, T. D., Forsyth, D. W., and Swain, C. J. 1987. Mechanisms of isostatic compensation in the vicinity of the East African Rift, Kenya. *Geophys. J. Roy. Astron. Soc.* 90:445–465.
Brace W. F., and Kohlstedt D. L. 1980. Limits on lithospheric stress imposed by laboratory experiments. *J. Geophys. Res.* 85:6248–6252.
Brown, C. D., and Grimm, R. E. 1995. Tectonics of Artemis Chasma: A Venusian "plate" boundary. *Icarus* 117:219–249.
Brown, C. D., and Grimm, R. E. 1996. Lithospheric rheology and flexure at Artemis Chasma, Venus. *J. Geophys. Res.* 101:12697–12708.
Burov, E. B., and Diament, M. 1995. The effective elastic thickness (T_e) of continental lithosphere: What does it really mean? *J. Geophys. Res.* 100:3905–3927.
Byerlee, J. D. 1978. Friction of rocks. *Pageoph.* 116:615–626.
Campbell, D. B., Head, J. W., Harmon J. K., and Hine, A. A. 1983. Venus: Identification of banded terrain in the mountains of Ishtar Terra. *Science* 221:644–647.
Campbell, D. B., Head, J. W., Harmon J. K., and Hine, A. A. 1984. Venus: Volcanism and rift formation in Beta Regio. *Science* 226:167–170.
Caristan, Y. 1982. The transition from high temperature creep to fracture in Maryland diabase. *J. Geophys. Res.* 87:6781–6790.
Cloetingh, S., and Wortel, R. 1986. Stress in the Indo-Australian plate. *Tectonophysics* 132:49–67.
Cloetingh, S., Wortel, R., and Vlaar, N. J. 1989. On the initiation of subduction zones. *Pageoph.* 129:7–25.
Courtney, R. C., and White, R. S. 1986. Anomalous heat flow and geoid across the Cape Verde Rise: Evidence for dynamic support from a thermal plume in the mantle. *Geophys. J. Roy. Astron. Soc.* 87:815–868.
DeBremaecker, J. C. 1977. Is the oceanic lithosphere elastic or viscous? *J. Geophys. Res.* 82:2001–2004.
Detrick, R. S., et al. 1986. Heat flow observations on the Bermuda Rise and thermal models of mid-plate swells. *J. Geophys. Res.* 91:3701–3723.
Dorman, L. M., and Lewis, B. T. R. 1970. Experimental isostasy 1. Theory of the determination of the Earth's isostatic response to a concentrated load. *J. Geophys. Res.* 70:3357–3365.
Ebinger, C. J., Bechtel, T. D., Forsyth, D. W., and Bowin, C. O. 1989. Effective elastic plate thickness beneath the East African and Afar Plateaus and dynamic compensation of the uplifts. *J. Geophys. Res.* 94:2883–2901.
Evans, S. A., Simons, M., and Solomon, S. C. 1992. Flexural analysis of uplifted rift flanks on Venus. In *International Colloquium on Venus*, LPI Contrib. No. 789, pp. 30–32 (abstract).
Ford, P. G., and Pettengill, G. H. 1992. Venus topography and kilometer-scale slopes. *J. Geophys. Res.* 97:13103–13114.
Forsyth, D. W. 1985. Subsurface loading and estimates of the flexural rigidity of continental lithosphere. *J. Geophys. Res.* 90:12623–12632.

Goetze, C. 1978. The mechanisms of creep in olivine. *Phil. Trans. Roy. Soc. London A* 288:99–119.
Goetze, C., and Evans, B. 1979. Stress and temperature in the bending lithosphere as constrained by experimental rock mechanics. *Geophys. J. Roy. Astron. Soc.* 59:463–478.
Grimm, R. E. 1994. Recent deformation rates on Venus. *J. Geophys. Res.* 99:23163–23171.
Grimm, R. E., and Solomon, S. C. 1988. Viscous relaxation of impact crater relief on Venus: Constraints on crustal thickness and thermal gradient. *J. Geophys. Res.* 93:11911–11929.
Head, J. W. 1990. Formation of mountain belts on Venus: Evidence for large-scale convergence, underthrusting, and crustal imbrication in Freyja Montes, Ishtar Terra. *Geology* 18:99–102.
Hirth, G., and Kohlstedt, D. L. 1995. Experimental constraints on the dynamics of the partially molten upper mantle: 2. Deformation in the dislocation creep regime. *J. Geophys. Res.* 100:15441–15449.
Janes, D. M., and Squyres, S. W. 1995. Viscous relaxation of topographic highs on Venus to produce coronae. *J. Geophys. Res.* 100:21137–21187.
Janes, D. M., et al. 1992. Geophysical models for the formation and evolution of coronae on Venus. *J. Geophys. Res.* 97:16055–16067.
Johnson, C., and Sandwell, D. T. 1994. Lithospheric flexure on Venus. *Geophys. J. Intl.* 119:627–647.
Karato, S., Paterson, M. S., and FitzGerald, J. D. 1986. Rheology of synthetic olivine aggregates: Influence of grain size and water. *J. Geophys. Res.* 91:8151–8176.
Karner, G. D., and Weissel, J. K. 1990. Factors controlling the location of compressional deformation of oceanic lithosphere in the central Indian Ocean. *J. Geophys. Res.* 95:19795–19810.
Kaula, W. M. 1990. Venus: A contrast in evolution to Earth. *Science* 247:1191–1196.
Kaula, W. M. 1995. Venus reconsidered. *Science* 270:1460–1464.
Kaula, W. M., et al. 1992. Styles of deformation in Ishtar Terra and their implications. *J. Geophys. Res.* 97:16085–16120.
Keep, M., and Hansen, V. L. 1994. Structural history of Maxwell Montes, Venus: Implications for Venusian mountain belt formation. *J. Geophys. Res.* 99:26015–26028.
Kiefer, W. S. 1995. Mantle plumes with temperature-dependent rheology and implications for the origin of volcanic rises on Venus. *Eos: Trans. AGU Suppl.* 76(46):F342 (abstract).
Kiefer, W. S., and Hager, B. H. 1991. A mantle plume model for the equatorial highlands of Venus. *J. Geophys. Res.* 96:20947–20966.
Kiefer, W. S., and Hager, B. H. 1992. Geoid anomalies and dynamic topography from convection in cylindrical geometry: Applications to mantle plumes on Earth and Venus. *Geophys. J. Intl.* 108:198–214.
Kirby, S. H., and Kronenberg, A. K. 1984. Deformation of clinopyroxenite: Evidence for a transition in flow mechanisms and semibrittle behavior. *J. Geophys. Res.* 89:3177–3192.
Kohlstedt, D. L., Evans B., and Mackwell, S. J. 1995. Strength of the lithosphere: Constraints imposed by laboratory experiments. *J. Geophys. Res.* 100:17587–17602.
Konopliv, A. S., and Sjogren, W. L. 1994. Venus spherical harmonic gravity model to degree and order 60. *Icarus* 112:42–54.
Kronenberg, A. K., and Shelton, G. L. 1980. Deformation microstructures in experimentally deformed Maryland diabase. *J. Struct. Geology* 2:341–353.
Lewis, B. T. R., and Dorman, L. M. 1970. Experimental isostasy 2. An isostatic

model for the U.S.A. derived from gravity and topography data *J. Geophys. Res.* 70:3367–3386.

Mackwell, S. J., Zimmerman, M. E., Kohlstedt, D. L., and Scherber, D. S. 1995. Experimental deformation of dry Columbia diabase: Implications for tectonics on Venus. In *Rock Mechanics: Proc. 35th U. S. Symposium*, eds. J. J. K. Daeman and R. A. Shultz (Brookfield, Vt.: A. A. Balkema), pp. 207–214.

Mackwell, S. J., Bystricky, M., White, J. C., Zimmerman, M. E., and Kohlstedt, D. L. 1996. High temperature deformation of dry diabase, with application to crustal deformation on Venus. *Lunar Planet. Sci. Conf.* XXVII:793–794 (abstract).

Masursky, H., et al. 1980. Pioneer Venus radar results: Geology from images and altimetry. *J. Geophys. Res.* 85:8232–8260.

McAdoo, D. C., and Sandwell, D. T. 1985. Folding of oceanic lithosphere. *J. Geophys. Res.* 90:8563–8569.

McKenzie D., and Bowin, C. 1976. The relationship between gravity and bathymetry in the Atlantic Ocean. *J. Geophys. Res.* 81:1903–1915.

McKenzie D., and Nimmo, F. 1996. Elastic thickness estimates for Venus from line of sight accelerations. *Icarus*, submitted.

McKenzie, D., et al. 1992. Features on Venus generated by plate boundary processes, *J. Geophys. Res.* 97:13533–13544.

McNutt, M. K. 1984. Lithospheric flexure and thermal anomalies. *J. Geophys. Res.* 89:11180–11194.

McNutt, M. K. 1988. Thermal and mechanical properties of the Cape Verde Rise. *J. Geophys. Res.* 93:2784–2794.

McNutt, M. K., and Shure, L. 1986. Estimating the compensation depth of the Hawaiian Swell with linear filters. *J. Geophys. Res.* 91:13915–13923.

Melosh, H. J. 1978. Dynamic support of outer rise topography. *Geophys. Res. Lett.* 5:321–324.

Moore, W., Schubert, G., and Sandwell, D. T. 1992. Flexural models of trench/outer rise topography of coronae on Venus with axisymmetric spherical shell thin elastic plates. In *International Colloquium on Venus*, LPI Contrib. No. 789, pp. 72–73 (abstract).

Moresi, L., and Parsons, B., 1995. Interpreting gravity, geoid, and topography for convection with temperature dependent viscosity: Application to surface features on Venus. *J. Geophys. Res.* 100:21155–21171.

Moresi, L. N., and Solomatov V. S. 1995. Numerical investigation of 2D convection with extremely large viscosity variations, *Phys. Fluids* 7:2154–2162.

Morgan, P., and Phillips, R. J. 1983. Hot spot heat transfer: Its application to Venus and implications to Venus and Earth. *J. Geophys. Res.* 88:8305–8317.

Mueller, S., and Phillips, R. J. 1991. On the initiation of subduction. *J. Geophys. Res.* 96:651–665.

Mueller, S., and Phillips, R. J. 1995. On the reliability of lithospheric constraints derived from models of outer-rise flexure. *Geophys. J. Intl.* 123:887–902.

Namiki, N., and Solomon, S. C. 1994. Impact crater densities on volcanoes and coronae on Venus: Implications for volcanic resurfacing. *Science* 265:929–933.

Neumann, G. A., and Zuber, M. T. 1995. A continuum approach to the development of normal faults. In *Rock Mechanics: Proc. 35th U. S. Symposium*, eds. J. J. K. Daeman and R. A. Shultz (Brookfield, Vt.: A. A. Balkema), pp. 191–198.

Phillips, R. J. 1986. A mechanism for tectonic deformation on Venus. *Geophys. Res. Lett.* 13:1141–1144.

Phillips, R. J. 1990. Convection-driven tectonics on Venus. *J. Geophys. Res.* 95:1301–1316.

Phillips, R. J. 1994. Estimating lithospheric properties at Atla Regio, Venus. *Icarus* 112:147–170.

Phillips, R. J., and Hansen, V. L. 1994. Tectonic and magmatic evolution of Venus. *Ann. Rev. Earth Planet. Sci.* 22:597–654.
Phillips, R. J., and Malin, M. C. 1983. The interior of Venus and tectonic implications. In *Venus*, eds. D. M. Hunten, L. Colin, T. M. Donahue and V. I. Moroz (Tucson: Univ. of Arizona Press), pp. 159–214.
Press, W. H., Teukolsky, S. A., Vetterling, W. T. and Flannery, B. P. 1992. *Numerical Recipes in FORTRAN* (Cambridge: Cambridge Univ. Press).
Price, M., and Suppe, J. 1994. Mean age of rifting and volcanism on Venus deduced from impact crater densities. *Nature* 372:756–759.
Rosenblatt, P., Pinet, P. C., and Thouvenot, E. 1994. Comparative hypsometric analysis of Earth and Venus. *Geophys. Res. Lett.* 21:465–468.
Sandwell, D. T., and Schubert, G. 1992a. Flexural ridges, trenches, and outer rises around coronae on Venus. *J. Geophys. Res.* 97:16069–16083.
Sandwell, D. T., and Schubert, G. 1992b. Evidence for retrograde lithospheric subduction on Venus. *Science* 257:766–770.
Saunders, R. S., et al. 1992. Magellan mission summary. *J. Geophys. Res.* 97:13067–13090.
Schatz, J. F., and Simmons, G. 1972. Thermal conductivities of Earth materials at high temperatures. *J. Geophys. Res.* 77:6966–6983.
Schubert, G., and Sandwell, D. T. 1995. A global survey of possible subduction sites on Venus. *Icarus* 117:173–196.
Schubert, G., Moore, W. B., and Sandwell, D. T. 1994. Gravity over coronae and chasmata on Venus. *Icarus* 112:130–146.
Senske, D. A., Head, J. W., Stofan, E. R., and Campbell, D. B. 1991. Geology and structure of Beta Regio, Venus: Results from Arecibo radar imaging. *Geophys. Res Lett.* 18:1159–1162.
Shelton, G., and Tullis, J. 1981. Experimental flow laws for crustal rocks. *Eos: Trans. AGU* 62:396.
Simons, M. 1996. Localization of Gravity and Topography: Constraints on the Tectonics and Mantle Dynamics of Earth and Venus. Ph.D. Thesis, Massachusetts Inst. of Technology.
Sleep, N. H. 1990. Hotspots and mantle plumes: Some phenomenology. *J. Geophys. Res.* 95:6175–6736.
Sleep, N. H. 1994. Lithospheric thinning by midplate mantle plumes and the thermal history of hot plume material ponded at sublithospheric depths. *J. Geophys. Res.* 99:9327–9343.
Smrekar, S. E. 1994. Evidence for active hotspots on Venus from analysis of Magellan gravity data. *Icarus* 112:2–26.
Smrekar, S., and Parmentier, E. M. 1996. The interaction of mantle plumes with surface thermal and chemical boundary layers: Applications to hotspots on Venus. *J. Geophys. Res.* 101:5397–5410.
Smrekar, S., and Phillips, R. J. 1988. Gravity-driven deformation of the crust on Venus. *Geophys. Res. Lett.* 15:693–696.
Solomatov, V. S. 1993. Parameterization of temperature-and stress-dependent viscosity convection and the thermal evolution of Venus. In *Flow and Creep in the Solar System: Observations, Modeling and Theory*, eds. D. B. Stone and S. K. Runcorn (Dordrecht: Kluwer), pp. 131–145.
Solomatov, V. S. 1995. Scaling of temperature- and stress-dependent viscosity convection, *Phys. Fluids* 7:266–274.
Solomatov, V. S., and Moresi, L. N. 1996. Stagnant lid convection on Venus. *J. Geophys. Res.* 101:4737–4753.
Solomon S. C., and Head, J. W. 1982. Mechanisms for lithospheric heat transport on Venus: Implications for tectonic style and volcanism. *J. Geophys. Res.* 87:9236.

Solomon, S. C., and Head, J. W, 1984. Venus banded terrain: Tectonic models for band formation and their relationship to lithospheric thermal structure. *J. Geophys. Res.* 89:6885–6897.

Solomon, S. C., and Head, J. W. 1990. Lithospheric flexure beneath the Freyja Montes foredeep, Venus: Constraints on lithospheric thermal gradient and heat flow. *Geophys. Res. Lett.* 17:1393–1396.

Solomon, S. C., et al. 1992. Venus tectonics: An overview of Magellan observations. *J. Geophys. Res.* 97:13199–13255.

Squyres, S. W., et al. 1992. Plains tectonism on Venus: The deformation belts of Lavinia Planitia. *J. Geophys. Res.* 97:13579–13599.

Stofan, E. R., et al. 1989. Geology of a rift zone on Venus: Beta Regio and Devana Chasma. *Geol. Soc. Amer. Bull.* 101:143–156.

Suppe, J. and Conners, C. 1992. Critical-taper wedge mechanics of fold-and-thrust belts on Venus: Initial results from Magellan. *J. Geophys. Res.* 97:13545–13561.

Surkov, Yu. A., et al. 1984. New data on the composition, structure and properties of Venus rock obtained by Venera 13 and 14. *J. Geophys. Res.* 89:B393–B402.

Surkov, Yu. A., et al. 1983. Determination of the elemental composition of rocks on Venus by Venera 13 and 14. *J. Geophys. Res.* 88:A481–A493.

Surkov, Yu. A., et al. 1986. Venus rock composition at the Vega 2 landing site. *J. Geophys. Res.* 91:E215–E218.

Turcotte, D. L. 1995. How does Venus lose heat? *J. Geophys. Res.* 100:16931–16940.

Turcotte, D. L., and Schubert, G. 1982. *Geodynamics: Applications of Continuum Physics to Geological Problems* (New York: J. Wiley).

Von Herzen, R. P., Cordery, M. J., Detrick, R. S., and Fang, C. 1989. Heat flow and the thermal origin of hot spot swells: The Hawaiian Swell revisited. *J. Geophys. Res.* 94:13783–13799.

Vorder Bruegge, R. W., and Fletcher, R. C. 1990. A model for the shape of overthrust zones on Venus. *Lunar Planet. Sci. Conf.* XXI:1278–1279 (abstract).

Wessel, P. 1992. Thermal stresses and the bimodal distribution of elastic thickness estimates of the oceanic lithosphere. *J. Geophys. Res.* 97:14177–14193.

Williams, C. A., Conners, C., Dahlen, F. A., Price E. J., and Suppe, J. 1994. Effect of the brittle-ductile transition on the topography of compressive mountain belts on Earth and Venus. *J. Geophys. Res.* 99:19947–19974.

Zuber, M. T. 1987. Constraints on the lithospheric structure of Venus from mechanical models and tectonic surface features. *J. Geophys. Res.* 92:E541–E551.

Zuber, M. T. 1994. Rheology, tectonics and the structure of the Venus lithosphere. *Lunar Planet. Sci. Conf.* XXV:1575–1576 (abstract).

Zuber, M. T., and Parmentier, E. M. 1990. On the relationship between isostatic elevation and the wavelengths of tectonic surface features on Venus. *Icarus* 85:290–308.

Zuber, M. T., and Parmentier, E. M. 1996. Formation of fold-and-thrust belts on Venus by thick-skinned deformation. *Nature* 377:704–707.

Zuber, M. T., Bechtel, T. D., and Forsyth, D. W. 1989. Effective elastic thicknesses of the lithosphere and mechanisms of isostatic compensation in Australia. *J. Geophys. Res.* 94:9353–9367.

THE CRUST OF VENUS

ROBERT E. GRIMM
Arizona State University

and

PAUL C. HESS
Brown University

Surface geochemical measurements and the morphology of volcanic structures imaged from orbit indicate that the surface of Venus is dominated by basalt. Even unusual geochemical signatures and volcanic landforms suggestive of silicic lavas are consistent with mafic to intermediate bulk compositions. Gravity data indicate that the mean thickness of the crust is 20 to 50 km; 30 km is the value preferred here. The planet's smooth hypsometry and the dominance of deep isostatic compensation argue that the crust has a comparatively uniform thickness, departing significantly only in the tessera highlands. The crust at present occupies 1 to 2% of the total volume of Venus. By contrast, Earth's present fractional volume of basaltic crust is ~0.5%, but may integrate to ~10% over the last 4 Gyr due to crustal recycling. In spite of the recent evidence from the Magellan mission that Venus underwent strong volcanotectonic resurfacing at or before several hundred million years ago, it is still unknown whether such resurfacing included significant crustal recycling. Recycling is required if the bulk of the present crust was generated by horizontal accretion analogous to seafloor spreading. However, full development of horizontal crustal accretion is difficult to envision because of the depth and temperature of melting necessary to generate basaltic crust so much thicker than terrestrial oceanic crust. In addition, the bulk composition of the crust must be picritic to komatiitic and a late global flooding event must conceal features diagnostic of earlier lithospheric recycling. Alternatively, a dry, stiff upper mantle may have allowed only limited horizontal movements, and vertical accretion dominated crustal generation. A tholeiitic to picritic crust could be formed at numerous regional centers of lithospheric extension during episodes of global activity. During periods of global quiescence, alkali basalts would be a significant melt component produced by minor stretching of a thick lithosphere or plume melting beneath a rigid lid. Recycling of crust is possible but not necessary in this model. The present crust is probably too thin to be globally limited by densification and detachment due to the granulite/eclogite phase transitions, but such controls, especially locally, cannot be ruled out. Efficient distribution of magma or subsequent solid-state creep is necessary to inhibit large permanent lateral variations in crustal thickness under vertical accretion. Contemporary crustal production is 0.01 to 0.4 km^{3}yr^{-1}, less than or equal to the intraplate crustal formation rate of Earth.

I. INTRODUCTION

The crust of a moon or planet is a thin outer layer that differs markedly in composition from the interior and from the composition of the primordial solar nebula (Taylor 1989). It further concentrates sizeable fractions of the planetary budget of incompatible elements. The crust is a chemical boundary layer formed by internal differentiation under a variety of melting processes and is distinct from the mechanical boundary layer or lithosphere. Because of the complexity of silicate phase relations, a range of crustal mineralogies is produced on rocky worlds. No such distinctions have yet been demonstrated for the icy moons.

Due to its similar bulk properties with Earth, Venus has been expected to show evidence for a similar thermal history and perhaps a comparable record of crustal differentiation. A variety of experimental and theoretical efforts based on Pioneer Venus (PV), Venera, and Vega orbital and surface data attempted to constrain the composition, volume, and differentiation history of the crust. Geochemical measurements made by a series of landers between 1973 and 1986 were largely consistent with a basaltic surface (Barsukov 1992). In the first Venus volume, McGill et al. (1982) explored differences in the rock cycle and hypsograms between Earth and Venus. They further determined that plate boundaries should be visible on Venus if present. Grimm and Solomon (1988) used geodynamic methods to determine the crustal thickness and inferred that the total crustal volumes of Earth and Venus were similar. They concluded that either Venus generated crust slowly or that some form of crustal recycling had occurred. Hess and Head (1990) discussed the compositions expected from a variety of crustal differentiation mechanisms. They found that, in spite of the hypothesized lack of water in the crust and mantle, no major Earth-like igneous suites could be excluded, although some systematic differences should be present. Head (1990) summarized many of these pre-Magellan ideas.

Perhaps the most startling result from Magellan is the implication from the cratering record that Venus underwent rapid resurfacing at or before several hundred million years ago, which then sharply declined (Schaber et al. 1992). Geodynamic models seek to understand the internal processes responsible for this abrupt transition, and to determine whether such changes are monotonic or episodic (Arkani-Hamed et al. 1993; Turcotte 1993; Parmentier and Hess 1992; chapter by Schubert et al.). Although Magellan gathered no primary geochemical data, these results have profoundly influenced geophysical concepts on the mechanisms, timing, and rate of crustal formation. In particular, the abandonment of a strictly uniformitarian perspective means that contemporary crustal production and recycling processes may be different, or least operate at a much reduced rate, than in the past.

In this chapter, we examine the crust of Venus in light of the Magellan results. As just mentioned, these data have a greater geophysical than geochemical impact, so our emphasis follows. We first review the mechanisms of crustal formation, both for terrestrial planets in general and Venus in particular. We re-assess the different kinds of evidence for crustal compositions on Venus. We analyze several methods for determining the crustal thickness (and volume) of the planet, and compare to the Earth. Lastly, we present three alternative hypotheses for crustal production and recycling on Venus in the past and estimate the recent rate of crustal formation.

II. CRUSTAL FORMATION ON THE TERRESTRIAL PLANETS

Taylor (1989) classified planetary crusts into three broad categories. A primary crust is formed by nearly complete melting and then subsequent crystal fractionation of the upper mantle during the late stages of accretion. For the Moon, the primary crust consists mostly of anorthosite which, according to the magma ocean hypothesis, formed by the flotation of calcic plagioclase over a global magma system (see Warren and Kallemeyn 1993, for an update). It is generally agreed that the magma system must have been large, as the Al_2O_3 content residing in the primary crust probably represents >70% of the global inventory (Wood 1986; O'Neill 1991). But how likely is it that the Earth or Venus once developed a stable primary crust? Taylor (1989) pointed out several reasons why Earth probably lacked an early anorthositic crust. The most relevant of these to the ensuing discussion of Venus is that plagioclase is not a liquidus phase in basalt melts above ~1 GPa. These pressures correspond to depths of <40 km on Venus and Earth but up to 200 km on the Moon. Thus, the depth range on Venus and Earth over which to form stable anorthosite crust by crystal flotation is very limited. Primary anorthositic crust has been confirmed only for the Moon. Such a crust is likely to have formed on Mercury and has been inferred from spectroscopic measurements (Jeanloz et al. 1995). At present there is no evidence for primary crust on Mars; spectroscopic observations suggest Mars is dominated by oxidized basalts and other alteration products (Soderblom 1992).

Secondary crusts are the result of regional, rather than global, melting and differentiation of the upper mantle. For silicate planets, this crust is overwhelmingly basaltic for the simple reason that basalt is the product of pressure-release partial melting of peridotite, which is expected to dominate the bulk composition of the mantles of the terrestrial planets. The exact composition is a function of the temperature at which melting is initiated and the degree of partial melting. Pressure-release melting is a consequence of mantle upwelling and/or lithospheric stretching. Terrestrial seafloor spreading forms prodigious quantities of secondary crust. Furthermore, seismic constraints on the continental crust (Christensen and Mooney 1995) and mass-balance calculations for island arcs (DeBari and Sleep 1991) indicate that upper intermediate-to-silicic rocks are in both cases underlain by mafic rocks, which could represent additional formation of secondary crust. Sample analyses have confirmed the presence of basaltic secondary crusts on the Moon,

Venus, and Mars; the morphology of plains regions on Mercury suggests that low-viscosity basalts may be present there also (Spudis and Guest 1988).

Tertiary crusts are formed by remelting pre-existing crust. Most terrestrial tertiary crust is formed at convergent margins where melting of the peridotite mantle wedge is initiated by the release of H_2O by dehydration of metamorphosed oceanic basaltic crust (Hess 1989). Underplating and intrusion of continental crust on Earth by basaltic magma has also been invoked as a mechanism to initiate magmatism in a variety of tectonic settings (Hildreth and Moorbath 1988). The remelting of subducted basaltic crust itself is a more contentious issue, although generation of trondhjemite-tonalite-dacite melts from relatively young subducted plates seems to be a reasonable hypothesis (Drummond and Defant 1990). Nonetheless, the key to the formation of andesitic tertiary crust on Earth is the ability to recycle H_2O-rich volatiles into the mantle. The existence of large volumes of tertiary crust on Venus and, indeed on other planets, will most likely depend on the same variables.

III. EXPECTED CRUSTAL COMPOSITIONS ON VENUS

Hess and Head (1990) gave a comprehensive summary of the petrogenetic constraints on crustal formation on Venus. Here we briefly summarize and expand upon this work, reorganizing the discussion to follow more closely Taylor's framework outlined above.

Venus does not display any elevated regions of distinctly higher crater density analogous to the lunar highlands that may indicate primary crust preserved since just after planetary formation. Furthermore, the young average impact crater retention age—comparable to the average age of the surface of the Earth—indicates a vigorous geological history. Therefore it is probable that primary crust has been completely recycled sometime over the last 4 Gyr. Alternatively, stable primary crust may never have formed on Venus, as Taylor argues for the Earth. However, we note that the present evidence cannot completely exclude pockets of primary crust; direct geochemical measurements are sparse, and crater retention ages do not correspond to crystallization ages where the surface has been strongly reworked, as in the highly deformed "complex ridged terrain" or tessera. We are not suggesting that tessera actually are primary crust, merely that crater retention ages are not indicators of crustal composition.

Bulk density and cosmogenic arguments predict that the mantle composition of Venus should be broadly similar to the Earth's (BVSP 1981) and hence the mantle should be peridotitic. Basalts are therefore expected as primary regional mantle melts forming secondary crust. A key parameter controlling the composition and amount of melt is the mantle potential temperature T_p, defined as the temperature that adiabatically upwelling material would reach at the planetary surface if it did not melt (see, e.g., McKenzie and Bickle 1988). The potential temperature is constant along an adiabat, whereas the actual temperature varies. Depending on T_p and the amount of upwelling, the full range of composition from quartz tholeiites ($T_p < 1400$ K) to komatiites ($T_p \sim 1800$ K) can be produced. The presence of volatiles, however, has significant effects on the melting regime. The role of H_2O is not only to lower the solidus temperature, but also to make near-solidus melts less olivine-normative compared to the dry system. The presence of CO_2 has the most dramatic effect at pressures greater than about 2 GPa. Near-solidus melts are more olivine-normative in a CO_2 atmosphere compared to the dry system. At very small degrees of melting, strongly SiO_2-undersaturated melts ranging in composition from alkali basalts to kimberlites and even carbonatites can be produced. Kaula (1990,1993) has argued that the lack of a moon-forming impact on Venus may have left it with an appreciable volatile content in the lower mantle.

Recycling of secondary basaltic crust can produce a range of tertiary crusts depending on the depth of melting and the volatile content. For shallow underthrusting and remelting of basalt under dry reducing conditions, ferrobasalt melts are expected for modest degrees of melting (pressures less than about 1.0 to 1.5 GPa and melt fractions greater than about 0.1). Silica-rich melts could result at smaller degrees of melting but these would be difficult to erupt. Such melts can be obtained at larger degrees of melting of basalt under more oxidizing conditions, however. At greater depths where basalt converts to eclogite ($P > 1.5$ GPa for most basalt compositions), more silica-rich melts such as trondjhemites, andesites, or even dacites are formed (Johnston 1986). Melting tholeiite basalt with H_2O or even a CO_2-H_2O fluid causes the solidus to be depressed by several hundreds of degrees and generates near-solidus SiO_2-enriched melts. Baker and Eggler (1987), for example, produced dacite melt with CO_2-H_2O fluids containing less than 25% H_2O at pressures less than 0.8 GPa. The effects of melting basalt in a pure CO_2 atmosphere are not well constrained, however. See Hess and Head (1990) for a more extensive discussion on the origin of silicic magmas.

IV. INFERENCES OF CRUSTAL COMPOSITION

There are two broad classes of evidence that have been used as indicators of crustal compositions on Venus. The first consists of direct measurements of elemental or oxide abundances from the Venera landers. The second class consists of indirect inferences from geomorphology, particularly the characteristics of lava flows. We will review these approaches in detail, as well as other suggestions derived from gravity and topography.

A. Lander Measurements

The most direct evidence for the compositions of crustal rocks on Venus has been provided by seven spacecraft with geochemical instruments that were successfully landed on Venus by the former Soviet Union between 1972 and 1986 (Vinogradov et al. 1973; Surkov 1977,1986; Surkov et al. 1984). Gamma-ray spectrometers (GRS) were carried on five spacecraft and X-ray flourescence (XRF) instruments were aboard three; the Vega 2 station had both. For convenience, the landing sites are shown in Fig. 1 and the principal results are summarized in Table I.

A major outstanding question for the surface materials sampled by the landers is what part of the rock cycle they represent. The K/U and K/Th ratios of the Venus samples are similar to those of terrestrial volcanics, but match the SNC meteorites and Martian samples even more closely. The K/U ratios are distinct, however, from those of lunar basalts and eucrites. A significant observation is that the K/U ratios are within a factor of 3 of each other. This relative constancy is important because near-surface metamorphic processes apparently have not metasomatically altered the relative abundances of K,U and Th by more than a factor of 3. If this conclusion is correct then the K/U ratios reflect the original bulk composition of the rocks. The ratios imply that the volatile-refractory element inventory of Venus is more like that of Mars and Earth than the Moon or the eucrite parent bodies. Evidence presented below supports this important conclusion.

While the major oxide compositions of the Venusian rocks are broadly basaltic (see below), stronger conclusions are difficult to make with confidence. Three serious uncertainties exist. First, the stated analytical errors are very large, and the disagreement in potassium abundance between the two instruments on Vega 2 suggests additional uncertainty (Table I). An important parameter describing basaltic rocks is the magnesium number, $Mg^* = MgO/(MgO+FeO)$. Typically, $Mg^* = 0.6$ to 0.7 for primitive terrestrial basalts. At face value, Mg^* from the Venera and Vega measurements is certainly consistent with terrestrial basalts, implying that Mg^* for the Venusian mantle is similar to terrestrial peridotite. But to calculate Mg^*, we need ferric as well as ferrous iron, i.e., the Fe_2O_3 concentration in addition to FeO. Given that the basaltic rocks are chemically weathered, the FeO/Fe_2O_3 ratio, even if known, would not constrain Mg^* of the Venusian mantle. We conclude, therefore, that such critical measures of igneous petrogenesis are not adequately constrained by the lander measurements.

A second uncertainty is that the samples analyzed are regolith and not intact bedrock. There is no guarantee that these two are equivalent in composition. Venera 9, 10, 13 and 14 panaramas show subhorizontally layered, slab-like rocks. Materials at the Venera 13 and 14 landing sites have low bearing capacity, low density (1.15–1.5 g cm^{-3}) and high porosity ($\sim 50\%$) (Zolotov and Volkov 1992). Evidence for physical weathering is widespread in the panoramas and includes the presence of detrital slabs, fractures, and fine-grained soils between the slabs and dust deposits (Zolotov and Volkov 1992). The distributed occurrence of wind streaks on Venus indicates that aeolian processes operate widely (Greeley et al. 1992). Basilevsky et al. (1985) concluded that the Venera 13 and 14 surface panoramas show volcaniclastic metasediments (cf., Garvin et al. 1984). Hence the samples may not represent pristine igneous rocks but may have been adulterated to various degrees by physical processes.

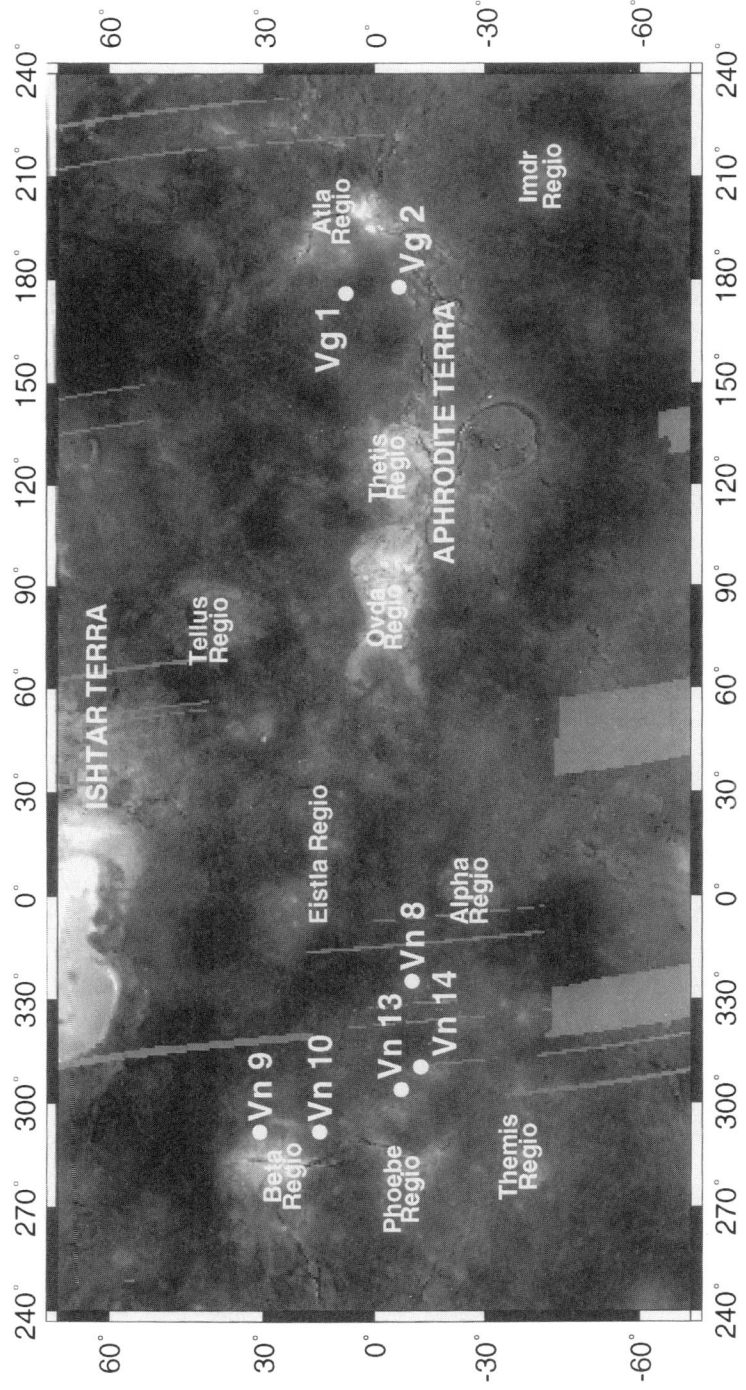

Figure 1. Venera (Vn) and Vega (Vg) landing sites with geochemical measurements. Magellan topographic base in Mercator projection. Nomenclature restricted to place names used in this chapter.

TABLE I
Surface Geochemical Measurements[a]

Constituent	Venera 8	Venera 9	Venera 10	Venera 13	Venera 14	Vega 1	Vega 2
SiO_2	—	—	—	45.1±3.0	48.7±3.6	—	45.6±3.2
TiO_2	—	—	—	1.59±0.45	1.25±0.41	—	0.2±10.1
Al_2O_3	—	—	—	15.8±3.0	17.9±2.6	—	16.0±1.8
FeO	—	—	—	9.3±2.2	8.8±1.8	—	7.74±1.1
MnO	—	—	—	0.2±0.1	0.16±0.08	—	0.14±0.12
MgO	—	—	—	11.4±6.2	8.1±3.3	—	11.5±3.7
CaO	—	—	—	7.1±0.96	10.3±1.2	—	7.5±0.7
K_2O	4.8±1.5[b]	0.6±0.1[b]	0.4±0.2[b]	4.0±0.63	0.2±0.07	0.54±0.27[b]	0.1±0.08
							0.48±0.24[b]
S	—	—	—	0.65±0.4	0.35±0.31	—	1.9±0.6
Cl	—	—	—	<0.3	<0.4	—	<0.3
U (ppm)	2.22±0.7	0.6±0.2	0.5±0.3	—	—	0.64±0.47	0.68±0.38
Th (ppm)	6.5±0.2	3.7±0.4	0.7±0.3	—	—	1.5±1.2	2.0±1.0

[a] wt. %. Table after Barsukov (1992).
[b] K converted to K_2O (K_2O wt. % = 1.21 K wt. %).

The last and perhaps most serious concern is that these sediments are not fragments of igneous rocks but are instead composed of metamorphic rocks. Thermodynamic calculations predict that basalts will react with the hot and caustic atmosphere to produce new secondary minerals (Volkov et al. 1986). Terrestrially weathered and metamorphosed basic to ultramafic rocks not only suffer a change in mineralogy but also in chemical composition. Note that the SO_3 content of the Venusian regolith is elevated, proving that sulfur has been added. What other elements have been added, mobilized, or subtracted?

It is well known from the comparison of metamorphosed and fresh terrestrial basalts and komatiites (Beswick 1982) that elements such as Ca, the alkalis, and Si are mobilized during even modest metamorphic processes. In contrast, elements such as Al or Ti appear to be unaffected by such metamorphic overprints. It is curious, therefore, that the basalts of Venus have low CaO contents and low CaO/Al_2O_3 ratios when compared to their terrestrial counterparts. There is a strong possibility that CaO has been lost relative to Al_2O_3 in the Venusian regolith. If CaO has been redistributed, other elements such as K and Si may have suffered a similar fate. Attempts to use these elements (and others) to constrain petrogenesis may be fraught with incalculable error. But do such processes operate on Venus?

The metasomatic effects observed in terrestrial komatiites and basalts are largely due to the transport of geochemical materials by H_2O-rich fluids. In contrast, Venus rocks in equilibrium with the atmosphere should contain relatively anhydrous minerals given the low H_2O contents ($\sim$100 ppm) and high temperatures of the atmosphere (Zolotov and Volkov 1992). But this conclusion does not necessarily apply to the interiors of volcanic flows or ash deposits. Bubble formation of H_2O-rich fluids is hindered on Venus because of the high atmospheric pressures. The solubility of H_2O in basalt melts is roughly 1% by weight at 100 bar, the atmospheric pressure on Venus. This constraint means that basalt melts with smaller amounts of dissolved H_2O (as is true for the vast majority of ocean floor and ocean island basalts on Earth) will not exsolve H_2O. The H_2O in these melts will be trapped in glass (quenched melt) or will exsolve from these melts after extended crystallization brings the melt to the H_2O solubility limit. Under the ambient conditions, the rock should crystallize to greenschist facies mineralogy (chlorite, epidote, micas). During this re-crystallization or during late stage exsolution of volatiles, H_2O-rich fluid is capable of migrating through the rock or ash column, producing just those metasomatic effects alluded to above. It should also be noted that Arvidson et al. (1992) have inferred from radar signatures that over 1 m of erosion or degradation has occurred in the oldest rocks. If correct, the rocks analyzed by Venera and Vega may have originally been in the interior of basalts or ash flows and would have recently been exposed to the atmosphere. Such rocks may not be in equilibrium with the atmosphere.

Nonetheless, a number of compositional inferences have been made assuming that the landers did sample relatively unaltered volcanic rocks representative of the crust. Whether these conclusions can stand close scrutiny depends on one's prejudices concerning the alteration of the analyzed samples.

Venera 14 is the closest analog to a tholeiite basalt notwithstanding that an analysis for Na_2O is not available. If we assume that Na_2O is 2 wt%, then the analysis sums to about 98%, a reasonable total given the analytical uncertainties. At first glance, the composition of the rock analyzed by Vega 2 appears anomalous in view of its low silica, TiO_2 and K_2O contents. Indeed, McKenzie et al. (1992a) suggest that the composition resembles more a cumulus gabbro than a tholeiite basalt. But if we add about 2 wt% Na_2O to the analysis, subtract the sulfur content and renormalize the analysis to 100%, the SiO_2 content becomes roughly 50% in line with typical tholeiite basalt (Barsukov 1992). The resulting CaO (8.3%) and TiO_2 contents remain on the low side compared to most terrestrial basalts, however. A plutonic origin for this rock cannot be discounted but considering the analytical difficulties this interpretation is highly unconstrained. The analysis of the composition of the rock at Venera 13 is characterized by comparatively low silica but very high K_2O contents. The absence of Na_2O in the analysis is critical to an understanding of the petrogenesis of this rock. The compositions of highly potassic mafic volcanic rocks on terrestrial continental crust typically have $K_2O/Na_2O>1$ (by weight), whereas ocean-island SiO_2-undersaturated basalts have $K_2O/Na_2O<1$. K_2O/TiO_2 ratios of 2.5 are more characteristic of ultrapotassic rocks in continental environments, however (see also McKenzie et al. 1992a). In detail the high MgO and Al_2O_3 contents and the low CaO contents are very unusual and do not characterize the vast majority of terrestrial rocks.

High K was also found by the Venera 8 GRS, as well as elevated U and Th. Nikolayeva (1990) inferred that this material is intermediate in silica content and stressed that this differs sharply from the Venera 13 sample, in spite of a similar potassium concentration. However, we see no reason to reject the hypothesis that the Venera 8 and Venera 13 sites are similar, as there were no U or Th measurements for the latter, and so it is possible that all three large-ion lithophile elements are partitioned similarly. In either case, an early suggestion that Venera 8 sampled granitic rocks (Vinogradov et al. 1973) now seems unlikely. We refer the reader to Barsukov (1992) and the references therein for more detailed speculations about petrogenesis.

B. Volcanic Morphology

It has long been recognized that the morphology of many volcanic structures on Venus, such as shield volcanoes and very long flows, indicate low-viscosity, effusive eruptions which are in turn characteristic of basalts (Masursky et al. 1980; Barsukov et al. 1986; Head et al. 1992). Such features account for the vast majority of cataloged volcanic landforms on the planet. However, two uncommon classes of structures suggest both extremely low-viscosity lavas and high-viscosity lavas: the canali for the former and the pancake domes and festoon flows for the latter.

The canali are very long (up to several thousand km), narrow, meandering channels found at many locations on Venus. Baker et al. (1992) considered three alternative hypotheses for the operating fluid (neglecting water): komatiitic lava, carbonatite lava, or liquid sulfur. They did not favor komatiite on the basis that its high eruption temperature would also promote rapid cooling and freezing of the lava, before it could produce a long channel. However, they did acknowledge that a combination of low viscosity, high discharge, warm atmospheric conditions, and thermal insulation by a solidified roof (now collapsed) might achieve low cooling rates. Gregg and Greeley (1993) have since shown that such insulating processes might act efficiently on Venus, obviating the problems of rapid cooling. Indeed, lunar rilles, which are comparable to if not as long as the Venusian canali, were formed from mare basalts, lavas which were very hot and fluid. Carbonatites are predicted as near-solidus melts for deep primary melts in the presence of CO_2 (Hess and Head 1990). Sulfur is also reasonably abundant in the Venera XRF analyses. However, Kaula (1993) has pointed out that it may be difficult to concentrate a large amount of low-melting point material. D. McKenzie (personal communication) has suggested that there is no reason to exclude fractionated tholeiite basalt as a channel-forming fluid. Such lava is a common product of differentiation and can travel large distances on Earth as surface flows or in dikes. But to our knowledge, these basalts rarely if ever cut channels into the continental crust, certainly not on the scale observed on Venus and the Moon. In contrast, ultramafic lavas on the Moon, the low and high TiO_2 mare basalts, produce very long rilles. The simplest hypothesis is that silicate magmas were responsible for the channels and these were magnesian in composition with affinities to picrite or komatiite magmas.

The pancake domes—formally called steep-sided domes—have been linked as analogs to terrestrial rhyolite domes and are presently considered the strongest morphological evidence for evolved lavas (Pavri et al. 1992). McKenzie et al. (1992*b*) mathematically modeled these structures as single effusions of a Newtonian fluid and concluded that the inferred high viscosity could be satisfied only by very silica-rich lavas such as dry rhyolite. There are several potential objections to this conclusion, however. Fink et al. (1993) compared the morphology of the steep-sided domes to several terrestrial analogs, and found that high aspect ratios are a feature of episodic rather than continuous effusions. Such episodic eruptions are, however, characteristic of silicic lavas. More recently, Bridges (1995) reported on large, steep-sided basaltic domes on the terrestrial seafloor, which suggests that such morphology can be dominated by eruption conditions rather than composition. Gregg and Fink (1995), scaling from laboratory analogs, concluded that episodic basaltic eruptions or continuous andesitic eruptions could produce the observed morphology. Magma crystallinity is another factor that could influence viscosity (Sakimoto and Zuber 1995). Lastly, Ford (1994) found that the radar-scattering properties of steep-sided domes were similar to those of the surrounding plains.

Even more rare than the pancake domes are the large, steep-sided, ridged, radar-bright "festoon" flows. Moore et al. (1992) mapped one such structure and reported that the distribution and morphology of its deposits suggested local magmatic differentiation. Using a variety of published scaling relations, they concluded that the breadth and thickness of the flow lobes and the large separations of regularly spaced ridges indicated an intermediate to high silica content. Subsequent laboratory analog experiments on flow morphology (Gregg and Fink 1995) and on fold spacing (Gregg et al. 1995) jointly point to an intermediate composition.

In summary, the morphology of the overwhelming majority of volcanic features on Venus indicate basaltic compositions. Although both exotic volatile fluids and highly silicic lavas have been invoked to explain the few unusual features, the possibility that all are silicate melts of ultramafic to intermediate composition cannot be excluded.

C. Linked Geochemistry and Morphology

Basilevsky et al. (1992) and Weitz and Basilevsky (1993) have examined the Venera and Vega landing sites in Magellan images. They found steep-sided domes at the Venera 8 and 13 sites, but not at the others. Although it is unlikely that these spacecraft actually landed on domes, this result does strongly suggest that non-tholeiitic compositions are spatially linked to volcanic structures indicative of somewhat evolved lavas. On a larger scale, however, there is no evident geological correlation in the locations of these sites: Veneras 13 and 14 landed within $\sim$900 km of each other on the eastern flank of Phoebe Regio. This a very heterogeneous region, composed of several tessera blocks separated by plains, with later uplift, rifting, and volcanism associated with the Beta-Themis-Aphrodite disturbances. In contrast, Venera 8, $\sim$2600 km farther to the east from Venera 14, lies in regionally unremarkable plains.

D. Other Approaches

One significant difference between Venus and Earth is their fractional distribution of elevation, or hypsometric curve; Venus' is smooth and unimodal, whereas Earth's is bimodal (Masursky et al. 1980). The split character of Earth's hypsogram is due of course to the distinct mean elevations of continents and ocean basins. However, this does not reflect so strongly the difference in density (composition) between oceanic and continental crust, but instead simply the thicker crust of the continents. The thickness of the oceanic crust is a near-constant 6 to 7 km as a consequence of the mechanics of decompression melting under the same repeating conditions at the mid-ocean ridge (see below). In contrast, the thickness of the continental crust is determined by the feedback between erosion and isostasy; erosion rapidly planes down continental elevations to near the erosional base level (sea level). Isostatic balance with the adjacent oceanic water and rock column (including potential differences between the thickness of oceanic and continental lithospheres) determines the mean thickness of the continents. This thickness has changed only slowly with time, as evinced by the near-constant continental freeboard (Wise 1974; Reymer and Schubert 1984). In the absence of oceans and an erosional base level, Earth's hypsogram might be unimodal (McGill et al. 1982). Alternatively, continental crust could still be magmatically and tectonically thickened and remain high-standing due to the preferential deformation of weaker silicic crust.

Gravity measurements are used to constrain planetary internal density differences, which might be used to search for light, felsic crust. Unfortunately, these differences are too small to detect in the presence of Airy isostatic compensation: the density difference between mafic and felsic crust on Earth (<0.3 Mg m^{-3}) is significantly smaller than the contrast between crust and mantle (>0.5 Mg m^{-3}). Given comparable vertical intervals representative of either the mean crustal thickness or boundary relief on the moho, the latter is likely to dominate any crustal gravity signal. The most likely scenario in which to detect a difference between mafic and felsic crust would be if differences in relief were entirely due to the density contrast between these materials, i.e., an intracrustal Pratt compensation. Such a mechanism is unlikely however; Airy isostatic adjustment is rapid on Earth ($\sim 10^4$ yr), and is controlled mainly by the viscosity of the upper mantle (Cathles 1975). Intracrustal isostatic adjustments could dominate only if the mantle could not respond over the lifetime of the topography, of order 10^8 yr for Venus. This would require an upper-mantle viscosity $\sim 10^4$ times higher than Earth's, which is still unlikely even if the mantle of Venus is drier (Kiefer and Hager 1991). Much of the long-wavelength topography on Venus appears to be dynamically supported (see, e.g., Phillips and Malin 1983), in which case gravity interpretations are even less sensitive to surficial density variations.

E. Summary

Detailed petrological inferences about the crust of Venus are hampered by concerns remaining about the accuracy of the Venera/Vega geochemical measurements and whether or not the surface materials are indeed relatively unaltered volcanic rocks. Nonetheless, mafic to intermediate silica contents are consistent with all geochemical and geomorphological data. This includes the two landing sites that apparently sampled more alkaline rocks and have nearby unusual volcanic structures, and for which no clear regional geological associations are evident. Any notably non-basaltic composition suggests that tertiary differentiation has occurred and been preserved. Although unusual volcanic structures appear to be volumetrically minor, it may be significant that over one-quarter of the landing sites suggest at least some tertiary crustal differentiation.

V. ESTIMATES OF CRUSTAL THICKNESS AND VOLUME

Although geophysical techniques have not been successful in determining crustal density (and hence composition), they have provided reasonable estimates of crustal thickness. This single measurement in turn places strong constraints on models of crustal generation. We therefore review these techniques in detail. There are three approaches: theoretical considerations based on the phase relations, geodynamic models, and gravity-topography relationships.

A. Theoretical Limits

The thickness of a gabbroic (basaltic) crust is limited by $P-T$ stability conditions (Fig. 2). Partially molten crust is not likely to be stable for hundreds of millions of years, and so equilibrium crustal thickness–temperature combinations to the right of the solidus may be ruled out. In addition, gabbroic crust will transform to denser garnet granulite and then eclogite with increasing depth and become gravitationally unstable. If the density increases approximately linearly from ~ 2.9 Mg m^{-3} to ~ 3.5 Mg m^{-3} across the granulite field (Namiki and Solomon 1993), then the density of the crust will not exceed the density of the mantle (3.3–3.4 Mg m^{-3}) until it is fairly near the eclogite boundary. These considerations limit the thickness of the crust to somewhere in the range 30 to 100 km, depending on temperature. Note, however, that these boundaries also depend on bulk composition. The selected curves (Ito and Kennedy, 1971) show the largest separation of the granulite and eclogite phase boundaries among the measurements summarized by Ringwood (1975). The full range of compositional variations (shaded region in Fig. 2) then allows the eclogite phase transition to lie practically anywhere between 10 and 100 km. Nonetheless, note also that these boundaries to gabbro stability are substantially shallower than those originally proposed by Anderson (1980).

Two thermal conditions serve as additional constraints on eclogite stability. First, this phase transition is unlikely to limit crustal thickness unless the temperature gradient is ≤ 10 to 15 K km^{-1}, as the crust would melt first. Second, the eclogite phase transition may be kinetically inhibited at low temperatures (Namiki and Solomon 1993; Jull and Arkani-Hamed 1995). The temperature T^* for which this restriction applies over the present surface age of the planet may be roughly quantified by a simple dimensional analysis. The expression for ionic diffusion (Namiki and Solomon 1993, their Eq. 2) has a characteristic time constant $\tau \sim r^2/D$, where r is the grain radius and D is the diffusion constant. Minimum and maximum values for $D(T)$ are given in Eq. (3) of Namiki and Solomon (1993) and a range of $r = 1$ to 10 mm was also chosen by these authors. A lower bound on T^* for $\tau = 400$ Myr follows by minimizing r and maximizing D, for which we find $T^* = 970$ K. The maximum blocking temperature that will achieve the same diffusion time (given by the maximum r and minimum D) is 1380 K. This simple dimensional analysis is in good agreement with Fig. 4 of Namiki and Solomon (1993). For comparison, Parmentier and Hess (1992) chose $T^* = 870$ to 1070 K.

Although the determination of reaction rate has been approached theoretically, quantitative experimental investigation has not followed, primarily because of the complexity of the reaction mechanisms and because of the slow rate of reaction. Observations of natural terrestrial eclogite settings may pro-

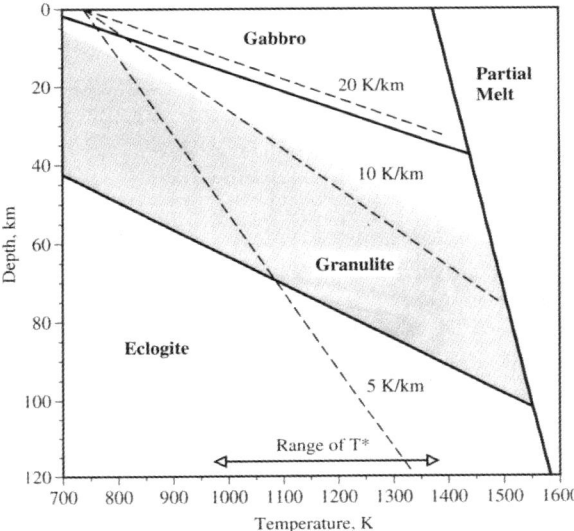

Figure 2. Phase boundaries for gabbro, garnet granulite, eclogite, and the gabbro solidus on Venus (Ito and Kennedy 1971). Shaded region shows range in granulite-eclogite phase boundary due to differences in composition (Ringwood 1975). Representative equilibrium conductive geotherms of 5, 10, and 20 K km^{-1} are shown as dashed lines. Granulite and eclogite phase transitions are inhibited at temperatures below T^* over the age of the present Venus surface, ~400 Myr. The range in T^* and in the positions of the phase boundaries do not allow controls on crustal thickness due to phase changes to be specified with confidence.

vide some insight (Herzog and Hess 1996). Terrestrial eclogite samples yield information about conditions of equilibration through equilibrium mineral compositions, and information about reaction mechanisms can be gleaned if textural features indicating partial reaction can be identified. Eclogite-forming reactions in natural rocks equilibrated at temperatures greater than 750 C usually do not show evidence for fluid infiltration (Rubie 1990). This may indicate lack of free fluid during reaction, or perhaps, expulsion of fluid during the early stages of reaction. Some field evidence shows that eclogite-bearing shear zones alternate on a scale of meters with the dry, unsheared granulites (Austrheim and Griffin 1985). These features demonstrate that granulite production is kinetically easier than eclogite because hydrous fluids, which facilitate rapid reaction, may be removed from the rock during metamorphism of basalt to granulite.

The question dealing with the gabbro-granulite-eclogite transition on Venus may therefore hinge on the question: how much H_2O is contained in the crust? We have argued earlier that volcanic rocks will immobilize H_2O within greenschist facies minerals provided that the H_2O content of the erupted liquids are less than about 1%. This constraint does not apply to plutonic melts where the solubility of H_2O increases with pressure. Because the intrusion/extrusion ratio is certainly greater than one and probably closer to about 10 on Venus (Crisp 1984), it follows that the crust is largely of a plutonic origin. If true, the H_2O content of the crust, particularly the lower crust, should be higher than implied by the relatively dry atmosphere. Studies of ^{40}Ar and H_2O loss are at present very model dependent, and no firm conclusions have been reached as to whether the interior of Venus is very dry (Namiki and Solomon 1995) or has simply not efficiently degassed (Kaula 1990,1993). As the crust passes into the granulite and eclogite regime a series of dehydration reactions ensue, releasing H_2O, enhancing the kinetics of solid state processes, and hence lowering T^*. Whether H_2O content of the Venus crust is sufficient to accelerate the kinetics of the granulite-eclogite transition is unknown. Other lines of reasoning including the apparent strength of the Venus crust point to a relatively dry crust (see below).

This discussion illustrates the strong dependence of crustal stability upon temperature and composition, particularly the H_2O content. If the granulite/eclogite boundary is deep and/or T^* is large, then the eclogite transition might never occur, regardless of crustal thickness. On the other hand, a shallow phase boundary and a low T^* could globally limit the crustal thickness to as little as 10 to 20 km. Intermediate values of all parameters lead to limiting crustal thicknesses $\sim$50 km for $dT/dz > 5$ to 10 K km^{-1}. We conclude that there is insufficient information at present to place a firm limit on crustal thickness due to phase transitions.

B. Geodynamic Models

These techniques seek to understand the forces that deform a planet's surface and the physical parameters that control the rate and style of deformation. In general, these models are responsive to the thickness of the elastic or mechanical lithosphere and not the crust. However, the "yield envelope" concept clearly shows the dependence of lithospheric yield strength upon crustal thickness and temperature gradient (Kohlstedt et al. 1995; chapter by Phillips et al.). Frictional sliding governs the yield strength of the uppermost lithosphere and is independent of rock composition. However, thermally activated creep dominates at greater depths and is a strong function of both temperature and composition. As the crust (even basaltic) contains more weak felsic minerals than the strong ultramafic mantle, increasing the thickness of the crust allows more rapid deformation at a fixed stress, or conversely, a smaller "creep strength" at a specified strain rate. Similarly, creep rates increase and rocks weaken under higher temperature gradients.

Two classes of geodynamic models have been used to estimate the thickness of the crust of Venus. In the first, Grimm and Solomon (1988) modeled the isostatic rebound of impact craters as either an elastic or viscous process. For both models, the topography of craters observed by Venera 15 and 16 constrained the crustal thickness H to <10 to 20 km for thermal gradients $dT/dz = 10$ to 20 K km^{-1}. In essence, the crustal rocks appeared to be too weak to support topography over geological time scales, and so the models required that strong mantle rocks be placed relatively close to the surface.

The second kind of approach was taken by Zuber (1987) and Banerdt and Golombek (1988). These workers used elastic, viscous, and plastic rheologies to model the characteristic spacings of tectonic structures. In general, the spacing of structures increases with lithospheric thickness. However, in some cases two superimposed wavelengths are apparent, which indicates two scales of strength in the lithosphere. The shorter spacing is controlled by the strong upper portion of the crust. The longer wavelength is thought to arise from the strong upper mantle or the suppression of intermediate wavelengths in the weak lower crust, depending on the model assumptions. In either case, a weak lower crust must separate strong upper crust and strong upper mantle. For such regions of multiple wavelengths, Zuber found $H = 5$ to 30 km and $dT/dz < 25$ K km^{-1}, whereas Banerdt and Golombek (1988) constrained H to 5 to 15 km and dT/dz to 10 to 15 K km^{-1}.

The crater rebound and tectonic wavelength techniques yielded remarkably similar results, thus mutually reinforcing each other. The conclusion that the crust of Venus was thin, <30 km, was a sharp departure from previous ideas, especially Anderson's (1980) suggestion that the lower uncompressed density of Venus than Earth could be explained by a very thick crust (>100 km).

Two recent developments call for revision of these results. First of course are the Magellan images and altimetry, which can be used to revise the crater isostatic rebound approach by searching for evidence of crater floor uplift and associated fracturing. In addition, various tectonic structures can be studied to update deformation wavelength studies. Second, Mackwell et al. (1995) have performed creep measurements on dry diabase, which may provide a better analog to the basaltic crust of Venus than previous flow laws. They found that the strength of crustal rocks under anhydrous conditions can approach that of the ultramafic mantle. In this case the strength stratification becomes less distinct, and so crustal thickness constraints cannot be derived (Hillgren and Melosh 1995; Brown and Grimm 1996a). The overall thickness of the mechanical lithosphere still depends on dT/dz, however. Future revision of the tectonic modeling will determine if the crust and mantle have been sufficiently decoupled during deformation to produce multiple wavelengths of structures and hence constrain crustal thickness. This will depend on both careful inspection of Magellan images to determine if multiple wavelengths are indeed contemporaneous, and the application of the anhydrous crustal flow law.

C. Gravity

The improved orbital geometry and tracking of Magellan have allowed gravity studies to replace geodynamic models as the principal way to infer crustal thickness on Venus. The main problem is to isolate the part of the gravity signal due to crustal thickness variations from other contributions to compensation.

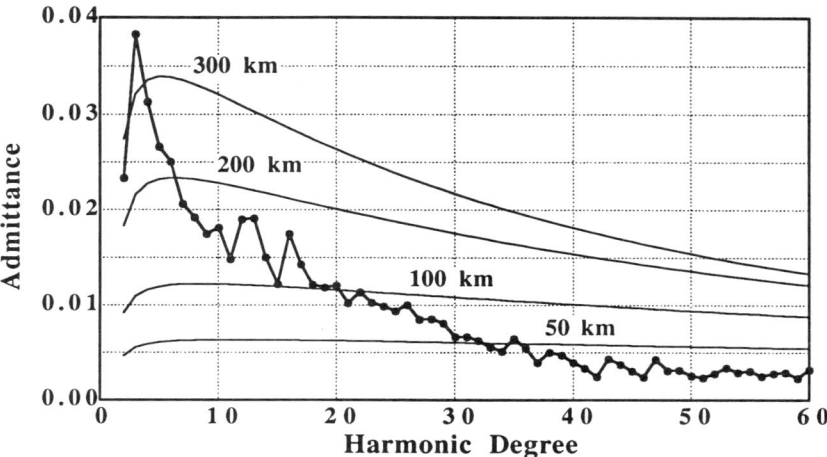

Figure 3. Evidence from gravity for a 20 to 50 km mean crustal thickness. Global spectral admittance from Konopliv and Sjogren (1994), showing short-wavelength asymptote ∼25 to 50 km.

Pioneer Venus showed that long-wavelength gravity and topography are highly correlated on Venus, which indicates an apparent depth of isostatic compensation (ADC) of 100 to 200 km (see, e.g., Kiefer et al. 1986). Phillips and Malin (1983) established that classical Airy isostasy could not support topography as any variations in crustal thickness at such great depths would quickly viscously relax and be eliminated (and the topography along with it). We note that this analysis has not been rigorously upheld in the post-Magellan view of Venus, in which the crust is thought to be significantly stronger (see above) and in which thermal gradients may be as low as 5 K km^{-1} in a thermal lithosphere >200 km thick (Sandwell and Schubert 1992; Turcotte 1993). However, simple calculations using the effective viscosity of the crust (Mackwell et al. 1995) and the relaxation time of compensated topography (Solomon et al. 1982) suggest that creep will still be activated at depths comparable to the ADC.

Rejection of crustal isostasy for the long-wavelength topography of Venus led Phillips and Malin (1983) to propose that compensation was due to thermal convection within Venus, and so the "hot-spot" model was born. Smrekar and Phillips (1991) further showed that the individual domal volcanic highlands are the most conspicuous class of deeply compensated features, which further reinforced the hot-spot model and concepts of thermal/dynamic compensation. However, these works and others that examined the role of varying both crustal and thermal lithospheric thickness in order to match gravity and topography (see, e.g., Banerdt 1986; Williams and Gaddis 1991; Grimm and Phillips 1992; Herrick and Phillips 1992) did not explicitly solve for mean crustal thickness, but specified it *a priori*.

Recent gravity models have exploited the higher resolution of the Magellan data to show reliably the spatial and/or spectral variations in compensation depth and in some cases to solve for crustal thickness explicitly. Konopliv and Sjogren (1994) synthesized Magellan line-of-sight spacecraft accelerations into a spherical harmonic model for vertical gravity. Using a spectral admittance technique (the ratio of gravity to topography in the wavenumber domain), they calculated the ADC as a function of wavelength (Fig. 3). The compensation depth is large at long wavelengths, showing the contribution of mantle density variations to the support of topography. However, the short-wavelength ADC has an asymptote at $\sim$25 km, which Konopliv and Sjogren interpreted as the average crustal thickness. They noted that improvements to the gravity models would boost the power at short wavelengths and increase the apparent crustal thickness; however, they claimed that it would be unlikely to exceed 50 km. This result should be interpreted with some additional caution, as mantle thermal and/or compositional compensation could still have an effect across the spectrum. We therefore interpret their 50-km value as an upper limit to the crustal thickness.

Simons et al. (1994) also computed spectral admittances from the spherical harmonic model of Konopliv and Sjogren, but used moving windows to highlight regional variations in compensation. They found short-wavelength ADCs of $\sim$50 km for several plateau highlands, confirming and extending earlier work by Smrekar and Phillips (1991). Kucinskas and Turcotte (1994) used a spatial admittance technique to compute a zero-elevation crustal thickness of 50 km for Ovda and 60 km for Thetis. We believe that the result for Thetis is influenced by deeper, noncrustal compensation (see below) and so take the 50-km value to be representative of the crustal thickness calculated by this method. McKenzie (1994) roughly estimated the crustal thickness at Beta Regio to be $\sim$30 km using the residual gravity from a simple space-domain filter and an estimate of upper crustal extension from radar imagery. However, this hybrid technique requires additional assumptions about extensional geometry, particularly that strain throughout the crust is vertically homogeneous.

Phillips (1994) performed a comprehensive study of the compensation of Atla Regio, a type example of a domal volcanic highland. He solved for the thicknesses of the crust, elastic lithosphere, and thermal lithosphere by minimizing misfits in the wavenumber domain. Phillips' best-fitting mean crustal thickness for the short-wavelength portion of the signal was 30 ± 13 km. The crustal thickness was unbounded for the long-wavelength band. Subsequent work using an improved gravity model (chapter by Phillips et al.) has failed to confirm these results, however.

Grimm (1994) focused on the gravity of four plateau-shaped highland regions. Compensation was partitioned into an upper layer for the crust and a lower layer representing mantle density variations (thermal or chemical). Alpha, Tellus, and Ovda Regiones were found to be well fit in the spatial domain by essentially complete compensation in a crust of thickness 35 to 45 km (Fig. 4). A deep component to the compensation of the fourth highland, Thetis

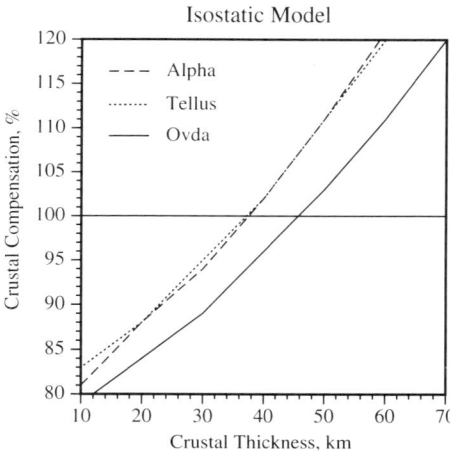

Figure 4. Evidence from gravity for a 20 to 50 km mean crustal thickness. Best-fitting crustal thickness beneath three plateau highlands from Grimm (1994). Isostatic correction to surrounding plains yields ~30 km for all three regions.

Regio, did not allow a unique solution for crustal thickness. Sensitivity studies examining the size of the study area vs the size of the highland showed that such results are biased away from the mean H of the study area and towards the crustal thickness existing below the highland. It is then a simple matter to correct the derived crustal thickness to that in the surrounding plains under conditions of Airy isostasy. For ~1 km of average long-wavelength (>1000 km) relief at Alpha and Tellus Regiones and ~2 km of such topography at Ovda Regio, the mean crustal thickness in the surrounding plains for all three regions is ~30 km. Errors and uncertainties allow a range in mean thickness of at most 20 to 40 km.

The full range of mean crustal thicknesses from these studies is 20 to 50 km. We prefer the 30-km best fit from Grimm (1994) because this study explicitly attempted to isolate crustal compensation for several widely spaced tessera, which even competing geological models suggest were formed by thickened crust (Bindschadler et al. 1992; Phillips et al. 1991). Also note that all of the quoted values are estimates of the mean, and not the actual range in crustal thickness: the latter is likely to be comparatively small over most of Venus, as is evident from the planet's smooth hypsogram, considering that long-wavelength, deeply compensated topography contributes substantially. However, several kilometers of relief is associated with some crustally compensated plateau highlands, which require a local thickening of the crust by ~20 to 40 km, i.e., roughly doubling it. The comparatively small area of crustal plateaus implies that the crustal thickness in the plains is close to the global average.

D. Summary: Venusian and Terrestrial Crustal Volumes

Improved gravity data from Magellan have allowed more robust estimates of the mean crustal thickness (~30 km) over previous geodynamic models. It is unlikely that this thickness is globally limited by the gabbro-granulite-eclogite phase transitions. However, the range of compositions and activation temperatures for the phase transition do not allow this hypothesis to be ruled out.

TABLE II
Comparison of Crustal Volumes of Venus and Earth[a]

	Venus		Earth	
	Secondary Crust	Tertiary Crust	Secondary Crust	Tertiary Crust
Present volume (% of planetary volume)	1–2	≪1 ?	0.5[b]	0.5[c]
Recent production rate ($km^3 \; yr^{-1}$)	<4[d]	?	20–30[e]	1–6[f]
Time-integrated volume (% of planetary volume)	?[g]	?	~10[h]	~1–2[i]

[a] Rounded to one significant figure.
[b] 61% of surface area occupied by oceanic crust (Turcotte and Schubert 1982) with mean thickness of 6 to 7 km (Reid and Jackson 1981), lower 15 km of 40-km thick continental crust (Christensen and Mooney 1995).
[c] Upper 25 km of continental crust.
[d] Head et al. (1992); Bullock et al. (1993); Strom et al. (1994).
[e] Mean seafloor spreading rate 3.45 $km^2 \; yr^{-1}$ (Parsons 1982), 0.3 to 4 $km^3 \; yr^{-1}$ combined oceanic and continental basaltic intraplate magmatism (Reymer and Schubert 1984; Crisp 1984; White and McKenzie 1989), mafic portion of arc magmatism 0.3 to 4 $km^3 \; yr^{-1}$ (prior references; one-third to one-half of arc magmatism from DeBari and Sleep [1991]).
[f] Reymer and Schubert (1984); Crisp (1984); less mafic portion arc magmatism above.
[g] Depends on presence of recycling and recurrence interval of "catastrophes."
[h] Same rate over 4 Gyr.
[i] Present volume plus recycled sediments, 0.6 $km^3 \; yr^{-1}$ (Reymer and Schubert 1984) to 2.5±1.2 $km^3 \; yr^{-1}$ (DePaolo 1983), over 4 Gyr.

The range of estimated mean crustal thicknesses reviewed here suggests that the basaltic secondary crust of Venus occupies 1 to 2% of the total planetary volume. This is rather greater than Earth's present basaltic crustal volume, but is significantly smaller than the amount of secondary crust that Earth may have produced by seafloor spreading throughout geologic time (Table II). Earth's time-integrated tertiary crustal production has probably been limited to a few percent or less of its total volume. The uncertainties in the composition of the Venus surface from both chemical and morphological studies (discussed above) preclude a robust estimate of the amount of tertiary crust on Venus.

VI. SPECULATIONS ON CRUSTAL HISTORY

We now review some key inferences about the crust of Venus, and examine how they constrain three alternative models of crustal production and recycling. Because the crust of Venus seems to be predominantly basaltic, much of it appears to have been produced by regional decompression of the mantle. On Earth, this occurs at mid-ocean ridges and hot spots. A combination of horizontal stretching and vertical upwelling emplaces new crust. Seafloor spreading involves infinite lateral strain, and therefore may be described as an environment of horizontal crustal accretion. In contrast, lateral strains are generally modest at hot spots (unless they occur on a spreading center, like Iceland), and so these may be generally characterized as environments of vertical crustal accretion. These two environments may generate broadly similar magmas, but they carry very different implications for the global geology of Venus.

Although basaltic melts are produced by stretching and upwelling on regional scales of ~ 1000 to 2000 km on Earth (White and McKenzie 1989), the apparently random distribution and relatively pristine morphology of impact craters on Venus indicates a global volcanic flooding event (Schaber et al. 1992). This synchroneity of plains emplacement does not require that the whole crust was generated in a short time, but rather that for some time much less than the crater retention age the entire planetary surface could be reached by flows and/or intrusions from a melt source. This may have taken the form of many plumes or sites of extension. At a minimum, the required extrusive volume is $\sim 10^8$ km^3, given by the thickness required to just obliterate any prior cratering record (a few hundred meters to cover crater rims). Given that intrusions likely outweigh extrusions by an order of magnitude (Crisp 1984), the total volume is probably ~ 100 times larger than individual major continental flood basalt provinces on Earth (Hess 1989; White and McKenzie 1989). The duration of major volcanic floods on Earth is typically only several Myr, so the number of active plumes or extension sites need only be ~ 10 at any time. Alternatively, plume heads on Venus might be larger than those on Earth by a factor of 2 to 3 if the core-mantle thermal boundary layer is similarly thickened (Herrick and Phillips 1990), which would also lead to ~ 10 sites.

The comparative uniformity of crustal thickness on Venus is also remarkable. On the Moon, the excavation and isostatic adjustment of impact basins has left many regional variations in crustal thickness of several tens of km (Bratt et al. 1985; Zuber et al. 1994). The hemispheric dichotomy of Mars has preserved comparable early-formed differences in crustal thickness on an even larger scale (Balmino et al. 1982). However, the larger, more active terrestrial planets Venus and Earth have several mechanisms with which to smooth crustal thickness variations. As discussed above, the thicknesses of Earth's oceanic and continental crusts are determined by the mechanics of decompression melting and by isostatic adjustment in the presence of an erosional base level, respectively. As there are no oceans on Venus, there is no fixed erosional isostatic adjustment. Seafloor spreading is therefore the only global process by which to attain uniformity of crustal thickness on Venus. However, the large size of plains-forming lava flows and the presence of long lava channels (Head et al. 1992; Baker et al. 1992) suggests that low-viscosity lavas traveled hundreds or even thousands of km, and lineaments interpreted as dike swarms cover comparable lengths (McKenzie et al. 1992*c*; Grosfils and Head 1994). Therefore the crust of Venus could have been built up close to true hydrostatic equilibrium. Alternatively, viscous relaxation may have acted to smooth out variations in crustal thickness caused by lateral variations in vertical crustal accretion (Masursky et al. 1980). Both lateral distribution of magma by dike swarms and collapse of topography by solid-state creep are effective on Earth as well.

Lastly, the mean thickness of the crust places some constraints on crustal origins. As just mentioned in the context of terrestrial oceanic crust, the thickness of primary melt and its approximate composition as a function of mantle temperature, thickness of the mechanical lithosphere, and amount of extension is now well understood (see, e.g., McKenzie 1984; McKenzie and Bickle 1988). These ideas have been previously applied to Venus (Sotin et al. 1989; Solomon and Head 1991), but the implications of a thicker crust have not been assessed.

These observations constrain three alternative end-member models for crustal creation and destruction on Venus: (1) horizontal accretion and recycling; (2) vertical accretion with no recycling; and (3) vertical accretion and recycling.

A. Horizontal Accretion and Recycling

Lateral movements of a sufficiently large scale to dominate crustal generation would have been associated with global lithospheric recycling (note that we distinguish lithospheric recycling in general from plate tectonics in particular; other workers have called for periods of "plate tectonics" on Venus, but the specific geometric form of any such lithospheric recycling is unknown). Secondary crust would dominate the crustal production budget, as on Earth. Also by analogy with the Earth, large-scale lateral recycling may have destroyed any primary crust, if such crust ever formed. Tertiary crust may be produced at convergent margins where secondary crust is remelted and therefore should be spatially distributed in some pattern marking former convergent zones.

Arkani-Hamed et al. (1993) have suggested that the present surface of Venus records a permanent end to lithospheric recycling, when the secular decline in heat flow changed the mode of mantle convection from oscillatory to steady. Turcotte (1993) has called for episodes of lithospheric recycling separating periods of one-plate quiescence on Venus. The active mode is brought on by overcooling and thickening of the thermal boundary layer, which subsequently detaches. In fact, Turcotte's one-dimensional model does not specify what the planform of surface movements would be, whether they would be confined to regional deformations associated with sinking lithospheric diapirs and the associated warm upwellings filling them, or whether large-scale lateral movements would truly develop. Head et al. (1994) have pointed out these same issues of horizontal scale in the context of the combined chemical/thermal boundary layer detachment model of Parmentier and Hess (1992). However, this latter model relies explicitly on vertical crustal accretion and will be discussed below.

The parameters required to produce 20 to 50 km of melt at spreading centers are fairly straightforward to calculate. We closely follow McKenzie's (1984) formulation and adopt the dry solidus curve (D16) and an entropy change during melting of 360 J kg^{-1} K^{-1} given in that paper. Upwelling mantle follows an adiabat defined by a specified value of the mantle potential temperature T_p (defined earlier), and this trajectory changes where the solidus is crossed. The total melt thickness is determined by integrating the melt weight fraction as a function of depth and converting to porosity. For infinite extension at a spreading center, there are no other major parameters; at intraplate hot spots, the thickness of the lithosphere and the amount of extension are also considered (see below). Several other factors that could affect the net crustal thickness are ignored by this model, including mechanisms of melt access to the surface, the role of a rigid lid, and equilibrium vs fractional melting.

The heavy line plotted in Fig. 5 agrees with previous work and illustrates the temperature dependence of crustal thickness produced by adiabatic decompression. The nominal 6 to 7 km thick terrestrial oceanic crust is produced by infinite extension of mantle at $T_p \sim 1600$ K (McKenzie and Bickle 1988; note that small differences of some tens of Kelvins exist between the referenced study and the present work due to differences in the solidus relation and because the gravity of Venus is used throughout). If the mantle of Venus were only 100 K hotter than Earth's, as predicted from parameterized convection theory (Kaula and Phillips 1981; Stevenson et al. 1983), then ~ 15 km of melt would be produced (Sotin et al. 1989). However, a 30-km mean crustal thickness on Venus requires $T_p = 1800$ K, about 100 K hotter than expected. The error bounds on mean crustal thickness allow a range of $T_p = 1700$ to 1900 K. As a rule of thumb, the crustal thickness produced by lithospheric divergence approximately doubles with each 100 K increment in T_p.

The compositional implications of this model agree well with gross volcanic morphology and are marginally consistent with lander geochemical data. The parameterized approach to predicting melt composition of McKenzie and Bickle (1988) suggests a range in bulk MgO of $\sim 17 \pm 4\%$, i.e., picritic to komatiitic. Certainly the low-viscosity lavas inferred from volcanic morphologies are consistent with highly mafic compositions (see above). The mean MgO measured by Venera 13, 14 and Vega 2 is $10 \pm 5\%$. Given the uncertainties in the experimental methods, in possible alteration of the surface rocks, and in the theory itself, we cannot rule out the possibility of more mafic compositions in agreement with this model.

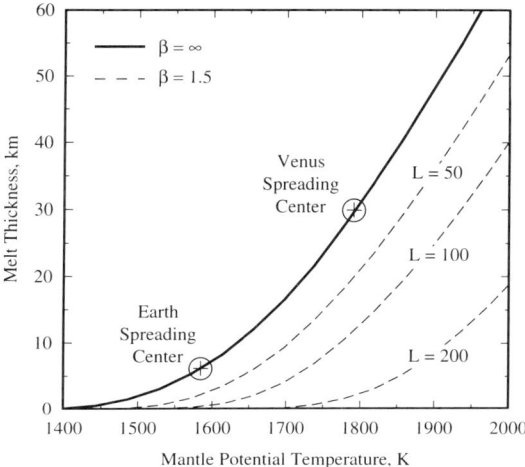

Figure 5. Basaltic crustal generation on Venus. For horizontal accretion at spreading centers ($\beta = \infty$), the best estimate of mean crustal thickness of Venus requires a mantle ~200 K hotter than Earth's. However, the depth at which melting begins must also be much larger, suggesting that spreading centers on Venus must be actively driven by hot plumes in this scenario. Alternatively, crust could be generated in multiple epsiodes of vertical accretion; melt thicknesses of several km at a time can be generated from modest thinning ($\beta = 1.5$) of mechanical lithospheres L initially 50 to 100 km thick at mantle potential temperatures only 100 K hotter than Earth.

The larger value of T_p implies not only a hotter mantle, but a greater depth at which melting begins (approximately 140 km at $T_p = 1800$ K). Such deep melting is indicative of hot jets or mantle plumes. Indeed, McKenzie and Bickle (1988) quote $H \sim 30$ km and $T_p \sim 1800$ K as characteristic of terrestrial plumes producing oceanic plateaus such as Iceland. Presumably, such plumes on Venus could be even hotter and form more melt (Fig. 6).

A potential flaw in this end-member model is that it requires *all* of the crust of Venus to be produced by plumes. All divergent margins would have to be underlain by hot upwelling sheets. As divergent margins on Earth are now understood to be the result of passive, or sink-driven, extension rather than as active upwellings (see, e.g., Forsyth and Uyeda 1975), this scenario seems unlikely for Venus.

A second problem for all models involving large-scale lateral movement is the lack of a globally interconnected network of structures indicative of even "fossil" lithospheric recycling (Solomon et al. 1992). Herrick (1994) has argued that the lock-up of lithospheric recycling would briefly raise upper mantle temperatures sufficiently to cause a global melting event and hence bury most pre-existing structures. We consider this mechanism still somewhat *ad hoc*, but there is no better explanation if large-scale lithospheric recycling or "plate tectonics" is invoked, such as in the models of Arkani-Hamed et

HORIZONTAL CRUSTAL ACCRETION AND RECYCLING

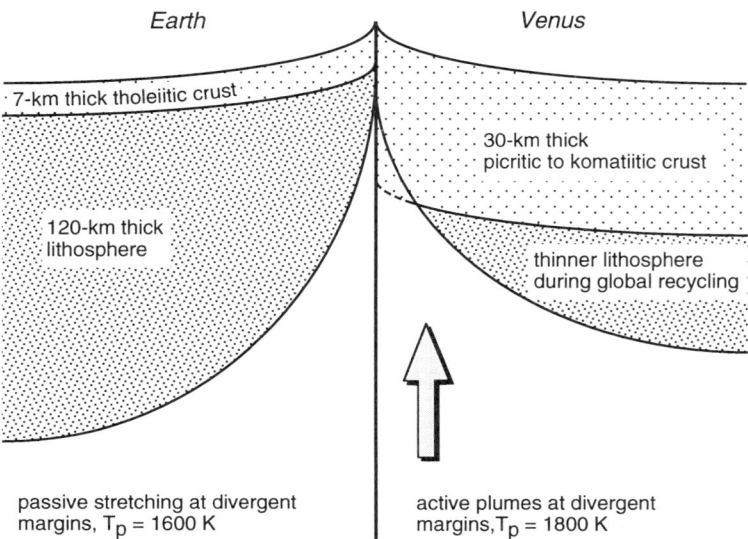

Figure 6. Crustal generation on Venus by horizontal accretion and comparison to Earth. Horizontal model implies crustal recycling.

al. (1993) and Turcotte (1993). We note that statistically significant lateral gradients in crater density diagnostic of horizontal recycling are not likely to be found if recycling was geologically rapid. In other words, if most lithosphere is recycled within 100 Myr, as on Earth, then there will not be a meaningful age gradient on a surface that is 400 Myr old.

B. Vertical Accretion With No Recycling

This end-member model represents another extreme view of the crust of Venus, one broadly more analogous to the Moon or Mars than Earth but which allows episodic global resurfacing (Fig. 7). In this scenario, hotspots formed by mantle plumes and/or modest lithospheric stretching are the agents of secondary crustal production. Tertiary crust forms by remelting of secondary crust, also at hotspots. Without substantial crustal recycling, there is nothing to eliminate a primary crust, and so we must invoke Taylor's (1989) arguments that Earth and Venus never formed sufficient quantities of this crust to stabilize and preserve it. We view doubtfully the possibility that primary crust could be buried and embedded in the secondary crust; a primary crust tens of km thick could be covered by only a veneer of basalt, which would leave Venus a thinly disguised Moon.

McKenzie and Bickle (1988) have examined the volume and composition of melt during lithospheric stretching as functions of the initial thickness of the

VERTICAL CRUSTAL ACCRETION WITHOUT RECYCLING

A. During Episodes of Global Resurfacing

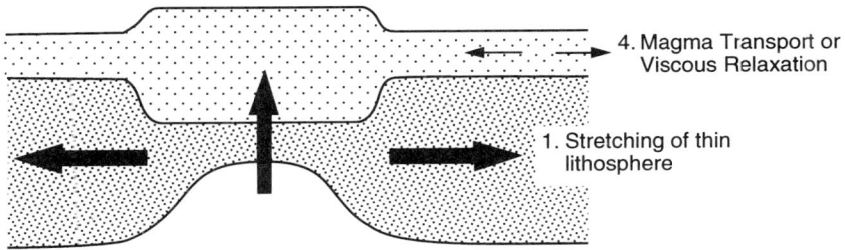

B. During Quiescent Intervals

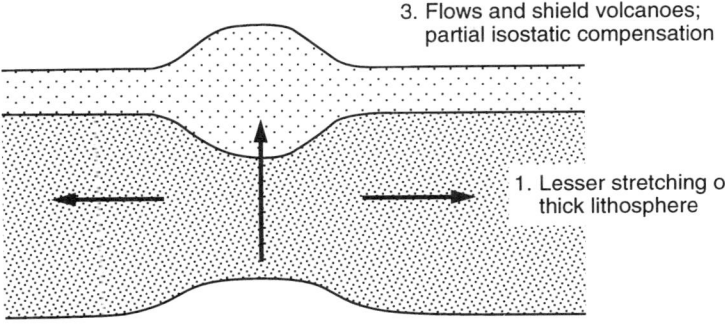

Figure 7.   Crustal generation on Venus by vertical accretion, but without recycling.

mechanical lithosphere, the ratio of final to initial surface area β (also given by strain+1), and T_p. Melt production increases with β, attaining a maximum when upwelling approaches the surface at a divergent margin ($\beta = \infty$). A thicker mechanical lithosphere results in lesser melt production.

On Earth, intraplate extension is commonly in the range $\beta = 1.2$ to 2, the larger values being typical of distributed stretching of the Basin and Range (McKenzie and Bickle 1988). Magee and Head (1995) have estimated a mean

$\beta \sim 1.2$ for several rift zones on Venus. This apparent stretching factor is a lower bound, as it has not been corrected for infilling due to rift-generated volcanism (McKenzie and Bickle 1988). If up to several km of melt were added to a crust already a few tens of km thick, then the true values of β are likely to lie in the range 1.2 to 2.0. We adopt $\beta = 1.5$ as exemplary (if not fully representative) of recent rifting on Venus.

The amount of melt generated at $\beta = 1.5$ as a function of the initial thickness of the mechanical lithosphere L is shown as a set of dashed lines in Fig. 5. The present initial thickness of the mechanical lithosphere prior to stretching could be 100 to 200 km (Turcotte 1993; Brown and Grimm 1996b). In this case, passive rifting at the nominal $T_p \sim 1700$ K for Venus will generate little or no melt. If T_p is raised 100 to 200 K due to upwelling plumes, several km or more of melt may be produced (note that the melt thicknesses over plumes are likely overestimated for the vertical recycling scenarios as there is no rigid lid in the model). In the upper limits of the melting model, episodes of global plume activity alone may be sufficient to resurface Venus. Several such events, say, one every several hundred million years, could vertically accrete the crust to its present thickness. In the lower limit, minor melt production might characterize Venus during quiescent periods. These conditions are broadly analogous to impingement of mantle plumes upon the terrestrial continental lithosphere, and therefore alkali basalts are expected to be included among the primary melts together with other primitive basaltic to picritic magmas (Fig. 7B). Alternatively, passive extension would dominate if global resurfacing was driven by lithospheric foundering (Parmentier and Hess 1992; Turcotte 1993). The "diapir" mode of sinking would be favored under a vertical crustal accretion model. Kaula (1993) has argued that the high viscosity of the upper mantle inferred from gravity data would lead to a predominance of regional- rather than global-scale mantle movements: a "distributed" convection. He further suggested that a more distributed upwelling of heat would lead to smaller melt production and a lower crustal production rate, obviating the need for crustal recycling.

If the mean thickness of the mechanical lithosphere was significantly less during global resurfacing, say $L < 100$ km, then extension events at $T_p = 1700$ K can each generate several km of melt (Fig. 7A). In comparision to terrestrial mid-ocean ridges, the hotter mantle temperature of Venus will partly offset the smaller amount of stretching, and so the comparatively shallow initiation of melting and similar net melt production should lead to tholeiitic to picritic compositions. A recurrence interval of several hundred million years would also be necessary to generate the full thickness of the Venusian crust.

Impingement of multiple plumes upon pre-existing crust, whether during the same global resurfacing event or in successive ones, may lead to remelting and the production of tertiary crust. Incompatible elements should be the first liberated by small degrees of partial melting, thus enhancing K, U, and Th in the melt products. In the vertical accretion model, tertiary crust should be randomly distributed across the planet and not confined to zones marking prior convergent margins. Crustal remelting could be fairly common, but might not proceed to high degrees of differentiation, if the lander geochemical measurements are representative.

Lastly, this model requires some mechanism to smooth out crustal thickness variations produced by a presumably arbitrary distribution of plumes during multiple crustal accretion events. As discussed above, this could be accomplished by lateral transport of magma by dikes, sills, and low-viscosity flows during crustal accretion, or subsequently by viscous relaxation. One problem with the latter idea is that there are no obviously relaxed impact craters, some of which should date to a time of high heat flow just after global resurfacing (Brown and Grimm 1996a). These objections might be removed by allowing impact craters to relax into an isostatic state, from which subsequent rebound slowed, or simply to invoke resurfacing the crater floors by lava. Nonetheless, lateral magma transport seems to be the simpler hypothesis.

C. Vertical Accretion With Recycling

This last end-member model differs from the previous model only in that large amounts of the crust are periodically stripped away and recycled into the mantle (Fig. 8). Primary crust may have been recycled. Secondary crust is generated by regional lithospheric stretching and tertiary crust by remelting of older secondary crust. Vertical recycling is accomplished either by detachment of eclogite or by lithospheric delamination.

In the model of Parmentier and Hess (1992), crustal thickness is controlled by both the depth of the eclogite phase transition and by a kinetic blocking temperature. Both the crust and buoyant melt residuum layer thicken with time, but the growing residuum layer further isolates the crust from the convecting mantle and lowers the thermal gradient, inhibiting the phase transition and/or solid-state creep in the lower crust. When the residuum itself is cool and dense enough to detach, the sudden heat pulse to the crust allows the eclogite to sink. Crustal recycling must occur in this model, as a sufficient thickness of crust must be produced so that its complementary residuum can thicken to the point of cooling and subsequent sinking. The eclogite phase boundary produces a natural horizon controlling crustal thickness variations, maintaining a comparatively uniform crustal thickness over time (Fig. 8A).

The predicted range of crustal thickness in the model of Parmentier and Hess (1992) is greater than that inferred from gravity data. In the model, crustal thickness oscillates between about 50 and 100 km, whereas the inferred thickness of the crust is only 20 to 50 km. Furthermore, the thinnest crust in the model occurs just after catastrophic overturn; the higher end of the range would be more appropriate at present. Recognizing that this preliminary model was intended only to illustrate a new concept, three alternatives exist to refine it and reconcile this disparity. First, the melting relation used was simplified: by not accounting for the increase in solidus temperature with melting, it over-estimates the amount of melt. Second, the granulite and

VERTICAL CRUSTAL ACCRETION AND RECYCLING

A. Controlled by Basalt-Granulite-Eclogite Phase Boundaries

1. Magmatic/tectonic thickening

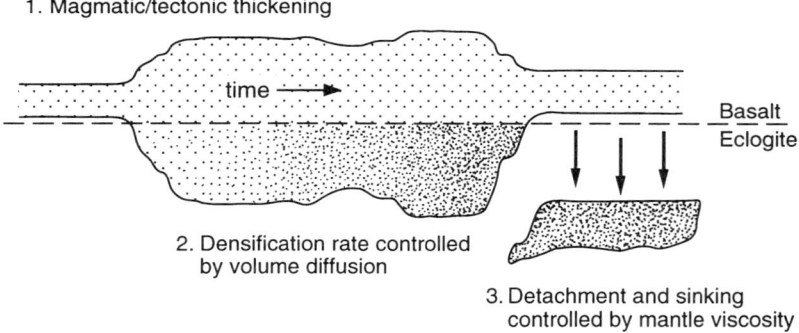

2. Densification rate controlled by volume diffusion

3. Detachment and sinking controlled by mantle viscosity

B. Controlled by Mantle Convective Stresses

1. Magmatic/tectonic thickening

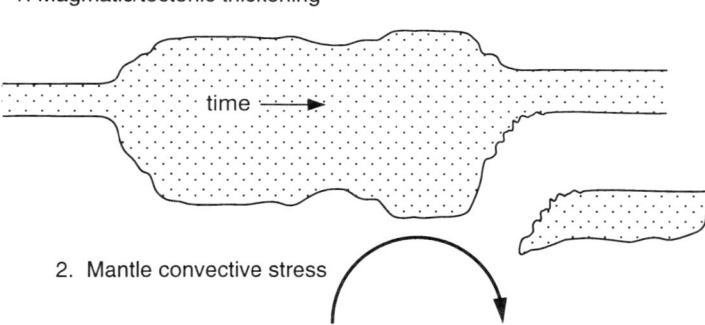

2. Mantle convective stress

Figure 8. Vertical accretion and recycling models for the crust of Venus.

eclogite phase boundaries could be much shallower than adopted in the model. As discussed above, the eclogite boundary in particular depends strongly on composition, and could lie anywhere in the range ~10 to 100 km. Third, in contrast to the basic model, the granulite and eclogite phases may have little to do with crustal recycling, which may be more strongly controlled by detachment of the chemical and thermal boundary layers.

Lithospheric delamination can include recycling of both chemical and thermal boundary layers. Detachment of dense mantle may place fresh, hot mantle in direct contact with the crust, causing a drastic reduction in viscosity and allowing rapid crustal flow. If sufficiently thinned and stretched by mantle flow, even buoyant crust may be recycled (Fig. 8B; Lenardic et al. 1993).

D. Summary

Of the three end-member models, only horizontal accretion and recycling can be ruled out, for several reasons. Hot plumes must actively drive global lithospheric spreading, in contrast to the passive or sink-driven divergent margins of Earth. The predicted crustal composition may be too mafic. Lastly, a "fossil" plate boundary network is predicted but not observed. It is also difficult to incorporate the horizontal model into a hybrid scenario, unless the dominant mode of crustal generation has changed over geologic time on Venus. For example, 15 to 20 km of crust could have been formed during a prior epoch of lithospheric recycling at $T_p = 1700$ K. During one or more episodes of more recent global resurfacing, vertical accretion could contribute the remaining 10 to 15 km of crust. Herrick's (1994) model is derived in the limit of a thick crust produced by horizontal accretion and only a thin veneer of vertically accreted volcanics. Unfortunately, such hypotheses are likely to remain untestable.

If, however, a uniformitarian perspective is adopted over the geologic history of Venus—where individual "catastrophic" episodes may be considered repetitions of a uniformitarian theme—then vertical crustal accretion is likely to have dominated. In this case, the "tie line" between the two remaining models depends only on the amount of crustal recycling. Due to the uncertainty in the position of the eclogite phase boundary, we cannot determine whether this transition is an important contributor to crustal recycling, and hence what role, if any, must be assumed by mantle convective flow in stripping the lower crust. There is no evidence at present to determine whether or not crustal recycling has occurred.

VII. CONTEMPORARY CRUSTAL FORMATION

A number of workers have attempted to estimate the recent rate of crustal formation on Venus. Pre-Magellan estimates, based on the rate of impact crater obliteration by lava infilling (Grimm and Solomon 1987) and equilibrium of SO_2 reaction with volcanic rocks (Fegley and Prinn 1989), allowed an extrusive flux up to 1 to 2 km^3 yr^{-1}. This corresponds to 5 to 20 km^3 yr^{-1} of total crustal production if 5 to 10% of the magma is erupted (Crisp 1984).

The Magellan images have provided much more stringent constraints, particularly in the pristine morphology and random distribution of impact craters. Bullock et al. (1993) found that these data could be matched in Monte Carlo resurfacing simulations only if the post-catastrophe volcanic flux was <0.4 $km^3 yr^{-1}$. The global reconnaissance mapping of Price and Suppe (1994) revealed that about 15% of the planet represented by volcanoes, flows, and coronae had a mean crater retention age of <100 Myr, or one-third to one-quarter of global value. This translates to a mean resurfacing rate of ~ 0.4 km^{2}yr^{-1} (assuming the crater retention age is one-half of the maximum age), which in turn implies an extrusive flux <0.4 km^{3}yr^{-1} for a maximum resurfacing depth of 1 km. Head et al. (1992) suggested that the flux necessary to form the estimated volumes of individual flows and edifices was <0.1 km^{3}yr^{-1}. Strom et al. (1994) favored a comparable upper limit to the extrusive flux, 0.15 km^{3}yr^{-1}, in the epoch following global resurfacing. However, these workers further suggested that preserved impact "splotches" could limit the resurfacing rate to as little as 0.01 km^{3}yr^{-1}. Using a maximum volcanic flux of 0.4 km^{3}yr^{-1}, the upper limit to the total crustal production rate is 4 km^{3}yr^{-1}; however, the net magmatic flux based on other estimates could be less than 0.5 km^{3}yr^{-1}.

These limits to the recent rate of crustal production on Venus may be compared to the Earth (Table II) and to model predictions. The rate of crustal accretion at Earth's divergent margins has been well constrained by the isochron patterns of the seafloor to 21 to 24 km^{3}yr^{-1} since the early Mesozoic (Parsons 1982). Estimates of the rates of continental, arc, and oceanic intraplate magmatism are subject to much greater uncertainty, however. The contribution of arc magmatism has been computed to be anywhere from 1.1 km^{3}yr^{-1} (Reymer and Schubert 1984) to 8.6 km^{3}yr^{-1} (Crisp 1984). As the lower one-third to one-half of the arc is mafic (DeBari and Sleep 1991), these volumes may be crudely partitioned into secondary and tertiary crust. Estimates for oceanic intraplate magmatism and continental magmatism are also variable, 0.2 to 2.4 km^{3}yr^{-1} and 0.1 to 1.6 km^{3}yr^{-1}, respectively (Crisp 1984; Reymer and Schubert 1984). White and McKenzie (1989) assessed the continental magmatic rate, which is dominated by flood basalts, at 0.4 km^{3}yr^{-1}. All told, the rate of intraplate basaltic magmatism on Earth lies in the range 0.3 to 4.0 km^{3}yr^{-1}.

This range for the intraplate flux of basaltic magma on Earth corresponds well to the estimated upper limits to recent crustal production rates on Venus; the total contemporary crustal production rate on Venus therefore is comparable to or less than the rate of intraplate magmatism on Earth. This conclusion had been suggested prior to Magellan by Phillips et al. (1991) and from the first global volcanic inventory from Magellan by Head et al. (1992). These fluxes may represent the similar control on the amount of melt generated by mantle plumes and/or lithospheric stretching in both intraplate and one-plate settings.

These estimates of the crustal production rate on Venus are significantly lower than the rates predicted by Parmentier and Hess (1992). In their model, crustal growth occurs at rates of 5 to 50 km^3 yr^{-1}, depending on the time within the cycle of episodic accretion and detachment. This implies 0.5 to 10 km^3 yr^{-1} of volcanism, which lies beyond the range accepted from resurfacing studies. As mentioned above, this model presently over estimates the melt production; a lower rate of magmatism predicted by the model might also be consistent with a thinner crust. The models of Turcotte (1993) and Arkani-Hamed et al. (1993) make no specific predictions about rates of contemporary crustal formation, but a terrestrial intraplate analog may be a likely upper bound.

VIII. CONCLUSION

Both geochemical and geomorphological data indicate that the crust of Venus is overwhelmingly basaltic, forming a volume that at present exceeds that of terrestrial oceanic crust and underplated basaltic rocks. Processes of secondary crustal generation thus appear to dominate the present surface of Venus more than known for any other planet. Some data have been used by other workers to indicate strongly non-basaltic compositions: the incompatible-element-enriched Venera 8 and 13 surface materials, the canali, the pancake domes, and the festoon flows. In contrast, we believe that all of these data are consistent with basaltic to intermediate compositions, or at least that such compositions cannot be ruled out at present. In particular, rhyolites and non-silicate fluids are nowhere required on the surface of Venus. Nonetheless, the unusual compositions and volcanic structures probably do indicate that some tertiary crustal differentiation has occurred. Unfortunately, further details on the volume and composition of tertiary crust still remain unanswered.

Analyses of Magellan gravity data give a range of 20 to 50 km for the mean crustal thickness; we suggest a value of 30 km. The planet's comparatively smooth hypsometry and the deep compensation of long-wavelength topography further indicate that this thickness is fairly uniform; significant thickening is present only beneath the large tessera plateau highlands and some minor thinning may exist beneath the domal volcanic highlands. Global control of crustal uniformity may be imposed through accretion at divergent margins, i.e., during a prior epoch of lithospheric recycling on Venus. However, the required mantle temperature is too high; even if mantle plumes lined divergent boundaries, temperature heterogeneities would probably lead to significant variations in crustal thickness. In addition, the crust would be more mafic than observed and a late *ad hoc* burst of global volcanism is necessary to completely bury structures diagnostic of lithospheric recycling.

Vertical crustal accretion is a more likely alternative. Major episodes of melting, crustal generation, and resurfacing occur in response to either upwelling mantle plumes or regional stretching, the latter linked to downwelling of an old lithosphere. Tholeiitic to picritic basalts are expected under these conditions. Heterogeneity in crustal thickness would be minimized by intrusive lateral transport by dikes and sills and extrusive flow by low-viscosity lavas. Minor addition to the crust takes place during quiescent periods at rates up to a few tenths of a cubic km per year, comparable to or less than the intraplate magmatic rate on Earth. Because of the thick lithosphere and modest stretching at present, alkali basalts are produced, analogous to terrestrial continental rifts.

We are unable to answer definitively whether significant crustal recycling has occured on Venus. This is partly due to the uncertainties in the nature of tertiary crust described above. It also follows from the view that densification due to the garnet granulite and eclogite phase transitions are the principal agents of vertical crustal recycling. Given evidence presented here that the crust may be too thin to be limited by these phase boundaries, such recycling is doubtful, or must be accomplished by other mechanisms.

Based on the inferences about its crustal history presented in this chapter, Venus has had periods of substantial surface mobility compared to one-plate planets, resulting in sufficient lithospheric stretching to produce batches of crust several kilometers or more in thickness at a time. The mechanisms of one or more such global catastrophes are the subjects of ongoing study. Nonetheless, it appears that a dry, stiff mantle did not allow the lithosphere sufficient freedom to produce an interconnected recycling network. Venus may be more of a "super Mars" and an ever more distant relative of Earth.

Acknowledgments. We thank N. Sleep and D. McKenzie for critical reviews and D. Brown for helpful comments. PCH was supported by a NASA grant. This chapter is dedicated to the memory of Robert S. Dietz, who would be shocked if he now knew whether or not seafloor spreading had occurred on Venus.

REFERENCES

Anderson, D. L. 1980. Tectonics and composition of Venus. *Geophys. Res. Lett.* 7:101–102.

Arkani-Hamed, J., Schaber, G. G., and Strom, R. G. 1993. Constraints on the thermal evolution of Venus inferred from Magellan data. *J. Geophys. Res.* 98:5309–5315.

Arvidson, R. E., et al. 1992. Surface modification of Venus as inferred from Magellan observations of plains. *J. Geophys. Res.* 97:13303–13318.

Austrheim, H., and Griffi, W. L. 1985. Shear deformation and eclogite formation within granulite-facies anorthosites of the Bergen Arcs, Western Norway. *Chem. Geol.* 50:267–281.

Baker, D. R., and Eggler, D. H. 1987. Compositions of anhydrous and hydrous melts coexisting with plagioclase, augite, and olivine or low Ca pyroxene from 1 atm to 8 kbar: Application to Aleutian volcanic center at Alka. *Amer. Mineral.* 72:12–28.

Baker, V. R., et al. 1992. Channels and valleys on Venus: Preliminary analysis of Magellan data. *J. Geophys. Res.* 97:13421–13444.

Balmino, G., Moynot, B., and Valés, N. 1982. Gravity field model of Mars in spherical harmonics up to degree and order eighteen. *J. Geophys. Res.* 87:9735–9746.

Banerdt, W. B. 1986. Support of long-wavelength loads on Venus and implications for internal structure. *J. Geophys. Res.* 91:403–419.
Banerdt, W. B., and Golombek, M. P. 1988. Deformational models of rifting and folding on Venus. *J. Geophys. Res.* 93:4759–4772.
Barsukov, V. L., et al. 1986. The geology and geomorphology of Venus as revealed by the radar images obtained by Veneras 15 and 16. *Proc. Lunar Planet. Sci. Conf.* 16, *J. Geophys. Res.* 91:378–398.
Barsukov, V. L. 1992. Venusian igneous rocks. In *Venus Geology, Geochemistry, and Geophysics: Research Results from the Soviet Union*, eds. V. L. Barsukov, A. T. Basilevsky, V. P. Volkov and V. N. Zharkov (Tucson: Univ. of Arizona Press), pp. 165–176.
Basilevsky, A. T., et al. 1985. The surface of Venus as revealed by the Venera landings: Part II. *Geol. Soc. Amer. Bull.* 96:137–144.
Basilevsky, A. T., Nikolaeva, O. V., and Weitz, C. M. 1992. Geology of the Venera 8 landing site region from Magellan data: Morphological and geochemical considerations. *J. Geophys. Res.* 97:16315–16336.
Beswick, A. E. 1982. Some geochemical aspects of alteration and genetic relations in komatiitic suites. In *Komatiites*, eds. N. T. Arndt and N. G. Nisbet (London: Allen and Unwin), pp. 211–214.
Bindschadler, D. L., Schubert, G., and Kaula, W. M. 1992. Coldspots and hotspots: Global tectonics and mantle dynamics of Venus. *J. Geophys. Res.* 97:13495–13532.
Bratt, S. R., Solomon, S. C., Head, J. W., and Thurber, C. H. 1985. The deep structure of lunar basins: Implications for basin formation and modification. *J. Geophys. Res.* 90:3049–3064.
Bridges, N. T. 1995. Submarine analogs to venusian pancake domes. *Geophys. Res. Lett.* 22:2781–2784.
Brown, C. D., and Grimm, R. E. 1996a. Floor subsidence and rebound of large Venus craters. *J. Geophys. Res.*, submitted.
Brown, C. D., and Grimm, R. E. 1996b. Lithospheric rheology and flexure at Artemis Chasma, Venus. *J. Geophys. Res.* 101:12697–12708.
BVSP (Basaltic Volcanism Study Project). 1981. *Basaltic Volcanism on the Terrestrial Planets* (New York: Pergamon Press).
Bullock, M. A., Grinspoon, D. H., and Head, J. W. 1993. Venus resurfacing rates: Constraints provided by 3-D Monte Carlo simulations. *Geophys. Res Lett.* 20:2147–2150.
Cathles, L. M. 1975. *The Viscosity of the Earth's Mantle* (Princeton, N. J.: Princeton Univ. Press).
Christensen, N. I., and Mooney, W. D. 1995. Seismic velocity structure and composition of the continental crust. *J. Geophys. Res.* 100:9761–9788.
Crisp, J. 1984. Rates of magma emplacement and volcanic output. *J. Volcanol. Geotherm. Res.* 20:177–211.
DeBari, S. M., and Sleep, N. H. 1991. High-Mg, Low-Al bulk composition of the Talkeetna island arc, Alaska: Implications for primary magmas and the nature of arc crust. *Geol. Soc. Amer. Bull.* 103:37–47.
DePaolo, D. J. 1983. The mean life of continents: Estimates of continent recycling rates from Nd and Hf isotopic data and implications for mantle structure. *Geophys. Res. Lett.* 10:705–708.
Drummond, M. S., and Defant, M. J. 1990. A model for trondhjemite-tonalite-dacite genesis and crustal growth via slab melting: Archean to modern compositions. *J. Geophys. Res.* 95::21503–21521.
Fegley, B., and Prinn, R. G. 1989. Estimation of the rate of volcanism of Venus from reaction rate measurements. *Nature* 337:55–58.

Fink. J. H., Bridges, N. T., and Grimm, R. E. 1993. Shapes of venusian "pancake" domes imply episodic emplacement and silicic composition. *Geophys. Res. Lett.* 20:261–264.
Ford, P. G. 1994. Radar scattering properties of steep-sided domes on Venus. *Icarus* 112:204–218.
Forsyth, D. W., and Uyeda, S. 1975. On the relative importance of driving forces of plate motion. *Geophys. J. Roy. Astron. Soc.* 43:163–200.
Garvin, J. B., Head, J. W., Zuber, M. T., and Helfenstein, P. 1984. Venus: The nature of the surface from Venera panoramas. *J. Geophys. Res.* 89:3381–3399.
Greeley, R., et al. 1992. Aeolian features on Venus: Preliminary Magellan results. *J. Geophys. Res.* 97:13319–13346.
Gregg, T. K. P., and Fink, J. H. 1995. Quantification of extraterrestrial lava flow effusion rates through experiments. *J. Geophys. Res.* 101:16891–16900.
Gregg, T. K. P., and Greeley, R. 1993. Formation of Venusian canali: Considerations of lava types and their thermal behaviors. *J. Geophys. Res.* 98:10873–10882.
Gregg, T. K. P., Fink, J. H., and Griffiths, R. W. 1995. Formation of multiple fold generations on lava flow surfaces: Influence of strain rate, cooling rate, and lava composition. *J. Volcanol. Geotherm. Res.*, submitted.
Grimm, R. E. 1994. The deep structure of Venusian plateau highlands. *Icarus* 112:89–103.
Grimm, R. E., and Phillips, R. J. 1992. Anatomy of a venusian hot spot: Geology, gravity, and mantle dynamics of Eistla Regio. *J. Geophys. Res.* 97:16035–16054.
Grimm, R. E., and Solomon, S. C. 1987. Limits on modes of lithospheric heat transport from impact crater density. *Geophys. Res. Lett.* 14:538–541.
Grimm, R. E., and Solomon, S. C. 1988. Viscous relaxation of impact crater relief on Venus: Constraints on crustal thickness and thermal gradient. *J. Geophys. Res.* 93:11911–11929.
Grosfils, E. B., and Head, J. W. 1994. The global distribution of giant radiating dike swarms on Venus: Implications for the global stress state. *Geophys. Res. Lett.* 21:701–704.
Hansen, V. L., and Phillips, R. J. 1995. Formation of Ishtar Terra, Venus: Surface and gravity constraints. *Geology* 23:292–296.
Head, J. W. 1990. Processes of crustal formation and evolution on Venus: An analysis of topography, hypsometry, and crustal thickness variations. *Earth Moon Planets* 50/51:25–55.
Head, J. W., et al. 1992. Venus volcanism: Classification of volcanic features and structures, associations, and global distribution from Magellan data. *J. Geophys. Res.* 97:13153–13197.
Head, J. W., Parmentier, E. M., and Hess, P. C. 1994. Venus: Vertical accretion of crust and depleted mantle and implications for geological history and processes. *Planet. Space Sci.* 42:803–811.
Herrick, R. R. 1994. Resurfacing history of Venus. *Geology* 22:703–706.
Herrick, R. R., and Phillips, R. J. 1990. Blob tectonics: A prediction for western Aphrodite Terra, Venus. *Geophys. Res. Lett.* 17:2129–2132.
Herrick, R. R., and Phillips, R. J. 1992. Geological correlations with the interior density structure of Venus. *J. Geophys. Res.* 97:16017–16034.
Herzog, S. G., and Hess, P. C. 1996. The role of water in the eclogite phase transition and crustal recycling on Venus. *Lunar Planet. Sci. Conf.* XXVII:533–534 (abstract).
Hess, P. C. 1989. *Origins of Igneous Rocks* (Cambridge, Mass.: Harvard Univ. Press).
Hess, P. C., and Head, J. W. 1990. Derivation of primary magmas and melting of crustal materials on Venus: Some preliminary petrogenetic considerations. *Earth Moon Planets* 50/51:57–80.

Hildreth, W., and Moorbath, S. 1988. Crustal contributions to arc magmatism in the Andes of central Chile. *Contrib. Mineral. Petrol.* 98:455–489.

Hillgren, V. J., and Melosh, H. J. 1995. Crater relaxation and the temperature gradient on Venus. *Icarus*, submitted.

Ito, K., and Kennedy, G. C. 1971. An experimental study of the basalt-garnet granulite-eclogite transition. In *The Structure and Physical Properties of the Earth's Crust*, ed. J. G. Heacock (Washington, D. C.: American Geophysical Union), pp. 303–314.

Jeanloz, R., Mitchell, D. L., Sprague, A. L., and de Pater, I. 1995. Evidence for a basalt-free surface on Mercury and implications for internal heat. *Science* 268:1455–1457.

Jull, M. G., and Arkani-Hamed, J. 1995. The implications of basalt in the formation and evolution of mountains on Venus. *Phys. Earth Planet. Inter.* 89:163–175.

Johnston, A. D. 1986. Anhydrous P-T relations of near-primary high alumina basalt from South Sandwich Islands. *Contrib. Mineral. Petrol.* 92:368–382.

Kaula, W. M. 1990. Venus: A contrast in evolution to the Earth. *Science* 247:1191–1196.

Kaula, W. M. 1993. Compositional evolution of Venus. In *Chemical Evolution of the Earth and Planets*, ed. E. Takahashi et al. (Washington, D. C.: American Geophysical Union).

Kaula, W. M., and Phillips, R. J. 1981. Quantitative tests for plate tectonics on Venus. *Geophys. Res. Lett.* 8:1187–1190.

Kiefer, W. S., and Hager, B. H. 1991. A mantle plume model for the equatorial highlands of Venus. *J. Geophys. Res.* 91:403–419.

Kiefer, W. S., Richards, M. A., and Hager, B. H. 1986. A dynamic model of Venus's gravity field. *Geophys. Res. Lett.* 13:14–17.

Kohlstedt, D. L., Evans, B., and Mackwell, S. J. 1995 Strength of the lithosphere: Constraints imposed by laboratory experiments. *J. Geophys. Res.*, in press.

Konopliv, A. S., and Sjogren, W. L. 1994. Venus spherical harmonic gravity model to degree and order 60. *Icarus* 112:42–54.

Kucinskas, A. B., and Turcotte, D. L. 1994. Isostatic compensation of equatorial highlands on Venus. *Icarus* 112:104–116.

Lenardic A., Kaula, W. M., and Bindschadler, D. L. 1993. A mechanism for crustal recycling on Venus. *J. Geophys. Res.* 98:18697–18705.

Mackwell, S. J., Zimmerman, M. E., Kohlstedt, D. L., and Scherber, D. S. 1995. Experimental deformation of dry Columbia diabase: Implications for tectonics on Venus. *Rock Mechanics: Proc. 35th U. S. Symposium*, eds. J. J. K. Daeman and R. A. Schultz (Brookfield, Vt.: A. A. Balkema), pp. 207–214.

Magee, K. P., and Head, J. W. 1995. The role of rifting in the generation of melt: Implications for the origin of the Lada Terra-Lavinia Plantia region of Venus. *J. Geophys. Res.* 100:1527–1532.

Masursky, H., et al. 1980. Pioneer Venus radar results: Geology from images and altimetry. *J. Geophys. Res.* 85:8232–8260.

McGill, G., et al. 1982. Topography, surface properties, and tectonic evolution. In *Venus*, eds. D. M. Hunten, L. Colin, T. M. Donahue and V. I. Moroz (Tucson: Univ. of Arizona Press), pp. 69–130.

McKenzie, D. 1984. The generation and compaction of partially molten rock. *J. Petrol.* 25:713–765.

McKenzie, D. 1994. The relationship between topography and gravity on Venus. *Icarus* 112:55–88.

McKenzie, D., and Bickle, M. J. 1988. The volume and composition of melt generated by lithospheric extension. *J. Petrol.* 29(3):625–679.

McKenzie, D., et al. 1992*a*. Features on Venus generated by plate boundary processes.

J. Geophys. Res. 97:13533–13544.
McKenzie, D., Ford, P. G., Liu, F., and Pettengill, G. H. 1992*b*. Pancakelike domes on Venus. *J. Geophys. Res.* 97:15967–15976.
McKenzie, D., McKenzie, J., and Saunders, R. S. 1992*c*. Dike emplacement on Venus and on Earth. *J. Geophys. Res.* 97:15977–15990.
Moore, H. J., Plaut, J. J., Schenk, P. M., and Head, J. W. 1992. An unusual volcano on Venus. *J. Geophys. Res.* 97:13479–13493.
Namiki, N., and Solomon, S. C. 1993. The gabbro-eclogite phase transition and the elevation of mountain belts on Venus. *J. Geophys. Res.* 98:15025–15032.
Namiki, N., and Solomon, S. C. 1995. Degassing of argon, helium, and water and the nature of crustal formation on Venus. *Lunar Planet. Sci. Conf.* XXVI:1029–1030.
Nikolayeva, O. 1990. Geochemistry of the Venera 8 material demonstrates the presence of continental crust on Venus. *Earth Moon Planet* 50/51:329–341.
O'Neill, H. St. C. 1991. The origin of the Moon and the early history of the Earth—A chemical model, Part 1: The Moon. *Geochem. Cosmochim. Acta* 55:1135–1157.
Parmentier, E. M., and Hess, P. C. 1992. Chemical differentiation of a convecting planetary interior: Consequences for a one-plate planet such as Venus. *Geophys. Res. Lett.* 19:2015–2018.
Parsons, B. 1982. Causes and consequences of the relation between area and age of the ocean floor. *J. Geophys. Res.* 87: 289–302.
Pavri, B., Head, J. W., Klose, K. B., and Wilson, L. 1992. Steep-sided domes on Venus: Characteristics, geologic setting, and eruption conditions from Magellan data. *J. Geophys. Res.* 97:13445–13478.
Phillips, R. J. 1994. Estimating lithospheric properties at Atla Regio, Venus. *Icarus* 112:147–170.
Phillips, R. J., and Malin, M. C. 1983. The interior of Venus and tectonic implications. In *Venus*, eds. D. M. Hunten, L. Colin, T. M. Donahue and V. I. Moroz (Tucson: Univ. of Arizona Press), pp. 159–214.
Phillips, R. J., Grimm, R. E., and Malin, M. C. 1991. Hot-spot evolution and the global tectonics of Venus. *Science* 252:651–658.
Price, M., and Suppe, J. 1994. Mean age of rifting and volcanism on Venus deduced from impact crater densities. *Nature* 372:756–759.
Reid, I., and Jackson, H.R. 1981. Oceanic spreading rate and crustal thickness. *Marine Geophys. Res.* 5:165–172.
Reymer, A., and Schubert, G. 1984. Phanerozoic addition rates to the continental crust and crustal growth. *Tectonics* 3:63–77.
Ringwood, A. E. 1975. *Composition and Petrology of the Earth's Mantle* (New York: McGraw-Hill).
Rubie, D. C. 1990. Role of kinetics in the formation and preservation of eclogites. In *Eclogite Facies Rocks*, ed. D. A. Carswell (Glasgow: Chapman and Hall).
Sakimoto, S. E. H., and Zuber, M. T. 1995. The spreading of variable-viscosity axisymmetric radial gravity currents: Application to the emplacement of Venusian "pancake" domes. *J. Fluid Mech.* 301:65–77.
Sandwell, D. T., and Schubert, G. 1992. Flexural ridges, trenches, and outer rises around coronae on Venus. *J. Geophys. Res.* 92:16069–16084.
Schaber, G. G., et al. 1992. Geology and distribution of impact craters on Venus: What are they telling us? *J. Geophys. Res.* 97:13257–13302.
Simons, M., Solomon, S. C., and Hager, B. H. 1994. Global variations in the geoid/topography admittance of Venus. *Science* 264:798–803.
Smrekar, S. E., and Phillips, R. J. 1991. Venusian highlands: Geoid to topography ratios and their interpretation. *Earth Planet. Sci. Lett.* 107:582–597.
Soderblom, L. 1992. The composition and mineralogy of the Martian surface from spectroscopic observations: 0.3 mm to 50 mm. In *Mars*, eds. H. H. Kieffer, B. W.

Jakosky, C. W. Snyder and M. S. Matthews (Tucson: Univ. of Arizona Press), pp. 557–593.
Solomon, S. C., and Head, J. W. 1991. Fundamental issues in the geology and geophysics of Venus. *Science* 252:297–312.
Solomon, S. C., Comer, R. P., and Head J. W. 1982. The evolution of impact basins: Viscous relaxation of topographic relief. *J. Geophys. Res.* 87:3975–3992.
Solomon, S. C., et al. 1992. Venus tectonics: An overview of Magellan observations. *J. Geophys. Res.* 97:13199–13255.
Sotin, C., Senske, D. A., Head, J. W., and Parmentier, E. M. 1989. Terrestrial spreading centers under Venus conditions: Evaluation of a crustal spreading model for Aphrodite Terra. *Earth Planet. Sci. Lett.* 95:321–333.
Spudis, P. D., and Guest, J. E. 1988. Stratigraphy and geologic history of Mercury. In *Mercury*, eds. F. Vilas, C. R. Chapman and M.S. Matthews (Tucson: Univ. of Arizona Press), pp. 118–164.
Stevenson, D. J., Spohn, T., and Schubert, G. 1983. Magnetism and thermal evolution of the terrestrial planets. *Icarus* 54:466–489.
Strom, R. G., Schaber, G. G., and Dawson, D. D. 1994. The global resurfacing of Venus. *J. Geophys. Res.* 99:10899–10926.
Surkov. Yu. A. 1977. Geochemical studies of Venus by the Venera 9 and 10 automatic interplanetary stations. *Proc. Lunar Planet. Sci. Conf.* 8:2665–2680.
Surkov, Yu. A. 1986. Venus rock composition at the Vega-2 landing site. *J. Geophys. Res.* 91:215–218.
Surkov, Yu. A., et al. 1984. New data on the composition, structure, and properties of Venus rock obtained by Venera 13 and Venera 14. *Proc. Lunar Planet. Sci. Conf.* 14. *J. Geophys. Res. Suppl.* 89:393–402.
Taylor, S. R. 1989. Growth of planetary crusts. *Tectonophysics* 161:147–156.
Turcotte, D. L. 1993. An episodic hypothesis for Venusian tectonics. *J. Geophys. Res.* 98:17061–17068
Turcotte, D. L., and Schubert, G. 1982. *Geodynamics: Applications of Continuum Physics to Geological Problems* (New York: J. Wiley).
Vinogradov, A. P., Surkov, Yu. A., and Kirnozov, F. F. 1973. The contents of uranium, thorium, and potassium in the rocks of Venus as measured by Venera 8. *Icarus* 20:253–259.
Volkov, V. P., Zolotov, M. Yu., and Khodakovsky, I. L. 1986. Lithosphere-atmospheric interaction on Venus. In *Chemistry and Physics of the Terrestrial Planets*, ed. S. K. Saxera (New York: Springer-Verlag), pp. 136–190.
Warren, P. H., and Kallemeyn, G. W. 1993. The ferroan anorthosite suite, the extent of primordial lunar melting, and the bulk composition of the Moon. *J. Geophys. Res* 98:5445–5455.
Weitz, C. M., and Basilevsky, A. T. 1993. Magellan observations of the Venera and Vega landing sites. *J. Geophys. Res.* 98:17069–17097.
White, R., and McKenzie, D. 1989. Magmatism at rift zones: The generation of volcanic continental margins and flood basalts. *J. Geophys. Res.* 94:7685–7729.
Williams, D. R., and Gaddis, L. 1991. Stress analysis of Tellus Regio, Venus, based on gravity and topography: Comparison with Venera 15/16 radar images. *J. Geophys. Res.* 96:18841–18860.
Wise, D. L. 1974. Continental margins, freeboard, and the volumes of continents and oceans through time. In *The Geology of Continental Margins* (New York: Springer-Verlag).
Wood, J. A. 1986. Moon over Mauna Loa: A review of hypotheses of formation of Earth's Moon. In *Origin of the Moon*, eds. W. K. Hartmann, R. J. Phillips and G. J. Taylor (Tucson: Lunar and Planetary Inst.), pp. 17–55.
Zolotov, M. Yu., and Volkov, V. P. 1992. Chemical processes on the planetary surface.

In *Venus Geology, Geochemistry, and Geophysics: Research Results from the Soviet Union*, eds. V. L. Barsukov, A. T. Basilevsky, V. P. Volkov and V. N. Zharkov (Tucson: Univ. of Arizona Press), pp. 177–199.

Zuber, M. T. 1987. Constraints on the lithospheric structure of Venus from mechanical models and tectonic surface features. *Proc. Lunar Planet. Sci. Conf.* 17, *J. Geophys. Res. Suppl.* 92:541–551.

Zuber, M. T., Smith, D. E., Lemoine, F. G., and Neumann, G. A. 1994. The shape and internal structure of the Moon from the Clementine mission. *Science* 266:1839–1843.

MANTLE CONVECTION AND THE THERMAL EVOLUTION OF VENUS

G. SCHUBERT
University of California at Los Angeles

V. S. SOLOMATOV
California Institute of Technology

P. J. TACKLEY
University of Calfornia at Los Angeles

and

D. L. TURCOTTE
Cornell University

Numerical models of mantle convection relevant to Venus are analyzed for their implications regarding the present state of Venus' interior and the planet's thermal evolution. The models are fully three-dimensional and account for strongly temperature-dependent viscosity and mantle phase transitions. Both modeling and observations suggest that the mantle of Venus is presently in a sluggish/stagnant-lid convective regime in contrast to Earth's mantle which is presently in the small viscosity contrast (plate tectonic-like) convective regime. The numerical models also suggest that layering of mantle convection by the endothermic phase change, at about 740 km depth in Venus and 660 km depth in the Earth, is more likely in Venus than Earth. The dependence of viscosity on temperature increases the propensity towards layering in the models. Realistically large geoid, dynamic topography, and admittance ratios can be obtained on a global scale in these fully dynamical calculations with a terrestrial-like heat flux and a thermal lithosphere less than 100 km thick. Whereas the numerical models of Venusian mantle convection obtain a quasi-steady balance between internal radiogenic heat generation and surface heat loss, inference of a thick lithosphere on Venus from Magellan observations of topography, gravity, and surface deformation on both large and small scales implies that the planet is not presently in thermal equilibrium with an Earth-like complement of radioactivity. Thermal history models of Venus are explored to explain how there might be a present imbalance between heat production and loss and what might account for the global resurfacing of the planet in a short period of time about 500 Myr ago implied by the cratering record. Possible explanations for this catastrophic event include episodic mantle overturning or a transition from a small viscosity contrast (plate tectonic-like) convective regime to the present sluggish/stagnant-lid convective regime. Mantle overturning might have been triggered by the foundering of a thick and heavy lithosphere or a global flush instability. Episodicity is inherent in the mantle overturning scenarios—the event of 500 Myr ago might recur.

I. INTRODUCTION

Magellan data have invigorated the study of Venusian mantle convection and thermal evolution (see, e.g., Phillips and Hanson 1994; Kaula 1995). Most important is the conclusion of Schaber et al. (1992) that the abundance and near random distribution of craters and their relative lack of degradation imply a global resurfacing event about 500 Myr ago on Venus. Further statistical studies have supported this hypothesis (Bullock et al. 1993; Strom et al. 1994). The event covered about 80 to 90% of the surface by volcanism during a relatively short period of time and there has been relatively little volcanic resurfacing since (Namiki and Solomon 1994; Price and Suppe 1994; Phillips and Izenberg 1995). Models of Venusian mantle convection and thermal evolution must provide an explanation for this catastrophic event.

Magellan observations of the topography and gravity of Venus have thrown open the question of whether the planet is presently able to rid itself of the heat it generates internally by radioactive decay of U, Th and K or it is now heating up inside because it is unable to lose this heat. This has become a major question about Venus because its lack of plate tectonics at present requires that near surface heat loss occur by conduction through a lithosphere (there is too little surface volcanism at present for this process to transfer much heat). If Venus contains heat producing elements in amounts comparable to Earth, and if Venus is losing this heat at about the rate of its production, then the lithosphere should be relatively thin. However, the strong correlation of gravity and topography and geologic structures that reflect brittle modes of lithospheric deformation suggest that Venus has a relatively thick lithosphere. The possible imbalance between present internal heat generation and surface heat loss on Venus may be connected with the event that caused the global resurfacing of the planet 500 Myr ago. Models of Venusian mantle convection and thermal evolution need to address the questions of how and at what rate Venus is presently losing its internal heat.

In the following, we discuss the results of recent attempts to model mantle convection in Venus and what these models have to contribute to our understanding of the global resurfacing event on Venus and the planet's present thermal state. We also discuss some thermal evolution scenarios that might explain the unusual geologic activity of the past and the relative lack thereof at present.

II. INTERNAL HEAT PRODUCTION

The amount of heat produced by the decay of radioactive elements within Venus is an essential input to all models of the present thermal state and thermal evolution of Venus. The usual basis for estimating heat production and surface heat loss for Venus is to assume an analogy with the Earth. The total heat loss at the surface of the Earth Q_E is close to 3.55×10^{13} W with an estimated error of less than 5%. The mean terrestrial surface heat flow q_E for a surface area of 5.1×10^8 km^2 is 70 mW m^{-2}. The origin of this heat is the decay of the radioactive isotopes of uranium, thorium, and potassium and the secular cooling of the Earth. The Urey number Ur is defined to be the ratio of radioactive heat generation to the total heat loss and $1 - Ur$ is the fraction of total heat loss attributed to secular cooling. Estimates of the Urey number for the Earth fall in the range $0.6 < Ur < 0.8$. Mean concentrations of the heat producing elements in the Earth's mantle are given in Table I for $\bar{U}r = 0.6$ and 0.8. Table I also includes the corresponding rates of heat generation.

The present surface heat flow on Venus can be estimated by scaling the Earth's heat loss ($Q_E = 3.55 \times 10^{13}$ W) to Venus using the masses of the two planets ($M_E = 5.97 \times 10^{24}$ kg, $M_V = 4.87 \times 10^{24}$ kg). The result is a present heat loss from Venus Q_V of 2.91×10^{13} W and a mean surface heat flow q_V of 63 mW m^{-2}. The assumption that Venus and the Earth have similar concentrations of the heat producing elements is reasonable in terms of present models of planetary accretion.

An independent constraint on the rate of heat production within Venus comes from data collected by the Vega and Venera landers. The concentrations of heat producing elements and rates of heat production for five landing sites on Venus (Vega 1, 2, Venera 8, 9, 10) are given in Table I (Surkov et al. 1987). Table I also lists potassium concentrations for two additional landing sites (Venera 13 and 14). Venera 8 and 13 sampled upland plains with weakly differentiated melanocratic alkaline gabbroids with high potassium content. Venera 14 sampled a lowland area covered with tholeiitic basaltic tuff with a low potassium content. Young shield structures were sampled by Venera 9 and 10 with lavas close in composition to tholeiitic basalts but with a calc-alkaline trend.

For comparison, data for several terrestrial basalt types are included in Table I. Mid-ocean ridge basalts (MORB) are taken as direct melt products of the Earth's mantle. However, the upper mantle is certainly depleted in incompatible elements relative to the bulk silicate Earth due to the concentration of these elements in the continental crust. Sun and McDonough (1989) argue that N-type MORB represents the melting of this depleted source region; as evidence they give the consistent depletion of the light rare Earth elements in these rocks relative to chondrites. On the other hand, they point out that E-type MORB have generally chondritic rare earth distributions and should represent the partial melting of a source region that has near bulk Earth concentrations. The concentrations of the heat producing elements in E-type MORB (Table I) are about 7 times the estimated bulk Earth values, leading to a reasonable 12% basaltic component in the undepleted mantle with complete transfer of incompatible elements to the liquid fraction. Also given in Table I are typical concentrations of heat-producing elements in ocean island basalts (OIB). Clearly these basalts have enriched concentrations of the incompatible elements.

The Venus data are generally consistent with values expected for basic rocks. Five landers give values that can be associated with moderately ra-

TABLE I
Concentrations of Heat Producing Elements and Rates of Heat Generation H for A Variety of Planetary Basalts and Source Rocks

			Uranium (U) (ppm)	Thorium (Th) (ppm)	Potassium (K) (ppm)	$\frac{Th}{U}$	$\frac{K}{U}$	Heat Production 10^{-12} W kg^{-1}
Chrondrite[a]			0.008	0.029	545	3.6	68,000	3.5
Bulk silicate Earth	Ur = 0.8[b]		0.029	0.116	290	4.0	10,000	7
	Ur = 0.6[a]		0.021	0.085	250	4.0	11,900	5.2
Basalts								
Earth	N type MORB[a]		0.047	0.12	600	2.6	12,800	10
	E type MORB[a]		0.18	0.60	2100	3.3	11,700	41
	OIB[a]		1.02	4.20	12000	4.1	11,800	255
Moon	Low-Ti olivine	12002[c]	0.22	0.75	415	3.4	1,900	43
		15545[c]	0.13	0.43	330	3.3	2,500	25
	Low-Ti pigeonite	12064[c]	0.22	0.84	580	3.8	2,600	46
		15597[c]	0.14	0.53	500	3.8	3,600	30
	High-Ti, low K	70215[c]	0.13	0.34	415	2.6	3,200	23
	High-Ti, high K	10049[c]	0.81	4.03	3000	5.0	3,700	197
	Low-Ti aluminous	14035[c]	0.59	2.1	830	3.6	1,400	117
Venus	Vega 1[d]		0.64	1.5	4,500	2.3	7,000	118
	Vega 2[d]		0.68	2.0	4,000	2.9	5,900	134
	Venera 8[d]		2.2	6.5	40,000	3.0	18,000	531
	Venera 9[d]		0.60	3.65	4,700	6.1	7,800	172
	Venera 10[d]		0.46	0.70	3,000	1.5	6,500	74
	Venera 13[e]		—	—	33,000	—	—	—
	Venera 14[e]		—	—	1,700	—	—	—

[a] Sun and McDonough (1989); [b] Turcotte and Schubert (1982); [c] Heiken et al. (1991), pp. 261–263; [d] Surkov et al. (1987); [e] Surkov et al. (1984).

diogenic basaltic rocks, two to three times higher than the E-type MORB but lower than typical OIB. Two landers give values more typical of silicic rocks on the Earth. In terms of modeling, an essential question is whether the surface values are typical of crustal values at depth. Certainly fractionation and crystallization are likely to lead to an upward concentration of the incompatible heat producing elements. In the continents of the Earth the concentrations of the heat producing elements decay exponentially with depth on a scale of about 10 km. However, it is impossible to estimate such variations in the crust of Venus. It seems reasonable to conclude that the direct scaling of terrestrial radiogenic element concentrations to Venus is appropriate.

III. OBSERVATIONAL CONSTRAINTS ON HEAT TRANSFER

At what rate and by what means is Venus presently losing its internal heat? We have no measurements of surface heat flow on Venus so we must look to other observations for some guidance in answering these questions.

There is no evidence that plate tectonics is presently active on Venus on a global scale (Kaula and Phillips 1981; Phillips et al. 1981; Grimm and Solomon 1988; Kaula 1990; Solomon et al. 1991,1992). Though some regions on Venus have features remarkably similar to those found at plate margins on Earth (Sandwell and Schubert 1992a,b; McKenzie et al. 1992; Schubert and Sandwell 1995), it is the quantitative insufficiency of these features that makes it improbable for plate tectonics to be a globally significant process on Venus. On Earth, mantle heat transfer is dominated by the reheating of subducted lithosphere. In the absence of global plate tectonics on Venus, lithospheric subduction cannot transfer a terrestrial-like heat flux through the Venusian mantle. However, it has been argued that retrograde lithospheric subduction is occurring along major segments of Venusian chasmata and some fraction of the heat flow on Venus might be carried by this process (Sandwell and Schubert 1992a,b; Schubert and Sandwell 1995).

Plate tectonics and lithospheric subduction contribute to the style of mantle convection on Earth and though these processes do not appear to have global significance for Venus, convection of a different sort is still the likely mode of heat transfer through the bulk of the Venus mantle. The style of Venusian mantle convection, as will be discussed below, is the sluggish/stagnant-lid type (Fig. 1). It involves motions beneath a "lithosphere" which is largely immobile in its upper part but sufficiently mobile in its lower part to serve as the source of relatively cold downflow for the deeper mantle. A broad upflow of relatively hot material and plume concentrations of even hotter material supply mass to the semi-rigid lower lithosphere.

The distinction between sluggish-lid convection and stagnant-lid convection lies mainly in the fraction of the heat flux removed by the moving lithosphere (Solomatov 1995). In sluggish-lid convection, heat is transferred to the surface by horizontally moving lithosphere, while in stagnant-lid convection most of the heat is transported to the surface by vertical conduction

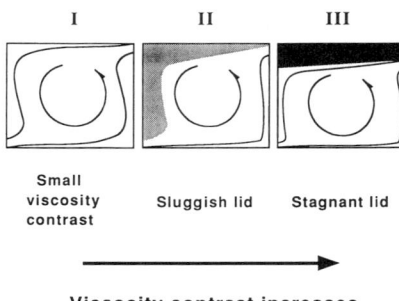

Convective regimes

Figure 1. A schematic representation of convective regimes. In the small viscosity contrast regime, the flow is similar to constant viscosity convection. In the sluggish-lid regime, the upper boundary layer is thicker than the lower boundary layer but it is still moving and is important for the heat transport. In the stagnant-lid regime, the upper bounary layer is essentially motionless and convection involves only the hottest part of the boundary layer. The dark region in the right panel shows the stagnant part of the cold boundary layer.

through effectively stationary lithosphere (Fig. 1). We describe Venus as being in the sluggish/stagnant-lid mode of convection because it is uncertain to what extent the motion of the lithosphere of Venus participates in heat removal. The possibility that the entire lithosphere is foundering along some segments of Venusian chasmata adds additional complexity to the nature of lithospheric heat transport on Venus (Schubert and Sandwell 1995).

What is really in question about heat transport on Venus is how much heat is transferred through the essentially rigid part of the lithosphere and what is the thickness of this nearly rigid layer. If the essentially rigid part of the lithosphere is very thick, as suggested by some interpretations of Venusian gravity, topography, and surface deformation discussed below, then magmatic and conductive heat transport are too inefficient to transfer heat through the lithosphere at a terrestrial-like rate of internal heat production. Magmatic heat transport through the lithosphere is likely unimportant because the present rate of volcanism on Venus, less than about 1 km^3 yr^{-1} (Fegley and Prinn 1989; Head et al. 1991,1992; Bullock et al. 1993; Strom et al. 1994; see the chapter by Grimm and Hess), is much smaller than the terrestrial rate of oceanic crust production, about 20 km^3 yr^{-1}. While not all magma can reach the surface (Greeley and Schneid 1991), this low rate of Venusian volcanism is indicative of either a low degree of melting in the upper mantle of Venus or difficulty in melt penetration through the lithosphere. Either way, the small rate of volcanism on Venus at present points to a minor role for magmatic heat transfer through the nearly rigid part of the lithosphere.

The efficacy of conductive heat transfer through the nearly rigid part of the lithosphere is inversely proportional to the thickness of the nearly rigid layer. For the terrestrial-like surface heat flux of 63 mW m^{-2} estimated above and a thermal conductivity of 3.3 W m^{-1} K^{-1}, the mean surface thermal gradient on Venus is 19 K km^{-1}. This thermal gradient implies a mean lithosphere thickness of 45 km on the assumption of a linear temperature profile through the lithosphere and a temperature drop of 850 K across it. This thickness is much less than the 200 to 300 km thickness suggested by many of the investigations discussed below. The lithosphere thickness estimate just above could be somewhat larger if a significant fraction of the heat producing elements were in the lithosphere or if the thermal conductivity was larger.

A number of observations argue against a Venusian lithosphere as thin as about 45 km.

1. The high topography of Ishtar Terra is largely compensated at depths indicative of a thick crustal root (Simons et al. 1994; Arkani-Hamed 1996; see the chapter by Kaula et al.). The 10 km of topography associated with Maxwell Montes implies a 50 to 75 km thick crustal root and therefore an even thicker lithosphere (see the chapter by Kaula et al.; Freed and Melosh 1995). Also, Ishtar Terra is bounded by escarpments several km high with slopes up to 30°. This topography must be supported by lithospheric stresses. Three-kilometer escarpments require stresses of the order of 100 MPa. There is no evidence that these surfaces are relaxing, so the stresses are most likely elastic.
2. The topography on Venus generally has large associated gravity anomalies. In a number of cases these anomalies are best explained by large depths of compensation that can be interpreted in terms of a thick lithosphere (Phillips 1994; Grimm 1994a; Kucinskas and Turcotte 1994; Schubert et al. 1994). The whole of Ishtar Terra is compensated at a depth of about 150 km (see the chapter by Kaula et al.).
3. Grimm and Solomon (1988) point out that unrelaxed craters imply a strong lithosphere. Banerdt et al. (1994) conclude that the isostatic compensation of Mead crater implies an elastic lithosphere with a thickness of at least 30 km; this suggests a thermal lithosphere at least twice as thick.
4. Sandwell and Schubert (1992a,b) and Schubert and Sandwell (1995) have also shown that the morphology of several coronae are in good agreement with lithospheric flexure models that have been successful in explaining the seafloor morphology at ocean trenches on Earth. Their flexural profiles yield elastic lithosphere thicknesses of 37 km for Artemis, 35 km for Latona, 15 km for Eithinoha, 40 km for Heng-o, and 18 km for Freyja, results which suggest even larger values for the mechanical thickness of the Venusian lithosphere.

If the lithosphere of Venus is as thick as these observations suggest then a terrestrial-like heat flow could not be conducted through the lithosphere.

Several alternatives are then consistent with a thick Venusian lithosphere: (1) Venus could have less radioactivity than Earth, a possibility we argued against earlier in this chapter; (2) Venus is presently not in an approximate steady state balance between internal heat production and surface heat loss, i.e., the present lithosphere is too thick to transport the required heat by conduction; (3) Venusian heat sources are concentrated close to the surface; (4) there are other lithospheric heat transfer mechanisms such as plume penetration of the lithosphere or lithospheric delamination. In the following sections we evaluate these possibilities.

IV. A MODEL OF VENUS WITH ALL THE HEAT SOURCES IN THE CRUST

One of the possibilities of reconciling a thick Venusian lithosphere with a terrestrial-like surface heat flux is for differentiation to have been so efficient on Venus that all the heat sources are concentrated in the crust. In this case, the internally generated heat only has to be conducted through the crust to reach the surface; the thick lithosphere does not interpose a thermal resistance between the region of heat generation and the surface.

In order to consider this possibility quantitatively, we assume a simple model in which the concentrations of heat-producing elements are uniform through a crust of thickness y_c and secular cooling can be neglected so that the heat flow to the base of the crust is zero. In the steady state, the temperature distribution in the crust of this model is (Turcotte and Schubert 1982, p. 145)

$$T = T_s + (T_m - T_s)\left(\frac{y}{y_c}\right)\left(2 - \frac{y}{y_c}\right) \tag{1}$$

where y is the depth, T_s is the surface temperature, and T_m is the uniform temperature of the mantle. Conservation of energy requires

$$q_s = \rho_c H_c y_c. \tag{2}$$

From Eqs. (1), (2) and Fourier's law of heat conduction

$$q_s = k_c \left(\frac{dT}{dy}\right)_{y=0} \tag{3}$$

we find

$$T_m - T_s = \frac{1}{2}\frac{q_s y_c}{k_c} \tag{4}$$

where H_c is the rate of heat production per unit mass of the crust and q_s is the surface heat flow.

For $Ur = 0.8$ the surface heat flow from radiogenic heating is $q_s = 50$ mW m^{-2} and with $Ur = 0.6$ it is $q_s = 38$ mW m^{-2}. We assume $T_s = 750$ K, $k_c = 2$ W m^{-1} K^{-1}, and $\rho_c = 2900$ kg m^{-3}. If the crust is thick the temperature within it will exceed its liquidus (assumed to be undesirable). If the crust is thin the heat production H_c will be large (exceeding the observed values given in Table I). Solutions for the two cases in which the basal temperature approaches the liquidus ($T_m = 1700$ K) are given in Fig. 2. The solution for $Ur = 0.8$ has $y_c = 75$ km and $H_c = 230 \times 10^{-12}$ W kg^{-1}; the solution for $Ur = 0.6$ has $y_c = 100$ km and $H_c = 130 \times 10^{-12}$ W kg^{-1}. Comparison of these rates of heat generation with the Venusian values given in Table I shows that the values for $Ur = 0.6$ are generally consistent with the Venus data.

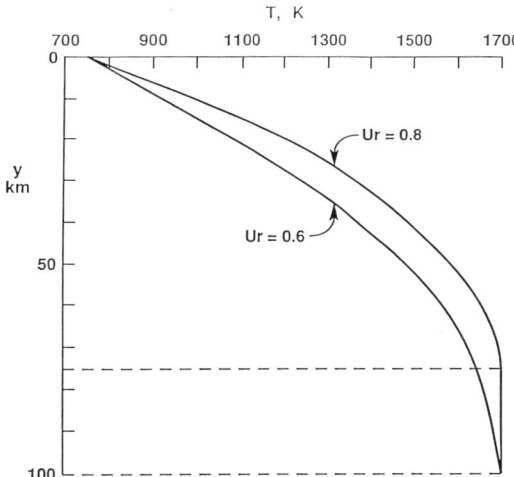

Figure 2. Temperature vs depth y for a model of a thick Venusian crust with a uniform concentration of heat producing elements and zero mantle heat flow from beneath the crust. The two values of the Urey number Ur, 0.6 and 0.8, correspond to surface heat flow values $q_s = 38$ mW m^{-2} and $q_s = 50$ mW m^{-2}, respectively.

Thus, it is possible to construct a model for the upward concentration of the heat producing elements that has a mantle temperature below the solidus and rates of heat generation compatible with the surface observations. However, this does require extreme assumptions: (1) almost complete transfer of the heat producing elements to the crust; (2) negligible secular cooling of Venus; and (3) uniform concentrations of the heat producing elements through the crust.

Presumably if this model is to be valid, the crust of Venus would have thickened with time with little crustal recycling. Expected consequences of this process would be the systematic depletion of the mantle heat producing elements with time leading to a reduction in the content of the radiogenic elements in the most recent volcanics and a gradual decay of volcanism with time. It is not clear if this type of decay could be consistent with the crater counts. Also, Kaula (1995) has pointed out that the low abundance of ^{40}Ar in the Venus atmosphere argues against a strong upward concentration of radioactive elements into the Venusian crust.

V. NONCONDUCTIVE MODES OF HEAT TRANSFER ACROSS A THICK VENUSIAN LITHOSPHERE

In this section we evaluate the possibility that a terrestrial-like heat flux could be transferred through a thick Venus lithosphere by plume penetration of the lithosphere or by lithospheric delamination.

A. Plume Penetration of the Lithosphere

The topography and associated gravity anomalies of volcanic rises on Venus have been attributed to mantle plumes by several authors (Phillips et al. 1991; Kiefer and Hager 1991,1992; Smrekar 1994; Smrekar and Parmentier 1996). Moore and Schubert (1995) have presented evidence from topography and gravity of lithospheric thinning beneath Beta Regio consistent with a mantle plume below this highland. It has been suggested that some coronae are also surface expressions of underlying mantle plumes (Stofan et al. 1991,1992). If Venusian mantle plumes could penetrate the lithosphere and if they are numerous enough and carry enough heat, they conceivably could transfer a terrestrial-like heat flow through the lithosphere. About eighty plumes with the strength of the Hawaiian plume (3.6×10^{11} W) would be required to transport a substantial fraction of the terrestrial-like heat flow of 2.9×10^{13} W. Such a large number of major plumes would effectively thin the lithosphere globally, in contradiction to the thick lithosphere dilemma we are attempting to reconcile, and would have surface signatures including extensive volcanism that are not seen. Estimates of the strength of active plumes on Venus suggest that the present plume flux is less than on the Earth (Smrekar and Phillips 1991). We will see in the models of Venusian mantle convection to be discussed later that plume-dominated convection is not the style of convection in a largely internally heated mantle. Instead, heat transport is primarily by downwellings originating from the near-surface cold boundary layer (see, e.g., Schubert et al. 1993; Parmentier et al. 1994; Travis and Olson 1994).

B. Lithospheric Delamination

Lithospheric delamination (Bird 1979) is the tearing away of the lithosphere from the crust and the sinking of the cold and heavy lithospheric slab into the hotter and lighter underlying mantle. The delaminated section of lithosphere is replaced by hot mantle with a net upward transfer of heat. If this process could occur sufficiently often and involve large volumes of lithosphere then the terrestrial-like heat flow could be transferred across the lithosphere. We distinguish lithospheric delamination from the downflow of cold lithospheric material originating in the mobile lower part of the lithosphere. The latter is the normal mode of convection in an internally heated medium and is presumably the process presently occurring on Venus as discussed in a later section.

A mechanism for lithospheric delamination on Venus has been proposed by Buck (1992). He decouples the upper crust from the upper mantle with a low viscosity lower crust, i.e., a lower crustal asthenosphere. The upper mantle convects, but the upper crust behaves as a scum that floats and does not participate in subduction. A similar model has been proposed by Arkani-Hamed (1993). However, the recent determination of the strength of dry diabase makes a weak lower crustal zone unlikely (Mackwell et al. 1995).

If the entire mantle heat flux is attributed to delamination, then it must occur sufficiently often to transmit $Q_D = 2.91 \times 10^{13}$ W. If the lithospheric temperature at which delamination occurs is $T_D = 1470$ K, the entire lithosphere would have to delaminate, on average, at intervals of $t_D = 1.2$ Myr. With $T_D = 1290$ K the lithospheric delamination time would have to be $t_D = 6$ Myr, and with $T_D = 1110$ K we would require $t_D = 19$ Myr. It seems unreasonable that global delamination events could take place at such short intervals. Also, global delamination events would be expected to disrupt the upper crust resulting in intensive volcanism that is not observed. Therefore, while delamination might occur on Venus, e.g., it may have played a role in creating the high plateau topography of Ishtar Terra, it is not likely that delamination could make a significant contribution to the global heat flow. This is consistent with our understanding of the role of delamination on the Earth.

Given that plume penetration of the lithosphere and lithospheric delamination are not plausible processes for the transfer of a terrestrial-like heat flow across a thick Venus lithosphere and given the implausibility of completely differentiating all Venusian heat sources into its crust, a nonsteady thermal state becomes an attractive hypothesis for reconciling a thick Venus lithosphere with a terrestrial-like complement of internal heat sources. We will discuss several scenarios for nonsteady thermal evolution of Venus later in this chapter after we describe some model results for the style of convection in the mantle.

VI. THE STYLE OF CONVECTION IN THE VENUSIAN MANTLE

In this section we discuss the results of numerical models of large-scale, three-dimensional thermal convection in the mantle of Venus.

A few numerical studies of global-scale mantle convection have specifically addressed Venus. Three-dimensional spherical models were presented by (Schubert et al. 1990), who showed that a rigid upper boundary (representing the base of Venus' lithosphere) decreases the temporal durability of downwellings compared to a free-slip boundary condition (representing an average boundary condition for Earth, where the plates are free to move). The internal temperature is also higher when a rigid lid is present. Similar models in more restrictive two-dimensional spherical axisymmetric geometry were presented by Leitch and Yuen (1991) and Leitch et al. (1992), who found that a rigid lid can decrease the width of convective cells in the upper mantle. Neither of these models includes the effects of an endothermic phase transition in the mantle of Venus. The endothermic phase change from γ-spinel to perovskite and magnesiowüstite occurs at a depth of about 660 km in the

Earth and is expected to occur at around 740 km depth in Venus. Studies both in two-dimensional geometry (Machetel and Weber 1991; Peltier and Solheim 1992; Solheim and Peltier 1994; Weinstein 1993; Zhao et al. 1992) and three-dimensional geometry (Honda et al. 1993; Tackley et al. 1993,1994; Yuen et al. 1994) indicate that the endothermic phase change in the Earth's mantle might impose a style of convection that lies between layered-and whole-mantle convection, by inhibiting material flow across itself, causing pooling of downwellings in the transition zone, followed by "avalanches" of this cold pooled material into the lower mantle (Tackley 1995).

A two-dimensional Cartesian model of phase-change modulated convection in Venus (Steinbach and Yuen 1992) found that the exothermic phase transition expected to occur at around 440 km depth in Venus increases the degree of layering; however, the greater depth of phase transitions in the Venus model compared with Earth and the presence of a rigid conductive lid in the Venus model decrease the degree of layering. Steinbach and Yuen (1992) also suggested that a global transition from whole mantle to layered convection, as first suggested by Christensen and Yuen (1985), might be responsible for the global resurfacing event 500 Myr ago on Venus. In contrast to the results of Steinbach and Yuen (1992), Solheim and Peltier (1994) and Tackley et al. (1994) find that the exothermic olivine-spinel transition decreases the tendency for layering in models with Earth-like parameters.

We have recently computed new models of three-dimensional, phase-change modulated convection in Venus' mantle. The model includes both an exothermic phase change at a depth of 440 km (410 km depth in the Earth) and an endothermic phase change at a depth of 740 km (660 km depth in the Earth); the Clapeyron slope of the exothermic phase change is 3 MPa K^{-1} and that of the endothermic phase change is -4 MPa K^{-1}. The computational domain is a spherical shell with isothermal boundaries; the inner boundary is free slip and the outer boundary is rigid (sometimes free slip to illustrate the effect of the Venus-like rigid upper boundary). The model is compressible, anelastic, and self-gravitating and thermodynamic and rheological properties vary with hydrostatic pressure or depth; parameter values and variations of properties with hydrostatic pressure are similar to those in the Earth's mantle. In particular, viscosity has a power-law dependence on density and increases by a factor of 17 over the depth of the model mantle. Heating is both from within and from below at realistic rates for Venus; other parameter values including volume-averaged Rayleigh numbers resulting from internal heating Ra_H and superadiabaticity Ra_T are also within the range thought to be appropriate for Venus. Results are presented for two cases. Case 1 has a rigid upper boundary and case 2, for contrast, has a free-slip upper boundary. For case 1, Ra_H is 1.4×10^8 and Ra_T is 4.8×10^6; for case 2, $Ra_H = 7 \times 10^7$ and $Ra_T = 2.4 \times 10^6$. The convective states discussed below represent statistical steady states.

The Venus mantle convection models have heat fluxes in the range 50 to 60 mW m^{-2}, with approximately 30% of the heat coming from the core and 70% from internal heating by radioactive decay. The heat flux is slightly lower than that of the Earth and may be realistic for Venus as discussed above. Maximum velocities are around 25 mm yr^{-1}, with values of 50 mm yr^{-1} being reached in the upper mantle of the free-slip boundary case 2.

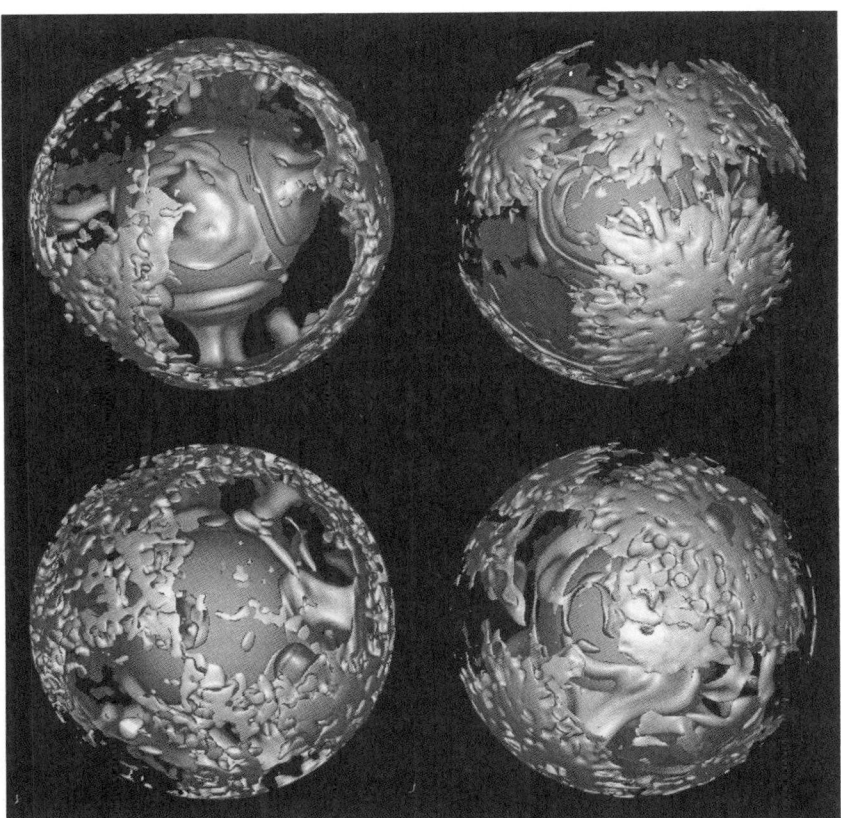

Figure 3. Cold downwellings (left) and hot upwellings (right) in the two spherical convection models of Venusian mantle convection. Illustrated are isosurfaces of residual temperature (temperature difference from the horizontally-averaged geotherm), showing where the temperature is 110 K lower (left) or 110 K higher (right) than the geotherm. The upper plots show case 1, with the rigid upper surface; the lower plots show case 2, with the free-slip upper surface. Convection in these models is time dependent; the time step shown is typical of the convective state.

Typical convective states in the two cases are illustrated in Fig. 3, which shows hot and cold residual temperature anomalies corresponding to upwellings and downwellings, respectively (residual temperatures are the temperature differences from the horizontally-averaged temperature at a particular radius). The broad-scale structure of convection is similar to that observed in previous three-dimensional spherical models of phase-change modulated convection (Tackley et al. 1993,1994). The upper mantle contains a network of interconnected downwelling sheets, which do not initially penetrate the endothermic phase transition, but are deflected, resulting in pools of cold material in the transition zone (left panels in Fig. 3). These pools build up until sufficient negative buoyancy has accumulated to overcome the buoyancy associated with phase change deflection, at which point cold material falls catastrophically into the lower mantle in the form of a broad cylindrical avalanche, which empties the local transition zone contents into broad cold pools at the base of the mantle. Avalanches at different places around the sphere overlap in time, so that the system is never completely layered or completely whole mantle on a global scale, as exhibited in some two-dimensional convection calculations (Weinstein 1993; Solheim and Peltier 1994). Globally synchronous transitions between layered and whole mantle convection appear to be an artifact of two-dimensional geometry, or small aspect ratio three-dimensional geometry—in three-dimensional spherical geometry the mantle is partially layered both spatially and temporally.

Comparison of these simulations to previous three-dimensional spherical Venus calculations (Schubert et al. 1990) shows that the characteristic horizontal wavelengths are significantly larger than observed when no phase transitions are present. This effect was previously observed in Earth-like calculations by Tackley et al. (1993). Particularly surprising in these results is that the rigid-boundary case 1 (upper panels in Fig. 3) displays greater wavelengths of flow than the free-slip boundary case 2 (lower panels in Fig. 3). This is opposite to previous calculations with no phase changes, in which a rigid upper boundary tends to induce shorter wavelengths of flow, with more closely spaced features (Schubert et al. 1990). It also appears that the rigid boundary case is more layered, an observation that is quantified in later sections.

Hot upwellings illustrated in the right panels of Fig. 3 display less clear structure than the cold downwellings. Near the core-mantle boundary, hot ridges are observed, separating the pools of cold material deposited by various avalanches. A few hot plumes penetrate the mid-mantle region and easily penetrate into the upper mantle. The upper mantle displays considerable long-wavelength hot structure.

Three distinct regions of the mantle can thus be identified on the basis of their thermal structure: (1) the upper mantle, containing downwelling sheets and pools of cold material in the transition zone; (2) the mid-mantle, between 740 and $\sim$2000 km depth, containing upwelling and broad downwelling plumes; and (3) the deep mantle, containing large pools of cold material above the core-mantle boundary separated by ridges of hot material.

Profiles of horizontally-averaged temperature are plotted for the two cases in Fig. 4, which also shows the reference state adiabat. In addition to the expected thermal boundary layers at the top and bottom of the mantle, another thermal boundary layer around the endothermic phase transition is clearly visible due to the partial layering of the flow by this phase transition. This

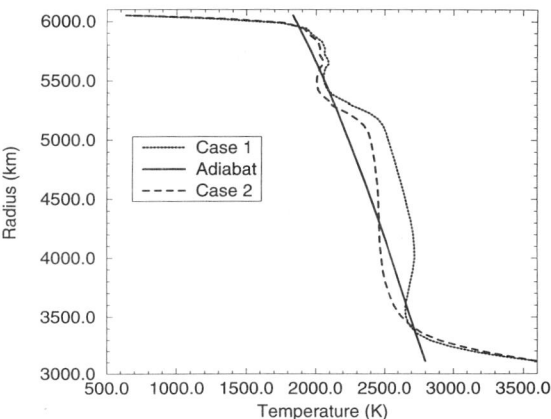

Figure 4. Horizontally-averaged temperature profiles for cases 1 and 2 in Fig. 3 and the reference state adiabat. The temperature profile is generally subadiabatic, with boundary layers visible at the surface, at the lower boundary and at 740 km depth, the location of the endothermic phase transition.

temperature drop is larger for case 1, implying a greater degree of layering in this case, and is around 300 to 400 K, compared to ~1300 K for the upper boundary layer and ~900 to 1000 K for the lower boundary layer.

Away from these boundary layers the temperature profile is generally subadiabatic, i.e., the temperature increase with depth is less than that displayed by an adiabat. This may seem surprising because it is commonly assumed that the interior of a convecting mantle is adiabatic; however, this assumption is only true for systems which are predominantly heated from below—subadiabatic temperature stratification is a fundamental property of convection which is heated predominantly from within (Schubert et al. 1993; Parmentier et al. 1994; Travis and Olson 1994). The mantles of Venus and Earth are thought to be predominantly heated from within by decay of radioactive elements and also by secular cooling which resembles internal heating both mathematically (in the energy equation) and physically (in its effect on convection) (Weinstein and Olson 1990).

In addition to the general subadiabaticity, a temperature minimum is displayed in the deep mantle, just above the lower thermal boundary layer. This feature, which is particularly pronounced in case 1, is caused by the pooling of cold avalanche material at this depth. Another feature is a temperature high in the mid-mantle, just below the endothermic phase transition. This reduces the temperature anomaly of upwelling plumes relative to the deep mantle, making it more difficult for them to penetrate into the upper mantle. Overall, departures from adiabaticity of up to 300 K, i.e., around 10% of the total temperature drop, are observed in the interior of the mantle.

A useful measure of the degree of flow stratification is the radial mass flux diagnostic (Peltier and Solheim 1992), a function of radius which is defined as the radial mass flux through a spherical shell at the radius in question, normalized so that the integral over nondimensional depth is equal to unity. This diagnostic for the two cases, averaged over a simulated period of several hundred million years, is illustrated in Fig. 5. The important feature here is the minimum in radial mass flux at 740 km depth caused by the inhibitive effect of the endothermic phase transition on radial flow. Although the curves in Fig. 5 are averaged over a long period of time, they do not change much with time because of the weakness of global episodicity, even though strong episodicity may be exhibited in local regions (Tackley et al. 1994).

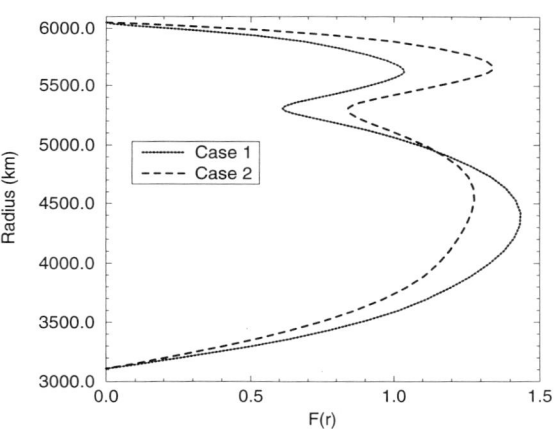

Figure 5. Radial mass flux diagnostics for the two cases of Fig. 3 averaged over a time of several hundred Myr.

Comparison of the two cases, shows that the degree of layering is greater with a rigid upper boundary (case 1) than with a free-slip upper boundary (case 2). Both the flow in the upper mantle and the flow through the phase boundary are more restricted in the presence of a rigid upper boundary than a free-slip upper boundary. This confirms the impression from visual examination of the temperature field (Fig. 3) and horizontally-averaged temperature profiles (Fig. 4) that the degree of layering is greater in case 1. If case 1 can be taken to be more representative of Venus, and case 2 more representative of the Earth, this implies that Venus' mantle may currently exhibit a greater degree of flow stratification than Earth's mantle.

The long-wavelength geoid and topography are perhaps the most important surface manifestations of Venusian mantle convection, offering strong, albeit ambiguous, constraints on the physical properties and dynamics of the mantle and lithosphere. On Venus (unlike the Earth) a strong positive correlation between geoid and topography is observed, admittance ratios are high, and inferred depths of compensation are large. As discussed above, large compensation depths can be taken to represent the thickness of the lithosphere. Such an interpretation leads to estimates for the thickness of Venus' lithosphere (200–300 km) much larger than expected for quasi-steady-state convection with a similar heat flux to that of the Earth (Turcotte 1993). The convection solutions discussed here allow us to evaluate whether compensation by deep dynamic heterogeneities that arise in a self-consistent, time-dependent, dynamically-convecting system with reasonable heat flux can provide an alternative explanation for the large admittance ratios (Kiefer and Hager 1992; McKenzie 1994).

The geoid, topography, and admittance ratios for the two models and Venus data are plotted in Fig. 6. The graphs show total amplitude for each spherical harmonic degree, not the rms amplitude of the coefficient as plotted in some other papers (Konopliv et al. 1993). The Venus data are a degree 75 version of the gravity model by Konopliv and Sjogren (1994) and the degree 360 topography model of Rappaport and Plaut (1994). Admittance ratios are calculated using the usual assumption that the total gravity for each spherical harmonic is given by a correlated and uncorrelated component (Kiefer et al. 1986; Simons et al. 1994)

$$G_{lm} = F_l T_{lm} + I_{lm} \tag{5}$$

where G_{lm} and T_{lm} are spherical harmonic coefficients of the geoid and topography, respectively, I_{lm} is the part of the geoid that is not correlated with topography, and F_l is the admittance ratio for each degree. The admittance ratios can be simply determined by

$$F_l = Re\left(\frac{\sigma_{gt}^2}{\sigma_{tt}^2}\right) \tag{6}$$

where σ_{gt} is the cross-covariance of geoid and topography, and σ_{tt} is the covariance of topography (Simons et al. 1994). The calculated admittance ratios are significantly different from those calculated using the earlier Pioneer data set (Kiefer et al. 1986), although the overall magnitude and trend are similar.

For the geoid and topography, the two simulations appear to straddle the Venus data for the long wavelengths at which these quantities are accurately known (Konopliv et al. 1993), i.e., spherical harmonic degrees less than about $l = 30$ to 40. The rigid boundary case (1) exhibits greater geoid and topography than the free-slip boundary case (2), probably because of the stronger coupling between convection and the surface. The admittance ratios straddle the Venus data for spherical harmonics less than $l = 7$, (except for $l = 2$ which is poorly determined), but are higher than Venus data, by around a factor of 2, for wavelengths below this.

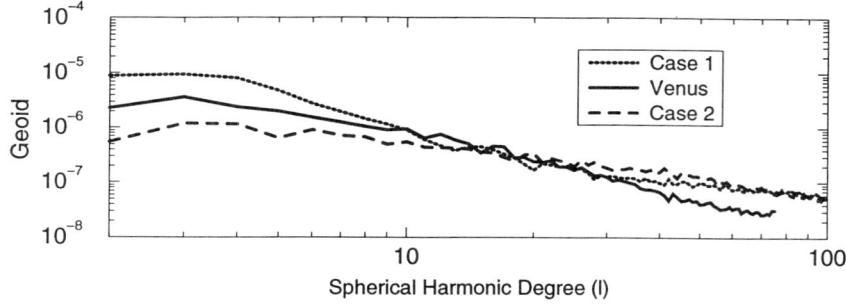
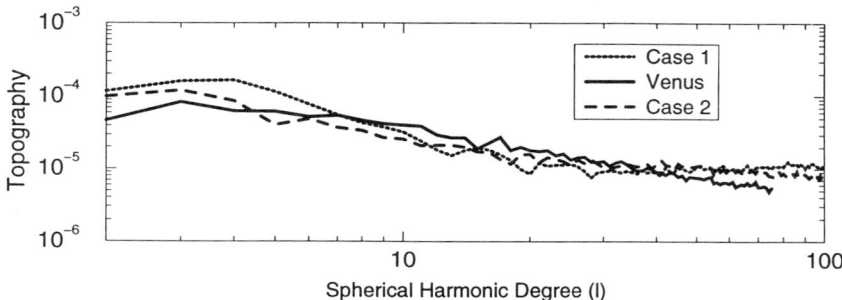
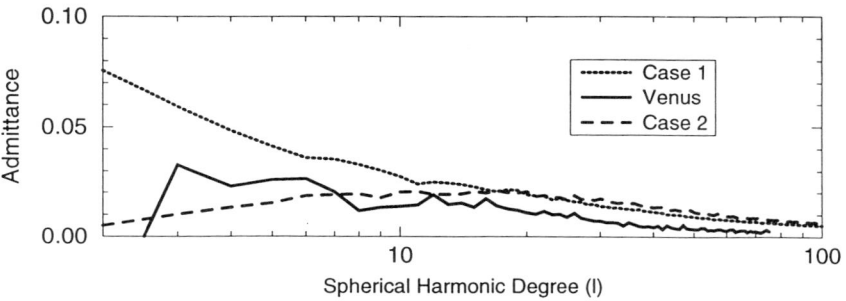

Figure 6. Frequency spectrum of geoid, topography and admittance ratios for the two cases of Fig. 3 plotted against Venus data. Geoid and topography are normalized by planetary radius and plotted as total amplitude for each spherical harmonic degree. The calculation of admittance ratios is described in the text.

These results show that realistically large geoid, dynamic topography and admittance ratios can be obtained on a global scale in a fully dynamical calculation even with a high heat flux thought to be appropriate to a steady-state Venus model. The upper boundary layer thickness, which would correspond to thermal lithospheric thickness if temperature-dependent viscosity were included, can be seen from the temperature profiles to be less than ~90 km, much smaller than the thickness required if admittance ratios are explained entirely by variations in lithospheric thickness. Thus, deeper mass anomalies associated with thermal convection in the interior are also important in determining large-scale geoid and topography. Because a rigid upper boundary condition gives a geoid signal which is too large, the results imply that the most appropriate upper boundary condition for Venus lies between the extremes of rigid and free slip.

Even though the calculations discussed above show that high admittance ratios are possible for the large scale of a convective model of Venus' mantle with a terrestrial-like complement of radiogenic heat sources and a relatively thin lithosphere ($\lesssim$100 km thick), the inference of a much thicker lithosphere (200–300 km) on Venus rests not only on large-scale admittance ratios but on the gravity and topography of many individual features of smaller scale (e.g., volcanic highlands, coronae, chasmata segments) and the deformation or lack thereof of small-scale structures (e.g., craters and flexural outer rises). Thus, a thick Venus lithosphere remains a viable interpretation of a large amount of data on the planet's gravity, topography, and geology.

Because of the stronger tendency for phase-change induced layering in Venus' mantle as compared with Earth's mantle exhibited by the above calculations, mantle overturn due to an avalanche of global scale is a credible explanation for the resurfacing event on Venus some 500 Myr ago. Although such a globally synchronous transition has not been observed in the fully three-dimensional spherical calculation, the present computations do not include the secular cooling of the planet which may enforce a globally synchronous transition from layered to whole mantle convection (Steinbach et al. 1993; Steinbach and Yuen 1994).

VII. EFFECTS OF TEMPERATURE-DEPENDENT VISCOSITY ON VENUSIAN MANTLE CONVECTION

The solutions discussed above do not take into account the strong dependence of the viscosity of mantle material on temperature, an effect likely to have a huge influence on the style of convection. When temperature-dependent viscosity is included in models, a rigid lithosphere arises naturally due to the very high viscosity in the upper boundary layer, provided the viscosity contrast is sufficiently large. Indeed, simple temperature-dependent viscosity models may resemble Venus much better than they resemble the Earth, where highly complex rheological behavior is required in order to describe the brittle failure that causes plate boundaries.

Numerical models of three-dimensional convection with temperature-dependent viscosity and free-slip boundaries have until recently been restricted to steady-state solutions in small boxes; notable findings include a preference for upwelling plumes and downwelling sheets at an aspect ratio of 1.5 (Christensen and Harder 1991), and the appearance of a rigid lid at large viscosity contrasts of greater than about 10^3 to 10^4 (Ogawa et al. 1991). Time-dependent, large aspect ratio calculations with strongly variable viscosity are only just coming into their own (Tackley 1993,1994,1996; Balachandar et al. 1995a,b; Ratcliff et al. 1995,1996a,b). Here we discuss two sets of wide-domain calculations in Cartesian geometry, one of which illustrates the dramatic effect of systematically increasing the viscosity contrast up to a factor of 10^5, the other of which illustrates the interaction of variable viscosity flow with mantle phase transitions (Tackley 1993,1996). We also discuss the first results of three-dimensional variable viscosity convection in a spherical shell with a viscosity contrast of 10^3 (Ratcliff et al. 1995).

Figure 7 shows the profound effect that viscosity variation has on the form of convection. The results are from Cartesian calculations in an $8 \times 8 \times 1$ box with periodic side boundaries and free-slip, isothermal top and bottom boundaries. Heating is from below and viscosity depends strongly on temperature according to an Arrhenius law. In these calculations the Rayleigh number based on the viscosity at the mean of the top and bottom boundary temperatures is 10^5. The viscosity contrasts are 1, 10, 10^2, 10^3, 10^4 and 10^5 in Figs. 7a–f, respectively. Three convective regimes are clearly visible: (i) the small viscosity contrast regime at viscosity contrasts of 1 to 10, characterized by a fairly symmetrical spoke pattern of upwelling and downwelling sheets which break up into plumes as they ascend or descend (Figs. 7a,b); (ii) the sluggish-lid regime at intermediate viscosity contrasts of 10^2 to 10^3, characterized by a very long-wavelength flow pattern consisting of a huge cylindrical downwelling plume surrounded by narrow upwelling sheets under a slowly deforming lithosphere (Figs. 7c,d); and (iii) the stagnant-lid regime at large viscosity contrasts of 10^4 to 10^5, characterized by a very small-wavelength pattern of upwelling plumes and downwelling sheets under an immobile stagnant lid (Figs. 7e,f).

With increasing viscosity contrast, Nusselt number (heat flux in units of the heat flux that would be conducted in the absence of convection) progressively decreased from 8.8 for the constant viscosity case to 4.8 for the case with a viscosity contrast of 10^5. Velocities, however, increased significantly with increasing viscosity contrast; rms velocity (nondimensionalized to the conductive velocity κ/D, where κ is the thermal diffusivity and D is the height of the box) increased from 197 to 293, and peak velocity increased from 763 to 1813. As the viscosity contrast is increased, convection below the lithosphere becomes more vigorous but with smaller temperature contrasts, so that heat flux decreases and velocity increases.

These two fundamental transitions in convective style have been discussed from a theoretical perspective by Solomatov (1995) and identified numerically in two-dimensional, small-domain experiments (Christensen 1984a; Moresi and Solomatov 1995) on the basis of changes in the Nusselt number-Rayleigh number relationship. The transition to the stagnant-lid regime was also identified in the three-dimensional simulations of Ogawa et al. (1991) on the basis of the velocity distribution, but the small aspect ratio of their com-

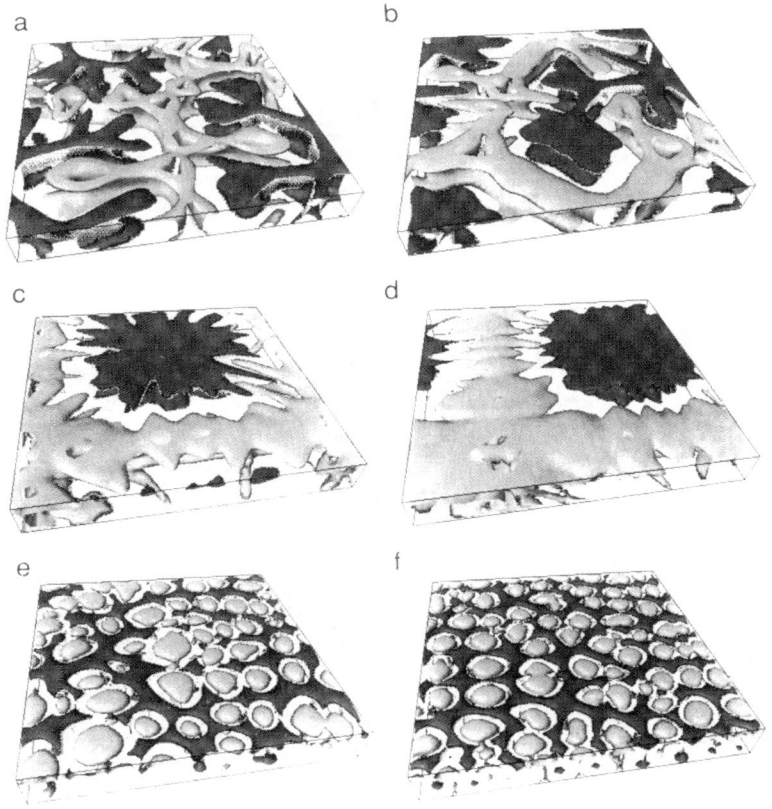

Figure 7. A series of Boussinesq calculations in an $8\times8\times1$ Cartesian box with fixed Rayleigh number based on the viscosity at the mean of the top and bottom boundary temperatures of 10^5 and increasing viscosity contrast. Figures show isosurfaces of residual temperature ΔT, i.e., temperature relative to a horizontally-averaged temperature profile, with dark surfaces representing cold (downwelling) material and light surfaces representing hot (upwelling) material. (a) constant viscosity, surfaces show $\Delta T = \pm 0.1$; (b) $\Delta\eta = 10$, $\Delta T = \pm 0.1$; (c) $\Delta\eta = 100$, $\Delta T = \pm 0.1$; (d) $\Delta\eta = 1000$, $\Delta T = \pm 0.1$; (e) $\Delta\eta = 10^4$, $\Delta T = \pm 0.075$; (f) $\Delta\eta = 10^5$, $\Delta T = \pm 0.05$.

putational domain precluded the change in preferred convective planform from occurring.

The transition from the small viscosity contrast mode of convection to the sluggish-lid mode of convection has also been found to occur in a numerical model of three-dimensional convection in a spherical shell (Ratcliff et al. 1995). In this model the upper and lower boundaries are isothermal and free

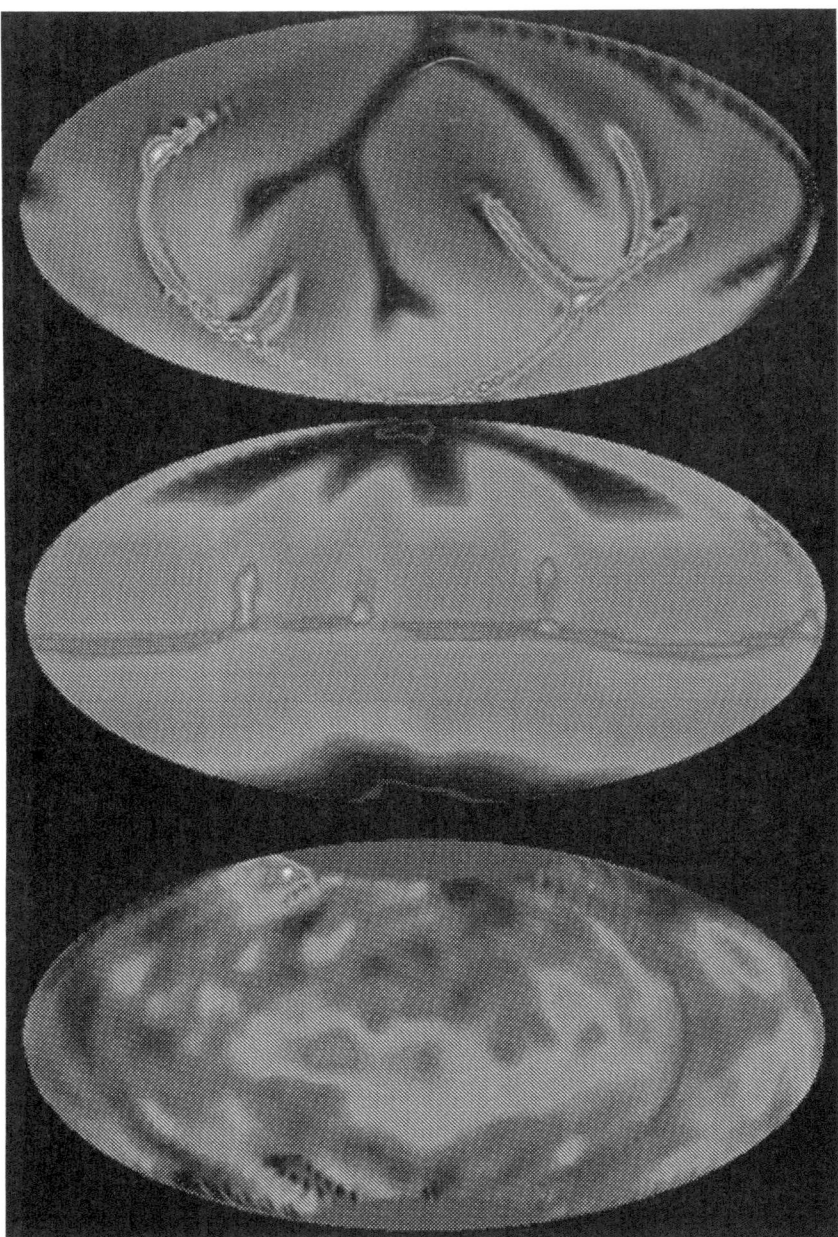

Figure 8. Patterns of convection in a spherical shell for constant viscosity (top), variable viscosity with a viscosity contrast of 10^3 (middle), and the topography of Venus (bottom). The convection patterns are illustrated by the temperature variations at mid-depth. Darker shades indicate either colder fluid or lower topography. The temperature fields and topography are displayed on a Hammer-Aitoff equal area projection (figure after Ratcliff et al. 1995).

slip and heating is from below. The Boussinesq approximation is made and the viscosity depends exponentially on temperature according to a linearized Arrhenius law. The Rayleigh number based on the mean of the top and bottom boundary temperatures is 10^5 and the viscosity varies by 10^3.

Figure 8 compares the spherical shell convection patterns for constant viscosity convection (also at a Rayleigh number of 10^5) and variable viscosity convection. The different convection patterns for the constant viscosity and variable viscosity cases show the transition to the sluggish-lid mode of convection. In the sluggish-lid mode that occurs for a viscosity ratio of 10^3, downwelling occurs as two large quasi-cylindrical structures. Finger-like spokes radiate asymmetrically from these features, reminiscent of the "spider planform" observed in the experimental work of Weinstein and Christensen (1991). Upwellings take the form of a nearly linear chain of partially connected plumes which encircle the sphere. The flow pattern has a dominant degree 2 spherical harmonic signature. Comparison of the horizontally averaged profiles of temperature, velocity and viscosity with those of the isoviscous case in Fig. 9 illustrates the warming of the interior due to formation of a cold and sluggish viscous lid adjacent to the upper boundary. The sluggish-lid style of convection in the spherical shell is quite similar to the sluggish-lid planform of convection in the Cartesian box discussed above. The transition from sluggish-lid to stagnant-lid convection in a spherical shell has recently been modeled by Ratcliff et al. (1996a,b).

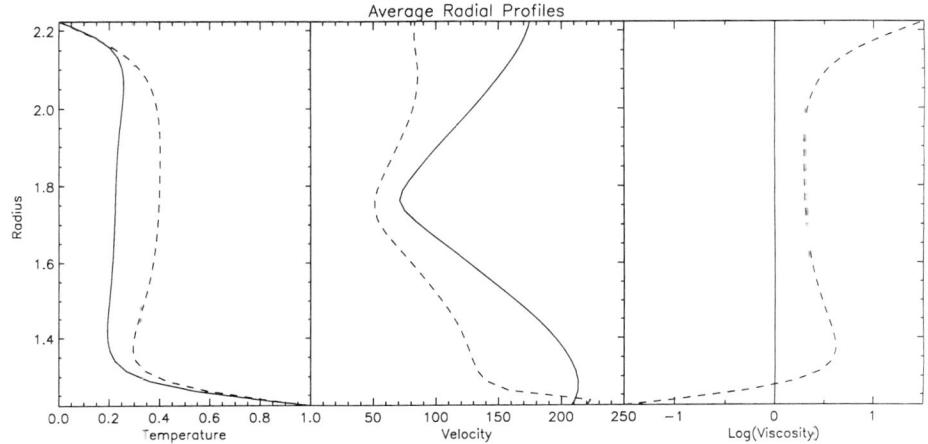

Figure 9. Horizontally-averaged profiles of temperature (left), velocity (center) and viscosity (right) in the spherical-shell variable viscosity convection models. Solid curves are for the constant viscosity case and the dashed curves represent the variable viscosity case with viscosity contrast 10^3 (figure after Ratcliff et al. 1995).

Figure 8 also compares the spherical shell convection patterns with the topography of Venus. The band of upwelling with its plume concentrations in the variable viscosity computation correspond to Venus' equatorial highlands while the polar downwellings in the computation may be analogous to the northern hemisphere high-latitude plateau highland Ishtar Terra on Venus (Bindschadler et al. 1990; see the chapter by Kaula et al.). This qualitative comparison of Venus' surface with the style of variable viscosity convection suggests that Venus may be in the sluggish-lid convection regime. The observed tectonic deformation rates on Venus are small (Grimm 1994b) and could be consistent with either the sluggish-lid or stagnant-lid regime because surface deformation occurs in both modes of convection. The viscosity contrast calculated from the activation energy and temperature drop across the upper boundary layer should place Venus in the stagnant-lid convection regime. The Venusian lithosphere may be weaker than predicted from a simple calculation because of nonviscous deformation of the lithosphere. The mode of Venusian convection has important implications for its thermal evolution as we discuss later in this chapter.

VIII. EFFECTS OF VARIABLE VISCOSITY AND PHASE TRANSITIONS ON VENUSIAN MANTLE CONVECTION

The convection models discussed in the previous section show how the strong temperature dependence of rock viscosity significantly influences the style of mantle convection on Venus. We have also discussed earlier in this chapter how phase transitions have a major effect on the nature of Venusian mantle convection. In this section we examine model results in which both variable viscosity and phase changes are simultaneously taken into account. Variable viscosity and phase changes, when acting together, may influence convection differently than they do when acting separately. Previous numerical studies on the interaction of temperature-dependent viscosity convection with mantle phase transitions have generally consisted of two-dimensional models which focus on individual features, particularly slabs (Christensen and Yuen 1984; Zhong and Gurnis 1994,1995) and plumes (Nakakuki and Fujimoto 1994; Nakakuki et al. 1994; Schubert et al. 1995).

We base the discussion of this section on new calculations of three-dimensional, variable viscosity, phase-change modulated convection in a $4 \times 4 \times 1$ Cartesian box heated from below with reflecting side boundaries and isothermal, free-slip upper and lower boundaries. The model uses the compressible anelastic liquid approximation and material properties are depth dependent except for viscosity which has an Arrhenius dependence on temperature. Both the exothermic and endothermic phase changes are included in the model with Clapeyron slopes of 2 and -4 MPa K^{-1}, respectively. Incorporation of a yield stress in some calculations allows comparison with an Earth-like model which simulates aspects of nonviscous lithospheric deformation involved in plate tectonics (Weinstein and Olson 1992; Bercovici 1993).

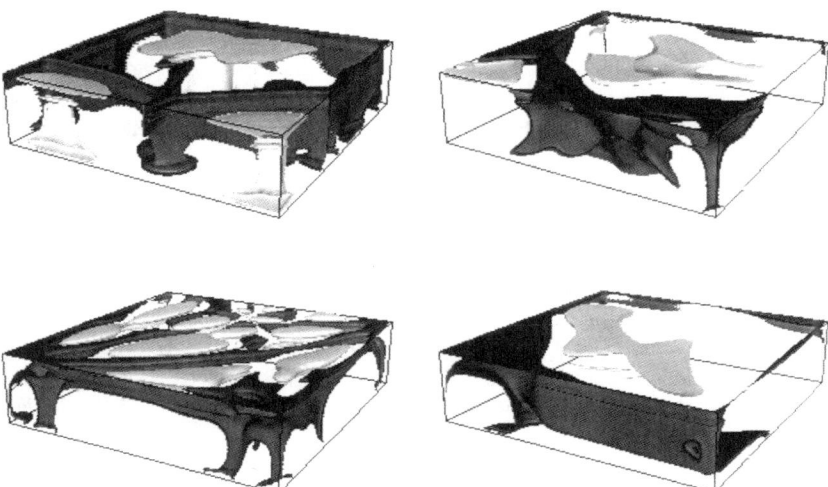

Figure 10. Cold downwellings (dark gray) and hot upwellings (light gray) for the three variable-viscosity phase change cases. Plotted are isosurfaces of residual temperature, i.e., temperature relative to the horizontally-averaged temperature profile, showing where the temperature is ±0.1 from the horizontal average. (a) (top left) constant viscosity case 1, (b) (lower left) variable viscosity case 2, (c) and (d) (right) variable viscosity with a yield stress case 3 at two different times.

Results for three cases—constant viscosity (case 1), variable viscosity (case 2), and variable viscosity with a yield stress (case 3)—are presented. In all cases the Rayleigh number, based on temperature, surface properties, and the viscosity of the reference adiabat at the surface, is 10^6. Viscosity varies by about a factor of 500 near the surface of the model and by about a factor of 2000 near the lower boundary. The patterns of convection in the three cases are shown in Fig. 10. Case 1 (with constant viscosity, top left in Fig. 10) displays a weakly time-dependent, whole-mantle mode of convection, with the broad, plume-like upwellings and interconnected sheet-like downwellings penetrating the endothermic phase transition with ease. This pattern is similar to previous convection calculations with no phase changes (Balachandar et al. 1992; Ratcliff et al. 1996b). The heat flux is 40.5 mW m^{-2}, about 20% lower than cases with similar parameters but no phase transitions reported in (Tackley 1994; Ratcliff et al. 1996b); this is due to the effect of the endothermic phase transition in inhibiting flow.

Case 2 (with temperature-dependent viscosity, lower left in Fig. 10) also exhibits only weak time-dependence, but in this case an appreciable degree of layering is apparent, with the sheet-like downwellings not penetrating the phase transition, but cold plumes at their intersections penetrating into the lower mantle. In contrast to the constant-viscosity calculations, these cold plumes do not display time-dependent "avalanche" behavior. This result is similar to a two-dimensional calculation by Zhong and Gurnis (1994) and corroborates the conclusion that a viscosity that is dependent on temperature only substantially increases the propensity of the mantle towards layering compared to constant viscosity. The heat flux, 22.8 mW m^{-2}, is substantially lower than in case 1 due to both the effect of a strong sluggish lithosphere in reducing heat flow, and to the greater degree of layering: layered convection is less efficient in transporting heat than whole-mantle convection.

Case 3 (similar to case 2 but with the addition of a yield stress, right panels in Fig. 10) exhibits very strong time dependence, with heat flux and velocity varying by more than a factor of 2 (heat flux varies between 20 and 45 mW m^{-2}). Two representative frames are illustrated in Fig. 10. The dominant feature in this case is a linear zone of downwelling, somewhat reminiscent of a terrestrial subduction zone, along the near and left edges of the box. Sections of this intermittently penetrate the phase transition into the lower mantle, with the form of penetration varying; the upper frame shows a large, substantially amorphous cold blob falling through the lower mantle, whereas the lower frame shows a coherent, linear slab-like feature extending from the surface to the core-mantle boundary. However, no cylindrical lower mantle avalanches, as observed in constant viscosity studies, occur.

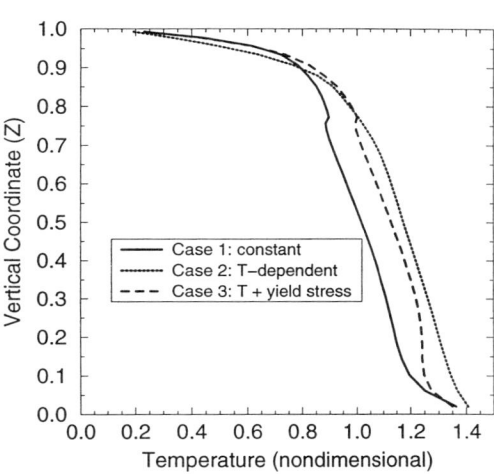

Figure 11. Profiles of horizontally-averaged temperature for the three variable viscosity phase change cases.

Profiles of the horizontally-averaged temperature for these cases are compared in Fig. 11. The important features in case 1 are the boundary layers near the upper and lower surfaces, the approximately adiabatic increase of temperature in the interior, and a jump in temperature at the depth of the endothermic phase transition due to the adiabatic release or absorption of latent heat by material passing through the transition. The lower boundary layer is much smaller than the surface boundary layer despite the 100% basal heating; this is a consequence of the increase of thermal diffusivity with depth in the model. In case 2, the interior temperature has increased greatly so that the lower boundary layer is almost imperceptible. The temperature step observed at the phase transition depth has disappeared because there is less vertical advection through the phase transition; much of the heat transport is now conducted. The yield stress introduced in case 3 has mitigated the effects of temperature-dependent viscosity and brought the temperature profile back in the direction of case 1, with a distinct lower boundary layer and kink in the profile at the phase transition.

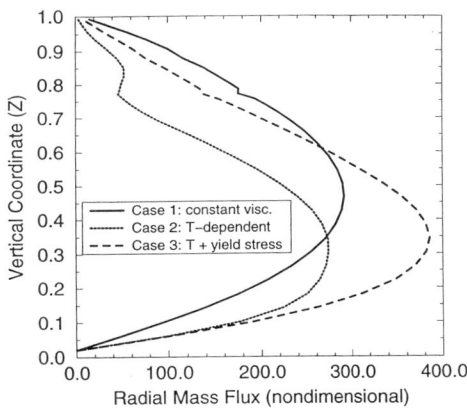

Figure 12. Profiles of horizontally-averaged vertical mass flux for the three variable viscosity phase change cases. Note that this is the mean of the absolute mass flux; conservation of mass requires the mean of the actual mass flux to be zero.

A more quantitative assessment of the degree of layering can be obtained from the radial mass flux (Fig. 12), here shown in absolute terms rather than normalized into the radial mass flux diagnostic. Case 1, constant viscosity, displays a smooth, symmetrical curve characteristic of whole mantle convection. The endothermic phase transition has a huge effect in the variable viscosity case 2; it reduces mass flux in the upper part of the mantle and causes a minimum in the mass flux at the endothermic phase transition depth, characteristic of partially-layered convection. In case 3, the signature of the phase transition is greatly reduced; only a slight change in the gradient of the curve is visible at the endothermic phase change depth. The curves for cases 1 and 2 are quite stable in time, whereas the curve for case 3 fluctuates by a factor of 2 to 3 in overall amplitude, though maintaining a similar profile.

These results illustrate the first-order effects of temperature-dependent viscosity on phase-change modulated mantle convection. Because the degree of layering increases with Rayleigh number (Christensen and Yuen 1985; Yuen et al. 1994), all cases would be more layered at the higher Rayleigh numbers realistic for Venus.

A comparison of cases 1 and 2 indicates clearly that inclusion of a viscosity which is dependent solely on temperature greatly increases the propensity of the system towards layering, a result previously indicated in two dimensions (Zhong and Gurnis 1994). An explanation for this is that downwellings in such temperature-dependent viscosity convection are broad and symmetric (i.e., two-sided), with considerable stress occurring where they leave the upper boundary layer (Tackley 1993; Steinbach and Yuen 1994). Thus, the negative buoyancy of the downwellings is largely used up in overcoming stress in the upper boundary layer, and is not available for overcoming the positive buoyancy associated with phase change deflection. This style of convection may be relevant to Venus, which does not display plate tectonics.

The Earth-like case (case 3) with a yield stress displays two main differences: (1) the degree of layering is reduced compared to that with a purely temperature-dependent viscosity; (2) the form of phase-change modulated convection was greatly changed compared to constant viscosity cases, with the high viscosity reducing the ability of cold material to spread in the transition zone, forcing avalanches to be linear. The variability in slab penetration both temporally and spatially and along strike of the model "subduction zone" is compatible with observations of subducted slabs in the Earth (Zhou 1990; van der Hilst et al. 1991; Fukao et al. 1992; van der Hilst and Seno 1993; Lay 1994).

Because plate tectonics is not observed on Venus it may be that case 2 is representative of Venus while case 3 is representative of the Earth. This implies that Venus' mantle may be more layered than Earth's mantle, including the possibility that Venus' mantle convection may currently be layered whereas Earth's mantle convection may be predominantly whole mantle. The spherical cases discussed above also suggest such a conclusion: the rigid-lid case was more layered than the free-slip case. A layered convection scenario for Venus would allow plumes from the upper/lower mantle interface to transport most of the heat (because they transport the heat from the lower mantle), which would be compatible with the plume-dominated tectonics favored by some researchers. The resurfacing event 500 Myr ago might correspond to a brief episode of whole-mantle convection.

IX. PARAMETERIZED CONVECTION MODEL OF THE THERMAL EVOLUTION OF VENUS

The numerical models of convection discussed above are probably most relevant to the present state of Venus' mantle although some of the results may be applicable to past states and the numerical results do suggest possible explanations for the global resurfacing event 500 Myr ago (mantle overturns due to a transition in convective mode or due to an avalanche). However, the realistic three-dimensional convection models are too computationally demanding to be run in a thermal evolution mode. Instead, the history of Venus is most readily explored in a quantitative manner using the parameterized convection approach. This technique accounts for the contribution of mantle convective heat transfer to the overall energy balance of a planet by a Nusselt number-Rayleigh number relation. The approach has been exploited for about two decades in studies of planetary thermal evolution (see, e.g., McKenzie and Weiss 1975; Sharpe and Peltier 1978,1979; Schubert et al. 1979; Schubert 1979; Turcotte et al. 1979; Daly 1980; McKenzie and Richter 1981; Spohn and Schubert 1982; Christensen 1985). Specific thermal evolution studies of Venus using the parameterized convection approach include those of Schubert et al. (1979), Stevenson et al. (1983), Solomatov et al. (1987); Solomatov and Zharkov (1990), Zharkov and Solomatov (1992), and Solomatov (1993). The thermal evolution study of Stevenson et al. (1983) is noteworthy in providing an explanation for the nonexistence of a Venusian intrinsic magnetic field—the likely absence of inner core solidification in Venus implies lack of an energy source to drive convection in the liquid part of the core. However, Arkani-Hamed's (1994) investigation of Venus' thermal history attributes the absence of a Venusian magnetic field to the freezing of Venus' core as a result of more efficient past heat removal on Venus compared with Earth. At present, with the Earth in the mobile-lid convective regime and Venus in the sluggish/stagnant lid regime, Earth is more efficient at heat removal.

Our discussion of Venus' thermal evolution in this section follows Solomatov (1995). We discuss the thermal evolution of a Venus model cooling from a hot initial state. Theories of the formation of the planets suggest that Venus must have been extensively heated by large impacts during its accretion (Safronov 1978; Wetherill 1990). This, and the addition of gravitational energy release due to core formation (Birch 1965; Flasar and Birch 1973), heating by adiabatic compression, heating by short-lived radioactive isotopes (Safronov 1978) and thermal blanketing by a greenhouse (Abe and Matsui 1985; Zahnle et al. 1988) all imply a very hot interior after the early formation period. The period of post-accretional cooling, melt differentiation and convective remixing is short compared to the entire evolution (Solomatov and Stevenson 1993). The thermal evolution controlled by solid state convection, which we discuss here, started around 4.5 Gyr ago and the initial potential temperature of Venus could not have been much higher than the solidus (Solomatov and Stevenson 1993).

The parameterized thermal evolution model assumes that the mantle of Venus contains a terrestrial-like complement of radiogenic heat sources, as discussed above. The influence of Venus' core on the thermal evolution of its mantle is minimal (Solomatov et al. 1987; Zharkov and Solomatov 1992) and the model does not include the core. The thermal evolution of Venus is found by integrating a simple energy balance equation for the average temperature

T of the mantle forward in time.

The appropriate form of the Nusselt number-Rayleigh number relation for convection with temperature-dependent viscosity depends on the regime of convection. In the small viscosity contrast regime, convection is controlled by the interior viscosity and is described by formulae identical to those for constant viscosity convection, provided the viscosity is calculated at the average temperature. This has been assumed in many parameterized convection calculations (see, e.g., Schubert et al. 1979; Stevenson et al. 1983).

In the sluggish-lid regime the heat transport depends almost entirely on the surface viscosity. Investigations in the parameter range corresponding to this regime led Christensen (1984b,1985) to conclude that the surface velocity and Nusselt number are almost constant during the evolution of the Earth provided the surface temperature and the surface viscosity are constant. The initial conditions control the entire evolution in this regime, while in traditional models the initial conditions are not important after about 1 Gyr of evolution. Christensen's (1985) conclusions are valid only for convection in this regime and it is not clear if they are applicable to the Earth and planets. In this regime, convection depends entirely on the rheology of the lithosphere.

In the stagnant-lid regime the efficiency of heat transport depends only on the rheology in the hot part of the cell where the viscosity variations do not exceed one order of magnitude, and convective heat transfer no longer depends on the rheology of the lid.

In order to calculate a thermal history for Venus with the parameterized convection approach we must first identify the thermal regime in which Venus is found at different stages of its evolution. On the basis of theoretical estimates of viscosity contrasts, mantle convection on any terrestrial planet should be in the stagnant-lid regime. This convective regime might be most consistent with a Venusian lithosphere as thick as 200 to 300 km and the small surface deformation rates on Venus (Grimm 1994b). However, as discussed earlier in this chapter, a comparison of features on the surface of Venus with numerically computed patterns of convection suggests that Venus may be in the sluggish-lid mode of convection.

The existence of plate tectonics on Earth suggests that the terrestrial lithosphere is weak due to brittle failure or other reasons and the effective viscosity contrast between the lithosphere and underlying mantle is not very large (Kaula 1980). The rheology of the Earth's lithosphere may have either no effect on mantle convection as in the small viscosity contrast regime (except, of course, for enabling this convective regime in the first place) or it may have a dominant effect as in the sluggish-lid regime. The answer is not known yet; the simplest assumption and the one usually adopted is that plate tectonics is similar to the small viscosity contrast regime.

As is the case for Earth, nonviscous modes of lithospheric deformation may control the convection regime on Venus and place Venus in the sluggish-lid regime. Venus might even be in the small viscosity contrast regime even though global plate tectonics is not now occurring on Venus. While global plate tectonics is equivalent to small viscosity contrast convection, the inverse is not necessarily true.

For purposes of "bounding" the thermal evolution of Venus we present results for two end-member thermal history models; one model assumes a thermal evolution controlled by small viscosity contrast convection and the other model assumes that rigid-lid convection dominates the thermal history. Details of the models and values of parameters can be found in Solomatov (1995). The thermal evolution of Venus models with mobile plates (small viscosity contrast regime) and with a stagnant lid are shown in Fig. 13. Three different initial temperatures are chosen to show the effect of initial conditions.

The regime with mobile plates gives evolution scenarios in which the present mantle temperature is close to that of the Earth. In the stagnant-lid regime, the temperature has to increase to more than 2000°C for the surface heat loss rate to become comparable with the radiogenic heating rate. This exceeds the melting temperature by several hundred degrees. A low rate of volcanism on Venus (at least 10 times smaller than on Earth) seems to be in contradiction with a higher temperature of the Venusian mantle.

Another problem with both solutions is that the lithospheric thickness (the thermal boundary layer) is much less than the 200 to 300 km value inferred from topography and gravity data. Even with a very depleted mantle, the lithosphere thickness would still be too small (Solomatov and Moresi 1996).

The problem with the lithosphere thickness prediction of these parameterized convection models has been discussed earlier in the chapter. The numerical convection models show that the global gravity and topography data can be consistent with a lithosphere only about 100 km thick overlying a convecting mantle and transporting a terrestrial-like heat flux. However, gravity and topography data on a smaller-scale and the nature of small-scale lithospheric deformation on Venus also suggest that Venus has a very thick lithosphere. In the following sections we discuss two *ad hoc* thermal evolution scenarios that might explain the thick lithosphere dilemma.

X. A THERMAL EVOLUTION SCENARIO INVOLVING CESSATION OF PLATE TECTONICS 500 MYR AGO

One possible explanation for both the global resurfacing event 500 Myr ago and a thick lithosphere at present is a change in convective style 500 Myr ago from plate tectonics (small viscosity contrast convection) to the stagnant-lid regime. The parameterized convection model of the previous section can be used to quantitatively evaluate this possibility. We first investigate the consequences of the hypothesized transition and then discuss why plate tectonics could have stopped on Venus 500 Myr ago. Cessation of plate tectonics 500 Myr ago as an explanation for the relative lack of resurfacing on Venus since that time has been suggested by Arkani-Hamed et al. (1993), Arkani-Hamed (1994), and Herrick (1994).

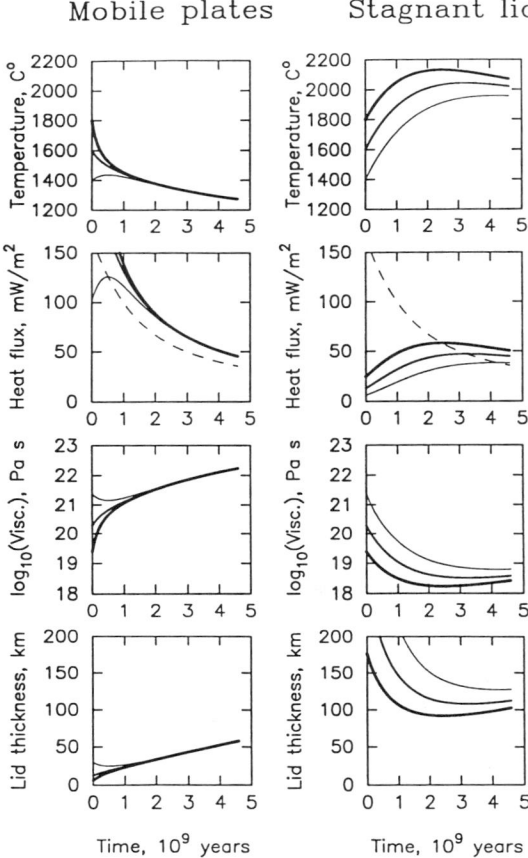

Figure 13. Parameterized convection calculations of the thermal evolution of Venus under the assumption that the convective regime on Venus was the same during the entire evolution. The regime with mobile plates (small viscosity contrast regime) is shown on the left, the stagnant-lid regime is on the right. The following convective parameters are calculated: the mantle temperature T, the surface heat flux, the viscosity at temperature T, and the thickness of the lithosphere (the cold thermal boundary layer). The dashed line shows the surface heat flux that would be in equilibrium with radiogenic heat production in the mantle. The three different curves correspond to different choices of initial temperature.

If a plate tectonic-like convective regime prevailed on Venus prior to 500 Myr ago and then ceased at that time, the deep interior of Venus would have heated up and a thicker lithosphere would have grown after the cessation of plate-like motions. Prior to the event of 500 Myr ago, the surface heat flux would have slightly exceeded the internal radiogenic heating as a result of a small contribution to the heat flux from secular cooling. After the plates stop, convective instabilities would begin to develop near the bottom of the lithosphere after a short incubation period (Davaille and Jaupart 1994). This is the stagnant-lid form of convective motion. It is much less efficient than plate-like convection and implies a substantial drop in the surface heat loss rate. The heat loss rate becomes smaller than the radiogenic heat production rate by almost an order of magnitude and the net result is that the interior of the planet heats up.

The cooling of the lithosphere is due to the difference between the large near-surface thermal gradients left over from the plate tectonics regime and the small heat flux carried by instabilities at the bottom of the lithosphere. As a result, the lithosphere is cooled from the top much faster than it is heated from below. The half-space cooling model adopted by (Turcotte 1993) can be used to calculate the thermal regime of the lithosphere for the first few hundred Myr after the cessation of plate tectonics (assuming the absence of secondary convection that would prevent continued lithospheric thickening). Accordingly, the surface heat flux F and the lithosphere thickness δ change with time as

$$F = \frac{k(T - T_s)}{\delta(t)}, \quad \delta(t) = [\pi \kappa (t - t_0)]^{1/2} \tag{7}$$

where κ is the thermal diffusivity, T_s is the surface temperature, and t_0 is a time slightly smaller than the time when plate motion stopped and is found from the requirement of a continuous change in the boundary layer thickness and the surface heat flux.

A sketch of this scenario is shown in Fig. 14 while Fig. 15 gives quantitative results from a thermal evolution calculation that *a priori* imposes the change in convective regime 500 Myr ago. The plots in Fig. 15 show an increase in interior temperature, a drop in surface heat flux, and a dramatic thickening of the lithosphere since the hypothesized change in convective regime 500 Myr ago. The magma production rate would also decrease after cessation of plate tectonics due to a drop in the convective velocity and an increase in the lithospheric thickness. A thicker lithosphere prevents convective flow from reaching close to the surface thereby substantially reducing melting due to adiabatic decompression (Fig. 14). Therefore, a decrease in the resurfacing rate on Venus would result not only from the change from mobile plates to a stagnant lid, but also from a large decrease in the rate of melt production. This is in agreement with the estimates of a very low level of present-day volcanism on Venus (Fegley and Prinn 1989; Head et al. 1991,1992). A continuing decay in the rate of volcanism due to gradual development of the stagnant lid is also in agreement with Price and Suppe's (1994) suggestion that limited volcanism continued after the resurfacing event.

Why would a plate tectonic-like regime on Venus cease to operate 500 Myr ago? A possible answer lies in the gradual thickening of the plates with time throughout the evolution of Venus. The thickening of the lithosphere with time increases its strength while the stresses in the lithosphere are decreasing with time (Solomatov and Moresi 1996). Early in the evolution of Venus the stresses could have been large enough to break the lithosphere,

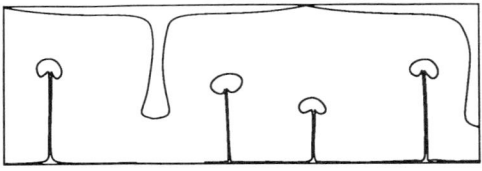

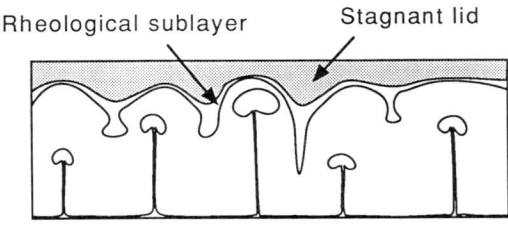

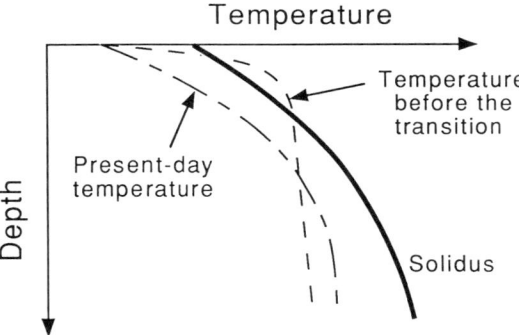

Figure 14. A schematic of convection in Venus before the cessation of the mobile plate regime around 0.5 Gyr ago and at present and a sketch of the depth profiles of temperature before and after the event.

perhaps by faulting, and result in a plate tectonic-like convective regime. When the stresses in the lid dropped below a critical value associated with the increasing strength of the lithosphere, perhaps 500 Myr ago, active surface motion would cease. This mechanical explanation of the cessation of subduction and initiation of the stagnant-lid convective regime is similar to a model suggested by Fowler (1993).

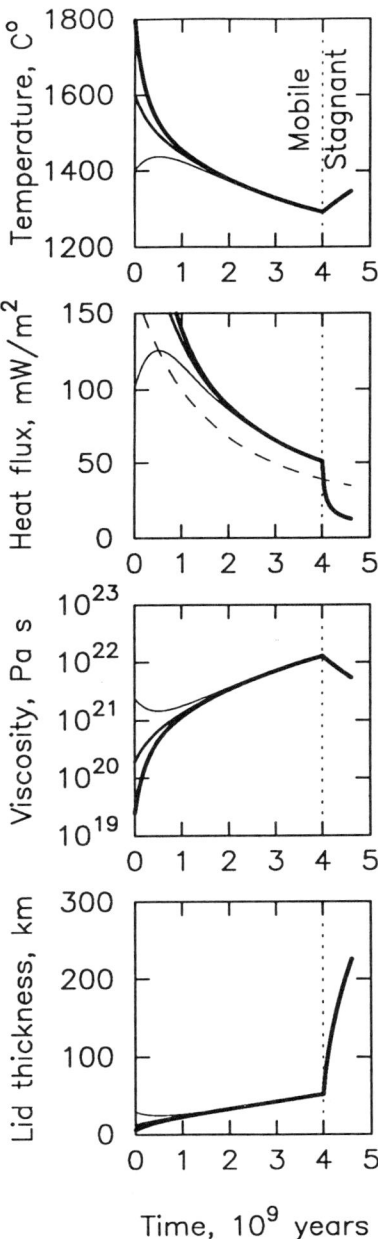

Figure 15. Parameterized convection calculations of the thermal evolution of Venus with the hypothesized change from the regime with mobile plates to the stagnant-lid regime 0.5 Gyr ago.

XI. A THERMAL EVOLUTION SCENARIO INVOLVING GLOBAL OVERTURN OF AN OVER-THICKENED LITHOSPHERE

In the scenario of the preceding section we have seen how the lithosphere would grow to a thickness of several hundred km (Fig. 15) following the event at 500 Myr ago. It is possible that this over-thickened, gravitationally unstable lithosphere would overturn, release heat stored in the interior during surface quiescence, and resurface the planet. If Venus is presently in a stagnant-lid, nonsteady convective regime with an already thick and thickening lithosphere, global lithospheric overturn by gravitational instability could lie in Venus' future. The event at 500 Myr could in fact have been such a lithospheric overturn if episodic mantle overturning interspersed with long periods of stagnant-lid convection characterized Venus' thermal evolution (Turcotte 1993). Mantle overturning and episodicity can also arise from effects of chemical differentiation (Parmentier and Hess 1992; Herrick and Parmentier 1994).

There is observational evidence from lava lakes that episodic subduction is a valid style of heat transfer in a convecting system with a very viscous (rigid) upper thermal boundary layer. Cooling to the atmosphere creates a "solid" crust on the top of a lava lake which is gravitationally unstable with respect to the molten magma beneath and undergoes episodic subduction. Wright et al. (1968) describe repeated crustal foundering events and spectacular crustal overturns of the Makaopuhi lava lake during the eruption of the Kilauea volcano in March 1965.

It is one thing to hypothesize the global overturn of a thick and gravitationally unstable lithosphere on Venus. However, a mechanism to initiate and facilitate such an overturn needs to be found. Before the strong, thick Venusian lithosphere can subduct it must be broken. Schubert and Sandwell (1995) have identified extension and rifting as the mechanism for initiating lithospheric foundering and subduction on Venus. Mantle plumes of unusual vigor could arise in the overheated deep interior of a planet with an over-thickened lithosphere and substantially weaken, thin, upwarp, and crack the lithosphere. Rifts propagating away from the center of such hotspot sites could break nearby lithosphere thick enough to be gravitationally unstable and initiate retrograde lithospheric foundering or subduction. If there are a sufficient number of such sites a global lithospheric overturn could occur. Venus may be on the verge of such an event. The evidence consists of the major volcanic rises such as Atla Regio and Beta Regio, the chasmata (rifts) emanating from the hotspot centers and extending into the lowlands away from the rises, and the possible occurrence of subduction along major segments of the chasmata far from the rises (Schubert and Sandwell 1995).

XII. SUMMARY AND CONCLUSION

We have discussed the possible style of mantle convection in Venus and suggested a number of scenarios for the planet's thermal evolution. Venus could be in any of three basic modes of convection—a small viscosity contrast regime, a sluggish-lid regime, and a stagnant-lid regime. The absence of global plate tectonics on Venus argues against the existence of the small viscosity contrast regime, but such a convective style cannot be totally ruled out because plate tectonics might not be the only surface manifestation of small viscosity contrast convection. Planforms of convection in the sluggish-lid regime resemble the distribution of major geologic features on Venus and inferences of surface deformation rates on Venus support the relevance to Venus of stagnant-lid convection. Stagnant-lid convection is a definite possibility if Venus has a lithosphere with globally-averaged thickness of 200 to 300 km. The inference of such a thick Venusian lithosphere is based in part on the strong correlation of gravity and topography and the large admittance ratios on the planetary scale. Numerical convection models demonstrate, however, that these attributes of the gravity and topography on the large scale are also characteristic of convection beneath a lithosphere only about 100 km thick on average. Even so, a several hundred kilometer thick Venusian lithosphere is also suggested by gravity and topography over individual features such as major highlands, and smaller features such as coronae. The observed lack of crater degradation and observations of flexural deformation lend further credence to the existence of a thick lithosphere on Venus. Simple arguments, the numerical convection calculations, and thermal history calculations all support the conclusion that if Venus indeed has a lithosphere as thick as 200 km to 300 km on average, then surface heat loss on Venus is much less than heat production inside Venus if the planet has a terrestrial complement of radiogenic elements and Venus cannot presently be in an approximate thermal steady state.

The premier event in the recent evolution of Venus is a major resurfacing of the planet about 500 Myr ago with little resurfacing since. There seems to be no lack of feasible scenarios to explain this inferred catastrophic event. Models of convection suggest that the endothermic phase change in Venus' mantle has a strong tendency to layer mantle convection. This suggests one possible explanation for the event 500 Myr ago—a transition via a global mantle avalanche from layered convection to a brief period of whole mantle convection followed again by layered convection (Steinbach et al. 1993; Steinbach and Yuen 1994). Other possible explanations for the event include a transition from a plate tectonic-like style of convection to stagnant-lid convection and episodic convection involving long periods of stagnant-lid convection interspersed with brief periods of mantle overturn by the gravitational instability of an overthickened lithosphere. Inherent in these mechanisms is the possibility of a repetition of the resurfacing event in the future. If subduction is occurring along segments of Venusian chasmata we may be witness to the beginning of the next catastrophe.

Acknowledgments. This research was supported in part by a grant from NASA and a grant from the NSF.

REFERENCES

Abe, Y., and Matsui, T. 1985. The formation of an impact generated H_2O atmosphere and its implications for the early thermal history of the Earth. *J. Geophys. Res.* 90:545–559.

Arkani-Hamed, J. 1993. On the tectonics of Venus. *Phys. Earth Planet. Int.* 99:2019–2033.

Arkani-Hamed, J. 1994. On the thermal evolution of Venus. *J. Geophys. Res.* 99:2019–2033.

Arkani-Hamed, J. 1996. Analysis and interpretation of high-resolution surface topography and gravity of Ishtar Terra, Venus. *J. Geophys. Res.* 101:4691–4710.

Arkani-Hamed, J., Schaber, G. G., and Strom, R. G. 1993. Constraints on the thermal evolution of Venus inferred from Magellan data. *J. Geophys. Res.* 98:5309–5315.

Balachandar, S., Yuen, D. A., and Reuteler, D. 1992. Time-dependent 3-dimensional compressible convection with depth-dependent properties. *Geophys. Res. Lett.* 19:2247–2250.

Balachandar, S., Yuen, D. A., and Reuteler, D. M. 1995a. Localization of toroidal motion and shear heating in 3-D high Rayleigh number convection with temperature-dependent viscosity. *Geophys. Res. Lett.* 22:477–480.

Balachandar, S., Yuen, D. A., Reuteler, D. M., and Lauer, G. S. 1995b. Viscous dissipation in 3-dimensional convection with temperature-dependent viscosity. *Science* 267:1150–1153.

Banerdt, W. B., et al. 1994. The isostatic compensation state of Mead crater. *Icarus* 112:117–129.

Bercovici, D. 1993. A simple-model of plate generation from mantle flow. *Geophys. J. Int.* 114:635–650.

Bindschadler, D. L., Schubert, G., and Kaula, W. M. 1990. Mantle flow tectonics and the origin of Ishtar Terra, Venus. *Geophys. Res. Lett.* 17:1345–1348.

Birch, F. 1965. Energetics of core formation. *J. Geophys. Res.* 70:6217–6221.

Bird, P. 1979. Continental delamination and the Colorado Plateau. *J. Geophys. Res.* 84:7561–7571.

Buck, W. R. 1992. Global decoupling of crust and mantle: Implications for topography, geoid, and mantle viscosity on Venus. *Geophys. Res. Lett.* 19:2111–2114.

Bullock, M. A., Grinspoon, D. H., and Head, J. W. 1993. Venus resurfacing rates: Constraints provided by 3-D Monte Carlo simulations. *Geophys. Res. Lett.* 20:2147–2150.

Christensen, U. R. 1985. Thermal evolution models for the Earth. *J. Geophys. Res.* 90:2995–3007.

Christensen, U. R. 1984a. Convection with pressure-dependent and temperature-dependent non-Newtonian rheology. *Geophys. J. Roy. Astron. Soc.* 77:343–384.

Christensen, U. R. 1984b. Heat-transport by variable viscosity convection and implications for the Earth's thermal evolution. *Phys. Earth Planet. Int.* 35:264–282.

Christensen, U. R., and Harder H. 1991. 3-d convection with variable viscosity. *Geophys. J. Int.* 104:213–226.

Christensen, U. R., and Yuen, D. A. 1984. The interaction of a subducting lithospheric slab with a chemical or phase boundary. *J. Geophys. Res.* 89:4389–4402.

Christensen, U. R., and Yuen, D. A. 1985. Layered convection induced by phase transitions. *J. Geophy. Res.* 90:291–300.

Daly, S. F. 1980. Convection with decaying heat sources: Constant viscosity. *Geophys. J. Roy. Astron. Soc.* 61:519–547.

Davaille, A., and Jaupart, C. 1994. Onset of thermal convection in fluids with temperature dependent viscosity: Application to the oceanic mantle. *J. Geophys. Res.* 99:19853–19,866.

Fegley, B., Jr., and Prinn, R. G. 1989. Estimation of the rate of volcanism on Venus from reaction rate measurements. *Nature* 337:55–58.

Flasar, F. M., and Birch, F. 1973. Energetics of core formation: A correction. *J. Geophys. Res.* 78:6101–6103.

Fowler, A. C. 1993. Boundary layer theory and subduction. *J. Geophys. Res.* 98:21997–22995.

Freed, A. M., and Melosh, H. J. 1995. Long term survival of the topography of Ishtar Terra, Venus. *Lunar and Planet. Sci. Conf.* XXVI:421–422 (abstract).

Fukao, Y., Obayashi, M., Inoue, H., and Nenbai, M. 1992. Subducting slabs stagnant in the mantle transition zone. *J. Geophys. Res.* 97:4809–4822.

Greeley, R., and Schneid, B. D. 1991. Magma generation on Mars: Amounts, rates, and comparison with Earth, Moon, and Venus. *Science* 254:996–998.

Grimm, R. E. 1994a. The deep structure of Venusian plateau highlands. *Icarus* 112:89–103.

Grimm, R. E. 1994b. Recent deformation rates on Venus. *J. Geophys. Res.* 99:23163–23171.

Grimm, R. E., and Solomon, S. C. 1988. Viscous relaxation of impact crater relief on Venus: Constraints on crustal thickness and thermal gradient. *J. Geophys. Res.* 93:11911–11929.

Head, J. W., et al. 1991. Venus volcanism: Initial analysis from Magellan data. *Science* 252:276–288.

Head, J. W., et al. 1992. Venus volcanism: Classification of volcanic features and structures, associations, and global distribution from Magellan data. *J. Geophys. Res.* 97:13153–13197.

Heiken, G., Vaniman, D., and French, B. M. 1991. *Lunar Source Book* (Cambridge: Cambridge Univ. Press).

Herrick, R. R. 1994. Resurfacing history of Venus. *Geology* 22:703–706.

Herrick, D. L., and Parmentier, E. M. 1994. Episodic large-scale overturn of two layer mantles in terrestrial planets. *J. Geophys. Res.* 99:2053–2062.

Honda, S., Yuen, D. A., Balachandar, S., and Reuteler, D. 1993. 3-dimensional instabilities of mantle convection with multiple phase transitions. *Science* 259:1308–1311.

Kaula, W. M. 1980. Material properties for mantle convection consistent with observed surface fields. *J. Geophys. Res.* 85:7031–7044.

Kaula, W. M. 1990. Venus: A contrast in evolution to Earth. *Science* 247:1191–1196.

Kaula, W. M. 1995. Venus reconsidered. *Science* 270:1460–1464.

Kaula, W. M., and Phillips, R. J. 1981. Quantitative tests for plate tectonics on Venus. *Geophys. Res. Lett.* 8:1187–1190.

Kiefer, W. S., and Hager, B. H. 1991. A mantle plume model for the equatorial highlands of Venus. *J. Geophys. Res.* 96:20947–20966.

Kiefer, W. S., and Hager, B. H. 1992. Geoid anomalies and dynamic topography from convection in cylindrical geometry: Applications to mantle plumes on Earth and Venus. *Geophys. J. Int.* 108:198–214.

Kiefer, W. S., Richards, M. A., and Hager, B. H. 1986. A dynamic model of Venus's gravity field. *Geophys. Res. Lett.* 13:14–17.

Konopliv, A. S., and Sjogren, W. L. 1994. Venus spherical harmonic gravity model to degree-60 and order-60. *Icarus* 112:42–54.

Konopliv, A. S., et al. 1993. Venus gravity and topography—60th degree and order model. *Geophys. Res. Lett.* 20:2403–2406.

Kucinskas, A. B., and Turcotte, D. L. 1994. Isostatic compensation of equatorial highlands on Venus. *Icarus* 112:104–116.

Lay, T. 1994. The fate of descending slabs. *Ann. Rev. Earth Planet. Sci.* 22:33–61.

Leitch, A. M., and Yuen, D. A. 1991. Compressible convection in a viscous Venusian mantle. *J. Geophys. Res.* 96:5551–5562.

Leitch, A. M., Yuen, D. A., and Lausten, C. L. 1992. Axisymmetrical spherical-shell models of mantle convection with variable properties and free and rigid lids. *J. Geophys. Res.* 97:20899–20923.

Machetel, P., and Weber, P. 1991. Intermittent layered convection in a model mantle with an endothermic phase change at 670 km. *Nature* 350:55–57.

Mackwell, S. J., Zimmerman, M. E., Kohlstedt, D. L., and Scherber, D. S. 1995. Experimental deformation of dry Columbia diabase: Implications for tectonics on Venus. In *Rock Mechanics: Proc. 35th U. S. Symp.*, eds. J. J. K. Daemen and R. A. Schultz (Brookfield, Vt.: A. A. Balkema), pp. 207–214.

McKenzie, D. 1994. The relationship between topography and gravity on Earth and Venus. *Icarus* 112:55–88.

McKenzie, D. P., and Richter, F. M. 1981. Parameterized thermal convection in a layered region and the thermal history of the Earth. *J. Geophys. Res.* 86:11667–11680.

McKenzie, D. P., and Weiss, N. O. 1975. Speculations on the thermal and tectonic history of the Earth. *Geophys. J. Roy. Astron. Soc.* 42:131–174.

McKenzie, D. P., et al. 1992. Features on Venus generated by plate boundary processes. *J. Geophys. Res.* 97:13533–13544.

Moore, W. B., and Schubert, G. 1995. Lithospheric thickness and mantle/lithosphere density contrast beneath Beta Regio, Venus. *Geophys. Res. Lett.* 22:429–432.

Moresi, L.-N., and Solomatov, V. S. 1995. Numerical investigation of 2D convection with extremely large viscosity variations. *Phys. Fluids* 7(9):2154–2162.

Nakakuki, T., and Fujimoto, H. 1994. Interaction of the upwelling plume with the phase and chemical-boundaries. 2. Effects of the pressure-dependent viscosity. *J. Geomag. Geoelec.* 46:587–602.

Nakakuki, T., Sato, H., and Fujimoto, H. 1994. Interaction of the upwelling plume with the phase and chemical-boundary at the 670 km discontinuity—Effects of temperature-dependent viscosity. *Earth Planet. Sci. Lett.* 121:369–384.

Namiki, N., and Solomon, S. C. 1994. Impact crater densities on volcanoes and coronae on Venus: Implications for volcanic resurfacing. *Science* 265:929–933.

Ogawa, M., Schubert, G., and Zebib, A. 1991. Numerical simulations of 3-dimensional thermal convection in a fluid with strongly temperature-dependent viscosity. *J. Fluid Mech.* 233:299–328.

Parmentier, E. M., and Hess, P. C. 1992. Chemical differentiation of a convecting planetary interior: Consequences for a one plate planet such as Venus. *Geophys. Res. Lett.* 19:2015–2018.

Parmentier, E. M., Sotin, C., and Travis, B. J. 1994. Turbulent 3-d thermal convection in an infinite Prandtl number, volumetrically heated fluid—Implications for mantle dynamics. *Geophys. J. Int.* 116:241–251.

Peltier, W. R., and Solheim, L. P. 1992. Mantle phase transitions and layered chaotic convection. *Geophys. Res. Lett.* 19:321–324.

Phillips, R. J. 1994. Estimating lithospheric properties at Atla Regio, Venus. *Icarus* 112:147–170.

Phillips, R. J., and Hansen, V. L. 1994. Tectonic and magmatic evolution of Venus. *Ann. Rev. Earth Planet. Sci.* 22:597–654.

Phillips, R. J., and Izenberg, N. R. 1995. Ejecta correlations with spatial crater density and Venus resurfacing history. *Geophys. Res. Lett.* 22:1517–1520.

Phillips, R. J., Kaula, W. M., McGill, G. E., and Malin, M. C. 1981. Tectonics and evolution of Venus. *Science* 212:879–887.

Phillips, R. J., Grimm, R. E., and Malin, M. C. 1991. Hot-spot evolution and the global tectonics of Venus. *Science* 252:652–658.

Price, M., and Suppe, J. 1994. Mean age of rifting and volcanism on Venus deduced from crater densities. *Nature* 372:756–759.

Rappaport, N., and Plaut, J. 1994. A 360-degree and -order model of Venus topography. *Icarus* 112:27–33.

Ratcliff, J. T., Schubert, G., and Zebib, A. 1995. Three-dimensional variable viscosity convection of an infinite Prandtl number Boussinesq fluid in a spherical shell. *Geophys. Res. Lett.* 22:2227–2230.

Ratcliff, J. T., Schubert, G., and Zebib, A. 1996a. Effects of temperature-dependent viscosity on thermal convection in a spherical shell. *Physica D*, in press.

Ratcliff, J. T., Tackley, P. J., Schubert, G., and Zebib, A. 1996b. Transitions in thermal convection with strongly variable viscosity. *Phys. Earth Planet. Int.*, submitted.

Safronov, V. S. 1978. The heating of the Earth during its formation, *Icarus* 33:3–12.

Sandwell, D. T., and Schubert, G. 1992a. Flexural ridges, trenches, and outer rises around coronae on Venus. *J. Geophys. Res.* 97:16069–16083.

Sandwell, D. T., and Schubert, G. 1992b. Evidence for retrograde lithospheric subduction on Venus. *Science* 257:766–770.

Schaber, G. G., et al. 1992. Geology and distribution of impact craters on Venus: What are they telling us? *J. Geophys. Res.* 97:13257–13301.

Schubert, G. 1979. Subsolidus convection in the mantles of terrestrial planets. *Ann. Rev. Earth Planet. Sci.* 7:289–342.

Schubert, G., and Sandwell, D. T. 1995. A global survey of possible subduction sites on Venus. *Icarus* 117:173–196.

Schubert, G., Cassen, P., and Young, R. E. 1979. Subsolidus convective cooling histories of terrestrial planets. *Icarus* 38:192–211.

Schubert, G., Bercovici, D., and Glatzmaier, G. A. 1990. Mantle dynamics in Mars and Venus—Influence of an immobile lithosphere on 3-dimensional mantle convection. *J. Geophys. Res.* 95:14105–14129.

Schubert, G., Glatzmaier, G. A., and Travis, B. 1993. Steady, 3-dimensional, internally heated convection. *Phys. Fluids A* 5:1928–1932.

Schubert, G., Moore, W. B., and Sandwell, D. T. 1994. Gravity over coronae and chasmata on Venus. *Icarus* 112:130–146.

Schubert G., Anderson, C., and Goldman, P. 1995. Mantle plume interaction with an endothermic phase change. *J. Geophys. Res.* 100:8245–8256.

Sharpe, H. N., and Peltier, W. R. 1978. Parameterized mantle convection and the Earth's thermal history. *Geophys. Res. Lett.* 5:737–740.

Sharpe, H. N., and Peltier, W. R. 1979. A thermal history model for the Earth with parameterized convection. *Geophys. J. Roy. Astron. Soc.* 59:171–205.

Simons, M., Hager, B. H., and Solomon, S. C. 1994. Global variations in the geoid/topography admittance of Venus. *Science* 264:798–803.

Smrekar, S. E. 1994. Evidence for active hotspots on Venus from analysis of Magellan gravity data. *Icarus* 112:2–26.

Smrekar, S. E., and Parmentier, E. M. 1996. The interaction of mantle plumes with surface thermal and chemical boundary layers: Applications to hotspots on Venus. *J. Geophys. Res.* 101:5397–5410.

Smrekar, S. E., and Phillips, R. J. 1991. Venusian highlands: Geoid to topography ratios and their implications. *Earth Planet. Sci. Lett.* 107:582–597.

Solheim, L. P., and Peltier, W. R. 1994. Avalanche effects in phase transition modulated thermal convection: A model of Earth's mantle. *J. Geophys. Res.* 99:6997–7018.

Solomatov, V. S. 1993. Parameterization of temperature- and stress-dependent viscosity convection and the thermal evolution of Venus. In *Flow and Creep in the Solar System: Observations, Modeling and Theory*, eds. D. B. Stone and S. K. Runcorn (Netherlands: Kluwer), pp. 131–145.

Solomatov, V. S. 1995. Scaling of temperature- and stress-dependent viscosity convection. *Phys. Fluids* 7:266–274.

Solomatov, V. S., and Moresi, L.-N. 1996. Stagnant lid convection on Venus. *J.*

Geophys. Res. 101:4737–4753.
Solomatov, V. S., and Stevenson, D. J. 1993. Nonfractional differentiation of a terrestrial magma ocean. *J. Geophys. Res.* 98:5391–5406.
Solomatov, V. S., and Zharkov, V. N. 1990. The thermal regime of Venus. *Icarus* 84:280–295.
Solomatov, V. S., Leontjev, V. V., and Zharkov, V. N. 1987. Models of thermal evolution of Venus in the approximation of parameterized convection. *Gerlands Beitr. Geophysik* 96:73–96.
Solomon, S. C., et al. 1991. Venus tectonics: Initial analysis from Magellan. *Science* 252:297–312.
Solomon, S. C., et al. 1992. Venus tectonics: An overview of Magellan observations. *J. Geophys. Res.* 97:13199–13255.
Spohn, T., and Schubert, G. 1982. Modes of mantle convection and the removal of heat from the Earth's interior. *J. Geophys. Res.* 87:4682–4696.
Stevenson, D. J., Spohn, T., and Schubert, G. 1983. Magnetism and thermal evolution of terrestrial planets. *Icarus* 54:466–489.
Steinbach, V., and Yuen, D. A. 1992. The effects of multiple phase transitions on Venusian mantle convection. *Geophys. Res. Lett.* 19:2243–2246.
Steinbach, V., and Yuen, D. A. 1994. Effects of depth dependent properties on the thermal anomalies produced in flush instabilities from phase transitions. *Phys. Earth Planet. Int.* 86:165–183.
Steinbach, V., Yuen, D.A., and Zhao, W. L. 1993. Instabilities from phase transitions and the timescales of mantle thermal evolution. *Geophys. Res. Lett.* 20:1119–1122.
Stofan, E. R., Bindschadler, D. L., Head, J. W., and Parmentier, E. M. 1991. Corona structures on Venus: Models of origin. *J. Geophys. Res.* 96:20933–20946.
Stofan, E. R., et al. 1992. Global distribution and characteristics of coronae and related features on Venus: Implications for origin and relation to mantle processes. *J. Geophys. Res.* 97:13347–13378.
Strom, R. G., Schaber, G. G., and Dawson, D. D. 1994. The global resurfacing of Venus. *J. Geophys. Res.* 99:10899–10926.
Sun, S. S., and McDonough, W. F. 1989. Chemical and isotopic systematics of oceanic basalts: Implications for mantle composition and processes. In *Magmatism in the Ocean Basins*, eds. A. D. Saunders and M. J. Norry (London: Geological Soc. Special Publ. 42), pp. 313–345.
Surkov, Y. A., et al. 1984. New data on the composition, structure, and properties of Venus rock obtained by Venera 13 and Venera 14. *J. Geophys Res.* 89:383–402.
Surkov, Y. A., et al. 1987. Uranium, thorium, and potassium in the Venusian rocks at the landing sites of Vega 1 and 2. *J. Geophys. Res.* 92:537–540.
Tackley, P. J. 1993. Effects of strongly temperature-dependent viscosity on time-dependent, 3-dimensional models of mantle convection. *Geophys. Res. Lett.* 20:2187–2190.
Tackley, P. J. 1994. Three-Dimensional Models of Mantle Convection: Influence of Phase Transitions and Temperature-Dependent Viscosity. Ph.D Thesis, California Inst. of Technology.
Tackley, P. J. 1995. Mantle dynamics: Influence of the transition zone. *Rev. Geophys. Suppl.* 33:275–282.
Tackley, P. J. 1996. Effects of strongly variable viscosity on three-dimensional compressible convection in planetary mantles. *J. Geophys. Res.* 101:3311–3332.
Tackley, P. J., Stevenson, D. J., Glatzmaier, G. A., and Schubert, G. 1993. Effects of an endothermic phase transition at 670 km depth in a spherical model of convection in the Earth's mantle. *Nature* 361:699–704.
Tackley, P. J., Stevenson, D. J., Glatzmaier, G. A., and Schubert, G. 1994. Effects of

multiple phase transitions in a 3-dimensional spherical model of convection in Earth's mantle. *J. Geophys. Res.* 99:15877–15901.
Travis, B., and Olson, P. 1994. Convection with internal heat sources and thermal turbulence in the Earth's mantle. *Geophys. J. Int.* 118:1–19.
Turcotte, D. L. 1993. An episodic hypothesis for Venusian tectonics. *J. Geophys. Res.* 98:17061–17068.
Turcotte, D. L., and Schubert, G. 1982. *Geodynamics* (New York: J. Wiley and Sons).
Turcotte, D. L., Cooke, F. A., and Willeman, R. J. 1979. Parameterized convection within the moon and the terrestrial planets. *Proc. Lunar Planet. Sci. Conf.* 10:2375–2392.
van der Hilst, R., and Seno, T. 1993. Effects of relative plate motion on the deep structure and penetration depth of slabs below the Izu-Bonin and Mariana island arcs. *Earth Planet. Sci. Lett.* 120:395–407.
van der Hilst, R., Engdahl, R., Spakman, W., and Nolet, G. 1991. Tomographic imaging of subducted lithosphere below northwest Pacific island arcs. *Nature* 353:37–43.
Weinstein, S. A. 1993. Catastrophic overturn of the Earth's mantle driven by multiple phase changes and internal heat generation. *Geophys. Res. Lett.* 20:101–104.
Weinstein, S., and Christensen, U. 1991. Convection planforms in a fluid with a temperature-dependent viscosity beneath a stress-free upper boundary. *Geophys. Res. Lett.* 18:2035–2038.
Weinstein, S., and Olson, P. 1990. Planforms in thermal convection with internal heat sources at large Rayleigh and Prandtl numbers. *Geophys. Res. Lett.* 17:239–242.
Weinstein, S. A., and Olson, P. 1992. Thermal convection with non-Newtonian plates. *Geophys. J. Int.* 111:515–530.
Wetherill, G. W. 1990. Formation of the Earth. *Ann. Rev. Earth Planet. Sci.* 18:205–256.
Wright, T. L., Kinoshita, W. T., and Peck, D. L. 1968. March 1965 eruption of Kilauea volcano and the formation of Makaopuhi lava lake. *J. Geophys. Res.* 73:3181–3205.
Yuen, D. A., et al. 1994. Various influences on 3-dimensional mantle convection with phase transitions. *Phys. Earth Planet. Int.* 86:185–203.
Zahnle, K. J., Kasting, J. F., and Pollack, J. B. 1988. Evolution of a steam atmosphere during Earth's accretion. *Icarus* 74:62–97.
Zhao, W. L., Yuen, D. A., and Honda, S. 1992. Multiple phase transitions and the style of mantle convection. *Phys. Earth Planet. Int.* 72:185–210.
Zharkov, V. N., and Solomatov, V. S. 1992. Models of thermal evolution of Venus. In *Venus Geology, Geochemistry and Geophysics*, eds. V. L. Barsukov, A. T. Basilevsky, V. P. Volkov and V. N. Zharkov (Tucson: Univ. of Arizona Press), pp. 280–319.
Zhong, S. J., and Gurnis, M. 1994. Role of plates and temperature-dependent viscosity in phase change dynamics. *J. Geophys. Res.* 99:15903–15917.
Zhong, S. J., and Gurnis, M. 1995. Mantle convection with plates and mobile, faulted plate margins. *Science* 267:838–843.
Zhou, H. W. 1990. Mapping of P-wave slab anomalies beneath the Tonga, Kermadec and New Hebrides arcs. *Phys. Earth Planet. Int.* 61:199–229.

APPENDIX

APPENDIX: README FILE FOR VENUS II CD-ROM

This README file contains installation instructions and release notes. Credit and copyright information is contained at the end of this file.

IMPORTANT

Images and QuickTime movies contained on this disk have been obtained from various sources. The Jet Propulsion Laboratory has been granted permission to distribute these images and movies for educational purposes only, on a non-profit bases. Credit is given to the image or movie owner in the accompanying captions and in the COPYRIGHT.TXT file. Copyright is non-transferable and is wholly retained by the owner. If you wish to use an image or movie, refer to the COPYRIGHT.TXT file and contact the owner directly.

OBTAINING ADOBE ACROBAT 3.0 READER™ SOFTWARE

Adobe Acrobat 3.0 software from Adobe Systems Incorporated can be obtained free of charge from the World Wide Web at http://www.adobe.com. This software for both Macintosh and PC platforms can be used to read PDF files found on this disk.

OBTAINING QUICKTIME™ SOFTWARE

QuickTime™ software from Apple Computer Company must be installed on your system in order to play many of the movies on this CD-ROM. QuickTime™ and QuickTime VR software for both Macintosh and PC platforms can be obtained free of charge from the World Wide Web (WWW) at: http://quicktime.apple.com. A version of QuickTime for both Macintosh and Windows PC is provided in the QTSW directory of this disk.

MACINTOSH SYSTEM REQUIREMENTS

- Macintosh computer with 68030 or greater processor
- MacOS 7.0 or above
- At least 8 MB of free RAM
- 640×480 pixel 256-color display
- 2X or faster CD-ROM player
- QuickTime 2.5™ or above. QuickTime 2.5™ is included on the CD-ROM in the "QTSW" folder

INSTALLING QuickTime 2.5™ MANUALLY

QuickTime 2.5™—A system extension needed to display QuickTime movies

Open the "QuickTime™ Install Disk 1" folder in the "QTSW" folder of the CD-ROM, and read the QuickTime™ READ ME file. Follow the directions. Then, launch or "click" the Install QuickTime™ application program to install QuickTime™. If you have downloaded these files from the WWW site you can then drag the "QuickTime" and "QuickTime PowerPlus" icons to your hard drive's System Folder icon (not into the open folder). Then re-start your Macintosh.

TO RUN VENUSII

Double-click on the "VENUS II" icon on the CD-ROM.

RECOMMENDED PC SYSTEM REQUIREMENTS

- 80386 (or greater) processor
- Windows 3.1 or above
- 16 MB RAM
- 256-color, 640×480 pixel monitor (24-bit color required for QuickTime VR™ movie only.)
- Audio capability
- QuickTime for Windows (included on CD)

INSTALLING QuickTime 2.5™ MANUALLY

A program to install QuickTime™ on your PC is included on this disk in the "QTSW" directory of the CD-ROM. Launch the qt16.exe program and the software will be loaded on your system. Then place the QTVRW.QTC file, also included on the disk, into the Windows System Directory on your PC. (Note: The QuickTime VR™ movie of the Venus rotating globe in Part 5 requires a 24-bit color system on PC's only.)

TO RUN VENUSII

Double-click on the "VENUS.EXE" icon in the "SOURCE" directory on the CD-ROM.

TROUBLESHOOTING

QuickTime™ Movies Do Not Play Back Smoothly

This CD-ROM requires at least a 2X CD-ROM player in order to play QuickTime™ movies properly. It is strongly recommended that a 4X or 6X CD-ROM player be used to achieve the full performance of this disk. If you have a 1X CD-ROM player, QuickTime™ playback from the CD-ROM will be problematic. QuickTime version 2.1™ is included on the CD-ROM in the "Software" folder.

The Color of Some Images Appears Weird on my Monitor

The program should run in 256, thousands or millions of colors. However, some images may not display properly in thousands of colors mode on some Macintosh models. This is a known bug in the application development software. To remedy the situation, switch to 256 colors mode.

TECHNICAL SUPPORT

If you have e-mail access to the Internet, you can request technical support from: pds_operator@jplpds.jpl.nasa.gov.

ADDITIONAL INFORMATION

The SPECIAL directory on this disk contains a scanned version of the Special Publication entitled "Magellan: Revealing the Face of Venus." This publication (JPL 400-494 3/93) is a production of the National Aeronautics and Space Administration and the Jet Propulsion Laboratory. It is stored in PDF format for viewing with Adobe Acrobat Reader™. The file name is magella.pdf.

The FILES directory contains all the data associated with the book, this disk and the multimedia presentation. Each one of the data files can be viewed externally with commercially available software. The formats used in the production of this disk are: GIF, PIC, Postscript (PS), Adobe PDF, TEXT and MOV for QuickTime™ movies. Please note: there are data files on this disk that were not able to be displayed in the multimedia presentation. However, these data files are all stored on the disk in this directory. Also, if a data file displayed in the multimedia presentation is difficult to read, check this directory for a higher resolution version.

The directories located on this CD-ROM are as follows: FILES contains all the data files submitted by the authors of the book. This directory is further broken down into directories such as ABSTRACT containing all the abstracts of the chapters in the book, BIBLIO containing all the submitted bibliographies, APPENDIX containing the appendix for the Fegley chapter (most if not all the text data are stored in both Postscript [PS] and PDF formats), and DATA which contains all the submitted data products related to the book chapters in GIF, PIC, MOV or TEXT format. However, not all chapters in the book have related data products. Only those authors who submitted data products for their chapters are represented here.

Navigation within the multimedia program is accomplished by use of the buttons at the bottom of your screen. There are multiple buttons, some of which are active only in certain screens (ZOOM and VOLUME). All visible buttons will be in a highlighted state when the mouse cursor is placed over them.

Starting from the bottom left there is the ZOOM button. For some chapter data, the ability to zoom in to see higher resolution images is available. When this is the case, this button will be visible. Next is the HOME button. Clicking this button will always take the user back to the first level menu or "Home." The BACK button acts as a hyperlink button taking the user back through the series of pages last visited, in reverse order. The UP button will allow the user to jump up one level in the program. The FORWARD button moves the user through a series of pages at the same level, if applicable. If there is only one entity at that level, this button will function the same as the UP button. The VOLUME UP and DOWN button adjusts the volume of your system in a series of steps. (Note: some movies have limited volume and are best heard using external speakers.) Finally, the QUIT button cancels the program and returns the user to the initiating system state.

The chapter abstracts that have more than a single page also have navigation arrows (purple) to allow the reader to jump to the next page or back to the beginning of the abstract. These arrows will be highlighted when the mouse cursor is placed over them. Clicking these arrows will take the user to the next page of the abstract or, if at the end of the abstract, back to the beginning.

More detailed information on the spacecraft and missions presented on this CD-ROM can be obtained from the Planetary Data System World Wide Web site located at http://pds.jpl.nasa.gov.

CREDITS

- VENUSII is a production of the Jet Propulsion Laboratory, California Institute of Technology. All rights reserved.
- CD-ROM DEVELOPMENT TEAM: Andrea Alazard, Steve Bougher, Adrian Godoy, David Hecox, Peter Kahn, Tom Kelly, Orlando Monroe, Randii Oliver, Amy Schumann Ruskin, Harman Smith and Sugi Sorensen.
- ADDITIONAL VENUSII TECHNICAL ADVISORS: Mike Martin and Yolanda Oliver.
- ADDITIONAL CONTRIBUTORS: Stephen Brecht.

SPECIAL THANKS

A special thanks to Sugi Sorensen, Harman Smith, Adrian Godoy, Amy Schumann Ruskin and Steve Bougher for all their help and support. Without their contributions, this disk would not have been possible. Thanks, Pete Kahn.

COPYRIGHTS AND ACKNOWLEDGMENTS

- VENUSII: Geology, Geophysics, Atmosphere and Solar Wind Environment, Copyright 1996 California Institute of Technology and its licensers. U.S. Government sponsorship under NASA Contract NAS-1270 is acknowledged. All rights reserved.
- DIRECTOR® COPYRIGHT©1994 Macromedia, Inc.
- QuickTime™ and the QuickTime™ logo are trademarks of Apple Computer Inc., used under license.
- Windows is a registered trademark of Microsoft Corporation.
- For specific image and movie credits, or re-use information, please check the COPYRIGHT.TXT file.

GLOSSARY

GLOSSARY*

admittance	used in gravity studies, it is the ratio of gravity or geoid to topography as a function of wavelength.
AAT	Anglo-Australian Telescope.
ADC	apparent depth of compensation. Refers to depth of a hypothetical surface where mass is in isostatic balance with the surface topography and the sum of the gravity effects of the topographic and compensating masses matches the observed gravity signal.
aeolian	literally, of the wind; wind-blown, as aeolian deposits.
aerobraking	a maneuver used by the Magellan spacecraft that used atmospheric drag to modify the spacecraft's elliptical orbit to one that was nearly circular.
airburst	refers to meteoroids that effectively explode before reaching the surface; the shock waves associated with airbursts may be responsible for the formation of several types of surface features, including dark splotches.
airglow	faint light emanated as a result of neutral or ion chemical reactions in the upper atmosphere.

---

* We have used some definitions from *Glossary of Astronomy and Astrophysics* by J. Hopkins (by permission of the University of Chicago Press, copyright 1980 by the University of Chicago), from *Glossary of Geology*, 3rd ed., edited by R. L. Bates and J. A. Jackson (Alexandria, Va.: American Geological Inst., 1987), from *Atmospheric Radiation: Theoretical Basis*, 2nd ed. (New York, N. Y.: Oxford University Press, 1989) and from *Astrophysical Quantities* by C. W. Allen (London: Athlone Press, 1973). We also acknowledge definitions and helpful comments from various chapter authors.

GLOSSARY

Airy compensation	a type of isostatic compensation whereby mountains are supported by roots, most often envisaged as less-dense crust protruding into more-dense mantle. Low-standing regions are compensated by "anti-roots."
albedo	ratio of reflected to incoming energy. Bond or spherical albedo: total over all directions. Geometric albedo: in backward direction. Monochromatic albedo: in some specific narrow wavelength band.
Alfvén Mach number	ratio of velocity in a plasma to the Alfvén speed. The Alfvén velocity is that speed at which a magnetic disturbance propagates along the background field direction.
altimeter	transmits radar waves and measures the time delay of radar echoes from the surface in vertical (near nadir) viewing geometries.
amu	atomic mass unit.
Aphrodite Terra	a continental size highland belt in the equatorial region of Venus.
apoapsis	that part in an elliptical orbit about a body which is the point of farthest distance from that body.
assemblage, mineral	the various mineral components of a rock.
asthenosphere	on Earth, a vertically confined "channel" in the mantle beneath the lithosphere that has a lower viscosity than both the mantle lithosphere above and "normal" mantle beneath. It has been argued that Venus does not have an asthenosphere.
Atla Regio	a large, rift-dominated rise, which includes the volcano Maat Mons.
atmosphere	as a unit, 1.013 bar, standard Earth surface pressure.
AU	astronomical unit; the mean distance from Earth to Sun, 1.496×10^8 km.

β	in plasma physics, the ratio of plasma and magnetic field energy densities (see chapters I.3 and I.4).
backscatter	also called backscatter cross section. On Magellan, this coefficient is a measure of the power of a radar echo from the Venusian surface.
bar	unit of pressure equal to 1.0×10^6 dyne cm^{-2} or 1.0×10^5 Pa (N m^{-2}).
basalt	a type of volcanic rock consisting mostly of the minerals pyroxene $(Ca,Mg,Fe)_2Si_2O_6$ and plagioclase feldspar, $CaAl_2Si_2O_8$-$NaAlSi_3O_8$. Basalt is created in abundance by partial melting in the mantles of terrestrial planets. It is by far the most common type of volcanic rock in the solar system.
bending moment	in the deformation of a thin lithospheric plate, it refers to the vertical integral, at any position along the plate, of horizontal normal stress times distance from a reference surface (the middle of the plate if it is elastic). That is, it is the net moment anywhere along the plate resulting from a load or moment applied at the surface or ends of the plate.
Beta Regio	a large, rift dominated volcanic rise, which includes the volcano Theia Mons.
BIMS	Pioneer Venus Multiprobe Bus Ion Mass Spectrometer.
bistatic radar	a radar experiment where the transmitting and receiving antennae are in different locations such that the angle of reflection is not zero.
BNMS	Pioneer Venus Multiprobe Bus Neutral Mass Spectrometer.

boundary layer (thermal)	in referring to mantle convection, it is a horizon where heat cannot be transported by mass advection, so it must be transported by thermal conduction. It is marked by high thermal gradients, and is found at the top of a convecting mantle, at the core-mantle boundary if heat is coming from the core, and possibly at phase boundaries within the mantle.
bow shock	a standing shock wave in the solar wind present upstream of an obstacle such as Venus. A shock wave is a discontinuity in a fluid flow across which the density, pressure, and temperature all increase. At a shock in a magnetized plasma, the magnetic field strength also increases.
bright halos	a surface deposit characterized by high radar backscatter and often surrounding a dark splotch.
Brunt-Väisälä frequency, N	frequency of buoyancy oscillations in an atmosphere (see chapter III.6).
buffer	a mineral or mineral system can control or buffer the composition of an atmosphere that reacts with it by withdrawing components from the atmosphere, or adding components to it, until the atmosphere is in chemical equilibrium with the mineral buffer.
Byerlee's law	an empirical result suggesting that a fixed relationship, nearly independent of rock type, exists between the normal and shear stress values required for frictional sliding along existing rock fractures; named after geophysicist J. D. Byerlee.

GLOSSARY

caldera
: a structural term referring to a circular to elongate depression and, on Venus, commonly characterized by concentric structural patterns of enveloping fractures; a caldera is distinguished from an eruptive volcanic vent crater; a caldera is larger than any vents from which eruptions may have occurred associated with caldera formation, and forms by collapse over an evacuated magma chamber.

canali
: channels having a constant width and depth. Canali are typically as wide as 3 km and up to 500 km in length. The longest canali-type channel known is 6800 km and it is the longest channel in the solar system. Canali may locally exhibit abandoned channel segments, cut-off meander bends, levees, and radar dark terminal deposits. Canali are hypothesized to have formed by large discharges of low viscosity lava over prolonged periods.

cauldron
: on Earth, a structural term referring to any circular or elongate volcanic subsidence regardless of size, shape, depth of exposure, or connection with surface volcanism.

CCD
: charged coupled device (an electronic imaging detector).

CFHT
: Canada-France-Hawaii Telescope.

Chandler wobble
: the Earth undergoes a free precession with a period of about 435 days about its spin axis. This is referred to as the Chandler wobble. Venus has a Chandler-like wobble.

chasma
: a geographic term for a broad valley, trough, or depression; plural: chasmata.

Cn-MIDRs	compressed <n times> resolution mosaicked image data records, denoting compressed-once MIDRs (C1-MIDRs: 15×15 degrees), twice-compressed MIDRs (C2-MIDRs: 45×45 degrees), and compressed-thrice MIDRs (C3-MIDRs: 120×80 degrees).
coagulation	growth of cloud particles by collisions due to their random (Brownian) motion.
coalescence	growth of large particles by sweeping up of smaller ones due to their different sedimentation speeds.
coherence	used in gravity studies, it is the normalized correlation of gravity or geoid to topography as a function of wavelength.
cold trap	the concept that a vapor (such as water) is kept out of higher atmospheric regions by condensation at some cold level.
continuous ejecta	relatively continuous, radar bright ejecta deposit, often with a lobate outer boundary, surrounding a Venusian impact crater and lying outward of hummocky ejecta, if it exists.
Coriolis force	a pseudo force that appears in the dynamical equations when set up for a rotating coordinate system. In the Earth's northern hemisphere, it tends to turn all velocity vectors to the right.
corona	(1) a geographic term for a raised, ovoid feature. On Venus, coronae form centers of tectonic deformation and volcanism; (2) faint glow about a planet due to scattering of light in the rarefied upper atmosphere or chemical reactions in that part of the atmosphere; (3) also used to describe an upper atmosphere of nonthermal (e.g., photochemical) origin.
Coulomb collisions	collisions between charged particles, mediated by the electrostatic force between them.

GLOSSARY

crater field	a collection of non-overlapping craters formed by simultaneous impact of several separated fragments of a single incoming meteoroid.
crater size–frequency distribution	the number of observed impact craters as a function of crater diameter; often normalized on a per unit area basis.
critical altitude	an altitude where, on many Venus mountains, the radar backscatter coefficient and radiothermal emissivity e (sometimes E) are seen to change abruptly. Beneath the critical altitude $e \sim 0.85$; above it e is much smaller, often ~ 0.4.
crustal (highland) plateau	large, relatively steep-sided elevated region found on Venus. The dominant example is Ovda Regio (approximately 3000×2000 km in size), found in the western portion of Aphrodite Terra. A crustal plateau typically has tesserae exposed at its surface.
current sheet	a narrow sheet, or layer, of plasma in which an electrical current flows and which separates regions with different (nearly opposite) magnetic field orientations and/or magnitudes.
cryosphere	the region of the upper atmosphere of Venus, above about 120 km on the night side, where the kinetic gas temperature is much lower than that in the atmosphere below it.
cyclostrophic balance	a condition in which the meridional atmospheric pressure gradient provides the horizontal component of centripetal force required to constrain the zonal flow circling the planet (see chapters II.4, III.1, and III.6).
cyclotron frequency (fc)	also known as gyrofrequency or Larmor frequency. A charged particle in a uniform magnetic field follows a circular, or helical path. The frequency of the periodic motion about this path is known as the cyclotron frequency.

Cytherean	Venusian.
dark halo	a diffuse, radar-dark zone surrounding the radar-bright ejecta of many impact craters. The most common explanation for dark haloes is pulverization of the surface by the downgoing atmospheric shock wave produced by the impactor.
dark margin	a radar-dark zone with a sharp boundary surrounding many impact craters. Possible explanations include deposition of fine-grained ejecta by a debris flow and embayment of a dark halo by subsequent volcanism.
delamination	tearing away from the crust and sinking of the dense mantle portion of the lithosphere.
D/H	deuterium to hydrogen ratio in water. For Venus, D/H is 150 ± 30 times that of terrestrial water.
diabatic	used to describe heating or cooling of an air parcel by radiation.
diamagnetic effect	when electrical currents, such as in a plasma, are generated so as to reduce the magnetic field in some region. High thermal pressure in a plasma can produce a diamagnetic effect.
dielectric constant	an intrinsic property measuring the electrical permittivity of a substance (denoted by ϵ).
diffusive equilibrium	a state, found in upper atmospheres, where each constituent takes up its own scale height.
diurnal variation	relating to the response of the atmosphere as a result of the apparent daily motion of the sun around the Earth; for non-rotating Venus it is, to first order, the response of the atmosphere as a function of angle between local nadir and the Sun; the day-to-night variation.

Doppler tracking	tracking of a spacecraft from Earth by measuring variations in the Doppler frequency shift of a communication signal. Subtle variations in the Doppler frequency correspond to small changes in spacecraft velocity; it is this information that is used to map planetary gravity fields.
drag force	the resistance offered to a satellite as a result of atmospheric particles impacting the surface.
eclogite	a basic igneous rock dominated typically by garnet and pyroxene. Can result from a high pressure phase change of basalt.
eddy conduction	heat conduction due to turbulent motion of the atmosphere.
eddy diffusion	diffusion due to turbulent motion of the atmosphere.
ejecta flows	also referred to as outflows, run-out flows, run-out ejecta, and fluidized ejecta in the literature. Extending beyond the nominal ejecta blankets of many impact craters are what appear to be ejecta deposits with a lava flow morphology. Some ejecta flows extend several hundred km from the crater rim.
emissivity (e, sometimes E)	the ratio of the radiance of a gray body and the radiance of a black body at the same temperature. For the hot Venusian surface and the frequency of Magellan radar, the ratio of the brightness temperature at 12.6-cm (T_b) and the physical temperature of the surface (T_p) is a close estimate of the emissivity. *See* critical altitude.
equatorial highlands	a geographic term referring to the high standing regions at low latitudes on Venus; dominated by Aphrodite Terra and Beta Regio.
EUV	extreme ultraviolet; radiation with wavelengths below about 100 nanometers.

exobase	altitude of the base of the exosphere, where the atmospheric scale height is equal to the mean free path, and above which particles can execute ballistic trajectories.
exosphere	the region of the upper atmosphere where particle-particle collisions are infrequent, with particles in ballistic orbits or escaping trajectories.
F10.7	10.7 cm radio flux, measured at Earth, normalized to 1 AU and measured in the units of 10^{-22} W m^{-2} Hz^{-1}; historically used as a proxy of EUV radiation incident upon the Earth's upper atmosphere.
F-BIDR	full-resolution basic data record (20 to 25 km wide); one F-BIDR was produced for each Magellan mapping orbit.
F-MIDR	full-resolution mosaicked image data record (5 × 5 deg).
festoon flow	on Venus, a large, radar-bright lava flow field characterized by organized patterns of internal ridges and flow bands.
fissure	also, fissure vent, or volcanic fissure; an elongate volcanic vent marking the site where a dike or dikes intersect the surface and volcanic materials have been erupted from a linear vent.
flexure	the bending of an elastic plate or shell in response to forces or moments; the term is used in conjunction with deformation of lithospheres.
flow field	a collection of individual, generally digitate, lava flows, erupted from the same source vent or source area and forming a spatially contiguous occurrence of numerous flows arranged on a pre-existing topographic gradient.
fluctus	on Venus, a large lava flow field.

GLOSSARY

flux ropes	magnetic flux ropes on Venus are narrow structures containing high magnetic field strength. Usually these ropes are force-free in which the field in the center is along the axis and the outer region has fields that are azimuthally-directed. Flux ropes are found in the Venus ionosphere.
foreshock	disturbed region upstream of a planetary bow shock. Foreshock is caused by plasma instabilities associated with plasma particles either reflected from or energized by the shock.
fractionation factor	the relative efficiency of atmospheric loss of one isotope over another (see chapter III.4).
fracture belt	linear to sinuous group of closely spaced, aligned fractures typically extending for hundreds of km.
FTS	Fourier Transform Spectrometer.
FWHM	full width of a spectral line or other feature at half-maximum intensity.
Gauss	CGS unit of magnetic field strength, 1.0×10^{-4} Tesla.
geoid	a surface that conforms to a gravitatonal-rotational equipotential. On Earth, it is tied to mean sea level; on Venus it is calculated, typically, at the planetary mean radius.
GHz	gigahertz (1.0×10^9 cycles per sec).

Gibbs free energy	thermodynamic property of chemical compounds (energy G of joules/mole). At constant temperature and pressure, chemical reactions among the compounds in a particular chemical system proceed in a direction that reduces the net G of the system. When net G reaches the lowest possible value, the system is said to have attained chemical equilibrium. $G(t)$ is not measured directly for compounds, but is formally derived from other measured quantities, chiefly the amount of heat required to raise the temperature of each compound, $C_p(t)$.
GM	gravitational constant times the mass of an object.
graben	a linear depression formed by a pair of inward-dipping normal faults.
gravity anomaly	departures of a planet's gravity from its average value.
gravity waves	waves in an atmosphere due to bouyancy effects (motion of fluid blobs) as opposed to sound waves.
gridded plains	plains characterized by lineations with a regular spacing of about a km.
GxDR	global x data record (x = altimeter, emissivity, radiometer, reflectivity, slope, or topography).
Gyr	one billion (1.0×10^9) yr.
Hadley cell	a simple, thermodynamically-direct, circulation cell with rising in warm regions and descending in cold ones. For Venus, a Hadley cell is manifested as a north-south (meridional) circulation cell, with rising motion at low latitudes and sinking motion at high latitudes. This flow pattern transports heat down the local thermal gradient, from the warmer tropics to the cooler poleward latitudes.
heat flux (flow)	the amount of heat, per unit area, escaping from a planet.

GLOSSARY

homopause	that altitude in a planet's upper atmosphere where the mean molecular mass of the atmosphere begins to decrease from its lower atmosphere mixed state; this is the altitude where the transition from mixing to molecular diffusion occurs. Often called the turbopause.
HST	Hubble Space Telescope.
hummocky ejecta	the inner ejecta zone found at many Venusian craters; characterized by a hummocky, blocky terrain that may represent the larger size fraction of material excavated during impact.
hydrodynamic escape	escape of a light atmospheric constituent with a very large flux, best described as a fluid flow rather than an atom-by-atom loss, called thermal or Jeans escape.
hypsometry	distribution of surface elevation. On Venus, most of the surface occurs within one km of the mean planetary radius.
IFSE	Venera 15 orbiter infrared Fourier spectrometer experiment.
IMF	interplanetary magnetic field.
in-situ measurements	direct measurements at a location as opposed to remote sensing radiation measurements at a distance.
intermediate volcano	on Venus, a volcano whose average edifice diameter is ≥ 20 km and <100 km (not to be confused with a volcano of intermediate composition).
intraplate	refers to tectonic and volcanic processes taking place on Earth well away from plate boundaries.
ionopause	abrupt termination of the ionosphere at the weakly magnetized planets (see chapters II.1 and II.2).

ionosheath	(also, magnetosheath). The region between a planetary obstacle and the bow shock, containing shocked solar wind plasma, is the magnetosheath. At Venus, where the obstacle is the ionosphere, this region (between the ionopause and the bow shock) is often called the ionosheath.
ionosphere	the ionized region of an upper atmosphere.
ionospheric holes	on Venus, large localized, depleted electron density regions in the antisolar sector of the ionosphere. Here the magnetic field is intensified and tends to be aligned parallel to the ecliptic plane.
ionotail	the region beyond the terminator at high altitudes in the shadow of Venus, where the ionosphere develops long comet-like tail rays that are aligned in the anti-sunward direction, with ion densities of about 50 to 500 cm^{-3}.
Ipe	photoelectron current from the OETP on-board the Pioneer Venus Orbiter.
IRHS	Infrared Heterodyne Spectrometer.
Ishtar Terra	continental size highland region in the northern hemisphere of Venus containing folded mountain belts
isostatic compensation	the mechanism whereby topographic mass is balanced at depth by a mass of the same magnitude and opposite sign.
IUE	International Ultraviolet Explorer.
J_2	the second zonal harmonic in a spherical harmonic expansion; "zonal" means that the harmonic does not depend on longitude. A second harmonic has two complete wavelengths around a sphere.
Jeans escape	thermal escape of light atoms from an upper atmosphere.

GLOSSARY

jelly sandwich	denotes a rheological stratification within the lithosphere wherein a weak layer is sandwiched between two strong layers. In most applications, the strong layers are the upper crust and upper mantle and the weak layer is the lower crust.
Joule heating	heating by dissipation of electric currents.
Kelvin-Helmholtz instability	tendency of waves to grow on a shear boundary between two regions in relative motion parallel to the boundary.
kHz	kilohertz (1.0×10^3 cycles per sec).
komatiite	magnesium rich, low silica volcanic rocks whose parent magmas form at very high temperatures. On Earth, most komatiites are of Archaean age.
KR	kilorayleigh (see R).
laminar flow	smooth, nonturbulent flow.
Landau damping	a form of plasma-wave absorption due to nonthermal electrons (see chapter I.4).
large volcano	on Venus, a volcano with a diameter of 100 km or greater as measured to the termini of its radially distributed lava flows; if a topographic edifice is present or identifiable, the diameter may be measured to the first prominent break in slope.
Larmor radius	also, gyroradius. Radius of gyration of a charged particle in the ambient magnetic field.
LCPS	Pioneer Venus Large Probe Cloud Particle Spectrometer.
line-of-sight (LOS) gravity	the component of a planet's spatially-varying gravity vector that is along the line of sight between a spacecraft and its tracking station on Earth.
LIR	Pioneer Venus Large Probe Infrared Radiometer.

lithosphere	the cool, relatively strong outermost shell of a planet; the definition is thermomechanical, as opposed to the crust, which is distinguished from the mantle beneath by compositional differences. There are three basic types of lithospheres: (1) elastic—rheology is elastic in the strict sense; (2) mechanical—an outer shell with a strength greater than a prescribed value, e.g., 50 MPa; and (3) thermal—outer shell that transports heat dominantly by conduction. *See* boundary layer.
LNMS	Pioneer Venus Large Probe Neutral Mass Spectrometer.
lossy	a reference to the loss of electrical energy as heat. In the radar context, a mineral is "electrically lossy" when its conductivity is relatively high so that propagating electromagnetic waves will attenuate significantly when they pass through the medium.
Love numbers	a set of dimensionless numbers that give the response of a planet to a tide raising potential. The Love number h characterizes the height of the solid body tide, while the Love number k characterizes the change in tidal potential due to mass redistribution caused by the tide raising potential.
lower cloud	Venus global cloud layer that extends from about 47 to 49 km altitude.
LSFR	Pioneer Venus Large Probe Solar Flux Radiometer.
LST	local solar time.
LTE	local thermodynamic equilibrium, the approximation that a radiation field can be computed as if all atomic and molecular energy states are in thermal equilibrium. Collisions dominate in maintaining energy states.
LVT	local Venus time.
Lyman-alpha	radiation emitted from the H atom at a wavelength 1216 Å.

GLOSSARY

magma	melts in the subsurface consisting of variable amounts of both solid and liquid silicates and volatiles.
magnetic barrier	the region just outside the ionosphere of Venus, containing enhanced magnetic field strength resulting from compressed interplanetary magnetic field.
magnetic pressure	equal to magnetic energy density. A magnetized plasma experiences a force directed from a region of high magnetic pressure to one with low magnetic pressure. Magnetic pressure is proportional to the square of the field magnitude.
magneto-hydrodynamics (MHD)	dynamics of a conducting medium coupled to a magnetic field (see chapter I.3).
magnetosonic Mach number	ratio of bulk flow speed to the magnetosonic speed in a magnetized plasma. The magnetosonic speed (the square root of the sum of the sound speed squared and the Alfvén speed squared) is the speed at which a pressure disturbance, in both thermal and magnetic pressure, propagates in a direction perpendicular to the magnetic field.
mantle convection	refers to transfer of heat in planetary mantles by the motion of mass in a continuous process where hot material rises and cold material sinks.
mass loading	the addition of mass to a plasma flow (due to the presence of an ion source), which results in the plasma slowing down.
Maxwellian distribution	speed distribution of the molecules or atoms of a gas in thermal equilibrium.
Maxwell Montes	highest mountain chain on Venus, on Ishtar Terra.
mbar	millibar, 1.0×10^{-3} bar; microbar, 1.0×10^{-6} bar.
meridional	north–south, parallel to longitude lines.

mesopause	the upper boundary of the mesosphere (100 to 120 km) where the local vertical temperature profile obtains a minimum before increasing into the dayside thermosphere. The nightside mesosphere does not exhibit a temperature minimum at this level, but rather reveals even colder temperatures above 120 km.
mesosphere	middle atmosphere of Venus, extending from the base of the upper cloud layer (58 km) to the base of the thermosphere (100–120 km). This region of the atmosphere is characterized by stable vertical temperature gradients and a reversed pole-to-equator temperature gradient, with temperatures that are warmer over the poles than over the equator. The atmospheric super-rotation also reaches is maximum amplitude (100 m s^{-1}) at the base of the mesosphere, at levels near the cloud tops (about 70 km).
metastable	an excited, long-lived state, of a atom or molecule that is chemically distinct from its ground state.
meteoroid	a piece of a comet or asteroid on a collision course with a planet or satellite.
mgal	milligal; unit of acceleration where 1 gal = 1 cm s^{-2} = 1^{-2} m s^{-2}; 100 mgal = 1 mm s^{-1}.
MGNP90LSAAP	the name for the gravity field described in the chapter by Sjogren et al., where MGN (Magellan) and P (Pioneer Venus Orbiter) data were used to produce a 90th degree and order spherical harmonic field version (L) with SAAP (surface acceleration *a priori*).
middle cloud	Venus global cloud layer that extends from about 49 to 58 km altitude.
MIDR	mosaicked image data record.

GLOSSARY

modified dome	also, fluted dome, scalloped-margin dome, or tick. On Venus, refers to steep-sided dome characterized by concave outer slopes and radial ridges and scallops occurring either over a small azimuth or around the entire periphery of the edifice.
Monte Carlo	a method of statistical modeling whereby a model is generated many times by randomly perturbing its parameters to find which sets of parameters are most consistent with observed data. Monte Carlo techniques have also been used to test for spatial randomness of impact craters on Venus by comparing the spatial relationships of the observed crater population with a large set of randomly generated populations.
mountain belt	mountainous terrain with large length to width ratio. Often contains compressional features such as fold belts. The dominant mountain belts on Venus surround the plateau Lakshmi Planum in the Ishtar Terra region (Maxwell, Freyja, Akna, and Danu Montes).
MR	megarayleigh (see R).
multiple-floored crater	a collection of overlapping craters formed by simultaneous impact of several separated fragments of a single incoming meteoroid.
multiple impact	a crater or craters that appears to be the result of simultaneous impact of several separated fragments of a single incoming meteoroid.
nadir	the downward direction; intensity seen looking downward.
nanotesla, nT	1.0×10^{-9} T.
nightglow	night airglow. *See* airglow.
nephelometer	instrument that measures the density of a cloud by measuring the intensity of radiation that is scattered by a cloud.

NIMS	Galileo near infrared mapping spectrometer.
NIR	near infrared.
NLTE	non-local thermodynamic equilibrium; collisions are no longer important in maintaining the energy states, rather the radiation field dominates. *See* LTE.
noble gases	He, Ne, Ar, Kr, Xe (also Rn); also called rare or inert gases.
nT	unit of magnetic field strength, also called gamma; 1.0×10^{-9} Tesla.
Nusselt number	a measure of heat transport in a planet given by the ratio of the total amount of heat transported to that transported by conduction alone. It is one measure of the efficiency of convection.
OAD	Pioneer Venus Orbiter atmospheric drag experiment.
obliquity	the angle between a planet's rotational axis and the normal to its orbital plane (ecliptic pole).
OCPP	Pioneer Venus Orbiter Cloud Photopolarimeter (and Imager).
OEFD	Pioneer Venus Orbiter Electric Field Detector (Plasma Wave Analyzer).
OETP	Pioneer Venus Orbiter Electric Temperature (Langmuir) Probe.
OIMS	Pioneer Venus Orbiter Ion Mass Spectrometer.
OIR	Pioneer Venus Orbiter Infrared Radiometer.
OMAG	Pioneer Venus Orbiter Magnetometer.
ONMS	Pioneer Venus Orbiter Neutral Mass Spectrometer.
OPA	Pioneer Venus Orbiter (solar wind) Plasma Analyzer.

ORAD	Pioneer Venus Orbiter Radar.
ORO	Pioneer Venus Orbiter radio occultation experiment.
ORPA	Pioneer Venus Orbiter Retarding Potential Analyzer.
OUVS	Pioneer Venus Orbiter Ultraviolet Spectrometer.
Ovda Regio	a large crustal plateau in the western portion of Aphrodite Terra.
pancake dome	steep-sided, flat-topped volcanic feature on Venus; typically a few hundred meters high, ranging from 10 to almost 100 km in diameter.
parabolas	about 10% of impact craters are associated with west-facing parabolas; they range up to ~1000 km in each dimension, and appear to be caused by windblown deposition of very fine-grained ejecta thrown up by the impact event.
parallax	relief technique to estimate relief of landforms. Magellan parallax is engendered by different geometric distortions of landforms in image pairs of the same scene acquired with different incidence angles or look directions. Such estimates of relief are particularly important for landforms and local relief at scale lengths that are too small for altimetry, and for terrain that is complicated and difficult to decipher with altimetry.
periapsis	that part in an elliptical orbit about a body which is the point of closest distance to that body.
phase angle	angle between Sun and observer, taken at the object observed.
photochemical	chemical reactions or processes initiated by light. On Venus, photon driven chemistry is important above 60 km.
photodissociation	dissociation of molecules induced by (light) photons.

photo-electrons	electrons formed as the by-product of ionization of atoms or molecules.
photoionization	ionization of atoms and molecules produced by (light) photons.
physiography	description of the topography and landforms.
pick-up ions	when ions are created within a moving plasma they are acted on by the Lorentz force associated with the plasma's large-scale electric and magnetic fields in such a way that they at least partially move together with the original plasma.
plains	the most widespread surface feature on Venus, relatively flat and of volcanic origin. They comprise over 80% of the surface and correspond to almost all elevations below 1.5 km; locally and regionally disturbed by tectonism and younger volcanism.
Planck function	energy distribution of blackbody radiation, usually in intensity units.
planetesimals	bodies of substantial size, believed to be an intermediate state in planetary accretion.
plasma	a gas containing significant abundances of charged particles, usually electrons and ions. Electric and magnetic fields play an important role in determining the behaviour of a plasma.
plasma frequency (f_p)	the natural frequency at which the electrons and ions in a plasma oscillate relative to one another.
plasma mantle	the lower portion of the magnetic barrier, just outside the Venus ionopause, containing a mixture of both solar wind and ionospheric plasma.
plateau-shaped highland	*see* crustal plateau.

GLOSSARY

plate tectonics	the paradigm for Earth's global tectonics—the lithosphere is broken into a finite number of plates in relative motion to one another. Plates are created at mid-ocean ridges and destroyed at subduction zones. Plate tectonics is thought not to occur on Venus.
plumes, mantle	rising buoyant material in the mantle; buoyancy is usually thermal, although it can also be compositional.
potential temperature	temperature that would be reached if air were compressed or expanded adiabatically to a standard pressure. Thus, the potential temperature, but generally not temperature, of air parcels displaced vertically and adiabatically remains constant.
ppb	parts per billion (1.0×10^9).
ppm	parts per million (1.0×10^6).
ppmv	parts per million by volume, a measure of gas concentration.
Pratt compensation	a type of isostatic compensation whereby mountains and low lying areas are supported by lateral changes in density down to a common depth of compensation. High-standing regions are associated with lower density columns beneath, and the opposite is true for low regions.
precession rate	the rate of motion of a planetary rotational axis about its ecliptic pole; the motion describes a cone.
production age	the age of a particular region of a planetary surface obtained by counting the number of impact craters on it and assuming that no craters have been destroyed since craters starting collecting after the surface was initially formed. Knowing the formation rate of craters on the surface will yield the age of surface formation, or production age.

prograde	rotating in the same sense as the orbital motions of the planets (also called direct).
propagating leader	part of the initiation of a lightning bolt.
PV	Pioneer Venus; PVO, Pioneer Venus Orbiter.
pyroclastics	particulate matter thrown into the air by volcanoes.
quenching	deactivation of an excited state in an atom or molecule.
R	Rayleigh; a unit of photon flux. 1 R = 10^6 photons emitted into 4π steradians per cm^2 vertical column per sec. It is most often used in measuring the luminous intensity of the airglow and aurora.
radar shadows	Magellan radar shadows are produced when no echoes are received because slopes are not illuminated by the SAR antenna.
radiogenic	produced by radioactive decay (e.g., ^{4}He, ^{40}Ar).
radiometer	Magellan instrument that passively measured the thermal emission from the surface at 12.6 cm by sampling the SAR receiver between radar echoes.
Rayleigh friction	drag or friction in a fluid that is proportional to the fluid's speed; typically used to approximate wave drag effects on the mean flow.
Rayleigh-Taylor instability	a "fingering" instability common when a dense fluid overlies (or is at higher pressure than) a less dense one.
RCE models	radiative-convective-equilibrium models. This type of thermal balance model determines the globally-averaged, altitude dependent equilibrium temperature profile in a planetary atmosphere that results from solar radiative heating, thermal radiative cooling, and vertical convective heat transport.

recycling	refers to the creation and destruction of crustal material on a planet by extracting crust from mantle and then returning it, for example, by subduction or by thickening, phase change, and delamination.
redox	contraction of the words reduction-oxidation; a reference to the oxidation state of a chemical system.
reflectivity	measure of the efficiency of a surface in reflecting electromagnetic radiation; ideally refers to a plane wave infringing on a flat surface, where the term Fresnel reflectively is used.
regolith	soils formed on the surface by weathering, impact, and aeolian activity.
residuum	magnesium-rich, low-density residue remaining after basalt has been extracted from a planetary mantle.
resurfacing, on Venus	the creation of a new surface in a region by volcanic or tectonic activity. By definition, there is no evidence of the previous surface.
retention age	on a planetary surface where craters are being destroyed, the number of impact craters divided by the crater formation rate gives not the age of the surface, but the size-dependent lifetime of a crater, known as the retention age.
retrograde	opposite of prograde; i.e. rotating in the opposite sense as the orbital motions of the planets.
rheology	deals with deformation and flow of rocks.
Richardson number (Ri)	non-dimensional number used to describe the onset of turbulence due to wind shear (see chapter III.6).
ridge belts	narrow to broad belts of mostly parallel ridges observed in the plains of Venus.
rifting	tectonic stretching of crustal rocks which results in linear, fault-bounded depressions (rift valleys).

rms slope	root mean square slope, derived from the Hagfors model, is an estimate of the slope distribution of the surface at scales of tens to hundreds of meters.
RSZ	retrograde super-rotating zonal; see superrotation.
runaway greenhouse	an effect that may have occurred on primitive Venus, if it had a global ocean. A greenhouse warming based on infrared absorption of water vapor could evaporate still more water, raising the temperature still further. Under certain conditions, this warming could continue to increase until the entire ocean evaporated into the atmosphere.
SAR	synthetic aperature radar; the SAR transmits radar waves and measures the radar echoes from the surface in oblique viewing geometries.
saturation vapor pressure	the vapor pressure of a chemical compound in equilibrium with the condensed form of the compound. A vapor pressure greater than the saturation value should promote additional condensation; less than the saturation value, additional evaporation of the condensed phase.
semi-empirical models	models of observations based, in part, on theory.
shield field	abbreviated term for "small shield volcano field". On Venus, an enhanced concentration of small volcanoes, generally over an area >100 km in diameter; a shield field may also include a concentrated occurrence of more widely scattered small volcanoes; although commonly consisting of small shield volcanoes, the term "shield field" also loosely refers to clusters that may consist of, or include, small conical volcanic edifices, small steep-sided domes, and small modified domes.
shield volcano	volcanic center characterized by low ($<10°$) angle and relatively straight slopes resulting in a conical edifice the relief of which is much less than its diameter.

GLOSSARY

shock front	a sudden jump in properties at a dynamical interface (e.g., a "sonic boom").
sinuous rilles	channels which enamate from distinct, circular or elongated regions of collapse. They are approximately 1 to 2 km wide and several tens of km long. They become narrower and shallower distally. They resemble lunar sinuous rilles. Their morphologies suggest an origin by the thermal erosion of flowing lava.
small dome	*see* small volcano.
small volcano	*also* small dome, small shield volcano. On Venus, refers to volcanic edifices <20 km in diameter; although shield volcano morphology is common, the edifice may be a steep-sided dome, small modified dome, or conical volcano with steep flank slopes.
SMOW	standard mean ocean water.
SNFR	Pioneer Venus Small Probe Net Flux Radiometer.
solar activity	any variation in the sun's radiation output with time; typically used in reference to the effects of such variations on the mean atmosphere of a planet.
solar cycle	fluctuation in the Sun's radiation output with a periodicity of 11 yr, as measured on Earth.
solar rotation	fluctuation in the Sun's radiation output with a periodicity of about 27 days, as measured on Earth.
solar wind dynamic pressure	the energy density associated with bulk solar wind flow. Also a momentum flux (equal to the mass density times the square of the bulk velocity). Dynamic pressure can be converted into thermal or magnetic pressure, such as at the magnetic barrier.
sonic Mach number	the ratio of a fluid flow speed to the speed of sound in that medium. Flows with sonic Mach number greater than unity are called supersonic.

spherical harmonics	a set of orthogonal functions used to represent any variable on a sphere. The functions are products of sines or cosines in longitude and Legendre polynomials in latitude.
splotches	circular, radar dark or bright patches on the Venusian surface thought to result from impact shocks of disintegrated asteroids and/or comets without formation of an attendant impact crater.
sprites	cloud to mesosphere-ionosphere lightning discharges on Earth. Sprites are one possibility for lightning discharges on Venus.
sputtering	release of atoms or molecules from a surface (or upper atmosphere) by the impact of ions; most efficient for impact energies of the order of kilo-eV.
SS–AS	subsolar-to-antisolar; the circulation cell that is the dominant feature in the thermospheric wind system.
stagnant (sluggish) lid convection	convection taking place beneath a largely immobile lithosphere. In stagnant lid convection, heat is transferred by vertical conduction through the lithosphere. In sluggish lid convection, some heat is transferred by horizontal motion of the lithosphere. Geologically, sluggish lid convection is equated with modest tectonics at the surface, while very little tectonics is associated with stagnant lid convection.
static stability	situation in which the temperature lapse rate in an atmosphere is less than the adiabatic lapse rate.
steep-sided	on Venus, generally refers to intermediate and small dome volcanoes characterized by convex slopes or convex basal margins; volcanoes of the latter description may also be referred to as "pancake" domes or pancake volcanoes.

stellate fracture center	also, "nova." On Venus, a concentrated radial or nearly radial pattern of fractures, grabens, and fissures, with or without surface evidence of volcanism or topographic expression.
stereoscopy	Magellan stereoscopic viewing of image pairs; used to enhance the ability to interpret Venus landforms, structures, and geologic relations between rock units.
subduction	the process on Earth whereby lithosphere is returned to the mantle at oceanic tenches.
superrotation	rotation of the atmosphere in excess of that due to solid body rotation alone. At Venus altitudes near the cloud tops, the entire atmosphere appears to rotate in almost solid body rotation at about 60 times the rotation rate of the surface. The maximum zonal wind speeds (100 m s^{-1}) are usually seen at the equator near 65 km, with the period of rotation being 4-days. The super-rotation is sometimes called the "4-day wind."
suprathermal electrons	hot electrons generated by non-thermal processes.
SZA	solar zenith angle.
tectonism	the process by which internally generated forces distort and reshape elements of a planet's crust.
temperature lapse rate	the vertical gradient of temperature (dT/Dz), usually expressed in degrees K per km (K/km).
terminator	the line dividing the illuminated and shaded regions of a planet.
Tesla (T)	MKS unit of magnetic field, 1.0×10^4 Gauss.
tessera	a radar-bright terrain made up of crosscutting sets of ridges and fractures. Crustal plateaus have tessera surfaces.

thermal tides	atmospheric variations with periods of 1 solar day and harmonics, induced by solar heating.
thermochemical	chemical reactions predominantly subject to thermal processes and vertical transport. On Venus, thermochemical reaction are important below the clouds.
thermosphere	the region of the upper atmosphere of Earth (and Venus) above about 120 km, where the kinetic gas temperature is much higher than that in the atmosphere below it.
tidal structure	any atmospheric variation having a period of 1 solar day.
trachyte	the volcanic (extrusive) equivalent of syenite; an evolved or differentiated lava composition characterized petrologically by adundant potassic to sodic feldspar and silica contents of up to 62 weight percent.
turbopause	see homopause.
upper cloud	Venus global cloud layer that extends from about 58 km to about 70 km altitude.
upper haze	Venus haze layer that extends from the cloud tops (70 km) into the upper mesosphere (90–100 km).
Urey number	the ratio of radioactive heat generation to total heat loss in a planet.
UV	ultraviolet radiation with wavelengths from about 100 nanometers to 380 nm.
UV absorber	an unknown material that absorbs ultraviolet radiation at altitudes within the Venus upper could layer.
VIRA	*Venus International Reference Atmosphere*.
viscosity	constant relating deviatoric stress to deviatoric strain rate in a fluid.
VOI	Venus orbit insertion.

GLOSSARY

volatiles	materials that accrete into a planet but subsequently degas as atmosphere or hydrosphere.
volcanic edifice	an accumulation of lavas and/or pyroclastic materials resulting in a topographic relief feature where magmas have been erupted onto the surface from a central vent or vent complex over 360 degrees of azimuth.
volcanic pit	a volcanic eruptive vent or non-eruptive pit, or pit crater collapse depression smaller than a caldera; frequently occurs at the summit of small and intermediate volcanoes; may also occur in a pre-existing surface or along a fissure vent.
volcanic rise	a broad topographic rise on Venus characterized by shield volcanism, extensional tectonics, and a large ADC value. Volcanic rises are interpreted as hotspots associated with mantle plumes.
volcano	a radial circular to elongate pattern of lavas and/or pyroclastic materials erupted from a central vent or vent complex; materials must have erupted over 360 degrees of azimuth implying the presence of a centralized topographic relief feature; measurable topographic evidence for an edifice may or may not be present.
VTGCM	Venus thermospheric general circulation model; three-dimensional, energy-hydrodynamic-chemical model, successfully used to simulate the Venus neutral upper atmosphere (above about 95 km) (see chapter II.4).
VTS3	Venus semi-empirical model of the neutral upper atmosphere, by Hedin et al. (1983) (see chapter II.3).
wavenumber	(1) for radiation, reciprocal of wavelength; (2) for planetary waves, the number of waves in one global circuit.
wave saturation	a condition where the local temperature lapse rate, produced by the sum of a wave and the mean state, is dry adiabatic.

weighting function	the kernel of an integral representing relative contributions of different layers to outgoing radiation.
whistler-mode waves	a kind of plasma wave, thought to be responsible for heating of Venus's dayside ionosphere. Also, the mode in which electromagnetic waves from lightning can penetrate the nightside ionosphere.
wind-induced diffusion	transport of light species by the thermospheric wind system. This occurs because light species have small atomic weights and large scale heights, thereby becoming increasingly dominant with altitude due to diffusion. For Venus, H and He atoms are transported to the nightside where they "pile-up" in a bulge region near 0400 LVT.
wind streaks	features of apparent wind origin that take on a variety of shapes and are often associated with impact craters and tectonically deformed areas. The deposition of fine particulate matter is responsible for many streaks, though wind scouring undoubtedly plays a role.
wrinkle ridges	long sinuous narrow features abundant on the plains of Venus (and on other planetary bodies) and interpreted to be compressional in origin.
yardang	streamlined hill formed by intense aeolian erosion of surface material.
yield strength envelope (YSE)	gives the strength of the lithosphere as a function of depth in both compression and extension. The envelope is determined by the weaker of viscous creep strength or frictional sliding strength. The frictional sliding strength is dependent on depth and column density. The ductile creep strength is strongly dependent on rock composition and temperature. Thus, the crust and mantle have different yield strength envelopes. *See* Byerlee's law.
zonal	east–west, parallel to the equator. The cloud-top superrotation is often referred to as a zonal wind.

INDEX

absorbing gases, 358
absorptivity, 532
acceleration mechanisms, 494
accretion
  horizontal, 1227
  vertical, 1230
ADC. *See* apparent depth of compensation.
adiabatic compressional heating, 213, 251, 273
  cooling, 274
  decompression, 1277
  expansion, 213
  expansional cooling, 374
Adivar Crater, 577, 1021
admittance ratios, 1261
aeolian features and processes, 492, 547, 585
  activity, 1023, 1077
  bedform, 556
  dune, 565
  Earth and Venus compared, 557
  microdunes, 570
  particle flux, 552
  particle threshold, 551
  resurfacing, 1062
  transport, 541
  weathering, 620
  wind abrasion, 554
  wind streaks, 578
  yardangs, 574
aerobraking, 228
aerodynamic erosion, 990
  stresses, 979
aerosols, 128, 326, 354, 367, 377
  composition, 436
  distribution, 375
  H_2SO_4, 354, 362, 376, 450
  profiles, 418
  sulfur, 444, 449
Aglaonice dune field, 565
Aino Planitia, 734, 774, 783, 905
  wrinkle ridges, 908
airburst, 993
airglow, 13, 448
  atomic oxygen, 268
  constraints, 264
  dayglow emissions, 242, 297
  mesosphere nightglows, 278, 314
  nitric oxide (NO) nightglow, 239, 242, 247, 264, 278, 282
  O_2
    Herzberg II nightglow, 265
    infrared nightglow, 268
    nightglow emission, 297, 319
AIRSAR data, 565

airshock, 992, 993
Airy compensation model, 939, 1153
  isostasy, 889, 1217, 1222, 1224
Akkruva Colles, 722
Akna Montes, 830, 831, 833, 879, 883
aluminum (Al), 1213
Al-Uzza Undae, 567. *See also* Fortuna-Meshkenet dune field.
albedo, 353, 450, 645, 990, 994, 1134
  features, 284
  modification, 470
  radiation force, 1132
albite, 604
alkalic rock, 504
alkaline igneous rocks, 597, 598, 605
Alpha Regio, 536, 542, 820, 825, 1224
Alpha-Lada fracture zone, 943
altimetry, 671, 892
ambipolar electric field, 114
Ampere's law, 68
amphiboles, 605
Ananke Tessera, 825, 836
anastomosing channels, 1033
  fractures, 1058
andalusite, 637
anemone, 720
angle of impact, 1038
Anglo-Australian Telescope (AAT) observations, 315, 328, 340
angular
  momentum, 480–482
    atmospheric, 1088
    flux, 485
  velocity, 480
anhydrite ($CaSO_4$), 621, 637–638, 640, 646, 650–651
  formation, 505, 618
anionic elements, 650
annuli, 811, 935, 938, 940, 942, 946, 1058
anomalous melting, 742
anorthite ($CaAl_2Si_2O_8$), 622
anorthosite, 896, 1207
anti-solar-to-subsolar flow, 319
Aphrodite Terra, 576, 578, 585, 667, 674, 735, 761, 774, 798, 822, 825, 905, 943, 1192
  Ovda Regio, 674
  Thetis Regio, 674
apparent depth of compensation (ADC), 802, 818, 823, 857–858, 863, 866, 868, 873, 887, 889, 1153, 1222, 1223
aprons, 730
arachnoids, 699, 700, 724, 726, 735, 773, 774

arches, 904, 911
arcuate
    features, 1186
    ridges, 804
Arecibo Observatory, 533, 820
    coronae data, 944
    images, 1006
    plateaus, crustal, 818
    system, 536
argon (Ar), 4, 387
Arrhenius relationship, 520, 1264, 1267
Arsia Mons, 727
Artemis Chasma, 774, 813–814, 905, 961, 1198
    model solutions, 1191
Artemis Corona, 810, 938–939, 961, 1187
Artemis Planitia, 759
Aruru Corona, 708
ash, 700, 717, 721
Asteria Regio, 846, 947
asteroids, 981
    collison rate, 1067
    C-type, 974, 975, 982
    impacts, 1041
    iron, 975
    M-type, 975
    S-type, 974, 975, 982
    Venus-crossing, 1067
asthenosphere, 798, 823, 857, 1008, 1185, 1255
Atalanta Planitia, 673, 734, 787, 808, 913, 915
Atalanta-Vinmara Planitiae, 836
Atla Group, 1053, 1055, 1059, 1061, 1062, 1065
    time, 1072, 1076
    volcanoes, 1070
Atla Regio, 657, 689, 731, 733, 744, 783, 846, 851, 856, 863, 866, 873, 879, 947, 949, 1140, 1177, 1194
    domal volcanic highland, 1223
    gravity model, 1174
atmosphere (deep), 331, 334, 336, 344, 348, 353, 470
    dynamics, 346
    momentum transport, 494
    near-infrared observations, 365
    radiative transfer modeling methods, 366
    spectrophotometric data, 363
    thermal balance, 354, 360
        unresolved issues, 362
    Vega data, 364
atmosphere (lower), 10, 325, 415. *See also* wind.
    carbon monoxide, 341
    chemistry, 417

circulation, 584, 585
clouds, 326, 438
composition, 326, 629
D/H ratio, 336
dynamics, 346, 493
element mixing ratio, 435
future study, 348, 449
hydrogen halides, 342
hydrostatic equilibrium, 235
local instability, 435
Mode 3 controversy, 452
near-infrared mapping, 335
photochemical models, 443
post-Pioneer Venus models, 443
redox state, 29
sulfur chemistry, 436
sulfur-bearing compounds, 343
superrotation, 251
thermochemical equilibrium, 434
water vapor, 336
wind dynamics, 327
atmosphere (upper), 214, 259, 282, 328. *See also* gravity, waves *and* near-infrared, windows.
    airglow constraints, 264
    dayside, 248
    eddy diffusion, 260
    global mean, 248
    gravity waves, 259, 260
    mesopause, 260
    retrograde superrotating zonal (RSZ), 259
    solar EUV, 259
    subsolar-to-antisolar (SS-AS), 259–260
    superrotating, zonal flow (RSZ), 260
    trace species, 301
    unresolved issues, 287
    water, 407
    wave drag, 260
    wind measurements, 270
atmosphere, 459, 591, 637, 775, 972, 1039. *See also* circulation of atmosphere; surface-atmosphere interactions; *and* weathering.
    angular momentum, 480
    boundary layer, 488
    chemistry, 434
    circulation, 549, 580, 583
        maintenance, 479
    clouds, 3
    composition, 3, 643, 650
    density at surface, 492
    deuterium, 17
    dry, 386
    flux calculations, 403
    hydrogen, 17
    hydrogen and deuterium escape, 400
    influence of impactors, 995

INDEX

ionosphere, magnetization, 24
layers, 488
lightning, 19
models, numerical, 492
near-surface, 347
observations
  radiation balance, 462
  structure, thermal, 460
optical properties, 358
oxygen, 17
parameters, surface pressure and temperature, 1077
properties, 394
rotation, 460, 473, 485, 489
solar wind, 26
spin, 468, 485, 493, 494
stratified, 487
sulfur, 8
superrotation, 494
thermal structure, 9, 353, 378, 461
  optical properties, 358
thermospheric composition, 14
tidal momentum transports, 493
water, 7
waves, 488
atmosphere-impactor interactions, 971, 984, 988, 995, 1006
atmospheric drag, 1132
  rotation, 237
  shielding, 972, 1066
Atropos Tessera, 828
Atropos/Itzpapaloti junction, 834
augen-shaped areas, 717
Aurelia Crater, 1049
Aurelia Group, 1049, 1054, 1055, 1061, 1071
  time, 1077
aureoles, 720. See also halos.
avalanche(s), 1258, 1263
axis, principal moment, 1089
axisymmetric advection, 480
azimuths, 580, 584
  wind streaks, 585

back venting, 994
backscatter, 512, 544, 572
  areas, 507
  cross section, 991
  data, 508, 511
  effects, 567
  functions, 534
  highlands, 519, 521
  measurements, 995
Baikal Rift, 873
Baltis Vallis (formerly Hildr Fossa), 783, 1064
baroclinic instabilities, 488
basalt, 516, 637, 644–645, 1179, 1237.
  See also crust; and weathering.

composition, 503, 554
  of surface, 504
crust, 830, 1207
diopside, 639
eclogite transition depth, 955
foams, 715–716
flows, weathering, 660
material, 544
melts, 1226
  H_2O, 1213
MORB, 1247
oxidation, 505, 616, 630
surface, 613
tholeitic, 599, 1053
basins
  and dome terrains, 804, 806, 823
  morphology, 884
bedforms, 556, 565. See also microdunes.
bedrock, 1210
  collapses, 506
Bell Regio, 731, 851, 856–858, 864, 866, 872, 873, 1140, 1194
belts, 804
  mountain, 881, 892
  ridge, 882
bending moment, 1190
Berkeley-Illinois-Maryland Array (BIMA), 311
Bermuda Rise, 866
beryllium (Be), 603
Beta-Atla-Themis Regiones (BAT) zone, 699, 730, 733, 735, 737, 745, 747, 942–943
Beta Regio, 533, 536, 657, 689, 731, 735, 744, 825, 846, 851, 856–858, 860, 863, 870, 873, 879, 944, 947, 1140
  crustal thickness, 1223
  extensional deformation, 1192
Beta-Themis-Aphrodite disturbances, 1216
Bezlea Dorsa, 911
bistatic radar observations, 519, 521, 536
blowoff, 408
blue absorption, 449
bolides, 512, 979, 1062
Boltzmann equation, 191, 199
boron (B), 603
boundary
  layer, 104, 488
    flows, 490
  margin, 1017
Boussinesq approximation, 1267
bow
  air pocket, 987
  shock, 27, 57, 62, 96, 988. See also Venus-solar wind interaction.
  position, 39

solar wind and, 40
bowls, 954
breakup, 987
Brewster angle, 519
bright halos, 993
brightness temperature, 528
brittle failure, 1166, 1274
brittle-ductile transition, 820, 824
brittle-plastic transition, 1182
broadband turbulence, 108
broad ridges, formation, 1057
Brunt frequency, 473, 476
buckling, 904, 949
buffer reactions, 603
buffering concept, 650
bulk density, 1208
buoyancy flux, 1195
　occurrence, 144
　rate approach, 140
Byerlee's law, 922, 1182

Cadiz dune field, 559
calcite ($CaCO_3$), 593, 595, 598, 620, 622, 651
　paragenesis, 596
calcium (Ca), 639, 1213
calderas, 700, 705, 724, 726, 730, 740, 743, 822, 828–829, 834, 880, 949, 960
　defined, 724
　downsag, 724, 727
　formation, catastropic callapse, 724
　Ishtar Terra, 827
　size, 737
　volcanism, 831
Canada-France-Hawaii Telescope (CFHT), 271, 340
　FTS observations, 315–316
canali, 758, 774, 785, 786, 790, 1214, 1215, 1237
　morphology, 759
　origin, 775
canali-type channels, 765, 772, 776, 783
　genetic processes, 774
carbon (C), 4
　density, atomic, 234
　oxidized, 387
carbon dioxide (CO_2), 3, 226, 592, 593, 650
　15 μm cooling, 247, 249, 276, 280
　absorption, 328
　atmosphere, 282
　bands, 337, 418
　density, 272
　gas transmission, 418
　in lower atmosphere, 326, 333
　in thermosphere, 283
　photolysis, 271, 314
　spectroscopy, 306–308

carbon monoxide (CO), 242, 319, 437, 627, 650
　absorption feature, 341
　abundance, 605, 630
　bulge, 273–274, 282, 319, 378
　density, 272
　distributions, 264
　emission, 378
　fluorescence, 271
　in atmosphere, 341, 443
　in mesosphere, 310, 318
　microwave emission measurements, 234
　mixing ratio, 278, 308, 341
　　in mesosphere, 318
　observations, in mesosphere, 311
carbonate stability, 593
carbonate-silicate equilibria, 593, 596
carbonates, 599
carbonatites, 596, 605, 1215
　flows, 776
　magmas, 597
carbonyl sulfide (OCS), 8
　mixing ratio, 343
Carson Crater, 660
Cartesian spectral analysis, 1177
catastrophic processes, 800
　convective episode, 1074
　resurfacing model, 1061
cauldrons, 726
central peaks, 1025
cerussite ($PbCO_3$), 596
chalogenides, 521, 627
channels (valles), 678, 757, 767, 770, 774, 785, 789, 924, 1215. See also Kallistos Vallis and Baltis Vallis.
　Baltis Vallis, 735, 1064
　classication, 758
　complex, 759, 761
　compound, 762
　distribution and settings, 772–773
　formation, 777, 790
　lava, 909, 910
　longitudinal profiles, 783
　meanders, 764, 766
　morphology, 758
　simple, 758
chaotic
　topography, 557
　zone, 1118
Chapman theory, 165–167
charge exchange, 41, 214
chasmata, 797, 798, 803, 806, 810, 811, 825, 835, 836, 939, 942, 953, 957, 961, 1199
　asymmetric, 811, 813
　characteristics, 811
　coronae, 943, 959

origin, 957
systems, 962
chemical equilibrium, 610
chemically weathered minerals, 649
Chenzy coefficient, 780
Chicxulub Crater, 999
chlorine (Cl), 430, 444–445, 603–604, 655
compounds, 282
chlorine-sulfur interactions, 448–449
chlorite, 1213
chondritic heat source, 1196
cinder, 721
circulation of atmosphere, 484, 549
experiments, 492
flow observations, 465
friction, 483
future study, 493
maintenance, 479
wave drag, 483
clasts, 1005
Cleopatra Crater, 539, 883, 895
Cleopatra Patera, 657
clinoenstatite ($MgSiO_3$), 640
Clotho Tessera, 882
clouds, 7, 9, 29, 282, 326, 415, 416, 464, 485, 495. See also atmosphere (lower).
aerosol profiles, 427
barrier, 394
chlorine, 443, 452
cloud-to-ground discharges, 128
cloud top
superrotation, 296, 306
ultraviolet absorber, 360
composition, 302, 342, 343, 361, 430
deck, meridional structure, 346
droplets, 328
ejecta, 506
O_2, 445
opacity, 327, 335
optical depths, 348
particles, 329, 364, 418
processes, 437
properties, data, 334
PV LNMS, 395
reflection, 268
SO_2, 421
spacecraft data, 427
spatial variations, 368
spectrophotometric data, 363
sulfur, 443, 452
types, 335
ultraviolet
features, 331
images, 470
markings, 451
waves, 284, 286
winds, 299–300, 305, 306, 317, 346

zonal, 287
CMB friction torque, 1120
CMF model, 1117
CO and CO_2
gas mixtures, 613
gas phase equilibration, 610
mechanism for equilibrating, 616
ratios, 607, 630, 643
cold
trap, 408
welding process, 586
coldspots, 821, 830. See also mantle, downwelling.
Colette Caldera, 880
Colette Patera, 735, 960
collapse pits, 777
collapsed terrain source, 769
collisonal momentum transfer, 283
collision-induced transitions, 367
collisions of electrons, 190
Columbia River flood basalts, 1076
comets, 394, 409, 973, 974, 981, 990
albedo, 974
comet Halley, 431
comet Shoemaker-Levy 9, 978, 985, 995
density, 974
Halley-family, 975
Jupiter-family, 974
comminution, 549
shock, 506
surface rocks, 504
compensation mechanisms, 1153
composite solid theory, 1092
compression, 1057
conductive heat transfer, 1250
cones, 719, 949
continental keel, 895
continuous wave (CW) downlink, 537
CONTRAST experiment, 609–610, 630
convection, 296, 1249. See also mantle, convection.
activity, 492
intensity, 489
models, 860, 1196
processes, 867
regime, 1274
core, 24, 1273
cooling, 1197
density, 1093
ellipticity, 1120
fluid, 1097
freezing, 1090
friction torque, 1088
liquid, 1093
magnetic field, 1090
mantle
boundary (CMB), 1088, 1090, 1258

1338 INDEX

fluid friction, 1105
Coriolis force, 585, 586
coronae, 676, 699–700, 726, 730, 735, 767, 773–774, 777, 797–798, 803, 806, 808, 810, 812–813, 825, 835–836, 846, 856, 868, 873–874, 902, 907, 942, 1059, 1069–1070, 1199, 1236, 1251
   asymmetric, 933
   centers, 705
   chains, 677
   characteristics, 811
   chasmata, 934, 943
   concentric, 933
   deformation, 1059
   early studies, 931
   Eistla Regio, 855
   evolution, subduction, 813
   evolutionary sequence, 940
   extension with upwelling model, 962
   flattening model, 953
   flexure model, 955
   formation, 811, 905, 1058, 1072, 1076
      mechanisms, 949
      models, 961
      time scales, 955
   gravity data, 938, 959
   Hecate Chasma, 944
   hotspots, 871
   isolated, 943
   moat, 811
   MPR, 811
   multiple, 933
   numbers, 932
   origin models, 949
   Parga Chasma, 947
   radial stress, 954
   radial/concentric, 933
   relaxation model, 953
   rim, 811
   rises, 942
   subduction/delamination model, 962
   tectonic characteristics, 935
   Themis Regio, 856
   topography, 935
   uplift model, 950
   Venera missions, 931
   volcanic characteristics, 935
corona-like interior, 708
Coulomb collisions, 191
crater densities, 687, 689. *See also* impact craters.
   Ishtar Terra, 827
cratering, 969, 972, 1015. *See also* atmosphere-impactor interactions; impact craters; *and* impactors.
   depth-diameter relations, 1000
   size and velocity, 970
   record, 848, 1206
craters, 512, 517, 571–572, 574, 660, 679, 985, 1001. *See also* ejecta.
   angle of impact, 1005
   basalt, 1005
   central uplift, 1002
   Cleopatra Crater, 883
   collapse, 1002
   complex, 1017
   cross-track distortion, 1000
   degradation, 1281
   density, 800
   deposits, 544
   depth-of-melting model, 1007
   depths, 679, 1025
   ejecta, 660, 997, 1000
      deposits, 540
      flows, 1004
   fields, 1040
   floors, 1023–1024
   flows, 800
   global survey, 1041
   hydrodynamic collapse model, 1007
creep, 1222
   behavior, 1179, 1180
   experiments, 1180
   measurements, 1221
   strength, 1220
critical elevation, 542
cross-terminator winds, 251
Crozet, 850
crust, 802, 922, 950, 1179, 1185, 1192, 1205, 1208
   basaltic, recycling, 1209
   carbon, 3
   compensation, 863
   composition, 1179, 1208–1209
   data results, 1237
   deformation, 1220
   dry, 1220
   felsic, 1217
   formation, 1207
   gabbroic (basaltic), 1218
   geochemistry and morphology, 1216
   geodynamic models, 1220
   gravity
      signal, 1217
      studies, 1222
   heat sources in, 1252
   history, 1226
   horizontal accretion and recycling, 1227
   intrusion, 817
   isostasy, 1222
   lander measurements, 1209
   layer, 786
   mafic, 1217

mantle boundary, lithosphere, 1172
material, 822
minerals, 625
models
  vertical accretion, no recycling, 1230
  vertical accretion, recycling, 1233
production rate, 1236
recent formation, 1235
recycling, 897, 1229, 1233, 1253
rheology, 950, 1183
rocks, 660
  anhydrous conditions, 1221
secondary, 1207, 1230, 1233
shortening, 737, 748
strength, 1029, 1044
structure, 902
tertiary, 1208
thickening
  compression, 916
  subsolidus flow, 823
  supersoldius flow, 823
thickness, 859, 889, 1217, 1223, 1226, 1232
thinning, 922
volcanic morphology, 1214
water, 1219
zone, 1255
cryosphere, 12, 226, 274
  nightside, 273, 276
crystal-rich models, 716, 717
crystallinity, 716
crystallization, 1249
Curie temperature, 520
cyclostrophic balance, 287, 470

Dali Chasma, 1199
Dali Chasmata, 851
Dali/Diana Chasma, 958
Danu Montes, 827, 833, 881
dayside ionosphere, 15, 161, 184
  chemical structure, 163
  ionopause, 171
  solar cycle variability, 165
  solar rotation, 203
Decan River flood basalts, 1076
decompression melting, 1226
Deep Space Network (DSN), 885, 1127
deformation, 802, 828, 832, 901, 944. See also Ishtar Terra and plains.
  coronae, 935
  craters, 1022
  extensional, 1192
  fragmentation and spreading, 985
  impactors, 978
  restricted spreading, 989
  spreading and breakup, 987
  structural, 1193
  zones, 813

degassing models, 411
Dekla Tessera, 825
delamination/subduction, 957
density perturbations, 237
depolarization, 541
  ratios, 543
depressions, 937
Derceto Corona, 767
desaturations, 1133
deuterated water vapor (HDO), 386
deuterium (D), 7, 17, 385, 397, 401, 403, 406. See also atmosphere.
  escape, 388, 400
  in atmosphere, 13, 395
  in thermosphere, 240
  loss mechanisms, 411
Devana Chasma, 707, 851, 859, 944
D/H ratios, 385, 386, 396, 598
  in atmosphere, 7, 336, 386, 388
  in future study, 411
  measurements, 394, 395, 396
diabase, 950, 1180, 1221, 1255
  Columbia, 802, 1180, 1182, 1184
  conditions, 1182
  dry, 810, 824, 1193
    measurements, 1180
  flow law, 954
  Maryland, 1180, 1182, 1184
Diana and Dali Chasmata, 1059
diapirs, 700, 745, 746, 811, 812, 950, 955, 961. See also coronae.
  in coronae formation, 953, 957
  lithosphere, 1228
  rise of, 950
  uplift, 954, 959
  upwellings, 962
dielectric
  constant, 544, 567, 653–654. See also surface.
  permittivity, 527, 653
  properties, 508, 527
  surface, 519
diffusion limit, 176
digital elevation models, 801
dikes, 907
Dione Regio, 851, 856, 870, 872
diopside ($CaMgSi_2O_6$), 622, 650–651
Direct Numerical Diagonalization (DND) method, 367
disappearing
  atmosphere, 247
  ionosphere, 182, 207
discharge, 766
disk-averaged observations
  in mesosphere, 308
disk-resolved observations, 311
dissociative recombination, 165, 183, 243
distal ejecta flows, 1033
disulfur monoxide (S_2O), 451

1340 INDEX

diurnal thermal forcing, 585
dolomite (CaMg(CO$_3$)$_2$), 595
domes, 574, 719, 846, 937, 949
   endogenic
      trachyte and alkalic termediate variants, 715
   fields, 720
   fluted, 715
   modified, 715
   pancake, 716
   scalloped-margin, 717
   silicic, 716
   trachyte, 716
domical uplift, 950
Doppler
   shifts, 149, 271, 299–300, 542
      CO lines, 306
      winds, 306
   tracking, 857
downwarping, 785
downwelling, 746, 825, 921, 1255, 1270, 1272
   example, Tibet, 890
   model (ADC), 892
   sheets, 1258
drag measurements, 228, 250, 263
dual-polarization emissivity data, 513
ductile flow, 1166
dunes, 506, 549, 565, 586
   fields, 547
   terrestrial, 557
   undae, 680
   wind formation, 995
dunite, 802
dust, 548, 586
dust clouds, 22
dynamo, magnetic, 23

Earth, 265, 337, 533, 907, 1152, 1227, 1229–1230
   aeolian
      features, 548, 556, 557
      transport, 551–552
   atmosphere, 285, 443, 483, 489, 591, 979
      wave drag, 483
   atmospheric effects, 747
   bulk properties, 1206
   calcite, 596
   calderas, 724, 726, 740
      sag-down, 727
   carbonate, 387
   circulation pattern, 493
   core, nonhydrostatic oblateness, 1120
   coronae, 810
   craters, 985
   crust, 1210, 1217, 1226, 1231, 1236
      seafloor spreading, 1225
   dunes, 574

   dynamo, 24
   exosphere, temperature, 245
   flexure, 1165
   flood basalts, 744
   fluorine, 603
   fluvial valleys, 764
   foreshock waves, 96
   fractionation, 4
   geologic history, 1061
   grabens, 907
   gravity, 1137
      data, 1151
   heat
      flow, 1194
      flux, 850
      production, 1246
   hotspots, 731, 849, 870
   hypsometry, 672
   impacts, 995
      craters, 975
      melts, 1004
   ionosphere, 218
   lava channels, 758, 774
   lavas, 776
   lightning, 126, 151
   magma, 748
      emplacement, 743
   magmatism, 1236
   mantle, 922, 1179, 1208, 1249, 1256, 1263
      cooling, 1198
      flow fields, 800
      plumes, 700
   mean heat flux, 1199
   meanders, 765, 766
   mineralogy, 593, 603–604
   oblateness, 1126
   ocean floor, 1198
   particle transport, 552, 554
   PEM, 1092
   plains, 902
   plate tectonics, 1274
   residuum, 833
   rock types, 637
   saltation threshold, 552
   satellites, 238
   spin, 1088
   spreading centers, 814
   stratosphere, equatorial, 484
   subduction zones, 814
   surface, 534, 540, 637, 655, 1074
   tectonics, 800, 1185
   temperature, global mean, 249
   thermal plumes, 849
   thermokarst landforms, 782
   thermosphere, 229, 261
   tropopause, 394
   valleys, 762, 764, 782
   volatiles, 387, 394

volcanic features, 957
volcanism, 1072, 1076
volcanoes, 741
weathering, 646
wrinkle ridges, 904
Earth and Venus compared
   atmosphere, 3, 15
   core, 1093
   crust, 1206, 1221, 1226
   gravity studies, 857
   hotspots, 873
   hypsometric curve, 1216
   interiors, 1179
   ion foreshock waves, 101
   lightning, 19
   lithosphere, 871
      deformation, 1274
      thinness, 1193
   magmatism, 741
   mass and radius, 797
   meanders, 765
   solar wind interaction, 26
   thermal evolution, 925
   volcanic rises, 849
   water, 7
eclogite, 1218–1220, 1233
   diapirs, 834
   phase
      boundaries, 1234
      transition, 1233
   sinkholes, 821
Eddington approximation, 375
eddy, 494
   activity, two-dimensional, 479
   conduction, 250
   diffusion, 252, 282
      coefficient, 250, 273, 279, 281–282, 286, 319
      formulations, 276
   flux, 493
   mechanisms, horizontal momentum transfer, 487
   momentum
      flux convergence, 482
      transfer processes, 484
   origin, 488
   Reynolds' stress, 484
   structures, 468
   thermal cooling, 248
   transport, 482
   types, 484
   viscosity, 274
edifices, classification, 719
Eistla Corona, 939
Eistla Regio, 731, 733, 737, 846, 851, 856–858, 863–864, 866, 872–873, 905, 938, 939, 942, 1194
   coronae, 854, 855
   volcanoes, 854

ejecta, 978, 1017, 1036, 1038 1043
   blankets, 512, 660, 680, 761, 991, 1005, 1017
   deposits, 506, 522, 577
   description, 1031
   fallback, 1002, 1024
   flows, 1005, 1017, 1033
elastic
   collisions, 190
   lithosphere, 1251. *See also* lithosphere *and* flexure.
   flexure, 1177
electrical conductivity, 543
electromagnetic
   radiation, 22
   waves, 144
electron
   cooling, 211
   density, 161, 180–181
   foreshock waves, 97
      measurements (ORO), 180
      profiles, 115, 167
   flux, 174, 216
   impact, 179, 181
      ionization, 168
   momentum equation, 67
   plasma oscillations, 99. *See also* Langmuir waves.
   precipitation, 174, 178, 179, 183, 184
   temperature, 170, 202
electrons, suprathermal, 174
elongated summit, 706
embayed corona-related flows, 856
embayments, 881, 882
emissivity, 528, 530, 995
   data, 541
   highlands, 543
   parabolas, 660
   values, 544
   vs altitude, 598
endothermic
   phase change, 1255, 1256, 1268, 1271, 1281
   phase transition, 1258, 1269
energy-minimization
   approach, 640
   calculations, 641
enstatite, 650
   ($MgSiO_3$), 639
ephemeris improvement, 1156
epidote, 1213
equator-to-pole temperature gradient, 296
equatorial highlands, 733
equilibrium
   minerals, 638. *See also* weathering.
   resurfacing model, 1062–1064
erosional mechanisms, 541, 549
escape flux, sputtering, 19
Eudocia Crater, 571, 572

EUV, 24
  flux conditions, 229, 276
  heating, 218, 248, 261
    efficiency, 250
  $I_{(pe)}$, 238
  solar, 15
evolution
  tectonic, 897
  thermal, 897
excitation, 249
exobase, 226
exogenic processes, 1077
exosphere, 225–226. *See also* cryosphere.
  escape rates, 7, 270
  models, 401
  nightside, 226
  temperature, 229, 231, 238, 243, 248, 250
  transport, 279, 287
exothermic phase change, 1256
extension, 1057, 1280
extrusive-to-intrusive ratio, 870

$F_{10.7}$ index, 165, 229, 238, 276
fan-type streaks, 577
farrum, 676
Fatua Corona, 938, 939
faulting, 904, 1056, 1072
faults, 822, 907, 944, 1057, 1059
fayalite, 1092
Fe mineral, 637
Fe^{3+}, 613
felsic minerals, 1220
ferric chloride ($FeCl_3$), 451, 655
ferrobasalt melts, 1209
ferroelectric
  minerals, 520, 597
  phases, 521, 598
  properties, 543
ferrous pyroxenes, 644
filaments, 172
fireball, 999
flank slopes, 719
flanking rift zones, 705
flattening, 973, 987
  of impactors, 978
  times, 956
flexural, 947
  deformation, 1281
  models, 864
flow
  deposits, 761
  fields, 703, 747
    large, 744
    morphology, 730
  laws, 1182–1183
    crustal, anhydrous, 1221
  lobes, 1216

  morphology, 1216
  observations, 470, 473
fluctūs, 674
fluid conservation equations, 67
fluid rings, 480
fluoramphiboles, 605
fluorapatite ($Ca_5(PO_4)_3F$), 640
fluorine (F), 604
fluorine-bearing minerals, 603
fluorite, 520
fluvial channels, 778
  systems, 766
  valleys, 764
flux, 175
  ropes, 26, 30, 75–89
  tube, 25
fold
  belts, 811, 818
  ridges, 828
  terrain, 804, 806
folding, 904
folds, 804, 805
  polyphase, 805
foreshock waves, 96. *See also* Earth *and* Earth and Venus compared.
forsterite (Mg_2SO_4), 1092
Forsyth-type model, 1173
Fortuna Group, 1049, 1051, 1057, 1061, 1062
Fortuna-Meshkenet dune field, 567. *See also* Al-Uzza Undae.
Fortuna Tessera, 657, 828, 884
Fortunian time, 1072, 1075
fractionation, 7, 401, 1249
fracture
  belts (lineae), 678, 688, 807, 808, 924
  system, 944
  zones, 772, 822
fractures, 804, 805, 856, 907
fragmentation, 979, 985
Fraunhofer lines, 306
free air gravity anomalies, 1143
  convection velocity, 477
  obliquity, 1093, 1120
  precession rate (FMN), 1088
free-slip
  boundaries, 1263
  boundary condition, 1255
  upper boundary, 1260
Fresnel
  equations, 530
  expressions, 513
  formulas, 528
  reflectivity, 512, 518–519
  transmission, 513
Freyja Montes, 833, 879, 881, 883, 889
friction, 492
frictional sliding, 1220

INDEX

fulgarites, 999
FWHM/outer rise amplitude region, 1191

gabbro
  cumulus, 1214
  stabilty, 1218
gabbro-granulite-eclogite transition, 894, 1219
Galileo mission, 394, 464, 470–471
  lightning observations, 136
  LNMS, data, 397
  Near Infrared Mapping Spectrometer (NIMS), 303, 305
    images, 315, 329, 347
    measurements, 398
  observations, 335
  plasma wave receiver, 22
  spectra, 341
  SSI, 305
  Ultraviolet Spectrometer instrument, 229
  Venus flyby, 467
gamma ray spectroscopy, 599
Ganis Chasmata, 851
Ganymede, target strength, 1025
gas chromatography, 629
gas-rock reactions, 618
gas-solid reactions, 629
Gaussian
  density anomaly, 950
  driving forces, 953
geochemistry, basaltic, 717
geoid 1137, 1260, 1262
  anomalies, 811, 849
geologic history, 1047–1048, 1059–1060, 1067. See also resurfacing.
  end-member models, 1060
  hypotheses regarding, 1074
  magma production rate, 749
geologic mapping, 667–668
  Arecibo Observatory, 681
  datum, 671
  feature-based, 681
  geographic information system (GIS), 682
  Magellan, 682
  scale, 685
  spacecraft data, 680
geomorphology
  channels, 678
  coronae, 676
  craters, 679
  dunes, 680
  fracture belts, 678
  mountain belts, 678
  plains, 673
  ridge belts, 677
  rift valleys, 677
  tessera, 674

  volcanoes, 675
  yardangs, 680
germanium (Ge), 543, 655
Gibbs free energy, 640, 651
  of formation, 639
global delamination events, 1255
  gridded altimetry data (GTNR), 1166
  magnetic field, 84
  mapping, 688
  resurfacing, 746
    event, 848, 1047, 1061, 1065
    model, 1061, 1063, 1064
global sulfuric acid, 622
GLORIA sonar images, 717
Goddard Space Flight Center (GSFC)
  Infrared Heterodyne Spectrometer (IRHS), 307
  model, spectral, 278, 282
Goldstone Deep Space Network antenna, 533, 536
grabens, 763, 764, 804, 805, 808, 820, 824, 856, 881, 907, 909–910, 915 924, 944, 947, 949
Grady-Kipp fragmentation, 985
Graham Crater, 1041
grains, 548
  windblown minerals, 649
granite, 831
granulite, 1218–1220
  phase boundaries, 1234
gravity, 892, 894, 1043, 1096. See also spherical harmonics and volcanic centers, rises.
  anomalies, 745, 1149, 1254
  constraint method, 1136
  data, 849, 857.
    LOS, 1142
  estimation software, JPL, 1133
  fields, 1126–1127
    data sources, 1127
    geodetic parameters, 1155
    geophysical implications, 1149
    line-of-sight model, 1142
    static, 1145
  measurements, 1217
  regional, 885
  relaxation, 955, 959
  signature, 800
  sliding stresses, 913
  slumping, 881
  target gravity, 1027
  waves, 471, 472, 481, 489
    breaking, 287
    dissipation, 285
    evidence, 282
    saturation theory, 285
    source, 286
gravity-to-topography ratios (GTRs), 802, 818, 823, 922, 1197. See

also plains.
  relationships, 810
  saturation processes, 282
greenhouse, 362
  agents, clouds, 364
  atmosphere, 8
  mechanism, 379
    solar-driven, 360
  models, 9, 354, 360, 495
  warming, 361
greenschist, 1213
ground shock, 993
groundbased observations, 747
  Arecibo data, 698, 717
Guan Daosheng Crater, 571, 572
Guinevere Planitia, 673, 774, 808, 809, 905, 918
Guinevere Supergroup, 1049, 1051
Guineverian period, 1076
Gula Mons, 706, 707, 851, 866, 1054, 1070
Guor Linea, 707, 851

H_2S, 627
H_2SO_4, 437
  aerosols, 378
Hadley
  cell, 364, 373, 477, 481
  circulation, 11, 465, 468, 481, 482, 487, 492, 494, 507, 549, 577, 583, 586
Hagfors
  behavior, 533
  formulation, 542
  scattering, 529
halides, 521, 627, 654
halos, 506–507, 512, 720, 990, 1073. *See also* aureoles.
  bright, 995
  dark, 992, 993, 995, 1017, 1022, 1032
halogen-bearing minerals, 593
Hathor Montes, 851
haze, 416
  aerosol profiles, 427
  layers, 326
  particles, 418
  spacecraft data, 427
HDO lines, 337
heat
  advection, 213
  flow, 192
  transport, 296
Hecate Chasma, 811, 812, 813, 851, 938, 939, 943, 944, 949, 958
Helen Planitia, 734, 774
helium (He), 261
  bulge, 233, 239, 273, 279
  density, 263, 279

  in atmosphere, 13
  in thermosphere, 283
hematite (Fe_2O_3), 592, 608–611, 613, 623, 637, 642, 644–645
  formation, 613
Heng-O Corona, 938–939
hercynite ($FeAl_2O_4$), 640
Hestia Rupes, 576
heterosphere, 283
highlands, 506, 542, 772, 806
HITEMP, data bases, 367
HITRAN data base, 337
homopause, 14, 226, 280–281, 396
horizontal momentum fluxes, 487
hotspots, 699, 821, 830, 835, 845, 872, 1196–1197. *See also* volcanic centers, rises.
  data, 1194
  elastic thickness, 871
  model, 823, 1222
  near infrared, 470
  plumes, 836
hot upwellings, 1258
Hubble Space Telescope (HST), 8, 415
Hugoniot elasic limit, 992
hydrochloric acid (HCl), 593
hydrodynamic escape, 387
hydrofluoric acid (HF), 520, 593
hydrogen (H), 7, 15, 17, 261, 385, 388, 403, 408, 445. *See also* atmosphere (lower), D/H ratio.
  atomic, neutral, 395
  bulges, 228, 233, 238, 269, 273
  densities, 263
  escape, 338, 388, 400
  halide buffers, 597
  halides, 342
  H-Lyman-α corona, 269
  H-polarized emissivity, 513
  in atmosphere, 13, 398, 443
  inventory, 411
  loss mechanisms, 411
  solar wind, 42
hydrostatic equilibrium, 1227
hypsometric curve, 1216
hypsometry, 667, 671, 1237

icy satellites, 1029, 1030, 1206
IFP, 421
IFSE
  data, 376
  temperatures, 375
ilmenite ($FeTiO_3$), 613
imbrication, 892
Imdr Regio, 855, 856, 872
impact craters, 504, 522, 541, 574, 783, 800, 821, 905, 990, 1008, 1015, 1031, 1062, 1073, 1233, 1235. *See also* craters, densities.

INDEX 1345

bright floors, 1023
complex, 1024
crater degradation, 1017
creep, 552
dark floors, 1024
density, 687
depths, 922
ejecta, 516
    deposits, 512, 581, 586, 649
    flows, 1033
formation, 506
highly oblique, 1041
interior morphometry, 1024, 1043
ionization, 41
melt, 1004, 1023, 1036
meteoroid disruption, 1039
modification, 1022
morphology, 1000, 1226
multi-asteroid impacts, 1041
multi-floored, 1040
peak-ring formation models, 1007
process, 1042
production rates, 1066
record, 387
retention ages, 1208
rims, 1006
ringed, 1017
scaling, 977
secondary, 1032
size and shape, 1016
size-frequency distribution, 1039
slumped, 977
terracing, 1028
wall slumping, 1002
impactor depression/imprint hypothesis, 1007
impactors, 971, 980. See also surface.
    airbursting, 992
    categories, 974
    debris, 995
    deceleration and deformation, 978
    deformation, 985
    ice, 987
    impact velocity, 973
    incidence angle, 973
    Jovian atmosphere, 987
    mass, 973
    population, 972
    rates, 973
    velocity distributions, 973
impacts, 387, 574
    oblique, 1041
infrared-active gases, 354
infrared spectroscopy, 629
Innini Montes, 851
interior, 1078, 1149, 1249. See also mantle, convection.
    outgassing, 409
intracloud discharges, 128

intracrustal differentiation, 890, 896
intraplate extension, 1231
intrusion/extrusion ratio, 1220
inversion
    procedure, 1144
    technique, 1029, 1030
ion
    acoustic waves, 104, 149, 151
    bulge, 17
    composition, 170
    density, 172
        profiles, dayside ionosphere, 163
    escape, 7
        flux, 179
        velocity, 401
    flux, 178, 179
    heating, 214
    kinetics, 57
    pickup, 34, 47, 50
    recombination, 179
    temperature, 177, 182, 207
    transport, 66, 174, 176, 179, 184
        density, 182
ionization, 16
ion-molecule reactions, 248
ion/neutral escape, 385
ionopause, 15, 64, 71, 88, 104, 171, 176, 180, 204, 210, 401
    current layer, 75, 89
    boundary, 34
    model, 216
    press balance, 71
    structure, 75, 86
ionosheath, 39. See also magnetosheath.
ionosphere, 15, 34, 64, 88, 401. See also atmosphere; dayside ionosphere; and nightside ionosphere.
    collision terms, 196
    dayside photochemical, 237
    electron fluxes, 201
    electrons, 401
    energy calculations, 197
    heat flow, 192
    heating, 147
    holes, 25, 182
    ion temperatures, 207
    low-solar activity, 162
    magnetization, 24, 204
    magnetized, 216
    magnetohydrodynamic (MHD) models, 78, 86
    microwave emission, 360
    model calculations, 209
    momentum balance, 68
    nighttime, 403
    photoelectron production rates, 196
    photoionization, 196
    plasma, 62

and field data, 71
temperatures, 203
properties, 396
pulses, 152
temperatures, 204
thermal conductivities, 192
transport equations, 191
unmagnetized ionosphere, 211
ionotail, 16
IRAM telescope, 312
iron (Fe), 655
Isabella Crater, 1008
ISAV-A, 421, 430
Ishtar Terra, 585, 667, 673–674, 735, 797, 803, 804, 806, 815, 818, 828, 830, 835, 837, 879, 887, 894, 896, 942, 1192, 1251
   Akna Montes, 674
   Danu Montes, 674
   data, imagery, 880
   formation, 829
   Fortuna Tessera, 674
   Freyja Montes, 674
   Lakshmi Planum, 674
     Colette Paterae, 674
     Sacajawea Paterae, 674
   Maxwell Montes, 674
   models
     dynamic, 890
     static, 889
   S-C terrain, 804
   topography, 1251
island model, 742
isostatic adjustment, 1226
   compensation, 1251
   models, 863
   rebound approach, 1221
isotropic blotches, 471
Itzpapalotl Tessera, 828, 883
   S-C terrain, 804

jadeite, 604
Jadwiga Crater, 567
Jeans escape flux, 400
jelly sandwich rheology, 1183, 1187, 1193. *See also* SWS rheology.
jets, 1229
   circumpolar, 470
   core, 470
   mid-latitude, 480
joint zones, 807
Jupiter, 985
   dynamo, 24
   fireballs, 996
   impacts, 995
   stratosphere, 996

K/Th ratios, 1210
K/U ratios, 1210
Kallistos Vallis, 762, 767, 769–771, 779

Karoo flood basalts, 850
Kaula power law, 1136–1137
Kelvin
   mode, 467
   waves, 305, 471
Kelvin-Helmholtz instability, 990
Kilauea volcano, 1280
Kilmogorov regime, 483
kimberlites, 596
kinematic processes, 823
kinetic blocking temperature, 1233
kipukas, 1051
Klenova Crater, 1008
komatiite, 1208, 1213, 1215
Kontrast experiment, 644
krypton (Kr), 29, 387

La Fayette Crater, 570
Lada Terra, 673, 958
Laima Tessera, 825, 836
Lakshmi Planum, 673, 735, 828, 829, 831, 833, 837, 880, 889, 894, 895. 1055
   Akna belt, 1050
   Colette Patera, 735
   Danu belt, 1050
   Freyja belt, 1050
   Maxwell belt, 1050
   plateau, 1192
   volcanism, 896
Lambert-law surface, 532
Landau damping, 78, 104, 210
   resonance waves, 149
lander measurements, crust, 1209
Langmuir waves, 99, 109
lateral autocorrelation, 541
Latona Corona, 813, 938, 939, 1199
lava, 1233
   carbonatite, 779, 782
     flows, 776
   channels, 766, 817
   fields, 1072
   flood, 637
   flows, 510, 516, 519, 540, 647, 699, 700, 721, 773, 880, 884, 920, 935, 1070, 1076, 1227, 1236
     amoeboid, 728
     digitate, 807, 909, 941, 946
     festoon, 728, 1215
     fluctus, 728
     lobate, 807, 1059
     morphology, 1069
     terrain, 804
   low-viscosity, 764, 777
   plains, 746
   sapping, 782
   silicate, 774, 779, 782
   silicic, 1215
   sulfur, 779, 782

flows, 776
    temperature, 778
    tholeiitic, 778
    tubes, 777
Lavinia Group, 1052, 1053, 1057, 1061, 1064
Lavinia Planitia, 730, 734, 761, 787, 808, 836, 905, 913, 916, 1052, 1058
    plains, 1057
    ridge belts, 1075
leaking greenhouses, 334
leucitite, 599
levees, 759
lightcurves, 431
lightning, 19, 125–128
    cloud-ionosphere, 128
    cloud-to-ground, 22
    electromagnetic wave observations, 135
    Galileo spacecraft, 136
    mechanisms, 127
    observations, Mt. Bigelow, Arizona, 133
    optical data inferences, 135
    optical measurements, 130
    Pioneer Venus Orbiter (PVO), 132, 138
    Vega Balloons, 133
    Venera missions, 130
Lindzen-Holton parameterization, 285
lineae, 678. *See also* fracture belts.
linears, 804, 805, 808, 907, 908, 915
liquid sulfur, 1215
lithium (Li), 603
lithosphere, 809–811, 814, 816, 824, 835, 846, 874, 890, 913, 924, 955, 1252. *See also* mantle, convection.
    convection modeling, 1196
    cooling, 1277
    coronae and, 934
    deformation, 1274
    delamination, 1233, 1254
    elasticity, 1223
    extension, 821
    flexural response, 915
    flexure, 857, 1177, 1199
        coronae/chasma results, 1198
        gravity data models, 1171, 1174
        models, 1251
        synthetic inversion model, 1173
        thin plate models, 1167
        topographic models, 1165
        viscous plate, 1169
    gravity studies, 857
    heat
        flux, 1194
        transport, 1250, 1254

inelastic flexure, 1185
    solutions, 1189
Magellan mission data, 1164
mean heat flow, 1195
mechanical, 802, 1220, 1232
moment-matching results, 1187
overturn, global, 1280
plume penetration, 1254
properties, 919, 1163
recycling, 1227, 1229, 1237
    delamination, 1234
rheologic properties, 1179
rigid, 1263
saturation moments, 1186
spacecraft data, 922
strength, 1165, 1179, 1184
stretching, 1207, 1230, 1236
structure, 809, 902, 1193
subduction, 814, 955, 1249
temperature gradient, 1194
thermal, 1197
thickening, 1281
    thermal evolution and 1280
thickness, 848, 868, 870, 873, 951, 1251, 1261, 1275, 1277
thinning, 836, 863, 959, 961
uplift, 817
vertical deformation, 909
LL/LR polarization ratio, 536
lobate flows, 1070
local fracture patterns, 902
Lorentzian profiles, 333
LOS accelerations, 1125, 1131, 1140
    gravity data, 1177
    wind velocities, 300
Love number, 1154
low-emissivity surface material, 658
low-viscosity channel, 1185
low-Z elements, 602
lower-hybrid-drift instability, 110
Lyman-α, 166
    airglow, 269
    resonance radiation, 15

Maat Mons, 519, 658, 851, 1054, 1055, 1070, 1140
Mach stem, 993
mafic lava flow, 775, 1217
Magellan (MGN) mission, 7, 167, 226, 228, 231, 394, 472, 492, 529, 698–699, 719, 747, 797, 845, 871–872, 885, 887, 902, 1077, 1127, 1131–1132, 1137, 1193, 1206, 1221, 1236, 1237
    aerobraking, data, mass density, 236
    altimetric data, 950
    altimetry, 671, 678, 797, 801, 935, 1001
    coronae, 932

crater data, 1008
crustal rheology, 922
data, 652, 718, 797, 820, 846, 863, 944, 1004, 1048
  aeolian features, 565
  emissivity, 801
  flow law, 802
  gravity, 797, 801, 811, 828, 846, 857, 938
  lithosphere, 1164
  synthetic-aperture radar (SAR) images, 798
  upwelling model, 822
  visualization techniques, 672
  volcanism, 699
drag measurements, 236, 237, 247, 263, 275, 284
images, 758, 776, 910–911, 918, 1066, 1235
  Venera and Vega landing sites, 1216
mapping, 681
nightside mass density data, 247
plateaus, crustal, 818
radar, 473
  images, 1006
  mapping, 757
radio occultations, 304
  measurements, 234
spacecraft, 536, 857
spreading ridges, 822
synthetic aperture radar (SAR), 532, 798, 894
  data, 533, 801, 950
  HH backscatter, 513
  images, 767, 783, 798, 804, 814, 830, 832, 903–904, 907
  Ishtar Terra, 880
  parabolas, 660
  plains data, 924
  resolution, 804, 809
tracking data, 815
wobble of Venus, 1090
Magellan Volcano Catalog (MVC), 703, 731
magma, 540, 544, 700, 741
  basaltic and picritic, 1232
  bodies, 834
  centers, 735
  chambers, 735
  crystallinity, 1215
  emplacement, 698, 748
  eruption, 743, 748
    frothy, 717
  mafic, 715
  oceans, 1207
  production, 738, 1277
  replenishment, 748
  reservoirs, 717, 735, 740, 742–744, 748, 821
  shallow, 715
  silicic, 716
  storage, 698, 743
  temperatures, 716
  transport, 698, 743, 1233
  trap, 744
magma-foam eruptions, 717
magmatic centers, 699, 700, 705, 746
  heat transport, 1250
magmatism, 703, 743, 812, 1208
magnesiowüstite, 1255
magnesite ($MgCO_3$), 595
magnesium number, 1210
magnetic barrier, 39
  convection-diffusion equation, 70
  field, 23, 39, 61, 171, 1273
    solar wind and, 27
  fluctuations, 210
  flux, 25
  turbulence, 104
magnetite (Fe_3O_4), 592, 608–611, 613, 623, 637, 640, 644, 645, 656
  oxidation, 623, 630
  ulvöspinel (Fe_2TiO_4), 642
magnetohydrodynamic (MHD) process. *See also* ionosphere.
  models, 44
  process, 61
    solar wind and, 62
  stability, 70
  theory, 66
  wave modes, 70
magnetosheath, 34, 39
  ions in, 41–42
  models, 36
magnetotail, 16, 27, 43, 86, 109
  formation, 50
  induced, 46, 50, 57
  lobes, 46
  wave activity, 119
Makaopuhi lava lake, 1280
Makh Dome, 951
MALAHIT, 421, 430
manganese (Mn), 602
Manning coefficients, 780
mantle, 104, 745, 786, 802, 871, 894, 950, 953–954, 1008, 1074, 1179, 1185, 1192, 1210, 1234. *See also* coldspots *and* hotspots.
  avalanche, 1246, 1281
  carbon, 3
  coldspots, 787, 822
  composition, 922, 1092, 1208
  convection, 811, 833, 1164, 1190, 1198, 1227, 1246, 1255–1280
    heat production, internal, 1246
    heat transfer, 1249
    models, heat transport, 1252

    variable viscosity and phase transition, 1268
    viscosity, temperature-dependent, 1263
cooling, 746, 814
debris, 1071
deep, 1259
D/H signature, 411
diapirs, 812
    upwelling, 934
downwelling, 737, 787, 821, 830, 835, 872, 1198
dynamics, 818
flow fields, 800
fluid, 1097
gravity studies, 857
outgassing source, 408
overturning, 1280
phenomena, 745
plasma, 107, 210, 216
plumes, 850, 857, 860, 867, 892, 1229, 1230, 1232, 1236–1237, 1254, 1280
return flow, 737
rigid upper boundary, 1260
solidus temperature, 812
temperature, 1229, 1237
ultramafic, 1220, 1221
uplift, 922
upwelling, 737, 816, 821, 835, 872, 874
    models, 830
viscosity, 1179, 1217
zone, 920
mantling, 513, 516
    materials, 508
mapping, 668. *See also* geologic mapping.
margin deposits, 1022
margins, 990, 1032
Mariner missions, 24, 161, 189, 235, 299, 305, 361, 592
    albedo, 465
    atmospheric waves, 284
Markham Crater, 516–517
Mars, 601, 1137, 1230
    aeolian transport, 554
    atmosphere, 15, 287
    bow shock, 104
    calderas, 724, 728, 740
    channels, 767, 769, 779
    chaotic terrains, 769
    composition, 4
    coronae, 810
    crust, 1002, 1208, 1210, 1226
    dunes, 574
    dynamo, 23
    fluvial
        channels, 757

    valleys, 764
    friable materials, 1074
    graben, 907
    gravity data, 1136, 1137
    hemispheric dichotomy, 1225
    ionosphere, 189, 209, 219
    lava channels, 758
    magmatism, 741
    particle transport, 552
    plains, 902
    planetosphere, 108
    saltation threshold, 552
    solar wind, 24
    surface, 522, 541
    target strength, 1025
    temperature, global mean, 249
    Tharsis volcanic region, 864
    thermokarst landforms, 782
    thermosphere, 229
    valley networks, 782
    valleys, 762, 764
    volatiles, 394
    volcanic rises, 817
    volcanism, 700, 705
    volcanoes, 731, 741
    water, 388
    wave activity, 108
    wrinkle ridges, 904, 911
mass
    density drag, 250
    loading, 53
    movement (slumping), 507
    spectroscopy, 629
    wasting, 574
master power law, 976
    simulation, 981
Maxwell Montes, 533, 536, 539, 540, 598, 652, 653, 657, 830, 831, 833, 834, 879, 881–883, 889
    mountain massif, 1051
Mead Crater, 574, 576, 580, 1008, 1017, 1024, 1149, 1251
meanders, 766
    bends, 759
    properties, 764
mean planetary radius (MPR), 798, 806, 811, 818
mechanical erosion and transport, 638
medial ejecta flows, 1033
Meitner Crater, 1008
melt, 894
    anomaly, 745
    ejecta, 1038
    products, 1247
Menat Undae, 565
Mercury, 522
    crater deposits, 1027
    craters, 1003, 1004
    crust, 1003, 1208

ejecta blankets, 1031
plains, 902
rotation rate, 1088
target strength, 1025
wrinkle ridges, 904, 911
mercury (Hg) compounds, 628
meridional (pole-to-equator)
circulation, 299, 346, 373
temperature gradients, 356, 372
velocities, 468
mesolands, 806
mesoscale dynamical cells, 471
mesosphere, 272, 295–296, 353
airglows, 314
circulation, 234
CO_2 observations, 301–302
cyclotropic balance, 317
dynamics, 296, 493
future research, 320
haze, 373
millimeter wave CO observations, 308–312
minor species, 318
polar, 379
poleward flow, 373
radiative cooling, 372
radio occultation observations, 303
structure, 296
temperature, 234
and wind, 317
measurements, 297–305
radio occultations, 297
temporal variations, 378
thermal
balance, 371
data, 355
winds, 298
winds, 252
above clouds, 299–300
measurements, 305–306, 308
zonal superrotation, 378
zonal wind, 305
thermal, 303
meteorites, 4
meteoroids, 1039–1040, 1043
fragmentation, 1016
meteorological activity, 348
methane, 397–398
Mg pyroxene, 637
MHD theory, 44. *See also* magnetohydrodynamic process.
micas, 1213
microcline ($KAlSi_3O_8$), 640
microdunes, 547, 557, 570
microwave data
observations, 353, 592
surface, 507
thermal emission, 533
MIDR, 578

mineralogy, 593, 629
Miranda, coronae, 810
Mnemosyne Regio, 942
Mössbauer (MB)
spectra, 611, 613
spectroscopy, 623, 629
moats, 954, 955
Mode 0 particles, 362
Mode 3 particles, 452
Mohr-Coulomb behavior, 985
Mokos Dome, 951
Mokosha Mons, 961
molecular diffusion coefficients, 281
moment
equations, 192
saturation, 1185
momentum
drag, 280
flux, 285, 482
transport, 346
equatorward, 493
monostatic
intensity observations, 532
polarization observations, 534
montes, 804, 827, 829, 831
monzonite, 599
Moon, 504, 522, 896, 1005, 1230
craters, 985, 1003, 1004, 1006, 1008, 1016, 1021, 1024, 1027, 1067
crust, 1003, 1207, 1208, 1210, 1226
dynamo, 23
ejecta blankets, 1031
grabens, 907
impact melt, 1004
lavas, ultramafic, 1215
plains, 902
regolith, 1074
rilles, 1215
sinuous, 758, 777, 778
rock types, 637
soil, 999
surface, 506, 541
target strength, 1025
wrinkle ridges, 904, 911
mountain belts, 678, 881
compression deformation, 1050
multiple and steep summits, 706
multiringed basins, 1006
Mylitta Fluctus, 533, 730, 761

narrowband waves, 108
NASA Infrared Telescope Facility (IRTF)
Cooled Grating Array Spectrograph, 338
NCAR VTGCM models, 280
near-Earth asteroids (NEAs), 975
near-infrared (NIR)
emission, 365
features, 331

spectra, 644
windows, 328, 629
  leaking greenhouse, 334
  models, 330
  remote sensing, 334
near-solidus melts, 1215
neon (Ne), 4
Nephele Dorsa, 911
nepheline ($NaAlSiO_4$), 597
nephelometer, 371
Neptune dynamo, 24
net global solar absorption, 463
neutral
  atmosphere composition, 230
  density
    measurements, 228
    profiles, 247
  mass spectrometers, 395
Newtonian fluid, 1215
nickel (Ni), 602
Nightingale Corona, 1169
nightside
  atmosphere, 247
  ionosphere, 15, 16, 96, 161, 213
    holes, 207
    ionotail, 171
    lower ionosphere, 172
    solar cycle changes in, 180
    sources, 174
Niobe Planitia, 905, 1053
nitrates, 603
nitrices, 603
nitrogen (N), 3–4, 387, 603
  atmospheric, 4, 443
  atomic, 242, 265
NNE trends, 585, 917
noble gases
  nonradiogenic, 3
  radiogenic, 6
nomenclature, geographic, 668
non-LTE (NLTE), 276, 297
non-Newtonian finite element code, 824
non-stagnant lid regime, 1197
noncatastrophic processes, 800
North Basin, 828, 834
North-Polar Control Network, 1157
nuclear magnetic resonance (NMR), 629
Nusselt number, 1196, 1197, 1264
  Rayleigh number relation, 1273–1274

oblateness, 1126
obliquity, 272, 372, 1093, 1120
  core friction equilibrium, 1111
  resonance, 1109
  torque balance, 1109
ocean, 411
  full terrestrial equivalent, 407
OCS, 437
  mixing ratio, 343, 625

Ohm's law, 67
olivine, 613, 950, 1179
  ductile rheology, 922
orbit, 247, 1127
  eccentricity, 229, 1094
  solar wind, 35
outgassing event, 8
Ovda Regio, 534, 598, 652, 820, 1223, 1224
  shape, 657
  tesserae, 820
Owens Valley Radio Observatory interferometer, 311
oxidation state of Venus surface, 592, 605–629. *See also* surface-atmosphere interactions.
  Venera spacecraft observations, 609
oxide redox, 616
oxygen (O), 17, 226, 603. *See also* atmosphere.
  abundance, 280
  atomic, 14, 18, 175, 239, 242, 267, 314
    collisions, 400
    hot, 234
  bulge, 273, 274, 282
  density, 272
  O^+
    flux, 178
    transport, 401
  fugacity, 605, 608–610, 613, 630
  in atmosphere, 443–444
  in clouds, 444
  in thermosphere, 283
  ion production, 43
  mixing ratios, 269, 273
oxygen (O_2), 319, 448
  Herzberg II nightglow, 265
  infrared intensity distribution, 278
  nightglow, infrared, 268
oxygen-carbon dioxide (O/CO_2)
  collisions, 248
  ratio, 234, 268, 273
  relaxation coefficient, 250
Ozza Mons, 851, 949, 1054, 1070

pahoehoe lava flows, 804, 991
paleoslopes, 817, 909
pancake domes, 719, 978, 987, 1215, 1237
  spreading model, 989, 990
parabolas, 473, 506–507, 517, 540, 997, 1017, 1020, 1032, 1043, 1077
  cavity, 1043
  dark, 1022, 1024
  deposits, 571, 997, 999
  ejecta deposits, 580, 581
  halos, age, 688
  radar-dark, 1071

parameterized convection model, 1272, 1275
Parga Chasma, 812, 813, 851, 856, 938, 939, 943, 944, 949, 958
Parga Corona, 811
parquet terrain, 803
particle, 548
 flux, 552
 movement, wind, 585
 threshold, 551
 transport, 549
Pavlova Corona, 942
peak
 heights, 1027
 ring basins, 1006
 rings, 1025, 1043
penumbrae, 991, 993, 995
peridotite, 1207–1208
permittivity, 652. *See also* dielectric constant.
perovskite (CaTiO$_3$), 520–521, 597, 1255
petrogenesis, 1214
phase-change modulated convection, 1256
Phoebe Regio, 735, 805, 824, 836, 991
 star terrain, 820
phosphorus, 430
photochemistry, 443, 449
photodissociation, 242–243, 248, 301
 of water molecules, 337
photoelectrons, 165, 190, 201, 216
 fluxes, 200
 impact, 242–243
 production rates, 196
 transport, 197, 211
photoionization, 41–43, 168, 170, 190, 196, 238, 248
photons, 444
 in deep atmosphere, 334
physiography, 667–668
 hypsometry, 667
pick-up ions, 47, 57
picrite, 1215, 1232
Pioneer Venus (PV) and Pioneer Venus Orbiter (PVO) missions, 8, 17, 24, 30, 132, 161, 169, 190, 226, 296, 299, 303, 337, 341, 343–344, 348, 355, 362, 385–386, 394, 396, 406, 436, 443, 452, 461, 470, 476, 492–493, 668, 822, 860, 931, 944, 1127–1128, 1193, 1222
 altimeter, 902
 atmosphere, 403
 atmospheric waves, 284
 data, gravity, 828, 855, 857
 Day and Night Probes, 461
 diurnal cycle, 238
 entry probes, 326

Large Probe, 397
 gas chromatograph, 432
Large Probe Cloud Particle Spectrometer (LCPS), 326, 362–363, 418
Large Probe Infrared Radiometer (LIR), 363
Large Probe Neutral Mass Spectrometer (LNMS), 340, 395, 398
Large Probe Solar Flux Radiometer (LSFR), 360, 362–363
lightning observations, 138
mapping, 680
measurements, 237
 solar flux, 495
meridional velocities, 474
North Probe Net Flux Radiometer, 336, 461
Orbiter Cloud Photopolarimeter (OCPP), 305, 326, 363
Orbiter Atmospheric Drag (OAD), 162, 170, 228
 mass density, 238, 240
 measurements, 230, 284
 nightside mass density data, 247
Orbiter Electric Field Detector (OEFD), 95, 100
Orbiter Electron Temperature Probe (OETP), 96, 162, 166, 171, 173, 182, 202, 228–229
Orbiter Infrared Radiometer (OIR), 228, 301, 363, 372
Orbiter Ion Mass Spectrometer (OIMS), 162–163, 169, 177, 179, 182, 228, 395
Orbiter Magnetometer (OMAG), 96
Orbiter Neutral Mass Spectrometer (ONMS), 162, 170, 228, 235, 237–238, 244, 260, 282
 density fluctuations, 238
 derived temperatures, 279
 measurements, 273
Orbiter Plasma Analyzer (OPA), 201
Orbiter Retarding Potential Analyzer (ORPA), 162–163, 169, 171–172, 174, 176–177, 180, 182, 201–202
Orbiter Radio Occultation (ORO), 162, 171–172, 180–181
 electron density profiles, 167
Orbiter Ultraviolet Spectrometer (OUVS), 162, 228, 268–269
 Lyman-α spectra, 231, 399
probes, 228, 468, 473
 data, 488
 entry, 461, 464, 468, 472
 flux measurements, 490
radio occultation data, 470
reflectivity data, 820

Small Probe Net Flux Radiometers
(SNFR), 363
Star Tracker, 284
topography and gravity data, 818
ultraviolet spectroscopy, 422
pits, 711, 725
plagioclase ($CaAlSi_2O_8$-$NaAlSi_3O_8$), 640
grains, 1182
plains, 504, 512, 516–517, 540, 544,
576, 651, 655, 734, 797, 798,
803, 806, 810, 820, 833, 836,
1049, 1056, 1057, 1074, 1076
age, 687
canali, 759
concentrated deformation, 910
craters, 1062
deformation, 807, 1064, 1075
belts, 808
distributed, 903
local, 918
water, 925
formation, 1072
fracture belts, 915
future studies, 925
geomorphology, 673
gravity and topography data, 920
gridded, 919
intratessera, 821
Lakshmi Planum, 880
linear
features, 907
ridges, 764
lobate, 1059
parallel fracture
patterns, 925
sets, 918
polygonal patterns, 920
regional, 1068
formation, 1075
ridge belts, 910
shield, 742
streaks, 580
tectonism, 901
lithosphere structure, 809
tension fractures, 925
vertical deformation, 909
volcanic, 803
wrinkle ridges, 903, 1068, 1069
Planck function, 334, 348
planetary-scale waves, 305
planetesimals, 386
icy, 4, 387
planetogenesis, 386
planimetric patterns, 907
planitiae, 673
plasma
density, 247
dynamics, 86, 88
mantle, 75

structure of, 86
motions, 162
pressure gradient force, 175
transport, 182–183
troughs, 172
wave phenomena, 95
electon foreshock waves, 96
ion foreshock waves, 100
ion populations, 100
ionospheric waves, 108
Langmuir waves, 119
mantle waves, 104
post-terminator waves, 116
wideband turbulence, 117
plastic hinge, 1185
plateaus, 835, 872, 874, 937, 954
basin-and-dome types, 820
crustal, 797, 798, 803, 804, 806, 836
characteristics, 818
craters, 821
downwelling model, 822, 823
formation, 825
formation models, 821
future studies, 826
tesserae, 820
deformed interior, 818
formation, 890
mantle upwellings, 824
gravity signatures, 818
lava-flow tessera types, 820
outboard, 883
surface age, 821
plate tectonics, 897, 1249, 1272, 1274,
1275, 1280
plumes, 745, 746, 797, 824, 825, 835,
851, 871, 890, 999. *See also*
thermal, plumes.
Alpha-type, 825
background, 1195
cold, 1269
convection models, 1197
crustal, 836
diapirs, 812
ejecta, 997
flow, 830
global activity, 1232
lithosphere, 1195
mantle plumes, 1228
penetration, 1254
temperatures, 871
thermal, 812
plutonic melts, 1219
poles, 303, 348, 1088
cloud top, 373
collar, 302–303, 305, 317, 336, 356,
379, 423, 464
orientation, stability, 1088
regions, 342, 348, 423, 671
aerosol distribution, 373

mesosphere, 379
vortex, 493
polarization, 528
  angle, 539
  dependence, 536
  dependent transmission, 535
  state, 532–533, 536
poleward drift, 480
polysulfur oxides, 451
ponded flows, 904
pores, super-pressurized, 994
potassium (K), 1214, 1247
  abundance, 1210
  feldspar, 637
  oxide (K_2O), 1214
Prandtl number, 778
pressure balances, 64, 71
pressure-release melting, 848, 864
primary crust, 1208, 1230, 1233
pristine craters, 1022, 1024
protobasins, 1007
proximal
  deposits, 1036, 1037
  ejecta flows, 1033
pyrite (FeS_2), 520, 592, 613, 622, 637, 640, 642, 646, 656
  decomposition, 505, 616, 622
pyrochlores, 597
pyroclastic eruptions, 649, 700
pyroxene, 613, 651, 1182
pyrrhotite, 622, 623, 625
  oxidation, 623, 625

quadrupole moment, 491
quartz (SiO_2), 593, 595, 637, 640, 650
quasi-specular, 529
quenching (relaxation), 249, 314

radar
  altimeter, 532
  backscatter, 533, 997
  echoes, intensity, 528
  emissivity, 592
  observation, 527
  reflectivity coefficient, 653
radar-bright flows, 949
radar-dark flows, 949
radial
  coronae, 961
  dike intrusion, 715
  flows, 703, 946
  gravity, 1137
    perturbation, 1143
  mass flux, 1271
    diagnostic, 1260
radially fractured
  exterior, 706
  interior, 707
radiative
  cooling, 372, 374, 377
  damping, 374, 485
  forcing, 379
  mesosphere, 372
  heating, 473, 484, 492
  rates, 377
  transfer modeling methods, 366
radiative-convective-equilibrium (RCE)
  experiments, 373
  model, 361
radioactive elements, 1259
radio emission measurements, 527
radiogenic
  elements, 1253
  heating, 1273, 1275, 1277
radiometer, 532
radio occultation measurements, 172, 180, 182, 297, 357
rainout, 995
Rayleigh
  drag, 285
  fractionation, 390–391, 407–409
  friction, 237, 247, 251, 274, 276, 278, 280, 285, 319
  number, 858, 860, 867, 1173, 1196, 1264, 1267
  scattering, 361
Rayleigh-Taylor instability, 953, 978, 987–988
reaction kinetics, 505
recycling, 1227. *See* crust.
redistribution of soil, 504
redox state, 8, 640
reflection coefficient, 528
regolith, 505, 522, 1210, 1213
relief features, 904
remnant thickness, 866
reservoirs, 700
residuum, 833, 834, 837, 870, 890, 1233
  layer, 873
  root, 864
resonance cone
  criterion, 146
  test, 144
resurfacing, 522, 549, 687, 745, 800, 836, 894, 1047, 1060, 1078, 1090, 1206, 1280–1281. *See also* geologic history.
  contemporary, 1236
  events, 7, 410
    global, 738, 1061, 1230, 1232, 1275
  exogenic, 1073
  history, 688, 871
  local zones, 1062
  rock-stratigraphic units, 1049
  stratigraphic
    analysis, 1048
    observations, 1063

structural deformation, 1055
  tectonic, 1072
  types of, 1071
  volcanic, 1072
retrograde
  rotation, 1088
  subduction, 955
Reynolds
  number, 778, 1105
  stresses, 480, 485
Rhea Montes, 858
rheology, 824, 889
  temperature-dependent, 867
rhodochrosite ($MnCO_3$), 596
rhyolite, 599, 1237
  domes, 1215
ribbon terrain, 804–806, 820, 823, 825, 828
  formation, 824
Richardson number, 473, 479, 488, 491
ridges, 504, 557, 1216, 1258. See also wrinkle ridges
  belts (dorsa), 516, 677, 688, 785, 804, 807–808, 810, 836, 895, 910, 916, 1057
    marginal, 820
  complexes, 883
  Rusalkian, 1076
  terrain, 1192
rifts, 707, 730, 747, 895, 958, 1057, 1059, 1070–1071, 1076, 1280
  age, 688
  valleys (chasmata), 677, 816
  zones, 715, 772, 1054, 1072, 1231
rilles, 1215
rims, 504, 947, 954, 955
ring massifs, 1025
ripples, 549, 556
rises, 937
rms
  slope, 541
  wave amplitudes, 283
rock rheology, 809, 1180
rocks
  crystalline metamorphic, 805
  gneissic, 805
  igneous, 1213
  mantle, 802
  metamorphic, 1213
  ultramafic, 1213
  ultrapotassic, 1214
rolling, 552
Rossby
  mode, 467
  number, 585
rotation
  atmosphere, 460
  axis, 342, 1089
  retrograde, 13

period, 272, 1155–1156
  seasonal, 1121
  synchronous, 1104
RSZ flow, 263, 270–272
runaway greenhouse, 408
Rusalka Group, 1052, 1053, 1059, 1061, 1064–1165
Rusalka Planitia, 735, 759, 774, 905, 1053
Rusalkian
  flows, 1070
  plains, 1057, 1075
rutile, 520, 637

Sacajawea Patera, 960
Sacajewea Caldera, 880
SALE hydrocode, 985
saltation, 548, 551
  flux, 553
  fully developed, 552
  threshold, 552
sand, 586
  dunes, 548
  supply, 557
Sapas Mons, 510, 512, 706
Sappho Corona, 942
saturation moment, 1198
Saturn, dynamo, 24
scaling, 1005
S/Ca ratios, 621
scarps, 671, 807, 808, 882, 907, 910, 915
  asymmetric, 1025
  fault, 907
  smooth, 884
scattering, 535
  diffuse, 512
  process, 536–537
S-C terrain, 804, 806
seafloor spreading, 1227
Seasat images, 559
sedimentation, 598
  budget, 549
  deposits, 586
Sedna Planitia, 734, 905, 911, 991, 1070
seismicity, 1078
Sekmet Mons, 706
Selu Corona, 953
semiconducting surface layer, 543
semi-diurnal boundary layer, 491
shadows, 990
shear failure, 985
shields, 719, 946, 949
  atmosphere, 1016
  edifices, 946
  fields, 676, 720, 722, 730, 744, 748
    apron, 722
    companions, 722
    dome, 688

origin and emplacement, 742
  plains, 722
shock
  heating, 996
  waves, 96, 992, 994
short wavelength convective signature, 867
Si, 1213
siderite ($FeCO_3$), 595–596
Sif Mons, 726, 851, 866
  volcano, 1070
SIGMA-3, 421, 430
Sigrun Fossae, 1051, 1056, 1057, 1059, 1061
Sigrunian time, 1072, 1076
Sikhote-Alin meteorite, 979
silica, 1216
silica-rich melts, 1209
silicate magmas, 1215
silicates, hydrated, 603
silicic, 896
sinuous
  lineament, 882
  rilles, 758, 763, 765, 767, 773, 777, 785, 790
slabs, 647
slope, 766
sluggish-lid
  convection, 1249
  mode, 1265, 1274
  regime, 1280
slumping, 895, 949
small cones, 946
SNC meteorites, 1210
SO_2, 444, 650
  measurements, 432
  mixing ratios, 378, 432
SO_3, 437
sodalite ($Na_4(AlSiO_4)_3Cl$), 597
sodium (Na), 601, 603
  oxide (Na_2O), 1214
  plagioclase, 637
  pyrovanadate, 645
solar. *See also* ionosphere; magneto-hydrodynamic process; and Venus-solar wind interaction.
  activity variations, 229. *See also* thermosphere.
    28-day rotation period, 238
    long term, 240
  cooling, 237
  cycle, 14, 27, 231
  emissions, scattering, 243
  EUV, 210
    radiation, 209
  EUV heating, 211, 261
    maximum (SMAX), 229–230, 275–276
    minimum (SMIN), 229, 275–276

moderate (SMED), 229–230, 276
  flux, 165, 170, 238, 306, 592
    absorption, 450
  heating, 12, 237, 238, 372, 375–375, 473, 483
  tides, 481
  maximum steady state flux, 406
  radiation, 62
    in lower atmosphere, 334
  resonance radiation, 15
  tide(s), 467, 484, 485
  wind, 15, 18, 25, 30, 33, 35, 61, 172, 199, 247
    dynamic pressure, 248
    electrons, 54, 216
    interaction with atmosphere, 3, 26
    ionospheric magnetization and, 24
    ions, 54
    plasma, 27
    protons, 54
    P_{sw}, 180
  zenith angle (SZA), 263
Sounder Probe, 461
South Scarp and Basin, 828
specific radar cross section, 541
Spectral Mapping Atmospheric Radiative Transfer (SMART) model, 366
sphene ($CaTiO_3$), 640
spherical harmonics, 1125, 1131
  gravity
    estimation, 1133
    field, 1127
    results, 1137
  solutions, 1126
spin dynamics, 1087–1121
  core mantle fluid friction, 1105
  equations, Euler, 1107
  eulerian wobble, 1116
  free obliquity, 1093
  groundbased radar, 1087
  Magellan (SAR) imaging, 1087
  models
    dynamical, 1097
    structure parameters, 1092
  obliquity
    resonance, 1109
    torque balance, 1109
  spin-orbit resonance, 1113
  tidal history, 1117
  tides
    atmospheric thermal, 1102
    solid body, 1097
splotches, 990–992, 1062
  dark, 1039
spreading, 985
  lateral, 987
  restricting, 989
sprites, 152
stability of pyrite, 643

stagnant-lid, 1277
    convection, 1164, 1249, 1281
        regime, 1197, 1264, 1274–1275, 1280
stagnation points, 318, 319
star terrain, 805–806, 820
static stability, 472
steep-sided domes, 1215. *See also* pancake domes.
stellate fracture centers, 705
Stokes
    flow, 863
    vector, 537
Stowe region, 572
stratigraphic
    analysis, 1047–1048. *See also* resurfacing.
    sequence, 1059
stratigraphy, 667, 688
    impact crater densities, 687
    local, 685
streak reflectivities, 578
streaks, 578. *See also* wind, streaks.
    type-P, 578, 580, 581, 583
stress-strain profiles, 649
sturzstroms, 1032
subduction, 1179, 1186, 1255, 1281
    episodic, 1280
    zones, 814, 955, 1192, 1272
subparallel, 1055
subsolar point, motion, 485
subsolar-to-antisolar (SSAS)
    circulation, 13, 318
        Pioneer Venus (PV), 272
    flows, 263, 270–271, 281, 300, 308
    mesospheric flow, 312
    thermospheric flow, 296
subsurface fluid flow, 763
sulfides, 627, 654
sulfur (S), 8, 430, 445, 450–451, 646, 1213, 1215
    allotropes, 450
        metastable, 450
    chemistry, 29, 436
    compounds, 343
    cycle, 436, 617
    flows, 776
sulfur dioxide (SO_2), 618, 622, 625
    enhancement, 128
    spacecraft data, 421
sulfuric acid (H_2SO_4), 326, 418, 430, 435, 448
    cloud, 437
    cloud droplets, 328
sulfuryl chloride (SO_2ClS_2), 448
summit pits, 719
Sun, 245, 248, 386, 407, 485, 487, 1088, 1102, 1132, 1153
    flux ropes, 26
    gravitational attraction, 1102, 1131

sunspots, 991
superadiabatic lapse rates, 365
superelastic collisions, 191
superheating, 778
superplumes, 1076
superrotation, 176, 263, 273, 275, 284, 346, 489, 491, 493, 495
    equatorial, 492
    zonal retrograde, 379
suprathermal electrons, 184, 199–201, 216
    fluxes, 181, 216
surface, 24, 343, 347, 527, 542, 547, 637, 1077. *See also* aeolian features and processes *and* dielectric properties.
    age, 981, 1066, 1067, 1071
    anomalous material, 652
    backscatter data, 513
    basalt flows, 991
    basalts, 1179
    bistatic observations, 536
    boundary layer flows, 490
    carbonates, 593, 599
    chemical analysis of material, 647
    coronae, 1069
    creep, 551
    crustal generation, secondary, 1237
    dating, 972
    deformation, 806, 1274
    dielectric constant, 512
    doming, 951
    dry, 386
    electrical resistivity of soil, 649
    features, 798
    future studies, 522
    geochemistry, 592
    halos, 990
    highland, 518
    hydrated silicates, 341
    lithosphere, 1172
    low-emissivity material, 658
    lowland roughness, 507
    mechanical and chemical weathering, 504
    microwave observations, 349
    mineralogy, 592, 595
    models of anomalous materials, 654
    monostatic observations, 532, 534
    morphology and mechanical properties, 647
    oxidation state, 592
    plains, regional, 1068
    radio waves, 529
    redox state, 643
    roughness, 567, 1106
    saltation, 492
    scattering, 527
    sediment, 505

simulations, 980
soil movement, 492
solar flux, 490
stresses, 1169
structure, 540
temperature, 593, 834
tesserae, 1068
volcanoes and lava flows, 1070
wind patterns, 580
surface-atmosphere interactions, 591
  carbon monoxide, 605
  equilibrating mechanisms, 616
  future study, 629
  hematite formation, 613
  hydrochloric acid (HCl), 603
  hydrofluoric (HF), 603
  sulfur cycle, 617
  thermochemical equilibrium models, 607, 610
  volatile element geochemistry, 627
  volcanism, 622
surficial units, 689
suspension, 551
swell topography, 860
SWS rheology, 1183–1184, 1192. *See also* jelly sandwich rheology.

tail rays, 50
target
  density, 1024
  strength, 1025
tectonics, 797, 798, 845, 848, 880, 887, 901, 1061, 1193. *See also* individual features; plains; *and* volcanic centers, rises.
  convective instabilities, 894
  decline, 889
  deformation, 783, 785
  emissivity, 801
  evolution, 834
  extensional, 882
  features
    concentrated deformation, 902
    distributed deformation, 902
  flow laws, 802
  gravity, 801
  Magellan mission, 798
  patterns, 893, 909
tektites, 999
tellurium (Te), 543, 655
Tellus Regio, 820, 1224
temperature, 418, 439, 484
Tepev Mons, 726, 851, 866
terminator, 13, 15, 178, 181
  waves, 75, 88, 207
terrace zones, 1044
terracing onset data, 1030
terrain
  plains, 1066

rifts and associated volcanics, 1066
  tesserae, 1066
tertiary crust, 1237
  differentiation, 1217
tesserae, 543, 574, 667, 674, 735, 748, 797–798, 804–806, 824, 827, 829, 831, 880, 1056, 1192, 1208
  age, 687
  deformation, 821, 1074
  extensional structures, 820
  flooded, 828
  formation, 1047, 1072, 1074, 1075
  highlands, 674
  inliers, 804, 806, 909
  terrain, 516,803, 851, 907, 1049
    deformation, 1049
  units, 859
Th, 1214
Tharsis region, 731
Theia Mons, 506, 536, 707, 851, 858
  volcano, 1070
Themis Regio, 731, 856, 872, 939, 942, 944
  coronae, 856
thermal heating. *See also* atmosphere; mesosphere; *and* plumes.
  balance, 353
  conduction, 296
  conductivities, 192, 210, 218
  convection, 10, 1222
  cooling, 375, 377
    decline, 889
  dynamic compensation, 1222
  electron heating rate calculations, 197
  electrons, 190, 216
  emission, 463, 464
    in lower atmosphere, 331, 333
  erosion, 778, 782
  evolution
    convection model, 1272
    lithosphere, over-thickened, 1280
    plate tectonics, 1275
  flux algorithm, 373
  heat conduction, 237
  opacity, 362, 367
  plumes, 797
  structure, 9
thermochemical equilibrium, 435, 619, 626, 627
thermosphere, 12, 14, 64, 176, 225, 282, 295–296. *See also* atmosphere; gravity, waves; *and* mesosphere.
  10-μm observations, 306–308
  circulation models, 272
  CO_2 cooling, 250
  data analysis, 230
  deuterium, 233, 240
  D/H ratios, 403

diurnal density variation, 261
drag force, 286
F2-ledge, 226
global
   circulation system, 268, 273
   mean eddy diffusion coefficient, 280
   mean heating rates, 250
   mean temperature, 239
   wind system, 276
hydrogen, 233, 240, 399
kinetic temperatures, 261
Lyman-α scattering, 399
mission observations, 228
models
   DRM model, 272, 273
   NCAR two-dimensional/VTGCM results, 245
   one-dimensional chemistry, 235
   photo-chemical diffusion, 236
   semi-empirical, VTS3, 235, 240
   spectral, 237
   Venus International Reference Atmosphere (VIRA) model, 236, 239, 1132
   Venus Thermosphere General Circulation Model (VTGCM), 237, 240
solar activity
   response, 229
   variations, 237
solar
   cycle variation, 240
   rotation variation, 238
   wind, 247
structure, 259
subsolar-to-antisolar flow, 296
superrotation, 234, 237, 260, 278, 279, 286
wind-induced diffusion, 261
winds, 270, 271
Thetis Regio, 516, 519, 657, 820, 825, 1223
   tesserae, 820
tholeiite, 1208
   basalt, 1214
   compositions, 1232
tholus, 676
thorium, 1247
thunder, 22
Ti, 1213
tides, 481, 1103. *See also* solar heating (tides).
   atmospheric, 1104
   atmospheric thermal, 1120
      torque, 1090
   bulge, 1153
   circulation, 465
   effect, 1132

forcing, 486
friction torque, 1088
   history, 1117, 1120
   solar-induced, 461
   solids, 1120
      body, 1097
      friction torque, 1090
   winds, 585
tilt patterns, 542, 544
TiO_2, 1214
topographic maps, 671
   swell, 850, 864
topography, 894, 1140. *See also* gravity, fields.
topside heat inflow, 211, 218
torque
   atmospheric, 1088
   core mantle friction, 1090
trace species, 301
transitional ejecta, 1033
transport
   equations, 191
   mechanical, 638
transterminator flow, 181
trench, 813
trondhjemite-tonalite-dacite melts, 1208
tropopause, 9, 335
troposphere, 9, 295–296. *See also* mesosphere.
   hydrogen, 394
troughs, 907, 937, 947
   formation, 940
   segments, 947
Tunguska event, 978, 1039
turbulence, 483
turbulent fluid friction, 1088
   torque, 1106
Tuulikki Mons, 705
twin planet model, 387
two-stream
   approach, 211
   method, 216
   transport equation, 200

Ulfrun Regio, 846
ULF waves, 100–101
ultraviolet
   absorber, 378
   photons, 451
   unidentified, 449
umbrae, 991–993, 995
underthrusting, 892
Uorsar Rupes, 830
upper mantle, 835
upwarps, 785
upwelling, 320, 358, 825, 851, 890, 921, 957, 1207, 1231
   anomalies, 746
   example, Iceland, 890

hypothesis, 959
   mantle, 1228
   models, earliest, 892
   radiation, 328
   sheets, 1229
uranium (U), 1214, 1247
Uranus, dynamo, 24
Urey number, 1247
Ushas Montes, 851
Uvaisi Crater, dark parabola, 1070

valleys
   networks, 789
   origins, 782
   rectangular, 763
   sapping, 782
   systems
      labyrinthic, 763
      pitted or irregular, 764
   types, 762
vapor-deposition mechanism, 656
vapor ejecta, 1038
Vega missions, 365, 415, 421, 599, 1210, 1216, 1217, 1247
   Ball nephelometer, results, 430
   balloons, 133, 284, 371, 461, 468, 471, 1214
   cloud observations, 417
   MgO, 1228
   probes, ISAV spectrometer, 431
   surface, 637
   water vapor, 427
velocity profiles, observed, 465
Venera missions, 22, 24, 130, 167, 174, 180, 234, 717, 1247
   atmosphere, 361, 363, 375
   atmospheric waves, 284
   cloud observations, 417
   FS, 422
   Infrared Fourier Spectrometer Experiment (IFSE), 302–303, 375
   lander, gas chromatograph, 432
   landers, 9 and 10, 326
   landing dynamics, 649
   lightning observations, 135
   lithospheric structure, 1193
   mapping, 680
   MgO, 1228
   mineralogy, 599
   plateaus, crustal, 818
   radar mappers, 991
   radio occultations, 303
   sulfur, 1215
   surface, 541, 548, 637, 644, 652
   water vapor, 425
vents, 700
Venus impact crater inventory (USGS), 1078
Venus-solar wind interaction, 33
   gas dynamic models, 35, 57
   hybrid models, 51
   ionosphere and, 62
   mass-loaded model, 40
   MHD models, 44, 57
   orbit and, 35
   test particle models, 47
Venusian Wind Tunnel (VWT), 548, 551
vertical gravity, 1137
   momentum transfer, 487
Very Large Array (VLA), 529, 536
Vesta Rupes, 830, 884
VEUV index, 238
Viking missions, 602
Vinmara Planitia, 734, 808, 913
VIRA reference profiles, 303
viscosity, 488
   contrast regime, 1280
   dissipation, 237, 481, 482
   relaxation, 1227
   temperature-dependent, 1263
   types of, 1269
visible spectra, 644
Vlasov theory, 104
VLF
   bursts, 152
   data, 108
   emissions, 106
   signals, 22, 108
   waves, 100
volatile
   elements, geochemistry, 627
   transport of minerals, 655
volatiles, 1209
volcanic centers, 699, 747–907
   calderas, 724
   characteristics, 700
   depressions, 674, 724
   enhanced concentration, 745
   flooding models, 825
   flow fields, 728
   geologic and tectonic distribution, 735
   intermediate volcanoes, 711
   large volcanoes, 703, 741
   morphologic classification, 705
   morphometry and classification, 705
   peaks, 846
   physiographic distribution, 733
   plains emplacement, 785
   rises, 675, 710, 797–798, 812, 815–816, 824, 836, 849, 856, 872, 939, 942, 959, 1254
      classification, 856
      corona-dominated, 816
      flexural models, 864
      gravity signatures, 815
      gravity studies, 857
      hotspot activity, 851, 868
      isostatic compensation models, 863

INDEX 1361

lithosphere thickness, 870
mantle upwelling models, 858
properties, 845
rift-dominated, 816
tectonics, 817
volcano-dominated, 816
size, 737
spatial distribution, 731
volcaniclastic meta-sediments, 1210
volcanism, 7, 128, 143, 152, 342, 410,
    507, 540, 544, 574, 622, 627,
    677, 697–698, 798, 800, 811,
    817, 822, 828, 848, 880, 887,
    940, 946, 949, 1061, 1076,
    1237, 1250, 1254, 1277
age, 688
basaltic, 1075, 1076
cones, 719
coronae, 777
corona-related, 1070
crater modification, 1022
crust formation, 1214
domes, 719
extrusive, 816
flood, 806
geomorphology, 676
global volcanic flooding event, 1226
intrusive, 850, 864
morphology, 1214
pyroclastic flows, 1034
shields, 719
syntectonic, 811
volcanoes, 338, 422, 512, 541, 580, 730,
    1070, 1236
  Atlian group, 1070
  domical, 676, 715
  materials, 734
  pancake domes, 676
    domes and steep-sided domes, 713
  Rusalkian group, 1070
  shield, 699, 711, 719, 807, 822, 902,
    1072
    construction, 1076
  small, 717, 720, 722
    characteristics, 719
    Magellan analysis, 722
    radial flow pattern, 720
volume-decreasing reactions, 505
volume-increasing reactions, 505
volume scattering, 532
vorticity transport, 487
VTGCM, 276, 281, 282
  model, 278, 319
    EUV temperature and $F_{(10.7)}$ index
      compared, 240
    simulations, 319
      SMAX conditions, 276, 277

water (H_2O), 7, 385, 407, 425, 443,
    757, 922, 1077, 1219 *See also*
    atmosphere *and* crust.
absorption, 328
accretion phase, 387
atmosphere, 333, 386, 394
cometary sources, 409
concentration profile, 336
crust, 802
deuterium, 407
evolution, 407
hydrogen, 407
mixing ratios, 340, 363
vapor, 326, 593
  bands, 329
  gradients, 336
  in atmosphere, 336, 396
  measurements, spacecraft data, 421
  mixing ratios, 362, 364
  Vega missions, 427
  Venera missions, 425
wave-mean-flow interaction, 274
wavelength-scale surface roughness, 534
waves, 318, 494, 557. *See also* meso-
    phere.
activity, 77, 95
drag, 275, 276, 282, 481, 483, 489
four-day, 467
gravity, 238
in mesosphere, 318
planetary scale, 238
weathering, 504, 522, 549, 637, 801,
    1023. *See also* surface.
as buffer, 650
chemical, 505, 522, 638, 1024
  basaltic rocks, 1210
  pyrite, 343
high altitude, 598
highlands, 504
mechanical, 510, 574
minerals produced, 638
primary basalt minerals, 645
processes, 511, 548
rock, 637
whistler
bursts, saucers, 149
mode, 139, 152
  attenuation, 146
  lightning and, 146
  propagation, 144, 148
waves, 78, 96, 104, 108, 147, 149,
    210
wind, 247, 298, 470, 492, 522, 547, 993,
    995. *See also* aeolian features
    and processes; mesosphere; *and*
    solar, wind.
abrasion, 554
crater ejecta, 999
equation, thermal, 475
fields, global-scale, 327

global system, 301
groundbased measurements, 270
impact-produced, 507
line-of-sight, 308
lower atmosphere, 10
meridional, 549
RSZ, 264–265, 270, 275, 278, 287
shear, 285, 471, 475
streaks, 473, 506, 547–548, 561–1210
  analysis, 576
  circulation, 580
subsolar-to-antisolar (SS-AS), 264, 279, 287, 319
  dawn and dusk, 286
  deceleration, 280
supersonic, 993
thermal, 298–299
transport, 242, 507
zonal, 252, 267–268, 272, 277, 287, 472, 479, 484, 489, 1033
  in troposphere, 298
  mesosphere, 312, 317
  retrograde, 306
  thermospheric, 265
  velocities, 10
windblown sediments, 586
WKB solution, 472
wobble, 552, 1121
  damping, 1089
  eulerian, 1116
wollastonite ($CaSiO_3$), 3, 593, 595, 596, 622, 651
wrinkle ridges, 785, 807, 810, 817, 821, 904, 924, 947, 1053, 1057
  age, 905
  formation, 1057
  morphology, 904
  patterns, 855
  properties, 903
wüstite ($Fe_{0.947}O$), 608

x-ray diffraction, 623, 629
  fluorescence, 599
xenon (Xe), 29, 387
  in atmosphere, 4
XRF data, 596, 602, 646

yardangs, 547, 561–680
yield strength envelope (YSE), 1182, 1184, 1186
  stress, 1268, 1271
Y pattern, 467

zebra pattern, reflectivity, 578, 580
Z elements, 602
zirconia oxygen fugacity sensors, 610
zonal mean thermal structure, 470
  superrotation period, 280
  wind, thermal, 303
zones of extension, 730

DATE DUE

BRODART, CO. Cat. No. 23-221-003